Lipo Wang Licheng Jiao
Guangming Shi Xue Li Jing Liu (Eds.)

Fuzzy Systems and Knowledge Discovery

Third International Conference, FSKD 2006
Xi'an, China, September 24-28, 2006
Proceedings

Springer

Series Editors

Jaime G. Carbonell, Carnegie Mellon University, Pittsburgh, PA, USA
Jörg Siekmann, University of Saarland, Saarbrücken, Germany

Volume Editors

Lipo Wang
Nanyang Technological University, Singapore
E-mail: elpwang@ntu.edu.sg

Licheng Jiao
Xidian University, Xi'an 710071, China
E-mail: lchjiao@mail.xidian.edu.cn

Guangming Shi
Xidian University, Xi'an, 710071 China
E-mail: gmshi@xidian.edu.cn

Xue Li
University of Queensland, Brisbane, Australia
E-mail: xueli@itee.uq.edu.au

Jing Liu
Xidian University, Xi'an, 710071, China
E-mail: neouma@mail.xidian.edu.cn

Library of Congress Control Number: 2006933055

CR Subject Classification (1998): I.2, F.4.1, F.1, F.2, G.2, I.2.3, I.4, I.5

LNCS Sublibrary: SL 7 – Artificial Intelligence

ISSN 0302-9743
ISBN-10 3-540-45916-2 Springer Berlin Heidelberg New York
ISBN-13 978-3-540-45916-3 Springer Berlin Heidelberg New York

Springer is a part of Springer Science+Business Media

springer.com

Printed in Germany

Typesetting: Camera-ready by author, data conversion by Scientific Publishing Services, Chennai, India
Printed on acid-free paper SPIN: 11881599 06/3142 5 4 3 2 1 0

Preface

This book, i.e., LNAI vol. 4223, is the proceedings of the Third International Conference on Fuzzy Systems and Knowledge Discovery (FSKD 2006), jointly held with the Second International Conference on Natural Computation (ICNC 2006, LNCS vols. 4221 and 4222) during September 24 – 28, 2006 in Xi'an, Shaanxi, China. FSKD 2006 successfully attracted 1274 submissions from 35 countries/regions (the joint ICNC-FSKD 2006 received 3189 submissions). After rigorous reviews, 165 high-quality papers, i.e., 115 long papers and 50 short papers, were included in the FSKD 2006 proceedings, representing an acceptance rate of 13.0%.

ICNC-FSKD 2006 featured the most up-to-date research results in computational algorithms inspired from nature, including biological, ecological, and physical systems. It is an exciting and emerging interdisciplinary area in which a wide range of techniques and methods are being studied for dealing with large, complex, and dynamic problems. The joint conferences also promoted cross-fertilization over these exciting and yet closely related areas, which had a significant impact on the advancement of these important technologies. Specific areas included neural computation, quantum computation, evolutionary computation, DNA computation, fuzzy computation, granular computation, artificial life, etc., with innovative applications to knowledge discovery, finance, operations research, and more. In addition to the large number of submitted papers, we were honored with the presence of six renowned keynote speakers.

On behalf of the Organizing Committee, we thank Xidian University for sponsorship, and the National Natural Science Foundation of China, the International Neural Network Society, the Asia-Pacific Neural Network Assembly, the IEEE Circuits and Systems Society, the IEEE Computational Intelligence Society, the IEEE Computational Intelligence Singapore Chapter, and the Chinese Association for Artificial Intelligence for technical co-sponsorship. We thank the members of the Organizing Committee, the Advisory Board, and the Program Committee for their hard work in the past 12 months. We wish to express our heartfelt appreciation to the keynote speakers, session chairs, reviewers, and student helpers. Our special thanks go to the publish, Springer, for publishing the FSKD 2006 proceedings as one volume of the *Lecture Notes in Artificial Intelligence* series (and the ICNC 2006 proceeding as two volumes of the Lecture Notes in Computer Science series). Finally, we thank all the authors and participants for their great contributions that made this conference possible and all the hard work worthwhile.

September 2006

Lipo Wang
Licheng Jiao

Organization

FSKD 2006 was organized by Xidian University and technically co-sponsored by the National Natural Science Foundation of China, the International Neural Network Society, the Asia-Pacific Neural Network Assembly, the IEEE Circuits and Systems Society, the IEEE Computational Intelligence Society, the IEEE Computational Intelligence Singapore Chapter, and the Chinese Association for Artificial Intelligence.

Organizing Committee

Honorary Conference Chairs	Shun-ichi Amari (RIKEN BSI, Japan)
	Xin Yao (University of Birmingham, UK)
General Co-chairs	Lipo Wang (Nanyang Technological University, Singapore)
	Licheng Jiao (Xidian University, China)
Program Committee Chairs	Guangming Shi (Xidian University, China)
	Xue Li (The University of Queensland, Australia)
Local Arrangement Chairs	Yuanyuan Zuo (Xidian University, China)
	Xiaowei Shi (Xidian University, China)
Proceedings Chair	Jing Liu (Xidian University, China)
Publicity Chair	Yuping Wang (Xidian University, China)
Sponsorship Chair	Yongchang Jiao (Xidian University, China)
Secretaries	Bin Lu (Xidian University, China)
	Tiantian Su (Xidian University, China)
Webmasters	Yinfeng Li (Xidian University, China)
	Maoguo Gong (Xidian University, China)

Advisory Board

Zheng Bao	Xidian University, China
Zixing Cai	Central South University, China
Guoliang Chen	University of Science and Technology of China, China
Huowang Chen	National University of Defense Technology, China
David Corne	The University of Exeter, UK
Dipankar Dasgupta	University of Memphis, USA
Kalyanmoy Deb	Indian Institute of Technology Kanpur, India
Baoyan Duan	Xidian University, China
Kunihiko Fukushima	Tokyo University of Technology, Japan
Tom Gedeon	The Australian National University, Australia

Aike Guo	Chinese Academy of Science, China
Yao Hao	Xidian University, China
Zhenya He	Southeastern University, China
Fan Jin	Southwest Jiaotong University, China
Yaochu Jin	Honda Research Institute Europe, Germany
Janusz Kacprzyk	Polish Academy of Sciences, Poland
Lishan Kang	China University of Geosciences, China
Nikola Kasabov	Auckland University of Technology, New Zealand
John A. Keane	The University of Manchester, UK
Soo-Young Lee	KAIST, Korea
Yanda Li	Tsinghua University, China
Zhiyong Liu	National Natural Science Foundation of China, China
Erkki Oja	Helsinki University of Technology, Finland
Nikhil R. Pal	Indian Statistical Institute, India
Yunhe Pan	Zhe Jiang University, China
Jose Principe	University of Florida, USA
Witold Pedrycz	University of Alberta, Canada
Marc Schoenauer	University of Paris Sud, France
Zhongzhi Shi	Chinese Academy of Science, China
Harold Szu	Office of Naval Research, USA
Shiro Usui	RIKEN BSI, Japan
Shoujue Wang	Chinese Academy of Science, China
Xindong Wu	University of Vermont, USA
Lei Xu	Chinese University of Hong Kong, Hong Kong, China
Bo Zhang	Tsinghua University, China
Nanning Zheng	Xi'an Jiaotong University, China
Yixin Zhong	University of Posts and Telecommunications, China
Syozo Yasui	Kyushu Institute of Technology, Japan
Jacek M. Zurada	University of Louisville, USA

Program Committee

Janos Abonyi	University of Veszprém, Hungary
Jorge Casillas	University of Granada, Spain
Pen-Chann Chang	Yuanze University, Taiwan
Chaochang Chiu	Yuanze Universizy, Taiwan
Honghua Dai	Deakin University, Australia
Suash Deb	National Institute of Science and Technology, India
Hepu Deng	RMIT University, Australia
Jun Gao	Hefei University of Technology, China
Saman Halgamuge	The University of Melbourne, Australia

Chongzhao Han	Xi'an Jiaotong University, China
Kaoru Hirota	Tokyo Institute of Technology, Japan
Frank Hoffmann	University of Dortmund, Germany
Dewen Hu	National University of Defense Technology, China
Jinglu Hu	Waseda University, Japan
Eyke Hüllermeier	University of Marburg, Germany
Hisao Ishibuchi	Osaka Perfecture University, Japan
Frank Klawoon	University of Applied Sciences, Germany
Naoyuki Kubota	Tokyo Metropolitan University, Japan
Sam Kwong	City University of Hong Kong, Hong Kong, China
Zongmin Ma	Northeastern University, China
Michael Margaliot	Tel Aviv University, Israel
Ralf Mikut	IAI, Germany
Pabitra Mitra	Indian Institute of Technology, India
Masoud Mohammadian	University of Canberra, Australia
Detlef Nauck	BT, UK
Hajime Nobuhara	Tokyo Institute of Technology, Japan
Andreas Nürnberger	University of Magdeburg, Germany
Quan Pan	Northwestern Polytechnical University, China
Da Ruan	SCK-CEN, Belgium
Thomas Runkler	Siemens AG, Germany
Yonghong Tan	Guilin University of Electronic Technology, China
Takao Terano	Tokyo Institute of Technology, Japan
Brijesh Verma	Central Queensland University, Australia
Guoyin Wang	Chongqing University of Posts and Telecommunications, China
Yugeng Xi	Shanghai Jiaotong University, China
Weixin Xie	Shenzhen University, China
Yiyu Yao	University of Regina, Canada
Gary Yen	Oklahoma State University, USA
Xinghuo Yu	RMIT, Australia
Shichao Zhang	University of Technology Sydney, Australia
Yanqing Zhang	Georgia State University, USA
Zhihua Zhou	Nanjing University, China

Reviewers

Aifeng Ren
Andrew Teoh
Andries Engelbrecht
Baolin Liu
Baolong Guo
Bekir Cakir
Ben-Shun Yi
Bernardete Ribeiro
Bin Xu
Bing Han

Binghai Zhou
Bo Liu
Bobby Gerardo
Bo-Qin Feng
Brijesh Verma
Byung-Joo Kim
Changjian Feng
Changyin Sun
Chao Deng
Chaojian Shi
Chen Yong
Chengxian Xu
Cheng-Yuan Chang
Cheol-Hong Moon
Chi Xie
Chijian Zhang
Ching-Hung Lee
Chor Min Tan
Chuanhan Liu
Chun Jin
Chunning Yang
Chunshien Li
Da-Kuan Wei
Daniel Neagu
Davide Anguita
Defu Cheng
Deok-Hwan Kim
Deqin Yan
Detlef Nauck
Dongbo Zhang
Donghu Nie
Dug Hun Hong
Du-Yun Bi
Ergun Eraslan
Eun-Jun Yoon
Fajun Zhang
Fangshi Wang
Fei Hao
Feng Chen
Feng Gao
Frank Klawoon
Fuyan Liu
Fu-Zan Chen
Gang Li
Gang Chen
Gaoshou Zhai
Gexiang Zhang
Golayoglu Fatullayev Afet
Guang Tian
Guang-Qiu Huang
Guangrui Wen
Guanjun Wang
Guanlong Chen
Gui-Cheng Wang
Guixi Liu
Guo-Chang Li
Guojian Cheng
Guoyin Wang
Gwi-Tae Park
H. B. Qiu
Hasan Cimen
He Jiang
Hengqing Tong
Hexiao Huang
Hisao Ishibuchi
Hong Jin
Hong Ge
Hongan Wang
Hongcai Zhang
Hongwei Li
Hongwei Si
Hong-Yong Yang
Huai-Liang Liu
Huaizhi Su
Huang Ning
Hua-Xiang Wang
Hui-Yu Wang
Hyun Chan Cho
Hyun-Cheol Jeong
I-Shyan Hwang
Ivan Nunes Da Silva
Jae Yong Seo
Jaeho Choi
Jae-Jeong Hwang
Jae-Wan Lee
Jeong Seop Sim
Jia Liu
Jia Ren
Jian Yu
Jian Xiao

Jian Yin
Jianbin Song
Jiancang Xie
Jiang Cui
Jianguo Ning
Jianhua Peng
Jianjun Xu
Jianjun Wang
Jianming Zhang
Jianqiang Yi
Jian-Rong Hou
Jianwei Yin
Jian-Yun Nie
Jiating Luo
Jie Hu
Jili Tao
Jimin Liang
Jin Yang
Jingwei Liu
Jinhui Zhang
Jinping Li
Jinwu Gao
Jiqing Qiu
Jiquan Shen
Jiulun Fan
Jiuying Deng
Jiyi Wang
John Q Gan
Jong Seo Park
Jong-Bok Kim
Joong-Hwan Baek
Jorge Casillas
Jorge Ropero
Jr-Syu Yang
Juan Liu
Juanying Xie
Jun Jing
Jun Meng
Jun Wu
Jun Zhao
Junping Zhang
Junping Wang
Junying Zhang
Kaoru Hirota
Kar-Ann Toh
Ke Lu
Kefeng Fan
Kejun Tan
Keon Myung Lee
Kongfa Hu
Kwan Houng Lee
Kyung Ha Seok
Lambert Spaanenburg
Lance Fung
Lean Yu
Lei Wang
Leichun Wang
Leslie Smith
Li Wu
Li Dayong
Li Sun
Li Zhang
Liang Gao
Liang Xiao
Liang Ming
Lianxi Wu
Licheng Jiao
Liguo Zhang
Lin Wang
Lin Gao
Lirong Jian
Longbo Zhang
Long-Shu Li
Manjaiah D h
Maoguo Gong
Maoyuan Zhang
Maurizio Marchese
Meihong Shi
Michael Margaliot
Ming Li
Ming Liu
Mingquan Zhou
Ming-Wen Shao
Mudar Sarem
Myo-Taeg Lim
Nagabhushan P
Naigang Cui
Nak Yong Ko
Nakaji Honda
Ning Xu

Ning Chen
Pei-Chann Chang
Peide Liu
Phill Kyu Rhee
Ping Guo
Ping Zhou
Ping Ji
Ping Yan
Pu Wang
Qiang Zhang
Qianjin Guo
Qinghe Ming
Qing-Ling Zhang
Qingqi Pei
Quan Zhang
Robo Zhang
Rongfang Bie
Ruijun Zhu
Ruiming Fang
Seok-Lyong Lee
Shangmin Luan
Shanmukhappa Angadi
Shaosheng Fan
Shi Zheng
Shigeo Abe
Shihai Zhang
Shi-Hua Luo
Shi-Jay Chen
Shitong Wang
Shuiping Gou
Shunman Wang
Shutao Li
Shuyuan Yang
Soo-Hong Park
Sung Kyung Hong
Sungshin Kim
Tae-Chon Ahn
Takao Terano
Tan Liu
Tan Ran
Tao Shen
Thi Ngoc Yen Pham
Tianli Li
Ting Sun
Tinghua Xu
Tinghuai Ma
Toly Chen
Toshio Eisaka
Wanchun Dou
Wang Lei
Wei Zou
Wei Xie
Wei Li
Wei Fang
Weida Zhou
Wei-Dong Kou
Weimin Ma
Wei-Ping Wang
Weixing Zhu
Wenbing Xiao
Wenchuan Yang
Wenpeng Lin
Wenqing Zhao
Wen-Shyong Tzou
Wen-Xiu Zhang
Wenxue Hong
Xian-Chuan Yu
Xianggang Yin
Xiangrong Zhang
Xianjin Zha
Xiaobing Liu
Xiaodong Liu
Xiaoguang Yang
Xiaoguang Liu
Xiaohui Yuan
Xiaohui Yang
Xiaoqing Lu
Xiaosuo Lu
Xin Li
Xinbo Gao
Xinbo Zhang
Xinghua Fan
Xingming Sun
Xinming Tang
Xinyu Yang
Xinyuan Lu
Xiongfei Li
Xiqin He
Xiu Jin
Xiuli Ma

Xiyue Huang
Xue-Feng Zhang
Xuejun Xu
Xueliang Bi
Xueling Ma
Xuemei Xie
Xueqin Feng
Xufa Wang
Yaguang Kong
Ya-Jun Du
Yan Li
Yan Liang
Yang Tao
Yangyang Wu
Yanmei Chen
Yanning Zhang
Yanning Zhang
Yanping Lv
Yanxia Zhang
Yan-Xin Zhang
Ye Du
Ye Bin
Yecai Guo
Yeon-Pun Chang
Yezheng Liu
Yidan Su
Yigong Peng
Yijun Yu
Yingfang Fan
Yinggang Xie
Yinghong Peng
Yong Fang
Yong Zhao
Yong Yang
Yong Fan
Yonggui Kao
Yongqiang Zhang
Yongqiang Zhao
Yongsheng Ding
Yon-Sik Lee
Young Chel Kwun
Young Hoon Joo
Young-Koo Lee
Yu Guo
Yuan Kang
Yuechao Ma
Yuehui Chen
Yufeng Liao
Yukun Bao
Yulong Lei
Yumin Tian
Yunjie Zhang
Yurong Zeng
Yutian Liu
Zhang Yang
Zhanhuai Li
Zhe-Ming Lu
Zhenbing Zeng
Zhengxing Cheng
Zhigang Xu
Zhigeng Fang
Zhi-Hong Deng
Zhihui Li
Zhiqiang Zuo
Zhiyun Zou
Zhonghua Li
Zixing Cai
Zongben Xu
Zong-Yuan Mao
Zoran Bojkovic

Table of Contents

Fuzzy Theory and Algorithms

Knowledge Discovery Theory and Algorithms

Fuzzy Applications

Knowledge Discovery Applications

Theory Research on a New Type Fuzzy Automaton

QingE Wu[1], Tuo Wang[1], YongXuan Huang[1], and JiSheng Li[1]

School of Electronic and Information Engineering Xi'an Jiao Tong University, Xi'an, Shaanxi, 710049, P.R. China
wqe969699@163.com

Abstract. For better solving some complicated problems in fuzzy automata hierarchy, simultaneously, in order to accomplish better task for fuzzy signal processing, this paper presents a kind of new automaton–fuzzy infinite-state automaton. The basic extracted frame of fuzzy infinite-state automaton is introduced by using neural networks. To the extracted fuzzy infinite-state automaton, this paper describes that it is equivalent to fuzzy finite-state automaton, and its convergence and stability on its hierarchy system will be discussed. Finally, the simulation is carried on and the simulation results show that the states of fuzzy infinite-state automaton converge to some stable states with extraction frame and training for weights what this paper provides at last. Finally, some problems of fuzzy infinite-state automaton and neural networks to be solved and development trends are discussed. These researches will not only extend further automata hierarchy, but also increase a new tool for application of fuzzy signal processing. It is an important base in the application of fuzzy automata theory.

1 Introduction

In previous work, we classify the fuzzy automata according to recognizing the type of the language. Accordingly, fuzzy automata have as well partition according to recognizing the feature of the language, and then the automaton is classified into deterministic automaton and non-deterministic automaton or fuzzy automaton (FA). The FA is classified into fuzzy finite-state automaton (FFA) and fuzzy infinite-state automaton (FIA). Non-deterministic automata and fuzzy automata can be transformed into deterministic automata.

Previously, fuzzy knowledge equivalence representations between neural networks, fuzzy systems and models of automata are discussed [1]. From a control point of view, fuzzy finite-state automata with recurrent neural networks [2-4] for often imitating fuzzy dynamical systems are very useful. Previously, works have been shown how FFA can be mapped into recurrent neural networks with second-order weights using a crisp representation of FFA states [5].

Until present, the automata of fuzzy or defuzzization what we study are all finite-state automata [6]. However, FIA is not introduced. In this paper, we will discuss extraction of the FIA by using recurrent neural networks. The equivalence

L. Wang et al. (Eds.): FSKD 2006, LNAI 4223, pp. 1–10, 2006.

of FIA and FFA is described. Since the FFA is equally powerful as the deterministic finite-state automaton (DFA), as well as FIA is equivalent to FFA, the FIA is equivalent to the DFA at last. The convergence and stability of FIA is also discussed. Some new definitions and theorems are given. Finally, the simulation results show that the states of FIA converge surely some stable points. Through these studies in this paper, it not only strengthens the relationship between the fuzzy systems and hierarchy of fuzzy automata, but also these problems to be solved will be directly impulse the development of theories and applications of fuzzy automata, and it will show the further and more spacious prospect in wide applications. Thus, there will be a theoretic base for extraction and application of any automata.

2 Preliminary FIA

According to [5], the definition of FFA is introduced as follows:

Definition 2.1. A fuzzy automaton (FA) is named for a fuzzy finite-state automaton (FFA) M if it consists of a six-tuple $M = (Q, \Sigma, F, Q_0, G, V)$. Each factor of the six-tuple denotes respectively as follows:

Where Q is a finite set of states; Σ is a finite set of input alphabet; $Q_0 \subseteq Q$ is a fuzzy set of initial states; $G \subseteq Q$ is a fuzzy set of final states; $V \subseteq [0,1]$ is a membership degree set of transition relation; and $F \in V : Q \times \Sigma \times Q \rightarrow V$ is a fuzzy relation between Q, Σ and Q, i.e., $F(q_i, \sigma, q_j) \in V$, where $q_i, q_j \in Q, \sigma \in \Sigma$. Then, the fuzzy automaton (FA) is called FFA.

Now, introduce how a FFA accepts the fuzzy language.

For $\sigma \in \Sigma$, denote $F_\sigma \in V$ by $F_\sigma(q_i, q_j) = F(q_i, \sigma, q_j)$. The degree $\big(L(FFA)\big)(\omega)$ that a FFA M accepts a word $\sigma_1 \cdots \sigma_n \in \Sigma^*$ is defined by:

$$\big(L(FFA)\big)(\sigma_1 \cdots \sigma_n) = P(q_0) \circ F_{\sigma_1} \circ \cdots \circ F_{\sigma_n} \circ G(q_n)$$

Where $P(q_0)$ and $G(q_n)$ are the membership degree in the initial state q_0 and the final state q respectively, and $\circ$ denotes the max-min composition of fuzzy relation, i.e.,

$$\big(L(FFA)\big)(\sigma_1 \cdots \sigma_n) = \bigvee_{q_0, q_1, \cdots, q_n \in Q} P(q_0) \bigwedge F_{\sigma_1}(q_0, q_1) \bigwedge \cdots \bigwedge F_{\sigma_n}(q_{n-1}, q_n) \bigwedge G(q_n)$$

$L(FFA)$ in Σ^* is called the fuzzy language accepted by FFA. $L(FFA)$ denotes the fuzzy language as follows:

$$L(FFA) = \Big\{(\omega, \mu) | \omega \in \Sigma, \mu = \bigvee_i \mu_i, F_\omega(q_0, q_i) = \mu_i \in V, \exists q_i \in G\Big\}. \quad (1)$$

Where $\mu = \bigvee_i \mu_i$ signifies μ is obtained by 'or' operator of μ_i.

Definition 2.2. A fuzzy automaton is called a fuzzy infinite-state automaton (FIA) M if it also consists of a six-tuple $M = (Q, \Sigma, \delta, Q_0, G, V)$. Each factor of the six-tuple also denotes respectively as follows:

Where Q is an infinite set of states, Σ is a set of input symbols, $Q_0 \subseteq Q$ is a fuzzy set of initial states, $G \subseteq Q$ is a fuzzy set of final states, $V \subseteq [0,1]$ is a membership degree set of transition function, and V is an infinite set, simultaneously, for any $\mu \in V$, $\delta : Q \times \Sigma \overset{\mu}{\rightarrow} Q$ is a transition function, i.e., $\delta(q_i, a, \mu) = \{q_j\}$, where $q_i, q_j \in Q, a \in \Sigma^*, \mu \in V$. Finally, the $G_\mu(q)$ denotes the fuzzy membership degree $\mu \in V$ at final state $q \in G$. Then, the FA is called the FIA.

Similarly, introduce how a FIA accepts the fuzzy language.

The degree $\big(L(FIA)\big)(\sigma_1 \cdots \sigma_n)$ that a FIA M accepts a word $\sigma_1 \cdots \sigma_n \in \Sigma^*$ is defined by $\big(L(FIA)\big)(\sigma_1 \cdots \sigma_n) = G_\mu\big(\delta(q_0, \sigma_1 \cdots \sigma_n, \mu)\big)$, where $q_0 \in Q_0, \mu \in V$, and where $\delta(q_0, \sigma_1 \cdots \sigma_n, \mu) = \delta\big(\delta(q_0, \sigma_1, \mu_1), \sigma_2 \cdots \sigma_n, \mu_2 \cdots \mu_n\big) = \cdots = \delta(q_{n-1}, \sigma_n, \mu_n) = \{q_n\}$, and $q_i \in Q$, $\mu_i \in V$, $i = 1, \cdots, n$.

$L(FIA)$ is called a fuzzy language accepted by FIA. The $L(FIA)$ is represented by the following set and where $\mu = \bigvee\limits_{ij} \mu_{ij}$ signifies μ is obtained by 'or' operator of μ_{ij}.

$$L(FIA) = \Big\{(\omega, \mu) | \omega \in \Sigma^*, \mu = \bigvee_{ij} \mu_{ij}, \delta(q_i, \omega, \mu_{ij}) = \{q_j\}, \\ \exists q_j \in G, \forall \mu_{ij} \in V, \forall q_i \in Q_0\Big\} \tag{2}$$

A fuzzy language L is acceptable by a FIA iff, $L = L(FIA)$, for some FIA.

3 Recurrent Neural Network Architecture for FIA

Based on a previous result that we encode FFA into recurrent neural networks [6], here we use the discrete recurrent neural network structure for mapping FIA into recurrent networks. The network architecture for extracting FIA is shown in Fig. 1.

3.1 Basic Structure of Recurrent Networks for FIA

The networks for FIA consist of two parts that are the trained networks and the extraction networks of FIA respectively. In training layer of networks, the recurrent neural networks are formed of N recurrent hidden neurons, and N output neurons, labeled $Y_j(t), j = 0, 1, \cdots, N-1$; M input neurons, labeled $x_l(t), l = 0, 1, \cdots, M-1$ with some weights ω_{jl}, associated to the links of these neurons. On extraction layer of networks for FIA, let neurons of extraction layer be a number of neurons and label L that are infinite, the L competitive neurons connect with the N output neurons by $N*L$ weights labeled $w_{ij}, i = 0, 1, \cdots, L-1, j = 0, 1, \cdots, N-1$.

The hidden unit activation function is the sigmoid function $f(x) = \frac{1}{1+e^{-x}}$. The output in training layer is discrete value that is determined by discretization function $D(x)$, and $D(x)$ is given by the following (3) or (4).

(I) When the membership degree is any variable value in interval [0,1], i.e., there is the infinite number of membership degrees: Then, we divide interval

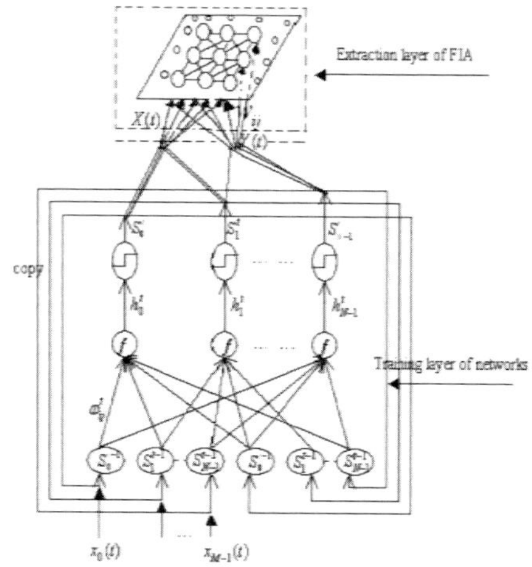

Fig. 1. Recurrent network architecture for FIA

[0,1]. Since the number of recurrent neuron is N, let us split the N intervals [0,1] into $n(n>1)$ coordinate subinterval, and the interval end-point value θ_s is obtained, where $s=0,1,\cdots,n$. Then, set

$$D(x)=\frac{\theta_i+\theta_{i+1}}{2} \quad if \;\; \theta_i<x\leqslant\theta_{i+1}, i=0,1,\cdots,n-1. \tag{3}$$

Where $\theta_0=0, \theta_n=1, \theta_i=\frac{i}{n}$.

(II) Based on statistic knowledge, when membership degree values are close to the corresponding finite real value $\{\theta_0,\theta_1,\cdots,\theta_{m-1}\}$, we set

$$D(x)=\theta_i, \;\; if \;\; |x-\theta_i|<\varepsilon, i=0,1,\cdots,m-1; x\in V. \tag{4}$$

Where ε is decided according to our demand, i.e., the final value x of neuron is close to θ_i after the whole string has been processed by the network.

For any values $\frac{\theta_i+\theta_{i+1}}{2}\neq 0$ and $\frac{\theta_i+\theta_{i+1}}{2}\neq 1$ are chosen instead of 0 and 1 here in order to give some power of influence to each of the current hidden unit values at the next time step, since a unit with value 0 would eliminate any influence of that unit.

We use h_i^t to denote the analog value of hidden unit i at the time step t, and S_i^t to denote the discretized value of hidden unit i at the time step t. ω_{ij}^t is the weight from unit j of layer 1 to unit i of layer 2 in training layer [6].

The dynamic process of network in training layer is described as follows:

$$h_i^t=f\Big(\sum_j\omega_{ij}^{t-1}S_j^{t-1}\Big), \forall i,t, \quad f(x)=\frac{1}{1+e^{-x}}, \quad S_i^t=D(h_i^t),$$

where $D(x)$ is obtained by the above equality (3) and (4).

3.2 Fuzzy States Representation for FIA

The current fuzzy state of FIA is a union of states $\{q_i\}$ with different fuzzy membership degrees. Consider the state q_j of FIA and the fuzzy state transition $\delta\Big(q_j, a_k, \{\theta_{ijk}\}\Big) = \{q_{i1}, \cdots, q_{ir}, \cdots\}$, where $a_k \in \Sigma$ is an input symbol and $\theta_{ijk} \in V$ is a fuzzy membership degree. We assign the corresponding recurrent state neuron S_j for state set $q_j = \{q_{j1}, \cdots, q_{jn} | n = 1, 2, \cdots\}$ of FIA, and the corresponding neurons $S_{i1}, \cdots, S_{ir}$ for states set $q_{i1}, \cdots, q_{ir}$ of FIA. The activation of recurrent state neuron S_i represents certainty θ_{ijk} with some state transition $\delta(q_j, a_k, \theta_{ijk}) = q_i$, i.e., $S_i^{t+1} \approx \theta_{ijk}$. If no state can reach q_i at time $t+1$, then let $S_i^{t+1} \approx 0$.

4 Extraction of FIA

To training of networks and extraction algorithm for FFA, see [6,7] for them in detail. It is similar to Kohonen's self-organizing feature map (SOFM)[7]. The extraction for FIA is similar to FFA, but there are some differences between the two. Here, a part of algorithm way that is different from the one of [6,7] is only given as follows:

(1) Input a sample signal to networks and train the networks, we obtain an output vector $Y(t)$, where $Y(t) = (S_0^t, S_1^t, \cdots, S_{N-1}^t)^T$, S_j^t is an input signal in extraction layer for $\forall t$ and $j \in \{0, 1, \cdots, N-1\}$. Let $w_{ij}(t)$ be the weights of connection from training layer, unit j to extracting layer, unit i in the case of binary inputs, and Let $W(t) = (w_{ij})$ be weights matrix. Regard the output vector $Y(t)$ as input vector $X(t)$ on extraction layer of FIA; the input $X(t)$ is obtained from $Y(t)$ with $X(t) = (X_{k_i}^i(t))_{L\times 1}$, where several pieces S_j^t in $Y(t)$ unite and achieve the vectors $X_{k_i}^i(t) = (S_0^t, S_1^t, \cdots, S_{k_i}^t)^T$, $0 \leqslant i \leqslant L-1$, $L = \{1, 2, \cdots\}$, $0 \leqslant j, k_i \leqslant N-1$.

(2) At first, in the region D of the large range, regulate the weights matrix $W(t)$. Parameter D will be obtained by trial and error, If the FIA has been extracted, next task is to check whether it recognizes all the training examples or not. If the answer is affirmative, we have found an optimal D^* that we are looking for. In the case of negative answer, the value of D is decreased for one unit. Otherwise, the procedure ends. Once the networks of extracting layer are trained, a unit will represent a state of the FIA. Regulate the weights of the connecting neurons from training layer to extraction layer with $w_{ij}(t+1) = w_{ij}(t) + \alpha(t)\big(S_j^t - w_{ij}(t)\big)$, where $j \in \{0, 1, \cdots, k_i\}$, $0 \leqslant i \leqslant L-1$, $0 \leqslant k_i \leqslant N-1$; $0 \leqslant \alpha(t) \leqslant 1$ is a kind of variable velocity of study, i.e., the more the difference of the fuzzy membership degree S_j^t and weight $w_{ij}(t)$ is at the moment t, the bigger the value $\alpha(t)$ is. Assume $B = \max_{i,j}\{|S_j^t - w_{ij}(t)|\}$, and here set,

$$\alpha(t) = \begin{cases} \frac{|S_j^t - w_{ij}(t)|}{B} & B \neq 0 \\ 0 & B = 0 \end{cases}.$$

(3) There is an index I set that is the number of neuron to participate in competition at a time in extraction layer, i.e., $I \subset \{0, 1, \cdots, L-1\}$. If $\forall s \in I$,

$\exists i \in I$, there is $||W^i_{k_i}(t) - X^i_{k_i}(t)||_2 = \min_{s\in I}\{||W^s_{k_s}(t) - X^s_{k_s}(t)||_2\}$, where $||\bullet||_2$ is an Euclidean 2-norm, $0 \leqslant k_i, k_s \leqslant N-1$, $W^i_{k_i}(t) = (w_{i0}, w_{i1}, \cdots, w_{ik_i})^T$, $0 \leqslant i \leqslant L-1$. We obtain the winner unit C_i, then the state of FIA that is extracted is q_i/x_i at the moment, where x_i is a fuzzy membership degree corresponding to the state q_i and is obtained by labeled $x_i = S^t_p$, with $|w_{ip}(t) - S^t_p| = \min_j\{|w_{ij}(t) - S^t_j|\}$ for any $j \in \{0, 1, \cdots, k_i\}$, $\exists p \in \{0, 1, \cdots, k_i\}$, $0 \leqslant k_i \leqslant N-1$.

(4) Regulate again the weights to connect the winner node C_i and the weights to connect the interior node in geometry neighborhood of C_i with $w_{ij}(t+1) = w_{ij}(t) + \alpha(t)\big(S^t_j - w_{ij}(t)\big)$. The $w_{ij}(t+1)$ has a larger or a smaller regulating until the $w_{ij}(t+1)$ approaches to the S^{t+1}_j in range of error. Therefore there is $||W^i_{k_i}(t+1) - X^i_{k_i}(t+1)||_2 \leqslant ||W^i_{k_i}(t) - X^i_{k_i}(t)||_2$. Assume $B_i = \max_j\big\{|S^t_j - w_{ij}(t)|\big\}$, and set,

$$\alpha(t) = \begin{cases} \frac{|S^t_j - w_{ij}(t)|}{B_i} & B_i \neq 0 \\ 0 & B_i = 0 \end{cases}.$$

The procedures of extracting FIA are shown as follows:

① At time $t = 0$, initialize S^0_0 to be 0.8 and all other S^0_j to be 0.2, $j \neq 0$ in order to give some power of influence for each of the current hidden unit values at the next time step. The network weights ω^0_{jl} are initialized randomly with a uniform distribution from -1 to 1 and ω^t_{jl} are given by trial and error later at time $t \neq 0$. Initialize $w_{ij}(0)$ randomly and let its value be in [0,1]. According to [6,7], by competition, the input $X(0)$ activates a unit (j_0, h_0) at extraction layer, which is taken as the initial state of the FIA, labeled $q_{(j_0,h_0)}/x_0$ that is determined by the vector $X^i_{k_i}(0) = (S^0_0, S^0_1, \cdots, S^0_{k_i})^T$, $0 \leqslant i \leqslant L-1$, $0 \leqslant k_i \leqslant N-1$, where $x_0 = S^0_p$ is a fuzzy membership degree.

② Starting out from the current activity unit (j, h) associated to state $q_{(j,h)}$ of FIA at time t. Introduce a previously unprocessed symbol $\xi_l \in \Sigma$ into the networks of training layer, and then an input vector $X(t)$ is obtained from producing an output vector $Y(t)$ and it activates a winner unit (m, n) that is taken as the corresponding state of the FIA. Now, a new state $q_{(m,n)}$ in FIA is or isn't created, but the associated transition, $\delta(q_{(j,h)}, \xi_l, \mu_{jm,hn}) = q_{(m,n)}$ is created. Calculate the membership degrees $\mu_{jm,hn}$ of state transitions by the above (3).

③ The following ξ_{l+1} is introduced into the networks at time $t+1$. Accordingly, it also obtains an active unit (m', n'). Thus, the transition has been created in the FIA from the activated unit (m, n) to the activated unit (m', n').

④ Repeat ②③ until all the symbols are processed.

5 Equivalence of FIA

The equivalence of FIA and FFA is discussed as follows:

Theorem 5.1. The FIA is equivalent to FFA.

Proof. Assume FIA $M_I = (Q, \Sigma, \delta, Q_0, G, V)$ accepts language $L(M_I)$, accordingly, a FFA $M_F = (Q_F, \Sigma, \delta_F, Q_{0F}, G_F, V_F)$ is made.

Since the V is a membership degree set of any transition and states of FIA, and the membership degree is from 0 to 1, choose $V = [0, 1]$ for general instance.

When the membership degree is any variable value in interval [0,1], i.e., there is the infinite number of membership degrees; let us divide the interval [0,1] into coordinate $n(n > 1)$ subinterval, and the interval end-point value θ_s is obtained, where $s = 0, 1, \cdots, n$. Then: $\mu_i = \frac{\theta_i + \theta_{i+1}}{2}$ if $\theta_i < x \leqslant \theta_{i+1}, i = 0, 1, \cdots, n-1$. Where $\theta_0 = 0, \theta_n = 1, \theta_i = \frac{i}{n}, x \in V$. We set $V_F = \{\mu_i | i = 0, 1, \cdots, n-1\}$.

$$\begin{aligned} q_i = \Big\{ q_x \Big| \exists q \in Q, \exists i, \delta(q_x, \sigma, x) = q, \sigma \in \Sigma^*, \forall q_x \in Q, \\ \forall x \in V, \theta_i < x \leqslant \theta_{i+1} \Big\}, i = 0, 1, \cdots, n-1. \end{aligned} \tag{5}$$

At the same time, we set $Q_F = \bigcup_{i=0}^{n-1} \Big\{ q_i \Big\}$ if there is a transition $\delta(q_j, \sigma, x_j) = \{q_i\}$, where $\forall \sigma \in \Sigma^*, \forall x_j \in V$.

It is obvious that the bigger n is, the more accurate FIA is equal to FFA.

Based on statistic knowledge, when membership degree values are close to the corresponding finite real value $\{\theta_0, \theta_1, \cdots, \theta_{m-1}\}$, i.e., $|x - \theta_i| < \varepsilon, x \in V$, let $\mu_i = \theta_i, i = 0, 1, \cdots, m-1$. At time, we set: $V_F = \{\mu_i | i = 0, 1, \cdots, m-1\}$.

$$\begin{aligned} q_i = \Big\{ q_x \Big| \exists q \in Q, \exists i, \delta(q_x, \sigma, x) = q, \sigma \in \Sigma^*, \forall q_x \in Q, \\ \forall x \in V, |x - \theta_i| < \varepsilon \Big\}, i = 0, 1, \cdots, m-1. \end{aligned} \tag{6}$$

At the same time, we set $Q_F = \bigcup_{i=0}^{m-1} \Big\{ q_i \Big\}$ if there is a transition $\delta(q_j, \sigma, x_j) = \{q_i\}$, where $\forall \sigma \in \Sigma^*, \forall x_j \in V$.

Assume $l = m$ or $l = n$, then the element of the Q_F is the $[q_0, q_1, \cdots, q_{l-1}]$; $Q_{0F} = [Q_0]$; $\bigcup_{i=0}^{l-1} q_i = Q$; $G_F \subseteq Q_F$ and each state of the G_F is one state subset of the final states of the M_I, i.e., the state of G_F is the following set:

$$\begin{aligned} q_G = \Big\{ q_x \Big| \exists q \in Q, \exists i, \delta(q, \sigma, x) = q_x, \sigma \in \Sigma^*, \forall q_x \in G, \forall x \in V, \\ \theta_i < x \leqslant \theta_{i+1} or |x - \theta_i| < \varepsilon \Big\}. \ i = 0, 1, \cdots, l-1. \end{aligned}$$

The δ_F is defined by $\delta_F\Big([q_0, q_1, \cdots, q_{l-1}], a, \mu_i\Big) = [p_0, p_1, \cdots, p_k]$ iff $\delta\Big(\{q_0, q_1, \cdots, q_{l-1}\}, a, x\Big) = \{p_0, p_1, \cdots, p_k\}$ is satisfied, where $\mu_i \in V_F, x \in V$. It manifests that δ_F is obtained by solving δ, i.e., $\bigcup_{i=0}^{l-1} \delta(q_i, a, x) = \{p_0, p_1, \cdots, p_k\}$, $p_i \subseteq Q$ $(i, k \in \{0, 1, \cdots, l-1\})$, the subset $\{p_0, p_1, \cdots, p_k\}$ implies $[p_0, p_1, \cdots, p_k]$, i.e., $\delta_F\Big([q_0, q_1, \cdots, q_{l-1}], a, \mu_i\Big) = [p_0, p_1, \cdots, p_k]$. It is obvious the $[q_0, \cdots, q_{l-1}]$ is the state of the FFA M_F.

Now, we prove the equality $L(FIA) = L(FFA)$.

With regard to the length of the string ω is proved as follows:

$$\delta_F(q_{0F}, \omega, \mu_i) = [q_0, q_1, \cdots, q_{l-1}] \Longleftrightarrow \delta(q_0, \omega, x) = \{q_0, q_1, \cdots, q_{l-1}\} \qquad (*)$$

where $\mu_i \in V_F, x \in V$.

If $|\omega| = 0$, i.e., $\omega = \varepsilon$, there is $\delta_F(q_{0F}, \varepsilon, 1) = q_{0F}$, $\delta(q_0, \varepsilon, 1) = \{q_0\}$, $\forall q_0 \in Q_0$, $q_{0F} \in Q_{0F}$. Since $Q_{0F} = [Q_0]$, the conclusion is affirmed.

If $|\omega| \leqslant k$, assume the above $(*)$ is true.

Then, if $|\omega| = k + 1$, i.e., $\omega = \omega_1 a$, $\omega_1 \in \Sigma^*$, $a \in \Sigma$, immediately, there is $\delta_F(q_{0F}, \omega_1 a, \mu_i) = \delta_F\big(\delta_F(q_{0F}, \omega_1, \mu_{i1}), a, \mu_{i2}\big)$ and $\delta_F(q_{0F}, \omega_1, \mu_{i1}) = [p_0, p_1, \cdots, p_i] \Leftrightarrow \delta(q_0, \omega_1, x_1) = \{p_0, p_1, \cdots, p_i\}$ is obtained by induction assumption.

Again, by the definition of the δ_F, $\delta_F\big([p_0, p_1, \cdots, p_i], a, \mu_{i2}\big) = [r_0, r_1, \cdots, r_j]$ is obtained and $\delta\big(\{p_0, p_1, \cdots, p_i\}, a, x_2\big) = \{r_0, r_1, \cdots, r_j\}$ is also satisfied. So, there is $\delta_F(q_{0F}, \omega_1 a, \mu_i) = [r_0, r_1, \cdots, r_j] \Longleftrightarrow \delta(q_0, \omega_1 a, x) = \{r_0, r_1, \cdots, r_j\}$ where $\mu_{i1}, \mu_{i2}, \mu_i \in V_F; x_1, x_2, x \in V$, $0 \leqslant i, j \leqslant l - 1$.

Finally, there must be $\delta_F(q_{0F}, \omega, \mu_i) \in G_F$ only if there is $\delta(q_0, \omega, x) \in G$. Thus, it proves that the equality $L(FIA) = L(FFA)$ holds.

Theorem 5.2. [8] The FFA is equally powerful as some L-nested system of DFA.

6 Stability and Convergence of FIA

Now, we discuss the stability of FIA. Let us divide the stability of FIA into two parts, which are the stability of the trained networks layer and the stability of the extraction layer of FIA respectively. For the stability of the trained networks, see [9][10]. Therefore, we now discuss only the stability of the extraction layer for FIA.

Definition 6.1. For the extraction FIA that has been obtained, assume the input vector to be $X(t)$ and the corresponding weights vector to be $w(t)$ in extraction layer. We call the fuzzy automaton to be stable, if there are always $||w(t) - X(t)|| < \varepsilon$ while $t > t_0$, for any $\varepsilon > 0$, $\exists t_0 > 0$.

Theorem 6.1. The extracted FIA is stable by the above extraction algorithm.

Proof. According to the above algorithm (4) for extraction of FIA and the definition 6.1, the conclusion is true.

Definition 6.2. Let V be a membership degree set of FIA. For $\mu_i \in V$, the neighborhood NV_i of μ_i is defined by: $NV_i = \big\{\mu_j | \mu_j \in V, |\mu_i - \mu_j| < \varepsilon\big\}$. There exists some μ_{ij} neighborhood NV_{ij}, for any $\mu_{lj} \in NV_{ij}$, if $\delta(q_i, \omega, \mu_{ij}) = q_j$ and $\delta(q_l, \omega, \mu_{lj}) = q_j$, then let q_l be the same as q_i, and μ_{lj} be the same as μ_{ij}, where ε is an error bound by requiring, $q_i, q_l, q_j \in Q$, $\omega \in \Sigma^*$.

Theorem 6.2. The states of FIA converge to some stable states.

Proof. According to the characteristics of the membership degree set V and the dividing algorithm of V in the above section 3.1, we can always obtain the finite membership degree values $\mu_i, i = 0, 1, \cdots, l-1$ for FIA. Thus, by the definition 6.2, the states of FIA can converge to some stable states.

(I) When the membership degree is any variable value in $V \subseteq [0, 1]$, let us divide the interval [0,1] into coordinate $n(n > 1)$ subinterval, and the interval end-point value θ_s is obtained, where $s = 0, 1, \cdots, n$.

Set $\mu_i = \frac{\theta_i + \theta_{i+1}}{2}$ if $\theta_i < x \leqslant \theta_{i+1}, i = 0, 1, \cdots, n-1$. ①

Where $\theta_0 = 0, \theta_n = 1, \theta_i = \frac{i}{n}, x \in V$. Let $\varepsilon_1 = \frac{1}{2n}$.

(II) When the membership degree values in V are close to the corresponding finite real value $\{\theta_0, \theta_1, \cdots, \theta_{m-1}\} \subseteq [0, 1]$, i.e., $|x - \theta_i| < \varepsilon_2$ for any $x \in V$, $i = 0, 1, \cdots, m-1$.

We set $\mu_i = \theta_i, i = 0, 1, \cdots, m-1$. ②

Assume $l = m$ or $l = n$, and let $\varepsilon = \varepsilon_1$ or $\varepsilon = \varepsilon_2$.

Thus, there are the corresponding l states q_i, $i = 0, 1, \cdots, l-1$.

By the definition 6.2, then, for $\forall \mu_i$, there exists its neighborhood NV_i, $i = 0, 1, \cdots, l-1$. Again, by $\theta_i < x \leqslant \theta_{i+1}$ in the above ① or $|x - \theta_i| < \varepsilon_2$, for any $x \in V$, there is always $x \in NV_i$, $i = 0, 1, \cdots, l-1$. Regulate the weights of the connecting neurons to make the state q_x of FIA satisfy $\delta(q_i, \omega, \mu_{ij}) = q_j$ and $\delta(q_x, \omega, \mu_{xj}) = q_j$, then q_x converge to q_i, $i = 0, 1, \cdots, l-1$.

So, the states of FIA converge to some stable states q_i, $i = 0, 1, \cdots, l-1$.

7 Simulation Results

In order to simplify in simulation, here we discuss the input $X(t)$ is a two-dimensional vector. The weight vector $w(t)$ is an eight-dimensional regulated vector in the networks of extraction layer. The activating function f is a gauss function. The simulation time T is 100 seconds.

When we calculate in the experiment, in order to make networks more quickly reflect the state distribution law of FIA on the whole, in general, the study speed α and the region D are chosen to be relatively bigger value at the beginning of training networks. Generally, the training time in the back period is 10-100 times that of training time of the fore period. The simulation results are shown in Fig.2 and Fig.3. From the Fig.2 known, the simulation results indicate that extraction

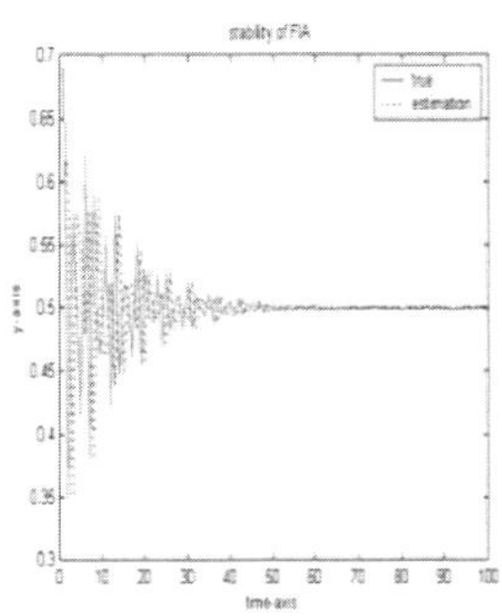

Fig. 2. Stability of FIA

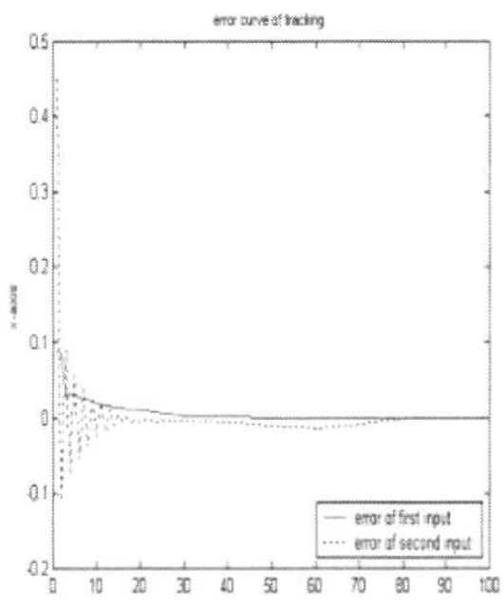

Fig. 3. Error curve of tracking

of FIA that has obtained is surely stable and convergent. By the Fig.3, the error curve for the difference of weights and inputs value in extraction layer reduces gradually and trends towards stability and convergence.

8 Conclusions

In this paper, these problems with respect to the definition, extraction, equivalence, convergence and stability of FIA are solved. The simulation results show that such extraction algorithm for FIA is surely stable and convergent. In conclusion, we have presented the basic ideas and algorithms for implementing stable recurrent networks and learning FIA in this paper. The network has similar capabilities for learning FIA as the analog FFA. These equivalent theorems imply that any two of FFA, FIA and L-nested systems of DFA are equally powerful. Then, the FIA is equivalent to the DFA at last.

Now, some questions require to be solved in the future: In order to learn better FIA, it is difficult how the states of FIA are minimized appropriate degree, i.e., how a new appropriate FFA will be obtained, and let it be equal to the FIA. It is difficult how the number of neuron and the layer of network are selected and designed for extracting the more stable FIA. These problems to be solved are worth being studied.

References

1. Blanco,A, Delgado,M, Pegalajar,M.C.: Identification of fuzzy dynamic systems using max-min recurrent neural networks. Fuzzy Set and Systems. 1 (2000) 63-70
2. Kosmatopoulos, E.B., Christodoulou, M.A.: Neural networks for identification of fuzzy dynamical systems: An Application to identification of vehicle highway systems. In Proc. 4th IEEE Mediterranean Symp. New Directions in control and Automation. 1 (1996) 23-38
3. Kosmatopoulos, E.B., Polycarpou, M.M., Christodoulou, M.A., etc.: High-order neural networks for identification of dynamical systems. IEEE Trans. Neural networks. 6 (1995) 422-431
4. Kosmatopoulos, E.B., Christodoulou, M.A.: Recurrent neural networks for approximation of fuzzy dynamical systems. Int.J. Intell. Control Syst. 1 (1996) 223-233
5. Omlin,C.W, Thornber,K.K, Giles,C.L.: Fuzzy finite state automata can be deterministically encoded into recurrent neural networks. IEEE Trans. Fuzzy Syst. 6 (1998) 76-89
6. Giles, C.L., Miller, C.B., Chen, D., etc.: Learning and extracting finite state automata with second-order recurrent neural networks. Neural Computation. 4 (1992) 393-405
7. Lalande,A, Jaulent,M.: A fuzzy automaton to detect and quantify fuzzy artery lesions from arteriograms. In: Proceedings of the Sixth International Conference IPMU'96. In Canada. 3 (1996) 1481-1487
8. Radim, Bělohlávek.: Determinism and fuzzy automata. Information Sciences. 143 (2002) 205-209
9. Zhang.Naiyao, Yan.Pingfan.: Neural Networks and Fuzzy Control. Tsing-Hua University Press (1998) (Chinese)
10. Yuan.Jiangwen, Shankou. Heng, Gu.Longsi.: Neural network and fuzzy signal processing. Science Press (2003) (Chinese)

Practical Stability Analysis and Synthesis of a Class of Uncertain T-S Fuzzy Systems

Linna Zhou, Qingling Zhang, and Chunyu Yang

Institute of Systems Science, Northeastern University,
Liaoning Shenyang 110004, China
ycyyyang@sina.com

Abstract. This paper is devoted to the investigation of practical stability of a class of uncertain T-S fuzzy systems. The uncertainties satisfy the so-called "matching conditions" and the bound is known. In this paper, a sufficient condition for practical stability of dynamic systems is first presented. Then a controller design method for uncertain T-S fuzzy systems is derived, which is based on the above sufficient condition and the quadratic stability condition for T-S fuzzy systems. The advantage of our approach is that the controller we present contains a tuning parameter which can improve systems performance and the main computation process can be conveniently performed by LMI toolbox. Finally, an example is given to illustrate the application of the obtained approach.

1 Introduction

During the past several years, fuzzy control has emerged as one of the most active and fruitful areas. Especially, since T-S model [1] was established by Takagi and Sugeno in 1985, more and more significative results and successful applications have been published. T-S fuzzy systems are nonlinear systems described by a set of IF-THEN rules which gives a local linear representation of an underlying system. Such models can approximate a wide class of nonlinear systems. Feng etc.[2], and Cao etc.[3] have proved that the T-S fuzzy system can approximate any continuous functions in R^n at any preciseness.

Stability is one of the most important concepts concerning the properties of control systems and some useful stability analysis techniques have come into being. These stability conditions of T-S fuzzy systems are all obtained from Lyapunov method. In [4] and [5], a positive definite matrix P must be found to satisfy all the subsystems' Lyapunov functions. In [2], [3] and [6], T-S models are reconstructed by the so-called "extreme" subsystems, then stability analysis is given via piecewise Lyapunov function. The above two classs of stability analysis methods did not consider sufficiently the interactions among the subsystems, so the stability conditions have more conservation. While in [7], the authors consider the interactions among the subsystems, and gave more simple and less conservative sufficient conditions in the terms of linear matrix inequalities (LMIs). All of these techniques are in the sense of Lyapunov stability, which qualitatively analysis systems performance. However, in practical engineering systems, uncertain external disturbance tends to introduce oscillation, even causes instability.

L. Wang et al. (Eds.): FSKD 2006, LNAI 4223, pp. 11–20, 2006.

In these cases, systems are usually not stable in the sense of Lyapunov stability, but sometimes their performance may be acceptable in practice just because they oscillate sufficiently near a mathematically unstable course. Furthermore, in some cases, though a system is stable or even asymptotically stable, it can not be acceptable in practice engineering just because the stable domain or the attraction domain is not large enough. To deal with these situations, the concept of practical stability [8, 9, 10], which is derived from the so called "finite time stability" and "technological stability" and quantitatively analysis systems performance, is more useful.

In this paper, we investigate a class of uncertain T-S fuzzy systems, for which the uncertainties satisfy "matching conditions" and the bound is known. Using Lyapunov functions, we first derive a sufficient condition for a nonlinear uncertain system to be practically stable. Then, by such a condition together with the quadratic stability condition in [7], we present a controller design method which can compensate the effect of the uncertainties. The controller contains a tuning parameter and we can improve systems performance by adjusting the parameter, which is illustrated by the example in the end of this paper.

The notation that is used here is standard in most respects. We use R to denote the set of real numbers. R^n and $R^{n1\times n2}$ are the obvious extensions to vectors and matrices of the specified dimensions. Let I or I_r denotes the identity matrix with appropriate dimension. M is a matrix with proper dimension, M^T stands for the transpose of M. For symmetric matrices P and Q, the notation $P > Q(P \geq Q)$ means that $P - Q$ is positive definite(positive semi-definite). $\|\cdot\|$ denotes the Euclidean norm of a vector or matrix. For given scalar functions $r_1(\cdot)$ and $r_2(\cdot)$, $(r_1 \circ r_2)(\cdot)$ denotes the function $r_1(r_2(\cdot))$.

2 Preliminary Knowledge

In this section, we first recall the definition of practical stability of dynamic systems and give a sufficient condition. Then we retrieve the continuous T-S fuzzy system and the related definitions and stability conditions.

2.1 Practical Stability Analysis of Dynamic Systems

Consider the following uncertain nonlinear dynamic system

$$\begin{aligned} \dot{x}(t) &= f(x,t) + B(t)(u(t) + w(t)) \\ x(t_0) &= x_0 \end{aligned} \tag{1}$$

where $t \in R$ is time, $x(t) \in R^n$ is the state, $u(t) \in R^m$ is the control, $w(t) \in R^m$ is uncertain element, $f(\cdot) : R^n \times R \rightarrow R^n$ and $B(\cdot) : R \rightarrow R^{n\times m}$ are known. $\|w(t)\| \leq \sigma$, $\sigma \in R^+$ is a constant.

Remark 1. The uncertainties are called to satisfy the so-called "matching conditions" if the input and the uncertain element have the same channel, that is, they have the same coefficient matrix.

Definition 1. *[11] The system (1) is said to be practically stable if, given* (λ, A) *with* $0 < \lambda < A$*, we have* $\|x_0\| < \lambda$ *implies* $\|x(t)\| < A$, $t > t_0$ *for some* $t_0 \in R^+$.

Assume there is a Lyapunov function $V(\cdot) \in C^1(R^n, R^+)$, such that for all $x \in R^n$

$$r_1(\|x\|) \leq V(x) \leq r_2(\|x\|)$$

$$\frac{d(V(x))}{dt}|_{(1)} \leq -r_3(\|x\|)$$

where $r_i(\cdot)$, $i = 1, 2, 3$ are continuous, strictly increasing functions, and

$$r_i(0) = 0,\ r_i(r) \to \infty,\ r \to \infty,\ i = 1, 2.$$

According to the above assumption, if

$$r_3(\eta) > 0,\ \forall \eta \in R^+$$

then system (1) is asymptotically stable obviously.

This paper addresses the case that

$$\exists \eta > 0,\ r_3(\eta) = 0$$

The following lemma will be used in the sequel.

Lemma 1. *If there exists a* $\eta \in R^+$ *such that*

$$\eta = R_3^{-1}(0)$$

then, for any $0 < l \leq \eta$*, we have*

$$\|x_0\| \leq l \Rightarrow \|x(t)\| \leq d(\eta),\ \forall t \in [t_0, \infty],$$

where $d(\eta) = (r_1^{-1} \circ r_2)(\eta)$.

Proof. Since $r_1(\cdot)$ and $r_2(\cdot)$ are increasing functions and $r_1(\cdot) \leq r_2(\cdot)$, we have

$$\eta \leq d(\eta).$$

For given $0 < l \leq \eta$, we have

$$l \leq \eta \leq d(\eta).$$

Now, suppose there is a $t_2 > t_0$ such that

$$\|x(t_2)\| > d(\eta).$$

Since $x(\cdot)$ is continuous and

$$\|x_0\| \leq l \leq d(\eta) < \|x(t_2)\|$$

there is a $t_1 \in [t_0, t_2)$ such that

$$\|x(t_1)\| = \eta \text{ and } \|x(t)\| \geq \eta, \forall t \in [t_1, t_2].$$

So, we have

$$\begin{aligned} r_1(\|x(t_2)\|) &\leq V(x(t_2)) \\ &= V(x(t_1)) + \int_{t_1}^{t_2} \dot{V}(x(\tau))d\tau \\ &\leq r_2(\|x(t_1)\|) + \int_{t_1}^{t_2} -r_3(\|x(\tau)\|)d\tau \\ &\leq r_2(\eta) + \int_{t_1}^{t_2} -r_3(\eta)d\tau \\ &= r_2(\eta). \end{aligned}$$

Hence,

$$\|x(t_2)\| \leq (r_1^{-1} \circ r_2)(\eta) = d(\eta)$$

which contradicts the above assumption. □

Remark 2. This lemma is a improvement on the theorem in reference [11].

Theorem 1. *System (1) is practically stable with respect to*(λ, A)*, if there exists* $\eta > 0$*, such that*

$$\lambda \leq \eta, \text{ and } A \geq d(\eta).$$

Theorem 1 is easy to prove by Lemma 1.

2.2 T-S Fuzzy Model and Its Stability Conditions

Consider the T-S fuzzy system described by the following IF-THEN rules:

$$IF\ \xi_1(t)\ is\ M_{1i}\ and\ \xi_2(t)\ is\ M_{2i} \ldots\ and\ \xi_p(t)\ is\ M_{pi},\ THEN$$

$$\dot{x} = A_i x(t) + B_i(u(t) + w(t)), \tag{2}$$

where $x(t) \in R^n$ is the state , $u(t) \in R^m$ is the input, $w(t) \in R^m$ is the uncertain element, $A_i \in R^{n\times n}$, $B_i \in R^{n\times m}$, $\xi_1(t),\ \xi_2(t), \ldots, \xi_p(t)$ are premise variables, we set $\xi = (\xi_1, \ldots, \xi_p)^T$. It's assumed that the premise variables do not depend on control variables and the uncertain element. Then, the state equation and the output are defined as follows:

$$\dot{x} = \sum_{i=1}^{r} h_i(\xi(t))(A_i x(t) + B_i(u(t) + w(t))), \tag{3}$$

where

$$h_i(\xi(t)) = \frac{\beta_i(\xi(t))}{\sum_{i=1}^{r} \beta_i(\xi(t))},\ \beta_i(\xi(t)) = \Pi_{j=1}^{p} M_{ij}(\xi_j(t)) \geq 0,\ \sum_{i=1}^{r} h_i(\xi(t)) = 1,$$

and $M_{ij}(\cdot)$ is the membership of fuzzy set M_{ij}.

Definition 2. *[12] For T-S fuzzy system (3), when $u(t) \equiv 0$, $w(t) \equiv 0$, if there exists $\alpha > 0$ and a positive-definite matrix X such that $\dot{V}(x(t)) \leq -\alpha x^T(t)x(t)$, where $V(x(t)) = x^T(t)Xx(t)$, then T-S fuzzy system (3) is called quadratically stable.*

The following lemma is Theorem 1 of [7].

Lemma 2. *[7] If there exist matrices M_i, Z, Y_{ij}, where Z is a positive-definite matrix, Y_{ii} are symmetric matrices, $Y_{ji} = Y_{ij}^T$, $i,\ j = 1, 2, \ldots, r$, $i \neq j$, satisfy the following LMIs:*

$$ZA_i^T + M_i^T B_i^T + A_i Z + B_i M_i < Y_{ii}, \tag{4}$$

$$\begin{aligned} &A_i Z + A_j Z + ZA_i^T + ZA_j^T + B_i M_j + B_j M_i + M_j^T B_i^T + M_i^T B_j^T \\ &\leq Y_{ij} + Y_{ij}^T,\ i \neq j, \end{aligned} \tag{5}$$

$$\begin{bmatrix} Y_{11} & \ldots & Y_{1r} \\ \vdots & \ddots & \vdots \\ Y_{r1} & \ldots & Y_{rr} \end{bmatrix} < 0. \tag{6}$$

Then, for T-S fuzzy system (3) when $w(t) \equiv 0$, the state feedback

$$u(t) = \sum_{i=1}^{r} h_i(\xi(t)) F_i x(t) \tag{7}$$

stabilizes the closed-loop system

$$\dot{x}(t) = \sum_{i=1}^{r} \sum_{j=1}^{r} h_i(\xi(t)) h_j(\xi(t)) (A_i + B_i F_j) x(t), \tag{8}$$

where $F_i = M_i Z^{-1}$, $i = 1, 2, \ldots, r$.

3 Fuzzy Controller Design

T-S fuzzy system (3) is rewritten as follows,

$$\dot{x}(t) = \sum_{i=1}^{r} h_i(\xi(t)) A_i x(t) + \sum_{i=1}^{r} h_i(\xi(t)) B_i (u(t) + w(t)). \tag{9}$$

Let the state feedback controller is

$$u = u_1 + u_2, \tag{10}$$

where u_1 is described by (7) and

$$u_2 = -\frac{r_0^2}{2} \sum_{i=1}^{r} h_i(\xi(t)) B_i^T X x(t), \tag{11}$$

where $X = Z^{-1}$, $r_0 \in R^+$ is a constant. Then we have the following closed-loop system

$$\dot{x}(t) = \sum_{i=1}^{r}\sum_{j=1}^{r} h_i(\xi(t))h_j(\xi(t))(A_i + B_iF_j)x(t) + \sum_{i=1}^{r}\sum_{j=1}^{r} h_i(\xi(t))h_j(\xi(t))B_i(-\frac{r_0^2}{2}B_j^T Xx(t) + w(t)). \quad (12)$$

Now, consider (4), (5), (6), pre- and postmultiply (4) and (5) with X and pre- and postmultiply (6) with $diag(X, \ldots, X)$, we get

$$A_i^T X + F_i^T B_i^T X + XA_i + XB_iF_i < XY_{ii}X, \quad (13)$$

$$X(A_i + B_iF_j + A_j + B_jF_i) + (A_i^T + F_j^T B_i^T + A_j^T + F_i^T B_j^T)X \leq XY_{ij}X + XY_{ij}^T X, \quad (14)$$

$$\begin{bmatrix} XY_{11}X & \ldots & XY_{1r}X \\ \vdots & \ddots & \vdots \\ XY_{r1}X & \ldots & XY_{rr}X \end{bmatrix} < 0. \quad (15)$$

By (15), we get

$$\begin{bmatrix} XY_{11}X & \ldots & XY_{1r}X \\ \vdots & \ddots & \vdots \\ XY_{r1}X & \ldots & XY_{rr}X \end{bmatrix} < -\alpha I,$$

where

$$\alpha = -\lambda_{min} \begin{bmatrix} XY_{11}X & \ldots & XY_{1r}X \\ \vdots & \ddots & \vdots \\ XY_{r1}X & \ldots & XY_{rr}X \end{bmatrix}$$

Note, $\lambda_{max}(\cdot)$ and $\lambda_{min}(\cdot)$ denote the maximal and minimal eigenvalue of the matrix separately. Then the time derivative of $V(x(t)) = x^T(t)Xx(t)$ along the state of system (9) is given as follows:

$$\begin{aligned}\dot{V}(x(t)) &= \dot{x}^T(t)Xx(t) + x^T(t)X\dot{x}(t) \\ &= \sum_{i=1}^{r}\sum_{j=1}^{r} h_i(\xi(t))h_j(\xi(t))x^T(t)(A_i^T X + F_j^T B_i^T X + XA_i + XB_iF_j)x(t) \\ &- \frac{r_0^2}{2}\sum_{i=1}^{r}\sum_{j=1}^{r} h_i(\xi(t))h_j(\xi(t))x^T(t)(XB_jB_i^i X + XB_iB_j^T X)x(t) \\ &+ \sum_{i=i}^{r} h_i(\xi(t))(w^T(t)B_i^T Xx(t) + x^T(t)XB_iw(t)) \\ &= \sum_{i=1}^{r} h_i^2(\xi(t))x^T(t)(A_i^T X + F_i^T B_i^T X + XA_i + XB_iF_i)x(t)\end{aligned}$$

$$
\begin{aligned}
&= \sum_{i=1}^{r} h_i^2(\xi(t))x^T(t)(A_i^T X + F_i^T B_i^T X + XA_i + XB_iF_i)x(t) \\
&+ \sum_{i=1}^{r}\sum_{i<j}^{r} h_i(\xi(t))h_j(\xi(t))((A_i^T + F_j^T B_i^T + A_j^T + F_i^T B_j^T)X \\
&+ X(A_i + B_iF_j + A_j + B_jF_i))x(t) \\
&- r_0^2 \sum_{i=1}^{r}\sum_{j=1}^{r} h_i(\xi(t))h_j(\xi(t))x^T(t)XB_iB_j^T Xx(t) \\
&+ w^T(t)(\sum_{i=1}^{r} h_i(\xi(t))B_i^T Xx(t)) + (\sum_{i=1}^{r} h_i(\xi(t))B_i^T Xx(t))^T w(t) \\
&= \sum_{i=1}^{r} h_i^2(\xi(t))x^T(t)(A_i^T X + F_i^T B_i^T X + XA_i + XB_iF_i)x(t) \\
&+ \sum_{i=1}^{r}\sum_{i<j}^{r} h_i(\xi(t))h_j(\xi(t))x^T(t)((A_i^T + F_j^T B_i^T + A_j^T + F_i^T B_j^T)X \\
&+ X(A_i + B_iF_j + A_j + B_jF_i))x(t) - (\frac{1}{r_0}w(t) - r_0\sum_{i=1}^{r} h_i(\xi(t))B_i^T X \times \\
&x\ (t))^T(\frac{1}{r_0}w(t) - r_0\sum_{i=1}^{r} h_i(\xi(t))B_i^T Xx(t)) + \frac{1}{r_0^2}w^T(t)w(t) \\
&\le -\alpha\|x(t)\|^2 + \frac{1}{r_0^2}w^T(t)w(t).
\end{aligned}
$$

We assume

$$
r_1(\|x\|) = \lambda_{min}(X)\|x\|^2,\ r_2(\|x\|) = \lambda_{max}(X)\|x\|^2,\ r_3(\|x\|) = \alpha\|x\|^2 - \frac{\sigma^2}{r_0^2}.
$$

Then according to Lemma 1 and Theorem 1,

$$
\eta = r_3^{-1}(0) = \frac{\sigma}{r_0\sqrt{\alpha}}. \tag{16}
$$

Based on the above discussion, we gain the following theorem.

Theorem 2. *For given $\lambda > 0$, system (9) is practically stabilized by (7) with respect to $(\lambda,\ d(\eta))$, if conditions in Lemma 2 hold and there exists $r_0 > 0$ such that*

$$
\lambda < \eta.
$$

Remark 3. By (16), η is strictly decreasing with respect to r_0, then $d(\eta)$ is strictly decrease with respect to r_0 since $d(\eta)$ is increasing with respect to η. In addition, the smaller $d(\eta)$ means smaller bound of the state. Thus, in practical engineering systems, we can improve systems performance by adjusting the parameter r_0, which can be illustrated by the following example.

4 An Example

To illustrate the application of the given approach, we consider a single pendulum system with uncertainty which is described as follows:

$$\begin{aligned}\dot{x}_1(t) &= x_2(t),\\ \dot{x}_2(t) &= -a\sin(x_1(t)) + u(t) + \Delta b\sin(t),\end{aligned}$$

where $a > 0$ is a constant, $\Delta b > 0$ is uncertain elements and Δb is bounded. Let $w(t) = \Delta b\sin(t)$, then we have

$$\begin{aligned}\dot{x}_1(t) &= x_2(t),\\ \dot{x}_2(t) &= -a\sin(x_1(t)) + u(t) + w(t).\end{aligned} \tag{17}$$

To design a fuzzy controller for the uncertain nonlinear system (17) we should construct a T-S model first. Using the idea of "sector nonlinearity" [10], the membership function can be calculated as

$$h_1 = \begin{cases} \dfrac{\sin(x_1(t)) - (2/\pi)x_1(t)}{(1-2/\pi)x_1(t)}, & x_1(t) \neq 0 \\ 1, & x_1(t) = 0 \end{cases}$$

$$h_2 = \begin{cases} \dfrac{x_1(t) - \sin(x_1(t))}{(1-2/\pi)x_1(t)}, & x_1(t) \neq 0 \\ 0, & x_1(t) = 0 \end{cases}$$

We name the membership functions "Zero" and "Not Zero" respectively. Then system (17) can represented by the following T-S model:

$$\begin{aligned}&Model\ \ Rule\ 1:\ \ IF\ x_1(t)\ is\ \text{"}Zero\text{"}\\ &THEN\ \dot{x}(t) = A_1x(t) + B_1(u(t) + w(t)),\\ &Model\ \ Rule\ 2:\ \ IF\ x_1(t)\ is\ \text{"}Not\ Zero\text{"}\\ &THEN\ \dot{x}(t) = A_2x(t) + B_2(u(t) + w(t)).\end{aligned}$$

Here,

$$A_1 = \begin{bmatrix} 0 & 1 \\ -a & 0 \end{bmatrix},\ A_2 = \begin{bmatrix} 0 & 1 \\ -(2/\pi)a & 0 \end{bmatrix},\ B_1 = B_2 = \begin{bmatrix} 0 \\ 1 \end{bmatrix}.$$

Let $a = 1,\ \Delta b = 1$. By Theorem 2, we have the following solution matrices:

$$Z = \begin{bmatrix} 22.8850 & -11.3012 \\ -11.3012 & 22.8850 \end{bmatrix},\ M_1 = \begin{bmatrix} 0.0000 & -23.7326 \end{bmatrix},$$

$$M_2 = \begin{bmatrix} -8.3160 & -19.6259 \end{bmatrix},\ Y_{11} = \begin{bmatrix} -11.3012 & 0 \\ 0 & -12.4314 \end{bmatrix},$$

$$Y_{12} = \begin{bmatrix} -8.1934 & 0 \\ 0 & -8.1934 \end{bmatrix},\ Y_{22} = \begin{bmatrix} -11.3012 & 0 \\ 0 & -12.4314 \end{bmatrix},$$

$$F_1 = \begin{bmatrix} -0.6773 & -1.3715 \end{bmatrix}, \ F_2 = \begin{bmatrix} -1.0407 & -1.3715 \end{bmatrix}, \ \alpha = 0.0031.$$

Then

$$\eta = \frac{17.9355}{r_0}, \ r_1(\|x\|) = 0.0293\|x\|^2, \ r_2(\|x\|) = 0.0863\|x\|^2,$$

$$d(\eta) = (r_1^{-1} \circ r_2)(\eta) = \sqrt{\frac{0.0863}{0.0293}\eta} = \frac{30.8115}{r_0}.$$

Set $\lambda = 0.1, \ x_0 = [-0.2 \ 0.2]^T$. Then

if $r_0 = 10$, we have, $\eta = 1.7953, \ d(\eta) = 3.0812$, while, if $r_0 = 50$, we have, $\eta = 0.3591, \ d(\eta) = 0.6162$.

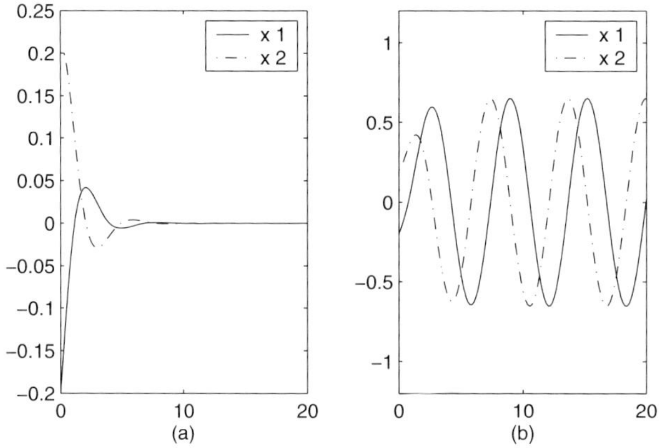

Fig. 1. (a)State responses with $x_0 = [-0.2 \ 0.2]^T, \ w(t) = 0, \ r_0 = 0$; (b)State responses with $x_0 = [-0.2 \ 0.2]^T, \ w(t) = sin(t), \ r_0 = 0$

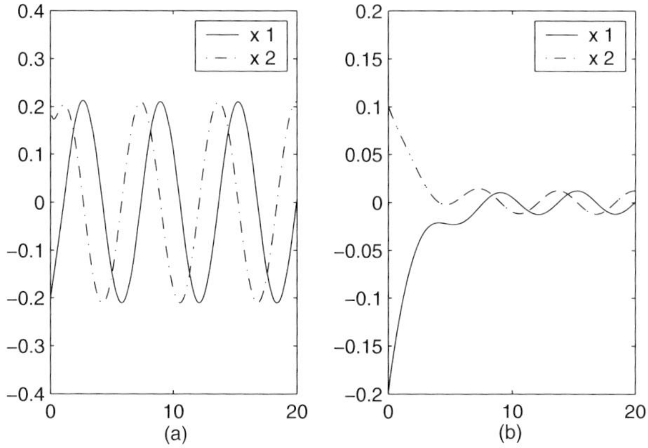

Fig. 2. (a)State responses with $x_0 = [-0.2 \ 0.2]^T, \ w(t) = \sin(t), \ r_0 = 10$; (b)State responses with $x_0 = [-0.2 \ 0.2]^T, \ w(t) = \sin(t), \ r_0 = 50$

Fig.1(a) shows the response of system (17) with $w(t) = 0$, $r_0 = 0$; Fig.1(b) shows the response of system (17) with $w(t) = \sin(t)$, $r_0 = 0$; Fig.2(a) shows the response of system (17) with $w(t) = \sin(t)$, $r_0 = 10$; Fig.2(b) shows the response of system 17 with $w(t) = \sin(t)$, $r_0 = 50$.

5 Conclusions

The problems of practical stability analysis and synthesis are considered for a class of uncertain T-S fuzzy system. Based on the well known PDC controller, a new type controller containing a tuning parameter is designed. By adjusting the parameter, the closed-loop systems' performance can be improved. The numerical example discussed in the end of this paper shows the validity of the given method.

References

1. Takagi, T., Sugeno, M.: Fuzzy identification of systems and its applications to modelling and control. IEEE Trans.on Systems, Man,and Cybernetics, **15** (1985) 116 - 132.
2. Feng, G., Cao, S. G., Rees, N. W., and Chak C.K.: Design of fuzzy control systems with guaranteed stability. Fuzzy Sets Syst., **85** (1997) 1-10.
3. Cao, S. G., Rees, N. W., and Feng, G.: Stability analysis and design for a class of continuous-time fuzzy control systems. Int.J. Control. **64** (1996) 1069-1087.
4. Wang, H. O., Tanaka, K., and Griffin, M.: Parallel distributed compensation of nonlinear systems by Takagi and Sugeno's Model. Proceedings of FUZZIEEE'95 (1995) 531-538.
5. Wang, H. O., Tanaka, K., and Griffin, M.: An approach to fuzzy control of nonlinear systems: stability and design issues. IEEE Trans. Fuzzy Syst., **4** (1996) 14-23.
6. Zhang, J. M., Li, R. H., and Zhang, P.A.: Stability analysis and systematic design of fuzzy control systems. Fuzzy sets and Systems, **120** (2001) 65-72.
7. Liu, X. D.,Zhang, Q. L.: An approaches to quadratic stability conditions and H controller designs for T-S fuzzy systems. Fuzzy Systems, **11** (2003) 830-839.
8. LaSalle, J. P., Lefschetz, S.: Stability by Lyapunov's direct method with applications, Academic Press New York (1961).
9. Lakshmikantham, V., Leela, S., and Martynyuk, A. A.: Practical stability of nonlinear systems, World Scientific Singapore (1990).
10. Lakshmikantham, V., Matrosov, V. M., and Sivasundaram, S.: Vector Lyapunov functions and stability analysis of nonlinear systems, Kluwer Academic Dordrecht Boston London (1991).
11. Corless, M., Leitmann, G.: Continuous state feedback guaranteeing uniform ultimate boundedness for uncertain dynamic systems.IEEE Tran Ac, **26** (1981) 1139-1144.
12. Taniguchi, T., Tanaka, K., Wang, H. O.: Fuzzy descriptor systems and nonlinear model following control.IEEE Trans. Fuzzy Syst., **8** (2000) 442-452.

Robust H_∞ Fuzzy Controller for Uncertain Nonlinear Systems with Time-Varying Delayed State

Taek Ryong Kim[1], Jin Bae Park[1], and Young Hoon Joo[2]

[1] Yonsei University, Seodaemun-gu, Seoul, 120-749, Korea
princebear@korea.com, jbpark@control.yonsei.ac.kr
[2] Kunsan National University, Kunsan, Chunbuk, 573-701, Korea
yhjoo@kunsan.ac.kr

Abstract. This paper addresses a robust H_∞ stabilization problem to uncertain nonlinear systems with time-varying delayed state via fuzzy static output feedback. We employ the Takagi–Sugeno (T–S) fuzzy model to represent the nonlinear system in both continuous-time and discrete-time cases. Sufficient conditions, which are in the format of bilinear matrix inequalities (BMI)s, are derived for robust stabilization and disturbance attenuation of the closed-loop fuzzy control system in both cases. Then, the BMIs are converted to linear matrix inequalities (LMI)s through some manipulations. The effectiveness of the proposed controller design methodology is demonstrated through a numerical simulation.

1 Introduction

In the industry, most plants have a strong nonlinearity and uncertainty. Thus, they post additional difficulties to the control theory of general nonlinear systems and the design of their controllers. Moreover, when nonlinear systems are controlled, time-varying delay is generally occurred and disturbance interrupts. Therefore to deal with these problems, many efforts have done.

There are many papers that propose the control methodology of the linear system with time-delay. But for the nonlinear system with time-delay, only few papers exist. This arises from the complexity of the nonlinear system. To overcome this difficulty, various schemes have been developed in the last two decades, among which a successful approach is fuzzy control.

Cao et al. first proposed the Takagi-Sugeno (T–S) fuzzy model with time-delay that represents the nonlinear system with time-delay and analyzed its stability [10]. Based on this, Lee et al. proposed a dynamic output feedback robust H_∞ control method for a class of uncertain fuzzy systems with time-varying delay [11]. In discrete-time case, Xu et al. proposed a dynamic output feedback control methodology for uncertain discrete-time fuzzy system with time-delay [13]. It should be noted that the problem of robust H_∞ control for nonlinear systems with time-delay via static output feedback is still open and remains unsolved, which motivates the present study.

L. Wang et al. (Eds.): FSKD 2006, LNAI 4223, pp. 21–30, 2006.

Lo et al. proposed the robust static output feedback control method of the nonlinear system without time-delay via fuzzy control approach [9]. In [9], controller can be easily designed by solving several linear matrix inequalities (LMIs). In this paper, we extend the methodology that proposed in [9] to the nonlinear system with time-varying delay for the robust stabilization of it via static output feedback strategy. In other words, we propose the methodology for designing the H_∞ fuzzy controller that can robustly control the nonlinear system with time-varying delay subject to external disturbances. To this end, we first represent the nonlinear system with time-delay to the T–S fuzzy model with time-delay as did in [10]. Then, parallel distributed compensation technique is applied for the design of the static output feedback fuzzy controller. After selecting one Lyapunov function, we derive some sufficient conditions for stability of the fuzzy system. But the conditions are composed of bilinear matrix inequalities (BMIs). By using similarity transform and congruence transform technique we convert it to LMIs . Therefore the H_∞ fuzzy controller can be easily designed by many current convex optimization algorithm tools.

The remainder of the paper is organized as follows: following the introduction, problem formulation is done in Section 2. In Section 3, the sufficient conditions for making the T–S fuzzy model with time-varying delay asymptotically stable in the H_∞ sense are derived for both the continuous-time case and discrete-time case. To show the effectiveness of the proposed method, two simulation examples are presented in Section 4. Finally, some conclusions are drawn in Section 5. In this paper, σx indicates a continuous-time state $\dot{x}(t)$ or discrete-time state $x(t+1)$ whenever appropriate. $C_{n,t} = C([-t,0], R^n)$ denotes the Banach space of continuous vector functions mapping the interval $[-t,0]$ into R^n with the topology of uniform convergence. The superscript $()^{'}$ denotes the transpose of a matrix and the symbol $*$ denotes a symmetric term in a block matrix representation.

2 Problem Formulation

The T–S fuzzy model is generally known as the universal approximator of nonlinear systems. We consider nonlinear systems represented by the following T–S fuzzy model with time-delay. *Plant Rule i:*

$$\begin{aligned}
&\text{IF } \theta_1(t) \text{ is } M_{i1} \ \cdots \text{ and } \theta_n(t) \text{ is } M_{in} \\
&\text{THEN } \sigma x(t) = (A_i + \Delta A(t))x(t) + (A_{di} + \Delta A_d(t))x(t-d(t)) \\
&\qquad\qquad + (B_{1i} + \Delta B_1(t))\omega(t) + (B_{2i} + \Delta B_2(t))u(t) \\
&\qquad\quad z(t) = (C_i + \Delta C(t))x(t) + (C_{di} + \Delta C_d(t))x(t-d(t)) \\
&\qquad\qquad + (D_{1i} + \Delta D_1(t))\omega(t) + (D_{2i} + \Delta D_2(t))u(t) \\
&\qquad\quad y(t) = Ex(t) \\
&\qquad\quad x(t) = \Lambda(t), \qquad t \in [-d_0,\ 0]
\end{aligned} \tag{1}$$

Where M_{ij} is the fuzzy set, $x(t) \in R^n$ is the state vector, $\omega(t) \in R^{q1}$ is unknown but the energy-bounded disturbance input, $u(t) \in R^{q2}$ is the controlled input,

$z(t) \in R^s$ is the controlled output, $(A_i, A_{di}, B_{1i}, C_i, C_{di}, D_{1i}, D_{2i})$ are some constant matrices of compatible dimensions, $i = 1, 2, \cdots, r$, in which r is the number of IF-THEN rules, $\Lambda(t) \in C_{n,d}$ is a vector-valued initial continuous function, and $\theta(t) = \left[\theta_1(t)\ \theta_2(t) \cdots \theta_n(t)\right]$ are the premise variables. It is assumed that the premise variables do not depend on the input variables $u(t)$ explicitly. The time-varying delay, $d(t)$, is assumed that

$$0 \le d(t) \le d_0, \qquad \dot{d}(t) \le \beta < 1. \tag{2}$$

The time-varying matrices are defined as follows:

$$\begin{pmatrix} \Delta A(t)\ \Delta A_d(t)\ \Delta B_1(t)\ \Delta B_2(t) \\ \Delta C(t)\ \Delta C_d(t)\ \Delta D_1(t)\ \Delta D_2(t) \end{pmatrix} = \begin{pmatrix} M\Delta(t) & (N_1\ N_2\ N_3\ N_4) \\ M_z\Delta_z(t) & (N_{z1}\ N_{z2}\ N_{z3}\ N_{z4}) \end{pmatrix} \tag{3}$$

where $(M, M_z, N_1, N_2, N_3, N_4, N_{z1}, N_{z2}, N_{z3}, N_{z4})$ are known real constant matrices, and (Δ, Δ_z) are unknown matrix functions with Lebesgue-measurable elements and satisfy $\Delta'(t)\Delta(t) \le I, \Delta_z'(t)\Delta_z(t) \le I$ in which I is the identity matrix of appropriate dimension.

The defuzzified output of (1) is represented as follows:

$$\begin{aligned} \sigma x(t) &= \sum_{i=1}^{r} \mu_i(\theta(t))[(A_i + \Delta A(t))x(t) + (A_{di} + \Delta A_d(t))x(t - d(t)) \\ &\quad + (B_{1i} + \Delta B_1(t))\omega(t) + (B_{2i} + \Delta B_2(t))u(t)] \\ z(t) &= \sum_{i=1}^{r} \mu_i(\theta(t))[(C_i + \Delta C(t))x(t) + (C_{di} + \Delta C_d(t))x(t - d(t)) \\ &\quad + (D_{1i} + \Delta D_1(t))\omega(t) + (D_{2i} + \Delta D_2(t))u(t)] \\ y(t) &= Ex(t) \end{aligned} \tag{4}$$

where

$$\mu_i(\theta(t)) = \frac{\omega_i(\theta(t))}{\sum_{i=1}^{r} \omega_i(\theta(t))}, \qquad \omega_i(\theta(t)) = \prod_{j=1}^{n} M_{ij}(\theta(t)), \qquad j = 1, 2, \cdots, n$$

$$\sum_{i=1}^{r} \mu_i(\theta(t)) = 1, \qquad \mu_i(\theta(t)) \ge 0, \qquad i = 1, 2, \cdots, r.$$

A static output feedback fuzzy controller is constructed by using parallel distributed compensation technique [7] and has the following defuzzified form:

$$u(t) = \sum_{i=1}^{r} \mu_i(\theta(t))K_i y(t), \tag{5}$$

where K_i's are constant control gains to be determined. For simplicity, we represent $\mu_i(\theta(t))$ as μ_i and abbreviate the time index, t, in time-varying matrices.

Substituting (5) into (4), the closed-loop system is obtained as follows:

$$\begin{aligned}\sigma x(t) &= [(A_\mu + B_{2\mu}K_\mu E + M\Delta(N_1 + N_4K_\mu E)]x(t) \\ &\quad + (A_{d\mu} + M\Delta N_2)x(t-d(t)) + (B_{1\mu} + M\Delta N_3)\omega(t) \\ z(t) &= [(C_\mu + D_{2\mu}K_\mu E + M_z\Delta_z(N_{z1} + N_{z4}K_\mu E)]x(t) \\ &\quad + (C_{d\mu} + M_z\Delta_z N_{z2})x(t-d(t)) + (D_{1\mu} + M_z\Delta_z N_{z3})\omega(t) \\ y(t) &= Ex(t) \end{aligned} \tag{6}$$

where

$$W_\mu = \sum_{i=1}^{r} \mu_i W_i, \quad W_i \in \{A_i, A_{di}, B_{1i}, B_{2i}, C_i, C_{di}, D_{1i}, D_{2i}\}.$$

The H_∞ performance considered here is the following:
In the continuous-time case

$$\int_0^\infty z'(t)z(t)dt < \gamma^2 \int_0^\infty \omega'(t)\omega(t)dt. \tag{7}$$

In the discrete-time case

$$\sum_{t=0}^{\infty} z'(t)z(t) < \gamma^2 \sum_{t=0}^{\infty} \omega'(t)\omega(t). \tag{8}$$

Definition 1. *If* (5) *satisfies the following two conditions, it is said to be H_∞ fuzzy controller.*

1) The controller makes the system (1) *or* (4) *robustly stable in the presence of $\omega(t)$*

2) Given γ, the closed-loop system (6) *must satisfy the criterion* (7) *in continuous-time case or* (8) *in discrete-time case, in which the initial condition is zero.*

3 Robust Stabilization by the H_∞ Fuzzy Controller

3.1 Continuous-Time Case

We first derive some sufficient conditions that guarantee the global asymptotic stability and disturbance attenuation of the closed-loop continuous-time T–S fuzzy system. Then, we convert these sufficient conditions to LMIs by using similarity transform and congruence transform. Therefore H_∞ fuzzy controller which satisfies Definition 1 can be easily designed.

Theorem 1. *Given a constant $\gamma > 0$, the system* (1) *is robustly stabilizable by the controller* (5) *if there exist the positive symmetric matrices P, S and the control gain K_j such that the following inequalities are satisfied. In other words,* (5) *is the H_∞ fuzzy controller.*

$$\begin{cases} M_{ii} < 0, & i = 1, \cdots, r \\ \frac{1}{r-1}M_{ii} + \frac{1}{2}(M_{ij} + M_{ji}) < 0, & 1 \le i \ne j \le r \end{cases} \tag{9}$$

where

$$M_{ij} = \begin{bmatrix} \Phi_{ij} & * & * & * & * & * & * & * \\ A'_{di}P & -S & * & * & * & * & * & * \\ B'_{1i}P & 0 & -\gamma^2 I & * & * & * & * & * \\ \epsilon_1 M'P & 0 & 0 & -\epsilon_1 I & * & * & * & * \\ \Psi_j & N_2 & N_3 & 0 & -\epsilon_1 I & * & * & * \\ \Gamma_{ij} & C_{di} & D_{1i} & 0 & 0 & -I & * & * \\ 0 & 0 & 0 & 0 & 0 & \epsilon_2 M'_z & -\epsilon_2 I & * \\ \Psi_{zj} & N_{z2} & N_{z3} & 0 & 0 & 0 & 0 & -\epsilon_2 I \end{bmatrix},$$

$$\Phi_{ij} = A'_i P + E'K'_j B'_{2i} P + PA_i + PB_{2i} K_j E + \frac{1}{1-\beta} S, \quad \Psi_j = N_1 + N_4 K_j E,$$

$$\Gamma_{ij} = C_i + D_{2i} K_j E, \quad \Psi_{zj} = N_{z1} + N_{z4} K_j E.$$

Proof. The proof is omitted due to the lack of the space.

The inequalities (9) are BMIs which are not solvable by the convex programming technique. In order to convert it to LMIs, we use the method proposed in [9]. If we define new state variables $x = T\tilde{x}$, then (6) is converted to the followings:

$$\begin{aligned} \sigma\tilde{x}(t) &= [(\tilde{A}_\mu + \tilde{B}_{2\mu} K_\mu \tilde{E} + \tilde{M}\Delta(\tilde{N}_1 + N_4 K_\mu \tilde{E})]\tilde{x}(t) \\ &\quad + (\tilde{A}_{d\mu} + \tilde{M}\Delta\tilde{N}_2)\tilde{x}(t - d(t)) + (\tilde{B}_{1\mu} + \tilde{M}\Delta N_3)\omega(t) \\ z(t) &= [(\tilde{C}_\mu + D_{2\mu} K_\mu \tilde{E} + M_z \Delta_z (\tilde{N}_{z1} + N_{z4} K_\mu \tilde{E})]\tilde{x}(t) \\ &\quad + (\tilde{C}_{d\mu} + M_z \Delta_z \tilde{N}_{z2})\tilde{x}(t - d(t)) + (D_{1\mu} + M_z \Delta_z N_{z3})\omega(t) \end{aligned} \tag{10}$$

where $\tilde{A}_\mu = T^{-1} A_\mu T$, $\tilde{B}_{2\mu} = T^{-1} B_{2\mu}$, $\tilde{A}_{d\mu} = T^{-1} A_{d\mu} T$, $\tilde{B}_{1\mu} = T^{-1} B_{1\mu}$, $\tilde{C}_\mu = C_\mu T, \tilde{C}_{d\mu} = C_{d\mu} T$, $\tilde{E} = ET$, $\tilde{M} = T^{-1} M$, $\tilde{N}_1 = N_1 T$, $\tilde{N}_2 = N_2 T$, $\tilde{N}_{z1} = N_{z1} T$, $\tilde{N}_{z2} = N_{z2} T$. Let $Q = P^{-1}$.

$$Q_{n\times n} = \begin{bmatrix} Q_{1p\times p} & 0 \\ 0 & Q_2 \end{bmatrix}$$

The transformation matrix, T, is selected in order to satisfy the following condition:

$$\tilde{E} = ET = \begin{bmatrix} I_p & 0 \end{bmatrix}$$

That is

$$T = \begin{bmatrix} E'(EE')^{-1}|_{n-p} \; ortc(E') \end{bmatrix}.$$

where $ortc(E')$ denotes orthogonal complement of E'. Applying Theorem 1 to (10), the sufficient condition to stabilize (10) is the following:

$$\begin{cases} \tilde{M}_{ii} < 0, & i = 1, \cdots, r \\ \frac{1}{r-1}\tilde{M}_{ii} + \frac{1}{2}(\tilde{M}_{ij} + \tilde{M}_{ji}) < 0, & 1 \le i \ne j \le r \end{cases} \tag{11}$$

where

$$\tilde{M}_{ij} = \begin{bmatrix} \tilde{\Phi}_{ij} & * & * & * & * & * & * & * \\ \tilde{A}'_{di}P & -S & * & * & * & * & * & * \\ \tilde{B}'_{1i}P & 0 & -\gamma^2 I & * & * & * & * & * \\ \epsilon_1 \tilde{M}'P & 0 & 0 & -\epsilon_1 I & * & * & * & * \\ \tilde{\Psi}_j & \tilde{N}_2 & N_3 & 0 & -\epsilon_1 I & * & * & * \\ \tilde{\Gamma}_{ij} & \tilde{C}_{di} & D_{1i} & 0 & 0 & -I & * & * \\ 0 & 0 & 0 & 0 & 0 & \epsilon_2 M'_z & -\epsilon_2 I & * \\ \tilde{\Psi}_{zj} & \tilde{N}_{z2} & N_{z3} & 0 & 0 & 0 & 0 & -\epsilon_2 I \end{bmatrix},$$

$$\tilde{\Phi}_{ij} = \tilde{A}'_i P + \tilde{E}' K'_j \tilde{B}'_{2i} P + P\tilde{A}_i + P\tilde{B}_{2i} K_j \tilde{E} + \frac{1}{1-\beta} S, \quad \tilde{\Psi}_j = \tilde{N}_1 + N_4 K_j \tilde{E},$$
$$\tilde{\Gamma}_{ij} = \tilde{C}_i + D_{2i} K_j \tilde{E}, \quad \tilde{\Psi}_{zj} = \tilde{N}_{z1} + N_{z4} K_j \tilde{E}.$$

Let $\Theta = diag \left[Q \; Q \; I \; I \; I \; I \; I \; I \right]$. Pre- and post-multiplying (11) by Θ, the inequality expounded is displayed as

$$\begin{cases} \bar{M}_{ii} < 0, & i = 1, \cdots, r \\ \frac{1}{r-1} \bar{M}_{ii} + \frac{1}{2}(\bar{M}_{ij} + \bar{M}_{ji}) < 0, & 1 \le i \ne j \le r \end{cases} \tag{12}$$

where

$$\bar{M}_{ij} = \begin{bmatrix} \bar{\Phi}_{ij} & * & * & * & * & * & * & * \\ Q\tilde{A}'_{di} & -X & * & * & * & * & * & * \\ \tilde{B}'_{1i} & 0 & -\gamma^2 I & * & * & * & * & * \\ \epsilon_1 \tilde{M}' & 0 & 0 & -\epsilon_1 I & * & * & * & * \\ \bar{\Psi}_j & \tilde{N}_2 Q & N_3 & 0 & -\epsilon_1 I & * & * & * \\ \bar{\Gamma}_{ij} & \tilde{C}_{di} Q & D_{1i} & 0 & 0 & -I & * & * \\ 0 & 0 & 0 & 0 & 0 & \epsilon_2 M'_z & -\epsilon_2 I & * \\ \bar{\Psi}_{zj} & \tilde{N}_{z2} Q & N_{z3} & 0 & 0 & 0 & 0 & -\epsilon_2 I \end{bmatrix},$$

$$\bar{\Phi}_{ij} = Q\tilde{A}'_i + \begin{bmatrix} F'_j \\ 0 \end{bmatrix} \tilde{B}'_{2i} + \tilde{A}_i Q + \tilde{B}_{2i} \left[F_j \; 0 \right] + \frac{1}{1-\beta} X, \quad \bar{\Psi}_j = \tilde{N}_1 Q + N_4 \left[F_j \; 0 \right],$$
$$\bar{\Gamma}_{ij} = \tilde{C}_i Q + D_{2i} \left[F_j \; 0 \right], \; \bar{\Psi}_{zj} = \tilde{N}_{z1} Q + N_{z4} \left[F_j \; 0 \right], \quad X = QSQ > 0, \quad F_j = K_j Q_1.$$

Note that positive definite matrix S can be always obtained and (12) are linear matrix inequalities that has following 5 variables: $(Q, X, F_j, \epsilon_1, \epsilon_2)$.

3.2 Discrete-Time Case

As we did for the continuous-time case, the same procedure is applicable to the discrete-time systems. The result is stated below.

Theorem 2. *Given a constant $\gamma > 0$, the system (1) is robustly stabilizable by the controller (5) if there exist the positive symmetric matrices P, S and the control gain K_j such that the following inequalities are satisfied. In other words, (5) is the H_∞ fuzzy controller.*

$$\begin{cases} M_{ii} < 0, & i = 1, \cdots, r \\ \frac{1}{r-1} M_{ii} + \frac{1}{2}(M_{ij} + M_{ji}) < 0, & 1 \le i \ne j \le r \end{cases} \tag{13}$$

where

$$M_{ij} = \begin{bmatrix} S-P & * & * & * & * & * & * & * & * \\ 0 & -S & * & * & * & * & * & * & * \\ 0 & 0 & -\gamma^2 I & * & * & * & * & * & * \\ \Phi_{ij} & A_{di} & B_{1i} & -P^{-1} & * & * & * & * & * \\ 0 & 0 & 0 & \epsilon_1 M' & -\epsilon_1 I & * & * & * & * \\ \Psi_j & N_2 & N_3 & 0 & 0 & -\epsilon_1 I & * & * & * \\ \Gamma_{ij} & C_{di} & D_{1i} & 0 & 0 & 0 & -I & * & * \\ 0 & 0 & 0 & 0 & 0 & 0 & \epsilon_2 M_z' & -\epsilon_2 I & * \\ \Psi_{zj} & N_{z2} & N_{z3} & 0 & 0 & 0 & 0 & 0 & -\epsilon_2 I \end{bmatrix}$$

$\Phi_{ij} = A_i + B_{2i} K_j E$, $\Psi_j = N_1 + N_4 K_j E$, $\Gamma_{ij} = C_i + D_{2i} K_j E$, $\Psi_{zj} = N_{z1} + N_{z4} K_j E$.

Proof. The proof is omitted due to the lack of the space.

Let $x = T\tilde{x}$, $Q = P^{-1}$, and $\Theta = diag\left[Q\ Q\ I\ I\ I\ I\ I\ I\ I\right]$. Applying Theorem 2 to (10) and pre- and post-multiplying by Θ, the sufficient condition to stabilize (10) is the following:

$$\begin{cases} \bar{M}_{ii} < 0, & i = 1, \cdots, r \\ \frac{1}{r-1} \bar{M}_{ii} + \frac{1}{2}(\bar{M}_{ij} + \bar{M}_{ji}) < 0, & 1 \le i \ne j \le r \end{cases} \tag{14}$$

where

$$\bar{M}_{ij} = \begin{bmatrix} X-Q & * & * & * & * & * & * & * & * \\ 0 & -X & * & * & * & * & * & * & * \\ 0 & 0 & -\gamma^2 I & * & * & * & * & * & * \\ \bar{\Phi}_{ij} & \tilde{A}_{di}Q & \tilde{B}_{1i} & -Q & * & * & * & * & * \\ 0 & 0 & 0 & \epsilon_1 \tilde{M}' & -\epsilon_1 I & * & * & * & * \\ \bar{\Psi}_j & \tilde{N}_2 Q & N_3 & 0 & 0 & -\epsilon_1 I & * & * & * \\ \bar{\Gamma}_{ij} & \tilde{C}_{di}Q & D_{1i} & 0 & 0 & 0 & -I & * & * \\ 0 & 0 & 0 & 0 & 0 & 0 & \epsilon_2 M_z' & -\epsilon_2 I & * \\ \bar{\Psi}_{zj} & \tilde{N}_{z2} Q & N_{z3} & 0 & 0 & 0 & 0 & 0 & -\epsilon_2 I \end{bmatrix}$$

$\bar{\Phi}_{ij} = \tilde{A}_i Q + \tilde{B}_{2i}\left[F_j\ 0\right]$, $\bar{\Psi}_j = \tilde{N}_1 Q + N_4 \left[F_j\ 0\right]$, $\bar{\Gamma}_{ij} = \tilde{C}_i Q + D_{2i}\left[F_j\ 0\right]$, $\bar{\Psi}_{zj} = \tilde{N}_{z1} Q + N_{z4}\left[F_j\ 0\right]$, $X = QSQ > 0$, $F_j = K_j Q_1$.

4 Computer Simulation

Consider the nonlinear system with time-varying delay, which is represented by the following T–S fuzzy model:

Plant Rule 1:

$$\begin{aligned} &\text{IF } x_2(t) \text{ is } M_{11} \\ &\text{THEN } \dot{x}(t) = (A_1 + \Delta A(t))x(t) + (A_{d1} + \Delta A_d(t))x(t - d(t)) \\ &\qquad\qquad + B_{11}\omega(t) + B_{21}u(t) \\ &\qquad\quad z(t) = C_1 x(t) \\ &\qquad\quad y(t) = Ex(t) \end{aligned}$$

Plant Rule 2:

$$\begin{aligned} &\text{IF } x_2(t) \text{ is } M_{21} \\ &\text{THEN } \dot{x}(t) = (A_2 + \Delta A(t))x(t) + (A_{d2} + \Delta A_d(t))x(t - d(t)) \\ &\qquad\qquad + B_{12}\omega(t) + B_{22}u(t) \\ &\qquad\quad z(t) = C_2 x(t) \\ &\qquad\quad y(t) = Ex(t) \end{aligned} \tag{15}$$

where

$$A_1 = \begin{bmatrix} -0.5 & 0.2 \\ 1 & 0 \end{bmatrix}, A_2 = \begin{bmatrix} -0.5 & -1.5 \\ 1 & 0 \end{bmatrix}, A_{d1} = \begin{bmatrix} -0.3 & -0.3 \\ 0 & 0 \end{bmatrix}, A_{d2} = \begin{bmatrix} -0.3 & -0.1 \\ 0 & 0 \end{bmatrix},$$

$$B_{11} = B_{12} = \begin{bmatrix} 1 \\ 0 \end{bmatrix}, B_{21} = B_{22} = \begin{bmatrix} 0.5 \\ 0 \end{bmatrix}, C_1 = C_2 = \begin{bmatrix} 0 & 1 \end{bmatrix}, E = \begin{bmatrix} 1 & 1 \end{bmatrix}.$$

where time-varying delay is $d(t) = 2 + 0.3\sin(t)$. The time-varying matrices $\Delta A(t)$ and $\Delta A_d(t)$ satisfies (3) and is defined by

$$M = \begin{bmatrix} 0.2 \\ 0 \end{bmatrix}, \quad N_1 = N_2 = \begin{bmatrix} 1 & 0 \end{bmatrix}.$$

By solving (12), we can obtain

$$Q = \begin{bmatrix} 9.7836 & 0 \\ 0 & 1.2778 \end{bmatrix}, \quad X = \begin{bmatrix} 227.7655 & 3.5372 \\ 3.5372 & 0.3782 \end{bmatrix}, \quad F_1 = \begin{bmatrix} -554.0317 & 0 \end{bmatrix},$$

$$F_2 = \begin{bmatrix} -529.6989 & 0 \end{bmatrix}, \quad \epsilon_1 = 185.5006, \quad \epsilon_2 = 209.1023.$$

From this, the values of S, K_1, and K_2 are

$$S = \begin{bmatrix} 2.3795 & 0.041 \\ 0.041 & 0.2316 \end{bmatrix}, \quad K_1 = -56.6286, \quad K_2 = -54.1415.$$

The simulation results for the nonlinear system are shown in Figs 1 and 2. In the simulation, the initial value is $x(t) = 0 \quad \text{for } t < 0, \quad x(0) = \begin{bmatrix} -1 \\ 1.2 \end{bmatrix}$ and the disturbance is defined by $\omega(t) = r/(9t + 1)$, where r is a random number taken from a uniform distribution over $[0, \ 3]$.

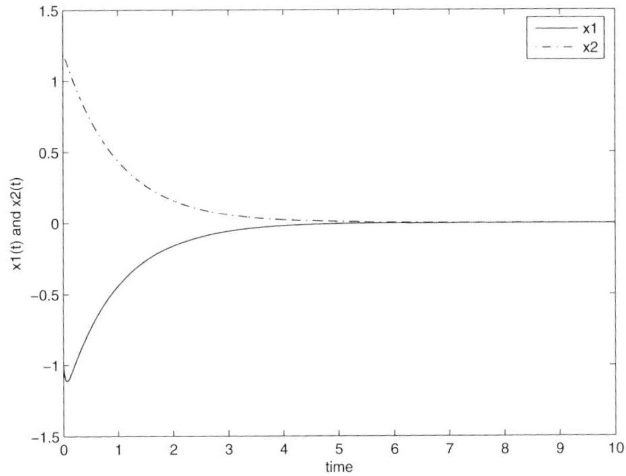

Fig. 1. Time response of (15) controlled by the H_∞ fuzzy controller

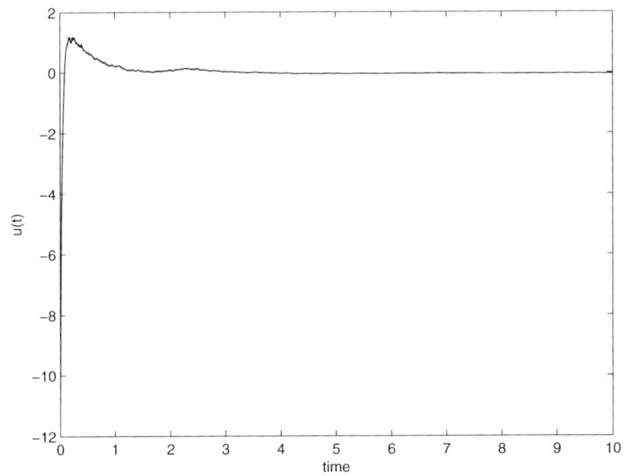

Fig. 2. The control input $u(t)$

5 Conclusion

We have developed and analyzed the H_∞ fuzzy static output feedback control method to robustly stabilize the nonlinear system with time-varying delay in both continuous-time and discrete-time cases. Using Lyapunov stability theory, the sufficient conditions, which are in the format of BMIs, were derived in both continuous-time case and discrete-time case. Through further manipulation, BMIs were converted to LMIs. Therefore, we can easily design the H_∞ fuzzy controller via current convex algorithm tools.

References

1. Y. H. Joo, L. S. Shieh, and G. Chen, "Hybrid state-space fuzzy model-based controller with dual-rate sampling for digital control of chaotic systems," *IEEE Trans. Fuzzy Syst.*, vol. 7, no. 4, pp. 394-408, 1999.
2. W. Chang, J. B. Park, Y. H. Joo, and G. Chen, "Design of sampled-data fuzzy-model-based control systems by using intelligent digital redesign," *IEEE Trans. Circuits Syst.-I*, vol. 49, no. 4, pp. 509-517, 2002.
3. H. J. Lee, Y. H. Joo, W. Chang, and J. B. Park, "A new intelligent digital redesign for TS fuzzy systems: Global approach," *IEEE Trans. Fuzzy Syst.*, vol. 12, no. 2, pp. 274-284, 2004.
4. K. Tanaka, T. Ikeda, and H. O. Wang, "A unified approach to controlling chaos via an LMI-based fuzzy control system design," *IEEE Trans. Circuits Syst.-I*, vol. 45, no. 10, pp. 1021-1040, 1998.
5. Y. C. Chang, S. S. Chen, S. F. Su, and T. T. Lee, "Static output feedback stabilization for nonlinear interval time-delay systems via fuzzy control approach," *Fuzzy Sets Syst.*, vol. 54, no. 6, 2004.
6. T. Takagi and M. Sugeno, "Fuzzy identification of systems and its applications to modeling and control," *IEEE Trans. Syst., Man, Cybern.*, vol. SMC-15, no. 1, pp. 116-132 ,1985.
7. H. O. Wang, K. Tanaka, and M. F. Griffin, "An approach to fuzzy control of nonlinear systems: Stability and design issues," *IEEE Trans. Fuzzy Syst.*, vol. 4, no. 1, pp. 14-23, 1996.
8. K. Tanaka and M. Sugeno, "Stability analysis and design of fuzzy control systems," *Fuzzy Sets Syst.*, vol. 45, no. 2, pp. 135-156, 1992.
9. J. C. Lo and M. L. Lin, "Robust H_∞ nonlinear control via fuzzy static output feedback," *IEEE Trans. Circuits Syst. I*, vol. 50, no. 11, pp. 1494-1502, 2003.
10. Y. Y. Cao and P. M. Frank, "Analysis and synthesis of nonlinear time-delay systems via fuzzy control approach," *IEEE Trans. Fuzzy Syst.*, vol. 8, no. 2, pp. 200-211, 2000.
11. K. R. Lee, J. H. Kim, E. T. Jeung, and H. B. Park, "Output feedback robust H^∞ control of uncertain fuzzy dynamic systems with time-varying delay," *IEEE Trans. Fuzzy Syst.*, vol. 8, no. 6, pp. 657-664, 2000.
12. F. Hao, T. Chu, L. Huang and L. Wang, "Non-fragile controllers of peak gain minimization for uncertain systems via LMI approach," *Dynamics of Continuous Discrete and Impulsive Systems.*, vol. 10, no. 5, pp. 681-693, 2003.
13. S. Xu and J. Lam, "Robust H_∞ Control for Uncertain Discrete-Time-Delay Fuzzy Systems Via Output Feedback Controllers," *IEEE Trans. Fuzzy Syst.*, vol. 13, no. 1, pp. 82-93, 2005.
14. H. J. Lee, J. B. Park, and G. Chen, "Robust fuzzy control of nonlinear systems with parametric uncertainties," *IEEE Trans. Fuzzy Syst.*, vol. 9, no. 2, pp. 369-379, 2001.
15. H. D. Tuan, P. Apkarain, T. Narikiyo, and Y. Yamamoto, "Parameterized linear matrix inequality techniques in fuzzy control system design," *IEEE Trans. Fuzzy Syst.*, vol. 9, no. 2, pp. 324-332, 2001.
16. S. Boyd, L. El Ghaoui, E. Feron, and V. Balakrishnan, *Linear Matrix Inequalities in System and Control theory.* Philadelphia, PA: SIAM, 1994.

Observer-Based H_∞ Controller Designs for T-S Fuzzy Systems

Jinping Wang, Shengjuan Huang, and Xiqin He*

The authors are with Institute of Applied Mathematics, AnShan University of Science and Technology, AnShan, LiaoNing Province, P.R. China, 114051
Tel.: 86 (0)412 5929509
xiqinhe@hotmail.com

Abstract. H_∞control designs for T-S fuzzy systems have been studied, based on the observers, the systems which composed of the fuzzy observers and the error systems are proposed. Some new sufficient conditions which guarantee the quadratical stability and the existence of the state feedback H_∞ control for the systems are proposed. The condition in the theorem 3 is in the form of a matrix inequality, which is simple and converted to the LMIs that can be solved by using MATLAB.

1 Introduction

T-S(Takagi-Sugeno)[1] fuzzy systems are nonlinear systems described by a set of IF-THEN rules which give a local linear representation of an underlining system. Feng et al. [2] and Cao et al. [3], [4] have proved that the T-S fuzzy system can approximate any continuous functions in a compact set of Rn at any preciseness, and that the method based on linear uncertain system theory can convert the stability analysis of a fuzzy control system to the stability analysis of linear time-varying extreme subsystems. This allows the designers to take advantage of conventional linear system to analyze and design the fuzzy control systems.

H_∞ control has been an attractive research topic since the last decade. Some papers have discussed the H_∞ feedback control for fuzzy systems. They deal with a state feedback control design that requires all system states to be measured. In many cases, this requirement is too restrictive. The existence of state feedback H_∞ control in some papers need to find a common symmetry and positive matrix satisfying the fuzzy subsystems. So the conditions are conservative. A new quadratically stable condition which is simple and relaxed is proposed in [5], and a new observer design for the T-S fuzzy system and two new conditions of the existence of H_∞ control based on the observers are also proposed, which are simple and in the forms of linear matrix inequalities which can be directly solved by using MATLAB. In this paper, a method which is different from that in [5] to deal with the control problems of T-S fuzzy systems is proposed.

The condition is in the form of a matrix inequality which can be converted to the LMIs and solved by using MATLAB, and the condition is relaxed.

* Corresponding author.

L. Wang et al. (Eds.): FSKD 2006, LNAI 4223, pp. 31–40, 2006.

The conditions of the existence of H_∞ control of T-S fuzzy systems are proposed in [6,7], considering the following T-S fuzzy systems:

$$\begin{aligned}\dot{x}(t) &= \sum_{i=1}^{r} \mu_i(\xi)(A_i x(t) + B_{1i}\omega(t) + B_{2i}u(t)) \\ z(t) &= \sum_{i=1}^{r} \mu_i(\xi)(C_i x(t) + D_i u(t))\end{aligned} \tag{1}$$

Theorem 1. *For a given constant $\gamma > 0$, if there exist matrices F_i, X, X_{ij}, where X is a positive-definite matrix, X_{ii} are symmetry matrices, $X_{ji} = X_{ij}^T, i = 1, ..., r$, satisfy the following matrix inequalities:*

$$A_i^T X + F_i^T B_{2i}^T X + X A_i + X B_{2i} F_i + \frac{1}{\gamma^2} X B_{1i} B_{1i}^T X < X_{ii} \tag{2}$$

$$\begin{aligned}&A_i^T X + F_j^T B_{2i}^T X + F_i^T B_{2j}^T X + A_j^T X + X A_i + X B_{2i} F_j + X B_{2j} F_i + X A_j \\ &+ \tfrac{1}{\gamma^2} X B_{1i} B_{1j}^T X + \tfrac{1}{\gamma^2} X B_{1j} B_{1i}^T X \le X_{ij} + X_{ij}^T, i \neq j\end{aligned} \tag{3}$$

$$H_k = \begin{bmatrix} X_{11} & \cdots & X_{1r} & C_1^T + F_k^T D_1^T \\ \vdots & \ddots & \vdots & \vdots \\ X_{r1} & \cdots & X_{rr} & C_r^T + F_k^T D_r^T \\ C_1 + D_1 F_k & \cdots & C_r + D_r F_k & -I \end{bmatrix} < 0, k = 1, ..., r \tag{4}$$

then the state feed-back

$$u(t) = \sum_{j=1}^{r} \mu_j(\xi(t)) F_j x(t) \tag{5}$$

makes the following system stable with the H_∞ performance bound with γ:

$$\begin{aligned}\dot{x}(t) &= \sum_{i=1}^{r} \mu_i(\xi)\mu_j(\xi)(A_i x(t) + B_{2i} F_j x(t) + B_{1i}\omega(t)) \\ z(t) &= \sum_{i=1}^{r} \mu_i(\xi)\mu_j(\xi)(C_i + D_i F_j) x(t)\end{aligned} \tag{6}$$

The conditions of existence of H_∞ control in theorem 1 are in the forms of matrix inequalities, but it is restricted by condition (3)(need to consider $i \neq j$), that is, a common symmetry positive matrix satisfying the fuzzy subsystems is still to be find, so the conditions of theorem 1 are still conservative. Based on the theorem 1, new conditions are proposed in [5]:

Theorem 2. *For a given constant $\gamma > 0$, if there exist matrices M_i, Z, Z_{ij}, where Z is a positeve-definite matrix, Z_{ii} are symmetric matrices, $Z_{ji} = Z_{ij}^T, i = 1, ..., r$, satisfy the following LMIs:*

$$Z A_i^T + M_i^T B_{2i}^T + A_i Z + B_{2i} M_i + \frac{1}{\gamma^2} B_{1i} B_{1i}^T < Z_{ii} \tag{7}$$

$$\begin{aligned}&ZA_i^T + M_j^T B_{2i}^T + M_i^T B_{2j}^T + ZA_j^T + A_i Z + B_{2i} M_j + B_{2j} M_i + A_j Z \\ &+\tfrac{1}{\gamma^2} B_{1i} B_{1j}^T + \tfrac{1}{\gamma^2} B_{1j} B_{1i}^T \leq Z_{ij} + Z_{ij}^T, i \neq j\end{aligned} \tag{8}$$

$$H_k = \begin{bmatrix} Z_{11} & \cdots & Z_{1r} & ZC_1^T + M_k^T D_1^T \\ \vdots & \ddots & \vdots & \vdots \\ Z_{r1} & \cdots & Z_{rr} & ZC_r^T + M_k^T D_r^T \\ C_1 Z + D_1 M_k & \cdots & C_r Z + D_r M_k & -I \end{bmatrix} < 0, k = 1, 2, ..., r \tag{9}$$

then the state feed-back (5) makes (6) stable with the H_∞ performance bound with γ.

The conditions of existence of H_∞ control in theorem 2 are in the forms of LMIs, but it is still restricted by (8) (as (3)), so the conditions of theorem 2 are also conservative.

In [8], first, the new systems are given based on the observers, and the error of the systems is considered, then the controllers are designed to obtain the H_∞control performance of the systems. But the conditions need to find a common symmetry and positive matrix P. So the conditions are still conservative.

In [9], the new systems are proposed based on fuzzy performance evaluator(FPE), and the disturbance rejection is added to the FPE, only the control performance of the error systems are considered, that is, the controllers are designed to obtain the H_∞ control performance of the error systems.

The paper is organized as follow: in section 2, we propose the systems based on the observers, and the error of systems is considered at the same time. In section 3, the controllers and the error matrices are designed to make the systems which composed of the observers and the error systems satisfy the given H1 control performance, especially, the condition in the theorem 3 is in the form of a matrix inequality, which is simple and does not need to find a common symmetry and positive definite matrix satisfying each subsystems(in fact, we consider the interactions among the fuzzy subsystems), so the condition is relaxed. In section 4, the designing approaches of the observers are propose. In section 5, An example is present to show the effectiveness of the results. The conclusion was made in section 6.

2 The Stability of T-S Fuzzy Systems

Consider the following T-S fuzzy systems:

$$\begin{aligned}\dot{x}(t) &= \sum_{i=1}^{r} \mu_i(\xi)(A_i x(t) + B_{1i}\omega(t) + B_{2i}u(t)) \\ y &= \sum_{i=1}^{r} \mu_i(\xi) C_i x(t)\end{aligned} \tag{10}$$

where $x(t) \in R^n$ is the state variable, $y(t) \in R^q$ is the output variable, $\omega(t) \in R^l$ is the disturbance variable, $u(t) \in R^m$ is the input variable, $A_i \in R^{n\times n}, B_{1i} \in R^{n\times l}, B_{2i} \in R^{n\times m}andC_i \in R^{q\times n}$ are premise variables. It is assumed that the premise variables do not depend on the control and disturbance variables. Where $\mu_i(\xi(t)) = \frac{\beta_i(\xi(t))}{\sum\limits_{j=1}^{r}\beta_j(\xi(t))}$,$\beta_i(\xi(t)) = \prod\limits_{i=1}^{p} M_{ij}(\xi(t))$,$M_{ij}(\cdot)$is the membership function of the fuzzy setM_{ij}. Obviously, we have $\sum\limits_{i=1}^{r}\mu_i(\xi(t)) = 1$,$\mu_i(\xi(t)) > 0, i = 1, ..., r, \forall t$.

Definition 1. *For (10), when $\omega(t) \equiv 0, u(t) \equiv 0$, if there exist $\alpha > 0$ and a symmetry positive definite matrix X such that*

$$\dot{V}(x(t)) \leq -\alpha x^T(t)x(t) \tag{11}$$

where $V(x(t)) = x^T(t)Xx(t)$, then (10) is called quadratically stable.

Based on the observers, we can obtain

$$\begin{aligned} \dot{\bar{x}} &= \sum_{i=1}^{r}\mu_i(\xi)(A_i\bar{x} + B_{2i}u + G_i(y - \bar{y})) \\ \bar{y} &= \sum_{i=1}^{r}\mu_i(\xi)C_i\bar{x} \end{aligned} \tag{12}$$

the state feedback is

$$u(t) = \sum_{j=1}^{r}\mu_j(\xi)K_j\bar{x} \tag{13}$$

in (12) and (13), $G_i \in R^{n\times q}, K_i \in R^{m\times n} (i = 1, 2, ..., r)$ are the feedback matrices of output error and the state feedback matrices.

Let $e(t) = x - \bar{x}$ ($e(t)$is the systems' error), then

$$\dot{\bar{x}} = \sum_{i=1}^{r}\sum_{j=1}^{r}\mu_i(\xi)\mu_j(\xi)\left[(A_i + B_{2i}K_j)\bar{x} + G_iC_j\ e(t)\right] \tag{14}$$

$$\dot{e}(t) = \dot{x} - \dot{\bar{x}} = \sum_{i=1}^{r}\mu_i(\xi)A_ie - \sum_{i=1}^{r}\sum_{j=1}^{r}\mu_i(\xi)\mu_j(\xi)G_iC_je + \sum_{i=1}^{r}\mu_i(\xi)B_{1i}\omega \tag{15}$$

and let $\tilde{x} = \begin{pmatrix} \bar{x} \\ e \end{pmatrix}$, $\tilde{A}_{ij} = \begin{pmatrix} A_i + B_{2i}K_j & G_iC_j \\ 0 & A_i - G_iC_j \end{pmatrix}$, $\tilde{B}_{1i} = \begin{pmatrix} 0\ 0 \\ 0\ B_{1i} \end{pmatrix}$, $\tilde{\omega} = \begin{pmatrix} 0 \\ \omega \end{pmatrix}$,then from (14) and (15), we can obtain

$$\dot{\tilde{x}} = \sum_{i=1}^{r}\sum_{j=1}^{r}\mu_i(\xi)\mu_j(\xi)(\tilde{A}_{ij}\tilde{x} + \tilde{B}_{1i}\tilde{\omega}) \tag{16}$$

Theorem 3. *If there exist matrices $K_i, G_i, i = 1, 2, ..., r$ and a symmetry positive definite matrix X such that*

$$Q = \begin{bmatrix} \Lambda_{11}^TP + P\Lambda_{11} \ldots & \Lambda_{1r}^TP + P\Lambda_{1r} \\ \ldots & \ddots\ \ldots \\ \Lambda_{r1}^TP + P\Lambda_{r1} \ldots & \Lambda_{rr}^TP + P\Lambda_{rr} \end{bmatrix} < -\alpha I, (\alpha > 0) \tag{17}$$

where $\Lambda_{ii} = \tilde{A}_{ii}, 2\Lambda_{ij} = \tilde{A}_{ij} + \tilde{A}_{ji}, \tilde{A}_{ij} = \begin{pmatrix} A_i + B_{2i}K_j & G_iC_j \\ 0 & A_i - G_iC_j \end{pmatrix}, i, j = 1, 2, ..., r$ $(\Lambda_{ij} = \Lambda_{ji})$, *then the state feedback (13) makes the closed-loop system (16) quadratically stable when* $\omega(t) \equiv 0$.

Proof. Let $\Lambda_{ii} = \tilde{A}_{ii}, 2\Lambda_{ij} = \tilde{A}_{ij} + \tilde{A}_{ji}, \omega(t) \equiv 0 (\tilde{\omega}(t) \equiv 0)$, we construct Lyapunov function $V(t) = \tilde{x}^T(t)P\tilde{x}(t)$, then

$$\begin{aligned}
\dot{V}(\tilde{x})) &= \sum_i^r \mu_i^2(\xi)\tilde{x}^T(\tilde{A}_{ii}^T P + P\tilde{A}_{ii})\tilde{x} \\
&+ 2\sum_{i=1}^r \sum_{i<j}^r \mu_i(\xi)\mu_j(\xi)\tilde{x}^T((\frac{\tilde{A}_{ij} + \tilde{A}_{ji}}{2})^T P + (\frac{\tilde{A}_{ij} + \tilde{A}_{ji}}{2})P)\tilde{x} \\
&= \begin{pmatrix} \mu_1\tilde{x} \\ \vdots \\ \mu_r\tilde{x} \end{pmatrix}^T \begin{pmatrix} \Lambda_{11}^T P + P\Lambda_{11} & \cdots & \Lambda_{1r}^T P + P\Lambda_{1r} \\ \vdots & \ddots & \vdots \\ \Lambda_{r1}^T P + P\Lambda_{r1} & \cdots & \Lambda_{rr}^T P + P\Lambda_{rr} \end{pmatrix} \begin{pmatrix} \mu_1\tilde{x} \\ \vdots \\ \mu_r\tilde{x} \end{pmatrix} \\
&\leq -\alpha\tilde{x}^T(t)\tilde{x}(t)
\end{aligned}$$

□

Corollary 1. *If there exist matrices* $K_i, G_i, i = 1, 2, ..., r$ *and a symmetry positive definite matrix* X *such that*

$$\begin{bmatrix} \Lambda_{11}^T P + P\Lambda_{11} & ... & \Lambda_{1r}^T P + P\Lambda_{1r} \\ ... & \ddots & ... \\ \Lambda_{r1}^T P + P\Lambda_{r1} & ... & \Lambda_{rr}^T P + P\Lambda_{rr} \end{bmatrix} < 0 \tag{18}$$

where $\Lambda_{ii} = \tilde{A}_{ii}, 2\Lambda_{ij} = \tilde{A}_{ij} + \tilde{A}_{ji}, \tilde{A}_{ij} = \begin{pmatrix} A_i + B_{2i}K_j & G_iC_j \\ 0 & A_i - G_iC_j \end{pmatrix}, i, j = 1, 2, ..., r$ $(\Lambda_{ij} = \Lambda_{ji})$, *then the state feedback (13) makes the closed-loop system (16) quadratically stable when* $\omega(t) \equiv 0$.

3 The H_∞ Controller Based on the Observer

In the closed-loop system (16), $\tilde{\omega}$ is the outside disturbance, it will destroy the robust performance of the control systems, even makes the systems unstable. So we use the H_∞ form to measure the robust performance of the systems, and the controllers $u(t)$ are designed to make the robust performance of the closed-loop system better.

Considering the following H_∞ control performance,

$$\int_0^{t_f} \tilde{x}^T(t)\tilde{Q}\tilde{x}(t)dt \leq \tilde{x}^T(0)P\tilde{x}(0) + \rho^2 \int_0^{t_f} \tilde{\omega}^T(t)\tilde{\omega}(t)dt \tag{19}$$

where t_f is the control final time, $\rho > 0$ is a constant, $P$and $\tilde{Q}$ is symmetry positive matrix. Our aim is to design fuzzy controllers to make the closed-loop system (16) satisfy the given H_∞ control performance (19). For the closed-loop system (16), we construct Lyapunov function $V(t) = \tilde{x}^T(t)P\tilde{x}(t)$, then we have

$$\dot{V}(t) \le \sum_{i=1}^{r}\sum_{j=1}^{r}\mu_i(\xi)\mu_j(\xi)\tilde{x}^T(\tilde{A}_{ij}^TP + P\tilde{A}_{ij} + \tfrac{1}{\rho^2}P\tilde{B}_{1i}\tilde{B}_{1i}^TP)\tilde{x} + \rho^2\tilde{\omega}^T\tilde{\omega} \quad (20)$$

Lemma 1. *For the closed-loop system (16), if there exist a common matrix* $P = P^T > 0$ *such that*

$$\tilde{A}_{ij}^TP + P\tilde{A}_{ij} + \frac{1}{\rho^2}P\tilde{B}_{1i}\tilde{B}_{1i}^TP + \tilde{Q} < 0, i, j = 1, ..., r \quad (21)$$

then for the given constant $\rho > 0$ *, system (16) can obtain the* H_∞ *control performance (19).*

Proof. from (20) and (21), we have $\dot{V}(t) < -\tilde{x}^TQ\tilde{x} + \rho^2\tilde{\omega}^T\tilde{\omega}$, and integrating both sides of the above inequalities from 0 to t_f, we have

$$\int_0^{t_f}\tilde{x}^T(t)\tilde{Q}\tilde{x}(t)dt \le \tilde{x}^T(0)P\tilde{x}(0) + \rho^2\int_0^{t_f}\tilde{\omega}^T(t)\tilde{\omega}(t)dt$$

□

Theorem 4. *For the closed-loop system (16), if there exist a constant* $\alpha > 0$ *and a common matrix* $P = P^T > 0$ *such that*

$$\begin{array}{l}\tilde{A}_{ij}^TP + P\tilde{A}_{ij} + \frac{1}{\rho^2}P\tilde{B}_{1i}\tilde{B}_{1i}^TP < -\alpha I \ \ (\alpha > 0) \\ i, j = 1, ..., r\end{array} \quad (22)$$

then for the given constant $\rho > 0$, *system (16) can obtain the* H_∞ *control performance(19).*

Proof. If we let $\tilde{Q} = \tilde{Q}^T > 0, \lambda_{\max}(\tilde{Q}) = \alpha$,then (22) is converted to (21). □

Theorem 5. *For the closed-loop system (16), if there exist* $K_i, G_i, i = 1, 2, ..., r$ *and* $P = P^T > 0$ *such that*

$$Q = \begin{bmatrix} Q_{11} & \cdots & Q_{1r} \\ \vdots & \ddots & \vdots \\ Q_{r1} & \cdots & Q_{rr} \end{bmatrix} < 0 \quad (23)$$

where $Q_{ij} = X\Lambda_{ij}^T + \Lambda_{ij}X + \frac{1}{\rho^2}B_{ij}$ *,* $i, j = 1, 2, ..., r,\ X = P^{-1}, \Lambda_{ij} = \frac{\tilde{A}_{ij}+\tilde{A}_{ji}}{2}$,

$$B_{ij} = \frac{\tilde{B}_{1i}\tilde{B}_{1i}^T + \tilde{B}_{1j}\tilde{B}_{1j}^T}{2}, \tilde{B}_{1i} = \begin{pmatrix} 0 & 0 \\ 0 & B_{1i} \end{pmatrix}$$

$$\tilde{A}_{ij} = \begin{pmatrix} A_i + B_{2i}K_j & G_iC_j \\ 0 & A_i - G_iC_j \end{pmatrix} , i, j = 1, 2, ..., r$$

then for the given constant $\rho > 0$, *system (16) can obtain the* H_∞ *control performance(19).*

Proof. From (20), we have

$$\begin{aligned}
\dot{V}(t) &\leq \sum_{i=1}^{r}\sum_{j=1}^{r}\mu_i(\xi)\mu_j(\xi)\tilde{x}^T(\tilde{A}_{ij}^T P + P\tilde{A}_{ij} + \frac{1}{\rho^2}P\tilde{B}_{1i}\tilde{B}_{1i}^T P)\tilde{x} + \rho^2\tilde{\omega}^T\tilde{\omega} \\
&= \sum_{i=1}^{r}\sum_{j=1}^{r}\mu_i(\xi)\mu_j(\xi)\tilde{x}^T(\frac{(\tilde{A}_{ij}+\tilde{A}_{ji})^T}{2}P + P\frac{(\tilde{A}_{ij}+\tilde{A}_{ji})}{2})\tilde{x} \\
&+ \sum_{i=1}^{r}\sum_{j=1}^{r}\mu_i(\xi)\mu_j(\xi)\tilde{x}(\frac{1}{\rho^2}P\frac{\tilde{B}_{1i}\tilde{B}_{1i}^T+\tilde{B}_{1j}\tilde{B}_{1j}^T}{2}P)\tilde{x} + \rho^2\tilde{\omega}^T\tilde{\omega} \\
&= \begin{pmatrix}\mu_1\tilde{x}\\ \vdots \\ \mu_r\tilde{x}\end{pmatrix}\widetilde{Q}\begin{pmatrix}\mu_1\tilde{x}\\ \vdots \\ \mu_r\tilde{x}\end{pmatrix} + \rho^2\tilde{\omega}^T\tilde{\omega}
\end{aligned}$$

where $\widetilde{Q} = \begin{bmatrix} \Lambda_{11}^T P + P\Lambda_{11} + \frac{1}{\rho^2}PB_{11}P & \cdots & \Lambda_{1r}^T P + P\Lambda_{1r} + \frac{1}{\rho^2}PB_{1r}P \\ \vdots & \ddots & \vdots \\ \Lambda_{r1}^T P + P\Lambda_{r1} + \frac{1}{\rho^2}PB_{r1}P & \cdots & \Lambda_{rr}^T P + P\Lambda_{rr} + \frac{1}{\rho^2}PB_{rr}P \end{bmatrix}$.

If $\widetilde{Q} < 0$,then from theorem 4 and corollary 2, there exist a symmetry positive matrix $\tilde{Z}$ such that

$$\widetilde{Q} < -\tilde{Z} \tag{24}$$

and for the given constant $\rho > 0$, system (16) can obtain the H_∞ control performance (19). Obviously, $\widetilde{Q} < 0$ is equivalent to the matrix inequality (23). □

4 Design of the H_∞ Controllers

In this section, we will deal with the design of the H_∞ controllers. We let $X = diag\{X_1\ X_2\}$, $X = X^T > 0$, and let

$$\begin{aligned}
N_{ij} &= X_1(A_i + A_j)^T + (A_i + A_j)X_1 + (B_{2i}F_j)^T + (B_{2i}F_j) + (B_{2j}F_i)^T \\
&+ (B_{2j}F_i)
\end{aligned}$$

$$\begin{aligned}
M_{ij} &= X_2(A_i + A_j)^T + (A_i + A_j)X_2 - X_2(G_iC_j + G_jC_i)^T \\
&- (G_iC_j + G_jC_i)X_2 + \frac{1}{\rho^2}(B_{1i}B_{1i}^T + B_{1j}B_{1j}^T)
\end{aligned}$$

$$H_{ij} = (G_iC_j + G_jC_i)X_2$$

(23) is equivalent to the following matrix inequality

$$Q = \begin{pmatrix} \begin{pmatrix} N_{11} & H_{11} \\ H_{11}^T & M_{11} \end{pmatrix} & \cdots & \begin{pmatrix} N_{1r} & H_{1r} \\ H_{1r}^T & M_{1r} \end{pmatrix} \\ \vdots & \ddots & \vdots \\ \begin{pmatrix} N_{r1} & H_{r1} \\ H_{r1}^T & M_{r1} \end{pmatrix} & \cdots & \begin{pmatrix} N_{rr} & H_{rr} \\ H_{rr}^T & M_{rr} \end{pmatrix} \end{pmatrix} < 0 \tag{25}$$

where $K_i = F_i X_1^{-1}, i = 1, 2, ..., r$ is the state feedback matrices of the observers, $G_i, i = 1, 2, ..., r$ is the error feedback matrices. (25) is not LMI. It can not be solved directly by using MATLAB. So we need to quote the following lemma proved in [10]:

Lemma 2. *For the given symmetrical matrix* $S = \begin{bmatrix} S_{11} & S_{12} \\ S_{21} & S_{22} \end{bmatrix}$ *where the dimension of* S_{11} *is* $r \times r$*. Then the following conditions are equivalent*
(1)$S < 0$;
(2)$S_{11} < 0, S_{22} - S_{12}^T S_{11}^{-1} S_{12} < 0$;
(3)$S_{22} < 0, S_{11} - S_{12} S_{22}^{-1} S_{12}^T < 0$.

We first solve the error feedback matrices $G_i, i = 1, ..., r$. From (25) and lemma 2, we have

$$M_{ii} = X_2 A_i^T + A_i X_2 - X_2 C_i^T G_i^T - G_i C_i X_2 + \frac{1}{\rho^2} B_{1i} B_{1i}^T < 0 \tag{26}$$

$$i = 1, 2, ..., r$$

from lemma 2 again, we have

$$\begin{pmatrix} (A_i^T X_3 + X_3 A_i - (H_i C_i)^T - (H_i C_i) & X_3 B_{1i} \\ B_{1i}^T X_3 & -\frac{1}{\rho^2} I \end{pmatrix} < 0, i = 1, 2, ..., r \tag{27}$$

where $X_3 = X_2^{-1}$, now (27) are LMIs. By using MATLAB, we have $X_3, G_i = X_3^{-1} H_i (i = 1, ..., r)$. With $G_i, i = 1, 2, ..., r$ already known, (25) become a LMI, and by using MATLAB, we can obtain $X_1, X_2, K_i = F_i X_1^{-1} (i = 1, ..., r)$. So the H_∞ controllers are obtained.

5 Simulation

We consider the following problem of balbancing an inverted pendulum on a cart. The equations for the pendulum are

$$\begin{aligned} \dot{x}_1 &= x_2 \\ \dot{x}_2 &= \frac{g \sin(x_1) - aml x_2^2 \sin(2x_1) - a \cos(x_1) u}{\frac{4l}{3} - aml \cos^2(x_1)} + \omega \end{aligned}$$

Where x_1 denotes the angle of the pendulum from the vertical, x_2 is the angular velocity, $g = 9.8 m/s^2$ is the gravity constant, ω is the external disturbance variable which is a sinusoidal signal, $\omega = \sin(2\pi t)$. m is the mass of the pendulum, M is the mass of the cart, $2l$ is the length of the pendulum, and u is the force applied to the cart. a=$1/(m + M)$. We choose m=2.0kg, M=8.0kg, $2l$=1.0m in the simulation. The following fuzzy model is used to design state feedback fuzzy controllers:

$$\begin{aligned} \dot{x}(t) &= \sum_{i=1}^{2} \mu_i(x(t))(A_i x(t) + B_{1i}\omega(t) + B_{2i} u(t)) \\ y(t) &= \sum_{i=1}^{2} \mu_i(x(t)) C_i x(t) \end{aligned}$$

where

$$A_1 = \begin{bmatrix} 0 & 1.0000 \\ 17.2941 & 0 \end{bmatrix}, B_{21} = \begin{bmatrix} 0 \\ -0.1765 \end{bmatrix}, B_{11} = \begin{bmatrix} 0 \\ 1 \end{bmatrix}, C_1 = [1, 0],$$

$$A_2 = \begin{bmatrix} 0 & 1.0000 \\ 12.6305 & 0 \end{bmatrix}, B_{22} = \begin{bmatrix} 0 \\ -0.0779 \end{bmatrix}, B_{12} = \begin{bmatrix} 0 \\ 1 \end{bmatrix}, C_2 = [1, 0].$$

We use the following membership functions:

$$\mu_1(x_1) = \left(1 - \frac{1}{1 + \exp(-7(x_1 + \frac{\pi}{4}))}\right), \mu_2(x_1) = 1 - \mu_1(x_1)$$

let $\rho = 1$, from (27) we have

$$G_1 = [3.2586, 26.9493]^T, G_2 = [3.2586, 22.2857]^T$$

Since $G_i, i = 1, 2$ are already known, (25) is a LMI, using MATLAB we have

$$X_1 = 1.0e + 003 * \begin{bmatrix} 1.1266 & -1.5615 \\ -1.5615 & 2.5377 \end{bmatrix}, X_2 = \begin{bmatrix} 4.3139 & 7.1103 \\ 7.1103 & 44.7897 \end{bmatrix},$$

$$F_1 = 1.0e + 005 * [1.3256, -0.7481], F_2 = 1.0e + 005 * [1.7632, -1.9974],$$

$$K_1 = [522.1859, 291.8400], K_2 = [322.3536, 119.6480].$$

Fig.1 shows the response of system (16) for an initial condition $x_1 = 70^\circ, x_2 = 1$, and $\omega(t) = \sin 2\pi t$.

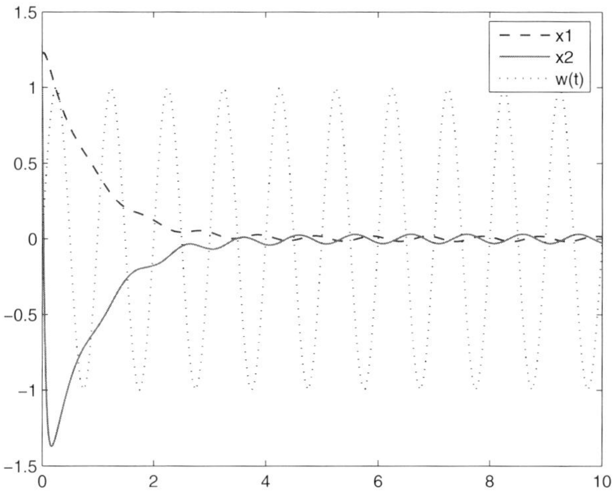

Fig. 1. $\omega(t) = \sin 2\pi t$ initial condition $x_1 = 70^\circ, x_2 = 1$

6 Conclusion

This paper mainly concerns about the systems which are composed of the observers and the error systems. Some new sufficient conditions which guarantee the quadratically stable and the existence of the state feedback H_∞ control for the systems are proposed. The condition in the theorem 3 is in the form of a matrix inequality, which is simple and holistic, and the condition is relaxed. Finally, the condition is converted to the LMIs which can be solved by using MATLAB.

References

1. Takagi, T., Sugeno, M.: Stability analysis and design of fuzzy control systems. Fuzzy Sets and Systems. **45**(2)(1993)135-156
2. Feng, G., Cao, G., Rees, W.: An approach to control of a class of nonlinear systems. Automatica. **32**(10)(1996)1469-1474
3. Cao, S. G., Rees, N. W., Feng, G.: Stability analysis and design for a class of continuous-time fuzzy control systems. International Journal of Control. **64**(6)(1996)1069-1087
4. Feng, G., Cao, S. G., Rees,N.W. Chak, C. K.: Design of fuzzy control systems with guaranteed stability. Fuzzy Sets Syst. **85**(1997)1-10
5. Liu,X. D., Zhang, Q. L.: Approaches to Quadratic Stability Conditions and H_∞ Control Designs for TS Fuzzy Systems. IEEE Trans On Fuzzy Systems. **11**(6)(2003)830-839
6. Tanaka, K., Ikede, T., Wang, H. O.: Fuzzy regulators and fuzzy observers: Relaxed stability conditions and LMI-baesd design.IEEE Trans on Fuzzy Systems. **6**(2)(1998)250-265
7. Tanaka, K., Ikede, T., Wang, H. O.: An LMI approach to fuzzy controller designs based on the relaxed stability conditions. Procof IEEE Int Conf Fuzzy System. Barcelona.(1997)171-176
8. Liu, G.R., Luo, Y.P., Wan, B.W.: H-two/H-innity fuzzy state feedback control based on fuzzy observer for nonlinear MIMO systems.Control Theory and Applications. **22**(1)(2005)12-16
9. Li, M., Zhang, H.G. Wang, C.H.: Fuzzy Tracking Control Design for Nonlinear Systems Via T-S Fuzzy Performance Evaluater.ACTA AUTOMATICA SINICA. **30**(4)(2004)578-582
10. Hale,J.: Theory of Functional Differential Equations. Springer. New York. 1977

New Robust Stability Criterion for Uncertain Fuzzy Systems with Fast Time-Varying Delays

Jiqing Qiu and Jinhui Zhang

College of Sciences, Hebei University of Science and Technology, Shijiazhuang, Hebei 050018, P.R. China
jinhuizhang82@gmail.com

Abstract. This paper considers the robust stability of uncertain T-S fuzzy system with time-varying delay. A new delay-dependent stability condition for the system is derived in terms of LMIs. The constraint on the time-varying delay function is removed, which means that a fast time-varying delay is allowed. Numerical example is given to illustrate the effectiveness and less conservativeness of the developed techniques.

1 Introduction

Recently, people have paid more and more attention on the robust stability of T-S fuzzy system with time-varying delay. Unfortunately, the existing results always assume that the time-varying delay function is continuously differentiable and its derivative is smaller than one, see [1] for example, which is a rigorous constraint. Therefore, it is interesting but challenging to develop the robust stability condition without any constraint on the time-varying delay.

In this paper, the problem of robust stability for uncertain T-S fuzzy system with time-varying delay is investigated. Based on the Lyapunov functional approach and Leibniz-Newton formula, a new delay-dependent criteria is presented in terms of linear matrix inequalities (LMIs) which can be easily solved by efficient interior-point algorithm. The derivative of the time-varying delay function may be larger than one. Numerical example is given to illustrate that the obtained results are less conservative than the existing results in the literature.

2 Problem Formulation

Consider the following uncertain T-S fuzzy system, the ith rule of this T-S fuzzy model is of the following form:

Plant Rule i:

IF $z_1(t)$ is M_{i1} and $\cdots$ and $z_p(t)$ is M_{ip} THEN

$$\dot{x}(t) = (A_i + \Delta A_i(t))x(t) + (B_i + \Delta B_i(t))x(t - \tau(t)), \tag{1}$$

$$x(t) = \varphi(t), \quad t \in [-\bar{\tau}, 0], \quad i = 1, 2, \cdots, r, \tag{2}$$

where $z_1(t)$, $z_2(t)$, $\cdots$, $z_p(t)$ are the premise variables, and M_{ij}, $j = 1, 2, \cdots, p$ are fuzzy sets, $x(t)$ is the state variable, r is the number of if-then rules, $\tau(t)$ is

L. Wang et al. (Eds.): FSKD 2006, LNAI 4223, pp. 41–44, 2006.

the time-varying delay satisfies $0 \leq \tau(t) \leq \bar{\tau}$ and $\dot{\tau}(t) \leq d$, $\varphi(t)$ is a vector-valued initial condition. The parametric uncertainties $\Delta A_i(t)$, $\Delta B_i(t)$ are time-varying matrices with appropriate dimensions, which are defined as follows:

$$\Delta A_i(t) = D_{1i}F_{1i}(t)E_{1i}, \quad \Delta B_i(t) = D_{2i}F_{2i}(t)E_{2i}, \quad i = 1, 2, \cdots, r, \tag{3}$$

where D_{1i}, E_{1i}, D_{2i}, E_{2i} are known constant real matrices with appropriate dimensions and $F_{1i}(t)$ and $F_{2i}(t)$ are unknown real time-varying matrices with Lebesgue measurable elements bounded by:

$$F_{1i}^T(t)F_{1i}(t) \leq I, \quad F_{2i}^T(t)F_{2i}(t) \leq I, \quad i = 1, 2, \cdots, r. \tag{4}$$

By fuzzy blending, the overall fuzzy model is inferred as follows:

$$\begin{aligned} \dot{x}(t) &= \frac{\sum_{i=1}^{r} \omega_i(z(t))[(A_i + \Delta A_i(t))x(t) + (B_i + \Delta B_i(t))x(t - \tau(t))]}{\sum_{i=1}^{r} \omega_i(z(t))} \\ &= \sum_{i=1}^{r} \mu_i(z(t))[\bar{A}_i(t)x(t) + \bar{B}_i(t)x(t - \tau(t))], \end{aligned} \tag{5}$$

with $\omega_i(z(t)) = \prod_{l=1}^{p} M_{il}(z_l(t))$, $\mu_i(z(t)) = \frac{\omega_i(z(t))}{\sum_{i=1}^{r} \omega_i(z(t))}$, $\bar{A}_i(t) = A_i + \Delta A_i(t)$, $\bar{B}_i(t) = B_i + \Delta B_i(t)$, $M_{il}(z_l(t))$ is the membership degree of $z_l(t)$ in M_{il}. It is assumed that $\omega_i(z(t)) \geq 0$, $i = 1, 2, \cdots, r$, $\sum_{i=1}^{r} \omega_i(z(t)) > 0$ for all t, so we have $\mu_i(z(t) \geq 0$ and $\sum_{i=1}^{r} \mu_i(z(t)) = 1$.

Remark 1. In many existing papers, the assumption $\dot{\tau}(t) \leq d < 1$ is needed, see [1] for example, but in this paper, this constraint is not necessary, which means that a fast time-varying delay is allowed.

3 Main Results

In this section, we will derive a delay-dependent robust stability condition for the following uncertain fuzzy system (6).

$$\begin{aligned} \dot{x}(t) &= \sum_{i=1}^{r} \mu_i(z(t))\bar{A}_i(t)x(t) + \sum_{i=1}^{r} \mu_i(z(t))\bar{B}_i(t)x(t - \tau(t)) \\ &= A(t)x(t) + B(t)x(t - \tau(t)) \end{aligned} \tag{6}$$

where $A(t) = \sum_{i=1}^{r} \mu_i(z(t))\bar{A}_i(t), B(t) = \sum_{i=1}^{r} \mu_i(z(t))\bar{B}_i(t)$.

Theorem 1. *The uncertain fuzzy system (6) is robustly asymptotically stable, if there exist symmetric positive definite matrices P, Q, R and real matrices P_1, P_2, N_1, N_2 and scalar $\varepsilon_{1i} > 0$, $\varepsilon_{2i} > 0$, $i = 1, 2, \cdots, r$, such that the following LMI holds:*

$$\Sigma = \begin{bmatrix} \Sigma_{11} & P - P_1^T + A_i^T P_2 & P_1^T B_i - N_1^T + N_2 & -N_1^T & P_1^T D_{1i} & P_1^T D_{2i} \\ \star & -P_2^T - P_2 + \bar{\tau}^2 R & P_2^T B_i & 0 & P_2^T D_{1i} & P_2^T D_{2i} \\ \star & \star & \Sigma_{33} & -N_2^T & 0 & 0 \\ \star & \star & \star & -R & 0 & 0 \\ \star & \star & \star & \star & -\varepsilon_{1i} I & 0 \\ \star & \star & \star & \star & \star & -\varepsilon_{2i} I \end{bmatrix} < 0, \tag{7}$$

where $\Sigma_{11} = P_1^T A_i + A_i^T P_1 + Q + N_1 + N_1^T + \varepsilon_{1i} E_{1i}^T E_{1i}$, $\Sigma_{33} = -(1-d)Q - N_2 - N_2^T + \varepsilon_{2i} E_{2i}^T E_{2i}$.

In all matrices, "$\star$*" denotes the symmetric terms in a symmetric matrix*

Proof. Choose the following positive definite Lyapunov functional:

$$V(t) = x^T(t)Px(t) + \int_{t-\tau(t)}^{t} x^T(s)Qx(s)ds + \bar{\tau}\int_{-\bar{\tau}}^{0}\int_{t+\theta}^{t} \dot{x}^T(s)R\dot{x}(s)dsd\theta. \quad (8)$$

Then, taking the derivative of $V(t)$ along the trajectory of system (6), and using the Lemma 1 in [2], we have

$$\begin{aligned}\dot{V}(t) &\leq 2x^T(t)P\dot{x}(t) + 2[x^T(t)P_1^T + \dot{x}^T(t)P_2^T] \times [-\dot{x}(t) + A(t)x(t) \\ &+ B(t)x(t-\tau(t))] + x^T(t)Qx(t) - (1-d)x^T(t-\tau(t))Qx(t-\tau(t)) \\ &+ \bar{\tau}^2\dot{x}^T(t)R\dot{x}(t) - \left[\int_{t-\tau(t)}^{t}\dot{x}(s)ds\right]^T R\int_{t-\tau(t)}^{t}\dot{x}(s)ds.\end{aligned}$$

Using Leibniz-Newton formula, we have

$$2[x^T(t)N_1^T + x^T(t-\tau(t))N_2^T] \times [x(t) - x(t-\tau(t)) - \int_{t-\tau(t)}^{t}\dot{x}(s)ds] = 0. \quad (9)$$

Then adding up (9) to $\dot{V}(t)$, we have

$$\dot{V}(t) = \eta^T(t)\begin{bmatrix} \Theta_{11} & P - P_1^T + A^T(t)P_2 & P_1^T B(t) - N_1^T + N_2 & -N_1^T \\ \star & -P_2^T - P_2 + \bar{\tau}^2 R & P_2^T B(t) & 0 \\ \star & \star & -(1-d)Q - N_2 - N_2^T & -N_2^T \\ \star & \star & \star & -R \end{bmatrix}\eta(t),$$

where $\Theta_{11} = P_1^T A(t) + A^T(t)P_1 + Q + N_1 + N_1^T$ and $\eta(t) = [x^T(t), \dot{x}^T(t), x^T(t-\tau(t)), (\int_{t-\tau(t)}^{t}\dot{x}(s)ds)^T]^T$.

Multiplying the left and the right sides of Σ by vector $\zeta^T(t)$ and $\zeta(t)$ respectively, we have $\zeta^T(t)\Sigma\zeta(t) < 0$, where

$$\begin{aligned}\zeta(t) = [x^T(t),\ \dot{x}^T(t),\ x^T(t-\tau(t)),\ &\int_{t-\tau(t)}^{t}\dot{x}^T(s)ds, \\ x^T(t)E_{1i}^T F_{1i}^T(t),\ x^T(t-\tau(t))&E_{2i}^T F_{2i}^T(t)]^T.\end{aligned}$$

Noting that, for any positive scalars $\varepsilon_{1i} > 0$ and $\varepsilon_{2i} > 0$, the following inequalities hold:

$$\begin{aligned}&\varepsilon_{1i}[F_{1i}(t)E_{1i}x(t)]^T[F_{1i}(t)E_{1i}x(t)] \leq \varepsilon_{1i}x^T(t)E_{1i}^T E_{1i}x(t), \\ &\varepsilon_{2i}[F_{2i}(t)E_{2i}x(t-\tau(t))]^T[F_{2i}(t)E_{2i}x(t-\tau(t))] \\ &\qquad \leq \varepsilon_{2i}x^T(t-\tau(t))E_{2i}^T E_{2i}x(t-\tau(t)).\end{aligned}$$

Based on the above two inequalities, we obtain $\eta^T(t)\Upsilon^{(i)}\eta(t) < 0$, where

$$\Upsilon^{(i)} = \begin{bmatrix} \Upsilon_{11}^{(i)} & P - P_1^T + \bar{A}_i^T(t)P_2 & P_1^T\bar{B}_i(t) - N_1^T + N_2 & -N_1^T \\ \star & -P_2^T - P_2 + \bar{\tau}^2 R & P_2^T\bar{B}_i(t) & 0 \\ \star & \star & -(1-d)Q - N_2 - N_2^T & -N_2^T \\ \star & \star & \star & -R \end{bmatrix} < 0, \quad (10)$$

and $\Upsilon_{11}^{(i)} = P_1^T\bar{A}_i(t) + \bar{A}_i^T(t)P_1 + Q + N_1 + N_1^T$.

Therefore, $\dot{V}(t) = \sum_{i=1}^{r}\mu_i(z(t))\eta^T(t)\Upsilon^{(i)}\eta(t) < 0$, from Lyapunov stability theorem, we can claim that if (7) holds, then system (6) is asymptotically stable.

4 Numerical Example

In this section, we borrow the example in [1] to illustrate the less conservativeness of our results.

Example 1. Consider the uncertain fuzzy system (6) with parameters:

$$A_1 = \begin{bmatrix} -3.2 & 0.6 \\ 0 & -2.1 \end{bmatrix},\ B_1 = \begin{bmatrix} 1 & 0.9 \\ 0 & 2 \end{bmatrix},\ A_2 = \begin{bmatrix} -1 & 0 \\ 1 & -3 \end{bmatrix},\ B_2 = \begin{bmatrix} 0.9 & 0 \\ 1 & 1.6 \end{bmatrix}$$

The membership function for Rule 1 and Rule 2 are

$$M_1(x_1(t)) = \frac{1}{1+\exp(-2x_1(t))}, \quad M_2(x_1(t)) = 1 - M_1(x_1(t)).$$

When $d = 0$, both Theorem 1 of [3] and Theorem 1 of [4] fail to verify that the system is asymptotically stable, and using Corollary 1 in [1], the upper bound of the time delay is $\bar{\tau}_{\max} = 0.58$, but using Theorem 1 in this paper, we have the upper bound of the time delay is $\bar{\tau}_{\max} = 0.6148$. Obviously, our result is less conservative than that obtained by the method in [1].

5 Conclusions

In this paper, we investigate the robust stability problem for uncertain T-S fuzzy system with time-varying delay. Based on the Lyapunov functional approach, a sufficient condition for the asymptotic stability of the uncertain fuzzy system is obtained. Numerical example illustrates the less conservativeness of our results.

References

1. Li, C., Wang, H., Liao, X.: Delay-Dependent Robust Stability of Uncertain Fuzzy Systems with Time-Varying Delays. IEE Proc. Control Theory Appl. **151** (2004) 417–421
2. Han, Q.: A Descriptor System Approach to Robust Stability of Uncertain Neutral Systems with Discrete and Distributed Delays. Automatica, **40** (2004) 1791–1796
3. Cao, Y., Frank, P.: Analysis and Synthesis of Nonlinear Time-Delay Systems via Fuzzy Control Approach. IEEE Trans. Fuzzy Syst. **8** (2000) 200–211
4. Zhang, Y., Heng, P.: Stability of Fuzzy Control Systems with Bounded Uncertain Delays. IEEE Trans. Fuzzy Syst. **10** (2002) 92–97

Stability Analysis and Controller Design of Discrete T-S Fuzzy System*

Jianxiang Yu, Songtao Zhang, and Guang Ren

Marine Engineering College, Dalian Maritime University, Dalian 116026, P.R. China
{yujianxiang, zst0626}@163.com, reng@dlmu.edu.cn

Abstract. For the conservative and difficulty of checking the stability of discrete T-S fuzzy control system with the common Lyapunov function approach and the fuzzy Lyapunov function approach, a fuzzy controller is designed to acquire globally asymptotical stability for discrete fuzzy system with the method of parallel distributed compensation (PDC) after the definition of a piecewise fuzzy Lyapunov function. Then a new sufficient condition to check the stability of closed-loop discrete T-S fuzzy system is proposed and proved. This condition is less conservative and difficult than above approaches. At last, a simulation example shows that the approach is effective.

1 Introduction

Recently, there has been a rapidly growing interest in the stability issues of T-S fuzzy systems. Most of stability conditions in terms of the common Lyapunov function [1] or the fuzzy Lyapunov function [2] are both conservative, since the common positive definite matrix $\mathbf{P}$ should satisfy r (rules' number) Lyapunov inequalities in the former, or a set of local matrices $\mathbf{P}_1, \mathbf{P}_2, \cdots \mathbf{P}_r$ should satisfy r^2 Lyapunov inequalities in the latter.

In order to overcome the shortcoming of the above two conditions, this paper proposes a new sufficient condition to check the stability of closed-loop discrete T-S fuzzy system based on the definition of a discrete piecewise fuzzy Lyapunov function. This condition only needs to satisfy the condition of fuzzy Lyapunov approach in each maximal overlapped-rule group. Therefore, the proposed condition is less conservative and difficult than former two approaches. A fuzzy controller is designed to acquire globally asymptotical stability for discrete fuzzy system with the method of parallel distributed compensation (PDC). A simulation example shows the approach is effective.

2 Main Result

A discrete T-S fuzzy model can be written as follows:

$$R_i : \text{IF } x_1(k) \text{ is } M_1^i, \text{and} \cdots, \text{and } x_n(k) \text{ is } M_n^i, \text{ THEN } \mathbf{X}(k+1) = \mathbf{A}_i \mathbf{X}(k) + \mathbf{B}_i \boldsymbol{u}(k), \tag{1}$$

* The work is supported by the Ministry of Communication of P.R. China (Grant #200332922505) and Doctoral Bases Foundation of the Educational Committee of P.R. China (Grant #20030151005).

L. Wang et al. (Eds.): FSKD 2006, LNAI 4223, pp. 45–48, 2006.

where i=1,2,⋯,r, r and n are the numbers of rules and state variables respectively, $\boldsymbol{X}^{\mathrm{T}}(k)=[x_1(k),x_2(k),\cdots,x_n(k)]$ is the state vector, M_j^i (j=1,⋯,n) is the fuzzy set. By the singleton fuzzifier, the product inference engine and center average defuzzification, the final output of (1) is inferred as:

$$\mathbf{X}(k+1)=\sum_{i=1}^{r} h_i(k)\mathbf{A}_i\mathbf{X}(k)+\mathbf{B}_i\boldsymbol{u}(k), \tag{2}$$

where $h_i(k)=\prod_{j=1}^{n} M_j^i(\mathbf{X}_j(k)) \Big/ \sum_{i=1}^{r}\prod_{j=1}^{n} M_j^i(\mathbf{X}_j(k))$.

Parallel distributed compensation (PDC) is a simple and natural design technique for a T-S fuzzy model (1). For (1), let $\boldsymbol{K}_i$ denote the state feedback gain of the ith local model, then the global model of a fuzzy controller can be inferred as follows:

$$\boldsymbol{u}(k)=-\sum_{i=1}^{r} h_i(k)\boldsymbol{K}_i\boldsymbol{X}(k) \tag{3}$$

In this paper, all of discussions and results are aimed at the prescribed the concepts of standard fuzzy partition (SFP) and maximal overlapped-rule group (MORG) as [3]. *Definition 1:* For a fuzzy system described by (2) with SFP inputs, if any of overlapped-rules groups is described as g_c (c=1,2,⋯,f), a discrete piecewise fuzzy Lyapunov function is defined as

$$V(\mathbf{X}(k))=\mathbf{X}^{\mathrm{T}}(k)\mathbf{P}(k)\mathbf{X}(k),\ \mathbf{P}(k)=\sum_{c=1}^{f}\lambda_c\mathbf{P}_c(k), \tag{4}$$

where $\lambda_c(\mathbf{X}(k))=\begin{cases}1 & \mathbf{X}(k)\in g_c\\ 0 & \mathbf{X}(k)\notin g_c\end{cases}$, $\sum_{c=1}^{f}\lambda_c(\mathbf{X}(k))=1$, $\mathbf{P}_c(k)=\sum_{i\in L_c} h_i(k)\mathbf{P}_i$, f denotes the number of overlapped-rules groups, and L_c={the sequence numbers of rules included in g_c }.

Theorem 1. For a fuzzy control system described by (2) and (3), if the input variables adopt SFPs, and let $\boldsymbol{G}_{ik}=\mathbf{A}_i-\mathbf{B}_i\boldsymbol{K}_k$, then the equilibrium of the closed-loop fuzzy control system is asymptotically stable in the large if there exist positive definite matrices $\mathbf{P}_i$ (or $\mathbf{P}_l$) in each MORG such that

$$\boldsymbol{G}_{ik}^{\mathrm{T}}\mathbf{P}_l\boldsymbol{G}_{ik}-\mathbf{P}_i<0,\ i,k,l\in \{\text{the sequence numbers of rules included in } G_q\}, \tag{5}$$

where G_q denotes the qth MORG, q=1,2,⋯,$\prod_{j=1}^{n}(m_j-1)$, and m_j denotes the number of fuzzy partition of the jth input variable.

Proof. The proof is similar to the proof of Theorem 4 in [3].

3 Numerical Example

In this section, a two-dimensional mapping Henon is chosen to illustrate the stability examination and the controller design for a T-S system in detail. The system state equation of the two-dimensional mapping Henon is following:

$$x(k+2)=1+bx(k)-ax^2(k+1) . \quad (6)$$

When $a=1$ and $b=-3$ the system takes on a chaos state[4].

Impose a force on the two-dimensional mapping Henon, then equation (7) is following:

$$x(k+2)=1+bx(k)-ax^2(k+1)+u . \quad (7)$$

In order to model (7), a T-S fuzzy system is considered as follows:

$$R_i : \text{IF } x_1(k) \text{ is } M_1^i \text{ and } x_2(k) \text{ is } M_2^i \text{ THEN } \boldsymbol{X}(k+1)=\mathbf{A}_i\boldsymbol{X}(k)+\mathbf{B}_i\boldsymbol{u} \quad (8)$$

where $i=1,2,\cdots,9$. Let $x_1(k)=x(k)$ and $x_2(k)=x_1(k+1)$. The fuzzy partitions of $x_1(k)$ and $x_2(k)$ shown in Fig. 1 are $F_1^t(x_1(k))$ and $F_2^s(x_2(k))$ $(t,s=1,2,3)$ respectively, and conform to the conditions of SFP, and

$$M_1^1=M_1^2=M_1^3=F_1^1,\ M_1^4=M_1^5=M_1^6=F_1^2,\ M_1^7=M_1^8=M_1^9=F_1^3,$$
$$M_2^1=M_2^4=M_2^7=F_2^1,\ M_2^2=M_2^5=M_2^8=F_2^2,\ M_2^3=M_2^6=M_2^9=F_2^3.$$

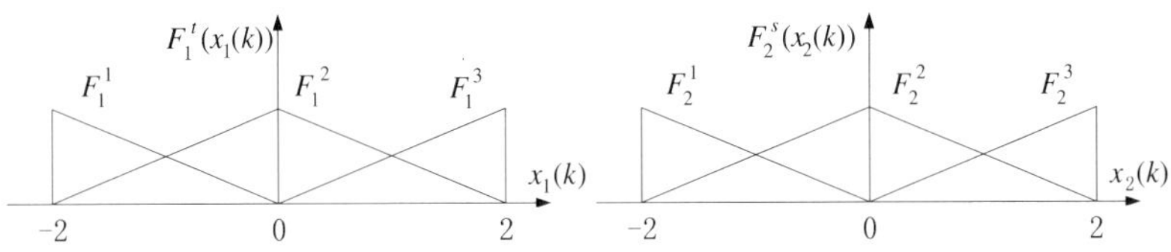

Fig. 1. The fuzzy partitions with $q_1=q_2=3$

We select the closed-loop eigenvalues of the 9 local linear subsystems via state feedback to be: $\mathbf{P}_1=\mathbf{P}_2=\mathbf{P}_3=\mathbf{P}_4=\mathbf{P}_5=\mathbf{P}_6=\mathbf{P}_7=\mathbf{P}_8=\mathbf{P}_9=[0.75 \quad 0.75]$. The state feedback gain of the local linear subsystems can be derived from Ackermann's formula.

We can conclude that this fuzzy system is stable by Theorem 1, for we have found 9 common positive definite matrices in the 4 MORGs satisfying the condition of Theorem 1 via the LMI approach.

We simulate the fuzzy system (7) using various initial conditions. The simulation result shows that this system is stable under all initial conditions. The system state responses under the initial condition of $\boldsymbol{X}_0(k)=[-1 \quad 1]^{\mathrm{T}}$ are shown in Fig. 2.

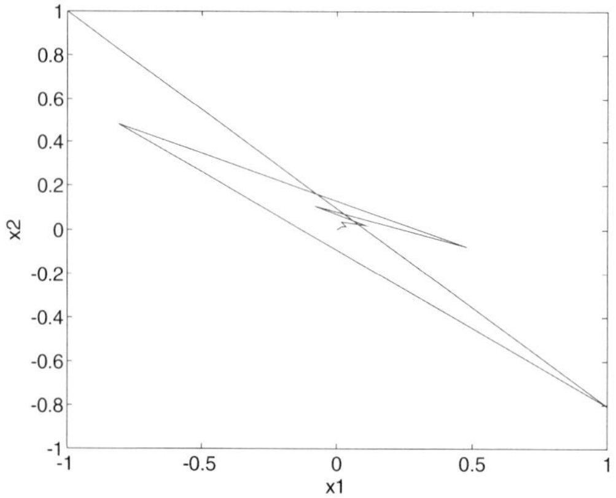

Fig. 2. The system state responses under the initial condition of $X_0(k) = [-1 \quad 1]^T$

4 Conclusions

This paper has contributed to the stability analysis and design of discrete T–S fuzzy control system. Based on the definition of a piecewise fuzzy Lyapunov function, a new stability condition of the closed-loop discrete T-S fuzzy control system is proposed. Our approach only needs to satisfy the condition of fuzzy Lyapunov approach in each MORG. This method can greatly reduce the conservatism and difficulty of the former stability analysis approaches. A design method of a T-S fuzzy controller is proposed by using the method of PDC. The simulation results show that this approach is effective.

References

1. Tanaka K., Sugeno M.: Stability Analysis and Design of Fuzzy Control System. Fuzzy Sets and Systems, 45 (1992) 135-156
2. Tanaka K., Hori T., Wang H. O.: A Multiple Lyapunov Function Approach to Stabilization of Fuzzy Control Systems. IEEE Trans. Fuzzy Systems, 11 (2003) 582-589
3. Xiu Z. H., Ren G.: Stability Analysis and Systematic Design of Takagi-Sugeno Fuzzy Control Systems. Fuzzy Sets and Systems, 151 (2005) 119-138
4. Henon M.: A Two Dimensional Map with A Strange Attractor. Math Phys. 50 (1976) 69-73

Stabilization of Multirate Sampled-Data Fuzzy Systems Based on an Approximate Discrete-Time Model

Do Wan Kim[1], Jin Bae Park[1], and Young Hoon Joo[2]

[1] Yonsei University, Seodaemun-gu, Seoul, 120-749, Korea
{dwkim, jbpark}@yonssei.ac.kr
[2] Kunsan National University, Kunsan, Chunbuk, 573-701, Korea
yhjoo@kunsan.ac.kr

Abstract. This paper studies a stabilization problem for a multirate digital control of fuzzy systems based on the approximately discretized model. In the multirate control scheme, a numerical integration scheme is used to approximately predict the current state from the state measured at the sampling points. It is shown that the multirate digital fuzzy controller stabilizing an approximate discrete-time fuzzy model would also stabilize the sampled-data fuzzy system in the sufficiently small control update time. Furthermore, some sufficient conditions for the stabilization of the approximate discrete-time fuzzy model are provided under the delta-operator frame work, which are expressed as the linear matrix inequalities (LMIs) and thereby easily tractable by the convex optimization techniques. A numerical example is demonstrated to visualize the feasibility of the developed methodology.

1 Introduction

Sampled-data systems are widespread more and more because most systems encountered in engineering applications are continuous while controls are implemented digitally using computers. Traditional analysis and design tools for continuous-time or discrete-time systems are unable to be directly used in the sampled-data systems. On way to address the sampled-data control is to develop a discrete-time model for the controlled, continuous-time plant and then pursue a digital controller based on the discretized model. This approach has been basically applicable for linear time-invariant (LTI) systems [8, 9, 12, 13, 14, 15, 16, 10, 11].

The nonlinear sampled-data control problem is difficult because exact discrete-time models of continuous-time processes are typically impossible to compute. From that reasons, there have been some researches focusing on the digital controller [1, 2, 3, 4, 5, 6] for Takagi–Sugeno (T-S) fuzzy systems based on their approximate discrete-time models. Although a great deal of effort has been made on digital control such as [1, 2, 3, 4, 5, 6], there still exists some matters that must be worked out. The first issue is how to efficiently tackle the stability preservation. It is a very important factor to preserve the stability in the digital controller,

L. Wang et al. (Eds.): FSKD 2006, LNAI 4223, pp. 49–58, 2006.

but the previous methods [1,2,3] do not only assure the stability of the sampled-data fuzzy closed-loop systems but also their approximately discretized model. At this point, the results [4, 5, 6] provided that sufficient conditions to stabilize the approximate discrete-time model of the sampled-data fuzzy system. However, they only show that the closed-loop sampled-data system is stable under the assumption that there exists no discretization error. Next, a considerable issue is about the multirate digital control. All of the previous results [1,2,3,4,5] are applicable only to a single-rate digital control in which the sampling and the control update periods are assumed to be equal. In practical applications, however, hardware restrictions can make two periods different essentially. There have been some investigations focusing on the multirate digital control of LTI systems from several disparate perspectives [12,13,14,15,16]. However, until now, no tractable method for the multirate digital fuzzy control has been proposed, with perhaps a few exceptions [6].

In this paper, we studies a multirate digital control of fuzzy systems based on the approximate discrete-time model. It is proved that the multirate digital fuzzy controller stabilizing an approximate discrete-time fuzzy model would also stabilize the sampled-data fuzzy system in the sufficiently small control update time. Some sufficient conditions for the stabilization of the approximate discrete-time fuzzy model are provided under the delta-operator framework, which are expressed as the linear matrix inequalities (LMIs) and thereby easily tractable by the convex optimization techniques. Furthermore, we show that the discretized error approach zero as increasing the input multiplicity. From this fact, we can design the digital controller stabilize the sampled-data fuzzy system in the wide range of the sampling period by increasing the input multiplicity.

The rest of this paper is organized as follows: Section 2 briefly reviews the T–S fuzzy system. In Section 3, the stability analysis and control synthesis of the multirate sampled-data fuzzy system is included. An example of a biodynamical system of human immunodeficiency virus type 1 (HIV-1) [21, 22] is provided in Section 4. Finally, Section 5 concludes this paper with some discussions.

2 Preliminaries

Consider the system described by the following T–S fuzzy model [17, 18]:

$$\dot{x}(t) = \sum_{i=1}^{r} \theta_i(z(t))(A_i x(t) + B_i u(t)) \tag{1}$$

where $x(t) \in \mathbb{R}^n$ and $u(t) \in \mathbb{R}^m$, r is the number of model rules, $z(t) = [z_1(t) \cdots z_p(t)]^T$ is the premise variable vector that is a function of states $x(t)$, and $\theta_i(z(t))$, $i \in \mathcal{I}_R(= \{1, 2, \cdots, r\})$ is the normalized weight for each rule, that is $\theta_i(z(t)) \geq 0$ and $\sum_{i=1}^{r} \theta_i(z(t)) = 1$.

We consider a multirate digital fuzzy system where $u(t)$ is held in constant between the (uniformly spaced) control update points. Let T and τ be the sampling and the control update periods, respectively, and assume $\tau = T/N$. The multirate digital fuzzy controller takes the following form:

$$u(t) = \sum_{i=1}^{r} \theta_i(z(kT + \kappa\tau))K_i x(kT + \kappa\tau) \tag{2}$$

for $t \in [kT + \kappa\tau, kT + \kappa\tau + \tau)$, $k \times \kappa \in \mathbb{Z}_{\geqslant 0} \times \mathbb{Z}_{[0,N-1]}$, where $x(kT + \kappa\tau)$, $\kappa \in \mathbb{Z}_{[1,N-1]}$ is predicted from $x(kT)$, and the subscript 'd' denotes the digital control. By substituting (2) into (1), the closed-loop sampled-data fuzzy system is obtained by

$$\dot{x}(t) = \sum_{i=1}^{r}\sum_{j=1}^{r} \theta_i(z(t))\theta_j(z(kT + \kappa\tau))(A_i x(t) + B_i K_j x(kT + \kappa\tau)) \tag{3}$$

for $t \in [kT + \kappa\tau, kT + \kappa\tau + \tau)$, $k \times \kappa \in \mathbb{Z}_{\geqslant 0} \times \mathbb{Z}_{[0,N-1]}$. A mixture of the continuous-time and discrete-time signals occurs in the above system (3). It makes traditional analysis tools for a homogeneous signal system unable to be directly used. It is found in [1, 2, 3, 4, 5, 6, 7] that the approximate discrete-time model of (3) takes the following form:

$$\begin{aligned} x(kT + \kappa\tau + \tau) \cong & \sum_{i=1}^{r}\sum_{j=1}^{r} \theta_i(z(kT + \kappa\tau))\theta_j(z(kT + \kappa\tau)) \\ & \times (G_i + H_i K_j)x(kT + \kappa\tau) \end{aligned} \tag{4}$$

where $G_i = \mathrm{e}^{A_i\tau}$ and $H_i = (G_i - I)A_i^{-1}B_i$.

3 Main Results

In this section, we show that the multirate digital fuzzy controller (2) stabilizing the approximate discrete-time fuzzy model (4) would also stabilize the multirate sampled-data fuzzy system (3) in the sufficiently small control update time. Furthermore, some sufficient conditions for the stabilization of the approximate discrete-time fuzzy model (4) are provided, which are expressed as the linear matrix inequalities (LMIs).

For the practical engineering approach, we consider the multirate control scheme that utilizes a numerical integration scheme to approximately predict the current state $x(kT+\kappa\tau)$ from the state $x(kT)$ measured at the sampling points, the delayed measurements. For more detail, see [19]. At this point, redefining (4) as $w(kT+\kappa\tau+\tau) \triangleq \mathcal{F}(w(kT+\kappa\tau))$, and rewriting (2) with $x(kT+\kappa\tau)$ replaced by $w(kT + \kappa\tau)$ leads

$$u(t) = \sum_{i=1}^{r} \theta_i(z(kT + \kappa\tau))F_i w(kT + \kappa\tau) \tag{5}$$

for $t \in [kT + \kappa\tau, kT + \kappa\tau + \tau)$, $k \times \kappa \in \mathbb{Z}_{\geqslant 0} \times \mathbb{Z}_{[0,N-1]}$, where $w(kT + \kappa\tau)$ is the approximate estimate of the state $x(kT + \kappa\tau)$ based on the measurements $x(kT)$, and $w(kT + \kappa\tau) = x(kT + \kappa\tau)$ if $\kappa = 0$.

Remark 1. Under the assumption that the premise variables vector $z(kT+\kappa\tau)$ can be computed from $x(kT+\kappa\tau)$, we can predict $w(kT+\kappa\tau)$ by the following recursive application of (4) defined as

$$\begin{aligned} w(kT+\tau) &\triangleq \mathcal{F}^1(w(kT)) \\ w(kT+2\tau) &= \mathcal{F}(w(kT+\tau)) \\ &= \mathcal{F}(\mathcal{F}^1(w(kT))) \\ &\triangleq \mathcal{F}^2(w(kT)) \\ w(kT+\kappa\tau) &\triangleq \mathcal{F}^\kappa(w(kT)). \end{aligned}$$

Redefining (1) as $\dot{x}(t) \triangleq f(x(t),u(t))$, and substituting (5) into (1) leads

$$\dot{x}(t) = f(x(t), w(kT+\kappa\tau)) \tag{6}$$

for $t \in [kT+\kappa\tau, kT+\kappa\tau+\tau)$, $k \times \kappa \in \mathbb{Z}_{\geqslant 0} \times \mathbb{Z}_{[0,N-1]}$.

Now, we show that the sampled-data system (6) is also asymptotically stable in the sufficiently small τ if the approximate discrete-time model (4) is asymptotically stable, which needs the following lemmas.

Lemma 1. *Let $f(x,u)$ be locally Lipschitz in their arguments. The exact discrete-time model of (6) takes the following form:*

$$\begin{aligned} x(kT+\kappa\tau+\tau) = &\mathcal{F}(x(kT+\kappa\tau), w(kT+\kappa\tau)) \\ &+ \tau^2\mathcal{E}(x(kT+\kappa\tau), w(kT+\kappa\tau)) \end{aligned} \tag{7}$$

Proof. The proof is omitted due to lack of space.

Lemma 2. *Let $\mathcal{F}(x,u)$ be locally Lipschitz in their arguments. Suppose that $\|\mathcal{E}(x(kT+\kappa\tau), u(kT+\kappa\tau))\| \leqslant \delta$ for some δ. Then,*

$$\|x(kT+\kappa\tau) - w(kT+\kappa\tau)\| \leqslant \frac{L_2^\kappa - 1}{L_2 - 1}\tau^2\delta \tag{8}$$

for any $k \times \kappa \in \mathbb{Z}_{\geqslant 0} \times \mathbb{Z}_{[0,N-1]}$.

Proof. The proof is omitted due to lack of space.

Remark 2. Note that the norm of the discretization error, $\|x(kT+\kappa_0\tau+\tau) - w(kT+\kappa_0\tau+\tau)\|$ will go to zero as τ approaches zero. Hence, the approximate discrete-time model can preserve the property and structure of (6) by increasing N.

Theorem 1. *The zero equilibrium $x_{eq} = [0]_{n\times 1}$ of (6) is asymptotically stable in the sufficiently small τ if the zero equilibrium $w_{eq} = [0]_{n\times 1}$ of the approximate discrete-time model (4) is asymptotically stable.*

Proof. The proof is omitted due to lack of space.

We now are in position to find some sufficient conditions such that the approximate discrete-time fuzzy model (4) is globally asymptotically stable in the sense of Lyapunov.

Remark 3. It is easy to see that $G_i \to I$ and $H_i \to [0]_{n\times m}$ as $\tau \to 0$, which signifies the eigenvalues of $G_i + H_iK_j$ gathers around one thereby weakens the numerical robustness of the related convex optimization problem.

To effectively tackle this problem, stability analysis technique based on the delta-operator is applied in this paper.

Remark 4. It has been shown that the delta-operator offers advantages over the shift operator, in terms of numerical robustness, i.e., lower coefficient sensitivity especially when the eigenvalues of a shift-operator-based discretized model are clustered around one, which corresponds to a fast sampling of the continuous-time representation of systems.

Theorem 2. *The system (4) is stabilizable by the controller (2) in the sufficiently small τ if there exist a matrix $Q = Q^T \succ 0$ and matrices $X_{ij} = X_{ij}^T = X_{ji} = X_{ji}^T$, M_i such that*

$$\begin{bmatrix} \left(\frac{QG_{\delta i}^T+M_j^TH_{\delta i}^T+QG_{\delta j}^T+M_i^TH_{\delta j}^T+G_{\delta i}Q+H_{\delta i}M_j+G_{\delta j}Q+H_{\delta j}M_i}{2}\right)+X_{ij} & (\bullet)^T \\ \left(\frac{\tau^{\frac{1}{2}}G_{\delta i}Q+\tau^{\frac{1}{2}}H_{\delta i}M_j+\tau^{\frac{1}{2}}G_{\delta j}Q+\tau^{\frac{1}{2}}H_{\delta j}M_i}{2}\right) & -Q \end{bmatrix} \prec 0 \tag{9}$$

$$[X_{ij}]_{r\times r} \succ 0, \quad 1 \leqslant i \leqslant j \leqslant r \tag{10}$$

where $(\bullet)^T$ denotes the transposed element in symmetric positions.

Proof. The proof is omitted due to lack of space.

Remark 5. The methodology in Theorems 2 for the state-feedback control can readily be modified to establish results for more general controls, which involve output feedback control, set-point regulation, robust control [23], and so on.

Corollary 1. *If $\tau \to 0$, then the following conditions are equivalent:*

(i) There exist $Q = Q^T \succ 0$ and matrices $X_{ij} = X_{ij}^T = X_{ji} = X_{ji}^T$, M_i such that LMIs (9) and (10)of Theorem 2.
(ii) There exist $Q = Q^T \succ 0$ and matrices $X_{ij} = X_{ij}^T = X_{ji} = X_{ji}^T$, M_i such that

$$\begin{aligned} &QA_{\delta i}^T + M_j^TB_{\delta i}^T + QA_{\delta j}^T + M_i^TB_{\delta j}^T \\ &\quad + A_iQ + B_{\delta i}M_j + A_{\delta j}Q + B_{\delta j}M_i + 2X_{ij} \prec 0 \end{aligned} \tag{11}$$

$$[X_{ij}]_{r\times r} \succ 0, \quad 1 \leqslant i \leqslant j \leqslant r \tag{12}$$

Proof. The proof is omitted due to lack of space.

Remark 6. Note that LMIs (11) and (12) is readily derived from the a continuous-time Lyapunov stability theorem by choosing $V = x(t)^T Px(t)$, and denoting $Q = P^{-1}$ and $K_i = M_i$. Hence, we conclude that the condition (i) of Corollary 1 converges a stabilizability condition [20] for the continuous-time fuzzy system as $\tau \to 0$.

4 Computer Simulations

We present in this section a numerical application in order to show the effectiveness of our approach. We wish to deign the multirate digital fuzzy controller (2) with $N = 5$ for the complex nonlinear systems. The comparisons of the recent method presented in [4] are provided.

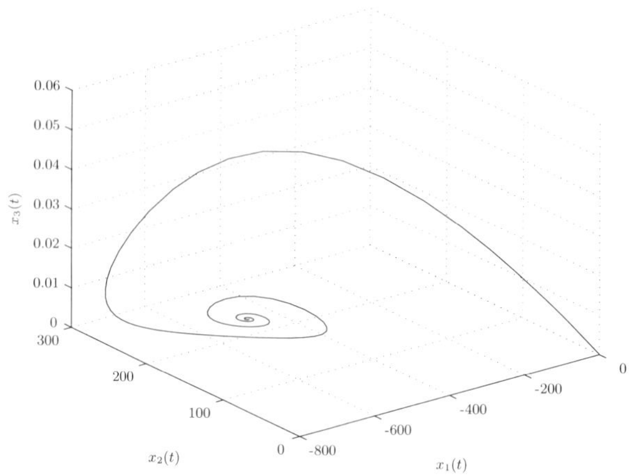

Fig. 1. Uncontrolled trajectory of the HIV-1 system

A biodynamic model of HIV-1 [21, 22] is given by

$$\begin{bmatrix} \dot{x}_1(t) \\ \dot{x}_2(t) \\ \dot{x}_3(t) \end{bmatrix} = \begin{bmatrix} -a_1 - a_2x_3(t) & 0 & -a_2b_1 \\ 0 & -a_3 + a_4x_3(t) & a_4b_2 \\ a_5x_3(t) & -a_6x_3(t) & a_5b_1 - a_6b_2 \end{bmatrix} \begin{bmatrix} x_1(t) \\ x_2(t) \\ x_3(t) \end{bmatrix} + \begin{bmatrix} 0 \\ 0 \\ 1 \end{bmatrix} u(t) \tag{13}$$

where $a_1 = 0.25$, $a_2 = 50$, $a_3 = 0.25$, $a_4 = 10.0$, $a_5 = 0.01$, $a_6 = 0.006$, $b_1 = 1000$ cells/mm^3, and $b_2 = 550$ cells/mm^3. As discussed in [21, 22], the HIV-1 system has two equilibrium points, where the desired equilibrium point is the origin. Fig. 1 shows the uncontrolled trajectory of this system. The initial conditions are $x_1 = x_2 = 0$ cells/mm^3 and $x_3 = 10^{-4}$ (corresponding ponding to k copies/ml).

To facilitate control design, we proceed to construct two-rule fuzzy model of HIV-1 system (13). To this end, the nonlinear terms x_3x_1 and x_3x_2 should be expressed as

$$x_3(t)x_1(t) = \theta_1(x_3(t)) \cdot x_{3min}x_1(t) + \theta_2(x_3(t)) \cdot x_{3max}x_1(t) \tag{14}$$
$$x_3(t)x_2(t) = \theta_1(x_3(t)) \cdot x_{3min}x_2(t) + \theta_2(x_3(t)) \cdot x_{3max}x_2(t) \tag{15}$$

where $\theta_1(x_3(t)) + \theta_2(x_3(t)) = 1$ and $x \in [x_{3min}, x_{3max}]$. Here, we can reasonably determine $[x_{3min}, x_{3max}]$ as $[-0.006, 0.006]$. Solving (14) or (15) for θ_1 and θ_2, and then using (14) and (15) to rewrite (13) as two-rule fuzzy model, we end up with

$$\dot{x}(t) = \sum_{i=1}^{2} \theta_i(x_3(t))(A_i x(t) + B_i u(t)) \tag{16}$$

where $\theta_1(x_3(t)) = \frac{-x_3(t)+x_{3max}}{x_{3max}-x_{3min}}$ and $\theta_2(x_3(t)) = \frac{x_3(t)-x_{3min}}{x_{3max}-x_{3min}}$, and the local system and input matrices are

$$\left[\begin{array}{c|c} A_1 & B_1 \\ \hline A_2 & B_2 \end{array}\right] = \left[\begin{array}{ccc|c} -a_1 - a_2 x_{3min} & 0 & -a_2 b_1 & 0 \\ 0 & -a_3 + a_4 x_{3min} & a_4 b_2 & 0 \\ a_5 x_{3min} & -a_6 x_{3min} & a_5 b_1 - a_6 b_2 & 1 \\ \hline -a_1 - a_2 x_{3max} & 0 & -a_2 b_1 & 0 \\ 0 & -a_3 + a_4 x_{3max} & a_4 b_2 & 0 \\ a_5 x_{3max} & -a_6 x_{3max} & a_5 b_1 - a_6 b_2 & 1 \end{array}\right]$$

This fuzzy model exactly represents the biodynamics of the nonlinear HIV-1 system under $x_{3min} \leq x_3 \leq x_{3max}$. Note that the fuzzy model does not has a common B, i.e., $B_1 = B_2$. In general, the fuzzy controller design of the common cases is simple. To show the effect of our approach, we consider a more difficult case, i.e., we change B_2 as follows:

$$B_2 = \begin{bmatrix} 0 \\ 0 \\ 5 \end{bmatrix}$$

We first seek to examine the convergence property of Theorem 2 for extremely small enough $T = 10^{-20}$ years. Using Theorem 2, we can find the multirate digital fuzzy gains

$$\left[\begin{array}{c} K_1 \\ \hline K_2 \end{array}\right] = \left[\begin{array}{ccc} 16 & -20 & -28878 \\ \hline 6 & -8 & -13004 \end{array}\right] \tag{17}$$

However, the LMIs given in [4] are infeasible due to the problem in Remark 3.

Next, we choose $T = 0.14$ years as the relatively large sampling time. Solving to Theorem 2 leads the following multirate digital gains:

$$\left[\begin{array}{c} K_1 \\ \hline K_2 \end{array}\right] = \left[\begin{array}{ccc} 0.0052 & -0.0070 & -25.6664 \\ \hline 0.0003 & -0.0010 & -8.4352 \end{array}\right] \tag{18}$$

However, the stability conditions in [4] are not strictly feasible, and then their digital gains are given by

$$\left[\frac{K_1}{K_2}\right] = \left[\frac{0.0008 \text{ -0.0009 -9.9082}}{\text{-0.0002 } 0.0001 \text{ -1.2297}}\right] \tag{19}$$

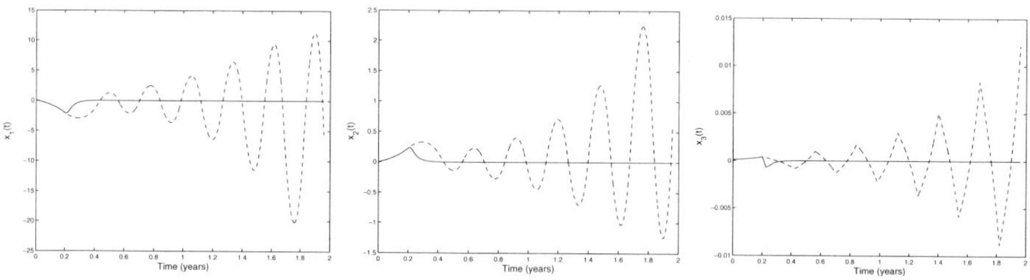

Fig. 2. Comparison of state responses of the controlled HIV-1 system (control input is activated at time $t = 0.2$ years): proposed (solid), [4] (dashed). The sampling period is $T = 0.14$ years.

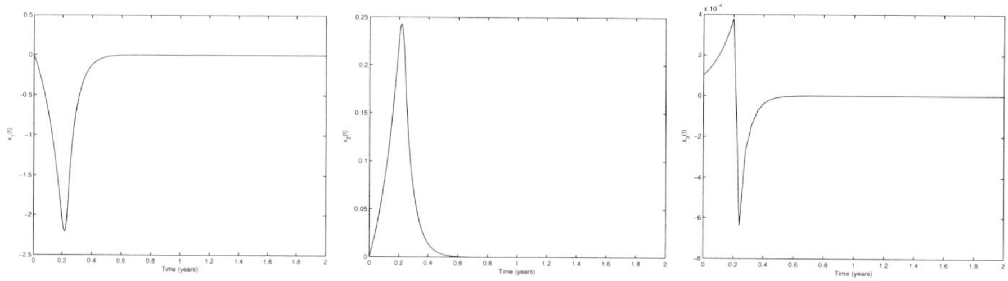

Fig. 3. State responses of the controlled HIV-1 system (control input is activated at time $t = 0.2$ years): proposed (solid) The sampling period is $T = 0.2$ years

Figs. 2 shows the time responses of two digitally controlled systems. As shown in the figures, the multirate digital control by the proposed method drives the trajectories to the equilibrium at the origin, while the other method fails to stabilize the system.

Another relatively longer sampling period $T = 0.2$ years is chosen. Applying Theorem 2 leads the following multirate digital gains are:

$$\left[\frac{K_1}{K_2}\right] = \left[\frac{0.0028 \; -0.0037 \; -18.8768}{0.0001 \; -0.0005 \; -5.6706}\right] \tag{20}$$

However, we cannot compute the feasible solution compute from the conditions in [4]. As shown in Fig. 3, the proposed controller well guarantee the stability preservation.

We emphasize that the proposed method guarantees the stability of the multirate sampled-data system in much wider range of sampling period than the previous method in which may fail to stabilize the system especially for relatively longer sampling period. This is because in the proposed method, the intersample behavior between sampling points can be considered, whereas the other approach does not.

5 Closing Remarks

In this paper, we have examined that a multirate digital controller that stabilize approximate discrete-time fuzzy model would also stabilize the resulting sampled-data fuzzy system in the sufficiently small control update time. To the authors' best knowledge, the proposed method is noble in several directions by considering: 1) the multirate digital control; 2) the stability of the multirate sample-data fuzzy system; 3) the stability analysis based on the delta operator. The simulation results on the HIV-1 convincingly demonstrated that it is possible to obtain the excellence performance through the proposed method.

Acknowledgment

This work was supported in part by the Korea Science and Engineering Foundation (Project number: R05-2004-000-10498-0).

References

1. Joo Y. H., Shieh L. S., and Chen G.: Hybrid state-space fuzzy model-based controller with dual-rate sampling for digital control of chaotic systems. IEEE Trans. Fuzzy Syst. **7** (1999) 394-408
2. Chang W., Park J. B., Joo Y. H., and Chen G.: Design of sampled-data fuzzymodel-based control systems by using intelligent digital redesign. IEEE Trans. Circ. Syst. I. **49** (2002) 509-517
3. Chang W., Park J. B., and Joo Y. H.: GA-based intelligent digital redesign of fuzzy-model-based controllers. IEEE Trans. Fuzzy Syst. **11** (2003) 35-44
4. Lee H. J., Kim H., Joo Y. H., Chang W., and Park J. B.: A new intelligent digital redesign for T.S fuzzy systems: global approach. IEEE Trans. Fuzzy Syst. **12** (2004) 274-284
5. Lee H. J., Park J. B., and, Joo Y. H.: Digitalizing a fuzzy observer-based output-feedback control: intelligent digital redesign approach. IEEE Trans. Fuzzy Syst. **13** (2005) 701-716
6. Kim D. W., Park J. B., and, Joo Y. H.: Intelligent digital control for nonlinear systems with multirate Sampling. LNAI **3613** (2005) 886-889
7. Kim D. W., Lee H. J., Park J. B., and, Joo Y. H.: Discretisation of continuous-time T–S fuzzy system: global approach. IEE Proc. Control Theory Appl. **153** (2006) 237-246
8. Chang W., Park J. B., Lee H. J., and Joo Y. H.: LMI approach to digital redesign of linear time-invarint systems. IEE Proc., Control Theory Appl. **149** (2002) 297-302
9. Lee H. J., Park J. B., and, Joo Y. H.: An efficient observer-based sampled-data control: Digital redesign approach. IEEE Trans. Circuits Syst. I **50**, (2003) 1595-1601

10. Shieh L. S., Wang W. M., and Tsai J. S. H.: Digital redesign of H8 controller via bilinear approximation method for state-delayed systems. Int. J. control **10** (1998) 665-683
11. Tsai J. S. H., Shieh C. S., and Sun Y. Y.: Digital modelling and robust digital redesign of sampled-data uncertain systems via the interval tuning bilinear approximation method. Journal of the Franklin Institute **338** (2001) 615-630
12. Rabbath C. A., Lechevin N., and Hori N.: Optimal dual-rate digital redesign with closed-loop order reduction. IEE Proc. Control Theory Appl. **152** (2005) 489-498
13. Chen Y. S., Tsai J. S. H., Shieh L. S., and Kung F. C.: New conditioning dual-rate digital-redesign scheme for continuous-time systems with saturating actuators. IEEE Trans. Circuits Syst. I **49** (2002) 1860-1870
14. Guo S. M., Shieh L. S., Chen G., and Lin C. F.: Effective chaotic orbit tracker: A prediction-based digital redesign approach. IEEE Trans. Circuits Syst. I **47** (2000) 1557-1570
15. Fujimoto H., Kawamura A., Tomizuka M.: Generalized digital redesign method for linear feedback system based on N-delay control. IEEE/ASME Trans. Mechatronics **4** (1999) 101-109
16. Shieh L. S., Wang W. M., Bain J., and Sunkel J. W.: Design of lifted dual-rate digital controllers for X-38 vehicle. Jounal of Guidance, Contr. Dynamics **23** (2000) 629-339
17. Wang H. O., Tananka K., and Griffin M. F.: An approach to fuzzy control of nonlinear systems: Stability and design issues. IEEE Trans. Fuzzy Syst. **4** (1996) 14-23
18. Tananka K., Ikeda T., and Wang H. O.: Fuzzy regulators and fuzzy obserbers: relaxed stability conditions and LMI-based designs," IEEE Trans. Fuzzy Syst. **6** (1998) 250-265
19. Polushin I. G. and Marquez H. J.: Multirate versions of sampled-data stabilization of nonlinear systems. Automatica **40** (2004) 1035-1041
20. Kim E.and Lee H.: New approaches to relaxed quadratic stability conditions of fuzzy control systems. IEEE Trans. Fuzzy Syst. **8** (2000) 523-534
21. Ge S. S., Tian Z., and Lee T. H.: Nonlinear control of a dynamic model of HIV-1. IEEE Trans. Biomed. Eng. **52** (2005) 353-361
22. Brandt M. E. and Chen G.: Feedback control of a biodynamical model of HIV-1. IEEE Trans. Biomed. Eng. **48** (2001) 754-759
23. Hao F., Chu T., Huang L. and Wang L.: Non-fragile controllers of peak gain minimization for uncertain systems via LMI approach. Dynamics of Continuous Discrete and Impulsive Systems **10** (2003) 681-693

An Algorithm for High-Dimensional Traffic Data Clustering

Pengjun Zheng and Mike McDonald

Transportation Research Group, University of Southampton,
Highfield, Southampton, SO17 1BJ, United Kingdom
{p.zheng, mm7}@soton.ac.uk

Abstract. High-dimensional fuzzy clustering may converge to a local optimum that is significantly inferior to the global optimal partition. In this paper, a two-stage fuzzy clustering method is proposed. In the first stage, clustering is applied on the compact data that is obtained by dimensionality reduction from the full-dimensional data. The optimal partition identified from the compact data is then used as the initial partition in the second stage clustering based on full-dimensional data, thus effectively reduces the possibility of local optimum. It is found that the proposed two-stage clustering method can generally avoid local optimum without computation overhead. The proposed method has been applied to identify optimal day groups for traffic profiling using operational traffic data. The identified day groups are found to be intuitively reasonable and meaningful.

1 Introduction

Data clustering is a process of finding natural groupings in a dataset so that data points belonging to same group are more similar than those belonging to different groups [1]. A number of clustering algorithms have been proposed in the past [2]. The most widely used clustering methods are c-means (or K-means) and fuzzy c-means algorithms ([3]-[5]).

The c-means (CM) is a hard clustering method where each point of the dataset belongs to one of clusters exclusively. The fuzzy c-means (FCM) allows for partial membership of belonging to several clusters, i.e. a data point can belong to more than one cluster with different degrees of memberships. This allows for ambiguity in the data and provides a natural partition method compatible with human inaccurate reasoning. The optimal partition is achieved by minimising a specified objective function, usually the weighted sum of squared Euclidean distances between data points and cluster centres [4]-[5].

Both CM and FCM find the optimal partition using iterative procedures. The dataset is initially partitioned into c clusters randomly. The algorithm then iteratively updates the c centres that implicitly represent the partition. The c-means algorithms have been successfully used in many data clustering applications, mainly for its simplicity and efficiency.

The objective function of CM and FCM is non-convex so that the algorithm may converge to a local optimal solution that is significantly inferior to the desirable

L. Wang et al. (Eds.): FSKD 2006, LNAI 4223, pp. 59–68, 2006.

global optimum [6]-[7]. To overcome this drawback, clustering methods using Genetic Algorithms have been proposed to find global optimums [8]-[9]. However, GA-based clustering methods are usually computational expensive [10] which may take thousands of iterations to find a global optimum. The approach may only be suitable for low-dimensional and small data sets.

Traffic data refer to time-series data collected using traffic monitoring equipments. With the rapid development in the Intelligent Transportation Systems, large scale traffic monitoring has now become more and more a commonplace. A typical regional traffic surveillance system usually has more than one thousand sensors collecting data at 1-minute interval round the clock. Time-series databases are often extremely large. As traffic conditions evolve on a daily basis (1440 minutes), 1-min traffic data collected in one day is a time series with a length of 1440, which can usually be considered as a point in 1440-dimensional space. In this sense, traffic databases are characterized by high-dimension and large size.

There has been much interest in the Knowledge Discovery in Data (KDD) from traffic data through data clustering, i.e. identification of natural day groups for traffic profiling purpose so that accurate historical traffic profiles can be constructed from those days with similar traffic conditions. This can theoretically be realised using time-series traffic data from a prolonged time period (e.g. 1 year) based on standard data clustering methods. However, the result may suffer from local optimum if CM and FCM are used, and may not be computationally feasible for GA-based algorithms because of the dimension and size of the data.

In this paper, a two-stage fuzzy clustering method is proposed. Dimensionality reduction is first applied on the original data so that a compact representation of the data is derived which contains only the main features of the original data. By clustering the low-dimensional compact data, optimal partition of the compact data is found. The identified partition is then used as the initial partition for the clustering of the complete data, thus effectively reduces the possibility of local optimum. The proposed method has been applied to identify optimal day groups for traffic profiling using operational traffic data. It is found that the two-stage algorithm is able to find the global optimum without increasing computation demands.

2 Fuzzy c-Means Clustering

The methodology for partition a data set into c fuzzy clusters, the fuzzy c-means (FCM) clustering algorithm, has been developed by Dunn and generalised by Bezdek [4]-[5]. FCM is an iterative clustering method that produces an optimal c partition by minimising the weighted sum objective function J_{FCM}

$$J_{FCM} = \sum_{k=1}^{n} \sum_{i=1}^{c} [A_i(x_k)]^q d^2(x_k, v_i) \qquad (1)$$

where $X=\{\boldsymbol{x}_1, \boldsymbol{x}_2, \ldots, \boldsymbol{x}_n\} \subset R^s$ is the data set in the s-dimension of the input variables, n is the number of data points, c is the number of clusters with $2 \le c < n$, $A_i(x_k) \in [0,1]$ is the degree of membership of $\boldsymbol{x}_k$ in the ith cluster, q is the weighing exponent, v_i is the prototype of the centre of cluster i, $d^2(x_j,v_i)$ is the Euclidean distance between object x_k and cluster centre v_i.

The optimal fuzzy set can be determined by an iterative process where J is successively minimised whilst $V=[v_1, v_2, ..., v_c]$ and $A=[A_1, A_2, ..., A_c]$ are updated using (2) and (3) at m^{th} iteration:

$$v^{(m)}{}_i = \frac{\sum_{k=1}^{n}[A_i^{(m)}(x_k)]^q x_k}{\sum_{k=1}^{n}[A_i^{(m)}(x_k)]^q} \tag{2}$$

$$A_i^{(m+1)}(x_k) = \frac{1}{\sum_{j=1}^{c}\left(\frac{d_{ik}}{d_{jk}}\right)^{2/(q-1)}}, \text{ if for } 1 \le i \le c,\ \left\|x_k - v_i\right\| \ne 0 \tag{3}$$
$$= 0 \qquad \text{otherwise}$$

The detailed iterative procedure can be described as follows:

1. initialise $A_i^{(m)}(x_j)$ for all j, $m=0$
2. calculate v_i based on $A^{(m)}$ using Eq. (2)
3. Compute $A^{(m+1)}(x_j)$ for all j, using Eq. (3)
4. If $A^{(m+1)}$ and $A^{(m)}$ are close enough, e.g. $\left\|A^{(m+1)} - A^{(m)}\right\| < \varepsilon$, stop. Else go to step 2.

The CM algorithm can be regarded as a special case of FCM only that the membership function A is a two-value function:

$$A_i(x_k) = 1, \text{ for } 1 \le i \le c, \left\|x_k - v_i\right\| \le \left\|x_l - v_i\right\|, l \ne k$$
$$= 0 \text{ , otherwise} \tag{4}$$

Some widely used implementations of FCM (e.g. Matlab) generate initial partition randomly. An outline of the FCM initialisation in Matlab is shown below:

```
function InitFCM(c,n)
     generate c(number of cluster) by n(number of data
     points) matrix of random number;
     calculate column sum;
     scale random numbers in each column by the column
     sum so that the column sum of the scaled partition
     matrix is always equal to one;
```

As FCM is sensitive to the initial partition, it may converge to a partition that is a local optimum under some initial partitions. This could be random in nature as the initial partition is generated randomly.

3 Dimensionality Reduction of High-Dimensional Data

A time series of length s can be considered as a point in s-dimensional space R^s. Dimensionality reduction is a technique of decomposition and representation of the data in a reduced parameter space R^S, where $S<s$. Major dimensionality reduction methods include Discrete Fourier Transform (DFT) [11], Singular Value Decomposition (SVD) and Discrete Wavelet Transform (DWT). Dimensionality reduction is achieved by ignoring some details in the source data. For instance, DFT decomposes a signal (time series) of length s into s sine/cosine waves that can be recombined into the original signal. Most of the Fourier coefficients have very low amplitude and can be discarded without much loss of information. The signal can still be approximately reconstructed from those few high amplitude Fourier coefficients. It is therefore possible to approximately represent the original time series using a few coefficients in another parameter space (e.g. frequency domain for DFT). A simple but rather useful technique for data dimensionality reduction is aggregate approximation [12], in which the consecutive data values within a time interval are collapsed by a single value:

$$\bar{x}_j = f(x \mid s_j, e_j) \tag{5}$$

where s_j and e_j be indices of dimensions in $\boldsymbol{x}$ for interval j such that $s_j \le e_j$. $f()$ is a mapping (e.g. the average and weighted average).

Depending on the choice of $f()$ and $[s_j, e_j]$, s-dimensional data $\boldsymbol{x}= \{x^1, x^2, \ldots, x^s\}$can be compacted into S-dimensional data $\bar{x} = \{\bar{x}^1, \bar{x}^2, \ldots, \bar{x}^S\}$, where $S<s$. In the simplest form, an equal-width aggregation scheme by averaging source data in equal aggregation intervals can effectively reduce data dimensionality by s/S times:

$$\bar{x}_j = \frac{S}{s} \sum_{i=s_j}^{e_j} x_i \tag{6}$$

The different between the original data and the compact data at any $[s_j, e_j]$ will be:

$$\Delta x_i = x_i - \bar{x}_j \tag{7}$$

Typical traffic flow data and its compact representation under reduced space are shown in Figure 1. The original data is in 1-min interval (1440 dimensional). By aggregating the data at one-hour interval, the compact data can be regarded as 24-dimensional. It can be observed that the compact data basically retain the global shape of the original data, while the differences are mainly short-term variations. It can also be found that variations within each time interval are significantly different; an indication that the equi-width aggregate approximation may not be optimal. It is therefore possible to aggregate data in varying intervals that can capture the global pattern of the data better. This is equivalent to finding optimal S pairs of aggregation intervals $\{(s_1, e_1), (s_2, e_2), \ldots, (sS, eS)\}$, to minimise the total:

$$SSE = \sum_{j=1}^{S} \sum_{i=sj}^{ej} (x_i - \bar{x}_j)^2 \tag{8}$$

where $s_{j+1} = e_j + 1$. This will result in an optimal aggregate approximation scheme in terms of approximation errors (by minimising SSE).

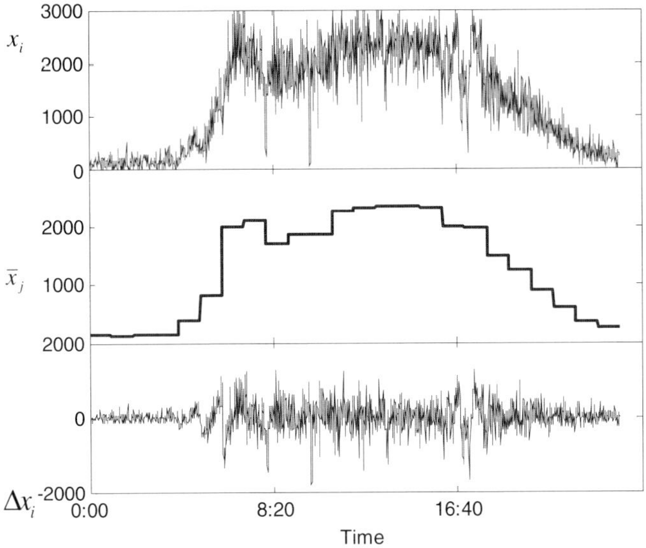

Fig. 1. Typical high-dimensional traffic data and its compact representation

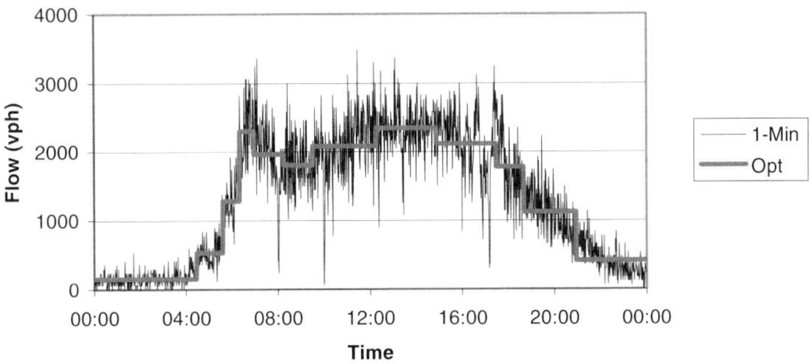

Fig. 2. Compact data based on optimal aggregation approximation

There exist many efficient algorithms to solve this kind of optimisation problem. In this research, the algorithm based on dynamic programming as proposed in [13] has been implemented to compute optimal aggregation intervals. The compact data at a reduced dimension of 12 is shown in Figure 2. It can be clearly observed that the compact data follow the global trends of the original data well. The SSE is even smaller than that based on equi-width aggregation approximation at dimension of 24.

4 Two-Stage Clustering Method

A two stage clustering method has been proposed in this research for clustering high-dimensional traffic data. The clustering is initially performed using low-dimensional data obtained through aggregation approximation. The identified optimal partition matrix is then used as the initial partition in the second stage clustering using full-dimensional data. The clustering based on compact data may be less likely to converge to local optimums. The second stage clustering based on full-dimensional data then looks for the further optimal partition by considering all details available in the data. An outline of the two-stage FCM is described as follows:

```
program 2stgFCM(c,n)
generate compact data by applying dimensionality reduc-
tion techniques;
apply FCM on the compact data (using random initial
partition matrix) to find the optimal partition matrix;
apply FCM on full-dimensional data using the optimal
partition matrix identified as the initial partition;
```

The optimal partition in data clustering is achieved by minimising the weighted sum of squared Euclidean distances between data points and cluster centres. It is not difficult to observe that the main components of Euclidean distance between two points in original space are preserved in the reduce space:

Let x and y are two points in s-dimensional space, S is the target dimensionality,

Let $\Delta x_i = x_i - \bar{x}$, $\Delta y_i = y_i - \bar{y}$ within each $[s_j, e_j]$, $j=1,2, \ldots, S$, the squared Euclidian distance between x and y will be:

$$\begin{aligned} \|x-y\|^2 &= \sum_{j=1}^{S} \sum_{i=sj}^{ej} (x_i - y_i)^2 = \sum_{j=1}^{S} \sum_{i=sj}^{ej} \left[(\bar{x}_j + \Delta x_i) - (\bar{y}_j + \Delta y_i)\right]^2 \\ &= \sum_{j=1}^{S} \left[(e_j - s_j + 1)(\bar{x}_j - \bar{y}_j)^2\right] + \sum_{j=1}^{S} \sum_{i=sj}^{ej} \left[(\Delta x_i - \Delta y_i)^2\right] \\ &= \|\bar{x}' - \bar{y}'\|^2 + \sum_{j=1}^{S} \sum_{i=sj}^{ej} \left[(\Delta x_i - \Delta y_i)^2\right] \end{aligned} \tag{9}$$

where $\bar{x}'_j = \sqrt{(e_j - s_i + 1)} \cdot \bar{x}_j$ and $\bar{y}'_j = \sqrt{(e_j - s_i + 1)} \cdot \bar{y}_j$ for $j=1,2,\ldots,S$

It can be found that the optimal partition found by clustering data at reduced dimensionality (based on $\|\bar{x}' - \bar{y}'\|$) is equivalent to clustering the full-dimensional data by ignoring local variations of $\sum_{j=1}^{S} \sum_{i=sj}^{ej} \left[(\Delta x_i - \Delta y_i)^2\right]$, thus is a high-level partition that is unlikely to converge to local optimum. In addition, clustering based on the compact data is usually computational inexpensive as the reduced dimensionality is often much lower compared with the original dimensionality. Repeated clustering based on the compact data is possible to reduce the chances of converging to a local optimum.

5 Application

The proposed two-stage clustering algorithm has been applied to find natural day groups for traffic profiling purposes. The traffic profile refers to the historically typical traffic conditions that represent the past experience and also the future expectations. The most common day groups for traffic profile are weekday, e.g. Monday - Sunday. This is intuitively reasonable as it can capture weekly patterns of traffic conditions. However, many other partitions are also possible, e.g. a partition between public holidays and working days. A promising approach of identifying these natural day groups is through learning from the operational data using data clustering techniques. Traffic flow data for each day can be regarded as a data point in s-dimensional space. By applying clustering on traffic data of more than one year (365 data points), the natural groups (clusters) can be identified from the optimal partition matrix using proper defuzzification process, e.g., the maximum membership procedure that assigns the data point (a day's traffic data) k to the cluster (day group) i with the highest membership

$$C_k = \arg_i\{\max(A_i(x_k)\}, \; i=1,\, 2,\, \ldots,\, c.$$

where $A_i(x_k) \in [0,1]$ is the degree of membership of $\boldsymbol{x_k}$ in the ith cluster.

5.1 The Data

Operational traffic data collected using loop detectors are used to identify natural day groups for traffic profiling. The database consists of 1-min traffic flow data collected over a period of 640 days from 1 April 2004 to 31 December 2005. Data collected on 26 days are incomplete and are not included. In total, traffic data collected on 614 days (884160 readings) are used in the clustering.

The compact data are generated using two aggregate approximation schemes:

- Optimal aggregate approximation to a reduced dimension of 12
- Equi-width aggregate approximation to a reduced dimension of 24

Both direct and the proposed two-stage fuzzy clustering methods are used to cluster data into 3-10 clusters. The two-stage clustering method has also been used in a repeated initial clustering way where clustering over the compact data is repeated more than one time to find the optimal initial partition matrix. This is intended to prevent converging to local optimums in the clustering process of the compact data and is computational affordable as clustering on low-dimensional data demands little CPU time. Thus, four clustering runs are performed for a given pre-defined number of clusters:

- Direct clustering using full-dimensional data
- Two-stage clustering using compact data generated from optimal aggregate approximation method
- Two-stage clustering using compact data generated from Equi-width aggregate approximation method
- Two stage clustering based on the best initial partition obtained by repeatedly clustering using compact data generated using both aggregate approximation methods

5.2 Results

The identified day groups based on a cluster number of 5 are shown in Figure 3. Five natural day groups are dominated by Mon-Tue, Wed-Thu, Fri, Sat and Sun

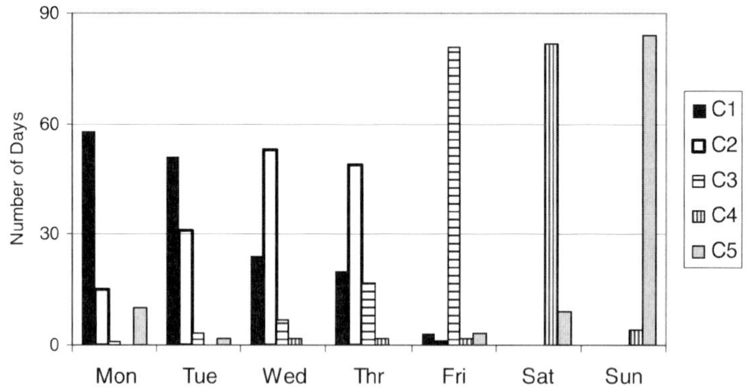

Fig. 3. Distribution of identified day groups over weekdays

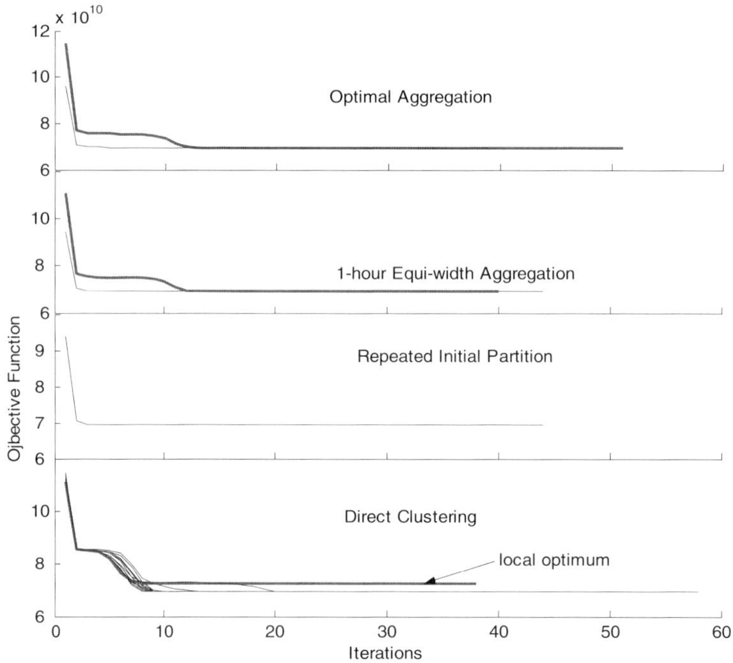

Fig. 4. Objective function values using different clustering procedures

respectively. This is intuitively reasonable as traffic patterns on Saturday and Sunday are usually different from working days. Traffic patterns on Friday have been identified as different from both other working days and non-working days. Traffic conditions on Monday and Tuesday are more alike, so do that on Wednesday and Thursday. It is worth noticing that 10 public holiday Mondays have all been successfully identified as belonging to Sunday group.

The corresponding objective function values evolving over each clustering iteration have been shown in Figure 4. This is based on 12 runs to find the optimal partition of 5-day groups. Convergences to an inferior local optimum have only been observed two times based on direct clustering using full-dimensional data.

The average CPU times based on 12 runs are shown in Table 1. It can be found that the direct clustering method actually takes slightly longer time. This is because the clustering requires more iterations to converge than that of other three methods. The proposed two-stage algorithm can achieve better optimum while not increasing computational demands.

Table 1. CPU times under different clustering schemes (Pentium4 3GHz CPU)

Schemes	1-hour Equi-width aggregation		Optimal Aggregation		Direct Clustering	Repeated Initial partitions	
CPU Time (s)	Initial Partition	Total	Initial Partition	Total	Total	Initial Partition	Total
AVG	0.58	11.82	0.36	11.73	12.53	0.89	12.14
STD	0.06	0.55	0.07	1.34	1.70	0.17	0.72

6 Conclusions

A two-stage algorithm for clustering high-dimensional traffic data has been described in this paper. It has been compared with direct clustering method. The performance of the proposed algorithm has been shown to outperform the standard implementation. The algorithm is less likely to converge to local optimums while the computational demands have not been increased (or even slightly reduced). The application of the algorithm to cluster operational traffic data has produced promising results. The identified natural day groups are intuitively reasonable and meaningful. However, one thing remains un-guaranteed, that is, the identified clusters using the proposed method could still not be the global optimum. The absolute optimum may only be guaranteed by exhaustive enumeration, which is not yet practically feasible for large database due to extremely high computational demands.

References

1. Jain, A. K., Dubes, R. C.: Algorithms for Clustering Data, Prentice-Hall, Englewood Cliffs, NJ (1988)
2. Jain, A. K., Murty, M. N., Flynn, P. J.: Data Clustering: A Review, ACM Computing Surveys, vol. 31(3) (1999) 264-323
3. Hand, D. J., Krzanowski, W. J.: Optimising k-means Clustering Results with Standard Software Packages, Computational Statistics & Data Analysis, vol. 49(4) (2005) 969-973
4. Dunn, J. C.: A fuzzy relative of the ISODATA process and its use in detecting compact well separated clusters, Journal of Cybernetics, Vol. 3 (1974) 32-57
5. Bezdek, J. C.: Pattern Recognition with Fuzzy Objective Function Algorithms, New York, Plenum Press, (1981)
6. Selim, S. Z. Ismail, M. A.: K-means Type Algorithms: A Generalised Convergence Theorem and Characterisation of Local Optimality. IEEE Transactions on Pattern Analysis and Machine Intelligence. vol. 6(1) (1984) 81-87.
7. Pollard, D.: A Central Limit Theorem for k-Means Algorithm, Annals of Probability, vol. 10 (1982) 919-926.
8. Murthy, C. A. Chowdhury, N.: In Search of Optimal Clusters using Genetic Algorithms. Pattern Recognition Letters, vol. 17 (1996) 825-832
9. Jones, D. Beltramo, M. A.: Solving Partitioning Problems with Genetic Algorithms. Proceedings of Fourth International Conference of Genetic Algorithms, (1991) 442-449.
10. Laszlo, M, Mukherjee, S. A Genetic Algorithm Using Hyper-Quadtrees for Low-dimensional K-means clustering, IEEE Transactions on Pattern Analysis and Machine Intelligence, 28 (4) (2006) 533-543
11. Vanloan, C. F.: Generalizing Singular Value Decomposition, SIAM Journal on Numerical Analysis, vol. 13 (1) (1976) 76-83
12. Keogh, E.J. Pazzani, M.J.: A simple dimensionality reduction technique for fast similarity search in large time series databases. Lecture Notes in Artificial Intelligence, Vol. 1805. Springer-Verlag, Berlin Heidelberg New York (2000), 122-133
13. Jagadish, H. V. Koudas, N., Muthukrishnan, S., Poosala, V., Sevcik, K. Suel, T.: Optimal Histograms with Quality Guarantees. Proceedings of the 24th VLDB Conference, New York, USA, (1998)

Hierarchical Clustering with Proximity Metric Derived from Approximate Reflectional Symmetry

Yong Zhang* and Yun Wen Chen

Department of Computer Science and Engineering
School of Information Science and Engineering
Fudan University
Shanghai, 200433, P.R. China
zhang_yong@fudan.edu.cn

Abstract. In order to address the problems arise from predefined similarity measure, learning similarity metric from data automatically has drawn a lot of interest. This paper tries to derive the proximity metric using reflectional symmetry information of the given data set. We first detect the hyperplane with highest degree of approximate reflectional symmetry measure among all the candidate hyper-planes defined by the principal axes and the centroid of the given data set. If the symmetry is prominent, then we utilize the symmetry information acquired to derive a retorted proximity metric which will be used as the input to the Complete-Link hierarchical clustering algorithm, otherwise we cluster the data set as usual. Through some synthetic data sets, we show empirically that the proposed algorithm can handle some difficult cases that cannot be handled satisfactorily by previous methods. The potential of our method is also illustrated on some real-world data sets.

1 Introduction

Cluster analysis, as one of the basic tools for exploring the underlying structure of a given data set, has been applied in a wide variety of fields. Actually, clustering (classification) plays an important and indispensable role in the long history of human development as one of the most primitive activities[7].

As pointed by most researchers, cluster analysis intends to partition a group of objects into a number of more or less homogeneous subgroups (clusters) such that patterns within a cluster are more similar to each other than patterns belonging to different clusters. Often, a clear distinction is made between supervised clustering and unsupervised clustering, the former involving only labeled data while the latter involving only unlabeled data in the process of learning.

Existing methods for clustering fall into two categories as hierarchical clustering and partitional clustering, based on the properties of clusters generated.

* Corresponding author.

L. Wang et al. (Eds.): FSKD 2006, LNAI 4223, pp. 69–78, 2006.

Hierarchical clustering groups data objects with a sequence of partitions, either from singleton clusters to a cluster including all individuals or vice versa, while partitional clustering directly divides data objects into some pre-specified number of clusters without the hierarchical structure.

Regardless of hierarchical or partitional clustering, they both rely on the definition of similarity (or dissimilarity) measure, which establishes a rule for assigning patterns to a particular cluster. Many algorithms adopt a predefined similarity measure based on some particular assumptions. These algorithms may fail to model the similarity correctly when the data distribution does not follow assumed scheme. Instead of choosing the similarity metric manually, a promising solution is to learn the metric from data automatically.

Usually, in order to extract appropriate metric from data, some additional background knowledge or supervisory information should be made available for unsupervised clustering. Supervisory information is often provided in the form of partial labeled data or pairwise similarity and/or dissimilarity constraints[4][5].

Despite the progress in semi-supervised clustering for similarity metric learning, metric learning for strict unsupervised clustering remains a challenge. An interesting trial is to extract similarity metric based on symmetry. In [6], for instance, a novel nonmetric distance measure based on the idea of point symmetry is proposed, where the point refers to the centroid of corresponding cluster.

Following a similar consideration, in this paper, we exploit the approximate reflectional symmetry of the given data set to induce the proximity metric such that the similarity between pairs of points is not strictly depend on their closeness in the feature space, but rather on their symmetrical affinity. Although this is similar to [6] to some extent, several important differences should be noticed. First, we don't assume compulsory symmetry exist in data set, we perform the symmetry detection as a preprocessing. If symmetry exist, we continue the clustering with the guidance of symmetry information detected, otherwise we process the data set as usual. Second, we consider reflectional symmetry, which is more common in abstract or in nature, rather than the point symmetry. Third, we mainly utilize the symmetry information acquired to deduce an appropriate proximity metric.

The remainder of this paper is organized as follows: Section 2 introduces the algorithm for reflectional symmetry measurement and detection. The proposed clustering algorithm, which can be divided into two stages: proximity metric construction and complete-link clustering, is described in section 3. Simulation results on both synthetic and real-world data are presented in section 4, comparing with some previous methods. In section 5 we review the related research briefly, and some concluding remarks are given in section 6.

2 Reflectional Symmetry Measurement and Detection

Symmetry is often described with symmetric transformation T, which is a transformation that when applied to all elements of a system S result in a system S' that is identical to the original system S.

Before we incorporate symmetry information into clustering, we must solve the following problem: How can we detect or measure the symmetry of the given system efficiently? If we only consider symmetry as a binary feature (i.e., a system is either symmetric or it is not symmetric), or exact symmetry, then the problem will be easier. However, even perfectly symmetric objects may lose their exact symmetry because of digitalization or quantification. Consider symmetry as an approximate feature[3], we can describe inexact symmetry more exactly. A system has approximate symmetry with respect to a transformation if it is "almost" invariant under that transformation. Obviously, the challenge is to interpret "almost" in an appropriate manner. Now, the problem can be presented as follows: How can we measure and/or detect approximate symmetry degree for a given system?

As in[3], this problem can be translated into two subproblems: First, how to measure the symmetry degree of a given data set with respect to any specified hyperplane, and second, how to find a hyperplane with highest degree of symmetry measure.

The reflectional symmetry about a hyperplane for a set of n-dimensional data points can be measured by reflecting each point through the hyperplane, and measuring the distance from the reflected point to the closest point in the original data set[3]. More accurately, the mean of these distances is a measure of reflectional asymmetry. In order to make the measure invariant to scale, we divide the measure by a scaling factor. We select the scaling factor as the maximum distance of any point from the centroid so that the asymmetry measure will be within the range zero to one for any hyperplane that contain the centroid. To create a symmetry measure from the asymmetry measure, the asymmetry measure can be subtracted from one. For perfectly symmetric data set, the symmetry measure will be one. When the degree of symmetry decreases, the symmetry measure also decreases.

Finding the most reflectional symmetric hyperplane for an n-dimensional point set requires the symmetry to be measured for every possible hyperplane. This is not practical because the number of possible hyper-planes is infinite. Even though we can narrow the search space by discretization, this approach is also computationally prohibitive especially for high dimension situations. A more practical approach is to utilize the principal axes of the data set to define n candidate hyper-planes, and choose the hyperplane associated with the highest symmetry measure as the approximate hyperplane of reflectional symmetry. The motivation for this approach is based on two theorems found in [2], which state that:

- Any plane of symmetry of a body is perpendicular to a principal axis
- Any axis of symmetry of a body is a principal axis

Fig. 1 shows the principal axes that would be found using the centroid and eigenvectors of the covariance matrix for 2D and 3D objects. A hyperplane can be uniquely defined using a point and a normal vector, so each principal axis can be used to define a hyperplane containing the centroid of the given data set.

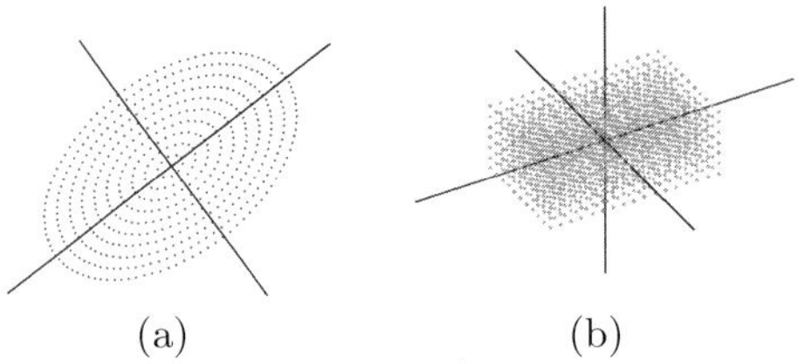

Fig. 1. The centroids and principal axes for a 2D ellipse (a) and 3D box (b)

The detailed algorithm description is as follows:

Input:
the data set $S = \{\overrightarrow{x_i}\}_{i=1}^{N}$, where $\overrightarrow{x_i}$ is represented as n-dimensional column vector

Output:
the hyperplane *symhp* with highest value of reflectional symmetry measurement, and the corresponding symmetry measure value *symv*

Steps:

Step 1 determine all the n candidate hyper-planes

1.a compute the centroid $\overrightarrow{m}$ of S

$\overrightarrow{m} = \frac{1}{|S|}\sum_{\overrightarrow{x}\in S}\overrightarrow{x}$

1.b estimate the covariance matrix C for S

$C = \frac{1}{|S|}\sum_{\overrightarrow{x}\in S}(\overrightarrow{x}-\overrightarrow{m})(\overrightarrow{x}-\overrightarrow{m})^T$

1.c compute all the n principal axis vectors $\overrightarrow{w_1}$, $\overrightarrow{w_2}$, $\cdots$, $\overrightarrow{w_n}$ based on C

1.d determine n candidate hyper-planes $\Pi_1, \Pi_2, \cdots, \Pi_n$ as:

$\Pi_i : \overrightarrow{w_i}\cdot(\overrightarrow{x}-\overrightarrow{m}) = 0$, where $\overrightarrow{x} \in R^n, i = 1, 2, \cdots, n$

Step 2 measure the symmetry degree w.r.t.every candidate hyperplane

2.a compute $d_{max} = max_{\overrightarrow{x}\in S} \parallel \overrightarrow{x}-\overrightarrow{m} \parallel$

2.b for each candidate hyperplane $\Pi_i, i = 1, 2, \cdots, n$

2.b.1 for each $\overrightarrow{x} \in S$, compute

$d(\overrightarrow{x}, \Pi_i) = min_{\overrightarrow{y}\in S} \parallel \overrightarrow{x^*} - \overrightarrow{y} \parallel$, where $\overrightarrow{x^*}$ is the reflection of $\overrightarrow{x}$ through Π_i

2.b.2 compute $\bar{d}(\Pi_i) = \frac{1}{|S|}\sum_{\overrightarrow{x}\in S} d(\overrightarrow{x}, \Pi_i)$

2.b.3 compute $symv(\Pi_i) = 1 - \frac{\bar{d}(\Pi_i)}{d_{max}}$

Step 3 find $\Pi_k (1 \leq k \leq n)$, that

$symv(\Pi_k) = max_{1\leq i\leq n} symv(\Pi_i)$,

and return Π_k as well as $symv(\Pi_k)$

The order of complexity of this algorithm is $O(nN^2)$, where N represents the number of points in the given data set and n is the dimension of data points. However, if the data are sorted along an axis, then the typical order of complexity can be reduced to $O(nNlogN)$ by using a binary search to find the nearest neighbor. If we want to find better symmetric hyperplane, we can also choose the principal axis as the initial candidate and optimize it further using some appropriate optimization methods[3].

3 The Proposed Clustering Algorithm

3.1 Proximity Metric Construction

We have acquired approximate reflectional symmetry information about the given data set using the algorithm presented in section 2. Then how can we incorporate this information into cluster analysis so as to improve the clustering results? The direct method is to induce approximate proximity metric based on the symmetry information available so that the similarity between pairs of points is not strictly depend on their closeness in the feature space, but rather on their symmetrical affinity.

Nevertheless, just as mentioned earlier, we cannot expect reflectional symmetry present for every data set. In the proposed algorithm, we introduce a pre-specified threshold ts. If the symmetry measure with respect to the acquired hyperplane is bigger than ts, then we modify the original proximity metric based on the symmetry information before clustering, otherwise we cluster the patterns directly as usual without any proximity modification.

If the symmetry information is available, we construct Approximate Reflection Pair Set (ARPS) with respect to the symmetric hyperplane. Concretely, let $\overrightarrow{x}$, $\overrightarrow{y}$ be two patterns in S, we will add $\overrightarrow{x}$ and $\overrightarrow{y}$ into ARPS if $\overrightarrow{y}$ is the nearest neighbor of $\overrightarrow{x^*}$, where $\overrightarrow{x^*}$ is the reflection of $\overrightarrow{x}$ through the specified hyperplane. If no extra constraint considered, all the patterns in S will be added into ARPS ultimately unless the cardinal of S is odd. This may be unreasonable, because the nearest neighbor of $\overrightarrow{x^*}$ may be very far apart from $\overrightarrow{x^*}$ especially when most of the patterns in S have been added into ARPS. In order to address this problem, we only consider a pattern within the specified neighborhood of $\overrightarrow{x^*}$, i.e., the distance between $\overrightarrow{x^*}$ and its nearest neighbor is smaller than a specified threshold tn, as the approximate reflection of $\overrightarrow{x}$, otherwise $\overrightarrow{x}$ is not considered.

Then we derive a new proximity metric on the basis of the ARPS by lowering the distances between the approximate reflectional pairs, i.e., pairs in ARPS which are mutual reflections, to zero. In this way, however, the points only satisfy the symmetric affinity imposed by ARPS, instead of the reflectional symmetry of the given data set as a whole and intrinsic feature. Hence, as in [5], we also require the points to satisfy the implied symmetric relation of ARPS. We interpret the proximity matrix as weights for a complete graph over the data points. Thus we obtain a new metric by running an all-pairs-shortest-paths algorithm on the modified proximity matrix. The concrete algorithm is presented as follows:

Input:

the data set $S = \{\overrightarrow{x_i}\}_{i=1}^{N}$, where $\overrightarrow{x_i}$ is represented as n-dimensional column vector

the symmetric hyperplane *symhp* and corresponding symmetry measure *symv*

the threshold ts and tn

Output:

the proximity matrix P

Steps:

Step 1 construct $P_{N \times N}$, where $P_{ij} = \| \overrightarrow{x_i} - \overrightarrow{x_j} \|, 1 \leq i, j \leq N$

Step 2 if $symv < ts$, goto step 5
Step 3 ARPS construction
3.a initialize $S^{'} = S$, $ARPS = \phi$
3.b repeat until $S^{'} = \phi$
3.b.1 select $\overrightarrow{x}$ from $S^{'}$, and find $\overrightarrow{y_0}$, that

$$\| \overrightarrow{x} - \overrightarrow{y_0} \| = min_{\overrightarrow{y} \in S^{'} - \{\overrightarrow{x}\}} \| \overrightarrow{x^*} - \overrightarrow{y} \|,$$

where $\overrightarrow{x^*}$ is the reflection of $\overrightarrow{x}$ through $symhp$
3.b.2 if $\| \overrightarrow{x} - \overrightarrow{y_0} \| < tn$, add $\overrightarrow{x}$ and $\overrightarrow{y_0}$ into $ARPS$, and let $S^{'} = S^{'} - \{\overrightarrow{x}, \overrightarrow{y_0}\}$, otherwise discard $\overrightarrow{x}$ from $S^{'}$, i.e., let $S^{'} = S^{'} - \{\overrightarrow{x}\}$
Step 4 proximity adaptation
4.a set $I = \{i | \overrightarrow{x_i} \in ARPS\}$
4.b for $k \in I$, for $i = 1$ to N, for $j = 1$ to N

$$P_{ij} = min\{P_{ij}, P_{ik} + P_{kj}\}$$

Step 5 return P matrix

3.2 Compete-Link Hierarchical Clustering

In this paper, we use complete-link hierarchical agglomerative clustering as the clustering approach. As a matter of fact, we can use any clustering algorithm provided that its input is a proximity matrix.

As a popular hierarchical clustering algorithm, complete-link agglomerative clustering is very familiar to most researchers. For the convenience, the algorithm is presented as follows[5]:

Input:
the proximity matrix $P_{N \times N}$
Output:
the linkage $link$
Steps:
Step 1 initialize $Clusters = \{c_i$ for each pattern $\overrightarrow{x} \in S\}$, $link = \phi$, distances $d(c_i, c_j) = P_{ij}, 1 \leq i, j \leq N$
Step 2 repeat until $|Clusters| = 1$
2.a choose closest $(c_1, c_2) = argmin_{c_i, c_j \in Clusters} d(c_i, c_j)$
2.b add (c_1, c_2) to $link$
2.c merge c_1 and c_2 into c_{new} in $Clusters$
2.d for $c_i \in Clusters$

$$d(c_i, c_{new}) = max\{d(c_i, c_1), d(c_i, c_2)\}$$

3.3 Algorithm Analysis

The proposed algorithm requires $O(nN^2)$ computations and $O(N^2)$ space which may be intractable for some large data sets. Notice that we allow a many-to-many correspondence for data points when we measure the degree of symmetry with respect to a specified hyperplane (section 2). However, for the construction of ARPS, we impose a one-to-one correspondence, i.e., each reflected point is only to find its nearest neighbor among the points that have not previously been matched. The reason is that we want to find more accurate reflectional symmetric pairs to describe the symmetry information.

4 Simulation Results

4.1 Evaluation Criteria

Rand Index[7] can be used to reflect the agreement of two clustering results. Let n_s, n_d be the number of point pairs that are assigned to the same/different cluster(s) in both partitions respectively. The Rand Index is defined as the ratio of (n_s+n_d) to the total number of point pairs, i.e., $N(N-1)/2$, where N represents the number of points in the given data set. The Rand Index lies between 0 and 1, and when the two partitions are consistent completely, the Rand Index will be 1.

In what follows, we use Rand Index to measure the agreement between the resultant partition and the ground truth.

4.2 Simulation with Synthetic Data

The synthetic data is designed to highlight the improvements brought by the proposed algorithm compared with other clustering algorithms like single-link, complete-link and SBKM[6]. The parameter ts and tn are chosen for 0.8 and 0.5, respectively, and this is kept the same irrespective of the data sets used. Fig.2 shows the target clustering and the corresponding clustering results generated by the proposed algorithm and other three algorithms.

Table 1 shows the degree of reflectional symmetry for every synthetic data set and the corresponding Rand Index values for the proposed algorithm as well as other three algorithms: single-link, complete-link, and SBKM.

Table 1. The reflectional symmetry degree of the three synthetic data sets and the Rand Index values for the proposed algorithm as well as other three algorithms: SL (Single-Link), CL (Complete-Link), SBKM (Symmetry Based K-Means) after applying to the three sets

Data Sets	Data set1	Data set2	Data set3
Symmetry Degree	0.9328	0.9171	0.9458
CL	0.5896	0.4982	0.5482
SL	**1.00**	0.7207	0.5619
SBKM	0.4962	0.5246	0.5118
The Proposed Algorithm	**1.00**	**1.00**	**0.9867**

Compared with other three algorithms, the proposed algorithm improves the clustering results substantially. Although single-link clustering can also detect some of these patterns, we should notice that: with the guidance of symmetry information, both single-link and complete-link algorithms are effective for these

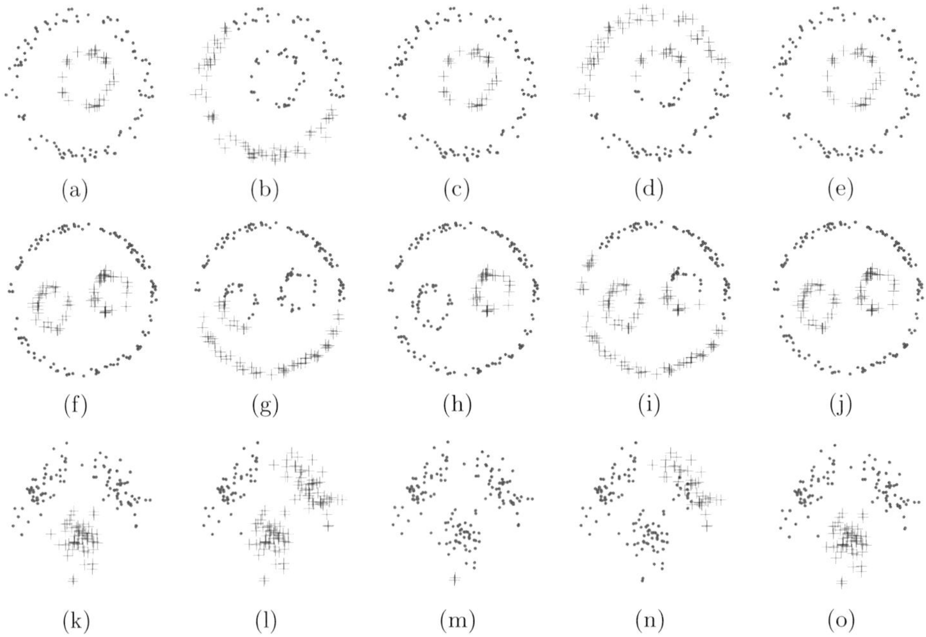

Fig. 2. Comparison of the proposed algorithm with Single-link, Complete-link, SBKM on four synthetic data sets. Subfigures in the first column show the target clusterings. Subfigures in the second, third, fourth and fifth column are the clustering results after applying Complete-link, Single-link, SBKM and the proposed algorithm, respectively.

data sets. This means the algorithm choice may be of less importance when symmetry information is available. Taken in this sense, we can indeed expect to improve the clustering results using acquired symmetry information even for a poorly-performing algorithm.

4.3 Simulation With Real-World Data

Simulations were conducted on three data sets from UCI repository: GLASS, PIMA, and BUPA[1]. Table 2 summarizes the properties of the three data sets: the number of instances, the number of dimensions (attributes), the number of classes as well as the symmetry degree acquired with the algorithm presented in Section 2.

Table 2 also shows the corresponding Rand Index values for the four algorithms considered. During the simulations, for SBKM, the parameter θ was chosen as 0.27, 0.3 and 0.18 for the three data sets respectively. And the results was averaged over 10 runs. As for the proposed algorithm, we set the parameter ts to 0.8 for all of the data sets to be studied, while the parameter tn was chosen as 0.16, 0.6 and 0.43 for the three data sets respectively.

Although the performance improvement compared with other three algorithms are not so substantial, the results are also encouraging since this indicates

Table 2. The Rand Index values for the proposed algorithm as well as other three algorithms: SL(Single-Link), CL(Complete-Link), SBKM(Symmetry Based K-Means) on the real-world data sets. The corresponding symmetry degrees(SD) are also presented. TPA is short for The Proposed Algorithm.

Dataset	#Classes	#Instances	#Attributes	#SD	CL	SL	SBKM	TPA
GLASS	7	214	9	0.9998	0.5822	0.2970	**0.6697**	0.6313
PIMA	2	768	8	0.9127	0.5466	0.5458	0.5185	**0.5559**
BUPA	2	345	6	0.9412	0.5056	**0.5104**	0.5026	**0.5104**

that our algorithm is not only confined to low-dimensional and two-category problems.

From the simulation results over synthetic and real-world data, we can confirm the effectiveness of the proposed algorithm. Different from previous algorithms, the proposed algorithm intends to learn the proximity metric using the symmetry information of the given data set. As we know, if reflectional symmetry is present in a candidate data set, then the symmetry should be considered a more significant feature or constraint. Hence, we can expect better results for the proposed algorithm when applied to data sets with approximate symmetry.

5 Related Work

Leaning metric from data has been an active research field. One traditional important family of algorithms that (implicitly) learn metrics is the unsupervised ones that take an input data set, find and embedding of it in some space. This includes algorithms such as Multidimensional Scaling (MDS) and Locally Linear Embedding (LLE) etc.

In the context of clustering, a promising approach was proposed by Wagstaff et al[8] for clustering with constraint (similarity) information. Pablo et al[4] also exploited some pairs of points with known dissimilarity value to teach a dissimilarity relation to a feed-forward neural network. Eric.P et al[9] learned a distance metric based on given examples of similar pairs of points in feature space.

The proposed algorithm is distinct from all the algorithms mentioned above. It exploits the reflectional symmetry information detected from data to derive an appropriate proximity metric over the input space. The motivation for the proposed algorithm is in accordance with the consideration in [6]to some extent.

Symmetry detection and measurement is also a key problem for the proposed algorithm. There are many literatures available about symmetry detection and/or measurement. In [2][3], the principal axes of the given data set were used as the initial setting for finding the symmetric plane(hyperplane) further. In this paper, however, we directly exploit the principal axes to determine the symmetric hyper-planes.

6 Conclusion

A method to derive the proximity metric based on the approximate reflectional symmetry information acquired from data has been proposed for complete-link hierarchical clustering. The results over several synthetic data sets demonstrate that the proposed algorithm can improve the clustering accuracy compared with other similar algorithms: single-link, complete-link and SBKM algorithm.

In this paper, we only consider to exploit the reflectional symmetry. In future, we intend to guide cluster analysis with more kinds of symmetry such as rotation, skew symmetry etc. Additionally, we are interested in extending the global symmetry detection to local symmetry detection.

Acknowledgments

This work is supported by Natural Science Foundation of China under grant No.60305002 and China/Ireland Science and Technology Collaboration Research Fund under grant No.CI-2004-09.

References

1. Blake, L., Merz, J.: UCI repository of machine learning databases. `http://www.ics.uci.edu/~mlearn/MLRepository.html` (1998)
2. Changming Sun., Sherrah, J.: 3D symmetry detection using the extended Gaussian image. Pattern Analysis and Machine Intelligence, IEEE Transactions on. **19(2)** (1997) 164–168
3. Colliot, O., Tuzikov, A., Cesar, R., Bloch, I.: Approximate reflectional symmetries of fuzzy objects with an application in model-based object recognition. Fuzzy Set and Systems. **147** (2004) 141–163
4. Corsini, P., Lazzerini, B., Marcelloni, F.: A fuzzy relational clustering algorithm based on a dissimilarity measure extracted from data. Systems, Man and Cybernetics, Part B, IEEE Transactions on. **34(1)** (2004) 775–781
5. Klein, O., Kamvar, S.O., Manning, C.: From instance-level constraints to space-level constraints: Making the most of prior knowledge in data clustering. Proceedings of the Nineteenth International Conference on Machine Learning. (2002) 307–314
6. Muchun Su., Chien-Hsing Chou.: A modified version of the k-means algorithm with a distance based on cluster symmetry. Pattern Analysis and Machine Intelligence, IEEE Transactions on. **23(6)** (2001) 674–680
7. Rui Xu., Wunsch, D.: Survey of clustering algorithms. Neural Networks, IEEE Transactions on. **16(3)** (2005) 645–678
8. Wagstaff, K., Cardie, C., Rogers, S., Schroedl, S.: Constrained k-means clustering with background knowledge. Proceedings of the Eighteenth International Conference on Machine Learning. (2001) 577–584
9. Xing, E.P., Ng, A.Y., Jordan, M.I., Russell, S.: Distance metric learning, with application to clustering with side-information. Advances in Neural Information Processing Systems **15** (2003) 505–512

Fuzzy Clustering Based on Vague Relations

Faxin Zhao, Z.M. Ma, and Li Yan

Northeastern University
Shenyang, Liaoning 110004, P.R. China
zongmin_ma@yahoo.com

Abstract. A vague relation, as well as an intuitionistic fuzzy relation, is a further generalization of a fuzzy relation. In fact there are situations where vague relations are more appropriate. In this paper, fuzzy clustering based on vague relations is investigate. On the basis of max-*t* & min-*s* compositions, we make a further extension of *n*-step procedure. With the proposed *n*-step procedure, a similarity vague relation matrix is obtained by beginning with a proximity vague relation matrix. Then a clustering algorithm is proposed for the similarity vague relation matrix.

1 Introduction

In the real world, there are many vaguely specified data values. In order to handle such vagueness, Zadeh introduced fuzzy set theory in 1965 [1]. In the fuzzy set theory, each object $u_i \in U$ is assigned a single value between 0 and 1, called the grade of membership, where U is a universe of discourse. As pointed out by Gau *et al.* in [2], the drawback of using the single membership value in the fuzzy set theory is that the evidence for $u_i \in U$ and the evidence against $u_i \in U$ are in fact mined together. They also pointed out that the single number reveals nothing about its accuracy. To tackle this problem, Gau *et al.* proposed the notion of vague set, which is regarded as a generalized version of fuzzy sets. They used a truth-membership function tr_V and a false-membership f_V to characterize the lower bound on μ_V, where μ_V is a membership function defined for a vague set V. These lower bounds are used to create a subinterval on [0, 1], namely, $[tr_V(u_i), 1 - f_V(u_i)]$, to generalize the $\mu_V(u_i)$ of fuzzy sets, where $tr_V(u_i) \leq \mu_V(u_i) \leq 1 - f_V(u_i)$. Being a further generalized version of fuzzy sets, vague set is not isomorphic to the *Intuitionistic Fuzzy Set* (IFS) [3], there are some interesting features for handling vague data that are unique to vague sets [4], and what's more, vague sets have been introduced to deal with imperfect information.

Cluster analysis is a very useful classification tool. It has been widely applied in taxonomy, business, engineering systems, pattern recognition, image processing, and etc. [5-7]. Based on the fuzzy set theory, fuzzy clustering has been widely studied in a variety of areas [8-10]. These fuzzy clusterings can be roughly divided into two categories: fuzzy clustering based on fuzzy relations and fuzzy clustering based on objective functions. Although the first type of clustering methods is eventually the novel method of agglomerative hierarchical clustering, they are simple and useful in application systems. For the first type of clustering methods, Tamura et al. [11] constructed an *n*-step procedure using max-min compositions by beginning with a reflexivity and symmetry fuzzy relation. A max-min similarity relation is obtained after the *n*-step

L. Wang et al. (Eds.): FSKD 2006, LNAI 4223, pp. 79–88, 2006.

max-min compositions are performed. Finally, one can achieve a multi-level hierarchical clustering based on this max-min similarity relation. However, because the max-min composition is composed of higher relation values, clustering results based on Tamura's n-step procedure of max-min compositions do not explore the data well. To tackle this problem, Recently, Yang et al. [12] extended Tamura's procedure to all types of max-t compositions, and then, based on this max-t similarity relation, proposed a clustering algorithm to obtain a clustering.

It is well known that a vague relation or an intuitionistic fuzzy relation is a generalization of fuzzy relation. Furthermore, De et al. [13] have applied intuitionistic fuzzy relation in medical diagnosis. They concluded that the non-membership (false-membership) functions have more important roles in comparison to the membership function corresponding to the complement of fuzzy sets because of the fact that in decision making problems, particularly in case of medical diagnosis. From [4], we can find that there are situations, such as analysis in data relationships and similarity measures, where vague relations are more appropriate. So, in this paper, we focus on fuzzy clustering based on vague relations. On the basis of max-t & min-s compositions, we make an extension of Yang and Shih's n-step procedure in [12]. With the proposed n-step procedure, a similarity vague relation matrix is obtained by beginning with a proximity vague relation matrix. Then a clustering algorithm is proposed for the similarity vague relation matrix.

The rest of this paper is organized as follows. In Section 2, we introduce some basic concepts and definitions. In Section 3, we propose an n-step procedure based on max-t & min-s compositions which is a further extension of Yang and Shih's n-step procedure in [12]. In section 4, a modified clustering algorithm is proposed based on max-t & min-s similarity vague relation constructed by the proposed n-step procedure. Finally, the conclusions are discussed in Section 5.

2 Basic Concepts and Definitions

Let U be a universe of discourse, with an element of U is denoted by u.

Definition 1 (Vague Set)**.** A vague set V in U is characterized by a *truth-membership* function tr_V and a *false-membership* function f_V. Here $tr_V(u)$ is a lower bound on the grade of membership of u derived from the evidence for u and $f_V(u)$ is a lower bound on the negation of u derived from the evidence against u. $tr_V(u)$ and $f_V(u)$ both associate a real number in the interval [0, 1] with each element in U, where $tr_V(u) + f_V(u) \leq 1$. Then

$$tr_V: U \to [0, 1] \text{ and } f_V: U \to [0, 1]$$

Suppose $U = \{u_1, u_2, \ldots, u_n\}$. A vague set V of the universe of discourse U can be represented by

$V = \sum_{i=1}^{n} [tr_V(u_i), 1 - f_V(u_i)] / u_i, \forall u_i \in U$, where $tr_V(u) \leq \mu_V(u) \leq 1 - f_V(u)$ and $1 \leq i \leq n$.

This approach bounds the grade of membership of u to a subinterval $[tr_V(u), 1 - f_V(u)]$ of [0, 1]. In other words, the exact grade of membership $\mu_V(u)$ of u may be unknown, but is bounded by $tr_V(u) \leq \mu_V(u) \leq 1 - f_V(u)$, where $tr_V(u) + f_V(u) \leq 1$. The precision of the knowledge about u is characterized by the difference $(1 - tr_V(u) - f_V(u))$. If the difference is small, the knowledge about u is relatively precise; if it is large, we know correspondingly little. If $tr_V(u)$ is equal to $(1 - f_V(u))$, the knowledge about u is exact, and the vague set reverts back to fuzzy set. If $tr_V(u)$ and $(1 - f_V(u))$ are both equal to 1 or 0, depending on whether u does or does not belong to V, the knowledge about u is very exact and the vague set reverts back to ordinary set. For example, the fuzzy set $\{0.6/u\}$ can be presented as the vague set $\{[0.6, 0.6]/u\}$, while the ordinary set $\{u\}$ can be presented as the vague set $\{[1, 1]/u\}$.

Let $VSs(X)$ denote the set of all the vague sets in X. Then for $\forall A, B \in VSs(X)$, we have the following expression.

$A \le B$ if and only if $tr_A(x) \le tr_B(x)$ and $1 - f_A(x) \le 1 - f_B(x)$, $\forall x \in X$.

Definition 2 (Vague Relation). Let X and Y be ordinary finite non-empty sets. We call a vague relation to be a vague subset of $X \times Y$, that is, to an expression R defined by:

$$R = \{<(x, y), tr_R(x, y), 1 - f_R(x, y)>| x \in X, y \in Y\}$$

where

$$tr_R: X \times Y \rightarrow [0, 1],\ f_R: X \times Y \rightarrow [0, 1],$$

which satisfies the condition $0 \le t_R(x, y) + f_R(x, y) \le 1$, $\forall (x, y) \in X \times Y$. We represent the set of all the vague subsets of $X \times Y$ as $VSR(X \times Y)$.

Definition 3 (max-t & min-s compositions). Given a t-norm, a s-norm and an initial vague-relation matrix, $R^{(0)} = [(tr_{ij}^{(0)}, 1 - f_{ij}^{(0)})]$, then $R^{(n)} = [(tr_{ij}^{(n)}, 1 - f_{ij}^{(n)})]$, with $tr_{ij}^{(n)} = \max_k \{t(tr_{ik}^{(n-1)}, tr_{kj}^{(n-1)})\}$ and $f_{ij}^{(n)} = \min_k \{s(f_{ik}^{(n-1)}, f_{kj}^{(n-1)})\}$, n = 1, 2, 3,... is called a max-t & min-s composition.

Definition 4 (Proximity relation). A vague relation R in X is called proximity relation if it satisfies

Reflexive, if $tr_R(x, x) = 1$ and $f_R(x, x) = 0$ $\forall x \in X$ and

Symmetry, if $R = R^{-1}$, i.e. $tr_R(x, y) = tr_R(y, x)$ and $f_R(x, y) = f_R(y, x)$ $\forall x, y \in X$.

Definition 5 (Similarity relation). A proximity vague relation R in X is called Similarity relation if it satisfies: (max-t & min-s) Transitive, if $R \ge R \circ R$, i.e. $t_R(x, z) \ge \vee_{y \in X} \{t[tr_R(x, y), tr_R(y, z)]\}$ and $1 - f_R(x, z) \ge 1 - \wedge_{y \in X} \{s[f_R(x, y), f_R(y, z)]\}$ for all x, z in X.

The t-norms or s-norms [14] were defined as a general class of intersection or aggregation operators. Two typical dual pairs of t-norms t and s-norms s are listed as follows, where $0 \le x, y \le 1$,

(1) $t_\nabla(x, y) = \max\{0, x + y - 1\}$, $s_\nabla(x, y) = \min\{1, x + y\}$.
(2) $t_m(x, y) = \min\{x, y\}$, $s_m(x, y) = \max\{x, y\}$.

3 An n-Step Procedure

A crisp relation R between two sets, say X and Y, denoted $R(X, Y)$, is defined as a subset of $X \times Y$. This relation is associated with an indicator function $\mu_R(x, y) \in \{0, 1\}$

for all $(x, y) \in X \times Y$. That is, if $(x, y) \in R(X, Y)$ then $\mu_R(x, y) = 1$, and if $(x, y) \notin R(X, Y)$ then $\mu_R(x, y) = 0$. Zadeh [15] defined a fuzzy relation R between X and Y as a subset of $X \times Y$ by allowing $\mu_R(x, y)$ to take value in [0, 1], which represents the strength of the relationship between x and y. Furthermore, he defined a fuzzy relation R in X, called a max-min similarity relation, if it satisfies $\mu_R(x, x) = 1$ for $\forall x \in X$ (reflexivity); $\mu_R(x, y) = \mu_R(y, x)$ for $\forall x, y \in X$ (symmetry); $\mu_R(x, z) \geq \max_{y \in X} \{\min\{\mu_R(x, y), \mu_R(y, z)\}\}$ for $\forall x, z \in X$ (transitivity). It is clear that the similarity relation R is a fuzzy-type generalization of the equivalence relation.

In real pattern recognition such as visual images, smells, and pictures, etc., human subjectivity provides important information. This subjective information may be represented by a proximity fuzzy relation. Based on fuzzy relations, fuzzy clustering was first proposed by Tamura et al.[11]. But proximity fuzzy relations do not have transitivity, they cannot be used in clustering directly. This is why Tamura et al. [11] constructed an n-step procedure by using the composition of a proximity fuzzy relation. Recently, Yang and Shih [8] extended Tamura's n-step procedure and proposed a clustering algorithm based on fuzzy relation. In this paper, we will extend Yang and Shih's algorithm to vague relations.

Theorem 1 (An n-step procedure). Let $R^{(0)} = [(tr_{ij}^{(0)}, 1 - f_{ij}^{(0)})]$ be a proximity vague relation matrix. Then, by the max-t & min-s compositions of vague relations, we have

$$I < R^{(0)} < R^{(1)} < \ldots < R^{(n)} = R^{(n+1)} = \ldots,$$

where $R^{(n)}$ is a similarity relation and $I = [(0, 0)]$. If n is infinite, then $\lim_{n\to\infty} R^{(n)} = R^{(\infty)}$ with $R^{(\infty)}$ a similarity relation, i.e.

$$I < R^{(0)} < R^{(1)} < \ldots < R^{(n)} < R^{(n+1)} < \ldots < R^{(\infty)}.$$

Proof. Since $R^{(0)}$ is a proximity vague relation matrix, $I < R^{(0)}$. Let $R^{(1)} = [(tr_{ij}^{(1)}, 1 - f_{ij}^{(1)})]$ with $tr_{ij}^{(1)} = \vee_k \{t(tr_{ik}^{(0)}, tr_{kj}^{(0)})\}$ and $f_{ij}^{(1)} = \wedge_k \{s(f_{ik}^{(0)}, f_{kj}^{(0)})\}$. Then

$$\left.\begin{array}{l} tr_{ij}^{(0)} = t(tr_{ij}^{(0)}, 1) = t(tr_{ij}^{(0)}, tr_{jj}^{(0)}) \leq \vee_k t(tr_{ik}^{(0)}, tr_{kj}^{(0)}) = tr_{ij}^{(1)} \\ f_{ij}^{(0)} = s(f_{ij}^{(0)}, 0) = s(f_{ij}^{(0)}, f_{jj}^{(0)}) \geq \wedge_k s(f_{ik}^{(0)}, f_{kj}^{(0)}) = f_{ij}^{(1)} \end{array}\right\} \Leftrightarrow R^{(0)} \leq R^{(1)}$$

Assume that $R^{(0)}$ does not have transitivity. Then, there exists i and j such that for some m we have

$$\left.\begin{array}{r} tr_{ij}^{(0)} < t(tr_{im}^{(0)}, tr_{mj}^{(0)}) \Rightarrow tr_{ij}^{(0)} < \vee_k t(tr_{ik}^{(0)}, tr_{kj}^{(0)}) = tr_{ij}^{(1)} \\ f_{ij}^{(0)} > s(f_{im}^{(0)}, f_{mj}^{(0)}) \Rightarrow f_{ij}^{(0)} > \wedge_k s(f_{ik}^{(0)}, f_{kj}^{(0)}) = f_{ij}^{(1)}, i.e. 1 - f_{ij}^{(0)} < 1 - f_{ij}^{(1)} \end{array}\right\} \Rightarrow R^{(0)} < R^{(1)}$$

If $R^{(1)}$ does not have transitivity either, then similarity one has $R^{(1)} < R^{(2)}$. If the transitivity is not reached after (n-1) compositions, then

$$I < R^{(0)} < R^{(1)} < \ldots < R^{(n)}.$$

Assume that $R^{(n)}$ has transitivity. Then for all i, j, we have $tr_{ij}^{(n)} \geq \vee_k t(tr_{ik}^{(n)}, tr_{kj}^{(n)}) = tr_{ij}^{(n+1)}$, and $t(tr_{ii}^{(n)}, tr_{ij}^{(n)}) = t(1, tr_{ij}^{(n)}) = tr_{ij}^{(n)} \Rightarrow tr_{ij}^{(n+1)} = \vee_k t(tr_{ik}^{(n)}, tr_{kj}^{(n)}) \geq t(tr_{ii}^{(n)}, tr_{ij}^{(n)}) = tr_{ij}^{(n)}$. So we can obtain $tr_{ij}^{(n+1)} = tr_{ij}^{(n)}$. Similarly, we can obtain $f_{ij}^{(n+1)} = f_{ij}^{(n)}$ either. Hence, we have $R^{(n+1)} = R^{(n)}$. Similarly, $R^{(n+2)} = R^{(n+1)}$. That is

$$I < R^{(0)} < R^{(1)} < \ldots < R^{(n)} = R^{(n+1)} = \ldots.$$

If there is not a finite n such that $R^{(n)} = R^{(n+1)} = \ldots$, then

$$I < R^{(0)} < R^{(1)} < \ldots < R^{(n)} < R^{(n+1)} < \ldots < R^{(*)},$$

Where $R^{(*)} = \begin{bmatrix} [1,1] & \ldots & [1,1] \\ \vdots & & \vdots \\ [1,1] & \cdots & [1,1] \end{bmatrix}$

It is known that $\{R^{(k)} | k = 0, 1, 2, \ldots\}$ is monotone and bounded. Then $R^{(\infty)} = \lim_{n\to\infty} R^{(n)}$ exists. Next, it is claimed that $R^{(\infty)}$ is a similarity relation.

Recall that $tr_{ij}^{(n)} = \vee_k t(tr_{ik}^{(n-1)}, tr_{kj}^{(n-1)})$ and $f_{ij}^{(n)} = \wedge_k s(f_{ik}^{(n-1)}, f_{kj}^{(n-1)})$. Now, two new terms are defined as $tr'^{(n)}_{ij} = \vee_k t(tr'^{(n-1)}_{ik}, tr'^{(n-1)}_{kj})$ and $f'^{(n)}_{ij} = \wedge_k s(f'^{(n-1)}_{ik}, f'^{(n-1)}_{kj})$. Although $R^{(n)}$ and $R'^{(n)}$ are different, $\lim_{n\to\infty} R^{(n)} = \lim_{n\to\infty} R'^{(n)} = R^{(\infty)}$.

$$tr'^{(2)}_{ij} = \vee_{k_1} t(tr'^{(1)}_{ik_1}, tr'^{(1)}_{k_1 j}),$$

$$tr'^{(3)}_{ij} = \vee_{k_2} t(tr'^{(2)}_{ik_2}, tr'^{(2)}_{k_2 j}) = \vee_{k_1,k_2} t(t(tr'^{(1)}_{ik_1}, tr'^{(1)}_{k_1k_2}), tr'^{(1)}_{k_2 j})$$

$$\vdots$$

$$tr'^{(n)}_{ij} = \vee_{k_1,\cdots,k_{n-1}} t(\cdots(t(t(tr'^{(1)}_{ik_1}, tr'^{(1)}_{k_1k_2}), tr'^{(1)}_{k_2k_3}),\cdots, tr'^{(1)}_{k_{n-1}j}))$$

$$tr'^{(m+n)}_{ij} = \vee_{k_1,\cdots,k_{m+n-1}} t(\cdots(t(t(tr'^{(1)}_{ik_1}, tr'^{(1)}_{k_1k_2}), tr'^{(1)}_{k_2k_3}),\cdots, tr'^{(1)}_{k_{m+n-1}j}))$$

$$\ge \vee_{k_1,\cdots,k_{m-1}} \vee_{k_{m+1},\ldots,k_{m+n-1}} t(\cdots(t(t(tr'^{(1)}_{ik_1}, tr'^{(1)}_{k_1k_2}), tr'^{(1)}_{k_2k_3}),\cdots, tr'^{(1)}_{k_{m+n-1}j}))$$

$$= t(\vee_{k_1,\cdots,k_{m-1}} t(\cdots t(tr'^{(1)}_{ik_1}, tr'^{(1)}_{k_1k_2}),\cdots, tr'^{(1)}_{k_{m-1}l}), \vee_{k_{m+1},\ldots,k_{m+n-1}} t(\cdots t(tr'^{(1)}_{lk_{m+1}}, tr'^{(1)}_{k_{m+1}k_{m+2}}),\cdots, tr'^{(1)}_{k_{m+n-1}j}))$$

$$= t(tr'^{(m)}_{il}, tr'^{(n)}_{lj})$$

and

Similarly, we can obtain $1 - f'^{(m+n)}_{ij} \ge 1 - s(f'^{(m)}_{il}, f'^{(n)}_{lj})$.

So, we have $tr'^{(m+n)}_{ij} \ge t(tr'^{(m)}_{il}, tr'^{(n)}_{lj})$ and $1 - f'^{(m+n)}_{ij} \ge 1 - s(f'^{(m)}_{il}, f'^{(n)}_{lj})$ for all l. As $m\to\infty$ and $n\to\infty$, we can obtain

$$tr'^{(\infty)}_{ij} \ge t(tr'^{(\infty)}_{il}, tr'^{(\infty)}_{lj}),\ 1 - f'^{(\infty)}_{ij} \ge 1 - s(f'^{(\infty)}_{il}, f'^{(\infty)}_{lj}), \text{ for all } l.$$

We also have $tr'^{(\infty)}_{ii} = 1, 1 - f'^{(\infty)}_{ii} = 1$ and $tr'^{(\infty)}_{ij} = tr'^{(\infty)}_{ji}, 1 - f'^{(\infty)}_{ij} = 1 - f'^{(\infty)}_{ji}$. That is, $R^{(\infty)}$ is a similarity relation.

Theorem 1 extends Yang and Shih's n-step procedure to the vague relation. That is, we can get a similarity vague relation matrix $R^{(n)}$ based on max-t & min-s compositions by beginning with a proximity vague relation matrix.

Example 1. Consider the proximity vague relation matrix $R^{(0)}$ with

$$R^{(0)} = \begin{bmatrix} [1,1] & & \\ [0.7,0.7] & [1,1] & \\ [0.6,0.65] & [0.1,0.15] & [1,1] \end{bmatrix}$$

(1) by max-t_m & min-s_m composition, we have

$$R^{(1)} = \begin{bmatrix} [1,1] & & \\ [0.7,0.7] & [1,1] & \\ [0.6,0.65] & [0.6,0.65] & [1,1] \end{bmatrix} = R^{(2)}$$

(2) by max-t_∇ & min-s_∇ composition, we have

$$R^{(1)} = \begin{bmatrix} [1,1] & & \\ [0.7,0.7] & [1,1] & \\ [0.6,0.65] & [0.3,0.35] & [1,1] \end{bmatrix} = R^{(2)}$$

A proximity relation only represents a subjective similarity relation. It is an initial relation which does not have transitivity. By Theorem 1, we can obtain a max-t & min-s transitivity. From Example 1, it is shown that [0.3, 0.35] obtained by max-t_∇ & min-s_∇ composition is relatively close to $[tr_{23}^{(0)}, 1 - f_{23}^{(0)}] = [0.1, 0.15]$, In some sense, max-t_∇ & min-s_∇ composition can attain transitivity and also remain nearest to the original subjective similarity. That is, the max-t_∇ & min-s_∇ composition seems to be more meaningful than the max-t_m & min-s_m composition.

Theorem 2. Let t_1 and t_2 be any two t-norms and s_1 and s_2 be any two s-norms. Let $R^{(0)}$ be a proximity vague relation matrix. Suppose that $I < R^{(0)} < R^{(1)} < \ldots < R^{(m)} = R^{(m+1)}$ by the n-step procedure based on max-t_1 & min-s_1 compositions and $I < R^{(0)} < R^{(1)} < \ldots < R^{(n)} = R^{(n+1)}$ by the n-step procedure based on max-t_2 & min-s_2 compositions. If t_1-norm $\leq t_2$-norm and s_1-norm $\geq s_2$-norm then $m \leq n$.

Proof. Since $R^{(n)}$ is a max-t_2 & min-s_2 similarity relation,

$tr_{ij}^{(n)} \geq \vee_k t_2(tr_{ik}^{(n)}, tr_{kj}^{(n)})$, $f_{ij}^{(n)} \leq \wedge_k s_2(f_{ik}^{(n)}, f_{kj}^{(n)})$.

but $t_2(tr_{ik}^{(n)}, tr_{kj}^{(n)}) \geq t_1(tr_{ik}^{(n)}, tr_{kj}^{(n)})$ and $s_2(f_{ik}^{(n)}, f_{kj}^{(n)}) \leq s_1(f_{ik}^{(n)}, f_{kj}^{(n)})$ for all k. Then

$tr_{ij}^{(n)} \geq \vee_k t_1(tr_{ik}^{(n)}, tr_{kj}^{(n)})$, $f_{ij}^{(n)} \leq \wedge_k s_1(f_{ik}^{(n)}, f_{kj}^{(n)})$.

Therefore, $m \leq n$.

By Theorem 2, we can find that the number of max-t_∇ & min-s_∇ compositions to reach its transitive closure $R^{(m)}$ is less than that of the max-t_m & min-s_m compositions to reach its transitive closure $R^{(n)}$, i.e. $m \leq n$.

Definition 6. for $0 \leq \alpha, \beta \leq 1$ with $\alpha + \beta \leq 1$, we define a α-β-cut of the vague relation R as follows:

$R_{(\alpha,\beta)} = \{(x, y) | tr_R(x, y) \geq \alpha, f_R(x, y) \leq \beta\}$

Thus, we have the proposition that if $0 \leq \alpha_1 \leq \alpha_2 \leq 1$, $0 \leq \beta_2 \leq \beta_1 \leq 1$ with $0 \leq \alpha_1 + \beta_1 \leq 1$, $0 \leq \alpha_2 + \beta_2 \leq 1$, then $R_{(\alpha_2,\beta_2)} \subset R_{(\alpha_1,\beta_1)}$.

Example 2. Consider the proximity vague relation matrix $R^{(0)}$ on $X = \{x_1, x_2, x_3\}$ with

$$R^{(0)} = \begin{bmatrix} [1,1] & & \\ [0.7,0.7] & [1,1] & \\ [0.6,0.65] & [0.1,0.15] & [1,1] \end{bmatrix}.$$

as in Example 1. Then

(1)

$$R^{(1)} = \begin{bmatrix} [1,1] & & \\ [0.7,0.7] & [1,1] & \\ [0.6,0.65] & [0.6,0.65] & [1,1] \end{bmatrix} = R^{(2)}$$

is a max-t & min-s similarity relation based on max-t_m & min-s_m composition. We have

(1.1) if $0 < \alpha \leq 0.6$, $0.35 \leq \beta < 1$(i.e. $0 < 1-f_R(x, y) \leq 0.65$) with $\alpha + \beta \leq 1$, one has $R^{(1)}_{(\alpha,\beta)} = \{(x_1,x_1),(x_1,x_2),(x_1,x_3),(x_2,x_1),(x_2,x_2),(x_2,x_3),(x_3,x_1),(x_3,x_2),(x_3,x_3)\}$, and then, the elements in X can be clustered into one class, i.e. $\{x_1, x_2, x_3\}$.

(1.2) if $0.6 < \alpha \leq 0.7$, $0.3 \leq \beta < 0.35$(i.e. $0.65 < 1-f_R(x, y) \leq 0.7$) with $\alpha + \beta \leq 1$, one has $R^{(1)}_{(\alpha,\beta)} = \{(x_1,x_1),(x_1,x_2),(x_2,x_1),(x_2,x_2),(x_3,x_3)\}$, and then, the elements in X can be clustered into two classes, i.e. $\{x_1, x_2\}$ and $\{x_3\}$.

(1.3) if $0.7 < \alpha \leq 1$, $0 \leq \beta < 0.3$(i.e. $0.7 < 1-f_R(x, y) \leq 1$) with $\alpha + \beta \leq 1$, one has $R^{(1)}_{(\alpha,\beta)} = \{(x_1,x_1),(x_2,x_2),(x_3,x_3)\}$, and then, the elements in X can be clustered into three classes, i.e. $\{x_1\}$, $\{x_2\}$ and $\{x_3\}$.

(2)

$$R^{(1)} = \begin{bmatrix} [1,1] & & \\ [0.7,0.7] & [1,1] & \\ [0.6,0.65] & [0.3,0.35] & [1,1] \end{bmatrix} = R^{(2)}$$

is a max-t & min-s similarity relation based on max-t_∇ & min-s_∇ composition. We have

(1.1) if $0 < \alpha \leq 0.3$, $0.65 \leq \beta < 1$(i.e. $0 < 1-f_R(x, y) \leq 0.35$) with $\alpha + \beta \leq 1$, one has $R^{(1)}_{(\alpha,\beta)} = \{(x_1,x_1),(x_1,x_2),(x_1,x_3),(x_2,x_1),(x_2,x_2),(x_2,x_3),(x_3,x_1),(x_3,x_2),(x_3,x_3)\}$, and then, the elements in X can be clustered into one class, i.e. $\{x_1, x_2, x_3\}$.

(1.2) if $0.6 < \alpha \leq 0.7$, $0.3 \leq \beta < 0.35$(i.e. $0.65 < 1-f_R(x, y) \leq 0.7$) with $\alpha + \beta \leq 1$, one has $R^{(1)}_{(\alpha,\beta)} = \{(x_1,x_1),(x_1,x_2),(x_2,x_1),(x_2,x_2),(x_3,x_3)\}$, and then, the elements in X can be clustered into two classes, i.e. $\{x_1, x_2\}$ and $\{x_3\}$.

(1.3) if $0.7 < \alpha \leq 1$, $0 \leq \beta < 0.3$(i.e. $0.7 < 1-f_R(x, y) \leq 1$) with $\alpha + \beta \leq 1$, one has $R^{(1)}_{(\alpha,\beta)} = \{(x_1,x_1),(x_2,x_2),(x_3,x_3)\}$, and then, the elements in X can be clustered into three classes, i.e. $\{x_1\}$, $\{x_2\}$ and $\{x_3\}$.

However,

(1.4) if $0.3 < \alpha \leq 0.6$, $0.35 \leq \beta < 0.65$(i.e. $0.35 < 1-f_R(x, y) \leq 0.65$) with $\alpha + \beta \leq 1$, one has $R^{(1)}_{(\alpha,\beta)} = \{(x_1,x_1),(x_1,x_2),(x_1,x_3),(x_2,x_1),(x_2,x_2),(x_3,x_1),(x_3,x_3)\}$. It shows that x_1 and x_2 or x_1 and x_3 can be in a class, but x_2 and x_3 cannot be in a class. That is, we cannot obtain a clustering result from $R^{(1)}_{(\alpha,\beta)}$. But from $R^{(1)}$, we can have $tr_{R^{(1)}}(x_1,x_2) = 0.7 > 0.6 = tr_{R^{(1)}}(x_1,x_3)$ and $f_{R^{(1)}}(x_1,x_2) = 0.3 < 0.35 = f_{R^{(1)}}(x_1,x_3)$. By the extension of maximum similarity principle (called max-min similarity principle) to have x_1 and x_2 in a class is preferred. Thus, we have the following clustering result:

$0.3 < \alpha \leq 0.7$, $0.3 \leq \beta < 0.65$ with $\alpha + \beta \leq 1 \Rightarrow \{x_1, x_2\}$, $\{x_3\}$.

4 Clustering Algorithm

There are many data which are presented in subjective relations and expressed in a proximity relation matrix. Proximity relations cannot be used in clustering because they do not have transitivity. A max-t & min-s similarity relation is obtained by an n-step procedure based on the max-t & min-s compositions given in Section 3. Then, this max-t & min-s similarity relation matrix is used for clustering. In Section 3, it is stated that the max-t_∇ & min-s_∇ composition is more meaningful and effective than the max-t_m & min-s_m composition when they are used in the n-step procedure. But we can not get a partition tree from max-t_∇ & min-s_∇ similarity matrix in Example 2. To tackle this problem, a clustering algorithm for any max-t & min-s similarity relation based on a modified maximum similarity principle is presented. The algorithm is given as follows:

Algorithm. Fuzzy Clustering Based on Vague Relations
Input: a proximity vague relation matrix: $R^{(0)} = [tr_{ij}^{(0)}, 1 - f_{ij}^{(0)}]_{n\times n}$
Output: class C
Begin

1. $I = \{1, 2, \ldots, n\}$; It is given that (α, β) with $0 < \alpha, \beta \le 1$ and $0 < \alpha + \beta \le 1$;
2. **for** $R^{(0)}$, evaluates its corresponding similarity-relation matrix $R = [tr_{ij}, 1 - f_{ij}]_{n\times n}$ using the n-step procedure of max-t & min-s compositions;
3. **for** all ($i = j$ or $tr_{ij} < \alpha$ or $f_{ij} > \beta$) set $[tr_{ij}, 1-f_{ij}] = [0, 0]$;
4. **Let** $[tr_{pq}, 1-f_{pq}] = [\max\{tr_{ij} | i < j,\ i, j \in I\}, 1 - \min\{f_{ij} | i < j,\ i, j \in I\}]$, $\forall p, q \in I$;
   if exists $[tr_{pq}, 1-f_{pq}] \ne [0,0]$ **then**
   $C = \{p, q\}$; **GOTO** 5;
   else
   GOTO 6;
5. **for each** $u \in I - C$
   if $\sum_{i\in C} tr_{iu} = \max\{ \sum_{i\in C} tr_{ij} \mid j \in I - C,\ tr_{ij} \ne 0 \text{ for all } i \in C\}$ and
   $\sum_{i\in C} f_{iu} = \min\{ \sum_{i\in C} f_{ij} \mid j \in I - C,\ f_{ij} \ne 0 \text{ for all } i \in C\}$ **then**
   $\{C = C + \{u\}$; GOTO 5;$\}$
   else
   {**OUTPUT** the cluster C; $I = I - C$; **GOTO** 4; }
6. **OUTPUT** all indices in I into separated cluster;

End

Example 3. Given a 10×10 proximity vague relation matrix, $R^{(0)}$, and given (α, β) = (0.55,0.3).

$$R^{(0)} = \begin{bmatrix}
[1,1] \\
[0.2,0.2] & [1,1] \\
[0.5,0.6] & [0.3,0.45] & [1,1] \\
[0.8,0.9] & [0.6,0.7] & [0.5,0.5] & [1,1] \\
[0.6,0.65] & [0.7,0.8] & [0.3,0.4] & [0.7,0.8] & [1,1] \\
[0.2,0.4] & [0.9,0.9] & [0.4,0.6] & [0.3,0.4] & [0.2,0.25] & [1,1] \\
[0.3,0.4] & [0.2,0.2] & [0.1,0.2] & [0.5,0.5] & [0.4,0.4] & [0.1,0.2] & [1,1] \\
[0.9,0.9] & [0.8,0.8] & [0.3,0.35] & [0.4,0.5] & [0.5,0.6] & [0.3,0.4] & [0.6,0.6] & [1,1] \\
[0.4,0.55] & [0.3,0.4] & [0.7,0.8] & [0.1,0.1] & [0.8,0.9] & [0.7,0.8] & [0.1,0.2] & [0.0,0.2] & [1,1] \\
[0.3,0.4] & [0.2,0.2] & [0.6,0.6] & [0.3,0.4] & [0.9,1] & [0.2,0.2] & [0.3,0.4] & [0.2,0.3] & [0.1,0.2] & [1,1]
\end{bmatrix}$$

Step 1. Let $I = \{1, 2, \ldots, 10\}$ and (α, β)=(0.55,0.3).

Step 2. evaluates similarity vague relation matrix $R = [tr_{ij}, 1-f_{ij}]_{10\times 10}$ using the n-step procedure of max-t_∇ & min-s_∇ compositions.

$$R=\left[\begin{array}{llllllllll}
[1,1] \\
[0.7,0.7] & [1,1] \\
[0.5,0.6] & [0.3,0.5] & [1,1] \\
[0.8,0.9] & [0.6,0.7] & [0.5,0.5] & [1,1] \\
[0.6,0.7] & [0.7,0.8] & [0.5,0.7] & [0.7,0.8] & [1,1] \\
[0.6,0.6] & [0.9,0.9] & [0.4,0.6] & [0.5,0.6] & [0.6,0.7] & [1,1] \\
[0.5,0.5] & [0.4,0.4] & [0.1,0.2] & [0.5,0.5] & [0.4,0.4] & [0.3,0.3] & [1,1] \\
[0.9,0.9] & [0.8,0.8] & [0.4,0.5] & [0.7,0.8] & [0.5,0.6] & [0.7,0.7] & [0.6,0.6] & [1,1] \\
[0.4,0.6] & [0.6,0.7] & [0.7,0.8] & [0.5,0.7] & [0.8,0.9] & [0.7,0.8] & [0.2,0.3] & [0.4,0.5] & [1,1] \\
[0.5,0.7] & [0.6,0.8] & [0.6,0.7] & [0.6,0.8] & [0.9,1] & [0.5,0.7] & [0.3,0.4] & [0.4,0.6] & [0.7,0.9] & [1,1]
\end{array}\right]$$

Step 3. for all ($i = j$ or $tr_{ij} < 0.55$ or $1-f_{ij} < 0.7$) set $[tr_{ij}, 1-f_{ij}] = [0, 0]$, then we have

$$R=\left[\begin{array}{llllllllll}
[0,0] \\
[0.7,0.7] & [0,0] \\
[0,0] & [0,0] & [0,0] \\
[0.8,0.9] & [0.6,0.7] & [0,0] & [0,0] \\
[0.6,0.7] & [0.7,0.8] & [0,0] & [0.7,0.8] & [0,0] \\
[0,0] & [0.9,0.9] & [0,0] & [0,0] & [0.6,0.7] & [0,0] \\
[0,0] & [0,0] & [0,0] & [0,0] & [0,0] & [0,0] & [0,0] \\
[0.9,0.9] & [0.8,0.8] & [0,0] & [0.7,0.8] & [0,0] & [0.7,0.7] & [0.0] & [0,0] \\
[0,0] & [0.6,0.7] & [0.7,0.8] & [0,0] & [0.8,0.9] & [0.7,0.8] & [0,0] & [0,0] & [0,0] \\
[0,0] & [0.6,0.8] & [0.6,0.7] & [0.6,0.8] & [0.9,1] & [0,0] & [0,0] & [0,0] & [0.7,0.9] & [0,0]
\end{array}\right]$$

Step 4. Since $[\max tr_{ij}, 1\text{-}\min f_{ij}] = [tr_{510}, 1-f_{510}] = [0.9, 1]$, so we have $p = 1$ and $q = 8$, then $C = \{5,10\}$.

Step 5. Since $(\max_{i\in I-C}(tr_{5i} + tr_{10i}), \min_{i\in I-C}(f_{5i} + f_{10i})) = [tr_{59}, f_{59}]+[tr_{109}, f_{109}]=(1.5, 0.2)$, so $C = \{5, 10, 9\}$.

Step 5. Since $(\max_{i\in I-C}(tr_{5i} + tr_{10i} + tr_{9i}), \min_{i\in I-C}(f_{5i} + f_{10i} + f_{9i})) = [tr_{52}, f_{52}]+[tr_{102}, f_{102}]+[tr_{92}, f_{92}]=(1.9, 0.7)$, so $C = \{5, 10, 9, 2\}$.

Step 5. There is no $u\in I-C$ so that $tr_{5u} + tr_{10u}+ tr_{9u} + tr_{2u}$ get a maximum and $f_{5u} + f_{10u}+ f_{9u} + f_{2u}$ get a minimum. So, $C = \{5, 10, 9, 2\}$ and $I = I - C = \{1, 3, 4, 6, 7, 8\}$.

Step 4. Since $[\max tr_{ij}, 1\text{-}\min f_{ij}] = [tr_{18}, 1-f_{18}] = [0.9, 0.9]$, so $C = \{1, 8\}$.

Step 5. Since $(\max_{i\in I-C}(tr_{1i} + tr_{8i}), \min_{i\in I-C}(f_{1i} + f_{8i})) = [tr_{14}, f_{14}]+[tr_{84}, f_{84}]=(1.5, 0.3)$, so $C = \{1, 8, 4\}$.

Step 5. There is no $u\in I-C$ so that $tr_{1u} + tr_{8u}+ tr_{4u}$ get a maximum and $f_{1u} + f_{8u}+ f_{4u}$ get a minimum. So $C = \{1, 8, 4\}$ and $I = I - C = \{3, 6, 7\}$.

Step 4. Since $[tr_{36}, 1 - f_{36}] = [tr_{37}, 1 - f_{37}] = [tr_{67}, 1 - f_{67}] = [0, 0]$ then go to Step 6.

Step 6. Output three separated clusters $\{3\}$, $\{6\}$ and $\{7\}$.

Thus, when $(\alpha, \beta) = (0.55, 0.3)$, we can get a partition $\{5, 10, 9, 2\}$, $\{1, 8, 4\}$, $\{3\}$, $\{6\}$, $\{7\}$.

5 Conclusions

In this paper, we extend Yang and Shih's n-step procedure to the vague relation. A max-t & min-s similarity vague relation matrix is obtained by beginning with a proximity vague relation matrix based on the extended max-t & min-s n-step procedure. Since a max-t_∇ & min-s_∇ similarity vague relation matrix may not have a resolution form of equivalence relations as a max-t_m & min-s_m similarity, a clustering algorithm for any max-t & min-s similarity vague relation matrix is proposed. It is shown that max-t_∇ & min-s_∇ compositions has better performance than that of max-t_m & min-s_m.

Acknowledgements

This work is supported by the National Research Foundation for the Doctoral Program of Higher Education of China under Grant No.20050145024 and partially by the Program for New Century Excellent Talents in University under Grant No.NCET-05-0288.

References

1. Zadeh, L.A.: Fuzzy Sets. Information and Control 8(3) (1965) 338–353
2. Gau, W.L., Buehrer, D.J.: Vague Sets. IEEE Transactions on Systems, Man, and Cybernetics 23(2) (1993) 610–614
3. Atanassov, K.T.: Intuitionistic Fuzzy Sets. Fuzzy Sets and Systems 20 (1986) 87–96
4. Lu A., Ng, W.: Vague Sets or Intuitionistic Fuzzy Sets for Handling Vague Data: Which One is Better? Lecture Notes in Computer Science 3716 (2005) 401–416
5. Aldenderfer, M., Blashfield, R.: Cluster Analysis. Beverly Hill Sage Publications (1984)
6. Arabie, P., Carroll, J.D., Desarbo, W.J., Wind: Overlapping Clustering: A New Method for Product Positioning. Journal of Marketing Research 18 (1981) 310–317
7. Punj, G.N., David, W.S.: Cluster Analysis in Marketing Research: Review and Suggestion for Application. Journal of Marketing Research 20 (1983) 135–148
8. Bezdek, J.C.: Pattern Recognition with Fuzzy Objective Function Algorithms. Plenum Press, New York (1981)
9. Trauwaert, E., Kaufman, L., Rousseeuw, P.: Fuzzy Clustering Algorithms Based on the Maximum Likelihood Principle. Fuzzy Sets and Systems 42 (1991) 213–227
10. Dave, R.N.: Generalized Fuzzy c-shells Clustering and Detection of Circular and Elliptical Boundaries. Pattern Recognition 25 (1992) 713–721
11. Tamura, S., Higuchi, S., Tanaka, K.: Pattern Classification Based on Fuzzy Relations. IEEE Trans. Systems Man Cybernet. 1 (1978) 61–66
12. Yang M.S., Shih, H.M.,: Cluster Analysis Based on Fuzzy Relations, Fuzzy Sets and Systems 120(2) (2001) 197–212
13. De, S.K., Biswas, R., Roy, A.R.: An Application of Intuitionistic Fuzzy Sets in Medical Diagnosis. Fuzzy Sets and Systems 117 (2001) 209–213
14. Zimmermann, H.J.: Fuzzy Set Theory and Its Applications, Kluwer, Dordrecht, 1991
15. Zadeh, L.A.: Similarity Relations and Fuzzy Ordering. Inform. Sci. 3 (1971) 177–200

The Fuzzy Clustering Algorithm Based on AFS Topology

Rui Ding[1], Xiaodong Liu[1,2], and Yan Chen[1]

[1] Economics and Management College, Dalian Maritime University, Dalian,116026, P.R. China
[2] Research Center of Information and Control, Dalian University of Technology, Dalian,116024, P.R. China
joking810908@163.com

Abstract. This paper establishes a new metric space for the clustering problems. The neighbors on the object set induced by the topology molecular lattice on $^*EI$ algebra are given and a new distance based on the neighbors is proposed. In the proposed clustering algorithm, the Euclidean metric is replaced by the new distance based on the order relationship of the samples on the attributes. As a result, using the method to Iris data we show it has a better result and clearer classification than the other clustering algorithm based on the Euclidean metric. This study shows that the AFS topology fuzzy clustering algorithm can obtain an high clustering accuracy according to order relationship.

1 Introduction

Fuzzy sets and systems has been developed rapidly and applied in many fields since it was proposed by Prof. Zadeh [1]. However, a fuzzy set is a rather abstract notion. Fuzzy sets are useful for many purposes, and membership functions do not mean the same thing at the operational level in each and every context. We are often perplexed by the problem how to properly determine the membership function according to the concrete situation. In order to deal with the above discussed problems, AFS (Axiomatic Fuzzy Set) theory was firstly proposed by Liu in 1995 [4]. In essence, the AFS framework provides an effective tool to convert the information in the training examples and databases into the membership functions and their fuzzy logic operations. AFS fuzzy logic can be applied to the data sets with various data types such as real numbers, Boolean value, partial order, even human intuition descriptions, which are very difficult or unsolved for other clustering algorithm such as the fuzzy c-means algorithm (FCMA) conceived by Dunn [8] and generalized by Bezdek [9] and the fuzzy k nearest neighbor algorithm, named as k-NN algorithm [10].

We know that human can classify, cluster and recognize the objects in the ordinary data set X without any metric in Euclidean space. What is human recognition based on if X is not a subset of some metric space in Euclidean space? For example, if you want to classify all your friends into two classes {close friends} and {common friends}. The criteria/metric you are using in the

L. Wang et al. (Eds.): FSKD 2006, LNAI 4223, pp. 89–98, 2006.

process is very important though it may not be based on the Euclidean metric. In [6, 11], using topological molecular theory, the topological structures on X induced by the topological molecular lattices generated by some fuzzy sets in EM have been obtained. This kind topology on X is determined by the original data and the chosen fuzzy sets in EM. It is an abstract geometry relations among the objects in X, the interpretations of the special topological structures on the AFS structures directly obtained by a given data set through the differential degrees between objects in X. With the topological space on X induced by the fuzzy concepts, the pattern recognition problems of ordinary data sets can be studied.

In this paper, we applied the topological structures induced by some concepts in EM to establish the metric for clustering problems. We study the topology molecular lattice on $^*EI$ algebra over some concepts, which based on AFS structure and AFS algebra, then give the neighbors on the object set reduced by the topology molecular lattice. We apply the neighbors of the topology to define a distance to study the fuzzy clustering analysis and the example shows that the new clustering algorithm is effective.

2 Preliminaries

In this section, we will recall the notations and definitions of AFS theory. AFS theory is made of AFS structures which is a special kind of combinatorics object [20] and AFS algebra which is a family of completely distributive lattices [7]. About the detail mathematical properties of AFS algebras please see [2-6, 11-19].

Definition 1 ([13]). *Let ζ be any concept on the universe of discourse X. R_ζ is called a binary relation (i.e., $R_\zeta \subset X \times X$) of ζ if R_ζ satisfies: $x, y \in X$, $(x, y) \in R_\zeta \Leftrightarrow x$ belongs to ζ at some degree and the degree of x belonging to ζ is larger than or equals to that of y, or x belongs to ζ at some degree and y does not at all.*

Definition 2 ([2,3]). *Let X be a set and R be a binary relation on X. R is called a sub-preference relation on X if for x, y, $z \in X$, $x \neq y$, R satisfies the following conditions:*

D5-1. If $(x, y) \in R$, then $(x, x) \in R$;
D5-2. If $(x, x) \in R$ and $(y, y) \notin R$, then $(x, y) \in R$;
D5-3. If (x, y), $(y, z) \in R$, then $(x, z) \in R$;
D5-4. If $(x, x) \in R$ and $(y, y) \in R$, then either $(x, y) \in R$ or $(y, x) \in R$.

In addition, ζ is called a simple concept or simple attribute on X if R_ζ is a sub-preference relation on X. Otherwise ζ is called a complex concept or a complex attribute on X.

Definition 3 ([2,3]). *Let X, M be two sets and 2^M be the power set of M, $\tau : X \times X \to 2^M$. (M, τ, X) is called an AFS structure if τ satisfies the following conditions:*

Table 1. Date Set

sample	age	weight	height	male	female	salary	fortune
x_1	21	50	1.69	yes	no	0	0.000
x_2	30	52	1.62	no	yes	120	200.000
x_3	27	65	1.80	yes	no	100	40.000
x_4	60	63	1.50	no	yes	80	324.000
x_5	45	54	1.71	yes	no	140	486.940.000

AX1: $\forall (x_1, x_2) \in X \times X,\ \tau(x_1, x_2) \subseteq \tau(x_1, x_1)$;
AX2: $\forall (x_1, x_2),\ (x_2, x_3) \in X \times X,\ \tau(x_1, x_2) \cap \tau(x_2, x_3) \subseteq \tau(x_1, x_3)$.
In addition, X is called universe of discourse, M is called an attribute set and τ is called a structure.

In practice, we always suppose that every concept in M is a simple concept on X. We can verify that (M, τ, X) is an AFS structure if τ is defined by

$$\tau(x_i, x_j) = \{m | m \subset M, (x_i, x_j) \in R_m\}, x_i, x_j \in X.$$

Example 1. Let $X = \{x_1, x_2, \ldots, x_5\}$ be a set of five persons. $M = \{m_1, m_2, \ldots, m_7\}$, where m_1=age, m_2=weight, m_3=height, m_4=male, m_5=female, m_6=salary, m_7=fortune, suppose there exists Table1:

According to Table1 and the preference relations, $\tau(x_1, x_1)$=\{age, height, weight, salary, male\}. This indicates that the person x_1 has the properties m_1, m_2, m_3, m_4. Similarly for $\tau(x_i, x_i)$, $i = 2, \ldots, 10$. $\tau(x_4, x_5) = \{m_1, m_2, m_5\}$. This implies that the degree of x_4 possessing properties m_1, m_2, m_5 is larger than that of x_5 or equal. Similarly for $\tau(x_i, x_j)$, $i, j = 1, 2, \ldots, 10$. It easily verifies that τ satisfies AX1, AX2 and (M, τ, X) is an AFS structure.

In order to study fuzzy concepts and their topological structure, we introduce $^*EI$ algebra($^*EI$ algebra is the opposite of EI algebra).

Definition 4 ([11]). *Let M be sets. In general, M is a set of fuzzy or crisp concepts,*

$$EM^* = \{\sum_{i \in I} A_i | A_i \subseteq M, i \in I, I \text{ is any no-empty indexing set}\}.$$

Each $\sum_{i \in I} A_i$ is an element of EM^*, where $\sum_{i \in I}$ is just a symbol meaning that element $\sum_{i \in I} A_i$ is composed of $A_i \subseteq M$, $i \in I$ separated by symbol "+". When I is a finite indexing set, $\sum_{i=1}^{n} A_i$ is also denoted as $A_1 + A_2 + \cdots + A_n$. $\sum_{i \in I} A_i$ represents the same element of EM^* when these $A_i (i \in I)$ are summed by different orders, for example, $\sum_{i \in \{1,2\}} A_i = A_1 + A_2 = A_2 + A_1$.

Definition 5 ([11]). *Let M be a non-vacuous set. We define a binary relation R on EM^* as follows: $\forall \sum_{i \in I} A_i$, $\sum_{j \in J} B_j \in EM^*$,*

$$(\sum_{i \in I} A_i) R (\sum_{j \in J} B_j) \Leftrightarrow \forall A_i (i \in I), \exists B_h (h \in J)$$

such that $A_i \supseteq B_h$ *and* $\forall\ B_j(j \in J), \exists A_u(u \in I)$ *such that* $B_j \supseteq A_u$. *It is obvious that* R *is an equivalence relation. We denote* EM^*/R *as* EM. *By* $\sum_{i\in I} A_i = \sum_{j\in J} B_j$, *we mean that* $\sum_{i\in I} A_i$ *and* $\sum_{j\in J} B_j$ *are equivalent under the equivalence relation* R.

Theorem 1 ([11]). *Let* $X_1, ..., X_n, M$ *be* $n+1$ *non-empty sets. Then* $(EM, \vee, \wedge)$ *forms a completely distributive lattice under the binary operations* $\vee, \wedge$ *defined as follows:* $\forall \sum_{i\in I} A_i, \sum_{j\in J} B_j \in EM$,

$$\sum_{i\in I} A_i \wedge \sum_{j\in J} B_j = \sum_{k\in I\sqcup J} C_k,$$

$$\sum_{i\in I} A_i \vee \sum_{j\in J} B_j = \sum_{i\in I, j\in J} (A_i \cup B_j),$$

where $\forall k \in I \sqcup J$ *(disjoin union of indexing sets* I *and* J*),* $C_k = A_k$ *if* $k \in I$ *and* $C_k = B_k$ *if* $k \in J$. $(EM, \vee, \wedge)$ *is called the* $^*EI$ *(expanding one set* M*) algebra over* M. $\emptyset$ *are the maximum and* M *is minimum element of* EM.

Theorem 2 ([15,16]). *Let* M *be a set.* $\forall \sum_{i\in I} A_i \in EM$, *if the operator "* $'$ *" is defined as follows*

$$(\sum_{i\in I} A_i)' = \wedge_{i\in I}(\vee_{a\in A_i}\{a'\}),$$

then " $'$ *" is an order-reversing involution on* $^*EI$ *algebra* EM.

3 Fuzzy Clustering Algorithmic Based on Topological Structure and $^*EI$ Algebra

In this section, we will discuss the topological molecular lattice structures on $^*EI$ algebras; and give the relations of these topological structures. As applications, we study the topology produced by a family of fuzzy concepts on $^*EI$ algebras and apply these to analyze relations among fuzzy concepts. Using these, we believe that we can study the law of human thinking. The most important fact is that all these can be operated by computers.

Definition 6 ([6,11]). *Let* X *and* M *be sets, and* (M, τ, X) *be an AFS structure.* $\eta \subseteq EM$ *, the* $^*EI$ *algebra over* M, η *is called a closed topology . if* $\sum_{m\in M}\{m\}, M \in \eta$, *and* η *is closed under finite unions (*$\vee$*or* $*$*) and arbitrary intersections (*$\wedge$*or*$+$*).* η *is called a topological molecular lattice on* $^*EI$ *algebra over* M *of AFS structure*(M, τ, X), *denoted as*(EM, η) *(* $\sum_{m\in M}\{m\}$ *is the minimal element, and* M *is the maximal element).*

Definition 7 ([6,11]). *Let* X *and* M *be sets, and* (M, τ, X) *be an AFS structure.* η *is a topological molecular lattice on* $^*EI$ *algebra over* M *of AFS structure* (M, τ, X). *For any* $x \in X$, $\sum_{i\in I} A_i \in EM$ *, and* $\sum_{i\in I} A_i \in \eta$ *we define*

$$N_{\sum_{i\in I} A_i}(x) = \{y|\tau(x,y) \geq \sum_{i\in I} A_i\}$$

this is called the neighborhood of x inducing by $\sum_{i\in I} A_i \in \eta$.

$$N_\eta(x) = \{N_{\sum_{i\in I} A_i}(x)| \sum_{i\in I} A_i \in \eta\}$$

is called the neighborhood of x inducing by η.

Theorem 3 ([6,11]). *Let X and M be sets, and (M,τ,X) be an AFS structure. η is a topological molecular lattice on $^*EI$ algebra over M of AFS structure (M,τ,X). if*

$$B = \{N_{\sum_{i\in I} A_i}(x)|x \in X, \sum_{i\in I} A_i \in \eta\}$$

then B is a base for some topology.

The topological space (X,T_η), in which B is a base for, T_η is called the topology induced by η.

As in example1, we consider the relations among age, height, and weight. Let η be the topological molecular lattice generated by $\{m_1\}, \{m_2\}, \{m_3\}$, which are elements in $^*EI$ algebra over M. $\eta(m_1,m_2,m_3)$ consists of the following:
$\alpha_1 = \{m_1\} + \{m_2\} + \{m_3\}$; $\alpha_2 = \{m_1\} + \{m_2\}$; $\alpha_3 = \{m_1\} + \{m_3\}$;
$\alpha_4 = \{m_2\} + \{m_3\}$; $\alpha_5 = \{m_1\}$; $\alpha_6 = \{m_2\}$; $\alpha_7 = \{m_3\}$;
$\alpha_8 = \{m_1,m_2\} + \{m_1,m_3\} + \{m_2,m_3\}$; $\alpha_9 = \{m_1,m_2\} + \{m_1,m_3\}$;
$\alpha_{10} = \{m_1,m_2\} + \{m_2,m_3\}$; $\alpha_{11} = \{m_1,m_3\} + \{m_2,m_3\}$; $\alpha_{12} = \{m_1,m_2\}$;
$\alpha_{13} = \{m_1,m_3\}$; $\alpha_{14} = \{m_2,m_3\}$; $\alpha_{15} = \{m_1\} + \{m_2,m_3\}$;
$\alpha_{16} = \{m_2\} + \{m_1,m_3\}$; $\alpha_{17} = \{m_3\} + \{m_1,m_2\}$; $\alpha_{18} = \{m_1,m_2,m_3\}$;
$\alpha_{19} = \emptyset$;

Now we consider the base of the topology for$\{x_1,x_2,x_3,x_4,x_5\}$:
$N_{\alpha 1}(x_1) = \{x_1,x_2,x_4\}$; $N_{\alpha 2}(x_1) = \{x_1,x_2,x_4\}$; $N_{\alpha 3}(x_1) = \{x_1\}$;
$N_{\alpha 4}(x_1) = \{x_1,x_2,x_4\}$; $N_{\alpha 5}(x_1) = \{x_1\}$; $N_{\alpha 6}(x_1) = \{x_1,x_2,x_4\}$;
$N_{\alpha 7}(x_1) = \{x_1\}$; $N_{\alpha 8}(x_1) = \{x_1\}$; $N_{\alpha 9}(x_1) = \{x_1\}$;
$N_{\alpha 10}(x_1) = \{x_1\}$; $N_{\alpha 11}(x_1) = \{x_1\}$; $N_{\alpha 12}(x_1) = \{x_1\}$;
$N_{\alpha 13}(x_1) = \{x_1\}$; $N_{\alpha 14}(x_1) = \{x_1\}$; $N_{\alpha 15}(x_1) = \{x_1\}$;
$N_{\alpha 16}(x_1) = \{x_1,x_2,x_4\}$; $N_{\alpha 17}(x_1) = \{x_1\}$; $N_{\alpha 18}(x_1) = \{x_1\}$;

Therefore the neighborhoods of x_1 induced by η is
$N\eta(x_1) = \{\{x_1,x_2,x_4\},\{x_1\}\}$;
Similarly, we get the neighborhoods of x_i $(i = 2....5)$ induced by η is
$N\eta(x_2) = \{\{x_1,x_2,x_3,x_4\},\{x_1,x_2,x_4\},\{x_1,x_2,x_3\},\{x_1,x_2\},\{x_2,x_4\},\{x_2\}\}$;
$N\eta(x_3) = \{\{x_1,x_2,x_3,x_4,x_5\},\{x_1,x_3\}\}$;
$N\eta(x_4) = \{\{x_1,x_2,x_3,x_4,x_5\},\{x_1,x_2,x_4,x_5\},\{x_4\}\}$;
$N\eta(x_5) = \{\{x_1,x_2,x_3,x_4,x_5\},\{x_1,x_2,x_3,x_5\},\{x_1,x_2,x_4,x_5\},\{x_1,x_2,x_5\}\}$;

Definition 8. *Let X and M be sets, and (M, τ, X) be an AFS structure. η is a topological molecular lattice on $^*EI$ algebra over M of AFS structure (M, τ, X). if only choose $\wedge$ operation i.e. η^* is closed under arbitrary intersections. η^* is called a intersectant topological molecular lattice on $^*EI$ algebra over M of AFS structure (M, τ, X), denoted as (EM, η^*)*

As in the example, let η^* be the intersectant topological molecular lattice generated by $\{m_1\}, \{m_2\}, \{m_3\}$, which are elements in $^*EI$ algebra over M. $\eta^*(m_1, m_2, m_3)$ consists of the following:

$\alpha_5 = \{m_1\}$; $\alpha_6 = \{m_2\}$; $\alpha_7 = \{m_3\}$; $\alpha_{12} = \{m_1, m_2\}$;
$\alpha_{13} = \{m_1, m_3\}$; $\alpha_{14} = \{m_2, m_3\}$; $\alpha_{18} = \{m_1, m_2, m_3\}$;

Definition 9. *Let $N_\eta(x)$ is the neighborhood of x induced by η .*

$$N_{x_i}^{x_j} = \{\delta \in N_\eta | x_i \in \delta, x_j \notin \delta\}.$$

then the distance from x_i to x_j is $d_{i-j} = \sum_{\delta \in N_{x_i}^{x_j}} |\delta|$. ($|\delta|$ is the length of the neighbor, i.e. the number of the topology base which produced the neighbor). Similarly $d_{j-i} = \sum_{\delta \in N_{x_j}^{x_i}} |\delta|$, so define the distance between x_i and x_j is $d(i,j) = (d_{i-j} + d_{j-i})/2$.

In example1 the neighbors including x_1 but not including x_2 are $\{x_1\}, \{x_1, x_3\}$, since the neighbor $\{x_1\}$ is produced by the following 13 base in $\eta(m_1, m_2, m_3)$:

$N_{\alpha 3}(x_1) = \{x_1\}$; $N_{\alpha 5}(x_1) = \{x_1\}$; $N_{\alpha 7}(x_1) = \{x_1\}$; $N_{\alpha 8}(x_1) = \{x_1\}$;
$N_{\alpha 9}(x_1) = \{x_1\}$; $N_{\alpha 10}(x_1) = \{x_1\}$; $N_{\alpha 11}(x_1) = \{x_1\}$; $N_{\alpha 12}(x_1) = \{x_1\}$;
$N_{\alpha 13}(x_1) = \{x_1\}$; $N_{\alpha 14}(x_1) = \{x_1\}$; $N_{\alpha 15}(x_1) = \{x_1\}$;$N_{\alpha 17}(x_1) = \{x_1\}$;
$N_{\alpha 18}(x_1) = \{x_1\}$;

So $|\{x_1\}| = 13$, similarly $|\{x_1, x_3\}| = 5$, the distance from x_1 to x_2 is $d_{1-2} = 18$. the neighbors including x_2 but not including x_1 are $\{x_2, x_4\}, \{x_2\}$, the sum of the length is 5, so the distance from x_2 to x_1 is $d_{2-1} = 5$, then $d(1,2) = (d_{1-2} + d_{2-1})/2 = 11.5$.

In the following, we describe the design method:

Let X be the universe of discourse, M be a set of simple features on X.

Step1: Consider the intersectant topological molecular lattice (EM, η^*) generated by all the correlative concepts $\Lambda \subseteq EM$, where Λ is a set of fuzzy sets which are selected to cluster the objects in X.

Step2: Establish AFS structure (M, τ, X) based on the original data and facts (refer to Example 1), and then get the neighborhoods $N_\eta(x)$ induced by the correlative intersectant topological molecular lattice (refer Definition 7).

Step3: For each $x \in X$, apply Definition 9 to calculate the distance between the objects, according to the neighborhoods.

Step4: Apply the distance between each $x \in X$ to establish the fuzzy relation matrix $M = (m_{ij})$ on $X = \{x_1, x_2, \ldots, x_n\}$. Since for any $i, j = 1, 2, ..., n$, $m_{ij} \leq m_{ii}$, hence $M^k \leq M^{k+1}$ for any $k = 1, 2, ...$ Therefore exists an integer r

such that $(M^r)^2 = M^r$, i.e., fuzzy relation matrix $R = M^r$ can yield a partition tree with equivalence classes.

Step5: Clustering analysis based on the fuzzy relation matrix R.

The distance between objects is defined acorrding to the neighbors induced by the concepts in Λ, which reflect the relation between the objects consider the concepts in Λ. We intend to cluster the objects based on the abstract geometrical relations determined by the selected concepts in Λ. The basic idea of the appraoch is based on the following observation:

(1) If none or few of the neighbors can separate the two objects x, y, then the distance of x, y is small;

(2) If any or a large numbers of the neighbors can separate the two objects x, y, then the distance of x, y is large;

(3) If the big neighbor can separate the two objects x, y, then the distance is large.

4 Example

In this section, we apply the AFS topology clustering algorithm to Fisher Iris data, which is well known to the pattern recognition community. The data set contains 150 patterns for 3 classes, each class has 50 instances, each class refers to a type of Iris plant. One class is linearly separable from other two but the latter are not linearly separable from each other. Patten classes are Iris-Setosa, Iris-Versicolor and Iris-Virginica. The four features of ordinal variables involved are the sepal length, sepal width, petal length, and petal width, respectively.

Let $X = \{x_1, x_2, \cdots, x_{150}\}$. From x_1 to x_{50} are "Iris-Setosa", from x_{51} to x_{100} are "Iris-Versicolor", and from x_{101} to x_{150} are "Iris-Virginica". In order to acquire more information from the original data, we expand the original four features to eight. Let M be a set of simple attributes on X, $M = \{m_1, m_1', m_2, m_2', m_3, m_3', m_4, m_4'\}$, where m_1 =sepal length, m_1' =sepal short, m_2 =sepal width, m_2' =sepal narrow, m_3 =petal length, m_3' =petal short, m_4 =petal width, m_4' =petal narrow.

Step1:

Let $\Lambda = \{m_1, m_1', m_2, \cdots, m_4'\} \subseteq EM$. Then calculate the intersectant topological molecular lattice (EM, η^*) generated by the concepts in Λ.

Since the intersect between the correlative concepts and their negation is near empty, hence we can ignore the neighbors induced by the concepts such as $m_i \wedge m_i'$, $i = 1, 2, 3, 4$. Thus the neighbor system of the topology induced by the concepts in Λ can be reduced greatly.

Step2:

Establish AFS structure (M, τ, X) based on $X = \{x_1, x_2, \cdots, x_{150}\}$ and $M = \{m_1, m_1', m_2, \cdots, m_4'\}$, and get the neighborhoods of x induced by $\eta^*(m)$, which denoted as $N_{\eta^*(m)}(x_i)$ (refer to Definition 7).

Step3:

Calculate the distance between $X = \{x_1, x_2, \cdots, x_{150}\}$, establish distance matrix:

$D = (d_{ij}) =$

$$\begin{bmatrix} 0\ 749 \cdots 1727\ 1577 \cdots 1967.5\ 2023.5 \cdots 1849 \\ 0 \ \cdots 1843\ 1707 \cdots 2152.5\ \ 2161 \ \cdots 1625 \\ \ddots \ \vdots \ \ \vdots \ \ \vdots \ \ \vdots \ \ \vdots \ \ \vdots \ \ \vdots \\ 0 \ \ 413 \ \cdots 971.5\ 1236.5 \cdots 992 \\ 0 \ \cdots 919.5\ 1156.5 \cdots 851 \\ \ddots \ \vdots \ \ \vdots \ \ \vdots \ \ \vdots \\ 0 \ \ 1146 \ \cdots 943.5 \\ 0 \ \cdots 517.5 \\ \ddots \ \vdots \\ 0 \end{bmatrix}$$

Step4:
Standardize these distance:

$$d^*(i,j) = d(i,j)/max(d(:,j)),$$

then get the similar relation:

$$N(x_i, x_j) = 1 - d^*(x_i, x_j).$$

Transforming the similarity matrix into its transitive closure, the fuzzy equivalent matrix as follows, which can yield a partition tree with equivalence classes:
$R =$

$$\begin{bmatrix} 1\ 0.8758 \cdots 0.7242\ 0.7242 \cdots 0.7242\ 0.7242 \cdots 0.7242 \\ 1 \ \cdots 0.7242\ 0.7242 \cdots 0.7242\ 0.7242 \cdots 0.7242 \\ \ddots \ \vdots \ \ \vdots \ \ \vdots \ \ \vdots \ \ \vdots \ \ \vdots \ \ \vdots \\ 1 \ \ 0.8568 \cdots 0.8408\ 0.8408 \cdots 0.8408 \\ 1 \ \cdots 0.8408\ 0.8408 \cdots 0.8408 \\ \ddots \ \vdots \ \ \vdots \ \ \vdots \ \ \vdots \\ 1 \ \ 0.8454 \cdots 0.8425 \\ 1 \ \cdots 0.8425 \\ \ddots \ \vdots \\ 1 \end{bmatrix}$$

We have validated that $R^2 = R$.

According to different thresholds, we get dynamic cluster results, and finally the most accurate result is when threshold $\lambda = 0.8409$, the result we got is accord with the nature the Iris have.

When threshold $\lambda = 0.8409$, cluster one is:
$x_1, \ldots, x_{22}, x_{24}, \ldots, x_{41}, x_{43}, \ldots, x_{50}$.
cluster two is:
$x_{51}, \ldots, x_{68}, x_{70}, x_{72}, x_{74}, \ldots, x_{77}, x_{79}, \ldots, x_{83}, x_{85}, \ldots, x_{87}, x_{89}, \ldots, x_{100}$.

cluster three is:
$x_{69}, x_{71}, x_{73}, x_{78}, x_{84}, x_{88}, x_{101}, \ldots, x_{106}, x_{108}, x_{111}, \ldots, x_{117}, x_{119}, \ldots, x_{131}, x_{133}, x_{134}, x_{136}, \ldots, x_{150}$.

There are two classifying errors in the class "Iris-Setosa"; there are six patterns in class "Iris-Versicolor" distributed to class "Iris-Virginica", and there are six patterns away from class "Iris-Virginica", i.e. total 14 classification errors. The clustering accurate rate is 90.67%.

We apply Euclidean metric for traditional algorithm to establish distance matrix[21] and transform it into its transitive closure, the most accurate result is when threshold $\lambda = 0.94182$, total 29 patterns were error, clustering accurate rate is 80.67%. Using the function kmeans in MATLAB toolbox for the iris-data, which is based on the well known k-mean clustering algorithm, the clustering accuracy rate is 89.33%. And Using the function fcm in MATLAB toolbox for the iris-data, which is based on the well known fuzzy c-mean clustering algorithm, the clustering accuracy rate is also 89.33%.

5 Conclusion

In this paper, we established metric space based on the topological structures induced by the involved fuzzy concepts in the AFS framework, proposed measure for membership functions and got the fuzzy similarity relations on X, then applied the measure to study the clustering analytic problems. The AFS topology clustering algorithm is applied to the well known iris-data, and an high clustering accurate rate is achieved. By the comparison of the accuracy with the current fuzzy clustering algorithm, such as c-means fuzzy algorithm, k-nearest-neighbor fuzzy algorithm, and Euclidean metric transitive closure algorithm, one can observe that:

(1) The performance of our algorithm is quite well;
(2) The clustering algorithms based on the topological distance are more simple and understandable, they needn't repeat to convergence;
(3) The attributes of objects in it can be various data types or sub-preference relations, even human intuition descriptions. But both k-mean, fuzzy c-mean algorithms and other current fuzzy clustering algorithm can only be applied to the data sets with numerical attributes;
(4) The cluster number or the class label need not be given beforehand.

From these results, we can conclude that the performance of our proposed algorithm is comparable with many other pattern clustering algorithms and can be treated as one of the most suitable clustering algorithm.

Acknowledgement

This work is supported in part by the National Natural Science Foundation of China under Grant 60575039 60534010 and in part by the National Key Basic Research and Development Program of China under Grant 2002CB312201-06.

References

1. Zadeh, L.A.: Fuzzy sets. Inf. Control **18** (1965) 338–353
2. Liu, X.D.: The fuzzy theory based on AFS algebras and AFS structure. J. Math. Anal. Appl. **217** (1998) 459–478
3. Liu, X.D.: The fuzzy sets and systems based on AFS structure, EI algebra and EII algebra. Fuzzy Sets Syst. **95** (1998) 179–188
4. Liu, X.D.: A new mathematical axiomatic system of fuzzy sets and systems. J. Fuzzy Math. **3** (1995) 559–560
5. Liu, X.D.: Two algebra structures of AFS structure. J. Fuzzy Math. **3** (1995) 561–562
6. Liu, X.D.: The topology on AFS algebra and AFS structure. J. Math. Anal. Appl. **217** (1998) 479–489
7. Wang, G.j.: Theory of topological molecular lattices. Fuzzy Sets Syst. **47** (1992) 351–376
8. J. C. Dunn.: A fuzzy relative of the ISODATA process and its use in detecting compact well-separated clusters. J. Cybern. vol. 3, no. 3. (1974) 32–57
9. J. C. Bezdek.: Fuzzy Mathematics in pattern classification. Ph.D. dissertation. Dept. Appl. Math. Cornell Univ. Ithaca. NY. (1973)
10. J. M. Keller., M. R. Gray. and J. A. Givens Jr.: A fuzzy K-nearest neighbors algorithm. IEEE Trans. Syst. Man, Cybern., vol. SMC-15, no. 4, Apr. (1985) 580–585
11. Liu, X.D., Zhang, Y.J.: The fuzzy theory based on AFS structure and AFS algebra. Dlian.: Dlian Maritime University Press (1998)
12. Liu, X.D., Chai, T.Y. and Wang, W.: AFS Fuzzy Logic Systems and Its Applications to Model and Control. International Journal of Information and Systems Sciences. vol. 2, no. 3. (2006) 1-21
13. Liu, X.D., Chai, T.Y. and Wang, W.: Approaches to the representations and logic operations of fuzzy concepts in the framework of axiomatic fuzzy set theory I, II. Information Sciences. Revised (2005)
14. Liu, X.D., Wang, W., and Chai, T.Y.: The Fuzzy Clustering Analysis Based on AFS Theory. IEEE Trans. Syst. Man, Cybern.-partB: Cybernetics **35** (3) (2005) 1013-1027
15. Liu, X.D., Zhu, K.J., Huang, H.Z.: The representations of fuzzy concepts based on the fuzzy matrix theory and the AFS theory. in Proc. IEEE Int. Symp. Intelligent Control, Houston, TX, Oct. (2003) 1006–1011
16. Liu, X.D., Witold, P., Zhang, Q.L.: Axiomatics fuzzy sets logic. in Proc. IEEE Int. Conf. Fuzzy Systems **1** St. Louis, MO (2003) 55–60
17. Liu, X.D., Witold, P.: The Development of Fuzzy Decision Trees in the Framework of Axiomatic Fuzzy Set Logic. Applied Soft Computing, accepted. (2005). available online.
18. Liu, X.D., Zhang, L.S., Zhu, K.J. and Zhang, Q.L.: The Structures of EI Algebras Generated by Information Attributes. Int. J. Intelligent Systems Technologies and Applications, in press.
19. Liu, X.D., Liu, W.Q.: Credit Rating Analysis with AFS Fuzzy Logic. Lecture Notes in Computer Science, LNCS 3612. (2005) 1198-1204
20. Graver J E., Watkins M E.: Combinatorics with Emphasis on the Theory of Graphs. New York. Inc.: Springer-Verlag.(1977)
21. Gao, X.B.: Fuzzy Clustering Analysis and its Applications. Xian.: Xidian University Press. (in chinese). (2004)

Fuzzy C-Means Algorithm with Divergence-Based Kernel

Young-Soo Song[1], Dong-Chul Park[1], Chung Nguyen Tran[1], Hwan-Soo Choi[1], and Minsoo Suk[2]

[1] Dept. of Information Engineering, Myong Ji University, Korea
{twobasehit, parkd, tnchung, hschoi}@mju.ac.kr
[2] School of Info. and Comm. Eng., Sungkyunkwan University, Korea
msu@ece.skku.ac.kr

Abstract. A Fuzzy C-Means algorithm with a Divergence-based Kernel (FCMDK) for clustering Gaussian Probability Density Function (GPDF) data is proposed in this paper. The proposed FCMDK is based on the Fuzzy C-Means algorithm and employs a kernel method for data transformation. The kernel method adopted in the proposed FCMDK is used to transform input data into a feature space of a higher dimensionality so that the nonlinear problems residing in input space can be linearly solved in the feature space. In order to deal with GPDF data, a divergence-based kernel employing a divergence distance measure for its similarly measure is used for data transformation. The proposed FCMDK is used for clustering GPDF data in an image classification model. Experiments and results on Caltech data sets demonstrate that the proposed FCMDK is more efficient than other conventional algorithms.

1 Introduction

Clustering algorithms have been successfully applied in many applications such as data analysis, image segmentation, pattern recognition, and speech recognition. Traditionally, conventional clustering algorithms such as the Self Organizing Map (SOM) [1] and the k-means [2] have been most widely used in practice. The SOM and the k-means algorithms execute a hard classification, in which each object is either assigned to a class or not. As a result, they assign an object to a single class and ignore the possibility that the object may also belong to other classes. This may lead to inaccuracies when boundaries among clusters are not sharp and severely overlapped.

More recently, fuzzy clustering techniques have been proposed for clustering problems. One of the most widely used algorithms employing fuzzy clustering techniques is the Fuzzy C-Means (FCM) algorithm. The FCM algorithm was originally introduced by Bezdek in 1981 [3] as an improvement on earlier clustering algorithms such as the SOM and the k-means. In the FCM, an object can belong to several classes at the same time but with different degrees of certainty, which are measured by the membership function [3]. The FCM algorithm has more robust abilities in comparison with the SOM and the k-means and has been successfully

L. Wang et al. (Eds.): FSKD 2006, LNAI 4223, pp. 99–108, 2006.

applied to many clustering applications. However, there remain problems with regard to clustering data where boundaries among clusters are nonlinear.

Recently, the kernel method has been used in various clustering algorithms [4,5,6]. The kernel method is based on mapping data from the input space to a feature space of a higher dimensionality, and then solving a linear problem in that feature space. One successful algorithm using the kernel method is the Support Vector Machine (SVM) [7]. The SVM has been successfully utilized in a number of applications. However, significant problems including choosing free parameters still exist. One alternative approache is to use the kernel method only for transforming the input data into the feature space and leave the clustering tasks to traditional clustering algorithms. This approach can thereby utilize advantageous features of the kernel method used in the SVM to obtain a nonlinear solution. It has been successfully employed in many traditional clustering algorithms such as Fuzzy Kernel Perceptron [8] and Kernel Fuzzy C-Means [9].

In this paper, a Fuzzy C-Means algorithm with a Divergence-based Kernel (FCMDK) for clustering GPDF data is proposed. The proposed FCMDK is based on the FCM algorithm and thus exploits advantageous features of fuzzy clustering techniques. Furthermore, transformation of data from the input space to the feature space of a higher dimensionality is adopted using the kernel method before clustering. Consequently, complex nonlinear problems in the original input space can be solved linearly in the feature space according to the well-known Mercer theorem [5,6]. In order to deal with GPDF data, a divergence-based kernel method that uses a divergence distance to measure the distance between two probability distributions is employed for data transformation. The FCMDK is applied for clustering GPDF data in an image classification model. The image classification model uses a localized image representation method to describe texture information of the image, where each image is represented by GPDF data.

The remainder of this paper is organized as follows. Section 2 presents Fuzzy C-Means and Kernel-based Fuzzy C-Means algorithms. Section 3 introduces the proposed Divergence-based Kernel Fuzzy C-Means algorithm. Section 4 presents experiments and results on Caltech data sets including comparisons with other conventional algorithms. Conclusions are presented in Section 5.

2 Kernel-Based Fuzzy C-Means Algorithm

2.1 Fuzzy C-Means Algorithm

The FCM algorithm has successfully been applied to a wide variety of clustering problems. The FCM algorithm attempts to partition a finite collection of elements $\boldsymbol{X} = \{\boldsymbol{x}_1, \boldsymbol{x}_2, ..., \boldsymbol{x}_N\}$ into a collection of C fuzzy clusters. Bezdek first generalized the *fuzzy ISODATA* by defining a family of objective functions $J_m, 1 < m < \infty$, and established a convergence theorem for that family of objective functions [3]. For the FCM, the objective function is defined as:

$$J_m(U, \boldsymbol{v}) = \sum_{i=1}^{C} \sum_{k=1}^{N} \mu_{ik}^m \|\boldsymbol{x}_k - \boldsymbol{v}_i\|^2 \tag{1}$$

where $\|.\|^2$ denotes Euclidean distance measure, $\boldsymbol{x}_k$ and $\boldsymbol{v}_i$ is the input data, k, and cluster prototype, i, respectively. μ_{ki} is the membership grade of the input data $\boldsymbol{x}_k$ to the cluster $\boldsymbol{v}_i$, and m is the weighting exponent, $m \in 1, \cdots, \infty$, while N and C are the number of input data and clusters, respectively.

The FCM objective function is minimized when high membership grades are assigned to objects which are close to their centroid and low membership grades are assigned when objects are far from their centroid.

By using the Lagrange multiplier to minimize the objective function, the center prototypes and membership grades can be updated as follows:

$$\mu_{ik} = \frac{1}{\sum_{j=1}^{C} \frac{\|\boldsymbol{x}_k - \boldsymbol{v}_i\|^2}{\|\boldsymbol{x}_k - \boldsymbol{v}_j\|^2}} \tag{2}$$

$$\boldsymbol{v}_i = \frac{\sum_{k=1}^{N} \mu_{ik}^m \boldsymbol{x}_k}{\sum_{k=1}^{N} \mu_{ik}^m} \tag{3}$$

The FCM finds the optimal values of group centers iteratively by applying Eq. (2) and Eq. (3) in an alternating fashion.

2.2 Kernel-Based Fuzzy C-Means Algorithm

Though the FCM has been applied to numerous clustering problems, it still suffers from poor performance when boundaries among clusters in the input data are nonlinear. One alternative approach is to transform the input data into a feature space of a higher dimensionality using a nonlinear mapping function so that nonlinear problems in the input space can be linearly treated in the feature space according to the well-known Mercer theorem [5,6]. One of the most popular data transformation methods adopted in recent studies is the kernel method [4]. One of the advantageous features of the kernel method is that input data can be implicitly transformed into the feature space without knowledge of the mapping function. Further, the dot product in the feature space can be calculated using a kernel function.

With the incorporation of the kernel method, the objective function in the feature space using the mapping function Φ can be rewritten as follow:

$$F_m = \sum_{i=1}^{C} \sum_{k=1}^{N} \mu_{ik}^m \|\Phi(\boldsymbol{x}_k) - \Phi(\boldsymbol{v}_i)\| \tag{4}$$

Through kernel substitution, the objective function can be rewritten as:

$$F_m = 2\sum_{i=1}^{C} \sum_{k=1}^{N} \mu_{ik}^m (1 - K(\boldsymbol{x}_i, \boldsymbol{v}_k)) \tag{5}$$

where $K(\boldsymbol{x}, \boldsymbol{y})$ is a kernel function used for calculating the dot product of vectors **x** and **y** in the feature space. To calculate the kernel between two vectors, the Gaussian kernel function is widely used:

$$K(\boldsymbol{x}, \boldsymbol{y}) = \exp\left(-\frac{\|\boldsymbol{x}-\boldsymbol{y}\|^2}{\sigma^2}\right) \tag{6}$$

By using the Lagrange multiplier to minimize the objective function, the cluster prototypes can be updated as follow:

$$\boldsymbol{v}_i = \frac{\sum_{k=1}^{N} \mu_{ik}^m K(\boldsymbol{x}_k, \boldsymbol{v}_i)\boldsymbol{x}_k}{\sum_{k=1}^{N} \mu_{ik}^m K(\boldsymbol{x}_k, \boldsymbol{v}_i)} \tag{7}$$

And the membership grades can be updated as follow:

$$\mu_{ik} = \frac{1}{\sum_{j=1}^{C} \left(\frac{1-K(\boldsymbol{x}_k,\boldsymbol{v}_i)}{1-K(\boldsymbol{x}_k,\boldsymbol{v}_j)}\right)^{\frac{1}{m-1}}} \tag{8}$$

3 Fuzzy C-Means Algorithm with Divergence-Based Kernel

Since conventional kernel-based clustering algorithms were designed for deterministic data, they cannot be used for clustering probability data. In this paper, we propose a Fuzzy C-Means algorithm with a Divergence-based Kernel (FCMDK) in which a divergence distance is employed to measure the distance between two probability distributions. The proposed FCMDK incorporates the FCM for clustering data and the divergence-based kernel method for data transformation.

For GPDF data, each cluster prototype is not represented by a deterministic vector in the input space but is represented by a GPDF with a mean vector and covariance matrix. In order to calculate the kernel between two GPDF data, a divergence-based kernel is employed. The divergence-based kernel is an extension of the standard Gaussian kernel. While the Gaussian kernel is the negative exponent of the weighted Euclidean distance between two deterministic vectors as shown in Eq. 6, the divergence-based kernel is the negative exponent of the weighted divergence measure between two GPDF data. The divergence-based kernel function between two GPDF data is defined as follows:

$$DK(g_{\boldsymbol{x}}, g_{\boldsymbol{y}}) = \exp\left(-\alpha D(g_{\boldsymbol{x}}, g_{\boldsymbol{y}}) + b\right) \tag{9}$$

where $DK(g_{\boldsymbol{x}}, g_{\boldsymbol{y}})$ is the divergence distance between two Gaussian distributions, $g_{\boldsymbol{x}}$ and $g_{\boldsymbol{y}}$. After evaluating several divergence distance measures, the popular

Bhattacharyya distance measure is employed. The similarity measure between two distributions using the Bhattacharyya distance measure is defined as follows:

$$D(G_i, G_j) = \frac{1}{8}(\boldsymbol{\mu}_i - \boldsymbol{\mu}_j)^T \left[\frac{\boldsymbol{\Sigma}_i + \boldsymbol{\Sigma}_j}{2}\right]^{-1} (\boldsymbol{\mu}_i - \boldsymbol{\mu}_j) + \frac{1}{2} \ln \frac{\left|\frac{\boldsymbol{\Sigma}_i + \boldsymbol{\Sigma}_j}{2}\right|}{\sqrt{|\boldsymbol{\Sigma}_i||\boldsymbol{\Sigma}_j|}} \tag{10}$$

where $\boldsymbol{\mu}_i$ and $\boldsymbol{\Sigma}_i$ denote the mean vector and covariance matrix of a Gaussian distribution G_i, respectively. T denotes the transpose matrix.

Similar to the cluster prototypes and membership grades in the kernel-based FCM, the cluster prototypes and membership grades in the FCMDK can be updated using a Lagrange multiplier to minimize its objective function. However, each cluster prototype representing a cluster in the FCMDK is a probability distribution with a mean vector and a covariance matrix. Therefore, cluster prototypes in each iteration are updated by modifying their mean vector and covariance matrix as follows:

$$m_{\boldsymbol{v}_i} = \frac{\sum\limits_{k=1}^{N} \mu_{ik}^m DK(\boldsymbol{x}_k, \boldsymbol{v}_i) m_{\boldsymbol{x}_k}}{\sum\limits_{k=1}^{N} \mu_{ik}^m DK(\boldsymbol{x}_k, \boldsymbol{v}_i)} \tag{11}$$

$$\Sigma_{\boldsymbol{v}_i} = \frac{\sum\limits_{k=1}^{N} \mu_{ik}^m DK(\boldsymbol{x}_k, \boldsymbol{v}_i) \Sigma_{\boldsymbol{x}_k}}{\sum\limits_{k=1}^{N} \mu_{ik}^m DK(\boldsymbol{x}_k, \boldsymbol{v}_i)} \tag{12}$$

where $m_{\boldsymbol{v}_i}$ and $m_{\boldsymbol{x}_k}$ are the mean of the cluster prototype $\boldsymbol{v}_i$ and the vector in input $\boldsymbol{x}_k$, respectively. $\Sigma_{\boldsymbol{v}_i}$ and $\Sigma_{\boldsymbol{x}_k}$ are the covariance of the cluster prototype $\boldsymbol{v}_i$ and the vector in input $\boldsymbol{x}_k$, respectively. $DK(\boldsymbol{x}_k, \boldsymbol{v}_j)$ is the divergence-based kernel function between two Gaussian distributions $\boldsymbol{x}_k$ and $\boldsymbol{v}_j$.

The membership grades are similar to those in the KFCM and can be updated as follows:

$$\mu_{ik} = \frac{1}{\sum\limits_{j=1}^{c} \left(\frac{1-DK(\boldsymbol{x}_k, \boldsymbol{v}_i)}{1-DK(\boldsymbol{x}_k, \boldsymbol{v}_j)}\right)^{\frac{1}{m-1}}} \tag{13}$$

where $\boldsymbol{x}_k$ and $\boldsymbol{v}_i$ are the probability distribution input vector and probability distribution cluster prototype, respectively. $DK(\boldsymbol{x}_k, \boldsymbol{v}_j)$ is the divergence-based kernel function between two Gaussian distributions, $\boldsymbol{x}_k$ and $\boldsymbol{v}_j$.

With the incorporation of the divergence-based kernel method and the FCM, the proposed FCMDK can be used for clustering GPDF data while utilizing the advantageous features of the fuzzy clustering techniques and the kernel method. Thus, it provides an efficient clustering algorithm for GPDF data.

4 Experiments and Results

The data set used for experiments in this paper is the Caltech image data set. The Caltech image data set consists of different image classes (categories) in which each class contains different views of an object. The Caltech image data were collected by the Computational Vision Group and are available at the following website: http://www.vision.caltech.edu/html-files/archive.html

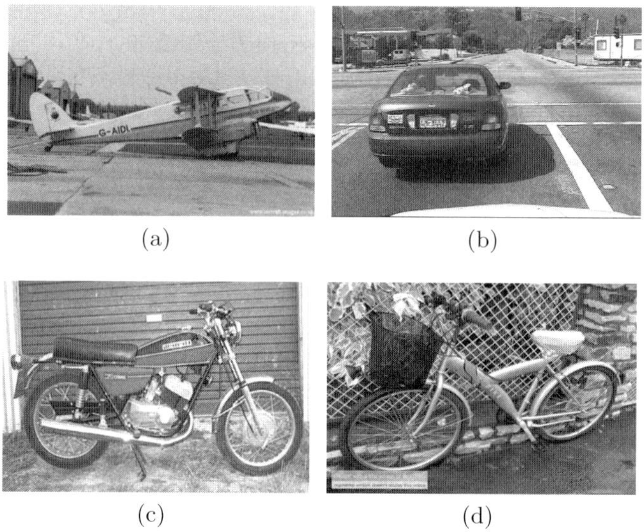

Fig. 1. (a) Airplane (b) Car (c) Motorbike (d) Bike

From these classes, we selected the 4 most easily confused classes: airplane, car, bike and motorbike for the experiments. Each class consists of 200 images with different views resulting in a total of 800 images in the data set. From this data set, 100 images were randomly chosen for training classifiers while the remaining images were used for testing classifiers. Before any further processing for feature extraction, the entire image sets were converted to grey scale with the same resolution. Fig. 1 shows examples of the 4 image categories used in the experiments. Figs. 1(a), 1(b), 1(c), and 1(d) are examples of car, airplane, motorbike, and bike, respectively.

In order to describe the texture information of images, the localized image representation method is employed. The localized representation method represents the content of the image by a collection of local features extracted from the image. These features are computed at different points of interest in the image. Afterwards, a Gaussian distribution, wherein the mean vector and covariance matrix are estimated from all local feature vectors obtained from the image, is used to represent the content of the image. Localized representation maintains the dimensions of the feature vectors tractable by adjusting the sizes of blocks. This method is consequently more robust to occlusions and clutter.

In order to obtain the texture information from the image, conventional texture descriptors based on a frequency domain analysis such as Gabor filters [10] and wavelet filters [11] are often used. However, these algorithms often require a high computational load for feature extraction and are not suitable for real-time applications. In this paper, the Discrete Cosine Transform (DCT) is adopted for extracting the texture information from each block of the image [12]. The DCT transforms the image from the spatial domain into the frequency domain.

For the localized representation, images are transformed into a collection of 8×8 blocks. Each block is then shifted by an increment of 2 pixels horizontally and vertically. The DCT coefficients of each block are then computed and return in 64 dimensional coefficients. Only the 32 lowest frequency DCT coefficients that are visible to the human eye are kept. Therefore, the feature vectors that are obtained from each block have 32 dimensions. In order to calculate the GPDF for the image, the mean vector and the covariance matrix are estimated from all blocks obtained from the image. Finally, a GPDF with a 32-dimensional mean vector and a 32×32 covariance matrix is used to represent the content of the image.

The performance of the proposed FCMDK algorithm is evaluated using the Caltech image data in an image classification model. The image data is modeled using the popular Gaussian Mixture Model (GMM). Each image class is modeled by a GMM. The proposed FCMDK is employed to obtain mixture components in each GMM for each image class.

After GMMs for all image classes are estimated, a minimum-likelihood (ML) classifier based on these GMMs is obtained. A class of a tested image using the ML classifier is determined using the following decision rule:

$$Class(x) = \arg\min_i D(\boldsymbol{x}, C_i) \tag{14}$$

$$D(G(\boldsymbol{x}; \mu, \Sigma), C_i) = \sum_{k=1}^{N_i} w_{ik} D(G(\boldsymbol{x}; \boldsymbol{\mu}, \boldsymbol{\Sigma}), G(\boldsymbol{x}; \boldsymbol{\mu}_{ik}, \boldsymbol{\Sigma}_{ik})) \tag{15}$$

where $\boldsymbol{x}$ is the tested image represented by a Gaussian distribution feature vector with a mean vector, $\boldsymbol{\mu}$, and a covariance matrix, $\boldsymbol{\Sigma}$. $\boldsymbol{\mu}_{ik}$ and $\boldsymbol{\Sigma}_{ik}$ represent the mean vector and covariance matrix of cluster k in class C_i, respectively. w_{ik} is the weight component of cluster k in class C_i and N_i is the number of clusters in class C_i.

In order to investigate the performance of the proposed FCMDK algorithm relative to that of other conventional algorithms, the SOM and the k-mean algorithms are also assessed for comparison. It should be noted that the SOM and the k-means algorithms used in the experiments are Divergence-based SOM (DSOM) and Divergence-based k-means (Dk-means). The divergence measure is employed to measure the distance between two probability distributions for the SOM and the k-means since the data used in our experiments are GPDF data.

Perhaps the most important parameter that has to be selected in most clustering algorithms is the number of clusters in the data. Most clustering algorithms partition data into a specified number of clusters, regardless of whether the clusters are meaningful. In the experiments, the number of code vectors is varied

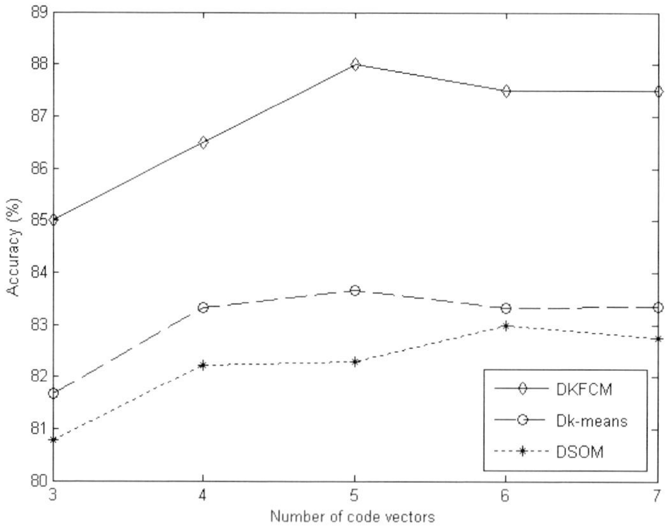

Fig. 2. Overall classification accuracies using different algorithms

from 3 to 7 in order to determine a sufficient number of code vectors to represent the number of mixture components in the GMMs. Fig. 2 shows the classification accuracy of the classification model using the Divergence-based SOM (DSOM), the Divergence-based k-means (Dk-means), and the proposed FCMDK, respectively. As can be seen from Fig. 2, the accuracies of all classification models are improved significantly when the number of code vectors is increased from 3 to 5 while they tend to saturate when the number of code vectors is greater than 5. This implies that using 5 code vectors for representing the mixtures of GPDF data is sufficient.

Table 1 shows the accuracy of the classification models using the DSOM, the Dk-means, and the proposed FCMDK. As can be seen from Table 1, the classification model using the proposed FCMDK always outperforms the models using the DSOM and Dk-means. Accuracy improvement of 4.93% and 6.50% is achieved over the Dk-means and DSOM algorithms, respectively.

A confusion matrix describing the classification results of the classification model using the proposed FCMDK is shown in Table 2. As can be inferred from Table 2, cars can be well discriminated from the others images while bikes and motorbikes are easily confused. These are logical results given that motorbikes

Table 1. Classification accuracy (%) of different algorithms using 5 code vectors

	Airplane	Car	Bike	Motorbike	Accuracy
DSOM	75.71	100	60.00	93.41	**82.28%**
Dk-means	82.66	100	57.33	94.65	**83.66%**
FCMDK	86.5	100	69.5	96.0	**88.00%**

Table 2. Confusion matrix of image categories, using 5 code vectors

	Airplane	Car	Bike	Motorbike	Accuracy
Airplane	**86**	2.0	10.0	2.0	**86.0%**
Car	0.0	**100**	0.0	0.0	**100%**
Bike	14.0	0.0	**70.0**	16.0	**70.0%**
Motorbike	0.0	0.0	4.0	**96.0**	**96.0%**

and bikes are quite similar to the human eye whereas the cars are significantly different.

5 Conclusion

A new Fuzzy C-Means algorithm with a Divergence-based Kernel (FCMDK) for clustering GPDF data is proposed. The proposed FCMDK is based on the Fuzzy C-Means (FCM) algorithm and the divergence-based kernel method. The divergence-based kernel method is used for data transformation in which the divergence measure is employed to measure the distance between two probability distribution. With the incorporation of the divergence-based kernel and the FCM, the proposed FCMDK can be used for clustering GPDF data. Furthermore, the kernel method adopted in the proposed FCMDK makes the overall approach more robust in dealing with complex nonlinear problems. Successful application in experiments of the proposed FCMDK to a GPDF data clustering problem in an image classification model provides motivation for utilizing the proposed FCMDK in other practical applications.

Acknowledgment

This work was supported by the Korea Research Foundation Grant funded by the Korea Government (MOEHRD-KRF-2005-042-D00265).

References

1. Kohonen,T.: The Self-Organizing Map. Proc. IEEE, Vol. 78 (1990) 1464-1480
2. Hartigan, J.: Clustering Algorithms. New York, Wiley (1975)
3. Bezdek, J.C.: Pattern Recognition with Fuzzy Objective Function Algorithms. New York(Plenum), (1981)
4. Muller, K.R., Mika, S., Ratsch, G., Tsuda, K., Scholkopf, B.: An Introduction to Kernel-Based Learning Algorithms. IEEE Transactions on Neural Networks 12(2) (2001) 181-201
5. Cover, T.M.: Geomeasureal and Statistical Properties of Systems of Linear Inequalities in Pattern Recognition. Electron. Computing, Vol. EC-14 (1965) 326-334.
6. Girolami,M.: Mercer Kernel-Based Clustering in Feature Space. IEEE Trans. on Neual Networks 13(3) (2002) 780-784.

7. Cristianini, N., Shawe-Taylor, J.: An Introduction to Support Vector Machine. Cambridge, Cambridge Univ. Press (2000)
8. Chen, J.H., Chen, C.S.: Fuzzy Kernel Perceptron. IEEE Trans. on Neural Networks 13(6) (2002) 1364-1373.
9. Chen, S., Zhang, D.: Robust Image Segmentation using FCM with Spatial Constraints Based on New Kernel-Induced Distance Measure. IEEE Trans. on Systems, Man and Cybernetics 34(4) 2004 1907-1916
10. Daugman, J.G.: Complete Discrete 2D Gabor Transform by Neural Networks for Image Analysis and Compression. IEEE Trans. on Acoust., Speech, and Signal Processing 36 (1988) 1169-11179
11. Pun, C.M., Lee, M.C.: Extraction of Shift Invariant Wavelet Features for Classification of Images with Different Sizes. IEEE Trans. on Pattern Analysis and Machine Intelligence 26(9) (2004) 1228-1233
12. Huang, Y.L., Chang, R.F.: Texture Features for DCT-Coded Image Retrieval and Classification. ICASSP Proc. IEEE Int. Conf. on Acoustics, Speech, and Signal Processing, Vol. 6 (1999) 3013-3016

Variable Threshold Concept Lattice and Dependence Space

Jian-Min Ma[1], Wen-Xiu Zhang[1], and Sheng Cai[2]

[1] Institute for Information and System Sciences, Faculty of Science,
Xi'an Jiaotong University, Xi'an, Shaan'xi 710049, P.R. China
cjm-zm@126.com, wxzhang@mail.xjtu.edu.cn

[2] School of Information Science and Engineering,
Fudan University, Shanghai, 200433, P.R. China
caisheng@fudan.edu.cn

Abstract. In this paper, a variable threshold concept lattice is introduced based on a fuzzy formal context with a threshold, and some properties are discussed. The number of concepts in a variable threshold concept lattice is far less than that in a fuzzy concept lattice. Then a dependence space is constructed according to the variable threshold concept lattice. Applying the congruences on the dependence space, a closed set is obtained. And a new approach is discussed by using the closed set to construct variable threshold formal concepts.

Keywords: Fuzzy formal context; Variable threshold concept lattice; Dependence space; Congruence; Closed set.

1 Introduction

Formal concept analysis, proposed by Wille in 1982 [3,7], is a kind of approaches to analyze data, study and represent implicit knowledge in data from different aspects. The basis of formal concept analysis are formal concepts and concept lattices. A concept lattice is an ordered hierarchical structure of formal concepts that are defined by a binary relation between a set of objects and a set of attributes. Each formal concept is an (objects, attributes) pair, which consists of two parts: the extent (objects covered by the concept) and the intent (attributes describing the concept). They uniquely determine each other [3,7]. As an effective tool for data analysis and knowledge processing, concept lattice has been successfully applied to various fields, such as data mining, information retrieval, software engineering, and so on.

The classical concept lattices reflect the accurate relationships between objects and attributes, while the fuzzy concept lattices show the uncertain relationships between concepts and attributes. Since lots of knowledge obtained from reality are vague and uncertain, it is important and interesting to study the fuzzy concept lattice. And more and more people have studied the fuzzy formal context and constructed all kinds of fuzzy concept lattices [1,2,4,8,9,10,12].

L. Wang et al. (Eds.): FSKD 2006, LNAI 4223, pp. 109–118, 2006.

Novotny thought that reductions of rough set theory and formal concept analysis could be depicted uniformly in mathematics and therefore he provided the theory of dependence space [5]. In this paper, we first propose a definition of a variable threshold concept lattice with a threshold δ. And we show that the variable threshold concept lattice has properties analogous to classical concept lattice. Then we define a dependence space based on the notions of the variable threshold concept lattices. For the congruence on the dependence space, we give another equivalent congruence and discuss their properties. Meanwhile, using each of these two congruence, we can get a partition of the set of attributes and then obtain a closed set, which can be used to easily construct variable threshold concepts from the variable threshold concept lattice.

This paper is organized as follows. Section 2 presents the definitions and properties about a variable threshold concept lattice. Section 3 shows the dependence space based on the variable threshold concept lattice, and gives a new approach to constructing variable threshold concepts of the variable threshold concept lattice. Finally, Section 4 concludes the paper.

2 The Variable Threshold Concept Lattice

A fuzzy formal context is a triplet $(U, A, \tilde{I})$, where U is a non-empty, finite set of objects called a universe, A is a non-empty, finite set of attributes, and $\tilde{I}: U \otimes A \rightarrow [0,1]$ is a fuzzy binary relation between U and A. Then for any $x \in U$ and $a \in A$, we have $\tilde{I}(x,a) \in [0,1]$. And $\tilde{I}(x,a)$ denotes the level that the object x has the attribute a, or the level that a is possessed by x. A fuzzy formal context is in fact an information table with the domain of attributes being $[0,1]$ in rough set theory.

With respect to a fuzzy formal context $(U, A, \tilde{I})$ and $\delta \in (0,1]$, we define a pair of dual operators for $X \subseteq U$, $B \subseteq A$ by:

$$X^{*\delta} = \{a \in A \mid \tilde{I}(x,a) \geq \delta, \text{ for all } x \in X\},$$
$$B^{*\delta} = \{x \in U \mid \tilde{I}(x,a) \geq \delta, \text{ for all } a \in B\}.$$

We write $\{x\}^{*\delta}$ as $x^{*\delta}$ for $x \in U$, and $\{a\}^{*\delta}$ as $a^{*\delta}$ for $a \in A$.

Definition 1. *Let $(U, A, \tilde{I})$ be a fuzzy formal context and $\delta \in (0,1]$. A pair (X, B) is referred to as a variable threshold formal concept, for short, a variable threshold concept, of $(U, A, \tilde{I})$, if and only if $X \subseteq U$, $B \subseteq A$, $X^{*\delta} = B$ and $X = B^{*\delta}$. X is referred to as the extent and B the intent of (X, B). We denote by $L_\delta(U, A, \tilde{I})$ the set of all variable threshold concepts of a fuzzy formal context $(U, A, \tilde{I})$. Note that*

$$Ext_\delta(U, A, \tilde{I}) = \{X \subseteq U \mid (X, B) \in L_\delta(U, A, \tilde{I})\},$$
$$Int_\delta(U, A, \tilde{I}) = \{B \subseteq A \mid (X, B) \in L_\delta(U, A, \tilde{I})\}.$$

For a fuzzy formal context $(U, A, \tilde{I})$, the following properties hold: for all X_1, X_2, $X \subseteq U$, $B_1, B_2, B \subseteq A$ and $\delta \in (0,1]$,

1. $X_1 \subseteq X_2 \Rightarrow X_2^{*\delta} \subseteq X_1^{*\delta}$, $B_1 \subseteq B_2 \Rightarrow B_2^{*\delta} \subseteq B_1^{*\delta}$.
2. $X \subseteq X^{*\delta *\delta}$, $B \subseteq B^{*\delta *\delta}$.

3. $X^{*\delta} = X^{*\delta*\delta*\delta}$, $B^{*\delta} = B^{*\delta*\delta*\delta}$.
4. $X \subseteq B^{*\delta} \Leftrightarrow B \subseteq X^{*\delta}$.
5. $(X_1 \cup X_2)^{*\delta} = X_1^{*\delta} \cap X_2^{*\delta}$, $(B_1 \cup B_2)^{*\delta} = B_1^{*\delta} \cap B_2^{*\delta}$.
6. $(X_1 \cap X_2)^{*\delta} \supseteq X_1^{*\delta} \cup X_2^{*\delta}$, $(B_1 \cap B_2)^{*\delta} \supseteq B_1^{*\delta} \cup B_2^{*\delta}$.
7. $(X^{*\delta*\delta}, X^{*\delta})$ and $(B^{*\delta}, B^{*\delta*\delta})$ are all variable threshold concepts.

For $\delta \in (0,1]$, (X_1, B_1), $(X_2, B_2) \in L_\delta(U, A, \tilde{I})$ are ordered by $(X_1, B_1) \leq_\delta (X_2, B_2) \Leftrightarrow X_1 \subseteq X_2 (\Leftrightarrow B_2 \subseteq B_1)$. Then " $\leq_\delta$ " is a partial relation on $L_\delta(U, A, \tilde{I})$. And the conjunction and disjunction are given by:

$$(X_1, B_1) \wedge_\delta (X_2, B_2) = (X_1 \cap X_2, (B_1 \cup B_2)^{*\delta*\delta}),$$
$$(X_1, B_1) \vee_\delta (X_2, B_2) = ((X_1 \cup X_2)^{*\delta*\delta}, B_1 \cap B_2).$$

Then $L_\delta(U, A, \tilde{I})$ is a complete lattice called the variable threshold concept lattice.

Definition 2. *Let $L_\delta(U, A_1, \tilde{I}_1)$ and $L_\delta(U, A_2, \tilde{I}_2)$ be two variable threshold concept lattices. If for any $(X, B) \in L_\delta(U, A_2, \tilde{I}_2)$ there exists $(X', B') \in L_\delta(U, A_1, \tilde{I}_1)$ such that $X = X'$, then $L_\delta(U, A_1, \tilde{I}_1)$ is said to be finer than $L_\delta(U, A_2, \tilde{I}_2)$, denoted by:*

$$L_\delta(U, A_1, \tilde{I}_1) \leq_\delta L_\delta(U, A_2, \tilde{I}_2).$$

If $L_\delta(U, A_1, \tilde{I}_1) \leq_\delta L_\delta(U, A_2, \tilde{I}_2)$ and $L_\delta(U, A_2, \tilde{I}_2) \leq_\delta L_\delta(U, A_1, \tilde{I}_1)$, then these two variable threshold concept lattices are said to be isomorphic to each other, denoted by:

$$L_\delta(U, A_1, \tilde{I}_1) \cong_\delta L_\delta(U, A_2, \tilde{I}_2).$$

For a fuzzy formal context $(U, A, \tilde{I})$ and $D \subseteq A$, $\tilde{I}_D : U \otimes D \to [0,1]$ is a fuzzy binary relation between U and D. And $(U, D, \tilde{I}_D)$ is also a fuzzy formal context. We denote by $X^{*\delta}$ the operator ${}^{*\delta}$ under $(U, A, \tilde{I})$, and denote by $X_D^{*\delta}$ the operator ${}^{*\delta}$ under $(U, D, \tilde{I}_D)$. Then $\tilde{I}_A = \tilde{I}$, $\tilde{I}_D \subseteq \tilde{I}_A$, $X_A^{*\delta} = X^{*\delta}$, $X_D^{*\delta} = X^{*\delta} \cap D$, and $X_D^{*\delta} \subseteq X^{*\delta}$.

Then for any $\delta \in (0,1]$, $D \subseteq A$ and $D \neq \emptyset$, we can easily get that

$$L_\delta(U, A, \tilde{I}) \leq_\delta L_\delta(U, D, \tilde{I}_D).$$

Definition 3. *Let $(U, A, \tilde{I})$ be a fuzzy formal context and $\delta \in (0,1]$. D is referred to as a $\delta-$consistent set of $(U, A, \tilde{I})$, if there exists an attribute set $D \subseteq A$ such that $L_\delta(U, D, \tilde{I}_D) \cong_\delta L_\delta(U, A, \tilde{I})$. And further, if $L_\delta(U, D - \{d\}, \tilde{I}_{D-\{d\}}) \ncong_\delta L_\delta(U, A, \tilde{I})$ for all $d \in D$, then D is called a $\delta-$reduct of $(U, A, \tilde{I})$.*

For any $\delta \in (0,1]$, $D \subseteq A$, and $D \neq \emptyset$, we have

$$D \text{ is a } \delta-\text{consistent set} \Leftrightarrow L_\delta(U,\ D,\ \tilde{I}_D) \leq_\delta L_\delta(U,\ A,\ \tilde{I}).$$

And the $\delta-$reduct exists for any fuzzy formal context.

Theorem 1. *Let $(U, A, \tilde{I})$ be a fuzzy formal context, $\delta \in (0,1]$, $D \subset A, D \neq \emptyset$, and $E = A - D$. Then*

$$D \textit{ is a } \delta - \textit{ consistent set} \Leftrightarrow \forall F \subseteq E, F \neq \emptyset, (F^{*\delta*\delta} \cap D)^{*\delta} = F^{*\delta}.$$

Proof. Suppose D is a $\delta-$consistent set, then $L_\delta(U,D,\tilde{I}_D) \leq_\delta L_\delta(U,A,\tilde{I})$. For any $F \subseteq E$ and $F \neq \emptyset$, we know that $(F^{*\delta}, F^{*\delta *\delta}) \in L_\delta(U,A,\tilde{I})$. Then there exists $C \subseteq D$ such that $(F^{*\delta}, C) \in L_\delta(U,D,\tilde{I}_D)$. Thus $C^{*\delta} = F^{*\delta}$. Combining this and $C = (F^{*\delta})^{*\delta}_D = F^{*\delta *\delta} \cap D$, we conclude that $(F^{*\delta *\delta} \cap D)^{*\delta} = C^{*\delta} = F^{*\delta}$.

Contrarily, take $(X,B) \in L_\delta(U,A,\tilde{I})$, we have $X^{*\delta}_D = X^{*\delta} \cap D = B \cap D$.

Obviously, $B = (B \cap D) \cup (B \cap E)$. If $B \cap E = \emptyset$, then $X = B^{*\delta} = (B \cap D)^{*\delta}$; If $B \cap E \neq \emptyset$, by $B \cap E \subseteq E$ we have $((B \cap E)^{*\delta *\delta} \cap D)^{*\delta} = (B \cap E)^{*\delta}$. Thus

$$\begin{aligned} B \cap E \subseteq B &\Rightarrow (B \cap E)^{*\delta *\delta} \subseteq B^{*\delta *\delta} = X^{*\delta} = B \\ &\Rightarrow (B \cap E)^{*\delta} = ((B \cap E)^{*\delta *\delta} \cap D)^{*\delta} \supseteq (B \cap D)^{*\delta}. \end{aligned}$$

Therefore $X = B^{*\delta} = (B \cap D)^{*\delta} \cap (B \cap E)^{*\delta} = (B \cap D)^{*\delta}$. From which we conclude that $L_\delta(U,D,\tilde{I}_D) \leq_\delta L_\delta(U,A,\tilde{I})$, and D is a $\delta-$consistent set.

Using Theorem 1 we can get for $\delta \in (0,1]$, $D \subset A, D \neq \emptyset$, and $E = A \setminus D$

$$D \text{ is a } \delta-\text{consistent set } \Leftrightarrow L_\delta(U,D,\tilde{I}_D) \leq_\delta L_\delta(U,E,\tilde{I}_E).$$

Property 1. Let $(U,A,\tilde{I})$ be a fuzzy formal context and $\delta_1, \delta_2 \in (0,1]$. Then

$$L_{\delta_1}(U,A,\tilde{I}) \leq_\delta L_{\delta_2}(U,A,\tilde{I}) \Leftrightarrow Ext_{\delta_2}(U,A,\tilde{I}) \subseteq Ext_{\delta_1}(U,A,\tilde{I}).$$

Property 2. Let $(U,A,\tilde{I})$ be a fuzzy formal context, $\delta_1, \delta_2 \in (0,1]$ and $\delta_1 < \delta_2$. Then for any $X \subseteq U$, $B \subseteq A$, we have

(1) $X^{*\delta_1 *\delta_2} \subseteq X^{*\delta_1 *\delta_1} \subseteq X^{*\delta_2 *\delta_1}$, $B^{*\delta_1 *\delta_2} \subseteq B^{*\delta_1 *\delta_1} \subseteq B^{*\delta_2 *\delta_1}$;
(2) $X^{*\delta_1 *\delta_2} \subseteq X^{*\delta_2 *\delta_2} \subseteq X^{*\delta_2 *\delta_1}$, $B^{*\delta_1 *\delta_2} \subseteq B^{*\delta_2 *\delta_2} \subseteq B^{*\delta_2 *\delta_1}$.

Proof. It is clear that $X^{*\delta_2} \subseteq X^{*\delta_1}$ and $B^{*\delta_2} \subseteq B^{*\delta_1}$ for $0 < \delta_1 < \delta_2 \leq 1$. Then $(X^{*\delta_1})^{*\delta_2} \subseteq (X^{*\delta_1})^{*\delta_1}$, and $(X^{*\delta_1})^{*\delta_1} \subseteq (X^{\delta_2})^{*\delta_1}$ by properties of $^{*\delta}$. Thus $X^{*\delta_1 *\delta_2} \subseteq X^{*\delta_1 *\delta_1} \subseteq X^{*\delta_2 *\delta_1}$. Likewise, we can prove the others.

For any fuzzy formal context $(U,A,\tilde{I})$ and $0 < \delta_1 < \delta_2 \leq 1$, we have

(1) If $X \in Ext_{\delta_1}(U,A,\tilde{I})$, then $X^{*\delta_1 *\delta_1} \subseteq X^{*\delta_2 *\delta_2}$;
(2) If $X \in Ext_{\delta_2}(U,A,\tilde{I})$, then $X^{*\delta_2 *\delta_2} \subseteq X^{*\delta_1 *\delta_1}$;
(3) If $X \in Ext_{\delta_1}(U,A,\tilde{I}) \cap Ext_{\delta_2}(U,A,\tilde{I})$, then $X^{*\delta_1 *\delta_1} = X^{*\delta_2 *\delta_2}$.

Actually, for any variable threshold concepts (X, B), $(Y, C) \in L_\delta(U,A,\tilde{I})$, we have $X \neq Y \Rightarrow X^{*\delta} \neq Y^{*\delta}$. However, for any $X, Y \subseteq U$, the result $X \neq Y \Rightarrow X^{*\delta} \neq Y^{*\delta}$ may not be true.

Example 1. Table 1 is an example of a fuzzy formal context with $U = \{1, 2, 3, 4\}$ and $A = \{a, b, c, d, e\}$.

Table 1. A fuzzy formal context of Example 1

U	a	b	c	d	e
1	0.7	0.6	0.0	0.9	1.0
2	1.0	0.9	1.0	0.4	0.0
3	0.3	0.0	0.5	1.0	0.3
4	0.6	1.0	0.8	0.0	0.4

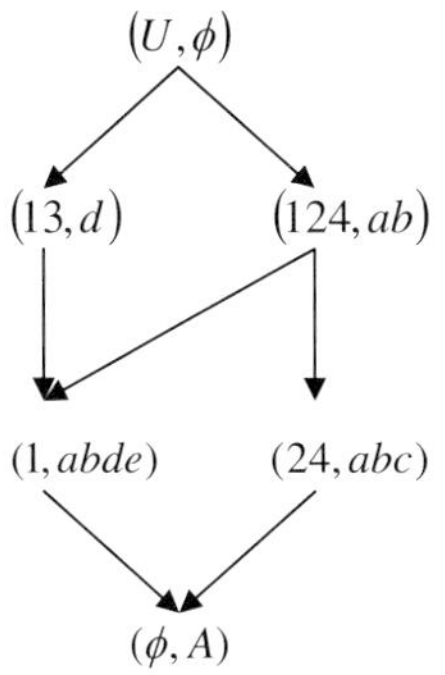

Fig. 1.

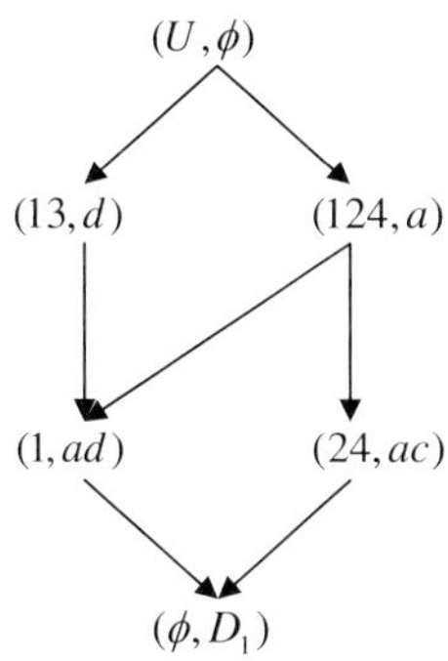

Fig. 2.

Fig.1. shows the variable threshold concept lattice of $(U, A, \tilde{I})$ for $\delta = 0.6$. There are 6 variable threshold concepts: $L_\delta(U, A, \tilde{I}) = \{(1, abde), (24, abc), (13, d), (124, ab), (U, \emptyset), (\emptyset, A)\}$. And there exist two $\delta-$reducts: $D_1 = \{a, c, d\}$ and $D_2 = \{b, c, d\}$. Fig. 2. shows the variable threshold concept lattice of $(U, D_1, \tilde{I}_{D_1})$. It is easy to see that the two variable threshold concept lattices are isomorphic to each other.

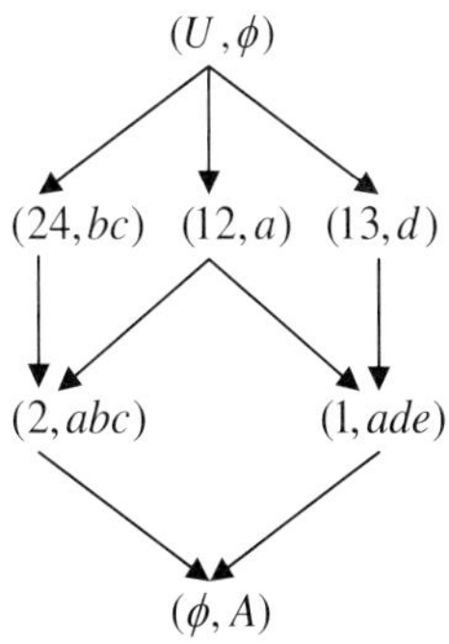

Fig. 3.

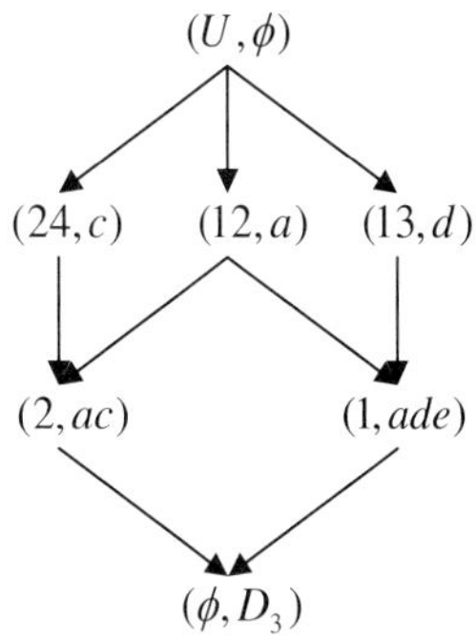

Fig. 4.

Fig. 3. shows the variable threshold concept lattice of $(U, A, \tilde{I})$ for $\delta = 0.7$. And $(U, A, \tilde{I})$ has two $\delta-$reducts: $D_3 = \{a, c, d\}$ and $D_4 = \{a, b, d\}$. Fig. 4. shows the variable precision concept lattice of $(U, D_3, \tilde{I}_{D_3})$. It is easy to see that the two variable precision concept lattices are isomorphic to each other.

Compared Fig.1. with Fig.3., and Fig.2. with Fig.4. respectively, we also can get that for $0 < \delta_1 < \delta_2 \leq 1$ and $(X, B) \in L_{\delta_1}(U, A, I)$, there exists $(X', B') \in L_{\delta_2}(U, A, I)$ such that $X' \subseteq X$.

In fact, for any $\delta \in (0, 1]$, a fuzzy formal context can be converted to a classical formal context. For any $\delta \in (0, 1]$, we can define $I_\delta \subseteq U \otimes A$ as follows: $I_\delta(x, a) = 1 \Leftrightarrow \tilde{I}(x, a) \geq \delta$. Then (U, A, I_δ) is a formal context with 0 and 1.

Thus properties of a classical concept lattice are analogous to those of a variable threshold concept lattice $L_\delta(U, A, \tilde{I})$ (see [3,7,8,9,10,11]).

3 Dependence Space Based on the Variable Threshold Concept Lattice

A triple (U, A, F) is called an information system, if U is a non-empty set of objects, A is a non-empty set of attributes, and $F = \{f_j \mid f_j : U \rightarrow V_j(a_j \in A)\}$, where f_j is an information function from U to V_j, and V_j is the domain of the attribute a_j. For $B \subseteq A$, we define a binary relation $R_B = \{(x_i, x_j) \mid f_l(x_i) = f_l(x_j)\ (\forall a_l \in B)\}$. R_B is an equivalence relation on U, and it determines a partition $U/R_B = \{[x_i]_B \mid x_i \in U\}$, where $[x_i]_B = \{x_j \mid (x_i, x_j) \in R_B\} = \{x_j \mid f_l(x_i) = f_l(x_j)\ (\forall a_l \in B)\}$. We denote by $\mathcal{P}(A)$ the power set of A (see [8,9,10,11]).

Definition 4. [5,10] *Let (U, A, F) be an information system and $\mathcal{R}$ a equivalence relation on $\mathcal{P}(A)$.*

(1) $\mathcal{R}$ is referred to as a congruence on $\mathcal{P}(A)$, if for any $(B_1,\ C_1) \in \mathcal{R}$, $(B_2,\ C_2) \in \mathcal{R}$, we have $(B_1 \cup B_2,\ C_1 \cup C_2) \in \mathcal{R}$.

(2) $(A, \mathcal{R})$ is referred to as a dependence space, if $\mathcal{R}$ is a congruence on $\mathcal{P}(A)$.

Theorem 2. *Let $(U, A, \tilde{I})$ be a fuzzy formal context and $\delta \in (0, 1]$. Note*

$$\mathcal{R}_\delta = \{(B, C) \in \mathcal{P}(A) \times \mathcal{P}(A) \mid B^{*\delta} = C^{*\delta}\}.$$

Then $(A, \mathcal{R}_\delta)$ is a dependence space.

Proof. It is clear that $\mathcal{R}_\delta$ is an equivalence relation on $\mathcal{P}(A)$. Suppose (B_1, C_1), $(B_2, C_2) \in \mathcal{R}_\delta$. Then $B_1^{*\delta} = C_1^{*\delta}$, $B_2^{*\delta} = C_2^{*\delta}$. By the property of $^{*\delta}$ we know

$$(B_1 \cup B_2)^{*\delta} = B_1^{*\delta} \cap B_2^{*\delta} = C_1^{*\delta} \cap C_2^{*\delta} = (C_1 \cup C_2)^{*\delta}.$$

Then $(B_1 \cup B_2, C_1 \cup C_2) \in \mathcal{R}_\delta$, and $\mathcal{R}_\delta$ is a congruence on $\mathcal{P}(A)$. Therefore, $(A, \mathcal{R}_\delta)$ is a dependence space.

Lemma 1. *Let $(U,\ A,\ \tilde{I})$ be a formal context and $\delta \in (0, 1]$. Then for any $x \in U$ and $B \in A$ we have*

$$x \in B^{*\delta} \Leftrightarrow B \subseteq x^{*\delta}.$$

Lemma 2. [10] *Let $(U,\ A,\ \tilde{I})$ be a fuzzy formal context and $\mathcal{H} \subseteq \mathcal{P}(A)$. Note that*

$$T(\mathcal{H}) = \{(B,\ C) \in \mathcal{P}(A)^2 \mid \forall D \in \mathcal{H},\ B \subseteq D \Leftrightarrow C \subseteq D\}.$$

Then $(A,\ T(\mathcal{H}))$ is a dependence space.

Proposition 1. *Let $(U, A, \tilde{I})$ be a fuzzy formal context and $\delta \in (0, 1]$. Note that*

$$\mathcal{R}_\delta = \{(B,\ C) \mid B^{*\delta} = C^{*\delta}\},$$
$$\mathcal{H}_\delta = \{x^{*\delta} \mid x \in U\}.$$

Then $T(\mathcal{H}_\delta) = \mathcal{R}_\delta$.

Proof. It is clear that $\mathcal{R}_\delta$ and $T(\mathcal{H}_\delta)$ are all congruences on $\mathcal{P}(A)$ in terms of Theorem 2 and Lemma 2.

If $(B,\ C) \in \mathcal{R}_\delta$, then $B^{*_\delta} = C^{*_\delta}$. By Lemma 1, for any $x^{*_\delta} \in \mathcal{H}_\delta$ we have

$$B \subseteq x^{*_\delta} \Leftrightarrow x \in B^{*_\delta} \Leftrightarrow x \in C^{*_\delta} \Leftrightarrow C \subseteq x^{*_\delta}$$

which leads to $(B,\ C) \in T(\mathcal{H}_\delta)$.

If $(B,\ C) \in T(\mathcal{H}_\delta)$, then $B \subseteq x^{*_\delta} \Leftrightarrow C \subseteq x^{*_\delta}$ for any $x^{*_\delta} \in \mathcal{H}_\delta$. Therefore,

$$x \in B^{*_\delta} \Leftrightarrow B \subseteq x^{*_\delta} \Leftrightarrow C \subseteq x^{*_\delta} \Leftrightarrow x \in C^{*_\delta}.$$

Thus $B^{*_\delta} = C^{*_\delta}$ and $(B,\ C) \in \mathcal{R}_\delta$. This completes the result.

Definition 5. [8] *Let A be a finite set of attributes and $\mathcal{P}(A)$ be the power set of A. Then $\mathcal{J}:\ \mathcal{P}(A) \to \mathcal{P}(A)$ is called a closure operator, if it satisfies:*

(1) $B \subseteq \mathcal{J}(B) \quad (\forall B \in \mathcal{P}(A))$;
(2) $B \subseteq C \Rightarrow \mathcal{J}(B) \subseteq \mathcal{J}(C) \quad (\forall B,\ C \in \mathcal{P}(A))$;
(3) $\mathcal{J}(\mathcal{J}(B)) = \mathcal{J}(B) \quad (\forall B \in \mathcal{P}(A))$.

Proposition 2. *Let $(U, A, \tilde{I})$ be a fuzzy formal context and $\delta \in (0,1]$. Note*

$$\mathcal{R}_\delta = \{(B,\ C) \mid B^{*_\delta} = C^{*_\delta}\},$$
$$[B]_\delta = \{C \mid (B, C) \in \mathcal{R}_\delta\},$$
$$\mathcal{J}_\delta(B) = \bigcup [B]_\delta = \bigcup \{C \mid (B, C) \in \mathcal{R}_\delta\}.$$

Then the following properties hold:

(1) $\mathcal{J}_\delta(B) \in [B]_\delta$, *and if* $B^{'} \in [B]_\delta$, *then* $B^{'} \subseteq \mathcal{J}_\delta(B)$;
(2) *If* $B_1, B_2 \in [B]_\delta$, *and* $B_1 \subseteq B^{'} \subseteq B_2$, *then* $B^{'} \in [B]_\delta$.

Proof. (1) Suppose $[B]_\delta = \{B_1, B_2, \cdots, B_k\}$. Since $(B, B_i) \in \mathcal{R}_\delta \quad (i \leq k)$, and $\mathcal{R}_\delta$ is a congruence, then $(B, \mathcal{J}_\delta(B)) = (B, \bigcup\limits_{i \leq k} B_i) \in \mathcal{R}_\delta$. Thus $\mathcal{J}_\delta(B) \in [B]_\delta$. Then for $B^{'} \in [B]_\delta$, $B^{'} \subseteq \mathcal{J}_\delta(B)$.

(2) Suppose $B_1, B_2 \in [B]_\delta$ and $B_1 \subseteq B^{'} \subseteq B_2$. Then $(B_1, B_2) \in \mathcal{R}_\delta$, $B_1 \cup B^{'} = B^{'}$ and $B^{'} \cup B_2 = B_2$. Since $\mathcal{R}_\delta$ is a congruence, we can conclude that $(B_1 \cup B^{'}, B_2 \cup B^{'}) \in \mathcal{R}_\delta$. Thus $(B^{'}, B_2) \in \mathcal{R}_\delta$. According to $(B_2, B) \in \mathcal{R}_\delta$, we have $(B^{'}, B) \in \mathcal{R}_\delta$. And then $B^{'} \in [B]_\delta$.

In fact, $[B]_\delta$ in Proposition 2 is an equivalent class of $\mathcal{P}(A)$. Since the congruence is first an equivalence relation on $\mathcal{P}(A)$, we can get

$$\mathcal{P}(A)/\mathcal{R}_\delta = \{[B]_\delta \mid \quad B \subseteq A\}.$$

That is, $\mathcal{R}_\delta$ determine a partition $\mathcal{P}(A)/\mathcal{R}_\delta$.

Proposition 3. *Let $(U, A, \tilde{I})$ be a fuzzy formal context and $\delta \in (0,1]$. Note*

$$\mathcal{R}_\delta = \{(B,\ C) \mid B^{*_\delta} = C^{*_\delta}\},$$
$$[B]_\delta = \{C \mid (B, C) \in \mathcal{R}_\delta\},$$
$$\mathcal{J}_\delta(B) = \bigcup [B]_\delta = \bigcup \{C \mid (B, C) \in \mathcal{R}_\delta\}.$$

Then $\mathcal{J}_\delta$ is a closure operator.

Proof. (1) Since $\mathcal{R}_\delta$ is reflexive, we have $B \in [B]_\delta$. Thus $B \subseteq \mathcal{J}_\delta(B)$.

(2) Suppose $B \subseteq C$. Since $\mathcal{R}_\delta$ is a congruence, by Proposition 2 we conclude that $(B, \mathcal{J}_\delta(B)) \in \mathcal{R}_\delta$, $(C, \mathcal{J}_\delta(C)) \in \mathcal{R}_\delta$, and then $(C, \mathcal{J}_\delta(B) \cup \mathcal{J}_\delta(C)) \in \mathcal{R}_\delta$. Therefore, $\mathcal{J}_\delta(B) \cup \mathcal{J}_\delta(C) \subseteq \mathcal{J}_\delta(C)$. Thus $\mathcal{J}_\delta(B) \subseteq \mathcal{J}_\delta(C)$.

(3) Since $\mathcal{R}_\delta$ is a congruence, we can observe that $(B, \mathcal{J}_\delta(B)) \in \mathcal{R}_\delta$ and $(\mathcal{J}_\delta(B), \mathcal{J}_\delta(\mathcal{J}_\delta(B))) \in \mathcal{R}_\delta$. Then $(B, \mathcal{J}_\delta(\mathcal{J}_\delta(B))) \in \mathcal{R}_\delta$. Thus $\mathcal{J}_\delta(\mathcal{J}_\delta(B)) \subseteq \mathcal{J}_\delta(B)$. By (1) we have $\mathcal{J}_\delta(B) \subseteq \mathcal{J}_\delta(\mathcal{J}_\delta(B))$. Therefore $\mathcal{J}_\delta(\mathcal{J}_\delta(B)) = \mathcal{J}_\delta(B)$.

Lemma 3. *Let $(U, A, \tilde{I})$ be a fuzzy formal context and $\delta \in (0, 1]$. Note that*

$$\mathcal{R}_\delta = \{(B,\ C) \mid B^{*\delta} = C^{*\delta}\},$$
$$[B]_\delta = \{C \mid (B, C) \in \mathcal{R}_\delta\},$$
$$\mathcal{J}_\delta(B) = \bigcup [B]_\delta = \bigcup \{C \mid (B, C) \in \mathcal{R}_\delta\}.$$

Then

$$\mathcal{J}_\delta(B) = B^{*\delta *\delta}.$$

Proof. Since $(B^{*\delta *\delta})^{*\delta} = B^{*\delta}$, we have $B^{*\delta *\delta} \subseteq \mathcal{J}_\delta(B)$. Take $a \in \mathcal{J}_\delta(B)$, then there exists $C \subseteq A$ such that $a \in C$ and $B^{*\delta} = C^{*\delta}$. Thus $B^{*\delta *\delta} = C^{*\delta *\delta}$ and $a \in C \subseteq C^{*\delta *\delta} = B^{*\delta *\delta}$ which leads to $\mathcal{J}_\delta(B) \subseteq B^{*\delta *\delta}$.

Let A be a non-empty finite set, and $\mathcal{P}(A)$ the power set of A. An element $B \in \mathcal{P}(A)$ is said to be $\mathcal{J}_\delta$−closed if $\mathcal{J}_\delta(B) = B$ (see [5,10]).

We denote by $\overline{\mathcal{J}}_\delta$ the set of all $\mathcal{J}_\delta$−closed elements in $\mathcal{P}(A)$, and call it $\mathcal{J}_\delta$−closed set, that is

$$\overline{\mathcal{J}}_\delta = \{B \mid \mathcal{J}_\delta(B) = B\}.$$

Theorem 3. *Let $(U,\ A,\ \tilde{I})$ be a fuzzy formal context and $\delta \in (0, 1]$. Note*

$$\mathcal{R}_\delta = \{(B,\ C) \mid B^{*\delta} = C^{*\delta}\},$$
$$[B]_\delta = \{C \mid (B, C) \in \mathcal{R}_o\},$$
$$\mathcal{J}_\delta(B) = \bigcup [B]_\delta = \bigcup \{C \mid (B, C) \in \mathcal{R}_\delta\}.$$

Then

$$(X,\ B) \in L_\delta(U, A, \tilde{I}) \Leftrightarrow B \in \overline{\mathcal{J}}_\delta.$$

Proof. We only need to prove

$$(X,\ B) \in L_\delta(U, A, \tilde{I}) \Leftrightarrow \mathcal{J}_\delta(B) = B.$$

Take $(X,\ B) \in L_\delta(U, A, \tilde{I})$, then $X^{*\delta} = B,\ X = B^{*\delta}$. And $B^{*\delta *\delta} = B$. By Lemma 3 we have $\mathcal{J}_\delta(B) = B$. On the other hand, if $\mathcal{J}_\delta(B) = B$, using Lemma 3 we can get $B = B^{*\delta *\delta}$. Let $X = B^{*\delta}$, then we have $B = X^{*\delta}$, and $(X,\ B) \in L_\delta(U, A, \tilde{I})$.

By Theorem 3 we can observe that $\overline{\mathcal{J}}_\delta = Int_\delta(U, A, \tilde{I})$.

Theorem 3 shows an approach to construct a variable threshold concept lattice by $\mathcal{J}_\delta$−closed set:

By Proposition 2 we know the greatest element of each equivalent class of $\mathcal{P}(A)/\mathcal{R}_\delta$ is $\mathcal{J}_\delta$−closed. While by Theorem 3 we have every $\mathcal{J}_\delta$−closed element is the intent of a variable threshold concept of $L_\delta(U, A, \tilde{I})$. Thus we can get the equivalent classes and $\mathcal{J}_\delta$−closed set according to $\mathcal{R}_\delta$ or $\mathcal{H}_\delta$, and then get the variable threshold concept lattice.

Example 2. The fuzzy formal context $(U,\ A,\ \tilde{I})$ is given as Example 1.

For $\delta = 0.6$, by Theorem 2 or Proposition 1 we can get the partition

$$\begin{aligned}\mathcal{P}(A)/\mathcal{R}_\delta &= \{[B]_\delta \mid B \subseteq A\}\\ &= \{\{\emptyset\},\ \{d\},\ \{a,b,ab\},\ \{c,\ ac,\ bc,\ abc\},\\ &\quad \{e,\ ad,\ ae,\ bd,\ be,\ de,\ abd,\ abe,\ bde,\ ade,\ abde\},\\ &\quad \{cd,\ ce,\ acd,\ ace,\ bce,\ cde,\ abcd,\ abce,\ acde,\ bcde,\ abcde\}\}.\end{aligned}$$

By Proposition 2 we can get that the greatest element of each equivalent class is $\mathcal{J}_\delta-$closed. Then the $\mathcal{J}_\delta-$closed set is

$$\overline{\mathcal{J}}_\delta = \{\emptyset,\ d,\ ab,\ abc,\ abde,\ abcde\}.$$

Compared with the variable threshold concept lattice $L_\delta(U, A, \tilde{I})$ obtained from Example 1 for $\delta = 0.6$, we can get that each element of $\overline{\mathcal{J}}_\delta$ is just the intent of a variable threshold concept of $L_\delta(U, A, \tilde{I})$, which is equal to Fig. 1.

By the dual of ${}^{*_\delta}$: $\mathcal{P}(A) \to \mathcal{P}(U)$ and ${}^{*_\delta}$: $\mathcal{P}(U) \to \mathcal{P}(A)$, we can define analogously the congruence $\mathcal{R}_\delta^{'} = \{(X,\ Y) \mid X^{*_\delta} = Y^{*_\delta}\}$ on U and construct a dependence space $(U, \mathcal{R}_\delta^{'})$. Then using $\mathcal{R}_\delta^{'}$ we have $\mathcal{P}(U)/\mathcal{R}_\delta^{'} = \{[X]_\delta \mid X \subseteq U\}$ and the $\mathcal{J}_\delta^{'}-$closed set $\overline{\mathcal{J}'}_\delta = \{X \mid \mathcal{J}_\delta^{'}(X) = X\}$. As a consequence, the greatest element of $\overline{\mathcal{J}'}_\delta$ is just the extent of a variable threshold concept of $L_\delta(U, A, \tilde{I})$. That is, $\overline{\mathcal{J}'}_\delta = Ext_\delta(U, A, \tilde{I})$.

4 Conclusion

Fuzzy set theory is to deal with uncertainty and vagueness. Then the fuzzy concept lattice is an effective tool for knowledge representation and knowledge discovery in uncertain fields. This paper introduced a variable threshold concept lattice based on a fuzzy formal context. And a dependence space is constructed based on the variable threshold concept lattice. Applying to the congruence on the dependence space, the equivalent classes and then a closed set are obtained. The introduced notions of the congruence and the closed set are useful for the variable threshold concept lattice. In fact, an approach to constructing variable threshold concepts of the variable threshold concept lattice is introduced.

In future, we will study a decision fuzzy formal context furthermore.

Acknowledgement

This work is supported by a grant from the Major State Basic Research Development Program of China (973 Program No. 2002CB312200).

References

1. Belohlavek, R.: Concept lattices and order in fuzzy logic. Annals of Pure and Applied Logic 128(2004) 277-298
2. Elloumi, S.: A multi-level conceptual data reduction approach based in the lukasiewicz implication. Information Sciences 163(4)(2004) 253-262

3. Ganter, B., Wille, R.: Formal Concept Analysis: Mathematical Foundations. Springer-Verlag, New York (1999)
4. Mieszkowicz-Rolka, A. Rolka, L.: Variable precision fuzzy rough sets model in the analysis of process data. In: *D.Ślezak* et al. (Eds.): RSFDGrC 2005, LNAI 3641 (2005) 354-363
5. Novotny, M.: Dependence Spaces of Information Systems. In : Ortowska E. (Ed.), Incomplete informations: Rough Set Analysis [M], Physica-Verlag (1998)
6. Saquer, J., Deogun, J.: Formal Rough Concept Analysis. New directions in Rough Sets, Data Mining, and Granular-Soft Computing. Lecture Notes in Computer Science 1711, Springer, Berlin (1999) 91-99
7. Wille, R.: Restructuring Lattice Theory: an Approach Based on Hierarchies of Concepts. In: Rival, I. (Ed.): Ordered Sets. Reidel,Dordrecht-Boston (1982) 445-470
8. Yao, Y.Y.: Concept lattices in rough set theory. Proceedings of 23rd International Meeting of the North American Fuzzy Information Processing Society (2004) 796-801
9. Yao, Y.Y.: A Comparative Study of Formal Concept Analysis and Rough set theory in Data Analysis, Rough Sets and Current Trends in Computing. Proceedings of 3rd International Conference, RSCTC'04 (2004) 59-68
10. Zhang, W.-X., Qiu G.-F.: Uncertain Decision Making Based on Rough Sets. Publishin of Tsinghua University, Beijing (2005)
11. Zhang, W.-X., Wei, L., Qi, J.-J.: Attribute Reduction in Concept Lattice Based on Discernibility Matrix. In: D. Ślezak et al. (Eds.): RSFDGrC 2005, LNAI 3642 (2005) 157-165
12. Ziarko, W.: Variable precision rough set model. Journal of Computer and System Sciences, 46(1993) 39-59

Non-fragile Robust H_∞ Fuzzy Controller Design for a Class of Nonlinear Descriptor Systems with Time-Varying Delays in States

Junsheng Ren

Key Laboratory of Marine Simulation & Control,
Dalian Maritime University, Liaoning Province, 116026, P.R. China
jsren@dlmu.edu.cn

Abstract. The controller fragility can cause the performance debasement of the closed-loop system due to small perturbations in the coefficients of the controller design, and is one of the most important factors to be considered during practical controller design. To take the controller fragility into consideration for a class of nonlinear time-delayed descriptor systems with norm-bounded time-varying uncertainties in the matrices of state, delayed state and control gain, we have proposed non-fragile robust H_∞ fuzzy control design via state feedback controllers in this paper. The nonlinear descriptor system is approximated by Takagi-Sugeno (T-S) fuzzy model. In combination of parallel-distributed compensation (PDC) scheme, sufficient conditions are derived for the existence of non-fragile robust H_∞ fuzzy controllers in terms of linear matrix inequalities (LMI). Finally, an example is given to demonstrate the use of the proposed controller design.

1 Introduction

Generally, the perturbations during the controller's implementation are quite difficult to avoid due to finite word length in digital systems, the imprecision inherent in analog systems, the need for additional tuning of parameters in the final controller implementation and other reasons. some examples in [1] had been presented to show that small perturbations in the coefficients of the controller designed by using robust H_2, H_∞, l_1 and μ approaches can destabilize the closed-loop control system. The authors therein had suggested to take into account uncertainties both in the controller structure and in the system structure. After [1], the research of non-fragile controller has been an active area during the past several years, see [2] and the references therein. However, the efforts therein were mainly focused on linear systems. The non-fragile controller for nonlinear system was discussed in [3]. The method therein needs positive-definite solution to a pair of coupled Hamilton-Jacobi inequalities, which are much complicated and only have solutions for a special kind of systems. Therefore, it's still an open problem to design non-fragile controller for nonlinear system.

Very recently, many effects [6,7] have been devoted to descriptor systems with time delay, because the descriptor system has a tighter representation for

L. Wang et al. (Eds.): FSKD 2006, LNAI 4223, pp. 119–128, 2006.

a wider class of systems for representing real independent parametric perturbations in comparison to traditional state-space representation [4]. Due to the difficulties of constructing Lyapunov function and the complexity of the existence and uniqueness of the solution, there still remain some difficulties in controlling the nonlinear descriptor systems with time delays. Recent studies [8,10,9] have shown Takagi-Sugeno (T-S) fuzzy model is a universal approximator of any smooth nonlinear systems having a first order that is differentiable. Therefore, it is meaningful to consider applying the fuzzy model to approximate the nonlinear descriptor system with time delays. To stabilize the nonlinear descriptor system with time delays, some researchers considered T-S fuzzy descriptor system with time delays [11,12].

Motivated by the aforementioned pioneering works, the goal of this paper is to propose non-fragile robust H_∞ fuzzy controller for a class of nonlinear descriptor systems with time-varying delays and norm-bounded uncertainties. First, the nonlinear descriptor system with time-varying delays is described by T-S fuzzy model. Then, the sufficient conditions for non-fragile robust H_∞ fuzzy controller are presented by use of PDC scheme. Finally, numerical example is given to illustrate the effectiveness of the controller design.

2 Problem Formulation and Some Preliminaries

We utilize T-S fuzzy system approximate the nonlinear time-delayed descriptor system with parametric uncertainties as follows

$$
\begin{aligned}
&\text{Plant Rule } k: \\
&\qquad \text{IF } \vartheta_1(t) \text{ is } N_{k1} \text{ and, } \ldots, \text{ and } \vartheta_g \text{ is } N_{kg}, \\
&\qquad \text{THEN} \\
&\quad \text{Right-Hand-Side Plant Rule } i: \\
&\qquad \text{IF } \vartheta_1(t) \text{ is } J_{i1} \text{ and, } \ldots, \text{ and } \vartheta_g \text{ is } J_{ig}, \\
&\qquad \text{THEN } E_k\dot{x}(t) = (A_i + \Delta A_i)\,x(t) + (A_{di} + \Delta A_{di})\,x(t-\sigma(t)) \qquad (1) \\
&\qquad\qquad + (B_i + \Delta B_i)\,u(t) + B_{2i}\omega(t), \\
&\qquad\quad z(t) = C_i x(t), \\
&\qquad\quad x(t) = \phi(t),\ t \in [-\sigma_0, 0], \\
&\qquad \text{for } k = 1,\ 2,\ \cdots,\ r,\ i = 1,\ 2,\ \cdots,\ r_e.
\end{aligned}
$$

where $\vartheta(t) = \{\vartheta_1(t), \vartheta_2(t), \cdots, \vartheta_g(t)\}$ denotes the variables of premise part, $A_i, A_{di} \in \mathbb{R}^{n\times n}$ and $B_i \in \mathbb{R}^{m\times n}$ are known real constant matrices, and N_{kl} and J_{il} denote fuzzy sets, the real-valued functional $\sigma(t)$ is the time-varying delay in the state and satisfies $\sigma(t) \le \sigma_0$, σ_0 is a real positive constant representing the upper bound of the time-varying delay. It is further assumed that $\dot{\sigma}(t) \le \beta < 1$ and β is a known constant. $\phi(t)$ are continuous vector-valued initial functions, and r and r_e denote the number of IF-THEN rules. We introduce the following assumption on E_i and C_i.

Assumption 1. *Without loss of generality, it is assumed that* $E_1 = E_2 = \cdots = E_g = E$ *and* $C_1 = C_2 = \cdots = C_g = C$ *in the fuzzy representation of the nonlinear descriptor system (1). Obviously, we have* $N_{kl} = J_{il}$ *and* $r = r_e$.

The assumption 1 will simplify our following discussions much but without loss of generality. In (1), $\Delta A_i,$, $\Delta A_{di} \in \mathbb{R}^{n\times n}$, $\Delta B_i(t) \in \mathbb{R}^{m\times n}$, are the system's uncertainty matrices and satisfy Assumption 2.

Assumption 2. *Uncertainty matrices* ΔA_i, ΔB_i *and* ΔA_{di} *in system (1) take the following structures*

$$\begin{bmatrix} \Delta A_i & \Delta B_i & \Delta A_{di} \end{bmatrix} = D_i F_i(\nu) \begin{bmatrix} E_{1i} & E_{2i} & E_{di} \end{bmatrix}, \tag{2}$$

where D_i, E_{1i}, E_{di} *and* E_{2i} *are constant real matrices of appropriate dimensions, and* $F_i(t) \in \mathbb{R}^{i\times j}$ *is unknown matrix-valued functions*

$$F_i^T(\nu) F_i(\nu) \le I, \tag{3}$$

where $\nu \in \Omega$, Ω *is a compact set in* $\mathbb{R}$. *and* I *is the identity matrix of appropriate dimensions.*

Based on Assumption 1, the final output of the T-S fuzzy model is inferred as follows, by using the fuzzy inference method with a singleton fuzzifier, product inference and center average defuzzifiers

$$\begin{aligned} E\dot{x}(t) = \sum_{i=1}^{r} h_i(\vartheta(t))[(A_i + \Delta A_i)x(t) + (A_{di} + \Delta A_{di})x(t - \sigma(t)) \\ + (B_i + \Delta B_i)u(t) + B_{2i}\omega(t)], \end{aligned} \tag{4}$$

where $h_i(\vartheta(t)) = w_i(\vartheta(t)) \Big/ \sum_{i=1}^{r} w_i(\vartheta(t))$, $w_i(\vartheta(t)) = \prod_{j=1}^{r} M_{ij}(\vartheta(t))$ and $J_{ij}(\vartheta(t))$ denotes the degree of membership of $z(t)$ on J_{ij}. It is assumed that the degree of membership satisfies $\sum_{i=1}^{r} w_i(\vartheta(t)) > 0, w_i(\vartheta(t)) \ge 0, i = 1,\ 2,\ \cdots,\ r$. Note that for all t, there exists $\sum_{i=1}^{r} h_i(\vartheta(t)) = 1, h_i(\vartheta(t)) \ge 0$.

For PDC scheme, non-fragile robust H_∞ fuzzy controller and the fuzzy model (1) possess the same premises. The resulting overall controller is nonlinear in general which is a fuzzy blending of each individual linear controller designed for each local linear model. Then, supposing that all the states are observable, the i-th controller rule can be expressed by

$$\begin{aligned} &\text{Controller Rule } i: \\ &\qquad \text{IF } \vartheta_1(t) \text{ is } N_{i1} \text{ and } \ldots, \text{ and } \vartheta_g \text{ is } N_{ig}, \\ &\qquad \text{THEN } u(t) = (K_i + \Delta K_i)x(t),\ i = 1,\ 2,\ \cdots,\ r. \end{aligned} \tag{5}$$

where $u(t)$ is the actually implemented local controller, K_i is the local nominal gain, ΔK_i represents drifting from the nominal solution. It has been proved that fuzzy logic controller in (5) is an approximator for any nonlinear state feedback controller [9]. The overall fuzzy controller can be represented as follows

$$u(t) = \sum_{i=1}^{r} h_i(\vartheta(t))(K_i + \Delta K_i)x(t). \tag{6}$$

Applying the controller (6) to the system (4) will result in the following closed-loop control system

$$\begin{cases} E\dot{x}(t) = \sum\limits_{i=1}^{r} h_i(\vartheta(t))\{[(A_i + \Delta A_i) + (B_i + \Delta B_i)(K_i + \Delta K_i)]x(t) \\ \qquad\qquad +(A_{di} + \Delta A_{di})x(t - \sigma(t)) + B_{2i}\omega(t)\}, \\ z(t) = Cx(t), \\ x(t) = \phi(t),\ t \in [-\sigma_0, 0], \end{cases} \tag{7}$$

In the following, we introduce some definitions and useful properties for the system (7).

Definition 1. *A pencil* $sE - \Sigma_{i=1}^{r} h_i(\vartheta(t))A_i$ *(or pair*$(E - \Sigma_{i=1}^{r} h_i\ (\vartheta(t))A_i)$*) is regular, if* $det(sE - \Sigma_{i=1}^{r} h_i(\vartheta(t))A_i)$ *is not identically zero;*

Fuzzy descriptor system (7) has no impulsive mode (or impulse free) if and only if $rank(E) = degdet(sE - \Sigma_{i=1}^{r} h_i(\vartheta(t))A_i)$.

Remark 1. The notations $det(\cdot)$, $rank(\cdot)$ and $deg(\cdot)$ denote determinant, rank and degree of a matrix, respectively. The property of regularity guarantees the existence and uniqueness of solution for any specified initial condition. The condition of impulse free ensures that singular system has no infinite poles.

Definition 2. *The closed-loop system 7 is asymptotically stable with disturbance attenuation* γ*, if the followings are fulfilled for time-varying delays and norm-bounded parametric uncertainties*

1). The closed-loop system (7) is asymptotically stable;

2). The closed-loop system guarantees, under zero initial conditions, $\|z(t)\|_2 \leq \gamma\,\|\omega(t)\|_2$*, for all non-zero* $\omega(t) \in L_2\,[0,\ \infty)$.

The objective of this paper is to design non-fragile robust H_∞ controller in the presence of time-varying delays, parameter uncertainties of system and additive uncertainty of controller. Also the controller guarantees disturbance attenuation of the closed-loop system from $\omega(t)$ to $z(t)$.

3 Non-fragile Robust H_∞ Fuzzy Controller Design

Now we are in a position to present the main result in this paper. Firstly, stability conditions are presented for the systems (7) without external disturbances.

Theorem 1. *Consider the uncertain nonlinear system with time delays (7) and suppose that the disturbance inputs are zero for all the time. The closed-loop system (7) is asymptotically stable if there exist positive definite matrix P, and controller gains K_i satisfying such that*

$$PE^T = EP \geq 0, \begin{bmatrix} \Pi_1 & * \\ A_{di}^T P & \Lambda_1 \end{bmatrix} < 0, \begin{bmatrix} \Pi_2 & * \\ A_{di}^T P + A_{dj}^T P & \Lambda_2 \end{bmatrix} < 0, \tag{8}$$

where

$$\begin{aligned} \Pi_1 &= PA_i + PB_iK_i + A_i^T P + K_i^T B_i^T P + \frac{R_1}{1-\beta} + (\varepsilon_{1i} + \varepsilon_{3i} \cdot \varepsilon_{2i} \\ &\quad + \varepsilon_{4i}) PD_iD_i^T P + \varepsilon_{2i} PB_i(I - \varepsilon_{3i}(E_{2i}H_i)^T(E_{2i}H_i))^{-1} B_i^T P + \varepsilon_{1i}^{-1}(E_{1i} \\ &\quad + E_{2i}K_i)^T(E_{1i} + E_{2i}K_i) + \varepsilon_{2i}^{-1} E_{Ki}^T E_{Ki}, \\ \Pi_2 &= PA_i + PB_iK_j + A_i^T P + K_j^T B_i^T P + PA_j + PB_jK_i + A_j^T P \\ &\quad + K_i^T B_j^T P + \frac{2R_1}{1-\beta} + (\varepsilon_{1ij} + \varepsilon_{2ij} \cdot \varepsilon_{3ij} + \varepsilon_{2ij}) PB_i(I - \varepsilon_{3ij}^{-1}(E_{2i}H_j)^T \\ &\quad \times (E_{2i}H_j))^{-1} B_i^T P + \varepsilon_{5ij} \cdot \varepsilon_{6ij} + \varepsilon_{4ij} + \varepsilon_{4i}) PD_iD_i^T P + \varepsilon_{4j} PD_jD_j^T P \\ &\quad + \varepsilon_{1ij}^{-1}(E_{1i} + E_{2i}K_j)^T(E_{1i} + E_{2i}K_j) + \varepsilon_{2ij}^{-1} E_{Kj}^T E_{Kj} \\ &\quad + \varepsilon_{5ij} PB_j(I - \varepsilon_{6ij}^{-1}(E_{2j}H_i)^T(E_{2j}H_i))^{-1} B_j^T P + \varepsilon_{5ij}^{-1} E_{Ki}^T E_{Ki} \\ &\quad + \varepsilon_{4ij}^{-1}(E_{1j} + E_{2j}K_i)^T(E_{1j} + E_{2j}K_i), \\ \Lambda_1 &= \varepsilon_{4i}^{-1} E_{di}^T E_{di} - \frac{R_1}{1-\beta}, \Lambda_2 = \varepsilon_{4i}^{-1} E_{di}^T E_{di} + \varepsilon_{4j}^{-1} E_{dj}^T E_{dj} - \frac{2R_1}{1-\beta}, \end{aligned}$$

where $1 \leq i < j \leq r$, ε_{ci} $(1 \leq c \leq 4)$, ε_{dij} $(1 \leq d \leq 6)$ *are arbitrary positive scalars,* $*$ *denotes the transposed element in the symmetric position.*

Proof. Define the following functional candidate for the system (7) as follows

$$V(x(t)) = x^T(t)E^T Px(t) + \frac{1}{1-\beta}\int_{t-\sigma(t)}^{t} x^T(s)R_1x(s)ds, \tag{9}$$

where P is a time-invariant, symmetric positive definite matrix. Then, the time derivative of the Lyapunov candidate $V(x(t))$ is given by

$$\begin{aligned} \frac{dV(x(t))}{dt} &= \dot{x}^T(t)E^T Px(t) + x^T(t)E^T P\dot{x}(t) + \frac{1}{1-\beta}x^T(t)R_1x(t) \\ &\quad - \frac{1-\sigma(t)}{1-\beta}x^T(t-\sigma(t))R_1x(t-\sigma(t)). \end{aligned}$$

After some manipulations, the above formulae can be rewritten as follows

$$\frac{dV(x(t))}{dt} = \sum_{i=1}^{r} h_i^2(\vartheta(t))x^T(t)(P((A_i + \Delta A_i) + (B_i + \Delta B_i)(K_i + \Delta K_i)$$
$$+ ((A_i^T + \Delta A_i^T) + (K_i^T + \Delta K_i^T)(B_i^T + \Delta B_i^T))P)x(t) + \sum_{i<j}^{r} h_i(\vartheta(t))h_j(\vartheta(t))$$
$$\times (P(A_i + \Delta A_i) + (B_i + \Delta B_i)(K_j + \Delta K_j)) + ((A_i^T + \Delta A_i^T)$$
$$+ (K_j^T + \Delta K_j^T)(B_i^T + \Delta B_i^T))P + P((A_j + \Delta A_j) + (B_j + \Delta B_j)(K_i + \Delta K_i))$$
$$+ ((A_j^T + \Delta A_j^T) + (K_i^T + \Delta K_i^T)(B_j^T + \Delta B_j^T))P)x(t)$$
$$+ x^T(t)P(A_{di} + \Delta A_{di})x(t - \sigma(t)) + x^T(t - \sigma(t))(A_{di}^T + \Delta A_{di}^T)Px(t))$$
$$+ \frac{1}{1-\beta}x^T(t)R_1x(t) - \frac{1-\sigma(t)}{1-\beta}x^T(t-\sigma(t))R_1x(t-\sigma(t)).$$

Applying Lemmas in [5] to the above formulae results in

$$\frac{dV(x(t))}{dt} \leq \Xi_1 + \Xi_2, \tag{10}$$

where

$$\Xi_1 = \sum_{i=1}^{r} h_i^2(\vartheta(t))x^T(t)[PA_i + PB_iK_i + A_i^TP + K_i^TB_i^TP + \varepsilon_{1i}PD_iD_i^T + \varepsilon_{1i}^{-1}$$
$$\times (E_{1i} + E_{2i}K_i)^T(E_{1i} + E_{2i}K_i) + \varepsilon_{2i}PB_i(I - \varepsilon_{3i}^{-1}(E_{2i}H_i)^T(E_{2i}H_i))^{-1}B_i^TP$$
$$+ \varepsilon_{3i} \cdot \varepsilon_{2i}PD_iD_i^TP + \varepsilon_{2i}^{-1}E_{Ki}^TE_{Ki}]x(t) + \varepsilon_{4i}^{-1}x^T(t-\sigma(t))E_{di}^TE_{di}x(t-\sigma(t))$$
$$+ \varepsilon_{4i}x^TPD_iD_i^TPx(t) + x^T(t)PA_{di}x(t-\sigma(t)) + x^T(t-\sigma(t))A_{di}^TPx(t)$$
$$+ \frac{1}{1-\beta}x^T(t)R_1x(t) - \frac{1}{1-\beta}x^T(t-\sigma(t))R_1x(t-\sigma(t))\},$$

$$\Xi_2 = \sum_{i<j}^{r} h_i(\vartheta(t))h_j(\vartheta(t))\{x^T(t)[PA_i + PB_iK_j + A_i^TP + K_j^TBP + \varepsilon_{1ij}^{-1}(E_{1i}$$
$$+ E_{2i}K_j)^T(E_{1i} + E_{2i}K_j) + \varepsilon_{1ij}PD_iD_i^TP + \varepsilon_{2ij}PB_i(I - \varepsilon_{3ij}^{-1}(E_{2i}H_j)^T$$
$$\times (E_{2i}H_j))^{-1}B_i^TP + \varepsilon_{3ij} \cdot \varepsilon_{2ij}PD_iD_i^TP + \varepsilon_{2ij}^{-1}E_{Kj}^TE_{Kj} + PA_j + PB_jK_i$$
$$+ A_j^TP + K_i^TB_j^TP + \varepsilon_{4ij}PD_iD_i^TP + \varepsilon_{4ij}^{-1}(E_{1j} + E_{2j}K_i)^T(E_{1j} + E_{2j}K_i)$$
$$+ \varepsilon_{5ij}PB_j(I - \varepsilon_{6ij}^{-1}(E_{2j}H_i)^T(E_{2j}H_i))^{-1}B_j^TP + \varepsilon_{5ij} \cdot \varepsilon_{6ij}PD_jD_j^TP$$
$$+ \varepsilon_{5ij}^{-1}E_{Ki}^TE_{Ki})x(t) + \varepsilon_{4i}x^T(t)PD_iD_i^TPx(t) + \varepsilon_{4j}PD_jD_j^TPx(t)$$
$$+ \varepsilon_{4i}^{-1}x^T(t-\sigma(t))E_{di}^TE_{di}x(t-\sigma(t)) + \varepsilon_{4j}^{-1}x^T(t-\sigma(t))E_{dj}^TE_{dj}x(t-\sigma(t))$$
$$+ x^T(t)PA_{di}x(t-\sigma(t))E_{dj}^TE_{dj}x(t-d(t)) + x^T(t-\sigma(t))A_{di}^TPx(t) + x^T(t$$
$$- \sigma(t))A_{dj}^TPx(t) + \frac{2}{1-\beta}x^T(t)R_1x(t) - \frac{2}{1-\beta}x^T(t-\sigma(t))R_1x(t-\sigma(t)).$$

From the properties of quadratic form, the above formulae will lead to

$$\frac{dV(x(t))}{dt} = \sum_{i=1}^{r} h_i^2(\vartheta(t)) \begin{bmatrix} x(t) \\ x(t-d(t)) \end{bmatrix}^T \begin{bmatrix} \Pi_1 & PA_{di} \\ A_{di}^T P & \Lambda_1 \end{bmatrix} \begin{bmatrix} x(t) \\ x(t-d(t)) \end{bmatrix}$$

$$+ \sum_{i<j}^{r} h_i(\vartheta(t))h_j(\vartheta(t)) \begin{bmatrix} x^T(t) & x^T(t-d(t)) \end{bmatrix}$$

$$\times \begin{bmatrix} \Pi_2 & PA_{di} + PA_{dj} \\ A_{di}^T P + A_{dj}^T P & \Lambda_2 \end{bmatrix} \begin{bmatrix} x(t) \\ x(t-d(t)) \end{bmatrix}.$$

So far, if inequalities (8) hold, there exists $dV(x(t))/dt < 0$, and the closed-loop control system (7) will asymptotically stable. This completes the proof.

Next, non-fragile robust H_∞ fuzzy controller is presented for the system (7) with external disturbances based on Theorem 1.

Theorem 2. *Consider uncertain nonlinear descriptor system with time-varying delays (7). (5) is non-fragile robust H_∞ fuzzy controller for the system (7), if there exist matrices M_i, symmetric positive definite matrix N, U such that*

$$NE^T = EN \geq 0, \begin{bmatrix} \Omega_{11} & \Omega_{12} & 0 \\ * & \Omega_{22} & \Omega_{23} \\ * & * & -\varepsilon_{4i} I \end{bmatrix} < 0, \begin{bmatrix} \Upsilon_{11} & \Upsilon_{12} & 0 \\ * & \Upsilon_{22} & \Upsilon_{23} \\ * & * & \Upsilon_{33} \end{bmatrix} < 0, \quad (11)$$

hold, where

$$\Omega_{11} = A_i N + B_i M_i + N A_i^T + M_i^T B_i^T + \frac{U}{1-\beta} + (\varepsilon_{1i} + \varepsilon_{3i} \cdot \varepsilon_{2i} + \varepsilon_{4i}) D_i D_i^T,$$

$$\Omega_{12} = \begin{bmatrix} A_{di} N & B_{2i} & N E_{1i}^T + M_i^T E_{2i}^T & N E_{Ki}^T & N C^T \end{bmatrix},$$

$$\Omega_{22} = -\mathrm{diag}\{ \tfrac{U}{1-\beta}, \ \gamma^2 I, \ \varepsilon_{1i} I, \ \varepsilon_{2i} I, \ I \},$$

$$\Omega_{23}^T = \begin{bmatrix} E_{di} N & 0 & 0 & 0 & 0 \end{bmatrix};$$

$$\Upsilon_{11} = A_i N + B_i M_j + N A_i^T + M_j^T B_i^T + A_j N + B_j M_i + N A_j^T + M_i^T B_j^T$$
$$+ \frac{2U}{1-\beta} + (\varepsilon_{1ij} + \varepsilon_{2ij} \cdot \varepsilon_{3ij} + \varepsilon_{5ij} \cdot \varepsilon_{6ij} + \varepsilon_{4ij} + \varepsilon_{4i}) D_i D_i^T + \varepsilon_{4j} D_j D_j^T,$$

$$\Upsilon_{12} = \big[\ (A_{di} + A_{dj}) N \quad B_{2i} + B_{2j} \quad N E_{1i}^T + M_j E_{2i} \quad N E_{1j}^T + M_i^T E_{2j}^T$$
$$N E_{Kj}^T \quad N E_{Ki}^T \quad NE \ \big],$$

$$\Upsilon_{22} = -\mathrm{diag}\left\{ \frac{2U}{1-\beta}, 2\gamma^2 I, \varepsilon_{1ij} I, \varepsilon_{4ij} I, \varepsilon_{2ij} I, \varepsilon_{5ij} I, \frac{1}{2} I \right\},$$

$$\Upsilon_{23}^T = \begin{bmatrix} E_{di} N & 0 & 0 & 0 & 0 & 0 & 0 \\ E_{dj} N & 0 & 0 & 0 & 0 & 0 & 0 \end{bmatrix}, \Upsilon_{33} = -\mathrm{diag}\left\{ \varepsilon_{4i} I, \varepsilon_{4j} I \right\}.$$

Proof. First, let

$$\Gamma = (A_i + \Delta A_i)x(t) + (A_{di} + \Delta A_{di})x(t - d(t)) + (B_i + \Delta B_i)(K_i + \Delta K_i)x(t),$$

then we have

$$\begin{aligned} J &= \int_0^\infty \{z^T(t)z(t) - \gamma^2\omega^T(t)\omega(t)\}dt \\ &\leq \int_0^\infty \{z^T(t)z(t) - \gamma^2\omega^T(t)\omega(t) + \frac{dV(x(t))}{dt}\}dt \\ &= \int_0^\infty \{\sum_{i<j}^r h_i(\vartheta(t))h_j(\vartheta(t))\xi^T(t)\Phi_2\xi(t) + \sum_{i=1}^r h_i^2(\vartheta(t))\xi^T(t)\Phi_1\xi(t)\}dt, \end{aligned}$$

where $\xi(t) = \begin{bmatrix} x^T(t) & x^T(t-\sigma(t)) & \omega^T(t) \end{bmatrix}^T$, $\Pi_1 = \tilde{\Pi}_1 + C^TC$ and $\Pi_2 = \tilde{\Pi}_2 + 2C^TC$.

If there exist $\Phi_1 < 0$ and $\Phi_2 < 0$, then $J \leq 0$, which implies that $\|z(t)\|_2 \leq \gamma\|\omega(t)\|_2$, for any $\omega(t) \in L_2[0, \infty)$. The closed-loop system (7) is asymptotically stable with disturbance attenuation γ according to definition 2 in section 2. Then, to make the make the results solvable by convex optimization method, we multiply the resulting inequalities $\Phi_1 < 0$ and $\Phi_2 < 0$ with $\Theta = \mathrm{diag}(P^{-1}, P^{-1}, I)$ both left and right side, respectively. Introduce new variables $N = P^{-1}$, $M_i = K_iP^{-1}$ and $U = NR_1N$. Then, we obtain

$$\begin{bmatrix} \tilde{\chi}_{ii} & * & * \\ NA_{di}^T & \hat{\Lambda}_1 & * \\ B_{2i}^T & 0 & -\gamma^2 I \end{bmatrix} < 0, \begin{bmatrix} \tilde{\chi}_{ij} & * & * \\ N(A_{di}^T + A_{dj}^T) & \hat{\Lambda}_2 & * \\ B_{2i}^T + B_{2j}^T & 0 & -2\gamma^2 I \end{bmatrix} < 0, \tag{12}$$

where

$$\begin{aligned} \tilde{\chi}_{ii} &= A_iN + B_iM_i + NA_i^T + M_i^TB_i^T + \frac{U}{1-\beta} + (\varepsilon_{1i} + \varepsilon_{3i}\cdot\varepsilon_{2i} \\ &\quad + \varepsilon_{4i})D_iD_i^T + \varepsilon_{2i}B_i(I - \varepsilon_{3i}(E_{2i}H_i)^T(E_{2i}H_i))^{-1}B_i^T + \varepsilon_{1i}^{-1}(E_{1i}N \\ &\quad + E_{2i}M_i)^T(E_{1i}N + E_{2i}M_i) + \varepsilon_{2i}^{-1}NE_{Ki}^TE_{Ki}N + NC^TCN, \\ \tilde{\chi}_{ij} &= A_iN + B_iM_j + NA_i^T + M_j^TB_i^T + A_jN + B_jM_i + NA_j^T \\ &\quad + M_i^TB_j^T + \frac{2U}{1+\beta} + (\varepsilon_{1ij} + \varepsilon_{2ij}\cdot\varepsilon_{3ij} + \varepsilon_{5ij}\cdot\varepsilon_{6ij} + \varepsilon_{4ij} \\ &\quad + \varepsilon_{4i})D_iD_i^T + \varepsilon_{4j}D_jD_j^T + \varepsilon_{2ij}B_i(I - \varepsilon_{3ij}^{-1}(E_{2i}H_j)^T(E_{2i}H_j))^{-1}B_i^T \\ &\quad + \varepsilon_{5ij}B_j(I - \varepsilon_{6ij}^{-1}(E_{2j}H_i)^T(E_{2j}H_i))^{-1}B_j^T + \varepsilon_{1ij}^{-1}(E_{1i}N + E_{2i}M_j)^T \\ &\quad \times (E_{1i}N + E_{2i}M_j) + \varepsilon_{4ij}^{-1}(\varepsilon_{1j}N + \varepsilon_{2j}M_i)^T(\varepsilon_{1j}N + \varepsilon_{2j}M_i) \\ &\quad + N(\varepsilon_{2ij}^{-1}E_{Ki}^TE_{Kj} + \varepsilon_{5ij}^{-1}E_{Ki}^TE_{Ki} + 2C^TC)N, \\ \hat{\Lambda}_1 &= \varepsilon_{4i}^{-1}NE_{di}^TE_{di}N - \frac{1}{1-\beta}NR_1N, \end{aligned}$$

$$\hat{A}_2 = \varepsilon_{4i}^{-1} N E_{di}^T E_{di} N + \varepsilon_{4j}^{-1} N E_{dj}^T E_{dj} N - \frac{2}{1-\beta} N R_1 N.$$

Then, multiply the resulting inequalities (12) with $\Theta = \mathrm{diag}(P^{-1}, P^{-1}, I)$ both left and right side, respectively. Introduce new variables $N = P^{-1}$, $M_i = K_i P^{-1}$ and $U = NR_1N$. However, the conditions are not jointly convex in M_is and N in Theorem 1. Therefore, Schur complement is applied to the obtained matrix inequalities. Then, the LMIs in 11 can be obtained. This completes the proof.

4 Numerical Example

To demonstrate the use of our method, we consider a nonlinear descriptor system with time-varying delays approximated by using the following IF-THEN fuzzy rules:

IF $x_1(t)$ is P, THEN
$E\dot{x}(t) = (A_1 + \Delta A_1)x(t) + (A_{d1} + \Delta A_{d1})x(t - \sigma(t)) + (B_1 + \Delta B_1)u(t) + B_{11}\omega(t);$
IF $x_1(t)$ is N, THEN
$E\dot{x}(t) = (A_2 + \Delta A_2)x(t) + (A_{d2} + \Delta A_{d2})x(t - \sigma(t)) + (B_2 + \Delta B_2)u(t) + B_{11}\omega(t);$

where the membership functions of 'P', 'N' are given as follows

$$M_1(x_1(t)) = 1 - \frac{1}{1 + \exp(-2x_1)}, \quad M_1(x_1(t)) = 1 - M_1(x_1(t)) \tag{13}$$

The uncertainties ΔA_i, ΔA_{di} and ΔB_i are assumed to have the form of (2). Then, the relevant matrices in the T-S fuzzy model are given as follows

$$E = \begin{bmatrix} 1 & 0 \\ 0 & 0 \end{bmatrix}, A_1 = \begin{bmatrix} -1 & 0.4 \\ 0 & -0.5 \end{bmatrix}, A_{d1} = \begin{bmatrix} 0.3 & -0.4 \\ 0 & 0 \end{bmatrix}, B_1 = \begin{bmatrix} 0 \\ 0.1 \end{bmatrix},$$
$$B_{11} = \begin{bmatrix} 0 \\ 1 \end{bmatrix}, A_2 = \begin{bmatrix} -0.5 & 0 \\ 0.5 & -1 \end{bmatrix}, A_{d2} = \begin{bmatrix} 0.4 & 0 \\ 0.4 & 0.3 \end{bmatrix}, B_2 = \begin{bmatrix} 0 \\ 0.5 \end{bmatrix},$$
$$B_{11} = \begin{bmatrix} 0 \\ 1 \end{bmatrix}, D_1 = \begin{bmatrix} 0.1 \\ 0.2 \end{bmatrix}, D_2 = \begin{bmatrix} 0.1 \\ 0.5 \end{bmatrix}, E_{11} = E_{12} = \begin{bmatrix} 1 & 0 \end{bmatrix},$$
$$E_{d1} = E_{d2} = \begin{bmatrix} 0.1 & 0 \end{bmatrix}, E_{21} = 0.3, E_{22} = 0.2, F_1(t) = F_2(t) = \sin(t),$$
$$H_1 = H_2 = 0.5, E_{K1} = E_{K2} = \begin{bmatrix} 0.5 & 0.5 \end{bmatrix}, \quad \phi(t) = \begin{bmatrix} e^{t+1} & 0 \end{bmatrix}^T,$$

and $\sigma(t) = h\sin t$. In Theorem 2, we choose the scalar coefficients $\varepsilon_{ci} = \varepsilon_{dij} = 1$, $1 \leq c \leq 4$, $1 \leq d \leq 6$. By using Matlab LMI Control Toolbox, positive definite matrices P, R_1 and feedback gain K_is can be obtained as follows

$$P = \begin{bmatrix} 5.8860 & 3.0870 \\ 3.0870 & 3.1162 \end{bmatrix}, \quad R_1 = \begin{bmatrix} 1.0412 & 0.7638 \\ 0.7638 & 1.5298 \end{bmatrix},$$
$$K_1 = [\text{-0.47855 -0.76740}], \quad K_2 = [\text{-0.97124 -1.1999}].$$

5 Conclusions

In this paper, non-fragile robust H_∞ fuzzy controller design has been addressed for a class of nonlinear descriptor systems with time-varying delays via fuzzy interpolation of a series of linear systems. The fuzzy controller is reduced to the solution of a set of LMIs, which make the design much more convenient. Furthermore, an example has demonstrated the use of the proposed fuzzy model-based controller.

Acknowledgements

This work was supported in part by National Natural Science Foundation of China under Grant No. 60474014 and the Research Fund for the Doctoral Program of Higher Education under Grant No. 20020151005.

References

1. Keel, L. H., Bhattacharryya, S. P.: Robust, fragile, or optimal? IEEE Trans. Automat. Contr. **42** (1997) 1098–1105
2. An, S., Huang, L., Gu, S., Wang, J.: Robust non-fragile state feedback control of discrete time-delay systems. In Proceedings of International Conference on Control and Automation, Budapest, Hungary (2005) 794–799
3. Yang, G. H., Wang, J. L.: Non-fragile nonlinear H_∞ control via dynamic output feedback. In Proc. Am. Control Conf., Denver, USA (2003) 2969–2972
4. Campbell, S. L.: Singular systems of differential equations. Pitman, Marshfield, Mass. (1980)
5. Li, X., De Souza, C. E.: Criteria for robust stability and stabilization of uncertain linear systems with state-delay. Automatica **33** (1997) 1657–1662
6. Chen, S. J., Lin, J. L.: Robust stability of discrete time-delay uncertain singular systems. IEE Proc.-Control Theory Appl. **150** (2003) 325-330
7. Ma, S.: Robust stabilization for a class of uncertain discrete-time singular systems with time-delays. In Proc. World Congr. Intelligent Control Autom., Hangzhou, P. R. China (2004) 970–974
8. Ying, H.: Sufficient conditions on uniform approximation of multivariate functions by general Tagaki-Sugeno fuzzy systems with linear rule consequence. IEEE Trans. Syst. Man, Cybern. **28** (1998) 515-521
9. Wang, H. O., Li, J., Tanaka, K.: T-S fuzzy model with linear rule consequence and PDC controller: A universal framework for nonlinear control systems. International Journal of Fuzzy Systems **5** (2003) 106–113
10. Wang, H. O., Tanaka, K., Griffin, M. F.: Parallel distributed compensation of nonlinear systems by Takagi-Sugeno fuzzy model. In Proc. IEEE Int. Conf. Fuzzy Syst., Yokohama, (1995) 531-538
11. Taniguchi, T., Tanaka, K., Yamafuji, K., Wang, H. O.: Fuzzy descriptor systems: Stability analysis and design via LMIs. In Proc. Am. Control Conf., San Deigo (1999) 1827-1831
12. Liu, G. Y., Zhang, Q. L., Yang, L., Zhai, D.: Quadratic stability study for a class of T-S fuzzy descriptor systems. Journal of Northeastern University, **25** (2004) 1131–1133

PWM Fuzzy Controller for Nonlinear Systems

Young Hoon Joo[1], Sung Ho Kim[1], and Kwang Baek Kim[2]

[1] Kunsan National University, Kunsan, Chonbuk, 573-701, Korea
{yhjoo, shkim}@kunsan.ac.kr

[2] Dept. of Computer Engineering, Silla University, Busan 370-701, Korea
gbkim@silla.ac.kr

Abstract. In this paper, we develop an intelligent digitally redesigned PAM fuzzy controller for nonlinear systems. Takagi-Sugeno fuzzy model is used to model the nonlinear systems and a continuous-time fuzzy-model-based controller is designed based on extended parallel-distributed-compensation method. The digital controllers are determined from existing analogue controllers. The proposed method provides an accurate and effective method for digital control of continuous-time nonlinear systems and enables us to efficiently implement a digital controller via pre-determined continuous-time TS fuzzy-model-based controller. We have applied the proposed method to the balancing problem of the inverted pendulum to show the effectiveness and feasibility of the method.

1 Introduction

Fuzzy logic control is one of most useful approaches for control of complex and ill-defined nonlinear systems. The main drawback of fuzzy logic control is the empirical design procedure, which are based on trial-and-error process. Therefore, recent trend in fuzzy logic control is to develop systematic method to design the fuzzy logic controller. The studies on the systematic design of fuzzy logic controller have largely been devoted to two approaches. One is based on soft-computing method [1]. This approach utilizes neural network theory and genetic algorithm, etc.. This method is quite efficient since the most appropriate and optimal fuzzy logic controllers can be designed without the aids of human experts. However, it may suffer difficulties in determining the overall stability of the system. The other is based on the well-established conventional linear system theory [2-3]. One popular method is the TS fuzzy model or dynamic fuzzy- model-based control theory, which combines the fuzzy inference rule with the local linear state model [2-3]. This kind of method has been widely used in the control of nonlinear systems since it is easy to incorporate the mathematical analysis developed in the linear control theory.

At the same time, we have been witnessed rapid development of flexible, low-cost microprocessors in the electronics fields. Therefore, it is desirable to implement the recent advanced controller in digital. There are three digital design approaches for digital control systems. The first approach, called the direct digital design approach, is to convert an analogue plant to a digital plant and then find the digital controller. The

L. Wang et al. (Eds.): FSKD 2006, LNAI 4223, pp. 129 – 138, 2006.

second is, which is named the digital redesign approach, is to find the analogue controller and then carry out the digital redesign. The other is to directly design a digital controller for the analogue plant, which is still under development [5-7]. In general, there exist two types of digital controllers: the pulse-amplitude-modulation (PAM) controller and the pulse-width-modulation (PWM) controller. The PAM controller, which is commonly used in digital control, produces a series of piecewise-constant continuous pulses having variable amplitude and variable or fixed width [5].

In this paper, we develop an intelligent digitally redesigned PAM fuzzy controller for digital control of continuous-time nonlinear systems. We first apply the digital redesign technique to each local linear model. By the proposed method, the conventional digital redesign method developed in linear system field can be then easily applied and extended to the control of nonlinear systems.

2 Fuzzy-Model-Based Controller

Consider a nonlinear dynamic system in the canonical form

$$x^{(n)}(t) = \mathbf{f}(\mathbf{x}) + \mathbf{g}(\mathbf{x})\mathbf{u}(t) \tag{1}$$

where, the scalar $x^{(n)}$ is the output state variable of interest, the vector $\mathbf{u}$ is the system control input, and $\mathbf{x} = \begin{bmatrix} x & \dot{x} & \ldots & x^{(n-1)} \end{bmatrix}^T$ is the state vector. In (1), the nonlinear function $\mathbf{f}(\mathbf{x})$ is a known nonlinear continuous function of $\mathbf{x}$, and the control gain $\mathbf{g}(\mathbf{x})$ is a known nonlinear continuous and locally invertible function of $\mathbf{x}$. This nonlinear system can be approximated by the TS fuzzy model. More specifically, the *i*th rule of the TS fuzzy model in the continuous time case is formulated in the following form:

$$\begin{aligned} &\text{Plant Rule } i\text{: IF } x(\text{t}) \text{ is } F_1^i \text{ and ... and } x^{(n-1)}(t) \text{ is } F_n^i, \\ &\text{THEN } \dot{\mathbf{x}}(t) = \mathbf{A}_i\mathbf{x}_c(t) + \mathbf{B}_i\mathbf{u}_c(t) \quad (i = 1, 2, ..., q) \end{aligned} \tag{2}$$

while the consequent part in the discrete-time case is represented by $\mathbf{x}_d(t+l) = \mathbf{F}_i\mathbf{x}_d(t) + \mathbf{G}_i\mathbf{u}_d(t)$ in (2).

Here, $\mathrm{F}_\mathrm{j}^\mathrm{i}$ $(j = 1, \ldots, n)$ is the fuzzy set, q is the number of rules, $\mathbf{x}_\mathrm{c}(\mathrm{t}) \in \Re^\mathrm{n}$ is the state vector, $\mathbf{u}_\mathrm{c}(\mathrm{t}) \in \Re^\mathrm{m}$ is the input vector, $\mathbf{A}_i \in \Re^{n\times n}$ and $\mathbf{B}_i \in \Re^{n\times m}$, $x_1(t)$, ..., $x_n(t)$ are the premise variables (which are the system states) and $(\mathbf{A}_i, \mathbf{B}_i)$ in the continuous-time case and $(\mathbf{F}_i, \mathbf{G}_i)$ in discrete-time case denote the *i*th local model of the fuzzy system, respectively. Subscript 'c' and 'd' represent continuous-time and discrete-time case, respectively. Using the center of gravity defuzzification, product inference, and single fuzzifier, the final output of the overall fuzzy system is given by

$$\dot{\mathbf{x}}_\mathrm{c}(\mathrm{t}) = \sum_{\mathrm{i=1}}^{\mathrm{q}} \mu_\mathrm{i}(\mathbf{x}_\mathrm{c}(\mathrm{t}))(\mathbf{A}_\mathrm{i}\mathbf{x}_\mathrm{c}(\mathrm{t}) + \mathbf{B}_\mathrm{i}\mathbf{u}_\mathrm{c}(\mathrm{t})) \tag{3}$$

where

$$w_i(\mathbf{x}(t)) - \prod_{j=1}^{n} F_j^i(x^{(j-1)}(t)), \quad \mu_i(\mathbf{x}(t)) = \frac{w_i(\mathbf{x}(t))}{\sum_{i=1}^{q} w_i(\mathbf{x}(t))}$$

Using the same premise as (2), the EPDC fuzzy controller in continuous-time model has the following rule structure:

$$\begin{aligned} &\text{Controller Rule } i\text{: IF } x(t) \text{ is } F_1^i \text{ and ... and } x^{(n-1)}(t) \text{ is } F_n^i, \\ &\text{THEN } \mathrm{u(t)} = -\mathbf{K}_i\mathbf{x}(\mathrm{t}) + \mathbf{E}_i(\mathrm{t})\mathbf{r}(\mathrm{t}) \quad (i = 1, 2, ..., q) \end{aligned} \tag{4}$$

where $\mathbf{K}_i = [k_1^i, \cdots, k_n^i]$ and $\mathbf{E}_i = [e_1^i \ \cdots \ e_n^i]$ are the feedback and feedforward gain vectors in ith subspace, respectively. $\mathbf{r}$(t) is the reference input.
The overall closed-loop fuzzy system becomes

$$\dot{\mathbf{x}} = \sum_{i=1}^{q} \sum_{j=1}^{q} \mu_i(\mathbf{x})\mu_j(\mathbf{x})((\mathbf{A}_i - \mathbf{B}_i\mathbf{K}_j)\mathbf{x}(t) + \mathbf{B}_i\mathbf{E}_j\mathbf{r}(t)) \tag{5}$$

3 PAM Fuzzy Controller

In general, when applying the dual-rate sampling method to a dynamic system, the fast sampling rate is used for the system parameter identification without losing the information of the dynamic system, and the slow sampling rate is used for the computation of advanced controllers in real time.

For the implementation of a digital control law, it needs to find a digital control law from the obtained optimal analogue control law. This can be carried out using the digital redesign technique, which is called the generalized digital redesign. This technique matches the continuous-time closed-loop state $\mathbf{x}_c(t)$ with the discrete-time closed-loop state $\mathbf{x}_d(t)$ at $t = kT_s$, where T_s is the slow sampling time. We can expect that this method consider the system responses only at the slow-rate sampling time, $t = k_sT_s$. If the slow sampling time T_s is not sufficiently small, then the control law cannot capture the system's behavior during the slow sampling time T_s. Thus, it needs to consider the system's behavior during the slow-rate sampling and reflect it to the control law. We adopt a new digital redesign method that considers the inter-sampling behaviors [6].

Subscript i representing ith subspace is omitted to avoid the complexity. Consider a controllable and observable analogue plant represented by

$$\dot{\mathbf{x}}_c(t) = \mathbf{A}\mathbf{x}_c(t) + \mathbf{B}\mathbf{u}_c(t), \quad \mathbf{x}_c(0) = \mathbf{x}_{c0} \tag{6}$$

$$\mathbf{y}_c(t) = \mathbf{C}\mathbf{x}_c(t) + \mathbf{D}\mathbf{u}_c(t) \tag{7}$$

The optimal state-feedback control law, which minimizes the performance index

$$J = \int_0^{\infty} \{[\mathbf{x}_c(t) - \mathbf{r}(t)]^T \mathbf{Q}[\mathbf{x}_c(t) - \mathbf{r}(t)] + \mathbf{u}_c(t)\mathbf{R}\mathbf{u}_c(t)\}dt \tag{8}$$

with $\mathbf{Q} \geq 0, \mathbf{R} > 0$ is

$$\mathbf{u}_c(t) = -\mathbf{K}_c\mathbf{x}_c(t) + \mathbf{E}_c\mathbf{r}(t) \tag{9}$$

where $\mathbf{r}(t)$ is a reference input vector and

$$\mathbf{K}_c = \mathbf{R}^{-1}\mathbf{B}^T\mathbf{P} \tag{10}$$

$$\mathbf{E}_c = -\mathbf{R}^{-1}\mathbf{B}^T(\mathbf{A} - \mathbf{B}\mathbf{K}_c)^{-T}\mathbf{Q} \tag{11}$$

where $\mathbf{P}$ is the solution to

$$\mathbf{A}^T\mathbf{P} + \mathbf{P}\mathbf{A} - \mathbf{P}\mathbf{B}\mathbf{R}^{-1}\mathbf{B}^T\mathbf{P} + \mathbf{Q} = 0 \tag{12}$$

The analogue control law $\mathbf{u}_c(t)$ in (7) can be approximated as

$$\mathbf{u}_c(t) \cong W_{k_f}\Phi_{k_f}(t) = \mathbf{u}_{dk_f}(k_f T_f) \tag{13}$$

for $\mathbf{r}(t) = \mathbf{r}(k_f T_f)$ with $k_f T_f \leq t < k_f T_f + T_f$, where

$$\begin{aligned} W_{k_f} &= \frac{1}{T_f}\int_{k_f T_f}^{K_f T_f + T_f} u_c(t)dt \\ &= -\frac{\mathbf{K}_c}{T_f}\int_{k_f T_f}^{K_f T_f + T_f} \mathbf{x}_c(t)dt + \mathbf{E}_c\mathbf{r}(k_f T_f) \end{aligned} \tag{14}$$

where $\Phi_{k_f}(t)$ is orthonormal series and T_f is the fast rate sampling time.

Consider and the closed-loop of the analogue system in (5) represented with piecewise constant input by

$$\dot{\mathbf{x}}_c(t) = \mathbf{A}_c\mathbf{x}_c(t) + \mathbf{E}_c\mathbf{r}(k_f T_f) \tag{15}$$

where $\mathbf{A}_c = \mathbf{A} - \mathbf{B}\mathbf{K}_c$.

The corresponding digital system is

$$\mathbf{x}_c(k_f T_f + T_f) = \mathbf{G}_{cN}\mathbf{x}_c(k_f T_f) + \mathbf{H}_{cN}\mathbf{E}_c\mathbf{r}(k_f T_f) \tag{16}$$

where $\mathbf{G}_{cN} = e^{\mathbf{A}_c\mathbf{T}_f}$, $\mathbf{H}_{cN} = [\mathbf{G}_{cN} - \mathbf{I}_n]\mathbf{A}_c^{-1}\mathbf{B}$.

Consider the digital system with a piecewise constant input $\mathbf{u}_{dk}(k_f T_f)$ as

$$\mathbf{x}_d(K_f T_f + T_f) = \mathbf{G}_N\mathbf{x}_d(k_f T_f) + \mathbf{H}_N\mathbf{u}_{dk}(k_f T_f) \tag{17}$$

where $\mathbf{G}_N = e^{\mathbf{A}T_f}$, $\mathbf{H}_N = [\mathbf{G}_N - \mathbf{I}_n]\mathbf{A}^{-1}\mathbf{B}$. Let the desired digitally redesigned control law be

$$\mathbf{u}_{\mathrm{dk}}(\mathrm{k_f T_f}) = -\mathbf{K}_{\mathrm{dk}}\mathbf{x}_{\mathrm{d}}(\mathrm{k_f T_f}) + \mathbf{E}_{\mathrm{dk}}\mathbf{r}(\mathrm{k_f T_f}) \tag{18}$$

Its digital closed-loop system becomes

$$\mathbf{x}_d(k_f T_f + T_f) = \hat{\mathbf{G}}_{cN}\mathbf{x}_d(k_f T_f) + \hat{\mathbf{H}}_{cN}\mathbf{r}(k_f T_f) \tag{19}$$

where $\hat{\mathbf{G}}_{cN} = \mathbf{G}_N - \mathbf{H}_N\mathbf{K}_{dk}$, $\hat{\mathbf{H}}_{cN} = \mathbf{H}_N\mathbf{E}_{dk}$.

The integration of the analogue closed-loop system in (13) is represented by

$$\begin{aligned} &\int_{k_f T_f}^{k_f T_f + T_f} \mathbf{x}_c(t)dt \\ &= \mathbf{A}_c^{-1}\left[\mathbf{x}_c(k_f T_f + T_f) - \mathbf{x}_c(k_f T_f) - T_f\mathbf{B}\mathbf{E}_c\mathbf{r}(k_f T_f)\right] \end{aligned} \tag{20}$$

Substituting (16) into (14), (20), and its result, $\mathbf{u}_c(\mathrm{t})$, into (8). Matching the resulting system's state with (17), we have

$$\mathbf{G}_N - \mathbf{H}_N\mathbf{K}_{dk} = \mathbf{G}_N - \mathbf{H}_N\mathbf{K}_c(\mathbf{A}_c T_f)^{-1}(\mathbf{G}_{cN} - \mathbf{I}_n) \tag{21}$$

$$\mathbf{H}_N\mathbf{E}_{dk} = \mathbf{H}_N\left[\mathbf{I}_m + (\mathbf{K}_c - \mathbf{K}_{dk})\mathbf{A}_c^{-1}\mathbf{B}\right]\mathbf{E}_c \tag{22}$$

The solutions to (21) and (22) become

$$\begin{aligned} \mathbf{K}_{dk} &= \mathbf{K}_c(\mathbf{A}_c\mathbf{T}_f)^{-1}(\mathbf{G}_{cN} - \mathbf{I}_n) \\ \mathbf{E}_{dk} &= \left[\mathbf{I}_m + (\mathbf{K}_c - \mathbf{K}_{dk})\mathbf{A}_c^{-1}\mathbf{B}\right]\mathbf{E}_c \end{aligned} \tag{23}$$

To improve the performance during the slow rate sampling time, it is desired to find the digitally redesigned control law that matches both the digital closed-loop state with the analogue closed-loop state, and has the slow sampling rate time. The slowly sampled digital system with the digitally redesigned fast rate sampling time control law $\overline{\mathbf{u}}_d^{(N)}$ can be described as

$$\begin{aligned} \mathbf{x}_d(k_s T_s + T_s) &= \mathbf{G}_N^N\mathbf{x}_d(k_s T_s) + \overline{\mathbf{H}}_N^{(N)}\overline{\mathbf{u}}_d^{(N)}(k_s T_s) \\ &= \mathbf{G}_N^N\mathbf{x}_d(k_s T_s) + \sum_{i=1}^{N}\overline{\mathbf{H}}_i\mathbf{u}_{di}(k_s T_s + (i-1)T_f) \end{aligned} \tag{24}$$

where,

$$\mathbf{G}_N^N = (\mathbf{G}_N)^N \tag{25}$$

$$\begin{aligned}\overline{\mathbf{H}}_N^{(N)} &= \begin{bmatrix}\overline{\mathbf{H}}_1 & \overline{\mathbf{H}}_2 & \cdots & \overline{\mathbf{H}}_{N-1} & \overline{\mathbf{H}}\end{bmatrix} \\ &= \begin{bmatrix}\mathbf{G}_N^{N-1}\mathbf{H}_N & \mathbf{G}_N^{N-2}\mathbf{H}_N & \cdots & \mathbf{G}_N\mathbf{H}_N & \mathbf{H}_N\end{bmatrix}\end{aligned} \tag{26}$$

$$\begin{aligned}\overline{\mathbf{u}}_d^{(N)}(k_sT_s) &= \begin{bmatrix}\overline{u}_{d1}^T(k_sT_s) & \overline{u}_{d2}^T(k_sT_s) & \cdots \end{bmatrix.\\ &= \qquad \left. \cdots \quad \overline{u}_{dN}^T(k_sT_s)\right]^T \\ &= \begin{bmatrix}\overline{u}_{d1}^T(k_sT_s) & \overline{u}_{d2}^T(k_sT_s+T_f) & \cdots \end{bmatrix. \\ &\qquad \left. \cdots \quad \overline{u}_{dN}^T(k_sT_s+(N-1)T_f)\right]^T \\ &= -\overline{\mathbf{K}}_d^{(N)}\mathbf{x}_d(k_sT_s) + \overline{\mathbf{E}}_d^{(N)}\mathbf{r}(k_sT_s)\end{aligned} \tag{27}$$

Considering the digital system in (19) with the reference input $\mathbf{r}(t)=\mathbf{r}(k_sT_s)$ for $k_sT_s \le t < k_sT_s + T_s$, the closed-loop state $\mathbf{x}_d$ can be represented at $t = k_fT_f + (i-1)T_f$ and $k_fT_f = k_sT_s$ by

$$\begin{aligned}\mathbf{x}_d(k_sT_s+(i-1)T_f) &= \mathbf{G}_{cN}^{i-1}\mathbf{x}_d(k_sT_s) \\ &\quad + \sum_{j=1}^{i-1} \hat{\mathbf{G}}_{cN}^{i-1-j}\hat{\mathbf{H}}_{cN}\mathbf{r}(k_sT_s)\end{aligned} \tag{28}$$

and by definition of (27), the digital control law with the dual rate sampling time can be written as

$$\begin{aligned}u_{di}(k_sT_s+(i-1)T_f) &= -\mathbf{K}_{dk}\mathbf{x}_d(k_sT_s+(i-1)T_f) \\ &\quad + \mathbf{E}_{dk}\mathbf{r}(k_sT_s)\end{aligned} \tag{29}$$

Substituting (28) into (29), its result is

$$\mathbf{x}_\mathrm{d}(\mathrm{K_f T_f + T_f}) = \mathbf{G}_\mathrm{N}\mathbf{x}_\mathrm{d}(\mathrm{k_f T_f}) + \mathbf{H}_\mathrm{N}\mathbf{u}_\mathrm{dk}(\mathrm{k_f T_f}) \tag{30}$$

$$\begin{aligned}\overline{\mathbf{E}}_d^{(N)} &= \begin{bmatrix}\overline{\mathbf{E}}_{d1}^T & \overline{\mathbf{E}}_{d2}^T & \cdots & \overline{\mathbf{E}}_{dN}^T\end{bmatrix} \\ &= \left[\mathbf{E}_{dk}^T \quad (\mathbf{E}_{dk} - \mathbf{K}_{dk}\hat{\mathbf{H}}_{cN})^T \quad \cdots \right. \\ &\qquad \left. \cdots \quad \left(\mathbf{E}_{dk} - \sum_{j=1}^{N-1}\mathbf{K}_{dk}\hat{\mathbf{G}}_{cN}^{N-1-j}\hat{\mathbf{H}}_{cN}\right)^T\right]^T\end{aligned} \tag{31}$$

4 Intelligent Digitally Redesigned Controller

The overall closed-loop system is obtained from the feedback interconnection of the nonlinear system (1) and the controller (5), resulting in the following equation:

$$\mathbf{x}^{(n)}(t) = \mathbf{F}_o(\mathbf{x}(t)) + \mathbf{G}_o(\mathbf{x}(t))\mathbf{r}(t) \tag{32}$$

where $\mathbf{F}_o(\mathbf{x}(t)) - \mathbf{f}(\mathbf{x}(t)) - \mathbf{g}(\mathbf{x}(t))\mathbf{K}(\mu)\mathbf{x}(t)$, $\mathbf{G}_o(\mathbf{x}(t)) = \mathbf{g}(\mathbf{x}(t))\mathbf{E}(\mu)$.

The following theorem is our stability result for the equilibrium state:

Theorem 1. Consider the following nonlinear system:

$$\dot{\mathbf{x}}(t) = \mathbf{A}\mathbf{x}(t) - \mathbf{K}(\mu)\mathbf{x}(t) + \mathbf{G}_o(\mathbf{x}(t), \mathbf{E}(\mu))\mathbf{r}(t) \tag{33}$$

where $\mathbf{r}(t)$ is a given reference signal, and $\mathbf{K}(\mu)$ and $\mathbf{E}(\mu)$ are control gain matrices with parameters $\mu \in \Re^q$, which can also be functions of $\mathbf{x}(t)$ and t in general. If $\mathbf{K}(\mu)$ and $\mathbf{E}(\mu)$ are designed such that

(i) the matrix $[\mathbf{A} - \mathbf{K}(\mu)]$ is stable uniformly for all $\mathbf{x} \in \Re^n$,

(ii) $\int_0^\infty \|\mathbf{G}_o(\mathbf{x}(t), \mathbf{E}(\mu)\mathbf{r}(t)\| \, dt < \infty$, or

(ii)' $\int_0^\infty \|\mathbf{G}_o(\mathbf{x}(t), \mathbf{E}(\mu)\mathbf{r}(t)\| \, dt \le \|\mathbf{C}(\mathbf{x}(t,t)\| \, \|\mathbf{x}(t)\|$ with $\int_0^\infty \|\mathbf{C}(\mathbf{x}(t), t)\| \, dt < \infty$,

where $\|\cdot\|$ is the Euclidean norm, then the controlled system (5) is stable in the sense of Lyapunov.

Proof. See [2].

Corollary 1. In the nonlinear control system (1) with a fuzzy controller (5), namely,

$$\mathbf{x}^{(n)}(t) = \mathbf{F}_o(\mathbf{x}(t), \mathrm{K}(\mu)) + \mathbf{G}_o(\mathbf{x}(t), \mathrm{E}(\mu))\,\mathbf{r}(t) \tag{35}$$

where,

$$\mathbf{F}_o(\mathbf{x}(t), \mathbf{K}(\mu)) = \mathbf{f}(\mathbf{x}(t)) - \mathbf{g}(\mathbf{x}(t))\,\mathbf{K}(\mu)\,\mathbf{x}(t), \quad \mathbf{G}_o(\mathbf{x}(t), \mathbf{E}(\mu)) = \mathbf{g}(\mathbf{x}(t))\,\mathbf{E}(\mu),$$

if the TS fuzzy model

$$\mathbf{x}^{(n)}(t) = [\mathbf{A} - \mathbf{B}\mathbf{K}(\mu)]\,\mathbf{x}(t) \tag{36}$$

is designed such that it can uniformly approximate the given uncontrolled system (35), namely, $\|\mathbf{F}_o(\mathbf{x}(t), \mathbf{K}(\mu)) - [\mathbf{A} - \mathbf{B}\mathbf{K}(\mu)]\|$ can be arbitrary small, and if the control gains $\mathbf{K}(\mu)$ and $\mathbf{E}(\mu)$ are designed such that the two conditions (i) and (ii) (or (ii)') of Theorem 1 are satisfied, then the fuzzy control system (35) is stable in the sense of Lyapunov.

Proof. See [2].

5 Simulation

To illustrate the proposed method, let us consider the problem of balancing of an inverted pendulum on a cart. The dynamic equation is in [9]

$$\dot{x}_1 = x_2$$
$$\dot{x}_2 = \frac{g\sin(x_1) - amlx_2^{\ 2}\sin(2x_1)/2 - a\cos(x_1)u}{4l/3 - aml\cos^2(x_1)}$$

where x_1 is the angle of the pendulum in radians from vertical axis, x_2 is the angular velocity, $g = 9.8m/s^2$ is the acceleration by the gravity with the mass of the pendulum $m = 2.0kg$ and the mass of the cart $M = 8.0kg$, $a = m + M$, $2l = 1.0m$ is the length of pendulum, and u is the force applied to the cart.
An approximated TS fuzzy model is as follows [9]:

Rule 1: IF x_1 is about 0 THEN $\dot{x} = A_1x + B_1u$

Rule 2: IF x_1 is about $\pm\pi$ THEN $\dot{x} = A_2x + B_2u$.

where

$$\mathbf{A}_1 = \begin{bmatrix} 0 & 1 \\ \dfrac{g}{4l/3 - aml} & 0 \end{bmatrix}, \qquad \mathbf{B}_1 = \begin{bmatrix} 0 \\ -\dfrac{a}{4l/3 - aml} \end{bmatrix}$$

$$\mathbf{A}_2 = \begin{bmatrix} 0 & 1 \\ \dfrac{2g}{\pi(4l/3 - aml\beta^2)} & 0 \end{bmatrix}, \quad \mathbf{B}_2 = \begin{bmatrix} 0 \\ -\dfrac{a\beta}{4l/3 - aml\beta^2} \end{bmatrix}$$

and $\beta = \cos(88^\circ)$. The membership function for Rules is shown in Fig. 1.

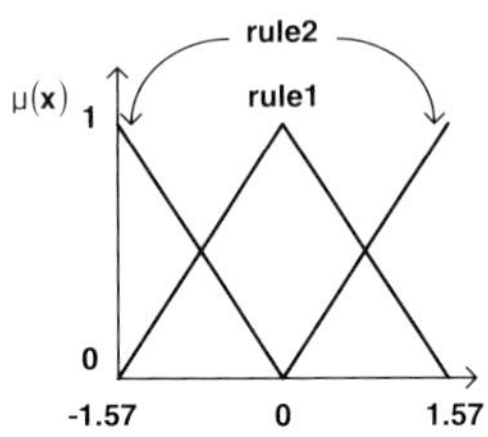

Fig. 1. Membership Functions

We choose the slow-rate sampling period T_s as 0.02 and the fast-rate sampling period T_f as 0.01, thus $N = T_s/T_f$ is 2. The initial conditions are $x_1 = 43^\circ$ (0.7505rad) and $x_2 = 0$. In order to check the stability of the global fuzzy control system, Based on LMI [3], we found the common positive definite matrix $\mathbf{P}_c$ to be

$$\mathbf{P}_c = \begin{bmatrix} 0.4250 & 0.0189 \\ 0.0189 & 0.0084 \end{bmatrix}$$

The other conditions are also satisfied. Therefore, the overall continuous-time fuzzy system is stable in the sense of Lyapunov. Figure 2–4 show the comparisons of the position angle $x_1(t)$, the angular velocity $x_2(t)$, and the control input of this example by the proposed method and the original analogue controller, respectively. As seen in these results, the proposed scheme is successful for digital control of nonlinear system.

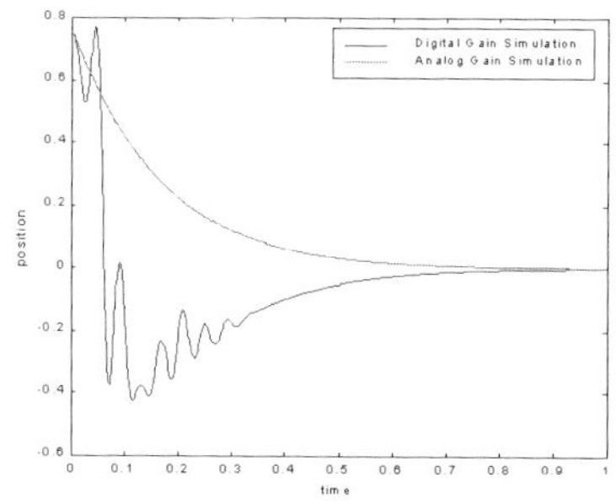

Fig. 2. Position angle $x_1(t)$

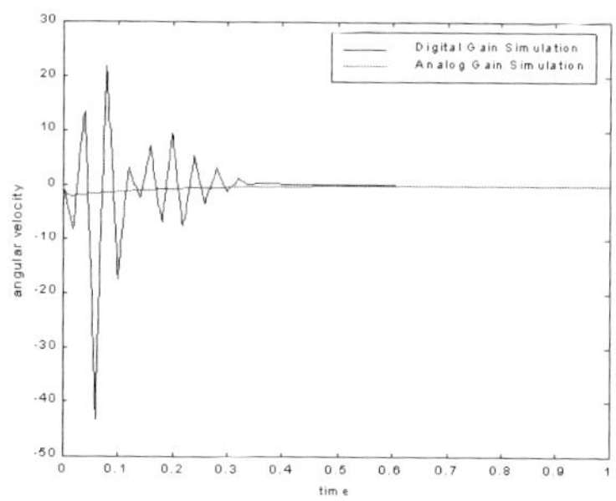

Fig. 3. Angular velocity $x_2(t)$

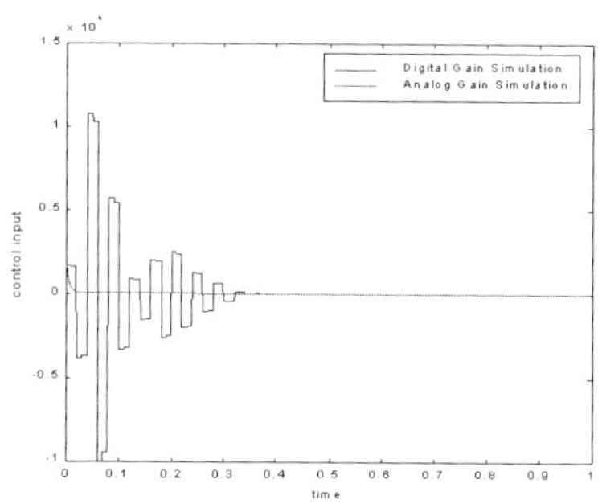

Fig. 4. Control input

6 Conclusions

In this paper, we have proposed the digitally redesigned PAM fuzzy-model-based controller design method for a nonlinear system. We represent the nonlinear system as a TS fuzzy-model-based system, and the EPDC technique is then utilized to design a fuzzy-model-based controller. In analogue control law design, the optimal regional pole assignment technique is adopted and extended with some new stability conditions to construct multiple local linear systems. The PAM digital redesign method is carried out to obtain the digital control law for control of each local analogue system, where a new state matching has been developed. The inverted pendulum balancing simulation has shown that the proposed method is very effective in controlling a nonlinear system.

Acknowledgement. This work has been supported by KESRI(R-2004-B-120), which is funded by MOCIE(Ministry of commerce, industry, and energy).

References

1. Joo, Y. H., Hwang, H. S., Kim, K. B., and Woo, K. B.: Fuzzy system modeling by fuzzy partition and GA hybrid schemes. Fuzzy Sets and Systems, **86** (1997) 279-288,
2. Joo Y. H., Shieh L. S., and Chen G.: Hybrid state-space fuzzy model-based controller with dual-rate sampling for digital control of chaotic systems. IEEE Trans. Fuzzy Syst. - -(1999) 394-408
3. Lee, H. J., Park, J.B., and Joo, Y.H., "Digitalizing a Fuzzy observer-based output-feedback control: intelligent digital redesign approach", IEEE Trans. on Fuzzy Systems, **13** (2005), 701-716
4. Kuo, B. C.: Digital Control Systems. Holt, Rinehart and Winston, N. Y., (1980)
5. Shieh, L. S., Wang, W. M., and Appu Panicker, M. K.: Design of lifted dual-rate PAM and PWM digital controllers for the X-38 Vehicle, IEE Proceedings(D), **148** (2001) 249-256
6. Shieh, L. S., Zhao, X. M., and Sunkel, J.: Hybrid state-space self-tuning control using dual-rate sampling. IEE Proceedings(D). **138** (1991) 50-58.
7. H. J. Lee, Y. H. Joo, W. Chang, J. B. Park, "A new intelligent digital redesign for TS fuzzy systems: global approach", IEEE Trans. on Fuzzy Systems, 12 (2004), 274-284
8. Tanaka, K. and Sugeno, M.: Stability analysis and design of fuzzy control systems. Fuzzy Sets and Systems. **45** (1992) 135-156
9. Wang, H. O., Tanaka, K., and Griffin, M. F.: Parallel distributed compensation of nonlinear systems by Takagi-Sugeno fuzzy model. Proc. of FUZZ-IEEE/IFES'95, (1995) 531-538.

Youla Parameterization and Design of Takagi-Sugeno Fuzzy Control Systems

Wei Xie[1] and Toshio Eisaka[2]

[1] College of Automation Science and Technology, South China University of Technology, Guangzhou, 510641, China
[2] Computer Sciences, Kitami institute of Technology, 165 Koencho, Kitami, Hokkaido, 090-8507, Japan
weixie@hotmail.co.jp, eisaka@cs.kitami-it.ac.jp

Abstract. A method of designing Takagi-Sugeno fuzzy control systems based on the parameterization of quadratically stabilizing controllers is presented. Conception of doubly coprime factorization and Youla parameterization of LTI systems are extended to T-S fuzzy system with respect to quadratic stability. The parameterization of the close-loop systems, which are affine with arbitrary stable Q-parameter, is then described. This description enables the application of the Q-parameter approach to various T-S fuzzy control-systems. Above all, a design scheme of Q to obtain L2-gain performance is clarified.

1 Introduction

Takagi-Sugeno (T-S) fuzzy systems can be formalized from a large class of nonlinear systems [1],[2]. Despite the fact that the global T-S models are nonlinear due to the dependence of the membership functions on the fuzzy variables, they are described as a special case of Polytopic Linear Differential Inclusions (PLDI) [3], in which the coefficients are normalized membership functions. That is, local dynamics in different state-space regions are represented by linear models; and the nonlinear systems are approximated by the overall fuzzy linear models. Most of the existing control techniques for T-S models utilized the parallel distributed compensation (PDC) law [4]. With quadratic Lyapunov functions and the PDC law, a great deal of attention has been focused on analysis and synthesis of these systems [4]-[9]. In particular, in [9], sufficient conditions described by linear matrix inequality (LMI) are provided for the existence of a quadratically stabilizing dynamic compensator or a performance-oriented controller based on the notion of dynamic parallel distributed compensator (DPDC).

On the other hand, it is well known that doubly coprime factorization of linear time invariant (LTI) systems is an essential tool for the analysis and design of control systems. Youla parameterization [10],[11], which constructs the set of output feedback stabilizing controllers, is also described with the coprime factorization of plants. The set of achievable closed-loop transfer matrices is affine and readily described with free Q-parameter. A control system design framework based on this description is referred to as "Q-parameter approach". Various control system designs including

L. Wang et al. (Eds.): FSKD 2006, LNAI 4223, pp. 139 – 148, 2006.

multi-objective problems resolves themselves to specify this Q-parameter. However, to the authors' knowledge, the general Youla parameterization control scheme including command tracking issue of T-S fuzzy systems has not yet been discussed explicitly. Since T-S fuzzy systems cannot be used to treat transfer functions or eigenvalues of state matrices, the above-mentioned Youla parameterization methodology for LTI system cannot be applied in a straightforward way to T-S fuzzy systems.

In the present paper, first, conceptions of doubly coprime factorization and Youla parameterization of LTI systems are extended to T-S fuzzy systems, with respect to quadratic stability using state space expression. Consequently, the parameterization of the close-loop state space expression, which is affine with a stable T-S fuzzy Q-parameter, is obtained. Accordingly, a T-S fuzzy control-system design resolves itself to specify this Q-parameter that satisfies the design specifications. Namely, we can apply this Q-parameter approach to a variety of T-S fuzzy control-system designs. Above all, a design scheme of Q to obtain L2-gain performance for T-S fuzzy systems is clarified with LMI methodology [12],[13].

2 Preliminary

In this section, some notation and assumptions regarding T-S fuzzy systems are introduced; and useful conceptions and a lemma are recapped.

Definition 1. Takagi-Sugeno Fuzzy Systems
The T-S fuzzy model G consists of a finite set of fuzzy IF…Then rules. Each rule has the following form:

Dynamic part:
Rule $i = 1, 2, \cdots, r$: IF $z_1(t)$ is $M_{i1}, \cdots$ and $z_p(t)$ is M_{ip},

$$\text{Then } \dot{x}(t) = A_i x(t) + B_{ui} u(t) + B_{wi} w(t)$$

Output part:
Rule $i = 1, 2, \cdots, r$: IF $z_1(t)$ is $M_{i1}, \cdots$ and $z_p(t)$ is M_{ip},

$$\text{Then } \begin{array}{l} y(t) = C_{yi} x(t) + D_{wyi} w(t) \\ z(t) = C_{zi} x(t) + D_{uzi} u(t) + D_{wzi} w(t) \end{array}$$

The vectors $x(t)$, $u(t), w(t), y(t)$ and $z(t)$ denote the state, input, disturbance and output vectors, respectively. Each variable $z_i(t)$ is a known parameter, which may be function of the state variables and/or external disturbances. The symbols M_{ij} represent membership functions for fuzzy sets.

Using the center of gravity method for defuzzification, we can express the aggregated fuzzy model $G(z)$ as shown in (1) where z denotes the vector containing all the individual parameters $z_i(t)$.

$$G(z) \triangleq \left[\begin{array}{c|cc} A(z) & B_u(z) & B_w(z) \\ \hline C_y(z) & 0 & D_{wy}(z) \\ C_z(z) & D_{uz}(z) & D_{wz}(z) \end{array}\right] = \sum_{i=1}^{r} h_i(z) \left[\begin{array}{c|cc} A_i & B_{ui} & B_{wi} \\ \hline C_{yi} & 0 & D_{wyi} \\ C_{zi} & D_{uzi} & D_{wzi} \end{array}\right]. \tag{1}$$

The $h_i(z)$ is normalized possibility for the ith rule to fire given by

$$h_i(z) = \frac{w_i(z)}{\sum_{i=1}^{r} w_i(z)}, \tag{2}$$

where the possibility for the ith rule to fire: $w_i(z)$ is given by the product of all the membership functions associated with the ith rule as

$$w_i(z) = \prod_{j=1}^{p} M_{ij}(z_j). \tag{3}$$

We will assume that at least one $w_i(z)$ is always nonzero so that $\sum_{i=1}^{r} w_i\left(z_j\right) \neq 0$.

It also should be noted that the normalized possibility $h_i(z)$ satisfies the conditions $h_i(z) > 0$ and $\sum_1^r h_i(z) = 1$.

We then consider the L2 gain and stability of T-S fuzzy systems.

Definition 2. L2 gain performance

The control system satisfying the property $\sup_{w\neq 0, \|w\|_2 < \infty} \frac{\|b\|_2}{\|a\|_2} < \gamma$ is said to have L2 gain performance with the bound γ related to signals a and b, where $\|c\|_2 = \sqrt{\int_0^\infty c^T(t)c(t)dt}$ is the L2-norm of the signal c.

Lemma 1. Quadratic stability

T-S fuzzy model (1) is said to be quadratically stable if and only if there exists $P > 0$ such that

$$A_i^T P + PA_i < 0 \tag{4}$$

3 Youla Parameterization for T-S Fuzzy Systems

In this section, the conception of Youla parameterization for LTI systems is extended to T-S fuzzy systems. Because we focus on feedback stability problem, T-S fuzzy feedback control system configured in Fig.1 is considered. Where, the $G_{22}(z)$ denotes the plant (1) around u and y, and $K(z)$ denotes a T-S fuzzy controller.

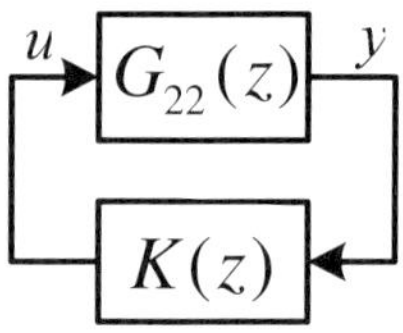

Fig. 1. T-S fuzzy feedback control system

Next, the conception of doubly coprime factorization of LTI systems is extended to T-S fuzzy systems. Then, based on the factorization, we introduce a systematic way of obtaining a set of quadratically stabilizing controller construction for T-S fuzzy systems.

3.1 Quadratically Stabilizing Observer-Based Controller for T-S Fuzzy Systems

The original T-S fuzzy plant $G_{22}(z)$ can be expressed as

$$\begin{aligned} \dot{x} &= A(z)x + B_u(z)u \\ y &= C_y(z)x \end{aligned} \tag{5}$$

Lemma 2. A quadratically stabilizing T-S fuzzy observer-based controller can be formulated as

$$\begin{aligned} \dot{\hat{x}} &= A(z)\hat{x} + B_u(z)u + L(C_y(z)\hat{x} - y) \\ u &= F\hat{x} \end{aligned} \tag{6}$$

where F, L are constant matrices. The $F = VP_f^{-1}$, $L = P_l^{-1}W$,the matrices V and W satisfy the following LMIs:

$$\begin{aligned} P_f > 0,\ A(z)P_f + P_f A^T(z) + B_u(z)V + V^T B_u^T(z) < 0, \\ P_l > 0,\ A^T(z)P_l + P_l A(z) + WC_y(z) + C_y^T(z)W^T < 0 \end{aligned} \tag{7}$$

Proof. Substituting (6) into (5), the closed-loop state matrix can be expressed as

$$A_{cl}(z) = \begin{bmatrix} A(z) + B_u(z)F & B_u(z)F \\ 0 & A(z) + LC_y(z) \end{bmatrix} \tag{8}$$

Based on (7) and lemma 1, we see that $A(z) + B_u(z)F$ and $A(z) + LC_y(z)$ are quadratically stable. Thus, the system (8) is quadratically stable.

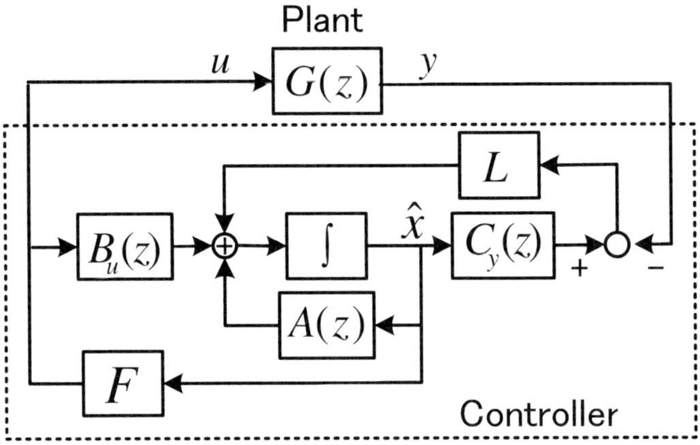

Fig. 2. Quadratically stabilizing T-S fuzzy observer-based controller

The quadratically stabilizing observer-based controller for T-S fuzzy systems can be shown as Fig.2.

Now we can define the coprime factorization of T-S fuzzy plant that resembles the case of LTI systems.

Definition 3. Coprime factorization for T-S fuzzy plant.

The T-S fuzzy plant $G(z)$ is said to have doubly coprime factorization if there exit right coprime factorization and left coprime factorization as:

$$G(z) = N_r(z)D_r(z)^{-1} = D_l(z)^{-1}N_l(z), \tag{9}$$

where a set of realizations for stable T-S fuzzy systems $N_r(z), D_r(z), N_l(z), D_l(z), X_l(z), Y_l(z), X_r(z)$ and $Y_r(z)$ can be chosen such that

$$\begin{bmatrix} D_r(z) & X_l(z) \\ N_r(z) & Y_l(z) \end{bmatrix} \begin{bmatrix} Y_r(z) & -X_r(z) \\ -N_l(z) & D_l(z) \end{bmatrix} = \begin{bmatrix} I & 0 \\ 0 & I \end{bmatrix}. \tag{10}$$

A particular set of realizations can be chosen such that

$$\begin{bmatrix} D_r(z) & X_l(z) \\ N_r(z) & Y_l(z) \end{bmatrix} \triangleq \left[\begin{array}{c|cc} A(z)+B_u(z)F & B_u(z) & -L \\ \hline F & I & 0 \\ C_y(z) & 0 & I \end{array}\right] = \sum_{i=1}^{r} h_i(z) \left[\begin{array}{c|cc} A_i+B_{ui}F & B_{ui} & L \\ \hline F & I & 0 \\ C_{yi} & 0 & I \end{array}\right],$$

$$\begin{bmatrix} Y_r(z) & -X_r(z) \\ -N_l(z) & D_l(z) \end{bmatrix} \triangleq \left[\begin{array}{c|cc} A(z)+LC_y(z) & -B_u(z) & L \\ \hline F & I & 0 \\ C_y(z) & 0 & I \end{array}\right] = \sum_{i=1}^{r} h_i(z) \left[\begin{array}{c|cc} A_i+LC_{yi} & B_{ui} & L \\ \hline F & I & 0 \\ C_{yi} & 0 & I \end{array}\right],$$

in which F and L are chosen such that both $A_i - B_iF$ and $A_i - LC_i$ are quadratically stable.

3.2 All Quadratically Stabilizing Controllers for T-S Fuzzy Systems

In this Subsection, the parameterization of quadratically stabilizing controllers for T-S fuzzy systems is investigated.

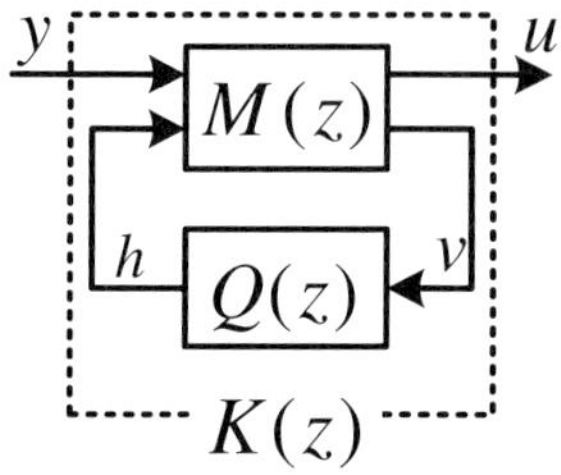

Fig. 3. Quadratically stabilizing controllers

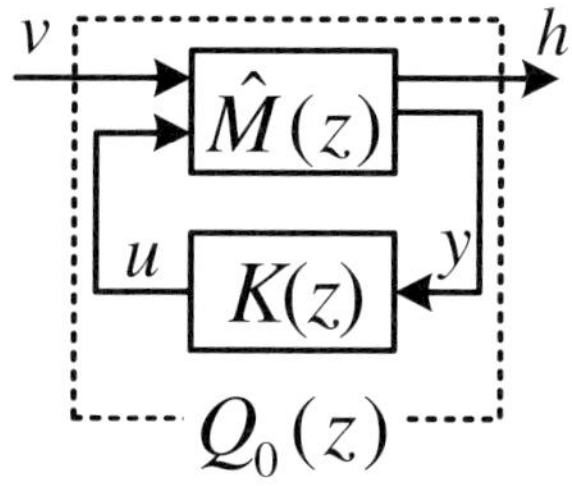

Fig. 4. A function $Q_0 = F_l(\hat{M}, K)$

Theorem 1
All quadratically stabilizing controllers for T-S fuzzy plant (1) can be described as lower LFT: $F_l(M(z), Q(z))$ shown in Fig. 3 with any quadratically stable $Q(z)$. Where, the $M(z)$ and $Q(z)$ are given by

$$M(z) \triangleq \left[\begin{array}{c|cc} A(z)+B_u(z)F+LC_y(z) & -L & B_u(z) \\ \hline F & 0 & I \\ -C_y(z) & I & 0 \end{array}\right] \tag{11}$$

$$Q(z) \triangleq \left[\begin{array}{c|c} A_Q(z) & B_Q(z) \\ \hline C_Q(z) & D_Q(z) \end{array}\right]; A_Q^T(z)P_Q + P_Q A_Q(z) < 0; P_Q > 0 \tag{12}$$

Proof
Sufficiency: Using (11) and (12), the controller $K(z)$ is derived as

$$K(z) \triangleq \left[\begin{array}{cc|c} A(z)+B_u(z)F+LC_y(z)-B_u(z)D_Q(z)C_y(z) & B_u(z)C_Q(z) & B_u(z)D_Q(z)-L \\ -B_Q(z)C_y(z) & A_Q(z) & B_Q(z) \\ \hline F-D_Q(z)C_y(z) & C_Q(z) & D_Q(z) \end{array}\right] \tag{13}$$

Connecting (13) to (1), the closed-loop system can be expressed as

$$\begin{bmatrix} \dot{x} \\ \dot{e} \\ \dot{x}_Q \end{bmatrix} = \begin{bmatrix} A(z)+B_u(z)F & B_u(z)F-B_u(z)D_Q(z)C_y(z) & B_u(z)C_Q(z) \\ 0 & A(z)+LC_y(z) & 0 \\ 0 & -B_Q(z)C_y(z) & A_Q(z) \end{bmatrix} \begin{bmatrix} x \\ e \\ x_Q \end{bmatrix} \tag{14}$$

where $e = \hat{x} - x$ and x_Q is the state variable of the $Q(z)$.

Using Lemma 1, there exists a quadratic Lyapunov function $V = x^T P_{cl} x$ that P_{cl} is positive such that $V = x^T P_{cl} x = \begin{bmatrix} x_f \\ x_l \\ x_q \end{bmatrix}^T \begin{bmatrix} \lambda_f P_f & 0 & 0 \\ 0 & P_l & 0 \\ 0 & 0 & \lambda_q P_q \end{bmatrix} \begin{bmatrix} x_f \\ x_l \\ x_q \end{bmatrix} > 0$, where λ_f, λ_q are positive numbers, this assures that the closed-loop system is quadratically stable.

Necessity: We will show that arbitrary quadratically stabilizing controller $K(z)$ can be expressed with a quadratically stable Q_0 as $K = F_l(M, Q_0)$.

Consider a function $Q_0 = F_l(\hat{M}, K)$ as shown in Fig.4, where

$$\hat{M} \triangleq \left[\begin{array}{c|cc} A(z) & -L & B_u(z) \\ \hline -F & 0 & I \\ C_y(z) & I & 0 \end{array}\right]. \tag{15}$$

Since (2,2) block of $\hat{M}(z)$ is the same as that of the original plant (1), then $K(z)$ also stabilized $\hat{M}(z)$. Accordingly, we see $Q_0 = F_l(\hat{M}, K)$ is quadratically stable. The substitution of this function Q_0 into $F_l(M, Q_0)$ yields $F_l(M, Q_0) = F_l(M, F_l(\hat{M}, K)) = F_l(J_{tmp}, K)$, where J_{tmp} can be obtained by using the sate space star product formula

$$J_{tmp}(z) \triangleq \left[\begin{array}{cc|cc} A(z)+LC_y(z)+B_u(z)F & -B_u(z)F & -L & B_u(z) \\ LC_y(z) & A(z) & -L & B_u(z) \\ \hline F & -F & 0 & I \\ -C_y(z) & C_y(z) & I & 0 \end{array}\right] \tag{16}$$

A similar transformation with $T = \begin{bmatrix} I & I \\ 0 & I \end{bmatrix}$, and eliminating stable uncontrollable and unobservable mode, gives $J_{tmp} \triangleq \left[\begin{array}{c|c} 0 & I \\ \hline I & 0 \end{array}\right]$. Consequently, the relation that we want to show is deduced:

$$F_l(M, Q_0) = F_l(J_{tmp}, K) = K. \tag{17}$$

The proof has been completed.

4 Design of Function Q with L2 Gain Performance

So far, the parameterization of the closed-loop T-S fuzzy systems with arbitrary stable Q has been obtained. Based on this parameterization, T-S fuzzy control-system designs resolve themselves to settle this Q that satisfies the design specifications. In this Section, a necessary and sufficient condition and also a design scheme of Q to obtain L2-gain performance is clarified with LMI methodology.

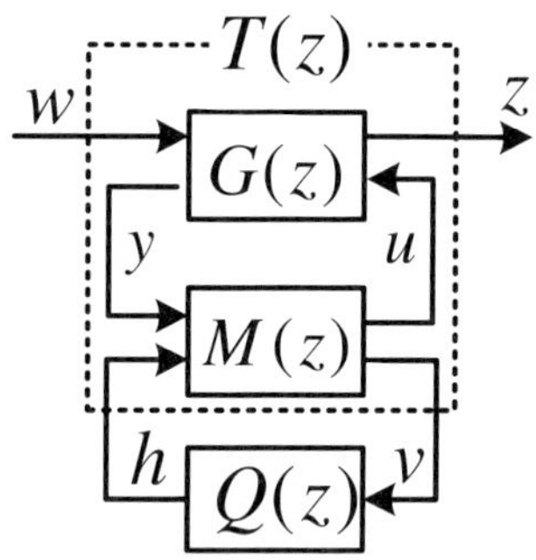

Fig. 5. Two-step design construction

Figure 5 shows a two-step design construction for the control performance design of T-S fuzzy systems, where $M(z)$ is an observer-based controller mentioned in lemma 2. The generalized plant $T(z)$ is then derived as

$$\left[\begin{array}{c|cc} A_p(z) & B_{p1}(z) & B_{p2}(z) \\ \hline C_{p1}(z) & D_{p11}(z) & D_{p12}(z) \\ C_{p2}(z) & D_{p21}(z) & 0 \end{array}\right] = \left[\begin{array}{c|cc} \begin{bmatrix} A(z)+B_u(z)F & B_u(z)F \\ 0 & A(z)+LC_y(z) \end{bmatrix} & \begin{bmatrix} B_w(z) \\ -B_w(z)-LD_{wy}(z) \end{bmatrix} & \begin{bmatrix} B_u(z) \\ 0 \end{bmatrix} \\ \hline \begin{bmatrix} C_z(z)+D_{uz}(z)F & D_{uz}(z)F \end{bmatrix} & D_{wz}(z) & D_{uz}(z) \\ \begin{bmatrix} 0 & -C_y(z) \end{bmatrix} & D_{wy}(z) & 0 \end{array}\right]. \tag{18}$$

For the simplicity, we assume constant B_Q, C_Q and $D_Q = 0$ as $Q(z)$ instead of general $B_Q(z), C_Q(z), D_Q(z)$. In this case, we have the following necessary and sufficient condition for the existence of γ-suboptimal compensators Q for the generalized plant (18).

Theorem 2
Consider above T-S fuzzy plant governed by (18), there exists a quadratically stable $Q = [A_Q(z), B_Q, C_Q, 0]$ guaranteeing quadratic $\|z\|_2 \le \gamma \|w\|_2$ along all parameter

trajectories, if and only if there exist two symmetric matrices X, Y and $\hat{A}_Q(z), \hat{B}_Q, \hat{C}_Q$ and satisfying the following LMIs,

$$\begin{pmatrix} X & I \\ I & Y \end{pmatrix} > 0,$$

$$\begin{bmatrix} XA_p(z)+\hat{B}_Q C_{p2}(z)+(*) & * & * & * \\ \hat{A}_Q^T(z)+A_p(z) & A_p(z)Y+B_{p2}(z)\hat{C}_Q+(*) & * & * \\ (XB_{p1}(z)+\hat{B}_Q D_{p21}(z))^T & (B_{p1}(z))^T & -\gamma I & * \\ C_{p1}(z) & C_{p1}(z)Y+D_{p12}(z)\hat{C}_Q & D_{p11}(z) & -\gamma I \end{bmatrix} < 0. \tag{19}$$

Terms denoted * is induced by symmetry.

Proof

According to the Theorem 1, it is noted that quadratic stability of Q is equivalent to quadratic stability of closed-loop control system including G, M and Q itself. Meanwhile, according to [12], [13], LMIs (19) above give the sufficient and necessary conditions that a compensator Q stabilizes the closed-loop system and guarantees the L2 gain performance. If the solution of Q function satisfying LMIs (19) exists, for closed-loop system is quadratically stable, Q itself is also a quadratically stable one.

It is well known that the above infinite LMIs problem is difficult to be solved, however, since this T-S fuzzy system has formulation as (18), the above problem can be transferred to finite vertex LMIs problems as

$$\begin{bmatrix} XA_{pi}+\hat{B}_Q C_{p2i}+(*) & * & * & * \\ \hat{A}_{Qi}^T+A_{pi} & A_{pi}Y+B_{p2i}\hat{C}_Q+(*) & * & * \\ (XB_{p1i}+\hat{B}_Q D_{p21i})^T & (B_{p1i})^T & -\gamma I & * \\ C_{p1i} & C_{p1i}Y+D_{p12i}\hat{C}_Q & D_{p11i} & -\gamma I \end{bmatrix} < 0 \tag{20}$$

The T-S fuzzy compensator $Q(z)$ can be derived through the following scheme.

1. Solve for H, J, the factorization problem $I - XY = HJ^T$.
2. Compute $A_Q(z), B_Q, C_Q$ with

$$A_Q(z) = \sum_{i=1}^{r} h_i(z) H^{-1}(\hat{A}_{Qi} - XA_{pi}Y - \hat{B}_Q C_{p2i}Y - XB_{p2i}\hat{C}_Q)J^{-T},$$
$$B_Q = H^{-1}\hat{B}_Q, \quad C_Q = \hat{C}_Q J^{-T}.$$

5 Conclusions

One methodology of designing T-S fuzzy control systems through the use of Youla-Parameterization has been proposed. A conception of doubly coprime factorization of LTI systems has been extended to T-S fuzzy control systems with respect to quadratic stability. In LTI systems, internal stability can be assured from the Bezout identity. However here, as for T-S fuzzy systems, stability is assured through the Lyapunov inequality.

Parameterization of the closed-loop systems with any quadratically stable Q-parameter in an affine fashion was then described. Consequently, the Q-parameter approach can be applicable to a variety of T-S fuzzy control-system designs. Among them, a necessary and sufficient condition and a design scheme of Q to obtain L2-gain performance was clarified with LMI methodology.

References

1. Castro, J.: Fuzzy Logic controllers are Universal Approximators, IEEE Transactions on Systems, Man and Cybernetics, Vol. 25, (1995) 629-635.
2. Tanaka.K, Sugeno. M.: Fuzzy identification of systems and its applications to modeling and control, IEEE Transactions on Systems, Man and Cybernetics, (1985) 15(1), 116-132.
3. Boyd, S., Ghaoui, L. E., Feron, E., Balakrishnan, V.: Linear Matrix Inequality in Systems and Control Theory. SIAM Studies in Applied Mathematics, Vol.15, SIAM, Philadelphia, (1994).
4. Wang, H. O., Tanaka, K., Griffin, M. F.: An Approach to Fuzzy Control of Nonlinear Systems: Stability and Design Issues, IEEE Trans. Fuzzy Sys, Vol.4, No.1, (1996) 14-23.
5. Johansson, M., Rantzer, A., Arzen, K. E.: Piecewise Quadratic Stability for Affine Sugeno Systems, Proc. 7th IEEE Int. Conf. Fuzzy Syst. Anchorage, AK, Vol. 1, (1998) 55-60.
6. Petterson, S., Lennartson, B.: An LMI Approach for Stability Analysis of Nonlinear Systems, Proc. Of European Control Conference, ECC'98, Brussels, Belgium, (1997).
7. Tanaka, K., Sugeno.M: Stability Analysis and Design of Fuzzy Control systems, Fuzzy Sets and systems, Vol.45, No.2, (1992) 135-156.
8. Tanaka,K., Ikeda, T., Wang, H. O.: Design of Fuzzy Control Systems Based on Relaxed LMI Stability Conditions, Proc.35th CDC, (1996) 598-603.
9. Li, J., Wang, H. O., Niemann, D., Tanaka, K.: Dynamic parallel distributed compensation for Takagi-Sugeno fuzzy systems: An LMI approach, Inf. Sci. 123(3-4), (2000) 201-221.
10. Youla, D.C., Jabr, H., Bongiorno.J.J.: Modern Wiener-Hopf design of optimal control controllers: part II:The multivariable case, IEEE Trans. Autom. Control., Vol. 21 (1976) 319-338.
11. Kučera, V.,: Discrete linear Control: the Polynomial Equation Approach, New York: Wiley, (1979).
12. Gahinet, P., Nemirovski, A., Laub, A. J., Chilali, M.: LMI control toolbox for use with matlab, User's guide, The Math Works Inc Natick, MA, USA, (1995).
13. Masubuchi, I., Ohara, A., Suda, N.: LMI based controller synthesis: A unified formulation and solution, International Journal of Robust and Nonlinear Control, Vol. 8, No. 8, (1998) 669-686.

On the Markovian Randomized Strategy of Controller for Markov Decision Processes

Taolue Chen[1,2,⋆], Tingting Han[2,3,4,⋆⋆], and Jian Lu[2,⋆⋆⋆]

[1] CWI, Department of Software Engineering, PO Box 94079, 1090 GB Amsterdam, The Netherlands
[2] Nanjing University, State Key Laboratory of Novel Software Technology, Nanjing, Jiangsu, P.R. China 210093
[3] RWTH Aachen, MOVES, Ahornstraße 55, D-52056 Aachen, Germany
[4] University of Twente, Faculty of EEMCS, FMT, PO Box 217, 7500 AE Enschede, The Netherlands

Abstract. This paper focuses on the so called *controller synthesis* problem, which addresses the question of how to limit the internal behavior of a given system implementation to meet its specification, regardless of the behavior enforced by the environment. We consider this problem in the probabilistic setting, where the underlying model has both probabilism and nondeterminism and the nondeterministic choices in some states are assumed to be controllable while the others are under the control of an unpredictable environment. As for the specification, it is defined by *probabilistic computation tree logic*. We show that under the restriction that the controller exploits only *Markovian randomized strategy*, the existence of such a controller is decidable, which is done by a reduction to the decidability of first-order theory for reals. This also gives rise to an algorithm which can synthesize the controller if it does exist.

1 Introduction

In this paper, we focus on *Markovian randomized* strategy of controllers for a *Markov Decision Process* (MDP for short) like model. Let us start by putting this problem in a more general setting: It is well-known that in the system design, to develop a system which satisfies the user requirement is one of the basic goals [1]. Undoubtedly, one hopes to automate this error-prone process as far as possible. To achieve this, one of the goals is to *synthesize* a system based on the requirements. However, this goal is usually too ambitious to be realized, and thus a more practical, but equally important task is to synthesize only a

⋆ This author is partially supported by the Dutch Bsik project BRICKS (Basic Research in Informatics for Creating the Knowledge Society).

⋆⋆ This author is partially supported by the Dutch NWO project QUPES (Verification of Quantitative Properties of Embedded Software).

⋆⋆⋆ This author is partially supported by 973 Program of China (No. 2002CB312002), NNSFC (No. 60233010, No. 60273034, No. 60403014), 863 Program of China (No. 2005AA113160, No. 2004AA112090).

L. Wang et al. (Eds.): FSKD 2006, LNAI 4223, pp. 149–158, 2006.

controller which can limit or control the behaviors of an existing system (in terms of control theory, this is usually called *plant*), to meet the given specification. This is often referred to as the *controller synthesis* problem.

In such a framework, the plant acts usually in an *environment*. The goal is to find a *schedule* for the controllable events that guarantees the specification to be satisfied considering all possible environmental behaviors. One can also understand the controller and environment as two *players*, and thus take a game-theoretical point of view on the controller synthesis problem (in the below, we take the liberty to switch between these two points of views). The plant constitutes the game board and controller synthesis becomes the problem of finding a *winning strategy* for the controller that satisfies the specification whatever move the environment does, or in other words, under any adversary [1]. Correspondingly, the requirement specification is the *winning condition* in terms of game theory [11], i.e. the controller wins once the specification is satisfied. We note that the winning condition can be given either internally or externally: the former often imposes restrictions, for instance, on the number of visits of a state in the plant, typical examples include Büchi or Muller condition. The latter is usually temporal logic formulas which are supposed to be satisfied by the controller plant.

As a fundamental problem in control theory and computer science, the controller synthesis problem has attracted a lot of attentions in recent years. For discrete systems, it is well understood [14]. Recently, it also has been studied for timed systems and probabilistic systems. In this paper, we shrink our attention to the probabilistic setting, following [1]. Our underlying model for the plant is Markov Decision Processes [7][12]. However, different from the standard MDP model [12], here we adopt the point of view of [7] and distinguish states that are under control of the plant from those that are under the control of the environment. To put this in the game theory, this model is also known as *turn-based stochastic $2\frac{1}{2}$-player games* [4] (we note that the normal MDP is basically, turn-based stochastic $1\frac{1}{2}$-player games, see [4] for details). And by translating the player to other terms, one can construct a lot of examples of the similar spirit. Actually it is a very popular model in planning, AI, control problems, etc. In particular, it has been extensively applied for natural computation, fuzzy systems and knowledge discovery.

In this paper, we study the problem of finding a strategy for the plant such that a given external specification formalized as a *probabilistic temporal logic* formula is fulfilled, no matter how the opponent (environment) behaves. Concretely speaking, here the specification is given by a formula of *probabilistic computation tree logic* ([9], PCTL for short), which is one of the most influential and widely used probabilistic temporal logics. As for strategy, we follow [1], i.e. we consider the following choices:

- The system or the opponent has to choose deterministically (D) or randomly (R); and
- The system or the opponent chooses according to the current state (M, also called stationary or Markovian) or the history of the game played so far (H).

The combination gives rise to (at least) four classes of strategies, i.e. MD, HD, MR and HR strategies. In [1], it is shown that any of the strategy-classes (HD, HR, MD and MR) requires its own synthesis algorithm. Moreover, in that paper, the problem is only solved for MD strategies w.r.t. PCTL properties. However, the other three are left open (We quote a sentence from [1] which says "For other strategy types (MR, HD or HR), the complexity or even the decidability of the controller synthesis problem for PCTL is an open problem"). The aim of this paper is to attack this problem by considering MR strategy. The importance of MR strategy is justified by the fact that based on it, the controller can use random number generator rather than memory to make decisions. In practice, especially for embedded systems, it often leads to some "cheap" solution. It turns out even under this restriction, the problem is non-trivial. In the below, we discuss the main difficulties and our solutions in brief, from both the strategy and the specification perspectives:

– For strategy. As for the MD strategy, the controller synthesis problem can be trivially reduced to the model checking problem for Markov Chain w.r.t. PCTL specification. This is because there are only finitely many MD strategies for a given MDP, and thus one can try out all possibilities. Actually, in [1], this brute forth approach is used. However, for MR strategy, some more sophisticated approach has to be exploited since the total number of MR strategies is infinite (even uncountable since one can easily show the cardinality of the set of MR strategy is $\aleph_1$). To overcome this problem, we appeal to the deep results on the complexity of decision procedures for the first-order theory of reals $(\mathbb{R}, +, \cdot, \leq)$, which is well-known to be decidable [13]. To be more precise, we encode the existence of a MR-controller to $(\mathbb{R}, +, \cdot, \leq)$, thus prove the decidability of the existence of MR-controller.
– For specification. As we suggested before, in this paper we consider external specification, which is expressed by PCTL. In the game theory, this is rather difficult and a common approach to deal with external specification (winning condition) is to turn this into some internal one and at the same time, to transform the underlying model. For example, for linear-time properties, one can "encode" the specification by deterministic Muller automata and then take a product with the game graph (here it is MDP), thus the problem can be reduced to the one with a new MDP w.r.t. Muller winning condition (which is an internal winning condition). However, in the case of branching-time properties considered in this paper, it is not obvious how to adapt this approach, since in order to "encode" PCTL, we have to introduce some notion like "probabilistic tree automata", which has not been well studied yet, due to the knowledge of the authors. Fortunately, we find that our idea to exploit the decidability of $(\mathbb{R}, +, \cdot, \leq)$ can also be used to circumvent this difficulty.

It is worth emphasizing that our underlying model essentially agrees with $2\frac{1}{2}$-player game. However, we allow the two players to use different sorts of strategies in the sense that on the controller's aspect, it is only allowed to take MR strategy while on the environment aspect, we do *not* exert any restriction on the strategy which it can take. We note that this situation is akin to the notion of "modulo checking" [10] which deals with the problem of checking whether a module behaves correctly no matter in which environment it is placed. Actually, the interested reader will detect that in the second step of our algorithm (see Section 3), we implicitly encode the "module checking" problem for MDP into $(\mathbb{R}, +, \cdot, \leq)$. We believe this kind of "asymmetry" makes our result stronger and more general, and it is more useful in some applications. Moreover, it should be noticed that some related work has appeared in [4][5]. In both of papers, essentially the same problem is considered. However, the difference lies in that they considered linear-time specification (which can be regarded as internal winning condition), while we consider branching time specification. As we have suggested before, for controller synthesis problem, branching time specification is much more involved.

The structure of this paper is as follows: Section 2 summarizes some background material on MDP, strategy and PCTL. Section 3 presents our algorithm for MR strategy, and a simple example from [1] is given. Due to space restriction, in this extended abstract, most of proofs and practical examples are omitted and we refer the interested readers to the technical report version of this paper for more details.

2 Preliminaries

Definition 1 (Distribution). A *distribution* on a countable set X denotes a function $\mu : X \to [0,1]$ such that $\sum_{x \in X} \mu(x) = 1$. We use $\text{Distr}(X)$ to denote the set of all distributions on X.

Definition 2 (Markov Decision Process). A *Markov Decision Process* is a tuple $\mathcal{M} = (S, Act, \mathbb{P}, s_{in}, AP, L)$ where

- S is a countable set of states;
- Act is a finite set of actions;
- $\mathbb{P} : S \times Act \times S \to [0,1]$ is a three-dimensional transition probability matrix such that for all states $s \in S$ and actions $a \in Act$, $\sum_{t \in S} \mathbb{P}(s, a, t) \in \{0, 1\}$;
- $s_{in} \in S$ is the initial state;
- AP denotes a finite set of atomic propositions;
- $L : S \to \wp(AP)$ is a labelling function which assigns to each state $s \in S$ the set $L(s)$ of atomic propositions that are (assumed to be) valid in s.

For technical reasons, we require that none of the states is terminal, i.e. for each state s, there exists an action a and a state s' with $\mathcal{P}(s, a, s') > 0$. $\mathcal{M}$ is called finite if the state space S is finite. For $T \subseteq S$, $\mathcal{P}(s, a, T) = \sum_{t \in T} \mathcal{P}(s, a, t)$ denotes the probability for s to move to a T-state, provided that action a has been selected in state s. We write $\mathcal{I}_{\mathcal{M}}(s)$ or $\mathcal{I}(s)$ (if $\mathcal{M}$ is clear from the context) for the action set $\{a \in Act \mid \mathcal{P}(s, a, S)\}$.

Definition 3 (Path and Trace). A *path* in $\mathcal{M}$ is a finite or infinite alternating sequence of states and actions $\sigma = s_1 a_1 \cdots a_n s_n$ or $\sigma = s_1 a_1 s_2 a_2 \cdots$ such that

$\mathcal{P}(s_i, a, s_{i+1}) > 0$. We use $\sigma[i]$ to denote the i-th state of σ. For finite path σ, we use $\sigma[\downarrow]$ to denote the last state of σ. Furthermore, we denote by $Path^*$ (resp. $Path^\omega$) the set of finite (resp. infinite) paths of a given MDP and by $Path^*(s)$ (resp. $Path^\omega(s)$) the set of finite (resp. infinite) paths of a given MDP starting at the state s.

For infinite path σ, we use $tr(\sigma)$ (*trace* of σ) to denote the infinite word over the alphabet $\wp(AP)$ which arises from σ by the projection of the induced state-sequence to the sequence of the labelings. For instance, if σ is defined as above, then $tr(\sigma) = L(s_1)L(s_2)\cdots \in (\wp(AP))^\omega$.

In the sequel, we assume that $\mathcal{M}$ is a finite MDP and S_0 a nonempty subset of S consisting of the states which are under the control of the system, i.e. where the system may decide which of the possible actions is executed. The states in $S \setminus S_0$ are controlled by the environment.

Definition 4 (Strategy). A *strategy* of $(\mathcal{M}, S_0)$ is an instance D that resolves the nondeterminism in the S_0 states. We distinguish four types of strategies:

- An MD-strategy is function $D : S_0 \to Act$ such that $D(s) \in \mathcal{I}(s)$;
- An MR-strategy is function $D : S_0 \to \mathrm{Distr}(Act)$ with $D(s) \in \mathrm{Distr}(\mathcal{I}(s))$;
- An HD-strategy is function D that assigns to any finite path σ in $\mathcal{M}$ with $\sigma[\downarrow] = s \in S_0$ and action $D(\sigma) \in \mathcal{I}(s)$;
- An HR-strategy is function D that assigns to any finite path σ in $\mathcal{M}$ with $\sigma[\downarrow] = s \in S_0$ and action $D(\sigma) \in \mathrm{Distr}(\mathcal{I}(s))$;

We note that the MD-strategy is often called *simple* or *purely memoryless* strategy. We refer to the strategies for the environment as *adversaries*. Formally, for $\mathrm{X} \in \{\mathrm{MD}, \mathrm{MR}, \mathrm{HD}, \mathrm{HR}\}$, an X-adversary for $(\mathcal{M}, S_0)$ denotes a X-strategy for $(\mathcal{M}, S \setminus S_0)$. The notion of *policy* will be used to denote a decision rule that resolves both the internal nondeterministic choices (to be resolved by the controller) and the nondeterministic choices (to be resolved by the environment). We will use D for strategy, E for adversary and C for policies. Policy will often be written as $C = (D, E)$.

MDPs and Markov Chains Induced by Strategies. Any strategy D for $(\mathcal{M}, S_0)$ induces a MDP $\mathcal{M}_D$ which arises through unfolding $\mathcal{M}$ into a tree-like structure where the nondeterministic choices in S_0-states are resolved according to D. If $S_0 = S$ and $D = C$ is a policy for $\mathcal{M}$ then *all* nondeterministic choices are resolved in $\mathcal{M}_C$. For any HR-policy C, the MDP $\mathcal{M}_C$ is an infinite-state discrete-time Markov chain. If C is a stationary policy, then $\mathcal{M}_C$ can be regarded as a *Discrete Time Markov Chain* (DTMC for short) with state space S. If C is a policy for $\mathcal{M}$, then we write $\mathbb{P}^C_{\mathcal{M}}$ or just $\mathbb{P}^C$ (if $\mathcal{M}$ is clear from the context) to denote the standard probability measure on $\mathcal{M}_C$. Moreover, we use $Path^*_C(s)$ (resp. $Path^\omega_C(s)$) to denote the set of finite (resp. infinite) paths of the Markov chain induced by MDP $\mathcal{M}$ and policy C. Similarly, $Path^*_C(s)$ (resp. $Path^\omega_C(s)$) the set of finite (resp. infinite) paths of the corresponding Markov chain starting at the state s.

2.1 Probabilistic Computation Tree Logic (PCTL)

PCTL, the probabilistic extension of CTL (computation tree logic), was defined by Hansson and Jonsson [9] and is one of the most widely used probabilistic temporal logics. Let $AP = \{p, q, \dots\}$ be a countable infinite set of *atomic propositions.* The syntax of PCTL is given by the following BNF:
State formula:

$$\Phi ::= \top \mid p \mid \neg\Phi \mid \Phi \wedge \Phi \mid [\varphi]_{\bowtie r}$$

Path formula:

$$\varphi ::= \bigcirc\Phi \mid \Phi \mathcal{U} \Phi$$

where $p \in AP$ and $r \in [0, 1]$.

The most interesting part is $[\varphi]_{\bowtie r}$, which intuitively asserts that the probability measure of the paths satisfying φ meets the bound given by $\bowtie r$. The path modalities $\bigcirc$ (next step) and $\mathcal{U}$ (until) have the same meaning as in CTL. Other boolean connectives and the temporal operations $\Diamond$ (eventually) and $\Box$ (always) can be derived as a standard way. In particular, we set $\bot \stackrel{\text{def}}{=} \neg(\top)$.

Semantics. Given a MDP $\mathcal{M}$ as before, the formal definition of the satisfaction relation $\models$ for PCTL path and state formulas is defined as

$$\begin{array}{lll}
\mathcal{M}, s \models \top & & \\
\mathcal{M}, s \models p & \Leftrightarrow & p \in L(s) \\
\mathcal{M}, s \models \neg\Phi & \Leftrightarrow & s \not\models \Phi \\
\mathcal{M}, s \models \Phi_1 \wedge \Phi_2 & \Leftrightarrow & s \models \Phi_1 \text{ and } s \models \Phi_2 \\
\mathcal{M}, s \models [\varphi]_{\bowtie r} & \Leftrightarrow & \text{for } all \text{ policies } C,\ \mathbb{P}^C_{\mathcal{M}}(\{\pi \in Path^\omega(s) \mid \pi \models \varphi\}) \bowtie r
\end{array}$$

$$\begin{array}{lll}
\mathcal{M}, \pi \models \bigcirc\Phi & \Leftrightarrow & \pi[1] \models \Phi \\
\mathcal{M}, \pi \models \Phi_1 \mathcal{U} \Phi_2 & \Leftrightarrow & \exists j \geq 1.(\pi[j] \models \Phi_2 \wedge \forall 1 \leq i < j. \pi[i] \models \Phi_1)
\end{array}$$

For any formula Φ (resp. φ), the (standard) *closure* of formula $cl(\Phi)$ (resp. $cl(\varphi)$) contains formulas whose truth values can influence the truth value of Φ (resp. φ).

3 Controller Synthesis for PCTL

Let us recall that the *controller synthesis* problem discussed in this paper is formalized by triples $(\mathcal{M}, S_0, Spec)$ where $\mathcal{M}$ is a finite MDP, S_0 a set of controllable states in $\mathcal{M}$ and $Spec$ a temporal logical formula. The question is to find a strategy D for $(\mathcal{M}, S_0)$ such that $Spec$ holds for the MDP $\mathcal{M}_D$, no matter how the environment (adversary) behaves. As shown in [1], the role of the strategy-type for controller synthesis is completely different from the situation in PCTL model checking, which has been addressed in [2]. While a single algorithm suffices for PCTL model checking, for controller synthesis, any strategy type requires its own synthesis algorithm. In the rest of this section, we focus on MR-strategy.

Definition 5. For any MDP $\mathcal{M}$, $s \in S$, and PCTL state-formula Φ, we define $Path^{\omega}_{C}(s, \Phi) = \{\sigma \in Path^{\omega}_{C}(s) \mid \sigma \models \Phi\}$ and furthermore

- $\mathbb{P}^{+}_{s}(\Phi) \stackrel{\text{def}}{=} \max\{\mathbb{P}(Path^{\omega}_{C}(s, \Phi)) \mid C \in \text{MD}\}$;
- $\mathbb{P}^{-}_{s}(\Phi) \stackrel{\text{def}}{=} \min\{\mathbb{P}(Path^{\omega}_{C}(s, \Phi)) \mid C \in \text{MD}\}$.

Where, C is ranged over by policies.

The satisfaction relation for PCTL does not depend on the chosen policy type because maximal and minimal probabilities for PCTL-path formulas under all HR-policies are reached with *simple* policies, as shown in [2]. This fact is stated by the following theorem, which is one of the cornerstones of our algorithm.

Theorem 1. *([2][6]) For any MDP $\mathcal{M}$, $s \in S$, and PCTL path-formula φ, the following properties hold:*

1. $s \models [\varphi]_{\bowtie r}$ *where* $\bowtie \in \{\leq, <\}$ *if and only if* $\mathbb{P}^{+}_{s}(\varphi) \bowtie r$*;*
2. $s \models [\varphi]_{\bowtie r}$ *where* $\bowtie \in \{\geq, >\}$ *if and only if* $\mathbb{P}^{-}_{s}(\varphi) \bowtie r$*.*

Before presenting our approach in a formal way, we give an overview of the main ideas. Basically, we will construct a closed formula of $(\mathbb{R}, +, \cdot, \leq)$ such that this formula is valid if and only of there exists an MR-controller for the given MDP $\mathcal{M}$ w.r.t. specification Φ, which is a PCTL state-formula. Let us denote the aforementioned MR-controller D, then the formula is of the form

$$\exists D.\mathcal{M}_{D}, s_{in} \models \Phi$$

Note here $\mathcal{M}_D$ is also an MDP with controllable sets attributed to the environment. Our main challenge is how to encode $\mathcal{M}_D \models \Phi$ in $(\mathbb{R}, +, \cdot, \leq)$, which turns out to be non-trivial and includes several subtle tricks. Now, let us start our construction in the following steps. Note that for convenience, we abuse the notation a little in the sense that for any first-order variable set $X = \{X_1, \cdots, X_n\}$ and formula ψ, we write $\exists X \psi$ as an abbreviation for $\exists X_1 \cdots \exists X_n \psi$.

***Step* 1**

For each state $s \in S_0$ and action $a \in \mathcal{I}(s)$, we introduce a first-order variable $X_{s,a}$. Intuitively, $X_{s,a}$ denotes the probability of choosing the action a in the state $s \in S_0$, thus it represents the strategy D. Then $\exists D.\mathcal{M}_D \models \Phi$ turns out to be

$$\begin{array}{l}\exists\{X_{s,a} \mid s \in S_0, a \in \mathcal{I}(s)\} \\ \quad \bigwedge_{X_{s,a}}(0 \leq X_{s,a} \leq 1) \wedge \bigwedge_{s \in S_0}(\sum_{a \in \mathcal{I}(s)} X_{s,a} = 1) \wedge \mathcal{M}_D, s_{in} \models \Phi\end{array} \tag{1}$$

***Step* 2**

The main goal of **Step 2** is to encode $\mathcal{M}_D, s_{in} \models \Phi$. To this end, for every $\psi \in cl(\Phi)$ and $s \in S$, we introduce a first-order variable $Y_{s,\psi}$, which ranges over $\{0, 1\}$. That is, these variables are essentially boolean values and we set

$$Y_{s,\psi} = 1 \text{ iff } s \models \psi$$

However, it is not easy to express $Y_{s,\psi} = 1$ in an inductive way. We then construct formula $W_{s,\psi}$ for every $s \in S$ and $\psi \in cl(\Phi)$, which can be defined inductively on the structure of ψ. Intuitively, $W_{s,\psi}$ denotes $s \models \psi$ and thus we have $Y_{s,\psi} = 1 \Leftrightarrow W_{s,\psi}$. It follows that (1) can be refined to

$$\begin{array}{l} \exists\{X_{s,a} \mid s \in S_0, a \in \mathcal{I}(s)\}. \\ \quad \bigwedge_{X_{s,a}} (0 \leq X_{s,a} \leq 1) \wedge \bigwedge_{s \in S_0} (\sum_{a \in \mathcal{I}(s)} X_{s,a} = 1) \Rightarrow \\ \quad\quad \exists\{Y_{s,\psi} \mid s \in S, \psi \in cl(\Phi)\}. \\ \quad\quad\quad \bigwedge_{Y_{s,\psi}} (Y_{s,\psi} = 0 \vee Y_{s,\psi} = 1) \wedge (Y_{s,\psi} = 1 \Leftrightarrow W_{s,\psi}) \wedge (Y_{s_{in},\Phi} = 1) \end{array} \tag{2}$$

Clearly, the remaining thing is to construct the formula $W_{s,\psi}$, which is the most difficult task. As we mentioned before, this will be done in an inductive way according to the structure of ψ. We proceed by a case analysis on the form of ψ.

1. $\psi = p$. If $p \in L(s)$, then $W_{s,\psi} \stackrel{\text{def}}{=} \top$ else $W_{s,\psi} \stackrel{\text{def}}{=} \bot$;
2. $\psi = \neg\psi_1$. Then $W_{s,\psi} \stackrel{\text{def}}{=} (Y_{s,\psi_1} = 0)$;
3. $\psi = \psi_1 \wedge \psi_2$. Then $W_{s,\psi} \stackrel{\text{def}}{=} (Y_{s,\psi_1} = 1) \wedge (Y_{s,\psi_2} = 1)$;
4. $\psi = [\bigcirc\psi_1]_{\bowtie r}$. We have two cases:
   - $s \in S_0$. Then
   $$W_{s,\psi} \stackrel{\text{def}}{=} \sum_{a \in \mathcal{I}(s), t \in S} X_{s,a} \cdot \mathbb{P}(s,a,t) \cdot Y_{t,\psi_1} \bowtie r$$
   - $s \in S \setminus S_0$. Then we have to distinguish the following two subcases according to $\bowtie$.
     - $\bowtie \in \{\leq, <\}$. Then by Theorem 1(1), $W_{s,\psi} \stackrel{\text{def}}{=} \mathbb{P}_s^+(\bigcirc\psi_1) \bowtie r$. According to Definition 5, it is easy to see that
     $$\mathbb{P}_s^+(\bigcirc\psi_1) = \max_{a \in \mathcal{I}(s)} \{\sum_{t \in S} \mathbb{P}(s,a,t) \cdot Y_{t,\psi_1}\}$$
     We can find a solution for the above equation by solving the following linear program:
     $$\begin{array}{l} \min x_s \\ \text{s.t. } \; x_s \geq \sum_{t \in S} \mathbb{P}(s,a,t) \cdot Y_{t,\psi_1} \text{ for any } a \in \mathcal{I}(s) \end{array}$$
     We then introduce first-order variables Z_s for $s \in S \setminus S_0$. It follows that $W_{s,\psi}$ is defined as
     $$\exists\{Z_s \mid s \in S \setminus S_0\}.(0 \leq Z_s \leq 1) \wedge \bigwedge_{a \in \mathcal{I}(s)} (Z_s \geq \sum_{t \in S} \mathbb{P}(s,a,t) \cdot Y_{t,\psi_1})$$
     $$\wedge (Z_s \bowtie r) \wedge (\forall\{Z'_s \mid s \in S \setminus S_0\}.(0 \leq Z'_s \leq 1) \wedge \bigwedge_{a \in \mathcal{I}(s)} (Z'_s \geq$$
     $$\sum_{t \in S} \mathbb{P}(s,a,t) \cdot Y_{t,\psi_1}) \Rightarrow (Z'_s \geq Z_s))$$
     - If $\bowtie \in \{\geq, >\}$. This is the duality of the previous subcase.

5. $\psi = [\psi_1 \mathcal{U} \psi_2]_{\bowtie r}$. This case is the most involved one. However, the basic idea remains similar as the previous case. We also have the following two cases:
   - $s \in S_0$. Then clearly it follows that

$$W_{s,\psi} \stackrel{\text{def}}{=} \begin{cases} \top & \text{if } Y_{s,\psi_2} = 1 \\ \bot & \text{if } Y_{s,\psi_1} = 0 \\ Y_{s,\psi_1} = 1 \wedge \sum_{a \in \mathcal{I}(s), t \in S} X_{s,a} \cdot \mathbb{P}(s,a,t) \cdot Y_{t,\psi} \bowtie r & \text{o.w.} \end{cases}$$

   It is easy to transform this definition into a normal first-order formula in $(\mathbb{R}, +, \cdot, \leq)$ as follows:

$$\begin{aligned} &(Y_{s,\psi_2} = 1 \Rightarrow \top) \wedge (Y_{s,\psi_1} = 0 \Rightarrow \bot) \wedge \\ &(Y_{s,\psi_2} = 0 \wedge Y_{s,\psi_1} = 1 \Rightarrow Y_{s,\psi_1} = 1 \wedge \sum_{a \in \mathcal{I}(s), t \in S} X_{s,a} \cdot \mathbb{P}(s,a,t) \cdot Y_{t,\psi} \bowtie r \end{aligned}$$

   - $s \in S \setminus S_0$. As in the previous case, we have to distinguish the following two subcases according to $\bowtie$.
     - $\bowtie \in \{\leq, <\}$. Then $W_{s,\psi} \stackrel{\text{def}}{=} \mathbb{P}^+_s(\psi_1 \mathcal{U} \psi_2) \bowtie r$. According to Definition 5, it is easy to see that

$$\mathbb{P}^+_s(\psi) = \begin{cases} 1 & \text{if } Y_{s,\psi_2} = 1 \\ 0 & \text{if } Y_{s,\psi_1} = 0 \\ \max_{a \in \mathcal{I}(s)} \{ \sum_{t \in S \setminus S_0} \mathbb{P}(s,a,t) \cdot \mathbb{P}^+_t(\psi) \\ \qquad + \sum_{t \in S_0} \mathbb{P}(s,a,t) \cdot Y_{t,\psi} \} & \text{o.w.} \end{cases}$$

       However, different from the previous case, now it is difficult to reduce this problem to solving a pure linear programming, since in the definition, the case distinctions have to be involved. Fortunately, we can still borrow the same idea, because actually what we need is to express the linear inequation in $(\mathbb{R}, +, \cdot, \leq)$ rather than to find the solution concretely. By this observation, we set

$$\begin{aligned} &\min x_s \\ &\text{s.t. for any } a \in \mathcal{I}(s), \\ &\qquad \begin{cases} x_s = 1 & \text{if } Y_{s,\psi_2} = 1 \\ x_s = 0 & \text{if } Y_{s,\psi_1} = 0 \\ x_s \geq \sum_{t \in S \setminus S_0} \mathbb{P}(s,a,t) x_t + \sum_{t \in S_0} \mathbb{P}(s,a,t) Y_{t,\psi} & \text{o.w.} \end{cases} \end{aligned}$$

       For simplicity, for Z_s $(s \in S)$, we define $\Im(Z_s, Z_t)$ as

$$\begin{aligned} \Im(Z_s, Z_t) \stackrel{\text{def}}{=} \; &(Y_{s,\psi_2} = 1 \Rightarrow Z_s = 1) \wedge \\ &(Y_{s,\psi_1} = 0 \Rightarrow Z_s = 0) \wedge \\ &(Y_{s,\psi_2} = 0 \wedge Y_{s,\psi_1} = 1 \\ &\quad \Rightarrow Z_s \geq \sum_{t \in S \setminus S_0} \mathbb{P}(s,a,t) Z_t + \sum_{t \in S_0} \mathbb{P}(s,a,t) Y_{t,\psi}) \end{aligned}$$

It follows that $W_{s,\psi}$ is defined as

$$\exists\{Z_s \mid S \setminus S_0\}.(0 \leq Z_s \leq 1) \wedge \Im(Z_s, Z_t) \wedge (Z_s \bowtie r) \\ \wedge(\forall\{Z'_s \mid S \setminus S_0\}.(0 \leq Z'_s \leq 1) \wedge \Im(Z_s, Z_t) \Rightarrow (Z'_s \geq Z_s))$$

- $\bowtie \in \{\leq, <\}$. This is the duality of the previous case.

This completes our construction for formula $W_{s,\psi}$.

With $W_{s,\psi}$ on hand, we can fill the gap in (2), which completes the construction of $\mathcal{M}_D, s_{in} \models \Phi$. □

The correctness of our algorithm can be ensured by the following theorem, whose proof is the "reverse" of our construction shown above and thus is omitted.

Theorem 2. *For MDP $\mathcal{M}$, PCTL formula Φ, there exists an MR-controller for $\mathcal{M}$ if and only if (2) holds.*

Thus, we have the following corollary concerning on the complexity of the algorithm, according to [8][3].

Corollary 1. *For MDP $\mathcal{M}$, specification Φ, the problem of deciding whether there exists an MR-controller is in EXPTIME. Moreover, if such a controller does exist, it can be effectively constructed.*

References

1. C. Baier, M. Größer, M. Leucker, B. Bollig, and F. Ciesinski. Controller synthesis for probabilistic systems. In *Proceeding of IFIP TCS'04*. Kluwer, 2004.
2. A. Bianco and L. de Alfaro. Model checking of probabilistic and nondeterministic systems. In *Proceeding of FSTTCS'95*. LNCS 1026, 499-513, Springer, 1995.
3. S. Basu, R. Pollack and M. Roy. On the combinatorial and algebraic complexity of quantifier elimination. *Journal of ACM*, 43(6):1002-1045, 1996.
4. K. Chatterjee, M. Jurdzinski and T. Henzinger. Simple stochastic parity games. In *Proceeding of CSL'03*. LNCS 2803, 100-113, Springer, 2003.
5. K. Chatterjee, M. Jurdzinski and T. Henzinger. Quantitative simple stochastic partity games. In *Proceeding of SODA'04*. SIAM.
6. C. Courcoubetis and M. Yannakakis. The compleixity of probabilistic verification. *Journal of ACM*, 42(4):857-907, 1995.
7. J. Filar and K. Vrieze. *Competitive Markov Decision Processes*. Springer, 1997.
8. D. Grigoriev. Complexity of deciding Tarski algebra. *Journal of Symbolic Computation*, 5(1-2):65-108, 1988.
9. H. Hansson and B. Jonsson. A Logic for reasoning about time and reliability. *Formal Aspects of Computing*, 6(5): 512-535, 1994.
10. O. Kupferman, M. Vardi and P. Wolper. Module checking. *Information and Computation*. 164(2): 322-344, 2001.
11. G. Owen. *Game Theory*. Academic Press, 1995.
12. M. Puterman. *Markov Decision Processes*. Wiley, 1994.
13. A. Traski. *A Decision Method for Elementary Algebra and Geometry*. Univ. of California Press, Berkeley, 1951.
14. W. Thomas. Infinite games and verification. In *Proceeding of CAV'03*, LNCS 2725, 58-64, Springer, 2003.

Improved Automatic Gain Control Circuit Using Fuzzy Logic

Jong-Won Kim, Liang Zhang, Jae-Yong Seo, Hyun-Chan Cho, Hwa-Il Seo, Tai-Hoon Cho, and Jong-Dae Jung

Depart of Electrical and Electronics Engineering,
Korea University of Technology and Education,
307, Gar-Jeon, Byeong-Cheon, Cheon-An, Chung-Nam, Korea
{Kamuiai, zzzlll, cholab, sjyong, hiseo, thcho, jungjd}@kut.ac.kr

Abstract. A problem that arises in most communication receivers concerns the wide variation in power level of the signals received at the antenna. These variations cause serious problems which are usually be solved in receiver design by using Automatic Gain Control (AGC). AGC is achieved by using an amplifier whose gain can be controlled by using external current or voltage. We have to note that the AGC circuit does not respond to rapid changes in the amplitude of input and multifrequency. Nowadays, with the development of the fuzzy theory, the advantages of the fuzzy logic are recognized widely and deeply. Applying fuzzy logic to AGC circuit is a way to enhance AGC circuit.

1 Introduction

A problem that arises in most communication receivers concerns the wide variation in power level of the signals received at the antenna. This variation is due to a variety of causes. For example, in a space-communications system the satellite or spaceship transmitter may be continuously altering its position with respect to the ground receiver. In receiver design these variations cause serious problems which can usually be solved by using automatic gain control (AGC)[1][2]. AGC is one method to adjust automatically the gain of the amplifier circuit according to the intensity of signal by an external current or voltage. The design is superior but it is not adapted to the fast, wide range changing signal and different frequency signal due to the existence of capacitor. For example, radio as one kind of receiver can receive different frequency signals. However, for the fixed capacitor, AGC circuit can not adapt itself to different frequency. We can solve the problem using fuzzy logic algorithm (FLA). Fuzzy Logic is a paradigm for an alternative methodology which can be applied in developing both linear and nonlinear systems for embedded control [3]. In this case, using FLA, it is not needed to calculate exactly the signal relation between input and output. This paper is organized as follows: In section 2, the configuration of AGC circuit is introduced. The improved AGC system using fuzzy logic is designed in section 3, simulation results are shown in section 4. Finally, conclusions are presented in section 5.

L. Wang et al. (Eds.): FSKD 2006, LNAI 4223, pp. 159 – 168, 2006.

2 AGC Circuit

2.1 Theory of Automatic Gain Control

Commercially available AGC circuits, such as the LM13600 AGC amplifier, employ a basic control current source within an OP-amp, as shown in figure 1. In this circuit transistor T3 act as the constant-current source supplying current I_{AGC} [1], where

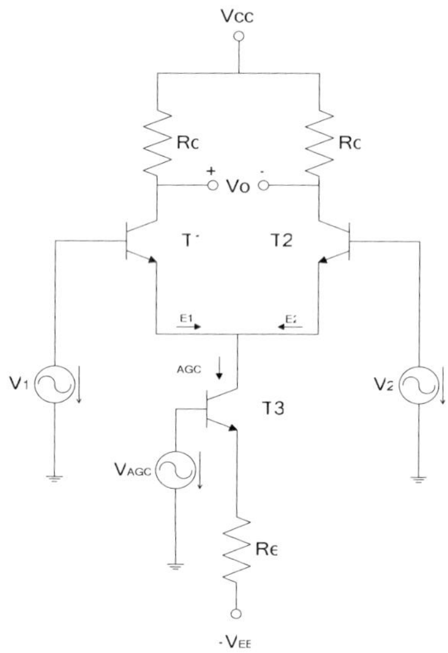

Fig. 1. A gain-control difference amplifier

$$I_{AGC} = \frac{V_{AGC} - 0.7 + V_{EE}}{R_e} \tag{2-1}$$

Using the small signal emitter currents analysis, we can know that the output voltage of v_o (setting $v_1 = 0$) is

$$v_o = (\frac{R_c}{2V_T} I_{AGC}) v_2(t) \tag{2-2}$$

If $v_2(t) = v_{2m}(t)\cos\omega_0 t$, then the amplitude voltage of the output is kept constant.

If we arrange to have I_{AGC} inversely proportional to the envelop of the input voltage,

$$I_{AGC} = \frac{K}{V_{2m}(t)} \tag{2-3}$$

Then (2-2) becomes

$$v_o = (\frac{R_c K}{2V_T})\frac{v_{2t}}{v_{2m}(t)} \tag{2-4}$$

From (2-4), we can know that output is not depended on input signal.

2.2 Calculation of the Output Voltage of the AGC System

The output voltage v_{o1} of the gain-controlled op-amp A_1 shown in figure 2 [1].

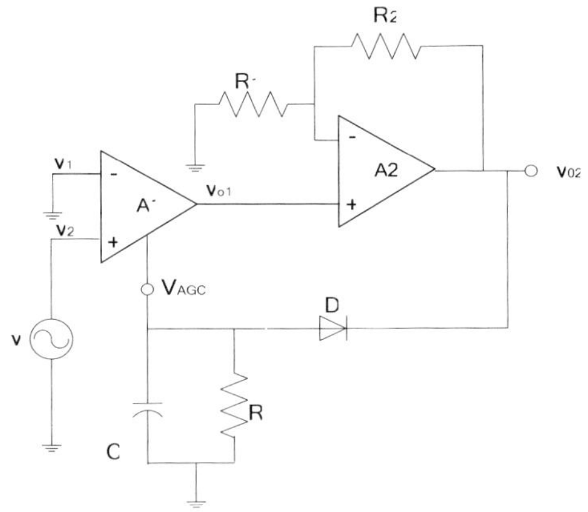

Fig. 2. Simple AGC system

$$v_{o1} = (\frac{R_c K_1}{2V_T} I_{AGC})v_i \tag{3-1}$$

The output voltage v_{o2} of the entire AGC amplifier is then

$$v_{o2} = (\frac{R_c}{2V_T} K_1 K_2 I_{AGC})v_i \tag{3-2}$$

where $K_2 = 1 + R_2 / R_1$, the gain of amplifier A_2. The output voltage $v_{o2}(t)$ is envelope detected so that the AGC voltage V_{AGC} is negative voltage. If $v_i(t) = V_{im}(t)\cos\omega_0 t$ and $v_o(t) = -V_{o2}(t)\cos\omega_0 t$, the AGC voltage can be shown as

$$V_{AGC} = -V_{o2m}(t) = -(\frac{R_c}{2V_T} K_1 K_2 I_{AGC})V_{im}(t) \tag{3-3}$$

The gain-control voltage and current are also related by (2-1). Substituting (3-3) into (2-1) and solving for I_{AGC} yields.

$$I_{AGC} = \frac{V_{EE} - 0.7}{R_e + (R_c / 2V_T) K_1 K_2 V_{im}(t)} \tag{3-4}$$

The gain-control voltage and current are also related by (2-1). Substituting (3-3) into (2-1) and solving for I_{AGC} yields.

$$v_{o2} = (\frac{R_c}{2V_T} K_1 K_2) \frac{V_{EE} - 0.7}{1 + (R_c / 2V_T R_e) K_1 K_2 V_{im}(t)} \tag{3-5}$$

The gain K_1 and K_2 are usually made large enough to ensure that

$$\frac{R_c}{2V_T} K_1 K_2 >> 1 \tag{3-6}$$

Finally

$$v_{o2} \approx (V_{EE} - 0.7) \frac{v_i(t)}{V_{im}(t)} \tag{3-7}$$

This important result indicates that v_{o2} is proportional to $v_i(t)/V_{im}(t)$. This ratio has a constant envelope since the envelope of $v_i(t)$ is $V_{im}(t)$. As a result of the peak-detector action the AGC circuit responds only to slowly varying changes in the envelope to changes in signal power. A typical value for the RC time constant of the peak detector is 1sec [1].

It is interesting to note that the AGC circuit does not respond to rapid changes in the amplitude of v_i. If the amplitude of v_i were to change instantaneously, then even if OP-amps could follow the change, the envelop detector capacitor could not, since the capacitor's voltage could not change instantaneously. Hence, in response to such a change, (3-6) no longer applies and $v_{o2}(t)$ is proportional to $v_i(t)$ until steady state is reached. Thus, an AGC circuit is considered a "slow-acting" limiter. Otherwise, the peak value detector is useful for just one frequency because of the fixed RC circuit. Therefore, the two disadvantages make the AGC system tremendously limited.

2.3 One Example of AGC System in Practice

Fig. 4 shows a good example of AGC circuit which gets the I_{AGC} by feeding back circuit. We can see that if the output is big enough the Q1 passes. Therefore, a corresponding I_{AGC} to get the voltage of the V_{AGC} point goes down to make the input point current decreases, so the output I_{AGC} keeps the same level. On the other hand, if the output decreases, the function can make the voltage of the V_{AGC} goes up to maintain the output the same level [3].

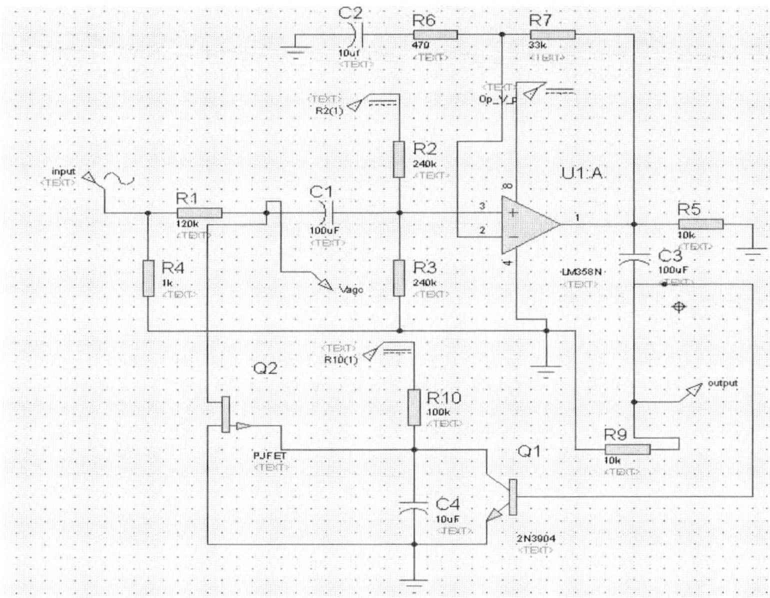

Fig. 4. Simulation of the AGC Circuit

2.4 Problem of the AGC Circuit

Referring to the figure 4, Q2, coupled with R2 and the equivalent resistance of R3 and R4, form a voltage divider to the input signal source. With input levels below 40mv p-p, the input is evenly divided between R2 (120k) and R3||R4 (120k). The output amplitude of LM358 isn't large enough to turn on Q2, which acts as a positive peak detector. The gate of the JFET is pulled to +5V, pinching its channel off and creating a very high resistance from drain to source. This essentially removes it from the circuit.

At input levels above 40mv p-p, Q1 is turned on at the positive peaks of the output of LM358, lowering the JFETs gate to source voltage. The channel resistance decreases and attenuates the input signal to maintain the output at approximately 1.2V p-p.

3 Improved AGC System Using Fuzzy Logic Algorithm (FLA)

3.1 Block Diagram of Conventional AGC System

Figure 5 shows the block diagram of the conventional AGC system which has been described in previous section and figure 6 shows the proposed AGC system. Fuzzy logic describes complex systems using knowledge and experience by fuzzy rules. It does not require system modeling or complex math equations governing the relationship between inputs and outputs. To overcome the problems of AGC circuit that discussed in previous section, we just need to change the envelope detector part by FLA. And we need not to calculate strictly the relation between the input and output voltage.

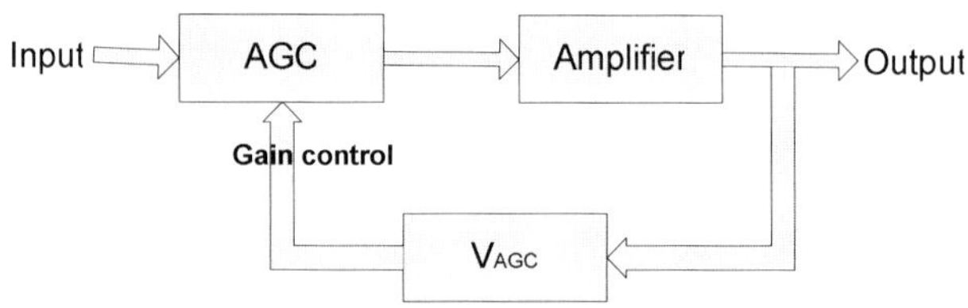

Fig. 5. Conventional AGC System

3.2 Improved AGC System

The example shown in figure 4 can be constructed as shown in figure 6. Use the FLA to replace the RC envelope detector part of conventional AGC.

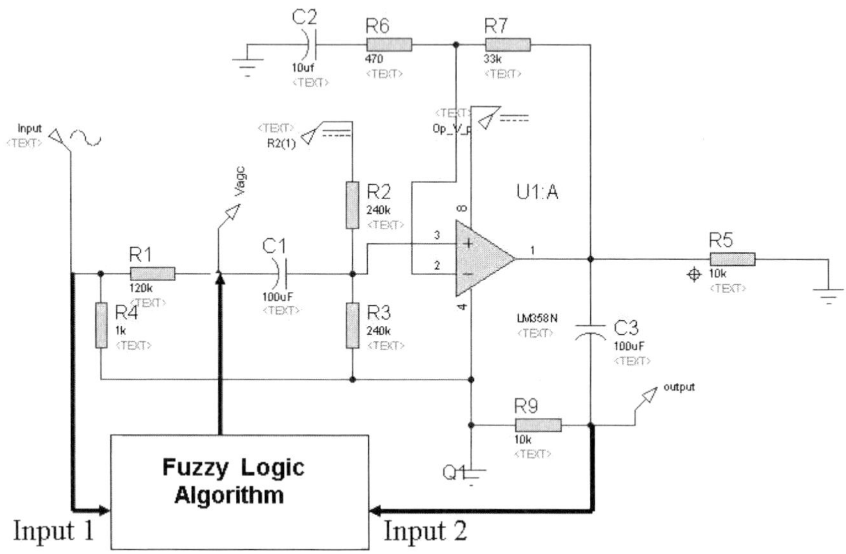

Fig. 6. Improved AGC circuit Using FLA

3.3 Construction of the Fuzzy Logic System

Determination of the state variables and control variables:

(a). state variables (the input variable of the FLC)
  -the input of the circuit, the output of the circuit
(b). control variable (the output variable of the FLC)
  -Vagc

Using Matlab 7.0 we can make the fuzzy rule and get the output. The input1 of the fuzzification is the input of AGC circuit, the input2 is the output of AGC [1],[2].

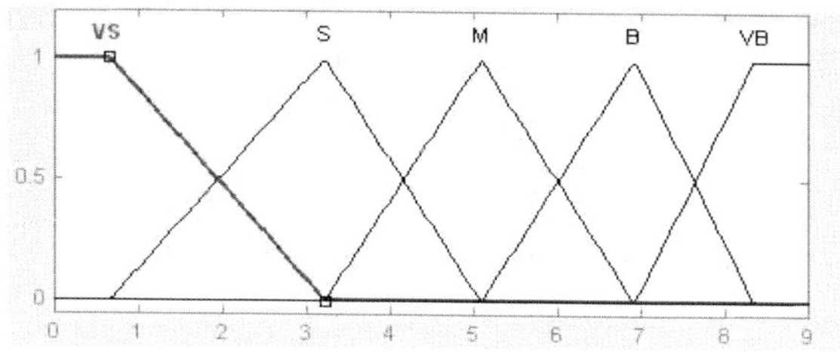

Fig. 7. Input1

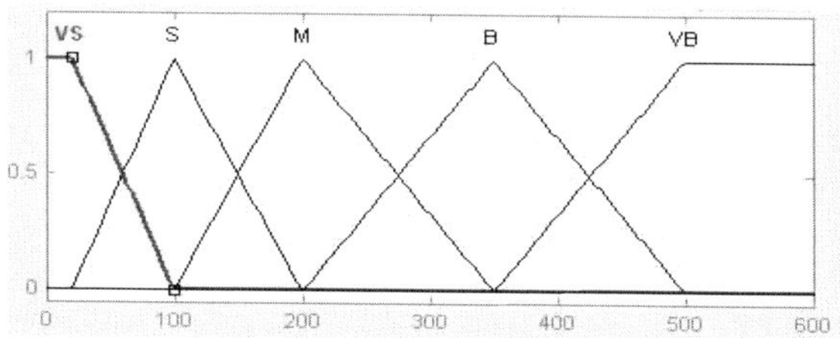

Fig. 8. Input2

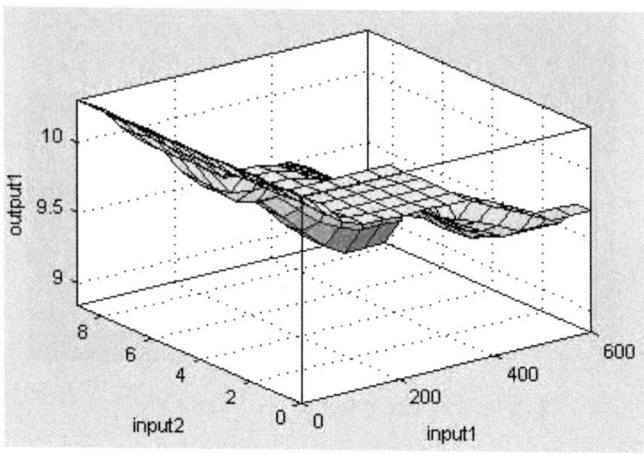

Fig. 9. Output surface

Use the input and output of the AGC circuit as the fuzzy input, and by correct fuzzy rules, we can get the Fuzzy Logic Controller output-gain control signal- Vagc.

4 Computer Simulation

4.1 Simulation of the Conventional AGC Circuit

In the Proteus 6 Professional which is a kind of circuit simulation tool, we can simulate conventional AGC circuit conveniently. Figure 4 is drawn using Proteus. After get the circuit pass, input a suitable signal and run the program, we can get the result as shown in figure 10.

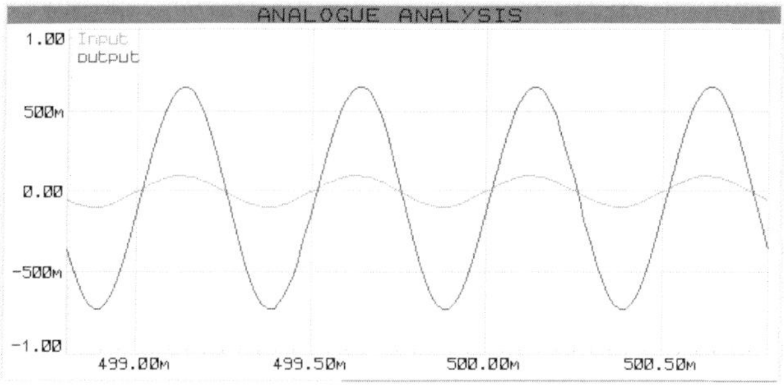

Fig. 10. Result of the Simulation

The green line (small signal) is the input signal and the red line (big signal) is the output signal. If we change the peak amplitude of the input signal at the period of 50mv ~ 1.2v, we can find the approximate output 650mv. When we apply the windows wave file (chord.wav) as the input signal, we get the following chart.

Fig. 11. The input of the Chord.wav

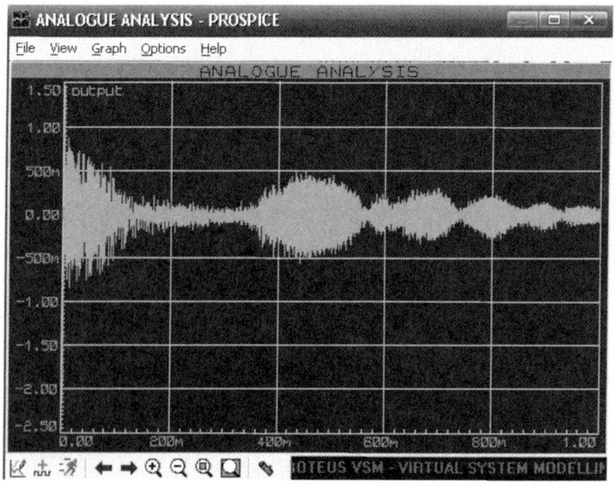

Fig. 12. The Output of the Chord.wav

From figure 12, we can see obviously that when the input signal is too small, the output can not be kept constant as what we have talked about before. So we can listen a short sound at the beginning of the signal but then there is no sound.

4.2 Simulation of the Improved AGC Circuit

Table 1 shows the data testing. Input one voltage signal and we can get the output in Proteus simulation. And then use the MATLAB fuzzy tool box what we have constructed fuzzy rule to calculate the Vagc control signal and then input the Vagc signal to the simulation circuit we can get the date testing table.

Table 1. Data testing

	V_{in}	V_{out1}	V_{AGC}	V_{out2}
1.	8mv	318mv	10.3mv	705mv
2.	70mv	2.47v	10.1mv	713mv
3.	230mv	5.55v	9.92mv	700mv
4.	379mv	7.28v	9.20mv	650mv
5.	463mv	8.01v	9.01mv	636mv

Then we simulate the No.3 data to get the figure following:

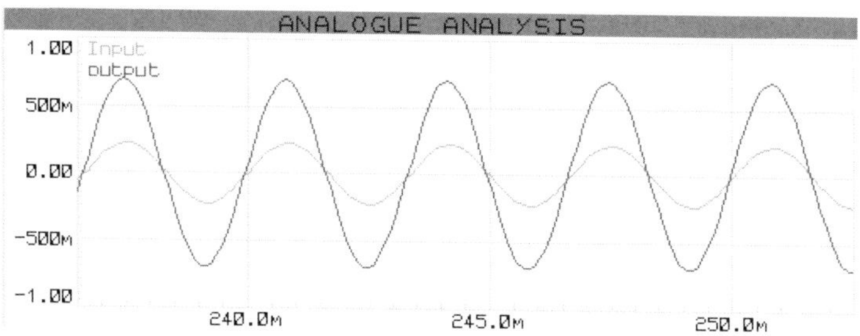

Fig. 13. Simulation Result

From this figure we can see that even the input signal is very small, the output is kept constant.

5 Conclusion

AGC circuit is an ingenious circuit and is used widely. In practice, capacitors are fixed to boards, so it is very hard to change the capacity of them. During the transmission of signals, there are too many unknown conditions affection. Therefore we have to find a way to increase the capacity of receiver part. When we add the FLA to AGC, we can make it more precise and efficient.

Acknowledgement

This work was supported by grant No. RTI04-01-02 from the Regional Technology Innovation Program of the Ministry of Commerce, Industry and Energy(MOCIE).

References

1. Donald L. Schilling. Charles Belove: Electronic Circuits. 3rd edn. McGraw-Hill (1989)
2. Adel S. Sedra, Kenneth C. Smith: Microelectronic Circuits. 3rd edn. (1991)
3. Zeungnam Bien: Fuzzy-Logic Control, Hong-Neung Science Books (1998)
4. Young Gu Kim: Apply Fuzzy Logic to Auto Gain Control. KUT press(2001) 12-21
5. Klir, G. J. and Yuan, B.: Fuzzy sets and Fuzzy Logic. Prentice Hall, Upper Saddle River, NJ(1995)
6. Robert T. Paynter: Electronic Devices And Circuits, Fifth Edition, Prentice-Hall (2000)

Reliable Control of Fuzzy Descriptor Systems with Time-Varying Delay*

Yuhao Yuan[1], Zhonghu Yuan[2], Qingling Zhang[1], Daqing Zhang[1], and Bing Chen[3]

[1] Institute of Systems Science, Northeastern University, Shenyang 110004
[2] School of Information Engineering, Shenyang University, Shenyang 110044
[3] Institute of Complexity Science, Qingdao University, Shandong 266071
yyhmds@sohu.com, qlzhang@mail.neu.edu.cn

Abstract. The reliable fuzzy controller design problem of T-S fuzzy descriptor systems with time-varying delay is introduced. Based on linear matrix inequality approach, a less conservative reliable controller design method is presented. The resulting fuzzy control systems are reliable in the sense that asymptotic stability is achieved not only when all control components are operating well, but also in the presence of some component failures. Moreover, the result is extended to the case of observer-based reliable fuzzy control.Two numerical examples are also given to illustrate the design procedures and their effectiveness.

1 Introduction

Reliable control is an effective approach to improve system reliability. The kernel idea of this approach is to design a fixed controller such that the closed-loop can maintain stability and performance, not only when all control components are operational, but also in the case of some admissible control component outages. In the past two decades, reliable control problems have been extensively studied by many researchers [1, 2, 3, 4].

On the other hand, many complex nonlinear systems can be expressed in a certain form of mathematical models locally. Takagi and Sugeno have proposed a fuzzy model to describe the complex systems [5]. In this T-S fuzzy model, local dynamics in different state space regions are represented by local linear systems. The overall model is obtained by 'blending' these linear models through membership functions. As a common belief, the control technique based on the T-S fuzzy model is conceptually simple and effective for the control of complex systems.

As to reliable control of T-S fuzzy systems, progress has been made in most recent years too [6, 7, 8]. In literature [6, 7], reliable controller design are based on a assumption that control component failures are modeled as outages, i.e., when a failure occurs, the signal (in the case of sensors) or the control action (in the case of actuators) simply becomes zero. In [8], a more general failure model

* Supported by Natural Science Foundation of Liaoning Province (No.20042001).

L. Wang et al. (Eds.): FSKD 2006, LNAI 4223, pp. 169–178, 2006.

is adopted for actuator failures, which studied problem that control components being failure to some extent, i.e., the failure coefficients take value in the interval [0, 1]. When the actuator is invalid but kept in the admissible area, the controller will stabilizes the system.

In 1999, Tanniguchi and Tanaka et al extended the T-S fuzzy model to descriptor nonlinear systems [9, 10]. They brought the concept of T-S fuzzy descriptor systems forward. But so far, the reliable control problems for T-S fuzzy descriptor system has scarcely been studied.

Time-delays often occur in many dynamic systems, it has been shown that the existence of delays usually becomes the source of instability and deteriorates performance of systems. So, it is worth to study a system with time-delay both theoretically and practicality.

In practical situations, failure of actuators often occurs. Thus, from a safety point as well as a performance point of view, an important requirement is to have a reliable control such that the stability and performance of the closed-loop system can tolerate actuator failures. In this paper, we will consider the reliable control problem of T-S fuzzy descriptor systems with time-delay.

The paper is organized as follows. Firstly, the problem is formulated. In section 3, based on the solvability of LMIs, taking account of affects of all subsystems, which gives the design method of the reliable state feedback controller. In section 4, the result obtained in section 3 is extended to the case of observer-based reliable fuzzy control. In section 5, numerical examples are used to illustrate the results. Finally, concluding remarks are made in section 6.

Notations: Matrix $X > 0$ $(X \geq 0)$ denotes that$X$ is a positive (semi-positive) definite matirx, $A > (\geq) B$ denotes $A - B > (\geq) 0$. Symbol I stands for the unit matrix with appropriate dimensions.

2 Problem Formulation and Failure Model

In this section, if the uncertain system parameter information is considered, the nonlinear descriptor system can be presented as an uncertain fuzzy descriptor model with time-varying delay. The ith fuzzy rule is of the following form:

$$\begin{aligned} R_i : \text{IF } z_1 \text{ is } & N_{i1} \text{ and} \cdots z_p \text{ is } N_{ip} \text{ , THEN} \\ E\dot{x}(t) &= (A_i + \Delta A_i(t)) x(t) + (A_{1i} + \Delta A_{1i}(t)) x(t - \tau(t)) \\ &\quad + (B_i + \Delta B_i(t)) u(t) \ , \\ y(t) &= C_i x(t) \ , \\ x(t) &= \varphi(t) \ , t \in [-\tau_0, 0] \ . \quad i = 1, 2, \ldots, r \ . \end{aligned} \tag{1}$$

Where N_{ij} are the fuzzy sets, $z_1, z_2, \cdots, z_p$ are premise variables. Scalar r is the number of IF-THEN rules. $E \in R^{n \times n}$ may be singular. $x(t) \in R^n$ is the state, $u(t) \in R^m$ is the control input and $y(t) \in R^l$ is the output. $A_i, A_{1i} \in R^{n \times n}, B_i \in R^{n \times m}, C_i \in R^{l \times n}, i = 1, 2, \cdots, r$. The $\tau(t)$ is the time-varying delay and satisfies $0 < \tau(t) \leq \tau_0 < \infty$, τ_0 is the upper bound of the delay. It is further assumed that $\dot{\tau}(t) \leq \beta < 1$ and β is a known constant. $\varphi(t)$ is initial vector. $\Delta A_i(t), \Delta A_{1i}(t), \Delta B_i(t)$ denote the uncertainties and take the form of $[\Delta A_i(t) \quad \Delta A_{1i}(t) \quad \Delta B_i(t)] = MF(t)[E_i \quad E_{1i} \quad E_{bi}]$, where M, E_i, E_{1i}, E_{bi} are known constant matrices and $F(t)$ is an unknown matrix function, $F(t)$ satisfying $F(t)^T F(t) \leq I$. The fuzzy descriptor model is assumed to be locally regular, i.e., exist $s_i \in C$, such that $\det(s_i E - A_i - \Delta A_i(t)) \neq 0 (i = 1, 2, \cdots, r)$.

By taking a standard fuzzy inference strategy, that is, using a singleton fuzzifier, procedure fuzzy inference and center average defuzzifier, the final fuzzy model of the systems is inferred as follows

$$E\dot{x}(t) = \sum_{i=1}^{r} \lambda_i(z)\left[(A_i + \Delta A_i(t))x(t) + (A_{1i} + \Delta A_{1i}(t))x(t - \tau(t)) + (B_i + \Delta B_i(t))u(t)\right], \tag{2}$$

where $\lambda_i(z) = \prod_{j=1}^{p} N_{ij}(z_j) / \sum_{i=1}^{r}\prod_{j=1}^{p} N_{ij}(z_j)$, $\sum_{i=1}^{r}\lambda_i(z) = 1$. $N_{ij}(z_j)$ is the grade of membership of z_j in N_{ij}. For simplicity, $x, x_\tau, u, \lambda_i, \Delta A_i, \Delta A_{1i}, \Delta B_i$ will be used instead of $x(t), x(t - \tau(t)), u(t), \lambda_i(z), \Delta A_i(t), \Delta A_{1i}(t), \Delta B_i(t)$, respectively.

Consider a nonlinear descriptor system $E\dot{x} = f(x, u)$, where $E \in R^{n\times n}, x \in R^n, u \in R^m, \det(E) = 0$. The following definition regarding the solvability of the nonlinear descriptor system was given in [11].

Definition 1. *If for any piecewise continuous input u and initial state x_0, there always exists a unique differentiable solution x with $x(0) = x_0$, then system $E\dot{x} = f(x, u)$ is called solvable.*

The purpose of this paper is to design a reliable controller. The analysis is developed under the assumption that the T-S fuzzy descriptor system is solvable.

Lemma 1. [12] *Let $M, E, F(t)$ be real matrices of appropriate dimensions with $F^T(t)F(t) \leq I$. Then for any scalar $\varepsilon > 0$,*

$$MF(t)E + E^T F^T(t) M^T \leq \varepsilon M M^T + \varepsilon^{-1} E^T E .$$

3 Reliable Control Via State Feedback for T-S Model

First, the reliable fuzzy controller will be designed to stabilize system (2). The fuzzy controller shares the same fuzzy sets with the fuzzy model in the premise parts and has local linear controller in the consequent parts. The ith fuzzy rule of the fuzzy controller is of the following form

$$R_i : \text{IF } z_1 \text{ is } N_{i1} \text{ and} \cdots z_p \text{ is } N_{ip}, \text{THEN}$$

$$u = K_i x , \quad i = 1, 2, \cdots, r .$$

For fuzzy reliable control problems, the following actuator fault model is used.

$$R_i : \text{IF } z_1 \text{ is } N_{i1} \text{ and} \cdots z_p \text{ is } N_{ip}, \text{THEN}$$

$$u = \Phi_{\omega i} K_i x \ , \ i = 1, 2, \cdots, r \ .$$

Where $K_i, i = 1, 2, \cdots, r$ are the local linear feedback gains. Hence, the overall fuzzy controller is given by

$$u_\omega = \sum_{i=1}^{r} \lambda_i \Phi_{\omega i} K_i x \ , \tag{3}$$

where $\Phi_{\omega i} = \text{diag} \ [\delta_{\omega i}(1), \delta_{\omega i}(2), \cdots, \delta_{\omega i}(m)]$, $\delta_{\omega i}(j) \in [0, 1]$, $i = 1, 2, \cdots, r$, $j = 1, 2, \cdots, m$. Matrix $\Phi_{\omega i}$ describes the fault extent. $\delta_{\omega i}(j) = 0$ means that the jth component in the ith local actuator is invalid, $\delta_{\omega i}(j) \in (0, 1)$ implies that the jth component is at fault in some extent and $\delta_{\omega i}(j) = 1$ denotes that the jth component operates properly. Thus, for a given diagonal matrix $\Phi_{\Omega i}, i = 1, 2, \cdots, m$. the set $\Omega = \{u_\omega = \sum_{i=1}^{r} \lambda_i \Phi_{\omega i} K_i x, \text{and } \Phi_{\omega i} \geq \Phi_{\Omega i}, i = 1, 2, \cdots, m\}$ is called an admissible set of actuator fault. Namely, symbol $\Phi_{\Omega i}, i = 1, 2, \cdots, m$ in set Ω describes the worst status of the scaling factor $\Phi_{\omega i}, i = 1, 2, \cdots m$. Once the scaling factor extent become smaller than $\Phi_{\Omega i}$, the reliable controller can not work properly anymore .

Remark 1. It is obvious that when $\Phi_{\omega i} = \Phi_\omega, i = 1, 2, \cdots, r$, and $\delta_\omega(j)$ takes only the values of 0 and 1, the actuator failure model is just the same as that in [6,7] and the references cited therein.From this point,the problem to be solved here is more general.

For the case of $u_\omega \in \Omega$, the closed-loop system is given by

$$E\dot{x} = \sum_{i=1}^{r} \sum_{j=1}^{r} \lambda_i \lambda_j \left[(A_i + \Delta A_i) x + (A_{1i} + \Delta A_{1i}) x_\tau + (B_i + \Delta B_i) \Phi_{\omega j} K_j x \right] \ . \tag{4}$$

Theorem 1. *Consider system (4), if there exist nonsingular matrices $P, S > 0, K_i, X_{ij}$, where $X_{ii} = X_{ii}^T, X_{ij} = X_{ji}^T, i \neq j, (i, j = 1, 2, \cdots, r)$, such that*

$$E^T P = P^T E \geq 0 \ , \tag{5}$$

$$\Psi_{ii} = \begin{bmatrix} \Theta_{ii} & P^T (A_{1i} + \Delta A_{1i}) \\ * & -S \end{bmatrix} < X_{ii} \ , \tag{6}$$

$$\Psi_{ij} = \begin{bmatrix} \Theta_{ij} + \Theta_{ji} & P^T (A_{1i} + \Delta A_{1i} + A_{1j} + \Delta A_{1j}) \\ * & -2S \end{bmatrix} \leq X_{ij} + X_{ij}^T \ , \ i < j, \tag{7}$$

$$X = (X_{ij})_{r \times r} = \begin{bmatrix} X_{11} & \cdots & X_{1r} \\ \vdots & \ddots & \vdots \\ X_{r1} & \cdots & X_{rr} \end{bmatrix} < 0 \ . \tag{8}$$

Where $\Theta_{ij} = P^T[(A_i + \Delta A_i) + (B_i + \Delta B_i)\Phi_{\omega j} K_j] + [(A_i + \Delta A_i) + (B_i + \Delta B_i)\Phi_{\omega j} K_j]^T P + \frac{1}{1-\beta} S$ and "$*$" denotes the transposed elements in the symmetric positions. Then the resultant closed-loop system (4) is asymptotically stable for any $u_\omega \in \Omega$.

Proof. Construct a Lyapunov function as

$$V(t) = x^T E^T P x + \frac{1}{1-\beta} \int_{t-\tau(t)}^{t} x^T(s) S x(s)\, ds ,$$

where $E^T P = P^T E \geq 0, S > 0$. Differentiating $V(t)$ along the trajectory of system (4) gives

$$\dot{V} \leq \sum_{i=1}^{r} \lambda_i^2 \xi^T \Psi_{ii} \xi + \sum_{i=1}^{r} \sum_{i<j}^{r} \lambda_i \lambda_j \xi^T \Psi_{ij} \xi \leq \begin{bmatrix} \lambda_1 \xi \\ \vdots \\ \lambda_r \xi \end{bmatrix}^T \begin{bmatrix} X_{11} & \cdots & X_{1r} \\ \vdots & \cdots & \vdots \\ X_{r1} & \cdots & X_{rr} \end{bmatrix} \begin{bmatrix} \lambda_1 \xi \\ \vdots \\ \lambda_r \xi \end{bmatrix} < 0 ,$$

where $\xi = \begin{bmatrix} x \\ x_\tau \end{bmatrix}$, This completes the proof. □

Remark 2. It is worthwhile to be pointed out that for a system with large dimension, more than two subsystems are often activated at the same time. So, the interactions of subsystems are taken into account in Theorem 1. The method to do this is to introduce the relaxation matrix X into Theorem 1, which was firstly utilized in [13] and improved by [14] .

We will give out the scheme of how to design the reliable controller. The main idea is to convert conditions in Theorem 1 into LMI conditions.

Theorem 2. *Consider system (4), if there exist* $\varepsilon_i > 0, \varepsilon_{ij} > 0$, *nonsingular matrix* X, $Y > 0, Z_{ij}$, *where* $Z_{ii} = Z_{ii}^T, Z_{ij} = Z_{ji}^T, i \neq j,\ i,j = 1,2,\cdots,r$ *such that the following LMIs are satisfied:*

$$X^T E^T = EX \geq 0 , \tag{9}$$

$$\begin{bmatrix} \begin{matrix} A_i X + X^T A_i^T - B_i \Phi_{\Omega i} B_i^T \\ -Z_{ii1} + \varepsilon_i M M^T \end{matrix} & A_{1i} Y - Z_{ii2} & 0 & X^T E_i^T & X^T \\ * & -Y - Z_{ii3} & 0 & Y E_{1i}^T & 0 \\ * & * & -I & E_{bi}^T & 0 \\ * & * & * & -\varepsilon_i I & 0 \\ * & * & * & * & -(1-\beta) Y \end{bmatrix} < 0 , \tag{10}$$

$$\begin{bmatrix} \Omega & (A_{1i} + A_{1j}) Y - Z_{ij2} - Z_{ij3}^T & 0 & X^T (E_i + E_j)^T & X^T \\ * & -2Y - Z_{ij4} - Z_{ij4}^T & 0 & Y (E_{1i} + E_{1j})^T & 0 \\ * & * & -I & [E_{bi} \ \ E_{bj}]^T & 0 \\ * & * & * & -\varepsilon_{ij} I & 0 \\ * & * & * & * & -\frac{(1-\beta)}{2} Y \end{bmatrix} < 0 , \tag{11}$$

$$\begin{bmatrix} Z_{11} & \cdots & Z_{1r} \\ \vdots & \ddots & \vdots \\ Z_{r1} & \cdots & Z_{rr} \end{bmatrix} < 0 . \tag{12}$$

Where $\Omega = (A_i + A_j)X + X^T(A_i + A_j)^T + 2(B_i - B_j)(B_i - B_j)^T - B_i\Phi_{\Omega j}B_i^T - B_j\Phi_{\Omega i}B_j^T - Z_{ij1} - Z_{ij1}^T + \varepsilon_{ij}MM^T$. $Z_{ij}s$ *are partitioned as* $Z_{ii} = \begin{bmatrix} Z_{ii1} & Z_{ii2} \\ * & Z_{ii3} \end{bmatrix}$, $Z_{ij} = \begin{bmatrix} Z_{ij1} & Z_{ij2} \\ Z_{ij3} & Z_{ij4} \end{bmatrix}$. *Then, the control gains are given by* $K_i = -B_i^T X^{-1}, i = 1, 2, \cdots, r$. *and the resultant closed-loop system (4) is asymptotically stable for any* $u_\omega \in \Omega$.

Proof. The proof is omitted because of the limited space. □

4 Observer-Based Reliable Control for T–S Model

In many cases, states are unknown or partly detected. Therefore, it is needed to estimate states. If the controller is designed with the effect of time-delay, the delay must be known exactly. But it is usually impossible to know the delay exactly. Here, we manage to design the state-observer not affected by the delay.

We consider system (2) without uncertainties. Construct the fuzzy observer

$$\begin{aligned} E\dot{\hat{x}} &= \sum_{i=1}^{r} \lambda_i \left[(A_i\hat{x} + B_i u) + L_i(y - \hat{y})\right] , \\ \hat{y} &= \sum_{i=1}^{r} \lambda_i C_i \hat{x} . \end{aligned} \tag{13}$$

where L_i is the observer gain. Define the estimation error as $e = x - \hat{x}$, then

$$E\dot{e} = E\dot{x} - E\dot{\hat{x}} = \sum_{i=1}^{r}\sum_{j=1}^{r} \lambda_i\lambda_j[(A_i - L_iC_j)e + A_{1i}x_\tau] .$$

Consider the following fuzzy controller $u = \sum_{i=1}^{r} \lambda_i K_i \hat{x}$. For any actuator failures $u_\omega \in \Omega$, the system can be expressed as follows:

$$\bar{E}\dot{\bar{x}} = \sum_{i=1}^{r}\sum_{j=1}^{r} \lambda_i\lambda_j[\bar{A}_{ij}\bar{x} + \bar{A}_{1i}\bar{x}_\tau + \bar{B}_i\Phi_{\omega j}\bar{K}_j\bar{x}] , \tag{14}$$

where $\bar{E} = \begin{bmatrix} E & 0 \\ 0 & E \end{bmatrix}$, $\bar{A}_{ij} = \begin{bmatrix} A_i & 0 \\ 0 & A_i - L_iC_j \end{bmatrix}$, $\bar{A}_{1i} = \begin{bmatrix} A_{1i} & 0 \\ A_{1i} & 0 \end{bmatrix}$, $\bar{B}_i = \begin{bmatrix} B_i \\ 0 \end{bmatrix}$, $\bar{K}_j = \begin{bmatrix} K_j & -K_j \end{bmatrix}$, $\bar{x} = \begin{bmatrix} x \\ e \end{bmatrix}$.

Theorem 3. *Consider system (14), if in Step 1 there exist nonsingular matrix* X_1, P_2, $Y_1 > 0$, $Y_2 > 0$, M_i, $\bar{Z}_{ii11} < 0$, $\tilde{Z}_{ii13} < 0$, $\bar{Z}_{ii31} < 0$, $\bar{Z}_{ii33} < 0$, $\bar{Z}_{ii21}$, $\tilde{Z}_{ii24}$, $\bar{Z}_{ij11}$, $\tilde{Z}_{ij14}$, $\bar{Z}_{ij21}$,, $\tilde{Z}_{ij24}$, $\bar{Z}_{ij31}^T$, $\tilde{Z}_{ij34}^T$, $\bar{Z}_{ij41}$, $\bar{Z}_{ij44}$, *such that*

$$X_1^T E^T = EX_1 \geq 0 , \quad E^T P_2 = P_2^T E \geq 0 , \tag{15}$$

$$\begin{bmatrix} \begin{matrix} A_iX_1 + X_1^TA_i^T \\ -2B_i\Phi_{\Omega i}B_i^T - \bar{Z}_{ii11} \end{matrix} & A_{1i}Y_1 - \bar{Z}_{ii21} & X_1^T \\ * & -Y_1 - \bar{Z}_{ii31} & 0 \\ * & * & -(1-\beta)Y_1 \end{bmatrix} < 0 , \tag{16}$$

$$\begin{bmatrix} \begin{array}{l}(A_i+A_j)X_1+X_1^T(A_i+A_j)^T\\ +2(B_i-B_j)(B_i-B_j)^T\\ -B_i\Phi_{\Omega j}B_i^T-B_j\Phi_{\Omega i}B_j^T\\ -\bar{Z}_{ij11}-\bar{Z}_{ij11}^T\end{array} & (A_{1i}+A_{1j})Y_1-\bar{Z}_{ij21}-\bar{Z}_{ij31}^T & X_1^T \\ * & -2Y_1-\bar{Z}_{ij41}-\bar{Z}_{ij41}^T & 0 \\ * & * & -\frac{(1-\beta)}{2}Y_1 \end{bmatrix}<0 \tag{17}$$

$$\begin{bmatrix} \begin{array}{l}P_2^TA_i-M_iC_i\\ +\left(P_2^TA_i-M_iC_i\right)^T-\tilde{Z}_{ii13}\end{array} & -\tilde{Z}_{ii24} & I \\ * & -Y_2-\bar{Z}_{ii33} & 0 \\ * & * & -(1-\beta)Y_2 \end{bmatrix}<0\ , \tag{18}$$

$$\begin{bmatrix} \begin{array}{l}P_2^T(A_i+A_j)-(M_iC_j+M_jC_i)\\ +(A_i+A_j)^TP_2-(M_iC_j+M_jC_i)^T\\ -\tilde{Z}_{ij14}-\tilde{Z}_{ij14}^T\end{array} & -\tilde{Z}_{ij24}-\tilde{Z}_{ij34}^T & I \\ * & -2Y_2-\bar{Z}_{ij44}-\bar{Z}_{ij44}^T & 0 \\ * & * & -\frac{(1-\beta)}{2}Y_2 \end{bmatrix}<0\ . \tag{19}$$

By computation, $\bar{Z}_{ii13}=P_2^{-T}\tilde{Z}_{ii13}P_2^{-1}$, $\bar{Z}_{ii24}=P_2^{-T}\tilde{Z}_{ii24}$. $\bar{Z}_{ij14}=P_2^{-T}\tilde{Z}_{ij14}P_2^{-1}$, $\bar{Z}_{ij24}=P_2^{-T}\tilde{Z}_{ij24}$, $\bar{Z}_{ij34}=P_2^{-T}\tilde{Z}_{ij34}^T$. *And (after solving the LMIs in Step 1) in Step 2 there exist matrices* $\bar{Z}_{ii12}$, $\bar{Z}_{ii22}$, $\bar{Z}_{ii23}$, $\bar{Z}_{ii32}$, $\bar{Z}_{ij12}$, $\bar{Z}_{ij13}$, $\bar{Z}_{ij22}$, $\bar{Z}_{ij23}$, $\bar{Z}_{ij32}$, $\bar{Z}_{ij33}$, $\bar{Z}_{ij42}$, $\bar{Z}_{ij43}$, *satisfying the following LMIs:*

$$Z=[\bar{Z}_{ij}]_{r\times r}=\begin{bmatrix}\bar{Z}_{11} & \cdots & \bar{Z}_{1r}\\ \vdots & \ddots & \vdots\\ \bar{Z}_{r1} & \cdots & \bar{Z}_{rr}\end{bmatrix}<0\ , \tag{20}$$

where

$$\bar{Z}_{ii}=\begin{bmatrix}\begin{bmatrix}\bar{Z}_{ii11} & \bar{Z}_{ii12}\\ * & \bar{Z}_{ii13}\end{bmatrix} & \begin{bmatrix}\bar{Z}_{ii21} & \bar{Z}_{ii22}\\ \bar{Z}_{ii23} & \bar{Z}_{ii24}\end{bmatrix}\\ \begin{matrix} * & * \\ * & * \end{matrix} & \begin{bmatrix}\bar{Z}_{ii31} & \bar{Z}_{ii32}\\ * & \bar{Z}_{ii33}\end{bmatrix}\end{bmatrix},\ \bar{Z}_{ij}=\begin{bmatrix}\begin{bmatrix}\bar{Z}_{ij11} & \bar{Z}_{ij12}\\ \bar{Z}_{ij13} & \bar{Z}_{ij14}\end{bmatrix} & \begin{bmatrix}\bar{Z}_{ij21} & \bar{Z}_{ij22}\\ \bar{Z}_{ij23} & \bar{Z}_{ij24}\end{bmatrix}\\ \begin{bmatrix}\bar{Z}_{ij31} & \bar{Z}_{ij32}\\ \bar{Z}_{ij33} & \bar{Z}_{ij34}\end{bmatrix} & \begin{bmatrix}\bar{Z}_{ij41} & \bar{Z}_{ij42}\\ \bar{Z}_{ij43} & \bar{Z}_{ij44}\end{bmatrix}\end{bmatrix}.$$

Then control gains are $K_i=-B_i^TX_1^{-1}$, *observer gains are* $L_i=P_2^{-T}M_i$, $i=1,2,\cdots,r$. *and the system (14) is asymptotically stable for any* $u_\omega\in\Omega$.

Proof. The proof is omitted because of the limited space. □

Remark 3. It is involved to find the state observer of time delay systems. [15] provided a one-step approach to design observer-based controller for T-S fuzzy system without time delay. However, for a T-S fuzzy system with time-delay, how to obtain a controller with less conservatism in one-step is still open.

5 Numerical Examples

In this section, two examples are employed to illustrate the validity and the effectiveness of the approaches proposed in this paper.

Example 1. Consider the following fuzzy model:

$$R_1 : \text{IF } x_1 \text{ is } P_1, \text{THEN} \quad E\dot{x} = (A_1 + \Delta A_1)x + (A_{11} + \Delta A_{11})x_\tau + B_1 u,$$

$$R_2 : \text{IF } x_1 \text{ is } P_2, \text{THEN} \quad E\dot{x} = (A_2 + \Delta A_2)x + (A_{12} + \Delta A_{12})x_\tau + B_2 u.$$

Where the membership function of $'P_1 ','P_2 '$ are given as following the effectiveness of the method $w_1(x_1) = 1 - \frac{1}{1+e^{-2x_1}}, w_2(x_1) = 1 - w_1(x_1)$. And

$$E = \begin{bmatrix} 1 & 0 \\ 0 & 0 \end{bmatrix}, A_1 = \begin{bmatrix} 3 & 1 \\ 1 & -2 \end{bmatrix}, A_2 = \begin{bmatrix} 2 & 1 \\ -2 & -0.5 \end{bmatrix}, A_{11} = \begin{bmatrix} 0 & 0 \\ 0.2 & 0.1 \end{bmatrix},$$

$$A_{12} = \begin{bmatrix} 0 & 0 \\ 0.1 & 0.5 \end{bmatrix}, B_1 = \begin{bmatrix} 3 \\ 0.1 \end{bmatrix}, B_2 = \begin{bmatrix} 5 \\ 0 \end{bmatrix}, M = \begin{bmatrix} 0.1 \\ 0.2 \end{bmatrix}, F(t) = \sin(t),$$

$$E_1 = E_2 = [0.1 \ \ 0], E_{11} = [1 \ \ 0], E_{12} = [0.1 \ \ 0], E_{b1} = 0.1, E_{b2} = 0.$$

Let $\tau(t) = 1 + 0.5 \sin t$. Consider $\Phi_{\omega 1} = 0.3 * I, \Phi_{\omega 2} = 0.01 * I$, if utilizing the method proposed in [16] to solve the problem, it is no way to design the controller, because the acurator failure was not taken into account in [16]. However, from Theorem 2, we get $P = \begin{bmatrix} 0.2842 & 0 \\ 0.3674 & 0.0281 \end{bmatrix}$, $S = \begin{bmatrix} 0.0063 & 0 \\ 0 & 0.0001 \end{bmatrix}$, state feedback gains are $K_1 = [-0.8893 \ \ -0.0028], K_2 = [-1.4210 \ \ 0]$.

Example 2. Let us consider the following T-S fuzzy system:

$$R_1 : \text{IF } x_1 \text{ is } N_1, \text{THEN} \quad E\dot{x} = A_1 x + A_{11} x_\tau + B_1 u, \quad y = C_1 x.$$

$$R_2 : \text{IF } x_1 \text{ is } N_2, \text{THEN} \quad E\dot{x} = A_2 x + A_{12} x_\tau + B_2 u, \quad y = C_2 x.$$

Where the membership function of $'N_1 ','N_2 '$ are given as following the effectiveness of the method $w_1(x_1) = 1 - \frac{1}{1+e^{-2x_1}}, w_2(x_1) = 1 - w_1(x_1)$. And

$$E = \begin{bmatrix} 1 & 0 \\ 0 & 0 \end{bmatrix}, A_1 = \begin{bmatrix} -2 & 1 \\ 1 & 2 \end{bmatrix}, A_2 = \begin{bmatrix} -2 & 2 \\ 1 & 2 \end{bmatrix}, A_{11} = A_{12} = \begin{bmatrix} 2 & 3 \\ 0 & 0 \end{bmatrix},$$

$$B_1 = \begin{bmatrix} 1 \\ 2 \end{bmatrix}, B_2 = \begin{bmatrix} 2 \\ 3 \end{bmatrix}, C_1 = [1 \ \ 2], C_2 = [0.5 \ \ 1].$$

Firstly, assume that $\Phi_{\Omega 1} = \Phi_{\Omega 2} = I$, that is, there is no any actuator failures occurring. Solving the LMIs (15)–(20) for a standard fuzzy controller produces

$$K_1 = [0.0731 \ \ 0.2175], \qquad K_2 = [0.0094 \ \ 0.2175],$$
$$L_1 = [0.9198 \ \ -0.1160]^T, \ L_2 = [1.0598 \ \ -0.1325]^T.$$

Assume the admissible set of actuator failures is given by

$$\Omega = \left\{ u_\omega = \sum_{i=1}^{2} \lambda_i \Phi_{\omega i} K_i x, \text{and } \Phi_{\omega 1} \geq 0.2 * I, \Phi_{\omega 2} \geq 0.3 * I \right\}.$$

Solving the LMIs (15)–(20) for a reliable fuzzy controller, we get

$$\begin{aligned} K_1 &= [0.2320 \ \ 0.7093], \qquad K_2 = [0.0154 \ \ 0.7093], \\ L_1 &= [3.1607 \ \ -0.3990]^T, L_2 = [3.6418 \ \ -0.4559]^T. \end{aligned}$$

Simulations were carried out for the delay $\tau(t) = 2$ and the initial conditions $\begin{bmatrix} x_1(0) & x_2(0) \end{bmatrix} = \begin{bmatrix} -1 & 0.5 \end{bmatrix}$. The responses of both design schemes of standard control design and reliable control design for the case without actuator failures are shown in Fig.1. It is obvious that both standard controller and reliable controller guarantee the asymptotic stabilities of the closed-loop system. When actuator failures with $\Phi_{\Omega 1} = 0.2 * I, \Phi_{\Omega 2} = 0.3 * I$ occurred, the state responses for the two cases are shown in Fig.2. It is observed that when actuator failures occur, the closed-loop system with the standard fuzzy controller is not even asymptotically stable, while the closed-loop system using the reliable fuzzy controller still operates well and maintains an acceptable level of performance.

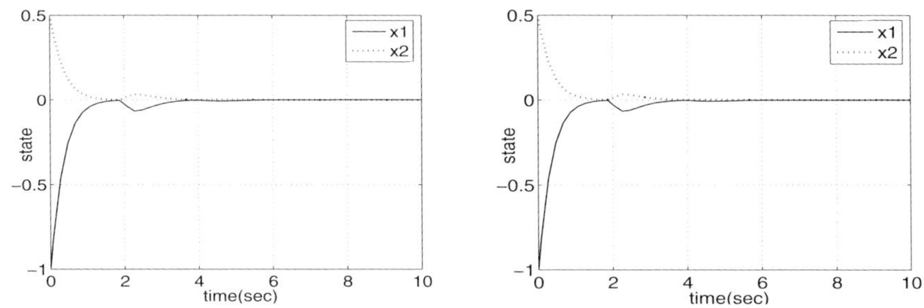

Fig. 1. standard control without failure and reliable control without failure

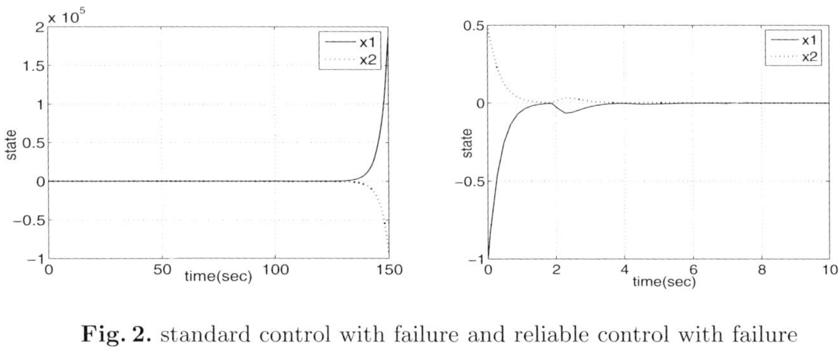

Fig. 2. standard control with failure and reliable control with failure

6 Conclusion

In this paper, we have considered the reliable fuzzy control design problem for nonlinear uncertain descriptor systems with state-delay. The effect among the subsystems are taken account of sufficiently. The less conservative reliable controller design schemes via state feedback and estimated state feedback are proposed.One of the future research topics is to extend the results developed in the present paper to the nonlinear systems, with more complex faults of actuator and sensor, to make the systems achieve appropriate performance.

References

1. R.J., V., Medanic, J., Perkins, W.: Design of reliable control systems. IEEE Trans. Automat. Control **37** (1992) 290–304
2. R.J., V.: Reliable linear-quadratic state-feedback control. Int. J. Control **31** (1995) 137–143
3. Yang, G.H., Lam, J., Wang, J.: Reliable H_∞ control for affine nonlinear systems. IEEE Trans. Automat. Control **43** (1998) 1112–1117
4. Liang, Y.W., Liaw, D., Lee, T.: Reliable control of nonlinear systems. IEEE Trans. Automat. Control **45** (2000) 706–710
5. Takagi, T., Sugeno, M.: Fuzzy identification of systems and its applications to modeling and control. IEEE Transactions on Systems,Man, and Cybernetics **15** (1985) 116–132
6. Wu, H.: Reliable LQ fuzzy control for continuous-time nonlinear systems with actuator faults. IEEE Transactions on Systems,Man, and Cybernetics **34** (2004) 1743–1752
7. Wu, H.: Reliable mixed $L2/H_\infty$ fuzzy static output feedback control for nonlinear systems with sensor faults. Automatica **41** (2005) 1925–1932
8. Chen, B., Liu, X.P.: Reliable control design of fuzzy dynamic systems with time-varying delay. Fuzzy Sets and Systems **146** (2004) 349–374
9. Taniguchi, T., Tanaka, K., Yamafuji, K., Wang, H.O.: Fuzzy descriptor systems: stability analysis and design via LMIs, San Diego (1999) 1827–1831
10. Taniguchi, T., Tanaka, K., Wang, H.O.: Fuzzy descriptor systems and fuzzy controller designs, Taipei, Taiwan (1999) 655–659
11. Liu, X.P.: Robust stabilization of nonlinear singular systems. (1995) 2375–2376
12. Wang, Y., Xie, L., De, S.: Robust control of a class of uncertain nonlinear systems. Systems Control Letter **19** (1992) 139–149
13. Kim, E., Lee, H.: New approaches to relaxed quadratic stability condition of fuzzy control systems. Control **8** (2000) 523–534
14. Liu, X.D., Zhang, Q.L.: New approaches to H_∞ controller designs based on fuzzy observers for T-S fuzzy systems via LMI. Automatica **39** (2003) 1571–1582
15. Lin, C., Wang, Q.G., Lee, T.H.: Improvement on observer-bsed H_∞ control for T-S fuzzy systems. Automatica **41** (2005) 1651–1656
16. Wang, Y., Sun, Z.Q., Sun, F.C.: Robust fuzzy control of a class of nonlinear descriptor systems with time-varying delay. International Journal of Control, Automation and Systems **2** (2004) 76–82

A Novel Fuzzy Approximator with Fast Terminal Sliding Mode and Its Application

Yunfeng Liu, Fei Cao, Yunhui Peng, Xiaogang Yang, and Dong Miao

303 Branch, Xi'an Research Inst. Of High-tech, Hongqing Town, 710025, China
footballliu@163.com

Abstract. A new learning algorithm for fuzzy system to approximate unknown nonlinear continuous functions is presented. Fast terminal sliding mode combining the finite time convergent property of terminal attractor and exponential convergent property of linear system is introduced into the conventional back-propagation learning algorithm to improve approximation ability. The Lyapunov stability analysis guarantees that the approximation is stable and converges to the unknown function with improved speed. The proposed fuzzy approximator is then applied in the control of an unstable nonlinear system. Simulation results demonstrate that the proposed method is better than conventional method in approximation and tracing control of nonlinear dynamic system.

1 Introduction

Fuzzy systems most often use linguistic information from experts. Functionally, a fuzzy system can be described as a function approximator. Theoretical investigations have revealed that fuzzy system is universal approximator [1, 2], i.e., which can approximate any continuous function to any prescribed accuracy provided that sufficient fuzzy rules. Some Researches have shown that fuzzy system could offer an approach for system modeling [3, 4]. Now there are varieties of learning algorithms available for fuzzy approximator, majority of them are of gradient descent type, such as the popular back-propagation learning algorithm [4]. It is well known that conventional gradient descent learning algorithm can be very slow. Several methods have been proposed for improving the speed of convergence of gradient descent, such as momentum [5], exponential gradient [6], etc.

In this paper, we present a novel learning algorithm for fuzzy system to approximate unknown nonlinear continuous function. To improve the learning rate, we use a particular kind of sliding mode control concept—the fast terminal sliding mode (TSM) [7]. The fast TSM has been recently developed based on the concept of terminal attractor, which exhibits some superior properties such as fast finite time convergence to the origin and less steady-state errors [8]. So we introducc a fast TSM into the conventional back-propagation learning algorithm to improve approximation ability based on fuzzy system proposed in [4]. The Lyapunov stability analysis

L. Wang et al. (Eds.): FSKD 2006, LNAI 4223, pp. 179 – 187, 2006.

guarantees that the approximation is stable and converges to the unknown function with improved speed. Simulation results verify the validity of the proposed learning algorithm.

2 Basic Concepts

2.1 Fuzzy Approximator

Fuzzy system is one of the function approximators. Wang described in his paper [4] as follows: The fuzzy logic systems with center average defuzzifier, product-inference rule equation, singleton fuzzifier are of the following form:

$$\hat{f}(\boldsymbol{x}|\boldsymbol{\theta}) = \frac{\sum_{l=1}^{M} \bar{y}^l (\prod_{i=1}^{n} \mu_{F_i^l}(x_i))}{\sum_{l=1}^{M} (\prod_{i=1}^{n} \mu_{F_i^l}(x_i))} \tag{1}$$

where $\hat{f} : U \subset R^n \to R$, $\boldsymbol{x} = (x_1, x_2, \cdots, x_n) \in U$, $\boldsymbol{\theta} = [\theta^1, \cdots, \theta^M]^T$. $\mu_{F_i^l}(x_i)$ is the Gaussian membership function defined by

$$\mu_{F_i^l}(x_i) = \exp\left[-\left(\frac{x_i - \bar{x}_i^l}{\sigma_i^l}\right)^2\right] \tag{2}$$

where $\theta^l = [\bar{y}^l, \bar{x}_i^l, \sigma_i^l]^T$,($i = 1, 2, \cdots, n$, $l = 1, 2, \cdots, M$). Fuzzy system as in the form of (1) is universal approximator, which can uniformly approximate any types of nonlinear function over U to any degree of accuracy if U is compact. By analyzing (1), we observe a fact: the mapping $\hat{f} : U \to R$ determined by (1) can be represented as a three-layer feed forward network, as shown in Fig. 1. Back-propagation learning algorithm can be applied to any types of feedforward networks. Therefore, we can train the fuzzy system using back-propagation learning algorithm.

2.2 Back-Propagation Learning Algorithm

After constructing the fuzzy system, and giving the training data ($\boldsymbol{x}$, d), the next step is to train the parameters $\bar{y}^l$, $\bar{x}_i^l$ and σ_i^l of the functional representation of the fuzzy system in (1)such that (3) is minimized.

$$E = \frac{1}{2}[\hat{f}(\boldsymbol{x}) - d]^2 \tag{3}$$

where d is the desired outputs. The error $e = \hat{f}(\boldsymbol{x}) - d$. Let us use f to denote $\hat{f}(\boldsymbol{x})$, and use $f = a/b$, $z^l = \prod_{i=1}^{n} \exp(-(\frac{x_i - \bar{x}_i^l}{\sigma_i^l})^2)$, $a = \sum_{l=1}^{M} \bar{y}^l z^l$, $b = \sum_{l=1}^{M} z^l$.

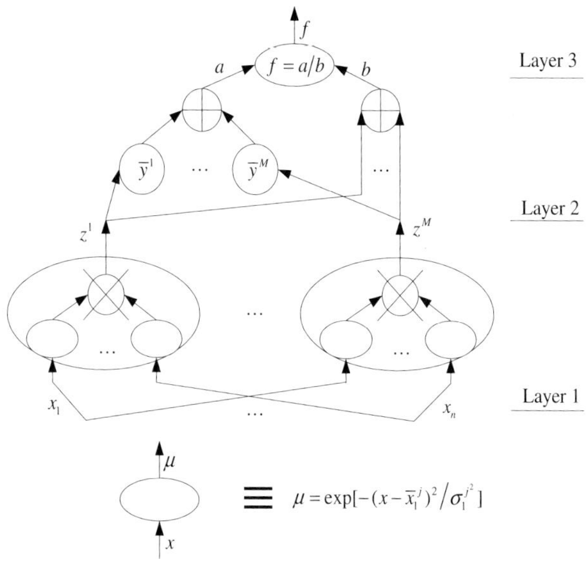

Fig. 1. Network representation of fuzzy systems

To train $\theta^l = [\bar{y}^l, \bar{x}_i^l, \sigma_i^l]^T$, we use

$$\theta^l(k+1) = \theta^l(k) - \alpha\frac{\partial E}{\partial \theta^l} = \theta^l(k) - \alpha e\frac{\partial \hat{f}(x|\theta^l)}{\partial \theta^l} \tag{4}$$

By using back-propagation and chain rules, we have:

$$\frac{\partial \hat{f}}{\partial \bar{y}^l} = \frac{z^l}{b}, \quad \frac{\partial \hat{f}}{\partial \bar{x}_i^l} = \frac{\bar{y}^l - f}{b} z^l \frac{2(x_i - \bar{x}_i^l)}{\sigma_i^{l^2}}, \quad \frac{\partial \hat{f}}{\partial \sigma_i^l} = \frac{\bar{y}^l - f}{b} z^l \frac{2(x_i - \bar{x}_i^l)^2}{(\sigma_i^l)^3} \tag{5}$$

So we obtain the following training algorithm for $\bar{y}^l$, $\bar{x}_i^l$ and σ_i^l.

$$\bar{y}^l(k+1) = \bar{y}^l(k) - \alpha e\frac{z^l}{b}, \quad \bar{x}_i^l(k+1) = \bar{x}_i^l(k) - \alpha e\frac{\bar{y}^l - f}{b} z^l \frac{2(x_i - \bar{x}_i^l(k))}{\sigma_i^{l2}(k)}$$
$$\sigma_i^l(k+1) = \sigma_i^l(k) - \alpha e\frac{\bar{y}^l - f}{b} z^l \frac{2(x_i - \bar{x}_i^l(k))^2}{\sigma_i^{l3}(k)} \tag{6}$$

By utilizing this learning algorithm, the error e will gradually decrease till the tolerance of approximation error is reached adaptively.

2.3 The Concept of Fast Terminal Sliding Mode

The fast TSM can be described by the following first-order dynamics [7]

$$s = \dot{x} + \alpha x + \beta x^{q/p} = 0 \tag{7}$$

where $x \in R$ is a scalar variable, $\alpha, \beta > 0$, $p, q(p > q)$ are the positive integers. Note that the parameter p must be an odd integer and only the real solution is considered so that for any real number $x^{q/p}$ is always a real number. Then we have

$$\dot{x} = -\alpha x - \beta x^{q/p} \tag{8}$$

For the properly chosen parameters, given an initial state $x(0) \neq 0$, the dynamics will reach $x = 0$ in finite time. The physical interpretation is: When x is far away from zero, the approximate dynamic become $\dot{x} = -\alpha x$ whose fast convergence is well understood. When close to $x = 0$, the approximate dynamics become $\dot{x} = -\beta x^{q/p}$ which is a terminal attractor. More precisely, we can solve the differential equation analytically. The exact time to reach zero, t_r is determined by

$$t_r = \frac{p}{\alpha(p-q)} \ln \frac{\alpha x(0)^{(p-q)/p} + \beta}{\beta} \tag{9}$$

and the equilibrium zero is a terminal attractor.

3 Learning Algorithm with Fast Terminal Sliding Mode

As we noted in the above section, back-propagation learning algorithm can minimize the approximation error. However, it can be very slow when the approximation error is very small according to (4). In order to increase the convergent speed, we propose to use a new form of learning algorithm based on the fast TSM concept. In this section, we will use a novel error criterion

$$E = \frac{1}{2} e^2 + \frac{p\eta_2}{(p+q)\eta_1} e^{(p+q)/p} \tag{10}$$

where $\eta_1, \eta_2 > 0$, $p, q(p > q)$ are odd integers, then $p+q$ is even, so (10) is a positive definite energy function. The corresponding learning algorithm becomes

$$\theta^l(k+1) = \theta^l(k) - \eta_1 \frac{\partial E}{\partial \theta^l} = \theta^l(k) - (\eta_1 e + \eta_2 e^{q/p}) \frac{\partial \hat{f}(x|\theta^l)}{\partial \theta^l} \tag{11}$$

where η_1, η_2 determine the learning rate together. So we obtain the following learning algorithm for $\bar{y}^l$, $\bar{x}_i^l$ and σ_i^l .

$$\bar{y}^l(k+1) = \bar{y}^l(k) - (\eta_1 e + \eta_2 e^{q/p}) \frac{z^l}{b} \tag{12}$$

$$\bar{x}_i^l(k+1) = \bar{x}_i^l(k) - (\eta_1 e + \eta_2 e^{q/p}) \frac{\bar{y}^l - f}{b} z^l \frac{2(x_i - \bar{x}_i^l(k))}{\sigma_i^{l2}(k)} \tag{13}$$

$$\sigma_i^l(k+1) = \sigma_i^l(k) - (\eta_1 e + \eta_2 e^{q/p}) \frac{\bar{y}^l - f}{b} z^l \frac{2(x_i - \bar{x}_i^l(k))^2}{\sigma_i^{l3}(k)} \tag{14}$$

Comparing with (6), only a nonlinearity item of approximation error is inserted into the conventional back-propagation learning algorithm, but it would make the learning faster. In the following, we will firstly prove that the proposed algorithm is Lyapunov stable, and then discuss its faster convergent property.

According to the universal approximation property of fuzzy system, any continuous function over a compact set Ω can be approximated by a fuzzy system to any accuracy as

$$d = \hat{f}(\boldsymbol{x}|\boldsymbol{\theta}^*) - \varepsilon(\boldsymbol{x}) \tag{15}$$

where $\boldsymbol{\theta}^* = [\theta^{1*}, \theta^{2*}, \cdots, \theta^{M*}]^T$ is an optimal constant weight vector, $\theta^{l*} = [\overline{y}^{l*}, \overline{x}_i^{l*}, \sigma_i^{l*}]^T$ ($i = 1,2,\cdots,n$, $l = 1,2,\cdots,M$) and $\varepsilon(\boldsymbol{x})$ is the optimal approximation error tolerance, and $|\varepsilon(\boldsymbol{x})| \le \varepsilon_M$, where ε_M is a small positive constant. In the case of three independent variables, using (3), (15) and Taylor formula, the approximation error can be expressed as

$$e = \hat{f}(\boldsymbol{x}|\boldsymbol{\theta}) - \hat{f}(\boldsymbol{x}|\boldsymbol{\theta}^*) + \varepsilon(\boldsymbol{x}) = \boldsymbol{\varphi}^T \left(\frac{\partial \hat{f}(\boldsymbol{x}|\boldsymbol{\theta})}{\partial \boldsymbol{\theta}} \right) + \hat{f}_0 + \varepsilon(\boldsymbol{x}) = \sum_{l=1}^{M} \varphi^l \frac{\partial \hat{f}(\boldsymbol{x}|\theta^l)}{\partial \theta^l} + \hat{f}_0 + \varepsilon(\boldsymbol{x}) \tag{16}$$

where $\varphi^l = \theta^l - \theta^{l*}$ is the estimation error of the optimal parameter θ^{l*} , $\boldsymbol{\varphi}$ is the vector form of φ^l . $\hat{f}_0$ is second-order approximation error of the Taylor series. Using the mean value theorem, we have

$$\hat{f}_0 \equiv \left(\frac{\partial \hat{f}(\boldsymbol{x}|\boldsymbol{\theta}_0)}{\partial \boldsymbol{\theta}} - \frac{\partial \hat{f}(\boldsymbol{x}|\boldsymbol{\theta})}{\partial \boldsymbol{\theta}} \right)^T \boldsymbol{\varphi} \tag{17}$$

where $\boldsymbol{\theta}_0 = \lambda\boldsymbol{\theta} + (1-\lambda)\boldsymbol{\theta}^*$, $\lambda \in [0,1]$. Because the optimal weight vector θ^{l*} does not vary with time, (i.e., $\dot{\theta}^{l*} = 0$), it follows from (11) and (16) that

$$\dot{\varphi}^l = \dot{\theta}^l - \dot{\theta}^{l*} = \dot{\theta}^l = (-\eta_1 e - \eta_2 e^{q/p}) \frac{\partial \hat{f}(\boldsymbol{x}|\theta^l)}{\partial \theta^l} \tag{18}$$

or in the vector form

$$\dot{\boldsymbol{\varphi}} = \dot{\boldsymbol{\theta}} = (-\eta_1 e - \eta_2 e^{q/p}) \frac{\partial \hat{f}(\boldsymbol{x}|\boldsymbol{\theta})}{\partial \boldsymbol{\theta}} \tag{19}$$

Theorem 1. If the parameter vector $\boldsymbol{\theta}$ of the fuzzy system (1) is continuously updated according to (19) to approximate the unknown continuous function(15)□then the fuzzy approximator is Lyapunov stable, and $\boldsymbol{\theta}$ will approximate $\boldsymbol{\theta}^*$ in the finite time.

Proof. From(16) and(19), we can have

$$\dot{\boldsymbol{\theta}} = -(\eta_1(\hat{f}(\boldsymbol{x}|\boldsymbol{\theta}) - \hat{f}(\boldsymbol{x}|\boldsymbol{\theta}^*) + \varepsilon(\boldsymbol{x})) + \eta_2(\hat{f}(\boldsymbol{x}|\boldsymbol{\theta}) - \hat{f}(\boldsymbol{x}|\boldsymbol{\theta}^*) + \varepsilon(\boldsymbol{x}))^{q/p}) \frac{\partial \hat{f}(\boldsymbol{x}|\boldsymbol{\theta})}{\partial \boldsymbol{\theta}} \tag{20}$$

where $\dot{\boldsymbol{\theta}} = [\dot{\boldsymbol{\theta}}^1, \dot{\boldsymbol{\theta}}^2, \cdots, \dot{\boldsymbol{\theta}}^M]^T$.

To explore the stability of (20), we choose the Lyapunov function candidate as

$$V(\theta) = \frac{1}{2}\tilde{\theta}^T \tilde{\theta} = \frac{1}{2}(\theta - \theta^*)^T (\theta - \theta^*) \tag{21}$$

then we have the first derivative of (21)as

$$\dot{V} = \tilde{\boldsymbol{\theta}}^T \dot{\tilde{\boldsymbol{\theta}}} = \tilde{\boldsymbol{\theta}}^T \dot{\boldsymbol{\theta}} = (-\eta_1 e - \eta_2 e^{q/p})\tilde{\boldsymbol{\theta}}^T \frac{\partial \hat{f}(\boldsymbol{x}|\boldsymbol{\theta})}{\partial \boldsymbol{\theta}} = (-\eta_1 e - \eta_2 e^{q/p})(e - \varepsilon(\boldsymbol{x}) - \hat{f}_0)$$

$$= -\eta_1 e^2 - \eta_2 e^{(q+p)/p} + \eta_1 e\varepsilon(\boldsymbol{x}) + \eta_2 e^{q/p}\varepsilon(\boldsymbol{x}) + \eta_1 e\hat{f}_0 + \eta_2 e^{q/p}\hat{f}_0$$

$$\le -\eta_1 |e|^2 - \eta_2 |e|^{(q+p)/p} + \eta_1 |e||\varepsilon(\boldsymbol{x})| + \eta_2 |e|^{q/p} |\varepsilon(\boldsymbol{x})| + \eta_1 |e||\hat{f}_0| + \eta_2 |e|^{q/p} |\hat{f}_0|$$

$$= -\eta_1 |e|(|e| - |\varepsilon(\boldsymbol{x})| - |\hat{f}_0|) - \eta_2 |e|^{q/p} (|e| - |\varepsilon(\boldsymbol{x})| - |\hat{f}_0|)$$

$$= (-\eta_1 |e| - \eta_2 |e|^{q/p})(|e| - |\varepsilon(\boldsymbol{x})| - |\hat{f}_0|) \tag{22}$$

Because $\partial \hat{f}(\boldsymbol{x}|\boldsymbol{\theta})/\partial \boldsymbol{\theta}$ is finite, so $|\hat{f}_0|$ is finite according to (17). While the approximation error $|e| > |\varepsilon(\boldsymbol{x})| + |\hat{f}_0|$, the above formula implies that $\dot{V} < 0$. So we conclude that $|e|$ is finite.

In the novel learning algorithm (18), the introduction of the nonlinearity item $e^{q/p}$ amplifies the approximation error's contribution to the convergent rate in the neighborhood of $e = 0$. For example, $q = 1$ □ $p = 3$, when the approximation error $e = 0.005$, the nonlinearity item $e^{q/p} = 0.1710$ .As a result, the convergent speed is increased in spite of the smaller approximation error. The closer to the minimum, the faster the convergent speed, resulting in the finite time convergence to $\varphi_l = 0$. This can be explained as follows. Consider the Jacobian around the minimum $\varphi_l = 0$, i.e.,

$$J = \frac{\partial \dot{\varphi}_l}{\partial \varphi_l} = \frac{\partial}{\partial \varphi_l}((-\eta_1 e - \eta_2 e^{q/p})\frac{\partial \hat{f}(x|\theta^l)}{\partial \theta^l}) = (-\eta_1 - \frac{q\eta_2}{p}e^{(q-p)/p})\left(\frac{\partial \hat{f}(x|\theta^l)}{\partial \theta^l}\right)^2 \tag{23}$$

We have $J \to -\infty$ when $e \to 0$, which indicates that at the minimum $e = 0$ the equivalent eigenvalue tends to negative infinity, and of course, the fuzzy approximator with such an infinitely negative eigenvalue, will converge to the minimum with an infinitely large speed which results in finite time convergence. This means that $\boldsymbol{\theta}$ will approximate $\boldsymbol{\theta}^*$ in the finite time.

Theorem 2. The proposed learning algorithm (11) guarantees that the fuzzy approximator converges to the global minimum of the approximation error e because the second partial derivative with respect to θ^l is positive.

Proof. From (11), we can have

$$\frac{\partial^2 E}{\partial^2 \theta^l} = \frac{\partial}{\partial \theta^l}\left(\frac{\partial E}{\partial \theta^l}\right) = \frac{\partial}{\partial \theta^l}\left(e\frac{\partial \hat{f}(x|\theta^l)}{\partial \theta^l} + \frac{\eta_2}{\eta_1}e^{q/p}\frac{\partial \hat{f}(x|\theta^l)}{\partial \theta^l}\right) = \left(\frac{\partial \hat{f}(x|\theta^l)}{\partial \theta^l}\right)^2 +$$

$$\frac{\eta_2 q}{\eta_1 p}e^{(q-p)/p}\left(\frac{\partial \hat{f}(x|\theta^l)}{\partial \theta^l}\right)^2 = (1+\frac{\eta_2 q}{\eta_1 p}e^{(q-p)/p})\left(\frac{\partial \hat{f}(\boldsymbol{x}|\theta^l)}{\partial \theta^l}\right)^2 > 0 \tag{24}$$

We note that p,q are odd, so $p-q$ is even. This means that the error criterion (10) is convex in the weight space θ^l and therefore possesses only single minimum.

4 Application of the Proposed Fuzzy Approximator

Consider a simple nonlinear system

$$\dot{x} = f(x) + u \tag{25}$$

where $f(x) = (1-e^{-x(t)})/(1+e^{-x(t)})$. It is clear that the plant is unstable without control input because if $u(t)=0$, $\dot{x}>0$ for $x>0$ and $\dot{x}<0$ for $x<0$.

We choose the initial state $x(0)=3$. The sampling period $t=0.001s$.The control objective is to design a continuous control strategy with a fuzzy approximator to compensate the unknown nonlinear function to drive the system state to a smaller neighborhood of the origin in faster speed. So a fast TSM is chosen as follows:

$$\dot{x} = -\alpha_1 x - \beta_1 x^{q_1/p_1} \tag{26}$$

and the corresponding control law is designed as

$$u = -\hat{f}(x) - \alpha_1 x - \beta_1 x^{q_1/p_1} \tag{27}$$

where the first item $\hat{f}(x)$ is the proposed fuzzy approximator for compensating the unknown nonlinear system function $f(x)$, and the other items aim at obtaining a fast TSM. $\hat{f}(x)$ is of the form (1) with $M=3$ and $n=1$. Suppose the nonlinear plant to be identified starts operation from $k=0$. Do not start the learning algorithm (12), (13) and (14) for the first M time points. According to M time points [4], set the initial parameters $\overline{y}(0)=[0.2449 \quad 0.4621 \quad 0.6351]$, $\sigma(0)=[0.1667 \quad 0.1667 \quad 0.1667]$, $\overline{x}(0)=[0.5 \quad 1 \quad 1.5]$. We start the learning from time point $k=4$, and train the parameter $\overline{y}^l$, $\overline{x}^l$ and σ^l for one cycle a each time point by using (12), (13) and (14) .

The parameters of learning parameter $\eta_1=10$, $\eta_2=10$, $q/p=1/3$ and the fast TSM parameters $\alpha_1=1$, $\beta_1=1$, $q_1/p_1=3/5$. The simulation results are shown in Fig. 2 to demonstrate the fast convergence of the proposed fuzzy approximator.

In order to demonstrate the faster convergence of the proposed algorithm (11) compared with the conventional algorithm (4), we only change the exponential power $q/p=1$ instead of $q/p=1/3$ with other parameters keeping invariant, and the novel learning algorithm is changed back to the conventional algorithm. The simulation results of approximation errors shown in Fig. 3 validate the improvements on the faster convergent speed and lower approximation error with the proposed algorithm.

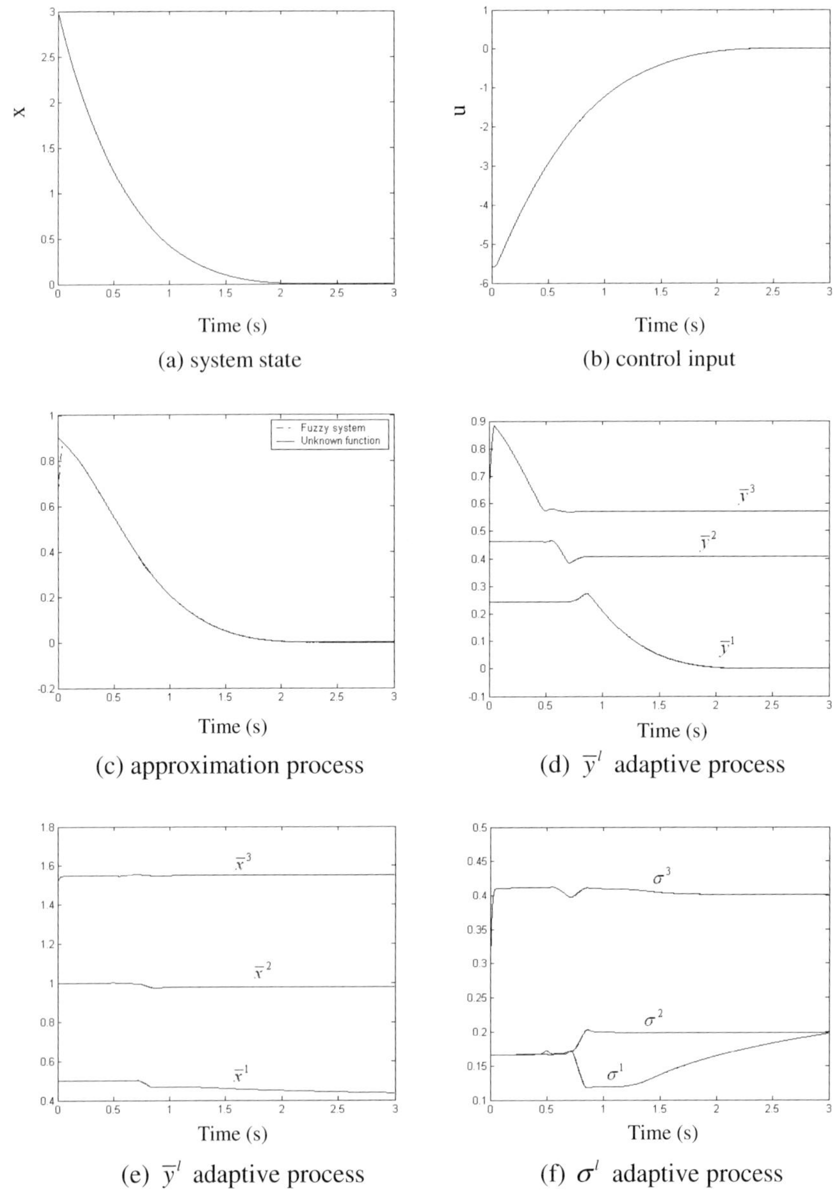

(a) system state (b) control input

(c) approximation process (d) $\overline{y}^l$ adaptive process

(e) $\overline{y}^l$ adaptive process (f) σ^l adaptive process

Fig. 2. Control with the proposed fuzzy approximator

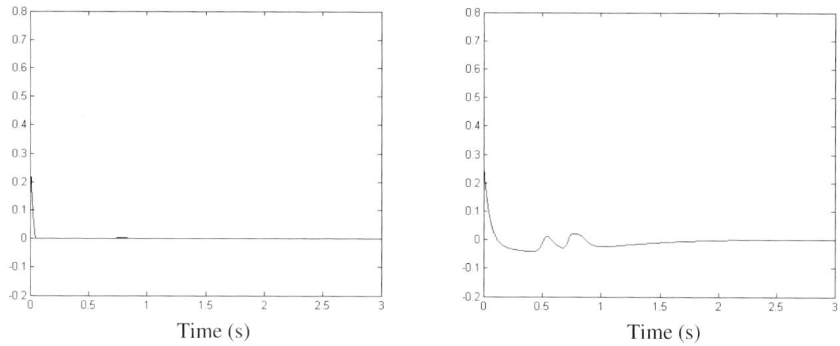

(a) $q/p = 1/3$ (b) $q/p = 1$

Fig. 3. The comparison of the approximation error

5 Conclusions

A modified learning algorithm that utilizes the concept of fast TSM for fuzzy systems to approximate unknown nonlinear continuous functions is presented. The Lyapunov stability analysis guarantees that the approximation is stable and converges to the unknown function with improved speed. The proposed fuzzy approximator is applied in the control of an unstable nonlinear system, and the effectiveness of the proposed algorithm has been verified by simulations.

References

1. Wang, L. X.: Fuzzy Systems are Universal Approximators. Proc. IEEE International Conf. on Fuzzy System, San Diego, (1992) 1163-1170
2. Zhang, T. P., Yang, Y. Q. and Zhang, H. Y.: Direct Adaptive Sliding Mode Control with Nonlinearly Parameterized Fuzzy Approximators. Proceedings of the 4th World Congress on Intelligent Control and Automation. Shanghai, China, (2002) 1915-1919
3. Yue, S. H., Zhang K. C.: An Optimal Fuzzy Approximator. Appl. Math. J. Chinese Univ. Ser. B, 17(2), (2002) 236-240
4. Wang, L. X., Mendel, J. M.: Back-propagation Fuzzy Systems as Nonlinear Dynamic System Identifiers. Proc. IEEE International Conf. on Fuzzy System, San Diego, (1992)1409-1418.
5. Qian, N.: On the Momentum Term in Gradient Descent Learning Algorithms. Neural Networks, 12 (2), (1999) 145-151
6. Hill, S. I., Williamson, R. C.: Convergence of Exponentiated Gradient Algorithm. IEEE Trans. Signal Process, 49 (2001) 1208-1215.
7. Yu, X., Man, Z. and Wu, Y.: Terminal Sliding Modes with Fast Transient Performance. Proceedings of the 36th IEEE CDC, San Diego, (1997) 962-963
8. Yu, X., Man, Z.: Fast Terminal Sliding Mode Control Design for Nonlinear Dynamic Systems. IEEE Trans. Circuits Systems Part I , 39 (2), (2002) 261-264

Robust Fuzzy Tracking Control of Nonlinear Systems with Uncertainty Via T-S Fuzzy Model

Jian Zhang[1], Minrui Fei[1], Taicheng Yang[2], and Yuemei Tan[1]

[1] Shanghai Key Laboratory of Power Station Automation Technology,
School of Mechatronics and Automation,
Shanghai University, Shanghai 200072, China
Jianzhang668@163.com, mrfei@staff.shu.edu.cn,
tanya@staff.shu.edu.cn

[2] Department of Engineering, University of Sussex, Brighton BN1 9QT, UK
taiyang@sussex.ac.uk

Abstract. This paper presents a novel robust fuzzy tracking control method for uncertain nonlinear systems. The Takagi-Sugeno fuzzy model is employed for fuzzy modeling of uncertain nonlinear system. Based on the fuzzy model, the internal model principle (IMP) is adopted to design the robust fuzzy tracking controller. Then the robust fuzzy observer is designed independently. Sufficient conditions are derived for stabilization of the robust fuzzy tracking controller and the robust fuzzy observer in the sense of Lyapunov asymptotic stability. The main contribution of this paper is the development of the robust fuzzy tracking control based on the internal model principle of uncertain nonlinear systems. A simulation example is given to illustrate the design procedures and asymptotic tracking performance of the proposed method.

1 Introduction

The task of tracking is a typical control problem in the industry. Most plants in the industry have severe nonlinearity and uncertainties such as the NOx emission in a power plant [1]. They post additional difficulties to the tracking control. Recently, there has been rapidly growing interest in fuzzy tracking control of nonlinear and there have been many successful applications [2], [3], [4], [5]. The most important issue for fuzzy tracking control systems is how to reduce the tracking error between the desired trajectory and the actual output values rapidly with the guaranteed stability. It has been extensively studied by a number of researchers [6], [7], [8], [9]. Some fuzzy control design methods for nonlinear systems with a guaranteed H_∞ model reference tracking performance is proposed [10], [11]. The tracking error only can be attenuated below some certain level. Therefore how to reduce the tracking error still need further study.

In this work, the fuzzy linear model of Takagi-Sugeno is used to approximate a uncertain nonlinear system. Then a hybrid robust fuzzy tracking controller based on the internal model principle is introduced to track the desired output trajectory. The robust fuzzy observer is also designed independently for the controller according to the

L. Wang et al. (Eds.): FSKD 2006, LNAI 4223, pp. 188 – 198, 2006.

separation property. The sufficient conditions for stabilization of the controller and the observer are both formulated as the linear matrix inequality problem.

It is believed that the main benefits of the proposed design method include: (1) The asymptotic tracking of the reference input signal and the disturbance attenuation are achieved by employing the internal model principle in the method proposed. (2) The issue of the robustness, i.e., the uncertainties in the nonlinear plant has been addressed in the design based on T-S fuzzy model easily.

2 Problem Formulation and Preliminaries

It is well known that a Takagi-Sugeno fuzzy model can be a universal approximation of a smooth nonlinear dynamic [5], [7], [12], [13]. Consider the following nonlinear system:

$$\dot{x}(t) = f(x(t)) + g(x(t))u(t) \tag{1}$$

where

$$x(t) = [x_1(t), x_2(t), \cdots, x_n(t)]^T \in R^{n\times 1}$$

denotes the state vector

$$u(t) = [u_1(t), u(t), \cdots, u_m(t)]^T \in R^{m\times 1}$$

denotes the control input.

In order to consider the parametric uncertainties in a continuous-time nonlinear system, the following fuzzy model with parametric uncertainties is employed [8], [14].

Plant Rule i:

IF $x_1(1)$ is M_{i1} and … and $x_n(1)$ is M_{in}

$$\begin{aligned} \text{THEN}\ \dot{x}(t) &= \left(A_i + \Delta A_i(t)\right)x(t) + (B_i + \Delta B_i(t))u(t) \\ y(t) &= C_i x(t), \qquad i = 1, 2, \cdots, g \end{aligned} \tag{2}$$

where M_{ij} is the fuzzy set, $A_i \in R^{n\times n}, B_i \in R^{n\times m}$, g is the number of IF-THEN rules, and $x_1(t), \cdots, x_g(t)$ are the premise variable. A_i, B_i, C_i are known constant matrixes that describe the nominal system. $\Delta A_i, \Delta B_i$ represent the time-varying parametric uncertainties having the following structure:

$$\left[\Delta A_i(t), \Delta B_i(t)\right] = H_i F_i(t)\left[E_{1i}, E_{2i}\right], \qquad i = 1, 2, \cdots, g \tag{3}$$

where H_i, E_{1i}, E_{2i}, are known constant matrix appropriate dimensions, $F_i(t)$ is unknown matrix which is bounded as follows:

$$F_i(t) \in \Omega_F := \{F(t) : \|F(t)\| \le 1, \textit{the elements of } F(t) \textit{ are Lebesgue measurable}\} \tag{4}$$

The overall fuzzy system is inferred as follows:

$$\dot{x}(t) = \frac{\sum_{i=1}^{g} \mu_i(x(t))\left\{\left(A_i + \Delta A_i(t)\right)x(t) + \left(B_i + \Delta B_i(t)\right)u(t)\right\}}{\sum_{i=1}^{g} \mu_i(x(t))}$$

$$= \sum_{i=1}^{g} h_i(x(t))\{(A_i + \Delta A_i(t))x(t) + (B_i + \Delta B_i(t))u(t)\}$$

$$y(t) = \sum_{i=1}^{g} h_i(x(t))C_i x(t) \tag{5}$$

where

$$x(t) = [x_1(t), x_2(t), \cdots, x_n(t)]^T, \quad \mu_i(x(t)) = \prod_{j=1}^{n} M_{ij}(x(t)), \quad h_i(x(t)) = \frac{\mu_i(x(t))}{\sum_{i=1}^{g} \mu_i(x(t))}$$

$M_{ij}(x_i(t))$ is the grade of membership of $x_j(t)$ in M_{ij}. It is assumed that

$$\mu_i(x(t)) \ge 0, \qquad i = 1, 2, \cdots, n; \qquad \sum_{i=1}^{n} \mu_i(x(t)) > 0$$

for all t.

Then the following conditions can be obtained:

$$h_i(x(t)) \ge 0, \qquad i = 1, 2, \cdots, g; \qquad \sum_{i=1}^{g} h_i(x(t)) = 1$$

for all t.

The objective of this paper is to design a T-S fuzzy tracking controller such that the closed-loop overall fuzzy system can track direct trajectory quickly with guaranteed asymptotic stability. Furthermore, the robust fuzzy tracking controller and the robust fuzzy observer are designed separately to reduce the difficulty of design.

3 Robust Fuzzy Tracking Controller

The robust fuzzy tracking controller is composed of a servo-compensator and a state feedback controller.

Consider a reference model as follows:

$$\begin{aligned} x_r(t) &= A_r x_r(t) \\ r(t) &= C_r x_r(t) \end{aligned} \tag{6}$$

where

$r(t)$ reference input
$x(t)$ reference state
A_r specific asymptotically stable matrix
$\{A_r \quad C_r\}$ are observable

It is assumed that $r(t)$, for all $t \ge 0$, represents a desired trajectory for y in equation (2) to follow.

According to the Internal Model Principle, we can design the following servo-compensator:

$$\begin{aligned} x_c &= A_c x_c + B_c(r - y) \\ y_c &= x_c \end{aligned} \tag{7}$$

Then the state equation of every fuzzy subsystem can be represented as follows:

$$\begin{bmatrix} \dot{x} \\ \dot{x}_c \end{bmatrix} = \begin{bmatrix} A_i & 0 \\ -B_c C_i & A_c \end{bmatrix} \begin{bmatrix} x \\ x_c \end{bmatrix} + \begin{bmatrix} B_i \\ 0 \end{bmatrix} u + \begin{bmatrix} 0 \\ B_c \end{bmatrix} r$$

$$\tilde{y} = \begin{bmatrix} C_i & 0 \end{bmatrix} \begin{bmatrix} x \\ x_c \end{bmatrix}, \qquad i = 1, 2, \cdots, g \tag{8}$$

Eigenvalues of the subsystem (8) can be placed randomly by the following state feedback controller:

$$u_i = \begin{bmatrix} K_i & K_{ci} \end{bmatrix} \begin{bmatrix} x \\ x_c \end{bmatrix} = K_i x + K_{ci} x_c \tag{9}$$

Let us denote

$$\tilde{x}(t) = \begin{bmatrix} x(t) \\ x_c(t) \end{bmatrix}, \ \tilde{A} = \begin{bmatrix} A_i & 0 \\ -B_c C_i & A_c \end{bmatrix}, \ \tilde{B} = \begin{bmatrix} B_i \\ 0 \end{bmatrix}, \ \tilde{B}_r = \begin{bmatrix} 0 \\ B_c \end{bmatrix}, \ \tilde{C} = \begin{bmatrix} C_i & 0 \end{bmatrix}, \ \tilde{K}_i = \begin{bmatrix} K_i & K_{ci} \end{bmatrix}$$

Therefore, the augmented system in (8) can be expressed as the following form:

$$\begin{bmatrix} \dot{x} \\ \dot{x}_c \end{bmatrix} = \begin{bmatrix} A_i & 0 \\ -B_c C_i & A_c \end{bmatrix} \begin{bmatrix} x \\ x_c \end{bmatrix} + \begin{bmatrix} B_i \\ 0 \end{bmatrix} \begin{bmatrix} K_j & K_{cj} \end{bmatrix} \begin{bmatrix} x \\ x_c \end{bmatrix} + \begin{bmatrix} 0 \\ B_c \end{bmatrix} r$$

$$= \tilde{A}_i \tilde{x} + \tilde{B}_i \tilde{K}_j \tilde{x} + \tilde{B}_r r = \left(\tilde{A}_i + \tilde{B}_i \tilde{K}_j \right) \tilde{x} + \tilde{B}_r r$$

$$\tilde{y} = \begin{bmatrix} C_i & 0 \end{bmatrix} \begin{bmatrix} x \\ x_c \end{bmatrix} = \tilde{C}_i \tilde{x}, \quad i, j = 1, 2, \cdots, g \tag{10}$$

that is

$$\begin{aligned} \dot{\tilde{x}} &= \left(\tilde{A}_i + \tilde{B}_i \tilde{K}_j \right) \tilde{x} + \tilde{B}_r r \\ \tilde{y} &= \tilde{C}_i \tilde{x}, \qquad i, j = 1, 2, \cdots, g \end{aligned} \tag{11}$$

The overall fuzzy tracking control system can be expressed as follows:

$$\begin{aligned} \dot{\tilde{x}}(t) &= \sum_{j=1}^{g} \sum_{i=1}^{g} h_i(\tilde{x}(t)) h_j(\tilde{x}(t)) \left(\tilde{A}_i + \tilde{B}_i \tilde{K}_j \right) \tilde{x}(t) + \tilde{B}_r r \\ \tilde{y}(t) &= \sum_{i=1}^{g} h_i(\tilde{x}(t)) \tilde{C} \tilde{x}(t) \end{aligned} \tag{12}$$

If the fuzzy system contains uncertainties, then the system can be expressed as follows:

$$\begin{aligned} \dot{\tilde{x}}(t) &= \sum_{i=1}^{g} \sum_{j=1}^{g} h_i h_j (\tilde{A}_i + \Delta \tilde{A}_i + (\tilde{B}_i + \Delta \tilde{B}_i) \tilde{K}_j) \tilde{x}(t) + \tilde{B}_r r \\ \tilde{y}(t) &= \sum_{i=1}^{g} h_i(\tilde{x}(t)) \tilde{C}_i \tilde{x}(t) \end{aligned} \tag{13}$$

If the system (13) is stable, then the output y can track the desired trajectory $r(t)$. The sufficient conditions that guarantee the global asymptotic stability of the controlled fuzzy system with parametric uncertainties are presented in terms of

Lyapunov's direct method. The fuzzy tracking controller based on the IMP with good tracking performance can be designed easily.

The closed-loop eigenvalues can be chosen according to the performance requirement of nominal closed-loop system. Then the controller $\tilde{K}_1, \tilde{K}_2, \cdots, \tilde{K}_g$ can be obtained. These controllers can guarantee the good tracking performance of controlled nonlinear systems. Furthermore, the following theorem can guarantee the asymptotically stable in the large.

Theorem 1: If there exist a symmetric and positive definite matrix P, and some scalars $\lambda_{ij}(i, j = 1, 2, \cdots, g)$ such that the following LMIs are satisfied, the T-S fuzzy tracking system (13) is asymptotically stable .

$$\begin{bmatrix} \tilde{A}_i^T P + P\tilde{A}_i + \tilde{K}_i^T \tilde{B}_i^T P + P\tilde{B}_i \tilde{K}_i & * & * \\ E_{1i} + E_{2i}\tilde{K}_i & -\lambda_{ii} I & * \\ H_i^T P & 0 & -\lambda_{ii}^{-1} I \end{bmatrix} < 0 \qquad (1 \le i \le g) \tag{14}$$

$$\begin{bmatrix} \begin{matrix} \tilde{A}_i^T P + P\tilde{A}_i + \tilde{A}_j^T P + P\tilde{A}_j + \tilde{K}_j^T \tilde{B}_i^T P \\ + P\tilde{B}_i \tilde{K}_j + \tilde{K}_i^T \tilde{B}_j^T P + P\tilde{B}_j \tilde{K}_i \end{matrix} & * & * & * & * \\ E_{1i} + E_{2i}\tilde{K}_j & -\lambda_{ii} I & * & * & * \\ E_{1j} + E_{2j}\tilde{K}_i & 0 & -\lambda_{ii} I & * & * \\ H_i^T P & 0 & 0 & -\lambda_{ii}^{-1} I & * \\ H_j^T P & 0 & 0 & 0 & -\lambda_{ii}^{-1} I \end{bmatrix} < 0 \; (1 \le i < j \le g) \tag{15}$$

where $*$ denotes the transposed elements in the symmetric positions.

Proof: Trivial.

4 Robust Fuzzy Observer

In practice, all of state variables are not fully measurable. It is necessary to design a robust fuzzy observer in order to implement the robust fuzzy tracking controller. To simplify the design procedure of fuzzy control systems, the robust fuzzy tracking controller and robust fuzzy observer can be designed independently according to separation property.

For the fuzzy observer design, it is assumed that the nominal fuzzy system is locally observable. The local state observers are designed as follows:

Observer Rule i:

IF $x_1(1)$ is M_{i1} and … and $x_n(1)$ is M_{in}

$$\begin{aligned} &\text{THEN}\; \dot{\hat{\tilde{x}}}(t) = \left(\tilde{A}_i + \Delta\tilde{A}_i(t)\right)\hat{\tilde{x}}(t) + (\tilde{B}_i + \Delta\tilde{B}_i(t))u(t) + G_i[\tilde{y}(t) - \hat{\tilde{y}}(t)] + \tilde{B}_r r \\ &\hat{\tilde{y}}(t) = C_i \hat{\tilde{x}}(t), \qquad i = 1, 2, \cdots, g \end{aligned} \tag{16}$$

where G_i ($i = 1, 2, \cdots, g$) are observation error matrices $y(t)$ and $\hat{y}(t)$ are the final output of the uncertain nonlinear system and the robust fuzzy observer respectively. Then the final estimated state of the robust fuzzy observer is:

$$\dot{\hat{\tilde{x}}}(t)=\sum_{i=1}^{g}h_i(\tilde{A}_i+\Delta\tilde{A}_i)\hat{\tilde{x}}(t)+\sum_{i=1}^{g}h_i(\tilde{B}_i+\Delta\tilde{B}_i)u(t)+\sum_{i=1}^{g}h_iG_i(\tilde{y}(t)-\hat{\tilde{y}}(t))+\tilde{B}_r r \tag{17}$$

The final output of the robust observer is

$$\hat{\tilde{y}}(t)=\sum_{i=1}^{g}h_iC_i\hat{\tilde{x}}(t) \tag{18}$$

By substituting (18) into (17), we obtain

$$\dot{\hat{\tilde{x}}}(t)=\sum_{i=1}^{g}h_i(\tilde{A}_i+\Delta\tilde{A}_i)\hat{\tilde{x}}(t)+\sum_{i=1}^{g}h_i(\tilde{B}_i+\Delta\tilde{B}_i)u(t)+\sum_{i=1}^{g}\sum_{j=1}^{g}h_ih_jG_i\tilde{C}_j[\tilde{x}(t)-\hat{\tilde{x}}(t)]+\tilde{B}_r r \tag{19}$$

The robust fuzzy tracking controller based on the robust fuzzy observer is as follows:

Controler Rule i:

IF $x_1(1)$ is M_{i1} and … and $x_n(1)$ is M_{in}

THEN $u(t)=K_i\hat{\tilde{x}}(t), \quad i=1,2,\cdots,g$ (20)

The final output of this fuzzy controller is

$$u(t)=\sum_{i=1}^{g}h_i(\tilde{x}(t))\tilde{K}_i\hat{\tilde{x}}(t) \tag{21}$$

By respectively substituting (21) into (13) and (19), we obtain

$$\tilde{x}(t)=\sum_{i=1}^{g}h_i(\tilde{A}_i+\Delta\tilde{A}_i)\tilde{x}(t)+\sum_{i=1}^{g}\sum_{j=1}^{g}h_ih_j(\tilde{B}_i+\Delta\tilde{B}_i)\tilde{K}_j\hat{\tilde{x}}(t)+\tilde{B}_r r \tag{22}$$

$$\dot{\hat{\tilde{x}}}(t)=\sum_{i=1}^{g}\sum_{j=1}^{g}h_ih_j(\tilde{A}_i+\Delta\tilde{A}_i+(\tilde{B}_i+\Delta\tilde{B}_i)\tilde{K}_j)\hat{\tilde{x}}(t)+\sum_{i=1}^{g}\sum_{j=1}^{g}h_ih_jG_i\tilde{C}_j[\tilde{x}(t)-\hat{\tilde{x}}(t)]+\tilde{B}_r r \tag{23}$$

Let $e(t)=\tilde{x}(t)-\hat{\tilde{x}}(t)$, then

$$e(t)=\sum_{i=1}^{g}\sum_{j=1}^{g}h_ih_j(\tilde{A}_i+\Delta\tilde{A}_i-G_i\tilde{C}_j)e(t) \tag{24}$$

The time derivative of $e(t)$ is

$$\dot{e}(t)=\sum_{i=1}^{g}\sum_{j=1}^{g}h_i(\tilde{x}(t))h_j(\tilde{x}(t))(\tilde{A}_i+\Delta\tilde{A}_i-G_i\tilde{C}_j)\tilde{x}(t)$$

$$=\sum_{i=1}^{g}h_i^2(\tilde{x}(t))(\tilde{A}_i+\Delta\tilde{A}_i-G_i\tilde{C}_i)\tilde{x}(t)+2\sum_{i<j}^{g}h_i(\tilde{x}(t))h_j(\tilde{x}(t))\times\left(\frac{\tilde{A}_i+\Delta\tilde{A}_i-G_i\tilde{C}_j+\tilde{A}_j+\Delta\tilde{A}_j-G_j\tilde{C}_i}{2}\right)\tilde{x}(t) \tag{25}$$

The following theorem provides a sufficient condition for robust stabilization of the above equation in the presence of parametric uncertainties.

Theorem 2: If there exists a symmetric and positive definite matrix P, some scalars λ_{ij}, ($i,j=1,2,\cdots,g$), such that the following LMIs are satisfied, the estimate error (24) is asymptotically stable.

$$\begin{bmatrix} \tilde{A}_i^T P + P\tilde{A}_i - (\tilde{K}_i^T \tilde{C}_i^T P + P\tilde{C}_i \tilde{K}_i) & * & * \\ E_{1i} & -\lambda_{ii} I & * \\ H_i^T P & 0 & -\lambda_{ii}^{-1} I \end{bmatrix} < 0 \quad (1 \le i \le g) \tag{26}$$

$$\begin{bmatrix} \tilde{A}_i^T P + P\tilde{A}_i + \tilde{A}_j^T P + P\tilde{A}_j - (\tilde{C}_j^T G_i^T P + PG_i\tilde{C}_j + \tilde{C}_i^T G_j^T P + PG_j\tilde{C}_i) & * & * & * & * \\ E_{1i} & \lambda_{ij}^{-1} I & * & * & * \\ E_{1j} & 0 & \lambda_{ij}^{-1} I & * & * \\ H_i^T P & 0 & 0 & \lambda_{ij} I & * \\ H_j^T P & 0 & 0 & 0 & \lambda_{ij} I \end{bmatrix} < 0 \quad (1 \le i < j \le g) \tag{27}$$

Proof: Trivial.

5 Numerical Example

For this example, a problem of balancing of an inverted pendulum on a cart to illustrate the performance of the robust fuzzy tracking controller. The state equation of the inverted pendulum is given by [12]:

$$\begin{aligned} \dot{x}_1 &= x_2 \\ \dot{x}_2 &= \frac{1.0}{\left[(M+m)(J+ml^2) - m^2 l^2 \cos^2 x_1\right]} \\ &\quad \times [-f_1 (M+m) x_2 - m^2 l^2 x_2^2 \sin x_1 \cos x_1 \\ &\quad + f_0 m l x_4 \cos x_1 + (M+m) m g l \sin x_1 - ml \cos x_1 u] \\ \dot{x}_3 &= x_4 \\ \dot{x}_4 &= \frac{1.0}{\left[(M+m)(J+ml^2) - m^2 l^2 \cos^2 x_1\right]} \\ &\quad \times [f_1 m l x_2 \cos x_1 + (J+ml^2) m l x_2^2 \sin x_1 \\ &\quad - f_0 (J+ml^2) x_4 - m^2 g l^2 \sin x_1 \cos x_1 + (J+ml^2) u] \\ y &= x_3 \end{aligned} \tag{28}$$

where x_1 denotes the angle(rad) of the pendulum from the vertical, x_2 is the angular velocity (rad/s), x_3 is the displacement(m) of the cart, and x_4 is the velocity (m/s) of the cart. In addition, g=9.8m/s2 is the gravity constant, m is the mass (kg) of the pendulum, M is the mass (kg) of the cart, f_0 is the friction factor (N/m/s) of the cart, f_1 is the frication factor (N/rad/s) of the pendulum, l is the length (m) from the center of mass of the pendulum to the shaft axis, J is the moment of inertia (kgm2) of the pendulum, u is the force (N) applied to the cart. The design parameters are M=2.68, m=0.23, f_0=18.46, f_1=0.00691, l=0.324, J=0.00501. The objective is to design a robust fuzzy tracking controller that is composed of a servo-compensator and a state feedback controller so as to the output y can track the desired trajectory $r(t)$.

The system (28) is approximated by the following two-rule fuzzy model:

Plant Rule 1:
IF $x_1(t)$ is about 0
THEN $\dot{x}(t) = A_1 x(t) + B_1 u(t)$
$y_1(t) = C_1 x(t)$

Plant Rule 2:
IF $x_1(t)$ is about $\pm\pi/3$
THEN $\dot{x}(t) = A_2 x(t) + B_2 u(t)$
$y_2(t) = C_2 x(t)$

where

$$A_1 = \begin{bmatrix} 0 & 1 & 0 & 0 \\ 26.8036 & -0.2536 & 0 & 17.3503 \\ 0 & 0 & 0 & 1 \\ -0.6864 & 0.0065 & 0 & -6.788 \end{bmatrix} \quad B_1 = \begin{bmatrix} 0 \\ -0.9399 \\ 0 \\ 0.3677 \end{bmatrix}$$

$$A_2 = \begin{bmatrix} 0 & 1 & 0 & 0 \\ 21.0601 & -0.2410 & 0 & 8.2422 \\ 0 & 0 & 0 & 1 \\ -0.2697 & 0.0031 & 0 & -6.4492 \end{bmatrix} \quad B_2 = \begin{bmatrix} 0 \\ -0.4465 \\ 0 \\ 0.3494 \end{bmatrix}$$

$$C_1 = C_2 = C_3 = C_4 = \begin{bmatrix} 0 & 0 & 1 & 0 \end{bmatrix}$$

The desired trajectory $r(t)$ is a step signal. Its Laplace transform of is as follows:

$$R(s) = \ell\left[1(t)\right] = \frac{1}{s}$$

We can design the servo-compensator

$$\dot{x}_c = A_c x_c + B_c e$$
$$y_c = x_c$$

where

$A_c = [0]$, $B_c = [1]$

It is assumed that all system parameters are uncertain but bounded within 20% of the maximal values. It is defined by

$$H_1 = H_2 = \begin{bmatrix} -0.2 & 0 & 0 & 0 & 0 \\ 0 & 0.2 & 0 & 0 & 0 \\ 0 & 0 & -0.2 & 0 & 0 \\ 0 & 0 & 0 & 0.2 & 0 \\ 0 & 0 & 0 & 0 & 0.2 \end{bmatrix}, \quad E_{11} = E_{12} = \begin{bmatrix} 0 & 1 & 0 & 0 & 0 \\ 23 & -0.246 & 0 & 12.8 & 0 \\ 0 & 0 & 0 & 1 & 0 \\ -0.38 & 0.0051 & 0 & -6.2 & 0 \\ 0 & 0 & -1 & 0 & 0 \end{bmatrix}$$

$$E_{21} = E_{22} = \begin{bmatrix} 0 & 0 & 0 & 0 & 0 \end{bmatrix}^T$$

Choosing the closed-loop eigenvalues [-5.0 -2.0 -3.0 -1.0 –2.0] for $\tilde{A}_1 - \tilde{B}_1 \cdot \tilde{K}_1$ and $\tilde{A}_2 - \tilde{B}_2 \cdot \tilde{K}_2$, we obtain the state feedback controller.

K_1=[-168.1311 -32.3925 -77.9398 -61.1534 73.1310]
K_2=[-356.5590 -76.3357 -99.3797 -73.7725 93.0752]

The robust fuzzy tracking controller is applied to the original system (28), then we obtain the simulation results as shown in Fig. 1 and 2. The initial condition is assumed

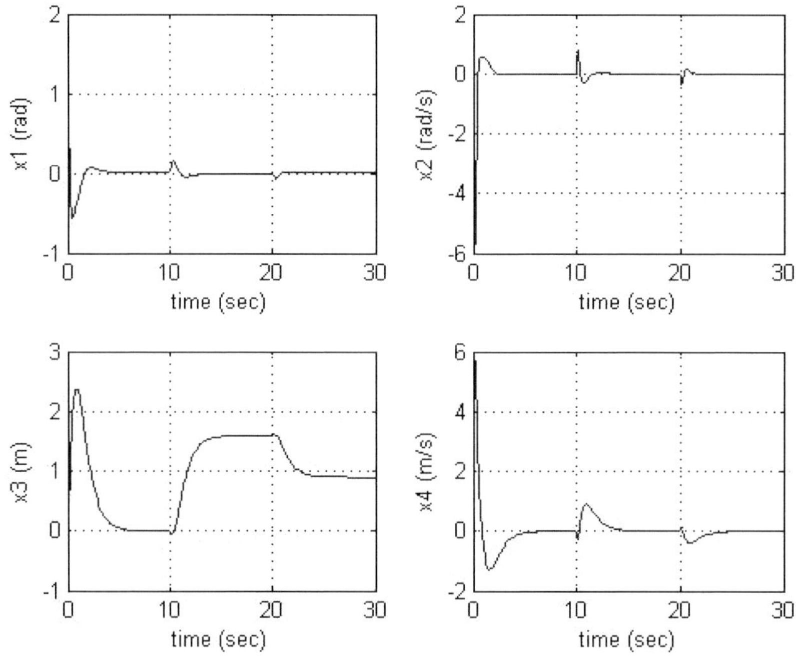

Fig. 1. x_1: the angle (rad) of the pendulum; x_2: the angular velocity (rad/s); x_3: the displacement (m) of the cart; x_4: the velocity (m/s) of the cart.

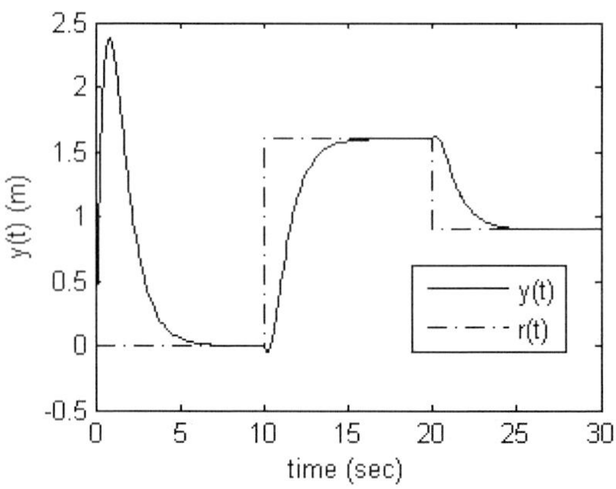

Fig. 2. The output of the cart

to be $(x_1(0), x_2(0), x_3(0), \ x_4(0)) = (\pi/4, 0, 0, 0)$. In the figure 1, the state variables verge equilibrium state quickly. It shows that the robust fuzzy tracking controller can balance the inverted pendulum with uncertainties. Fig. 2 presents the simulation results for the proposed robust fuzzy tracking control. The desired trajectory $r(t)$ is tracked by output $y(t)$ without error. The tracking performance is obviously better than that of other fuzzy tracking controller.

6 Conclusion

In this paper, we have proposed the robust fuzzy controller based on the IMP. Then the robust fuzzy observer is proposed for the controller. Sufficient conditions for robust stabilization of uncertain nonlinear plants with robust fuzzy tracking controller and robust fuzzy observer are also developed and analyzed. These conditions are formulated in the LMI format so that they can be solved easily. Because the internal model of the desired trajectory is embedded in the control system, the robust fuzzy tracking controller can track the desired trajectory without error.

Furthermore, we propose the separation design of robust fuzzy tracking controllers and robust fuzzy observers. Thus we can design robust fuzzy controllers and robust fuzzy observers separately, which reduces the difficulty of design. Simulation results on the inverted pendulum system show that the desired trajectory for uncertain nonlinear systems can be tracked via the proposed method without error.

Acknowledgement

This work is supported by Key Project of Science & Technology Commission of Shanghai Municipality (04JC14038 and 04DZ11008), Doctoral Program Foundation of Science & Technology Special Project in University (20040280017), Program for New Century Excellent Talents in University (NCET-04-0433), and Shanghai Leading Academic Disciplines (T0103).

References

1. Li, K., Thompson, S., Peng, J.: Modelling and prediction of NOx emission in a coal-fired power generation plant. Control Engineering Practice. 12 (2004) 707-723
2. Chiang, C.-L.; Su, C.-T.: Tracking control of induction motor using fuzzy phase plane controller with improved genetic algorithm. Electric Power Systems Research, 2 (2005) 239-247
3. El, H. A., Bentalba, S.: Fuzzy path tracking control for automatic steering of vehicles. Robotics and Autonomous Systems, 4 (2003) 203-213
4. Marseguerra, M., Zio, E.: Model-free fuzzy tracking control of a nuclear reactor. Annals of Nuclear Energy, 9 (2003) 953-981
5. Park, C.-W., Cho, Y.-W.: Adaptive tracking control of flexible joint manipulator based on fuzzy model reference approach. IEE Proceedings: Control Theory and Applications, 2 (2003) 198-204

6. Tseng, C.-S.; Chen, B.-S.; Uang, H.-J.: Fuzzy tracking control design for nonlinear dynamic systems via T-S fuzzy model. IEEE Transactions on Fuzzy Systems, 3 (2001) 381-392
7. Kim, E.: Output feedback tracking control of robot manipulators with model uncertainty via adaptive fuzzy logic. IEEE Transactions on Fuzzy Systems, 3 (2004) 368-378
8. Tong, S.-C.; Wang, T.; Li, H.-X.: Fuzzy robust tracking control for uncertain nonlinear systems. International Journal of Approximate Reasoning, 2 (2002) 73-90
9. Guan, X., Chen, C.: Adaptive fuzzy control for chaotic systems with H infinity tracking performance. Fuzzy Sets and Systems, 1 (2003) 81-93
10. Yu, W.-S.: H infinity tracking-based adaptive fuzzy-neural control for MIMO uncertain robotic systems with time delays. Fuzzy Sets and Systems, 3 (2004) 375-401
11. Chang, Y.-C.: Robust tracking control for nonlinear MIMO systems via fuzzy approaches. Automatica, 10 (2000) 1535-1545
12. Ma, X. J., Sun, Z. Q., He, Y. Y.: Analysis and design of fuzzy controller and fuzzy observer. IEEE Transactions on Fuzzy Systems, 1 (1998) 41-51
13. Zhang J., Fei M.R.: Analysis and Design of Robust Fuzzy Controllers and Robust Fuzzy Observers of Uncertain Nonlinear Systems. The 6th World Congress on Intelligent Control and Automation, Dalian, China (2006) 3767-3771
14. Lee H. J., Park J. B., Chen G.: Robust fuzzy control of nonlinear systems with parametric uncertainties. IEEE Transactions on Fuzzy Systems, 2 (2001) 369-379

Adaptive Dynamic Surface Fuzzy Control for a Class of Uncertain Nonlinear Systems

Chen Gang

College of Automation, Chongqing University,
Chongqing 400044, China
chengang@cqu.edu.cn

Abstract. Adaptive dynamic surface fuzzy control approach is proposed for a class of uncertain nonlinear systems in strict-feedback form. The dynamic surface control technique is introduced to overcome the problem of explosion of terms associated with backstepping design method. Fuzzy logic system is used as a universal approximator to approximate unstructured uncertain functions and the bounds of the reconstruction error is estimated online. The algorithm has the adaptive mechanism with minimum learning parameterizations. Furthermore, all the signals in the closed-loop systems are guaranteed to be semi-globally uniformly ultimately bounded and the output of the system is proved to converge to a small neighborhood of the desired trajectory. The control performance can be guaranteed by an appropriate choice of the design parameters. Simulation results demonstrate the effectiveness of the proposed control method.

1 Introduction

Control of uncertain nonlinear systems for regulation and tracking has been one of the research focuses in the past decade. Robust control and adaptive control are two popular methods to tackle this kind of questions. Recently, adaptive nonlinear control has made significant progress after introduction of the adaptive backstepping design procedure [1]. However, we will find that the design procedure of the traditional backstepping design is lengthy and tedious. Because of the repeated differentiations of certain nonlinear functions in the design, the complexity of controller grows drastically with the increase of the system order. In [2], the dynamic surface control technique was introduced to overcome this problem. Low-pass filters are integrated into the design to avoid explicit differentiation of the nonlinear functions.

The challenge of controlling complex plant without model based knowledge has motivated the application of fuzzy set theory. Parallel to the development of robust control and adaptive control, fuzzy logic control has been widely used in many engineering applications. Wang has proven that a fuzzy logic system(FLS) can approximate uniformly any nonlinear continuous function under a compact set[4]. Fuzzy set theory provides a way to the modeling and design of nonlinear control system. In order to improve the performance and stability of control system, the synthesis approach to construct adaptive fuzzy controllers has received much attention[4-14]. However,

L. Wang et al. (Eds.): FSKD 2006, LNAI 4223, pp. 199 – 207, 2006.

when many rules are used in the fuzzy control, the learning time tend to become very large. Thus, the computation burden of the control system is very high.

In this note, we present an adaptive dynamic surface fuzzy control method for a class of strict-feedback nonlinear systems. The algorithm not only avoids the explosion of complexity, but also has the adaptive mechanism with minimal learning parameterizations.

2 Problem Formulation

2.1 System Description

Consider the n-order nonlinear system of the form

$$\begin{aligned} \dot{x}_i &= f_i(x_1,\cdots,x_i)+x_{i+1},\ x(0)=x_0,\ 1\le i\le n-1 \\ \dot{x}_n &= f_n(x)+u, \\ y &= x_1, \end{aligned} \tag{1}$$

where $u\in R$ is the input, $y\in R$ is the output, $x(t)\in R^n$ is the state, and $f_i(x_1,\cdots,x_i)$, $i=1,\cdots,n$, are unknown functions satisfying $f_i(0,\cdots,0)=0$. The control objective is to make the output y track the desired smooth trajectory $y_r(t)$ with any prescribed small error.

2.2 Fuzzy Logic System

In this note, we consider a FLS with the product-inference rule, singleton fuzzifier, center average defuzzifier, and Gaussian membership function [4]. The fuzzy rules are in the following form:

RULE $i(i=1,\cdots,m)$: IF x_1 is F_{i1} and$\cdots$, and x_n is F_{in}, THEN y is w_i, where $x=(x_1,\cdots,x_n)^{\mathrm{T}}\in X=X_1\times\cdots X_n\subset R^n$ and $y\in Y\subset R$ are the input and the output of the FLS, respectively. n is the number of the premise variables. m is the number of IF-THEN rules. w_i is the fuzzy singleton in the i-th rule. F_{ij}, $j=1,\cdots,n$, are the labels of fuzzy sets in the universes of discourse X_j with Gaussian membership function

$$\mu_{F_{ij}}(x_j)=\exp\left(-\left(\frac{x_j-\bar{x}_{ij}}{\sigma_{ij}}\right)^2\right) \tag{2}$$

where $\bar{x}_{ij},\sigma_{ij}$ are design parameters.

The overall fuzzy model is achieved by fuzzy aggregation of each individual rules as

$$y(\boldsymbol{x})=\frac{\sum_{i=1}^{m} w_i\left(\prod_{j=1}^{n}\mu_{F_{ij}}(x_j)\right)}{\sum_{i=1}^{m}\left(\prod_{j=1}^{n}\mu_{F_{ij}}(x_j)\right)} \tag{3}$$

where $y: X \subset R^n \to R$. If the membership functions (i.e., $\bar{x}_{ij}, \sigma_{ij}$) are fixed, and the fuzzy singleton $w_i, i=1,\cdots,m$,are adjustable parameters, then (3) can be rewritten as

$$y(x|W)=\sum_{i=1}^{m} w_i \xi_i(x) = W^{\mathrm{T}} \xi_0(x)$$
$$\xi_i(\boldsymbol{x}) = \frac{\prod_{j=1}^{n} \mu_{F_{ij}}(x_j)}{\sum_{i=1}^{m}\left(\prod_{j=1}^{n} \mu_{F_{ij}}(x_j)\right)} \tag{4}$$

where $\xi_0(x) = (\xi_1(x),\cdots,\xi_m(x))^{\mathrm{T}} \in R^m$ is called the fuzzy basis function vector, and $W=(w_1,\cdots,w_m)^{\mathrm{T}} \in R^m$ is called the parameter vector.

Lemma 1 [4]: For any given real continuous function y on a compact set $X \subset R^n$ and arbitrary $\varepsilon > 0$, there exits a FLS y^* in the form of (4) with (2) such that $\sup_{x\in X}\left|y^*(x|W) - y(x)\right| < \varepsilon$.

Based on Lemma 1, for the unknown nonlinear function $f_0(x)$, we have the result over the compact set Ω_0: $f_0(x) = w_0^{*T}\xi_0(x) + v_0(x)$. Where $v_0(x)$ is the reconstruction error. w_0^* is an optimal weight vector and is chosen as $w_0^* = \arg\min_{w_0\in R^n}\left\{\sup_{x\in\Omega_0}\left|f_0(x) - w_0^{\mathrm{T}}\xi_0(x)\right|\right\}$.

3 Design of Adaptive Fuzzy Tracking Controller

In this note, FLSs are employed to approximate the uncertain functions. The design procedure consists of n steps. At each step i, the virtual controller α_i will be developed by employing an appropriate Lyapunov function V_i. The detailed design procedure is given below.

Step 1. Consider the first function $\dot{x}_1 = f_1(x_1) + x_2$ in system (1). Since $f_1(x_1)$ is unknown, we employ a FLS to approximate it, that is $f_1(x_1) = \theta_1^{*\mathrm{T}}\xi_1(x_1) + \delta_1^*$. Based on a priori knowledge, the premise parts of the FLS as well as the nominal vector $\bar{\theta}_1$ are designed first and fixed. Thus, there exist positive constants ρ_{11} and ρ_{12} such that $\left\|\theta_1^* - \bar{\theta}_1\right\| \le \rho_{11}$, $\left|\delta_1^*\right| \le \rho_{12}$. Define the first error surface $s_1 = x_1 - y_r$. The time derivative of s_1 is

$$\dot{s}_1 = x_2 + f_1 - \dot{y}_r\,. \tag{5}$$

To make the system (5) stable, we choose the following fictitious controller

$$\alpha_1 = -\bar{\theta}_1^{\mathrm{T}}\xi_1(x_1) - k_1 s_1 + \dot{y}_r - \varphi_1\,, \tag{6}$$

where $\varphi_1 = \dfrac{\hat{\rho}_{11}^2 \|\xi_1\|^2 s_1}{\hat{\rho}_{11}|s_1|\|\xi_1\| + \varepsilon_{11}\exp(-ct)} + \dfrac{\hat{\rho}_{12}^2 s_1}{\hat{\rho}_{12}|s_1| + \varepsilon_{12}\exp(-ct)}$, c is a positive design parameter. Choosing the following adaptive laws

$$\begin{aligned} \dot{\hat{\rho}}_{11} &= -\sigma_{w1}\hat{\rho}_{11} + r_{w1}|s_1|\|\xi_1\|, \\ \dot{\hat{\rho}}_{12} &= -\sigma_{v1}\hat{\rho}_{12} + r_{v1}|s_1|, \end{aligned} \tag{7}$$

where σ_{w1}, σ_{v1}, r_{w1}, and r_{v1} are positive constants. Let α_1 pass through a first-order filter with time constant τ_2 to get

$$\tau_2 \dot{z}_2 + z_2 = \alpha_1, \; z_2(0) = \alpha_1(0). \tag{8}$$

Step i $(2 \le i \le n)$. Consider the i-th function $\dot{x}_i = f_i(x_1, \cdots, x_i) + x_{i+1}$ in system (1). Let $u = x_{i+1}$. According to the universal approximation property of the FLS, we have $f_i = \theta_i^{*\mathrm{T}} \xi_i + \delta_i^*$. Based on a priori knowledge, the premise parts of the FLS as well as the nominal vector $\overline{\theta}_i$ are designed first and fixed. Thus, there exist positive constants ρ_{i1} and ρ_{i2} such that $\|\theta_i^* - \overline{\theta}_i\| \le \rho_{i1}$, $\|\delta_i^*\| \le \rho_{i2}$. Define the i-th error surface $s_i = x_i - z_i$. The time derivative of s_i is

$$\dot{s}_i = x_{i+1} + f_i - \dot{z}_i, \tag{9}$$

To make the system (9) stable, we choose the following fictitious controller

$$\alpha_i = -\overline{\theta}_i^{\mathrm{T}} \xi_i - k_i s_i + \dot{z}_i - \varphi_i, \tag{10}$$

where $\varphi_i = \dfrac{\hat{\rho}_{i1}^2 \|\xi_i\|^2 s_i}{\hat{\rho}_{i1}|s_i|\|\xi_i\| + \varepsilon_{i1}\exp(-ct)} + \dfrac{\hat{\rho}_{i2}^2 s_i}{\hat{\rho}_{i2}|s_i| + \varepsilon_{i2}\exp(-ct)}$. Choosing the following adaptive laws

$$\begin{aligned} \dot{\hat{\rho}}_{i1} &= -\sigma_{wi}\hat{\rho}_{i1} + r_{wi}|s_i|\|\xi_i\|, \\ \dot{\hat{\rho}}_{i2} &= -\sigma_{vi}\hat{\rho}_{i2} + r_{vi}|s_i|, \end{aligned} \tag{11}$$

where σ_{wi}, σ_{vi}, r_{wi}, and r_{vi} are positive constants. Let α_i pass through a first-order filter with time constant τ_{i+1} to get

$$\tau_{i+1} \dot{z}_{i+1} + z_{i+1} = \alpha_i, \; z_{i+1}(0) = \alpha_i(0). \tag{12}$$

According to (5), (6), (9), and (10), we have

$$\begin{aligned} \dot{s}_1 &= s_2 - k_1 s_1 + (\theta_1^* - \overline{\theta}_1)^{\mathrm{T}} \xi_1 + \delta_1^* - \varphi_1 + y_2, \\ \dot{s}_i &= s_{i+1} - k_i s_i + (\theta_i^* - \overline{\theta}_i)^{\mathrm{T}} \xi_i + \delta_i^* - \varphi_i + y_{i+1}, \; i = 2, \cdots, n-1 \\ \dot{s}_n &= -k_n s_n + (\theta_n^* - \overline{\theta}_n)^{\mathrm{T}} \xi_n + \delta_n^* - \varphi_n. \end{aligned} \tag{13}$$

Define $y_j = z_j - \alpha_{j-1}, 2 \le j \le n$. Differentiating y_j yields

$$\begin{aligned}
\dot{y}_2 &= \dot{z}_2 + \bar{\theta}_1^{\mathrm{T}} \frac{\partial \xi_1}{\partial x_1} + k_1 \dot{s}_1 + \dot{\varphi}_1 - \ddot{y}_r , \\
&= -\frac{y_2}{\tau_2} + B_2\left(s_1, s_2, y_2, \hat{\rho}_{11}, \hat{\rho}_{12}, y_r, \dot{y}_r, \ddot{y}_r\right), \\
\dot{y}_{i+1} &= -\frac{y_{i+1}}{\tau_{i+1}} + \bar{\theta}_i^{T} \frac{\partial \xi_i}{\partial (x_1, \cdots, x_i)} \begin{bmatrix} \dot{x}_1 \\ \vdots \\ \dot{x}_i \end{bmatrix} + k_i \dot{s}_i + \dot{\varphi}_i + \frac{\dot{y}_i}{\tau_i} , \\
&= -\frac{y_{i+1}}{\tau_{i+1}} + B_{i+1}\left(s_1, \cdots, s_{i+1}, \cdots, y_2, \cdots, y_{i+1}, \hat{\rho}_{11}, \hat{\rho}_{12}, \cdots, \hat{\rho}_{i1}, \hat{\rho}_{i2}, y_r, \dot{y}_r, \ddot{y}_r\right), \\
& \qquad 2 \le i \le n-1
\end{aligned} \tag{14}$$

Define $\tilde{\rho}_{i1} = \hat{\rho}_{i1} - \rho_{i1}$, $\tilde{\rho}_{i2} = \hat{\rho}_{i2} - \rho_{i2}$. For any $D_1 > 0$, $D_2 > 0$, the sets $A_1 = \left\{(y_r, \dot{y}_r, \ddot{y}_r): \ y_r^2 + \dot{y}_r^2 + \ddot{y}_r^2 \le D_1\right\}$ and $A_2 = \left\{\sum_{i=1}^{n}\left(\varepsilon_i^2 + \frac{1}{r_{wi}}\tilde{\rho}_{i1}^2 + \frac{1}{r_{vi}}\tilde{\rho}_{i2}^2\right) + \sum_{i=1}^{n-1} y_{i+1}^2 \le 2D_2\right\}$ are compact in R^3 and R^{4n-1}, respectively. Thus, $A_1 \times A_2$ is compact in R^{4n+2}. Therefore, $|B_{i+1}|$ has a maximum M_{i+1} on $A_1 \times A_2$.

Consider the Lyapunov function $V = \frac{1}{2}\sum_{i=1}^{n}\left(s_i^2 + \frac{1}{r_{wi}}\tilde{\rho}_{i1}^2 + \frac{1}{r_{vi}}\tilde{\rho}_{i2}^2\right) + \frac{1}{2}\sum_{i=1}^{n-1} y_{i+1}^2$. According to (7), (11), (13), and (14), the time derivative of V is

$$\begin{aligned}
\dot{V} \le & \sum_{i=1}^{n-1}\left(-k_i s_i^2 + s_i s_{i+1} - \tilde{\rho}_{i1}|s_i|\|\xi_i\| - \tilde{\rho}_{i2}|s_i| + \varepsilon_{i1}e^{-ct} + \varepsilon_{i2}e^{-ct} + y_{i+1}s_i + \frac{1}{r_{wi}}\tilde{\rho}_{i1}\dot{\hat{\rho}}_{i1} + \frac{1}{r_{vi}}\tilde{\rho}_{i2}\dot{\hat{\rho}}_{i2}\right) \\
& -k_n s_n^2 - \tilde{\rho}_{n1}|s_n|\|\xi_n\| - \tilde{\rho}_{n2}|s_n| + \varepsilon_{n1}e^{-ct} + \varepsilon_{n2}e^{-ct} + \frac{1}{r_{wn}}\tilde{\rho}_{n1}\dot{\hat{\rho}}_{n1} + \frac{1}{r_{vn}}\tilde{\rho}_{n2}\dot{\hat{\rho}}_{n2} + \sum_{i=1}^{n-1}\left(-\frac{y_{i+1}^2}{\tau_{i+1}} + |y_{i+1}B_{i+1}|\right) \\
\le & \sum_{i=1}^{n-1}\left(-k_i s_i^2 + s_i s_{i+1} + s_i y_{i+1}\right) - k_n s_n^2 + \sum_{i=1}^{n}\left(-\frac{\sigma_{wi}}{r_{wi}}\tilde{\rho}_{i1}\hat{\rho}_{i1} - \frac{\sigma_{vi}}{r_{vi}}\tilde{\rho}_{i2}\hat{\rho}_{i2}\right) + \sum_{i=1}^{n}\left(\varepsilon_{i1}e^{-ct} + \varepsilon_{i2}e^{-ct}\right) \\
& + \sum_{i=1}^{n-1}\left(-\frac{y_{i+1}^2}{\tau_{i+1}} + |y_{i+1}B_{i+1}|\right).
\end{aligned}$$

Noticing that $s_i^2 + s_{i+1}^2/4 \ge s_i s_{i+1}$, $s_i^2 + y_{i+1}^2/4 \ge s_i y_{i+1}$, $\tilde{\rho}_{i1}\hat{\rho}_{i1} \ge \left(\hat{\rho}_{i1}^2 - \rho_{i1}^2\right)/2$, $|y_{i+1}B_{i+1}| \le \delta/2 + y_{i+1}^2 B_{i+1}^2/(2\delta)$ $(\delta > 0)$, we further have

$$\begin{aligned}
\dot{V} \le & \sum_{i=1}^{n-1}\left(-k_i s_i^2 + 2s_i^2 + \frac{s_{i+1}^2}{4} + \frac{y_{i+1}^2}{4}\right) - k_n s_n^2 + \sum_{i=1}^{n}\left(\frac{\sigma_{wi}}{2r_{wi}}\left(\rho_{i1}^2 - \hat{\rho}_{i1}^2\right) + \frac{\sigma_{vi}}{2r_{vi}}\left(\rho_{i2}^2 - \tilde{\rho}_{i2}^2\right)\right) + \sum_{i=1}^{n}\left(\varepsilon_{i1}e^{-ct} + \varepsilon_{i2}e^{-ct}\right) \\
& + \sum_{i=1}^{n-1}\left(-y_{i+1}^2/\tau_{i+1} + y_{i+1}^2 B_{i+1}^2/(2\delta) + \delta/2\right).
\end{aligned}$$

Let $k_1 = 2 + b_0$, $k_i = 2.25 + b_0$ $(i = 2, \cdots, n-1)$, $k_n = 0.25 + b_0$, $b_0 = \min\left(\frac{\sigma_{w1}}{2}, \cdots, \frac{\sigma_{wn}}{2}, \frac{\sigma_{v1}}{2}, \cdots, \frac{\sigma_{vn}}{2}\right)$, $\frac{1}{\tau_{i+1}} = 0.25 + \frac{M_{i+1}^2}{2\delta} + b_0$, $e_0 = \sum_{i=1}^{n}\left(\frac{\sigma_{wi}}{2r_{wi}}\rho_{i1}^2 + \frac{\sigma_{vi}}{2r_{vi}}\rho_{i2}^2\right) + \sum_{i=1}^{n}\left(\varepsilon_{i1}e^{-ct} + \varepsilon_{i2}e^{-ct}\right)$.

It follows that

$$\begin{aligned}\dot{V} &\leq \sum_{i=1}^{n}\left(-b_0\left(s_i^2 + \frac{\tilde{\rho}_{i1}^2}{r_{wi}} + \frac{\tilde{\rho}_{i2}^2}{r_{vi}}\right)\right) + \sum_{i=1}^{n-1}\left(\frac{y_{i+1}^2}{4} - \left(\frac{1}{4} + \frac{M_{i+1}^2}{2\delta} + b_0\right)y_{i+1}^2 + \frac{y_{i+1}^2 B_{i+1}^2}{2\delta}\right) \\ &\qquad + \frac{(n-1)\delta}{2} + e_0 \\ &\leq -2b_0 V + \frac{(n-1)\delta}{2} + e_0 .\end{aligned} \tag{15}$$

Solving inequality (15) yields

$$V(t) \leq ((n-1)/2\delta + e_0)/(2b_0) + (V(0) - ((n-1)/2\delta + e_0)/(2b_0))e^{-2b_0 t} \tag{16}$$

By increasing the value of b_0, the quantity $((n-1)/2\delta + e_0)/(2b_0)$ can be made arbitrarily small. Based on the previous analysis, we get the following theorem.

Theorem 1. Consider the closed-loop system consisting of (1), controller (6), (10), and parameter adaptive laws (7), (11). Assume that all the initial conditions are on the compact set A_2. All the signals in the closed-loop system are uniformly bounded. Furthermore, the tracking error can be made arbitrarily small by adjusting the design parameters suitably.

Compared with the fuzzy control methods presented in [4-9, 13, 14], the proposed control scheme in this note has the following features.

Remark 1. No matter how many rules are used in the FLS, our algorithm only requires $2n$ parameters to be updated online, where n denotes the number of the state variables in the designed system. The online computation burden is reduced drastically. This feature is particularly different from most fuzzy control methods reported in the literature.

Remark 2. The algorithm can easily incorporate a priori information of the plant into the controller design. Based on the priori knowledge, we can first design the nominal fuzzy controller. In control engineering, the fuzzy control is very useful when the plants are too complex for analysis using conventional techniques, and have available qualitative knowledge from domain experts for the controller design.

Remark 3. The algorithm is more suitable for practical implementation. The controllers and the parameter adaptive laws are highly structural. Such a property is particularly suitable for parallel processing and hardware implementation in the practical applications.

Remark 4. There are two important advantages associated with dynamic surface control technique. It prevents the problem of explosion of terms and allows the design where the model is not differentiated.

4 Simulation Example

The dynamics for a single-link manipulator with the inclusion of motor dynamics is described by [10-12]

$$D\ddot{q} + B\dot{q} + N\sin(q) = \tau,$$
$$M\dot{\tau} + H\tau = u - k_m\dot{q}, \tag{17}$$

where q, $\dot{q}$, $\ddot{q}$ denote the link angular position, velocity, acceleration, respectively. τ is the motor shaft angle. u represents the motor torque. Let $x_1 = q$, $x_2 = \dot{q}$, $x_3 = \tau$. Then equations (17) equal to the following state space representation

$$\dot{x}_1 = x_2,$$
$$\dot{x}_2 = \frac{1}{D}x_3 + f_1(x_1, x_2),$$
$$\dot{x}_3 = \frac{1}{M}u + f_2(x_2, x_3), \tag{18}$$

where $f_1(x_1, x_2) = -\frac{B}{D}x_2 - \frac{N}{D}\sin(x_1)$, $f_2(x_2, x_3) = -\frac{k_m}{M}x_2 - \frac{H}{M}x_3$. We define five fuzzy sets for each variable x_1, x_2, and x_3 with labels negative big(NB), negative small(NS), zero(ZO), positive small(PS), positive big(PB). The fuzzy membership functions are chosen as $\mu_{NB}(x_1) = e^{-0.01(x_1+1)^2}$, $\mu_{NS}(x_1) = e^{-0.01(x_1+0.5)^2}$, $\mu_{ZO}(x_1) = e^{-0.01x_1^2}$, $\mu_{PS}(x_1) = e^{-0.01(x_1-0.5)^2}$, $\mu_{PB}(x_1) = e^{-0.01(x_1-1)^2}$. If $x_1 < -1$, then $\mu_{NB} = 1$. If $x_1 > 1$, then $\mu_{PB} = 1$. $\mu_{NB}(x_j) = e^{-0.02(x_j+1)^2}$, $\mu_{NS}(x_j) = e^{-0.02(x_j+0.5)^2}$, $\mu_{ZO}(x_j) = e^{-0.02x_j^2}$, $\mu_{PS}(x_j) = e^{-0.02(x_j-0.5)^2}$, $\mu_{PB}(x_j) = e^{-0.02(x_j-1)^2}$, $j = 2,3$. If $x_j < -1$, then $\mu_{NB} = 1$. If $x_j > 1$, then $\mu_{PB} = 1$. The desired output is assumed to be $y_r = \sin(\pi t)$. In the simulation study, the model parameter values are $D = 1$, $M = 0.05$, $B = 1$, $k_m = 10$, $H = 0.5$, $N = 10$. The controller parameters chosen for simulation are $c = 0.2$, $k_1 = k_2 = 5$, $k_3 = 1$, $\varepsilon_{i1} = \varepsilon_{i2} = 0.1$, $\tau_1 = \tau_2 = 0.02$. The parameter values in the adaptive laws are $r_{w1} = 15$, $\sigma_{w1} = 5$, $r_{v1} = 15$, $\sigma_{v1} = 5$, $r_{w2} = 20$, $\sigma_{w2} = 10$, $r_{v2} = 20$, $\sigma_{v2} = 10$. Two FLSs are constructed in the controller design procedure. The nominal vector for the first FLS is $w_1 = (-66\ -28\ -0.1\ 28.5\ 66\ -66\ -28\ 0.1\ 28.5\ 66)^{\mathrm{T}}$. The nominal vector for the second FLS is $w_2 = (-70\ -35\ 0\ 35\ 70\ -70\ -35\ 0\ 35\ 70)^{\mathrm{T}}$. The initial parameter values chosen for simulation are $x_1(0) = 0$, $x_2(0) = \pi$, $x_3(0) = \pi$,

$z_1(0)=\pi$, $z_2(0)=\pi$, $\hat{\rho}_{11}(0)=\hat{\rho}_{12}(0)=\hat{\rho}_{21}(0)=\hat{\rho}_{22}(0)=0$. Fig.1 shows that the proposed algorithm achieves good tracking performance. The control input is given in Fig.2. Fig. 3 and Fig.4 show that the adaptive parameters $\hat{\rho}_{11}(t)$, $\hat{\rho}_{12}(t)$, $\hat{\rho}_{21}(t)$, and $\hat{\rho}_{22}(t)$ are bounded.

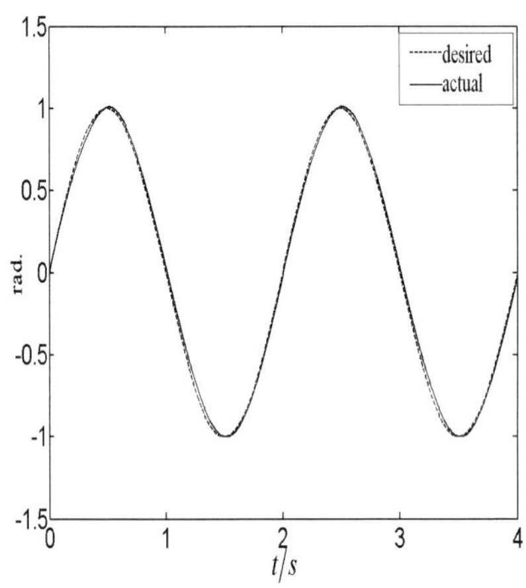

Fig. 1. Tracking performance

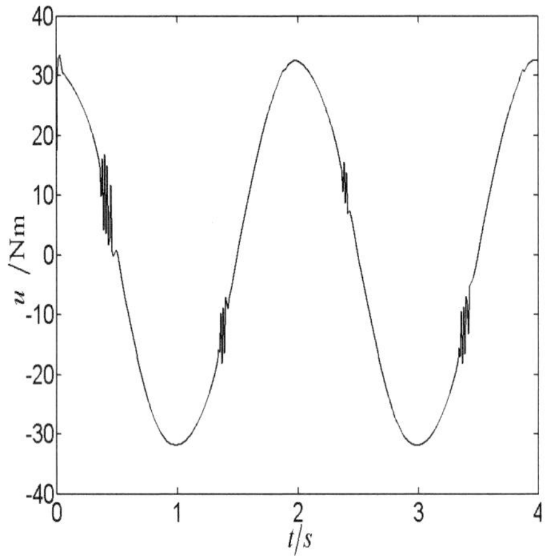

Fig. 2. Control input

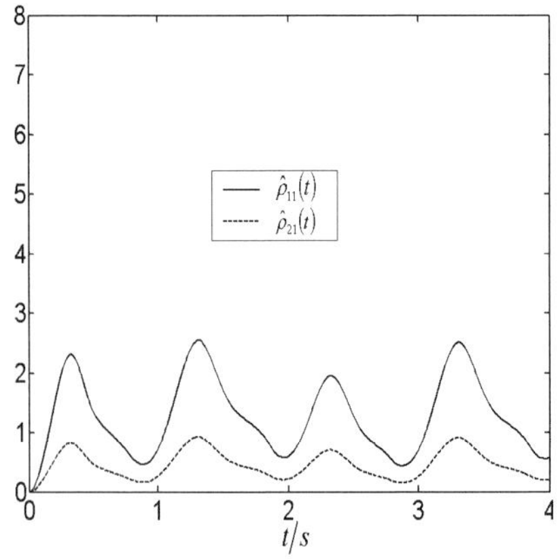

Fig. 3. Adaptive parameters $\hat{\rho}_{11}(t)$and $\hat{\rho}_{21}(t)$

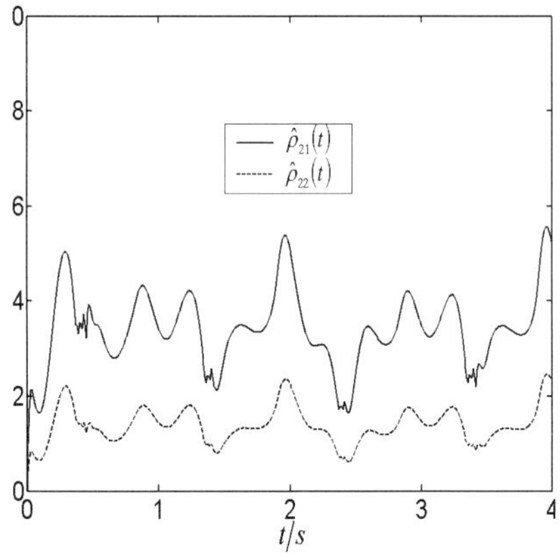

Fig. 4. Adaptive parameters $\hat{\rho}_{21}(t)$and $\hat{\rho}_{22}(t)$

5 Conclusions

We present an adaptive dynamic surface fuzzy control method for a class of uncertain nonlinear systems in strict-feedback form. By introducing dynamic surface control and fuzzy control techniques, the explosion of terms caused by the traditional backstepping approach is avoided and the expert knowledge can be incorporated into the controller design. Furthermore, the algorithm has the adaptive mechanism with minimum learning parameterizations. The computation burden of the control system is reduced drastically. It is proved that all the signals in the closed-loop system are uniformly ultimately bounded and the tracking error can be made arbitrarily small. Simulation results demonstrate the effectiveness of the proposed algorithm.

Acknowledgement

This paper is supported by the Scientific Research Foundation of Chongqing University.

References

1. Kokotovic, P. V.: The Joy of Feedback. IEEE Contr. Syst. Mag. 3(1992)7-17
2. Swaroop, D., Gerdes, C., Yip, P. P., Herdrick, J. K.: Dynamic Surface Control for a Class of Nonlinear Systems. IEEE Trans. Autom. Control. 45(2000)1893-1899
3. Wang, D., Huang, J.: Neural Network-Based Adaptive Dynamic Surface Control for a Class of Uncertain Nonlinear Systems in Strict-Feedback Form. IEEE Trans. Neural Networks. 16(2005)195-202
4. Wang,L. X.: Adaptive Fuzzy Systems and Control: Design and Stability Analysis, Prentice-Hall, Englewood Cliffs, NJ(1994)
5. Chen,G., Wang, S., Zhang, J.: Robust and Adaptive Backstepping Control for Nonlinear Systems Using Fuzzy Logic Systems. Lecture Notes in Artificial Intelligence. 3614(2005) 452-461
6. Chen, G, Wang, S. Robust Control of Time-Varying Nonlinear Systems with Unknown Control Directions. Control and Decision. 20(2005)1397-1400
7. Yang, Y., Feng, G., Ren, J.: A Combined Backstepping and Small-Gain Approach to Robust Adaptive Fuzzy Control for Strict-Feedback Nonlinear Systems. IEEE Trans. Syst. Man, Cybern. 34(2004)406-420
8. Chen,G., Wang, S., Zhang, J.: Robust Adaptive Fuzzy Control for Uncertain Nonlinear Systems. Lecture Notes in Artificial Intelligence. 3613(2005) 841-850
9. Jagannathan, S., Lewis,F. L.: Robust Backstepping Control of a Class of Nonlinear Systems Using Fuzzy Logic. Inform. Sci.123(2000) 223-240
10. Chen,G., Wang, S.: Robust Control For Multi-fingered Hand Manipulation in a Constrained Environment. Acta Automatica Sinica. 31(2005)901-906
11. Chen, G, Wang, S.: Dynamic Control of Four-Fingered Hand Manipulation with Sliding. Proceedings of the World Congress on Intelligent Control and Automation. 6(2004)4855-4859
12. Chen,G., Wang, S., Zhang, J.: Adaptive Neural Network Control for Multi-fingered Robot Hand Manipulation in the Constrained Environment. Lecture Notes in Computer Science. 3611(2005) 270-273
13. Zheng,F., Wang, Q. G., Lee, T. H.: Adaptive and Robust Controller Design for Uncertain Nonlinear Systems via Fuzzy Modeling Approach. IEEE Trans. Syst. Man, Cybern. 34(2004) 166-178
14. Lee, H. J., Park, J. B., Chen, G.: Robust Fuzzy Control of Nonlinear Systems with Parameter Uncertainties. IEEE Trans. Fuzzy Syst. 9(2001)369-379

Fuzzy Nonlinear Regression Model Based on LS-SVM in Feature Space

Dug Hun Hong[1] and Changha Hwang[2]

[1] Department of Mathematics, Myongji University, Yongin Kyunggido, 449-728, South Korea
dhhong@mju.ac.kr

[2] Corresponding Author, Division of Information and Computer Science, Dankook University, Yongsan Seoul, 140-714, South Korea
chwang@dankook.ac.kr

Abstract. This paper presents a new method of estimating fuzzy multivariable nonlinear regression model for fuzzy input and fuzzy output data. This estimation method is obtained by constructing a fuzzy linear regression based on least squares support vector machine(LS-SVM) in a high dimensional feature space for the data set with fuzzy inputs and fuzzy output. Experimental results are then presented which indicate the performance of this algorithm.

1 Introduction

A fuzzy regression model is used in evaluating the functional relationship between the dependent and independent variables in a fuzzy environment. In many cases of fuzzy regression, the linear regression is recommended for practical situations when decisions often have to be made on the basis of imprecise and/or partially available data. Several methods have been presented to estimate fuzzy regression models. Fuzzy regression, as first developed by Tanaka et al.[15] for linear case, is based on the extension principle.

Tanaka et al.[15] initially applied their fuzzy linear regression procedure to non-fuzzy experimental data. In the experiments that followed this pioneering effort, Tanaka et al.[15] used fuzzy input experimental data to build fuzzy regression models. Fuzzy input data used in these experiments were given in the form of triangular fuzzy numbers. The process is explained in more detail by Dubois and Prade[5]. Hong et al.[6][7] proposed the fuzzy linear regression model using shape-preserving fuzzy arithmetic operations based on Tanaka's approach. A technique for linear least squares fitting of fuzzy variable was developed by Diamond[4] giving the solution to an analog of the normal equation of classical least squares.

Fuzzy linear regression assumes the linear fuzzy model for describing the functional relationship between data pairs. In nonlinear case, this assumption may lead to large modeling errors. Thus fuzzy nonlinear regression methods were suggested to overcome the disadvantages of fuzzy linear regression methods. In practice, the nonlinear functional relationship between input-output data pairs

L. Wang et al. (Eds.): FSKD 2006, LNAI 4223, pp. 208–216, 2006.

is frequently unknown. For this reason we need model-free fuzzy nonlinear regression. There have been a few articles on fuzzy nonlinear regression. The most ofen used nonlinear regression methods for fuzzy input-output data are the fuzzified neural networks. This kind of method was studied by Ishibuchi et al.[11], Ishibuchi and Nii[12], Cheng and Lee[3] and Zhang et al.[16]. On the other hand, for crisp input-fuzzy output data the kernel-based nonlinear regression methods were studied by Hong and Hwang[8][9][10]. Buckley and Feuring[1] proposed a nonlinear regression method for fuzzy input-fuzzy output. However they prespecified regression model functions such as linear, polynomial, exponential and logarithmic, which look somewhat unrealistic for the application.

In this paper we concentrate on the nonlinear regression analysis for the data set with fuzzy input-fuzzy output. We want a model-free method suitable for fuzzy nonlinear regression model. For this purpose, we consider the least squares support vector machine(LS-SVM, Suykens and Vanderwalle[14]). It allows us to derive a computationally simple and easy fuzzy nonlinear regression.

The rest of this paper is organized as follows. Section 2 illustrates LS-SVM approach to fuzzy linear regression for fuzzy input-fuzzy output. Section 3 describes fuzzy nonlinear regression achieved by LS-SVM in a high dimensional feature space. Section 4 illustrates how to determine the important parameters associated with the proposed method. Section 5 illustrates numerical studies. Finally, Section 6 gives the conclusions.

2 Fuzzy Linear LS-SVM Regression

In this section we will modify the underlying idea of LS-SVM for the purpose of deriving the convex optimization problems for fuzzy multivariable linear regression models for fuzzy input-fuzzy output. We can use SVM in Hong and Hwang[8]. However, the basic idea of LS-SVM gives computational simplicity and efficiency in finding solutions of fuzzy regression models. We will focus on fuzzy regression models based on triangular fuzzy number since this type of fuzzy number is mostly used in practice. Fuzzy regression models based on trapezoidal and Gaussian fuzzy numbers can be constructed in a similar manner.

Suppose we are given the training data $\{\boldsymbol{X}_i, Y_i\}_{i=1}^{l} \subset T(R)^d \times T(R)$, where $\boldsymbol{X}_i = ((m_{X_{i1}}, \alpha_{X_{i1}}, \beta_{X_{i1}}), \cdots, (m_{X_{id}}, \alpha_{X_{id}}, \beta_{X_{id}}))$ and $Y_i = (m_{Y_i}, \alpha_{Y_i}, \beta_{Y_i})$. Here $T(R)$ and $T(R)^d$ are the set of triangular fuzzy numbers and the set of d-vectors of triangular fuzzy numbers, respectively. Let $\boldsymbol{m}_{\boldsymbol{X}_i} = (m_{X_{i1}}, \cdots, m_{X_{id}})$, $\boldsymbol{\alpha}_{\boldsymbol{X}_i} = (\alpha_{X_{i1}}, \cdots, \alpha_{X_{id}})$, $\boldsymbol{\beta}_{\boldsymbol{X}_i} = (\beta_{X_{i1}}, \cdots, \beta_{X_{id}})$, $B = (m_B, \alpha_B, \beta_B)$ and $\boldsymbol{w} = (w_1, \cdots, w_d)$.

For the fuzzy inputs and fuzzy output we consider the following model $H2$:

$$\begin{aligned} H2 : Y(\mathbf{X}) &= \langle \boldsymbol{w}, \boldsymbol{X} \rangle + B, \ \ B \in T(R), \boldsymbol{w} \in R^d \\ &= (\langle \boldsymbol{w}, \boldsymbol{m}_{\boldsymbol{X}} \rangle + m_B, \langle |\boldsymbol{w}|, \boldsymbol{\alpha}_{\boldsymbol{X}} \rangle + \alpha_B, \langle |\boldsymbol{w}|, \boldsymbol{\beta}_{\boldsymbol{X}} \rangle + \beta_B, \end{aligned}$$

where $|\boldsymbol{w}| = (|w_1|, |w_2|, \cdots, |w_d|)$. We arrive at the following convex optimization problem for the model $H2$ by modifying the idea for crisp multiple linear regression:

$$\text{minimize } \frac{1}{2}\|\boldsymbol{w}\|^2 + \frac{C}{2}\sum_{k=1}^{3}\sum_{i=1}^{l} e_{ki}^2 \tag{1}$$

$$\text{subject to } \begin{cases} m_{Y_i} - \langle \boldsymbol{w}, \boldsymbol{m}_{\boldsymbol{X}_i}\rangle - m_B = e_{1i}, \\ (m_{Y_i} - \alpha_{Y_i}) - (\langle \boldsymbol{w}, \boldsymbol{m}_{\boldsymbol{X}_i}\rangle + m_B - \langle |\boldsymbol{w}|, \boldsymbol{\alpha}_{\boldsymbol{X}_i}\rangle - \alpha_B) = e_{2i} \\ (m_{Y_i} + \beta_{Y_i}) - (\langle \boldsymbol{w}, \boldsymbol{m}_{\boldsymbol{X}_i}\rangle + m_B + \langle |\boldsymbol{w}|, \boldsymbol{\beta}_{\boldsymbol{X}_i}\rangle + \beta_B) = e_{3i}. \end{cases}$$

Here, the parameter C is a positive real constant and should be considered as a tuning parameter in the algorithm. This controls the smoothness and degree of fit. The cost function with squared error and regularization corresponds to a form of ridge regression.

The optimal values of $B = (m_B, \alpha_B, \beta_B)$ and Lagrange multipliers α_{1i}, α_{2i} and α_{3i} can be obtained by the optimality conditions, which lead to the optimal value of $\boldsymbol{w}$. Introducing Lagrange multipliers α_{1i}, α_{2i} and α_{3i}, we construct a Lagrange function as follows:

$$\begin{aligned} L = {} & \frac{1}{2}\|\boldsymbol{w}\|^2 + \frac{C}{2}\sum_{k=1}^{3}\sum_{i=1}^{l} e_{ki}^2 + \sum_{i=1}^{l} \alpha_{1i}(e_{1i} - m_{Y_i} + \langle \boldsymbol{w}, \boldsymbol{m}_{\boldsymbol{X}_i}\rangle + m_B) \\ & - \sum_{i=1}^{l} \alpha_{2i}(e_{2i} - (m_{Y_i} - \alpha_{Y_i}) + (\langle \boldsymbol{w}, \boldsymbol{m}_{\boldsymbol{X}_i}\rangle + m_B - \langle |\boldsymbol{w}|, \boldsymbol{\alpha}_{\boldsymbol{X}_i}\rangle - \alpha_B)) \quad (2) \\ & - \sum_{i=1}^{l} \alpha_{3i}(e_{3i} - (m_{Y_i} + \beta_{Y_i}) + (\langle \boldsymbol{w}, \boldsymbol{m}_{\boldsymbol{X}_i}\rangle + m_B + \langle |\boldsymbol{w}|, \boldsymbol{\beta}_{\boldsymbol{X}_i}\rangle + \beta_B)). \end{aligned}$$

Then, the conditions for optimality are given by

$$\begin{aligned} \frac{\partial L}{\partial \boldsymbol{w}} = \boldsymbol{0} \to \boldsymbol{w} = {} & \sum_{i=1}^{l} \alpha_{1i}\boldsymbol{m}_{\boldsymbol{X}_i} + \sum_{i=1}^{l} \alpha_{2i}(\boldsymbol{m}_{\boldsymbol{X}_i} - sgn(\boldsymbol{w}) \cdot \boldsymbol{\alpha}_{\boldsymbol{X}_i}) \\ & + \sum_{i=1}^{l} \alpha_{3i}(\boldsymbol{m}_{\boldsymbol{X}_i} + sgn(\boldsymbol{w}) \cdot \boldsymbol{\beta}_{\boldsymbol{X}_i}) \end{aligned} \tag{3}$$

$$\frac{\partial L}{\partial m_B} = 0 \to \sum_{k=1}^{3}\sum_{i=1}^{l} \alpha_{ki} = 0 \tag{4}$$

$$\frac{\partial L}{\partial \alpha_B} = 0 \to \sum_{i=1}^{l} \alpha_{2i} = 0 \tag{5}$$

$$\frac{\partial L}{\partial \beta_B} = 0 \to \sum_{i=1}^{l} \alpha_{3i} = 0 \tag{6}$$

$$\frac{\partial L}{\partial e_{ki}} = 0 \to e_{ki} = \frac{\alpha_{ki}}{C}, \quad k = 1, 2, 3 \tag{7}$$

$$\frac{\partial L}{\partial \alpha_{1i}} = 0 \to m_{Y_i} - \langle \boldsymbol{w}, \boldsymbol{m}_{\boldsymbol{X}_i}\rangle - m_B - e_{1i} = 0 \tag{8}$$

$$\frac{\partial L}{\partial \alpha_{2i}} = 0 \to m_{Y_i} - \alpha_{Y_i} - \langle \boldsymbol{w}, \boldsymbol{m}_{\boldsymbol{X}_i}\rangle - m_B + \langle |\boldsymbol{w}|, \boldsymbol{\alpha}_{\boldsymbol{X}_i}\rangle + \alpha_B - e_{2i} = 0 \tag{9}$$

$$\frac{\partial L}{\partial \alpha_{3i}} = 0 \rightarrow m_{Y_i} + \beta_{Y_i} - \langle \boldsymbol{w}, \boldsymbol{m_{X_i}} \rangle - m_B - \langle |\boldsymbol{w}|, \boldsymbol{\beta_{X_i}} \rangle - \beta_B - e_{3i} = 0 \quad (10)$$

where $sgn(\boldsymbol{w}) = (sgn(w_1), \cdots, sgn(w_d))$ and the '$\cdot$' represents the component-wise product. Here $sgn(t) = 1$ or -1 depending on whether $t > 0$ or $t < 0$. Note that we have used $\frac{\partial}{\partial t}|t| = sgn(t)$. We notice that we can tell $sgn(\boldsymbol{w})$ by performing regression in advance for model values of fuzzy variables $\boldsymbol{m_{X_i}}, i = 1, \ldots, l$. There could be other different ways to tell their signs.

Therefore, the optimal values of $B = (m_B, \alpha_B, \beta_B)$ and Lagrange multipliers $\alpha_{1i}, \alpha_{2i}, \alpha_{3i}$ can be obtained from the linear equation as follows:

$$\begin{pmatrix} 0 & 0 & 0 & \mathbf{1}' & \mathbf{1}' & \mathbf{1}' \\ 0 & 0 & 0 & \mathbf{0}' & \mathbf{1}' & \mathbf{0}' \\ 0 & 0 & 0 & \mathbf{0}' & \mathbf{0}' & \mathbf{1}' \\ \mathbf{1} & \mathbf{0} & \mathbf{0} & \boldsymbol{S}_{11} & \boldsymbol{S}_{12} & \boldsymbol{S}_{13} \\ \mathbf{1} & -\mathbf{1} & \mathbf{0} & \boldsymbol{S}'_{12} & \boldsymbol{S}_{22} & \boldsymbol{S}_{23} \\ \mathbf{1} & \mathbf{0} & \mathbf{1} & \boldsymbol{S}'_{13} & \boldsymbol{S}'_{23} & \boldsymbol{S}_{33} \end{pmatrix} \begin{pmatrix} m_B \\ \alpha_B \\ \beta_B \\ \boldsymbol{\alpha}_1 \\ \boldsymbol{\alpha}_2 \\ \boldsymbol{\alpha}_3 \end{pmatrix} = \begin{pmatrix} 0 \\ 0 \\ 0 \\ \boldsymbol{m_Y} \\ \boldsymbol{m_Y} - \boldsymbol{\alpha_Y} \\ \boldsymbol{m_Y} + \boldsymbol{\beta_Y} \end{pmatrix} \quad (11)$$

with

$$\begin{aligned}
\boldsymbol{S}_{11} &= [\langle \boldsymbol{m_{X_i}}, \boldsymbol{m_{X_j}} \rangle] + \boldsymbol{I}/C \\
\boldsymbol{S}_{12} &= [\langle \boldsymbol{m_{X_i}}, \boldsymbol{m_{X_j}} - sgn(\boldsymbol{w}) \cdot \boldsymbol{\alpha_{X_j}} \rangle] \\
\boldsymbol{S}_{13} &= [\langle \boldsymbol{m_{X_i}}, \boldsymbol{m_{X_j}} + sgn(\boldsymbol{w}) \cdot \boldsymbol{\beta_{X_j}} \rangle] \\
\boldsymbol{S}_{22} &= [\langle \boldsymbol{m_{X_i}} - sgn(\boldsymbol{w}) \cdot \boldsymbol{\alpha_{X_i}}, \boldsymbol{m_{X_j}} - sgn(\boldsymbol{w}) \cdot \boldsymbol{\alpha_{X_j}} \rangle] + \boldsymbol{I}/C \\
\boldsymbol{S}_{23} &= [\langle \boldsymbol{m_{X_i}} - sgn(\boldsymbol{w}) \cdot \boldsymbol{\alpha_{X_i}}, \boldsymbol{m_{X_j}} + sgn(\boldsymbol{w}) \cdot \boldsymbol{\beta_{X_j}} \rangle] \\
\boldsymbol{S}_{33} &= [\langle \boldsymbol{m_{X_i}} + sgn(\boldsymbol{w}) \cdot \boldsymbol{\beta_{X_i}}, \boldsymbol{m_{X_j}} + sgn(\boldsymbol{w}) \cdot \boldsymbol{\beta_{X_j}} \rangle] + \boldsymbol{I}/C,
\end{aligned}$$

where $\boldsymbol{\alpha}_1$, $\boldsymbol{\alpha}_2$, $\boldsymbol{\alpha}_3$, $\boldsymbol{m_Y}$, $\boldsymbol{\alpha_Y}$ and $\boldsymbol{\beta_Y}$ are the $l \times 1$ vectors of α_{1i}, α_{2i}, α_{3i}, m_{Y_i}, α_{Y_i} and β_{Y_i}, respectively, and $[a_{ij}]$ represents the $l \times l$ matrix with elements a_{ij}.

Hence, the prediction of $Y(\boldsymbol{X}_q)$ given by the LS-SVM on the new unlabeled data $\boldsymbol{X}_q$ is

$$\hat{Y}(\boldsymbol{X}_q) = (\langle \boldsymbol{w}, \boldsymbol{m_{X_q}} \rangle + m_B, \langle |\boldsymbol{w}|, \boldsymbol{\alpha_{X_q}} \rangle + \alpha_B, \langle |\boldsymbol{w}|, \boldsymbol{\beta_{X_q}} \rangle + \beta_B). \quad (12)$$

3 Fuzzy Nonlinear LS-SVM Regression

In this section, we study a new method of estimating fuzzy multivariable nonlinear regression model for fuzzy input and fuzzy output data. This method is obtained by constructing a fuzzy linear regression based on LS-SVM in a high dimensional feature space of fuzzy inputs. To do this, we need to briefly look at again the idea used in LS-SVM for crisp nonlinear regression. This could be achieved by simply preprocessing input patterns $\boldsymbol{x}_i$ by a map $\boldsymbol{\Phi} : \boldsymbol{R}^d \rightarrow \boldsymbol{\mathcal{F}}$ into some feature space $\boldsymbol{\mathcal{F}}$ and then applying the standard LS-SVM regression algorithm. First notice that the only way in which the data appears in algorithm is in the form of dot products $\langle \boldsymbol{x}_i, \boldsymbol{x}_j \rangle$. The algorithm would only depend on the data through dot products in $\mathcal{F}$, i.e. on functions of the form $\langle \boldsymbol{\Phi}(\boldsymbol{x}_i), \boldsymbol{\Phi}(\boldsymbol{x}_j)\rangle$. Hence it suffices to know and use $K(\boldsymbol{x}_i, \boldsymbol{x}_j) = \langle \boldsymbol{\Phi}(\boldsymbol{x}_i), \boldsymbol{\Phi}(\boldsymbol{x}_j)\rangle$ instead of $\boldsymbol{\Phi}(\cdot)$ explicitly. The well used kernels for regression problem are given below.

$$K(\boldsymbol{x}, \boldsymbol{y}) = (1 + \langle \boldsymbol{x}, \boldsymbol{y}\rangle)^p : \text{ Polynomial kernel}$$
$$K(\boldsymbol{x}, \boldsymbol{y}) = e^{-\frac{\|\boldsymbol{x}-\boldsymbol{y}\|^2}{2\sigma^2}} : \text{ Gaussian kernel}$$

Consider for example a feature space with two inputs and a polynomial kernel of degree 2. Then

$$\begin{aligned} K(\boldsymbol{x}, \boldsymbol{y}) &= (1 + \langle \boldsymbol{x}, \boldsymbol{y}\rangle)^2 \\ &= (1 + x_1 y_1 + x_2 y_2)^2 \\ &= 1 + 2x_1 y_1 + 2x_2 y_2 + (x_1 y_1)^2 + (x_2 y_2)^2 + 2x_1 y_1 x_2 y_2. \end{aligned}$$

Thus, if we choose $\boldsymbol{\Phi}(\boldsymbol{x}) = (1, \sqrt{2}x_1, \sqrt{2}x_2, x_1^2, x_2^2, \sqrt{2}x_1 x_2)$, then $K(\boldsymbol{x}, \boldsymbol{y}) = \langle \boldsymbol{\Phi}(\boldsymbol{x}), \boldsymbol{\Phi}(\boldsymbol{y})\rangle$.

Unlike the case of numerical inputs, the function $\boldsymbol{\Phi}$ should be increasing for fuzzy inputs, since the shape of triangular fuzzy number should be preserved after preprocessing fuzzy inputs with $\boldsymbol{\Phi}$. When $m_{X_{ij}} - \alpha_{X_{ij}} \geq 0$, the feature mapping function $\boldsymbol{\Phi}$ associated with polynomial kernel is increasing and thus preserves the shape of triangular fuzzy number. However, the feature mapping function associated with Gaussian kernel does not suffice this condition. Hence, throughout the paper we assume that $m_{X_{ij}} - \alpha_{X_{ij}} \geq 0$ by a simple translation of all data, and consider only the feature mapping function associated with polynomial kernel. In fact, we use the feature mapping function rather than the kernel function. Let us define a function $\boldsymbol{\Phi}^*$ by $\boldsymbol{\Phi}^* : \boldsymbol{T}(\boldsymbol{R})^d \rightarrow \boldsymbol{T}(\boldsymbol{\mathcal{F}})$ such that $\boldsymbol{\Phi}^*((\boldsymbol{m}_{\boldsymbol{X}}, \boldsymbol{\alpha}_{\boldsymbol{X}}, \boldsymbol{\beta}_{\boldsymbol{X}})) = (\boldsymbol{\Phi}(\boldsymbol{m}_{\boldsymbol{X}}), \boldsymbol{\Phi}(\boldsymbol{m}_{\boldsymbol{X}}) - \boldsymbol{\Phi}(\boldsymbol{m}_{\boldsymbol{X}} - \boldsymbol{\alpha}_{\boldsymbol{X}}), \boldsymbol{\Phi}(\boldsymbol{m}_{\boldsymbol{X}} + \boldsymbol{\beta}_{\boldsymbol{X}}) - \boldsymbol{\Phi}(\boldsymbol{m}_{\boldsymbol{X}}))$, and let us define $\boldsymbol{\alpha}^{\Phi}_{\boldsymbol{X}_i}, \boldsymbol{\beta}^{\Phi}_{\boldsymbol{X}_i}$ by

$$\begin{aligned} \boldsymbol{\alpha}^{\Phi}_{\boldsymbol{X}_i} &= \boldsymbol{\Phi}(\boldsymbol{m}_{\boldsymbol{X}_i}) - \boldsymbol{\Phi}(\boldsymbol{m}_{\boldsymbol{X}_i} - \boldsymbol{\alpha}_{\boldsymbol{X}_i}) \\ \boldsymbol{\beta}^{\Phi}_{\boldsymbol{X}_i} &= \boldsymbol{\Phi}(\boldsymbol{m}_{\boldsymbol{X}_i} + \boldsymbol{\beta}_{\boldsymbol{X}_i}) - \boldsymbol{\Phi}(\boldsymbol{m}_{\boldsymbol{X}_i}). \end{aligned}$$

Then, similar to the linear case in Section 2, we have

$$\begin{aligned} \boldsymbol{w}^{\Phi} = &\sum_{i=1}^{l} \alpha_{1i} \boldsymbol{\Phi}(\boldsymbol{m}_{\boldsymbol{X}_i}) + \sum_{i=1}^{l} \alpha_{2i} (\boldsymbol{\Phi}(\boldsymbol{m}_{\boldsymbol{X}_i}) - sgn(\boldsymbol{w}^{\Phi}) \cdot \boldsymbol{\alpha}^{\Phi}_{\boldsymbol{X}_i}) \\ &+ \sum_{i=1}^{l} \alpha_{3i} (\boldsymbol{\Phi}(\boldsymbol{m}_{\boldsymbol{X}_i}) + sgn(\boldsymbol{w}^{\Phi}) \cdot \boldsymbol{\beta}^{\Phi}_{\boldsymbol{X}_i}). \end{aligned} \tag{13}$$

Furthermore, replacing $\boldsymbol{w}, \boldsymbol{m}_{\boldsymbol{X}_i}, \boldsymbol{\alpha}_{\boldsymbol{X}_i}, \boldsymbol{\beta}_{\boldsymbol{X}_i}$ with $\boldsymbol{w}^{\Phi}, \boldsymbol{\Phi}(\boldsymbol{m}_{\boldsymbol{X}_i}), \boldsymbol{\alpha}^{\Phi}_{\boldsymbol{X}_i}, \boldsymbol{\beta}^{\Phi}_{\boldsymbol{X}_i}$ in the linear equation (11) and solving it, we have the nonlinear prediction of $Y(\boldsymbol{X}_q)$ on the new unlabeled data $\boldsymbol{X}_q$, which is given by

$$\hat{Y}(\boldsymbol{X}_q) = (\langle \boldsymbol{w}^{\Phi}, \boldsymbol{\Phi}(\boldsymbol{m}_{\boldsymbol{X}_q})\rangle + m_B, \langle |\boldsymbol{w}^{\Phi}|, \boldsymbol{\alpha}^{\Phi}_{\boldsymbol{X}_q}\rangle + \alpha_B, \langle |\boldsymbol{w}^{\Phi}|, \boldsymbol{\beta}^{\Phi}_{\boldsymbol{X}_q}\rangle + \beta_B). \tag{14}$$

4 Model Selection

When we use LS-SVM for fuzzy linear regression, we must determine an optimal choice of the regularization parameter C. But for the fuzzy nonlinear regression, we have to determine one more parameter, which is polynomial degree p. In this paper we use cross-validation method for parameter selection. If data is not scarce then the set of available input-output measurements can be divided into two parts - one part for training and one part for testing. In this way several different models, all trained on the training set, can be compared on the test set. This is the basic form of cross-validation. A better method is to partition the original set in several different ways and to compute an average score over the different partitions. In this paper the average score is computed by using the squared error based on the following distance between two outputs.

$$\begin{aligned} d^2(Y,Z) = (m_Y - m_Z)^2 + ((m_Y - \alpha_Y) - (m_Z - \alpha_Z))^2 \\ + ((m_Y + \beta_Y) - (m_Z + \beta_Z))^2. \end{aligned} \tag{15}$$

An extreme variant of this is to split the measurements into a training set of size and a test set of size 1 and average the squared error on the left-out measurements over the possible ways of obtaining such a partition. This is called leave-one-out cross-validation. In the leave-one-out cross-validation method, we train using all but one training measurement, then test using the left out measurement. We repeat this, leaving out another single measurement. We do this until we have left out each example. Then we average the results on the left out measurements to assess the generalization capability of our fuzzy regression procedure.

$$\begin{aligned} CV(C,p) = \frac{1}{l}[\sum_{i=1}^{l}(m_{Y_i} - \hat{m}_{Y_i}^{(-i)})^2 + \sum_{i=1}^{l}((m_{Y_i} - \alpha_{Y_i}) - (\hat{m}_{Y_i}^{(-i)} - \hat{\alpha}_{Y_i}^{(-i)}))^2 \\ + \sum_{i=1}^{l}((m_{Y_i} + \beta_{Y_i}) - (\hat{m}_{Y_i}^{(-i)} + \hat{\beta}_{Y_i}^{(-i)}))^2], \end{aligned} \tag{16}$$

where $(\hat{m}_{Y_i}^{(-i)}, \hat{\alpha}_{Y_i}^{(-i)}, \hat{\beta}_{Y_i}^{(-i)})$ is the predicted values of $Y_i = (m_{Y_i}, \alpha_{Y_i}, \beta_{Y_i})$ obtained from training data without X_i.

5 Numerical Studies

In contrast to fuzzy linear regression, there have been only a few articles on fuzzy nonlinear regression. What researchers in fuzzy nonlinear regression were concerned with was data of the form with crisp inputs and fuzzy output. Some papers(Buckley and Feuring[1], Celmins[2]) are concerned with the data set with fuzzy inputs and fuzzy output. However, we think those fuzzy nonlinear regression methods look somewhat unrealistic and treat the estimation procedures of some particular models. In this paper we treat fuzzy nonlinear regression for data of the form with fuzzy inputs and fuzzy output, without assuming the underlying model function.

In order to illustrate the performance of the nonlinear regression prediction for fuzzy inputs and fuzzy outputs, two examples are considered. In examples, centers of X_i's were randomly generated in $[0, 0.25, \cdots, 10.0]$, and spreads were randomly generated in $[0.3, 0.4, \cdots, 1.0]$ and some of them were modified so that $m_{X_{ij}} - \alpha_{X_{ij}} \geq 0$ are satisfied. The centers of Y_i's of the first and second examples were generated by

$$m_{Y_i} = 2.1 + \exp(0.2 m_{X_i}) + \epsilon_i,$$
$$m_{Y_i} = 1.1 + 2.5 \log(1 + m_{X_i}) + \epsilon_i,$$

respectively, where $\epsilon_i, i = 1, 2, \cdots, 25$, is a random error from the normal distribution with mean 0 and variance 0.01.

By the leave-one-out cross-validation method, we selected $(5000, 3)$ as the value of (C, p) for both examples. In figures four corners of each solid box - the lower left, the lower right, the upper left, and the upper right - represent $(m_{X_i} - \alpha_{X_i}, m_{Y_i} - \alpha_{Y_i})$, $(m_{X_i} + \beta_{X_i}, m_{Y_i} - \alpha_{Y_i})$, $(m_{X_i} - \alpha_{X_i}, m_{Y_i} + \beta_{Y_i})$, and $(m_{X_i} + \beta_{X_i}, m_{Y_i} + \beta_{Y_i})$, respectively, four corners of each dotted box represent $(m_{X_i} - \alpha_{X_i}, \hat{m}_{Y_i} - \hat{\alpha}_{Y_i})$, $(m_{X_i} + \beta_{X_i}, \hat{m}_{Y_i} - \hat{\alpha}_{Y_i})$, $(m_{X_i} - \alpha_{X_i}, \hat{m}_{Y_i} + \hat{\beta}_{Y_i})$, and $(m_{X_i} + \beta_{X_i}, \hat{m}_{Y_i} + \hat{\beta}_{Y_i})$. And '·' represents (m_{X_i}, m_{Y_i}) and the dashed line is a connection between $(m_{X_i}, \hat{m}_{Y_i})$'s. From figures we can see that the proposed model derives good results on the nonlinear regression for fuzzy inputs and fuzzy outputs.

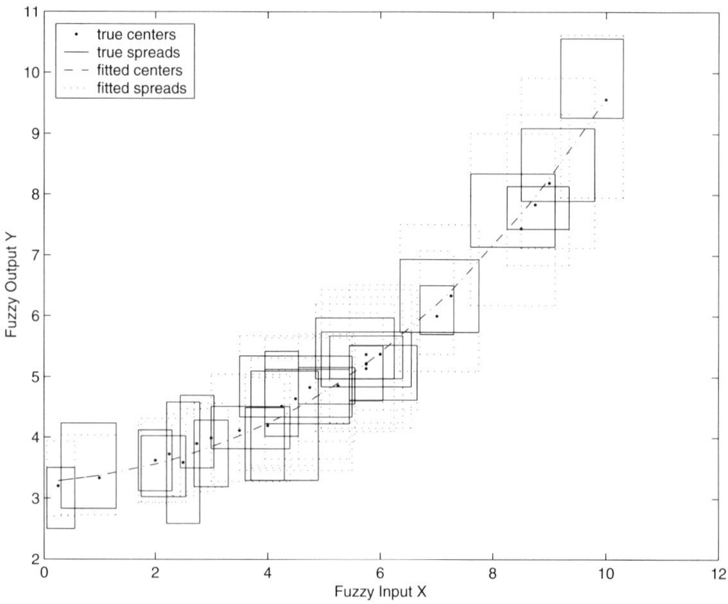

Fig. 1. Fuzzy nonlinear regression model for thr first example

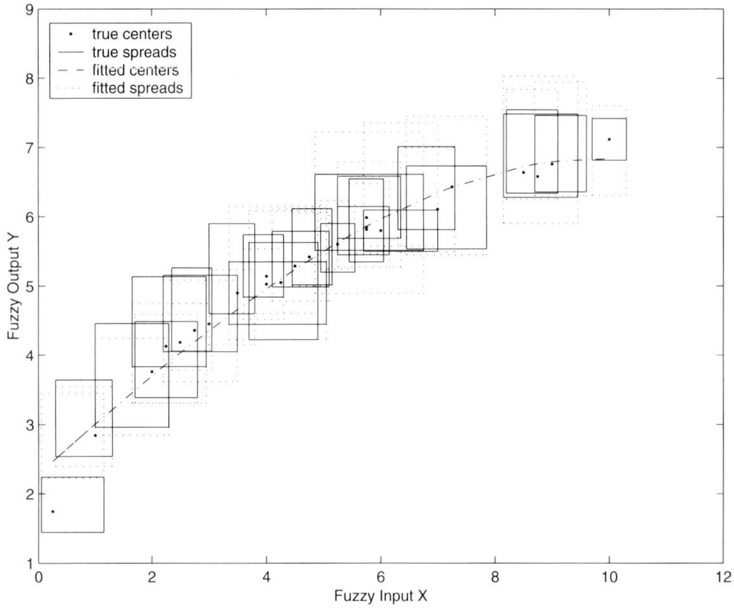

Fig. 2. Fuzzy nonlinear regression model for the second example

6 Conclusions

In this paper we have presented an estimation strategy based on LS-SVM for fuzzy multivariable nonlinear regressions. The experimental results show that the proposed fuzzy nonlinear regression model derives the satisfying solutions and is an attractive approach to modelling fuzzy data.

There have been some papers treat fuzzy nonlinear regression models. They usually assume the underlying model functions or are computationally expensive. The proposed algorithm here is a model-free method in the sense that we do not have to assume the underlying model function. This model-free method turned out to be a promising method which has been attempted to treat fuzzy nonlinear regression model with fuzzy inputs and fuzzy output. The main formulation results in solving a simple matrix inversion problem. Hence, this is a computationally simple and efficient way. The hyperparameters of the proposed model can be tuned using cross-validation method.

Acknowledgement

This work was supported by the Korea Research Foundation Grant(KRF-2004-042-C00020).

References

1. Buckley, J. and Feuring, T. : Linear and non-linear fuzzy regression: evolutionary algorithm solutions. Fuzzy Sets and Systems **112** (2000) 381–394
2. Celmins, A. : A practical approach to nonlinear fuzzy regression. SIAM Journal of Scientific and Statistical Compututing **12** (1991) 521–546
3. Cheng, C. B. and Lee, E. S. : Fuzzy regression with radial basis function network. Fuzzy Sets and Systems **119** (2001) 291–301
4. Diamond, P. : Fuzzy least squares. Information Sciences **46** (1988) 141–157
5. Dubois, D. and Prade, H. : Theory and applications, fuzzy sets and systems. Academic Press, New York (1980)
6. Hong, D. H., Song, J. K. and Do, H. Y. : Fuzzy least-squares linear regression analysis using shape preserving operations. Information Sciences **138** (2001) 185–193
7. Hong, D. H., Lee, H. and Do, H. Y. : Fuzzy linear regression analysis for fuzzy input-output data using shape-preserving operations. Fuzzy Sets and Systems **122** (2001) 513–526
8. Hong, D. H. and Hwang, C. : Support vector fuzzy regression machines. Fuzzy Sets and Systems **138** (2003) 271–281
9. Hong, D. H. and Hwang, C. : Extended fuzzy regression models using regularization method. Information Sciences **164** (2004) 31–46
10. Hong, D. H. and Hwang, C. : Ridge regression procedures for fuzzy models using triangular fuzzy numbers. International Journal of Uncertainty, Fuzziness and Knowledge-Based Systems **12** (2004) 145–159
11. Ishibuchi, H. , Fujioka, R. and Tanaka, H. : Fuzzy regression analysis using neural networks. Fuzzy Sets and Systems **50** (1992) 257–265
12. Ishibuchi, H. and Nii, M. : Fuzzy regression using asymmetric fuzzy coefficients and fuzzified neural networks. Fuzzy Sets and Systems **119** (2001) 273–290
13. Saunders, C., Gammerman, A. and Vork, V. : Ridge regression learning algorithm in dual variable. Proceedings of the 15th International Conference on Machine Learning (1998) 515–521
14. Suykens, J. A. K. and Vandewalle, J. : Recurrent least squares support vector machines. IEEE Transactions on Circuits and Systems-I **47** (2000) 1109–1114
15. Tanaka, H., Uejima, S. and Asia, K. : Linear regression analysis with fuzzy model. IEEE Transactions on Systems, Man and Cybernetics **12** (1982) 903-907
16. Zhang, D., Deng, L., Cai, K. and So, A. : Fuzzy nonlinear regression with fuzzified radial basis function network. IEEE Transactions on Fuzzy Systems **13** (2005) 742–760

Ranking Fuzzy Variables in Terms of Credibility Measure

Jin Peng, Huanbin Liu, and Gang Shang

College of Mathematics and Information Sciences
Huanggang Normal University, Huanggang City, Hubei 438000, China
pengjin01@tsinghua.org.cn

Abstract. Fuzzy variables are used for representing imprecise numerical quantities in a fuzzy environment, and the comparison of fuzzy variables is considered an important and complicated issue in fuzzy logic theory and applications. In this paper, we propose a new type of method for ranking fuzzy variables in the setting of credibility measure. Some basic properties of this type of ranking fuzzy variable in terms of credibility measure are investigated. As an illustration, the case of ranking rule for typical trapezoidal fuzzy variables is examined.

Keywords: fuzzy theory, ranking fuzzy variable, credibility measure.

1 Introduction

Fuzzy variables are used for representing imprecise numerical quantities in a fuzzy environment, and their comparison is very important and fundamental for theory and application purposes. The importance of fuzzy ranking is becoming more and more paramount in the current information age. Many fuzzy ranking methods have been suggested in the literature so far. To name a few, see the articles Bortolan and Degani [1] and Facchinetti et al. [6] and the references therein for a review of some methods. Wang and Kerre [20] presented a comprehensive survey of the available ranking methods more than 35 indices in the fuzzy literature.

Researchers interested in fuzzy ranking have proposed different ranking methods from different angles. Lee and Li [10] provided comparison method of fuzzy numbers based on the probability measure of fuzzy events. Liou and Wang [11] proposed a method of ranking fuzzy numbers with integral value. Fortemps and Roubens [7] presented some interesting properties related to the area compensation procedure to compare fuzzy numbers. Yoon [23] and Yager et al. [21] suggested some approaches to comparing fuzzy numbers motivated by a probabilistic view of the underlying uncertainty. Modarres and Sadi-Nezhad [16] proposed a ranking method in which a preference function is defined and fuzzy numbers are ranked on the basis of their preference ratio. Detyniecki and Yager [4] studied the ranking process based on the valuations methods that was introduced initially by Yager and Filev [22]. Chu and Tsao [3] in 2002 proposed ranking fuzzy numbers with the area between the centroid point and original point to

L. Wang et al. (Eds.): FSKD 2006, LNAI 4223, pp. 217–220, 2006.

improve the distance method presented by Cheng [2] in 1998. A new approach for ranking fuzzy numbers based on a distance measure is introduced by Tran and Duckstein [19]. A characterization of the consistency property defined by the additive transitivity property of the fuzzy preference relations is presented by Herrera-Viedma et al. [8] in 2002. Facchinetti and Ricci [5] discussed some properties for ranking fuzzy numbers. Lee et al. [9] in 2004 proposed a method for ranking the sequences of fuzzy values that assigns a preference degree to each ranked sequence. A new approach for ranking fuzzy numbers based on a fuzzy simulation analysis method has been recently proposed by Sun and Wu [18].

The purpose of this paper is to provide a new type of ranking fuzzy variable based on credibility measure proposed by Liu and Liu [15]. Some preliminary concepts, such as fuzzy variable, mmembership function, credibility measure, credibility distribution, optimistic and pessimistic values, expected value and variance of fuzzy variable, can be consulted in Liu [12,13,14].

2 Ranking Fuzzy Variables Via Credibility Measure

We begin with introducing basic concepts of ranking fuzzy variables based on credibility measure and then focus on some basic properties related to the introduced fuzzy ranking method.

Definition 1. *Let ξ and η be two fuzzy variables defined on a credibility space $(\Theta, \mathcal{P}(\Theta), \mathrm{Cr})$. We say that ξ is almost sure greater than η in credibility, denoted by $\xi \succeq_{\mathrm{Cr}} \eta$, if and only if $\mathrm{Cr}\{\xi \geq \eta\} = 1$.*

Theorem 1. *Let ξ, η and ζ be fuzzy variables. Then*
(a) Reflexivity: $\xi \succeq_{\mathrm{Cr}} \xi$;
(b) Antisymmetry: $\xi \succeq_{\mathrm{Cr}} \eta$ and $\eta \succeq_{\mathrm{Cr}} \xi$ implies ξ and η are identical in credibility, i.e., $\mathrm{Cr}\{\xi = \eta\} = 1$;
(c) Transitivity: $\xi \succeq_{\mathrm{Cr}} \eta$ and $\eta \succeq_{\mathrm{Cr}} \zeta$ implies $\xi \succeq_{\mathrm{Cr}} \zeta$.

Theorem 2. *Let ξ and η be two fuzzy variables, $\Phi(x)$ and $\Psi(x)$ denote the credibility distribution functions of ξ and η, respectively. Then $\xi \succeq_{\mathrm{Cr}} \eta$ implies $\Phi(x) \leq \Psi(x)$ for all $x \in \Re$.*

Remark: The above theorem means that fuzzy ranking in credibility here implies the first fuzzy dominance introduced by Peng et al. [17].

Theorem 3. *Let ξ and η be two fuzzy variables, $\alpha \in (0, 1]$. Let $\xi_{\inf}(\alpha)$ and $\eta_{\inf}(\alpha)$ denote the α-pessimistic values of ξ and η, respectively. Then $\xi \succeq_{\mathrm{Cr}} \eta$ implies both $\xi_{\inf}(\alpha) \geq \eta_{\inf}(\alpha)$ and $\xi_{\sup}(\alpha) \geq \eta_{\sup}(\alpha)$ for all $\alpha \in (0, 1]$.*

Theorem 4. *Let ξ and η be two fuzzy variables and $\{\xi_n\}$ be a fuzzy variable sequence. If $\xi_n \succeq_{\mathrm{Cr}} \eta$ for all $n = 1, 2, \cdots$ and $\{\xi_n\}$ converges uniformly to ξ, then $\xi \succeq_{\mathrm{Cr}} \eta$.*

Theorem 5. *Let ξ and η be two fuzzy variables, $U(x)$ is a strictly increasing continuous function defined on $\Re$. Then $\xi \succeq_{\mathrm{Cr}} \eta$ iff $U(\xi) \succeq_{\mathrm{Cr}} U(\eta)$.*

Theorem 6. *Let ξ and η be two fuzzy variables, $a, b \in \Re$ and $a > 0$. If $\xi \succeq_{\mathrm{Cr}} \eta$, then $a\xi + b \succeq_{\mathrm{Cr}} a\eta + b$.*

Theorem 7. *Let ξ and η be two fuzzy variables, $U(x)$ is a strictly decreasing continuous function defined on $\Re$. Then $\xi \succeq_{\mathrm{Cr}} \eta$ iff $U(\eta) \succeq_{\mathrm{Cr}} U(\xi)$.*

Theorem 8. *Let ξ and η be two fuzzy variables, $a, b \in \Re$ and $a < 0$. If $\xi \succeq_{\mathrm{Cr}} \eta$, then $a\eta + b \succeq_{\mathrm{Cr}} a\xi + b$. Especially, if $\xi \succeq_{\mathrm{Cr}} \eta$, then $-\eta \succeq_{\mathrm{Cr}} -\xi$.*

Theorem 9. *Let ξ and η be two fuzzy variables with $E[\xi] < \infty$ and $E[\eta] < \infty$. If $\xi \succeq_{\mathrm{Cr}} \eta$, then $E[\xi] \geq E[\eta]$.*

Theorem 10. *Let ξ and η be two fuzzy variables with $E[\xi] < \infty$ and $E[\eta] < \infty$. If $\xi \succeq_{\mathrm{Cr}} \eta$, then $E[U(\xi)] \geq E[U(\eta)]$ for all strictly increasing continuous function U for which these expected values are finite.*

As an illustration, the case of fuzzy ranking rule for typical trapezoidal fuzzy variables is documented as follows.

Theorem 11. *Let $\xi = (r_1, r_2, r_3, r_4)$ and $\eta = (s_1, s_2, s_3, s_4)$ be two trapezoidal fuzzy variables. Then $\xi \geq_{\mathrm{Cr}} \eta$ iff $r_4 \geq r_3 \geq r_2 \geq r_1 \geq s_4 \geq s_3 \geq s_2 \geq s_1$.*

3 Conclusions

This short paper is concerned with the new approach of ranking fuzzy variables by means of credibility measure. Some basic properties of fuzzy ranking in the setting of credibility measure are investigated.

It should be pointed out that there exist different routes for dealing with ranking fuzzy variables within the axiomatic framework of credibility theory. Different ranking methods can produce different order for the same sample of fuzzy variables. Future research including comparing the introduced method to other existing standard ranking methods will be deeply continued.

Acknowledgments

This work was supported by the Natural Science Foundation No. 2006ABA185 of Hubei Province, the Significant Project No. Z200527001 of Hubei Provincial Department of Education, the Scientific and Technological Innovative Group Project of Hubei Provincial Institutions of Higher Education, China.

References

1. Bortolan, G., Degani, R.: A review of some methods for ranking fuzzy subsets. Fuzzy Sets and Systems **15** (1985) 1–19
2. Cheng, C.-H., A new approach for ranking fuzzy numbers by distance method. Fuzzy Sets and Systems **95** (1998) 307–317

3. Chu, T.-C., Tsao, C.-T.: Ranking fuzzy numbers with an area between the centroid point and original point. Computers & Mathematics with Applications **43** (2002) 111-117
4. Detyniecki, M., Yager, R. R.: Ranking fuzzy numbers using alpha-weighted valuations. International Journal of Uncertainty, Fuzziness and Knowledge-Based Systems **8** (2001) 573–592
5. Facchinetti, G., Ricci, R. G.: A characterization of a general class of ranking functions on triangular fuzzy numbers. Fuzzy Sets and Systems **146** (2004) 297–312
6. Facchinetti, G., Ricci, R. G., Muzzioli, S.: Note on ranking fuzzy triangular numbers. International Journal of Intelligent Systems **13**(1998) 613–622
7. Fortemps, P., Roubens, M.: Ranking and defuzzification methods based on area compensation. Fuzzy Sets and Systems **82** (1996) 319–330
8. Herrera-Viedma, E., Herrera, F., Chiclana, F., and Luque, M.: Some issues on consistency of fuzzy preference relations. European Journal of Operational Research **154** (2004) 98–109
9. Lee, S., Lee, K. H., and Lee, D.: Ranking the sequences of fuzzy values. Information Sciences **160** (2004) 41–52
10. Lee, E. S., Li, R. J.: Comparison of fuzzy numbers based on the probability measure of fuzzy events. Computer and Mathematics with Applications **15** (1987) 887–896
11. Liou, T.-S., Wang, M. J.: Ranking fuzzy numbers with integral value. Fuzzy Sets and Systems 50 (1992) 247-255
12. Liu, B.: Theory and Practice of Uncertain Programming. Physica-Verlag, Heidelberg (2002)
13. Liu, B.: Uncertainty Theory: An Introduction to its Axiomatic Foundations. Springer-Verlag, Heidelberg (2004)
14. Liu, B.: A survey of credibility theory. Fuzzy Optimization and Decision Making **5**(2006) 1–19
15. Liu, B., Liu, Y.-K.: Expected value of fuzzy variable and fuzzy expected value model. IEEE Transactions on Fuzzy Systems **10**(2002) 445–450
16. Modarres, M., Sadi-Nezhad, S.: Ranking fuzzy numbers by preference ratio. Fuzzy Sets and Systems **118** (2001) 429–436
17. Peng, J., Mok, H. M. K., Tse, W.-M.: Fuzzy dominance based on credibility distributions. Lecture Notes in Artificial Intelligence (LNAI 3613) (2005) 295–303
18. Sun, H., Wu, J.: A new approach for ranking fuzzy numbers based on fuzzy simulation analysis method. Applied Mathematics and Computation **174** (2006) 755–767
19. Tran, L., Duckstein, L.: Comparison of fuzzy numbers using a fuzzy distance measure. Fuzzy Sets and Systems **130** (2002) 331–341
20. Wang, X., Kerre, E. E.: Reasonable properties for the ordering of fuzzy quantities (I)(II). Fuzzy Sets and Systems **118** (2001) 375–385, 387–405
21. Yager, R. R., Detyniecki, M., Bouchon-Meunier, B.: A context-dependent method for ordering fuzzy numbers using probabilities. Information Sciences **138** (2001) 237–255
22. Yager, R. R., Filev, D.: On ranking fuzzy numbers using valuations. International Journal of Intelligent Systems **14**(1999) 1249–1268
23. Yoon, K. P.: A probabilistic approach to rank complex fuzzy numbers. Fuzzy Sets and Systems **80**(1996) 3167–176

Controllability for the Semilinear Fuzzy Integrodifferential Equations with Nonlocal Conditions

Jin Han Park[1], Jong Seo Park[2], and Young Chel Kwun[3,⋆]

[1] Division of Math. Sci., Pukyong National University, Pusan 608-737, South Korea
jihpark@pknu.ac.kr
[2] Department of Math. Education, Chinju National University of Education, Chinju 660-756, South Korea
parkjs@cue.ac.kr
[3] Department of Mathematics, Dong-A University, Pusan 604-714, South Korea
yckwun@dau.ac.kr

Abstract. In this paper, we study the controllability for the semilinear fuzzy integrodifferential control system with nonlocal condition in E_N by using the concept of fuzzy number whose values are normal, convex, upper semicontinuous and compactly supported interval in E_N.

1 Introduction

Many authors have studied several concepts of fuzzy systems. Kaleva [3] studied the existence and uniqueness of solution for the fuzzy differential equation on E^n where E^n is normal, convex, upper semicontinuous and compactly supported fuzzy sets in R^n. Seikkala [6] proved the existence and uniqueness of fuzzy solution for the following equation:

$$\dot{x}(t) = f(t, x(t)) \ , \quad x(0) = x_0,$$

where f is a continuous mapping from $R^+ \times R$ into R and x_0 is a fuzzy number in E^1. Diamond and Kloeden [2] proved the fuzzy optimal control for the following system:

$$\dot{x}(t) = a(t)x(t) + u(t), \quad x(0) = x_0$$

where $x(\cdot), u(\cdot)$ are nonempty compact interval-valued functions on E^1. Kwun and Park [4] proved the existence of fuzzy optimal control for the nonlinear fuzzy differential system with nonlocal initial condition in E^1_N using by Kuhn-Tucker theorems. Recently, Balasubramaniam and Muralisankar [1] proved the existence and uniqueness of fuzzy solutions for the following semilinear fuzzy integrodifferential equation ($u(t) = 0$) with nonlocal initial condition:

$$\frac{dx(t)}{dt} = A\left[x(t) + \int_0^t G(t-s)x(s)ds\right] + f(t,x) + u(t), \quad t \in I = [0,T], \tag{1}$$

$$x(0) + g(t_1, t_2, \cdots, t_p, x(\cdot)) = x_0 \in E_N, \tag{2}$$

⋆ Corresponding author.

L. Wang et al. (Eds.): FSKD 2006, LNAI 4223, pp. 221–230, 2006.

where $A : I \to E_N$ is a fuzzy coefficient, E_N is the set of all upper semicontinuous convex normal fuzzy numbers with bounded α-level intervals, $f : I \times E_N \to E_N$ is a nonlinear continuous function, $G(t)$ is $n \times n$ continuous matrix such that $\frac{dG(t)x}{dt}$ is continuous for $x \in E_N$ and $t \in I$ with $\|G(t)\| \leq k$, $k > 0$, $u : I \to E_N$ is control function and $g : I^p \times E_N \to E_N$ is a nonlinear continuous function. In the place of $\cdot$ we can replace the elements of the set $\{t_1, t_2, \cdots, t_p\}$, $0 < t_1 < t_2 \cdots < t_p \leq T$, $p \in N$, the set of all natural numbers.

In this paper, we find the sufficient conditions of controllability for the control system (1)-(2).

2 Preliminaries

A fuzzy subset of R^n is defined in terms of membership function which assigns to each point $x \in R^n$ a grade of membership in the fuzzy set. Such a membership function $m : R^n \to [0, 1]$ is used synonymously to denote the corresponding fuzzy set.

Assumption 1. m maps R^n onto $[0, 1]$.
Assumption 2. $[m]^0$ is a bounded subset of R^n.
Assumption 3. m is upper semicontinuous.
Assumption 4. m is fuzzy convex.

We denote by E^n the space of all fuzzy subsets m of R^n which satisfy assumptions 1-4; that is, normal, fuzzy convex and upper semicontinuous fuzzy sets with bounded supports. In particular, we denoted by E^1 the space of all fuzzy subsets m of R which satisfy assumptions 1-4 (see [2]).

A fuzzy number a in real line R is a fuzzy set characterized by a membership function m_a as $m_a : R \to [0, 1]$. A fuzzy number a is expressed as $a = \int_{x \in R} m_a(x)/x$, with the understanding that $m_a(x) \in [0, 1]$ represent the grade of membership of x in a and $\int$ denotes the union of $m_a(x)/x$'s [5].

Let E_N be the set of all upper semicontinuous convex normal fuzzy number with bounded α-level intervals. This means that if $a \in E_N$ then the α-level set

$$[a]^\alpha = \{x \in R : m_a(x) \geq \alpha,\ 0 < \alpha \leq 1\}$$

is a closed bounded interval which we denote by

$$[a]^\alpha = [a_l^\alpha, a_r^\alpha]$$

and there exists a $t_0 \in R$ such that $a(t_0) = 1$ (see [4]).

The support Γ_a of a fuzzy number a is defined, as a special case of level set, by the following

$$\Gamma_a = \{x \in R : m_a(x) > 0\}.$$

Two fuzzy numbers a and b are called equal $a = b$, if $m_a(x) = m_b(x)$ for all $x \in R$. It follows that

$$a = b \Leftrightarrow [a]^\alpha = [b]^\alpha \quad \text{for all } \alpha \in (0, 1].$$

A fuzzy number a may be decomposed into its level sets through the resolution identity

$$a = \int_0^1 \alpha[a]^\alpha,$$

where $\alpha[a]^\alpha$ is the product of a scalar α with the set $[a]^\alpha$ and $\int$ is the union of $[a]^\alpha$'s with α ranging from 0 to 1.

We denote the supremum metric d_∞ on E^n and the supremum metric H_1 on $C(I : E^n)$.

Definition 1. Let $a, b \in E^n$.

$$d_\infty(a,b) = \sup\{d_H([a]^\alpha, [b]^\alpha) : \alpha \in (0,1]\}$$

where d_H is the Hausdorff distance.

Definition 2. Let $x, y \in C(I : E^n)$

$$H_1(x,y) = \sup\{d_\infty(x(t), y(t)) : t \in I\}.$$

Let I be a real interval. A mapping $x : I \to E_N$ is called a fuzzy process. We denote

$$[x(t)]^\alpha = [x_l^\alpha(t), x_r^\alpha(t)],\ t \in I,\ 0 < \alpha \leq 1.$$

The derivative $x'(t)$ of a fuzzy process x is defined by

$$[x'(t)]^\alpha = [(x_l^\alpha)'(t), (x_r^\alpha)'(t)], \quad 0 < \alpha \leq 1$$

provided that is equation defines a fuzzy $x'(t) \in E_N$.

The fuzzy integral

$$\int_a^b x(t)dt, \quad a, b \in I$$

is defined by

$$\left[\int_a^b x(t)dt\right]^\alpha = \left[\int_a^b x_l^\alpha(t)dt, \int_a^b x_r^\alpha(t)dt\right]$$

provided that the Lebesgue integrals on the right exist.

Definition 3. [1] The fuzzy process $x : I \to E_N$ is a solution of equations (1)-(2) without the inhomogeneous term if and only if

$$(\dot{x}_l^\alpha)(t) = \min\Big\{A_l^\alpha(t)\big[x_j^\alpha(t) + \int_0^t G(t-s)x_j^\alpha(s)ds\big],\ i,j = l,r\Big\},$$
$$(\dot{x}_r^\alpha)(t) = \max\Big\{A_r^\alpha(t)\big[x_j^\alpha(t) + \int_0^t G(t-s)x_j^\alpha(s)ds\big],\ i,j = l,r\Big\},$$

and

$$(x_l^\alpha)(0) = x_{0l}^\alpha - g_l^\alpha(t_1, t_2, \cdots, t_p, x(\cdot)),$$
$$(x_r^\alpha)(0) = x_{0r}^\alpha - g_r^\alpha(t_1, t_2, \cdots, t_p, x(\cdot)).$$

Next hypotheses and existence result are Balasubramaniam and Muralisakar's results (see [1]).

(H1) The nonlinear function $g : I^p \times E_N \to E_N$ is a continuous function and satisfies the inequality

$$d_H([g(t_1, t_2, \cdots, t_p, x(\cdot))]^\alpha, [g(t_1, t_2, \cdots, t_p, y(\cdot))]^\alpha) \leq c_1 d_H([x(\cdot)]^\alpha, [y(\cdot)]^\alpha),$$

for all $x(\cdot), y(\cdot) \in E_N$, c_1 is a finite positive constant.

(H2) The inhomogeneous term $f : I \times E_N \to E_N$ is a continuous function and satisfies a global Lipschitz condition

$$d_H([f(s, x(s))]^\alpha, [f(s, y(s))]^\alpha) \leq c_2 d_H([x(s)]^\alpha, [y(s)]^\alpha),$$

for all $x(\cdot), y(\cdot) \in E_N$, and a finite positive constant $c_2 > 0$.

(H3) $S(t)$ is a fuzzy number satisfying for $y \in E_N$, $S'(t)y \in C^1(I : E_N) \bigcap C(I : E_N)$ the equation

$$\begin{aligned} \frac{d}{dt} S(t)y &= A\left[S(t)y + \int_0^t G(t-s)S(s)y ds\right] \\ &= S(t)Ay + \int_0^t S(t-s)AG(s)y ds, \quad t \in I, \end{aligned}$$

such that

$$[S(t)]^\alpha = [S_l^\alpha(t), S_r^\alpha(t)],$$

and $S_i^\alpha(t)$ $(i = l, r)$ is continuous. That is, there exists a constant $c > 0$ such that $|S_i^\alpha(t)| \leq c$ for all $t \in I$.

Theorem 1. [1] *Let $T > 0$, and hypotheses* (H1)-(H3) *hold. Then for every x_0, $g \in E_N$, the fuzzy initial value problem* (1)-(2) *without control function has a unique solution $x \in C(I : E_N)$.*

3 Nonlocal Controllability

In this section, we show the nonlocal controllability for the control system (1)-(2).

The control system (1)-(2) is related to the following fuzzy integral system:

$$\begin{aligned} x(t) = S(t)(x_0 - g(t_1, t_2, \cdots, t_p, x(\cdot))) &+ \int_0^t S(t-s)f(s, x(s))ds \\ &+ \int_0^t S(t-s)u(s)ds, \end{aligned} \tag{3}$$

where $S(t)$ is satisfy (H3).

Definition 4. The equation (3) is nonlocal controllable if, there exists $u(t)$ such that the fuzzy solution $x(t)$ of (3) satisfies $x(T) = x^1 - g(t_1, t_2, \cdots, t_p, x(\cdot))$ (i.e., $[x(T)]^\alpha = [x^1 - g(t_1, t_2, \cdots, t_p, x(\cdot))]^\alpha$) where x^1 is target set.

We assume that the linear fuzzy control system with respect to nonlinear fuzzy control system (3) is nonlocal controllable. Then

$$\begin{aligned} x(T) &= S(T)(x_0 - g(t_1, t_2, \cdots, t_p, x(\cdot))) + \int_0^T S(T-s)u(s)ds \\ &= x^1 - g(t_1, t_2, \cdots, t_p, x(\cdot)) \end{aligned}$$

and

$$\begin{aligned} [x(T)]^\alpha &= \left[S(T)(x_0 - g(t_1, t_2, \cdots, t_p, x(\cdot))) + \int_0^T S(T-s)u(s)ds \right]^\alpha \\ &= \Big[S_l^\alpha(T)(x_{0l}^\alpha - g_l^\alpha(t_1, t_2, \cdots, t_p, x(\cdot))) + \int_0^T S_l^\alpha(T-s)u_l^\alpha(s)ds, \\ &\qquad S_r^\alpha(T)(x_{0r}^\alpha - g_r^\alpha(t_1, t_2, \cdots, t_p, x(\cdot))) + \int_0^T S_r^\alpha(T-s)u_r^\alpha(s)ds \Big] \\ &= \left[(x^1 - g(t_1, t_2, \cdots, t_p, x(\cdot)))_l^\alpha, (x^1 - g(t_1, t_2, \cdots, t_p, x(\cdot)))_r^\alpha \right] . \end{aligned}$$

Defined the fuzzy mapping $G : \widetilde{P}(R) \to E_N$ by

$$G^\alpha(v) = \begin{cases} \int_0^T S^\alpha(T-s)v(s)ds, \ v \subset \overline{\Gamma_u} \ , \\ 0, \qquad\qquad\qquad\qquad \text{otherwise.} \end{cases} \tag{4}$$

Then there exists $G_i^\alpha \quad (i = l, r)$ such that

$$\begin{aligned} G_l^\alpha(v_l) &= \int_0^T S_l^\alpha(T-s)v_l(s)ds \ , \ v_l(s) \in [u_l^\alpha(s), u^1(s)] \ , \\ G_r^\alpha(v_r) &= \int_0^T S_r^\alpha(T-s)v_r(s)ds \ , \ v_r(s) \in [u^1(s), u_r^\alpha(s)] \ . \end{aligned}$$

We assume that G_l^α, G_r^α are bijective mappings.

Hence α-level of $u(s)$ are

$$\begin{aligned} [u(s)]^\alpha &= [u_l^\alpha(s), u_r^\alpha(s)] \\ &= \Big[(\widetilde{G}_l^\alpha)^{-1} \big((x^1)_l^\alpha - g_l^\alpha(t_1, t_2, \cdots, t_p, x(\cdot)) \\ &\qquad\qquad -S_l^\alpha(T)(x_{0l}^\alpha - g_l^\alpha(t_1, t_2, \cdots, t_p, x(\cdot))) \big), \\ &\quad (\widetilde{G}_r^\alpha)^{-1} \big((x^1)_r^\alpha - g_r^\alpha(t_1, t_2, \cdots, t_p, x(\cdot)) \\ &\qquad\qquad -S_r^\alpha(T)(x_{0r}^\alpha - g_r^\alpha(t_1, t_2, \cdots, t_p, x(\cdot))) \big) \Big]. \end{aligned}$$

Thus we can be introduced $u(s)$ of nonlinear system

$$\begin{aligned} [u(s)]^\alpha &= [u_l^\alpha(s), u_r^\alpha(s)] \\ &= \Big[(\widetilde{G}_l^\alpha)^{-1} \big((x^1)_l^\alpha - g_l^\alpha(t_1, t_2, \cdots, t_p, x(\cdot)) - S_l^\alpha(T)(x_{0l}^\alpha \end{aligned}$$

$$-g_l^\alpha(t_1,t_2,\cdots,t_p,x(\cdot))) - \int_0^T S_l^\alpha(T-s)f_l^\alpha(s,x(s))ds\big),$$
$$(\widetilde{G}_r^\alpha)^{-1}\big((x^1)_r^\alpha - g_r^\alpha(t_1,t_2,\cdots,t_p,x(\cdot)) - S_r^\alpha(T)(x_{0r}^\alpha$$
$$-g_r^\alpha(t_1,t_2,\cdots,t_p,x(\cdot))) - \int_0^T S_r^\alpha(T-s)f_r^\alpha(s,x(s))ds\big)\Big].$$

Then substituting this expression into the equation (3) yields α-level of $x(T)$.

$$\begin{aligned}
&[x(T)]^\alpha \\
&= \Big[S_l^\alpha(T)(x_{0l}^\alpha - g_l^\alpha(t_1,t_2,\cdots,t_p,x(\cdot))) + \int_0^T S_l^\alpha(T-s)f_l^\alpha(s,x(s))ds \\
&\quad + \int_0^T S_l^\alpha(T-s)(\widetilde{G}_l^\alpha)^{-1}\big((x^1)_l^\alpha - g_l^\alpha(t_1,t_2,\cdots,t_p,x(\cdot)) \\
&\quad -S_l^\alpha(T)(x_{0l}^\alpha - g_l^\alpha(t_1,t_2,\cdots,t_p,x(\cdot))) - \int_0^T S_l^\alpha(T-s)f_l^\alpha(s,x(s))ds\big)ds, \\
&\quad S_r^\alpha(T)(x_{0r}^\alpha - g_r^\alpha(t_1,t_2,\cdots,t_p,x(\cdot))) + \int_0^T S_r^\alpha(f_r^\alpha(s,x(s))ds \\
&\quad + \int_0^T S_r^\alpha(T-s)(\widetilde{G}_r^\alpha)^{-1}\big((x^1)_r^\alpha - g_r^\alpha(t_1,t_2,\cdots,t_p,x(\cdot)) \\
&\quad -S_r^\alpha(T)(x_{0r}^\alpha - g_r^\alpha(t_1,t_2,\cdots,t_p,x(\cdot))) - \int_0^T S_r^\alpha(T-s)f_r^\alpha(s,x(s))ds\big)ds\Big] \\
&= \Big[S_l^\alpha(T)(x_{0l}^\alpha - g_l^\alpha(t_1,t_2,\cdots,t_p,x(\cdot))) + \int_0^T S_l^\alpha(T-s)f_l^\alpha(s,x(s))ds \\
&\quad +G_l^\alpha\cdot(\widetilde{G}_l^\alpha)^{-1}\big((x^1)_l^\alpha - g_l^\alpha(t_1,t_2,\cdots,t_p,x(\cdot)) \\
&\quad -S_l^\alpha(T)(x_{0l}^\alpha - g_l^\alpha(t_1,t_2,\cdots,t_p,x(\cdot))) - \int_0^T S_l^\alpha(T-s)f_l^\alpha(s,x(s))ds\big), \\
&\quad S_r^\alpha(T)(x_{0r}^\alpha - g_r^\alpha(t_1,t_2,\cdots,t_p,x(\cdot))) + \int_0^T S_r^\alpha(T-s)f_r^\alpha(s,x(s))ds \\
&\quad +G_r^\alpha\cdot(\widetilde{G}_r^\alpha)^{-1}\big((x^1)_r^\alpha - g_r^\alpha(t_1,t_2,\cdots,t_p,x(\cdot)) \\
&\quad -S_r^\alpha(T)(x_{0r}^\alpha - g_r^\alpha(t_1,t_2,\cdots,t_p,x(\cdot))) - \int_0^T S_r^\alpha(T-s)f_r^\alpha(s,x(s))ds\big)\Big] \\
&= [(x^1)_l^\alpha - g_l^\alpha(t_1,t_2,\cdots,t_p,x(\cdot)), (x^1)_r^\alpha - g_r^\alpha(t_1,t_2,\cdots,t_p,x(\cdot))] \\
&= [x^1 - g(t_1,t_2,\cdots,t_p,x(\cdot))]^\alpha.
\end{aligned}$$

We now set

$$\begin{aligned}
\varPhi x(t) &= S(t)(x_0 - g(t_1,t_2,\cdots,t_p,x(\cdot))) + \int_0^t S(t-s)f(s,x(s))ds \\
&\quad + \int_0^t S(t-s)\widetilde{G}^{-1}\big(x^1 - g(t_1,t_2,\cdots,t_p,x(\cdot))
\end{aligned}$$

$$-S(T)(x_0 - g(t_1, t_2, \cdots, t_p, x(\cdot))) - \int_0^T S(T-s)f(s, x(s))ds\big)ds$$

where the fuzzy mappings $\widetilde{G}^{-1}$ satisfied above statements.

Notice that $\Phi x(T) = x^1 - g(t_1, t_2, \cdots, t_p, x(\cdot))$, which means that the control $u(t)$ steers the equation(3) from the origin to $x^1 - g(t_1, t_2, \cdots, t_p, x(\cdot))$ in time T provided we can obtain a fixed point of the nonlinear operator Φ.

Assume that the following hypotheses:

(H4) Linear system of equation (3) ($f = 0$) is nonlocal controllable.
(H5) $(1 + 2c)c_1 + 2cc_2T < 1$.

Theorem 2. *Suppose that hypotheses* (H1)-(H5) *are satisfied. Then the equation(3) is nonlocal controllable.*

Proof. We can easily check that Φ is continuous function from $C([0, T] : E_N)$ to itself. For $x, y \in C([0, T] : E_N)$,

$$\begin{aligned}
& d_H\big([\Phi x(t)]^\alpha, [\Phi y(t)]^\alpha\big) \\
&= d_H\Big(\big[S(t)(x_0 - g(t_1, t_2, \cdots, t_p, x(\cdot))) + \int_0^t S(t-s)f(s, x(s))ds \\
&\quad + \int_0^t S(t-s)\widetilde{G}^{-1}\big(x^1 - g(t_1, t_2, \cdots, t_p, x(\cdot)) \\
&\qquad -S(T)(x_0 - g(t_1, t_2, \cdots, t_p, x(\cdot))) - \int_0^T S(T-s)f(s, x(s))ds\big)ds\big]^\alpha, \\
&\quad \big[S(t)(x_0 - g_l^\alpha(t_1, t_2, \cdots, t_p, y(\cdot))) + \int_0^t S(t-s)f(s, y(s))ds \\
&\quad + \int_0^t S(t-s)\widetilde{G}^{-1}\big(x^1 - g(t_1, t_2, \cdots, t_p, y(\cdot)) \\
&\qquad -S(T)(x_0 - g(t_1, t_2, \cdots, t_p, y(\cdot))) - \int_0^T S(T-s)f(s, y(s))ds\big)ds\big]^\alpha\Big) \\
&\le d_H\Big(\big[S(t)g(t_1, t_2, \cdots, t_p, x(\cdot))\big]^\alpha, \big[S(t)g(t_1, t_2, \cdots, t_p, y(\cdot))\big]^\alpha\Big) \\
&\quad + d_H\Big(\big[\int_0^t S(t-s)f(s, x(s))ds\big]^\alpha, \big[\int_0^t S(t-s)f(s, y(s))ds\big]^\alpha\Big) \\
&\quad + d_H\Big(\big[\int_0^t S(t-s)\widetilde{G}^{-1}(-g(t_1, t_2, \cdots, t_p, x(\cdot)) \\
&\qquad + S(T)g(t_1, t_2, \cdots, t_p, x(\cdot)) - \int_0^T S(T-s)f(s, x(s))ds)\ ds\big]^\alpha, \\
&\quad \big[\int_0^t S(t-s)\widetilde{G}^{-1}(-g(t_1, t_2, \cdots, t_p, y(\cdot)) \\
&\qquad + S(T)g(t_1, t_2, \cdots, t_p, y(\cdot)) - \int_0^T S(T-s)f(s, y(s))ds)\ ds\big]^\alpha\Big)
\end{aligned}$$

$$\le (1+2c)c_1 d_H([x(\cdot)]^\alpha,[y(\cdot)]^\alpha)$$
$$+cc_2\Big(\int_0^t d_H([x(s)]^\alpha,[y(s)]^\alpha)ds+\int_0^T d_H([x(s)]^\alpha,[y(s)]^\alpha)ds\Big).$$

Therefore

$$d_\infty\big(\Phi x(t),\Phi y(t)\big)=\sup_{\alpha\in(0,1]} d_H\big([\Phi x(t)]^\alpha,[\Phi y(t)]^\alpha\big)$$
$$\le (1+2c)c_1 d_\infty(x(\cdot),y(\cdot))+cc_2\Big(\int_0^t d_\infty(x(s),y(s))ds+\int_0^T d_\infty(x(s),y(s))ds\Big).$$

Hence

$$\begin{aligned}H_1\big(\Phi x,\Phi y\big)&=\sup_{t\in[0,T]} d_\infty\big(\Phi x(t),\Phi y(t)\big)\\&\le((1+2c)c_1+2cc_2T)H_1(x,y).\end{aligned}$$

By hypotheses (H5), Φ is a contraction mapping. By the Banach fixed point theorem, (3) has a unique fixed point $x\in C([0,T]:E_N)$.

4 Example

Consider the semilinear one-dimensional heat equation on a connected domain $(0,1)$ for a material with memory, boundary conditions $x(t,0)=x(t,1)=0$ and with initial condition $x(0,z)+\sum_{k=1}^p c_k x(t_k,z)=x_0(z)$ where $x_0(z)\in E_N$. Let $x(t,z)$ be the internal energy and $f(t,x(t,z))=\widetilde{2}tx(t,z)^2$ be the external heat. Let $A=\widetilde{2}\frac{\partial^2}{\partial z^2}$, $\sum_{k=1}^p c_k x(t_k,z)=g(t_1,t_2,\cdots,t_p,x(\cdot))$ and $G(t-s)=e^{-(t-s)}$ then the balance equation becomes

$$\frac{dx(t)}{dt}=\widetilde{2}\left[x(t)-\int_0^t e^{-(t-s)}x(s)ds\right]+\widetilde{2}tx(t)^2+u(t), \tag{5}$$
$$x(0)+g(t_1,t_2,\cdots,t_p,x(\cdot))=x_0 \tag{6}$$

The α-level set of fuzzy number $\widetilde{2}$ is $[\widetilde{2}]^\alpha=[\alpha+1,3-\alpha]$ for all $\alpha\in[0,1]$. Then α-level set of $f(t,x(t)$ is

$$[f(t,x(t)]^\alpha=t[(\alpha+1)(x_l^\alpha(t))^2,(3-\alpha)(x_r^\alpha(t))^2]. \tag{7}$$

Further, we have

$$\begin{aligned}&d_H\left([f(t,x(t))]^\alpha,[f(t,y(t))]^\alpha\right)\\&=d_H\left(t[(\alpha+1)(x_l^\alpha(t))^2,(3-\alpha)(x_r^\alpha(t))^2],t[(\alpha+1)(y_l^\alpha(t))^2,(3-\alpha)(y_r^\alpha(t))^2]\right)\\&=t\max\{(\alpha+1)|(x_l^\alpha(t))^2-(y_l^\alpha(t))^2|,(3-\alpha)|(x_r^\alpha(t))^2-(y_r^\alpha(t))^2|\}\\&\le 3T|x_r^\alpha(t)+y_r^\alpha(t)|\max\{|x_l^\alpha(t)-y_l^\alpha(t)|,|x_r^\alpha(t)-y_r^\alpha(t)|\}\\&=c_2 d_H([x(t)]^\alpha,[y(t)]^\alpha),\end{aligned}$$

where c_2 is satisfies the inequality in hypothesis (H5), and also

$$\begin{aligned}
&d_H\left([g(t_1,t_2,\cdots,t_p,x(\cdot))]^\alpha,[g(t_1,t_2,\cdots,t_p,y(\cdot))]^\alpha\right)\\
&=d_H\left(\sum_{k=1}^p c_k[x(t_k)]^\alpha,\sum_{k=1}^p c_k[y(t_k)]^\alpha\right)\\
&\le|\sum_{k=1}^p c_k|\max_k d_H([x(t_k)]^\alpha,[y(t_k)]^\alpha)\\
&=c_1 d_H\left([x(\cdot)]^\alpha,[y(\cdot)]^\alpha\right),
\end{aligned}$$

where c_1 is satisfies the inequality in hypothesis (H5). Therefore f and g satisfy the global Lipschitz condition. Let initial value x_0 is $\widetilde{0}$. Target set is $x^1=\widetilde{2}$. The α-level set of fuzzy numbers $\widetilde{0}$ is $[\widetilde{0}]^\alpha=[\alpha-1,1-\alpha]$. We introduce the α-level set of $u(s)$ of equation (5)-(6).

$$\begin{aligned}
[u(s)]^\alpha&=[u_l^\alpha(s),u_r^\alpha(s)]\\
&=\Bigg[\widetilde{G}_l^{-1}\Bigg((\alpha+1)-\sum_{k=1}^p c_k x_l^\alpha(t_k)-S_l^\alpha(T)((\alpha-1)\\
&\qquad-\sum_{k=1}^p c_k x_l^\alpha(t_k))-\int_0^T S_l^\alpha(T-s)s(\alpha+1)(x_l^\alpha(s))^2ds\Bigg),\\
&\quad\widetilde{G}_r^{-1}\Bigg((3-\alpha)-\sum_{k=1}^p c_k x_r^\alpha(t_k)-S_r^\alpha(T)((1-\alpha)\\
&\qquad-\sum_{k=1}^p c_k x_r^\alpha(t_k))-\int_0^T S_r^\alpha(T-s)s(3-\alpha)(x_r^\alpha(s))^2ds\Bigg)\Bigg].
\end{aligned}$$

Then substituting this expression into the integral system with respect to (5)-(6) yields α-level set of $x(T)$.

$$\begin{aligned}
[x(T)]^\alpha=&\Bigg[S_l^\alpha(T)((\alpha-1)-\sum_{k=1}^p c_k x_l^\alpha(t_k))+\int_0^T S_l^\alpha(T-s)s(\alpha+1)(x_l^\alpha(s))^2ds\\
&+\int_0^T S_l^\alpha(T-s)(\widetilde{G}_l^\alpha)^{-1}\Bigg((\alpha+1)-\sum_{k=1}^p c_k x_l^\alpha(t_k)-S_l^\alpha(T)((\alpha-1)\\
&-\sum_{k=1}^p c_k x_l^\alpha(t_k))-\int_0^T S_l^\alpha(T-s)s(\alpha+1)(x_l^\alpha(s))^2ds\Bigg)ds,\\
&S_r^\alpha(T)((1-\alpha)-\sum_{k=1}^p c_k x_r^\alpha(t_k))+\int_0^T S_l^\alpha(T-s)s(3-\alpha)(x_r^\alpha(s))^2ds\\
&+\int_0^T S_l^\alpha(T-s)(\widetilde{G}_r^\alpha)^{-1}\Bigg((3-\alpha)-\sum_{k=1}^p c_k x_r^\alpha(t_k)-S_r^\alpha(T)((1-\alpha)
\end{aligned}$$

$$-\sum_{k=1}^{p} c_k x_r^\alpha(t_k)) - \int_0^T S_l^\alpha(T-s)s(3-\alpha)(x_r^\alpha(s))^2 ds\Bigg) ds\Bigg]$$

$$= \left[(\alpha+1) - \sum_{k=1}^{p} c_k x_l^\alpha(t_k), (3-\alpha) - \sum_{k=1}^{p} c_k x_r^\alpha(t_k)\right]$$

$$= \left[\widetilde{2} - \sum_{k=1}^{p} c_k x(t_k)\right]^\alpha .$$

Then all the conditions stated in Theorem 2 are satisfied, so the system (5)-(6) is nonlocal controllable on $[0, T]$.

5 Conclusion

In this paper, by using the concept of fuzzy number in E_N, we study the controllability for the semilinear fuzzy integrodifferential control system with nonlocal condition in E_N and find the sufficient conditions of controllability for the control system (1)-(2).

References

1. P. Balasubramaniam and S. Muralisankar, Existence and uniqueness of fuzzy solution for semilinear fuzzy integrodifferential equations with nonlocal conditions, *An International J. Computer & Mathematics with applications*, 47(2004), 1115–1122.
2. P. Diamand and P. E. Kloeden, Metric space of Fuzzy sets, *World scientific*, (1994).
3. O. Kaleva, Fuzzy differential equations, *Fuzzy set and Systems*, 24(1987), 301–317.
4. Y. C. Kwun and D. G. Park, Optimal control problem for fuzzy differential equations, *Proceedings of the Korea-Vietnam Joint Seminar*, (1998), 103–114.
5. M. Mizmoto and K. Tanaka, Some properties of fuzzy numbers, *Advances in Fuzzy Sets Theory and applications, North-Holland Publishing Company*,(1979), 153–164.
6. S. Seikkala, On The Fuzzy Initial Value problem,*Fuzzy Sets and Systems*, 24(1987), 319–330.

The Analysis and Design of IG_gHSOFPNN by Evolutionary Optimization

Ho-Sung Park and Tae-Chon Ahn

School of Electrical Electronic and Information Engineering, Wonkwang University, 344-2, Shinyong-Dong, Iksan, Chon-Buk, 570-749, South Korea
{neuron, tcahn}@wonkwang.ac.kr

Abstract. In this paper, we introduce the analysis and design of Information granulation based genetically optimized Hybrid Self-Organizing Fuzzy Polynomial Neural Networks (IG_gHSOFPNN) by evolutionary optimization. The architecture of the resulting IG_gHSOFPNN results from a synergistic usage of the hybrid system generated by combining fuzzy polynomial neurons (FPNs)-based Self-Organizing Fuzzy Polynomial Neural Networks(SOFPNN) with polynomial neurons (PNs)-based Self-Organizing Polynomial Neural Networks(SOPNN). The augmented IG_gHSOFPNN results in a structurally optimized structure and comes with a higher level of flexibility in comparison to the one we encounter in the conventional HSOFPNN. The GA-based design procedure being applied at each layer of IG_gHSOFPNN leads to the selection of preferred nodes available within the HSOFPNN. The obtained results demonstrate superiority of the proposed networks over the existing fuzzy and neural models.

1 Introduction

When the dimensionality of the model goes up (say, the number of variables increases), so do the difficulties. In particular, when dealing with high-order nonlinear and multi-variable equations of the model, we require a vast amount of data to estimate all its parameters. GMDH-type algorithms have been extensively used since the mid-1970's for prediction and modeling complex nonlinear processes [1]. While providing with a systematic design procedure, GMDH comes with some drawbacks. To alleviate the problems associated with the GMDH, Self-Organizing Polynomial Neural Networks (SOPNN)[2] Self-Organizing Fuzzy Polynomial Neural Networks (SOFPNN)[3], and Hybrid Self-Organizing Fuzzy Polynomial Neural Networks (HSOFPNN)[4] introduced. In this paper, to address the above design problems coming with thc development of conventional self-organizing neural networks, in particular, HSOFPNN, we introduce the IG_gHSOFPNN with the aid of the information granulation [5] as well as the genetic algorithm [6]. The determination of the optimal values of the parameters available within an individual PN(viz. the number of input variables, the order of the polynomial and the collection of preferred nodes) and FPN(viz. the number of input variables, the order of the polynomial, the collection of preferred nodes, the number of MFs for each input variable, and the selection of MFs) leads to a structurally and parametrically optimized network.

L. Wang et al. (Eds.): FSKD 2006, LNAI 4223, pp. 231 – 240, 2006.

2 The Architecture and Development of HSOFPNN

2.1 Fuzzy Polynomial Neuron : FPN

This neuron, regarded as a generic type of the processing unit, dwells on the concept of fuzzy sets. When arranged together, FPNs build the first layer of the HSOFPNN. As visualized in Fig. 1, the FPN consists of two basic functional modules. The first one, labeled by **F**, is a collection of fuzzy sets that form an interface between the input numeric variables and the processing part realized by the neuron. The second module (denoted here by **P**) concentrates on the function – based nonlinear (polynomial) processing.

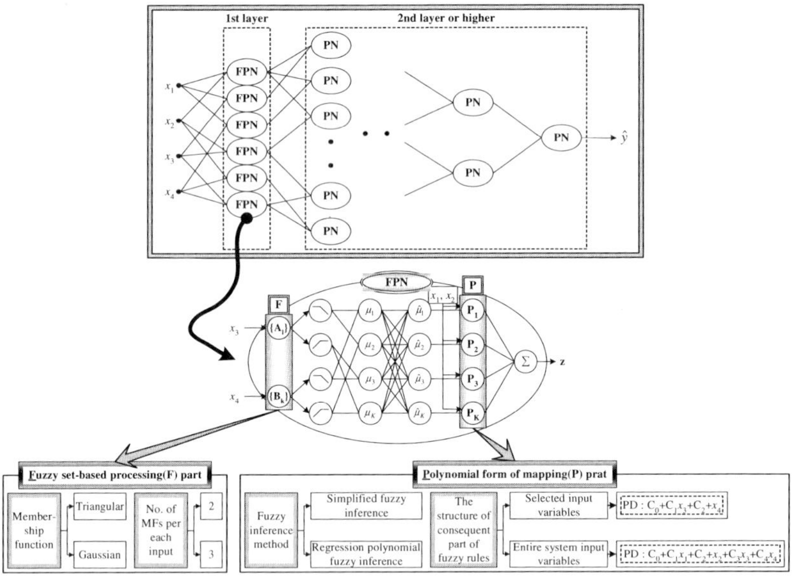

Fig. 1. Magnetization as a function of applied field

In other words, FPN realizes a family of multiple-input single-output rules. Each rule, refer again to Fig. 1, reads in the form

$$\text{If } x_p \text{ is } \boldsymbol{A}_l \text{ and } x_q \text{ is } \boldsymbol{B}_k \text{ then } z \text{ is } P_{lk}(x_i, x_j, \boldsymbol{a}_{lk}) \tag{1}$$

The activation levels of the rules contribute to the output of the FPN being computed as a weighted average of the individual condition parts (functional transformations) P_K.

$$z = \frac{\sum_{K=1}^{all\ rules} \mu_K P_K(x_i, x_j, \boldsymbol{a}_K)}{\sum_{K=1}^{all\ rules} \mu_K} = \sum_{K=1}^{all\ rules} \tilde{\mu}_K P_K(x_i, x_j, \boldsymbol{a}_K), \quad \tilde{\mu}_K = \mu_K \Big/ \sum_{L=1}^{all\ rules} \mu_L \tag{2}$$

2.2 Polynomial Neuron : PN

The SOPNN algorithm in the PN based layer of HSOFPNN is based on the GMDH method and utilizes a class of polynomials such as linear, quadratic, modified quadratic, etc. to describe basic processing realized there.

The input-output relationship for the above data realized by the SOPNN algorithm can be described in the following manner

$$y=f(x_1, x_2, \ldots, x_N) \tag{3}$$

The estimated output $\hat{y}$ reads as

$$\hat{y} = c_0 + \sum_{i=1}^{N} c_i x_i + \sum_{i=1}^{N}\sum_{j=1}^{N} c_{ij} x_i x_j + \sum_{i=1}^{N}\sum_{j=1}^{N}\sum_{k=1}^{N} c_{ijk} x_i x_j x_k + \cdots \tag{4}$$

Table 1. Different forms of the regression polynomials building a FPN and PN

Order of the polynomial: Order	Type: FPN	Type: PN	No. of inputs: 1	2	3
0	Type 1		Constant	Constant	Constant
1	Type 2	Type 1	Linear	Bilinear	Trilinear
2	Type 3	Type 2	Quadratic	Biquadratic-1	Triquadratic-1
	Type 4	Type 3		Biquadratic-2	Triquadratic-2

1: Basic type, 2: Modified type

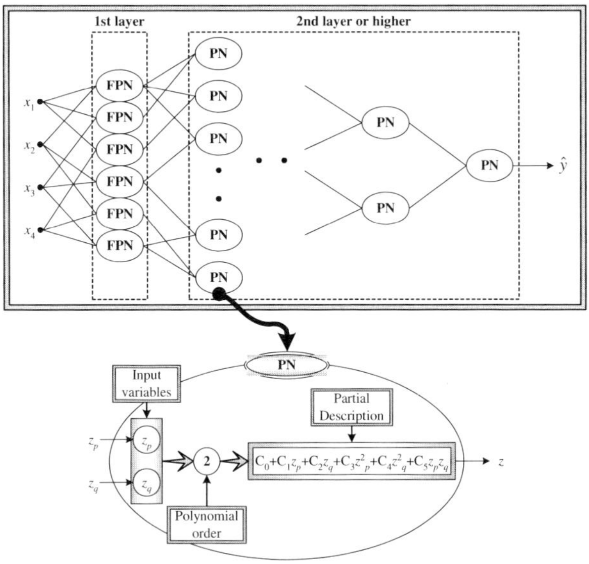

Fig. 2. A general topology of the PN based layer of HSOFPNN

The detailed PN involving a certain regression polynomial is shown in Table 1. The architecture of the PN based layer of HSOFPNN is visualized in Fig. 2.

3 Optimization of HSOFPNN by Information Granulation and Genetic Algorithms

3.1 Optimization of HSOFPNN by Information Granulation

3.1.1 Definition of the Premise Part of Fuzzy Rules Using IG

We assume that given a set of data $X=\{x_1, x_2, \ldots, x_n\}$ related to a certain application, there are some clusters revealed by the HCM[7]. Each cluster is represented by its center and all elements, which belong to it. Each membership function in the premise part of the rule is assigned to be complementary with neighboring ones.

3.1.2 Restructure of the Consequence Part of Fuzzy Rules Using IG

Here, let us concentrate on building the consequent part of the fuzzy rule. Each cluster (and the resulting rule) can be regarded as a sub-model of the overall system. The premise parts of the fuzzy rules help quantify how much overlap occurs between the individual sub-models. The consequent part of the fuzzy rule is a polynomial with independent variables for which the center point on this cluster (that is the sub-model) is mapped onto the point of origin. Therefore, all data belonging to the cluster are mapped into new coordinates. This is done by subtracting the value of the center point from all data belonging to the corresponding cluster.

3.2 Optimization of HSOFPNN by Genetic Algorithm

GAs has been theoretically and empirically demonstrated to provide robust search capabilities in complex spaces thus offering a valid solution strategy to problems

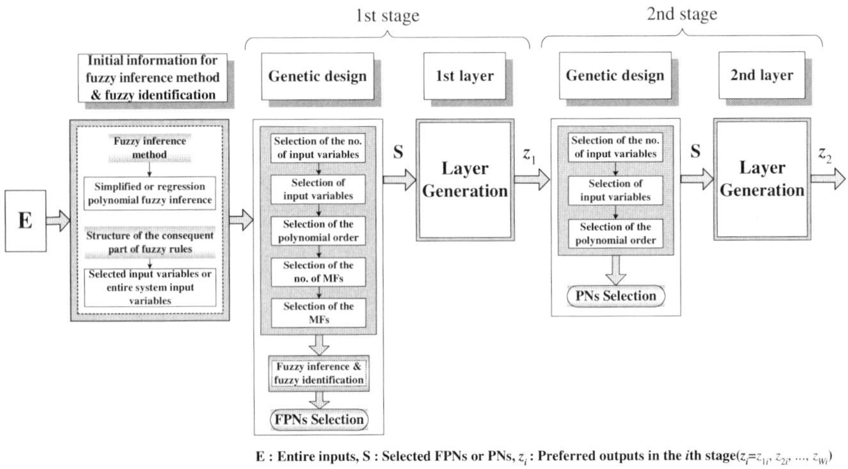

Fig. 3. Overall optimization process of HSOFPNN using GAs

requiring efficient and effective searching. It is eventually instructive to highlight the main features that tell GA apart from some other optimization methods: (1) GA operates on the codes of the variables, but not the variables themselves. (2) GA searches optimal points starting from a group (population) of points in the search space (potential solutions), rather than a single point. (3) GA's search is directed only by some fitness function whose form could be quite complex; we do not require it need to be differentiable.

In this study, for the optimization of the HSOFPNN model, GA uses the serial method of binary type, roulette-wheel used in the selection process, one-point crossover in the crossover operation, and a binary inversion (complementation) operation in the mutation operator. To retain the best individual and carry it over to the next generation, we use elitist strategy [5].

4 The Algorithm and Design Procedure of IG_gHSOFPNN

Overall, the framework of the design procedure of the IG_gHSOFPNN architecture comprises the following steps.

[Step 1] *Determine system's input variables.*
Define system's input variables $\boldsymbol{x}_i(i=1, 2, \ldots, n)$ related to the output variable y.
[Step 2] *Form a training and testing data.*
The input-output data set $(\boldsymbol{x}_i, y_i)=(x_{1i}, x_{2i}, \ldots, x_{ni}, y_i)$, $i=1, 2, \ldots, N$ is divided into two parts, that is, a training and testing dataset.
[Step 3] *Decision of axis of MFs by Information granulation*
As mentioned in '3.1.1 Definition of the premise part of fuzzy rules using IG', we obtained the new axis of MFs by information granulation.
[Step 4] *Decide initial information for constructing the HSOFPNN structure.*
Here we decide upon the essential design parameters of the HSOFPNN structure. Those include
a) Initial specification of the fuzzy inference method and the fuzzy identification
b) Initial specification for decision of HSOFPNN structure
[Step 5] *Decide a structure of the PN and FPN based layer of HSOFPNN using genetic design.*
This concerns the selection of the number of input variables, the polynomial order, the input variables, the number of membership functions, and the selection of membership functions to be assigned at each node of the corresponding layer.

In nodes (PN and FPNs) of each layer of HSOFPNN, we adhere to the notation of Fig. 4.
[Step 6] *Estimate the coefficient parameters of the polynomial in the selected node (PN or FPN).*
[Step 6-1] *In case of a PN (PN-based layer)*

The vector of coefficients $\mathbf{C}_i$ is derived by minimizing the mean squared error between y_i and z_{mi}.

$$E=\frac{1}{N_{tr}}\sum_{i=0}^{N_{tr}}(y_i-z_{mi})^2 \tag{5}$$

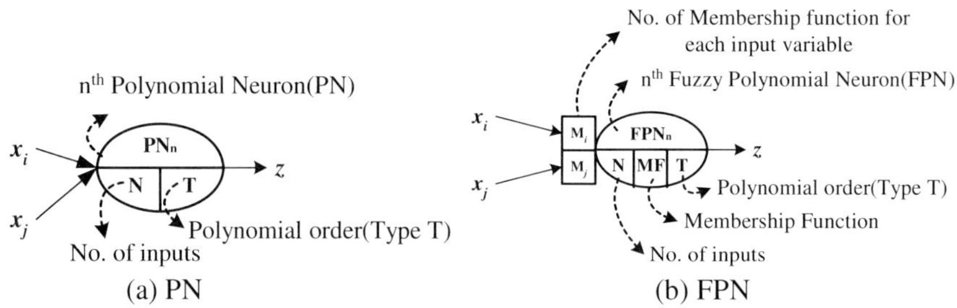

Fig. 4. Formation of each PN or FPN in HSOFPNN architecture

Using the training data subset, this gives rise to the set of linear equations

$$\mathbf{Y}=\mathbf{X}_i\mathbf{C}_i \tag{6}$$

Evidently, the coefficients of the PN of nodes in each layer are expressed in the form

$$y=f(x_1, x_2, \ldots, x_N) \quad \mathbf{C}_i=(\mathbf{X}_i^T\mathbf{X}_i)^{-1}\mathbf{X}_i^T\mathbf{Y} \tag{7}$$

[Step 6-2] *In case of a FPN (FPN-based layer)*
i) Simplified inference
The consequence part of the simplified inference mechanism is a constant. Using information granulation, the new rules read in the form

$$R^n : If\ x_1\ is\ A_{n1}\ and\ \cdots\ and\ x_k\ is\ A_{nk}\ then\ y_n - M_n = a_{n0} \tag{8}$$

$$\hat{y} = \frac{\sum_{j=1}^{n}\mu_{ji}y_i}{\sum_{i=1}^{n}\mu_{ji}} = \frac{\sum_{j=1}^{n}\mu_{ji}(a_{j0}+M)}{\sum_{i=1}^{n}\mu_{ji}} = \sum_{j=1}^{n}\hat{\mu}_{ji}(a_{j0}+M_j) \tag{9}$$

$$\mu_{ji} = A_{j1}(x_{1i}) \wedge \cdots \wedge A_{jk}(x_{ki}) \tag{10}$$

The consequence parameters (a_{j0}) are produced by the standard least squares method.
ii) Regression polynomial inference
The use of the regression polynomial inference method gives rise to the expression.

$$\begin{aligned} &R^n : If\ x_1\ is\ A_{n1}\ and \cdots and\ x_k\ is\ A_{nk}\ then \\ &\quad y_n - M_n = f_n\{(x_1 - v_{n1}), (x_2 - v_{n2}), \cdots, (x_k - v_{nk})\} \end{aligned} \tag{11}$$

$$\begin{aligned} \hat{y}_i &= \frac{\sum_{j=1}^{n}\mu_{ji}y_i}{\sum_{j=1}^{n}\mu_{ji}} = \frac{\sum_{j=1}^{n}\mu_{ji}\{a_{j0}+a_{j1}(x_{1i}-v_{j1})+\cdots+a_{jk}(x_{ki}-v_{jk})+M_j\}}{\sum_{i=1}^{n}\mu_{ji}} \\ &= \sum_{j=1}^{n}\hat{\mu}_{ji}\{a_{j0}+a_{j1}(x_{1i}-v_{j1})+\cdots+a_{jk}(x_{ki}-v_{jk})+M_j\} \end{aligned} \tag{12}$$

The coefficients of consequence part of fuzzy rules obtained by least square method(LSE) as like a simplified inference.

[Step 7] *Select nodes (PNs or FPNs) with the best predictive capability and construct their corresponding layer.*

All nodes of this layer of the IG_gHSOFPNN are constructed genetically. To evaluate the performance of PNs or FPNs constructed using the training dataset, the testing dataset is used. Based on this performance index, we calculate the fitness function. The fitness function reads as

$$F(\text{fitness Function}) = \frac{1}{1+EPI} \tag{13}$$

where *EPI* denotes the performance index for the testing data (or validation data).

[Step 8] *Check the termination criterion.*

As far as the depth of the network is concerned, the generation process is stopped at a depth of less than three layers. This size of the network has been experimentally found to build a sound compromise between the high accuracy of the resulting model and its complexity as well as generalization abilities.
In this study, we use a measure (performance indexes) that is the Mean Squared Error (MSE)

[Step 9] *Determine new input variables for the next layer.*

The outputs of the preserved nodes (z_{1i}, z_{2i}, …, z_{Wi}) serves as new inputs to the next layer (x_{1j}, x_{2j}, …, x_{Wj})($j=i+1$). This is captured by the expression

$$x_{1j} = z_{1i},\ x_{2j} = z_{2i},\ \ldots,\ x_{Wj} = z_{Wi} \tag{14}$$

The IG_gHSOFPNN algorithm is carried out by repeating steps 4-9 of the algorithm.

5 Simulation Study

We consider a nonlinear static system [8] with two inputs, x_1, x_2 and a single output that assumes the following form

$$y = (1 + x_1^{-2} + x_2^{-1.5})^2, \quad 1 \le x_1, x_2 \le 5 \tag{15}$$

This nonlinear static equation is widely used to evaluate modeling performance of the fuzzy modeling. Using (15), 50 input-output data are generated: the inputs are generated randomly and the corresponding output is then computed through the above relationship. We consider the MSE to serve as a performance index. Table 2 summarizes the list of parameters used in the genetic optimization of the networks.

Table 3 summarizes the results: According to the information of Table 2, the selected input variables (Node), the selected polynomial type (T), the selected no. of MFs (M), and its corresponding performance index (PI and EPI) were shown when the genetic optimization for each layer was carried out.

Table 2. Computational overhead and a list of parameters of the GAs and the HSOFPNN

	Parameters	1st layer	2nd layer	3rd layer
GAs	Maximum generation	150	150	150
	Total population size	100	100	100
	Selected population size	30	30	30
	Crossover rate	0.65	0.65	0.65
	Mutation rate	0.1	0.1	0.1
	String length	3+3+30+5+1	3+3+30+5	3+3+30+5
HSOFPNN	Maximal no. of inputs to be selected(Max)	$1 \leq l \leq$ Max(2~3)	$1 \leq l \leq$ Max(2~3)	$1 \leq l \leq$ Max(2~3)
	Polynomial type (Type T) of the consequent part of fuzzy rules	1≤T_F≤4	1≤T_P≤3	1≤T_P≤3
	Membership Function (MF) type	Triangular Gaussian		
	No. of MFs per each input(M)	2 or 3		

l, T_F, T_P : integer, T_F : Type of SOFPNN, T_P : Type of SOPNN.

Table 3. Performance index of IG_gHSOFPNN for nonlinear function process

Max	1st layer Node(M)	T	MF	PI	2nd layer Node	T	PI	3rd layer Node	T	PI
	(a) In case of selected input									
2	1(3) 2(3)	3	T	2.0176e-24	17 21	1	1.8799e-24	3 25	3	1.8787e-24
3	1(3) 2(3) 0	3	T	2.0176e-24	10 19 21	1	1.7301e-24	2 22 30	2	1.7297e-24
	(b) In case of entire system input									
2	1(3) 2(3)	3	T	2.0176e-24	17 22	1	1.8799e-24	2 27	2	1.8787e-24
3	1(3) 2(3) 0	3	T	2.0176e-24	3 10 19	3	1.8016e-24	11 16 30	2	1.8013e-24

In case of selected input, the result for network in the 3rd layer is obtained when using Max=3 with Type 2 polynomials (quadratic functions) and 3 node at input (node numbers are 2, 22, 30); this network comes with the value of PI=1.7297e-24.

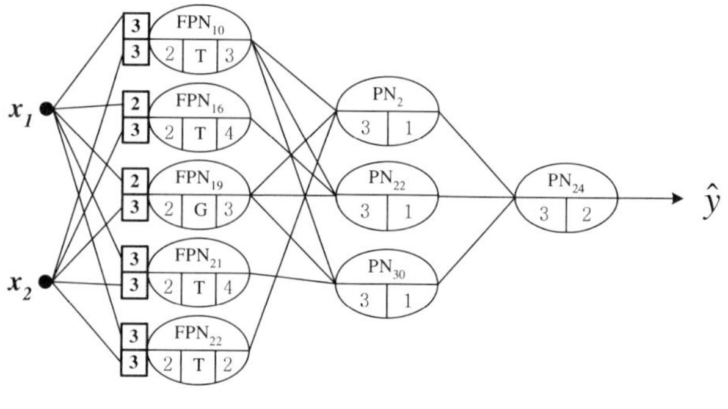

Fig. 5. Optimal IG_gHSOFPNN architecture

Fig. 6 illustrates differences in learning observed between selected input and entire system input by visualizing the values of the performance index obtained in successive generations of GA when using Max=2 and Max=3.

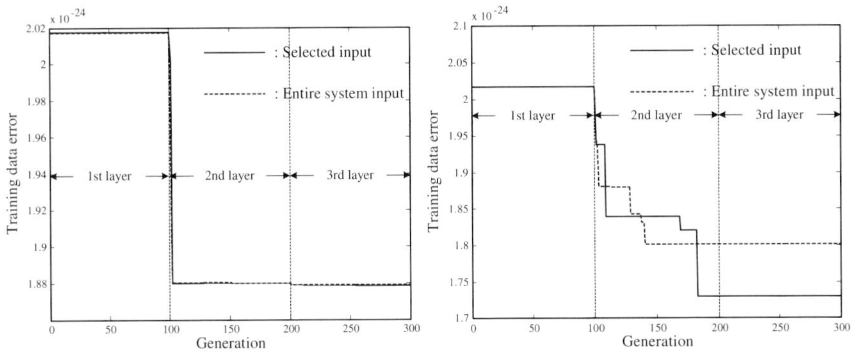

(a) Max=2 (b) Max=3

Fig. 6. Optimal procedure by IG and GAs

Table 4 includes a comparative analysis of the performance of the proposed network with other models.

Table 4. Comparative analysis of the performance of the network; considered are models reported in the literature

Model			Performance Index
Sugeno and Yasukawa's model [8]			0.0790
Gomez-Skarmeta et al.'s model [9]			0.0700
Kim et al.'s model [10]			0.0190
Kim et al.'s model [11]			0.0089
PNN [12] (5th layer)	CASE I	Basic	0.0212
		Modified	0.0212
	CASE II	Basic	0.0041
		Modified	0.0105
FPNN [13] (5th layer)	Case 1(BFPNN)		0.0033
	Case 2(MFPNN)		0.0023
ANFPN[14] (5th layer)	Basic	Case 1	0.0105
		Case 2	0.0153
	Modified	Case 1	0.0081
		Case 2	0.0082
Our model	Selected input	3rd layer(Max=2)	1.8787e-24
		3rd layer(Max=3)	1.7297e-24
	Entire system input	3rd layer(Max=2)	1.8787e-24
		3rd layer(Max=3)	1.8013e-24

6 Concluding Remarks

In this paper, we have introduced and investigated a class of Information Granulation based genetically optimized Hybrid Self-Organizing Fuzzy Polynomial Neural Networks (IG_gHSOFPNN) driven to genetic optimization and information granulation regarded as a modeling vehicle for nonlinear and complex systems.

The GA-based design procedure applied at each stage (layer) of the HSOFPNN driven to information granulation leads to the selection of the preferred nodes (or FPNs and PNs) with optimal local. These options contribute to the flexibility of the resulting architecture of the network.

Through the proposed framework of genetic optimization we can efficiently search for the optimal network architecture (being both structurally and parametrically optimized) and this design facet becomes crucial in improving the overall performance of the resulting model.

Acknowledgement. This work was supported by the Korea Research Foundation Grant funded by the Korean Government(MOEHRD)(KRF-2005-041-D00713).

References

1. Ivakhnenko, A.G.: Polynomial theory of complex systems. IEEE Trans. on Systems, Man and Cybernetics. SMC-1 (1971) 364-378
2. Oh, S.K., Pedrycz, W.: The design of self-organizing Polynomial Neural Networks. Information Science. 141 (2002) 237-258
3. Park, H.S., Park, K.J., Lee, D.Y, Oh, S.K.: Advanced Self-Organizing Neural Networks Based on Competitive Fuzzy Polynomial Neurons. Transactions of The Korean Institute of Electrical Engineers. 53D (2004) 135-144
4. Oh, S.K., Pedrycz, W., Park, H.S.: Multi-layer hybrid fuzzy polynomial neural networks: a design in the framework of computational intelligence. Neurocomputing. 64 (2005) 397-431
5. Jong, D.K.A.: Are Genetic Algorithms Function Optimizers?. Parallel Problem Solving from Nature 2, Manner, R. and Manderick, B. eds., North-Holland, Amsterdam (1992)
6. Zadeh, L.A., et al.: Fuzzy Sets and Applications: Selected Paper. Wiley, New York (1987)
7. Bezdek, J.C.: Pattern Recognition with Fuzzy Objective Function Algorithms. New York. Plenum (1981)
8. Sugeno, M., Yasukawa, T.: A Fuzzy-Logic-Based Approach to Qualitative Modeling. IEEE Trans. Fuzzy Systems. 1 (1993) 7-31
9. Gomez-Skarmeta, A.F., Delgado. M, Vila, M.A.: About the use of fuzzy clustering techniques for fuzzy model identification. Fuzzy Sets and Systems. 106 (1999) 179-188
10. Kim, E.T., et al.: A new approach to fuzzy modeling. IEEE Trans. Fuzzy Systems. 5 (1997) 328-337
11. Kim, E.T., et al.: A simple identified Sugeno-type fuzzy model via double clustering. Information Science. 110 (1998) 25-39
12. Oh, S.K., Kim, D.W., Park, B.J., Hwang, H.S.: Advanced Polynomial Neural Networks Architecture with New Adaptive Nodes. Transactions on Control, Automation and Systems Engineering of Korea. 3 (2001) 43-50
13. Park, B.J, Pedrycz, W., Oh, S.K.: Fuzzy Polynomial Neural Networks : Hybrid Architectures of Fuzzy Modeling. IEEE Trans. Fuzzy Systems. 10 (2002) 607-621
14. Park, B.J, Oh, S.K.: The Analysis and Design of Advanced Neurofuzzy Polynomial Networks. Journal of the Institute of Electronics Engineers of Korea. 39-CI (2002) 18-31

A Note on the Handling of Fuzziness for Continuous-Valued Attributes in Decision Tree Generation

Dug Hun Hong[1], Sungho Lee[2], and Kyung Tae Kim[3]

[1] Department of Mathematics, Myongji University, Kyunggido 449-728, South Korea
dhhong@mju.ac.kr

[2] Department of Statistics, Daegu University, Kyungbuk 712-714, South Korea
shlee1@taegu.ac.kr

[3] Department of Electronics and Electrical Information Engineering
Kyungwon University, Sungnam Kyunggido, South Korea
ktkim@kyungwon.ac.kr

Abstract. Recently, Xizhao and Hong [Fuzzy Sets and Systems 99(1998), 283-290] proposed to revise the cut-point in a decision tree algorithm as the cross-point between two symmetric fuzzy membership functions. In this note we show that in the general class of non symmetric membership function, the cross-point depend on the precise form of the membership function.

1 Introduction

There are many algorithms based on decision tree generation, e.g., ID3 [4], GID3 [1], N2 [2], and C4 [5]. These algorithms have the most powerful heuristics in inductive learning. In [3], a result about the information entropy minimization heuristic used is discretizing continuous-valued attributes is derived. A new technique, soft thresholds, has been presented by Quinlan [5]. Recently Xizhao and Hong [6] proved the fact that selection of membership functions in a class of symmetric distribution does not influence the decision tree generation on the basis of likelihood possibility maximization. In this note, we further discuss this problem and show that the selection of membership functions in a class of non-symmetric distribution influences the decision tree generation.

2 Parameter Estimation

μ is said to be a fuzzy number if it is a convex, closed fuzzy set on R (the real line). For a given fuzzy number μ with the property that the set $\{x|\mu(x) = 1\}$ consists of only one point, we denote $\{Q_\mu(x; a, b) = \mu((x - a)/b),\ a \in R, b > 0\}$ by Ω_μ. Ω_μ is called a family of fuzzy numbers, generated by μ, where a and b are location parameter and scale parameter, respectively.

A fuzzy number (possibility distribution) has membership function v which belongs to Ω_μ and the parameters of v, a and b remain to be determined. A crisp

L. Wang et al. (Eds.): FSKD 2006, LNAI 4223, pp. 241–245, 2006.

sample from the distribution $v, (x_1, x_2, \cdots, x_m)$, is known. The problem is how to reasonably estimate parameters a and b by using the sample. We consider maximum scale likelihood method.

The integral of a membership function μ on R is called fuzzy entropy of the Possibility distribution μ, denoted by $E[\mu]$.

Let the membership function be $Q_u(x : a, b)$, and $(x_1, \cdots, x_m)$ a sample. We denote by

$$L(a,b) = \frac{1}{E[Q_\mu]} T(Q_\mu(x_1, : a, b), Q_\mu(x_2 : a, b), \cdots, Q_\mu(x_m : a, b))$$

the likelihood possibility with which the T-related sample appears. The Sup-T estimators of a and b are defined to be $\hat{a}$ and $\hat{b}$ satisfying $L(\hat{a}, \hat{b}) = \sup_{a \in R, b > 0} L(a,b)$.

In the following, we take a continuous , not necessarily symmetric fuzzy number μ whose support is assumed to be R and strictly increasing on $(-\infty, 0]$ and strictly decreasing on $[0, \infty)$. Let $x^{(1)} = \min_{1 \le i \le m} x_i$, $x^{(m)} = \max_{1 \le i \le m} x_i$, and $\mu_L = \mu \wedge I_{(-\infty,0]}$, $\mu_R = \mu \wedge I_{[0,\infty)}$, where we denote $\wedge$ by "min" and $\vee$ by "max", then there exists unique $\alpha \in [0,1]$ such that $x^{(m)} - x^{(1)} = \mu_R^{-1}(\alpha) - \mu_L^{-1}(\alpha)$.

Theorem 1. *The Maxmin estimator of the parameter* $\theta = (a, b)$ *is* $(\hat{a}, \hat{b}) = (x^{(m)} - \mu_R^{-1}(\alpha)(= x^{(1)} - \mu_L^{-1}(\alpha)), \mu_R^{-1}(\alpha) \vee (-\mu_L^{-1}(\alpha))/c)$, *where* α *is the number satisfying* $x^{(m)} - x^{(1)} = \mu_R^{-1}(\alpha) - \mu_L^{-1}(\alpha)$, *and* c *is real number at which the function* $g(t) = t(\mu(t) \wedge \mu(-t))$ $(t \ge 0)$ *attains its maximum.*

Proof. Let $\int_{-\infty}^{\infty} \mu(t)dt = l$, then $E[Q_\mu] = lb$. Since $T = min$,

$$\begin{aligned} L(a,b) &= \min_{1 \le i \le m} Q_\mu(x_i : a, b)/E[Q_\mu] \\ &= \min_{1 \le i \le m} \mu(\frac{x_i - a}{b})/(lb) \\ &= [\mu(\frac{x^{(1)} - a}{b}) \wedge \mu(\frac{x^{(m)} - a}{b})]/(lb). \end{aligned}$$

Let $\alpha \in [0,1]$ is the number satisfying $x^{(m)} - x^{(1)} = \mu_R^{-1}(\alpha) - \mu_L^{-1}(\alpha)$, and let $a^* = x^{(m)} - \mu_R^{-1}(\alpha) = x^{(1)} - \mu_L^{-1}(\alpha)$. Then, for any given $b > 0$,

$$\begin{aligned} \sup_{a \in R} L(a,b) &= \frac{1}{lb} \sup_{a \in R} [\mu(\frac{x^{(1)} - a}{b}) \wedge \mu(\frac{x^{(m)} - a}{b})] \\ &= \frac{1}{lb} [\mu(\frac{\mu_L^{-1}(\alpha)}{b}) \wedge \mu(\frac{\mu_R^{-1}(\alpha)}{b})] \\ &= L(x^{(m)} - \mu_R^{-1}(\alpha), b) \\ &= L(a^*, b), \end{aligned}$$

where the second equality comes from the fact that, for any $a \in R$, either $x^{(1)} - a \le \mu_L^{-1}(\alpha)$ or $x^{(m)} - a \ge \mu_R^{-1}(\alpha)$ and the monotonicity property of μ. Hence, we have $\hat{a} = a^*$. Next, we consider that

$$\begin{aligned}\sup_{a\in R,b>0} L(a,b) &= \sup_{b>0} L(x^{(m)} - \mu_R^{-1}(\alpha), b)\\ &= \sup_{b>0}[\frac{\mu(\frac{\mu_R^{-1}(\alpha)}{b})}{lb} \wedge \frac{\mu(\frac{\mu_L^{-1}(\alpha)}{b})}{lb}]\\ &= \sup_{t>0}\frac{t\mu(t)}{l\mu_R^{-1}(\alpha)} \wedge \sup_{t>0}\frac{t\mu(-t)}{l(-\mu_L^{-1}(\alpha))}\\ &= \frac{1}{l(\mu_R^{-1}(\alpha)\vee(-\mu_L^{-1}(\alpha)))}\sup_{t>0} t(\mu(t)\wedge\mu(-t))\\ &= \frac{1}{l(\mu_R^{-1}(\alpha)\vee(-\mu_L^{-1}(\alpha)))} c\mu^*(c),\end{aligned}$$

where $\mu^*(t) = \mu(t) \wedge \mu(-t)$ for $t \in [0, \infty)$. Let $c = \frac{\mu_R^{-1}(\alpha)}{b_R} = \frac{-\mu_L^{-1}(\alpha)}{b_L}$ for some $b_R > 0$ and $b_L > 0$. Since

$$\mu^*(c) = \mu^*(\frac{\mu_R^{-1}(\alpha)}{b_R}) = \mu^*(\frac{x^{(m)} - a^*}{b_R})$$

and

$$\mu^*(c) = \mu^*(\frac{-\mu_L^{-1}(\alpha)}{b_L}) = \mu^*(\frac{x^{(1)} - a^*}{-b_L}),$$

we have that

$$\begin{aligned}&\sup_{a\in R,b>0} L(a,b) = \frac{1}{l[(\mu_R^{-1}(\alpha)\vee(-\mu_L^{-1}(\alpha)))/c]}\mu^*(c)\\ &= \frac{1}{l[(\mu_R^{-1}(\alpha)\vee(-\mu_L^{-1}(\alpha)))/c]}[\mu(\frac{x^{(m)} - a^*}{b_R}) \wedge \mu(\frac{x^{(1)} - a^*}{-b_L})]\\ &= \frac{1}{l[(\mu_R^{-1}(\alpha)\vee(-\mu_L^{-1}(\alpha)))/c]}[\mu(\frac{x^{(m)} - a^*}{\mu_R^{-1}(\alpha)/c}) \wedge \mu(\frac{x^{(1)} - a^*}{-\mu_L^{-1}(\alpha)/c})].\end{aligned}$$

Now from the decreasing property of μ^* on $[0, \infty)$, we have that

$$\begin{aligned}\mu(\frac{x^{(m)} - a^*}{\mu_R^{-1}(\alpha)/c}) \wedge \mu(\frac{x^{(1)} - a^*}{-\mu_L^{-1}(\alpha)/c}) &= \mu(\frac{x^{(m)} - a^*}{(\mu_R^{-1}(\alpha)\vee(-\mu_L^{-1}(\alpha)))/c})\\ &\quad \wedge\mu(\frac{x^{(1)} - a^*}{(\mu_R^{-1}(\alpha)\vee(-\mu_L^{-1}(\alpha)))/c}),\end{aligned}$$

and hence we have

$$\sup_{a\in R,b>0} L(a,b) = L(a^*, \frac{\mu_R^{-1}(\alpha)\vee(-\mu_L^{-1}(\alpha))}{c})$$

Therefore, the Maxmin estimation of parameter (a, b) is

$$(\hat{a}, \hat{b}) = (x^{(m)} - \mu_R^{-1}(\alpha), \frac{\mu_R^{-1}(\alpha)\vee(-\mu_L^{-1}(\alpha))}{c}),$$

which completes the proof.

Example 1. Let $\mu_L(x) = e^{-|x|}$ and $\mu_R(x) = e^{-x^2}$ and let $x^{(1)} = 1, x^{(m)} = 13$. Let $\alpha = e^{-9}$, then $x^{(m)} - x^{(1)} = 13 - 1 = 3 - (-9) = \mu_R^{-1}(e^{-9}) - \mu_L^{-1}(e^{-9})$. Hence, by Theorem 2, $(\hat{a}, \hat{b}) = (10, \frac{9}{c})$.

3 Steadiness of Decision Tree

The revised value of the best cut point can be determined by computing the cross point of two membership function which describe the family F_1 and F_2. In previous section, we have used a family Ω_μ which is generated by a non-symmetric fuzzy number μ. But, not like the case that μ is 0-symmetric fuzzy numbers, the cross point of two membership functions depend on the selection of μ. we see the following discussion.

Proposition 1. *Let μ be a given non-symmetric fuzzy numbers with continuous membership function and R-support, and let $Q_\mu(x : a_1, b_1)$ and $Q_\mu(x : a_2, b_2)$, which are two fuzzy numbers describing a successive pair of families of attribute values, be generated by using Maxmin μ/E estimation. Then the cross point of these two membership function T of the first sample $(x^{(1)}, x^{(2)}, \cdots, x^{(m)})$ and the second sample $(y^{(1)}, y^{(2)}, \cdots, y^{(n)})$ is*

$$T = \begin{cases} \textit{solution of } \mu_R(\frac{x-\hat{a}_1}{\hat{b}_1}) = \mu_L(\frac{x-\hat{a}_2}{\hat{b}_2}) & \textit{if } \hat{a}_1 \leq \hat{a}_2 \\ \textit{solution of } \mu_L(\frac{x-\hat{a}_1}{\hat{b}_1}) = \mu_R(\frac{x-\hat{a}_2}{\hat{b}_2}) & \textit{if } \hat{a}_1 > \hat{a}_2 \end{cases},$$

where $x^{(m)} - x^{(1)} = \mu_R^{-1}(\alpha) - \mu_L^{-1}(\alpha)$, $y^{(n)} - y^{(1)} = \mu_R^{-1}(\beta) - \mu_L^{-1}(\beta)$, $(\hat{a}_1, \hat{b}_1) = (x^{(m)} - \mu_R^{-1}(\alpha), \frac{\mu_R^{-1}(\alpha) \vee (-\mu_L^{-1}(\alpha))}{c})$, *and* $(\hat{a}_2, \hat{b}_2) = (y^{(n)} - \mu_R^{-1}(\beta), \frac{\mu_R^{-1}(\beta) \vee (-\mu_L^{-1}(\beta))}{c})$.

4 Conclusion

In this paper, we considered the parameter estimation problems on the basis of likelihood possibility maximization where the possibility distribution family is generated by non-symmetric fuzzy number μ. But, not like the symmetric min-related case of Wang and Hong [6], we show that the selection of membership function μ does influence decision tree generation.

References

1. J. Cheng, U.M. Fayyad, K.B. Irani and Z. Qian: Improved decision trees : a generalized version of ID3, Proc. 5th Int. Conf. on Machine Learning (Morgan Kaufmann, San mateo, CA, 1988) 100-108.
2. P. Clark and T. Niblett: The CN2 induction algorithm, Machine Learning **3** (1989) 261-284.

3. U.M. Fayyad, K.B. Irani: On the handing of continuous-valued attributes in decision tree generation, Machine Learning **8** (1992) 87-102.
4. J.R. Quinlan: Probabilistic decision trees, in: Y. Kodratoff and R. Michalski, eds., Machine Learning: An Artificial Intelligence Approach, **3** (Morgan Kaufmann, San Mateo, CA, 1990).
5. J.R. Quinlan: C4.5: Programs for Machine Learning(Morgan Kaufmann, San Mateo, CA, 1993).
6. W. Xizhao and J. Hong: On the handling of fuzziness for continuous-valued attributes in decision tree generation, Fuzzy Sets and Systems **99** (1998) 283-290.

Weighted Reduction for Decision Tables

Changzhi Xu[1] and Fan Min[2]

[1] The State Key Laboratory of Information Security,
Graduate School of the Chinese Academy of Sciences, Beijing 100049, China
xchzh04B@mails.gucas.ac.cn
[2] School of Computer Science and Engineering,
University of Electronic Science and Technology of China, Chengdu 610054, China
minfan@uestc.edu.cn

Abstract. As a classical problem of Rough Sets, the reduct problem still attracts research interests recently. Many existing works on the reduct problem are devoted into finding optimal reducts where the optimal metrics is the number of attributes. In reality, however, this optimal metrics is not fair since some attributes may have much larger domains than others, and they tend to have better discernibility thus more likely to be included in optimal reducts. To cope with this fairness problem, in this paper we propose the concept of *average discernibility* which takes into consideration the cardinality of the attribute domain. Attribute reduction based on average discernibility can be implemented through assigning each attribute an appropriate weight in the reduction process to adjust attribute significance. We point out further that some human experts knowledge can also be expressed by the weight vector formed by weights of all attributes. Then we propose a weighted reduction algorithm based on discernibility, and analyze the usefulness the weight vector along with its setting policies. This algorithm is consistent with the existing reduction algorithm based on discernibility in that the former contains the latter as a special case when all elements of the weight vector are equal and non-zero. Experiment results of the Bridges dataset in the UCI library validate the usefulness of our algorithm.

Keywords: Rough Sets, attribute reduction, average discernibility, weight vector and decision rule.

1 Introduction

As a classical problem of Rough Sets[1], the reduct problem still attracts research interests recently (see, e.g., [2][3][4][5]). Objectives of attribute reduction include rejecting unimportant and/or superfluous information, facilitating learning process among others.

Most existing works are devoted into finding minimal reducts, i.e., reducts with least attributes, because those reducts tend to give better generalization ability. It has been proven that finding the whole reduct set or a minimal reduct is NP complete [6]. In many applications, people firstly find out a small set

L. Wang et al. (Eds.): FSKD 2006, LNAI 4223, pp. 246–255, 2006.

of (possibly minimal) reducts using heuristic algorithms such as genetic algorithm [7], entropy based algorithm [2] or discernibility based algorithm [4], then decide which ones to use by human experts.

Unfortunately, minimal reducts do not necessarily give good generalization ability. In many applications, some attributes may have much larger domains than others, and they tend to have good discernibility and high attribute significance. If such attributes are essentially unimportant and we use minimal reducts containing them for rule generation, we will obtain rule sets with both bad generalization ability and bad predication accuracy. We call this the fairness problem of the reduct problem.

In order to cope with the fairness problem, we propose the concept of *average discernibility*. Attribute reduction based on average discernibility can be implemented through employing a weight vector. In fact, some human experts knowledge can also be expressed by the weight vector. We propose a weighted reduction algorithm based on discernibility, and analyze the usefulness the weight vector along with its setting policies. We use the Bridges dataset in the UCI library to test the validity of our algorithm.

The rest of the paper is organized as follows: Section 2 lists relative concepts. Section 3 gives the definition of the *average discernibility* and an algorithm which is a variation of discernibility based reduction algorithm. Then we analyze this algorithm in detail, with emphasizing on the usefulness of the weight vector and its setting. Section 4 lists some experimental results of our algorithm on the Bridges dataset.

2 Preliminaries

In this section we enumerate some concepts introduced by Pawlak [8].

2.1 Decision Tables and Reducts

Formally, a *decision table* is a triple $S = (U, C, \{d\})$ where $d \notin C$ is the decision attribute and elements of C are called *conditional attributes* or simply *conditions*. Table 1 lists a decision table S_1 where $U = \{x_1, x_2, x_3, x_4, x_5, x_6, x_7\}$, $C = \{\text{Number, Muscle-pain, Headache, Leucocyte, Temperature, Tonsil}\}$ and $d =$ Flu. The *domain* of attribute $a \in C$ is denoted by $DOM(a)$, and the cardinality of $DOM(a)$ is denoted by $|DOM(a)|$, or $|a|$ for briefness. For example, $DOM(\text{Headache}) = \{\text{yes, no}\}$ and $|\text{Number}| = 7$.

Let $\underline{B}X$ denote $B-$*lower approximation* of X, the positive region of $\{d\}$ with respect to $B \subseteq C$ is denoted by

$$POS_B(\{d\}) = \bigcup_{X \in U/\{d\}} \underline{B}X \tag{1}$$

A reduct is the minimal subset of attributes that enables the same classification of elements of the universe as the whole set of attributes. This can be formally defined as follows:

Table 1. An exemplary decision table

Patient	Number	Muscle-pain	Headache	Leucocyte	Temperature	Tonsil	Flu
x_1	1052	no	yes	high	high	swell	yes
x_2	1066	no	yes	high	high	swell	yes
x_3	1074	no	yes	high	high	swell	yes
x_4	1125	no	no	high	normal	swell	no
x_5	1258	no	yes	normal	low	normal	no
x_6	1357	no	yes	normal	high	normal	no
x_7	1388	yes	yes	normal	high	normal	yes

Definition 1. *Any $B \subseteq C$ is called a* reduct *of $S = (U, C, \{d\})$ iff:*

1. $POS_B(\{d\}) = POS_C(\{d\})$;
2. $\forall a \in B, POS_{B-\{a\}}(\{d\}) \subset POS_C(\{d\})$.

Let $Red(S)$ denotes the set of all reducts of $S = (U, C, \{d\})$, the *core* of S is given by

$$Core(S) = \bigcap Red(S). \tag{2}$$

2.2 Discernibility

Given a decision table $S = (U, C, \{d\})$, the discernibility of any $a \in C$ is defined by the set of object pairs with different decision attribute values it can discern. This is formally given by

$$DP(d|a) = \{(x_i, x_j) \in U \times U | d(x_i) \neq d(x_j), i < j, a(x_i) \neq a(x_j)\}, \tag{3}$$

where $i < j$ is required to ensure that no duplicate pairs exist.

For example, in S_1 listed in Table 1, $DP(d|\text{Leucocyte}) = \{(x_1, x_5), (x_1, x_6), (x_2, x_5), (x_2, x_6), (x_3, x_5), (x_3, x_6), (x_4, x_7)\}$.

The discernibility of a set of attributes is defined by the union of discernibility of all attributes, i.e.,

$$DP(d|B) = \bigcup_{a \in B} DP(d|a). \tag{4}$$

For example, $DP(d|\{\text{Leucocyte, Temperature}\}) = \{(x_1, x_5), (x_1, x_6), (x_1, x_7), (x_2, x_5), (x_2, x_6), (x_2, x_7), (x_3, x_5), (x_3, x_6), (x_4, x_7), (x_1, x_4), (x_2, x_4)\}$.

According to Theorem 4 in [3], the following theorem is obvious:

Theorem 1. *Given $S = (U, C, \{d\})$ and $B \subseteq C$,*

$$DP(d|B) = DP(d|C) \Rightarrow POS_B(\{d\}) = POS_C(\{d\}). \tag{5}$$

It should be noted that the reverse of this theorem does not hold in inconsistent decision tables.

3 Weighted Reduction

In this section we firstly propose the concept of average discernibility, then we propose a weighted reduction algorithm based on discernibility, finally we analyze the algorithm in detail with the focus on the usefulness of the weight vector.

3.1 Average Discernibility

Nguyen S.H. [9] proposed to discern objects by triples (a, v_1, v_2) called *chains*, where $v_1, v_2 \in DOM(a)$. The discernibility of any chain (a, v_1, v_2) is defined by the set of object pairs with different decision attribute values it can discern. This is formally given by

$$DP(d|(a, v_1, v_2) = \{(x_i, x_j) \in U \times U | d(x_i) \neq d(x_j), a(x_i) = v_1, a(x_j) = v_2\}. \quad (6)$$

Clearly, for any attribute a, there are $C_a = C(|a|, 2)$ chains. For example, in Table 1, Number has 21 chains: (Number, 1052, 1066), (Number, 1052, 1074), ..., (Number, 1357, 1388); while Headache has only 1 chain: (Headache, Yes, No). The following two theorems are straightforward:

Theorem 2. *The discernibility of an attribute is the union of the discernibility of chains of the attribute:*

$$DP(d|a) = \bigcup_{v_1, v_2 \in DOM(a)} DP(d|(a, v_1, v_2)). \quad (7)$$

Theorem 3. *The number of object pairs discerned by an attribute is the sum of the number of object pairs discerned by all chains of the attribute:*

$$|DP(d|a)| = \sum_{v_1, v_2 \in DOM(a)} |DP(d|(a, v_1, v_2))|. \quad (8)$$

It is then natural for us to define the *average discernibility* of attribute a as:

$$ADP(d|a) = \frac{|DP(d|a)|}{C_a}. \quad (9)$$

For example, in Table 1, $|DP(\text{Flu}|\text{Number})| = 12$ and $|DP(\text{Flu}|\text{Headache})| = 5$, while $ADP(\text{Flu}|\text{Number}) = \frac{12}{21} = \frac{4}{7}$ and $ADP(\text{Flu}|\text{Headache}) = 5$. As an heuristic information, average discernibility seems to be more fair thus more appropriate than discernibility.

3.2 A Weighted Reduction Algorithm

Many existing reduction algorithms use the bottom-up approach. They start from the core and then add attribute according to certain heuristic information such as conditional entropy [2], discernibility [4] until the positive region is equal to that of S.

Given a decision table $S = (U, C, \{d\})$, let $C = \{a_1, a_2, \ldots, a_{|C|}\}$, the weight vector $W = \{w_1, w_2, \ldots, w_{|C|}\}$. Our algorithm is listed in Fig. 1.

WeightedReduction $(S = (U, C, \{d\}))$
{input: A decision table S.**}**
{output: A reduct B.**}**
Step 1. $B = \emptyset$, $Att = C$, $disPairs = 0$; //Initialize
Step 2. Move attributes from Att to B according to their (weighted) significance;
Step 2.1 For any $a_i \in Att$, $newDispairs_i = |DP(d|B \cup \{a_i\})|$,
$SGF_i = w_i \times (newDispairs_i - dispairs)$;
Step 2.2 If for all $a_i \in Att$ $SGF_i = 0$, goto **Step 3**;
Step 2.3 Let the attribute with the largest SGF_i be a_j, $Att = Att - \{a_j\}$,
$B = B \cup \{a_j\}$, $disPairs = newDispaires_j$;
Step 2.4 If $Att = \emptyset$ goto **Step 3**, else goto **Step 2.1**;
Step 3. For any $a \in B$, if $POS_{B-\{a\}}(\{d\}) = POS_B(\{d\})$, $B = B - \{a\}$;

Fig. 1. A weighted reduction algorithm based on discernibility

3.3 Analysis of the Algorithm

The major characteristic of this algorithm is the introduction of the weight vector W. If we set elements of W to be all the same, this algorithm reduces to a traditional reduction algorithm. From this point of view, this algorithm is more general than traditional ones. If we set $w_i = 1/C_{a_i}$ for any $1 \leq i \leq |C|$, the heuristic information is essentially the average discernibility.

In more general cases, we only require that w_i to be a non-negative value. This is because that in many applications human experts are taking part in the learning process and his/her knowledge may be incorporated in the weight vector. We will analyze this in more detail in the next subsection.

According to Step 2.1, if $w_i = 0$ then $SGF_i = 0$, which in turn indicates that a_j would never be included in B. Now a problem arises: Can this algorithm find a reduct for any given W? Let C' be the set of attributes with respective weight set to 0, namely, $C' = \{a_i \in C | w_i = 0\}$, we have the following property:

Property 1. Let the output of the algorithm be B,

$$B \in Red(S) \Leftrightarrow POS_{C-C'}(\{d\}) = POS_C(\{d\}). \tag{10}$$

Proof. Denote B in the algorithm before executing Step 3 by B', according to Step 2.1 of the algorithm, we have

$$B' \cap C' = \emptyset, \tag{11}$$

$$B' \subseteq C - C'. \tag{12}$$

While according to the stopping criteria (Step 2.2 and 2.4),

$$DP(d|B') = DP(d|C - C'). \tag{13}$$

From Theorem 1 we know that

$$POS_{B'}(\{d\}) = POS_{C-C'}(\{d\}). \tag{14}$$

Step 3 ensure that

$$POS_B(\{d\}) = POS_{B'}(\{d\}), \tag{15}$$

and for any $a \in B$,

$$POS_{B-\{a\}}(\{d\}) = POS_B(\{d\}). \tag{16}$$

($\Leftarrow$) Integrate $POS_{C-C'}(\{d\})$=$POS_C(\{d\})$ with equations (14) and (15), we obtain

$$POS_B(\{d\}) = POS_C(\{d\}). \tag{17}$$

According to Definition 1 and equations (16) and (17), $B \in Red(S)$.

($\Rightarrow$) Assume that $POS_{C-C'}(\{d\}) \neq POS_C(\{d\})$, according to equations (14),(15),

$$POS_B(\{d\}) = POS_{C-C'}(\{d\}) \neq POS_C(\{d\}), \tag{18}$$

which indicates $B \notin Red(S)$.

In fact, the introduction of W incurred the deletion of C', and the reduction process is essentially on the attribute set $C - C'$. It is easy to obtain the following property:

Property 2. Let the output of the algorithm be B, and $S' = (U, C - C', \{d\})$, then

$$B \in Red(S'). \tag{19}$$

The following proposition can then be obtained.

Proposition 1. *If $w_i > 0$ for all $i \in \{1, 2, \ldots, |C|\}$, then $B \in Red(S)$.*

3.4 The Usefulness of the Weight Vector

Now we investigate the usefulness of the weight vector. First of all, the use of the weight vector is to assist or make balance rather than *control* the reduction process. During the reduction process, the weight vector is helpful in the following aspects:

1. Cope with the fairness problem. This is the initial motivation of this work and we have analyzed it.
2. Delete useless attributes. For many datasets, there is an object ID attribute, such as Number in Table 1, ID number in the WDBC (see breast-cancer-wisconsin of the UCI library [10]) dataset, Names in the zoo dataset (also in UCI library). On one hand, this attribute has the best discernibility since it can discern any object pairs; while on the other hand, it is totally useless in generating decision rules since respective rules cannot be used to predicate for any *new* instance. In applications this kind of attributes are deleted during the preprocess stage. While in the weighted reduction, we simply set the weight of respective attributes to 0.

3. Reduce the influence of noise. Noise exists in many applications. It is often impossible to eliminate noise. Determining which data are noise is also hard. In fact, reduction is also a process of eliminating noise. Sometimes we know which attributes are more likely to be influenced by noise. In such cases we can assign relatively small weights for respective attributes, and the possibilities of these attributes being chosen is expected to decrease. It should be noted that we cannot simply delete such attributes since they may be quite important, or even irreducible (core attributes).
4. Reduce the influence of missing values. In many datasets, there are some missing values. Their influence is quite similar with that of noise, so we should also assign relatively small weights for them.
5. Reduce the influence of potential useless attributes. According to some human knowledge, it is quite suspectable if some given conditional attributes have any causation relationship with the decision attribute. Again, we can use smaller weight to reduce the possibility of such attributes being chosen.
6. Increase the significant of easy to obtain attributes. In many applications, the cost of obtaining different attributes many vary. For example, in a clinic, it is much more easier and cheaper to obtain the Tonsil attribute than that of Leucocyte (blood test is required). In this case we can set the weight of Tonsil a little higher than that of Leucocyte. If their discernibility have no much difference, Tonsil would be chosen. While if the discernibility of Leucocyte is much stronger (contains more elements), Leucocyte would still be chosen.

3.5 The Setting of the Weight Vector

In the last subsection we have mentioned some general policies in setting the weight vector. We use S_1 listed in Table 1 to explain them in more detail.

Now we assume human expert are involved. Obviously, the weight corresponds to Number should be 0. Since Headache is a commonly used attribute, we can take it as a benchmark attribute, and set its weight as 1. Tonsil is also a easy to obtain attribute, we set its weight as 0.9. The cost of Leucocyte is relatively high, and its weight is set to 0.6. Muscle-pain is potentially useless, so its weight is set to 0.4. Temperature varies according to the current status of the patient, or the time (some patients may have very high temperature in the midnight or noon). So we take it as an attribute with noise, and its weight is set to 0.3. In this example, no attribute seems very important or unreplaceable, so we do not have very high weight.

In order to make comparison, we consider three other settings. The first one is set elements of W to be all 1s. This scheme corresponds with non-weighted reduction without deleting useless attributes. The second one only take into consideration of the average discernibility. The third one take into consideration of human experts, and the last one also use the same policies mentioned above, while the setting changed a little.

Settings of W and respective results are listed in Table 2.

Table 2. Settings of W and respective reduct

Scheme #	W	B
1	[1, 1, 1, 1, 1, 1]	{Number}
2	$[\frac{1}{21}, 1, 1, 1, \frac{1}{3}, 1]$	{Muscle-pain, Headache, Leucocyte}
3	[0, 0.4, 1, 0.6, 0.3, 0.9]	{Muscle-pain, Tonsil, Headache}
4	[0, 0.5, 1, 0.5, 0.3, 0.9]	{Muscle-pain, Tonsil, Headache}

From Table 2 we know that the result of Scheme #1 is meaningless. The result of Scheme #2 is much better, but it still contain hard to obtain attribute (Leucocyte). The result of Scheme #3 also contains three attributes, but these attributes are easier to obtain, and less influenced by noise.

Based on Scheme #3, Scheme #4 increased weights of Muscle-pain, and Temperature. Their results are the same. This shows that the setting of the weight vector does not have to be very *accurate*. It is quite enough just to reflect respective policies.

4 Experiments

We choose the Bridges dataset from the UCI library [10] to test the validity of our algorithm.

The Bridges dataset contains 13 attributes: 7 (including River, Location, Erected, Purpose, Length, Lanes and Clear-G) for specifications and 5 (including T-or-D, Material, Span, Rel-L and Type) for design descriptions. We use version2 rather than version1 so that no more discretization work is required. Taking all specifications as conditions and one design description at a time as the decision attribute, we obtain 5 decision tables. Because their purposes are similar, we use the same weight vector. The Purpose attribute of a bridge is relatively important in designing issues, so we set the corresponding weight to 1.5. The Location attribute seems not quite relevant with the design issues and could be viewed as potentially useless, so we set the corresponding weight to 0.4. The Length attribute has many missing values (27 out of 108), so we set the corresponding weight to 0.6. Weights of all other attributes are set to 1.

In order to obtain more meaningful results, we use RSES 2.2 [11] to obtain rule sets of the reduced decision table, and then test their coverage and accuracy. Specifically, we use the *Classify* → *Cross-validation method* function where parameters are set as follows: Number of Fold: 10, Discretization: NO, Algorithm of rules computation: LEM2 with cover parameter 0.95, Shortening ratio: 0.95, Conflicts resolve by: Standard Voting.

The experimental results are shown in Table 3, where $W_1 = [1, 1, 1, 1, 1, 1, 1]$, $W_2 = [1/C_{a_1}, \ldots, 1/C_{a_7}]$ and $W_3 = [1, 0.4, 1, 1.5, 0.6, 1, 1]$. W_1 is the degenerated setting and W_2 is an auto setting. We take the experiment 10 times and each data of *coverage* and *accuracy* is the average value.

Table 3. Experiment results of the Bridges dataset

d	W	B	coverage	accuracy
T-or-D	W_1	{Location, Erected, Lanes}	0.647	0.7517
	W_2, W_3	{Location, Purpose, Lanes, Clear-G}	0.719	0.7936
Material	W_1	{Location, Erected, Length, Clear-G}	0.767	0.8345
	W_2, W_3	{Location, Erected, Purpose, Length}	0.763	0.8849
Span	W_1	{Location, Erected, Purpose, Length}	0.491	0.5839
	W_2, W_3	{Location, Erected, Purpose, Length}	0.491	0.5839
Rel-L	W_1	{Location, Erected, Purpose, Length}	0.579	0.7389
	W_2, W_3	{Location, Erected, Purpose, Length}	0.579	0.7389
Type	W_1	{Location, Erected, Length, Lanes, Clear-G}	0.575	0.5209
	W_2, W_3	{Location, Erected, Purpose, Length, Clear-G}	0.606	0.5343

Next we analyze the experimental results. In all 5 decision tables W_2 and W_3 produces the same result, indicating that the auto setting is valid when no expert is involved. Clearly this does not hold for any decision table.

Although we assign a relatively low weight to the Location attribute, it still appears in all reducts. That is because it is a core attribute. Deleting this attribute (assigning 0 as its weight) would give better results, but it is rather artificial thus out of the scope of this paper.

The result of the decision table taking T-to-D as the decision attribute is especially interesting. Although the reduct obtained using our algorithm contains more (4 vs. 3) attributes, it is helpful for obtaining decision rule set with higher coverage and accuracy. Results of the decision tables taking Material or Type as the decision attribute also validates the usefulness of our algorithm and weight vector setting policy.

For the other two decision tables our algorithm produces the same reduct set as that of the traditional one.

5 Conclusions and Further Works

In this paper we proposed the concepts of weighted reduction. We pointed out that this algorithm is more general than traditional ones and discussed the usefulness of the weight vector in detail. We also discussed the weight vector setting policies. Experimental results of the Bridges dataset indicate that both the average discernibility based weight vector and the user specified weight vector are valid in searching for better reducts than that of the traditional discernibility based approach.

It should be noted that the idea of using weight vector is also applicable to many other kinds of reduction algorithm, such as entropy-based reduction algorithm.

We are applying this idea to real applications, such as natural language (especially Chinese) processing, to test its validity. The setting of W deserves more detailed investigation in such applications.

References

1. Pawlak, Z.: Rough sets. International Journal of Computer and Information Sciences **11** (1982) 341–356
2. Wang, G., Yu, H., Yang, D.: Decision table reduction based on conditional information entropy. Chinese Journal of Computers **25**(7) (2002) 1–8
3. Zhang, W., Mi, J., Wu, W.: Knowledge reductions in inconsistent information systems. Chinese Journal of Computers **26**(1) (2003) 12–18
4. Xu, Y., Peng, H., Wang, Z.: Reduction algorithm based on discernibility and its applications. Chinese Journal of Computers **26**(1) (2003) 97–103
5. Min, F., Bai, Z., He, M., Liu, Q.: The reduct problem with specified attributes. In: Rough Sets and Soft Computing in Intelligent Agent and Web Technology, International Workshop at WI-IAT 2005. (2005) 36–42
6. Wong, S.K.M., Ziarko, W.: On optimal decision rules in decision tables. Bulletin of polish academy of sciences **33** (1985) 693–696
7. Wróblewski., J.: Finding minimal reducts using genetic algorithms. In Wang, P.P., ed.: JCIS'95, Wrightsville Beach, North Carolina (1995) 186–189
8. Pawlak, Z.: Some issues on rough sets. In Peters, J.F., Skowron, A., Grzymała-Busse, J.W., Kostek, B., Świniarski, R.W., Szczuka, M.S., eds.: Transactions on Rough Sets I. LNCS 3100. Springer-Verlag, Berlin Heidelberg (2004) 1–58
9. Nguyen, S.H.: Regularity Analysis And Its Applications In Data Mining. PhD thesis, Warsaw University, Warsaw, Poland (1999)
10. Blake, C.L., Merz, C.J.: UCI repository of machine learning databases, http://www.ics.uci.edu/˜mlearn/mlrepository.html (1998)
11. Bazan, J., Szczuka, M.: The RSES homepage, http://alfa.mimuw.edu.pl/˜rses (1994–2005)

On Rough Fuzzy Set Algebras

Wei-Zhi Wu and You-Hong Xu

Information College, Zhejiang Ocean University,
Zhoushan, Zhejiang, 316004, P. R. China
wuwz@zjou.net.cn, xyhong160@sina.com

Abstract. A rough fuzzy set is a pair of fuzzy sets resulting from the approximation of a fuzzy set in a crisp approximation space. A rough fuzzy set algebra is a fuzzy set algebra with added dual pair of rough fuzzy approximation operators. In this paper, structures of rough fuzzy set algebras are studied. It is proved that if a system $(\mathcal{F}(U), \cap, \cup, \sim, L, H)$ is a (a serial, a reflexive, a symmetric, a transitive, an Euclidean, a Pawlak, respectively) rough fuzzy set algebra then the derived system $(\mathcal{F}(U), \cap, \cup, \sim, LL, HH)$ is a (a serial, a reflexive, a symmetric, a transitive, an Euclidean, a Pawlak, respectively) rough fuzzy set algebra. Properties of rough fuzzy approximation operators in different types of rough fuzzy set algebras are also examined.

Keywords: Approximation operators; Fuzzy sets; Rough fuzzy set algebras; Rough fuzzy sets; Rough sets.

1 Introduction

The basic notions in rough set theory are the lower and upper approximation operators [11]. There are at least two methods to define approximation operators, the constructive and axiomatic approaches. In the constructive approach, binary relations on a universe of discourse, partitions of the universe of discourse, neighborhood systems, and Boolean algebras are all primitive notions. The lower and upper approximation operators are constructed by means of these notions [1,4,5,7,8,9,10,12,13,14,15,19,21,22,23,25,27,28]. On the other hand, the axiomatic approach takes the lower and upper approximation operators as primitive notions and a set of axioms is used to characterize approximation operators that are the same as the ones produced by using the constructive approach. Under this point of view, a rough set algebra is a set algebra with two additional approximation operators and rough set theory may be interpreted as an extension theory with two additional unary operators [3,6,8,9,13,14,19,21,23,24,26,27,29]. The lower and upper approximation operators are related to the necessity (box) and possibility (diamond) operators in modal logic, the interior and closure operators in topological space [2,16,17,18,27], the belief and plausibility functions in the Dempster-Shafer theory of evidence [20,30]. Thus the axiomatic approach helps us to gain much more insights into the mathematical structures of rough set approximation operators.

L. Wang et al. (Eds.): FSKD 2006, LNAI 4223, pp. 256–265, 2006.

In [26], Yao defined different types of crisp rough set algebras by using axiomatic approach and examined connections between a rough set algebra and its derived systems. In this paper, we mainly focus on the study of rough fuzzy set algebras. We first define rough fuzzy approximation operators via axiomatic approach. We then discuss mathematical structures of various types of rough fuzzy set algebras and examine their relationships with the deriving systems.

2 Definitions of Rough Fuzzy Set Algebras

Let X be a finite and nonempty set called the universe of discourse. The class of all subsets (resp., fuzzy subsets) of X will be denoted by $\mathcal{P}(X)$ (resp., by $\mathcal{F}(X)$). For any $A \in \mathcal{F}(X)$, we denote by $\sim A$ the complement of A. 1_y denotes the fuzzy singleton with value 1 at y and 0 elsewhere. $\widehat{\alpha}$ denotes the constant fuzzy set, i.e., $\widehat{\alpha}(x) = \alpha$ for all $x \in X$, where $\alpha \in I = [0, 1]$, the unit interval. For $A \in \mathcal{P}(X)$, we denote by 1_A the characteristic function of A, i.e., $1_A(x) = 1$ for $x \in A$ and 0 elsewhere.

Let U be a finite and nonempty universe of discourse. A subset $R \in \mathcal{P}(U \times U)$ is referred to as a binary relation on U. Let

$$R_s(x) = \{y \in U : (x, y) \in R\}, \quad x \in U,$$

$R_s(x)$ is referred to as the successor neighborhood of x with respect to R [22,28]. The relation R is referred to as serial if for all $x \in U$ there exists $y \in U$ such that $(x, y) \in R$, i.e., $R_s(x) \neq \emptyset$ for all $x \in U$; R is referred to as reflexive if for all $x \in U$, $(x, x) \in R$, i.e., $x \in R_s(x)$ for all $x \in U$; R is referred to as symmetric if for all $x, y \in U$, $(x, y) \in R$ implies $(y, x) \in R$, i.e., $y \in R_s(x)$ implies $x \in R_s(y)$ for all $(x, y) \in U \times U$; R is referred to as transitive if for all $x, y, z \in U$, $(x, y) \in R$ and $(y, z) \in R$ imply $(x, z) \in R$, i.e., $y \in R_s(x)$ and $z \in R_s(y)$ imply $z \in R_s(x)$; R is referred to as Euclidean if for all $x, y, z \in U$, $(x, y) \in R$ and $(x, z) \in R$ imply $(y, z) \in R$, i.e., $y \in R_s(x)$ and $z \in R_s(x)$ imply $z \in R_s(y)$; R is referred to as equivalent if R is a reflexive, symmetric, and transitive relation.

Definition 1. *A fuzzy unary operator $L : \mathcal{F}(U) \to \mathcal{F}(U)$ is referred to as a rough fuzzy lower approximation operator iff it satisfies axioms:*

(FLC) $L(1_{U-\{y\}}) \in \mathcal{P}(U), \quad \forall y \in U,$

(FL1) $L(A \cup \widehat{\alpha}) = L(A) \cup \widehat{\alpha}, \quad \forall A \in \mathcal{F}(U), \quad \forall \alpha \in I;$

(FL2) $L(A \cap B) = L(A) \cap L(B), \ \forall A, B \in \mathcal{F}(U).$

A fuzzy unary operator $H : \mathcal{F}(U) \to \mathcal{F}(U)$ is referred to as a rough fuzzy upper approximation operator iff it satisfies axioms:

(FHC) $H(1_y) \in \mathcal{P}(U), \quad \forall y \in U,$

(FH1) $H(A \cap \widehat{\alpha}) = H(A) \cap \widehat{\alpha}, \quad \forall A \in \mathcal{F}(U), \ \forall \alpha \in I,$

(FH2) $H(A \cup B) = H(A) \cup H(B), \ \forall A, B \in \mathcal{F}(U).$

The lower and upper operators $L, H : \mathcal{F}(U) \to \mathcal{F}(U)$ are referred to as dual operators iff

(FD) $H(A) = \sim L(\sim A), \quad \forall A \in \mathcal{F}(U).$

According to [21,23] we can conclude that for the dual rough fuzzy lower and upper operators $L, H : \mathcal{F}(U) \to \mathcal{F}(U)$ there exists a binary relation R on U such that

$$\underline{R}(A) = L(A), \quad \overline{R}(A) = H(A), \quad \forall A \in \mathcal{F}(U), \tag{1}$$

where

$$\underline{R}(A)(x) = \bigwedge_{y \in R_s(x)} A(y), \quad \overline{R}(A)(x) = \bigvee_{y \in R_s(x)} A(y), x \in U. \tag{2}$$

$\underline{R}(A)$ and $\overline{R}(A)$ defined by Eq.(2) are resp. called the lower and upper approximations of fuzzy set A, in such a case, (U, R) is called a crisp approximation space and $(\underline{R}(A), \overline{R}(A))$ is called the rough fuzzy set of A w.r.t. (U, R) [23].

Definition 2. *If $L, H : \mathcal{F}(U) \to \mathcal{F}(U)$ are dual rough fuzzy lower and upper operators, i.e., L satisfies axioms* (FLC), (FL1), (FL2) *and* (FD), *or equivalently, H satisfies axioms* (FHC), (FH1), (FH2), *and* (FD). *Then the system $S_L =: (\mathcal{F}(U), \cap, \cup, \sim, L, H)$ is referred to as a rough fuzzy set algebra (RFSA). Moreover, if there exists a serial (resp. a reflexive, a symmetric, a transitive, an Euclidean, an equivalence) relation R on U such that $L(A) = \underline{R}(A)$ and $H(A) = \overline{R}(A)$ for all $A \in \mathcal{F}(U)$, then S_L is referred to as a serial (resp. a reflexive, a symmetric, a transitive, an Euclidean, a Pawlak) RFSA.*

Axiom (FD) implies that operators L and H in a RFSA S_L are dual. Axiom (FL2) implies the equivalent axioms (FL3) and (FL4), and axiom (FH2) implies the equivalent axioms (FH3) and (FH4):

(FL3) $L(A \cup B) \supseteq L(A) \cup L(B), \quad \forall A, B \in \mathcal{F}(U)$,
(FH3) $H(A \cap B) \subseteq H(A) \cap H(B), \forall A, B \in \mathcal{F}(U)$,
(FL4) $A \subseteq B \Longrightarrow L(A) \subseteq L(B), \quad \forall A, B \in \mathcal{F}(U)$,
(FH4) $A \subseteq B \Longrightarrow H(A) \subseteq H(B), \ \forall A, B \in \mathcal{F}(U)$.

Axiomatic characterizations of other special rough fuzzy operators are summarized in the following Theorem 1 [23]:

Theorem 1. *Suppose that S_L is a RFSA. Then*

(1) *it is a serial RFSA iff one of following equivalent axioms holds:*
(FL0) $L(\widehat{\alpha}) = \widehat{\alpha}, \quad \forall \alpha \in I$, (FH0) $H(\widehat{\alpha}) = \widehat{\alpha}, \quad \forall \alpha \in I$,
(FL0)$'$ $L(\emptyset) = \emptyset$, (FH0)$'$ $H(U) = U$,
(FLH0)$'$ $L(A) \subseteq H(A), \ \forall A \in \mathcal{F}(U)$.

(2) *it is a reflexive RFSA iff one of following equivalent axioms holds:*
(FL5) $L(A) \subseteq A, \quad \forall A \in \mathcal{F}(U)$, (FH5) $A \subseteq H(A), \quad \forall A \in \mathcal{F}(U)$.

(3) *it is a symmetric RFSA iff one of the following equivalent axioms holds:*
(FL6)$'$ $L(1_{U-\{x\}})(y) = L(1_{U-\{y\}})(x), \ \forall (x, y) \in U \times U$,
(FH6)$'$ $H(1_x)(y) = H(1_y)(x), \ \forall (x, y) \in U \times U$,
(FL6) $A \subseteq L(H(A)), \ \forall A \in \mathcal{F}(U)$, (FH6) $H(L(A)) \subseteq A, \ \forall A \in \mathcal{F}(U)$.

(4) *it is a transitive RFSA iff one of following equivalent axioms holds:*
(FL7) $L(A) \subseteq L(L(A)), \quad \forall A \in \mathcal{F}(U)$,
(FH7) $H(H(A)) \subseteq H(A), \forall A \in \mathcal{F}(U)$.

(5) *it is an Euclidean RFSA iff one of following equivalent axioms holds:*
(FL8) $H(L(A)) \subseteq L(A), \quad \forall A \in \mathcal{F}(U)$,
(FH8) $H(A) \subseteq L(H(A)), \quad \forall A \in \mathcal{F}(U)$.

In sequel, we will denote

$$LL(A) = L(L(A)), \quad HL(A) = H(L(A)), \quad A \in \mathcal{F}(U),$$
$$HH(A) = H(H(A)), LH(A) = L(H(A)), \quad A \in \mathcal{F}(U).$$

Similarly, we can define other composed rough fuzzy approximation operators.

Theorem 2. *If S_L is a RFSA, then $S_{LL} =: (\mathcal{F}(U), \cap, \cup, \sim, LL, HH)$ is also a RFSA.*

Proof. It is only to prove that HH satisfies axioms (FHC), (FH1), (FH2), and (FD). For any $A \in \mathcal{F}(U)$ and $\alpha \in I$, by axiom (FH1) we have

$$HH(A \cap \widehat{\alpha}) = H(H(A \cap \widehat{\alpha})) = H(H(A) \cap \widehat{\alpha}) = H(H(A)) \cap \widehat{\alpha} = HH(A) \cap \widehat{\alpha}. \quad (3)$$

That is, operator HH satisfies axiom (FH1). Since H obeys axioms (FH2) and (FD), it is easy to verify that HH also satisfies axioms (FH2) and (FD). For any $y \in U$, let $X = H(1_y)$, since H obeys axioms (FHC) and (FH2), we conclude that $X \in \mathcal{P}(U)$ and

$$HH(1_y) = H(H(1_y)) = H(X) = H(\bigcup_{x \in X} 1_x) = \bigcup_{x \in X} H(1_x). \quad (4)$$

Since $H(1_x) \in \mathcal{P}(U)$ for all $x \in X$, we conclude that $HH(1_y) \in \mathcal{P}(U)$. Thus we have proved that HH satisfies axiom (FHC).

3 Special Classes of RFSAs

In this section we discuss properties of approximation operators in special classes of RFSAs. We will discuss the relationships between a RFSA S_L and its producing two systems S_{LL} and $S_{HL} := (\mathcal{F}(U), \cap, \cup, \sim, HL, LH)$.

3.1 Serial RFSAs

In a serial RFSA S_L, $L(A)$ is a subset of $H(A)$ for all $A \in \mathcal{F}(U)$ and L and H map any constant fuzzy set into itself. We then have the following relationships between the approximation operators:

$$LL(A) \subseteq LH(A)(\text{and } HL(A)) \subseteq HH(A), \quad A \in \mathcal{F}(U). \quad (5)$$

By Theorem 2 we can obtain the following Theorem 3.

Theorem 3. *If S_L is a serial RFSA, then S_{LL} is also a serial RFSA.*

3.2 Reflexive RFSAs

In a reflexive RFSA S_L, L and H resp. satisfy axioms (FL5) and (FH5). It is easy to observe that LL and HH also obey axioms (FL5) and (FH5) resp., thus by using Theorem 2 we can establish following Theorem 4.

Theorem 4. *If S_L is a reflexive RFSA, then S_{LL} is also a reflexive RFSA.*

We have the following relationships between the composed approximation operators which are stronger than those in Eq.(5):

$$\begin{aligned} &LL(A) \subseteq L(A) \subseteq A \subseteq H(A) \subseteq HH(A), \quad A \in \mathcal{F}(U), \\ &LL(A) \subseteq L(A) \subseteq LH(A) \subseteq H(A) \subseteq HH(A), \quad A \in \mathcal{F}(U), \\ &LL(A) \subseteq L(A) \subseteq HL(A) \subseteq H(A) \subseteq HH(A), \quad A \in \mathcal{F}(U). \end{aligned} \tag{6}$$

Together with the monotonicity of L and H, axioms (FL5) and (FH5) imply the following properties: $\forall A, B \in \mathcal{F}(U)$,

$$\begin{aligned} A \subseteq L(B) \quad &\Longrightarrow L(A) \subseteq L(B), \\ H(A) \subseteq B \quad &\Longrightarrow H(A) \subseteq H(B), \\ L(A) \subseteq B \quad &\Longrightarrow L(A) \subseteq H(B), \\ A \subseteq H(B) \quad &\Longrightarrow L(A) \subseteq H(B), \\ L(A) \subseteq L(B) &\Longrightarrow L(A) \subseteq B, \\ H(A) \subseteq H(B) &\Longrightarrow A \subseteq H(B), \\ H(A) \subseteq L(B) &\Longrightarrow A \subseteq L(B),\ H(A) \subseteq B. \end{aligned}$$

3.3 Symmetric RFSAs

In a symmetric RFSA S_L, approximation operators L and H resp. obey (FL6) and (FH6). From [23] we know that operator L in a symmetric RFSA can be equivalently characterized by axioms (FD), (FL1), (FL2), and (FL6) (or equivalently, H can be characterized by axioms (FD), (FH1), (FH2), and (FH6)), that is, axioms (FLC) and (FHC) can be omitted.

Property 1. If S_L is a symmetric RFSA, then

$$HLH(A) = H(A), \quad LHL(A) = L(A),\ \forall A \in \mathcal{F}(U). \tag{7}$$

Proof. Since S_L is a symmetric RFSA, $A \subseteq LH(A)$ holds for all $A \in \mathcal{F}(U)$ and in terms of the monotonicity of H we then obtain

$$H(A) \subseteq HLH(A),\ \forall A \in \mathcal{F}(U). \tag{8}$$

On the other hand, replacing A with $H(A)$ in $HL(A) \subseteq A$ we have

$$HLH(A) \subseteq H(A),\ \forall A \in \mathcal{F}(U). \tag{9}$$

Thus

$$HLH(A) = H(A),\ \forall A \in \mathcal{F}(U). \tag{10}$$

Likewise, we can conclude

$$LHL(A) = L(A),\ \forall A \in \mathcal{F}(U). \tag{11}$$

Property 2. If S_L is a symmetric RFSA, then, $\forall A, B \in \mathcal{F}(U)$,

$$H(A) \subseteq B \Longleftrightarrow A \subseteq L(B). \tag{12}$$

Proof. Since S_L is a symmetric RFSA, for any $A, B \in \mathcal{F}(U)$, by the monotonicity of H and the duality of L and H we have

$$\begin{aligned} H(A) \subseteq B &\Longleftrightarrow \sim L(\sim A) \subseteq B \Longleftrightarrow \sim B \subseteq L(\sim A) \\ &\Longrightarrow H(\sim B) \subseteq HL(\sim A) \subseteq \sim A \Longrightarrow A \subseteq \sim H(\sim B) = L(B). \end{aligned}$$

Hence

$$H(A) \subseteq B \Longrightarrow A \subseteq L(B). \tag{13}$$

On the other hand, we have

$$A \subseteq L(B) \Longleftrightarrow A \subseteq \sim H(\sim B) \Longleftrightarrow H(\sim B) \subseteq \sim A \Longrightarrow LH(\sim B) \subseteq L(\sim A)$$
$$\Longleftrightarrow \sim HL(B) \subseteq L(\sim A) = \sim H(A) \Longrightarrow H(A) \subseteq HL(B) \subseteq B.$$

Thus we conclude Eq.(12).

Theorem 5. *If S_L is a symmetric RFSA, then S_{LL} is also a symmetric RFSA.*

Proof. From Theorem 2 we see that S_{LL} is a RFSA. By axiom (FL6) we have

$$H(A) \subseteq LHH(A), \quad \forall A \in \mathcal{F}(U). \tag{14}$$

By the monotonicity of L, it follows that

$$LH(A) \subseteq LLHH(A), \quad \forall A \in \mathcal{F}(U). \tag{15}$$

Since $A \subseteq LH(A)$, we have

$$A \subseteq LLHH(A), \quad \forall A \in \mathcal{F}(U). \tag{16}$$

Consequently, by the duality of L and H, we conclude that

$$HHLL(A) \subseteq A, \quad \forall A \in \mathcal{F}(U). \tag{17}$$

Thus operators LL and HH resp. obey axioms (FL6) and (FH6). Therefore, S_{LL} is a symmetric RFSA.

3.4 Transitive RFSAs

In a transitive RFSA S_L, L and H resp. obey axioms (FL7) and (FH7). We then have

$$HHHH(A) \subseteq HHH(A) \subseteq HH(A), \quad \forall A \in \mathcal{F}(U). \tag{18}$$

That is, HH obeys axiom (FH7), thus in terms of Theorem 2 we obtain following Theorem 6.

Theorem 6. *If S_L is a transitive RFSA, then S_{LL} is also a transitive RFSA.*

By monotonicity and axioms (FL7) and (FH7) we have properties:

$$\begin{array}{ll} L(A) \subseteq B \Longrightarrow L(A) \subseteq L(B), & A, B \in \mathcal{F}(U), \\ A \subseteq H(B) \Longrightarrow H(A) \subseteq H(B), & A, B \in \mathcal{F}(U). \end{array} \tag{19}$$

3.5 Euclidean RFSAs

In an Euclidean RFSA S_L, L and H resp. obey axioms (FL8) and (FH8). By axiom (FL8) we have

$$HLL(A) \subseteq L(A), \quad \forall A \in \mathcal{F}(U). \tag{20}$$

Then by the monotonicity of H and axiom (FL8), it follows that

$$HHLL(A) \subseteq HLL(A) \subseteq LL(A), \quad \forall A \in \mathcal{F}(U). \tag{21}$$

Similarly, we have

$$HH(A) \subseteq LLHH(A), \quad \forall A \in \mathcal{F}(U). \tag{22}$$

Therefore, in terms of Theorem 2 we can obtain following Theorem 7.

Theorem 7. *Suppose that S_L is an Euclidean RFSA. Then S_{LL} is also an Euclidean RFSA.*

It is easy to verify that approximation operators L and H in an Euclidean RFSA S_L have the following properties:

$$\begin{array}{l} HL(A) \subseteq LL(A), \quad HH(A) \subseteq LH(A), \quad A \in \mathcal{F}(U), \\ A \subseteq L(B) \Longrightarrow H(A) \subseteq L(B), \quad A, B \in \mathcal{F}(U), \\ H(A) \subseteq B \Longrightarrow H(A) \subseteq L(B), \quad A, B \in \mathcal{F}(U). \end{array} \tag{23}$$

3.6 Serial and Symmetric RFSAs

Operator L in a serial and symmetric RFSA S_L can be characterized by axioms (FD), (FL1), (FL2), (FL0), and (FL6). It is easy to verify that

$$LL(A) \subseteq HL(A) \subseteq A \subseteq LH(A) \subseteq HH(A), \quad \forall A \in \mathcal{F}(U). \tag{24}$$

Thus from Theorems 2 and 5 we conclude that a serial and symmetric RFSA S_L produces a reflexive and symmetric RFSA S_{LL}. Moreover, if S_L is a reflexive RFSA, we have the following relationship: $\forall A \in \mathcal{F}(U)$,

$$LL(A) \subseteq L(A) \subseteq HL(A) \subseteq A \subseteq LH(A) \subseteq H(A) \subseteq HH(A). \tag{25}$$

In such a case, for each fuzzy set $A \in \mathcal{F}(U)$ three systems S_{HL}, S_L, and S_{LL} produce a nested family of approximations.

3.7 Serial and Transitive RFSAs

Operator L in a serial and transitive RFSA S_L is characterized by axioms (FLC), (FD), (FL0), (FL1), (FL2), and (FL7), thus S_{LL} is also a serial and transitive RFSA. It is easy to verify that

$$\begin{array}{ll} L(A) \subseteq LL(A) \subseteq HL(A) \subseteq HH(A) \subseteq H(A), & A \in \mathcal{F}(U), \\ L(A) \subseteq LL(A) \subseteq LH(A) \subseteq HH(A) \subseteq H(A), & A \in \mathcal{F}(U). \end{array} \tag{26}$$

Moreover, if S_L is a reflexive RFSA, L and H obey axioms:

$$\text{(FL9) } L(A) = LL(A), \ \forall A \in \mathcal{F}(U), \quad \text{(FH9) } H(A) = HH(A), \ \forall A \in \mathcal{F}(U).$$

In such a case, two systems S_L and S_{LL} become the same one. A reflexive RFSA is a serial one and thus operators L and H resp. obey axioms (FL0) and (FH0). It should be noted that axioms (FL0), (FL2), (FL5), and (FL9) of L, and (FH0), (FH2), (FH5), and (FH9) of H are the axioms of interior and closure operators of a fuzzy topological space. Such an algebra is thus referred to as a topological RFSA. With the topological RFSA, a fuzzy set A is said to be open if $L(A) = A$, and closed if $H(A) = A$. It follows from axioms (FL9) and (FH9) that L and H resp. map any fuzzy set into an open fuzzy set and a closed fuzzy set.

Operators L and H in a reflexive and transitive RFSA S_L have relationships:

$$\begin{array}{ll} L(A) = LL(A) \subseteq A \subseteq HH(A) = H(A), & A \in \mathcal{F}(U), \\ L(A) = LL(A) \subseteq HL(A) \subseteq HH(A) = H(A), & A \in \mathcal{F}(U), \\ L(A) = LL(A) \subseteq LH(A) \subseteq HH(A) = HA, & A \in \mathcal{F}(U). \end{array} \tag{27}$$

3.8 Serial and Euclidean RFSAs

Operator L in a serial and Euclidean RFSA S_L is characterized by axioms (FLC), (FD), (FL0), (FL1), (FL2), and (FL8), then S_{LL} is also a serial and Euclidean RFSA. In terms of Eq.(23) it is easy to verify that

$$LL(A) = HL(A) \subseteq L(A) \subseteq H(A) \subseteq LH(A) = HH(A), \quad \forall A \in \mathcal{F}(U). \quad (28)$$

In such a case, two systems S_{LL} and S_{HL} become the same one and thus S_{HL} is also a serial and Euclidean RFSA.

Moreover, if S_L is a reflexive rough set algebra, then

$$HL(A) \subseteq L(A) \subseteq A, \quad A \subseteq H(A) \subseteq LH(A), \quad \forall A \in \mathcal{F}(U). \quad (29)$$

From Eq.(29) we see that L and H resp. obey axioms (FL6) and (FH6), thus S_L is a symmetric RFSA. On the other hand, by the monotonicity of H, axioms (FH8), (FL8), and (FL5) we have

$$HH(A) \subseteq HLH(A) \subseteq LH(A) \subseteq H(A), \quad \forall A \in \mathcal{F}(U). \quad (30)$$

That is, H obeys axiom (FH7), therefore S_L is a transitive RFSA. Thus we conclude that a reflexive and Euclidean RFSA is a Pawlak RFSA. It is easy to observe

$$LL(A) = HL(A) = L(A) \subseteq A \subseteq H(A) = LH(A) = HH(A), \forall A \in \mathcal{F}(U). \quad (31)$$

In such a case, three systems S_L, S_{LL}, and S_{HL} become the same RFSA.

3.9 Symmetric and Transitive RFSAs

Operator L in a symmetric and transitive RFSA S_L is characterized by axioms (FD), (FL1), (FL2), (FL6), and (FL7), the system S_{LL} is also a symmetric and transitive RFSA. By axioms (FL6) and (FL7) we have

$$H(A) \subseteq LHH(A) \subseteq LH(A), \ \forall A \in \mathcal{F}(U). \quad (32)$$

That is, H obeys axiom (FH8), therefore S_L is an Euclidean RFSA. Hence we have following relationships:

$$HL(A) \subseteq L(A) \subseteq LL(A), \quad HH(A) \subseteq H(A) \subseteq LH(A), \ \forall A \in \mathcal{F}(U). \quad (33)$$

Moreover, if S_L is a serial rough fuzzy algebra, we have

$$A \subseteq LH(A) \subseteq HH(A) \subseteq H(A), \ \forall A \in \mathcal{F}(U). \quad (34)$$

That is, S_L is a reflexive RFSA. Therefore, S_L is a Pawlak RFSA and three systems S_L, S_{LL}, and S_{HL} become the same one.

3.10 Symmetric and Euclidean RFSAs

Operator L in a symmetric and Euclidean RFSA S_L is characterized by axioms (FD), (FL1), (FL2), (FL6), and (FL8), the system S_{LL} is also a symmetric and Euclidean RFSA. By axioms (FL6), (FH6), and (FL8) we have

$$L(A) \subseteq LHL(A) \subseteq LL(A), \quad \forall A \in \mathcal{F}(U). \quad (35)$$

That is, S_L is a transitive RFSA. Moreover, if S_L is a serial RFSA, then from Subsection 3.9 we see that S_L is a Pawlak RFSA and the three systems S_L, S_{LL}, and S_{HL} become a same RFSA.

3.11 Transitive and Euclidean RFSAs

Operator L in a transitive and Euclidean RFSA S_L is characterized by axioms (FLC), (FD), (FL1), (FL2), (FL7), and (FL8), the system S_{LL} is also a transitive and Euclidean RFSA. Obviously, Eq.(33) holds. Moreover, if S_L is a serial RFSA, it is not difficult to prove

$$LL(A) = L(A) = HL(A), \quad HH(A) = H(A) = LH(A), \quad A \in \mathcal{F}(U). \qquad (36)$$

Therefore, three systems S_L, S_{LL}, and S_{HL} become the same RFSA. Only if S_L is a reflexive or a symmetric RFSA can S_{LL} be a Pawlak RFSA.

4 Conclusion

A rough set algebra is a set algebra with two additional approximation operators. Under this point of view, rough set theory may be treated as an extension of set theory. In this paper we have studied various types of RFSAs and examined their properties. We have proved that if $(\mathcal{F}(U), \cap, \cup, \sim, L, H)$ is a (a serial, a reflexive, a symmetric, a transitive, an Euclidean, a Pawlak, resp.) RFSA then $(\mathcal{F}(U), \cap, \cup, \sim, LL, HH)$ is also a (a serial, a reflexive, a symmetric, a transitive, an Euclidean, a Pawlak, resp.) RFSA. We have also investigated compositions of different types of RFSAs. This work may be viewed as an extension of Yao [26] and it may also be treated as a completion of Wu and Zhang [23]. The analysis will facilitate further research in uncertain reasoning under fuzziness.

Acknowledgement

This work was supported by a grant from the National Natural Science Foundation of China (No. 60373078).

References

1. Boixader, D., Jacas, J., Recasens, J.: Upper and lower approximations of fuzzy sets. International Journal of General Systems **29**(2000) 555–568
2. Chuchro, M.: On rough sets in topological Boolean algebras. In: Ziarko, W.(ed.): Rough Sets, Fuzzy Sets and Knowledge Discovery. Springer-Verlag, Berlin, 1994, pp.157–160
3. Comer, S.: An algebraic approach to the approximation of information. Fundamenta Informaticae **14**(1991) 492–502
4. Dubois, D., Prade, H.: Rough fuzzy sets and fuzzy rough sets. International Journal of General Systems **17**(1990) 191–209
5. Inuiguchi, M.: Generalizations of rough sets: from crisp to fuzzy cases. In: Tsumoto, S., Slowinski, R., Komorowski, J., et al.(eds.): RSCTC 2004, LNAI 3066, pp.26–37
6. Lin, T.Y., Liu, Q.: Rough approximate operators: axiomatic rough set theory. In: Ziarko, W.(ed.): Rough Sets, Fuzzy Sets and Knowledge Discovery. Springer, Berlin, 1994, pp.256–260
7. Mi, J.-S., Leung, Y., Wu, W.-Z.: An uncertainty measure in partition-based fuzzy rough sets. International Journal of General Systems **34**(2005) 77–90

8. Mi, J.-S., Zhang, W.-X.: An axiomatic characterization of a fuzzy generalization of rough sets. Information Sciences **160**(2004) 235–249
9. Morsi, N.N., Yakout, M.M.: Axiomatics for fuzzy rough sets. Fuzzy Sets and Systems **100**(1998) 327–342
10. Nanda, S., Majumda, S.: Fuzzy rough sets. Fuzzy Sets and Systems **45**(1992) 157–160
11. Pawlak, Z.: Rough Sets: Theoretical Aspects of Reasoning about Data. Kluwer Academic Publishers, Boston, 1991
12. Pei, D.W., Xu, Z.-B.: Rough set models on two universes. International Journal of General Systems **33**(2004) 569–581
13. Pomykala, J.A.: Approximation operations in approximation space. Bulletin of the Polish Academy of Sciences: Mathematics **35**(1987) 653–662
14. Radzikowska, A.M., Kerre, E.E.: A comparative study of fuzzy rough sets. Fuzzy Sets and Systems **126**(2002) 137–155
15. Slowinski, R., Vanderpooten, D.: A generalized definition of rough approximations based on similarity. IEEE Transactions on Knowledge and Data Engineering **12**(2000) 331–336
16. Vakarelov, D.: A modal logic for similarity relations in Pawlak knowledge representation systems. Fundamenta Informaticae **15**(1991) 61–79
17. Wiweger, R.: On topological rough sets. Bulletin of Polish Academy of Sciences: Mathematics **37**(1989) 89–93
18. Wu, W.-Z.: A study on relationship between fuzzy rough approximation operators and fuzzy topological spaces. In: Wang, J., Jin, Y.(eds.): FSKD 2005, LNAI 3613, pp.167–174
19. Wu, W.-Z., Leung, Y., Mi, J.-S.: On characterizations of $(\mathcal{I}, \mathcal{T})$-fuzzy rough approximation operators. Fuzzy Sets and Systems **154**(2005) 76–102
20. Wu, W.-Z., Leung, Y., Zhang, W.-X.: Connections between rough set theory and Dempster-Shafer theory of evidence. International Journal of General Systems **31**(2002) 405–430
21. Wu, W.-Z., Mi, J.-S., Zhang, W.-X.: Generalized fuzzy rough sets. Information Sciences **151**(2003) 263–282
22. Wu, W.-Z., Zhang, W.-X.: Neighborhood operator systems and approximations. Information Sciences **144**(2002) 201–217
23. Wu, W.-Z., Zhang, W.-X.: Constructive and axiomatic approaches of fuzzy approximation operators. Information Sciences **159**(2004) 233–254
24. Yao, Y.Y.: Two views of the theory of rough sets in finite universes. International Journal of Approximate Reasoning **15**(1996) 291–317
25. Yao, Y.Y.: Combination of rough and fuzzy sets based on α-level sets. In: Lin, T.Y., Cercone, N.(eds.): Rough Sets and Data Mining: Analysis for Imprecise Data. Kluwer Academic Publishers, Boston, 1997, pp.301–321
26. Yao, Y.Y.: Constructive and algebraic methods of the theory of rough sets. Journal of Information Sciences **109**(1998) 21–47
27. Yao, Y.Y.: Generalized rough set model. In: Polkowski, L., Skowron, A.(eds.): Rough Sets in Knowledge Discovery 1. Methodology and Applications. Physica-Verlag, Heidelberg, 1998, pp.286–318
28. Yao, Y.Y.: Relational interpretations of neighborhood operators and rough set approximation operators. Information Sciences **111**(1998) 239–259
29. Yao, Y.Y., Lin, T.Y.: Generalization of rough sets using modal logic. Intelligent Automation and Soft Computing: An International Journal **2**(1996) 103–120
30. Yao, Y.Y., Lingras, P.J.: Interpretations of belief functions in the theory of rough sets. Information Sciences **104**(1998) 81–106

On Reduction of Morphological Covering Rough Sets

Tingquan Deng[1] and Yanmei Chen[2]

[1] College of Science, Harbin Engineering University, Harbin 150001 P.R. China
[2] Department of Mathematics, Harbin Institute of Technology, Harbin 150001 P.R. China
{Tq_Deng, Chen.Yanmei}@163.com

Abstract. Both rough set theory and mathematical morphology originated from the classical set theory, and are characterized by using dual operators sharing similar properties. In this paper, a notion of morphological covering rough set is introduced by means of combining rough set with fundamental morphological operators. The focus of this paper is on constructing a covering and a pair of morphological rough approximation operators. The issue on reduction of a morphological covering is explored and the procedure for generating the reduction is presented. Necessary and sufficient conditions on characterization of the reduction are proved.

Keywords: Rough set, morphological operators, morphological covering, adjunction, reduction.

1 Introduction

Rough set theory [5] was introduced, as an extension of classical set theory, to deal with incomplete data in intelligent information systems. The main purpose of rough set is to mine much more deterministic information by locating and classifying objects in a knowledge representation system with an indiscernible relation on it. Many researchers successively extended the classical definition of rough set (Pawlak rough set) from an equivalence relation to a general mathematical object, a similarity relation, a topological neighborhood structure or a covering, and developed various generalizations of rough set [8,10,11,12].

Mathematical morphology [4] was originally developed as a tool for studying porous media. Later, it evolved to a general theory of shape analysis, and was subsequently enriched and popularized by [7] for image analysis, in particular for those applications where geometric aspects are relevant. Its basic principle is probing and analyzing spatial structure and characteristic of an object with a small geometric template known as the structuring element. Mathematical morphology has been extended from the classical set theory to grey-scale images and multi-scale images [3,9].

[1] This work was supported by the postdoctoral science-research developmental foundation of Heilongjiang province (LBH-Q05047) and the fundamental research foundation of Harbin Engineering University (2005).

L. Wang et al. (Eds.): FSKD 2006, LNAI 4223, pp. 266–275, 2006.

Rough set and mathematical morphology share their corresponding essentials (a binary relation and a structuring element). Although both of them are distinct techniques in data analysis and processing, there are many analogies in their appearances and algebraic properties. The combined study of them is an appealing topic and can find broad applications for the two theories. To our knowledge, the only work that puts them together owns to Polkowski LT [6] and Bloch I [1]. This paper combines the foundations of mathematical morphology with rough set and introduces a concept called morphological covering rough set in a general universe. A covering is generated from a morphological dilation, which is used for replacing the indiscernible relation in Pawlak rough approximation space. The reduction of the morphological covering is studied. Necessary and sufficient conditions on characterization of the reduction are proved.

2 Pawlak Rough Approximations

Let E be a nonempty set (finite or infinite), called the universe of discourse, and R be an equivalence relation on it. For $x \in E$, $[x]_R = \{y \in E \mid (x\,, y) \in R\}$ is the equivalence class of x with respect to R, in which all elements are indiscernible. R is therefore called an indiscernible relation and $(\mathcal{P}(E)\,, R)$ is called a Pawlak rough approximation space, where $\mathcal{P}(E)$ is the power set of E. The definition of rough set was introduced by Pawlak as follows.

Definition 1. *Assume that $(\mathcal{P}(E)\,, R)$ is a Pawlak rough approximation space, for $X \in \mathcal{P}(E)$, let $\underline{R}(X) = \{x \in E \mid [x]_R \subseteq X\}$ and $\overline{R}(X) = \{x \in E \mid [x]_R \cap X \neq \emptyset\}$, then $\underline{R}(X)$ and $\overline{R}(X)$ are called the Pawlak rough lower approximation and Pawlak rough upper approximation, respectively. $(\underline{R}(X)\,, \overline{R}(X))$ characterizes the rough approximation of X in $(\mathcal{P}(E)\,, R)$. If $\underline{R}(X) = \overline{R}(X)$, X is called definable.*

The Pawlak rough approximations of $X \in \mathcal{P}(E)$ can also be rewritten as $\underline{R}(X) = \cup\{[x]_R \in \mathcal{P}(E) \mid [x]_R \subseteq X\}$ and $\overline{R}(X) = \cup\{[x]_R \in \mathcal{P}(E) \mid [x]_R \cap X \neq \emptyset\}$, which fulfill many algebraic properties.

Theorem 1. *Let $(\mathcal{P}(E)\,, R)$ be a Pawlak rough approximation space, then*

(1) $\underline{R}(\emptyset) = \overline{R}(\emptyset) = \emptyset$, $\underline{R}(E) = \overline{R}(E) = E$;
(2) $\underline{R}(X^c) = (\overline{R}(X))^c$, *where Y^c denotes the complement of Y;*
(3) $\underline{R}$ *and* $\overline{R}$ *are increasing;*
(4) $\underline{R}$ *is anti-extensive, whereas* $\overline{R}$ *is extensive, i.e.,* $\underline{R} \subseteq \mathrm{id} \subseteq \overline{\mathrm{R}}$, *where* id *denotes the identity operator on* $\mathcal{P}(E)$;
(5) $\underline{R}$ *and* $\overline{R}$ *are idempotent, i.e.,* $\underline{R}\underline{R} = \underline{R}$ *and* $\overline{R}\overline{R} = \overline{R}$;
(6) $\overline{R}\underline{R} = \underline{R}$, $\underline{R}\overline{R} = \overline{R}$;
(7) $\underline{R}(X \cap Y) = \underline{R}(X) \cap \underline{R}(Y)$, $\overline{R}(X \cup Y) = \overline{R}(X) \cup \overline{R}(Y)$;
(8) *For any* $x \in E$, $\underline{R}([x]_R) = \overline{R}([x]_R) = [x]_R$;
(9) *For any* $x \in E$, $\underline{R}(\{x\}^c) = ([x]_R)^c$ *and* $\overline{R}(\{x\}) = [x]_R$.

Theorem 1 lays down the foundations of rough set theory and its applications. In practice, an equivalence relation limits applications of Pawlak rough set. In the study of image segmentation, for example, the transitivity of a relation is not satisfied when the relation is determined by the compatibility of pixels. Neither the symmetry nor the transitivity of a relation is easily realized when the relation is concerned with the characterization of topological neighborhood structures in image analysis. Several extensions of rough set have been proposed by means of substituting a non-equivalence relation for the indiscernible relation in Pawlak rough approximation space [2,8,10,11,12]. The following sections focus on the combined study of rough set with basic morphological operators.

3 Rough Set Based on Morphology

3.1 Fundamental Morphological Operators

Definition 2. *Let E be a nonempty set and let $\mathcal{L} = \mathcal{P}(E)$, if there are two operators $\delta : \mathcal{L} \to \mathcal{L}$ and $\varepsilon : \mathcal{L} \to \mathcal{L}$ satisfying*

$$\delta(X) \subseteq Y \iff X \subseteq \varepsilon(Y) \tag{1}$$

for arbitrary $X, Y \in \mathcal{L}$, the pair (ε, δ) is called an adjunction on $\mathcal{L}$.

Definition 3. *An operator $\delta : \mathcal{L} \to \mathcal{L}$ is called a dilation on $\mathcal{L}$ if*

$$\delta(\cup_i X_i) = \cup_i \delta(X_i) \tag{2}$$

for any family $\{X_i\} \subseteq \mathcal{L}$. An operator $\varepsilon : \mathcal{L} \to \mathcal{L}$ is called an erosion on $\mathcal{L}$ if for any family $\{Y_i\} \subseteq \mathcal{L}$,

$$\varepsilon(\cap_i Y_i) = \cap_i \varepsilon(Y_i)\,. \tag{3}$$

It is easily verified that if a pair (ε, δ) is an adjunction on $\mathcal{L}$, δ is a dilation and ε is an erosion. On the other hand, given a dilation δ (an erosion ε, resp.) on $\mathcal{L}$, there is a unique erosion ε (a unique dilation δ, resp.) on $\mathcal{L}$ such that the pair (ε, δ) forms an adjunction on $\mathcal{L}$. Furthermore, the compositions of δ and ε, separatively, generate a morphologically algebraic opening (an increasing, idempotent and anti-extensive operator) $\delta\varepsilon$ and a morphologically algebraic closing (an increasing, idempotent and extensive operator) $\varepsilon\delta$ on $\mathcal{L}$ [3].

In what follows, every dilation δ on $\mathcal{L}$ is supposed to satisfy the premise $\cup_{X \in \mathcal{L}} \delta(X) = \mathcal{L}$.

Assume that δ is a set mapping on $\mathcal{L}$, for $X \in \mathcal{L}$, let $\delta^*(X) = \cup_{x \in X} \delta^*(\{x\})$, where $\delta^*(\{x\}) = \{y \in E \mid x \in \delta(\{y\})\}$, δ^* is called the inverse of δ. If $\delta^* = \delta$, δ is called symmetric. δ is called extensive if $X \subseteq \delta(X)$ for any $X \in \mathcal{P}(E)$. Note the definition of δ^*, the following results can be proved easily.

Proposition 1. *(1) For any set mapping δ on $\mathcal{L}$, $(\delta^*)^* = \delta$.*
(2) δ is a dilation on $\mathcal{L}$ if and only if δ^* is a dilation on $\mathcal{L}$.

Example 1. Let E be a nonempty universe, δ be a mapping on $\mathcal{P}(E)$, and let $A(x) = \delta(\{x\})$ and $A^*(x) = \delta^*(\{x\}) = \{y \in E \mid x \in \delta(\{y\})\}$, $x \in E$, then for any $X \in \mathcal{P}(E)$, $\delta(X) = \cup_{x \in X} A(x) = \{x \in E \mid A^*(x) \cap X \neq \emptyset\}$ is a dilation of

X on $\mathcal{P}(E)$. Meanwhile, an erosion ε on $\mathcal{P}(E)$ can be given by $\varepsilon(X) = \{x \in E \mid A(x) \subseteq X\}$, which shares the adjunction relation with δ. Clearly, δ is symmetric (extensive, resp.) if and only if $A^*(x) = A(x)$ ($x \in A(x)$, resp.) for all $x \in E$. If δ is extensive on $\mathcal{P}(E)$, the equation $\cup_{X \in \mathcal{P}(E)} \delta(X) = E$ holds naturally.

Example 2. Let $E = \mathbf{R}^n$, where $\mathbf{R}^n$ denotes the n-dimensional Euclidean space, given a nonempty set $A \in \mathcal{P}(E)$, it is true that $\delta(X) = X \oplus A = \{x + a \mid x \in X, a \in A\} = \cup_{a \in A} X_a$ is a dilation, whereas $\varepsilon(X) = X \ominus A = \cap_{a \in A} X_{-a}$ is an erosion, and (ε, δ) is an adjunction on $\mathcal{P}(E)$, where $X_a = \{x + a \mid x \in X\}$ is the translation of X along $a \in E$, $\oplus$ and $\ominus$ denote the Minkowski set addition and subtraction, respectively. The A is called a structuring element in mathematical morphology or a template in image processing. As a rule, there are many choices for the structuring element such an area in $\mathbf{R}^2$ with the structure as a square, a rectangle, a cross, a circle, or a segment with a determined direction. Let o be the origin of $\mathbf{R}^n$, if $o \in A$, δ is extensive. δ is symmetric if $\check{A} = A$, where $\check{A} = \{-a \mid a \in A\}$ is the reflection of A.

Example 3. Let $E = \{a_1, a_2, a_3\}$, define an operator δ on $\mathcal{P}(E)$ as: $\delta(\{a_1\}) = \delta(\{a_1, a_2\}) = \{a_1, a_2\}$, $\delta(\{a_2\}) = \{a_1\}$, $\delta(\{a_3\}) = \{a_2, a_3\}$, and $\delta(\{a_1, a_3\}) = \delta(\{a_2, a_3\}) = \delta(E) = E$, then δ is a non-extensive and non-symmetric dilation on $\mathcal{P}(E)$. An erosion ε on $\mathcal{P}(E)$ corresponding to δ is given by $\varepsilon(\emptyset) = \varepsilon(\{a_2\}) = \varepsilon(\{a_3\}) = \emptyset$, $\varepsilon(\{a_1\}) = \varepsilon(\{a_1, a_3\}) = \{a_2\}$, $\varepsilon(\{a_1, a_2\}) = \{a_1, a_2\}$, $\varepsilon(\{a_2, a_3\}) = \{a_3\}$, which satisfies the adjunction with δ on $\mathcal{P}(E)$.

To conclude this subsection, it is interesting to investigate the relationship between a dilation, a binary relation and a covering.

Let E be a nonempty universe and δ be a dilation on $\mathcal{P}(E)$, then $R = \{(x, y) \mid y \in \delta(\{x\}), x \in E\}$ is a binary relation on E, called the induced relation of δ. Evidently, R is reflexive if and only if δ is extensive, R is symmetric when and only when δ is symmetric, and R is transitive if and only if $\delta^2(X) \subseteq \delta(X)$ for every $X \in \mathcal{P}(E)$. Let $C(x) = \{y \in E \mid (x, y) \in R\}$ and $C^*(x) = \{y \in E \mid (y, x) \in R\}$, then $C(x) = \delta(\{x\})$, $C^*(x) = \delta^*(\{x\})$, and both $\mathcal{C} = \{C(x) \mid x \in E\}$ and $\mathcal{C}^* = \{C^*(x) \mid x \in E\}$ are the coverings of E from the premise $\cup_{X \in \mathcal{P}(E)} \delta(X) = E$.

On the other hand, assume that R is a binary relation on E (R in this paper is always supposed to satisfy the assumption that for every $x \in E$, there is $y \in E$ and for every $y \in E$, there is $x \in E$ satisfying $(x, y) \in R$)), we have

Theorem 2. *$\delta(X) = \cup_{x \in X} C(x)$ is a dilation, $\varepsilon(X) = \{x \in E \mid C(x) \subseteq X\}$ is an erosion, and the pair (ε, δ) is an adjunction on $\mathcal{P}(E)$.*

Proof. It is sufficient to prove the statement that (ε, δ) is an adjunction.

Let $X, Y \in \mathcal{P}(E)$, then $\delta(X) \subseteq Y \iff \cup_{x \in X} C(x) \subseteq Y \iff \forall x \in X, C(x) \subseteq Y \iff \forall x \in X, x \in \varepsilon(Y) \iff X \subseteq \varepsilon(Y)$.

3.2 Morphological Covering Rough Set

Let E be a nonempty universe and let δ be a dilation on $\mathcal{P}(E)$, the pair $(\mathcal{P}(E), \delta)$ is called a morphological rough approximation space.

Definition 4. *Let $(\mathcal{P}(E), \delta)$ be a morphological rough approximation space, for $X \in \mathcal{P}(E)$, its morphological rough lower approximation and morphological rough upper approximation are defined by*

$$\underline{\delta}(X) = \cup_{x \in E}\{\delta(\{x\}) \mid \delta(\{x\}) \subseteq X\} \tag{4}$$

and

$$\overline{\delta}(X) = \{x \in E \mid x \in \delta^*(\{y\}) \Rightarrow \delta^*(\{y\}) \cap X \neq \emptyset , y \in E\} , \tag{5}$$

respectively. The pair $(\underline{\delta}(X), \overline{\delta}(X))$ characterizes the rough approximation of X in $(\mathcal{P}(E), \delta)$. If $\underline{\delta}(X) = X$ ($\overline{\delta}(X) = X$, resp.), X is called lower (upper, resp.) definable, and if $\underline{\delta}(X) = \overline{\delta}(X)$, X is called definable.

From Definition 4, it is trivial that both the empty set $\emptyset$ and the whole universe E are definable for any dilation δ on $\mathcal{P}(E)$.

In Example 3, a straightforward computation shows that $\underline{\delta}(\emptyset) = \underline{\delta}(\{a_2\}) = \underline{\delta}(\{a_3\}) = \emptyset$, $\underline{\delta}(\{a_1\}) = \underline{\delta}(\{a_1, a_3\}) = \{a_1\}$, $\underline{\delta}(\{a_1, a_2\}) = \{a_1, a_2\}$, $\underline{\delta}(\{a_2, a_3\}) = \{a_2, a_3\}$, $\underline{\delta}(E) = E$; $\overline{\delta}(\emptyset) = \emptyset$, $\overline{\delta}(\{a_1\}) = \overline{\delta}(\{a_1, a_2\}) = \{a_1, a_2\}$, $\overline{\delta}(\{a_2\}) = \{a_2\}$, $\overline{\delta}(\{a_3\}) = \{a_3\}$, and $\overline{\delta}(\{a_1, a_3\}) = \overline{\delta}(\{a_2, a_3\}) = \overline{\delta}(E) = E$. Thus $X = \{a_1, a_2\}$ is a definable set and all of the other nontrivial subsets of E are not definable in $(\mathcal{P}(E), \delta)$. Both $\{a_1\}$ and $\{a_2, a_3\}$ are lower definable, whereas $\{a_2\}$ and $\{a_3\}$ are upper definable.

The following proposition shows the relationship between the morphological rough lower approximation and the rough upper approximation.

Proposition 2. *Let $(\mathcal{P}(E), \delta)$ be a morphological rough approximation space, for any $X \in \mathcal{P}(E)$, $\underline{\delta}(X^c) = (\overline{\delta^*}(X))^c$. In particular, if δ is symmetric, $\underline{\delta}$ and $\overline{\delta}$ are dual.*

Proof. Let $X \in \mathcal{P}(E)$ and $x \in E$, then $x \in (\overline{\delta^*}(X))^c \iff \exists y \in E, \ni x \in (\delta^*)^*(\{y\}) = \delta(\{y\})$ and $\delta(\{y\}) \cap X = \emptyset \iff \exists y \in E, \ni x \in \delta(\{y\})$ and $\delta(\{y\}) \subseteq X^c \iff x \in \underline{\delta}(X^c)$.

Proposition 3. *Let $(\mathcal{P}(E), \delta)$ be a morphological rough approximation space, for any $x \in E$, $\underline{\delta}(\delta(\{x\})) = \delta(\{x\})$, $\overline{\delta}(\delta^*(\{x\})^c) = \delta^*(\{x\})^c$; $\underline{\delta}(\{x\}^c) = \{y \in E \mid \delta^*(\{y\}) \subseteq \delta^*(\{x\})\}^c$, and $\overline{\delta}(\{x\}) = \{y \in E \mid \delta(\{y\}) \subseteq \delta(\{x\})\}$.*

Proof. The inclusion $\underline{\delta}(\delta(\{x\})) \subseteq \delta(\{x\})$ is obvious from the definition of $\underline{\delta}$. The reverse inclusion $\underline{\delta}(\delta(\{x\})) = \cup_{y \in E}\{\delta(\{y\}) \mid \delta(\{y\}) \subseteq \delta(\{x\})\} \supseteq \delta(\{x\})$ is also clear.

For the second equality, $\overline{\delta}(\delta^*(\{x\})^c) = \{y \in E \mid \exists z \in E, y \in \delta^*(\{z\}), \delta^*(\{z\}) \cap (\delta^*(\{x\}))^c = \emptyset\}^c = \{y \in E \mid \exists z \in E, y \in \delta^*(\{z\}), \delta^*(\{z\}) \subseteq \delta^*(\{x\})\}^c = \{y \in E \mid y \in \delta^*(\{x\})\}^c = \delta^*(\{x\})^c$.

$\underline{\delta}(\{x\}^c) = \cup_{y \in E}\{\delta(\{y\}) \mid \delta(\{y\}) \subseteq \{x\}^c\} = \cup_{y \in E}\{\delta(\{y\}) \mid x \notin \delta(\{y\})\} = \cup_{y \in E}\{\delta(\{y\}) \mid y \notin \delta^*(\{x\})\} = \cup\{\delta(\{y\}) \mid y \in \delta^*(\{x\})^c\} = \delta(\delta^*(\{x\})^c) = \{y \in E \mid \delta^*(\{y\}) \subseteq \delta^*(\{x\})\}^c$.

$\overline{\delta}(\{x\}) = \{y \in E \mid y \in \delta^*(\{z\}) \Rightarrow \delta^*(\{z\}) \cap \{x\} \neq \emptyset, \forall z \in E\} = \{y \in E \mid y \in \delta^*(\{z\}) \Rightarrow x \in \delta^*(\{z\}), \forall z \in E\} = \{y \in E \mid z \in \delta(\{y\}) \Rightarrow z \in \delta(\{x\}), \forall z \in E\} = \{y \in E \mid \delta(\{y\}) \subseteq \delta(\{x\})\}$.

Proposition 3 shows that every granule $\delta(\{x\})$ is lower definable, whereas the complement of every granule $\delta^*(\{x\})$ is upper definable in $(\mathcal{P}(E), \delta)$. These properties of granules are very useful in data analysis, especially in image data analysis for locating isolated points and apertures. Besides them, the morphological rough approximation operators have the following algebraic characterizations.

Theorem 3. *Let $(\mathcal{P}(E), \delta)$ be a morphological rough approximation space, then*

(1) $\underline{\delta}$ *and* $\overline{\delta}$ *are increasing;*
(2) $\underline{\delta}$ *is anti-extensive, whereas* $\overline{\delta}$ *is extensive, i.e.,* $\underline{\delta} \subseteq \mathrm{id} \subseteq \overline{\delta}$*;*
(3) $\underline{\delta}$ *and* $\overline{\delta}$ *are idempotent, i.e.,* $\underline{\delta}\underline{\delta} = \underline{\delta}$ *and* $\overline{\delta}\overline{\delta} = \overline{\delta}$*.*

Proof. (1) The monotone of $\underline{\delta}$ and of $\overline{\delta}$ is clear.

(2) The anti-extension property of $\underline{\delta}$ is obvious from the definition of $\underline{\delta}$.

To prove the extension property of $\overline{\delta}$. Let $X \in \mathcal{P}(E)$, if $x \notin \overline{\delta}(X)$, there exists $y \in E$ such that $x \in \delta^*(\{y\})$ and $\delta^*(\{y\}) \cap X = \emptyset$. Therefore $x \notin X$.

(3) From the duality relation between $\underline{\delta}$ and $\overline{\delta}$, it suffices to prove $\underline{\delta} \subseteq \underline{\delta}\underline{\delta}$.

Let $X \in \mathcal{P}(E)$ and $x \in \underline{\delta}(X)$, there is $y \in E$ such that $x \in \delta(\{y\})$ and $\delta(\{y\}) \subseteq X$. Note the monotone of $\underline{\delta}$ and the equalities in Proposition 3, we have $\delta(\{y\}) = \underline{\delta}(\delta(\{y\})) = \underline{\delta}\underline{\delta}(\delta(\{y\})) \subseteq \underline{\delta}\underline{\delta}(X)$, which implies $x \in \underline{\delta}\underline{\delta}(X)$.

Proposition 4. *Let $(\mathcal{P}(E), \delta)$ be a morphological rough approximation space, for all $X, Y \in \mathcal{P}(E)$,*

(1) $\underline{\delta} \subseteq \overline{\delta}\underline{\delta} \subseteq \overline{\delta}$, $\underline{\delta} \subseteq \underline{\delta}\overline{\delta} \subseteq \overline{\delta}$*;*
(2) $\overline{\delta}\underline{\delta}$ *and* $\underline{\delta}\overline{\delta}$ *are increasing and idempotent;*
(3) $\underline{\delta}(X \cap Y) \subseteq \underline{\delta}(X) \cap \underline{\delta}(Y) \subseteq X \cap Y \subseteq \overline{\delta}(X \cap Y) \subseteq \overline{\delta}(X) \cap \overline{\delta}(Y)$ *and* $\underline{\delta}(X) \cup \underline{\delta}(Y) \subseteq \underline{\delta}(X \cup Y) \subseteq X \cup Y \subseteq \overline{\delta}(X) \cup \overline{\delta}(Y) \subseteq \overline{\delta}(X \cup Y)$*;*
(4) *If δ is an extensive dilation on $\mathcal{P}(E)$,* $\varepsilon(X) \subseteq \underline{\delta}(X) \subseteq X \subseteq \overline{\delta}(X) \subseteq \delta(X)$.

Proof. (1) It is straightforward from Theorem 3.

(2) The monotone of $\overline{\delta}\underline{\delta}$ and of $\underline{\delta}\overline{\delta}$ follows from that of $\underline{\delta}$ and of $\overline{\delta}$.

From (1) and the statements in Theorem 3, on the one hand, $\overline{\delta}\underline{\delta} = \overline{\delta}\underline{\delta}\underline{\delta} \subseteq \overline{\delta}\underline{\delta}\overline{\delta}\underline{\delta}$, on the other hand, $\overline{\delta}\underline{\delta}\overline{\delta}\underline{\delta} \subseteq \overline{\delta}\overline{\delta}\underline{\delta} = \overline{\delta}\underline{\delta}$. Therefore, $\overline{\delta}\underline{\delta}$ is idempotent.

The idempotence of $\underline{\delta}\overline{\delta}$ can be proved in the same way.

(3) They are true from the definitions of rough approximations.

(4) For $X \in \mathcal{P}(E)$, the inclusions $\underline{\delta}(X) \subseteq X \subseteq \overline{\delta}(X)$ hold. According to the extension of δ, the inclusions $\varepsilon(X) \subseteq X \subseteq \delta(X)$ hold for every $X \in \mathcal{P}(E)$. Let $x \in \varepsilon(X)$, then $\delta(\{x\}) \subseteq X$, which implies $x \in \underline{\delta}(X)$. Thus $\varepsilon(X) \subseteq \underline{\delta}(X)$.

If $x \notin \delta(X)$, $\delta^*(\{x\}) \cap X = \emptyset$, and so $x \notin \overline{\delta}(X)$ since $x \in \delta^*(\{x\})$. Therefore $\overline{\delta}(X) \subseteq \delta(X)$.

Theorem 4. *Let $(\mathcal{P}(E), \delta)$ be a morphological rough approximation space, then*

(1) $\underline{\delta} = \delta\varepsilon$ *and* $\overline{\delta} = \varepsilon\delta$*;*
(2) *If δ is extensive and transitive,* $\overline{\delta}\underline{\delta} = \underline{\delta}$ *and* $\underline{\delta}\overline{\delta} = \overline{\delta}$*;*
(3) *If δ is extensive, symmetric and transitive, then* $\underline{\delta} = \underline{R}$ *and* $\overline{\delta} = \overline{R}$*, where* $R = \{(x, y) \mid y \in \delta(\{x\}), x \in E\}$.

Proof. (1) They can be proved from straightforward computations.

(2) It is sufficient to prove that $\overline{\delta}\underline{\delta} \subseteq \underline{\delta}$ and $\overline{\delta} \subseteq \underline{\delta}\overline{\delta}$.

From (2) in Theorem 3, we have $\varepsilon\underline{\delta} \subseteq \varepsilon$ and $\varepsilon \subseteq \varepsilon\overline{\delta}$, which imply $\varepsilon\delta\varepsilon = \varepsilon$.

The inclusion $\varepsilon \subseteq \underline{\delta}$ is obvious from the extension of δ. Therefore, $\overline{\delta}\underline{\delta} = \varepsilon\delta\delta\varepsilon \subseteq \varepsilon\delta\varepsilon = \varepsilon \subseteq \underline{\delta}$.

According to the transitivity of δ and the adjunction between δ and ε, it is true that $\delta^2(X) \subseteq \delta(X) \iff \delta(X) \subseteq \varepsilon\delta(X) \iff X \subseteq \varepsilon\varepsilon\delta(X)$ for any $X \in \mathcal{P}(E)$. Therefore, $\underline{\delta}\overline{\delta} = \delta\varepsilon\varepsilon\delta \supseteq \delta$, and so $\underline{\delta}\overline{\delta} \supseteq \delta \supseteq \overline{\delta}$.

(3) Clearly, $R = \{(x\,,y) \mid y \in \delta(\{x\})\,, x \in E\}$ is an equivalence relation on E. Thus for all $x \in E$, $\delta(\{x\}) = [x]_R$, so $\underline{\delta}(X) = \cup\{\delta(\{x\}) \mid \delta(\{x\}) \subseteq X\,, x \in E\} = \underline{R}(X)$.

By using the duality relations between $\underline{R}$ and $\overline{R}$, and between $\underline{\delta}$ and $\overline{\delta}$, $\overline{\delta}(X) = (\underline{\delta}^*(X^c))^c = (\underline{\delta}(X^c))^c = (\underline{R}(X^c))^c = \overline{R}(X)$ for every $X \in \mathcal{P}(E)$.

Theorem 4 shows a straightforward link of rough approximation operators with fundamental morphological operators. It also indicates that both $\underline{\delta}(X)$ and $\overline{\delta}(X)$ for arbitrary set X are definable if δ is an extensive and transitive dilation.

Example 4. Let $E = \{a_1\,, a_2\,, a_3\,, a_4\}$, and let δ be a dilation on $\mathcal{P}(E)$ given by $\delta(\{a_1\}) = \{a_1\,, a_2\}$, $\delta(\{a_2\}) = \{a_2\}$, $\delta(\{a_3\}) = \{a_2\,, a_3\}$, and $\delta(\{a_4\}) = \{a_4\}$, then δ is extensive and transitive (the induced relation $R = \{(a_1\,, a_1)$, $(a_1\,, a_2)$, $(a_2\,, a_2)$, $(a_3\,, a_2)$, $(a_3\,, a_3)$, $(a_4\,, a_4)\}$ is therefore reflexive and transitive), so $\overline{\delta}\underline{\delta} = \underline{\delta}$ and $\underline{\delta}\overline{\delta} = \overline{\delta}$. If δ is defined by $\delta(\{a_1\}) = \delta(\{a_3\}) = \{a_1\,, a_3\}$, $\delta(\{a_2\}) = \{a_2\}$, and $\delta(\{a_4\}) = \{a_4\}$, then the corresponding induced relation $R = \{(a_1\,, a_1)$, $(a_1\,, a_3)$, $(a_2\,, a_2)$, $(a_3\,, a_1)$, $(a_3\,, a_3)$, $(a_4\,, a_4)\}$ is an equivalence one, and therefore, $\overline{\delta}\underline{\delta} = \underline{\delta} = \underline{R}$ and $\underline{\delta}\overline{\delta} = \overline{\delta} = \overline{R}$. In the latter case, the morphological rough approximation space $(\mathcal{P}(E)\,, \delta)$ is identical with the Pawlak one $(\mathcal{P}(E)\,, R)$.

4 Reduction of a Morphological Covering

As shown in Section 3, given a dilation δ on $\mathcal{P}(E)$, the family $\mathcal{C} = \{C(x) = \delta(\{x\}) \mid x \in E\}$ is a covering of E, called a morphological covering. The pair $(\mathcal{P}(E)\,, \mathcal{C})$ is referred to as a covering rough approximation space. In $(\mathcal{P}(E)\,, \mathcal{C})$ let $\underline{\mathcal{C}}(X) = \cup_{x\in E}\{C(x) \in \mathcal{C} \mid C(x) \subseteq X\}$ and $\overline{\mathcal{C}}(X) = \{x \in E \mid x \in C^*(y) \Rightarrow C^*(y) \cap X \neq \emptyset\}$, then $\underline{\mathcal{C}}(X)$ and $\overline{\mathcal{C}}(X)$ are, respectively, called the covering rough lower approximation and covering rough upper approximation of X in $(\mathcal{P}(E)\,, \mathcal{C})$, which are indeed identical with $\underline{\delta}$ and $\overline{\delta}$.

Example 5. Let $E = \{a_1\,, a_2\,, a_3\,, a_4\}$, define a dilation δ on $\mathcal{P}(E)$ as: $\delta(\{a_1\}) = \{a_1\,, a_3\}$, $\delta(\{a_2\}) = \{a_2\,, a_3\}$, $\delta(\{a_3\}) = \{a_1\,, a_2\,, a_3\}$, and $\delta(\{a_4\}) = \{a_4\}$, then $\mathcal{C} = \{C_1\,, C_2\,, C_3\,, C_4\}$ is a covering of E, where $C_i = \delta(\{a_i\})$, $i = 1\,, 2\,, 3\,, 4$. $\mathcal{C}_1 = \{C_1\,, C_2\,, C_4\}$, $\mathcal{C}_2 = \{C_1\,, C_3\,, C_4\}$, $\mathcal{C}_3 = \{C_2\,, C_3\,, C_4\}$, and $\mathcal{C}_4 = \{C_3\,, C_4\}$ are all the coverings of E as well. But not all of the rough universes $(\mathcal{P}(E)\,, \mathcal{C}_i)\,(i = 1\,, 2\,, 3\,, 4)$ are equivalent to $(\mathcal{P}(E)\,, \delta)$. That is, there are some $i \in \{1\,, 2\,, 3\,, 4\}$ and $X \in \mathcal{P}(E)$ such that $\underline{\delta}(X) \neq \underline{\mathcal{C}_i}(X)$ or $\overline{\delta}(X) \neq \overline{\mathcal{C}_i}(X)$, which is true when $i = 2\,, 3$ or 4.

Let δ be a dilation on $\mathcal{P}(E)$, if there are $x\,, y \in E$, $x \neq y$, satisfying $\delta(\{x\}) = \delta(\{y\})$, that is, $\mathcal{C} = \{\delta(\{x\}) \mid x \in E\}$ is a multi-set, then all such $\delta(\{x\})$ but one are superfluous when computing $\underline{\mathcal{C}}$ and $\overline{\mathcal{C}}$ for a set. We may retain any one of them and remove the others until $\mathcal{C}$ is changed into a common set (a non-multi-set), which is also denoted by $\mathcal{C}$. Therefore, without loss of generality, the covering $\mathcal{C}$ is supposed to be a common set.

If $C_1 \in \mathcal{C}, \mathcal{C} - \{C_1\}$ is still a covering of E, and the same rough approximations of arbitrary set $X \in \mathcal{P}(E)$ can be generated in $(\mathcal{P}(E)\,, \mathcal{C} - \{C_1\})$ as those in $(\mathcal{P}(E)\,, \delta)$, C_1 is called a reducible set in $\mathcal{C}$. In the same way, if there is a set $C_2 \in \mathcal{C} - \{C_1\}$ such that $\mathcal{C} - \{C_1\,, C_2\}$ is a covering of E and the same rough approximations can be developed in $(\mathcal{P}(E)\,, \mathcal{C} - \{C_1\,, C_2\})$ as those in $(\mathcal{P}(E)\,, \delta)$, C_2 is called reducible in $\mathcal{C}$ as well. These steps are continued until there are no reducible sets in $\mathcal{C} - \{C_1\,, C_2\,, \cdots, C_k\}$ anymore, C_1, C_2, $\cdots$, C_k are reducible sets, and every element in $\mathcal{C} - \{C_1\,, C_2\,, \cdots, C_k\}$ is called irreducible in $\mathcal{C}$.

In fact, if $\mathcal{C}$ is a multi-set, all repeated elements in $\mathcal{C}$ are reducible. It is also clear that the gain of reducible sets of a morphological covering is independent of the order of priorities.

Definition 5. *If C_1, C_2, $\cdots$, C_k are reducible sets of a covering $\mathcal{C}$, and there are no other reducible sets in $\mathcal{C}$ anymore, then $\mathcal{C} - \{C_1\,, C_2\,, \cdots, C_k\}$ is called a reduction of $\mathcal{C}$, and is denoted by* $\mathrm{Red}(\mathcal{C})$.

$\mathrm{Red}(\mathcal{C})$ is composed of all irreducible sets of $\mathcal{C}$, and is also a covering of E. It is a minimum covering that ensures $\underline{\delta}(X) = \underline{\mathrm{Red}(\mathcal{C})}(X)$ and $\overline{\delta}(X) = \overline{\mathrm{Red}(\mathcal{C})}(X)$ for every $X \in \mathcal{P}(E)$. If every element in $\mathcal{C}$ is irreducible, $\mathrm{Red}(\mathcal{C}) = \mathcal{C}$.

Theorem 5. *Assume that δ is a dilation on $\mathcal{P}(E)$, for $x \in E$, let $I_x = \{y \in E \mid \delta(\{y\}) \subseteq \delta(\{x\})\}$, then $\delta(\{x\})$ is a reducible set of $\mathcal{C} = \{\delta(\{x\}) \mid x \in E\}$ if and only if $\delta(\{x\}) = \cup_{y \in I_x - \{x\}} \delta(\{y\})$.*

Proof. $\Leftarrow$: If there is $x_0 \in E$ satisfying $\delta(\{x_0\}) = \cup_{y \in I_{x_0} - \{x_0\}} \delta(\{y\})$, then $\mathcal{C} - \{\delta(\{x_0\})\}$ is a covering of E. It suffices to prove the equations $\underline{\mathcal{C} - \{\delta(\{x_0\})\}}(X) = \underline{\delta}(X)$ and $\overline{\mathcal{C} - \{\delta(\{x_0\})\}}(X) = \overline{\delta}(X)$ for all $X \in \mathcal{P}(E)$.

For $X \in \mathcal{P}(E)$, if $\delta(\{x_0\}) \cap X^c \neq \emptyset$, it is true that $\underline{\delta}(X) = \underline{\mathcal{C} - \{\delta(\{x_0\})\}}(X)$; otherwise, $\delta(\{x_0\}) \subseteq X$, and so $\underline{\mathcal{C} - \{\delta(\{x_0\})\}}(X) \subseteq \underline{\delta}(X)$.

Let $x \in \underline{\delta}(X)$, there is $z \in E$ satisfying $x \in \delta(\{z\}) \subseteq X$. If $z \neq x_0$, then $x \in \delta(\{z\}) \subseteq X$ and $\delta(\{z\}) \in \mathcal{C} - \{\delta(\{x_0\})\}$. Thus $x \in \underline{\mathcal{C} - \{\delta(\{x_0\})\}}(X)$; otherwise, there is $y \in I_{x_0} - \{x_0\}$ satisfying $x \in \delta(\{y\}) \subseteq X$, $\delta(\{y\}) \subseteq \delta(\{x_0\})$, and $\delta(\{y\}) \in \mathcal{C} - \{\delta(\{x_0\})\}$. Therefore $x \in \underline{\mathcal{C} - \{\delta(\{x_0\})\}}(X)$.

The result on upper approximations can be proved from the duality principle.

$\Rightarrow$: If there is $x_0 \in E$ such that $\delta(\{x_0\})$ is reducible in $\mathcal{C}$ but $\delta(\{x_0\}) \neq \cup_{y \in I_{x_0} - \{x_0\}} \delta(\{y\})$, by the definitions and properties of the morphological rough approximations and of the covering ones, we have $\underline{\delta}(\delta(\{x_0\})) = \delta(\{x_0\})$. But $\underline{\mathcal{C} - \{\delta(\{x_0\})\}}(\delta(\{x_0\})) = \cup\{\delta(\{y\}) \in \mathcal{C} - \{\delta(\{x_0\})\} \mid \delta(\{y\}) \subseteq \delta(\{x_0\})\} = \cup_{y \in I_{x_0} - \{x_0\}} \delta(\{y\}) \neq \delta(\{x_0\})$, which contradicts the reducible set $\delta(\{x_0\})$ of $\mathcal{C}$.

Theorem 5 shows that every reducible set of morphological covering $\mathcal{C}$ is certainly a union of some irreducible sets of $\mathcal{C}$. For instance in Example 5, C_3 is reducible in $(\mathcal{P}(E), \mathcal{C})$ since $C_3 = C_1 \cup C_2$, and there are no other reducible sets in $\mathcal{C}_1 = \{C_1, C_2, C_4\}$, so $\mathrm{Red}(\mathcal{C}) = \mathcal{C}_1$.

Proposition 5. *For a dilation δ on $\mathcal{P}(E)$, the reduction* $\mathrm{Red}(\mathcal{C})$ *of morphological covering $\mathcal{C} = \{\delta(\{x\}) \mid x \in E\}$ is unique.*

Proof. Suppose that $\mathcal{C}_1$ is also a reduction of $\mathcal{C}$, for each $C \in \mathcal{C}_1$, there is $x_0 \in E$ satisfying $C = \delta(\{x_0\})$. If $\delta(\{x_0\}) \notin \mathrm{Red}(\mathcal{C})$, then $\delta(\{x_0\}) = \cup_{y \in I_{x_0} - \{x_0\}} \delta(\{y\})$. For every $y \in I_{x_0} - \{x_0\}$, there are irreducible sets $\delta(\{y_1\})$, $\delta(\{y_2\})$, $\cdots$, $\delta(\{y_{k_y}\})$ satisfying $\delta(\{y\}) = \cup_{i=1,2,\cdots,k_y} \delta(\{y_i\})$, so $\delta(\{x_0\}) = \cup_{y \in I_{x_0} - \{x_0\}} \cup_{i=1,2,\cdots,k_y} \delta(\{y_i\})$, meaning that $\delta(\{x_0\})$ is an reducible set of $\mathcal{C}_1$, which contradicts $\delta(\{x_0\}) \in \mathcal{C}_1$. Therefore $\mathcal{C}_1 \subseteq \mathrm{Red}(\mathcal{C})$.

The reverse inclusion can be proved analogously.

From Proposition 5, the following result holds.

Theorem 6. *Let δ be a dilation on $\mathcal{P}(E)$, then the morphological covering rough set are identical with the Pawlak one if and only if the reduction* $\mathrm{Red}(\mathcal{C})$ *of the morphological covering $\mathcal{C} = \{\delta(\{x\}) \mid x \in E\}$ is a partition of E.*

It is interesting to investigate the reconstruction of a dilation from the reduction of morphological covering. Here the universe E is supposed to be a countably finite set, $\{a_1, a_2, \cdots, a_n\}$, for example. The following propositions hold.

Proposition 6. *Suppose that δ is a dilation on $\mathcal{P}(E)$, and that $\delta(\{a_{i_1}\})$, $\delta(\{a_{i_2}\})$, $\cdots$, $\delta(\{a_{i_k}\})$ are the irreducible sets of $\mathcal{C} = \{\delta(\{x\}) \mid x \in E\}$. Let $\Delta : \mathcal{P}(E) \to \mathcal{P}(E)$, $\Delta(X) = \cup_{x \in X} \Delta(\{x\})$ for $X \in \mathcal{P}(E)$, where $\Delta(\{x\})$ is defined as follows. For every $j \in \{1, 2, \cdots, n\}$, if $j \in \{i_1, i_2, \cdots, i_k\}$, $\Delta(\{a_j\}) = \delta(\{a_j\})$; otherwise, $\Delta(\{a_j\}) = \cup\{\delta(\{a_m\}) \mid a_j \in \delta(\{a_m\}), m \in \{i_1, i_2, \cdots, i_k\}\}$, then Δ is a dilation on $\mathcal{P}(E)$.*

Proposition 7. *In the morphological rough approximation spaces $(\mathcal{P}(E), \delta)$ and $(\mathcal{P}(E), \Delta)$, $\underline{\delta}(X) = \underline{\Delta}(X)$ and $\overline{\delta}(X) = \overline{\Delta}(X)$ for all $X \in \mathcal{P}(E)$.*

5 Conclusions

This paper studies a generalization of Pawlak rough set based on a morphological covering. It has been shown that rough sets and mathematical morphology can be integrated provided that the covering is generated by a morphological dilation. Similar algebraic properties of morphological covering rough set to those of Pawlak rough set have been preserved.

Another main contribution of this paper is on the study of reduction of a morphological covering. We have pointed out that a reduction of morphological covering is a unique minimum covering that generates the same rough approximations. The gain of reduction is independent of the order of priority of irreducible sets. The equivalence of the morphological covering rough set with the Pawlak one has been established.

This paper provides one an idea to do data analysis by using morphological techniques. We will in the future work on algorithms and applications of morphological covering rough set in data mining and reduction.

References

1. Bloch, I.: On links between mathematical morphology and rough sets. Pattern Recognition **33** (2000) 1487–1496
2. Gomolinska, A.: A comparative study of some generalized rough approximations. Fundamenta Informaticae **51** (2002) 103–119
3. Heijmans, H.: Morphological Image Operators. Academic Press, Boston (1994)
4. Matheron, G.: Random Sets and Integral Geometry. John Wiley & Sons, New York (1975)
5. Pawlak, Z.: Rough sets. International Journal of Computer and Information Sciences **11** (1982) 341–356
6. Polkowski, L.T.: Rough set approach to mathematical morphology: approximate compression data. In Proceedings of Information Processing and Management of Uncertainty, Paris (1998) 1183–1189
7. Serra, J.: Image Analysis and Mathematical Morphology. Academic Press, London (1982)
8. Slowinski, R., Vanderpooten, D.: A generalized definition of rough approximations based on similarity. IEEE Transactions on Knowledge and Data Engineering **12** (2000) 331–336
9. Soille, P.: Morphological Image Analysis. 2nd edn. Springer-Verlag, Heidelberg (2003)
10. Wu, W.-Z., Zhang, W.-X.: Neighborhood operator systems and approximations. Information Sciences **144** (2002) 201–217
11. Yao, Y.Y.: Relational interpretations of neighborhood operators and rough set approximation operators. Information Sciences **111** (1998) 239–259
12. Zhu, W., Wang, F.-Y.: Reduction and axiomization of covering generalized rough sets. Information Sciences **152** (2003) 217–230

Binary Relation Based Rough Sets*

William Zhu[1,2,3] and Fei-Yue Wang[1,4]

[1] The Key Laboratory of Complex Systems and Intelligent Science, Institute of Automation, The Chinese Academy of Sciences, Beijing 100080, China
[2] Department of Computer Science, University of Auckland, Auckland, New Zealand
[3] Computer Information Engineering College, Jiangxi Normal University, China
[4] Systems and Industrial Engineering Department, The University of Arizona, Tucson, AZ 85721, USA
fzhu009@ec.auckland.ac.nz, feiyue@sie.arizona.edu

Abstract. Rough set theory has been proposed by Pawlak as a tool for dealing with the vagueness and granularity in information systems. The core concepts of classical rough sets are lower and upper approximations based on equivalence relations. This paper studies arbitrary binary relation based generalized rough sets. In this setting, a binary relation can generate a lower approximation operation and an upper approximation operation. We prove that such a binary relation is unique, since two different binary relations will generate two different lower approximation operations and two different upper approximation operations. This paper also explores the relationships between the lower or upper approximation operation generated by the intersection of two binary relations and those generated by these two binary relations, respectively.

Keyword: Rough set, Lower approximation, Upper approximation, Binary relation, Fuzzy set, Granular computing.

1 Introduction

At the Internet age, more and more data are being collected and stored, thus, how to extract the useful information from such enormous data becomes an important issue in computer science. In order to cope with this issue, researchers have developed many techniques such as fuzzy set theory [40], rough set theory [18], computing with words [27,41,42,43,44], computational theory for linguistic dynamic systems [28], etc.

Rough set theory has been proposed by Pawlak [18] as a tool to conceptualize, organize and analyse various types of data in data mining. This method is especially useful for dealing with uncertain and vague knowledge in information systems. Many examples of applications of the rough set method to process control, economics, medical diagnosis, biochemistry, environmental science,

* The first author is in part supported by the New Economy Research Fund of New Zealand and this work is also in part supported by two 973 projects (2004CB318103) and (2002CB312200) from the Ministry of Science and Technology of China.

L. Wang et al. (Eds.): FSKD 2006, LNAI 4223, pp. 276–285, 2006.

biology, chemistry psychology, conflict analysis and other fields can be found in [1,5,6,7,8,12,13,15,16,17,19,20,21,22,23,26,29,30,31,32,33,34,45,51,52].

The classical rough set theory is based on equivalent relations, but in some situations, equivalent relations are not suitable for coping with the granularity, thus classical rough set method is extended to similarity relation based rough set [10,11,25], covering based rough sets [2,48,49,50], etc [4].

Papers [3,9,35,36,37,38,39] have done extensive research on binary relation based rough sets. In this paper, we also study general binary relation based rough sets. Our focus is on relationships between two lower approximation operations generated by two binary relations, and relationships between two upper approximation operations generated by two binary relations.

The other parts of this paper are organized as follows: In Section 2, we present the fundamental concepts and properties of the Pawlak's rough set theory, and basic definitions and properties of binary relations. Section 3 discusses binary relation based rough sets in literature. Section 4 is the major contribution of this paper. We explore the relationships between rough set generated by two relations on a universe and claim that two different binary relations will generate two different lower approximation operations and two different upper approximation operations. This paper concludes in section 5.

2 Background

2.1 Fundamentals of the Pawlak's Rough Sets

Let U be a finite set, the domain of discourse, and R an equivalent relation on U. R is generally called an indiscernability relation in rough set theory [18]. R will generate a partition $U/R = \{Y_1, Y_2, \ldots, Y_m\}$ on U where $Y_1, Y_2, \ldots, Y_m$ are the equivalent classes generated by the equivalent relation R, and, in the rough set theory, they are also called elementary sets of R. For any $X \subseteq U$ we can describe X by the elementary sets of R and the two sets

$$R_*(X) = \cup\{Y_i \in U/R | Y_i \subseteq X\}$$
$$R^*(X) = \cup\{Y_i \in U/R | Y_i \cap X \neq \phi\}$$

are called the lower and the upper approximation of X, respectively.

Let ϕ be the empty set, $-X$ the complement of X in U, from the definition of approximation sets, we have the following conclusions about them.

The properties of the Pawlak's rough sets: (1L) $R_*(U) = U$
(1H) $R^*(U) = U$
(2L) $R_*(\phi) = \phi$
(2H) $R^*(\phi) = \phi$
(3L) $R_*(X) \subseteq X$
(3H) $X \subseteq R^*(X)$
(4L) $R_*(X \cap Y) = R_*(X) \cap R_*(Y)$
(4H) $R^*(X \cup Y) = R^*(X) \cup R^*(Y)$
(5L) $R_*(R_*(X)) = R_*(X)$

(5H) $R^*(R^*(X)) = R^*(X)$
(6L) $R_*(-X) = -R^*(X)$
(6H) $R^*(-X) = -R_*(X)$
(7L) $X \subseteq Y \Rightarrow R_*(X) \subseteq R_*(Y)$
(7H) $X \subseteq Y \Rightarrow R^*(X) \subseteq R^*(Y)$
(8L) $R_*(-R_*(X)) = -R_*(X)$
(8H) $R^*(-R^*(X)) = -R^*(X)$
(9L) $\forall K \in U/R, R_*(K) = K$
(9H) $\forall K \in U/R, R^*(K) = K$

The (3L), (4L), and (8L) are characteristic properties for the lower approximation operations [14,46,47], i.e., all other properties of the lower approximation operation can be deduced from these three properties. Correspondingly, (3H), (4H), and (8H) are characteristic properties for the upper approximation operation.

2.2 Relations on a Set

In this subsection, we present some basic concepts and properties of binary relations to be used in this paper. For detailed description and proof of them, please refer to [24].

Definition 1. *(Relations) Let U be a set, $U \times U$ the product set of U and U. Any subset R of $U \times U$ is called a relation on U. For any $(x, y) \in U \times U$, if $(x, y) \in R$, we say x has relation R with y, and denote this relationship as xRy.*

For any $x \in U$, we call the set $\{y \in U | xRy\}$ the right neighborhood of x in R and denote it as $RN_R(x)$.

For any $x \in U$, we call the set $\{y \in U | yRx\}$ the left neighborhood of x in R and denote it as $LN_R(x)$.

When there is no confusion, we omit the lowercase R.

Definition 2. *(Reflexive relations) Let R be a relation on U. If for any $x \in U$, xRx, we say R is reflexive. In another word, If for any $x \in U$, $x \in RN(x)$, R is reflexive.*

Definition 3. *(Symmetric relations) Let R be a relation on U. If for any $x, y \in U$, $xRy \Rightarrow yRx$, we say R is symmetric. In another word, If for any $x, y \in U$, $y \in RN(x) \Rightarrow x \in RN(y)$, R is symmetric.*

Definition 4. *(Transitive relations) Let R be a relation on U. If for any $x, y, z \in U$, xRy, and $yRz \Rightarrow xRz$, we say R is transitive.*

Definition 5. *(Equivalent relations) Let R be a relation on U. If R is reflexive, symmetric, and transitive, we say R is a equivalent relation on U.*

3 Binary Relation Based Generalized Rough Sets

An extensive research on algebraic properties of rough sets based on binary relations can be found in paper [3,9,35,36,37,38,39]. They proved the existence of a certain binary relation for an algebraic operator with special properties, but they did not consider the uniqueness of such a binary relation. Furthermore, we consider the relationships between rough sets generated by the join of two binary relations and rough sets generated by these two binary relations, respectively. We also discuss the above issue for the intersection of two binary relations.

Definition 6. *(Rough set based on a relation [38]) Suppose R is a binary relation on a universe U. A pair of approximation operators, $L(R), H(R) : P(U) \rightarrow P(U)$, are defined by:*

$$L(R)(X) = \{x|\forall y,\ xRy \Rightarrow y \in X\}{=}\{x|RN(x) \subseteq X\},$$
$$H(R)(X) = \{x|\exists y \in X,\ s.t.\ xRy\}{=}\{x|RN(x) \cap X \neq \phi\}.$$

They are called the lower approximation operation and the upper approximation operation, respectively. The system $(P(U), \cap, \cup, -, L(R), H(R))$ is called a rough set algebra, where $\cap, \cup$, and - are set intersection, union, and complement.

Example 1. (A relation and its lower and upper approximation operations) Let $U = \{a, b, c\}$ and $R = \{(a, a), (b, b), (b, c), (c, a), (c, b), (c, c)\}$, then
$RN(\{a\}) = \{a\}$, $RN(\{b\}) = \{b, c\}$, $RN(\{c\}) = \{a, b, c\}$.
$L(R)\{a\} = \{a\}$, $L(R)\{b\} = \{\phi\}$, $L(R)\{c\} = \{\phi\}$,
$L(R)\{a, b\} = \{a\}$, $L(R)\{a, c\} = \{a\}$, $L(R)\{b, c\} = \{b\}$,
$L(R)\{a, b, c\} = \{a, b, c\}$.
$H(R)\{a\} = \{a, c\}$, $H(R)\{b\} = \{b, c\}$, $H(R)\{c\} = \{b, c\}$,
$H(R)\{a, b\} = \{a, b, c\}$, $H(R)\{a, c\} = \{a, b, c\}$, $H(R)\{b, c\} = \{b, c\}$,
$H(R)\{a, b, c\} = \{a, b, c\}$.

Proposition 1. *(Basic properties of lower and upper approximation operations [38]) Let R be a relation on U. $L(R)$ and $H(R)$ satisfy the following properties: $\forall X, Y \subseteq U$,*
(1) $L(R)(U) = U$
(2) $L(R)(X \cap Y) = L(R)(X) \cap L(R)(Y)$
(3) $H(R)(\phi) = \phi$
(4) $H(R)(X \cup Y) = H(R)(X) \cup H(R)(Y)$
(5) $L(R)(-X) = -H(R)(X)$

Proposition 2. *[38] Let R be a relation on U. If operation $L : P(U) \rightarrow P(U)$ satisfies the following properties:*
(1)$L(U) = U$ (2)$L(X \cap Y) = L(X) \cap (Y)$
then there exists a relation R on U such that $L = L(R)$.

Proposition 3. *[38] Let R be a relation on U. If operations $H : P(U) \rightarrow P(U)$ satisfies the following properties:*
(1)$H(\phi) = \phi$ (2)$H(X \cup Y) = H(X) \cup H(Y)$
then there exists a relation R on U such that $H = H(R)$.

Proposition 4. *[38] Let U be a set. If an operator $L : P(U) \rightarrow P(U)$ satisfies the following properties:*
(1)$L(U) = U$ (2)$L(X \cap Y) = L(X) \cap (Y)$ (3)$L(X) \subseteq X$
then there exists one reflexive relation R on U such that $L = L(R)$.

Proposition 5. *[38] Let U be a set. If an operator $H : P(U) \to P(U)$ satisfies the following properties:*
(1) $H(\phi) = \phi$ (2) $H(X \cup Y) = H(X) \cup H(Y)$ (3) $X \subseteq H(X)$
then there exists one reflexive relation R on U such that $H = H(R)$.

Proposition 6. *[38] Let U be a set. If an operator $L : P(U) \to P(U)$ satisfies the following properties:*
(1)$L(U) = U$ (2)$L(X \cap Y) = L(X) \cap (Y)$ (3)$L(X) \subseteq L(-L(-X))$
then there exists one symmetric relation R on U such that $L = L(R)$.

Proposition 7. *[38] Let U be a set. If an operator $H : P(U) \to P(U)$ satisfies the following properties:*
(1) $H(\phi) = \phi$ (2) $H(X \cup Y) = H(X) \cup H(Y)$ (3) $H(-H(X)) \subseteq H(-X)$
then there exists one symmetric relation R on U such that $H = H(R)$.

Proposition 8. *[38] Let U be a set. If an operator $L : P(U) \to P(U)$ satisfies the following properties:*
(1)$L(U) = U$ (2)$L(X \cap Y) = L(X) \cap (Y)$ (3)$L(X) \subseteq L(L(X))$
then there exists one transitive relation R on U such that $L = L(R)$.

Proposition 9. *[38] Let U be a set. If an operator $H : P(U) \to P(U)$ satisfies the following properties:*
(1) $H(\phi) = \phi$ (2) $H(X \cup Y) = H(X) \cup H(Y)$ (3) $H(H(X)) \subseteq H(X)$
then there exists one transitive relation R on U such that $H = H(R)$.

4 Uniqueness of Binary Relations to Generate Rough Sets

For two relations R_1 and R_2 on a set U, R_1 and R_2 will generate their respective lower approximation operations and the upper approximation operations. $R_1 \cup R_2$ is also a relation on U, so it will generate its own lower approximation operation and the upper approximation operation. Then, what is the relationships among these lower approximation operation and upper approximation operations? How about the relation $R_1 \cap R_2$? We start to answer these questions. Firstly, we consider the situation for $R_1 \cup R_2$.

Theorem 1. *Let R_1 and R_2 be two relations on U and $X \subseteq U$. $L(R_1 \cup R_2)(X) = L(R_1)(X) \cap L(R_2)(X)$ and $H(R_1 \cup R_2)(X) = H(R_1)(X) \cup H(R_2)(X)$.*

Proof. $\forall X \subseteq U$, $L(R_1 \cup R_2)(X) = \{x | \forall y \in U, x(R_1 \cup R_2)y \Rightarrow y \in X\}$
$= \{x | \forall y \in U, xR_1y \text{ or } xR_2y \Rightarrow y \in X\}$
$= \{x | \forall y \in U, xR_1y \Rightarrow y \in X\} \cap \{x | \forall y \in U, xR_2y \Rightarrow y \in X\}$
$= L(R_1)(X) \cap L(R_2)(X)$.
$H(R_1 \cup R_2)(X) = \{x | \exists y \in X, x(R_1 \cup R_2)y\}$
$= \{x | \exists y \in X, xR_1y \text{ or } xR_2y\}$
$= \{x | \exists y \in X, xR_1y\} \cup \{x | \forall y \in X, xR_2y\}$
$= H(R_1)(X) \cup H(R_2)(X)$.

Proposition 10. *Let R_1 and R_2 are two relations on U. If $R_1 \subseteq R_2$, then $L(R_2) \subseteq L(R_1)$ and $H(R_1) \subseteq H(R_2)$.*

Then, we consider the situation for $R_1 \cap R_2$.

Theorem 2. *Let R_1 and R_2 be two relations on U and $X \subseteq U$. $L(R_1)(X) \cup L(R_2)(X) \subseteq L(R_1 \cap R_2)(X)$ and $H(R_1 \cap R_2)(X) \subseteq H(R_1)(X) \cap H(R_2)(X)$.*

Proof. It is easy to prove this theorem by Proposition 10.

Example 2. (Equalities in Theorem 2 do not hold generally)

Let $U = \{a, b, c\}$, $R_1 = \{(a, a), (a, b), (b, b)\}$, and $R_2 = \{(a, a), (a, c), (c, a), (c, b), (c, c)\}$, we have

$RN_{R_1}(\{a\}) = \{a, b\}$, $RN_{R_1}(\{b\}) = \{b\}$, $RN_{R_1}(\{c\}) = \phi$.
$RN_{R_2}(\{a\}) = \{a, c\}$, $RN_{R_2}(\{b\}) = \phi$, $RN_{R_2}(\{c\}) = \{a, b, c\}$,
$R_1 \cap R_2 = \{(a, a)\}$, and
$RN_{R_1 \cap R_2}(\{a\}) = \{a\}$, $RN_{R_1 \cap R_2}(\{b\}) = \phi$, $RN_{R_1 \cap R_2}(\{c\}) = \phi$.

For $X = \{a\}$ and $Y = \{b\}$, we have
$L(R_1)(X) = \{c\}$, $H(R_1)(Y) = \{a, b\}$,
$L(R_2)(X) = \{b\}$, $H(R_2)(Y) = \{c\}$,
and
$L(R_1 \cap R_2)(X) = \{a, b, c\}$, $H(R_1 \cap R_2)(Y) = \phi$.

Thus, $L(R_1)(X) \cup L(R_2)(X) \subset L(R_1 \cap R_2)(X)$ and
$H(R_1 \cap R_2)(Y) \subset H(R_1)(Y) \cap H(R_2)(Y)$.

A relation on U will generate a lower approximation operation and an upper approximation operation, then is it possible for two different relations on U to generate the same lower approximation operation and the same upper approximation operation? We start to study this problem.

Proposition 11. *Let R_1 and R_2 are two relations on U. If $H(R_1) \subseteq H(R_2)$, then $R_1 \subseteq R_2$.*

Proof. $\forall x, y \in U$, if $(x, y) \in R_1$, $y \in RN_{R_1}(x)$, $x \in H(R_1)\{y\} \subseteq H(R_2)\{y\}$, so $RN_{R_2}(x) \cap \{y\} \neq \phi$, that means $(x, y) \in R_2$, thus $R_1 \subseteq R_2$.

Corollary 1. *Let R_1 and R_2 are two relations on U. If $H(R_1) = H(R_2)$, then $R_1 = R_2$.*

Theorem 3. *Let R_1 and R_2 are two relations on U. If $H(R_1) = H(R_2)$ if and only if $R_1 = R_2$.*

Proof. It comes from Proposition 10 and Corollary 1.

By the duality between $H(R)$ and $L(R)$, we have the following result about $L(R)$.

Proposition 12. *Let R_1 and R_2 are two relations on U. If $L(R_1) \subseteq L(R_2)$, then $R_2 \subseteq R_1$.*

Corollary 2. *Let R_1 and R_2 are two relations on U. If $L(R_1) = L(R_2)$, then $R_1 = R_2$.*

Theorem 4. *Let R_1 and R_2 are two relations on U. If $L(R_1) = L(R_2)$ if and only if $R_1 = R_2$.*

Theorem 3 and 4 show that two different binary relations will certainly generate two different lower approximation operations and two different lower approximation operations. Recall that Proposition 2 and 3 show an operator on U with two certain properties can be generated by a binary relation, we actually have proved the uniqueness of such a binary relation.

Theorem 5. *Let R be a relation on U. If operation $L : P(U) \to P(U)$ satisfies the following properties:*
(1)$L(U) = U$
(2)$L(X \cap Y) = L(X) \cap L(Y)$
then there exists one and only one relation R on U such that $L = L(R)$.

Theorem 6. *Let R be a relation on U. If operations $H : P(U) \to P(U)$ satisfies the following properties:*
(1)$H(\phi) = \phi$
(2)$H(X \cup Y) = H(X) \cup H(Y)$
then there exists one and only one relation R on U such that $H = H(R)$.

Theorem 7. *Let U be a set. If an operator $L : P(U) \to P(U)$ satisfies the following properties:*
(1)$L(U) = U$
(2)$L(X \cap Y) = L(X) \cap (Y)$
(3)$L(X) \subseteq L(-L(-X))$
then there exists one and only one symmetric relation R on U such that $L = L(R)$.

Proof. It comes from Proposition 2 and Theorem 4.

Theorem 8. *Let U be a set. If an operator $H : P(U) \to P(U)$ satisfies the following properties:*
(1) $H(\phi) = \phi$
(2) $H(X \cup Y) = H(X) \cup H(Y)$
(3) $H(-H(X)) \subseteq H(-X)$
then there exists one and only one symmetric relation R on U such that $H = H(R)$.

Proof. It comes from Proposition 3 and Theorem 3.

Theorem 9. *Let U be a set. If an operator $L : P(U) \to P(U)$ satisfies the following properties:*
(1)$L(U) = U$
(2)$L(X \cap Y) = L(X) \cap (Y)$
(3)$L(X) \subseteq L(L(X))$
then there exists one and only one transitive relation R on U such that $L = L(R)$.

Proof. It comes from Proposition 2 and Theorem 4.

Theorem 10. *Let U be a set. If an operator $H : P(U) \to P(U)$ satisfies the following properties:*
(1) $H(\phi) = \phi$
(2) $H(X \cup Y) = H(X) \cup H(Y)$
(3) $H(H(X)) \subseteq H(X)$
then there exists one and only one transitive relation R on U such that $H = H(R)$.

Proof. It comes from Proposition 3 and Theorem 3.

5 Conclusions

In this paper we have studied relationships between generalized rough sets generated by two binary relations. We proved that two different binary relations will generate two different lower approximation operations and two different upper approximation operations. As for the applications of binary relation based rough sets to knowledge discovery from database, please refer to paper [10,11,25].

We will explore the relationships between binary relation based rough sets and covering based rough sets [48] in our future works. Another future research topic is to apply binary relation based rough set theory to the computational theory for linguistic dynamic systems [28] and security [52].

Acknowledgments

The first author thanks Prof. Thomborson from the University of Auckland for his support for this work and thanks Prof. Yiyu Yao from the University of Regina for his in-depth suggestions and comments on this paper. Discussions with Prof. JingTao Yao and Prof. Guilong Liu are also appreciated.

References

1. Angiulli, F., Pizzuti, C.: Outlier mining in large high-dimensional data sets. IEEE Trans. On Knowledge and Data Engineering **17** (2005) 203–215
2. Bonikowski, Z., Bryniarski, E., Wybraniec-Skardowska, U.: Extensions and intentions in the rough set theory. Information Sciences **107** (1998) 149–167
3. Cattaneo, G., Ciucci, D.: Algebraic structures for rough sets. In: LNCS. Volume 3135. (2004) 208–252
4. Dai, J.: Logic for rough sets with rough double stone algebraic semantics. In: RSFDGrC 2005. Volume 3641 of LNCS. (2005) 141–147
5. Dong, G., Han, J., Lam, J., Pei, J., Wang, K., Zou, W.: Mining constrained gradients in large databases. IEEE Trans. On Knowledge and Data Engineering **16** (2004) 922–938
6. Hall, M., Holmes, G.: Benchmarking attribute selection techniques for discrete class data mining. IEEE Trans. On Knowledge and Data Engineering **15** (2003) 1437–1447

7. Hu, F., Wang, G., Huang, H., et al: Incremental attribute reduction based on elementary sets. In: RSFDGrC 2005. Volume 3641 of LNCS. (2005) 185–193
8. Jensen, R., Shen, Q.: Semantics-preserving dimensionality reduction: Rough and fuzzy-rough-based approaches. IEEE Trans. On Knowledge and Data Engineerin **16** (2004) 1457–1471
9. Kondo, M.: On the structure of generalized rough sets. Information Sciences **176** (2005) 589–600
10. Kryszkiewicza, M.: Rough set approach to incomplte information systems. Information Sciences **112** (1998) 39–49
11. Kryszkiewicza, M.: Rule in incomplte information systems. Information Sciences **113** (1998) 271–292
12. Leung, Y., Wu, W.Z., Zhang, W.X.: Knowledge acquisition in incomplete information systems: A rough set approach. European Journal of Operational Research **168** (2006) 164–180
13. Li, D.G., Miao, D.Q., Yin, Y.Q.: Relation of relative reduct based on nested decision granularity. In: IEEE GrC 2006. (2006) 397–400
14. Lin, T.Y., Liu, Q.: Rough approximate operators: axiomatic rough set theory. In Ziarko, W., ed.: Rough Sets, Fuzzy Sets and Knowledge Discovery, Springer (1994) 256–260
15. Liu, Q.: Semantic analysis of rough logical formulas based on granular computing. In: IEEE GrC 2006. (2006) 393–396
16. Liu, G.: The transitive closures of matrices over distributive lattices. In: IEEE GrC 2006. (2006) 63–66
17. Pal, S., Mitra, P.: Case generation using rough sets with fuzzy representation. IEEE Trans. On Knowledge and Data Engineering **16** (2004) 292–300
18. Pawlak, Z.: Rough sets: Theoretical aspects of reasoning about data. Kluwer Academic Publishers, Boston (1991)
19. Polkowski, L., Skowron, A., eds.: Rough sets and current trends in computing. Volume 1424. Springer (1998)
20. Polkowski, L., Skowron, A., eds.: Rough sets in knowledge discovery. Volume 1. Heidelberg: Physica–Verlag (1998)
21. Polkowski, L., Skowron, A., eds.: Rough sets in knowledge discovery. Volume 2. Heidelberg: Physica–Verlag (1998)
22. Qin, K., Pei, Z., Du, W.: The relationship among several knowledge reduction approaches. In: FSKD 2005. Volume 3613 of LNCS. (2005) 1232–1241
23. Qin, K., Pei, Z.: On the topological properties of fuzzy rough sets. Fuzzy Sets and Systems **151** (2005) 601–613
24. Rajagopal, P., Masone, J.: Discrete Mathematics for Computer Science. Saunders College, Canada (1992)
25. Slowinski, R., Vanderpooten, D.: A generalized definition of rough approximations based on similarity. IEEE Trans. On Knowledge and Data Engineering **12** (2000) 331–336
26. Su, C., Hsu, J.: An extended chi2 algorithm for discretization of real value attributes. IEEE Trans. On Knowledge and Data Engineering **17** (2005) 437–441
27. Wang, F.Y.: Outline of a computational theory for linguistic dynamic systems: Toward computing with words. International Journal of Intelligent Control and Systems **2** (1998) 211–224
28. Wang, F.Y.: On the abstraction of conventional dynamic systems: from numerical analysis to linguistic analysis. Information Sciences **171** (2005) 233–259
29. Wang, G., Liu, F.: The inconsistency in rough set based rule generation. In: Rough Sets and Current Trends in Computing. Volume 2005 of LNCS. (2000) 370–377

30. Wu, W.Z., Zhang, W.X.: Rough set approximations vs. measurable spaces. In: IEEE GrC 2006. (2006) 329–332
31. Wu, W.Z., Leung, Y., Mi, J.S.: On characterizations of (i, t) -fuzzy rough approximation operators. Fuzzy Sets and Systems **154** (2005) 76–102
32. Wu, W.Z., Zhang, W.X.: Constructive and axiomatic approaches of fuzzy approximation operator. Information Sciences **159** (2004) 233–254
33. Yang, X.P., Li, T.J.: The minimization of axiom sets characterizing generalized approximation operators. Information Sciences **176** (2006) 887–899
34. Yao, J., Liu, W.N.: The STP model for solving imprecise problems. In: IEEE GrC 2006. (2006) 683–687
35. Yao, Y.: A comparative study of fuzzy sets and rough sets. Information Sciences **109** (1998) 227–242
36. Yao, Y.: On generalizing pawlak approximation operators. In: LNAI. Volume 1424. (1998) 298–307
37. Yao, Y.: Relational interpretations of neighborhood operators and rough set approximation operators. Information Sciences **101** (1998) 239–259
38. Yao, Y.: Constructive and algebraic methods of theory of rough sets. Information Sciences **109** (1998) 21–47
39. Yao, Y., Chen, Y.: Subsystem based generalizations of rough set approximations. In: LNCS. Volume 3488. (2005) 210–218
40. Zadeh, L.A.: Fuzzy sets. Information and Control **8** (1965) 338–353
41. Zadeh, L.: The concept of a linguistic variable and its application to approximate reasoning – I. Information Sciences **8** (1975) 199–249
42. Zadeh, L.: The concept of a linguistic variable and its application to approximate reasoning – II. Information Sciences **8** (1975) 301–357
43. Zadeh, L.: The concept of a linguistic variable and its application to approximate reasoning – III. Information Sciences **9** (1975) 43–80
44. Zadeh, L.: Fuzzy logic = computing with words. IEEE Transactions on Fuzzy Systems **4** (1996) 103–111
45. Zhong, N., Yao, Y., Ohshima, M.: Peculiarity oriented multidatabase mining. IEEE Trans. On Knowledge and Data Engineering **15** (2003) 952–960
46. Zhu, F., He, H.C.: The axiomization of the rough set. Chinese Journal of Computers **23** (2000) 330–333
47. Zhu, F., He, H.C.: Logical properties of rough sets. In: Proc. of The Fourth International Conference on High Performance Computing in the Asia-Pacific Region. Volume 2., IEEE Press (2000) 670–671
48. Zhu, W., Wang, F.Y.: Reduction and axiomization of covering generalized rough sets. Information Sciences **152** (2003) 217–230
49. Zhu, W., Wang, F.Y.: A new type of covering rough sets. In: IEEE IS 2006, 4-6 September 2006. (2006)(to appear)
50. Zhu, W., Wang, F.Y.: Relationships among three types of covering rough sets. In: IEEE GrC 2006. (2006) 43–48
51. Zhu, W., Wang, F.Y.: Axiomatic systems of generalized rough sets. In: RSKT 2006. Volume 4062 of LNAI. (2006) 216–221
52. Zhu, W., Wang, F.Y.: Covering based granular computing for conflict analysis. In: IEEE ISI 2006. Volume 3975 of LNCS. (2006) 566–571

On the Categorizing of Fully Symmetric Relations in Partial Four-Valued Logic*

Renren Liu and Ting Wang

College of Information Engineering, Xiangtan University, Xiangtan, Hunan, 411105, China
renrenliu@hotmail.com

Abstract. In multiple-valued logic theories, the decision and construction for Sheffer functions is an important problem. The decision for Sheffer functions is interrelated to the decision for completeness of functions set, and the solution can be reduced to determining the minimal coverings of precomplete. It's well known that each precomplete set is a function set, $T(G_m)$, preserving the relation G_m, therefore, the categorizing of this relation has provided the determination of precomplete set's minimal covering with more convenient ways. In this paper, fully symmetric relations in partial four-valued logic are categorized by similar relation.

1 Introduction

Multiple-valued logic is an important branch of computer science. The structure theory of multiple-valued logic functions is an important research field in multiple-valued logic theory. One of the most important and fundamental problems is the completeness of function sets. The solution of this problem depends on determination of all the precomplete categoryes in multiple-valued logic function sets.

Another important problem in multiple-valued logic completeness theory is the decision on Sheffer [1] functions, which depends on the deciding of the minimal covering of precomplete sets of all precomplete sets. For the partial multiple-valued logic function, the author has concisely decided the minimal covering of precomplete sets in 3-valued by using the similar relation among the precomplete sets [2]. In addition, the author proved that T_E , $P_K \cup \{*\}, L_P, L_{G_{4,2}}$ are included in the minimal covering for any *k*. For the complexity of full symmetric function sets, simply separable function sets and regularly separable function sets, the problem has not been completely solved yet [2-4].

It's well known that each precomplete set is a function set, $T(G_m)$, preserving the relation G_m. The author has already proved that [17], if G_m is similar to G'_m, $T(G_m)$ and $T(G'_m)$ are either within or without the minimal covering. Therefore, the categorizing of this relation has provided the determination of precomplete set's minimal covering with more convenient ways. In this paper, fully symmetric relations in P_4^* are categorized by the similar relation.

* This work is supported by the National and Hunan Province Natural Science Foundation of China (60083001,60375021, 60433020, 03JJY3099).

L. Wang et al. (Eds.): FSKD 2006, LNAI 4223, pp. 286–289, 2006.

G_m is said to be the fully symmetric relation if

$$G_m = \bigcup_{\substack{i,j=1 \\ i \neq j}}^{m} G_m(\{i,j\}) \cup G_m^*$$

where $G_m(\{i,j\}) = \{< a_1, a_2, \cdots, a_m > | \, a_i = a_j, a_r \in E_K, 1 \le r \le m\}$, G_m^* is an empty set (except m=2) or only includes m-tuples whose elements are different from each other, and G_m is symmetric about S_m, i.e., $G_m = G_m^{\sigma} = \{< a_{\sigma(1)}, \cdots, a_{\sigma(m)} > | < a_1, \cdots, a_m > \in G_m\}$, for arbitrary $\sigma \in S_m$, where S_m is a symmetric group over {1, 2, ⋯, m}. The set of all functions preserving G_m is denoted by $F_{s,m} = T(G_m)$ and called the Fully Function Set.

2 Main Results

There are 62 binary fully symmetric relations such as $F_{S,2} = T(G_2)$, where $G_2 =$ {<0,0>,<1,1>,<2,2>,<3,3>} $\cup \, G_2^*$, where G_2^* is one of these as follows:

(1) The relations that include two binary couples is one category as follows:

{<0,1>,<1,0>} $\overset{(12)}{\sim}$ {<0,2>,<2,0>} $\overset{(23)}{\sim}$ {<0,3>,<3,0>} $\overset{(01)}{\sim}$ {<1,3>,<3,1>} $\overset{(12)}{\sim}$ {<2,3>,<3,2>} $\overset{(13)}{\sim}$ {<2,1>,<1,2>} .

(2) The relations that include four binary couples are 2 categories as follows:

Category 1 :

{ <0,1>,<1,0> , <2,3>,<3,2> } $\overset{(12)}{\sim}$ { <0,2>,<2,0> , <1,3>,<3,1> } $\overset{(23)}{\sim}$ {<0,3>, <3,0>, <1,2>,<2,1>} ;

Category 2 :

{ <0,1>,<1,0> , <0,2>,<2,0> } $\overset{(23)}{\sim}$ { <0,1>,<1,0> , <0,3>,<3,0> } $\overset{(01)}{\sim}$ { <0,1>, <1,0> , <1,3>,<3,1> } $\overset{(13)}{\sim}$ { <0,3>,<3,0> , <1,3>,<3,1> } $\overset{(123)}{\sim}$ { <0,1>,<1,0> , <1,2>,<2,1> } $\overset{(12)}{\sim}$ { <0,2>,<2,0> , <1,2>,<2,1> } $\overset{(02)\,(13)}{\sim}$ { <0,2>,<2,0> , <0,3>,<3,0> } $\overset{(02)}{\sim}$ { <0,2>,<2,0> , <2,3>,<3,2> } $\overset{(23)}{\sim}$ { <0,3>,<3,0> , <2,3>,<3,2> } $\overset{(01)}{\sim}$ { <1,3>,<3,1> , <2,3>,<3,2> } $\overset{(23)}{\sim}$ {<1,2>,<2,1>, <2,3>,<3,2>} $\overset{(12)}{\sim}$ {<1,2>,<2,1>, <1,3>,<3,1>} .

(3) The relations that include six binary couples are 3 categories as follows :

Category 1 :

$$\{\langle 0,1\rangle,\langle 1,0\rangle, \langle 0,2\rangle,\langle 2,0\rangle, \langle 0,3\rangle,\langle 3,0\rangle\} \overset{(01)}{\sim} \{\langle 0,1\rangle,\langle 1,0\rangle, \langle 1,2\rangle,\langle 2,1\rangle,\langle 1,3\rangle,\langle 3,1\rangle\} \overset{(12)}{\sim} \{\langle 0,2\rangle,\langle 2,0\rangle, \langle 1,2\rangle,\langle 2,1\rangle, \langle 2,3\rangle,\langle 3,2\rangle\} \overset{(23)}{\sim} \{\langle 0,3\rangle,\langle 3,0\rangle, \langle 1,3\rangle,\langle 3,1\rangle, \langle 2,3\rangle,\langle 3,2\rangle\};$$

Category 2:

$$\{\langle 0,1\rangle,\langle 1,0\rangle, \langle 0,2\rangle,\langle 2,0\rangle, \langle 1,2\rangle,\langle 2,1\rangle\} \overset{(23)}{\sim} \{\langle 0,1\rangle,\langle 1,0\rangle, \langle 0,3\rangle,\langle 3,0\rangle,\langle 1,3\rangle,\langle 3,1\rangle\} \overset{(12)}{\sim} \{\langle 0,2\rangle,\langle 2,0\rangle, \langle 0,3\rangle,\langle 3,0\rangle, \langle 2,3\rangle,\langle 3,2\rangle\} \overset{(01)}{\sim} \{\langle 1,2\rangle,\langle 2,1\rangle, \langle 1,3\rangle,\langle 3,1\rangle, \langle 2,3\rangle,\langle 3,2\rangle\};$$

Category 3:

$$\{\langle 0,2\rangle,\langle 2,0\rangle, \langle 1,2\rangle,\langle 2,1\rangle, \langle 1,3\rangle,\langle 3,1\rangle\} \overset{(13)}{\sim} \{\langle 0,2\rangle,\langle 2,0\rangle, \langle 1,3\rangle,\langle 3,1\rangle, \langle 2,3\rangle,\langle 3,2\rangle\} \overset{(23)}{\sim} \{\langle 0,3\rangle,\langle 3,0\rangle, \langle 1,2\rangle,\langle 2,1\rangle, \langle 2,3\rangle,\langle 3,2\rangle\} \overset{(12)}{\sim} \{\langle 0,3\rangle,\langle 3,0\rangle, \langle 1,2\rangle,\langle 2,1\rangle, \langle 1,3\rangle,\langle 3,1\rangle\} \overset{(01)}{\sim} \{\langle 0,2\rangle,\langle 2,0\rangle, \langle 0,3\rangle,\langle 3,0\rangle, \langle 1,3\rangle,\langle 3,1\rangle\} \overset{(23)}{\sim} \{\langle 0,2\rangle,\langle 2,0\rangle, \langle 0,3\rangle,\langle 3,0\rangle, \langle 1,2\rangle,\langle 2,1\rangle\} \overset{(12)}{\sim} \{\langle 0,1\rangle,\langle 1,0\rangle, \langle 0,3\rangle,\langle 3,0\rangle, \langle 1,2\rangle,\langle 2,1\rangle\} \overset{(23)}{\sim} \{\langle 0,1\rangle,\langle 1,0\rangle, \langle 0,2\rangle,\langle 2,0\rangle, \langle 1,3\rangle,\langle 3,1\rangle\} \overset{(12)}{\sim} \{\langle 0,1\rangle,\langle 1,0\rangle, \langle 0,2\rangle,\langle 2,0\rangle, \langle 2,3\rangle,\langle 3,2\rangle\} \overset{(01)}{\sim} \{\langle 0,1\rangle,\langle 1,0\rangle, \langle 1,2\rangle,\langle 2,1\rangle, \langle 2,3\rangle,\langle 3,2\rangle\} \overset{(23)}{\sim} \{\langle 0,1\rangle,\langle 1,0\rangle, \langle 1,3\rangle,\langle 3,1\rangle, \langle 2,3\rangle,\langle 3,2\rangle\} \overset{(01)}{\sim} \{\langle 0,1\rangle,\langle 1,0\rangle, \langle 0,3\rangle,\langle 3,0\rangle, \langle 2,3\rangle,\langle 3,2\rangle\}.$$

(4) The relations that include eight binary couples are 2 categories as follows:

Category 1:

$$\{\langle 0,1\rangle,\langle 1,0\rangle, \langle 0,2\rangle,\langle 2,0\rangle, \langle 0,3\rangle,\langle 3,0\rangle, \langle 1,2\rangle,\langle 2,1\rangle\} \overset{(23)}{\sim} \{\langle 0,1\rangle, \langle 1,0\rangle, \langle 0,2\rangle, \langle 2,0\rangle, \langle 0,3\rangle,\langle 3,0\rangle, \langle 1,3\rangle,\langle 3,1\rangle\} \overset{(12)}{\sim} \{\langle 0,1\rangle,\langle 1,0\rangle, \langle 0,2\rangle,\langle 2,0\rangle, \langle 0,3\rangle,\langle 3,0\rangle, \langle 2,3\rangle,\langle 3,2\rangle\} \overset{(013)}{\sim} \{\langle 0,1\rangle,\langle 1,0\rangle, \langle 0,2\rangle,\langle 2,0\rangle, \langle 1,2\rangle,\langle 2,1\rangle, \langle 1,3\rangle,\langle 3,1\rangle\} \overset{(23)}{\sim} \{\langle 0,1\rangle,\langle 1,0\rangle, \langle 0,3\rangle,\langle 3,0\rangle, \langle 1,2\rangle,\langle 2,1\rangle, \langle 1,3\rangle,\langle 3,1\rangle\} \overset{(02)}{\sim} \{\langle 0,1\rangle,\langle 1,0\rangle, \langle 1,2\rangle,\langle 2,1\rangle, \langle 1,3\rangle,\langle 3,1\rangle, \langle 2,3\rangle,\langle 3,2\rangle\} \overset{(1203)}{\sim} \{\langle 0,1\rangle,\langle 1,0\rangle, \langle 0,2\rangle,\langle 2,0\rangle, \langle 1,2\rangle,\langle 2,1\rangle, \langle 2,3\rangle,\langle 3,2\rangle\} \overset{(13)}{\sim} \{\langle 0,2\rangle,\langle 2,0\rangle, \langle 0,3\rangle,\langle 3,0\rangle, \langle 1,2\rangle,\langle 2,1\rangle, \langle 2,3\rangle,\langle 3,2\rangle\} \overset{(01)}{\sim} \{\langle 0,2\rangle,\langle 2,0\rangle, \langle 1,2\rangle,\langle 2,1\rangle, \langle 1,3\rangle,\langle 3,1\rangle, \langle 2,3\rangle,\langle 3,2\rangle\} \overset{(23)}{\sim} \{\langle 0,3\rangle,\langle 3,0\rangle, \langle 1,2\rangle,\langle 2,1\rangle, \langle 1,3\rangle,\langle 3,1\rangle, \langle 2,3\rangle,\langle 3,2\rangle\} \overset{(01)}{\sim} \{\langle 0,2\rangle,\langle 2,0\rangle,$$

<0,3>,<3,0>, <1,3>,<3,1>, <2,3>,<3,2>} $\overset{(12)}{\sim}$ {<0,1>,<1,0>, <0,3>,<3,0>, <1,3>,<3,1>, <2,3>,<3,2>} ;

Category 2 :

{<0,1>,<1,0>, <0,2>,<2,0>, <1,3>,<3,1>, <2,3>,<3,2>} $\overset{(23)}{\sim}$ {<0,1>,<1,0>, <0,3>,<3,0>, <1,2>,<2,1>, <2,3>,<3,2>} $\overset{(12)}{\sim}$ {<0,2>,<2,0>, <0,3>,<3,0>, <1,2>,<2,1>, <1,3>,<3,1>} ;

(4) The relations that include ten binary couples are 1 category as follows:

{ <0,1>,<1,0>,<0,2>,<2,0>,<0,3>,<3,0>,<1,2>,<2,1>,<1,3>,<3,1> } $\overset{(13)}{\sim}$
{ <0,1>,<1,0>,<0,2>,<2,0>,<0,3>,<3,0>,<1,3>,<3,1>,<2,3>,<3,2> } $\overset{(23)}{\sim}$
{ <0,1>,<1,0>,<0,2>,<2,0>,<0,3>,<3,0>,<1,2>,<2,1>,<2,3>,<3,2> } $\overset{(01)}{\sim}$
{ <0,1>,<1,0>,<0,2>,<2,0>,<1,2>,<2,1>,<1,3>,<3,1>,<2,3>,<3,2> } $\overset{(23)}{\sim}$
{ <0,1>,<1,0>,<0,3>,<3,0>,<1,2>,<2,1>,<1,3>,<3,1>,<2,3>,<3,2> } $\overset{(12)}{\sim}$
{<0,2>,<2,0>,<0,3>,<3,0>,<1,2>,<2,1>,<1,3>,<3,1>,<2,3>,<3,2>} .

References

1. Sheffer, H.M., A Set of Five Independent Postulates for Boolean Algebras with Application to Logical Constants, Trans. Am. Math. Soc., 14, (1913) 481-488.
2. Renren Liu, Some Results on the Decision for Sheffer Functions in Partial K-Valued Logic, Proceedings of the 28th International Symposium on Multiple-Valued Logic (II), (1998) 77-81.
3. Renren Liu, Research on the Similarity among Precomplete Sets Preserving m-ary Relations in Partial K-Valued Logic, Proceedings of the 29th International Symposium on Multiple-Valued Logic, (1999) 136-139.
4. Renren Liu et al, On the Categorization of Simply Separable Relations in Partial Four-Valued Logic, Lecture Notes in Computer Science 3612, Springer, Berlin, 2005, 1251-1256.

Concept Lattice and AFS Algebra

Lishi Zhang[1,2] and Xiaodong Liu[1]

[1] Research Center of Information and Control, Dalian University of Technology, Dalian,116024, P.R. China

[2] School of Science, Dalian Fisheries College , Dalian,116023, P.R. China

lishizhangcc@163.com, xdliuros@dlut.edu.cn

Abstract. Formal concept analysis is a field of applied mathematics based on mathematical analysis of concept and conceptual hierarchy,AFS algebra was first proposed and studied by Xiaodong Liu 1998 (Journal of Mathematical Analysis and Applications, vol. 217, Fuzzy Sets and Systems, vol. 95)[1,2]. In this paper, we explore the relationships between concept lattices, the AFS algebra, we analyze concept from the point of AFS algebra , We pave a way to explore concept lattice with AFS theory.

1 Introduction

Concept Lattices are used to represent conceptual hierarchies which are inherent in data. Introduced in the early 1980s,they are the core of the mathematical theory of Formal Concept Analysis (FCA)[10], FCA has over the years grown to a powerful theory for data analysis, and knowledge discovery[11],knowledge representation mechanism[12],conceptual clustering method [9] for crisp concepts. FCA is a branch of lattice theory motivated by the need for a clear mathematization of the notions of concept and conceptual hierarchy[6],FCA arouse the interests of practitioners from many fields such as pattern classification, data mining[7], knowledge acquisition[9], class hierarchy design and management[14].

AFS structure and AFS algebra were first proposed and studied by Liu [1, 2, 3, 4, 5], a new axiomatic system relating to the theory is introduced. In this procedure, the original data is the only source of all concerns, the membership functions and the fuzzy logic operations are automatically and impersonally obtained by the consistent mathematical methods according to the original data, a correspondence between a fuzzy concept and an element of AFS algebra is established. Such algebra is a completely distributive lattice, i.e., molecular lattices. Recently, AFS theory has been developed further and applied to fuzzy clustering analysis[5], fuzzy cognitive maps[8, 17, 18], concept representations[13] and fuzzy decision trees[15].

2 Previews About AFS Theory and Concept Lattice Theory

Definition 1. *(Liu [1]) Let* X, M *be sets,* 2^M *be the power set of* $M, \tau{:}X \times X \to 2^M$ (M, τ, X) *is called an AFS structure if* τ *satisfies the followings*

L. Wang et al. (Eds.): FSKD 2006, LNAI 4223, pp. 290–299, 2006.

$AX1: \forall(x_1, x_2) \in X \times X, \tau(x_1, x_2) \subseteq \tau(x_1, x_1)$;

$AX2: \forall(x_1, x_2), (x_2, x_3) \in X \times X, \tau(x_1, x_2) \cap \tau(x_2, x_3) \subseteq \tau(x_1, x_3)$

X is called the universe of discourse, M is called the attribute set and τ is called the structure relating to discourse X. If τ satisfies

$AX3: \forall(x_1, x_2) \in X \times X, \tau(x_1, x_2) \cup \tau(x_2, x_1) = M$

Then (M, τ, X) is called a strong relative AFS structure. X is called the universe of discourse, M is called the attribute set and τ is called the structure, structure τ gives us a vivid description about the system (M, τ, X),In the real world applications, if M is a set of simple attributes on X, then τ can be defined as $\tau(x, y) = \{m|m \in M, (x, y) \in R_m\}$ for any $x, y \in X$,where R_m is the binary relation on $X, (x, x) \in R_m$ means that x belong to m at some degree and for any $(x, y) \in R_m$,$x \neq y$ implies that x belong to m at a larger or equal degree than that of y.

Example 1. Let $X = \{x_1, x_2, ..., x_{10}\}$ be a set of ten persons. $M = \{m_1, m_2, ..., m_{10}\}$,and $m_1 = age, m_2 = height, m_3 = weight, m_4 = salary, m_5 = fortune, m_6 = male$, $m_7 = female, m_8 = black, m_9 = white, m_{10} = yellow$.

	m_1	m_2	m_3	m_4	m_5	m_6	m_7
x_1	20	1.9	90	1	0	y	n
x_2	13	1.2	32	0	0	n	y
x_3	50	1.7	67	140	34	n	y
x_4	80	1.8	73	20	80	y	n
x_5	34	1.4	54	15	2	y	n
x_6	37	1.6	80	80	28	n	y
x_7	45	1.7	78	268	90	y	n
x_8	70	1.65	70	30	45	y	n
x_9	60	1.82	83	25	98	n	y
x_{10}	3	1.1	21	0	0	n	y

Fig. 1. The Attributive Descriptions

About dermal color black, white and yellow, we have the preference relations

black: $x_7 > x_{10} > x_4 = x_8 > x_2 = x_9 > x_5 > x_6 = x_3 = x_1$;
white: $x_6 = x_3 = x_1 > x_5 > x_2 = x_9 > x_4 = x_8 > x_{10} > x_7$;
yellow: $x_2 = x_9 > x_4 = x_8 = x_5 > x_{10} > x_6 = x_3 = x_1 = x_7$;

where $x = y$ does not mean that x, ybe the same element in X, it denotes $(x, y) \in R$ and $(y, x) \in R$, that is to say the intensity of x is equal to the intensity of y.

According to table 1 and the preference relations of dermal color, construct τ: $X \times X \rightarrow 2^M$ as followings,$\forall x, y \in X, m \in \tau(x, y) \Leftrightarrow (x, y) \in R_m$ and $m \in M$. We learn $\forall m \in M$, R_m is a sub-preference relation. So τ suffices definition 1. For example the person x_2 whose fortune is zero, salary is zero, and is female, she has not the properties m_4, m_5, m_6,that is $m_6, m_5, m_4 \notin \tau(x_2, x_2) = \{m_1, m_2, m_3, m_7, m_8, m_9, m_{10}\}$.

An AFS structure (M, τ, X) is mathematical abstract of complex relations between the universe of discourse X and the attributive set M, membership functions and logic operations can be defined from AFS structure.

Now we explore the properties of structure τ,we give the following results.

Definition 2. *(Liu [1]) Let X, M be sets. Set $EXM^*$is defined as following*

$$EXM^* = \{\sum_{i\in I} a_iA_i | a_i \subseteq X, A_i \subseteq M, i \in I\}$$

Where I is any no empty indexing set, $\sum\limits_{i\in I}$ is just a symbol means that as an element of set EXM^* is composed of items a_iA_i, separated by symbol "+". When I is a finite index-ing set, is also denoted as $a_1A_1+a_2A_2+...+a_nA_n$.. $\sum\limits_{i\in I} a_iA_i$and $\sum\limits_{i\in I} a_{p(i)}A_{p(i)}$ are the same elements of if p is a bijection from I to I (e.g. ,$\sum\limits_{i=1}^{2} a_iA_i$ $a_1A_1 + a_2A_2$ and $a_2A_2 + a_1A_1$ are the same elements of EXM^*.

In [2], the author has defined an equivalence relation R on EXM^*as follows: $(\sum\limits_{i\in I} a_iA_i \sum\limits_{j\in J} b_jB_j) \in R$,$\forall i \in I$,$\exists k \in J$ such that $A_i \subseteq B_k$ and $\forall j \in J, \exists q \in I$ such that $B_j \supseteq A_q$,If $A_u \subseteq A_v$ and $a_u \supseteq a_v$, $u, v \in I, u \neq v$, then $\sum\limits_{i\in I} a_iA_i = \sum\limits_{j\in J, i\neq v} b_jB_j$

In the following, EXM^*/R (i.e. the quotient set) is denoted as EXM and $\sum\limits_{i\in I} a_iA_i = \sum\limits_{j\in J} b_jB_j$ implies that $\sum\limits_{i\in I} a_iA_i$, $\sum\limits_{j\in J} b_jB_j$ are equivalent. In [3], the author has proved that $(EXM, \vee, \wedge)$ are completely distributive lattices if the lattice operators $\vee, \wedge$ are defined as following: $\forall \sum\limits_{i\in I} a_iA_i, \sum\limits_{j\in J} b_jB_j \in EXM$,

$$\sum_{i\in I} a_iA_i \vee \sum_{j\in J} b_jB_j = \sum_{k\in I\sqcup J} c_kC_k$$

$$\sum_{i\in I} a_iA_i \wedge \sum_{j\in J} b_jB_j = \sum_{i\in I, j\in J} (a_i \cap b_j)(A_i \cup B_j)$$

where $k \in I \sqcup J$ is the disjoint union set of indexing sets I and J, $c_k = a_k$, $C_k = A_k$, when $k \in I$ and when $k \in J$. $c_k = b_k$,$C_k = B_k$. For the sake of simple, we define

$$\sum_{k\in I\sqcup J} c_kC_k = \sum_{i\in I} a_iA_i + \sum_{j\in J} b_jB_j$$

$(EXM, \vee, \wedge)$ is called the EII (expanding two sets X, M) algebra over X and $M.X\emptyset$ is the maximum element of EXM and $\emptyset M$ is the minimum element of EXM. In AFS theory, EXM are applied to represent the degrees of membership for fuzzy sets and $\sum\limits_{i\in I} a_iA_i = \sum\limits_{j\in J} b_jB_j$ implies that the membership degrees represented by $\sum\limits_{i\in I} a_iA_i$ and $\sum\limits_{j\in J} b_jB_j$ are equal.

Remark 1. In fact,we can define expand EII algebra when we choose sets $X_1, X_2, \cdots, X_n, M$ to construct a $EX_1X_2\cdots X_nM$,the operations are similar to the case as $n = 1$,thus,we obtain $EI, EII, \cdots, EI^n$ algebra.

Definition 3. *(Liu [2]) Let (M,τ,X) be an AFS structure,$x\in X, A\subseteq X, B\subseteq M$, we define the symbol*

$$\begin{aligned}\underline{B}(\{x\}) &= \{y \mid y\in X, (x,y)\supseteq B\}.\\ B(\{x\}) &= \{y \mid y\in X, (y,x)\supseteq B\}.\\ \underline{B}(A) &= \{y \mid y\in X, (x,y)\supseteq B, \forall x\in A\}.\\ B(A) &= \{y \mid y\in X, (y,x)\supseteq A, \forall x\in A\}.\end{aligned}$$

Definition 4. *Let (M,τ,X) be an AFS structure, $x\in X, A\subseteq M$, we define the membership of $\sum\limits_{i\in I} A_i\in EM$ as this*

$$\mu_{\sum\limits_{i\in I} A_i}(x)=\sum_{i\in I}\underline{A_i}(\{x\})A_i$$

For more detail,authors can refer [2, 3, 4, 5]

Lemma 1. *Let (M,τ,X) be an AFS structure*

$A_i\subseteq X, i\in I, B_j\subseteq M, j\in J, A\subseteq X, B\subseteq M$,
then we have that
(1) $\underline{(\bigcup\limits_{i\in I} A_i)}(B)=\bigcap\limits_{i\in I} A_i(B)$; $\underline{A}(\bigcup\limits_{j\in J} B_j)=\bigcap\limits_{j\in J}\underline{A}(B_j)$.
(2) $\underline{\overline{(\bigcup\limits_{j\in J} B_j)}}(A)=\bigcap\limits_{j\in J} B_j(A)$; $\underline{B}(\bigcup\limits_{i\in I} A_i)=\bigcap\limits_{i\in I} B(A_i)$.

Proof: From definition 4, it is straightforward to obtain them.

Proposition 1. *Let (M,τ,X) be an AFS structure, $A\subseteq X, B\subseteq M$,then we have that*

1) $\underline{A}(B)=M \Longleftrightarrow \forall g\in A$ and $m\in B, \underline{\{g\}}(\{m\})=M$
2) $\underline{B}(A)=G \Longleftrightarrow \underline{A}(B)=M$.

Proof: From Lemma 2 it is easy to obtain them.

Definition 5. *(B.Ganter,R.Wille,Formal [6]) A formal context is a triple concept (G,M,I) where G is a set of objects, M is a set of attributes, and I is a binary relation from G to M, i.e. $I\subseteq G\times M$, gIm is also written as $(g,m)\in I$, means that the object g possesses the attribute $m, g\in G, m\in M$. For a set of objects $A\subseteq G, \beta(A)$ is defined as the set of attributes shared by all objects in A, that is*

$$\beta(A)=\{m\in M|(g,m)\in I\ ,g\in A\}$$

Similarly, for $B\subseteq M, \alpha(B)$ is defined as the set of objects possesses all attributes in B, that is

$$\alpha(B)=\{g\in G|(g,m)\in I,\ m\in B\}$$

Definition 6. *(B.Ganter,R.Wille,Formal [6]) A formal concept of the context (G,M,I) is a pair (A,B) with $A\subseteq G, B\subseteq M$ and $\beta(A)=B, \alpha(B)=A$. We call A the extent of and B the intent of the concept (A,B).$\mathcal{B}(G,M,I)$ denote the set of all concepts of the context (G,M,I).*

	m_1	m_2	m_3	m_4	m_5
g_1	0	0	1	1	1
g_2	1	1	1	1	0
g_3	1	1	1	1	0
g_4	1	0	0	0	1
g_5	1	0	0	0	1
g_6	1	0	0	0	1

Fig. 2. Example of a Context

Example 2. Let $M = \{m_1, m_2, m_3, m_4, m_5\}, G = \{g_1, g_2, g_3, g_4, g_5, g_6\}$

As $\alpha(m_1, m_2) = \{g_2, g_3\}$, and $\beta\{g_2, g_3\} = \{m_1, m_2,\ m_3, m_4\}$,thus $(\{g2, g3\}, \{m1, m2\})$ is not a concept of $\mathcal{B}(G, M, I)$,and the pair $(\{g_1\}, \{m_3, m_4, m_5\})$ is a concept.

Lemma 2. *(B. Ganter, R.Wille, Formal [6]) Let (G, M, I) be a context, The following assertions hold*

1) For every $A_1, A_2 \subseteq G, A_1 \subseteq A_2$, implies that $\beta(A_2) \subseteq \beta(A_1)$,and for every $B_1, B_2 \subseteq M, B_1 \subseteq B_2$,implies $\alpha(B_2) \subseteq \alpha(B_1)$.

2) For any $A \subseteq G, A \subseteq \alpha(\beta(A))$ and $\beta(A) = \beta(\alpha(\beta(((A)))$; For any $B \subseteq M$, and $B \subseteq \beta(\alpha((B)))$ and $\beta(B) = \alpha(\beta(\alpha(B)))$.

Theorem 1. *(B. Ganter, R.Wille, Formal [6]) If T, S is an index set and, for every $t \in T, A_t \subseteq G$ is a set of objects, for every $s \in S, B_s \subseteq M$ is a set of attribute, then*

$$\beta(\bigcup_{t\in T} A_t) = \beta(\bigcap_{t\in T} A_t)\ ;\ \alpha(\bigcup_{s\in S} B_s) = \alpha(\bigcap_{s\in S} B_s)$$

Lemma 3. *Let (G, M, I) be a context, G and M be the object set and attribute set respectively. If τ:$G\times G \to 2^M$ is defined as following: For any $(g_1, g_2) \in G\times G$*

$$\tau(g_1, g_2) = \beta(g_1) \cup (\beta(g_2))^c$$

Then(G, M, I) is an AFS structure.

Proof: For any $(g_1, g_2), (g_2, g_3) \in G \times G$

$\tau(g_1, g_2) = \beta(g_1) \cup (\beta(g_2))^c \subseteq G = \beta(g_1) \cup (\beta(g_1))^c = \tau(g_1, g_1)$

$\tau(g_1, g_2) \cap \tau(g_2, g_3) \subseteq (\beta(g_1) \cup (\beta(g_2))^c) \cap (\beta(g_2) \cup (\beta(g_3))^c)$

$\subseteq \beta(g_1) \cup (\beta(g_3))^c \subseteq \tau(g_1, g_3)$.

Consequently, $AX1$ and $AX2$ of Definition 1 are satisfied.

3 Concept Lattice and AFS Algebra

In this section, we illustrate that the concept lattice is a special case of AFS algebra when M, the attribute set, contains all crisp attributes.

From Lemma 4, we know that (G, M, I) can be put as an AFS algebra. The key point is that whether or not the two structures coordinate in meaning. In what follows, we show that when M, the attribute set, contains all crisp attributes, it is indeed a concept lattice. The data in structure τ can be represented by a cross table, i.e.,by rectan-gular table the rows of which are headed by the object names and the column headed by the object names. A cross in row x_i and column m_j means that x_i has value v_{ij} in column m_j,where v_{ij} has value 1 or 0.If $m_k \in \beta(x_i) \cup (\beta(x_j))^c$,then we have that

If $m_k \in \beta(x_i) \cap (\beta(x_j))^c$,then $v_{ik} = 1$ and $v_{jk} = 0$.

If $m_k \in \beta(x_i), m_k \notin (\beta(x_j))^c$,then $v_{ik} = 1$ and $v_{jk} = 1$.

If $m_k \notin \beta(x_i), m_k \in (\beta(x_j))^c$,then $v_{ik} = 0$ and $v_{jk} = 0$.

Therefore, we have that $v_{ik} \geq v_{jk}$ holds in any cases, this implies that the structure τ keeps the meaning proposed by Liu[1],now we consider the operations defined on concept lattice and AFS algebra, we have that

Theorem 2. *Let (G, M, I) be a context, G and M be the object set and attribute set respectively. If $\tau : G \times G \to 2^M$ is defined as following: $Forall(g_1, g_2) \in G \times G$*

$$\tau(g_1, g_2) = \beta(g_1) \cup (\beta(g_2))^c$$

$A \subseteq G, B \subseteq M$, then we have that

1) For all $y \in G, x \in \alpha(B) \Rightarrow \mu_B(y) \leq \mu_B(x)$.

2) (A, B) is a concept of $\mathcal{B}(G, M, I) \Leftrightarrow \underline{A}(B) = M$ (or equivalently $\underline{B}(A) = G$) and $A \subseteq A_1, B \subseteq B_1$,such that $M = \underline{A_1}(B_1)$(or equivalently $G = \underline{B_1}(A_1)$) stands if and only if $A_1 = A$ and $B_1 = B$.

Proof: 1) From the definition of membership, we get that $\mu_{B(x)} = \underline{B}(\{x\})B$ and $\mu_B(y) = \underline{B}(\{y\})B$, If $\underline{B}(\{y\}) \subseteq \underline{B}(\{x\})$,then for $y \in G.y \in \underline{B}(\{y\}) \subseteq \underline{B}(\{x\})$.

Suppose $m_k \in B$ such that $v_{ik} = 0$,where $x = x_i$,thus as $m_k \in \tau(x_i, y)$,as the arbitrariness of y, then we get for all $i = 1, 2 \cdots\cdots |G|$,$v_{ik} = 0$.As a general assumption of G and M, full rows and full columns are always reducible,thus the fact that for all $i = 1, 2 \cdots\cdots, |G|$,$v_{ik} = 0$ leads to a contradict. Therefore we get that for all $m_k \in B, v_{ik} = 1$,thus we get that $B \subseteq \beta(x)$.

On the other hand, if $x \in \alpha(B)$,then we have that $B \subseteq \beta(x)$,as $\tau(x, y) = \beta(x) \cup (\beta(y))^c$,thus, for all $y \in G, \tau(x, y) \supseteq B, \forall s \in \underline{B}(\{y\})$,as $\tau(y, s) \supseteq B$, and $B \subseteq \tau(x, y) \cap \tau(y, s) \subseteq \tau(x, s)$,this implies that $s \in \underline{B}(\{x\})$,that is,$\underline{B}(\{y\}) \subseteq \underline{B}(\{x\})$.

2) we now in a position to show that if $\underline{B}(A) = G$ then $B = \beta(A)$.From the definition of $\underline{B}(A)$,we know that for all $g \in G$ and $a \in A, \beta(a) \cup (\beta(g))^c \supseteq B$, As

$$B \subseteq \beta(\alpha(B)) = \beta(\bigcup_{g \in \alpha(B)} \{g\}) = \bigcap\nolimits_{g \in \alpha(B)} \beta(g).$$ Choose one $g \in \alpha(B)$,

then $B \subseteq \beta(g)$,from $\beta(a) \cup (\beta(g))^c \supseteq B$, we get that for all $a \in A$, $B \subseteq \beta(a)$,thus $B \subseteq \bigcap_{a \in A} \beta(a) = \beta(\bigcup\limits_{a \in A} \{a\}) = \beta(A)$.$\forall m \in M$ and $\forall b \in B$, $\alpha(b) \cup \alpha((m))^c \supseteq \alpha(B)$,this implies that $M \subseteq \overline{\alpha(B)}(B)$,it follows that $M = \overline{\alpha(B)}(B)$,from the assumption,we have that $A = \alpha(B)$, equivalently,we can get that $B = \beta(A)$, thus (A, B) is a concept of $\underline{\mathcal{B}}(G, M, I)$. Conversely, Suppose (A, B) is a concept of $\underline{\mathcal{B}}(G, M, I)$,for all $b \in B$ and $m \in M$, $\tau(b, m) = \alpha(b) \cup \alpha((m))^c$,as $b \in B = \alpha(A)$,thus $\alpha(b) \supseteq \alpha(\beta(A)) = A$, this implies that $m \in \underline{A}(B)$.therefore, $M =$

$\underline{A}(B)$,it is easy to verify that $\underline{B}(A) = G$ .If there exist $A \subseteq A_1$, $B \subseteq B_1$,such that $M = \underline{A_1}(B_1)$ then $G = \underline{B_1}(A_1)$,from the above proof ,we know that $A_1 \subseteq \alpha(B_1)$ thus it follows that $A_1 \subseteq \alpha(B_1) \subseteq \alpha(B) = A$, that is $A_1 = A$,equivalently,we get and $B_1 = B$, this complete our proof.

Remark 2. 1) Theorem (2) and Proposition (1) provide an algorithm for the determination of concepts, if for $m \in B \subseteq M$and $g \in A \subseteq G$, we have that $\underline{m}(g) \neq G$ or $\underline{g}(m) \neq M$ then (A, B) is by no means a concept.

2) In AFS theory,$\underline{B}(A)$ and $\underline{A}(B)$ can be obtained with procedures.

Now we give an example to show the methods Now we manifest the table, the

	m_1	m_2	m_3	m_1, m_2	m_2, m_3	m_1, m_3	m_1, m_2, m_3
x_1	**1**	**1**	**0**	√	×	×	×
x_2	**1**	**0**	**1**	×	×	√	×
x_3	**0**	**1**	**1**	×	√	×	×
x_1, x_2	√	×	×	×	×	×	×
x_1, x_3	×	√	×	×	×	×	×
x_2, x_3	×	×	√	×	×	×	×
x_1, x_2, x_3	×	×	×	×	×	×	×

Fig. 3. Example of a Way to Find Concepts

top left table (with the red letters) is the original data, from the original table, we get that as $\underline{\{x_1\}}(\{m_3\}) = M$(or equivalently $\underline{\{m_3\}}(\{x_1\}) = G$, thus for all $m_3 \in B \subseteq M$ and $x_1 \in A \subseteq G, (A, B)$ is not a concept, therefore, a " × " is put in row which corresponds to set A and column which corresponds to set B, we can check them from lines to lines, or from columns to columns,When all the " × " are put in due places, the left corresponds to the concepts. in this example, the concepts are listed as the follows

$$(\{x_1\}, \{m_1, m_2\}), (\{x_2\}, \{m_1, m_3\}), (\{x_3\}, \{m_2, m_3\}),$$
$$(\{x_1, x_2\}, \{m_1\}), (\{x_1, x_3\}, \{m_2\}), (\{x_2, x_3\}, \{m_3\}).$$

With the Matlab procedures, we can easily handle them. Apart from this, the following distance function is useful for calculating the concepts in the context.

Definition 7. *(G, M, I) is a formal context, for $x, y \in G, B \subseteq M$,distance function d, induced by B, is defined like this;*

$$d(x, y) = |(\tau(x, y)^c \cup \tau(y, x)^c) \cap B|/|B|.$$

where $\tau(x, y)$ is defined as Lemma 4.

Lemma 4. *The function defined above is a distance function which satisfies;*

$\forall\ x, y, z \in G$

(1)$1 \geq d(x, y) \geq 0$ and $d(x, y) = 0 \Rightarrow x \in \alpha(B)$ if and only if $y \in \alpha(B)$.For $x, y \in \alpha(B), d(x, y) = 0$.

(2)$d(x, y) = d(y, x)$.
(3)$d(x, z) \leq d(x, y) + d(y, z)$.

Proof: By the definition of $\tau(x, y)$,it is a simple matter to verify them.
Now, we give an example about this.

	m_1	m_2	m_3	m_4
g_1	1	1	0	0
g_2	0	1	1	0
g_3	1	1	1	0
g_4	0	0	0	1

Fig. 4. Example of a Context

Let $B = \{m_2, m_3\}$,By simple computing, we can get that
$d(g_1, g_2) = 1/2, d(g_1, g_3) = 1/2, d(g_1, g_4) = 1/2, d(g_2, g_3) = 0, d(g_2, g_4) = 1$
$d(g_3, g_4) = 1$,as $(\{g_2, g_3\}, \{m_2, m_3\})$ is a concept, thus $d(g_2, g_3) = 0$.In fact, we can get all subsets of M easily, if for one set $N \subseteq M$, there exists an element $s \in \alpha(B)$ such that $d(s, t) = 0$ for some $t \in G$, then all such t and s form the extent with N as the intent, if $d(s, t) \neq 0$ for all $t \in G$, then N is not a intent at all.

In the follows, we find some concepts in AFS structure, most of all, we will give their implications in the fuzzy data, now we establish some definitions

Let (M, τ, X) is an AFS structure, for all $x \in X$,
$\beta(x) = \bigcap_{y \in X} \tau(y, x)$ and $\alpha(m) = \bigcap_{y \in X} \underline{m}(\{y\})$

Lemma 5. *Let (M, τ, X) be a structure, The following assertions hold:*

1) For every $A_1, A_2 \subseteq X, A_1 \subseteq A_2$, implies that $\beta(A_2) \subseteq \beta(A_1)$, and for every B_1, $B_2 \subseteq M, B_1 \subseteq B_2$, implies $\alpha(B_2) \subseteq \alpha(B_1)$.

2) For any $A \subseteq G, A \subseteq \alpha(\beta(A))$ and $\beta(A) = \beta(\alpha(\beta(A)))$ stands; For any $B \subseteq M$, and $B \subseteq \beta(\alpha(B))$ and $\alpha(B) = \alpha(\beta(\alpha(B)))$ holds.

3) $x \in \alpha(m) \Longleftrightarrow m \in \beta(x)$.

4) If T, S is an index set and, for every $t \in T, A_t \subseteq X$ is a set of objects, for every $s \in S$, $B_s \subseteq M$ is a set of attribute, then

$$\beta(\bigcup_{t \in T} A_t) = \beta(\bigcap_{t \in T} A_t) \ ; \ \alpha(\bigcup_{s \in S} B_s) = \alpha(\bigcap_{s \in S} B_s)$$

Proof: By the definition of $\beta(x)$ and $\alpha(m)$ it is easy to verify.

Remark 3. If we define $\beta(x) = \bigcap_{y \in X} \tau(x, y)$ and $\alpha(m) = \bigcap_{y \in X} m(\{y\})$,then we can get a result similar to Lemma 6.the corresponding concept is called max-concept.

Definition 8. *Let (M, τ, X) be a AFS structure, $A \subseteq X, B \subseteq G, (A, B)$ is called a min-concept of (M, τ, X) if $\beta(A) = B, \alpha(B) = A$.*

objects	attributes
x_1	$\{m_5, m_7\}$
x_2	$\{m_4, m_5, m_6\}$
x_3	$\{m_6\}$
x_4	$\{m_7\}$
x_5	$\{m_7\}$
x_6	$\{m_6\}$
x_7	$\{m_7\}$
x_8	$\{m_7\}$
x_9	$\{m_6\}$
x_{10}	$\{m_1, m_2, m_3, m_4, m_5, m_6\}\}$

attributes	objects
m_1	$\emptyset$
m_2	$\emptyset$
m_3	$\emptyset$
m_4	$\{x_2, x_{10}\}$
m_5	$\{x_1, x_2, x_{10}\}$
m_6	$\{x_2, x_3, x_6, x_9, x_{10}\}$
m_7	$\{x_1, x_4, x_5, x_7, x_8\}$

Fig. 5. Example of Attributes Objects

Example 3. Now we only list the min-concepts in example 1.

It is clear that $\{\{x_1, x_2, x_{10}\}, \{m_5\}\}$ is a concept, it corresponds to the fact "misfortune", $\{\{x_2, x_{10}\}, \{m_4, m_5\}\}$is a concept too, it corresponds to the fact "misfortune and the lowest salary",

Now we show the role of the concept in EI algebra.

Lemma 6. *Let X, M be sets. Set EXM^* is defined as following*
$A \subseteq X, B \subseteq G, (A, B)$ is a min-concept of (M, τ, X).
1) For all $y \in X$, the membership of $\mu_{B(x)} \leq \mu_{B(y)} \Longleftrightarrow x \in \alpha(B)$.
2) For all $B \subseteq M, y \in X, \mu_{A(x)} \leq \mu_{A(y)} \Longleftrightarrow x \in \beta(A)$.

Proof: From the definition of membership,we can get them. For the max-concept, the corresponding lemma can be easily obtained.

References

1. Xiaodong Liu.The Fuzzy Theory Based on AFS Algebras and AFS Structure, Journal of Mathematical Analysis and Applications, vol. 217(1998) 459-478
2. Xiaodong Liu.The Topology on AFS Algebra and AFS Structure, Journal of MathematicalAnalysis and Applications, vol. 217(1998) 479-489
3. Xiaodong Liu.A New Mathematical Axiomatic System of Fuzzy Sets and Systems,Journal of Fuzzy Mathematics, vol. 3(1995) 559-560
4. Xiaodong Liu.The Fuzzy Sets and Systems Based on AFS Structure, EI Algebra and EII algebra, Fuzzy Sets and Systems, vol. 95(1998)179-188
5. Xiaodong Liu,Wei Wang,Tianyou Chai.The Fuzzy Clustering Analysis Based on AFS Theory,EEE Transactions on Systems, Man and Cybernetics Part B, vol.35,No 5,October,(2005)1013-1027 No 5,October, (2005)1013-1027
6. Ganter B,Wille,R.Formal Concept Analysis: Mathematical Foundations, Springer, Berlin, (1999).
7. Pasquier N,Bastide Y,Taouil R, Lakhal T.Efficient mining of association rules using closed itemset lattices, Inform Systems 24(1)(1999)25-46.
8. Xiaodong Liu,Wanquan Liu.Credit Rating Analysis with AFS Fuzzy Logic,"Lecture Notes in Computer Science, LNCS 3612, 1198-1204, 2005.

9. Mineau G W,Godin R.Automatic structureing of knowledge bases by conceptual clustering,IEEE Trans.knowledge Data Eng.7(5) (1995) 824-828
10. Wille R.Restructuring lattice theory: an approach based on hierarchies of concepts, in: R. Ivan Rival (Ed.), Ordered Sets, Reidel, Dordecht,Boston,1982(445–470)
11. Stumme G,Wille R.(Eds.), Begriffliche Wissensverarbeitung—-Methoden und Anwendungen, Springer, Heidelberg (2000).
12. Wille R.Concept lattices and conceptual knowledge systems, Computers and Mathematics with Applications, vol. 23(1992)493—515.
13. Xiaodong Liu,Lishi Zhang, Zhou Jun, Kejiu Zhu and Qingling Zhang.The Structures of EI Algebras Generated by Information Attributes, Int. J. Intelligent Systems Technologies and Applications, in press
14. Godin R, Mili H, Mineau G, Missaoui R, Arfi A,ChauT.Design of class hierarchies based on concept Galois lattices, TAPOS, vol.4(2)(1998)117–134
15. Xiaodong Liu, Witold Pedrycz.The Development of Fuzzy Decision Tree in the Framework of Axiomatic Fuzzy Set Logic, Applied Soft Computing, accepted(2005)available online.
16. Vogt F, Wille R.TOSCANA—-A graphical tool for analyzing and exploring data, LNCS 894, Springer,Heidelberg(1995)226–233.
17. Xiaodong Liu, Qingling Zhang.The Fuzzy Cognitive Maps Based on AFS Fuzzy Logic,Dynam-ics of Continuous, Discrete and Impulsive Systems, Volume11, Number 5-6(2004)787-796.
18. Xiaodong Liu, Tianyou Chai, Wei Wang.AFS Fuzzy Logic Systems and Its Applicationsto Model and Control, " International Journal of Information and Systems Sciences,vol. 2, no. 3, 1-21, 2006.

Integrating Multiple Types of Incomplete Linguistic Preference Relations in Multi-person Decision Making

Zeshui Xu

Department of Management Science and Engineering
School of Economics and Management
Tsinghua University, Beijing 100084, China
Xu_zeshui@263.net

Abstract. In this paper, the multi-person decision making problems with various different types of incomplete linguistic preference relations are studied. Some new concepts, including incomplete uncertain linguistic preference relation, incomplete triangular fuzzy linguistic preference relation, incomplete trapezoid fuzzy linguistic preference relation, expected incomplete linguistic preference relation and acceptable expected incomplete linguistic preference relation, are defined. Based on some transformation functions, all these types of incomplete linguistic preference relations are transformed into the expected incomplete linguistic preference relations. By using the additive consistency property, the acceptable expected incomplete linguistic preference relations are then extended to the complete linguistic preference relations. Moreover, an approach is proposed for multi-person decision making based on multiple types of incomplete linguistic preference relations including incomplete traditional linguistic preference relations, incomplete uncertain linguistic preference relations, incomplete triangular fuzzy linguistic preference relations and incomplete trapezoid fuzzy additive linguistic preference relations.

1 Introduction

In the real world, many decision making problems involve choices from a finite discrete set of alternatives $X=\{x_1,x_2,...,x_n\}$, where each decision maker (DM) provides his/her preference information by using linguistic labels [1-10]. Suppose that $S=\{s_\alpha \mid \alpha=-t,...,t\}$ is a finite and totally ordered discrete linguistic label set, whose cardinality value is odd one, such as 7 and 9 [11-13], where s_α represents a possible value for a linguistic variable. For example, S can be assumed as

$$S=\{s_{-4}=extremely\ low,\ s_{-3}=very\ low,\ s_{-2}=low,\ s_{-1}=slightly\ low$$
$$s_0=fair,\ s_1=slightly\ high,\ s_2=high,\ s_3=very\ high,\ s_4=extremely\ high\}$$

where $s_\alpha < s_\beta$ iff $\alpha < \beta$. Normally, the mid label s_0 represents an assessment of "indifference", and with the rest of the linguistic labels being placed symmetrically around it.

L. Wang et al. (Eds.): FSKD 2006, LNAI 4223, pp. 300 – 309, 2006.

Usually, in the actual decision making process, some results may do not exactly match any linguistic labels in S. To preserve all the given linguistic preference information, Xu [13] extended the discrete linguistic label set S to a continuous label set $\overline{S} = \{s_\alpha \mid \alpha \in [-q,q]\}$, where $q(q > t)$ is a sufficiently large positive integer. If $s_\alpha \in S$, then s_α is termed an original linguistic label; otherwise, s_α is termed a virtual linguistic label. In general, the DM uses the original additive linguistic labels to evaluate alternatives, and the virtual additive linguistic labels can only appear in the actual calculations.

Linguistic preference relation is a common form used to express the DM's preference (judgment) for each pair of alternatives by means of linguistic labels. A number of studies have been conducted on multi-person decision making problems with linguistic preference relations [9,10,14-20]. All of these attempts focus on a single type of representation format of linguistic preference relations, and all the preferences are assumed to be available (for convenience, we call this type of linguistic preference relations the traditional linguistic preference relations). However, each DM is characterized by his/her own personal background and experience of the problem to be solved, the DMs' opinions may differ substantially, which could lead to the situations where some of them would not be able to efficiently express any kind of preference degree between two or more of the available options [21]. Thus, the DMs generally use different representation formats to express their linguistic preferences for each pair of alternatives in a multi-person decision making problem, and sometimes, some of preferences would be missing. Therefore, it is necessary to investigate this issue.

In this paper, we shall study the multi-person decision making problems with various different types of incomplete linguistic preference relations. We define the concepts of incomplete uncertain linguistic preference relation, incomplete triangular fuzzy linguistic preference relation, incomplete trapezoid fuzzy linguistic preference relation, expected incomplete linguistic preference relation and acceptable expected incomplete linguistic preference relation, etc. To make all these various different types of incomplete linguistic preference relations uniform, we transform them into the expected incomplete linguistic preference relations by using some transformation functions. We then extend all the acceptable expected incomplete linguistic preference relations to the complete linguistic preference relations by using the additive consistency property, and finally, we develop a practical approach to multi-person decision making based on multiple types of incomplete linguistic preference relations.

2 Definitions

Definition 1. *Let* $\tilde{s} = [s_\alpha, s_\beta] \in \tilde{S}$, *where* $s_\alpha, s_\beta \in \overline{S}$, s_α *and* s_β *are the lower and upper limits, respectively, then* $\tilde{s}$ *is called an uncertain linguistic variable ïïïï, where* $\tilde{S}$ *is the set of all the uncertain linguistic variables.*

Definition 2. *Let* $s_\alpha, s_\beta \in \overline{S}$, *then* $d(s_\alpha, s_\beta) = |\alpha - \beta|$ *is called the distance between* s_α *and* s_β.

$d(s_\alpha, s_\beta)$ reflects the similarity measure between s_α and s_β. It is clear that the greater the value of $d(s_\alpha, s_\beta)$, the closer s_α to s_β. Especially, if $d(s_\alpha, s_\beta)=0$, then $s_\alpha = s_\beta$.

3 Incomplete Linguistic Preference Relations

Consider a decision making problem, the DM compares each pair of alternatives in X by the linguistic labels in the set S, and may provide his/her linguistic preference information for each pair of alternatives by using linguistic preference relation, which is characterized by a function $\mu_A : X \times X \to D$, where D is the domain of representation of preference degrees [21]. A complete linguistic preference relation of order n necessitates the completion of all $n(n-1)/2$ judgments in its entire top triangular portion, Sometimes, however, the DM may develop a linguistic preference relation with incomplete information because of time pressure, lack of knowledge, and the DM's limited expertise related with problem domain, that is, some of preferences in a linguistic preference relation would be missing. In the following, we shall define the concepts of some incomplete linguistic preference relations.

Definition 3. *Let $A = (a_{ij})_{n \times n}$ be a linguistic preference relation, then A is called an incomplete traditional linguistic preference relation, if some of its elements can not be given by the DM, which we denote by the unknown variable "x", and the others can be provided by the DM, which satisfy*

$$a_{ij} \in \bar{S}, \; a_{ij} \oplus a_{ji} = s_0, \; a_{ii} = s_0$$

Definition 4. *Let $\tilde{A} = (\tilde{a}_{ij})_{n \times n}$ be a linguistic preference relation, then $\tilde{A}$ is called an incomplete uncertain linguistic preference relation, if some of its elements can not be given by the DM, which we denote by the unknown variable "x", and the others can be provided by the DM, which satisfy*

$$\tilde{a}_{ij} = [a_{ij}^{(l)}, a_{ij}^{(u)}] \in \tilde{S}, \; a_{ij}^{(l)} \oplus a_{ji}^{(u)} = s_0, \; a_{ij}^{(u)} \oplus a_{ji}^{(l)} = s_0, \; a_{ii}^{(l)} = a_{ii}^{(u)} = s_0$$

Definition 5. *Let $\hat{A} = (\hat{a}_{ij})_{n \times n}$ be a linguistic preference relation, then $\hat{A}$ is called an incomplete triangular fuzzy linguistic preference relation, if some of its elements can not be given by the DM, which we denote by the unknown variable "x", and the others can be provided by the DM, which satisfy*

$$\hat{a}_{ij} = [a_{ij}^{(l)}, a_{ij}^{(m)}, a_{ij}^{(u)}] \in \hat{S}, \; a_{ij}^{(l)} \oplus a_{ji}^{(u)} = s_0, \; a_{ij}^{(m)} \oplus a_{ji}^{(m)} = s_0$$

$$a_{ij}^{(u)} \oplus a_{ji}^{(l)} = s_0, \; a_{ii}^{(l)} = a_{ii}^{(m)} = a_{ii}^{(u)} = s_0$$

Definition 6. *Let* $\widehat{A} = (\widehat{a}_{ij})_{n\times n}$ *be a linguistic preference relation, then* $\widehat{A}$ *is called an incomplete trapezoid fuzzy linguistic preference relation, if some of its elements can not be given by the DM, which we denote by the unknown variable " x ", and the others can be provided by the DM, which satisfy*

$$\widehat{a}_{ij} = [a_{ij}^{(l)}, a_{ij}^{(m_1)}, a_{ij}^{(m_2)}, a_{ij}^{(u)}] \in \widehat{S},\ a_{ij}^{(l)} \oplus a_{ji}^{(u)} = s_0,\ a_{ij}^{(m_1)} \oplus a_{ji}^{(m_2)} = s_0$$

$$a_{ij}^{(m_2)} \oplus a_{ji}^{(m_1)} = s_0,\ a_{ij}^{(u)} \oplus a_{ji}^{(l)} = s_0,\ a_{ii}^{(l)} = a_{ii}^{(m_1)} = a_{ii}^{(m_2)} = a_{ii}^{(u)} = s_0$$

For convenience, we let Ω, $\widetilde{\Omega}$, $\hat{\Omega}$, and $\widehat{\Omega}$ be the sets of all the known elements of incomplete traditional linguistic preference relation A, incomplete uncertain linguistic preference relation $\widetilde{A}$, triangular fuzzy linguistic preference relation $\hat{A}$ and trapezoid fuzzy linguistic preference relation, respectively.

4 Expected Incomplete Linguistic Preference Relations

In order to make all the above types of incomplete linguistic preference relations uniform, below we shall define the concepts of expected incomplete linguistic preference relations of incomplete uncertain linguistic preference relation, incomplete triangular fuzzy linguistic preference relation and incomplete trapezoid fuzzy linguistic preference relation, respectively.

Definition 7. *Let* $\widetilde{A} = (\widetilde{a}_{ij})_{n\times n}$ *be an incomplete uncertain linguistic preference relation, then we define its expected linguistic preference relation as* $E(\widetilde{A}) = (E(\widetilde{a}_{ij}))_{n\times n}$, *where*

$$E(\widetilde{a}_{ij}) = \frac{1}{2}(a_{ij}^{(l)} \oplus a_{ij}^{(u)}),\ \textit{for all}\ i, j \in \widetilde{\Omega} \tag{1}$$

Clearly, by Definitions 4 and 7, we have

$$E(\widetilde{a}_{ij}) \oplus E(\widetilde{a}_{ji}) = s_0,\ E(\widetilde{a}_{ii}) = s_0,\ \textit{for all}\ i, j \in \widetilde{\Omega}$$

and thus, $E(\widetilde{A})$ is an incomplete traditional linguistic preference relation.

Definition 8. *Let* $\hat{A} = (\hat{a}_{ij})_{n\times n}$ *be an incomplete triangular fuzzy linguistic preference relation, then we define its expected linguistic preference relation as* $E(\hat{A}) = (E(\hat{a}_{ij}))_{n\times n}$, *where*

$$E(\hat{a}_{ij}) = \frac{1}{4}a_{ij}^{(l)} \oplus \frac{1}{2}a_{ij}^{(m)} \oplus \frac{1}{4}a_{ij}^{(u)},\ \textit{for all}\ i, j \in \hat{\Omega} \tag{2}$$

By Definitions 5 and 8, we have

$$E(\hat{a}_{ij}) \oplus E(\hat{a}_{ji}) = s_0,\ E(\hat{a}_{ii}) = s_0,\ \textit{for all}\ i, j \in \hat{\Omega}$$

and thus, $E(\hat{A})$ is an incomplete traditional linguistic preference relation.

Definition 9. *Let* $\widehat{A} = (\widehat{a}_{ij})_{n\times n}$ *be an incomplete trapezoid fuzzy linguistic preference relation, then we define its expected linguistic preference relation as* $E(\widehat{A}) = (E(\widehat{a}_{ij}))_{n\times n}$, *where*

$$E(\widehat{a}_{ij}) = \frac{1}{4}(a_{ij}^{(l)} \oplus a_{ij}^{(m_1)} \oplus a_{ij}^{(m_2)} \oplus a_{ij}^{(u)}),\ \textit{for all}\ i, j \in \widehat{\Omega} \tag{3}$$

By Definitions 6 and 9, we have

$$E(\widehat{a}_{ij}) \oplus E(\widehat{a}_{ji}) = s_0,\ E(\widehat{a}_{ii}) = s_0,\ \textit{for all}\ i, j \in \widehat{\Omega}$$

and thus, $E(\widehat{A})$ is also an incomplete traditional linguistic preference relation.

5 Acceptable Incomplete Traditional Linguistic Preference Relations

Definition 10. *Let* $A = (a_{ij})_{n\times n}$ *be an incomplete traditional linguistic preference relation, then* A *is called a consistent incomplete traditional linguistic preference relation, if*

$$a_{ij} = a_{ik} \oplus a_{kj},\ \textit{for all}\ i, j, k \in \Omega$$

Definition 11. *Let* $A = (a_{ij})_{n\times n}$ *be an incomplete traditional linguistic preference relation, the elements* a_{ij} *and* a_{kl} *are called adjoining, if* $(i, j) \cap (k, l) \neq \phi$. *For the unknown element* a_{ij}, *if there exist two adjoining known elements* a_{ik}, a_{kj}, *then* a_{ij} *is called available. Here,* a_{ij} *can be obtained indirectly by using* $a_{ij} = a_{ik} \oplus a_{kj}$.

Definition 12 [22]. *Let* $A = (a_{ij})_{n\times n}$ *be an incomplete traditional linguistic preference relation, if every unknown element can be obtained by its adjoining known elements, then* A *is called acceptable*; otherwise, A *is called unacceptable*.

Definition 13. *Let* $A = (a_{ij})_{n\times n}$ *be an acceptable incomplete traditional linguistic preference relation, if we replace each unknown* a_{ij} *in* A *with*

$$\overline{a}_{ij} = \frac{1}{p_{ij}} \sum_{k \in N_{ij}} (a_{ik} \oplus a_{kj}) \tag{4}$$

where $N_{ij} = \{k \mid a_{ik} \in \Omega, a_{kj} \in \Omega\}$, p_{ij} *is the number of the elements in* N_{ij}, *then we can get an extended complete traditional linguistic preference relation* $\overline{A} = (\overline{a}_{ij})_{n\times n}$ *of* $A = (a_{ij})_{n\times n}$.

6 Linguistic Aggregation Operators

As is well known, in multi-person decision making, all the individual decision information is needed to be aggregated into the collective decision information. To do so, in the following, we introduce some operational laws of linguistic labels and linguistic aggregation operators.

Definition 14 [13,22]. *Consider any two linguistic labels* $s_\alpha, s_\beta \in \overline{S}$, *we define their operational laws as follows:*

1) $s_\alpha \oplus s_\beta = s_\beta \oplus s_\alpha = s_{\alpha+\beta}$;
2) $\lambda s_\alpha = s_{\lambda\alpha}$, $\lambda \in [0,1]$;
3) $\lambda(s_\alpha \oplus s_\beta) = \lambda s_\alpha \oplus \lambda s_\beta$, $\lambda \in [0,1]$.

Definition 15 [22]. *Let* $LWAA : \overline{S}^n \to \overline{S}$, *if*

$$LWAA_\omega(s_{\alpha_1}, s_{\alpha_2}, \ldots, s_{\alpha_n}) = \omega_1 s_{\alpha_1} \oplus \omega_2 s_{\alpha_2} \oplus \cdots \oplus \omega_n s_{\alpha_n}$$

where $\omega = (\omega_1, \omega_2, \ldots, \omega_n)^T$ *is the weighting vector of the* s_{α_j}, *and* $\omega_j \in [0,1]$, $\sum_{j=1}^{n} \omega_j = 1$ $s_{\alpha_j} \in \overline{S}$, *then* $LWAA$ *is called a linguistic weighted arithmetic averaging (LWAA) operator. Especially, if* $\omega = (1/n, 1/n, \ldots, 1/n)^T$, *then LWAA operator is reduced to a linguistic arithmetic averaging (LAA) operator.*

Definition 16 [13]. *A linguistic ordered weighted aggregating (LOWA) operator of dimension* n *is a mapping* $LOWA: \overline{S}^n \to \overline{S}$ *that has an associated* n *vector* $w = (w_1, w_2, \ldots, w_n)^T$ *such that* $w_j \in [0,1]$, $\sum_{j=1}^{n} w_j = 1$. *Furthermore*

$$LOWA_w(s_{\alpha_1}, s_{\alpha_2}, \ldots, s_{\alpha_n}) = w_1 s_{\beta_1} \oplus w_2 s_{\beta_2} \oplus \cdots \oplus w_n s_{\beta_n}$$

where s_{β_j} is the jth largest of the s_{α_j}. *The LOWA operator is an extension of the ordered weighted operator (OWA) [23].*

It is clear that the LWAA operator weights the linguistic argument, and the LOWA operator weights the ordered position of the linguistic argument instead of weighting the argument itself, and therefore, weights represent different aspects in both the LWAA and LOWA operators. Xu [24] introduced another linguistic aggregation operator called linguistic hybrid arithmetic averaging (LHAA) operator, which generalizes both the LWAA and LOWA operators, and reflects the importance degrees of both the given argument and its ordered position.

Definition 17 [24]. *A linguistic hybrid arithmetic averaging (LHAA) operator is a mapping* $LHAA:\bar{S}^n \to \bar{S}$ *, which has associated with it a weighting vector* $w = (w_1, w_2, ..., w_n)^T$ *, with* $w_j \in [0,1]$, $\sum_{j=1}^{n} w_j = 1$ *, such that*

$$LHAA_{\omega,w}(s_{\alpha_1}, s_{\alpha_2}, ..., s_{\alpha_n}) = w_1 s_{\beta_1} \oplus w_2 s_{\beta_2} \oplus \cdots \oplus w_n s_{\beta_n}$$

where s_{β_j} *is the* j *th largest of the linguistic weighted argument* $\bar{s}_{\alpha_i}$ ($\bar{s}_{\alpha_i} = n\omega_i s_{\alpha_i}$, $i = 1,2,..., n$) , $\omega = (\omega_1, \omega_2, ..., \omega_n)^T$ *is the weighting vector of the* s_{α_i} *, with* $\omega_j \in [0,1]$, $\sum_{j=1}^{n} \omega_j = 1$ *, and* n *is the balancing coefficient.*

7 An Approach to Multi-person Decision Making Based on Multiple Types of Incomplete Linguistic Preference Relations

Based on the LAA and LHAA operators, in the following, we shall develop an approach to the multi-person decision making problem with various different types of incomplete linguistic preference relations.

Step 1. For a multi-person decision making problem with incomplete linguistic preference relations, let $X = \{x_1, x_2, ..., x_n\}$ be a finite set of alternatives, and $E = \{e_1, e_2, ..., e_t\}$ be a finite set of DMs. Let $v = (v_1, v_2, ..., v_p)^T$ be the weight vector of DMs, where $v_k \in [0,1]$, $\sum_{k=1}^{p} v_k = 1$. Each DM $e_k \in E$ provides his/her preference for each pair of alternatives by using the linguistic label set $S = \{s_\alpha \mid \alpha = -t, ..., t\}$, and constructs an incomplete linguistic preference relation $A_k = (a_{ij}^{(k)})_{n\times n}$ using one of the following representation formats: the incomplete traditional linguistic preference relation, incomplete uncertain linguistic preference relation, incomplete triangular fuzzy linguistic preference relation and incomplete trapezoid fuzzy additive linguistic preference relation.

Step 2. Utilize the expressions (1)-(3) to transform all the incomplete uncertain additional linguistic preference relations, incomplete triangular fuzzy linguistic preference relations and incomplete trapezoid fuzzy linguistic preference relations into the corresponding expected incomplete linguistic preference relations. We denote the expected incomplete linguistic preference relations of all the incomplete linguistic preference relations $A_k = (a_{ij}^{(k)})_{n\times n}$ $(k = 1,2,..., p)$ by $E(A_k) = (E(a_{ij}^{(k)}))_{n\times n}$

$(k = 1,2,..., p)$ (clearly, the expected incomplete traditional linguistic preference relation equals the incomplete traditional linguistic preference relations). If there exists an unacceptable expected incomplete linguistic preference relation, then it should be returned to the DM for reconstruction till an acceptable incomplete linguistic preference is reached.

Step 3. Utilize the expression (4) to extend all the expected incomplete linguistic preference relations $E(A_k) = (E(a_{ij}^{(k)}))_{n\times n}$ $(k = 1,2,..., p)$ into the expected complete linguistic preference relations $E(\overline{A}_k) = (E(\overline{a}_{ij}^{(k)}))_{n\times n}$ $(k = 1,2,..., p)$.

Step 4. Utilize the LAA operator

$$E(\overline{a}_i^{(k)}) = LAA\ (E(\overline{a}_{i1}^{(k)}), E(\overline{a}_{i2}^{(k)}),..., E(\overline{a}_{in}^{(k)})), \text{ for all } i, k$$

to aggregate the linguistic preference $E(\overline{a}_{ij}^{(k)})\,(j = 1,2,..., n)$ in the i th line of the $E(\overline{A}_k)$, and then get the averaged linguistic preference degree $E(\overline{a}_i^{(k)})$ of the i th alternative over all the other alternatives.

Step 5. Utilize the LHAA operator

$$E(a_i) = LHAA_{v,w}\ (E(\overline{a}_i^{(1)}), E(\overline{a}_i^{(2)}),..., E(\overline{a}_i^{(p)})), \text{ for all } i$$

to aggregate $E(\overline{a}_i^{(k)})$ $(k = 1,2,..., p)$ corresponding to the alternative x_i, and then get the collective averaged linguistic preference degree $E(\overline{a}_i)$ of i th alternative over all the other alternatives, where $w = (w_1, w_2,..., w_n)^T$ is the weighting vector of LHAA operator, with $w_j \in [0,1]$, $\sum_{j=1}^{n} w_j = 1$, and $v = (v_1, v_2,..., v_p)^T$ is the weight vector of DMs, with $v_k \in [0,1]$, $\sum_{k=1}^{p} v_k = 1$.

Step 6. Rank all the alternatives $x_i\,(i = 1,2,..., n)$ and select the best one(s) in accordance with the values of $E(\overline{a}_i)\,(i = 1,2,..., n)$.

Step 7. End.

8 Concluding Remarks

In many multi-person decision making problems, the DMs usually provide their preference information for each pair of alternatives with linguistic preference relations. These linguistic preference relations generally take different representation formats, and sometimes, some of their elements are missing. In this paper, we have focused on the multi-person decision making problems with various different types of incomplete linguistic preference relations. We have defined some new concepts such as incomplete uncertain linguistic preference relation, incomplete triangular fuzzy linguistic preference relation and incomplete trapezoid fuzzy linguistic preference relation, etc.

We have utilized some transformation functions to make all these incomplete linguistic preference relations uniform, and then utilized the additive consistency property to extend the acceptable incomplete linguistic preference relations to the complete linguistic preference relations, in which all the missing elements have been estimated by a simple procedure. We have also utilized some aggregation operators to fuse all the individual preference information into the collective preference information, by which the optimal alternative(s) can be derived.

Acknowledgement

The work was supported by the National Natural Science Foundation of China under Grant (70571087).

References

1. Zadeh, L.A.: The concept of a linguistic variable and its application to approximate reasoning. Part 1,2 and 3, Information Sciences 8(1975) 199-249, 301-357; 9(1976) 43-80.
2. Degani, R., Bortolan, G.: The problem of linguistic approximation in clinical decision making. International Journal of Approximate Reasoning 2(1988) 143-162.
3. Yager, R.R.: An approach to ordinal decision making. International Journal of Approximate Reasoning 12(1995) 237-261.
4. Torra, V.: Negation functions based semantics for ordered linguistic labels. International Journal of Intelligent Systems 11(1996) 975-988.
5. Bordogna, G., Fedrizzi, M., Passi, G.: A linguistic modeling of consensus in group decision making based on OWA operator. IEEE Transactions on Systems, Man, and Cybernetics 27(1997) 126-132.
6. Herrera, F., Martínez, L.: An approach for combining linguistic and numerical information based on 2-tuple fuzzy linguistic representation model in decision-making. International Journal of Uncertainty, Fuzziness, Knowledge-based Systems 8(2002) 539-562.
7. Xu, Z.S., Da, Q.L.: An overview of operators for aggregating information. International Journal of Intelligent Systems 18(2003) 953-969.
8. Herrera-Viedma, E., Peis, E.: Evaluating the informative quality of documents in SGML-format using fuzzy linguistic techniques based on computing with words. Information Processing and Management 39(2003) 195-213.
9. Xu, Z.S.: Deviation measures of linguistic preference relations in group decision making. Omega 33(2005) 249-254.
10. Xu, Z.S.: An approach based on the uncertain LOWG and the induced uncertain LOWG operators to group decision making with uncertain multiplicative linguistic preference relations. Decision Support Systems 41(2006) 488-499.
11. Kacprzyk, J., Fedrizzi, M.: Developing a fuzzy logic controller in case of sparse testimonies. International Journal of Approximate Reasoning 12(1995) 221-236.
12. Herrera, F., Martínez, L.: A model based on linguistic 2-tuples for dealing with multigranular hierarchical linguistic contexts in multi-expert decision-making. IEEE Transactions on Systems, Man, and Cybernetics 31(2001) 227-234.
13. Xu, Z.S.: Uncertain Multiple Attribute Decision Making: Methods and Applications. Tsinghua University Press, Beijing (2004).

14. Herrera, F., Herrera-Viedma, E., Verdegay, J.L.: A model of consensus in group decision making under linguistic assessments. Fuzzy Sets and Systems 78(1996) 73-87.
15. Herrera, F., Herrera-Viedma, E., Verdegay, J.L.: Direct approach processes in group decision making using linguistic OWA operators. Fuzzy Sets and Systems 79(1996) 175-190.
16. Herrera, F., Herrera-Viedma, E., Verdegay, J.L.: A rational consensus model in group decision making using linguistic assessments. Fuzzy Sets and Systems 88(1997) 31-49.
17. Herrera, F., Herrera-Viedma, E.: Choice functions and mechanisms for linguistic preference relations. European Journal of Operational Research 120(2000) 144-161.
18. Herrera, F., Herrera-Viedma, E.: Linguistic decision analysis: steps for solving decision problems under linguistic information. Fuzzy Sets and Systems 115(2000) 67-82.
19. Xu, Z.S.: A method based on linguistic aggregation operators for group decision making with linguistic preference relations. Information Sciences 166(2004) 19-30.
20. Xu, Z.S.: EOWA and EOWG operators for aggregating linguistic labels based on linguistic preference relations. International Journal of Uncertainty, Fuzziness and Knowledge-Based Systems 12(2004) 791-810.
21. Alonso, S., Chiclana, F., Herrera, F., Herrera-Viedma, E.: A Learning Procedure to Estimate Missing Values in Fuzzy Preference Relations Based on Additive Consistency. Lecture Notes in Computer Science 3131(2004) 227-238.
22. Xu, Z.S.: An approach to group decision making based on incomplete linguistic preference relations. International Journal of Information Technology and Decision Making 4(2005) 153-160.
23. Yager, R.R.: On ordered weighted averaging aggregation operators in multicriteria decision making. IEEE Transactions on Systems, Man, and Cybernetics 18(1988) 183-190.
24. Xu, Z.S.: A note on linguistic hybrid arithmetic averaging operator in group decision making with linguistic information. Group Decision and Negotiation, in press (2006).

Fuzzy Dual Ideal in a BCK-Algebra

Wenbo Qu[1,⋆], Zhihao Ma[2,⋆⋆], and Hao Jiang[3]

[1] Department of Mathematics, Shanghai Business School, Shanghai 200235 P.R. China
[2] Department of Mathematics, Shanghai Jiaotong University, Shanghai 200240 P.R. China
[3] Department of Mathematics, Zhejiang University, Hangzhou 310027 P.R. China
mamitli@zju.edu.cn

Abstract. In this paper, we study the fuzzy dual ideal of a BCK-algebra, and get the connection between dual ideals and fuzzy dual ideals.

Keywords: BCI-algebra, BCK-algebra, dual ideal, fuzzy dual ideal.

1 Introduction

The concept of fuzzy sets was introduced by Zadeh ([2]). Since then these ideas have been applied to other algebraic structures such as semigroups, groups, rings, modules, vector spaces and topologies. In 1991, Xi ([1]) applied the concept of fuzzy sets to BCK-algebras which are introduced by Imai and Iseki [3]. We know that BCK/BCI algebras play an important role in information science(see [6],[7],[8]). In 2001, Liu YongLin and Meng Jie studied Fuzzy ideals in BCI-algebras([5]). In 2002, C.Lele, C.Wu, and T.Mamadou studied the fuzzy filters in BCI-algebra[4].

In this paper, we study the fuzzy dual ideal of BCK-algebras, and get some interesting results.

2 Preliminaries

Let $(X, *, 0)$ be an algebra of $(2, 0)$ type, where X is a set and 0 is a constant. Then $(X, *, 0)$ is called a BCI-algebra, if the following four conditions hold:

(BCI-1).$((x * y) * (x * z)) * (z * y) = 0$. (BCI-2). $(x * (x * y)) * y = 0$. (BCI-3). $x * x = 0$. (BCI-4). $x * y = 0$ and $y * x = 0 \Rightarrow x = y$.

$(X, *, 0)$ is called to be a BCK-algebra, if it also satisfies the following condition: (BCK-5). $0 * x = 0$. And in a BCK-algebra, we can define a partial order as follows:$x \leq y$ if and only if $x * y = 0$.

⋆ The corresponding author.

⋆⋆ This work was supported by the education research foundation of Jiangsu province.

L. Wang et al. (Eds.): FSKD 2006, LNAI 4223, pp. 310–313, 2006.

3 Bounded BCK-Algebra, Dual Ideal, Fuzzy Dual Ideal

Definition 1. ([7]) *Let X be a BCK-algebra, if there is an element 1 of X satisfying $x \leq 1$ for all $x \in X$, then the element 1 is called unit of X. A BCK-algebra with unit is called to be bounded. In a bounded BCK-algebra, we denote $1 * x$ by Nx. In the following, X always means a bounded BCK-algebra.*

Definition 2. ([7]) *Let X be a bounded BCK-algebra, a nonempty subset D of X is called to be a dual ideal, if the following hold:*

(D1): $1 \in D$; (D2): $N(Nx * Ny) \in D$ and $y \in D$ imply $x \in D$ for any $x, y \in X$.

Recall that we have an equivalent definition of dual ideal:

Definition 3. ([7]) *Let X be a bounded BCK-algebra, a nonempty subset D of X is called to be a dual ideal, if the following hold:*

(D3): If $y \leq x$ and $y \in D$, then $x \in D$; (D2): $N(Nx * Ny) \in D$ and $y \in D$ imply $x \in D$ for any $x, y \in X$.

Let X be a bounded BCK-algebra, recall that a fuzzy set A in X is a map $A : X \to [0, 1]$.

Now we will introduce the concept of fuzzy dual ideal.

Definition 4. *Let X a bounded BCK-algebra, a fuzzy subset A of X is said to be a fuzzy dual ideal, if the follows hold:*

(FD1) If $x \leq y$, then $A(x) \leq A(y)$.
(FD2) For any $x, y \in X$, $A(x) \geq \min(A(N(Nx * Ny)), A(y))$.

Definition 5. ([4]) *Let X a bounded BCK-algebra, and let D be a subset of X and $\lambda \in (0, 1]$, we define a fuzzy set λ_D as following:*

If $x \in D$, then $\lambda_D(x) = \lambda$, otherwise $\lambda_D(x) = 0$.

So we know λ_D is similar to the character function on the set D.

Definition 6. ([9]) *Let A be a fuzzy subset of X, for $t \in [0, 1]$, define $A_t := \{x \in X | A(x) \geq t\}$, and is called the t-level subset of A.*

Theorem 1. *A nonempty subset D of X is a dual ideal if and only if λ_D is a fuzzy dual ideal for any $\lambda \in (0, 1]$.*

Proof. Suppose that D is a dual ideal. We want to prove that λ_D is a fuzzy dual ideal.

(a) First, let $x, y \in X$.

1. Assume $N(Nx * Ny) \in D$ and $y \in D$. We have $\lambda_D(N(Nx * Ny)) = \lambda$ and $\lambda_D(y) = \lambda$. Since D is a dual ideal, we have that $x \in D$, so $\lambda_D(x) = \lambda$. And we get that $\lambda_D(x) \geq \min(\lambda_D(N(Nx * Ny)), \lambda_D(y))$.

2. Assume $N(Nx * Ny) \in D$ and $y \notin D$. We have $\lambda_D(N(Nx * Ny)) = \lambda$ and $\lambda_D(y) = 0$. So $\min(\lambda_D(N(Nx*Ny)), \lambda_D(y)) = 0$. Since $\lambda_D(x) = \lambda$ or $\lambda_D(x) = 0$, we get that $\lambda_D(x) \geq \min(\lambda_D(N(Nx * Ny)), \lambda_D(y))$.

3. Assume $N(Nx * Ny) \notin D$ and $y \in D$. We have $\lambda_D(N(Nx * Ny)) = 0$ and $\lambda_D(y) = \lambda$. So $\min(\lambda_D(N(Nx*Ny)), \lambda_D(y)) = 0$. Since $\lambda_D(x) = \lambda$ or $\lambda_D(x) = 0$, we get that $\lambda_D(x) \geq \min(\lambda_D(N(Nx * Ny)), \lambda_D(y))$.

4. Assume $N(Nx * Ny) \notin D$ and $y \notin D$. We have $\lambda_D(N(Nx * Ny)) = 0$ and $\lambda_D(y) = 0$. So $\min(\lambda_D(N(Nx*Ny)), \lambda_D(y)) = 0$. Since $\lambda_D(x) = \lambda$ or $\lambda_D(x) = 0$, we get that $\lambda_D(x) \geq \min(\lambda_D(N(Nx * Ny)), \lambda_D(y))$.

(b) Let $x, y \in X$ and $x \leq y$, we will show that $\lambda_D(x) \leq \lambda_D(y)$. If $x \in D$, we have $\lambda_D(x) = \lambda$, since D is a dual ideal, we get that $y \in D$, so $\lambda_D(y) = \lambda$, and $\lambda_D(x) \leq \lambda_D(y)$. If $x \notin D$, we have $\lambda_D(x) = 0$, Since $\lambda_D(y) = \lambda$ or $\lambda_D(y) = 0$, so we get that $\lambda_D(x) \leq \lambda_D(y)$.

Conversely, assume that λ_D is a fuzzy dual ideal, we will prove that D is a dual ideal.

(a). Let $x \in D$ and $x \leq y$. Then we get that $\lambda_D(x) = \lambda$. Since λ_D is a fuzzy dual ideal, we get that $\lambda_D(x) \leq \lambda_D(y)$, so $\lambda_D(y) = \lambda$. So we get $y \in D$.

(b). Let $N(Nx * Ny) \in D$ and $y \in D$. We get that $\lambda_D(N(Nx * Ny)) = \lambda$ and $\lambda_D(y) = \lambda$. Since λ_D is a fuzzy dual ideal, we get that$\lambda_D(x) \geq \min(\lambda_D(N(Nx * Ny)), \lambda_D(y)) = \lambda$.

So we get that $\lambda_D(x) = \lambda$, and $x \in D$. Theorem is proved.

Theorem 2. *A fuzzy subset A of X is a fuzzy dual ideal if and only if for any $t \in [0, 1]$, the t-level subset $A_t := \{x \in X | A(x) \geq t\}$ is a dual ideal when $A_t \neq \emptyset$.*

Proof. Suppose that A_t is a dual ideal for any $t \in [0, 1]$. We want to prove that A is a fuzzy dual ideal.

(a) First, let $x, y \in X$ and $t = \min(A(N(Nx * Ny)), A(y))$. Then $N(Nx * Ny), y \in A_t$ and because A_t is a dual ideal, we get that $x \in A_t$, so we get $A(x) \geq t$, and $A(x) \geq t = \min(A(N(Nx * Ny)), A(y))$.

(b) Let $x, y \in X$ and $x \leq y$, we will show that $A(x) \leq A(y)$. Let $t = A(x)$. Then $x \in A_t$ and because A_t is a dual ideal, we get that $y \in A_t$, so $A(y) \geq t = A(x)$.

Conversely, assume that A is a fuzzy dual ideal, we will prove that A_t is a dual ideal.

(a). Let $x \in A_t$ and $x \leq y$. Then we get that $A(x) \geq t$. Since A is a fuzzy dual ideal, we get that $A(y) \geq A(x) \geq t$. So we get $y \in A_t$.

(b). Let $N(Nx * Ny) \in A_t$ and $y \in A_t$. We get that $A(N(Nx * Ny)) \geq t$ and $A(y) \geq t$. Since A is a fuzzy dual ideal, we get that $A(x) \geq \min(A(N(Nx * Ny)), A(y)) \geq t$.

So we get that $A(x) \geq t$, and $x \in A_t$. Theorem is proved.

Suppose A is a fuzzy set, we can define a new fuzzy set as following:

Definition 7. ([4]) *If A is a fuzzy subset of X, and $\alpha \in [0, 1]$, let $A^\alpha : X \to [0, 1]$ be given by $A^\alpha(x) = (A(x))^\alpha$.*

Theorem 3. *If A is a fuzzy dual ideal of X, and $\alpha \in [0, 1]$, then A^α is also a fuzzy dual ideal.*

Proof. (a) First, let $x, y \in X$. $A^\alpha(N(Nx * Ny)) = (A(N(Nx * Ny)))^\alpha$. Since A is a fuzzy dual ideal, we have: $A(x) \geq \min(A(N(Nx * Ny)), A(y))$. So we get that: $A^\alpha(x) = (A(x))^\alpha \geq (\min(A(N(Nx * Ny)), A(y)))^\alpha = \min((A(N(Nx * Ny)))^\alpha, (A(y))^\alpha) = \min(A^\alpha(N(Nx * Ny)), A^\alpha(y))$. So we get $A^\alpha(x) \geq \min(A^\alpha (N(Nx * Ny)), A^\alpha(y))$.

(b) Let $x, y \in X$ and $x \leq y$, we will show that $A^\alpha(x) \leq A^\alpha(y)$. Let $t = A(x)$. Because A is a fuzzy dual ideal, we get that $A(x) \leq A(y)$. So we get that $(A(x))^\alpha \leq (A(y))^\alpha$, that is, $A^\alpha(x) \leq A^\alpha(y)$. Theorem is proved.

Definition 8. ([4]) *Let X, Y be two bounded BCK algebras. Let $f : X \to Y$ be a mapping. Let B be a fuzzy subset of $f(X)$. Then $f^{-1}(B)(x) = B(f(x))$ is a fuzzy subset.*

A mapping f is called a BCK- homomorphism if $f(x * x') = f(x) * f(x')$. It is easy to get that $f(0) = 0$ and $f(x) \leq f(x')$ when $x \leq x'$.

Theorem 4. *Let f be an onto BCK- homomorphism. If B is a fuzzy dual ideal, then $f^{-1}(B)$ is also a fuzzy dual ideal.*

Proof. (a) First, let $x, y \in X$, we have the following: $f^{-1}(B)(N(Nx * Ny)) = B(f(N(Nx * Ny))) = B(N(Nf(x) * Nf(y)))$. Since B is a fuzzy dual ideal, we have, $B(f(x)) \geq \min(B(f(N(Nx * Ny))), B(f(y)) = \min(f^{-1}(B)(N(Nx * Ny)), f^{-1}(B)(y))$. So we proved that $f^{-1}(B)(x) \geq \min(f^{-1}(B)(N(Nx * Ny)), f^{-1}(B)(y))$.

(b) Let $x, y \in X$ and $x \leq y$, so we get that $f(x) \leq f(y)$. Because B is a fuzzy dual ideal, we get that $B(f(x)) \leq B(f(y))$. So we get that $f^{-1}B(x) \leq f^{-1}B(y)$. Theorem is proved.

References

[1] Xi, O.: Fuzzy BCK-algebra. Math. Japon. 36(1991) 935–942

[2] Zadeh, L.A.: Fuzzy sets. Information and Control 8(1965) 338–353

[3] Imai, Y., Iseki, K.: On axiom systems of propositional calculi. Proc. Japan Academy 42(1966) 19–22

[4] Lele, C., Wu, C.X., Mamadou, T.: Fuzzy filters in BCI-algebras. Int. J. Math. Math. Sci. 29(2002) 47–54

[5] Liu, Y.L., Meng, J.: Fuzzy ideals in BCI-algebras. Fuzzy Sets and Systems 123(2001) 227–237

[6] Meng, J., Jun, Y.B.: BCK-algebras. Kyung Moon Sa Co., Seoul, 1994

[7] Jun, Y.B., Xin, X.L.: On derivations of BCI-algebras. Inform. Sci. 159(2004) 167–176

[8] Zimmermann, H.J.: Fuzzy set theory–and its applications. Kluwer Academic Publishers, 2001

Theory and Practice on Information Granule Matrix*

Ye Xue[1,2] and Chongfu Huang[1]

[1] Institute of Disaster and Public Security, College of Resources Science, Beijing Normal University, No.19 Xinjiekouwai Street, Beijing 100875, China
[2] Department of Mathematics, Taiyuan University of Technology, Taiyuan, Shanxi, 030024, China
{jdjyxy, nortzw}@ires.cn

Abstract. In this paper, a new framework called information granule matrix is suggested to illustrate a given granule sample for showing its information structure. The new framework does not any extra condition but the observations. An information granule matrix can be turned into a fuzzy relation matrix for fuzzy inference. The concept of information granule matrix is firstly formulated. Then information granule matrix is shown by a simple example and discussed from the meaning of mechanism. To display the advantage of the new framework, it is compared with some existed methods. We also use our suggested framework to illustrate the relationship between earthquake magnitude M and isoseismal area S. The result shows that the new model is better than both Linear Regression and BP Network.

1 Introduction

Information exists everywhere all the time. Information is philosophically defined to be the reflection of motion state and existential fashion of objective reality. For example, a plane in the sky is an object. By using radar, we can monitor its motion and shape. A series of radar images are information. On another hand, information is technically defined as the object coding that can provide knowledge to the receiver. For example, speed and geometry of the monitored plane are the object codings. In fact, any information possesses its fixed information structure. In order to be satisfied with the need of research, many scholars explore the method of illustrating information structure.

In the 18th century, researchers found that numerous phenomena of physics and mechanics can be described by boundary value problems for differential equations. Then, equations of mathematical physics were developed into a popular approach to describe the relationships in physical systems. However, the equations of mathematical physics can describe physical phenomena with functions (i.e., relationships) in terms of partial differential equations, presuming that the basic laws of physics are known.

The traditional regression with respect to a given sample is to estimate a conditional expectation $E(Y|X=x)$ versus x with the given sample $\{(x_i, y_i)|i=1,2\cdots,n\}$. The regression result can describe the relationship between input and output, presuming that the shape of the population from which observations are drawn has been known

* Project Supported by National Natural Science Foundation of China, No. 40371002.

L. Wang et al. (Eds.): FSKD 2006, LNAI 4223, pp. 314–323, 2006.

and the size of the given sample is sufficiently large. But, for a small sample without any information about the population shape, it is very difficult to obtain a reasonable regression result.

The function approximate described by a trained neural network can be regarded as the estimation for the relationship we want to know. However, when a trained neural network is performing as a mapping from input space to output space, it is a black box. This means it is not possible to understand how a neural system works, and it is very hard to incorporate human a priori knowledge into a neural network. Furthermore, the well-known back-propagation (BP) algorithm has the problem of getting trapped in local minima, by which BP Network may lead to failure in finding a global optimal solution [1]. Besides, the convergence rate of BP Network is still too slow even if learning can be achieved.

The main advantage of the fuzzy graph concept [2] is the very compact and easy to understand representation of a function with if-then rules. Some fuzzy rule generators from training data by neural networks is more automatic. However, if the first-cut fuzzy approximation is far away the real relationship, the training data can do nothing. The method does not ensure that we can always find the rules with a given pool of experts or with a fixed set of data. Another problem is that fuzzy graphs suffer from the curse of dimensionality: rule explosion. They need too many rules to approximate a nonlinear function. The number of rules grows exponentially with the number of input and output variables.

In this paper, we suggest a new framework which does not any extra condition but the observations. It is made up of three kinds of information granule matrixes based on the types of information granule, which are called Type I-II Information Granule Matrix, Type III Information Granule Matrix, Type IV Information Granule Matrix, respectively. We discuss their difference by a simple example and from the meaning of mechanism. What's more, we understand that Type IV Information Granule Matrix is best of all. Finally, Type IV Information Granule Matrix can be compared with some existed methods. An application of illustrating the relationship between earthquake magnitude and isoseismal area proves its benefits.

2 Basic Concepts

In most real-world applications, the information would relate to continuous mathematical models. However, in many cases, the information is received piece by piece. The concept of a piece of information occurs under various evidences for perceiving any object, in broad sense, any observation is called a piece of information. In narrow sense, a sample point x_i of a given sample $W = \{x_1, x_2, \cdots, x_n\}$ drawn from population Ω is called a piece of information. In this case, sample point x_i provides an evidence in the following proposition, $g \triangleq (X \text{ is } x_i)$ is $1/n$. What's more, this proposition is perfectly identical to the definition of information granule provided by Zadeh [3].

Definition 1. *Let X be a variable taking values in U and G be a crisp subset of U . An information granule, g, in U is induced (or characterized) by a proposition of the form,* $g \triangleq X$ *is G.*

For example, a proposition, $g \stackrel{\Delta}{=} X$ is 2, is a granule.

Remark 1. If G be a fuzzy subset of U , then g is called fuzzy information granulation.

For example, the proposition, $g \stackrel{\Delta}{=} X$ is "around 2", is a fuzzy granulation.

Therefore, a sample point is an information granule. Or, an observation is an information granule. It is common, at least for real systems, there are four kinds of information granules [4]. Type I Information Granule: An observation with a crisp value from observing, experiments and data. For example, tossing a coin n times, we obtain n information granules. Type II Information Granule: An observation with a crisp vector which is more than 1-dimension from observing, experiment and data. For example, the scores of a student in mathematics, physics and chemistry. Type III Information Granule: An observation with a crisp set from observing, experiment and data. For example, a contour map includes a series of Type III Information Granules. Type IV Information Granule: An observation with a fuzzy set from observing, experiment and data. For example, proposition "Mary is young" is a Type IV Information Granule.

In fact, the relationship between observations from observing, experiments and data is actually the relationship between input information granules and output information granules. In the following section, we introduce a new approach called information granule matrix to describe this relationship. In addition, it is noted that a sample made up of information granules is called a granule sample, and that an element of the granule sample is called a sample granule.

3 Information Granule Matrix

Let $X = \{(x_i, y_i)|i = 1, 2, \cdots, n\}$ be a 2-dimension granule sample including an input granule $\{x_1, x_2, \cdots, x_n\}$ and output granule $\{y_1, y_2, \cdots, y_n\}$.

Let U be the domain of the input granule and V be the range of the output granule, respectively. The granule of U will be denoted by u, the same by v for V.

Let $A_j, j = 1, 2, \cdots, t$ and $B_k, k = 1, 2, \cdots, l$ be granules of U and V, respectively.

Let $U = \{A_j|j = 1, 2, \cdots, t\}$ and $V = \{B_k|k = 1, 2, \cdots, l\}$, their Cartesian product $U \times V$ is called an illustrating space. (A_j, B_k) is called an illustrating granule.

Unless stated otherwise, it is always assumed that X, A_j, B_k, U and V are nonempty.

(1) A_j, B_k are Type I Information Granule, i.e., $A_j \stackrel{\Delta}{=} u_j, B_k \stackrel{\Delta}{=} v_k$.

Definition 2. *Let*

$$g_{jk}(x_i, y_i) = \begin{cases} 1, \text{ if } x_i = u_j \text{ and } y_i = v_k, \\ 0, \text{ otherwise.} \end{cases} \quad \text{and } G_{jk} = \sum_{i=1}^{n} g_{jk}(x_i, y_i). \tag{3.1}$$

Then

$$G = \begin{array}{c} \\ u_1 \\ u_2 \\ \vdots \\ u_t \end{array} \begin{array}{c} \begin{array}{cccc} v_1 & v_2 & \cdots & v_l \end{array} \\ \begin{pmatrix} G_{11} & G_{12} & \cdots & G_{1l} \\ G_{21} & G_{22} & \cdots & G_{2l} \\ \vdots & \vdots & \vdots & \vdots \\ G_{t1} & G_{t2} & \cdots & G_{tl} \end{pmatrix} \end{array}. \tag{3.2}$$

G is called Type I Information Granule Matrix of X on $U \times V$, g_{jk} is called information gain of sample granule (x_i, y_i) at granule (u_j, v_k).

(2) A_j, B_k are Type III Information Granule, i.e., $A_j = [a_{1j}, a_{2j}] \triangleq U_j$, $B_k - [b_{1k}, b_{2k}] \triangleq V_k$.

Definition 3. *Let the information gain be*

$$e_{jk}(x_i, y_i) = \begin{cases} 1, \text{ if } (x_i, y_i) \in U_j \times V_k, \\ 0, \text{ otherwise.} \end{cases} \quad \text{and } E_{jk} = \sum_{i=1}^{n} e_{jk}(x_i, y_i). \tag{3.3}$$

Then

$$E = \begin{array}{c} \\ U_1 \\ U_2 \\ \vdots \\ U_t \end{array} \!\!\begin{array}{c} \begin{array}{cccc} V_1 & V_2 & \cdots & V_l \end{array} \\ \begin{pmatrix} E_{11} & E_{12} & \cdots & E_{1l} \\ E_{21} & E_{22} & \cdots & E_{2l} \\ \vdots & \vdots & \vdots & \vdots \\ E_{t1} & E_{t2} & \cdots & E_{tl} \end{pmatrix} \end{array}. \tag{3.4}$$

E is called Type III Information Granule Matrix of X on $U \times V$.

(3) A_j, B_k are Type IV Information Granule, i.e. A_j, B_k are fuzzy sets, respectively, written as $\widetilde{A}_j$ and $\widetilde{B}_k$. Let their membership functions be $\mu_j(u), u \in U$ and $\mu_k(v), v \in V$.

Definition 4. *Let the information gain be*

$$q_{jk}(x_i, y_i) = \mu_j(x_i) \times \mu_k(y_i) \text{ and } Q_{jk} = \sum_{i=1}^{n} q_{jk}(x_i, y_i) \tag{3.5}$$

Then

$$Q = \begin{array}{c} \\ \widetilde{A}_1 \\ \widetilde{A}_2 \\ \vdots \\ \widetilde{A}_t \end{array} \!\!\begin{array}{c} \begin{array}{cccc} \widetilde{B}_1 & \widetilde{B}_2 & \cdots & \widetilde{B}_l \end{array} \\ \begin{pmatrix} Q_{11} & Q_{12} & \cdots & Q_{1l} \\ Q_{21} & Q_{22} & \cdots & Q_{2l} \\ \vdots & \vdots & \vdots & \vdots \\ Q_{t1} & Q_{t2} & \cdots & Q_{tl} \end{pmatrix} \end{array}. \tag{3.6}$$

Q is called Type IV Information Granule Matrix of X on $U \times V$.

Remark 2. we don't mention Type II Information Granule Matrix, for it can be obtained by extending Type I Information Granule Matrix. Therefore, in the following, Type I Information Granule Matrix is expressed as Type I-II Information Granule Matrix.

4 Difference Among Them

4.1 A Simple Example

Example 1. There are 6 students in a group. Measuring their height, x_i in meters, and weight, y_i in kilograms, we obtain a granule sample

$$\begin{aligned} X &= \{(x_i, y_i) | i = 1, 2, \cdots, 6\} \\ &= \{(1.60, 50), (1.70, 65), (1.65, 55), (1.70, 60), (1.66, 62), (1.60, 55)\}. \end{aligned}$$

(1) Taking $U=\{u_1,u_2,u_3\}=\{1.60,1.65,1.70\}$ and $V=\{v_1,v_2,v_3,v_4\}=\{50,55,60,65\}$ by calculating with formula (3.1), we have got Type I-II Information Granule Matrix G,

$$G=\begin{array}{c} \\ u_1 \\ u_2 \\ u_3 \end{array}\begin{array}{c} \begin{array}{cccc} v_1 & v_2 & v_3 & v_4 \end{array} \\ \begin{pmatrix} 1 & 1 & 0 & 0 \\ 0 & 1 & 0 & 0 \\ 0 & 0 & 1 & 1 \end{pmatrix} \end{array}. \tag{4.1}$$

(2) Taking $U=\{U_1,U_2,U_3\}=\{[1.60,1.65),[1.65,1.70),[1.70,1.75)\}$ and $V=\{V_1,V_2,V_3,V_4\}=\{[50,55),[55,60),[60,65),[65,70)\}$ by calculating with formula (3.3), we have got Type III Information Granule Matrix E,

$$E=\begin{array}{c} \\ U_1 \\ U_2 \\ U_3 \end{array}\begin{array}{c} \begin{array}{cccc} V_1 & V_2 & V_3 & V_4 \end{array} \\ \begin{pmatrix} 1 & 1 & 0 & 0 \\ 0 & 1 & 1 & 0 \\ 0 & 0 & 1 & 1 \end{pmatrix} \end{array}. \tag{4.2}$$

If we slightly change U_j, V_k, then we get the following result.
Taking $U=\{U_1,U_2,U_3\}=\{[1.60,1.65],(1.65,1.70],(1.70,1.75)\}$ and $V=\{V_1,V_2,V_3,V_4\}=\{[50,55],(55,60],(60,65],(65,70)\}$ by calculating with formula (3.3), we have got Type III Information Granule Matrix E',

$$E'=\begin{array}{c} \\ U_1 \\ U_2 \\ U_3 \end{array}\begin{array}{c} \begin{array}{cccc} V_1 & V_2 & V_3 & V_4 \end{array} \\ \begin{pmatrix} 3 & 0 & 0 & 0 \\ 0 & 1 & 2 & 0 \\ 0 & 0 & 0 & 0 \end{pmatrix} \end{array}. \tag{4.3}$$

(3) Taking $U=\{\widetilde{A}_1,\widetilde{A}_2,\widetilde{A}_3\}=\{$"*around* 1.60", "*around* 1.65", "*around* 1.70"$\}$ and $V=\{\widetilde{B}_1,\widetilde{B}_2,\widetilde{B}_3,\widetilde{B}_4\}=\{$"*around* 50", "*around* 55", "*around* 60", "*around* 65"$\}$ by calculating with formula (3.5), we have got Type IV Information Granule Matrix Q,

$$Q=\begin{array}{c} \\ \widetilde{A}_1 \\ \widetilde{A}_2 \\ \widetilde{A}_3 \end{array}\begin{array}{c} \begin{array}{cccc} \widetilde{B}_1 & \widetilde{B}_2 & \widetilde{B}_3 & \widetilde{B}_4 \end{array} \\ \begin{pmatrix} 1 & 1 & 0 & 0 \\ 0 & 1 & 0.48 & 0.32 \\ 0 & 0 & 1.12 & 1.08 \end{pmatrix} \end{array}. \tag{4.4}$$

From the formulas (4.1), (4.2), (4.3) and (4.4), we have the following conclusions: (i) Type I-II Information Granule Matrix cannot include all sample granules, because sample granule (1.66, 62) hasn't been illustrated in it. (ii) Type III Information Granule Matrix can include all sample granules, but all sample granules falling into the same $U_j \times V_k$ are considered to play a same role. Thus, it neglects the difference between them. Neglecting the position difference implies that we throw away some information. (iii) If E is compared with E', then there exists very much stronger change although intervals of U_j, V_k have a little change in the boundary for Type III Information Granule Matrix. (iv) Type IV Information Granule Matrix fills up these gaps, that is, it picks up the information about the differences, what's more, it includes all sample granules.

4.2 The Meaning of Mechanism

When we use an information granule matrix to illustrate a given granule sample X for its information structure, the matrix plays a role to collect the observation's information by some means. An sample granule can be regarded as a small "ball". The information of the ball is transmitted to sensors located on the matrix when the ball falls.

Type I-II Information Granule Matrix framework is special which consists of a number of nodes (u_j, v_k) with sensors. Fig. 1 shows the dynamics model of Type I-II Information Granule Matrix. Because the balls are small, some of them may escape from the framework when they are falling on it. Therefore, the model, in many cases, cannot collect all the information carried by a given granule sample.

Type III Information Granule Matrix framework can be regarded as a group of boxes, with interval length and width. The cover of a box is formed by four nodes (u, v) where u, v are end points of the intervals. Suppose that a box has an inverted cone bottom, and its cover has been taken off. We place these boxes in order, one by one without leaving any empty space to fill in the square area encircled by the framework. Then, for each box, we set a sensor at the center of the bottom. Certainly, the box-matrix can catch all balls and every ball can touch one of the sensors located in the bottoms. However, the sensors cannot identify the first position where a ball falls. Hence, this model cannot totally show all gradients. In other words, some information cannot be grabbed by the model. Fig. 2 shows the dynamics model of Type III Information Granule Matrix.

Type IV Information Granule Matrix framework is a more intelligent structure. Firstly, to each node (u_j, v_k), where u_j, v_k are the gravity of fuzzy sets $\widetilde{A}_j, \widetilde{B}_k$, we set a sensor at it. Then, for a group of four sensors, we image a thin wood board, with length and width, covering on the sensors. We can put $(t-1)\times(l-1)$ boards on all groups. These boards are independent.

Now, when a ball falls, it must strike one of the boards. If its first position is not just in one of the supporting nodes, all four sensors under the board can detect it. However, the messages received by these sensors are not as strong as that when a ball falls straightly

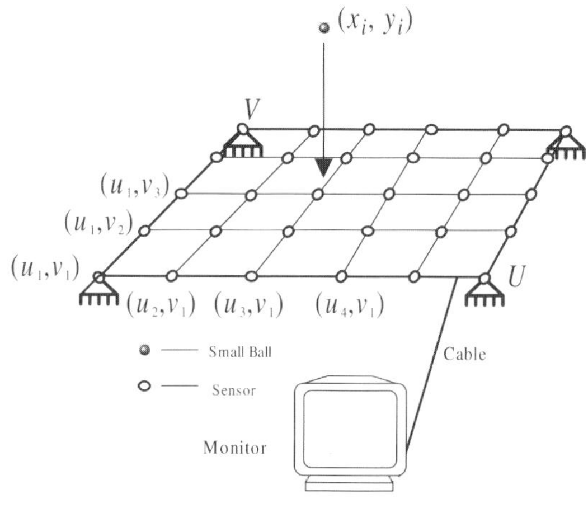

Fig. 1. Dynamics model of a Type III Information Granule Matrix

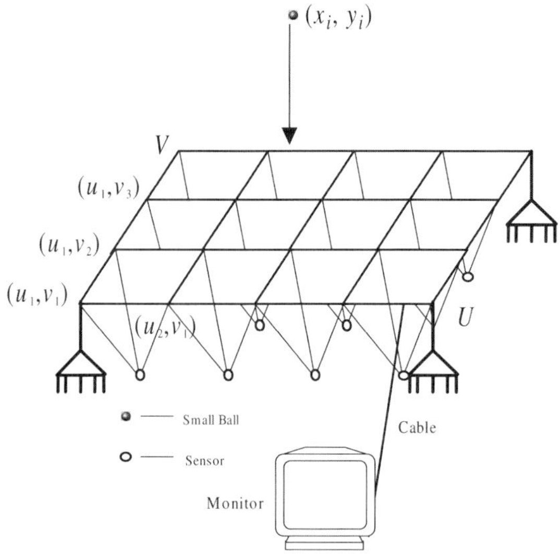

Fig. 2. Dynamics model of a Type III Information Granule Matrix

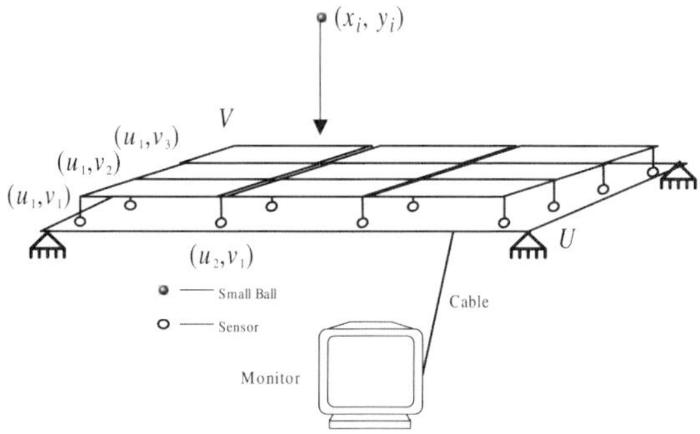

Fig. 3. Dynamics model of a Type IV Information Granule Matrix

on one of sensors. In other words, the striking force caused by a ball is shared by the four supporting nodes, and, at the same time, the information carried by a ball is distributed among the four sensors. Particularly, when a ball falls straightly on one of sensors, the other three sensors receive zero message. By this way, not only no ball can be lost, but also we can identify the first positions by four sensors. Fig. 3 shows the dynamics model of Type IV Information Granule Matrix.

5 Comparison with Other Existed Methods

In this section, we theoretically compare information granule matrix model with equations of mathematical physics, regression methods, artificial neural networks and fuzzy graphs.

Equations of mathematical physics would be the most ideal, but when it is used, we must meet the following natural requirements: (i) a solution must be exist. (ii) the solution must be unique. (iii) the solution must depend continuously on the data of the problem.

Regression methods in terms of mathematical statistics would be the most common, but when it is used, we must presume that the type of the population from which observations are taken is known and the size of the given granule sample is sufficiently large.

Artificial neural networks could be considered as the latest fashion, but when it is performing as a mapping from input space to output space, it is very hard to incorporate human a prior knowledge into it. Besides, it only lead to local minima and the convergence rate is still too slow.

Fuzzy graphs might be the most visual, but it not only is too rough, but also it depends on expert's experience.

Therefore, relatively speaking, information granule matrix is more visual and precise on the conditions as the following: (i) We don't know the basic laws of physics with respect to the given observations. (ii) We don't know the shape of the population from which observations are drawn. (iii) The patterns (granules (x_i, y_i), $i = 1, 2, \cdots, n$) may be contradictory (same input but different output). (iv) There isn't any given pool of experts. (v) There does not involve any complex operation.

6 Illustrating the Relationship Between Earthquake Magnitude and Isoseismal Area

In Yunnan Province of China, there is a data set of strong earthquakes consisting of 25 records from 1913 to 1976 with magnitude, M, and isoseismal area, S, of intensity, $I \geq VII$. Obviously, M and S belong to Type I information granule. Therefore, the granule sample with magnitude information granule m and logarithmic isoseismal area information granule $y = \log S$ is in (6.1).

$$\begin{aligned} W = & \{(x_1, y_1), (x_2, y_2), \cdots, (x_{25}, y_{25})\} \\ = & \{(6.5, 3.455), (6.5, 3.545), (7, 3.677), (5.75, 2.892), (7, 3.414), \\ & (7, 3.219), (6.25, 3.530), (6.25, 3.129), (5.75, 2.279), (6, 1.944), \\ & (5.8, 1.672), (6, 3.554), (6.2, 2.652), (6.1, 2.865), (5.1, 1.279), \\ & (6.5, 3.231), (5.4, 2.417), (6.4, 2.606), (7.7, 3.913), (5.5, 2.000), \\ & (6.7, 2.326), (5.5, 1.255), (6.8, 2.301), (7.1, 2.923), (5.7, 1.996)\}. \end{aligned} \tag{6.1}$$

To calculate, we employ the following universes U and V for the granule sample W.

$$\begin{aligned} U &= \{\text{“}\mathit{around}\ 5.010\text{”}, \text{“}\mathit{arond}\ 5.106\text{”}, \cdots, \text{“}\mathit{around}\ 7.790\text{”}\} \\ V &= \{\text{“}\mathit{around}\ 1.164\text{”}, \text{“}\mathit{around}\ 1.262\text{”}, \cdots, \text{“}\mathit{around}\ 4.004\text{”}\} \end{aligned}$$

Let $\mu_j(u), u \in U$ and $\mu_k(v), v \in V$ be triple fuzzy numbers in (3.5), we can obtain Type IV Information Granule Matrix Q by (3.5).

$$Q = \begin{array}{c} \\ \widetilde{A}_1 \\ \widetilde{A}_2 \\ \widetilde{A}_3 \\ \widetilde{A}_4 \\ \widetilde{A}_5 \\ \widetilde{A}_6 \\ \widetilde{A}_7 \\ \vdots \\ \widetilde{A}_{29} \\ \widetilde{A}_{30} \end{array} \begin{array}{c} \begin{array}{cccccc} \widetilde{B}_1 & \widetilde{B}_2 & \widetilde{B}_3 & \cdots & \widetilde{B}_{29} & \widetilde{B}_{30} \end{array} \\ \begin{pmatrix} 0.000000 & 0.051658 & 0.010842 & \cdots & 0.000000 & 0.000000 \\ 0.000000 & 0.774872 & 0.162628 & \cdots & 0.000000 & 0.000000 \\ 0.000000 & 0.000000 & 0.000000 & \cdots & 0.000000 & 0.000000 \\ 0.000000 & 0.000000 & 0.000000 & \cdots & 0.000000 & 0.000000 \\ 0.000000 & 0.000000 & 0.000000 & \cdots & 0.000000 & 0.000000 \\ 0.063988 & 0.831845 & 0.000000 & \cdots & 0.000000 & 0.000000 \\ 0.007440 & 0.096726 & 0.000000 & \cdots & 0.000000 & 0.000000 \\ \vdots & \vdots & \vdots & \vdots & \vdots & \vdots \\ 0.000000 & 0.000000 & 0.000000 & \cdots & 0.573342 & 0.030825 \\ 0.000000 & 0.000000 & 0.000000 & \cdots & 0.375638 & 0.020196 \end{pmatrix} \end{array} . \tag{6.2}$$

In order to compare the results of Type IV Information Granule Matrix model with ones of Linear Regression model and BP Network model [5], firstly, we obtain fuzzy relation matrix R_f of Q by use the formula (6.3).

$$\begin{cases} R_f = \{r_{ij}\}_{m\times t} \\ r_{ij} = Q_{jk}/s_k \\ s_k = \max_{1\le j\le m} Q_{jk} \end{cases} \tag{6.3}$$

$$R_f = \begin{array}{c} \\ \widetilde{A}_1 \\ \widetilde{A}_2 \\ \widetilde{A}_3 \\ \widetilde{A}_4 \\ \widetilde{A}_5 \\ \widetilde{A}_6 \\ \widetilde{A}_7 \\ \vdots \\ \widetilde{A}_{29} \\ \widetilde{A}_{30} \end{array} \begin{array}{c} \begin{array}{cccccc} \widetilde{B}_1 & \widetilde{B}_2 & \widetilde{B}_3 & \cdots & \widetilde{B}_{29} & \widetilde{B}_{30} \end{array} \\ \begin{pmatrix} 0.000000 & 0.062101 & 0.066667 & \cdots & 0.000000 & 0.000000 \\ 0.000000 & 0.931510 & 1.000000 & \cdots & 0.000000 & 0.000000 \\ 0.000000 & 0.000000 & 0.000000 & \cdots & 0.000000 & 0.000000 \\ 0.000000 & 0.000000 & 0.000000 & \cdots & 0.000000 & 0.000000 \\ 0.000000 & 0.000000 & 0.000000 & \cdots & 0.000000 & 0.000000 \\ 1.000000 & 1.000000 & 0.000000 & \cdots & 0.000000 & 0.000000 \\ 0.116279 & 0.116299 & 0.000000 & \cdots & 0.000000 & 0.000000 \\ \vdots & \vdots & \vdots & \vdots & \vdots & \vdots \\ 0.000000 & 0.000000 & 0.000000 & \cdots & 1.000000 & 1.000000 \\ 0.000000 & 0.000000 & 0.000000 & \cdots & 0.655173 & 0.655183 \end{pmatrix} \end{array} . \tag{6.4}$$

When x_0 is given, by using max – min composition fuzzy inference we obtain $\widetilde{y}_0$ based on R_f.

$$\widetilde{y}_0 = \max\{\min\{\mu_{\text{``}around\ m_0\text{''}}(u_i), R_f\}\}. \tag{6.5}$$

In order to defuzzify $\widetilde{y}_0$, we compute the gravity center of fuzzy set [6] in Eq.(6.6).

$$y_0 = \sum_{i=1}^{25}(\mu_{\widetilde{y}_0}(v_k)\times v_k)/\sum_{i=1}^{25}\mu_{\widetilde{y}_0}(v_k). \tag{6.6}$$

For the granule sample W, the results obtained by using Type IV Information Granule Matrix model are shown in Fig. 4 with a thick solid curve, where thin solid curve is from Linear Regression model, $y = -2.61 + 0.85x$, and dash curve from an BP Network model with momentum rate $\eta = 0.9$, learning rate $\alpha = 0.7$.

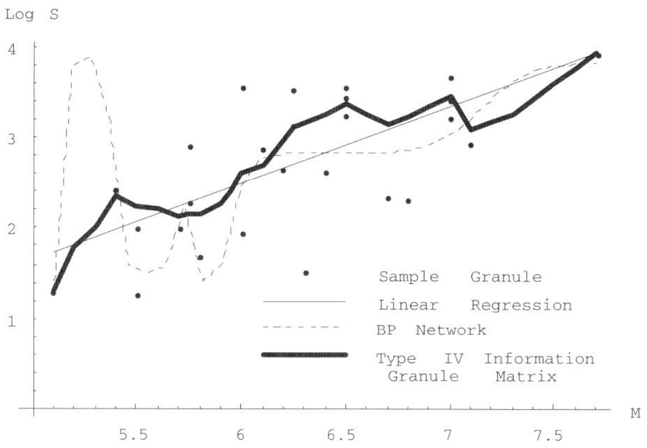

Fig. 4. Relationship between earthquake magnitude and logarithmic isoseismal area estimated by Linear Regression (thin solid curve), BP Network(dash curve) and Type IV Information Granule Matrix (thick solid curve)

Obviously, as far as the incomplete granule sample is concerned, Type IV Information Granule Matrix model is best to represent the relationship between earthquake magnitude and logarithmic isoseismal area. Because its curve is slightly surging, but BP Network curve is strongly surging and Linear Regression curve can not capture the observed relationship. In other words, seen from Fig. 4, the values calculating with Type IV Information Granule Matrix is the nearest to true value although its curve is not too smooth.

References

1. Gori, M., Tesi, A.: On the Problem of Local Minima in Back-propagation. IEEE Transactions on Pattern Analysis and Machine Intelligence, Vol. 14. (1992) 76-86
2. Kosko, B.: Fuzzy Engineering. Prentice-Hall, Upper Saddle River, New Jersey (1997)
3. Da Ruan and Chongfu Huang: Fuzzy Sets and Fuzzy Information Granulation Theory — Key Selected by Zadeh, L.A.. Beijing: Beijing Normal University Press (2000)
4. Chongfu Huang, Ye Xue: Some Concepts and Methods of Information Granule Diffusion. In: Xiaohua Hu, Qing Liu, Andrzej Skowron, Tsau Young Lin, Yager, Ronald R., Bo Zhang(eds.): Proceedings of the 2005 IEEE International Conference on Granular Computing, Vol. I. (2005) 28-33
5. Huang, C.F., Moraga, C.: A Diffusion-neural-network for Learning From Small Samples. International Journal of Approximate Reasoning, 35 (2004) 137-161
6. Chongfu Huang and Yong Shi: Towards Efficient Fuzzy Information Processing — Using the Principle of Information Diffusion. Physica-Verlag (Springer), Heidelberg, Germany (2002)

Fuzzy Topological Relations Between Fuzzy Spatial Objects*

Xinming Tang[1], Yu Fang[2], and Wolfgang Kainz[3]

[1] Key Laboratory of Geo-informatics of State Bureau of Surveying and Mapping
Chinese Academy of Surveying and Mapping
16 Beitaiping Rd., Haidian Dist, Beijing, China,100039
Tang@casm.ac.cn
[2] Institute of Remote Sensing and Geographic Information System
Peking University, Beijing, China,100871
csfang@pku.edu.cn
[3] Cartography and Geoinformation
Department of Geography and Regional Research
University of Vienna, Universitätsstrasse 7, A-1010 Vienna, Austria
wolfgang.kainz@univie.ac.at

Abstract. Fuzziness is an internal property of spatial objects. How to model fuzziness of a spatial object is a main task of next generation GIS. This paper proposes basic fuzzy spatial object types based on fuzzy topology. These object types are the natural extension of current non-fuzzy spatial object types. A fuzzy cell complex structure is defined for modeling fuzzy regions, lines and points. Furthermore, fuzzy topological relations between these fuzzy spatial objects are formalized based on the 9-intersection approach. This model can be implemented for GIS applications due to its scientific theory basis.

1 Introduction

In Geographic Information System (GIS) natural phenomena are usually modeled as spatial features including points, arcs or polygons. Generally these features are modeled based on the cell complex or simplicial complex structure, mathematically. This structure has been applied in commercial GIS software widely and successfully (such as ArcGIS etc). Currently these features are represented in crisp (non-fuzzy) points, arcs or polygons. That is, these features have a determinate boundary. However, natural phenomena are not always as crisp as spatial objects described in conventional GIS data models. Many objects have a fuzzy boundary in nature, such as downtown areas, mountains, soil distribution, grasslands and forests These objects have a common characteristic that they have no clear boundaries[1][9] . How to model these fuzzy objects is a significant topic in GIS applications. Several efforts have been made for representation of fuzzy objects, such as [3][4] [13].

* This research is funded by Natural Science Foundation of China (Project No. 40571127) and Key Laboratory of Geo-informatics of SBSM, Chinese Academy of Surveying and Mapping.

L. Wang et al. (Eds.): FSKD 2006, LNAI 4223, pp. 324–333, 2006.

On topological relations between fuzzy spatial objects, several models have been proposed to tackle the topological relations between fuzzy regions [5][6] [14][15][16][18]. Clementini and Di Felice identified 44 relations algebraically between fuzzy objects by use of the well-known 9-intersection approach proposed by Egenhofer and Franzosa[7], Egenhofer and Sharma[8]. 46 relations were identified by use of Cohn and Gotts' egg-yolk model, which is based on logic. Tang and Kainz [16] identified 152 relations between simple fuzzy regions in a fuzzy topological space of R^2 by use of the 4*4-intersection approach, which is the extension of the 9-intersection approach. They also identified 44 relations between fuzzy simple regions in general fuzzy topological space of R^2 by use of 9-intersection matrix.

However, fuzzy objects include not only regions but also lines and points. The topological relations between different kinds of features should be revealed.

This paper establishes a fuzzy topological structure for fuzzy spatial objects and investigates their topological relations. The structure of the paper is as follows. After the introduction of crisp cell complex, fuzzy cell is defined. Section 3 is the formalism of a fuzzy cell complex structure in which the primitives are fuzzy 0-cells, 1-cells and 2-cells. Section 4 is the model for accommodation of fuzzy spatial objects. Section 5 analyses the relations between fuzzy regions, lines and points. Section 6 is the conclusion.

2 Cell Complex and Topological Relation Models

2.1 Cell and Cell Complex

The structures of simplex and simplicial complex, and (crisp) cell and cell complex in algebraic topology have been adopted by almost all GIS software for modeling spatial objects. We briefly review the cell and cell complex structure for discussion. An *n-cell* e^n is a space that is homeomorphic to an (open) unit n-disk of the Euclidean space R^n, where an *open unit n-disk* D^n is the set of all points x of R^n for which the norm $||x||<1$ [10] [12]. By definition, a *0-cell* is a node (the minimal 0-dimensional object), a *1-cell* is the link between two distinct 0-cells, and a *2-cell* is the area described by a closed sequence of non-intersecting 1-cells.

A *cell complex* is defined by an inductive way: Start from a discrete set X^0, whose points are regarded as 0-cells. Inductively, form the *n-skeleton* X^n from X^{n-1} by attaching n-cells via maps $\varphi_\alpha : S^{n-1} \rightarrow X^{n-1}$. This means that X^n is the quotient space of the disjoint union $X^{n-1} \coprod_\alpha D^n_\alpha$ of X^{n-1} with a collection of n-disks D^n_α under the identifications $x \sim \varphi_\alpha(x)$ for $x \in \partial D^n_\alpha$. Thus as a set, $X^n = X^{n-1} \coprod_\alpha D^n_\alpha$, where each D^n_α is an open n-disk. One can either stop this inductive process at a finite stage, setting $X = X^n$, for some $n < \infty$ or one can continue infinitely, setting $X = \bigcup_n X^n$. In the latter case X is given the weak topology: A set $A \subset X$ is open (or closed) iff $A \cap X^n$ is open (or closed) in X^n for each n. A space X constructed in this way is called a *cell complex*.

2.2 Topological Relation Models

The intersection matrix is a well-known approach to identifying topological relation models between two subsets in topological space. In crisp topological space (cts), the 9-intersection matrix has been widely adopted for identification of topological relations between crisp regions. The idea is to adopt the topological notions of subsets in cts: the interior, boundary and exterior, to formalize an intersection matrix. By use of some topological invariants of the intersection such as the empty/non-empty contents, the topological relations between two subsets can be identified. This approach implies the following facts in cts:

1. The interior, boundary and the exterior of a subset are topological invariants;
2. These topological invariants are mutually disjoint in cts;
3. The empty/non-empty contents of the intersections between these three topological parts of two subsets are topological invariants.

However, fact (2) cannot hold in fuzzy topological space (fts). That is, the interior, boundary and exterior of a fuzzy set may not be disjoint with each other although they are topological properties. Therefore the 9-intersection approach cannot be directly applied for the identification of relations between two fuzzy sets. Tang and Kainz [17] investigated a special space and formalized the 9-intersection in this fts. They proved that the fts C whose open sets are crisp is able to meet the above conditions. The 9-intersection matrix can be formalized as:

$$I_9(A,B) = \begin{pmatrix} A^o \cap B^o & A^o \cap \partial B & A^o \cap B^e \\ \partial A \cap B^o & \partial A \cap \partial B & \partial A \cap B^e \\ A^e \cap B^o & A^e \cap \partial B & A^e \cap B^e \end{pmatrix} \quad (1)$$

(9-intersection matrix in the fts C)

By use of 9-intersection matrix, 44 topological relations are identified based on their defined simple fuzzy regions.

3 Fuzzy Cell and Fuzzy Cell Complex

In order to identify the topological relations between fuzzy spatial objects with the same or different dimensions, we now extend the cell and cell complex into the fuzzy domain. We adopt Chang's definition of fuzzy topological space [2][11]. Define the Euclidean distance between two fuzzy points [20] $p_a(x_1, x_2, \ldots x_n)$ and $q_b(y_1, y_2, \ldots, y_n)$ in fuzzy n-dimensional space $\tilde{R}^n$ by: $d(p_a, q_b) = \sqrt{\sum_{i=1}^{n}(x_i - y_i)^2}$. Define a fuzzy closed disk $\overline{D}_p^n$ of a fuzzy point p_a in $\tilde{R}^n$ by: $\overline{D}_p^n = \{q_b : d(p_a, q_b) \le r\}$. Define a fuzzy (open) disk D_p^n of a fuzzy point p_a in $\tilde{R}^n$ by: $D_p^n = \{q_b : d(p_a, q_b) < r\}$. Let $\tilde{R}^n$ be equipped with fuzzy topology δ that is induced from the Euclidean topology of R^n. Then a fuzzy (open) disk D_p^n is open if it is a lower semicontinuous mapping from R^n to [0,1]. A fuzzy closed disk $\overline{D}_p^n$ is closed if it is an upper semicontinuous mapping from R^n to [0,1]. $\tilde{R}^n$ is normal, p-normal and connected.

We define a *fuzzy closed n-cell* $\tilde{\tilde{e}}_n$ by a fuzzy subset of $\tilde{R}^n$, such that its space is homeomorphic to a fuzzy closed disk and it is an upper semicontinuous mapping from R^n to [0,1]. A *fuzzy (open) n-cell* $\tilde{e}_n$ is a fuzzy subset of $\tilde{R}^n$, such that its space is homeomorphic to a fuzzy (open) disk and it is a lower semicontinuous mapping from R^n to [0,1].

By definition, *a fuzzy 0-cell* is a fuzzy point. *A fuzzy 1-cell* is a fuzzy line segment such that its support is a crisp 1-cell, and the membership function is a lower semicontinuous mapping from *R* to [0,1]. *A fuzzy 2-cell* is a fuzzy disk such that its support is a crisp 2-cell, and the membership function is a lower semicontinuous mapping from R^2 to [0,1]. Fig.1 shows these fuzzy cells.

A fuzzy *n*-cell is a fuzzy subset of $\tilde{R}^n$. We adopt the dimension of its support in $\tilde{R}^n$. That is, a fuzzy cell has dimension *n* if the dimension of its support is *n*.

A fuzzy *n*-cell can be regarded as a subset of fuzzy Euclidean space $\tilde{R}^n$. In general fuzzy topological space, there are many topological properties, such as closure, interior, boundary. We describe the topological properties of a fuzzy *n*-cell in $\tilde{R}^n$. That is, the *closure* of a fuzzy *n*-cell $\tilde{e}_n$ is just the fuzzy closed *n*-disk; the *interior* of a fuzzy *n*-cell is the interior of the fuzzy closed *n*-cell in $\tilde{R}^n$, which is a fuzzy open *n*-cell $\tilde{e}_n$; the *core* $\tilde{e}_n^{\oplus}$ of a fuzzy *n*-cell is the crisp subset (where membership values equal to 1) of the fuzzy *n*-cell in $\tilde{R}^n$; the *boundary* $\partial\tilde{e}_n$ is the difference between the fuzzy closed *n*-cell [19] and the core of the fuzzy *n*-cell. These topological properties are illustrated in Fig. 2.

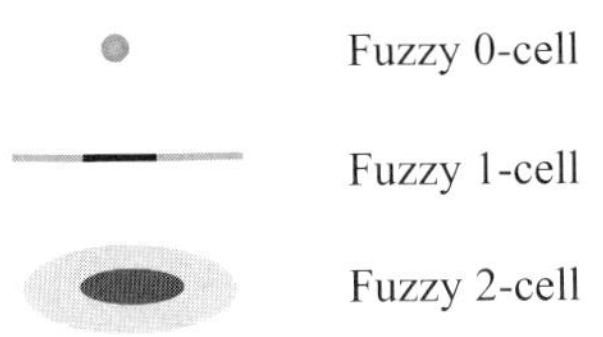

Fig. 1. Fuzzy 0-cell,1-cell and 2-cell

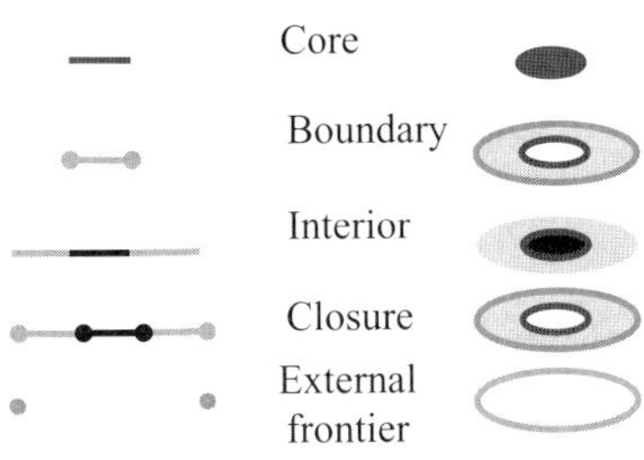

Fig. 2. Topological properties of the fuzzy 1- and 2-cells

In non-fuzzy cell complex structure, the boundary of a cell has always lower dimension than that of the cell. For example, the boundary of a 2-cell is always one-dimensional. The boundary of a 1-cell is always composed of two end points, which are zero-dimensional. In fuzzy cell structure, the boundary of an *n*-cell may have the same dimension as the *n*-cell. Fig. 3 shows some forms of a 1-cell and 2-cell.

It should be noted that these topological properties are defined based on the fact that a fuzzy n-cell has the same dimension as the fuzzy Euclidean space. For example, the topological properties of a 2-cell are defined based on the fact that the dimension of a 2-cell is equal to the dimension of $\tilde{R}^2$. This is because these topological properties are different in different fuzzy Euclidean spaces. For example, the interior of a 2-cell is a fuzzy open disk in $\tilde{R}^2$, but the interior of a 2-cell is empty in $\tilde{R}^3$.

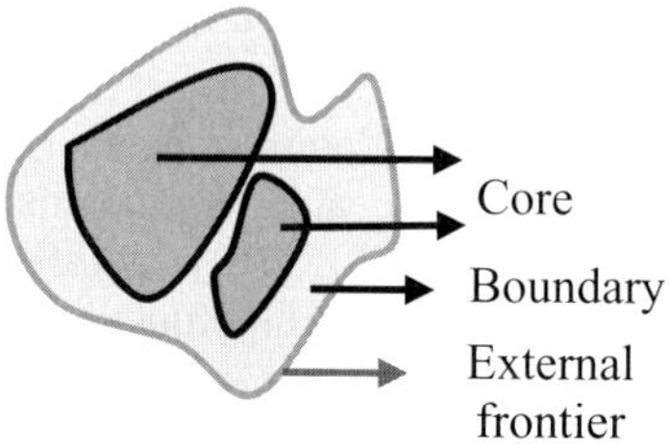

Fig. 3. Different forms of a fuzzy 1-cell and 2-cell

In non-fuzzy cell complex, each n-cell is attached onto p-cells where $p < n$, and the dimension of the boundary of a crisp n-cell is always less than the dimension of the n-cell and that for fuzzy cells this does not hold. The boundary of a fuzzy n-cell may have the same dimension as the fuzzy n-cell. In order to define a fuzzy cell complex based on fuzzy cells, we define the external frontier of a fuzzy n-cell. The *external frontier* $\partial^{ex}\tilde{e}_n$ of a fuzzy n-cell is difference between the fuzzy closed n-cell and its open n-cell in $\tilde{R}^n$. The external frontier is one-dimension less than the fuzzy n-cell since it is homeomorphic to the boundary of the support of a fuzzy n-cell in $\tilde{R}^n$. Fig. 3 shows the external frontier and other topological properties.

A fuzzy cell complex X is a collection of disjoint fuzzy (open) cells $\tilde{e}_i^n$ whose union is X, such that:

1. The support of X is an induced space of a crisp Hausdorff topological space;
2. For each open fuzzy n-cell $\tilde{e}_i^n$ of the collection, there exists a continuous mapping $f_i : \text{supp}(D^n) \rightarrow \text{supp}(X)$ that maps the interior of a fuzzy closed disk $(\overline{D}^n)^o$ (fuzzy) homeomorphically onto $\tilde{e}_i^n$ and carries $\partial^{ex}(D^n)$ into a finite union of fuzzy open cells, each of dimension less than n;
3. A set A is closed in X if $A \cap \overline{\tilde{e}}_i^n$ is closed in $\overline{e}_i^n$ for each i.

A *finite fuzzy cell complex X* is a fuzzy cell complex such that the cells are finite. Call each 0,1,…,n-cell in X a *face* of X. A *proper face* is any 0,1,…,(n-1)-cell. A fuzzy cell complex has the structure that each fuzzy n-cell is attached onto a p-cell where $p < n$ along the external frontier of each n-cell (n>0). Therefore a proper face is in the external frontier of fuzzy 1,2,…,n-cells.

We simply call X a fuzzy complex. A subset of a complex is called a *fuzzy subcomplex* if it is a subset of the complex and still a fuzzy complex. A finite fuzzy cell complex X can be formed in terms of faces. X is called a fuzzy cell complex if (1) every face of a fuzzy cell of X is in X, and (2) the intersection of two cells is either empty or a common face of both fuzzy cells.

4 Data Model for Fuzzy Spatial Objects

The finite combinatorial fuzzy cell complex structure actually constitutes a data model for fuzzy spatial objects. A 2-dimensional fuzzy spatial object abstracted from real world can be easily expressed by a cell complex. Fig. 4 represents a mountain with two cores.

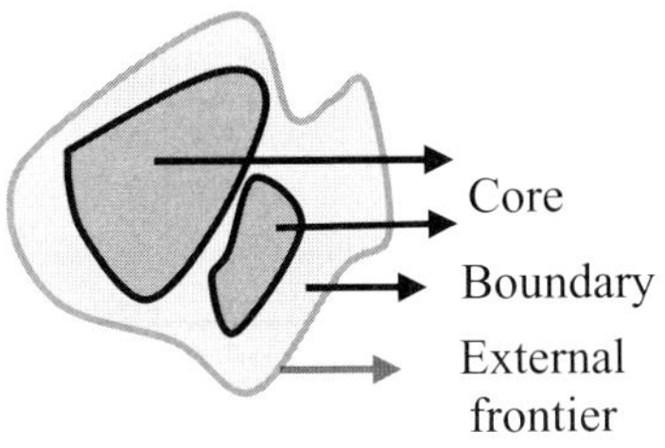

Fig. 4. Fuzzy spatial object *mountain*

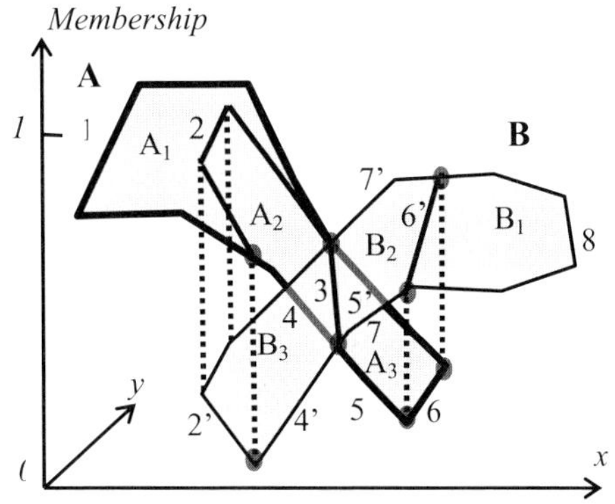

Fig. 5. Fuzzy spatial data model

When more fuzzy objects involve, we can identify cells according to the cell complex structure. For example there are two fuzzy objects: forest (A) and grassland (B), showing in Fig. 5. There are totally 6 fuzzy 2-cells (A_1, A_2, A_3, B_1, B_2 and B_3). *2* and *2'* are two common faces of A_1, A_2 and B_3, in which *2'* is the projection of one part of outer boundary (*1*) of B_3 onto A_1 and A_2, and *2* is the projection of the outer boundary A_1 and A_2 (*2'*) onto B_3. *3* is the common face between A_2, A_3, B_2 and B_3. Totally there are 6 2-cells, 13 1-celles and 8 0-cell fuzzy cells. The union (say B_{12}) of B_1 with B_2 is not a single cell since the intersection between B_{12} with A_2 is B_3, which is neither A_2 nor B_{12}. The combination set $\{A_1, A_2, A_3\}$ of A_1, A_2 and A_3, and the combination set $\{B_1, B_2, B_3\}$ of B_1, B_2 and B_3 constitute two fuzzy sets representing forest and grassland. The combination of A_2 with B_3 is not a fuzzy object since there exist two membership values in the universe.

5 Topological Relations Between Fuzzy Regions, Lines and Points

By use of fuzzy cell complex, fuzzy spatial objects can be structured. Since the interior, boundary and exterior[1] of fuzzy cells are mutually disjoint, we can adopt the 9-intersection matrix to identify the topological relations. We omit the formal

[1] The exterior is re-defined as the complement of the support of a fuzzy cell.

definition of fuzzy point, (simple) fuzzy line and (simple) fuzzy regions due to the page limit, instead readers can refer to Fig. 1 to get an intuitive meaning.

5.1 Relations Between Two Simple Fuzzy Regions

According to Tang and Kainz [17], 12 limitations hold between two simple fuzzy regions. We list 4 of 12 limitations:

1. The exteriors of two fuzzy regions intersect with each other;
2. If one fuzzy region is a subset of the core of the other, then its exteriors must intersect with the other's core, and vice versa;
3. If both cores are disjoint, then one fuzzy region's core intersects with the other's boundary, or with the other's outer, and vice versa;
4. Any part of one fuzzy region must intersect with at least one part of the other fuzzy region, and vice versa;

44 topological relations can be realized between two simple fuzzy regions when the empty/no-empty contents are applied as the topological invariants of the intersection (Table 1 lists 3 of 44 relations).

Table 1. 3 of 44 relations between two simple regions

Illustration	Matrix	Illustration	Matrix	Illustration	Matrix
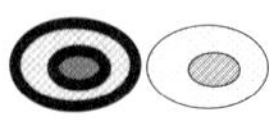	(1) $\begin{bmatrix} 0 & 0 & 1 \\ 0 & 0 & 1 \\ 1 & 1 & 1 \end{bmatrix}$	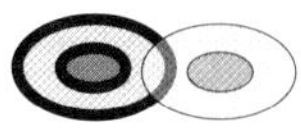	(2) $\begin{bmatrix} 0 & 0 & 1 \\ 0 & 1 & 1 \\ 1 & 1 & 1 \end{bmatrix}$	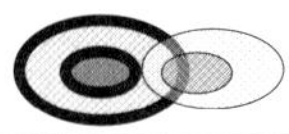	(3) $\begin{bmatrix} 0 & 1 & 1 \\ 0 & 1 & 1 \\ 1 & 1 & 1 \end{bmatrix}$

5.2 Relations Between Two Simple Fuzzy Lines

Limitations (1)-(4) between two regions still hold between two simple fuzzy lines. Furthermore, there are other limitations between two simple fuzzy lines:

5. If one's boundary does not intersect with the boundary of the other, and its core intersect with the core and boundary of the other, then the core must intersect with the exterior of the other;
6. If one's core does not intersect with the boundary of the other, and its boundary does not intersect with the core of the other, if both cores intersect with each other and both boundaries intersect with each other, then either both cores intersect with the exterior of the other, or both cores cannot intersects with the exterior of the other;
7. If both cores do not intersect with each other, and both boundaries do not intersect with each other, then their cores intersect with the exterior of the opposite;
8. If both boundaries do not intersect with the exterior of the opposite, then either both cores intersect with the exterior of the other, or both cores cannot intersect with the exterior of the other;
9. If one's core intersects with all parts except the boundary of the other, then the exterior must intersect with the core of the other.

Table 2. 3 of 97 relations between two simple lines

Matrix	Illustration	Matrix	Illustration	Matrix	Illustration
(1) $\begin{bmatrix} 0 & 0 & 1 \\ 0 & 0 & 1 \\ 1 & 1 & 1 \end{bmatrix}$	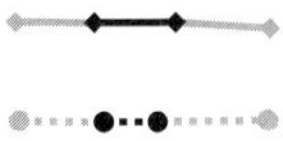	(2) $\begin{bmatrix} 1 & 0 & 1 \\ 0 & 0 & 1 \\ 1 & 1 & 1 \end{bmatrix}$	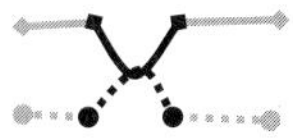	(3) $\begin{bmatrix} 1 & 0 & 0 \\ 0 & 0 & 1 \\ 0 & 1 & 1 \end{bmatrix}$	

By use of the 9-intersection matrix, 97 relations can be identified between two simple fuzzy lines (Table 2 lists 3 of 97 relations).

5.3 Relations Between a Simple Fuzzy Region and a Simple Fuzzy Line

Besides limitations (1)-(4), there are five more limitations between a fuzzy region and a fuzzy line.

10. The region's core and boundary always intersects with the line's exterior;
11. If the line's core is a subset of the support of the region's boundary, then the boundaries must intersect with each other;
12. If both boundaries do not intersect with each other, and the line's core is a subset of the region's core, then the line's closure is a subset of the region's core;
13. If both cores do not intersect with each other, and the line's boundary intersects with the region's core, then it also intersects with the region's boundary;
14. If both cores intersect with each other, and the line's core does not intersect with the region's boundary, then it does not intersect with the region's exterior.

30 relations between a simple fuzzy region and a simple fuzzy line are identified based on these limitations (Table 3 lists 3 of 30 relations).

Table 3. 3 of 30 relations between a simple fuzzy region and a simple fuzzy line

Matrix	Illustration	Matrix	Illustration	Matrix	Illustration
(1) $\begin{bmatrix} 0 & 0 & 1 \\ 0 & 0 & 1 \\ 1 & 1 & 1 \end{bmatrix}$	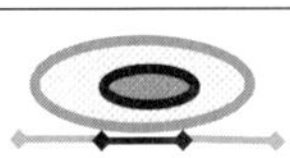	(2) $\begin{bmatrix} 0 & 0 & 1 \\ 0 & 1 & 1 \\ 1 & 1 & 1 \end{bmatrix}$		(3) $\begin{bmatrix} 0 & 0 & 1 \\ 0 & 1 & 1 \\ 1 & 0 & 1 \end{bmatrix}$	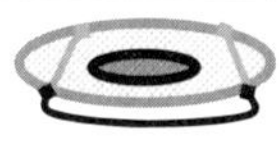

5.4 Relations Between a Fuzzy Point and a Fuzzy Line/Fuzzy Region

Since the boundary of a fuzzy point is empty, there is a strong limitation between a simple fuzzy point and a simple fuzzy line or a simple fuzzy region, that is, a fuzzy point is contained in only one support of parts of the line/region. There are three relations between a point and a line and three relations between a point and a region.

Table 4. 3 relations between a simple fuzzy point and a simple fuzzy line/a simple fuzzy region

Matrix	Illustration	Matrix	Illustration	Matrix	Illustration
(1) $\begin{bmatrix} 1 & 0 & 1 \\ 0 & 0 & 1 \\ 0 & 0 & 1 \end{bmatrix}$		(2) $\begin{bmatrix} 0 & 0 & 1 \\ 1 & 0 & 1 \\ 0 & 0 & 1 \end{bmatrix}$		(3) $\begin{bmatrix} 0 & 0 & 1 \\ 0 & 0 & 1 \\ 1 & 0 & 1 \end{bmatrix}$	
(1) $\begin{bmatrix} 1 & 0 & 1 \\ 0 & 0 & 1 \\ 0 & 0 & 1 \end{bmatrix}$	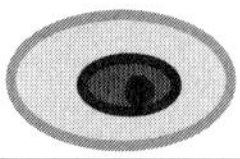	(2) $\begin{bmatrix} 0 & 0 & 1 \\ 1 & 0 & 1 \\ 0 & 0 & 1 \end{bmatrix}$	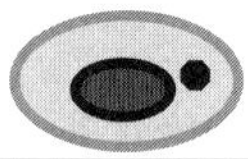	(3) $\begin{bmatrix} 0 & 0 & 1 \\ 0 & 0 & 1 \\ 1 & 0 & 1 \end{bmatrix}$	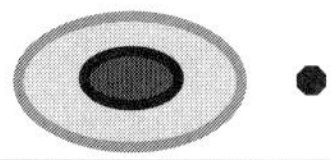

6 Conclusions and Discussions

This paper proposes a framework for dealing with fuzzy spatial objects theoretically. The data model is built on the fuzzy cell complex structure, which is a natural extension of crisp data model into fuzzy domains. The extension of cell complex to fuzzy cell complex keeps many properties of cell complex. It is also compatible with non-fuzzy spatial object. When a spatial object is crisp, it will be turned to be a crisp cell complex. By this way, the abnormalities existing in the boundary of a fuzzy spatial object are solved.

The fuzzy topological relations are investigated since they are basic to spatial data modeling. The idea for the identification is based on the 9-intersection matrix, which results from fuzzy topology. It can be easily perceived that the relations between fuzzy spatial objects are also just an extension of those between crisp spatial objects. If all fuzzy objects are crisp, the relations between these objects turn out to be crisp relations, just as the relations identified by Egenhofer and Franzosa [7]. Therefore the framework of fuzzy cell complex is also compatible with data models for crisp spatial objects.

The above relations are all based on the 9-intersection matrix. More topological relations can be identified if the 4*4-intersection matrix is adopted [14][15][16]. This is useful when more planar relations between two fuzzy spatial objects should be differentiated. These relations can also be tuned if other topological invariants are adopted. This method can be adopted for more relations between fuzzy lines and spatial features of other dimensions.

References

1. Burrough, P.A.: Natural Objects with Indeterminate Boundaries. In: P. Burrough and A.U. Frank (eds), Geographic Objects with Indeterminate Boundaries, Taylor & Francis, London (1996) 1-8
2. Chang, C.L.: Fuzzy Topological Space. Journal of Mathematical Analysis and Applications 24(1968) 182-190
3. Cheng, T., Molenaar, M., Bouloucos, T.: Identification of Fuzzy Objects from Field Objects. Spatial Information Theory, A theoretical Basis for GIS, Lecture Notes in Computer Sciences 1327, COSIT'97, Springer-Verlag, Germany (1997) 241-259
4. Cheng, T., Molenaar, M., Lin, H.: Formalizing Fuzzy Objects from Uncertain Classification Results. International Journal of Geographical Information Science 15(1) (2001) 27- 42

5. Clementini, E., Di Felice, P.: An Algebraic Model for Spatial Objects with Indeterminate Boundaries. In: P. Burrough and A.U. Frank (eds), Geographic Objects with Indeterminate Boundaries, Taylor & Francis, London (1996) 155-169
6. Cohn, A.G., Gotts, N.M.: The `Egg-Yolk' Representation of Regions with Indeterminate Boundaries. In: P. Burrough and A.U. Frank (eds), Geographic Objects with Indeterminate Boundaries, Taylor & Francis, London (1996) 171-187
7. Egenhofer, M. J., Franzosa, R.: Point-set Topological Spatial Relations. International Journal of Geographic Information Systems 5(2) (1991) 161-174
8. Egenhofer, M.J., Sharma, J.: Assessing the Consistency of Complete and Incomplete Topological Information. Geographic Systems 1(1993) 47-68
9. Fisher, P.: Boolean and Fuzzy Region. In: P. Burrough and A.U. Frank (eds), Geographic Objects with Indeterminate Boundaries, Taylor & Francis, London (1996) 87-94
10. Hatcher, A.: Algebraic Topology. Cambridge University Press, London (2002)
11. Liu, Y.M., Luo, M.K.: Fuzzy Topology. World Scientific, Singapore (1997)
12. Munkres, J.R.: Elements of Algebraic Topology. Addison-Wesley Publishing Company, California (1984)
13. Schneider, M.: Uncertainty Management for Spatial Data in Databases: Fuzzy Spatial Data Types. The 6th Int. Symp. on Advances in Spatial Databases (SSD), LNCS 1651, Springer Verlag (1999) 330-351
14. Tang, X.M., Kainz, W.: Analysis of Topological Relations between Fuzzy Regions in General Fuzzy Topological Space. In the Proceedings of Canadian Geomatics Conference, Ottawa, Canada (2002) 114-129
15. Tang, X.M., Kainz, W., Fang, Y.: Reasoning about Changes of Land Cover Objects with Fuzzy Settings. International Journal of Remote Sensing 26(14) (2005) 3025-3046
16. Tang, X.M., Fang, Y., Kainz, W.: Topological Relations between Fuzzy Regions in a Special Fuzzy Topological Space. Geography and Geo-information Science (in Chinese) (2003) 19(2): 1-10
17. Tang, X.M., Fang, Y., Kainz, W.: Topological Matrices for Topological Relations between Fuzzy Regions. In the Proceedings of the 4th International Symposium on Multispectral Image Processing and Pattern Recognition (SPIE), Wuhan, China (2005)
18. Tang, X.M., Kainz, W., Fang, Y.: Modeling of Fuzzy Spatial Objects and their Topological Relations. In: Proceedings of the 2nd Symposium on Spatial Data Quality (SDQ) W.Z. Shi, M.F. Goodchild and P. Fisher (eds), Hong Kong (2003) 34-50
19. Warren, H.R.: Boundary of a Fuzzy Set. Indiana University Mathematics Journal 26 (2)(1977) 191-197
20. Wong, C.K.: Fuzzy Points and Local Properties of Fuzzy Topology. Journal of Mathematical Analysis and Applications 46 (1974) 316-328

On Properties and the Corresponding Problems of Triangular Fuzzy Number Complementary Preference Relations*

Zaiwu Gong and Sifeng Liu

College of Economics and Management, Nanjing University of Aeronautics and Astronautics, Nanjing 210016, China

Abstract. Properties of fuzzy complementary preference relation have been studied. Based on the transformation between fuzzy complementary preference relation and reciprocal fuzzy preference relation, this paper proposes definitions of additive consistency for fuzzy complementary preference relation, gives the concepts of restricted max-min transitivity, restricted max-max transitivity and weak monotonicity for fuzzy complementary preference relation. Using the comparative method of triangular fuzzy number, the inherent relationships between the consistent fuzzy complementary preference relation and their properties are studied. It is also given an aggregation method for fuzzy complementary preference relation based on OWA operation. At the same time, this paper also proposes an algorithm that can judge whether a fuzzy complementary preference relation has the property of satisfied transitivity. Finally, it is illustrated by a numerical example that this method is feasible and effective.

1 Introduction

In the process of multiple attribute decision making, the pairwise comparison method may be used to rank a finite number of alternatives. Usually, decision makers(DMs) express their pairwise comparison information in two formats: multiplicative preference relations or fuzzy preference relations, of which entries are crisp numbers. However, in some practical situations, due to either the uncertainty of objective things or the vague nature of human judgment, DMs often provide imprecise judgment[1,2,3,4,5,6]. Vas Laarhoven and Pedrycy[1]proposed that these judgment information can be represented by triangular fuzzy numbers. In[2], Chang applies row mean method to obtain priorities for fuzzy preference relation with triangular fuzzy numbers.

Recently, people have paid much attention to study the properties of fuzzy preference relation, of which the consistency is the most important property. Traditionally, the research on the consistency of fuzzy preference relation is associated with transitivity, in the sense that if an alternative A is preferred to

* This work was supported by National Natural Science Foundation of China (70473037).

L. Wang et al. (Eds.): FSKD 2006, LNAI 4223, pp. 334–343, 2006.

or equivalent to B, and B to C, then A must be preferred to or equivalent to C. Tanino et al.[6,7,8,9] propose several important definitions of consistency such as weak transitivity, max-min transitivity, max-max transitivity, restricted max-max transitivity, multiplicative consistency and additive consistency. These properties tell us a minimal logical requirement and a fundamental principle that fuzzy preference relations should meet the psychological characteristics of human being. In [8], Herrera-Viedma et al. present a new characterization of the consistency property defined by the additive consistency of fuzzy preference relations. Gogus et al.[6] define strong transitivity and weak monotonicity for fuzzy reciprocal preference relations with triangular numbers, and demonstrate that such a fuzzy preference relation which satisfies the strong transitivity will satisfies the weak monotonicity theorem as well. In this paper, we will research the properties of fuzzy complementary preference relations with triangular fuzzy numbers.

The paper is organized as follows. In section 2, we will review briefly some operations of triangular fuzzy numbers, the definitions of fuzzy reciprocal preference relations and fuzzy complementary preference relations. In section 3, we will define some consistency concepts such as satisfied transitivity, general transitivity, restricted max-min transitivity, restricted max-max transitivity and weak monotonicity for fuzzy complementary preference relation with additive consistency; meanwhile, we will develop the relations among these properties, and propose an algorithm that can judge whether a fuzzy complementary preference relation has the property of satisfied transitivity. In section 4, we will give some aggregation approaches for fuzzy complementary preference relations. Finally, it is illustrated by a numerical example that the algorithm proposed is feasible and effective.

2 Some Concepts

Consider two positive triangular fuzzy numbers M_1 and M_2, $M_1 = (l_1, m_1, u_1)$, $M_2 = (l_2, m_2, u_2)$. Their operational laws are as follows[1,2,3]:

1) $(l_1, m_1, u_1) + (l_2, m_2, u_2) = (l_1 + l_2, m_1 + m_2, u_1 + u_2)$,
2) $(l_1, m_1, u_1) \odot (l_2, m_2, u_2) \approx (l_1 l_2, m_1 m_2, u_1 u_2)$,
3) $(\lambda, \lambda, \lambda) \odot (l_1, m_1, u_1) = (\lambda l_1, \lambda m_1, \lambda u_1), \lambda > 0, \lambda \in R$,
4) $(l_1, m_1, u_1)^{-1} \approx (1/u_1, 1/m_1, 1/l_1)$,
5) $(l_1, m_1, u_1) \div (l_2, m_2, u_2) \approx (l_1/u_2, m_1/m_2, u_1/l_2)$,
6) $ln(l, m, u) \approx (lnl, lnm, lnu)$,
7) $e^{(l,m,u)} \approx (e^l, e^m, e^u)$,

Definition 1. *Let $\widetilde{M} = (M_l, M_m, M_u)$ and $\widetilde{N} = (N_l, N_m, N_u)$ be two triangular fuzzy numbers. The degree of possibility of $\widetilde{M} \geq \widetilde{N}$ [2,6] is defined as*

$$v(\widetilde{M} \geq \widetilde{N}) = \sup_{x \geq y} min(\mu_M(x), \mu_N(y)) \tag{1}$$

By definition 1, we can see that when a pair(x, y) exists such that $x \geq y$ and $\mu_M(x) = \mu_M(y) = 1$, then we have $v(\widetilde{M} \geq \widetilde{N}) = 1$. So we get[10]:

$$v(\widetilde{M} \geq \widetilde{N}) = 1 \iff M_m \geq N_m \tag{2}$$

$$v(\widetilde{N} \geq \widetilde{M}) = height(\widetilde{M} \cap \widetilde{N}) = \frac{M_l - N_u}{(N_m - N_u) - (M_m - M_l)} \tag{3}$$

where height $(\widetilde{M} \cap \widetilde{N})$ is the membership value of the intersection point of the two numbers.

Theorem 2. *Buckley [10] defines a subjective θ such that*

$$\widetilde{M} > \widetilde{N} \Leftrightarrow v(\widetilde{M} \geq \widetilde{N}) = 1, v(\widetilde{N} \geq \widetilde{M}) < \theta \tag{4}$$

$$\widetilde{M} \approx \widetilde{N} \Leftrightarrow min(v(\widetilde{M} \geq \widetilde{N}), v(\widetilde{N} \geq \widetilde{M})) \geq \theta \tag{5}$$

Definition 3. *Let $A = (\widetilde{a_{ij}})_{n \times n}$ be a preference relation, then A is called a fuzzy reciprocal preference relation[1,3], if*

$$\widetilde{a_{ii}} = \widetilde{1} \tag{6}$$

$$a_{ijl}a_{jiu} = a_{ijm}a_{jim} = a_{iju}a_{jil} = 1 \tag{7}$$

where $\widetilde{a_{ij}} = (a_{ijl}, a_{ijm}, a_{iju}), i, j \in N (N = 1, 2, \cdots, n)$.

Definition 4. *Let $R = (\widetilde{r_{ij}})_{n \times n}$ be a preference relation, then R is called a fuzzy complementary preference relation[4,5], if*

$$\widetilde{r_{ii}} = \widetilde{0.5} \tag{8}$$

$$r_{ijl} + r_{jiu} = r_{ijm} + r_{jim} = r_{iju} + r_{jil} = 1 \tag{9}$$

where $\widetilde{r_{ij}} = (r_{ijl}, r_{ijm}, r_{iju}), i, j \in N$.

3 On Fuzzy Complementary Preference Relations with Additive Consistency

3.1 Additive Consistency

Theorem 5. *A fuzzy complementary preference relation $R = (\widetilde{r_{ij}})_{n \times n}$ and a fuzzy reciprocal preference relation $A = (\widetilde{a_{ij}})_{n \times n}$ can be transformed each other by the following formulae:*

$$\widetilde{r_{ij}} = 0.5 + 0.2 log_3^{\widetilde{a_{ij}}} \tag{10}$$

$$\widetilde{a_{ij}} = 3^{5(\widetilde{r_{ij}} - 0.5)}, \ i, j \in N \tag{11}$$

Proof. Let A be a fuzzy reciprocal preference relation. By(10), if $\widetilde{a_{ii}}=\widetilde{1}$, then $\widetilde{r_{ii}}=\widetilde{0.5}$. And if $a_{ijl}a_{jiu} = a_{ijm}a_{jim} = a_{iju}a_{jil} = 1$, then $r_{ijl} + r_{jiu} = 0.5 + 0.2log_3^{a_{ijl}} + 0.5 + 0.2log_3^{a_{jiu}} = 1$. It is also easily to get $r_{ijm} + r_{jim} = r_{iju} + r_{jil} = 1$. Thus R is complementary.

Using the same method, we can transform R into A by(11). □

Definition 6. *A fuzzy reciprocal preference relation $A = (\widetilde{a_{ij}})_{n\times n}$ is completely consistent(strong transitivity)[6], if the following equation satisfies:*

$$\widetilde{a_{ij}}\widetilde{a_{jk}}=\widetilde{a_{ik}}, i,j,k \in N \tag{12}$$

Theorem 7. *A fuzzy reciprocal preference relation $A = (\widetilde{a_{ij}})_{n\times n}$ is completely consistent if and only if the following equations hold[6]:*

$$a_{ijm}a_{jkm} = a_{ikm} \tag{13}$$

$$\sqrt{a_{iju}a_{ijl}}\sqrt{a_{jku}a_{jkl}} = \sqrt{a_{iku}a_{ikl}}, \ \forall\ i,j,k \in N \tag{14}$$

In the following, we will give the definition of the fuzzy complementary preference relation with additive consistency.

By(13), we have $0.2log_3^{a_{ijm}} + 0.2log_3^{a_{jkm}} = 0.2log_3^{a_{ikm}}$, then $(0.5+0.2log_3^{a_{ijm}}) + (0.5 + 0.2log_3^{a_{jkm}}) = 1 + 0.2log_3^{a_{ikm}}$. That is

$$r_{ijm} + r_{jkm} = r_{ikm} + 0.5 \tag{15}$$

Meanwhile, for $r_{ikm} = 1 - r_{kim}$, equation (15) can be rewritten as follows:

$$r_{ijm} + r_{jkm} + r_{kim} = 1.5 \tag{16}$$

By(14), the following equation can be got easily:

$$r_{iju} + r_{ijl} + r_{jku} + r_{jkl} + r_{kiu} + r_{kil} = 3 \tag{17}$$

Definition 8. *A complementary fuzzy preference relation $R = (\widetilde{r_{ij}})_{n\times n}$ satisfying equations (16) and (17) has additive consistency.*

Definition 9. *Let $R = (\widetilde{r_{ij}})_{n\times n}$ be a fuzzy complementary preference relation, for any $i,j,k \in N, i \neq j \neq k$, if*

1 *when $0.5 \leq \lambda \leq 1$. if $\widetilde{r_{ij}} \geq \widetilde{\lambda}, \widetilde{r_{jk}} \geq \widetilde{\lambda}$, then $\widetilde{r_{ik}} \geq \widetilde{\lambda}$;*

2 *when $0 \leq \lambda \leq 0.5$. if $\widetilde{r_{ij}} \leq \widetilde{\lambda}, \widetilde{r_{jk}} \leq \widetilde{\lambda}$, then $\widetilde{r_{ik}} \leq \widetilde{\lambda}$.*

then R has general transitivity.

Theorem 10. *A fuzzy complementary preference relation with additive consistency has general transitivity.*

Proof. Let $R = (\widetilde{r_{ij}})_{n\times n}$ be a fuzzy complementary preference relation, and when $0.5 \leq \lambda \leq 1$, let $\widetilde{r_{ij}}\widetilde{\geq}\widetilde{\lambda}, \widetilde{r_{jk}}\widetilde{\geq}\widetilde{\lambda}$, then $v(\widetilde{r_{ij}}\widetilde{\geq}\widetilde{\lambda}) = 1, v(\widetilde{r_{jk}}\widetilde{\geq}\widetilde{\lambda}) = 1$. Thus $r_{ijm} \geq \lambda, r_{jkm} \geq \lambda$.

For $r_{ijm}+r_{jkm} = r_{ikm}+0.5$, then $r_{ikm}+0.5 \geq 2\lambda$, we have $r_{ikm} \geq 2\lambda-0.5 \geq \lambda$. That is, $v(\widetilde{r_{ik}}\widetilde{\geq}\widetilde{\lambda}) = 1$, then $\widetilde{r_{ik}}\widetilde{\geq}\widetilde{\lambda}$.

Similarly, when $0 \leq \lambda \leq 0.5$, if $\widetilde{r_{ij}}\widetilde{\leq}\widetilde{\lambda}, \widetilde{r_{jk}}\widetilde{\leq}\widetilde{\lambda}$, we can also get $\widetilde{r_{ik}}\widetilde{\leq}\widetilde{\lambda}$. □

Definition 11. *Let $R = (\widetilde{r_{ij}})_{n\times n}$ be a fuzzy complementary preference relation, for any $i, j, k \in N, i \neq j \neq k$, if the following conditions hold:*

1) *if $\widetilde{r_{ij}}\widetilde{\geq}\widetilde{0.5}, \widetilde{r_{jk}}\widetilde{\geq}\widetilde{0.5}$, then $\widetilde{r_{ik}}\widetilde{\geq}\widetilde{0.5}$;*

2) *if $\widetilde{r_{ij}}\widetilde{\leq}\widetilde{0.5}, \widetilde{r_{jk}}\widetilde{\leq}\widetilde{0.5}$, then $\widetilde{r_{ik}}\widetilde{\leq}\widetilde{0.5}$.*

then R has satisfied consistency.

In the light of theorem 10 and definition 11, we get the following lemma.

Lemma 12. *A fuzzy complementary preference relation with additive consistency has satisfied consistency.*

The equivalent definition 11 is definition 13.

Definition 13. *If a fuzzy complementary preference relation has satisfied consistency, then the corresponding preference of the alternatives $X = \{x_1, x_2, \cdots, x_n\}$ has transitivity property. That is, there exists a priority chain in $X = \{x_1, x_2, \cdots, x_n\}$ satisfying $x_{u_1} \geq x_{u_2} \geq \cdots \geq x_{u_n}$, where x_{u_i} denotes the ith alternative in the priority chain, and $x_{u_1} \geq x_{u_2}$ represents x_{u_1} is preferred (superior) to x_{u_2}. If there exists a circulation $x_{u_1} \geq x_{u_2} \geq \cdots \geq x_{u_n} \geq x_{u_1}$, then the corresponding preference relation on the set of alternatives $X = \{x_1, x_2, \cdots, x_n\}$ has not transitivity property, and the fuzzy complementary preference relation is inconsistent.*

According to this definition, we can see that satisfied consistency is the minimal logical requirement and a fundamental principal of fuzzy preference relations, which reflects the thinking characteristic of human being[11]. Therefore, it is very important to set up a approach that can judge whether a fuzzy complementary preference relation has satisfied consistency. In the following, we will give a definition of preference matrix.

Definition 14. *Let $R = (\widetilde{r_{ij}})_{n\times n}$ be a fuzzy complementary preference relation. $P = (p_{ij})_{n\times n}$ is called preference matrix of R, where*

$$p_{ij} = \begin{cases} 1 & \widetilde{r_{ij}} > \widetilde{0.5}, \\ 0 & otherwise \end{cases}$$

Theorem 15. *Let $P = (p_{ij})_{n\times n}$ be a preference matrix of $R = (\widetilde{r_{ij}})_{n\times n}$, P^i be the ith sub-matrix of P, where $i(i = 0, 1, \cdots, n)$.(That is, deleting one 0 row* *vector and a corresponding column vector, we get a sub-matrix P^1 of $P^0 = P$;$\cdots$; deleting one 0 row vector and the corresponding column vector of P^i, we get a sub-matrix P^{i+1} of P^i;$\cdots$; deleting one 0 row vector and the corresponding column vector of P^{n-1}, we get a sub-matrix P^{n-2} of P^{n-1}, $P^n = (0)$.)*

For any $i(i = 0, 1, \cdots, n)$, R is satisfied consistent if and only if there is a 0 row vector in P^i.

Proof. Necessity.

If $R = (\widetilde{r_{ij}})_{n\times n}$ has satisfied consistency, suppose that we have a set of alternatives, $X = \{x_1, x_2, \cdots, x_n\}$, which is associated with a priority chain $x_{u_1} \geq x_{u_2} \geq \cdots \geq x_{u_n}$, where x_{u_i} denotes the ith alternative in the priority chain.

For being x_{u_n} the most inferior alternative, we have that $\widetilde{r_{u_nj}} \leq \widetilde{0.5}, j = 1, 2, \cdots, n$, then $p_{u_nj} = 0, j = 1, 2, \cdots, n$. That is, the u_nth row with entries are all 0. Deleting the u_nth row and the u_nth column, we get a sub-matrix P^1. At this time, the priority relations of the rest alternatives have no change, thus $x_{u_{n-1}}$ is the most inferior alternative of the rest. Obviously, in P^1, the entries of row represented by $x_{u_{n-1}}$ are all 0. Deleting the u_nth row and the u_nth column, the u_{n-1}th row and the u_{n-1}th colum of P, we get P^2. According to this method, at last we have a $n-1$th sub-matrix

$$P^{n-1} = \begin{pmatrix} 0 & 1 \\ 0 & 0 \end{pmatrix} or \begin{pmatrix} 0 & 0 \\ 1 & 0 \end{pmatrix} or \begin{pmatrix} 0 & 0 \\ 0 & 0 \end{pmatrix}$$

In P^{n-1}, the 0 row vector is represented by x_{u_2}. Deleting the 0 row vector and the corresponding column, we have $P^n = (0)$, then the most superior alternative x_{u_1} is gotten.

Sufficiency.

Let the entries of the u_nth row vector in P be 0, it is obviously that x_{u_n} is the most inferior alternative. Now deleting the u_nth row and the u_nth column in P, we get a sub-matrix P^1. Let the 0 row vector be represented $x_{u_{n-1}}$, then $x_{u_{n-1}}$ is superior to x_{u_n}. According to this method, at last we get the most superior alternative x_{u_1}. Thus we have a priority chain $x_{u_1} \geq x_{u_2} \geq \cdots \geq x_{u_n}$, where x_{u_i} denotes the ith superior alternative in the set of alternatives, $X = \{x_1, x_2, \cdots, x_n\}$. Then R has satisfied consistency. □

Actually, according to theorem 15, we can get a priority algorithm of the satisfied consistent fuzzy complementary preference relation.

step1. Construct preference matrix.

step2. Let $i = 0$.

step3. Search the 0 row vector in sub-matrix P^i, if the 0 row exists, then the alternative represented this row is denoted $x_{u_{n-i}}$, and go to step4. Otherwise go to step 5.

step4. Delete the 0 row in P^i (if there are more than 1 such rows then select a 0 row random)and the corresponding column, set $i = i + 1$. If $i = n$, then the alternative represented this row is denoted x_{u_1}(That is, R has satisfied consistency). End. otherwise, go to step 3.

step5. R is inconsistent. End.

Definition 16. *Let* $R = (\widetilde{r_{ij}})_{n\times n}$ *be a fuzzy complementary preference relation, for any* $i, j, k \in N, i \neq j \neq k$*,* R *has restricted max-max transitivity, if* $\widetilde{r_{ij}} \geq \widetilde{0.5}$*,* $\widetilde{r_{jk}} \geq \widetilde{0.5} \Rightarrow \widetilde{r_{ik}} \geq max\{\widetilde{r_{ij}}, \widetilde{r_{jk}}\}$*.*

Definition 17. *Let* $R = (\widetilde{r_{ij}})_{n\times n}$ *be a fuzzy complementary preference relation, for any* $i, j, k \in N, i \neq j \neq k$*,* R *has restricted max-min transitivity, if* $\widetilde{r_{ij}} \geq \widetilde{0.5}$*,* $\widetilde{r_{jk}} \geq \widetilde{0.5} \Rightarrow \widetilde{r_{ik}} \geq min\{\widetilde{r_{ij}}, \widetilde{r_{jk}}\}$*.*

Theorem 18. *If a fuzzy complementary preference relation satisfies additive consistency, then it verified restricted max-max transitivity.*

Proof. By $\widetilde{r_{ij}} \geq \widetilde{0.5}, \widetilde{r_{jk}} \geq \widetilde{0.5}$, we get $v(\widetilde{r_{ij}} \geq \widetilde{0.5}) = 1, v(\widetilde{r_{jk}} \geq \widetilde{0.5}) = 1$. So $r_{ijm} \geq 0.5, r_{jkm} \geq 0.5$. by equation (15), $r_{ijm} + r_{jkm} = r_{ikm} + 0.5$, we get $r_{ikm} \geq r_{ijm}; r_{ikm} \geq r_{jkm}$, that is $v(\widetilde{r_{ik}} \geq \widetilde{r_{ij}}) = 1, v(\widetilde{r_{ik}} \geq \widetilde{r_{jk}}) = 1$. Thus $\widetilde{r_{ik}} \geq \widetilde{r_{ij}}, \widetilde{r_{ik}} \geq \widetilde{r_{jk}}$. □

Lemma 19. *A fuzzy complementary preference relation with additive consistency has restricted max-min transitivity.*

Definition 20. *Let* $R = (\widetilde{r_{ij}})_{n\times n}$ *be a fuzzy complementary preference relation, for any* $i, j, p, l, s, n \in N$*,* R *has restricted weak monotonicity if* $\widetilde{r_{ij}} > \widetilde{r_{pl}}$*,* $\widetilde{r_{jn}} > \widetilde{r_{ls}}$ $\Rightarrow \widetilde{r_{in}} > \widetilde{r_{ps}}$*.*

Theorem 21. *A fuzzy complementary preference relation with additive consistency has restricted weak monotonicity.*

Proof. Let $R = (\widetilde{r_{ij}})_{n\times n}$ be a fuzzy complementary preference relation satisfying completely consistency, and assume that $\widetilde{r_{ij}} > \widetilde{r_{pl}}$, $\widetilde{r_{jn}} > \widetilde{r_{ls}}$. Then

1) by $\widetilde{r_{ij}} > \widetilde{r_{pl}}$, we have $v(\widetilde{r_{ij}} \geq \widetilde{r_{pl}}) = 1$, hence $r_{ijm} \geq r_{plm}$;
2) by $\widetilde{r_{jn}} > \widetilde{r_{ls}}$, we have $v(\widetilde{r_{jn}} \geq \widetilde{r_{ls}}) = 1$, hence $r_{jnm} \geq r_{lsm}$.

By 1) and 2), we get $r_{ijm} + r_{jnm} \geq r_{plm} + r_{lsm}$, and by $r_{ijm} + r_{jnm} = r_{inm} + 0.5, r_{plm} + r_{lsm} = r_{psm} + 0.5$, hence $r_{inm} \geq r_{psm}$. That is

$$v(\widetilde{r_{in}} \geq \widetilde{r_{ps}}) = 1 \tag{18}$$

In the following, we should prove only

$$v(\widetilde{r_{ps}} > \widetilde{r_{in}}) < \theta \tag{19}$$

Supposed that $v(\widetilde{r_{ps}} > \widetilde{r_{in}}) < \theta$ holds, by equation(4), we have $\widetilde{r_{in}} > \widetilde{r_{ps}}$. Otherwise, $v(\widetilde{r_{ps}} > \widetilde{r_{in}}) \geq \theta$, then

$$\widetilde{r_{in}} \approx \widetilde{r_{ps}} \tag{20}$$

by equation(5), we have $\tilde{r}_{in} \geq \tilde{r}_{ps}$. By the choice of a suitable $\theta \in [0,1]$[6], we get $v(\tilde{r}_{ps} > \tilde{r}_{in}) < \theta$. By(18)and(19), we have $\tilde{r}_{in} > \tilde{r}_{ps}$. □

Definition 22. *Let* $R = (\tilde{r}_{ij})_{n\times n}$ *a fuzzy complementary preference relation, for any* $i, j, p, l, s,\ n \in N$*,* R *has weak monotonicity if* $\tilde{r}_{ij} \geq \tilde{r}_{pl}$*,* $\tilde{r}_{jn} \geq \tilde{r}_{ls} \Rightarrow \tilde{r}_{in} \geq \tilde{r}_{ps}$*.*

According to the proof of theorem 21, we can easily get the following lemma.

Lemma 23. *The fuzzy complementary preference relation with additive consistency has weak monotonicity.*

3.2 The Aggregation of Fuzzy Complementary Preference Relations

Theorem 24. *The arithmetic average combination of fuzzy complementary preference relations is still complementary.*

Proof. Let $R^{(k)} = (\tilde{r}_{ij}^{(k)})_{n\times n}$ be a fuzzy complementary preference relation, $\sum_{k=1}^{l} \omega_k = 1, \omega_k \geq 0, k = 1, 2, \cdots, l$. And let $\tilde{r}_{ij} = \sum_{k=1}^{l} \omega_k\ \tilde{r}_{ij}^{(k)}$. Obviously, $\tilde{r}_{ii} = \sum_{k=1}^{l} \omega_k\ \tilde{r}_{ii}^{(k)} = \sum_{k=1}^{l} \omega_k\ \tilde{0.5} = \tilde{0.5}$, $r_{ijm} + r_{jim} = \sum_{k=1}^{l} \omega_k (r_{ijm}^{(k)} + r_{jim}^{(k)}) = \sum_{k=1}^{l} \omega_k = 1$; $r_{ijl} + r_{jiu} = \sum_{k=1}^{l} \omega_k (r_{ijl}^{(k)} + r_{jiu}^{(k)}) = \sum_{k=1}^{l} \omega_k = 1$; $r_{iju} + r_{jil} = \sum_{k=1}^{l} \omega_k (r_{iju}^{(k)} + r_{jil}^{(k)}) = \sum_{k=1}^{l} \omega_k = 1$. □

Theorem 25. *The arithmetic average combination of additive consistent fuzzy complementary preference relations still has additive consistency.*

Proof. Let $R^{(k)} = (\tilde{r}_{ij}^{(k)})_{n\times n}$ be the kth additive consistent fuzzy complementary preference relation, ω_k be the relative importance weight of $R^{(k)}, k = 1, 2, \cdots, l$. satisfying the condition $\sum_{k=1}^{l} \omega_k = 1, \omega_k \geq 0$. Let $\tilde{r}_{ij} = \sum_{k=1}^{l} \omega_k\ \tilde{r}_{ij}^{(k)}$. According to theorem 24, We should only prove that $\tilde{R}$ satisfies the additive consistent property. $r_{ijm} + r_{jkm} + r_{kim} = \sum_{k=1}^{l} \omega_k r_{ijm}{}^{(k)} + \sum_{k=1}^{l} \omega_k r_{jkm}{}^{(k)} + \sum_{k=1}^{l} \omega_k r_{kim}{}^{(k)}$ $= \sum_{k=1}^{l} \omega_k (r_{ijm}{}^{(k)} + r_{jkm}{}^{(k)} + r_{kim}{}^{(k)}) = \sum_{k=1}^{l} 1.5\omega_k = 1.5$. Using the same method, we have $r_{ijl} + r_{jkl} + r_{kil} + r_{iju} + r_{jku} + r_{kiu} = 3$. □

Let $\tilde{r}_{ij}^{(k)}$ be the ith row, the jth column and the kth entry of fuzzy complementary preference $R^{(k)} = (\tilde{r}_{ij}^{(k)})_{n\times n}, i, j \in N, k = 1, 2, \cdots, l$. $P = (\tilde{r}_{ij}^{C})_{n\times n}$ is called a collective preference relation which aggregated by OWA operator[12]. In this case, $\tilde{r}_{ij}^{C} = \Phi_Q(\tilde{r}_{ij}^{1}, \tilde{r}_{ij}^{2}, \cdots, \tilde{r}_{ij}^{l}) = \sum_{k=1}^{l} \omega_k\ \tilde{t}_{ij}^{k}$. where $\tilde{t}_{ij}^{k}$ is the kth largest value in the set $\{\tilde{r}_{ij}^{\,1}, \tilde{r}_{ij}^{\,2}, \cdots, \tilde{r}_{ij}^{\,l}\}$, Q is a relative non-decreasing quantifier [12] with a membership function:

$$Q(x) = \begin{cases} 0 \; 0 \leq x \leq a \\ \frac{x-a}{b-a} \; a \leq x \leq b \\ 1 \; b \leq x \leq 1 \end{cases}$$

$a, b \in [0, 1], \omega_k = Q(k/l) - Q(k - 1/l)$.

Theorem 26. *Let $R^{(k)} = (\tilde{r}_{ij}^{\,(k)})_{n\times n}$ be a fuzzy complementary preference, and $P = (\tilde{r}_{ij}^{\,C})_{n\times n}$ be a collective preference relation which aggregated by OWA operator. P is a fuzzy complementary preference if and only if $a + b = 1$. (If ${r_{ijm}}^s = {t_{ijm}}^k$ is the sth largest value in the set $\{{r_{ijm}}^i, i = 1, 2, \cdots, l\}$, then we consider ${r_{ijl}}^s = {t_{ijl}}^k$ and ${r_{iju}}^s = {t_{iju}}^k$ are the sth largest value in the set $\{{r_{ijl}}^i, i = 1, 2, \cdots, l\}$ and the set $\{{r_{ijl}}^i, i = 1, 2, \cdots, m\}$, respectively).*

Being very similar to the method proposed in literature[12], we omit the process of proof.

4 Numerical Example

Suppose that a DM gives a fuzzy complementary preference relation on an alternatives set $X = \{x_1, x_2, x_3, x_4\}$ as follows:

$$R = \begin{pmatrix} (0.5, 0.5, 0.5) & (0.5, 0.7, 0.8) & (0.3, 0.6, 0.8) & (0.2, 0.8, 0.9) \\ (0.2, 0.3, 0.5) & (0.5, 0.5, 0.5) & (0.3, 0.4, 0.6) & (0.6, 0.6, 0.7) \\ (0.2, 0.4, 0.7) & (0.4, 0.6, 0.7) & (0.5, 0.5, 0.5) & (0.5, 0.7, 0.7) \\ (0.1, 0.2, 0.8) & (0.3, 0.4, 0.4) & (0.3, 0.3, 0.5) & (0.5, 0.5, 0.5) \end{pmatrix}$$

Step1: Construct preference matrix. The preference matrix of P is as follows:

$$P = \begin{pmatrix} 0 & 1 & 1 & 1 \\ 0 & 0 & 0 & 1 \\ 0 & 1 & 0 & 1 \\ 0 & 0 & 0 & 0 \end{pmatrix}$$

Step2: Let $P^0 = P$.

Step3: Search the 0 row vector in P^0. Obviously, the entries of the fourth row are 0, then x_4 is the most inferior alternative.

Step4: Delete the fourth row and the fourth column in P^0, we get P^1:

$$P^1 = \begin{pmatrix} 0 & 1 & 1 \\ 0 & 0 & 0 \\ 0 & 1 & 0 \end{pmatrix}$$

Step5: Search the 0 row vector in P^1. Obviously, the entries of the second row are 0, this row is also the second row of P, so x_2 is superior to x_4.

Step6: Delete the fourth row and the fourth column, the second row and the second column in P, we get P^2:

$$P^2 = \begin{pmatrix} 0 & 1 \\ 0 & 0 \end{pmatrix}$$

Step7: Search the 0 row vector in P^2. Obviously, the entries of the second row are 0, this row is the third row of P, so x_3 is superior to x_2.

Step8: Delete the fourth row and the fourth column, the second row and the second column, the third row and the third column in P, we get P^3, where $P^3 = (0)$.

Step9: In the light of theorem 17 and the corresponding algorithm, the fuzzy preference relation has satisfied transitivity, and we get the most superior alternative x_1.

Therefore, the priority chain of the alternatives set$\{x_1, x_2, x_3, x_4\}$ is $x_1 \succ x_3 \succ x_2 \succ x_4$.

References

1. Laarhoven, P.J.M, Pedrycs, w.: A fuzzy extension of Saaty' priority theory.Fuzzy Sets and Systems. **11** (1983)229-241
2. Chang D Y: Applications of the extent analysis method on fuzzy AHP. European Journal of Operational Research. **95** (1996)649-655
3. Kwiesielewicz M.: A note on the fuzzy extension of Saaty's priority theory. Fuzzy Sets and Systems. **95** (1998)161-172
4. Xu Z S: A method for priorities of triangular fuzzy number complementary judgement matrices. Fuzzy Systems and Mathematics. **16** (2003)47-50
5. Xu Z S: Priority method of triangular fuzzy number of complementary judgement matrix. Journal of systems engineering. **19** (2004)85-88
6. Gogus O.,Boucher T. O.: Strong transitivity, rationality and weak monotonicity in fuzzy pairwise comparisons. Fuzzy Sets and Systems. **94** (1998)133-144
7. Tanono T.: Fuzzy Preference orderings in Group Decision Making. Fuzzy Sets and Systems. **12** (1984)117-131
8. Herrera E., Herrera F., Chiclana F., Luque M.: Some issues on consistency of fuzzy preference relations. European Journal of Operational Research. **154** (2004)98-109
9. Leung L C, Cao D: On consistency and ranking of alternatives in fuzzy AHP. European Journal of Operational Research. **124** (2000)102-113
10. Buckley J. J.: Fuzzy hierarchical analysis. Fuzzy Sets and Systems. **17** (1985) 233-247
11. Gonzalez-Pachon J., Rodriguez-Galiano M. I., Romero C: Transitive approximation to pairwise comparison matrices by using interval goal programming. J. Oper. Res,Soc. **54** (2003)532-538
12. Chiclana F.,Herrera F.,Herrera-Viedma E., Martinez L.: A note on the reciprocity in the aggregation of fuzzy preference relations using OWA operations. Fuzzy Sets and Systems. **137** (2003) 71-83

Knowledge Acquisition in Vague Objective Information Systems*

Lin Feng[1,2,3], Guoyin Wang[1], Yong Liu[1], and Zhenguo Zhu[1]

[1] Institute of Computer Science and Technology
Chongqing University of Posts and Telecommunications
Chongqing, 400065, P.R. China
wanggy@cqupt.edu.cn

[2] School of Information Science and Technology, Southwest Jiaotong University
Chengdu, 610031, P.R. China
mgyfl@tom.com

[3] Department of Engineering and Technology, Sichuan Normal University
Chengdu, 610072, P.R. China
mgyfl@tom.com

Abstract. Vague set is a new theory in the field of fuzzy information processing. In order to extract vague knowledge from vague information systems effectively, a generalized rough set model, rough vague set, is proposed. Its algebra properties are discussed in detail. Based on rough vague set, the approaches for low approximation and upper approximation distribution reductions are also developed in vague objective information systems(VOIS). Then, the method of knowledge requisition from VOIS is developed. These studies extended the corresponding methods in classical rough set theory, and provide a new approach for uncertain knowledge acquisition.

Keywords: Rough Set, Vague Set, Rough Vague Set, Knowledge Acquisition.

1 Introduction

Knowledge acquisition from vague information systems is one of the key problems in the area of intelligence information processing. How to extract knowledge from incomplete, imprecise, and vague information systems has gained much attention among researchers[1,2].

The uncertainty of knowledge acquisition from decision table is caused by two major reasons, indiscernibility of knowledge and vagueness of data[3]. The former has been solved effectively in rough set theory[1]. Unfortunately, vague data can't be dealt with classical rough set theory directly. In order to extract fuzzy

* This paper is supported by National Natural Science Foundation of P.R. China (No.60373111, No.60573068), Program for New Century Excellent Talents in University (NCET), Natural Science Foundation of Chongqing of China, and Research Program of the Municipal Education Committee of Chongqing of China (No.040505).

L. Wang et al. (Eds.): FSKD 2006, LNAI 4223, pp. 344–354, 2006.

knowledge from fuzzy information systems with rough set theory, many authors have generalized rough set model into a fuzzy environment. The result of these studies lead to the introduction of notions of rough fuzzy set and fuzzy rough set[4,5,6]. Professor Zhang defined lower approximation distribution reduction, upper approximation distribution reduction, maximum lower approximation distribution reduction, maximum upper approximation distribution reduction in fuzzy objective information systems, and developed theories and approaches for generating reductions[7,8]. These studies extended the corresponding methods in Pawlak information systems.

A fuzzy set $\tilde{F}$ is a class of objects U along with a grade of membership function[9]. This membership function $\mu_{\tilde{F}}(x), x \in U$, assigns to each object x a grade of membership ranging between zero and one. In[10], Gau pointed out that this single value combines the evidence for x belonging to $\tilde{F}$ and the evidence against x belonging to $\tilde{F}$. He also pointed out that the single number tells us nothing about its accuracy. Therefore, in [10], Gau proposed the concept of vague set. All membership function values of a vague set are a subinterval of $[0, 1]$. This subinterval contains three kinds of information of an element x belonging to $\tilde{F}$, i.e., support degree, negative degree and ignorance degree. Vague set are more accurate to describe some vague information than fuzzy set[11,12,13,14]. Although vague set generalized fuzzy set, it is a pity that the existing theories and approaches of fuzzy knowledge acquisition are not applied to vague information systems. So, it is necessary to develop extended rough set model and extract knowledge from vague information systems. With respect to studies of vague knowledge acquisition, Professor Xing proposed a method for extracting knowledge form vague decision table[15,16]. He analyzed the degree between containment and intersection of condition vague sets and decision vague sets. Furthermore, association rules could be generated in this way.

In this paper, a generalized rough set model by combining rough set theory and vague set theory, rough vague set, is proposed. Then, its algebra properties are discussed in detail. Finally, theories and methods for attribute reduction and knowledge acquisition in vague objective information system(VOIS) are developed. A new approach for uncertain knowledge acquisition from Vague Information System is proposed.

The rest of this paper is organized as follows. In section 2, we briefly review theoretical foundations of vague set and rough set. In section 3, we discuss rough vague set and its algebra properties. In section 4, we develop approaches for attribute reduction and knowledge acquisition in VOIS. In section 5, we conclude our studies and future work.

2 Theoretical Foundations of Rough Set and Vague Set

For the convenience of later discussion, we introduce some basic notions of vague set and rough set at first.

2.1 Basic Concepts of Vague set

Definition 2.1[10]. Let U be a finite and non-empty set of objects called universe, for any $x \in U$, A vague set $\hat{V}$ in U is characterized by a true-membership function $t_{\hat{V}}(x)$ and a false-membership function $f_{\hat{V}}(x)$. $t_{\hat{V}}(x)$ is a lower bound on the grade of membership of x derived from the evidence for x, and $f_{\hat{V}}(x)$ is lower bound on the negation of x derived from the evidence against x. $t_{\hat{V}}(x)$ and $f_{\hat{V}}(x)$ both associate a real number in the interval [0,1] with each point in U, where $t_{\hat{V}}(x) + f_{\hat{V}}(x) \leq 1$. That is $t_{\hat{V}} : U \to [0,1], f_{\hat{V}} : U \to [0,1]$.

Definition 2.2[10]. Let $\hat{A}, \hat{B}$ be a vague set in U, for any $x \in U$, operator of containment, union, intersection between vague set, and complement of vague set, is defined as follows:

(1)A vague set $\hat{A}$ is contained in vague set $\hat{B}, \hat{A} \subseteq \hat{B}$, if and only if $t_{\hat{A}}(x) \leq t_{\hat{B}}(x), 1 - f_{\hat{A}}(x) \leq 1 - f_{\hat{B}}(x)$;

(2)The union of two vague sets $\hat{A}$ and $\hat{B}$ with respect to truth-membership and false-membership function $t_{\hat{A}}(x)$, $f_{\hat{A}}(x)$, $t_{\hat{B}}(x)$, and $f_{\hat{B}}(x)$, is a vague set $\hat{X}$, written as $\hat{X} = \hat{A} \bigcup \hat{B}$, whose truth-membership and false-membership function are related to those of $\hat{A}$ and $\hat{B}$ by $t_{\hat{X}}(x) = max(t_{\hat{A}}(x), t_{\hat{B}}(x))$, $1 - f_{\hat{X}}(x) = max(1 - f_{\hat{A}}(x), 1 - f_{\hat{B}}(x))$.

(3)The intersection of two vague sets $\hat{A}$ and $\hat{B}$ with respect to truth-membership and false-membership function $t_{\hat{A}}(x)$, $f_{\hat{A}}(x)$, $t_{\hat{B}}(x)$,and $f_{\hat{B}}(x)$ is a vague set $\hat{X}$, written as $\hat{X} = \hat{A} \bigcap \hat{B}$, whose truth-membership and false-membership function are related to those of $\hat{A}$ and by $\hat{B}$ by $t_{\hat{X}}(x) = min(t_{\hat{A}}(x), t_{\hat{B}}(x))$, $1 - f_{\hat{X}}(x) = min(1 - f_{\hat{A}}(x), 1 - f_{\hat{B}}(x))$.

(4)The complement of vague set $\hat{A}$ is denoted by $\hat{A}^c$ and is defined by $t_{\hat{A}^c}(x) = f_{\hat{A}}(x)$ and $1 - f_{\hat{A}^c}(x) = 1 - t_{\hat{A}}(x)$.

2.2 Basic Concepts of Rough Set

Difinition2.3[18]. Let U be a finite and non-empty set of objects called universe, and let R be an equivalence relation on universe U, i.e., R is reflexive, symmetric and transitive. The pair (U, R) is called Pawlak approximation space. The equivalence relation R partitions U into disjoint subsets called equivalence classes. If two objects $x, y \in U$ belong to the same equivalence classes, we say that x and y are indiscernible. Given an arbitrary set $X \subseteq U$, it may be impossible to describe X precisely using the equivalence classes of R. In this case, one may characterize X by a pair of lower and upper approximations:

$$R_{-}(X) = \bigcup\{[x]_R | [x]_R \subseteq X\} = \{x \in U | [x]_R \subseteq X\},$$
$$R^{-}(X) = \bigcup\{[x]_R | [x]_R \bigcap X \neq \emptyset\} = \{x \in U | [x]_R \bigcap X \neq \emptyset\}\}.$$

Where $[x]_R = \{(x, y) \in R, \forall y \in U\}$ is the $R-$equivalence class containing x. The pair $(R_{-}(X), R^{-}(X))$ is called a rough set of X in (U, R).

Definition 2.4[18]. An information system S is a quadruple $S = (U, R, V, f)$, where U is a finite non-empty set of objects, R is a finite non-empty set of attributes, $V = \bigcup_{r \in R} V_r$, V_r is a non-empty set of values for r, $f : U \times R \rightarrow V$ is an information function that maps an object of U to exactly one value in V_r, i.e., for any $x \in U$, $r \in R$, has $f(x, r) \in V_r$. if $R = C \bigcup \{d\}, C \bigcap \{d\} = \emptyset$,$S$ is called a decision information system or decision table, where C is a set of condition attributes, d is a decision attribute.

Definition 2.5[18]. Let $S = (U, R, V, f)$ be an information system, each subset of $B \subseteq R$ determines an indiscernibility relation $IND(B)$ as follows:
$IND(B)=\{(x, y)|(x, y) \in U \times U, \forall b \in B(b(x) = b(y))\}$.

3 Rough Vague Set and Its Algebra Properties

In the section 2, we reviewed a rough set X, $X \subseteq U$, which can be represented by $(R_-(X), R^-(X))$ on the Pawlak approximation space (U, R). Suppose X is a vague set in U, how to describe X in (U, R)? In order to solve this problem, we develop rough vague set in this section.

Definition 3.1. Let $\hat{X}, \hat{Y}$ be a vague set in U, for any $x \in U$,
$\hat{I}(\hat{X}, \hat{Y}) = [\inf_{x \in U} \max(f_{\hat{X}}(x), t_{\hat{Y}}(x)), \inf_{x \in U} \max(1 - t_{\hat{X}}(x), 1 - f_{\hat{Y}}(x))]$;
$\hat{T}(\hat{X}, \hat{Y}) = [\sup_{x \in U} \min(t_{\hat{X}}(x), t_{\hat{Y}}(x)), \sup_{x \in U} \min(1 - f_{\hat{X}}(x), 1 - f_{\hat{Y}}(x))]$.
$\hat{I}$ is called the degree of $\hat{Y}$ contains $\hat{X}$, $\hat{T}$ is called the degree of $\hat{X}$ intersects $\hat{Y}$.

Definition 3.2. Let $AS = (U, R)$ be a Pawlak approximation space, $\hat{X}$ is a vague set in U, for each $x \in U$, a membership function of lower approximation $R_-(\hat{X})$ and upper approximation $R^-(\hat{X})$ of $\hat{X}$ in AS , are defined:

$$\mu_{R_-(\hat{X})}(x) = [\inf_{y \in U} \max(f_{[x]_R}(y), t_{\hat{X}}(y)), \inf_{y \in U} \max(1 - t_{[x]_R}(y), 1 - f_{\hat{X}}(y))];$$
$$\mu_{R^-(\hat{X})}(x) = [\sup_{y \in U} \min(t_{[x]_R}(y), t_{\hat{X}}(y)), \sup_{y \in U} \min(1 - f_{[x]_R}(y), 1 - f_{\hat{X}}(y))].$$

Therefore, a rough vague set is a vague set. The following formulas could be induced from Definitions 2.2 and 3.2:

$$\mu_{R_-(\hat{X})}(x) = [\min\{t_{\hat{X}}(y)|y \in [x]_R\}, \min\{1 - f_{\hat{X}}(y)|y \in [x]_R\}];$$
$$\mu_{R^-(\hat{X})}(x) = [\max\{t_{\hat{X}}(y)|y \in [x]_R\}, \max\{1 - f_{\hat{X}}(y)|y \in [x]_R\}].$$

If $R_-(\hat{X}) = R^-(\hat{X})$, then $\hat{X}$ is R definable, otherwise, $\hat{X}$ is an R rough vague set. If vague set $\hat{X}$ degenerates a classical set, namely, for any $x \in U, \mu_{\hat{X}}(x) = [1, 1]$ or $\mu_{\hat{X}}(x) = [0, 0]$, $R_-(\hat{X})$ and $R^-(\hat{X})$ degenerate lower approximation and upper approximation of Pawlak rough sets respectively.

Example 1. Let $AS = (U, R)$ be a Pawlak approximation space, suppose $U/R = \{\{x_1, x_3\}, \{x_2, x_5\}, \{x_4\}\}$, $\hat{X} = [0.6, 0.7]/x_1 + [0.4, 0.5]/x_3 + [0.5, 0.8]/x_4$. Lower and upper approximation sets $(R_-(\hat{X}), R^-(\hat{X}))$ of $\hat{X}$ in AS can be calculated, that is,

$R_{-}(\hat{X}) = [0.4, 0.5]/\{x_1, x_3\} + [0.5, 0.8]/\{x_4\};$
$R^{-}(\hat{X}) = [0.6, 0.7]/\{x_1, x_3\} + [0.5, 0.8]/\{x_4\}.$

Definition 3.3. Given a Pawlak approximation space (U, R), $\hat{X}$, $\hat{Y}$ are vague sets in U,

(1) if $R_{-}(\hat{X}) \subseteq R_{-}(\hat{Y})$, we say that $\hat{X}$ is R lower contained by $\hat{Y}$, or $\hat{Y}$ R lower contains $\hat{X}$, we denote $\hat{X} \subseteq_{-} \hat{Y}$;

(2) if $R^{-}(\hat{X}) \subseteq R^{-}(\hat{Y})$, we say that $\hat{X}$ is R upper contained by $\hat{Y}$, or $\hat{Y}$ R upper contains $\hat{X}$, we denote $\hat{X} \subseteq^{-} \hat{Y}$;

(3) if $\hat{X} \subseteq_{-} \hat{Y}$ and $\hat{X} \subseteq^{-} \hat{Y}$, we say that $\hat{X}$ is R rough vague contained by $\hat{Y}$, or $\hat{Y}$ R rough vague contains $\hat{X}$, we denote $\hat{X} \subseteq_{-}^{-} \hat{Y}$.

Theorem 3.1. The containment relation between rough vague sets has the following properties:

(1) $\hat{X} \subseteq \hat{Y} \Rightarrow (\hat{X} \subseteq_{-} \hat{Y}) \bigwedge (\hat{X} \subseteq^{-} \hat{Y}) \bigwedge (\hat{X} \subseteq_{-}^{-} \hat{Y});$
(2) $(\hat{X}^c \subseteq^{-} \hat{X}) \bigwedge (\hat{Y}^c \subseteq^{-} \hat{Y}) \Rightarrow (\hat{X}^c \bigcup \hat{Y}^c) \subseteq^{-} (\hat{X} \bigcup \hat{Y});$
(3) $(\hat{X}^c \subseteq_{-} \hat{X}) \bigwedge (\hat{Y}^c \subseteq^{-} \hat{Y}) \Rightarrow (\hat{X}^c \bigcup \hat{Y}^c) \subseteq_{-} (\hat{X} \bigcup \hat{Y}).$

Proof. According to the Definitions 2.2, 3.2 and 3.3, the Theorem holds.

Definition 3.4. Given a Pawlak approximation space (U, R), $\hat{X}, \hat{Y}$ are vague sets in U,

(1) if $R_{-}(\hat{X}) = R_{-}(\hat{Y})$, we say that $\hat{X}$ is R lower equal to $\hat{Y}$, we denote $\hat{X} =_{-} \hat{Y}$;

(2) if $R^{-}(\hat{X}) = R^{-}(\hat{Y})$, we say that $\hat{X}$ is R upper equal to $\hat{Y}$, we denote $\hat{X} =^{-} \hat{Y}$;

(3) if $\hat{X} =_{-} \hat{Y}$ and $\hat{X} =^{-} \hat{Y}$, we say that $\hat{X}$ is R rough vague equal to $\hat{Y}$, we denote $\hat{X} =_{-}^{-} \hat{Y}$.

Theorem 3.2. The equal relation between rough vague sets has the following properties:

(1) $(\hat{X} \bigcap \hat{Y} =_{-} \hat{X}) \bigwedge (\hat{X} \bigcap \hat{Y} =_{-} \hat{Y}) \Rightarrow \hat{X} =_{-} \hat{Y};$
(2) $(\hat{X} \bigcup \hat{Y} =^{-} \hat{X}) \bigwedge (\hat{X} \bigcup \hat{Y} =^{-} \hat{Y}) \Rightarrow \hat{X} =^{-} \hat{Y};$
(3) $(\hat{X} =^{-} \hat{X}^c) \bigwedge (\hat{Y} =^{-} \hat{Y}^c) \Rightarrow (\hat{X} \bigcup \hat{Y}) =^{-} (\hat{X}^c \bigcup \hat{Y}^c);$
(4) $(\hat{X} =_{-} \hat{X}^c) \bigwedge (\hat{Y} =_{-} \hat{Y}^c) \Rightarrow (\hat{X} \bigcap \hat{Y}) =_{-} (\hat{X}^c \bigcap \hat{Y}^c);$
(5) $(\hat{X} \subseteq \hat{Y}) \bigwedge (\hat{Y} =^{-} \emptyset) \Rightarrow \hat{X} =^{-} \emptyset;$
(6) $(\hat{X} \subseteq \hat{Y}) \bigwedge (\hat{Y} =_{-} U) \Rightarrow \hat{X} =_{-} U;$
(7) $(\hat{X} =_{-} \emptyset) \bigvee (\hat{Y} =_{-} \emptyset) \Rightarrow (\hat{X} \bigcap \hat{Y}) =_{-} \emptyset;$
(8) $(\hat{X} =^{-} U) \bigvee (\hat{Y} =^{-} U) \Rightarrow (\hat{X} \bigcup \hat{Y}) =^{-} U;$
(9) $\hat{X} =_{-} U \Rightarrow \hat{X} = U;$
(10) $\hat{X} =^{-} \emptyset \Rightarrow \hat{X} = \emptyset.$

Proof. According to the Definition 2.2, 3.2 and 3.4, the Theorem holds.

We can draw a conclusion from Theorem 3.1, and 3.2 that although rough vague set is a vague set, the operations of intersection, union, complementation and equation between rough vague sets are not same as the corresponding operation laws of vague sets. Because rough vague set is defined on the special knowledge space, it shows many properties that vague set don't own. Therefore, rough vague set has special method of knowledge acquisition in the field of data mining with these properties. Next, we will use model of rough vague set to study approach for attributes reduction and knowledge discovery in vague objective information systems.

4 Attribute Reduction and Knowledge Acquisition in Vague Objective Information Systems(VOIS)

4.1 Distribution Reduction in VOIS

Attribute reduction is one of the core notions in rough set. Many approaches for attribute reduction have been developed in rough sets[17,18]. Because the decision space in the VOIS isn't crisp, the existing methods of attribute reduction could not be applied to VOIS. In this section, we will discuss theories and methods of attribute reduction in VOIS based on the conceptions of distribution reduction in [7,8].

Definition 4.1. Let $\hat{G} =< U, C \bigcup \{d\}, V, f >$ be a decision information system, especially when $V_d = \{\hat{D}_1, \hat{D}_2, ..., \hat{D}_{|V_d|}\}$, we call $\hat{G}$ is a vague object information system(VOIS), where $\hat{D}_r$ is a vague set in $U(1 \leq r \leq |V_d|)$.

For any $x \in U$, a membership function of lower approximation $A_-(\hat{D}_r)$ and upper approximation $A^-(\hat{D}_r)$ are defined:

$$\mu_{A_-(\hat{D_r})}(x) = [\inf_{y \in U} \max(f_{[x]_A}(y), t_{\hat{D}_r}(y)), \inf_{y \in U} \max(1 - t_{[x]_A}(y), 1 - f_{\hat{D}_r}(y))];$$
$$\mu_{A_-(\hat{D_r})}(x) = [\sup_{y \in U} \min(t_{[x]_A}(y), t_{\hat{D}_r}(y)), \sup_{y \in U} \min(1 - f_{[x]_A}(y), 1 - f_{\hat{D}_r}(y))].$$

At the same time, for any $x \in U$, we denote:

$$\xi_A(x) = (\mu_{A_-(\hat{D}_1)}(x), \mu_{A_-(\hat{D}_2)}(x), ..., \mu_{A_-(\hat{D}_{|V_d|})}(x));$$
$$\eta_A(x) = (\mu_{A^-(\hat{D}_1)}(x), \mu_{A^-(\hat{D}_2)}(x), ..., \mu_{A^-(\hat{D}_{|V_d|})}(x));$$
$$\gamma_A(x) = \{\hat{D}_k | max(\underline{t}_{\hat{D}_k}(x) - \underline{f}_{\hat{D}_k}(x))\};$$
$$\delta_A(x) = \{\hat{D}_k | max(\overline{t}_{\hat{D}_k}(x) - \overline{f}_{\hat{D}_k}(x))\};$$

where, $\mu_{A_-(\hat{D}_k)}(x) = [\underline{t}_{\hat{D}_k}(x), 1 - \underline{f}_{\hat{D}_k}(x)]$, $\mu_{A^-(\hat{D}_k)}(x) = [\overline{t}_{\hat{D}_k}(x), 1 - \overline{f}_{\hat{D}_k}(x)]$ $(1 \leq k \leq |V_d|)$.

$\xi_A(x), \gamma_A(x)$ is called the lower approximation distribution, maximum lower approximation distribution of x with respect to d on A. $\eta_A(x), \delta_A(x)$ is called the upper approximation distribution, maximum upper approximation distribution of x with respect to d on A. For any an element $\mu_{A_-(\hat{D}_k)}(x)$ in $\xi_A(x)$, if $\underline{t}_{\hat{D}_k}(x) \geq \underline{f}_{\hat{D}_k}(x)$, we call that $[x]_A$ lower belongs to vague decision conception $\hat{D}_k$, otherwise, we call that $[x]_A$ doesn't lower belong to $\hat{D}_k$. Similarly, for any an element $\mu_{A^-(\hat{D}_k)}(x)$ in $\eta_A(x)$, if $\bar{t}_{\hat{D}_k}(x) \geq \overline{f}_{\hat{D}_k}(x)$, we call that $[x]_A$ upper belongs to vague decision conception $\hat{D}_k$, otherwise, we called that $[x]_A$ doesn't upper belong to $\hat{D}_k$.

Definition 4.2. Let $\hat{G} =< U, C \bigcup \{d\}, V, f >$ be a VOIS, $B \subset A, \forall x \in U$, for any vague decision conception $\hat{D}_k (1 \leq k \leq |V_d|)$, if

(1)$\xi_B(x)$ doesn't change that $[x]_A$ lower belongs to $\hat{D}_k$, or $\xi_B(x)$ doesn't change that $[x]_A$ doesn't lower belong to $\hat{D}_k$, then we call $\xi_A(x) = \xi_B(x)$;

(2)$\eta_B(x)$ doesn't change that $[x]_A$ upper belongs to $\hat{D}_k$, or $\eta_B(x)$ doesn't change that $[x]_A$ doesn't upper belong to $\hat{D}_k$, then we call $\eta_A(x) = \eta_B(x)$.

Definition 4.3. Let $\hat{G} =< U, C \bigcup \{d\}, V, f >$ be a VOIS, $B \subset A, \forall x \in U$,

(1)If $\xi_B(x) = \xi_A(x)$, then B is called a lower approximation distribution consistent set, if $\forall b \in B(\xi_{B-\{b\}}(x)) \neq \xi_A(x))$, then B is called a lower approximation distribution reduction;

(2)If $\eta_B(x) = \eta_A(x)$, then B is called an upper approximation distribution consistent set, if $\forall b \in B(\eta_{B-\{b\}}(x)) \neq \eta_A(x))$, then B is called a upper approximation distribution reduction;

Now, we develop an approach to compute the lower approximation distribution reduction in VOIS based on the former discussion.

Algorithm 1. *low approximation distribution reduction in* VOIS.

Input: $\hat{G} =< U, C \bigcup \{d\}, V, f >$, where condition attributes $C = \{a_i | i = 1, 2, ..., |U|\}$, vague decision attribute $d = \{\hat{D}_k | k = 1, 2, ..., |V_d|\}$;

Output: A lower approximation distribution reduction, $\underline{RED}$.

Step 1. $P \leftarrow A$, $\underline{RED} \leftarrow A$;

Step 2. $\forall x \in U$, compute $\xi_A(x)$;

Step 3. For any $a_i \in C$, $\forall x \in U$, compute $\xi_{A-\{a_i\}}(x)$, According to Definition 4.2, if $\xi_{A-\{a_i\}}(x) = \xi_A(x)$, We say that a_i is dispensable, $\underline{RED} \leftarrow \underline{RED} \backslash \{a_i\}$;

Step 4. output $\underline{RED}$, namely, $\underline{RED}$ is a lower approximation distribution reduction;

Step 5. end.

Similarly, we can get upper approximation distribution reduction in VOIS.

4.2 Attribute Reduction Based on Discernibility Matrix in VOIS

Let $\hat{G} =< U, C \bigcup \{d\}, V, f >$ be a VOIS, $\forall x, y \in U, k = 1, 2, ..., |V_d|$, we denote:

$\xi_A(x) \Theta \xi_A(y) = [\hat{D}_1(x) \bigcap \hat{D}_1(y), \hat{D}_2(x) \bigcap \hat{D}_2(y), ..., \hat{D}_{|V_d|}(x) \bigcap \hat{D}_{|V_d|}(y)]$,

$\kappa(\xi_A(x) \Theta \xi_A(y)) = \{\hat{D}_k | (t_{\hat{D}_k(x) \cap \hat{D}_k(y)}(z) - f_{\hat{D}_k(x) \cap \hat{D}_k(y)}(z)) \geq 0\}$. Similarly, we can define $\eta_A(x) \Theta \eta_A(y)$,$\kappa(\eta_A(x) \Theta \eta_A(y))$.

Definition 4.4. Let $\hat{G} =< U, C \bigcup \{d\}, V, f >$ be a VOIS, for any $x_i, x_j \in U$,

(1)An element of the lower approximation distribution discernibility matrix $C_{\underline{d}}$ is defined as:

$$C_{\underline{d}}(i,j) = \begin{cases} \{a | a \in C \wedge a(x_i) \neq a(x_j)\}, & \kappa(\xi_A(x) \Theta \xi_A(y)) = \emptyset \\ \emptyset & \kappa(\xi_A(x) \Theta \xi_A(y)) \neq \emptyset \end{cases};$$

(2)An element of the upper approximation distribution discernibility matrix $C_{\overline{d}}$ is defined as:

$$C_{\overline{d}}(i,j)=\begin{cases}\{a|a\in C\wedge a(x_i)\neq a(x_j)\}, & \kappa(\eta_A(x)\Theta\eta_A(y))=\emptyset\\ \emptyset & \kappa(\eta_A(x)\Theta\eta_A(y))\neq\emptyset\end{cases}.$$

Definition 4.5. Let $\hat{G}=<U,C\bigcup\{d\},V,f>$ is a VOIS, for any $x_i,x_j\in U$,

(1) $LA=\bigwedge(\bigvee\{a|a\in C_{\underline{d}}(i,j)\})$ is called the discernibility formula of lower approximation distribution;

(2) $HA=\bigwedge(\bigvee\{a|a\in C_{\overline{d}}(i,j)\})$ is called the discernibility formula of upper approximation distribution.

Theorem 4.4. The disjunction normal formula of LA is the lower approximation distribution reduction sets, the disjunction normal formula of HA is called upper approximation distribution reduction sets.

Proof. It is easy to prove this theorem with logic concepts and so omitted here.

Algorithm 2. *Attribute reduction based on discernibility matrix in*VOIS

Input: $\hat{G}=<U,C\bigcup\{d\},V,f>$, where condition attributes $C=\{a_i|i=1,2,...,|U|\}$, vague decision attribute $d=\{\hat{D}_k|k=1,2,...,|V_d|\}$;

Output: Lower (or upper) approximation distribution reduction in VOIS.

Step 1. compute the lower (or upper) approximation discernibility matrix in VOIS, $C_{\underline{d}}(C_{\overline{d}})$;

Step 2. compute $\underline{L}_{ij}=\bigvee_{a_k\in C_{\underline{d}}(i,j)}a_k(\overline{L}_{ij}=\bigvee_{a_k\in C_{\overline{d}}(i,j)}a_k)$, where $C_{\underline{d}}(i,j)\neq\emptyset(C_{\overline{d}}(i,j)\neq\emptyset)$;

Step 3. compute $\underline{L}=\bigwedge_{i,j}\underline{L}_{ij}(\overline{L}=\bigwedge_{i,j}\overline{L}_{ij})$;

Step 4. convert $\underline{L}(\overline{L})$ to the disjunction normal formula $\underline{L}=\bigvee_i\underline{L}_i(\overline{L}=\bigvee_i\overline{L}_i)$;

Step 5. each conjunctive item in the disjunction normal formula corresponds to a lower (or upper) approximation reduction.

4.3 Knowledge Acquisition in VIOS

Proposition 4.1. Let $\hat{G}=<U,C\bigcup\{d\},V,f>$ be a VOIS, B, $E\subseteq A$. B, E is lower and upper approximation distribution reduction respectively, $x\in U$,

(1) Lower approximation distribution decision value of $[x]_B$ is $\gamma_B(x)$, namely, $\bigwedge_{a\in B}(a=a(x))\rightarrow\gamma_B(x)$. The confidence $con=\mu_{B\text{-}(\hat{D}_i)}(x)\bigcap...\bigcap\mu_{B\text{-}(\hat{D}_j)}(x)$, the support $sup=|[x]_B|/|U|$, where $\hat{D}_i,...,\hat{D}_j$ are values of $\gamma_B(x)$;

(2) Upper approximation distribution decision value of $[x]_E$ is $\delta_E(x)$, namely, $\bigwedge_{a\in E}(a=a(x))\rightarrow\delta_E(x)$. The confidence $con=\mu_{E\text{-}(\hat{D}_i)}(x)\bigcap...\bigcap\mu_{E\text{-}(\hat{D}_j)}(x)$, the support $sup=|[x]_E|/|U|$, where $\hat{D}_i,...,\hat{D}_j$ are values of $\delta_E(x)$.

Example 2. We discuss the knowledge acquisition from Table 1.

Table 1. A vague objective information system

U	Condition attributes A				Decision attribute d		
	a_1	a_2	a_3	a_4	$\hat{D}_1$	$\hat{D}_2$	$\hat{D}_3$
x_1	1	0	2	1	[0.8,0.9]	[0.1,0.2]	[0.1,0.2]
x_2	0	0	1	2	[0.2,0.2]	[0.7,0.8]	[0.2,0.3]
x_3	2	0	2	1	[0.7,0.9]	[0.2,0.3]	[0.4,0.5]
x_4	0	0	2	2	[0.1,0.2]	[0.2,0.4]	[0.8,0.9]
x_5	1	1	2	1	[0.6,0.8]	[0.3,0.5]	[0.2,0.5]
x_6	0	0	1	2	[0.2,0.5]	[0.6,0.8]	[0.2,0.3]
x_7	0	0	2	2	[0.2,0.4]	[0.3,0.4]	[0.7,0.9]
x_8	0	0	1	2	[0.7,0.8]	[0.1,0.2]	[0.2,0.4]

First, we compute the lower and upper approximation distribution of A, maximum lower and upper approximation distribution of A. The result is shown in Table 2. From Table 2, with Definition 4.4, the lower approximation discernibility matrix is induced.

$$\begin{pmatrix} \emptyset & a_1a_3a_4 & \emptyset & a_1a_4 & \emptyset & a_1a_3a_4 & a_1a_4 & a_1a_3a_4 \\ & \emptyset & a_1a_3a_4 & a_3 & a_1a_2a_3a_4 & \emptyset & a_3 & \emptyset \\ & & \emptyset & a_1a_4 & \emptyset & a_1a_3a_4 & a_1a_4 & a_1a_3a_4 \\ & & & \emptyset & a_1a_2a_4 & a_3 & \emptyset & a_3 \\ & & & & & a_1a_2a_3a_4 & a_1a_2a_4 & a_1a_2a_3a_4 \\ & & & & & \emptyset & a_3 & \emptyset \\ & & & & & & \emptyset & a_3 \\ & & & & & & & \emptyset \end{pmatrix}$$

Table 2. Lower and upper approximation distribution, maximum lower and upper approximation distribution

U/A	$\xi_A(x_i)$			$\eta_A(x_i)$			$\gamma_A(x_i)$	$\delta_A(x_i)$
	$\hat{D}_1$	$\hat{D}_2$	$\hat{D}_3$	$\hat{D}_1$	$\hat{D}_2$	$\hat{D}_3$		
x_1	[0.8,0.9]	[0.1,0.2]	[0.1,0.2]	[0.8,0.9]	[0.1,0.2]	[0.1,0.2]	$\hat{D}_1$	$\hat{D}_1$
$x_2x_6x_8$	[0.2,0.2]	[0.1,0.2]	[0.2,0.3]	[0.7,0.8]	[0.7,0.8]	[0.2,0.4]	$\hat{D}_3$	$\hat{D}_1\hat{D}_2$
x_3	[0.7,0.9]	[0.2,0.3]	[0.4,0.5]	[0.7,0.9]	[0.2,0.3]	[0.4,0.5]	$\hat{D}_1$	$\hat{D}_1$
x_5	[0.6,0.8]	[0.3,0.5]	[0.2,0.5]	[0.6,0.8]	[0.3,0.5]	[0.2,0.5]	$\hat{D}_1$	$\hat{D}_1$
x_4x_7	[0.1,0.2]	[0.2,0.4]	[0.7,0.9]	[0.2,0.4]	[0.3,0.4]	[0.8,0.9]	$\hat{D}_3$	$\hat{D}_3$

Next, from Algorithm 2, the disjunction normal formula of lower approximation distribution will be got, that is, $(a_1 \bigwedge a_3) \bigvee (a_3 \bigwedge a_4)$. So, the lower approximation distribution reductions of Table 1 are $\{a_1, a_3\}$ or $\{a_3, a_4\}$. Similarly, the upper approximation distribution reductions of Table 1 are $\{a_1, a_3\}$ or $\{a_3, a_4\}$.

Let lower approximation distribution reduction be $B = \{a_1, a_3\}$, we can have lower approximation distribution of B:

(1)$\xi_B(x_1) = \xi_B(x_5) = ([0.6, 0.8], [0.1, 0.2], [0.1, 0.2]), \gamma_B(x_i) = \{\hat{D_1}\}(i = 1, 5)$,
(2)$\xi_B(x_2) = \xi_B(x_6) = \xi_B(x_8) = ([0.2, 0.2], [0.1, 0.2], [0.2, 0.3]), \gamma_B(x_i) = \{\hat{D_3}\}$ $(i = 2, 6, 8)$,
(3)$\xi_B(x_3) = ([0.7, 0.9], [0.2, 0.3], [0.4, 0.5]), \gamma_B(x_3) = \{\hat{D_1}\}$,
(4)$\xi_B(x_4) = \xi_B(x_7) = ([0.1, 0.2], [0.2, 0.4], [0.7, 0.9]), \gamma_B(x_i) = \{\hat{D_3}\}(i = 4, 7)$.

Therefore, we can extract decision rules from lower approximation distribution in Table 1 as follows:

R_1: IF $a_1 = 1 \bigwedge a_3 = 2$ THEN $d = \hat{D_1}, con = [0.6, 0.8], sup = 0.25$,
R_2: IF $a_1 = 0 \bigwedge a_3 = 1$ THEN $d = \hat{D_3}, con = [0.2, 0.3], sup = 0.375$,
R_3: IF $a_1 = 2 \bigwedge a_3 = 2$ THEN $d = \hat{D_1}, con = [0.7, 0.8], sup = 0.125$,
R_4: IF $a_1 = 0 \bigwedge a_3 = 2$ THEN $d = \hat{D_3}, con = [0.7, 0.9], sup = 0.25$.

Regarding to R_2, since *con*=[0.2,0.3], its true-membership t is less than false-membership f, we can delete it. Similarly, we can extract decision rules from upper approximation distribution in Table 1.

5 Conclusion and Future Work

Currently, vague set is one of the hot spots in the field of vague information processing. In this paper, we studied the vague knowledge acquisition using rough set. From the definition of the low approximation and upper approximation distribution, we proposed approaches for distribution reductions in VOIS. This studies provide a new avenue to vague information processing. Some efficient heuristic algorithms, as well as the information systems which conditional attribute values are vague, can be further developed based on these result. This is our future research work.

References

1. Pawlak,Z.: Rough Sets. *International Journal of Computer and Information Sciences.* 11(1982)341-356
2. Li,D.Y., Liu,C.Y., et al.: Artificial Intelligence with Uncertainty. *Journal of Software.* 11(2004)1583-1594
3. Ma,Z.F., Xing,H.C., et al.: Research on the Uncertainty of Rule Acquisition from Decision Table. *Control and Decision.* 6(2000)704-707
4. Dubois,D., Prade,H.: Putting Rough Sets and Fuzzy Sets together. *Intelligent Decision Support: Handbook of Applications and Advances of the Rough Sets Theory.* Slowinski, R., (Ed.), Kluwer Academic Publishers, Boston, (1992)203-222
5. Wu,W.Z., Zhang,W.X., et al.: Characterizating Rough Fuzzy Sets in Constructive and Axiomatic Approaches. *Chinese Journal of Computer.* 2(2004)197-203
6. Wu,W.Z., Mi,J.S., Zhang,W.X.: Generalized Fuzzy Rough sets. *Information Science.* 151(2003)263-282
7. Zhang,W.X., Qiu,G.F.: *Uncertainty Decision Based on Rough Sets.* Tsinghua University Press, Beijing(2005)

8. Yuan,X.J., Zhang,W.X.: The Inclusion Degree and Similarity Degree of Fuzzy Rough Sets. *Fuzzy Systems and Mathematics.* 1(2005)111-115
9. Zadeh,L.A.: Fuzzy sets.*Information and Control.* 3(1965)338-353
10. Gau,W.L., Daniel,J.B.: Vague Sets. *IEEE Transactions on Systems, Man and Cybemetics.* 2(1993)610-614
11. Li,F.: Measures of Similarity between Vague Sets. *Journal of Software.* 6(2001) 922-927
12. Yan,D.Q., Chi,Z.X.: Similarity Measure Between Vague sets. *pattern recognition and artificial intelligence.* 1(2004)22-26
13. Xu,C.Y.: The Universal Approximation of A Class of Vague Systems. *Chinese Journal of Computer.* 9(2005)1508-1513
14. Chen,S.M.: Analyzing Fuzzy System Reliability Using Vague Set Theory. *International Journal of Applied and Engineering.* 1(2003)82-88
15. Ma,Z.F., Xing,H.C., et al.: Strategies of Ambiguous Rule Acquisition from Vague Decision Table. *Chinese Journal of Computer,* 4(2001)382-389
16. Ma,Z.F., Xing,H.C., et al.: Approximations Based Machine Learning Approaches in Incomplete Vague Decision Table. *Journal of Computer Research and Development.* 9(2000)1051-1057
17. Wang,G.Y., Yu,H., Yang,D.C.: Decision Table Reduction based on Conditional Information Entropy. *Chinese Journal of Computer.* 7(2002)759-766
18. Wang,G.Y.:*Rough Sets Theory and knowledge acquisition.* Xi'an Jiaotong University Press, Xi'an(2001)

Multiple Documents Summarization Based on Genetic Algorithm

Derong Liu[1,2], Yongcheng Wang[1], Chuanhan Liu[1], and Zhiqi Wang[1]

[1] Dept. of Comp. Sci. and Engineering, Shanghai Jiao Tong University
{drliu, ycwang, chliu, shrimpwang}@sjtu.edu.cn
[2] Merchant Marine College, Shanghai Maritime University

Abstract. With the increasing volume of online information, it is more important to automatically extract the core content from lots of information sources. We propose a model for multiple documents summarization that maximize the coverage of topics and minimize the redundancy of contents. Based on Chinese concept lexicon and corpus, the proposed model can analyze the topic of each document, their relationships and the central theme of the collection to evaluate sentences. We present different approaches to determine which sentences are appropriate for the extraction on the basis of sentences weight and their relevance from the related documents. A genetic algorithm is designed to improve the quality of the summarization. The experimental results indicate that it is useful and effective to improve the quality of multiple documents summarization using genetic algorithm.

1 Introduction

Automatic document summarization is a system whose goal is to produce a condensed representation of document information content for the benefit of the reader and task [1]. More and more information became accessible, so information overload became a serious problem to challenge the researcher. With the explosive growth of online information, it is more important and necessary to generate useful summarization from a range of information sources. Multiple documents summarization is a summarization of collections of related documents. It helps reader to grasp key issues or important insights in short time. Differing with single-document summarization, it has more coverage of content and a higher compression ratio. Its goal is to extract the core content while removing redundancy by analyzing similarities and differences in information content.

Sentence extraction is one of main methods for multiple documents. Mani and Jing employed the method to extract salient sentences from related documents. It was one robust and domain-independent approach. Gregory [2] proposed using lexical chain for the efficient text summarization. Michael [3] combined information extraction and natural language generation for summarization in specific domain. These applications were limited to small and selected domains. Pascal Fang [4] combined optimal clustering and probabilistic model for extractive

L. Wang et al. (Eds.): FSKD 2006, LNAI 4223, pp. 355–364, 2006.

summarization. However, a vocabulary switching problem exists in natural language. People often use different terms to describe one subject. The clustering algorithm based on words could not get satisfactory performance especially for Chinese multi-documents summarization.

In this paper, we propose a new approach based on the theme concept to develop the summarization system, which maximize the coverage of topic and minimize the redundancy of contents. We extract a set of concepts combining semantic analysis and statistic techniques to represent the information content of related news documents. The concepts extracting for topic detection got the satisfied results in the evaluation of National '863' project at the 2004 HTRDP [5]. Information entropy of each sentence is obtained according to the theme concept space of the collection. The system calculates the relevance between sentences based on concept cohesion. Special equations are designed to extract the characteristic sentences considering information entropy, relevance and the other synthetic features of sentences from the related documents.

2 Multi-document Analysis Modeling

The collection of multiple documents is composed of many types of elements such as documents, paragraphs, sentences etc. Mann presented Rhetorical Structure Theory (RST) to describe single text organization. Radev presented Cross-Document Structure Theory (CST) to extend RST techniques for multi-document summaries [6]. We present Multi-Document Analysis Theory (MAT) to improve CST for representing sets of related documents. The elements and their relationships in MAT can be structured as a network. To illustrate the inter-relationship within the text collection, we define the network G as a set of elements and their relationships. We can configure the elements and their relationships in multi-document graph (Fig 1).

These elements are documents, paragraphs, sentences, topics and central theme of the collection. The model includes four types of relationship links. The first type represents inheritance relationships among elements within a single

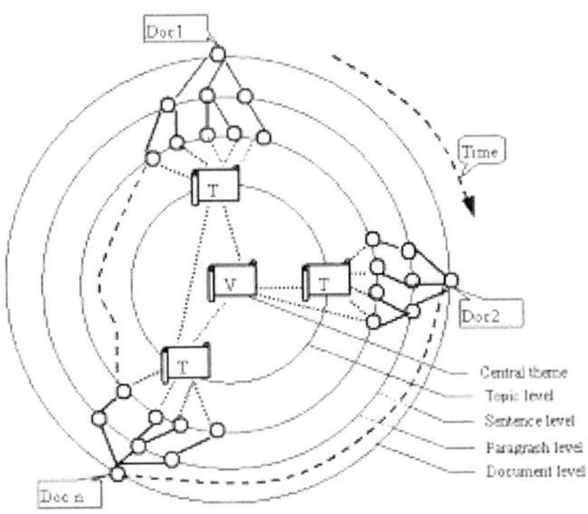

Fig. 1. Multi-document Graph

document. These links are drawn using thicker lines. The second type represents mapping links from documents or sentences to topic within a document and central theme in the entire collection. These links are drawn using dotted lines. The third type represents semantic relevance based topic and central theme space among textual units. These links are drawn using dashed lines. The fourth type represents chronological ordering of documents. These links are drawn using the clockwise arrow.

The follows are the description of elements and relationships.

The collection includes many members of documents: $M = \{D_1, D_2, \ldots, D_n\}$. Where D_n denotes the document n; M denotes the collection;

Each document includes many sentences: $D_i = \{S_{i1}, S_{i2}, \ldots, S_{il_i}\}$. Where S_{il_i} denotes the sentence l_i in document i;

Paragraphs are intermediate units in our processing model. So we define the direct relations of documents and sentences to simplify the modeling descriptions.

A set of concept spaces represents the topic of the document:

$$T_i = \{C_{i1}, C_{i2}, \ldots, C_{it_i}\}$$

Where C_{it_i} denotes the concept t_i in document i; T_i denotes the topic vector of document i;

The central theme of documents collection can be defined as following:

$$V = \{A_1, A_2, \ldots, A_m\}$$

Where A_m is the central concept in the collection; V denotes the central theme vector of documents collection.

We define concept as a lingual unit representing semantic object. Concept can be denoted by words, terms and phrases. Concept cohesion is defined to represent the relationship of concept semantic relevance. We explore the mixture strategies to calculate the concept cohesion based on the Chinese concept lexicon and corpus. The extraction and function of concepts will be described in the following section.

The network G is represented as graph characterized by 7-tuple:

$$G = (D, S, T, V, X, R, C) \tag{1}$$

where: D- document; S- sentence; T- topic of the document; V - central theme of the collection; X- relevance of documents; R- relevance of sentences; C- concept cohesion;

D, S, T and V are nodes in graph and the edges can be used to express their relationships. These relations include the relevance of documents, the relevance of sentences and the concept cohesion of theme.

The sentence is a short and explicit basic semantic unit in document. Our system tends to extract the sentence series from the related documents based MAT theory. Referring to the human consideration, the system had emphasized the following regulations for generating the summarization.

(1)The maximum of information entropy ($Max\ \Sigma\ I(S_i)$):
$I\ (S_i)$ denotes the information entropy of sentence i. we expect the output summary with maximum information entropy.

(2)The minimum of information redundancy ($Min\ \Sigma\ R(S_i,S_j)$):
$R(S_i,S_j)$ denotes the relevance between sentence i and sentence j. We hope to keep the diversity of target sentences.

(3)The more reasonable coherence:
It is expected that output summary have good coherence.

3 Implementation of the System

Our system includes several steps similar to human summarization. First, documents which have different formats are converted to the uniform system text. the system scans full text to detect the topic and identify some features of each document. Then, synthetic theme is obtained from the common topic of related documents. We analyze information entropy and similarity of each sentence based on concept space. Last, the system can create a summary which synthesizes core information across documents and constrains the repetitions.

3.1 Features Mining

Topic concept detection. It is necessary to identify the topic automatically for the large amount of text analysis. It is more difficult to segment Chinese words than English words from documents, because there is no inherent partition tag for automatic text processing. We use the concept as a lingual unit representing semantic object. Concept maybe includes more than one word or phrases from text. We utilize Chinese characters co-occurrences based on Markov statistics model to obtain topic concept. First, semantic strings are extracted from documents based on term frequency (TF) and inverse document frequency (IDF). Furthermore, base on concept knowledge base, our system take into account some conditions such as title, headline and position, to extract a group of subject concepts for expressing the topic. Our extracting technique had been proven in '863'evaluation.

Thus, each document can be represented by a set of concept space.

$$T_i = (\omega_{i1}(C_{i1}), \omega_{i2}(C_{i2}), \ldots, \omega_{ik}(C_{ik})) \tag{2}$$

where T_i denotes the topic of document i; C_{ik} denotes topic concept; $\omega_{ik}(C_{ik})$ is the weight parameter of the concept;

We explore concept cohesion to express the semantic relevance between concepts to improve the text analysis performance. Concept cohesion is obtained by semantic analysis and statistic technique based on concept lexicon and corpus [7]. Our system combines multiple resources including HowNet[8], Synonym Thesaurus[9] and news corpus to calculate Chinese concept cohesion.

Central theme. Each document has a set of concept. We cluster topic concepts and select a set of synthesis concepts for representing the theme of the collection.

If $\forall\ C_{ij} \in A_k\ Weight(C_{ij}, T) > L$;

Then extract $A = \Sigma\ A_k$ to represent the synthesis theme.

T denotes the collection of topic concepts; L denotes the threshold; $Weight$ (C_{ij}, T) denotes the weight of the topic concept in the collection.

The theme concepts of the collection are denoted as the following:

$$A = (\omega_{A_1}, \omega_{A_2}, \ldots, \omega_{A_m}) \tag{3}$$

ω_{A_m} denotes the weight parameter of theme concept A_m;

External features. Electronic document maybe contain some rich features. Some external features besides content can be extracted to assist the text analysis. For example, system can get the event time and news source from the news articles.

$TM(D_i)$ represents the stamp time of events for document i. We can capture the time according the special format of text, such as, "BEIJING, Nov. 21, (Xinhua News Agency)". In our MAT model, the $max(TM(D_i))$ equals to $TM(D_n)$ and the $min(TM(D_i))$ equal to $TM(D_1)$.

$SR(D_i)$ denotes the source ranking of document i. The authoritative news societies have higher-ranking.

3.2 Methods of Scoring Sentence Weight

Sentence weight is one of important parameters for extracting sentences from the text collection. Our system computes the weight of all sentences by three layers factors. The factors include information entropy in document sets,the weight within the document and the influence from the viewpoint of document level.

Information entropy. A set of theme concepts is obtained from the given collection. We can calculate information entropy of each sentence based on Shannon information theory. Sentence is the basic unit expressing semantic content in text. Information entropy expresses the information content covering theme concepts. The Shannon information entropy of sentence in the collection is defined as

$$I(S_{ij}) = \sum_{k=1}^{m} -N(A_k) * \omega_{A_k} * \log P(A_k) \tag{4}$$

Where $N(A_k)$ is the count of theme concepts in sentence; ω_{A_k} is the weight coefficient of theme concepts; $P(A_k)$ is the probability distribution of theme concepts in corpus.

Weight within the document. We take similar approaches in single document summarization to compute the scores of sentences. The main scoring methods include topic concept method, indicator phrase method, location method and title (or headline) method. Our system relies on these methods to determine the importance of sentence within in the document.

$$W(S_{ij}) = k_1 * N(C_i) + k_2 * Title() + k_3 * Ind() + k_4 * Loc() \tag{5}$$

Where $N(C_i)$ denotes the sum of topic concepts scores from the sentence $S_(ij)$. The function $Title()$ computes that sentence contains the concepts existing in title and headline. The function $Ind()$ represents that the sentence which contain an indicating phrase have scores boosted or penalized depending on the phrase. The function $Loc()$ describes the position score of the sentence in corresponding paragraphs structure. $k_1, k_2, k_3, andk_4$ are the coefficients, respectively.

Documents evaluation. As the container of sentences, the importance of document can be estimated for information content of sentences in some extent. We investigate that it is useful to evaluate sentence from the view of the document.

Each document has a set of concept for representing the topic.

$$T_i = (\omega_{i1}(C_{i1}), \omega_{i2}(C_{i2}), \ldots, \omega_{it_k}(C_{it_k}))$$

The relevance of two documents can be induced by topic concept:

$$X_{ij} = X(D_i, D_j) = \sum_{p=1}^{k}\sum_{q=1}^{k} \omega_{ip} * \omega_{jq} * Concept\_cohesion(C_{ip}, C_{jq}) \tag{6}$$

X_{ij} denotes the relevance between document D_i and document D_j. $Concept\_cohesion(C_{ip}, C_{jq})$ denotes the semantic relevance between concepts. We apply mixed strategies by lingual lexicon and corpus methods to get it [7]. If the total number of documents in the collection is N, the importance of each document can be obtained from the sum of relevance.

$$Q_i = P_i / \max_{j=1}^{n}(P_j); P_i = \sum_{j=1, j \neq i}^{N} X_{ij} \tag{7}$$

Q_i is the importance of document D_i in the collection.

Calculation of Sentences Weight. The scoring of a sentence is a weighted function of these parameters.

$$Score(S_{ij}) = [\beta_1 * I(S_{ij}) + \beta_2 * W(S_{ij})] * Q_i * TM_i * SR_i \tag{8}$$

$$TM_i = 1 + (TM(D_i) - TM(D_1))/(TM(D_n) - TM(D_1)) \tag{9}$$

$$SR_i = SR(D_i) \tag{10}$$

β_1,β_2 denotes the coefficients being established through experiments, respectively.

In formula (8), the calculation of sentences weight mainly depends on the information entropy and the weight within the document, but the document ranking, time and news source also are the important parameters.

3.3 Relevance Between Sentences

Each sentence can be expressed by vector space with theme concepts. The system can compute the relevance according to the relations of theme concepts.

Two sentence S_{ij} and S_{pq}:
$\boldsymbol{S_{ij}} = (t_1, t_2, \ldots, t_m)$
$\boldsymbol{S_{pq}} = (t'_1, t'_2, \ldots, t'_m)$
$t_k = N(A_k)$, represents the count of theme concept (A_k) in the sentence.

It is easy to compute the relevance through cosine similarity in a vector space. To improve the analysis precision, the system will calculate sentences relevance based on concept cohesion. The relevance of S_{ij} and S_{pq} is defined as the following:

$$R(S_{ij}, S_{pq}) = \sum_{k=1}^{m}\sum_{h=1}^{m} t_k * t'_h * Concept\_cohesion(A_k, A_h) \tag{11}$$

$Concept\_cohesion(A_k, A_h)$ denotes the semantic relevance between concepts.

3.4 Extraction Algorithms

Sentence is the basic unit of extraction in our system. Our goal is that candidates should contain more important information and less redundancy.

***CNS_MMR* Algorithm.** Carbonell has presented maximal margin relevance (MMR) to reorder the documents with a query in a retrieval system [10]. We develop the extraction algorithm similar to MMR. Our algorithm (CNS_MMR) utilizes the balance of sentence weight and relevance to evaluate the candidates. The sentence weight is obtained by considering some features such as information entropy, sentence position and document importance etc.

$$CNS\_MMR = \arg \max_{S_{ij} \in M \setminus U} [\alpha(Score(S_{ij})) - (1-\alpha) \max_{S_{pq} \in U} Sim(S_{ij}, S_{pq}, U)] \tag{12}$$

$$Sim(S_{ij}, S_{pq}, U) = \gamma * R(S_{ij}, S_{pq}) \tag{13}$$

In equation (12), M is the whole collection; U is the selected sentences collection; S_{ij} is the sentence to be selected; S_{pq} is the selected sentence; $Score(S_{ij})$ is the weight metric of importance; Sim is the anti-redundancy metric; α, γ denote the coefficient, respectively.

Genetic Algorithm (GA). It is possible that the best summarization is not produced by CNS_MMR algorithm, because the inclusion of a sentence in summarization depends on the sentences which were included before. Genetic Algorithms perform search tasks based on mechanisms similar to those of natural selection and genetics [11]. It is initialized with a set of solutions (represented by chromosomes) called a population. Solutions which are selected to form new solutions (offspring) are selected according to their fitness - the more suitable they are, the more chances they have to reproduce. The new population is created by repeating the process of selection, crossover and mutation.

Our algorithm is described as follows.

Coding
We decided that the genes (chromosome) take integer values, each value (g) representing the orders of a sentence (S_g) to be included in summarization. The g can be defined by i and j for the sentence (S_{ij}).

$$g = \sum_{k=0}^{i-1} S\_Count(D_k) + j \tag{14}$$

The function $S\_Count()$ denotes the count of sentences within a document. The length (L) of chromosome (H) is the desired summarization.

Fitness function
This is the heart of the GA. In our case fitness function ($f(H)$) represents the sum of the scores of sentences weight subtracts the value of relevance between sentences.

$$f(H) = \alpha * \sum^{L} Score(S_g) - \theta * \sum^{C_L^2} Sim(S_p, S_q, H) \tag{15}$$

$Sim(S_p, S_q, H)$ denotes the relevance between sentences by gene p and q of the chromosome H. α and θ denote the coefficient, respectively.

Genetic Operators
We used weighted roulette wheel selection to select chromosomes. The algorithm used the classical single point crossover operator and two mutation operators. The first one replaces the value of a gene with a randomly generated integer value. If a duplication is found in H, the gene's value is incremented by one. The second mutation operator replaces the values of a gene with the value of the sentence number of maximal weight from the corresponding document.

3.5 Reordering Strategies

Differing with single document summarization, the output sentences are extracted from different documents. Multiple documents summarization system has to design strategy to reorder the output sentences. The system employs the following reordering strategies based time and space information:

Begin
If $TM(S_{ij}) > TM(S_{pq})$, Then $Rank(S_{ij}) > Rank(S_{pq})$;
If $i = p, j > q$, Then $Rank(S_{ij}) > Rank(S_{pq})$;
If $Score(S_{ij}) > Score(S_{pq})$ Then $Rank(S_{ij}) < Rank(S_{pq})$;
End

Where $TM(S_{ij})$ denotes the release time of sentence; it can be obtained from the external feature of document. $Rank(S_{ij})$ denotes the output order of sentence S_{ij}. $Score(S_{ij})$ denote the information weight of the sentence.

4 Experiments

4.1 Evaluation Method

Many research have proposed various evaluation methods for summarization [12]. Borrowed from information retrieval research, precision and recall are used to measure how effectively our system generates a good summary. Supposing there is a system summary S and a baseline summary M. the number of sentences in S is N_s, and the number of sentences in M is N_m. the number of sentence occurring in both S and M is N_o. So the precision and recall can be denoted as follows. $Precision(P_c) = N_o/N_m$; $Recall(R_c) = N_o/N_s$;If $N_m = N_s$, so precision is equal to recall.

In practice, it is rough to calculate precision only by the number of same sentences in both S and M. McKeown treated the completeness of coverage between two sentences as the threshold. We take the completeness of coverage as the similarity score(C), 1 for all, 2/3 for most, 1/3 for some, 0 for hardly or none. Our system experiments employed similarity measure to evaluate the precision of performance.

$$P_c = \sum_{i=1}^{N_o} C_i/N_m \tag{16}$$

4.2 Results Analysis

Chinese news articles were collected from the website of $XinHua$ news agency for the experiments. The news articles cover the range of terrorist attack, flood disaster and meeting report etc. we prepared 4 sets of articles, about 100 articles in the experiments. A professional language teacher was invited to make manually 4 pieces of summaries as baseline standard. We adjust the compression rate to return the same number of sentences in system summary for each standard summary. Two graduate students were asked to compare the system summary against the baseline standard summary.

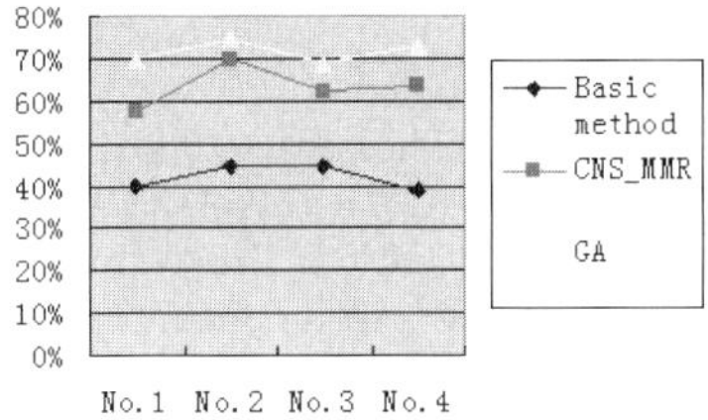

Fig. 2. The precisions by different algorithms

The system generates three kinds of summaries by basic algorithm, CNS_MMR and genetic algorithm. Measurement of precision is utilized to determine how comprehensive the summary covers the topics by different methods (in Fig.2).

The first average score of precision is 44.8% for 4 pieces of the basic summaries. The basic summaries only depend on the weight of sentences not considering the redundancy. The second average score of precision is 63.4% for the summaries based on CNS_MMR algorithm. The third precision is 71.8% for the summaries based on genetic algorithm against the standard summaries. Our results indicate that the system performance based on genetic algorithm is more robust than the two other methods referring to human judge.

5 Conclusions

In this paper we proposed a summarization prototype system based on genetic algorithm for multiple documents. The system can analyze the link relationships of sentences and documents based on concept space through Multi-Document Analysis Theory.Two algorithms were presented to extract the summary sentences by balancing information weight and redundancy. Our experiments indicate such approaches can be used to extract essential information effectively.

References

1. Iderjeet Mani. Automatic Summarization.. John Benjamins Publishing Company. 2001
2. H.Gregory, An Efficient Text Summarizer using Lexical Chains. NAACL-ANLP 2000 Workshop.
3. Michael White et al. Multi-document Summarization via Information Extraction. First International Conference on Human Language Technology Research (HLT), 2001.
4. Pascale Fung et al. Combining Optimal Clustering and Hidden Markov Models for Extractive Summarization. Proceedings of the ACL 2003.
5. 2004 HTRDP Evaluations. http://www.863data.org.cn/
6. Dragomir R. A common theory of information fusion from multiple text sources, step one: Crossdocument structure. Proceedings of the 1st ACL SIGDIAL 2000.
7. Derong Liu et al. Study of concept cohesion based on lexicon and corpus. the 1st National Conference on Information Retrieval and Content Security. 2004.
8. Dong zengdong, Dong qiang. Hownet http://www.keenage.com
9. Mei Jiaju, Zhu Yiming. Synonym Thesaurus 1983.
10. Jaime Carbonell et al. The use of MMR, diversity-based reranking for reordering documents and producing summarization. Proceedings of SIGIN-98.
11. Holland, JH. Adaptation in Natural and Artificial Systems. University of Michigan Press. 1975.
12. Dragomir R. et al. Evaluation challenges in large-scale document summarization. Proceedings of the 41st Annual Meeting of the ACL 2003.

Dynamic K-Nearest-Neighbor Naive Bayes with Attribute Weighted

Liangxiao Jiang[1], Harry Zhang[2], and Zhihua Cai[1]

[1] Faculty of Computer Science, China University of Geosciences
Wuhan, Hubei, P.R. China, 430074
ljiang@cug.edu.cn
[2] Faculty of Computer Science, University of New Brunswick
P.O.Box 4400, Fredericton, NB, Canada E3B 5A3
hzhang@unb.ca

Abstract. K-Nearest-Neighbor (KNN) has been widely used in classification problems. However, there exist three main problems confronting KNN according to our observation: 1) KNN's accuracy is degraded by a simple vote; 2) KNN's accuracy is typically sensitive to the value of K; 3) KNN's accuracy may be dominated by some irrelevant attributes. In this paper, we presented an improved algorithm called Dynamic K-Nearest-Neighbor Naive Bayes with Attribute Weighted (DKNAW) . We experimentally tested its accuracy, using the whole 36 UCI data sets selected by Weka[1], and compared it to NB, KNN, KNNDW, and LWNB[2]. The experimental results show that DKNAW significantly outperforms NB, KNN, and KNNDW and slightly outperforms LWNB.

1 Introduction

Classification is a fundamental issue in data mining. Typically, given a set of training instances with corresponding class labels, a classifier is learned form these training instances and used to predict the class of a test instance. In this paper, an instance x is described by an attribute vector $< a_1(x), a_2(x), \ldots, a_n(x) >$, where $a_i(x)$ denotes the value of the ith attribute A_i of x, and we use C and c to denote the class variable and its value respectively. The class of the instance x is denoted by $c(x)$.

K-Nearest-Neighbor (KNN) has been widely used in classification problems. KNN is based on a distance function that measures the difference or similarity between two instances. The standard Euclidean distance $d(x, y)$ is often used as KNN's distance function. Given an instance x, KNN assigns the most common class label of x's K nearest neighbors to x.

KNN is a typical example of lazy learning which simply stores training data at training time and delays its learning until classification time. Although KNN has been widely used as classifiers for decades, there exist three main problems confronting KNN according to our observation: 1) KNN's accuracy is degraded by a simple vote; 2) KNN's accuracy is typically sensitive to the value of K; 3) KNN's accuracy may be dominated by some irrelevant attributes.

L. Wang et al. (Eds.): FSKD 2006, LNAI 4223, pp. 365–368, 2006.

Motivated by these three main problems, researchers have made a substantial amount of effort to improve KNN's accuracy. An obvious approach to improving KNN is to weight the vote of k nearest neighbors according to their distance to the test instance x, by giving greater weight to closer neighbors. The resulting classifier is K-Nearest-Neighbor with Distance Weighted (KNNDW).

Recently, researchers have paid considerable attention to investigate the approach to combining KNN with naive Bayes. For example, Locally Weighted Naive Bayes[2] (LWNB). In LWNB, each of nearest neighbors is weighted in terms of its distance to the test instance. Then a local naive Bayes is built from the weighted training instances. Their experiments show that LWNB significantly outperforms naive Bayes and is not particularly sensitive to the value of K as long as it is not too small. Surely, it is a K-related algorithm.

In order to address K-related algorithm's sensitivity to the value of K, Xie et al.[3] proposed an improved algorithm called Selective Neighborhood Naive Bayes (SNNB). SNNB firstly constructs multiple Naive Bayes classifiers on multiple neighborhoods by using different radius values for a test instance. Then, it selects the most accurate one to classify the test instance. Their experimental results show that SNNB significantly outperforms naive Bayes. However, this process of searching for the best K value is very time-consuming.

Besides, KNN's accuracy may be dominated by some irrelevant attributes, referred as the curse of dimensionality. A major approach to deal with it is to weight attributes differently when calculating the distance between two instances. Motivated by this idea, Han et al.[4] present an improved algorithm called Weight Adjusted K-Nearest-Neighbor (WAKNN). WAKNN weight the importance of discriminating words using mutual information[5] between each word and the class variable in order to build more accurate text classifiers.

The rest of the paper is organized as follows. In Section 2, we present our improved algorithm simply called DKNAW. In Section 3, we describe the experimental setup and results. In Section 4, we make a conclusion.

2 Dynamic K-Nearest-Neighbor Naive Bayes with Attribute Weighted

Our motivation is to improve KNN's accuracy by synchronously using three improved approaches to deal with KNN's three corresponding problems: 1) Deploy a local naive Bayes on the K nearest neighbors of a test instance; 2) Learn a best value of K for each training data in training time; 3) Weight each attribute's contribution to the distance function using mutual information between each attribute and the class attribute. We call our improved algorithm Dynamic K-Nearest-Neighbor Naive Bayes with Attribute Weighted (DKNAW).

DKNAW combines eager learning and lazy learning. At training time, a best K value $bestK$ is eagerly learned to fit the training data. At classification time, for each given test instance, a local naive Bayes within $bestK$ nearest neighbors is lazily built. So, the whole algorithm of DKNAW can be partitioned into an eager algorithm (DKNAW-Training) and a lazy algorithm (DKNAW-Test). They are depicted below in detail. In our experiment, the minimum K value is set to 1, and the maximum K value of K is set to 50 if the number of training instances is below 100, otherwise to 100.

Algorithm. DKNAW-Training ($\mathbf{T}$, $maxK$, $minK$)
Input: a set $\mathbf{T}$ of training instances, the maximum K value $maxK$, and minimum K value $minK$
Output: the value of $bestK$

```
Let count[K] is the number of instances correctly classified
For K = minK to maxK
    coun[K] = 0
For each intance e in T
    T = T- {e}
    Find the maxK nearest neighbors of e from T
    For K = maxK to minK
        Train a local naive Bayes using the K nearest neighbors
        Classify e using the local naive Bayes
        If the classification of e is correct, count[K]++
        Remove the instance with the greatest distance
    T = T+ {e}
bestK = minK
maxCount = count[minK]
For K = minK + 1 to maxK
    If count[K] > maxCount
        bestK = K
        maxCount = count[K]
Return the value of bestK
```

Algorithm. DKNAW-Test ($\mathbf{T}$, x, $bestK$)
Input: a set $\mathbf{T}$ of training instances, a test instance x, and the value of $bestK$
Output: the class label c of x

```
Find the bestK nearest neighbors of x from T
Deploy a local naive Bayes on the bestK nearest neighbors of x
Use naive Bayes to produce the class label c of x
Return the class label c of x
```

3 Experimental Methodology and Results

We conduct the experiments to compare DKNAW with others using 36 UCI data sets selected by Weka[1] after the following three preprocessing steps. The value of K in all K-related algorithms is 10. In our all experiments, the classification accuracy of each classifier on each data set was obtained via 1 run ten-fold cross validation. Runs with the various classifiers were carried out on the same training sets and evaluated on the same test sets. Finally, we conducted a two-tailed t-test with a 95% confidence level to compare each pair of classifiers.

1. Missing values: We used the unsupervised filter *ReplaceMissingValues* in Weka to replace the missing values in each data set.
2. Discretization of numeric attributes: We used the unsupervised filter *Discretize* in Weka to discretize all the numeric attributes.
3. Removal of useless attributes: We used the unsupervised filter *Remove* in Weka to remove three useless attributes.

Table 1 shows the compared results of two-tailed t-test with a 95% confidence level between each pair of algorithms, each entry $w/t/l$ in Table 1 means that the classifier at the corresponding row wins in w data sets, ties in t data sets, and loses in l data sets, compared to the classifier at the corresponding column. From Table 1, we can see that DKNAW significantly outperforms NB, KNN, and KNNDW and slightly outperforms LWNB.

Table 1. Summary of experimental results: accuracy comparisons

	NB	KNN	KNNDW	LWNB
KNN	5/23/8			
KNNDW	9/22/5	11/25/0		
LWNB	11/20/5	13/20/3	2/32/2	
DKNAW	10/25/1	16/20/0	7/29/0	6/28/2

4 Conclusions

In this paper, we present an improved algorithm called Dynamic K-Nearest-Neighbor Naive Bayes with Attribute Weighted (DKNAW), to upgrade KNN's classification accuracy, in which three improved approaches have been used. The experimental results show that DKNAW significantly outperforms NB, KNN, and KNNDW and slightly outperforms LWNB.

References

1. Witten, I. H., Frank, E.: data mining-Practical Machine Learning Tools and Techniques with Java Implementation. Morgan Kaufmann (2000). http://prdownloads.sourceforge.net/weka/datasets-UCI.jar
2. Frank, E., Hall, M., Pfahringer, B.: Locally Weighted Naive Bayes. Proceedings of the Conference on Uncertainty in Artificial Intelligence (2003). Morgan Kaufmann(2003), 249-256.
3. Xie, Z., Hsu, W., Liu, Z., Lee, M.: SNNB: A Selective Neighborhood Based Naive Bayes for Lazy Learning. Proceedings of the Sixth Pacific-Asia Conference on KDD. Springer (2002) 104-114
4. K. Kumar Han. Text categorization using weight adjusted k-nearest neighbour classification. Technical report, Dept. of CS, University of Minnesota, 1999.
5. Friedman, Geiger, and Goldszmidt. "Bayesian Network Classifiers", Machine Learning, Vol. 29, 131-163, 1997.

Efficiently Mining Both Association and Correlation Rules

Zhongmei Zhou[1,2], Zhaohui Wu[1], Chunshan Wang[1], and Yi Feng[1]

[1] College of Computer Science and Technology, Zhejiang University, China
[2] Department of Computer Science, Zhangzhou Normal University, China
{zzm, wzh, cswang, fengyi}@zju.edu.cn

Abstract. Associated and correlated patterns cannot fully reflect association and correlation relationships between items like both association and correlation rules. Moreover, both association and correlation rule mining can find such type of rules, "the conditional probability that a customer purchasing A is likely to also purchase B is not only greater than the given threshold, but also significantly greater than the probability that a customer purchases only B. In other words, the sale of A can increase the likelihood of the sale of B." Therefore, in this paper, we combine association with correlation in the mining process to discover both association and correlation rules. A new notion of a both association and correlation rule is given and an algorithm is developed for discovering all both association and correlation rules. Our experimental results show that the mining combined association with correlation is quite a good approach to discovering both association and correlation rules.

1 Introduction

Data mining aims to discover useful patterns or rules in large data sets. Although association mining [1][2][5] can find many interesting rules, the following kind of rules is sometimes meaningless in some applications. "A and B are associated but not correlated, that is, although the conditional probability that a customer purchasing A is likely to also purchase B is great enough, it is not significantly greater than the probability that a customer purchases only B. For instance, if $P(B)=88\%$ and $P(B/A)=90\%$, the sale of A cannot increase the likelihood of the sale of B, even though the conditional probability $P(B/A)=90\%$ is much higher than the given threshold. It is the case that A and B are associated but not correlated."

To overcome this difficulty, correlation mining has been adopted [3][4][6][7]. However, such kinds of correlation rules are misleading on some occasions, especially on making business decisions. For example, if $P(B)=90\%$ and $P(B/A)=20\%$, the sale of A cannot increase the likelihood of the sale of B either, even if the purchase of B is influenced by the purchase of A. It is the case that A and B are correlated but not associated, i.e. the conditional probability $P(B/A)=20\%$ is not high enough. Based on these reasons, in this paper, we combine association with correlation in the mining process to discover both association and correlation rules.

L. Wang et al. (Eds.): FSKD 2006, LNAI 4223, pp. 369 – 372, 2006.

2 Mining Both Association and Correlation Rules

We use the measure all-confidence [5] as an association interestingness measure. In terms of the definition of all-confidence, if a pattern has all-confidence greater than or equal to a given minimum all-confidence, any two sub-patterns X, Y of this pattern have conditional probabilities $P(X/Y)$ and $P(Y/X)$ greater than or equal to the given minimum all-confidence, in other words X and Y are associated.

On the other hand, by statistical theory, $A_1, A_2, \cdots, A_n$ are independent, if $\forall k$ and $\forall 1 \le i_1 < i_2 < \cdots < i_k \le n$, $P(A_{i_1} A_{i_2} \cdots A_{i_k}) = P(A_{i_1})P(A_{i_2}) \cdots P(A_{i_k})$.

The definition of a both association and correlation rule is given as follows using the notions of association and independence:

Definition (a both association and correlation rule). Let η be the minimum correlation-confidence and ξ be the minimum confidence. Rule $X \leftrightarrow Y$ is called a both association and correlation rule if

$$\rho(XY) = P(XY) - P(X)P(Y) \big/ P(XY) + P(X)P(Y) \ge \eta$$

and the conditional probabilities $P(X/Y)$ and $P(Y/X)$ greater than or equal to the minimum confidence ξ.

According to the definition, if XY has all-confidence greater than or equal to the given minimum threshold and has correlation-confidence $\rho(XY)$ greater than or equal to the given minimum correlation-confidence, then rule $X \leftrightarrow Y$ is a both association and correlation rule.

Algorithm. Mining both association and correlation rules
Input: a transaction database TDB, a minimum support ξ, a minimum correlation-confidence η and a minimum all-confidence λ.
Output: the complete set of both association and correlation rules.

C_k : Candidate patterns of size k

L_k : Frequent associated patterns of size k

M_k : both association and correlation rules of size k

L_1 = {frequent items}

For ($k = 1; M_k \,!\neq \phi; k{+}{+}$) do begin

  C_{k+1} = candidates generated from $L_k * L_k$

   For each transaction t in database do

    increment the count of all candidates in C_{k+1} that are contained in t

  L_{k+1} = candidates in C_{k+1} with minimum support and minimum all-confidence

```
For each pattern l_{k+1} in L_{k+1}
    derive all both association and correlation rules
from l_{k+1} and insert them into M_{k+1}
Return ∪M_{k+1}
```

3 Experiments

All experiments are performed on two kinds of datasets: 1. Mushroom characteristic dataset, 2. Traditional Chinese Medicine (TCM) formula dataset, which consists of 4,643 formulas with 21689 kinds of medicine involved.

We compare both association and correlation rule mining with associated-correlated pattern mining [8] by the experimental results.

Table 3 shows the number of associated-correlated patterns, associated but not correlated patterns, both association and correlation rules generated in mushroom dataset when the minimum all-confidence increases with the fixed minimum support 1% , minimum correlation-confidence 1% , minimal pattern length 2 and maximal pattern length 5 . Table 4 and Table 5 shows the number of associated-correlated patterns, associated but not correlated patterns, both association and correlation rules generated in TCM dataset and mushroom dataset respectively as the minimum correlation-confidence varies. Because the TCM dataset is very sparse and a great number of patterns generated on TCM data have only two or three items, the number of both association and correlation rules in TCM dataset is significantly less than the number in mushroom dataset. We conclude from our experimental results that both association and correlation rule mining is quite a good method for exploring all both association and correlation relationships between items in a pattern.

Table 3. Num. of patterns in mushroom data (min_sup 1%, min_len 2, max_len 5, c_conf 1%)

All-connfidence	Independent	correlated	Correlation rule
30	112	3678	24501
40	90	1012	3937
50	61	279	618
60	31	83	150
70	12	36	74
80	12	16	28
90	7	8	15

Table 4. Num. of patterns in TCM data (min_sup 1%, min_len2 max_len5, all_conf 10%)

Corr-confidence	independent	correlated	Correlation rule
5	3	1058	1922
10	7	1054	1918
15	16	1045	1909
20	31	1030	1894
25	55	1006	1868
30	76	985	1845

Table 5. Num. of patterns in mushroom (support1%, min_len 2, max_len 5, all_conf 30%)

Corr-confidence	independent	correlated	Correlation rule
5	603	3187	20690
10	1066	2724	18502
15	1367	2423	16622
20	1613	2177	14193
25	1875	1915	11578
30	2100	1690	9349
35	2262	1528	6849
40	2423	1367	5005

4 Conclusions

Both association and correlation rule mining can discover rules which are extraordinary useful for making business decisions. In this paper, a notion of a both association and correlation rule is proposed. We combine association with correlation in the mining process to develop an algorithm for discovering all both association and correlation rules. Experimental results show that the techniques developed in the paper are feasible.

Acknowledgments. The work is funded by subprogram of China 973 project (NO. 2003CB317006), China NSF program (No. NSFC60503018).

References

1. R. Agrawal, T. Imielinski, A. Swami. Mining Association Rules Between Sets of Items in Large Databases. In Proc. 1993 ACM SIGMOD Int. Conf. Management of Data (SIGMOD'93), pp. 207-216.
2. R. Agrawal, R. Srikant. Fast Algorithms for Mining Association Rules in Large Databases. In Proc. 1994 VLDB Int. Conf. Very Large Databases (VLDB'94), pp. 487-499.
3. S. Brin, R. Motwani, C. Silverstein. Beyond Market Basket: Generalizing Association Rules to Correlations. In Proc. 1997 ACM SIGMOD Int. Conf. Management of Data (SIGMOD'97), pp. 265-276.
4. H. Liu, H. Lu, L. Feng, F. Hussain. Efficient Search of Reliable Exceptions. In Proc. Pacific-Asia Conference on Knowledge Discovery and Data Mining (PAKDD'99), pp. 194-203.
5. E. Omiecinski. Alternative interesting measures for mining associations. *IEEE Trans. Knowledge and Data Engineering*, 2003(15): 57-69.
6. G. Piatetsky-Shapiro. Discovery, Analysis and Presentation of Strong Rules. Knowledge Discovery in Databases, AAAI/MIT Press, 1991. pp. 229-248.
7. Y.-K. Lee, W.-Y. Kim, Y. D. Cai, J. Han. CoMine: Efficient Mining of Correlated Patterns. In Proc. 2003 Int. Conf. Data Mining (ICDM'03), pp.581-584.
8. Zhongmei Zhou, Zhaohui Wu, Chunshan Wang, Fengyi. Mining both associated and correlated patterns. In Proc. 2006 ICCS Int. Conf. Computational Science (ICCS'06), pp.468-475.

Estimating the Number of Clusters Using Multivariate Location Test Statistics

Kyungmee Choi[1], Deok-Hwan Kim[2,⋆], and Taeryon Choi[3]

[1] College of Science and Technology,
Hongik University at Jochiwon, Korea
kmchoi@hongik.ac.kr

[2] Department of Electronics Engineering,
Inha University at Incheon, Korea
deokhwan@inha.ac.kr

[3] Department of Mathematics and Statisics,
University of Maryland at Baltimore, U.S.A.
tchoi@math.umbc.edu

Abstract. In the cluster analysis, to determine the unknown number of clusters we use a criterion based on a classical location test statistic, Hotelling's T^2. At each clustering level, its theoretical threshold is studied in view of its statistical distribution and a multiple comparison problem. In order to examine its performance, extensive experiments are done with synthetic data generated from multivariate normal distributions and a set of real image data.

Keywords: Information Retrieval, Clustering, p-values, Multiple Comparison Procedures.

1 Introduction

Recently the cluster analysis has been popularized in the Information Retrieval. It groups or segments a collection of data into clusters so that those within each cluster are more similar to one another than data assigned to different clusters. However most of the clustering algorithms force experts to determine the unknown number of clusters which is to be estimated. In order to avoid this contradictory process, we propose to use some useful statistics which can estimate the number of clusters.

In this paper, to estimate the unknown number of clusters we explore Hotelling's T^2 [7], which has been known as the most powerful test of the two-population problem for normally distributed multivariate data. Since Hotelling's T^2 compares only two clusters in terms of their locations, we have to deal with a multiple comparison problem if there are more than two clusters. So, at each clustering level, we study its theoretical threshold in view of a multiple comparison problem and we obtain its upper bound based on its distribution. We compare this to some other criteria developed from classical location test statistics

⋆ Corresponding author.

L. Wang et al. (Eds.): FSKD 2006, LNAI 4223, pp. 373–382, 2006.

for the given data collection generated from multivariate normal distributions. With this threshold we analyze a real set of multidimensional data to seek the number of its latent clusters.

Estimating the number of optimum clusters has lead to a variety of different clustering methods. Early works include those by Ward [17] and Mojena [13], Miligan and Cooper [12] and Gordon [5]. Latest works include Rousseeuw and Driessen [15], Duda et al. [3], Hastie et al. [6], and Gallegos [4]. Yet, Duda et al. [3] pointed out that in general the number of optimum clusters is not known. So, instead of defining and seeking optimum clusters we will try to determine the number of latent clusters at the points where the provided criteria show significant changes. In order to evaluate its performance we conduct experiments with synthetic data generated from the pre-determined clusters and examine how well it figures out the original number of the clusters.

Most previous methods have used, as criteria, functions of variances within each cluster (within-variances). Smaller within-variances tend to provide well separated clusters, and bigger variances between clusters (between-variances) also imply that clusters are well separated. If the between-variance is big compared to the within-variance, it is very clear that the clusters are well separated. However if the between-variance is small compared to the within-variance, the clusters are often unlikely to be well separated even with the small within-variance. Therefore it is more reasonable to use the criteria which consider both the within-variance and the between-variance.

For comparison, we review Sum-of-Squared Error (SSE) and Wilks' Λ. All of them are well known multivariate statistics which test the different locations of the given clusters or groups. See Mardia et al. [11], Duda et al. [3], Hastie et al. [6], and Rencher [14]. SSE is one of the most popular criteria based on the within-variance while SSE does not consider the between-variance, the rest two consider a ratio of the between-variance to the within-variance. Wilks' Λ tests whether locations of more than two clusters differ or not. Hotelling's T^2 is a generalized t statistic in high dimensions to test whether locations of two clusters differ or not. When the data follow a normal distribution, SSE approximately follows a χ^2 distribution, Wilks' Λ follows a Wilks' Λ distribution and also approximately follows a χ^2 distribution. Hotelling's T^2 follows an F distribution. Thus as long as the data size is big enough, their distributions provide us clear thresholds for the given significance levels (significant error rates) along with nice statistical interpretation.

The hierarchical clustering is used because it reproduces the hierarchical structure of the data or the underlying structure of the data [13]. However all the criteria mentioned in this paper can be also used with the K-means clustering. These criteria will be calculated at each hierarchical level and presented in graphs to depict the estimates of the number of clusters. However the partitions are not necessarily optimal [17].

In section 2, two criteria are reviewed and their thresholds are sought for the given significance levels (significant error rates). In section 3, an approach using Hotelling's T^2 is presented. Section 4 contains an example using a set of real image data and various simulations with synthetic data for the evaluation of the proposed method. We make a conclusion in Section 5.

2 The Notations and the Criterion Functions

Suppose that for $x_i \in R^p$, $i = 1, \cdots, n$, let the data be a set of $D = \{x_1, x_2, \cdots, x_n\}$ and cluster them into the c disjoint clusters, $D_1, D_2, \cdots, D_c$. Let n_i be the size of D_i. For each cluster D_i, let us define the mean and variance, $m_i = \sum_{x \in D_i} x/n_i$ and $S_i = \sum_{x \in D_i} (x - m_i)(x - m_i)^T$. The grand mean is $m = \sum_{x \in D} x/n$. Then the total scatter matrix S_T is obtained as the sum of the within-cluster scatter matrix, S_W, and the between-cluster scatter matrix, S_B, that is, $S_T = S_W + S_B$. Here,

$$S_W = \sum_{i=1}^{c} S_i, \text{ and } S_B = \sum_{i=1}^{c} n_i(m_i - m)(m_i - m)^T.$$

At the hierarchy of clusters, the level c corresponds to c clusters. Let the given significance level (significant error rate) at each clustering level be α, which is controlled by the thresholds.

2.1 The Sum-of-Squared-Error

Let us define the Sum-of-Squared-Error as follows:

$$SSE = \sum_{i=1}^{c} \sum_{x \in D_i} ||x - m_i||^2.$$

Note that SSE is the trace of the within-variance. Since $tr[S_T] = tr[S_B] + tr[S_W]$ and $tr[S_T]$ is fixed, minimizing S_W implies maximizing S_B. Duda, Hart, and Stork [3] suggested to find the number of clusters by minimizing SSE and pointed out that SSE worked best when the clusters are compact and well-separated. They also mentioned that when there was no minimum, the natural number of clusters was determined at the big gap. However often SSE decreases monotonically in c and tends to converge, so that there is not always the minimum. Also there could be multiple big gaps.

Ward [17] tried to estimate the number of clusters by minimizing increase of SSE, which lead to the use of both the within-variance and the between-variance. Mojena [13] evaluated Ward's Incremental Sum of Squares as the best among seven criteria studied at that time. On the other hand Rousseeuw and Driessen [15] used $\prod_{i=1}^{c} det(S_i)^{|S_i|}$, where $|S_i|$ is the cardinality of ith cluster. Gallegos [4] used $\prod_{i=1}^{c} det(S_i)$ as a criterion, and showed that m_i and S_i were Maximum Likelihood Estimators of means and variances of each clusters when data were generated from normal distributions. Using the trace considers only diagonals of the variance matrices, while using the determinant considers correlations, too.

2.2 Wilks' Λ

Wilks' Λ is one of the traditional statistics which test whether the locations of more than two groups are equal. This measure can be expressed as a function of the ratio of the between-variance to the within-variance, which is defined by

$$\Lambda = \frac{det(S_W)}{det(S_B + S_W)} = \frac{1}{det(S_W^{-1} S_B + I)}.$$

See Mardia [11]. The number of clusters is sought where Λ is minimized. However like SSE, Λ decreases monotonically in c. When the data follow a multivariate normal distribution, this Statistic follows a Wilks' Λ distribution $\Lambda(p, n-c, c-1)$. When the sample size is large enough, its log transformation approximately follows a χ^2 distribution.

To obtain the statistically meaningful threshold which controls the significant level (significant error rate), let us define the p-value of a given value Λ_o at the cth clustering level as follows : $p = P(\Lambda \leq \Lambda_o)$. A small p-value provides a strong evidence of two separate clusters. If there is not a significant decrease in p-value from the cth clustering level to the $(c-1)th$ clustering level, then c is closer to the optimal number of clusters, where the criteria reaches the minimum.

The related statistics have been introduced by Pillai, Lawley-Hotelling, and Roy. $tr(S_W^{-1} S_B)$ and $tr(S_T^{-1} S_B)$ are also closely related to Λ. See Hastie [6], Duda, Hart, and Stork [3], and Rencher [14]. Rencher [14] introduced an analog of the univariate analysis of variance, $[tr(S_B)/(c-1)]$ / $[tr(S_W)/(n-c)]$, which has a local maximum.

3 Hotelling's T^2

The classical Hotelling's T^2 tests whether the locations of two clusters are equal or not. For D_i and D_j clusters with $i \neq j$, it is defined by

$$T_{ij}^2 = \frac{n_i n_j (n-2)}{(n_i + n_j)^2} (m_i - m_j)^T S_{pij}^{-1} (m_i - m_j),$$

where $S_{pij} = (S_i + S_j)/(n_i + n_j - 2)$. When the data follow a multivariate normal distribution, $(n_i + n_j - p - 1)/p(n_i + n_j - 2)T^2$ follows an $F(p, n_i + n_j - p - 1)$. This statistic can be interpreted as the Mahalanobis distance between the centers of two clusters. See Mardia [11].

To start finding the number of clusters, let us consider two clustering levels with $(c-1)$ and c clusters. It is necessary to decide which level is more optimal than the other. At the cth clustering level, there are $\binom{c}{2}$ of Hotelling's T^2s used to decide which pair of clusters to be merged. Note that this leads to a classical multiple comparison (multiple-inference) procedure. If no significant merging occurs, then c is closer to the optimal than $(c-1)$. Otherwise $(c-1)$ is closer to the optimal.

More precisely, let us assume the value T_o be Hotelling's T^2_{ij} for the pair of D_i and D_j clusters at the cth clustering level. Then the corresponding p-value, p_{ij}, is defined by

$$p_{ij} = P((n_i+n_j-p-1)/p(n_i+n_j-2)T^2_{ij} \geq T_o) = P(F(p, n_i+n_j-p-1) \geq T_o),$$

where $F(p, n_i+n_j-p-1)$ is the F distribution with p and n_i+n_j-p-1 degrees of freedom. A small p_{ij} is a good evidence of two separate clusters. Especially if $\max_{1\leq i\neq j\leq c} p_{ij}$ (MPH) is less than the given threshold, all c clusters are separated and so c is closer to the optimal than $(c-1)$. Thus in order to obtain a meaningful threshold, the bound of MPH should be studied. See Proposition 1 below.

Traditional multiple comparison controlled the significance level (significant error rate) α by controlling the probability of committing any falsely declared significant inference under simultaneous consideration of $\binom{c}{2}$ multiple inferences. Yet, Kim et al. [10] have used Hotelling's T^2 as an individual inference to decide whether each pair of clusters were to be merged, so that each individual inference used α as a threshold. Ignoring the multiplicity of the inference, however, leads to a greatly increased false significant error rate [1]. Choi and Jun [2] did not figure out how to control the multiplicity effect even though they noticed it. In this paper we will consider two competitive ways of controlling the multiplicity effect. They are Bonferroni-Type Significance Level procedure $(BSLP)$[14] and Benjamin and Hochberg's False Discovery Rate procedure $(FDRP)$[1].

Let R be the number of pairs of clusters which are declared to be separated. Let V be the number of pairs of clusters which are falsely declared to be separated. $BSLP$ tests individually each pair of clusters at level $\alpha_s = \alpha/\binom{c}{2}$, which guarantees the probability of at least one falsely declared significant to be less than α. That is, $P(V \geq 1) \leq \alpha$. Since α_s gets usually very small as c grows, $BSLP$ is known to be very conservative and relatively lose its power.

On the other hand, $FDRP$ controls $E(V/R) \leq \alpha$, the expected proportion of errors committed by declaring the pairs of clusters to be separated. According to the $FDRP$, α_s changes for each pair. Let $p_{(k)}$ be the kth smallest p-value among $\binom{c}{2}$ p_{ij}s. If $p_{(k)} \leq k\alpha/\binom{c}{2}$, then k pairs of clusters corresponding to the smallest p-values are separated. So if $\max p_{ij} \leq \alpha\binom{c}{2}/\binom{c}{2} = \alpha$, then all pairs of clusters are separated. This means that c is closer to the optimal number of clusters than $(c-1)$. The $FDRP$ is known to be more powerful than the $BSLP$.

For example, let us assume that $\alpha = 0.05$ as the total significance level (significant error rate) at the $4th$ clustering level. There are $\binom{4}{2}$ pairs of clusters and p-values, and they are ordered from the smallest to the largest. In the $BSLP$, $\alpha_s = (0.0083, \cdots, 0.0083)$ for all $\binom{4}{2}$ pairs. So $\max_{1\leq i\neq j\leq c} p_{ij}$ is compared to 0.0083. In the $FDRP$, α_s's are $(0.0083, 0.0167, \cdots, 0.05)$. So $\max_{1\leq i\neq j\leq c} p_{ij}$ is compared to 0.05.

Proposition 1. Let α be the given significance level (significant error rate) at each clustering level and p_{ij} for $i \neq j$ be p-value of the individual test T^2_{ij}. Then,

$$\textit{in the BSLP}: P\left(\max_{1\leq i\neq j\leq c} p_{ij} \leq \alpha/\binom{c}{2} \text{ occurs falsely}\right) \leq \alpha$$

$$\textit{in the FDRP}: P\left(\max_{1\le i\ne j\le c} p_{ij} \le \alpha \text{ occurs falsely}\right) \le \alpha.$$

Proofs follow directly from definitions of both the $BSLP$ and the $FDRP$. Therefore using MPH guarantees α as the significance level (significant error rate). So the algorithm follows right away with two different thresholds α_s based on both the $BSLP$ and the $FDRP$.

We now propose an algorithm to estimate the number of clusters using MPH as follows:

Algorithm. Estimating the Number of Clusters
begin
1. Set the number of clusters as c.
2. Do clustering the given data with c clusters.
3. Make pairwise comparisons with $\binom{c}{2}$ pairs of two clusters.
4. If all of the corresponding $\binom{c}{2}$ p-values are less than the given significance level α, at each clustering level, then accept c as the number of clusters.
5. If not, consider the number of clusters as $c - 1$ and iterate procedures from (2) to (4).

end

4 Experiments and Examples

4.1 An Example with Real Data

A collection of Corel photo image data is used as the test data set. Images in this collection have been classified into distinct categories like sunset, sea, flower etc. by experts, and each category includes about 100 images. Fig. 1 shows some representative images of the sunset class and the sea class, respectively. Here, visual features are extracted from the images. The feature values of three color moments are extracted from each channel of HSV (hue, saturation, value) color space. We conduct experiments with images in two pre-determined clusters and examine whether the algorithm can figure out the original number of clusters or not.

In Fig. 2, 100 images in the Sunset category are labeled as 1 and another 100 images in the Sea category are labeled as 2. The top left shows x and y components of the feature data, the top right shows x and z components, the bottom left shows y and z components. From these graphs we see that two clusters are well separated in directions of both x-axis vs y-axis and x-axis vs z-axis, while they are not in directions of y-axis vs z-axis.

We started with initial values of the number of clusters and a threshold as $c = 9$ and $\alpha = 1.0e^{-10}$, respectively. The bottom right of Fig. 2 shows a significant change of MPH between $c = 2$ and $c = 3$ and the MPH becomes less than α at

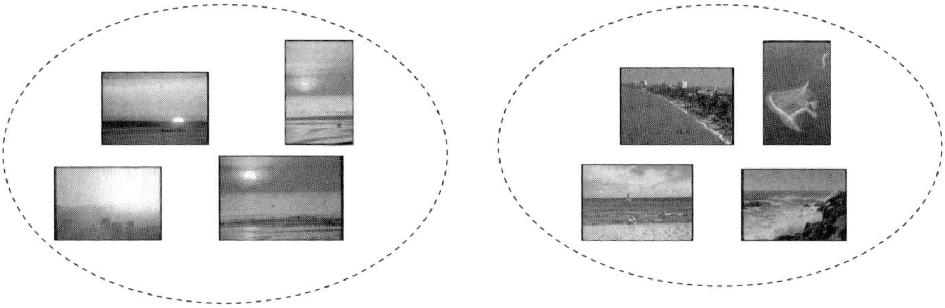

Fig. 1. Some representative images in the sunset cluster and the sea cluster

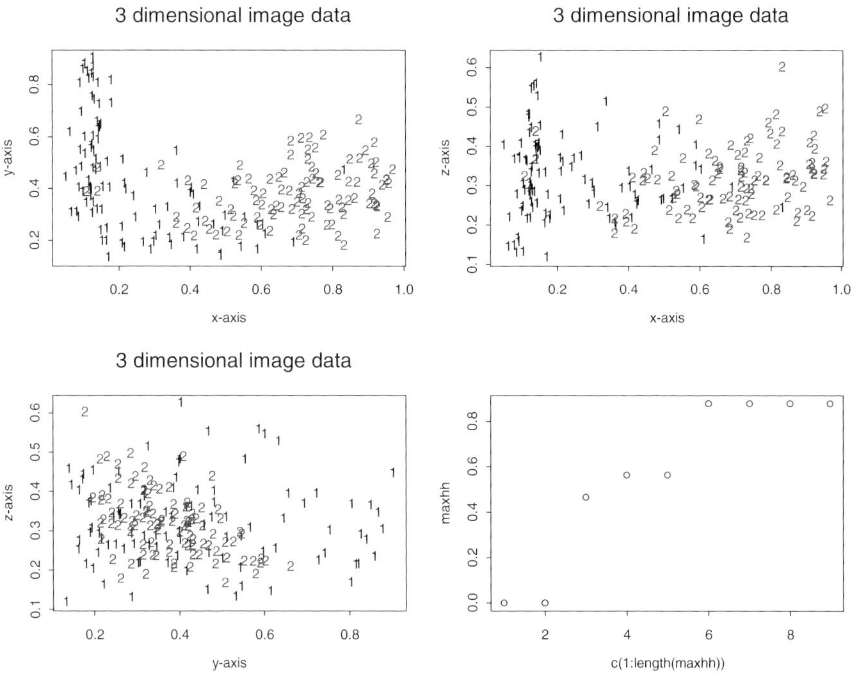

Fig. 2. Estimation of the number of the clusters in the image data: The images in the Sunset cluster are labeled as 1 and those in the Sea cluster are labeled as 2. The top left shows x and y components of the data, the top right shows x and z components of the data, and the bottom left shows y and z components of the data. The bottom right shows the jump at the clustering level 2.

$c = 2$. So, the number of clusters of this data is estimated as two. This example demonstrates that the proposed method can find the proper number of clusters even when the data are not in well separated in certain directions. Note that the proposed method works well even with some outliers in these image clusters.

4.2 Performance Evaluation

In order to evaluate the performance of the proposed method, we used two sets of simulated data. The first data set has three clusters in two dimensions, where the clusters are made up of standard bivariate normal variables with 50 observations in each cluster, centered at (0,0), (0,5) and (5,-3). The second data set has four clusters in three dimensions, where the clusters consist of standard tri-variate normal variables with 50 observations in each cluster, centered at (0,0,0), (0,5,0), (5,-3,1) and (-3,2,5). In this experiment, we use a statistical package R [8] and one thousand simulations were performed at both settings, and the results are summarized in Table 1. From Table 1, we can observe that the proposed method estimates the number of clusters well in these simulation studies. As you can see, the results from two-dimensional cases are very promising; The algorithm selects the correct number of clusters more than 97% of the cases. The results from three-dimensional cases didn't perform as well but it was still more than 85%. These results are comparable to Table 1 in Tibshrani et al. [16], in which they made comparisons of the performance of gap statistic with other existing methods. The idea of gap statistic is as follows: If we plot the graph of within cluster dissimilarity along y-axis and number of clusters along x-axis then this graph generally takes a form of a elbow depending upon the measure on the y-axis. The gap statistic seeks to locate this elbow because the value on the x-axis at this elbow is the correct number of clusters for the data. Although our method does not outperform that of gap statistic from Tibshrani et al. [16], it gives nearly comparable performance. In addition, our method is easy to implement and works very well when the normality assumption holds or the sample size increases.

Table 1. Result of performance evaluation

Case	Estimates of number of clusters : $\hat{K}$									
	1	2	3	4	5	6	7	8	9	10
Three clusters in 2 dimensions	0	11	973	3	0	0	0	0	0	0
Four clusters in 3 dimensions	0	4	126	870	0	0	0	0	0	0

5 Conclusion

Clustering methods in information retrieval systems enables the retrieval algorithms to be more efficient by forming clusters which maximize the degree of similarity within clusters, or maximize the degree of dissimilarity between clusters. However to make the clustering methods more efficient, experts have to determine the unknown number of clusters which is to be estimated. We reviewed some useful criteria which have been developed from location test statistics. Especially we used p-values of Hotelling's T^2 as a criterion to estimate the number of clusters. At the notable big jump in a series of p-values, the number of clusters was estimated. This point corresponded to the threshold of this statistic and was studied theoretically in view of a multiple comparison problem.

The proposed Hotelling's T^2 showed good performances with both the synthetic data and the real image data. Also its hit ratio to estimate the correct number of latent clusters was 97% in the bivariate normal data, and 85% in the trivariate normal data. Even though the experiments were done with normal data only, it is not very hard to predict its performance on non-normal data. Hotelling's T^2 has been known as the most powerful and popular multivariate statistic to test the location difference between two groups or two clusters. Its performance has been intensively studied by many statisticians. It has been well known that it outperforms other statistics when the data has light tails than the normally distributed data, while it does not when the data have heavy tails. However its performance was not too bad even with non-normal data [14].

There are a few issues that we would like to consider in the future. Since experiments were confined to only two and three dimensional cases, we can consider more general situations such as high-dimensional settings later on. Secondly, it was a challenge to decide a threshold for p-values since they were very small values. Nevertheless, it returned us satisfactory empirical results.

References

1. Y. Benjamin, and Y. Hochberg. Controlling the False Discovery Rate: a practical and powerful approach to multiple testing. J. R. Statist. Soc. B, Vol.57, No.1, pages 289-300, 1995.
2. K. Choi, and C. Jun. A systematic approach to the Kansei factors of tactile sense regarding the surface roughness. Applied Economics, in Press, 2006.
3. R.D. Duda, P. E. Hart, and D.G. Stork. Pattern Classification. John Wiley & Sons, Inc. New York, 2001.
4. M. T. Gallegos. Maximum likelihood clustering with outliers, Classification, Clustering, and Data Analysis. Jajuga et al Ed., Springer, 2002.
5. A. Gordon. Classification, 2nd ed. London, Chapman and Hall-CRC, 1999.
6. T. Hastie, R. Tibshirani, and J. Friedman. The elements of statistical learning. Data Mining, Inference, and Prediction. Springer, 2001.
7. H. Hotelling. Multivariate Quality Control. In C. Eisenhart, M. W. Hastay, and W. A. Wallis, eds. Techniques of Statistical Analysis. N.Y., McGraw-Hill, 1947.
8. R. Ihaka, and R. Gentleman. R: A language for data analysis and graphics. Journal of Computational and Graphical Statistics, Vol.5, No.3, pages 299-314, 1996.
9. K. Jajuga, A. Sokolowski, and H.-H. Bock. (Eds.) Classification, Clustering, and Data Analysis. Springer, 2002.
10. D. H. Kim, C. W. Chung, and K. Barnard. Relevance Feedback using Adaptive Clustering for Image Similarity Retrieval, The Journal of Systems and Software, Vol.78, pages 9-23, 2005.
11. K.V. Mardia, J.T. Kent, and J.M. Bibby. Multivariate Analysis. Academic Press, 1979.

12. G.W. Miligan, and M.C. Cooper. An examination of procedure for determining the number of clusters in a data set. Psychometrika, Vol.50, pages 159-179, 1985.
13. R. Mojena. Hierarchical grouping methods and stopping rules: An evaluation. The Computer journal, Vol.20, No.4, 1975.
14. A.C. Rencher. Methods of Multivariate Analysis. John Wiley and Sons, 2002.
15. P. J. Rousseeuw, and K. Van Driessen. A first algorithm for the minimum covariance determinant estimator. Technometrics, Vol.41, pages 212-223, 1999.
16. R. Tibsirani, G. Walther, and T. Hasite. Estimating the number of clusters in a data set via the gap statistic. J.R. Statist. Soc. B, Vol.63, pages 411-423, 2001.
17. J. H. Ward. Hierarchical Grouping to optimize an objective function. J. of Amer. Stat. Assoc., Vol.58, pages 236-244, 1963.

Some Comments on Error Correcting Output Codes

Kyung Ha Seok[1,⋆] and Daehyeon Cho[2]

[1] Department of Data Science, Institute of Statistical Information, Inje University, Kyungnam 621-749, Korea
skh@stat.inje.ac.kr
[2] Department of Data Science, Institute of Statistical Information, Inje University, 621-749, Kyungnam, Korea
cho@stat.inje.ac.kr

Abstract. Error Correction Output Codes (ECOC) can improve generalization performance when applied to multiclass problems. In this paper, we compared various criteria used to design codematrices. We also investigated how loss functions affect the results of ECOC. We found that there was no clear evidence of difference between the various criteria used to design codematrices. The One Per Class (OPC) codematrix with Hamming loss yields a higher error rate. The error rate from margin based decoding is lower than from Hamming decoding. Some comments on ECOC are made, and its efficacy is investigated through empirical study.

1 Introduction

For some classification problems, both two class and multiclass, it is known that the lowest error rate is not always reliably achieved by trying to design a single best classifier. An alternative approach is to employ a set of relatively simple sub-optimal classifiers and to determine a strategy that combines the results. This strategy, a method proposed by Diettrich and Bakiri [4][5], is known as ECOC. ECOC, because it offers good performance, has been widely used. ECOC is popular, too, because a good binary classifier is not necessarily a good multiclass classfier; binary problems are simpler to handle, and repeated computation, provided by ECOC, can reduce error.

ECOC was initially developed for pattern recognition, but these days it is used widely in machine learning, especially text classification (Kittler et al. [8]), Aha and Bankert [1], Windeatt and Ghaderi [13], Berger [2]). Diettrich and Bakiri [4][5] proposed ECOC and compared some codematrix design methods. They showed that ECOC performs well with multiclass problems. Allewein et al. [6] studied various codematrices and loss functions. We cannot agree with their results, however, since they ignored the choice of optimal parameters. Windeatt and Ghaderi [13] and Kuncheva [10] proposed an efficient ECOC design method

⋆ Corresponding author.

L. Wang et al. (Eds.): FSKD 2006, LNAI 4223, pp. 383–392, 2006.

and introduced various loss functions. However, though much research has dealt with ECOC and the criteria to design ECOC, there is still no unified standard for comparison of codematrix design criteria and loss functions. This paper introduces a unified criterion for codematrix design, and makes a case for the superiority of that criterion. The relationship between codematrices and loss functions is investigated, as well as the degree to which loss functions affect the result. Above all, a rationale for using the All Pairwise Comparison (APC) codematrix is provided. The paper is organized as follows. The output coding concept and original ECOC are briefly reviewed in Section 2. In Section 3 we give some comment on the ECOC. Experimental results with natural and synthetic data are given in Section 4. Finally, concluding remarks are given in Section 5.

2 ECOC

2.1 Encoding

ECOC consists of two steps, encoding and decoding. In the encoding step, a multiclass problem is reduced to multiple binary classification problems. After the binary classification problems are solved, the resulting set of binary classifiers must be combined. This is the decoding step. Now, we will introduce the codematrix design method, which is the core of encoding. Let $\Omega = \{1, ..., 4\}$ be a set of class labels with class number $c = 4$. Using a codematrix, we can reduce the multiclass problem to multiple binary classification problems. Note that the target of the binary classification problems is -1 or 1. Suppose that we use the codematrix shown Table 1. Then $L = 7$ binary classifiers f_i, $i = 1, .., 7$ are required.

Table 1. A codematrix with $c = 4$ and $L = 7$

Class Label	f_1	f_2	f_3	f_4	f_5	f_6	f_7
1	-1	-1	-1	1	-1	1	1
2	-1	-1	1	-1	-1	-1	-1
3	-1	1	-1	-1	1	-1	1
4	1	-1	-1	-1	1	1	-1

The ith column, denoted as the ith classifier, determines the target of the ith classifier. For example, the target of the first classifier takes -1 when the original class label is 1 or 2 or 3, and takes 1 otherwise. After training the 7 classifiers, we have to combine them to estimate the class label of the original multiclass problem. The rows of the codematrix are called codewords, denoted by $C_i = (C(i, 1), ..., C(i, L))$, $i = 1, ..., c$. Codewords represent the target of each classifier for a class label. For example, class label 1 is reduced to -1 for the first classifier, -1 for the second, -1 third,.., and 1 for the seventh.

A simple codematrix, called OPC, was designed to compare each class to all of the others. Accordingly, the OPC codematrix has its value 1 for the diagonal element, and -1 elsewhere. The APC codematrix designed for all pairs of classes are compared with each other. The unused class labels are denoted by 0. Table 2 shows an APC codematrix with $c = 4$.

Table 2. APC codematrix with $c = 4$

Class Label	f_1	f_2	f_3	f_4	f_5	f_6
1	1	1	1	0	0	0
2	-1	0	0	1	1	0
3	0	-1	0	-1	0	1
4	0	0	-1	0	-1	-1

The Exhaustive codes(EC) codematrix is generated by all possible different $2^{(c-1)} - 1$ classifiers for c classes. When $c = 3$, the EC codematrix is the same as the OPC codematrix. This means that when $c = 3$, the effect of ECOC is small. When $4 \leq c \leq 11$, the EC codematrix is mainly used. When $c > 11$, there are many experimental results, so the random selection of codewords is possible (Diettrich and Bakiri [5], Windeatt and Ghaderi [13], Schapire [11]). Contrarily, the effect of ECOC is not efficient, though when $4 \leq c \leq 11$, there is research that shows that a smaller number of classifiers with a well designed codematrix is more efficient than with an EC codematrix (Allwein et al.[6]).

The Hamming distance between codewords C_i and C_j is defined as follows:

$$HC_{ij} = \sum_{k=1}^{L} |C(i,k) - C(j,k)|/2, \quad i,j = 1,...,L. \tag{1}$$

The codematrix must be designed so as to make the Hamming distance large. In a binary classifier, when the maximum of HC_{ij} is d, the number of errors $d_1 = [(d-1)/2]$ can be overcome (Diettrich and Bakiri [5]). Here, $[x]$is the largest integer that is not larger than x. We denote d_1 as the error correcting power. If the number of the binary classifiers is not greater than d_1, the combined classifier classifies correctly. Accordingly, a codematrix must be designed so as to make d as large as possible. Allwein et al. [6] showed that a small value of d is related to a small codematrix training error, and proposed to design a codematrix to make d large. When the classifiers are independent of each other, the codematrix is very meaningful. So, the Hamming distance between classifiers f_i and f_j, for $i,j = 1,...,L$, becomes

$$HB_{ij} = min_{ij} min \sum_{k=1}^{c} |C(k,i) - C(K,j)|, \sum_{k=1}^{c} |2 - C(k,i) - C(K,j)|. \tag{2}$$

There are some other methods as well, such as Bagging (Breiman [3]) and Boosting (Freund and Schipire [7]), to render the correlation between classifiers low. As we discussed above, in designing code matrix, it was our object to lengthen the distances between columns and between rows. For this purpose, let AHC and AHB be the means of HC_{ij} and HB_{ij}, respectively for $i, j = 1, ..., L$. And let LHC and LHB be the minimum of HC_{ij} and HB_{ij}, respectively for $i, j = 1, ..., L$. And let us define A_α and L_α as follows.

$$A_\alpha = \alpha AHC + (1 - \alpha)AHB \tag{3}$$

$$L_\alpha = \alpha LHC + (1 - \alpha)LHB, \quad 0 \leq \alpha \leq 1 \tag{4}$$

Here, A_α indicates the weighed mean of the average distance, and L_α, the weighed mean of the minimum distance. Accordingly, we will determine which criteria can produce a better codematrix. Lately, there have been published some papers concerning codematrix design criteria. But we could not find any result with which we can agree. For example, Kuncheva [10] suggested that $A_{0.5}$ should be a codematrix design criterion because a codematrix from $A_{0.5}$ provides better performance than an OPC codematrix. But it is well known that an OPC codematrix yields a poor result, and so we cannot agree that $A_{0.5}$ should be the criteria of codematrix design. And neither can we accept the result of Allwein et al. [6], because they ignored the selection of the parameter of the binary classifiers, which is crucial to the resulting performance.

2.2 Decoding

Once we have an output $f(x) = (f_1(x), .., f_L(x))$ from x, we have to calculate the distance between $f(x)$ and $C_i, i = 1, ..., c$ to classify x into $k = \text{argmin}_i L(f(x), C_i)$, where L is a loss function that indicates a distance between $f(x)$ and C_i. Just as it is very important to use a good criterion for designing a codematrix in the encoding step, so it is also meaningful to consider which loss function to use in the decoding step. The simplest method of combining binary classifiers is Hamming decoding, which uses Hamming distance as a loss function. It is notable that Hamming decoding ignores the loss function that was used during training as well as the confidences attached to predictions made by classifiers. To overcome the weak points of Hamming decoding, margin-based decoding is used. The margin of an example (x, y) with respect to a real valued function $f(x)$ is $yf(x)$. Note that the margin is positive if and only if the sign of $f(x)$ agrees with y. For margin-based decoding, we consider the three following loss functions.

$$\text{Linear Loss} : L_{lin}(f, C_i) = \frac{1}{L} \sum_{k=1}^{L} \{f_k(x) - C(i, k)\} \tag{5}$$

$$\text{Exponential Loss} : L_{exp}(f, C_i) = \frac{1}{L} \sum_{k=1}^{L} exp\{-f_k(x)C(i, k)\} \tag{6}$$

$$\text{Positive Linear Loss} : L_{plin}(f, C_i) = \frac{1}{L} \sum_{k=1}^{L} [1 - f_k(x)C(i, k)]_+, \tag{7}$$

where $[x]_+ = max(0, x)$.

3 Some Comments on ECOC

To this point we have briefly introduced encoding and decoding. Since the error rate depends on encoding and decoding, we should select the criteria carefully. If the performance of APC is not worse than those of the others, we recommend its use, because if we use APC, we save computation time and we do not need to design a codematrix. Note that in the APC codematrix, unused class labels are denoted by 0. This means that we would use only a part of the training data set in each classifier. Since the margin is interpreted as the prediction confidence of classifiers, we can use it with ECOC. For example, in support vector machine (SVM, Vapnik[12]), if the margin of an input x_0 equals to C (regularization parameter), then the confidence that x_0 can be classified correctly is very high. And if the margin of x_0 is at or near 0, the confidence that x_0 can be classified correctly is very low. Therefore, input points with small margins have a large probability of misclassification. If the error rate of a training data set is 0 and the margin of a new input x_0 is very small, we can infer that x_0 has a different class label from that of the training data set. That is, it is reasonable to classify x_0 into neither the label 1 class nor the label -1 class. Therefore we conclude that the APC codematrix reduces the multiclass classification problem to 3 class (1, -1 , not 1 and -1) problems. For example, in Table 2, data sets with label 1 and 2 classes are used with the classifier f_1. The classifier f_1 trains the data in order to classify it as label 1 to 1, 2 to -1 and 3, 4 to 0. Accordingly, the result also depends on the loss function. Note that when $C(i,k) = 0$, $exp\{-f_k(x)C(i,k)\}$ and $[1 - f_k(x)C(i,k)]_+$ are constant, so the f_k is no more useful in these two loss functions, L_{plin} and L_{exp} . However, the loss L_{lin} reflects the magnitude of f_k. Therefore we can expect a better result if we use the linear loss function to design the APC codematrix. From a training data set, we can obtain $f(x) = (f_1(x), ..., f_L(x))$ and error rates e_k, $k = 1, .., L$, from the classifier $f_k(x)$ in the training data set. Then we can make a loss function using error rate e_k. The idea in this method is similar to that of boosting. For example, we can write the error rate weighted L_{exp} as follows:

$$EL_{exp}(f, C_i) = \sum_{k=1}^{L} (1 - e_k) exp\{-f_k(x)C(i,k)\} . \tag{8}$$

4 Experiment Results

In this section, we investigate the comments we made in Section 3, using both the natural data set and the synthetic data set. Accordingly, the objects of our simulation are 1) The differences of codematrix design criteria 2) The differences of loss functions 3) The efficiency of the error-rate-weighted loss function. 4) The relationship between linear loss and the APC codematrix

In our simulations we used SVM as a base binary classifier, radial basis function as a kernel function, and the natural data sets that we obtained in the UCI repository (ftp://ftp.ics.uci.edu/pub/machine-learning-databases), which are

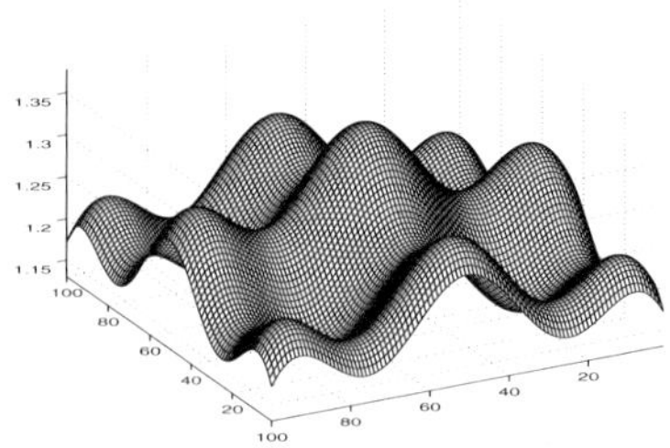

Fig. 1. The plot of the density function $p(x)$ without $C(x)$ in (9)

the dermatology-, glass-, ecoli-, and vowel data sets. These data sets are illustrated in Table . Except the OPC, the number of classifiers L of each codematrix set to be equal. The synthetic data set is a 9 class problem in 2-dimensional input

Table 3. Description the datasets used in the experimets

Name	Train	Test	Inputs	Classes
Glass	214	-	9	7
Ecoli	336	-	8	8
Dermatology	366	-	34	6
Vowel	528	462	10	11

space. We generated uniform random vectors $x_i = (x_{i1}, x_{i2})$ on the unit square. Then, we assigned class labels to each x_i according to the following conditional probabilities:

$$p(x) = \sum_{i,j=1}^{3} C(x) exp\left\{-10[(x_1 - \frac{(i-1)}{2})^2 + (x_2 - \frac{(j-1)}{2})^2]\right\}, \qquad (9)$$

where $C(x)$ is a normalizing function at x such that $\int C(x)p(x) = 1$. The density function without $C(x)$ is shown in Fig 1. We generated a data set of size $n = 300$. Note that the class numbers of all of the data sets were greater than 6. If the test set is not available, the original data set is divided into a training set and a test data set with the ratio of 7:3. The kernel parameter and the regularization parameter in SVM are obtained through the cross-validation method. Fig. 2 and Table 4 show the experimental results. In Fig. 2, there are five panels that represent the results from the glass-, ecoli-, dermatology-, vowel-, and synthetic data sets, respectively.

For the first object of our simulation, we considered three codematrix design criteria: $A_{0.5}$(average of averages of between-rows distance and between-classifiers distance, ARC), $L_{0.5}$ (average of minima of between-rows distance and between-classifiers distance, MRC), and L_1 (minimum of between-rows distance, MR). Because Kuncheva [10] proposed the use of $A_{0.5}$, we included $A_{0.5}$ in the

Table 4. Experiment results with various codematrices and loss functions. Natural datasets from UCI and a synthetic dataset are used. For each dataset five codematrices and four loss functions are used.

Dataset	Loss function	OPC	APC	MR(L_1)	MRC($L_{0.5}$)	APC($A_{0.5}$)
Glass	Hamming	0.4213	0.3550	0.3469	0.3409	0.3200
	Linear	0.3463	0.3417	0.3416	0.3369	0.3306
	Exp.	0.3466	0.3459	0.3422	0.3378	0.3281
	Pos. Lin.	0.3466	0.3414	0.3400	0.3378	0.3300
Ecoli	Hamming	0.1912	0.1810	0.1410	0.1470	0.1342
	Linear	0.1576	0.1646	0.1364	0.1390	0.1358
	Exp.	0.1574	0.1686	0.1376	0.1390	0.1382
	Pos. Lin.	0.1574	0.1693	0.1362	0.1380	0.1354
Dermatology	Hamming	0.0761	0.0350	0.0411	0.0404	0.0411
	Linear	0.0391	0.0327	0.0407	0.0378	0.0357
	Exp.	0.0391	0.0327	0.0411	0.0374	0.0357
	Pos. Lin.	0.0391	0.0327	0.0407	0.0376	0.0359
Vowel	Hamming	0.7987	0.5141	0.5238	0.4913	0.4762
	Linear	0.4784	0.4535	0.5390	0.4913	0.4654
	Exp.	0.4740	0.4562	0.5476	0.4848	0.4654
	Pos. Lin.	0.4740	0.4459	0.5390	0.4827	0.4719
Synthetic	Hamming	0.7867	0.2933	0.2000	0.2733	0.2100
	Linear	0.4100	0.3133	0.2033	0.2600	0.2000
	Exp.	0.4100	0.2833	0.2000	0.2500	0.2000
	Pos. Lin.	0.4100	0.3100	0.2033	0.2533	0.2000

criteria; and because Allwein et al. [6] proposed L_0, we included $A_{0.5}$; L_0 does not take into consideration the independence of classifiers, and so we included $L_{0.5}$; the OPC and APC codematrices are commonly used, and so we included them as well. Because EC has a long running time and is considered in many papers, we did not use it as a criterion in our simulation. These criteria appear in the order of OPC, APC, L_1(MR), $L_{0.5}$(MRC) and $A_{0.5}$(ARC) in the x axis. Among 10,000 random matrices, we selected codematrices maxmizing each criterion as optimal. In each panel, the different marker types of stem plot show different types of loss function. The circle represents the result from the Hamming loss, the square from the linear loss, the diamond from the exponential Loss, and the star from the positive linear loss. The result of the simulation shows that using a different codematrix does not affect the error rate significantly. Though the APC criterion yields a good result in the dermatology and vowel data sets, it does have a high error rate in the ecoli data. In synthetic data, the codematrices from the $L_1$and $A_{0.5}$ criteria produce low error rates. Especially, we noticed that the OPC result is not bad if Hamming loss is not used. This fact differs from the results of Allewin et al. [6] and Kuncheva [10], and might have resulted from the more considered selection of parameters in each classifier.

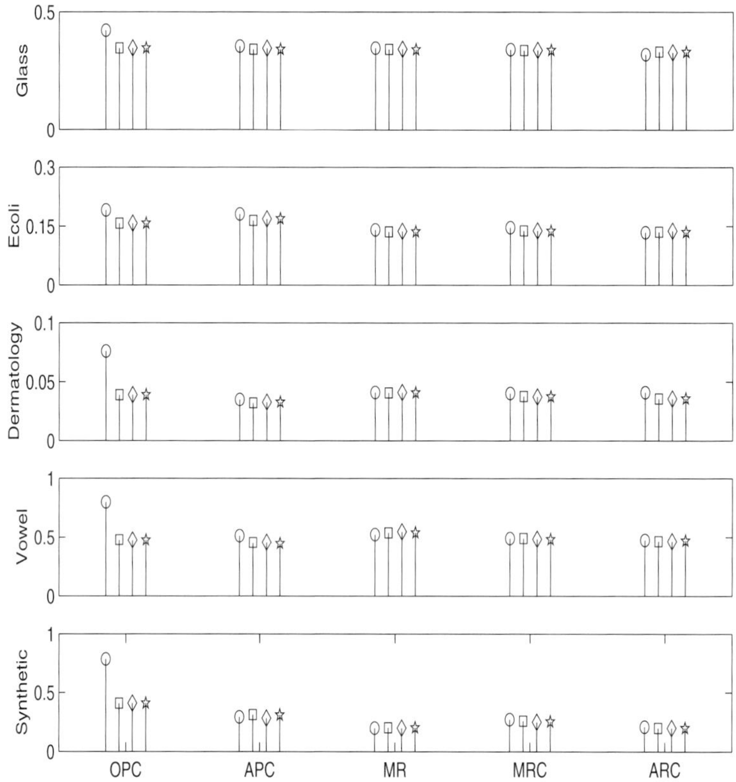

Fig. 2. Stem plot of experiment results with various codematrices and loss functions. Natural datasets from UCI and a synthetic dataset are used. For each dataset five codematrices and four loss functions are used. circle: Hamming loss, square: linear loss, diamond: exponential loss, star: pos. lin. loss.

In Fig 2, we can see that the OPC error rate using Hamming loss grows significantly larger. In the dermatology data set, the error rates using other loss functions averaged about 0.039, but using Hamming loss, the error rate was 0.076. But in the cases besides OPC, a different loss function did not produce a significantly different error rate. And the error rate in APC using linear loss was not as small as we expected. But we can see that in the ecoli-, dermatology- and vowel data sets, using linear loss, the error rate is relatively small.

Fig. 3 shows a stem plot of the error rate when we used error-rate-weighted loss(EL_{exp} and EL_{lin}) and L_{lin} and L_{exp} with APC. The above panel represents the results using L_{lin}, and the other panel shows the results from using L_{exp}. The x axis represents, in order, the glass-, ecoli-, dermatology-, vowel- and synthetic data sets. The diamond marker indicates the results from the error rate weighted loss and circle from L_{lin} or L_{exp}. In Fig. 3, we can see that there is no significant

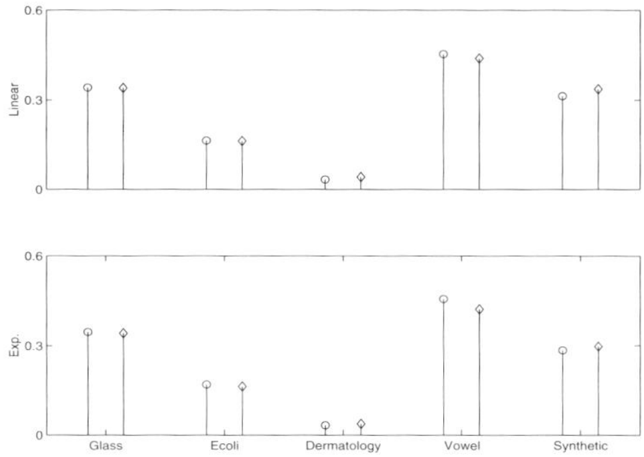

Fig. 3. Stem plot of error rate from EL_{exp}, EL_{lin}(circle), L_{lin} and L_{exp}(diamond) losses for 5 datasets

difference in error rate between L_{lin} and L_{exp}. Actually, error rate weighted exp. loss gives lower error rate for vowel dataset, higher error rate for synthetic dataset. The idea of boosting, which combines classifiers of higher weight with those of lower error rate, is not suitable for ECOC.

In summary, use of the APC codematrix is recommended, since the error rate in APC is relatively not so bad. Significantly, when we used APC, we could reduce the computation time by about 90%. Furthermore, with APC, the effort of designing a codematrix is unnecessary. The three loss functions L_{lin}, L_{plin}, and L_{exp}, do not significantly affect the error rate. As was in boosting, we used the error-weighted loss function to expect a good result, but found that the error rate was not significantly improved.

5 Conclusion

In this paper, we compared the codematrix design criteria and loss functions. From simulations, we showed that the loss functions, besides Hamming loss, are not significantly different. Thus, we recommend the use APC as a codematrix. APC saves computation time and, with it, design of a codematrix is unnecessary. And if the size of label class is given, APC is also fixed. So we believe that APC helps greatly to estimate parameters. Having studied the parameter selection that minimizes the error rates of each classifiers, in future work we will study the parameter selection that minimizes the error rates of the ensemble.

Acknowledgement

This work was supported by a Korea Research Foundation Grant funded by the Korean Government (MOEHRD, Basic Research Promotion Fund; KRF-2005-015-C00097).

References

1. Aha, D. W. and Bankert, R. L.: Cloud classification using error correcting output codes. Artificial Intelligence Applications : Natural Resources, Agriculture and Environmental Science, **11** (1997) 13-28
2. Berger, A.: Error-correcting output coding for text classification. In Proceedings of Int. Joint Conf. Artificial Intelligence, IJCAI'99 (1999) Stockholm, Sweden
3. Breiman, L.: Bagging predictors. Machine Learning, **24**(1997) 123-140
4. Dietterich, T. G. and Bakiri, G.: Error-correcting output codes : A general method for improving multi-class inductive learning programs. In Proceedings of the Ninth National Conference on Artificial Intelligence (AAAI-91), (1991) 572-577
5. Dietterich, T. G. and Bakiri, G.: Solving multi-class learning problems via error-correcting output codes. Journal of Artificial Intelligence Research, **2** (1995) 263 -286
6. Allwein, E. L., Schapire, R. E. and Singer, Y.: Reducing multi-class to binary : A unifying approach for margin classifiers. Machine learning research, **1** (2000) 113-141
7. Freund, Y. and Schapire, R. E.: A decision-theoretic generalization of on-line learning and application to boosting. Journal of computer and system science, bf55 (1997) 119 -139
8. Kittler, J., Ghaderi, R., Windeatt, T. and Matas, G.: Face verification using error correcting output codes. In Computer Vision and Pattern Recognition CVPR01, (2001) Hawaii, IEEE Press
9. Kuncheva, L. I. and Whitaker, C. J.: Measures of diversity in classifier ensembles. Mach. Learn. **51** (2003) 181-207
10. Kuncheva, L. I.: Using diversity measures for generating error -correcting output codes in classifier ensembles. Pattern Recognition Letters **26** (2005) 83 -90
11. Schapire, R. E.: Using output codes to boost multi-class learning problems. In 14th International Conf. on Machine Learning, (1997) 313-321, Morgan Kaufman
12. Vapnik, V.: Statistical Learning Theory. Springer, New York, 1998
13. Windeatt, T. and Ghaderi, R.: Coding and decoding strategies for multi-class learning problems. Information Fusion **4** (2003) 11-21

Pattern Recognition Using Evolutionary Classifier and Feature Selection

Mi Young Nam and Phill Kyu Rhee

Dept. of Computer Science & Engineering, Inha University
253, Yong-Hyun Dong, Incheon, South Korea
rera@im.inha.ac.kr, pkrhee@inha.ac.kr

Abstract. In this paper, we propose face feature selection and classifier selection method for face image group according illuminant. In knowledge based, we stored context and weight for feature points and selected classifier for context. This context is distinguished the face images having varying illumination. This context knowledge can be accumulated and used later. Therefore we designed the face recognition system by using evolution method and efficient face feature point selection. It can improve its performance incrementally using proposed algorithm. And we proposed efficient context modeling method by using SOM. For context awareness, we made artificial face images from FERET fa dataset and divided several group. Therefore we improved face recognition ratio using adaptable classifier, feature and weight for feature points.

1 Introduction

In this paper, we discuss about a context based classifier selection and feature points selection. The difficulties of object recognition vision are caused by the variations in internal or external illumination[1]. Thus, it can hardly be used for mobile applications due to uneven environments. We employs the concept of context-awareness and the genetic algorithm and determines (selects) a most effective structure of vision system for a given input data. In this paper, we will focus on object recognition system under varying illumination environment [2, 3, 4].

The knowledge of an individual context category and its associated chromosomes of effective classifiers are stored in the context knowledge base. Similar research can be found by [5]. We will deal with image objects the spatial boundaries of which can be well estimated in prior [6] without loss of generality. Recently, face recognition becomes a popular task in object recognition area [1, 2, 3, 4].

In classifier selection, the selection of a proper classifier system that is most likely to produce an accurate output for a given environment is attempted. In classifier selection approaches, input data are partitioned onto several regions, and the best classifier is assigned to each partitioned region [7].

In the section 2, we present context-aware evolutionary computation and the overview of the proposed object recognition scheme. In the section 3, we discuss about evolution system. Finally, we give the experimental results and the concluding remarks in the section 4 and 5, respectively.

L. Wang et al. (Eds.): FSKD 2006, LNAI 4223, pp. 393 – 399, 2006.

2 Data Context-Awareness and Context Knowledge Accumulation

2.1 An Efficient Context-Aware Definition

The set of context data is clustered into data context categories. Each data context category denotes the characteristics of context data that affect the performance of classifier system. Data context-awareness is carried out by modeling and identification of context data. The context data set should be modeled in association with context modeling. An input context data need be identified and used to select a most effective classifier system for an associated action data. We use the same input image as the context data as well as the action data here. In our system, we decide context awareness module by using SOM and FuzzyART [8]. Experimental result,

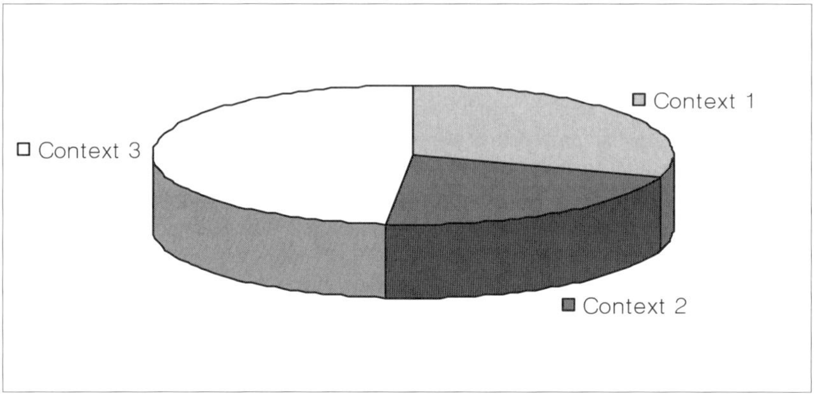

Fig. 1. Face image clustering by using SOM

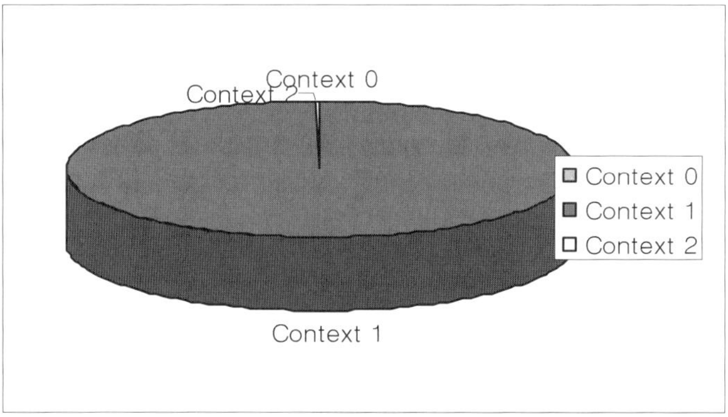

Fig. 2. Face images clustering by using FART

SOM is established for divide the face image's group. Fig 1 and Fig2 show clustering result by using SOM and FART. In the vertical scanning method, the input context data, i.e. 128 x 128 face image is reduced into 6 x 6 images, and the reduced image is scanned in the vertical direction first (from top to bottom) and from left to right.

3 Classifier Optimization for Face Recognition

3.1 Knowledge-Base Creation

Context knowledge describes a trigger of system action (combined classifier system output) in association with an identified context stored in the context knowledge base over a period of time [8]. Initially, the evolutionary weight learns and accumulates the knowledge of context-action configuration chromosome associations, and stores them in the knowledge base. The knowledge of context-action association denotes that of a most effective classifier system for an identified context. The AM configuration is encoded by the action configuration chromosome, and the fitness of the GA in the evolution system is decided.

Various approaches using GA for feature selection problem can be found in [9]. GA based optimization of classifier components can also be found in adaptive preprocessing based recognition [10], and neural network classifier systems [5]. In this paper, all possible classifier system combinations are encoded as artificial chromosomes. However, GA can hardly be used under dynamically changing environment alone since they usually consume too much time to evaluate the population of chromosomes in the genospace until finding an optimal solution. The knowledge of an individual context category and its associated chromosomes of effective classifier systems are stored in the context knowledge base. In addition, once the context knowledge is constructed, the system can react to changing environments at run-time.

Evolution face recognition consists of three stages: the preprocessing, feature representation, and class decision. In this paper, our major points are weight for

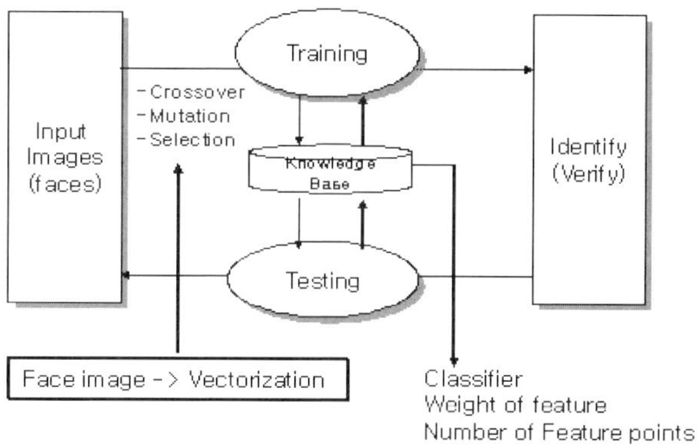

Fig. 3. The proposed evolution system : weight of feature points, number of feature and classifier selection by using evolution system

feature, number of feature selection, a classifier selection by using evolution. We adopt Gabor vectors with different weight values of individual fiducial points as the action primitives of feature representation. We made efficient chromosome and evolution parameters. The details of generation of Gabor vector will be discussed later. We use cosine distance for each classifier's identify. The architecture of face recognition using the evolutionary system is shown in Fig. 3.

In evolutionary based face recognition, the input images are used as the trained data as well as the context data. We assume that the training set of input face image is provided. In the testing step, and the data context module are identified by the context identifier module. In our system, we evolve weight of feature, feature points selection and classifier selection for contexts. It's different for context.

3.2 The Chromosome Encoding

The ECM is implemented by employing the genetic algorithm. The design of the chromosome and the fitness function of the GA are discussed in the followings. The GA explores the huge search space of action reconfiguration chromosomes to find an optimal classifier system structures for an identified data context. The optimality of the chromosome, i.e. fitness, is defined by classification accuracy and generalization capability. Fig.4 shows a possible encoding of the chromosome description.

CS_1	CS_2	CS_2	...	CS_n

Fig. 4. Chromosome description of the proposed scheme

CS_1, CS_2, . . ., and CS_n denote feature points selection and classifier selection, and they are represented by 32 bits.

As the GA searches the genospace, the GA makes its choices via genetic operators as a function of probability distribution driven by fitness function. The genetic operators used here are selection, crossover, and mutation.

The GA needs a salient fitness function to evaluate current population and chooses offspring for the next generation. Evolution or adaption will be guided by a fitness function defined in terms of the system accuracy and the class scattering criterion. The evolutionary module derives the classifier being balanced between successful recognition and generalization capability. The fitness function adopted here is defined as follows[6]:

$$\eta(V) = \lambda_1 \eta_s(V) + \lambda_2 \eta_g(V) \tag{1}$$

where $\eta_s(V)$ is the term for the system correctness, i.e., successful recognition rate and $\eta_g(V)$ is the term for class generalization. λ_1 and λ_2 are positive parameters that indicate the weight of each term, respectively.

In this paper, first, we decide weight of the feature point and we made the weight for context. The evolution method is used weight evolution and this weight is constructed classifier. Second we made multi classifier from number of feature. In evolution, we used chromosome following:

Table 1. Feature weight chromosome example

Feature 1	Feature 2	Feature 3	Feature 4	Feature 5
0.05	0.9	0.8	0.6	0.3

Table 2. Feature point selection

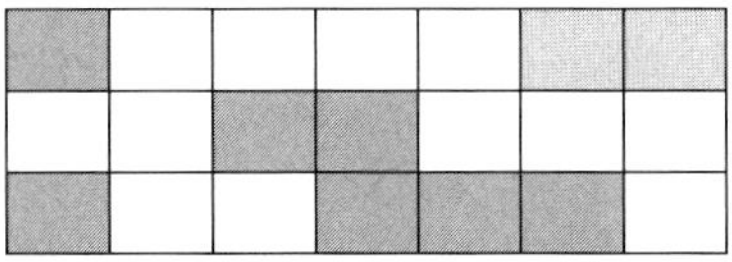

In this paper, we used classifier Gabor28, Gabor13 and weighted Gabor wavelet. Gabor28 classifier is used face feature point 28, and Gabor13 is used feature points, 13. And weight Gabor classifier is made proposed evolution method for decide weight of feature. The kernels of the Gabor wavelets show biological relevance to 2-D receptive field profiles of mammalian cortical cells [6]. The feature vector thus includes all the Gabor transform at the fiducial point $\vec{x}$, $V = (\overrightarrow{F(x_1)}\overrightarrow{F(x_2)}\cdots\overrightarrow{F(x_n)})$.

4 Experimental Results

In this paper, we used FERET dataset[11] for experiment. We used 2418 images from 1195 persons in FERET data set. The above data sets are merged for the context

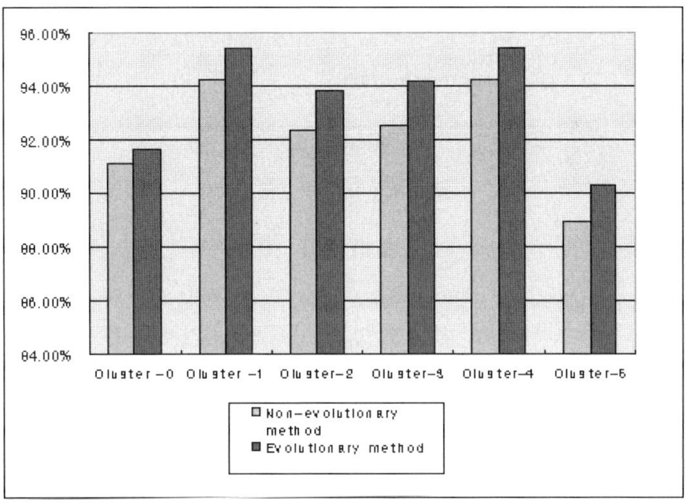

Fig. 4. The comparison of successful recognition rates between the non evolutionary method and weighted feature evolutionary methods

modeling. First, we clustered the images into context models by SOM. Second, we evolve classifier systems for individual context models. Evolution topic is feature points, weight for feature and classifier selection one of classifier combination. In this paper we divided three groups. Tables 1 shows the experimental results of six context models and weight for feature points selection.

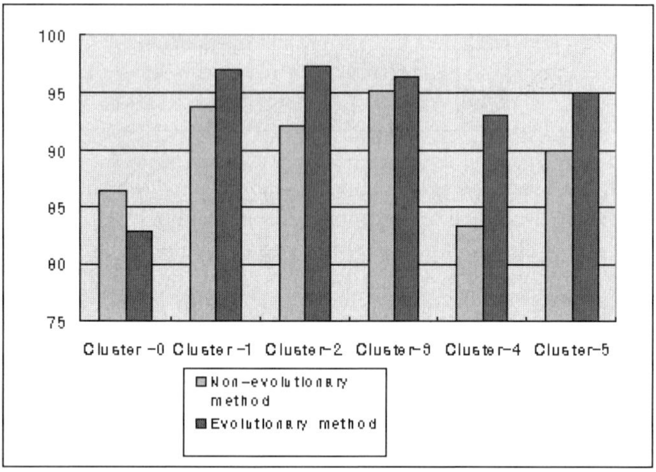

Fig. 5. The comparison of successful recognition rates between the non-evolutionary method and classifier selection and feature selection evolutionary method

Table 3. Performance comparison of the proposed system comparing with other approaches [12]

Algorithm/Method	FERET fafb
excalibur	0.794
mit_mar_95	0.834
mit_sep_96	0.948
umd_mar_97	0.962
usc_mar_97	0.95
Proposed method	0.9624

From tables, we can know that the classifier selection and feature selection is higher performance than weighted feature. The best recognition can be found in the proposed face recognition with the hybrid scanning when the twelve context model is adopted.

5 Conclusion

In this paper, we proposed method, the classifier and a feature selection using data context-awareness. In knowledge base, we stored chromosome for evolutionary result for each context. We store its experiences, face image illumination classification information, the best classifier information and feature point information. The proposed method tries to distinguish its input data context and evolves the classifier combination and feature selection. The main difference of the proposed classifier selection method from other methods is that it can select classifiers in accordance with the identified context. In experimental results, classifier selection and feature selection is efficient for face recognition.

References

1. M. Potzsch, N. Kruger, and C. Von der Malsburg, Improving Object recognition by Transforming Gabor Filter reponses, Network: Computation in Neural Systems, Vol.7, No.2 (1996) 341-347.
2. X. Wang and X. Tang, Random Sampling LDA for Face Recognition, *in Proceedings of CVPR*, Vol.2 (2004).
3. X. Wang and X. Tang, A Unified Framework for Subspace Face Recognition, *IEEE Trans. on PAMI,* Vol. 26, No. 9 (2004) 1222- 1228.
4. H. Kim, D. Kim, and S. Y. Bang, Face Recognition Using LDA Mixture Model, in *Proceedings of ICPR*, (2002) 486-489.
5. Ludmila I. Kuncheva, Switching Between Selection and Fusion in Combining Classifiers: An Experiment, IEEE Transactions on Systems, Man, and Cybernetics - part B: cybernetics, Vol.32, No.2 (2002) 146-156.
6. In Ja Jeon, Ki Sang Kwon, and Phill Kyu Rhee, Optimal Gabor Encoding Scheme for Face Recognition Using Genetic Algorithm, KES 2004, (2004) 227-236.
7. C. Liu and H. Wechsler, Evolutionary Pursuit and Its Application to Face recognition, IEEE Trans. on PAMI, Vol. 22, No. 6 (2000) 570-582.
8. M.Y. Nam and P.K. Rhee, An Efficient Face Recognition for Variant Illumination Condition, ISPACS2005, Vol.1 (2004) 111-115.
9. L. Kuncheva, K Andreeva, DREAM: a shell-like software system for medical data analysis and decision support, *Computer Methods and Programs in Biomedicine,* Vol.40, No.2 (1993) 73-81.
10. M.Y. Nam and P.K. Rhee, A Novel Image Preprocessing by Evolvable Neural Network, LNAI3214, Vol.3, (2004) 843-854.
11. P. Phillips, The FERET database and evaluation procedure for face recognition algorithms, Image and Vision Computing, Vol.16, No.5 (1999) 295-306.
12. http://www.nist.gov/

Robust Discriminant Analysis of Latent Semantic Feature for Text Categorization

Jiani Hu, Weihong Deng, and Jun Guo

Beijing University of Posts and Telecommunications, 100876, Beijing, China
{cughu, cvpr_dwh}@126.com, guojun@bupt.edu.cn

Abstract. This paper proposes a Discriminative Semantic Feature (DS F) method for vector space model based text categorization. The DSF method, which involves two stages, first reduces the dimension of the document vector space by Latent Semantic Indexing (LSI), and then applies a Robust linear Discriminant analysis Model (RDM), which improves the classical LDA by a energy-adaptive regularization criteria, to extract the discriminative semantic feature with enhanced discrimination power. As a result, DSF method can not only uncover latent semantic structure but also capture the discriminative feature. Comparative experiments on various state-of-art dimension reduction schemes such as our DSF, LSI, orthogonal centroid, two-stage LSI+LDA, LDA/QR and LDA/GSVD, are also performed. Experiments using the Reuters-21578 text collection show the proposed method performs better than other algorithms.

1 Introduction

Text categorization (TC) is a supervised learning task for assigning text documents to one or more pre-defined categories. It is used to find valuable information from a huge collection of text documents available in digital libraries, knowledge databases, the world wide web, and so on. The application area of interest in this paper is vector space based text categorization, in which documents are represented as column vectors in a *term-document* matrix [1]. Several characteristics have been observed in the vector space model based TC, such as very high dimensionality, data sparseness, high number of relevant features, and high level of redundancy [6]. Based on such properties, it is readily to reduce the dimensionality of the vector space for fast and robust categorization.

Latent Sematic Indexing (LSI) is a well-known dimension reduction technique for text data [2][7]. Since LSI is unsupervised, the LSI-based techniques do not preserve the cluster structure for discrimination in the low dimensional space. Thus, it is not able to enhance class separability. Linear discriminant analysis (LDA) is a supervised feature extraction method, which is capable of deriving compact and well-separated clusters [4]. It is commonly defined as an optimization problem involving covariance matrices that represent the scatter within and between clusters. However, the requirement that one of these matrices be nonsingular limits its application to the high dimensional text data. To deal with

L. Wang et al. (Eds.): FSKD 2006, LNAI 4223, pp. 400–409, 2006.

this problem, Torkkola [12] implemented LSI+LDA for document classification, which applies LSI to obtain a more compact representation so that the singularity of the scatter matrix is decreased. However, the overall performance of this two-stage approach is sensitive to the reduced dimensionality in the first stage.

In this paper, we propose a novel Discriminative Semantic Feature (DSF) method for vector space model based text categorization. The DSF method, which involves two stages, first reduces the dimension of the document vector space by LSI, and then applies a Robust linear Discriminant analysis Model (RDM) to extract the discriminative semantic feature with enhanced discrimination power. The RDM improves the classical LDA by a energy-adaptive regularization criteria in order to restrain the noise captured by LSI. Comparative experiments on various state-of-art dimension reduction schemes are also performed and the results show our method outperforms other algorithms generally.

Section 2 of this paper reviews classical LDA and introduces the proposed RDM. The two-stage DSF algorithm is presented in Section 3, where the detail steps of the TC system are also described. Experimental studies are presented in Section 4. Finally, conclusions are given in Section 5.

2 Robust Linear Discriminant Analysis Model

In this section, we first explain the reason classical LDA tends to overfit, and then introduce a new Robust linear Discriminant analysis Model (RDM) to improve the generalization ability of classical LDA.

2.1 Classical LDA Procedure and Its Limitation

Linear discriminant analysis is a widely used discriminant criterion, which defines a projection that makes the within-class scatter small and the between-class scatter large. For simplicity of discussion, we will assume the training data vectors $a_1, \ldots, a_n$ form columns of a matrix $A \in \mathbb{R}^{m \times n}$ and are grouped into k class as $A = [A_1, A_2, \ldots, A_k]$, where $A_i \in \mathbb{R}^{m \times n_i}$ and $\sum_{i=1}^{k} n_i = n$. Let N_i denote the set of column indices that belong to class i. The class centroid $c^{(i)}$ is computed by taking the average of columns in the class i, i.e., $c^{(i)} = \frac{1}{n_i} \sum_{j \in N_i} a_j$, and the global centroid c is defined as $c = \frac{1}{n} \sum_{j=1}^{n} a_j$. Then the within-class and between-class matrices are defined as

$$S_w = \frac{1}{n} \sum_{i=1}^{k} \sum_{j \in N_i} (a_j - c^{(i)})(a_j - c^{(i)})^T, \text{and} \tag{1}$$

$$S_b = \frac{1}{n} \sum_{i=1}^{k} n_i (c^{(i)} - c)(c^{(i)} - c)^T, \tag{2}$$

respectively [4]. LDA wants to find a linear transformation W, $W \in \mathbb{R}^{m \times l}$, with $l \ll m$ that makes the within-class scatter small and the between-class scatter large in the projected space. This simultaneous optimization can be approximated the criterion as follows:

$$J(W) = \arg\max_{W} \frac{|W^T S_b W|}{|W^T S_w W|}, \tag{3}$$

where W is the optimal discriminant projection. The W can be readily computed by calculating the eigenvector of the matrix $S_w^{-1} S_b$ [4]:

$$S_w^{-1} S_b W = W \Delta. \tag{4}$$

For the high dimensional problems such as text categorization, the inverse of S_w is especially problematic if for m-dimensional data vectors less than $m+1$ training vectors are available. Because S_w is a singular matrix if the sample size n is less than the dimension of the feature space m.

Another problem related to the S_w is its instability due to limited samples. This effect can be explicitly seen by writing the S_w^{-1} matrix in its spectral decomposition form, that is

$$S_w^{-1} = \sum_{i=1}^{m} \frac{\phi_i \phi_i^T}{\lambda_i}, \tag{5}$$

where λ_i is the ith eigenvalue of S_w^{-1} and ϕ_i is its corresponding eigenvector. It can be observed that the inverse of S_w is heavily weighted by the trivial eigenvalues and the directions associated with their respective eigenvectors. Hence, a poor or unreliable estimation of S_w^{-1} tends to exaggerate the importance of the low-variance information and consequently distorts the LDA algorithms [13]. In order to improve the ill-conditioned problem and enhance the generalization capability, a decent approximation of S_w^{-1} must be computed.

2.2 Robust Linear Discriminant Analysis Model

The Robust linear Discriminant analysis Model (RDM) improves the classical LDA by 1) decomposing the LDA procedure into a simultaneous diagonalization of both the within- and between- class scatter matrices [4], and 2) compute a decent approximation of S_w^{-1} by a energy-adaptive regularization criteria. Specifically, we first factorize the within-class scatter matrix, S_w, into following form using the principal components analysis (PCA):

$$S_w = \Phi \Lambda \Phi^T, \text{ with } \Lambda = diag\{\lambda_1, \lambda_2, \ldots \lambda_N\}, \tag{6}$$

where $\Phi \in \mathbb{R}^{N \times N}$ is an orthogonal eigenvector matrix and $\Lambda \in \mathbb{R}^{N \times N}$ is a diagonal eigenvalue matrix with diagonal elements in decreasing order $\lambda_1 \geq \lambda_2 \geq \ldots \geq \lambda_N$. To deal with the ill-condition problem of S_w, the RDM performs a form of regularization by adding a multiple of identity matrix to Λ, as $\hat{\Lambda} = \Lambda + \sigma I$, for some $\sigma > 0$. The determination of σ is not a trivial problem: When $\sigma \to \infty$, we will lose the information on S_w, while very small values of σ may not be sufficient effective. Cross-validation can be applied to estimating the optimal

σ as Regularized LDA (RLDA) does [14], but this solution is computationally expensive and the optimal value varies in different training data.

The RDM circumvents this problem by taking into account both the spectral energy and the magnitude requirement. The eigenvalue spectral of S_w provides a good criterion for estimating the optimal σ. The RDM determines the optimal σ^* as

$$\sigma^* = \lambda_m, \text{ with } J(m) = \min_m \left\{ m \middle| \frac{\sum_{i=1}^m \lambda_i}{\sum_{i=1}^N \lambda_i} \geq E(\theta) \right\} \tag{7}$$

where $J(m)$ determines the value of m according to the proportion of energy captured in the first m eigenvectors. In the paper, we empirically set the threshold $E(\theta) = 0.95$ for all the experiments. It is evident that with a settled $E(\theta)$, σ^* is self-adaptive to different training data according to the eigenvalue spectral of S_w.

The inverse of within-class scatter matrix can now be estimated as

$$\hat{S}_w^{-1} = \Phi\hat{\Lambda}^{-1}\Phi^T, \text{ with } \hat{\Lambda} = \Lambda + \sigma^* I. \tag{8}$$

With the decent estimate in (8), we can rework the LDA procedure in (4) as following:

$$(\Phi\hat{\Lambda}^{-1/2})^T S_b (\Phi\hat{\Lambda}^{-1/2})(\Phi\hat{\Lambda}^{-1/2})^{-1} W = (\Phi\hat{\Lambda}^{-1/2})^{-1} W \Delta. \tag{9}$$

Let $K_b = (\Phi\hat{\Lambda}^{-1/2})^T S_b (\Phi\hat{\Lambda}^{-1/2})$, $Y = (\Phi\hat{\Lambda}^{-1/2})^{-1} W$, then we can simplify the above equation to a eigenvalue problem as follows:

$$K_b Y = Y\Delta. \tag{10}$$

Finally, the projection matrix for RDM can be derived as:

$$W = \Phi\hat{\Lambda}^{-1/2} Y \tag{11}$$

3 Discriminant Semantic Feature Based Text Categorization System

This section presents a text categorization system which is based on the new DSF algorithm. Preprocessing and indexing are performed to transform a document into a document vector. Latent semantic analysis is used to uncover the latent semantic structure among index terms. The DSF algorithm is then applied to extract the low dimensional discriminative semantic feature. Finally, the K-Nearest Neighbor (K-NN) classifier is used for classification.

3.1 Preprocessing and Indexing

The preprocessing stage uses two techniques, *stopping* and *stemming*, to remove information that is supposed to be category neutral. Stopping is the removal of all words expected to be poor index terms (stopword). The application of stemming is based on the hypothesis that different inflected forms of a certain word do not carry category dependent information. In our system, the stemming procedure is performed by the Porter's suffix stripping algorithm [9].

After the preprocessing, the original documents are available as streams of terms containing only the information supposed to be useful. The well-known vector space model (VSM) [10] is used in our system. In VSM, the documents are represented as vectors where each component accounts for the term belonging to the dictionary. The indexing is typically based on a *term by document* matrix A, where each column j corresponds to a document in the text corpus and each row i corresponds to a term of the dictionary. The element A_{ij} is the component of document vector j related to term i and can be express as:

$$A_{ij} = tf(i,j) \cdot log(\frac{N}{N_i}), \tag{12}$$

where $tf(i,j)$ is the frequency of term i in document j, N is the total number of the documents, and N_i is the number of documents containing term i. The entity $log(\frac{N}{N_i})$ is called *inverse document frequency*, which gives more weight to the term when it is assumed to be more discriminative [11]. Moreover, to account for documents of different length, the length of each document vector (each column of A) is normalized to unit length.

3.2 Latent Semantic Analysis

As document vectors in the VSM reside in a space of very high dimensionality, it is necessary to find a low-dimensional representation for the robust machine learning and fast categorization. Latent semantic analysis, or LSI, seeks to represent the data in a lower dimensional space in the mean square error sense. It is fundamentally based on SVD (Singular Value Decomposition). Suppose the rank of A is r, LSI decomposes A as follows:

$$A = U\Sigma V^T, \tag{13}$$

where $\Sigma = diag(\sigma_1, \cdots, \sigma_r)$ and $\sigma_1 \geq \sigma_2 \cdots \geq \sigma_r$ are the singular values of A. $U = [\phi_1, \cdots, \phi_r]$ and ϕ_i is called the left singular vector. $V = [\upsilon_1, \cdots, \upsilon_r]$ and υ_i is called the right singular vector. It can be easily checked that the column vector of U are the eigenvectors of covariance matrix AA^T. Since the entries in the covariance matrix represent co-occurring terms in the documents, the eigenvectors with largest eigenvalues are directions related to dominant combinations of terms occurring in the corpus (i.e., "topic", "semantic concepts"). Following this property, for a document vector $a \in \mathbb{R}^{m\times 1}$, the low dimensional semantic feature vector, $\tilde{a}$, can be extracted as

$$\tilde{a} = U_l^T a, \tag{14}$$

where the column of $U_l \in \mathbb{R}^{m\times l}$ are the first l left singular vectors. The lower dimensional vector $\tilde{a}$ captures semantic features of the original document vector a.

3.3 Discriminative Semantic Feature Extraction and Classification

Given a training corpu with n documents corresponding to k classes, the TC system first performs preprocessing and indexing to produce the data vectors, and forms a training data matrix $A = [a_1 \dots a_n] \in \mathbb{R}^{m \times n}$, where m is the number of index terms.

The DSF algorithm is then developed to extract low dimensional features with enhanced discrimination power. The DSF algorithm first applies LSI for dimension reduction before proceeding with discriminant analysis. Unlike the previous "LSI+LDA" approach [12], DSF algorithm uses the proposed RDM, described in section 2.2, to perform discriminant analysis. The detail algorithm is as follows.

DSF algorithm
Input: Data matrix $A \in \mathbb{R}^{m \times n}$ with k class and an input vector $a \in \mathbb{R}^{m \times 1}$
Output: Optimal transformation matrix $G \in \mathbb{R}^{m \times (k-1)}$ and the $k-1$ dimensional discriminative semantic feature vector y of a

---

LSI stage:

1. Compute the SVD on A, as $A = U \Sigma V^T$
2. $\tilde{A} \leftarrow U_l^T A$, where $U_l \in \mathbb{R}^{m \times l}$ is defined in (14), $\tilde{A} \in \mathbb{R}^{l \times n}$, $l \leq rank(A)$.

LDA stage:

3. Construct the matrices $\tilde{S}_w$ and $\tilde{S}_b$ based on $\tilde{A}$.
4. Apply the RDM on $\tilde{S}_w$ and $\tilde{S}_b$, and compute W as (11)
5. $G \leftarrow U_l W$
6. $y \leftarrow G^T a$.

According to the DSF algorithm, the system projects the training data matrix into the discriminative semantic feature space as $F = G^T A$, where $F = [f_1 \dots f_n]$ and the ith column of F is the discriminative semantic feature vector of the training document a_i.

The classification stage is performed in the discriminative semantic feature space. When a novel document q is presented to the system, its discriminative semantic feature vector can be derived by $f_q = G^T q$. The class to which f_q belongs can be found by the K-NN algorithm [3] as follows:

1) From the similarity measure $sim(f_q, f_j)$ for $1 \leq j \leq n$, find the K nearest neighbors of f_q.
2) Among these K vectors, count the number belonging to each class.
3) Assign f_q to the class with the greatest count in the previous step.

The similarity measure used in our experiments to determine the closeness is the Euclidean distance, which is defined as $Ed(f_q, f_j) = (f_q - f_j)^T (f_q - f_j)$.

4 Experimental Results

4.1 Data Sets and Experimental Methodology

In order to assess the feasibility and performance of our new DSF method on text categorization task, experiments are carried out on two subsets of the *Reuters-21578* text categorization test collection distribution [8]. Specifically, they are the *re0* and *re1* text data sets [15]. The *re0* contains 1,504 documents corresponding to 13 classes, and the *re1* contains 1,657 documents of 25 classes. After the preprocessing and indexing procedures described in section 3.1, there are 2,886 and 3,758 index terms respectively.

We compare our DSF method with other five dimension reduction methods: LSI, two-stage LSI+LDA [12], Orthogonal Centroid (OC) [5], LDA/QR [14], LDA/GSVD [5], in term of the classification accuracy, which is defined as the number of correctly classified documents divided by the total number of documents. The K-NN algorithm (for K=1,5,10,30,50) based on the Euclidean distance is used as the classifier. In all the experiments, the classification accuracies are estimated by *5-fold cross-validation*. In 5-fold validation, we divide the data into 5 subsets of (approximately) equal size and each time leave out one subset for testing. The classification accuracy is the average of the 5 runs.

4.2 Results

Comparison of Classification Accuracies. In the following experiment, we evaluate our DSF method and compare it with other competing algorithms (LSI, OC, LDA/QR, LDA/GSVD) based on the classification accuracies, using the two data sets *re0* and *re1*. For LSI and DSF method, the results depend on the choice of reduced dimension l by LSI. 40% of the total semantic features derived from LSI are retained in this experiment. In other word, we set $l = 0.4n$ for LSI and DSF. The results on the two data sets are summarized in Table 1, which reports results with five different choices of K ($K = 1, 10, 15, 30, 50$) used in K-NN classifier. We also report the results on the original document vector space without dimension reduction, denoted as "FULL".

The main observations from these experiments are:

1. LDA-based dimension reduction algorithms, like LDA/QR, LDA/GSVD and our DSF algorithms do improve the performance of classification. In particular, the 10-NN classification accuracy of *re0* data set increases from 78.59% to 87.96% when reducing the original 2886 dimensional document vector into mere 12-dimension discriminative semantic features;
2. Supervised algorithms, like the the proposed DSF, OC, LDA/QR, and LDA/GSVD, have better performance than the one that does not use the label information, i.e. LSI;
3. The performance of the OC and LDA/QR is generally close to LDA/GSVD in all experiments. However, their computational costs are greatly cheaper than LDA/GSVD;

4. The proposed DSF performs generally better than all the other dimension reduction algorithms, which shows that the DSF algorithm can extract discriminative semantic features with low dimensionality ($k-1$) and enhanced discrimination power (higher accuracies).

Table 1. Comparative performance of the different dimension reduction algorithms on the *re0* and *re1* data sets. The mean and standard deviation of accuracies (%) from five runs are shown.

Dataset	KNN	FULL	LSI	OC	LDA/QR	LDA/GSVD	DSF
	1	77.06(2.85)	77.66(2.05)	83.98(2.14)	83.71(1.64)	**85.57**(1.95)	84.51(2.66)
	10	78.59(3.45)	79.05(3.26)	84.31(1.90)	84.57(1.88)	86.10(2.16)	**87.96**(2.49)
re0	15	78.79(2.73)	78.72(2.48)	84.37(1.76)	84.37(1.75)	86.30(2.57)	**87.50**(2.49)
	30	79.25(3.19)	79.98(3.45)	83.44(2.33)	83.64(2.18)	86.24(2.18)	**87.43**(1.95)
	50	79.52(2.60)	80.19(3.50)	82.71(2.91)	82.91(2.12)	85.17(2.02)	**86.50**(1.87)
	1	80.27(1.11)	81.35(1.24)	86.36(1.49)	85.70(2.03)	86.66(1.68)	**87.39**(2.17)
	10	83.22(0.95)	83.71(1.00)	86.54(2.45)	86.30(1.79)	86.36(1.54)	**87.93**(2.07)
re1	15	82.68(1.70)	82.86(1.10)	86.60(2.53)	86.18(2.29)	86.36(1.56)	**87.51**(1.70)
	30	81.53(1.93)	81.59(1.98)	85.16(2.31)	85.03(2.33)	86.42(1.69)	**87.63**(2.17)
	50	80.87(1.41)	80.57(2.11)	83.41(1.71)	83.10(1.90)	85.04(2.15)	**85.64**(2.04)

Effect of the Intermediate Dimension after LSI (l). The DSF method first introduces LSI to reduce the dimension of document vectors to l before the RDM is applied. In this experiment, we test the performance of our DSF method on different values of l, ranging from $0.02n$ to n, where n is the number of training documents. Specifically, 14 values of l is tested, namely $0.02n$, $0.04n$, $0.06n$, $0.08n$, $0.1n$, $0.2n$, ..., $0.9n$, n. For the LSI+LDA, the maximal value of l is $n-k$ rather than n, because in classical procedure of LDA the rank of S_w should be not large than $n-k$. Note that n is also the maximum dimension retained by LSI. $K = 10$ nearest neighbors algorithm is used for classification. The effectiveness of the DSF is shown in term of the comparative accuracies against the LSI and LSI+LDA with the same l. The comparative results are shown in Fig 1.

The main observations from these experiments are:

1. The performance of LSI is similar to the "FULL" (without dimension reduction) method. With the increasing of dimension l, its performance only varies slightly. Less than 10% of total semantic features are enough to win considerably high accuracy for LSI;
2. By performing discriminant analysis on the semantic features, LSI+LDA and DSF method both achieve much higher performance than LSI;
3. Within a appropriate range of $l \in [0.2n\ 0.4n]$, LSI+LDA performs slightly better than other LDA-based algorithms. However, the performance of LSI+

LDA deteriorates with further increase of l, especially when l gets close to $n-k$. This is because the $\tilde{S}_w$ becomes to be singular when l gets close to $n-k$, and the classical LDA is broken down, as described in section 2.1;

4. The performance of the proposed DSF method do not fluctuate very much with varying intermediate dimension l, which shows that the DSF method can get rid of the limitation of previous LSI+LDA method involving the selection of l. This result shows that the proposed RDM is effective to improve the generalization capability of the classical LDA. Furthermore, its classification accuracies remains higher than all the other algorithms when $l > 0.2n$.

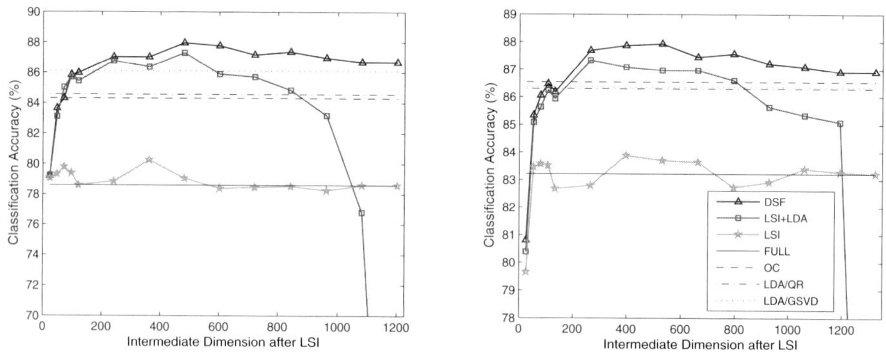

Fig. 1. Comparative performance of the proposed DSF method against other state-of-art methods under various LSI dimensions. The mean of accuracies (%) from five runs are shown. Left: results on *re0* data set, Right: results on *re1* data set.

5 Conclusions

This paper proposes a novel discriminative semantic feature extraction method and applies it successfully to the vector space model based text categorization. The discriminative semantic features, with low dimensionality and enhanced discrimination power, are derived from applying a robust linear discriminant model to the latent semantic features. The contribution of this paper lies in:

1. The RDM improves the classical LDA by 1) decomposing the LDA procedure into a simultaneous diagonalization of the within- and between- class scatter matrices, and 2) computing a decent approximation of S_w^{-1} by a new energy-adaptive regularization criteria;
2. The DSF method, whose performance does not fluctuate with varying intermediate dimension l, improves the limitation of previous LSI+LDA methods involving the selection of l;
3. Extensive evaluation on various state-of-art dimension reduction algorithms is performed for text categorization, including our new DSF, LSI, LSI+LDA, OC, LDA/QR, and LDA/GSVD. Experimental results show the DSF method performs better than other algorithms in general.

Acknowledgements

This research was sponsored by NSFC (National Natural Science Foundation of China) under Grant No.60475007 and the Foundation of China Education Ministry for Century Spanning Talent.

References

[1] Baeza-Yates, R., Ribeiro-Neto, B.: Modern Information Retrieval. Addison Wesley. (1999)
[2] Deerwester, S., Dumais, S.T., Furnas, G.W., and Landauer, T.K.: Indexing by latent semantic analysis. Journal of the American Society for Information Science. **41** (1990) 391–407
[3] Duda, R.O., Hart, P.E., and Stork, D.: Pattern Classification. Wiley. (2000)
[4] Fukunaga, K.: Introduction to Statistical Pattern Recognition, second ed. Academic Press. (1990)
[5] Howland, P. and Park, H.: Generalizing Discriminant Analysis Using the Generalized Singular Value Decomposition. IEEE Trans. Pattern Anal. Machine Intell. **26** (2004) 995–1006
[6] Joachims, T.: Learning to Classify Text Using Support Vector Machines. Kluwer. (2002)
[7] Landauer, T.K., Foltz, P.W., and Laham, D.: An introduction to latent semantic analysis. Discourse Processes. **25** (1998) 259–284
[8] Lewis, D.D.: Reuters-21578 text categorization test collection. http://www.daviddlewis.com/resources/testcollections/reuters21578/
[9] Porter, M.F.: An Algorithm for Suffix Stripping. Program. **14** (1980) 130–137
[10] Salton, G., Wong, A., Yang, C.S.: A vector space model for automatic indexing. Communications of the ACM. **18** (1975) 613–620
[11] Salton, G., and Buckley, C.: Term-Weighting Approaches in Automatic Text Retrieval. Information Processing and Management. **24** (1988) 513–523
[12] Torkkola, K.: Linear discriminant analysis in document classification. In IEEE International Conference on Data Mining (ICDM) Workshop on Text Mining. (2001)
[13] Thomaz, C.E., Gillies, D.F., and Feitosa, R.Q.: A New Covariance Estimate for Bayesian Classifier in Biometric Recognition. IEEE Trans. CSVT. **14** (2004) 214-223
[14] Ye, J., and Li, Q.: A Two-Stage Linear Discriminant Analysis via QR-Decomposition. IEEE Trans. Pattern Anal. Machine Intell. **27** (2005) 929–941
[15] Zhao, Y., and Karypis, G.: Empirical and Theoretical Comparisons of Selected Criterion Functions for Document Clustering. Machine Learning. **55** (2004) 311–331

Self-organizing Isometric Embedding Based on Statistical Criterions

Ruiguo Yu, Yuexian Hou, and Pilian He

School of Computer Science and Technology, Tianjin University, Weijin Road 92, 300072 Tianjin, China
{rgyu, yxhou, plhe}@tju.edu.cn

Abstract. Popular nonlinear dimensionality reduction algorithms, e.g., LLE, Isomap and SIE suffer a difficulty in common: neighborhood parameter has to be configured in advance to gain meaningful embedding results. Simulation shows that embedding often loses relevance under improper parameters configures. But current embedding residual criterions of neighborhood parameters selection are not independent to neighborhood parameters. Therefore it cannot work universally. To improve the availability of nonlinear dimensionality reduction algorithms in the field of self-adaptive machine learning, it is necessary to find some transcendent criterions to achieve unsupervised parameters selection. This paper begins with a discussion of optimal embedding principles and proposes a statistics based on spatial mutual information and normalized dependency index spectrum to determine reasonable parameters configuration. The simulation supports our proposal effectively.

1 Introduction

Automatic dimensionality reduction has been the main research issue of machine learning since its application in feature extraction, data visualization and data compression. Generally, automatic dimensionality reduction algorithm has the characters of biologic perception and thinking activity: Visual perception is considered a process of redundancy reduction, the outputs of visual neural cells should be independent of each other[1]. The expression mode of reduction is considered the principal qualification in the formation of thinking and meaning in mind[2][3]. The process that a new theory comes into being and purely understanding of wisdom in daily life and science activity is actually a quest for constringent expression and synopsis explanation of complicated relations based on statistical dependence of experiential data.

The existing linear dimensionality reduction algorithm such as PCA and CMDS[4] are easy to realize, but they can't reflect the nonlinear manifold structure[5]. Now, more attention is paid to nonlinear dimensionality reduction algorithm. Popular nonlinear dimensionality reduction algorithms such as LLE[6], Isomap[5][7] and SIE[9] suffer a difficulty in common: neighborhood parameter has to be configured in advance to gain meaningful embedding results, while embedding often loses relevance under improper parameters configures. But current embedding residual criterions of neighborhood parameters selection[14] are not independent to neighborhood

L. Wang et al. (Eds.): FSKD 2006, LNAI 4223, pp. 410 – 419, 2006.

parameters that cannot work universally. Simulation shows that current embedding residual criterions based neighborhood selection algorithm couldn't confirm the rational neighborhood parameters of data containing noise and couldn't get reasonable embedding results.

To improve the availability of nonlinear dimensionality reduction algorithms in the field of self-adaptive machine learning. This paper begins with a discussion of optimal embedding principles and proposes a statistics based on spatial mutual information and normalized dependency index spectrum to determine reasonable parameters configuration. The simulation supports our proposal effectively.

2 The Statistical Criterions of Optimal Embedding

We hope that the definitions of statistical criterions achieve quantitative evaluation results of embedding results of several different nonlinear reduction algorithms in different neighbors. To facilitate description, there are a number of agreed terms: X_D is the point sets in D dimensional space, the dimensionality reduction can be defined as the problem of seeking the mapping from X_D to a $d\,(d < D)$ dimensional embedding set X_d while maintaining the topological and geometric characteristics of X_D. X_D and X_d are known as the original data sets and embedding data sets respectively, the mapping f, which is from X_D to X_d, is called embedding mapping, and d is called embedding dimension. Correspondingly, given X_D and its embedding X_d, the problem that calculates the reverse embedding mapping f^{-1} from X_d to X_D is known as fitting restoration issue or compressed expression issue.

The statistical criterions of optimal embedding derived from three simple principles:

***a*. Simple principle:** From the perspective of reducing redundancy, optimal embedding should be the simple embedding.
***b*. Visual principle:** From the perspective of data visualization, optimal embedding should be more intuitive.
***c*. Accuracy principle:** Under reasonable precision, the original data sets X_D can be restored from the embedding data sets X_d.

Firstly to clarify the content of the above-mentioned three principles, then give the formal definition. There are intrinsic link and consistency between the three principles.

The purpose of dimensionality reduction is to gain the compressed form of the original data sets. Simple principle requires that the description complexity of X_d and $f^{-1}(\cdot)$ should be as far as possible close to the Kolmogorov complexity[9] of X_D. Here, the description complexity of data sets X_d and function $f^{-1}(\cdot)$ should be defined separately. The direct way is to regard X_d as a bit string sets, then define the description complexity of X_d as the minimum bits required to describe the bit string sets. But the above defined description complexity of the point sets often can not truly reflect the topological and geometric nature of X_d. Besides, dimensionality reduction algorithm is assumed to work in a potential continuous space. So, description complexity of point sets is agreed to be the number of bits required to describe the spatial distribution of point sets. Assuming that $f^{-1}(\cdot)$ is achieved by the given universal fitting model, then its description complexity can be agreed to be the minimum bits needed to describe the free parameters of the fitting model[1]. Above all, the simple principle can be further broken down into the simple point sets principle and the simple model principle.

Visual principle requires that the embedding results should be more intuitive. With the nature of rules, the various dimensionality of X_d can be abstractly interpreted as the hidden variables to determine X_D and can naturally become the basis for the formation of abstract theory. Visual principle and simple principle have obvious links: The spatial distribution of X_d corresponds to the description complexity of small point sets and simple border fitting conditions, and the latter is relevant to the described small function complexity.

Accuracy principle requires that the fitting of $f^{-1}(X_d)$ for X_D should meet reasonable accuracy. The assumption that the underlaying manifold of the original data set is no dramatic curvature fluctuations naturally corresponds to the requirements that the point-to-point distance in the embedding data sets is maintained in proportion to the original data sets. This will ensure that the dramatic changes in local curvature will not lead to excessive complexity and precision decline of fitting. For those nonlinear dimensionality reduction algorithms that maintain geodesic distance effectively, they can approximately confirm the reasonable smoothness of the fitting function. The regulation degree of domain X_d also has important implications for fitting precision. It is obvious that the more irregular X_d is, the more fitting bases is needed to achieve a precise fitting.

We use the spatial mutual information and normalized dependency index spectrum as the quantitative indicators of the three principles. The spatial mutual information describes the spatial distribution of point sets and can be used as the measure of description complexity and visibility of X_d. Furthermore, spatial mutual information and geodesic distance criterion can be the approximate measure of fitting precision. The normalized dependency index spectrum reflects the dependence among the freedom of the embedded sets, meanwhile, this dependence also reflects the redundancy of X_d's description complexity and the redundancy of $f^{-1}(\cdot)$'s description complexity. The latter is the redundancy of model's free parameters sets.

2.1 Spatial Mutual Information

Traditionally, the mutual information [10] of a time-series $X_t, t = 1,2,...$ can be defined as follows:

$$I(X_{t+1} \mid X_t) \equiv H(X_{t+1}) - H(X_{t+1} \mid X_t) = H(X_{t+1}) + H(X_t) - H(X_{t+1}, X_t)$$

[1] Obviously, there is an assumption that the fit model meets certain requirements of precision.

Where, $I(X_{t+1} | X_t)$ reflects the nonlinear dependency between adjacent sequence value of time-series [13]. The definition of mutual information can be naturally extended to spatial domain. Assuming that the points set X is in certain spatial domain, divide this domain equally into a number of hypercube and X_i is the number of points in i-cube. The regulations degree of spatial distribution of point sets X can be achieved by calculating spatial mutual information defined as follows:

$$\begin{aligned} I(X) &\equiv I(X_i | \{X_{N(i)}\}) \\ &\equiv H(X_i) - H(X_i | \{X_{N(i)}\}) = H(X_i) + H(\{X_{N(i)}\}) - H(X_i, \{X_{N(i)}\}) \end{aligned} \tag{1}$$

Here, i is the index of any hypercube in the domain. $N(i)$ is the neighborly hypercube of i-cube. For two-dimensional case, every non-border square is adjacent to eight squares. The mutual information $I(X)$ measured the nonlinear dependency between the number of points in one hypercube and the points number of its adjacent hypercube, that is the nonlinear dependency of point sets' spatial distribution.

2.2 Normalized Dependency Index

Simple principle sets the focus for statistical independence between dimensions of the embedded sets. We hope that the definition of a statistical criterion can effectively reflect the nonlinear dependency between various dimensions of X_d. The direct idea is to use the mutual information as the measure. But the mutual information of dimensions x_i and x_j depends not only on their dependency but also on the absolute information capacity of x_i and x_j. According to the definition of mutual information, if the absolute information capacity $H(x_i)$ and $H(x_j)$ is small, the absolute value of their mutual information cannot exceed $\max\{H(x_i), H(x_j)\}$. So, the mutual information cannot accurately reflect the dependency between various dimensions of X_d.

We rely on the definition of the normalized dependence index q to measure the dependence between systems. There is no dependency between two systems when q=0 where q lies in real interval [0,1]. When q=1, there is a determined function relationship between two systems. The information entropy has the following attributes: for independent system A and B, the entropy of their additive system $A+B$ reaches maximum which is equivalent to entropy of A plus entropy of B:

$$H(A+B) = H(A) + H(B) \tag{2}$$

On the other side, if there is determined dependence relation between system A and B, the entropy of system $A+B$ reaches minimum.

$$H(A+B) = \max\{H(A), H(B)\} \tag{3}$$

For those systems that are neither determined dependent nor complete independent, the entropy of their additive system lies between the above two extreme situations. Generally, $H(A+B)$ may be expressed as a function of $H(A)$ and $H(B)$.

$$H(A+B) = H(A) + H(B) + (1-q')H(A)H(B) \quad q' \geq 1 \tag{4}$$ [1]

Assuming that $H(A)$ and $H(B)$ is not zero:

$$q' = (H(A) + H(B) - H(A+B))/(H(A) \cdot H(B)) + 1 \tag{5}$$

From (2) and (3), q' lies in $[1, \min\{H(A), H(B)\}/(H(A) \cdot H(B)) + 1]$:

$$q'_{\max} \equiv \min\{H(A), H(B)\}/(H(A) \cdot H(B)) + 1 \tag{6}$$

Then we can define the normalized dependence index between A and B :

$$q_{AB} \equiv (q'-1)/(q'_{\max} - 1) \tag{7}$$

Actually, when $H(A)$ or $H(B)$ is 0, we let $q_{AB} \equiv 1$.

For point sets X_d with d dimensions, its normalized dependence index spectrum is the matrix below used to measure the dependency between various dimensions:

$$Q(X_d) \equiv \begin{bmatrix} q_{11} & q_{12} & \cdots & q_{1d} \\ q_{21} & q_{22} & \cdots & q_{2d} \\ \vdots & \vdots & \ddots & \vdots \\ q_{d1} & q_{d2} & \cdots & q_{dd} \end{bmatrix} \tag{8}$$

Where q_{ij} is the normalized dependence index between dimension i and j.

2.3 Statistical Criterion of Optimized Embedding

Define the statistical criterion c of optimized embedding to measure the embedding quality of X_d:

$$c(X_d) \equiv I(X_d) / sum_U(Q(X_d)) \tag{9}$$

Where, $I(X_d)$ is the spatial mutual information of X_d, $Q(X_d)$ is theindependency index spectrum of X_d, $sum_U(Q(X_d))$ is the summation of the first half diagonal elements of matrix $Q(X_d)$. High value of $c(X_d)$ corresponds to high spatial dependence and low dimension redundancy, that is a high quality embedding.

3 Statistical Criterions Based Self-organizing Isometric Embedding

Compared to the globally analytic algorithms, SIE has a speedup of $n/\log n$ [8]. So it is more applicable to the use of searching the optimized embedding in the great parameters range. The SIE* algorithm, which is based on c criterion is given below:

[1] The form of formula (4) is derived from the expression of Tsallis[11] entropy.

1. Give a sufficiently large neighbors parameters range.
2. For each parameter in the above range, implement operating 3 to 5.
3. Randomly select a few anchor point sets.
4. Embed the anchor point sets with current neighbors parameters using SIE [8].
5. Evaluate the embedded results based on all anchor point sets and choose the one with maximal $c(X_d)$.
6. Choose the one with the largest $c(X_d)$ as the optimized embedding.

The framework of SIE is directly applicable to the Isomap. For those nonlinear dimension reduction algorithms that maintain a topological relation such as Laplacian Eigenmap[12], we guess the best principles based on statistical criterion c can be also applied.

4 Simulation Results

From the definition of spatial mutual information and normalized dependency index spectrum, the statistical criterion naturally reflected the simple principle of the point sets of the embedding set. The following will show the reflection of other optimized embedding principles by the statistical criterion c. The test is based on clean and noisy Swiss roll data[5] (figure 1 and figure 2) data sets of 500 points.

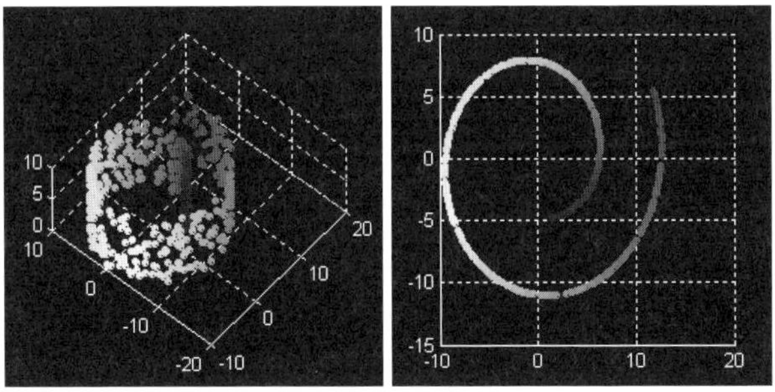

Fig. 1. Noise-free swiss roll data sets, left is the original data sets and right is projection of original data sets on X-Y plane

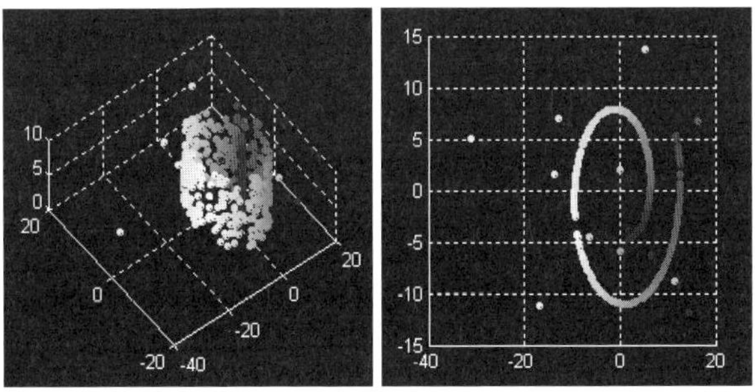

Fig. 2. Noise-polluted swiss roll data sets, left is the original data sets and right is projection of original data sets on X-Y plane

To evaluate the statistical criterion c's reflection of the visual principle, embed two data sets with different k values from 5 to 30 with the interval of 5 using Isomap and SIE algorithm. The calculation results of c are in table 1. Figure 3 illustrate the embedding results of noise-free data sets using Isomap. The visual regularity of the embedding sets is effectively reflected by the statistical criterion c. This result is also applicable to data sets containing noise and the SIE algorithm. More, statistical criterion c can also be applied for embedding results selection of different algorithm. Figure 4 shows the embedding results of SIE and Isomap when k=10,15 and 20. As showed in table 1 and figure 4, statistical criterion c effectively reflects the embedding quality. Because the embedding algorithm of Isomap is more sensitive to noise points, the global features of embedding manifold may distort[8] because of several noise points which leads to a worse embedding quality than SIE.

Table 1. The c value of embedding results of both noise-polluted and noise-free data sets using Isomap and SIE algorithm with various k values, I,S,C and N stand for Isomap, SIE, noise-free data sets and noise polluted data sets ,I-C represent the embedding of noise-free data sets using Isomap, So are I-N, S-C and S-N , k is the number of points in the neighborhoods

k	5	10	15	20	25	30
I-C	0.4808	3.1423	7.5668	0.5590	0.5501	1.1057
I-N	0.6882	0.2460	0.8751	0.8876	0.8108	0.8229
S-C	1.9434	2.9295	7.1690	1.1409	2.5376	1.4634
S-N	0.8835	3.6430	1.3001	1.9222	1.8199	2.0691

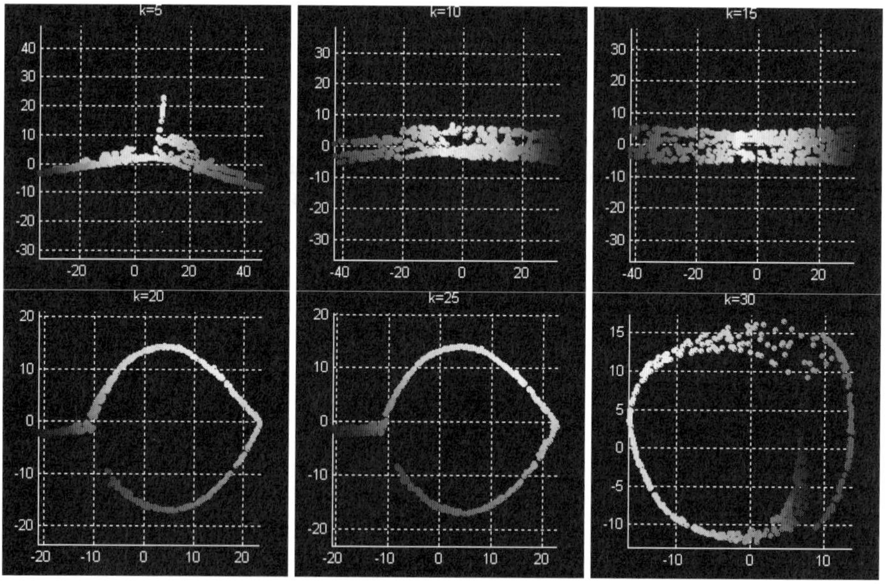

Fig. 3. Embedding results of noise-free data sets using Isomap algorithm with various k values(5-30)

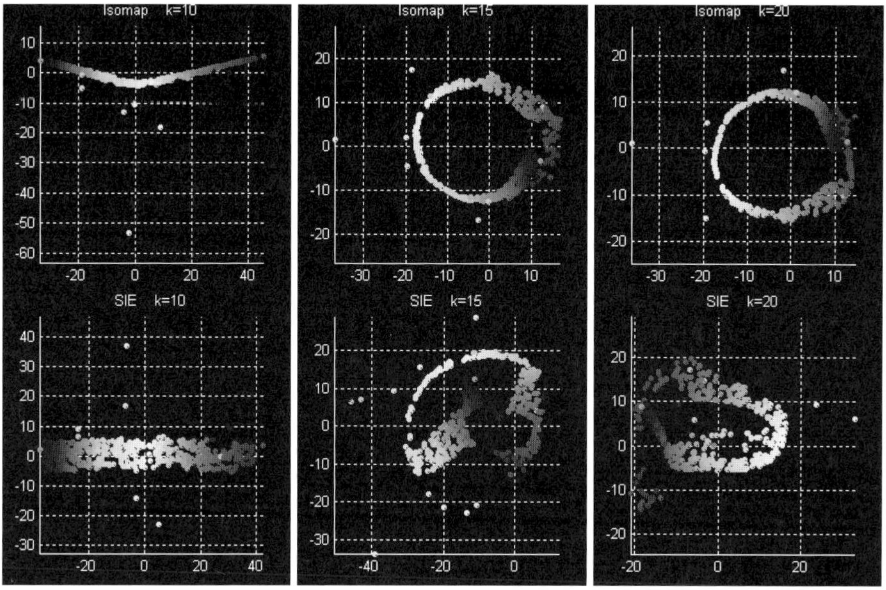

Fig. 4. Embedding results of noise-polluted data sets using Isomap and SIE algorithm with various k values(10-20)

For embedding algorithms maintaining geodesic distance, i.e., Isomap, a direct idea is to apply the maintenance extent of geodesic distance to evaluate the embedding quality. For example, define a normalized cost index[8]:

$$NS \equiv \sum_{1\le i<j\le n} |Dis_D(i,j) - Dis_d(i,j)| \Big/ \sum_{1\le i<j\le n} Dis_D(i,j)$$

Here, $Dis_D(i,j)$ is the geodesic distance in the original D dimensions data space of points i and j. And $Dis_d(i,j)$ is the Euclidean distance in the embedding d dimensions space. But the calculation of geodesic distance depends on the number of points in the neighborhoods. Using the value of NS for the selection of embedding sets with different k value is not applicable. Simulation shows that for noise-free data sets, there is a negative correlation between the optimized extent and the value of NS. This attribute of relevance is not applicable to datasets containing noise (figure 5). Actually, the algorithm in reference 14 is not applicable too.

Figure 6 shows the fitting and restoring precision of the original data sets using optimized embedding selected by the statistical criterion c. The fitting model is RBF network having 25 Gauss kernels. Embed the noise-polluted data sets with different k values from 7 to 17 with the interval of 2 using SIE algorithm and repeat the operation 10 times for each value. Select those with the highest c value and lowest NS value from the embedding results for different k and regard them as the optimized embedding for certain value of k. The optimized value of c and NS for different k is showed in the subfigure of figure 6 during which the red points represent the global optimized value. The blue points line and black points line in the upper subfigure of figure 6 represent the relative fitting error of embedding algorithms based on value c and value NS for different k value. The definition of relative fitting error is the sum of standard deviation of various dimensions in original data sets divided by error mean of fitting. There is a significant negative correlation between the c value of the embedding sets and relative fitting error. In the contrast, the value of NS cannot effectively reflect the trends of fit precision and model complexity of fitting.

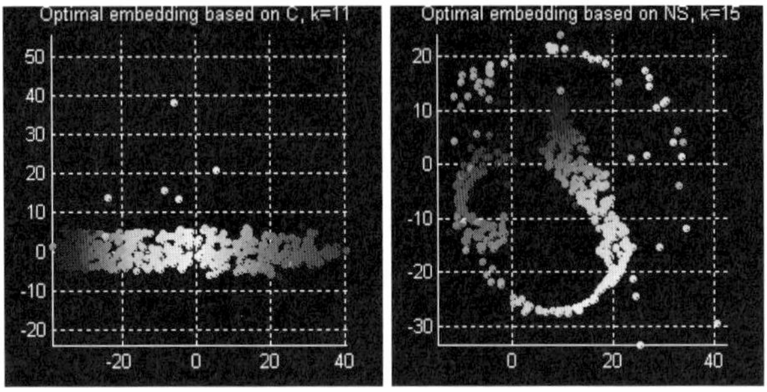

Fig. 5. Embedding results of noise-polluted data sets using SIE, The left represents the embedding with maximum c value(k=11), the right corresponds to the minimized normalization cost index(k=15)

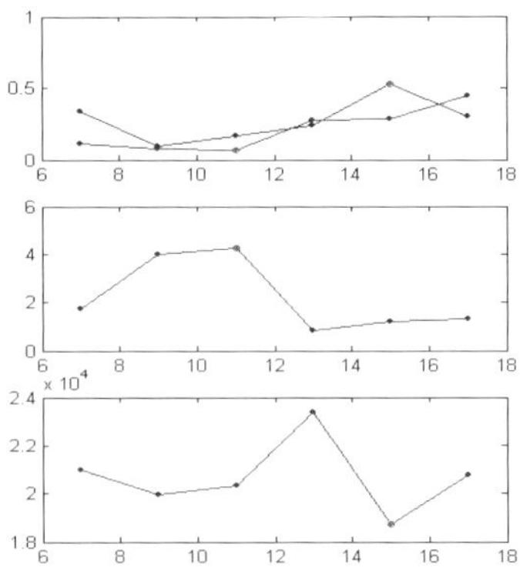

Fig. 6. The statistical value of embedding results of noise polluted data sets using SIE algorithm with various k

5 Conclusion

This paper proposed the evaluation principles of optimal embedding, defined the statistical criterion reflecting the principles and illustrated the effectiveness of statistical criterion with simulation results. The statistical criterion is applicable to the quantitative evaluation of embedding results of geodesic distance based SIE algorithm with different domain parameters. Simulations demonstrated the robustness and efficiency of the combination of SIE and the statistical criterion, i.e., SIE*. More work need to be down to test the effect of SIE* on a high capacity data sets, e.g., multimedia database, text base, biological DNA database, structure of web pages, etc.

References

1. H. Barlow, Unsupervised learning, Neural Computation, vol 1, pp. 295-311, 1989.
2. Gary Marcus, Programs of the Mind, Science 304: 1450-1451.
3. Eric Baum, What Is Thought?, MIT Press, Cambridge, MA, 2004.
4. K. V. Mardia, J. T. Kent, J. M. Bibby, Multivariate Analysis, Academic Press, London, 1979.
5. Joshua B. Tenenbaum et al. A Global Geometric Framework for Nonlinear Dimensionality Reduction. Science, 2000, 290: 2319-2323.
6. Sam T. Roweis et al. Nonlinear Dimensionality Reduction by Locally Linear Embedding. Science, 2000,290 : 2323-2326.
7. Vin De Silva and Joshua Tenenbaum, Global versus local methods in nonlinear dimensionality reduction, V. de Silva et al, NIPS'2002.
8. Hou Yuexian, et. al., Robust Nonlinear dimension reduction: a self-organizing approach, FSDK'05.
9. Ming Li and P. M. B. Vitanyi, An Introduction to Kolmogorov Complexity and Its Applications (Second Edition), Springer Verlag, New York, 1997.
10. R. Ash, Information Theory, John Wiley and Sons Press, Indianapolis, 1965.
11. C. Tsallis, Possible generalization of Boltzmann-Gibbs statistics. J. Stat. Phys., 1988, 52 (1/2): 479-487.
12. Mikhail Belkin and Partha Niyogi, Laplacian Eigenmaps and Spectral Techniques for Embedding and Clustering, NIPS'2001.
13. Hou Yuexian, He Pilian, Identification of Neural Network Predictor by Means of Prediction Complexity, Information and control, 2001, 30(1).
14. The ISOMAP algorithm and topological stability (by M. Balasubramanian and E.L. Schwartz, and response by J.B. Tenenbaum, V. de Silva, J.C. Langford). Science, 2002, vol.295, 7a.

Intra-pulse Modulation Recognition of Unknown Radar Emitter Signals Using Support Vector Clustering*

Gexiang Zhang, Haina Rong, and Weidong Jin

School of Electrical Engineering, Southwest Jiaotong University,
Chengdu 610031 Sichuan, China
gxzhang@ieee.org

Abstract. Unknown radar emitter signal (RES) recognition is an important issue in modern electronic warfare because the enemy's RESs are usually uncertain in the battlefield. Although unsupervised classifiers are used generally in many domains, few literatures deal with applications of unsupervised classifiers to RES recognition. In this paper, three unsupervised classifiers including competitive learning neural network (CLNN), self-organizing feature map neural network (SOMNN) and support vector clustering (SVC) are used to recognize unknown RESs. 135 RESs with 7 intra-pulse modulations are used to test the performances of the three classifiers. Experimental results show that SVC is only slightly superior to CLNN and is greatly inferior to SOMNN.

1 Introduction

With rapid development of radar technology, advanced radars become more and more popular in modern electronic warfare. Because of military purpose, the intra-pulse modulation types of enemy's radars are usually uncertain. Radar emitter signal (RES) recognition is the precondition and foundation of electronic jamming. The more detailed information of enemy's radars is obtained, the more definite aim the electronic jamming has and consequently the more effective the electronic jamming is [1-2]. To identify the enemy's unknown RESs, unsupervised classifiers, instead of supervised classifiers, should be used. However, comparing with supervised classifiers, there has been relatively little work done in pattern recognition and machine learning.

Unsupervised classifiers use unsupervised learning algorithm, in which unlabelled samples are classified into different classes only in terms of the number of classes. Because unsupervised classifiers need much smaller information about classes than supervised classifiers, they are more suitable for recognizing RESs in electronic intelligence system, electronic support measure system and radar warning receiver. At present, unsupervised classifiers include mainly competitive

* This work was supported by the National Natural Science Foundation of China (60572143), Science Research Foundation of SWJTU (2005A13) and National EW Lab Pre-research Foundation (NEWL51435QT220401).

L. Wang et al. (Eds.): FSKD 2006, LNAI 4223, pp. 420–429, 2006.

learning neural network (CLNN), self-organizing feature map neural network (SOMNN) and support vector clustering (SVC). Although CLNN, SOMNN and SVC are used generally in many domains [3-15], such as meteorological situation clustering [3], fault diagnosis [4], oil-well monitoring [6], market segmentation [7], nucleic acid clustering [8], location-allocation problems [9] and other applications [10-15], few literatures deal with applications of unsupervised classifiers to RES recognition.

This paper discusses the applications of unsupervised classifiers to intra-pulse modulation recognition of advanced RESs. First of all, the problem of RES recognition is described. Then, the design procedures of three unsupervised classifiers including CLNN, SOMNN and SVC are presented. Thirdly, CLNN, SOMNN and SVC are applied to recognize unknown RESs. According to experimental results, the performances of CLNN, SOMNN and SVC are compared. Finally, conclusions are drawn and future work is discussed.

2 Problem Description

In traditional recognition methods of RESs, one and more of 5 conventional parameters including carrier frequency, time of arrival, direction of arrival, pulse width and pulse amplitude are used to be inputs of classifiers. The methods are only suitable for conventional RESs of which the 5 parameters keep unchanging. In modern electronic warfare, plenty of advanced RESs are put into service, how to recognize them effectively is an emergent issue. In [13], intelligent recognition method (IRM) was presented to recognize advanced RESs. The structure of this method is shown in Fig.1.

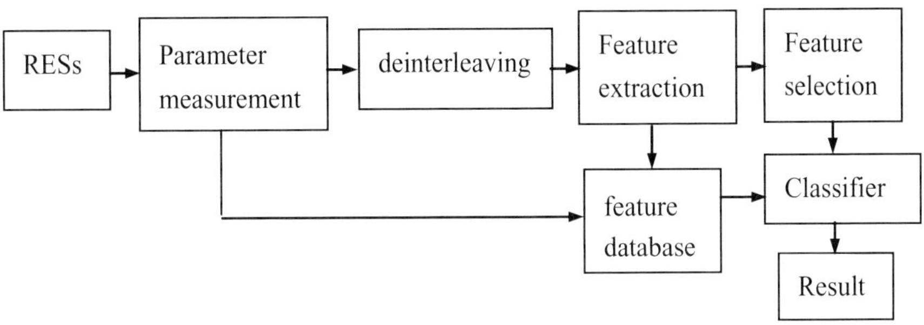

Fig. 1. Intelligent recognition method of RESs

In Fig.1, parameter measurement is used to obtain conventional 5 parameters. Deinterleaving is employed to transform the interleaved RES pulse train to several pulse trains, in which each pulse train only represents one RES. Feature extraction is used to extract valid features from advanced RESs. Because RESs have many changes and plenty of noise, the best feature that can identify all RESs cannot be found easily. For the difficult problem of RES recognition, multiple features need be extracted from RESs using multiple feature extraction methods. Feature selection is used to select the most discriminatory features from multiple extracted features so as to simplify the classifier structure and to decrease recognition error rate. Thus, feature database reserves conventional parameters and the extracted features. Some machine learning methods are used to design classifiers to fulfill automatic recognition of RESs.

The main difference between IRM and traditional methods is that IRM uses new features to construct parameter set for recognizing RESs. In [13, 16-17], supervised classifiers were discussed. In this paper, unsupervised classifiers are used to fulfill automatic recognition of RESs.

3 Unsupervised Classifiers

3.1 CLNN

Except for a kind of supervised neural networks, there is another kind of unsupervised neural networks. CLNN and SOMNN are two typical types of unsupervised neural networks. This subsection only discusses CLNN and the next subsection will deal with SOMNN.

The structure of CLNN is shown in Fig.2. CLNN is a two-layer neural network composed of input layer and output layer [3-6]. The number of neurons in the input layer depends on the dimensionality of input samples. The number of neurons in the output layer is the same as the number of classes. Full interconnection is adopted between the input layer and the output layer. The competitive transfer function is chosen as the transfer functions in the neurons of output layer. CLNN uses competitive learning algorithm, in which the competitive transfer function accepts a neuron input vector for a layer and returns neuron outputs of 0 for all neurons except for the winner, the neuron associated with the most positive element of neuron input I_i $(i = 1, 2, \cdots, n)$. The winner's output is 1. If all biases are 0, then the neuron whose weight vector is closest to the input vector has the least negative net input and, therefore, wins the competition to output O_j $(j = 1, 2, \cdots, m)$. The training process is accomplished by adjusting constantly the weights between the input layer and the output layer.

CLNN has some advantages of simple architecture and simple learning algorithm. However, there are two disadvantages in CLNN. One is that the number of classes need be set before training CLNN. In most cases, the number of classes is uncertain. The other is that the classification result is sensitive to the initial weight values. If the initial values of weights are chosen improperly, it is possible that one of the classes has no samples, or two of the classes are merged into one class, or one of the classes is classified into two classes. The shortcomings usually bring much effect on the classification results to CLNN.

3.2 SOMNN

SOMNN, developed by Kohonen in the early 1980s', is a kind of unsupervised learning neural network. SOMNN is also a two-layer neural network composed

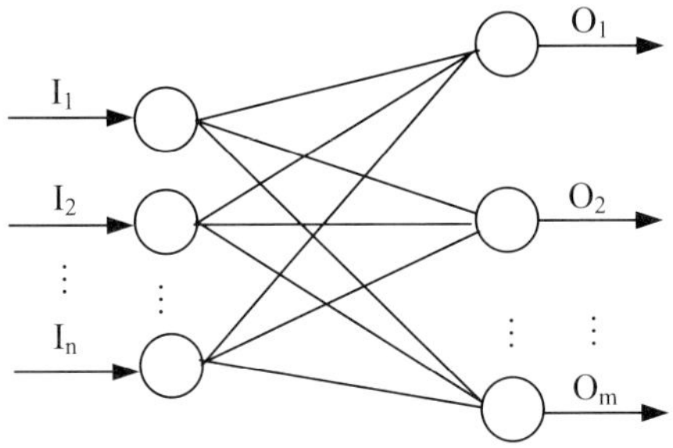

Fig. 2. The structure of CLNN

of input layer and output layer [7-10]. The number of neurons in the input layer depends on the dimensionality of input samples. The output layer is a competitive layer composed of m^2 neurons that form a two-dimensional plane array. Like CLNN, full interconnection is adopted between the input layer and the output layer. Different from CLNN, SOMNN uses side-suppressive connection among the neurons of the competitive layer. The structure of SOMNN is given in Fig.3. SOMNN can generate different responses to different inputs and have classification capability, which is similar to CLNN. Also, SOMNN can cluster the neurons with homologous function in the space. In the training process, SOMNN adjusts not only the weight of winner neuron, but the weights of all neurons in the neighboring domain of the winner neuron, which makes the neighboring neurons have homologous functions.

The main characteristics of SOMNN are as follows. The network weights are the memory of input samples. In the process of weight adjustment in the learning algorithm of SOMNN, not only the weights of the excited neuron, but also the weights of the neighboring neurons of the excited neuron, will be adjusted, which results in the abnormality of samples and large noise tolerance. The learning of CLNN makes the locations of the neighboring samples be very near in the output two-dimensional plane.

3.3 SVC

SVC is a non-parametric clustering algorithm based on the support vector approach [11-15]. In this algorithm, data points are mapped by means of a Gaussian kernel to a high dimensional feature space, where the minimal enclosing sphere is searched. This sphere, when mapped back to data space, can separate into several components, each enclosing a separate cluster of points. Based on a kernel method, support vector clustering is efficient algorithm because it avoids explicit calculations in the high-dimensional feature space [11-15]. Moreover, relying on the SVM quadratic optimization, SVC can obtain one global solution. The following description gives brief introduction of SVC [11-12].

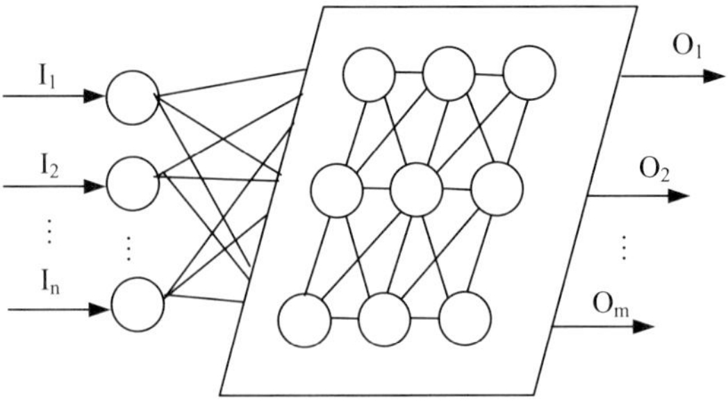

Fig. 3. The structure of SOMNN

Let $\{\mathbf{x}_i\} \subseteq \chi$ be a data set of N points, with $\chi \subseteq \mathbf{R}^d$, the data space. Using a nonlinear transformation Φ from χ to some high dimensional feature space. Looking for the smallest enclosing sphere of radius R is described as

$$\|\Phi(\mathbf{x}_j) - a\|^2 \leq R^2, \ \forall j \ . \tag{1}$$

Where $\|\cdot\|$ is the Euclidean norm and a is the center of the sphere. Soft constraints are incorporated by adding slack variables $\xi_j(\xi_j \geq 0)$. The following Lagrangian is introduced.

$$L = R^2 - \sum_j (R^2 + \xi_j - \|\Phi(\mathbf{x}_j) - a\|^2)\beta_j - \sum \beta_j \mu_j + C \sum \xi \ . \tag{2}$$

Where $\beta_j \geq 0$ and $\xi_j \geq 0$ are Lagrange multipliers. C is a constant and $C \sum \xi_j$ is a penalty term. According to KKT complimentary conditions, we can obtain the Wolfe dual form of the Lagrangian (2).

$$W = \sum_j \Phi(\mathbf{x}_j)^2 \beta_j - \sum_{ij} \beta_i \beta_j \Phi(\mathbf{x}_i) \Phi(\mathbf{x}_j) \ . \tag{3}$$

Subject to

$$0 \leq \beta_j \leq C, \ \sum_j \beta_j = 1, \ \ j = 1, 2, \cdots, N \ . \tag{4}$$

By introducing Mercer kernel $K(\mathbf{x}_i, \mathbf{x}_j)$, the Lagrangian W can be written as:

$$W = \sum_j K(\mathbf{x}_i, \mathbf{x}_j)\beta_j - \sum_{ij} \beta_i \beta_j K(\mathbf{x}_i, \mathbf{x}_j) \ . \tag{5}$$

The distance of the image of the point $\mathbf{x}$ in feature space from the center of the sphere is defined as

$$R^2(\mathbf{x}) = \|\Phi(\mathbf{x}) - a\|^2 \ . \tag{6}$$

According to the definition of kernel, the distance can be rewritten as

$$R^2(\mathbf{x}) = K(\mathbf{x}, \mathbf{x}) - 2 \sum_j \beta_j K(\mathbf{x}_j, \mathbf{x}) + \sum_{ij} \beta_i \beta_j K(\mathbf{x}_i, \mathbf{x}_j) \ . \tag{7}$$

Thus, the radius of the sphere R is $R(\mathbf{x}_i)$, where $\mathbf{x}_i$ is a support vector.

If $\beta_i = C$ in (4), the corresponding point $\mathbf{x}_i$ is mapped to the inside or to the surface of the feature space sphere and is called a bounded support vector (BSV). If $0 < \beta_i < C$, the image $\Phi(\mathbf{x}_i)$ of the corresponding point $\mathbf{x}_i$ lies on the surface of the feature space sphere and such a point is regarded as a support vector (SV). BSVs lie outside the boundaries, SVs lie on the cluster boundaries and all other points lie inside them [11, 12]. If $C \geq 1$, there are no BSVs. So C is usually set to 1.

4 Experiments

In our experiments, 135 RESs with different parameters are used to compare the performances of CLNN, SOMNN and SVC. Each of the 135 RESs has one of 7 intra-pulse modulations. The 7 modulations include CW, BPSK, MPSK, LFM, NLFM, FD and IPFE. Because of different parameters, CW has 15 different RESs and the rest 6 modulations have 20 RESs respectively. In the experiment, BPSK, MPSK and IPFE use Barker code and L-sequence pseudo-random code. Carrier frequencies of the 135 RESs vary from 400MHz to 750MHz. Pulse repetition frequency varies from 300 Hz to 1500 Hz. Pulse width varies from $0.25\mu s$ to $22.5\mu s$. Considering measure error, the signal parameters vary randomly in a certain range instead of fixed value.

In our prior work, 16 features including two resemblance coefficient features, information dimension, box dimension, correlation dimension, Lempel-Ziv complexity, approximate entropy, norm entropy and 8 wavelet packet decomposition features are extracted from the 135 RESs [1,2,18,19]. For every RES, 50 feature samples are extracted in each signal-to-noise rate (SNR) point of 5dB, 10dB, 15dB and 20dB. Thus, when SNR varies from 5 dB to 20 dB, every RES has 200 feature samples. CW has 3000 feature samples and other 6 modulations have 4000 feature samples respectively. The total feature samples of 135 RESs are 27000. These samples are classified equally into two groups: training group and testing group.

A supplementary explanation for our experiments must be given. RES recognition is usually used for military purpose. There is no benchmark dataset for testing the validity of the presented method. Also, there is no experimental result to be benchmark of the method.

Feature selection algorithm described in [18,19] is used to select the most discriminatory features from the 16 RES features. The samples in training group are applied to make the feature selection experiment. The experimental result is that the feature set composed of two feature including one of resemblance coefficient features and one of wavelet packet decomposition features has the most discriminatory and the lowest computational complexity. Thus, the selected features constitute input vector of unsupervised classifiers. The feature distribution graph is shown in Fig.4.

CLNN, SOMNN and SVC are used to recognize the 7 intra-pulse modulations of RESs, respectively. The three unsupervised classifiers are evaluated by using computing time and error rate. The structure of CLNN is shown in Fig.2.

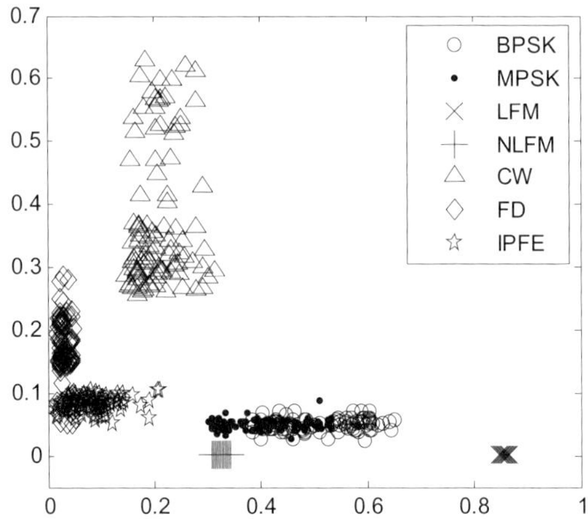

Fig. 4. Feature distribution graph of RESs

The number of neurons in the input layer and in the output layer is 2 and 7, respectively. The Kohonen learning rate, conscience learning rate, performance goal and the maximal training generation are set to 0.01, 0.001, 0 and 10000, respectively. The average time and average error rate of 20 tests are given in Table 1. In SOMNN, the performance goal and the maximal training generation are the same as those in CLNN. The ordering phase learning rate, ordering phase steps, tuning phase learning rate and tuning phase neighborhood distance are set to 0.9, 1000, 0.02 and 1, respectively. *hextop* is chosen as the topology function. Because the size of the output layer dimension has a directly effect on

Table 1. Comparisons of SVC, SOMNN and CLNN (%)

Methods	SVC	SOMNN	CLNN
BPSK	33.33	40.83	19.72
MPSK	33.33	64.72	94.79
LFM	0.00	0.00	0.00
NLFM	33.33	0.00	24.65
CW	0.00	0.28	34.52
FD	50.00	12.78	53.82
IPFE	50.00	5.28	100.00
Average error rate	42.86	17.70	46.79
Computing time (s)	42391.59	3090.19	3655.00

recognition result, different cases are tested and experimental results are shown in Table 2, in which the values are statistical data of 20 experiments and the error rate is computed from 7 modulations. The best results in Table 2 are chosen to compare with CLNN and SVC.

Table 2. Experimental results of SOMNN in different sizes of the output layer dimension

Diffrent sizes ([row, column])	Average error rate (%)	Computing time (s)
[2, 4]	28.02	3091.36
[3, 3]	17.70	3090.19
[3, 4]	34.29	3112.92
[4, 4]	31.59	3206.39
[4, 5]	40.04	3237.36
[5, 6]	45.91	3242.69
[5, 8]	51.75	3689.53
[6, 8]	57.54	3855.83
[7, 8]	63.13	4049.14
[8, 8]	71.19	4383.75
[9, 10]	75.67	5302.81
[14, 16]	82.42	13860.19
[20, 20]	87.30	35673.34

Table 3. Experimental results of SVC in different q and C

C	q	Average error rate (%)	Computing time (s)
1	50	85.71	16649.95
1	100	71.43	26923.75
1	200	42.86	42391.59
1	300	42.86	43040.08
1	400	43.14	44167.70
1	500	43.14	44167.70
0.8	500	45.71	44335.53
0.5	500	43.36	44011.14
1	1000	47.57	47705.16

When SVC is used to recognize the 7 modulations, we choose Gaussian kernel function [11]

$$K(\mathbf{x}_i, \mathbf{x}_j) = e^{-q|\mathbf{x}_i - \mathbf{x}_j|^2} \quad . \tag{8}$$

Where $\mathbf{x}_i$ and $\mathbf{x}_j$ are the ith and the jth feature vectors respectively. In the tests, the soft margin constant C is set to 1 and the scale parameter of the Gaussian kernel q varies from 1 to 1000. Some of experimental results are given in Table 3. The best results in the experiment are chosen to compare with CLNN and SOMNN.

As can be seen from Table 1, Table 2 and Table 3, some conclusions can be drawn. The size of the output layer dimension in SOMNN has much effect on recognition error rates. In this experiment, in general, the small size of the output layer dimension in SOMNN has low error rate. In SVC, the setting of parameter q is very important to the performance of classifier. While the changing of parameter C is little effect on experimental results. Experimental results also show that SVC is only slightly superior to CLNN and is greatly inferior to SOMNN, in terms of error rates and computing time.

5 Conclusions

In RES recognition, there are often some unknown radars of the enemies. This requires electronic reconnaissance systems, including electronic intelligence system, electronic support measure system and radar warning receiver, have the capability of identifying unknown RESs. By introducing unsupervised classifiers into RES recognition, this paper discusses the issue of unknown RES recognition. After CLNN, SOMNN and SVC are introduced briefly, extensive experiments are conducted to analyze the performances of the three methods comparatively. Experimental results show that SOMNN is a better unsupervised classifier for recognizing unknown RESs than CLNN and SVC. Though, how to choose the proper parameters in SOMNN and SVC is still an ongoing issue. Especially, the computational complexity of SVC is still very large. We have to do further research to quicken the algorithm and decrease the error rates of SVC. Also, the validity of the unsupervised method is proven further by using factual RESs in modern electronic warfare. These issues are our further work.

References

1. Zhang, G.X., Rong, H.N., Jin, W.D., Hu, L.Z.: Radar Emitter Signal Recognition Based on Resemblance Coefficient Features. In: Tsumoto,S., et al. (eds.): Rough Sets and Current Trends in Computing. Lecture Notes in Artificial Intelligence, Vol. 3066. Springer-Verlag, Berlin Heidelberg New York (2004) 665-670
2. Zhang, G.X., Hu, L.Z., Jin, W.D.: Intra-pulse Feature Analysis of Radar Emitter Signals. Journal of Infrared and Millimeter Waves, **23** (2004) 477-480
3. Turias, I.J., Gonzalez, F.J., Martin, M.L., Galindo, P.L.: A Competitive Neural Network Approach for Meteorological Situation Clustering. Atmospheric Environment, **40** (2006) 532-541
4. Khanmohammadi, S., Hassanzaseh, I., Zarei Poor H.R.: Fault Diagnosis Competitive Neural Network with Prioritized Modification Rule of Connection Weights. Artificial Intelligence in Engineering, **14** (2000) 127-132
5. Meyer-Baese, A., Pilyugin, S.S., Chen, Y.: Global Exponential Stability of Competitive Neural Networks with Different Time Scales. IEEE Transactions on Neural Networks, **14** (2003) 716-719
6. Simoes, M.G., Furukawa, C.M., Mafra, A.T., Adamowski, J.C.: A Novel Competitive Learning Neural Network Based Acoustic Transmission System for Oil-Well Monitoring. IEEE Transactions on Industry Applications, **36** (2000) 484-491
7. Kuo, R.J., An, Y.L., Wang, H.S., Chung, W.J.: Integration of Self-Organizing Feature Maps Neural Network and Genetic K-Means Algorithm for Market Segmentation. Expert Systems with Applications, **30** (2006) 313-324
8. Beckers, M.L.M., Melssen, W.J., Buydens, L.M.C.: A Self-Organizing Feature Map for Clustering Nucleic Acids Application to a Data Matrix Containing A-DNA and B-DNA Dinucleotides. Computer Chemistry, **21** (1997) 377-390

9. Hsieh, K.H., Tien, F.C.: Self-Organizing Feature Map for Solving Location-Allocation Problems with Rectilinear Distances. Computers and Operation Research, **31** (2004) 1017-1031
10. Sangole, A., Knopf, G.K.: Visualization of Randomly Ordered Numeric Data Sets Using Spherical Self-Organizing Feature Maps. Computers and Graphics, **27** (2003) 963-976
11. Ben-Hur, A., Horn, D., Siegelmann, H.T., Vapnik, V.: Support Vector Clustering. Journal of Machine Learning Research, **2** (2001) 125-137
12. Horn, D.: Clustering via Hilbert space. Physica A, **302** (2001) 70-79
13. Zhang, G.X.: Intra-pulse Modulation Recognition of Advanced Radar Emitter Signals Using Intelligent Recognition Method. In: Wang, G., et al. (eds.): Rough Set and Knowledge Technology. Lecture Notes in Artificial Intelligence, Vol. 4062. Springer-Verlag, Berlin Heidelberg New York (2006) 707-712
14. Lee, J., Lee, D.: An Improved Cluster Labeling Method for Support Vector Clustering. IEEE Transactions on Pattern Analysis and Machine Intelligence, **27** (2005) 461-464
15. Saketha Nath, J., Shevade, S.K.: An Efficient Clustering Scheme Using Support Vector Methods. Pattern Recognition, **39** (2006) 1473-1480
16. Zhang, G.X., Cao, Z.X., Gu, Y.J.: A Hybrid Classifier Based on Rough Set Theory and Support Vector Machines. In: Wang, L.P., Jin, Y.C. (eds.): Fuzzy Systems and Knowledge Discovery. Lecture Notes in Artificial Intelligence, Vol. 3613. Springer-Verlag, Berlin Heidelberg New York (2005) 1287-1296
17. Zhang, G.X.: Support Vector Machines with Huffman Tree Architecture for Multi-Class Classification. In: Lazo, M., Sanfeliu, A. (eds.): Progress in Pattern Recognition, Image Analysis and Applications. Lecture Notes in Computer Science, Vol. 3773. Springer-Verlag, Berlin Heidelberg New York (2005) 24-33
18. Zhang, G.X., Hu, L.Z., Jin, W.D.: Discretization of Continuous Attributes in Rough Set Theory and Its Application. In: Zhang, J., He, J.H., Fu, Y.X. (eds.): Computational and Information Science. Lecture Notes in Computer Science, Vol. 3314. Springer-Verlag, Berlin Heidelberg New York (2004) 1020-1026
19. Zhang, G.X., Hu, L.Z., Jin, W.D.: Radar Emitter Signal Recognition Based on Feature Selection Algorithm. In: Webb, G.I., Yu, X.H. (eds.): Advances in Artificial Intelligence. Lecture Notes in Artificial Intelligence, Vol. 3339. Springer-Verlag, Berlin Heidelberg New York (2004) 1108-1114

Difference Similitude Method in Knowledge Reduction

Ming Wu, Delin Xia, and Puliu Yan

School of Electronic Information, Wuhan University, P.R. China 430079
ming-wu@vip.sina.com

Abstract. An intergraded reduction method, which includes attributes reduction and rules induction, is proposed in this context. Firstly, U/C is calculated for reducing the complexity of the reduction. Then, difference and similitude sets, of the reduced information system, are calculated. The last, the attributes are selected according to their abilities for giving high accurate rules. The time complexity of the reduction, including attributes reduction and rules induction, is $O(|C|^2|U/C|^2)$.

1 Introduction

Because the real-world databases often include superfluous attributes or values, knowledge reduction is necessary before data mining. Usually, the reduction process is divided into two separate parts, *attributes reduction* and *values reduction*. After the reduction, an induction algorithm of rules will be employed to construct the simplest rules. In this paper, we will present a reduction method which not only reduces the conditional attributes and but also induces the decision rules.

Knowledge reduction includes attribute reduction and value reduction[1-3]. *AQ11*[4], *ID3*[5], *C4.5*[6] and *rough sets*[7] are the representative algorithms in value reduction and rules induction. There are three broad classes' reduction methods in the frame of rough sets: 1) based on *positive region*[8], 2) based on *information entropy*[9], 3) based on *discernibility matrix*[10]. If there are superfluous objects in decision table, we need not compare all the pairs of the objects. Thus, we reduce the information system by calculating *U/C* first. The superfluous objects are rejected from information system, and the time complexity is cut down to $O(|C|^2|U/C|^2)$.

2 Calculation of *U/C*

Let $IS=<U,C,D,V,f>$ be a given Information System. Herein, U is a non-empty finite set of objects (or instances) called universe; C is a non-empty finite set of conditional attributes; D is a non-empty finite set of decisional attributes; V is a non-empty finite set of all the conditional attribute values; f is an information function which identify the attribute values of each object.

Suppose A is a subset of C and $U/A=\{X_1,X_2,\cdots X_{|X|}\}$has been calculated. It is easy to obtain equation (1) and U/C can be calculated by employing it.

L. Wang et al. (Eds.): FSKD 2006, LNAI 4223, pp. 430–433, 2006.

$$U/\{A \cup a' \notin A\} = X_1/\{a'\} \cup X_2/\{a'\} \cdots \cup X_{|X|}/\{a'\} \\ = \bigcup_i X_i/\{a'\} \quad (1)$$

3 Difference Similitude Based Reduction

Let ***IS*=<*U*,*C*,*D*,*V*,*f*>** be a given Information System. Suppose there are $\boldsymbol{n}$ objects in ***IS***, and they are categorized into $\boldsymbol{m}$ categories. Let us denotes the ***i***th object as $\boldsymbol{x_i}$, the union of objects in the ***m***th category as $\boldsymbol{U_m}$, the value of attribute $\boldsymbol{a}$ of $\boldsymbol{x_i}$ as $\boldsymbol{f(a,x_i)}$.

Core: The conditional attributes, which can not be rejected from ***IS***, are called core attributes. The union of all the core attributes is denoted as $\boldsymbol{C_{core}}$.

Difference Sets. The difference set of $\boldsymbol{x_i}$, denoted as $\boldsymbol{DS(x_i)}$, is defined as:

$$DS(x_i) = \left\{ ds \,\middle|\, ds = \bigcup \left\{ a \,\middle|\, \begin{matrix} f(d,x_i) \neq f(d,x_j) \wedge d \in D \\ f(a,x_i) \neq f(a,x_j) \wedge a \in C \end{matrix} \right\}, j = 1,2,\cdots n \right\} \quad (2)$$

Similitude Sets. The similitude set of $\boldsymbol{x_i}$, denoted as $\boldsymbol{SS(x_i)}$, is defined as:

$$SS(x_i) = \left\{ ss \,\middle|\, ss = \bigcup \left\{ a \,\middle|\, \begin{matrix} f(d,x_i) = f(d,x_j) \wedge d \in D \\ f(a,x_i) = f(a,x_j) \wedge a \in C \end{matrix} \right\}, j = 1,2,\cdots n \right\} \quad (3)$$

In general, the original decision rule could be written in the following format:

$$if\ f(C_i^r,x)=f(C_i^r,x_i)\ then\ f(d,x)=f(d,x_i) \quad (4)$$

Suppose $\boldsymbol{A}$ is the set of the selected attributes, define:

$$\boldsymbol{frq(A,a,i)=frq_1(a,i)+frq_2(A,a,i)} \quad (5)$$

Where: $frq_1(a,i) = \left|\{ds \,|\, ds \in DS(x_i) \wedge a \in ds\}\right|$

$frq_2(A,a,i) = \left|\{ss \,|\, ss \in SS(x_i) \wedge ss \supseteq (A \cup \{a\})\}\right|$

The following is our reduction algorithm.

Rules induction and Attributes Reduction:

Input: universe ***U***, condition attribute set: ***C***, decision attribute set: ***D***, attribute value set: ***V***; information function: ***f***.

Output: ***Rules***, ***AttributeRed***

Steps:

(1) ***Calculate U/C,U/C={X$_1$,X$_2$···X$_S$}***
As in Section 2

(2) ***Reject the superfluous objects in the information system***

(3) ***Compute difference sets and similitude sets***
As in Equation (2) and (3)

(4) ***Rules induction and attributes reduction***
a)For each object $\boldsymbol{x_i}$ in ***U***
$C_i^r \leftarrow \emptyset$

For k=1,|C|
  Recount $frq_1(c_k,i)$
  Recount $frq_2(C_i^r,c_k,i)$
  $frq(c_k,i)=frq_1(c_k,i)+frq_2(C_i^r,c_k,i)$
  Choose attribute c_k with the greatest $frq(C_i^r,c_k,i)$
  $C_i^r \leftarrow C_i^r \cup \{c_k\}$
  $DS(x_i) \leftarrow DS(x_i) - \{ds|ds \cap \{a\} \wedge ds \in DS(x_i)\}$
  If $DS(x_i)=\emptyset$ then
    $R_i \leftarrow$ "if $f(C_i^r,x)=f(C_i^r,x_i)$ then $f(d,x)=f(d,x_i)$"
    Remove the instances which could be covered by R_i
    goto step a)

(5) $Rules \leftarrow \{ R_i \}$
  $AttributeRed \leftarrow \cup C_i^r$

3.1 Time Complexity of the Algorithm

The cost of Step(1) is $|C||U|+\sum|V_a|+|C|\cdot|U|$. The cost of Step (2) is $|U/C|$. Because there are only $|U/C|$ objects in the reduced system, the cost of Step (3) is $|C||U/C|^2$. In Step (4.a), the cost of calculating the $frq(c_k,i)$ is $|C||U/C|$. Suppose there is no superfluous attribute in the information system, there will be $|C||U/C|$ loops. Hence, the cost in step 4 is $|C|^2|U/C|^2$.The time complexity of the algorithm is $O(|C|^2|U/C|^2.)$

4 Experiment Evaluation

We used some databases in UCI repository of machine learning to evaluate our algorithm. The experimental results are listed in the following table.

Table 1. Reduction Results of several UCI databases

Database	|U|	|C|	|U/C|	$\|C_{red}\|$	Reduct	Rules	Simple
Adult+Stretch	20	4	20	2	Yes	3	Yes
Adult–Stretch	20	4	20	2	Yes	3	Yes
Balance	625	4	625	4	Yes	303	Yes
Ballon(YS)	20	4	20	2	Yes	3	Yes
Ballon(YSAS)	16	4	16	4	Yes	6	Yes
Breast-Cancer	699	9	463	4	Yes	73	Yes
Lenses	24	4	24	4	Yes	9	Yes
Hayes	117	4	69	3	Yes	24	Yes
Post-operative	90	8	74	8	Yes	33	Yes
Tic-Tac-Toe	958	9	958	8	Yes	33	Yes
Vote	435	16	342	8	Yes	48	Yes
Zoo	101	16	59	6	No	10	Yes

In table 1, the first column is the database's name. The 2nd,3rd,4th,5th,7th columns are the number of the objects, the conditional attribute number, the number of objects in the reduced system after step (1), thc attributes number after reduction and the number of the generated rules. The values in the sixth column mean whether the result of attribute reduction is the shortest. The values in the last column mean whether the rules are the simplest.

5 Conclusion

An intergraded algorithm, including feature selection and rule induction, is proposed in this paper. Firstly, ***U*/*C*** is calculated for reducing the complexity of reduction. Then, difference and similitude sets, of the reduced information system, are calculated. The last, the attributes are selected according to their abilities for giving high accurate rules. The time complexity of the reduction, includes attributes reduction and rules induction, is $\mathbf{O}(|C|^2|U/C|^2)$. The experiments show that the algorithm could extract the simplest rules from the information system.

Acknowledgement. This project is funded by Wuhan Sunlight Foundation (Project No.20055003059-3).

References

1. Blum Avrim L., Langley Pat. Selection of relevant features and examples in machine learning. Artificial Intelligence vol.97(1-2),(1997)245-271.
2. Kohavi R., John G. H. Wrappers for feature subset selection. Artificial Intelligence vol.97(1-2),(1997)273-324.
3. Isabelle Guyon, Elisseeff Andre. An Introduction to Variable and Feature Selection. Journal of Machine Learning Research vol.3,(2003)1157-1182.
4. Michalski R. S ., Chilausky R. L. Learning by being told and learning from examples: An experimental comparison of two methods of knowledge acquisition in context of developing on expert system for soybean disease diagnosm. Policy Analysis and Information Systems vol.4(2),(1980)125-150.
5. Quinlan J. Induction of decision trees. Machine Learning vol.1,(1986)81-106.
6. Liu H., Setiono R. Dimensionality reduction via discretization. Knowledge-Based Systems vol.9(1),(1996)67-72.
7. Pawlak Zdzislaw. Rough sets. International Journal of Parallel Programming vol.11(5),(1982)341-356.
8. Hu X. H., Nick C. Learning in relational databases: A rough set approach. International Journal of Computational Intelligence vol.11(2),(1995)323-338.
9. Liang Ji-Ye, Xu Zong-Ben. The algorithm on knowledge reduction in incomplete information systems. International Journal of Uncertainty Fuzziness and Knowledge-Based Systems vol.10(1),(2002)95-103.
10. Skowron A., C. Rauszer. The discernibility matrices and functions in information systems. In: R. Slowinski, editor. Intelligent Decision Support: Handbook of applications and advances of rough set theory. Dordrecht, Kluwer Academic Publishers, (1992)331-362.

An Approach for Reversely Generating Hierarchical UML Statechart Diagrams

Hua Chu, Qingshan Li, Shenming Hu, and Ping Chen

Xidian University, Software Engineering Institute,
P.O. Box 168, 710071, Shaan Xi, Xian, China
{hchu, qshli, shmhu}@mail.xidian.edu.cn,
chenping@mail.xidian.edu.cn

Abstract. One of the most crucial and complex steps of object-oriented system design lies in the transition needed to go from system behavior (defined by means of scenario models) to component behavior (described by means of communicating hierarchical state machine models).This paper presents a re-verse approach for generating hierarchical UML statechart diagrams. Firstly, we put forward a generation algorithm for a flat statechart diagram based on the BK-algorithm, which is validated useful in our reverse engineering tool XDRE by generating UML statechart diagrams from a set of UML sequence diagrams. Secondly, according to UML composite state, an automatic approach of introducing hierarchy to the generated flat statechart diagrams is proposed and implemented. Finally, systematic experiment is conducted in the paper in order to verify the validity of this approach.

1 Introduction

Scenario models and state machine models play central roles in current object-oriented system modelling processes. They provide two orthogonal views of systems. The former view describes system behavior as sequences of responsibilities that need to be executed by components in order to achieve overall system requirements, while the latter view addresses complete component behavior in terms of states and transitions.

The first work of this paper is to put forward a reverse approach for state machine from scenario models based on the concept of BK-algorithm[1]. And then we introduce hierarchy to the generated flat state machine in terms of the UML composite state[2]. Both approaches are validated useful in our reverse engineering tool, XDRE[4].

2 Reversely Generating Statechart Diagram from Scenario Models

UML[2] provides a standardized collection of notations for describing artifacts in a software-intensive system. Each UML notation represents a particular aspect of a software system from a particular viewpoint. However, there exists a good deal

L. Wang et al. (Eds.): FSKD 2006, LNAI 4223, pp. 434–437, 2006.

of overlap between many notations. This overlap can be exploited, in the form of automatic translations between notations, to reduce the time spent in design and to help maintain consistency between the models of different developers.

2.1 Approach for Statechart Diagrams Generation Based on BK-Algorithm

Biermann and Krisnashwamy present their algorithm(BK-algorithm) as to the synthesizing programs from their traces[1]. From the standpoint of the particular object, similarities between the concepts of a sequence diagram and a program trace(the input of the BK-algorithm) can be found. Both of them are sets of elements having partial relation arranged in time sequence. There is also a correspondence between a state diagram and a program (the output of the BK-algorithm). A program can be represented as a directed graph, so dose the statechart diagrams. Due to the similarities, the BK-algorithm can also be applied to generate the statechart diagram.

The process of our approach is divided into two steps:

1. Extracting the trace of an object concerned from UML sequence diagrams, which is a set of trace items in terms of (a_i, e_i). As for a trace item(a_i, e_i), a_i corresponds to the sending message of the object concerned and the receiving message e_i.
2. Recovering a statechart diagram mirroring dynamic behaviors of the object from the traces. A trace item (a_i, e_i) corresponds to such a state s : its action is $do : a_i$ and the trigger in the transition is e_i. If a_i is NULL, which represents this state has no do action. If e_i is VOID, which represents there has an unlabelled transition. If e_i is NULL, which represents this state is the last state of (referring to the set of states in state machine). An important work in this period is to identifying the similar states and merging them. States in the generated statechart diagram are represented by their $do : action$. Hence, two states are similar when their $do : action$ are completely same. If there exits a state whose $do : action$ is null, it is different from any state.

2.2 Hierarchical Statechart Diagrams Generation

Harel has introduced some concepts for raising the expressive power of flat state transition diagrams[3]. One of them is a superstate notation-a way to cluster and refine states, which is called a composite state in UML. The semantics of a superstate is an exclusive-or (XOR) of its substates; to be in a superstate an object must be in exactly one of its substates. The basic idea of our approach is to generate a hierarchical statechart diagram by introducing composite states to the flat statechart diagram.

The set of states S will be scanned according the order during the execution of the approach. For each state s in S, the transition function of s is checked in order to find states and events satisfying the semantics of XOR. The output of the approach is all composite states in the state machine.

3 A Case Study

In order to prove the correctness and validity of the state machine generation approach, a kernel communication subsystem of the call-center software system[5] is used as our test case. There are 129 C++ source files, 93 classes (including template class), 47,296 lines of source code in the system.

A part of one of the generated UML statechart diagrams is shown in Fig.1.

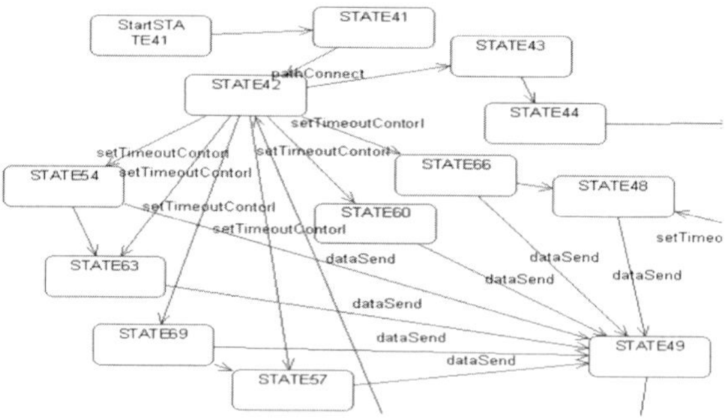

Fig. 1. Hierarchical statechart diagram with composite state

The hierarchical statechart shown in Fig.2 is the result of introducing hierarchy to Fig.1.

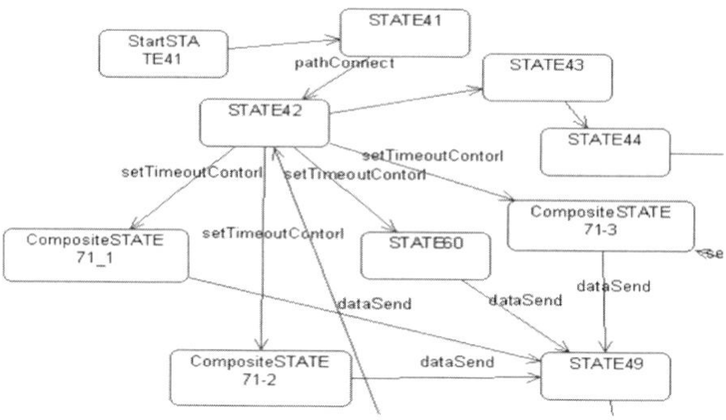

Fig. 2. Hierarchical statechart diagram with composite state

4 Related Research

In the current literature, several papers address the problem of defining state machines from different types of interaction diagrams (i.e., scenario models).

Whittle et al.[6] also provides an algorithm for generating UML statechart diagrams from scenario. They stress the important of obtaining a specification which can be read, understood and modified by a designer. In contrast, our approach emphasizes the dynamic behaviors of an object.

Minimally Adequate Synthesizer(MAS)[7]is an interactive algorithm that synthesizes UML statechart diagrams from sequence diagrams. It follows Angluin's framework of minimally adequate teacher to infer the desired statechart diagram by consulting the user. In contrast, our approach centers on the dynamic information of classes/objects. The sequence diagrams (input of our approach) is generated from the result of dynamic analysis[8]. Therefore, information displayed in the generated statechart diagrams represents the execution behavior of an object.

5 Conclusion

This paper presents a reverse generation approach for state machines from scenario models based on the concept of BK-algorithm and an approach for introducing hierarchy to the statechart machines.

In the future work, the presented technique may be extended. The next step is to adding some semantics during the transforming approach in order to making the generated state machine more readable.

Acknowledgements. This work is partially supported the National Research Foundation for the Doctoral Program of Higher Education of China and the Graduate Innovation Foundation of Xidian University.

References

1. Biermann A. W., Krishnaswamy R.: Constructing Programs from Example Computations. IEEE Trans, Software Engineering, vol.2 (1976)141–153
2. Booch G., Jacobson I., Rumbaugh J.: The Unified Modeling Language User Guide. Addison- Wesley(1999)
3. Harel D.: Statecharts: A Visual Formalism for Complex System. Science of Computer Programming8(1987)231–274
4. Li Qingshan, Chen Ping: XDRE 2.0 Reference Manual Software Engineering Institute. Xidian University(2004)
5. Chu Hua, Chen Ping: A Method of Designing Software Architecture Based on Process Group Component Model. Computer Science, vol.31 (2004)118–119,137
6. Whittle J., Schumann J.: Generating Statechart Designs from Scenarios. In Proceedings of the 22rd International Conference on Software Engineering, IEEE Computer Society(2000)314-323
7. Erkki M., Tarja S.: MAS-an Interactive Synthesizer to Support Behavioral Modeling in UML. In Proceedings of the 23rd International Conference on Software Engineering, IEEE Computer Society(2001)15–24
8. Li Qingshan, Chen Ping, Wang Wei: A C++ Instrumental Mechanism based on Reflection and Open Compile. Systems Engineering and Electronics,vol.25(2003) 851–855

A Novel Approach for Computing Partial Similarity Between 3D Models*

Wei Chen

State Key Lab of CAD&CG, Zhejiang University
chenwei@cad.zju.edu.cn

Abstract. In this paper, we present our initial solution to partial similarity computation between arbitrary 3D polygon models. The task is considered as the estimation of similarity transformations between the query pattern and target object. Two steps accounting for the scaling and rotation/translation parts are carried out, facilitated by applying EMD (earth mover's distance) to search the correspondence between focused point sets. In order to reduce the computation complexity involved in the second step, we use K-means algorithm to cluster the vertices of each model. We report our early experiments testing the efficiency of the proposed method on a small database as well as detailed discussions and the outline for the future work.

1 Introduction

Normally, shapes of 3D models are compared based on a *global* similarity measure. By *global*, we mean that the 3D query model is fully matched to a candidate from a 3D model database, penalizing any form of mismatch in indexing features. While the assumption of complete matching is convenient, it is not the right case where semantic relation is manifested only through a partial matching between objects.

So far as we know, there are little work on partial similarity computing of 3D models. Nevertheless, *Partial* matching is a useful technique for finding some components in 3D objects. We divide the partial matching problem into two sub-categories, namely, "one-to-one" and "one-to-many":

– **One-to-One:** An example of "one-to-one" partial matching is to find a car by its hood ornament. Obviously, they are semantically related objects. The characteristic of this search technique is that the query model is a part of expected returned results.
– **One-to-Many:** This case refers to finding an object within a large-scale scene, e.g. the $\langle chair, office \rangle$ pair. The presence of chairs usually suggests an office scene. We can take advantage of the assumption that all objects

* This paper is supported by National Science Fund Key Project of China under grant No. 60533050, National Science Fund Project of China under grant No. 60503056 and National Natural Science Funds of China for Innovative Research Groups under grant No.60021201.

L. Wang et al. (Eds.): FSKD 2006, LNAI 4223, pp. 438–447, 2006.

in a scene are semantically related with others, which is dominant in most situations.

We address the problem of "one-to-one" type of partial matching. Without ambiguity, we simply use the term "partial matching" in the follow-up sections. Section 2 contains a review of the related work. Section 3 presents our basic approach, while an improved version appears in Section 4. Finally, a brief summary appears in Section 5, followed by a discussion of topics for future work.

2 Research Background

The main challenges in supporting partial 3D shape-based similarity queries are **a well-defined representation of shape**: Allowing partial matching means that we are unaware of the location and the scaling of the query object in the 3D model. Known shape representations can be conceptually categorized into two classes concerning whether they require a preprocess for normalization of 3D models in a canonical coordinate system. Normalization-free representations might be extended to partial similarity computing. They are further classified as "statistical" or "topological":

- The main advantage of a "statistical" shape representation is its simplicity. It can be computed efficiently and is usually general enough, regardless of the geometric representation, topology, or application domain. However, further improvement in recognizing subtle distinctions between objects is still required.
- The skeletal and topological structure is the most intuitive notion of a 3D shape. However, actual models are often topologically incorrect, e.g, overlapping boundaries, gaps, and other topological degeneracies. Even for topologically valid models, problems still arise when mapping from a continuous model representation (e.g. surface representation) to a discrete one (e.g. volumetric representation), or from a continuous function to a discrete graph. This results in numerous graph structures for one 3D model that are entirely dependent on the choice of the sampling interval size.

Scaling/Pose estimation
To our best knowledge, there are few preliminary research work on partial similarity computing till now. To find partially similar 3D triangular faces, Motofumi *et al.* [1] propose to first divide 3D polygonal models into sets of triangular face sets by using binary trees. The angles created by normal vectors of triangular faces and the triangle areas are used as shape feature descriptors for scale invariant similarity computing. To speedup the algorithm, a dynamic programming approach is further introduced [2]. However, this method only works for 3D polygonal model databases of the same classification.

Recently, Funkhouser *et al.* [3] present a part-in-whole shape matching approach by representing a single model with many different descriptors. The final used descriptor is chosen by users and the distance between two models is defined as the sum of the squares of the distances from every point on one surface to the closest point on the other, and vice-versa.

It must be emphasized that our method concentrates on arbitrary 3D polygon model databases for models with the similar parts as the query pattern, which is different from above two algorithms.

3 Basic Approach

We choose EMD as our measure of dissimilarity since it can be efficiently computed by an algorithm with polynomial time of complexity. In addition, it is valid to all problems where objects defined in the problem can be expressed by a distribution in the form in Section 3.1.

The difficulty in making EMD amenable to partial matching of 3D models lies in representing an object as an invariant distribution under similarity transforms. As an initial attempt, we define the shape signature as a statistical distribution of shape index over the entire 3D model. The confidence of existence of the query pattern within the target object is measured by a scale estimation procedure built upon the EMD.

3.1 The Earth Mover's Distance

The EMD measures distance between two discrete distributions X and Y:

$$X = \{(x_1, w_1), (x_2, w_2), \cdots, (x_m, w_m)\}, Y = \{(y_1, u_1), (y_2, u_2), \cdots, (y_n, u_n)\} \tag{1}$$

where x_i, y_j are point locations in R^k space, w_i, u_j are nonnegative point weights, and m, n denote the number of points of each distribution respectively.

The EMD is the minimum amount of work to morph X into Y, normalized by the total weight of the lighter distribution. It can be quickly solved by transportation simplex algorithm:

$$\mathrm{EMD}(X, Y) = \frac{\min_{F \in \mathcal{F}(X,Y)} \mathrm{WORK}(F, X, Y)}{\min(\sum w_i, \sum u_j)} \tag{2}$$

where,

$$\mathrm{WORK}(F, X, Y) = \sum_i \sum_j f_{ij} d(x_i, y_j), for \qquad F = (f_{ij}) \in \mathcal{F}(X, Y). \tag{3}$$

Subject to:

1. $f_{ij} \geq 0$;
2. $\sum_{j=1}^{n} f_{ij} \leq w_i$;
3. $\sum_{i=1}^{m} f_{ij} \leq u_i$;
4. $\sum_{i=1}^{m} \sum_{j=1}^{n} f_{ij} = \min(\sum w_i, \sum u_i)$.

A *flow* between X and Y is an arbitrary matrix $F = (f_{i,j}) \in R^{m \times n}$, with f_{ij} denotes the amount of x_i matched to y_j. $\mathcal{F}(X, Y)$ is the set of all *feasible flows* between X and Y, where a *feasible flow* is the *flow* satisfying conditions 1–4. $d(x_i, y_j)$ computes the distance between point x_i and y_j.

3.2 Shape Index over Polyhedral Surface

Let $\langle k_1, k_2 \rangle$ denotes the principle curvature vector associated with a point p on a regular 3D surface, the shape index si at p is the angular coordinate of the polar representation of $\langle k_1, k_2 \rangle$:

$$si = \frac{2}{\pi} \arctan \frac{k_2 + k_1}{k_2 - k_1}, \qquad k_2 \geq k_1. \tag{4}$$

All surface patches, except planar surfaces where $k_1 = k_2 = 0$, are mapped onto the interval [-1, 1].

The computation of shape index depends on accurate estimation of the principle curvatures of a subjacent smooth surface from its polyhedral approximation. We use the linear complexity algorithm described in [4] for estimating the principle curvatures and principle directions at the vertices of a triangulated surface.

3.3 Scale Estimation Using EMD

The shape representation of a 3D model is the shape index distribution, computed from the shape index value at every vertex of the triangulated surface. The distribution is approximated by a discrete histogram with fixed sized bins. The contribution from each vertex is weighted by the normalized summed area of its neighboring triangles:

$$weight(v_i) = \frac{\sum_{f_k \in F^i} S(f_k)}{3 \times \sum_i S(f_i)}, \tag{5}$$

where F^i is the set of triangles containing the vertex v_i, S is the surface area of f_k. We require the polygonal model to be triangulated so that every facet is taken into consideration exactly three times. The total weight under the distribution is less than or equals to 1 as shape index has no definition on some vertices.

Let $X = (x, w)$ and $Y = (y, u)$ denote the normalized shape index distributions of a 3D model and the query respectively. Considering the ideal case of an exact pattern occurrence, where the query occurs in the 3D model as a fraction $c^* \in (0, 1]$ of the total surface area.

Practical scaling estimation differs from the ideal case in the following aspects: (1) Although $E(c)$ becomes constant as c approaches zero, the constant may not necessarily be zero. $E(c)$ can level off at whatever value, where small value suggests that the pattern may occur in the target. (2) c_0 is an estimation of c^*, and may be different from c^* depending on the model properties. Figure 1 shows an example of overestimation. The estimated c_0 for a mouse's ear is approximately twice of the c^*, since the normalized shape index distributions for one and two ears are nearly the same.

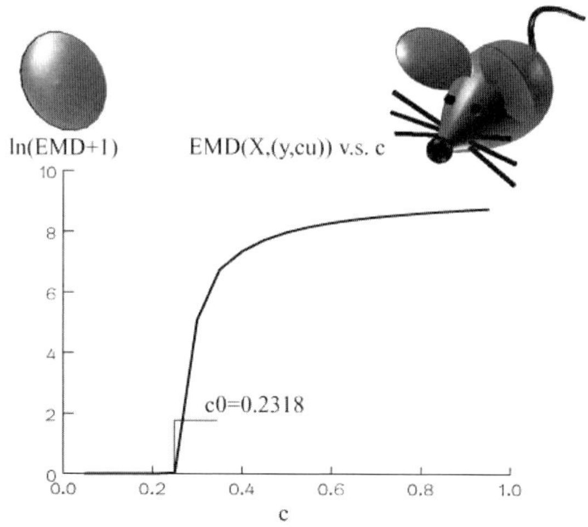

Fig. 1. An example of overestimated scale. $c_0 = 0.2318$, approximately twice the $c^* = 0.1231$.

The basic approach of partial similarity computing is carried out in three phases. First, $\text{EMD}(X, (y, c_0 u))$ and c_0 are estimated by means of a binary search along the c-axis. Second, if c_0 is less than c_θ and $\text{EMD}(X, (y, c_0 u))$ is larger than EMD_θ, it is concluded that the pattern does not occur in the model. Third, models are ranked according to their EMDs. If two EMDs are nearly identical, e.g. their difference is within a small ϵ, the corresponding two models are compared by the value of c_0, or say, the larger the c_0 is, the higher the rank is.

In practice, c_θ is set to 0.05. This accounts for the fact that if a model's representation spans all the range of shape index distribution, EMD will become zero upon whatever query pattern when c approaches zero.

3.4 Initial Results

We test on a small database of 30 models accessed from the World Wide Web and grouped into 3 classes: decoration, lamps, and toys. Objects from the same group have common parts, which can be chosen as the query pattern. Our primary testing patterns include slab, lamp-chimney and abdomen. The most similar models returned from the database are shown in Figure 2. We will defer our discussion of the experimental results until Sect. 4.5, along with the comparison between the basic and the improved approach.

4 Integrating Position as an Improvement

The initial results are encouraging despite that we only use the shape index information at each vertex. The shape index is a local geometry attribute while the position information introduced in this section is a global geometry attribute.

The position $(x_{pos}, y_{pos}, z_{pos})$ of each vertex depends on the scaling, orientation, and translation parameters of the model. The key to the solution lies in eliminating the influence posed by the two local coordinate systems. The method used here is the EMD under transformation sets.

4.1 EMD Under Transformation Sets

The EMD between two distributions X and Y under a transformation set $\mathcal{G}$ is defined as:

$$\mathrm{EMD}_{\mathcal{G}}(X,Y) = \min_{g \in \mathcal{G}} \mathrm{EMD}(X, g(Y)), \tag{6}$$

where,

$$\mathrm{EMD}(X, g(Y)) = \frac{\min_{F \in \mathcal{F}(X,g(Y))} \mathrm{WORK}(F, X, g(Y))}{\min(\sum w_i, \sum u_i)}. \tag{7}$$

For a distribution $Y = (y, u)$, $\mathcal{G}$ can be one of the following three types:

1. $c(Y) = c(y, u) = (y, cu)$;
2. $g(Y) = g(y, u) = (g(y), u)$;
3. $g_c(Y) = g_c(y, u) = (g(y), cu)$.

We have already encountered the first type of transformation in Sect. 3.3, when estimating the relative scaling between the query and the target. It changes a distribution's weights, but leaves its points fixed. The second type of transformation changes the points of a distribution, but leaves its weights fixed. It can be any of the following: Translation, Euclidean, Similarity, Linear or Affine. The last type of transformation allows changes to both the weights and points of a distribution.

We restate our problem as follows: Let $X = (x, w)$ and $Y = (y, u)$ denote two distributions representing the target and the query respectively. x_i and y_j are in the form of a four-tuple $\langle x_{pos}, y_{pos}, z_{pos}, s\rangle$, where s is the corresponding shape index value. Our aim is to compute the EMD between the query and the target under a best estimation of scaling, rotation and translation.

Note that the scaling changes the weights in the distribution representing the query, and the weights determine the set of feasible flows. For the sake of the simplification, we use an approximation method to estimate the transformation parameters in two steps: (1) determine the relative scaling between the target and the query. (2) determine the Euclidean transformation between the target and the scaled query. With the above parameters, we are able to compute the EMD as a measure of dissimilarity.

The solution to the first step is described in Sect. 3.3. The set of feasible flows is then fixed, which is independent of the rotation and translation $\langle R, T\rangle$. For the 3D model $\{X\}$ and the scaled query $\{Y\}$, we approximate the optimal solution for R and T iteratively:

– find the best flow for a fixed transformation:

$$F^{(k)} = \arg\min_{F} \mathrm{WORK}(F, X, g^{(k)}(Y)) for \qquad F \in \mathcal{F}(X, Y);$$

– find the best transformation for the previously computed flow:

$$g^{(k+1)} = \arg\min_{g} \mathrm{WORK}(F^{(k)}, X, g(Y)) for \qquad g \in \mathcal{G}.$$

Using the definition of EMD, it is easily proved that the iteration converges, though to a locally optimal transformation[5]. The iteration should be run with different initial conditions in search of the global optima.

4.2 Vertex Classification

A typical 3D model in our data set contains over 2000 vertices. Due to the following two reasons, we believe that the representation of a 3D model $\langle x_{pos}, y_{pos}, z_{pos}, s\rangle$ defined on every vertex can and should be reduced. First, vertices that are close enough in spatial coordinates also exhibit strong resemblance in their shape index values. This is because the surface we deal with is usually smooth, with continuously changing curvatures. Second, the transportation simplex algorithm has a polynomial time of complexity. Thus restricting the size of distribution will efficiently reduce the problem size.

We use K-means least-squares partitioning method to divide the cloud of vertices into K groups. It needs no prior knowledge about 3D models, and allows us to control the size of distributions efficiently. The distance from a sample to each cluster is a weighted sum of spatial distance and shape index difference. For each model, we execute the algorithm several times and select the best partition with the least squared error to avoid getting stuck at local optima.

After clustering, the signature of a 3D model is reduced to K components: $X = \{(x_1, w_1), \cdots, (x_K, w_K)\}$:

$$w_i = \sum_j w_{ij}, x_i = \sum_j w_{ij} x_{ij}, \tag{8}$$

where x_{ij} is the property (position or shape index) of the jth vertex from the ith cluster, and w_{ij} is the corresponding weight.

4.3 The Choice of Ground Distance

We choose the ground distance $d(x_i, y_j)$ to be a weighted sum of two items:

$$d(x_i, y_j) = (1 - \delta) d_{pos} + \delta d_{attr}, \tag{9}$$

where d_{pos} is the squared Euclidean distance between two points, and $d_{attr} = \|s_{x_i} - s_{y_j}\|^2$. Since the shape index is a local attribute, d_{attr} is independent of the Euclidean transformation.

With the above definition of ground distance, and the assumption of Euclidean transformation, the estimation of the optimal transformation under a fixed flow can be further simplified:

$$\begin{aligned} g^{(k+1)} &= \arg\min_{g\in\mathcal{G}} \mathrm{WORK}(F^{(k)}, X, g(Y)) \\ &= \arg\min_{g\in\mathcal{G}} \sum_i \sum_j f_{ij}^k d(x_i, g(y_j)) \\ &= \arg\min_{g\in\mathcal{G}} \sum_i \sum_j f_{ij}^k d_g(i,j) \\ &= \arg\min_{R,t} \sum_i \sum_j f_{ij}^k d_{R,t}(i,j), \end{aligned} \tag{10}$$

where,

$$\begin{aligned} d_g(i,j) &= (1-\delta)d_{pos} + \delta d_{attr})(x_i, g(y_j)), \\ d_{R,t}(i,j) &= \|x_i^{pos} - \langle R, T\rangle y_j^{pos}\|_2^2. \end{aligned}$$

For fixed R, the optimal translation must be $T(R) = \overline{x}_{pos} - R\overline{y}_{pos}$, where the overline means "average operation". The best rotation problem in Eq. 10 is solved completely in [6].

4.4 An Improved Searching Strategy

For each query pattern, we propose the following strategy to find objects that contain the pattern, and list them by the likelihood of pattern occurrence:

1. Compute the shape index distribution of the query pattern.
2. Compare it with the shape index distribution of every 3D model from the database, decide the relative scaling c_0 and the corresponding $\mathrm{EMD}_{scaling}$.
3. If $c_0 < c_\theta$ and $\mathrm{EMD}_{scaling} > \mathrm{EMD}_\theta$, we assert that the pattern doesn't occur in the 3D model, and quit the estimation procedure.
4. If $c_0 > c_\tau$, $d(pattern, model) = \mathrm{EMD}_{scaling}$, record it and go to 7.
5. Obtain a compact representation of the query by K-means clustering.
6. Estimate EMD_{trans}, as well as the best $\langle R, T\rangle$ from the scaled query to the target, go to 8.
7. These models are returned as the most similar objects, ranked by their EMDs. For those with nearly identical EMDs, ranking is determined by the magnitude of c_0.
8. These models are among the second class of retrieved objects. They are simply ranked by EMD_{trans}.

Note that steps 1–4,7 constitute our basic approach which is based only on the scaling estimation.

4.5 Experimental Results

We use the same data set and query pattern from the previous experiment. We partition the vertices into $K = 128$ groups, and set $c_\tau = 0.1$. Table 1 compares the experimental results of the two methods side by side. The terms including

FT("First Tier"), ST("Second Tier") and NN("Nearest Neighbor") are defined in [7].

Figure 2.(a) shows the first five 3D models returned if we use "slab" as the query. (b) and (c) illustrate different results by the basic and improved approach respectively. Since we set $c_{\mathcal{T}} = 0.1$, the sixth object (a lamp) will be regarded as among the most similar group of objects. That's why we return the wrong result. In this example, the EMD under Euclidean transformation performs better in that it lows the rank of two lamps and retrieves all the remaining correct objects.

Fig. 2. Query by slab. (a) five most similar objects from database, using the basic/improved query method. The estimated scale c_0 for each object is greater than $c_{\mathcal{T}}$. (b) second five most similar objects from database, using the basic query method. (c) second five most similar objects from database. The (7)–(10) are retrieved using the improved one.

Table 1. The evaluation of similarity computing results

Query Method	FT	ST	NN
Basic	86.7%	100%	100%
Improved	93.3%	100%	100%

5 Conclusion and Future Work

The main properties of our approach are

- **Simplicity.** We have used a statistical approach instead of extracting topological structure of 3D models to meet the requirements of both time and generality.

- **Efficiency.** Efficiency is guaranteed by the following aspects: (1) EMD can be quickly calculated by the transportation simplex algorithm. (2) K-means algorithm is used to control the problem size. (3) We either accept or reject some candidates in the basic step of the search strategy, reducing the number of 3D models to be processed by the second step.

- **Transformation Invariance.** We are aware that models may undergo free rigid movements and can have different scales, and have taken the similarity transformation invariance into consideration.

References

1. T.-S. Motofumi and Y.-S. Yuji. A search method to find partially similar triangular faces from 3D polygonal models. In *Proceedings of the IASTED International Conference Modeling and Simulation 2003*, pages 323–328. Palm Springs, 2003.
2. T.-S. Motofumi. A dynamic programming approach to search similar portions of 3D models. *The World Scientific Engineering Academy and Society Transaction on Systems*, 3(1):125–132, 2004.
3. T. Funkhouser, M. Kazhdan, P. Shilane, P. Min, W. Kiefer, A. Tal, S. Rusinkiewicz, and D. Dobkin. Modeling by example. *ACM Transactions on Graphics 2004 (SIGGRAPH'2004)*, 23(3):652–663, 2004.
4. G. Taubin. Estimating the tensor of curvature of a surface from a polyhedral approximation. In *The fifth International Conference on Computer Vision*, pages 902–907, June 1995.
5. S. Cohen. Finding color and shape patterns in images. Technical Report STAN-CS-TR-99-1620, Stanford University, May 1999.
6. S. Umeyama. Least-squares estimation of transformation parameters between two point patterns. *IEEE Transactions on Pattern Analysis and Machine Intelligence*, 13(4):376–380, April 1991.
7. R. Osada, T. Funkhouser, B. Chazelle, and D. Dobkin. Shape distributions. *ACM Transactions on Graphics*, 21(4):93–101, October 2002.

A New and Fast Method of Image Indexing

Lina-Huang and Zhijing-Liu

Dept of Computer Science Xidian University, Xi'an, 710071
huangnanadreams@163.com

Abstract. Traditional indexing methods face the difficulty of " curse of dimensionality" at high dimensionality. In this paper, the traditional vector approximation is improved. Firstly, it decreases the dimensions by LLE (locally-linear-embedding). As a result of it, a set of absolute low dimensions are gotten. Then, this paper uses the Gaussian mixture distribution and estimates the distribution through EM (expectation-maximization) method. The original data vectors are replaced by vector approximation. This approach gains higher efficiency and less run time. The experiments show a remarkable reduction of I/O. They also show an improvement on the indexing performance and then speed the image retrieval.

1 Introduction

Generally, CBIR must deal with the large image database and provides appropriate querying method to retrieval images efficiently. For high-dimensional vectors always express the image features, it is necessary to bring in an effective image base indexing method to speed up the indexing.

The performance of indexing methods degrades drastically as the dimensionality increases. The indexing method based on Vector approximation (VA) is a unique precious one which better than precious search. A sample distribution (the irritated vectors and the sample Gaussian distribution) of data is applied to VA. It is not a good selection to describe the holistic distribution. And the partition of scalar quantity leads the VA with low precision. [Ferhatomanoglu 2000] put forward VA+, which was used to deal with the symmetrical datasets. It used the Karhunen-Loeve to destroy the linear pertinence among the dimensions. The effect is weaker for non-linear dimensions in Gaussian distribution.

Considering the bugs of two methods, this paper put forwards a new and fast method. The detailed procession will be described in hinder chapters.

2 The Original Indexing Method Based on VA

In the VA-file method, the space of a dataset is partitioned into equal cells by a uniform partition strategy. A position code is then generated for all the vectors in a cell and the code is stored in an approximation file, the index file used in the VA-file method. During the search phase, using the k -nearest neighbor query algorithm, it

L. Wang et al. (Eds.): FSKD 2006, LNAI 4223, pp. 448 – 451, 2006.

reads a relatively smaller approximation file instead of the whole dataset and tries to filter the vectors, so that only a small fraction(usually 0.1％～10％)of them will be read. The partition method will decide the positions of the cells on that dimension.

The VA-file method divides that data space into 2^b rectangular cells, where $b=\sum_{i=1}^{d} b_i$ is a user-defined number of bits for each cell, d is the dimension and b_i is the number of bits assigned for the i -th dimension. Every dimension is partitioned in order to make each interval equally full. The figure1 explains it well. Based on an assumption that the dataset is of even-distribution, the method divides each dimension equally with 4 intervals, with the same number of vector components in each interval. Each data vector is expressed by a 4 –bits vector, it is far shorter than original length.

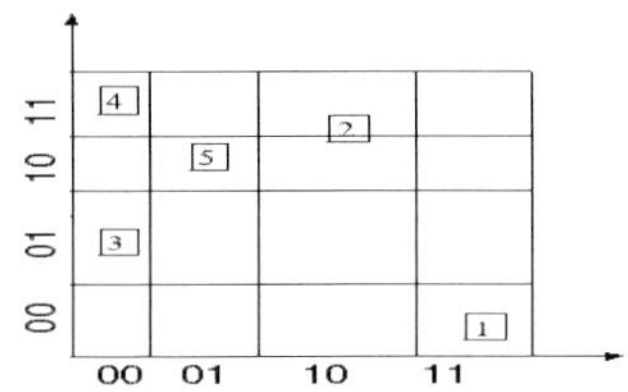

	原始向量	近似向量
1	(0.9, 0.1)	(11 00)
2	(0.6, 0.8)	(10 11)
3	(0.1, 0.4)	(00 01)
4	(0.1, 0.9)	(00 11)
5	(0.3, 0.7)	(01 10)

Fig. 1. The Original Vector and the Approximate Vector

3 The Improved Arithmetic

The partition method of the original VA-file is sensitive to the distribution of data and needs to be improved when it faces non-uniform data sets. To tackle the problem of processing datasets with non-uniform distribution in the VA-file, VA$^+$-file method uses the KL-transformation to reduce the correlation of data in different dimensions. But it only can deal with the relativity of the linear dimensions. In this paper, firstly, it uses LLE to reduce dimensions to solve the problem. Then it uses the Gaussian mixture distribution and estimates the distribution through EM (expectation-maximization). After two steps below, to produce vector approximate and then indexes the image dataset by $k-NN$ .The detailed arithmetic is showed as follows.

3.1 LLE

Locally Linear Embedding (LLE) is a nonlinear feature extraction algorithm that maps high-dimensional into a low-dimensional Euclidean space while preserving local topological structures. A data point in the high-dimensional input dataset is thought located in a local patch that can be linearly reconstructed by its closest neighboring points, one patch for each data point. Global data topological structures in the high-dimensional space can be approximated and preserved by stitching up all the patches. Piecewise linear curve fitting is an analogy of this approximation in 2-dimensional space. A weight matrix is constructed by grouping all the coefficients for reconstructing each point from its nearest neighbors. Based on the weight matrix, which contains the local topological information of the input data, a low-dimensional embedding can be derived through optimization. LLE incorporates three steps:

Step1-searching for its nearest neighbors

Where, d is the Euclidian distance between x_i and x_j in column vectors, for I, j=1,2,…,n, and n is the number of data points. Here, set n =10.

$$d = \sqrt{(x_i - x_j)}^T \sqrt{(x_i - x_j)} \tag{1}$$

Step2- calculating reconstruction weights

Where, k is the number of nearest neighbors used for reconstructing each vector. This is a constrained least squares problem that has the following closed-form solution. v = $[1,1,\ldots 1]^T$; $s_i = (X_i - N_i)^T (X_i - N_i)$, X_i is a $d \times k$ matrix whose columns are duplicates of x_i ,and N_i is also a $d \times k$ matrix whose columns are the k nearest neighbors of x_i determined in Step1, d is the dimensionality of the original space.

$$\min \varepsilon(W) = \sum_{i=1}^{N} | x_i - \sum_{j=1}^{k} w_j{}^i x_{ij} |^2 \quad \sum_{j=1}^{k} w_j{}^i = 1 \quad w_i = s_i{}^{-1} \mathrm{v} \tag{2}$$

Step3- determining low-dimensional embedding

Where, y_i is the coordinate of the data point x_i in the low-dimensionality of m. It is proved that solving(3) for y_i is equivalent to calculate the eigenvectors corresponding to the m smallest nonzero eigenvectors of $(\mathrm{I}-\mathrm{W})^{\mathrm{T}}(\mathrm{I}-\mathrm{W})$, where I is a $n \times n$ identity matrix and W is also a $n \times n$ matrix, each of whose rows is composed of a data point's reconstruction weights.

$$\min \varepsilon(Y) = \sum_{i=1}^{N} | y_i - \sum_{j=1}^{k} w_j{}^i y_{ij} |^2 \quad \frac{1}{N}\sum_{i=1}^{N} y_i y_i{}^T = \mathrm{I} \quad \sum_{i=1}^{N} y_i = 0 \tag{3}$$

Because the dimensionality of the low-dimensional embedding, m , is far smaller than the dimensionality of the original space, d , data dimensionality reduction is effectively achieved.

3.2 The EM Partition Based on Mixing Gaussian Model

In this paper, a mixing Gaussian Model is used for the distribution of the space of the dataset. The probability function of a mixture sample y defined by

$$\mathrm{p}(\mathrm{y}|\theta) = \sum_{m=1}^{k} a_m p(y|\theta_m) \quad a_m \geq 0, \quad m = 1, \cdots, k, \quad \sum_{m=1}^{k} a_m = 1 \tag{4}$$

Where, a_m is the mixing proportion, the $\mathrm{p}(\mathrm{y}|\theta_m)$ is the densities of each component , it according to the Gaussian model, it is written as

$$p(y|\theta_m) = 2\pi^{-d/2} | \Sigma_m |^{-1/2} \bullet \exp\{-\frac{1}{2}(y_n - \mu_m)^T \Sigma_m^{-1} (y_n - \mu_m)\} \quad (5)$$

Where, μ_m is mean square value, Σ_m is covariance matrix.

In this paper, EM is used to part the space of dataset. Firstly, it initializes the value of μ_m and Σ_m. And then, it uses the iterative method to estimate the values of parameters. The arithmetic would be stopped when the results vary lightly. Through it, the space of dataset is divided to k clusters.

4 The Results

The dataset has 275500 images. Each texture feature with 60 dimensions is reduced to 10 ones.

In this paper, the experiments are carried out based on VA-file□VA+-file and the new improved method respectively. Then the sample sets used in the proposed partition methods are generated by randomly selecting a certain percentage of feature vectors from the whole datasets. The number of it in this paper is 1000. The final result is the average of the 1000 random vectors.

In the experiments, the bits which is assigned to each dimension is 1,2,and 3 respectively. The K-NN arithmetic is used to index images in the text. Here, set k=5. In K-NN, there are two moments. The first moment is filtering vectors, while the second moment is dealing with original data. The comparisons in filtering efficiency are shown in Fig 2 below:

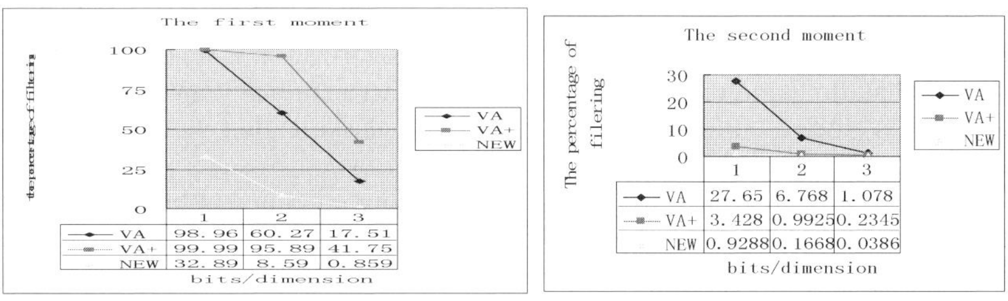

Fig. 2. The comparisons of filtering efficiency

In the case of few bits each dimension, VA^+-file has no effect in the first moment. But in the whole process, it has a high efficiency than VA-file. The results show that the new method is better than others . The smaller instruction of indexing is, the less run time for it runs. It provides a promising method for high-dimension indexing.

Reference

1. S T Roweis, L K Saul. Nonlinear Dimensionality Reduction by Locally Linear Embedding [J].Scientist

A Novel Algorithm for Text Categorization Using Improved Back-Propagation Neural Network

Cheng Hua Li and Soon Cheol Park

Division of Electronics and Information Engineering, Chonbuk National University
Jeonju, Jeonbuk, 561-756, Korea
lchjk@msn.com, scpark@chonbuk.ac.kr

Abstract. This paper describes a novel adaptive learning approach for text categorization based on a Back-propagation neural network (BPNN). The BPNN has been widely used in classification and pattern recognition; however it has some generally acknowledged defects, which usually originate from some morbidity neurons. In this paper, we introduce an improved BPNN that can overcome these defects and rectify the morbidity neurons. We tested the improved model on the standard Reuter-21578, and the result shows that the proposed model is able to achieve high categorization effectiveness as measured by the precision, recall and F-measure.

1 Introduction

With the current explosive growth of internet usage, extracting the accurate information that people need quickly is becoming harder and harder. The demand for fast and useful access to online data is increasing. Text categorization is an efficient technology for the handling and organizing of text data. There are many applications for text categorization, such as information retrieval, news classification, the sorting of email and text filtering.

The task of text categorization is to decide whether a document belongs to a set of pre-specified classes of documents. Automatic classification schemes can greatly facilitate this process, and many different approaches have been attempted, including the K-Nearest Neighbor [1, 2], Rocchio [3, 4], Decision Tree [5], and Neural Network [6, 7]. BPNNs have many advantages compared with other networks, so they can be used very widely, however they also have their limitations. The main defects of the BPNN can be described as: slow convergence; difficulty in escaping from local minima; easily entrapped in network paralyses; uncertain network structure. In previous experiments, it was demonstrated that these limitations are all related to the morbidity neurons. Therefore, we propose an improved model called MRBP (Morbidity neuron Rectify Back-Propagation neural network) to detect and rectify the morbidity neurons; this reformative BPNN divides the whole learning process into many learning phases. It evaluates the learning mode used in the phase evaluation after every learning phase. This can improve the ability of the neural network, making it more adaptive and robust, so that the network can more easily escape from a local minimum, and be able to train itself more effectively.

L. Wang et al. (Eds.): FSKD 2006, LNAI 4223, pp. 452 – 460, 2006.

This paper consists of 5 sections. In section 2, we describe the theory of back-propagation neural networks, including the basic theory and improved method. The experiments are discussed in section 3. The evaluation results are given in section 4. Finally, the conclusion and a discussion of future work are given in section 5.

2 Theory of Back-Propagation Neural Networks

2.1 Basic Theory of the BPNN

The back-propagation neural network is a generalization of the delta rule used for training multi-layer feed-forward neural networks with non-linear units. It is simply a gradient descent method designed to minimize the total error (or mean error) of the output computed by the network. Fig. 1 shows such a network.

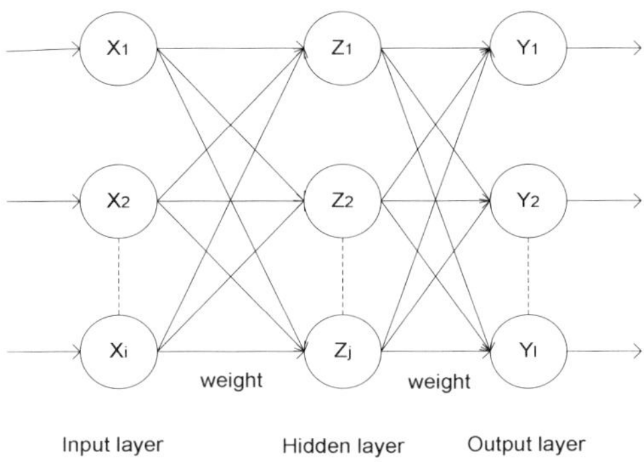

Fig. 1. Typical three layer BP network

In such a network, there is an input layer, an output layer, with one or more hidden layers in between them.

During training, an input pattern is given to the input layer of the network. Based on the given input pattern, the network will compute the output in the output layer. This network output is then compared with the desired output pattern. The aim of the back-propagation learning rule is to define a method of adjusting the weights of the networks. Eventually, the network will give the output that matches the desired output pattern given any input pattern in the training set.

The training of a network by back-propagation involves three stages: the feed-forward of the input training pattern, the calculation and back-propagation of the associated error, and the adjustment of the weight and the biases.

2.2 BPNN Defects Analysis and Commonly Used Improved Methods

The three main defects of the BPNN and some commonly used improved methods are as follows:

Slow convergence. In the beginning, the learning process proceeds very quickly, in each epoch, and can make rapid progress, however it slows down in the later stages [8]. There are two commonly used methods of improving the speed of training for BPNNs. *a) Introduce momentum into the network.* Convergence is sometimes faster if a momentum term is added to the weight update formulas. The weight update formula for a BPNN with momentum is

$$W_{ij}(k+1)=W_{ij}(k)+\eta\delta_i x_j+u\left(W_{ij}(k)-W_{ij}(k-1)\right) \tag{1}$$

where momentum parameter u is constrained to be in the range from 0 to 1. The new weights for the training step t+1 are based on the weights at training steps t and t-1. *b) Using the adaptive learning rate to adjust the learning rate.* The role of the adaptive learning rate is to allow each weight to have its own learning rate, and to let the learning rates vary with time as training progresses. The formulas for a BPNN with an adaptive learning rate is

$$\eta^{(n+1)}=\eta^{(n)}\times\frac{E^{n-1}}{E^{n}} \tag{2}$$

where n is the epoch during the training process, and E is the absolute error in each epoch. When E decreases, the learning effect will increase (the weight may change to a greater extent). Otherwise, the learning effect will decrease.

These two kinds of methods accelerate the convergence of the BPNN, but they can not solve other problems associated with the BPNN, especially when the size of the network is large.

Local minimum. When training a BPNN, it is easy to enter into a local minimum, and usually the GA and simulated annealing algorithms have been used to solve this problem. These algorithms can prevent the problem of entering into a local minimum, but they cannot ensure that the network will not enter into a global minimum, and they are even slower than the traditional BPNN.

Network paralyses. During training, the value of the weights may be very large and, consequently, the input of the network will be very large. Thus, the output value of the activation functions, O_j (or O_l), tends to 1, according to the formula of error back propagation, and the back propagation error will tend to 0. This phenomenon is referred to as saturation. The speed of training becomes very slow when saturation occurs. Finally it will cause the weight not to change any more, and this will lead to network paralysis. P.D. Wasserman [9] provided the suggested formula to limit the weight between (-a, a), but it is only used for weight initialization. It cannot prevent the value of the weight increasing during training, and it also has the possibility of leading to network paralysis.

2.3 MRBP Algorithm

The defects mentioned above are all related to saturation, the convergence will become slower and the system will change to a higher learning rate. Also, the weight becomes larger due to the larger learning rate, and this will cause the output value of the activation function to tend to 1. Under this situation, the network can easily enter into a local minimum and ultimately become entrapped by network paralysis. Based on our experience with such problems, we also found that there is another phenomenon which can cause such defects. For some of the neurons, the range of input values is restricted to a small range during each epoch, and this causes the values of the output to be extremely close to each other at each epoch, while the error during each epoch changes slowly. In other words, the speed of convergence is slow. Finally, this situation causes a local minimum or even network paralysis. In this paper, we refer to these two kinds of phenomena as neuron overcharge and neuron tiredness respectively. We call these neurons morbidity neurons. In general, if some morbidity neurons occur within it, then the network cannot function effectively.

The MRBP improved method: During the learning process, neurons face two kinds of morbidity: overcharge and tiredness. If we avoid the appearance of morbidity neurons during the learning phase or rectify the problem in time, then the networks can train and evolve effectively.

[Definition 1]: Neuron overcharged. If the input value of the neuron is very big or very small, it will cause the output value to tend to -1 or 1, and cause the back-propagation error to tend to 0. We refer to such a neuron as being overcharged. That is, for the activation function,

$$f(net_j + \theta_j) = \frac{2}{(1 + e^{-\lambda(net_j + \theta_j)})} - 1 \cdot \tag{3}$$

If $f(net_j + \theta_j) \to 1$ or $f(net_j + \theta_j) \to -1$, then $\delta_j \to 0$. When this happens, we refer to neuron j as being overcharged.

[Definition 2]: Neuron tiredness. If a certain neuron always receives the similar stimulation, then its response to this stimulation will be very similar, so that it is difficult to distinguish different stimulations by its response. We refer to such a neuron as being tired. That is, when neuron j during one learning phase (defined as follows) obeys,

$$\left(\underset{k}{MAX} f(net_j^k + \theta_j^k) - \underset{k}{MIN} f(net_j^k + \theta_j^k) \right) \to 0 \cdot \tag{4}$$

When this happens, we refer to the neuron j as being tired.

[Definition 3]: Learning phase. Choosing N iterations (or leanings) as a period, during this period we record some important data, and calculate the effect of the learning process, as the direction for the next period. We called this period the learning phase and, based on our experience, we use 50 epochs as the learning phase.

According to the definition of an overcharged neuron and a tired neuron, we know that they are directly related to the activation function. In the conventional activation function $f(x) = \frac{2}{(1+e^{-\lambda x})} - 1$, λ using 1 or other constants, whereas in our model, λ is an adjustable variable. V.P. Plagianakos [10] tried to use an adjustable value of λ in his paper. Actually, different combination of λ corresponds to different learning models.

The determinant rule of the morbidity neuron is: If $f(net_j + \theta_j) >= 0.9$ or $f(net_j + \theta_j) <= -0.9$, then neuron j is overcharged. And if $\underset{k}{MAX} f(net_j^k + \theta_j^k) - \underset{k}{MIN} f(net_j^k + \theta_j^k) <= 0.2$, then the neuron j is tired.

The formulae used to rectify the morbidity neuron are

$$\theta_j = \theta_j - \frac{\underset{k}{MAX} f(net_j^k + \theta_j^k) + \underset{k}{MIN} f(net_j^k + \theta_j^k)}{2} \tag{5}$$

and

$$\lambda_j = -\frac{Ln\left(\frac{2}{1.9} - 1\right)}{\underset{k}{MAX} f(net_j^k + \theta_j^k) - \underset{k}{MIN} f(net_j^k + \theta_j^k)}. \tag{6}$$

Formula (5) is used to normalize the maximum and minimum input values in the previous phase in order to make them symmetric with respect to the origin. Formula (6) is used to limit the maximum and minimum output values to the normal range. In our experiments, the range is (-0.9, 0.9). In our study, the morbidity neurons were rectified in each phase after their evaluation.

3 Experiments

3.1 Text Representation

In order to use an automated learning approach, we first need to transform the text into a feature vector representation. This involves several steps: word extraction, stop words removal, word stemming, and term weighting. In our experiment, we employ Porter's stemming algorithms [11] for word stemming, and a logarithmic function as term weight,

$$W_{i,j} = \log(1 + tf_{i,j}) \tag{7}$$

where $W_{i,j}$ is the term weight of the ith indexing term in document j, and $tf_{i,j}$ is the term frequency of the ith indexing term in document j.

3.2 Reuters Test Corpus

In order to measure its performance of our system, we tested the system on a standard test collection designed for text categorization used in the literature. This collection is known as Reuters-21578. We chose 1600 documents belonging to the Reuters data set with ten frequent categories. 600 documents were used for training and 1000 documents for testing.

After word stemming, we merged the sets of stems from each of the 600 training documents and removed the duplicates. This resulted in a set of 6122 indexing terms in the vocabulary.

In order to create the set of initial feature vectors to represent the 600 training documents, we measured the term weight for each of the 6122 indexing terms. The feature vectors were then formed from the term weights, and each of the feature vectors was of the form,

$$D_j = \langle W_{1,j}, W_{2,j}, \dots\dots\dots W_{6122,j} \rangle \tag{8}$$

For each training and testing document, we created the feature vectors corresponding to the 600 training documents, where each feature vector had a dimensionality of 6122.

3.3 Dimensional Reductions

The main difficulty in the application of a neural network to text categorization is the high dimensionality of the input feature space which is typical for textual data. This is because each unique term in the vocabulary represents one dimension in the feature space, so that the size of the input of the neural network depends upon the number of stemmed-words. In [12], the authors introduce four kinds of methods designed to reduce the dimensional of the feature space. Based on our experience, we reduced this size by choosing the highest term weights. We chose 1000 terms as the neural network's input since it offers a reasonable reduction neither too specific nor too general.

3.4 Experimental Results

The number of output nodes is equal to the number of pre-defined categories. In order to determine the number of the hidden nodes, we used the following rule [13],

$$\log_2^n < n_1 < \sqrt{n+m} + a \tag{9}$$

where n_1 is the number of hidden nodes; n is the number of input nodes; m is the number of output nodes and a is a constant ranging from (1, 10). In our experiments, the number of hidden nodes is 15 referred by the above rule. Our network has three layers consisting of 1000, 15 and 10 nodes respectively.

All of our experiments were conducted on a Pentium personal computer. We compared the mean absolute error using three different methods. The first method is the conventional BP network, which we refer to as the traditional BPNN. The second method is the commonly used improved method, which we call the Modified BPNN. The third method is our proposed method, which we call the MRBP network.

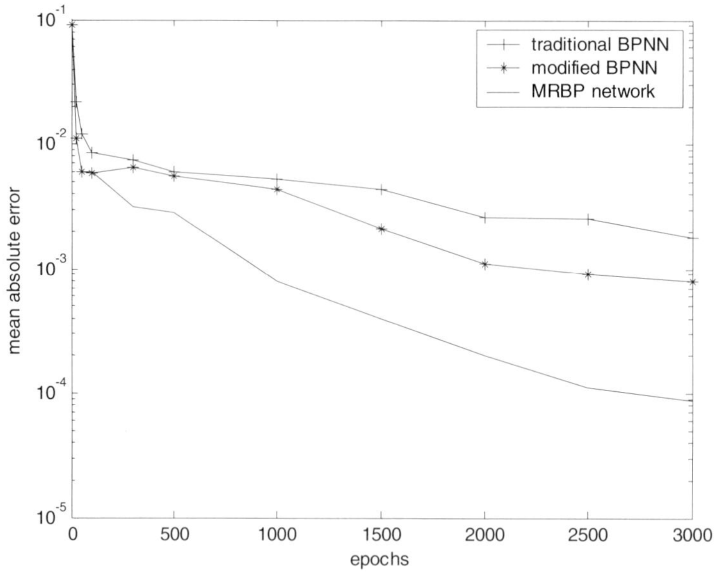

Fig. 2. Mean absolute error reduction during training with the three methods

In the case of the traditional BPNN, we can see in Fig. 2 that, at the beginning of the training phase, the error is reduced very rapidly. But this reduction slows down after a certain number of epochs, and then levels off. The Modified BPNN is 2-3 times faster than the first method at the beginning of the training phase. It slows down when the error reduction reaches a certain value and then fluctuates. However, in the case of our method, no morbidity neurons are produced in the first learning phase, with the result that the next learning phase is very similar to that of the Modified BPNN. However, from the third learning phase, our method progresses more rapidly than the Modified BPNN. It also has a good tendency in the later part of the training.

4 Evaluations

The performance of text categorization systems can be evaluated based on their categorization effectiveness

The effectiveness measures of recall, precision and F-measure are defined as [14]:

$$\text{Precision} = \frac{\text{Number of test set category members assigned to category}}{\text{Total number of test set members assigned to category}} \tag{10}$$

$$\text{Recall} = \frac{\text{Number of test set category members assigned to category}}{\text{Number of category members in test set}} \tag{11}$$

$$\text{F-measure} = \frac{2 \times (\text{Precision} \times \text{Re}\,call)}{\text{Precision} + \text{Re}\,call} \tag{12}$$

We used the macro-average method to obtain the average value of the precision and recall. The F-measure is based on the micro-average value. The performance results are given in table1.

Table 1. Comparison of the performances of the three kinds of networks

Category	Traditional BPNN		Modified BPNN		MRBP Networks	
	Preci-sion	Recall	Preci-sion	Recall	Preci-sion	Recall
Money-supply	0.832	0.913	0.913	0.916	0.938	0.946
coffee	0.847	0.901	0.882	0.900	0.929	0.933
gold	0.943	0.914	0.944	1.000	0.955	1.000
sugar	0.952	0.884	0.954	0.883	0.927	0.914
trade	0.725	0.786	0.766	0.836	0.824	0.895
crude	0.944	0.896	0.945	0.916	1.000	0.932
grain	0.937	0.928	0.928	0.934	0.948	0.924
Money-fx	0.890	0.855	0.908	0.877	0.918	0.912
Acq	0.937	0.873	0.943	0.893	0.943	0.903
earn	0.932	0.923	0.957	0.934	0.967	0.952
micro-average	0.892	0.887	0.914	0.908	0.935	0.931
F-measure	0.889		0.911		0.933	

The size of the network and some parameters used in our experiments are given in table 2.

Table 2. The network size and parameters

#Input Nodes	#Hidden Nodes	#Output Nodes	Learning Rate	Momentum
1000	15	10	0.01	0.8

5 Conclusion and Future Works

This paper proposes an algorithm for text categorization using an improved Back-propagation neural network. The MRBP detects and rectifies the morbidity neuron in each learning phase. This method overcomes the network paralysis problem and has good ability to escape from the local minima. The speed of the convergence is also increased and the results of our experiments show that the proposed method outperforms both the traditional BPNN and the commonly used improved BPNN. The superiority of the MRBP is obvious, especially when the size of the networks is large.

Even though the MRBP does not solve the problems associated with the structure of the BPNN, it provides a rule for adjusting the neurons and training the network effectively. In a future study, we intend to analyze and generalize the previous training as a direction for the next trainings.

References

1. Yang, Y. and Liu, X. A Re-examination of Text Categorization Methods. In Proceedings of SIGIR-99, 22nd ACM International Conference on Research and Development in Information Retrieval, (1999) 42-49
2. Mitchell, T.M. Machine Learning. McGraw Hill, New York, NY, (1996)
3. Rocchio, Jr. J. JRelevance Feedback in Information Retrieval. The SMART Retrieval System: Experiments in Automatic Document Processing, editor: Gerard Salton, Prentice-Hall, Inc., Englewood Cliffs, News Jersey, (1971)
4. Joachims, T. A Probabilistic Analysis of the Rocchio Algorithm with TFIDF for Test Categorization. In Proceedings of ICML-97, 14th International Conference on Machine Learning, pages 143-151, (1997)
5. Cohen, W. W. and Singer, Y. Context–Sensitive Learning Methods for Text Categorization. ACM Trans. Inform. Syst. 17, 2, (1999) 141-173
6. Ruiz, M. E. and Srinivasan, P. Hierarchical Neural Networks for Text Categorization. In Proceedings of SIGIR-99, 22nd ACM International Information Retrieval, (1999) 281-282
7. David A. Grossman, Ophir Frieder Information Retrieval: Algorithms and Heuristics Kluwer Academic Publishers, (2000)
8. Wei Wu, Guorui Feng, Zhengxue Li, and Yuesheng Xu Deterministic Convergence of an Online Gradient Method for BP Neural Networks IEEE TRANSACTIONS ON NEURAL NETWORKS, VOL. 16, NO. 3, (2005)
9. P.D.Wasserman.Neural Computing: Theory and Practice [M].New York: Van Nostrand Reinhold. (1989)
10. V.P. Plagianakos, M.N. Vrahatis, Training Neural Networks with Threshold Activation Functions and Constrained Integer Weights, *ijcnn*, p.5161, IEEE-INNS-ENNS International Joint Conference on Neural Networks (IJCNN'00)-Volume 5,(2000)
11. M. F. Porter. An algorithm for suffix stripping. Program, Vol.14 no. 3 (1980) 130-137
12. Savio L.Y. Lam, Dik Lun Lee, Feature Reduction for Neural Network Based Text Categorization, 6th International Conference on Database Systems for Advanced Applications (DASFAA '99), (1999) 195
13. Liming, Zhang. Models and applications of Artificial Neural Networks. Fudan University. Shanghai. (1993) 50
14. David D. Lewis and William A. Gale, A Sequential Algorithm for Training Text Classifiers SIGIR '94 Proceedings of the 17th Annual International ACM SIGIR Conference, London, (1994) 3-12

Image Retrieval Based on Similarity Score Fusion from Feature Similarity Ranking Lists

Mladen Jović, Yutaka Hatakeyama, Fangyan Dong, and Kaoru Hirota

Dept. of Computational Intelligence and Systems Science
Interdisciplinary Graduate School of Science and Engineering
Tokyo Institute of Technology
G3-49, 4259 Nagatsuta, Midori-ward, Yokohama 226-8502, Japan
{jovic, hatake, tou, hirota}@hrt.dis.titech.ac.jp

Abstract. An image similarity method based on the fusion of similarity scores of feature similarity ranking lists is proposed. It takes an advantage of combining the similarity value scores of all feature types representing the image content by means of different integration algorithms when computing the image similarity. Three fusion algorithms for the purpose of fusing image feature similarity scores from the feature similarity ranking lists are proposed. Image retrieval experimental results of the evaluation on four general purpose image databases with 4,444 images classified into 150 semantic categories reveal that a proposed method results in the best overall retrieval performance in comparison to the methods employing single feature similarity lists when determining image similarity with an average retrieval precision higher about 15%. Compared to two well-known image retrieval system, SIMPLicity and WBIIS, the proposed method brings an increase of 4% and 27% respectively in average retrieval precision. The proposed method based on multiple criteria thus provides better approximation of the user's similarity criteria when modeling image similarity.

1 Introduction: Image Similarity Computation

One of the most important issues in the present image retrieval is modeling image similarity [1], [12]. Image is typically modeled as a collection of low-level image features [11], [1]. Image similarity computation involves the application of different *feature similarity measures* on the extracted *image features*. Based on the overall image similarity to the (user's) query image, the database images are ranked in a *single* similarity ranking list. Finally, the most similar images from the similarity ranking list are presented to the end user [14]. In the recent survey of 56 Content Based Image Retrieval (CBIR) systems [13], most of the systems are employing a *single* similarity values ranking list.

As an alternative approach to this one, the application of the three different feature similarity rank's score fusion algorithms(cf. 2) when ranking overall image similarity in terms of partial *feature similarities* without using human relevance judgments is approached. The **focus** on *how* the creation of the final similarity values ranking list between a query image q and database images from

L. Wang et al. (Eds.): FSKD 2006, LNAI 4223, pp. 461–470, 2006.

the low-level image feature's similarity ranking lists is emphasized. Combining different feature similarity score ranking using data fusion methods based on multiple criteria in a *rank aggregation* [15] manner is done. Fusion of feature similarity value scores in case of two algorithms is derived in a non-heuristical manner. The third approach is done in a heuristical manner. The empirical evaluation is done on four general purpose image databases containing 4,444 images in 150 semantic categories. In total, more than 66,000 queries are executed, based on which several performance measures are computed.

In 2, proposed feature similarity ranking lists fusion algorithms are described. An empirical evaluation and comparison of the proposed algorithms is demonstrated in 3.

2 Feature Similarity Ranking Score Fusion Algorithms

The employment of a *single* similarity ranking list is done by most of the CBIR systems surveyed in [13](Fig. 1(a)). As an alternative to this, a calculation of the overall image similarity between the query image q and all database images based on a *multi* feature similarity ranking lists is proposed. Image similarity computation involves the application of different *feature similarity measures* on the extracted *image features*. Based on the feature similarity values(feature similarities) between query image q and all database images, with a help of image retrieval techniques, the *feature similarity ranking lists* are determined. When calculating the *feature similarity ranking lists*, a *ranking position*(feature similarity scores) of each database image i with respect to the query image q is

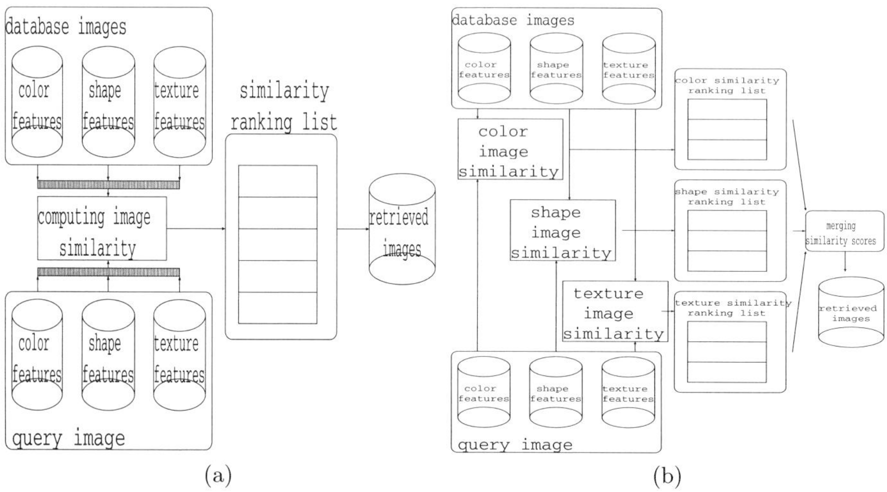

Fig. 1. (a) Computing the overall image similarity based on low-level image features in most of the CBIR systems surveyed in [13]; **(b)** A new approach of computing image similarity based on low-level image features by employing multi feature similarity ranking lists

uniquely determined. Number of features used for the image representation will determine the number of *feature similarity lists*. Integrated ranking of the *multi* feature similarity ranking lists is then determined by the the fusion algorithms. Such a framework is illustrated on the Fig. 1(b). The fusion is done in such a way to optimize retrieval performance. Three feature similarity score fusion algorithms are proposed. As for the image feature representation, color, shape and texture image features are chosen. Color feature is represented by the *color moments* [3]. Shape feature is represented by the *edge–direction histogram* [4]. Texture feature is represented by the *texture neighborhood* [9]. Color feature similarity in is measured by the *weighted Euclidean* distance [2], while shape and texture feature similarity are measured with a help of *city-block* distance.

Let us for a given query image q, with respect to all database images, define three feature similarity ranking lists: *color feature similarity ranking list* (**CFSRL**), *shape feature similarity ranking list*(**SFSRL**) and *texture feature similarity ranking list*(**TFSRL**). Next, let us assume that at **CFSRL**, **SFSRL** and **TFSRL** top five positions, the images with identifiers {a, b, c, d, e} are ordered as following:

$$\mathrm{CFSRL} = (\mathrm{a,b,c,e,d})\,; \quad \mathrm{SFSRL} = (\mathrm{d,a,c,e,b})\,; \quad \mathrm{TFSRL} = (\mathrm{b,a,c,e,d})\,. \tag{1}$$

Inverse Rank Position Algorithm(IRP) is a first algorithm to merge the *multi* feature similarity lists into a *single* overall similarity ranking list. The inverse of the sum of inverses of the feature similarity rank scores for each individual feature for a given image from relevant feature similarity ranking lists is used(3).

$$\mathrm{IRP(q,i)} = \frac{1}{\sum_{\text{feature similarity}=1}^{n} \frac{1}{\text{rank position}_{\text{feature similarity}}}}\,. \tag{2}$$

$$\text{feature similarity} \in \{\mathrm{CFSRL, SFSRL, TFSRL}\}; \quad \mathrm{i} \in \{\mathrm{a,b,c,d,e}\};\ n = 3. \tag{3}$$

Example. According to the sample feature similarity ranking lists given in 2, the overall similarity ranking of the images {a, b, c, d, e} with respect to the query image q is calculated as following:

$$\mathrm{IRP}\,(a) = \frac{1}{2}; \quad \mathrm{IRP}\,(b) = \frac{10}{19}; \quad \mathrm{IRP}\,(c) = 1; \quad \mathrm{IRP}\,(d) = \frac{5}{7}; \quad \mathrm{IRP}\,(e) = \frac{4}{3}. \tag{4}$$

$\Longrightarrow$ e > c > d > b > a, meaning that image a is the most relevant image to the query q, image b is the next relevant etc(Fig. 2).

Borda Count Algorithm(BC) taken from social theory in voting [16] is a second algorithm to merge the *multi* feature similarity lists into a final overall similarity ranking list. An image with the highest rank on each of the feature similarity ranking lists (in an n–way vote) gets n votes. Each subsequent image gets one vote less (so that the number two gets *n-1* votes, number three *n-2*

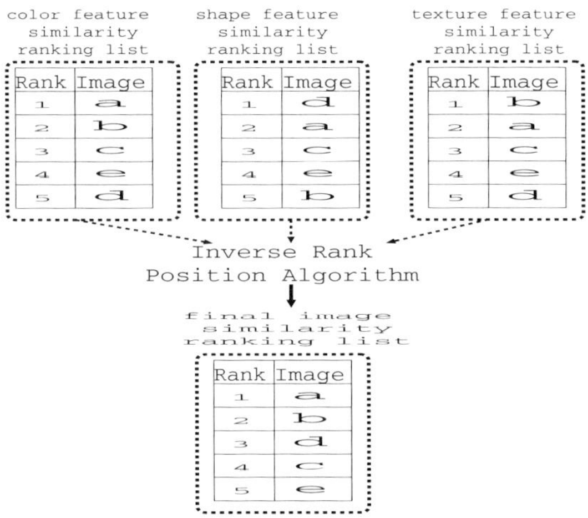

Fig. 2. Ordering of the first five retrieved images based on the color-shape-texture features merged by Inverse Rank Position Algorithm

votes etc.). Finally, for each database image, all the votes from all of the three feature similarity ranking lists are summed up and the image with the highest number of votes is ranked as the most relevant to the query image, winning the election.

$$\mathrm{BC(q,i)} = \sum_{\text{feature similarity}=1}^{n} \text{rank position}_{\text{feature similarity}}. \tag{5}$$

$$\text{feature similarity} \in \{\mathrm{CFSRL}, \mathrm{SFSRL}, \mathrm{TFSRL}\}; \quad \mathrm{i} \in \{\mathrm{a,b,c,d,e}\};\ n = 3. \tag{6}$$

Example. According to the sample feature similarity ranking lists given in 2, the overall similarity ranking of the images {a, b, c, d, e} with respect to the query image q is calculated as following:

$$\mathrm{BC}\,(a) = 5; \quad \mathrm{BC}\,(b) = 8; \quad \mathrm{BC}\,(c) = 9; \quad \mathrm{BC}\,(d) = 11; \quad \mathrm{BC}\,(e) = 12. \tag{7}$$

$\Longrightarrow$ e > d > c > b > a, meaning that image a is the most relevant image to the query q, image b is the next relevant etc(Fig. 3).

Leave Out Algorithm(LO) is a third algorithm to merge the *multi* feature similarity lists into a single overall similarity ranking list. The elements are inserted into the final similarity ranking list circularly from three feature similarity ranking lists(see Algorithm 1). Repeating elements from feature similarity ranking lists are not inserted into the final similarity ranking list if already appeared there. Order of the next selected element from the feature similarity ranking lists to be inserted into the final similarity ranking list can be arbitrary and will therefore influence on the retrieval precision. In the experimental part, the

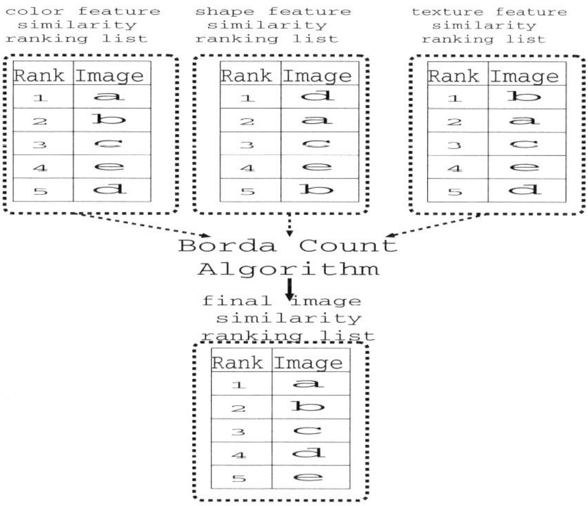

Fig. 3. Ordering of the first five retrieved images based on the color-shape-texture features merged by Borda Count Algorithm

---

Algorithm 1. Leave Out Algorithm

Input: Q, I : images (e.g., a query image and a database image); $\{Q[q] : 1 \le q \le N\}$, $\{I[i] : 1 \le i \le N\}$, respectively; feature similarity ranking lists: *feature similarity* $\in \{\text{CFSRL}, \text{SFSRL}, \text{TFSRL}\}$;

Output: overall image similarity ranking list for $\{(Q[q], I[i]) : 1 \le q, i \le N\}$;

for $(q \leftarrow 1$ **to** $N)$ **do**
  for $(i \leftarrow 1$ **to** $N)$ **do** {3 feature similarity values computed for a pair of images }
    compute Feature Similarities$(Q[q], I[i])$;
  end for
end for
for $(iteration \leftarrow 1$ **to** $N)$ **do**
  get the image[I] from **CFSRL** with highest rank $\notin$ {final similarity list}
  image[I] $\leftarrow$ insert final image similarity list;
  get the image[I] from **SFSRL** with highest rank $\notin$ {final similarity list}
  image[I] $\leftarrow$ insert final image similarity list;
  get the image[I] from **TFSRL** with highest rank $\notin$ {final similarity list}
  image[I] $\leftarrow$ insert final image similarity list;
end for

---

order *CFSRL*, *SFSRL*, *TFSRL* is chosen, as comparing to the other permutations of the feature similarity ranking lists when employing Leave Out Algorithm it provides the optimal retrieval precision. Therefore, this ranking score fusion algorithm is rather heuristical compared to the previous two. In such a way, in each of the similarity score merging iteration, only one element is inserted from each feature similarity list, as illustrated on Fig. 4.

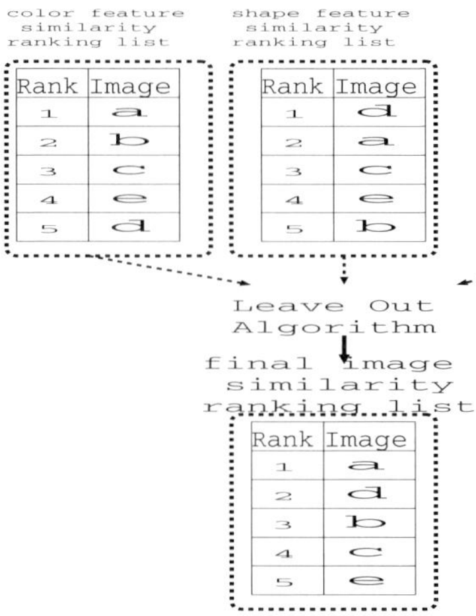

Fig. 4. An example of the ordering of the first five retrieved images based on the color-shape-texture features merged by Leave Out Algorithm

Example. According to the sample feature similarity ranking lists given in 2, the overall similarity ranking of the images {a, b, c, d, e} with respect to the query image q is calculated as following:

$$\mathrm{LO}\,(iter\#1) - > \mathrm{a}; \quad \mathrm{LO}\,(iter\#2) - > \mathrm{d}; \quad \mathrm{LO}\,(iter\#3) - > \mathrm{b}; \tag{8}$$

$$\mathrm{LO}\,(iter\#4) - -\mathrm{c}; \quad \mathrm{LO}\,(iter\#5) - -\mathrm{e}; \tag{9}$$

$\Longrightarrow$ a < d < b < c < e, meaning that image a is the most relevant image to the query q, image d is the next relevant etc(Fig. 4).

3 Experimental Evaluation

All the experiments are performed on AMD Athlon Processor Machine 64-bit Processor Machine, with 1 GByte RAM Memory. Four standard test databases are used when conducting experiments, containing 4,444 images, divided into 150 semantic categories **C-1000-A database** [5]; **C-1000-B database** [5]; **V-668 database** [8] and **B-1776 database** [10]. All the four test databases originate from the well-known image collections, used for the evaluation of the image retrieval systems [6], [7]. Partitioning of each database into semantic categories is determined by the creators of the database, and reflects the human perception of image similarity. The semantic categories define the **ground truth**. For a given query image, the relevant images are considered only those belonging to the same semantic category as the query image. This implies that the number of relevant images for a given query image equals the number of images in the category to which that image belongs. The performance measures are: (1) precision –*[P.]*, (2) weighted precision–*[W. P.]* and (3) average rank–*[A.R.]*.

These the most frequently used measures of the image retrieval performance [1]. All the performance measures are computed for each query image, based on the given ground truth. Since each image in each test database is used as a query, all the performance measures are averaged for each test database. For each algorithm, average values of retrieval precision (P.), weighted precision (W.P.) and average rank (A.R.) are provided in Table 1 and Table 2. For a given query image, precision is computed as the fraction of the relevant images that are retrieved. Weighted precision is computed in a similar way, however, the higher a relevant image is ranked, the more it contributes to the overall weighted precision value [7]. This means that, unlike the precision, weighted precision takes into account the rank of the retrieved relevant images as well. Average rank is simply the average of the rank values for the relevant images. In addition to the *global* image representation, five region-based image similarity representations are also experimentally evaluated. Image is initially divided into $N \times N, (N \in \{1, 2, 3, 4, 5\})$ non-overlapping image regions. From each region color, shape and texture feature are extracted, as described in 2. Therefore, each resolution is uniquely determined by N, that is the number of image regions.

3.1 Experiment Results and Discussion

Initially, three proposed fusion algorithms are compared to each other. Next, a comparison to six conventional image similarity models employing a *single* overall similarity ranking list of color, shape, texture, color-shape, color-texture and shape-texture image features, respectively, is done. As among image similarity

Table 1. Comparison of the **IRP**, **BC** and **LO** algorithms on C-1000-A, C-1000-B, B-1776 and V-668 test databases in the resolutions providing optimal retrieval performance(5×5 resolution, except for the **LO** algorithm on B-1776 and V-668 test databases – where 1×1 resolution provides the optimal performance)

	C-1000-A			*C-1000-B*			*B-1776*			*V-668*		
	P.[%]	*W.P.[%]*	*A.R.*	*P.[%]*	*W.P.[%]*	*A.R.*	*P.[%]*	*W.P.[%]*	*A.R.*	*P.[%]*	*W.P.[%]*	*A.R.*
IRP	49.04	61.72	203.8	39.13	52.88	246.0	73.89	85.69	40.06	41.23	63.4	162.6
BC	44.34	56.5	217.9	36.63	48.42	252.5	69.12	82.00	46.0	38.04	58.28	170.2
LO	20.18	31.57	435.0	18.09	26.57	449.7	30.30	45.56	158.0	20.98	38.02	256.6

Table 2. Comparison of the average retrieval precision to the system employing *single* overall similarity ranking lists: color-texture image feature combination (*Resolution 5*) as well as two advanced well known image retrieval systems: SIMPLicity and WBIIS

Image Database	*Number of Images*	*Color Texture*	*SIMPLicity*	*WBIIS*	*IR*	*BC*	*LO*
C-1000-A	1000	43.00	45.3	22.6	49.04	44.34	20.18
C-1000-B	1000	22.05	–	–	39.13	36.63	18.09
B-1776	1776	50.09	–	–	73.89	69.12	30.3
V-668	668	25.01	–	–	41.23	38.04	20.98

models mentioned color-texture feature based image similarity model performed optimally, only results for this model are reported. Next, proposed fusion algorithms are compared to two representative well-known image retrieval systems, WBIIS and SIMPLicity, also employing a *single* similarity ranking list. Finally the conclusions are drawn.

Evaluation of the Fusion Algorithms. A shown in Tables 1 and 2, with respect to the retrieval effectiveness measured by average retrieval precision, the Inverse Rank Position algorithm performs optimally on all databases. The difference between Inverse Rank Position and Borda Count algorithm is smaller than the difference between Borda Count and Leave Out algorithms. As seen in Tables 1 and 2, the retrieval performance of the Borda Count is about twice higher than Leave Out algorithm, independently from the database. With respect to the region-based modeling, in most cases, partitioning the image into more regions(*Resolution 5*) improves the retrieval performance in case of all fusion algorithms on all databases, according to the expectations. However, this is not the case for the Brodatz-1776 and Vistex-668 databases, where in case of *Leave Out* algorithm global image representation provides the optimal retrieval performance. Comparing Inverse Rank Position and Leave Out algorithms, the highest difference in average precision is in the case of Brodatz-1776 database, reaching the difference about 43%. On other three databases, the difference in average precision is about 25%. Next, as shown in Table 2, when compared to the best out of the six conventional image similarity methods– employing color-texture image features – while Inverse Rank Position and Borda Count have higher average retrieval precision on all databases, this is not the case for Leave Out algorithm. This fact might be explained to the heuristical nature of feature similarity ranking lists fusion into the final similarity ranking list. Finally, as shown in Table 2, on C-1000-A testing database, when compared to two state-of-art CBIR systems, Inverse Rank Position algorithm reaches the values of SIMPLicity system, while WBIIS lags behind both Inverse Rank Position and Borda Count algorithms. Among three proposed algorithms, Inverse Ranking provides the optimal retrieval performance. Possible explanation is that equal emphasis is put on any of the three image features – as important image similarity modeling elements, compared to other two algorithms, which are more heuristics based. In particular to the Leave Out algorithm, which in some cases might be only based either on color, or shape or texture image feature.

4 Conclusion

Image similarity model based on the fusion of the feature similarity ranking list scores is proposed. It takes an advantage of combining them by means of the data fusion algorithms when computing the overall image similarity. Three fusion algorithms are proposed for this reason. The evaluation on the four test databases, containing 4,444 images, in 150 semantic categories is done, based on which the (weighted) precision and average rank are computed.

Effectiveness of the three fusion algorithms measured by the retrieval performance is compared to the conventional systems using a *single* similarity ranking list. In addition, the effectiveness of two advanced CBIR systems(SIMPLicity and WBIIS) is compared, too. As shown in the experiments, data fusion methods based on the *multi* feature similarity lists provides better approximation of the user's similarity criteria than a *single* feature similarity list. The out performance is with an average retrieval precision higher about 15%. Compared to SIMPLicity and WBIIS, this is also the case. Possible explanation of this is that the *combining different rankings using data fusion methods in general based on* **multi** *criteria(feature similarity lists) provides better approximation of the user's similarity criteria than a single feature similarity list.* Thus the assessment of such approach in ranking image similarity in terms of partial *feature similarities* without using human relevance judgments should be carefully considered when modeling image similarity.

Reported improvements might be of the significance for the various application domains covering different image domain(s). The experimental results are only reported for images covering (1) color homogeneous structures and (2) unconstrained color photographs image domains. Large variance among the visual characteristics of the images in all testing databases allows for the general conclusions about the performance of the proposed algorithms. However, the applicability is not strictly concerned to two above mentioned domains. Validity of the results is also applicable to any databases containing images with large variance among the visual characteristics. These could be e.g. medical image or finger prints databases. Additionally, an investigation of extending testing data sets from the image to video data are already going on.

Acknowledgments

The authors would like to thank Dr Zoran Stejić, from Ricoh Co., Ltd., Japan and Thomas Seidl from RWTH Aachen University, for their comments, constructive suggestions and valuable research discussions on content based image retrieval as well as the source codes of the image features.

References

1. A. W. M. Smeulders, M. Worring, S. Santini, A. Gupta, R. Jain, Content-based image retrieval at the end of the early years, in: IEEE Transactions in Pattern Analysis and Machine Intelligence, 22(12), (2000) 1349–1380.
2. M. Stricker, M. Orengo, Similarity of color images, in: Storage and Retrieval for Image and Video Databases, Proc. SPIE 2420, (1995) 381-392.
3. M. Stricker, M. Orengo, Similarity of color images, in: Proc. of IS&T and SPIE Storage and Retrieval of Image and Video Databases III, San Jose, CA, USA, (1995) 381-392.
4. S. Brandt, J. Laaksonen, E. Oja, Statistical shape features in content-based image retrieval. in: Proc. of 15th Int. Conf. on Pattern Recognition (ICPR-2000), Vol. 2. Barcelona, Spain, (2000) 1066-1069.

5. Corel Corporation, Corel Gallery 3.0., 2000 Available: http://www3.corel.com/
6. Z. Stejić, Y. Takama, K. Hirota, Genetic algorithm-based relevance feedback for image retrieval using Local Similarity Patterns. in: Information Processing and Management, 39(1), (2003) 1-23.
7. J. Z. Wang, J. Li, G. Wiederhold, SIMPLIcity: Semantics-sensitive Integrated Matching for Picture LIbraries, in: IEEE Transactions on Pattern Analysis and Machine Intelligence, 23(9), (2001) 947-963.
8. Massachusetts Institute of Technology, Media Lab, Vision Texture Database, 2001, Available: ftp://whitechapel.media.mit.edu/pub/VisTex/.
9. J. Laaksonen, E. Oja, M. Koskela, S. Brandt, Analyzing low-level visual features using content-based image retrieval, in: Proc. 7th Int. Conf. on Neural Information Processing (ICONIP'00), Taejon, Korea, (2000) 1333-1338.
10. P. Brodatz, Textures: a photographic album for artists and designers, New York: Dover Publications, (1966) 40-46.
11. J.M. Jolion, Feature Similarity, in: M.S. Lew (Ed.), Principles of Visual Information Retrieval, Springer, London, (2001) 121-143.
12. V. Castelli, L.D. Bergman: Digital imagery: fundamentals, in V. Castelli, L.D. Bergman(Eds.), Image Databases: Search and Retrieval of Digital Imagery, Wiley, New York, USA, (2002) 1-10.
13. R.C. Veltkamp, M. Tanase Content-Based Image Retrieval Systems: A Survey Technical Report UU-CS-2000-34, (2000).
14. Y. Rui, T.S. Huang, M. Ortega, S. Mehrotra: Relevance feedback: a power tool for interactive content-based image retrieval in: *IEEE Trans. on Circuits Syst. Video Technol.* 8 (5), (1998) 664 – 655.
15. Dwork, C., Kumar, R., Naor, M., Sivakumar, D.: Rank aggregation methods for the Web, in: *Proceedings of 10th International World Wide Web conference* , Hong Kong, (2001) 613 – 622.
16. Roberts, F. S. *Discrete mathematical models with applications to social, biological, and environmental problems*, Englewood Cliffs, NJ: Prentice Hall

A Novel Feature Weighted Clustering Algorithm Based on Rough Sets for Shot Boundary Detection*

Bing Han, Xinbo Gao, and Hongbing Ji

School of Electronic Engineering, Xidian Univ., Xi'an 710071, China
hanbing@lab202.xidian.edu.cn, xbgao@lab202.xidian.edu.cn,
hbji@xidian.edu.cn

Abstract. Shot boundary detection as the crucial step attracts much more research interests in recent years. To partition news video into shots, many metrics were constructed to measure the similarity among video frames based on all the available video features. However, too many features will reduce the efficiency of the shot boundary detection. Therefore, it is necessary to perform feature reduction before shot boundary detection. For this purpose, the classification method based on clustering algorithm of Variable Precision Rough-Fuzzy Sets and Variable Precision Rough Sets for feature reduction and feature weighting is proposed. According to the particularity of news scenes, shot transition can be divided into three types: *cut transition*, *gradual transition* and *no transition*. The efficiency of the proposed method is extensively tested on UCI data sets and more than 3 ***h*** of news programs and 96.2% recall with 96.3% precision have been achieved.

1 Introduction

With the increasing proliferation of digital video contents, efficient techniques for analysis, indexing, and retrieval of videos according to their contents have become important. A common first step for most content-based video analysis techniques available is to segment a video into elementary shots, each comprising a continuous in time and space. These elementary shots are composed to form a video sequence during video sorting or editing with either cut transitions or gradual transitions of visual effects such as fades, dissolves, and wipes.

In recent years, a large number of metrics have been proposed to segment a video into shots by measuring the dissimilarity, or distance, between two or a short sequence of adjacent frames [1-3]. These metrics make use of such frames or video features as pixel values, statistic features, intensity and color histogram and *etc*. If the measured dissimilarity is greater than some predetermined threshold, the shot boundary is assumed. How to adequately use features available is becoming the hot topic on shot boundary detection to improve the detection efficiency with keeping the detection accuracy, it is necessary to perform feature reduction of shot boundary.

* This work was supported by the program for New Century Excellent Talents in University of China(NCET-04-0948), National Natural Science Foundation of China (No.60202004) and the Key Project of Chinese Ministry of Education (No.104173).

L. Wang et al. (Eds.): FSKD 2006, LNAI 4223, pp. 471–480, 2006.

Nowadays, some researchers use the Rough Sets theory and fuzzy clustering to reduct feature and detect shot boundary [3]. But, these method is efficient for disperse or categorical data. For the video data during shot boundary detection, the feature value is continuous. So, the first step is to disperse the data. Therefore, the performance of disperse affects directly the efficiency of the method. To this end, a novel fuzzy *c*-prototype algorithm based on the Variable Precision Rough-Fuzzy Sets (VPFRS) and Variable Precision Rough Sets (VPRS), which is called as Hybrid Rough Sets algorithm based clustering method (HRS-FWFCP), to deal with categorical, numerical and mixed data. The most important difference between the proposed algorithm and the method in reference [3] is the Rough-Fuzzy Sets is syncretized into clustering algorithm and the feature attribute is weighted in new method. Li[4] proposed a feature weighted clustering method based on ReliefF algorithm. Since this method chooses the samples at random, the optimal weights are obtained by multi-step iteration. However, new algorithm need not extract the samples. So, the optimal weights can be achieved by only once iteration and the computational complexity is decreased. In addition to, this method can deal with continuous attribute value, which is suitable for video data.

2 Fuzzy *c*-Prototypes Algorithm and Rough Sets

The rough sets theory introduced by Pawlak in the early 1980s[5] is an effective mathematical analysis tool to deal with vagueness and uncertainty.

Let R be an equivalence relation on a universal set X. Moreover, let X/R denote the family of all equivalence classes introduced on X by R. One such equivalence class in X/R, which contains $x \in X$, is designated by $[x]_R$. For any output class $A \subseteq X$, we can define the lower and upper approximations, denoted as $\underline{R}(A)$ and $\overline{R}(A)$, which approach A as closely as possibly from inside and outside respectively.

Definition 1: Let S and Q be the attribute sets (such as condition attributes and decision attributes). $S \subseteq R, Q \subseteq R : U/S = \{X_1, X_2, \cdots, X_l\}$, $U/Q = \{Y_1, Y_2, \cdots, Y_l\}$. We say that Q depends on S in the degree of k on S if

$$k = \gamma_S(Q) = |\mathrm{POS}_S(Q)| \big/ |U| = \bigcup_{X \in U/IND(Q)} \underline{S}(X) . \quad (1)$$

where $|\cdot|$ denotes the cardinality of a set. Thus, the coefficient k expresses the ratio of all elements of the universe which can be properly classified into blocks of the partition $U/I(Q)$, employing attributes S. It can be dealt with consistency of information.

Definition 2[6]: Let X be a set, R be an equivalence relation defined on X and the output class A be a fuzzy set. A rough-fuzzy set is a tuple ($\underline{A}$, $\overline{A}$) where the lower approximation $\underline{A}$ and the upper approximation $\overline{A}$ of A are fuzzy sets of X/R, with membership functions defined by

$$\underline{A}(x) = \inf\{A(y) \mid y \in [x]_R\}, \quad x \in U \quad ; \quad \overline{A}(x) = \sup\{A(y) \mid y \in [x]_R\}, \quad x \in U \quad (2)$$

Definition 3: Let X be a set, R be an equivalence relation defined on X and the output class A be a fuzzy set. A rough-fuzzy set is a tuple $(\underline{A}_\alpha, \overline{A}_\beta)$ where the lower approximation $\underline{A}_\alpha$ and the upper approximation $\overline{A}_\beta$ of A are fuzzy sets of X/R depending coefficients $0 \le \beta \le \alpha \le 1$, with membership functions defined by

$$\underline{A}_\alpha = \{x \in U \mid \underline{A}(x) \ge \alpha\} \ ; \ \overline{A}_\beta = \{x \in U \mid \underline{A}(x) \ge \beta\}. \tag{3}$$

Definition 4: Let X be a set, R be an equivalence relation defined on X, $A \in F(U)$, the definition of class accuracy $\eta_R(A)$ of A is

$$\eta_R(A) = {|\underline{A}_\alpha|} \Big/ {|\overline{A}_\beta|}. \tag{4}$$

where $|\underline{A}_\beta| = 0$, then $\eta_R(A) = 0$.

3 Feature Weighted Clustering Algorithm Based on VPFRS and VPRS(Hybrid Rough Sets: HRS)

Corresponding to the fuzzy c-prototypes algorithm(FCP), the feature-weighted FCP (FWFCP) algorithm also represents the cluster analysis as the following mathematical programming problem.

Let $X = \{x_1, x_2, \cdots, x_n\}$ be a given set of objects to be clustering processed, and $x_i = [x_i^r, x_i^c]^T$ denotes the m features of the i-th object (sample), in which $x_i^r = [x_{i1}^r, \cdots, x_{it}^r]$ indicates the numerical features and $x_i^c = [x_{i,t+1}^c, \cdots, x_{i,m}^c]$ stands for the categorical features. Let $P = \{p_1, p_2, \cdots, p_c\}$, $p_i = [p_{i1}^n, \cdots, p_{it}^n, p_{i,t+1}^c, \cdots, p_{im}^c]^T$ represent the prototype of the i-th class. The fuzzy c-prototypes objective function is modified as [4]

$$J(W,P) = \sum_{i=1}^{c} \Big(\sum_{j=1}^{n} \mu_{ij}^2 \sum_{l=1}^{t} |x_{jl}^r - p_{jl}^r|^2 + \lambda \sum_{j=1}^{n} \mu_{ij}^2 \sum_{l=1}^{t} \delta(x_{jl}^c, p_{jl}^c) \Big). \tag{5}$$

In the right hand of (5), the first term is the squared Euclidean distance in numerical feature space, and the second term is a simple dissimilarity matching measurement. Here $\delta(\cdot)$ is defined as

$$\delta(a,b) = \begin{cases} 0 & a = b \\ 1 & a \ne b \end{cases}. \tag{6}$$

Let ω_l denote the weights for numerical features and σ_l stand for the weights for categorical features. Then the clustering objective function is modified as

$$J(W,P) = \sum_{l=1}^{c} \Big(\sum_{i=1}^{n} \mu_{li}^2 \sum_{j=1}^{m} \omega_l |x_{jl}^r - p_{jl}^r|^2 + \sum_{i=1}^{n} \mu_{li}^2 \sum_{j=1}^{m} \sigma_l \delta(x_{jl}^c, p_{jl}^c) \Big). \tag{7}$$

In which $\mu_{li} \in [0,1]$ indicates the membership degree of sample x_j to the l-th cluster. ω_l and σ_l denote the weights for the numerical features and the categorical features respectively.

By minimizing the objective function $J(W,P)$, the optimal clustering result can be achieved. Note that since all the weights can be classified into two groups, one for numerical features and another for categorical features, the two groups weights will be updated with VPFRS and VPRS algorithm respectively. The weights for numerical features and for for categorical features will be calculated by Variable Precision Rough-Fuzzy Sets and Variable Precision Rough Sets as Eq.(8) and Eq.(9) according to the Eq.(1) and Eq.(4) respectively

$$\sigma = \gamma_S(Q) = |\mathrm{POS}_S(Q)| \big/ |U| \tag{8}$$

$$\omega = \eta_S(Q) = |\underline{Q}_\alpha| \big/ |\overline{Q}_\beta| \tag{9}$$

where S and Q are conditional attributes and decision attribute respectively. $\underline{S}_\beta(X)$ is the β ($0 \le \beta \le 1$)lower approximation of S depending on Q.

For the categorical features, we use the classical Variable Precision Rough Sets to update the weights in the second term of Eq.(9) and shown in the Eq.(10), due to the effectiveness of classical Variable Precision Rough Sets dealing with the discrete features (categorical features). For the continuous values, we first normalize these values and regard them as the membership. Then, the Variable Precision Rough-Fuzzy Sets can be applied into calculating the feature weights in the first term of Eq.(9) and given in the Eq.(11).

Then the parameter ω_l and σ_l have different value, the algorithm is degrade into several clustering method, shown in the Table 1.

Table 1. The different detection algorithm according to different coefficient

Algorithms	λ^r	λ^c
Fuzzy *c*-means method	$\omega_l^2 = 1$	$\sigma_l^2 = 0$
Fuzzy *c*-modes method	$\omega_l^2 = 0$	$\sigma_l^2 = 1$
Fuzzy *c*-prototypes method	$\omega_l^2 = 1$	$\sigma_l^2 = 1$
Feature weighted fuzzy *c*-means method based on VPFRS(VPFRS-FWFCM)	$\omega_l^2 \neq 0,1$	$\sigma_l^2 = 0$
Feature weighted fuzzy *c*-modes method based on VPRS(VPRS-FWCM)	$\omega_l^2 = 0$	$\sigma_l^2 \neq 0,1$
Feature weighted fuzzy *c*-prototypes method based on HRS(HRS-FWFCP)	$\omega_l^2 \neq 0,1$	$\sigma_l^2 \neq 0,1$

The processed samples are unlabelled in cluster analysis. In our algorithm, we have to use the traditional clustering algorithm to label the samples. Then the HRS-FWFCP algorithm will be able to obtain the proper weight for each feature and the proposed new clustering algorithm will achieve the final optimal clustering result. So, the proposed method is a semi-supervised clustering method.

4 Shot Boundary Detection Scheme Based on HRS-WFCM

First, features of video sequences used as conditional attributes are extracted and the initialization decision attributes (the types of shot boundary) are given. Then, by calculating the correlation between conditional attributes, the importance of conditional attributes can be obtained. The final features can be achieved by clustering feature attributes with our algorithm.

4.1 The Feature Extraction of Video Data

To detect the video shot boundaries, 12 candidate features are usually extracted for common use [1-3,7,8]. The component in RGB model and HSV model respectively(*R,G,B*;*H,S,V*), Gray-histogram(*G-H*), Color-histogram (the color histogram of RGB model and the color histogram of HSV model respectively: *RGB-H* and *HSV-H*) and Statistic features(mean, variance and skewness: *M*, *St*, *P*).

4.2 The Detection Scheme

According to the characteristics of news scenes, shot transition can be divided into three types: *cut transition*, *gradual transition* and *no transition*. Due to the great capacity of video data, the computer cannot deal with a lot of data once. So, the video is partitioned into several clips. During a mount of observation and experiments, a little news unit often lasts less than 150 seconds and the shot transition is no more than 5 seconds. Therefore, we select 150 frames in length and deal with 300 units video clips at random to select optimal feature for shot boundary detection and generate general rules for different shot transition types firstly. That is to say, the 300 units with each of including 150 frames are selected to perform feature selection. Therefore, the number of condition attributes, the number of samples and the number of decisions in proposed method are 12,300 and 3 respectively, where the original decisions of RS is maken by Twin Comparison method [7]. The detail procedure is given in the Ref. [8].

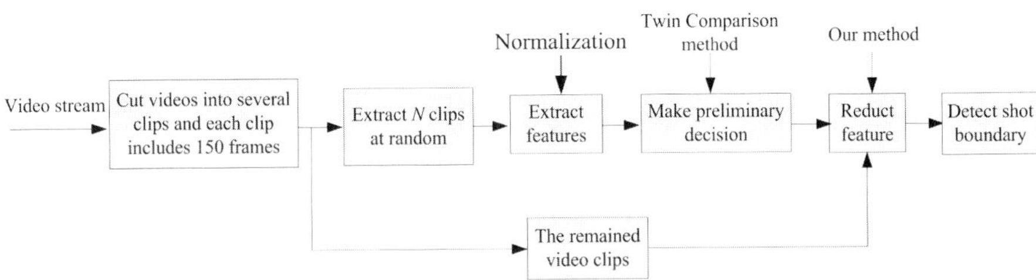

Fig. 1. The scheme of shot boundary detection

4.3 Feature Reduction

The importance of feature obtained from VPFRS-WFCM is arranged in descending ordered, If the ratio of the importance for the first *k* feature value to the whole feature value is higher than a predetermined threshold *T*, that is $\sum_{j=1}^{k} \kappa_j \Big/ \sum_{j=1}^{n} \kappa_j \geq T$, the first k feature is the main attribute for shot boundary detection. So, we define a adaptive threshold as $T = c \cdot \sum_{j=1}^{n} \kappa_j$, c is a constant and the corresponding discussion is given later.

According to the analysis above, the dissimilarity function for shot boundary detection is defined as follows,

$$S = \sum_{j=1}^{m} \omega_j \left(I_j(t+1) - I_j(t) \right) \tag{10}$$

Where m is the number of selected condition attributes (features).$I_j(t)$ is the jth attribute of the tth frame.

5 Experimental Results

To verify the classification performance of the proposed feature weighted clustering algorithm, some preliminary experiments are conducted for comparing proposed algorithm with traditional fuzzy c-types algorithm on UCI real data sets and video data. The experimental results illustrate the good performance of the new algorithm.

5.1 Experiments on the UCI Real Data

1) Experiment with numerical data set

We employ the real data set of *Iris* as tested data [9] to verify the effectiveness and robustness of our proposed algorithm for numerical data set. The IRIS data set contains 150 samples in 4-dimensional feature space, and the 4 components of each sample represent the petal length, petal width, sepal length and sepal width of IRIS. The whole data set is often divided into 3 categories, i.e., Setosa, Versicolor and Virginica(These are denoted as S, Ve and Vi respectively), each of which is composed of 50 samples. In feature space, the samples of the Setosa are separated from the other 2 categories, while there exists overlapping between the Versicolor and the Virginica. We employ the traditional fuzzy c-means algorithm (FCM) and the proposed feature weighted algorithm (VPFRS-FWFCM) to classify the *Iris* data set shown in the Table 2 in $\alpha = 0.5$ which is obtained from experiments. And the wrong classified number (WCN) of samples and the wrong classification rate (WCR) are used as criteria for comparing the performance of the 2 clustering algorithms.

Table 2. The classification results of *Iris* data sets

	The classified results by our method			
	VPFRS-FWFCM /FCM	S	Ve	Vi
Actual Class	S	50/50	0/0	0/0
	Ve	0/0	49/45	4/12
	Vi	0/0	1/5	46/38

There are two coefficient α and β in Rough-Fuzzy Sets. According to the need of this paper, we set β is one. Then the variation of WCN and WCR with the coefficient α is shown in Fig 2.The optimal coefficient α can be obtained by our proposed method during clustering. That is, the best performance of clustering is in $\alpha = 0.5$.

In addition, the obtained feature weight is ω = [0, 0, 0.2835, 0.2365] in $\alpha = 0.5$, which implies that the forth features have the bigger contribution and the second features have the smaller contribution for classification, which is accord with the real data.

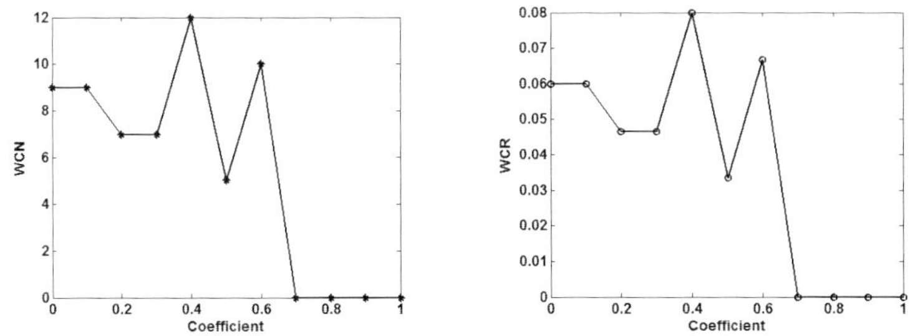

(a) (b)

Fig. 2. The variation of WCN and WCR with the coefficient α by our method

2) Experiment with categorical data set

To verify the classification performance of the proposed feature weighted clustering algorithm for the categorical data set, we employ the real data set of *bean diseases* as tested data set [9].The bean disease data set contains 47 recorders, and each of which is described with 35 features. Each recorder is labeled as one of the following 4 kinds of diseases, i.e., Diaporthe stem canker, Charcoal rot, Rhizoctonia root rot, and Phytophthora rot. Except the Phytophthora rot with 17 recorders, all the other categories have 10 recorders. The traditional *c*-modes (CM) and the proposed feature weighted CM (VPRS-FWCM) algorithms are used to classify the data set of bean disease. Table 4 shows the classification results of the 4 types in the form of a confusion matrix, in which D,C,R,P denote one of disease types respectively.

Table 4. The classification results of 2 clustering algorithms on *bean disease* data set

	The classified results by our method				
	VPRS- FWCM/CM	D	C	R	P
	D	10/10	0/6	0/0	0/0
Actual Class	C	0/0	10/4	0/2	0/1
	R	0/0	0/0	10/8	2/0
	P	0/0	0/0	0/0	15/16

During the test, we obtain that WCN and WCR of our proposed method (VPRS-FWCM) are 2 and 4%, those of CM are 12 and 26%, which shows that our algorithm is outperformed than fuzzy *c*-modes algorithm (CM) in effectiveness and robustness.

Fig.3 draws the obtained weights for features of bean disease data set by our proposed algorithm (VPRS-FWCM). It can be found that the weights of the 5-th, 6-th, from 9-th to 11-th, from the 13-th to the 19-th and from the 29-th to the 34-th features are zero, which implies these 18 features having no contribution for classification. By checking the original data set, all the recorders have the same values in the 11-th, from the 13-th to the 19-th and from the 29-th to the 34-th features. While the distributions of the value for 5-th,6-th, 9-th and 10-th features in the original data set are unorderly, the orderliness of sample distribution is not taken on. So, these features have less contribution for classification and the weights of them are zero. For this data sets, the weights by our algorithm accord with for the real situation. So the proposed algorithm obtains the optimal weights for every feature.

3) Experiment with mixed data set

As well known, the mixed data sets with numerical and categorical attributes are often encountered in data mining and other applications. To test the performance of the proposed algorithm to such mixed data set, we select the real data set of zoo as testbed [9], which contains 101 recorders, and each recorder including 15 categorical attributes and 1 numerical attribute.The fuzzy *c*-prototypes (FCP) and our feature weighted FCP (FRS-FWFCP) algorithms are adopted to classify the data set of zoo. The traditional *c*-prototypes algorithm makes 19 mistakes in classification, while the proposed algorithm makes 3 mistakes. The classification result of our algorithm is shown in Table 5, in which the number with '*' denotes the amount of the wrong classified samples. Since the number of mammals is greater than others, it is partitioned into 2 classes falsely, i.e., class 1 and class 7. The crawlers and amphibians are merged into one class. The other classes are achieved correct classification.

Table 5. The classification results of 2 clustering algorithms on zoon data set

Standard categories	Class1 (30)	Class2 (20)	Class3 (14)	Class4 (10)	Class5 (8)	Class6 (8)	Class7 (11)
Mammals(41)	30						11*
Birds(20)		20					
Fish(13)			12	1*			
Insectology (8)				8			
Molluscs(10)			1*	1*	8		
Crawlers(5)						5	
Amphibians(4)			1*			3	

Fig.4 shows the obtained weights for features of zoo data set. It is obvious that the forth feature has the biggest weight, which implies this feature has the biggest contribution for classification. In fact, the forth feature is the key attribute for distinguishing the mammals from others. While the fourteenth feature takes on the domestication character of animals and it has the smallest weight, which implies the smallest contribution of this feature for classification.

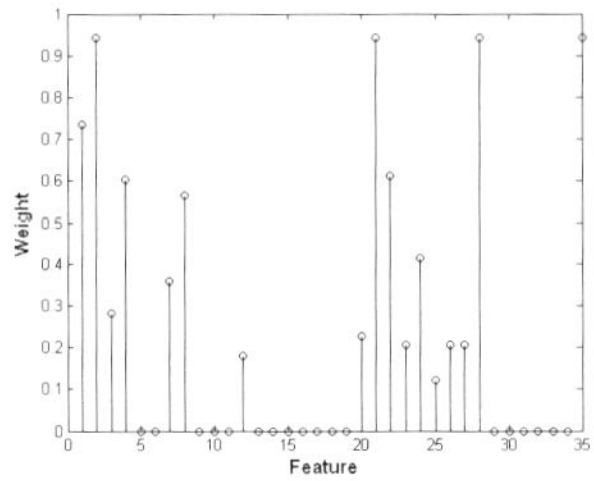

Fig. 3. The obtained weights of bean data set

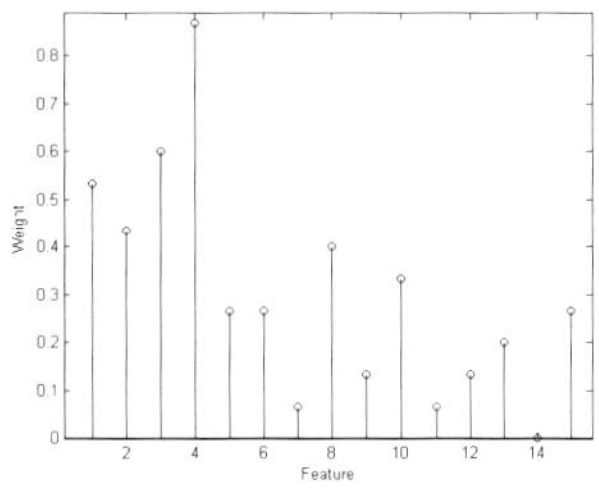

Fig. 4. The obtained weights of zoo data set

5.2 Experiments on the News Video Real Data

To verify the proposed method on large data sets, we apply our algorithm into video data from CCTV to detect shot boundary transition.

The method described above is applied into 7 news from CCTV lasting over 3*h*, whose frame size is 352×240 and frame rate is 30 frames per second, which include cut, fade and dissolve, as well as zoom, pan and other camera motions and object motions, including 2112 shots, where there are 1999cuts and 113 gradual transitions. We conduct an experiment with Twin Comparison method [7] on the same video clips. The experimental results are summarized in Table 6. We use the standard *recall* and *precision* criteria, shown in reference[8].H,M,F,R and P are denoted Hits, Misses, False alarms, Recall, Precision respectively. And the number of right hits is hits minus false alarms.

Table 6. The comparison of our method with the histogram method (DOH)

Program video	Proposed method					Twin Comparison[7] method				
	H	M	F	R	P	H	M	F	R	P
News 1	372	12	10	96.8%	97.3%	380	30	36	92.0%	90.5%
News 2	322	9	8	97.2%	97.5%	350	15	42	95.4%	88.0%
News 3	400	16	20	96.0%	95.0%	400	46	50	88.4%	87.5%
News 4	245	13	11	94.7%	95.5%	250	22	25	91.1%	90.0%
News 5	187	10	10	94.7%	94.7%	190	17	20	90.9%	89.5%
News 6	246	7	8	97.1%	96.8%	255	18	28	92.7%	89.0%
News 7	338	14	12	95.9%	96.4%	358	22	40	93.5%	88.9%
Total	2110	81	79	96.2%	96.3%	2183	170	241	92.0%	89.0%

5 Conclusions

The fuzzy *c*-means, *c*-modes and *c*-prototypes algorithms were designed for numerical, categorical and mixed data sets respectively. However, all the above algorithms assume that each feature of the samples plays a uniform contribution for cluster analysis. To consider the particular contributions of different features, this paper presents a novel feature weighted clustering algorithm based on Variable Precision Rough Sets and Variable Precision Rough-Fuzzy Sets. This method can obtain better classification performance than traditional clustering algorithms. In addition, the new algorithm can be used to analyze the different contributions among the features for classification, which is suitable for feature optimal choice in pattern recognition.

The experimental results on UCI data sets demonstrate the effectiveness of the proposed feature weighted clustering algorithm for numerical data, categorical data, even for mixed data. In real application, the proposed algorithm is complemented into shot transition detection in video indexing and browsing. The experimental results with real news totaled over 190 minutes in length from CCTV show that our method is reasonable and effective.

In the Future, our attention is focused on the automatic selection of coefficient α and β .

References

1. JOHN S. Boreczky, Lawrence A. Rowe. Comparison of video shot boundary detection techniques. In SPIE Conf. Storage & Retrieval for Image & Video Databases, Vol. 2670, (1996) 170-179
2. GARGI, U., KASTURI, R., STRAYER, S.H.: Performance characterization of video-shot-change detection methods. IEEE Trans. Circuits Syst. Video Technol., Vol.10(1) (2000) 1-13
3. Gao Xin-bo, Han Bing, Ji Hong-bing: Shot boundary detection Method for News video based on Rough sets and fuzzy clustering. Lecture Notes in Computer Sciences, 3656, (2005) 231-238
4. Li Jie, Gao Xinbo, Jiao Licheng: A new feature weighted fuzzy clustering algorithm. The Tenth International Conference on Rough Sets, Fuzzy Sets, Data Mining, and Granular Computing, Lecture Notes in Artificial Intelligence, (3641)(2005) 412-420 (2005)
5. Z. Pawlak,: Rough Set. International Journal of Computer and Information Science. Vol. 11(5) (1982) 341-356
6. D. Dubois, H. Prade: Rough fuzzy sets and fuzzy rough sets. International journal of general systems, (17), (1990) 191-209
7. Zhang H J, et al.: Automatic partitioning of full motion video. Multimedia Systems, 1(1) (1993) 10-28
8. HanBing, Gao Xin-bo, Ji Hong-bin.: An efficient algorithm of gradual transition for shot boundary segmentation. SPIE on MIPPR, Vol. 5286(2) (2003) 956-961
9. UCI Repository of Machine Learning Databases and Domain Theories. ftp://ftp.ics.uci.edu/pub/machine-learning- databases

An Effective Combination of Multiple Classifiers for Toxicity Prediction

Gongde Guo[1,2], Daniel Neagu[2], Xuming Huang[1,2], and Yaxin Bi[3]

[1] Dept. of Computer Science, Fujian Normal Univ., Fuzhou, 350007, China
[2] Dept. of Computing, Univ. of Bradford, Bradford, BD7 1DP, UK
{G.Guo, D.Neagu, X.Huang5}@bradford.ac.uk
[3] School of Computing and Mathematics, Univ. of Ulster, BT37 0QB, UK
Y.Bi@ulster.ac.uk

Abstract. The performance of individual classifiers applied to complex data sets has for predictive toxicology a significant importance. An investigation was conducted to improve classification performance of combinations of classifiers. For this purpose some representative classification methods for individual classifier development have been used to assure a good range for model diversity. The paper proposes a new effective multi-classifier system based on Dempster's rule of combination of individual classifiers. The performance of the new method has been evaluated on seven toxicity data sets. The classification accuracy of the proposed combination models achieved, according to our initial experiments, 2.97% better average than that of the best individual classifier among five classification methods (Instance-based Learning algorithm, Decision Tree, Repeated Incremental Pruning to Produce Error Reduction, Multi-Layer Perceptrons and Support Vector Machine) studied.

1 Introduction

Multiple Classifier System (MCS) has been widely applied to various fields of pattern recognition, including character recognition [1], speech recognition [2], text categorization [3],[4] and toxicity prediction [5], [6]. The idea of combination of classifiers is motivated by the observation of their complementary characteristics. It is desirable to take advantage of the strengths of individual classifiers and to avoid their weaknesses, resulting in the improvement of classification accuracy [4]. The work presented here is inspired by an idea from common sense reasoning and also from artificial intelligence research, i.e. a decision made on the basis of the multiple pieces of evidence should be more effective than one based on single piece of evidence. A classification problem is seen as a process of inferences about class concepts from concrete examples [7]. The inference process can be modeled as forward reasoning under uncertainty, as in production rule systems, which allows prior knowledge (prior performance assessments of classifiers) to be incorporated and multiple pieces of evidence from the classifiers to be combined to achieve precise classification decisions [4].

In the context of combining multiple classifiers for applications of toxicity prediction of chemical compounds, a number of researchers [5],[6] have shown that

L. Wang et al. (Eds.): FSKD 2006, LNAI 4223, pp. 481–490, 2006.

combining different classifiers can improve classification accuracy. Guo et al. [5] studied four similarity-based classifier combination methods which include Majority Voting-based combination (MV), Maximal Similarity-based Combination (MSC), Average Similarity-based Combination (ASC) and Weighted Similarity-based Combination (WSC). MV is the simplest approach, where the classification decision on each class is made on the basis of majority classifiers being in favor of that class for a given input [9]. MSC is based on the local highest similarity among a set of individual classifiers for combination. The classifier with highest local similarity will be dynamically selected for classifying the instances. ASC is a global combination method, where the similarities to each class are determined by individual classifiers and averaged together. The averaged similarities are then used for class label assignment to each test instance [5]. WSC is an intermediate approach between MSC and ASC, where instead of selecting the best classifier with the highest local similarity or considering all the classifiers' similarities to each class into account, WSC uses a control parameter α, where $0 < \alpha < 1$, to control the balance between the local optimization and global optimization [5].

In this paper, we propose to use Dempster's rule of combination to combine multiple classifiers for toxicity prediction of chemical compounds. Dempster's rule of combination provides a theoretical underpinning for achieving more accurate prediction through aggregating the majority voting principle and the belief degrees of decisions. The work presented in this paper mainly focuses on combining the outputs from different classifiers at the measurement level and incorporating the prior performance (prior knowledge) of each classifier into the definition of the mass functions, which is different from the work done by Xu et al. [1] and Bi et al. [4]. Xu et al aimed at combining the outputs from classifiers at the label level, and Bi et al. incorporate the prior performance of each classifier into the classification decision process.

2 Background Knowledge

Consider a number of exhaustive and mutually exclusive propositions $h_i, i = 1, \ldots, m$, which form a universal set Θ, called the frame of discernment. For any subset $H_i = \{h_{i1}, \ldots, h_{ik}\} \subseteq \Theta$, h_{ij} $(0 < j \leq k)$ represents a proposition, called a focal element, and when H_i is one element subset, i.e. $H_i = \{h_i\}$, it is called a singleton. All the subsets of Θ constitute a powerset 2^Θ, i.e. for any subset $H \subseteq \Theta$, $H \in 2^\Theta$. The D-S theory uses a numeric value in a range [0, 1] to represent the strength of some evidence supporting a proposition $H \subseteq \Theta$ based on a given evidence, denoted by $m(H)$, called the mass function, and uses a sum of strength for all subsets of H to indicate a belief degree to the proposition H on the basis of the same evidence, denoted by *bel(H)*, often called belief function. The formal definitions for these functions are given below [10]:

Definition 1. *Let Θ be a frame of discernment, given a subset* H $\subseteq \Theta$*, a mass function is defined as a mapping* $m : 2^\Theta \to [0, 1]$*, and satisfies the following conditions:*

$m(\phi) = 0$
$\sum_{H \subseteq \Theta} m(H) = 1$

Definition 2. *Let Θ be a frame of discernment and* m *be a mass function on Θ, the belief of a subset $H \subseteq \Theta$ is defined as*

$$bel(H) = \sum_{B \subseteq H} m(B) \tag{1}$$

and satisfies the following conditions:
$bel(\phi) = 0$
$bel(\Theta) = 1$

When H is a singleton, $m(H) = bel(H)$. It can be seen that a belief function gathers all of the support that a subset H gets from all of the mass functions of its subsets.

Definition 3. *Let m_1 and m_2 be two mass functions on the frame of discernment Θ, and for any subset $H \subseteq \Theta$, the orthogonal sum of two mass functions on H is defined as:*

$$m(H) = m_1 \oplus m_2(H) = \frac{\sum_{X,Y \subseteq \Theta, X \cap Y = H} m_1(X) \times m_2(Y)}{1 - \sum_{X,Y \subseteq \Theta, X \cap Y = \phi} m_1(X) \times m_2(Y)} \tag{2}$$

This formula is also called Dempster's rule of combination. It allows two mass functions to be combined into a third mass function, pooling pieces of evidence to support propositions of interest.

3 Proposed Combination Technique

3.1 Definition of Mass Function

Let φ be a classifier, $C = \{c_1, c_2, \ldots, c_{|C|}\}$ be a list of class labels, and d be any test instance, an assignment of class labels to d is denoted by $\varphi(d) = \{s_1, s_2, \ldots, s_{|C|}\}$, where $s_i \geq 0, i = 1, 2, ..., |C|$ represents the relevance of the instance d to the class label c_i. The greater the score assigned to a class, the greater the possibility of the instance being under this class. For convenience of discussion, we define a function ϖ, $\varpi(c_i) = s_i + \delta$ for all $c_i \in C$, where $1 > \delta > 0$ represents the prior knowledge of classifier φ. It is clear that $\varpi(c_i) > 0, i = 1, 2, ..., |C|$. Alternatively, $\varphi(d)$ is written as $\varphi(d) = \{\varpi(c_1), \varpi(c_2), \ldots, \varpi(c_{|C|})\}$ which is treated as a general form of the output information at the measurement level.

A formal definition of mass function in this context is described as follows:

Definition 4. *Let C be a frame of discernment, where each class label $c_i \in C$ is a proposition that the instance d is of class label c_i, and $\varphi(d)$ be a piece of evidence that indicates a possibility that the instance comes from each class label*

$c_i \in C$, then a mass function is defined as a mapping, m: $2^C \rightarrow [0,1]$, i.e. mapping a basic probability assignment (bpa) to $c_i \in C$ for $1 \leq i \leq |C|$ as follows:

$$m(\{c_i\}) = \frac{\varpi(c_i)}{\sum_{j=0}^{|C|} \varpi(c_j)} \quad where \quad 1 \leq i \leq |C| \tag{3}$$

This expresses the degrees of belief in propositions of each class label to which a given instance should belong. The mass function defined in this way satisfies the conditions given in Definition 1.

With formula (3), the expression of the output information $\varphi(d)$ can be rewritten as $\varphi(d) = \{m(\{c_1\}), m(\{c_2\}), \ldots, m(\{c_{|C|}\})\}$. Therefore two or more outputs derived from different classifiers as pieces of evidence can be combined by using formula (2) to obtain a combined output as a new piece of evidence, forming a combined classifier for classification tasks.

3.2 Combination Method

Given a group of learning algorithms and a training data set, each of learning algorithms can build one or more classifiers (models) based on different subsets, e.g., feature subsets, of training data set. Moreover, different classification algorithms can build different classifiers on the same subsets. The combination task of multiple classifiers, in this context, is to summarize the classification results by the classifiers derived from diverse learning algorithms on different feature subsets.

Let ψ be a group of L learning algorithms, $\varphi_1^k, \varphi_2^k, \ldots, \varphi_n^k$ be a group of classifiers associated with learning algorithm L_k, where $1 \leq k \leq L$ and n is a parameter that is related to the number of feature subsets, then each of the classifiers, φ_i^k assigns an input instance d to Y_i^k, i.e. $\varphi_i^k(d) = Y_i^k$ and $1 \leq i \leq n$. The results output by multiclassifiers are represented as a matrix:

$$\begin{bmatrix} Y_1^1 & Y_2^1 & \ldots & Y_n^1 \\ Y_1^2 & Y_2^2 & \ldots & Y_n^2 \\ \ldots & \ldots & \ldots & \ldots \\ Y_1^L & Y_2^L & \ldots & Y_n^L \end{bmatrix} \tag{4}$$

where Y_i^k is a vector denoted as $(m_i^k(c_1), m_i^k(c_2), ..., m_i^k(c_{|C|})$. Each row in the matrix corresponds to one of learning algorithms, and each column corresponds to one of the feature subsets, i.e. Y_i^k is the result yielded by the classifier φ_1^k - a classifier built by L_k learning algorithm on i feature subset. If the number of classification algorithms $L = 5$, and the number of feature subsets is 5, 5 classifiers will be generated by each of the classification algorithms, denoted by $\{\varphi_1^k, \varphi_2^k, \ldots, \varphi_5^k\}_{k=1}^5$. Thus the combination task based on this matrix is made both on the columns and rows, i.e. for each column, all the rows will be combined using formula (5), and the combined results in each column will be combined again using formula (6), thereby producing a new mass distribution over all the class labels that represents the consensus of the assignments of the multiple classifiers to test class labels. The final classification decision will be made by using the decision rule of formula (7).

$$m_i^{'}(c_i) = m_i^1 \oplus m_i^2 \oplus \ldots \oplus m_i^L = [\ldots[[m_i^1 \oplus m_i^2] \oplus m_i^3] \oplus \ldots \oplus \; m_i^L](c_i) \quad (5)$$

$$bel(c_i) = m_1^{'} \oplus m_2^{'} \oplus \ldots \oplus m_K^{'} = [\ldots[[m_1^{'} \oplus m_2^{'}] \oplus m_3^{'}] \oplus \ldots \oplus \; m_K^{'}](c_i) \quad (6)$$

With all belief values of class labels to which class labels could belong obtained by using Equation (5) and (6), we can define a decision rule for determining a final class label in general cases below:

$$\varphi_{DRC}(d) = c_i \; if \; bel(c_i) = argmax_{c_i \in C} \; \{bel(c_i) | i = 1, 2, \ldots, |C|\} \quad (7)$$

In Equation (7) the abbreviation *DRC* stands for Dempster's rule of combination.

4 Experiments and Evaluation

4.1 Data Sets

To evaluate the effectiveness of our proposed classifier combination method, seven toxicity data sets: Trout, Bee, Daphnia, Dietary_Quail, Oral_Quail, APC and Phenols from the real-world applications have been collected for evaluation. Among these data sets five of them, i.e. Trout, Bee, Daphnia, Dietary_Quail and Oral_Quail come from DEMETRA project [11], each of them contains all the descriptors from both 2D_MDL_ABLegend and 2D_Pallas subsets; APC data set is proposed by CSL [12]; Phenols data set comes from TETRATOX database [13]. Some general characteristics of the data sets are given in Table 1.

Table 1. General information about the data sets

Data set	NF	NFFS	NN	NO	NB	NC	NI	CD
Trout	248	22	0	22	0	3	282	129:89:64
Bee	252	11	0	11	0	5	105	13:23:13:42:14
Daphnia	182	20	0	20	0	4	264	122:65:52:25
Dietary_Quail	254	12	0	12	0	5	123	8:37:34:34:10
Oral_Quail	253	8	0	8	0	4	116	4:28:24:60
APC	248	6	0	6	0	4	60	17:16:16:11
Phenols	173	11	0	11	0	3	250	61:152:37

Titles of columns in Table 1 have the following meanings: NF - Number of Features; NFFS - Number of Features after Feature Selection; NN - Number of Nominal features; NO - Number of Ordinal features; NB - Number of Binary features; NC - Number of Classes; NI - Number of Instances; CD - Class Distribution.

4.2 Classifiers

Five classification methods involved in generating classifiers for combination are chosen in terms of their representability and diversity which include the Instance-based Learning algorithm (IBL), Decision Tree learning algorithm (DT), Repeated Incremental Pruning to Produce Error Reduction (RIPPER), Multi-Layer Perceptrons (MLPs) and Support Vector Machine (SVM). The IBL, DT, RIPPER, MLPs, and SVM used in our experiments are from the Weka software package [14]. A brief introduction of the five classifiers applied in this study is given below:

Instance Based Learners: IBLs classify an instance by comparing it to a set of pre-classified instances and choose a dominant class of similar instances as the classification result.
Decision Tree: DT is a widely used classification method in machine learning and data mining. The decision tree is grown by recursively splitting the training set based on a locally optimal criterion until all or most of the records belonging to each of the leaf nodes bear the same class label.
Repeated Incremental Pruning to Produce Error Reduction: RIPPER is a propositional rule learning algorithm that performs efficiently on large noisy data sets. It induces classification (if-then) rules from a set of pre-labeled instances and looks at the instances to find a set of rules that predict the class of earlier instances. It also allows users to specify constraints on the learned if-then rules to add prior knowledge about the concepts, in order to get more accurate hypothesis.
Multi-Layer Perceptrons: MLPs are feedforward neural networks with one or two hidden layers, trained with the standard backpropagation algorithm. They can approximate virtually any input-output map and have been shown to approximate the performance of optimal statistical classifiers in difficult problems.
Support Vector Machine: SVM is based on the Structural Risk Minimization principle from statistical learning theory. Given a training set in a vector space, SVM finds the best decision hyperplane that separates the instances in two classes. The quality of a decision hyperplane is determined by the distance (referred as margin) between two hyperplanes that are parallel to the decision hyperplane and touch the closest instances from each class.

4.3 Combination Schemes

(1) Majority Voting-based Combination (MVC)
Given x a new instance to be classified with true class label t_x and k predefined classifiers $A_1, A_2, \cdots, A_k$ respectively, where classifier A_i approximates a discrete-valued function $f_{A_i} : \Re^n \longrightarrow C$, then the final class label of x is:

$$f(x) \leftarrow argmax_{c \in C} \sum_{i=1}^{k} \delta(c, f_{A_i}(x)) \tag{8}$$

where $\delta(a, b) = 1$ if $a{=}b$, and $\delta(a, b) = 0$ otherwise.

Based on the hypothesis above, the classification result of x classified by A_j is a vector of probabilities of x to each class $P =< P_{j1}, P_{j2}, \cdots, P_{jm} >$, where $j = 1, 2, \cdots, k$ and m is the number of predefined classes. The final class label of x can be obtained either as:

(2)Maximal Probability-based Combination (MPC)

$$f_1(x) \leftarrow argmax_{c_v \in C}\{max_u\{P_{uv}|u = 1, 2, \cdots, k\}|v = 1, 2, \cdots, m\} \quad (9)$$

(3) Average Probability-based Combination (APC)

$$f_2(x) \leftarrow argmax_{c_v \in C}\{\sum_{u=1}^{k}(P_{uv}/k)|v = 1, 2, \cdots, m\} \quad (10)$$

4.4 Statistical Tool for Comparison

There are many approximate statistical tests for determining whether one learning method outperforms another on a particular learning task. Among these the *Signed Test* [15] is commonly used. Here we give a brief description of this method which will be used to measure the statistical difference between the performances of two classification methods in the next section.

The Signed Test [15] is a general statistical tool for comparing the performance of different classification methods. Given n data sets, let n_A (n_B, respectively) be the number of data sets in which classification method A does better (worse respectively) than classification method B in terms of the classification accuracy. Then we have:

$$z = \frac{\frac{n_A}{n_A+n_B} - p}{\sqrt{\frac{p \times q}{n_A+n_B}}} \approx N(0, 1) \quad (11)$$

where p is the probability that classification method A does better than classification method B; and q=1-p. Under the null hypothesis, p=0.5, so

$$z = \frac{\frac{n_A}{n_A+n_B} - 0.5}{\sqrt{\frac{0.5 \times 0.5}{n_A+n_B}}} \approx N(0, 1) \quad (12)$$

which has (approximately) a standard normal distribution $N(0, 1)$. We can reject the null hypothesis that two classification methods are the same in terms of performance if $|Z| > Z_{\infty,0.975} = 1.96$.

4.5 Evaluation

[Experiment 1]. In this experiment, we test both five classification methods, i.e. IBL, DT, RIPPER, MLPs and SVM, and four combination methods, i.e. MVC, MPC, APC and DRC (the abbreviation DRC here stands for the proposed combination method which is based on Dempster's rule of Combination), over seven toxicity data sets using a ten-fold cross validation. The class distribution

Table 2. Performance of individual classifiers evaluated on seven data sets

Data set	IBL	k	DT	RIPPER	MLPs	LR	SVM
TROUT	59.93	5	55.32	56.74	58.16	0.9	62.06
ORAL_QUAIL	57.76	5	62.93	60.34	51.72	0.3	65.52
DAPHNIA	54.17	5	50.38	50.00	53.41	0.3	54.55
DIETARY_QUAIL	48.78	10	45.53	39.84	55.28	0.3	48.78
BEE	58.09	5	45.71	46.67	51.43	0.3	53.33
PHENOLS	74.80	10	74.40	76.40	78.40	0.3	80.00
APC	43.33	5	43.33	40.00	40.00	0.3	43.33
Average	56.69	/	53.94	52.86	55.49	/	58.22

Table 3. Performance of different combination methods evaluated on seven data sets

Data set	MVC	MPC	APC	DRC
TROUT	63.12	56.38	59.22	64.93
ORAL_QUAIL	62.93	56.03	60.34	63.34
DAPHNIA	54.17	53.78	53.78	54.92
DIETARY_QUAIL	53.66	43.90	52.03	53.78
BEE	58.10	42.86	55.24	60.29
PHENOLS	80.40	79.20	82.40	82.40
APC	38.33	40.00	36.67	40.00
Average	58.67	53.16	57.10	59.95

of each data set is presented in Table 1. The experimental results are presented in Table 2 and 3.

In Table 2, each row recorded the best performances of different classification methods evaluated on a feature subset of the leftmost data set by CfsSubsetEval method which is implemented in the Weka software package [14]. Parameter k stands for the number of nearest neighbors chosen for IBL, which is tuned from 1 to 10 with step 1; LR represents the learning rate set for MLPs, which is tuned from 0.1 to 0.9 with step 0.1.

Table 3 reported the experimental results of different classifier combination methods carried out on the seven aforementioned data sets. The performances of MVC, MPC and APC in Table 3 are based on the results reported in Table 2. The performance of DRC is calculated on a $L \times n$ performance matrix by using Dempster's rule of combination where L stands of the number of classifiers and n stands for the number of feature subsets for each toxicity data set.

Eight feature selection methods are involved in extracting different subsets for each original toxicity data set, which are: Correlation-based Feature Selection; Chi-Chi squared ranking filter; Consistency Subset evaluator; Gain Ratio feature evaluator; Information Gain ranking filter; kNNMFS Feature Selection [8]; ReliefF ranking filter; SVM feature evaluator. All the feature selection methods except kNNMFS are implemented in the Weka software package [14], where kNNMFS is implemented in our own prototype system.

From Table 2 and 3 it is clear that the average classification accuracy of DRC based combination method over seven data sets is better than that of any other classification methods. Moreover, DRC based combination method performs best compared to other classifier combination methods.

[Experiment 2]. The goal of this experiment is to measure the statistical difference between the performances of any two methods studied. We compare the performance of any two classification methods based on the results obtained in Table 2 and 3. The statistical difference between the performances of any two methods is calculated using the signed test and is given in Table 4.

Table 4. The signed test of different classifiers

Signed Test	IBL	DT	RIPPER	MLPs	SVM	MVC	MPC	APC
$n_A : n_B$	6:1	6:1	6:0	5:1	5:2	7:0	6:0	6:0
DRC	1.89(+)	1.89(+)	2.45(+)	1.63(+)	1.13(-)	2.65(+)	2.45(+)	2.45(+)

In Table 4, the item 1.63(+) in cell (3, 5), for example, means DRC is better than MLPs in terms of performance over the seven data sets. That is, the corresponding $|Z| > Z_{0.90} = 1.415$. The item 1.13(-) in cell (3, 5) means there is no significant difference in terms of performance between DRC and APC over seven data sets as the corresponding $|Z| < Z_{0.90} = 1.415$. From the statistical point of view the proposed DRC classifier combination algorithm outperforms individual classification algorithms and other combination systems with an exception of SVM. Although there is no significant difference in terms of performance between DRC and SVM, the average classification accuracy of DRC is still 2.97% better than that of SVM.

5 Conclusions

In this work, we proposed an approach for combining multiple classifiers using Dempster's rule of combination. Various experiments have been carried out on seven collected toxicity data sets from real-world applications to evaluate the performance of classification algorithms individually and in combination. Based on our experimental results, it is fairly to draw a conclusion: the performance of the combination method based on Dempster's rule of combination is better than that of any other combination method studied, i.e. MVC, MPC and APC, and is 2.97% on average, better than the best individual classification method SVM. The experimental results have shown the promise of the proposed approach. However more experiments both on toxicity data sets and also benchmark data are necessary for a full evaluation of the approach proposed.

Acknowledgments

This work is partially funded by the EPSRC project PYTHIA: GR/T02508/01. The authors acknowledge also the support of the EU EP5 project DEMETRA (www.demetra-tox.net).

References

1. L.Xu, A.Krzyzak, C.Y.Suen. Several Methods for Combining Multiple Classifiers and Their Applications in Handwritten Character Recognition, IEEE Trans. on System, Man and Cybernetics, Vol. 22 (3), 1992, pp.418-435.
2. T.Denoeux. A neural Network Classifier based on Dempster-Shafer Theory. IEEE transactions on Systems, Man and Cybernetics A, Vol. 30(2), 2000, pp.131-150.
3. Y.Yang, T.Ault, T.Pierce. Combining Multiple Learning Strategies for Effective Cross Validation. Proc. of The Seventeenth International Conference on Machine Learning (ICML'00), 2000, pp.1167-1182.
4. Y.Bi, D.Bell, H.Wang, G.Guo et al. Combining Multiple Classifiers for Text Categorization using Dempster-Shafer Theory of Evidence. In Proc. of the 1st International Conference on Modeling Decisions for Artificial Intelligence (MDAI'2004), 2004, pp.127-138.
5. G.Guo, D.Neagu. Similarity-based Classifier Combination for Decision Making. Proc. of IEEE International Conference on Systems, Man and Cybernetics, 2005, pp.176-181.
6. E.Benfenati, P.Mazzatorta, D.Neagu, G.Gini. Combining Classifiers of Pesticides Toxicity through a Neuro-fuzzy Approach, Lecture Notes in Computer Science, Volume 2364, 2002, pp. 293.
7. T.K.Ho. Multiple Classifier Combination: Lessons and Next Steps, Tin Kam Ho, in A. Kandel, H. Bunke, (eds.), Hybrid Methods in Pattern Recognition, World Scientific, 2002, pp.171-198.
8. G.Guo, D.Neagu, M.T.D.Cronin. Using kNN Model for Automatic Feature Selection. In Proc. of ICAPR 2005, LNCS 3686, pp. 410-419, 2005.
9. Y.H.Li and A.K.Jain. Classification of Text Documents. The Computer Journal, Vol. 41(8), 1998, pp.537-546.
10. J.W.Guan, and D.Bell. Evidence Theory and its applications. 1991, Vol.1 & 2.
11. EU FP5 Quality of Life DEMETRA QLRT-2001-00691: Development of Environmental Modules for Evaluation of Toxicity of pesticide Residues in Agriculture.
12. CSL: Development of artificial intelligence-based in-silico toxicity models for use in pesticide risk assessment, 2004-2007.
13. T.W.Schultz. TETRATOX: Tetrahymena pyriformis population growth impairment endpoint-A surrogate for fish lethality. Toxicol. Methods 7: 289-309, 1997.
14. I.H.Witten and G.Frank. Data Mining: Practical Machine Learning Tools with Java Implementations, Morgan Kaufmann, San Francisco, 2000.
15. G.W.Snedecor and W.G.Cochran. Statistical Methods. Iowa State University Press, 1989.

A Contourlet Transform Based Fusion Algorithm for Nighttime Driving Image

Shengpeng Liu, Min Wang, and Yong Fang*

School of Communication and Information Engineering,
Shanghai University, Shanghai (200072), China
yfang@staff.shu.edu.cn

Abstract. A novel contourlet transform based fusion algorithm for nighttime driving image is proposed in this paper. Because of advantages of the contourlet transform in dealing with the two or higher dimensions singularity or the image salient features, such as line, curve, edge and etc., each of the accurately registered images is decomposed into a low frequency subband image and a sets of high frequency subband images with various multiscale, multidirectional local salient features. By using different fusion rules for the low frequency subband image and high frequency subband images, respectively, the fused coefficients are obtained. Then, the fused image is generated by the inverse contourlet transform. The simulation results indicate that the proposed method outperforms the traditional wavelet packet transform based image fusion method.

1 Introduction

Visibility is very important for the safe driving. While it reduces sharply in the nighttime with limited illumination provided by the vehicle headlights and the environment, so obstacle such as pedestrian may be hard to see in the light. Though the infrared camera can easily identify and locate the obstacle, it misses the key street feature such as landmark and traffic lights. Based on this, image fusion was used to perform this task. Image fusion generates a single image, which contains more accurate description of the scene and less artifacts or distortion than any of the individual source images. It has emerged as a new promising research area in recent years.

In the past, many techniques have been proposed. Conventional methods can be generally classified into four groups: classical Intensity-Hue-Saturation (IHS) transform [1], Principal Component Analysis (PCA) [2][3], statistical and arithmetic combination [4], and recently popular multiscale fusion. The multiscale transforms, such as High-Pass Filtering (HPF) method [5], Laplacian Pyramid (LP) [6], gradient pyramid[7], morphological pyramid[8], and wavelet transform[9], are very useful for analyzing the information content of images for the purpose of fusion and gain tense interests from many researchers. Comparing with other multiscale transforms, wavelet transform is more compact, and able to provide directional information in the

* IEEE Senior Member.

L. Wang et al. (Eds.): FSKD 2006, LNAI 4223, pp. 491–500, 2006.

low-low, high-low, low-high, and high-high bands, and contains unique information at different resolutions. Image fusion based on wavelet transform can provide better performance than those based on other multiscale methods as [10]. Wavelet packet transform (WPT), compared with WT, is a finer decomposition method and offers a richer range of possibilities for image processing because it can improve both high-frequency and low-frequency information in all wavelet packet fields. It has become one of the major fusion techniques.

Although the 2-D wavelet packet transform is able and optimal to represent the discontinuities at edge points, it will not "see" the smoothness along the contours which universally exist in the natural images [11]. It provides a poor directional selectivity and not optimal to extract the salient features in image. As a result, it limits the image fusion performance in some extent. Recently, the contourlet transform has been shown as a powerful method to preserve the spectral characteristics of the multipolarization and multifrequency images [12], allowing decomposition of each image into channels based on their local frequency content.

In this paper, a new multiscale transform, contourlet transform (CT), is used to improve the fusion performance. CT was introduced by Minh N. Do and Martin Vetterli in [12], [13], It meets the image representation "wish list": multiresolution, localization, critical sampling, directionality and anisotropy, while the separated wavelet packet transform provides only the first three, it provides an abundant directional selectivity and can represent the singularity in two or higher dimensions especially represent various directional smooth contours in natural images. The proposed technique firstly decomposes the image into a set of multisacle and different directional subband images by the CT, secondly applies different fusion rules to the low frequency and high frequency subband coefficients, then reconstructs the fusion image by performing backward CT to the gotten coefficients beforehand. The CT based fusion technique outperforms the WPT Based Image Fusion Algorithm (WPTBIFA), and the simulation results demonstrate it.

The rest of this paper is organized as follow. Section 2 presents the contourlet transform and the motivation for utilizing new transform to improve the fusion quality. Section 3 discusses the contourlet transform based fusion technique. Simulation setups in Section 4 and the results showed in this section confirm the validity of the proposed method. The concluding remarks in Section 5 summarize the advantages of the proposed algorithm.

2 Contourlet Transform

CT is a new multiscale, directional selectivity transform. It bases on an efficient two-dimensional non-separable filter banks and provides a flexible multi-resolution, local and directional approach for image processing [13]. It meets the image representation "wish list": multiresolution, localization, critical sampling, directionality and anisotropy, while the separated wavelet packet transform provides only the first three. CT is better than wavelet transform (or wavelet packet transform) in dealing with the singularity in two or higher dimensions, it provides an abundant directional selectivity and can represent various directional smooth contours in natural images [14], as shown in Fig. 1.

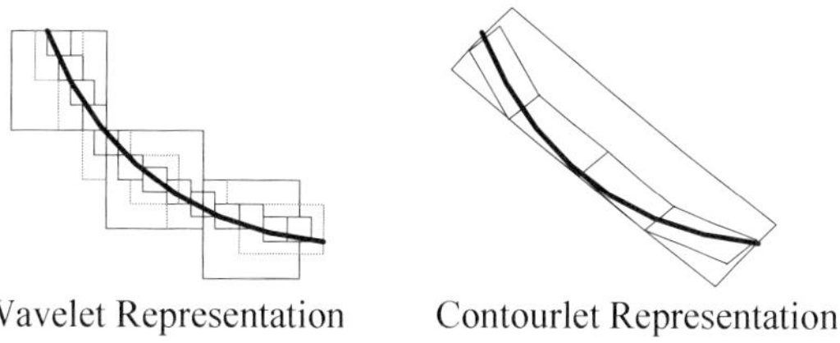

Fig. 1. Comparison of the curves representations with the Wavelet and the Contourlet

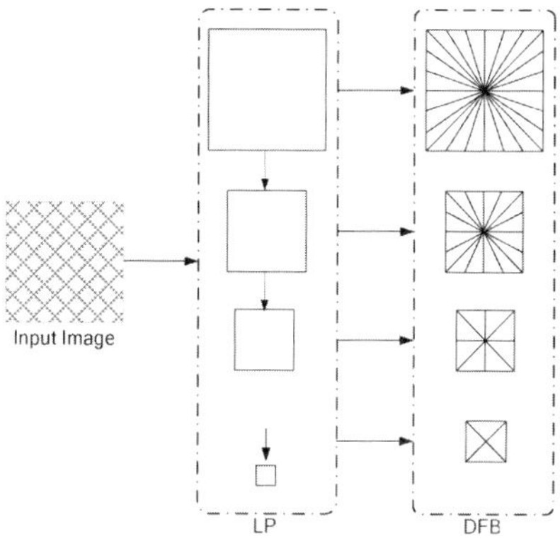

Fig. 2. Contourlet Transform Diagram.

There are two stages in the CT: multi-scale analysis stage and directional analysis stage, as shown in Fig. 2. The first stage is used to capture the point discontinuities. A Laplacian pyramid decomposes the input image into a detail sub-image and band-pass image which is difference between the input image and the prediction image. In the second stage, the band-pass image is decomposed into 2^{ld} ($ld = 1,2,...,n$) wedge shape sub-image by the directional filter banks (DFB), and the detail sub-image is then decomposed by the LP for the next loop, this stage to link point discontinuities into linear structures. The whole loop can be done lp ($lp = 1,2,...,n$) iteratively, and the number of direction decomposition at each level can be different, which is much more flexible than the three directions in wavelet. The overall result is an image expansion by using basic elements like contour segments. Fig. 3 shows the decomposition results of the standard image 'Lena (512x512)' by a two levels LP decomposition and 8 levels DFB decomposition in the finest level CT.

From above, it is clear that it can represent the two or higher dimension singularity or the salient features with abundant directional coefficients. All these two or higher dimension singularity or the salient features, such as line, curve and edge, are vital to the image fusion performance.

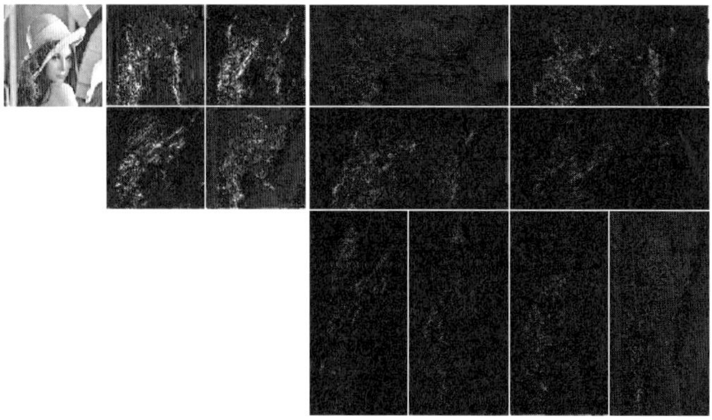

Fig. 3. Decomposition results of the standard image 'Lena (512x512)'

3 Fusion Algorithm Based on Contourlet Transform

In this paper, the image fusion is performed in contourlet domain. By using the CT's predominance in dealing with the two or higher dimensions singularity, the abundant directional selectivity and better representing various directional smooth contours in natural images are provided. Therefore, the fusion image generated by the proposed technique is more suitable for the purpose of human visual perception, object detection, target recognition and other further image processing.

3.1 Fusion Algorithm

To perform image fusion successfully, the proposed technique needs a prerequisite that the input images have to be accurately registered. Fig. 4 illustrates the framework of the new image fusion algorithm based on contourlet transform. In the figure, $\boldsymbol{I}_1, \boldsymbol{I}_2, \cdots, \boldsymbol{I}_n$ are the n accurately registered input images, and $\boldsymbol{F}$ is the fused image.

Suppose that there are just only two input images, $\boldsymbol{I}_1$ and $\boldsymbol{I}_2$ in the fusion approach, without loss the generality. Firstly, the CT decomposes the input images into a set of multisacle and different directional subband images, the number of the decomposition scale $\boldsymbol{N}_s$ and the number of the directional subbands $\boldsymbol{N}_{md}$ in scale m are decided before contourlet decomposition. Suppose that (m,n) indicate one of the subband image in the decomposition subband set, where m and n is the scale level and the n directional subband in scale m, respectively. Secondly, according to the subband's characteristics, the proposed approach applies different fusion rules. The low frequency subbands $\boldsymbol{I}_1(m_l, n_l)$ and $\boldsymbol{I}_2(m_l, n_l)$ hold the most of low frequency information, the variation of these coefficients is very small, a well-dealing low frequency fusion rule, fusion rule A, will be chosen to fuse these coefficients from the both two input images. While the high frequency directional subband $\boldsymbol{I}_1(m_i, n_i)$ and $\boldsymbol{I}_2(m_i, n_i)$ $(i \neq l)$ contain highly fluctuated coefficients, the large absolute decomposition coefficients correspond to shaper brightness changes in different direction and thus to the salient features in the image such as edges, lines and region boundaries. These features are the key information of image and fit to the strong human visual effects. The proposed technique chooses another corresponding fusion rule, rule B, to fully use these salient information. All these two fusion rules will be descript in the next section. Then, the inverse contourlet transform is applied to the gotten fused coefficients, and the fused image is generated.

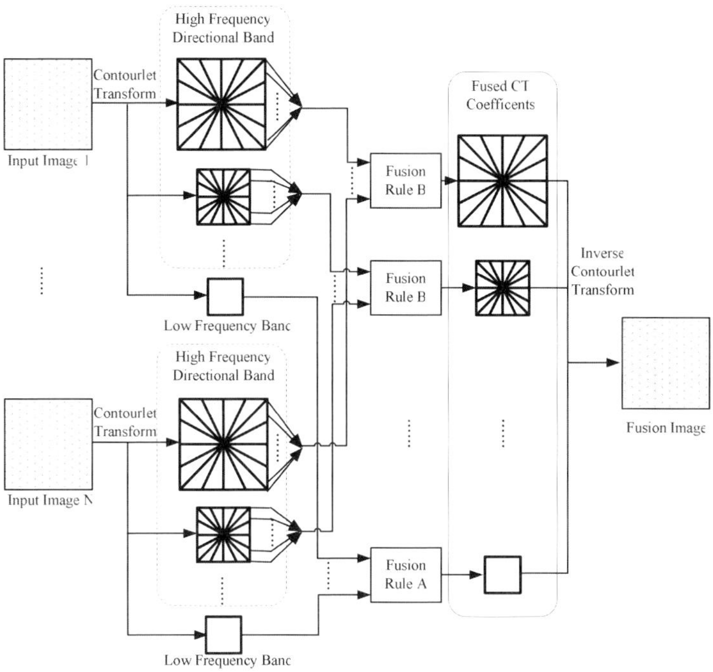

Fig. 4. Procedure of contourlet transform based image fusion

3.2 Fusion Rules

In the image fusion, the selection of fusion rules is vital for improving fusion quality. Different fusion rules will generate different fusion qualities and their quality varies heavily. Based on the characteristics of the CT decomposition coefficients, two type of fusion rules are applied to the different kinds of subbands. Fusion rule **A** is used to fuse the low frequency subband $\boldsymbol{I}_1(m_l,n_l)$ and $\boldsymbol{I}_2(m_l,n_l)$, while the fusion rule **B** is put forward to generate the high frequency subbands from the corresponding parts in both the two input images.

Fusion Rule A. Considering the low frequency subband is the images' approximate information, the coefficients are much similar and vary very slowly. So, here we choose the average of the corresponding low frequency subband as the fusion rule, it defines as,

$$F(m_l,n_l)=\frac{(I_1(m_l,n_l)+I_2(m_l,n_l))}{2}\,. \tag{1}$$

Fusion Rule B. As described above, the CT is predominant in dealing with the two or higher dimensions singularity, it provides an abundant directional selectivity and various directional smooth contours representing in natural images. The directional features in each high frequency subbands should be fully used to improve the fusion quality. In this paper, the proposed approach selects the feature based salience measure as the fusion rule in high frequency subbands. The salience computation estimates the evidence of a pattern related to the information contained in the neighboring area. It can be defined as the local energy of the incoming pattern within neighborhood $\boldsymbol{p}$:

$$\boldsymbol{S}(i,j,m,n)=\sum_{i'=1}^{r}\sum_{j'=1}^{h}\boldsymbol{p}(i',j')[\boldsymbol{D}(i+i',j+j',m,n)]^2\,, \tag{2}$$

where $\boldsymbol{S}$ is the salience measure, $\boldsymbol{p}$ is a small $r\times h$, centered on the considered pixel (i, j) window, typically is 3×3 or 5×5 window. $\boldsymbol{D}$ is the pyramid.

From above, it is clear that the salience of a particular component pattern is high if that pattern plays a role in representing important information in a scene, while it is low if the pattern represents unimportant information, or if it represents corrupted image data. So the criterion is

$$\boldsymbol{F}(i,j,m,n)=\begin{cases}\boldsymbol{I}_1(i,j,m,n) & \text{if}\quad \boldsymbol{S}_{I_1}(i,j,m,n)>\boldsymbol{S}_{I_2}(i,j,m,n)\\ \dfrac{\boldsymbol{I}_1(i,j,m,n)+\boldsymbol{I}_2(i,j,m,n)}{2} & \text{if}\quad \boldsymbol{S}_{I_1}(i,j,m,n)=\boldsymbol{S}_{I_2}(i,j,m,n)\\ \boldsymbol{I}_2(i,j,m,n) & \text{if}\quad \boldsymbol{S}_{I_1}(i,j,m,n)<\boldsymbol{S}_{I_2}(i,j,m,n)\end{cases}\,. \tag{3}$$

So, after the salience computation step, the fused high frequency coefficients will be generated according to the above criterion.

4 Simulations and Results

In this simulation, the proposed algorithm was applied to image fusion for two groups of images: road-scene images [15] and pedestrian on the road images [16]. The wavelet packets transform based image fusion algorithm (WPTBIFA) had also been realized for comparison.

These two approaches performed with the same fusion rules. The CT was two levels LP decomposition and 32 levels DFB decomposition in the finest level, contrast to the two levels, 'db4' based wavelet packets transform.

4.1 Evaluation Criteria

Performance measures are essential to determine the possible benefits of fusion as well as to compare results obtained with different algorithms. Visual inspection provided an overall impression of the detail information and the similarity of the original and resultant images. However, it is difficult to evaluate image fusion result objectively. Therefore, in addition to the visual analysis, three evaluation criteria are used for quantitatively assessing the performance of the fusion in this paper.

Correlation Coefficient. The correlation coefficient is the most popular similarity metric in image fusion. In this paper, the following Conversion Correlation Coefficient (CCC) criteria was used,

$$CCC = \sqrt{\frac{corr(F,I_1)^2 + corr(F,I_2)^2}{2}} . \tag{4}$$

Mutual Information. Mutual information has also selected as a means of assessing image fusion quality. The mutual information between source image and the fused image is described as [6],

$$MI_{FI_1}(f,i_1) = \sum_{f,i_1} p_{FI_1}(f,i_1)\log_2\left(\frac{p_{FI_1}(f,i_1)}{p_{FI}(f)\, p_{I_1}(i_1)}\right), \tag{5}$$

$$MI_{FI_2}(f,i_2) = \sum_{f,i_2} p_{FI_2}(f,i_2)\log_2\left(\frac{p_{FI_2}(f,i_2)}{p_F(f)\, p_{I_2}(i_2)}\right). \tag{6}$$

In this paper, the fusion performance was evaluated by the Compound Mutual Information (CMI), which defined as

$$CMI = MI_{FI_1}(f,i_1) + MI_{FI_2}(f,i_2) . \tag{7}$$

From above description, the bigger one implies better performance.

Standard Deviation. Suppose that the size of the fused image was $N \times M$, the standard deviation was estimated by

$$SD = \sqrt{\frac{1}{MN}\sum_{i=1}^{M}\sum_{j=1}^{N}(f(i,j)-\mu)^2} . \tag{8}$$

where $f(i,j)$ was the (i, j) th pixel intensity value and μ was the sample mean of all pixel values of the image. As we know, the high contrast image will have high standard deviation, and a low contrast image will have a low standard deviation. So, the higher the fused image have, the better fusion performance is.

4.2 Simulation Results

The simulation was carried out under the procedure described in Fig. 4. The two simulation results are shown in Fig. 5 and Fig. 6, respectively.

Fig. 5 describes the fusion result of the group 1, road-scene images: (a) and (b). The former is the intensified visible image and the latter is the thermal IR (FLIR) image. (c) is the WPTBIFA fused result, and (d) is the fused result of the proposed CTBIFA over through this paper.

(a) Intensified Visible Image

(b) Thermal IR (FLIR) Image

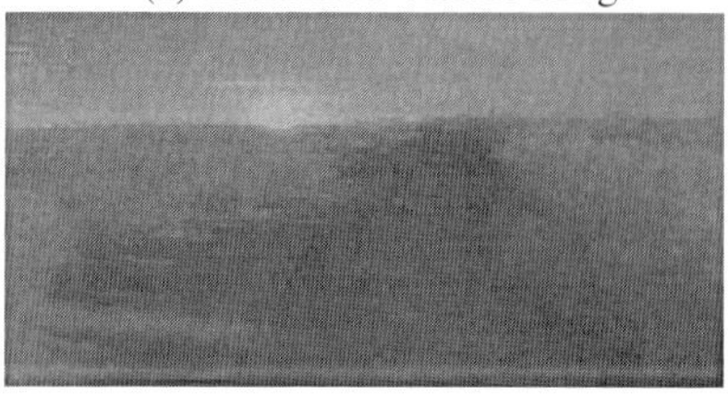

(c) WPTBIFA Result

(d) The Proposed CTBIFA Result

Fig. 5. Group 1 Input images and the fused results of WPTBIFA and the proposed CTBIFA

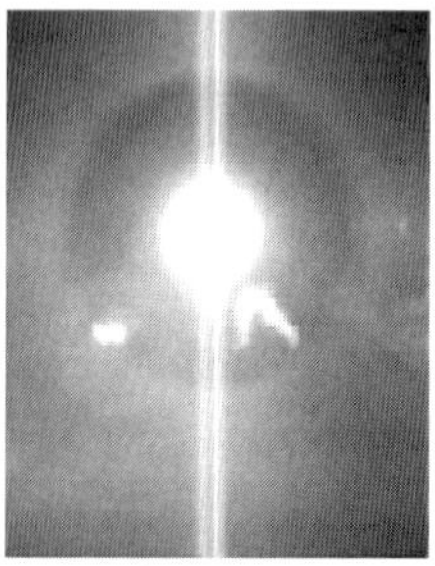

(a) Visible Image

(b) Infrared Image

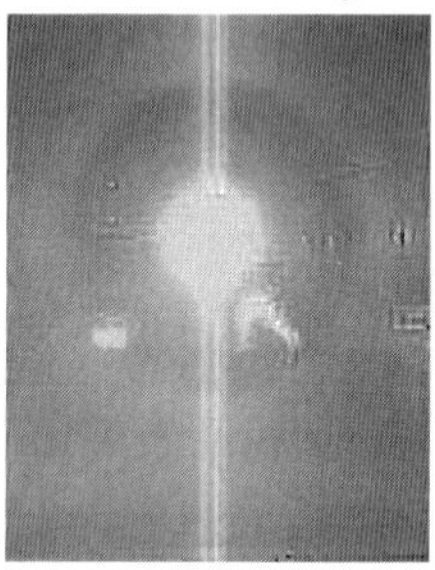

(c) WPTBIFA Result

(d) The Proposed CTBIFA Result

Fig. 6. Group 2 input images and the fused results of WPTBIFA and the proposed CTBIFA.

While the Fig. 6 shows the fusion results of group 2, pedestrian on the road. (a) and (b) are the intensified visible image and the short-wave infrared image. (c) and (d) are the fused results of the WPTBIFA and the proposed CTBIFA, respectively.

From the above two figures and under the view of visual inspection, it is clear that the fused image of the proposed CTBIFA in this paper contains the both features of the visible image and the infrared image. It is much clearer than any of the input images and the WPTBIFA fused image. So, it can be concluded that the proposed CTBIFA outperforms the WPTBIFA. This conclusion also can be gotten from Table 1. Table 1 indicates the performance index valued by the three criteria described just now.

Table 1. Performance index of the WPTBIFA and the proposed CTBIFA

	WPTBIFA			The Proposed CTBIFA		
	CCC	CMI	SD	CCC	CMI	SD
Group 1	1.3371	1.7203	33.1607	1.3804	1.7641	46.2392
Group 2	1.3307	1.9798	22.3054	1.3711	2.0232	61.3482

5 Conclusions

This paper proposes a new fusion technique for nighttime driving image, which based on the new multiscale, direction selectivity contourlet transform. The contourlet tranform, better than wavelet packet transform, in dealing with the two or higher dimensions singularity, such as line, curve, edge and etc. It provides an abundant directional selectivity and can represent various directional smooth contours in natural images. The proposed CTBIFA makes fully use of the multiscale, multidirectional information in the contourlet decomposition coefficients and gets the fused information according to the feature based fusion rules. The simulation results indicate that the proposed CTBIFA outperforms the WPTBIFA and verify the validity of the proposed CTBIFA. So, from this paper, it is obviously that the proposed CTBIFA is a useful image fusion approach.

Acknowledgments

This research has been supported in part by the National Natural Science Foundation of China (60472103), Shanghai Excellent Academic Leader Project (05XP14027), and Shanghai Leading Academic Discipline Project (T0102).

References

1. Carper, W.J., Lillesand, T.M., Kiefer, R.W.: The use of Intensity-Hue-Saturation transform for merging SPOT panchromatic and multispectral image data. Photogrammetric Engineering and Remote Sensing, Vol. 56. (1990) 459-467
2. Chavez, P.S., Sides, S.C., Anderson, J.A.: Comparison of three different methods to merge multi resolution and multi-spectral data: Landsat TM and SPOT panchromatic. Photogrammetric Engineering and Remote Sensing, Vol 57. (1991) 295-303
3. Zhang, Y.: Problems in the Fusion of Commercial High-Resolution Satellite Images as well as Landsat 7 Images and Initial Solutions. ISPRS, CIG, SDH Joint International Symposium on GeoSpatial Theory, Processing and Applications, Ottawa, Canada, (2002) 9-12

4. Vrabel, J.: Multispectral imagery band sharpening study. Photogrammetric Engineering and Remote Sensing, Vol. 62. (1969) 1075-1083
5. Sheffigara, V.K.: A Generalized Component Substitution Technique for Spatial Enhancement of Multispectral Images Using A Higher Resolution Data Set. Photogrammetric Engineering and Remote Sensing. Vol. 58. (1992) 561-567
6. Burt, P.J., Adelson, E.H.: The laplacian pyramid as a compact image code. IEEE Trans. On Communications. Vol. 31. (1983) 523-540
7. Petrovic, V.S., Xydeas, C.S.: Gradient-Based Multi-resolution Image Fusion. IEEE Transactions on Image Processing, Vol. 13.(2004) 228-237
8. Toet,A.: A Morphological Pyramid Image Decomposition. Pattern Recognition Letters. Volume 9. (1989) 255-261
9. Chipman, L., Orr, T.: Wavelets and image fusion. In: IEEE International Conference on Image Processing. Volume 3. (1995) 248-251
10. Wang, H.H., Peng, J.X., Wu, W.: A fusion algorithm of remote sensing image based on discrete wavelet packet. Proceedings of the Second International Conference on Machine Learning and Cybernetics. (2003) 2557-2562
11. Do, M., Vetterli, M.: The Contourlet Transform: An efficient directional multiresolution image representation. IEEE Transactions on Image Processing. (2003) 1-16
12. Do, M., Vetterli, M.: Contourlets, in: J. Stoeckler, G. V. Welland, Beyond Wavelets, Academic Press. (2002) 1-27
13. He, Z.H., Bystrom, M.: Reduced feature texture retrieval using contourlet decomposition of luminance image component. 2005 Int. Conf. on Communications, Circuits and Systems, (2005) 878-882
14. Chen, Y., Rick, S.B.: Experimental tests of image fusion for night vision. 7th international Conf. on information fusion. (2005) 491-498
15. Waxman, A.M., Aguilar, M., et al.: Solid-state color night vision: fusion of low-light visible and thermal infrared imagery. Lincoln Laboratory Journal. (1998) 41-60
16. Krebs, W.K., McCarley, J.S., et al.: An evaluation of a sensor fusion system to improve drivers' nighttime detection of road hazards. Proceedings of the Human Factors and Ergonomics Society 43rd Annual Meeting. (1999) 1333-1337.

Consistency Measures of Linguistic Preference Relations and Its Properties in Group Decision Making*

Yucheng Dong[1] and Yinfeng Xu[1,2]

[1] School of Management, Xi'an Jiaotong University, Xi'an, 710049, China
[2] The State Key Lab for Manufacturing Systems Engineering, Xi'an, 710049, China
{ycdong, yfxu}@mail.xjtu.edu.cn

Abstract. This paper first introduces the concept of consistent linguistic preference relations. By defining the distance of a linguistic preference relation to the set of consistent linguistic preference relations, we present a consistency index of linguistic preference relations and develop a consistency measure method for linguistic preference relations. This method is performed to ensure that the decision maker is being neither random nor illogical in his or her pairwise comparisons using the linguistic label set. At last, we discuss two consistency properties on group decision making using linguistic preference relations. These properties are of vital importance for group decision making with linguistic preference relations.

1 Introduction

Everybody makes decisions all the time, educated or uneducated, young or old, with ease or with difficulty. In the multiple attribute decision making (MADM), decision makers supply their preferences on alternatives in different preference representation structures. Among these structures, the preference relation is the most common one. According to element forms in preference relations, there are often two kinds of preference relations: numerical preference relation (i.e., multiplicative preference relation and fuzzy preference relation) [1, 2] and linguistic preference relation [3-6].

Recently, many researchers pay attention to group decision making (GDM) using linguistic preference relations. Herrera et al. [3] introduce a framework to reach consensus in GDM under linguistic assessments. Herrera et al. [4, 6-7] present the linguistic ordered weighted averaging (LOWA) operators to aggregate linguistic preference relations. Herrera and Herrera-Viedma [8] analyze the problem of finding a solution set of alternatives from a collective linguistic preference relation, following two research lines: choice functions and mechanisms. Xu [9, 10] proposes the linguistic OWG operators. Fan et al. [11, 12] discuss the satisfying consistency of linguistic preference relations, basing on three-way transitivity.

* This research is supported by NSF of China under Grants 10371094 and 70471035.

L. Wang et al. (Eds.): FSKD 2006, LNAI 4223, pp. 501–511, 2006.

In this paper, we focus our attention on consistency measures of linguistic preference relations. At the same time, we also discuss the consistency properties on GDM using linguistic preference relations. In order to do so, the rest of this paper is organized as follows. In section 2, we introduce some basic notations and operational laws of linguistic variables. In section 3, we develop a consistency measure of linguistic preference relations. In section 4, we discuss the consistency properties on GDM using linguistic preference relations. In section 5, an illustrative example is pointed out. Concluding remarks and future research are included in section 6.

2 Basic Notations and Operational Laws [13]

Xu introduces some basic notations and operational laws of linguistic variables in [13]. Let $S = \{s_\alpha | \alpha = -t, ..., -1, 0, 1, ..., t\}$ be a linguistic label set with odd cardinality. Any label, s_α, represents a possible value for a linguistic variable, and it is required that the linguistic label set should satisfy the following characteristics: (1) The set is ordered: $s_\alpha > s_\beta$ if and only if $\alpha > \beta$; (2) There is the negation operator: neg $(s_\alpha)=s_{-\alpha}$.

We call this linguistic label set S as the linguistic scale. For example, S can be defined as:

$$\begin{aligned} S = \{ & s_{-4} = extremely\ poor, s_{-3} = very\ \ poor, s_{-2} = poor \\ & s_{-1} = slightly\ poor, s_0 = fair, s_1 = slightly\ good \\ & s_2 = good, s_3 = very\ good, s_4 = extremely\ good\}. \end{aligned}$$

To preserve all the given information, we extend the discrete linguistic label set S to a continuous linguistic label set $\overline{S} = \{s_\alpha | \alpha \in [-q, q]\}$, where $q (q \geq t)$ is a sufficiently large positive integer. If $s_\alpha \in S$, then we call s_α the original linguistic label; otherwise, we call the virtual linguistic label. In general, the decision maker uses the original linguistic labels to evaluate alternatives, and the virtual linguistic labels can only appear in operations. Consider any two linguistic terms $s_\alpha, s_\beta \in \overline{S}$, and $\mu, \mu_1, \mu_2 \in [0, 1]$, we introduce some operational laws as follows: (1)$s_\alpha \oplus s_\beta = s_{\alpha+\beta}$; (2)$s_\alpha \oplus s_\beta = s_\beta \oplus s_\alpha$; (3)$\mu s_\alpha = s_{\mu\alpha}$; (4)$(\mu_1 + \mu_2) s_\alpha = \mu_1 s_\alpha \oplus \mu_2 s_\alpha$; (5)$\mu(s_\alpha \oplus s_\beta) = \mu s_\alpha \oplus \mu s_\beta$.

Let $s \in S$, we denote $I(s)$ as the lower indices of s, and call it as the gradation of s in S.

3 Consistency Measures of Linguistic Preference Relations

There is a finite set of alternatives, $X = \{x_1, x_2, ..., x_n\} (n \geq 2)$, as well as a preestablish label set, $S = \{s_\alpha | \alpha = -t, ..., -1, 0, 1, ..., t\}$. The decision maker provides his/her opinions on X as a linguistic preference relation, $P = (p_{ij})_{n \times n} \subseteq X \times X$, with membership function $u_p : X \times X \rightarrow S$, where $u_p(x_i, x_j) = p_{ij}$ denotes the linguistic preference degree of the alternative x_i over x_j. We assume, without loss of generality, that P is reciprocal in the sense that $p_{ij} \oplus p_{ji} = s_0$($s_0$ is the median label in S).

Definition 1: $P = (p_{ij})_{n \times n}$ is called a transitive linguistic preference relation if there exists $p_{ik} > s_0$, $p_{kj} > s_0$, then $p_{ij} > s_0$ for $i, j, k = 1, 2, ..., n$.

For a linguistic label set S, if the preference intensities of all gradations in S form an arithmetic progression, we call S as the arithmetic progression linguistic label set. In other words, we called S as the arithmetic progression linguistic label set if there exists the alternative F_1 is "s_α" over the alternative F_2 , the alternative F_2 is "s_β" over the alternative F_3, then the alternative F_1 is "$s_{\alpha+\beta}$" over the alternative F_3 for any $s_\alpha, s_\beta \in S$. In this paper, we suppose that S is an arithmetic progression linguistic label set. Basing on this, we define consistent linguistic preference relations.

Definition 2: $P = (p_{ij})_{n \times n}$ is called a consistent linguistic preference relation if there exists $p_{ik} \oplus p_{kj} = p_{ij}$ for $i, j, k = 1, 2, ..., n$.

Theorem 1: $P = (p_{ij})_{n \times n}$ is a transitive linguistic preference relation under the condition that P is a consistent linguistic preference relation.

Proof: Since $P = (p_{ij})_{n \times n}$ is a consistent linguistic preference relation, it follows that

$$p_{ij} = p_{ik} \oplus p_{kj}, \qquad i, j, k = 1, 2, ..., n. \tag{1}$$

Without loss of generality, we suppose that

$$p_{ik} \geq s_0 \tag{2}$$

and

$$p_{kj} \geq s_0. \tag{3}$$

By (1), (2) and (3), we have

$$I(p_{ij}) = I(p_{ik}) + I(p_{kj}) \geq 0 \tag{4}$$

i.e.

$$p_{ij} \geq s_0 \tag{5}$$

which completes the proof of theorem 1.

Let $s_\alpha, s_\beta \in S$ be two linguistic variables, Xu [13] defines the distance (namely deviation degree) between s_α and s_β as follows:

$$d(s_\alpha, s_\beta) = \frac{|s_\alpha - s_\beta|}{T} \tag{6}$$

where T is the number of linguistic terms in the set S.

Definition 3: Let $A = (a_{ij})_{n \times n}$ and $B = (b_{ij})_{n \times n}$ be two linguistic preference relations, then we define the distance between A and B as follows:

$$d(A, B) = \sqrt{\frac{2}{n(n-1)} \sum_{j=i+1}^{n} \sum_{i=1}^{n} \left(d\left(a_{ij}, b_{ij}\right)\right)^2} \tag{7}$$

Xu [13] defines a different distance metric (namely deviation degree) between two linguistic preference relations, and obtains a useful property on GDM using linguistic preference relations. When using our distance metric, we can obtain the same property (see theorem 5). Moreover, this distance metric will bring us the additional advantages in establishing the consistency thresholds (see theorem 3). This is the reason that we adopt the new distance metric.

Definition 4: Let $A=(a_{ij})_{n\times n}$ be a linguistic preference relation and M_{p_n} be the set of the $n\times n$ consistent linguistic preference relations, then we define the distance between A and M_{p_n} as follows:

$$d(A,M_{p_n})=\min_{P\in M_{pn}}\left(d\left(A,P\right)\right). \tag{8}$$

We set $d(A,M_{p_n})$ as the consistency index (CI) of the linguistic preference relation A, namely

$$CI(A)=d(A,M_{p_n}). \tag{9}$$

This consistency index has a definite physical implication and reflects deviation degree between the linguistic preference relation A and consistent linguistic preference relations. Obviously, the smaller the value of $CI(A)$, the more consistent the linguistic preference relation A. If $CI(A)=0$, then A is a consistent linguistic preference relation.

Let $A=(a_{ij})_{n\times n}$ be a linguistic preference relation.

Lemma 1: $d(A,M_{p_n})\geq\frac{1}{T}\sqrt{\frac{2}{n(n-1)}\sum\limits_{j=i+1}^{n}\sum\limits_{i=1}^{n}\left(I(a_{ij})-\frac{1}{n}\sum\limits_{k=1}^{n}\left(I(a_{ik})+I(a_{kj})\right)\right)^2}$

Proof: Since

$$d(A,M_{p_n})=\frac{1}{T}\min_{P\in M_{p_n}}\sqrt{\frac{2}{n(n-1)}\sum_{j=i+1}^{n}\sum_{i=1}^{n}\left(I(a_{ij})-I(p_{ij})\right)^2} \tag{10}$$

Let $c>1$, then we have

$$d(A,M_{p_n})=\frac{1}{T}\min_{P_n\in M_{p_n}}\sqrt{\frac{2}{n(n-1)}\sum_{j=i+1}^{n}\sum_{i=1}^{n}(\log_c(c^{I(a_{ij})})-\log_c(c^{I(p_{ij})}))^2}. \tag{11}$$

Let $A'=(a'_{ij})_{n\times n}=(c^{I(a_{ij})})_{n\times n}$ and $P'=(p'_{ij})_{n\times n}=(c^{I(p_{ij})})_{n\times n}$. Since A is a linguistic preference relation and P is a consistent preference relation, we obtain that

$$a'_{ij}\times a'_{ji}=c^{I(a_{ij})}\times c^{I(a_{ji})}=c^{I(a_{ij})+I(a_{ji})}=1 \qquad i,j=1,2,...,n \tag{12}$$

and

$$p'_{ik}\times p'_{kj}=c^{I(p_{ik})}\times c^{I(p_{kj})}=c^{I(p_{ij})}=p'_{ij} \qquad i,j,k=1,2,...,n. \tag{13}$$

By (12) and (13), we have that A' is a AHP judgement matrix and P' is a consistent AHP judgement matrix. Let $M'_{p_n}=\{P'=(p'_{ij})_{n\times n}|p'_{ij}=c^{I(p_{ij})}\ and\ P\in$

$M_{p_n}\}$, and $M_{R_n^+}$ is the set of $n\times n$ consistent AHP judgement matrices, obviously we have

$$M'_{p_n} \subseteq M_{R_n^+}. \tag{14}$$

Let $E = (e_{ij})_{n\times n}$ and $F = (f_{ij})_{n\times n}$ are two AHP judgement matrices, we define the distance between E and F as follows

$$D(E,F) = \sqrt{\frac{2}{n(n-1)}\sum_{j=i+1}^{n}\sum_{i=1}^{n}(\log_c(e_{ij}) - \log_c(f_{ij}))^2}. \tag{15}$$

By (11),(14) and (15), it can be got that

$$d(A, M_{p_n}) = \min_{P\in M_{p_n}} d(A,P) = \frac{1}{T}\min_{P'\in M'_{p_n}} D(A',P') \geq \frac{1}{T}\min_{W\in M_{R_n^+}} D(A',W). \tag{16}$$

By the theory of logarithmic least squares method in AHP [1], we have

$$\min_{W\in M_{R_n^+}} D(A',W) = \sqrt{\frac{2}{n(n-1)}\sum_{j=i+1}^{n}\sum_{i=1}^{n}\left(I(a_{ij}) - \frac{1}{n}\sum_{k=1}^{n}(I(a_{ik}) + I(a_{kj}))\right)^2}. \tag{17}$$

Thus

$$d(A, M_{p_n}) \geq \frac{1}{T}\sqrt{\frac{2}{n(n-1)}\sum_{j=i+1}^{n}\sum_{i=1}^{n}\left(I(a_{ij}) - \frac{1}{n}\sum_{k=1}^{n}(I(a_{ik}) + I(a_{kj}))\right)^2} \tag{18}$$

which completes the proof of lemma 1.

Lemma 2: Let $\overline{P} = (\overline{p_{ij}})_{n\times n}$, where $\overline{p_{ij}} = \frac{1}{n}\sum_{c=1}^{n}(a_{ic}\oplus a_{cj})$, then P is a consistent linguistic preference relation.

Proof: Since

$$\begin{aligned}\overline{p_{ik}}\oplus\overline{p_{kj}} &= (\tfrac{1}{n}\textstyle\sum_{c=1}^{n}(a_{ic}\oplus a_{ck}))\oplus(\tfrac{1}{n}\sum_{c=1}^{n}(a_{kc}\oplus a_{cj}))\\ &= (\tfrac{1}{n}\textstyle\sum_{c=1}^{n}(a_{ic}\oplus a_{ck}\oplus a_{kc}\oplus a_{cj}))\\ &= \tfrac{1}{n}\textstyle\sum_{c=1}^{n}(a_{ic}\oplus s_0\oplus a_{cj})\\ &= \tfrac{1}{n}\textstyle\sum_{c=1}^{n}(a_{ic}\oplus a_{cj}) = \overline{p_{ij}},\end{aligned} \tag{19}$$

we complete the proof of lemma 2.

Theorem 2: $d(A, M_{p_n}) = \frac{1}{T}\sqrt{\frac{2}{n(n-1)}\sum\limits_{j=i+1}^{n}\sum\limits_{i=1}^{n}\left(I(a_{ij}) - \frac{1}{n}\sum\limits_{k=1}^{n}(I(a_{ik}) + I(a_{kj}))\right)^2}$

Proof: Since $\overline{p_{ij}} = \frac{1}{n}\sum_{k=1}^{n}(a_{ik}\oplus a_{kj})$, from definition 3, we have

$$d(A,\overline{P}) = \frac{1}{T}\sqrt{\frac{2}{n(n-1)}\sum_{j=i+1}^{n}\sum_{i=1}^{n}\left(I(a_{ij}) - \frac{1}{n}\sum_{k=1}^{n}(I(a_{ik}) + I(a_{kj}))\right)^2}. \tag{20}$$

Since $\overline{P}$ is a consistent linguistic preference relation, it can be got that

$$d(A,\overline{P}) \geq \min_{P\in M_{p_n}} d(A,P) = d(A,M_{p_n}). \tag{21}$$

By Lemma 1 and (20), we also have

$$d(A,\overline{P}) \leq d(A,M_{p_n}). \tag{22}$$

So

$$d(A,M_{p_n}) = d(A,\overline{P}). \tag{23}$$

i.e.

$$d(A,M_{p_n}) = \frac{1}{T}\sqrt{\frac{2}{n(n-1)}\sum_{j=i+1}^{n}\sum_{i=1}^{n}\left(I(a_{ij}) - \frac{1}{n}\sum_{k=1}^{n}(I(a_{ik})+I(a_{kj}))\right)^2}. \tag{24}$$

which completes the proof of theorem 2.

Corollary 1: $CI(A) = d(A,\overline{P})$.

By corollary 1, we find that $CI(A)$ virtually reflects deviation degree between A and $\overline{P}$. We may approximately regard $\overline{P}$ as the impersonal linguistic preference relation. Let $\varepsilon_{ij} = I(a_{ij}) - I(p_{ij})$, then we have $CI(A) = \frac{1}{T}\sqrt{\frac{2}{n(n-1)}\sum_{j=i+1}^{n}\sum_{i=1}^{n}(\varepsilon_{ij})^2}$. The decision makers often have certain consistency tendency in making pairwise comparisons [9], and the values of ϵ_{ij} relatively centralizes the domain close to zero. Thus, we assume that $\varepsilon_{ij}(i<j)$ is independent normally distributed with a mean of 0 and a standard deviation of σ.

Theorem 3: $\frac{n(n-1)}{2}(T\times\frac{1}{\sigma}\times CI(A))^2$ is chi-square distribution with $\frac{n(n-1)}{2}$ degree of freedom, namely $\frac{n(n-1)}{2}(T\times\frac{1}{\sigma}\times CI(A))^2 \sim \chi^2(\frac{n(n-1)}{2})$, on the condition that $\varepsilon_{ij}(i<j)$ is independent normally distributed with a mean of 0 and a standard deviation of σ, namely $\varepsilon_{ij}\sim N(0,\sigma^2)$.

Proof: Since $\frac{n(n-1)}{2}(T\times\frac{1}{\sigma}\times CI(A))^2 = \sum_{j=i+1}^{n}\sum_{i=1}^{n}(\frac{\varepsilon_{ij}}{\sigma})^2$, and $\frac{\varepsilon_{ij}}{\sigma}(i<j)$ is independent normally distributed with a mean of 0 and a standard deviation of 1, we have that $\frac{n(n-1)}{2}(T\times\frac{1}{\sigma}\times CI(A))^2 \sim \chi^2(\frac{n(n-1)}{2})$. This completes the proof of theorem 3.

If we further suppose that $\sigma^2=\sigma_0^2$, namely $\varepsilon_{ij}\sim N(0,\sigma_0^2)$, then the consistency measure is to test hypothesis H0 versus hypothesis H1:

Hypothesis H0: $\sigma^2 \leq \sigma_0^2$;
Hypothesis H1: $\sigma^2 \geq \sigma_0^2$.

The degree of freedom of the estimator $\chi^2 = \sum_{j=i+1}^{n}\sum_{i=1}^{n}(\frac{\varepsilon_{ij}}{\sigma_0})^2$ is $\frac{n(n-1)}{2}$. By right one-sided test , we can get the critical value λ_α of χ^2 at the significance level α. In this way, we have that

$$\overline{CI} = \frac{\sigma_0}{T}\sqrt{\frac{2}{n(n-1)}\lambda_\alpha} \tag{25}$$

Table 1. The values of $\overline{CI}$ for different n and T when $\alpha = 0.1$ and $\sigma_0 = 2$

	n=3	n=4	n=5	n=6	n=7	n=8	n=9
T=5	0.1765	0.2424	0.2790	0.3019	0.3176	0.3290	0.3376
T=9	0.0980	0.1347	0.1550	0.1677	0.1765	0.1828	0.1876
T=17	0.0519	0.0713	0.0821	0.0888	0.0934	0.0968	0.0993

If $CI(A) \leq \overline{CI}$, we consider that A is of acceptable consistency; otherwise, we consider A is of unacceptable consistency and hope the decision maker adjust the elements in A for better consistency. According to the actual situation, the decision makers can set different values for α and σ_0^2. Table 2 shows the values of $\overline{CI}$ for different n and T when setting $\alpha = 0.1$ and $\sigma_0 = 2$.

4 Consistency Properties on GDM Using Linguistic Preference Relations

Consider a group decision making problem with linguistic preference relations. Let $D = \{d_1, d_2, ..., d_m\}$ be the set of decision makers, and $\lambda = \{\lambda_1, \lambda_2, ..., \lambda_m\}$ be the weight vector of decision makers, where $\lambda_k > 0, k = 1, 2, ..., m, \sum_{k=1}^{m} = 1$. Let $A_1, A_2, ..., A_m$ be the linguistic preference relations provided by m decision makers $d_k(k = 1, 2, ..., m)$, where $A_k = (a_{ij}^k)_{n\times n}$, $a_{ij}^k \in S(k = 1, 2, ..., m; i, j = 1, 2, ..., n)$, then denote $\overline{A} = (\overline{a_{ij}})_{n\times n} = \lambda_1 A_1 \oplus \lambda_2 A_2 \oplus ... \oplus A_m$ as the weight combination of $A_1, A_2, ..., A_m$, where

$$\overline{a_{ij}} = \lambda_1 a_{ij}^1 \oplus \lambda_2 a_{ij}^2 \oplus ... \oplus \lambda_m a_{ij}^m, \quad i, j = 1, 2, ..., n. \tag{26}$$

Xu [13] proves that $\overline{A}$ is a linguistic preference relation. Now, we first discuss a property on the consistency of $\overline{A}$.

Theorem 4: $CI(\overline{A}) \leq \overline{CI}$ under the condition that $CI(A_k) \leq \overline{CI}(k = 1, 2, ..., m)$.

Proof: Let

$$y_{ij}^k = \left(I(a_{ij}^k) - \frac{1}{n}\sum_{c=1}^{n}(I(a_{ic}^k) + I(a_{cj}^k))\right)^2 \tag{27}$$

then $CI(A_k) = \frac{1}{T}\sqrt{\frac{2}{n(n-1)}\sum_{j=i+1}^{n}\sum_{i=1}^{n} y_{ij}^k}$. Because $CI(A_k) \leq \overline{CI}(k = 1, 2, ..., m)$, it can be got that

$$\sum_{j=i+1}^{n}\sum_{i=1}^{n} y_{ij}^k \leq \frac{n(n-1)}{2}(T \times \overline{CI})^2 \tag{28}$$

Since (26) and

$$CI(\overline{A}) = \frac{1}{T}\sqrt{\frac{2}{n(n-1)}\sum_{j=i+1}^{n}\sum_{i=1}^{n}\left(I(\overline{a_{ij}}) - \frac{1}{n}\sum_{c=1}^{n}(I(\overline{a_{ic}}) + I(\overline{a_{cj}}))\right)^2} \tag{29}$$

we have that

$$CI(\overline{A})=\frac{1}{T}\sqrt{\frac{2}{n(n-1)}\sum_{j=i+1}^{n}\sum_{i=1}^{n}\left(\sum_{k=1}^{n}\left(\lambda_k(I(a_{ij}^k)-\frac{1}{n}\sum_{c=1}^{n}(I(a_{ic}^k)+I(a_{cj}^k)))\right)\right)^2} \tag{30}$$

By (27) and (30), it can be got that

$$\begin{aligned}CI(\overline{A})&=\tfrac{1}{T}\sqrt{\tfrac{2}{n(n-1)}\textstyle\sum_{j=i+1}^{n}\sum_{i=1}^{n}(\sum_{k=1}^{m}(\lambda_k^2y_{ij}^k)+2\sum\limits_{k<l}(\lambda_k\lambda_l\sqrt{y_{ij}^ky_{ij}^l}))}\\&\le\tfrac{1}{T}\sqrt{\tfrac{2}{n(n-1)}\textstyle\sum_{j=i+1}^{n}\sum_{i=1}^{n}\left(\sum_{k=1}^{m}(\lambda_k^2y_{ij}^k)+\sum\limits_{k<l}(\lambda_k\lambda_l(y_{ij}^k+y_{ij}^l))\right)}\\&=\tfrac{1}{T}\sqrt{\tfrac{2}{n(n-1)}\left(\sum\limits_{k=1}^{m}(\lambda_k^2\sum\limits_{j=i+1}^{n}\sum\limits_{i=1}^{n}y_{ij}^k)+\sum\limits_{k<l}(\lambda_k\lambda_l(\sum\limits_{j=i+1}^{n}\sum\limits_{i=1}^{n}y_{ij}^k+\sum\limits_{j=i+1}^{n}\sum\limits_{i=1}^{n}y_{ij}^l))\right)}\end{aligned} \tag{31}$$

From (28) and (31), we know that

$$CI(\overline{A})\le\overline{CI}\sqrt{\textstyle\sum_{k=1}^{m}\lambda_k^2+2\sum\limits_{k<l}(\lambda_k\lambda_l)}=\overline{CI}\sqrt{\textstyle\sum_{k=1}^{m}\lambda_k}=\overline{CI} \tag{32}$$

which completes the proof of theorem 4.

Corollary 2: $\overline{A}$ is a consistent linguistic preference relation on the condition that all of $A_1, A_2, ..., A_m$ are consistent linguistic preference relation.

Xu [13] defines a different distance metric (or deviation degree) between two linguistic preference relations and obtains a useful property on GDM using linguistic preference relations. When using our distance metric, we also can obtain the same property. The proof of this property (see theorem 5) is similar to theorem 4.

Theorem 5: Let $A_1, A_2, ..., A_m$ and B be $m+1$ linguistic preference relations. If $d(A_k, B) < \alpha (k = 1, 2, ..., m)$, then $d(\overline{A}, B) < \alpha$.

Proof: Let $z_{ij}^k = (I(a_{ij}^k) - I(b_{ij}^k))^2$, then $d(A_k, B) = \frac{1}{T}\sqrt{\frac{2}{n(n-1)}\sum\limits_{j=i+1}^{n}\sum\limits_{i=1}^{n} z_{ij}^k}$.
Because $d(A_k, B) < \alpha(k = 1, 2, ..., m)$, it can be got that $\sum_{j=i+1}^{n}\sum_{i=1}^{n} z_{ij}^k \le \frac{n(n-1)}{2}(T\times\alpha)^2$. Since

$$\begin{aligned}d(\overline{A},B)&=\tfrac{1}{T}\sqrt{\tfrac{2}{n(n-1)}}\sqrt{\sum\limits_{j=i+1}^{n}\sum\limits_{i=1}^{n}(I(\overline{a_{ij}})-I(b_{ij}))^2}\\&=\tfrac{1}{T}\sqrt{\tfrac{2}{n(n-1)}}\sqrt{\sum\limits_{j=i+1}^{n}\sum\limits_{i=1}^{n}(\sum\limits_{k=1}^{m}(\lambda_k(I(a_{ij}^k)-I(b_{ij}^k))))^2}\end{aligned}$$

$$
\begin{aligned}
&= \frac{1}{T}\sqrt{\frac{2}{n(n-1)}}\sqrt{\sum_{j=i+1}^{n}\sum_{i=1}^{n}(\sum_{k=1}^{m}(\lambda_k^2 z_{ij}^k) + 2\sum_{k<l}(\lambda_k\lambda_l\sqrt{z_{ij}^k z_{ij}^l}))} \\
&\leq \frac{1}{T}\sqrt{\frac{2}{n(n-1)}}\sqrt{\sum_{j=i+1}^{n}\sum_{i=1}^{n}(\sum_{k=1}^{m}(\lambda_k^2 z_{ij}^k) + \sum_{k<l}(\lambda_k\lambda_l(z_{ij}^k + z_{ij}^l)))} \\
&= \frac{1}{T}\sqrt{\frac{2}{n(n-1)}}\sqrt{\sum_{k=1}^{m}(\lambda_k^2\sum_{j=i+1}^{n}\sum_{i=1}^{n} z_{ij}^k) + \sum_{k<l}(\lambda_k\lambda_l(\sum_{j=i+1}^{n}\sum_{i=1}^{n} z_{ij}^k + \sum_{j=i+1}^{n}\sum_{i=1}^{n} z_{ij}^l))} \\
&\leq \alpha\sqrt{\sum_{k=1}^{m}\lambda_k^2 + 2\sum_{k<l}(\lambda_k\lambda_l)} = \alpha\sqrt{\sum_{k=1}^{m}\lambda_k} = \alpha
\end{aligned}
\tag{33}
$$

This complete the proof of theorem 5.

5 Illustrative Example

In order to show how these theoretical results work in practice. Let us consider the example used by Xu [13]. In the example, there are five decision makers $d_k(k = 1, 2, ..., 5)$. The decision makers compare five alternatives with respect to certain criterion by using the linguistic scale

$$
\begin{aligned}
S = \{&s_{-4} = \textit{extremely poor}, s_{-3} = \textit{very poor}, s_{-2} = \textit{poor} \\
&s_{-1} = \textit{slightly poor}, s_0 = \textit{fair}, s_1 = \textit{slightly good} \\
&s_2 = \textit{good}, s_3 = \textit{very good}, s_4 = \textit{extremely good}\}
\end{aligned}
$$

and construct, respectively, the linguistic preference relations. Suppose that the linguistic preference relation B is given by a leading decision maker, and the linguistic preference relations A_1, A_2, A_3 and A_4 are given by the other four decision makers respectively. They are listed as follows:

$$
B = \begin{pmatrix} s_0 & s_0 & s_2 & s_{-1} & s_4 \\ s_0 & s_0 & s_{-1} & s_0 & s_3 \\ s_{-2} & s_1 & s_0 & s_2 & s_1 \\ s_1 & s_0 & s_{-2} & s_0 & s_2 \\ s_{-4} & s_{-3} & s_{-1} & s_{-2} & s_0 \end{pmatrix} \qquad
A_1 = \begin{pmatrix} s_0 & s_{-1} & s_3 & s_{-1} & s_3 \\ s_1 & s_0 & s_1 & s_0 & s_2 \\ s_{-3} & s_{-1} & s_0 & s_{-1} & s_2 \\ s_1 & s_0 & s_1 & s_0 & s_0 \\ s_{-3} & s_{-2} & s_{-2} & s_0 & s_0 \end{pmatrix}
$$

$$
A_2 = \begin{pmatrix} s_0 & s_1 & s_2 & s_0 & s_4 \\ s_{-1} & s_0 & s_{-1} & s_0 & s_0 \\ s_{-2} & s_1 & s_0 & s_{-1} & s_3 \\ s_0 & s_0 & s_1 & s_0 & s_1 \\ s_{-4} & s_0 & s_{-3} & s_{-1} & s_0 \end{pmatrix} \qquad
A_3 = \begin{pmatrix} s_0 & s_0 & s_3 & s_1 & s_3 \\ s_0 & s_0 & s_{-2} & s_2 & s_2 \\ s_{-3} & s_2 & s_0 & s_1 & s_1 \\ s_{-1} & s_{-2} & s_{-1} & s_0 & s_{-1} \\ s_{-3} & s_{-2} & s_{-1} & s_1 & s_0 \end{pmatrix}
$$

$$
A_4 = \begin{pmatrix} s_0 & s_2 & s_0 & s_{-1} & s_2 \\ s_{-2} & s_0 & s_{-1} & s_1 & s_0 \\ s_0 & s_1 & s_0 & s_{-1} & s_2 \\ s_1 & s_{-1} & s_1 & s_0 & s_1 \\ s_{-2} & s_0 & s_{-2} & s_{-1} & s_0 \end{pmatrix}
\overline{A} = \begin{pmatrix} s_0 & s_{0.5} & s_2 & s_{-0.25} & s_3 \\ s_{-0.5} & s_0 & s_{-0.75} & s_{0.75} & s_1 \\ s_{-2} & s_{0.75} & s_0 & s_{-0.5} & s_2 \\ s_{0.25} & s_{-0.75} & s_{0.5} & s_0 & s_{0.25} \\ s_{-3} & s_{-1} & s_{-2} & s_{-0.25} & s_0 \end{pmatrix}
$$

$$\overline{\overline{A}} = \begin{pmatrix} s_0 & s_{0.1} & s_{2.4} & s_{-0.3} & s_{3.2} \\ s_{-0.1} & s_0 & s_{-0.4} & s_{0.5} & s_{1.2} \\ s_{-2.4} & s_{0.4} & s_0 & s_{-0.6} & s_{2.1} \\ s_{0.3} & s_{-0.5} & s_{0.6} & s_0 & s_{0.2} \\ s_{-3.2} & s_{-1.2} & s_{-2.1} & s_{-0.2} & s_0 \end{pmatrix}$$

Without loss of generality, supposing $\lambda_1 = \lambda_2 = \lambda_3 = \lambda_4 = \frac{1}{4}$, the collective linguistic preference relation of A_1, A_2, A_3 and A_4 is $\overline{A}$. We also take $\lambda_1 = 0.4, \lambda_2 = 0.3, \lambda_3 = 0.2, \lambda_4 = 0.1$, then the collective linguistic preference relation of A_1, A_2, A_3 and A_4 is $\overline{\overline{A}}$. Table 2 shows the results of the illustrative example. We can find that all of A_1, A_2, A_3, A_4 are of acceptable consistency(see table 1). At the same time, the conclusions in this example are in accordance with theorem 4 and theorem 5.

Table 2. The results of the illustrative example

	A_1	A_2	A_3	A_4	$\overline{A}$	$\overline{\overline{A}}$
$d(\bullet, B)$	0.1648	0.1757	0.1648	0.2018	0.1438	0.1421
$CI(\bullet)$	0.1176	0.1030	0.1277	0.0981	0.0885	0.0965

6 Conclusions

This paper first defines the distance between two linguistic preference relations, and the distance of a linguistic preference relation to the set of consistent linguistic preference relations. Basing on this, a consistency measure method of linguistic preference relations is proposed to test whether a linguistic preference relation is of acceptable consistency. This paper also discusses the consistency properties on GDM using linguistic preference relations, and shows that the weight combination of linguistic preference relations $A_1, A_2, ..., A_m$ is of acceptable consistency under condition that each of $A_1, A_2, ..., A_m$ is of acceptable consistency. Suppose that the linguistic preference relation B is given by a leading decision maker, and the linguistic preference relations $A_1, A_2, ..., A_m$ are given by the other decision makers, we prove that the distance between B and the weight combination of linguistic preference relations $A_1, A_2, ..., A_m$ is no greater than the largest distances between B and each of linguistic preference relations $A_1, A_2, ..., A_m$. These results are very important for the application of linguistic preference relations in group decision making. As future research, we plan to discuss how to deal with inconsistency in linguistic preference relations, basing on our consistency measure.

References

1. Saaty T L. The analytic hierarchy process. New York: McGraw-Hill (1980).
2. Orlorski S A. Decision-making with a fuzzy preference relation. Fuzzy Sets and Systems, 3(1978) 155-167.

3. Herrera F, Herrera-Viedma E, Verdegay J L. A Model of Consensus in group decision making under linguistic assessments. Fuzzy Sets and Systems, 78(1996) 73-87.
4. Herrera FHerrera-Viedma E, Verdegay J L. Direct approach processes in group decision making using linguistic OWA operators, Fuzzy Sets and Systems, 79 (1996) 175-190.
5. Herrera-Viedma E, Herrera F, Chiclana F. A Consensus Model for multiperson decision making with different preference structures. IEEE Transactions on Systems, Man and Cybernetics,32(2002) 394-402.
6. Herrera F, Martinez L, A 2-tuple fuzzy linguistic representation model for computing with words, IEEE Transactions on fuzzy systems 8 (2000) 746-752.
7. Herrera F, A sequential selection process in group decision making with linguistic assessment, Information Sciences 85 (1995) 223-239.
8. Herrera F, Herrera-Viedma E, Linguistic decision analysis: steps for solving decision problems under linguistic information, Fuzzy Sets and Systems 115 (2000) 67- 82.
9. Xu Z S, A method based on linguistic aggregation operators for group decision making with linguistic preference relations, Information Sciences 166 (2004) 19-30.
10. Xu Z S, Uncertain linguistic aggregation operators based approach to multiple attribute group decision making under uncertain linguistic environment, Information Sciences 168 (2004) 171-184.
11. Fan Zhi-Ping and Chen X, Consensus measures and adjusting inconsistency of linguistic preference relations in group decision making, Lecture Notes in Artificial Intelligence, Germany: Springer-Verlag, 3613(2005):130-139.
12. Fan Zhi-Ping, Jiang Y P, A judgment method for the satisfying consistency of linguistic judgment matrix, Control and Decision 19(2004):903-906.
13. Xu Z S, Deviation measures of linguistic preference relations in group decision making, Omega 33 (2005) 249-254.
14. P. de Jong. A Statistical approach to Saaty's scaling method for priorities. Journal of Mathematical Psychology 28(1984) 467-478.

Adapting OLAP Analysis to the User's Interest Through Virtual Cubes *

Dehui Zhang, Shaohua Tan, Shiwei Tang, Dongqing Yang, and Lizheng Jiang

School of Electronics Engineering and Computer Science, Peking University, Beijing 100871, China
{dhzhang, tan, lzjiang}@cis.pku.edu.cn, {tsw, dqyang}@pku.edu.cn

Abstract. The manually performing of the operators turns OLAP analysis a tedious procedure. The huge user's exploration space is the major reason of this problem. Most methods in the literature are proposed in the data perspective, without considering much of the users' interests. In this paper, we adapt the OLAP analysis to the user's interest on the data through the virtual cubes to reduce the user's exploration space in OLAP. We first extract the user's interest from the access history, and then we create the virtual cube accordingly. The virtual cube allows the analysts to focus their eyes only on the interesting data, while the uninteresting information is maintained in a generalized form. The Bayesian estimation was employed to model the access history. We presented the definition and the construction algorithm of virtual cubes. We proposed two new OLAP operators, through which the whole data cube can be obtained, and we also prove that no more response delay is incurred by the virtual cubes. Experiments results show the effectiveness and the efficiency of our approach.

1 Introduction

OLAP systems enable powerful analysis of large amounts of summary data and the data are often organized in multidimensional cubes. Through the pre-aggregations, the analysts can get quick answers of analysis queries.

However, in order to find some valuable knowledge from the data cubes, there are too many paths and too huge volume of data to be examined during the analysis, i.e. the analysts have to face a large exploration space. The explosion on the size of user exploration space is a real problem, making manual inspection of the data a daunting and overwhelming task.

The reasons for huge exploration space can be roughly divided into two categories: the huge data volume and lack of navigations when analysis. The researchers have made a lot of effort to reduce user's exploration space in OLAP from different perspectives. One method to reduce exploration space is to shrink the stored data size by approximation or compressing. The approximation methods [1, 2, 3] are to find some concise representations of the data cube, so the examined dataset is reduced

* This work is supported by the National Natural Science Foundation of China under Grant No.60473072.

L. Wang et al. (Eds.): FSKD 2006, LNAI 4223, pp. 512 – 521, 2006.

when analysis. The challenge is to find a way to estimate the data as precisely as possible. Approximation is working at the cost of precision, so the researchers try to find some compress methods without precision lost [4, 5, 6]. These works are to divide the cube cells into some equivalent classes, so both the examined paths and the size of dataset are reduced. However, the effectiveness depends on the existence of the *single tuple*[6] in the data. Another method is to provide navigation when analysis. The navigation in the literature is either by automatizing OLAP operations [7, 8] or by extracting characteristics of the data, such as frequency [9], correlations [10], clusters [11], exceptions [12] and mini-classes [13]. The navigation methods are effective that they reduce exploration space and find knowledge simultaneously.

However, most existing works seldom take the users' analysis behaviors into consideration. If we can find some approaches to adapt the analysis only to the data where he is interested, then the user will have a much small data cube to surf. So, how to adapt the OLAP analysis to the user's interest is a valuable and challenge issue.

In this paper, we propose a new mechanism called *virtual cubes* based on the access history to adapt the OLAP analysis to the user's interest, and hence to reduce the exploration space. The modified dimension-member-tree will serve as the user's access history. We employ the Bayesian method to model the access history. In order to update the access history, we present the analysis-tracking algorithm.

A virtual cube is a cube derived from a subset of a cuboid of the original one, with some dimension members being noted as 'interesting'. In order to maintain the uninteresting information, we aggregate all the original cells that include uninteresting dimension members into some special cells, called virtual cells. By querying upon these special cells, the whole detail of the original cube can be obtained. The cells in the virtual cube are divided into two categories: all the dimension members of a cell are interesting, or it includes uninteresting dimension members, i.e. a virtual cell.

We propose the construction algorithm of virtual cube in this work. The algorithm is effective that it only takes one original cuboid as input and using 'interesting' members as the pruning strategy in one pass scanning.

Addition to the basic OLAP operators, two new operators on the virtual cubes are presented to query the uninteresting information, and we also prove that no more response delay is incurred due to the virtual cube mechanism.

We implement consolidated experiments in this paper. The experiments results prove that the virtual cube mechanism is effective and efficient.

The remainder of the paper is organized as follows. The problem will be formulated in section 2. In section 3 we will propose the Bayesian estimation model, the corresponding algorithm and the evaluation of the model. We will discuss the virtual cube construction algorithm in section 4. Two new operators on the virtual cube and the analysis tracking method are also proposed there. In section 5, experiments results will be presented. At last, conclusions are presented in section 6.

2 Problem Formulation

The users' exploration space is the data volume and the paths that the users have to examine during the analysis process. To reduce the users' exploration space is to find ways that can help the users reach their analysis destinations with as little data and few paths examined as possible. In this section, we give the definitions of the virtual cubes and the user's access history as follows.

2.1 Virtual Cubes

We first give the definition of Dimensions, and then we introduce the virtual cube. When a dimension has several aggregation paths, each one can be viewed as a separate structure. For the convenience of narration, we suppose that each dimension have only one aggregation path.

Definition 2.1: A dimension D can be shown as $D = (L_0, L_1,..., L_{k-1})$, member(D) representing the members set of D, and M_i=member (L_i) is the members set of L_i. Let '$\leqq$' be the partial order of aggregation, then $L_i \leqq L_j$, for $i \geqslant j$. For each member $x \in M_i$, *children(x)*=$\{y \mid y \in M_{i+1} \wedge y \leqq x\}$, and for each $y \in$children(x), *parent(y)*=x.

Definition 2.2: Given a dimension $D = (L_0, L_1, ..., L_{k-1})$, the *interesting members* of D is a set $I_D = \bigcup_{i=0}^{k-1} M'_i$, where $M'_i \subseteq M_i$. For each member $x \in M'_i$, if children(x)$\cap$ I_D $\neq \Phi$ and children(x)$\cap$ I_D $\neq$children(x), we define a *virtual member* O_x, where parent(O_x)=x, and children(O_x)=Φ. A *virtual dimension* from D is $V_D=(V_0, V_1,..., V_{k-1})$, where $V_i = M'_i \cup O_{Li}$ and $O_{Li}=\{ O_x \mid x \in M'_i \}$.

Property-1: Given two dimension members x, y of dimension D, and $y \leqq x$, then $y \in I_D \Rightarrow x \in I_D$.

Without loss of generality, we use 'sum' as the aggregation function, and based on which we define the virtual cubes.

Definition 2.3: A *virtual cell* c: $(d_1, d_2,..., d_n; v)$ of a virtual cube is defined as:

1) There exists an 'i', where d_i is a virtual member;
2) If d_i is a virtual member, $c.v=z.v-\Sigma y.v$, where $z= (d_1, d_2,...,d_{i-1}, parent(d_i), d_{i+1},..., d_n; v)$ and $y=(d_1, d_2,...,d_{i-1}, t, d_{i+1},..., d_n; v) \wedge t \in (children(x) \cap I_{Di})$.

Definition 2.4: Given an n-dimension dataset $R=(D_1, D_2,..., D_n; Measure)$, the data cube derived from R is called the *original cube*. Let $I=(I_{D1}, I_{D2},..., I_{Dn})$ be the interesting members of the dimensions, and $V_D=(V_{D1}, V_{D2},..., V_{Dn})$ is the virtual dimensions. The *virtual cube* C_V of R is defined as: $C_V= C_1 \cup C_2$, where $C_1=\{(d_1, d_2,..., d_n; v) \mid \forall d_i \in I_{Di}\}$, $C_2=\{(d_1, d_2,..., d_n; v) \mid \exists d_i \in (V_{Di}- I_{Di})\}$.

2.2 User's Access History

If a dimension has only one aggregation path, the dimension members will be organized as a tree, called the dimension member tree (DMT), where each node represents a dimension member and the edge represents the aggregation relationship between members.

Definition 2.5: Given an n-dimension dataset $R=(D_1, D_2,..., D_n; Measure)$, an *access history H* has the form $H =(T_1, T_2,..., T_n; N)$, N is the accumulative total of the requested members, T_i is a modified DMT of D_i. A node of T_i has the form (key, flag, count; pointers), where the *key* is the label of this member, the *flag* represents that whether this entry has been modified or not till now, the *count* is the member's requested times and the *pointers* point to its children nodes in T_i.

3 User's Interest Extracting

In this paper, we employ the Bayesian estimation to extract the users' interests on the data, and in this section we will introduce the details.

3.1 A Bayesian Model to Extract User's Interest from the Access History

Bayesian estimation is a probability method and is widely used in many industrial fields. Given a variable x, whose distribution function is f $(x|\theta)$, where θ is an unknown parameter. The parameter θ can be viewed as another variable, and its distribution function is $\pi(\theta)$, which is called the prior distribution. The Bayesian estimation is to estimate the parameter θ using observation samples X, and the result is the function $\pi(\theta|X)$, which is called the posterior distribution.

From probability perspective, each dimension member is either accessed during analysis or not in Bernoulli distribution. And the history information forms an observation of this member. Let the parameter of Bernoulli distribution be *p*, we can use Bayesian method to estimate the value of p. Given large enough observations, i.e. sufficient history information, the Bayesian estimation can get optimized estimated value of p. We then use the value of p to estimate user interest on the data.

Given a dimension member *x*, x is a Bernoulli variable, if x is requested then x=1 else x=0. The cumulative distribution function is: $f(x|\theta)=\theta^x(1-\theta)^{1-x}$. Let $\pi(\theta)$ be the prior distribution of θ, although $\pi(\theta)$ can be acquired according to Jeffreys rules [14], in real applications, $\pi(\theta)$ is usually represented using the Beta distribution function. That is:

$$\pi(\theta)=\mathrm{Be}(\theta;a,b)=\frac{\Gamma(a+b)}{\Gamma(a)\Gamma(b)}\theta^{a-1}(1-\theta)^{b-1},\ \text{where } 0<\theta<1.$$

For the above formulation, if the real value of θ is close to 0, then a>b; and if it is close to 1, then a<b. If we have no experience about θ in advance, then a=b=1.

Definition 3.1: Suppose that $\mathbf{X}=(X_1, X_2, \ldots, X_n)$ be the observation samples of X, $f(\mathbf{X}|\theta)=\prod_{i-1}^{n}\theta^{X_i}(1-\theta)^{n-\sum X_i}=\theta^Y(1-\theta)^{n-Y}$, where $Y=\sum_{i=1}^{n}X_i$, then the Bayesian estimation of posterior distribution of θ is:

$$\pi(\theta|X)=\pi(\theta)f(X|\theta)/\int_0^1\pi(\theta)f(X|\theta)d\theta$$

3.2 The Algorithm for Extracting Interest from the Access History

Given an access history H =(T_1, T_2,..., T_n; N), we use it as the observation samples to estimate distribution parameter θ for each dimension member and N is the capacity of the samples.

```
Algorithm-1: extractInterest Algorithm
Input: H=(T_1,T_2,…,T_n;N), threshold γ
Output: user's interest I=( I_D1, I_D2,…, I_Dn)
{  For each T_i
        travel(root(T_i),I_Di);
}
travel(e_x,I_D)
{   P(e_x.key=1)=θ; f(x|θ) =θ^(e_x.count) (1−θ)^(N−e_x.count) ;
    Compute π(θ|H);
    if P(e_x.key=1)<γ, return;
    else I_Di = I_Di ∪ {e_x.key};
    for each child node e_y of e_x, travel(e_y,I_D);
}
```

Suppose the dataset have n dimensions and λ members of each dimension, then the cost of algorithm-1 is O($n\log\lambda$). In fact, according to the property-1, not all the members are taken into computation, so the real cost is less than O($n\log\lambda$).

3.3 Model Evaluation

It is obvious that, if the uninteresting members are not requested during the analysis, the model will be completely accurate. In fact, we can use the probability of this random event to represent the precision of our model. Let Ω be the set of all the dimension members, Θ be the interesting members set, and μ be the confidence of the model. Then μ can be noted as the following formulation:

$$\mu = \prod_{x\in\Omega-\Theta}(P(x=0)) \geq (1-\gamma)^{|\Omega-\Theta|}$$

The above formulation shows that, the model will be more precise if there are fewer members in $\Omega-\Theta$ or γ is getting little enough. However, in order to reduce user exploration space, we expect that there are as many members in $\Omega-\Theta$ as possible. Fortunately, when the user's interest becomes stable, the dimension members can be roughly divided into two categories: the interesting members whose probabilities are close to 1 and the uninteresting members whose probabilities are close to 0. So a threshold γ close to 0 must be enough, and thus, the γ guarantees the precision of our model.

4 Virtual Cubes Mechanism

In section-3, we have discussed how to extract the user's interest from his access history. We then create a virtual cube for this user according to his interest. When the user logs in, we track the user's analysis process to update his access history. After the user logs out, the virtual cube is incrementally maintained offline if needed.

4.1 Virtual Cubes Construction

The virtual cube is consisted of two categories of cells: virtual cells or not. If a cell of virtual cube is not a virtual cell, its value can be acquired from the original cube directly. So, only the virtual cells should be stored on the disk to speed the analysis. In this paper, we choose Dwarf [15] as the storage structure for these cells.

The virtual cube construction algorithm is presented in *Algorithm-2*. The input for the construction of a virtual cube is a cuboid of original cube. However, the input is possible not the base dataset, e.g. for level L_i of a dimension D, if $M'_i=\Phi$, then all the levels L_k that $k \geqslant i$ need not be included in the input cuboid.

```
Algorithm-2: creatVirtualCube Algorithm
Input: the original cube C, interest I=(I_D1, I_D2,…, I_Dn)
Output: the virtual cube C_V
{   Choose cuboid p of C as the input.
    Call the CreateDwarfCube[15] algorithm, while for
cell x=(x_1, x_2,…, x_n; v) of the Dwarf, x.v=vcell(x);
}
Function: vcell(x)
{   If x has been computed, then return x.v;
    If x includes none virtual member
        Get x.v from original cube, and return x.v;
    Let x_i be the first virtual member
    s=parent (x_i);
    z= (x_1, x_2,…, x_i-1, s , x_i+1,…, x_n; v);
    y=(x_1,…,x_i-1, t, x_i+1,…, x_n; v) and t∈(children(s)∩I_Di);
    x.v= vcell (z)-Σvcell(y);
    Fill the aggregated value of x;
}
```

For the step of choosing a correct cuboid as the input of the Dwarf construction, all the I_Ds should be examined that whether it is equal to {all}. The number of I_Ds is n, which is the number of dimensions. Since n is a constant for a given dataset, so the cost of this step is O(1).

The CreateDwarfCube algorithm in [15] needs a sorted dataset as its input. In the virtual cube circumstance, the input is a cuboid, and it usually is not the base dataset. So a memory sort algorithm might be enough. Let m be the size of input cuboid, then the upper-bound of overall I/O cost is O (m*log*m) and the lower-bound is O (m).

4.2 Operations on Virtual Cubes

Analysis using virtual cubes is straightforward that the users can just view the virtual cubes as the normal ones, and all the OLAP operators are available.

Theorem 4.1: The OLAP operators (slice, dice, drill and pivot) of virtual cubes are achieved with **no more** response than that with the normal cubes.

We will give the proof for drill-down operator, and for other operations the Theorem 4.1 can be proved analogously.

Proof: Given an n-dimension dataset R=(D_1, D_2,..., D_n; Measure). Let I=(I_{D1}, I_{D2},..., I_{Dn}) be the user interest of the dataset R. Suppose that G=(L_1, L_2,..., L_n; Measure) be the currently shown dataset, where L_i is a dimension level of ith dimen- sion. Let $x \in$ member(L_i), the drill-down operation on the original cube from 'x' is to fetch the cells set G_o={c|c=(x_1, x_2,..., x_{i-1},t, x_{i+1},..., x_n; v), $x_i \in L_i \wedge t \in$ children(x)}, and the drill-down operation on the virtual cube from 'x' is to fetch the cells set G_v={c|c=(x_1, x_2,..., x_{i-1},t, x_{i+1},..., x_n; v), $x_i \in V_{Di} \wedge t \in$ children(x)}.

For the virtual cube, if children(x)$\cap I_{Di}=\Phi$, let $G_v=G_1 \cup G_2$, where G_1={c| c is not a virtual cell} and G_2={c| c is a virtual cell}. All the cells in G_1 can be gotten from the original cube, so no more delay is incurred. For each cell c=(x_1, x_2,..., x_n; v) in G_2, there should be a $j \neq i$ that x_j is a virtual member, then c.v=z.v-Σy.v, where z= (x_1, x_2,...,x_{j-1}, parent(x_j), x_{j+1},..., x_n; v) and y=(x_1, x_2,...,x_{j-1}, r, x_{j+1},..., x_n; v)$\wedge$ $r \in$(children(parent(x_j))$\cap I_{Dj}$). In fact, z is a cell in Q=(L_1, L_2,...,L_i+1,...,L_j-1,..., L_n; Measure), and $y \in G_1$, thus, c.v can be computed in a constant time. Since $|G_2| \leq |G_o - G_1|$, to compute G_2 is not slower than to compute G-G_1.

If children(x)$\cap I_{Di} \neq \Phi$, $G_v=G_1 \cup G_2$, where G_1={c| c is not a virtual cell} and G_2={c| c is a virtual cell}. G_1 does not lead to more delay as explained former. While for each cell in G_2, its value is stored in the Dwarf, which supports faster point queries [15]. ■

In order to get the uninteresting information, we propose a pair of new operators operating on the virtual cube, called *wrap* and *unwrap*.

Definition 4.1: The operator *unwrap* operates on the ith member of a cell x=(x_1, x_2,..., x_n; v), noted as unwrap(x, i), where unwrap(x, i)={y|y=(x_1, x_2,...,x_{i-1},t,x_{i+1}..., x_n) $\wedge t \in$children(parent(x_i))-I_{Di}}. The *wrap* operator is the anti-operation of unwrap.

The result of unwrap is a cells set, and each cell of this set is either a normal cell or a virtual cell not existed in the virtual cube. For a normal cell, the value can be obtained from the original cube. For a virtual cell, if children(x_i)$\cap I_{Di}=\Phi$, the result of unwrap can be gotten from definition 2.3, and if children(x_i)$\cap I_{Di} \neq \Phi$, the result is stored in the Dwarf of the virtual cube.

4.3 Analysis Tracking

Analysis tracking is to update the access history of this user, by increasing the count of requested times of this member, according to the operations on the virtual cube. The analysis tracking algorithm is presented in *Algorithm-3*.

```
Algorithm-3: analysisTracking Algorithm
Input: operations, access history H =( T1, T2,…, Tn; N);
{ While (not log out)
  { Let o be the operation on i^th dimension;
    Let m be the members set requested by o;
    If o is in (slice, dice, drill-down, unwrap):
      N=N+|m|;
      for each node t of Ti, and t.key∈m
         t.count++; t.flag=true;
  }
}
```

After the user left the system, we update the Bayesian estimation of the members whose flags are 'true', then set them back to 'false'. It is apparently that the cost of the algorithm is O(1). The updated Bayesian estimation for dimension members might result in the changing of I_Ds, so we compute new I_Ds needed based on which the virtual cube should be maintained.

5 Experiments

There are three key factors that decided the effectiveness of our mechanism: the precision of Bayesian estimation, the reduction of exploration space and the disk space for virtual cube. In this section, we perform a thorough analysis of these factors.

Let R be the dataset, C_o be the original cube from R, and C_v represent the virtual cube. Suppose that the user's real interest be the cells set I_R, and $C_v=C_1 \cup C_2$, where C_1={c| c is a virtual cell} and C_2={c| c is not a virtual cell}.

The experiments were conducted on a real data warehouse application of the taxation bureau of Hubei province of China. We take two typical users in to our consideration. One is the executive, and the other is the officer. Suppose that the size of access history be Φ, i.e. the analysis has been recorded for the Φ times logging in for the user, we then regard the next logging in be his real interest of this user.

5.1 Precision of Bayesian Estimation

We use the coverage ratio $\beta = |C_2| / | I_R |$ and the out range times of the analysis request to evaluate the precision of our mechanism. The results can be shown as follows.

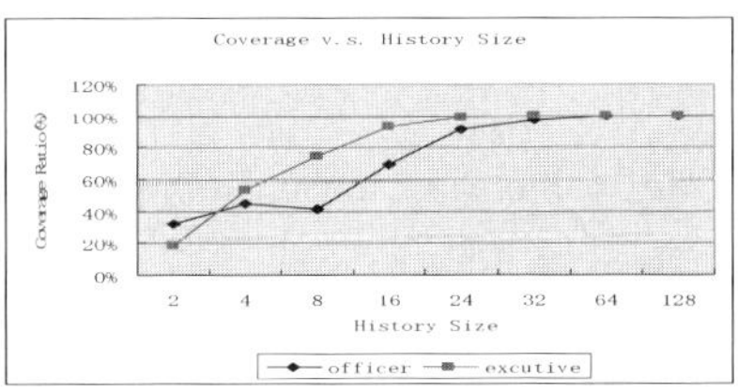

Fig. 1. Coverage Ratio vs. History Size

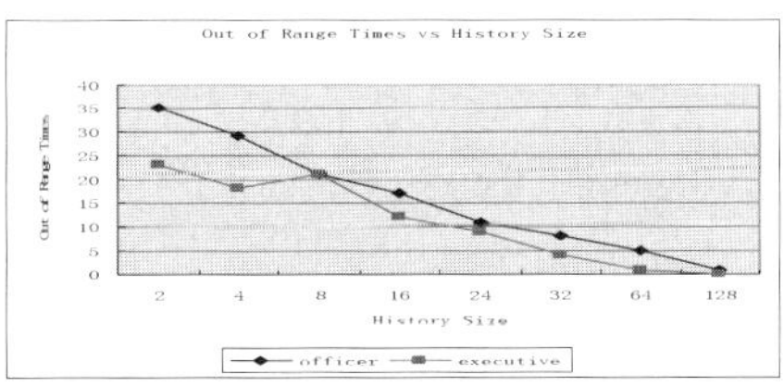

Fig. 2. Out of Range Times vs. History Size

The Fig.1 shows that the coverage ratio will increase along with the history size, while the out range times drops along with the history size in the Fig.2. For the executive, the coverage ratio increases faster and the out range times drops faster. The reason is obvious that there are less interesting cells for the executive than that for the officer. The figures above show that when the history size is greater than 24, the precision of our mechanism is effective—coverage is above 90% and the out range request is less than 10.

5.2 Exploration Space Reduction

We use the ratio $\beta = |C_v|/|C_o|$ to evaluate the exploration space reduction. Fig. 3 and Fig. 4 show the results.

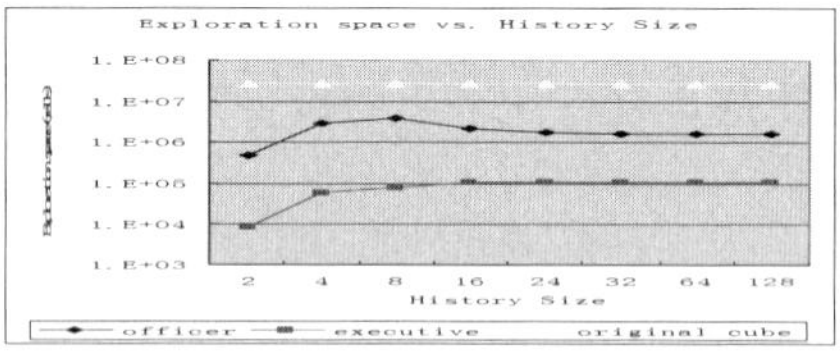

Fig. 3. Exploration Space vs. History Size

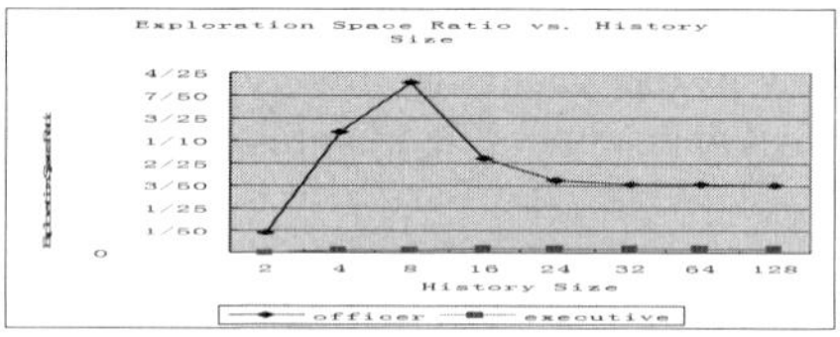

Fig. 4. Exploration Space Ratio vs. History Size

For the executive, the number of interesting cells is becoming stable soon. While for the officer, because there are more cells, so the number is fluctuant that it might greater than the real value. However, when the history size is getting greater than 15, the cells number will be nearly fixed. From figure-4, we can learn that the reduction ratios are 1/17 and 1/250 for the two users respectively.

5.3 Disk Requirement of Virtual Cube

The original cube needs 7.63GB to store it, while the disk requirement for virtual cube is much less. The experiments results can be shown in Fig. 5 and Fig. 6.

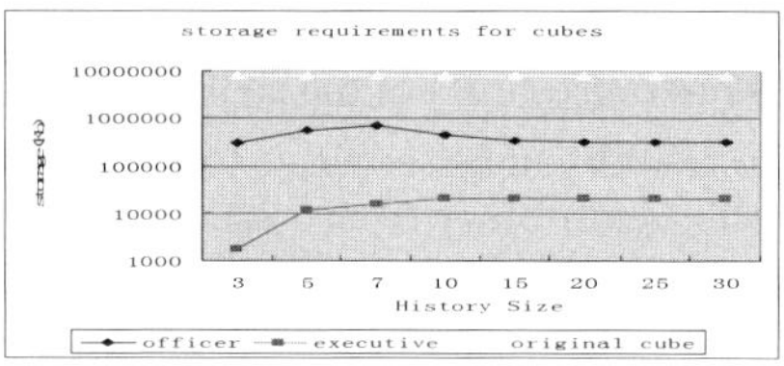

Fig. 5. Storage of the Virtual Cubes

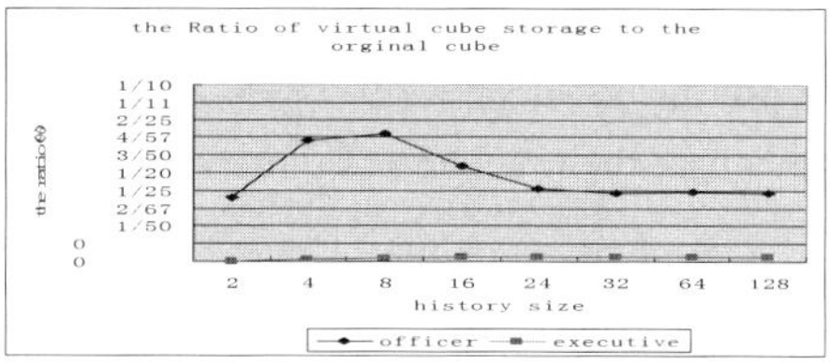

Fig. 6. Storage Ratio to the Original Cube

The lines in Fig.5 and Fig.6 have almost the same shape as those in Fig.3 and Fig.4 respectively. This is just what we expect because given the structure of dimension, it is the number of cells that decides the disk requirement if the data is not skewed. But in Fig.6, the ratio is smaller than that of the Fig.4, and the reasons are that the Dwarf eliminates the suffix and prefix redundancies and only the virtual cells are stored.

6 Conclusions

In this paper, we propose the virtual cube mechanism to reduce exploration space in OLAP. We extract the users' interests from their access histories, and effective Bayesian estimation method was employed. We prove that no more response delay is incurred due to the virtual cube mechanism. Through the experiments results, we can draw the following conclusions: the precision of the Bayesian model work well in the real application, the virtual cubes mechanism can dramatically reduce the exploration space and it only take a negligible disk requirement.

References

1. J.Vitter, M.Wang and B.Lyer.Data cube approximation and histogram via wavelets. CIKM98
2. Daniel Barbará, Mark Sullivan: Quasi-Cubes: Exploiting Approximations in Multidimensional Databases. SIGMOD Record 26(3): 12-17 (1998)
3. Jayavel Shanmugasundaram, Usama M. Fayyad, Paul S. Bradley: Compressed Data Cubes for OLAP Aggregate Query Approximation on Continuous Dimensions. KDD 1999: 223-232
4. Ying Feng, Divyakant Agrawal, Amr El Abbadi, Ahmed Metwally: Range CUBE: Efficient Cube Computation by Exploiting Data Correlation. ICDE 2004: 658-670
5. Laks V. S. Lakshmanan, Jian Pei, Jiawei Han: Quotient Cube: How to Summarize the Semantics of a Data Cube. VLDB 2002: 778-789
6. W.Wang, J.Feng, H.Lu and J.Yu. Condensed Cube: An Effective Approach to Reducing Data Cube Size. ICDE 2002.
7. Sunita Sarawagi. Explaining differences in multidimensional aggregates. VLDB'99
8. G. Sathe & S. Sarawagi. Intelligent Rollups in Multidimensional OLAP Data. VLDB'01:531-540
9. G. Cormode, F. Korn, S. Muthukrishnan, and D. Srivastava. Diamond in the Rough: Finding Hierarchical Heavy Hitters in Multi-Dimensional Data. ACM SIGMOD 2004.
10. G. Dong, J. Han, J. Lam J. Pei and K. Wang. Mining Multi-Dimensional Constrained Gradients in Data Cubes. VLDB 2001.
11. Y.Shi, Y. Song, A. Zhang. A Sringking-Based Approach for Multi-Dimensaional data Analysis.VLDB2003.
12. LI Cui-Ping, LI Sheng-En, WANG Shan, DU Xiao-Yong. A Constraint-Based Multi-Dimensional Data Exception Mining Approach. Journal of Software. 2003.
13. Hui Yu, Jian Pei, Shiwei Tang, Dongqing Yang. Mining The Most General Multidimensional Summarization of "Probable Groups" in Data Warehouses. SSDBM2005.
14. G.E.P. Box, G.C. Tiao. Bayesian Inference in Statistical Analysis. Addision-Wesley, Reading. MA, 1973
15. Y. Sismanis et al. Dwarf: Shrinking the Petacube. SIGMOD 2002.

Computational Grid-Based 3-tier ART1 Data Mining for Bioinformatics Applications*

Kyu Cheol Cho, Da Hye Park, and Jong Sik Lee

School of Computer Science and Engineering
Inha University
#253, YongHyun-Dong, Nam-Ku,
Incheon 402-751, South Korea
{landswell, audrey57}@empal.com, jslee@inha.ac.kr

Abstract. Computational Grid technology has been noticed as an issue to solve large-scale bioinformatics-related problems and improves data accuracy and processing speed on multiple computation platforms with distributed bioDATA sets. This paper focuses on a GPCR data mining processing which is an important bioinformatics application. This paper proposes a Grid-based 3-tier ART1 classifier which operates an ART1 clustering data mining using grid computational resources with distributed GPCR data sets. This Grid-based 3-tier ART1 classifier is able to process a large-scale bioinformatics application in guaranteeing high bioDATA accuracy with reasonable processing resources. This paper evaluates performance of the Grid-based ART1 classifier in comparing to the ART1-based classifier and the ART1 optimum classifier. The data mining processing time of the Grid-based ART1 classifier is 18% data mining processing time of the ART1 optimum classifier and is the 12% data mining processing time of the ART1-based classifier. And we evaluate performance of the Grid-based 3-tier ART1 classifier in comparing to the Grid-based ART1 classifier. As data sets become larger, data mining processing time of the Grid-based 3-tier ART1 classifier more decrease than that of the Grid-based ART1 classifier. Computational Grid in bioinformatics applications gives a great promise of high performance processing with large-scale and geographically distributed bioDATA sets.

1 Introduction

BioDATA is continuously increased and demands of high performance system for solving these data. GPCR (G-Protein Coupled Receptor) [1], which takes a major part at recent biology and industry, is one of the receptor-ligand interactions in signal transduction. These sequence data have been identified in current genome research and have been focused in demand of classifier for protein data mining. We induct to a data mining method for identifying GPCR with ART1 [2][3] (Adaptive Resonance Theory 1) clustering in Grid environment.

* This research was supported by the MIC(Ministry of Information and Communication), Korea, under the ITRC(Information Technology Research Center) support program supervised by the IITA(Institute of Information Technology Assessment).

L. Wang et al. (Eds.): FSKD 2006, LNAI 4223, pp. 522–531, 2006.

Recently, construction of high performance system has been noticed with a Grid-based construction [4] which is with cooperation of human and resource through data management system. The Grid computing leaps forward the next generation infrastructure. The Grid computing improves system performance and reduces the data mining processing time and required cost. Also, the Grid computing enables to reuse BioDATA. This paper designs the Grid-based ART1 classifier which supplies resources in the Grid environment. We design the ART1-based classifier and the ART1 optimum classifier for comparing improved performances of the Grid-based ART1 classifier using component units in the Grid environment.

This paper is organized as follows: Section 2 describes Grid computing and bioinformatics and bioinformatics applications with neural network. Section 3 designs the ART1-based classifier, the ART1 optimum classifier, and the Grid-based ART1 classifier with the ART1 clustering. Section 4 demonstrates accuracy improvement and data mining processing time reduction of pattern classifier through preprocess and pattern data mining and evaluates the performance of the Grid-based ART1 classifier. The conclusion is in Section 5.

2 Related Work

2.1 Grid Computing and Bioinformatics

Recently, bioinformatics demands and encompasses a high performance computing for performance evaluation and data mining with increasing of biology data processing. These demands expands a high performance computing concept to Grid computing which is focused on computational operation among computing resources on high performance networking. A combination of Grid computing and bioinformatics is used to bioinformatics application such as the GPCR data analysis. OBIGrid (Open Bioinformatics Grid) [5] is designed bioinformatics and initiative for parallel bioinformatics processing and emphasized the importance of network transparency and security policy issues. And, the bioGrid [6] fast calculates biology data through computer resources in the Grid environment. Also, it is more effective than parallel computing. The bioDATA applies the Grid computing for solving problems of real-world. And, it designs model [7] managing the bioDATA tree structure. This model is based on interface about processing result and statistical information. The bioDATA tree structure uses meta-data for management, and manages raw data and processing results. The Grid computing is used by bioinformatics implementation of the whole world. For example, UK BioGrid is called MyGrid [8], and is e-Science Grid project for automatic job flow. And, North Carolina BioGrid is called NC Grid [9][10], and provides data store and network for understanding genome variation. And, Singapore BioGrid applies Clustal-G [11] for implementing a sequence alignment. Also, the other BioGrid projects are Asia-Pacific BioGrid [12], EuroGrid [13] and so forth.

2.2 Bioinformatics Applications with Neural Network

ART1 as one of the Neural Network solves nerve network problems. Neural Network [14] is the more effective method than standard statistics methods in path physiologic system and is used treatment prediction in bioinformatics. It solves a problem by prediction and decreases error by repetitive training. The Artificial Neural Network (ANN) is a simple model operating by practice in biological nerve system. The ANN is used in various medicine diagnosis fields and is compared to diagnosis of doctor and the existing consultation fractionation [15]. The ANN training network repeats output and obtains accurate result. First ANN with false-positive detection tool [16] applied digital chest radiation to searching tumor of lung cancer. After, the ANN is widely used to minimum of chest tumor false-positive detection [17] from digital chest radiation. And, the ANN applies survival data (recurrence of death and disease) of lung cancer data to censor [18]. The Genetic Algorithm Neural Network (GANN) applying genetic algorithm [19] is the most known recognizer and is used to searching in clinically variable set. In case of train, data mining processing time increases, but the GANN gets accurate result [20] because gradient descent optimization doesn't fall a minimum. And, the Fuzzy Neural Network (FNN) [21] is relatively advanced artificial Neural Network model and is used to fast gene differentiation.

3 ART1 Classifier on Computational Grid

3.1 ART1-Based Classifier

The ART1 automatically integrates new learning knowledge into all knowledge base and solves nerve network stability and plasticity. And, the ART1 makes learned clusters with new learning data and selects new cluster for new data mining prediction category learning and keeps that the existing contents is deleted by new excess input. Therefore, the ART1 is a stable structure which is quick at learning in constantly varying environments. This result ascertains structure and function of protein, and plays an important part in constituting new protein family. The ART1 is advantage which can identify data without trained data pattern group in online state. Also the ART1 is advantage which can divide the GPCR classes of two or above.

The ART1-based classifier is a module processing classifier with the ART1 clustering algorithm for GPCR data mining. This classifier is available when user knows an optimized threshold value for high success rate of data mining prediction. Data mining prediction success rate and data mining processing time of the ART1-based classifier are set by Threshold Value Definition. And, threshold value for proper performance is known by user continuously training. But, this classifier does not assure expands of data. Fig. 1 shows function of module in the ART1-based classifier.

Threshold Value Definition processing predicts number of creating cluster providing stable data mining success rate and sets threshold value. And, the ART1-based classifier can provide stable performance by proper threshold value. But, inputting value of user does not always provide stable data mining success rate, because number of cluster about inputting threshold value varies with data. Input Data Adjustment processing is data reading and readjusting module for the ART1 data mining. This processing needs high costs because it is data input step. Training processing makes cluster which is based on threshold value definition and has unsupervised property. So, this processing makes cluster creating and training through inputted

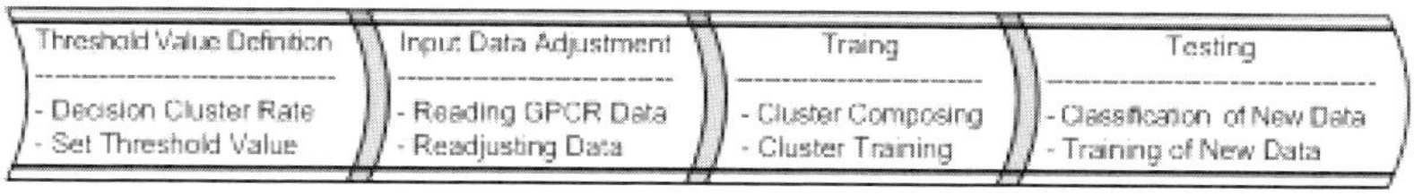

Fig. 1. Job flow of ART1-based classifier

data. Testing processing input a new data based on composing information of cluster into making cluster. And, this processing belongs to proper cluster through vigilance value.

3.2 ART1 Optimum Classifier

The ART1 optimum classifier works on a single machine for the GPCR data mining. This classifier leads in the optimal data mining processing time and assures the optimal data mining success rate. This classifier adjusts clusters, data mining processing time, and success rate through continuously data-set training, and controls threshold value. But, the ART1 optimum classifier does not assure expand for a large amount of data mining. Fig. 2 shows component progress in the ART1 optimum classifier.

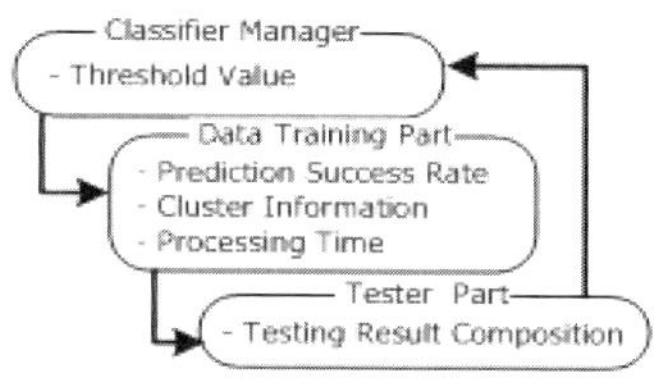

Fig. 2. Components of ART1 optimum classifier

Classifier Manager component searches for proper threshold value through data mining success rate and data mining processing time. And, searching for proper threshold provides low cost and data mining processing time because number of cluster is different according to data. Also, this component assures data mining success rate and can find out appropriate threshold using binary pattern that half value about a previous value. After Data Training Part component reads data for the ART1 data mining, this component assures cluster about data and progresses continuously training. When this component is connected, this component progresses training based on the existing data and sends composed cluster information, data mining processing time, and data mining success rate to Tester Part. Tester Part component progresses test through composed cluster information, data mining processing time, and data mining success rate, and sends results to Classifier Manager component.

In fig. 3, it is a flowchart which assures the optimal data mining success rate (95%) and uses binary pattern that half value of a previous value for getting the optimal threshold value. Threshold value assures the optimal data mining success rate and finds out threshold value providing the optimal data mining processing time. If data mining success rate is high, this module adds binary value to threshold value otherwise deducts binary value from threshold value. This processing applies threshold value to data training machine and progresses same processing.

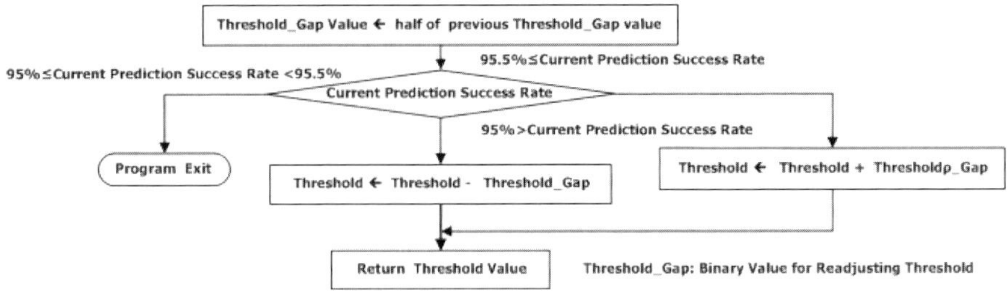

Fig. 3. Flowchart of threshold value readjustment

3.3 Grid-Based ART1 Classifier

The Grid-based ART1 classifier uses the ART1 data mining for the GPCR data mining, and leads in the optimal data mining processing time to assure the optimal data mining success rate. Each component matches resource in grid environment. This classifier applies in different data and transformed data and assures expand about the large-scale of data, because this classifier readjusts cluster information, data mining processing time, and data mining success rate through data-set training, and controls threshold value. And, this classifier does not depend on a system and divides jobs and enables to reuse, share data and makes automatically a back-up system for data access.

The Grid-based ART1 classifier usefully applies the Grid resource, and reduces total data mining processing time and provides low cost. Fig. 4 shows component progress in the Grid-based ART1 classifier.

Classifier management part is manager for applying the grid resource. Clustering rate manager adjusts result and readjusts threshold value. And, this component does broadcasting to the Grid resource launching data training parts. After tester part receives cluster information, data mining processing time and data mining success rate from data training part, this component merges results and sends test result to clustering rate readjustment part. Data Training Part receives broadcasted message from the classifier rate readjustment part, and runs clustering processing about applicable data-set. This component usefully uses the Grid resource, because this component needs processing division for data mining processing time in the Grid environment.

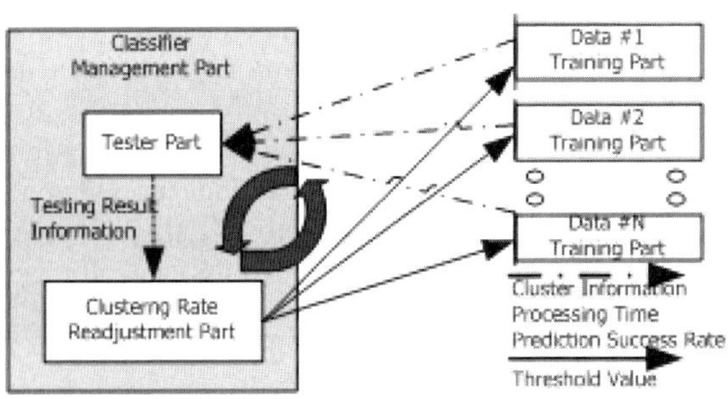

Fig. 4. Architecture of Grid-based ART1 classifier

3.4 Grid-Based 3-tier ART1 Classifier

The Grid-based 3-tier ART1 classifier expands the Grid-based ART1 classifier for the GPCR data advanced data mining, and leads in data mining processing time to assure the grid-based data mining success rate. Each component matches resource in grid environment. This classifier applies in different data and transformed data and assures expand about the large-scale of data, because this classifier readjusts cluster information, data mining processing time, and data mining success rate through transformed to agents, data-set training, and controls threshold value.

The Grid-based 3-tier ART1 classifier usefully applies the Grid resource, and reduces total data mining processing time and provides low cost for large-scale data. Fig. 5 shows component progress in the Grid-based 3-tier ART1 classifier.

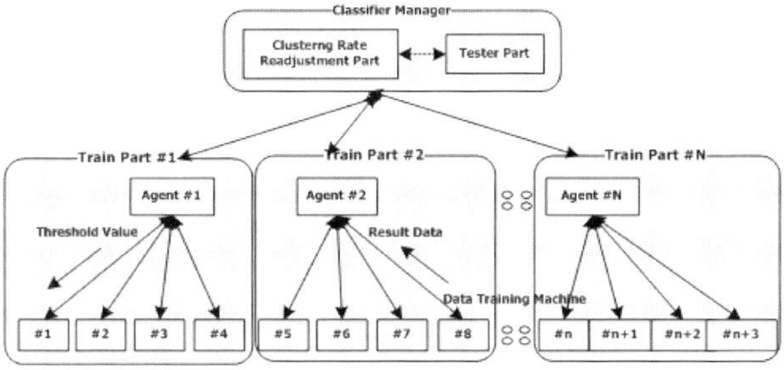

Fig. 5. Architecture of Grid-based 3-tier ART1 classifier

Classifier management is manager for applying the grid resource. Clustering rate manager adjusts result, readjusts threshold value and find out appropriate threshold using test. And, this component does broadcasting to the Grid resource launching agents. After classifier manager receives results information of clusters, data mining processing time and data mining success rate from data training part, this component merges results and sends test result to clustering rate readjustment part. Each agent receives broadcasted message from the classifier manager and broadcast threshold information message to their placed data training machine and send result information to classifier manager when received message and merge result data of their placed data training machine. And, data training mart receives broadcasted message from the agent, and runs clustering processing about applicable data-set. This component usefully uses the Grid resource, because this component needs processing division for data mining processing time in the Grid environment.

4 Experiment and Result

4.1 Grid-Based 3-tier ART1 Classifier Implementation

To simulate the Grid-based 3-tier ART1 data mining, we develop a Grid test-bed using the HLA (High-Level Architecture) [22] middleware specification and RTI (Run-Time Infrastructure) [23] implementation. As Figure 6 illustrates, the inter-federate communication works on the Grid-based 3-tier ART1 data mining federation. The federation includes one rate readjust federate and twelve training federates. The RTI message passing for data management among federates depends on the inter-federate communication inside the federation. In the platform setting, we develop the Grid system using RTI implementation which operates on Windows operating systems. The total of thirteen federates are allocated to thirteen machines, respectively, and they are connected via a 10 Base T Ethernet network.

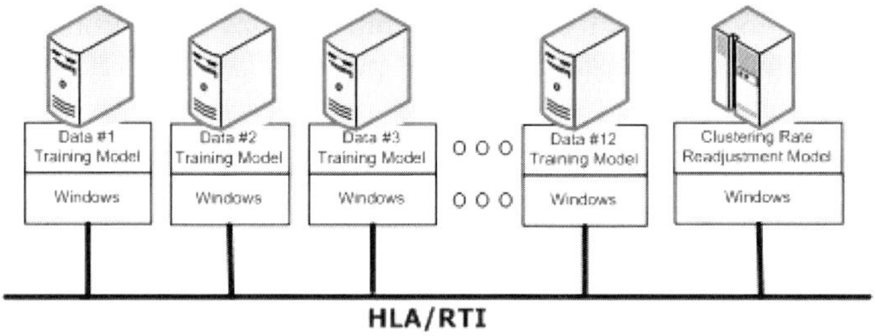

Fig. 6. Grid-based 3-tier ART1 data mining federation

4.2 Experiment 1: Data Mining Processing Time (ART1-Based Classifier vs. ART1 Optimum Classifier vs. Grid-Based ART1 Classifier)

This experiment compares data mining processing time of the Grid-based ART1 classifier to the ART1-based classifier and the ART1 optimum classifier. This experiment adjusts data-set number of three classifiers and assures the optimal data mining rate(95%). And, this experiment 1 measures data mining processing time until threshold value uses the optimal data mining processing time. We demonstrate that any classifier is useful through this experiment, because it enables to measure a possible data-set number and can estimate data mining processing cost and system stability. We compose GPCR data into 1, 6, 12, 25, 50 and 100 data-set through measure data mining processing time of classifier. Fig. 7 shows data mining processing time about three classifiers.

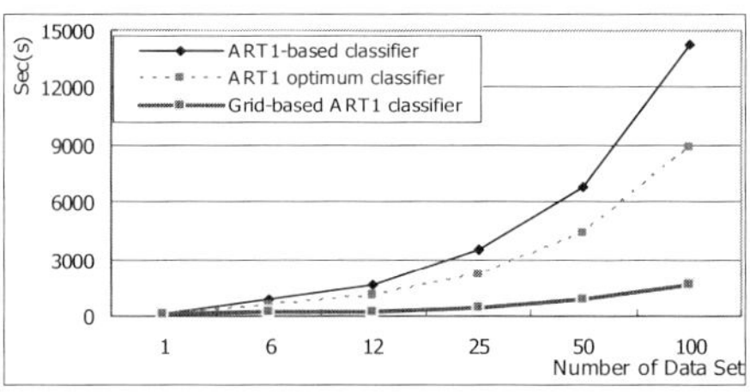

Fig. 7. Data mining processing time (ART1-based classifier vs. ART1 optimum classifier vs. Grid-based ART1 classifier)

Data mining processing time by data-set number is on the increase. If the Grid-based ART1 classifier uses 100 data-set, driving time records about 1615 seconds. And, if the ART1 optimum classifier uses 100 data-set, driving time records about 8877 seconds. This result shows that driving time of the Grid-based ART1 classifier is the 82% less than that of the ART1 optimum classifier. Also, if the ART1-based classifier uses 100 data-set, driving time records about 13850 seconds. This result shows that driving time of the Grid-based ART1 classifier is the 88% less than that of the ART1-based classifier. These results demonstrate that the Grid-based ART1 classifier reduces cost and data mining processing time than the two classifiers.

4.3 Experiment 2: Data Mining Processing Time (Grid-Based ART1 Classifier vs. Grid-Based 3-tier ART1 Classifier)

This experiment compares data mining processing time of the Grid-based ART1 classifier to the Grid-based 3-tier ART1 classifier. This experiment operates to measure data mining processing time of ART1 classifiers systems though grid data mining system composition using several grid computation resources. In large-scale data, data mining system for data needs high performance and absurdly data mining processing time. To be short data mining processing time means that data mining system makes high performance, makes efficient and makes low cost through properly grid resources management for GPCR data. This experiment adjusts data-set number of three classifiers and assures the optimal data mining rate (95%). And, this experiment 3 measures data mining processing time until threshold value uses the optimal data mining processing time. We demonstrate that any classifier is useful through this experiment, because it enables to measure a possible data-set number and can estimate data mining processing cost and system stability. We compose GPCR data into 1, 6, 12, 25, 50 and 100 data-set through measure data mining processing time of classifier.

Fig. 7 evaluates data mining processing time performance of the ART1-based classifier, the ART1 optimum classifier and the Grid-based ART1 classifier. Fig. 8 evaluates data mining processing time performance of the Grid-based ART1 classifier and the Grid-based 3-tier ART1 classifier.

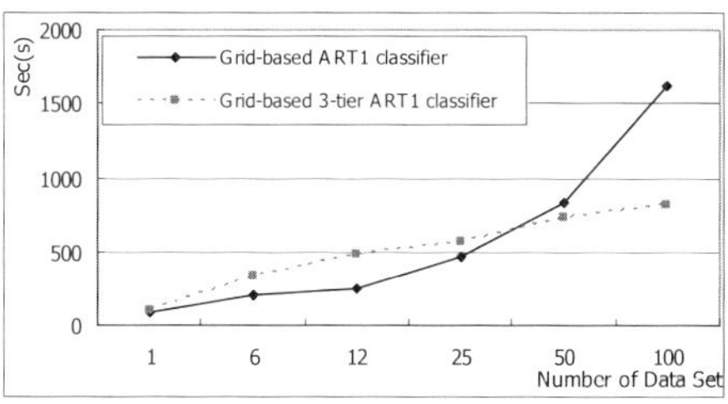

Fig. 8. Data mining processing time of classifiers (Grid-based ART1 classifier vs. Grid-based 3-tier ART1 classifier)

Data mining processing time by data-set number is on the increase. If the Grid-based 3-tier ART1 classifier uses 100 data-set, driving time records about 814 seconds. And, if the Grid-based ART1 classifier uses 100 data-set, driving time records about 1615 seconds. This result shows that driving time of the Grid-based ART1 classifier is about two times than that of the Grid-based 3-tier ART1 classifier. As more and more increasing data, calculate increasing rate becomes gradually decrease. Increasing rate of classifiers shows that the Grid-based 3-tier ART1 classifier need short time than the Grid-based ART1 classifier when large-scale data set. This result shows that driving time of the Grid-based ART1 classifier is the 88% less than that of the ART1-based classifier. These results demonstrate that the Grid-based ART1 classifier reduces cost and data mining processing time than the two classifiers.

5 Conclusion

Grid technology has been noticed as an issue to solve large-scale bioinformatics-related problems and improves data accuracy and processing speed on multiple computation platforms with distributed bioDATA sets. This paper demonstrates the Grid-based 3-tier ART1 classifier which enables to process high performance data in the Grid environment and improve accuracy and reduce data mining processing time using grid resources. We design the Grid-based 3-tier ART1 classifier and compare data mining processing time of the Grid-based ART1 classifier to those of the ART1-based classifier and the ART1 optimum classifier. The data mining processing time of the Grid-based ART1 classifier is 18% data mining processing time of the ART1 optimum classifier and is the 12% data mining processing time of the ART1-based classifier. And we evaluate performance of the Grid-based 3-tier ART1 classifier in comparing to the Grid-based ART1 classifier. As data-sets is large-scale, data mining processing time of the Grid-based 3-tier ART1 classifier more effect than the Grid-based ART1 classifier. These results demonstrate that the Grid-based 3-tier ART1 classifier tremendously reduce the GPCR data mining time. The Grid-based 3-tier ART1 classifier provides high performance data mining processing in saving overall execution time with large-scale GPCR data separated on geographically distributed computing resources.

References

1. Watson, S. and Arkinstall, S.: The G-protein Linked Receptor Facts Book. Academic Press, Burlington, MA. (1994)
2. Grossberg, S.: Neural Networks and Natural Intelligences, The definitive collection of papers from Grossberg's group. MIT Bradford Press (1988)
3. Carpenter, G.A. & Grossberg, S.: Adaptive resonance theory: Stable self-organization of neural recognition codes in response to arbitrary lists of input patterns. Proceedings of the 8th Conference of the Cognitive Science Society, Hillsdale, NJ: Erlbaum Associates (1988) 45-62
4. Foster, I. and Kesselman, C.: The Grid: Blueprint for a New Computing Infrastructure. Morgan Kaufmann, San Francisco. (1998)

5. Fumikazu, K., Hiroyuki, U., Kenji, S., and Akihiko, K.: A network design for Open Bioinformatics Grid(GBIGrid), Proc. The 3rd Annual Meeting, Chem-Bio Informatics Society, (2002) 192-193
6. Stevens, R.D., Robinson, A.J., Goble, C.A.: myGrid: personalised bioinformatics on the information grid. Bioinformatics. (2003) 302-304
7. Karo, M., Dwan, C., Freeman, J., Weissman, J., Livny, M., Retzel, E.: Applying Grid technologies to bioinformatics. 10th IEEE International Symposium on High Performance Distributed Computing. (2001) 441 –442
8. http://www.myGrid.org.uk/
9. http://www.ncbioGrid.org/
10. http://www.ncgbc.org/
11. Li, K.B.: ClustalW-MPI:ClustalW Analysis Using Distributed and Parallel Computing. Bioinformatics 19 (2003) 1585-1586
12. http://www.apbionet.org/grid/
13. http://www.eurogrid.org/
14. Baxt WG.: Application of neural networks to clinical medicine. Lancet, 346 (1995) 1135-8
15. Finne P, Finne R, Stenman UH.: Neural network analysis of clinicopathological factorss in urological disease: a critical evaluation of available techniques. BJU Int 88 (2001) 825-831
16. Lin JS, Ligomenides PA, Freedman MT, et al.: Application of artificial neural networks for reduction of false-positive detections in digital chest radiographs. Proc Annu Symp Comput Appl Med Care (1993) 434-438
17. Wu YC, Doi K, Giger ML, et al.: Reduction of false positives in computerized detection of lung ndodules in chest radiographs using artificial neural networks, discriminant analysis, and a rule-based scheme. J Digit Imaging 7 (1994) 196-207
18. Biganzoli E, Boracchi P, Mariani L, et al.: Feed forward neural networks for the analysis of censored survival data: a partial logistic regression approach. Stat Med 17 (1998) 1169-1186
19. Goldberg DE.: Genetic algorithms in search, optimization and machine learning. 1st edition. Reading, MA: Addison-Welsey Publishing Co. (1989)
20. Jefferson MF, Narayanan MN, Lucas SB.: A neural network computer method to model the INR response of individual patients anticoagulated with warfarin. Br J Haematol 89(1):29, (1995)
21. Noguch H, Hanai T, Honda H, Harrison LCm Kobayashi T.: Fuzzy neural network-based prediction of the motif for MHC class II binding peptides. J Biosci Bioeng 92 (2001) 227-31
22. Kapolka, A.: The Extensible Run-Time Infrastructure (XRTI): An Experimental Implementation of Proposed Improvements to the High Level Architecture. Master's Thesis, Naval Postgraduate School. (2003)
23. Defense Modeling and Simulation Office(DMSO). Runtime Infrastructure (RTI). http://www.dmso.mil/public/transition/hla/rti/ (2003)

Parallel Computing for Optimal Genomic Sequence Alignment

Zhihua Du[1,2], Zhen ji[1], and Feng Lin[2]

[1] Faculty of Information Engineering, Shenzhen University, Shenzhen 518060, China
[2] School of computer engineering, Nanyang Technological University, Singapore

Abstract. This paper presents a new parallel algorithm, called "block-based wavefront", to produce optimal pairwise alignment for biological sequences with reliable output and reasonable cost. It takes advantage of dynamic programming and parallel computing to produce optimal results in reasonable time. More importantly, the algorithm makes it possible for biologists to analyze datasets that were previously considered too long, often leading to memory overflow or prohibitively long time for computation.

1 Introduction

Pairwise alignment aims to find the best match between two DNA or protein sequences. In generally, two categories of methods have been recognized. The first category is the dynamic programming based technique, such as the Needleman-Wunsch algorithm [1] and the Smith-Waterman algorithm [2] [3]. The key idea is that the best alignment that ends at the positions of a given pair in two sequences is the best alignment previous to the two positions plus the score for aligning the two positions. It works by forming a dynamic programming table to characterize scores of all possible alignments, then tracing back through the table to find the maximal scoring alignment. FASTA [4] and BLAST [5] [6] are based on the secondary category: heuristic sequence comparison. Heuristic methods can only provide sub-optimal solutions in which some good answers may be left out by trading speed for precision.

2 Block-Based Wavefront

Mathematically, the dynamic programming based method is to construct an $m \times n$ matrix F, where m and n are the length of the two sequences. There are data dependencies between the matrix elements in directions of left-to-right, top-to-down and main-diagonal. These dependencies imply a particular order of computation of the matrix. The matrix F is filled from top left to bottom right with i going from 1 to m and j from 1 to n. The similarity matrix can be computed in parallel by distributing the computation along anti-diagonals because elements which can be computed independently of each other are located on a so-called *wavefront*. However, such a *wavefront* computation mode has a

L. Wang et al. (Eds.): FSKD 2006, LNAI 4223, pp. 532–535, 2006.

few problems. One problem is each parallel "wavefront" leads to lots of communications among processes. For example, after process 1 computes the top-left element, it has to send the result to process 2 and 3. Therefore, this mode demands an extremely fast inter-process communication such as on systolic arrays. The other problem is that it requires a very large number of processors if real biological data is to be considered.

To solve these problems, we proposed a "block-based wavefront" algorithm to compute blocks instead of individual elements. The algorithm divides the similarity matrix by column into p groups (p is the number of processors) evenly with a number of complete columns, and assigns each processor one such group. The columns in each processor are grouped into blocks with the height of B (B is the height of block that would be adjusted according to the number of rows). Therefore, the computation of a given block requires only the column segment of the block to its immediate left, and the main-diagonal element, a total $B+1$ elements. The parallel alignment is executed in a block-based wavefront such that computing nodes will first calculate the blocks along the first anti-diagonal in parallel, then along the second diagonal in parallel, the third, the fourth,..., until the last diagonal.

Figure 1 shows an example of computing a 16×16 matrix on 4 processors. The horizontal sequence x with 16 columns is distributed evenly to 4 processors. In each round, processes compute a 4×4 block of matrix. Initially process $p1$ starts computing block 1 in round 1. Then processes $p1$ and $p2$ can work in round 2, processes $p1$, $p2$ and $p3$ at round 3 and so on.

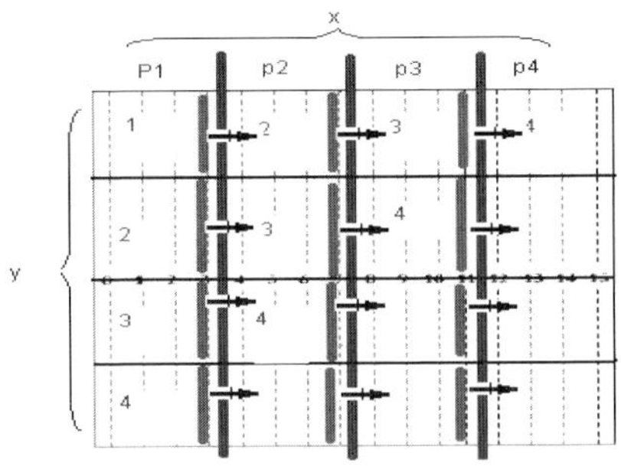

Fig. 1. Block-based wavefront with 4×4 block size

In a *wavefront* computation mode, each element on different processes can be computed only after receiving two input values (elements of the left-to-right and the main diagonal). However, in the proposed algorithm, if a block has 4 rows and 4 columns, 16 elements will be computed after receiving 5 input values. Thus, communication-to-computation ratio drops from 2:1 to 5:16, an 84% reduction. Even though this algorithm will increase some serial computations when computing elements within a block, it decreases the communication load dramatically. The computing time complexity of each process is $\frac{m \times n}{p}$ when distributing the whole work load to p processes.

This algorithm does not need a global control; that is, once it is started it continues to complete the whole matrix over the cluster of compute nodes without the external intervention. We would step back from global update and ask only what each process need to do its job and what it must pass on to other processes. In order to ensure that no data value is updated at step $t + 1$ until the data values in neighboring blocks have been updated at step t, we use block communication. Process will be stopped until the **Send** application buffer is free or **Recv** application buffer is written. That is, each block will not begin computing until its previous division has completed computing.

3 Experimental Results

In order to evaluate the performance of the proposed algorithm a dataset consisting of sequences that range from 100k to 900k nucleotides was used. The algorithm was run on 4 to 48 processors to study the execution time and speedup. For uniprocessor performance, the serial version is used as a baseline. During the experiments, the height of block (parameter "B") is assigned 100. In order to remove the unpredicted noise generated by the operating system, five consecutive runs for each pair of sequences were performed. The average results as from the five runs were used.

Table 1 lists the execution time for different problem sizes. If we use the largest data as an example, we can notice that the execution time is dramatically reduced from more than four days when running a serial program to about 2 hours when running the parallelized program on 48 processors.

Table 1. Execution time (sec)

Proc No.	1	4	8	12	16	20	24	28	32	36	40	44	48
$100k \times 100k$	3824	1159	605	487	318	254	210	188	168	162	156	144	132
$300k \times 300k$	38000	10200	5400	3999	2805	2200	1700	1500	1383	1222	1188	1061	952
$900k \times 900k$	390000	103660	53000	35000	27753	21400	17300	15300	13266	12600	11700	10200	8700

Figure 2 displays the speedup. It can be found that, for $100k \times 100K$ sequences, there is a little drop in speedup when more processors are added to the task. However, as the sequence sizes increase the speedup approaches the optimal linear speedup. The lack of speedup for the smaller dataset is a result of there not being enough jobs to fully exploit all the 48 processors' computing power. According to Amdahl's Law [7], as the problem size increases, the opportunity for parallelism grows, and the serial fraction shrinks in its importance for speedup. Thus, the best speedup curve is obtained for the largest sequences that are aligned. The granularity of work is more reasonable and the speedup becomes linear for multiple processors because of the large sequence size.

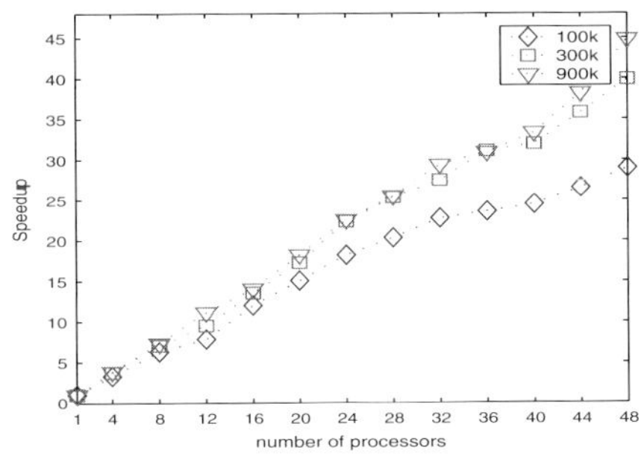

Fig. 2. Speedup of the "block-based wavefront" algorithm for optimal pairwise alignment

Acknowledgment

This work was partially supported by the National Natural Science Foundation of China under grant no. 60572100, Foundation of State Key Laboratory of Networking and Switching Technology (CHINA) and Science Foundation of Shenzhen City under grant no. 200408.

References

1. Needleman,S.B., Wunsch,C.D.: A general method applicable to the search for similarities in the amino acid sequence of two proteins. J.Mol.Biol. **48** (1970) 443–453.
2. Smith,T.F. , Waterman,M.S.: Identification of common molecular subsequences. J.Mol.Biol. **147** (1981) 195–197.
3. Smith,T.F. , Waterman,M.S.: Comparison of biosequences. Adv.Appl.Math. **2** (1981) 482–489.
4. Pearson, Lipman: Improved tools for biological sequence comparison. Proc.Natl.Acad.Sci.USA. **85** (1988) 2444–2448.
5. Altschul,S.F., Gish,W., Miller,W., Myers,E.W., Lipman,D.J.: Basic local alignment search tool. J. Mol. Biol. **215** (1990) 403–410.
6. Gish,W., States,D.J.:Identification of protein coding regions by database similarity search. Nature Genet. **3** (1993) 266–272.
7. Amdahl,G.M.: Validaity of the single processor approach achieving large-scale computing capabilities.In AFIPS Conference Proceedings. AFIPS Press,Reston. **30** (1967) 483–485.

Several Speed-Up Variants of Cascade Generalization

Zhipeng Xie

Department of Computing and Information Technology
Fudan University, Shanghai, 200433, China
xiezp@fudan.edu.cn

Abstract. Cascade generalization sequentially composes different classification methods into one single framework. However, the latter classification method has to be up against a larger attribute set. As a result, its learning process will slow down, especially for the data sets that have many class labels, and for the learning algorithms whose computational complexity is high with respect to the number of attributes, among others. This paper is devoted to propose several variants of the original cascade generalization, where the basic idea is to reduce the number of augmented attributed each time in the Cascade framework. Extensive experimental results manifest that all the variants are much faster than the original one as supposed. In addition, all the variants have achieved a little reduction of error rates, compared with the original Cascade framework.

1 Introduction

Cascade generalization, presented in [2], is a sequential composition of several classification algorithms into one single framework. At each step, a classification algorithm is applied to the data set to generate a base classifier, and then the probability class distributions generated by the base classifier are used as the new attributes that are inserted into the data set. Thereafter, the subsequent classification method will be applied to this larger data set (with more attributes, but same number of examples) and insert its own class distributions as new attributes. Experimental results provide in [2] manifested that Cascade generalization outperforms substantially Stack generalization and Boosting.

Consider a training set $D = \{ (x_i, y_i) \mid i=1, \ldots, n \}$, where $x_i = [a_1(x_i), a_2(x_i), \ldots, a_m(x_i)]$ is an instantiation of the conditional attribute set $\{a_1, a_2, \ldots, a_m\}$, and y_i takes value from a set of predefined class labels, that is $y_i \in \{Cl_1, \ldots, Cl_c\}$, where c is the number of classes. Let $\mathcal{M}$ be a classification algorithm, we use $\mathcal{M}(D)$ to denote the classifier learned by applying $\mathcal{M}$ on the data set D. This classifier, also represented by $\mathcal{M}(x, D)$, assigns a class label to an example x. This is the typical description for classification task.

Howbeit, in the framework of cascade generalization, it is required that each classifier $\mathcal{M}(x, D)$ should output a probability distribution, $[p_1, p_2, \ldots, p_c]$, over the possible class labels, where p_i ($1 \le i \le c$) represents the probability with which the example x belongs to the i-th class label Cl_i. The class label with the highest probability is assigned to the example as the predicted value.

L. Wang et al. (Eds.): FSKD 2006, LNAI 4223, pp. 536–540, 2006.

Given a classifier $\mathcal{M}(D)$, each example x in the instance space could be concatenated with the output probability class distribution. Here, the example after concatenation is denoted as $x'=\varphi(x, \mathcal{M}(x, D))$. If the operator φ is applied to all the examples in a data set D_1, we could derive a new data set $\Phi(D_1, \mathcal{M}(D_1, D)) = \{\varphi(x, \mathcal{M}(x, D)) \mid x \in D_1\}$. This new data set contains all the examples that appear in D_1 augmented with the $\#c$ new attributes that generated as the probability class distribution by $\mathcal{M}(D)$.

Given training set D, and two classification algorithms $\mathcal{M}_1$ and $\mathcal{M}_2$. Cascade generalization will use $\mathcal{M}_1$ to generate the $Level_1$ training data

$$Level_1train = \Phi(D, \mathcal{M}_1(D, D))$$

Classification algorithm $\mathcal{M}_2$ is applied on $Level_1$ training data, with $\mathcal{M}_2(Level_1train)$ as the learned classifier. After this learning process, if an unlabelled example x is given, it will be firstly concatenated with the probability class distribution generated by $\mathcal{M}_1(D)$. We use $\varphi(x, \mathcal{M}_1(x, D))$ to denote it after concatenation. Then, $\mathcal{M}_2(Level_1train)$ is used to make the decision, with $\mathcal{M}_2(\varphi(x, \mathcal{M}_1(x, D)), Level_1train)$ as the output.

2 Several Variants of Cascade Generalization

In Cascade generalization, the $Level_i$ training set is produced by augmenting $\#c$ new continuous attributes to the $Level_{i-1}$ training set. When the classification problem at hand has many class labels, the learning process of the latter classifier will be slowed down a lot, especially for those classifiers with high computational complexity with respect to the number of attributes. In addition, continuous attributes are normally more time-consuming than nominal attributes, for most learning algorithm. For example, C4.5, a famous decision-tree algorithm, will spend $O(n \times \log n)$ for each continuous attribute at each node, where n is the number of instances at this node, while it will only spend $O(n)$ for each nominal attribute.

The aim of this section is to develop several variants of the original cascade generalization, which are expected to be faster than and also to be comparable in accuracy to the original one.

The first variant: class label vs. class distribution

The first variant, called CASC-A, is to extend the data set with one nominal attribute each time instead of $\#c$ continuous attributes in original CASC. This nominal attribute carries the information of class label predicted by the current base classifier. Although it was stated in [2] that combining classifiers by means of categorical classes looses the strength of the classifier in its prediction, our experiments in section 5 show that such loss is none of importance in the framework of cascade.

Why we could do that? The reason is that the $\#c$ continuous attributes appended by the original Cascade contain a lot of insignificant information and redundant information. The $\#c$ continuous attributes form a probability class distribution. For a given example, its values on these $\#c$ attributes sum up to be 1. Thus, there is one redundant attribution among the $\#c$ attributes. In addition, Due to the fact that the class distribution is generated by a classifier, it is very common that only one of them has a big value (probability) while the others take small values.

Formally, the learning process is depicted as follows: Given training set D with attribute set $\{a_1, a_2, \ldots, a_m\}$, and two classification algorithms $\mathcal{M}_1$ and $\mathcal{M}_2$. Let $\mathcal{M}_1(D)$ be the classifier generated by applying $\mathcal{M}_1$ on D, which assigns a class label to an example. We could define a new attribute a_{meta} that $a_{meta}(x) = \mathcal{M}_1(x, D)$ for each $x \in D$. After that, each example $x \in D$ can be transformed into a newly generated example $x'=[a_1(x), a_2(x), \ldots, a_m(x), a_{meta}(x)]$. Let us denote the set of all the generated examples as D', which has the attribute set $\{a_1, a_2, \ldots, a_m, a_{meta}\}$. Algorithm $\mathcal{M}_2$ is then applied on D' to get the classifier $\mathcal{M}_2(D')$.

Because CASC-A uses one nominal attribute representing the class labels predicted by the current base classifier, it makes CASC-A look like Stacked generalization. But CASC-A still belongs to the cascade family, because all classifiers have access to the original attributes of the data sets.

The second variant: binarization of class label

The second variant, CASC-B, is similar to the first variant in that they do not require the base classifiers output probability class distribution, but different in that #c binary attributes are generated each time, which are the corresponding binarization of the nominal attribute in CASC-A.

Let $a_{meta,1}, a_{meta,2}, \ldots, a_{meta,c}$ denote the #c binary attributes. For a given example x, $a_{meta,i}(x) = 1$ if and only if $\mathcal{M}_1(x, D) = Cl_i$, otherwise $a_{meta,i}(x) = 0$.

The third variant: class label plus probability

We have to admit that using the predicted class label instead of the probability class distribution means discarding some information. As remedy, we could add another continuous attribute, which records the corresponding conditional probability of the predicted class label. The value of this attribute is the maximum value in the vector of the probability class distribution. This attribute represents in some degree the correctness possibility of the prediction stored in the first nominal attribute.

This variant, called CASC-APROB, augments 2 attributes (1 nominal attribute and 1 continuous attribute) each time. Here, it is required that base classifiers could output the probability class distribution. For a given example x, let $[p_1, p_2, \ldots, p_c]$ be the output class distribution of $\mathcal{M}_1(x, D)$. If $t = \arg\max_{1 \le i \le c} p_i$, the predicted class label of $\mathcal{M}_1(x, D)$ should be Cl_t, and the maximum value in the vector of class distribution is p_t. Here, if we use a_{meta1} and a_{meta2} to denote the augmented nominal attribute and continuous attribute respectively, then we have $a_{meta1}(x)=Cl_t$ and $a_{meta2}(x)=p_t$.

3 Experimental Results and Summary

We used 30 datasets from UCI machine learning repository [1] to evaluate the algorithms in this paper. Table 1 lists all the datasets used and the related characteristics. Ten-fold cross validation was executed on each dataset to obtain the results. For comparison of these algorithms, we made sure that the same cross-validation folds were used for all the different learning algorithms involved in the comparison. All the algorithms were coded in Visual C++ 6.0 with the help of standard template library.

Because the experimental results in [2] have already shown that the most promising combination is to use a decision tree as the high-level classifier and naïve Bayes or Discrim as low-level classifiers, naïve Bayes and a decision tree algorithm (C4.5)

Table 1. Datasets Information

DATA SET	#EXMP	#CLS	DATA SET	#EXMP	#CLS	DATA SET	#EXMP	#CLS
ADULT	48842	2	HEART	270	2	SATIMAGE	6435	6
ANNEAL	898	5	HEPATITIS	155	2	SEGMENT	2310	7
AUSTRALIAN	690	2	HORSE	368	2	SHUTTLE-SMALL	5800	6
BREAST	699	2	HYPO	3163	2	SICK	2800	2
CHESS	3196	2	IONO	351	2	SOLAR	323	6
CLEVE	303	2	LETTER	20000	26	SOYBEAN-LARGE	683	19
CRX	690	2	MUSHROOM	8124	2	TIC-TAC-TOE	958	2
DIABETES	768	2	NURSERY	12960	5	VEHICLE	846	4
GERMAN	1000	2	PENDIGITS	10992	10	VOTE	435	2
GLASS	214	7	PIMA	768	2	WAVEFORM	5000	3

are here used as the low-level classifier and high-level classifier respectively. The algorithms evaluated here include: (1) CASC (the original Cascade generalization algorithm), (2) CASC-A, CASC-B, CASC-APROB (the algorithms described in this paper), (3) C4.5 (the well-known decision tree algorithm).

Error rate comparison

The mean error rates across all the experimental domains are presented in Table 2. In addition, with the significance level set at 95%, statistical comparisons between algorithms were conducted for each data set, using the paired sample t-test. For the space limitation, the detailed information of error rates is omitted.

Table 2. Error rates on the experimental domains

	C4.5	CASC	CASC-A	CASC-APROB
MEAN	13.35	12.60	12.31	12.39

Table 3. Summary of Comparisons with CASC

	CASC-A VS. CASC	CASC-APROB VS. CASC
Number of wins	16/12	19/9
Significant wins	5/3	2/1

Computational efficiency comparison

Let us have a look at those data sets whose attributes are all nominal and those that contains at least 10 class labels. The information about training time on those domains is displayed in table 4. Conclusions can be drawn easily that both CASC-A and CASC-APROB run much faster than CASC. For the other datasets, the variants proposed in this paper are also faster, but we do not list all of them here.

Table 4. Computational efficiency comparison (in seconds)

DATA SET	CASC	CASC-A	CASC-APROB	DATA SET	CASC	CASC-A	CASC-APROB
LETTER	16.72	4.59	4.75	NURSERY	1.405	0.086	1.472
PENDIGITS	1.87	0.62	0.63	SOYBEAN-L	0.117	0.037	0.0391
CHESS	0.195	0.044	0.091	TICTACTOE	0.061	0.009	0.0218
MUSHROOM	0.247	0.047	0.07				

Acknowledgements. This work was funded in part by National Natural Science Foundation of China under grant number 60503025, and the Science & Technology Commission of Shanghai Municipality under grant number 03ZR14014.

References

[1] Blake, C. L., & Merz, C. J.: UCI repository of machine learning databases. University of California, Irvine, CA (1998) http://www.ics.uci.edu/~mlearn/MLRepository.html

[2] Gama, J., and Brazdil, P.: Cascade Generalization. Machine Learning, 41 (2000) 315-343.

An Improvement of Posteriori

Zhi-Hong Deng

National Laboratory of Machine Perception, School of Electronics Engineering and Computer Science, Peking University, Beijing 100871, China
zhdeng@cis.pku.edu.cn

Abstract. Interactive mining, which is the problem of mining frequent itemsets in a database under different thresholds, is becoming one of the interesting topics in frequent itemsets mining because of needs of practical application. In this paper, we propose a heuristic method for greatly decreasing the number of possible candidates in Poteriori, which is an algorithm based on Apriori for interactive mining. Fewer possible candidates make Poteriori more efficient. Analysis based on an example shows the advantage of our method.

1 Introduction

Data mining, or knowledge discovery in databases (KDD) has attracted tremendous amount of attention in the database research community due to its wide applicability in many areas. Frequent itemsets mining plays an essential role in many important data mining tasks [1]. Since the first introduction of mining of frequent itemsets in [2], various algorithms have been proposed to discover frequent itemsets efficiently.

However, most of the previous work has focused on mining frequent itemsets under specified minimum support threshold (or minsup for short) as soon as possible, and very little work has been done on the mining problem where minsup may change. As stated in [3], users are often unsure about their requirements on the minimum support at first due to the lack of knowledge about the application domains or the outcomes resulting from different threshold settings. Therefore, they may have to re-execute the mining procedure many times with varied thresholds in order to get satisfied results in real-world applications. This simple method of re-execution is clearly inefficient because all the computations done initially for finding the frequent itemsets under old thresholds are wasted. As a result, it is both desirable and imperative to develop effective approaches for interactive mining, which is the problem of mining frequent itemsets in a database under different thresholds. To deal with such kind of mining problem, Liu and Yin proposed an algorithm called Posteriori [3], which is based on Apriori.

By analyzing Poteriori, we find that the way of joining itemsets to generate possible candidates is inefficient. Therefore, we propose a new method for joining itemsets in this paper. Analysis based on an example shows the advantage of our method.

The remaining of the paper is organized as follows. Section 2 gives a detailed problem description. Section 3 presents our heuristic method. Section 4 summarizes our study and points out some future research issues.

L. Wang et al. (Eds.): FSKD 2006, LNAI 4223, pp. 541 – 544, 2006.

2 Problem Description

Let $I = \{a_1, a_2, \ldots, a_m\}$ be a set of items. Let $DB = \{T_1, T_2, \ldots, T_n\}$ be a transaction database ,where T_k ($k \in [1..n]$) is a transaction which has a unique identifier and contains a set of items in I. Given an itemset P ($\subseteq I$), which is a set of items, a transaction T contains P if and only if $P \subseteq T$. The support of P is the number of transactions containing P in DB. An itemset P is a frequent itemset if P's support is no less than a predefined minsup ξ.Given a transaction database DB and a minsup ξ, the problem of finding the complete set of frequent itemsets is called the frequent itemsets mining problem.

Let FI be the set of frequent itemsets in the database DB, and ξ be the minimum support. After users have found some frequent itemsets, they may be unsatisfied with the mining results and want to try out new results with certain changes on the minimum support thresholds, such as from ξ to ξ'. We call mining frequent itemsets under different minsups is interactive mining of frequent itemsets. The essence of the problem of interactive mining is to find the set FI' of frequent itemsets under a new minsup ξ'.

When the minsup is changed, two cases may happen:

1. $< \xi'$: some frequent itemsets in FI will become infrequent under ξ'. Therefore, these frequent itemsets don't belong to FI'.
2. $\xi > \xi'$: all frequent itemsets in FI will still be frequent under ξ'. Therefore, FI is a subset of FI'. At the same time, some itemsets, which don't belong to FI, will become frequent under ξ' and become an element of FI'.

For the first case, the finding of frequent itemsets is simple and intuitive. Just select those frequent itemsets in FI with support no less than ξ', and put them to FI'. In the paper, we concentrate on the second case.

3 A Heuristic Method for Joining Itemsets

Limited by space, we omit the description of Posteriori. Please refer to [3] for the details of Posteriori.

The method adopted by Posteriori to generate C_k[3] obtains all candidates by emulating all kinds of possible itemsets. This emulating method may generate a large number of possible candidates. Let n_i be the number of itemsets in $L_i[1]$ and m_i be the number of itemsets in $L_i[2]$. The number of possible candidates with length of k will be $\sum_{i=1}^{k-1} n_i m_{k-i}$, which is often a very big number. For ensuring a possible candidate C to be a candidate, we must check whether each $(k-1)$-subset S of C is in L_{k-1}'. As we know, the checking process is time-consuming. Therefore, a large number of possible candidates may result in consuming too much time. This will finally affect the efficiency of Posteriori. In fact, we can decrease the number of possible candidates dramatically by some heuristic rule. Before describing our method, we first give one concept and two lemmas.

Let $L_1 = \{x_1, x_2, \ldots, x_s\}$, $A_1 = \{y_1, y_2, \ldots, y_t\}$, and L_2' be the set of all frequent 2-itemsets under the new minsup ξ'.

Definition 1 (antipathic set): Let x be a 1-itemset in L_1. The antipathic set of x is defined as $\{y\in A_1 \mid \{x, y\} \notin L_2'\}$.
The antipathic set of x is also denoted as $x^{\sim}$ for short.

Theorem 1 (Apriori property): All the subsets of a frequent itemset must also be frequent.

Proof. This is the basic property of frequent itemsets. Its proof can be found in [4].

Lemma 1: if an itemset is not frequent, its superset can never be frequent.

Proof. Let P be an itemset that is not frequent and Q be a superset of P. If Q is frequent, we have P is frequent according to Theorem 1. This conflicts with the fact that P is not frequent. Therefore, Q is not frequent.

Lemma 2: Let P be a k-itemset ($k \geq 3$). If $\exists\, x, y\in P \wedge y\in x^{\sim}$, then P must be an infrequent itemset under minsup ξ'.

Proof. Assume that there exist $x, y\in P$ and $y\in x^{\sim}$. According to the definition of antipathic set, we know that 2-itemsets $\{x, y\}$ is not frequent. Because of $x, y\in P$, we know that $\{x, y\}$ is a subset of P. According to lemma 1, we have that P is not frequent.

By exploring lemma 2 in the process of generating $C_k[3]$, we can greatly decrease the number of possible candidates. The method is as follows. Before combining a frequent itemset P in $L_i[1]$ with a frequent itemset Q in $L_{k-i}[2]$ to generate a possible candidate, we first check whether elements in Q are in antipathic set of elements in P. If $y\in Q$ is in the antipathic set of $x\in P$, $P \cup Q$ can not be frequent according to lemma 2. Therefore, $P \cup Q$ does not need to be regarded. The following procedure shows the details of our heuristic method.

```
Procedure Heuristic_Generate_3(k(≥3): the length of
itemsets)

for i = 1 to k-1 {

    for each frequent itemset P in L_i[1] {

        P~ = ∅;

        for each element x in P {

            P~ = P~ ∪ x~; }

        for each frequent itemset Q in L_{k-i}[2] {

            if Q ∩ P~ ≠ ∅ then do next;

            else {

                C = P ∪ Q;
```

```
                    if each (k–1)-subset S of C is in L'_{k-1} then
                    add C to C_k [3];}}
        }
    }
    return C_k [3];
```

For depicting the efficiency of Heuristic_Generate_3 more distinctly, Let's consider an example as follows.

Let $L_1 = L_1[1] = \{x_1, x_2, x_3\}$, $A_1 = L_1[2] = \{y_1, y_2, y_3\}$, $L_2[1] = \{\{x_1, x_2\}, \{x_1, x_3\},\{x_2, x_3\}\}$, $L_2[2] = \{\{y_1, y_2\}, \{y_1, y_3\},\{y_2, y_3\}\}$, $L_2[3] = \{\{x_1, y_1\}, \{x_1, y_2\}\}$. Now, we consider the generation of $C_3[3]$. By Generate_3, we know the number of possible candidates is equal to 18(=3×3+3×3). According to the definition of antipathic set, we have $x_1^{\sim} = \{y_3\}$, $x_2^{\sim} = x_3^{\sim} = \{y_1, y_2, y_3\}$. By Heuristic_Generate_3, the only possible candidate is $\{x_1, y_1, y_2\}$. The radio is 18:1. It shows that Heuristic_Generate_3 is much better than Generate_3.

4 Conclusions

In this paper, we presented a heuristic method for efficiently decrease the number of possible candidates in Posteriori. By filtering those infrequent itemsets in advance, this method can increase the efficiency of Posteriori.

Recently, there have been some interesting studies at mining maximal frequent itemsets [5, 6] and closed frequent itemsets [7, 8]. The extension of our technique for interactive mining of these special frequent itemsets is an interesting topic for future research.

References

1. J. Han, J. Pei, and Y. Yin. Mining frequent patterns without candidate generation. In SIGMOD'00, pp. 1-12.
2. R. Agrawal, T. Imielinski, and A. Swami. Mining Association Rules between Set of Items in Large Databases. In SIGMOD'93, pp. 207-216.
3. J. Liu and J. Yin. Towards efficient data re-mining (DRM). In PAKDD'01, pp. 406-412.
4. R. Agrawal and R.Srikant. Fast algorithm for mining Association rules. In VLDB'94, pp. 487-499.
5. R. J. Bayardo. Efficiently mining long patterns from databases. In SIGMOD'98, pp. 85-93.
6. D. Burdick, M. Calimlim, and J. Gehrke. MAFIA: A maximal frequent itemset algorithm for transactional databases. In ICDE'01, pp. 443-452.
7. M. Zaki and C. Hsiao. CHARM: An efficient algorithm for closed itemset mining. In SDM'02, pp. 12-28.
8. J. Y. Wang, J. Han, and J. Pei. CLOSET+: Searching for the Best Strategies for Mining Frequent Closed Itemsets. In SIGKDD'03, PP. 236-245.

An Approach Based on Wavelet Analysis and Non-linear Mapping to Detect Anomalies in Dataset*

Song Yanpo[1,2], Tang Ying[3], Peng Xiaoqi[1,2], Wang Wen[2], and Tang Lu[2]

[1] School of Information Science and Engineering, Central South University, Changsha, 410083, China
[2] School of Energy Science and Engineering, Central South University, Changsha, 410083, China
[3] School of Physics Science and Technology, Central South University, Changsha, 410083, China

Abstract. An approach based on wavelet analysis and non-linear mapping is proposed in this paper. Using the non-linear mapping to decrease the dimensions of data, taking full advantage of wavelet analysis' superiority in local analysis, the approach is able to detect anomalies accurately. The experiments show that the approach is accurate and practical.

1 Introduction

Anomalies in dataset are the samples that are considerably different from the remainders as if they are generated by a different mechanism. The anomalies in numeric dataset can be classified into two classes: outliers and violators. Mapping all the samples in dataset to a multi-dimensional space, some points probably are apart from any cluster, the corresponding samples are defined as outliers; while some samples' relation among some attributes (for example, among the independent variables and the dependent variables) is very different from others, they are defined as violators. Anomalies analysis has been seen as one of the radical tasks of data mining. There are many methods to detect outliers, such as the methods based on statistics[1], distance[2], density[3], or clustering information[4]. But, there are not so many effective methods to detect violators. The method integrating semantic knowledge[5] can be seen as a method to detect violators, but it can only deal with the data samples which decision attribute (dependent variable) can provide clear information for classifying. Simon Hawkins et al proposed to detect anomaly using RNN (Replicator Neural Networks), this method detects anomaly by its reconstruction. In theory, the method can detect both the outliers and violators, but in fact, its performance is in close relation to the structure and train algorithm of RNN, which are not easy to be decided.

Wavelet analysis is an effective mathematic tool which can detect the high frequency transformation of signal, while the violators can be seen as the points where high frequency transformation happened in a sense, so it can be used to detect anomalies in dataset. But because multi-dimensional wavelet analysis is difficult to understand and realize, dimensions reducing is often need before wavelet analysis for

* This research is supported by National Natural Science Foundation of China (50374079) and Specialized Research Fund for the Doctoral Program of Higher Education(20030533008).

L. Wang et al. (Eds.): FSKD 2006, LNAI 4223, pp. 545 – 548, 2006.

some multi-dimensional data set. Therefore, an approach based on wavelet analysis and non-linear mapping (*NLM*) to detect anomalies in data set with multi independent variables and single variable is proposed, this approach is shorten as "*WANLM*" in this paper. *WANLM* includes three primary phases: 1) transform the independent variables of the sample *p* and its *K* nearest neighbors to two-dimensional data using *NLM*; 2) compute the Wavelet Transformation Coefficient (short as *WTC*) of the sample *p* based on the transformed data; 3) judge the sample *p* is whether anomaly or not by comparing with a threshold *T*.

2 Description for *WANLM*

2.1 Revised *NLM*

NLM transforms the *K*+1 data with more than 2 dimensions into two-dimensional data and keeps the distances between them as much as possible[6]. In this paper, it is used to reduce the dimensions of data and prepare for the following wavelet analysis, so keeping the distances from sample *p* to its neighbors is more important than keeping the distances among its neighbors. Whereas, error function is revised as follow:

$$E=\left(\sum_{i<j}^{K+1}(d_{ij}^{*}-d_{ij})^{2}/d_{ij}^{*}+\sigma\sum_{j\neq p}^{K+1}(d_{pj}^{*}-d_{pj})^{2}/d_{pj}^{*}\right)\Big/\left(\sum_{i<j}^{K+1}d_{ij}^{*}+\sigma\sum_{j\neq p}^{K+1}d_{pj}^{*}\right) \quad (1)$$

Where, d^{*}_{ij}, d_{ij}, are the distances between the sample *i* and sample *j* before and after the transformation. σ is a revised coefficient bigger than 0.

2.2 Two-Dimensional Wavelet Transformation

One kind of two-dimensional wavelet transformation is defined as follow:

$$WT_{f}(a,b_{1},b_{2})dx_{2}=\frac{1}{a}\iint f(x_{1},x_{2})\varphi\left((x_{1}-b_{1})/a,(x_{2}-b_{2})/a\right)dx_{1} \quad (2)$$

Two-dimensional Marr wavelet is often used as core function of wavelet transformation, its analytic expression is $\varphi(x_1,x_2)=(2-x_1^{2}-x_2^{2})e^{-(x_1^2+x_2^2)/2}$, it also can be expressed in form of polar coordinate: $\Phi(r,\theta)=(2-r^{2})e^{-r^{2}/2}$. For the sake of computing convenience, Marr wavelet function is revised so that $\Phi(r,\theta \mid r=1.4 \ or \ r\geq 3.5)=0$, which is used as the core function of wavelet transformation in this paper.

2.3 Main Steps of *WANlM*

The main steps of *WANLM* to detect anomalies from data set $\{(x_{i1}, x_{i2},\ldots\ldots, x_{im}, y_i) \mid i=1,2,\ldots\ldots, N\}$ as follow:

1. Set the parameters relational to *WANLM* such as the scale coefficient *a* of wavelet analysis, the total amount *K* of neighbors.

Scale coefficient *a* decides the size of computational scope of wavelet transformation, the smaller scale coefficient *a* is, the smaller the domain is. In order to set a proper value for *a*, calculate the average distance among samples in data set as: $d_m=\sum_{i=1}^{N-1}dn_i/N$, where, dn_i is the least distance from sample *i* to other sample in dataset.

Then, set :

$$a=ka*d_m/2.8 \quad (3)$$

where, *ka* is a undetermined coefficient, which reflect approximately the ratio of radius of computational scope to average distance. In general, it can be set as 1~10. It maybe lose some anomalies using only one scale coefficient, therefore multi-scale analysis is often needed.

K should be big enough so that the domain of neighbors can cover the computational scope, but too big *K* will aggravate the computing burden and error of *NLM*. Then, set:

$$K=\lceil (2*ka+0.5)*k_c \rceil \quad (4)$$

where, k_c is the total amount of independent variables, $\lceil x \rceil$ is the minimal integer which is not smaller than x.

2. Set $j = 1$.
3. Searching *K* neighbors for sample *j*.
4. Using the revised *NLM* proposed in section 2.1, transform the independent variables of sample *j* and its *K* neighbors, keep the dependent variables unchanged.
5. According to section 2.2, compute the *WTC* of sample *j*, marked it as wt_j.
6. If $j< N$, let $j = j+1$ and go step 4; otherwise, go step 7.
7. Standardize every sample's *WTC* using following expression: $wt'_p = (wt_p-M)/S$, where, $p(1\leqslant p\leqslant N)$ is the serial number of a sample, wt_p is *WTC* of this sample, *M* is criterion value which can be set as the average or 0, *S* is the standard deviation.
8. According to the threshold which has been set for detecting, sign the samples whose standardized *WTC*s are larger than threshold as anomalies. In general, the threshold *T* can be set as 2.0~4.0.
9. Output the serial number of anomalies, end.

3 Simulated Experiments and Result Analysis

Produce 400 data with five independent variables and single dependent variable $\{(C_{i1}, C_{i2}, C_{i3}, C_{i4}, C_{i5})|i=1,......,400\}$ as experimental data, the independent variables of every sample is random numbers varying from –2 to 2, the corresponding decision attribute D_i is:

$$D_i = \sin(C_{i1}) + \sin(C_{i2}) + 0.3*C_{i3} + 0.3*C_{i4} + 0.4*C_{i5} \quad (5)$$

Give every sample a unique serial number from 1 to 400, add noise to the sample which serial number is integer multiple of 20, that is:

$$D_{j\times 20} = D_{j\times 20} \pm 1.0\text{,}\ j=1, 2, \ldots\ldots, 20 \quad (6)$$

Therefore, 20 abnormal samples are made. The original experimental data is shown in Fig.1, vertical ordinate shows the value of decision attribute of every sample, and horizontal ordinate shows the serial number of every sample. Obviously, it is impossible to detect anomalies in data set according to Fig.1 Set ka = 1, compute the scale coefficient of wavelet analysis a and the total amount of neighbors K according to expressions (3) and (4), then compute *WTC*s of the samples and standardize them, results are show in Fig.2.

As the shown in Fig.2, *WTCs* of the polluted samples are relatively bigger than others. If set the threshold T=2.5, 18 samples would be detected as anomalies by *WANLM*, 3 real anomalies would be missed, 1 normal sample would be misjudged as anomaly. If set the threshold T=2.0, 2 real anomalies would be missed and 2 normal samples would be misjudged. If set T =3.0, 10 real anomalies would be missed and no anomaly would be misjudged. Therefore, *WANLM* can detect the anomalies effectively in multi-dimensional dataset with appropriate threshold. In general, setting threshold should accord with two key principles: 1) Ensuring anomalies are minority; 2) Making the difference between anomalies and normal samples distinct.

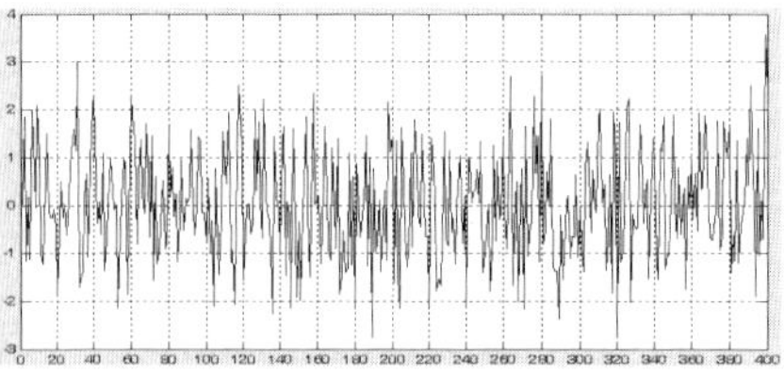

Fig. 1. Original experimental data

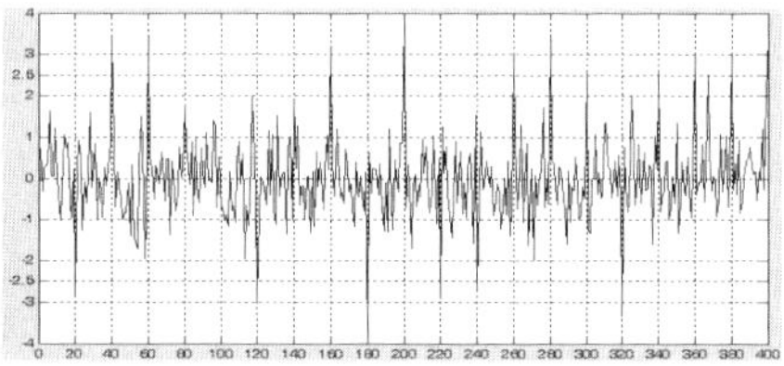

Fig. 2. *WTC*s of samples

4 Conclusions

1. *WANLM* is accurate and practical in detecting the anomalies in multi-dimensional data set. Above all, only the result of single scale *WANLM* (that is, using only one scale coefficient) is shown in this paper. Using multi-scale *WANLM*, the result of detecting would be better.
2. In practical application, the value spans of independent variables often are very different, therefore, the independent variables often should be standardized firstly to avoid effect of some variables with short value span being ignored.

References

1. Eleazar Eskin. Anomaly Detection over Noisy Data using Learned Probability Distributions[A]. Proc. 17th International Conf. on Machine Learning[C]. Morgan Kaufmann Publishers Inc. San Francisco, CA, USA, 2000: 255~262
2. Knorr E.M., Ng R.T.. Finding intensional knowledge of distance-based outliers[A]. Proceedings of the 25th International Conference on Very Large Data Bases. Edinburgh[C]. Scotland, Morgan Kaufmann,1999: 211~222
3. Breunig M. M., Kriegel H. P., Ng R. T., et al. LOF: identifying density-based local outliers[A]. Proceedings of the ACM SIGMOD International Conference on Management of Data. Dallas[C]. Texas, ACM Press, 2000: 93~104
4. Zengyou He, Xiaofei Xu, Shengchun Deng. Discovering cluster-based local outliers[J]. Pattern Recognition Letters, 2003, 24(9-10):1641~1650
5. Zengyou He, Shengchun Deng, Xiaofei Xu. Outlier Detection Integrating Semantic Knowledge[A]. In: Proceeding of the 3rd International Conference on Web-Age Information Management[C], Beijing, China, 2002: 126~131
6. Sammon J. W.. A nonlinear mapping for data structure[J]. IEEE Transactions on Computers, 1969, 18(4): 401~409

Classifying Noisy Data Streams*

Yong Wang[1], Zhanhuai Li[1], and Yang Zhang[2,**]

[1] Dept. Computer Science & Software, Northwestern Polytechnical University, P.R. China
{Wangyong, Lizhanhuai}@nwpu.edu.cn
[2] School of Information Engineering, Northwest A&F University, P.R. China
Zhangyang@nwsuaf.edu.cn

Abstract. The two main challenges associated with mining data streams are concept drifting and data noise. Current algorithms mainly depend on the robust of the base classifier or learning ensembles, and have no active mechanisms to deal noisy. However, noise still can induce the drastic drops in accuracy. In this paper, we present a clustering-based method to filter out hard instances and noise instances from data streams. We also propose a trigger to detect concept drifting and build *RobustBoosting*, an ensemble classifier, by boosting the hard instances. We evaluated *RobustBoosting* algorithm and *AdaptiveBoosting* algorithm [1] on the synthetic and real-life data sets. The experiment results show that the proposed method has substantial advantage over *AdaptiveBoosting* algorithm in prediction accuracy, and that it can converge to target concepts efficiently with high accuracy on datasets with noise level as high as 40%.

1 Introduction

A difficult problem for learning from data streams is that the concept of interest may depend on some hidden context, which is not given explicitly in the form of predictive features. Changes in the hidden context can induce more or less radical changes in the target concept, which is generally known as concept drifting. An effective learner should be able to track such changes and adapt to them quickly.

Noise is another difficult problem for mining data streams. Firstly, the noisy instances have a detrimental effect on the classification performance before concept drifting. Secondly, when concept drifts, on one hand, if a learning algorithm is highly robust to noise, it may adjust to concept drifting too slowly, or even neglects the concept drifting; on the other hand, if a learning algorithm is too sensitive to noise, it may interpret the noise instances as the instances from the new concept wrongly, which will furthermore cause overfitting.

In this paper, we propose *RobustBoosting*, a novel ensemble classifier to solve the above problems. In our method, we identify not only noisy instances, but also hard instances. Then, we depress the impact of noisy instances for mining noisy data streams by boosting the hard learned instances. Furthermore, we propose a concept drifting detection technique to actively adaptive to concept drifting.

* This research is supported by NSF 60373108.

** Corresponding author.

L. Wang et al. (Eds.): FSKD 2006, LNAI 4223, pp. 549–558, 2006.

This paper is organized as follows. Section 2 reviews related work. Section 3 analyzes the noise instances distribution in data streams, followed by a density-based clustering algorithm to identify noise instances and hard learned instances in section 4. An active concept drifting detection method and our *RobustBoosting* algorithm is presented in section 5 and section 6, respectively. Section 7 contains experimental results and we conclude this paper in section 8.

2 Related Work

A substantial amount of recent work has focused on mining data streams. However, few classifiers have mechanisms to deal noisy data, with most of them depending on the robust of the base classifier or learning ensembles. The use of pruning and learning ensembles partially addresses the problem. However, noise can still drastically affect the accuracy in mining data streams. Dominges et al. [2][3] devised Hoeffding tree, a novel decision tree algorithm, which performances asymptotically the same as or better than its batch version. Last [4] presented an online classification system, which use IFN (Info-Fuzzy Network) as a base classifier and can dynamically adjusts the size of the training window. Kunchera [5] pointed out that the ensemble methods are accurate, flexible and sometimes more efficient than single classifiers. Street er al. [6] gave an ensemble algorithm that builds one classifier per data block independently. Adapting relies solely on retiring old classifier one at a time. Wang [7] used a similar ensemble building method. Their algorithm tries to adapt to concept drifting by assigning weights to classifiers proportional to their accuracy on the most recent data block. Fang Chu [1] proposed *AdaptiveBoosting*, a novel boosting ensemble method, which is based on a dynamic sample-weight assignment scheme and achieves the accuracy of traditional boosting. This approach aims at significant data changes that could cause serious deterioration of the ensemble performance. As *AdaptiveBoosting* is the most related, we compare our work with it in our experiment.

Although there have been a great deal of off-line algorithms [8, 9] for noise identification, these type of algorithms can not deal with noise in data stream mining with concept drifting. FLORA4 [10] can detect concept drifting and have mechanism to deal with noise. However, it is oriented to small-sized dataset instead of data streams [3]. It represents concepts by conjunctions of attribute values, which is less applicable for data streams.

3 Noise and Hard Instances

The incoming data stream is partitioned into sequential chunks $s_1, s_2, \cdots s_j \cdots$, $j \geq 1$ with s_j being the most up-to-data chunk, and each chunk being of the same size. We learn a classifier C_i from each s_j, and then these classifiers can form an ensemble classifier E_c. When learning classifier C_i, traditional boosting technical pay more attention to the instances which have been misclassified by E_c, namely hard instances. Hard instance is not noise, but instances that are hard to be classified correctly by previous classifiers. For noisy data chunk s_j, focusing on misclassified instances may cause classifier C_i to focus on boosting the noisy instances with high probability. And this will go on cause the decreasing in the accuracy of C_i and E_c.

The core idea of our *RoustBoosting* is that we boost hard instance as well as filter the noisy data so as to increase the accuracy. To clearly define our research scope, we make some reasonable assumptions. Firstly, we make the same assumption as the one used in [11] by FanWei. The assumption assumes that "training data is collect without any known prior bias". In other words, if instance x has probability of τ to be seen in the universe of valid instance, then it has the same probability τ to be sampled without replacement from the universe to form the training instance in the data stream. Secondly, we also assume that when labeling instance in data streams, mislabeled instances (class noise) are generated without any known prior bias, namely, random class noise. Specifically, for a binary classification task, the label of each instance is independently switched from the true label $f(x)$ with probability θ, ($0<\theta<\frac{1}{2}$). Here, the value θ is referred as noise rate. Random class noise is a most standard assumption in learning theory [13]. From the experiment in real-life dataset in section 7, it is shown that our algorithm get good results under these assumptions. This proves that these assumptions are reasonable and incarnate the character of data streams, and that they just exclude rare and unrealistic situation.

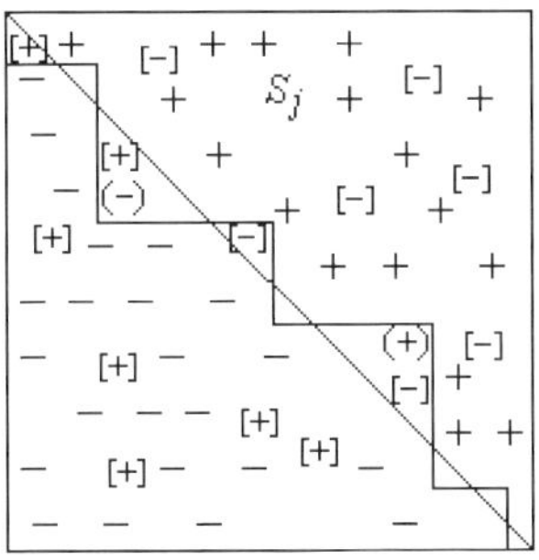

Fig. 1. True concept and its ensemble classifier E_c

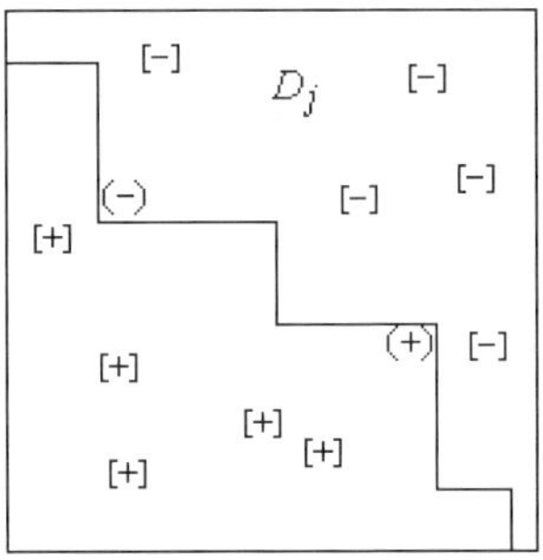

Fig. 2. Dataset D_j consists of instances misclassified by E_c

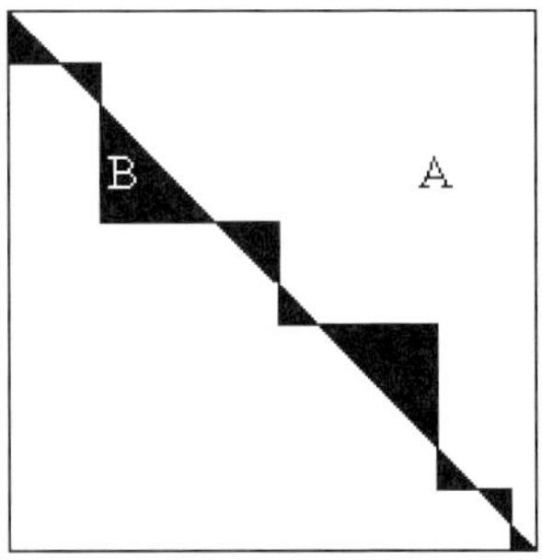

Fig. 3. Area A and area B

We illustrate our idea by a simple hyperplane example. In Fig.1, the hyperplane stands for a true concept and the interpolated straight line for the hyperplane is optimal model E_c of the true concept. If an instance is above the hperplane, it is labeled "+" (positive) by the true concept; otherwise, it is labeled negative "-". Noisy instances are denoted by "[+]" and "[-]". And hard instances are denoted by "(+)" and "(-)". E_c is used to classify the instances in data chuck s_j so as to identify the noise in s_j. All instances which do not satisfy the optimal model E_c constitute the data set D_j in figure 2. Please refer to Fig.3, all hard instances, which lie inside area B (between the hperplane and the interpolated straight lines), fail to satisfy the optimal model E_c. Parts of the noisy instances of s_j, which dissatisfy optimal model E_c, locate in area A. In other words, D_j consists of all hard instances and parts of noisy instances in data chunk s_j. When learning classifier for data chunk s_j, if we can focus the hard instances in area B and reduce the effect of noise in area A, then the accuracy of learned classifier could be improved. However, we do not know and usually will never know the true concept.

4 Identify Noise by Clustering Algorithm

According our assumptions, the instances in data chunk s_j follow uniform distribution in the universe of valid instances, and the noisy instances in data chunk s_j follow a different parameter uniform distribution. Under this assumption, for data chunk s_j, the number of noise in area A, say, μ, is a random variable; and the number of hard instances in area B, ν, is a random variable too. If the noise rate is q, the clean rate is p, then $p=1-q$. $A=|\mathbf{A}|$ and $B=|\mathbf{B}|$ stand for measure of area A and B, respectively. Since the noise rate q in data stream satisfy $q \ll 50\%$ generally, we get $E(\frac{\nu}{B}) \Big/ E(\frac{\mu}{A}) = \frac{np}{A+B} \Big/ \frac{nq}{A+B} = \frac{p}{q} = \frac{1-q}{q} = \frac{1}{q} - 1 > 1$. In other words, for data chunk s_j, the density of hard instances in area B is at least twice as large as the density of noise in area A.

DBSCAN [12] is a density-based clustering algorithm, which define a cluster as a set of density-connected points. We modified this algorithm so as to ensure that the instances which have been assigned to some clusters have high probability to be hard instances, and that the instances which have not been assigned to any clusters have high probability to be noise.

```
Function  SearchNoise(D_j ,r,MinPts)
Input:     D_j = {(x_1,y_1),(x_2,y_2),...,(x_m,y_m)} // y_i ∈ {+,-}  is class lable
Output: HardPointSet NoisePointSet
begin
FOR i FROM 1 TO m{
  Get point(x_i,y_i) from D_j;
  seeds=SetofPoints.RegionQuary(point , r, y_i );
  IF seeds.size<MinPts NoisePointSet.add((x_i,y_i));
  ELSE{
    HardPointSet.add(seeds);
    Seeds.delete((x_i,y_i));
    WHILE seeds≠ Ø {
      currentP=seeds.first();
```

```
          result=SetofPoints.RegionQuery(currentP, r, yi);
          IF result.size≥MinPts{
            FOR j FROM 1 TO result.size{
              resultP=result.get(j);
              IF(resultP∉ HardPointSet)&&
              (resultP∉ NoisePointSet){
                 seeds.append(resultP);
              }
              IF resultP∈ NoisePointSet{
                NoisePointSet.delete(resultP);
                HardPointSet.add(resultP);
              }
            }
          }
          seeds.delete(currentP);
        }
      }
    }
    end.
```

Alg. 1. Identifing noise and hard instances

Algorithm 1 is used to identify the noise and hard instances in data chunk s_j. In this algorithm, the call of $SetofPoints.RegionQurery(point, r, y_i)$ returns a set of instances in which all instances lie in the r-neighborhood of $point$ and have class label y_i. If the set contains more than $MinPts$ instances, then the density in r-neighborhood of point is believed to be large enough. Thus, according the foregoing statement the instance point is believed by our algorithm to be a hard instance, otherwise, noise.

5 Detecting Concept Drifting

For data chunk s_j, we estimate the probability of $E_c(x) \neq y$. For this purpose, we construct binary random variable φ, with $\varphi = 1$ denoting $E_c(x) \neq y$ happens and $\varphi = 0$ denoting $E_c(x) = y$ happens. If the mathematical expectation of φ is p, say, $E(\varphi) = p$, then the variance of p is $\sigma_\varphi^2 = p(1-p)$. If k denotes the number of times that $E_c(x) \neq y$ happens in data chunk s_j, then for enough large $n = |s_j|$, the sample mean $\bar{\varphi}$ follows the normal distribution $N(p, \sqrt{p(1-p)/n})$ approximately. That is to say

$$P(|\bar{\varphi} - p| < z_\alpha \sqrt{\frac{p(1-p)}{n}}) = 2\alpha - 1 \quad . \tag{1}$$

By solving the equation $(\frac{k}{n} - p)^2 = z_\alpha^2 \frac{p(1-p)}{n}$, the upper endpoint, p_U^j, and the lower endpoint, p_L^j, of $2\alpha - 1$ confidence intervals for the p are given by

$$p_U^j = \frac{k}{n+z_\alpha^2} + \frac{z_\alpha^2}{2(n+z_\alpha^2)} + \frac{z_\alpha}{n+z_\alpha^2}\sqrt{z_\alpha^2 + 4k - \frac{4k^2}{n}} . \quad (2)$$

$$p_L^j = \frac{k}{n+z_\alpha^2} + \frac{z_\alpha^2}{2(n+z_\alpha^2)} - \frac{z_\alpha}{n+z_\alpha^2}\sqrt{z_\alpha^2 + 4k - \frac{4k^2}{n}} . \quad (3)$$

Here, Z_α is the upper α percentile of the standard normal distribution, $k = |D_j|$ $n = |s_j|$. If $\frac{|D_j|}{n} > p_U^{j-1}$ or $\frac{|D_j|}{n} < p_L^{j-1}$, we believe the concept drifting occurs.

6 Classification Algorithm

Based on the above discussion about identifying noise and hard instances, and the algorithm of concept drifting detection, we present our *RobustBoosting* algorithm, a boosting ensemble classifier.

```
Function RobustBoosting(data stream, M )
Input:data stream s_1,s_2,···s_j··· and s_j ={(x_1,y_1),(x_2,y_2),···(x_n,y_n)}
M//The maximal number of classifiers in ensemble E_c
Output: an ensemble classifier E_c
begin
read a data chunk s_j from the stream;
IF E_c ≠∅ {
  FOR i FROM 1 TO n{
     E_c(x_i) =round(1/m Σ_{k=1}^{m} C_k(x_i)) ;
     IF  E_c(x_i) ≠ y_i    D_j ← (x_i,y_i) ;
  }
  e_j = |D_j| / n ;
  compute p_L^j and p_U^j by formulae (2) and (3);
  IF p_L^{j-1} <e_j < p_U^{j-1} {
     SearchNoise( D_j, r ,MinPts );
     FOR i FROM 1 TO n{
        IF (x_i,y_i) ∈ HardPointSet     ω_i=(1-e_j)/e_j ;
        ELSE IF (x_i,y_i) ∈ NoisePointSet    ω_i =0 ;
        ELSE   ω_i =1 ;
     }
  }
  ELSE{
     E_c =∅ ;
```

```
        FOR i FROM 1 TO n   ω_i =1 ;
    }
}
ELSE{
    FOR i FROM 1 TO n   ω_i =1 ;
    p_L^1 =0 ;  p_U^1 =1 ;
}
Train a classifier C from weighted dataset s_j ;
IF |E_c| =M{
    E_c.add(C) ;
    E_c.deleteFirst() ;
}
ELSE  E_c.add(C) ;
end.
```

Alg. 2. *RobustBoosting* : Robust boosting ensemble algorithm

As a data stream continuously flows in, it is broken into chunks of equal size. Individual classifiers are built from each data chunk. Then, these classifiers are added into a fix-sized ensemble E_c in sequence. Once the ensemble is full, we remove the first existing classifier. Before training a classifier on the up-to-date data chunks s_j, we determine whether a concept drifting has occurred by the approach described in section 5. If there is no concept drifting, we use the ensemble E_c to classify the data chunks s_j. The weights of the identified noise are set to 0 and the weights of correctly classified instances are left unchanged. The weights are normalized to be a valid distribution, and then a classifier is constructed from the re-weighted data chunk. When concept drifting is detected, we clear ensemble E_c and just add the classifier trained on the up-to-date data chunk into ensemble E_c . The reason for clearing ensemble E_c is that obsolete classifiers have bad effects on overall ensemble performance.

7 Empirical Study and Results

7.1 Moving Hyperplane Dataset

The *Moving Hyperplane* dataset is widely used for experiment [7,11]. A hyperplane in a d-dimensional space is denoted by equation: $\sum_{i=1}^{d} a_i x_i = a_0$. All the instances which satisfy $\sum_{i=1}^{d} a_i x_i \geq a_0$ are labeled positive and otherwise negative. Weights a_i are initialized randomly in the range of $[0,1]$. The value of a_0 is always set as $a_0 = \frac{1}{2}\sum_{i=1}^{d} a_i$, so that roughly half of the instances are positive, and the other half are negative. We simulate concept drifting by a series of parameters. Parameter K

specifies the total number of dimensions whose weights are changing. Parameter $t \in R$ specifies the magnitude of the change (every N instances) for weights a_i $(1 \le i \le K)$. Weights change continuously, i.e., a_i is adjusted by t/N after every N instances being generated. Furthermore, there is a possibility 10% that the change would reverse direction after every N instance. Also, each time the weights are updated, we recomputed a_0 so that the class distribution is not disturbed. In this experiment, we choose Parameter $d = 10$, N =1000, $K = 7$, and $t = 5$ to generate data stream. The predictive performance was tested on clean testing instances which were also generated randomly, according to the same underlying concept. The experiment results shown below were obtained when the chunk size equals 1000, and similar results could be obtained for other chunk sizes, such as 2000, 3000, 4000, etc.

We introduced noise by randomly switching the labels of p% of the instances. To better evaluate algorithm 1, we adopt three factors: $R1$, $R2$ and ER for each data chunk s_j. They can be defined as: $R1 = \frac{|F_{noise} \cap G|}{|G|}$, $R2 = \frac{|F_{hard} \cap H|}{|H|}$, and $ER = \frac{|F_{hard} \cap G|}{|H|}$. Here G is the set of noisy instances in s_j; H is the set of hard instances in s_j; F_{noise} is the set of instances that are identified as noise in s_j; F_{hard} is the set of instances that are identified as hard instances in s_j. At different noise rate, we checked these three factors. Figure 4, figure 5 and figure 6 show a typical run when the noise rate is around 20%.

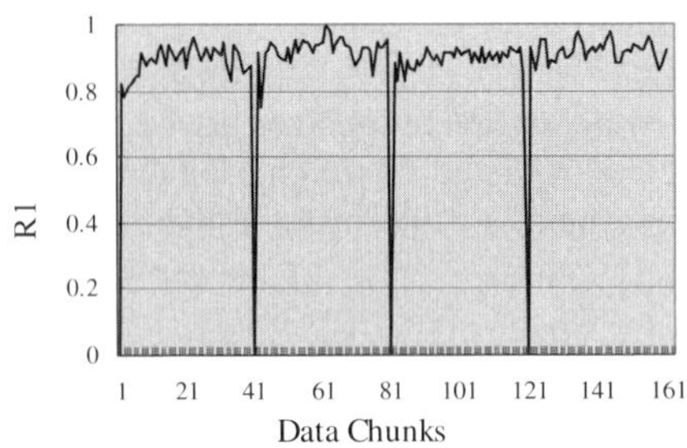

Fig. 4. Noise detection results with noise rate around 20%

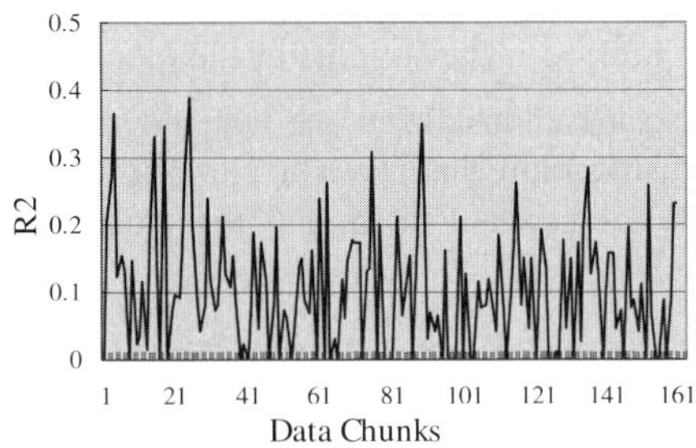

Fig. 5. Hard instances detection result with noise rate around 20%

Figure 4 shows that before concept drifting, most noise in each data chunk could be founded ($R1$). The reason for sudden drops in data chunk 41, 81 and 121 is that the system detects the concept drifting on these chunks and there is no enough information to estimate whether instances in these chunks comes from a new concept or noise. Figure 5 presents the results of identified hard instances in each data chunk ($R2$). Figure 6 presents the proportion that noisy instance has been mistakenly identified as a hard instances in H (ER). This portion of instances has very crucial effects to predictive performance in a boosting ensemble classifier. From figure 6, we can see that the maximal value is 6%. Figure 7 presents the result of *RobustBoosting* compared with *AdaptiveBoosting* on the *Moving Hyperplane* dataset set when noise rate is 40%. Obviously, our *RobustBoosting* always performs better and can converge to the new concept more quickly.

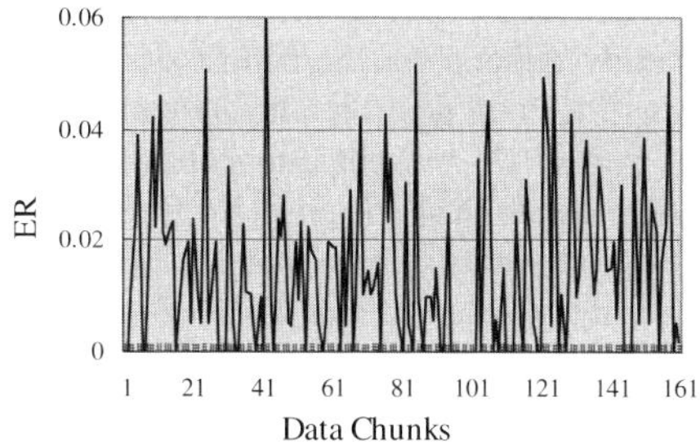

Fig. 6. Proportion of noise instances which are wrongly identified as hard instances with noise rate around 20%

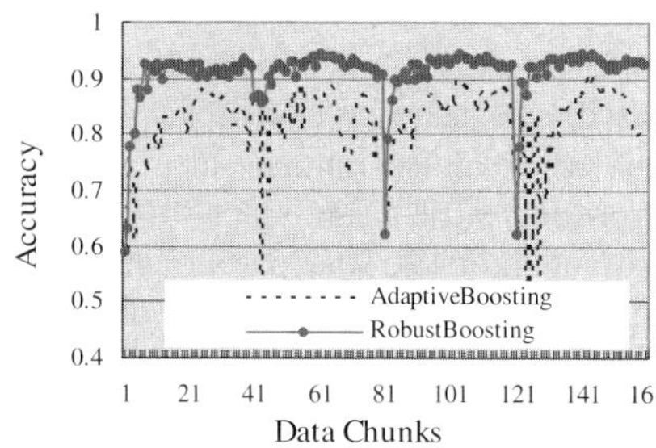

Fig. 7. Predictive accuracy for *RobustBoosting* and *AdaptiveBoosting* on the synthetic dataset with 40% noise rate

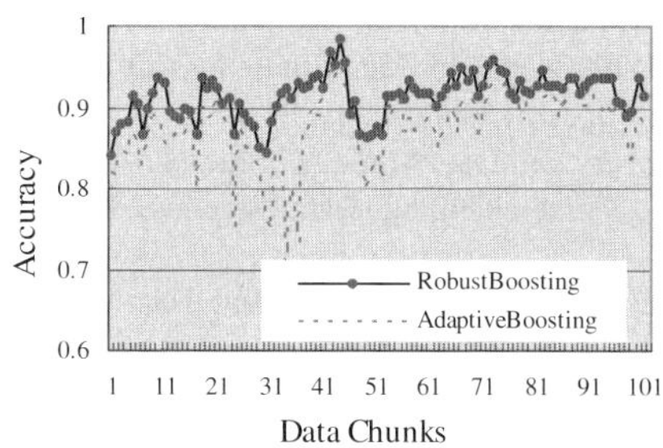

Fig. 8. Predictive accuracy for *RobustBoosting* and *AdaptiveBoosting* on credit card fraud dataset

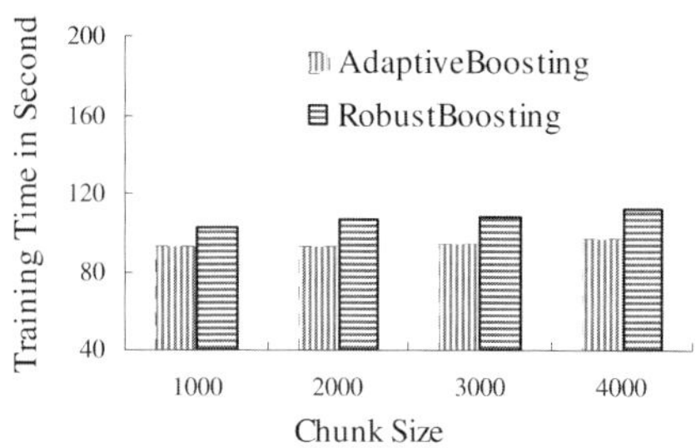

Fig. 9. Comparison of *AdaptiveBoosting* and *RobustBoosting* in training time

7.2 Credit card Fraud Data

This data contains 100k credit card transactions. Concept drifting is simulated by sorting transactions by changes by the transaction amount [1]. We study the ensemble performance using various chunk size (1k, 2k, 4k). The experiment result with chunk size set to 1000 is shown in Fig.8. It is obvious that *RobustBoosting* performs fewer and smaller drops in accuracy than *AdaptiveBoosting* does.

Fig.9 compares the running time for training *RobustBoosting* and *AdaptiveBoosting* on the credit card fraud dataset. The x-axis represents the chunksize and y-axis represents the training time. The experiment is conducted on a PC with 2.4M HZ CPU and 256M memory. The result shows that it takes more time to train *RobustBoosting*. This is because *RobustBoosting* runs algorithm 1 on each data chunk to identify noise and hard instances. In general, *RobustBoosting* is 12% slower than *AdaptiveBoosting*, and this is compensated by 30% less error rate than *AdaptiveBoosting*.

8 Conclusion

Current research on mining data streams mainly depend on the robust of the base classifier or learning ensembles, and few works focus on dealing noisy. However, the problem of noise remains un-solved. In this paper, we present our *RobustBoosting* algorithm, which can adapt to concept drifting and robust to noise by filtering out most of the noise and parts of hard instances from data chunks. We also propose a trigger to actively adaptive to changes in the noisy data stream. *RobustBoosting* is compared with *AdaptiveBoosting* by extensive experiments. The experiment results show that the proposed method has substantial advantage over *AdaptiveBoosting* algorithm in prediction accuracy, and that it can converge to target concepts efficiently with high accuracy on datasets with noise level as high as 40%.

References

1. F.Chu , C.Zaniolo: Fast and Light Boosting for Adaptive Mining of Data Streams. In Proc. of the 5th Pacific-Asic Conference on Knowledge Discovery and Data Mining PAKDD (2004)
2. Pedro Domingos, Geoff: Hulten: Mining High-Speed Data Streams. In Proc. of the Sixth International Conference on Knowledge Discovery and Data Mining (2000) 71-80
3. G.Hulten, L.Spencer, P. Domingos: Mining Time-changing Data Streams. In ACM SIGKDD (2001)
4. M. Last: Online Classification of Nonstationary Data Stream. Intelligent Data Analysis Vol.6 (2002) 129-147
5. Ludmila I. Kuncheva. Classifier Ensembles for Changing Environments. In Proc.5th Int. Workshop on Multiple Classifier Systems.(2004) 1-15
6. W.Street, Y., Kim: A Streaming Ensemble Algorithm(sea) for Large–Scale Classification. In Int'l Conf. on Knowledge Discovery and Data Mining (2001)
7. H.Wang, Wei Fan, P., Yu, J., Han: Mining Concept-Drifting Data Streams Using Ensemble Classifiers In int'l Conf. on Knowledge Discovery and Data Mining (2003)
8. J., Kubica, A., Moore: Probabilistic Noise Identification and Data Cleaning. In Int'l Conf. Data Mining (2003)
9. Xingquan Zhu,Xindong Wu, Qijun Chen: Eliminating Class Noise in Large Datasets. In the Proc. of the 20th International Conf. On Maching Learning (2003)
10. G.,Widmer, M., Kubat: Learning in the Presence of Concept Drift and Hidden Contexts. Machine learning 23 (1996) 69-101
11. Wei Fan: Systematic Data Selection to Mine Concept-Drifting Data Streams. In the Proceeding of the Conf. KDD (2004) 128-137
12. M., Ester, H., P., Kriegel, J., Sander, X., Xu: A Density-Based Algorithm for Discovering Clusters in Large Spatial Database with Noise. In Proc.of Int. Conf. on Knowledge Discovering and Data Mining (1996) 226-231
13. Adam Kalai, Rocco A., Servedio: Boosting in the Presence of Noise. Journal of Computer and System Science, Vol.71,Issue3 (2003) 226-290

FCM-Based Clustering Algorithm Ensemble for Large Data Sets

Jie Li, Xinbo Gao, and Chunna Tian

School of Electronic Engineering, Xidian Univ., Xi'an 710071, P.R. China

Abstract. In the field of cluster analysis, most of the available algorithms were designed for small data sets, which cannot efficiently deal with large scale data set encountered in data mining. However, some sampling-based clustering algorithms for large scale data set cannot achieve ideal result. For this purpose, a FCM-based clustering ensemble algorithm is proposed. Firstly, it performs the atom clustering algorithm on the large data set. Then, randomly select a sample from each atom as representative to reduce the data amount. And the ensemble learning technique is used to improve the clustering performance. For the complex large data sets, the new algorithm has high classification speed and robustness. The experimental results illustrate the effectiveness of the proposed clustering algorithm.

1 Introduction

Cluster analysis is to partition an unlabelled sample set into subsets according to some certain criteria. It is one of the multivariate statistical analysis methods and an important branch of unsupervised pattern recognition [1]. In cluster analysis, homogeneous samples are aggregated into same cluster, and vice versa. So it can be used to determine the closeness quantificationally among objects under study. Thus, valid classification and analysis can be achieved.

Among the traditional cluster methods, the objective function based clustering algorithms become more and more popular for converting the cluster analysis into an optimization problem. As one of the most typical approaches for this purpose, Fuzzy c-means (FCM) [2] has been widely used in various application fields. As well known, the computational complexity of the FCM algorithm is $O(ncl)$, where n is the sample number of the data set, c is the category number and l is the iteration generation. Compared with other clustering algorithms, the FCM has a higher efficiency. Whereas, with the augment of data amount or cluster number, the complexity of the FCM algorithm will be increased gradually, which leads to difficulty of real-time data analysis.

With the rapid development of the computer technology, the capability of data acquisition is improved greatly, and the capacity of information is increased even faster. Facing the large data set, data mining techniques emerges as time requires. As one of effective analysis tools, clustering algorithm attracts much more attention. In data mining research area, the clustering algorithm is a powerful tool in processing large amount of high-dimensional data set [3]. To this

L. Wang et al. (Eds.): FSKD 2006, LNAI 4223, pp. 559–567, 2006.

end, some fast clustering algorithms are presented, such as CURE, CLARA etc. [4]. Those algorithms improve the computation efficiency by randomly selecting a small part of sample set for cluster analysis, which cannot always guarantee to obtain the optimal result.

The ensemble learning technique [5] uses several versions of component learners to solve the same problem, which can evidently improve the generalization ability of the learning system. In recent years, it has been widely used in machine learning, neural network and statistics and other fields, and become an important research issue. For this purpose, a FCM-based clustering algorithm ensemble is proposed for large data sets. The new algorithm pre-processes the data set with atom clustering to divide the whole data set into a series of atom clusters. Then, randomly select a sample from each atom as representative to reduce the data amount. Finally, the ensemble learning technique is used to improve the clustering performance. It has approximate linear time complexity and fits for large data sets analysis.

The rest of this paper is organized as follows. The FCM algorithm is briefly introduced in next section. Section 3 describes the FCM-based clustering algorithm ensemble method. The proposed the FCM-based clustering algorithm ensemble for large data sets is presented in Section 4. Section 5 provides the experimental results comparison among the proposed algorithm, the traditional FCM algorithm and the sampling-based FCM algorithm. Concludes and several topics for further study are drawn in Section 6.

2 Fuzzy *c*-Means Clustering Algorithm

Let $X = \{x_1, x_2, \cdots, x_n\}$ be a set of n observation samples in feature space, in which $x_i = [x_{i1}, x_{i2}, \cdots, x_{im}]^T$ denotes a feature vector of the sample x_i, and x_{ij} is the j-th attribute value of vector x_i. For a given integer $c(2 \leq c \leq n)$, the fuzzy c-means (FCM) clustering of the set X can be represented as the following mathematical programming problem.

$$min\Big\{J_2(U,P) = \sum_{i=1}^{c}\sum_{j=1}^{n}(\mu_{ij})^2 \cdot d^2(x_j,p_j)\Big\} \qquad s.t. U \in M_{fc} \tag{1}$$

Where, $U = [\mu_{ij}]_{c\times n}$ is fuzzy partition matrix [6]. We have

$$M_{fc} = \Big\{U \in R^{c\times n} \big| \mu_{ij} \in [0,1]; \quad \sum_{i=1}^{c}\mu_{ij} = 1, \forall j; \quad 0 < \sum_{j=1}^{n}\mu_{ij} < n, \forall i\Big\} \tag{2}$$

which is the fuzzy c-partition space of sample set X. $P = \{p_1, p_2, \cdots, p_c\}$ denotes the clustering prototype set with c clustering center vectors. $d(\cdot)$ is defined as Euclidean distance for measuring the dissimilarity between sample and clustering prototype. For $X \subset R^m$, we have

$$d^2(x_j,p_i) = (x_j - p_i)^T \cdot (x_j - p_i) \tag{3}$$

In 1981, Bezdek presented the iterative formulas of FCM algorithm as follows [7].

$$\mu_{ij} = \left(\sum_{l=1}^{c} \left(d(x_j, p_l)\right)^{-2}\right)^{-1} \Big/ \left(d(x_j, p_i)\right)^2, \quad \forall i, j \tag{4}$$

$$p_i = \sum_{j=1}^{n} \mu_{ij}^2 x_j \Big/ \sum_{j=1}^{n} \mu_{ij}^2 \tag{5}$$

However, the objective functions of c-means type algorithms are nonlinear and multimode. The traditional gradient-based optimization methods are easy to get trap into local optima. On the other hand, users require the clustering algorithm with not only high efficiency but also strong generalization ability. Therefore a FCM-based fuzzy clustering algorithm ensemble is proposed in next section.

3 FCM-Based Clustering Algorithm Ensemble

The ensemble learning generally consists of two stages, i.e., component learner generation and component learner aggregation. For the given data set $X = \{x_1, x_2, \cdots, x_n\}$, the FCM algorithm is employed to perform k times cluster analysis on X, and cluster labels of the m-th time clustering is denoted as $\{L_1^m, L_2^m, \cdots, L_c^m\}$ $m = 1, 2, \cdots, k$. Note that without priori knowledge of categories those obtained label vectors cannot be directly used for the following conclusion synthesization.

For example, although the label vectors $[2, 3, 3, 1, 4, 4, 5]^T$ and $[1, 2, 2, 3, 5, 5, 4]^T$ are different in expression, they stand for the same clustering result. To combine the different clustering results, the cluster label vectors should be aligned. That is to say, the corresponding relationships among the label vectors should be established by matching operation defined as follows. If

$$S_i^m(n, l) = L_i^m \bigcap L_l^n \qquad n = 1, 2, \cdots, k, \quad l = 1, 2, \cdots, c \tag{6}$$

$$j = \arg\left\{\max_l |S_i^m(n, l)|, \quad l = 1, 2, \cdots, c\right\} \tag{7}$$

Then $L_i^m = L_j^n$ which means that the i-th cluster in the m-th clustering corresponds to the j-th cluster in the n-th clustering.

After all the cluster results are aligned, the conclusion combination can be achieved. It is assumed that the obtained sample membership functions in k times cluster analysis are indicated as $\{\mu_{1l}^m, \mu_{2l}^m, \cdots, \mu_{nl}^m\}, m = 1, 2, \cdots, k, l = 1, 2, \cdots, c$. Then, the ensemble of final membership function of each sample is defined as

$$\mu_{il} = \frac{1}{k} \sum_{j=1}^{k} \mu_{il}^j \qquad l = 1, 2, \cdots, c \tag{8}$$

Finally, according to the obtained membership function, the category label of each sample can be determined.

4 FCM-Based Clustering Algorithm Ensemble for Large Data Set

As we all know, cluster analysis often encounter large data set in data mining. Like the traditional FCM algorithm, the FCM-based clustering algorithm ensemble also involves the real-time problem when applied to large data set. Moreover, the CPU time increases with the increase of the number of component learner. When the sample amount is huge and the number of category is too large, the proposed FCM-based clustering algorithm ensemble is also time-consuming, which cannot satisfy the requirement of real-time. To efficiently analyze large scale data set, a FCM-based clustering ensemble for large data set is proposed as follows.

4.1 Atom Clustering Algorithm

The ensemble learning generally consists of two stages, i.e., component learner generation and component learner aggregation. For the given data set , the FCM algorithm is employed to perform k times cluster analysis on X, and cluster labels of the m-th time clustering is denoted as . Note that without priori knowledge of categories those obtained label vectors cannot be directly used for the following conclusion synthesization.

Definition 1. For a given arbitrary point x in data space and a distance e, all the points within the circle region with center at x and radii of e form an atom with nucleus x. The other points within an atom is called as electrons around nucleus x.

The atom clustering algorithm first selects one of samples as nucleus, and then computes distance between the rest samples and the nucleus. If the distance is less than the threshold e, these samples are defined as electrons of the nucleus. Otherwise, the sample will be looked as a new nucleus. Till all the samples are processed, the atom clustering algorithm stops and outputs the atom clusters, $C = c_1, c_2, \cdots, c_t$, where t is the total number of nucleus.

The atom clustering algorithm can be described as the following pseudo code.

```
Input: the data set X containing n points in m-dimensional space
        e-the distance threshold
Output: C-a set a atomic clusters
  NumberOfAtom = 1
  C[1] = X[1]
  FOR i = 1 TO NumberOfObjects
    FOR j = 1 TO NumberOfAtom
      distance= Distance(X[i],C[j])
        IF distance < e
          NumberOfElectron[j] + 1
          Atomship[j] = j
        ELSE
```

```
            NumberOfAtom + 1
            C[NumberOfAtom] = X[i]
            Atomship[i] = NumberOfAtom
        ENDIF
    ENDFOR
ENDFOR
```

The atom clustering algorithm is a partition-based method, which classifies the n samples into some atom clusters. This algorithm only needs pre-setting a threshold, i.e., the radii of atom e, and performs once iteration for obtaining the data partition. The number of atoms will be much less than that of samples.

4.2 FCM-Based Clustering Ensemble for Large Data Set

For the obtained atom clusters $C = \{c_1, c_2, \cdots, c_t\}$ from the data set, we had even take the center of atom, i.e., *nucleus*, as representative of each cluster to reduce the data amount [8]. However, when the electrons around nucleus do not distribute uniformly, the nucleus cannot stand for the atom cluster completely, which influences the clustering result.

Here, we select one of *electrons* from each atom randomly to form training data subset. Several training subsets can be obtained by repeating the above process. For the atom clusters $C = \{c_1, c_2, \cdots, c_t\}$, randomly generate k subsets $s_i = \{x_1^i, x_2^i, \cdots, x_t^i\}, i = 1, 2, \cdots, k$, in which

$$x_j^i = random\{c_j\} \tag{9}$$

Where, *random* is a function that randomly select a sample from c_j. Then, the FCM algorithm is performed on each subset s_i. The corresponding results are combined by the ensemble learning method as Section 3 mentioned.

4.3 Category Labeling of Large Data Set

After performing the FCM algorithm on each atom representative set, a partition matrix of atoms can be obtained. The label of the corresponding *electrons* (samples) in each atom is labeled the same as the category of atom. By combining all the labeling results from different training subsets, the final labels of all the samples in the large data set are obtained.

In this way, through classifying the atom clusters, each sample of large data set can be classified. Since the number of atom clusters is much smaller than that of samples in large data set, the CUP time of cluster analysis is reduced greatly.

5 Experimental Results

To verify the effectiveness of the proposed FCM-based clustering algorithm ensemble method, some preliminary experiments are conducted to compare the proposed algorithm, the traditional FCM algorithm and sampling-based FCM algorithm. The experimental results illustrate the effectiveness of our algorithm.

5.1 Classification Performance Test for Large Data Set

For the sake of visualization, we synthesize 9 Guassian distributed subsets in 2D plane, each of which consists of 10000 samples, as shown in Fig.1 (a). Fig.1(b) gives the result of atom clustering, in which each symbol "o" denotes a nucleus. Here we take , and the number of atom nucleus is 162. It is clear that the number of nucleus is much smaller than that of samples. Performing the FCM algorithm on the atom clusters, the result is shown in Fig.1(c), which presents clustering centers of the data set. Fig.1(d) shows the clustering centers obtained by the proposed clustering algorithm ensemble method. It is obvious that the clustering result is acceptable.

Fig.1 (e) shows the nucleuses of the obtained 21 atom clusters with the distance threshold of $e = 0.04$. Performing the FCM algorithm on the 21 nucleuses with the cluster number of 9, we get the cluster centers shown in Fig.1 (f). It is clear that since the number of nucleuses is too few, the obtained cluster centers cannot reflect the real distribution of samples. While, Fig.1 (g) shows the cluster centers obtained by the proposed FCM-based clustering algorithm ensemble method. The new algorithm randomly selects an *electron* from an atom rather than the *nucleus* as representative. By repeating such process several times, the distribution of *electrons* in each atom cluster can be estimated more and more accuracy. So, the proposed algorithm can obtain good classification performance.

For the large scale data set, to obtain good clustering performance, the distance threshold e should take a small value, which leads to much more atom clusters and CPU time for further cluster analysis. While the new algorithm can obtain good performance with a large distance threshold e, thus much CPU time can be saved.

To test the classification performance, we perform the new algorithm on the data set shown in Fig.1 (a) 20 times independently and count the correct classification rates as Fig.1 (h), in which the ensemble scale takes 5. Contrasting to the sampling-based FCM algorithm, the new algorithm employs the ensemble technique to improve the generalization ability, and thus achieving higher correct classification rate and robustness.

Fig.1 (i) presents the relationship between the correct classification rate and the ensemble scale of the new algorithm, in which the correct classification rate is the average result of running the algorithm 10 times independently. It can be seen that the average correct classification rate is raised with the increase of the ensemble scale. With the increase of the ensemble scale, the sampled *electrons* can estimate the distribution of *electrons* in atom clusters more and more accurate, which of course result in the higher and higher correct classification rate.

5.2 The Scalability of the Proposed Algorithm on Large Data Sets

To test the scalability of the proposed clustering algorithm ensemble method, we design two groups of experiments, one for sample number scalability and another for cluster number scalability. Both experiments are conducted on the PC with Intel Pentium 4 CPU (2.4GHz) and memory of 512MB. Fig.2(a) shows the

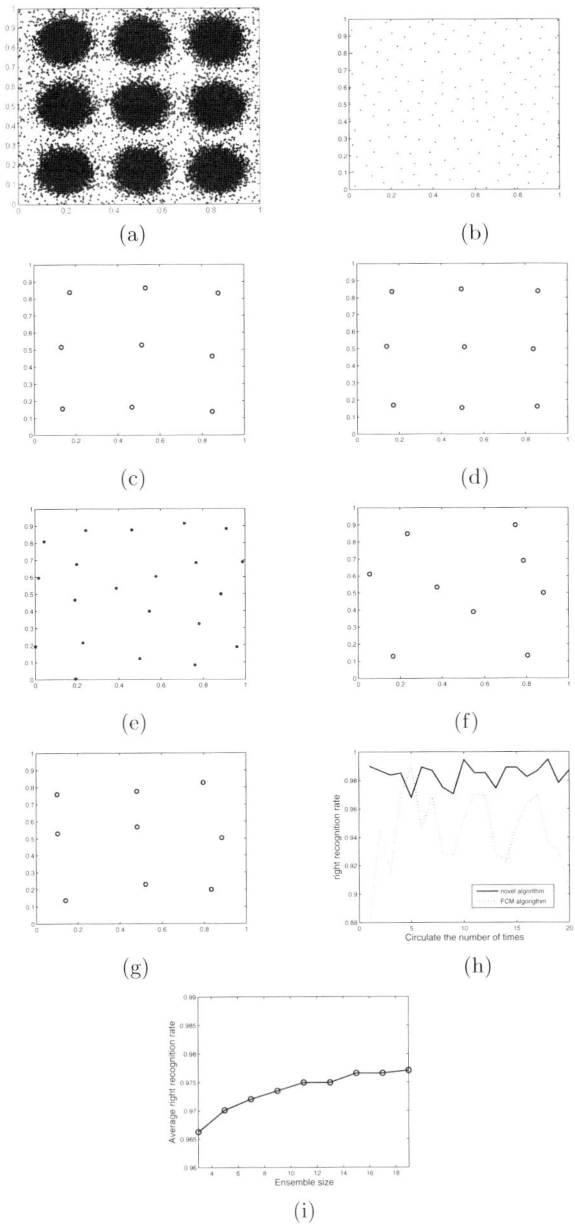

Fig. 1. Experimental results of classification performance test for large data set(a)A large scale test data set (b)Nucleuses obtained by atom clustering with $e = 0.002$ (c)Clustering centers obtained by FCM algorithm (d)Clustering centers obtained by our algorithm (e)Nucleuses obtained by atom clustering with $e = 0.04$ (f)Clustering centers obtained by FCM algorithm (g)Clustering centers obtained by our algorithm (h)Comparison of correct classification rates of 2 algorithms (i)The correct classification rate vs. the scale of ensemble

variation of the CPU time with the increasing of the data amount, in which the solid line and dashed line denote the variation curves of the FCM algorithm and the proposed algorithm respectively with the data amount ranging from 90 thousands to 1.8 millions. Each curve is an average of 10 independent experiments. Fig.2(b) presents the variation of the CPU time of the two algorithms with the increase of the cluster number, in which the data set contains 1.8 millions of samples.

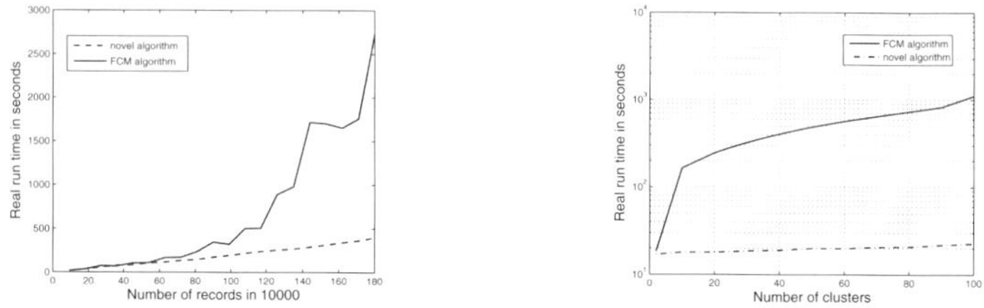

(a)The CPU time vs. data amount (b)The CPU time vs. cluster number

Fig. 2. The scalability comparison of FCM and novel algorithms

Since the new algorithm employs the atom clustering to reduce the data amount, it is faster than the standard FCM algorithm for the large data set. Moreover, the component clustering learners in the new algorithm can be implemented in parallel fashion, which can further save CPU time. Therefore, the proposed algorithm suits for the cluster analysis of large data sets very well.

6 Conclusions

This paper proposes a FCM-based clustering algorithm ensemble for large data set. The new algorithm first performs the atom clustering on the data set. Then randomly select a sample from each atom as representative to reduce the data amount, which is repeated several times to generate training subsets. The FCM algorithm is employed on theses training subsets, and the results are aggregated by the ensemble learning to improve the generalization ability of learning system and the accuracy of clustering result. Moreover, the proposed algorithm is essential parallel, and it can be implemented in various parallel machines for processing large data sets in real-time.

Acknowledgment

This work was supported by the National Natural Science Foundation of China (No.60202004), the Key Project of Chinese Ministry of Education (No.104173), China Postdoctoral Science Foundation, and the Program for New Century Excellent Talents in University (NCET-04-0948), China.

References

1. He Qing: Advance of the theory and application of fuzzy clustering analysis, Fuzzy System and Fuzzy Mathematics, **12(2)** (1998) 89–94. (In Chinese)
2. Xinbo Gao: Optimization and Applications Research on Fuzzy Clustering Algorithms, Doctoral Thesis, Xidian University, Xi'an 710071, China, (1999).
3. Anderberg, M.R. Cluster Analysis for Applications. Academic Press. (1973).
4. Kaufman L. and Rousseeuw P. J. Finding groups in data: an introduction to cluster analysis. (1990).
5. Dietterich T.G, Machine learning research: Four current directions. AI Magazine, **18(4)** (1997) 97–136.
6. B. Everitt. Cluster Analysis. New York, Heinemann Educational Books Ltd., (1974) 45–60.
7. Bezdek J. C., Pattern Recognition with Fuzzy Object Function Algorithms, Plenum. New York, (1981).
8. Xinbo Gao, Jie Li, Hongbing Ji, An automatic multi-threshold image segmentation algorithm based on weighting FCM and statistical test, Acta Electronica Sinica, **32(4)** (2004) 661–664.

Time Series Subsequence Searching in Specialized Binary Tree

Tak-chung Fu[1,2,*], Hak-pun Chan[1], Fu-lai Chung[1], and Chak-man Ng[2]

[1] Department of Computing, The Hong Kong Polytechnic University, Hong Kong
{cstcfu, 03500210t, cskchung}@comp.polyu.edu.hk
[2] Department of Computing and Information Management,
Hong Kong Institute of Vocational Education (Chai Wan), Hong Kong
cmng@vtc.edu.hk

Abstract. Subsequence searching is a non-trivial task in time series data analysis and mining. In recent years, different approaches are published to improve the performance of subsequence searching which based on index the time series and lower bound the Euclidean distance. In this paper, the problem of applying Euclidean distance on time series similarity measure is first reviewed. Previous approaches to align time series for similarity measure are then adopted for subsequence searching, they include: dynamic time warping (DTW) and perceptually important point (PIP). Furthermore, a tree data structure (SB-Tree) is developed to store the PIP of a time series and an approximate approach is proposed for subsequence searching in the SB-Tree. The experimental results performed on both synthetic and real datasets showed that the PIP approach outperformed DTW. The approximate approach based on SB-Tree can further improve the performance of the PIP-based subsequence searching while the accuracy can still be maintained.

1 Introduction

Recently, the increasing use of temporal data has initiated various research and development attempts in the field of data analysis and mining. Temporal data analysis, in particular time series analysis, has long been an active research area. Indeed, one of the hottest topics is subsequence searching [1]. Given a query pattern Q, the task is to identify the most similarity subsequence in a time series P, of length n and m respectively, where $m>>n$, $P=(p_1,\cdots,p_m)$ and $Q=(q_1,\cdots,q_n)$. Another variant is to identify all subsequences in the time series P with distance less than ε where ε can be user specified or determined automatically.

In traditional databases, similarity search are exact based. However, in time series database, which is characterized by its numerical and continuous nature, similarity search is typically carried out in an approximate manner. Consider the stock time series, one may expect to search pattern like:

* Corresponding author.

L. Wang et al. (Eds.): FSKD 2006, LNAI 4223, pp. 568–577, 2006.

***Q1:** Find all "head-and-shoulder" patterns last for a month in the closing prices of all high-tech stocks.*

The simplest method to identify those patterns is sequential scanning by placing the query pattern on every offsets of the time series and *m-n*+1 subsequences will be compared. Euclidean distance is always adopted for the distance measure. Different indexing approaches, which lower bounding the Euclidean distance, are proposed [2-5] to improve the performance of the subsequence searching process. However, Euclidean distance is not always being the suitable distance function in some domains [6-8]. For example, stock time series has its own characteristics over other time series data (e.g. data from scientific areas like electrocardiogram ECG) which the salient points are important. As shown in Fig.1, although the Euclidean distances among these patterns are large, they should be considered as similar.

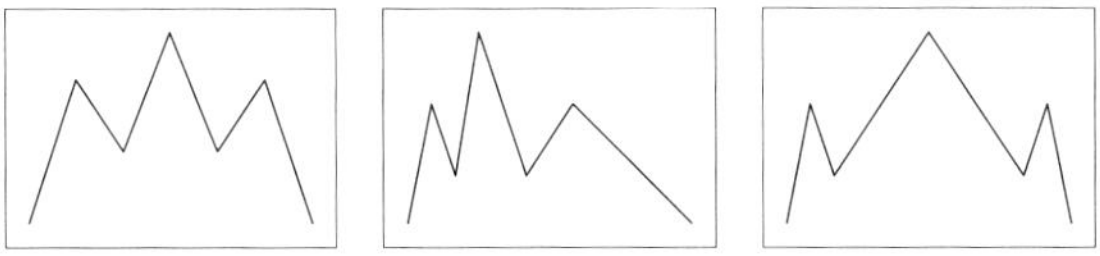

Fig. 1. Head-and-shoulder time series patterns

To fill this gap, other distance measures are proposed like [7-9] and the most common one among them is dynamic time warping (DTW) [9]. This approach can align the time series of similar lengths during the matching process. However, the required length of the subsequences is specified by the user and in most of the cases, it is longer than the length of the query pattern, that is, $w>n$. For example, the length of the subsequences looking for in **Q1** is 1 month while only 7 points are needed to describe the query pattern. As shown in our previous work [10], DTW is not a suitable distance measure method in this case due to the accumulated error along the warping path.

Therefore, instead of further investigate on improving the performance of the subsequence searching process, it is necessary to review the accuracy of the subsequence searching. In this paper, our previous proposed flexible time series pattern matching scheme based on perceptually important point (PIP) [10] is studied and applied to subsequence searching. Emphasis of this work is on stock pattern matching and hence the query pattern is generally referred to the technical (analysis) patterns, e.g., head-and-shoulder or double top. The proposed scheme adopts the time domain approach, which is more intuitive to stock market analysts and investors and will not smooth out those salient points as in most of the existing approaches. It is able to match the time series subsequences and the query pattern of different lengths in an effective manner. Moreover, an approximate subsequence searching approach based on a Specialized Binary (SB) Tree data structure is proposed to improve the performance while the accuracy can still be maintained. The paper is organized into four sections. The proposed PIP-based subsequence searching approach together with the approximate approach is introduced in section 2. The simulation results are reported in section 3 and the final section concludes the paper.

2 Our Approach

In this section, our previous proposed time series pattern matching approach based on Perceptually Important Point (PIP) and a tree data structure, SB-Tree, for storing the reordered time series data based on the concept of PIP identification process are first revisited. Then, the proposed approximate subsequence searching approach based on the SB-Tree is proposed.

2.1 Time Series Pattern Matching Based on Perceptually Important Point

Time series pattern matching based on Perceptually Important Point (PIP) identification is firstly introduced by the authors in [10]. Like DTW, the proposed scheme adopts the time domain approach. As to the technical analysis of stock data in financial domain, the frequently used stock patterns are typically characterized by a few salient points. For example, the head-and-shoulder pattern should at least consist of a head point, two shoulder points and a pair of neck points. These points are perceptually important in the human identification process and should also be taken into accounts in the pattern matching process. The proposed scheme follows this idea by locating those PIPs in the time series *P* in accordance with the query pattern *Q*. The location process works as follows.

With sequences *P* and *Q* being normalized to a unit square (for shifting and uniform amplitude scaling invariant), the PIPs are located in order according to Fig.2. Currently, the first two PIPs will be the first and last points of *P*. The next PIP will be the point in *P* with maximum distance to the first two PIPs. The fourth PIP will then be the point in *P* with maximum distance to its two adjacent PIPs, i.e., in between either the first and second PIPs or the second and the last PIPs. The PIP location process continues until the length of *SP* is equal to that of query pattern *Q*.

```
Procedure PIPIdentification (P,Q)
   Input: sequence P[1..m], template Q[1..n]
   Output: pattern SP[1..n]
Begin
   Set SP[1]=P[1], SP[n]=P[m]
   Repeat until SP[1..n] are all filled
   begin
      Select point P[j] with maximum distance to the adjacent points in
      SP (SP[1] and SP[n] initially)
      Add P[j] tO SP
   end
   Return SP
end
```

Fig. 2. Pseudo code of the perceptually important point identification process

To determine the maximum distance to the two adjacent PIPs, as depicted in Fig.3, is the vertical distance (VD) between the test point p_3 and the line connecting the two adjacent PIPs, i.e.,

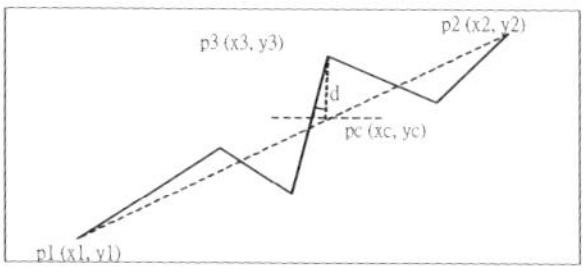

$$VD(p_3, p_c) = |y_c - y_3| = \left| \left(y_1 + (y_2 - y_1) \cdot \frac{x_c - x_1}{x_2 - x_1} \right) - y_3 \right|$$

where $x_c = x_3$.

Fig. 3. Vertical distance measure: PIP-VD

It is intended to capture the fluctuation of the sequence and the highly fluctuated points would be considered as PIPs. As different sequences may have different "amplitudes", after identifying the PIPs in the data sequence, it is necessary to re-scale (normalize) the points so that the comparison between sequences in different "amplitude" ranges can be facilitated. This is typically addressed by normalizing all the sequence values to a given range (e.g. –1 to 1). Then, the simplest way to measure the distance between the sequences is the amplitude distance (*AD*) and the distance between two sequences *P* and *Q* can be computed using direct point-to-point difference, i.e.,

$$AD(SP,Q) = \sqrt{\frac{1}{n}\sum_{k=1}^{n}(sp_k - q_k)^2} \tag{1}$$

for a query pattern *Q*, *SP* and (sp_k) denote the PIPs found in *P*. More suggested similarity measure methods can be found in [10].

Given with a query pattern Q_n and the required length of the subsequence *w*, the primitive way to apply the PIP-based pattern matching approach for subsequence searching is using sequential scanning. That is, divides the time series P_m into subsequences using sliding window of size *w* and compares the query pattern with the subsequences. Therefore, the total number of candidate subsequences need to compare is *m*-*w*+1.

2.2 Specialized Binary Tree Data Structure

Based on the PIP identification process, a time series reordering approach based on the data point importance and a binary tree (B-tree) structure is proposed in [11] for storing the time series data. It is called Specialized Binary (SB) Tree.

To create a SB-Tree, the overall PIP identification process is adopted. The first and last data points in the sequence are the first two nodes of the SB-Tree. The node, which represents the last data point of the time series, becomes the root of the tree for easy updating and the node, which represents the first data point, becomes the child on the left-hand-side of the root. The third PIP identified becomes the child on the right-hand-side of the second node. The tree can then be built recursively as follow. Starting from the parent node *pnode* (the third PIP initially) and the current/child node *cnode* (the fourth PIP initially), we have

- If *cnode.x* < *pnode.x* then goto the left arc of *pnode*
    - If *pnode.left* is empty, add *cnode* to this position
    - Else *pnode* = *pnode.left* and next iteration start
- Else goto the right arc of *pnode*
    - If *pnode.right* is empty, add *cnode* to this position
    - Else *pnode* = *pnode.right* and next iteration start

A sample time series and the corresponding SB-Tree built are shown in Fig.4. The arc of the tree represents the VD of the corresponding node (PIP). The major contribution of the SB-Tree representation is its ability to support incremental updating of the time series [12] and multi-resolution visualization [13].

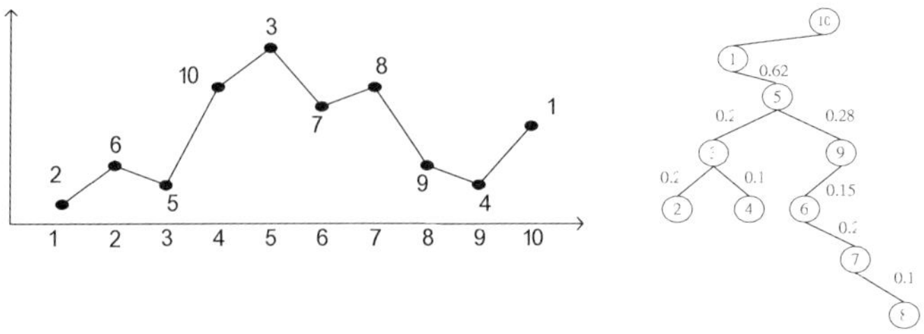

Fig. 4. Sample time series and the SB-Tree built

2.3 Subsequence Searching Based on SB-Tree

It is reasonable to identify the necessary PIPs in each subsequence formed during subsequence searching because the overall shape of a subsequence is different from its whole sequence, the data points of the subsequence should be retrieved in an exact order based on their importance according to this subsequence.

However, it is time consuming to identify the PIPs of every subsequence during subsequence searching. An approximate approach is therefore proposed. When a subtree in the SB-Tree is considered, it represents a subsequence of the time series. The candidate subsequences are retrieved according to the data point importance in the whole SB-Tree of the corresponding time series. It is generally faster than the exact approach as no re-calculation is necessary. It is only based on accessing the SB-Tree with additional criteria. Given the starting and ending points of a subsequence, the subsequence can be retrieved from the SB-Tree which built from the time series it belongs to. The data points within that subsequence can be identified based on their positions in the SB-Tree. A procedure for retrieving a subsequence from a SB-Tree is shown in Fig.5. Furthermore, the number of PIPs needs to retrieve from the SB-Tree is controlled by the parameter *no_pip* which should be equal to the length of the query pattern during subsequence searching.

For example, if the subsequence from point 5 to point 11 of a sample time series in Fig.6a is needed to retrieve, the corresponding nodes need to retrieve are shown in Fig.6b. The order of retrieving the data points is: 5→9→11→10→6→7→8 according to the procedure in Fig.5.

After retrieving the necessary number of PIPs for representing a subsequence, the same distance measure as Eqt.1 can be conducted. As shown in next section, although the set of PIPs retrieved based on the proposed approximate approach may not be exactly the same as the set identified by conducting the PIP identification process in the subsequence, similar shape can always be represented by the PIPs retrieved from the approximate approach.

```
Function Retrieve_Subsequence(root, begin, end, no_pip)
   Input:      SBTree root, Point begin, Point end, Length no_pip
   Output:     List L[1..no_pip]
Begin
   NodeType node
   Repeat until L[1..no_pip] all filled
   Begin
      node.dist = -1
      Find_Next_PIP_Constraint(root, begin, end, node)
      Append node.x TO L
      Marked node as USED
   End
   Return L
End

Function Find_Next_PIP_Constraint(cur_node, begin, end, next_node)
   Input:      NodeType cur_node
               Point begin
               Point end
               NodeType next_node
Begin
   If (cur_node NOT marked USED) and (begin <= cur_node.x <= end) Then
      If (cur_node.dist > next_node.dist) Then
       next_node = cur_node
      End If
   Else
      If (cur_node.left <> NULL) Then
       Find_Next_PIP_Constraint(cur_node.left, begin, end, next_node)
      End If
      If (cur_node.right <> NULL) Then
       Find_Next_PIP_Constraint(cur_node.right, begin, end, next_node)
      End If
   End If
End
```

Fig. 5. Pseudo code of retrieving time series subsequence from the SB-Tree

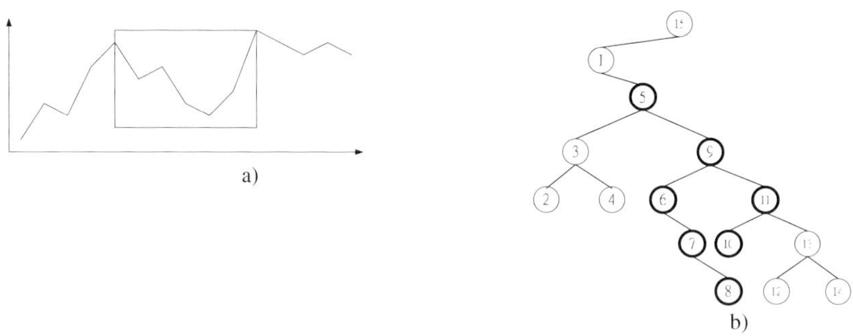

Fig. 6. Example of retrieving a time series subsequence from the SB-Tree

3 Experimental Results

In this section, we evaluate the accuracy as well as the speed of the approximate approach for subsequence searching in SB-Tree. The accuracy is measured by the number of correctly best-matched subsequences and the targeted pattern is retrieved within the top 3 queried subsequences. The other two sequential scanning methods based on PIP and DTW distance measures are also implemented for comparison. The two parameters tested are the length of the time series and the preferred length of the subsequence.

Both the synthetic and real datasets are used in the experiment. The real dataset is the Hong Kong Heng Sang Index (HSI) from 1989 to 2006 (4214 data points). For the synthetic dataset, the time series is generated by two steps. First, a set of patterns with different lengths was generated. Each of them belongs to one of the five technical patterns (with the corresponding number of PIP in the bracket), head-and-shoulder (7), double tops (5), triple tops (7), rounded top (4 or 5) and spike top (5) as shown in Fig.7. The patterns are uniformly time scaled. Then, each salient point of the patterns can be warped between its previous and next salient points. Finally, noise is added to the set of patterns. Noise adding is controlled by two parameters, namely, the probability of noise adding for each data point and the level of noise being added to such point.

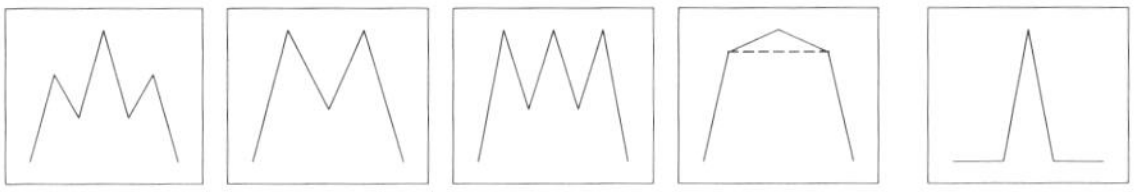

Fig. 7. The five technical patterns: head-and-shoulder (7 PIPs), double tops (5 PIPs), triple tops (7 PIPs), rounded top (4 or 5 PIPs) and spike top (5 PIPs)

The second step is to insert the generated pattern to a long time series. The time series is either generated by random walk function (Fig.8a) or a real time series (Fig.8b). By doing this way, an objective evaluation on the accuracy of the methods can be achieved by identifying the inserted pattern from the time series. The technical patterns in Fig.7 will be served as the query patterns.

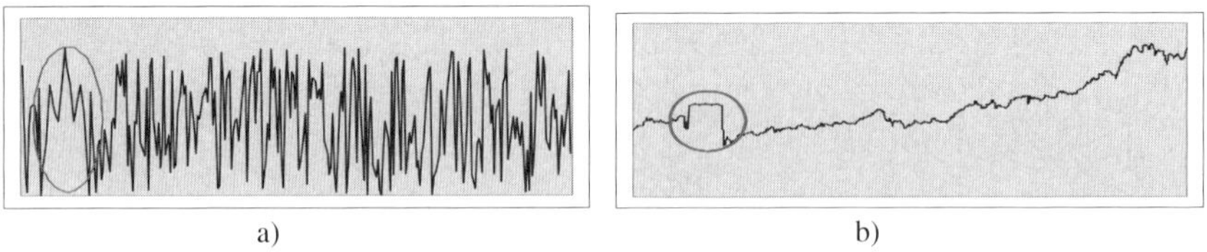

Fig. 8. Sample synthetic time series (a) pattern inserted to a random walk time series and (b) pattern inserted to a real (stock) time series

3.1 Synthetic Dataset

First, the effect of increasing the length of the time series is tested. The length of time series is varied from 500 to 4000. The length of the subsequence is fixed to 61. Fig.9a shows the accuracy of the three methods. The performance of DTW is surprisingly bad to retrieve the inserted pattern. As stated in section 1, DTW can only be worked when the different between the lengths of the matching time series is not large. However, in our case, the query pattern is only 4 to 7 data points (Fig.7) but the length of the preferred subsequence is 61 data points. The great difference between the length of the query pattern and query subsequence leads to the poor performance of the DTW method. On the other hand, the accuracy by using PIP and the proposed SB-Tree is similar. The result also shows that the accuracy of subsequence searching is independent to the length of time series. As shown in Fig.9b, the time of query is greatly increased with the increasing of the length of time series when sequential scanning is based on PIP. But the speed on subsequence searching by scanning the SB-Tree can be kept in a constant level.

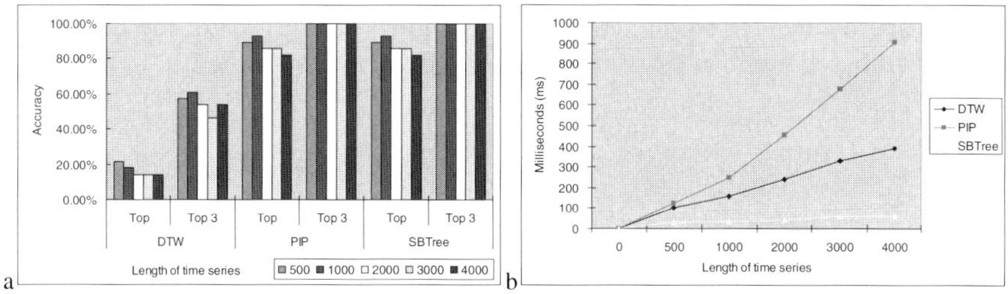

Fig. 9. (a) The accuracy of subsequence searching versus the length of time series and (b) the speed of subsequence searching versus the length of time series

Then, the effect on different preferred subsequence lengths (the sliding window size w) is evaluated. The length of time series used is fixed to 2000 data points. The length of the preferred subsequence used is varied from 25 to 61. As shown in Fig.10a, the accuracy is also independent to the subsequence length. As the differences between the lengths of the query pattern and the lengths of subsequence are large in all cases (i.e. 4 to7 compare to 25-61), the DTW method resulted in bad performance. The PIP and SB-Tree methods are outperformed the DTW method. Similar result is also obtained by varying the length of time series, the time of query is greatly increased with the further increasing of the length of subsequence when sequential scanning is based on PIP. But the speed on subsequence searching by scanning the SB-Tree can be kept in a constant level (Fig.10b).

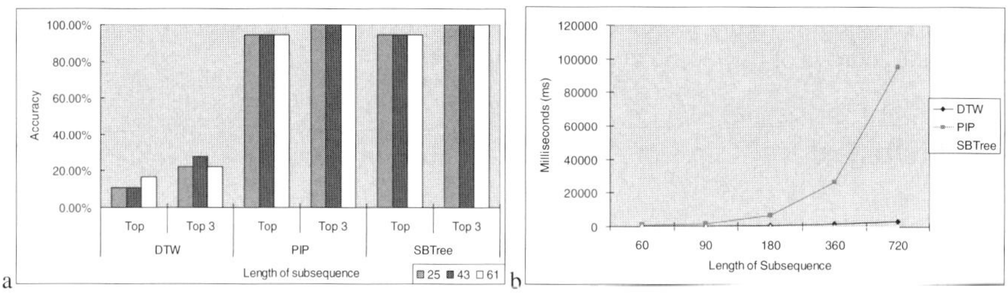

Fig. 10. (a) The accuracy of subsequence searching versus the preferred length of subsequence (i.e. the window size) and (b) the speed of subsequence searching versus the preferred length of subsequence

3.2 Real Dataset

Finally, sample query results on the real time series (Hong Kong HSI, 4214 data points) are shown in this subsection. The head-and-shoulder pattern is served as the query pattern and the preferred length of subsequence is set to 90. As shown in Fig.11, the first circled pattern is the query result obtained by using DTW method and second circled pattern is the result obtained by using PIP-based approaches (i.e. both the PIP and SB-Tree approaches). Fig.12a shows the zoom in of the first circled pattern and Fig.12b shows the corresponding shape after identified the 7 PIPs. Comparing the shape of the second circled pattern as shown in Fig.12c&d, the query result from the PIP-based approaches is much better than the DTW method.

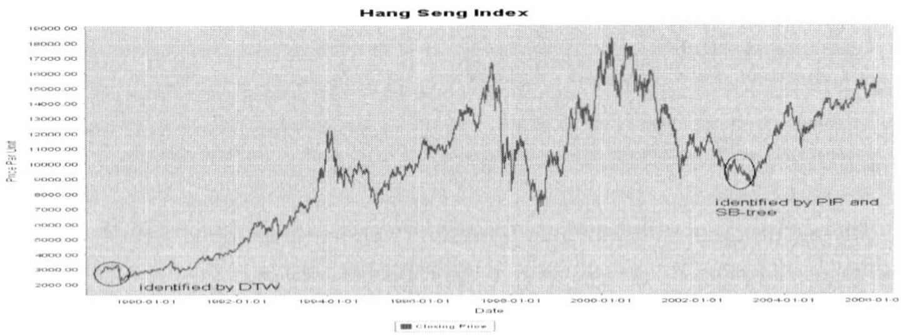

Fig. 11. Sample stock time series (i.e. Hong Kong Heng Sang Index with 4214 data points), the first circled pattern is the query result from DTW method and the second circled pattern is the query result from both PIP and SB-Tree approaches

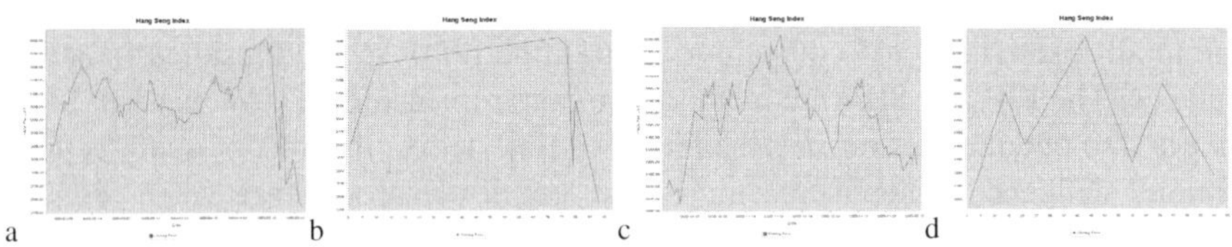

Fig. 12. (a) The original pattern of the zoom in of the first circled pattern in Fig.11, (b) the shape represented by 7PIPs of the zoom in of the first circled pattern in Fig.11, (c) the original pattern of the zoom in of the second circled pattern in Fig.11 and (d) the shape represented by 7PIPs of the zoom in of the second circled pattern in Fig.11

4 Conclusion

In this paper, the current works on subsequence matching in time series database are revisited. The limitation on the similarity measure approaches which based on preserving the Euclidean distance and the problem of using dynamic time warping during pattern matching are discussed. Our previous proposed pattern matching method based on Perceptually Important Point identification shows its strength on subsequence searching especially when the different of the lengths of the query pattern and the time series is large. Furthermore, an approximate subsequence searching approach which based on a tree data structure is proposed to speedup the PIP identification process while the accuracy can still be maintained. One may find it particularly attractive in financial applications like stock data analysis. To further develop the proposed approach, an indexing mechanism is now investigating like most of the similar researches.

References

1. Agrawal, R., Faloutsos, C. and Swami, A.: Efficient Similarity Search. In Sequence Databases. In Proceedings of the 4th International Conference on Foundations of Data Organization and Algorithms (1993) 69-84
2. Faloutsos, C., Ranganathan, M. and Manolopoulos, Y.: Fast Subsequence Matching in Time-Series Databases. In Proceedings of the 1994 ACM SIGMOD International Conference on Management of Data (1994) 419-429
3. Morinaka, Y., Yoshikawa, M., Amagasa, T. and Uemura, S.: The L-index: An Indexing Structure for Efficient Subsequence Matching in Time Sequence Databases. In Proceedings of the 5th Pacific-Asia Conference on Knowledge Discovery and Data Mining (2001) 51-60
4. Keogh, E., Lin, J. and Fu, A.: HOT SAX: Efficiently Finding the Most Unusual Time Series Subsequence. In Proceedings of the 5th IEEE International Conference on Data Mining (2005) 226-233
5. Wu, H., Salzberg, B., Sharp, G., Jiang, S., Shirato, H. and Kaeli, D.: Subsequence Matching on Structured Time Series Data. In Proceedings of the 2005 ACM SIGMOD International Conference on Management of Data (2005) 682-693
6. Keogh, E.: A Fast and Robust Method for Pattern Matching in Time Series Databases. In Proceedings of the 9th IEEE International Conference on Tools with Artificial Intelligence (1997) 578-584
7. Megalooikonomou, V., Wang, Q., Li, G. and Faloutsos, C.: A Multiresolution Symbolic Representation of Time Series. In Proceedings of the 21st IEEE International Conference on Data Engineering (2005) 668-679
8. Perng, C.S., Wang, H., Zhang, R. and Parker, D.: Landmarks: A New Model for Similarity-Based Pattern Querying in Time Series Databases. In Proceedings of the 16th IEEE International Conference on Data Engineering (2000) 33-42
9. Berndt, D.J. and Clifford, J.: Using Dynamic Time Warping to Find Patterns in Time Series. In AAAI Working Notes of the Knowledge Discovery in Databases Workshop (1994) 359-370
10. Chung, F.L., Fu, T.C., Luk, R. and Ng, V.: Flexible Time Series Pattern Matching Based on Perceptually Important Points. International Joint Conference on Artificial Intelligence Workshop on Learning from Temporal and Spatial Data (2001) 1-7
11. Fu, T.C., Chung, F.L., Luk, R. and Ng, C.M.: A Specialized Binary Tree for Financial Time Series Representation. The 10th ACM SIGKDD International Conference on Knowledge Discovery and Data Mining Workshop on Temporal Data Mining (2004) 96-103
12. Fu, T.C., Chung, F.L., Tang, P.Y., Luk, R. and Ng, C.M.: Incremental Stock Time Series Data Delivery and Visualization. In Proceedings of The ACM 14th Conference on Information and Knowledge Management (2005) 279-280
13. Fu, T.C., Chung, F.L., Lam, C.F., Luk, R. and Ng, C.M.: Adaptive Data Delivery Framework for Financial Time Series Visualization. In Proceedings of the 4th International Conference on Mobile Business (2005) 267-273

Research of Local Co-location Pattern in Spatial Event Sequences

Wang Zhanquan[1], Yu Huiqun[1], and Chen Haibo[2,*]

[1] East China University of Science and Technology, Department of Computer Science and Engineering, Postfach 20 02 37, Shanghai, China
{zhqwang, yqh}@ecust.edu.cn
[2] Zhejiang Sci-tech University, College of Science, Postfach 310018, Hangzhou, China
chen_hb@sohu.com

Abstract. The present works are focusing on the discovery of global co-location patterns. It is a challenging problem to find the local co-location patterns. A novel method was presented to find local co-location patterns in an event sequence. The local co-location patterns were found by using an effective multi-layer index in a given time window (win) and local neighbor domain set. The experiment was done to prove algorithm effective and feasible.

1 Introduction

The spatial database is used widely to lead to an increasing interest in mining interesting and useful but implicit spatial patterns [1],[2],[3],[4]. Reference [5] proposed the reference feature centric model and can effectively deal with spatial association rule. But may yield duplicate count for candidate associations. The data-partition approach [6] defines transactions by dividing spatial datasets into disjoints partitions. However imposing artificial disjoint transactions may undercount instances of tuples co-located together. Spatial co-location method [7] defines co-location patterns based on spatial correlation. It can obtain the good results because of considering spatial correlation property, but can't process categorical data. Reference [8] can effectively process the categorical data, but it can't obtain local co-location patterns. To resolve these defects, spatially local co-location pattern in event sequences is proposed.

2 Model Building

Spatially local co-location pattern in event sequences is described as follows: there are N spatial events whose number of spatial features is less than K, the inherent patterns of spatial features would be mined after its neighbor correlation is completely considered in win. Spatially local co-location pattern can obtain the effective rules, for example in win, the rule is $A \xrightarrow{win} B$, fox example, A is occurrences of nutria, B

This work is partially supported by the NSF of China under grant No. 60473055 and 60373075, Shanghai Pujiang Program under grant No. 05PJ14030.

* Corresponding author.

L. Wang et al. (Eds.): FSKD 2006, LNAI 4223, pp. 578 – 581, 2006.

is appearance of SARS's patient. If the rule is confident, it has the probability to find patients of SARS in a nearby region whenever the nutria is found in win, besides A, B has the order of time. Some relevant definitions are given as follows:

Definition 1: A spatial framework S is quintuple, $S = \{SE, F, V, R, sf\}$, where SE is sequence sets, $SE = \{E_i \mid 0 \le i \le M, T\}$, E_i is a set of spatial event sequence. Limited by space, the other symbols are shown in [8].

Definition 2: Spatially local co-location patterns C is a set of spatial feature patterns which are composed of spatial features, the value of spatial features, and the win: $c_{i1} \wedge c_{i2} \wedge \ldots \wedge c_{ik} \wedge win$, where c_{ij} is $f_{ij} = v_{ij} (0 < j \le K)$.

Definition 3: Local neighbor domain set L is a set of instances such that all pairwise locations in L are neighbors in win. If every feature pattern c_i in pattern C appears in an instance of L in order and there exists no proper subset of L does so, then a neighbor-set L is said to be an instance of local co-location pattern C in win. We denote all instances of local co-location pattern C as $\text{instancet}(C)_{win}$.

Definition 4: To measure the implication strength of a spatial feature in a local co-location pattern in win, a participation ratio: $pr(C, f, win) = (\pi_f(\text{instancet}(C, win)))/(\text{instancet}(f, win))$, where π is the relational projection operation with duplication elimination. A feature f has a partition ratio $pr(C, f, win)$ means wherever f is observed, with probability $pr(C, f, win)$, all other features in C can be observed. For a pattern C in win, participation index $PI(C, win) = \min_{f \in C}\{pr(C, f, win)\}$. Given threshold min_sup, a pattern is called prevalent if $PI(C, win) \ge \text{min\_sup}$. For a rule: $c_1 \xrightarrow{win} c_2$, the conditional of rule is defined as: $p(c_2 | c_1, win) = (\pi_{c_1}(\text{instancet}(c_2, win)))/(\text{instancet}(c_1, win))$, where $c_1 \le c_2$. p is the probability that a neighbor-set in $\text{instancet}(c_2)_{win}$ is a part of a neighbor-set in $\text{instancet}(c_1 \cup c_2)_{win}$.

Lemma 1: Let C, C' be two local co-location patterns in win such $C \le C'$, then $PI(C, win) \ge PI(C', win)$.

Proof: Given every spatial feature $c, c \in C \cap C'$, $pr(C, c, win) \ge pr(C', c, win)$, thus $PI(C, win) \ge PI(C', win)$.

Let us describe algorithm, Limited by space, the other symbols are shown in [8].

Input: spatial framework S, win, min_sup, min_conf;

1.To initialize feature itemsets C_1 which $k = 1$, make $L_1 = C_1$, and $PI = 1$;

2. IS_1=Gen_Instance_Set(C_1, S);

3. for($k = 2; (L_k \neq \phi) and (k \leq K); k++$){

4. C_{k+1}=Gen_Local_Co-location(L_k, k);

5. IS_{k+1}=Gen_Instance_Set(C_{k+1}, IS_k, R ,win);

6. P_{k+1}=Select_fre_Co-location(min_sup, C_{k+1}, IS_{k+1});

7. R_{k+1}=Gen_ Local_Colocation_Rule(min_conf, L_{k+1}, IS_{k+1}); }

8. Return $L = \cup_k L_k, R = \cup_k R_k$;

3 Analysis

Completeness: The spatial join produces all pair (p', p'') of instances where $p'.feature \neq p''.feature$, and p', p'' are neighbors. Any now instance of any size 2 co-location satisfying these two conditions in the join predicate will be generated [2]. Thus the algorithm is complete; Correctness: the row instance of each local co-location is correct, as that will imply the correctness of the participation index values and that of each local co-location meeting the user specified threshold. Thus the local co-location algorithm is correct; Time Complexity: Cost is a paramount factor in spatial data set. Its time complexity is: o(KxMxN), where K is number of spatial features, M denotes number of spatial instances, N denotes number of nodes which meet requirements. The time cost is effectively reduced as the dynamic structure R-tree is used in data set. The search depth of method is controlled by pruning algorithm if there are too many nodes.

4 Experiments

All the experiments were performed on a Pentium 4 1.8GMHz PC machine with 384M main memory. All methods were implemented using C++ language. The totally 7320 data records were collected. The fire levels:{0,I,II,III}, the loss increases in turn.

Limited by space, we report only the results on some representative datasets. The symbol of (win, 1440) denotes that the win is 1440 minutes; (min_sup, 25) denotes that min_sup is 25. A value of min_sup is max, and regularly decrease with the increase of layers; (m, 2K) denotes that number of spatial objects is 2K. From Fig.1, it was found that increase of number of spatial objects would lead to enhancing of frequent item number. In any other unchanging case, the larger the win, the more number of frequent items and vice versa; In Fig.2, number of frequent items fall with increase of the participation index value. The frequent item number increases with number of total nodes if the other case is invariable. As can be seen in fig.3, as the win advances, the number of frequent items was the same. But there are obvious changes if the win is larger than some value because the length of total item set is dominated by number of spatial features in spatial event sequence.

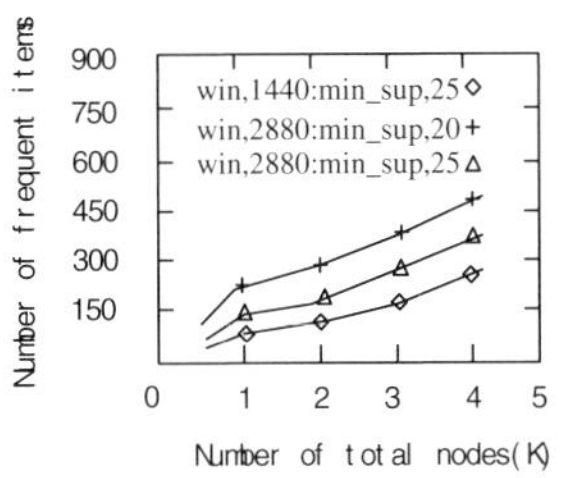

Fig. 1. Relation between total node number and frequent items

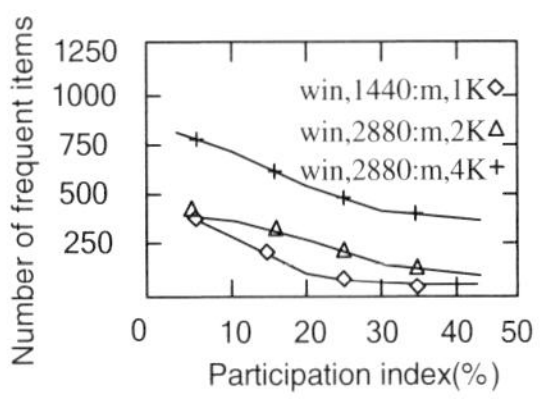

Fig. 2. Relation between participation index and frequent items

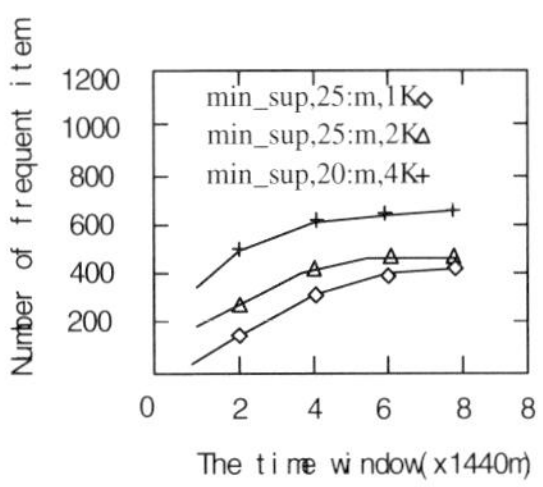

Fig. 3. Relation between the win and frequent items

5 Conclusions

A novel approach was presented to obtain local co-location patterns. It is effective in this study.In the present study, three interesting directions were opened for future explorations. First, it would be applied to other fields (city's layout, and so on). Second, the desirable results were obtained effectively from large data set. Third, for moving objects, it was highly interesting to mine temporal-spatial co-location patterns in three-dimensional (X-Y-TIME) coordinate.

References

[1] Agarwal, R., Srikant,R. Fast Algorithms for Mining Association Rules. In Proc. 20th Int. VLDB, Santiago, Chile, September 12-15.(1994)487-499

[2] Yan Huang.Mining Co-Location Patterns from Large Spatial Datasets ,PhD dissertation of the University of Minnesota, (2003)

[3] S.Chawla, S.Sekhar, W.Wu, U.Ozesmi. Extending Data Mining for Spatial Applications A Case Study in Predicting Nest Locations. In Conf. on ACM SIGMOD, (2000)

[4] R.H. Guting. An introduction to Spatial Database Systems. In VLDBJ, (1994)

[5] Koperski, K., Han, J.W.. Discovery of Spatial Association Rules in Geographic Information Database. In Proc 4th Int. Large Spatial Database, Maine, August .(1995)47-66

[6] Morimoto, Y. Mining Frequent Neighboring Class Sets in Spatial Databases. In Proc 7th int. Knowledge discovery and data mining, San Francisco, California, (2001)353-358

[7] Shekhar, S., Huang,Y. Discovering Spatial Co-location Patterns: A Summary of Results. In Proc 7th Int. Spatial and Temporal Databases, Redondo Beach, CA, USA, (2001)236-256.

[8] Wang, Z.Q, Spatial Co-location Rule Mining Algorithm in Categorical Data, Journal of Computer-aided Design & Computer Graphics, 17(10),(2005)320-327

Adaptive Nearest Neighbor Classifier Based on Supervised Ellipsoid Clustering

Guo-Jun Zhang[1,2], Ji-Xiang Du[1,2], De-Shuang Huang[1,2], Tat-Ming Lok[3], and Michael R. Lyu[4]

[1] Intelligent Computing Lab, Hefei Institute of Intelligent Machines, Chinese Academy of Sciences, P.O. Box 1130, HeFei Anhui 230031, China
{zhanggj, dshuang}@iim.ac.cn
[2] Department of Automation, University of Science and Technology of China, Hefei, China
[3] Information Engineering Dept., The Chinese University of Hong Kong, Hong Kong
[4] Computer Science & Engineering Dept., The Chinese University of Hong Kong, Hong Kong

Abstract. Nearest neighbor classifier is a widely-used effective method for multi-class problems. However, it suffers from the problem of the curse of dimensionality in high dimensional space. To solve this problem, many adaptive nearest neighbor classifiers were proposed. In this paper, a locally adaptive nearest neighbor classification method based on supervised learning style which works well for the multi-classification problems is proposed. In this method, the ellipsoid clustering learning is applied to estimate an effective metric. This metric is then used in the *K*-NN classification. Finally, the experimental results show that it is an efficient and robust approach for multi-classification.

1 Introduction

One of most popular classification approaches is the Nearest Neighbor (NN) method. However, the NN rule becomes less appealing in the case of limited training samples and high dimensional feature space due to the curse of dimensionality. Severe bias can be caused in such situations. Recently, several methods [1-4] characterized by query-based local distance functions were proposed to reduce this bias. However, the computation seems not to be efficient and there are too many parameters despite of the relatively high accuracy they achieve. Are there any better alternatives which could estimate the local relevance efficiently?

Our method uses the supervised ellipsoid clustering(SEC) to produce boundary and utilizes its boundary to estimate the local features relevance of the query with a scheme provided in LAMANN[1]. The relevance is then used in the weighted Euclidean distance during the *K*-NN classifications. The algorithm is referred to as Ellipsoids-boundary weighting adaptive Nearest Neighbor Algorithm (EWANN Algorithm) because the class boundary is constructed according to the surface of ellipsoids.

L. Wang et al. (Eds.): FSKD 2006, LNAI 4223, pp. 582 – 585, 2006.

This paper is organized as follows. Section 2 presents the feature relevance theory. Some empirical evaluation of our method is given in Section 3. Finally, a concluding remark is included in Section 4 .

2 Ellipsoids-Boundary Weighting Adaptive Nearest Neighbor Algorithm

It is often noted that different classes may have different discriminant features. This means that distance computation does not always have equal strength on features in the feature space. Considering this, many locally adaptive nearest neighbor classifiers are designed to find these relevant dimensions.

Supervised ellipsoid clustering (SEC) is to cover each class region with ellipsoids. Each ellipsoid will represent a set of points [5]. Ellipsoids can be generated by an incremental learning procedure. The ellipsoids are created, constricted, or enlarged gradually at the present of each training sample. After SEC, each class is stuffed with a set of ellipsoids. In other views, the outer surface of the ellipsoids of a certain class composes the boundary of that class, i.e. the boundary of the class against the other classes is formed by the surface of its ellipsoids.

The gradient vector of the point on the boundary identifies a direction along which nearby data points are well separated. We use this gradient vector to measure local feature relevance and weighting features accordingly.

The feature relevance $R(x)$ can be given as $R_i(x) = |N_d \bullet u_i| = |N_{di}|$, where d is the nearest point on the boundary to the query, N_d the gradient vector at point d , u_i the vector unit. After r is transformed from $R(x)$ by a scheme given by Jing Peng [1], it could be applied in the weighted distance computation during the NN classification. The resulting algorithm is summarized in Fig. 1.

```
INPUT: Ellipsoids of all classes produced by SEC, query q and
parameter K for K-NN
Step 1. Find ellipsoid E_n which is nearest to query q
Step 2. Find the point d which is the nearest point to q on the E_n .
Step 3. Compute the gradient vector N_d
Step 4. Compute R(x) and transform it to r by the scheme given by
Jing Peng
Step 5. Use r in weighted distance computation and apply K-NN rule
```

Fig. 1. Ellipsoids-boundary Weighting Adaptive Nearest Neighbor Algorithm

3 Empirical Evaluation

In the following, we compare several competing classification methods using a number of data sets. The classifiers are 5NN which uses the simple five NN rule, DANN using the discriminant adaptive NN rule.[2], Morph which uses he morphing rule [3],ADAMENN using the adaptive Metric Nearest Neighbor Algorithm [4] and LAMANN using the large margin nearest neighbor.[1]. The data sets were taken from the UCI Machine Learning Database Repository. We randomly select 60% of samples of each class as training samples and other 40% for testing.

Table 1. Average classification error rate

	Iris	Heart	Diabetes	Cancer
EWANN	**5.8**	24.4	**24.3**	23.6
5NN	**5.8**	24.8	25.4	24.1
DANN	**5.8**	23.7	24.8	22.7
Morph	**5.8**	**22.7**	25.7	**22.8**
ADAMENN	**5.8**	22.9	25.0	25.0
LAMNN	**5.8**	24.0	24.8	23.1

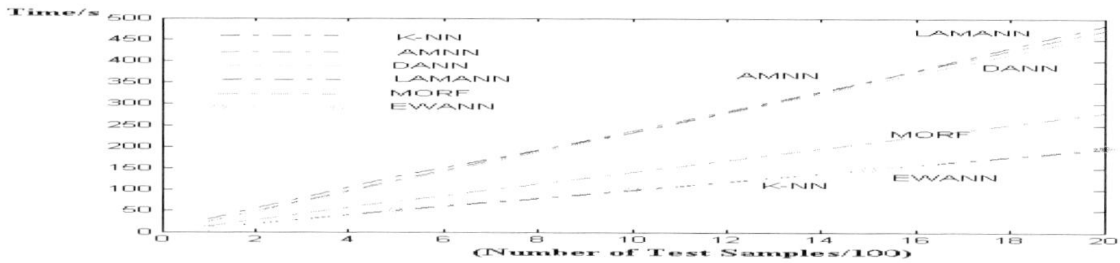

Fig. 2. On-line computing time performance

From Table 1, it can be found that EWANN achieved the best performance of the 2/4, followed closely by LAMANN. The result shows that our method is as competing as other adaptive classifiers. What's more, as it is shown in Fig. 2, the proposed method is superior to other adaptive nearest neighbor methods in terms of online computing.

4 Conclusions

This paper presents a new flexible metric method for effective and efficient pattern classification. It employs SEC to generate class boundary, and use its information for the following nearest neighbor classification. The experimental results show clearly that the proposed algorithm can potentially improve the performance of K-NN in several classification problems.

References

1. Domeniconi, C., Gunopulos ,D., and Peng, J.: Large Margin Nearest Neighbor Classifiers, IEEE Transaction on Neural Networks, VOL 16, No. 4 , (2005), 899-909
2. Hastie, T., Tibshirani, R.: "Discriminant Adaptive Nearest Neighbor Classification", IEEE Trans. on PAMI, Vol. 18,No. 6, (1996), 607-615
3. Peng, J., Heisterkamp, D. R., Dai, H.K.: LDA/SVM Driven Nearest Neighbor Classifiers, IEEE Transaction on Neural Networks, VOL 14, No. 4, (2003), 940-942
4. Domeniconi, C., Peng, J., Gunopulos, D.: Locally Adaptive Metric Nearest Neighbor Classification, IEEE Transaction on PAMI, (2002), 1281-1285
5. Kositsky, M., Ullman, S.: Learning Class Regions by the Union of Ellipsoids, Proceedings of the 13th ICPR, IEEE Computer Society Press, (1996), 750-757

Mining Temporal Patterns from Sequence Database of Interval-Based Events

Yen-Liang Chen and Shin-Yi Wu

Department of Information Management, National Central University,
300, Jungda Road, Chung-Li, Taiwan 320, China
{ylchen, sywu}@mgt.ncu.edu.tw

Abstract. Sequential pattern mining is one of the important techniques of data mining to discover some potential useful knowledge from large databases. However, existing approaches for mining sequential patterns are designed for point-based events. In many applications, the essence of events are interval-based, such as disease suffered, stock price increase or decrease, chatting etc. This paper presents a new algorithm to discover temporal pattern from temporal sequences database consisting of interval-based events.

1 Introduction

Sequential pattern mining, which is an important data mining problem, was first proposed by Agrawal and Srikant in [1]. Given a sequence database, which consists of three attributes: sequence-id, event-id, time-stamp, sequential pattern mining is to discover all sequential patterns with *supports* that are no less than the user defined minimum support, *min_sup*. In a supermarket, the sequence database is a set of transaction records, and the three attributes correspond to customer-id, purchased items, and transaction time. A sequential pattern discovered from such a transactional database might be "30% of customers bought 'milk and diapers', then 'toys', and then 'cosmetics'.

Sequential pattern mining can be applied in many domains [5, 8], for example, in retailing industry, it can be used to find purchasing habit; in finance, it can be applied to discover the trends of stock prices; in biology, it can be utilized to analyze the DNA sequence; and many other applications. However, in the existing studies of sequential pattern mining, the events in a sequence are point-based, and the sequential patterns represented only the chronological order of event points. Undoubtedly, events in many daily applications are interval-based, since the starting time and the ending time of each event have been recorded. For example, in a clinical dataset, diseases suffering are interval-based events; in a library, books loaning are interval-based events; in a web chat room, users staying are interval-based events; and many others. From sequences of interval-based events, we can discover patterns, called temporal patterns, more diversiform.

Since essences of interval-based and point-based events are different, the relationships between two interval-based events are more complicated than those between two

L. Wang et al. (Eds.): FSKD 2006, LNAI 4223, pp. 586–595, 2006.

point-based events. While relationships among point-based events possess only three possibilities, "X *before* Y", "X *equal* Y", and "X *after* Y", relationships among two intervals have been classified into 13 classes in Allen's taxonomy [2], including "X *before* Y", "X *equal* Y", "X *meets* Y", "X *overlaps* Y", "X *during* Y", "X *starts* Y", "X *finishes* Y" and six inverse relationships except "*equal*". Due to this difference, temporal pattern mining would discover some interesting patterns which are never found from sequences of point-based events. For example, in medical application, a temporal pattern might be "patients frequently start 'fever' when they start to 'cough' and these symptoms all happened during they caught 'flu'". In other application such as web chat room, "'David' often login during the duration when 'Peter' and 'Mary' are in chat" might be a frequent temporal pattern.

The existing sequential pattern mining algorithms, including apriori-like algorithms [1, 6], data projection based algorithms [3, 5], lattice based algorithm [7], or memory indexing based algorithm [4], were unable to discover temporal patterns directly or with slight modification. In this paper, we first formally define the problem of temporal pattern mining for interval-based events in Section 2. Besides, because the patterns to be discovered are totally different from those in the original problem of sequential pattern mining, we also define the representations for temporal patterns and temporal sequences. After defining the standard format to represent temporal sequences, we propose an algorithm *T-Apriori*, in Section 3, to discover temporal pattern for interval-based events. *T-Apriori*, which is modified from the well-known *GSP* algorithm [6], is demonstrated by explaining a complete simulated case. Finally, the conclusions are drawn in Section 4.

Table 1. Original temporal database *D*, consists of three attributes: sequence id, event id, and time period

Sid	Eid	Interval	Sid	Eid	Interval
01	a	{5, 10}	02	c	{16, 18}
01	b	{8, 12}	02	c	{19, 21}
01	c	{14, 18}	02	d	{16, 22}
01	d	{14, 20}	03	a	{12, 20}
01	b	{17, 22}	03	b	{22, 25}
02	a	{8, 14}	03	c	{29, 31}
02	b	{9, 15}	03	d	{28, 32}

2 Problem Definition

Let *D* denote a temporal database with three attributes: sequence-id, event-id, and time period. For example, a list of clinical records may contain the attributes: patient-id, disease the patient suffered, duration that the patient suffered from this disease. Table 1 is an example of temporal database *D*. The duration of happened each event is recorded by a pair of timestamps t^+ and t^-, where $t^+ \leq t^-$, which are noted for the starting time and the ending time. Suppose there are totally *u* event types that could happen in *D*, called *event types* 1, 2, …, *u*. Then an event e_i can be represented as a triple (ty_i, t_i^+, t_i^-), where

$ty_i \in \{1, \ldots, u\}$ is an event type, t_i^+ and t_i^- are the starting time and the ending time of e_i. Based on these notations, we describe how the temporal sequences are represented and define the problem of temporal pattern mining.

Definition 1 (Event end points and order relations). An event $e_i = (ty_i, t_i^+, t_i^-)$ can be represented as two end points ty_i^+ and ty_i^-,called *event end points*, where ty_i^+ and ty_i^- are the *starting point* (*esp* for short) and *ending point* (*eep* for short) of e_i, respectively. Let the time of an event end point u be denoted as $time(u)$. Then the *order relation* $Rel(u, v)$ of two event end points u and v can be defined as "<" if $time(u) < time(v)$; and as "=" if $time(u) = time(v)$.

In Fig. 1, we have six event end points a^+ at time 5, a^- at time 9, b^+ at time 5, b^- at time 11, c^+ at time 6 and c^- at time 13. Some order relations among them are $Rel(a^+, b^+)$ = "=", $Rel(a^-, b^-)$ = "<", and $Rel(a^+, c^+)$ = "<".

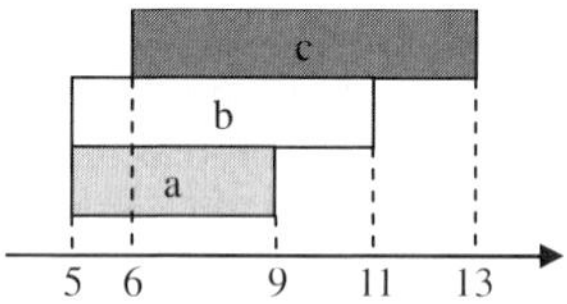

- $e_1 = (a, 5, 9)$
- $e_2 = (b, 5, 11)$
- $e_3 = (c, 6, 13)$

Fig. 1. Three events e_1, e_2, and e_3 consist of six event end points. Six event end points possess 15, C_2^6, order relations.

Definition 2 (Arrangement of event end points in a temporal sequence). In a temporal sequence to be defined below, end point u must be placed before end point v if the following conditions are satisfied.

1. $time(u) < time(v)$.
2. If $time(u) = time(v)$ but u is an *esp* and v is an *eep*.
3. Conditions 1 and 2 are in ties but the event type of u alphabetically precedes that of v.

Base on Definition 2, the six points of the temporal sequence in Fig. 1 can be arranged in the following order: a^+, b^+, c^+, a^-, b^- and c^-.

Definition 3 (Temporal sequence). A *temporal sequence* can be defined constructively as follows.

1. A temporal sequence of one event, called 1-event temporal sequence, can be written as $(e_i^+ \oplus e_i^-)$, where $\oplus \in \{<, =\}$ is the order relation of e_i^+ and e_i^-.
2. Let $p = (p_1 \oplus_1 p_2 \oplus_2 \ldots \oplus_{i-1} p_i \oplus_i p_{i+1} \oplus_{i+1} \ldots \oplus_{j-1} p_j \oplus_j p_{j+1} \oplus_{j+1} \ldots \oplus_{2k-1} p_{2k})$ denote a temporal sequence of k events, called k-events temporal sequence, where p_i is an end point and $\oplus_i \in \{<, =\}$, $1 \le i \le (2k-1)$. Suppose p' is the temporal sequence obtained by inserting event $e_a = (e_a^+ \oplus_a e_a^-)$into p. Assume that e_a^+ must be placed between p_i and p_{i+1} and e_a^- between p_j and p_{j+1} according to the rules given in Definition 2. Then, we have

$p' = (p_1 \oplus_1 p_2 \oplus_2 \ldots \oplus_{i-1} p_i \oplus_{i'} e_a^+ \oplus_{a+} p_{i+1} \oplus_{i+1} \ldots \oplus_{j-1} p_j \oplus_{j'} e_a^- \oplus_{a-} p_{j+1} \oplus_{j+1} \ldots \oplus_{2k-1} p_{2k})$, where $Rel(p_i, e_a^+) = \oplus_{i'}$, $Rel(e_a^+, p_{i+1}) = \oplus_{a+}$, $Rel(p_j, e_a^-) = \oplus_{j'}$, and $Rel(e_a^-, p_{j+1}) = \oplus_{a-}$.

Example 1. $p = (b^+ < a^+ = c^+ < a^- < b^- < c^-)$, $e_a = (d^+ < d^-)$, where $time(c^+) < time(d^+) < time(a^-)$ and $time(b^-) < time(d^-) < time(c^-)$. Fig. 2 shows that after adding e_a into p, we have p' = $(b^+ < a^+ = c^+ < d^+ < a^- < b^- < d^- < c^-)$.

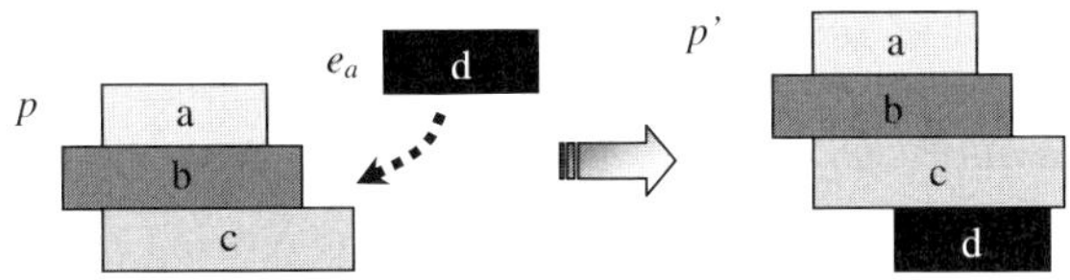

Fig. 2. Illustration for construction of a temporal sequence

We use an occurrence number attached to an *esp* or *eep* to distinguish the multiple occurrences of the same event type. For example, $(a^+ < b^{+1} < a^- < b^{-1} < c^+ = d^+ < b^{+2} < c^- < d^- < b^{-2})$ and $(a^+ < b^+ < a^- < b^- < c^{+1} = d^+ < c^{-1} < c^{+2} < c^{-2} < d^-)$ are two temporal sequences with occurrence numbers, where event type b and event type c occur twice in the first and second sequences, respectively. Throughout this paper, when we write $p_i = p_j$, we mean these two *esps* (or *eeps*) are of the same event type.

Basically, all the end points in a temporal sequence are arranged according to the rules in Definition 2. To deal with multiple occurrences of events, the following condition should be added into Definition 2: "4. Conditions 1, 2 and 3 are in ties but the occurrence number of *u* is smaller than that of *v*". For example, we may have a temporal sequence like $(a^+ < b^+ < a^- < b^- < c^{+1} = c^{+2} = d^{+1} = d^{+2} < c^{-1} < c^{-2} = d^{-1} < d^{-2})$.

Table 2. Temporal Sequences *S* constructed from temporal database *D*

Sid	Temporal Sequences																				
		0		1		2		3		4		5		6		7		8		9	
01	(	a^+	<	b^{+1}	<	a^-	<	b^{-1}	<	c^+	=	d^+	<	b^{+2}	<	c^-	<	d^-	<	b^{-2}	)
02	(	a^+	<	b^+	<	a^-	<	b^-	<	c^{+1}	=	d^+	<	c^{-1}	<	c^{+2}	<	c^{-2}	<	d^-	)
03	(	a^+	<	a^-	<	b^+	<	b^-	<	d^+	<	c^+	<	c^-	<	d^-	)				

Suppose we are given a temporal database like the one shown in Table 1. Then since an event sequence is formed by a set of events, we can transform it into a temporal sequence according to the method given in Definition 3. Doing so repeatedly can transform an event sequence database into a corresponding temporal sequence database. For example, the set of temporal sequences, *S*, in Table 2 is constructed from the temporal database *D* in Table 1. From the constructed temporal sequence database, the patterns we are looking for are those subsequences occurred frequently. The following formally describes these ideas.

Definition 4 (*Small* operation). Function $Small(\oplus_r, \oplus_{(r+1)}, \ldots, \oplus_q)$, where $\oplus_i \in \{<, =\}$, is defined as follows. If any $\oplus_i$, $r \le i \le q$, is "<", then the output of *Small* is "<". If all $\oplus_i$, $r \le i \le q$, are "=", then the output of *Small* is "=".

Let $p = (p_1 \oplus_1 p_2 \oplus_2 \ldots \oplus_{2k-1} p_{2k})$ denote a temporal sequence of k events. The order relation of any two end points in p can be derived by *Small*. Let p_r, p_q be two end points in p. Then the order relation of p_r and p_q, denoted as $Rel(p_r, p_q) \in \{<, =\}$, is $Small(\oplus_r, \ldots, \oplus_{q-1})$. For example, if $p = (a^+<b^+=a^-=c^+<c^-<b^-)$, then $Rel(a^+, c^+)$ = "<" = $Small(<, =, =)$.

Property 1 (Property of *Small*). Let $p = (p_1 \oplus_1 p_2 \oplus_2 \ldots \oplus_{2k-1} p_{2k})$ be a temporal sequence of k events. For any two end points p_r and p_{q+1} in p, the order relation of p_r and p_{q+1}, $Rel(p_r, p_{q+1})$, is $Small(\oplus_r, \oplus_{(r+1)}, \ldots, \oplus_q)$.

Proof. The temporal sequence is a list of end points ordered by their timestamps in non-decreasing order. Even if we insert new end points into the list, the list must keep sorted in this way. Because of this ordering, the property holds.

Definition 5 (Containment). Suppose we have two temporal sequences $ts = (s_1 \oplus_1 s_2 \oplus_2 \ldots \oplus_{n-1} s_n)$ and $p = (p_1 \otimes_1 p_2 \otimes_2 \ldots \otimes_{r-1} p_r)$, $r \le n$. We declare that temporal sequence p is *contained* in temporal sequence ts, denoted $p \subseteq ts$, if all the following conditions are satisfied.

1. We have r indexes in ts, denoted $1 \le w_1 < w_2 < \ldots < w_r \le n$, satisfying the condition of $p_1 = s_{w_1}$, $p_2 = s_{w_2}$, …, and $p_r = s_{w_r}$.
2. $\otimes_i = Small(\oplus_{w_i}, \ldots, \oplus_{w_{i+1}-1})$, for i=1 to r-1.
3. For every event e in ts, the two end points of e are either both in p or none in p.

***Example* 2.** Suppose we have temporal sequences $ts = (a^+<b^{+1}<a^-<b^{-1}<c^+=d^+<b^{+2}<c^-<d^-<b^{-2})$, $p_1 = (a^+<a^-<d^+<d^-)$, $p_2 = (d^+<d^-<a^+<a^-)$, $p_3 = (a^+<a^-<e^+<e^-)$ and $p_4 = (b^+<c^+<c^-<b^-)$. Then p_1 is contained in ts, but p_2, p_3 and p_4 are not contained in ts.

Definition 6 (Support). The *support* of a temporal sequence p in a temporal sequence database S is defined as Equation (1).

$$Support(p, S) = \frac{|\{ts_i \mid ts_i \in S,\ p \text{ is contained in } ts_i\}|}{|S|} \tag{1}$$

A frequent temporal k-pattern is a temporal sequence of k events with a support grater than the threshold, denoted as *min_sup*, which is given by the user,. The objective of this research is to find all the frequent temporal k-patterns for all possible k.

3 Mining Temporal Patterns

The proposed algorithm *T-Apriori* is based on the well-known algorithm *GSP* [6]. The differences between *T-Apriori* and *GSP* are (1) *GSP* is designed for mining sequential patterns for point-based events, but *T-Apriori* is developed for discovering temporal patterns for interval-based events; (2) Candidates generated from *T-Apriori* are different from *GSP*, because the generated patterns are different; (3) In *T-Apriori*, the occurring positions of each pattern in every temporal sequence are recorded for speeding time of pattern searching and frequency counting. Before introducing algorithm *T-Apriori*, we have to explain the format of candidates and large patterns generated by this algorithm.

Definition 7 (Candidates). A candidate c in C_k is noted as $c = \{cp, PE\}$, where cp is the candidate pattern and PE is a set of the position entries associated with cp. Candidate pattern $cp = (p_1, p_2, \ldots, p_{2k})$, where $p_i \in \{ty^+, ty^-\}$, $ty \in \{event\ type\ 1, event\ type\ 2, \ldots, event\ type\ u\}$, $1 \le i \le 2k$. Position entries $PE = \{en_1, en_2, \ldots, en_m\}$, where m is the number of temporal sequences in S, and each en_j, $1 \le j \le m$, is defined in

$$en_j = \begin{cases} < app_1, app_2, \ldots, app_x >, \text{if } cp \text{ can be mapped into } s_j \in S \\ \qquad\qquad \text{with } x \text{ combinations.} \\ \phi, \text{if } cp \text{ can't be mapped into } s_j \in S \end{cases} \tag{2}$$

Since, a candidate cp may be mapped into several position combinations in each $s_i \in S$, all possibilities are recorded in en_j. Each $app_{ii} = (pos_1, pos_2, \ldots, pos_{2k})$, which means the positions where each point of cp is mapped in s_i.

Example 3. Given a temporal sequence database S in Table 2, some examples of candidate are shown as follows. $c_1 = \{(a^+, a^-, c^+, c^-), \{<(0, 2, 4, 7)>, <(0, 2, 4, 6), (0, 2, 7, 8)>, <(0, 1, 5, 6)>\}\}$, $c_2 = \{(b^+, b^-, d^+, c^+, c^-, d^-), \{\phi, <(1, 3, 5, 7, 8, 9)>, <(2, 3, 4, 5, 6, 7)>\}\}$

Definition 8 (Large temporal patterns). A large temporal pattern ltp is noted as $ltp = \{ts, PE\}$, where ts is a temporal sequence as defined in Definition 2, and Definition 3, and PE, refer to that in Definition 7, is a set of position entries associated with tp.

Example 4. Continuing from Example 3, after checking each order relation between adjacent elements and the frequency, the two candidate c_1, and c_2 are frequent, then they become large temporal patterns $ltp_1 = \{(a^+<a^-<c^+<c^-), \{<(0, 2, 4, 7)>, <(0, 2, 4, 6), (0, 2, 7, 8)>, <(0, 1, 5, 6)>\}\}$, and $ltp_2 = \{(b^+<b^-<d^+<c^+<c^-<d^-), \{\phi, <(1, 3, 5, 7, 8, 9)>, <(2, 3, 4, 5, 6, 7)>\}\}$.

Here, the temporal pattern ts in each large temporal pattern is the output knowledge discovered by *T-Apriori*. Having defined the notations of candidates and large temporal patterns, we can now propose the algorithm *T-Apriori*.

```
Method: Call T-Apriori(S, min_sup)
Input: S: Temporal sequence database; min_sup: A threshold
of support given by the user
Output: LTS: The set of all large temporal patterns

Procedure T-Apriori(S, min_sup) {
   L1= FindLOne(S, min_sup);
   For (k=2; L(k-1) ≠ ϕ; k++) {
      Ck = CandidateGen(L(k-1), min_sup); //Generate candidates
```

```
        Pk = CheckRel(Ck, min_sup); //Check the relations between
any two adjacent points of each candidate in Ck from S
        Lk = ComputeLarge(Pk, min_sup); //Count support
        LTP = LTP ∪ Lk;
    }
    Output LTP;
}
```

Given a temporal sequence database *S* in Table 2, and a minimum support threshold, setting to *min_sup* = 50% or 2, *T-Apriori* first discovers L_1 from the known set of event types by the subroutine *FindLOne*. Second, the candidate set C_2 is generated from L_1 by the subroutine *CandidateGen*. Third, *CheckRel* and *ComputeLarge* will check the order relations and compute the frequency for each candidate to generate large temporal patterns, such as the steps presented in Example 4. Repeat step 2 and 3, until L_k generated by *ComputeLarge* contains no patterns.

Having known the basic process of *T-Apriori*, we are now able to see the detail of each subroutine utilized in *T-Apriori*. Suppose there are four event types, a, b, c, and d in the temporal sequence database *S* in Table 2. The set of event type is supposed to be known before mining, for example, the types of diseases recorded in a clinical database are known. From the four event types, we can generate eight C_1, candidates with length 1, that are (a^+<a^-), (a^+=a^-), (b^+<b^-), (b^+=b^-), (c^+<c^-), (c^+=c^-), (d^+<d^-), and (d^+=d^-). The process of *FindLOne* is to scan the temporal sequence database once, then check the frequency of each candidate *c* in C_1. If *Support*(*c*) ≥ 50%, put *c* into L_1. After execute *FindLOne* in this example, we got L_1 = {(a^+<a^-), (b^+<b^-), (c^+<c^-), (d^+<d^-)}, the position entries are omitted.

Next, we explain how to generate candidates from a set of large temporal patterns, i.e. the subroutine *CandidateGen*. The basic idea of *CandidateGen* is that, if two temporal patterns ltp_1 = {ts_1, PE_1} and ltp_2 = {ts_2, PE_2} with length *l* share the same (*l* - 1) patterns, join the position entries PE_1 and PE_2, and then generate candidates.

```
Subroutine CandidateGen(L(k-1), min_sup) {
    For each two patterns ltp1, ltp2 in L(k-1){
        Let α1, α2 as the last event of ltp1 and ltp2;
        ltp1' = ltp1 - α1; ltp2' = ltp2 - α2
        If ((ltp1' = ltp2') & (∃ltp* | ltp*∈L(k-1), ltp* ⊆ α1,
ltp* ⊆ α2))
        { Join position entries of tp1 and tp2 and generate
the candidate sets;
            Add each candidate with no less than min_sup
entries of PE in the candidate sets into Ck;
        }
    }
   Return Ck
}
```

Example 5. Continuing the example above, L_1 = {(a^+<a^-), (b^+<b^-), (c^+<c^-), (d^+<d^-)}. For each two temporal pattern, ltp_1 and ltp_2, join their position entries. Suppose ltp_1 = {(a^+<a^-), {<(0, 2)>, <(0, 2)>, <(0, 1)>}}, and ltp_2 = {(b^+<b^-), {<(1, 3), (6, 9)>, <(1, 3)>, <(2, 3)>}}. Refer to Fig 3, after joining position entries, we obtained (0, 1, 2, 3) and (0,

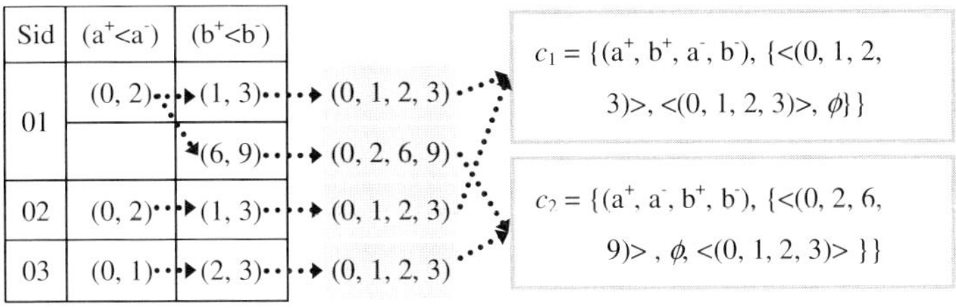

Fig. 3. Generating C_2 from L_1

2, 6, 9) in s_{01}, (0, 1, 2, 3) in s_{02}, (0, 1, 2, 3) in s_{03}. From these position entries, two candidates c_1 = {(a^+, b^+, a^-, b^-), {<(0, 1, 2, 3)>, <(0, 1, 2, 3)>, ϕ}} and c_2 = {(a^+, a^-, b^+, b^-), {<(0, 2, 6, 9)> , ϕ, <(0, 1, 2, 3)>}} are generated by checking temporal sequences database S.

If we repeat the process done in Example 5 for all pairs of large one patterns, we may generate C_2 = {(a^+, b^+, a^-, b^-), (a^+, a^-, b^+, b^-), (a^+, a^-, c^+, c^-), (a^+, a^-, d^+, d^-), (b^+, b^-, c^+, c^-), (b^+, b^-, d^+, d^-), (c^+, d^+, c^-, d^-), (d^+, c^+, c^-, d^-)}, of which position entries are omitted. The next problem, how to generate C_k from $L_{(k-1)}$ when $k > 2$, is explained in Example 6.

Example 6. Given two large temporal patterns ltp_1 = {($b^+<b^-<c^+<c^-$), {<(1, 3, 4, 7)>, <(1, 3, 4, 6), (1, 3, 7, 8)>, <(2, 3, 5, 6)> }} and ltp_2 = {($b^+<b^-<d^+<d^-$), {<(1, 3, 5, 8)>, <(1, 3, 5, 9)>, <(2, 3, 4, 7)>}} with length 2, we now explain how the candidates with length 3 are generated. Because ts_1 = ($b^+<b^-<c^+<c^-$) and ts_2 = ($b^+<b^-<d^+<d^-$) share the common prefix ($b^+<b^-$), and we can find ($c^+<d^+<c^-<d^-$), ($d^+<c^+<c^-<d^-$) in L_2, ltp_1 and ltp_2 are appropriate pair to generate candidates. After joining the position entries, we got (1, 3, 4, 5, 7, 8) in s_{01}, (1, 3, 4, 5, 6, 9) and (1, 3, 5, 7, 8, 9) in s_{02}, and (2, 3, 4, 5, 6, 7) in s_{03}. From these position entries, we may generate two candidates c_1 = {(b^+, b^-, c^+, d^+, c^-, d^-), {<(1, 3, 4, 5, 7, 8)>, <(1, 3, 4, 5, 6, 9)>, ϕ}} and c_2 = {(b^+, b^-, d^+, c^+, c^-, d^-), {ϕ, <(1, 3, 5, 7, 8, 9)>, <(2, 3, 4, 5, 6, 7)>}} after checking temporal sequences database S.

After generating the candidates, we are now ready to consider the process to generate large temporal patterns. The process to generate L_k from C_k can be divided into two parts. The first part is checking the relations between any two adjacent points of each candidate in C_k, and then generating the set of potential large patterns P_k, i.e. *CheckRel*. The second part is counting support of each pattern in P_k, and then the set of large temporal pattern L_k is obtained, i.e. *ComputeLarge*. We explain the two subroutines *CheckRel* and *ComputeLarge* together in the Example 7 and Example 8.

```
Subroutine CheckRel(S, C_k, min_sup) {
  For each c={cp, PE_c} in C_k {
    For each en_i=(pos_1, pos_2, …, pos_l) in PE {
      Check ⊕ between each pair of pos_j and pos_(j+1) to get
temporal sequence ts*.
      if ts* is exist in a potentially large pattern plp*
in P
         add en_i into the PE_plp*
      else {  construct a new plp = {ts, PE_plp}
              P = P ∪ plp }
```

```
            }
        }
        Return P
    }

    Subroutine ComputeLarge(Pk, min_sup) {
        For each candidate plp ∈ Pk {
            Add plp into Lk if support(plp) ≥ min_sup;
        }
    }
```

Table 3. Another temporal sequence database *S'*

Sequence-id	Temporal Sequences																
		0		1		2		3		4		5		6		7	
01	(	a^+	<	a^-	=	b^+	<	b^-	)								
02	(	a^+	<	a^-	=	b^{+1}	<	b^{-1}	<	b^{+2}	<	b^{-2}	<	b^{+3}	=	b^{-3}	)
03	(	a^+	<	a^-	<	b^+	=	b^-	)								

Example 7. Given another temporal sequence database *S'* in Table 3, we now discuss how *CheckRel* generates potential large patterns from the candidate $c = \{(a^+, a^-, b^+, b^-), \{<(0, 1, 2, 3)>, <(0, 1, 2, 3), (0, 1, 4, 5), (0, 1, 6, 7)>, <(0, 1, 2, 3)>\}\}$. When taking s'_{02} as example, after checking each of the three position entries from s'_{02}, we will get (0<1=2<3), (0<1<4<5), and (0<1<6<7). After checking the order relations for each position entries of c, three potential large patterns $plp_1 = \{(a^+<a^-=b^+<b^-), \{<(0, 1, 2, 3)>, <(0, 1, 2, 3)>, \phi\}\}$, $plp_2 = \{(a^+<a^-<b^+<b^-), \{\phi, <(0, 1, 4, 5)>, \phi\}\}$, and $plp_3 = \{(a^+<a^-<b^+=b^-), \{\phi, <(0, 1, 6, 7)>, <(0, 1, 2, 3)>\}\}$ are generated. Since *min_sup* is set to 2, only plp_1 and plp_3 are large.

Table 4. The frequent temporal patterns discovered from *S*

LTP	Pattern	Illustration	*LTP*	Pattern	Illustration
L_1	$(a^+<a^-)$	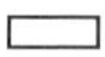	L_2	$(c^+=d^+<c^-<d^-)$	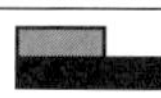
	$(b^+<b^-)$			$(d^+<c^+<c^-<d^-)$	
	$(c^+<c^-)$		L_3	$(a^+<b^+<a^-<b^-<c^+<c^-)$	
	$(d^+<d^-)$			$(a^+<b^+<a^-<b^-<d^+<d^-)$	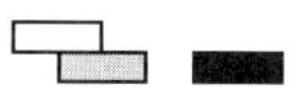
L_2	$(a^+<b^+<a^-<b^-)$	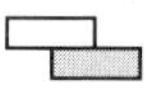		$(a^+<a^-<c^+=d^+<c^-<d^-)$	
	$(a^+<a^-<b^+<b^-)$			$(a^+<a^-<d^+<c^+<c^-<d^-)$	
	$(a^+<a^-<c^+<c^-)$			$(b^+<b^-<c^+=d^+<c^-<d^-)$	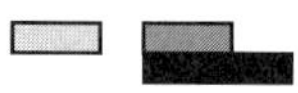
	$(a^+<a^-<d^+<d^-)$			$(b^+<b^-<d^+<c^+<c^-<d^-)$	
	$(b^+<b^-<c^+<c^-)$		L_4	$(a^+<b^+<a^-<b^-<c^+=d^+<c^-<d^-)$	
	$(b^+<b^-<d^+<d^-)$				

Example 8. Returning to Example 5, from $c_1 = \{(a^+, b^+, a^-, b^-), \{<(0, 1, 2, 3)>, <(0, 1, 2, 3)>, \phi\}\}$ and $c_2 = \{(a^+, a^-, b^+, b^-), \{<(0, 2, 6, 9)>, \phi, <(0, 1, 2, 3)>\}\}$, after checking the order relations for each position entries, we got two potential large patterns $plp_1 = \{(a^+<b^+<a^-<b^-), \{<(0, 1, 2, 3)>, <(0, 1, 2, 3)>, \phi\}\}$ and $plp_2 = \{(a^+<a^-<b^+<b^-), \{<(0, 2, 6, 9)>, \phi, <(0, 1, 2, 3)>\}\}$. Since both the number of position entries of plp_1 and plp_2 are 2, plp_1 and plp_2 are large temporal patterns.

In the example of temporal sequence database S in Table 2, after repeating the process of generating candidates and constructing large temporal patterns, the output patterns are shown in Table 4.

4 Conclusion

Existing researches of sequential pattern mining discovered patterns from sequences of point-based events. In this paper, we proposed a new mining problem of discovering temporal patterns for interval-based events. For representing this new kind of pattern, we designed a standard format of temporal sequences. Based on this standard format, we develop the algorithm *T-Apriori* to discover temporal patterns for interval-based events.

References

1. Agrawal, R., Srikant, R.: Mining Sequential Patterns. In: Proceedings of the 11th International Conference on Data Engineering (1995) 3-14
2. Allen, J. F.: Maintaining knowledge about temporal intervals. Communications of ACM, 26 (11) (1983) 832-843.
3. Han, J., Pei, J., Mortazavi-Asl, B., Chen, Q., Dayal, U., Hsu, M.–C.: Freespan: frequent pattern-projected sequential pattern mining. In: Proceedings of the International Conference on Knowledge Discovery and Data Mining (2000) 355-359
4. Lin, M. –Y., Lee, S.- Y.: Fast discovery of sequential patterns by memory indexing. In: Proceedings of 2002 DaWak (2002) 150-160
5. Pei, J., Han, J., Mortazavi-Asl, B., Pinto, H., Chen, Q., Dayal, U., Hsu, M.-C.: PrefixSpan: mining sequential patterns efficiently by prefix-projected pattern growth. In: Proceedings of the 17th International Conference on Data Engineering (2001) 215-224
6. Srikant, R., Agrawal, R.: Mining sequential patterns: generalizations and performance improvements. In: Proceedings of the Fifth International Conference on Extending Database Technology (EDBT) (1996) 3-17
7. Zaki, M. J.: SPADE: an efficient algorithm for mining frequent sequences. Machine Learning Journal, 42(1/2) (2001) 31-60
8. Zhao, Q., Bhowmick, S. S.: Sequential pattern mining: a survey. Technical Report, CAIS, Nanyang Technological University, Singapore (2003)

Ontology-Based Framework of Robot Context Modeling and Reasoning for Object Recognition

Wonil Hwang[1], Jinyoung Park[1], Hyowon Suh[1], Hyungwook Kim[2], and Il Hong Suh[2]

[1] Department of Industrial Engineering, KAIST, Daejeon, Korea
{onil, zknowledge, hw_suh}@kaist.ac.kr
[2] Graduate school of Information and Communications, Hanyang University, Seoul, Korea
{hwkim, ihsuh}@incorl.hanyang.ac.kr

Abstract. This paper introduces Multi-layered Context Ontology Framework (MLCOF) for comprehensive, integrated robot context modeling and reasoning for object recognition. MLCOF consists of six knowledge layers (KLayer) including rules such as an image layer, three geometry (1D, 2D and 3D geometry) layers, an object layer, and a space layer. For each KLayer, we use a 6-tuple ontology structure including concepts, relations, relational functions, concept hierarchies, relation hierarchies and axioms. The axioms specify the semantics of concepts and relational constraints between ontological elements at each KLayer. The rules are used to specify or infer the relationships between ontological elements at different KLayers. Thus, MLCOF enables to model integrated robot context information from a low level image to high level object and space semantics. With the integrated context knowledge, a robot can understand objects not only through unidirectional reasoning between two adjacent layers but also through bidirectional reasoning among several layers even with partial information.

1 Introduction

Intelligent service robot has to understand and carry out user's requirements. For this, object recognition is essential to robot. In the early researches of image processing, objects are recognized using pattern matching with low level image data extracted from sensors. However these approaches are limited in object recognition. To solve this problem, some researchers have been focusing on context based object recognition. They modeled context information such as objects and their geometric shape. But, the approaches do not have concrete reasoning mechanism. In addition, the context information is limited in integrating low level information. Moreover, the context information describing robot environment is very complex and not well organized. Thus systematic framework is needed to represent it.

This paper introduces Multi-layered Context Ontology Framework (MLCOF) for comprehensive integrated context modeling and reasoning for object recognition in a robot environment. This proposed approach is based on ontology. By constructing ontology for context information such as image, geometries, objects and spaces, the robot environment knowledge can be managed comprehensively and synthetically, and object reasoning with partial context information is possible.

L. Wang et al. (Eds.): FSKD 2006, LNAI 4223, pp. 596 – 606, 2006.

In this research, context information includes geometries, objects and spaces. In a robot environment, the use of multi-layered context information can enable the accuracy of object recognition. This proposed framework is represented with first-order logic to maintain integrated uniform representation.

In section 2, we discuss previous researches, and basic concept of MLCOF is provided in section 3. The details of MLCOF based object ontology model is discussed in section 4. Finally in section 5, a summary and consideration for future research are provided.

2 Related Works

There are several knowledge-based image retrieval systems related to object recognition such as a computer system for interpreting scenes [2], a knowledge based image understanding system [4], a knowledge based galaxy classification system [6]. These systems use data schemas for context information, which are limited in containing sufficient and structured context information. So, ontology based systems have been developed.

Ontology-based image processing: S Liu et al. [5] introduced a framework for building ontology to provide semantic interpretations for image contents. Their ontology is focused on visual information such as color, texture and edge features which are extracted from an image. They define general concepts and properties in the ontology. However their ontology is developed only for visual concepts. N Maillot et al. [3] proposed to use visual concept ontology to hide the low-level image layer. The mapping between objects and image is based on the visual concept ontology. These ontological concepts are linked both to object knowledge and to low-level vision numerical descriptors. This ontology can be considered as a guide that provides the vocabulary for the visual description of objects. J Vompras [7] also used ontology to reduce the gap between low-level visual features of images and high-level human perception of inferred semantic contents. The ontology framework contains image primitives as well as semantic information. However, the above studies are limited in using full ontology, and their reasoning algorithms are not well organized.

Ontology-based Context Modeling: Go and Sohn [9] studied how to model context information in a robot environment. They developed the context model using rules and ontology. Rules are used for modeling dynamic information such as current user's location, current robot dictator and so on. Ontology is mainly used for describing static information about most parts of device, space, person and artifact. Their research focuses on modeling object, space and activities. They do not concern low-level context which is generally used to recognize their context of interest in practical application. CONON [8] proposes OWL encoded context ontology for modeling context and supporting logic based context reasoning. The context ontology includes location, person, activity and composite entity. Furthermore they developed context reasoning process. Using this reasoning mechanism, implicit context (activities such as sleeping, cooking) can be deduced from explicit context (space, person, device and so on). However their approach does not provide sensor-driven image or geometry information. They only modeled objects, spaces and activities, and defined rules for reasoning between them.

3 Basic Concept of MLCOF

MLCOF consists of six knowledge layers (KLayer): image, 1D geometry, 2D geometry, 3D geometry, object and space. For managing integrity and consistency of this framework, geometry, object and space layer have meta-ontology, ontology and ontology-instance layer while image layer has ontology and ontology instance layer. We use a 6-tuple ontological structure including concepts, relations, concept hierarchies, relation hierarchies, relation functions and axioms. The rules are used to specify or infer the relationships between ontological elements at different KLayers. In Appendix, for the formality of MLCOF, seven ontological definitions are defined based on KAON ontology [1].

MLCOF enables us to model integrated context information from a low level image to high level object and space semantics. The uses of various forms of context information such as geometry, object, and space enable more accurate object recognition. Thus, if we use multi-layered approach containing several forms of context information, systematic information management and effective reasoning process are possible. Figure 1 shows overall architecture of MLCOF.

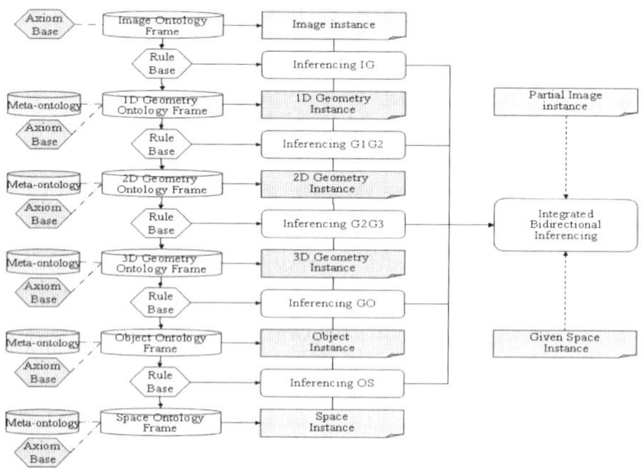

Fig. 1. Architecture of MLCOF (Dynamic View)

4 MLCOF-Based Context Ontology Modeling

MLCOF provides a framework for integrated and systematic context ontology modeling and reasoning. We can model all the context information related to visual objects and infer an object with an MLCOF based context ontology. For MLCOF-based context ontology modeling and reasoning, we need to define usage mechanism of MLCOF. First, we need to describe role and usage of each layer. Second, building rules and axioms has to be discussed. Finally, we need to describe not only unidirectional reasoning between two adjacent layers, but also bidirectional reasoning among several layers with partial information.

4.1 Descriptions of KLayers

We divide KLayer into a meta-ontology layer, an ontology layer and an ontology instance layer. Meta ontology layers capture general features of context entities. They support sharable and reusable framework to represent the entities such as geometries and objects. We construct meta-ontology layers using the 6-tuple ontology structure. Ontology layers are for describing domain specific entities like a refrigerator in the object layer. When we construct an ontology layer, we use a meta-ontology layer. Ontology instance layers are the instances of ontology layers, which are recognized results by reasoning process based on MLCOF from an input image.

Image Layer: it represents information of an image including pixel information as shown in figure 2. In addition, an image layer may include low-level features extracted from an image such as edge features, SIFT feature points and etc. Thus, an image layer needs to have additional entities to represent them. In the example in figure 2 (a), an image layer has SIFT feature points as well as its pixel information. A set of SIFT feature points in an image layer is directly matched to an object in an object layer. In this case, a geometry layer is not necessary. However, a set of pixels is recognized as a set of edges, and the edges are then recognized as lines or curves in geometry layer by edge detection rules and geometry reasoning rules. Thus, geometry layer are generally necessary.

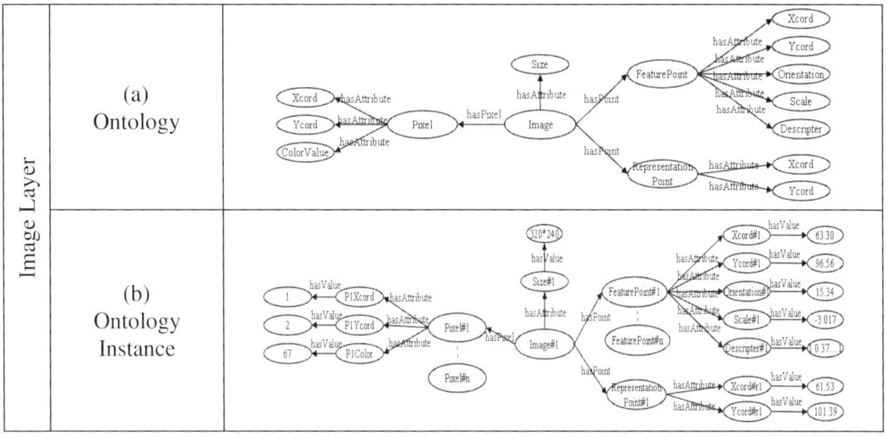

Fig. 2. Example of image layer

Geometry Layer: it represents geometries obtained from a set of pixels in an image layer. Geometry meta-ontology layer describes representative geometry concepts, attributes, and their relations as shown in figure 3 (a). Geometry concept is divided into 1-dimensional geometry, 2-dimensional geometry, 3-dimensional geometry, and spatial relations between the geometries are defined. Geometry concepts also have attributes and geometry reference image to represent a related image in the image layer. Geometry ontology layer represents specific geometry such as 1D geometry-line, 2D geometry-rectangle, 3D geometry-hexahedron, etc. Thus, we classify geometry ontology layer into three KLayers such as 1D, 2D and 3D geometry ontology layer as shown in figure 3 (b). This classification supports flexible definitions of geometries. Geometry instance layer represents an instance of a geometry ontology layer. Figure 3 (c) shows an example of geometry instances for a quadrangle example. This instance can be obtained from an image instance by edge detection and other rules, which will be discussed later.

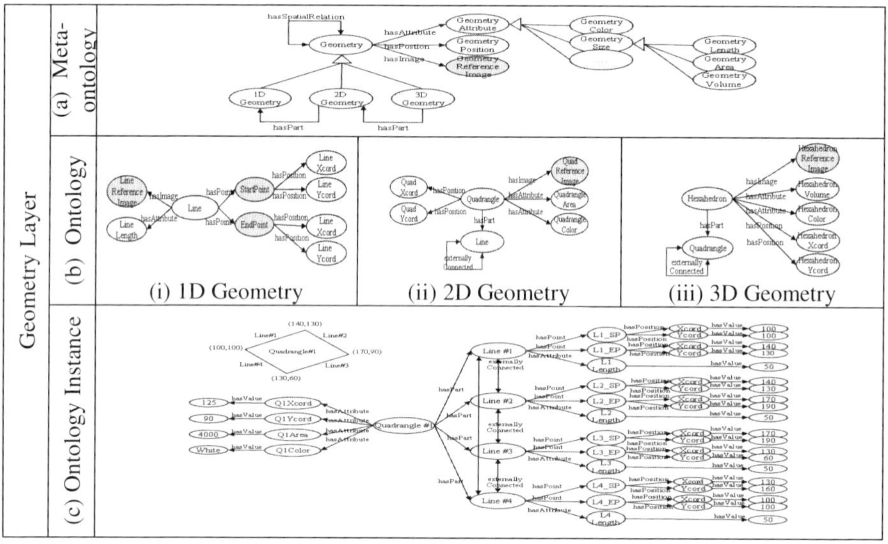

Fig. 3. Example of geometry layer

Object Layer: it represents object and its properties. Objects are divided into top objects and child objects. Top object is the target object which we want to recognize, and it usually consists of several child objects. For example, we regard a refrigerator as a top object while we consider its door and knob as child objects. Child object is a functionally significant subset of top object or another child object. Object meta-ontology layer defines object, object position, object reference image, object function and object attribute as concepts. In object ontology layer, we represent specific real world object such as refrigerator, chair, etc. All concepts and relations in this layer are defined according to the object meta-ontology layer, and an object ontology instance layer includes specific instance of a refrigerator obtained from input geometry instances. Figure 4 shows an example of an object layer.

Space Layer: it represents space and its properties. We deemed space as a set of functionally related objects. So space has several objects in it. Space Layer consists of meta-ontology, ontology and ontology instance layer, and they are similar to those in geometry and object layer. Figure 5 shows an example of a space layer.

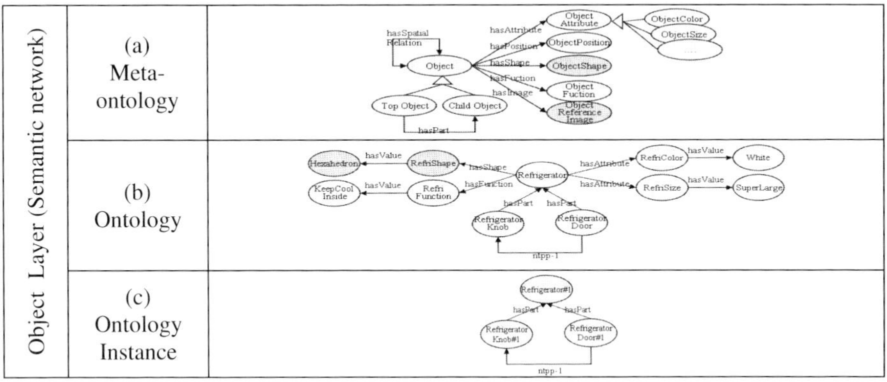

Fig. 4. Example of object layer

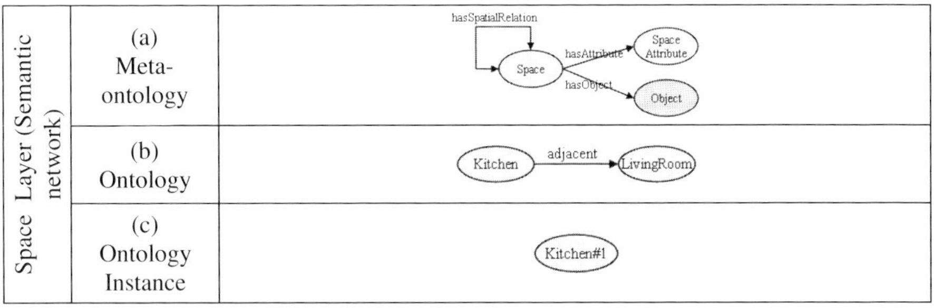

Fig. 5. Example of space layer

4.2 Axioms for Each KLayer

According to definition 4 and 5, we can define axioms. The axioms are categorized into two types. One is to explicitly specify the semantics of concepts and relations so that the ambiguity of context knowledge can be alleviated. The other is the relational constraints between the concepts and the relations. Relational constrains are for both spatial and hierarchical constraints. Table 1 shows the examples of axioms. The axioms are generally represented with logical language so that they can be a basis for additional reasoning.

4.3 Rules Between KLayers

According to definition 6 and 7, we can define rules. While axioms are defined in each KLayer, rules are defined between KLayers. The rules are used for reasoning the information which is necessary for recognizing instances in one layer from thosc in another layer. These rules are generally represented in the form of IF-THEN. And we semantically classify rules into two categories such as common rules and integrated rules which are discussed in the followings.

Unidirectional reasoning can be achieved by rules which are defined between two adjacent layers. Using these rules, image, geometry and object instances are inferred orderly. Table 3 describes rough examples which show how the rules are used to reason geometries from an image and object from those geometries. First, we derive lines from an image by an edge detection rule. Quadrangles are then inferred from the lines by a 'Rule_G1G2'. After that, we can reason a hexahedron based on quadrangles by a 'Rule_G2G3'. Lastly, we can recognize a refrigerator with the hexahedron and some additional information. In real situation, rules are defined more complicatedly. In these examples, we assume that a refrigerator has three types as shown in table 3.
Bidirectional reasoning is represented by integrated rules which are defined among several layers. Integrated rules are used to infer objects or some necessary information for object recognition when related information is given. A simple example of integrated rules in table 2 presents how we can recognize an unknown object as a refrigerator (or a computer) with object candidates and space information such as the kitchen (or the office).

Table 1. Examples of axioms

Layer	Type	Meaning	[Identifier] / FOL representation
I.L[1]	S[4]	Range of row value.	[AIS_Row] $\forall$x, c Xcord(x) $\wedge$ hasValue(x, c) $\wedge$ ($-2^{63} \leq c \leq +2^{64}$)
	S	Range of column value.	[AIS_Col] $\forall$y, c Xcord(y) $\wedge$ hasValue(y, c) $\wedge$ ($-2^{63} \leq c \leq +2^{64}$)
G.L[2]	S	Commutativity of 'Externally connected' relation	[AGS_EC] $\forall$x,y Geometry(x) $\wedge$ Geometry(y) $\wedge$ externallyConnected(x,y) $\Rightarrow$ externallyConnected(y,x)
	S	Inverse relation of 'TPP' and 'TPP-1'	[AGS_NTTP] $\forall$x,y Geometry(x) $\wedge$ Geometry(y) $\wedge$ TPP (x,y) $\Rightarrow$ TPP-1(y,x)
O. L[3]	S	Transitivity of 'hasPart' relation	[AOS_Transitivity] $\forall$x,y,z Object(x) $\wedge$ Object(y) $\wedge$ Object(z) $\wedge$ hasPart(x,y) $\wedge$ hasPart(y,z) $\Rightarrow$ hasPart(x,z)
	R[5]	Spatial relation between object	[AOR_DoorKnob] $\forall$x,y Knob(x) $\wedge$ Door(y) $\wedge$ hasPart(y,x) $\wedge$ $\neg$ inCenterOf(x, y)

Abbreviated terms:
(1: Image layer, 2 Geometry layer, 3: Object layer,
4: Specification of the semantics for concepts and relation, 5: Relational constraints)

Table 2. Example of rules for recognizing refrigerator (bidirectional reasoning with partial space information)

Rule Type	Meaning	[Identifier] / FOL representation
Integrated Rule	IF X's candidates are refrigerator and cabi net AND X is in space: kitchen THEN X is refrigerator	[IR_ref] $\forall$x,h,k (refrigerator(x) $\vee$ cabinet(x)) $\wedge$ kitchen(h) $\wedge$ hasObject(k,x) $\Rightarrow$ Refrigerator(x)
Integrated Rule	IF X's geometry is hexahedron AND X is in office AND There is keyboard and mouse ne ar X THEN X is computer	[IR_com] $\forall$x,h,k,m,o Object(x) $\wedge$ Hexahedron(h) $\wedge$ Keyboar d(k) $\wedge$ Mouse(m) $\wedge$ Office(o) $\wedge$ hasGeometry(x,h) $\wedge$ adjacent(x,k) $\wedge$ adjacent(x,m) $\wedge$ hasObject(o,x) $\Rightarrow$ Computer (x)

Table 3. Example of rules for recognizing refrigerator (unidirectional reasoning from image layer to object layer)

Rule Type	Meaning	[Identifier] / FOL representation
Rule_IG	IF S is a set of edge pixel. AND orientations of all the ele ments in S are almost same. AND Y refers S THEN Y is line	[RIG_Line] S1: ∀p,x,y,r pixel(p)∧xcord(x)∧ycord(y)∧hasXcord(p,x)∧hasYcord(p,y) ∧R(r)∧hasR(p,r)⇒(fv(x,y)=multiply(differential(Gaussian(y)),Gaussian (x))∧(fh(x,y)=multiply(differential(Gaussian(x)),Gaussian(y))∧(Rv(x,y) =convolution(I(x,y),fv(x,y)))∧(Rh(x,y)=convolution(I(x,y),fg(x,y))) (R(x,y)=square(Rv(x,y))+square(Rh(x,y)))∧(r = R(x,y)) S2: ∀p,r,t pixel(p) ∧ R(r) ∧ hasR(p,r) ∧(r>T) ⇒ edgepixel(p) S3: ∀s,e2 ∃ e1, lineSet(s) ∧ edgepixel(e1) ∧ (e1∈s) ∧ edgepixel(e2) ∧n eighbor(e1,e2)∧direction(d1)∧direction(d2)∧hasDirection(s,d1)∧hasDir ection(e2,d2) ∧(abs(d1-d2) < D) ⇒ (e2 ∈ s) S4: ∀s,e1,e2,x1,y1,x2,y2,a,p1,p2 lineSet(s)∧refer(a,s)∧hasStart(s,e1)∧ h asEnd(s,e2)∧hasXcord(e1,x1)∧hasYcord(e1,y1)∧hasXcord(e2,x2)∧has Ycord(e2,y2) ⇒ line(a) ∧ hasStartPoint(p1) ∧ hasEndPoint(p2) ∧ hasXc ord(p1,x1) ∧hasYcord(p1,y1)∧hasXcord(p2,x2)∧hasYcord(p2,y2) hasle ngth(a, squreroot(square(x1-x2)+square(y1-y2))
Rule_ G1G2	IF Y is consisted of 4 lines. AND Each line's end point is c onnected to other line's start po int. AND All lines don't intersect e ach other. THEN Y is quadrangle.	[RGG_Quad] ∀a1,a2,a3,a4,sp1,ep1,sp2,ep2,sp3,ep3,sp4,ep4,y Line(a1) ∧ Line(a2) ∧ Line(a3) ∧ Line(a4) ∧ StartPoint(sp1) ∧ EndPoint(ep1) ∧ StartPoint(sp2) ∧ EndPoint(ep2) ∧ StartPoint(sp3) ∧ EndPoint(ep3) ∧ StartPoint(sp4) ∧ EndPoint(ep4) ∧ hasPoint(a1,sp1) ∧ hasPoint(a1,ep1) ∧ hasPoint(a2,s p1) ∧ hasPoint(a2,ep2) ∧ hasPoint(a3,sp3) ∧ hasPoint (a3,ep3) ∧ hasPoi nt(a4,sp4) ∧ hasPoint(a4,ep4) ∧ samePoint(sp1,ep4) ∧samePoint(sp2,ep 1) ∧samePoint(sp3,ep2) ∧samePoint(sp4,ep3) ∧ hasPart(y,a1) ∧ hasPart(y,a2) ∧ hasPart(y,a3) ∧ hasPart(y,a4) ∧ ¬ intersect(a1,a2) ∧ ¬ intersect(a 2,a3) ∧ ¬ intersect(a3,a4) ∧ ¬ intersect(a1,a4) ∧ ¬ intersect(a1,a3) ⇒ Qu adrangle(y)
Rule_ G2G3	IF Y is consisted of 3 quadrangles. AND Each quadrangle shares t wo lines with other two quadra ngles. AND The number of shared lin e is two. THEN Y is Hexahedron	[RGG_Hexahedron] ∀a1,b1,a2,b2,a3,b3,a4,b4,r1,r2,r3,r4,h Line(a1)∧Line(b1)∧Line(a2)∧Li ne(b2)∧Line(a3)∧Line(b3)∧Line(a4)∧Line(b4)∧Quadrangle(r1)∧Quadr angle(r2)∧Quadragle(r3)∧Quadrangle(r4)∧hasPart(r1,a1)∧hasPart(r1,b1)∧hasPart(r2,a2)∧hasPart(r2,b2)∧hasPart (r3,a3) ∧hasPart(r3,b3) ∧hasPa rt(r4,a4) ∧hasPart(r4,b4) ∧hasPart(h,r1)∧ hasPart(h,r2)∧hasPart(h,r3)∧h asPart(h,r4)∧sameLine(a1,a2)∧sameLine (b2,b3)∧sameLine(b1,a3)∧diff erentLine(a1,b1)∧differentLine(a2,b2)∧differentLine(a3,b3) => Hexahe dron(h)
Rule_ GO	IF X 'shape is hexahedron. AND X's vertical side length is larger than its horizontal size l ength. AND X' color is white. AND it has one door. THEN X is refrigerator	[RGO_refrigerator1] ∀r,d,c,h,v,w Door(d) ∧ Hexahedron(h) ∧ ObjectColor(c) ∧ hasPart(r,d) ∧ hasValue(c,White) ∧ hasAttribute(r ,c) ∧ hasShape(r,h) ∧ hasHeight(r,v) ∧ hasWidth(r,w) ∧ (w>v) ⇒ Refrigerator(r) ∧ hasType(r, T1) Type1 
Rule_ GO	IF X has two doors which are plac ed above and below. AND Other conditions equal to 'Refrigerator (Type1)' THEN X is also refrigerator	[RGO_refrigerator2] ∀r,d1,d2,h,c,v,w Door(d1) ∧ Door(d2) ∧ Hexahedron (h) ∧ ObjectColor(c) ∧ hasPart(r, d1) ∧ hasPart(r, d1) ∧ ec(d1,d2) ∧ verticalWith(d1,d2) ∧ hasValue(c,White) ∧ hasAttribute(r,c) ∧ hasShape(r,h) ∧ hasHeight(r,v) ∧ hasWidth(r,w) ∧ (w>v) ⇒ Refrigerator(r) ∧ hasTyp e(r, T2) Type2 
Rule_ GO	IF X has two doors which are plac ed which are placed left and rig ht. AND Other conditions equal to 'Refrigerator (Type1)' THEN X is also refrigerator	[RGO_refrigerator3] ∀r,d1,d2,h,c,v,w Door(d1)∧Door(d2) ∧ Hexahedron(h) ∧ ObjectColor(c) ∧ hasPart(r, d) ∧ ec(d1,d2) ∧ horiz ontalWith(d1,d2) ∧ hasValue(c, White) ∧ has Attribute (r, c) ∧ hasShape(r,h) ∧ hasHeight(r,v) ∧ hasWidth(r, w) ∧ (w>v)⇒ Refrigerator(r) ∧ hasType(r, T3) Type3 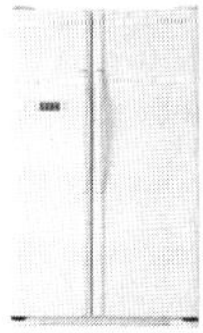

5 Conclusions and Further Study

This paper introduces Multi-layered Context Ontology Framework (MLCOF) for comprehensive, integrated object modeling and reasoning in a robot environment. MLCOF consists of five knowledge layers (KLayer) including axioms and rules. Five KLayers with axioms and rules enable to model integrated context information from sensor driven low level image to high level object semantics. With the integrated context information, a robot can understand objects through not only unidirectional reasoning between two adjacent layers but also bidirectional reasoning among several layers even with partial information. This research makes integrated robot-environment-recognition possible. However, further researches are necessary for practical application. The rich context vocabulary, axioms and rules need to be defined. In addition, going ahead of current object recognition level, the recognition of space and dynamic situation should be realized. An XML-based representation of the model is also necessary for sharing the environment knowledge with other agents.

Acknowledgement

This work is supported by the Intelligent Robotics Development Program, one of the 21st Century Frontier R&D Programs funded by the Korea Ministry of Commerce, Industry and Energy.

References

1. Bozsak, E., Ehrig M., Handschuh S., Hotho A., Maedce A., Motik B., Oberle D., Schmitz C., Staab S. and Stojanovic L., (2002), "KAON – Towards a Large Scale Semantic Web," Lecture notes in computer science, No. 2455, pp. 304-313.
2. Hanson**,** AR**, Riseman EM,** (1978), "VISIONS: a computer system for interpreting scenes", Computer vision systems. Academic, New York**, pp. 303-333**.
3. Maillot, N, Thonnat M, Boucher A, (2004), "Towards ontology-based cognitive vision", Machine Vision and Applications, pp. 33-40.
4. Matsuyama**,** T, Hwang V, (1990), "SIGMA – a knowledge-based aerial image understanding system", Plenum, New York**.**
5. Song, L, Liang-Tien C, Syin C, (2004), "Ontology for Nature-Scene Image Retrieval", CoopIS/ DOA/ ODBASE 2004, LNCS 3291, pp1050-1061.
6. Thonnat, M, Bijaoui, (1989), "A Knowledge-based galaxy classification systems. Knowledge-based systems in astronomy.", Lecture Notes Phys , pp. 121-159.
7. Vompras, J., (2005), "Towards Adaptive Ontology-Based Image Retrieval", 17th GI-Workshop on the Foundations of Databases, pp148-152.
8. Wang, X.H. , Zhang, D.Q. , Gu, T. and Pung, H.K., (2004), "Ontology based context modeling and reasoning using OWL", Pervasive Computing and Communications Workshops, 2004. Proceedings of the Second IEEE Annual Conference on, pp.18-22.
9. Young Cheol Go, Joo-Chan Sohn, (2005), "Context Modeling for Intelligent Robot Services using Rule and Ontology", The 7th international conference on, Vol. 2, pp. 813-816.

Appendix: Formal Definition of MLCOF

Definition 1

Multi-layered Context Ontology Framework

$MLCOF := (KLayer, R^0)$

Such that KLayers are knowledge layers and R^0 is a finite set of rules.

Definition 2

A set of Knowledge Layers of MLCOF consists of 17 Layers;

$KLayers := \{KLayer_{ij} \mid 1 \leq i \leq 6, 1 \leq j \leq 3\} \setminus \{KLayer_{11}\}$

We define a knowledge layer $KLayer_{ij}$ for $i,j \in N$ (set of natural numbers), $1 \leq i \leq 6$, $1 \leq j \leq 3$.

$KLayer_{1j}$ is a knowledge layer for the image layer,
$KLayer_{2j}$ is a knowledge layer for the 1D geometry layer,
$KLayer_{3j}$ is a knowledge layer for the 2D geometry layer,
$KLayer_{4j}$ is a knowledge layer for the 3D geometry layer,
$KLayer_{5j}$ is a knowledge layer for the object layer,
$KLayer_{6j}$ is a knowledge layer for the space layer,
$KLayer_{i1}$ is a knowledge layer for the meta-ontology layer,
$KLayer_{i2}$ is a knowledge layer for the ontology layer,
$KLayer_{i3}$ is a knowledge layer for the ontology instance layer.

Specially,
$KLayer_{31}$ and $KLayer_{41}$ are equal to $KLayer_{21}$.

Definition 3

A ij-th knowledge layer in MLCOF consists of 6-tuples;

$KLayer_{ij} := (C_{ij}, R_{ij}, Rel_{ij}, H_{ij}^{C}, H_{ij}^{R}, A_{ij}^{0})$

Such that

C_{ij} is a set of concepts in $KLayer_{ij}$
R_{ij} is a set of relations in $KLayer_{ij}$
Rel_{ij} is a set of relation functions in $KLayer_{ij}$
H_{ij}^{C} is a set of concept hierarchies in $KLayer_{ij}$
H_{ij}^{R} is a set of relation hierarchies in $KLayer_{ij}$
A_{ij}^{0} is a set of axioms

Definition 4

The set of axioms of each KLayer is a set of sentences Λ which follows the representation of logical language. Λ is represent d by 3-tuples of KLayer (elements of C_{ij}, R_{ij} and Rel_{ij} of $KLayer_{ij}$), which are the elements in the same KLayer. Also, a sentence of Λ specifies the meaning of the elements by describing the relationship of the elements in a KLayer. Any sentence in Λ can not be entailed by other sentences in Λ.

Definition 5

Axioms are based on a structure of 3-tuples;

$A^0 = \{AI, \Lambda, \alpha\}$

(i) AI is a set of axiom identifiers
(ii) Λ is a set of logical sentences, and
(iii) α is a set of axiom mapping functions: α: AI $\Rightarrow$ Λ

Definition 6
The set of rules is a set of sentences Ɗ which follows the representation of logical language. Ɗ is represented by 3-tuples of KLayer (elements of C_{ij}, R_{ij} and Rel_{ij} of $KLayer_{ij}$). A sentece of Ɗ represents the relationship between the three elements of KLayers (C_{ij}, R_{ij} and Rel_{ij}), and is used to entail other concept or relation. The rule should include at least two elements, one from a KLayer and the other from another KLayer.

Definition 7
Rules are a structure of 3 tuples:

$$R^0 = \{RI, Ɗ, \beta\}$$

(i) RI is a set of rule identifiers
(ii) Ɗ is a set of logical sentences, and
(iii) β is a set of rule mapping functions: β: RI $\Rightarrow$ Ɗ

Extended Ontology Model and Ontology Checking Based on Description Logics*

Yu Changrui[1], Wang Hongwei[2], and Luo Yan[1]

[1] School of Management, Shanghai Jiao Tong University , 200052 Shanghai, China
{yucr, yanluo}@sjtu.edu.cn
[2] School of Economic and Management, Tongji University, 200092 Shanghai, China
hwwang@tongji.edu.cn

Abstract. Ontology formal model and ontology checking recently are still under hot discussion. In this paper, an extended ontology model is constructed using Description Logics. Based on the extended model, the issue on ontology checking is studied with the conclusion that the four kinds of term checking, including term satisfiability checking, term subsumption checking, term equivalence checking and term disjointness checking, can be reduced to the satisfiability checking, and satisfiability checking can be transformed into instantiation consistence checking.

1 Introduction

Currently, the formal ontology model suiting to consistency checking is still under hot discussion. Some researches of ontology are based on first order logic (FOL), e.g. Ontolingua, CycL, LOOM. Although FOL have a more expressive power, the reasoning process are complex and most of them are even undecidable, which is not suitable for ontology model checking [1]. Description Logics is equipped with a formal and logic-based semantics allowing inferring implicitly represented knowledge from the knowledge that is explicitly contained in the knowledge base [2]. Although DL has a less expressive power than FOL, its inference procedures are more efficient and decidable, which is more suitable for ontology checking: description logic (DL) is used in ConsVISor tool for consistency checking of ontologies [3][4],and temporal description logic (TDL)in [5].

Ref. [6] proposed a naive ontology model *O=<T, X, TD, XD>*, consisting of Term Set, Individual Set, Term Definition Set, and Instantiation Assertion Set in DL. In this paper, an extended ontology model *O=<T, X, TD, XD, TR>* is established in description logics through introducing Term Restriction Set to the naive model in Ref. [6]. Based on the extended model, we study the issue of how to check the ontology model for preserving the consistence of ontology.

2 Extended Ontology Model

Definition 1: Given a terminology description language *L*, an extended ontology model is a 5-tuples,

* This research is supported by the Natural Science Fund of China (#70501022) and (#70501024).

L. Wang et al. (Eds.): FSKD 2006, LNAI 4223, pp. 607 – 610, 2006.

$$O=<T, X, TD, XD, TR> \tag{1}$$

Where T is a Term Set, X is an Individual Set, TD is a Term Definition Set, XD is an Instantiation Assertion Set, and TR is a Term Restriction Set.

Term Set comprises a group of atomic terms, and we adopt the term constructors from Description Logic to build term formulas for the expression of more complex contents. Given L, we call the expression, satisfying the syntax rule below, an L-based term formula.

$$D, E \rightarrow C \mid \top \mid \bot \mid \neg C \mid D \sqcap E \mid \forall P.D \mid \exists P.\top \tag{2}$$

Eq. (2) shows that the expressive power of term formulas strongly depends on the type of term constructors.

Definition 2 (Subsumption, Equivalence, Disjointness): Given L and $O=<T, X, TD, XD, TR>$, D and E are two L-based term formulas:

(i) D is subsumed by E, if $D^I \subseteq E^I$ for any an interpretation I;
(ii) D and E are equivalent, if $D^I = E^I$ for any an interpretation I;
(iii) D and E are disjointness, if $D^I \cap E^I = \varnothing$ for any an interpretation I.

Given $O=<T, X, TD, XD, TR>$, Term Definition Set TD is such a set that consists of term definition items subject to the following restrictions, written as $TD=\{C_1 \equiv D_1, C_2 \equiv D_2, \ldots, C_n \equiv D_n\}$. Where, $C_i \in T$, D_i is a term formula, and every term in D_i is from T.

(i) for any i, j $(i \neq j, 1 \leq i \leq n, 1 \leq j \leq n)$, $C_i \neq C_j$ holds.
(ii) if there exist $C_1' \equiv D_1'$, $C_2' \equiv D_2', \ldots, C_m' \equiv D_m'$ in TD, and C_i' occurs in D_{i-1}' $(1 < i \leq m, m \leq n)$, then C_1' must not occur in D_m'.

Individual Set is a set of individuals whose names are denoted as a, b, c. Instantiation Assertion Set consists of a group of class instantiation assertions, property instantiation assertions and individual inequality assertions.

Definition 3 (Model of Instantiation Assertion): Given $O=<T, X, TD, XD, TR>$, if there exists an ontology interpretation I making an instantiation assertion α holds, then I is said to be a model of α. If I is model of all the instantiation assertions in XD, then I is called a model of XD.

A class instantiation assertion, written as $C(a)$, states that individual a belongs to class C. A property instantiation assertion, written as $P(a, b)$, states that there exists a relation P between a and b, and b is called the value of a about property P.

Given $O=<T, X, TD, XD, TR>$, the Term Restriction Set TR is a set of term relationships in the form of subsumption, equivalence or disjointness, which is intended to restrict the logical relationship between terms in T.

Definition 4 (Expansion of Term Restriction Set): Given $O=<T, X, TD, XD, TR>$. For convenience, we assume $TR=\{D_1 \mathrm{M} E_1, D_2 \equiv E_2, D_3 ⩀ E_3\}$. If each term relation in TR has been transformed into the expansion form, a new Term Restriction Set $TR'=\{e(D_1)\mathrm{M}e(E_1), e(D_2) \equiv e(E_2), e(D_3) ⩀ e(E_3)\}$ is obtained. TR' is said to be the expansion of TR, written as $e(TR)$.

3 Checking of Ontology Model

Ontology model is semantic-oriented. Semantic consistency checking of ontology model can be divided into two kinds of problems, term checking and Instantiation checking, and these problems can transform each other.

3.1 Term Checking

Term checking of ontology model includes four sub-problems: term satisfiability checking, term subsumption checking, term equivalence checking and term disjointness checking.

Proposition 1 (Reduction to Subsumption): Given O=<T, X, TD, XD, TR>, for any two term formulas D, E:

(i) D is unsatisfiable with respect to TR and TD, iff $(TR+TD)\models D\mathrm{M}\bot$;
(ii) $(TR+TD)\models D\equiv E$, iff $(TR+TD)\models D\mathrm{M}E$ and $(TR+TD)\models E\mathrm{M}D$;
(iii) $(TR+TD)\models D\pitchfork E$, iff $(TR+TD)\models (D\sqcap E)\mathrm{M}\bot$.

Proposition 2 (Reduction to Unsatisfiability): Given O=<T, X, TD, XD, TR>, for any two term formulas D, E:

(i) $(TR+TD)\models D\mathrm{M}E$, iff $D\sqcap\neg E$ is unsatisfiable with respect to TR and TD;
(ii) $(TR+TD)\models D\equiv E$, iff both $D\sqcap\neg E$ and $\neg D\sqcap E$ are unsatisfiable with respect to TR and TD;
(iii) $(TR+TD)\models D\pitchfork E$, iff $D\sqcap E$ is unsatisfiable with respect to TR and TD.

Proposition 1 and 2 imply that all the four kinds of term checking can be reduced to the (un)satisfiability or subsumption.

Proposition 3: Given O=<T, X, TD, XD, TR>. D and E are two term formulas. $e(D)$ is the expansion of D with respect to TD, and $e(E)$ is the expansion of E with respect to TD. We have:

(i) D is satisfiable with respect to TR and TD, iff $e(D)$ is satisfiable with respect to $e(TR)$;
(ii) $(TR+TD)\models D\mathrm{M}E$, iff $e(TR)\models e(D)\mathrm{M}e(E)$;
(iii) $(TR+TD)\models D\equiv E$, iff $e(TR)\models e(D)\equiv e(E)$;
(iv) $(TR+TD)\models D\pitchfork E$, iff $e(TR)\models e(D)\pitchfork e(E)$.

Proposition 3 states that problems of term checking can be solved through checking the expansion of the terms.

3.2 Instantiation Checking

Definition 5 (Consistence of Instantiation Assertion): Given O=<T, X, TD, XD, TR> and an instantiation assertion α. If there exists such an ontology interpretation I that is a model of α, then we say α is consistent, and inconsistent otherwise. If I is not only a model of α, but also a common model of both TD and TR, then we say α is consistent with respect to TD and TR. If I is a model of XD, then we say XD is consistent. If *I* is not only a model of *XD*, but also a common model of both *TD* and *TR*, then we say *XD* is consistent with respect to *TD* and *TR*.

Proposition 4: Given *O=<T, X, TD, XD, TR>* and a class instantiation assertion *C*(*a*), we have:

(i) *C*(*a*) is consistent with respect to *TD* and *TR*, iff *e*(*C*)(*a*) is consistent with respect to *e*(*TR*).
(ii) *XD* is consistent with respect to *TD* and *TR*, iff *e*(*XD*) is consistent with respect to *e*(*TR*).

Proposition 5: Given an ontology *O=<T, X, TD, XD, TR>* and a term formula *C*, *C* is satisfiable with respect to *TD* and *TR*, iff *C*(*a*) is consistent with respect to *TD* and *TR*, where *a* is an arbitrarily chosen individual name.

Proposition 4 states that the consistence of an instantiation assertion is equivalent to the consistence of the expansion of the instantiation assertion. Proposition 5 states that the consistence of instantiation assertion can be determined through checking the satisfiability of the term.

4 Conclusions

In this paper, an extended ontology model is constructed using Description Logics, which is a 5-tuples including Term Set, Individual Set, Term Definition Set, Instantiation Assertion Set and Term Restriction Set. Based on the extended model, issues of ontology checking, including term checking and instantiation checking, are studied with the conclusion that: the four kinds of term checking, including term satisfiability checking, term subsumption checking, term equivalence checking and term disjointness checking, can be reduced to satisfiability checking, and satisfiability checking can be transformed into instantiation checking. The problem of instantiation consistence can be decided by Tableau Algorithm (see Ref. [7] for details).

References

1. Anderson, C.A., Alonzo Church's contributions to philosophy and intensional logic, Bull. Symbolic Logic 4 (2) (1998) 129-171.
2. F. Baader, D. Calvanese, D. McGuinness, and D. Nardi, editors. The Description Logic Handbook - Theory, Implementation and Applications. Cambridge University Press (2003).
3. K. Baclawski, M. Kokar, and J. Smith. Consistency Checking of RM-ODP Specifications. International Conference on Enterprise Information Systems, Setúbal, Portugal (2001).
4. K. Baclawski, M. Kokar, et al. Consistency Checking of Semantic Web Ontologies. Lecture Notes in Computer Science, Vol. 2342, Springer-Verlag (2002) 454—459.
5. F. Weitl and B. Freitag. Checking semantic integrity constraints on integrated web documents. Lecture Notes in Computer Science, Vol.3289. Springer-Verlag(2004)198–209.
6. Wang Hongwei, Wu Jiachun, Jiang Fu. Study on naive ontology model based on Description Logics. System Engineering 21(3) (2003) 101-106.
7. Baader F, Sattler U. Tableau algorithm for description logics. Lecture Notes in Artificial Intelligence, Vol.1847. Springer-Verlag (2000) 1-18.

A General Fuzzy-Based Framework for Text Representation and Its Application to Text Categorization

Son Doan[1], Quang-Thuy Ha[2], and Susumu Horiguchi[1]

[1] Graduate School of Information Science
Tohoku University, Aoba 09, Sendai, 980-8579, Japan
{s-doan, susumu}@ecei.tohoku.ac.jp
[2] College of Technology
Vietnam National University, Hanoi
144 Xuan Thuy, Cau Giay, Hanoi, Vietnam
thuyhq@vnu.edu.vn

Abstract. In this paper we develop the general framework for text representation based on fuzzy set theory. This work is extended from our original ideas [5],[4], in which a document is represented by a set of fuzzy concepts. The importance degree of these fuzzy concepts characterize the semantics of documents and can be calculated by a specified aggregation function of index terms. Based on this representation, a general framework is proposed and applied to text categorization problem. An algorithm is given in detail for choosing fuzzy concepts. Experiments on the real-world data set show that the proposed method is superior to the conventional method for text representation in text categorization.

1 Introduction

Text representation is one of the most important steps in text processing such as information retrieval, text routing, topic detection, text management, particularly in text categorization. In general, there are two common framework for text representation: a vector space model [18] and a probabilistic model [7],[14]. In the vector space model, a document is considered as a vector in vector space in which each dimension corresponds to a term. Apart from the vector space model, each document is considered as a event in the probabilistic model in which a document is represented as a "bag-of-words". In both methods, each term is weighted by an indexing function characterizing the importance of that term in the document. Naturally, there are two ways of indexing functions [8]: human indexing and automated indexing. Human indexing often produces a high performance but is very time consuming. Thus, in practice automated indexing is often used.

Representation in automated indexing can be implemented in two main steps [13]: standard representation and dimension reduction. In the first step, it takes all terms existing in documents to represent text. In the second step, high frequency words (also called "stop words") like *the, a, an,...* are eliminated. After eliminating stop words, the remaining terms is used for text representation.

L. Wang et al. (Eds.): FSKD 2006, LNAI 4223, pp. 611–620, 2006.

Conventional representations based on frequency of terms do not entirely reflect importance of terms in the document. Some terms with low frequencies can play importance in topic of sentences or the document. Otherwise, some terms with high frequencies can be irredundant or noisy. In addition, in real-world applications, users play a important role, particularly in specifying topics and content of documents based on existing terms in documents.

To solve such problems, several automated approaches have been proposed. Deerwester et al. [3] proposed Latent Semantic Indexing (LSI) and Billhardt et al. [1] proposed the Context Vector Model. Both methods reduce the vector space of terms by calculating co-occurrences of index terms in documents. However, both methods require many operations when calculating a terms matrix.

Fuzzy set proposed by L.A. Zadeh [24] is a mathematical tool for dealing with uncertainty, vagueness and ambiguity. Its application in text representation for information retrieval was first proposed by Buell [2], in which a document can be represented as a fuzzy set of terms. Murai et. al. [15] also investigated a fuzzy document retrieval method, in which a keyword is represented as the extension of set of keywords with fuzzy relations. Molinari and Pasi [12] proposed the representation of structured documents, e.g., HTML documents, based on fuzzy set theory in information retrieval. Miyamoto investigated thoroughly applications of fuzzy set theory in information retrieval and cluster analysis in [11]. Other recent studies have looked at some real applications of fuzzy sets in information retrieval [9] and text summarization [21]. Basic idea in text representation for text categorization problem was first presented in [5] by Doan and Horiguchi, and an algorithm based on k-fuzzy set of a document was proposed in [4].

In this paper, we extend the work presented in [5],[4]. A general framework for text representation based on fuzzy set theory for text categorization is proposed. In the proposed method, A document can be represented as a fuzzy subset in which each element called a fuzzy concept. The semantics of documents are defined by the importance degree of fuzzy concept. This degree can be calculated by a fuzzy aggregation function of index terms.

The rest of this paper is organized as follows. Section 2 introduces vector space model and document indexing function. Section 3 describes the proposed general framework of document representation based on fuzzy sets. An algorithm applied to text categorization is given in Section 4. Experimental results are presented in Section 5. Conclusions are drawn in Section 6.

2 Vector Space Model and Document Indexing

The vector space model for document representation was first proposed by Salton et al. [18]. In this model, each document is represented in a vector space in which each dimension corresponds to a term. Let a database $\mathcal{D}$ consist of m documents $\mathcal{D} = \{d_1, d_2, \dots d_{|\mathcal{D}|}\}$ and a vocabulary $\mathcal{T} = \{t_1, t_2, \dots t_{|\mathcal{T}|}\}$. The importance of a term t_i in a document d_j is characterized by a term weight w_{ij}. First we consider the vector space model, and the next is document indexing functions.

2.1 Vector Space Model

In the vector space model, a document d_j is represented as a vector of $|\mathcal{T}|$ dimensions as follows,

$$d_j = (w_{1j}, w_{2j}, \ldots w_{|\mathcal{T}|j}), \tag{1}$$

where w_{ij} characterizes the importance of a term t_i in the document d_j. Mathematically, w_{ij} can be expressed by a function w as follows,

$$\begin{aligned} w : \mathcal{T} \times \mathcal{D} &\rightarrow [0, 1] \\ (t_i, d_j) &\mapsto w(t_i, d_j) = w_{ij} \end{aligned} \tag{2}$$

where w_{ij} can be determined based on indexing functions TF, IDF, or $TFIDF$ which indicated in the next section. w_{ij} can be normalized into interval [0,1] as follows:

$$w_{ij} := \frac{w_{ij}}{\max_{j=1}^{|\mathcal{T}|} w_{ij}}, \tag{3}$$

or

$$w_{ij} := w_{ij} / \sqrt{\sum_{i=1}^{|\mathcal{T}|} w_{ij}^2}. \tag{4}$$

2.2 Document Indexing

Traditional methods for document indexing are Boolean method and frequency methods as follows.

Boolean method. The Boolean method is the simplest method for document representation [8]. Each term weight is 0 or 1; 0 means the term is not present and 1 means that it is present in the document.

$$w_{ij} = \begin{cases} 0, & \text{if } t_i \in d_j, \\ 1, & \text{otherwise.} \end{cases} \tag{5}$$

Frequency methods. Frequency methods assign non-negative values to the weight matrix, instead of binary values as in the Boolean method. According to term frequency and document frequency, there are three indexing methods: term frequency (TF); inverse document frequency (IDF); and term frequency/inverse document frequency ($TFIDF$), which is the combination of the first two methods [8,10].

Term Frequency (TF) method

Term weights are calculated based on the frequency of the terms in the document. Let f_{ij} be the number of occurrences of term t_i in document d_j, w_{ij} may be a function of f_{ij},

$$w_{ij} = TF(f_{ij}). \tag{6}$$

TF should be a monotonic function of f_{ij}, indicating that the importance of a term is increased when its frequency is high; it was introduced in [10] as the following

$$TF(f_{ij}) = \sqrt{f_{ij}} \text{ or } w_{ij} = 1 + \log_2(f_{ij}) \text{ or } w_{ij} = f_{ij}. \tag{7}$$

Inverse Document Frequency (IDF) method

The term's weight matrix is calculated as follows:

$$w_{ij} = \log_2 \frac{m}{df_i}, \tag{8}$$

where m is the number of documents in the database, df_i is the document frequency of terms t_i, the number of documents in which the term t_i occurs. The reason for using IDF weights is that terms which occur in very few documents are better discriminators than those occurring in many documents, and therefore should be weighted higher [10].

TFIDF method

The third method combines term frequency, f_{ij}, and document frequency, df_i. It is a very common method in which w_{ij} is defined by multiplying TF and IDF indexing methods.

$$w_{ij} = \begin{cases} TF(f_{ij}) \log_2 \frac{m}{df_i} \text{ if } f_{ij} \geq 1, \\ 0 \text{ if } f_{ij} = 0 \end{cases} \tag{9}$$

In the $TFIDF$ method, a term in a document is combined by term frequency by inverse document frequency. The advantage of $TFIDF$ term weighting was analyzed by Sparck Johnes [20], and Salton and Buckley [17].

There are other extensions of these document indexing methods. For example, Latent Semantic Indexing [3] and Contex Vector Model [1].

3 A General Fuzzy-Based Framework for Text Representation

Let a database $\mathcal{D} = \{d_1, d_2, \ldots, d_{|\mathcal{D}|}\}$, a vocabulary $\mathcal{T} = \{t_1, t_2, \ldots, t_{|\mathcal{T}|}\}$ as mentioned in the previous section. A document d can be represented in vector space as follows:

$$d_j = (w_{1j}, w_{2j}, \ldots, w_{|\mathcal{T}|j}). \tag{10}$$

In terms of fuzzy set theory, it can be re-written as :

$$d_j = \mu(t_{1j})/t_1 + \mu(t_{2j})/t_2 + \ldots + \mu(t_{|\mathcal{T}|j})/t_{|\mathcal{T}|}, \tag{11}$$

where $\mu(t_{ij}) = w_{ij}$.

We present a fuzzy concept $\tilde{t}$ as follows.

Definition 1. *A fuzzy concept $\tilde{t}$ is a subset of T consisting terms that have a relation to each other. Mathematically,*
$\tilde{t} \in T = \{t_1, t_2, \ldots, t_{|\tilde{t}|}\}$, where $t_i R t_j$, R is a binary relation between t_i and t_j.

A binary relation R between two terms t_i and t_j can be Boolean, frequency, etc.

Definition 2. *A membership function of a fuzzy concept is an aggregation function of terms existing in that fuzzy concept.*

$$\mu : \tilde{t} \rightarrow [0,1],$$
$$\tilde{t} \mapsto \mu(\tilde{t}) = aggre(\mu(t_1), \mu(t_2), \ldots, \mu(t_{|\tilde{t}|})),$$

where $aggre(\mu(t_1), \mu(t_2), \ldots, \mu(t_{|\tilde{t}|}))$ is an aggregation function of $\mu(t_i)$, $\mu(t_i)$ is determined by document indexing function as described in Section 2.2.

An aggregation function can be Max, Min, OWA operators. It characterizes the semantics of of fuzzy concepts in the text document. We then represent a document based on fuzzy set theory as follows.

Proposition 1. *A document can be represented by a set of fuzzy concept in vector space model. Denote $\tilde{\mathcal{T}}$ a set consisting of fuzzy concepts, and $\tilde{t}_{ij} \in \tilde{\mathcal{T}}$ a fuzzy concept $\tilde{t}_i$ belonging to a document d_j, we have:*

$$\mu : \tilde{T} \times D \rightarrow [0,1]$$
$$(\tilde{t}_i, d_j) \mapsto \mu(\tilde{t}_{ij})$$

A document d_j can be then represented as a set of fuzzy concepts as follows:

$$d_j = \mu(\tilde{t}_{1j})/\tilde{t}_{1j} + \mu(\tilde{t}_{2j})/\tilde{t}_{2j} + \ldots + \mu(\tilde{t}_{|\tilde{T}|j})/\tilde{t}_{|\tilde{T}|j} \tag{12}$$

4 A Fuzzy-Based Algorithm for Text Representation in Text Categorization

In this Section, we propose an algorithm for choosing fuzzy concepts and specifying aggregation function in text categorization problem. Given a set of predefined classes $\mathcal{C} = (c_1, c_2, \ldots, c_{|\mathcal{C}|})$. The problem of text categorization is to assign a document into one or more predefined classes.

We define the fuzziness of a document by terms frequency in the vocabulary as follows.

$$d_j = \mu(t_{1j})/t_1 + \mu(t_{2j})/t_2 + \ldots + \mu(t_{|\mathcal{T}|j})/t_{|\mathcal{T}|}, \tag{13}$$

where $\mu(t_{ij}) \equiv$ the frequency of a terms t_i in a document d_j.

Then, we define the fuzziness of a set of categories with respect to a set of terms as the following.

$$\mathcal{C} = \mu'(t_{1j})/t_1 + \mu'(t_{2j})/t_2 + \ldots + \mu'(t_{|\mathcal{T}|j})/t_{|\mathcal{T}|}, \tag{14}$$

where,

$$\mu' : T \times \mathcal{C} \rightarrow [0,1]$$
$$(t_i, \mathcal{C}) \mapsto \mu'(t_i, \mathcal{C}),$$

where $\mu'(t_i, \mathcal{C})$ can be determined by information measure, such as mutual information, information gain, etc [23]. They are calculated as follows.

Information gain measure: The information gain of term t is given by [19],[23]:

$$IG(t,\mathcal{C}) = -\sum_{i=1}^{|\mathcal{C}|} P(c_i)\log P(c_i) + P(t)\sum_{i=1}^{|\mathcal{C}|} P(c_i|t)\log P(c_i|t) + P(\bar{t})\sum_{i=1}^{|\mathcal{C}|} P(c_i|\bar{t})\log P(c_i|\bar{t}). \tag{15}$$

Mutual information measure: Mutual information of term t in class c is given by [19],[23].

$$MI(t,c) = \log\frac{P(t \wedge c)}{P(t).P(c)}, \tag{16}$$

for a set of categories $\mathcal{C} = \{c_i\}_{i=1}^{|\mathcal{C}|}$, mutual information of each term t can be calculated by,

$$MI(t,\mathcal{C}) = \sum_{i=1}^{|\mathcal{C}|} \log\frac{P(t \wedge c_i)}{P(t).P(c_i)}, \tag{17}$$

where $P(.)$ is the notation of probability.

Then, $\mu'(t_i, \mathcal{C})$ of a term t_i in $\mathcal{C}$ can be determined by $\mu'(t_i, \mathcal{C}) := IG(t_i, \mathcal{C})$ or $\mu'(t_i, \mathcal{C}) := MI(t_i, \mathcal{C})$.

Practically, we choose fuzzy concepts as terms in the vocabulary and Max operator as the aggregation function *aggre*. A fuzzy concept is then the term that has the highest membership function in the vocabulary. Based on the fuzziness of documents and categories with respect to terms in Equation (13) and Equation (14), we select fuzzy concepts as in Algorithm 1.

Algorithm 1. Choose fuzzy concepts from text

1: Select p fuzzy concepts from Equation (13).
2: Select q fuzzy concepts from Equation (14).
3: Combine fuzzy concepts from Step 1 and Step 2.

5 Experiments

5.1 Experimental Methodology

In order to investigate the proposed method, we apply it to text categorization as follows. We use the real-world data set 20Newsgroups as the benchmark data set. It consists of 20 categories, each has 1000 documents. The details of data collection can be found in [6].

Performance of text categorization can be evaluated by two basic measures: precision and recall. In addition, BEP and F_1 measures which are combination of both precision and recall are often used in text categorization. To evaluate the whole system, there are some measures, e.g., micros and macros. In this paper

Table 1. micro F1 performance of 20Newsgroups, Rocchio algorithm

Parameters		Trials										
p	q	1	2	3	4	5	6	7	8	9	10	ave.
2500	2	82.0	79.4	80.0	80.8	79.6	80.5	79.1	79.9	79.6	80.8	80.2
2500	4	80.4	79.2	81.4	80.7	79.7	81.0	79.5	80.0	79.6	81.0	80.3
2500	6	80.6	81.5	81.0	80.3	80.6	81.0	80.8	79.9	80.1	81.3	80.7
2500	8	80.7	81.6	81.9	81.4	80.4	80.2	80.7	80.4	80.3	80.3	**80.8**
10000	2	80.1	81.5	79.1	79.5	80.6	80.2	80.3	80.7	79.7	81.4	80.9
10000	4	81.0	80.2	80.1	81.3	82.0	81.8	81.3	80.8	80.3	81.1	**81.0**
10000	6	81.2	81.1	81.8	80.3	80.4	79.1	80.1	79.7	81.6	81.2	80.6
10000	8	80.2	81.3	80.1	81.3	79.7	79.9	81.5	83.4	81.4	80.0	80.1
baseline 1		74.9	75.3	76.3	77.2	76.8	75.8	75.4	77.0	75.2	76.7	**76.1**
baseline 2		79.6	80.1	80.3	81.0	78.8	79.9	80.5	79.5	79.9	76.7	**80.0**
all-term		81.6	80.9	80.6	81.1	77.8	81.8	79.9	81.5	82.1	81.6	**80.8**

we consider the micro $F1$ which was mostly used to evaluate text categorization system [23],[22].

We use two baseline algorithms, Rocchio in vector space model and naive Bayes in probabilistic model, to classify documents. The Rocchio algorithm built each category as a vector and calculated distance between a document and a category by their cosine. A document will be assigned into the category which has the closest distance. For the lack of paper, we do not give the details of Rocchio algorithm; it can be found in [16]. In the experiments we set two parameters α and β to 1 as the original algorithm. The naive Bayes algorithm is based on Bayes theorem to assign the probability of document into each category, the details of the algorithm can be found in [23].

In order to compare to conventional methods in text representation, we choose three baseline methods: all-term method uses total vocabulary with 114444 terms, baseline 1 uses only 2500 terms, baseline 2 uses 1000 terms ($\approx$ 1/10 vocabulary), both have highest mutual information measure.

To investigate the effectiveness of parameters in the proposed method, parameters p and q in Algorithm 1 are set as follows. Parameter p is chosen as $2, 4, 6$ and 8, q is chosen as 2500 and 10000, respectively. Fuzzy membership function in Equation (13) is frequency of terms in a document and is normalized into [0,1] and one in Equation (14)is mutual information measure. The total number of terms when $q = 2500, q = 2, 4, 6, 8$ are $9751, 14676, 18912, 22599$, respectively, and when $q = 10000, q = 2, 4, 6, 8$ are $14022, 17605, 21041, 24207$, respectively.

Our experiments were done using 10-fold cross-validation in Unix environment, C and Perl programming language.

5.2 Experimental Results

The experimental results are shown in Table 1 and in Table 2. Table 1 indicate micro-F_1 performances for Rocchio algorithm and Table 2 for naive Bayes algorithm. Each algorithm was implemented on the same data set. The three

Table 2. micro F1 performance of 20Newsgroups, naive Bayes algorithm

Parameters		Trials										
p	q	1	2	3	4	5	6	7	8	9	10	ave.
2500	2	81.2	80.3	80.3	81.0	79.8	80.8	81.3	79.9	79.1	79.9	80.4
2500	4	80.2	81.0	81.5	80.6	81.0	81.1	80.9	81.1	81.5	81.3	81.0
2500	6	82.5	82.0	83.2	81.2	81.9	81.8	81.9	82.1	79.4	81.8	81.8
2500	8	82.2	82.3	83.0	81.8	81.9	83.0	82.2	81.2	82.8	82.6	**82.3**
10000	2	80.9	80.9	81.7	82.5	81.1	81.6	80.1	82.7	82.4	81.0	81.5
10000	4	80.9	81.7	82.9	81.2	81.5	79.9	80.9	82.0	82.5	81.3	81.6
10000	6	82.0	81.7	81.3	82.9	81.2	81.5	82.6	81.0	82.3	82.1	81.8
10000	8	82.5	81.3	81.7	82.5	82.6	81.1	82.2	81.5	81.1	81.6	**81.9**
baseline 1		78.7	77.8	71.1	79.0	78.4	75.9	76.5	78.3	78.7	78.0	**77.3**
baseline 2		79.5	79.6	80.9	80.2	80.4	80.4	78.8	80.5	79.1	79.7	**79.9**
all-term		82.8	83.6	82.3	82.6	81.3	81.9	83.7	83.9	81.5	83.6	**82.7**

last rows describe the results of three standard methods: baseline 1, baseline 2 and all-term methods. The remaining are the proposed methods.

Firstly, we consider only three standard methods. Is is easily to observe that the all-term method archieve the highest performance and the baseline 1 achieve the worst one. Moreover, naive Bayes algorithm achieves better performance than Rochio algorithm. In more details, all-term method achieve 82.7% in naive Bayes algorithm versus 80.8% in Rochio algorithm, baseline 1 and baseline 2 using naive Bayes algorithm achieve 77.3% and 79.9% versus 76.1% and 80.0% using Rochio algorithm. These results confirmed some experiments reported in [23],[22].

Secondly, we can easily observe that almost cases using the proposed method outperformed baseline 1 and baseline 2, and comparable to all-term method. In details, baseline 1 has the lowest performance for both Rocchio and naive Bayes algorithms. In details, with $p = 2500$ (as the same number of terms as baseline 1) and $q = 2, 4, 6, 8$ the results of both Rochio and naive Bayes achieve performances much more than baseline 1. Rochio algorithm in baseline 1 achieves only 76.1% while the proposed methods achive 80.2%, 80.3%, 80.7%, and 80.8%. For naive Bayes algorithm, baseline 1 achieve only 77.3% while the proposed methods achieve 80.4%, 81.0%, 81.8%, and 82.3%, respectively. The results are the same as the baseline 2 and other cases of proposed methods.

In Table 1 and Table 2, we also observe that with appropriate parameters p and q we can achieve the highest performance. In this experiment it is the case of $p = 2500$ and $q = 8$ with 80.8% for Rochio and 82.3% for naive Bayes algorithms. The reduced number of terms is only 22599, approximately $22599/114444 \approx$ 19.8% total number of terms in vocabulary. The results also show that selected terms are independent of classification algorithms.

In summary, the experimental results show that the proposed method of document representation achieves better performance than conventional representation methods; and significantly reduces number of terms in vocabulary.

6 Conclusions

In this paper we developed the general framework for text representation based on fuzzy set theory. Based on this framework, an algorithm for choosing fuzzy concepts was proposed and applied to text categorization problem. Experiments on the real-world data set 20Newsgroups showed that the proposed method outperformed the baseline methods, including text representation using selected terms based on mutual information measure, and comparable to one that used all vocabulary in text categorization.

Acknowledgements

We thank to Doctor Van-Nam Huynh, Japan Advanced Institute of Science and Technology for his helpful comments on the manuscript.

This work is supported in partly the Grand-in-Aid of Scientific Research, JSPS, Japan and the project "Information Extraction Models for discovering entities and sematic relations from Vietnamese Web pages" in College of Technology, Vietnam National University, Hanoi.

References

1. H. Billhardt, D. Bonajo, and V. Maojo. A context vector model for information retrieval. *Journal of the American Society for Information Science and Technology (JASIST)*, 53(3):236–249, 2002.
2. D.A. Buell. An analysys of some fuzzy subsets application to information retrieval systems. *Fuzzy Sets and Systems*, 7(1):35–42, 1982.
3. S. Deerwester, G.W. Furnas S. Dumais, and T.K. Landauer. Indexing by latent semantic indexing. *Journal of the American Society for Information Science and Technology (JASIST)*, 41(6):391–407, 1990.
4. S. Doan. A fuzzy-based approach to text representation in text categorization. In *Proceeding of 14th IEEE Int'l Conference onn Fuzzy Systems - FUZZ-IEEE 2005*, pages 1008–1013, Nevada, U.S., 2005.
5. S. Doan and S. Horiguchi. A new text representation using fuzzy concepts in text categorization. In *Proceeding of 1st International Conference on Fuzzy Set and Knowledge Discovery (FSKD)*, volume 2, pages 514–518, Singapore, 2002.
6. CMU Text Learning Group. 20newsgroups dataset. `http://www.cs.cmu.edu/~textlearning`.
7. T. Joachims. Text categorization with support vector machines: Learning with many relevant features. In *Proceedings 10th European Conference on Machine Learning(ECML)*, pages 137–142, 1998.
8. D. Lewis. *Representation and Learning in Information Retrieval.* PhD thesis, Graduate School of the University of Massachusetts, 1991.
9. D. Lucarella and R.Marara. First: fuzzy informatioon retrieval system. *Journal of Information Science*, 17(2):81–91, 1991.
10. C.D. Manning and H. Schutze. *Foundations of Statistical Natural Language Processing.* The MIT Press, 1999.

11. S. Miyamoto. *Fuzzy Sets in Information Retrieval and Cluster Analysis.* Kluwer Academic Publishers, 1990.
12. A. Molinari and G. Pasi. A fuzzy representation of html document for information retrieval system. In *Proceeding of 5th IEEE Int't Conference on Fuzzy Systems*, pages 107–112, 1996.
13. I. Moulinier. A framework for comparing text categorization approaches. In *AAAI Symposium on Machine Learning and Information Access.* Stanford University, 1996.
14. I. Moulinier and J.G. Ganascia. Applying an existing machine learning algorithm to text categorization. In S. Wermter, E. Riloff, and G. Schaler, editors, *Connectionist, Statistical and Symbolic Approaches to Learning for Natural Language Processing*, pages 343–354. Springer-Verlag, Heidelbeg, 1996.
15. T. Murai, M. Miyakoshi, and M. Shimbo. A fuzzy document retrieval method based on two-valued indexing. *Fuzzy Sets and Systems*, 30(2):103–120, 1989.
16. J. Rocchio. Relevance feedback in information retrieval. In G. Salton, editor, *The SMART retrieval system: Experiments on Automatic Document Processing*, chapter 14, pages 313–323. Prentice Hall, 1971.
17. G. Salton and C. Buckley. Term weighting approaches in automatic text retrieval. *Information Processing and Management*, 24(5):513–523, 1988.
18. G. Salton, A. Wong, and C.S. Yang. A vector space model for automatic indexing. *Communications of the ACM*, 18(11):613–620, 1975.
19. F. Sebastiani. Machine learning in automated text categorization. *ACM computing survey*, 34(1):1–47, 2002.
20. K.A. Sparck-Jones. A statistical interpretation of term specifility and its application in retrieval. *Journal of Documentation*, 28(1):11–20, 1972.
21. R. Witte and S. Bergler. Fuzzy coreference resolution for summarization. In *Proceedings of 2003 International Symposium on Reference Resolution and Its Applications to Question Answering and Summarization (ARQAS)*, pages 43–50, Venice, Italy, June 23–24 2003. Università Ca' Foscari. `http://rene-witte.net`.
22. Y. Yang. An evaluation of statistical approaches to text categorization. *Information Retrieval Journal*, 1:69–90, 1999.
23. Y. Yang and J.O. Pedersen. A comparative study on feature selection in text categorization. In *Proceeding of the 14th International Conference on Machine Learning (ICML97)*, pages 412–420, 1997.
24. L.A. Zadeh. Fuzzy sets. *Information Control*, 8:338–353, 1965.

Risk Assessment of E-Commerce Projects Using Evidential Reasoning

Rashid Hafeez Khokhar, David A. Bell, Jiwen Guan, and QingXiang Wu

The School of Electronics, Electrical
Engineering and Computer Science
Queen's University Belfast
Belfast, BT7 1NN, N.I. UK
Tel.: (44)2890974783, (44)2890974165; Fax: (44)2890975666
{r.khokhar, da.bell, j.guan, q.wu}@qub.ac.uk

Abstract. The purpose of this study is to develop a decision making system to evaluate the risks in E-Commerce (EC) projects. Competitive software businesses have the critical task of assessing the risk in the software system development life cycle. This can be conducted on the basis of conventional probabilities, but limited appropriate information is available and so a complete set of probabilities is not available. In such problems, where the analysis is highly subjective and related to vague, incomplete, uncertain or inexact information, the Dempster-Shafer (DS) theory of evidence offers a potential advantage. We use a direct way of reasoning in a single step (i.e., extended DS theory) to develop a decision making system to evaluate the risk in EC projects. This consists of five stages 1) establishing knowledge base and setting rule strengths, 2) collecting evidence and data, 3) determining evidence and rule strength to a mass distribution for each rule; i.e., the first half of a single step reasoning process, 4) combining prior mass and different rules; i.e., the second half of the single step reasoning process, 5) finally, evaluating the belief interval for the best support decision of EC project. We test the system by using potential risk factors associated with EC development and the results indicate that the system is promising way of assisting an EC project manager in identifying potential risk factors and the corresponding project risks.

Keywords: Dempster-Shafer Theory, Evidential Reasoning, Software Performance Evaluation, Financial Engineering, E-Commerce Application.

1 Introduction

Electronic Commerce (EC) is possibly the most promising information technology application that enterprises have seen in recent years. EC addresses the needs of organizations, suppliers and costumers to reduce costs while improving the quality of goods and services, and increasing the speed of service delivery [8]. The current highly competitive business environment needs a good quality EC system, but EC development is subject to various kinds of risk such as malicious code attacks [2], uncertain legal jurisdiction [7], absence of firewall [8], and lack of using cryptography [10].

L. Wang et al. (Eds.): FSKD 2006, LNAI 4223, pp. 621 – 630, 2006.

A comprehensive collection of potential risk factors associated with EC development can be found in Refs [11].

Leung et al., [9] have developed an integrated knowledge-based system that assists project managers to determined potential risk factors and the corresponding project risks. According to this knowledge-base system, most project managers worry about the time involved in risk management when it comes to identifying and assessing risks. However, with the aid of computers and the use of software systems, the time for risk analysis can be significantly reduced. Addison [1] used a Delphi technique to collect the opinion of 32 experts and proposed 28 risks for EC projects. Meanwhile, Carney et al. [3] designed a tool called COTS Usage Risk Evaluation (CURE) to predict the risk areas of COTS products in which he identified four categories comprising 21 risk areas. Ngai and Wat [12] have developed a web-based fuzzy decision support system (FDSS) to assist project manager in identifying potential EC risk factors. However, FDSS has not been tested with real life EC projects, and it can only handle the available risk variables. Also the various membership functions need to be estimated to be as realistically as possible. Cortellessa et al., [4] have introduced a methodology which elaborates annotated UML diagrams to estimate the performance failure probability and combines it with the failure severity estimate which is obtained using the Functional Failure Analysis. This methodology is still have some limitation and only suitable for the analysis of performance-based risk in the early phases of the software life cycle.

In this paper, a direct way of reasoning in a single step is presented to develop our evidential reasoning based system to assess the risk in EC projects. A direct way of reasoning in a single step is actually an extended DS theory of evidence. It is a generalization of Yen's [13, 14] model from Bayesian probability theory to the DS theory of evidence.

Section 2 presents the system development methodology involved in the system for the risk assessment of EC projects. In section 3, a direct way of reasoning in a single step is described and the experimental results are presented in section 4. Finally, conclusions are given in section 5.

2 System Development Methodology

In this paper, a decision making system is developed to assist EC project managers in specifying potential risk factors and evaluating the corresponding EC development risks. Figure 1 presents the methodology involved in the system. Before applying the proposed methodology, it is important to conduct risk identification and compile a list of the most significant uncertainty factors and their descriptions. For this purpose, we use the results of exploratory factor analysis (EFA) by Wat et al., [11] to identifying potential risks associated with EC development. Wat et al., [11] used a source-based approach to categorizing EC development risks is initially used with technical, organizational, and environmental risks as the three primary source categories. Then potential risks associated with EC development was identified with 51 risk items associated with EC development based on a comprehensive literature review and interviewed with EC practitioners.

The project manager first inputs all risk factor and different pieces of evidence. Then system will search for a rule from the existing rules database and get a rule strength for the selected evidence and hypothesis. The rules database is developed according to the support strengths of different pieces of evidence for different conclusions. The user can edit the rules strengths dynamically, this step will help us to handle uncertain situation. The next step is to combine the outputs from the different rules for all risk factors and its corresponding pieces of evidence using the extended DS engine, and to evaluate the belief intervals for the risk assessment of EC projects.

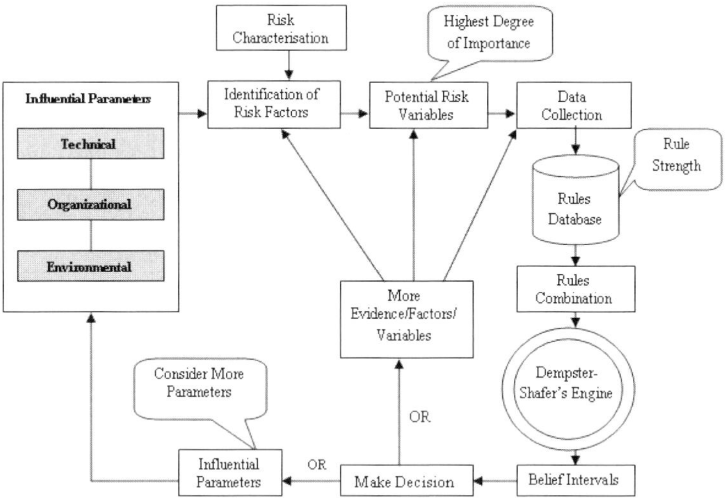

Fig. 1. System development methodology for risk assessment of EC projects

3 The Direct Way of Reasoning in Single Step

Guan and Bell [5, 6] extended the DS theory by introducing an evidential mapping that uses mass functions as used in the DS theory to express uncertain relationships. It is a generalization of Yen's [13, 14] model from Bayesian probability theory to the DS theory of evidence. The sections below describe the direct way of reasoning in a single step process through an example of its use.

3.1 Rule Strengths in Expert System

In the first stage, we describe the knowledge base and rule strengths. Instead of a mass function being used to express strengths for each element of the evidence space, a mass function is used to express strength for each subset of the evidence space.

Consider a frame of discernment with 5 hypotheses $\Theta = \{h_1, h_2, h_3, h_4, h_5\}$:
$h_1 = "VeryLow", h_2 = "Low", h_3 = "Medium", h_4 = "High", h_5 = "VeryHigh"$.

Consider a particular piece of evidence e_1 e.g., "*wrong schedule estimation*", which comes from results of an exploratory factor analysis (EFA) [11] for "*resource risk factor*". We can obtain a general rule which uses this evidence, when present indicated by $\{e_1\}$, strongly supports $h = \{h_1, h_2, h_3\}$ of Θ and refutes $\bar{h} = \{h_4, h_5\}$. When the evidence is not present, indicated by $\{\bar{e}_1\}$, the support strengths are divided between $\bar{h}$ and Θ. More specifically, we say here that there is an evidence space $\Xi_1 = \{e_1, \bar{e}_1\}$ and mass functions $s_{11}, s_{12}, s_{13} : 2^{\Theta} \rightarrow [0,1]$ such that

$$s_{11}(\{h_1,h_2,h_3\} \mid \{e_1\}) = 0.64, s_{11}(\{h_4,h_5\} \mid \{e_1\}) = 0.34, s_{11} = (\{\Theta\} \mid \{e_1\}) = 0.02;$$
$$s_{12}(\{h_1,h_2,h_3\} \mid \{\bar{e}_1\}) = 0.00, s_{12}(\{h_4,h_5\} \mid \{\bar{e}_1\}) = 0.50, s_{12} = (\{\Theta\} \mid \{\bar{e}_1\}) = 0.50;$$
$$s_{13}(\{h_1,h_2,h_3\} \mid \{\Xi_1\}) = 0.25, s_{13}(\{h_4,h_5\} \mid \{\Xi_1\}) = 0.45, s_{13} = (\{\Theta\} \mid \{\Xi_1\}) = 0.30.$$

Guan and Bell [5, 6] used mass function $m(X) = s(X \mid E)$ on the power set of hypothesis space Θ to express the rule strength for each subset E of the evidence space Ξ. Yen [13, 14] used $m(X) = s(X \mid e)$ for each element e of the evidence space to express the rule strength. This means that Guan and Bell [5, 6] have generalized Yen's *subset-probability-pair-collection-valued (s-p-p-c-v)* mapping to a *subset-mass-pair-collection-valued (s-m-p-c-v)* mapping.

The s-m-p-c-v mapping Γ from the power set 2^{Ξ} of evidence space Ξ to $2^{2^{\Theta} \times [0,1]}$ is

$$\Gamma : (2^{\Xi} - \{\phi\}) \rightarrow 2^{2^{\Theta} \times [0,1]} \tag{1}$$

such that for every non-empty $E \subseteq \Xi$

$$\Gamma(E) = \{(A_{E1}, s_E(A_{E1} \mid E)), (A_{E2}, s_E(A_{E2} \mid E)), \ldots, (A_{En_E}, s_E(A_{En_E} \mid E))\} \tag{2}$$

Where $A_{E1}, A_{E2}, \ldots, A_{En_E} \in 2^{\Theta}$; *i.e.*, $A_{E1}, A_{E2}, \ldots, A_{En_E} \subseteq \Theta$ are the focal elements of mass function $m_E(X) = s_E(X \mid E)$ on 2^{Θ}:

$$0 < s_E(A_{E1} \mid E), s_E(A_{E2} \mid E), \ldots, s_E(A_{En_e} \mid E) \leq 1 \tag{3}$$

and

(1) $A_{Ei} \neq \phi$ for $i = 1, \ldots\ldots.n_E$;
(2) $s_E(A_{Ei} \mid E) > 0$ for $i = 1, \ldots\ldots.n_E$;
(3) $\sum_{i=1}^{nE} s_E(A_{Ei} \mid E) = 1$

Then a rule is a collection $RULE =< E, \Theta, \Gamma >$, where Ξ is an evidence space, Θ is a hypothesis space, and Γ is a s-m-p-c-v mapping from the power set 2^{Ξ} of evidence space Ξ to hypothesis space Θ (more precisely, to $2^{2^{\Theta} \times [0,1]}$).

Also, a rule can be expressed by a collection of $|2^{\Xi}| - 1$ "strength" mass functions $m_E(A) = s_E(A \mid E)$ for $A \subseteq \Theta$,

$$RULE = \{s_E(A \mid E) \mid E \in (2^{\Xi} - \{\phi\})\} \tag{4}$$

$$= \{s_1(A \mid E_1), s_2(A \mid E_2), ..., s_{|2^{\Xi}-1|}(A \mid E_{|2^{\Xi}|-1})\},$$

$$2^{\Xi} - \{\phi\} = \{E_1, E_2, ..., E_{|2^{\Xi}-1|}\}; \; s_i = s_{Ei}, E_i \neq \phi$$

for $i = 1, ... | 2^{\Xi} | - 1$.

Consider source of evidence e_2 e.g., "*project over budget*" which comes from the same results of EFA [11]. This evidence when present indicated by $\{e_2\}$, strongly support subset $h = \{h_3, h_4, h_5\}$ of Θ, and refutes $\bar{h} = \{h_1, h_2\}$. When the evidence is not present, indicated by $\{\bar{e}_2\}$, the support strengths are divided between $\bar{h}$ and Θ. More specifically, we say here that there is an evidence space $\Xi_2 = \{e_2, \bar{e}_2\}$ and mass functions $s_{21}, s_{22}, s_{23} : 2^{\Theta} \rightarrow [0,1]$ such that

$$s_{21}(\{h_3, h_4, h_5\} \mid \{e_2\}) = 0.76, s_{21}(\{h_1, h_2\} \mid \{e_2\}) = 0.20, s_{21} = (\{\Theta\} \mid \{e_2\}) = 0.04;$$
$$s_{22}(\{h_3, h_4, h_5\} \mid \{\bar{e}_2\}) = 0.00, s_{22}(\{h_1, h_2\} \mid \{\bar{e}_2\}) = 0.50, s_{22} = (\{\Theta\} \mid \{\bar{e}_2\}) = 0.50;$$
$$s_{23}(\{h_3, h_4, h_5\} \mid \{\Xi_2\}) = 0.65, s_{23}(\{h_1, h_2\} \mid \{\Xi_2\}) = 0.20, s_{23} = (\{\Theta\} \mid \{\Xi_2\}) = 0.15.$$

Summarizing, following the method in [5], the knowledge base includes the following rules:

RULE-1
IF EVIDENCE $\{e_1\}$ THEN
HYPOTHESIS $\{h_1, h_2, h_3\}$ WITH STRENGTH $s_{11}(\{h_1, h_2, h_3\} \mid \{e_1\}) = 0.64$
HYPOTHESIS $\{h_4, h_5\}$ WITH STRENGTH $s_{11}(\{h_4, h_5\} \mid e_1) = 0.34$
HYPOTHESIS $\{\Theta\}$ WITH STRENGTH $s_{11}(\{\Theta\} \mid \{e_1\}) = 0.02$
ELSE IF EVIDENCE $\{\bar{e}_1\}$ THEN
HYPOTHESIS $\{h_1, h_2, h_3\}$ WITH STRENGTH $s_{12}(\{h_1, h_2, h_3\} \mid \{\bar{e}_1\}) = 0.00$
HYPOTHESIS $\{h_4, h_5\}$ WITH STRENGTH $s_{12}(\{h_4, h_5\} \mid \{\bar{e}_1\} = 0.50$
HYPOTHESIS $\{\Theta\}$ WITH STRENGTH $s_{12}(\{\Theta\} \mid \{\bar{e}_1\} = 0.50$
ELSE IF EVIDENCE $\{\Xi_1\}$ THEN
HYPOTHESIS $\{h_1, h_2, h_3\}$ WITH STRENGTH $s_{13}(\{h_1, h_2, h_3\} \mid \Xi_1) = 0.25$
HYPOTHESIS $\{h_4, h_5\}$ WITH STRENGTH $s_{13}(\{h_4, h_5\} \mid \Xi_1) = 0.45$
HYPOTHESIS $\{\Theta\}$ WITH STRENGTH $s_{13}(\{\Theta\} \mid \{\Xi_1\}) = 0.30$

Here $\Xi_1 = \{e_1, \bar{e}_1\}$ is an evidence space and

$$m_{11}(X) = s_{11}(X \mid e_1) \tag{5}$$
$$m_{12}(X) = s_{12}(X \mid \bar{e}_1) \tag{6}$$
$$m_{13}(X) = s_{13}(X \mid \Xi_1) \tag{7}$$

are mass functions $2^{\Theta} \rightarrow [0,1]$; i.e., they are the functions $m : 2^{\Theta} \rightarrow [0,1]$ such that

$$m(\phi) = 0, \sum_{X \subseteq \Theta} m(\Theta) = 1. \tag{8}$$

RULE-2
IF EVIDENCE $\{e_2\}$ THEN
HYPOTHESIS $\{h_3,h_4,h_5\}$ WITH STRENGTH $s_{21}(\{h_3,h_4,h_5\}|\{e_2\})=0.76$
HYPOTHESIS $\{h_1,h_2\}$ WITH STRENGTH $s_{21}(\{h_1,h_2\}|\{e_2\}=0.20$
HYPOTHESIS $\{\Theta\}$ WITH STRENGTH $s_{21}(\{\Theta\}|e_2)=0.04$
ELSE IF EVIDENCE $\{\bar{e}_2\}$ THEN
HYPOTHESIS $\{h_3,h_4,h_5\}$ WITH STRENGTH $s_{22}(\{h_3,h_4,h_5\}|\{\bar{e}_2\}=0.00$
HYPOTHESIS $\{h_1,h_2\}$ WITH STRENGTH $s_{22}(\{h_1,h_2\}|\{\bar{e}_2\}=0.50$
HYPOTHESIS $\{\Theta\}$ WITH STRENGTH $s_{22}(\{\Theta\}|\{\bar{e}_2\}=0.50$
ELSE IF EVIDENCE $\{\Theta\}$ THEN
HYPOTHESIS $\{h_3,h_4,h_5\}$ WITH STRENGTH $s_{23}(\{h_3,h_4,h_5\}|\{\Xi_2\}=0.65$
HYPOTHESIS $\{h_1,h_2\}$ WITH STRENGTH $s_{23}(\{h_1,h_2\}|\{\Xi_2\})=0.20$
HYPOTHESIS $\{\Theta\}$ WITH STRENGTH $s_{23}(\{\Theta\}|\{\Xi_2\}=0.15$

Here $\Xi_2=\{e_2,\bar{e}_{12}\}$ is an evidence space and

$$m_{21}(X)=s_{21}(X\mid e_2) \tag{9}$$

$$m_{22}(X)=s_{22}(X\mid \bar{e}_2) \tag{10}$$

$$m_{23}(X)=s_{23}(X\mid \Xi_2) \tag{11}$$

are mass functions $2^{\Theta}\rightarrow[0,1]$

3.2 Data and Evidence

Suppose from the above rules and given a particular confidence in the presence of the data items, we can derive pieces of evidence which are in the conventional DS format. The confidence c_1, we have that e_1 evidence is in fact present is as follows:
i.e., we have data strength:

$$c_1(\{e_1\})=0.70, c_1(\{\bar{e}_1\})=0.20, c_1(\Xi_1)=0.10$$

Here c_1 is a mass function over the evidence space Ξ_1, intuitively representing the confidence we have that e_1 is present.

Similarly, evidence e_2 is present into the following data strengths:

$$c_2(\{e_2\})=0.75, c_2(\{\bar{e}_2\})=0.20, c_2(\Xi_2)=0.05$$

Here c_2 is again a mass function over Ξ_2.

Generally, there is a mass function $c_i:2^{\Theta}\rightarrow[0,1]$ over the evidence space Ξ_i for $i=1,2,........n$.

3.3 Hypothesis Strength

In this stage we present the procedure to get from evidence and rule strength to a mass distribution for each rule; i.e., the first half of a single step reasoning process.

Now, for each rule we can get a hypothesis strength mass function from the evidence strength and the rule strength.

For RULE-1; i.e., for rule strengths s_{11}, s_{12}, s_{13} and from evidence c_1, the risk variable mass distribution $r_1 : 2^{\Theta} \rightarrow [0,1]$ is obtained as follows.

$r_1(\{h_1, h_2, h_3\}) = 0.47, r_1(\{h_4, h_5\}) = 0.38, r_1(\{\Theta\}) = 0.15$

By RULE-2; i.e., for rule strengths s_{21}, s_{22}, s_{23} and from evidence c_2 we get the following mass distribution $r_2 : 2^{\Theta} \rightarrow [0,1]$ for the other node:

$r_2(\{h_3, h_4, h_5\}) = 0.60, r_2(\{h_1, h_2\}) = 0.26, r_2(\{\Theta\}) = 0.14$

This is the first half of our reasoning process.

3.4 Combining Prior Mass and Different Rules

Now, let us discuss the second half of the single step reasoning process. If μ_1 and μ_2 are two mass functions corresponding to two independent evidential sources, then the combined mass function $\mu_1 \otimes \mu_2$ is calculated according to Dempster rule of combination:

1. $(\mu_1 \otimes \mu_2)(\phi) = 0$;
2. For every $A \subseteq \Theta, A \neq \phi$,

$$(\mu_1 \otimes \mu_2)(A) = \frac{\sum_{X \cap Y = A} [P(A) \frac{\mu_1(X)}{P(X)} \frac{\mu_2(Y)}{P(Y)}]}{\sum_{\theta \subseteq \Theta, \theta \neq \phi} (\sum_{X \cap Y = \theta} [P(\theta) \frac{\mu_1(X)}{P(X)} \frac{\mu_2(Y)}{P(Y)}])} \tag{12}$$

For our example, the intersection table of $\mu_1 \otimes \mu_2$ for RULE-1 and RULE-2 is shown in the following table 1.

Table1. Intersection table to combine two rules

$\mu_1 \otimes \mu_2$	$\{h_3, h_4, h_5\}(0.60)$	$\{h_1, h_2\}(0.26)$	$\{\Theta\}(0.14)$
$\{h_1, h_2, h_3\}(0.47)$	$\{\{h_3\}(0.20)$	$\{h_1, h2\}(0.19)$	$\{h_1, h_2, h_3\}(0.07)$
$\{h_4, h_5\}(0.38)$	$\{h_4, h_5\}(0.36)$	$\phi(0)$	$\{h_4, h_5\}(0.05)$
$\{\Theta\}(0.15)$	$\{h_3, h_4, h_5\}(0.09)$	$\{h_1, h_2\}(0.04)$	$\Theta(0.02)$

We get the normalization constant (required to discount for mass committed to ϕ, the empty set)

$$N = \sum_{X \cap Y \neq \phi} P(X \cap Y) \frac{\mu_1(X)}{P(X)} \frac{\mu_2(Y)}{P(Y)} = 1.0083 \tag{13}$$

3.5 Belief Intervals and Ignorance

Finally, we can establish the belief interval for our conclusion after applying this reasoning process. We convert the above results to a set of beliefs for the respective conclusions by adding the masses of all subsets of each conclusion to get the belief allocated to it, and then we get the belief intervals for risk assessment in the EC project using 2 pieces of evidence. So the conclusion is: "The '*resource risk*' is '$\{h_3\} = Medium$' using the two pieces of evidence, '*wrong schedule estimation*' and '*project over budget*' with the belief intervals $[bel_{\mu}(A), pls_{\mu}(A)] = [0.2007, 0.3657]$ and $ignorance(A) = 0.1657$".

4 Case Study

The design of the ease of use interface is a key element for the risk assessment of EC developments. Therefore, we design an interface that can be used by any user in EC environments. In this paper, we describe our results with the help of three general steps including: 1) input risk factors and different pieces of evidence 2) edit the rules strengths if required and 3) finally evaluate the belief intervals for the best supported decision. The risk evaluation form is presented in figure 2 to input all potential risk factors, pieces of evidence and data associated with EC project. In this form the project manager/evaluator first inputs all potential risk factors and different pieces of evidence using the results of an EFA [11]. The project manager/evaluator then selects the appropriate hypothesis among five given hypothesis (e.g. *VeryLow, Low, Medium, High, and VeryHigh*) and assign the data with the help of slider for each evidence. The next step is to determine the rule strengths to get a mass distribution for each rule.

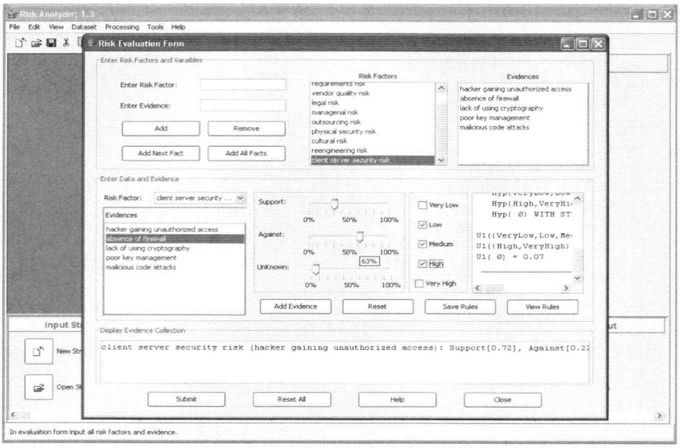

Fig. 2. Risk evaluation form to input potential risk associated with EC projects

The system will search for the rules from the existing rules database for the particular piece of evidence involved. The user can also modify the rule strengths to handle uncertainty. Similarly the user inputs all data/evidence and determines the rules for every risk factor. Finally, we apply the DS engine to evaluate the degrees of belief for these risk factors. The risk assessment results using 10 important risk factors from [11] with belief intervals are presented in table 2. In this table different pieces of evidence (in column 3) are presented for the corresponding risk factors. The next columns demonstrate the assessment results with belief intervals and these risk assessment results are explained in figure 3. For example the first result is described as: The 'resource risk' is '*Low*' with support (37%), against (46%) and uncertain (17%).

Table 2. Belief intervals and ignorance for potential risk associated with EC projects

No	Risk factors	Variables (pieces of evidence)	Conclusion	$[bel_l(A), pls_l(A)]$	$ign(A)$
1	Resources risk	$V_{21}, V_{22}, V_{23}, V_{24}, V_{25}, V_{27}$	*Low*	*[0.3700, 0.5357]*	*0.1656*
2	Requirements risk	$V_{14}, V_{15}, V_{16}, V_{17}, V_{19}, V_{20}$	*Very Low*	*[0.1393, 0.2088]*	*0.0795*
3	Vendor quality risk	$V_{46}, V_{47}, V_{48}, V_{49}$	*Very Low*	*[0.0598, 0.1866]*	*0.1367*
4	Client-server security risk	V_1, V_2, V_3, V_4, V_5	*Low*	*[0.2736, 0.6100]*	*0.3364*
5	Legal risk	V_{38}, V_{39}, V_{40}	*Very Low*	*[0.0244, 0.0836]*	*0.0592*
6	Managerial risk	$V_{28}, V_{29}, V_{30}, V_{31}, V_{32}$	*Very Low*	*[0.0815, 0.1111]*	*0.0296*
7	Outsourcing risk	$V_{40}, V_{41}, V_{42}, V_{43}, V_{45}$	*Very Low*	*[0.3091, 0.7453]*	*0.4363*
8	Physical security risk	V_7, V_8, V_9, V_{10}	*Low*	*[0.5284, 0.7513]*	*0.2229*
9	Cultural risk	V_{50}, V_{51}	*Low*	*[0.3958, 0.6727]*	*0.2769*
10	Reengineering risk	V_{33}, V_{34}	*Low*	*[0.0554, 0.2333]*	*0.1779*

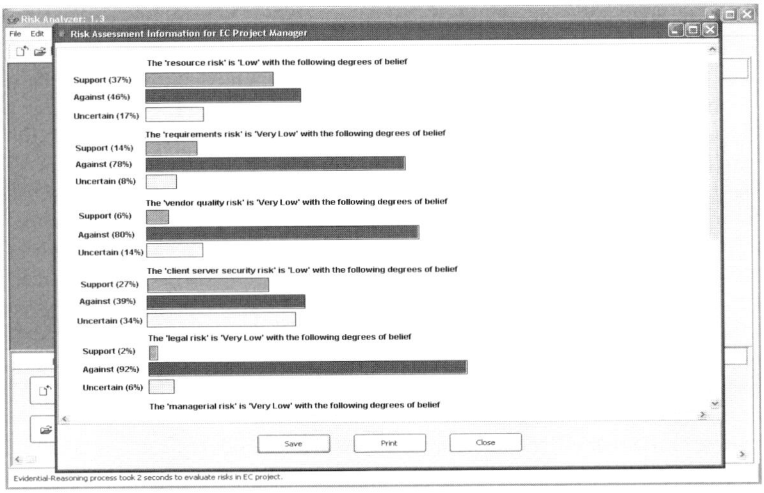

Fig. 3. Risk assessment results using direct way of reasoning

5 Conclusions

This paper has outlined an approach to the assessment of the risks associated with EC development using a direct way of evidential reasoning with data plus general rule. This method of evidential reasoning has been proposed to assist EC project managers and decision makers in formalizing the types of thinking that are required when assessing the current risk environment of their EC development in a systematic manner.

A system has been designed and developed to incorporate is risk analysis model. System evaluation was performed to ascertain whether the system achieved its designed purpose, and the results are satisfactory. The results of the evaluation strongly support the viability of this approach to risk analysis using a direct way of evidential reasoning, and it is demonstrated to be feasible for evaluating EC project risk.

References

1. Addison. T.: E-commerce project development risks: evidence from a Delphi survey, International Journal of Information Management. 1 (2003) 25–40.
2. Bandyopadhyay, K., Mykytyn, P, P., Mykytyn, K.: A framework for integrated risk management in information technology. Management Decision, Vol. 37. No. 5. (1999) 437-444.
3. Carney, D., Morris, E., Patrick, R.: Identifying commercial off-the-shelf (COTS) product risks: the COTS usage risk evaluation. Technical Report, CMU/SEI (2003)-TR-023.
4. Cortellessa, V., Goseva-Popstojanova, K., Kalaivani Appukkutty, Guedem, A.R., Hassan, A., Elnaggar, R., Abdelmoez, W., Ammar, H.H., Model-based performance risk analysis Software Engineering, IEEE Transactions on Vol. 31. Issue 1. (2005) 3 - 20
5. Guan, J. W., Bell, D. A.: Evidence theory and its applications. Vol.1, Studies in Computer Science and Artificial Intelligence 7, Elsevier, The Netherlands, (1991).
6. Guan, J. W., Bell, D. A.: Evidence theory and its applications. Vol.2, Studies in Computer Science and Artificial Intelligence 8, Elsevier, The Netherlands, (1992).
7. Halpern, J., Mehrotra, A. K.: The tangled web of e-commerce: identifying the legal risks of online marketing. The computer Lawyer, Vol. 17, No. 2, 8-14.
8. Kalakota, R., Whinston, A.B.: Frontiers of the Electronic Commerce. , Addison-Wesley, Reading, MA (1996).
9. Leung, H.M., Chuah, K.B., Tummala, V.M.R.: A knowledge-based system for identifying potential project risks. OMEGA: International Journal of Management Science. Vol. 26, Issue. 5. (1998) 623–638.
10. Treese, G. W., Stewart, L. C.: Designing Systems for Internet Commerce, Addison Wesley, Massachusetts.
11. Wat, F. K. T., Ngai, E.W.T., Cheng, T.C.E.: Potential risks to e-commerce development using exploratory factor analysis. International Journal of Services Technology and Management (2004), Vol. 6. Part 1, Pages 55-71.
12. Ngai, E.W.T., Wat, F.K.T.: Fuzzy decision support system for risk analysis in e-commerce development. Journal of Decision Support Systems. Vol. 40. Issue 2. (2005) 235-255
13. Yen, J.: A reasoning model based on an extended Dempster-Shafer theory. Proceedings AAAI-(1986) 125-131.
14. Yen, J.: GERTIS: A Dempster-Shafer Approach to Diagnosing Hierarchical Hypotheses", Communications of the ACM 5 Vol. 32, (1989), 573-585.

A Service-Oriented Modeling Approach for Distributed Management of Multidisciplinary Design Knowledge in the Semantic Grid

W.Y. Zhang[1], L. Zhang[2], and Y. Xie[3]

[1] School of Information, Zhejiang University of Finance & Economics, Hangzhou 310018, China
wyzhang@pmail.ntu.edu.sg
[2] School of Computer Science, Zhejiang University, Hangzhou 310027, China
Zhangli_zju@163.com
[3] School of Mechatronic Engineering, China Jiliang University, Hangzhou 310018, China
xieyue@cjlu.edu.cn

Abstract. This paper describes a preliminary attempt at using Semantic Grid paradigm, especially service-oriented modeling approach, for distributed management of multidisciplinary design knowledge, enabling to add semantics to grid services to endow them with capabilities needed for their successful deployment and reuse in multidisciplinary collaborative design. Domain resource ontology is identified as the key enabler to a meaningful agent communication for collaborative work among multidisciplinary organizations. A semantic-driven workflow representation adopted in the proposed service-oriented modeling framework can speed up the design process of a complex engineering system by composing the process with existing multidisciplinary design resources intelligently. The proposed approach has been evaluated with a multidisciplinary collaborative design example of metal stamping progressive die.

1 Introduction

Today's engineering design community is exhibiting a growing trend towards design processes that are more knowledge-intensive, distributed and collaborative. The increasing complexity of engineering systems, coupled with the fact that disparate design knowledge is often scattered among multidisciplinary organizations and lacks consistency, makes effective capture, retrieval, reuse, sharing and exchange of multidisciplinary design knowledge through knowledge management a critical issue in collaborative product development.

Although the current model-based knowledge management technologies have laid the foundation for the emerging fields of multidisciplinary collaborative design, the heterogeneity of multidisciplinary design knowledge representation is still a major obstacle to sharing and exchanging multidisciplinary design knowledge among multidisciplinary organizations that collaborate over internet.

The recent popularity of Semantic Grid [1] has renewed people's interest in building open, dynamic and adaptive knowledge management systems, with a high

L. Wang et al. (Eds.): FSKD 2006, LNAI 4223, pp. 631 – 640, 2006.

degree of automation, which supports flexible coordination and collaboration on a global scale. Aiming at representing multidisciplinary design knowledge explicitly and formally and sharing it among multiple design agents, this paper describes a preliminary attempt at using Semantic Grid paradigm, especially service-oriented modeling approach for distributed management of multidisciplinary design knowledge, enabling to add semantics to grid services to endow them with capabilities needed for their successful deployment and reuse in multidisciplinary collaborative design. The service-oriented modeling process evolves along five consecutive layers, i.e., knowledge elicitation, product modeling, ontology modeling, workflow planning and knowledge application layers, with diverse knowledge assets wrapped up as grid services to facilitate knowledge consumption and supply in the Semantic Grid. Domain resource ontology is identified as the key enabler to a meaningful agent communication for collaborative work among multidisciplinary organizations. Formal knowledge representation in OWL (Web ontology language) format [2] extends traditional product modeling with capabilities of knowledge sharing and distributed problem solving, and is used as a content language within the FIPA ACL (agent communication language) [3] messages to support cooperation among multiple design agents. Owing to a semantic-driven workflow representation in the proposed service-oriented modeling framework, a workflow of grid services can be constructed to compose a complex engineering design process with existing multidisciplinary design resources that are arranged sequentially, parallelly or iteratively. The feasibility of the semantic-driven workflow planning strategy is manifested using a multidisciplinary collaborative design example of metal stamping progressive die.

The effort in service-oriented modeling will remove one of the most commonly stated criticisms of the suitability of knowledge modeling used for knowledge management, due to the difficulty in building a single flawless model that contains all contextual information at different levels of abstraction. The proposed approach is viewed as a promising knowledge management method that facilitates the implementation of computer supported cooperative work (CSCW) in multidisciplinary design by allowing multiple design agents to share a clear and common understanding to the definition of multidisciplinary design problem and the semantics of exchanged multidisciplinary design knowledge.

2 Related Work

The knowledge management research community has come a long way towards taking a modeling perspective on knowledge engineering. The modeling approach represents an effort to obtain a better understanding, description and representation of the problem. With the modeling approach, development of knowledge management systems can be faster and more efficient through the re-use of existing models for different areas of the same domain. Specifically, the effort at knowledge modeling usually proceeds along mediating representation and ontology modeling, to which the science of knowledge engineering has much to contribute.

The importance of knowledge modeling in knowledge management has been identified in CommonKADS [4], which provides tools for corporate knowledge management and includes methods that perform a detailed analysis of knowledge intensive tasks and processes. A suite of mediating models including organization model, task model, agent model, communication model, expertise model and design model form the core of its systematic knowledge management methodology.

On the other hand, research in the growing field of ontology modeling offers a firm basis for solving knowledge modeling problems. The main motivation behind ontology is to establish standard models, taxonomies, vocabularies and domain terminologies, and use them to allow for sharing and reuse of knowledge bodies in computational form. For example, an ontology-based knowledge modeling system is proposed by Chan [5] to facilitate building an application ontology of a domain by explicitly structuring and formalizing both domain and task knowledge of any industrial problem domain. The system also supports knowledge sharing by converting the ontology into the extensible mark-up language (XML).

Ontologies are also expected to play a major role in the emerging Semantic Web [6]. The Semantic Web possesses a huge potential to overcome knowledge modeling difficulties over the web, by modeling the concepts in a knowledge domain with a high degree of granularity and formal structure including references to mutually agreed-on semantic definitions in ontologies. An example of the use of Semantic Web in knowledge modeling is configuration knowledge representations [7], which compares the requirements of a general configuration ontology with the logics chosen for the Semantic Web, and describes the specific extensions required for the purpose of communicating configuration knowledge between state-of-the-art configurators via Semantic Web languages OIL and DAML+OIL. Because Semantic Web languages are relatively new languages – having only become official W3C standards since 2001 – their use in the engineering field, in particular, multidisciplinary design area has not yet reached the pervasive level that has been seen in the information technology world.

Convergence between the Semantic Web and another recent development in grid computing technologies [8] has seen grid technologies evolving towards the Semantic Grid [1]. The Semantic Grid is an extension of the current grid in which knowledge and services are given well-defined meaning, better enabling intelligent agents to work in cooperation. Ontologies serve as a best vehicle to formally hold a formal, explicit specification (of the knowledge assets) that can be shared within the virtual organizations, and also to enable semantic-driven knowledge modeling on the Semantic Grid. Chen et al. [9] proposed a distributed knowledge management framework for semantics and knowledge creation, population and reuse in the Engineering Grid, better facilitating problem solving in computation and data intensive engineering design optimization involving fluid dynamics.

Notwithstanding the promising results reported from existing research work for model-based knowledge management, there has been little research using the service-oriented modeling approach to support the management of multidisciplinary design knowledge, especially, for Semantic Web or Semantic Grid applications. In addition, most existing approaches lack an ontology-based collaborative product modeling framework that supports a meaningful agent communication for multidisciplinary collaborative design.

3 Distributed Management of Multidisciplinary Design Knowledge in the Semantic Grid

Multidisciplinary collaborative design is a very complex process, which involves plenty of product modeling tools and engineering knowledge from various disciplines at different design phases. However, these resources are often located geographically and represented in heterogeneous formats, makes effective capture, retrieval, reuse, sharing and exchange of knowledge a critical issue in a collaborative design development. The Semantic Grid infrastructure is utilized in this work to enable designers to carry out multidisciplinary collaborative design by seamless access to a state-of-the-art collection of product modeling tools and other knowledge resources around the internet.

In order to manage the multidisciplinary design knowledge in a manner that is explicit, formal, modular, extensible, interoperable, and yet comprehensible, an ontology-based service-oriented modeling approach to the multidisciplinary design knowledge in the Semantic Grid is proposed. It evolves along five consecutive layers, i.e., knowledge elicitation, product modeling, ontology modeling, workflow planning and knowledge application layers, with diverse knowledge assets wrapped up as grid services to facilitate knowledge consumption and supply in the Semantic Grid (Figure 1).

3.1 Knowledge Elicitation Layer

Knowledge elicitation covers the interactions with various knowledge sources such as application database systems, legacy systems, and documents through a set of generic knowledge acquisition services such as data mining service, data conversion service and information extraction service in order to elicit multidisciplinary design knowledge of the domain and produce a federated, distributed description of it, i.e., multidisciplinary design knowledge warehouse.

3.2 Product Modeling Layer

The initially elicited multidisciplinary design knowledge is analyzed in the product modeling layer in order to structure it and develop product models that are used as a communication between domain experts and knowledge engineers, as an aid in structuring and describing the domain-specific multidisciplinary design knowledge independently of any particular implementation. The development of multidisciplinary models will be facilitated by a set of generic product modeling services provided by various product modeling tools and technologies.

3.3 Ontology Modeling Layer

Though various standalone product modeling services are able to describe and distinguish involved disciplinary-specific design knowledge while maintaining efficiency and computability in standalone, one-off product modeling environment, it cannot rigorously and unambiguously capture the semantics of exchanged multidisciplinary design knowledge, therefore prohibiting automated reasoning in multidisciplinary design environments. Towards composing and configuring various

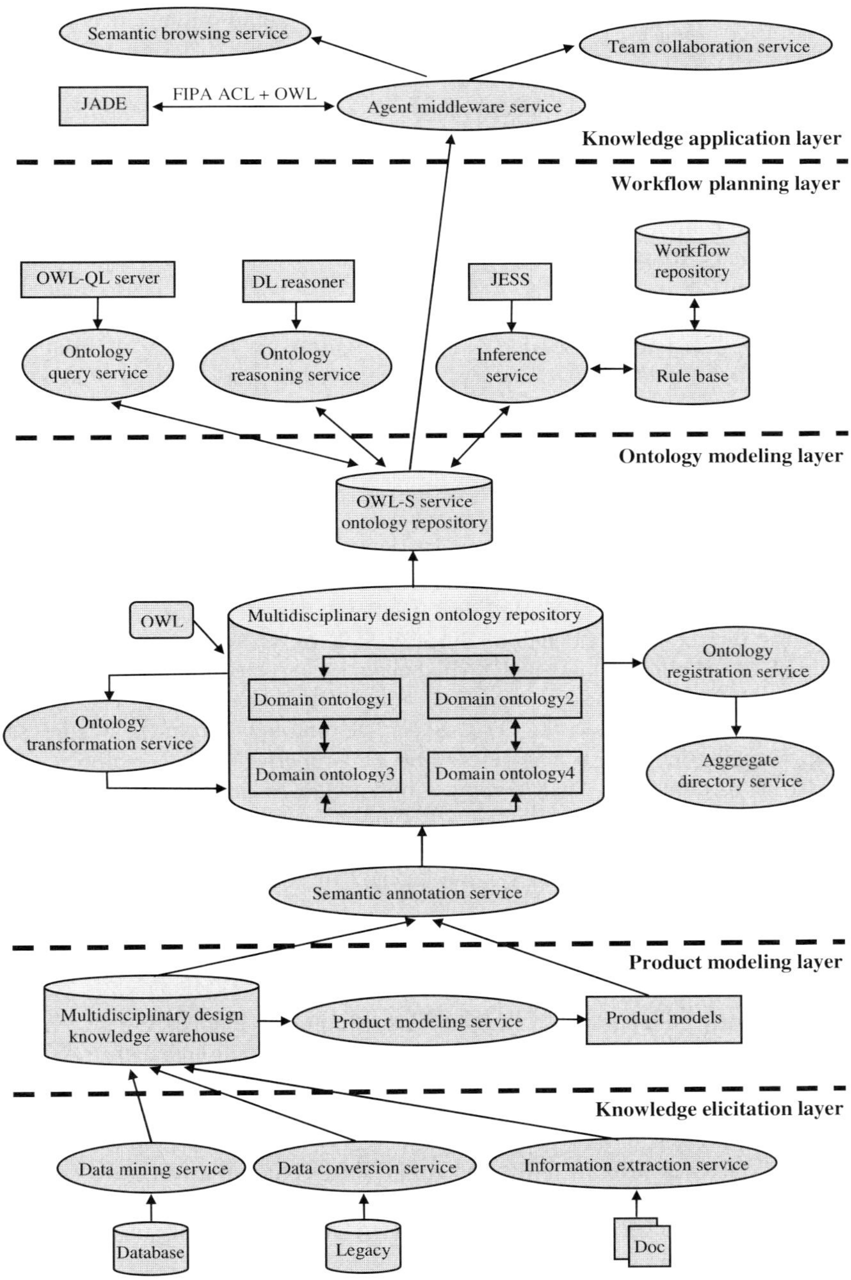

Fig. 1. Service-oriented modeling to the multidisciplinary design knowledge in the Semantic Grid

product modeling services to support distributed, multidisciplinary collaborative design, an ontological description to the multidisciplinary design knowledge is necessary to be exploited in the ontology modeling layer.

The need for rigorous and unambiguous description of multidisciplinary design knowledge can be summarized as a common ontological foundation that supports consistent conceptualization of distributed product design models. The key concepts of multidisciplinary design knowledge are represented as different domain ontologies through a semantic annotation service at different design phases. Further, the multidisciplinary design ontology is built with the formal representation language OWL that is the most expressive semantic markup language up to date in the Semantic Grid.

An ontology registration service is used to register the ontology to an aggregate directory and to notify the directory service of the availability of the required ontology. An ontology transformation service is used to offer the grid infrastructure the capabilities to translate or map information from one ontology to another, and to negotiate meaning or otherwise resolve differences between ontologies.

In the Semantic Grid infrastructure, all resources including a collection of product modeling services, program modules and other knowledge resources act as grid services. Each grid service has a WSDL (Web service description language) interface for service description, is registered in the UDDI (universal description, discovery and integration) service repository, and has a SOAP (simple object access protocol) listener for service implementation. Since WSDL only describes Web services as collections of operation names and XML Schema data type at a syntactic level, it is enriched by adding semantic information with OWL-S [10] service ontology in the proposed modeling framework. OWL-S service ontology provides a core set of markup language constructs for describing the properties and capabilities of grid services in unambiguous and computer-interpretable form. As OWL-S service ontology does not provide complete vocabulary sets for describing specific grid services in various engineering domains, the domain-specific terms and concepts used in OWL-S to describe grid services are defined in domain resource ontology, i.e., multidisciplinary design ontology.

3.4 Workflow Planning Layer

Multidisciplinary collaborative design process of complex engineering systems often involves construction of a workflow either manually or automatically to realize a series of simple design activities that may be heterogeneous and belong to different disciplines. In the Semantic Grid infrastructure this process amounts to discovering existing multidisciplinary design resources, i.e., grid services, each with semantically defined interface and responsibility, and composing them into a workflow of services in a workflow planning layer.

The OWL-S service ontology repository is used to discover and retrieve a semantically matching service available in the network according to the request of the wanted grid service specification. An ontology query service and an ontology reasoning service, which are realized through OWL-QL [11] and DL (description logic) reasoner respectively, control the whole process of service discovery and matchmaking. The ontology query service provides query to the multidisciplinary design concepts, their properties and relationships in the underlying ontology repository, e.g., by returning the properties and relationships (such as parents or children) of a concept using OWL-QL. The DL reasoner provides reasoning capabilities over various knowledge entities in the ontology repository. Any practical DL reasoner such as Racer [12] can be applied to perform common ontological operations such as terminological and assertion reasoning, subsumption checking, navigating concept hierarchies, and so on.

A rule-based inference engine JESS (Java expert system shell) [13] is employed to provide an inference service to compose retrieved grid services into a workflow that is then stored in a workflow repository for workflow reuse. JESS is a rule engine and scripting environment written entirely in Sun's Java language. A rule base is used to contain the workflow planning skills, which may be acquired through experiences in manual workflow planning. The rules include task decomposition rules, service configuration rules, service iteration rules, etc. For example, the task decomposition rules may be used to decompose a complex design process into a series of simple design activities; the service configuration rules may be used to retrieve a coupled grid service from a retrieved grid service; and the service iteration rules may be used to iterate the design process if the executed workflow or partial workflow doesn't satisfy the problem specifications fully. All rules are formulated in the form of IF-THEN formats in CLIPS language [14].

3.5 Knowledge Application Layer

The development of knowledge applications will be facilitated by a set of generic application-level services such as agent middleware service, semantic browsing service and team collaboration service.

A meaningful communication for multi-agent distributed collaborative design is implemented upon a FIPA-compliant Java agent development environment (JADE) [15]. JADE provides an agent middleware service to support the agent representation, agent management and agent communication on the top of OWL-S service ontology. FIPA ACL enables agents to collaborate with each other by setting out the encoding, semantics and pragmatics of the communicating messages. The query request for multidisciplinary design ontology can be transformed from FIPA ACL messages into OWL-QL format, while the multidisciplinary design ontology with OWL format can be encapsulated into FIPA ACL messages to facilitate communication and sharing among multiple agents.

The semantic browsing service allows user to explore the multidisciplinary design ontology and OWL-S service ontology at the semantic level. A widely accepted ontology editor Protégé-2000 with the OWL Plugin [16] is used as the semantic browser to browse ontology, generate ontology graph, and classify new multidisciplinary design concepts.

The team collaboration service is used to provide collaborative work such as the status of collaborative team members, discussion minutes, meeting status, things to do list, project status, etc.

4 An Illustrative Design Example

To demonstrate the feasibility of the proposed approach for multidisciplinary collaborative design, a Java-based software prototype is implemented in a network of PCs with Windows NT/2000. Tomcat™ is adopted to host the XML database endpoint, and Globus Toolkit 4.0 is employed to realize the grid service registry.

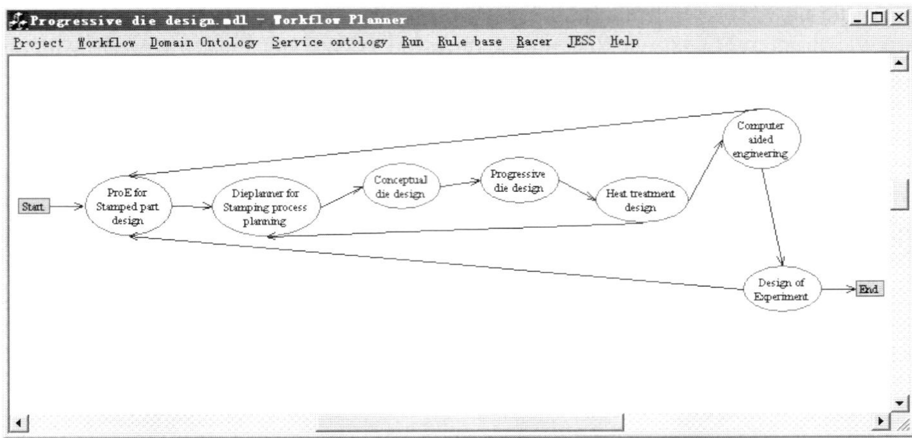

Fig. 2. Workflow of the design example

An illustrative design example of metal stamping progressive die is elaborated, which needs to compose and instantiate a set of grid enabled design resources into an executed workflow. A typical workflow of multidisciplinary collaborative design of metal stamping progressive die consists of stamped part design, stamping process planning, conceptual die design, progressive die design, heat treatment design, computer-aided engineering (CAE) and design of experiment (DOE). These steps might be performed iteratively in order to design a qualified progressive die. Each partial design problem may be solved with different design resources, and each design resource may have different performances with different configuration and instantiation. For example, there are over a hundred CAD tools for stamped part design such as ProE, Autocad and CATIA, each of which is geared to dealing with specific type of engineering circumstance. Even with a selected CAD tool, different instantiations of control parameters may produce different external interfaces and semantics. It is up to the ontology query service, ontology reasoning service and inference service in the workflow planner which CAD service to choose and how to instantiate it. According to a certain design specification of the metal stamping progressive die and based on the existing design resources, figure 2 shows the automatically generated workflow of instantiated grid services. If the CAE or DOE result doesn't satisfy the design specification, the design process will be iterated to the previous steps for redesign. The resulted workflow will be stored in the workflow repository for workflow reuse. If more domain-specific multidisciplinary design ontologies are loaded into the domain resource ontology, the existing workflow needs to be updated. Moreover, when new grid services appear or old ones become unavailable, the existing workflow should be updated with new services too.

5 Conclusion

This paper has presented a service-oriented modeling approach for distributed management of multidisciplinary design knowledge in the Semantic Grid, which aims to populate the multidisciplinary design knowledge with ubiquitous semantic content and facilitate semantic interoperability among multiple agents. Domain resource ontology is identified as the key enabler to a meaningful agent communication for collaborative work among currently available product modeling tools and technologies because of its modularity, extensibility and interoperability. Formal knowledge representation in OWL format extends standalone product modeling with capabilities of knowledge sharing and distributed problem solving, and is used as a content language within the FIPA ACL messages in the multi-agent distributed design environment. A workflow of grid services can be constructed to compose a complex engineering design process with existing multidisciplinary design resources that are arranged sequentially, parallelly or iteratively. The semantic-driven workflow planning strategy was demonstrated by application to the multidisciplinary collaborative design of metal stamping progressive die. However it is not limited to this task domain only; the methodology is also applicable to other task domains.

Our future work will look into developing and publishing diverse domain-specific multidisciplinary design ontologies in OWL format using the proposed approach, in order to capture an extensive set of annotations of general multidisciplinary design knowledge with a community-wide agreement. As a result, more and more standalone, one-off, locally stored product modeling frameworks can be federated, integrated, and consumed by multiple agents in the Semantic Grid.

Acknowledgement. This work was supported in part by Zhejiang Natural Science Fund of China (ZJNSF) (Y105003) and Zhejiang Provincial Education Department's Specialized Project of China (20051056).

References

1. de Roure D. and Jennings, N.: The Semantic Grid: Past, Present, and Future. Proceedings of the IEEE, 93 (2005) 669-681
2. McGuinness, D.L. and Harmelen, F.V.: OWL Web Ontology Language Overview. http://www.w3.org/TR/2004/REC-owl-features-20040210/, March 20 (2004)
3. Foundation for Intelligent Physical Agents: FIPA Specifications. http://www.fipa.org/specifications/ (2002)
4. Schreiber, G., Akkermans, H., Anjewierden, A., et al.: Knowledge Engineering and Management: The CommonKADS Methodology. MIT Press, Massachusetts (1999)
5. Chan, C.W.: From Knowledge Modeling to Ontology Construction. International Journal of Software Engineering and Knowledge Engineering 14 (2004) 603-624
6. Berners-Lee, T., Hendler, J. and Lassila, O.: The Semantic Web. Scientific American 284 (2001) 34-43

7. Felfernig, A., Friedrich, G., Jannach, D., et al.: Configuration Knowledge Representations for Semantic Web Applications. Artificial Intelligence for Engineering Design, Analysis and Manufacturing: Aiedam 17 (2003) 31-50
8. Foster, I. and Kesselman, C.: The Grid: Blueprint for a New Computing Infrastructure. Morgan Kaufmann (1999)
9. Chen, L., Shadbolt, N., Goble, C., et al.: Semantic-assisted Problem Solving on the Semantic Grid. Journal of Computational Intelligence, Special Issue 21 (2005) 157-176
10. OWL-S: http://www.daml.org/services/owl-s/1.1/ (2004)
11. Fikes, R., Hayes, P. and Horrocks, I.: OWL-QL – A Language for Deductive Query Answering on the Semantic Web. Knowledge Systems Laboratory, Stanford University, Stanford, CA (2003)
12. Haarslev, V. and Moller, R.: Racer: A Core Inference Engine for the Semantic Web. In: Proceedings of the 2nd International Workshop on Evaluation of Ontology-Based Tools (2003) 27-36
13. Rriedman-Hill, E. J.: JESS, the Expert System Shell for the Java Platform. http://herzberg.ca.sandia.gov/jess (2002)
14. Giarratano, J. and Riley, G.: Expert Systems: Principles and Programming, 3rd ed. PWS, Boston (1998)
15. Bellifemine, F., Poggi, A. and Rimassa, G.: Developing Multi Agent Systems with a FIPA-compliant Agent Framework. Software Practice & Experience 31 (2001) 103-128
16. Knublauch, H., Musen, M.A. and Rector, A.L.: Editing Description Logics Ontologies with the Protégé OWL Plugin. In: International Workshop on Description Logics. Whistler, BC, Canada (2004)

Batch Scheduling with a Common Due Window on a Single Machine

Hongluan Zhao[1,⋆], Fasheng Hu[1], and Guojun Li[1,2]

[1] School of Math. and Sys. Sci., Shandong University, Jinan, China
honglzhao@163.com
[2] Computational Systems and Biology Lab., Department of Biochemistry and Molecular Biology, The University of Georgia, USA

Abstract. A common due window scheduling problem with batching on a single machine is dealt with to minimize the total penalty of weighted earliness and tardiness. The processing time of a batch is defined as the longest processing time among the jobs in the batch. This model is motivated by applications in the manufacturing of integrated circuits. Based on several optimal properties, a polynomial algorithm is proposed for given due window.

Keywords: Scheduling; Batch; Due Window; Earliness; Tardiness.

1 Introduction

In the recent years, JIT (Just-In-Time) sequencing and scheduling problems have attracted the attention of many researchers. With the development of manufacturing, the conception of due window is important since most due dates are specified with some tolerances. It is a time interval defined by an early due date and a tardy due date. In the other hand, there has been significant interest in scheduling problems that involve an element of batching. The motivation is mainly for efficiency, since it may be cheaper or faster to process jobs in a batch than to process them individually. A batch is a set of jobs processed simultaneously and completed together when the processing of all jobs in the batch is finished. Therefore, the processing time of each batch equals to the longest processing time among all the products assigned into it.

In this paper, we extend due window scheduling to the situation with jobs processed in batches. A single machine batching scheduling problem is considered to minimize the total penalty of the weighted earliness and tardiness.

In the recent twenty years, many results are concerned with earliness and tardiness penalty, but most of them on due date constraint surveyed in [2]. Articles on window scheduling problems are limited and almost about common due window. Further, some batching scheduling papers are effused in the last decade reviewed by [5], but most is about regular objective function and only a few of them consider the existence of due dates.

⋆ Corresponding author.

L. Wang et al. (Eds.): FSKD 2006, LNAI 4223, pp. 641–645, 2006.

2 Description and Optimal Algorithm

Given is a set $J = \{J_1, J_2, \cdots, J_n\}$ of n jobs processed on a batch processing machine with batch capacity unlimited. Let p_i be the processing time of job $J_i \in J$. Once the processing of a batch is initiated, it cannot be interrupted, whose processing time equals to the largest processing time of the jobs assigned to it. For batch B, its processing time $p(B) = \max\{p_i | \ J_i \in B\}$.

A common due window is given by an early due date e and a tardy due date d with the window size $K = d - e$. Let S_i, C_i be the starting time and completion time of J_i respectively, $i = 1, 2, \cdots, n$. Then for the batch B including J_i, its start time is $S(B) = S_i$ and the completion time is $C(B) = C_i$. The earliness and tardiness of job J_i are $E_i = \max\{0, e - C_i\}$ and $T_i = \max\{0, C_i - d\}$ respectively. The objective function for schedule σ is defined as

$$Z(\sigma) = \sum_{i=1}^{n} (\alpha E_i + \beta T_i),$$

where α and β are penalty coefficients and job-independent. Assume that all parameters are positive integers. Our goal is to partition the jobs into batches and schedule the batches to minimize $Z(\sigma)$. In schedule σ, the early set, window set and tardy set are defined respectively as:

$$E(\sigma) = \{J_i | \ C_i < e\}, \quad W(\sigma) = \{J_i | \ e \le C_i \le d\}, \quad T(\sigma) = \{J_i | \ C_i > d\}.$$

They are denoted by E, W and T respectively when without confusion. Since the jobs in one batch have the same completion time, the batch is also defined in the corresponding set of its jobs for simplicity. Obviously, in an optimal schedule, no idle time is inserted between the starting time of the first batch and the completion time of the last batch. Several properties of an optimal schedule are presented in the following.

Property 1. *There exists an optimal schedule σ, where the batches in $W(\sigma)$ contain the smallest jobs among all n jobs.*

Property 2. *In any optimal schedule σ, $W(\sigma) = \emptyset$ or it contains only one batch, whose processing time is the smallest of all batches.*

Property 3. *In an optimal schedule σ, if $W(\sigma) \neq \emptyset$ or the processing time of the first batch in $T(\sigma)$ is not larger than d, then $E(\sigma) = \emptyset$.*

Property 3 implies that whenever $E(\sigma) \neq \emptyset$, we have $W(\sigma) = \emptyset$ and the processing time of the first batch in $T(\sigma)$ is larger than d. Keep in mind that the processing time of the only batch in $E(\sigma)$ is smaller than that of the first batch in $T(\sigma)$. For an optimal schedule, denote the batch in window set as B_W and that of early set as B_E. Further, we have

Property 4. *In any optimal schedule σ, if $E(\sigma) \neq \emptyset$, then the start time of B_E is 0 or $e - p(B_E) - 1$.*

Proof. Since $E(\sigma) \neq \emptyset$, there must have $W(\sigma) = \emptyset$. Let $\{B_1, B_2, \cdots, B_m\}$ be the batch sequence in $T(\sigma)$. For a set A, $\mid A \mid$ denotes how many jobs are contained in A. Then

$$\begin{aligned}Z(\sigma) &= \alpha(e - C(B_E)) \times |B_E| + \textstyle\sum_i \beta T_i \\ &= \alpha(e - C(B_E)) \times |B_E| + \beta\{[C(B_E) + p(B_1) - d] \times |B_1| + \\ &\quad [C(B_E) + p(B_1) + p(B_2) - d] \times |B_2| + \cdots + \\ &\quad [C(B_E) + p(B_1) + \cdots + p(B_m) - d] \times |B_m|\} \\ &= \beta\{[|T| \times p(B_1)] + [(|T| - |B_1|) \times p(B_2)] + \cdots + [|B_m| \times p(B_m)] \\ &\quad + |T| \times C(B_E)\} + \alpha(e - C(B_E)) \times |B_E| - |T|\beta d \\ &= \beta\{[|T| \times p(B_1)] + [(|T| - |B_1|) \times p(B_2)] + \cdots + [|B_m| \times p(B_m)]\} \\ &\quad + \alpha e \times |B_E| - |T|\beta d + (|T|\beta - |B_E|\alpha) \times (S(B_E) + p(B_E))\end{aligned}$$

Thus, it is linear about the start time $S(B_E)$. Additionally, when the early batch B_E is determined, we have $C(B_E) \leq e - 1$. In a result, $S(B_E)$ is equal to 0 if $|T|\beta \geq |B_E|\alpha$ and is $e - p(B_E) - 1$ otherwise. □

For the situation without batch processing, the jobs in tardy set are sequenced in a nondecreasing order of processing times (SPT order) in an optimal schedule. Similarly, the following extended argument occurs.

Property 5. *In an optimal schedule, the batches in tardy set are sequenced in nondecreasing order of their processing times.*

Assume that the jobs are indexed according to the SPT order such that $p_1 \leq p_2 \leq \cdots \leq p_n$. An *SPT-batch schedule* is one in which adjacent jobs in the sequence $\{J_1, J_2, \cdots, J_n\}$ may be grouped to form batches. Obviously, an optimal schedule is SPT-batch rule. It is specified by the jobs that start intuitively their batches. That is to say, if we know the start job of each batch, the schedule is determined. For an optimal schedule, let the batch sequence in tardy set be $\{B_1, B_2, \cdots, B_m\}$.

Even if $p_i \leq d$, J_i is not necessarily in B_W since it influences the tardy penalties of $\mid T \mid$ jobs. Once the elements of batch B_W are confirmed, shift it to left most possibly only if $C(B_W) \geq e$ in order to reduce the tardiness mostly of tardy jobs. If $W \neq \emptyset$, there must have $E = \emptyset$ and the objective function is

$$Z(\sigma) = \sum_{i=1}^{n} (\alpha E_i + \beta T_i) = \beta \sum_{J_i \in J \setminus W} (C_i - d).$$

On the other hand, if $E \neq \emptyset$, there must have $W = \emptyset$ and the penalty function becomes

$$\begin{aligned}Z(\sigma) &= \sum_{i=1}^{n} (\alpha E_i + \beta T_i) \\ &= \alpha(e - S(B_E) - p(B_E)) \times |B_E| + \beta \sum_{J_i \in J \setminus E} (C_i - d).\end{aligned}$$

Recall that the start time of B_E is 0 or $e - p(B_E) - 1$ in an optimal schedule and that the batch B_E contains the smallest jobs.

Above all, the objective penalty is to minimize the total completion time of tardy jobs after batch B_E or B_W is fixed. Assume that the processing of the first tardy batch starts at time t. Suppose that a batch $\{J_j, \cdots, J_{k-1}\}$ is inserted at the start of tardy set for jobs $J_k, \cdots, J_n$. The total completion time of jobs $J_k, \cdots, J_n$ increases by $(n-k+1)p_{k-1}$, while the total completion time for jobs $J_j, \cdots, J_{k-1}$ is $(k-j)(t+p_{k-1})$. Thus, the overall increase in total completion time is $(n-j+1)p_{k-1}+(k-j)t$. Then we will use a dynamic programming to determine the tardy batches for an optimal schedule as [1]. Without loss of generality, assume that $p_1 \leq d$. Suppose that job J_0 is a fictitious job and $p_0 = 0$.

Algorithm

Step 1.1. Sort the jobs in SPT order such that $p_1 \leq p_2 \leq \cdots \leq p_n$.

Step 1.2. Suppose that $E = \emptyset$ and initialize $i = 0$. Let $B_W = \{J_0, J_1, \cdots, J_i\}$ where $p_i \leq d$ and shift it to the left mostly only if $C(B_W) \geq e$. Set $t = C(B_W)$.

Step 1.3. Let G_j be the minimum total completion time for SPT-batch schedule containing jobs $J_j, \cdots, J_n$. Taking as initialization $G_{n+1} = t$, we calculate G_{i+1} by the following recursion for $j = n, n-1, \cdots, i+1$,

$$G_j = \min_{k=j+1,\cdots,n+1}\{G_k + (n-j+1)p_{k-1} + (k-j)t\}.$$

Step 1.4. If $i+1 \leq n$ and $p_{i+1} \leq d$, set $i = i+1$ and go to Step 1.2. Otherwise, find the smallest index $s \in \{1, 2, \ldots, i\}$ such that the total penalty $Z = \beta G_{s+1} - \beta(n-s)d$ is minimized.

Step 1.5. Suppose that $E \neq \emptyset$. In this case we have $W = \emptyset$. Let $B_E = \{J_1, \cdots, J_i\}$ for $i = 1, \cdots, n$ where $p_i < e$. Then set $S(B_E) = 0$ if $(n-i)\beta \geq i\alpha$ and $S(B_E) = e - p_i - 1$ otherwise. Initialize $G'_{n+1} = t = S(B_E) + p_i$ and calculate G'_{i+1} by the below recursion for $j = n, n-1, \cdots, i+1$,

$$G'_j = \min_{k=j+1,\cdots,n+1}\{G_k + (n-j+1)p_{k-1} + (k-j)t\},$$

until $p_{i+1} \geq e$. The index $s \in \{1, 2, \ldots, i\}$ is chosen such that the total penalty $Z' = \alpha s(e-t) + \beta G'_{s+1} - \beta(n-s)d$ minimized.

Step 1.6. The optimal value is the smaller one of Z and Z', and the corresponding optimal schedule is found by backtracking. □

By simple computationally time argument, we see that the time complexity is $O(n^2)$ in Step 1.3 and Step 1.5. Thus the total time complexity can be upbounded by $O(n^3)$.

Theorem 1. *For a given due window, Algorithm always outputs an optimum schedule in time* $O(n^3)$.

However, since the dynamic program of our algorithm has a structure that admits geometric techniques, the time complexity can be reduced to $O(n^2 \log n)$.

3 Conclusion

We have investigated a common due window scheduling problem on a batching machine to minimize the total penalty of weighted earliness and tardiness. Based on some properties, an optimal algorithm is presented.

References

[1] P. Brucker, A. Gladky, H. Hoogeveen, et al. Scheduling a batching machine. *Journal of Scheduling*, 1, (1998), 31-54.

[2] V. Gordon, J. M. Proth, C. Chu. A survey of the state-of-the-art of common due date assignment and scheduling research. *European Journal of Operational Research*, 139, (2002), 1-25.

[3] M.Y. Kovalyov. Batch scheduling and common due date assignment problem:An NP-hard case. *Discrete Applied Mathematics*, 80, (1997), 251-254.

[4] F. J. Kramer, C. Y. Lee. Common due window scheduling. *Production and Operations Management*, 2, (1993), 262-275.

[5] C.N. Potts, M.Y. Kovalyov. Scheduling with batching: a review. *European Journal of Operational Research*, 120, (2000) 228-249.

A Secure and Efficient Secret Sharing Scheme with General Access Structures

Liao-Jun Pang[1], Hui-Xian Li[2], and Yu-Min Wang[*]

[1] The Ministry of Edu. Key Lab. of Computer Networks and Information Security, Xidian Univ., Xi'an 710071, China
ljpang@mail.xidian.edu.cn, ymwang@xidian.edu.cn

[2] Department of Computer Science and Engineering, Dalian Univ. of Technology, Dalian, 116024, China
hxli@student.dlut.edu.cn

Abstract. A new secret sharing scheme with general access structures was proposed, which is based on Shamir's secret sharing scheme and the discrete logarithm problem. In this scheme, the dealer need not send any secret information to participants. And the shared secret, the participant set and the access structure can be changed dynamically without updating any participant's secret shadow. The degree of the used Lagrange interpolation polynomial is only one, which makes the computational complexity of the proposed scheme very low. The proposed scheme has advantages over the existing schemes and thus provides great capabilities for many applications.

1 Introduction

The notion of secret sharing was introduced by Shamir [1] and Blakley [2], and their schemes are called (*k*, *n*)-threshold schemes. Benaloh *et al.* [3] pointed out that a threshold scheme could only handle a small fraction of the secret sharing idea, and proposed a secret sharing scheme with general access structures. Hwang *et al.* [4] also proposed a secret sharing scheme with general access structures, in which each participant's secret shadow is selected by the participant himself, and the shared secret, the participant set and the access structure can be changed dynamically without updating any participant's secret shadow. This scheme is very useful, but its computational complexity is too large, which will have a negative effect on its practical application. In this paper, we shall propose a secret sharing scheme with general access structures, which is also based on Shamir's secret sharing. Because the used Lagrange interpolation polynomial is only of degree one, compared with Hwang *et al.*'s scheme, the proposed scheme is more efficient and easier to implement.

The rest of this paper is organized as follows. In Section 2, we shall present our secret sharing scheme. In Section 3, we shall analyze the security of the proposed scheme. Finally, we shall come to our conclusion in Section 4.

[*] This work is supported by the National Key 973 Project of China (G1999035805).

L. Wang et al. (Eds.): FSKD 2006, LNAI 4223, pp. 646 – 649, 2006.

2 The Proposed Scheme

The propose scheme can be used to share multiple secrets without updating participants' secret shadows. The process of sharing each secret is similar, so here we only introduce our scheme for sharing a single secret. Assume that $P = \{P_1, P_2, \ldots, P_n\}$ is a set of n participants and $\Gamma = \{\gamma_1, \gamma_2, \cdots, \gamma_t\}$ is the access structure. Additionally, a public bulletin [5] is required in the proposed scheme.

2.1 Initialization Phase

At first, the dealer selects two strong prime numbers [6], p_1 and p_2. Let m denote the multiplication of p_1 and p_2. In succession, randomly select an integer g from $[m^{1/2}, m]$ such that $g \neq p_1$ or p_2, and another prime q larger than m. Then, the dealer publishes g, m and q on the public bulletin.

After g, m and q are published, each participant P_i randomly selects an integer x_i from $[2, m]$ and computes $y_i = g^{x_i} \bmod m$. Keep x_i secretly and deliver y_i to the dealer.

At last, the dealer should publish each y_i on the public bulletin.

2.2 Secret Distribution Phase

The dealer can perform as follows to share a secret s among these participants:

Firstly, randomly select an integer x_0 from $[2, m]$ such that x_0 is relatively prime to (p_1-1) and (p_2-1), and compute $y_0 = g^{x_0} \bmod m$. Then, find another integer h such that $x_0 \times h = 1 \bmod \phi(m)$, where $\phi(m)$ is the Euler function [18]. In succession, select an integer a from $[1, q-1]$ randomly, and construct a $1st$ degree polynomial $f(x)=s+ax$. At the same time, select t distinct random integers, $d_1, d_2, \ldots, d_t$, from $[1, q-1]$, to denote these t qualified subsets in Γ, respectively. Compute $f(1)$, and for each subset $\gamma_j = \{P_{1j}, P_{2j}, \ldots, P_{dj}\}$ in Γ, compute $H_j = f(d_j) \oplus (y_{1j}^{x_0} \bmod m) \oplus (y_{2j}^{x_0} \bmod m) \oplus \ldots \oplus (y_{dj}^{x_0} \bmod m)$. At last, publish y_0, h, $f(1)$, $H_1, H_2, \ldots, H_t$, $d_1, d_2, \ldots, d_t$ on the public bulletin.

2.3 Secret Reconstruction Phase

All participants of any subset γ_j can cooperate to reconstruct the secret s. Without loss of generality, we assume that the participants of $\gamma_j = \{P_{1j}, P_{2j}, \ldots, P_{dj}\}$ want to reconstruct s. The reconstruction procedure is showed as follows:

Firstly, each participant P_{ij} computes $x_{ij}' = y_0^{x_{ij}} \bmod m$ by using the public information y_0 and his secret shadow x_{ij}, and then delivers it to the designated combiner. With these values, the combiner can computes $H_j' = H_j \oplus x_{1j}'' \oplus x_{2j}' \oplus \ldots \oplus x_{dj}'$. Using $f(1)$, d_j and H_j', he can reconstruct $f(x) = xf(1) - xH_j' - d_i f(1) + H_j'(1-d_i)^{-1}$. Finally, the shared secret can be recovered by computing $s = f(0) \bmod q$.

In this scheme, the combiner can check whether x_{ij}' is true or not by the equation $x_{ij}'^{h}=y_{ij} \bmod m$, because $x_{ij}'^{h}=(y_0^{x_{ij}})^h=(g^{x_0 x_{ij}})^h=(g^{x_0 h})^{x_{ij}}=g^{x_{ij}}=y_{ij} \bmod m$.

3 Analyses and Discussions

3.1 Security Analyses

In order to get the secret, the $1st$ polynomial $f(x)$ should be reconstructed firstly, which needs two distinct points satisfying $y=f(x)$. Using the public information, one can easily obtain the point $(1, f(1))$. All participants of each qualified subset γ_j can cooperate to get another point $(d_j, H_j')=(d_j, f(d_j))$, but the participants of any unqualified subset cannot. Using only one point to reconstruct the $1st$ polynomial $f(x)$ will face the difficulty to break Shamir's secret sharing scheme.

Additionally, in the secret reconstruction phase, each participant P_{ij} in γ_j only provides a public value x_{ij}' generated by the formula $x_{ij}'=y_0^{x_{ij}} \bmod m$, so he does not have to disclose his secret shadow x_{ij}. Anyone who wants to get the participant's secret shadow x_{ij} from x_{ij}' will face the difficulty in solving the discrete logarithm problem [6]. The reuse of the secret shadow in our scheme is secure. Similarly, it is computationally impossible to derive the secret shadow x_i of each participant P_i from his public information y_i generated by the formula $y_i=g^{x_i} \bmod m$.

Through the security analyses discussed above, it is concluded that our scheme is a computationally secure one.

3.2 Renew Process

The shared secret, the participant set and the access structure can be changed dynamically without updating any participant's shadow. We shall discuss this problem in the following.

Firstly, to alter the shared secret, all the dealer needs to do is to update some public information of the shared secret on the public bulletin. If a new qualified subset γ_{d+1} needs to be added, the dealer should randomly select an integer d_{d+1} for γ_{d+1} and to compute H_{d+1} in the secret distribution. Then, publish the information d_{d+1} and H_{d+1}. However, if an old qualified subset γ_j needs to be cancelled, the only thing the dealer needs to do is to delete the information d_j and H_j. If a new participant P_{n+1} needs to be absorbed, P_{n+1} should select a random integer x_{n+1} and then compute $y_{n+1}=g^{x_{n+1}} \bmod m$. In succession, he/she keeps x_{n+1} secretly and delivers y_{n+1} to the dealer for publication. On the contrary, if an old participant P_i needs to be disenrolled, then the dealer should remove the information about P_i from the public bulletin.

3.3 Performance Analyses

In shamir's secret sharing scheme and those schemes based on shamir's secret sharing, the most time-consuming operation in the proposed scheme is the polynomial interpolation computation. In the proposed scheme, the degree of the used Lagrange polynomial $f(x)$ is only 1, and we can construct $f(x)$ only by four multiplication operations and four addition (or subtraction) operations. Therefore, the proposed scheme is very efficient and easy to implement.

4 Conclusions

In this paper, we propose a new secret sharing scheme with general access structures, in which the dealer need not send any secret information to each participant, and the shared secret, the participant set and the access structure can be changed dynamically without updating any participant's shadow. Analyses show that the proposed scheme has the advantages of the existing schemes and overcomes their disadvantages, which should make this scheme find more extensive applications than other schemes do.

References

1. A. Shamir. How to share a secret. Communications of the ACM 22 (1979), pp.612-613.
2. G. Blakley. Safeguarding cryptographic keys. Proc. AFIPS 1979 Natl. Conf., New York, 1979, pp. 313-317.
3. J. Benaloh, J. Leichter. Generalized secret sharing and monotone functions. Advance in Cryptology-Crypto'88, Springer-Verlag, Berlin, 1990, pp.27-35.
4. Hwang Ren-Junn, Chang Chin-Chen. An on-line secret sharing scheme for multi-secrets. Computer Communications, 21(13), 1998, pp. 1170-1176.
5. Pang Liao-jun, Wang Yu-min. A new (t, n) multi-secret sharing scheme based on Shamir's secret sharing. Applied Mathematics and Computation, 167(2), 2005, pp. 840-848.
6. R.L. Rivest, A. Shamir, L. Adleman. A method for obtaining digital signatures and public key cryptosystem. Communication of ACM, 1978, 21, pp. 120-126
7. K.H. Rosen. Elementary Number Theory and Its Applications. Addison-Wesley, MA, 1993.

Content-Based Information Security Technique for Chinese Text

Wenyin Zhang

College of Information, Linyi Normal University,
Shandong Linyi 276005, China
zwenyin@126.com

Abstract. With the development of E-government and E-commerce in China, more and more attention has been paid to the protection of Chinese information security such as authenticity, integrality, confidentiality as well as copyrights. In this paper, a new technique for Chinese text information security protection is provided based on the thought of the mathematical expression of a Chinese character. The proposed method embeds the watermarking signals into some Chinese characters with occlusive components by readjusting the size of the closed rectangular regions in these components. The algorithm is very simple and totally based on the content. Experiments show that the proposed text watermarking technique is more robust and transparent than the counterpart methods. It will play an important role in protecting the security of Chinese documents over Internet.

Keyword: Information security, Copyright Protection,Chinese Text Watermarking.

1 Introduction

With the fast development of E-government and E-commerce in China, large amount of Chinese information is published and distributed over Internet. How to protect these Chinese information such as the authenticity, integrality and confidentiality of government documents or commerce bills has attracted much concerns. Copyright protection of Chinese Texts is also of importance. In order to solve these problem, nowadays more and more attention has been paid to digital watermarking technology.

Digital watermarking, formed in 1990s, is a new research field in information security [1,2,3]. As an important approach to ensure the security of information, digital watermarking can overcome many disadvantages of encryption. For example, if the encrypted messages are decrypted, the decrypted messages will be open to everyone, and will be short of protection from unauthorized copy, unauthorized publication, unauthorized access, and vicious tampering. So, digital watermarking has gained a large international interest and may be widely used in attestation, copyright protection, labelling, monitoring, tamper proofing, conditional access, national defense, national security, and etc.

L. Wang et al. (Eds.): FSKD 2006, LNAI 4223, pp. 650–659, 2006.

Chinese Text watermarking techniques are used to protect the Chinese information security. Unlike other media watermarking which make use of the redundant information of their host media as well as the characteristics of human perceptual system, text watermarking techniques are different from that of none-text watermarking because text has no redundant information. Therefore, text watermarking algorithm is very difficult to satisfy the requirements of transparency (invisibility or imperceptibility), robustness and etc.

The research of text watermarking may be dated to 1993. The IEEE Journal on Selected Areas in Communication was issued on Internet by embedding text watermark in its articles to protect its copyright in September of 1995 [4]. There were 1,200 registered readers of this issue in the first month. A special issue of IEEE Journal on Selected Areas in Communication on copyright protection was issued in 1998. Since 1993, some text watermarking techniques has been put forward. Line-shift coding, word-shift coding, and character coding are three main approaches adopted in text watermarking [4,5,6]. In line-shift coding approach[4], a mark is embedded on a page by vertically displacing an entire text line. A line is moved up or down, while the line immediately above or below (or both) are left unmoved. These unmoved adjacent lines can be served as reference locations in the decoding process. In word-shift coding approach [5], a mark is embedded by horizontally shifting the location of a word within a text line. A word is displaced left or right, while the words immediately adjacent are left unmoved. These unmoved words can then serve as reference locations in the decoding process. Character coding approach [6] is a class of techniques which embed a mark by altering a particular feature of an individual character. Examples of possible feature alterations include a change to an individual characters' height or its position relative to other characters. Once again, some character features are left unchanged to facilitate decoding. For example, a detection algorithm might compare the height of a hypothetically altered character with that of another unmodified instance of the same character elsewhere on the page. Since line-shift, word-shift, and character coding can not be implemented in unformatted (plain) text file such as TXT, these techniques can not be used to embed a watermark into plain text file. Other techniques such as adding some blanks at the end of a line and so on have been developed to embed a watermark into plain text file [4]. Each technique enjoys certain advantages or applicability, but some of them are not adaptive to Chinese Text, because Chinese characters are different from western letters. In the same time, since all the marks embedded by these techniques are embedded outside the true content of the text, the marks can be attacked or deleted easily, these marking techniques are not very robust.

Unlike the techniques discussed above which embed watermarking signals outside the content of a text, the semantics-based word replacement technique can embed the watermark signals into the true content of a text. But the semantics-based word replacement technique is mainly on replacing words with their synonyms. In the research of semantics-based word replacement techniques, M.J. Atallah proposed a technique based on TMR (Text meaning representation)[7].

Though TMR-based technique is a good idea, it is hard to be implemented since a computer can not understand the meaning of a text fully correctly. Furthermore, Chinese semantics and word parsing are more difficult than that of western language. Therefore, generally speaking, Chinese text watermarking is more difficult than English text watermarking.

Through deep analysis on the structure of Chinese characters, the literature [8] presented a new Chinese text watermarking algorithm based on the thought of the mathematical expressions [9]. The method choose left-right structured Chinese characters and automatically adjust the width of the two components and the space between the two components to embed watermarking information into the true content of a formatted Chinese text documents. Compared with other approaches such as [4,5,6], it is totally based on the text content, and achieve more robustness and transparency. To some extent, it breaks through the difficulties of Chinese text watermarking, though it maybe affects the aesthetic feeling of Chinese characters because of the changes to characters.

In this paper, we continue to explore the Chinese text watermarking techniques based on the mathematical expressions and its automatic generation [10]. Here, we want to embed the watermarking information into Chinese text without affecting the integrality and aesthetic feeling of Chinese characters. According to the structure knowledge of all the Chinese characters mined by mathematical expressions, we know that among the most commonly used about 6000 Chinese characters in the National Standard GB2312-80 of China, there are more than 2800 characters whose components have one or more occlusive components with closed rectangular regions, and most of them are in common use. By experiments, we have found that properly adjusting the size of these regions is suited for embedding the watermarking information with much more robustness and transparency than the method [8]. In this paper, we will describe this new watermarking technique in detail.

This paper is organized as follows: In Section 2, we will present the thought of the mathematical expression of a Chinese character and its automatic generation algorithm briefly. Section 3 will expound the principle and algorithm of our text watermarking technique. The experiment results will be presented in Section 4. Finally, Section 5 concludes the paper.

2 The Mathematical Expressions of Chinese Characters and Their Automatic Generation

The mathematical expression of Chinese characters is a novel mathematical method to express Chinese characters based on deep analysis of knowledge about character structure. In this method, a Chinese character is expressed into a mathematical expression in which the components of Chinese characters act as operands and the position relationships between two components act as operators which satisfy some certain operation laws, just like general math expressions.

In [9], based on the statistical analysis, 505 basic components are selected from the Chinese characters in Class 1 and Class 2 which are authorized by the

National Standard GB2312-80 of China, and six spatial relations of two components are needed to be defined as the operators in the mathematical expression. These six operators are lr, ud, ld, lu, ru and we which represent respectively the spatial relation of left-right, up-down, left-down, left-upper, right-upper, and whole-enclosed defined strictly in [9]. Some of the selected basic components and their serial numbers are shown in Figure 1, and the intuitive explanation of six operators according to the component positions is shown by the Figure 2.

Fig. 1. Some of the basic components and their serial numbers

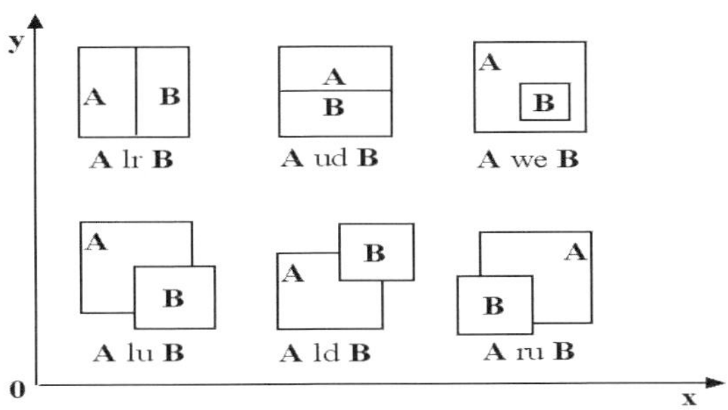

Fig. 2. Intuitive explanation of six operators

The method is simple and suitable for Chinese characters. With this method, Chinese characters can be processed like English letters. Some applications have been achieved, such as platform-spanning transmission of Chinese information, knowledge mining of character structure and automatic generation of character shapes.

In [10], based on the work [9], we presented the method of automatic generation of mathematical expressions by image processing. In order to the convenience of computer processing, we selected subjectively about 500 components and extracted 9 features for every component: Connectivity number, Occlusiveness number, Extremity number, Inflexion number, Joint number, Cross number and other three statistic features. According to these features, we introduced the approach to separate a Chinese character into its components, and realized the automatic generation of mathematical expressions by the components recognition. We did experiments to more than 6000 Chinese characters in GB2312-80, and correct generation ratio is 92.7 percent.

The work on the mathematical expressions facilitates the processing of Chinese information better than ever in the fields of Internet transmission and has been applied to the mobile communication of Chinese information. The structure knowledge and beautiful shapes of Chinese characters can be obtained. In addition, present Chinese literature can be automatically translated into mathematical expressions for storage and transmission. Furthermore, some encryption of Chinese information may be developed.

In this paper, we use some of the automatic generation algorithms of mathematical expressions to embed watermark into Chinese text. In order to make the following sections easily understood, some of the definitions are selected here.

Definition 1. *Component*
A basic component is composed of several strokes, and it may be a Chinese character or a part of a Chinese character.

Definition 2. *Component relations*
Let A,B be two components, B lr A , B ud A , B ld A , B lu A , B ru A and B we A represent that A and B have the spatial relation of lr, ud, ld, lu, ru *and we respectively.*

Definition 3. *Occlusive component*
A component is occlusive when it has one or more hollow closing regions. The number of hollow occlusive regions (holes) is defined as occlusiveness number.

Definition 4. *Minimal Rectangle of a component*
A rectangle is defined as a minimal rectangle of a component, if it is a minimal one and can can envelop the component.

3 The Proposed Watermarking Technique of Chinese Text

Through deep analysis on the components of Chinese characters, we find that the components with one or more occlusive regions is suited for embedding watermarking information. In this section, we present the principle and algorithm of the Chinese text watermarking technique.

3.1 Principle

Since Chinese character is an ideographic and pictographic character, and all the Chinese characters can be expressed by components in their mathematical forms, we may separate a Chinese character into its components and then we select the components with occlusive regions as potential watermarking embedding positions. We readjust the occlusive region size of a component to embed watermarking information.

According to the occlusive region number of a component, we can classify the components with occlusive rectangular regions into four types. Different type leads to different watermarking embedding scheme which is listed as follows:

(1) The first type of component (S_1) has one occlusive rectangular region, we adjust the aspect ratio (T_1) of the occlusive region to embed the watermarking information. A parameter α_1 is used to control the aspect ratio. The sketch map is shown in Fig.3.T_1. Here, S_1 is regarded as a set containing all the components of first type.
(2) The second type of component (S_2) has two occlusive rectangular regions which adjoin each other vertically or horizontally, we change the ratio of the widths or heights (T_2) of the two regions to embed the watermarks. A parameter α_2 is used to control the ratio. The sketch map is shown in Fig.3.T_2.
(3) The third type of component (S_3) includes three occlusive rectangular regions adjacent with each other vertically or horizontally. We rearrange the width or height ratio (T_3) between the upper or right two regions. The control parameter is α_3. The sketch map is Fig.3.T_3.
(4) The last type of component (S_4) has four or more adjacent occlusive region. We also use the width or height ratio (T_4) to embed information by moving central line left or right or up or down. The parameter is α_4. The sketch map is Fig.3.T_4.

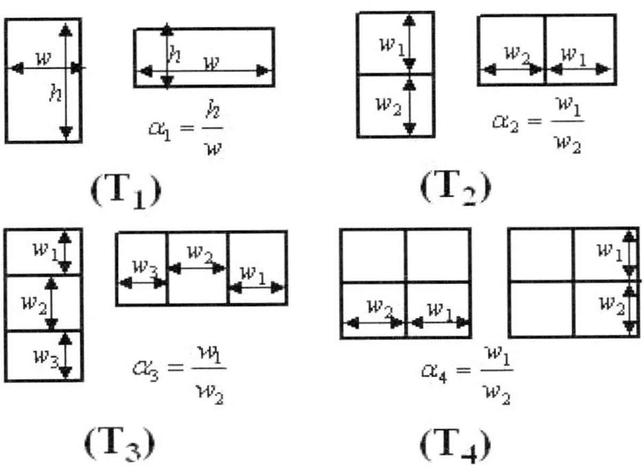

Fig. 3. The sketch map of embedding methods of four types of components

Many Chinese characters may have more than one component which has occlusive regions, the following priority is used to decide which component be used to embed information. (Let c_i be a component, N_{c_i} be the occlusive rectangular region number of c_i and P_{c_i} be the priority of c_i.)

(1) $N_{c_i} < N_{c_j} \Rightarrow P_{c_i} > P_{c_j}$;
(2) $(N_{c_i} = N_{c_j}) \wedge (c_j\ we\ c_i \vee c_i\ lr\ c_j \vee c_i\ ud\ c_j) \Rightarrow P_{c_i} > P_{c_j}$;

In order to keep the integrality and aesthetic feeling of Chinese characters, we sum up some rules as follows: (Let C be a Chinese character, $C^{'}$ be the watermarked character of C, and $c_i^{'}$ be the watermarked component of c_i, R_c, R_C be the minimal rectangle of c and C.)

(1) $R_{c_i^{'}} \subseteq R_{c_i}$;
(2) $R_{C'} \subseteq R_C$;

3.2 Algorithms

To describe the principle clearly, some definitions and backgrounds should be presented formally at first.

$$\Omega = \{C|C \in GB2312-80\};$$
$$\Phi = \{c|c \in \cup S_j, j \in [1..4]\};$$
$$\Theta = \{C|C \in \Omega, \exists c \subset C, c \in \Phi\};$$

Ω is a set which comprises all the Chinese characters in GB2312-80, Φ contains all the components with occlusive regions which can be used to embed watermarking information, and Θ includes all the Chinese characters which have one or more component belonging to Φ.

Embedding Algorithm

Input: a hosted Chinese text document H_t, watermark W and the embedding parameters $\alpha_1, \alpha_2, \alpha_3, \alpha_4$.

Output: the watermarked hosted Chinese text document $H_t^{'}$ in which the watermark W is embedded.

```
begin
generate the bit stream W' from W;
determine the embedding positions for watermarking signals.
while (C ∈ H_t ⋂ Θ) do
{
  get a component c ∈ Φ from C;
  if the corresponding bit of W' is 1, then
  {
    determine the type of the component c to obtain the value i ∈ [1..4];
    if c ∈ S_i, then change T_i according to α_i and reconstruct the C;
  }
}
end.
```

Extracting Algorithm

Input: a watermarked Chinese text document $H_t^{'}$ and the parameters $\alpha_1, \alpha_2, \alpha_3, \alpha_4$;

Output: watermark W;
begin
while $(C \in H_t^{'} \bigcap \Theta)$ do
{
get a component $c \in \Phi$ from C;
determine the type of the component c to obtain the value $i \in [1..4]$;
if $c \in S_i$ and $|T_i - \alpha_i| < \varepsilon$, then the bit of $W^{'}$ is 1;
else the bit of $W^{'}$ is 0;
}
map $W^{'}$ to watermark W.
end.

4 Experimental Results

We have implemented the watermarking technique in the Automatic Generating System of Mathematical Expressions of Chinese Characters (AGS-MECC)[10]. In the experiments, we only process the bold-face font because of its good properties [10]. The process of other fonts will be the our future work, because it is more difficult in processing them than processing bold-face font.

In the experiment, we embed a watermark "SEI OF ECNU" into a text in BMP format whose length is of 5475 Chinese characters and whose font is bold-face. By statistic, There are 2517 embedding positions in this hosted Chinese text. We regard the parameters as secret keys, and use '00000000' to denote the beginning and end of watermarking bit-stream so that we can locate the watermarks. We transform the watermark "SEI OF ECNU" into binary ASCII code bit-stream and embed it to the source text. In this paper, we don't discuss the technique of mapping W to $W^{'}$.

The experimental results show that the proposed method is more transparent and robust than the existing techniques such as line-shift coding, word-shift coding, character coding. Compared with the method [8], our method has more powerful ability of conceal the changes to Chinese characters, though it has less capacity for watermarking information. Fig.4 shows a part of original text and watermarked text.

With the extracted watermarking, we can make sense of whether the host Chinese text is changed. For example, by comparison of original watermarking and extracted watermarking, we can decide where the host text is tampered so that we can obtain the authenticity, integrality and confidentiality of the host data, in addition to the tamper proof. Copyright owners can use the watermarks as proof to protect their benefits in the copyright violation. Another merit of the proposed method is that it can resist geometrical transformation such as enlarging, condensing or distortion with little effect on the characters.

To strength the performance of the proposed technique, we take the following measures: (1)Regarding the parameters as secret keys. With the secret keys, it is easy to extract the watermark exactly, but it is difficult in removing the

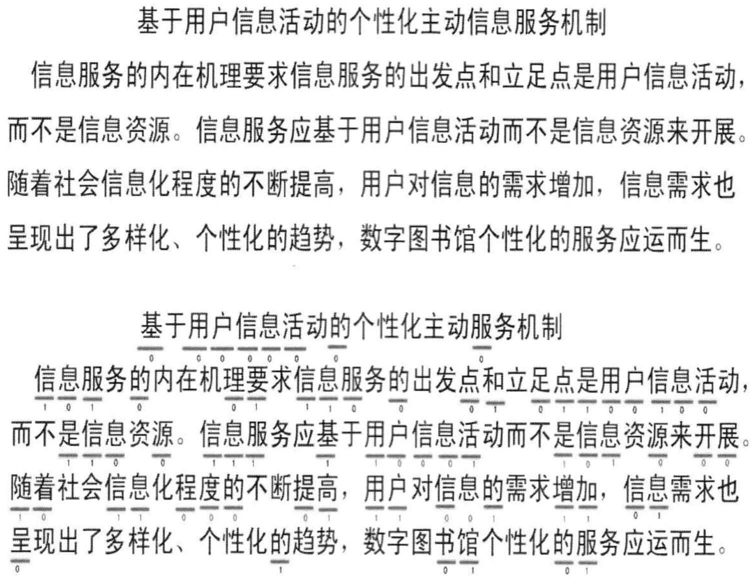

基于用户信息活动的个性化主动信息服务机制

信息服务的内在机理要求信息服务的出发点和立足点是用户信息活动，而不是信息资源。信息服务应基于用户信息活动而不是信息资源来开展。随着社会信息化程度的不断提高，用户对信息的需求增加，信息需求也呈现出了多样化、个性化的趋势，数字图书馆个性化的服务应运而生。

基于用户信息活动的个性化主动服务机制

信息服务的内在机理要求信息服务的出发点和立足点是用户信息活动，而不是信息资源。信息服务应基于用户信息活动而不是信息资源来开展。随着社会信息化程度的不断提高，用户对信息的需求增加，信息需求也呈现出了多样化、个性化的趋势，数字图书馆个性化的服务应运而生。

Fig. 4. A part of original text and watermarked text. The upper part is the original hosted text, and the under part is the watermarked hosted text. The embedding positions (marked by red underline) and embedded bits are displayed for comparison.

watermarks without the keys. (2)Embedding the watermarking information more times as long as there are enough embedding positions, in order to improve the watermarking robustness.

A direct attack method to the proposed technique is to use the unmarked characters to replace the Characters with occlusive regions, but it costs too much. Another attack method is to destroy the occlusive rectangular regions of Chinese characters, but it will do harm to the beauty of Chinese characters.

5 Conclusions

With the development of the Internet, there are more and more Chinese text transmitted over it. It is very necessary to take the Chinese information security into account. Chinese Text watermarking can compensate for the disadvantage of encryption. The proposed scheme based on the unique characteristic of Chinese characters in this paper will play an important role in protecting the security of Chinese documents over Internet because of its good performance such as good robustness and transparency. The technique will play an important role in the text security and will be widely applied in fields such as national defense, information hiding, secret transmission , copyright protection, e-government and e-commerce and so on.

References

1. Tanaka, K., Nakamura, Y., and Matsui, K. Embedding secret information into a dithered multi-level image. In Proc of 1990 IEEE Military Communications Conference, 1990:216- 220.
2. Cox, I.J., Miller, M.L. The First 50 Years of Electronic Watermarking. J. of Applied Signal Processing. 2002, 2:126- 132.
3. Fridrich, J., and Goljan, M. On Estimation of Secret Message Length in LSB Steganography in Spatial Domain. In Proc. of SPIE Symposium on Electronic Imaging, USA, 2004.
4. Brassil, J. , Low, S., Maxemchuk, N. F., and O'Gorman, L. Electronic Marking and Identification Techniques to Discourage Document Copying. IEEE Journal on Sel. Areas in Commun.. 1995,13(8):1495-1504.
5. Brassil, J., Low, S.,.and Maxemchuk, N.F. Copyright Protection for the Electronic Distribution of Text Documents. Proceedings of the IEEE, 1999, 87(7).
6. Huang, D., Yan, H. Interword distance changes represented by sine waves for watermarking text images. IEEE Trans. On Circuits and Systems for Video Technology, 2001, 11(12):1237-1245.
7. Atallah, M.J., Raskin, V., Hempelmann, C.F., et al. Natural Language Watermarking and Tamperproofing. In Proc. of 5th International Information Hiding Workshop (IHW), The Netherlands, 2002.
8. Sun, X.M., Luo, G., Huang H.J. Component-based Digital Watermarking of Chinese Texts. In ACM International Conference Proceeding Series, Vol.85, Proc. of the 3rd international conference on Information security, November, 2004.
9. Sun, X.M., Chen, H.W., Yang, L.H., and Tang, Y.Y. Mathematical Representation of a Chinese Character and its Applications. International Journal of Pattern Recognition and Artificial Intelligence. 2002, 16(8):735-747.
10. Wu, J.Z., Zhang, W.Y., Zeng Z.B. Automatic generation of mathmatical expressions of Printed Chinese Characters. In Proc. of the IASTED International Conference on Computational Intelligence, Calgary, Alberta, Canada, July 4-6, 2005

Global Transaction Control with Multilevel Security Environments

Hyun-Cheol Jeong

Dept. of Medical Engineering, Kwangju Health College
683-3 Shinchang-Dong, Gwangsan-Gu Gwangju 506-701 Korea
hcjeong@kjhc.ac.kr

Abstract. It is the most important issue that the restrictive security policy and correct concurrency control is guaranteed. This paper considers the security of heterogeneous system with replicated data. The existed read-from relationship in the existed serializability is improper. So, we define secure read-from relationship and propose secure one-copy quasi-serializability by utilizing this relationship and display some examples. The proposed secure one-copy quasi-serializability is very proper for global transactions in that this serializability doesn't violate security autonomy and prevents covert channel between global transactions.

1 Introduction

Security policy is essentially supported in heterogeneous system to protect data. Security manager assigns security level to user and data. BLP[1] assures secure property to prevent data with high security level from being directly disclosed to unauthorized users. One-copy serializability(1SR) is utilized as the correctness criteria of replicated data. W. Du[2] suggested the one-copy quasi-serialzability(1QSR) of global transaction(GT) that has weaker constraint than 1SR. This paper proposes the secure one-copy quasi-serializability (S1QSR) to effectively manage GT in MDBS with security manager(MDBS/SM) and displays some examples. MDBS/SM consists of multilevel local systems with security manager that are heterogeneous and autonomous. Section 2 describes related works. Section 3 presents a model, MDBS/SM. Section 4 describes S1QSR and presents some examples for S1QSR. Section 5 discusses the S1QSR. Section 6 describes our conclusion and future work.

2 Previous Researches

Many researches were proposed to manage replicated data. In W. Du[2], 1QSR easily preserves the serializability of GTs because it does not consider local indirect conflicts. S. Jajodia[3] solved covert channel. But, He didn't suggest the secure propagation mechanism. M.H. Kang[4] proposed new transaction model for multilevel secure database and some techniques. O. Costich[5] suggested multilevel one-copy serializability. I.E. Kang[6] considered local autonomy and security constraints in multilevel secure federated database system. Security level in each site must be globally ordered. Because each transaction manager is in every security level, there is considerable overhead in system construction. R. Sandhu[7] restricted write operation when transaction level is the same as data level.

L. Wang et al. (Eds.): FSKD 2006, LNAI 4223, pp. 660 – 663, 2006.

3 System Model

Global transaction manager(GTM) decomposes GT into some subtransactions and decides their global order to execute the physical operation in each site. Next, GTM submits them to local execution monitor(LEM) and plays a role of coordinator. But, GTM not recognizes information for LT and can not control LT. GSM manages the security level of GT according to security criteria standardized in MDBS/SM. When primary copy(Pc) is updated, GSM submits ST to local execution security module(LESM) to write secondary copy(Sc) in every site. To decide primary copy, first, GSM assigns security level to GT and GTM submits subtransactions of GT to each site. Then, finds out the read set and write set of the submitted GT. Second, every data must be carefully updated in order to maintain the consistency and security of replicated data. Primary copy is {each replicated $D \in {}^{\forall}$ (site)|highest $SL(D)_{D \in writeset}$}. The transaction with the same as the security level of primary copy executes write operation. After global security manager(GSM) consults the information window(IW) in global information dictionary(GID), GSM decides the site that Pc virtually is in by the Pc decision rule. IW preserves some information that are data item, security level, Pc and Sc for data in every sites, update bit, Before Value(BeV) and After Value(AeV) for a data. Security transactions(ST) executes only update for Sc in all sites. We assume that updating Sc is executed after Pc is updated. Each Sc has BeV and AeV when it is updated. ST submitted to each site is $W^S(AeV:S_i)$. GSM refers to IW and decides the security level of ST by security function (SF). SF assigns security level that satisfies the restricted *-property for every Sc to ST. So, ST is able to access each Sc. Local execution manager(LEM) plays a role of a participant for GT and takes subtransactions for GTM to submit.

4 Transaction Processing with Security Constraints

4.1 Problem Definition

If global order is $T_i \rightarrow T_j$, every site executes operation in ${}^{\forall}op_i \rightarrow {}^{\forall}op_j$. For example, $GT_1=\{G_{1i}, G_{1j}\}$; $G_{1i}:w_{g1}(x_i)$, $G_{1j}:w_{g1}(x_j)$, $GT_2=\{G_{2i}, G_{2j}\}$, $G_{2i}:r_{g2}(x_i)w_{g2}(y_i)$, $G_{2j}:w_{g2}(y_j)$, $RT_1=\{RT_{1i}, RT_{1j}\}$; $RT_{1i}:w_{r1}(x_i)$, $RT_{1j}:r_{r1}(y_j)r_{r1}(z_j)$, $RT_2=\{RT_{2i}, RT_{2j}\}$; $RT_{2i}:w_{r2}(x_i)$, $RT_{2j}:w_{r2}(y_j)$, $LT_1:r_{l1}(x_i)r_{l1}(y_i)$, $LT_2:w_{l2}(z_j)r_{l2}(y_j)$, Then, $E_i:r_{l1}(x_i)w_{g1}(x_j)w_{r1}(x_i)r_{g2}(x_i)w_{g2}(y_i)w_{r2}(y_i)$ $r_{l1}(y_i)$, $E_j:w_{g1}(x_j)w_{r1}(x_j)r_{r1}(z_j)w_{l2}(z_j)r_{l2}(y_j)r_{g2}(x_j)w_{r2}(x_j)$, $E_i:w_{g2}(y_i)w_{r2}(y_i)r_{l1}(x_i)r_{l1}(y_i)w_{g1}(x_j)w_{r1}(x_i)r_{g2}(x_i)$ $E_j:w_{g1}$ $(x_j)w_{r1}(x_j)r_{r1}(z_j)w_{l2}(z_j)r_{l2}(y_j)$ $r_{g2}(x_j)$ $w_{r2}(x_j)$. Then, E_i and E_j are $GT_1 \rightarrow RT_1 \rightarrow GT_2 \rightarrow RT_2$, $GT_1 \rightarrow RT_1 \rightarrow GT_2$, $GT_2 \rightarrow RT_2$ in E'_i, $GT_1 \rightarrow RT_1 \rightarrow GT_2 \rightarrow RT_2$ in E'_j. So, E' is 1QSR because it is equivalent to E. But, this situation violates serializability and causes the covert channel between transactions because security manager assigns level into transactions and those transactions utilize the exited 1QSR.

4.2 Secure Serializability for Transaction

For For $T=\{T_1, T_2, ..., T_n\}$, $D=\{x,y,..., \}$, $SL(T) \in \{T, S, C, U\}$, and $SL(D) \in \{T, S, C, U\}$, $U <_H C <_H S <_H T$. LSM in each site assigns level to local data and transaction and maintains correct value by protecting local data from being insecure. Therefore, we consider security autonomy together with the existed local autonomy in MDBS/SM. Each LSM independently assigns security level to transaction and data belonging to self-site. So, he has privileges to protect his database. Because the read-from relationship of the existed serializability is improper for MDBS/SM, new secure read-from relationship is requested. This relationship considers both security and read-from relationship. When security level is $B <_H A$, $SL(T_i) \leq_H SL(T_j) = SL(x)$, if there is the relation of $R^A{}_j[W^B{}_i(X^B)]$ or $R^B{}_j[W^B{}_i(X^B)]$ between two transactions, T_i and T_j, there is secure read-from relationship between them. To define S1QSR, we consider some relevant matters. First, the transaction to be executed according to global order must satisfy secure read-from relationship in behalf of preventing direct information efflux. Second, when transactions act in collusion to establish covert channel, this indirect information efflux is prevented. Also, we maintained the BeV and AeV for each data and preserved the most recent value. When global order is $T_i \rightarrow T_j$, $SL(T_j) <_H SL(T_i)$, $T_i, T_j \in$ {only GTs}, $w_i(y)r_i(x)w_j(x)$, indirect information efflux is occurred. GSM assigns security level to ST. ST that is transaction with reliable level is submitted to update secondary copy. Covert channel is not established because there is no conflict between ST and GT. Read operation and write operation have the same level for $R^B{}_j\,[W^B{}_i\,(X^B)]$. For global order $T_i \rightarrow T_j$, $SL(T_j)\ {}_H{>}\ SL(T_i)$, if $T^T{}_j{:}w^T{}_j(y^T)r^T{}_j(x^S)$, $T^S{}_i{:}w^S{}_i(x^S)$ and E_1: $\underline{w^S{}_i(x^S)}w^T{}_j(y^T)\ \underline{r^T{}_j(x^S)}$, there is no covert channel because T_i with lower level is preceded according to global order. Namely, It is a reason that the high and the low level for two transactions not violate global execution. But, if $SL(T_j) <_H SL(T_i)$, $T^T{}_i{:}w^T{}_i(y^T)r^T{}_i(x^S)$, $T^S{}_j$: $w^S{}_j(x^S)$, E_2: $w^T{}_i(y^T)\ \underline{r^T{}_i(x^S)\ w^S{}_j(x^S)}$, there is covert channel because they are executed according to global order and violate level. According to global order $T_i \rightarrow T_j$, covert channel is established for x. The conditions for secure one-copy quasi-serial execution are as follows. If a set of transactions $\{E_1, E_2, ..., E_n\}$ satisfies the following conditions, it is S1QSE. 1. Each site execution, E_i, is serializable without direct or indirect information efflux. 2. If T_i precedes T_j, all operations of T_i is executed prior to the operations of T_j through the global order of transaction. 3. for T_i, T_j {(GT) $\cup$(ST)}, if T_i is last transaction to update data and secure read-from relationship, T_j read AeV. If there is the secure read-from relationship with covert channel, T_i precedes T_j in the order and T_j reads BeV. The first two conditions of secure one-copy quasi-serial executions are similar to those of 1QSR. ST executes write operation to update copies in each local site. The global order between ST and GT must be assured to maintain data consistency. The last condition means that the last write operation and secure read-from relationship is equivalent to last execution without indirect information efflux, called covert channel. Let $SL(GT_1)=T$, $SL\ (GT_2)=S$, $SL(LT_1)=T$, $SL(LT_2)=S$; $SL(x_i)=S$, $SL(x_j)=S$, $SL(y_i)=C$, $SL(y_j)=T$, $SL(z_i)\ =T$, $SL(z_j)=S$, $SL(k_i)=S$, $SL(k_j)=U$. Primary copies are z, k in $site_i$, x, y in $site_j$. Then, global transactions and local transactions are submitted as follows. $T_1=\{G_{1i}{:}\ w_{g1}(z_i)r_{g1}(y_i)r_{g1}(k_i), G_{1j}{:}r_{g1}(z_j)\ w_{g1}(y_j)r_{g1}(x_j)\}$, $GT_2=\{G_{2i}{:}r_{g2}(x_i)w_{g2}(k_i), G_{2j}{:}w_{g2}(x_j)r_{g2}(k_j)\}$, $ST_1=\{S_{1i}{:}w_{s1}(x_i)w_{s1}(y_i), S_{1j}{:}w_{s1}(z_j)w_{s1}(k_j)\}$, $LT_i{:}w_{l1}(z_i)r_{l1}(y_i)$, LT_j: $r_{l1}(x_j)\ r_{l1}(k_j)$. The executions in each site are as follows. $E_i{:}w^T{}_{g1}(z^T{}_i)w^T{}_{li}(z^T{}_i)w_{s1}(x_i)r^T{}_{g1}(y^C{}_i)\ r^T{}_{g1}(k^S{}_i)r^T{}_{li}(y^C{}_i)w_{s1}(y_i)\ r^S{}_{g2}(x^S{}_i)w^S{}_{g2}(k^S{}_i)$. $E_j{:}r^T{}_{g1}(z^C{}_j)\underline{w^T{}_{g1}(y^T{}_j)\ r^T{}_{g1}(x^S{}_{jBeV})}r^T{}_{lj}(x^S{}_j)w_{s1}(z_j)w_{s1}(k_j)r^T{}_{lj}(k^U{}_j)\underline{w^S{}_{g2}(x^S{}_j)}\ r^S{}_{g2}(k^U{}_j)$. For primary copy, updating data by ST_{1i} and ST_{1j} creates new value in each site. So, secondary copies have BeV and AeV. If the underlined part of E_j is $w^T{}_{gl}(y^T{}_j)r^T{}_{gl}(x^S{}_j)$ .... $w^S{}_{g2}(x^S{}_j)$, covert channel between GT_1 and GT_2 is established. To avoid covert channel GT_1 reads $x^S_{j/BeV}$.E is S1QSE since $GT_1 \rightarrow ST_1 \rightarrow GT_2$, $GT_1 \rightarrow GT_2$ in E_i and E_j. So, the secure one-copy quasi-serializability (S1QSR) is considered as follows. A global execution of a set of transactions is secure one-copy quasi-serializability if it is equivalent to a secure one-copy quasi-serial execution of the same set of transaction. An example of S1QSR is presented. For the same situation in an example of 1QSR, let $E'=\{E'_i,E'_j\}$. Where, E'_i:$r^T_{g1}(k^S_i)w^T_{li}(z^T_i)w^S_{g2}(k^S_i)w^T_{g1}(z^T_i)r^T_{li}(y^C_i)w_{s1}(x_i)r^T_{g1}(y^C_i)w_{s1}(y_i)r^S_{g2}(x^S_i)$. E'_j:$r^T_{g1}(z^C_j)w_{s1}(z_j)w^T_{g1}(y^T_j)r^T_{g1}(x^S_{j/BeV})w^S_{g2}(x^S_j)w_{s1}(k_j)r^T_{lj}(x^S_j)r^T_{lj}(k^U_j)r^S_{g2}(k^U_j)$. $GT_1 \rightarrow GT_2$, $GT_1 \rightarrow ST_1$, $ST_1 \rightarrow GT_2$ in E'_i, $GT_1 \rightarrow ST_1$, $GT_1 \rightarrow GT_2$, $ST_1 \rightarrow GT_2$ in E'_j. Therefore, E'_i is S1QSR because it is equivalent to E in an example of S1QSE.

4 Discussions

There are various serializabilities[2,5,6]. The ML-1SR introduced security policies into the existed one-copy serializability in homogeneous database system with replicated data. In ML-1SR, the higher security level database is, the more memory is required. The proposed S1QSR introduces security policy into one-copy quasi-serializability in heterogeneous system. Security manager is in global and local module. S1QSR has no problem for indirect conflict because this considers global order. ML-1SR has transaction manager in each security level. This is considerable overhead. Also, ML-1SR has strong constraint that security level is globally ordered. But, S1QSR mitigates the strong constraints of ML-1SR.

5 Conclusion

We must strongly recognize the importance of security. This paper proposes secure one-copy quasi-serializability and presents the examples. Each site keeps security autonomy and prevents both direct and indirect information efflux and assures serializability. In the future work, we will develop and prove the algorithms for assuring S1QSR. Also, we will research the secure recovery mechanism concering the S1QSR.

References

1. C. P. Pfleeger, Security in Computing, Prentice Hall (1989)
2. W. Du, et al, "Supporting Consistent Updates in Replicated Systems" VLDB (1993)
3. S. Jajodia, B. Kogan , "Transaction Processing in Multilevel Secure Databases Using Repli cated Architecture", Symposium on Security and Privacy (1990)
4. M. H.Kang , et al., "A Practical Transaction Model and Untrusted Transaction Manager for a Multilevel Secure Database System", Database Security VI IFIP(1993)
5. O. Costich , " Transaction Processing Using an Untrusted Scheduler in a Multilevel Database with Replicated Architecture", Database Security V IFIP (1992)
6. I. E. Kang,T. F. Keefe, "Concurrency Control for Federated Multilevel Secure Database Sys-tems", 8th IEEE Computer Security Foundations Workshop(1995)
7. R. Sandhu, "Lattice-Based Access Control Models", IEEE Computer (1993)

A Privacy Preserving Mining Algorithm on Distributed Dataset*

Shen Hui-zhang, Zhao Ji-di, and Yang Zhong-zhi

Aetna School of Management, Shanghai Jiao Tong University, Shanghai, P.R. China, 200052
{hzshen, judyzhao33}@sjtu.edu.cn

Abstract. The issue of maintaining privacy in data mining has attracted considerable attention over the last few years. The difficulty lies in the fact that the two metrics for evaluating privacy preserving data mining methods: privacy and accuracy are typically contradictory in nature. This paper addresses privacy preserving mining of association rules on distributed dataset. We present an algorithm, based on a probabilistic approach of distorting transactions in the dataset, which can provide high privacy of individual information and at the same time acquire a high level of accuracy in the mining result. Finally, we present experiment results that validate the algorithm.

1 Introduction

In the light of developments in technology to analyze personal data, public concerns regarding privacy of personal information are rising. Data mining, with its objective to efficiently discover valuable and inherent information from large databases, is particularly sensitive to misuse [1] [2]. Therefore an interesting new direction for data mining research is the development of techniques that incorporate privacy concerns[2] and to develop accurate models without access to precise information in individual data records [3]. The difficulty lies in the fact that these two metrics for evaluating privacy preserving data mining methods: privacy and accuracy, are typically contradictory in nature, with the consequence that improving one usually incurs a cost in the other [4].

In this paper, we assume such a scenario: a transaction database DB is horizontally partitioned among n sites which are $S_1, S_2, \cdots, S_n$, $DB = DB_1 \cup DB_2 \cup \cdots \cup DB_n$ and DB_i locates at site S_i ($1 \le i \le n$). The itemset X has *local* support count of $X_{.\sup i}$ at site S_i if $X_{.\sup i}$ of the transactions contains X. The *global* support count of X is given as $X_{.\sup} = \sum_{i=1}^{n} X_{.\sup i}$. An itemset X is globally supported if $X_{.\sup} \ge S_{\min} * \sum_{i=1}^{n} |DB_i|$, where $S_{\min}$ is the user-defined

---

* Supported by IBM SUR (SURTHU5).

L. Wang et al. (Eds.): FSKD 2006, LNAI 4223, pp. 664–673, 2006.

minimum support. Global confidence of a rule $X \Rightarrow Y$ can be given as $\{X \cup Y\}_{.sup} / X_{.sup}$.

The set of frequent itemsets $L_{(k)}$ consists of all k-itemsets that are globally supported. The set of locally frequent itemsets $LL_{i(k)}$ consists of all k-itemsets supported locally at site S_i. $GL_{i(k)} = L_{(k)} \cap LL_{i(k)}$ is the set of globally large k-itemsets locally supported at site S_i. The objective of distributed association rule mining is to find the itemsets $L_{(k)}$ for all $k > 1$ and the support counts for these itemsets and, based on this, generate association rules with the specified minimum support and minimum confidence.

2 Related Work

The issue of maintaining privacy in data mining has attracted considerable attention over the last few years. Previous work can be broadly classified into cryptographic approach and distortion approach. In the first approach, privacy preserving is achieved using cryptographic methods[1][6]. The other approach uses distortion methods to build privacy preserving data mining solutions. The basic idea in distortion approach is to modify data values such that reconstruction of the values for any individual transaction based on the distorted data is difficult and thus is safe to use for mining, while on the other hand, the distorted data and information on the distribution of the random data used to distort the data, can be used to generate valid rules and simulate an approximation to the original data distribution. Randomization is done using a statistical method of value distortion [5] that returns a value $x_i + r$ instead of x_i, where r is a random value drawn from a specific distribution. A Bayesian procedure for correcting perturbed distributions is proposed and three algorithms for building accurate decision trees that rely on reconstructed distributions are presented in [6]. [3] studied the feasibility of building accurate classification models using training data in which the sensitive numeric values in a user's record have been randomized so that the true values cannot be estimated with sufficient precision. More Recently, the data distortion method has been applied to Boolean association rules [7][8]. [8] investigate whether users can be encouraged to provide correct information by ensuring that the mining process cannot violate their privacy on some degree and put forward an algorithm named MASK(Mining Associations with Secrecy Konstraints) based on Bernoulli probability model. [7] presented a framework for mining association rules from transactions consisting of categorical items and proposed a class of randomization operators for maintaining data privacy.

In this paper, we follow the distortion approach, and however, address the problem of mining association rules with a Markov distortion approach in the context of distributed database environments, that is, all sites have the same schema, but each site has information on different entities. Thus a classical association mining rule such as Apriori should be extended to distributed environments to generate association rules. An algorithm Count Distribution (CD), which is an adaptation of the Apriori algorithm, has been proposed for the same parallel mining environment[9]. While a thorough study of distributed association rule mining in a distributed environment can be found in[10]. The goal of this paper is to accomplish the distributed mining process as accurately as possible without compromising the private information of local large itemsets for all the sites.

3 Markov Chain Model and Itemset Transition Probability

A discrete Markov chain model can be defined by the tuple $\langle S, P, \lambda \rangle$, where S corresponds to the state space, P is a matrix representing transition probabilities from one state to another, and λ is the initial probability distribution of the states in S. The fundamental property of Markov model is the dependency on the previous state. If the vector $S(t)$ denotes the probability vector for all the states at time t, then

$$\hat{S}(t) = \hat{S}(t-1) * P \tag{1}$$

If there are n states in our Markov chain, then the matrix of transition probabilities P is of size $n \times n$. Markov chains can be applied to privacy preserving data mining. In this formulation, a Markov state can correspond to a frequent k-itemset.

Let $\mathrm{M} = \{X_1, X_2, \cdots, X_m\}$ be a set of itemsets where an itemset X_i is a k-itemset. Let P_M be the itemset transition probability matrix of M subject to (1) $p_{kl} \geq 0$, (2) $\forall k\,(1 \leq k \leq m), \sum_{l=1}^{m} p_{kl} = 1$. Where $p_{kl}(1 \leq k \leq m, 1 \leq l \leq m)$ is the probability with which itemset X_k transits to itemset X_l.

3.1 Quantifying Privacy and Privacy Measure

As mentioned earlier, the mechanism adopted in this paper for achieving privacy is to distort the user data before it is subject to the mining process. Accordingly, we measure privacy with regard to the probability with which the user's distorted items can be reconstructed.

In the algorithm we present in the later section, an itemset X_k is transited to an itemset X_l randomly with probability p_{kl}. We generate the distorted data value from a transaction by randomizing each given itemset in the transaction.

Denoting an original itemset as X^M and the distorted itemset as X^D, the probability of correct reconstruction of itemset X_i is given by

$$R(X_i) = \sum_{j=1}^{m} p_{ij}^2 * \frac{P(X^M = X_i)}{\sum_{k=1}^{m} p_{kj} P(X^M = X_k)}$$

Where $P(X^M = X_i)$ is given by the support of X_i in the original dataset.

And the probability of correct reconstruction of the set of itemsets $\mathbf{M}$ is given by

$$R(M) = \prod_{i=1}^{m} \sum_{j=1}^{m} p_{ij}^2 * \frac{P(X^M = X_j)}{\sum_{k=1}^{m} p_{kj} P(X^M = X_k)}$$

After computing the reconstruction probability of the set of itemsets, we can define user privacy as the following percentage:

$$P(M) = (1 - \overline{R}(M)) * 100 \tag{2}$$

Where $\overline{R}(M)$ is the average reconstruction probability in a mining process.That is, when the reconstruction probability is 0, the privacy is 100%, whereas it is 0 if the $\overline{R}(M) = 1$.

3.2 Support Recovery

We denote the original true set of k-itemsets by $\mathbf{M}$ and the distorted set of k-itemsets, obtained with a distortion probability matrix P_M, as D. Let $S^M(X_1, X_2, \cdots, X_m)$ be the vector of expected support of itemsets on M and $S^D(X_1, X_2, \cdots, X_m)$ be the vector of support of itemsets on D. P_M is the itemset transition probability matrix of M as defined above. We have the following theorem based on the property of Markov chain.

Theorem

$$S^D(X_1, X_2, \cdots, X_m) = S^M(X_1, X_2, \cdots, X_m) * P_M \tag{3}$$

And

$$S^M(X_1, X_2, \cdots, X_m) = S^D(X_1, X_2, \cdots, X_m) * P_M^{-1} \tag{4}$$

Where P_M^{-1} is the inverse matrix of P_M.

Proof

Let $C^M(X_i)$ be the support count of itemset X_i on the original dataset ($1 \le i \le m$). After the distortion procedure, approximately $C^M(X_1) * p_{11}$ of original X_1 will remain X_1 on the distorted dataset, $C^M(X_2) * p_{21}$ of original X_2 will transit to X_1 on the distorted dataset, $C^M(X_3) * p_{31}$ of original X_3 will transit to X_1, …, $C^M(X_m) * p_{m1}$ of original X_m will transit to X_1, given $C^M(X_i)$ is large enough.

Thus the overall support count of X_1 on the distorted dataset is $C^D(X_1)=\sum_{i=1}^{m} C^M(X_i)*p_{i1}$, and $S^D(X_1)=\sum_{i=1}^{m} S^M(X_i)*p_{i1}$.

Similarly, the overall support of X_k ($2\le k\le m$)on the distorted dataset is $S^D(X_k)=\sum_{i=1}^{m} S^M(X_i)*p_{ik}$. That is,

$$S^D(X_1,X_2,\cdots,X_m)=S^M(X_1,X_2,\cdots,X_m)*P_M$$

Remember there are two constraints that must be satisfied when generating P_M. From these two constraints, it is easy to derive that the inverse matrix of P_M, denoted as P_M^{-1}, exists. Therefore, we have

$$S^M(X_1,X_2,\cdots,X_m)=S^D(X_1,X_2,\cdots,X_m)*P_M^{-1}$$

3.3 Recovery Error

As[7][8] pointed out in their studies, in a probabilistic distortion approach, fundamentally we cannot expect the reconstructed support values to coincide exactly with the actual supports. This means that we may have errors in the estimated supports of frequent itemsets with the reported values being either larger or smaller than the actual supports. This kind of error is qualified as the metric of *Support Error*, ρ, which reflects the average relative error in the reconstructed support values for those itemsets that are correctly identified to be frequent. Let *r_sup* be the reconstructed support and *a_sup* be the actual support, the support error is computed over all frequent itemsets as

$$\rho=\frac{100*\sum_f \left|r\_sup_f-a\_sup_f\right| / a\_sup_f}{|f|} \tag{5}$$

Errors in support estimation can also result in errors in the identification of the frequent itemsets. It is quite likely that for an itemset slightly above *Smin* that one of its subsets will have recovered support below *Smin*. The itemset will be discarded from the candidate set due to a key property of Apriori algorithm that if itemsets is a frequent itemset, that all of its subsets must have support larger than *Smin*. It will become especially an issue when the *Smin* setting is such that the support of a number of itemsets lies very close to this threshold value. This kind of error is measured by the metric of *Identification Error*, which reflects the percentage error in identifying frequent itemsets and has two components: δ^+, indicating the percentage of false positives, and δ^- indicating the percentage of false negatives. Denoting the reconstructed set of frequent itemsets with R and the correct set of frequent itemsets with F, these two metrics are computed as:

$$\delta^{+} = \frac{|R - F|}{|F|} * 100 \quad (6)$$

$$\delta^{-} = \frac{|F - R|}{|F|} * 100 \quad (7)$$

Where $|F|$ indicates the number of frequent itemsets at k-itemset.

Hence, to reduce such errors in frequent itemset identification, we discard only those itemsets whose recovered support is smaller than a Candidate Limit, given as *Smin* *(1- σ), for candidate set generation in the algorithm we will give in Section 4. Here σ is a reduction coefficient. However, because the distortion procedure is executed in each iteration during the mining process, the error in identifying an itemset of small size will not have a similar ripple effect in terms of causing errors in identifying itemsets of longer size as presented in [8].

4 Privacy Preserving Mining Algorithm of Global Association Rules on Distributed Data

We will now use above distortion procedure to construct a distributed association rule mining algorithm to preserve the privacy of individual sites. The goal of the algorithm is to find the global frequent itemsets and discover the global association rules satisfying the predefined thresholds while not, with some reasonable degree of certainty, disclosure the local frequent itemsets and the local support count of each individual site, thus protect its privacy.

The distributed association rule mining algorithm, given the global minimum support S*min*, the global minimum confidence and the reduction coefficient σ , works as follows.

1. Let k=1, let the candidate set be all the single items included in the dataset.
   Repeat the following steps until no itemset left in the candidate set. {

(1) The common site broadcasts the transition probability matrix P_k for k-itemsets included in the candidate set.

(2) Each site generates its local frequent k-itemsets $LL_{i(k)}$ using Apriori-like algorithm. Here the local minimum support is $S_{\min} * \frac{n_i}{N}$, where n_i is the transactions on site S_i and N is the number of overall transactions on all the sites. To each transaction located in its local dataset, it finds out the frequent itemsets included in this transaction and distorts them with the distortion approach given above and sends the distorted data to the common site.

(3) The first site S_1 gets some "fake" itemsets randomly choosing from the predefined set of fake itemsets, adds them to its local frequent set of k-itemsets, and then sends its local frequent set (not the real one at this time) to the second site. The second site adds its local frequent set, the real one without fake itemsets,

to the set it gets, deletes the duplicated itemsets, and sends the set to the third site. The third and other sites do the similar work till the set is sent back to the first site. The first site removes the random fake itemsets from the set and thus gets the global candidate set. And then the global candidate set is sent to the common site for data mining.

(4) The common site reads the distorted dataset from all the sites and scans it to compute all the supports ($S^D(X_i)$) for each k-itemset in the global candidate set.

(5) The common site recovers the supports on the original dataset ($S^T(X_i)$) for each k-itemset in the global candidate set using the formulae from the above theorem.

(6) The common site discards every itemset whose support is below its candidate limit(*Smin* *(1- σ)).

(7) The common site saves for output only those k-itemsets $L_{(k)}$ and their supports whose recovered support is at least *Smin*.

(8) The common site forms all possible (k+1)-itemsets such that all their k-subsets are among the remaning itemsets generated in step(6). Let these (k+1)-itemsets be the new candidate set.

(9) Let k=k+1. }

2. The common site finds out all the association rules with all the saved itemsets and their supports, given the user-specified global minimum confidence. This step is straightforward and need not to be described in detail.

5 Experiments

We carry out the evaluation of the new algorithm on a synthetic dataset. This dataset was generated from the IBM Almaden generator[11]. We took the parameters T10.I4.D1M.N1K resulting in a million customer tuples with each customer purchased about ten items on average. These tuples were distributed on 10 sites with a hundred thousand tuples on each site. We took a variety of *Smin* and P_k values to evaluate the privacy and accuracy of our algorithm. [12]shows that providing high accuracy and at the same time preventing exact or partial disclosure of individual information are conflicting objectives. Therefore, in this experiment, we plan to check this point on balancing accuracy and privacy. In order to simplify the experiments, we focus the evaluation of generating frequent itemsets. The presented value of *Smin* in this paper is 0.2%. Because the transition probability matrixs P_k are different in each iteration in our experiments, we present them in the tables. We use two different values of the reduction coefficient, namely $\sigma = 0$ and $\sigma = 10\%$ to evaluate the difference on recovery errors discussed in Section 3.5.

The results for the experiments are shown in Table 1 and Table 2. In the tables, the level indicates the length of the frequent itemset, $|F|$ indicates the actual number of frequent itemsets generated by Apriori algorithm at this level, $|F^1|$ indicates the number of frequent itemsets correctly generated by our algorithm at this level, P_k indicates the transition probability matrix simply presented in its size, ρ indicates the support error, δ^+ indicates the percentage error in identifying false positive frequent itemsets, and δ^- indicates the percentage error in false dropped frequent itemsets.

Table 1. *Smin* = 0.2%, $\sigma = 0$

Level	$\|F\|$	$\|F^1\|$	P_k	ρ	δ^+	δ^-
1	863	850	858×858	3.57	0.93	1.51
2	6780	6619	6691×6691	4.02	1.063	2.37
3	1385	1345	1359×1359	1.61	1.013	2.89
4	890	874	882×882	1.63	0.90	1.80
5	392	378	381×381	1.65	0.77	3.57
6	150	145	145×145	1.53	0	3.33
7	47	45	45×45	1.50	0	4.26
8	10	10	10×10	1.02	0	0

The user privacy value under these two conditions is, computed from formula (2), 67%, we adjust the elements of P_k in order to take the comparison under the same privacy value for these two conditions. The results in both Tables show that, even for a low minimum support of 0.2%, most of the itemsets are mined correctly from the distorted dataset. The support error is less than 5% at all levels. Note that the percentage error in identifying false positive frequent itemsets is significantly small at all levels, partially due to each iteration in the algorithm, the global candidate set is sent to the common site in an anonymous way. And because the distortion procedure is executed on the original data transactions in each iteration, the percentage error in both false positive frequent itemsets and false negative frequent itemsets is not accumulated during the mining process.

In Table 2, the value of *Smin* is relaxed with a reduction of 10%. We can see from the results that the negative identifying error goes down significantly, at an average

Table 2. *Smin* = 0.2%, $\sigma = 10\%$

Level	$\|F\|$	$\|F^1\|$	P_k	ρ	δ^+	δ^-
1	863	855	859×859	3.59	0.46	0.93
2	6780	6659	6711×6711	3.97	0.77	1.78
3	1385	1367	1371×1371	2.63	0.29	1.30
4	890	879	885×885	1.79	0.67	1.24
5	392	392	393×393	2.01	0.26	0
6	150	150	150×150	1.45	0	0
7	47	47	47×47	1.48	0	0
8	10	10	10×10	0.98	0	0

reduce of 0.5%, while the positive identifying error remains almost the same. The results confirm the correctness of reducing identification error through a reduction of the minimum support.

Now we increase the privacy value to 86% by changing the element values of P_k and evaluate this change on the accuracy of the mining results which are shown in Table 3($\sigma = 0$).

Table 3. *Smin* = 0.2%, $\sigma = 0$ with different P_k values from experiment 1

Level	$\|F\|$	$\|F^1\|$		P_k	ρ	δ^+	δ^-
1	863	754	769	769×769	8.6	1.74	12.63
2	6780	5872	5941	5941×5941	8.78	1.02	13.39
3	1385	1185	1203	1203×1203	6.65	1.30	14.44
4	890	738	746	746×746	6.43	0.90	17.08
5	392	303	307	307×307	5.98	1.02	22.70
6	150	121	129	129×129	6.5	5.33	19.33
7	47	39	40	40×40	5.87	2.13	17.02
8	10	9	9	9×9	4.77	0	10

Table 3 shows that the support error ρ and the two identification errors all become much higher. For example, the false negative errors at all levels become higher than 10 with the highest at level 5. This experiment implies that the tradeoff between privacy and accuracy are very sensitive to the transition matrix. We should choose appropriate transition matrix to achieve the goal of acquiring plausible values for both privacy level and accuracy.

6 Conclusions and Discussions

In this paper, we focus on the task of finding frequent itemsets and extend the problem of mining association rules to distributed environments. We develop formal notions of privacy obtained from the distortion procedure and show that our distortion approach can provide guarantees against privacy disclosure. We propose a distributed privacy preserving mining algorithm and present the full process. We also evaluate the tradeoff between privacy guarantees and reconstruction accuracy and show the practicality of our approach. But because of time limitation, we have only completed the experiments on a synthetic dataset. The experimental results on a real dataset are expected to be reported soon.

In our future work, we plan to combine the probabilistic approach and the cryptographic approach to get better performance for privacy preserving mining. Another issue of future research involves extending the probabilistic approach to other applications.

References

1. Clifton, C. and Marks, D. *Security and privacy implications of data mining*. in *ACM SIGMOD Workshop on Research Issues on Data Mining and Knowledge Discovery*. May 1996.
2. Agrawal, R. *Data Mining: Crossing the Chasm*. in *the 5th International Conference on Knowledge Discovery in Databases and Data Mining*. August 1999. San Diego, California, an invited talk at SIGKDD.
3. Agrawal, R. and Srikant, R. *Privacy preserving data mining*. in *Proc. of the ACM SIGMOD Conference on Management of Data*. May 2000. Dallas, Texas.
4. Agrawal, D. and Aggarwal, C. *On the Design and Quantification of Privacy Preserving Data Mining Algorithms*. in *Proc. of 20th ACM Symp. on Principles of Database Systems (PODS)*. 2001.
5. Conway, R. and Strip, D. *Selective partial access to a database*. in *Proc. ACM Annual Conf.* 1976.
6. Breiman, L., et al. *Classification and Regression Trees*. 1984. Wadsworth, Belmont.
7. Evfimievski, A., et al., *Privacy Preserving Mining of Association Rules*. Information Systems, Jun 2004. **29**(4): p. 343-364.
8. Rizvi, S.J. and Haritsa, J.R. *Maintaining Data Privacy in Association Rule Mining*. in *Proc. 28th International Conf. Very Large Data Bases*. 2002.
9. Agrawal, R. and Shafer, J.C., *Parallel Mining of Association Rules: Design, Implementation, and Experience*, in *IBM Research Report*. 1996.
10. Cheung, David W., et al., *Efficient mining of association rules in distributed databases*. IEEE Transactions on Knowledge and Data Engineering, 1996. **8**(6): p. 911 - 922.
11. Agrawal, R. and Srikant, R., *Fast Algorithms for Mining Association Rules*. June 1994, IBM Almaden Research Center: San Jose, California.
12. Adam, R. and Wortman, J. C., *Security-control methods for statistical databases*. ACM Computing Surveys, Dec. 1989. **21**(4): p. 515-556.

Improvement of Decision Accuracy Using Discretization of Continuous Attributes

QingXiang Wu[1,2,3], David Bell[2], Martin McGinnity[3], Girijesh Prasad[3], Guilin Qi[2], and Xi Huang[1]

[1] School of Physics and OptoElectronic Technology, Fujian Normal University
Fujian, Fuzhou, China
{qxwu, xihuang}@fjnu.edu.cn

[2] School of Computer Science, Queen's University, Belfast, BT7 1NN, UK
{Q.Wu, DA.Bell}@qub.ac.uk

[3] School of Computing and Intelligent Systems, University of Ulster at Magee
Londonderry, BT48 7JL, N.Ireland, UK
{Q.Wu, G.Prasad, TM.McGinnity}@ulster.ac.uk

Abstract. The naïve Bayes classifier has been widely applied to decision-making or classification. Because the naïve Bayes classifier prefers to dealing with discrete values, an novel discretization approach is proposed to improve naïve Bayes classifier and enhance decision accuracy in this paper. Based on the statistical information of the naïve Bayes classifier, a distributional index is defined in the new discretization approach. The distributional index can be applied to find a good solution for discretization of continuous attributes so that the naïve Bayes classifier can reach high decision accuracy for instance information systems with continuous attributes. The experimental results on benchmark data sets show that the naïve Bayes classifier with the new discretizer can reach higher accuracy than the C5.0 tree.

Keywords: Decision-making, Classification, Naive Bayes Cassifier, Discretizer.

1 Introduction

The naïve Bayes classifier [1,2] is a highly practical machine learning method, and has been widely applied to decision-making and classification. Since the original naïve Bayes classifier encounters problem when sample set is small, the virtual samples have been applied in [1,3]. Based on the experiments on benchmark data sets from the UCI Machine Learning Repository, a modified naïve Bayes classifier with an empirical formula is proposed in this paper. Because this classifier prefers to dealing with symbolic data, a transformation from continuous data to symbolic data is required. This transformation is also called continuous attribute discretizer. Two classes of approaches (unsupervised discretizers and supervised discretizers) have been surveyed and proposed in [4,5,6]. Based on statistical information used in the naïve Bayes classifier, a new adaptive discretizer is proposed to solve the data type transformation problem. For this, a *dichotomic entropy* is defined and applied to determine splitting point within an interval. Based on the decision distribution and the value distribution, a *compound distributional index* is defined to guide to an interval

L. Wang et al. (Eds.): FSKD 2006, LNAI 4223, pp. 674 – 683, 2006.

that should be split. The discretizer can adaptively discretize any continuous attribute according to the adaptive rules based on *minimal dichotomic entropy* and *maximal compound distributional index.* As a set of optimal value intervals are obtained by the discretizer, decision accuracy can be improved when the naïve Bayes classifier is used. The discretizer can be also applied to other knowledge discorvery approaches for discretization of continuous attributes.

In section 2, a modified naïve Bayes classifier is proposed. In section 3, a *dichotomic entropy,* a *compound distributional index,* and *adaptive rules* are defined. An example is applied to illustrate the algorithm. Experimental results and analysis are given in Section 4. Section 5 presents conclusions.

2 Modified Naïve Bayes Classifier

Following the notations in [3,8], let $H = <U, A>$ represent *an information system,* where $U = \{o_1, o_2, \ldots, o_i, \ldots, o_n\}$ is a finite non-empty set, called an object space or universe; o_i called an object. Each object has a finite non-empty set of attributes $A = \{ a_1, a_2, a_3, \ldots, a_i, \ldots, a_m\}$, where m is the number of attributes. An *instance information system* is defined to distinguish an information system with decision attributes from a general information system. An *instance* is defined to distinguish an object with decision values from general objects. Let $I = < U, A \cup D >$ represent an *instance information system*, where $U = \{u_1, u_2, \ldots, u_i, \ldots, u_n\}$ is a finite non-empty set, called an instance space or universe, where u_i is called an *instance* in U, and n is number of instances. Each instance has a set of attributes A and decision values D. D is a non-empty set of decision attributes or class attributes, and $A \cap D = \varnothing$.

As each instance u has a set of attribute values, $a(u)$ represents a value of attribute a by applying an operation on instance u. In other words, $a(u)$ is the value of attribute a of instance u. Domain V_a is defined as

$$V_a = \{a(u) : u \in U\} \text{ for } a \in A. \tag{1}$$

For a given universe U, all attribute domains can be obtained according to (1). For an instance information system, the domain of decision attribute or class attribute is defined as

$$V_d = \{d(u) : u \in U\} \text{ for } d \in D. \tag{2}$$

The *condition vector space*, which is generated from attribute domain V_a, is denoted by

$$\begin{aligned} V_{\times A} &= \mathop{\times}_{a \in A} V_a = V_{a1} \times V_{a2} \times \ldots \times V_{a|A|} \\ |V_{\times A}| &= \prod_{i=1}^{|A|} |V_{a_i}| \end{aligned} \tag{3}$$

where $|V_{\times A}|$ is the size of the *condition vector space*.

The *decision vector space*, which is generated from decision domain (or class domain) V_d, is denoted by

$$V_{\times D} = \bigtimes_{d \in D} V_d = V_{d1} \times V_{d2} \times \ldots \times V_{d|D|}$$
$$|V_{\times D}| = \prod_{i=1}^{|D|} |V_{d_i}| \qquad (4)$$

where $|V_{\times D}|$ is the size of the *decision vector space*. A conjunction of attribute values for an instance corresponding to a *condition vector* in the *condition vector space* is denoted by

$$\vec{A}(u) = [a_1(u), a_2(u), \ldots, a_{|A|}(u)] \qquad (5)$$

Let AU represent a set of *condition vectors* which exist in the instance information system.

$$AU = \{\vec{A}\ (u) : u \in U\} \qquad (6)$$

If $|AU| = |V_{\times A}|$, the system is called a *complete instance system*. In the real world, training sets for decision-making or classification are rarely complete instance systems. In order to illustrate algorithms in this paper, Table 1 is taken as an example of instance information system.

Table 1. Example instance information system

U	*a1*	*a2*	*a3*	*a4*	*d*
u1	1	1	1	4	+
u2	1	2	3	3	-
u3	2	3	1	4	+
u4	2	4	2	1	-
u5	3	4	2	2	-
u6	4	4	2	3	+
u7	4	3	3	3	-
u8	5	2	2	4	+
u9	6	1	1	4	+
u10	7	1	2	3	+
u11	7	2	3	1	-
u12	7	3	3	2	-

In this instance information system, there are 4 Attributes *A*={a1,a2,a3,a4}, 12 instances *U*={*u1*,*u2*,...,*u12*}, and one decision attribute with 2 values V_d = {+,-}. The value domains for each attribute are as follows:

V_{a1} ={1,2,3,4,5,6,7}, $|V_{a1}|$ =7. V_{a2}={1,2,3,4}, $|V_{a2}|$ =4. V_{a3} ={1,2,3}, $|V_{a3}|$ =3.
V_{a4} ={1,2,3,4}, $|V_{a4}|$ =4.

The size of the condition vector space: $|V_{\times A}| = 7 \times 4 \times 3 \times 4 = 336$. The number of condition vectors appearing in the table: $|AU| = 12$. Clearly, 324 possible condition vectors (or conjunctions of attribute values) do not appear in Table 1. Table 1 is not a complete instance information system with 324 unseen instances. A classifier can extract rules from such incomplete training set and classify all instances including 324 unseen instances. According to the Bayes classifier, the most probable decision can be expressed as follows:

$$d_{MP} = \arg\max_{d_i \in V_d} P(d_i \mid \vec{A}) \tag{7}$$

This expression can be rewritten by means of the Bayesian theorem.

$$d_{MP} = \arg\max_{d_i \in V_d} \frac{P(\vec{A} \mid d_i)P(d_i)}{P(\vec{A})} = \arg\max_{d_i \in V_d} P(\vec{A} \mid d_i)P(d_i) . \tag{8}$$

Unseen instances cannot be classified by rules based on Equation (8) because condition vector $\vec{A}$ for an unseen instance does not appear in the training set and thus $P(\vec{A} \mid d_i)$ cannot be obtained from the training set. In order to classify unseen instances, it is assumed that attribute values are conditionally independent given the decision value. i.e.

$$P(\vec{A} \mid d_i) = P(a_1, a_2, \ldots, a_{|A|} \mid d_i) = \prod_j P(a_j \mid d_i) . \tag{9}$$

And so the naïve Bayes classifier [7] is obtained.

$$d_{MP} = \arg\max_{d_i \in V_d} P(d_i) \prod_j P(a_j \mid d_i) . \tag{10}$$

In order to represent Equation (10) with distribution numbers and share the statistical information with a discretizer, a set of statistical numbers is defined as follows. Suppose that there is an instance information system $I = < U, A \cup D >$. Let N_{d_k} represent the number of instances with decision value d_k .

$$N_{d_k} = |\{u : d(u) = d_k \;\; for\ all\ u \in U\}| . \tag{11}$$

Let N_{d_k, a_i, v_x} represent the number of instances with decision value d_k and attribute value $v_x \in V_{a_i}$.

$$N_{d_k, a_i, v_x} = |\{u : d(u) = d_k \;\; and\ a_i(u) = v_x \;\; for\ all\ u \in U \}| \tag{12}$$

Let N_{a_i, v_x} represent the number of instances for all decisions $d_x \in V_d$ and attribute value $v_x \in V_{a_i}$.

$$N_{a_i, v_x} = |\{u : a_i(u) = v_x \;\; for\ all\ u \in U \}| . \tag{13}$$

For example, value number distribution for Table 1 is shown in Table 2.

The number N_{d_k, a_i, v_x} is a basic distribution number. Based on the number N_{d_k, a_i, v_x} , the numbers N_{d_k} and N_{a_i, v_x} can be calculated by following expression.

$$N_{d_k} = \sum_{v_x \in V_{a_i}} N_{d_k, a_i, v_x} \ \ for\ any\ a_i\ . \tag{14}$$

$$N_{a_i, v_x} = \sum_{d_x \in V_d} N_{d_x, a_i, v_x}\ . \tag{15}$$

Table 2. Distribution of value numbers

Decision	Attribute		N_{d_k,a_i,v_x} for $v_x \in V_{ai}$							N_{d_k}
	Name	Domain V_a	v_{x1}	v_{x2}	v_{x3}	v_{x4}	v_{x5}	v_{x6}	v_{x7}	
d1='+'	a_1	{1, 2, …, 7}	1	1	0	1	1	1	1	6
	a_2	{1, 2, 3, 4}	3	1	1	1	---	---	---	
	a_3	{1, 2, 3}	3	3	0	---	---	---	---	
	a_4	{1, 2, 3, 4}	0	0	2	4	---	---	---	
d2='-'	a_1	{1, 2, …, 7}	1	1	1	1	0	0	2	6
	a_2	{1, 2, 3, 4}	0	2	2	2	---	---	---	
	a_3	{1, 2, 3}	0	2	4	---	---	---	---	
	a_4	{1, 2, 3, 4}	2	2	2	0	---	---	---	

Based on these distributional numbers, Equation (10) is rewritten as follows.

$$d_{MP} = \arg\max_{d_k \in V_d} \frac{N_{d_k}}{|U|} \prod_i \frac{N_{d_k, a_i, a_i(u)}}{N_{d_k}}, \tag{16}$$

where $|U|$ is total number of instances in information system. Consider the virtual samples [1]. A modified naïve Bayes classifier is proposed as follows.

$$d_{mp} = \arg\max_{d_k \in V_d} \frac{N_{d_k}}{|U|} \prod_i \frac{N_{d_k, a_i, a_i(u)} + \beta \bullet |U|}{N_{d_k} + \beta \bullet |U| \bullet |V_a|}, \tag{17}$$

where β is a constant with small number and a typical value $\beta = 0.02$ is chosen in our experiments. By tuning this constant, a high accuracy can be obtained. Note that Equation (17) is different from the formula in [1]. The value number $|V_a|$ and instance number $|U|$ are considered here so that high accuracy can be obtained. Here, distribution numbers are used instead of probabilities because the numbers can be updated easily when a new instance is added to the instance information system. Suppose that the new instance *u13* = (*a1*=7, *a2*=3, *a3*=3, *a4*=1, *d* = '-') is added to the instance information system. Only 5 numbers need to be updated as follows.

$$N_{d\_, a_1, v_{x7}} \Rightarrow 3;\ N_{d\_, a_2, v_{x3}} \Rightarrow 3;\ N_{d\_, a_3, v_{x3}} \Rightarrow 5;\ N_{d\_, a_4, v_{x1}} \Rightarrow 3;\ N_{d\_} \Rightarrow 7.$$

By means of this update approach, any number of new instances can be added online. Equation (17) and distribution of value numbers as shown in Table 2 can be regarded as a type of knowledge for decision-making. The advantage of this representation is that it enables an intelligent system to update new knowledge in a changing environment.

3 Discretization of Continuous Attributes

3.1 Definition of Dichotomic Entropy

In order to share the statistical information with the naïve Bayes classifier, a discretizer, which is based on distributional index and minimum of a dichotomic entropy, is proposed to discretize the continuous attributes, instead of existing discretization approaches [5-7]. A compound index is obtained from distribution numbers for guiding to split intervals.

Let $v_x \in V_{a_i}$ is a value of continuous attribute a_i and N_{d_k,a_i,v_x} represent the number of instances with decision value $d_k \in V_d$ and value v_x for attribute a_i. Suppose that continuous attribute a_i is split by border value v_{bd}. The number of instances with decision d_k and value $a_i(u) \le v_{bd}$ is represented by $N_{d_k,left}$.

$$N_{d_k,left} = \sum_{v_x \le v_{bd}} N_{d_k,a_i,v_x} . \tag{18}$$

The number of instances with value $a_i(u) \le v_{bd}$ for all decision $d_k \in V_d$ is represented by $N_{a_i,left}$.

$$N_{a_i,left} = \sum_{d_k \in V_d} N_{d_k,left} . \tag{19}$$

The number of instances with decision d_k and value $a_i(u) > v_{bd}$ is represented by $N_{d_k,right}$.

$$N_{d_k,right} = \sum_{v_x > v_{bd}} N_{d_k,a_i,v_x} . \tag{20}$$

The number of instances with value $a_i(u) > v_{bd}$ for all decision $d_k \in V_d$ is represented by $N_{a_i,right}$.

$$N_{a_i,right} = \sum_{d_k \in V_d} N_{d_k,right} . \tag{21}$$

In order to indicate instance number and homogeneous degree over decision space within an attribute value interval, a *decision distributional index* is defined as follows.

$$E_d(v_{start} \to v_{end}) = \sum_{d_k \in V_d} -N_{d_k,v_{start} \to v_{end}} \log_2\left(\frac{N_{d_k,v_{start} \to v_{end}}}{N_{a_i,v_{start} \to v_{end}}}\right) , \tag{22}$$

where $N_{d_k,v_{start} \to v_{end}}$ represents the number of instances with decision value d_k and attribute value within v_{start} and v_{end}, and $N_{a_i,v_{start} \to v_{end}}$ represents the number of instances with attribute value within v_{start} and v_{end} for all decision values. This *decision distributional index* indicates that the larger the number of instances within the interval, the larger the index. The more homogeneous the distribution is, the larger the index. Based on this concept, two *decision distributional indexes* can be obtained when an interval is split. A *left* decision distributional index can be represented by following expression.

$$E_{left}(v_x \le v_{bd}) = \sum_{d_k \in V_d} -N_{d_k,left} \log_2(\frac{N_{d_k,left}}{N_{a_i,left}}) . \quad (23)$$

A *right* decision distributional index can be represented by following expression.

$$E_{right}(v_x > v_{bd}) = \sum_{d_k \in V_d} -N_{d_k,right} \log_2(\frac{N_{d_k,right}}{N_{a_i,right}}) . \quad (24)$$

A *dichotomic entropy* for splitting point v_{bd} is defined as

$$E(v_{bd}) = \frac{1}{|U|} E_{left}(v_x \le v_{bd}) + \frac{1}{|U|} E_{right}(v_x > v_{bd}) , \quad (25)$$

where $|U|$ is total number of instances. According to information entropy principles in machine learning theory [1,4,5,7,8], the smaller the entropy is, the better the attribute discretization. Applying Equation (25), a border value v_{border} can be obtained by minimization of dichotomic entropy.

$$v_{border} = \arg\min_{v_{bd} \in V_{a_i}} \frac{1}{|U|} E_{left}(v_x \le v_{bd}) + \frac{1}{|U|} E_{right}(v_x > v_{bd}) . \quad (26)$$

In other words, the minimal entropy can be obtained if v_{border} is applied to split attribute into two intervals.

3.2 Continuous Attribute Discretizer

Applying Equation (26), a continuous attribute can be split into two intervals. The two intervals can be selected one interval to split into 2 intervals by analogy. The total number of intervals becomes 3. Therefore, a continuous attribute can be split into any desired number of intervals. Here two questions have to be answered.

Which interval needs to be split further?

How many intervals are best for decision-making?

In principle, an interval with inhomogeneous value distribution and large number of instances should be split. In order to get a quantitative criterion, a distributional index for instance number distribution over the attribute values and the decision space within an interval, which is called a *value distributional index,* is defined as follows.

$$E_v(v_{start} \to v_{end}) = \sum_{v_{start} \le v_x < v_{end}} \sum_{d_k \in V_d} -N_{d_k,a_i,v_x} \log_2(\frac{N_{d_k,a_i,v_x}}{N_{a_i,v_x}}) . \quad (27)$$

It is obvious that $E_v(v_{start} \to v_{end})$ is small if the distribution varies with value v_x at high frequency. $E_v(v_{start} \to v_{end})$ is large if the distribution varies with value v_x very slow, in other words, the distribution over value v_x is homogeneous. Based on the difference of $E_d(v_{start} \to v_{end})$ - $E_v(v_{start} \to v_{end})$, a *compound distributional index* is defined as follows.

$$\Delta E(v_{start} \to v_{end}) = \frac{E_d(v_{start} \to v_{end}) - E_v(v_{start} \to v_{end})}{|U|} , \quad (28)$$

where $|U|$ is total number of instances in the instance information system. Dividing by $|U|$ ensures that $\Delta E(v_{start} \to v_{end})$ is a real value within [0,1] and $\Delta E(v_{start} \to v_{end})$ decreases monotonously when instance number of the interval decreases i.e. the interval becomes smaller. This *compound distributional index* can be applied as a criterion to determine whether an interval is to be split further. An interval with largest $\Delta E(v_{start} \to v_{end})$ is selected for splitting further in an adaptive discretizer. If $\Delta E(v_{start} \to v_{end}) = 0$ i.e. the value distribution is *homogeneous distribution with respect to* v_x, it is not needed to split the interval further. In most cases, 5 intervals are enough for reaching high decision accuracy in our experiments. Therefore, adaptive discretizer stops when the number of intervals reaches five or $\Delta E(v_{start} \to v_{end})$ is less than a threshold. These adaptive rules are thus very different from the approach in [4]. The value of $\Delta E(v_{start} \to v_{end})$ is applied to select an interval that should be split. The formal algorithm is as follows.

A1. Algorithm for discretization of continuous attribute

1 Calculate the distribution numbers according to Equations (11)-(15) and get the distributional numbers over sampled values and over decision space. The results are shown as Table 2.

2 Calculate dichotomic entropy and determine the splitting point

2.1 Initial values

Interval control number: n = 0;

$v_{start} = v_{min}$; $v_{end} = v_{max}$

Splitting point sequence list $S\_list = [v_{min}, v_{max}]$

2.2 Determine the splitting point

$$v_{border-n} = \arg\min_{v_{bd} \in V_{a_i}} E(v_{bd})$$

$$= \arg\min_{v_{bd} \in V_{a_i}} \frac{1}{N_{a_i, v_{start} \to v_{end}}} E_{left}(v_x \le v_{bd}) + \frac{1}{N_{a_i, v_{start} \to v_{end}}} E_{right}(v_x > v_{bd})$$

2.3 Add the splitting point into splitting point sequence list

S_list =[v_{min}, $v_{border-n}$, v_{max}]

3 Select an interval for splitting further

3.1 Calculate compound index

$$E_v(v_{start} \to v_{end}) = \sum_{v_{start} \le v_x < v_{end}} \sum_{d_k \in V_d} -N_{d_k, a_i, v_x} \log_2\left(\frac{N_{d_k, a_i, v_x}}{N_{a_i, v_x}}\right)$$

$$\Delta E(v_{start} \to v_{end}) = \frac{E_d(v_{start} \to v_{end}) - E_v(v_{start} \to v_{end})}{|U|}$$

3.2 Record compound index for each interval

C-index-list =[$\Delta E(v_{min} \to v_{border-n})$, $\Delta E(v_{border-n} \to v_{max})$]

Adaptive rule control

n=n+1;

If $\Delta E_{max} < 0.0001$ then end program

If n > 5 then end program

3.4 Select the interval with maximal ΔE_{max} and set the parameters

$v_{start} = v_{start\text{-}with\text{-}max}$; $v_{end} = v_{end\text{-}with\text{-}max}$

Goto 2.2

//After running this program, a splitting point sequence list can be obtained

4 Experimental Results

In order to test the improvement of the improved naïve Bayes classifier with adaptive discretizer, 16 benchmark data sets from the UCI Machine Learning Repository were applied. The decision accuracies under ten-fold cross validation standard are given in column *Bayes* in Table 3. Sub column *Org* lists decision accuracies for the classifier without the adaptive discretizer. Sub column *Dich* lists decision accuracies for the classifier with the adaptive discretizer. Sub column *D5* lists decision accuracies for the classifier with a 5-identical-interval discretizer. It can be seen that the modified naïve Bayes classifier with the adaptive discretizer improved decision accuracies for 14 data sets. The average accuracy over 16 data sets is better than that approaches without the adaptive discretizer and with a 5-identical-interval discretizer. The adaptive discretizer was applied to the C5.0 tree. The results are shown in column *C5.0* tree. The average accuracy is still improved when this adaptive discretizer is attached for data preparation. The modified naïve Bayes classifier combined with the discretizer can obtain higher average accuracy than the C5.0 tree for the 16 data sets. Column *Att* is for attribute numbers in data sets. The string '60c60' indicates that there are 60 attributes and 60 attributes are continuous attributes. Column *N* is for instance numbers in data sets. The names with '♣' indicate that some attribute values

Table 3. Comparable results for the Naïve Bayes classifier with the discretizer

Data Name	Att	N	Bayes			C5.0 tree		
			Org	Dich	D5	Org	Dich	D5
Sonar	60c60	208	64.0	89.9	81.7	71.0	74.0	73.0
Horse-colic ♣	27c7	300	73.7	75.0	73.7	78.3	80.0	80.7
Ionosphere	34c34	351	82.6	90.3	88.6	88.3	91.5	89.5
Wine	13c13	178	98.3	98.3	96.1	93.2	97.7	95.5
Crx_data♣	15c6	690	81.0	86.5	85.0	85.1	86.4	84.1
Heart	13c6	270	76.3	84.4	84.8	78.1	76.3	79.3
Hungarian♣	13c6	294	83.3	83.7	83.7	79.2	80.9	78.2
SPECTF	44c44	80	66.3	88.8	75.0	70.0	76.2	68.8
Astralian	14c6	690	85.2	85.5	86.5	83.2	85.9	84.5
Echocard♣	12c8	132	60.6	73.5	68.0	68.2	57.5	69.0
Bupa	6c6	345	64.4	70.2	67.0	67.2	65.2	66.4
Iris_data	4c4	150	94.0	96.7	82.0	96.7	96.7	96.0
Ecoli	6c6	336	71.5	75.3	75.0	71.5	83.6	82.4
Anneal♣	38c6	798	85.4	86.7	86.2	98.6	98.6	98.6
Hepatitis♣	19c6	155	69.6	70.8	69.6	62.0	68.5	67.3
Bands♣	39c20	540	64.8	66.3	68.5	68.0	68.3	69.4
Average			76.9	**82.6**	79.5	78.7	**80.5**	80.2

Table 4. Comparision with other existing approaches

Data/ Approach	Bayes-Dich	CLIP4-CAIM	C4.5-MChi	Chi-C4.5	MDLPC-C4.5
Iris	96.7	92.7	94.7	94.0	94.0
Heart	84.4	79.3	32.9	55.1	54.5

are missing in the data set. As it is a NP-hard problem to find the best solution for discretization, the proposed discretizer gives a good solution with a low time complexity. Therefore, for some data sets, the decision accuracies are slightly lower than 5-identical-interval discretizer, for example, data sets Heart and Bands. Results for comparison with other existing approaches [5] are shown in Table 4.

5 Conclusion

Dichotomic entropy is defined and minimal dichotomic entropy can be applied to determine the border value for splitting an interval. A compound index, which composes of decision distributional index and value distributional index, is defined and applied for guiding to the interval that should be split during the discretization. Based on these concepts, a continuous attribute can be split in two intervals at the border point with minimal dichotomic entropy, and then the compound index is applied to select an interval to split further until the desired number of intervals is reached or the compound index is small enough. Applying the improved naïve Bayes classifier with adaptive discretizer to 16 benchmark data sets, experimental results show that the average accuracy has been improved with different scales. The improved naïve Bayes classifier with adaptive discretizer obtained higher average accuracy than the C5.0 tree.

References

1. Mitchell, T.: Machine Learning, McGraw Hill, Co-published by the MIT Press Companies, Inc. (1997).
2. Rish, I.: An empirical study of the naive Bayes classifier, IJCAI-01 Workshop on Empirical Methods in Artificial Intelligence, (2001).
3. Wu, Q. X., Bell, D.A. and McGinnity, T.M.: Multi-knowledge for Decision Making. International Journal of Knowledge and Information Systems, Springer-Verlag, 2 (2005) 246 – 266.
4. Wu, X.: A Bayesian Discretizer for Real-Valued Attributes. The Computer J., 39(8) (1996)688-691.
5. Kurgan, L.A., and Cios, K. J.: CAIM Discretization Algorithm. IEEE Transactions on Knowledge and Data Engineering, 16(2) (2004) 145-153.
6. Dougherty, J., Kohavi, R., and Sahami, M.: Supervised and Unsupervised Discretization of Continuous Features. Proc. of International Conference on Machine Learning, (1995)194-202.
7. Wu, Q.X. and Bell, D.A.: Multi-Knowledge Extraction and Application. In Rough Sets, Fuzzy Sets, Data Mining, and Granular Computing, (eds) Wang, G.Y., Liu, Q., Yao Y.Y. and Skowron, A., LNAI 2639, Springer, Berlin, (2003) 274-279.
8. Quinlan, J. R.: Induction of Decision Trees. Machine Learning, 1(1) (1986) 81 - 106.

Model Inference of a Dynamic System by Fuzzy Learning of Geometric Structures

Kaijun Wang[1], Junying Zhang[1], and Jingxuan Wei[2]

[1] School of computer science and engineering, Xidian University, Xi'an 710071, P.R. China
kjwang@mail.xidian.edu.cn, jyzhang@xidian.edu.cn
[2] Dept of applied mathematics, Xidian University, Xi'an, P.R. China
wjxjingxuan@yahoo.com.cn

Abstract. One of difficult tasks on dynamic systems is the exploration of connection models of variables from time series data. Reasonable time regions for constructing the models are crucial to avoid improper models or the loss of important information. We propose fuzzy learning of geometric structures to find reasonable time regions and proper models to reveal varying laws of system. By comparing values of fuzzy merging function for shorter time regions and fuzzy unmerging function for larger varying actions, reasonable model regions are inferred. Experimental results (for both simulated and real data) show that the proposed method is very effective in finding connection models adaptive to the evolution of a dynamic system, and it detected large varying actions in the regions below preset minimal region length, whereas the non-fuzzy learning method failed.

1 Introduction

Knowledge discovery of a dynamic system is a difficult task, since connections between variables (or connection model) often vary with evolution of the system. For this unsupervised task, reasonable time regions for model construction (or *model regions*) are crucial to avoid improper models. Nevertheless, few literatures refer to the problem of finding reasonable model regions. Some important approaches for dynamic systems like Dynamic bayesian network, which represents connections of a variable set by joint probability distribution, have not concerned this problem.

We propose fuzzy learning of geometric structures to find reasonable model regions and a series of proper connection models adaptive to evolution of a dynamic system. The geometric structures (as possible model regions) from time series data are used as guidance to reasonable model regions. To avoid unreasonable models in small time regions and information losing of large varying actions, flexible fuzzy techniques [1] are employed: a fuzzy function of region merging for small model regions and a fuzzy function of region unmerging for larger varying actions, and their fuzzy values are compared to reason reasonable model regions.

L. Wang et al. (Eds.): FSKD 2006, LNAI 4223, pp. 684 – 687, 2006.

2 Model Inference Based on Geometrization of Data

The introduction of geometrization of time series data into continuous binomial curves to form curve manifolds and model inference for curve manifolds are from [2].

Let a dynamic system have m continuous variables, and let d time series data $y_j \in R$ of variable Y be given, and the fitting of y_j be by a binomial curve $y=f_i(z)$ (z is time variable). Let $f_i(z)=c_0+c_1z+c_2z^2$, then its parameters c_k can be found under least square error criterion. Thus, A curve manifold [3] is formed when all the local curves $y=f_i(z)$ are joined, i.e. $M_Y=\bigcup\{y=f_i(z)\}$. When parameter u is used, the local region U_i of M_Y is $r_i(u)=(y=f_i(u), z=u)$ (or parameter curves) instead of $y=f_i(z)$, and then any tangent vector on U_i is $(f_i'(u),1)$ or only $f_i'(u)$ for convenience.

Definition 2.1 (Geometric frame)**.** Let H be a set with k ordered geometric entities on curve manifold M, which consist of local curves or tangent vectors, then H is called a geometric frame (noted as $H(M, f_i)=\{f_i(u)\}$ when $f_i(u)$ is used).

The model inference is to find parameters of linear regression model from given curve manifolds under least square error criterion when model structure is known. Let model G_i be $M_Y=\phi_0 I+\phi_i M_X$, n pairs of geometric frames $\{p_l, h_l\}$ for M_Y and M_X respectively be given, we set $\overline{p}=\frac{1}{n}\sum_l p_l$, $\overline{h}=\frac{1}{n}\sum_l h_l$, and choose uniform local scope $u\in[u_1, u_2]$. Then, we can infer $\phi_0=\overline{p}-\phi_i\overline{h}$ and

$$\phi_i=\left(\sum_{l=1}^{n}\int_{u_1}^{u_2} f_l(u)g_l(u)du-n(u_2-u_1)\overline{p}\overline{h}\right)\Big/\left(\sum_{l=1}^{n}\int_{u_1}^{u_2} g_l(u)g_l(u)du-n(u_2-u_1)\overline{h}\,\overline{h}\right)$$

When model structure is unknown, structure reasoning is to find a G_k fitting curve manifolds most from hypothesis model space $S=\{G_i\}$. To prevent over-fitting, we need a criterion like AIC [4] to evaluate which model is optimal (at minimal AIC).

3 Finding Model Regions by Fuzzy Learning

We will define waves to describe geometric structures (rise and descent of curves) of curve manifolds, which give guide information for reasonable model regions.

Definition 3.1 (Wave)**.** Let U and V are two domains of curve manifold M, and U_1 and U_2 be arbitrary small proximate regions in both sides of U. If tangent vectors on U are all positive or negative (called wave condition), but the same wave condition is not satisfied in U_1 and U_2, then U is called an ascending wave or descending wave of M. If V contains both ascending and descending waves, and all the ascending waves are in front of descending waves under time direction, then V is called a wave of M.

Let a geometric unit be a wave or a region above preset minimum length on a curve manifold, and then a geometric unit is a possible model region. In order to avoid unreasonable models in too small time regions or the loss of large varying information, we adopt fuzzy technique and set a threshold, a minimum length of model region. Fuzzy technique can make a balance between the catching of larger varying actions of system and the preferred larger model regions.

We design a fuzzy function to measure merging necessity of small model regions and a fuzzy function of region unmerging for larger varying actions, and compares values of two fuzzy functions to determine whether a geometric unit remains as a model region or merges with its neighbor.

Definition 3.2 (fuzzy merging/unmerging function)**.** Let L_k be length of a geometric unit k. Let *Lmin* be lower bound of a model region, or preset minimal length of a model region. And let *Lm* be merging threshold, below which a region must be merged. Then, $S_{merg}(k) = 1 - \min\{(L_k - L_m)/(L_{min} - L_m), 1\}$ is the fuzzy merging function. Let V_k be data variance of a variable in a geometric unit k, *Vm* be lower bound of V_k for a region k entitled by a "large varying action", and *Vmax* be up bound of V_k. Then, $S_{unm}(k) = \min\{(V_k - V_m)/(V_{max} - V_m), 1\}$ is fuzzy unmerging function for large varying actions.

Remarks. Generally, the *Lm* may be set as half *Lmin*, and *Vm* and *Vmax* as a median and maximum of data variances among all variables of system respectively. Thus, only *Lmin* needs presetting.

4 Experimental Results

The simulated data with random white Gaussian noise of variance 0.2 are from model $Z(t)=2.5Y(t)-5$, where $X(t)=14.5+0.5\sin(0.28t+4)$ and $Y(t)=10+5\sin(0.031t)$, $t\in[1,100]$; and $Z(t)=Y(t)+10$, where $X(t)=14.5+0.5\sin(0.28t+4)$ and $Y(t)=10.5+0.5\sin(0.18t)$, $t\in[101,300]$; but a big pulse $X(t)=14+5\sin(0.105t-21.06)$ and $Z(t)=2X(t)-8$ in $t=[201,230]$. The minimal length of model regions is set to be 50.

The real data Bupa from UCI repository of machine learning databases (Blake, C. L. and Merz, C. J.) contains instances in 345 days for 5 blood variables sensitive to liver disorders. The possible relation between alamine aminotransferase (or *Z*) and aspartate aminotransferase (or *X*) are evaluated. The first 190 data are used and normalized to have variance of 1 for each variable. The minimal region length is set to be 30 days.

The results (including found models from non-fuzzy method) are listed in Table 1 and 2. It can be seen that the fuzzy learning method found reasonable model regions at relatively low error, whereas the non-fuzzy method could not reveal a large varying action of system in T3 for simulated data and in T2 & T4 for Bupa. This suggests that the fuzzy learning method can find proper models adaptive to system variation, and detect larger varying actions of system in shorter regions too.

Table 1. Models found by fuzzy learning and non-fuzzy method for simulated data

Time region	T1=1−104	T2=105−194	T3=195−238	T4=239−300
correct model	Z=2.5Y−5	Z=1Y+10	Z=2X−8	Z=1Y+10
Fuzzy learning	2.502Y−5.14	1.082Y+9.10	2.014X−8.61	0.985Y+10.09
coefficient error	0.08%	8.2%	0.7%	1.5%
Non-fuzzy	2.502Y−5.14	1.855X+0.936Y−16.09		

Table 2. Models found by fuzzy learning and non-fuzzy method for Bupa data

Time region	Fuzzy	Time region	Non-fuzzy
T1=1−30	Z=1.417X+0.37	T1=1−190	Z=0.765X+0.001
T2=31−48	Z=1.135X+0.09		
T3=49−140	Z=0.483X−0.15		
T4=141−164	Z=0.936X+0.06		
T5=165−190	Z=0.248X+0.43		

5 Conclusion

Exploring connection models of dynamic systems from time series data is a difficult task. The proposed fuzzy learning method can find appropriate connection models adaptive to the evolution of system and catch larger varying actions in shorter regions, which was demonstrated in the experiments.

Geometric structure information inherent in data can help find reasonable model regions, and fuzzy technique can make a balance between the catching of larger varying actions of system and the preferred larger model regions. Their integration leads to the proposed method with a better ability for this task.

Acknoledgement

This work is supported by the National Science Fund of China under Grant No. 60574039 and 60371044, and by the national visiting scholar fund of China.

References

1. Honga, T., Lin, K., Wang, S.: Fuzzy data mining for interesting generalized association rules, Fuzzy Sets and Systems 138 (2003), 255–269.
2. Wang, K., Zhang, J., Guo, L.: Geometric Frame Network: a Geometrical Learning Theory for Knowledge Discovery. Technical report (2006), Xidian University.
3. Chen, W.: An introduction to Differential Manifold. High education Press 2001.
4. Akaike, H.: A new look at the statistical model identification. IEEE Trans. Automatic Control, Vol. AC-19 (1974), 716 ~ 723.

Context Modeling with Bayesian Network Ensemble for Recognizing Objects in Uncertain Environments

Seung-Bin Im, Youn-Suk Song, and Sung-Bae Cho

Department of Computer Science
Yonsei University
134 Shinchon-dong, Sudaemoon-ku, Seoul 120-749, Korea
{envymask, corlary}@sclab.yonsei.ac.kr, sbcho@cs.yonsei.ac.kr

Abstract. It is difficult to understand a scene from visual information in uncertain real world. Since Bayesian network (BN) is known as good in this uncertainty, it has received significant attention in the area of vision-based scene understanding. However, BN-based modeling methods still have the difficulties in modeling complex relationships and combining several modules, as well as the high computational complexity of inference. To overcome them, this paper proposes a method to divide and select the BN modules for recognizing the objects in uncertain environments. The method utilizes the behavior selection network to select the most appropriate BN modules. Several experiments are performed to verify the usefulness of the proposed method.

1 Introduction

Scene understanding, recognizing contexts to understand the class of scene or situation, is a very difficult and largely unsolved problem. When running a scene understanding process, visual information may be uncertain due to motion blur, bad lighting condition, etc. We need the recognition methods that are robust in uncertainty.

Since Bayesian network (BN) is good for modeling uncertain domain, it has received significant attention in the area of vision-based scene understanding [1, 2]. However, if we adopt BN techniques for real world scene understanding, there are some problems we must solve. At first, modeling complex relationships of real world is difficult and the nodes that represent relationships require large amount of computation [3]. Moreover, it is difficult to combine various goal-specific BN modules. Therefore, a novel method is required for modeling and combining BNs without any significant increase of computation.

This paper proposes a practical method of objects recognition for service robots. The method utilizes the behavior selection network (BSN) to effectively combine the pre-designed BN modules and the activity-object context BN modules for the prediction of target objects and the control of robot behaviors to find objects in several places.

L. Wang et al. (Eds.): FSKD 2006, LNAI 4223, pp. 688 – 691, 2006.

2 Context Modeling and BN Ensemble Based on Domain Knowledge

2.1 BN Design Method Based on Domain Knowledge of Occluded Object

We propose a method of BN design for reasoning the occluded objects using the object relationships based on activity. Fig.1 shows the process of designing them as tree structure based on common-cause principle. Constructed BN modules are composed of activity, class and primitive nodes. The BN is a singly-connected structure like tree, and common-cause structures are used for sub-trees: it is a sort of causal relationships among three nodes. It has two nodes that have another node as a parent in common. This structure allows us to represent the relationship of objects more easily and precisely [4]. The process of designing BNs for occluded objects is a simple causal chain because it is possible to assign the parameters to each node for representing the relationship between them.

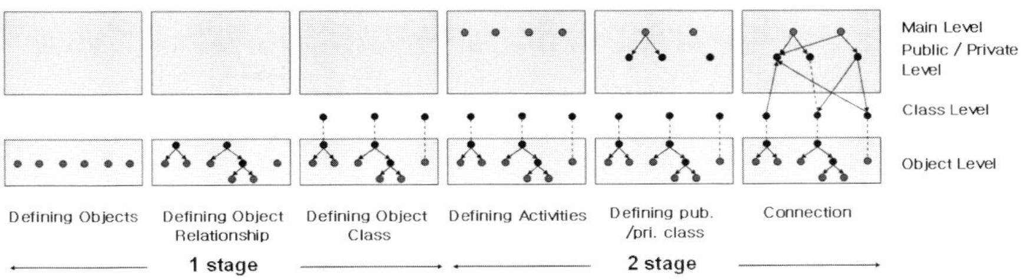

Fig. 1. The process of designing BN modules for occluded objects

2.2 Behavior Selection Network for BN Ensemble

Behavior selection network accomplishes the objectives through action selection and state transition. Usual input for BSN is sensory information, but we connect the BN inference results to sensor input for more accurate environmental perception. We have modified the original definition of BSN inference [5] with the following equations.

At the forward propagation, activation a_1 is

$$\Delta a_1 = \sum_{i=1}^{n} f(a_{s_i}) + \sum_{k=1}^{m} f(a_{B_k}), \text{ where } f(a_{B_k}) = \phi \times a_{B_k}$$
$$f(a_{B_k}) = P(y_k \mid Parents(y_k), Child(y_k)), \text{ where } y_k \text{ is the query node of each BN} \quad (1)$$

Here, n means the number of sensors, m means the number of inferential results that can be the input of BSN, and $f(a)$ is the Bayesian inference function of each BN. BSN module combines the pre-designed BNs (for recognizing places and objects) and activity BN that was constructed by the proposed method.

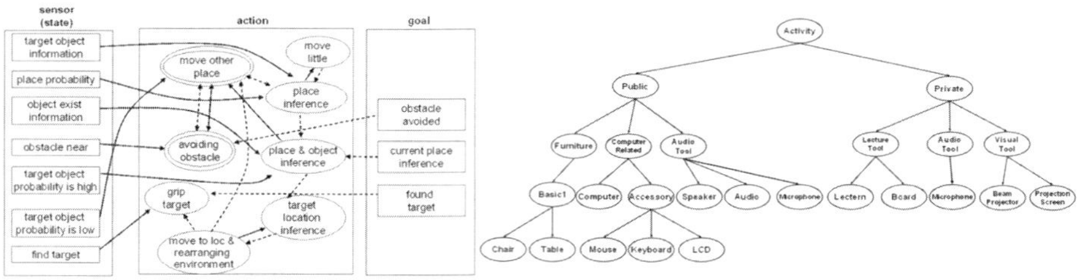

(a) (b)

Fig. 2. (a) Structured behavior selection network, (b) Constructed BN for objects that have the relationship with the activity of 'presentation'

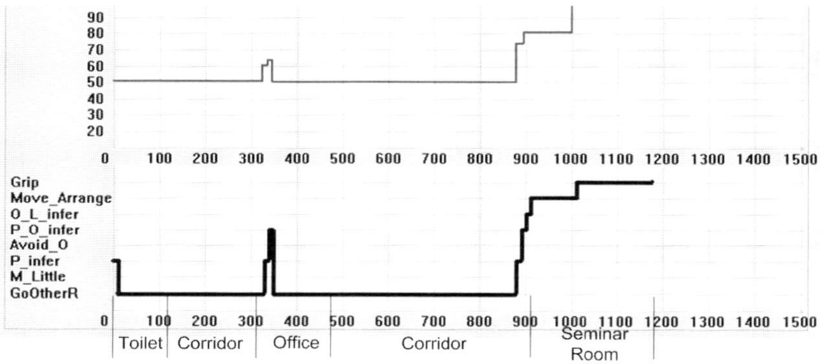

Fig. 3. Experimental results of finding beam-projector, occluded target object. Upper thin line shows the object existence probability and under thick line shows the state transition.

3 Experimental Results

The task of the experiments is to let the service robot find occluded objects in indoor environments and design Bayesian networks according to the proposed method. In the experiment, the robot wanders around four places (toilet, corridor, office, seminar room) and finds a beam projector, the target object occluded. In this paper, we have used simulated experimental environment, and randomly generated place probabilities in each places according to real world probability distribution that extracted from vision data. Fig. 3 shows one of the experimental processes that were executed in the webot simulator. Upper graph shows the change of object existence probability and lower graph depicts the states selected by BSN. If the inferred probability is lower than threshold of 70%, deeper-level inference is triggered. We confirmed that BSN module could find target objects precisely and efficiently: The BSN produced 100% accuracy in ten repeated runs.

Fig. 4 shows the result of the second experiment: prediction of beam-projector existence in contextually similar places. That is shown under threshold of 70% until the robot finds five objects with the activity-object context BN in the BSN module.

Predictions seem reasonable except one case. It can be seen that false-positive error is likely to occur in the similar environment from the result of accumulating evidences.

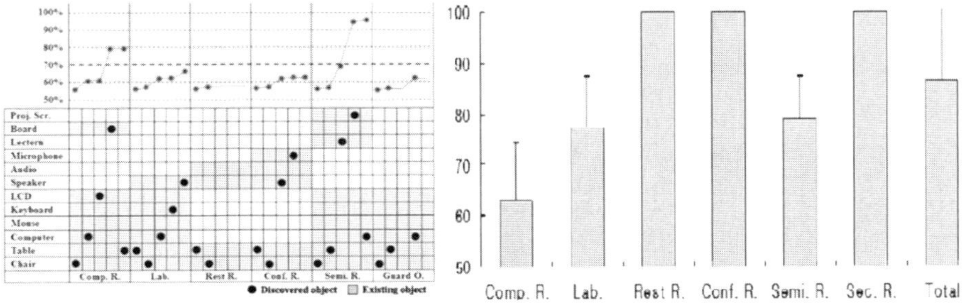

Fig. 4. The result of predicting an occluded-object (beam- projector) in six places (computer room, laboratory, rest room, conference room, seminar room, and security room)

4 Concluding Remarks

In this paper, we proposed a context modeling method for scene understanding (especially, object recognition) and intelligent service of robot based on image. We use the BSN for properly combining several types of BN modules used for inferring places and occluded objects as well as controlling robot behavior for its goal. In addition, we proposed a BN modeling method for inferring place and occluded objects. The experiments show the usefulness of the proposed method. For the future works, we will extend the scope of domains and apply the method to the real robots.

Acknowledgments. This work was supported by the Korea Science and Engineering Foundation (KOSEF) through the Biometrics Engineering Research Center (BERC) at Yonsei University.

References

1. B. Neumann, *A Conceptual Framework for High-Level Vision*, Bericht, FB Informatik, FBI-HH-B245/03, 2003.
2. M. Marengoni, A. Hanson, S. Zilberstein and E. Riseman, "Decision making and uncertainty management in a 3D reconstruction system," *IEEE Trans. Pattern Analysis and Machine Intelligence*, vol. 25, no. 7, pp. 852-858, 2003.
3. M. Neil, N. Fenton and L. Nielson, "Building large-scale Bayesian networks," *The Knowledge Engineering Review*, vol. 15, no. 3, pp. 257-284, 2000.
4. Y.-S. Song, S.-B. Cho, and I.-H. Suh, "Activity-object Bayesian networks for detecting occluded objects in uncertain indoor environment," *Int. Conf. on Knowledge-Based Intelligent Information & Engineering Systems*, vol. 3683, pp. 937-944, 2005.
5. K.-J. Kim, S.-B. Cho, "BN+BN: Behavior network with Bayesian network for intelligent agent," *Lecture Notes in Artificial Intelligence*, vol. 2903, pp 979-991, 2003.

Mining Sequential Patterns in Large Datasets

Xiao-Yu Chang[1], Chun-Guang Zhou[1,*], Zhe Wang[1], Yan-Wen Li[1,2], and Ping Hu[1]

[1] College of Computer Science and technology, Jilin University, Key Laboratory of Symbol Computation and Knowledge Engineering of the Ministry of Education, Changchun 130012, P.R. China
[2] Department of Computer Science, Northeast Normal University Changchun 130012, P.R. China
*cgzhou@jlu.edu.cn

Abstract. A novel algorithm FFSPAN (**F**ast **F**requent **S**equential **Pa**tter**n** mining algorithm) is proposed in this paper. FFSPAN mines all the frequent sequential patterns in large datasets, and solves the problem of searching frequent sequences in a sequence database by searching frequent items or frequent itemsets. Moreover, the databases that FFSPAN scans keep shrinking quickly, which makes the algorithm more efficient when the sequential patterns are longer. Experiments on standard test data show that FFSPAN is very effective.

1 Introduction

Sequential pattern mining [1] is a very important field in data mining with broad applications, including the analysis of market baskets, Web access patterns, disease treatments, natural disasters, DNA sequences, and so on. Most existing algorithms are variants of Apriori [2], which employ a breadth-first approach, i.e. finding all k-length sequences before considering (k+1)-length sequences. This approach limits the effectiveness of them, since useful longer frequent patterns can not been discovered. Recently, the merits of the depth-first approach have been recognized [3]. To mine long patterns, we propose a novel algorithm FFSPAN which mines the complete set of sequential patterns but greatly reduces the size of database to be scanned. The search strategy of FFSPAN integrates a depth-first traversal of the search space with effective pruning mechanisms. Traditionally, to judge whether a sub-sequence is frequent in a database, it is needed to compare the whole sub-sequence with every sequence in the original database, however FFSPAN only needs to compare the last itemset of the sub-sequence with every sequence in a database that keeps shrinking. So it is much more efficient when patterns become long.

2 Preliminaries and Related Work

Given a sequence $s = \{t_1,t_2,\ldots,t_m\}$ and an item α, $s \lozenge \alpha$ means that s concatenates with α. A new sequence s' generated by $s \lozenge \alpha$ can be classified to two classes: (1) α is an **SES** (a sequence-extended step [4][5][6]): $s \lozenge_s \alpha = \{t_1,t_2,\ldots,t_m,(\alpha)\}$; (2) α is an **IES** (an itemset-extended step): $s \lozenge_i \alpha = \{t_1,t_2,\ldots,t_m \cup (\alpha)\}$, $\forall\, k \in t_m,\ k < \alpha$. Also, given two

L. Wang et al. (Eds.): FSKD 2006, LNAI 4223, pp. 692–695, 2006.

sequences $s = \{t_1,t_2,\ldots,t_m\}$ and $p = \{t_1',t_2',\ldots,t_m'\}$, $s \Diamond p$ means s concatenates with p, and it can be classified to two classes: p is an **IES**, or p is an **SES**. If $s = p' \Diamond s'$, $p \subseteq p'$, and the last item of p is the same as the last item of p', then p is a *prefix* of s, s' is a *suffix* of s with respect to (w.r.t.) p. For example, {(a,d)(b)},{(a)(b)} are two prefixes of {(a,d)(b,c,d)}, and {(c,d)} is a suffix of {(a,d)(b,c,d)} w.r.t. prefix {(a,d)(b)} and {(a)(b)}. We can get a suffix s' of s by removing a prefix p of s, denoted as $s' = s - p$.

3 The FFSPAN Algorithm

Given a sequence s, let *lastItemset*(s) and *firstItemset*(s) represent the last itemset and the first itemset of s respectively. Given a prefix sequence p and a database D, the new database D' consists of all of the sequences in D whose prefixes are p, and corresponding postfixes are not Φ. D_p is the projected database of D w.r.t. p, which consists of all of the postfixes of each sequence in D' w.r.t. the prefix p. For each sequence, s_j', in D', let β_j represent the itemset of s_j' which *lastItemset*(p) occurs in; e.g. in D' of Table 1, β_2 of s_2' is (ab), and β_4 of s_4' is (abd). In D_s, let s_j^P represent the corresponding postfix sequence of s_j' w.r.t. the prefix p. (1) If β_j - *lastItemset*(p) $\neq \Phi$ (e.g. (ab) - (b) = Φ, but (abc) - (b) = (c)), remove *firstItemset*(s_j^P) of s_j^P in D_p. The new database is denoted as D_{p-}. (2) If β_j - *lastItemset*(p) $\neq \Phi$, replace *firstItemset*(s_j^P) of s_j^P with β_j; else replace s_j^P with $\beta_j \Diamond_s s_j^P$ in D_p. The new database is denoted as D_{p+}. Table 1 is an example to illustrate the definitions of the new databases above.

Table 1. Databases w.r.t. prefix sequence p = {(a)(b)}

	D	D'	D_p	D_{p-}	D_{p+}
ID	Sequence	Sequence	Sequence	Sequence	Sequence
1	{(a,b)(c,d)(a,d)}	Φ	Φ	Φ	Φ
2	{(*a*,c)(a,*b*)(d)}	{(*a*,c)(a,*b*)(d)}	{(d)}	{(d)}	{(a,*b*)(d)}
3	{(*a*)(a)(a,*b*)}	Φ	Φ	Φ	Φ
4	{(*a*,c)(a,*b*,d)(c,d)}	{(*a*,c)(a,*b*,d)(c,d)}	{(d)(c,d)}	{(c,d)}	{(a,*b*,d)(c,d)}

Theorem 1. *Given a prefix sequence p, let supD(s) represent the support of s in database D, then for any item α*
(1) $sup_D(p \Diamond_s \alpha) = sup_{Dp-}(\alpha)$; *(2)* $sup_D(p \Diamond_i \alpha) = sup_{Dp+}(lastItemset(p) \Diamond_i \alpha)$.

Theorem 1 can be easily proved and, we can propose the algorithm as follow:

Algorithm1: FFSPAN(t, *offset*[], S_n, I_n, *minsup*)
Input: t, an itemset, t = *lastItemset*(p); *offset*[], an array of pointers to distinguish D_{p-} and D_{p+}; I_n and S_n, the candidate sets; *minsup*, minimum support.
Output: The complete set of frequent sequences.

$S_{temp} = \Phi$; $I_{temp} = \Phi$;
For each $\alpha \in S_n$
  if $\sup_{Dp-}(\alpha) \geq minsup$
  //based on Theorem 1, we improve the step: "**if** $\sup_D(p \Diamond_s \alpha) \geq minsup$"
  then $S_{temp} = S_{temp} \cup \{\alpha\}$

```
For each (α∈ S_temp)
    FFSPAN ((α), offset[] w.r.t. α, S_temp, all items in S_temp greater than α, minsup)
For each α∈ I_n
    if sup_Dp+ (t ◊_i α) ≥ minsup
    //based on Theorem 1, we improve the step: "if sup_D (p ◊_i α) ≥ minsup"
    then I_temp = I_temp ∪ {α}
For each (α∈ I_temp)
    FFSPAN (t ◊_i α, offset[] w.r.t. α, S_temp, all items in I_temp greater than α, minsup)
```

4 Experimental Results

The experiments are implemented on a 2.7GHZ Intel PC with 512MB main memory, running Windows XP. All the codes are written by C++.The test datasets are generated by the IBM Quest Synthetic Data Generator [1]. It can be downloaded in the following website: http://www.almaden.ibm.com/software/quest/Resources/datasets/syndata. html#assocSynData. Now let us examine the effectiveness of the application of the idea in Theorem 1. The algorithm that is not improved by the idea is called "SPAM". Figures 1-6 demonstrate the differences between FFSPAN and SPAM. Figures 1-3 vary minimum supports for different size datasets. We can see that when min_sup becomes lower, FFSPAN is much faster than SPAM. Because when min_sup becomes lower, more patterns will be mined and the times of scanning a database will be more and more, and the cost of scanning a shrinking database and finding a frequent item or itemset for FFSPAN is much less than that of scanning the original database and finding a frequent sequence for SPAM. Figures 4-6 vary the parameters of the IBM data generator for large datasets, the results show that FFSPAN is faster than SPAM.

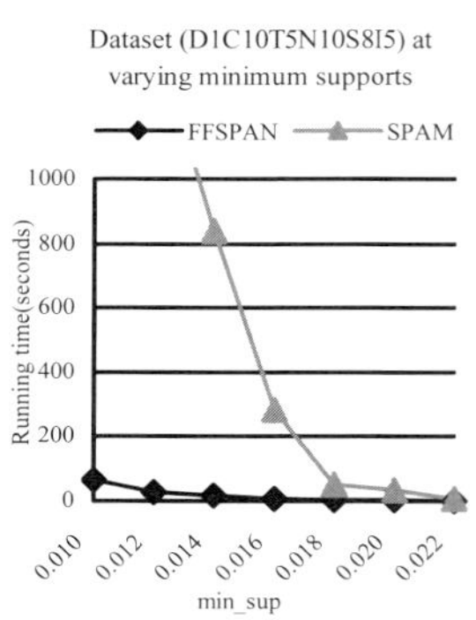

Fig. 1. Varying support for small dataset

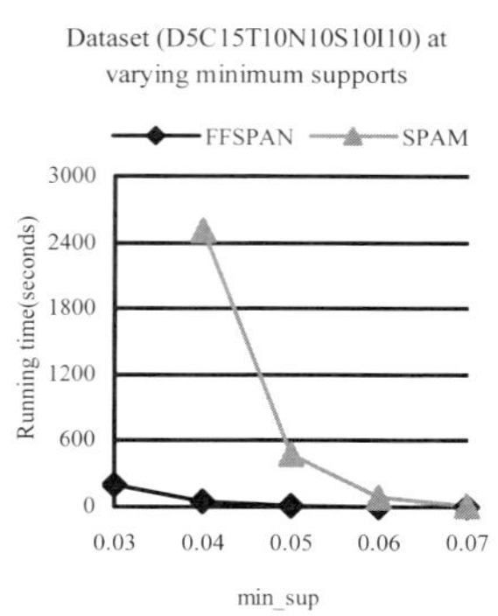

Fig. 2. Varying support for medium-sized dataset

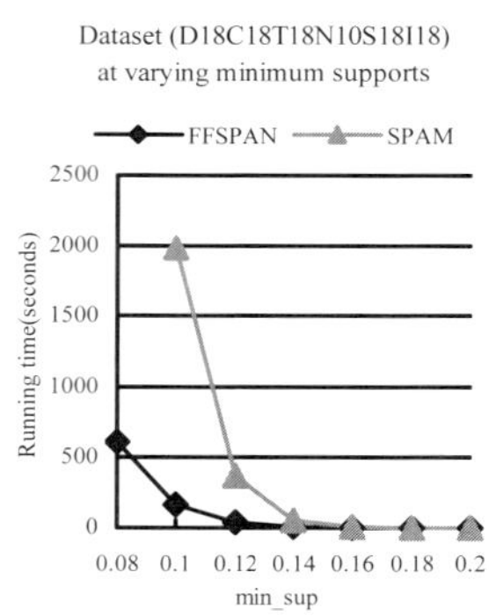

Fig. 3. Varying support for large dataset

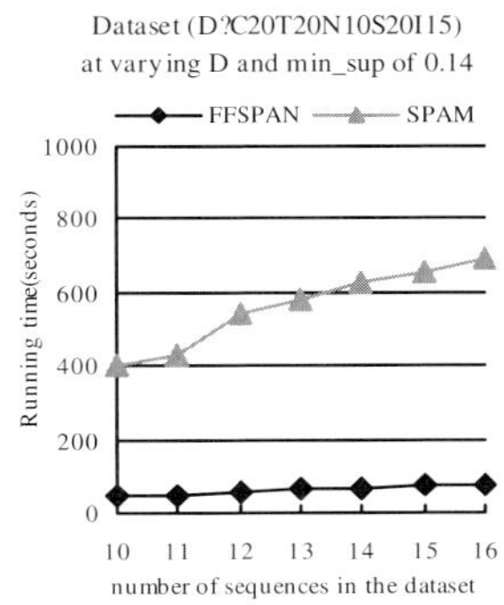

Fig. 4. Varying number of customers

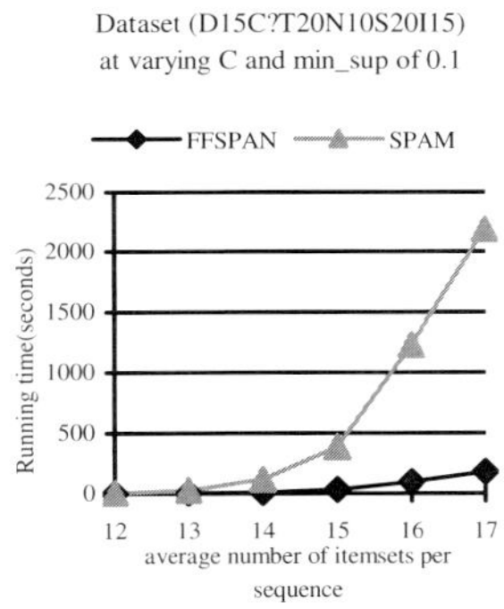

Fig. 5. Varying number of itemsets per sequence

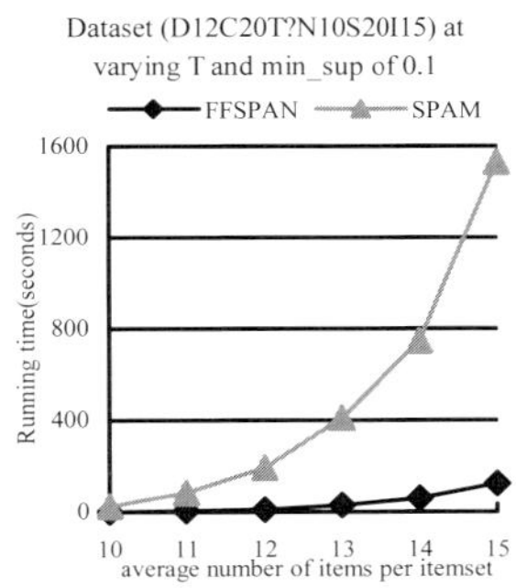

Fig. 6. Varying number of items per itemset

5 Conclusions and Discussions

In this paper, we propose a novel algorithm to mine the set of all the frequent sequential patterns in a database. FFSPAN mines a frequent sequence by judging whether the last itemset of that sequence is frequent in a database that keeps shrinking, which makes it especially efficient when the sequential patterns are long. Both theory and experiments examine the effectiveness of FFSPAN.

Acknowledgement

This work is supported by the National Natural Science Foundation of China under Grant No. 60433020 and the Key Laboratory of Symbol Computation and Knowledge Engineering of the Ministry of Education, and "985" Project of Jilin University.

References

1. R. Agrawal and R. Srikant: Mining Sequential Patterns. In ICDE 1995, Taipei, Taiwan (1995)
2. R. Agrawal and R. Srikant: Fast algorithms for mining association rules. In VLDB'94, pages 487–499, Santiago, Chile (1994)
3. D. Burdick, M. Calimlim, and J. Gehrke: Mafia: A maximal frequent itemset algorithm for transactional databases. In ICDE 2001, Heidelberg, Germany (2001)
4. X. Yan, J. Han, and R. Afshar: CloSpan: Mining Closed Sequential Patterns in Large Databases. In SDM 2003, San Francisco, California, USA (2003)
5. J. Ayres, J. Flannick, J. Gehrke, et al: Sequential pattern mining using a bitmap representation. In SIGKDD2002, pages 429–435, Edmonton, Alberta, Canada (2002)
6. J. Pei, J. Han, B. Mortazavi-Asl, H. Pinto, Q. Chen, U. Dayal and M-C. Hsu: PrefixSpan: Mining Sequential Patterns Efficiently by PrefixProjected Pattern Growth. In. Proc. 2001 Int. Conf. Data Engineering (ICDE'01), pages 215-224, Heidelberg, Germany (April 2001)

EAST: Energy Alignment Search Tool

Dariusz Mrozek, Bożena Małysiak, and Stanisław Kozielski

Silesian University of Technology, Department of Computer Science, Akademicka 16,
44-100 Gliwice, Poland
{Dariusz.Mrozek, Bozena.Malysiak, Stanisław.Kozielski}@polsl.pl

Abstract. The inspection of structural changes (in selected proteins' regions) that take place in particular reactions of signal pathways is very important for their analysis. Conformational modifications as an effect of biochemical reactions or environmental influences cause changes in the conformational energy distributions. Observations of the energy characteristics allow to detect structural changes of selected proteins. In the paper, we present the EAST – method of similarity searching based on energy profiles. This method can be employed to find similar proteins during the analysis of substrates of signal paths' reactions and to detect regions of protein structure modifications.

1 Introduction

Signal transduction is a process of passing signal inside and between cells in living organisms [1], [2]. The process is realized through many consecutive reactions that occur one-by-one. Signal is passed by the activation of appropriate biological molecules in given cellular reaction causing changes in their chemical construction and/or spatial conformation. Each component reaction of the signal path may involve many molecules on all pathway's levels [3]. Reactions in the particular pathway occur according to the "key-lock" paradigm [4] involving several substrates, enzymes, coenzymes and other cofactors. These are usually proteins – biological molecules that play a main role in all biochemical processes in living cells [5], [6].

In the analysis of biochemical reactions and proteins' activity in signal pathways, protein structures and protein chemical constructions are much more descriptive than protein sequence. Changes of proteins' chemical activity are usually reached by their conformational changes in particular reaction. In the living cells, proteins usually switch their conformation as an effect of the molecular interactions with other molecules, e.g. in the phosphorylation process [7], or environment factors, e.g. temperature or pH changes of the vicinity.

The analysis of molecular regions, where proteins change their conformation in particular reaction, is very important for the analysis of the entire signal path. Some of these places are favorable from the biological point of view. The biological behavior of proteins is determined by their chemical features that can be difficult to extract by visual inspection of series of active and inactive compounds [16]. For this reason, in our work we use energy characteristics to compare modifications in protein structures. We use the theory of molecular mechanics [8] to determine the potential energy for

L. Wang et al. (Eds.): FSKD 2006, LNAI 4223, pp. 696 – 705, 2006.

the particular protein conformation. The total potential energy of the protein structure is a summary result of the component energies that comes from different interactions between atoms in the protein. Therefore, according to the molecular mechanics, changes of the protein conformation result in alteration of the potential energy value. And conversely, observation of the potential energy changes may reveal the conformational modifications. Knowledge of proteins' activity and its regulation on the molecular level can give us additional information about interactions regulating the cell cycle and potential targets for the treatment of some dangerous diseases like: cancer or diabetes mellitus.

In the paper, we describe EAST (Energy Alignment Search Tool) – the system and algorithm of similarity searching of proteins in different biological states. The EAST is based on energy profiles computed for particular protein structures. We benefit from the method during the analysis of substrates of signal paths' reactions. However, one can find it useful in other situations.

2 Related Works

There is plenty of work made in the area of similarity searching for biological molecules. All these works use various representations of molecules. Existing retrieval algorithms for biological databases are grounded in principles of approximate retrieval methods and heuristics. In the biological databases two trends can be distinguished: (1) similarity searching based on protein sequences alignment, (2) similarity searching based on the alignment of three-dimensional protein structures.

Similarity searching by protein sequence is usually one of the first steps in many studies on biological molecules. In the approach, there are two leading competitive methods – FASTA [9] and BLAST [10] – that have many implementations.

In the second approach, protein structures, originally represented by atomic coordinates, are first represented in the simpler form in order to reduce the search space. The most characteristic representatives of the group of methods are: VAST [11], DALI [12], CE [13], CTSS [14], and others. The reduced representation of proteins depends on the method. All mentioned methods of structural similarity searching were considered in our research. However, they are much better for homology modeling or protein structure prediction based on threading experiments.

There is also a group of algorithms that use Molecular Interaction Potentials (MIPs) or Molecular Interaction Fields (MIFs), like [15], [16], [17]. MIP/MIFs are results of interaction energies between the considered compounds and relevant probes. MIPs are frequently calculated with the popular GRID [18] program and are used for the comparison of series of compounds displaying related biological behavior [16]. The group of algorithms based on MIPs use atomic coordinates of biological molecules to calculate component, nonbonded interaction energies. This group of algorithms is usually used in the process of drug design.

Finally, in the work [19] authors examine the correlation between energy and different, native distance measures that are frequently used in many algorithms of similarity searching on the structural level. This work inspired us in our research.

3 Energy Profiles and Their Component Characteristics

The distribution of energy of various types along the protein/enzyme polypeptide chain may be very descriptive for protein function, activity and may reflect some distinctive properties. We define a single **protein energy profile E_P** as a set of energy characteristics of various type of energy, determined for a given protein structure.

$$E_P = \{E_{st}, E_{ben}, E_{tor}, E_{vdw}, E_{el}, E_{tot}\} \tag{1}$$

In the expression above: E_{st} denotes a distribution of the bond stretching energy which is a result of a deformation of the optimal bond length, E_{ben} denotes a distribution of the angle bending component energy which is a consequence of changes of the optimal angles between each pair of adjacent covalent bonds, E_{tor} is a distribution of the torsional angle energy, E_{vdw} is a distribution of the van der Waals energy that is caused by correlated motion of the electron clouds of interacting atoms, E_{el} denotes a distribution of the electrostatic energy which is represented by the Coulomb's law. E_{tot} is a distribution of the total energy.

Each component energy characteristic that constitutes the energy profile is determined by energy values calculated for consecutive residues of the polypeptide chain and, in some cases, for their neighborhood. Let the R be an ordered set of m residues in the polypeptide chain $R = \{r_1 r_2 r_3 ... r_m\}$, and X_i^n be a set of atomic coordinates building the ith residue (n is a number of atoms of the residue), then single value of energy E_i^t of the type t can be expressed as:

$$E_i^t = f^t(r_i) = f^t(X_i^n), \tag{2}$$

and generally, a distribution (characteristic) of the energy of type t as:

$$E^t = f^t(R) = f^t(X^k), \tag{3}$$

where k is a number of all atoms in the biological molecule.

Distributions of energy over the peptide chains can be computed based on the protein atomic coordinates stored in public, macromolecular structure databases, like Protein Data Bank (PDB) [20]. Entire energy profiles can then be built, stored in a separate database and compared to each other in order to discover some interesting features. The Energy Distribution Data Bank (EDB) is a special database that we established to store information about energy profiles. At present, EDB consists of profiles for about 2.000 proteins from PDB. Moreover, additional features may be extracted during the process of profiles calculation – they can be: protein sequences, descriptions and annotations, values of torsion angles, secondary structure information, and others. These additional features may be used later in the comparison process.

Single energy profile is specific for one protein structure, not for one protein. It implies that proteins in different spatial conformations may result with different or modified energy profiles. In our research, we deal with such cases while we test our new methods of conformation simulation [21], [21].

Calculating each of the component energy characteristics we use TINKER program [23], atomic coordinates and one of several standard force fields, usually Amber [24].

4 Similarity Searching with Energy Profiles

In the EAST method (Energy Alignment Search Tool), we concentrate on strong structural similarity between molecules or their structural modifications. Our method benefits from the fact that conformational changes of protein structures have an effect on their energy distributions. Even mere differences in some small fragments of the entire construction can be caught on the energy level. Detection of the modified parts can now be accomplished with the use of profiles stored in the EDB. We developed appropriate analysis tools in order to analyze energy profiles. They allow to compare two juxtaposed structures or search similar proteins through the batch comparison of selected types of energy.

The simple profile search process can be performed in the following steps:

1. A user specifies a query protein by entering the PDB ID for the molecule that has an existing energy profile in the EDB or by the PDB molecular structure file. In the first case, the query profile is retrieved from the EDB and takes part in the comparison process. In the second case, the energy profile is computed by the Feature Extraction module (FEm) and additional features are retrieved from the PDB file.
2. Additional features of the query protein constitute a base for the preselection process. For profiles existing in the EDB, these features are stored in the Additional Features Database (AFDB) and were extracted during an early profile computation process (offline process). They can be just retrieved from the AFDB based on given PDB ID. In other cases, they should be extracted from the PDB file by FEm. The preselection module returns *id*s of these proteins that are similar with respect to the given additional feature (e.g. a sequence).
3. The Energy Profile Comparison module (EPCm) compares query profile to each energy profile from the EDB that *id* was returned by the Preselection module. It is a pairwise comparison process which consists of the following operations:
    a. the distance matrix is created for the query energy characteristic and the candidate characteristic from database. The distance matrix is transformed to binary matrix or cut-off matrix – depending on type of the Smith-Waterman algorithm that is used in the next step.
    b. standard [25] or author's, fuzzy version of the Smith-Waterman algorithm is run in order to find the optimal alignment. We can parameterize the algorithm using different penalties of gap opening and gap extension.
    c. the similarity measures: *RMSD* and *SCORE* are calculated and show the similarity between two compared energy profiles. The *Smith-Waterman Score* is returned as well.
4. The EPCm returns the best k solutions with respect to the *SCORE* similarity measure. The k is determined at the beginning of the search process. Now, it is possible to juxtapose the best matches considering consecutive energy characteristics from the energy profile.

4.1 Preselection Phase

Preslection phase is the part of the algorithm that is crucial for the efficiency of the entire method. Good preselection process distinguishes a group of possible best molecules for which distance matrices are built in the next step. At present, in the preselection phase of the EAST we use BLAST method [10] to retrieve molecules with the same or similar amino acid sequence. This is because we often search for proteins that have the same sequence but differ with structure (conformation switching in particular reactions of the signal cascades) or have similar sequence (as an effect of point mutation) and similar structure. Certainly, there may be some cases of similar sequence and completely different structures.

4.2 Distance Matrices

For molecules that *id*s were returned in the preselection, we perform a pairwise comparison with the query protein. The comparison is carried on the energy level. In the version of our algorithm only one, selected at the beginning, type of energy from the profile is considered (e.g. torsion angle energy, bond stretching energy, or other).

The entry d^{AB}_{ij} of the energy distance matrix D denotes the distance between energy values of the ith residue of protein A and the jth residue of protein B:

$$d_{ij}^{AB} = \left| e_{t,i}^{A} - e_{t,j}^{B} \right|, \tag{4}$$

where: t is a type of energy that is considered in the searching.

In Fig. 1a we present the energy distance matrix for query protein 1b38 and candidate protein 1w98 from the EDB. The component energy considered in the profile while searching is set to *Torsional angle* and we use *Amber94* [24] force field to compute energies (in fact, they were computed earlier and are now retrieved from the proper EDB set – see section 4.5). The 1w98 was one of the molecules found in the result set of the searching process. Darker regions indicate lower energetic distances and higher similarity between compared residues.

a) b)

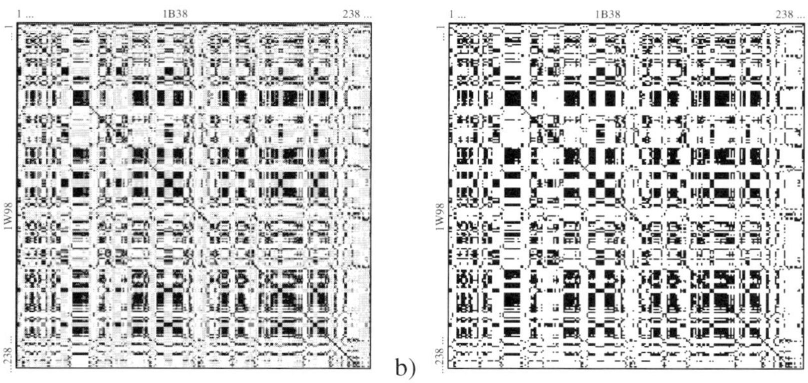

Fig. 1. Distance matrix a) for the compared pair of molecules: 1b38 vs. 1w98, energy type: *torsional angle*, b) the same matrix after transformation to binary matrix B

Before the analysis of possible alignment paths we use threshold values (cut-off values) to qualify given fragments of energy characteristics as similar or not. In the version of the algorithm we have two alternative possibilities how we can transform distance matrix D:

1. we can transform it to binary matrix B according to the following expression:

$$b_{ij} = \begin{cases} 1 & \text{if } d_{ij} \le d_0 \\ 0 & \text{otherwise} \end{cases}, \tag{5}$$

where d_0 is cut-off energy distance (cut-off value),
2. we can transform it to cut-off matrix C according to the following expression:

$$c_{ij} = \begin{cases} d_{ij} & \text{if } d_{ij} \le d_0 \\ +\infty & \text{otherwise} \end{cases}, \tag{6}$$

where d_0 is cut-off energy distance (cut-off value).

Further processing of the matrix B or C to find optimal alignment with the use of the Smith-Waterman algorithm depends on the selected approach.

The cut-off value is selected arbitrary and depends on the energy characteristic that is considered in the comparison process. E.g. in our work, for *torsion angle* energy characteristics it is between 0,5 and 1,5 (Kcal/mole) and for *total* energy it can be from 1 to 10. The cut-off matrix for the same pair of molecules and *torsional angle* energy type, obtained by the transformation of the distance matrix from Fig. 1a, is presented in Fig. 1b. The cut-off value was established to 1 Kcal/mole.

4.3 Optimal Alignment

In order to find the optimal alignment with the Smith-Waterman algorithm we use binary matrix B or cut-off matrix C presented in the previous section. The first approach is faster and more tolerant for energetic discrepancies. In the case, we use only cut-off value (e.g. d_0=1,2 Kcal/mole) and standard settings for the Smith-Waterman method – e.g. similarity value $\mu(b_{ij})=1$ for $b_{ij}=1$ (match), $\mu(b_{ij})=-1/3$ for $b_{ij}=0$ (mismatch), the deletion weight $W_k=1+1/3*k$, where k is a number of deletion and b_{ij} is single cell value of the binary matrix B.

In the second approach, we modified Smith-Waterman algorithm using fuzzy term to define the similarity value for matching energy points (low cell values of the C matrix). Therefore, we don't use the constant value to award the similarity. Except, we define the similarity value using a kind of membership function $\mu(c)$ that is given by the expression (7) and presented in Fig. 2a.

$$\mu(c) = \begin{cases} 1 & 0 \le c \le d_{id} \\ \dfrac{d_0 - c}{d_0 - d_{id}} & d_{id} < c \le d_0 \\ 0 & c > d_0 \end{cases}. \tag{7}$$

In the fuzzy scoring function we have to define two values: a cut-off value d_0, which was set to 1 Kcal/mole in the example presented in Fig. 1b, and identity threshold d_{id}, e.g. 0,2 Kcal/mole. Similarity calculation for the single cell value of the C matrix is presented in Fig. 2b. The values of the cut-off and identity threshold were chosen based on the *a priori* observations of energy characteristics for many proteins. They may vary for different types of energy.

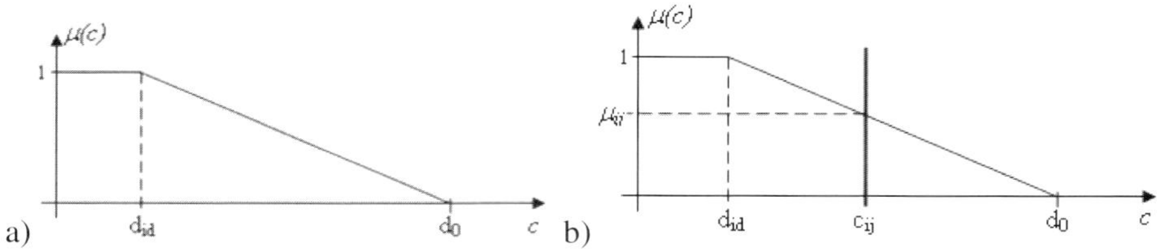

Fig. 2. Fuzzy similarity award for matching energy points of compared molecules (a), determined on the basis of distance value in the cut-off matrix *C* (b)

The approach with the fuzzy similarity award is a little bit slower but gives more accurate results than first approach. The *Smith-Waterman Score* obtained from standard Smith-Waterman algorithm does not take into account the quality of the alignment. It just returns the cumulated award for residues matching on the energy level. The Smith-Waterman algorithm with fuzzy similarity award considers both, number of matching energy points and the quality of the match. Additional measures are calculated as well (see next section).

4.4 Scoring System

In the result set a user obtains the list of proteins having selected energy characteristic similar to the proper energy characteristic of the query protein. However, in the juxtaposition, during detailed analysis of entire profiles, the user can compare any energy type from the energy profiles of the query and resultant proteins. In order to evaluate the similarity degree between selected energy characteristics of two proteins we calculate the *SCORE* measure using the following expression:

$$SCORE = Length_a \,/\, RMSD_a \qquad (8)$$

where:

$Length_a$ – is a length of the alignment (in residues)
$RMSD_a^t$ – is a *Root Mean Square Deviation* [19] of the alignment

4.5 Off-Line Backstage of the Method

Specifying query protein as a PDB ID we must be sure the EDB stores the molecule's energy profile. In example presented in Fig. 1, it was true and thus, feature extraction process was avoided as all needed data were stored in the EDB/AFDB. This makes the searching process much faster on demand. However, feature extraction and energy profiles calculation had to be performed earlier in the 'off-line' process. The computation is time consuming but fortunately, it should be done only once and profiles are inserted into the EDB. Calculation of the energy profile for single biological molecule takes usually from 1 minute to 30 minutes. The time depends on the number of residues and number of polypeptide chains.

5 Experiments

For our experiments we used the EDB/AFDB databases containing about 2.000 proteins from our mirror of the Protein Data Bank (containing more than 33.000 proteins). The subset was chosen arbitrary.

In Fig. 3 we can observe comparison of the *torsion angle* energy characteristics from energy profiles of the Cyclin-Dependent Protein Kinase 2 (CDK2) molecules found in the searching process with EAST. These structures were best results of the similarity searching with the use of our EAST method for the given query PDB ID = 1B38. Retrieved entries are very similar in general shape and show strong energetic similarity. In the comparison we use the analysis toolkit that we built as a graphical supplement of the EAST method. The small changes in the conformation of the CDK2 can be difficult to observe in the graphical tools that present a protein structure (like *RasMol* [26]), unless, we zoom-in to appropriate fragments. However, the clear difference can be caught in the energy characteristics may be observed between residues 157 and 164 which may indicate the conformation modification in the part.

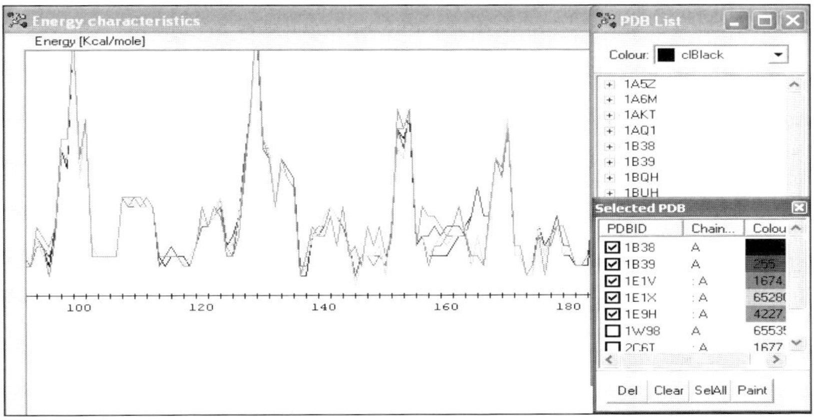

Fig. 3. Juxtaposition of the *torsion angle* component energy characteristics of different forms of the Cyclin-Dependent Kinase CDK2 (x axe – residue no., y axe – energy value)

Results of the searching process for query PDB ID = 1B38 are presented in Table 1. Searching was performed for: the *Torsional angle* component energy, *Amber94* force field, with BLAST preselection and modified version of the Smith-Waterman algorithm, a cut-off value d_0 was set to 1 Kcal/mole and identity threshold d_{id}=0,2 Kcal/mole. Other parameters, e.g. BLAST settings, were set to defaults.

In the result set presented in Table 1, we can observe very good alignment of the 1B39 molecule, with 97% of matching positions, high value of the *Score* measure = 486,41, and good quality of alignment (*RMSD*=0,592). The *Smith-Waterman Score* is also the highest (255,76). For other molecules *Score* value and *Smith-Waterman Score* decrease and rise values of the *RMSD*. There are some molecules, like 1W98 or 2C6T, which create long alignments with the query protein and have relatively high percentage of matching positions. However, their alignments have poor quality – the

Table 1. Results of the searching process for 1B38 molecule with fuzzy *S-W Score*

PDB ID	Length of alignment	Matches	RMSD	Score	Matches %	S-W Score
1B39	288	278	0,592	486,41	97%	255,76
1E1X	285	246	1,239	229,94	86%	214,89
1E1V	288	250	1,336	215,52	87%	202,50
2EXM	286	234	1,508	189,61	82%	185,65
1E9H	277	187	1,791	154,64	68%	136,51
2C6T	281	200	1,929	145,61	71%	124,00
1W98	279	220	2,135	130,65	79%	103,87

value of the *RMSD* is higher than other molecules in the set. The poor quality is reflected in their *Score* and fuzzy *Smith-Waterman Score*.

6 Performance Evaluation and Concluding Remarks

Presented EAST method of similarity searching is relatively efficient – response time is in the level of minutes, in the case we get the query profile from the EDB. The time covers retrieving the query data from the EDB/AFDB, preselection by sequence, computation of the distance matrices and optimal alignment. However, we must remember that profiles have to be computed earlier in the offline process – fortunately, only once. The searching time grows when the query protein is passed as a PDB file. Then, its profile has to be computed online which costs 2-30 minutes, depending on the protein length (computation made on Intel Pentium IV CPU 3,20GHz, 1GB RAM).

In comparison to other similarity searching methods our algorithm can be considered as a high-performance method, even with online profile computation. However, it is based on different paradigm, structure representation and has different purpose.

The EAST method can be placed on the borderline of the sequence/structure similarity. Molecules found with the use of EAST usually indicate high sequence similarity and structural similarity. They can differ in some regions which are caught by the method and are reported in the final result set. Thus, the method is very good in observation of conformational switching in molecular structures associated with their activation or inactivation. The analysis of the protein energy profiles provides additional information about studied protein molecules and supplies our elementary knowledge beyond sequence and 3D structure. This supplementary information may lead to discover interesting regularities or anomalies. We developed the method in our research on molecules' activity in cellular processes, especially signal pathways, and we are going to continue our work. We believe all this works are worth our efforts.

References

1. Ray, L.B.: The Science of Signal Transduction. Science, 284, (1999) 755-756
2. Berridge, M.J.: The Molecular Basis of Communication within the Cell. Scientific American, 253 (4), 142-152

3. Mrozek, D., Małysiak, B., Frączek, J., Kasprowski, P.: Signal Cascades Analysis in Nanoprocesses with Distributed Database System. Springer, *LNCS*, 3516/3 (2005) 334-341
4. Lodish, H., Berk, A., Zipursky, S.L., et al.: Molecular Cell Biology. Fourth Edition. W. H. Freeman and Company, NY (2001)
5. Branden, C., Tooze, J.: Introduction to Protein Structure. Garland (1991)
6. Cantor, C.R., Schimmel, P.R.: Biophysical Chemistry. W.H. Freeman (1980)
7. Brown, N.R., Noble, M.E.M., et al.: Effects of Phosphorylation of Threonine 160 on Cyclin-dependent Kinase 2 Structure and Activity. J. Biol. Chem., 274(13), (1999) 8746-8756
8. Burkert, U., Allinger, N.L.: Molecular Mechanics. American Chemical Society, Washington D.C. (1980)
9. Pearson, W.R., Lipman, D.J.: Improved Tools for Biological Sequence Analysis. PNAS 85 (1988) 2444–2448
10. Altschul, S.F., Gish, W., Miller, W., Myers, E.W., Lipman, D.J.: Basic local alignment search tool. J Mol Biol, 215, (1990) 403-10
11. Gibrat, J.F., Madej, T., Bryant, S.H.: Surprising similarities in structure comparison. Curr Opin Struct Biol, 6(3), (1996) 377–385
12. Holm, L, Sander, C.: Protein structure comparison by alignment of distance matrices. J Mol Biol. 233(1), (1993) 123-38
13. Shindyalov, I.N., Bourne, P.E.: Protein structure alignment by incremental combinatorial extension (CE) of the optimal path. Protein Engineering, 11(9), (1998) 739-747
14. Can, T., Wang, Y.F.: CTSS: A Robust and Efficient Method for Protein Structure Alignment Based on Local Geometrical and Biological Features. Proceedings of the 2003 IEEE Bioinformatics Conference (CSB 2003) 169–179
15. Thorner, D.A., Wild, D.J., Willett, P., Wright, P.M.: Similarity Searching in Files of Three-Dimensional Chemical Structures: Flexible Field-Based Searching of Molecular Electrostatic Potentials. J. Chem. Inf. Comput. Sci. (1996) 900-908
16. Rodrigo, J., Barbany, M., et al.: Comparison of Biomolecules on the Basis of Molecular Interaction Potentials, J. Braz. Chem. Soc., Vol. 13, No. 6, (2002) 795-799
17. Ji, H., Li, H., Flinspach, M., Poulos, T.L., Silverman, R.B.: Computer Modeling of Selective Regions in the Active Site of Nitric Oxide Synthases: Implication for the Design of Isoform-Selective Inhibitors. J. Med. Chem. (2003) 5700-5711
18. GRID, version 19; Molecular Discovery Ltd. (20 A Bearkeley Street): Mayfair, London, England (2001)
19. Wallin, S., Farwer, J., Bastolla, U.: Testing Similarity Measures with continuous and discrete Protein Models. Proteins 50, (2003) 144-157
20. Berman, H.M., Westbrook, J., Feng, Z., Gilliland, G., Bhat, T.N., Weissig, H., et al.: The Protein Data Bank. Nucleic Acids Res., 28, (2000) 235–242
21. Znamirowski, A.W., Znamirowski, L.: Two-Phase Simulation of Nascent Protein Folding, Proc. of the 4th IASTED Inter. Conference on Modelling, Simulation, and Optimization 2004, Kauai, Hawaii, ACTA Press (2004) 293-298
22. Znamirowski, L., Zukowska, E.D.: Simulation of Post-translational Conformations in the Ribosomal Polypeptide Synthesis. Proc. of the IASTED Intern. Conf. on Modelling and Simulation, Marina del Rey, California, ACTA Press (2002) 97-102
23. Ponder, J.: TINKER – Software Tools for Molecular Design, Dept. of Biochemistry & Molecular Biophysics, Washington University, School of Medicine, St. Louis (2001)
24. Cornell, W.D., Cieplak, P., et al.: A Second Generation Force Field for the Simulation of Proteins, Nucleic Acids, and Organic Molecules. J.Am. Chem. Soc., 117, (1995) 5179-5197
25. Smith, T.F., Waterman, M.S.: Identification of common molecular Subsequences. J. Mol. Biol. 147, (1981) 195-197
26. Sayle, R., Milner-White, E.J.: RasMol: Biomolecular graphics for all. Trends in Biochemical Sciences (TIBS), Vol. 20, No. 9, (1995) 374

A Fuzzy Advance Reservation Mechanism of Network Bandwidth in Video Grid

Xiaodong Liu, Qionghai Dai, and Chuang Lin

Broadband Network and Multimedia Research Center
Graduate School at Shenzhen, Tsinghua University
Shenzhen, 518055, P.R. China
liuxd@sz.tsinghua.edu.cn

Abstract. Next generation grid networks will be required to support video delivering services which need bandwidth is reserved at an earlier time so that desired QoS requirement could be ensured in the future. In this paper, we propose a new intelligent admission control algorithm for advance reservation. The main design principle underlying our algorithm is delay and jitter are playing more and more important roles in providing high quality video to users, but they are often fuzzy and indefinite expression from clients in advance and it is therefore hard to get the satisfied advance reservations for network resources requirements using the accurate mathematic models like traditional schemes. So, we propose a fuzzy neural network model to cluster priority of advance reservation requests according to delay and jitter variables and then allocate the bandwidth based on priority preempting principle. Experimental results show our algorithms achieve well-balanced performance.

1 Introduction

A Grid is a technology that utilizes a network according to the requests from users by utilizing various kinds of resources such as computers, storage devices, and observation devices, which are geographically distributed. The Grid uses these resources in a flexible, simple, integrated, and efficient manner, and it represents an infrastructure. In most Grid scheduling systems, submitted jobs are initially placed into a queue if there are no available resources. Therefore, there is no guarantee as to when these jobs will be executed. This causes problems in most grid applications [1]. Advance reservations are a useful method to allocate resources of various kinds in many different environments. Among others, a major advantage of this kind of reservation is the improved admission probability for requests that are made sufficiently early [2]. Advance reservation for global grids becomes an important research area as it allows users to gain concurrent access for their applications to be executed in parallel, and guarantees the availability of resources at specified future times [3].

Grid computing applications require quality-of-service (QoS) guarantees on various fields. Recently, delivering live video streams across the Internet is becoming more and more popular as computers gradually replace other media and entertainment channels such as radio, television, newspapers etc. [4], and some real-time synchronous collaboration applications such as videoconference and sensor nets are developing on

L. Wang et al. (Eds.): FSKD 2006, LNAI 4223, pp. 706 – 715, 2006.

Internet. However, Nowadays Internet, based on "best effort" and "non-status" can not provide real-time, high-performance, reliable, and secure transport fashion for these time-sensitive applications. The network bandwidth and other parameters such as delay and jitter have a crucial impact on the quality of a media stream delivered over IP connection. The grid contains many high-performance core services to support the above QoS requirements. For example, the allocation of computing resources (processor nodes) is made using advance reservation mechanisms, i.e., the processors are allocated a long time before they are actually used. It also implements the advance reservation mechanisms on computer networks in the same way.

Advance reservations are especially useful for grid computing but also for a variety of other applications that require network quality-of-service, such as content distribution networks or even mobile clients, which need advance reservation to support handovers for streaming video. Some of the important multimedia applications of integrated services networks require that advance reservations be possible. The clients who wish to set up video conferencing need to schedule those meetings in advance to make sure that all or most of the participants will be able to attend; at the time the meeting is scheduled, they must also be certain that the network connections and the other resources required will be available when needed and for the entire duration of the meeting. Unfortunately, distributed multimedia applications must be supported by real-time communication services, which are to provide the necessary QoS guarantees, and these services cannot admit an arbitrary number of connections. Thus, there is no guarantee that the resources for a pre-scheduled meeting will be available at the time the meeting is expected to start, unless they can be reserved in advance. In general, two types of resource reservations in computer networks can be distinguished: immediate reservations which are made in a just-in-time manner and advance reservations which allow reserving resources a long time before they are actually used. The advanced reservation feature makes it possible to obtain a guaranteed start time in advance. A guaranteed start time brings two advantages. It makes it possible to coordinate the job with other activities, and resource selection can be improved as the resource comparison is based on a guaranteed start time rather than on an estimate.

In the field of advance reservations in grid networks, some work has been carried out. In the following section, we give a brief description of the advance reservation environment and discuss the pros and cons of two important evolutions.

2 Advance Reservations

In general, advance reservations differ from immediate reservations only by the time a request is submitted to the network management thus decoupling the submission of the request completely from the usage of the resources. In order to perform reliable admission control, status information about the currently known future utilization of links is required. The period for which requests can be submitted is called *book-ahead period*. Usually, this period is divided into slots of fixed size. Access to the network is controlled by a management system usually called *bandwidth broker*. A reservation request is submitted to the bandwidth broker. The bandwidth broker admits only those requests for which sufficient bandwidth can be guaranteed during the requested transmission period. [5] proves such advance reservation mechanisms can lead to a reduced performance of the network with respect to the amount of admitted requests and the allocated bandwidth. Because in advance reservation scenarios, requests with a large book-ahead can be issued, i.e. requests can be made a long time before the transmission actually takes place. This results in peaks which occur at different times within the book-ahead period and block the bandwidth at those times. These peaks lead to a fragmentation of the available network resources such that gaps appear which cannot be filled with requests since these gaps are too short for additional requests to fit in. So, the concept of malleable reservations to bandwidth reservation is proposed in [5] in order to overcome the performance degradation of advance reservations. The idea is to implement a new service for clients that allows defining requests which do not have fixed constraints in terms of start time, stop time, and bandwidth requirement. These reservations are an opportunity to fill the gaps and thus to improve the network performance. But it is only useful for a certain type of reservations, e.g. only the total amount of data to be transmitted might be of interest and perhaps a deadline until the transmission must be finished. Examples for such reservations are transmissions of large amounts of data such as backups of data sent to a storage server where the data is written to tapes and the backup must be finished at a certain time. It is not suitable for real-time streaming video delivery which is often affected by network jitter and need a constant bandwidth.

[6] formally defines and analyzes advance reservations scheme and presents a constrained mathematical model for advance reservations. It provides us with two useful metrics, Rejection Probability and Reservation Probability. The rejection probability is a measure of the capacity that the underlying network can handle. The reservation probability is a measure of the utilization of the network based on the pattern of user requests and can be used to define the appropriate number of network resources, which would keep the rejection probability down to an acceptable level. But its analytical model is based on eleven assumptions (including each request targets reservation for only one time slot which is unpractical) and end-to-end lightpaths which is essentially a connection-oriented setup. These presuppositions are not suitable for common grid conditions based on Internet.

So, we propose an intelligent advance reservations mechanism based on fuzzy neural network for video grid to resolve the above limitations.

3 A Fuzzy Neural Network Architecture for Advance Reservations

Network delay and jitter requests play the important role for video delivering and reconstruction quality. They are often fuzzy and indefinite expression from clients in advance. It is therefore hard to get the satisfied advance reservations for network resources requirements using the accurate mathematic models based on delay and jitter variables. So, we propose an advance reservation requirements priority prediction model based on fuzzy neural network that merge a neural network with a fuzzy system into one integrated system to combine the real time parallel computation and learning abilities of neural network with the human-like knowledge representation and explanation abilities of fuzzy systems.

There are two input signals x_1 and x_2 corresponding to delay and jitter and one output signal y representing the priority fuzzy sets about advance reservation requests. For each input variable, we define three fuzzy sets A_{1j}, A_{2j}, A_{3j} ($j=1,2$) to represent the large, medium and small value of corresponding network parameters. The membership function of every fuzzy set is defined as follows:

$$\mu_{A_{1j}} = -1/(1+e^{-w_g(x_j-w_c^1)})+1 \ . \tag{1}$$

$$\mu_{A_{2j}} = 1/(1+e^{-w_g(x_j-w_c^1)})-1/(1+e^{-w_g(x_j-w_c^2)}) \ . \tag{2}$$

$$\mu_{A_{3j}} = 1/(1+e^{-w_g(x_j-w_c^2)}) \ . \tag{3}$$

Where w_c^1, w_c^2 are the center of Sigmoid function, and w_g deciding the pitch of Sigmoid function.

The common form of fuzzy rules is as follows:

$$R_k^l: \quad (if\ x_1\ is\ A_{p1}^k\ and\ x_2\ is\ A_{q2}^k,\ then\ y\ is\ B_l)\ is\ \tau_k^l \ . \tag{4}$$

Where $p,q=1,2,3;\ \ k=1,2,\cdots 9$. B_l ($l=1,\ 2$) are the fuzzy sets of output y and represent the high priority and low priority of network resources requirements. The membership function of B_l is triangular and are defined as follows:

$$\mu_{B_1(y)} = (y - w_{c1}^{'})/w_{g1}^{'} \ . \tag{5}$$

$$\mu_{B_2(y)} = (y - w_{c2}^{'})/w_{g2}^{'} \ . \tag{6}$$

Where $w_{c1}^{'}$ and $w_{c2}^{'}$ are the values of independent variable y when function value is equal to zero. $1/w_{g1}^{'}$ and $1/w_{g2}^{'}$ are the slopes of the function. Because the membership functions are monotone, so their inverse functions are defined as follows:

$$B_1^{-1}(\mu_0) = w_{g1}^{'}\mu_0 + w_{c1}^{'} \ . \tag{7}$$

$$B_2^{-1}(\mu_0) = w_{g2}^{'}\mu_0 + w_{c2}^{'} \ . \tag{8}$$

τ_k^l is the truth value when the conclusion of rule k is B_l . The rule is absolutely correct when τ_k^l is equal to one and is absolutely false when τ_k^l is equal to zero. The output of the fuzzy system is defined as follows:

$$y = \sum_{l=1}^{2} \alpha_l^{'} \mu_{B_l}^{-1}(\alpha_l^{'}) \Big/ \sum_{l=1}^{2} \alpha_l^{'} = \sum_{l=1}^{2} \overline{\alpha}_l^{'} \mu_{B_l}^{-1}(\alpha_l^{'}) . \tag{9}$$

Where $\overline{\alpha}_l^{'} = \alpha_l^{'} \Big/ \sum_{i=1}^{2} \alpha_i^{'}$, $\alpha_l^{'} = \sum_{k=1}^{9} \overline{\alpha}_k \tau_k^l$, $\overline{\alpha}_k = \alpha_k \Big/ \sum_{m=1}^{9} \alpha_m$, and $\alpha_k = \mu_{A_{p1}^k}(x_1) \bullet \mu_{A_{q2}^k}(x_2)$.

According to the above definition, we designed a ten layers feedforward fuzzy neural network architecture for getting the priority of network resources requests. (See Fig. 1).

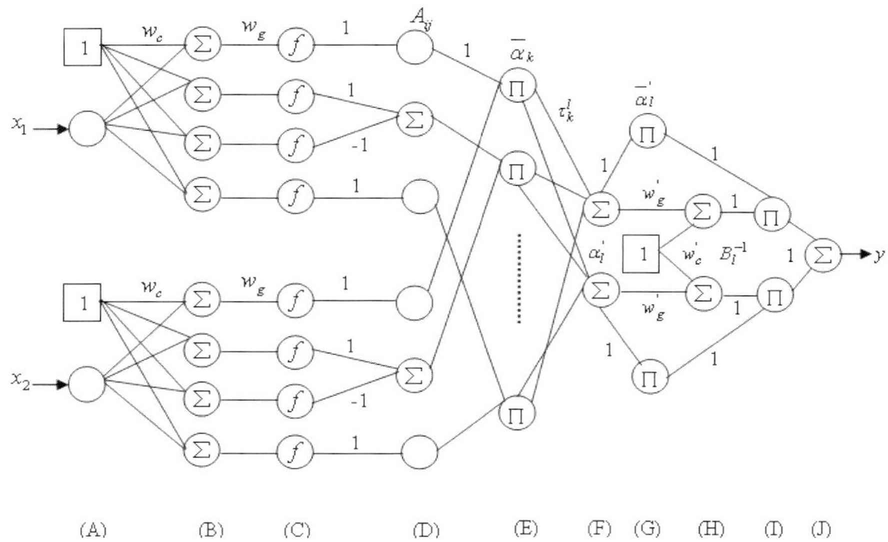

Fig. 1. A ten layers feedforward fuzzy neural network architecture for getting the priority of advance reservation requests

In input layer A of Fig. 1, blank circle denote that the neuron has no processing for input information, one in rectangular denote a fixed input. The weights from rectangular to neurons belong to the second layer B are w_c (w_c^1 or w_c^2). The weights from circle to the neurons of layer B are 1. There are eight neurons in layer B, each neuron receives two inputs: x_j and w_c and then does $x_j - w_c$ operation. The weights between layer B and layer C are w_g . There are also eight neurons in layer C. Each neuron receives one input $w_g(x_j - w_c)$ and then calculates the Sigmoid function $\{1+\exp(-w_g(x_j - w_c))\}$. The weights between layer C and layer D are 1 or -1. There are six neurons in layer D, among them only the second and the fifth neuron do sum operation and the others do nothing. The outputs of these six neurons are membership function $\mu_{A_{ij}}(x_j)$. So far, the fuzzification processing is completed. The weights between layer D and layer E is one. The layer E is a fuzzy rule layer and has nine neurons corresponding to nine fuzzy rules. Each neuron does calculation to produce $\overline{\alpha}_k$. The weights between layer E and layer F are τ_k^l ($k = 1, 2, \cdots 9$, $l = 1, 2$). There are only two neurons in layer F and they are corresponding to fuzzy sets B_1 and B_2 separately. These two neurons do calculations to produce α_l'. The layer G produces $\overline{\alpha}_l'$ and layer H gets $B_l^{-1}(\alpha_l')$ according to equation (7) and (8). Layer I and J produce output of the system according to equation (9). The output value y represents the priority of the network resources requests. Through the process of the fuzzy neural network, we get two clustering: one is advance reservation requests with high priority and the other is low priority. Then we can make advance reservation according to the priority and other parameters using the admission control algorithm introduced in next section.

4 An Intelligent Admission Control Algorithm

Usually, in an advance reservation environment a request *r* can be defined as *r* = (*u, v*, t_{start}, t_{stop}, *dl, jt, bw*), where *u, v* denote the start and destination node, t_{start}, t_{stop} denote the start and stop time of the transmission and *dl, jt, bw* the requested delay, jitter and bandwidth. Admission control in this case is to find a feasible path from *u* to *v* such that the bandwidth *bw* is available during the requested transmission period [t_{start}, t_{stop}]. An admission control algorithm for the case of using these parameters with priority getting from the above fuzzy neural network is described as follows:

```
Algorithm:
  Input: G(V,E) /*network graph*/;
         r /*advance reservation requests*/;

  Begin:
    for each r do
        x1:= dl;
        x2:= jt;
        y=FNN(x1,x2);
    endfor.
    for each r in high priority do
      if(find_path(u,v,t_start,t_stop,bw)==success) then
        reservation;
        break;
      else
        preempt the bandwidth allocated to others;
      endif.
```

```
  (13)endfor.
  (14)for each r in low priority do
  (15) if(find_path(u,v,t_start,t_stop,bw)==success) then
  (16)   reservation;
  (17)   break;
  (18) endif.
  (19)endfor.
End.
```

The algorithm takes as the input the network graph G(V, E) and the requests *r*. The first "for" loop cluster the requests into high priority and low priority sets. Line (2) and (3) take delay and jitter as inputs and line (4) compute the priority of the request according to the ten layers feedforward fuzzy neural network architecture introduced in above section. The second "for" loop in line (6) allocate network bandwidth for high priority requests in advance. Line (7) tests whether a feasible path, i.e. a path with sufficient bandwidth, exists. If such a path is found, the bandwidth is reserved (line (8)) and the loop exit (line (9)). The purpose of the function find_path is to determine a path from *u* to *v* with sufficient bandwidth *bw* within the interval $[t_{start}, t_{stop}]$. If path is not found, we need to release the bandwidth occupied by the requests of low priority or later t_{start} and assign it to the higher priority request(line (11)). The third "for" loop in line (14) process the low priority requests. If the path is not found for a request, it simply stops.

During the actual running time, assign the immediate requests middle priority, so they can preempt the bandwidth reserved for low priority requests in advance.

5 Experimental Results

To evaluate the performance of our proposed intelligent scheme, we employ a simulation model. Based on this simulation model, we compare the performance of our scheme with fixed advance reservation scheme.

We used the eqos network topology depicted in [5]. a set of requests was generated for the network topology. Each request is defined by its start time, stop time, delay, jitter and bandwidth requirement. For the case of immediate reservations, requests are issued at the start time. In order to examine advance reservations, a varying number of requests was randomly chosen from the set and issued a certain amount of time before the corresponding transmission is to start. The book-ahead period had a length of 2^{15} = 32768 slots. Each link of the networks was assumed to have a bandwidth capacity of 100Mbps. Bandwidth requirements used in the simulation experiment range from 32Kbps simulating email transmission to 1Mbps simulating live video delivering.

Performance measures obtained through simulation are immediate call acceptance rate and Peak Signal Noise Ratio(PSNR) of video playing.

In Fig. 2, the influence of a varying percentage of advantage reservations on the immediate call acceptance rate is outlined. It can be observed that the immediate call acceptance rate of Fixed Advance Reservation(FAR) decreases sharply when the percentage of the advance reservation increase. But our Intelligent Advance

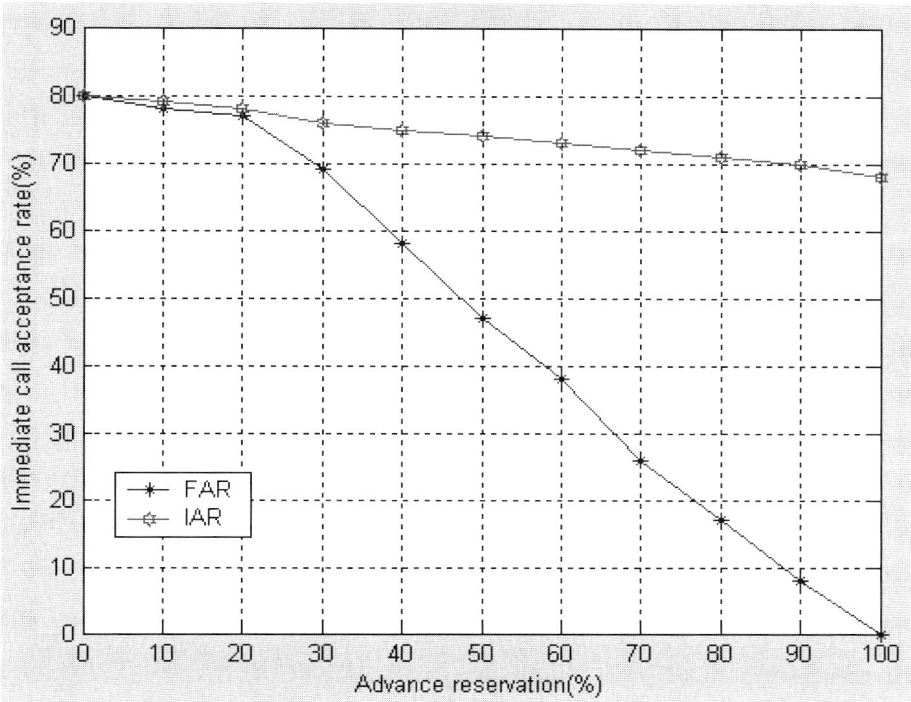

Fig. 2. Relationship between Immediate call acceptance rate and Advance reservation

Reservation(IAR) keeps steady. Because even if the bandwidth had been assigned to advance reservation requests, the immediate requests assigned middle priority in our algorithm could preempt the advance reservation requests with low priority to acquire the bandwidth. Thus our algorithm ensures the smooth running of some real-time immediate requests(for example, video-on-demand).

In order to verify the effectiveness of our algorithm in video grid, we request a MPEG4 VOD video sequence based on LSMP(a streaming media system developed by ourselves) randomly. The video sequence encoded by 25 frames per second and the rate is 450Kbps. From Fig. 3, we can see that the PSNR of our algorithm(IAR) remain high with the percentage of advance reservation increase. However, the PSNR of fixed advance reservation algorithm(FAR) decrease sharply with the percentage of advance reservation increase and this result in the poor video playing quality. The main reason lies in our algorithm pay more attention to the effect of the delay and jitter which are important to video playing. So, the VOD video sequence request gets high priority through fuzzy neural network inference and network bandwidth is allocated to it preferentially according to our intelligent admission control algorithm. Thus ensure the VOD video sequence request get enough bandwidth to satisfy its delay and jitter requests so as to obtain the good playing quality.

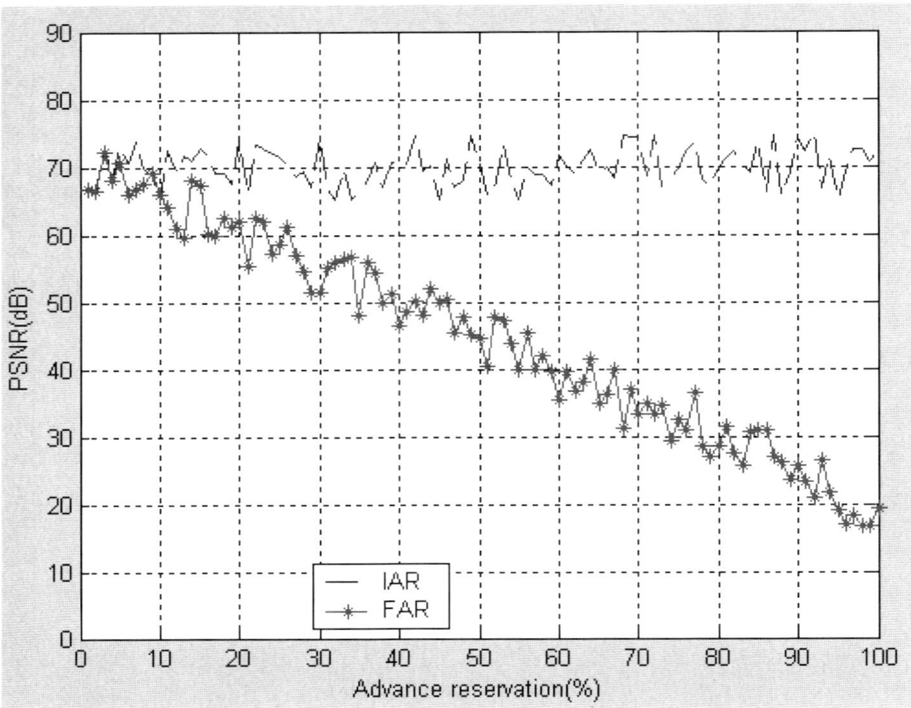

Fig. 3. Relationship between video PSNR and percentage of advance reservation

6 Conclusions

In order to provide advance reservation in QoS sensitive video grid networks, an intelligent admission control algorithm based on ten layers feedforward fuzzy neural network architecture is proposed. An important design principle underlying our scheme is intelligent bandwidth allocation based on real time priority clustering estimation according to fuzzy delay and jitter requests. Our approach is able to resolve conflicting QoS performance criteria while ensuring efficient network performance. In addition, proposed algorithm has low complexity making it practical for video transmission networks. Performance evaluation results clearly indicate that our scheme maintains well-balanced immediate call acceptance rate and PSNR value of video signal with the percentage of advance reservation increase while other schemes can not offer such an attractive performance. The main concept of our scheme can also be applied beyond bandwidth reservation in video grid networks to other resource management algorithms, e.g., CPU scheduling, disk and memory management, etc.

Acknowledgement

This work is partially supported by CNSF Grant #60432030 and #90412012.

References

1. Anthony Sulistio, Rajkumar Buyya: A Grid Simulation Infrastructure Supporting Advance Reservation. In Proc. of the 16th International Conference on Parallel and Distributed Computing and Systems (PDCS'04), Cambridge, USA, November 9-11 (2004)
2. Lars-Olof Burchard: Analysis of Data Structures for Admission Control of Advance Reservation Requests, IEEE Transactions on Knowledge and Data Engineering, vol. 17, no. 3, Mar., (2005) 413-424
3. Libing Wu, Jianbing Xing, Chanle Wu, Jianqun Cui: An Adaptive Advance Reservation Mechanism for Grid Computing, Sixth International Conference on Parallel and Distributed Computing Applications and Technologies (PDCAT'05), (2005) 400-403
4. A. Henning D. Raz: Efficient Management of Transcoding and Multicasting Multimedia Streams, The Ninth IFIP/IEEE International Symposium on Integrated Network Management (IM 2005), Nice, France. 15-19 May (2005)
5. Lars-Olof Burchard, Hans-Ulrich Heiss: Performance Issues of Bandwidth Reservations for Grid Computing, Proceedings of the 15th Symposium on Computer Architecture and High Performance Computing (SBAC-PAD'03), IEEE, (2003)
6. Sumit Naiksatam, Silvia Figueira, Stephen A. Chiappari, Nirdosh Bhatnagar: Analyzing the Advance Reservation of Lightpaths in Lambda-Grids, Proceedings of the 5th International Symposium on Cluster Computing and Grid, Cardiff, Wales, UK, May 9~12, (2005)

Design and Implementation of a Patch Management System to Remove Security Vulnerability in Multi-platforms

Jung-Taek Seo[1], Yun-ju Kim[1], Eung-Ki Park[1], Sang-won Lee[2], Taeshik Shon[2], and Jongsub Moon[2]

[1] National Security Research Institute
463-1 Jeonmin-dong, Yu-seong-gu, Daejeon, 305-811, Republic of Korea
{seojt, zzuya99, ekpark}@etri.re.kr
[2] *CIST, Korea University* 1-Ga, Anam-dong, Sungbuk-Gu, Seoul, Republic of Korea
{a770720, 743zh2k,jsmoon}@korea.ac.kr

Abstract. Because of worms which make use of vulnerability of computer systems, computer incidents are increasing. Although there is an opportunity to defend these attacks at an earlier stage, people undergo several serious disturbances because many administrators and users didn't realize the gravity of the patch management, that is, the patch management system (PMS) prevents the intrusion. Prevention with "Prevention is better than cure." is very important. Therefore we need installing an effective patch management system. We designed and implemented the proposed framework which provides solutions for managing and distributing critical patches that resolve known security vulnerabilities and other stability issues with various platforms.

1 Introduction

Worms have been the bane of computer systems since Code Red and Nimda hit computers in summer 2001, which put not only Microsoft but also other OS Vendors on the path toward making security to No. 1 priority [1]. In January 2003, the Slammer worm spreads to corporate networks worldwide, causing databases to go down, bank teller machines to stop working and some airline flights to be canceled. At six months earlier before the worm spread, a researcher had released codes that exploited the major Microsoft SQL vulnerability abused by the worm to spread [2]. According to Symantec Internet Security Threat Report, time is decreasing between vulnerability disclosure and widespread exploitation [3]. As a result, we should have installed the security patches at the earlier opportunity. We can easily find that the time gap between the announcement of vulnerability and the release of an associated exploit is reduced recently as shown in Table 1 [4], [5].

Thereafter, the reasons both that an attack using vulnerability of system is protected just if the patch for the attack is installed and that the costs and business interruptions related to security incidents have been more increased make an automatic patch distribution process be very important [6].

L. Wang et al. (Eds.): FSKD 2006, LNAI 4223, pp. 716 – 724, 2006.

Table 1. Vulnerabilities posted date and worm issue date

Name	Microsoft Security Bulletin Code	Vulnerability posted date	Worm issue date
Nimda	MS00-078	2000 / 10 / 17	2001 / 09 / 18
CodeRed	MS01-033	2001 / 06 / 18	2001 / 07 / 16
SQL Slammer	MS02-039	2002 / 07 / 24	2003 / 01 / 25
Blaster	MS03-026	2003 / 07 / 16	2003 / 08 / 11
Sasser	MS04-011	2004 / 04 / 13	2004 / 05 / 01

In this paper, we propose a Patch Management System (PMS), after researching on existing patch management systems. The rest of this paper is organized as follows: Our approach and design of the PMS is described in Section 3. The result of the implementation is explained in Section 4, followed by conclusion in Section 5.

2 A Survey of Patch Management Product Trends

Several companies or institutions released solutions for the patch management. We analyzed functions of the announced patch management systems in table 2 [7, 8, 9, 10, 11, 12, 13, 14, 15, 16]. The functions of patch management systems are analyzed below.

A hierarchically structured patch distribution is a solution for expanding a network domain. As network scale is larger, the need for an integrated management is higher. In a near future many products may support this facility. There were five products to support this function.

The function for analyzing vulnerability of a target is desired for patch management system in order for the most appropriate patch version to be installed to the target.

There can be attacks such as Replay, Man in the middle or forgery of patch during downloading a patch through communication network. patch management system have to provide some protecting method against the attacks. This function is essential. Nevertheless, only four products offer confidentiality, authentication, and integrity.

Since an organization equips with many computers, all computers in the organization do not install only one operating system. To maintain all computers consistently safe in the organization, patch management system should support all platforms. Four companies support multi platform function.

Now a day, a software program is usually coded with reusable method. This method reduces the time to develop new software and the cost for producing. So, a software package doesn't consist of all modules, but requires other modules like dynamic linking library when it is installed. That is, a dependency problem can be occurred when a patch management system is installed unless pre-required modules are already resided within the target computer. A patch management system must support to resolve the dependency problem.

Most products offer central management. This function supports so that an administrator may grasp present status of the patch installed in its network domain and take action about each circumstance.

Table 2. A survey of Patch Management Product Trends

	Patch Link	Ecora	Big Fix	Security Profiling	St. Bernard Software	GFI	Gravity Storm Software	Sha Vli k	Soft Run	MS
Extension	O	X	O	O	O	X	X	X	X	O
Analysis of Vulnerability	O	O	O	O	O	O	O	O	O	O
Secure Transfer	O	X	O	X	X	X	O	X	X	O
Multi Platform	O	O	O	O	X	X	X	X	X	X
Solution of Dependency	O	O	O	O	O	O	O	O	O	O
Central Management	O	O	O	O	X	O	X	O	O	O

3 A Proposed Approach

A proposed patch management system (PMS) can extend the structure of previous PMS's. The existing PMS's may provide services in large scale network environments, and analyze vulnerability and solve dependency [17]. In addition the proposed PMS supports secure transmission during patch.

3.1 Overall Structure

3.1.1 Basic Framework

PMS is a system that distributes, installs and manages patches in multi platforms, and has components as shown in figure 1; a patch server, patch client agents, a patch server manager, patch client managers, data storage and patch storage.

- **Patch Server:** Patch Server communicates with an agent of patch clients using patch distribution protocols, distributes a ScanList to a client, offers patch files to a client's agent, and reconstructs information in relation with a present status of patch installation of the client.
- **Patch Client Agent:** It is responsible for communicating with patch server, scanning the system's information in which it resides, detecting the installed patches and the uninstalled patches in the system, and installing patches into the system in which it is. Patch client agent communicates with the Patch server periodically to find new patch information. It can be resided in multi platforms such as Windows, Linux, and Solaris.
- **Patch Server Manager:** It manages a patch server and provides web-based user interfaces to administrator. In case of new patch releasing from a vendor, it adds and modifies new patch information and acquires the system's information from each client.
- **Patch Client Manager:** It provides web-based user interfaces to end users of each client, and manages patch information about the installed patches and the uninstalled patches of each client.

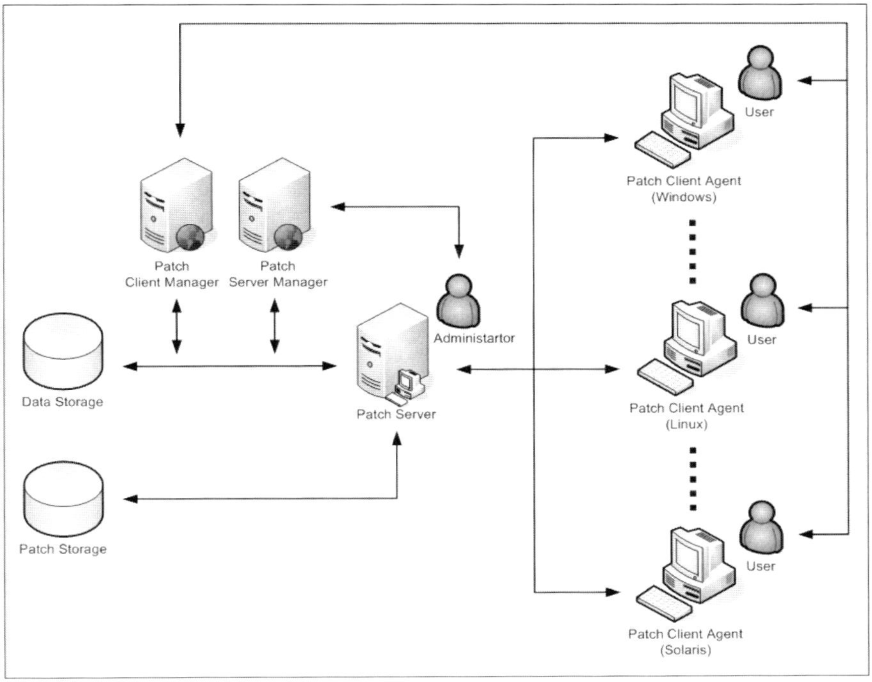

Fig. 1. Basic Framework of PMS

- **Data Storage:** It keeps important data of PMS such as ScanList, information of user, system's information of clients, low-ranking server, and log. ScanList is a criterion of detecting vulnerability and contains information of patches.
- **Patch Storage:** It has patch files.

3.1.2 Framework for Extending the Network Domain

In case of managing patches in the distributed and heterogeneous networks, we use many servers and many client systems. This situation causes an administrator to manage a lot of patch servers and in consequence, increases the expenses and the time for the patch collection, the patch distribution and the test. The framework shown in figure 2 can easily manage many groups and reduce the overload of the patch server for managing patch.

- **Primary Server:** It is not only security patch server but also manages the secondary server. It can become a secondary server of the other server.
- **Secondary Server:** It is both security patch server and managed by primary server. It can become a primary server of the other server.
- **Distribution Server:** It is not security patch server. It simply functions distributing patch files.

A primary server transmits a new patch package and its information to secondary servers when an administrator registers new patch from vendor. The secondary servers distribute the patch package to client agents.

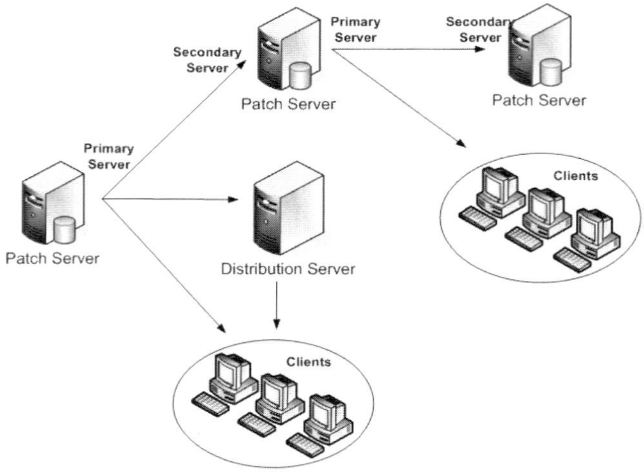

Fig. 2. Framework for extending the management network

3.2 The Vulnerability Analysis of Client Systems

Database access method used in previous study [17] can cause server's overload because the patch server analyzes vulnerability of all client systems. Also, because server extracts much information from clients in order to analyze vulnerability, and/or if the user of the client does not want, it becomes serious privacy problem. The proposed method of vulnerability analysis of client is to use XML. We call it "ScanList". The XML document contains both a criterion for detecting vulnerability and information about all patches for target operating system's version. A patch server searches relevant ScanList using basic information of client system and transmits it to agent. This scheme prevents server's overload because each agent analyzes its own vulnerability. This method escape from privacy violation because the server needs only basic data of client system such as type, version and build number of OS. Also, server's construction is easy because it doesn't use database and nor update the database. Finally, the most significant advantage of this scheme is to solve dependency problem between patches as ScanList includes information for deciding patches excluded. A ScanList contains the following components as shown in table 3.

Table 3. Tag and Description of ScanList

Tag	depth	Description
os	1	Type and version of operating system - In case of Windows, it includes version, product type, su ite mask, build number, service pack information, archite cture and etc. - In case of Linux and UNIX, it includes version, architec ture, kernel version etc.
patch	2	Identification of each patches

Table 3. (*continued*)

Detection	3	Criterion of detecting for installed patch and uninstalled pat ch of system yet.
installed	4	Information for verifying installation - It includes version of system files, registry key value and etc.
exception	4	Criterion for deciding whether the patch is excluded or not. - It solves a problem of dependency.
install	3	Information for installing security patch
description	4	Patch description to notify user
installation	5	Information for installation process - It includes patch file name, file size, installer type, installat ion command, necessity of reboot , necessity of individual i nstallation and etc.

ScanList consists of information to verify installation, to decide whether system is exceptional or not and for installation process, and depends on each operating system vendor because of security patch characteristics. We can make the ScanList automatically tied up with operation system vendors or manually build with open information from them.

The ScanList had better be encrypted to prevent an attacker from abusing the information.

3.3 Secure Communication Mechanism

Previous study [22] of patch distribution generated a new session key in each patch distribution processing. PMS supports transfer of ScanList and the present status of patch installation as well as patch distribution. Transmitting information with one session key for a long time is unreliable. Therefore, we have to guarantee safety communication as the proposed protocol changes session key periodically.

3.3.1 Advanced Authentication and Key Agreement Protocol

A modified Needham-Schroeder protocol [20, 21] is used in order for both a client and a server to authenticate each other based on certificates [22]. This protocol enables an agreed session key between a client and a server. Secure communication channels are established based on this agreed key of the protocol.

It is assumed that certificates of a client and the patch server are exchanged during a user registration process. Advanced authentication and key agreement protocols are followed.

client → server : "Hello", Epub_s(K1|Client Name|R1|T)
server → client : Epub_c(K2|R1|R2|V)
client → server : R2

Where E is encryption, pub_s and pub_c are public keys of server and client each, R1 and R2 are random numbers, K1 and K2 are keys to be used to generate a session key, and T and V are each timestamp and validity time of session key each.

The client sends a key, a random number and timestamp encrypted with the server's public key to the server. The server replies with another key, the random number received from the client, a random number, and expiration time of session key encrypted with client's public key. The client sends the received random number to the server. Upon completion of these protocols, a server and a client are mutually authenticated.

A session key is computed as *f*(K1, K2) using an appropriated publicly known non-reversible function *f*.

Once the authentication is completed, the session key is used to encrypt data between the server and the client. This mechanism changes the session key periodically as the validation time of the session key expired.

3.3.2 The Integrity of Patches

The patch server offers patches and ScanList that contains patch name, description of patch, information about installation, and size of file. Also ScanList includes MD5 checksum of patch file. The integrity of the patch is proved as comparing checksum in ScanList with computed one before agent installs patch.

4 Implementation Results

The proposed PMS is implemented by Java (JDK 1.4.2_03) in Windows Server 2000, RedHat Linux 9, and Solaris 8 platform. The program is coded with C++, MFC and Windows API for supporting the scanning Window systems and user interface. Also

Table 4. Remote Interface of the server and client

Remote Interface of the server	
boolean AuthenticateClient (String strID, String strPW)	It authenticates clients.
FILE_BUFFER DownloadFile (String strFilePathName, int lLength, int lFilePointer)	It transfer files to clients.
void SetClientTimer(String strUserId, String strTime)	It sets up the time of schedule.
void Connection (ClientInterface cinterface, String strKey)	It registers client's interface object.
boolean SendSystemInfo (String filename, String User_id)	It receives the information about client system.
boolean SendResultofInstallation(boolean bInstalled, boolean bNeedInstallation, String User_id)	It receives present status of patch installation.
void InsertNewPatchfile()	It registers new patch.
void ServerSynchronization(String[] filenames)	It synchronizes with the low ranking server.
boolean ClientRequestPatchfile(String strUserID)	It requests the installation of patch.
boolean CheckConnection(String UserId)	It checks a condition of connection
Remote Interface of the client	
FILE_BUFFER RequestProfile (String strProfileName, int lLength, int lFilePointer)	It transfer files to server.
boolean SendScanList (String[] strfileName, String strDistributionServerIP)	It receives the ScanList.

server manager and client manager are implemented by ISS and PHP (version 4.3.8). The communication between Server and client agent is activated under JAVA RMI. We define remote interfaces of a server and a client as in Table 4, and they acquire necessary events through these interfaces.

5 Conclusion

The proposed Patch Management System (PMS) for removing security vulnerability in various platforms was designed with hierarchical structure so that it could manage a large network efficiently and cope actively with change of various networks. Because each client agent instead of server analyzed vulnerability of its own system by using ScanList, we could reduce server's overload, solve privacy problem as it extracts system information, and cope dependency. We presented security protocols for safe patch distribution, prevented that a session was being kept for a long time without changing key. Moreover we guaranteed a stable communication between server and client and eased extension of platform or function by implementation using distributed object.

In a future, standardization for ScanList form and methodlogy to offer patch information for efficient management in multi-platform environment will be studied.

References

1. Robert Lemos : Microsoft publishes program to blast MSBlast, CNET News.com (January 6, 2004)
2. Robert Lemos : Hacker code could unleash Windows worm, CNET News.com (July 25, 2003)
3. Symantec Internet Security Threat Report Volume V, Symantec Corporation (March 2004)
4. CERT/Coordination Center : CERT Advisory, www.cert.org/advisories/CA-2001-26.html
5. Microsoft Corporation : Microsoft Security Bulletin, www.microsoft.com/technet/security /bulletin/ms00-078.mspx,
6. Bashar, M.A., Krishnan, G., Kuhn, M.G.: Low-threat security patches and tools. In: Proceedings of the International Conference on Software Maintenance (1997)
7. PatchLink Corporation, www.patchlink.com
8. Ecora Software Corporation, www.ecora.com
9. BigFix, Inc, www.bigfix.com
10. Security Profiling, www.securityprofiling.com
11. St. Bernard Software, www.stbernard.com
12. GFI, ww.gfi.com
13. Gravity Storm Software, www.securitybastion.com
14. Shavlik, www.shavlik.com
15. SoftRun, www.softrun.com
16. Microsoft SUS, www.microsoft.com/windowsserversystem/sus/default.mspx
17. Seo J.T., Yoon J.B., Choi D.S., Park E.K., Sohn T.S., Moon J.S.: A study on the Patch Management System In Intranet, PACO 2004
18. Seo J.T., Choi D.S., Park E.K., Shon T.S., Moon J.S.:Patch Management System for Multiplatform Environment, PDCAT 2004

19. Sohn T.S., Moon J.S., Seo J.T, Im E.K., Lee C.W.:Safe Patch Distribution Architecture in Intranet Environments, SAM 2003
20. Needham, R.M., Schroeder, M.D.:Using encryption for authentication in large networks of computers. Communications of the ACM 21 (1978) 993-999
21. Menezes, A.J., van Oorschot, P.C., Vanstone,S.A.: Handbook of Applied Cryptography. CRC press (1996)
22. Rivest, R., Shamir, A., Adleman, L.: A method for obtaining digital signature and public-key cryptosystems. Communications of the ACM 21 (1978) 120-126

Fuzzy Logic Anomaly Detection Scheme for Directed Diffusion Based Sensor Networks*

Sang Hoon Chi and Tae Ho Cho

School of Information and Communication Engineering, Sungkyunkwan University, 300 Cheoncheon-dong, Jangan-gu, Suwon, Gyeonggi-do, 440-746, Korea
{craken78, taecho}@ece.skku.ac.kr

Abstract. The wireless sensor network is rapidly emerging as an important tool for various applications. In this network, where a large subset of the network applications monitors and protects critical infrastructures, security is a particularly important issue. However, ensuring the security in the sensor network is complicated due to limited energy resources. In this paper, a fuzzy logic based anomaly detection scheme which provides the security to the directed diffusion protocol is proposed. The scheme is effective in preventing Denial-of-Service type attacks, which drain the energy resources within the nodes. Fuzzy logic is exploited in order to obtain a high detection rate by considering factors such as the node energy level, neighbor nodes list, message transmission rate, and error rate in the transmission.

1 Introduction

Recent advance in micro electronics and wireless communications has enabled the development of low cost sensor networks. A wireless sensor network (WSN) is composed of small nodes with sensing, computation, and wireless communication capabilities. Sensor nodes are usually scattered in a sensor field, which is an area where sensor nodes are deployed. The sensor nodes have the ability to communicate either with each other or directly to an external Base Station (BS) [1]. Generally, the WSN is composed of hundreds or thousands of sensor nodes, and a BS, or BSes. Each sensor node collects environmental information using a sensing unit and then transmits sensed data to neighbor nodes or to a BS. The sensor network is promising technology, and enables economically practical solutions for a variety of applications [2,3]. Figure 1 presents the entire architecture of WSNs.

A large subset of sensor network applications requires security, particularly if the sensor network monitors critical infrastructures. As a result, many security systems have emerged as important countermeasures against various attacks [4]. However, these systems are complicated, due to the broadcast nature of wireless communication. In addition, most of these systems cannot be directly applied to the

* This research was supported by the MIC (Ministry of Information and Communication), Korea, under the ITRC (Information Technology Research Center) support program supervised by the IITA (Institute of Information Technology Assessment).

L. Wang et al. (Eds.): FSKD 2006, LNAI 4223, pp. 725 – 734, 2006.

sensor network, due to the resource constraints on sensor nodes such as limited energy [5,6]. Therefore, security systems must be designed to operate with simple algorithms, minimizing resource utilization. Intrusion detection is an important solution, capable of preventing serious damage from various attacks. Although other security techniques defend outsiders from injecting packets, these techniques do not solve the problem of Denial-of-Service (DoS) type attacks. However, intrusion detection is able to solve this problem by using simple cryptographic techniques [7,8].

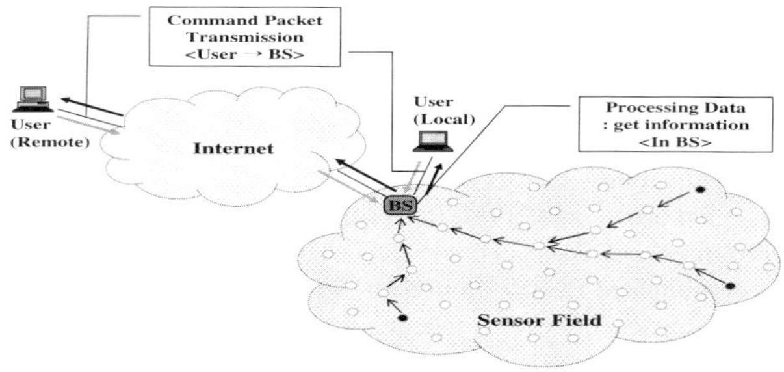

Fig. 1. Architecture of the wireless sensor networks

In this paper, a fuzzy logic based anomaly detection scheme, which provides the security to the directed diffusion protocol in WSN, is proposed. A BS or Master Node (MN) collects data about abnormal behaviors from neighbor nodes, for detecting whether attacks are being conducted. In order to perform this process, each sensor node monitors the condition of nodes within the transmission range, and itself. The condition indicates that abnormal behaviors of a node or neighbor nodes, is presented as four factors, such as the node energy level, neighbor node list, message transmission rate, and error rate in the transmission. Limited energy resources on sensor nodes are one of the major issues in WSN. The proposed scheme is effective in preventing DoS type attacks, which drain energy resources within nodes. Fuzzy logic is used, in order to obtain high detection rate, by considering various factors.

2 Background

This section introduces the directed diffusion as a routing protocol for this work, and attack models that focus on the directed diffusion, briefly describing other research relating to intrusion detection in WSNs.

2.1 Directed Diffusion Overview

Generally, sensor network routing protocols can be categorized into three types, flat-based, hierarchical-based, and location-based routing. In flat-based routing, each sensor node typically plays the same role. This routing is an efficient way to reduce the amount of energy within the specific application, and performs data aggregation

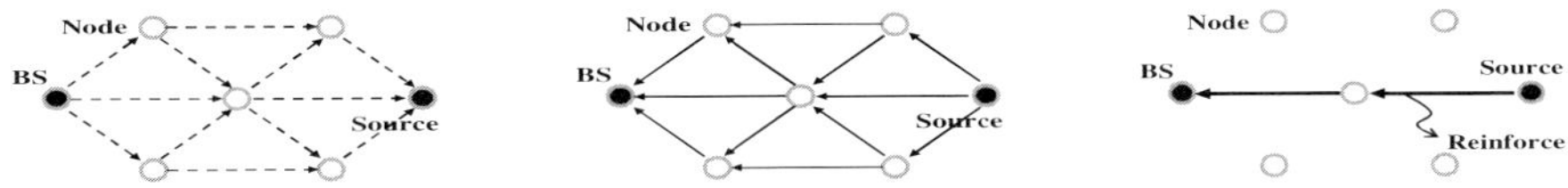

(a) Interest message propagation (b) Gradients setup (c) Data delivery along reinforced path

Fig. 2. The simplified process of directed diffusion

and elimination of redundant data [2,3]. In particular, directed diffusion, a component of flat-based routing, is a data-centric routing algorithm, where the BS transmits queries to specific regions and waits for data from nodes in the specific regions. An interest message is a query specifying what a user wants. In order to create a query, an interest message is defined using a list of attribute-value pairs [9].

Figure 2 (a) presents the interest message propagation. After the interest message is propagated throughout the sensor field, gradients are setup to draw data satisfying the query, towards the requesting node. The gradient specifies the direction of data flow. This process continues until the gradients are set up from the target regions back to the BS. Figure 2 (b) presents the setup of gradients. When a sensor node in the target region receives the message, it activates its sensors and initiates collection of events. If the node discovers a query event, then the sensed data is returned in the reverse path of the interest message. The best paths are reinforced to prevent further flooding. An example of the data delivery along reinforced paths is presented in Figure 2 (c).

2.2 Attack Models

A variety of attacks often result in a loss of availability of the sensor network. In particular, the directed diffusion is vulnerable to DoS type attacks, due to the robust nature of flooding. The concern here is the following types of attacks, which influence the directed diffusion [6,8,10].

- Path Influence and Selective Forwarding: An adversary can influence the routing path taken by a message flow by spoofing positive and negative reinforcements to control the routing path in directed diffusion.
- Resource Depletion: An adversary can create a high volume of data and routing messages that can quickly deplete the batteries of many nodes, because of the large-scale, multi-hop and cooperative nature of sensor networks.
- Radio Interference: An adversary conducts attacks through transmitting malicious messages into a common channel. These malicious messages collide with normal messages.
- Hello Flood: A malicious node can send or record or replay the interest message at a high transmission rate. This creates the illusion of being a neighbor to many nodes in the sensor network. In addition, it can confuse network routing.
- Node Impersonation: In order to disrupt the sensor network, an adversary must establish itself as a legitimate node, most likely by spoofing the ID of another node. The adversary may begin to deplete the resources or propagate false alarms.

2.3 Related Works

The sensor network commonly uses cryptography for security against unauthorized nodes. However, cryptography can only protect the network against external nodes. In addition, it does not solve the problem of compromised nodes, physical destruction, and DoS type attacks. Therefore, the sensor network requires an intrusion detection scheme. Existing researches related to intrusion detection are introduced as follows.

The paper [11] proposes an intrusion detection scheme to detect localization anomalies caused by adversaries. A number of ways to detect malicious attacks in localizations, in which sensor nodes can verify whether their derived locations are consistent with the deployment knowledge, are proposed. Another paper [12] presents three schemes to find the most vulnerable node in the sensor network, for defense against intrusions. In the first scheme it formulates the attack-defense problem between an attacker and sensor network. In the second scheme it uses the Markov Decision Process, in order to predict the most vulnerable sensor node. In the third scheme it uses an intuitive metric of the node's traffic, and protects the node with the highest value of this metric. In [8], in order to make a sensor node capable of detecting an intruder, a dynamic statistical model of the neighboring nodes is built in conjunction with a detection algorithm, by monitoring received packet power levels and arrival rates. The paper [10] proposes a secure MAC protocol which is based on the RTS/CTS mechanisms. It utilizes probabilistic intrusion indicators: analyzing the security problems of RTS/CTS based MAC protocols, and designs the intrusion detection method using soft decision theory. More research relating to intrusion detection for WSN not introduced in this section can be found in [13-15].

3 Fuzzy Logic Anomaly Detection Scheme (FLADS)

In this paper, DoS type attacks that compromise the directed diffusion are focused on, where an adversary's purpose is to cause false alarms or deplete resources of sensor nodes. As a result, network lifetime and availability is considerably reduced. The goal is to design the FLADS to defend against malicious message injection attacks and resource depletion attacks launched by the adversary.

The FLADS uses factors such as the node energy level, neighbor nodes list, message transmission rate, and error rate in the transmission, in order to monitor abnormal sensor node behavior. The BS or MNs generate a detection value to determine whether the attacks exist. The detection value is determined by a fuzzy logic controller, by consideration of the four factors. In order to archive this value, the BS or MNs collect the inform message about the factors from neighbor nodes. The MN creates a new advertisement (ADV) message for notification of its location, and the sensor node creates a new inform message for reporting abnormal behaviors.

3.1 Assumption

- All nodes are peer entities. They use the same hardware with constant transmission power, and run the same protocol stack.
- BS and MNs are considerably more powerful than ordinary sensor nodes in terms of computational power, energy, storage, and so on.

- MNs know their locations (e.g., GPS receivers, etc), and other nodes discover their locations, based on the beacons provided by the MN [11].
- The sensor nodes cannot always transmit the correct values and the communication errors are considered as a kind of random noise.
- Sensor networks may be deployed in hostile environments, where sensor nodes can be compromised and destructed. However it is assumed that base station and beacon nodes cannot be compromised.

3.2 Fuzzy Logic Based Anomaly Detection Architecture

In directed diffusion, the function of each sensor node is equal to that of other nodes with the exception of the BS. However, in the FLADS, the sensor network is composed of a hierarchical structure in which MNs are locally deployed in the subset region of the network. The MN is used for detecting attacks from a fuzzy logic based controller. Figure 3 presents the architecture of the FLADS with a hierarchical structure based on directed diffusion.

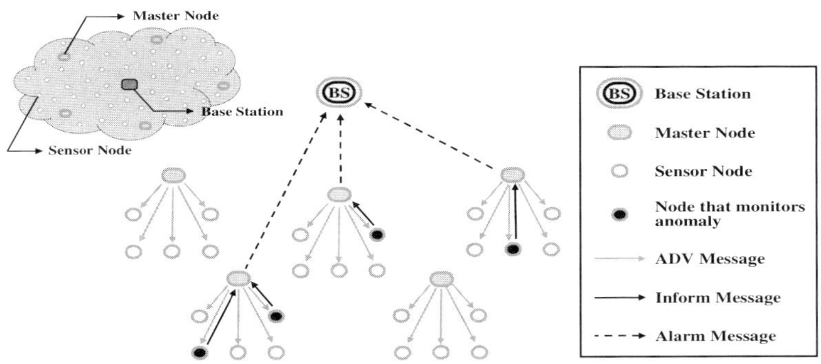

Fig. 3. Architecture of the fuzzy logic based anomaly detection scheme in directed diffusion routing protocol

Each MN transmits an ADV message to neighbor nodes with location information. Since the ADV message includes a hop count considering the network size, this message is transmitted until the hop count is permitted. In the event that hop counts of the received messages are same, a sensor node selects an ADV message, and includes a low hop count value. It means that the MN limits the number of neighbor nodes. The sensor node conducts a monitoring task using a local anomaly detection module, and the sensor node creates an inform message in the event that it detects abnormal behavior. In this distributed fashion, sensor nodes have the ability to record simple statistics regarding their neighbors' behavior and anomalies. An inform message that is generated by the sensor node is delivered to the MN. Then, the fuzzy logic controller of MN determines whether the nodes have come under attack. If the attack is real, the MN transmits an alarm message to the BS with detection information.

3.3 Detection Entities and Factors That Affect the Anomaly Detection

Detection entities of the FLADS are divided into two parts: local detection modules and global detection modules. In case of the local detection modules, each sensor node equipped with an Anomaly Detection Module (ADM), is capable of monitoring the behaviors of its system and neighbor nodes. The global detection modules are used to monitor attacks by collecting the inform messages from sensor nodes. In this case, the BS and MNs equipped with fuzzy controllers are capable of determining whether attacks exist. The ADM uses four factors to monitoring anomalies of the sensor node. These factors can be exploited as key enablers for providing fuzzy input variables. The components of this scheme are presented in Figure 4.

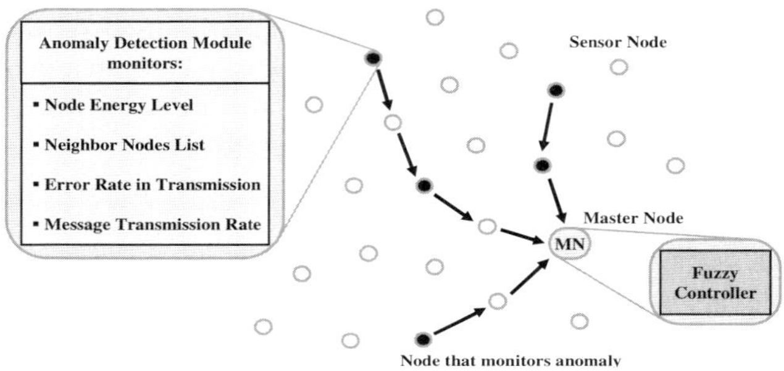

Fig. 4. The components for launching anomaly detection scheme against attacks

The node energy level presents the remaining energy of a sensor node. A rapid decrease in the energy level indicates a resource depletion attack, for example, resulting from a high volume of data and large number of routing messages [5,7]. In the interest message propagation and gradients setup process, each sensor node exchanges routing messages with neighboring nodes, then constructs a neighbor node list. The neighbor node list presents the relationship with neighbor nodes within the radio transmission range [11]. In the event that the number of neighbor nodes increases or decreases the initial number in the neighbor node list, this may indicate attacks such as the node impersonation and physical attacks [4-7]. The message transmission rate is an average message arrival rate representative of the sensor node or neighbor activities. At every node, the last messages received from each neighbor are used to calculate the statistics for that neighbor and each arriving message is compared against these values. Each node records the message arrival time of each incoming message [8,10]. This factor indicates a notable feature of attacks such as hello flood, node impersonation, and resource depletion. The adversary could randomly access the link and interfere with messages from the channel. More seriously, this node may inject and alter transmitted data. These types of attacks can be detected by path influence and selective forwarding or radio interference. The error rate in message transmission presents a feature of these attacks. To monitor the attacks, each node performs error checking of the following: CRC, fixed packet size, destination field, handler field, and application data unit.

Figure 5 presents the process of ADM for monitoring behaviors of a sensor node. In the event that messages such as the interest message and data message are delivered to the sensor node, the ADM of the sensor node monitors anomalies according to each layer [15]. If the ADM detects anomalies, the sensor node creates an inform message, in which the factors for monitoring anomalies are included. The sensor node transmits an inform message toward the MN. In [8,10-14], the utilization and features of these factors used in the ADM are described in detail.

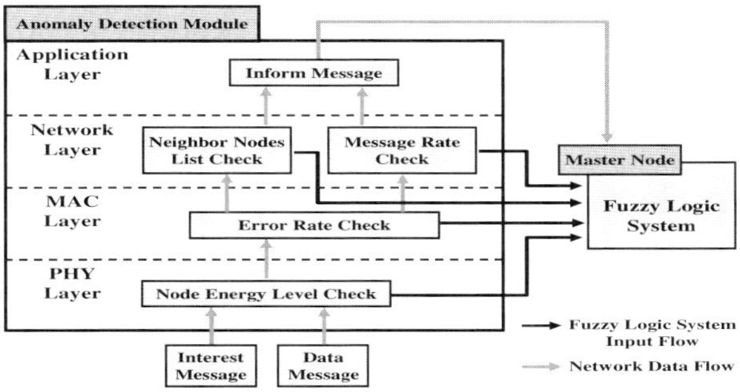

Fig. 5. The process of anomaly detection module for monitoring behaviors of the sensor node and neighbor

3.4 Fuzzy Logic Design

In FLADS, the MN can detect an attack by exploiting the four factors of the fuzzy logic based controller. The fuzzy controller has a hierarchical structure composed of four input variables and three output variables. Figure 6 presents the architecture of the fuzzy controller in determining whether the attack is real.

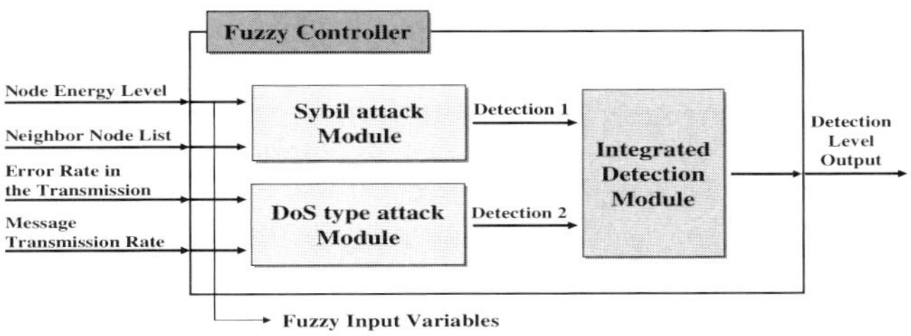

Fig. 6. Architecture of the fuzzy logic based controller

This controller determines a detection value using the four input parameters of the energy level, neighbor node list, error rate, and message rate included in the inform message. A final output variable among these output variables presents the detection level. The detection level for the final output is divided to two detection modules. These sub-detection modules are output from four input variables. Since the node energy level and neighbor node list is related to each other, these parameters are exploited to generate the detection 1. Also, the detection 2 is generated by the error rate and message rate in the transmission.

Figure 7 illustrates the membership functions of four input parameters of the fuzzy logic. Three output parameters of the fuzzy logic are detection 1, detection 2, and integrated detection level, which are also represented by the membership functions as shown in Figure 7. Some of the fuzzy rules are shown below.

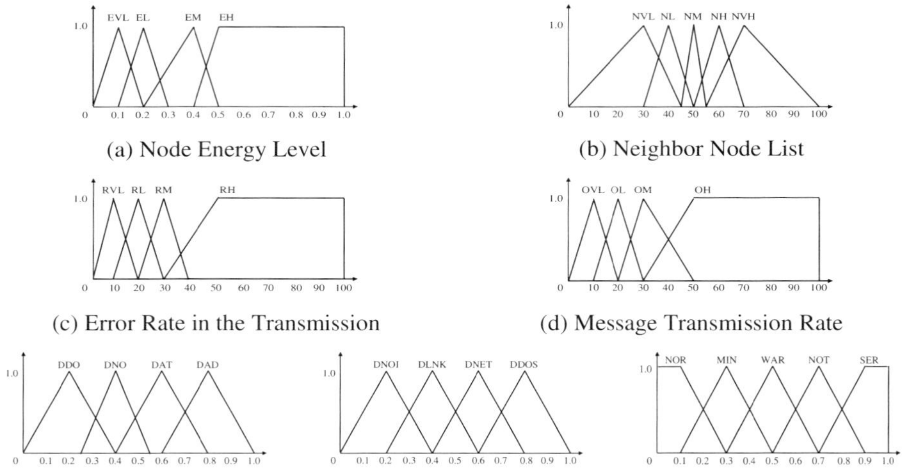

(a) Node Energy Level
(b) Neighbor Node List
(c) Error Rate in the Transmission
(d) Message Transmission Rate
(e) Detection 1
(f) Detection 2
(g) Integrated Detection Level

Fig. 7. Membership functions for input and output variables in fuzzy controller

```
R09: IF (Node_Energy_Level IS EL)
     AND (Neighbor_Nodes_List IS NH)
     THEN (Detection_1 IS DNO)

R10: IF (Node_Energy_Level IS EL)
     AND (Neighbor_Nodes_List IS NVH)
     THEN (Detection_1 IS DAD)

R11: IF (Node_Energy_Level IS EM)
     AND (Neighbor_Nodes_List IS NVL)
     THEN (Detection_1 IS DDO)
```

4 Simulation Result

In order to show the performance of the FLADS, the FLADS is evaluated through simulation. The sensor network is composed of 600 sensor nodes deployed in a rectangular area of 400×400 m^2. It is assumed that each node's initial energy level is 1 J, the transmission and reception cost for a message is 60.48 μJ and 14.4 μJ, respectively. The transmission and reception cost for the data message is 107.52 μJ

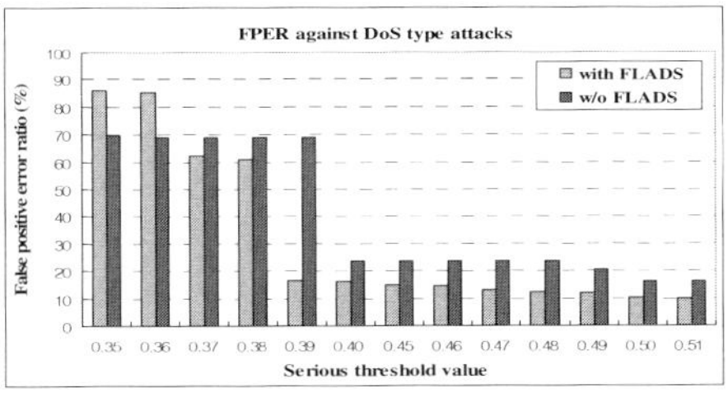

Fig. 8. False Positive Error Ratio (FPER) against DoS type attacks

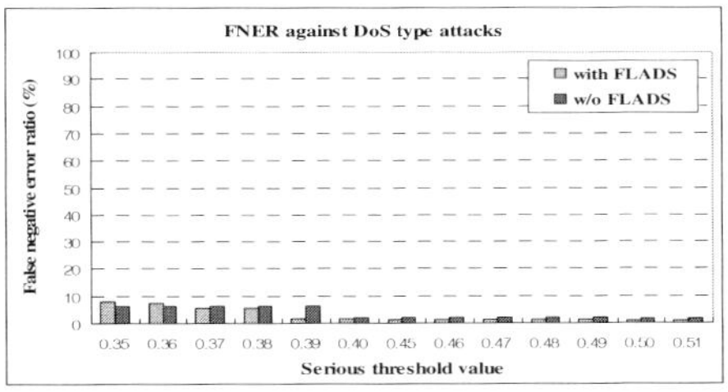

Fig. 9. False Negative Error Ratio (FNER) against DoS type attacks

and 25.6 μJ, respectively. Ratio of energy spent for packet reception to packet transmission is kept at approximately 1:4 [9]. A sensor node has a transmission range of 40 meters. Following four graphs are used for evaluating the FLADS.

The above mentioned attack models, are used, and new nodes are randomly deployed in sensor fields at an interval of 300 simulation times, to measure the FPER and FNER. Figure 8 presents the result of the FPER against directed diffusion DoS type attacks. A serious threshold value presents the sensitivity of a detection module. As presented in the graph, the proposed FLADS has a lower FPER, than the scheme without the FLADS. The average FPER of this scheme is reduced by approximately 8.4%. It is important to note that the FNER of the system with the FLADS is similar to that of the system without the FLADS. This result is presented in Figure 9.

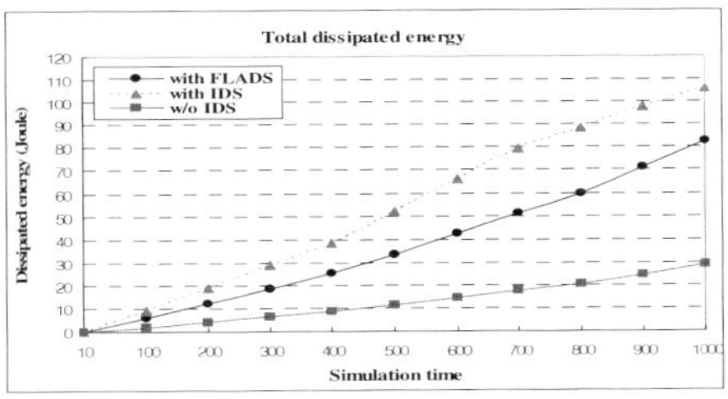

Fig. 10. Total dissipated energy of the sensor network

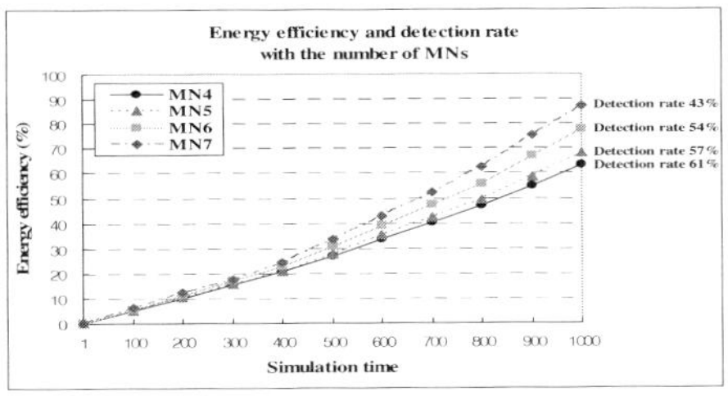

Fig. 11. Energy efficiency and detection rate with the number of MNs

In Figure 10, the simulation results demonstrate that the system with intrusion detection (an energy level of approximately 44.3%) consumes more energy than the system with the FLADS (an energy level of approximately 30.4%), whereas the system without intrusion detection (about 10.5% energy level) consumes less energy than the system with the FLADS. The result of energy efficiency and detection rate with the number of MNs is presented in Figure 11. The more MNs deployed in the sensor field, the more energy efficiency is increased, since the overhead of messages between MN and nodes is reduced. However, the detection rate in this case decreases faster than the rate when a small number of MNs exist.

5 Conclusion

The proposed FLADS exploits the fuzzy logic system in detecting attacks within WSN. The work is motivated by the directed diffusion routing protocol, which is vulnerable to DoS type attacks that drain the energy resources within sensor nodes. The fuzzy logic system determines whether an attack is being conducted using four factors as inputs, these are the node energy level, neighbor node list, message transmission rate, and error rate of the transmission. The FLADS demonstrates high detection rate and energy efficiency through the simulation in which the FPER and FNER are measured. As future work, the response mechanism to various attacks will be investigated and included in the FLADS with the aid of blackboard architecture.

References

1. Akyildiz, I.F., Weilian, S., Sankarasubramaniam, Y., Cayirci, E.: A Survey on Sensor Networks. IEEE Commun. Mag. 40(8) (2002) 102-114
2. Akkaya, K., Younis, M.: A Survey of Routing Protocols in Wireless Sensor Networks. Ad Hoc networks Journal 3(3) (2004) 325-349
3. Al-Karaki, J. N., Kamal, A. E.: Routing Techniques in Wireless Sensor Networks: A Survey. IEEE Wirel. Commun. 11(6) (2004) 6-28
4. Shi, E., Perrig, A.: Designing secure sensor networks. IEEE Wirel. Commun. 11(6) (2004) 38-43
5. Perrig, A., Stankovic, J., Wagner, D.: Security in Wireless Sensor Networks. Communication of the ACM 47(6) (2004) 53-57
6. Karlof, C., Wagner, D.: Secure Routing in Wireless Sensor Networks: Attacks and Countermeasures. Proc. of the IEEE SNPA (2003) 113-127
7. Wood, A.D., Stankovic, J.A.: Denial of Service in Sensor Networks. IEEE Computer 35(10) (2002) 54-62
8. Onat, I, Miri, A.: An Intrusion Detection System for Wireless Sensor Networks. IEEE Int'l Conf. on WiMob 3 (2005) 253-259
9. Intanagonwiwat, C., Govindan, R., Estrin, D., Heidemann, J., Silva, F.: Directed Diffusion for Wireless Sensor Networking. IEEE ACM T. Network 11(1) (2003) 2-16
10. Ren, Q., Liang, Q.: Secure Medial Access Control (MAC) in Wireless Sensor Networks: Intrusion Detection and Countermeasures. Proc. of the IEEE PIMRC 4 (2004) 3025-3029
11. Wenliang, D., Lei, F., Peng, N.: LAD: Localization Anomaly Detection for Wireless Sensor Networks. Proc. of the IEEE IPDPS (2005)
12. Agah, A., Das, S.K., Basu, K., Asadi, M.: Intrusion Detection in Sensor Networks: A Non-cooperative Game Approach. Proc. of the IEEE NCA (2004) 343-346
13. Su, C.C., Chang, K.M., Kuo, Y.H., Horng, M.F.: The New Intrusion Prevention and Detection Approaches for Clustering-based Sensor Networks. IEEE Conf. on WCNE 4 (2005) 1927-1932
14. Zhou, G., He, T., Stankovic, J.A., Abdelzaher, T.: RID: Radio Interference Detection in Wireless Sensor Networks. Proc. of the IEEE INFCOM 2 (2005) 891-901
15. Roman, R., Zhou, J., Lopez, J.: Applying intrusion detection systems to wireless sensor networks. Proc. of the IEEE CCNC (2006) 640-644

An Entropy-Based Stability QoS Routing with Priority Scheduler in MANET Using Fuzzy Controllers*

Baolin Sun[1,2,3], Chao Gui[1,3], Hua Chen[2], and Yue Zeng[3]

[1] College of Computer & Technology, Hubei University of Economics
Wuhan 430205, P.R. China
blsun@163.com

[2] Department of Mathematics and Physics, Wuhan University of Science and Engineering
Wuhan 430073, P.R. China

[3] School of Computer Science and Technology, Wuhan University of Technology
Wuhan 430063, P.R. China

Abstract. Due to the dynamic nature of the network topology and restricted resources, quality of service (QoS) and stability routing in mobile ad hoc network (MANET) is a challenging task. This paper presents an Entropy-based Stability QoS Routing with Priority scheduler in MANET using fuzzy controllers (ESQRP). The key idea of ESQRP algorithm is to construct the new metric-entropy and select the stability path with the help of entropy metric to reduce the number of route reconstruction so as to provide QoS guarantee in the ad hoc network. The simulation results shows that the ESQRP approach provide an accurate and efficient method of estimating and evaluating the route stability in dynamic mobile networks.

1 Introduction

A Mobile Ad Hoc Network (MANET) is an autonomous system of mobile nodes connected by wireless links. There is no static infrastructure such as base station as that was in cell mobile communication. All the nodes are free to move around randomly, thus changing the network topology dynamically. For such networks, an effective routing algorithm is critical for adapting to node mobility as well as possible channel error to provide a feasible path for data transmission [1-8].

Fuzzy logic based decision algorithm influences caching decisions of multiple paths uncovered during route discovery and avoids low quality paths [5-7]. Differentiated resource allocation considering message type and network queue status is evaluated using fuzzy logic scheme [4,6,7]. Entropy [2,8] presents the uncertanity and a measure of the disorder in a system.

In this paper, we present an Entropy-based Stability QoS Routing with Priority scheduler in MANET using fuzzy controllers (ESQRP).

* This work is supported by National Natural Science Foundation of China (No. 90304018), NSF of Hubei Province of China (No. 2005ABA231, 2006ABA301) and Key Scientific Research Project of Hubei Province Education Department (No. D200617001).

L. Wang et al. (Eds.): FSKD 2006, LNAI 4223, pp. 735 – 738, 2006.

2 Entropy Metric

We also associate each node m with a set of variable features denoted by $a_{m,n}$ where node n is a neighbor of node m. Any change of the system can be described as a change of variable values $a_{m,n}$ in the course of time t such as $a_{m,n}(t) \rightarrow a_{m,n}(t+\Delta_t)$. Let us denote by $v(m, t)$ the velocity vector of node m and by $v(n, t)$ the velocity vector of node n at time t. The relative velocity $v(m, n, t)$ between nodes m and n at time t is defined as: $v(m, n, t)= v(m, t) - v(n, t)$, Let us also denote by $p(m, t)$ the position vector of node m and by $p(n, t)$ the position vector of node n at time t. The relative position $p(m, n, t)$ between nodes m and n at time t is defined as: $p(m, n, t) = p(m, t) - p(n, t)$, Then, the relative mobility between any pair (m, n) of nodes during some time interval is defined as their absolute relative speed and position averaged over time. Therefore, we have:

$$a_{m,n} = \frac{1}{N}\sum_{i=1}^{N} \frac{|\, p(m,n,t_i) + v(m,n,t_i) \times \Delta_{t_i} \,| - |\, p(m,n,t_{i+1}) \,|}{R}$$

where N is the number of discrete times t_i that velocity information can be calculated and disseminated to other neighboring nodes within time interval Δ_t. R is radio range of nodes. In general the entropy $H_m(t, \Delta_t)$ at mobile is calculated as follows:

$$H_m(t,\Delta_t) = \frac{-\sum_{k\in F_m} P_k(t,\Delta_t)\log P_k(t,\Delta_t)}{\log C(F_m)}, \qquad F'(s,d) = \prod_{i=1}^{N_r} H_i(t,\Delta_t)$$

where $P_k(t, \Delta_t) = (a_{m,k} \,/\, \sum_{i\in F_m} a_{m,i})$, N_r denotes the number of intermediate mobile nodes over a route between the two end nodes (s, d).

$$F(s,d) = -\ln F'(s,d) = -\sum_{i=1}^{N_r} \ln H_i(t,\Delta_t)$$

We are computing $F(s, u)$, and queuing it from the smallest to the biggest, namely, $F(s,u_1) \leq F(s, u_2) \leq \ldots \leq F(s, u_m)$, then, the min value is the best stability path.

3 Fuzzy Controller QoS Routing

The controllable elements in the different priority architecture are shown in Fig. 1. In this architecture, all nodes have a separate queue for each service class; a classifier places the packets into the respective queue and the scheduler selects packets from these queues for transmission in the output links.

Fuzzy systems reason with multi-valued fuzzy sets instead of crisp sets. The output of the fuzzy routing in ad hoc networks in Fig. 2 is used to tune the controlled system's parameters based on the state of the system. Rule base is an IF-THEN rule group with fuzzy sets that represents the desired behavior of a fuzzy system.

Ri: **IF** x_1 is A_{i1} and … and x_n is A_{in} **THEN** y is C_i, $i = 1, 2, \ldots, L$

where L is the number of fuzzy rules, $x_j \in U_j$, $j = 1, 2, \ldots, n$, are the input variables, y is the output variable, and A_{ij} and C_i are linguistic variables or fuzzy sets for x_j and y

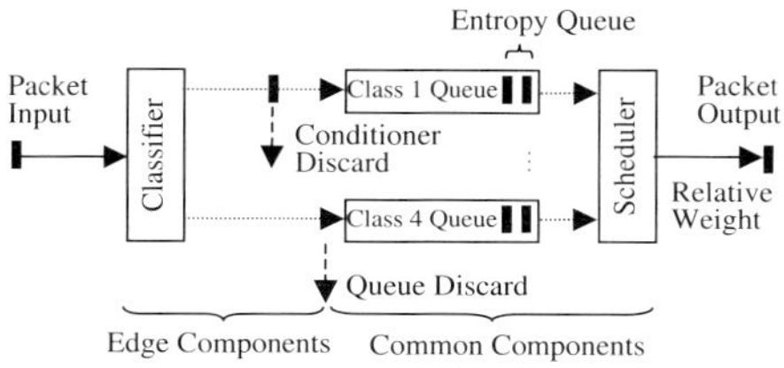

Fig. 1. Different priority controllable architecture

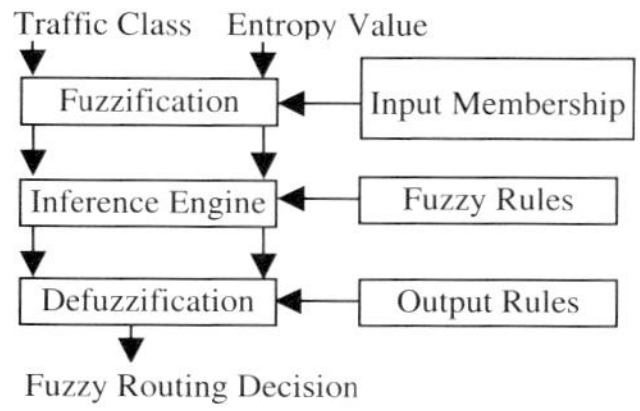

Fig. 2. The fuzzy routing in MANET

Table 1. Fuzzy rule base

Rule No.	Entropy Value	QoS Class	Priority Index
1	very low	high	very high
2	low	high	high
3	medium	medium	medium
4	high	low	low

respectively. A_{ij} and C_i are characterized by both membership functions. The table 1 shows the fuzzy conditional rules for the fuzzy scheduler.

4 Simulation

We simulated the proposed scheduler in OPNET [10] and conducted experiments to evaluate the effectiveness of the proposed scheduler. In this performance evaluation the following performance metrics were evaluated: percentile of data transmission rate, path success ratio, and end-to-end delay.

Fig. 3 depicts a comparison of data transmission rate AODV and ESQRP scheduler. The data transmission rate is still higher than that of AODV, which means it is more suitable for the routing choosing under timely data transmission application and dynamic network structure. The average end-to-end delay performance as shown in

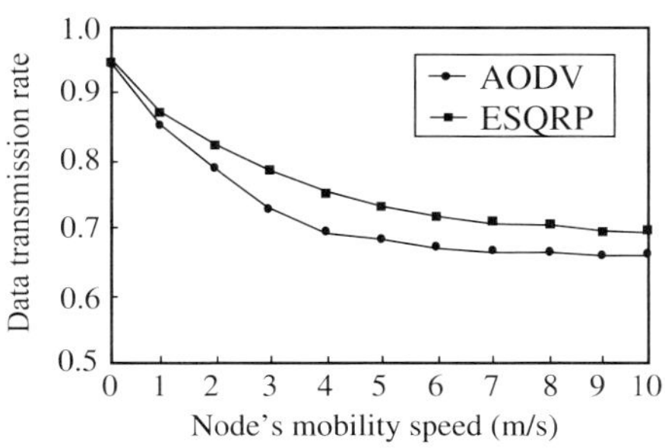

Fig. 3. Data transmission rate vs Node's mobility speed

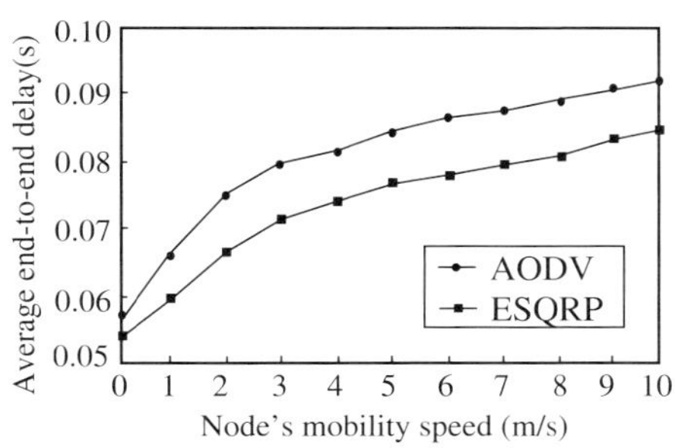

Fig. 4. Average end-to-end delay vs. Node's mobility speed

the Fig. 4, proves that the end-to-end delay improves when scheduler is included. As the mobility varies from 0-10 m/s, the fuzzy controllers scheduler provides an end-to-end delay reduced by around 0.005 sec. to 0.01 sec.

5 Conclusion

In this paper, we present an entropy-based stability QoS routing with priority scheduler in MANET using fuzzy controllers (ESQRP). Our entropy-based stability QoS routing algorithm with priority scheduler has produced significant improvements in data transmission rate, and average end-to-end delay. Fuzzy logic implementation relates input and output in linguistic terms, the overlap composition of many input variables in taking a single output decision shows the robustness of the system in adapting to constantly changing mobile scenario. The membership functions and rule bases of the fuzzy scheduler are carefully designed. The use of fuzzy logic improves the handling of inaccuracy and uncertainties of the ingress traffic into the domain.

References

1. Sun, B. L., and Li, L. Y.: A QoS Multicast Routing Optimization Algorithms Based on Genetic Algorithm. Journal of Communications and Networks, Vol. 8, No.1, (2006) 116-122
2. Sun, B. L., Li, L. Y., Yang, Q., and Xiang, Y.: An Entropy-Based Stability QoS Multicast Routing Protocol in Ad Hoc Network. Advances in Grid and Pervasive Computing (GPC 2006). Lecture Notes in Computer Science, Vol. 3947, Springer-Verlag Berlin Heidelberg, (2006) 217-226
3. Sun, B. L., Yang, Q., Ma J. and Chen H.: Fuzzy QoS Controllers in Diff-Serv Scheduler using Genetic Algorithms. Computational Intelligence and Security (CIS2005). Lecture Notes in Artificial Intelligence, Vol. 3801, Springer-Verlag Berlin Heidelberg, (2005) 101-106
4. Perkins, C., Royer, E. B., and Das, S.: Ad hoc On Demand Distance Vector (AODV) Routing. RFC 3561, July (2003)
5. Rea, S., and Pesch, D.: Multi-metric routing decisions for ad hoc networks using fuzzy logic. In Proceedings of 1st International Symposium on Wireless Communication Systems, Mauritius, 20- 22 September, (2004) 403-407
6. Alandjani, G., and Johnson, E.: Fuzzy Routing in ad hoc networks. In Proceedings of the IEEE International Conference on Performance, Computing and Communications, Phoenix, Arizona, April, (2003) 525-530
7. Gomathy, C., and Shanmugavel, S.: An Efficient Fuzzy Based Priority Scheduler for Mobile Ad hoc Networks and Performance Analysis for Various Mobility Models. 2004 IEEE Wireless Communications and Networking Conference (WCNC 2004), Atlanta, USA, Vol. 2, March (2004) 1087-1092
8. An, B., and Papavassiliou, S.: An Entropy-Based Model for Supporting and Evaluating Route Stability in Mobile Ad hoc Wireless Networks. IEEE Communications Letters, Vol. 6, No. 8, (2002) 328-330
9. Sun, Q., Li, L. Y.: An Efficient Distributed Broadcasting Algorithm for Ad Hoc Networks. Advanced Parallel Processing Technologies (APPT 2005), Lecture Notes in Computer Science, Vol. 3756, Springer Verlag Berlin Heidelberg, (2005) 363-372
10. OPNET Technologies Inc., Bethesda, MD, USA, available at http://www.opnet.com

Design of a Multi-model Fuzzy Controller for AQM

Ming Liu, Wen-hua Dou, and Rui Xiao

Computer College, National University of Defense Technology
Hunan, 410073, P.R. China
liutomorrow@hotmail.com

Abstract. Given the fact that the current Internet is getting more difficult in handling the traffic congestion control, new techniques are required. This paper proposes a multi-model fuzzy controller for AQM routers which consolidates the advantages of FLC and PID controller. Simulation results show that the proposed AQM scheme does improve the end-to-end performance.

1 Introduction

The traditional technique to manage router queue lengths is to drop incoming packets when the queue is full (tail drop). Although tail drop is widely adopted in the Internet, it presents important drawbacks, such as long delays experienced by packets and the lock-out phenomena. To eliminate the tail drop disadvantages and to anticipate the source answers to incipient congestion situations, AQM policies have been proposed. The main objective of AQM is to make TCP choose source rates so as to maximize the aggregate utility subject to the constraint that the total source rate at any link does not exceed the link capacity.

Since the IETF recommendation[1], the main focus of the research community has been on the development of new AQM schemes. In recent study, it was noted that more than 50 new algorithms have been proposed since 1999. Several AQM schemes are proposed and analyzed using either control theoretic model or optimization model. Despite the research efforts spanning several years, there are no universally acceptable control solutions in this area. However, these AQM algorithms show weaknesses to detect and control congestion under dynamically changing network situations.

Internet is a rather complex huge and nonlinear dynamic system. Fuzzy logic controllers have been developed and applied to nonlinear system for the last two decades. Recently, some research papers use fuzzy logic investigating solutions to congestion control issues. Fuzzy control is suitable to the dynamical network environment without precise model. The approaches in [2]-[4] are fuzzy logic algorithms for AQM under IP networks. The major problem is it's hard to configure the fuzzy rule and parameters in membership function for dynamical environments under IP networks, to debug them is difficult and the steady state performance are not desirable.

2 Multi-model Fuzzy Controller for AQM

Although the fuzzy logic is suitable for the changeable network, it is extremely difficult to have a traditional fuzzy controller with good performance due to the complexity of

L. Wang et al. (Eds.): FSKD 2006, LNAI 4223, pp. 739 – 742, 2006.

the applications. It must define more subtle fuzzy sets, more accurate membership functions and fuzzy control rules to improve the performance (precision) of fuzzy logic controller, but it is hard to design and test in complex internet. Although the linear PI controller has slow response, it is easy to implement and have good steady state performance. In this section, we propose a multi-model fuzzy controller for AQM which consolidates the advantages of FLC and PID controller. The block diagram can be depicted in figure 1.

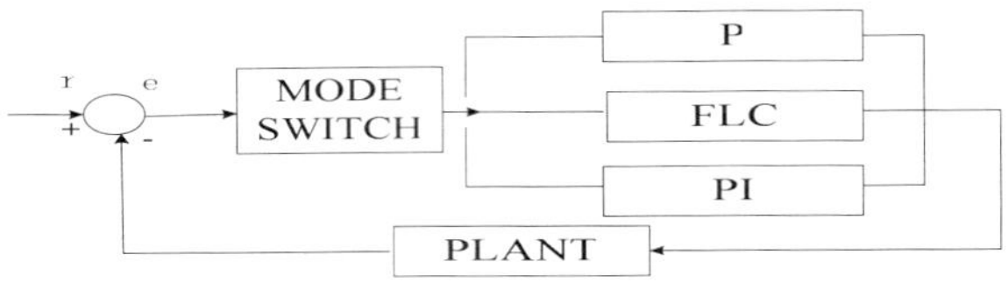

Fig. 1. Block diagram of P-Fuzzy-PI

e is the error between the current router queue size and the target queue size. In the discrete form, the error is expressed as: $e(kT) = q(kT) - q\_ref$

We define $E = K_i e(kT)$, $EC = K_p \Delta e(kT)$, where K_p and K_i are the proportional and integral gains. EP and ZE are max and min threshold provided by user. p is the packet drop probability in router. The algorithm can be depicted as follows:

1) if $E \geq EP$, we use P controller: $C(s) = K'_p$, the purpose is to get fast response.

2) if ZE<E<EP, we use fuzzy logic controller[4], we define $U = f(E, EC)$, $p(kT) = p(kT - T) + K_f f(E, EC) = p(kT - T) + K_f U$

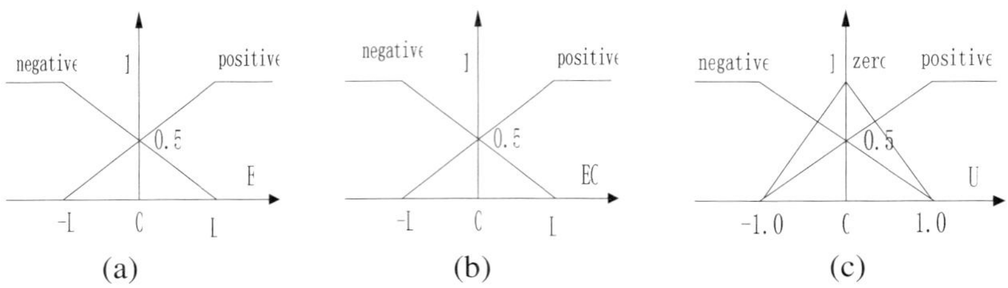

Fig. 2. Membership functions for E, EC and U

Figure2(a)(b) show the membership functions for E and EC. Both of the input variables have two fuzzy term sets: negative(N) and positive(P). Here L is a constant. Inference produces a new fuzzy set from the result of the fuzzification using a set of rules. Figure2 (c) shows the membership function for the output U. the output have three fuzzy term sets, negative(N), zero(Z) and positive(P). The rule used in the fuzzy controller is as follows: if E is negative and EC is negative then U is negative; if E is negative and EC is positive then U is zero; if E is positive and EC is negative then U is zero; if E is positive and EC is positive then U is positive.

The defuzzification process is based on the center of mass formula: $U = \sum b_i \mu_i / \sum \mu_i$, here μ_i is the membership value of input, b_i is the corresponding output, $b_i = -1,0,0,1 (i = 1,2,3,4)$.

$\mu_1 = \min(\mu_N(E), \mu_N(EC))$ $\mu_2 = \min(\mu_N(E), \mu_P(EC))$

$\mu_3 = \min(\mu_P(E), \mu_N(EC))$ $\mu_4 = \min(\mu_P(E), \mu_P(EC))$

Then the controller output can be depicted as: $p(kT) = p(kT - T) + K_f U$

3) if $E \leq ZE$, then we use PI controller $C(s) = K_p + K_i / s$, this can be converted into a difference equation, at time $t = kT$ where $T = 1/f_s$:

$$p(kT) = p(kT - T) + K_I T * (q(kT) - q\_ref) + K_P * (q(kT) - q(kT - T))$$

Our purpose is to get good performance in steady state and benefits the merits from both fuzzy controller and PI controller.

3 Simulation Results

In this section we study the performance of P-Fuzzy-PI in various traffic conditions and compare it with PI and ARED. We evaluate them by NS2 simulator and use common topology shown in figure 3. Link bandwidth C is 3750pkt/s. RTT ranges between 40ms and 220ms. Buffer size is 800 packets. We use ns default parameters set in PI and ARED scheme, and P-Fuzzy-PI accord with reference[4]:

$K_p = 1.771e-5$ $K_I = 9.643e-7$ $T = 0.1$ $ZE = 20$ $L = K_i * 100$ $EP = 400$ $K_f = 0.01$.

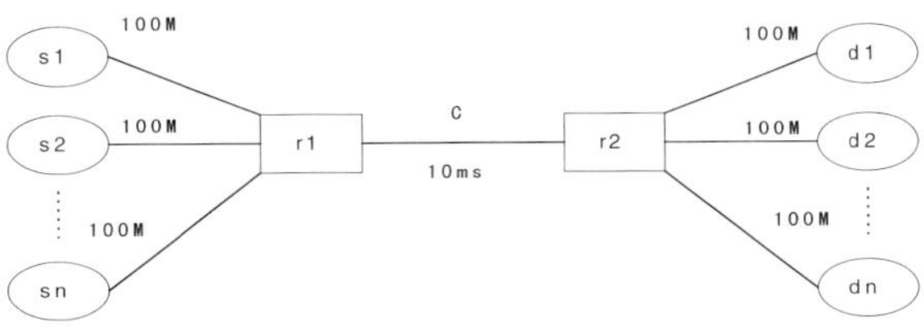

Fig. 3. Network topology

Firstly, we analyze the performance of the AQM schemes under varying traffic load. We compare the responsiveness and queue size of P-Fuzzy-PI, PI and ARED in the presence of long-lived FTP flows only. The number of FTP flows is 200 at the beginning, 100 FTP flows leave the link after 100 seconds, they join the link again when t=200s. The total simulation lasted for 300s. The queue lengths for the three algorithms are depicted in figure 4(a). As shown in figure 4(a), under varying traffic load, P-Fuzzy-PI can regulate queue length to the desired reference value quickly while PI converges slowly and ARED keeps the queue length at the desired value with oscillations.

Then we analyze the performance of the AQM schemes when unresponsive flows exist, we use two mixtures: FTP and ON/OFF flows. The burst and idle times of the ON/OFF service model are 0.5s and 1s respectively, and the sending rate during "on" duration is 200Kbps. The number of FTP flows is 100 at the beginning, 50 ON/OFF

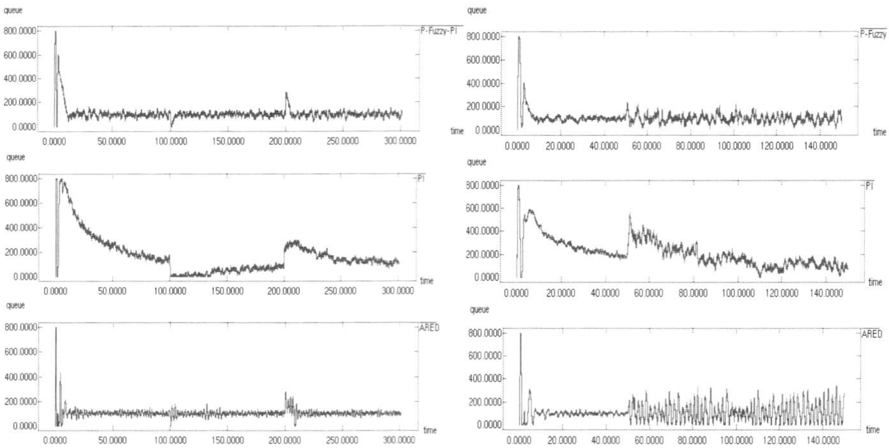

(a) (b)

Fig. 4. Evolution of the queue lengths with FTP flows and ON/OFF flows

flows arrive at the link 50 seconds later. The queue lengths, plotted in Figure 4(b), show that P-Fuzzy-PI reaches the steady state in a short time, whereas PI takes longer time to stabilize. ARED keeps the queue length at the desired value with large oscillations.

The results of experiment show that P-Fuzzy-PI is more robust under changefully traffic load and when unresponsive flows exist.

4 Conclusions

Fuzzy logic will offer significant improvements on congestion control in TCP/IP networks although there are still issues required investigation. In this paper, we present a multi-model fuzzy controller for AQM. The main idea is using multi-model control and fuzzy logic to improve the robustness and performance of AQM controller. The performance of Multi-model Fuzzy Controller is evaluated by simulations and compared with PI and ARED. The results indicate that the proposed scheme exhibits more robust congestion control behavior than PI and ARED.

References

1. B. Braden, D. Clark, J. Crowcroft, B. Davie, etc, Recomedations on Queue Management and Congestion Avoidance in the Internet, RFC2309, April 1998.
2. F. Ren, Y. Ren, and X. Shan, Design of a fuzzy controller for active queue management, Computer Communications, vol. 25, pp. 874-883, 2002.
3. Di Fatta.G, Hoffmann.F, Lo Re.G, Urso.A, A genetic algorithm for the design of a fuzzy controller for active queue management, IEEE Transactions on Systems, Man, and Cybernetics, Part C: Applications and Reviews, Volume: 33 , Issue: 3 , Aug. 2003
4. Jalili-Kharaajoo, Active queue management by means of fuzzy sliding mode control theory, Advanced Communication Technology, 2004. The 6th International Conference on , Volume: 2 , 2004 Pages:670 – 674

Fuzzy Optimization for Security Sensors Deployment in Collaborative Intrusion Detection System

Chengchen Hu, Zhen Liu, Zhen Chen, and Bin Liu

Dept. of Computer Science and Technology
Tsinghua University, Beijing, China, 100084

Abstract. This paper argues about the deployment positions of Network-based Intrusion Detection System and suggests the "Distributed Network Security Sensors" distributed among the nodes of the internal network to monitor traffic. We study the tradeoff between cost and monitoring coverage to determine the positions and processing rates of the sensors. To handle the uncertain nature of flow, we build fuzzy expected value optimization models and develop a hybrid intelligent algorithm to obtain the deployment strategy. From the experiments in actual and synthesized network topologies, we observe that a small number of low-speed sensors are sufficient to maintain a high monitoring coverage. It also depicts that deploying DSS is much more efficient in larger topologies.

1 Introduction

Network-based intrusion detection system (NIDS) performs security analyses on packets captured by eavesdropping on a network link/node. NIDS has been widely studied since Heberlein etal. published their pioneering work on *Network Security Monitor (NSM)* [1]. The current generation of NIDS, referred to "centralized NIDS", typically places the sensors at the gateway nodes between internal and external networks, with the idea of monitoring all the traffic at the entries of the internal network. However, the ever-increase in line speed and ever-demanding computation intensive processing pose great challenges to NIDS. The aggregation points where intrusion sensors are located tend to become congestion points. In addition, it becomes ineffective totally when the attacks are initiated by internal hosts, and when there exists other entrances to the internal network, like dial-up and wireless access.

We address ourselves to the problem of the sensors locations and suggest a different architecture for NIDS that a number of sensors should be distributed among the nodes of the internal network – hereafter referred to as *"Distributed Network Security Sensors" (DNSS)*. We may deploy more than one low speed sensor to monitor a flow, since a flow can be monitored at any node of its routing path. When the flow rate is faster than the processing rate, the low speed sensor samples the traffic to control the resource consumption. Although a single sensor may miss attacks, the cooperation of all the sensors can protect a high-speed network. The architecture of DNSS is scalable because the processing rate in a single sensor is not the bottleneck any longer. In addition, since the sensors are distributed among the internal network, intrusions initiated from internal network or backdoor accesses can be detected and timely dealt with. In

L. Wang et al. (Eds.): FSKD 2006, LNAI 4223, pp. 743 – 752, 2006.

this paper, we study the deployment strategy of DNSS, which refers to the positions and processing rates of the distributed security sensors. First, we strike a right tradeoff between deployment cost and effective protection. One can simply believe that the protection is more effective if more traffic can be monitored. It is straightforward that the more sensors and/or faster processing rates, the larger fraction of traffic can be monitored. However, locating a sensor incurs a deployment cost, and selecting faster processing sensors requires more budgets. Second, we introduce the concept of "uncertain optimization" [2] to handle the flow fluctuation. One may estimate a certain value of the flow rate as most related work on different placement problems did [3-5], but such deterministic optimizations are not suitable for our application in real uncertain environment. Many stochastic traffic characterization problems/methods are widely studied, however, rare Internet traffic invariants were found for this complex system. Thus, we treat flow rates as fuzzy variables.

A "clustered sensors" approach for network security analyses was proposed in [6] to support in-depth intrusion detection on high-speed links. As the number of streams increases, the slicing and reassembly operations will incur heavy implementation overhead. A number of products in industrial world could also work in high-speed environment [7-9], but as well as [6], attacks from backdoor accesses and internal hosts may be missed. It was suggested in [10] that NIDS should be deployed on end hosts with specialized hardware. This solution required bulk of hardware-based sensors to be deployed on the end hosts, whose implementation cost is huge. A peer-to-peer approach for collaborative intrusion detection was proposed in [11]. This work addressed the challenges of scalability and failure avoidance in central point and show better performance in latency and load balancing compared with centralized NIDS. Nevertheless, this work did not mention how to place the peers.

2 Fuzzy Expected Value Optimization Models

The network can be modeled as a graph $G(V,E)$, where V is the set of nodes and $E \subseteq V \times V$ is the set of links. Table 1 summarizes the notations used in this paper. A flow is defined as a collection of packets with the same *Source and Destination address (SD)* pairs. We consider the node-based sensors implemented inside routers/switches, so all the links connected to the nodes can be monitored. We assume the sensor i operates at a rate of q_i Mb/s for all the incoming flows, which should be lower than a maximum acceptable value L. The processing rate at each sensor can be adjusted independently.

We assume that placing a sensor incurs two parts of cost. One is a fixed cost component such as a space cost (or/and a maintenance cost), which is represented as a constant C. The other is the hardware (or/and software) cost, which should be relevant to a non-decreasing convex function of its processing rate. Therefore, $c_i = D(q_i) = C + k(\frac{q_i}{L})^n$, $n > 1$, where k, n are two constant parameters. The total budget should be the sum of the deployment cost of all the sensors, $\sum_{i \in V} c_i x_i$. It is straightforward that if the processing rate in node i is faster than the flow rate, the

Table 1. Notation descriptions

Notation	Description
S	Set of all flows
R_j	Monitored fraction of flow j
r_{ij}	Monitored fraction of flow j in node i
y_{ij}	$y_{ij}=1$ if node i is in the routing path of flow j, else $y_{ij}=0$
c_i	Cost of placing a sensor in node i
C	Fixed cost component of placing a sensor
q_i	Processing rate in the node i
$\mathbf{q}$	Vector of q_i, $\mathbf{q}=\{q_i\}, i\in V$
x_i	$x_i=1$ if a sensor is placed in node i, else $x_i=0$
$\mathbf{x}$	Vector of x_i, $\mathbf{x}=\{x_i\}, i\in V$
f_j	Traffic rate of flow j (fuzzy parameters)
$\mathbf{f}$	Fuzzy vector of f_j, $\mathbf{f}=\{f_j\}, j\in S$
L	Maximum acceptable processing rate
B	Total budget
T	Predefined coverage threshold
N	Number of flows

fraction should be 1; otherwise, the fraction should be the ratio between processing rate and flow rate. Monitored fraction of flow j in node i can be achieved by

$$r_{ij}=\begin{cases}\dfrac{q_i x_i y_{ij}}{f_j}, & \text{if } \dfrac{q_i x_i y_{ij}}{f_j}\le 1 \;\; i\in V, j\in S\\[2ex] 1, \;\; \text{if } \dfrac{q_i x_i y_{ij}}{f_j}>1 \;\; i\in V, j\in S\end{cases}$$

Suppose the processing in each sensor to be independent, therefore, the monitored fraction of each flow can be determined by $R_j=1-\prod_{i\in V}(1-r_{ij}), j\in S$. We calculate the monitoring coverage as the mean of monitoring fraction of each flow, i.e., $F(\mathbf{x},\mathbf{q},\mathbf{f})=[\sum_j R_j]/N$. As mentioned above, the flow rates are uncertain parameters and we consider $\mathbf{f}$ as a fuzzy vector. Therefore, the monitored faction of each flow and the monitoring coverage obtained by $\mathbf{f}$ are also fuzzy variables. In order to quantitate the monitoring coverage and further rank it, we employ the fuzzy expected value operator, $E[F(\mathbf{x},\mathbf{q},\mathbf{f})]=\frac{1}{N}E[\sum_j R_j]$.

There are several possible ways to define a mean value for fuzzy variables. We borrow the more general definition in [12] that is applicable to continuous fuzzy variables, discrete fuzzy variables, and a function of multiple fuzzy variables.

Maximizing the monitoring coverage and minimizing the implementation cost are two conflicting objectives. We introduce two problems to consider the tradeoff between these two objectives: Budget Constrained Problem (BCP) and Coverage Constrained Problem (CCP). BCP is included when the total budget is limited, whose objective is to maximize the monitoring coverage without violating the budget constraint for the sensors' deployment cost. The objective is to maximize the monitoring coverage and the constraint indicates that we have limited budgets to cover the deployment cost. The dual of BCP is CCP, whose objective is to minimize the implementation cost when a certain monitoring coverage is achieved. Thus, the optimization model, whose decision variables are also **x** and **q**. The objective function minimizes the total deployment cost and the constraint guarantees that the coverage should not be less than a predefined coverage level T.

3 Hybrid Intelligent Algorithm

The existence of the fuzzy expected value function $E[F(\mathbf{f})]$ makes the solution more complicated. If the uncertain function is determined, there is no essential difference between the solutions of our models and classical optimization models. We propose a *Hybrid Intelligent (HI)* algorithm, which is composed of an uncertain function approximation part to determine the uncertain function and an optimization method part to find the optimal solution after the uncertain function is approximated.

We first describe a fuzzy simulation method to estimate $E[F(\mathbf{f})]$. As mentioned above, let $\mathbf{f}=\{f_1,f_2,\cdots,f_N\}$ be a fuzzy vector with a membership function $\mu:\Re\rightarrow[0,1]$, where flow rates $f_i(i=1,2,\cdots,N)$ are fuzzy variables. First, randomly generate vectors $\xi_j=\{\xi_{1j},\xi_{2j},\cdots,\xi_{Nj}\}$, $j=1,\cdots,m$ from the ε-cut of $\mathbf{f}$, where ε is a sufficiently small number. We can estimate $E[f(\xi)]$ by the following algorithm.

Step 1: $E=0$, $i=1$;
Step 2: randomly generate $\xi_j=\{\xi_{1j},\xi_{2j},\cdots,\xi_{nj}\}$, $j=1,\cdots,m$;
Step 3: $a=f(\mathbf{x}_1)\wedge f(\mathbf{x}_2)\wedge\cdots\wedge f(\mathbf{x}_n)$, $b=f(\mathbf{x}_1)\vee f(\mathbf{x}_2)\vee\cdots\vee f(\mathbf{x}_n)$;
Step 4: generate r from [a, b] uniformly;
Step 5: if $r\geq 0$, $Cr\{F(\mathbf{f})\geq r\}=\frac{1}{2}(\max_{j=1,\cdots,m}\{\mu_j \mid F(\xi_j)\geq r\}+1-\max_{j=1,\cdots,m}\{\mu_j \mid F(\xi_j)<r\}),\forall r\geq 0$,

if $r<0$, $Cr\{F(\mathbf{f})\leq r\}=\frac{1}{2}(\max_{j=1,\cdots,m}\{\mu_j \mid F(\xi_j)\leq r\}+1-\max_{j=1,\cdots,m}\{\mu_j \mid F(\xi_j)>r\}),\forall r<0$,

Step 6: $E\leftarrow E-Cr\{F(\mathbf{f})\leq r\}$;
Step 7: if $i<m$, return to step 4 and $i\leftarrow i+1$;
Step 8: $E[F(\mathbf{f})]=a\vee 0+b\wedge 0+(b-a)E/m$.

Fuzzy simulation is sufficient to approximate the fuzzy expected value function; however, fuzzy simulation is required whenever the uncertain function is used. To reduce the computational time, we introduce the *Artificial Neural Networks (ANN)* [13] to speedup the approximation. ANN could approximate the uncertain function and has high operation speed after it is trained. Our motivation is to train the ANN once, and use it many times. In addition, ANN has the ability to compensate for the error of training data though fuzzy simulation is not very precise.

Genetic Algorithm (GA) is an effective optimization algorithm inspired by nature selection and population genetics [14], which is selected as the basic optimization

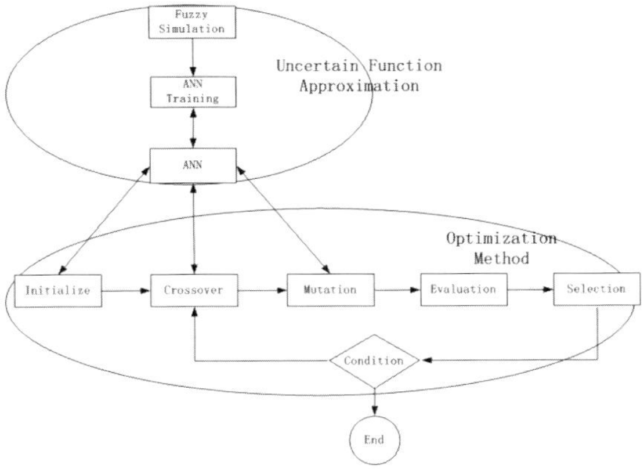

Fig. 1. Hybrid intelligent algorithm

method of HI algorithm. Most of the classical optimization methods, e.g. gradient methods, Hessian methods and heuristic greedy, are much dependent upon the objectives or/and may stop search at a local optimal solution. To use GA, we first code the chromosomes and design the operations for our problem. We use a $2n$-dimensional vector $V=(x_1,x_2,\cdots,x_n,q_1,q_2,\cdots,q_n)$ as a chromosome to represent candidate decision variables, where n is the number of nodes. A chromosome is initialized by setting x_i as a random integer from {0, 1} and setting q_i as a random number between 0 and 1. Two pairs of crossover positions (n_1, n_2) and (n_3, n_4) are randomly generated with crossover probability such that $n_1<n_2$ and $n_3<n_4$. The genes between n_1 and n_2, n_3 and n_4, are exchanged. To mutate the chromosome, a position pair (m_1, m_2) is randomly generated ($m_1<m_2$). Sequences $\{x_{m_1},x_{m_1+1},\cdots,x_{m_2}\}$ and $\{q_{m_1},q_{m_1+1},\cdots,q_{m_2}\}$ are regenerated randomly to form a new chromosome with mutation probability. The chromosome is replaced if it is feasible. The evaluation operation use rank-based evaluation function to evaluate the chromosomes. And the selection operation is a fitness-proportional selection according to the fitness value obtained in the evaluation operation. Note that, a chromosome can be selected more than one times.

The entire algorithm can be summarized in Fig. 1 and by the following algorithm.

Step 1: Use fuzzy simulation to generate sufficient groups of training data;

Step 2: Train an ANN using the training data generated in Step 1;

Step 3: Initialize the chromosomes, and check their feasibility (by ANN for CCP);

Step 4: Update the chromosomes by crossover and mutation operations, and check the feasibility of offspring (by ANN for CCP);

Step 5: Calculate the objective values for all chromosomes (by ANN to solve BCP) and the fitness of each chromosome according to the objective values;

Step 6: Select the chromosomes according to the fitness;

Step 7: To step 4 until pre-defined generations are produced.

---

4 Experiments and Evaluation

We first evaluate the proposed DNSS deployment algorithm on a 10-node real topology. The deployment cost function is defined specifically as $c_i = D(q_i) = 1 + \frac{1}{2}(q_i / 500)^2$. Since the flow rates are "uncertain parameters", we generate two types of synthetic traffic flow with different fuzzy membership functions. More specifically, trapezoidal scenario[1] and triangle scenario[2] are considered. In trapezoidal scenario, the rate of flow j is a trapezoidal variable $(100, \lambda_j^1, \lambda_j^2, 500)$, where λ_j^1 and λ_j^2 are generated uniformly from the interval (100, 300) and (300, 500) respectively. In triangle scenario, the rate of flow j is a triangle variable $(100, \lambda_j, 500)$, where λ_j is generated uniformly from the interval (100, 500). It has been shown that the Internet has "hot spot" behavior, i.e., a few *POP (Point-of-Presence)* pairs have very large flows, while the majority of POP pairs have substantially less flows between them. For this reason, we first generate flows between any two nodes and the number of flows is 90, then we randomly select 22% of the SD pairs to generate flows and the number of flows is 20.

Fig. 2 - Fig. 5 show the results obtained by BCP. Fig. 2 plots the monitoring coverage as a function of the number of sensors in triangle traffic scenario, and Fig. 3 plots the function for BCP in trapezoidal traffic scenario. Only 4 sensors can monitor almost all of the traffic in the experimental network topology. It is indicated in the figures that more sensors are required to keep the same monitoring coverage if number of flows is increased from 20 to 90. This phenomenon is more obvious in triangle traffic scenario. Fig. 4 and Fig. 5 show the monitoring coverage under different budget constraints in triangle traffic scenario and trapezoidal scenario respectively. Monitoring coverage increases as the budget increases, however, we observe that it has the trend of a diminishing coverage gain. Therefore, blindly increasing budget will lead to ineffectiveness in sense of incurring disproportion penalty in implementation cost vis-à-vis the gain in increased coverage. If the number of flows is decreased from 90 to 20, same budget can deploy a DNSS system monitoring more traffic.

Fig. 6 - Fig. 9 illustrate the results obtained by CCP. Fig. 6 and Fig. 7 represent the relationship between the monitoring coverage and the number of sensors in triangle traffic scenario and trapezoidal scenario respectively. In the worst case that each SD pair has a connection (i.e., there are 90 flows inside the network), two figures indicate the results that only 2-3 sensors can monitor 80% of the traffic and only 4 sensors can monitor 95% of the traffic. As mentioned above, there will be less flow in the network

[1] A trapezoidal fuzzy variable can be represented by (r_1, r_2, r_3, r_4) with the membership function of $\mu(x) = \begin{cases} (x - r_1)/(r_2 - r_1) & r_1 \le x \le r_2 \\ 1 & r_2 \le x \le r_3 \\ (x - r_3)/(r_4 - r_3) & r_3 \le x \le r_4 \\ 0 & otherwise \end{cases}$.

[2] When $r_2 = r_3$, a trapezoidal fuzzy variable becomes a triangle fuzzy variable.

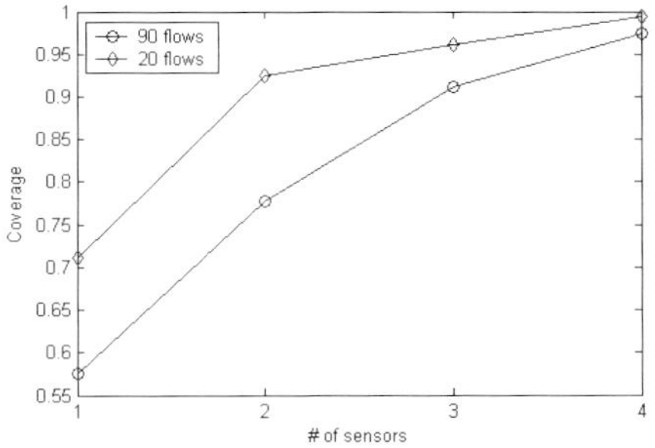

Fig. 2. # of sensors vs. Coverage, triangle, BCP

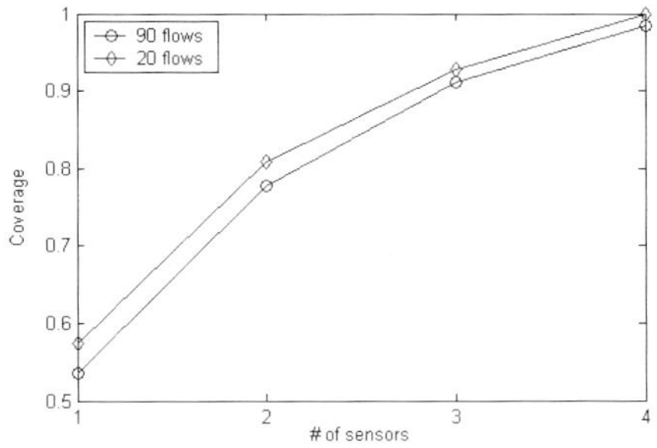

Fig. 3. # of sensors vs. Coverage, trapezoidal, BCP

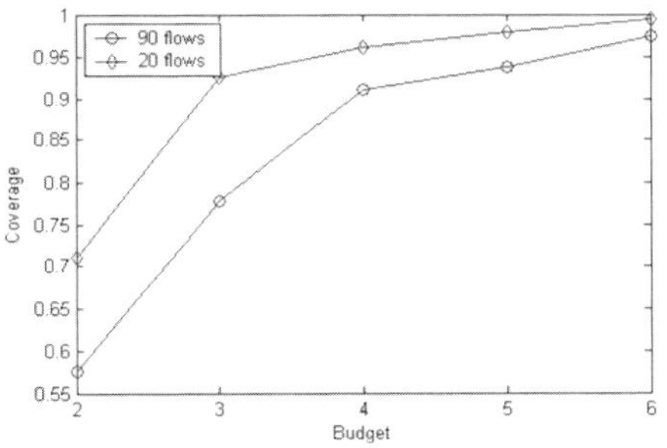

Fig. 4. Budget vs. Coverage, triangle, BCP

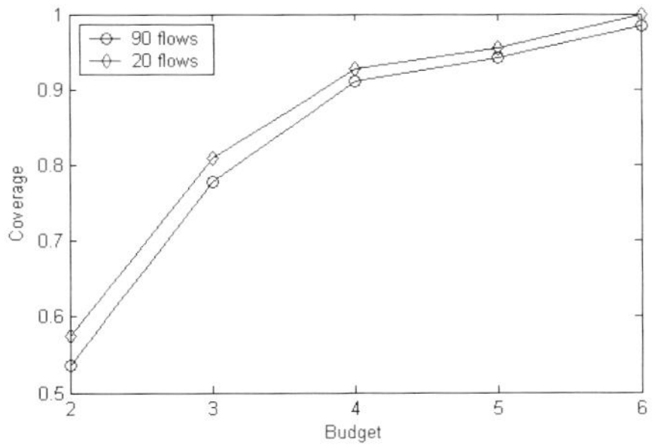

Fig. 5. Budget vs. Coverage, trapezoidal, BCP

and we reduce the number of flows to 20 by randomly selecting 22% of the total SD pairs. The results are much better, especially in the triangle traffic scenario. Note that, the number of the sensors is same under some coverage constraint, but the deployment strategies are different. It can be revealed in Fig. 8 and Fig. 9 that the costs are not the same under different coverage constraints. Fig. 8 and Fig. 9 depict the results of the monitoring coverage vs. the deployment cost in triangle traffic scenario and trapezoidal scenario respectively. With the increase of coverage constraint, the cost also increases. We also observe that the curves of implementation cost are approximately the convex functions of coverage. It means that the differential coefficient of cost is also increased with the increase of coverage. Furthermore, less cost is required if the number of flows is reduced from 90 to 20.

We set up simulations to assess the calculation, which are done under a same topology and generated traffic scenarios. By configuring the virtual sensors in the calculated places with the calculated processing rates, the comparison results are revealed in Fig. 10. The results demonstrate that the calculation meets the simulation quite well.

Using the Waxman model [15], we randomly generate larger topologies to investigate the effectiveness. Table 2 lists the number of security sensors required to obtain a monitoring coverage larger than 90% under different topologies. It depicts that deploying DSS is much more efficient in larger topologies, since the increase in the number of monitors is much less than the increase in the number of topology nodes.

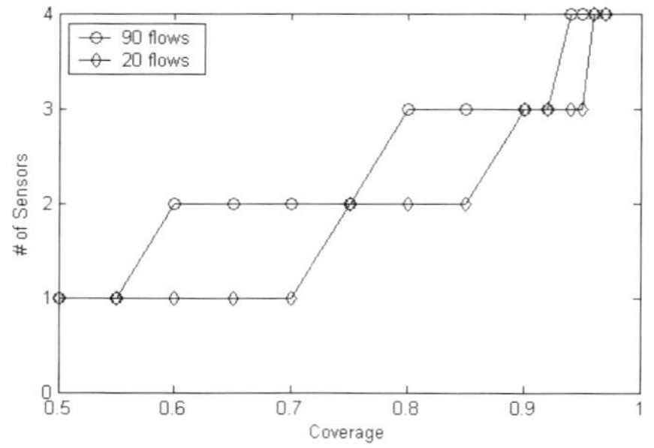

Fig. 6. Coverage vs. # of sensors, triangle, CCP

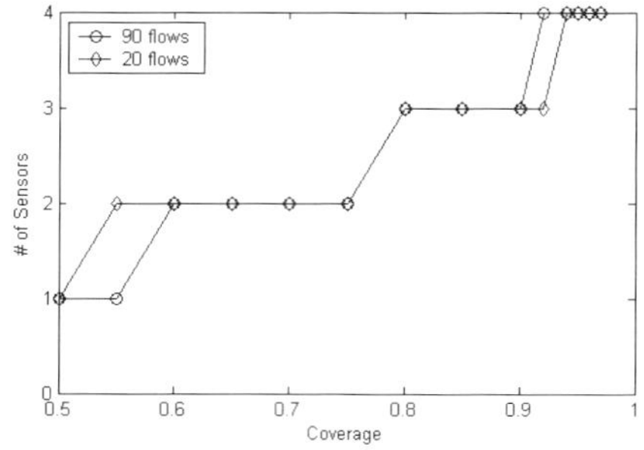

Fig. 7. Coverage vs. # of sensors, trapezoidal, CCP

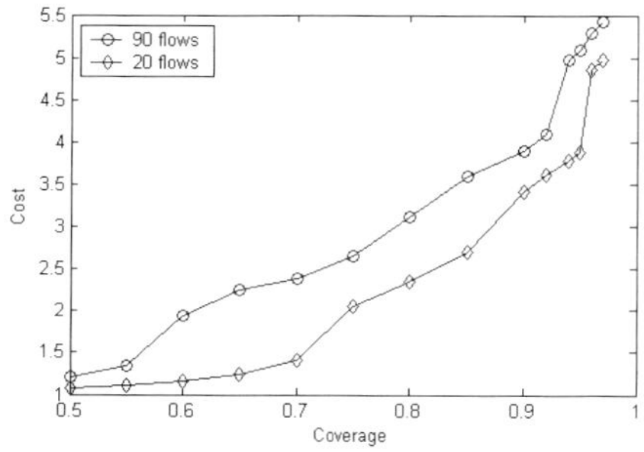

Fig. 8. Coverage vs. Cost, triangle, CCP

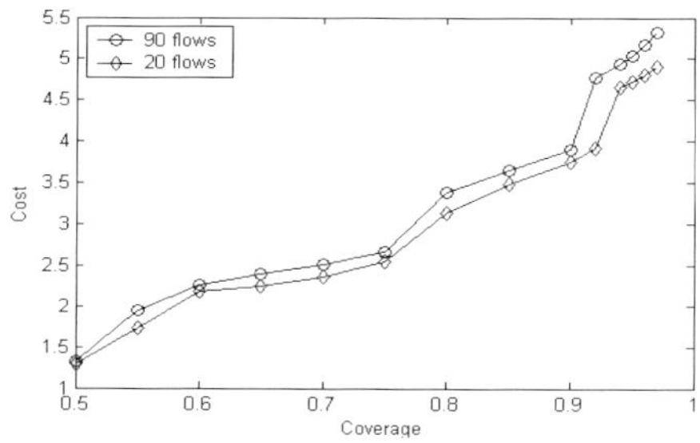

Fig. 9. Coverage vs. Cost, trapezoidal, CCP

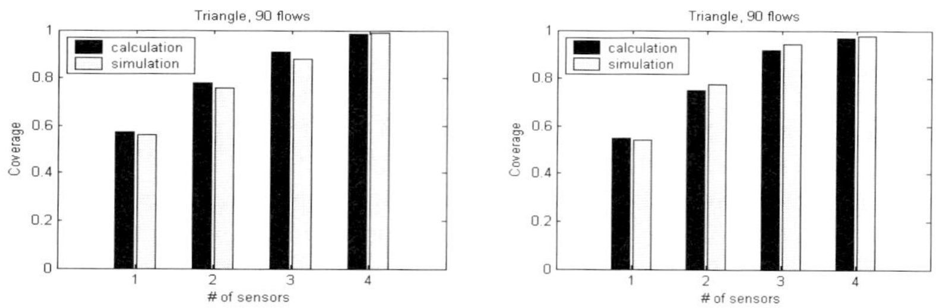

Fig. 10. Simulation evaluation. The left one is for BCP and the right one is for CCP

Table 2. Number of sensors in different topologies

# of nodes	10	20	30	40	50	60	70	80	90	100
# of sensors	3	5	8	9	11	13	16	17	20	22

The experiment results evaluate the model and HI algorithm to deploy the sensors, and demonstrate the effectiveness of the proposed architecture of DNSS. Only a small number of sensors can monitor almost all the traffic and thus protect the high-speed internal network effectively. Furthermore, it is much more efficient in larger topologies.

5 Discussions

The calculation time of HI may be a little long when it is applied to large-scale topologies. In fact, the algorithm can be further optimized in three aspects. First, the inherent parallelism of GA makes it attractive. Second, the fuzzy simulation of each training data is also independent and can be handled in parallel. As the parallel and distributed computing becomes more readily available, the computational time complexity can be greatly amortized in aforementioned two aspects. Third, ANN has the high speed of operation after it is trained, but the training of ANN utilizing classical gradient-based method (e.g. back-propagation algorithm) is a time consuming task when it is applied to a large topology. Nevertheless, recent breakthroughs in ANN researches shorten the training time thousands of times. Both in theory and in experiments, it is demonstrated that a novel training algorithm called *Extreme Learning Machine (ELM)* provides the good generalization performance at extremely fast learning speed. Such outcomes are easy to use (no parameters setting problem) and have been utilized by us to speed up the training process of the HI algorithm.

In order to obtain the input parameters of HI algorithm, we should first have some prior knowledge of flow rates distributions of the network. For a specific network without firstly deploying a monitoring system to obtain the specific distribution, we can trigger the following iteration to tackle this difficulty.

Step 1: Calculate a rough deployment strategy assuming the flow rate distributions;
Step 2: Configure the monitors with the deployment strategy and collect traffic;
Step 3: Use the collected information as the input and get a new deployment;
Step 4: Go to step 2.

Note that, in Step 3, we do not need to know the exact flow distribution to generate training data. We could directly input the sampled information to train a ANN and further achieve the deployment strategy.

6 Conclusion

In this paper, we propose the architecture of NSS that several low speed sensors are distributed among the nodes of the internal networks. We study the tradeoff between deployment cost and monitoring coverage to decide the deployment strategy of DNSS, which determines the locations and processing rates of the sensors. The DNSS is scalable in high-speed network and can detect the attacks initiated from backdoors or the internal network. Our work takes into account the uncertain factors of flow rates and constructs fuzzy optimization models, which make the solution more valid in practice. In addition, we propose the hybrid intelligent algorithm to achieve the optimal solution. From the experiments in actual and synthesized network topologies, we observe that a small number of low-speed sensors are sufficient to maintain high monitoring coverage. The experiments also depict that deploying DSS is much more efficient in larger topologies.

References

1. L. T. Heberlein, G. V. Dias, K. N. Levitt, B. Mukherjee, J. Wood, and D. Wolber, "A Network Security Monitor," IEEE Symposium on Research on Security and Pri-vacy, 1990.
2. B. Liu, "Theroy and Practice of Uncertain Programming", Heidelberg: Physica-Verlag, 2002.
3. S. Jamin, C. Jin, Y. Jin, D. Raz, Y. Shavitt, and L. Zhang, "On the Placement of Internet Instrumentation," INFOCOM, 2000.
4. K. Suh, Y. Guoy, J. Kurose, and D. Towsley, "Locating Network Monitors: Com-plexity, Heuristics, and Coverage," INFOCOM, 2005.
5. X. Tang and J. Xu, "On Replica Placement for QoS-aware Content Distribution," INFOCOM, 2004.
6. C. Kruegel, F. Valeur, G. Vigna, and R. A. Kemmerer, "Stateful Intrusion Detection for High-Speed Networks," IEEE Symposium on Research on Security and Privacy, 2002.
7. ISS, "RealSecure Network Gigabit." http://www.iss.net/products_services/enterprise_ protection /rsnetwork/gigabitsensor.php.
8. T. Networks, "Attack Mitigator IPS 5500." http://www.toplayer.com/content/products/ intrusion_detection/attack_mitigator.jsp.
9. E. Carter, Cisco Intrusion Detection System, 1st ed: Cisco Press, 2001.
10. C. Clark, W. Lee, D. Schimmel, D. Contis, M. Kone, and A. Thomas, "A Hardware Platform for Network Intrusion Detection and Prevention," In Proceedings of The 3rd Workshop on Network Processors and Applications (NP3), 2004.
11. C. V. Zhou, S. Karunasekera, and C. Leckie, "A Peer-to-Peer Collaborative Intru-sion Detection System," International Conference on Networks 2005, Kuala Lumpur, Malaysia, 2005.
12. B. Liu and Y.-K. Liu, "Expected Value of Fuzzy Variable and Fuzzy Expected Value Models," IEEE Transaction on Fuzzy System, vol. 10, 2002.
13. S. Haykin, Neural Networks - A Comprehensive Foundation. New York: Macmillan College Publishing Company, 1994.
14. D. A. Coley, An Introduction to Genetic Algorithms for Scientists and Engineers. Singapore: World Scientific, 1999.
15. B. M. Waxman, "Routing of Multipoint Connections," IEEE Journal on Selected Areas in Communications, vol. 6, pp. 1617 - 1622, 1988.
16. J. P. Cohoon, S. U. Hedge, W. N. Martin, and D. Richards, "Punctuated Equilibria: A Parallel Genetic Algorithm," Second Internatioanal Conference on Genetic Algo-rithms, 1987.
17. M. Tomassini, "Parallel and Distributed Evolutionary Algorithms," in Evolutionary Algorithms in Engineering and Computer Science. Chichester: John Wiley & Sons, 1999.
18. G.-B. Huang, Q.-Y. Zhu, and C.-K. Siew, "Extreme learning machine: a new learn-ing scheme of feedforward neural networks," 2004 IEEE International Joint Confer-ence on Neural Networks, 2004.
19. M.-B. Li, G.-B. Huang, P. Saratchandran, and N. Sundararajan, "Fully Complex Extreme Learning Machine," Neurocomputing, 2005.

Objective Evaluation for Compressed Video Quality Based on Fuzzy Synthetic Judgment

Wen Ji, Haoshan Shi, and Ying Wang

Dept. Electronic Engineering, Northwestern Polytechnical University
710072, Xi'an, Shaanxi, P.R. China
{jill, shilaoshi}@nwpu.edu.cn

Abstract. In this paper, a new effective and reliable objective video evaluation model is proposed based on fuzzy synthetic judgment. Firstly, the essential design of the model is described systemically. Then the fuzzy membership functions of all factors which affect compressed video quality are determined. And the factor weights are also estimated by approximate algorithm. This model highlights a comprehensive evaluation by taking account of multiple properties of the compressed video which affect the whole video sequence, such as quality, fluency and motion info. On the other side, this model has well capability of compatibility with other evaluation models or standards. Simulation results demonstrate the advantages of the fuzzy objective video quality evaluation model especially in comprehensive judgment of the whole video.

1 Introduction

Many different quality assessment methods have been proposed in order to evaluate the reconstructed quality of compressed video sequences. The most used measures are quantitative metrics based on a simple difference between frames, like MSE (Mean Square Error) and PSNR (Peak Signal to Noise Ratio). These are appealing because they are simple to calculate, have clear physical meanings, and are mathematically convenient in the context of optimization. But they are not very well matched to perceived visual quality of whole video sequence. These years, great effort has plunged into the development of quality assessment methods that take advantage of known characteristics of the human visual system (HVS) and other video features[1]-[6]. But these models are usually complicated and less compatible with each other so as to inadequate for practical applications. The majority of the proposed fuzzy[7] systematic quality evaluation model is centralized in giving a comprehensive assessment including most video affecting factors such as motion info, fluency and quality. At the same time, well compatibility with other evaluation models or standards is considered thoughtfully. Section 2 provides the deduction process of fuzzy video quality evaluation model. In section 3, we describe paradigms for quality assessment through comparing the test results of different standardized video sequences. Finally section 4 takes some conclusions.

L. Wang et al. (Eds.): FSKD 2006, LNAI 4223, pp. 753 – 761, 2006.

2 The Objective Video Evaluation Based on Fuzzy Synthetic Judgment

2.1 The Description of Fuzzy Synthetic Evaluation Model

Definition 1. Let $U = \{u_1, u_2, ..., u_n\}$ be the set of video quality affecting factors, evaluation results set is $V = \{v_1, v_2, ..., v_m\}$, $R = (r_{ij})_{n\times m}$ is a fuzzy relation from U to V ,and $0 < r_{ij} < 1$. If the fuzzy subset $A = \forall U$ and fuzzy subset $B = \forall V$ meet $B = A \circ R$, then there is a fuzzy evaluation from fuzzy set U to fuzzy set V.

Theorem 1. (Extension theorem) Given $R \in \mathcal{F}(U \times V)$, a mapping T_R : $\mathcal{F}(U) \to \mathcal{F}(V)$,and meet $A \mapsto T_R(A) = A \circ R$, then fuzzy evaluation is unique.

Then finding an objective video quality evaluation model can be equivalent to solve a fuzzy synthetic evaluation problem. This model can be expressed as follows:

Case 1: A known video quality effecting factors set $U = \{u_1, u_2, ..., u_n\}$, corresponding to judgment objects;

Case 2: A known evaluation-making set $V = \{v_1, v_2, ..., v_m\}$ according to different affecting factors;

Case 3: A known weight vector assignment of each factors, affiliation with a fuzzy set in U , noted $A = \{a_1, ... a_n\}$, where a_i is the weight value of i th factor u_i ;

Case 4: Get judgment matrix $R = (r_{ij})_{n\times m}$, it includes n fuzzy sets in V and judgment vectors, where $R_i = (r_{i1}, r_{in}, ..., r_{im})$;

Case 5: Finally, a fuzzy synthetic video quality evaluation matrix can be formularized as $B = A \circ R$, where B is a fuzzy subset in V .

2.2 The Building of Single- and Multi- affecting Factors

Video quality affecting factors set include subset U_i of l, ($i = 1, 2, ..., l$), and $\bigcup_{i=1}^{l} U_i = U$. Then according to the fuzzy model , evaluation vectors corresponding to each factors constitute the evaluation matrix $R = \begin{pmatrix} B_1 \\ B_2 \\ \vdots \\ B_L \end{pmatrix} = \begin{pmatrix} b_{11} & \cdots & b_{1m} \\ \vdots & \ddots & \vdots \\ b_{l1} & \cdots & b_{lm} \end{pmatrix}$.The weights vector of single- factor is $A = (a_1, a_2, ..., a_l)$. Consequently, the results of the fuzzy synthetic evaluation can be written as follows:

$$B = A \circ R = (a_1, a_2, ..., a_l) \circ \begin{pmatrix} b_{11} & \cdots & b_{1m} \\ \vdots & \ddots & \vdots \\ b_{l1} & \cdots & b_{lm} \end{pmatrix} = (\sum_{i=1}^{l} a_1 b_{l1}, \sum_{i=1}^{l} a_i b_{l2}, ..., \sum_{i=1}^{l} a_i b_{im}) .$$

For building multi- affecting factors, the weights vector is written as A_j = ($a_1, a_2, ..., a_l$) according to each U_j, the evaluation results are educed from the second fuzzy transform, described as $B = A_j \circ R$. If A_j is still a subset, then from recursion method, we can have the final evaluation result $B = A \circ R$. So the evaluation model has well capability of compatibility.

2.3 Getting Weights Assignment Matrix

Before achieving the results, the most difficult is to decide the weights. The weight vectors are numbers assigned to some objects that express their relative importance. It may happen that the objects whose weight vectors are to be determined are elements of a hierarchy, because the evaluation results are judged separately for each criterion. So the criteria themselves have various weight vectors. In this paper, firstly we decided the initial weight vectors based on aggregation of preferences, and then optimized adaptation of assessment is performed in accordance with statistic contributions of video quality affecting factors.

2.4 The Judgment of Objective Video Quality Based on Motion Info, Fluency and Quality

Definition 2. Disjunctive operation: A or B, noted $A \vee B$; Conjunctive operation: A and B, noted $A \wedge B$; Implicational operation: A implicate B, noted $A \rightarrow B$; incompatibly operation: A is incompatibly to B, noted $A \wedge B = \phi$

Definition 3. The reliability of $A, B, C...$ is expressed as $P(A)$, $P(B)$, $P(C)$ ... , respectively.

Property 1. $0 \le P(A) \le 1$, $P(\phi) = 0$, $P(X) = 1$.

Property 2. $P(A \vee B) = P(A) \vee P(B)$.

Property 3. $P(A_1 \wedge A_2 \wedge ... \wedge A_n) = P(A_2 \mid A_1) \cdots P(A_n \mid A_1 A_2 A_{n-1})$.

Property 4. if $A \rightarrow B$, then $P(B \mid A) = 1$, and $P(A \wedge B) = P(A) \ge P(A \wedge \overline{B})$

Theorem 2. (Inductive inference) If $A \rightarrow B_1$, and $A \rightarrow B_2, \cdots$, and $B_1, B_2, \cdots$ is easy to be conformed. Then the larger n, the more reliability of A, after $B_1, B_2, \cdots$, B_n are conformed respectively.

Proof. When $P(B_n / A) = 1$, $P(A \wedge B_1 \wedge \cdots \wedge B_n) \ne 0 (n \ge 1)$, as shown in property 1,2 and 3, then we can get $P(A) \le P(A / B_1 \wedge B_2 \wedge \cdots \wedge B_n) \le P(A / B_1 \wedge B_2 \wedge \cdots \wedge B_{n+1})$.

Q.E.D.

Theorem 3. (Converse inference) If B is mutually exclusive to A, then the reliability of A can be enhanced by negation of B.

Proof. $A \wedge B = \phi$, $P(A \wedge B) = P(\phi) = 0$, and $P(A \wedge B) = P(A) \cdot P(B/A)$, hence $P(A) \leq P(A/\overline{B})$. Q.E.D.

Corollary 1. If the factors sets are the subset of all video quality affecting factors set, then the evaluation result is the subset of synthetic evaluation results.

Proof. The factors are implicated to all affecting factors set, as shown in theorem 3 and theorem 4, proposition is affirmative. Q.E.D

Definition 4. $U = \{u_1, u_2, u_3, u_4\}$ is the set of video quality affecting factors, where u_1 is MSE, u_2 is the interval between adjacent frames, u_3 conclude motion characters, u_4 means the quantization parameter.

For a fair performance evaluation of compressed video, video characteristics including quality, motion info, interval and sensitivity should be involved in the criteria affecting the finial performance of judgment. Consequently, the factor vectors of the fuzzy synthetic evaluation can be taken as below.

$R_1 = \{ MSE_1, ... MSE_n \}$: $MSE = \frac{1}{M \times N} \sum_{i=0}^{M-1} \sum_{j=0}^{N-1} [x(i,j) - \hat{x}(i,j)]^2$, which is mean-squared-error of each encoded picture. Formula $PSNR = 10\lg \frac{255^2}{MSE}$ can be adopted to convert the MSE value to $PSNR$. This judgment vector is complied with the conventional video quality measurement.

$R_2 = \{ \Delta t_1, ..., \Delta t_n \}$:The intervals of adjacent frames. The objective $PSNR$ or MSE results do not always provide a reliable metric for the continuity performance evaluation of a video sequence. Since video signals are three-dimensional, the temporal factor plays a major role in the overall assessment, especially in wobble and fluency aspects.

$R_3 = \{ motion_1, ..., motion_n \}$:The motion vectors from motion estimation in each block. The main limitation of $PSNR$ or MSE is that they do not correlate well with subjective quality evaluations. For a fair evaluation of a compressed video sequence, motion information should also be included. The sequences have a strong relativity with one or several preceding images referred. Different prediction modes affect the quality of coded video. Therefore, motion estimation info is taken into account as a factor in this model.

$R_4 = \{ \sigma_1, ..., \sigma_n \}$:The parameters set of quantization. The compression process in block-based video coders is mainly attributed to the quantization of the transformed coefficients. The quantizer is regarded as the most important component of the video encoder since is controls both the coding efficiency and quality of the reconstructed video sequence. Most video compressed algorithms include many techniques such as flow control, optimal search of motion etc which can be adapted simply by these coefficients. The finial compressed video quality is closely related to the quantization, so the reliability of evaluation could be improved when the quantization step size is involved in factor vectors.

The four video quality affecting factors are all within 'or ' or 'implication' relations each other. According to theorem 2, the reliability of synthetic evaluation results can be enhanced. The variability range of factors value is in inverse proportion to value of judgment result, that consistent with the tendency of MSE . The best performance of video sequence will be occurred when the judgment value is closely to 0 in theory.

When weight assignment vector of affecting factor becomes $A = (1,0,0,0)$, the video quality evaluation degenerates to $PSNR$ measurement system which being widely used. When $A = (0,1,0,0)$ or $(0,0,1,0)$, it degenerates to judgment of fluency and wobble of video sequence. As a result, many evaluations in research can be included to the subset of the fuzzy video quality evaluation model.

Model extension: This model presents a new evaluation method with emphasis on the systematization. Highlight the multi- affecting factors for compressed video sequence and address for video characters including video quality, fluency and motion info. Most other evaluation methods can be annexed to the fuzzy systematic video evaluation model as multi-affecting factors, such as multi-channel properties of the HVS[1], perceptual distortion assessment[2], 3D-WSNR[4], automatic video quality assessment on the metric[5], perceptual prefiltering model[6]. According to theorem2, the reliability of the fuzzy systematic video evaluation model will be enhanced when the relation of these models belongs to implication. Therefore the model has a widely compatibility.

3 Video Quality Evaluation Results and Analysis

For these experiments, six video sequences were used. All sequences are in a temporal resolution of 30Hz and in QCIF of 176 pixels by 144 lines. The selected material covered a broad range of image complexity in terms of spatial detail and motion. The following provides a brief description of the test material[8]:

Foreman sequence, it is the most active scene of all since it includes a shaky background, high noise and a fair amount of bi-directional motion of the foreground object.

Carphone sequence, it shows a moving background with fair details, include a talking head in a moving vehicle with more motion in the foreground object and a non-uniform changing background.

Suzie sequence, it is with high contrast and moderate noise. It contains a fast head motion with the subject, being the foreground, holding a telephone handset with a stationary and plain-textured background.

Miss_am sequence, it is rather more active than Claire and Grandma with the subject more moving her shoulders before a static camera.

Claire and Grandma are both typical head-and-shoulder sequences with uniform and stationary background and minimal amount of activity confined to moving lips and flickering eyelids. Both are low motion video sequences with moderate contrast and noise and a uniform background. According to 2.3, weights assignment vector is $A = (0.02,0.8,0.1,0.001)$.All sequences can be contrasted in GOP units, where 1 intra and 39 inter frames in each GOP.

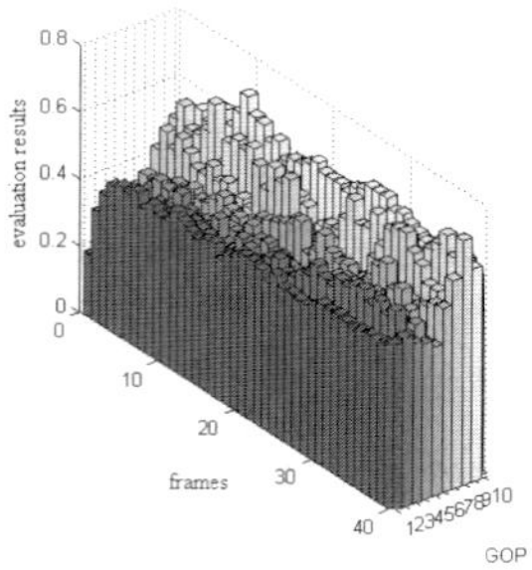

(a) evaluation result of foreman sequence

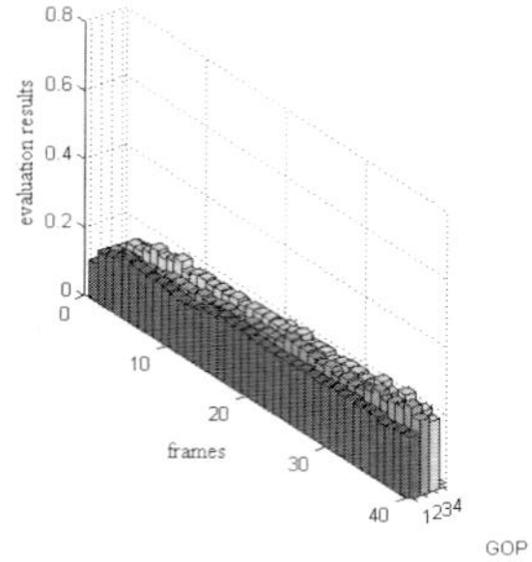

(b) evaluation result of miss_am sequence

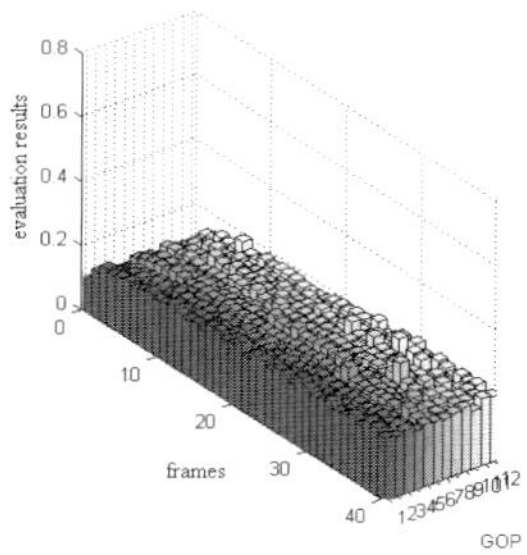

(c) evaluation result of claire sequence

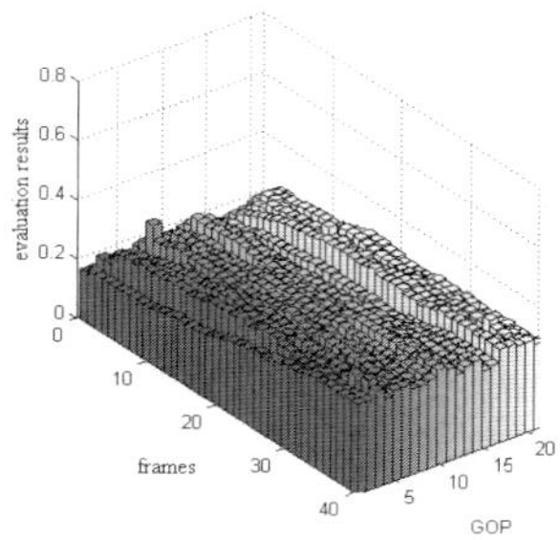

(d) evaluation result of grandma sequence

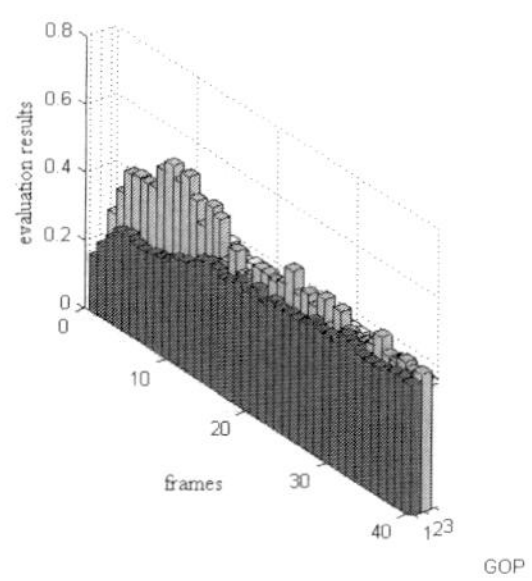

(e) evaluation result of suzie sequence

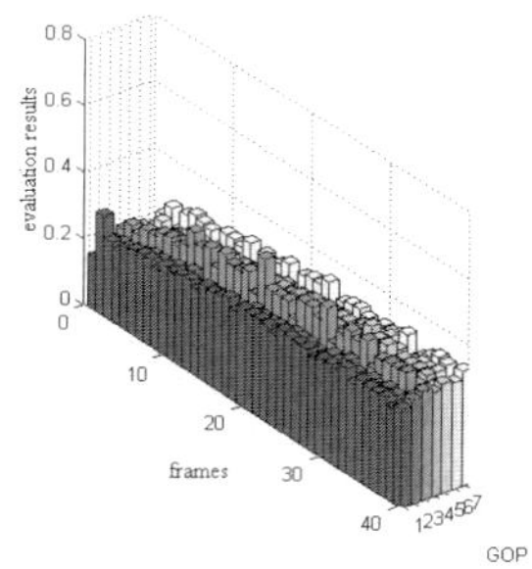

(f) evaluation result of news sequence

Fig. 1(a)-(f). The overall evaluation results of each standard sequence

The overall fuzzy symmetric video quality evaluation results are constructed in three-dimensional histogram (Fig. 1). It is clear that the results are consistent with the judgments of subjective DSCQS. Fig. 3 and Fig. 4 provide the weights assignment in affecting factors and their results.

The judgment results of video continuity and fluency features are illustrated in Fig. 2 –Fig. 4 by foreman sequence example. The original video frequency is 30Hz, while contrastive video frequency is 10Hz. Corresponding frame skip mode is adopted in the contrastive sequence so as to test influence on temporal interval. The evaluation of video fluency feature in this model can be shown in Fig. 2 and Fig. 1 (a), with a well demonstration on both motility and relativity of video sequences features. Fig. 3 and

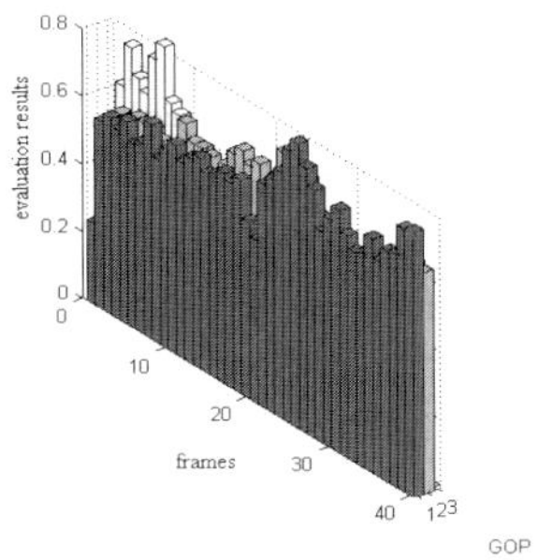

Fig. 2. evaluation result of foreman sequence of 10Hz

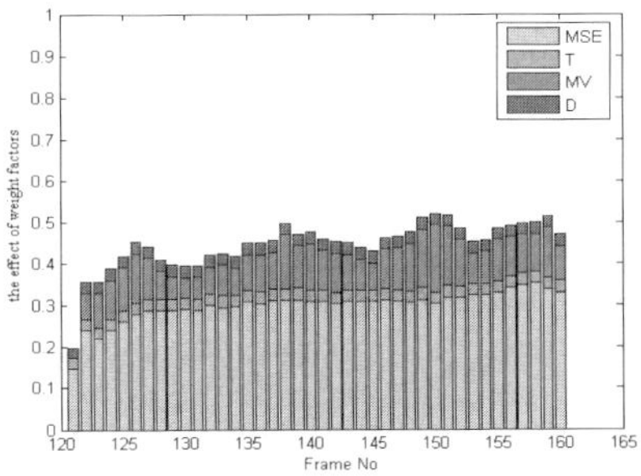

(a) effect of weight factors in GOP4

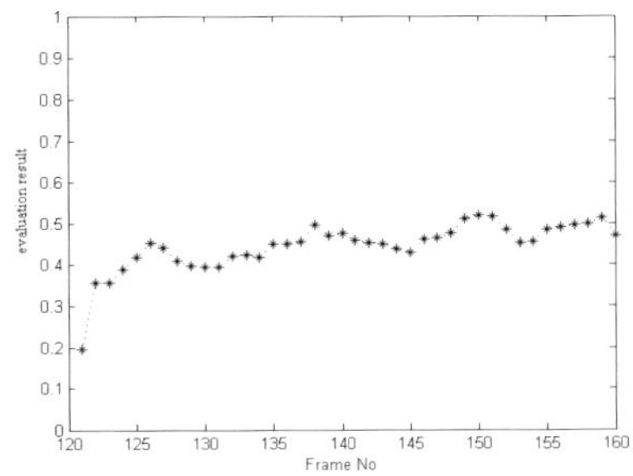

(b) evaluation result of GOP4

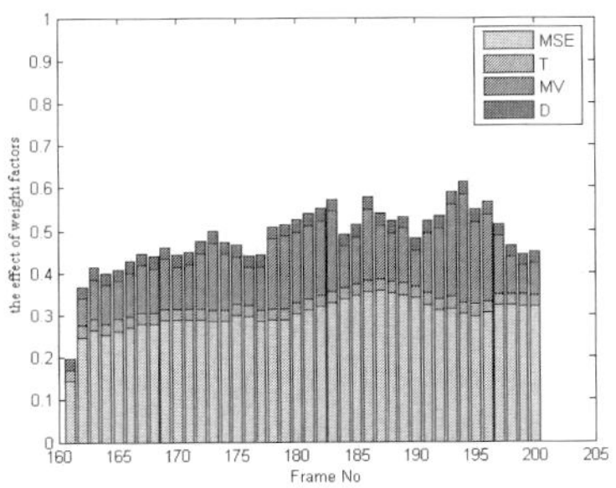

(c) effect of weight factors in GOP5

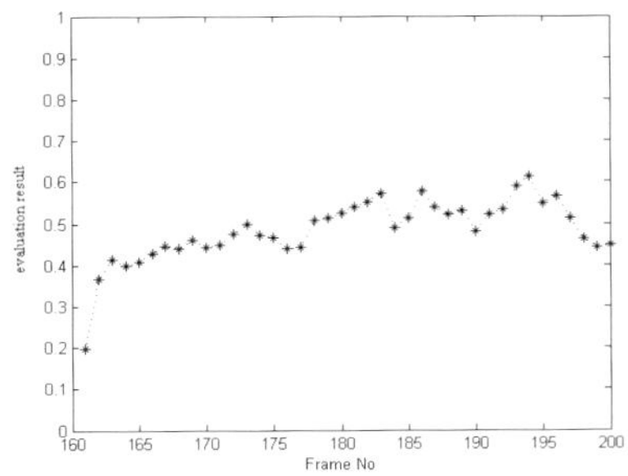

(d) evaluation result of GOP5

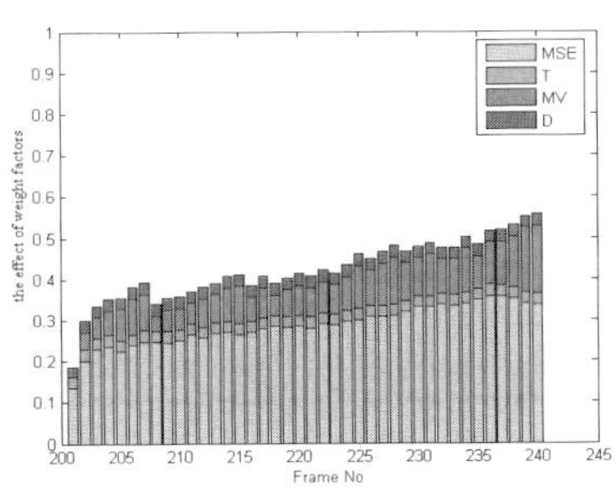

(e) effect of weight factors in GOP6

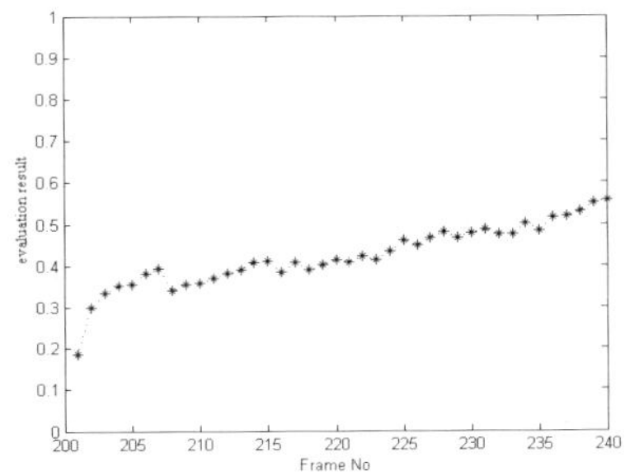

(f) evaluation result of GOP6

Fig. 3. 120-240 frames evaluation results and weight factor affects of foreman sequence of 30Hz

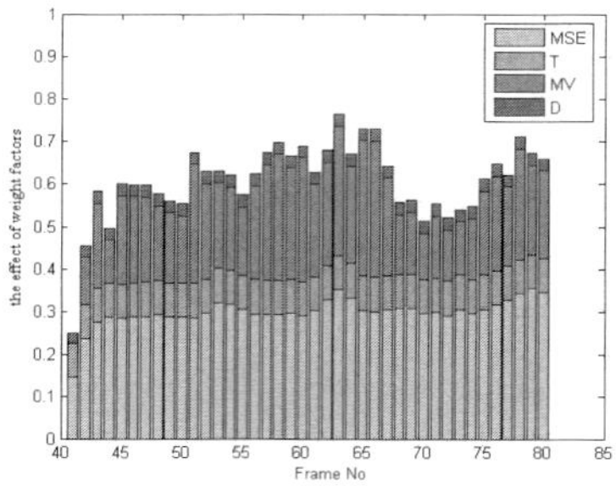

(a) effect of weight factors

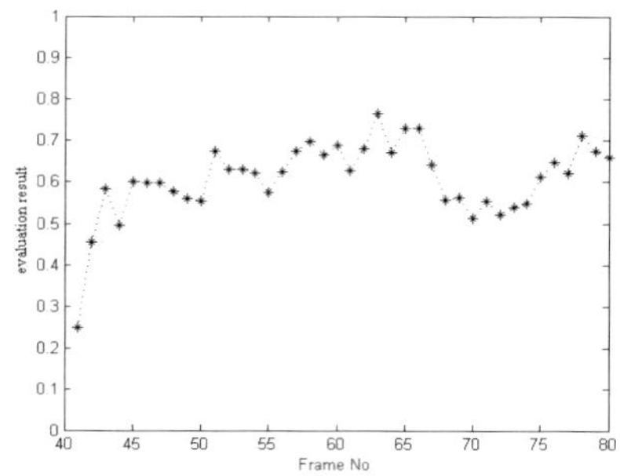

(b) evaluation result

Fig. 4. corresponding 41-80 frames evaluation results and weight factor affects of foreman sequence of 10Hz

Fig. 4 provide the testing result of the most acute segment of foreman. The quality values of 10Hz segment have not been obviously affected by the varieties of *PSNR* (*MSE*), but the systematic evaluation results are not as good as the ones of 30Hz due to increasing in both sampling interval and motility which lead to worse fluency and wobble. It is illustrated in Fig. 4 (a) and Fig. 3 (a)(c)(e) through affects of the factor weights assignment. Therefore, these results demonstrate the advantages of the fuzzy objective video quality evaluation model especially in integrative video judgment.

4 Conclusion

The primary focus of this work has been to develop effective and systematic methods for assessing the quality of compressed video. Most current objective assessment methods for video quality are complicated and unilateral so as to be inadequate for practical application. Objective metric based on fuzzy systematic evaluation is firstly proposed. The paper presents a way to evaluate the objective quality of video sequences by using multiple video affecting factors including motion info, fluency and quality features, as well as other evaluation models. Subjective and objective experiments have confirmed the feasibility, which is important for communication applications such as error resilience, flow control and video transcoding so as to make communication more efficiently.

References

1. Jing Guo, Van Dyke-Lewis, et al.: Gabor Difference Analysis of Digital Video Quality. Broadcasting, IEEE Transactions on Volume 50, Issue 3, Sept 2004, Page(s): 302 – 311
2. Ong E., Lin W., Lu Z., et al.: Quality Evaluation of MPEG-4 and H.26l Coded Video for Mobile Multimedia Communications. Signal Processing and Its Applications. 2003. Proceedings. Seventh International Symposium on Volume 1, 1-4 July 2003 Page(s): 473 - 476

3. Masry M., Hemami, et al.: Perceived Quality Metrics for Low Bit Rate Compressed Video. Image Processing. Proceedings 2002 International Conference on Volume 3, 24-28 June 2002 Page(s): III-49 - III-52
4. Nannan Wang, Qiang Zhang, et al.: Objective Quality Evaluation of Digital Video. Circuits and Systems. The 2000 IEEE Asia-Pacific Conference on 4-6 Dec. 2000 Page(s):791 - 794
5. Kuhmanch C., Schremmer C .: Mpirical Evaluation of Layered Video Coding Schemes. Image Processing, Proceedings 2001 International Conference on Volume 2,7-10 Oct. 2001 Page(s): 1013 – 1016
6. Steiger O., Ebrahimi T.,et al.: Evaluating Perceptually Prefitered Video, Multimedia And Expo. ICME2005. IEEE International Conference on 06-06, July 2005 Page(s): 1290 - 1293
7. Xiedong Cao.: Fuzzy Information Processing and Application, Science Press. 2003
8. Sadka A.H.: Compressed Video Communications.2002.

Appendix

Wen Ji is with the Department of Electronic Engineering, Northwestern Polytechnical University, Xi'an, Shaan xi 710072, China. Now she is a graduate student for her Ph.D. degree. Her current research topics include wireless networking and multimedia communication techniques. E-mail: jill@nwpu.edu.cn

HaoShan Shi is currently a professor in the Department of Electronic Engineering, Northwestern Polytechnical University, Xi'an, Shaan xi 710072, China. He is the director of the Multimedia Communication and Computer Networks Lab, where he is involved in research and development projects for the application of multimedia technologies to the communication field.

Ying Wang is with the Department of Electronic Engineering, Northwestern Polytechnical University, Xi'an, Shaan xi 710072, China. Now she is a graduate student for her Ph.D. degree. Her current research topics include wireless networking and coding techniques. E-mail: coolman-coolman@163.com

The Generalization of λ-Fuzzy Measures with Application to the Fuzzy Option*

Liyan Han[1] and Wenli Chen[2]

School of Economics and Management, Beihang University,
Beijing, 100083, China

Abstract. In this paper, we present the definition of λ-fuzzy signed measure, its generalized transform function, and the distribution properties of λ-fuzzy measures. Then investors' heterogeneity is revealed with a family of λ-fuzzy measures. The approach to asset pricing based on λ-fuzzy measures breaks through the traditional "identical rationality" assumption, and provides an useful analytical tool for recent researches of behavioral economics and finance.

Keywords: fuzzy; fuzzy measures; heterogeneity; asset pricing; fuzzy options.

1 Introduction

It is known that information can be represented by measures based on sets and risk is the evaluative form of information. In 1970s, fuzzy measures have been applied to the nonadditivity of information in many fields. They can represent the overlap, complement and conflict among information in real life. Yager *et al.* adapt the fuzzy measures to solve nonadditivity of information (see[1]-[2]).

In economics, "identical rationality" assumption is not for abstracting real world but just for convenience. Fuzziness and uncertainty of information can be used to express people's different attitude which is the imperfect characteristic of financial market. However, few quantified models are available in economics and financial analysis. In 1994, Cherubini fuzzified Black-Schole model by λ-fuzzy measures, where λ is the indicator of the uncertainty (see[3]). In 2002, Han and Zheng used the family of fuzzy measures to exploit diversification of rationality and constructed non-identical rationality frame (see[4]). Hence we can overcome the limitation of representative agent in classical economics and finance with the above method. But the distribution properties of λ-fuzzy measures are not available so far.

This paper is devoted to the original definitions, the relationship between λ-fuzzy measures and probability concerning the parametric changes, their distribution properties and application to option pricing. In section 2, we provide

* The research is supported by National Natural Science Foundation of China, titled as "Research on the pricing methods of options based on the fuzzy measures" (70271010).

L. Wang et al. (Eds.): FSKD 2006, LNAI 4223, pp. 762–765, 2006.

some definitions, generalization and properties of λ-measures. Their application to option pricing is in section 3. Finally, the section 4 is a summary of this study.

2 Related Definitions, Construction and Distribution Properties

The key properties of a fuzzy measure are semi-continuous and monotonicity with respect to set inclusion. λ-fuzzy measures satisfy λ-rule (see[5]) and they are essentially nonadditive measures corresponding to real information.

Let Ω be a nonempty set and Σ be the algebra of Ω.

Definition λ-fuzzy measure [6]**:** μ is called a λ-fuzzy measure on Σ iff it satisfies the $\sigma-\lambda$-rule (see[5]) on Σ and there exists at least one set $E \in \Sigma$ such that $\mu(E) < \infty$.

Usually the λ-fuzzy measure is denoted by g_λ. When Σ is a σ-algebra, $g_\lambda(\Omega) = 1$ and $\lambda \in (-1, +\infty)$, the λ-fuzzy measure is also called a Sugeno measure, noted as S_λ-measure.

Signed measure [8] is an extended real valued, countably additive set function μ on the class of all measurable sets of a measurable space $<\Omega, \Sigma>$, such that $\mu(0) = 0$, and such that μ assumes at most one of the values $+\infty$ and $-\infty$.

Definition 2.1. Fuzzy signed measure: μ is a fuzzy signed measure on Σ iff it is an extended real valued fuzzy measure.

Definition 2.2. λ-fuzzy signed measure: μ is a fuzzy signed measure on Σ iff it is a fuzzy signed measure and satisfies the λ-rule.

The λ-fuzzy measure on σ-algebra is a fuzzy measure with λ-additive, so it can be constructed by means fuzzy measures constructed (see[9]). In this paper, we use transform functions to construct λ-fuzzy signed measures and λ-fuzzy measures.

Theorem 1. *Let $<\Omega, \Sigma, P>$ be a measurable space, if P is an arbitrary additive measure, and $\mu(A) = \frac{1}{\lambda}[f(\lambda)^{P(A)} - 1]$, where $f(\lambda)$ is an arbitrary function with respect to λ, then μ is a λ-fuzzy signed measure.*

Proof. From the theorem in [5], we must show that it satisfies λ-rule, so the proof is trivial.

At the same time, it must note that μ is a regular fuzzy measure iff $f(\lambda) = 1+\lambda$. For example, Wang and Klir (see[5]) give out a concrete transform function of cumulative distributions and construct the new continuous distribution function.

We call normal, lognormal, Chi square exponential and student distribution as classical distributions. And select $f(\lambda) = 1 + \lambda$, then $\mu_\lambda = \frac{(1+\lambda)^P - 1}{\lambda}$ is a S_λ-measure. The following conclusions can be drawn by numerical analysis, drawing and observation.[1] Here we only display three representative graphs.

[1] Want to know the details, please write to the author.

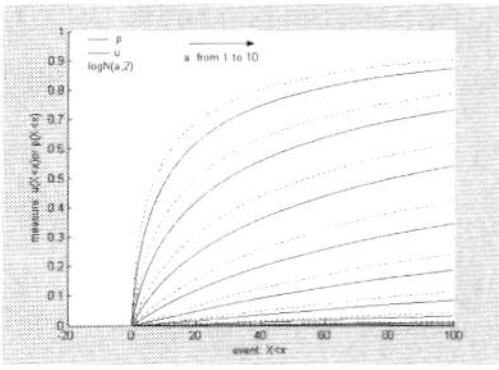

Fig. 1. images with the changes in lognormal means from 1 to 10, where lognormal variance is 2, when $\lambda = 0.9$

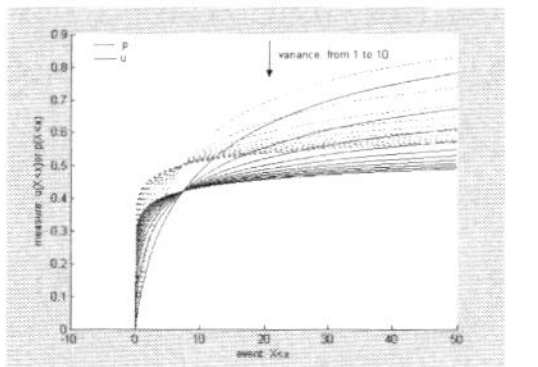

Fig. 2. images with the changes in lognormal variance from 1 to 10, where the lognormal mean is 2, when $\lambda = 0.9$

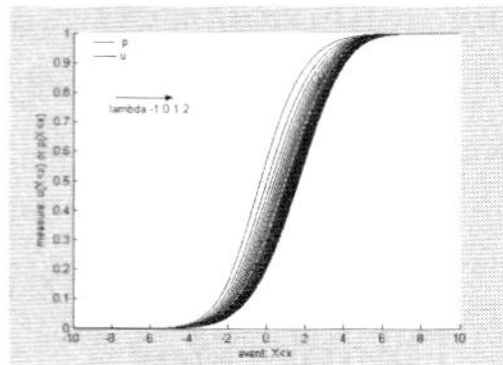

Fig. 3. images with the changes in λ, where the lognormal mean is 1, and variance is 2

Proposition 1. *a: If $\lambda > 0$, then μ_λ is an increasing function with respect to $f(\lambda)$; if $\lambda < 0$, then μ_λ is a decreasing function.*
b: μ_λ is increasing in λ is increasing, when $\lambda < 0$. On the contrary, μ_λ is decreasing in λ is increasing.

Proof. From the monotonicity of generalized transform function to $f(\lambda)$ and λ, the proof is obvious. So it is omitted.

Proposition 2. *a: The distribution function of μ_λ, where $\mu_\lambda = \frac{(1+\lambda)^{F(x)}-1}{\lambda}$, satisfies the following properties, for any λ: (1) $\mu_\lambda(X)$ is a nondecreasing function, that is, if $x_1 \leq x_2$, then $\mu_\lambda(x_1) \leq \mu_\lambda(x_2)$; (2) $0 \leq \mu_\lambda(X) \leq 1$ and $\lim_{x\to-\infty} \mu_\lambda(x) = 0$, $\lim_{x\to+\infty} \mu_\lambda(x) = 1$; (3) $\forall x_0$,$\mu_\lambda(X)$ is a right continuous function; (4) $\forall a, b, a < b$, $\mu_\lambda(a < x < b) = \frac{\mu_\lambda(x\leq b)-\mu_\lambda(x\leq a)}{1+\lambda\mu_\lambda(x\leq a)}$.*
b: With the changes in classical distribution parameters, the relationships among λ-fuzzy measures are similar to ones of classical probability distributions when λ is fixed.

Proof. a: 1. $\frac{\partial \mu_\lambda}{\partial x} = \frac{\partial \mu_\lambda}{\partial F(x)}\frac{\partial F(x)}{\partial x} = \frac{(1+\lambda)^{F(x)}\ln(1+\lambda)}{\lambda}\frac{\partial F(x)}{\partial x}$, $\forall \lambda \in [-1, \infty]$, $\frac{(1+\lambda)^{F(x)}\ln(1+\lambda)}{\lambda} > 0$, hence $\frac{\partial \mu_\lambda}{\partial x}$ and $\frac{\partial F(x)}{\partial x}$ have the same signal, and $\mu_\lambda(X)$ and $F(X)$ have the same monotonicity with respect to x; 2. $\mu_\lambda(X)$ and $F(X)$ have the same asymptotic lines: $\lim_{x\to-\infty} \mu_\lambda(x) = \lim_{x\to-\infty} F(x) = 0$, $\lim_{x\to+\infty} \mu_\lambda(x) = \lim_{x\to+\infty} F(x) = 1$; 3. $\forall x_0$, $\lim_{x\to(x_0)^+} \mu_\lambda(x) = \frac{(1+\lambda)^{\lim_{x\to(x_0)^+} F(x)} - 1}{\lambda} = \mu_\lambda(x_0)$; 4. It is trivial from property in [6]. b: The proof is omitted.

So the structure of the lambda fuzzy measures can character the heterogeneous information in financial market, especially for overactive behaviors of investors.

3 Application to the Fuzzy Option

In the classical finance theory, risk neutral probability measure stands for the foreseeing towards the natural states by the unique rational representative agent in the economy. While in order to breaking through the homogeneity in real market, we take a family of λ-fuzzy measures in stead of the unique probability measure, and substitute Choquet expectation for expectation under risk neutral probability. In this framework, with a given λ-fuzzy measure and its dual measure the λ-interval of the fuzzy price can be deduced, and then a fuzzy set can be constructed with expression theorem in fuzzy set theory. In paper [4], the fuzzy price of options based on λ-fuzzy measure is presented.

Therefore, the price of the option is a fuzzy number not a determinate number, which covered the classical option value by Black-Scholes model. Then the option price with nature language can be estimated in advance, and the outcome is stable. So The fuzzy prices contain more information than that in classical situation. Based on this fuzzy price, some sensitivities of options can be deduced, for example $\delta, \theta, \rho, \lambda, \gamma$. So the fuzzy hedge ratios and hedge strategies can contribute to both measuring of the default risk and portfolio insurance.

4 Conclusion

In this paper, we show that λ-fuzzy measures are fit for the real information. Especially in financial market, they display heterogeneous behaviors of individuals and they are useful to explain the occurrences of impossible events. The research of this paper solves the choice of λ-fuzzy measures in the pricing of the fuzzy option, and provides the analytical tool for measure of heterogeneity adapting for variation of different risk utility.

References

1. Yager, R.R: Measuring the information and character of a fuzzy measure. IEEE. **3** (2001) 1718–1722
2. Grabish, M.Murofushi, T. and Sugeno: Fuzzy measures and integrals: theory and applications. New York: Phusica Verlag. (2000)
3. Umberto Cherubini: Fuzzy measures and asset prices: accounting for information ambiguity. Applied Mathematical Finance. **4** (1997)
4. Han Liyan and Zheng Chengli: Fuzzy options with application to default risk analysis for municipal bonds in China. Nonlinear Analysis, Theory, Methods and Applications. J. Diff. Eq. **63**(Issue 5-7) (2005) e2353-e2365 (SCI)
5. Wang, Z. and Klir, G.J.: Fuzzy Measure Theory. New York: Plenum Press. (1992)
6. Liu Yingming: Liu Yingming. Sichuan education press, Sichuan, China, (1992)
7. M.Sugeno: Theory of fuzzy integrals and its applications. Ph.D. Thesis, Tokyo Institute of Technology (1974)
8. Paul R. Halmos: Measure Theory, Springer-Verlag New York Inc. (1974)
9. Klir George J., Zhenyuan Wang, and Harmanec, David: Constructing fuzzy measures in expert systems. In Fuzzy Sets Syst., **92** (1974) 251–264

An Interval Semi-absolute Deviation Model For Portfolio Selection*

Yong Fang and Shouyang Wang

Institute of Systems Science,
Academy of Mathematics and Systems Science,
Chinese Academy of Sciences, Beijing 100080, China
yfang@amss.ac.cn, swang@iss.ac.cn

Abstract. Interval number is a kind of special fuzzy number and the interval approach is a good method to deal with some uncertainty. The semi-absolute deviation risk function is extended to an interval case. Based on the extended semi-absolute deviation risk function, an interval semi-absolute deviation model for portfolio selection is proposed. By introducing the concepts of pessimistic satisfactory index and optimistic satisfactory index of interval inequality relation, an approach to compare interval numbers is given. The interval portfolio selection problem is converted to two parametric linear programming problems. A numerical example is given to illustrate the behavior of the proposed portfolio selection model.

1 Introduction

Since Markowitz[5] published his pioneering work which established the foundation of modern portfolio analysis, the mean variance methodology for portfolio selection has served as a basis for the development of modern financial theory over the past five decades. However, Markowitz's standard model is not used extensively to construct large-scale portfolios. Konno and Yamazaki[4] used the absolute deviation risk function to replace the risk function in Markowitz's model to formulate a mean absolute deviation portfolio optimization model. It turns out that the mean absolute deviation model maintains the nice properties of Markowitz's model and removes most of the main difficulties in solving Markowitz's model. Furthermore, a mean semi-absolute deviation portfolio selection model was proposed by Speranza and Mansini[8].

Though probability theory is one major tool used for analyzing uncertainty in finance, it cannot describe the uncertainty completely since there are many other uncertain factors that differ from the random ones found in financial markets. Some other techniques can be applied to handle uncertainty of financial markets, for instance, fuzzy set theory [10] and possibility theory. Recently, some authors, such as Ramaswamy[6], Tanaka and Guo[9], Inuiguchi and Ramik[2] have studied

* Supported by the National Natural Science Foundation of China under Grant No. 70221001.

L. Wang et al. (Eds.): FSKD 2006, LNAI 4223, pp. 766–775, 2006.

portfolio selection models based on possibility theory. It is always assumed that the distribution functions of possibility returns are known in these portfolio selection models. But, in reality, it is not always easy for an investor to specify them in financial markets. Therefore, in some cases, for instance, the historical data of stocks are lacking, it is a good idea that the uncertain returns of assets are determined as interval numbers through making use of experts' knowledge.

In this paper, we propose an interval semi-absolute deviation model for portfolio selection where the expected return of securities are treated as interval numbers. This paper is organized as follows. In Section 2, we give some notations for interval numbers and briefly introduce some interval arithmetics. An order relation over intervals is introduced. The concepts of pessimistic satisfactory index and optimistic satisfactory index of interval inequality relation are given. Based on these concepts, an approach to compare interval numbers is proposed. In Section 3, an interval semi-absolute deviation model for portfolio selection is proposed. In Section 4, an example is given to illustrate our approach by using real data from the Shanghai Stock Exchange. A few concluding remarks are finally given in Section 5.

2 Notations and Definitions

Denote the set of all the real numbers by R. An order pair in a bracket defines an interval

$$a = [\underline{a}, \overline{a}] = \{x : \underline{a} \leq x \leq \overline{a}, x \in R\}$$

where $\underline{a}$ is the lower bound and $\overline{a}$ is the upper bound of interval a respectively. The center and the width of a can be easily calculated as

$$m(a) = \frac{1}{2}(\underline{a} + \overline{a}) \text{ and } w(a) = \frac{1}{2}(\overline{a} - \underline{a}).$$

a can also be denoted by its center and width as

$$a = \langle m(a), w(a) \rangle = \{x : m(a) - w(a) \leq x \leq m(a) + w(a), x \in R\}.$$

The extension of ordinary arithmetic to closed intervals is known as interval arithmetic. For a detailed discussion, one can refer to [1]. First, we quote a basic concept as follows.

Definition 1. ([1]) Let $\circ \in \{+, -, \times, \div\}$ be a binary operation on R. If a and b are two closed intervals, then

$$a \circ b = \{x \circ y : x \in a, y \in b\}$$

defines a binary operation on the set of all the closed intervals. In the case of division, it is always assumed that 0 is not in b.

An interval number can be viewed as a special fuzzy number whose membership function takes value 1 over the interval, and 0 anywhere else. It is clear that the above three operations of intervals are equivalent to the operations of addition, subtraction and scalar multiplication of fuzzy numbers via the extension principle. Rommelfanger, Hanscheck and Wolf [7] investigated the interval programming problem as a fuzzy programming problem. Ishibuchi and Tanaka[3] suggested an order relation $\preceq$ between two intervals as follows.

Definition 2. If intervals a and b are two profit intervals, the order relation $\preceq$ between a and b is defined as

$$a \preceq b \text{ if and only if } \underline{a} \leq \underline{b} \text{ and } \overline{a} \leq \overline{b};$$

$$a \prec b \text{ if and only if } a \preceq b \text{ and } a \neq b.$$

For describing the interval inequality relation in detail, we give three new concepts in the following.

Definition 3. For any two interval numbers $a = [\underline{a}, \overline{a}]$ and $b = [\underline{b}, \overline{b}]$, there is an interval inequality relation $a \leq b$ between the two interval numbers a and b if and only if $m(a) \leq m(b)$. Furthermore, if $\overline{a} \leq \underline{b}$, we say the interval inequality relation $a \leq b$ between a and b is optimistic satisfactory; if $\overline{a} > \underline{b}$, we say the interval inequality relation $a \leq b$ between a and b is pessimistic satisfactory.

Definition 4. For any two interval numbers $a = [\underline{a}, \overline{a}]$ and $b = [\underline{b}, \overline{b}]$, if the interval inequality relation between them is pessimistic satisfactory, the pessimistic satisfactory index of the interval inequality relation $a \leq b$ can be defined as

$$PSD(a \leq b) = 1 + \frac{\underline{b} - \overline{a}}{w(a) + w(b)}$$

Definition 5. For any two interval numbers $a = [\underline{a}, \overline{a}]$ and $b = [\underline{b}, \overline{b}]$, if the interval inequality relation between them is optimistic satisfactory, the optimistic satisfactory index of the interval inequality relation $a \leq b$ can be defined as

$$OSD(a \leq b) = \frac{\underline{b} - \overline{a}}{w(a) + w(b)}.$$

3 Model Formulation

Assume that an investor wants to allocate his wealth among n risky assets offering random rates of returns and a risk-free asset offering a fixed rate of return. We use V shape function to express the transaction costs, so the total transaction costs of the portfolio $x = (x_1, x_2, \cdots, x_n, x_{n+1})$ can be denoted by

$$C(x) = \sum_{i=1}^{n+1} C_i(x_i) = \sum_{i=1}^{n+1} k_i |x_i - x_i^0|,$$

where k_i is the rate of transaction costs for the asset i $(i = 1, 2, \cdots, n+1)$ and x_i^0 is the proportion of the risky asset i $(i = 1, 2, \cdots, n)$ or risk-free asset $n+1$ owned by the investor.

It is well known that future returns of securities cannot be accurately predicted in any emerging securities market. Traditionally, researchers consider the arithmetic mean as the expected return of the security according to historical observation. So the expected return of the security is a crisp value in this way. However, the later historical data of a security most often indicate that the performance of a corporation is more important than that of the earlier historical data. In addition, if the historical data of a security are not enough, one cannot accurately estimate the statistical parameters due to data scarcity. Considering these problems, perhaps it is good idea that the expected return of a security may be considered as an interval number rather than a crisp value based on the arithmetic mean of historical data. Investors may make use of corporation's financial report and the security's historical data to determine the expected return interval's range. The uncertain expected return of the risky asset i $(i = 1, 2, \cdots, n)$ can be represented as the following interval number:

$$\tilde{r}_i = [\underline{r}_i, \overline{r}_i].$$

After removing the transaction costs, the net expected return interval of portfolio $x$can be represented as

$$\tilde{r}(x) = \sum_{i=1}^{n} \tilde{r}_i x_i + r_{n+1} x_{n+1} - \sum_{i=1}^{n+1} k_i |x_i - x_i^0|.$$

If the expected returns of securities are crisp values, the semi-absolute deviation of the return of portfolio x below the expected return at the past period t, $t = 1, 2, \cdots, T$ can be represented as

$$w_t(x) = |\min\{0, \sum_{i=1}^{n} (r_{ti} - r_i) x_i\}| = \max\{0, \sum_{i=1}^{n} (r_i - r_{ti}) x_i\}$$

where r_{ti} is the historical rate of return of risky asset i and r_i is the expected returns of security i. For a detailed discussion, one can refer to [8].

Because the expected returns on securities are considered as interval numbers, we may consider the semi-absolute deviation of the rates of return on portfolio x below the expected return over all the past periods as an interval number too.

Since the expected return interval on portfolio x is

$$\hat{r}(x) = [\sum_{i=1}^{n} \underline{r}_i x_i + r_{n+1} x_{n+1}, \sum_{i=1}^{n} \overline{r}_i x_i + r_{n+1} x_{n+1}],$$

we can get the semi-absolute deviation interval of return of portfolio x below the expected return over the past period t, $t = 1, 2, \cdots, T$. It can be represented as

$$\tilde{w_t}(x) = [\max\{0, \sum_{i=1}^{n} (\underline{r}_i - r_{ti}) x_i\}, \max\{0, \sum_{i=1}^{n} (\overline{r}_i - r_{ti}) x_i\}].$$

Then the average value of the semi-absolute deviation interval of return on portfolio x below the uncertain expected return over all the past periods can be represented as

$$\begin{aligned}\tilde{w}(x) &= \tfrac{1}{T}\sum_{t=1}^{T}\tilde{w}_t(x)\\ &= \tfrac{1}{T}\sum_{t=1}^{T}[\max\{0,\sum_{i=1}^{n}(\underline{r}_i - r_{ti})x_i\}, \max\{0,\sum_{i=1}^{n}(\overline{r}_i - r_{ti})x_i\}]\end{aligned}$$

We use $\tilde{w}(x)$ to measure the risk of portfolio x. Suppose that the investor wants to maximize the return on a portfolio after removing the transaction costs within some given level of risk. If the risk tolerance interval $\tilde{w}$ is given, the mathematical formulation of the portfolio selection problem is

$$\begin{aligned}\text{(ILP)}\ \max\ \tilde{r}(x) &= \sum_{i=1}^{n}\tilde{r}_i x_i + r_{n+1}x_{n+1} - \sum_{i=1}^{n+1}k_i|x_i - x_i^0|\\ \text{s.t.}\quad & \tilde{w}(x) \le [\underline{w}, \overline{w}],\\ & \sum_{i=1}^{n+1}x_i = 1,\\ & 0 \le x_i \le u_i, i = 1, 2, \cdots, n+1.\end{aligned}$$

where $\underline{w}$ and $\overline{w}$ are two given constants, $\underline{w}$ represents the pessimistic tolerated risk level, $\overline{w}$ represents the optimistic tolerated risk level.

We introduce the order relation $\preceq$ in the interval objective function of (ILP). Based on the concepts of the pessimistic satisfactory index and optimistic satisfactory index proposed by us in Section 2, the interval inequality relation $\tilde{w}(x) \le [\underline{w}, \overline{w}]$ in (ILP1) is sure to be expressed by one of two crisp equalities. The two crisp equivalent equalities of the interval constraint condition $\tilde{w}(x) \le [\underline{w}, \overline{w}]$ can be represented as follows:

$$PSD(\tilde{w}(x) \le [\underline{w}, \overline{w}]) = \alpha$$

and

$$OSD(\tilde{w}(x) \le [\underline{w}, \overline{w}]) = \beta.$$

Then the interval linear programming problem (ILP1) can be decomposed into two interval linear programming problems in which the objective functions are interval numbers and the constraint conditions are crisp equalities and inequalities. The two interval objective function linear programming problems are represented as follows:

$$\begin{aligned}\text{(PO1)}\ \max_{\preceq}\ \tilde{r}(x) &= \sum_{i=1}^{n}\tilde{r}_i x_i + r_{n+1}x_{n+1} - \sum_{i=1}^{n+1}k_i|x_i - x_i^0|\\ \text{s.t.}\quad & PSD(\tilde{w}(x) \le [\underline{w}, \overline{w}]) = \alpha,\\ & \sum_{i=1}^{n+1}x_i = 1,\\ & 0 \le x_i \le u_i, i = 1, 2, \cdots, n+1.\end{aligned}$$

where α is given by the investor.

$$
\begin{aligned}
\text{(PS1)}\ \max_{\preceq}\ & \tilde{r}(x) = \sum_{i=1}^{n} \tilde{r}_i x_i + r_{n+1}x_{n+1} - \sum_{i=1}^{n+1} k_i|x_i - x_i^0| \\
\text{s.t.}\quad & OSD(\tilde{w}(x) \leq [\underline{w}, \overline{w}]) = \beta, \\
& \sum_{i=1}^{n+1} x_i = 1, \\
& 0 \leq x_i \leq u_i, i = 1, 2, \cdots, n+1.
\end{aligned}
$$

where β is given by the investor.

We can find that the constraint conditions of (PO1) are stricter than those of (PS1). Hence, we can get an optimistic investment strategy by solving (PO1), and a pessimistic investment strategy by solving (PS1).

Denote F_1 as the feasible set of (PO1) and F_2 as the feasible set of (PS1).

Definition 6. $x \in F_1$ is a satisfactory solution of (PO1) if and only if there is no other $x' \in F_1$ such that $\tilde{r}(x) \prec \tilde{r}(x')$; $x \in F_2$ is a satisfactory solution of (PS1) if and only if there is no other $x' \in F_2$ such that $\tilde{r}(x) \prec \tilde{r}(x')$.

By Definition 6, the satisfactory solution of (PO1) is equivalent to the non-inferior solution set of the following bi-objective programming problem:

$$
\begin{aligned}
\text{(PO2)}\ \max\ & \sum_{i=1}^{n} \underline{r}_i x_i + r_{n+1}x_{n+1} - \sum_{i=1}^{n+1} k_i|x_i - x_i^0| \\
\max\ & \sum_{i=1}^{n} \overline{r}_i x_i + r_{n+1}x_{n+1} - \sum_{i=1}^{n+1} k_i|x_i - x_i^0| \\
\text{s.t.}\ & \text{and all constraints of (PO1).}
\end{aligned}
$$

The satisfactory solution of (PS1) is equivalent to the non-inferior solution set of the following bi-objective programming problem:

$$
\begin{aligned}
\text{(PS2)}\ \max\ & \sum_{i=1}^{n} \underline{r}_i x_i + r_{n+1}x_{n+1} - \sum_{i=1}^{n+1} k_i|x_i - x_i^0| \\
\max\ & \sum_{i=1}^{n} \overline{r}_i x_i + r_{n+1}x_{n+1} - \sum_{i=1}^{n+1} k_i|x_i - x_i^0| \\
\text{s.t.}\ & \text{and all constraints of (PS1).}
\end{aligned}
$$

Introducing a new variable x_{n+2} such that $x_{n+2} \geq \sum_{i=1}^{n+1} k_i|x_i - x_i^0|$. Let

$$
d_i^+ = \frac{|x_i - x_i^0| + (x_i - x_i^0)}{2}; d_i^- = \frac{|x_i - x_i^0| - (x_i - x_i^0)}{2};
$$

$$
\underline{y}_t^+ = \frac{|\sum_{i=1}^{n}(r_{ti} - \underline{r}_i)x_i| + \sum_{i=1}^{n}(r_{tj} - \underline{r}_j)x_j}{2};
$$

$$
\underline{y}_t^- = \frac{|\sum_{i=1}^{n}(r_{tj} - \underline{r}_i)x_i| - \sum_{i=1}^{n}(r_{ti} - \underline{r}_i)x_i}{2};
$$

$$\overline{y}_t^+ = \frac{|\sum_{i=1}^{n}(r_{ti}-\overline{r}_i)x_i| + \sum_{j=1}^{n}(r_{ti}-\overline{r}_i)x_i}{2};$$

$$\overline{y}_t^- = \frac{|\sum_{i=1}^{n}(r_{ti}-\overline{r}_i)x_i| - \sum_{j=1}^{n}(r_{tj}-\overline{r}_j)x_j}{2}.$$

Thus, by multi-objective programming theory, the non-inferior solution to (PO2) can be generated by solving the following parametric linear programming problem:

$$\begin{aligned}
\text{(PO3)}\ \max\ & \sum_{i=1}^{n}[\lambda \underline{r}_i + (1-\lambda)\overline{r}_i]x_i + r_{n+1}x_{n+1} - x_{n+2} \\
\text{s.t.}\ & \tfrac{1}{T}\sum_{t=1}^{T}(1+\alpha)\overline{y}_t^- + (1-\alpha)\underline{y}_t^- = (1-\alpha)\overline{w} + (1+\alpha)\underline{w}, \\
& \sum_{i=1}^{n+1} k_i(d_i^+ + d_i^-) \le x_{n+2}, \\
& \overline{y}_t^- + \sum_{i=1}^{n}(r_{ti}-\overline{r}_i)x_i \ge 0, \\
& \underline{y}_t^- + \sum_{i=1}^{n}(r_{ti}-\underline{r}_i)x_i \ge 0, \\
& d_i^+ - d_i^- = x_i - x_i^0, i = 1, 2, \cdots, n+1, \\
& \sum_{i=1}^{n+1} x_i = 1, \\
& d_i^+ \ge 0, d_i^- \ge 0, i = 1, 2, \cdots, n+1, \\
& \underline{y}_t^- \ge 0, \overline{y}_t^- \ge 0, t = 1, 2, \cdots, T, \\
& x_i \ge 0, i = 1, 2, \cdots, n+1.
\end{aligned}$$

The non-inferior solution to (PS2) can be generated by solving the following parametric linear programming problem:

$$\begin{aligned}
\text{(PS3)}\ \max\ & \sum_{i=1}^{n}[\lambda \underline{r}_i + (1-\lambda)\overline{r}_i]x_i + r_{n+1}x_{n+1} - x_{n+2} \\
\text{s.t.}\ & \tfrac{1}{T}\sum_{t=1}^{T}(2+\beta)\overline{y}_t^- - \beta\underline{y}_t^- = (2+\beta)\underline{w} - \beta\overline{w}, \\
& \sum_{i=1}^{n+1} k_i(d_i^+ + d_i^-) \le x_{n+2}, \\
& \overline{y}_t^- + \sum_{i=1}^{n}(r_{ti}-\overline{r}_i)x_i \ge 0, \\
& \underline{y}_t^- + \sum_{i=1}^{n}(r_{ti}-\underline{r}_i)x_i \ge 0, \\
& d_i^+ - d_i^- = x_i - x_i^0, i = 1, 2, \cdots, n+1, \\
& \sum_{i=1}^{n+1} x_i = 1, \\
& d_i^+ \ge 0, d_i^- \ge 0, i = 1, 2, \cdots, n+1, \\
& \underline{y}_t^- \ge 0, \overline{y}_t^- \ge 0, t = 1, 2, \cdots, T, \\
& x_i \ge 0, i = 1, 2, \cdots, n+1.
\end{aligned}$$

(PO3) and (PS3) are two standard linear programming problems. One can use several algorithms of linear programming to solve them efficiently, for example, the simplex method. So we can solve the original portfolio selection problem (ILP) by solving (PO3) and (PS3).

4 Numerical Example

In this section, we suppose that an investor chooses twelve componential stocks of Shanghai 30 index and a risk-less asset for his investment. We use a kind of saving account as the risk-less asset and the term of the saving account is three months. So the rate of return on the risk-less asset is 0.0014 per month. We collect historical data of the twelve kinds of stocks from January, 1999 to December, 2002. The data are downloaded from the web site www.stockstar.com. Then we use one month as a period to obtain the historical rates of return during forty-eight periods. We obtain the expected rate of return interval according to experts' knowledge. The intervals are given in Table 1.

Suppose the investor gives the risk level interval $\tilde{w} = [0.015, 0.040]$, By the method proposed in above section, we can solve the portfolio selection problem by solving (PO3) and (PS3).

For the given risk level interval $\tilde{w}$, more optimistic portfolios can be generated by varying the values of the parameters λ and α in (PO3); more pessimistic portfolios can be generated by varying the values of the parameters λ and β in (PS3). The return intervals, the risk intervals and the values of parameters of optimistic portfolios are listed in Table 2. The optimistic portfolios are listed in Table 3. The return intervals, the risk intervals and the values of parameters of

Table 1. The expected rates of returns intervals

Exchange Code	600001	600002	600009
Return Interval	[0.0060, 0.0068]	[0.0062, 0.0069]	[0.0104, 0.0114]
Exchange Code	600058	600068	600072
Return Interval	[0.0231, 0.0238]	[0.0067, 0.0078]	[0.0089, 0.0098]
Exchange Code	600098	600100	600104
Return Interval	[0.0164, 0.0173]	[0.0261, 0.0268]	[0.0078, 0.0087]
Exchange Code	600115	600120	600631
Return Interval	[0.0156, 0.0167]	[0.0223, 0.0229]	[0.0120, 0.0128]

Table 2. The return intervals, the risk intervals and the values of parameters of optimistic portfolios

	Return Interval	Risk Interval	λ	α
Portfolio 1	[0.0145, 0.0149]	[0.0273, 0.0276]	0.60	0
Portfolio 2	[0.0140, 0.0145]	[0.0248, 0.0251]	0.50	0.2
Portfolio 3	[0.0106, 0.0110]	[0.0178, 0.0180]	0.30	0.8

Table 3. The allocation of Portfolio 1, 2, 3

Exchange Code	600001	600002	600009	600058	600068
Portfolio 1	0.0000	0.0000	0.0000	0.0000	0.0000
Portfolio 2	0.0000	0.0000	0.0000	0.0730	0.0000
Portfolio 3	0.0000	0.0000	0.0000	0.0572	0.0000
Exchange Code	600072	600098	600100	600104	600115
Portfolio 1	0.0000	0.0000	0.4146	0.0000	0.0000
Portfolio 2	0.0000	0.0000	0.2886	0.0000	0.0000
Portfolio 3	0.0000	0.0000	0.1825	0.0000	0.0000
Exchange Code	600120	600631	Saving		
Portfolio 1	0.3078	0.0000	0.2776		
Portfolio 2	0.3610	0.0000	0.2774		
Portfolio 3	0.2938	0.0000	0.4665		

Table 4. The return intervals, the risk intervals and the values of parameters of pessimistic portfolios

	Return Interval	Risk Interval	λ	β
Portfolio 4	[0.0091, 0.0094]	[0.0148, 0.0150]	0.60	0
Portfolio 5	[0.0066, 0.0068]	[0.0144, 0.0147]	0.50	0.8
Portfolio 6	[0.0049, 0.0052]	[0.0132, 0.0138]	0.30	1.5

Table 5. The allocation of Portfolio 4, 5, 6

Exchange Code	600001	600002	600009	600058	600068
Portfolio 4	0.0000	0.0000	0.0189	0.0472	0.0000
Portfolio 5	0.0000	0.0000	0.0000	0.0000	0.0000
Portfolio 6	0.0000	0.0000	0.0000	0.0000	0.0000
Exchange Code	600072	600098	600100	600104	600115
Portfolio 4	0.0090	0.0000	0.0839	0.0000	0.0000
Portfolio 5	0.0000	0.0000	0.2662	0.0000	0.0000
Portfolio 6	0.0000	0.0000	0.1850	0.0000	0.0000
Exchange Code	600120	600631	Saving		
Portfolio 4	0.3274	0.0000	0.5136		
Portfolio 5	0.0000	0.0000	0.7338		
Portfolio 6	0.0000	0.0000	0.8150		

pessimistic portfolios are listed in Table 4. The pessimistic portfolios are listed in Table 5.

The investor may choose his own investment strategy from the portfolios according to his attitude towards the securities' expected returns and and the degree of portfolio risk with which he is comfortable. If the investor is not satisfied with any of these portfolios, he may obtain more by solving the two parametric linear programming problems (PO3) and (PS3).

5 Conclusion

The semi-absolute deviation risk function is extended to an interval case. An interval semi-absolute deviation model with no short selling and no stock borrowing in a frictional market is proposed for portfolio selection. By introducing the concepts of pessimistic satisfactory index and optimistic satisfactory index of interval inequality relation, an approach to compare interval numbers is given. By using the approach, the interval semi-absolute deviation model can be converted into two parametric linear programming problems. One can find a satisfactory solution to the original problem by solving the corresponding parametric linear programming problems. An investor may choose a satisfactory investment strategy according to an optimistic or pessimistic attitude. The model is capable of helping the investor to find an efficient portfolio that is in the closest possible accord with his goals.

References

1. Alefeld, G., Herzberger, J.: Introducing to Interval Computations. Academic Press New York (1983).
2. Inuiguchi, M., Ramik, J.: Possibilistic Linear Programming: A Brief Review of Fuzzy Mathematical Programming and a Comparison with Stochastic Programming in Portfolio Selection Problem. Fuzzy Sets and Systems 111 (2000) 3–28.
3. Ishibuchi, H., Tanaka, H.: Formulation and Analysis of Linear Programming Problem with Interval Coefficients. Journal of Japan Industrial Management Association 40 (1989) 320–329.
4. Konno, H., Yamazaki, H.: Mean Absolute Portfolio Optimization Model and Its Application to Tokyo Stock Market. Management Science 37(5) (1991) 519–531.
5. Markowitz, H.M.: Portfolio Selection. Journal of Finance 7 (1952) 77–91.
6. Ramaswamy, S.: Portfolio Selection Using Fuzzy Decision Theory. Working Paper of Bank for International Settlements No.59 (1998).
7. Rommelfanger, H., Hanscheck, R., Wolf, J.: Linear Programming with Fuzzy Objectives. Fuzzy Sets and Systems 29 (1989) 31–48.
8. Mansini, R., Speranza, M.G.: Heuristic Algorithms for the Portfolio Selection Problem with Minimum Transaction Lots. European Journal of Operational Research 114 (1999) 219–233.
9. Tanaka, H., Guo, P.: Portfolio Selection Based on Upper and Lower Exponential Possibility Distributions. European Journal of Operational Research 114 (1999) 115–126.
10. Zadeh, L.A.: Fuzzy Sets. Information and Control 8 (1965) 338–353.

A New Dictionary Learning Method for Kernel Matching Pursuit

Shuiping Gou, Qing Li, and Xiangrong Zhang

Institute of Intelligent Information Processing, Xidian University Xi'an, 710071, China

Abstract. This paper presents a method for dictionary training of Kernel matching pursuits (KMP) [1] applied in large size data classification. This algorithm uses the existing fuzzy clustering technique to design function dictionary from a set of training data. The motivation is to enhance the local and quickly searching ability of basic matching pursuit algorithm. As a result, this method reduces computation complexity of matching pursuit algorithm. Simulation results show the proposed algorithm is efficacious for training large size pattern space.

1 Introduction

The matching pursuit (MP) algorithm is adaptive signal decomposing method, and any signal is decomposed into a linear expansion of waveforms that are selected from a redundant dictionary of functions. On the basis of the MP algorithm, a matching pursuit method using kernel function sets to optimize was presented [1]. Theoretically the MP algorithm is an excellent method but its implement is a greedy algorithm [3]. So, training dictionary function of the MP has arisen more interesting for many researchers. Recently, several improved MP algorithm was suggested to construct optimal dictionary function [4]. Further, optimal size of dictionary was also researched in [5-6]. More recently, the MP has been successfully applied in many areas [6, 7]. Hence, a pattern recognition algorithm of Kernel matching pursuits based on fuzzy clustering is given by us, which overcomes large computational number of the basic matching pursuit algorithm. The presented algorithm is effective on the image recognition.

2 Kernel Matching Pursuit

Kernel matching pursuit (KMP) is simply the idea of applying the Matching Pursuit (MP) family of algorithms to problem in machine learning, using a kernel-based dictionary [1]. Given a kernel function *K*, we construct the basis dictionary of MP by the kernel centered on the training data: $D = \{d_i = K(x, x_i) | i = 1 \cdots l\}$. There exist a lot of commonly used Mercer kernels, such as polynomial kernel with the form of $K(x, x_i) = [(x, x_i) + 1]^d$ and RBF kernel with the form of $K(x, x_i) = \exp(-\|x - x_i\| / 2p)$. Running matching pursuit algorithm to get the approximation functions in regression or the decision function in classification.

L. Wang et al. (Eds.): FSKD 2006, LNAI 4223, pp. 776 – 779, 2006.

$$f_N(x) = \sum_{k=1}^{N} \beta_k g_k(x) = \sum_{k=1}^{N} \beta_k K(x, x_k) . \tag{1}$$

$$f_N(x) = \operatorname{sgn}\left(\sum_{k=1}^{N} \beta_k g_k(x) \right) = \operatorname{sgn}\left(\sum_{k=1}^{N} \beta_k K(x, x_k) \right). \tag{2}$$

Where $\{x_k | k = 1 \cdots N\} \in \{x_1, \cdots, x_l\}$ is support point. There is two ways to stop the algorithm. One is that the basis functions reach the maximum N, the other is that the error goes below a predefined given threshold. More details about KMP can be found in [1].

3 Dictionary Learning of Kernel Matching Pursuit Algorithm

KMP requires every step of searching process be global optimal searching in the redundant dictionary of function in order to select best matching signal structure. Namely, the dictionary learning time of the KMP was too long. The fuzzy clustering algorithm was used widely data compression and model constructing. The fuzzy c-mean clustering (FCM) is the typical algorithm based object function. The FCM modifies clustering center V and partition U many times to carry out dynamic iteration process. So FCM has greatly local search performance. We use the FCM to divide rough dataset into parts and get some small size dictionaries.

In *KMP* algorithm, learning is also the training and modification process on dictionary function. Its learning problem can be reduced to how to get optimal dictionary partition d_i, *i*=1…M and the size of dictionary [8-9]. We adopt the fuzzy clustering algorithm to speed up the process of learning dictionaries. It can be realized that partition object datasets into subsets with clustering center using the *FCM*. And then we use *KMP* algorithm to classify objection for every clustering partition (subset). As compared with original datasets as the presented method can reduce the complexity of the problem for object recognition. And this method can also keep intact dataset information compared with independent factor analysis [4] and vector quantization [5] in training dictionary.

The *KMP* based on *FCM* can reduce greatly training time of the dataset on the premise of the decreasing indistinctively accurate recognition rate. Then, a family of fuzzy clustering algorithm for dictionary learning of KMP to classify large size datasets is as follows:

Step1. Given the datasets of training classifier $S=\{(x_1, y_1),\ldots,(x_l, y_l)\}$, kernel function K, here we adopt for RBF kernel and Set a positive number p as kernel parameter, the number of fuzzy partition C. Set a small positive number,ε, Set t is 1,the number of iteration L.

Step2. For datasets X={ $x_1 \cdots x_l$ }, repeat to Calculate *U* and *V* of using the FCM to get dictionary D={ $d_1 \cdots d_C$ }, then compute $g_i(x) = K(x, x_i)$, for each dictionary d_j, *j*=1…*C*.

Step3. Based on KMP algorithm to obtain weight coefficient with $\omega_i = g_i^T(x)\cdot y \Big/ \|g_i(x)\|^2$, $x\in d_j$. Select the anterior minimal residue corresponding vectors $\overline{x}_j$ and ω_j from kernel function sets.

Step4. Calculate the decision function: $f_t(\omega_j, x_j) = \sum_{t=1}^{L} \omega_j^t g_j^t(x)$, j=1...C.

Step5. Let $y = y - f_t$, if $\|y\| >= \varepsilon$ go to step3 and largen the number of iteration L for each d_j.

Step6. get classifier f_t, subsequently object is recognized by equation (2).

4 Experiment Results and Analysis

In the first case, we used datasets for UCI repository from the Web page http://www.ics.uci.edu/~mlearn/MLRepository html. On the Pima and Cancer datasets we used the same kernel function parameters P setting as the one in [7] and the Chess (KR-vs-KP), the termination condition is specified the residue R be 0.01. The average test rates of 50 times are shown in Table 1, in which between parentheses is indicated the error recognition rate, substitutes UP for Uniformly Partition.

Table 1. Comparison of error recognition results on 4 UCI datasets

	KMP	KMP on UP	KMP on FCM
Pima Indians （768）	0.9840 （23.9% ）	0.1250(26.33%)	0.340 （24.13%)
Wisc.Cancer （699）	0.8910 （3.4%)	0.1410(4.862%)	0.2030 （3.13%)
Chess(KR-vs-KP)3196	38.1090 (0.5%)	1.2660 (3.0%)	1.5630 (1.3%)
Waveform(5300)	11.0859(11.61%)	0.2190(15.20%)	0.2660 (10.89%)
Mushrooms(8124)	105.9037 （1.1%)	3.3910 （2.3%)	5.4340(0.91%)

Firstly, the results of table 1 are all based on limited iteration number of the KMP algorithm and the KMP based on FCM has much more shorter training time than the KMP, especially for the large size dataset. Secondly, recognition accuracy rate of the presented algorithm is close to the KMP. This is the fact that the FCM method can compress data, which may improve classification a certain extent. The presented method has a slight long training time than the KMP on UP but the error recognition rate increased a lot than the KMP and the KMP on FCM. In fact, the KMP on UP divides into several subsets without decrease the relativity of the initial datasets.

The second set of tests was Image collection is composed of 1064 binary value remote sense image. The collection contains 608 planes and 456 marines. A part of them is shown in Fig.1. The termination condition is specified the residue R be 0.01. The average error recognition results of 20 times are shown in Table 2.

Fig. 1. Part of planes and marines images

Table 2. Comparison of recognition results by several methods

	KMP	KMP on FCM
training time (s)	87.63	6.14
error recognition rate (%)	2.36	1.21

Time of training data is very important to classify image object. In Table2 we can see that the KMP based on FCM method has much shorter training time while higher recognition rate than the KMP. So the presented method is satisfied whether recognition times or recognition accuracy rate for image object.

In this paper, the proposed approach overcomes large computational time of the basic matching pursuit algorithm and it is effective on the UCI database and remote sense image recognition. In addition, we argue that convergence decision and selection of the parameters of the kernel function in KMP deserve more research.

References

1. Pascal, V., Yoshua,B.: Kernel Matching Pursuit. Machine Learning. 48(2002)165-187
2. Mallat, S., Zhang, Z.: Matching Pursuits with Time-frequency Dictionaries. IEEE Trans. Signal Process. 12 (41) (1993) 3397-3415
3. Davis, G.M., Mallat, S., Avelanedo, M.: Greedy adaptive approximations. J. Constr. Approx. 13 (1997) 57-98
4. Olshausen, B.A., Millman, K.J. Learning sparse codes with a mixture of Gaussian prior. Advances in Neural Information Processing Systems. 12(2000)
5. Liu, Q.S., Wang, Q., Wu, L.: Size of the Dictionary in Matching Pursuit Algorithm. IEEE Transactions on Signal Processing. 52(2004)3403-3408
6. Popovici, V., Thiran, J.P.: Adaptive Kernel Matching Pursuit for Pattern Classification. Artificial Intelligence and Applications. 411-094(2004)235-239
7. Yaghllbi, M., Rabiee, H.R., Ghanbari, M., Shamsollahi, M.B.: Anew Image Texture Extraction Algorithm Based on Matching Pursuit Gabor Wavelets. 2005 IEEE International Conference on Acoustics, Speech, and Signal Processing (2005)741-744

Facial Expression Recognition Using Fuzzy Kernel Discriminant Analysis

Qingjiang Wu[1], Xiaoyan Zhou[2,3], and Wenming Zheng[2]

[1] Department of Computer Science, Huaqiao University, Quanzhou, 362011, China
[2] Research Center for Learning Science, Southeast University, Nanjing, 210096, China
[3] Department of Electrics Engineering, Nanjing University of Information Science and Technology, Nanjing, 210044, China
wenming_zheng@seu.edu.cn

Abstract. In this paper, we address the facial expression recognition task using fuzzy kernel discriminant analysis (Fuzzy KDA) method. The Fuzzy KDA method improves the performance of kernel discriminant analysis (KDA) by considering the class membership of each training sample. We conduct the facial expression recognition on two well-known facial expression databases to demonstrate the better performance of the proposed method.

1 Introduction

Kernel discriminant analysis (KDA) was first proposed by Mika et al. [1] as the non-linear extension of the classical Fisher discriminant analysis, and had demonstrated the powerful discriminant ability in many nonlinear pattern recognition problems. However, the traditional KDA algorithm is always derived under the assumption that each training sample belongs to a unique class, which is not appropriate for some recognition problems such as facial expression recognition (FER) because each facial image may contain several basic expression categories. For this reason, we propose a fuzzy KDA (Fuzzy KDA) method for the FER problem. The idea of using fuzzy method for linear discriminant analysis (LDA) was first introduced by Kwak et al.[2] for face recognition.

Facial expression recognition (FER) has become a very hot research topic in computer vision and pattern recognition. During the last decade, various approaches have been proposed to this goal. General speaking, the FER task can be divided into the facial feature extraction part and the facial expression classification part, in which the feature extraction plays a very important role. We adopt our preliminary work [3] by using Gabor wavelet transformation method to extract some facial expression features from each facial image, and then use the Fuzzy KDA method to reduce the dimensionality of the expression feature vector for classification.

2 Fuzzy Kernel Discriminant Analysis

KDA had demonstrated the powerful discriminant ability in many pattern recognition problems. However, this method is derived under the assumption that

L. Wang et al. (Eds.): FSKD 2006, LNAI 4223, pp. 780–783, 2006.

each training sample belongs to a unique class, which may not be appropriate for FER, where each facial image may contain more than one of the six basic facial expressions. To overcome the drawbacks of KDA while make use of its advantages, we adopt the fuzzy method on KDA and apply it to FER in this section.

Let $X = \{x_i^j | j = 1, \cdots, N_i; i = 1, \cdots, c\}$ be an n-dimensional sample set with N elements, c is the number of the total classes, and N_i is the number of the samples in ith class. Let Φ be a nonlinear mapping that maps X from the input space into a Hilbert space F, i.e., $\Phi : X \rightarrow F, x \rightarrow \Phi(x)$, where the inner product of any two points, $\Phi(x)$ and $\Phi(y)$, can be computed via the kernel function $k(x, y) = (\Phi(x))^{\mathsf{T}}\Phi(y)$. Let $\mu_{ij}(i = 1, \cdots, c; j = 1, \cdots, N)$ denote the membership grade of the jth sample belonging to the ith class. For simplicity, we simply denote all the training samples in F by $\Phi(x_1), \Phi(x_2), \cdots, \Phi(x_N)$. Then, according to the fuzzy Fisherfaces method [2] , the expressions of the ith class mean, the mean of all training samples, the between-class matrix, and the within-class scatter matrix in the feature space F can be respectively formulated by: $u_i^{\Phi} = \sum_{j=1}^{N} \mu_{ij}\Phi(x_j) / \sum_{j=1}^{N} \mu_{ij}$, $u^{\Phi} = \sum_{i=1}^{N} \Phi(x_i)/N$, $S_B^{\Phi} = \sum_{i=1}^{c} N_i(\mu_i^{\Phi} - \mu^{\Phi})(\mu_i^{\Phi} - \mu^{\Phi})^{\mathsf{T}}$, $S_W^{\Phi} = \sum_{i=1}^{c} \sum_{j=1}^{N_i} (\Phi(x_i^j) - \mu_i^{\Phi})(\Phi(x_i^j) - \mu_i^{\Phi})^{\mathsf{T}}$. Let $S_T^{\Phi} = S_B^{\Phi} + S_W^{\Phi}$. Define the Fisher discriminant criterion by $J(\omega) = S_B^{\Phi}/S_T^{\Phi}$. Let

$$\Phi(X) = [\Phi(x_1^1), \Phi(x_1^2), \cdots, \Phi(x_1^{N_1}), \cdots, \Phi(x_c^1), \cdots, \Phi(x_c^{N_c})] \tag{1}$$

Then we have $S_B^{\Phi} = \Phi(X)(W_\mu - M)(W_\mu - M)(\Phi(X))^{\mathsf{T}}$, $S_B^{\Phi} = \Phi(X)(I - W_\mu)(I - W_\mu)(\Phi(X))^{\mathsf{T}}$, where I is the $N \times N$ identity matrix, $M = (m_{ij})_{i=1,\cdots,N; j=1,\cdots,N}$ is an $N \times N$ matrix with all terms equal to $1/N$, and $W_\mu = (W_l)_{l=1,\cdots,c}$ is an $N \times N$ matrix, where W_l is an $N_l \times N_l$ matrix with the jth element of each column equals to $\mu_{ij} / \sum_{j=1}^{N} \mu_{ij}$. Moreover, it is easy to check that the optimal discriminant vectors of Fuzzy KDA, denoted by ω, can be written as the following forms:

$$\omega = \sum_{p=1}^{c} \sum_{q=1}^{N_p} \alpha_{pq}(\Phi(x_p^q) - \mu^{\Phi}) = \Phi(X)(I - M)\alpha \tag{2}$$

Therefore, we obtain that the Fisher discriminant function $J(\omega)$ can be rewritten as $J(\alpha) = \alpha^{\mathsf{T}} B\alpha / \alpha^{\mathsf{T}} T\alpha$, where $B = (I - M)K(W_\mu - M)(W_\mu - M)K(I - M)$, $T = (I - M)K[(W_\mu - M)(W_\mu - M) + (I - W_\mu)(I - W_\mu)]K(I - M)$. In this case, solving the optimal discriminant vectors of Fuzzy KDA is equivalent to solving the discriminant coefficient vectors that maximize the Fisher discriminant criterion $J(\alpha)$, which can be implemented using the traditional KDA algorithm. Suppose that $\omega_i = \Phi(X)(I - M)\alpha_i$ $(i = 1, \cdots, c - 1)$ are the $c - 1$ optimal discriminant vectors of Fuzzy KDA. Let $W_{FKDA} = [\omega_1, \cdots, \omega_{c-1}] = \Phi(X)[\alpha_1, \cdots, \alpha_{c-1}]$ be the transformation matrix. Then, the projection of a test sample $\Phi(x_{test})$ onto W_{FKDA} can be computed by $y_{test} = W_{FKDA}^{\mathsf{T}}\Phi(x_{test})\kappa$, where $\kappa = [k(x_1, x_{test}, \cdots, k(x_N, x_{test})]$. Let $y_i^j(j = 1, \cdots, N_i; i = 1, \cdots, c)$ are the projections of $\Phi(x_i^j)(j = 1, \cdots, N_i; i = 1, \cdots, c)$ onto W_{FKDA}, respectively. Then the classification of $\Phi(x_{test})$ can be conducted using the following formula

$$i^* = \arg\min_i \|y_{test} - y_i^j\| \tag{3}$$

3 Facial Expression Recognition Using Fuzzy KDA

In this section, we will apply the Fuzzy KDA method to the facial expression recognition task. Similar with the method in [3], we manually locate 34 landmark points from each facial image by referring to the landmark points used in [6], and then use the Gabor wavelet representation of each facial image at the landmark points to represent the facial features of each image, where all of the wavelet convolution values (magnitudes) at these landmark points are combined into a 1020 dimensional vector. The Gabor kernel is defined as follows:

$$\Psi_{u,v} = \frac{\|k_{u,v}\|^2}{\sigma^2} \exp(-\frac{\|k_{u,v}\|^2 \|z\|^2}{2\sigma^2}) \left[\exp(ik_{u,v} \cdot z) - \exp(-\frac{\sigma^2}{2})\right] \quad (4)$$

where u and v represent the orientation and scale of the Gabor kernels, and $k_{u,v}$ is defined as: $k_{u,v} = k_v \exp(i\phi_u)$, where $k_v = \pi/2^v$ ($v \in \{1, 2, \cdots, 5\}$) and $\phi_u = \pi u/6$ ($u \in \{0, 1, \cdots, 5\}$).

To use the Fuzzy KDA method for FER, we use the Fuzzy K-nearest neighbor (Fuzzy K-NN) method proposed by Keller et al.[7] to compute the class membership of each training sample. Consider that we are conducting the FER task in the feature space F, thus the distance metric in Fuzzy K-NN should be modified as: $d(\Phi(x_p), \Phi(x_q)) = \|\Phi(x_p) - \Phi(x_q)\|^2 = k(x_p, x_p) - 2k(x_p, x_q) + k(x_q, x_q)$.

4 Experiments and Conclusions

We use the Japanese Female Facial Expression (JAFFE) database [5] and the Ekman's 'Pictures of Facial Affect' database [4], respectively, to test the performance of the proposed method. The JAFFE facial expression database contains 213 facial images covering all the 7 facial expressions posed by 10 Japanese female.The Ekman's facial expression database contains 110 images consisting of 6 male and 8 female subjects. After removing the neutral images, the JAFFE database contains 183 images and the Ekman's database contains 96 iamges. We adopt the "leave-one-class-out" cross-validation strategy to conduct this experiment, where the nearest neighbor classifier is used in the experiment. Moreover, we use the monomial kernel $k(x, y) = (x^\top y)^d$ and the gaussian kernel $k(x, y) = \exp\{\frac{\|x-y\|^2}{\sigma}\}$ as the kernel function. Table 1 and 2 show the experimental results on the JAFFE database and the Ekman's database, respectively.

Table 1. Comparison of Average Recognition Rate on JAFFE database

Methods	Recognition Rate (%)
Fuzzy KDA(Gaussian kernel with $\sigma = 2e6$)	**78.14**
KDA(Gaussian kernel with $\sigma = 2e6$)	77.05
Fuzzy KDA (Monomial kernel with $d = 3$)	76.50
KDA (Monomial kernel with $d = 3$)	68.85
KCCA[3]	77.05
LDA [5]	75.00

Table 2. Comparison of Average Recognition Rate on Ekman database

Methods	Recognition Rate (%)
Fuzzy KDA(Gaussian kernel with $\sigma = 7e6$)	**82.29**
KDA(Gaussian kernel with $\sigma = 7e6$)	78.13
Fuzzy KDA (Monomial kernel with $d = 2$)	79.17
KDA (Monomial kernel with $d = 2$)	76.04
KCCA[3]	77.08
LDA [5]	82.00

From table 1 and 2, we can see that the Fuzzy KDA method achieves the recognition rate as high as 78.14% and 82.29%, respectively, which is much better than those of the KDA method.

Acknowledgments

This work was partly supported by the NSF of China under grant 60503023, and partly supported by the NSF of Jiangsu province under the grant BK2005407, partly supported by the key laboratory of image processing and image communication of Jiangsu province under the grant ZK205013, and partly supported by Program for New Century Excellent Talents in University (NCET).

References

1. Mika S., Rätsch G., Weston J., Schölkopf B., and Müller K.-R.: Fisher discriminant analysis with kernels. Neural Networks for Signal Processing IX, Y.-H. Hu, J. Larsen, E. Wilson, and S. Douglas, Eds. Piscataway, IEEE, 41-48, 1999.
2. Kwak K.-C., and Pedrycz W.: Face Recognition Using a Fuzzy Fisherface Classifier. Pattern Recognition, Vol.38, pp. 1717-1732, 2005.
3. Zheng W., Zhou X., Zou C. and Zhao L.: Facial Expression Recognition Using Kernel Canonical Correlation Analysis (KCCA). IEEE Transactions on Neural Networks, Vol.17, No.1, pp.233-238, 2006.
4. Ekman P., and Friesen W. V.: Pictures of Facial Affect. Human Interaction Laboratory, Univ. of Calification Medical Center, San Francisco, 1976.
5. Lyons M., Budynek J., and Akamatsu S.: Automatic Classification of Single Facial Images. IEEE Trans. On Pattern Analysis and Machine Intelligence, Vol. 21, No. 12, pp.1357-1362, 1999.
6. Zhang Z., Lyons M., Schuster M., and Akamatsu S.: Comparison between geometry based and Gabor wavelets based facial expression recognition using multilayer perceptron. In Proceedings of Third IEEE International Conference on Automatic Face and Ges-ture Recognition, pp.454-459, 1998.
7. Keller J. M., Gray M.R., and Givens J.A.: A Fuzzy K-Nearest Neighbor Algorithm. IEEE Transactions on Systems, Man, and Cybernetics, Vol.15, No.4, pp.580-585, 1985.

A Classifier Ensemble Method for Fuzzy Classifiers

Ai-min Yang [1], Yong-mei Zhou [1,2], and Min Tang[1]

[1] Department of Computer Science, Hunan University of Technology, ZhuZhou, 412008, China
[2] College of Information Science & Engineering,Central South University, ChangSha, 410083, China
Amyang18@163.com

Abstract. In this paper, a classifier ensemble method based on fuzzy integral for fuzzy classifiers is proposed. The object of this method is to reduce subjective factor in building a fuzzy classifier, and to improve the classification recognition rate and stability for classification system. For this object, a method of determining fuzzy integral density based on membership matrix is proposed, and the classifier ensemble algorithm based on fuzzy integral is introduced. The method of selecting classifier sets is also presented. The proposed method is evaluated by the comparison of experiments with standard data sets and the existed classifier ensemble methods.

1 Introduction

Fuzzy Classification is an important application of Fuzzy Set. Fuzzy classification rule is widely considered a well-suited representation of classification knowledge, and is readable and interpretable. Fuzzy classification has been widely applied in many fields, such as image processing, words recognition, voice recognition etc.

The auto-generation of fuzzy partition and fuzzy classification rules is a key problem for the fuzzy classification research, along with expressions and adjustments of classification rules and the improvement of the classification recognition rate. Although a single fuzzy classifier has implemented the auto-generation of fuzzy Partition and fuzzy classification rules with good classification performance to some extent, it needs to select the type of membership function and parameters and to take some time to learn these parameters for a good classifier. This paper proposed a classifier ensemble method with fuzzy integral density[11] which can generate fuzzy classification rules automatically and can decrease subjective factors during training classifier.And the method of measuring generalization difference(GD) for classifier sets is also introduced. The proposed methods are evaluated by the experiments.

2 Related Works

(1) Fuzzy Classifier Rules
The typical fuzzy classification IF-THEN rules[1-2] have the form as Eq.(1).

$$\begin{aligned} &R_k : IF \quad x_1 \quad is \quad A_{1,i(1,k)} \quad AND \quad \ldots A_{j,i(j,k)} \quad \ldots \quad AND \quad x_n \quad is \quad A_{n,i(n,k)} \\ &THEN \quad g_{k,1} = z_{k,1} \quad AND \ldots AND \quad g_{k,M} = z_{k,M} \end{aligned} \quad . \tag{1}$$

L. Wang et al. (Eds.): FSKD 2006, LNAI 4223, pp. 784 – 793, 2006.

In Eq.(1), $x=[x_1,\ldots,x_n]^T \in R^n$ is input pattern, x_i is feature property, $\Omega=\{C_1,\ldots, C_m,\ldots, C_M\}$ is the set of class label. FSN_j is the number of linguistic label of the j-th feature, $A_{j,i}$ is the i-th fuzzy set in x_j feature axis (i =1,..., FSN_j, j =1,...,n), $g_{k,m}$ is the discriminant function of C_m related with rule R_k, suffix i(j,k) is the function of fuzzy set serial number describing x_j in rule R_k, $z_{k,m} \in R$ can be seen as the support degree of C_m(m=1,2,...,M) for R_k rule. $z_{k,m} \in [0,1]$, $[z_{k,1},\ldots,z_{k,M}]^T$ is soft classification output.

(2) Fuzzy Integral

Definition 1. Suppose g_λ is Fuzzy measure[11], and has property as follows.

If $A, B \subset X \quad and \quad A \cap B = \Phi$, then Eq.(2)

$$g_\lambda(A \cup B) = g_\lambda(A) + g_\lambda(B) + \lambda g_\lambda(A) g_\lambda(B) \qquad \lambda > -1. \tag{2}$$

So, g_λ is called as λ fuzzy measure. g_λ has the following properties:

Suppose $X=\{x_1,\ldots,x_n\}$ is a finite set, and $g^i=g_\lambda(\{x_i\})$, then $\{g^i,\ldots, g^n\}$ is called as g_λ fuzzy density function. So, for arbitrary subset of X, $A=\{x_{i_1},\cdots,x_{i_m}\} \subseteq X$, the measure value of g_λ can be got from fuzzy density function, as Eq.(3)

$$\begin{aligned} g_\lambda(A) &= \sum_{j=1}^{m} g^{i_j} + \lambda \sum_{j=1}^{m-1} \sum_{k=j+1}^{m} g^{i_j} g^{i_k} + \cdots + \lambda^{m-1} g^{i_m} \cdots g^{i} \\ &= \left[\prod_{x_i \in A} (1+\lambda g^i) - 1 \right] \Big/ \lambda, \qquad \lambda \neq 0 \end{aligned} \tag{3}$$

λ is calculated according to Eq.(4).

$$X = \bigcup_{i=1}^{n} \{x_i\}, \quad g(X) = 1 \quad \text{i.e., } \lambda + 1 = \prod_{i=1}^{n} (1 + \lambda g^i). \tag{4}$$

For a set $\{g^i\}(0<g^i< 1)$, the above equation has a solution satisfying the following form:$\lambda \in (-1,+\infty)$, and $\lambda \neq 0$.

So,if fuzzy density g^i(i=1,2,...,n) is known, g_λ can be constructed. For information integral, the description of fuzzy density g^i can be as the important degree of final decision from information source x_i. The fuzzy measure of arbitrary set A expresses the important degree of final decision for A.

Definition 2. Assume (X, Ψ) is a measure space, h: X→[0,1] is a measure function, then the fuzzy integral of h about fuzzy measure g_λ in $A(A \subseteq X)$ is Eq.(5).

$$\int_A h(x) \circ g_\lambda(\cdot) = \sup_{\alpha \in [0,1]} [\min(\alpha, g_\lambda(A \cap F_\alpha))] \tag{5}$$

Where, $F_\alpha = \{x \in A \mid h(x) \geq \alpha\}$.

If X is a finite set, Fuzzy Integral[3] can be calculated. Suppose $X=\{x_1,x_2,\ldots,x_n\}$ is a finite set, h:X→[0,1] is a function, and $h(x_1) \geq h(x_2) \geq \ldots \geq h(x_n)$, then the value of fuzzy integral h(x) for fuzzy measure g_λ can be computed using Eq.(6).

$$\int_X \mathrm{h(x)} \circ \mathrm{g}_\lambda = \max_{i=1}^{n} [\min(h(x_i), g_\lambda(A_i))]. \tag{6}$$

Where, $A_i=\{x_1,\ldots,x_i\}$, and $g_\lambda(A_i)$ can be determined by Eq.(7).

$$
\begin{aligned}
&g_\lambda(A_1)=g_\lambda(\{x_1\})=g^1 \\
&g_\lambda(A_i)=g^i+g_\lambda(A_{i-1})+\lambda g^i g_\lambda(A_{i-1}),1<i\le n
\end{aligned}
\quad (7)
$$

(3)Three Types of Fuzzy Classification Model

The propposed ensemble method uses three types of fuzzy classification models[12] which were proposed in our early research.

Model I. Fuzzy Classification Model Based on Fuzzy Kernel Hypersphere Perception (FCMBFKHP). For this model, firstly the input patterns in the initial input space are mapped to high dimensional feature space by selecting a suitable kernel function. In the feature space, the hypersphere which covers all training patterns of the class is founded for every class by the algorithm of FKHP. A hypersphere is regarded as a fuzzy partition and a IF-THEN rule is created for a fuzzy partition. A hyper-cone membership function is defined with regarding the center and radius as parameters. Fuzzy classification rule is as Eq.(8).

$$R_m:\ IF\ \Phi(x)\ is\ around\ C_m\ \ THEN\ x\in C_m\ with\ CF=\alpha_m\ . \quad (8)$$

Where, R_m denotes labels of rule, created by the m-th class, CF denotes the degree of pattern $\Phi(x)$ belonged to this rule, $\alpha_m\in[0,1]$,Φ is a kernel function.

Model II. Fuzzy Classification Model Based on Evolving Kernel Clustering (FCMBEKC). For this model, firstly the patterns in the initial input space are mapped to high dimensional feature space by selecting a suitable kernel function. In the feature space, several hyperspheres are got by clustering for each class training patterns by the algorithm of EKC(Evolving Kernel Clustering). A hypersphere is regarded as a cluster which corresponds to a fuzzy partition that creates a IF-THEN rule. A hyper-ellipse membership function is defined with the center of each cluster as parameters. Fuzzy Classification rule is as Eq.(9).

$$R_{mj}:\ IF\ \Phi(x)\ is\ around\ C_{mj}\ \ THEN\ x\in C_m\ with\ CF=\alpha_{mj}\ . \quad (9)$$

Where, R_{mj} denotes labels of rule, created by the j-th cluster of the m-th class, CF denotes the degree of pattern $\Phi(x)$ belonged to this rule, Φ is a kernel function.

Model III. Fuzzy Classification Model Based on Support Vector Machine (FCMBSVM). In the initial stage of the model construction, the center around of each training pattern is regarded as a fuzzy partition.Each training pattern corresponds to a fuzzy partition which creates a IF-THEN rule. Kernel function is constructed by selecting suitable membership function. The parameters of SVM and rules are gained using SVM learning method. This model can automatically generate fuzzy partition and fuzzy classification rule. Classification rule is as Eq.(10).

$$
\begin{aligned}
&R_k: IF\ \ x_1\ \ is\ \ A_{1,k}\ \ AND\ \ \ldots A_{j,k}\ \ \ldots\ \ AND\ \ x_n\ \ is\ \ A_{n,k} \\
&THEN\ \ Class\ \ is\ \ C_m\ \ with\ \ \alpha_{k,m}
\end{aligned}
\quad (10)
$$

Where, R_k is the k-th rule(k=1,2,…,l) , $A_{j,k}$ is fuzzy subset from the projection on the j-th axis (feature) using the i-th training pattern as center, is also the subset of the k-th rule in the j-th axis. $C_m \in \{-1,1\}$(m=1,2) represents the class, $\alpha_{k,m}$(k=1,2,…,l) can be seen as the support degree of class C_m for rule R_k.

3 The Method of Classifier Ensemble Based on Fuzzy Integral

During classifier ensemble with Fuzzy Integral, there are two factors for a pattern evaluation. One is the individual classifier evaluation. In this paper,this evaluation is membership degree, i.e. measure function h in Fuzzy Integral theory. The other is dependability degree of each classifier, i.e. fuzzy integral density g. These factors can be expressed by classification precision for each class.

3.1 The Generation of Individual Classifier

The most important technique in individual classifier generation[10] is Boosting[4] and Bagging (Bootstrap Aggregating) algorithm[5] . In our research, the following aspects are considered.

① According to different classification models,in 2 section, three classification models are introduced,FCMBFKHP, FCMBEKC and FCMBSVM. So, individual classifiers can be generated from the three models.

② By selecting different types of kernel functions and parameters, in the proposed models, the initial model spaces are mapped into high dimensional feature spaces with kernel function. So, different individual classifier can be got by selecting different kernel functions and parameters, such as radial basis kernel function, polynomial kernel function. And different individuals can be created from the different parameters, such as δ parameter of radial basis kernel function, penalty parameter C of support vector machine etc.

3.2 Selection of Classifier Sets

After selection of individual classifier, an important question in classifier ensemble system is how to construct classifier sets in order to decrease the relativity of classifiers. A common method is firstly constructing N individual classifiers, then a classifier set is built with K(K<N) classifiers selected, and then several classifier sets and their relativity are got through repetition by defining the method of computing total relativity of a group of individual classifiers. At last, a criterion of selecting classifier set is defined to select classifier sets for classifier ensemble system.

Turner and Gosh [6] point out that improvement of multi-classifier ensemble performance depends on the speciality that wrong decision patterns for each classifier. That is, the less the patterns in which each classifier makes wrong decision at the same time, the higher the recognition performance.So according to this idea and the different influences on ensemble decision by different number of classifiers making wrong decision, a measuring generalization difference method (GD)for individual classifiers is introduced as the criterion of selecting classifier set. This method is similarity to the idea of generalization difference among neural network ensemble individuals proposed by Partridge [7-8].

First, define the wrong probability of an arbitrary individual classifier as Eq.(11).

Definition 3. $p(\text{arbitrary misclassification}) =$

$$\sum_{k=1}^{K} p(\text{selected misclassifications} | \text{k misclassifications}) \times p(\text{k misclassifications})$$

$$= \sum_{k=1}^{K} \frac{k}{K} \times p_k \tag{11}$$

(p_k is the probability that k classifiers are misclassifications at the same time)

The following defines the wrong probability of two arbitrary classifier on randomly selecting test patterns as Eq.(12).

Definition 4. $p(\text{two arbitrary misclassifications}) =$

$$\sum_{k=2}^{K} p(\text{selected misclassifications} | \text{k misclassifications}) \times \text{p}(\text{k misclassifications})$$

$$= \sum_{k=2}^{K} \frac{k}{K} \frac{k-1}{K-1} p_k \tag{12}$$

(p_k is the probability that k classifiers are misclassifications at the same time)

A misclassification table can be built for several classifiers. p (one misclassification) is the wrong probability of an arbitrary classifier selected from K classifiers on test patterns, while p (two misclassifications) is the wrong probability of two arbitrary classifiers on test set. Like this generalization, the generalization difference of different individual classifier sets can be determined. The following defines a calculation method for generalization difference(GD) of classification set.

Definition 5. GD of classifier set in some test set is defined as Eq.(13).

$$GD = \frac{p(\text{one misclassification}) - p(\text{two misclassifications at the same time})}{p(\text{one misclassification})} \tag{13}$$

According to Definition 5, the classifier set with the max GD is selected to ensemble.

3.3 Determination of Fuzzy Integral Density Based on Membership Degree Matrix

The following is the way to determine fuzzy integral density, firstly, membership degree matrix (MDM) is got from the given test set, then, using membership degree matrix, confusion matrix (CM) is got, which can be used to calculate fuzzy integral density g.

Supposes $C_m (m \in \Lambda = \{1,...,M\})$ denotes M different classes , $e_k (k=1,...,K)$ denotes K different classifiers respectively, then the output of classifier e_k for pattern x can be expressed by $\mu_k(x) = (\mu_{k1},..., \mu_{km},..., \mu_{kM})$, where $0 \le \mu_{km} \le 1$ means membership degree of classifier e_k for x, and then select the label of the maximal μ_{km} as the class label of pattern x, thus pattern x belongs to the corresponding class.

Definition 6. $MDM(x) = [\mu_1(\mathrm{x})^T, \mu_2(\mathrm{x})^T, ..., \mu_K(\mathrm{x})^T]^T = \begin{bmatrix} \mu_{11} & \cdots & \mu_{1\mathrm{M}} \\ \vdots & \mu_{km} & \vdots \\ \mu_{K1} & \cdots & \mu_{K\mathrm{M}} \end{bmatrix}$ is called membership degree matrix of pattern x for multi classifiers $\{e_k, k=1,\ldots,K\}$.

Definition 7. For pattern set S=$\{x_i,\ i=1,\ldots,L\}$, MDM(S)=[MDM (x_1),…, MDM (x_i),…, MDM (x_L)] is membership degree matrix of pattern set S with multi classifiers $e_k(k=1,2,\ldots,K)$.

Membership degree matrix includes all classification results of each pattern in pattern or pattern set, which can be used to statistic and analyze classification precision, relativity of classes and so on. The following will analyze how to get confusion matrix from membership degree function.

Calculation of Confusion Matrix(CM),CM for e_k is Eq.(14).

$$\mathrm{CM}_k = \begin{bmatrix} r_{11}^{(k)} & r_{12}^{(k)} \cdots & & r_{1\mathrm{M}}^{(k)} \\ r_{21}^{(k)} & \ddots & & \vdots \\ & & r_{ml}^{(k)} & \vdots \\ r_{\mathrm{M}1}^{(k)} & \cdots & \cdots & r_{\mathrm{MM}}^{(k)} \end{bmatrix}. \quad (14)$$

Where, $r_{ml}^{(k)}$ is the probability that $\mathrm{e_k}$ judges the pattern belonging to $\mathrm{C_m}$ as C_l. CM can be calculated by MDM, and MDM of pattern set S of multi-classifier $\mathrm{e_k}$(k =1,2,..., K) is MDM(S)=[MDM $(\mathrm{x_1})$,…, MDM $(\mathrm{x_i})$,…, MDM $(\mathrm{x_L})$].

The k-th row of MDM(S) includes the classification output of all patterns of pattern set S with classifier e_k, CM of pattern set S with classifier e_k can be got with statisticing these cases.

The algorithm of calculating CM of classifier e_k is as follows.

Algorithm 1: Calculating CM of classifier e_k
Input: Pattern set S and corresponding MDM(S)
Output: CM of classifier e_k
Step1: Initialize R_k=M×M matrix, and let R_k=0.
Step2: Select x from pattern set S, determine its real class label (the class of training patterns or test patterns is given), assume it is m, the classification output μ=[μ_{k1},…, μ_{km} ,…, μ_{kM}]of x is obtained from MDM(S) with classifier e_k, and add μ to row m-th of matrix R_k, and remove pattern x from S.
Step3: Judge S whether is null or not, if not null, the algorithm goes to Step1, else,go to **Step4.**
Step4: Normalize each row of matrix R_k with Eq.(15).

$$r_{ml} = R_k(m,l) \Big/ \sum_{q=1}^{M} R_k(m,q) \quad \text{where, } l=1,2,\ldots,\mathrm{M}\,. \quad (15)$$

Step5: Output confusion matrix R_k of classifier e_k.

Confusion matrix $R_k(k=1,2,\ldots,K)$ corresponding to other classifiers can be got, according to the above algorithm.Belief degree of classifier e_k as fuzzy integral density g^k can be got by Eq.(16) with confusion matrix.

$$g^k = \sum_{m=1}^{M} r_{mm}^{(k)} \Big/ \sum_{m=1}^{M} \sum_{l=1}^{M} r_{ml}^{(k)} . \tag{16}$$

3.4 Classifier Ensemble Method

For a given multi-classifier ensemble question, individual classifier is e_k, $k=1,2,...,K$, K is the number of classifiers, $\Omega=\{C_1,C_2,\ldots,C_M\}$ is class label set, M is number of classes, $\mu^k{}_m$, $m=1,2,\ldots,M$ represents output of each individual classifier. For fuzzy integral, $\mu^k{}_m$ is the evaluation of classifier e_k for input pattern belonging to the *m-th* class, that is h_k. The performance of the current classifier shows the evaluation reliability, i.e. fuzzy integral density $g^{k.}$ The method of calculating h_k and g^k has been introduced before.

Suppose τ: $\{\mu_1, \mu_2,\ldots, \mu_K\}$ is a finite set, h: $\tau\to[0,1]$ is a function, and $h(\mu_1)\geq h(\mu_2)\ldots\geq h(\mu_K)$, fuzzy integral is Eq.(17) according to Eq.(6).

$$FI = \max_{k=1}^{K}\left[\min\left(h(\mu_k), g_\lambda(A_k)\right)\right]. \tag{17}$$

Where, $A_k=\{\mu_1, \mu_2,\ldots, {}_{ik}\}$.

g_λ can be calculated with Eq.(7).

$$\begin{aligned} &g_\lambda(A_1) = g_\lambda(\{\mu_1\}) = g^1 \\ &g_\lambda(A_k) = g^k + g_\lambda(A_{k-1}) + \lambda g^k g_\lambda(A_{k-1}), \quad 1 < k \leq K \end{aligned} . \tag{18}$$

Where, λ is calculated by Eq.(4).

$$\lambda + 1 = \prod_{k=1}^{K}(1+\lambda g^k) \qquad \text{Where, } \lambda\in(-1,+\infty), \text{and } \lambda\neq 0.$$

Fuzzy integral of a input pattern x for a certain class can be got with Eq.(17), and fuzzy integral of this pattern for other classes can be calculated by the same way. If $FI_m(x)(m=1,2,\ldots,M)$ is fuzzy integral of input pattern x for each class, the decision model of multi-classifier ensemble system is as follows.

$$\text{Class}(x) = \arg\max_{m=1}^{M} FI_m(x) . \tag{19}$$

The classifier ensemble algorithm is as follows.

Algorithm 2: Classifier Ensemble Algorithm
Input: pattern x;**Output:** class of x

Step1: For input pattern x, each individual classifier outputs the membership degree of x corresponding to each class.

Step2: For each class C_m, each classifier e_k calculates $h_m(\mu_k)$ and $g_\lambda(\mu_k)$, and fuzzy integral FI_m corresponding to C_m.

Step3: Judge the class for the pattern x with Eq.(19).

4 Analysis of Experiment Results

Wine data set and waveform data set are adopted for the experiment analysis, which come from UCI machine learning database[9]. wine data set has 13 features, the number of classes is 3, the number of training patterns is 118, and the number of test pattern is 60. Waveform data set has 21 features, number of classes m=3, training pattern 300, and test pattern 4700.

Individual classifier can be generated with fuzzy classification model introduced in section 2. Each model creates 10 classifiers with the strategy of selecting different kernel functions and parameters, and different membership functions and parameters.

So, the total number of individual classifiers is 30, which means there are 30 individual classifiers for selection. Each classifier set includes 6 individual classifiers to ensemble for constructing pattern classification system, which is selected by its generalization difference (GD). Wine data set is to train and test 30 individual classifiers, and randomly select 6 classifiers to compose an ensemble classifier. The recognition rate is got by the method of classifier ensemble based on fuzzy integral with test pattern testing. Calculating GD of individual classifier in ensemble classifier, the relationship of the recognition rate and generalization difference of classifier sets is got as Fig.1(some points are got rid off in convenient to observe).

In Fig.1, the whole trend is that system recognition rate improves with GD increasing. So, the method based on GD for selection classifier is feasible.

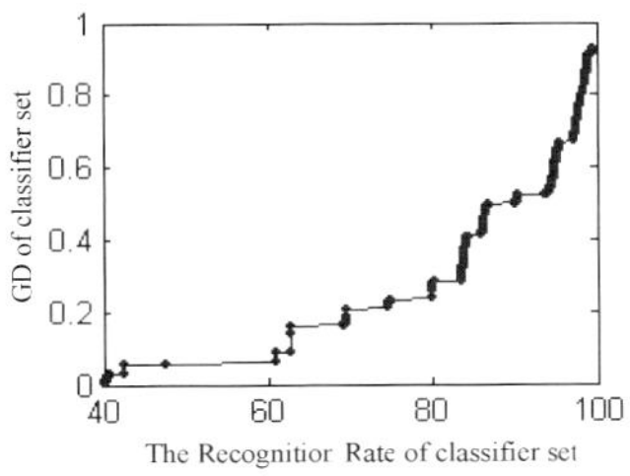

Fig. 1. Relationship of the recognition rate and generalization difference of classifier sets

Table 1. Comparison of recognition rate between the system of classifier ensemble and individual classifier

Classifier	wine (%)	waveform (%)
Classifier 1	91.67	74.65
Classifier 2	93.33	72.82
Classifier 3	90.00	80.0
Classifier 4	808.33	82.49
Classifier 5	95.00	79.1
Classifier 6	93.33	83.75
Classifier ensemble system	96.67	85.5

Table 2. Comparison of recognition rate between the proposed method and the others classifier ensemble methods

Ensemble method		Wine (%)	Waveform (%)
Max	means	93.85	84.83
	Bayesian	95.21	86.02
	vote	94.58	85.73
	proposed method	96.66	86.31
Average	means	93.25	83.39
	Bayesian	94.92	84.89
	vote	94.19	84.10
	proposed method	95.50	85.5

After selecting individual classifier, the performance of classifier system is analyzed though experiment which is constructed by multi-classifier ensemble. The comparison is done in two aspects, one is the recognition rate of classifier ensemble system and that average of individual classifier, the other is recognition rate between the proposed method and others classifier ensemble methods.

The experiment data set is also wine data set and waveform data set. Table 1 is comparison of recognition rate between the system of classifier ensemble and individual classifier. Table 2 is comparison of recognition rate between the proposed method and the others classifier ensemble methods.

From the experiment, the recognition rate of the classification system based on fuzzy integral classifier ensemble is obviously higher than the average recognition rate of individual classifier. So, how to determine parameters in individual classifier is solved by multi-classifier ensemble method.

5 Conclusions

This paper proposes the method of multi-classifier ensemble based on fuzzy integral and introduces the method of getting fuzzy integral density with membership degree matrix, the method of selecting individual classifier with GD of classifier set, and the algorithm of classifier ensemble based on fuzzy integral. We validate the efficiency of these methods and the performance of the classifier ensemble system with typical data set. The experiment shows it obviously improves the performance of the classification system based on classifier ensemble. This paper suggests a method how to select the parameters and optimize the performance for a single fuzzy classifier.

Acknowledgements

This paper is supported by the Hunan Natural Science Fund of China (05JJ40101). And supported by Scientific Research Fund of Hunan Provincial Education Department(03C597).

References

1. Ludmila I, Kuncheva.:How good are fuzzy If-Then classifiers?.IEEE Transactions on Systems, Man, and Cybernetics, ,30(4) (2000) 501-509
2. O. Cord_on, M.J. del Jesus, F. Herrera.: A proposal on reasoning methods in fuzzy rule-based classification systems[J].Int. J. of Approximate Reasoning, 20(1) (1999)21-45
3. Wierzechon.S.T.:On fuzzy measure and fuzzy integral. Fuzzy information and decision processes, New York: North-Holland,(1982)78-86.
4. Schapire R E.:The strength of weak learn ability.Machine Learning, 5(2) (1990) 197 -227
5. Breiman L:Bagging predictors.Machine Learning, 24(2) (1996)123-140
6. K. Turner,J.Gosh.:Error correlation and error reduction in ensemble classifiers.Texas: Dept. of ECE, University of Texas,(1996)
7. Partridge D.:Network generalization differences quantified. Neural Networks, (9) (1996) 263-271.
8. D.Partridge,W.B.Yates. Engineering Multiversion Neural-Net Systems.Neural Computation, (8) (1998)869-893
9. Blake,C.L.Merz, C.J. UCI Repository of machine learning databases http://www.ics.uci.edu/~mlearn/MLRepository.html,(1998)
10. Zhou Zhi-Hua,Cheng Shi-Fu. Neural Network Ensemble.Chinese Journal of Computers, 25(1) (2005)1-8
11. ZaoXie-Dong.Fuzzy Information Managing and Application.BeiJing:Science Publishing Company,(2003)174-176
12. Yang Ai-Min.The Model Research on Fuzzy Classification.Doctor's Degree Paper of Fudan University,(2005)

A Hybrid Soft Computing Approach to Link Travel Speed Estimation*

Seung-Heon Lee, M. Viswanathan, and Young-Kyu Yang

College of Software, Kyungwon University, Bokjeong-Dong,
Sujung-Gu, Seongnam-Si, Gyeonggi-do, South Korea 461-701
shleejj@gmail.com, {murli, ykyang}@kyungwon.ac.kr

Abstract. The field of dynamic vehicle routing and scheduling is growing at a strong pace nowadays due to the many potential applications in urban traffic management. In recent times there have been many attempts to estimate the vehicle travel times over congested links. As opposed to the previous decade where traffic information was collected mainly by fixed devices with high maintenance costs, the advent of GPS has resulted in data being progressively collected using probe cars equipped with GPS-based communication modules. Typically traditional methods used for analyzing the data collected using fixed devices need to be extended. The aim of this research is to propose a hybrid method for estimating the optimal link speed using the data acquired from probe cars using combination of the fuzzy c-means (FCM) algorithm with multiple regression analysis. The paper describes how the probe data are analyzed and automatically classified into three groups of speed patterns and the link speed is predicted from these clusters using multiple regression. In performance tests, the proposed method is robust and is able to provide accurate travel time estimates.

1 Introduction

Public agencies everywhere spend public funds on transport infrastructure in the hope of providing opportunities for faster and more efficient movement of traffic. The vital need of accurate real time traffic data for supporting ITS applications has resulted in decades of research with a strong focus to improve the efficiency of data acquisition techniques and data analysis techniques. Among this data, travel time has been recognized to be one of the most valuable ones, particularly for ATIS (Advanced Traveller Information System) or ATMS (Advanced Traffic Management System). There are several systems commercially available that are capable of estimating real-time travel times. These can be broadly classified into spot speed measurement systems, spatial travel time systems, and probe vehicle technologies. Spot speed measurement systems, specifically inductance loop detectors, have been the main source of real-time traffic information for the past two decades. Other technologies for measuring spot

* This research was supported by the MIC (Ministry of Information and Communication), Korea, under the ITRC (Information Technology Research Center) support program supervised by the IITA (Institute of Information Technology Assessment), and the BK21 project.

L. Wang et al. (Eds.): FSKD 2006, LNAI 4223, pp. 794 – 802, 2006.

speeds have also evolved, such as infrared and radar technologies. Regardless of the technology, the spot measurement approaches only measure traffic stream speeds over a short roadway segment at fixed locations along a roadway. These spot speed measurements are used to compute spatial travel times over an entire trip using space-mean-speed estimates [10]. With the wide spread availability of GPS receivers in car navigation systems traffic information can now be easily collected by probe cars equipped with GPS receivers.

There are some companies trying to service the link speed (i.e. vehicle speed between a node to node) estimated solely by GPS data due to its low cost and data availability in real time fashion. This requires developing novel analytical algorithms which are different from the algorithms used for the conventional fixed detectors. The conventional methods use the statistical and the simple neural net algorithms to calculate the link speed. The data set collected by the fixed detection devices are speed, road occupancy rate, and the amount of traffic flow and other data for the fixed locations where the devices are installed. But, data acquired from GPS probe cars are only vehicle speeds for specified time and location. And thus traffic information using GPS requires a new analysis method.

Many researchers have endeavored to develop reliable travel time forecasting models using various methods including historical profile approaches, time series models, neural networks, nonparametric regression models, traffic simulation models, and dynamic traffic assignment (DTA) models. The aim of this research is to propose and test a novel algorithm for calculating the link speed for the collected information from the probe cars equipped with GPS receivers. The algorithm in this paper has been specifically targeted towards data collected from GPS devices. Since the use of GPS in urban traffic systems is a relatively recent phenomenon the primary advantage itself is the development of a system which is applicable in the GPS framework. In support of this objective, we automatically cluster the GPS data into three levels of speed layers using the Fuzzy C-means (FCM) algorithm [9] and apply these layers of speed patterns to predict the link speed. These are defined as slow, middle, and high. But in the clusters produced by the FCM, there exists some gaps between the three clusters and the maximum speed sometimes falls into the gap instead of one of the cluster because analyzed data does not include all speed regions. In order to eliminate the gap between speed layers, the boundaries of the three speed layers are expanded to form a continuous chain of speed values. Estimated link speed is mapped into final link travel speed using multiple regression. Finally, we generate speed patterns for all of the links using three expanded speed layers in a fast and efficient manner. The comparison between the actual recorded speeds and the predicted speeds of our algorithm suggests that our approach is plausible and offers optimal results in the performance testing.

2 Related Work

Models about link travel time estimation has been ADVANCE project, Ali-Scout dynamic route-guidance system, hybrid forecasting model, time series models, Kalman filtering model, historical and real-time profiles and neural network model, etc [3][4][7]. This chapter describes for ADVENCE project, Ali-Scout, hybrid forecasting model. The Advanced Driver and Vehicle Advisory Navigation ConcEpt

(ADVANCE) was an in-vehicle advanced traveler information system (ATIS) that operated in the northwestern portion and northwest suburbs of Chicago, Illinois. ADVANCE, which pioneered many of the public/private organizational arrangements and technical features now considered commonplace for Intelligent Transportation System deployments across the United States, was unique in its use of "probe" vehicles to dynamically generate travel time information about arterial and local streets. ADVANCE project for the express highway and arterial by Boyce and Rouphail used the traffic volume and occupancy ratio collected from the fixed detectors and used link-specific models and a generalized model for estimation of link travel time [1][7]. Ali-Scout dynamic route-guidance system is a second generation product developed by Siemens, which provides real-time, turn-by-turn guidance to drivers who have units installed in their vehicles. Ali-Scout vehicles communicate with infrared roadside beacons, which send travel times to the traffic control center and receive sequential routing instructions from the center [3]. A hybrid forecasting model for travel time forecasting has been developed and tested by deploying GIS technologies in the following areas: ① storing, retrieving, and displaying traffic data to assist in the forecasting procedures, ② building road network data, and ③ integrating historical databases and road network data [6].

Our research uses fuzzy c-mean and multiple regression algorithms for estimation of link travel time. A clustering approach that involves minimization of some objective function, or error criterion, belongs to a family of objective function clustering algorithms. The purpose of these algorithms is to partition the space of a given data samples. When the algorithms minimizes an error function it is often called C-Means being c the number of classes or clusters. If the classes are allowed to be fuzzy, the Fuzzy C-Means (FCM) clustering algorithm may be used [2][8]. Multiple regression is linear regression model using independent variable above two things for explaining variation of dependent variable. Multiple regression reduces error variance and is able to explain or estimate for dependent variable. Also, it is able to analyze of a causal relationship and compares effect which is attained to dependent variable from independent variable because we can measure effect which is independently attained to dependent variable from some independent variable as controlled state of value of another independent variable.

3 Link Speed Estimation System Framework

We generate patterns for the high, middle and low speed level using FCM and estimate link travel speed using multiple regression. The final link speed is calculated by means of smoothing using the center value of each cluster after confirming of cluster membership of collected data. And then link speed is recalculated by multiple regression. Figure 1 shows structure of system for calculating the link speed.

The data collection interval is of 5 minute durations and the probe cars represented by taxis using GPS devices. The data collected from the probe taxis have many patterns such as vehicle stopping and running, slow and fast running and searching running. Due to the various traveling patterns, time estimation using the mean speed has many gaps on comparisons with actual speed. Therefore, this research proposes a novel speed calculation method for getting speed near to the actual speed.

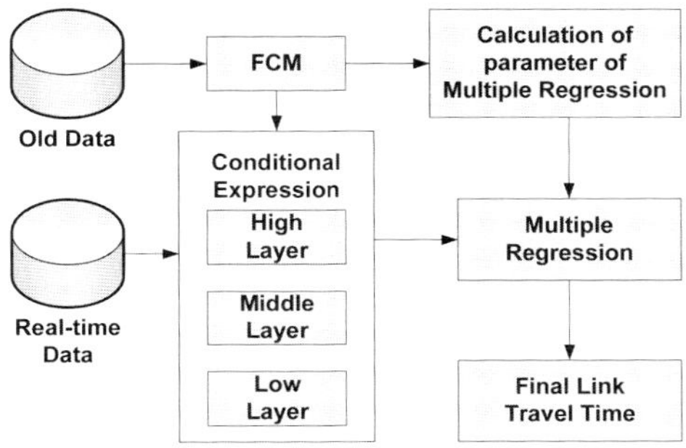

Fig. 1. Structure model of link speed estimation

3.1 Generation of Speed Pattern Layers

In the calculation of vehicle speed V using GPS, we divide length of link by time using pass times of previous and current nodes. Data from similar time intervals of 5 minute durations is used to calculate three patterns (high, middle, low speed) using FCM. FCM recognizes spherical clusters of points in p-dimensional space. Having a finite set of objects $X=\{x_1,\dots,x_2\}$ and the number of cluster centers c to be calculated, the assignment of the n objects to the c clusters is represented by the proximity matrix $U=[U_{ik}]$. With $k=1,\dots,n$ and $i=1,\dots,c$, $u_{ik} \in [0,1]$ expressing the fuzzy proximity or affiliation of object x_k to cluster center v_i. Matrix U is used to calculate region of speed and v_i is used to calculate link speed.

After generating the clusters, we obtain three important quantities. The first is the center value of a cluster. The second is min-max value of cluster. Min-max values are used to decide boundary of speed layer from GPS data for calculation of link speed. Also, center values are used to make from low-speed to high speed. Finally, the third is number of objects in a speed cluster. These values are used to make a center of speed which is not including speed domain U.

In the above case, maximum speed needs to be included in one of the patterns during calculation of link travel time. But as the application of the FCM, results in some gaps between three clusters and the maximum speed sometimes falls into the gap instead of one of the cluster. Therefore, we need to recalculate speed domain U using (1).

$$v_i^{non\ cluster} = w \bullet v_i^{upper\ cluster} + (1-w) \bullet v_i^{lower\ cluster} \tag{1}$$

$v^{non\ cluster}$ is calculated by providing weight on center of upper and lower cluster of collected maximum speed. ω is weight for connection between upper cluster and lower cluster and is distributed over a region from 0 to 1. The weight is calculated using (2).

$$w = \frac{N_{upper}}{N_{upper} + N_{lower}} \tag{2}$$

N is number of object including upper and lower cluster and is used to calculate weight. The weights are higher for clusters that have many objects in comparison with another cluster. Once the weight is computed, the base speed for non-pattern area and link speed can be computed by using (1) and (3) respectively. The estimated link speed is usually larger than the mean speed because the highest speed data of the link is used for the calculating the link speed. Finally, if calculated speed is faster than the speed limit, it can not be used due to the traffic regulations. Therefore, we will offer users the regulation speed in those cases.

3.2 Multiple Regression for Travel Time Estimation

Using matrix U, we derive a conditional expression for pattern of speed. Each cluster has maximum value, minimum value and center value for the speed. These values are used for calculating the link speed using the real time collection speed. V_T is time mean speed and can be calculated by dividing number of collected data by sum of all speeds. V_T is difficult to use for the link speed production which uses GPS since the volume of collected data using GPS is small. Also, this expression has a tendency to ignore the high speed data and calculates to speed of middle inside putting first. Therefore, we propose a new speed calculation method for the GPS data.

$$V_T = \frac{\sum_{j=1}^{n} V_{t1} + \sum_{k=1}^{m} \left(\frac{n}{n+m} V_{t2} + (1 - \frac{n}{n+m}) v_i \right)}{N} \tag{3}$$

As v_i is center value of cluster which includes maximum speed, V_{tl} is all of speed values which is included in the cluster. And V_{t2} is mean low speed that is little higher than center value of cluster. $n/(n+m)$ is threshold value of speed for partitioning near into high-speed layer. n is number of cluster object including maximum speed and m is number of the rest. Since in this research the traffic information is collected by probe taxis, the data has many outliers on low speed data than high speed. This is result of slowing down and stopping frequently in order to look for prospective passengers. Therefore, we calculate near maximum speed by considering a higher weightage to high speed data than low speed.

Estimated link travel speed by (3) is not robust due to insufficient number of data. For solving problem, we adapt to previous speed into current speed using multiple regression on same link. Multiple regression is applied using (4).

$$y = a_0 + a_1 x_1 + a_2 x_2 \tag{4}$$

But, case of adapting (4) produces false relations of two speeds since previous speed and current speed on same link have different number of speed data. Therefore we solve problem using (5) which applies weighting according to number of data.

$$y = a_0 + a_1 x_1 \beta + a_2 x_2 (1 - \beta) \tag{5}$$

$$\beta = \frac{N_{nowspeed}}{N_{prev} + N_{now}}$$

Weight β use number of now and previous speed data. Also, coefficient (a_0, a_1, a_2) is calculated by X (which is adapting weights) and y (which is actual speed). Final link Travel speed is calculated by (5) using coefficient and weights.

4 Estimation of Link Speed Using GPS Data

In order to generate patterns regarding the traffic speed, as mentioned earlier, we used taxi based probe car like for collection of data. The data acquisition for link speed analysis was performed using these probe cars which were equipped with GPS devices. In this experimental analysis the collected information of actual speed was for the time duration from 10am to 11am. We analyzed the data collected thus using FCM. Figure 2 and 3 showed variation of speed about center value of each cluster.

Speed patterns are described based on the traffic density. Therefore, quantity of collected GPS data is important in pattern analysis. Although it is not mentioned in

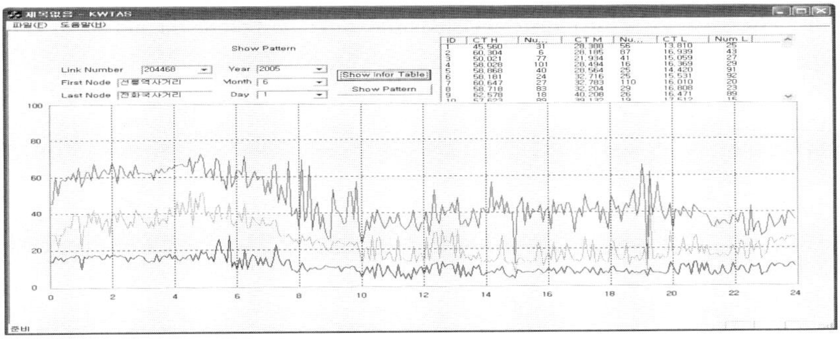

Fig. 2. Variation in speed patterns during 1 month from 0:00 to 24:00

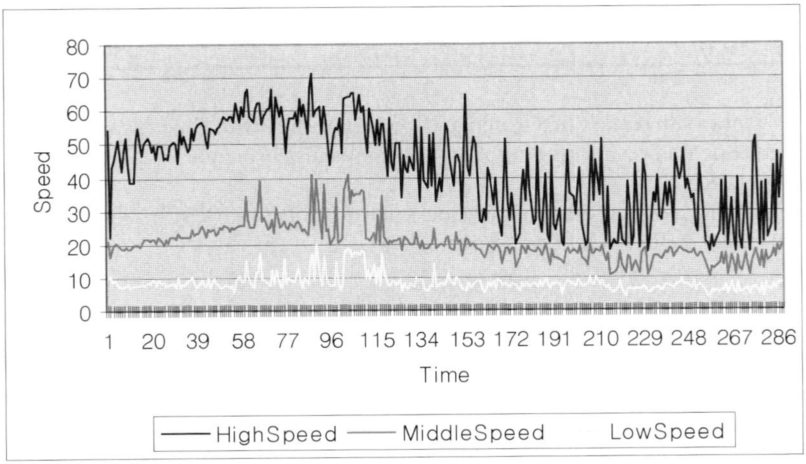

Fig. 3. Variation in speed patterns

this research, it is necessary to study the quantity of data required to estimate the link speed properly. It can be observed that at the peak office hours and closing times from PM 2:00, there is a sudden lowering of speed and at the other times speed is rising as in figure 4. However, not all links have the same pattern and each one varies according to state of road and environment. Table 1 showed center and min-max value of each cluster by getting result of analysis at 5 minute interval until 11:00 from 10:00 using FCM and multiple regression.

Table 1. Center of calculated speed pattern

interval	1	2	3	4	5	6	7	8	9	10	11	12
H_Speed	36	30	40	34	42	38	34	31	34	38	34	34
M_Speed	21	14	27	14	26	27	14	15	17	18	14	13
L_Speed	11	7	11	4	12	11	5	7	11	8	6	4

Table 2. Speed result of each link calculated using mean, test probe car and our algorithm

5 interval	Mean		Actual		New Algorithm	
	1st	2nd	1st	2nd	1st	2nd
10:00-10:05	17.800	29.750	28	34	30.899	37.245
10:05-10:10	4.000	22.000	4	30	2.016	25.615
10:10-10:15	15.000	20.000	21	25	17.982	25.391
10:15-10:20	13.286	26.286	28	50	29.728	35.018
10:20-10:25	18.600	29.250	31	42	33.738	39.083
10:25-10:30	14.250	16.000	22	19	20.216	19.494
10:30-10:35	20.333	31.000	31	39	30.899	38.246
10:35-10:40	19.000	30.400	30	51	26.337	37.136
10:40-10:45	10.667	27.000	14	37	9.083	31.349
10:45-10:50	13.000	18.400	17	23	20.154	17.322
10:50-10:55	13.750	34.333	20	42	17.206	44.367
10:55-11:00	3.000	12.500	15	16	3.459	7.640

Table 2 presents results that compared our algorithm with time mean speed and actual speed. Such as result, our algorithm is showed near actual speed more than time mean speed.

For verifying compatibility of model, our research used RMSE formula for comparing of mean, actual and new algorithm.

$$RMSE = \left[\frac{1}{N} \sum_{n=1}^{N} \left(x_n' - x_n \right)^2 \right]^{1/2} \tag{6}$$

RMSE for speed estimation of our algorithm is 4.31726 and 7.044891 and RMSE for the time mean speed is 9.19839 and 11.26033.

Table 3. Speed result of each link that calculated using mean, test probe car and our algorithm

	RMSE	
	1st	**2nd**
Between Mean and Actual	9.19839	11.26033
Between New Algorithm and Actual	4.31726	7.044891

This case is estimated for reason of providing near speed which is running to high speed in our algorithm and being mean for all of speed of passed probe car in time mean speed. Also, it is estimated for reasons of small number of collected data and difference of collected data from existing fixed detector and GPS. Our algorithm has high performance to analyze collected data using GPS and easy to analyze by small number of collected data. Therefore, it is able to provide correct speed through much supply of GPS.

5 Conclusion

Travel time data is vital to consumers and governments alike. This paper identifies the ubiquity of GPS probe data in estimating travel times under congested traffic conditions and proposes a new travel time estimation approach without any additional data requirements or infrastructure investment. One of the main problems in analyzing the traffic patterns of all roads and applying it to estimate the link speeds is the huge computational burden. In attempting to solve these issues, we used the FCM which shows very high efficiency in clustering the GPS data and calculating the link speed in conjunction with multiple regression. By the application of this methodology we are able to provide estimation close to near real speed even when the amount of collected data is small and the speed deviation is significant.

Through performance tests we observe that our proposed algorithm is robust and performs significantly well. For example, in our empirical analysis the average rmse in travel speed estimation between our algorithm and the mean speed can be observed in Table 3. The overall analysis and result shows that our approach has a higher performance in comparison to the time mean speed. The accuracy in calculation of link speed is enhanced by this research, however we hope that more accurate link speeds can be computed by considering the time delay occurring in cross roads when vehicles are waiting to turn left or doing u-turns.

References

1. Boyce, D., Rouphail, N., Kirson, A.: "Estimation and measurement of link travel times in ADVANCE project," Proceedings of Vehicle Navigation and Information Systems Conference, IEEE, 1993, pp.62–66
2. Lázaro,J., Arias, J., Martín, J. L., Cuadrado, C., Astarloa, A.: "Implementation of a Modified Fuzzy C-Means Clustering Algorithm for Real-Time Applications," Microprocessors and Microsystems. Vol. 29 (8-9), 2005, pp. 375-380

3. Dharia, A. Adeli, H.: “Neural network model for rapid forecasting of freeway link travel time,” Engineering Applications of Artificial Intelligence, Vol. 16, No. 7, 2003, 607-613
4. Coifman, B.: “Estimating Travel Times and Vehicle Trajectories on Freeways Using Dual Loop Detectors,” Transportation Research: Part A, Vol. 36, No. 4, 2002, pp. 351-364
5. Cortes, C.E., Lavanya, R. Jun-Seok Oh, Jayakrishnan, R.: “A General Purpose Methodology for Link Travel Time Estimation Using Multiple Point Detection of Traffic,” Transportation Research Record, vol. 1802, 2002, pp. 181-189
6. You, J., Kim, T.J.: "Development and evaluation of a hybrid travel time forecasting model," Transportation Research Part C: Emerging Technologies, Vol. 8. Issues 1-6, 2000, pp. 231-256
7. Kim, Y.C., Choi, K.J., Kim, D.K., Oh, K.D.:“Estimation of Link Travel Speed Using Single Loop Detector Measurements for Signalized Arterials,” Journal of Transportation Research Society of Korea, Vol.15, No. 4, 1997, pp.53-71
8. Goktepe, A.B., Altun, S., Sezer, A.: “Soil clustering by fuzzy c-means algorithm,” Advances in Engineering Software, Vol. 36. Issue 10, 2005, pp. 691-698
9. J. C. Bezdek: "Pattern Recognition with Fuzzy Objective Function Algoritms", Plenum Press, New York, 1981.
10. Dion F., and, Rakha H. Estimating dynamic roadway travel times using automatic vehicle identification data for low sampling rates, Transportation Research Part B: Methodological, In Press.

Wigner-Ville Distribution Based on EMD for Faults Diagnosis of Bearing

Hui Li[1], Haiqi Zheng[2], and Liwei Tang[3]

First Department, Shijiazhuang Mechanical Engineering College, Shijiazhuang 050003, People's Republic of China
[1]Huili68@163.com, [2]jxgcxy@heinfo.net, [3]Tangliwei@163.com

Abstract. Wigner-Ville distribution (WVD) is a joint time-frequency analysis for non-stationary signals. The main difficulty with the WVD is its bilinear characteristic which leads to cross terms in the time-frequency domain. Recently the technique of empirical mode decomposition (EMD) has been proposed as a novel tool for the analysis of nonlinear and non-stationary data. In this paper, key elements of the numerical procedure and principles of EMD are introduced. Wigner-Ville distribution based on EMD is applied in the research of the faults diagnosis of the bearing. Firstly, the original time series data is decomposed in intrinsic mode functions (IMFs) using the empirical mode decomposition. Then, the Wigner-Ville distribution for selected IMF is calculated. The signal simulation and experimental results show that Wigner-Ville distribution based on EMD can not only successfully eliminate the cross terms but also effectively diagnosis the faults of the bearing.

1 Introduction

A local component of rotating machine fault maybe produce successive periodical impacts and the signature of a defective component is spread across a wide frequency band and hence can be easily masked by noise and low frequency effects. Thus, in order to obtain the useful information that is hidden in the noisy signal, an effective method for feature extraction has to be used. Time-frequency analysis method has been used for non-stationary signal feature extraction. Time-frequency dependent analysis method is a novel approaches in the fields of applied mathematics and has rapidly emerged as a common subject of research and application in scientific and engineering investigations in the last two decades. Therefore, a number of time-frequency analysis technology have been developed for analyzing non-stationary signals. Among those, the Short Time Fourier Transform (STFT) [1], Wavelet Transform (WT) [2] and Wigner-Ville distribution (WVD) [3] are widely used. The STFT [1] uses sliding windows in time to capture the frequency characteristics as functions of time. Therefore, spectrum is generated at discrete time instants. Three-dimensional display is required to describe frequency, magnitude, and time. An inherent drawback with the STFT is the limitation between time and frequency resolutions. A finer frequency resolution can only be achieved at the expense of time resolution and vice-versa. Furthermore, this method requires large amounts of computation and storage for display. The Wavelet Transform (WT), on the other hand, is similar to the SHFT

L. Wang et al. (Eds.): FSKD 2006, LNAI 4223, pp. 803 – 812, 2006.

in that it also provides a time-frequency map of the signal being analyzed. The improvement that the WT makes over the STFT is that it can achieve high frequency resolution with sharper time resolutions. A very appealing feature of the wavelet analysis is that it provides a uniform resolution for all the scales. Limited by the size of the basic wavelet function, the downside of the uniform resolution is uniformly poor resolution. Moreover, a important limitation of the wavelet analysis is its non-adaptive nature. Once the basic wavelet is selected, one will have to use it to analyze all the data. The Wigner-Ville distribution is a basic time-frequency representation, which is part of the Cohen class of distribution. Furthermore, it possesses a great number of good properties and is of popular interest for non-stationary signal analysis. Therefore, the Wigner-Ville distribution has received considerable attention in recent years as an analysis tool for non-stationary or time-varying signals. It has been widely used in the areas of structure-bone noise identification, optics, machinery condition monitoring and so on. The difficulty with this method is the severe cross terms as indicated by the existence of negative power for some frequency ranges. In addition, the WVD of discrete time signals suffers from the aliasing problem, which may be overcome by employing various approaches.

In this work, we introduce a novel approach for nonlinear, non-stationary data analysis and apply it to eliminating the cross terms of the WVD. This new method is a recently developed Empirical Mode Decomposition (EMD)[4] applied to non-stationary phenomena. The EMD is included in the so called Hilbert-Huang Transform based on the direct extraction of the energy associated with the intrinsic time scales in the signal. This process generates a set of components, called the intrinsic modes functions (IMF). Each intrinsic modes function is a mono-component function respect to time. The WVD of mono-component has no cross term. Therefore, the WVD based on EMD can eliminate the cross terms effectively.

This paper has been organized as follows: Section 1, gives a brief introduction of the time-frequency analysis technology. Section 2, gives a brief description of the Empirical Mode Decomposition (EMD), of the original signal in intrinsic modes or IMFs, by means of the sifting process. Section 3, presents the method and procedure of the WVD based on EMD. Section 4, gives an example of the WVD based on Empirical Mode Decomposition to a simulation signal, and some special facts which we have for the correct to the method. Section 5, gives the applications of the WVD based on Empirical Mode Decomposition to faults diagnosis of the inner race faults of the bearing. Section 6, gives the main conclusions of this paper.

2 Brief Description of the EMD Method [4]

Empirical Mode Decomposition is an emerging novel technique of signal decomposition having many interesting properties. In particular, EMD has been applied to numerous scientific investigations, such as biomedical signals processing, geophysics, civil engineering and so on. In order to facilitate the reading of this paper we will introduce in detail the Empirical mode decomposition, which is a relatively novel technique.

To extract the IMF from a given data set, the sifting process is implemented as follows. First, identify all the local extrema, and then connect all of the local maxima by a cubic spline line as the upper envelope. Then, repeat the procedure for the local minima to produce the lower envelope. The upper and lower envelopes should cover all the data between them. Their mean is designated $m_1(t)$, and the difference between the data and $m_1(t)$ is $h_1(t)$, i.e.:

$$x(t)-m_1(t)=h_1(t). \tag{1}$$

Ideally, $h_1(t)$ should be an IMF, for the construction of $h_1(t)$ described above should have forced the result to satisfy all the definitions of an IMF by construction. The sifting process has to be repeated as many times as it is required to reduce the extracted signal to an IMF. In the subsequent sifting process steps, $h_1(t)$ is treated as the data; then:

$$h_1(t)-m_{11}(t)=h_{11}(t). \tag{2}$$

where $m_{11}(t)$ is the mean of the upper and lower envelopes of $h_1(t)$. This process can be repeated up to k times; $h_{1k}(t)$ is then given by

$$h_{1(k-1)}(t)-m_{1k}(t)=h_{1k}(t). \tag{3}$$

After each processing step, checking must be done on whether the number of zero crossings equals the number of extrema.

Let the sifting stop at the k th sifting, then

$$c_1(t)=h_{1k}(t). \tag{4}$$

It is designated as the first IMF component from the data. Overall, $c_1(t)$ should contain the finest-scale or the shortest-period component of the signal corresponding to this stopping condition.

Once $c_1(t)$ is obtained, it can be separated from the rest of the data by using

$$x(t)-c_1(t)=r_1(t). \tag{5}$$

Since the residue, $r_1(t)$, still contains longer-period components, it is treated as the original complete input data and subjected to the same sifting process as described above. This procedure can be repeated for all the subsequent $r_j(t)$ values, and the result is

$$\begin{cases} r_1(t)-c_2(t)=r_2(t) \\ \qquad\vdots \\ r_{n-1}(t)-c_n(t)=r_n(t) \end{cases}. \tag{6}$$

The sifting process should continue until the residue, $r_n(t)$, becomes a constant value, a monotonic function or a function with only one extremum from which no more IMFs can be extracted. Even for data with a zero mean, the final residue can still be different from zero. If the data have a trend, the final residue will be that trend. By summing equations (5) and (6), it follows that

$$x(t) = \sum_{j=1}^{n} c_j(t) + r_n(t) . \quad (7)$$

The sifting processes are finished and a set of IMFs have been produced. Thus, one achieves a decomposition of the data into n-empirical IMF modes, plus a residue, $r_n(t)$, which can be either the mean trend or a constant.

3 Proposed Wigner-Ville Distribution Based on the EMD

The procedure of proposed Wigner distribution based on the EMD is given as follows

1) To decompose the analyzed signal using EMD and to obtain IMFs;
2) To select the interested IMF according to the objective of fault diagnosis,;
3) To calculate the Wigner-Ville distribution of selected IMF;
4) To synthesize the Wigner-Ville distribution of selected IMF;
5) To analyze the Wigner-Ville distribution of selected IMF and to draw a diagnostic conclusion.

4 Signal Simulation of Wigner-Ville Distribution Based on the EMD

In the following section, the results of a simple signal simulation studying on the performance of the EMD are presented to real a better understanding of this numerical method. We present the performance of the EMD on time variant frequency modulated data series. For example, Equation (10) gives one possible mathematical description of a frequency-modulated signal.

$$x_1(t) = \cos[2\pi 40t + 0.5\sin(2\pi 20t)] . \quad (8)$$

$$x_2(t) = \sin(2\pi 160t) . \quad (9)$$

$$x(t) = x_1(t) + x_2(t) . \quad (10)$$

Signal $x(t)$ is composed of a carrier frequency is 40Hz, frequency-modulated is 20 Hz and constant or time-independent frequency of 160 Hz sine wave.

Therefore, the frequency-modulated $f(t)$ can be written as

$$f(t) = 40 + 10\cos(40\pi t) . \quad (11)$$

The variation range of the frequency-modulated $f(t)$ is given as

$$30 \le f(t) \le 50 . \quad (12)$$

Fig.1 displays a graphical sketch of the signal $x(t)$ which was generated over a total time span T=0.2s with a sampling frequency $f_s = 2560$ Hz. Fig.2 shows the

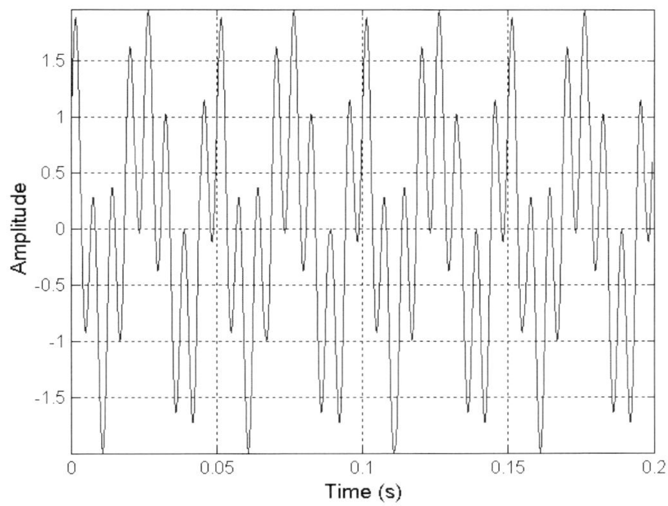

Fig. 1. Time histories of the simulation signal $x(t)$

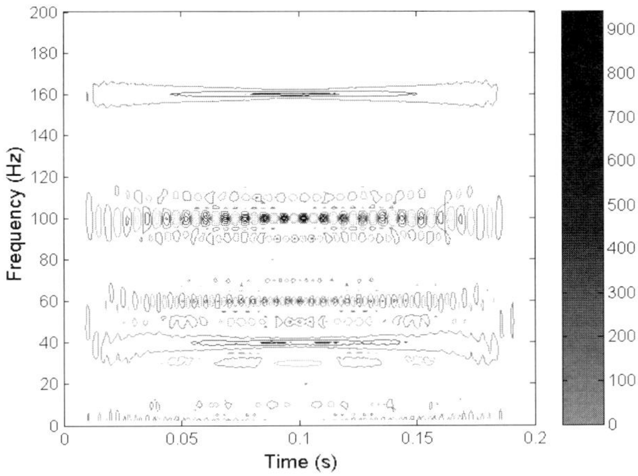

Fig. 2. WVD of the recording in Fig.1

Wigner-Ville Distribution of the signal $x(t)$, in which the cross terms is clearly evident as an oscillatory component midway between the two auto-terms. Moreover, the characteristics of frequency-modulated for $x_1(t)$ is not evident.

Fig.3 shows the empirical mode decomposition in IMF of the signal $x(t)$. With the help of the sifting algorithm explained in Section 2, we carried out this decomposition. The decomposition identifies two modes: c_1 represents the sine wave of $x_2(t)$, c_2 represents the frequency-modulated signal, $x_1(t)$, c_3 is the residue, respectively. By virtue of EMD method, signal can be decomposed into two complete and orthogonal intrinsic mode functions. Therefore, we can know not only the frequency components of the signal, but also the variation of the amplitude and period. These IMFs component can reflect the actual physical meaning of the signal.

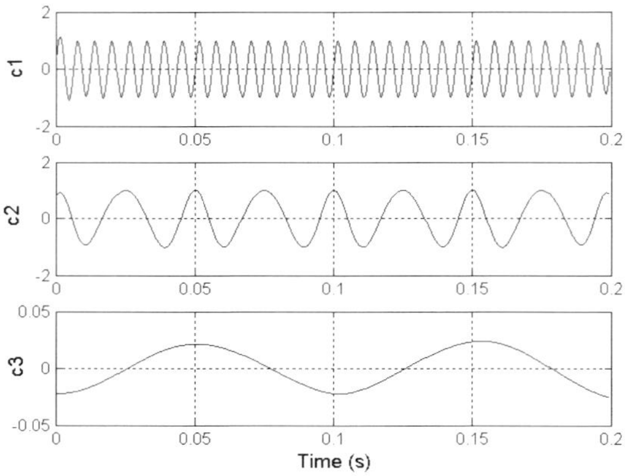

Fig. 3. The three IMF components of the recording in Fig.1

Fig.4 displays the Wigner-Ville Distribution based on the empirical mode decomposition. The Wigner-Ville Distribution spectrum in Fig.4 shows a clear picture of temporal frequency energy distribution of the data, i.e., the linear response has constant frequency at 160 Hz, the non-linear response has frequency dependence modulated around 40Hz and bounded by 30 and 50 Hz, and the decaying energy of the non-linear response with the color changing from the white at the beginning to dark blue at the end of the record. Therefore, the Wigner-Ville distribution based on the empirical mode decomposition is better to description the characteristics of the time-frequency distribution. In contrast, the Wigner-Ville Distribution in Fig.2 not only has the cross terms, but also can not express the non-linear response of the modulated frequency.

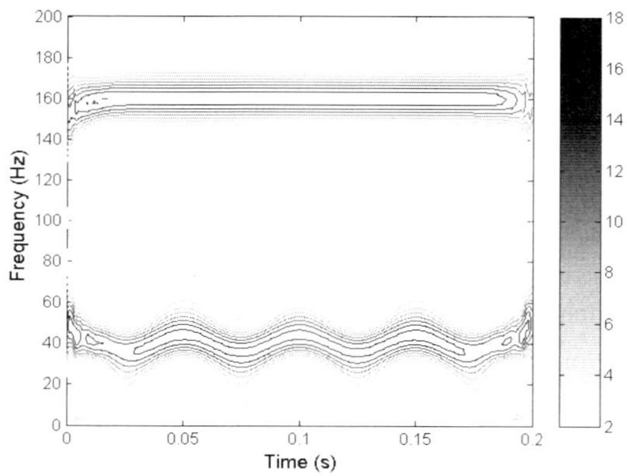

Fig. 4. WVD of $c_1 + c_2$ components

5 Bearing Faults Diagnosis Using WVD Based on EMD

For many mechanical dynamic signals, impulses usually indicate the occurrence of faults. In the following, one example is given to show the WVD based on EMD is applied to bearing faults monitoring.

Rolling bearings are installed in many kinds of machinery. A lot of problem of those machines may be caused by defects of the rolling bearing. Generally, local defects may occur on inner race, outer race or rollers of bearing. A local fault may produce periodic impacts, the size and the repetition period which are determined by the shaft rotation speed, the type of fault and the geometry of the bearing. The successive impacts produce a series of impulse response, which maybe amplitude modulated because of the passage of fault through the load zone. The spectrum of such a signal would consist of a harmonics series of frequency components spaced at the component fault frequency with the highest amplitude around the resonance frequency. These frequency components are flanked by sidebands if there is an amplitude modulation due to the load zone. According to the period of the impulse, we can judge the location of the defect using characteristic frequency formulae. Because inner race defect has more transfer segments when transmitting the impulse to the outer surface of the case, usually the impulse components are rather weak in the vibration signal. The tested bearing was used to study only one kind of surface failure: the bearing was damaged on the race. The rolling bearing tested has a groove on inner race. Localized defect was seed on the inner race by an electric-discharge machine to keep their size and depth under control. The size of the artificial defect was 1mm in depth and the width of the groove was 1.5mm. The speed of the spindle is 1500r/min, that is, the rotating frequency f_r is 25 Hz. The type of the ball bearing is 208. There are 10 rollers ($Z=10$) in a bearing and the contact angle $\alpha = 0°$, roller diameter d= $55/3$ mm, bearing pitch diameter D=97.5mm. Then the characteristic frequency of the inner race defect can be calculated by the equation (13).

$$f_{inner} = \frac{z}{2}\left(1+\frac{d}{D}\cos\alpha\right)f_r . \tag{13}$$

Therefore, according to equation (13), the characteristic frequency of the inner race defect is calculated to be at 148.5Hz.

The vibration signals of the inner defects are sampled on a gearbox. The motion is produced by a DC motor, with which it is easy to adjust the rotation speed. The monitoring and diagnostic system is composed of four accelerometers, amplifiers, B&K 3560 spectrum analyzer and a computer. The sampling span is 12.8 kHz and the sampling frequency is 32768 Hz.

The original vibration signal of inner race defect is displayed in Fig.5. It is clear that there are periodic impacts in the vibration signal. There are significant fluctuations in the peak amplitude of the signal, and there are also considerable variations of frequency content. Fig.6 shows the Wigner-Ville distribution of the vibration signal, in which we can hardly find the characteristic frequency component of the inner race defect. Because the vibration signal is generally the high frequency modulated signal,

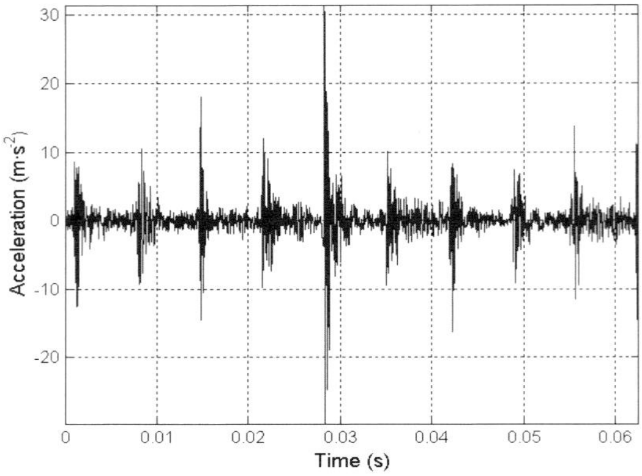

Fig. 5. Original vibration signal of the bearing with inner race fault

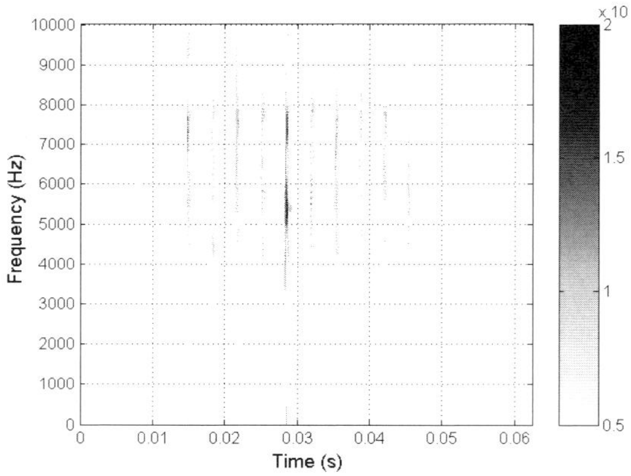

Fig. 6. WVD of the vibration signal shown in Fig.5

the Wigner-Ville distribution is concentrated on high frequency section. Therefore, it is hardly to distinguish the type of the defect in Fig.6.

Fig.7 displays the empirical mode decomposition in eleven IMFs of the vibration signal in Fig.5. The decomposition identifies eleven modes: $c_1 \sim c_{10}$ represents the frequency components excited by the inner race defects, c_{11} is the residue, respectively. From Fig.7, it can be easily proven that the EMD decomposes vibration signal very effectively on an adaptive method. Because the IMF mode, which is related to the inner race defect of the bearing, is the c_7, and on account of the fact that the sifting procedure select first modes that display higher frequencies, we conclude that we can use c_7 mode in the analysis of the inner race defect. Fig.8 shows the WVD of c_7. It is very evident that the WVD plot is able to clearly show not only the characteristic frequency of the inner race fault, but also occur time of the fault.

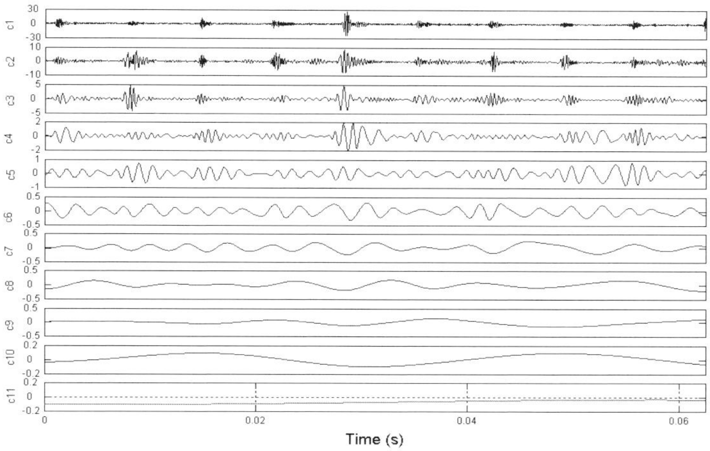

Fig. 7. IMFs of the signal shown in Fig.5

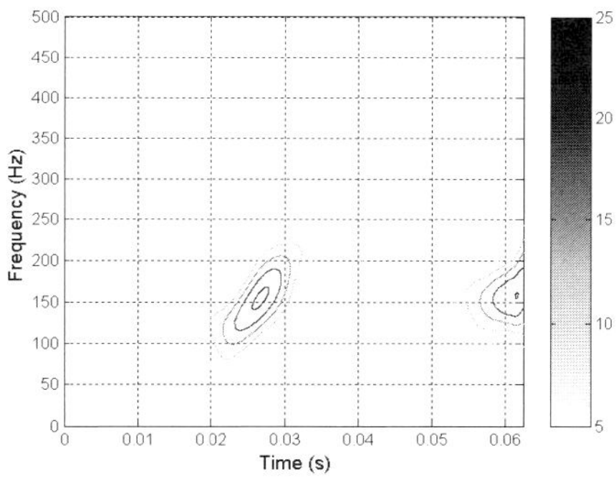

Fig. 8. WVD of c_7 component

6 Conclusions

The technique of empirical mode decomposition (EMD) has been proposed as a novel tool for the analysis of nonlinear and non-stationary data. The application of Wigner-Ville distribution based on the EMD to the faults diagnosis of ball bearing is presented. The EMD is applied to decompose the fault signal, derived from a tested ball bearing, into their components in order to recognize the fault type associated with the characteristic frequency of the inner race fault. It has been shown that the WVD based on the EMD is capable of detecting local faults in ball bearing. Dark zones and cured bands in the contour plots are the main feature of an impulse caused by a local fault. It has been shown that the Wigner-Ville distribution based on the EMD is an effective method for elimination of cross terms. Simulation and bearing faults diagnosis using WVD based on EMD verify the presented method.

Acknowledgement

The authors are grateful to the National Natural Science Foundation of China for supporting this research under grant No. 50375157.

References

1. Cohen L.: Time-frequency Analysis. Prentice-Hall, Englewood Cliffs, NJ. (1995)
2. James Li C., Ma Jun: Wavelet decomposition of vibration for detection of bearing-localized defects. NDT&E International. 30(1997)143-149
3. Staszewski W.J., Worden K., Tomlinson G.R.: The-frequency analysis in gearbox fault detection using the Wigner-Ville distribution and pattern recognition. Mechanical Systems and Signal Processing. 11(5)(1997)673-692
4. Huang N .E, et al :The empirical mode decomposition and the Hilbert spectrum for nonlinear and non-stationary time series analysis. Proceeding of Royal Society London. A(1998)454:903-995

Active Learned Multi-view Face Detection Tree Using Fuzzy Cluster Validity Analysis*

Chunna Tian, Xinbo Gao, and Jie Li

School of Electronic Engineering, Xidian University, Xi'an 710071, China
tianchunna@lab202.xidian.edu.cn, xbgao@lab202.xidian.edu.cn,
leejie@mail.xidian.edu.cn

Abstract. An active learned face detection tree based on FloatBoost method is proposed to accommodate the in-class variability of multi-view faces. To handle the computation resource constraints to the size of training example set, an embedded Bootstrap example selection algorithm is proposed, which leads to a more effective predictor. The tree splitting procedure is realized through dividing face training examples into the optimal sub-clusters using the fuzzy c-means algorithm together with a new cluster validity function based on the modified partition fuzzy degree. Then each sub-cluster of face examples is conquered with the FloatBoost learning to construct branches in the node of the detection tree. During training, the proposed algorithm is much faster than the original detection tree. The experimental results illustrate that the proposed detection tree is more efficient than the original one while keeping its detection speed. And the E-Bootstrap strategy outperforms the Bootstrap one in selecting relevant examples.

1 Introduction

In the recent years, face and facial expression recognition have attracted much more attention though they have been studied for more than 20 years by psychophysicists, neuroscientists, and engineers. A great deal of research demonstrations and commercial applications have been developed from these efforts. The first step of any face processing system is detecting the locations of faces in scene. Such a problem is challenging because faces are non-rigid and have a high degree of variability in size, color, pose and sampling noise *etc* [1]. So far, numerous algorithms have been proposed for this purpose, which can be broadly classified into two main groups according to the style of detection foundation. They are feature-based face detection and template-based face detection.

Usually, these feature-based methods are not robust in complex cases. Template-based methods recognize human faces as a kind of model. Actually, the face detection problem is translated into a two-class problem of statistical pattern recognition. The template-based methods, which attract more and more attention nowadays, are more robust and superior in accuracy. Approaches to template-based

* This work was partially supported by the National Natural Science Foundation of China (No.60202004) the Key Project of Chinese Ministry of Education (No.104173), and the Program for New Century Excellent Talents in University (NCET-04-0948), China.

L. Wang et al. (Eds.): FSKD 2006, LNAI 4223, pp. 813 – 821, 2006.

face detection may be further divided into two subgroups, template matching methods and template learning methods. As to the former methods, several standard patterns of faces are stored to describe the face as a whole or the facial features separately. The correlations between an input image and the stored patterns are computed for detection [2]. In the latter case, templates or classifiers are learned from a large set of training positive/negative examples, which should capture the representative variability of facial/ non-facial appearance. These learned classifiers are then used for detection. Thus the example-based learning algorithms have two essential issues, relevant example selection and feature selection. Most of the available approaches for this purpose use Bootstrap [3] to select relevant examples, which goes a ways toward letting the learning system select or focus on training examples by themselves to reduce its current prediction error. However, there is still much redundant information in the training data sampled by Bootstrap. In order to reduce this redundancy and solve the problem of computation resource constraints, an embedded Bootstrap (E-Bootstrap) strategy is proposed, which sieving elaborately through extremely large training data sets for more typical examples relevant to the learning problem.

For feature selection, it is difficult to obtain one feature based good classifier at once, but it is very easy to get weak classifiers whose performances may only be better than random guess. So Viola *et al.* adopted AdaBoost learning algorithm [4] to boost those weak classifiers into a strong one of higher accuracy, which has been successfully validated for frontal upright face detection. However, statistics show that approximately 75% of the faces in home photos are non-frontal. So Stan *et al.* proposed a detector-pyramid for multi-view face detection [5], which divides the object patterns manually into several more homogeneous sub-pattern classes firstly, then constructs multiple parallel cascade classifiers each handling a specific sub-pattern. Finally, they merge their individual results. Besides, they present a FloatBoost learning algorithm to select more effective weak classifiers and solve the monotonicity problem encountered in the AdaBoost. There are still two challenges: (1) It is empirical and difficult work to determine the right sub-pattern class for each object in most cases; (2) Multiple specialized classifiers increase the computational complexity.

To solve these problems, Rainer *et al.* presented a tree of boosted classifiers to detect mouths in images, which employed a clustering-and-splitting step to construct branches in the tree nodes recursively [6]. The splitting principle they chose is to split with lower computational complexity. In order to find the optimal splitting style, they have to train every possible splitting manner, which leads to a really time consuming training process. Besides, when applied to face detection task, such a reduction of the computational complexity is slight and the splitting style they chose does not agree with the most valid clustering result. So the sub-patterns they classified are not the most homogeneous ones according to the clustering validity, which is often converted to the optimal classification of data. Therefore we adopt the fuzzy c-means (FCM) algorithm with a new cluster validity function based on the modified partition fuzzy degree [7] to choose the optimal clustering result, which has been proved more efficient and accurate.

2 Training of Face Detection Tree

As typical machine learning based pattern recognition problem, face detection in this paper has two essential issues. One issue is E-Bootstrap based example selection, which is realized through embedding another Bootstrap process into each iteration of Bootstrap strategy to reduce the redundant information in the training set thus get more typical and representative training examples. Another issue is classifier design based on the selected examples, which will be discussed in details as follows.

2.1 Embedded Bootstrap Strategy for Example Selection

The eventual performance of a face detection system depends heavily on the quality of examples it receives during training. For "face" patterns, the task at hand seems rather straight forward. We simply collect all the frontal views of faces we can find in all kinds of image sources. Because we do not have access to achieve a large positive example set, we even artificially enlarged our data set by randomly mirroring, rotating, and re-scaling these obtained face images. For "non-face" patterns, the task we have seems quite different. There are simply too many possibilities to consider. How to build a comprehensive but tractable database of "face" and "non-face" patterns? Most of the available methods adopt Bootstrap strategy [3-5], which is difficult to converge in this task. During training, the partially-trained system is applied to a sequence of scenery images, which do not contain faces. Any regions in the image detected as faces are errors, which can be added into the set of negative training examples. Note that some of the newly collected examples may be taken from the same image resource, so those examples may have similar texture, and they may represent each other very well. Therefore we presented an embedded Bootstrap example selection strategy to improve the utility of our negative example set. This idea is formulated as follows, that is illustratively demonstrated in Fig.1.

Initialize: Give a candidate example set S . The false classification rate of classifier P_i on the extended example set S_i is denoted as f_i. Specify the maximum acceptable false prediction rate f_{th}. Initialize the iteration time $i = 1$.

(E1) Start with a random selected small training example set S^r , which may be highly non-representative in S, as the initial training dataset T_1;

(E2) Train a classifier P_i on the current training example set T_i ;

(E3) Run P_i on an extended example set $S_i \subset S$. Collect the examples that P_i falsely predicted in S_i into a subset S_i^f . Calculate current false prediction rate $f_i = |S_i^f|/|S_i|$, where $|\cdot|$ denotes the set cardinality;

(E4) If $f_i \leq f_{th}$, then goto (E5); Else: windowing select a random subset of S_i^f denoted as S_i^{fr} , then define $T_i' = T_i \cup S_i^{fr}$, with which another classifier $P_i' = P(T_i')$ is induced. Run P_i' on S_i^f and collect the false predicted examples into a subset $S_i^{ff} = \mathcal{F}(S_i^f \mid P_i')$. Specify $S_i^{\hat{f}} = S_i^{fr} \cup S_i^{ff}$. Usually $S_i^{\hat{f}}$ is much smaller than S_i^f . Let i=i+1, $T_i = T_{i-1} \cup S_{i-1}^{\hat{f}}$, go to step (E2);

(E5) After i^E iterations of E-Bootstrap, a high utility example dataset T^E defined as Eq. (1) can be achieved. Then perform training procedure on it, and we get the detector $P^E = P(T^E)$.

$$T^E = T_{i^E} = \left(\bigcup_{j=1}^{i^E-1} S_j^{\hat{f}}\right) \cup S^r = \left(\bigcup_{j=1}^{i^E-1} (S_j^{fr} \cup S_j^{ff})\right) \cup S^r \,, \tag{1}$$

where

$$S_j^{\hat{f}} = S_j^{fr} \cup \mathcal{F}(S_j^f \mid P_j') = S_j^{fr} \cup \mathcal{F}[S_j^f \mid P_j(T_j')] \ . \tag{2}$$

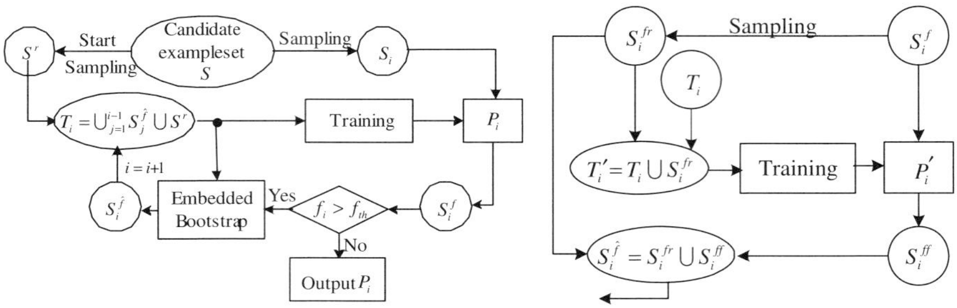

Fig. 1. Sketch map of the E-Bootstrap sampling process. (a) E-Bootstrap sampling process. (b) The embedded Bootstrap procedure.

The training cycle in Fig.1 can be realized with PAC learning model, ANN or SVM learning algorithms. In this paper we adopt the FloatBoost learning to build a detection tree, in which the tree splitting is determined by a new cluster validity function based on the modified partition fuzzy degree.

2.2 FloatBoost Based Detection Tree Design

Our object-detection procedure classifies images based on the value of simple features. We use three types of Haar wavelet-like features as reference [5] does. These features can be non-symmetrical to cater to non-symmetrical characteristics of non-frontal faces. The value of one Haar wavelet-like feature can be calculated through "integral image" technique[4]. Due to the complexity of non-frontal faces, we divide the positive training examples into more homogeneous sub-pattern classes, then training each sub-pattern with the learning algorithm. Thus the detection tree is formed as Fig. 2 demonstrates. Recall that the set of Haar wavelet-like feature is over complete. Even though each feature can be computed very efficiently, computing the complete set is prohibitively expensive. So we adopt the cascade FloatBoost to select these essential features for face detection. The result of each stage FloatBoost training is a strong classifier with the given false alarm rate (say, 50%) and hit rate (say, 99.5%). And those stage strong classifiers are cascaded into a detector.

The training process of such a tree is a recursive procedure as Fig. 3 shows, which starts with the root tree node. At each node all positive and negative training examples, see Fig. 4 for example, specified by the parent node (denoted as S_+ and S_- respectively) are used for training a stage classifier that consists of p features [5]. The essential problems are when and how to split the node of the tree into branches. To the end, we adopt the FCM algorithm to classify the positive training example set S_+ into c

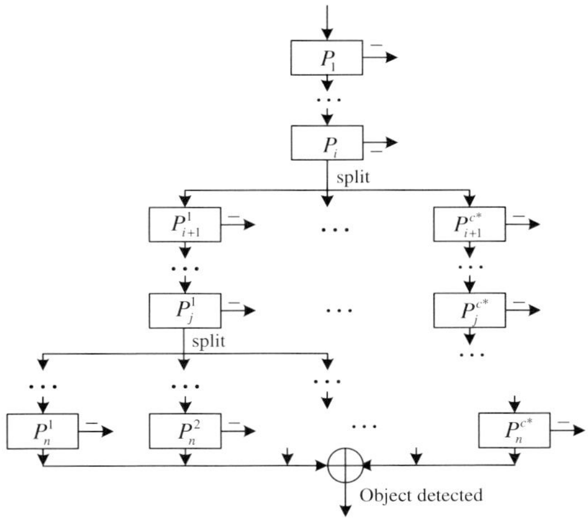

Fig. 2. Tree-like clssifier

sub-clusters. Let $X = \{x_1, x_2,..., x_n\}$ denote a positive data set S_+, and $x_i = \{x_{i1}, x_{i2},..., x_{ip}\}^T$ represent the p features of the i-th face example. The fuzzy clustering can be described the following mathematical programming problem,

$$\min J_m(U,c) = \sum_{i=1}^{c}\sum_{j=1}^{n} u_{ij}^m d_{ij}^2 \quad , \tag{3}$$

where m is the weight parameter (m >1), $d_{ij} = \| x_j - V_i \|$ is the Euclidean distance between example x_j to the cluster center V_i. u_{ij} is the membership of x_j belongs to cluster i. Whereas the FCM algorithm asks the specification of cluster number c in advance. Thus questions arise from the rationality of the clustering result and the determination of the optimal cluster number c^*, which is equivalent to obtain the optimal partition matrix $U = [u_{ij}]$ of the given positive example set.

To evaluate the partition effect, the fuzzy partition entropy and partition fuzzy degree (PFD) were proposed as criteria of cluster validity defined as Eq.(4) and Eq.(5) respectively. Bezdek used the fuzzy partition entropy to construct a cluster validity criterion [8] as Eq.(6) for determining the optimal cluster number c^*.

$$H(U;c) = -\frac{1}{n}\sum_{i=1}^{c}\sum_{j=1}^{n} u_{ij} \cdot \log_a(u_{ij}) \quad , \tag{4}$$

$$P_f(U;c) = \frac{1}{n}\sum_{i=1}^{c}\sum_{j=1}^{n} \| u_{ij} - (u_{ij})_H \| \quad , \tag{5}$$

where $(u_{ij})_H$ is the defuzzifying result of the fuzzy partition matrix.

$$H(U^*, c^*) = \min_{c}\left\{ \min_{\Omega_c} H(U;c) \right\} . \tag{6}$$

Unfortunately, functions of $H(U;c)$ and $P_f(U;c)$ have a increasing tendency with the augment of cluster number c, which will impact the detection of the optimal number. So we adopt the modified partition fuzzy degree (MPFD) defined as Eq.(7) to avoid those problems,

$$M_{pf}(U;c)=\frac{P_f(U;c)}{\tilde{H}(U;c)}, \quad (7)$$

where $\tilde{H}(U;c)$ is the smoothed $H(U;c)$ by the 3-point smoothing operator or median filter. In this way, by compensating the increasing tendency of PFD function with the increasing of cluster number, the MPFD function can be easily used to select the optimal cluster number c^* as Eq. (8).

$$M_{pf}(U,c^*)=\min_c\left\{\min_{\Omega_c} MPF(U;c)\right\}. \quad (8)$$

If the $M_{pf}(U,c^*)$ is less than the given threshold (say, 0.5), the current node of cascade will be replaced by the divided c^* branches. Each branch receives only the corresponding subset of positive samples together with a new filtered set of negative samples, which will be trained with FloatBoost respectively. Otherwise, the monolithic classifier will be used preserving the cascade structure at this node. This recursive procedure does not stop until a given target depth of the tree is reached.

1. Load all positive and negative training examples, and then start to train the root node of the detection tree with FloatBoost learning;
2. Load all positive training examples S_+ and negative training examples S_- assigned by the parent node, with which to train the standard stage classifier P. If the given performance of the detection tree achieve go to step 5;
3. For $c = 2, 3, \cdots, 2\ln(|S_+|)$, do FCM clustering with all the p features composing P. Then determine the optimal clustering result according to the MPFD $M_{pf}(U, c^*)$;
4. If $M_{pf}(U, c^*) > 0.5$, then go to step 2 to train the tree node of next level; Else: set S_+^i (i=1, 2, $\cdots$, c^*) of the i-th positive example sub-cluster of S_+, Train c^* standard stage classifier P_i on S_+^i plus S_-, replace P with nodes $\{P^1, P^2, \cdots, P^{c^*}\}$, thus the node P is divided into c^* branches; then go to step 2 to train each node of the new split branches;
5. Output the detection tree.

Fig. 3. Detection tree training algorithm

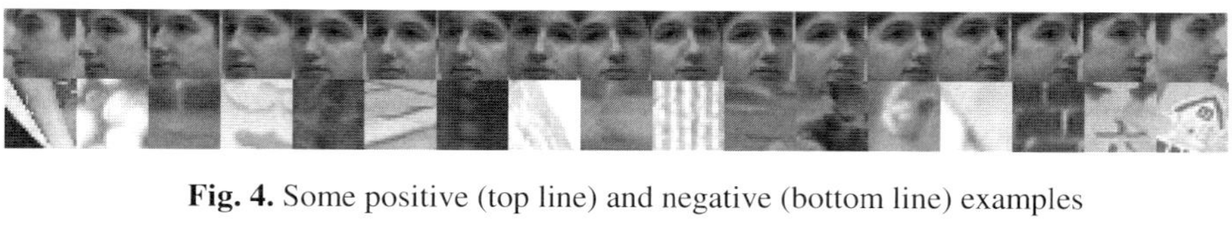

Fig. 4. Some positive (top line) and negative (bottom line) examples

When such a constructed tree is applied to detection, the candidate image regions passed through the tree detector in a depth-first way from the root to a terminal node. If the input pattern is rejected by a tree node, the search will trace back to the nearest upper split point and try another branch, until a path can accept the input pattern or all the possible paths have been searched without acceptances. In the former case, the input pattern will be labeled positive, and the later negative. Since facial regions in an image contains edges of eyes, mouth *etc.*, we can use Canny edge detector to reject some sub-windows of image that contain too few or too much edges, which are impossible of including faces.

3 Experimental Results and Analysis

In this section, two experiments are conducted to compare the performance of Bootstrap and E-Bootstrap algorithms and test the performance of the proposed detector tree, respectively. To show the validity of E-Bootstrap, we adopt the Adaboost learning based method [4] to train classifier for frontal face detection. We simply collect about 6000 frontal face examples from all kind of available dataset. Together with the same positive examples, databases of "non-face" collected by Bootstrap and E-Bootstrap strategies are used to train detectors P^B and P^E individually. Both P^B and P^E are applied to three different test databases to show their validity. The first dataset is MIT + CMU test set, which consists of 125 images containing 481 frontal faces [5]. The second one contains 236 images of Champions database [9], and each image involves one face. The last one is a home-brewed database including 150 images (involving 563 faces) with cluttered background. The performances of face classifiers on those test sets are measured by *Recall* and *Precision* [10].

Experiment 1: In this experiment, the size of extended example set in Bootstrap is smaller than that in E-Bootstrap, viz. $|S_i^B|<|S_i^E|$. Thus we enable $|S^B| \approx |S^E| \approx 15000$ with $i^B > i^E$. Since E-Bootstrap converges faster in this case, the training time of E-Bootstrap and Bootstrap is comparative though we embed a training procedure in each iteration of E-Bootstrap. However, the *Precision* of P^E is almost two times higher than that of P^B, which can be seen in Table 1. During training the positive training sets of Bootstrap and E-Bootstrap are hold constant while the negative set are augmented respectively with the increase of iteration time, thus the *Recall* of P^E and P^B are parallel, bus the *Precision* of P^E is higher. So we will adopt the E-Bootstrap for our negative example selection in the training of multi-view detection tree.

Table 1. Performances comparison between detectors P^B and P^E

Test datasets		MIT + CMU	Champions	Home-brew	Average
Recall	P^B	83.6%	98.1%	93.2 %	91.6%
	P^E	82.1%	98.6%	90.7%	90.5%
Precision	P^B	34.2%	82.9%	14.2%	43.8%
	P^E	83.4%	97.1%	80.8%	87.1%

Experiment 2: In this experiment, about 3000 multi-view face examples and abundant images without faces are collected from all kind of available dataset before training. Through random mirroring, slightly rotating, and re-scaling these obtained face images, about 18000 multi-view faces examples were generated. With the all positive multi-view face samples and negative examples collected by E-Bootstrap we built three face detectors by the same FloatBoost learning algorithm: *Detector* 1 is a single cascade detector with 21 stages; *Detector* 2 is a detection tree trained by the method proposed in reference [6]. And *Detector* 3 is a detection tree trained with the proposed method. We apply the three detectors to two test databases to validate the performance of our algorithm. One dataset is the CMU profile face set [5], which consists of 208 images with 441 faces of which 347 are profile views. Another one is a home-brewed database contains 452 faces in 100 images with cluttered background. The performance of the three detectors on the two test sets is shown in Table 2.

We can see that *Detector* 1 has the lowest detection rate because it can not deal with the variety of multi-view faces, the tree detector proposed by Lienhart *et al* though can deal with the face variety better, the sub-patterns they divide is not quite homogeneous according to the clustering validity. So *Detector* 3 trained with the optimal partition method of the multi-view face examples has the highest detection rate. However, the

Table 2. Performance comparison of three detectors

Test datasets		CMU	Home-brew	Average
Recall	*Detector* 1	71.1%	65.5%	68.3%
	Detector 2	78.0%	80.6%	79.3%
	Detector 3	87.8%	89.2%	88.5%

Fig. 5. Some multi-view face detection results

detection speed of *Detector* 1 is the fastest. And the speed of *Detector* 2 and 3 are comparative but slightly lower than that of *Detector* 1. Fig. 5 partially demonstrates the test results of our multi-view face detection tree.

4 Conclusions

In this paper, an E-Bootstrap example selection algorithm for active learning is proposed. In comparison with the traditional Bootstrap sampling algorithm, experimental results show that the proposed one, with almost the same training time, improves the diversity and typicality of sampled example set. Besides a doable face detection algorithm using a novel FloatBoost classifier tree has been presented. Dissimilar to the widely used cascade classifier, the tree classifier allows the stages of a cascade to split into several branches in order to deal with the potential diversity of multi-view faces. The optimal splitting is realized by the FCM clustering algorithm and the cluster validity function of MPFD, which improve the discriminative power compared to the original detection tree [6] with slightly lower detection speed. Experimental results show that the proposed algorithm has a better detection performance. Although the proposed approach has been applied only to multi-view human face detection, it can be applied to other complex object detection problems as well.

References

1. Yang, M.H., Kriegman, D.J., Ahuja, N.: Detecting Faces in Images: A Survey. IEEE Trans. on Pattern Analysis and Machine Intelligence, Vol. 24 (2002) 34–58
2. Lanitis, A., Taylor, C.J., Cootes, T.F.: An Automatic Face Identification System Using Flexible Appearance Models. Image and Vision Computing, Vol. 13 (1995) 393–401
3. Sung, K. K., Poggio,T.: Example-based Learning for View-based Human Face Detection. IEEE Trans. on Pattern Analysis and Machine Intelligence, Vol. 20 (1998) 39–51
4. Viola, P., Jones, M.: Rapid Object Detection Using a Boosted Cascade of Simple Features. Proc. IEEE Conf. on Computer Vision and Pattern Recognition, Kauai, Hawaii USA, Vol. 1 (2001) 511–518
5. Li, S.Z., Zhang, Z.Q.: FloatBoost Learning and Statistical Face Detection. IEEE Trans. on Pattern Analysis and Machine Intelligence, Vol. 26 (2004) 1112–1122
6. Lienhart,. R., Liang, L.H., Kuranov, A.: A Detector Tree of Boosted Classifiers for Real-time Object Detection and Tracking. IEEE Conf. International Conference on Multimedia and Expo, Baltimore, MD, USA, Vol. 2 (2003) 6–9
7. Li, J., Gao, X.B., Jiao, L.C.: A New Cluster Validity Function Based on the Modified Partition Fuzzy Degree. LNAI, RSCTC, Vol. 3066 (2004) 586–591
8. Bezdek, J.C.: Pattern Recognition with Fuzzy Objective Function Algorithms. Plenum Press, New York (1981)
9. http://www.libfind.unl.edu/alumni/events/breakfast_for_champions.htm
10. Gao, X. B., Tang, X.: Unsupervised Video Shot Segmentation and Model-free Anchorperson Detection for News Video Story Parsing, IEEE Trans. on Circuits Systems for Video Technology. Vol. 12 (2002) 765–776

A Novel Fourier Descriptor for Shape Retrieval

Bin Wang[1] and Chaojian Shi[1,2]

[1] Department of Computer Science and Engineering, Fudan University, Shanghai, 200433, P.R. China
[2] Merchant Marine College, Shanghai Maritime University, Shanghai, 200135, P.R. China
wangbin.cs@fudan.edu.cn, cjshi@shmtu.edu.cn

Abstract. A novel Fourier descriptor (FD), which is derived from chord-length functions (CLF) obtained through equal-arc-length partitions of a contour, is proposed. The proposed FD is tested on a standard shape database and experimental results show that it outperforms the existing FDs which are derived from other shape signatures.

1 Introduction

Shape-based image retrieval is a hot topic in image processing and pattern recognition. Its applications can be found in many areas, such as meteorology, medicine, space exploration, manufacturing, entertainment, education, law enforcement and defense. Shape retrieval includes three primary issues: shape description, shape similarity measure and shape indexing. Among them, shape description is the most important issue.

Fourier descriptor (FD) [1,2] is one of the widely used shape descriptors. In general, the FD is obtained by applying a Fourier transform on a shape signature. A shape signature is any 1D function representing 2D areas or boundaries. Since different shape signature will lead to different FD, the performance of FD method is affected by the shape signature. Till now, many shape signatures, such as complex coordinates, centroid distance, tangent angle, curvature, cumulative angle and so on, have been proposed for deriving FD. Zhang et al.[3] compared six different FDs which are derived from different shape signatures. They claim that the FD derived from centroid distance signature is significantly better than those derived from the other shape signatures.

To further improve the performance of FD, we develop a novel shape signature, chord-length functions (CLF), which is obtained by partitioning the contour into arcs of the same length. The advantage of CLF is that it can capture both the global and local shape features. Therefore, CLF can characterize the shape more accurately. Experimental results show that our method can achieve higher retrieval performance than the existing FDs.

2 Brief Review of Fourier Descriptor

A contour C can be denoted as an ordered sequence of N coordinate points, $C = \{\lambda_t = (x(t), y(t)), t = 0, 1, \ldots, N-1\}$, where C is closed, i.e. $\lambda_{i+N} = \lambda_i$. Suppose

L. Wang et al. (Eds.): FSKD 2006, LNAI 4223, pp. 822–825, 2006.

that $r(t)$ is a shape signature derived from contour C. One dimensional Fourier transform is then applied on $r(t)$ to obtain the Fourier transform coefficients

$$U(n) = \frac{1}{N}\sum_{t=0}^{N-1} r(t)exp(\frac{-j2\pi nt}{N}),\ n = 0, 1, \ldots, N-1. \tag{1}$$

Since the signal, $r(t)$, is real, there are only $N/2$ different frequencies in the result of Fourier transformation (magnitude of the frequency response is symmetric) and the feature vector is formed by the first $N/2$ coefficients corresponding to the low frequency components. Scale invariance is achieved by dividing the magnitude values of FD, rotation invariance is achieved by taking only the magnitude values of the FD. The invariant feature vector used to describe the shape consists of the first $N/2$ coefficients, that is,

$$f = \left[\frac{|U(1)|}{|U(0)|}, \frac{|U(2)|}{|U(0)|}, \ldots, \frac{|U(N/2)|}{|U(0)|}\right]^T \tag{2}$$

3 The Proposed Fourier Descriptor

3.1 Chord-Length Function (CLF)

Let $L = \sum_{i=0}^{N-1} d(\lambda_i, \lambda_{i+1})$ be the perimeter of the contour C, where $d(\lambda_i, \lambda_{i+1})$ is Euclidean distance between points λ_i and λ_{i+1}, Let us start from a point $\lambda_i \in C$ and follow the contour anti-clockwise to equally divide it into k sections $\widehat{\lambda_i s_1}, \widehat{s_1 s_2}, \ldots, \widehat{s_{k-1}\lambda_i}$, and obtain $k-1$ chords $\overline{\lambda_i s_1}, \overline{\lambda_i s_2}, \ldots, \overline{\lambda_i s_{k-1}}$, where s_j is the jth division point and $k > 1$ is a pre-specified parameter. We now have $k-1$ chord lengths $L_1^{(i)}, L_2^{(i)}, \ldots, L_{k-1}^{(i)}$, where $L_j^{(i)}$ is the length of the chord $\overline{\lambda_i s_j}$, i.e. the Euclidean distance between the points λ_i and s_j.

As point λ_i moves along the contour, the chord lengths $L_j^{(i)}, j = 1, \ldots, k-1$, vary accordingly. In other words, $L_j^{(i)}$ are functions of λ_i. Without loss of generality, we specify λ_0 as the reference point. Then each point λ_i can be uniquely identified with the length $l_i \in [0, L]$ of arc $\widehat{\lambda_0\lambda_i}$. Therefore each chord length $L_j^{(i)}$ can be considered as a function of arc length l_i. Then we obtain a set of chord length functions $\Phi = \{L_1, L_2, \ldots, L_{k-1}\}$. Since $L_j(l) = L_{k-j}(l + j \cdot L/k), j = 1, 2, \ldots, k-1$, only half of the set of chord-length functions are needed for description shape, i.e. $\Phi_h = \{L_1, L_2, \ldots, L_{k/2}\}$.

From the definition of the chord-length functions, we can see that the value of function L_j is the length of the chord corresponding to the arc whose length is $j \cdot L/k$. Different level chords which correspond to arcs with different length are used to characterize the shape and in these chord-length functions, both the global feature and local feature can be reflected. Therefore, CLF descriptor is superior to the existing shape signatures such as centroid distance, curvature function and so on. It should be pointed out that k is the only parameter of CLF. The larger the k is, the smaller the partitions will be and the more details of the boundary will be described. So if we expect higher accuracy in shape distinction, k will be set larger.

3.2 The FD Using CLF for Shape's Difference Measure

We have proposed a shape description CLF $\Phi_h = \{L_1, L_2, \ldots, L_{k/2}\}$. One dimensional Fourier transformation is then applied on each chord-length function L_i, and an invariant feature vector which is similar with Eq. 2 is obtained as follows

$$f_i = [\mu_1^{(i)}, \mu_2^{(i)}, \ldots, \mu_{N/2}^{(i)}]^T \tag{3}$$

The set of feature vectors $\{f_1, f_2, \ldots, f_{k/2}\}$ which is invariant to translation, scaling and rotation is then used to describe the shape.

Suppose $F^{(A)} = \{f_1^{(A)}, f_2^{(A)}, \ldots, f_{k/2}^{(A)}\}$ and $F^{(B)} = \{f_1^{(B)}, f_2^{(B)}, \ldots, f_{k/2}^{(B)}\}$ are FDs of shape A and shape B derived from CLF. The difference between shape A and shape B is then defined as follows

$$d(A, B) = (\sum_{i=1}^{k/2} | f_i^{(A)} - f_i^{(B)} |^2)^{1/2} \tag{4}$$

where $|\cdot|$ denotes the Euclidean distance between two feature vectors.

4 Experimental Results and Discussions

To evaluate the retrieval performance of the proposed FD method, we use a standard shape database, MPEG-7 Part B [4], which consists 1400 images: 70 shape

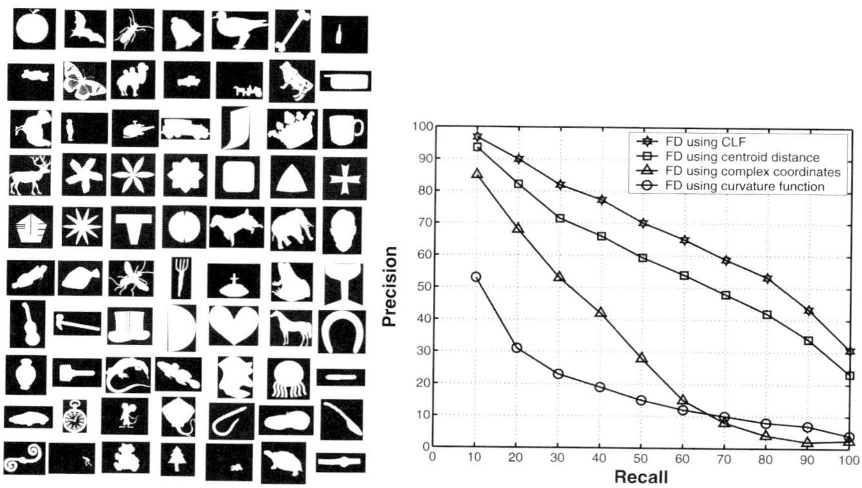

Fig. 1. Left: MPEG-7 Part B, 70 categories with 20 shapes each. Right: Precision-Recall plot of the proposed FD and other three classical FDs.

categories as shown in Fig. 1(Left), 20 images per category. Three widely used FDs which are derived from centroid distance signature, complex coordinates and curvature function, respectively, are selected for comparison.

The commonly used retrieval performance measurement, precision and recall [5] are adopted as evaluation of the query results. The precision P and recall R are calculated as $P = r/n$ and $R = r/m$, where r is the number of retrieved relevant shapes, n is the total number of retrieved shapes and m is the total number of relevant shapes in the whole database. For each shape in the database, take it as a query to match all the shapes in the database. The precision at each level of recall is recorded. The final precision of the retrieval for a certain FD is the average precision of all the queries in the query set. The resulting precision and recall for different FDs are plotted in Fig. 1(Right), where the parameter for CLF is set to $k = 8$.

From the precision and recall plot, we can see that the proposed FD achieves higher precision at each level of recall than the other three FDs.

5 Conclusion

We have presented a FD which is derived from a novel shape signature: chord-length functions (CLF). CLF is obtained through equal-arc-length partitions of a contour. The proposed FD has been tested on a standard shape database and the experimental results show that it outperforms other existing FDs.

Acknowledgement

The research work in this paper is partially sponsored by Shanghai Leading Academic Discipline Project, T0603.

References

1. Chellappa, R., Bagdazian, R.: Fourier Coding of Image Boundaries. IEEE Tans. Pattern Anal. Mach. Intell. **6(1)** (1984) 102-105
2. Wallace, T. P., Wintz, P. A.: An Efficient Three-Dimensional Aircraft Recognition Algorithm Using Normalized Fourier Descriptors. Computer Graphics and Image Processing. **13** (1980) 99-126
3. Zhang, D., Lu, G.: Study and Evaluation of Different Fourier Methods for Image Retrieval. Image and Vision Computing. **23** (2005) 33-49
4. Latecki, L. J., Lakamper, R., Eckhardt, U.: Shape Descriptors for Non-Rigid Shapes with A Single Closed Contour. IEEE Conf. on Comp. Vis. and Patt. Recog. (2000) 424-429
5. Bimbo, A. Del.: Visual Information Retrieval. Morgan Kaufmann Publishers. Inc, San Francisco. CA. (1999) 56-57

Tracking Control of a Nonholonomic Mobile Robot Using a Fuzzy-Based Approach

An-Min Zou, Zeng-Guang Hou, Min Tan, and Zeng-Shun Zhao

Laboratory of Complex Systems and Intelligence Science,
Institute of Automation, The Chinese Academy of Sciences,
P.O. Box 2728, Beijing 100080, China
amzou@126.com, {zengguang.hou, min.tan, zengshun.zhao}@ia.ac.cn

Abstract. This paper investigates the tracking problem of nonholonomic mobile robots. A control structure combining a kinematic controller and a dynamic controller based on nonlinear feedback control plus fuzzy compensator is presented. The fuzzy compensator, whose parameters are tuned on-line, is employed to approximate the total uncertainty including the structured and unstructured uncertainties due to the universal approximation property of fuzzy logic systems. The stability of the proposed approach is guaranteed by the Lyapunov theory. Simulation results show the efficiency of the proposed approach.

1 Introduction

In the last few years, control of a nonholonomic system such as a nonholonomic mobile robot has received wide attention. However, most of the existing methods for solving the control problem of a nonholonomic mobile robot only consider the kinematic model and neglect the vehicle dynamics. Recently, some approaches have been proposed to integrate a kinematic controller and a torque controller for a nonholonomic mobile robot with kinematics and dynamics [1-3]. In [1], Fierro and Lewis studied a control structure that integrated a kinematic controller and a feedforward neural network computed-torque controller for a nonholonomic mobile robot, and the neural network weights are tuned on-line, with no "off-line learning phase" needed. In [2], the kinematic and dynamic controllers are designed by using polar coordinates and wavelet networks, respectively, and the parameters and structure of wavelet networks are adapted on-line. The adaptive control methods, which linearly parameterize the motion model of the nonholonomic mobile robot, have been employed to solve the tracking problem of nonholonomic mobile robots with kinematics and dynamics [3]. However, the external disturbances have not been considered, and computation of the "regression matrix" needs tedious analysis. Fuzzy logic systems (FLS) with product inference, center-average defuzzifier, and Gaussian membership function are capable of approximating any real continuous functions on a compact set to arbitrary accuracy [4]. Neural networks and wavelet networks have the capability to approximate continuous functions, and have been used for the control of a nonholonomic mobile robot with dynamics and kinematics [1] [2]. However, only

L. Wang et al. (Eds.): FSKD 2006, LNAI 4223, pp. 826–835, 2006.

the FLS is constructed from a set of IF-THEN rules, and linguistic information from human experts can be incorporated into the FLS.

In this paper, a control structure combining a kinematic controller and a dynamic controller based on nonlinear feedback control plus fuzzy compensator is presented for a nonholonomic mobile robot. First, the backsteping method is employed to design a kinematic controller to give the velocity control inputs in order to make the pose (position and orientation) error asymptotically stable. Second, a torque controller based on nonlinear feedback control plus fuzzy compensator is designed such that the mobile robot's velocities converge to the given velocity inputs obtained by the kinematic controller, and the stability of the proposed approach is guaranteed by the Lyapunov theory.

This paper is organized as follows. The motion model of nonholonomic systems is described in Section 2. Section 3 discusses the kinematic controller and dynamic controller based on nonlinear feedback control plus fuzzy compensator with applications to the tracking problem. Stability is proven by the Lyapunov stability theory. Some simulation results are presented in Section 4 and conclusions are given in Section 5.

2 Motion Model of a Nonholonomic Mobile Robot

In a 2-dimensional (2D) Cartesian space, the pose of a nonholonomic mobile robot is represented by

$$q = (x, y, \theta)^T, \tag{1}$$

where $(x, y)^T$ is the position of the robot in a reference coordinate system, and the heading direction θ is taken counterclockwise from the OX-axis.

In thisi paper, we consider a class of nonholonomic mobile robots with s constraints whose motion model including kinematics and dynamics can be described by [1]

$$M(q)\ddot{q} + V(q, \dot{q})\dot{q} + F(\dot{q}) + G(q) + \tau_d = B(q)\tau - A^T(q)\lambda, \tag{2}$$

$$\dot{q} = J(q)v, \tag{3}$$

where $M(q) \in R^{3\times3}$ is a symmetric, positive definite inertia matrix; $V(q, \dot{q}) \in R^{3\times3}$ is the centripetal and Coriolis matrix; $F(\dot{q}) \in R^{3\times1}$ denotes the surface friction; $G(q) \in R^{3\times1}$ is the gravitational vector; τ_d denotes bounded unknown disturbances including unmodeled dynamics; $B(q) \in R^{3\times r}$ with $r = 3 - s$ is the input transformation matrix; $\tau \in R^{r\times1}$ is the input vector; $A(q) \in R^{s\times3}$ is the matrix associated with the constraints, and $\lambda \in R^{s\times1}$ is the vector of constraint forces. $v = (v_0, \omega)^T$, v_0, ω are the linear and angular velocities of the robot respectively, and $J(q)$ is given by

$$J(q) = \begin{pmatrix} \cos\theta & -d\sin\theta \\ \sin\theta & d\cos\theta \\ 0 & 1 \end{pmatrix}. \tag{4}$$

It is assumed that the wheels of the robot do not slide. This is expressed by the nonholonomic constraint

$$\dot{x}\sin\theta - \dot{y}\cos\theta - d\dot{\theta} = 0. \tag{5}$$

We consider that all kinematic equality constraints are independent of time, and can be expressed as follows

$$A(q)\dot{q} = 0. \tag{6}$$

Since the trajectory of the mobile robot base is constrained to the horizontal plane, $G(q) = 0$. The kinetic energy K is given by [1]

$$K = \frac{1}{2}\dot{q}^T M(q)\dot{q}. \tag{7}$$

Differentiating (3), substituting this result in (2), and then multiplying by J^T, we can eliminate the constraint matrix $A^T(q)\lambda$ as follows [1]

$$J^T M J\dot{v} + J^T(M\dot{J} + VJ)v + J^T F + J^T\tau_d = J^T B\tau. \tag{8}$$

By appropriate definitions we can rewrite (8) as follows

$$\bar{M}\dot{v} + \bar{V}v + \bar{F} + \bar{\tau}_d = \bar{B}\tau. \tag{9}$$

The parameters $\bar{M}$ and $\bar{V}$ in (9) are functions of physical parameters of the nonholonomic mobile robot such as mass of the robot and wheels, radius of the wheel, moments of inertia and so on. Because the measurements may be inaccurate and the environment and payloads may be variable, the precise values of these parameters are difficult to acquire. Therefore, it is assumed that actual values of $\bar{M}$ and $\bar{V}$ can be separated as nominal and uncertain parts as follows

$$\begin{aligned} \bar{M} &= \bar{M}_0 + \Delta\bar{M}, \\ \bar{V} &= \bar{V}_0 + \Delta\bar{V}, \end{aligned} \tag{10}$$

where $\bar{M}_0$ and $\bar{V}_0$ are known precisely, and $\Delta\bar{M}$ and $\Delta\bar{V}$ denote the structured uncertainty.

By defining $\bar{\zeta} = \Delta\bar{M}\dot{v} + \Delta\bar{V}v + \bar{F} + \bar{\tau}_d$, we have

$$\bar{M}_0\dot{v} + \bar{V}_0 v + \bar{\zeta} = \bar{B}\tau, \tag{11}$$

where $\bar{\zeta}$ denotes the total uncertainty including the structured and unstructured uncertainties. Note that $\bar{\zeta}$ is a function of $\dot{v}$ and v that can be measured.

3 Controller Design

The trajectory tracking problem for a nonholonomic mobile robot is described as follows: given the reference pose $q_r = (x_r, y_r, \theta_r)^T$ and the reference velocities

$v_r = (v_{0r}, \omega_r)^T$ with $v_{0r} > 0$ for all time, find a smooth velocity v_c such that $\lim_{t\to\infty}(q_r - q) = 0$ and $\lim_{t\to\infty}(v_r - v_c) = 0$. Then compute the torque input τ for (2), such that $v \to v_c$ as $t \to \infty$.

The tracking error vector is expressed in the robot coordinate system as

$$e_r = T_e(q_r - q),$$
$$\begin{pmatrix} e_1 \\ e_2 \\ e_3 \end{pmatrix} = \begin{pmatrix} \cos\theta & \sin\theta & 0 \\ -\sin\theta & \cos\theta & 0 \\ 0 & 0 & 1 \end{pmatrix} \begin{pmatrix} x_r - x \\ y_r - y \\ \theta_r - \theta \end{pmatrix}. \quad (12)$$

Using the backstepping approach, the following control law can be obtained [6]

$$v_c = \begin{cases} v_{0c} = v_{0r}\cos e_3 + k_1 e_1 \\ \omega_c = \omega_r + k_2 v_{0r} e_2 + k_3 v_{0r} \sin e_3 \end{cases}. \quad (13)$$

with $k_i > 0 (i = 1, 2, 3)$.

Let's define the auxiliary velocity error $e_c = v_c - v$. So, by considering $\bar{\zeta} = 0$, a nonlinear feedback control for (11) can be obtained by

$$\tau = \bar{B}^{-1}(\bar{M}_0 u + \bar{V}_0 v), \quad (14)$$

where $u = \dot{v}_c + k_4 I_{2\times 2} e_c$, k_4 is a positive constant, and I is the identity matrix.

Substituting (14) into (11), we have

$$\dot{e}_c + k_4 I_{2\times 2} e_c = 0, \quad (15)$$

i.e., v converges to v_c with an exponential ratio k_4.

However, the control law (14) includes only known parts of the systems. Therefore, the structured uncertainty, disturbance, unmodeled dynamic and surface friction will hinder the performance of the closed system.

Motivated by the capability of FLS as universal approximators [4], we can design a controller that combines the controller (14) and an FLS-based controller for the total uncertainty of the system including the structured and unstructured uncertainties. In this way, applying (14) to (11) yields

$$\dot{e}_c + k_4 I_{2\times 2} e_c = \rho, \quad (16)$$

where $\rho = -\bar{M}_0^{-1}\bar{\zeta}$.

Then, the overall control law becomes

$$\tau = \tau_0 + \tau_f, \quad (17)$$

where τ_0 is defined by (14) and τ_f is a compensating torque to be determined below.

ρ in (16) can be approximated by FLS as follows [4], [5]

$$\rho = W^*\xi(z) + \varepsilon(z), \quad (18)$$

where $\varepsilon(z)$ is the approximation error. $\xi(z) = (\xi^1(z), \xi^2(z), ..., \xi^N(z))^T$ is the fuzzy basis function vector fixed by the designer, and z is given by

$$z = (\dot{v}^T, v^T)^T, \tag{19}$$

and $W^* = (\omega_{ij}) \in R^{m \times N}$ in (18) is an optimal weight matrix satisfying that

$$W^* = \arg\min_{\hat{W}} \{ \sup_{z \in D_z} |\hat{\rho}(z|\hat{W}) - \rho(z)| \}, \tag{20}$$

where D_z denotes the sets of suitable bounds of z, and $\hat{\rho}(z|\hat{W})$, which is an estimation of ρ, is given by

$$\hat{\rho}(z|\hat{W}) = \hat{W}\xi(z), \tag{21}$$

with $\hat{W}$ an adjustable weight matrix. Define the compensating torque τ_f in (17) as

$$\tau_f = -\bar{B}^{-1}\bar{M}_0\hat{\rho}(z|\hat{W}). \tag{22}$$

Substituting the control law (17) including the nonlinear feedback control (14) and the fuzzy part (22) into (11), we have

$$\dot{e}_c + k_4 I_{2\times 2} e_c + \tilde{\rho}(z) = 0, \tag{23}$$

where

$$\tilde{\rho}(z) = \rho(z) - \hat{\rho}(z|\hat{W}) = \varepsilon(z) + \tilde{W}\xi(z), \tag{24}$$

with $\tilde{W} = W^* - \hat{W}$ denoting estimation error of the weight matrix.

Equation (23) can be rewritten in the state-space equation as follows

$$\dot{e}_c = Ae_c + C\tilde{\rho}(z), \tag{25}$$

where $A = -k_4 I_{2\times 2}$ and $C = -I_{2\times 2}$.

Assumption 1: The desired reference trajectory is continuous and bounded so that $\|q_r\| \le q_M$ with q_M a known scalar bound, and the total uncertainty $\bar{\zeta}$ is bounded so that $\|\bar{\zeta}\| \le \zeta_M$.

Assumption 2: The reference linear velocity v_{0r} is bounded, and $v_{0r} > 0$ for all $t \ge 0$, and the angular velocity ω_r is bounded.

Assumption 3: The norm of the optimal weight matrix of the FLS is bounded so that $\|W^*\| \le w_m$, and the approximation error is bounded so that $\|\varepsilon\| \le \varepsilon_m$.

Theorem 1. *Consider the nonholonomic system (11), and supposed Assumptions 1-3 are satisfied. Given the kinematic control laws (13), the torque control (17), which consists of the nonlinear feedback control (14) and the fuzzy part (22), and the FLS adaptation laws*

$$\dot{\hat{W}} = \Lambda^{-1}C^T P e_c \xi^T, \tag{26}$$

where $\Lambda = diag(\Lambda_1, \Lambda_2, ..., \Lambda_n)$ with $\Lambda_i > 0$ is a gain matrix, and P is the symmetric positive definite solution of the following Riccati equation

$$A^T P + PA + P^T CC^T P + Q = 0, \tag{27}$$

where Q is a constant positive definite matrix with appropriate dimensions given in advance. Then $\|e_c\|$ and $\|\tilde{W}\|$ are uniformly ultimately bounded.

Proof. Consider the following Lyapunov function candidate

$$V = \frac{1}{2}(e_1^2 + e_2^2) + \frac{1}{k_2}(1 - \cos e_3) + V_1, \tag{28}$$

where

$$V_1 = e_c^T P e_c + tr(\tilde{W} \Lambda \tilde{W}). \tag{29}$$

The time derivative of V is

$$\dot{V} = \dot{e}_1 e_1 + \dot{e}_2 e_2 + \frac{1}{k_2}\dot{e}_3 \sin e_3 + \dot{V}_1, \tag{30}$$

and differentiating V_1 along (25) results in

$$\begin{aligned} \dot{V}_1 =& e_c^T(A^T P + PA)e_c + \tilde{\rho}^T C^T P e_c + e_c^T P C \tilde{\rho} + 2tr(\tilde{W}^T \Lambda \dot{\hat{W}}) \\ =& e_c^T(A^T P + PA)e_c + 2e_c^T P C \tilde{\rho} - 2tr(\dot{\hat{W}}^T \Lambda \tilde{W}). \end{aligned} \tag{31}$$

Applying repeatedly the properties of trace of matrix and substituting (25) and (26) into (31) yields

$$\begin{aligned} \dot{V}_1 &= -e_c^T(P^T C C^T P + Q)e_c + 2e_c^T P C(\varepsilon + \tilde{W}\xi) - 2tr(\tilde{W}^T \Lambda \dot{\hat{W}}) \\ &= -e_c^T(P^T C C^T P + Q)e_c + 2e_c^T P C\varepsilon + 2e_c^T P C \tilde{W}\xi - 2tr(\tilde{W}^T \Lambda \dot{\hat{W}}) \\ &= -e_c^T(P^T C C^T P + Q)e_c + 2e_c^T P C\varepsilon + 2tr(e_c^T P C \tilde{W}\xi) - 2tr(\tilde{W}^T \Lambda \dot{\hat{W}}) \\ &= -e_c^T(P^T C C^T P + Q)e_c + 2e_c^T P C\varepsilon + 2tr(\xi^T \tilde{W}^T C^T P e_c) - 2tr(\tilde{W}^T \Lambda \dot{\hat{W}}) \\ &= -e_c^T(P^T C C^T P + Q)e_c + 2e_c^T P C\varepsilon + 2tr(\tilde{W}^T C^T P e_c \xi^T) - 2tr(\tilde{W}^T \Lambda \dot{\hat{W}}) \\ &= -e_c^T(P^T C C^T P + Q)e_c + 2e_c^T P C\varepsilon + 2tr(\tilde{W}^T (C^T P e_c \xi^T - \Lambda \dot{\hat{W}})) \\ &= -e_c^T(P^T C C^T P + Q)e_c + 2e_c^T P C\varepsilon \\ &= -e_c^T Q e_c - (C^T P e_c - \varepsilon)^T (C^T P e_c - \varepsilon) + \varepsilon^T \varepsilon \\ &\le -e_c^T Q e_c + \varepsilon^T \varepsilon. \end{aligned} \tag{32}$$

The velocity tracking error is

$$\begin{aligned} e_c = v_c - v = \begin{pmatrix} e_4 \\ e_5 \end{pmatrix} = \begin{pmatrix} v_{0c} - v_0 \\ \omega_c - \omega \end{pmatrix} \\ = \begin{pmatrix} v_{0r} \cos e_3 + k_1 e_1 - v_0 \\ \omega_r + k_2 v_{0r} e_2 + k_3 v_{0r} \sin e_3 - \omega \end{pmatrix}, \end{aligned} \tag{33}$$

i.e.,

$$v = \begin{pmatrix} v_{0r} \cos e_3 + k_1 e_1 - e_4 \\ \omega_r + k_2 v_{0r} e_2 + k_3 v_{0r} \sin e_3 - e_5 \end{pmatrix}. \tag{34}$$

By substituting (32) and the derivatives of the pose error into (30), we obtain

$$\dot{V} \leq e_1(\omega e_2 - v_0 + v_{0r}\cos e_3) + e_2(-\omega e_1 + v_{0r}\sin e_3) + \frac{1}{k_2}(\omega_r - \omega)\sin e_3 - e_c^T Q e_c + \varepsilon^T \varepsilon. \quad (35)$$

Substituting (34) into (35) results in

$$\begin{aligned}\dot{V} &\leq -k_1 e_1^2 - \frac{k_3}{k_2} v_{0r}\sin^2 e_3 + e_1 e_4 + \frac{1}{k_2} v_{0r} e_5 \sin e_3 - e_c^T Q e_c + \varepsilon^T \varepsilon \\ &\leq -k_1(e_1 - \frac{e_4}{2k_1})^2 - \frac{k_3 v_{0r}}{k_2}(\sin e_3 - \frac{e_5}{2k_3})^2 - e_c^T Q e_c + \varepsilon^T \varepsilon \\ &\leq -e_c^T Q e_c + \varepsilon^T \varepsilon \\ &\leq -\lambda_{min}(Q)\|e_c\|^2 + \varepsilon^T \varepsilon, \end{aligned} \quad (36)$$

where $\lambda_{min}(Q)$ is the minimum eigenvalue of matrix Q, thus $\dot{V}$ is negative outside the following compact set Σ_{e_c}

$$\Sigma_{e_c} = \{e_c(t) | 0 \leq \|e_c(t)\| \leq \sqrt{\frac{1}{\lambda_{min}(Q)}}\|\varepsilon\|\}. \quad (37)$$

According to the Lyapunov theory and LaSalle extension [7], this demonstrates that both $\|e_c\|$ and $\|\tilde{W}\|$ are uniformly ultimately bounded.

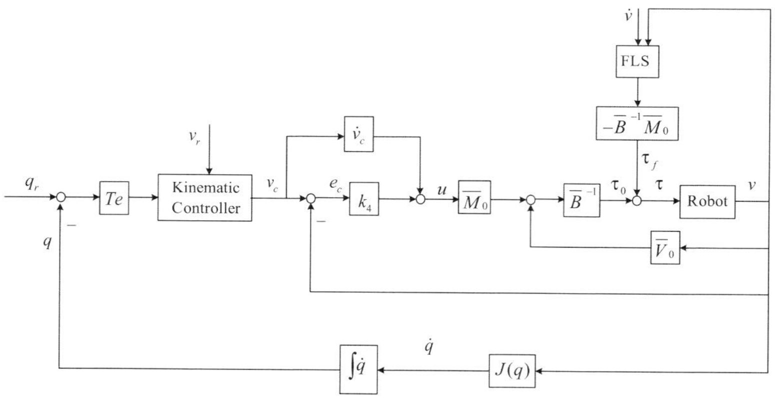

Fig. 1. FLS control to mobile robot

4 Simulation Results

In this section, we will provide some simulation results to show the effectiveness of our proposed methods. The nominal parameters of the robot are: $m_0 = 10$kg, $I_0 = 5$kg.m^2, $R_0 = 0.5$m, $r_0 = 0.05$m, $d = 0$; the uncertain parts of the robot parameters are $\Delta m = 0.5$kg, $\Delta I = 0.2$kg.m^2, $\Delta R = 0$m, $\Delta r = 0$m. The sampling time is 0.01s. The parameters used in the controller are: $k_1 = 10$, $k_2 = 5$, $k_3 = 2$, $k_4 = 4$. The initial pose of the desired virtual cart is $(0, -1, 0)^T$. The actual robot starts at $(-1.8, -1.8, 0)^T$, this means that the initial error is $(1.8, 0.8, 0)^T$. The reference linear and angular velocities are given as follows

$$v_{0r} = \begin{cases} 0.5 + 0.5\sin t, & 0 \le t < 4 \text{ s} \\ 1.5, & 4 \le t < 6 \text{ s} \\ 1 + \cos t, & 6 \le t < 8 \text{ s} \\ 1, & 8 \le t \le 10 \text{ s} \end{cases} \tag{38}$$

and

$$\omega_r = \begin{cases} \cos t, & 0 \le t < 4 \text{ s} \\ 0, & 4 \le t < 6 \text{ s} \\ 1 + 1.5\sin t, & 6 \le t < 8 \text{ s} \\ 1, & 8 \le t \le 10 \text{ s} \end{cases} . \tag{39}$$

The FLS control scheme is shown in Fig. 1. The input vectors of the FLS are $z = (z_1, z_2, z_3, z_4)^T$, where $z_1 = \dot{v}_0$, $z_2 = \dot{\omega}$, $z_3 = v_0$ and $z_4 = \omega$. For each input vector, five Gaussian membership functions are defined as

$$\mu_{F_i^1}(z_i) = \exp\left(-(\frac{z_i + 2}{0.5})^2\right), \ \mu_{F_i^2}(z_i) = \exp\left(-(\frac{z_i + 1}{0.5})^2\right),$$

$$\mu_{F_i^3}(z_i) = \exp\left(-(\frac{z_i}{0.5})^2\right), \ \mu_{F_i^4}(z_i) = \exp\left(-(\frac{z_i - 1}{0.5})^2\right),$$

$$\mu_{F_i^5}(z_i) = \exp\left(-(\frac{z_i - 2}{0.5})^2\right), \ i = 1, 2, 4$$

and

$$\mu_{F_i^1}(z_i) = \exp\left(-(\frac{z_i}{0.5})^2\right), \ \mu_{F_i^2}(z_i) = \exp\left(-(\frac{z_i - 0.5}{0.5})^2\right),$$

$$\mu_{F_i^3}(z_i) = \exp\left(-(\frac{z_i - 1}{0.5})^2\right), \ \mu_{F_i^4}(z_i) = \exp\left(-(\frac{z_i - 1.5}{0.5})^2\right),$$

$$\mu_{F_i^5}(z_i) = \exp\left(-(\frac{z_i - 2}{0.5})^2\right), \ i = 3.$$

To reduce the number of fuzzy rules, like [8], $20(4 \times 5)$ fuzzy rules are adopted as follows

$$R_i^k : \text{ if } z_i \text{ is } F_i^k, \text{ then } \rho_1 \text{ is } B_1^k \text{ and } \rho_2 \text{ is } B_2^k,$$

where $i = 1, 2, 3, 4$ and $k = 1, 2, ..., 5$. ρ_1 and ρ_2 are two outputs of the FLS satisfying that $\rho = (\rho_1, \rho_2)^T$. The initial values of the weight matrix $\hat{W}(0)$ are set equal to zero matrix. The positive definite matrix Q in Riccati equation (31) and the gain matrix Λ are selected as $Q = diag(15, 15)$ and $\Lambda = diag(4, 4)$, respectively. It is assumed that the external disturbance and the surface friction are $\tau_d + F = (5\sin t, 5\cos t, 5\sin t)^T$. The simulation results are shown in Figs. 2 and 3. The results show that the proposed control scheme using fuzzy compensator greatly improves the performances of robot tracking with model uncertainty.

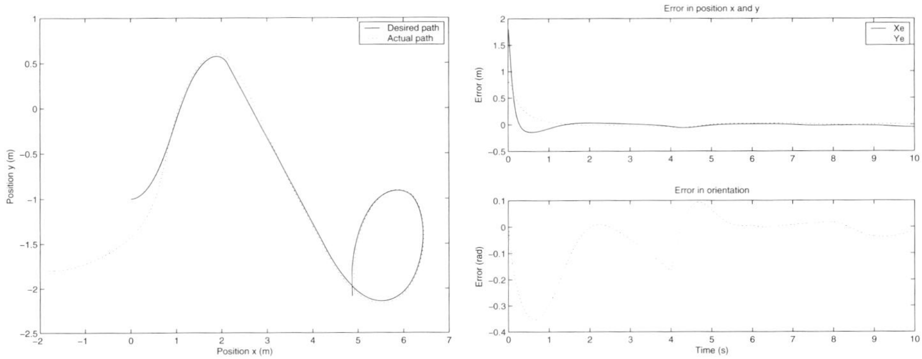

(a) (b)

Fig. 2. Tracking with uncertainty and feedback control: (a) mobile robot trajectory; (b) position and orientation errors

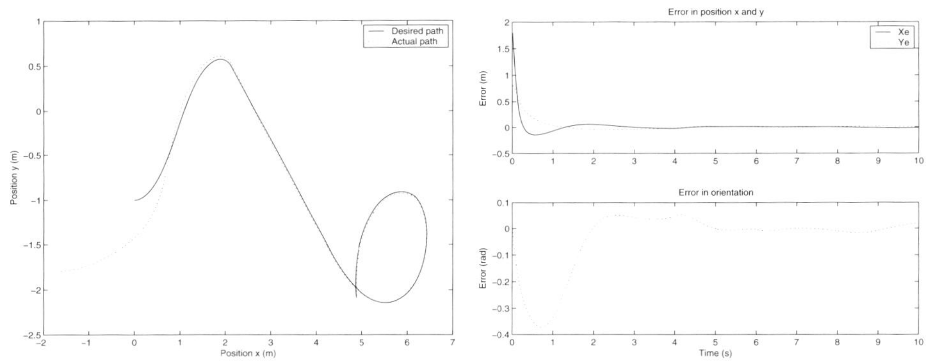

(a) (b)

Fig. 3. Tracking using FLS control with uncertainty: (a) mobile robot trajectory; (b) position and orientation errors

5 Conclusions

A stable control algorithm capable of dealing with the total uncertainty including the structured and unstructured uncertainties for path tracking of a nonholonomic mobile robot is proposed in this paper. The control structure combines a kinematic controller obtained by the backstepping approach and a dynamic controller based on nonlinear feedback control plus fuzzy compensator. In the proposed scheme, the fuzzy part with a set of "on-line" tunable parameters with no "off-line learning phase" needed is employed to approximate the total uncertainty due to the universal approximation property of the FLS. Stability of the proposed method is proven by Lyapunov theory. Some simulations are provided in order to illustrate the feasibility of the proposed method.

Acknowledgments

This research was supported in part by the National Natural Science Foundation of China (Grants 60205004, 50475179 and 60334020), the National Basic Research Program (973) of China (Grant 2002CB312200), and the Hi-Tech R&D Program (863) of China (Grants 2002AA423160 and 2005AA420040).

References

1. Fierro, R., Lewis, F. L.: Control of a Nonholonomic Mobile Robot Using Neural Networks. IEEE Transactions on Neural Networks, **9** (4) (1998) 589-600
2. de Sousa, C. Jr., Hemerly, E. M., Galvao, R. K. H.: Adaptive Control for Mobile Robot Using Wavelet Networks. IEEE Transactions on Systems, Man and Cybernetics, Part B, **32** (4) (2002) 493-504
3. Fukao, T., Nakagawa, H., Adachi, N.: Adaptive Tracking Control of a Nonholonomic Mobile Robot. IEEE Transactions on Robotics and Automation, **16** (5) (2000) 609-615
4. Wang, L. X.: Fuzzy Systems are Universal Approximators. Proceedings of the IEEE International Conference on Fuzzy Systems, (1992) 1163-1170
5. Wang, L. X., Mendel, J. M.: Fuzzy Basis Function, Universal Approximation, and Orthogonal Least-Squares Learning. IEEE Transactions on Neural Networks, **3** (5) (1992) 807-814
6. Kanayama, Y., Kimura, Y., Miyazaki, F., Noguchi, T.: A Stable Tracking Control Method for an Autonomous Mobile Robot. Proceedings of IEEE International Conference on Robotics and Automation, **1** (1990) 384-389
7. LaSalle, J. P.: Some Extensions of Lyapunov's Second Method. IRE Transactions Circuit Theory, **7** (4) (1960) 520-527
8. Chan, P. T., Rad, A. B., Wang, J.: Indirect Adaptive Fuzzy Sliding Mode Control: Part II: Parameter Projection and Supervisory Control. Fuzzy Sets and Systems, **122** (1) (2001) 31-43

Implementation of the Avoidance Algorithm for Autonomous Mobile Robots Using Fuzzy Rules

Jang Hyun Kim[1], Jin Bae Park[2], and Hyunseok Yang[3]

[1] Department of Electrical and Electronic Engineering, Yonsei University,
134 Shinchon-Dong Seodaemun-ku, Seoul, Korea
jhkim@control.yonsei.ac.kr
http://control.yonsei.ac.kr
[2] Professor, Department of Electrical and Electronic Engineering, Yonsei University,
134 Shinchon-Dong Seodaemun-ku, Seoul, Korea
jbpark@yonsei.ac.kr
http://control.yonsei.ac.kr
[3] Professor, Department of Mechanical Engineering, Yonsei University,
134 Shinchon-Dong Seodaemun-ku, Seoul, Korea
hsyang@yonsei.ac.kr
http://mservo.yonsei.ac.kr

Abstract. Complex "lifelike" behaviors are composed of local interactions of individuals under fundamental rules of artificial life. In this paper, fundamental rules for cooperative group behaviors, "Avoidance" of multiple autonomous mobile robots are represented by a small number of fuzzy rules. Fuzzy rules in Sugeno type and their related parameters are automatically generated from clustering input-output data obtained from the algorithms for the group behaviors. Simulations demonstrate the fuzzy rules successfully realize group intelligence of mobile robots.

1 Introduction

Study of artificial life is one of the advanced technologies intelligence control. Research concerned with the artificial life is devoted to understanding life by attempting to abstract the fundamental dynamical principles underlying biological phenomena, and recreating these dynamics in other physical media including computers. In addition to providing new ways to study the biological phenomena associated with biological life(B-life), artificial life(A-life) allows us to extend our concepts to the larger domain of "life-as-it-could-be"[1][2]. Group intelligence in complex lifelike behaviors is considered to be composed of local interactions of individuals with simple functions under fundamental rules of artificial life. In this paper, fundamental rules in cooperative behaviors of multiple autonomous mobile robots(AMRs) are represented by small number of fuzzy system rules. Group intelligence can be observed in such cases as in obstacle avoidance of birds and in forming a swarm of insects[3][4][5]. One type of group behavior, obstacle avoidance is considered in this paper as simulations. Avoidance refers to

L. Wang et al. (Eds.): FSKD 2006, LNAI 4223, pp. 836–845, 2006.

the behavior that robots scattered in a space gather to make a squad. Our goal is to generate a small number of fuzzy rules as fundamental rules that govern complex lifelike group behaviors of mobile robots. Generating the fuzzy rules consists of two steps. First, we realize mathematical algorithms responsible for the group behaviors and collect numerical input-output data from the process. Second, We cluster the input-output data to identify fuzzy system structure and parameters to construct fuzzy rules from the clustering result. The performance of group behaviors governed by the fuzzy rules is compared with those by the algorithm[6]. Section 2 describes the algorithms for group behaviors of avoidance of autonomous mobile robots. Section 3 shows the method of generating fuzzy rules from clustering the input-output data. In Section 4, simulation results show the fuzzy system with a small number of fuzzy rules successfully realizes cooperative behaviors of autonomous mobile robots. Final section describes conclusion and future plan for artificial life.

2 Alogrithm for the Obstacle Avoidance Behavior

2.1 Design Autonomous Mobile Robot Using Infrared Sensors

Autonomous Mobile Robot is independently, but it limit in our environment and that is not communicate with another robot. We design several Autonomous Mobile Robot because we explain Autonomous Mobile Robot based on artificial intelligence[1][2][15]. The characteristics of that include the velocity and the direction. We propose that define group behavior of obstacle avoidance Autonomous Mobile Robots based on the algorithm and fuzzy rules[1][15].

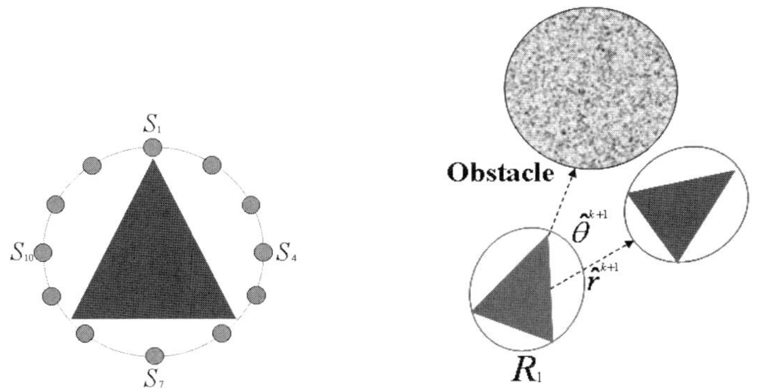

(a) Architecture of autonomous mobile robot (b) Algorithm for obstacle avoidance

Fig. 1. Design Autonomous Mobile Robot using infrared sensors and Obstacle avoidance algorithm

Table 1. Infrared Sensors of Autonomous Mobile Robots

Sensor number	Angle
S_1	0°
S_2	30°
S_3	60°
S_4	90°
S_5	120°
S_6	150°
S_7	180°
S_8	210°
S_9	240°
S_{10}	270°
S_{11}	300°
S_{12}	330°

The obstacle avoidance algorithm is based on the relative angle and distance between a robot and other robots. Each time a robot calculates the relative angle and distance by 12 sensor of robots. Autonomous mobile robot has 12 infrared sensors and Each time a robot calculates the relative angle and distance by 12 infrared sensors. This calculative values are important factors of algorithm. In the future, we define AMRs for Autonomous Mobile Robots[1][2][15].

Fig. 1 shows the definitions of relative distances and angles for the obstacle avoidance group behaviors. Each step, a robot calculates the relative rotation angle and the distance. The movement of a robot in the obstacle avoidance behavior requires recognition of the obstacle. When a robot moves straight, the rotation of the robot is divided into three grades to avoid radical rotation of the robot. Such robot moves repeatedly to make arrangement as a circle. In the obstacle avoidance procedures, a robot moves in three different distance steps and in the angle range between 0 and 30 degrees to avoid abrupt movement which is not practically realizable[15].

2.2 Obstacle Avoidance According to Fuzzy Rules

Complex group behaviors are considered as simple local interactions between individuals under fundamental rules of artificial life. Fuzzy system can handle uncertainty involved in natural phenomena in linguistic terms. Fundamental rules involved in the group behavior are represented by a set of fuzzy system rules. If the fuzzy system is modeled as in Sugeno form[7], adaptive training algorithms such as clustering can identify the structure and its parameters of the fuzzy system. Flocking refers to the procedure that gather robots scattered in an open space to form a group of robots. The goal of arrangement is to form a circular shape of robots. Each time a robot calculates the relative distances and angles with other robots. Fig. 2 shows the input and the output variables for the avoidance blocks. If robots locate in a wide place and approach random obstacle, robots avoid random obstacle that use fuzzy rules.

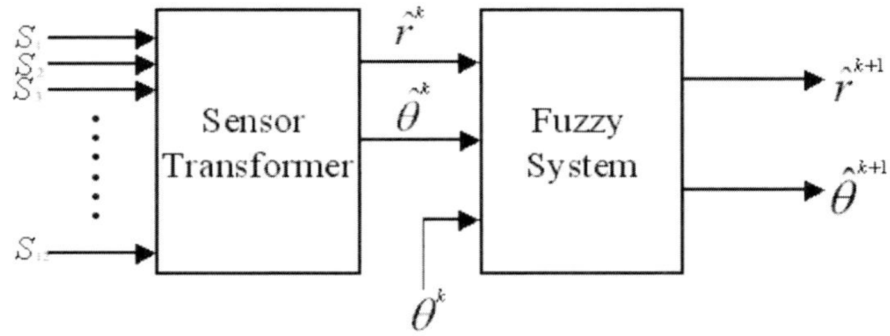

Fig. 2. Fuzzy system for obstacle avoidance behaviors and input-output data parameters

Each fuzzy system consists of 3 inputs and 2 outputs. The input variables of the avoidance block are the distance $\hat{r}^k$, and the angles $\hat{\theta}^k$ for the obstacle, and robot's angle. The outputs of the fuzzy system are the distance $\hat{r}^{k+1}$ and the rotation angle $\hat{\theta}^{k+1}$ of an individual at the next move. We denotes the on-going direction of the individual. In order to avoid abrupt movement which cannot be implemented practically, angle of robot rotation and distance of robot movement are set to have values from 0 to 30 degrees and from 0.005 to 0.2 at 0.025 distance. Positive values of indicate the rotation of counterclockwise direction.

3 Fuzzy Rules Generate from Clustering

3.1 Clustering Algorithm

Subtractive clustering[5] can identify fuzzy system models by to determining cluster centers from the numerical input-output data. The number of cluster centers corresponds to the number of fuzzy rules. If we consider the Sugeno-type fuzzy model, the parameters are also determined from the clustering algorithm. The clustering algorithm calculates the potential values P_i from N normalized data obtained from the input-output product-space.

$$P_i = \sum_{k=1}^{N} exp(-\alpha \|x_k - x_i\|^2) \tag{1}$$

$$P_1^* = \max P_i \tag{2}$$

Here, $i = 1, \cdots, N$ and γ_a is a positive constant to set data far apart from a cluster center so as not to influence on the potential value[5][7][8]. The first cluster center x_i^* corresponds to the largest potential value P_i^*. The second cluster center is calculated after removing the effect of the first cluster center. Eq'n 3 shows how to remove the effect of the first cluster center. The second cluster center X_2^* corresponds to the largest potential value of $P_i^{'}$.

$$P_i^{'} = P_i - P_i^* exp(-\beta \|x_i - x_1^*\|^2), \ \beta = 4/\gamma_a^2 \tag{3}$$

$$P_2^* = \max P_i^{'} \tag{4}$$

The positive constant γ_b presents cluster centers from assembling to close. This process repeats until potential values reach a fixed limit $(\underline{\varepsilon}, \bar{\varepsilon})$. Cluster centers $\{x_1^*, x_2^*, \cdots, x_M^*\}$ determine M fuzzy rules. They also determine the center position of input membership functions. Widths of the membership functions are fixed based on experience. The parameters $\alpha_{i1}, \alpha_{i2}, \cdots, \alpha_{in}$ can be optimized by linear least squares estimation or adaptive training algorithms.

3.2 Generating Fuzzy Rules for Autonomous Mobile Robots

Sugeno fuzzy system model[5] is used to represent fundamental rules of group intelligence. The MISO type fuzzy rules are of the form given in (5)

$$\begin{aligned} &IF\ x_1\ is\ A_{i1}\ and\ x_2\ is\ A_{i2}\ and\ \cdots\ x_n\ is\ A_{in} \\ &THEN\ y_1 = a_{0i} + a_{1i}x_1 + \cdots + a_{ni}x_i \end{aligned} \tag{5}$$

A_{ij} is Gaussian membership functions for input fuzzy variables, coefficients a_{0i}, a_{1i}, $\cdots$ a_{ni} determine the output of the fuzzy system. Fuzzy model process based on the clustering of input-output data determines the centers of the membership functions for antecedent fuzzy variables. In order to develop the fuzzy model, input-output data are obtained from the group behavior algorithms. Fuzzy rules for avoidance behaviors are generated from clustering the input-output data. We obtained input-output data for 90 times with 5 mobile robots, and for 10 times with 10 mobile robots. The parameters are set to $\gamma_a = 0.5$, $\gamma_b = 2.0$, $\bar{\epsilon} = 0.2$, and $\underline{\epsilon} = 0.1$.

Table 2. Parameter of the input membership functions for avoidance

Rules \ Variables	a_{0i}	a_{1i}	a_{2i}	a_{3i}	b_{0i}	b_{1i}	b_{2i}	b_{3i}
1	0.3628	-0.005	-0.0018	-0.0358	-0.0273	0.4025	1.6860	7.3220
2	0.0164	-0.0003	0.0006	-0.0235	-22.7885	0.1088	0.8482	4.8559
3	-0.0223	-0.0003	0.0005	-0.0266	-76.9204	-0.0150	0.3701	3.7974
4	0.4144	-0.0018	-0.0046	0.0357	-222.9720	1.2176	2.6398	-23.4144
5	0.0155	-0.0002	-0.0001	0.0084	18.4710	-0.0707	0.0359	-3.1046
6	-0.3336	-0.0007	0.0028	-0.0422	0.0246	0.6163	-1.6640	3.6929
7	1.2459	-0.0025	-0.0034	-0.0576	-0.0499	1.0968	1.5311	5.4053
8	0.0000	-0.0002	0.0000	0.0026	0.0000	-0.0137	0.0000	0.2917
9	0.0366	-0.0008	0.0030	0.0495	4.7194	0.2357	-1.0777	-17.7164
10	0.0338	0.0004	-0.0003	0.0203	-26.0544	-0.4205	1.4586	-21.1434
11	0.2010	-0.0001	-0.0006	0.0008	-64.2467	-0.1078	0.1749	5.5097
12	0.1183	0.0001	-0.0002	0.0187	-60.3032	0.0186	0.1041	18.9206
13	0.0229	-0.0006	0.0003	0.0102	73.5939	-0.0914	-0.3023	-2.9522
14	0.0867	-0.0006	-0.0005	0.0221	-3.6213	0.3328	0.5946	-14.1480
15	0.1036	-0.00002	-0.0001	0.0083	-26.1434	0.0260	0.0528	2.2723
16	0.1267	0.00005	-0.0001	0.0067	-97.0739	0.0317	0.2886	7.2859

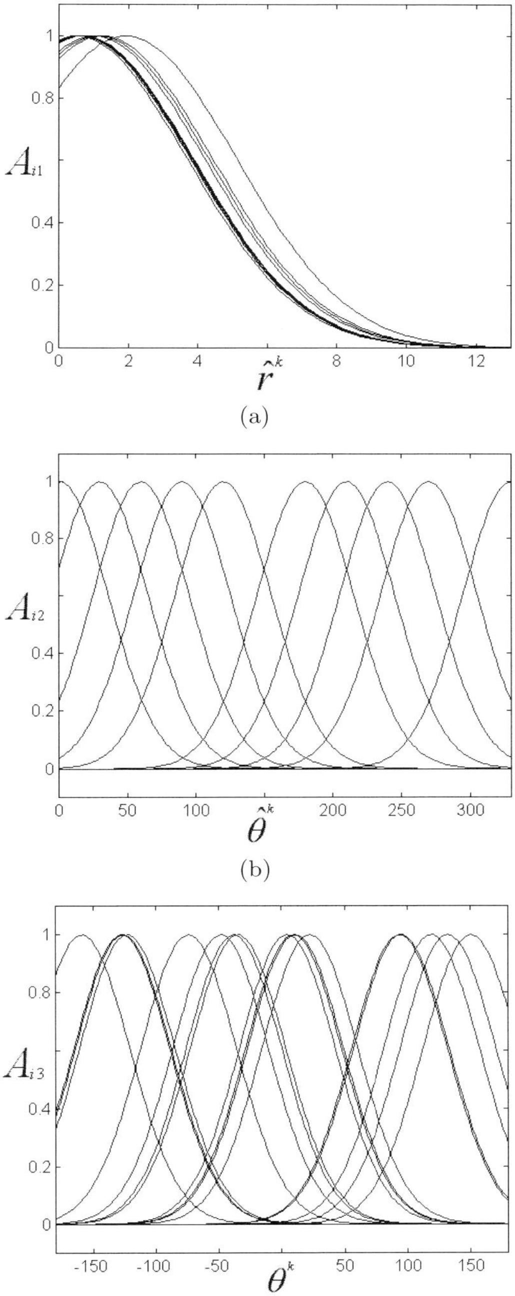

Fig. 3. Membership function for avoidance

After clustering the data, 16 cluster centers and therefore 16 fuzzy rules are obtained for the avoidance block. The 16 fuzzy rules for avoidance are of the form:

$$\begin{aligned} &IF\ r^k\ is\ A_{i1}\ and\ \hat{\theta}^k\ is\ A_{i2}\ and\ \theta^k\ is\ A_{i3} \\ &THEN\ \hat{r}_{k+1} = a_{0i} + a_{1i}\hat{r}^k + a_{2i}\hat{\theta}^k + a_{3i}\theta^k, \\ &\qquad b_{0i} + b_{1i}\hat{r}^k + b_{2i}\hat{\theta}^k + b_{3i}\theta^k \\ &\qquad (i = 1, 2, 3, \cdots, 16) \end{aligned} \tag{6}$$

Initial position of robots are randomly selected. Table 2 shows the center locations of the 16 fuzzy rules for avoidance. Fig. 3 shows the membership functions for avoidance. Center locations of the membership functions for $\hat{r^k}$ is close, but are not redundant since in the 3 dimensional input space, the centers are apart enough.

4 Computer Simulations

According to the obstacle avoidance algorithm, An autonomous mobile robot avoids sporadic obstacle. Fuzzy rules are generated from the input-output data by the algorithm. Group behaviors governed by the fuzzy rules are compared with those by the algorithm. From random initial positions, in case of avoidance, robots avoid sporadic obstacle. A robot calculates the distance and the angle to the obstacle.

Fig. 4 shows an ideal simulation of avoidance of Autonomous Mobile Robot and a simulation by fuzzy system of avoidance.

The AMRs behavior algorithms produced ideal movements of avoidance of 3 mobile robots without collision from random initial positions. Fig. 5 shows an ideal simulation of avoidance of 3 robots. Fig. 5 shows robot trace and demonstrates the final location of last step.

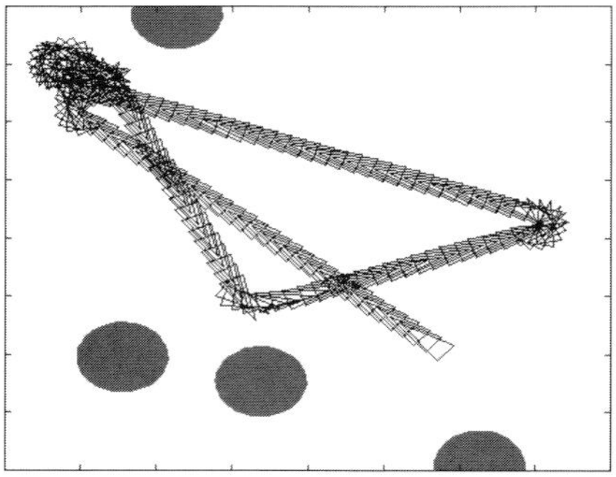

(a) Algorithm

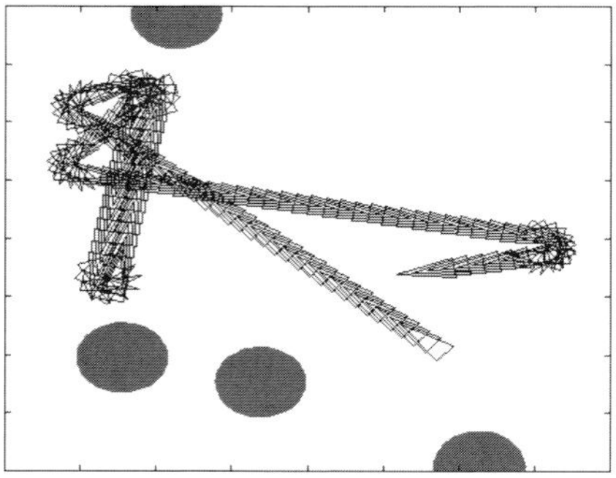

(b) Fuzzy rules

Fig. 4. Simulation result

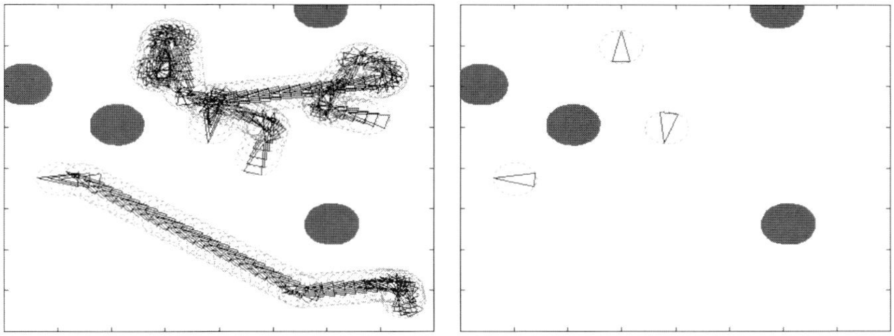

(a) Trace of AMRs (b) Location of AMRs

Fig. 5. Simulation result by the algorithm

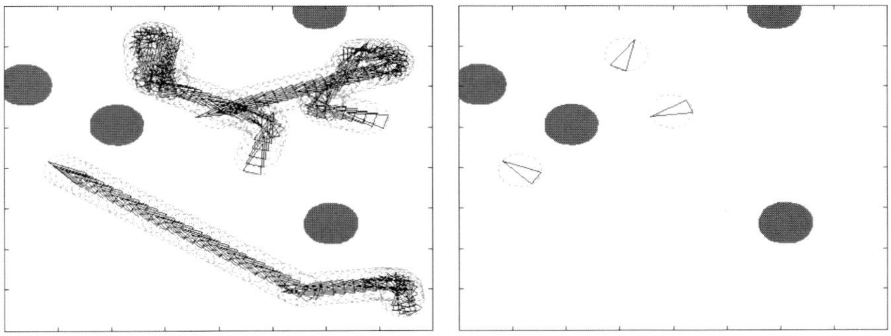

(a) Trace of AMRs (b) Location of AMRs

Fig. 6. Simulation result by fuzzy rules

Fig.7(a) and Fig.7(b) shows performance measurement for robots by algorithms and fuzzy system

5 Conclusions

In this paper, cooperative behaviors of autonomous mobile robots are investigated. Avoidance behaviors are implemented by the group behavior algorithms. The algorithm produces ideal trace and movement of multiple robots in the obstacle avoidance. A complex lifelike behavior is believed to consist of a set of simple fundamental rules. The fundamental rules are represented by the fuzzy rules, generated from clustering input-output data obtained from the movement by the algorithms. The fuzzy systems with a small number of fuzzy rules successfully realize cooperative behavior of mobile robots. Future plan of study includes minimization of the number of rules for group high intelligence control of artificial life, and other intelligent cooperative behaviors.

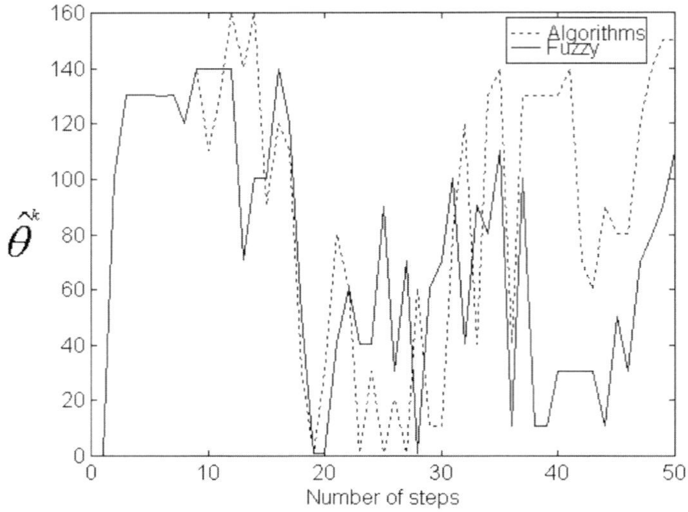

(a) Average of input angle variation

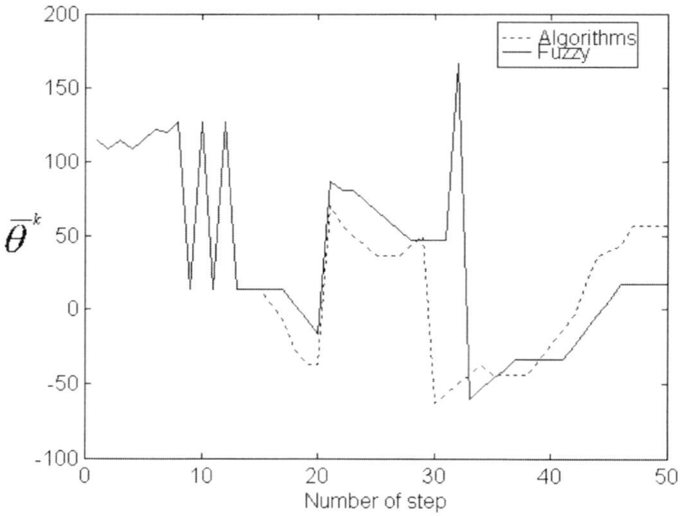

(b) Average of robot angle variation

Fig. 7. Performance Measurement

References

1. C. G. Langton, (ed.): Artificial Life II. Addison Wesley, 1992.
2. M. Sipper: An Introduction to Artificial Life. Explorations in Artificial Life, pp.4-8, 1995.
3. C. W. Reynolds: Flocks, Herds, and Schools: A Distributed Behaviorsal Model. Computer Graphics, 21(4) pp.25-34, July, 1987.

4. D.Teraopoulos, X, Tu, and R. Grzeszczuk: Artificial Fishes: Autonomous Locomotion, Perception, Behavior, and Learning in a Simulated Physical World. Artificial Life I, pp.337-351, 1994.
5. R.Yager and D.Filev: Essentials of Fuzzy Modeling and Control. John Wiley and Sons, 1994.
6. Won Nah, JoongHwan Baek: Classification of C.elegans Behavioral Phenotypes Using Clustering. Proceedings of the IEEK Conference 2003, 7 v.2003, n.IV, pp.1743-1746.
7. L. A. Zadeh: Fuzzy Sets. Information Control, Vol. 8, pp.338-353, 1965.
8. L. X. Wang and J. Mendel: Fuzzy Basic Functions, Universal Approximation, and Orthogonal Least Square Learning. IEEE Trans. On Neural Networks, Vol. 3, pp. 807-874, 1992.
9. Y. Davidor: A Natural Occurring Niche and Species Phenomenon: The Model and First Results. Proc. Of The Fourth Int. Conf. On Genetic Algorithms, pp.257-263, 1991.
10. D. E. Goldberg: Genetic Algorithms in Search, Optimization and Machine Learning. Addison-Wesley, 1989.
11. J. Liska and S. S. Melsheimer: Complete Design of Fuzzy Logic Systems Using Genetic Algorithms. Proc. Of the third IEEE Conf. Fuzzy Systems, pp. 1377-1382, 1994.
12. C. W. Xu: Fuzzy Model Identification and Self-Learning for Dynamic Systems. IEEE Trans. On Syst. Man. Cybern., Vol. **17**, pp. 683-689, 1987.
13. M. Sugeno and T. Yasukawa: A Fuzzy-Logic-Based Approach to Qualitative Modeling. IEEE Trans. On Fuzzy System, Vol. **1**, pp. 7-31, 1993.
14. J. S. R. Jang, C.T.Sun, and E. Mizutani: Neuro-Fuzzy and Soft Computing, Prentice Hall pp. 353-360, 1997.
15. J. H. Kim, J. B. Park, H. S. Yang, Y. P. Park: Generation of Fuzzy Rules and Learning algorithms for Cooperative Behavior of Autonomous Mobile Robots(AMRs). LECTURE NOTES IN ARTIFICIAL INTELLIGENCE, vol. **3613**, pp. 1015, 2005.

Fuzzy Likelihood Estimation Based Map Matching for Mobile Robot Self-localization*

Jinxia Yu[1,2], Zixing Cai[1], and Zhuohua Duan[1,3]

[1] College of Information Science & Engineering, Central South University, 410083 Changsha Hunan, China

[2] College of Computer Science & Technology, Henan Polytechnic University, 454003 Jiaozuo Henan, China
melissa2002@163.com

[3] Department of Computer Science, Shaoguan University 512003 Shaoguan Guangdong, China

Abstract. Reliable self-localization is a key issue in mobile robot navigation techniques under unknown environment. Aimed at an experimental platform of mobile robot with two rocker-bogie suspensions and four drive wheels, the dead-reckoning error of the proprioceptive sensors (odometry, fiber optic gyros) and the ranging performance of the exteroceptive sensor (2D time of fight laser scanner) are analyzed in this paper. Then, the environmental map using occupancy grids is adopted to fuse the information of the robot's pose by dead-reckoning method and the range to obstacles by laser scanner. In this condition, the map matching method, combined fuzzy logic and maximum likelihood estimation, is presented to improve mobile robot self-localization. By experiments of the robot platform, the effectiveness of this method is validated and the self-localization performance of mobile robot is enhanced.

1 Introduction

The localization of mobile robot, the "where am I" problem, is a fundamental issue of robot navigation[1]. Because of the scarcity of a priori knowledge and the uncertainty of the physical environment, a mobile robot exploring under an unknown environment has no absolute frame of reference for its position, other than the information it detects through its sensors[2]. However, sensor affected by the limited performance may obtain the imprecise measurement information and accordingly it is difficult to realize the accurate self-localization. Therefore, it has become an urgent research project for mobile robot self-localization under unknown environment[3].

Under unknown environment, mobile robot uses map matching to realize its self-localization, which substantively reduces the localization uncertainty caused by dead-reckoning errors by establishing the corresponding relationship between the current local map and the previous global map[4]. The algorithm of map matching is based on icon or feature and the special realized techniques mainly include EKF, Markov localization, Monte Carlo localization (MCL), maximum likelihood estimation (MLE), etc[5]. In relation to other methods, MLE is more fit for realizing the reliable

* This work is supported by the National Natural Science Foundation of China (No. 60234030).

L. Wang et al. (Eds.): FSKD 2006, LNAI 4223, pp. 846 – 855, 2006.

self-localization as it is dependent of some100 measurement accuracy and able to generate a consistent map even if all features look alike and cannot be distinguished perceptually. At the same time, fuzzy techniques have been proved it is effective in addressing the uncertainty of the self-localization problem[6]. So, fuzzy logic, together with MLE, is employed into map matching to deal with such problem in order to improve the self-localization performance of mobile robot. With an experimental platform of mobile robot, which utilizes odometry, fiber optic gyros (FOG) etc as its proprioceptive sensors and a 2D laser scanner based on the ranging principle of time of flight as its exteroceptive sensor, it analyzes the measurement errors of these sensors and uses occupancy grids as the environment map to fuse the information of the robot's pose by dead reckoning and the range to obstacles by laser scanner in this paper. In this condition, the map matching method, combined fuzzy logic and maximum likelihood estimation, is presented to improve the mobile robot self-localization. By experiments of the robot platform, the effectiveness of this method is validated and the self-localization performance of mobile robot is enhanced.

2 Errors Analysis of Sensors

With the research requirement for the theories and methods of mobile robot navigation, Cai et al design and develop an experimental platform of mobile robot (see figure 1). It adopts the architecture of two rocker-bogie suspensions; four drive wheels and an omnidirectional wheel. The proprioceptive sensors such as odometry, FOG are used to measure the pose of mobile robot and the exteroceptive sensors such as laser scanner, cameras are used to realize the environment perception where the study is only focus on laser scanner in the 2D plane.

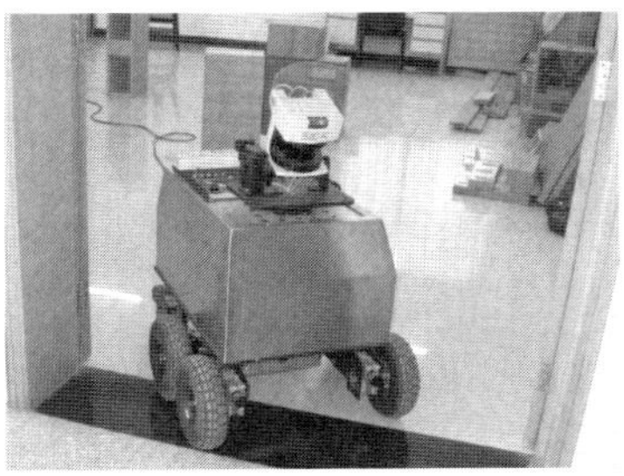

Fig. 1. Mobile robot MORCS-1

2.1 Sensors for Dead-Reckoning

The pose of mobile robot in the 2D plane can be expressed by the status $X = (x_r, y_r, \theta_r)$ where (x_r, y_r) is the position of mobile robot in the world coordinate system and θ_r is its heading. The pose is measured by odometry and FOG. Odometry is implemented by means of incremental optical encoders that monitor the wheel revolutions or steering angle. The encoder data are then used to compute the robot's offset from a known starting position. FOG is used to measure the heading of mobile robot. FOG used in our system is KVH E-Core RD1100 gyros.

UMBmark (University of Michigan Benchmark) test is utilized to evaluate the dead-reckoning performance of the robot platform based on the proprioceptive sensors. The measurement errors by experiments are shown in figure 2. From it, we can see that the accumulated dead-reckoning errors increase over time and range that are inevitable on account of odometry and FOG are both based on relative position measurements. Dead-reckoning errors are the main source of mobile robot self-localization.

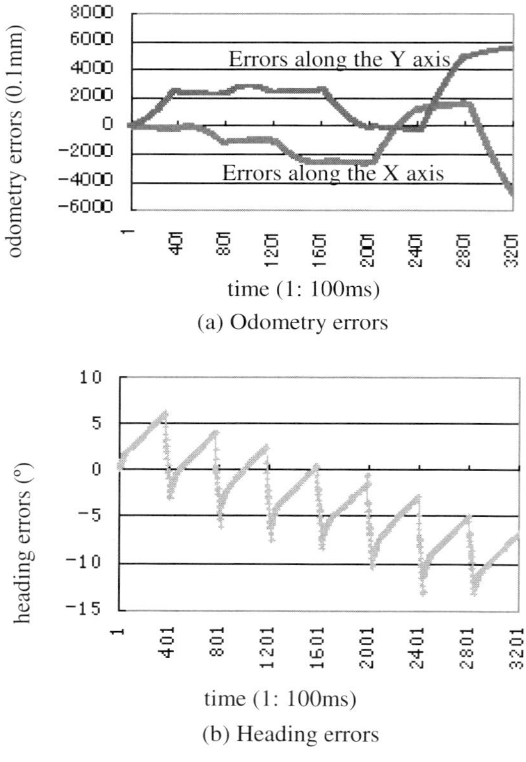

Fig. 2. Dead-reckoning errors of mobile robot

2.2 Sensor for Environment Perception

Laser scanner LMS291 made by Sick is adopted to provide 2D scanning angle (180 degrees). It is based on a time-of-fight (TOF) measurement principle. The angular resolution of LMS291 is selectable at 1°, 0.5° or 0.25° and the scanning angle is selectable at 180° or 100°. In our research, LMS291 would get 361 measurement data throughout 180° scanning angle with the 0.5° resolution and the time is 26.67ms. The laser scanner's range is basically dependent on the object's surface. At the same time, it is also affected by the incidence angle of the laser beam.

The performance specifications of the measurement accuracy for laser scanner are mainly the minimum resolution, system error are statistical error. In order to verify the measurement accuracy of LMS291, we aligned the range from 20 to 248 cm. with the test platform in figure 3(a) and moved the target every 3 cm at which the target is measured 100 times. It can get 7700 measurement data as shown in figure 3(b) in which the error mean of each measured point and the upper limit and the lower limit of error of all measurement data are painted. This test discovered that the maximum deviation of the target with fixed reflectivity (dust color carton) is between 3 and 33 cm. In addition, the standard deviation of measuring error is 0.94 cm in this test.

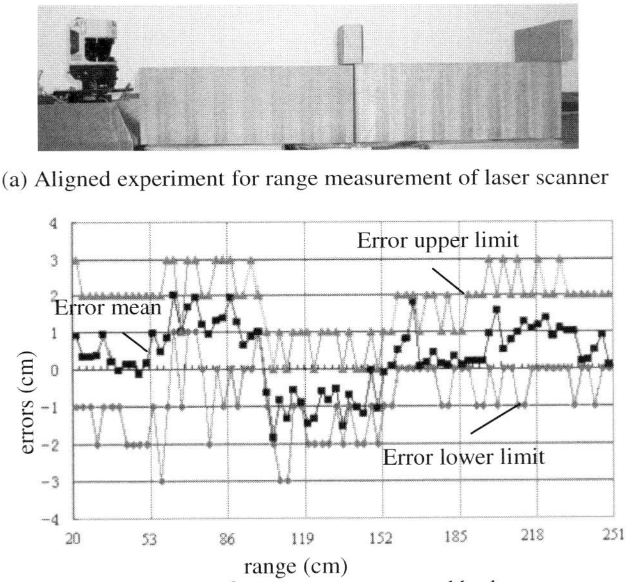

(a) Aligned experiment for range measurement of laser scanner

(b) Distribution curve of range errors measured by laser scanner

Fig. 3. Experiment for range measurement of laser scanner

3 Self-localization Based on Fuzzy Likelihood Estimation

3.1 Map Building

Kinematics model is displayed as figure 4 after reducing the driver wheel on the same side to a wheel. On the assumption that the coordinate origin *O* be the start point of mobile robot in the world coordinate system, *X* axis be the original motion direction at the start point of mobile robot, *Z* axis be vertical to the horizontal plane and *Y* axis be determined according to the rule of right hand. If the sampling time is short enough, the practical operation of mobile robot is approximate to the real kinematics model. The pose of the robot platform can be calculated by accumulating its increment in a sampling time ΔT as shown in formula 1.

$$\begin{cases} x_r(k+1) = x_r(k) + \Delta d(k)\cdot\cos(\theta_r(k) + \Delta\theta(\mathrm{k})) \\ y_r(k+1) = y_r(k) + \Delta d(k)\cdot\sin(\theta_r(k) + \Delta\theta(\mathrm{k})) \\ \theta_r(k+1) = \theta_r(k) + \Delta\theta(\mathrm{k}) \end{cases} \tag{1}$$

Where, $(x_r(k+1), y_r(k+1))$ and $(x_r(k), y_r(k))$ is the current and previous position of robot center, $\theta_r(k+1)$ and $\theta_r(k)$ is the current and previous heading of robot center, $\Delta d(k)$ is the estimated values of translation displacement provided by odometry and $\Delta\theta(k)$ is the estimated values of rotation displacement provided by FOG at time k in a sampling period ΔT.

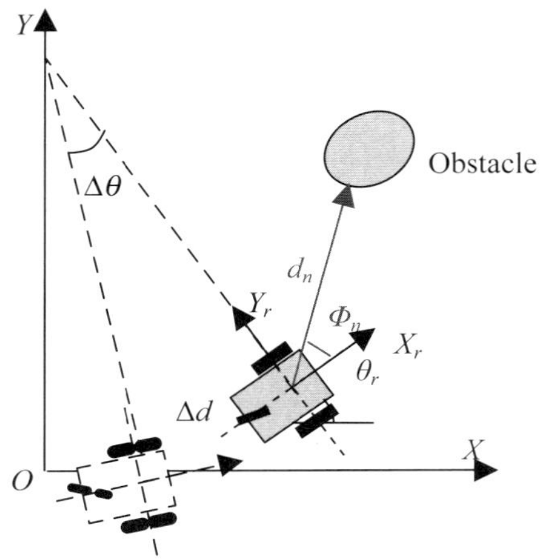

Fig. 4. Environment mapping based on laser scanner

The environmental information is the snapshot from LMS291, and the operating environment of mobile robot is described by 2D Cartesian grids in this paper. We take a 2D array as the environment map model and an array value corresponding to its each grid cell is recorded according to the times that the obstacle is observed. It assumed each grid cell can only be in one of two states, empty or occupied, and that has to be estimated from sensor observations. The knowledge that robot at time k about occupancy of a given cell is stored as probability of these two states, given all the prior sensor observations. When an obstacle is observed at such position, the array value is added by 1. Since the size of grids directly makes an effect on the resolution of control algorithms, each grid cell corresponding to 5×5cm^2 under real environment is adopted considering that LMS291 has the advantage of high measurement resolution and quick response time. If doing so, the environmental model of mobile robot can be created in real time and keep the high accuracy. The conversion from the polar to Cartesian coordinate system is written in equation (2).

$$\begin{cases} x_n' = x_r + \rho_n \cdot \cos(\theta_r + \phi_n) \\ y_n' = y_r + \rho_n \cdot \sin(\theta_r + \phi_n) \end{cases} \tag{2}$$

Where, ρ_n is the distance from the origin of LMS291 to the n-th obstacle, ϕ_n is the angle of the n-th laser beam relative to the home orientation, n is the number of scanning data from LMS291, (x_n', y_n') is the coordinate of the n-th obstacle in the world coordinate system; (x_r, y_r) and θ_r is the current center coordinate of mobile robot in the world coordinate system and its heading obtained from the equation (1), respectively.

3.2 Map Matching Based on MLE

Since most objects in operating environment of mobile robot are relatively stationary, a few dynamic obstacles can be treated as the environmental noise, while the inherent uncertainty of environment grids is mainly derived from the accumulated pose errors. The errors between the estimated and the real values of robot's pose are represented by $\delta X = (\delta x, \delta y, \delta\theta)$ and statistical characteristics of these parameters can be gathered by dead-reckoning sensors (odometry, FOG) during robot motion. By searching for the optimal parameters δx, δy, $\delta\theta$ under scan matching, we can find the most likely position of mobile robot in the world coordinate system. Therefore the estimation for the parameters δx, δy, $\delta\theta$ is a maximum likelihood estimation problem.

$$(\delta x^*, \delta y^*, \delta\theta^*) = \arg\max_{\delta x, \delta y, \delta\theta} l(\delta x, \delta y, \delta\theta) \tag{3}$$

For estimating the parameters δx, δy, $\delta\theta$, the map matching is formulated in terms of maximum likelihood estimation using the distance from the occupied cells in the current local map to their closest occupied cells in the previous map with respect to some relative position between the maps as measurements. We denote the distances for these occupied grid cells at some position of mobile robot by $D_1, \cdots, D_n$, thus the likelihood function for the parameters δx, δy, $\delta\theta$ is formulated as follows.

$$l(\delta x, \delta y, \delta\theta) = \sum_{i=1}^{n} \ln p(D_i; \delta x, \delta y, \delta\theta) \tag{4}$$

In doing so, we adopted the following criterion for the probability distribution function (PDF) of grid-grid matching $p(D_i; \delta x, \delta y, \delta\theta)$. If an occupied grid cell in the current local map can fit with the corresponding grid cell in the previous map then the PDF values of this cell is evaluated to be 1; if it border on the corresponding grid cell in the top, down, left and right direction then the PDF values are evaluated to be 0.6, or 0.3 in the diagonal direction; if it doesn't map to any grid cells above then the PDF

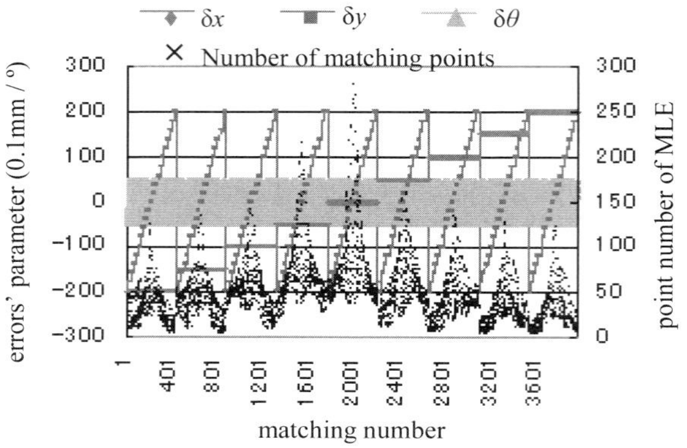

Fig. 5. Map matching based on MLE

values is evaluated to be 0. That is to say, we limit the map matching in a 3×3 grid area centered by the matching grid cell. With the UMBmark test, map matching based on MLE is applied to estimate the parameters and the experimental result is shown in figure 5.

3.3 Fuzzy Likelihood Estimation

It can realize self-localization to a certain extent by applying map matching based on MLE to estimate the parameters δx, δy, $\delta\theta$ and by using these to correct the robot's pose and the obstacle's position. However, the ranging data of laser scanner include noisy disturbance from environment such as ambient light, mixed pixel or exist data losing for the smaller reflectivity and the bigger incidence angle. Moreover, due to the existence of random noise, dynamic disturbances, obstacles obstructing or the scanning gap at some fixed scanning resolution. All these may produce the ambiguity information about the ambient environment. Hence, it may be impossible to find the most likely position of mobile robot under unknown environment. For this reason, fuzzy logic is employed into local map matching based on MLE to deal with such uncertainty in order to improve the robot localization performance.

In order to be robust to deal with this uncertainty and to introduce fuzzy logic, the scan area of LMS291 from –90 to 90 degrees can be divided into 3 equal sectors A (left), B(center), C(right). On account of the analysis above, we can know it is different for the results of maximum likelihood estimation in various sector whether the robot moves or not. Hence, we assign a fuzzy degree of membership to the number of likelihood estimation of each sector between scans $m_i (m_i <= 120, i = A, B, C)$ and define its fuzzy set is {S, M, B} (see figure 6a). For the result of likelihood estimation of each sector, it can also define the corresponding fuzzy set as {S, PS, M, PB, B} (see figure 6b).

In order to feedback roundly the inference information, the centroid average method is used to do a fuzzy decision-making. For the requirement of real-time

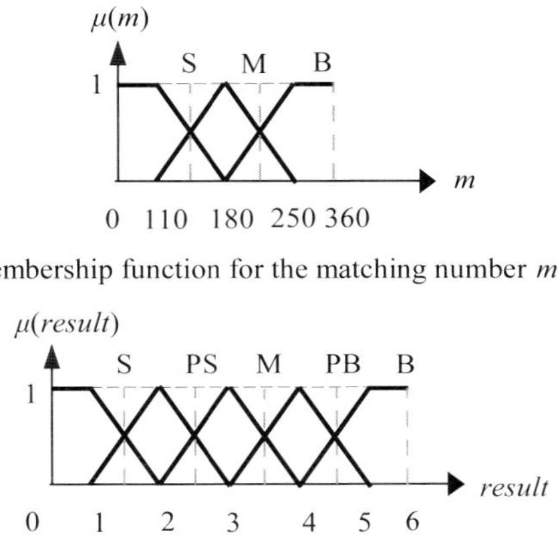

(a) Membership function for the matching number m

(b) Membership function for the fuzzy matching result

Fig. 6. Fuzzy membership function

correspondence, the scanning data of LMS291 is only dealt with map matching at a fixed time, and the preferable matching results is defuzzificated and used to correct the pose of mobile robot in the world coordinate system.

4 Experimental Analysis

To validate the effectiveness of this method in a larger and complex environment, the self-localization algorithm is integrated with other navigation methods such as exploring strategy, path planning and locomotion control and the following experiment is carried out. First, mobile robot MORCS-1 explores the operating environment at some strategy, and builds the environmental map based on the information stemmed from dead-reckoning and laser scanner, and uses the map matching method based on fuzzy logic and MLE to realize its self-localization. Then it implement the path planning assumed that it starts at the beginning point A and moves at the ending point B, which is handed over the system of locomotion control to carry out. In experiment, MORCS-1 starts from the beginning point A whose velocity is selectable from 20cm/s to 60cm/s, detects and avoids the obstacles under its operating environment successfully and reaches the ending point B lastly. Figure 7a shows the self-localization experiment of mobile robot MORCS-1 from the beginning point A to the ending point B by map matching using fuzzy logic and MLE, which is randomly run at 10 times to compare the self-localization by different method (see figure 7b and 7c). Experimental results discovered that MORCS-1 could fulfill the navigation task by its self-localization performance. At the same time, a fine environment map can be built and the self-localization errors of both odometry and heading are reduced distinctly. The effectiveness of this method under rather complex environment and the adaptability of that to the dynamic environment are also validated.

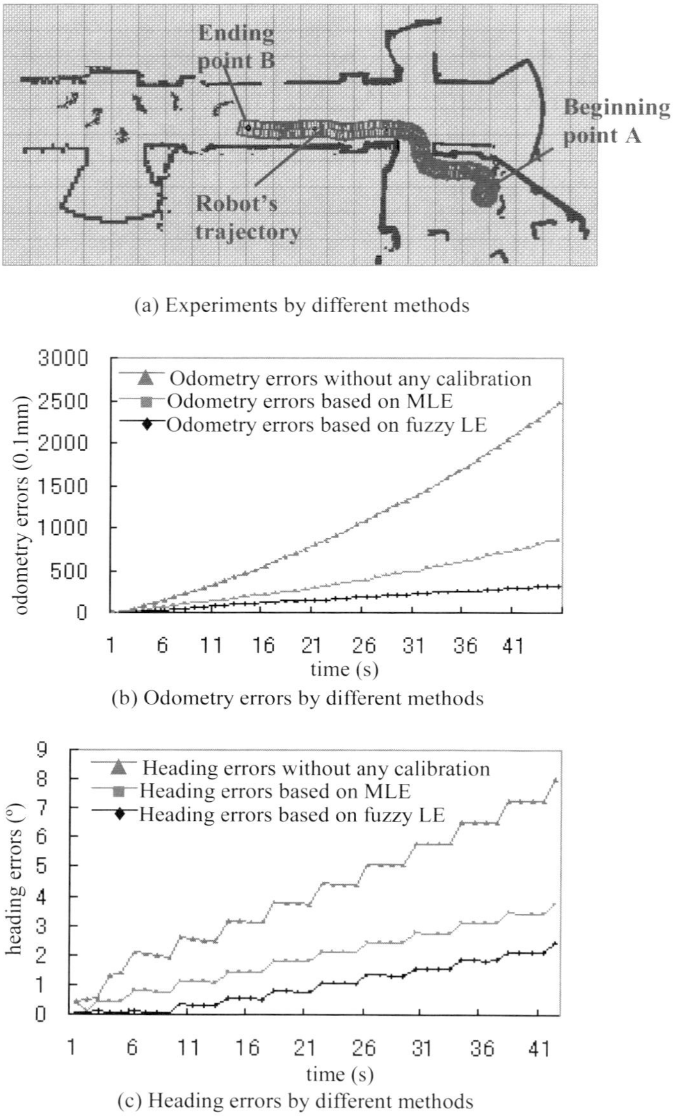

Fig. 7. Self-localization experiments with MORCS-1

5 Conclusions

Aimed at the self-localization problem of mobile robot navigation and combined with an experimental platform MORCS-1, it present a method by integrating fuzzy logic into map matching based on MLE to reduce robot's errors and improve its precision.

First, the dead-reckoning performance is estimated by analyzing the accumulated errors of the proprioceptive sensors (odometry, FOG) and the ranging performance is simultaneously estimated by analyzing the error disturbance of the exteroceptive sensor (2D laser scanner). Then, the pose of mobile robot is reckoned according to its kinematics model and the environmental map is built based on the information of the laser scanner and the pose of mobile robot. The environmental map using occupancy grids is adopted to fuse the information of the robot's pose by dead-reckoning method and the range to obstacles by laser scanner. After analyzing the shortage of map matching based on MLE, the map matching method, combined fuzzy logic and maximum likelihood estimation, is presented to improve the mobile robot self-localization.

In order to verify the effectiveness of this method, we use mobile robot MORCS-1 as the experimental platform under an office environment. Experiments under the large and complex environment are implemented to validate this method. Experimental results discovered that this method could build more accurate environmental map, reduce sensors' noise and the uncertainty from the ambient environment.

References

1. Leonard, J.J., Durrant-Whyte, H.F.: Mobile Robot Localization by Tracking Geometric Beacons. IEEE Transactions on Robotics and Automation, 7(5) (1991) 376-382
2. Lu, F., Milios, E.E.: Robot Pose Estimation in Unknown Environments by Matching 2D Range Scans. In: Proceedings of the IEEE Conference on Computer Vision and Pattern Recognition (CVPR94), Seattle, WA, USA, (1994) 935-938
3. Cai, Z.X., He, H.G., Chen, H.: Some issues for mobile robot navigation under unknown environments (in Chinese). Control and Decision 17(4) (2002) 385-391
4. Lu, F., Milios, E.E.: Globally Consistent Range Scan Alignment for Environment Mapping. Autonomous Robots, 4(4) (1997) 333-349
5. Robin, R.M.: Introduction to AI Robotics, 1st ed. Beijing: Publishing House of Electronics Industry, (2004)
6. Gasós, J., Rosetti, A.: Uncertainty representation for mobile robots: perception, modeling and navigation in unknown environments. Fuzzy Sets and Systems, 107(1) (1999) 1-24

Research on Attitude Law of Mass Moment Missile

Qing Guo, Ming Yang, and Zi-cai Wang

Control & Simulation Center, Harbin Institute of Technology,
150001 Harbin, P.R. China
guoqinghit@126.com

Abstract. The ability of moving mass attitude control is investigated for mass moment missile with three-axial stabilization. On basis of constructing the attitude model of mass moment missile with three moving masses, the coupled influence to this system is described which is caused by the relative movement between the moving masses and the cartridge case. Then an attitude control law based on fuzzy logic is presented for the mass moment missile and the fuzzy rule tables corresponding to the moving mass position in respective axial are constituted. According to analysis of the relation between angular acceleration and searching condition, three moving mass positions are regulated in phase. Simulation results show that the system stabilization can be satisfied to realize efficiently attitude adjustment for missile.

1 Attitude Mathematical Models

To carry out attitude-variant control quickly and furthest decrease coupling, one mass is fixed at x-axis. Another two masses are fixed at y-axis and z-axis in tail and pass through interceptor's axis. These are shown in Fig. 1.

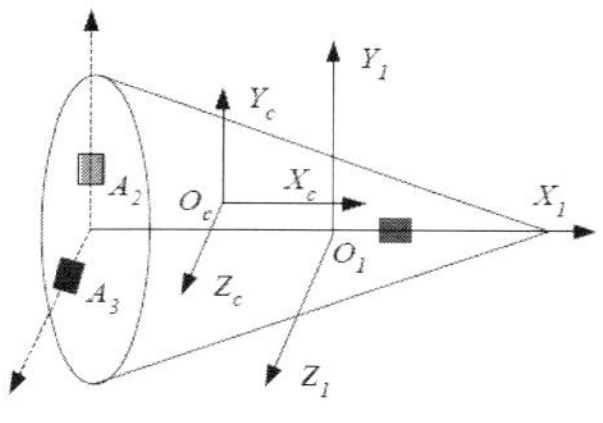

Fig. 1. Layout of Three Moving Masses in Respective Axis

According to the dynamics theory of Rigid Bodies, the momentum equation of system centroid is shown in equation (1).

$$d\vec{H} / dt = \sum \vec{M} \ . \tag{1}$$

Foundation Item. This work was also supported by National 863 Project under Contract 2004AA811052.

L. Wang et al. (Eds.): FSKD 2006, LNAI 4223, pp. 856 – 859, 2006.

The total moment vector ($\sum\vec{M}$) forces on the interceptor. Due to passing the center of mass, the G-force can't affect the variety of attitude. So $\sum\vec{M}$ viz. the aerodynamic moment vector (M_R).

$$\boldsymbol{M_R} = \vec{r}\times\vec{F} = \begin{bmatrix} i & j & k \\ x_f - x_c & y_f - y_c & z_f - z_c \\ R_{cx} & R_{cy} & R_{cz} \end{bmatrix} = \begin{bmatrix} M_{Rx} \\ M_{Ry} \\ M_{Rz} \end{bmatrix}. \tag{2}$$

In the case of the instantaneous center of press $O_f(x_f, y_f, z_f)^T$, center of mass $O_c(x_c, y_c, z_c)^T$ in reference frame $(O_1X_1Y_1Z_1)$ and aerodynamic coordinate $(R_{cx}, R_{cy}, R_{cz})^T$, the aerodynamic moment vector (M_R) is shown in equation (2).

Then the dynamic equation of attitude is given by

$$\begin{bmatrix} \dot{\omega}_x \\ \dot{\omega}_y \\ \dot{\omega}_z \end{bmatrix} = \begin{bmatrix} I_{11} & I_{12} & I_{13} \\ I_{12} & I_{22} & I_{23} \\ I_{13} & I_{23} & I_{33} \end{bmatrix}^{-1} \begin{bmatrix} M_{Rx} - M_{Ax} \\ M_{Ry} - M_{Ay} \\ M_{Rz} - M_{Az} \end{bmatrix} \tag{3}$$

where M_{Ax}, M_{Ay} and M_{Az} represent the additive moments engendered by coupled system dynamics.

2 Fuzzy Attitude Law

Equation (3) shows that three angular accelerations of missile are relative to the moments of aerodynamic M_x, M_y, M_z. So moments of aerodynamic are used for fuzzy variable to infer. According to PD-control idea, corresponding equivalent-required moment in three respective axis is shown in Equation (4).

$$\begin{cases} M_r = k_1\Delta\gamma + k_2\Delta\omega_x \\ M_\psi = k_3\Delta\psi + k_4\Delta\omega_y \\ M_\vartheta = k_5\Delta\vartheta + k_6\Delta\omega_z \end{cases}. \tag{4}$$

In Equation (4), k_i is PD parameter of respective channels.

Based on above analysis, self-revised-fuzzy attitude law is designed. The principle of this system is shown in Fig.2.

1 If $|M_r| < 0.5M_{r\max}$, it is shown that roll channel is regulated well. So we only control another two channels. The format of fuzzy rule tables is given by

if $|M_\psi| = B_i$ and $|M_\vartheta| = C_i$ then $x_{A1/p1} = D_i$

where B_i, C_i, D_i represent input and output fuzzy sets. Fuzzy rule table for mass A_1 in x-axis is shown in Tab.1.

For mass A_2 and mass A_3 in y-axis and z-axis, one-dimensional fuzzy rule is adopted to judge. It is shown in Tab.2.

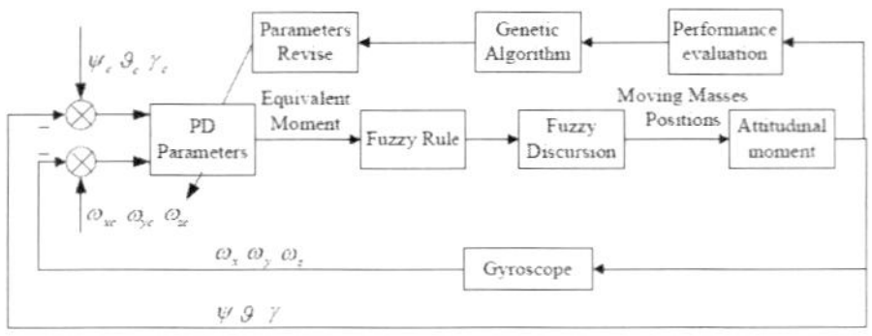

Fig. 2. The Principle of Self-Revised-Fuzzy Attitudinal System

Table 1. Rule Table for Mass A_1 in x-axis

$x_{A1/p1}$	$\lvert M_{\vartheta}\rvert$			
$\lvert M_{\psi}\rvert$	B	MB	M	S
B	NM	NM	NM	NM
MB	NM	ZE	ZE	ZE
M	NM	ZE	PM	PM
S	NM	ZE	PM	PB

Table 2. Rule Table for Mass A_2 and Mass A_3 in Radial Axes

M_{ϑ} or M_{ψ}	NB	NM	ZE	PM	PB
$x_{A2/p2}$ or $x_{A3/p3}$	PB	PM	ZE	NM	NB

2 If $\lvert M_r\rvert \geq 0.5M_{r\max}$, it is shown that roll channel is not controlled well. Here we should process to search three moving mass positions. Ternary array *nSeekCond*[3] is defined as current searching condition. The elements of this array deposit integers -2, -1, 0, 1, 2. We establish the fuzzy rule table as Table 3.

The searching condition *nSeekCond*[0] is relative to M_{ψ} and the *nSeekCond*[1] is relative to M_{ϑ}. There fuzzy rules are similar to *nSeekCond*[2].

Table 3. Rule Table for Current Searching Condition

M_{γ}	NB	NM	ZE	PM	PB
nSeekCond[2]	-2	-1	0	1	2

When the roll channel is not be controlled well, mass A_1 in x-axis should keep current position steady. The fuzzy rule table of A_2 and A_3 is given by

Table 4. Rule Table for Radial Masses A_2 and A_3

i or *j*	0	1	2	3	4
$x_{A2/p2}$ or $x_{A3/p3}$	NB	NM	ZE	PM	PB

where i and j are integral variables which satisfy the computation as follows.

$$i = nSeek \,/\, 5,\ \ j = nSeek \ \% \ 5 \tag{5}$$

In equation (5), $nSeek$ denote the searching times that is noted ultimately.

3 Simulation Results

In Fig. 3, the attitude angles and angular velocities are regulated from initial value to anticipant value 5.5 seconds later. Fig. 4 shows the current positions of moving masses in reference frame of path.

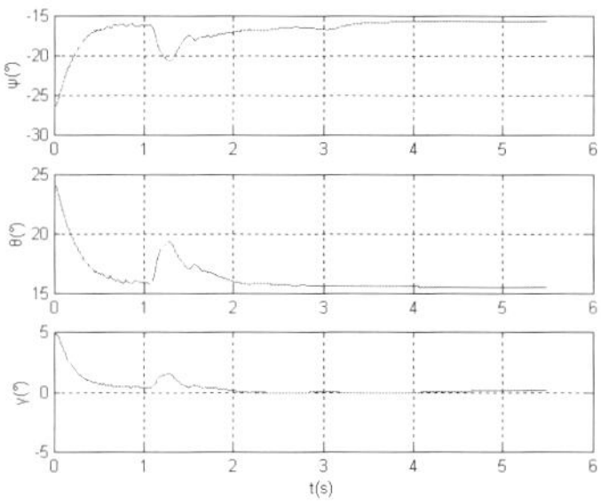

Fig. 3. Attitude Angles of Missile

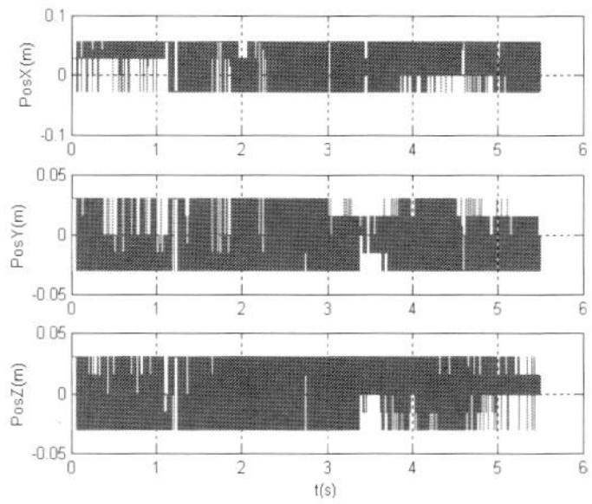

Fig. 4. Positions of Moving Mass in 3 Axes

4 Conclusions

An attitude control law based on fuzzy logic which used in missile's terminal phase has been studied after the model of attitudinal movement with three-axial stabilization is constructed. Three equivalent-required moments and three moving masses positions are used for fuzzy variables to control positional trend in phase. Simulation results show that the system static stabilization can be satisfied to regulate the attitude of missile optionally since this attitude law is adopted.

References

1. P. K. Menon, G. D. Sweriduk, E. J. Ohlmeyer, "Integrated Guidance and Control of Moving Mass Actuated Kinetic Warheads". Journal of Guidance, Control and Dynamics, 2004, Vol. 27, No. 1, January, pp: 118-126.
2. Raymond H B, Rush D R and Beverly R S, "Moving Mass Trim Control System Design," AIAA-96-3826- CP, 1996.
3. R. D. Robinet, B. Rainwater, and S. A. Kerr, "Moving Mass Trim Control for Aerospace Vehicles," Journal of Guidance, Control and Dynamics, 1996, 19(5), pp. 1064-1079.
4. You-zhi. He, Xiao-yu. Zhang, "Design of Three Moving Masses Missile System Based on the Fuzzy Variable Structure Control," Systems Engineering and Electronics, vol. 27, no. 2, pp. 292-295, Feb. 2005.

Multiple Models Fuzzy Decoupling Controller for a Nonlinear System*

Xin Wang[1,2], Hui Yang[2], and Bing Wang[3]

[1] Center of Electrical & Electronic Technology, Shanghai Jiao Tong University, Shanghai, P.R. China, 200030
wangxin26@sjtu.edu.cn
[2] School of Electrical & Electronic Engineering, East China Jiaotong University, Jiangxi, P.R. China, 330013
[3] School of Information Engineering, Shandong University at Weihai, 264209

Abstract. For a discrete-time nonlinear MIMO system, a multiple models fuzzy decoupling controller is designed. At each equilibrium point, the system is expanded into linear and nonlinear terms. The linear term is identified using one FLSs, while nonlinear term using other FLSs, which compose one system model. Then, all models got at all equilibrium points compose the multiple models set. At each instant, the best model is chosen out according to the switching index. Accordingly, the nonlinear term of the best model is viewed as measurable disturbance and eliminated using the feedforward strategy.

1 Introduction

In recent years, for nonlinear MIMO systems, few works have been observed. Ansari *et al.* simplified a nonlinear system into a linear system by using Taylor's expansion at the equilibrium point and controlled it using a linear adaptive decoupling controller accordingly [1]. However, for a system with strong nonlinearity, it can not get good performance. In [2], an exact linear system can be produced utilizing a feedback linearization approach. But accurate information must be known precisely. Furthermore, a variable structure controller with sliding mode was proposed [3], which requires the system an affine system. Although the design methods above can realize nonlinear decoupling control, there were too many assumptions required. To solve this problem, intelligent decoupling controller was proposed. A hierarchical fuzzy sliding mode controller was designed for SIMO nonlinear system [4]. In [5], two NNs were needed to identify the linear and nonlinear terms expanded using Taylor's formula at the origin. Unfortunately, when the equilibrium point was far from the origin, the system lost its stability.

In this paper, a Multiple Models Fuzzy Decoupling Controller (MMFDC) is designed. Utilizing Taylor's formula, the linear term is the first-order derivative of the

* This work is supported by National Natural Science Foundation (No. 60504010, 50474020) and Shanghai Jiao Tong University Research Foundation.

L. Wang et al. (Eds.): FSKD 2006, LNAI 4223, pp. 860 – 863, 2006.

system and identified using Fuzzy Logic Systems (FLSs) while the nonlinear term using other FLSs, which compose one system model. All models compose the multiple models set. At each instant, one best model is chosen out. The nonlinear term of the above model is viewed as measurable disturbance and eliminated.

2 Description of the System

The system is a nonlinear MIMO system of the form $\boldsymbol{y}(t+1)=\boldsymbol{f}[\boldsymbol{y}(t),\cdots,\boldsymbol{u}(t),\cdots]$, where $\boldsymbol{u}(t)$, $\boldsymbol{y}(t)$ are $n\times 1$ input, output vectors respectively and $\boldsymbol{f}[\cdot]$ is a vector-based nonlinear function which is continuously differentiable and Lipshitz.

Suppose that $(\boldsymbol{u}_1,\boldsymbol{y}_1),\cdots(\boldsymbol{u}_l,\boldsymbol{y}_l),\cdots(\boldsymbol{u}_m,\boldsymbol{y}_m)$ are m equilibrium points. At each equilibrium point $(\boldsymbol{u}_l,\boldsymbol{y}_l)$, using Taylor's formula, it obtains

$$\boldsymbol{y}(t+1)=\boldsymbol{y}_l+\sum_{n_1=1}^{n_a}\boldsymbol{f}'_{n_1}\Big|_{\substack{u=u_l\\ y=y_l}}\cdot[\boldsymbol{y}(t-n_a+n_1)-\boldsymbol{y}_l]+\sum_{n_2=0}^{n_b}\boldsymbol{f}'_{n_2}\Big|_{\substack{u=u_l\\ y=y_l}}\cdot[\boldsymbol{u}(t-n_b+n_2)-\boldsymbol{u}_l]+\boldsymbol{o}[\boldsymbol{x}(t)], \tag{1}$$

where $\boldsymbol{f}'_{n_1}=\dfrac{\partial \boldsymbol{f}}{\partial \boldsymbol{y}(t-n_a+n_1)}$, $\boldsymbol{f}'_{n_2}=\dfrac{\partial \boldsymbol{f}}{\partial \boldsymbol{u}(t-n_b+n_2)}$, $\boldsymbol{x}(t)=[\boldsymbol{y}(t)-\boldsymbol{y}_l,\cdots;\boldsymbol{u}(t)-\boldsymbol{u}_l,\cdots]$, $\boldsymbol{o}[\boldsymbol{x}(t)]$ satisfies $\lim_{\|\boldsymbol{x}(t)\|\to 0}\dfrac{\|\boldsymbol{o}[\boldsymbol{x}(t)]\|}{\|\boldsymbol{x}(t)\|}=0$, where $\|\cdot\|$ is the Euclidean norm operator.

Define $\bar{\boldsymbol{y}}(t)=\boldsymbol{y}(t)-\boldsymbol{y}_l$, $\bar{\boldsymbol{u}}(t)=\boldsymbol{u}(t)-\boldsymbol{u}_l$, $\boldsymbol{v}(t)=\boldsymbol{o}[\boldsymbol{x}(t)]$, $\boldsymbol{A}^l_{n_1}=(-1)\cdot\boldsymbol{f}'_{n_1}\Big|_{\substack{u=u_l\\ y=y_l}}$, $\boldsymbol{B}^l_{n_2}=\boldsymbol{f}'_{n_2}\Big|_{\substack{u=u_l\\ y=y_l}}$, $\boldsymbol{A}^l(z^{-1})=\boldsymbol{I}+\cdots+\boldsymbol{A}^l_{n_a}z^{-n_a}$, $\boldsymbol{B}^l(z^{-1})=\boldsymbol{B}^l_0+\cdots+\boldsymbol{B}^l_{n_b}z^{-n_b}$, the system (1) is

$$\boldsymbol{A}^l(z^{-1})\bar{\boldsymbol{y}}(t+1)=\boldsymbol{B}^l(z^{-1})\bar{\boldsymbol{u}}(t)+\boldsymbol{v}(t). \tag{2}$$

3 Design of MMFDC

At each equilibrium point $(\boldsymbol{u}_l,\boldsymbol{y}_l)$, one group of FLSs is utilized to approximate the system's first-order derivative offline. So $\hat{\boldsymbol{A}}^l(z^{-1})$ and $\hat{\boldsymbol{B}}^l(z^{-1})$ are obtained. The other is employed to approximate the nonlinear term $\boldsymbol{v}(t)$ online.

For the multiple models set, the switching index is chosen as $J_l=\|\boldsymbol{e}^l(t)\|^2=\|\boldsymbol{y}(t)-\boldsymbol{y}^l(t)\|^2$, where $\boldsymbol{e}^l(t)$ is the output error between the real system and the model l. $\boldsymbol{y}^l(t)$ is the output of the model l. Let $j=\arg\min(J_l)$ correspond to the model whose output error is minimum, then it is chosen as the best.

For the system (2), like the conventional controller design, the cost function is as $J_c=\|\boldsymbol{P}(z^{-1})\bar{\boldsymbol{y}}(t+k)-\boldsymbol{R}(z^{-1})\boldsymbol{w}(t)+\boldsymbol{Q}(z^{-1})\bar{\boldsymbol{u}}(t)+\boldsymbol{S}(z^{-1})\boldsymbol{v}(t)\|^2$, where $\boldsymbol{w}(t)$ is the known reference signal, $\boldsymbol{P}(z^{-1}), \boldsymbol{Q}(z^{-1}), \boldsymbol{R}(z^{-1})$, $\boldsymbol{S}(z^{-1})$ are weighting polynomial matrices respectively. Introduce $\boldsymbol{P}(z^{-1}) = \boldsymbol{F}(z^{-1})\boldsymbol{A}(z^{-1}) + z^{-1}\boldsymbol{G}(z^{-1})$, the control law is

$$\boldsymbol{G}(z^{-1})\overline{\boldsymbol{y}}(t) + [\boldsymbol{F}(z^{-1})\boldsymbol{B}(z^{-1}) + \boldsymbol{Q}(z^{-1})]\overline{\boldsymbol{u}}(t) + [\boldsymbol{F}(z^{-1}) + \boldsymbol{S}(z^{-1})]\boldsymbol{v}(t) = \boldsymbol{R}\boldsymbol{w}(t). \tag{3}$$

Combing (3) with (2), the closed loop system equation is obtained as follows

$$\begin{aligned}\left[\boldsymbol{P}(z^{-1}) + \boldsymbol{Q}(z^{-1})\boldsymbol{B}^{-1}(z^{-1})\boldsymbol{A}(z^{-1})\right]\overline{\boldsymbol{y}}(t+1) &= \boldsymbol{R}(z^{-1})\boldsymbol{w}(t) \\ &+ \left[\boldsymbol{Q}(z^{-1})\boldsymbol{B}(z^{-1})^{-1} - \boldsymbol{S}(z^{-1})\right]\boldsymbol{v}(t).\end{aligned} \tag{4}$$

To eliminate the nonlinear form and the interactions of the system exactly, let $\boldsymbol{Q}(z^{-1}) = \boldsymbol{R}_1\boldsymbol{B}(z^{-1})$, $\boldsymbol{S}(z^{-1}) = \boldsymbol{R}_1$, $\boldsymbol{P}(z^{-1}) + \boldsymbol{R}_1\boldsymbol{A}(z^{-1}) = \boldsymbol{T}(z^{-1})$, $\boldsymbol{R}(z^{-1}) = \boldsymbol{T}(1)$, where $\boldsymbol{T}(z^{-1})$ is a stable diagonal polynomial matrix decided by the designer and $\boldsymbol{R}_1$ is a constant matrix. Then the closed loop system is derived as $\boldsymbol{T}(z^{-1})\boldsymbol{y}(t+k) = \boldsymbol{T}(1)\boldsymbol{w}(t)$. By the choice of weighting polynomial matrixes, it not only decouples the system dynamically but also places poles arbitrarily.

4 Simulation Studies

A discrete-time nonlinear multivariable system is described as follows

$$\begin{cases} y_1(t+1) = \dfrac{-0.2y_1(t)}{1+y_1^2(t)} + \sin[u_1(t)] - 0.5\sin[u_1(t-1)] + 1.5u_2(t) + 0.2u_2(t-1) \\ y_2(t+1) = 0.6y_2(t) + 0.2u_1(t) + 1.3u_1(t-1) + u_2(t) + u_2^2(t) + \dfrac{1.5u_2(t-1)}{1+u_2^2(t-1)} \end{cases} \tag{5}$$

which is the same as the simulation example in [5]. The known reference signal $\boldsymbol{w}$ is set to be a time-varying signal. When $t = 0$, w_1 equals to 0 and when t is 40, 80, 120, 160, 200, it changed into 0.05, 0.15, 0.25, 0.35, 0.45 respectively.

In Fig.1 and 2, the system (5) is expanded only at the original point (0,0) and a Fuzzy Decoupling Controller (FDC) is used. In Fig.3 and 4, the system is expanded at six equilibrium points, *i.e.* $[0, 0]^T$, $[0.1, 0]^T$, $[0.2, 0]^T$, $[0.3, 0]^T$, $[0.4, 0]^T$.

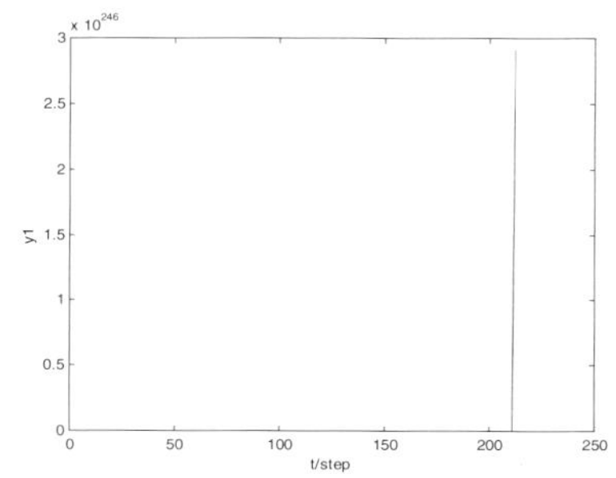

Fig. 1. The output $y_1(t)$ of FDC

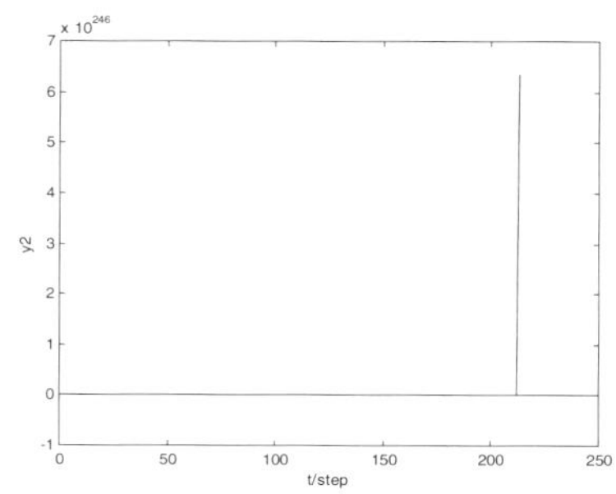

Fig. 2. The output $y_2(t)$ of FDC

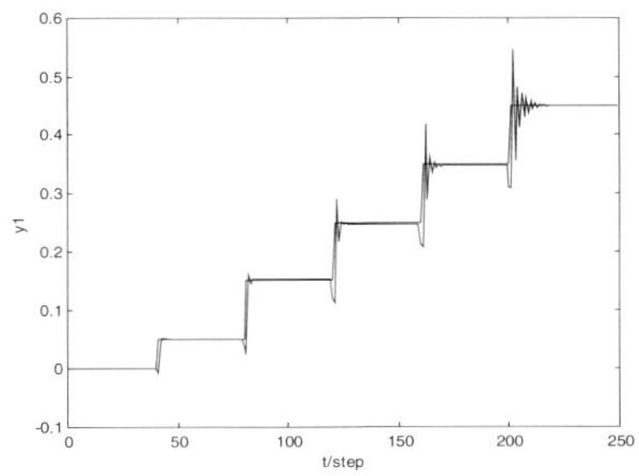

Fig. 3. The output $y_1(t)$ of MMFDC

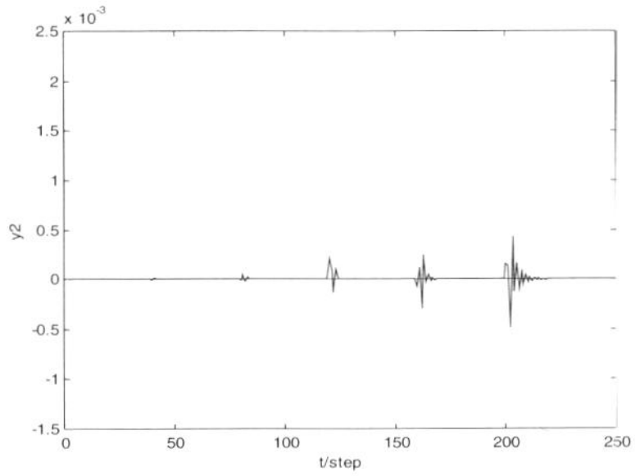

Fig. 4. The output $y_2(t)$ of MMFDC

Note that the equilibrium points are far away from the set points. The results show that although the same FDC method is adopted, the system using FDC loses its stability (see Fig.1 and 2), while the system using MMFDC not only gets the good performance but also has good decoupling result (see Fig.3 and 4).

5 Conclusion

A MMFDC is designed to control the discrete-time nonlinear multivariable system. At each equilibrium point, one group of FLSs is used offline to identify the linear term of the nonlinear system and the other is trained online to identify the nonlinear term. The multiple models set is composed of all models, which are got from all equilibrium points. According to the switching index, the best model is chosen as the system model. The nonlinear term is viewed as measurable disturbance and eliminated using feedforward strategy. The simulation example shows that the effectiveness of the controller proposed.

References

1. Ansari R.M., Tade M.O.: Nonlinear Model-based Process Control: Applications in Petroleum Refining. Springer, London (2000)
2. Wang W.J., Wang C.C.: Composite Adaptive Position Controller for Induction Motor Using Feedback Linearization, IEE Proceedings D Control Theory and Applications, 45 (1998) 25–32
3. Wai R.J., Liu W.K.: Nonlinear Decoupled Control for Linear Induction Motor Servo-Drive Using The Sliding-Mode Technique. IEE Proceedings D Control Theory and Applications, 148 (2001) 217–231
4. Lin C.M., Mon Y.J.: Decoupling Control by Hierarchical Fuzzy Sliding-Mode Controller. IEEE Transactions on Control Systems Techonology. 4 (2005) 593–598
5. Yue H., Chai T.Y.: Adaptive Decoupling Control of Multivariable Nonlinear Non-Minimum Phase Systems Using Neural Networks. Proceedings of the American Control Conference. (1998) 513–514

EWFCM Algorithm and Region-Based Multi-level Thresholding

Jun-Taek Oh and Wook-Hyun Kim

School of EECS, Yeungnam University, 214-1 Dae-dong,
Gyeongsan, Gyeongbuk 712-749, South Korea
{ohjuntaek, whkim}@yumail.ac.kr

Abstract. Multi-level thresholding is a method that is widely used in image segmentation. However, most of the existing methods are not suited to be directly used in applicable fields, and moreover they are not extended into a step of image segmentation. This paper proposes region-based multi-level thresholding as an image segmentation method. At first, we classify pixels of each color channel to two clusters by using EWFCM algorithm that is an improved FCM algorithm with spatial information between pixels. To obtain better segmentation results, a reduction of clusters is then performed by a region-based reclassification step based on a similarity between regions existing in a cluster and the other clusters. We finally perform a region merging by Bayesian algorithm based on Kullback-Leibler distance between a region and the neighboring regions as a post-processing method, as many regions still exist in image. Experiments show that region-based multi-level thresholding is superior to cluster-, pixel-based multi-level thresholding, and an existing method and much better segmentation results are obtained by the proposed post-processing method.

1 Introduction

Image segmentation plays an important role in understanding and analyzing image. In particular, region segmentation and object detection in image are both essential procedures for practical applications. Methods for image segmentation[1] include texture analysis-based methods, histogram thresholding-based methods, clustering-based methods, and region-based split and merging methods, among which threshold-based image segmentation[1, 2, 6-9] is widely used in many applications, such as document processing and object detection, as it is simple and efficient as regards dividing image into the foreground and background. Histogram thresholding-based methods use various criteria, such as Otsu's method[8], entropy-based method[2, 9], minimum error thresholding[10], and etc. However, none of these histogram thresholding-based methods include spatial information, which can lead to serious errors in the case of image segmentation. Plus, the selection of a threshold is very difficult, as the histograms of most real-images have an ambiguous and indistinguishable distribution. To

L. Wang et al. (Eds.): FSKD 2006, LNAI 4223, pp. 864 – 873, 2006.

solve the problem, FCM(fuzzy c-mean algorithm)[3-5], as a representative fuzzy clustering algorithm, has become a powerful tool that has been successfully applied to image thresholding to segment image into meaningful regions. However, certain problems like noise still remain, as no spatial information is included. In segmenting image, spatial information is an essential part since a pixel in real-image has a relation with the neighbors. In this paper, we propose EWFCM(entropy-based weighted FCM) algorithm using classification information between a pixel and the neighbors for classifying pixels in each color channel.

Threshold-based methods segment image using thresholds extracted from a brightness distribution of the image and many methods concerning them are being proposed at present. However, most of them focused on selecting the optimal thresholds for segmenting image. In case of segmenting image only using thresholds, they are not suited to be directly used in applicable fields, as image is segmented into very many regions. And most of the existing methods left the extension into image segmentation as a future work, otherwise they were proposed as a pre-processing method to obtain a finally segmented image. Y. Du[6, 7] used a histogram thresholding-based method for each color component in color image, and then multi-level image thresholding is performed by the optimal clusters determined by the within-class and between-class distance of the clusters, which are classified for each color component. Yet, this method is difficult to extend to multi-level thresholding for each color component and a reclassification of cluster-units to detect the optimal clusters leads to incorrect image segmentation. In this paper, we propose region-based multi-level thresholding as an extended method for image segmentation, which is performed to obtain better segmentation results by reducing clusters. Region-based multi-level thresholding is performed by a reclassification step based on similarities between the reclassified regions in a cluster and the other clusters. However, many similar or small regions still exist in image. To remove these regions, we perform a region merging using Bayesian algorithm based on Kullback-Leibler distances between regions.

2 EWFCM Algorithm and Region-Based Multi-level Thresholding

This paper consists of three steps, including EWFCM algorithm for classifying pixels in each color channel, region-based multi-level thresholding for efficiently reducing the clusters, and a region merging by Bayesian algorithm based on Kullback-Leibler distances between regions to obtain better segmentation results.

2.1 EWFCM(Entropy-Based Weighted FCM) algorithm

FCM(fuzzy c-means) algorithm[3-5] is widely used in image segmentation as an unsupervised segmentation algorithm. The objective function $J_m(U,V)$ in FCM algorithm is given by :

$$J_m(U,V)=\sum_{i=1}^{c}\sum_{j=1}^{n}(u_{ij})^m \left\| v_i - x_j \right\|^2 \tag{1}$$

where x_j is the gray-level value of j'th pixel and v_i is the mean value of i'th cluster. A solution of the objective function $J_m(U,V)$ can be obtained via an iterative process, where the degrees of membership and the mean value of cluster are updated via:

$$u_{ij} = \frac{1}{\sum_{k=1}^{c}\left(\| v_i - x_j \| / \| v_k - x_j \| \right)^{2/m-1}} \qquad v_i = \sum_{j=1}^{n}(u_{ij})^m x_j \Big/ \sum_{j=1}^{n}(u_{ij})^m \tag{2}$$

where u_{ij} is the degree of membership between the j'th pixel and i'th cluster, v_i is the mean value of i'th cluster, c is the number of clusters, and m is an arbitrarily chosen FCM weighting exponent that must be greater than one.

FCM algorithm can classify most of noise-free real-images, which have an uncertain and complex data distribution. However, as FCM algorithm does not incorporate spatial information, it may fail to segment image corrupted by noise and other imaging artifacts.

The neighborhood information is incorporated into FCM algorithm to remove any noise. In image, since the center pixel has a relationship with its neighbors, the probability that the center pixel and its neighbors will be classified in the same cluster is high. As such, Y. Yang[5] proposed a spatially weighted FCM algorithm using k-NN(nearest neighbor) algorithm, which is based on a distance between a mean gray-value of a cluster and a gray-value of a current pixel. However, Y. Yang's method may be lead to an incorrect classification if the histogram distributions of clusters are different. And it needs to define a parameter in advance. Therefore, this paper proposes an improved entropy-based weighted FCM(EWFCM) algorithm, where a weight based on entropy that takes into account the spatial relationship between the current pixel and its neighbors is applied to FCM algorithm.

The improved degrees of membership u_{ij}^* and v_i^* are given by:

$$u_{ij}^* = \frac{w_{ij}}{\sum_{k=1}^{c}\left(\| v_i - x_j \| / \| v_k - x_j \| \right)^{2/m-1}} \qquad v_i^* = \sum_{j=1}^{n}(u_{ij}^*)^m x_j \Big/ \sum_{j=1}^{n}(u_{ij}^*)^m \tag{3}$$

For i'th cluster, j'th pixel possessed a high ratio of belonging to i'th cluster when many neighbors of j'th pixel belong to i'th cluster. Then w_{ij} possesses a high weight as regards belonging to i'th cluster, and w_{ij} is calculated as:

$$w_{ij} = 1 - \frac{e_i}{e_i + e_k} = 1 - \frac{p_i \log(p_i)}{p_i \log(p_i) + p_k \log(p_k)} \tag{4}$$

$$p_i = \frac{1 + \#\text{of } x_{en} \text{ in the set } N_j^i}{1 + \#\text{of } x_{en} \text{ in the set } N_j} \qquad p_k = \frac{1 + \#\text{of } x_{en} \text{ in the set } N_j^k}{1 + \#\text{of } x_{en} \text{ in the set } N_j} \tag{5}$$

where x_{en} is the neighboring pixel of j'th pixel that is the center pixel, N_j is the set of the neighbors nearest to the center pixel, N_j^i is the subset of N_j composed of the pixels belonging to i'th class, N_j^k is the subset of N_j except N_j^i, p_i is the ratio that the neighbors of the j'th pixel belong to the same cluster, and p_k is the ratio that the neighbors of the j'th pixel do not belong to the same cluster. w_{ij} is then obtained by the entropy of Shannon based on those ratios. EWFCM algorithm can correctly classify pixels by only using a classification index between a current pixel and the neighbors and is performed faster than Y. Yang's method. And code image that is based on the cluster number extracted by EWFCM algorithm for each color component is created by $c_j = r_j level^0 + g_j level^1 + b_j level^2$ where, for j'th pixel, c_j is the combined cluster numbers in code image and (r_j, g_j, b_j) is the cluster number extracted by EWFCM algorithm for each color channel. And $level$ is the number of clusters for each color channel. If $level$ is set to 2, code image consists of all 8 clusters, and each cluster is assigned a cluster number from 0 to 7. If $level$ increases, the clusters in the code image abruptly increases. Therefore the clusters need to be reduced in the reclassification step.

2.2 Region-Based Multi-level Thresholding

Based on classification results for each color channel obtained by EWFCM algorithm, a pixel, region, or cluster that exists in code image is used as a reclassification unit. By reducing the clusters in code image by a reclassification step, we can obtain better segmentation results. In this paper, we describe region-based multi-level thresholding and pixel- and cluster-based multi-level thresholding are performed by the same procedure as region-based multi-level thresholding.

Region-based multi-level thresholding is performed by distances between regions segmented from a reclassified cluster and the other clusters. At first, a selection of the reclassified cluster is given by :

$$\max_{k \in all\_cluster} \left(\frac{\mathrm{var}_k}{1.0 + \mathrm{var}_{all}} \times \frac{size_{all} - size_k}{size_{all}} \times \frac{\max\_dis_{all} - \min\_dis_k}{\max\_dis_k} \right) \tag{6}$$

where var_k, var_{all}, $size_k$, and $size_{all}$ are variances and sizes of k'th cluster and image, respectively. And $\max\_dis_k$ and $\min\_dis_k$ are maximum and minimum distances between k'th cluster and the other clusters, respectively. For all clusters($all\_cluster$), a cluster is selected when its variance is large while its size is small and distances between it and the others are short. Regions existing in the reclassified cluster are then reclassified into most similar clusters by :

$$\min_{c \in all\_cluster - k_{index}} \left(\frac{\mathrm{var}_{r,c}}{\mathrm{var}_r + \mathrm{var}_c} \right) \tag{7}$$

where var_r and var_c are variances of r'th region in the reclassified cluster(k_{index}) and c'th cluster among the other clusters ($all\_cluster - k_{index}$), respectively. And $\mathrm{var}_{r,c}$ is variance after r'th region and c'th cluster are merged. A cluster is selected as a most similar cluster when a ratio of a sum of each variance before merging them to variance after merging them is the lowest.

A reclassification step is repeatedly performed until the number of clusters is the same as a pre-defined number. If the number of clusters is not defined, the optimal number of clusters is selected when an average within-class distance for the clusters in process of reclassifying all clusters into 2 clusters is minimal. The optimal number of clusters is selected by:

$$opt_{cluster} = \min_{all\_cluster}\left(\sum_{i=1}^{size_{all\ cluster}} wd_i \Big/ size_{all\_cluster} \right) \qquad all\_clsuter \geq 2 \tag{8}$$

$$wd_i = \frac{\sqrt{\sum_{j=1}^{size_i}\left((r_j - m\_r_i)^2 + (g_j - m\_g_i)^2 + (b_j - m\_b_i)^2\right)}}{size_i} \tag{9}$$

where $opt_{cluster}$ is a minimum average within-class distance for all clusters, $size_{all\_cluster}$ is the number of clusters, wd_i and $size_i$ are a within-class distance and size of i'th cluster, respectively. (r_j, g_j, b_j) is gray-values for red, green, blue color channel of j'th pixel existing in i'th cluster and (m_r_j, m_g_j, m_b_j) is average gray-values for red, green, blue color channel of i'th cluster.

2.3 Region Merging Using Bayesian Algorithm

As a post-processing method that is performed to obtain better segmentation results, regions that size is small are merged into the most similar neighboring regions. And similar regions are merged by Bayesian algorithm based on Kullback-Leibler distances between a merged region and the neighbors. The process for a region merging is as follows:

① A region that has the largest variance among all regions is selected as a merged region by :

$$\max_{r \in all\_region}\left(\mathrm{var}_r \times \frac{size_r}{size_{image}} \right) \tag{10}$$

where var_r and $size_r$ are variance and size of r'th region, respectively, $size_{image}$ is size of image.

② Kullback-Leibler distances between the region selected at ① step and its neighbors are measured by :

$$d(h_c, h_j) = \sum_{g=0}^{255}\left(h_c(g) - h_j(g)\right)\log\frac{h_c(g)}{h_j(g)} \tag{11}$$

where $d(h_c, h_j)$ is a distance between j'th region and c'th region and $h()$ is a function that has probability values obtained from a histogram distribution of the region.

③ After a most similar neighboring region for the region selected at ① step is selected by Bayesian algorithm, the regions are merged if their similarity is larger than a given threshold(0.7).

$$\max_{j \in nr}\left(\frac{P(r_c|r_j)}{\sum_{i=1}^{nr} P(r_c|r_i)}\right) \quad P(r_c|r_j) = \frac{dis_{c\_j}}{\sum_{k=1}^{nr} dis_{c\_k}} \quad dis_{c\_k} = \frac{1.0}{1.0 + d(h_c, h_k)} \tag{12}$$

where, on the assumption that $P(r_j)$ is same, $P(r_c|r_j)$ is a probability value based on a distance(dis_{c_j}) between the region(r_c) selected at ① step and j'th region(r_j) among the neighboring regions(nr) and it is a similarity that takes account of distances of the neighboring regions.

④ ① ~ ③ steps are repeatedly performed until the clusters are not reduced.

3 Experiment

All the algorithms in this paper were coded using SDK Version 1.4.1 in Window XP. And a function developed by M. Borsotti[11] was used for the performance evaluation.

$$Q(I) = \frac{\sqrt{R}}{10000\,(N \times M)} \times \sum_{i=1}^{R}\left(\frac{e_i^2}{1 + \log(A_i)} + \left(\frac{R(A_i)}{A_i}\right)^2\right) \tag{13}$$

where I is a segmented image, N and M are the width and height of the image, respectively, R is the number of regions in a segmented image, A_i and e_i are the area and average color error for the i'th region, respectively. $R(A_i)$ represents the number of regions with an area equal to A_i. The smaller the value of $Q(I)$, the better the segmentation result.

Fig. 1 shows code images for comparing performances of each method for noise removal. Fig. 1(a) is noisy images with added 5% salt & pepper noise. Figs. 1(b) ~ (d) are code images that consist of 8 clusters after classifying pixels for each color channel into 2 clusters by FCM algorithm, Y. Yang method, and EWFCM algorithm. Y. Yang method and EWFCM algorithm effectively removed noise while FCM algorithm left noise as it is. And as compared to Y. Yang method, EWFCM algorithm obtained a little better result. Moreover, the computational time of EWFCM algorithm showed approximately 59% from Y. Yang method for Fig. 1(a). After creating code image, a reclassification step is performed to reduce clusters existing in code image. Table 1 shows performance evaluations and the number of regions for the segmented image when a pixel, region, and cluster are used as an unit of the reclassification. Pixel- and cluster-based multi-level thresholding are performed by the same process as region-based multi-level thresholding. Bold and slant letters in table 1 show performance evaluations obtained from the optimal clusters. As be seen in table 1, the optimal number of clusters selected by a minimum average within-class distance doesn't accord with that by M. Borsotti, since a minimum average within-class distance only depends on a difference between average gray-values of a cluster and gray-value of pixels that are classified into the cluster. However, we used a minimum average within-class distance for more quickly selecting the optimal clusters. Region-based multi-level thresholding showed the best performance although the number of regions is smaller. Pixel- and cluster-based multi-level thresholding showed that the number of regions is either same or more although the number of clusters is reduced. This means that a pixel or regions in a cluster are reclassified into the remaining clusters as an independent region and a reclassification of a pixel or a cluster has no effect on more improvement of segmentation results. That is to say, this shows that a region is a more important factor than a pixel and a cluster in segmenting an image. And all methods showed that performance evaluations have a high value if too many clusters are reduced in a reclassification step, since a brightness error in a region is as much larger as reducing the number of regions.

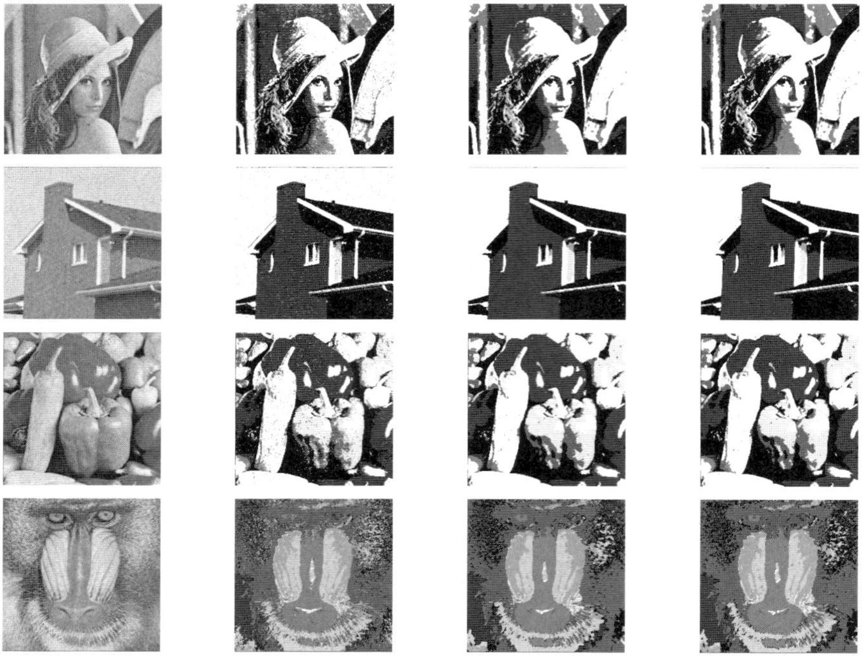

(a) (b) (c) (d)

Fig. 1. Performance evaluation comparison of the proposed method and the existing methods for removal of noise. (a) Noisy images with added 5% salt & pepper noise. (b) Code images by FCM algorithm. (c) Code images by Y. Yang method. (d) Code images by EWFCM algorithm.

Table 1. Comparison of performance evaluations by region-, cluster-, and pixel-based multi-level thresholding. Row : Number of clusters. Column : Experimental images and reclassification methods, including region-based, cluster-based, and pixel-based reclassification.

		8	7	6	5	4	3	2
Lena	Region	2212.9 (668)	2210.0 (666)	2180.2 (648)	2118.6 (610)	1954.0 (499)	***2578.9 (254)***	5100.7 (61)
	Cluster	2212.9 (668)	2212.9 (668)	2212.9 (668)	2207.9 (665)	2100.1 (600)	***2769.5 (300)***	5821.9 (77)
	Pixel	2212.9 (668)	2212.9 (668)	2240.6 (684)	2099.9 (599)	2383.8 (740)	***3523.6 (507)***	6072.4 (89)
House	Region	313.5 (427)	311.1 (420)	307.9 (411)	306.7 (407)	241.0 (212)	***245.3 (198)***	2307.5 (42)
	Cluster	313.5 (427)	313.1 (426)	307.8 (409)	310.8 (407)	255.8 (230)	263.6 (223)	***1586.2 (18)***
	Pixel	313.5 (427)	313.8 (428)	310.5 (419)	310.2 (417)	311.1 (419)	***267.6 (246)***	1791.4 (24)
Peppers	Region	1218.4 (607)	1204.1 (591)	1153.8 (531)	1124.8 (481)	***1005.7 (286)***	5918.4 (194)	3669.3 (57)
	Cluster	1218.4 (607)	1218.4 (607)	1201.8 (590)	1164.8 (519)	1456.8 (313)	7143.4 (253)	***3464.1 (52)***
	Pixel	1218.4 (607)	1220.4 (607)	1235.1 (618)	1209.1 (557)	1676.1 (443)	5347.6 (254)	***6349.2 (83)***
Baboon	Region	3024.9 (1234)	3022.6 (1232)	2997.9 (1207)	3755.4 (1022)	***3365.5 (749)***	6733.9 (502)	31568.2 (216)
	Cluster	3024.9 (1234)	3024.9 (1234)	3020.0 (1230)	3855.6 (1095)	3479.1 (817)	***6470.0 (440)***	16550.8 (55)
	Pixel	3024.9 (1234)	3037.5 (1244)	3194.7 (1329)	3274.6 (1340)	***3377.9 (1356)***	5435.7 (507)	34327.2 (660)

Fig. 2 shows the segmented images and performance evaluations by the proposed method and Y. Du's method. Y. Du segmented image using the optimal clusters that are selected by a within-class and between-class distances of clusters after creating code image and classifying pixels into 2 clusters for each color channel by Otsu and Kapur methods. In Fig. 2, the proposed method was superior to Y. Du's method despite having less clusters and regions for all experimental images except that Y. Du(Kapur) showed the best performance evaluation for 'lena' image. However, as be seen in Fig. 2, the proposed method showed the best segmentation results in point of visual view. And the optimal number of clusters that is selected by the proposed method was less than that by Y. Du's method. That is to say, this means that the proposed method can display image using less color information than Y. Du's method. Fig. 3 shows the resulting images that are finally segmented by a region merging for

Fig. 2(c). A region merging is performed by Bayesian algorithm based on Kullback-Leibler distances between regions. As compared with the results that are only obtained by region-based multi-level thresholding, better segmentation results were showed although many regions were reduced. Therefore, it shows that the proposed region merging algorithm is valid and effective as a post-processing method for image segmentation.

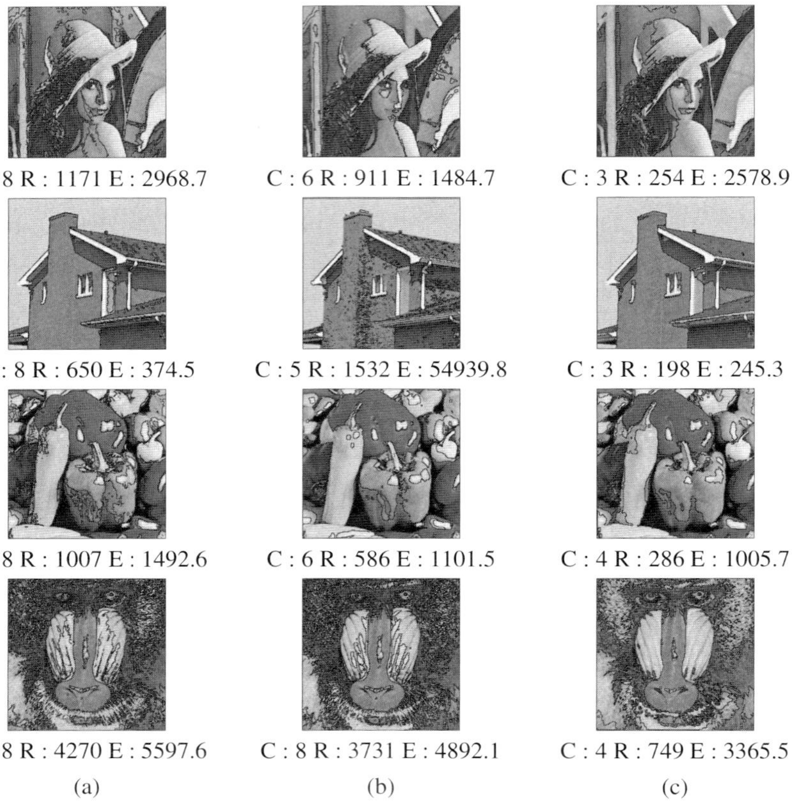

Fig. 2. Segmented images and performance evaluations (a) : by Y. Du(Otsu) (b) : by Y. Du(Kapur) (c) : by the proposed method. C : Number of clusters. R : Number of regions. E : Performance evaluation

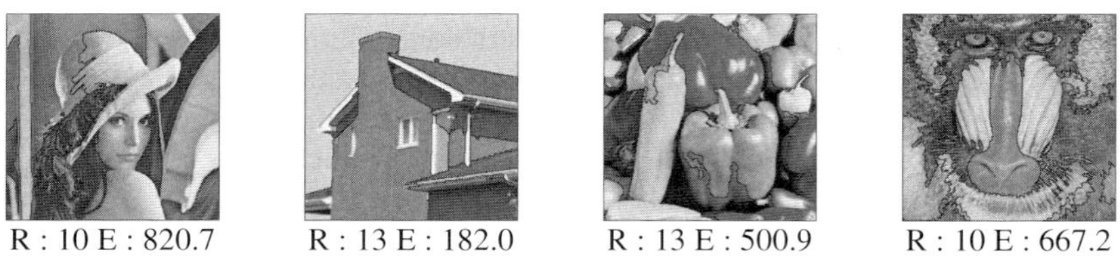

Fig. 3. Finally segmented images and performance evaluations by region merging. R : Number of regions. E : Performance evaluation.

4 Conclusion

This paper proposes region-based multi-level thresholding for color image segmentation. EWFCM algorithm that is used in classifying pixels for each color channel into 2 clusters effectively removed noise and was faster performed than an existing method. And as a multi-level thresholding method that is extended into image segmentation, region-based reclassification showed better segmentation results than a pixel- and a cluster-based reclassification as well as an existing method. In image segmentation, this means that a region is a more important factor than a pixel and a cluster. In addition, by performing a region merging using Bayesian algorithm based on Kullback-Leibler distances between a region and the neighbors, we obtained more accurate segmentation results than those that are obtained by only using region-based multi-level thresholding. The proposed method is possible to be applied into various fields, including extraction of principal color information, object detection, image retrieval, and so on. And an application of the proposed method with reducing the computational time is areas under further study.

References

1. Sezgin, M. and Sankur, B.: Survey over image thresholding techniques and quantitative performance evaluation. Journal of Electronic Imaging, Vol.13, No.1, (2004) 146-165
2. Brink, A. D.: Minimum spatial entropy threshold selection. IEE Proc. Vis. Image Signal Process., Vol.142, No.3, (1995) 128-132
3. Pal, N., and Bezdek, J.: On cluster validity for the fuzzy c-means model. IEEE Trans. Fuzzy Syst., Vol.3, No.3, (1995) 370-379
4. Pham, D. L.: Fuzzy clustering with spatial constraints. Proc. of IEEE Conf. on Image Process., Vol.2, (2002) 65-68
5. Yang, Y., Zheng, C., and Lin, P.: Image thresholding based on spatially weighted fuzzy c-means clustering. Proc. of IEEE Conf. on Computer and Information Technology, (2004) 184-189
6. Du, Y., Chang, C., and Thouin, P. D.: Unsupervised approach to color video thresholding. Opt. Eng. Vol.32, No.2, (2004) 282-289
7. Du, Y., Change, C. I., and Thouin, P. D.: An unsupervised approach to color video thresholding. Proc. of IEEE Conf. on Acoustics, Speech and Signal Processing, Vol.3, (2003) 373-376
8. Otsu, N.: A threshold selection method from gray level histograms. IEEE Trans. Syst. Man Cybern. Vol.9, No.1, (1979) 62-66
9. Kapur, J. N., Sahoo, P. K., and Wong, A. K. C.: A new method for gray level picture thresholding using the entropy of the histogram. Graph. Models Image Process., Vol.29, (1985) 273-285
10. Lloyd, D. E.: Automatic target classification using moment invariant of image shapes. Technical Report, RAE IDN AW 126, Farnborough, UK, (1985)
11. Borsotti, M., Campadelli, P., and Schettini, R.: Quantitative evaluation of color image segmentation results. Patt. Recogn. Lett. Vol.19, No.8, (1998) 741-747

Feature-Oriented Fuzzy Shock-Diffusion Equation for Adaptive Image Resolution Enhancement*

Shujun Fu[1,2,**], Qiuqi Ruan[2], Wenqia Wang[1], and Jingnian Chen[3]

[1] School of Mathematics and System Sciences, Shandong University, Jinan, 250100, China
[2] Institute of Information Science, Beijing Jiaotong University, Beijing, 100044, China
[3] School of Arts and Science, Shandong University of Finance, Jinan, 250014, China
**shujunfu@163.com

Abstract. Image resolution enhancement is receiving a great deal of attention in the wide increasing use of digital imaging technologies recently. This paper presents a feature-oriented fuzzy shock-diffusion equation, where the shock term is used to sharpen edges along the normal direction to the isophote line (edge), while the diffusion term is used to remove artifacts ("jaggies") along the tangent direction. A fuzzy decision mechanism is used to preserve image features such as edge, texture and fine part. Finally, a shock capturing scheme with a special limiter function is developed to speed the process with numerical stability. Experimental results on real images demonstrate that our algorithm substantially improves the subjective quality of the enhanced images over conventional interpolations and some related equations.

1 Introduction

The recent increase in the wide use of digital imaging technologies in various markets has brought a simultaneous demand for higher-resolution images. The demand for such high-resolution (HR) images can be met by algorithmic advances in super-resolution (SR) technology in place of–or in tandem with–hardware development [1]. Instead of using a sequence of video frames or multiple exposure, we elaborate an approach suitable for SR based on a single image, exploiting the properties common to a wide range of natural images.

Image interpolation (magnification) is a common image resolution enhancement process to gain a high-resolution image from its low-resolution version for a single image [2-6]. However, most interpolation algorithms in existence to enhance image resolution suffer visually the effects of blurred edges and annoying artifacts ("jaggies" and "mosaics") to some extent. And, they may also lose some features of the image.

In the past decades there has been a growing amount of research concerning partial differential equations in image enhancement, such as anisotropic diffusion filters [7-9] for edge preserving noise removal, and shock filters [10-11] for edge sharpening.

* This work is supported by the national natural science fund, China (No. 60472033); the open project of the national laboratory of pattern recognition at the institute of automation of the chinese academy of sciences, China; and the technological innovation fund of excellent doctorial candidate of Beijing Jiaotong university, China (No. 48007).

L. Wang et al. (Eds.): FSKD 2006, LNAI 4223, pp. 874 – 882, 2006.

Image interpolation means "to read between the original pixels", which also can be considered as a diffusion process: "to diffuse gray levels from pixels of the original image to the blank interpolated pixels between them". Therefore, we extend the nonlinear PDE-based flow method, and apply it to image interpolation.

As the extension of conventional (crisp) set theory, L. A. Zadeh put forward the fuzzy set theory to model the vagueness and ambiguity in complex systems which is a useful tool for handling the uncertainty associated with vagueness and/or imprecision. Image and its processing bear some fuzziness in nature. Moreover, some definitions, such as edges, boundaries and even the definition of contrast, are fuzzy. Therefore, fuzzy set theory has been successfully applied to image processing and computer vision [12]. In this paper, incorporating anisotropic diffusion with shock filter, we present feature-oriented fuzzy shock-diffusion equation (FFSE) for adaptive image resolution enhancement, where a fuzzy decision mechanism is used to preserve image features such as edge, texture and fine part.

In numerical implementation, in order to solve effectively the nonlinear equation to obtain discontinuous solution with numerical instability, a shock capturing scheme is developed with a special limiter function to speed the process.

This paper is organized as follows. In section 2, some related equations are introduced for enhancing images: anisotropic diffusions and shock filters. Then, a feature-oriented fuzzy shock-diffusion equation is proposed to enhance adaptively image resolution. In section 3, we implement the proposed method and test it on real images. Conclusions are presented in section 4.2.

2 Feature-oriented Fuzzy Shock-diffusion Equation

2.1 Adaptive Image Resolution Enhancement

The FFSE process has two steps. First, the image is interpolated to the new desired size. We use bilinear interpolation. The first step provides good results over smooth areas, but edges are smeared, and artifacts ("jaggies") are introduced. In Fig.1, local parts of results obtained by the bilinear interpolation on images, such as the Lena and the medical MRI, are shown respectively. One can see obviously blurred edges and annoying artifacts in the interpolated image.

Secondly, we perform the nonlinear FFSE process to enhance the edges and to smooth the interpolation byproducts, where a fuzzy decision mechanism is used to preserve image features.

2.2 Some Related Work

One of most influential work in using partial differential equations (PDEs) in image processing is the anisotropic diffusion (AD) filter, which was proposed by P. Perona and J. Malik [8] for image denoising, enhancement, etc. Let $(x, y) \in \Omega \subset R^2$, and $t \in [0, +\infty)$, a multi-scale image $u\,(x, y, t): \Omega \times [0, +\infty) \to R$, is evolved according to the following equation:

$$\frac{\partial u(x,y,t)}{\partial t} = div(g(|\nabla u(x,y,t)|)\nabla u(x,y,t)), \quad g(|\nabla u|) = 1/(1+(|\nabla u|/K)^2) \qquad (1)$$

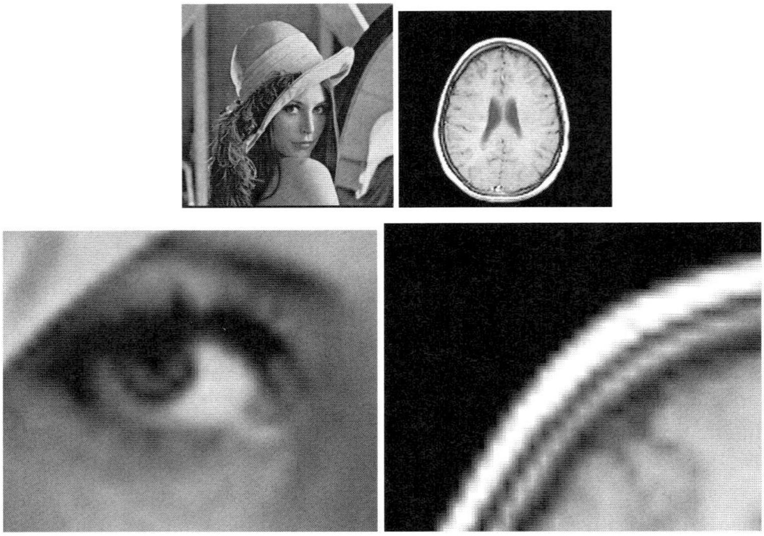

Fig. 1. Top: original images; Bottom: local parts of results by the bilinear interpolation

where K is a gradient threshold. The scalar diffusivity $g(|\nabla u|)$, chosen as a non-increasing function, governs the behaviour of the diffusion process. By formally developing the divergence term, (1) can be put in term of second order derivatives taken in the directions of the gradient vectors ($\vec{N}$) and in the orthogonal tangent ones ($\vec{T}$):

$$\frac{\partial u}{\partial t} = (K^2(K^2 - |\nabla u|^2)/(K^2 + |\nabla u|^2)^2)u_{NN} + (K^2/(K^2 + |\nabla u|^2))u_{TT} \tag{2}$$

This expression allows an easier interpretation of the original equation: (1) acts like a low pass filter diffusing along the edge directions and, selectively, for diffusion functions as (1), can preserve edges without diffusing across edges or even enhance them, provided that their gradient value is greater than K.

Different from the nonlinear parabolic diffusion process, L. Alvarez and L. Mazorra [11] proposed an anisotropic diffusion with shock filter (ADSF) equation by adding a hyperbolic equation, called shock Filter which was introduced by S.J. Osher and L.I. Rudin [10], for noise elimination and edge sharpening:

$$\frac{\partial u}{\partial t} = -\text{sign}(G_\sigma * u_{NN})\text{sign}(G_\sigma * u_N)|\nabla u| + cu_{TT} \tag{3}$$

where G_σ is a Gaussian function with standard deviation σ, c is a positive constant.

2.3 Fuzzy Shock-Diffusion Equation

In equation (3), however, to enhance an image using the symbol function sign(x) is a binary decision process. This is a hard partition without middle transition. Unfortunately, the obtained result is a false piecewise constant image, where a bad visual quality is produced in some areas. There are two reasons for the failure of the binary shock filter. One is that the binary shock filter is a hard sharpening process, and it does not sharpen differently the image according to the different characteristics of image areas. As a result, too many shocks are formed, which make the result look unreal. Another is that the binary shock filter sharpens the image equally without adjusting its behaviour according to image features. Some artifacts and false textures appear inevitably in the image.

Fuzzy techniques are powerful tools for knowledge representation and processing. In image processing applications, many difficulties arise because the data / tasks / results are uncertain. This uncertainty, however, is not always due to the randomness but to the ambiguity and vagueness. Beside randomness which can be managed by probability theory, we can distinguish between three other kinds of imperfection in the image processing: grayness ambiguity, geometrical fuzziness and vague (complex / ill-defined) knowledge. These objects are fuzzy in nature [12]. Denote the fuzzy set S on the region R as:

$$S = \int \frac{\mu_S(x)}{x}, \quad x \in R \tag{4}$$

where $\mu_s(x) \in [0, 1]$ is called the membership function of S on R. Chen, etc [13] extended further above set to the generalized fuzzy set, where they denoted the generalized membership function (GMF) $\mu_s(x) \in [-1, 1]$ to substitute $\mu_s(x) \in [0, 1]$.

An image comprises regions with different features, such as edges, textures and details, and flat areas, which should be treated differently to obtain a better result in an image processing task. We divide an image into two-type regions by its smoothed gradient magnitude: big gradients (such as boundaries of different objects, textures and details), and small gradients (such as smoother segments inside different areas). In our algorithm, for edges, textures and details, shock filters with the sign function enhance image features in a binary decision process, which produce unfortunately a false piecewise constant result. We notice that the variation of texture and detail is fuzzy in these areas. In order to approach this variation, we extend the binary decision to a fuzzy one substituting sign(x) by a hyperbolic tangent membership function th(x), which guarantees a natural smooth transition in these areas, by controlling softly changes of gray levels of the image. As a result, a fuzzy shock-type backward diffusion is introduced to enhance these features while preserving a natural transition in these areas. The normal derivative of the smoothed image is used to detect image feature. A forward diffusion in the isophote line direction is performed to smooth artifacts and jaggies. Finally, an isotropic diffusion is used to smooth flat areas simultaneously.

Thus, incorporating anisotropic diffusion with shock filter, we develop a feature-oriented fuzzy shock-diffusion equation (FFSE) process to enhance image resolution while preserving image features simultaneously:

$$\begin{cases} v = G_\sigma * u \\ \dfrac{\partial u}{\partial t} = c_N L_N(u) + c_T L_T(u) \end{cases} \tag{5}$$

where

$$L_N(u) = \begin{cases} -\mathrm{th}(l_1 v_{NN})\,|u_N|, & |v_N| \geq \mathrm{T} \\ 0, & \text{else} \end{cases}, \quad L_T(u) = \begin{cases} u_{TT}, & |v_N| \geq \mathrm{T} \\ \Delta u, & \text{else} \end{cases}$$

with Neumann boundary condition, where l_1 is a parameter to control the gradient of the membership function th(x); T is a threshold to divide the image into two-type different regions; c_N and c_T are the backward and forward flow control coefficients respectively. We choose $c_N = 1$ and the following forward flow control coefficient to prevent excess smoothness to smaller details:

$$c_T = 1/(1 + l_2 u_{TT}^2) \tag{6}$$

where l_2 is a constant.

An image magnification method using Level-Set Reconstruction (we call it LSR) is presented in [14], where instead of assuming a smoothness prior for the underlying intensity function, it assumes smoothness of the level curves, and produces appealing visually images. However, with $c_N = 0$ and $c_T = 1$, it does not sharpen edges in the gradient direction, and may smooth away corners and small details at the same time.

3 Numerical Implementation and Experimentals

3.1 A Shock Capturing Scheme

When solving numerically a nonlinear convection-diffusion equation like (5) using a difference scheme, the hyperbolic term must be discretized carefully because discontinuity solutions, numerical instability and spurious oscillation may appear [15]. Here, we develop a speeding shock capturing scheme by using a proper limiter function.

An xplicit Euler method with central difference scheme is used to approximate (5) except the gradient term $|u_N|$. Below we detail a numerical approach to it. On the image grid, the approximate solution is to satisfy:

$$u_{ij}^n \approx u(ih, jh, n\Delta t), \; i, j, n \in Z^+ \tag{7}$$

where h and $\triangle t$ are the spatial and temporal step respectively. Let $h = 1$, $\delta^+ u_{ij}^n$ and $\delta^- u_{ij}^n$ areforward and backward difference schemes of u_{ij}^n respectively. A limiter function MS is used to approximate the gradient term:

$$|u_N| = \sqrt{(MS(\delta_x^+ u_{ij}^n, \delta_x^- u_{ij}^n))^2 + (MS(\delta_y^+ u_{ij}^n, \delta_y^- u_{ij}^n))^2} \tag{8}$$

where

$$MS(x, y) = \begin{cases} x, & |x| < |y| \\ y, & |x| > |y| \\ x, & |x| = |y| \text{ and } xy > 0 \\ 0, & |x| = |y| \text{ and } xy \le 0 \end{cases} \tag{9}$$

The MS function gets fewer 0 in value than the *minmod* function does in the x-y plane, which also make the scheme satisfy the numerical instability. Because the gradient term represents the transport speed of the scheme, the MS function makes our scheme evolve faster with a bigger transport speed than those with the *minmod* function.

3.2 Experiments

We present results obtained by using our scheme (5), and compare its performance with above related methods. A number of images have been used to test our scheme. Examples shown in Fig.2-4 are the Lena and the MRI, where we interpolate them by a factor 2 with the parameters: $[l_1, l_2]$=[300, 7×10^{-4}], $T=10$ and 14 respectively.

It is commonly agreed that the peak signal-to-noise ratio (PSNR) of the original image to the interpolated result does not always provide an accurate measure of the visual quality for natural images [4,5,16]. Therefore, we only rely on subjective evaluation to assess the visual quality of the enhanced images in this paper. The proposed FFSE is compared with conventional interpolation methods: the nearest, the bilinear and the bicubic, the LSR, and the ADSF respectively. In Fig.2 and Fig.3, the resolution of the Lena image has been increased by a factor 2. Conventional interpolation methods: the nearest, the bilinear and the bicubic, result in blurred edges and annoying jaggies, specially the nearest. The level-set reconstruction (LSR) can effectively smooth jaggies and obtain pleasing visually contours. But it does not sharpen edges in the gradient direction, and the edges remain blurry. At the same time, it smooths away some fine part. As for the ADSF, it produces sharp edges and smooth contours. However, indicating the zero-crossing of the edge by the sign function is a binary decision process, by which, unfortunately, the obtained result is a false piecewise constant image. With a discontiguous transition between two different areas, the result of image looks unnatural. Finally, it can be seen that the best visual quality is obtained by enhancing the image resolution using the proposed method, which preserves most features of the image with a natural transition between two different areas, and produces pleasing sharp edges and smooth contours (see Lena's brim, cheek and eyeballs in Fig.2 and Fig.3).

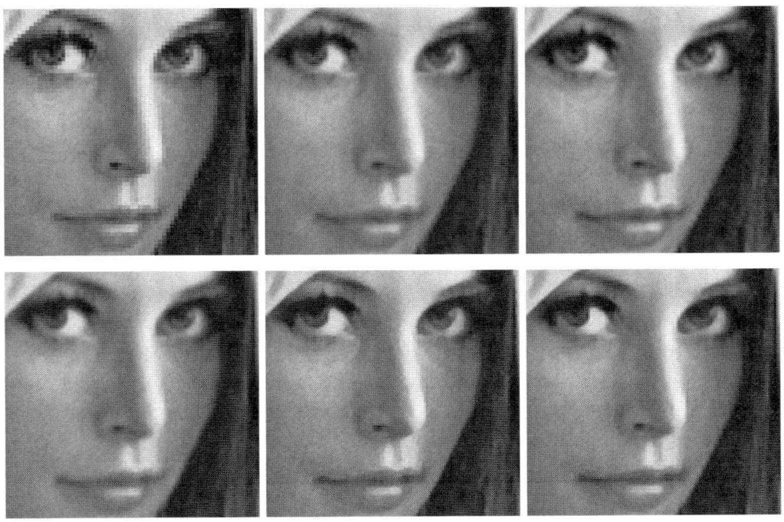

Fig. 2. Enhancing the Lena image (top-left to bottom-right): the nearest, the bilinear, the bicubic, the LSR, the ADSF, and the FFSE

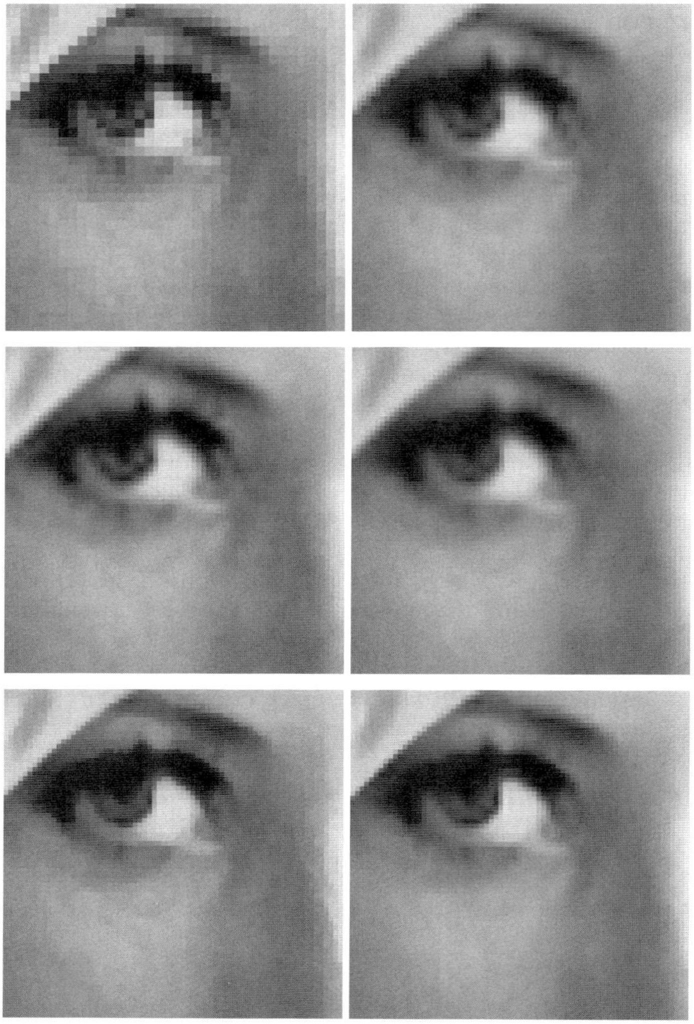

Fig. 3. Local parts of above results (top-left to bottom-right): the nearest, the bilinear, the bicubic, the LSR, the ADSF, and the FFSE

The next experiment is to enhance the medical MRI image by above methods. In Fig.4, we show the results by the bilinear, the ADSF and the FFSE respectively. It can be seen that the bilinear method produces blurry edges and jagged contours; while the ADSF method yields unnatural transition in some regions. The best visual quality is obtained by enhancing the image resolution using the proposed method.

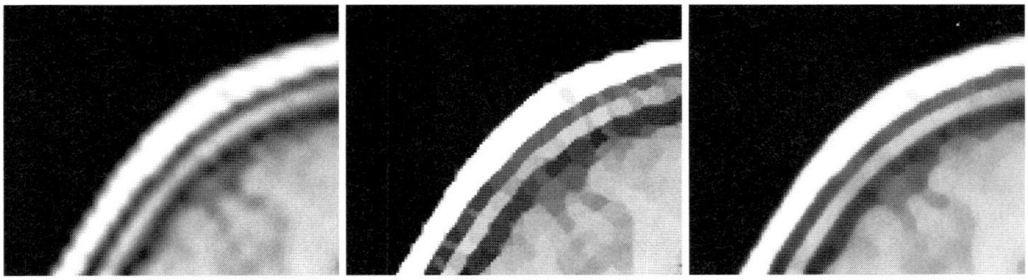

Fig. 4. Local parts of results by enhancing the MRI image using different methods (left to right): the bilinear, the ADSF, and the FFSE

4 Conclusions

This paper presents an adaptive image resolution enhancement scheme using fuzzy shock-diffusion equation, by which we not only can effectively sharpen edges, but also can smooth contours of the interpolated image. Preserving image features such as edges, corners and fine part with a natural transition between two different areas, this method produces better visual results of the interpolated images than conventional interpolations and some related equations. In numerical implementation, a shock capturing scheme speeds the process.

References

1. Castleman, K.R.: Digital Image Processing, Prentice Hall (1995).
2. Jensen K., Anastassiou D.: Subpixel edge localization and the interpolation of still images. IEEE Trans. on Image Processing, 4(3)(1995) 285–295.
3. Allebach J., Wong P. W.: Edge-directed interpolation. Proc. IEEE Int. Conf. Image Processing, 3(1996) 707–710.
4. Battiato S., Gallo G., Stanco F.: A locally adaptive zooming algorithm for digital images. Image Vision and Computing, Elsevier Science. Inc., 20(2002) 805-812.
5. Li Xin, Orchard M. T.: New Edge-Directed Interpolation. IEEE transactions on image processing, 10(10)(2001) 1521-1527.
6. Chang-Qing Zhu, Qian Wang, et al: Image Magnification Based on Multi-Band Wavelet Transformation. China Journal of Image and Graphics, 7(A) (3)(2003) 653-656.
7. Aubert, G., Kornprobst, P.: Mathematical Problems in Image Processing: Partial Differential Equations and the Calculus of Variations, vol.147 of Applied Mathematical Sciences, Springer-Verlag (2001).
8. Perona, P., Malik, J.: Scale-space and edge detection using anisotropic diffusion. IEEE Trans. Pattern Anal. Machine Intell., 12(7)(1990) 629-639.
9. You, Y.L., Xu, W., Tannenbaum, A., Kaveh, M.: Behavioral analysis of anisotropic diffusion in image processing. IEEE Trans. on Image Processing, 5(11)(1996) 1539-1553.
10. Osher, S.J., Rudin, L.I.: Feature-oriented image enhancement using shock filters. SIAM J. Numer. Anal., 27(1990) 919-940.
11. Alvarez, L., Mazorra, L.: Signal and image restoration using shock filters and anisotropic diffusion. SIAM J. Numer. Anal., 31(2)(1994) 590-605.

12. Tizhoosh Hamid R.: *Fuzzy Image Processing: Introduction in Theory and Applications*, Springer-Verlag (1997).
13. Chen W.F., Lu X.Q., Chen J.J., Wu G.X.: A new algorithm of edge detection for color image: Generalized fuzzy operator: *Science in China (Series A)*, 38(10)(1995) 1272-1280.
14. Morse B.S., Schwartzwald D.: Image magnification using level-set reconstruction. Proceedings of IEEE Computer Society Conference on Computer Vision and Pattern Recognition, 1(2001) 333-340.
15. Liu, R.X., Shu, Q.W.: Some new methods in Computing Fluid Dynamics, Science Press of China, Beijing (2004).
16. Muresan D. D., Parks Thomas W.: Adaptively quadratic (aqua) image interpolation. IEEE Transactions on Image Processing, 13(5)(2004) 690-698.

Fuzzy Edge-Sensitive Filter for Noise Reduction

Zheng-fang Li, Qing-jun Yu, and Wei-jun Li

[1] College of Electronics & Information Engineering, South China University of Technology
Guangzhou, China, 510641
adshlxie@scut.edu.cn

Abstract. In this paper, a fuzzy edge-direction detector is introduced firstly, and then a fuzzy edge-sensitive noise reduction filter is presented in the basis of edge-sensitive noise reduction algorithm proposed by Adelmann (1999). The experimental results show that the method can obtain better results than Adelmann's and other schemes, such as Russo's FIRE filter (1997), in respect of smoothing impulse, Gaussian and mixed noise and preserving edges.

1 Introduction

Noise smoothing and image enhancement are very popular in most of image processing applications. Because smoothing a region might destroy an edge and sharpening edge might lead to unnecessary noise, the key of the problem lies on how to seek a trade-off between noise reduction and detail preserving. Many different techniques have been proposed to address this problem, and have got many promising results [1]~[3]. Xie [4] separated the homogeneous blocks from inhomogeneous ones to prevent edges from blurring by blocking artifacts reduction. Xu [5] used some characteristics of human visual system to discriminate edges and blocking artifacts. However, during the image processing, finding a good trade-off between noise reduction and preservation of details is a rather difficult work, and still need to be investigated further.

Adelmann (1999)[6] proposed an edge-sensitive noise reduction algorithm for digital image processing. However, this algorithm does not fit for natural scenes. In [7], Russo proposed some fuzzy-rule-based operator for smoothing, sharpening, and edge detection. In this paper, we incorporate heuristic fuzzy rules into edge-sensitive noise reduction algorithm, and propose an edge-sensitive noise reduction method based on fuzzy-logic-control, which can not only smooth Gaussian noise, but also remove impulsive noise, while preserve edges and image details efficiently.

2 Fuzzy Edge-Direction Detector

Adelmann's [6] scheme is not robust against noise, especially impulsive noise. Therefore, it cannot provide good performance of impulsive noise reduction, and its ability

[1] The work is supported by the National Natural Science Foundation of China for Excellent Youth (Grant 60325310), the Guangdong Province Science Foundation for Program of Research Team (Grant 04205783), the Specialized Prophasic Basic Research Projects of Ministry of Science and Technology, China (Grant 2005CCA04100).

L. Wang et al. (Eds.): FSKD 2006, LNAI 4223, pp. 883 – 886, 2006.

of edge preserving is also limited. To address this problem, we introduce a fuzzy edge-direction detector based on [8] in this section.

Firstly, we define the input of this fuzzy detection system as follow

$$\Delta_{k,\theta}(i,j) = \left|P(i+k+\sigma_\theta, j-1) - P(i+k-\sigma_\theta, j+1)\right| \tag{1}$$

$$\Delta_{\beta,k}(i,j) = \left|P(i-1, j+k+\sigma_\beta) - P(i+1, j+k-\sigma_\beta)\right| \tag{2}$$

where $P(i,j)$ denotes the pixel luminance at location (i,j), and $\Delta_{k,\theta}(i,j)$ denotes the pixel variation in direction θ at the pixel $P(i+k,j)$ (k is the horizontal offset), and σ_θ is the orientation offset. Similarly, $\Delta_{\beta,k}(i,j)$ and σ_β are defined. Here we only consider four possible edge directions for θ and β, i.e., 0°, 45°, 90° and 135°, and let $\sigma_{0^o} = 0$, $\sigma_{45^o} = 1$, $\sigma_{90^o} = 0$ and $\sigma_{135^o} = -1$.

Secondly, we converted these inputs into fuzzy variables represented by the name of the associated fuzzy sets and a membership value. For this fuzzy detector, we use four fuzzy sets, named $SMALL_{0^o}$, $SMALL_{45^o}$, $SMALL_{90^o}$ and $SMALL_{135^o}$, respectively, to characterize small pixel variation. These membership functions have all the same shape, which are defined differently by two parameters: a_θ and b_θ.

Thirdly, a rule base is constructed for the fuzzy detector. To be simplified, only one rule is used to detect each of the four directions. For example, the second rule can be read as

R2:

$$\begin{aligned}&\text{If } \Delta_{-1,45^o}(i,j) \text{ is } SMALL_{45^\circ} \text{ and } \Delta_{1,45^o}(i,j) \text{ is } SMALL_{45^\circ} \text{ and } \Delta_{0,45^\circ}(i,j) \text{ is } SMALL_{45^\circ}\\ &\quad\text{and } \Delta_{45^o,-1}(i,j) \text{ is } SMALL_{45^\circ} \text{ and } \Delta_{45^o,1}(i,j) \text{ is } SMALL_{45^\circ} \text{ then } dir(i,j) = 45^\circ\end{aligned} \tag{3}$$

Finally, to make the final decision about the edge-direction at the pixel $P(i,j)$, defuzzification for our fuzzy detector takes the direction with the maximum membership value, as described by

$$Direction(i,j) = \arg\max_{i=0,1,2,3,4} (\lambda_i) \tag{4}$$

where λ_i is the activity degree of rule R_i $(i = 1,2,3,4)$, can be computed by

$$\begin{aligned}\lambda_1 &= \min\{SMALL_{0^\circ}(\Delta_{k,0^o});\, k = -2,-1,0,1,2\}\\ \lambda_2 &= \min\{SMALL_{45^\circ}(\Delta_{k,45^o}), SMALL_{45^\circ}(\Delta_{45^\circ,k});\, k = -1,0,1\}\\ \lambda_3 &= \min\{SMALL_{90^\circ}(\Delta_{90^o,k});\, k = -2,-1,0,1,2\}\\ \lambda_4 &= \min\{SMALL_{135^\circ}(\Delta_{k,135^\circ}), SMALL_{135^\circ}(\Delta_{135^\circ,k});\, k = -1,0,1\}\end{aligned} \tag{5}$$

For the ELSE-rule, R0, we apply the following formula to calculate its activity degree

$$\lambda_0 = \max\left\{0, 1 - \sum_{i=1}^{4} \lambda_i\right\} \tag{6}$$

3 Edge-Sensitive Median Filter

To improve the performance in impulsive noise reduction and edge-preserving, we choose a 3×3 window to calculate the updated value of the objective pixel, and consider a modification of the fuzzy rule base of the above section as following

$$\begin{aligned}
&\text{R1: if } dir(i,j)=0^\circ \text{ then } y_1 = median_{0^\circ}\\
&\text{R2: if } dir(i,j)=45^\circ \text{ then } y_2 = median_{45^\circ}\\
&\text{R3: if } dir(i,j)=90^\circ \text{ then } y_3 = median_{90^\circ}\\
&\text{R4: if } dir(i,j)=135^\circ \text{ then } y_4 = median_{135^\circ}\\
&\text{R0: else } y_0 = median_{+}
\end{aligned} \tag{7}$$

where $median_\theta$ denotes the median value of all the pixels lying on the central line in the direction θ, and $median_+$ denotes the median value of all the pixels in the cross neighborhood. Similarly, the activity degree of every rule is calculated through (5) ~ (6), and the output is obtained by means of (7). As in the previous case, the value y is assigned to the objective pixel $P(i, j)$ and re-used for further processing. Here we call this filter fuzzy edge-sensitive median filter.

4 Experimental Results

The visual comparisons among Adelmann's [6] algorithm, Russo's FIRE filter [7] and the proposed algorithm against impulsive noise are shown in Fig.1. It can be found that our filter can provide better performance.

(a) Noisy image (b) Adelmann's (c) Russo's (d) Proposed

Fig. 1. Comparison results among different scheme on image contaminated by impulsive noise

This numerical comparison is based on two quantitative and objective measures: the mean square error (MSE) and the mean absolute error (MAE).

The capabilities of all algorithms for reducing impulsive noise, Gaussian noise are demonstrated in Table 1, 2, respectively. According to quantitative comparison, we can conclude that our method can provide better performance than Adelmann's algorithm and Russo's FIRE filter in respect of noise reduction and edge preservation.

However, from the computer complexity, our filter needs more time than Adelmann's scheme, but less than Russo's filter.

Table 1. Error obtained from removing impulsive noise by different algorithms

Impulsive noise density	MSE			MAE		
	10%	20%	30%	10%	20%	30%
Adelmann	0.0074	0.0168	0.0279	0.0406	0.0743	0.1079
Russo	0.0031	0.0062	0.0112	0.0235	0.0350	0.0492
Proposed	0.0021	0.0048	0.0092	0.0143	0.0198	0.0304

Table 2. Error obtained from removing Gaussian noise by different algorithms

Gaussian white noise variance	MSE			MAE		
	0.05	0.08	0.1	0.05	0.08	0.1
Adelmann	0.0147	0.0214	0.0254	0.0948	0.1142	0.1250
Russo	0.0229	0.0318	0.0363	0.1176	0.1394	0.1492
Proposed	0.0092	0.0122	0.0149	0.0681	0.0856	0.0948

5 Conclusions

In this paper we present a new edge-sensitive noise reduction method based on the use of fuzzy edge-direction detector. The goal of the filtering process is edge preservation as well as noise reduction. Experiment results show that the proposed filter can provide better performance in smoothing noise than other algorithms.

References

1. Chan, T.F., Osher S., Shen J.H.: The Digital TV Filter and Nonlinear Denoising. IEEE Transactions on Image Processing, vol.10, no.2 (2001) 231-241
2. Zhou, Z.H., Xie, S.L.: Error concealment based on adaptive MRF-MAP framework. Advances in Machine Learning and Cybernetics, Lecture Notes in Artificial Intelligence 3930 (2006) 1025-1032
3. Xu, Z.L, Xie, S.L.: An adaptive deblocking algorithm based on MRF. Journal of South China University of Technology (Natural Science Edition), Vol.33, No.7 (2005) 15-19
4. Xie, S.L., Zhou, Z.H.: New adaptive reduction algorithm of blocking artifacts. Acta Electronica Sinica, Vol.33, No.10 (2005) 1897-1900
5. Xu, Z.L., Xie, S.L.: A deblocking algorithm based on HVS. Journal of Electronics and Information Technology, Vol.27, No.11 (2005) 1717-1721
6. Adelmann, H.G.: An edge-sensitive noise reduction algorithm for image processing. Computers in Biology and Medicine, Vol.29 (1999) 137-145
7. Russo,F. : Edge-Detection in Noisy Image Using Fuzzy Reasoning. IEEE Transaction on Instrumentation and Measurement, Vol.47, No.5 (1998) 1102-1105
8. Michaud,F., Chon, T.L.D., Lachiver,G. : Fuzzy detection of edge-direction for video line doubling. IEEE Transactions on Circuits and Systems for Video Technology, Vol.7, No.3 (1997) 539-542

Fusing Color and Texture Features for Background Model

Hongxun zhang and De xu

School of Computer and Information Technology,Beijing Jiaotong University
No.3 of Shangyuan Residence Haidian District, Beijing, 100044, China
{zhx, xd}@computer.njtu.edu.cn

Abstract. Background subtraction is a method typically used to segment moving regions in image sequences taken from a static camera by comparing each new frame to a model of the scene background. We present a novel approach that uses fuzzy integral to fuse the texture and color features for background subtraction. The method could handle various small motions of background objects such as swaying tree branches and bushes. Our method requires less computational cost. The model adapts quickly to changes in the scene that enables very sensitive detection of moving targets. The results show that the proposed method is effective and efficient in real-time and accurate background maintenance in complex environment.

1 Introduction

Background subtraction is often one of the first tasks in machine vision applications, making it a critical part of the system. The output of background subtraction is an input to a higher-level process that can be, for example, the tracking of an identified object. The performance of background subtraction depends mainly on the background modeling technique used to model the scene background. Especially natural scenes put many challenging demands on background modeling since they are usually dynamic in nature including illumination changes, swaying vegetation, rippling water, flickering monitors etc. A robust background model algorithm should handle situations where new stationary objects are introduced to or old ones removed from the scene. Even in a static scene frame-to-frame changes can occur due to noise and camera jitter. Moreover, the background model algorithm should operate in real-time. In this paper, we propose a novel approach to background subtraction. We integrate the texture and color features for background subtraction. The goal of the new approach was to address all of the above-mentioned difficulties. The remainder of this paper is organized as follows. Section 2 outlines the related work. Section 3 describes color and texture similar measure methods. Section 4 discusses the background to be updated. Section 5 and 6 contain the experimental results and conclusion.

L. Wang et al. (Eds.): FSKD 2006, LNAI 4223, pp. 887–893, 2006.

2 Related Works

Different kinds of methods for detecting moving objects have been proposed in many literatures. The most commonly used cue is pixel intensity. One very popular technique is to model each pixel in a video frame with a Gaussian distribution. This is the underlying model for many background subtraction algorithms. A simple technique is to calculate an average image of the scene, to subtract each new video frame from it and to threshold the result. The adaptive version of this algorithm updates the model parameters (mean and covariance) recursively by using a simple adaptive filter. This approach is used in [1]. The previous model does not work well in the case of dynamic natural environments since they include repetitive motions like swaying vegetation, rippling water, flickering monitors, camera jitter etc. This means that the scene background is not completely static. By using more than one Gaussian distribution per pixel it is possible to handle such backgrounds [2, 3, 4]. Stauffer and Grimson [4] modeled each pixel as a mixture of weighted Gaussian distributions and on-line approximation to update the model. The weights were related to the persistence of the distributions in the model. In [2], the mixture of Gaussians approach was used in a traffic monitoring application. The model for pixel intensity consisted of three Gaussian distributions corresponding to the road, the vehicle and the shadow distributions. Adaptation of the Gaussian mixture models can be achieved using an incremental version of the EM algorithm. The Gaussian assumption for pixel intensity distribution does not always hold. To deal with the limitations of parametric methods, a non-parametric approach to background modeling was proposed in [5]. The proposed method utilizes a general non-parametric kernel density estimation technique for building a statistical representation of the scene background. The probability density function for pixel intensity is estimated directly from the data without any assumptions about the underlying distributions. In [6, 7], each image pixel was modeled with a Kalman filter. The proposed method can adapt to changes in illumination, but has problems with complex dynamic backgrounds. This approach was used in the automatic traffic monitoring application presented in [8]. In [9], the dynamic texture (background) is modeled by an ARMA model. A robust Kalman filter algorithm is used to iteratively estimate the intrinsic appearance of the dynamic texture as well as the regions of the foreground objects. In [10], the background model was constructed from the first video frame of the sequence by dividing it into equally sized blocks and calculating an edge histogram for each block. The histograms were constructed using the pixel-specific edge directions as bin indices and incrementing the bins with the corresponding edge magnitudes. In [11], a fusion of edge and intensity information was used for background subtraction. Motion based approaches have also been proposed for background subtraction. The algorithm presented in [12] detects salient motion by integrating frame-to-frame optical flow over time. Salient motion is assumed to be motion that tends to move in a consistent direction over time. The saliency measure used is directly related to the distance over which a point has traveled with a consistent direction. Region-based algorithms usually divide an image into blocks and calculate block specific features. Change detection is achieved via block matching. In [13], the block correlation is measured using the normalized vector distance measure. In [10], an edge histogram calculated over the block area is used as a feature vector describing the block. The region-based approaches allow only coarse detection of the moving objects unless a multi-scale approach is used.

3 Color and Texture Similar Measure

3.1 Color Feature Similar Measure

We transform the individual RGB frames into the Ohta color space in which the three features are approximately irrelevance [14]. The three orthogonal color features of Ohta color space, $I_1 = (R+G+B)/3$, $I_2 = (R-B)/2$ or $(B-R)/2$, and $I_3 = (2G - R - B)/4$, are important components for representing color information. These three color features are significant in many cases. A good result of color image processing can be achieved by using only the first two, as we use in our background model. To test for similarity between the background and a current frame we apply uncertainty color features $\triangle I_k^t$ which is defined as

$$\triangle I_k^t = \begin{cases} I_k^t(x,y)/I_k^B(x,y) & \text{if } I_k^t(x,y) < I_k^B(x,y) \\ 1 & \text{if } I_k^t(x,y) = I_k^B(x,y) \\ I_k^B(x,y)/I_k^t(x,y) & \text{if } I_k^t(x,y) > I_k^B(x,y) \end{cases} \tag{1}$$

Where $k(=1,2,3)$ is one of the three color features, t represents time t video frame, B represents the background which is used at time t.

3.2 Texture Feature Similar Measure

The degree of computational complexity of most texture models is too high. Therefore, we choose a simple texture model that is the local binary pattern operator (LBP) [15]. The operator labels the pixels of an image block by setting threshold of the neighborhood of each pixel with the center value and considering the result as a binary number (LBP code):

$$LPB(x_c, y_c) = \sum_{p=0}^{P-1} s(g_p - g_c)2^p \tag{2}$$

where g_c corresponds to the value of the center pixel (x_c, y_c) and g_p represents the grey values of the P neighborhood pixels(Fig. 1). The function $s(x)$ is defined as follows:

$$s(x) = \begin{cases} 1 & \text{if } x > 0 \\ 0 & \text{if } x \leq 0 \end{cases} \tag{3}$$

To test for similarity between an image block texture and the background texture we apply uncertainty $h_{texture}$ which is defined as

$$h_{texture}(x_c, y_c) = \begin{cases} G_t(x_c,y_c)/G_B(x_c,y_c) & \text{if } G_t(x_c,y_c) < G_B(x_c,y_c) \\ 1 & \text{if } G_t(x_c,y_c) = G_B(x_c,y_c) \\ G_B(x_c,y_c)/G_t(x_c,y_c) & \text{if } G_t(x_c,y_c) > G_B(x_c,y_c) \end{cases} \tag{4}$$

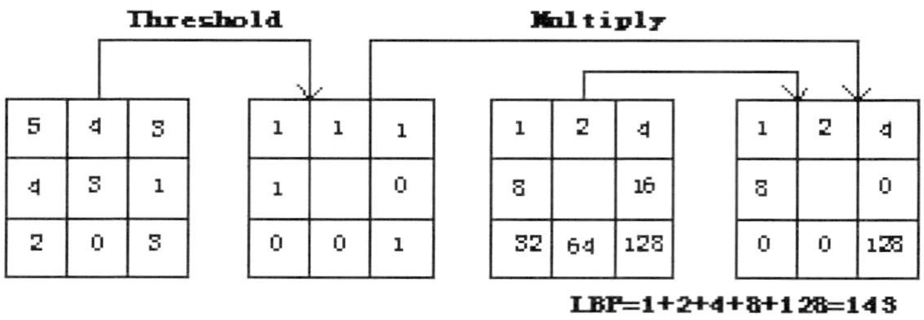

Fig. 1. The values of neighbors that are estimated by bilinear interpolation

where $G_B(x_c, y_c)$ is the texture LBP of pixel (x_c, y_c) in background and $G_t(x_c, y_c)$ is the texture LBP of pixel (x_c, y_c) in time t video frame. $h_{texture}$ is close to one if $G_B(x_c, y_c)$ and $G_t(x_c, y_c)$are very similar.

4 Fusion of Texture and Color Feature Using Fuzzy Integral

Definition 1. *Let X be a finite set and $h : X \mapsto [0, 1]$be a fuzzy subset of X. The fuzzy integral over X of the function h with respect to a fuzzy measure g is defined by*

$$\oint_{\mathcal{X}} h(x) \circ g(x) = \sup_{\alpha \in [0,1]} [min(\alpha, g(X \bigcap H_\alpha))] \tag{5}$$

Where H_α is the α-level set of h,

$$H_\alpha = \{x : h(x) \geq \alpha\}$$

Suppose that an object is evaluated from the point of view of a set of sources X. Let denote the decision for the object when source is considered and let $g(\{x\})$ denote the degree of importance of this source. We can adopt g_λ -fuzzy measure calculate the values of $g(X)$. For background subtraction problem, we use fuzzy integral to integrate color and texture features of two pixels of same position in the background and the current frame respectively and decide whether two pixels are similar or not. Certainly, the feature sets is $X = \{x_1, x_2, x_3\}$. One element is $x_1 = \{texture\}$ and the others are $x_2 = \{I_1\}$ and $x_3 = \{I_2\}$. Now corresponding to each x_i the degree of important, $g_\lambda(x_i)$, of how important x_i is in background subtraction problem. The values $g_\lambda(X)$ of can be determined as:

$$g_\lambda(x_1) = g(\{x_1\}), g_\lambda(x_2) = g(\{x_2\}), g_\lambda(x_3) = g(\{x_3\}) \tag{6}$$

$$g_\lambda(X) = g(\{x_1, x_2, x_3\}) = [\prod_{\lambda \in X} (1 + \lambda g_\lambda(x_i)) - 1]/\lambda \tag{7}$$

Let $h_i : X \mapsto [0, 1]$ be a fuzzy function. Fuzzy function $h_1 = h(x_1) = h_{texture}$ is the evaluation of texture feature. Fuzzy function $h_2 = h(x_2) = h_{\triangle I_1}$ is the evaluation of color feature I_1. Fuzzy function $h_3 = h(x_3) = h_{\triangle I_2}$ is the evaluation of color feature I_2. Then a fuzzy integral, S, with respect to a fuzzy measure g over X can be computed by

$$\oint_{\mathcal{X}} h(x) \circ g(.) = \sup_{\alpha \in [0,1]} [min(\alpha, g(H_\alpha))] \tag{8}$$

The calculation of the fuzzy integral is as follows: suppose $h(x_1) \geq h(x_2) \geq h(x_3)$, if not, X is rearranged so that this relation holds. Then a fuzzy integral, S, with respect to a fuzzy measure g over X can be computed by

$$S = \max_{i=1}^{n}[min(h(x_i), g(X_i))] \tag{9}$$

where $X = \{x_1, x_2, x_3\}$.

We set a threshold value to decide whether processing pixels are similar or not. If the fuzzy integral value is larger than the threshold value, we consider two processing pixels are similar.

5 Updating the Background

Background subtraction, the process of subtracting the current image from a reference one, is a simple and fast way to obtain the moving object in the foreground region and has been employed by many surveillance systems. In our background model, the pixel motion character is used to decide which pixel should be updated, and the subtraction results between the input video and background are used to make a decision. We present a dynamic matrix $D(X)$ to analyzing the changes detection result, where the motion state of each pixel is stored in the matrix. Only those pixels whose values do not change much can be updated into the background. At first, we choose the first video frame to be our original background. According to the matrix $D(X)$, we update the background dynamically. Let denotes the input frame at time t. Equation (10) and (11) show the expression of the dynamic matrix $D(X)$ and the pixel x_t is a moving pixel or not at time t

$$D(x_t) = \begin{cases} D(x_{t-1}) & \text{if } m(x_t) = 0 \quad and \quad D(x_{t-1}) \neq 0 \\ \omega & \text{if } m(x_t) \neq 0 \end{cases} \tag{10}$$

Where

$$m(x_t) = \begin{cases} 0 & \text{if } S \geq \delta \\ 1 & \text{if } otherwise \end{cases} \tag{11}$$

where S represent the fuzzy integral value with respect to a fuzzy measure g over X, δ is the threshold to make a decision whether the pixel is changing at time t or not, ω is the time length to record the pixel's moving state, once the $D(x_t)$ equates to zero, the pixel update method will make a decision that this pixel should be updated into the background.

6 Experiments

The performance of the proposed method was evaluated using several video sequences. We take $\omega = 5$ and $\delta = 0.85$. Table 1. shows the fuzzy density values that we chose. For computing conveniently and efficiently, we use the fuzzy density function $g_\lambda(x_1) + g_\lambda(x_2) + g_\lambda(x_3) = 1$. When we choose {0.7,0.2,0.1} or {0.6,0.3,0.1} to use as fuzzy density values, results of our experiments have good effect. Figure 2. shows the results of our algorithm for the outdoor test sequence, which contains relatively small moving objects. The original sequence has been taken from the PETS database (ftp://pets.rdg.ac.uk/). The proposed algorithm successfully handles this situation. For the parameter values used in the tests, a frame rate of 25 fps was achieved. We used a standard PC with a 1.8 GHz Pentium CPU processor and 256 MB of memory in our experiments. The image resolution was 352*288 pixels.

Table 1. The fuzzy density values

No.	$\{x_1\}$	$\{x_2\}$	$\{x_3\}$	$\{x_1, x_2\}$	$\{x_1, x_3\}$	$\{x_2, x_3\}$	$\{x_1, x_2, x_3\}$
1	0.7	0.2	0.1	0.9	0.8	0.3	1
2	0.6	0.3	0.1	0.9	0.7	0.4	1
3	0.5	0.4	0.1	0.9	0.6	0.5	1
4	0.5	0.3	0.2	0.8	0.7	0.5	1

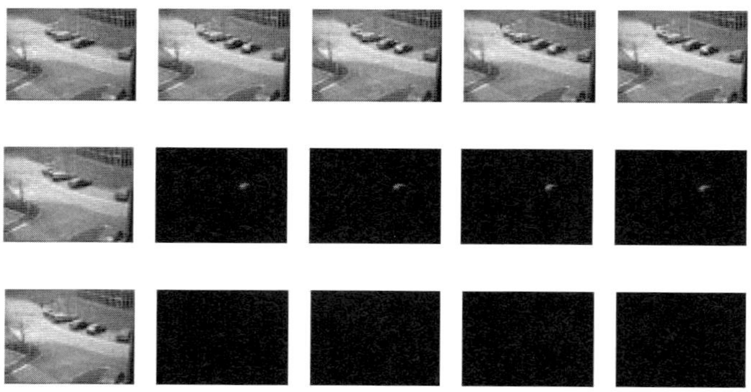

Fig. 2. Results of our algorithm for the outdoor test sequence

7 Conclusion

In this paper, we presented a new tool for constructing and maintaining the scene background in video surveillance. The model is based on fusing color and texture features. The model achieves sensitive detection of moving targets against cluttered backgrounds. It can handle situations where the scene background is not completely static but contains small motions such as moving tree branches and bushes. The model is also adaptive to changes in the scene illumination. It is able to suppress false detections that arise due to small camera displacements. Experiments on complex environments show that the system can deal with difficult situations. While the system requires measure densities, these densities must be subjectively assigned by expert. Future work includes using uncertainty theory or multiple neural networks to determinate these densities.

References

1. C. R. Wren, A. Azarbayejani, T. Darrell, and A. P. Pentland: Pfinder: Real-time tracking of the human body. Vol.19. IEEE Transactions on Pattern Analysis and Machine Intelligence(1997)780–785.
2. N. Friedman and S. Russell: Image segmentation in video sequences: A probabilistic approach. In 13th Conference on Uncertainty in Artificial Intelligence(1997)175–181.]
3. W. E. L. Grimson, C. Stauffer, R. Romano, and L. Lee: Using adaptive tracking to classify and monitor activities in a site. In IEEE Computer Society Conference on Computer Vision and Pattern Recognition(1998)22–29.
4. C. Stauffer and W. E. L. Grimson: Adaptive background mixture models for real-time tracking. Vol.2. In IEEE Computer Society Conference on Computer Vision and Pattern Recognition(1999)246–252.
5. A. Elgammal, R. Duraiswami, D. Harwood, and L. S. Davis: Background and foreground modeling using nonparametric kernel density estimation for visual surveillance. Vol.90. Proceedings of the IEEE(2000)1151–1163.
6. K. P. Karmann and A. Brandt: Moving object recognition using an adaptive background memory. In Cappellini V. (ed.) Vol.2. Time-Varying Image Processing and Moving Object Recognition(1990).
7. C. Ridder, O. Munkelt, and H. Kirchner: Adaptive background estimation and foreground detection using Kalman-filtering. In International Conference on Recent Advances in Mechatronics(1995) 193–199.
8. D. Koller, J. Weber, T. Huang, J. Malik, G. Ogasawara, B. Rao, and S. Russell: Towards robust automatic traffic scene analysis in real-time. Vol.1. In 12th IAPR International Conference on Pattern Recognition(1994)126–131.
9. J. Zhong and S. Sclaroff: Segmenting foreground objects from a dynamic textured background via a robust Kalman filter. Vol.1. In 9th IEEE International Conference on Computer Vision(2003)44–50.
10. M. Mason and Z. Duric: Using histograms to detect and track objects in color video. In 30th Applied Imagery Pattern Recognition Workshop(2001)154–159.
11. S. Jabri, Z. Duric, H. Wechsler, and A. Rosenfeld: Detection and location of people in video images using adaptive fusion of color and edge information. Vol.4. In 15th International Conference on Pattern Recognition(2000)627–630.
12. L. Wixon: Detecting salient motion by accumulating directionally-consistent flow. Vol.22. IEEE Transactions on Pattern Analysis and Machine Intelligence(2000)774–780.
13. T. Matsuyama, T. Ohya, and H. Habe: Background subtraction for non-stationary scenes. In 4th Asian Conference on Computer Vision(2000)622–667.
14. Y-I Ohta, T. Kanade, and T. Sakai: Color Information for Region Segmentation. Vol.13. Comp. Graphics & Image Processing(1980)222–241.
15. PIETIKINEN M., OJALA T., AND Z. XU: Rotation-invariant texture classification using feature distributions. Vol.33. Pattern Recognition(2000)43–52.

Texture Segmentation by Fuzzy Clustering of Spatial Patterns

Yong Xia[1,3], Rongchun Zhao[1], Yanning Zhang[1], Jian Sun[2], and Dagan Feng[3,4]

[1] School of Computer, Northwestern Polytechnical University, Xi'an 710072, China
{YongXia, RCZhao, YNZhang}@nwpu.edu.cn
[2] School of Automatic Control, Northwestern Polytechnical University, Xi'an, China
[3] School of Information Technologies, F09, University of Sydney, NSW 2006, Australia
[4] Dept. of Electronic and Information Engineering, Hong Kong Polytechnic University

Abstract. An approach to perceptual segmentation of textured images by fuzzy clustering of spatial patterns is proposed in this paper. The dissimilarity between a texture feature, which is modeled as a spatial pattern, and each cluster is calculated as a combination of the Euclidean distance in the feature space and the spatial dissimilarity, which reflects how much of the pattern's neighborhood is occupied by other clusters. The proposed algorithm has been applied to the segmentation of texture mosaics. The results of comparative ex-periments demonstrate that the proposed approach can segment textured im-ages more effectively and provide more robust segmentations.*

1 Introduction

Texture segmentation, which has long been an important and challenging topic in image processing, can be achieved by adopting two independent sub-processes: texture feature extraction and feature clustering [1]. However, texture segmentation is different from traditional clustering problem in that each texture feature implies the spatial information of the texture patch it represented. Features of the same texture region are not only numerically similar, but spatially compact. Therefore, some spatial constraints must be incorporated into the clustering algorithm.

There are many methods to utilize the spatial information. A straightforward one is to include the coordinates as features [2]. Many other approaches adopt the Markov random field (MRF) model and interpret the spatial constraint in terms of the potential of each pixel clique [3]. A recent approach uses a linear filter to explore the spatial constraint [4]. If all pixels in a patch belong to the same class, the center pixel will be smoothed by its neighbors so that eventually all pixels in the window have high and similar membership values in one of the clusters. Although outperforming conventional algorithms, those methods often suffer from various inaccuracies.

In this paper, we solve the texture segmentation problem from the point of view of fuzzy clustering of spatial patterns. To incorporate the spatial information into the

* This research is partially supported by the HK-RGC grant, the ARC grant, the NSFC under Grant No. 60141002, and the ASFC under Grant No. 02I53073.

L. Wang et al. (Eds.): FSKD 2006, LNAI 4223, pp. 894 – 897, 2006.

object function of fuzzy clustering, we define a novel metric of dissimilarity between a feature and a cluster to reflect not only the distance in feature space, but the location of the feature. We present the results of our approach when used to segment the mosaics of Brodatz textures [5]. We also compare them with the results obtained by using an MRF-based algorithm [3] and the spatial fuzzy clustering algorithm [4].

2 Segmentation Algorithm

An image is a 2D array of pixels defined on a $W \times H$ rectangular lattice. Each pixel can be represented by a feature vector, which is named as a pattern in the terminology of clustering. The pattern corresponding to the pixel lying on a site $s \in S$ is denoted as x_s. The value of pattern x_s indicates its position in feature space, and the subscript s specifies its position on the lattice. Pattern x_s implies both the feature information and the spatial information, and hence is called a spatial pattern in this paper. Accordingly, a textured image can be modeled as a spatial pattern set $X = \{x_s : s \in S\}$. Texture segmentation is equivalent to clustering of the spatial pattern set X, which can be achieved by minimizing the following sum of dissimilarity

$$J_m(U,V) = \sum_{s \in S} \sum_{r=1}^{C} u_{rs}^m d_{rs}^2 \tag{1}$$

where C is the desired number of texture patterns, m is a fuzzy factor $(m > 1)$, u_{rs} is the membership of the pattern x_s to the r-th cluster, and d_{rs} is the dissimilarity between x_s and the prototype v_r. Similar to traditional clustering algorithms, a local minimum can be reached by performing the Picard iteration [6].

Generally, the dissimilarity d_{rs} is computed by using a distance measure defined in the feature space, which, however, is not fully competent for clustering of spatial patterns because of the lack of the spatial constraints. Here, we define the dissimilarity d_{rs} as a combination of the feature distance and the spatial dissimilarity with respect to the position of pattern x_s on the lattice

$$d_{rs} = d_{rs}^F + \alpha \cdot d_{rs}^S , \tag{2}$$

Where d_{rs}^F is the Euclidean distance, d_{rs}^S is the normalized spatial dissimilarity, and the coefficient α presents a trade-off between them.

Our philosophy of defining the spatial dissimilarity is that if a pixel s lies in the r-th textured region, the pattern x_s and the prototype v_r must have a small dissimilarity. The normalized spatial dissimilarity d_{rs}^S is defined as follows, reflecting how much of the pattern x_s' neighbourhood is occupied by the r-th cluster

$$d_{rs}^S = 1 - \sum_{t \in \eta_s} u_{rt} \beta_t \Big/ \sum_{c=1}^{C} \sum_{t \in \eta_s} u_{ct} \beta_t , \tag{3}$$

where η_s is the set of sites that are contained in site s ' neighbourhood and the factor

$$\beta_t = h\left(\sqrt{(t_i - s_i)^2 + (t_j - s_j)^2}\right) \tag{4}$$

characterizes the contribution of each neighbor to the overall dissimilarity. In our experiments, we choose $h(.)$ as a sigmoid function.

Apparently, the importance of the feature distance d_{rs}^F and the spatial dissimilarity d_{rs}^S is not invariant during the clustering process. In the early stage, the spatial information implied in the fuzzy partition is not reliable, and the clustering should be dominated by d_{rs}^F. As the partition gradually approaching a convergence, the spatial dissimilarity should play an increasingly important role so that the misclassification can be corrected. Therefore, a variable weight factor $\alpha(n)$, which satisfies a sigmoid function, is used in our experiments to substitute for the constant α.

3 Experimental Results

To assess its ability to segment textured images, the proposed algorithm has been compared with two commonly used segmentation approach, one is based on the MRF model [3] and the other is based on spatial fuzzy clustering (SFC) [4]. The comparative experiments have been carried out on a set of four-class texture mosaics, which are generated by using twelve natural textures chosen from the Brodatz album. With the purpose of comprehensive investigation, the test image set MⅣ is made of ${}_{12}C_4 = 495$ samples. The 6-dimensional feature [7] derived from Conditional Markov (CM) model is uniformly used by all three approaches to make a fair comparison.

Two test cases, together with their corresponding segmentations, are presented in Fig. 1. Both the percentage of incorrectly classified pixels and the time cost

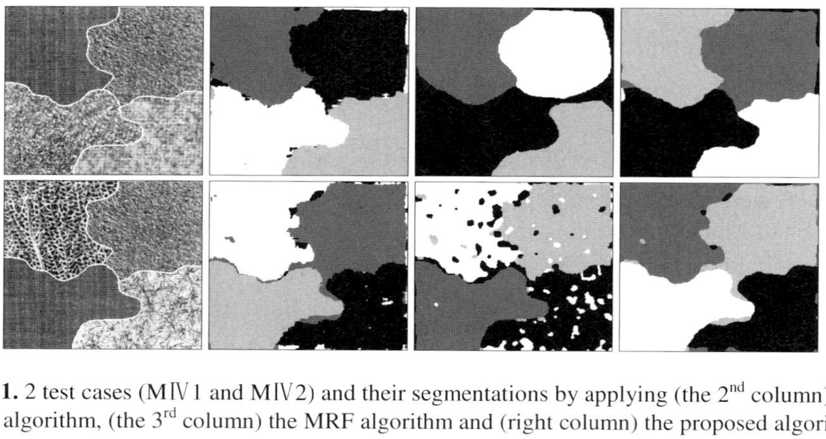

Fig. 1. 2 test cases (MⅣ1 and MⅣ2) and their segmentations by applying (the 2nd column) the SFC algorithm, (the 3rd column) the MRF algorithm and (right column) the proposed algorithm

Table 1. Performance of three segmentation algorithms

Image Index	Error Percentage			Time Cost		
	SFC	MRF	Proposed	SFC	MRF	Proposed
MIV1	4.95 %	9.94 %	4.23 %	3.50 s	5.58 s	4.92 s
MIV2	6.98 %	12.18 %	5.30 %	4.09 s	5.70 s	5.24 s
Average	12.10 %	17.38 %	10.26 %	5.11 s	5.78 s	6.44 s

(Intel Pentium IV 4.0 GHz Processor, 2G Memory) of those two cases are given in Table 1. The average performance of those three approaches over the entire image set is also listed in Table 1. It is obvious that the proposed approach can achieve more accurate segmentation, especially in suppressing small mis-segmented regions, but at a cost of the slightly increased computational complexity.

4 Conclusions

In this paper, a textured image is modeled by a set of spatial patterns and texture segmentation can be achieved by from the point of view of fuzzy clustering of spatial patterns. The dissimilarity between a spatial pattern and each cluster is defined by using both the feature value and the spatial information. Comparative experiments on texture mosaics have demonstrated that the novel algorithm is more effective.

References

1. Reed, T. R., Hans du Buf, J. M.: A Review of Recent Texture Segmentation and Feature Extraction Techniques, CVGIP: Image Understanding, Vol. 57, No. 3 (1993) 359-372
2. Krishnapuram, R., Freg, C.P.: Fitting an Unknown Number of Lines and Planes to Image Data through Compatible Cluster Merging, Pattern Recogn., Vol. 25 (1992) 385–400
3. Deng, H., Clausi, D.A.: Unsupervised Segmentation of Synthetic Aperture Radar Sea Ice Imagery Using a Novel Markov Random Field Model, IEEE Trans. Geosci. Remote Sensing, Vol. 43 (2005) 528-538
4. Liew, A.W.C., Leung, S.H., Lau, W.H.: Segmentation of Color Lip Images by Spatial Fuzzy Clustering, IEEE Trans. Fuzzy Systems, Vol. 11 (2003) 542-549
5. Brodatz, P.: Texture: A Photographic Album for Artists and Designers, Dover, New York, 1966
6. Bezdek, J.C.: Pattern Recognition with Fuzzy Objective Function Algorithms, Plenum Press, New York, 1981
7. Manjunath, B.S., Chellappa, R.: Unsupervised Texture Segmentation Using Markov Random Fields, IEEE Trans. Pattern Anal. Machine Intell., Vol. 13 (1991) 478–482

Uncertainty Analysis Using Geometrical Property Between 2D-to-3D Under Affine Projection

Sungshik Koh[1] and Phil Jung Kim[2]

[1] Insan Innovation Telecom Co., Ltd.,
524-19, Uunnam-dong, Gwangsan-gu, Gwnagju, 506-812, Korea
phdkss@chosun.ac.kr
[2] Dept. of IT, Sunghwa College,
224, Wolpyeong-ri, Seongjeon-myeon, Gangjin-gun, Jeollanam-do, 527-812, Korea
philjung@hanmail.net

Abstract. In this paper, we propose uncertainty analysis using geometrical property between 2D-to-3D under affine reconstruction. In situations when are no missing data in an observation matrix, the accurate solution is known to be provided by Singular Value Decomposition (SVD). However, when converting image sequences to 3D, several entries of the matrix have not been observed and other entries have been perturbed by the influence of noise. In this case, there is no simple solution. In this paper, a new approach is applied for recovering missing data using geometrical properties between a 2D image plane and 3D shape and for estimating noise level in an observation matrix using ranks of SVD. This paper consists of four main phases: geometrical properties between 2D image plane and 3D error space, geometrical recovering of missing data, and noise level estimation in the observation matrix.

1 Introduction

In order to reconstruct a 3D shape and camera motion from a 2D observation matrix, the matrix factorization methods have been widely used in computer vision. To be factored out, it is essential that all of feature points (FPs) are visible in the observed matrix. In real video clips, these projections however are not visible along the entire image sequence due to the occlusion and the limited fields of view.

Many researchers have been developed 3D reconstruction algorithms with recovering the missing data. Filling of missing data was first realized by Tomasi and Kanade [1] for affine camera. However in their method there was the NP-hard problem of finding the largest full sub-matrix of a matrix with missing elements. Jacobs [2] improved their method by fitting an unknown matrix of a certain rank to an incomplete noisy matrix resulting from measurements in images, which we call Linear Fitting method (LF). Guerreiro and Aguiar [3] proposed Expectation-Maximization (EM) and Two-Step (TS) iterative algorithm. The algorithms converged to the global optimum in a very small number of iterations. However, the performance of them is influenced by noisy observation. Moreover, although there have been quite a few studies that addressed error analysis for Shape from Motion (SfM) algorithms [4]-[8], the

L. Wang et al. (Eds.): FSKD 2006, LNAI 4223, pp. 898 – 907, 2006.

perturbation of the 2D observation matrix is usually estimated by correspondence covariance which is obtained from image gradients. Due to the complexity and nonlinearity of the factorization method, the above methods may face with a difficulty to derive the exact analysis.

In order to solve the problem of missing data, this paper offers uncertainty analysis using geometry properties between 2D image plane and 3D error space and applies it to recovery of missing data. Moreover this paper introduces a new noise level estimation algorithm without helping specific techniques such as covariance, information matrix, and so on.

2 Geometrical Properties and Its Application

In this section, a new algorithm for solving the problem of missing data from noisy observations of a subset of entries is introduced. The primary objective of this method is to obtain more accurate 3D reconstruction by recovering the missing data. In this section, some specific geometry between the noise of the 2D observation matrix and the error of 3D shape is described.

2.1 2D Image Plane and 3D Error Space

To accurately evaluate the precision of the 3D shape reconstructed from a noisy observation matrix is extremely challenging. In order to recover and evaluate missing data, at first, 3D error space is introduced as the following:

(a) For recovering a missing FP (unobserved), the position of the missing FP ($\boldsymbol{p}_m$) is first fit roughly, and three FPs are randomly selected ($\boldsymbol{p}_a$, $\boldsymbol{p}_b$, and $\boldsymbol{p}_c$), which are called bias FPs, on neighbors of the missing FP ($\boldsymbol{p}_m$). Next, new FPs ($\boldsymbol{q}_i$) are added, which are called Reference Points (RPs), on a circular pattern centering on the missing FP ($\boldsymbol{p}_m$) (see Fig. 1(a)).

Π_2: reference plane composed of $\boldsymbol{q}_i$ on 2D image plane,

$\overrightarrow{\boldsymbol{p}_m\boldsymbol{q}_i}$: reference vector (RV) composed of $\boldsymbol{p}_m$ and $\boldsymbol{q}_i$ on 2D image plane.

(b) Using affine SVD factorization, the roughly fitted FP ($\boldsymbol{p}_m$), three bias FPs ($\boldsymbol{p}_a$, $\boldsymbol{p}_b$, and $\boldsymbol{p}_c$), and the RPs ($\boldsymbol{q}_i$) are reconstructed to $\boldsymbol{P}_m^*$, ($\boldsymbol{P}_a^*$, $\boldsymbol{P}_b^*$, and $\boldsymbol{P}_c^*$), and $\boldsymbol{Q}_i^*$ on 3D, respectively (see Fig. 1(b)).

(c) A new 3D error space is defined as follows (see Fig. 1(c)). A point vector $\mathbf{L}$ on 3D error space, which corresponds to any point ($\boldsymbol{P}^*$) on 3D, is obtained by the distances from bias FPs ($\boldsymbol{P}_a^*$, $\boldsymbol{P}_b^*$, and $\boldsymbol{P}_c^*$).

$$\mathbf{L} = (L_a, L_b, L_c),$$

$$L_a = \left\|\overrightarrow{\boldsymbol{P}_m^*\boldsymbol{P}_a^*}\right\|,\ L_b = \left\|\overrightarrow{\boldsymbol{P}_m^*\boldsymbol{P}_b^*}\right\|,\ \text{and}\ L_c = \left\|\overrightarrow{\boldsymbol{P}_m^*\boldsymbol{P}_c^*}\right\|.$$

Hence, $\boldsymbol{P}_m$ and $\boldsymbol{Q}_i$ can be expressed as (L_a, L_b, L_c) and (L_{ai}, L_{bi}, L_{ci}) on the 3D error space, respectively.

$$L_j = \left\| \overrightarrow{\boldsymbol{P}_m^*, \boldsymbol{P}_j^*} \right\|,$$

$$L_{ji} = \left\| \overrightarrow{\boldsymbol{Q}_i^*, \boldsymbol{P}_j^*} \right\|, \; j = a,b,c \in \{1,2,\cdots,N\}, \; i = 1,2,...,Z .$$

Here,

Π_3: reference plane composed of $\boldsymbol{Q}_i$ on the 3D error space,

$\overrightarrow{\boldsymbol{P}_m\boldsymbol{Q}_i}$: Reference Vector (RV) composed of $\boldsymbol{P}_m$ and $\boldsymbol{Q}_i$ on the 3D error space.

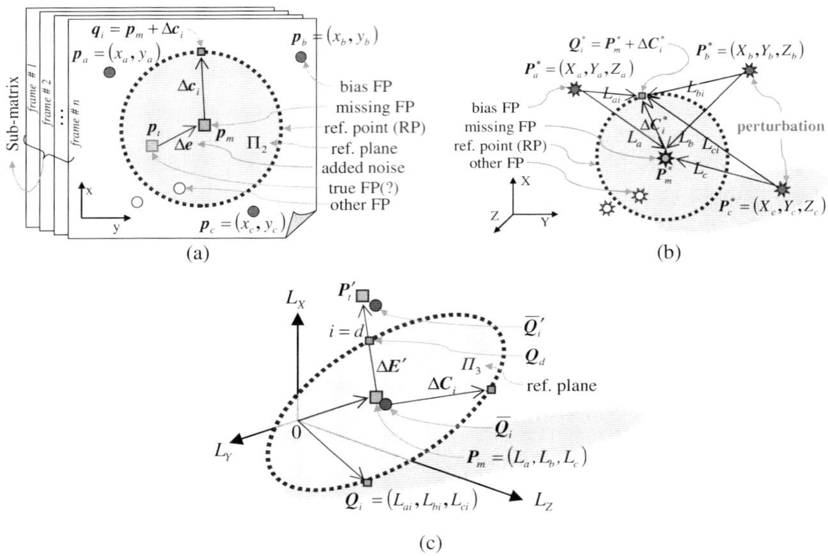

Fig. 1. Comparison of some parameters on 2D image plane, 3D reconstruction space, and 3D error space. (a) noisy FP, three bias FPs and circular RPs on 2D image plane. (b) parameters reconstructed from (a) on 3D reconstruction space. (c) parameters transformed from (b) on 3D error space.

2.2 Recovery of Missing Data

When a FP deviates from its observation matrix position, its reconstructed 3D point, by affine SVD factorization is also misaligned. In this section, the error vector is recovered using the geometrical property facts described previously. The procedure of the proposed algorithm is as follows. In general, a noisy FP on a sub-matrix is expressed by

$$\boldsymbol{p}_m = \boldsymbol{p}_t + \Delta \boldsymbol{e} , \tag{3}$$

where $\boldsymbol{p}_t$ is a true FP and $\Delta \boldsymbol{e}$ is white Gaussian noise. In order to fit the unknown $\Delta \boldsymbol{e}$, First, three bias FPs ($\boldsymbol{p}_j$) are selected randomly from the observed entries except for $\boldsymbol{p}_t$ on the same frame. Next, circular RPs around a central $\boldsymbol{p}_m$ are imputed.

$$\boldsymbol{q}_i = \boldsymbol{p}_m + \Delta \boldsymbol{c}_i,\ i = 1,2,\cdots,Z, \tag{4}$$

where, $\boldsymbol{q}_i$ and $\Delta \boldsymbol{c}_i$ are the RPs and circle radius on the 2D image plane. The reconstructed 3D points of (4) are

$$\boldsymbol{Q}_i^* = \boldsymbol{P}_m^* + \Delta \boldsymbol{C}_i^*, \tag{5}$$

where, $\boldsymbol{P}_m^*$ and $\Delta \boldsymbol{C}_i^*$ are the reconstructed RPs and its circular parameter on 3D. In this case, $\boldsymbol{P}_m(L_a, L_b, L_c)$ and $\boldsymbol{Q}_i(L_{ai}, L_{bi}, L_{ci})$ on the 3D error space are transformed to (5) (see Section 3.1). The error can be expressed as

$$\Delta \boldsymbol{E} = \overrightarrow{\boldsymbol{P}_m \boldsymbol{P}_t}. \tag{6}$$

An approximate $\boldsymbol{P}_t'$ is substituted, obtained from a sub-matrix without missing FP, since $\boldsymbol{P}_t$ is also unknown. The relationship is satisfied with $\boldsymbol{P}_t' \cdot \overline{\boldsymbol{Q}}_i \cong \boldsymbol{P}_t \cdot \overline{\boldsymbol{Q}}_i'$. Where $\overline{\boldsymbol{Q}}_i$ is the mean of $\boldsymbol{Q}_i$s. Hence, the approximate error can be represented as

$$\Delta \boldsymbol{E}' = \overrightarrow{\boldsymbol{P}_m \boldsymbol{P}_t'} = \boldsymbol{P}_t' \langle \times \rangle \left(\overline{\boldsymbol{Q}}_i \langle / \rangle \overline{\boldsymbol{Q}}_i' - (1,1,1) \right), \tag{7}$$

where, $\begin{cases} \boldsymbol{P}\langle\times\rangle\boldsymbol{Q} \equiv (x_1x_2, y_1y_2, z_1z_2) \\ \boldsymbol{P}\langle/\rangle\boldsymbol{Q} \equiv \left(\dfrac{x_1}{x_2}, \dfrac{y_1}{y_2}, \dfrac{z_1}{z_2}\right) \end{cases}$, for $\begin{cases} \boldsymbol{P} = (x_1, y_1, z_1) \\ \boldsymbol{Q} = (x_2, y_2, z_2) \end{cases}$.

If $\|\Delta \boldsymbol{E}'\| \neq 0$, this means some noise exists in $\boldsymbol{P}_m$ on 3D error space and also in $\boldsymbol{p}_m$ of (3). In order to obtain the error vector ($\Delta \boldsymbol{e}'$) on Π_2, it is derived from the parameters represented on Π_3 of the 3D error space.

$$\Theta_d = \underset{d \in i}{\mathbf{argmin}} \|f_i\|^2, \quad \text{for } f_i = \mathbf{cos}^{-1}\left(\frac{\left\|\overrightarrow{\boldsymbol{P}_m \boldsymbol{P}_t'}\right\| \cdot \left\|\overrightarrow{\boldsymbol{P}_m \boldsymbol{Q}_i}\right\|}{\overrightarrow{\boldsymbol{P}_m \boldsymbol{P}_t'} \cdot \overrightarrow{\boldsymbol{P}_m \boldsymbol{Q}_i}} \right), \tag{8}$$

$$A_d = \frac{\left\|\overrightarrow{\boldsymbol{P}_m \boldsymbol{P}_t'}\right\|}{\left\|\overrightarrow{\boldsymbol{P}_m \boldsymbol{Q}_d}\right\|},\ d \in \{1,2,\cdots,Z\}, \tag{9}$$

where Θ_d is the minimum angle between $\overrightarrow{\boldsymbol{P}_m \boldsymbol{P}_t'}$ and $\overrightarrow{\boldsymbol{P}_m \boldsymbol{Q}_d}$, and A_d is the ratio of the size of $\overrightarrow{\boldsymbol{P}_m \boldsymbol{P}_t'}$ based on $\overrightarrow{\boldsymbol{P}_m \boldsymbol{Q}_d}$. In using geometrical properties, the error vector on Π_2 from (8) and (9) is derived by

$$\theta_d = \mathbf{cos}^{-1}\left(\frac{\left\|\overrightarrow{\boldsymbol{p}_m \boldsymbol{q}_d}\right\| \cdot \left\|\overrightarrow{\boldsymbol{p}_m \boldsymbol{q}_1}\right\|}{\overrightarrow{\boldsymbol{p}_m \boldsymbol{q}_d} \cdot \overrightarrow{\boldsymbol{p}_m \boldsymbol{q}_1}} \right), \tag{10}$$

$$\alpha_d = A_d \left\|\overrightarrow{\boldsymbol{p}_m \boldsymbol{q}_d}\right\|. \tag{11}$$

Therefore, the missing FP (p_m) can be updated as

$$\tilde{p}_t = p_m - \Delta e', \tag{12}$$

where $\Delta e' \equiv f(\alpha_d, \theta_d)$, which is a vector with magnitude α_d and angle θ_d. If $\|\Delta E'\|$ is larger than the predefined threshold, then $\tilde{p}_t$ is set up to p_m and the above procedure is repeated until the position converses sufficiently close to the true position.

3 Estimation of Noise Level

Given an image sequence, suppose there are a set of P tracked feature points over F frames. We then obtain trajectories of image coordinate. The affine projection can be written with the all views and object points as follows:

$$\underbrace{\begin{bmatrix} m_{11} \cdots m_{1P} \\ \vdots \quad \ddots \quad \vdots \\ m_{F1} \cdots m_{FP} \end{bmatrix}}_{\mathbf{W}_{2FxP}} = \underbrace{\begin{bmatrix} \mathbf{R}_1^T \\ \vdots \\ \mathbf{R}_F^T \end{bmatrix}}_{\mathbf{R}_{2Fx3}^T} \underbrace{[\mathbf{S}_1 \cdots \mathbf{S}_P]}_{S_{3xP}}. \tag{13}$$

The $2F$xP matrix $\mathbf{W}$ is the measurement matrix and (13) means that this matrix can be decomposed into a $2F$x3 matrix $\mathbf{R}$ representing camera motion and a 3xP matrix $\mathbf{S}$ representing the object shape. This factorization can be performed through SVD as follows: Let

$$\begin{aligned} W_{2F\times P} &= V\, diag(\sigma_1, \cdots, \sigma_N) U^T \\ &= \underbrace{V_1 diag(\sigma_1, \sigma_2, \sigma_3) U_1^T}_{\mathbf{W}^*_{2F\times P}} + \underbrace{V_2 diag(\sigma_4, \cdots, \sigma_N) U_2^T}_{\Delta\mathbf{W}_{2F\times P}} \end{aligned} \tag{14}$$

be SVD of the measurement matrix $\mathbf{W}$ with $N \equiv \mathbf{min}(2F, P)$ and singular values $\sigma_1 \geq \sigma_2 \geq \cdots \geq \sigma_N \geq 0$. The column vectors of the PxN ($2F$x3) matrix $\mathbf{U} \equiv [\mathbf{U}_1, \mathbf{U}_2] (\mathbf{V} \equiv [\mathbf{V}_1, \mathbf{V}_2])$ are mutually orthogonal, which means $\mathbf{U}^T\mathbf{U} = \mathbf{I}_N (\mathbf{V}^T\mathbf{V} = \mathbf{I}_N)$. Clearly, if there is no noise in the 2D image coordinates, $\mathbf{W}^*$ is rank 3 and $\Delta\mathbf{W}$ in (14) is zero. Tomasi and Kanade [1] perform a SVD of $\mathbf{W}$ and use the 3 largest eigenvalues to construct to construct camera motion and object shape. If the SVD returns a rank greater than 3, then the affine projection model is invalid. It is well known that the 3D affine structure cannot be reconstructed completely and the degenerated 3D shape is relative to the number of image sequences [9]-[13]. We use this as a check point of noise estimation.

Suppose that unknown noise is distributed in observation matrix $\mathbf{W}$. We can describe the singular values of $\Delta\mathbf{W}$ presented by (14) as follows:

$$\Delta\sigma_i^a = diag(\sigma_4, \cdots, \sigma_n), \ \textit{for only 3 frames}, \ n = \textbf{\textit{min}}\{6, P\}, \tag{15}$$

$$\Delta\sigma_i^b = diag(\sigma_4, \cdots, \sigma_m), \ \textit{for all frames}, \ m = \mathbf{min}\{2\times F, P\}. \tag{16}$$

A error of 3D shape can be then evaluated by using the singular values of $\Delta\mathbf{W}$.

$$E_W = \sum_{i=3}^{n} \Delta\sigma_i^a + \sum_{i=3}^{m} \Delta\sigma_i^b \cdot \tag{17}$$

Finally, to estimate unknown noise level in **W**, we calculate again E_{WN_k} using (17) after embedding known noise N_k into **W**. Hence, the estimation equation of the noise level can be derived.

$$N_u = \frac{N_k \cdot E_W}{E_W - E_{WN_k}}, \tag{18}$$

where N_u is a unknown noise level included within 2D observation matrix.

4 Experimental Results

In this section, experiments are described that illustrate the behavior of the proposed algorithm using synthetic and real video sequences. The proposed system confirms the wide application of the proposed method and also certifies how much accurately recover the missing data. Simulation was undertaken using Matlab7.04.

4.1 Synthetic Video Sequence

In Fig. 2(a), a cube is set in a 3D world with a set of cameras. The size of the box is 1 [unit]^3, and the box contains eight 3D corner points. All points are tracked from 20 image frames taken to cover 180 degrees with the same intrinsic parameters. The cube is placed with its centroid at 2.5 [unit] from the first camera. The cameras are pointed towards the centroid of the cube. Fig. 2(b) presents a pattern of the missing data. The positions of its entries in the observation matrix are corrupted with holes. Yellow squares represent successfully tracked entries. Red circles denote missing entries that are not received initially, and these results are some holes in the observation matrix.

For confirming the convergence of the proposed geometrical approach, the following reference points are set. The RPs ($\boldsymbol{q}_i$, $i = 1,2,\cdots,10$) around a missing FP ($\boldsymbol{p}_m$) are located at the constant interval angles on a circle and its radius is set to $\Delta \boldsymbol{c}_i$ =0.3

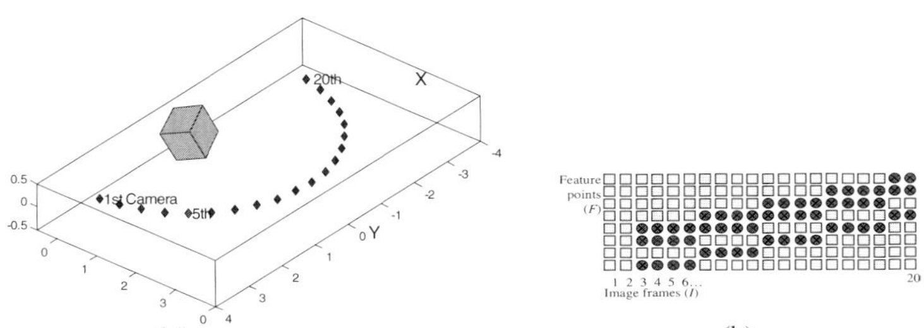

(a) (b)

Fig. 2. Synthetic sequence. (a) the view of a 3D cube and its camera configuration. (b) the pattern of missing data (Red) in a cube sequence.

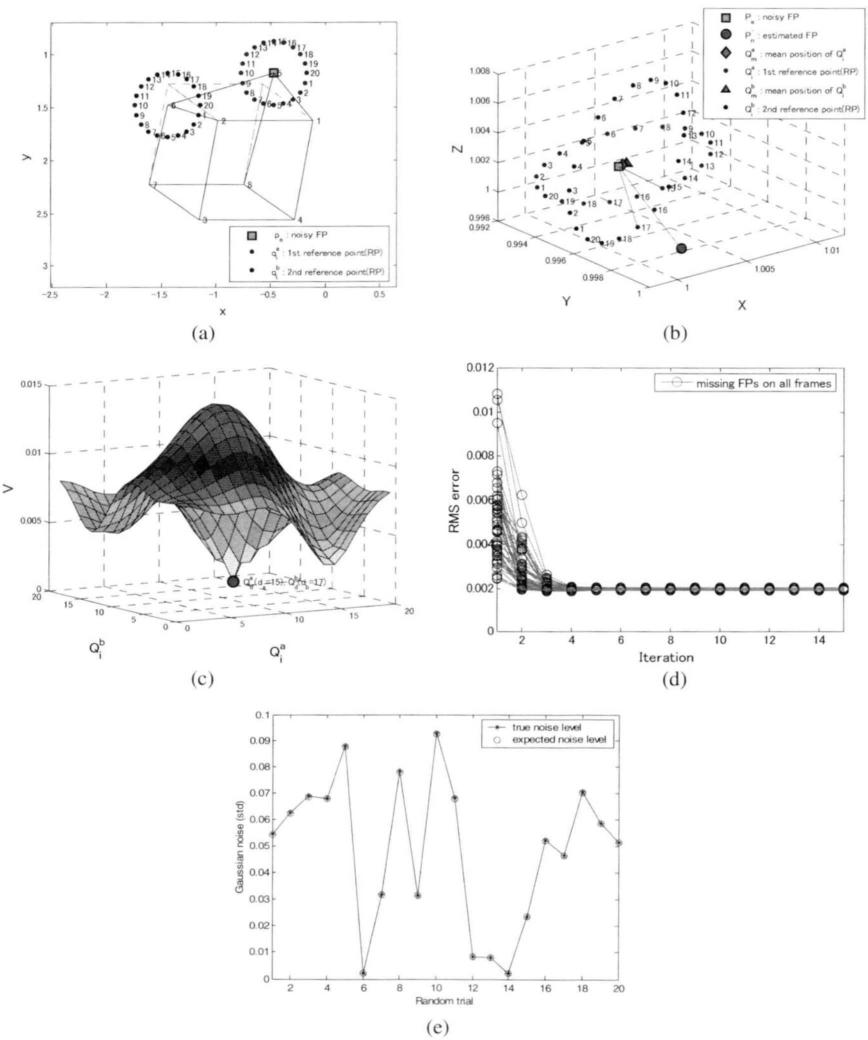

Fig. 3. An example for two missing FPs on the 12th frame. (a) missing FP and RPs on Π_2. (b) missing FP and RPs on Π_3. (c) orientation for recovering missing FP on Π_3. (d) *RMS* errors of the recovered missing FPs on observation matrix with Gaussian noise (*Δe*= 0.002 [unit]). (e) the results of estimation for Gaussian noise level (*Δe*) from zero to 0.1 [unit].

[unit]. The number of RPs has no limitation because the proposed algorithm is able to use the geometrical properties if the number of reference points is greater then three (see Fig. 1). Recovering missing data using geometrical properties between Π_2 and Π_3 is shown in Fig. 3. Fig. 3(a) and (b) illustrate a pattern of missing FP and RPs on

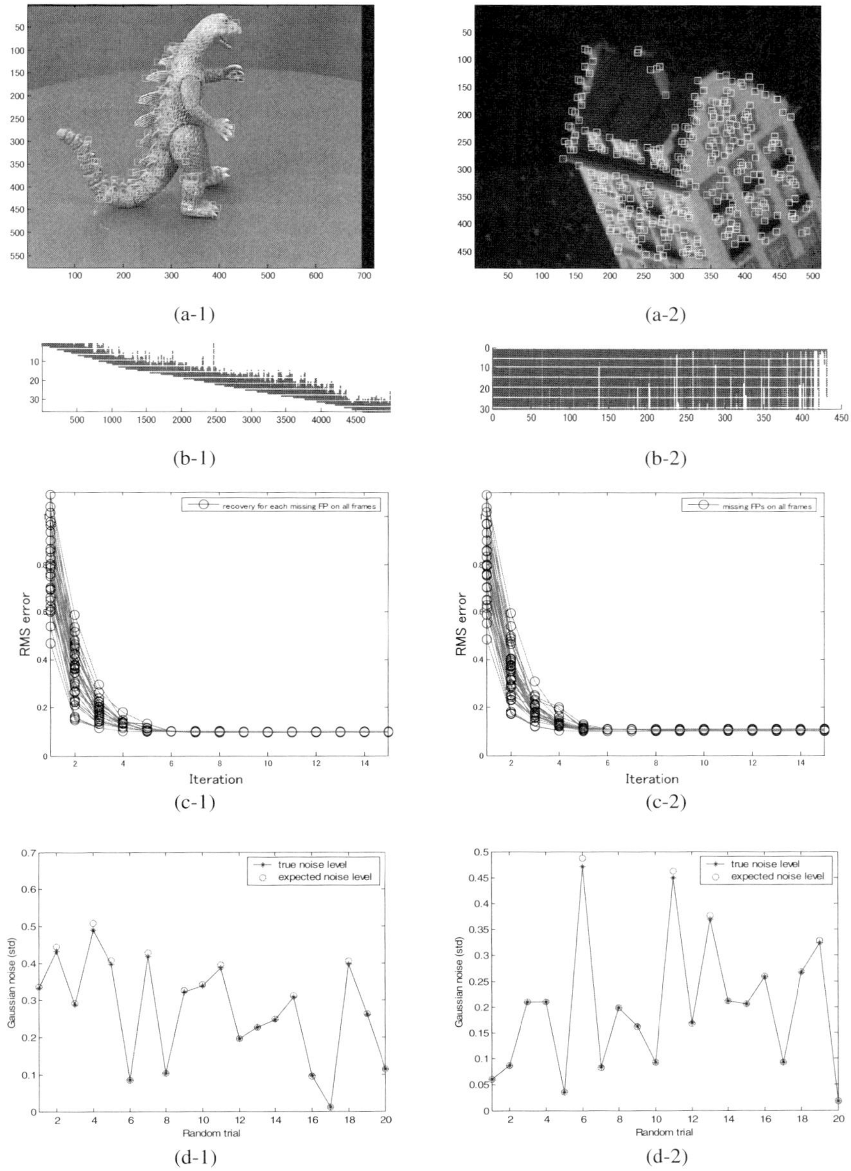

Fig. 4. Two examples of 'dinosaur (720x576)' (left) and 'hotel (512x480)' (right) image sequence. (a) 3rd frames with tracked FPs. (b) observation matrices. (c) *RMS* errors of the recovered missing FPs when added noise (Δe =0.1[pixel]). (d) the results of estimation for Gaussian noise level(Δe) from zero to 0.5 [pixel].

Π_2 and Π_3 respectively and Fig. 3(c) an expected orientation for recovering missing FP on Π_3. The final simulation results are presented in Fig. 3(d). We can confirm that the geometrical approach is superior to noise, as the proposed method is designed to minimize the influence of noise. Improved results can be achieved after greater iterations. After fitting the missing data, the estimation of unknown noise level included within the 2D observation matrix is shown in Fig. 3(e), where our algorithm is simulated for various Gaussian noise levels from zero to 0.1 [unit]. It is easy to see the strong correlation between the two results.

4.2 Real Video Sequence

In order to test the proposed algorithm on 36-frame 'dinosaur' and 30-frame 'hotel' real sequence, respectively, the entries of an observation matrix is tracked over the sequence. The 3rd frames are presented in Fig. 4(a), where the tracked FPs are denoted by square symbol. The FPs were extracted by the Harris interest operator [14]. The observation matrices of the video frames are presented in Fig. 4(b), where red points are the observed entries. Fig. 4(c) illustrates the results of geometrical fitting of missing data. The estimated results of unknown noise level is shown in Fig 4(d), where our algorithm is simulated for various noise levels from Gaussian noise zero to 0.1 [pixel]. The proposed method leads to results of greater accuracy in various levels of noise. In addition, the trend of the *RMS* errors of recovered missing data and the estimation errors of various noise levels for real sequence closely follow that of the synthetic sequence. Therefore, it can be confirmed that the proposed method provides more satisfied results not only for the synthetic sequences, but also for real sequences.

5 Conclusions

In this paper, we have presented the recovery algorithm of missing data using geometrical properties between 2D image plane and 3D error space and the noise level estimation algorithm for evaluating noise level in 2D observation matrix using ranks of SVD. Because of handling directly the orientation and distance of missing data by geometrical properties under wide ranges of noise levels, the proposed geometrical approach is presented by minimizing the influence of noise, which differs substantially from previous methods. The achievements of the recovery performances can give the satisfied results for synthetic or real sequences as the missing data are recovered accurately within only 10 times iterative processing. Also the estimation of noise level leads to results of greater accuracy in various levels of noise. According to the simulation, the results are presented to validate the claims about the performance of our approach.

References

1. Tomasi and T. Kanade : Shape and motion from image streams under orthography: A factorization method. Int. J. of Computer Vision, vol.9-2 (1992) 137-154
2. Sturm, P., Triggs, B.: A factorization based algorithm for multi-image projective structure and motion. In Proc. ECCV, vol. 2, Cambridge, UK (1996) 709-720

3. Poelman, C.J., Kanade, T. : A paraperspective factorization method for shape and motion recovery. IEEE Trans. on PAMI, vol.19-3 (1997) 206-218
4. Morita, T., Kanade, T. : A sequential factorization method for recovering shape and motion from image streams. IEEE Trans. on PAMI, vol.19-8 (1997) 858-867
5. Aguiar, P.M.Q., Moura, J.M.F. : Rank 1 Weighted Factorization for 3D Structure Recovery: Algorithms and Performance Analysis. IEEE Trans. on PAMI, vol.25-9 (2003) 1134-1149
6. Irani, M., Anandan, P. : Factorization with uncertainty. In Proc.ECCV, Dublin, Ireland (2000) 539-553
7. Shapiro, L., : Affine Analysis of Image Sequences. Cambridge University Press (1995)
8. Quan, L., Kanade, T. : A factorization method for affine structure from line correspondences. In Proc. IEEE CVPR, San Francisco CA, USA (1996) 803-808
9. Zhaohui Sun, V. Ramesh, and A. Murat Tekalp : Error characterization of the factorization method. Computer Vision and Image Understanding, vol.82 (2001) 110-137
10. K. Daniilidis and M. Spetsakis : Understanding noise sensitivity in structure from motion, Visual Navigation (1996) 61–88
11. D. D. Morris, K. Kanatani, and T. Kanade : Uncertainty modeling for optimal structure from motion. In Vision Algorithms: Theory and Practice, Lecture Notes in Computer Science, Springer-Verlag, Berlin/New York, vol.1883 (2000) 200–217
12. R. Szeliski and S. B. Kang : Shape ambiguities in structure-from-motion. IEEE Trans. on PAMI. vol.19 (1997) 506–512
13. R. Szeliski, and S.B. Kang : Shape Ambiguities in Structure From Motion. Technical Report 96/1, Digital Equipment Corporation, Cambridge Research Lab, Cambridge, Mass. (1996)
14. C. J. Harris and M. Stephens : A Combined Corner and Edge Detector. In Proc. Alvey Vision Conf. (1998) 147-151

Novel Prediction Approach – Quantum-Minimum Adaptation to ANFIS Outputs and Nonlinear Generalized Autoregressive Conditional Heteroscedasticity

Bao Rong Chang*

Department of Computer Science and Information Engineering
National Taitung University, Taitung, Taiwan 950
phone: +886-89-318855 ext. 2607; fax: +886-89-350214
brchang@nttu.edu.tw

Abstract. Volatility clustering degrades the efficiency and effectiveness of time series prediction and gives rise to large residual errors. This is because volatility clustering suggests a time series where successive disturbances, even if uncorrelated, are yet serially dependent. To overcome volatility clustering problems, an adaptive neuro-fuzzy inference system (ANFIS) is combined with a nonlinear generalized autoregressive conditional heteroscedasticity (NGARCH) model that is adapted by quantum minimization (QM) so as to tackle the problem of time-varying conditional variance in residual errors. The proposed method significantly reduces large residual errors in forecasts because volatility clustering effects are regulated to trivial levels. Two experiments using real financial data series compare the proposed method and a number of well-known alternative methods. Results show that forecasting performance by the proposed method produces superior results, with good speed of computation. Goodness of fit of the proposed method is tested by Ljung-Box Q-test. It is concluded that the ANFIS/NGARCH composite model adapted by QM performs very well for improved predictive accuracy of irregular non-periodic short-term time series forecast and will be of interest to the science of statistical prediction of time series.

1 Introduction

In practice, predictions are obtained by extrapolating a value at the next time instant based on a prediction algorithm [1]. The autoregressive moving-average (ARMA) is a traditional method very suitable for forecasting regular periodic data like seasonal or cyclical time series [2]. On the other hand, ARMA does not work well on irregular or non-periodic data sequences such as international stock prices or future volume indices [3]. This is because ARMA lacks a learning mechanism and cannot tackle large fluctuations in a complex time series. In particular, the back-propagation neural

* Corresponding author.

L. Wang et al. (Eds.): FSKD 2006, LNAI 4223, pp. 908 – 918, 2006.

network (BPNN) [4] and radial basis function neural network (RBFNN) [5] has been successfully applied to time series forecasting but requires a large amount of pattern/target training data to capture the dynamics of the time series. An alternate predictor, the grey model [6], has been widely applied to non-periodic short-term forecasts and however commonly encounters the overshoot phenomenon [1] whereby huge residual errors emerge at the inflection points of a data sequence. The adaptive neuro-fuzzy inference system (ANFIS) [7] has been widely applied to random data sequences with highly irregular dynamics [8] [9], e.g. forecasting non-periodic short-term stock prices [1]. However, volatility clustering effects [10] in the data sequence prevent ANFIS from reaching desired levels of accuracy. Further, a revised version of GARCH called a nonlinear generalized autoregressive conditional heteroscedasticity (NGARCH) [11] was presented for resolving volatility clustering effects. To do so, an adaptation called quantum minimization (QM) [12] is applied to adapt the coefficients of a linear combination of ANFIS and NGARCH so that large residual error is compensated by NGARCH and near-optimal solutions can be obtained.

2 ANFIS/NGARCH Composite Model Resolving Volatility Clustering

2.1 NGARCH Resolving Volatility Clustering

ARMAX(r,m,Nx) [13] encompasses autoregressive (AR), moving-average (MA) and regression (X) models, in any combination, as expressed below

$$y_{armax}(t) = C^{armax} + \sum_{i=1}^{r} R_i^{armax}\, y(t-i) + e_{resid}(t) + \sum_{j=1}^{m} M_j^{armax}\, e_{resid}(t-j) + \sum_{k=1}^{N_x} \beta_k^{armax}\, \mathbf{X}(t,k) \quad (1)$$

where C^{armax} = a constant coefficient, R_i^{armax} = autoregressive coefficients, M_j^{armax} = moving average coefficients, $e_{resid}(t)$ = residuals, $y_{armax}(t)$ = responses, β_k^{armax} = regression coefficients, $\mathbf{X}$ = an explanatory regression matrix in which each column is a time series and $X(t,k)$ denotes a element at the t th row and k th column of input matrix.

NGARCH(p,q) [11] describes nonlinear time-varying conditional variances and Gaussian residuals $e_{resid}(t)$. Its mathematical formula is

$$\sigma_{ntvcv}^2(t) = K^{ng} + \sum_{i=1}^{p} G_i^{ng}\sigma_{ntvcv}^2(t-i) + \sum_{j=1}^{q} A_j^{ng}\sigma_{ntvcv}^2(t-j)\left[\frac{e_{resid}(t-j)}{\sqrt{\sigma_{ntvcv}^2(t-j)}} - C_j^{ng}\right]^2 \quad (2)$$

with constraints

$$\sum_{i=1}^{p} G_i^{ng} + \sum_{j=1}^{q} A_j^{ng} < 1,\ K^{ng} > 0,\ G_i^{ng} \geq 0,\ i = 1,\dots,p,\ A_j^{ng} \geq 0,\ j = -1,\dots,q$$

where K^{ng} = a constant coefficient, G_i^{ng} = linear-term coefficients, A_j^{ng} = nonlinear-term coefficients, C_j^{ng} = nonlinear-term thresholds, $\sigma^2_{ntvcv}(t)$ = a nonlinear time-varying conditional variance and $e_{resid}(t-j)$ = j-lag Gaussian distributed residual in ARMAX.

In the presence of conditional heteroscedasticity, this composite model can perform ARMAX and NGARCH separately over every period in a time series. For simplicity as employed in [14], it is possible to merge the outputs of ARMAX and NGARCH linearly to attain better results as

$$y_{Composite\ Model}(t) = f(y_{ARMAX}(t),\ y_{NGARCH}(t)) = Cf_1 \cdot y_{ARMAX}(t) + Cf_2 \cdot y_{NGARCH}(t) \tag{3}$$

where f is defined as a linear function of ARMAX and NGARCH outputs, $y_{ARMAX}(t)$ and $y_{NGARCH}(t)$. Cf_1 and Cf_2 in Eq. (3) are the coefficients of a linear combination of ARMAX and NGARCH outputs. The resulting residual $y_{NGARCH}(t)$ at time t is obtained from a product of $\sqrt{\sigma^2_{ntvcv}(t)}$ in Eq. (2) and a normalized random number $randn(1)$ where $0 \le randn(1) \le 1$.

2.2 ANFIS Coordinated with NGARCH to Improve Regression

ARMAX cannot fit data sequences very well for irregular or non-periodic time series due to the lack of a dynamic learning mechanism. So, we propose an improved approach, i.e. to replace ARMAX with ANFIS for the conditional mean component of composite model because ANFIS has its own self-adaptive learning ability to fit irregular or non-periodic time series. This proposed composite model is rewritten as ANFIS/NGARCH. Formulation of the linear combination [14] is expressed as

$$y_{Proposed\ Composite\ Model}(t) = g(y_{ANFIS}(t),\ y_{NGARCH}(t)) = Coef_1 \cdot y_{ANFIS}(t) + Coef_2 \cdot y_{NGARCH}(t) \tag{4}$$

where g is defined as a linear function of the ANFIS and NGARCH outputs, respectively, $y_{ANFIS}(t)$ and $y_{NGARCH}(t)$, while $Coef_1$ and $Coef_2$ are respectively the coefficients of the linear combination of the ANFIS and NGARCH outputs.

A novel adaptation mechanism, called quantum minimization (QM) [12], is presented in the next section and will be exploited to search for optimal or near-optimal coefficients $Coef_1$ and $Coef_2$ in Eq. (4).

3 Quantum Minimization Adapting ANFIS/NGARCH

3.1 Quantum Exponential Searching Algorithm

As reported in [15], we assume in this section that the number t of solutions is known and that it is not zero. Let $A = \{i \mid F(i) = 1\}$ and $B = \{i \mid F(i) = 0\}$.

Step 1: For any real numbers k and l such that $tk^2 + (N-t)l^2 = 1$, redefine

$$|\Psi(k,l)\rangle = \sum_{i\in A} k\,|i\rangle + \sum_{i\in B} l\,|i\rangle .$$

A straightforward analysis of Grover's algorithm shows that one iteration transforms $|\Psi(k,l)\rangle$ into

$$\left|\Psi\left(\frac{N-2t}{N}k + \frac{2(N-t)}{N}l, \frac{N-2t}{N}l - \frac{2t}{N}k\right)\right\rangle .$$

Step 2: This gives rise to a recurrence similar to the iteration transforms in Grover's algorithm [16], whose solution is that the state $|\Psi(k_j,l_j)\rangle$ after j iterations is given by

$$k_j = \frac{1}{\sqrt{t}}\sin((2j+1)\theta) \quad and \quad l_j = \frac{1}{\sqrt{N-t}}\cos((2j+1)\theta) .$$

where the angle θ is so that $\sin^2\theta = t/N$ and $0 < \theta \le \pi/2$.

3.2 Quantum Minimum Searching Algorithm

We second give the minimum searching algorithm [12] in which the minimum searching problem is to find the index i such that $T[i]$ is minimum where $T[0,...,N-1]$ is to be an unsorted table of N items, each holding a value from an ordered set.

Step 1: Choose threshold index $0 \le i \le N-1$ uniformly at random.
Step 2: Repeat the following stages (2a and 2b) and interrupt it when the total running time is more than $22.5\sqrt{N} + 1.4\lg^2 N$. Then go to stage (2c).

(a) Initialize the memory as $\sum_j \frac{1}{\sqrt{N}}|j\rangle|i\rangle$. Mark every item j for which $T[j] < T[i]$.

(b) Apply the quantum exponential searching algorithm [15].

(c) Observe the first register: let i' be the outcome. If $T[i'] < T[i]$, then set threshold index i to i' .

Step 3: Return i

This process is repeated until the probability that the threshold index selects the minimum is sufficiently large.

3.3. QM-AFNG Forecasting Based on Signal Deviation

Single-step-look-ahead prediction, as shown in Fig. 1 and Fig. 2, can be arranged by adding the most recent predicted signal deviation $\delta\hat{o}(k+1)$ of Eq. (5) to the observed current output $o(k)$.

$$\delta\hat{o}(k+1) = h(o(k), o(k-1),..., o(k-s), \delta o(k), \delta o(k-1),..., \delta o(k-s)) \tag{5}$$

$$\hat{o}(k+1) = o(k) + \delta\hat{o}(k+1) \tag{6}$$

Based on the QM-AFNG structure, one can form the function p of the ANFIS output, $\delta\hat{o}_{anfis}(k+1)$, and the square-root of NGARCH's output, $\hat{\sigma}_{\delta\hat{o}}(k+1)$, as presented below and shown in Fig. 1.

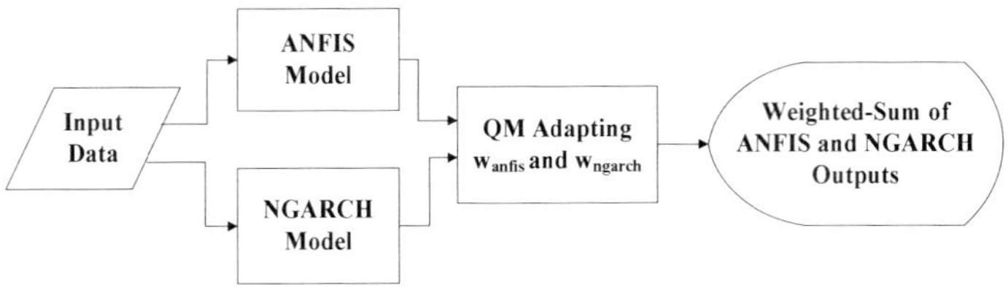

Fig. 1. Diagram of QM adapting ANFIS/NGARCH outputs (denoted as QM-AFNG)

$$\delta\hat{o}_{qm-afng}(k+1) = p(\delta\hat{o}_{anfis}(k+1), \hat{\sigma}_{\delta o}(k+1)) \tag{7}$$

A weighted-average function is assumed to combine both $\delta\hat{o}_{anfis}(k+1)$ and $\hat{\sigma}_{\delta o}(k+1)$ to attain a near-optimal result $\delta\hat{o}_{qm-afng}(k+1)$.

$$\begin{aligned} \delta\hat{o}_{qm-afng}(k+1) &= w_{anfis} \cdot \delta\hat{o}_{anfis}(k+1) + w_{ngarch} \cdot \hat{\sigma}_{\delta o}(k+1) \\ s.t. \quad & w_{anfis} + w_{ngarch} = 1 \end{aligned} \tag{8}$$

Here, the linear combination of two nonlinear functions, $\delta\hat{o}_{anfis}(k+1)$ and $\hat{\sigma}_{\delta o}(k+1)$, can also optimally approximate an unknown nonlinear target $\delta\hat{o}_{qm-afng}(k+1)$. Let $W_{afng} = [w_{anfis}\ w_{ngarch}]^T$ denote a weight-vector of w_{anfis} and w_{ngarch}. A digital cost-function (DCF) [17] is defined as

$$DCF = \frac{\left\|W_{afng}\right\|^2}{2} + Const. \sum_{k=0}^{l-1} \left\| y(k+1) - y(k) - o(k) - \delta\hat{o}_{qm-afng}(k+1) \right\|^2, \tag{9}$$

$Const.$: a constant,

which can be used for measuring the accuracy when the respected cost is minimized. Quantum minimization mentioned above is employed for adapting the appropriate weights, w_{anfis} and w_{ngarch}, for the forecast $\delta\hat{o}_{anfis}(k+1)$ and $\hat{\sigma}_{\delta o}(k+1)$ as per Eq. (8), respectively. Quantum minimization gives an order of computational cost as $O(\sqrt{N})$.

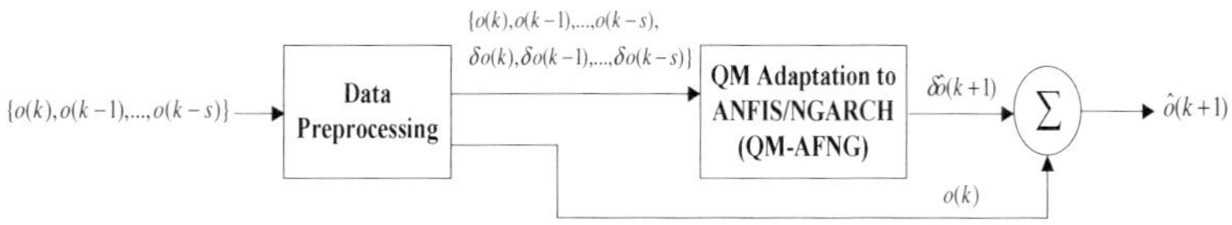

Fig. 2. Prediction using QM-AFNG system

4 Experimental Results and Discussions

In order to justify reasonable accuracy for a time series forecast, four well-known criteria [18] are commonly utilized. The terminology of these criteria is indicated as: (a) mean absolute deviation (MAD); (b) mean absolute percent error (MAPE); (c) mean squared error (MSE); (d) Theil'U inequality coefficient (Theil'U).

$$Theil'U = \sqrt{\frac{MSE}{MS}} = \sqrt{\frac{\sum_{t=1}^{l}\left(y_{t_c+t} - \hat{y}_{t_c+t}\right)^2 \Big/ l}{\sum_{t=1}^{l} y_{t_c+t}^2 \Big/ l}} \tag{10}$$

where l = the number of periods in forecasting, t_c = the current period, y_{t_c+t} = a desired value at the $t_c + t$ th period and $\hat{y}_{t_c+t}$ = a predicted value at the $t_c + t$ th period.

As shown in Figs. 3 to 8, with a sliding widow size of 7 data points, the forecasting abilities of our proposed method and several alternative methods are compared in experiments The alternative methods used are grey model (GM), auto-regressive moving-average (ARMA), back-propagation neural network (BPNN), ARMA/NGARCH composite model (ARMAXNG), adaptive neuro-fuzzy inference system (ANFIS), and the ANFIS/NGARCH composite model adapted by quantum minimization (QM-AFNG). Single-step-look-ahead prediction methodology is employed in all experiments. In single-step-look-ahead design, a small number of the most recent observed data are collected as a sliding window (i.e. data queue) for modeling an intermediate predictor to predict the next period output. Once the next period's sampled datum is obtained, we drop a datum at the bottom of the data queue and add the most recent sampled datum into the data queue at the top position, thereby forming the new data queue used for the next prediction. This process continues until the task is terminated. To simplify comparison of the tested methods as plotted curves, only the three most representatives are shown in the figures. Thus GM, ARMA and the proposed QM-AFNG are illustrated in Figs. 3 to 8, where "•" represents the sequential output of GM prediction, "∘" represents the sequential output of ARMA prediction and "$-*-$" represents the sequential output of QM-AFNG prediction. All six methods, however, are compared for goodness-of-fit in Tables 1 to 8.

First, the forecast of international stock price indices of four markets (New York Dow-Jones Industrials Index, London FTSE-100 Index, Tokyo Nikkei Index, and Taipei Taiex Index) [19] are shown in Figs. 3 to 6. In addition, this study shows performance evaluation based on (a) mean absolute deviation (MAD), (b) mean absolute percent error (MAPE) ×100, (c) mean squared error (MSE) (unit=10^5), and (d) Theil'U inequality coefficient (Theil'U) between the actual sampled values and the predicted results of international stock price monthly indices over 48 months from Jan. 2002 to Dec. 2005. Forecasting performance of all six methods is summarized in Tables 1 to 4, showing QM-AFNG obtains the best prediction results. The goodness of fit of QM-AFNG prediction modeling for the four markets is tested by Ljung-Box Q-test [20] with p-values of 0.5082, 0.3239, 0.4751 and 0.3702, where each p-value is greater than the level of significance (0.05).

Second, Figs 7 and 8 show the comparative forecasts of the equity volume index futures and options over 24 months (Jan. 2001 to Dec. 2002) as quoted from the London International Financial Futures and Options Exchange (LIFFE) [21]. Performance evaluation is again made on the basis of MAD, MAPE, MSE, and Theil'U between the actual and predicted values. Tables 5 to 8 summarize prediction performance of our alternative methods and shows that QM-AFNG achieves superior results. The goodness of fit of QM-AFNG prediction modeling for futures and options is also tested by Ljung-Box Q-test with p-values of 0.2677 and 0.1523, in which each p-value is greater than level of significance (0.05).

Table 1. The comparison between different prediction models based on Mean Absolute Deviation (MAD) on international stock price monthly indices

Methods	Mean Absolute Deviation				
	New York D.J. Industrials Index	London FTSE-100 Index	Tokyo Nikkei Index	Taipei TAIEX Index	Average
GM	340.5970	153.8277	477.2157	355.1361	331.6941
ARMA	339.7215	153.7628	439.8190	321.1152	313.6046
BPNN	279.1350	134.5064	453.7069	277.5879	286.2341
ARMAXNG	320.7695	152.3504	437.0319	317.9291	307.0202
ANFIS	284.5725	145.3118	441.5919	296.1719	291.9120
QM-AFNG	274.8238	125.3910	430.0475	269.3103	274.8932

Table 2. The comparison between different prediction models based on Mean Absolute Percent Error (MAPE) on international stock price monthly indices

Methods	Mean Absolute Percent Error (unit=10^{-2})				
	New York D.J. Industrials Index	London FTSE-100 Index	Tokyo Nikkei Index	Taipei TAIEX Index	Average
GM	3.65	3.54	4.49	6.49	4.54
ARMA	3.61	3.53	4.14	5.81	4.27
BPNN	2.98	3.06	4.19	5.05	3.82
ARMAXNG	3.52	3.50	4.12	5.77	4.23
ANFIS	3.06	3.31	4.13	5.40	3.98
QM-AFNG	2.83	2.97	4.05	4.93	3.70

Table 3. The comparison between different prediction models based on Mean Squared Error (MSE) on international stock price monthly indices

Methods	Mean Squared Error (unit=10^{5})				
	New York D.J. Industrials Index	London FTSE-100 Index	Tokyo Nikkei Index	Taipei TAIEX Index	Average
GM	1.9582	4.0063	3.2209	1.7472	2.7332
ARMA	1.8230	3.8832	2.9384	1.4737	2.5296
BPNN	1.2652	3.0656	3.0189	1.0461	2.0990
ARMAXNG	1.8170	3.8527	2.9193	1.4772	2.5166
ANFIS	1.3550	3.8494	2.8912	1.1683	2.3160
QM-AFNG	1.1784	2.9536	2.7689	1.0113	1.9781

Table 4. The comparison between different prediction models based on Theil'U Inequality Coefficient (Theil'U) on international stock price monthly indices

Methods	Theil'U Inequality Coefficient				
	New York D.J. Industrials Index	London FTSE-100 Index	Tokyo Nikkei Index	Taipei TAIEX Index	Average
GM	0.0435	0.0414	0.0501	0.0721	0.0518
ARMA	0.0420	0.0408	0.0479	0.0662	0.0492
BPNN	0.0349	0.0362	0.0485	0.0558	0.0439
ARMAXNG	0.0409	0.0406	0.0477	0.0663	0.0489
ANFIS	0.0362	0.0411	0.0475	0.0590	0.0460
QM-AFNG	0.0331	0.0351	0.0463	0.0545	0.0423

Table 5. The comparison between different prediction models based on Mean Absolute Deviation (MAD) on futures and options volumes monthly indices of equity products

Methods	Mean Absolute Deviation		
	Futures Index of Equity Products	Options Index of Equity Products	Average
GM	0.2607	0.0957	0.1782
ARMA	0.1935	0.1198	0.1567
BPNN	0.1022	0.0746	0.0884
ARMAXNG	0.1803	0.0722	0.1263
ANFIS	0.0851	0.0713	0.0782
QM-AFNG	0.0704	0.0668	0.0686

Table 6. The comparison between different prediction models based on Mean Absolute Percent Error (MAPE) on futures and options volumes monthly indices of equity products

Methods	Mean Absolute Percent Error		
	Futures Index of Equity Products	Options Index of Equity Products	Average
GM	0.0441	0.0168	0.0305
ARMA	0.0328	0.0210	0.0269
BPNN	0.0172	0.0131	0.0152
ARMAXNG	0.0305	0.0127	0.0216
ANFIS	0.0144	0.0125	0.0135
QM-AFNG	0.0132	0.0109	0.0121

Table 7. The comparison between different prediction models based on Mean Squared Error (MSE) on futures and options volumes monthly indices of equity products

Methods	Mean Squared Error		
	Futures Index of Equity Products	Options Index of Equity Products	Average
GM	0.0945	0.0138	0.0542
ARMA	0.0547	0.0114	0.0331
BPNN	0.0196	0.0087	0.0142
ARMAXNG	0.0507	0.0096	0.0302
ANFIS	0.0112	0.0092	0.0102
QM-AFNG	0.0093	0.0071	0.0082

Table 8. The comparison between different prediction models based on Theil'U Inequality Coefficient (Theil'U) on futures and options volumes monthly indices of equity products

Methods	Theil'U Inequality Coefficient		
	Futures Index of Equity Products	Options Index of Equity Products	Average
GM	0.0483	0.0191	0.0337
ARMA	0.0303	0.0332	0.0318
BPNN	0.0220	0.0152	0.0186
ARMAXNG	0.0368	0.0183	0.0276
ANFIS	0.0166	0.0155	0.0161
QM-AFNG	0.0152	0.0137	0.0145

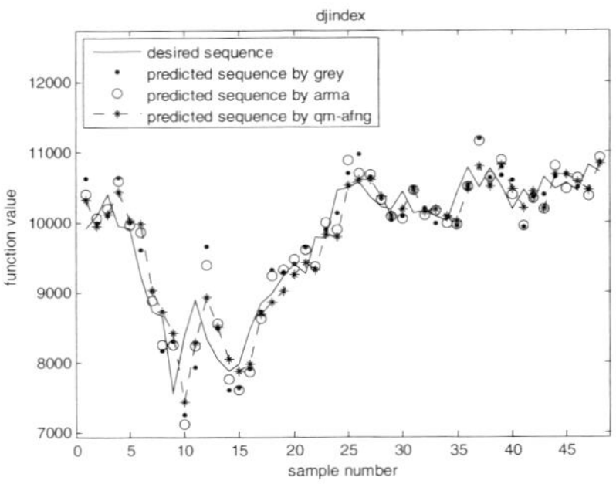

Fig. 3. Forecasts of monthly New York D.J. industry index

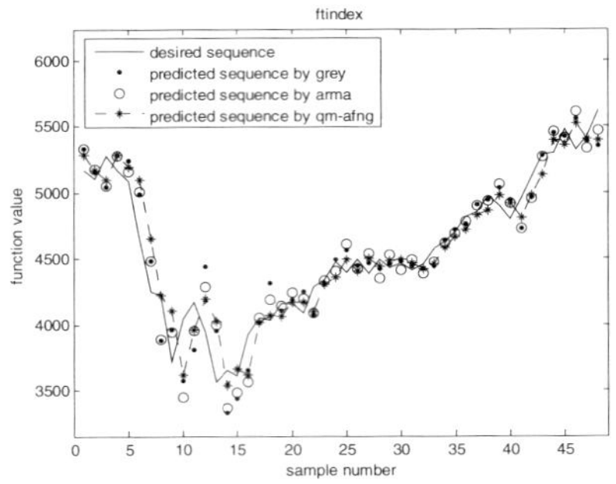

Fig. 4. Forecasts of monthly London FTSE-100 index

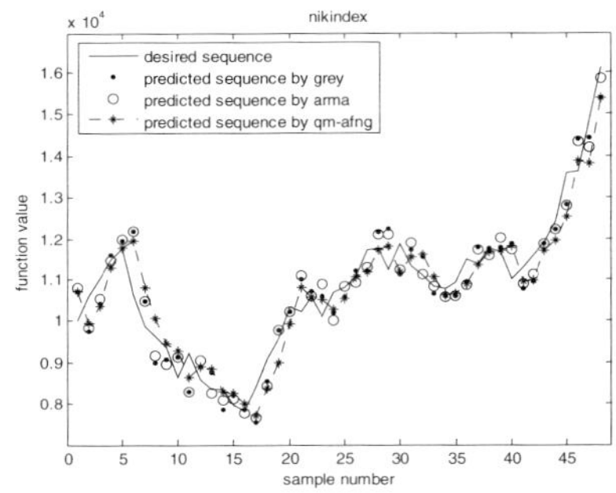

Fig. 5. Forecasts of monthly Tokyo Nikkei index

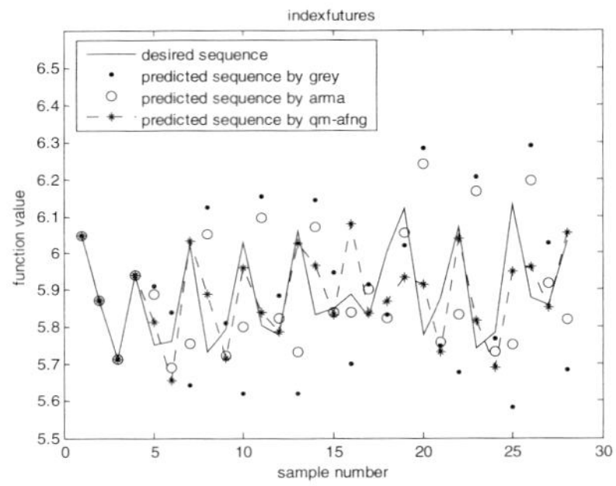

Fig. 7. Forecasts of monthly equity volume index futures

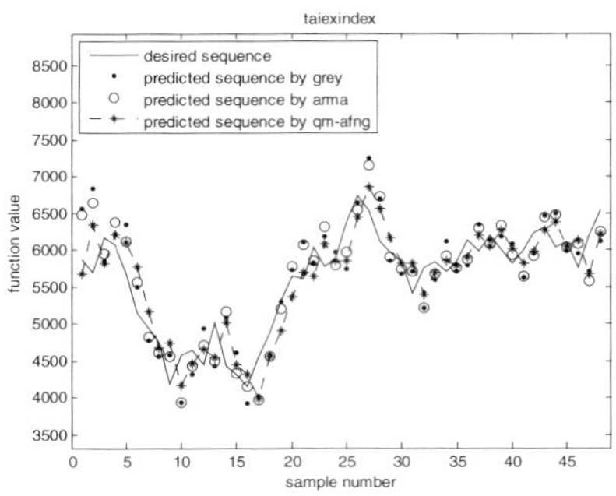

Fig. 6. Forecasts of monthly Taipei Taiex index

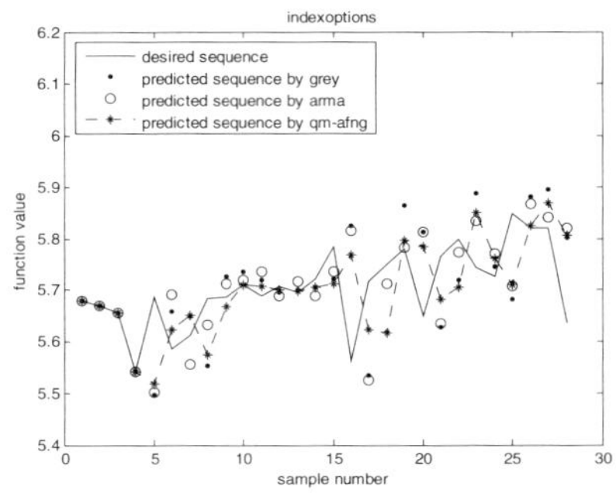

Fig. 8. Forecasts of monthly equity volume index options

5 Concluding Remarks

This study has proposed a method that incorporates a nonlinear generalized autoregressive conditional heteroscedasticity (NGARCH) into an ANFIS approach so as to correct the crucial problem of time-varying conditional variance in residual errors. In this manner, large residual error is significantly reduced because the effect of volatility clustering is regulated to a trivial level. Experimental comparison of a range of systems shows that the ANFIS/NGARCH composite model adapted by QM provides superior prediction accuracy and good computation speed for irregular non-periodic short-term time series forecast.

Acknowledgements

This work is fully supported by the National Science Council, Taiwan, Republic of China, under grant number **NSC 94-2218-E-143-001**.

References

1. Chang, B. R.: Advanced Hybrid Prediction Algorithm for Non-Periodic Short-Term Forecasting. International Journal of Fuzzy System. 5 3 (2003) 151-160
2. Box, G. E. P., Jenkins, G. M., Reinsel, G. C.: Time Series Analysis: Forecasting & Control. Prentice-Hall, New Jersey (1994)
3. Chang, B. R.: Tuneable Free Parameters C and Epsilon-Tube in Support Vector Regression Grey Prediction Model -SVRGM(1,1|C,ε) Approach. In Proc. IEEE International Conference on Systems, Man, and Cybernetics. (2004) 2431-2437
4. Castillo, O., Melin, P.: Simulation and Forecasting Complex Economic Time Series Using Neural Network and Fuzzy Logic. In Proc. International Joint Conference on Neural Network, (2001) 1805-1810
5. Thomson, R., Hodgman, T. C., Yang, Z. R., Doyle, A. K.: Characterizing Proteolytic Cleavage Site Activity Using Bio-Basis Function Neural Networks. Bioinformatics. 19 14 (2003) 1741–1747.
6. Chang, B. R.: Hybrid BPNN-Weighted Grey-CLMS Forecasting. Journal of Information Science and Engineering. 21 1 (2005) 209-221
7. Jang, J.-S. R.: ANFIS: Adaptive-Network-based Fuzzy Inference Systems. IEEE Transactions on Systems, Man, and Cybernetics. 23 3 (1993) 665-685
8. Wang, J.-S.: An Efficient Recurrent Neuro-Fuzzy System for Identification and Control of Dynamic Systems. In Proc. IEEE International Conference on Systems, Man, and Cybernetics. (2003) tracking #: 146
9. Neter, J., Wasserman W., and Kutner M.H.: Applied Linear Statistical Models. 2nd Ed. Homewood, IL: Irwin, (1985)
10. Bellerslve, T.: Generalized Autoregressive Conditional Heteroscedasticity. Journal of Econometrics. 31 (1986) 307-327
11. Gourieroux, C.: ARCH Models and Financial Applications, Springer-Verlag, New York (1997)
12. Durr, C., Hoyer, P.: A Quantum Algorithm for Finding the Minimum. (1996) http://arxiv.org/abs/quant-ph/9607014

13. Hamilton, J. D.: Time Series Analysis. Princeton University Press, New Jersey (1994)
14. Pshenichnyj, B. N., Wilson, S. S.: The Linearization Method for Constrained Optimization. Springer, New York (1994)
15. Boyer, M., Brassard, G., Hoyer, P., Tapp, A.: Tight Bounds on Quantum Searching. Fortschritte Der Physik (1998)
16. Grover, L. K.: A Fast Quantum Mechanical Algorithm for Database Search. In Proc.28th Ann. ACM Symp, Theory of Comp., ACM Press (1996) 212-219
17. Anguita, D., Ridella, S., Rivieccio F., Zunino R.: Training Support Vector Machines: a Quantum- Computing Perspective. In Proc. IEEE IJCNN. (2003) 1587-1592
18. Diebold, F. X.: Elements of Forecasting. South-Western, Cincinnati (1998)
19. FIBV FOCUS MONTHLY STATISTICS, International Stock Price Index (2005)
20. Ljung, G. M., Box, G. E. P.: On a Measure of Lack of Fit in Time Series Models. Biometrika. 65 (1978) 67-72
21. London International Financial Futures and Options Exchange (LIFFE). http://www.liffe.com/, 2002

Parallel-Structure Fuzzy System for Sunspot Cycle Prediction in the Railway Systems

Min-Soo Kim

Maglev Train System Research Team, Korea Railroad Research Institute
360-1 Woulam-Dong, Uiwang-City, Kyonggi-Do, Korea
Ms_kim@krri.re.kr

Abstract. This paper presents a parallel-structure fuzzy system (PSFS) for prediction of sunspot cycle in the railway communication and power systems based on smoothed sunspot number time series. The PSFS consists of a multiple number of fuzzy systems connected in parallel. Each component fuzzy system in the PSFS predicts the same future data independently based on its past time series data with different embedding dimension and time delay. According to the embedding dimension and the time delay, the component fuzzy system takes various input-output pairs. The PSFS determines the final predicted value as an average of all the outputs of the component fuzzy systems excluding the predicted data with the minimum and the maximum values in order to reduce error accumulation effect.

1 Introduction

Predicting future behavior of a system based on the knowledge regarding its previous behavior is one of the essential and ultimate objectives of science[1]. There are two basic approaches to prediction: model-based approach and nonparametric method. Model-based approach assumes that sufficient prior information is available with which one can construct an accurate mathematical model for prediction. Nonparametric approach, on the other hand, directly attempts to analyze a sequence of observations produced by a system to predict its future behavior. Though nonparametric approaches often cannot represent full complexity of real systems, many contemporary prediction theories are developed based on the nonparametric approach because of difficulty in constructing accurate mathematical models.

Sunspots are dark areas that grow and decay on the lowest level of the Sun that are visible from the Earth. Short-term predictions of solar activity are essential to help operating and to design railway communication / power systems that will survive for their useful lifetimes. The most visible appearance of solar activity is sunspots on the photosphere. Sunspots are magnetic regions on the Sun with magnetic field strengths thousands of times stronger than the Earth's magnetic field and appear as dark spots on the surface of the Sun and typically last for several days. Sometimes magnetic fields change rapidly releasing huge amounts of energy in solar flares and ejection of material from and through the corona. Solar activity like sunspots tends to vary from a

L. Wang et al. (Eds.): FSKD 2006, LNAI 4223, pp. 919 – 928, 2006.

minimum to a maximum and back again in a solar cycle of about 11 years. Planning for satellite orbits and ground/space missions often require knowledge of levels years in advance through prediction of sunspot time series [2]~[5].

This paper presents a parallel-structure fuzzy system (PSFS) for prediction of smoothed sunspot data which show characteristic of chaotic time series. This approach corresponds to a nonparametric approach since it generates a prediction result based on past observations of the system output. The PSFS consists of a multiple number of component fuzzy systems connected in parallel. Each component fuzzy system in the PSFS predicts future data independently based on its past time series data with different embedding dimension and time delay. The embedding dimension determines the number of inputs of each component fuzzy system. According to the time delay, the component fuzzy system takes inputs at different time intervals. Each component fuzzy system produces separate prediction results for a future data at a specific time index. The PSFS determines the final predicted value as an average of all the outputs of the component fuzzy systems. Each component fuzzy system contains a small number of multiple-input single-output (MISO) Sugeno-type fuzzy rules[6], which are generated by clustering input-output training data. Fuzzy systems can represent uncertainties involved in the behaviors of complex physical systems easier than conventional prediction algorithms. In many cases, fuzzy rules are determined according to experience of human experts or engineering common sense. When a structured knowledge is not available and the information is given in the form of numerical input-output data as in many real-world cases, adaptive clustering algorithms can automatically produce fuzzy rules.

2 Time Series Prediction

Nonparametric approach of time series prediction is based on the assumptions that future behavior of a time series can be represented by a functional relationship of its previous observations. If k previous input data are given at the k th time step, for a time series data $x(k)$, the τ-step-ahead value $\hat{x}(k+\tau)$ can be expressed as

$$\hat{x}(k+\tau) = P[x(k), x(k-1), x(k-2), \cdots, x(2), x(1)] \tag{1}$$

where $P[\cdot]$ denotes a function that represents input-output relationship of nonparametric time series prediction process and positive integer τ is called time delay.

Time series prediction methods can be classified into either one-step-ahead prediction or short-term prediction depending on the fact that predicted values are again used as input values. In one-step-ahead prediction, the predicted value of future data $\hat{x}(k+\tau)$ is expressed by its previous m inputs with time delay τ of the data sequence as in Eq. (2).

$$\hat{x}(k+\tau) = P[x(k), x(k-\tau), x(k-2\tau), \cdots, x(k-(m-1)\tau)] \tag{2}$$

where positive integer m is time delay. Future value of a time series can be predicted by an output of a linear or nonlinear function $P[\cdot]$ of $m\tau$ previous input data.

In short-term or long-term prediction of time series data, predicted values of the data are again used as inputs for prediction of future data. Eq. (3) shows short-term or long-term prediction of time series. The predicted value of future data $\hat{x}(k+\tau)$ is expressed according to the data previously predicted $\hat{x}(k), \hat{x}(k-\tau), \cdots, \hat{x}(k-(m-1)\tau)$. Therefore, long-term prediction of chaotic time series based on the data previously predicted is a difficult task since small initial error causes enormous error accumulation effects in future values.

$$\hat{x}(k+\tau) = P[\hat{x}(k), \hat{x}(k-\tau), \hat{x}(k-2\tau), \cdots, \hat{x}(k-(m-1)\tau)] \quad (3)$$

The optimal embedding dimension m is determined for a specific time delay τ. For given τ, error performance measures are calculated for the training data and for the validation data. Validation data is a data set not used in training phase for checking if training result is acceptable. Error performance measures are defined as mean-square error and maximum absolute error calculated from the difference between the one-step-ahead prediction results and real data. For a given time series data, MSE and MAE are computed as one increases possible embedding dimension values for a fixed time delay. This process is repeated for training data and for validation data. The optimal embedding dimension corresponds to the embedding dimension whose error measure is minimized both for training and validation data.

3 Parallel-Structure Fuzzy System

3.1 Configuration of a Parallel-Structure Fuzzy System

A parallel-structure fuzzy system (PSFS) predicts future data according to several prediction mechanisms based on different number and different samples of previous data. The PSFS consists of a multiple number of component fuzzy systems connected in parallel for predicting time series. Fig. 1 shows the structure of the parallel-structure fuzzy system. The PSFS contains N component fuzzy systems, $Fuzzy\ System_1$, $Fuzzy\ System_2, \ldots, Fuzzy\ System_N$ connected in parallel. Each component fuzzy system produces independently predicted values of same future data $\hat{x}(k+1)$ at a time index $k+1$ based on previous data. The PSFS produces the final predicted value according to the N prediction results $\hat{x}_1(k+1), \hat{x}_2(k+1), \cdots, \hat{x}_N(k+1)$ of the N component fuzzy systems.

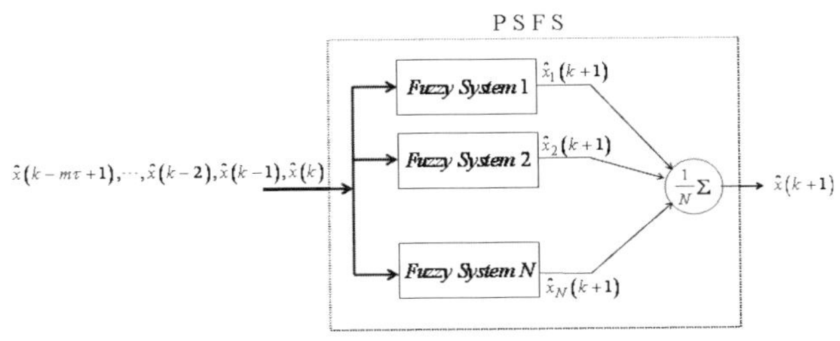

Fig. 1. Structure of the parallel-structure fuzzy system(PSFS)

Time series prediction with the PSFS is characterized by the two parameters τ and m. The embedding dimension m defines the number of inputs to each component fuzzy system, and the time delay τ defines the time interval of input data to component fuzzy systems.

For the PSFS with N component fuzzy systems in general, each fuzzy system produces prediction results $\hat{x}_1(k+1), \hat{x}_2(k+1), \cdots, \hat{x}_N(k+1)$ based on previous data $\hat{x}(k), \hat{x}(k-1), \hat{x}(k-2), \cdots$, and the final predicted data $\hat{x}(k+1)$ as in Eq. (4) becomes an average of all the prediction results by component fuzzy systems.

$$\hat{x}(k+1) = \frac{1}{N}\left[\sum_{i=1}^{N} \hat{x}_i(k+1)\right] \tag{4}$$

In short-term or long-term prediction, a small amount of prediction error is accumulated to become a big error after several iterations. The PSFS reduces error accumulation effect by averaging the prediction results of all the component fuzzy systems after removing the extreme values of prediction results.

Modeling fuzzy systems involves identification of the structure and the parameters with given training data. In the Sugeno fuzzy model[6], unlike the Mamdani method[7], the consequent part is represented by a linear or nonlinear function of input variables. The Sugeno model can represent nonlinear input-output relationships with a small number of fuzzy rules. Each rule in the Sugeno model corresponds to an input-output relationship of a fuzzy partition. Fuzzy rules in the MISO Sugeno model with n inputs $x_1, \cdots, x_n$ and an output variable y_i of the ith fuzzy rule is of the form:

$$IF\ x_1\ is\ A_{i1}\ and \cdots and\ x_n\ is\ A_{in}, THEN\ y_i = f_i(x_1, \cdots, x_n) \tag{5}$$

where $i = 1, \cdots, M$ and A_{ij} is a linguistic label represented by the membership function $A_{ij}(x_j)$, and $f_i(\cdot)$ denotes a function that relates input to output. M denotes the number of rules in the fuzzy system. The Sugeno fuzzy model can easily generates fuzzy rules from numerical input-output data obtained from an actual process. The function $A_{ij}(x_j)$ defines a membership function assigned to the input variable x_j in the ith fuzzy rule. In this paper, Gaussian membership functions and a linear function are used for simplicity.

$$A_{ij}(x_j) = \exp\left(-\frac{1}{2}\left((x_j - c_{ij})/w_{ij}\right)^2\right) \tag{6}$$

$$f_i(x_1, \cdots, x_n) = a_{0i} + a_{1i}x_1 + \cdots + a_{ni}x_n \tag{7}$$

where the parameters c_{ij} and w_{ij} define center and width of the Gaussian membership function $A_{ij}(x_j)$. The coefficients $a_{01}, a_{1i}, \cdots, a_{ni}$ are to be determined from input-output training data. In the simplified reasoning method, the output y of the fuzzy system with M rules is represented as

$$y = \sum_{i=1}^{M} \mu_i f_i(x_1, \cdots, x_n) \Big/ \sum_{i=1}^{M} \mu_i \tag{8}$$

where μ_i is a degree of relevance. In the product implication method, the degree of relevance is defined as

$$\mu_i = \prod_{j=1}^{n} A_{ij}(x_j) \tag{9}$$

$$= \exp\left(-\frac{1}{2} \sum_{j=1}^{n} \left(\frac{x_j - c_{ij}}{w_{ij}} \right)^2 \right) \tag{10}$$

In the Sugeno fuzzy model, the time-consuming rule extraction process from experience of human experts or engineering common sense reduces to a simple parameter optimization process of coefficients $a_{01}, a_{1i}, \cdots, a_{ni}$ for a given input-output data set since the consequent part is represented by a linear function of input variables. Using linear function in the consequent part, a group of simple fuzzy rules can successfully approximate nonlinear characteristics of practical complex systems.

Construction of the parallel-structure fuzzy system is basically off-line. The parameters of the PSFS must be determined by the training data before time series prediction operation.

3.3 Modeling of Component Fuzzy System Based on Clustering

The parameters of the component fuzzy systems are characterized using clustering algorithms. The subtractive clustering algorithm[14] finds cluster centers $x_i^* = (x_{1i}^*, \cdots, x_{ni}^*)$ of data in input-output product space by computing the potential values at each data point. The potential value is inversely proportional to distance between data points, which means densely populated data produces large potential values and therefore more cluster centers.

In the Subtractive clustering algorithm, the first cluster center corresponds to the data with the largest potential value. After removing the effect of the cluster center just found, the next cluster center becomes the data with the largest potential value, and so on. This procedure is repeated until the potential value becomes smaller than a predetermined threshold. There are n -dimensional input vectors $x_1, x_2, \cdots, x_m$ and 1-dimensional outputs $y_1, y_2, \cdots, y_m$ forming $(n+1)$ -dimensional space of input-output data. For data $X_1, X_2, \cdots, X_N$ in $(n+1)$ -dimensional input-output space, the subtractive clustering algorithm produces cluster centers as in the following procedure:

Step 1: Normalize given data into the interval [0,1].
Step 2: Compute the potential values at each data point. The potential value P_i of the data X_i is computed as

$$P_i = \sum_{j=1}^{N} \exp(-\alpha \|X_i - X_j\|^2), \quad i = 1, 2, \cdots, N \tag{11}$$

where a positive constant $\alpha = 4 / r_a^2$ determines the data interval which affects the potential values. Data outside the circle with radius over positive constant $r_a < 1$ do not substantially affect potential values.

Step 3: Determine the data with the largest potential value P_1^* as the first cluster center X_1^*.

Step 4: Compute the potential value $P_i^{'}$ after eliminating the influence of the first cluster center.

$$P_i^{'} = P_i - P_1^* \sum_{j=1}^{N} \exp(-\beta \left\| X_i - X_1^* \right\|^2) \tag{12}$$

where positive constant $\beta = 4/r_b^2$ prevents the second cluster center from locating close to the first cluster center. If the effect of potential of the first cluster center is not eliminated, second cluster center tends to appear close to the first cluster center, since there are many data concentrated in the first cluster center. Taking $r_b > r_a$ makes the next cluster center not appear near the present cluster center.

Step 5: Determine the data point of the largest potential value P_2^* as the second cluster center X_2^*. In general, compute potential values $P_i^{'}$ after removing the effect of the k th cluster center X_k^*, and choose the data of the largest potential value as the cluster center X_{k+1}^*

$$P_i^{'} = P_i - P_k^* \exp(-\beta \left\| X_i - X_k^* \right\|^2) \tag{13}$$

Step 6: Check if we accept the computed cluster center. If $P_k^* / P_1^* \geq \bar{\varepsilon}$, or $P_k^* / P_1^* > \underline{\varepsilon}$ and $d_{\min}/r_a + P_k^*/P_1^* \geq 1$, then accept the cluster center and repeat step 5. Here $d_{\min}$ denotes the shortest distance to the cluster centers $X_1^*, X_2^*, \cdots, X_k^*$ determined so far. If $P_k^* / P_1^* > \underline{\varepsilon}$ and $d_{\min}/r_a + P_k^*/P_1^* < 1$, then set the X_k^* to 0 and select the data of the next largest potential. If $d_{\min}/r_a + P_k^*/P_1^* \geq 1$ for the data, choose this data as the new cluster center and repeat step 5. If $P_k^* / P_1^* \leq \underline{\varepsilon}$, terminate the iteration.

When determining cluster centers, upper limit $\bar{\varepsilon}$ and lower limit $\underline{\varepsilon}$ allows the data of lower potential and of larger distance $d_{\min}$ between cluster centers to be cluster centers. Step 6 is the determining process of the calculated cluster center according to $d_{\min}$, the smallest distance to the cluster centers $X_1^*, X_2^*, \cdots$ calculated so far. When determining the cluster centers, data with low potential value can be chosen as a cluster center if $d_{\min}$ is big enough due to upper limit $\bar{\varepsilon}$ and lower limit $\underline{\varepsilon}$.

Fuzzy system modeling process using the cluster centers $X_1^*, X_2^*, \cdots, X_M^*$ in input-output space is as follows. The input part of the cluster centers corresponds to antecedent fuzzy sets. In $(n+1)$-dimensional cluster center X_i^*, the first n values are n-dimensional input space $x_i^* = (x_{i1}^*, \cdots, x_{in}^*)$. Each component determines the center of membership functions for each antecedent fuzzy sets. The cluster centers become the center of the membership functions $c_{ij} = x_{ij}^*$. The width of the membership function w_{ji} is decided as

$$w_{ij} = r_a \left\| \max_i(x_i^*) - \min_i(x_i^*) \right\| / \sqrt{M} \tag{14}$$

where M denotes the number of cluster centers, $\left\|\max_i(x_i^*) - \min_i(x_i^*)\right\|$ denotes the difference between the maximum and the minimum distances between cluster centers. The number of cluster centers corresponds to the number of fuzzy rules. The next process is to compute optimal consequent parameters $a_{0i}, a_{1i}, \cdots, a_{ni}$ in order to produce output y_j of the y_jth rule in the Sugeno fuzzy model. The number of centers equals the number of fuzzy rules. The output of the fuzzy system is defined as a linear function of input variables.

$$y_i = a_{0i} + a_{1i}x_1 + a_{2i}x_2 + \cdots + a_{ni}x_n \tag{15}$$

$$= a_i^T x + a_{0i} \tag{16}$$

Compute parameters g_i through linear least-squares estimation, the final output y of the Sugeno fuzzy model is given as

$$y = \sum_{i=1}^{M} \mu_i (a_i^T x + a_{0i}) \Big/ \sum_{i=1}^{M} \mu_i \tag{17}$$

This is the final output of the fuzzy system.

4 Simulations

4.1 Sunspot Time series

Time series data used in this paper is the sunspot number data that is monthly averaged of the number of individual spots through solar observation and consists of monthly sample collected from 1749/1 to 2005/9 like Table 1.

The sunspot number is computed as

$$R = k\left(10g + s\right) \tag{18}$$

where g is the number of sunspot regions, s is the total number of individual spots in all the regions, and k is a scaling factor (usually < 1).

And the sunspot number is represented as

$$\hat{R}_n = \frac{1}{24}\left(\sum_{i=-6}^{5} R_{n+i} + \sum_{i=-5}^{6} R_{n+i} \right) \tag{19}$$

Table 1. The sunspot time series

Index	Year / Month	Sunspot Number(R)
1	1749 / 1	58.0
2	1749 / 2	62.6
…	…	…
3080	2005 / 8	36.4
3081	2005 / 9	22.1

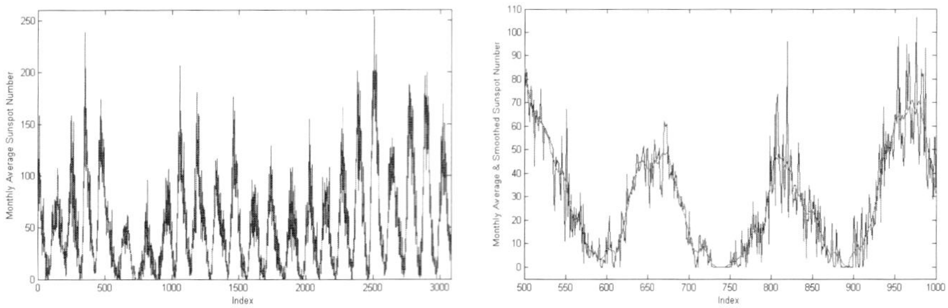

(a) Monthly average sunspot number (b) Sample view

Fig. 3. The smoothed sunspot time series data

Fig. 3 shows the sunspot time series data (an approximate 11-year cycle) and smoothed sunspot number time series used in the prediction with the PSFS. Each data set contains 3,081 samples.

The subtractive clustering algorithm with the parameters $r_a = 0.3, r_b = 0.75, \overline{\varepsilon} = 0.3$, and $\underline{\varepsilon} = 0.1$ generates the cluster centers. In this case, the PSFS with 5 component fuzzy systems ($N = 5$) is applied to time series prediction with 3,081 data. For the modeling of PSFS we use the first 2,921 data samples, except the next 160 test data samples, which divide two parts: one is training data (2,337 samples= $2,921 \times 0.8$) and the other is validation data (584 samples = $2,921 \times 0.2$).

In order to configure the PSFS for time series prediction, several embedding dimensions m must be determined for a specific time delay τ. For given τ, error performance measures are calculated from the difference between the one-step ahead prediction results trained with training data and validation data. The optimal value of m at a fixed τ corresponds to an integer for which the performance measures MSE and MAE is the smallest value of one-step ahead prediction for both training and validation data. Training data are used in constructing the fuzzy system based on the clustering algorithm. Validation data, which is not used to construct the fuzzy system, determines the optimal m according to τ when applied with the one-step-ahead prediction method.

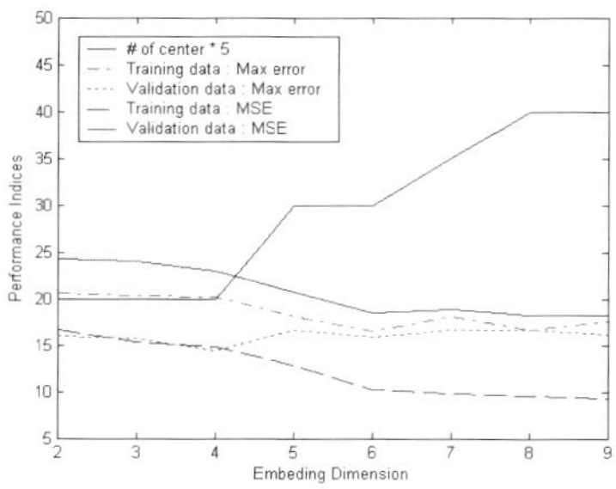

Fig. 4. Determination of embedding dimension when time delay is 3

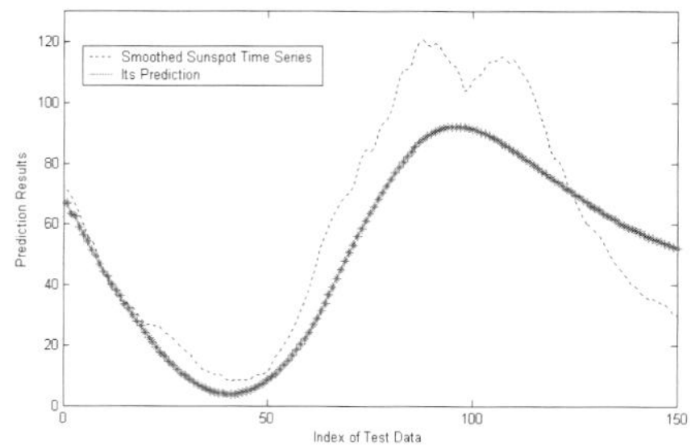

Fig. 5. Prediction result of the PSFS to test data

Fig. 4 shows how to find the optimal embedding dimension m for given time delay $\tau = 3$. The optimal value of m becomes 4. The smoothed sunspot time series data is characterized by the five (τ, m) pairs of (1,5), (2,3) , (3,4) ,(4,3), and (5,4) because the PSFS is $N = 5$.

4.2 Prediction with Parallel-Structure Fuzzy Systems

The PSFS contains three component fuzzy systems ($N = 5$) where τ is changed 1 to 5 ($\tau = 1, 2, \cdots, 5$). Each component fuzzy system is characterized by several embedding dimensions for fixed time delay. Three prediction results produced by the component fuzzy systems are averaged at each step. Fig. 5 shows the prediction result excluding the initial data by the PSFS.

Next the PSFS is applied to the pure future data represented by the period between 2005/10 and 2018/1. Fig. 6 shows the prediction result of the PSFS.

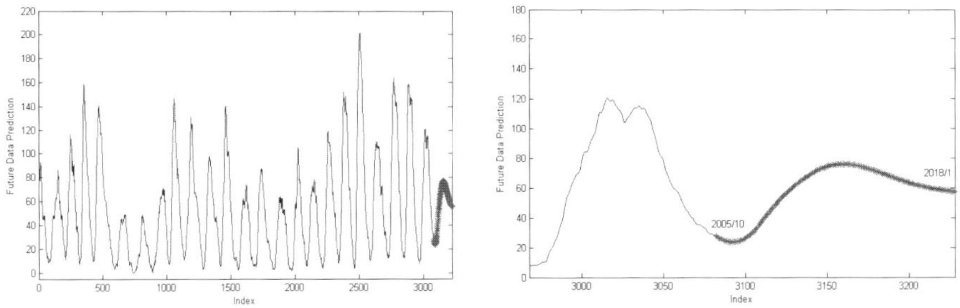

(a) Total view (b) Closed view

Fig. 6. Prediction result to the future samples

5 Conclusions

This paper presents a parallel-structure fuzzy system(PSFS) for predicting smoothed sunspot cycle in the railway communication and power systems. The PSFS corresponds to a nonparametric approach of time series prediction. The PSFS consists of a multiple number of component fuzzy systems connected in parallel. Each component fuzzy system is characterized by multiple-input single-output Sugeno-type fuzzy rules, which are useful for extracting information from numerical input-output training data. The component fuzzy systems for time series prediction are modeled by clustering input-output data. Each component fuzzy system predicts the future value at the same time index with different values of embedding dimension and time delay. The PSFS determines the final prediction value by averaging the results of each fuzzy system excluding the minimum and the maximum values out of N outputs in order to reduce error accumulation effect. Model-based approach was not considered since it has a difficulty in that it is not always possible to construct an accurate model. Performance comparison with the model-based approach will depend on modeling accuracy.

Computer simulations show that the PSFS trained with training and validation data successfully predicts the smoothed sunspot number data. The embedding dimension determines the number of inputs of a component fuzzy system, and the inputs to each component fuzzy system are characterized by the time delay. The number of component fuzzy systems chosen is five in this simulation. At each choice of time delay, optimal embedding dimension is determined by the value at which both the mean-square prediction error and the maximum absolute prediction error are constant. The PSFS produces the final prediction result by averaging the outputs of the component fuzzy systems after removing the maximum and the minimum prediction values.

References

1. A. S. Weigend and N. A. Gershenfeld, eds. Time Series Prediction: Forecasting the Future and Understanding the Past, Addison-Wesley Pub., pp.175-193, 1994.
2. M. S. Kim, H. S, Lee, C. H. You, and C. S. Chung, "Chaotic Time Series Prediction using PSFS2," 41st Annual Conference on SICE, August 2002.
3. Thompson, R. J., "A Technique for Predicting the Amplitude of the Solar Cycle," Solar Physics 148, 1993.
4. Hathaway, D. H., Wilson, R. M., and Reichmann, E. J., "A Synthesis of Solar Cycle Prediction Techniques," Journal Geophys. Res., 104, No. A10, 1999.
5. K. J. Li, H.S. Yun, H. F. Liang, and X. M. Gu, "Solar activity in Extended Cycles," Journal Geophys. Res., 107, No. A7, 2002.
6. M. Sugeno, Industrial Applications of Fuzzy Control, Elsevier Science Pub., 1985.
7. E. H. Mamdani, and S. Assilian, "An Experiment in Linguistic Synthesis with a Fuzzy Logic Conroller," Int. J. of Man Machine Studies, Vol. 7, No. 1, pp.1-13, 1975.
8. M. Casdagal, "Nonlinear Prediction of Chaotic Time Series," Physica D, pp.335-356, 1989.
9. D. Lowe and A. R. Webb, "Time Series Prediction by Adaptive Networks: A Dynamical Systems Perspective," Artificial Neural Networks, Forecasting Time Series, V. R. Vemuri, and R. D. Rogers (ed.), IEEE Computer Society Press, pp. 12-19, 1994.
10. D. S. Broomhead and D. Lowe, "Multi-variable Functional Interpolation and Adaptive Networks," Complex Systems, pp.262-303, 1988.
11. M.-S. Kim and S.-G. Kong, "Time Series Prediction using the Parallel-Structure Fuzzy System," 1999 IEEE Int. Fuzzy Systems Conference Proceedings, Vol. 2, 934-938, August 1999.
12. M. S. Kim and S. G. Kong, "Parallel Structure Fuzzy Systems for Time Series Prediction," Int. Jurnal of Fuzzy Systems, Vol. 3, No1, March 2001.
13. J.-S. R. Jang, and C.-T. Sun, "Neuro-Fuzzy Modeling and Control," Proceedings of the IEEE, March 1995.
14. S. Chiu, "Fuzzy Model Identification Based on Cluster Estimation," Journal of Intelligent & Fuzzy Systems, Vol. 2, No. 3, Sept. 1994.
15. R. R. Yager and D. P. Filev, Essentials of Fuzzy Modeling and Control, John Wiley & Sons, pp.246-264, 1994.

An Improved Fuzzy Approach to Planning and Scheduling Problems in Hybrid Distributed MES*

Xiaobing Liu, Hongguang Bo, Yue Ma, and Qiunan Meng

CIMS Centre, Dalian University of Technology, Postfach 11 60 23, Dalian, China
xbliu@dlut.edu.cn, bohongguang@sohu.com,
mayue_dalian@sina.com, code_name@163.com

Abstract. This paper proposes an improved approach which based on the model dealing with uncertain processing times, flexible due-dates and input-output radio. Associating with fuzzy set, classic scheduling rules, multi-closed loop control and user cooperating, hybrid system problems are hierarchically decomposed into the planning level problems, the scheduling level problems and material tracking feedback level problems.

1 Introduction

A multi-location iron and steel enterprise production management system is a representative hybrid distributed manufacturing execution system. In HDMES, planning and scheduling problems and material tracking problems have been considered the typical hybrid distributed computing problems.

Iron and steel product has the features of multi- variety, diversification and small batch. The make-to- order (MTO) production positioning strategy is widely adopted by iron and steel enterprises. Controlling the expenses and cost is core target for improving competition abilities of any enterprises. Due to continuous casting and hot rolling, planning & scheduling and material tracking in the integrated process are the combined lot scheduling and tracking problems integrating multiple production stages [1]. Mould casting and cold rolling technologies are still execute production planning and material tracking management independently.

2 Fuzzy Scheduling Model

Production scheduling problems in iron and steel enterprises are classified into static and dynamic scheduling problems. Static problems are such types in which the information is previously known. On the other hand, dynamic problems are such types in which production tasks are inserted into randomly over a scheduling period and the scheduler has no information on the recent production tasks prior to scheduling them [2]. The marked characteristics of a hybrid-scheduling problem is semi-continuum and semi-discretisation, and the primitive elements of iron and steel enterprises' scheduling problems are a collection of productions and a set of machines to be arranged. Aiming at MTO production pattern, we discuss production scheduling problems in HDMES

* This work was supported by NHTDP for CIMS (2003AA414044) and NSFC (70471057).

L. Wang et al. (Eds.): FSKD 2006, LNAI 4223, pp. 929 – 932, 2006.

using the concept of fuzzy due date [3]. Productions P_i (i = 1, 2… n), have fuzzy due dates, *Di*, whose membership functions are:

$$u_{Di}(x)=\begin{cases}1\left(d_i^{L1}\le x<d_i^{L2}\right)\\ \frac{1}{2}+\frac{1}{2}\frac{x-d_i^{M1}}{d_i^{L1}-d_i^{M1}}\left(d_i^{M1}\le x<d_i^{L1}\right)\\ \frac{1}{2}+\frac{1}{2}\frac{d_i^{M2}-x}{d_i^{M2}-d_i^{L2}}\left(d_i^{M2}\le x<d_i^{L2}\right)\\ \frac{1}{2}\frac{x-d_i^{S1}}{d_i^{M1}-d_i^{S1}}\left(d_i^{S1}\le x<d_i^{M1}\right)\\ \frac{1}{2}\frac{d_i^{S2}-x}{d_i^{S2}-d_i^{M2}}\left(d_i^{M2}\le x<d_i^{S2}\right)\\ 0\left(d_i^{S2}\le x, x<d_i^{S1}\right)\end{cases} \tag{1}$$

Where $u_{Di}(x)$ is a strictly decreasing function (satisfying range is from 1 to 0). The fuzzy due date corresponds to the satisfaction level for the production task completion time. In this model, as we dislike 'tardiness', so the membership function is defined as a hexagon form [4]. Their processing times are trapeziform fuzzy numbers, distributed by the following membership functions, respectively:

$$u_{Tmn}(x)=\begin{cases}1\left(p_{mn}^{L1}\le x<p_{mn}^{L2}\right)\\ \frac{x-p_{mn}^{S1}}{p_{mn}^{L1}-p_{mn}^{S1}}\left(p_{mn}^{S1}\le x<p_{mn}^{L1}\right)\\ \frac{p_{mn}^{L2}-x}{p_{mn}^{S2}-p_{mn}^{L2}}\left(p_{mn}^{L2}\le x<p_{mn}^{S2}\right)\\ 0\left(p_{mn}^{S2}\le x, x<p_{mn}^{S1}\right)\end{cases} \tag{2}$$

In order to schedule production tasks, we must synthetically consider the processing times and the due dates. We discuss the description approach of scheduling problems in HDMES mentioned above. In this paper, owing to the processing times and the due dates are all fuzzy numbers; we adopt fuzzy satisfaction level as one of optimal guideline. Fuzzy satisfaction level is defined according to the characters of fuzzy processing times and fuzzy due dates:

$$W_{FSLi}=\frac{areaD_i\cap areaP_{mn}}{areaD_i} \tag{3}$$

And $areaD_i$ is fuzzy processing time; $areaP_{mn}$ is fuzzy due date; $areaD_i \cap areaP_{mn}$ is the intersection of fuzzy processing time and fuzzy due date. W_{FSL} is fuzzy satisfaction level.

The possibility measure aggregate $\prod$ is defined to feasible scheduling aggregate, element $\sigma\in\prod$, optimizing goal function $f(\sigma^*)$, and satisfaction level S :

$$\max_{\sigma\in\prod} S=\max_{\sigma\in\prod} f\left(\sigma^*\right)=\frac{1}{m}\sum_i^m W_{FSLi} \tag{4}$$

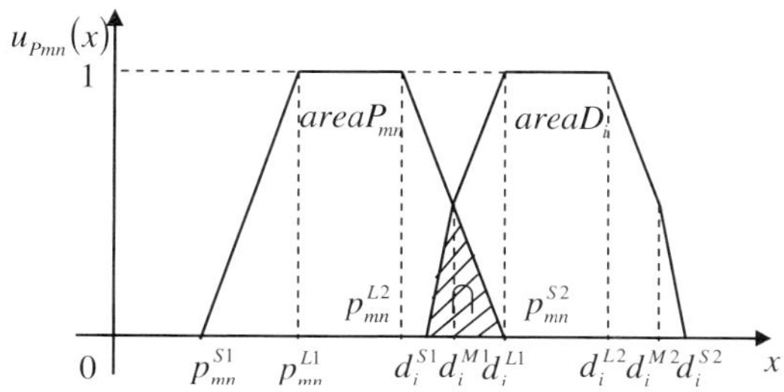

Fig. 1. Fuzzy satisfaction level

Product cost is an important economic index in reflecting the operation of iron and steel enterprises generally. In cost management, input-output ratios control is considered as an important method. There are several ways to handle input-output ratios plan problems in connection with data envelopment analysis [11]. Based on input-output ratios and fuzzy set, scheduling model appending input-output ratios is brought forward. Production tasks are denoted: $P = \{P_1, P_2, ..., P_n\}$, and each production task possibly transforms a finite number of inputs $\{I_1, I_2, ..., I_j\}$ into a finite number of outputs $\{O_1, O_2, ..., O_{j'}\}$, i.e. $P_i = (I_i, O_i)$, for $i = 1,2,..., n$. Associating with fuzzy theory, each input and output is given by a triangular fuzzy number: $I = (x; a, b, c)$ where $0 < a \le b \le c$, and $O = (y; u, v, w)$ where $0 < u \le v \le w$. Let each *input* $in = 1,2,..., j$ and *output* $out = 1,2,..., j'$ can be denoted by triangular fuzzy number [5] $I_{in} = (x; a_{in}, b_{in}, c_{in})$ where $0 < a_{in} \le b_{in} \le c_{in}$; $O_{out} = (y; u_{out}, v_{out}, w_{out})$ where $0 < u_{out} \le v_{out} \le w_{out}$. So the model appending input-output ratio is given by $R_{in-out}^{i} = \dfrac{O_{out}^{i}}{I_{in}^{i}} = \dfrac{(y; u_{out}^{i}, v_{out}^{i}, w_{out}^{i})}{(x; a_{in}^{i}, b_{in}^{i}, c_{in}^{i})}$.

3 Integrated Scheduling Algorithm

According as Earliest Due Date (EDD) rule, Shortest Processing Time (SPT) rule, First In First Out (FIFO) rule and basal sorting principle, scheduling algorithm arrange production tasks' sequences repetitiously [6]. The start working time is computed by due date and process time, start working time matrix is acquired:

$$T_{is}^{(0)} = \begin{matrix} p_{i1}^{(1)} \\ p_{i2}^{(1)} \\ \vdots \\ p_{in_1}^{(1)} \end{matrix} \overset{\begin{matrix} R_{i1} & R_{i2} & \cdots & R_{im_1} \end{matrix}}{\begin{bmatrix} t_{s11}^{(0)} & t_{s12}^{(0)} & \cdots & t_{s1m_1}^{(0)} \\ t_{s21}^{(0)} & t_{s22}^{(0)} & \cdots & t_{s2m_1}^{(0)} \\ \vdots & \vdots & & \vdots \\ t_{sn_11}^{(0)} & t_{sn_12}^{(0)} & \cdots & t_{sn_1m_1}^{(0)} \end{bmatrix}} \tag{5}$$

R_{ij} $(j = 1, 2, \cdots, m_1)$—all correlative resource m_1—resource gross

Scheduling process instance is computed according as formula (5), and Gantt chart as:

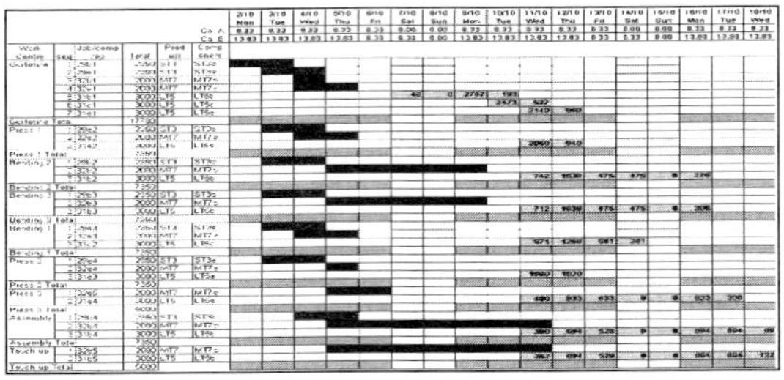

Fig. 2. Gantt chart of multi-job scheduling results

4 Conclusions

The model for planning and scheduling problems in HDMES with special restrict conditions was build up and solved by improved fuzzy algorithms. It is deployed and applied in Dongbei Special Steel Group. Cross-referring this application, it is confirmed that the system is beneficial for the enterprise to arrange the relationship of the departments and raise its production ratios and competitive abilities.

References

1. Lixin Tang, Jiyin Liu, Aiying Rong, Zihou Yang, A.: A review of planning and scheduling systems and methods for integrated steel production. European Journal of Operational Research, Vol. 133. (2000) 1-20
2. B.S.Gershwin: Hierarchical flow control, a framework for scheduling and planning Discrete Events in Manufacturing System. Proc. of the IEEWE, Vol. 77, No1, (1989) 195-209
3. C. Kao, S.-T. Liu, Fuzzy efficiency measures in data envelopment analysis, Fuzzy Sets and Systems 113 (2000) 427–437
4. K. Triantis, S. Sarangi, D. Kuchta, Fuzzy pairwise dominance and fuzzy indices: an evaluation of productive performance, Eur. J. Oper. Res. 144 (2003) 412–428
5. Brucker, P., 1998. Scheduling algorithms, 2nd ed. Springer, Heidelberg
6. Sadeh, N., D.W. Hildum, T.J. LaLiberty, J. McAnulty, D. Kjenstad and A. Tseng: A Blackboard Architecture for Integrating Process Planning and Production Scheduling. Concurrent Engineering, Research & Applications, 6(2), 1998

Fuzzy Modeling Technique with PSO Algorithm for Short-Term Load Forecasting

Sun Changyin[1,2], Ju Ping[1], and Li Linfeng[1]

[1] College of Electric Engineering, Hohai University, Nanjing 210098, P.R. China
[2] Research Institute of Automation, Southeast University, Nanjing 210096, P.R. China
cysun@hhu.edu.cn

Abstract. This paper proposes a new modeling approach for building TSK models for short-term load forecasting (STLF). The approach is a two-stage model building technique, where both premise and consequent identification are simultaneously performed. The fuzzy C-regression method (FCRM) is employed at stage-1 to identify the structure of the model. The resulting model is reduced in complexity by selection of the proper model inputs which are achieved using a Particle Swarm Optimization algorithm (PSO) based selection mechanism at stage-2. To obtain simple and efficient models we employ two descriptions for the load curves (LC's), namely, the feature description for the premise part and the cubic B-spline curve for the consequent part of the rules. The proposed model is tested using practical data, while load forecasts with satisfying accuracy are reported.

1 Introduction

Short-term load forecasting (STLF) plays an important role in power systems. Accurate short-term load forecasting has a significant influence on the operational efficiency of a power system, such as unit commitment, and interchange evaluation [1-5, 7-10].

The PSO algorithm [11] is a new evolutionary computation stochastic technique. In this paper, a selection mechanism is suggested based on PSO algorithms. This tool provides a means to selecting the past daily LC's that should be considered in the premise part of the model obtained at the previous stage. Since the selection of the most significant past inputs is of great importance in STLF, PSO helps establishing a correct mapping between the past LC's and the LC of the day to be forecasted. At this stage we obtain a reduced fuzzy model having a simple structure and small number of parameters. The simplicity and flexibility of PSO helps not only to simplify the implementation but also to combine with any kinds of estimators easily; in addition, it reduces the time cost of model selection a lot and has superior performance.

In this paper, the entire load curve (LC) of a day is considered as a unique load datum. Our intention is to create a fuzzy model mapping the LC's of past input days to the LC of the day to be predicted. The paper tackles all problems related to the structure and parameter identification of the model with TSK Fuzzy modeling and PSO.

L. Wang et al. (Eds.): FSKD 2006, LNAI 4223, pp. 933 – 936, 2006.

2 Model Identification

The model building method is a two-stage procedure, dealing simultaneously, with two important issues relevant to fuzzy modeling, namely, structure identification, input selection.

Stage-1: At this stage the structure identification problem is tackled. It comprises the following two tasks that are related to each other: (a) partitioning the input space, that is, determining the fuzzy sets of the model inputs (premise identification), and (b) calculating the parameters of the consequent regression models (consequent identification). The modeling algorithm suggested in this paper is based on the fuzzy C-regression model (FCRM) method. The FCRM is a modified version of the FCM clustering algorithm, FCRM develops clusters whose prototypes are regression models, hyper-planes etc. Hence, FCRM method suits the type of fuzzy models considered here. In our case, the cluster prototypes are LC-shaped CBS curves. The identification objective is to separate the daily load data into c fuzzy clusters and determine the LC-shaped prototypes. The resulting fuzzy partition is then assigned to the premise variables, which permits defining the premise fuzzy sets. Note that each cluster corresponds to a fuzzy rule.

Stage-2: Based on the initial fuzzy model generated at stage-1 and a candidate input set, a PSO is developed in this stage. The goal of PSO is to select a small subset comprising the most significant model inputs. At the end of this stage we obtain a reduced fuzzy model with simple structure and small number of parameters.

3 Load Forecasting Fuzzy Models

A. Framework of the fuzzy model

The load of next day is output of the model and the corresponding load influencing factors such as history load data, temperature information are the input data of the model. The training data is supplied by history database. The final target is to find an enough simple and accurate mapping function from influencing factors to future load with a good generalization.

B. Definitions and notation

The suggested method is employed to develop TSK fuzzy models for the forecasting of the next day's hourly loads of the Chinese Henan Interconnected Power System. To obtain an economical forecast model with reduced parameter complexity, we considered four day types: the Weekday (Tuesday, Wednesday, Thursday, Friday), Saturday, Sunday and Monday. So fuzzy models are generated for the forecasting of the whole week. For each day type a separate fuzzy model is generated to perform hourly based load forecasting for the time period of interest.

For the identification of a fuzzy model we employ two data sets, the training and the checking set. The training data set contains historical load and weather data from a time period of six months, starting from 1st May and ending at 31st October and is used to develop the fuzzy models. The forecasting capabilities of the obtained models are evaluated by means of the checking data set that contains load and weather data of the year 2001.

4 Test Results and Conclusions

The fuzzy models obtained by the suggested method are employed for STLF of the Chinese Henan interconnected power system. Table 1 summarizes the four day types' forecast APE and weekly forecast APE. The suggested modeling method generated fuzzy models with three or four clusters (rules). In the great majority of cases, the CBS curves are described by eight control points with the interior knots set at the time instants where the load extremals occur. The forecasting results on May 23 2001 are shown in Table 2.

Table 1. Simulation results for the fuzzy models developed by our method: Four day types' forecast APE and weekly forecast APE

Day Type	Monday	Weekday	Saturday	Sunday	Week
APE(%)	2.01	2.26	1.95	2.32	2.14

Table 2. The forecasting results on May 23 2001

Hour	Actual load /MW	forecasting load /MW	Error/%
1	5051	5125.9	1.48
2	4905	5123.0	4.44
3	4883	4899.9	0.35
4	4884	4712.1	-3.52
5	4781	4758.7	-0.47
6	4988	5012.0	0.48
7	5314	5388.0	1.39
8	5677	5802.4	2.21
9	6425	6171.0	-3.95
10	6591	6409.8	-2.75
11	6646	6434.6	-3.18
12	6581	6420.1	-2.44
13	6419	6377.2	-0.65
14	6296	6221.0	-1.19
15	6238	6047.1	-3.06
16	6105	6296.1	3.13
17	6193	6408.8	3.48
18	6982	6925.9	-0.80
19	7725	7588.1	-1.77
20	7862	7636.3	-2.87
21	7628	7305.8	-4.22
22	7187	7011.1	-2.45
23	6168	6361.5	3.14
24	5482	5266.1	-3.94
APE(%)			1.98

From the above discussion, the resulting model is reduced in complexity by discarding the unnecessary input variable and is optimized using a richer training data set. This method is used to generate fuzzy models for the forecasting of the Chinese power system. The simulation results demonstrate the effectiveness of the suggested method.

Acknowledgements

This work was supported by the National Natural Science Foundation of China under Grant 50595412, also supported by the Natural Science Foundations of Jiangsu Province under Grant BK2006564.

References

1. Papadakis S.E., Theocharis J.B. , Bakirtzis A.G., A load curve based fuzzy modeling technique for short-term load forecasting, Fuzzy Sets and Systems 135 (2003) 279–303.
2. Paris A. Mastorocostas, John B. Theocharis, Vassilios S. Petridis, A constrained orthogonal least-squares method for generating TSK fuzzy models: Application to short-term load forecasting, Fuzzy Sets and Systems 118 (2001) 215-233.
3. Paris A. Mastorocostas, John B. Theocharis, An orthogonal least-squares method for recurrent fuzzy- neural modeling, Fuzzy Sets and Systems 140 (2003) 285–300.
4. Liao G., Tsao T., Application of fuzzy neural networks and artificial intelligence for load forecasting, Electric Power Systems Research 70 (2004) 237–244.
5. Xie H., Niu D., Zhang G., Yang W. A hybrid fuzzy modeling and its application in short term load forecast.. Proceedings of the CSEE, 2005, 25(8)□17□22 .
6. Wang H., Xiao J., T-S fuzzy system based on multi-resolution analysis. Control Theory and Applications, 2005, 22(2):325-329.
7. Mara L., Lopes M., Carlos R., Anna Diva P., Electric load forecasting using a fuzzy ART & ARTMAP neural network. Applied Soft Computing, 5 (2005) 235–244.
8. Al-Kandaria A.M.,, Solimanb S.A., Fuzzy short-term electric load forecasting. Electrical Power and Energy Systems 26 (2004) 111–122.
9. Song K., Baek Y., Hong D., Jang G. Short-Term Load Forecasting for the Holidays Using Fuzzy Linear Regression Method. IEEE Transactions on Power Systems. 20, (2005), 96-101
10. Agnaldo J. Rocha Reis, Alexandre P. Alves da Silva. Feature Extraction via Multiresoltion Analysis for Short-Term Load Forecasting. IEEE Transactions on Power Systems, 20(2005), 189-198.
11. Zeng J., Jie J., Cui Z. Particle Swarm Optimization. Beijing□Scientific Publishing Company, 2004.
12. Wang L. Fuzzy System and Fuzzy Control. Beijing□TsingHua University Publishing Company, 2003.

A Fuzzy Symbolic Inference System for Postal Address Component Extraction and Labelling

P. Nagabhushan[1], S.A. Angadi[2], and B.S. Anami[3]

[1] Department of Studies in Computer Science, University of Mysore, Mysore
[2] Department of Studies in Computer Science, University of Mysore, Mysore and Basaveshwar Engineering College, Bagalkot
[3] Department of Computer Science and Engineering, BEC, Bagalkot
vinay_angadi@yahoo.com

Abstract. It is important to properly segregate the different components present in the destination postal address under different labels namely addressee name, house number, street number, extension/ area name, destination town name and the like for automatic address reading. This task is not as easy as it would appear particularly for unstructured postal addresses such as that are found in India. This paper presents a fuzzy symbolic inference system for postal mail address component extraction and labelling. The work uses a symbolic representation for postal addresses and a symbolic knowledge base for postal address component labelling. A symbolic similarity measure treated as a fuzzy membership function is devised and is used for finding the distance of the extracted component to a probable label. An alpha cut based de-fuzzification technique is employed for labelling and evaluation of confidence in the decision. The methodology is tested on 500 postal addresses and an efficiency of 94% is obtained for address component labeling.

Keywords: Postal address component labelling, Fuzzy methodology, Symbolic similarity measure, alpha cut based de-fuzzification, Inference System.

1 Introduction

Efforts to make postal mail services efficient are seen the world over. There is a spurt of activity in postal automation area in recent times. [1] enlists the computer vision tasks in postal automation. Delivery of mail to the addressee at the destination place requires sorting for onward dispatch at the origin post office and re-sorting if needed at intermediate post offices and lastly sorting for distribution. Hence mail sorting is a very important and skilled task which should be made efficient to improve the quality of mail services. It can be made efficient by devising tools for the automation of various sub tasks of mail sorting. Towards this end, tools/ techniques from different domains such as pattern recognition, image processing, graph theory, optimization, soft computing etc need to be applied.

The different aspects of postal services that need to be automated are discussed in [2]. The literature survey reveals that researchers around the world are addressing various issues required for postal automation especially contributing to mail sorting,

L. Wang et al. (Eds.): FSKD 2006, LNAI 4223, pp. 937 – 946, 2006.

but there is little effort found in simulating the human expertise required for postal mail handling, a few of them are described here. An algorithmic prototype for automatic verification and validation of postal addresses is presented in [3]. [4] proposes a methodology for truthing, testing and evaluation of postal address components. A formal method for information theoretic analysis of postal address components is given in [5]. The address component identification, required for postal automation in India and other countries, which do not have structured address formats, is not attempted. The task of address component labelling is similar to text/ word categorization. Literature is abound with general text categorization works applied to other domains [6].

In this work a fuzzy symbolic inference system for extraction and labelling of postal address components is presented. A Symbolic similarity measure is devised for identifying the address component labels using the symbolic representation of the postal address and a symbolic knowledge base. The similarity measure is a fuzzy membership function as it gives approximate nearness to various possible labels. This necessitates the disambiguation of the similarity formulation and is carried out by an inference mechanism using fuzzy alpha cut methodology. The alpha cut set is further used in defining a confidence value for the decision made. The methodology has given a labelling accuracy of 94%.

The remaining part of the paper is organized into five sections. Section 2 presents a discussion on the postal mail address component labelling problem. Section 3 gives the symbolic representation of the postal address and the symbolic knowledge base employed. Section 4 describes the fuzzy symbolic inference system for address component labelling. It elaborates the similarity formulation and alpha cut based de-fuzzification technique used for disambiguation and confidence evaluation. Section 5 gives the results and provides critical comments. Section 6 presents the conclusion.

2 Postal Mail Address Component Labelling Problem

The structure of postal addresses in developed countries like USA, UK etc is fairly standardized [7,8] as brought out by the examples in Figure 1, and this is facilitated by the structured layout of the localities. The addresses are always written using the same structure hence the line of occurrence is sufficient to identify the address component such as street name, postal code etc. The same standardization though is not found in a country like India and it is difficult to devise a standard address format for the postal addresses in India. Indian postal addresses generally give a description of the geographical location of the delivery point of the addressee, for example, Near Playground, Behind CTO, Besides City Hospital etc. A typical set of examples of UK, USA and Indian addresses are given in Figure 1.

As indicated by address-3 in Figure 1, the postal addresses in the Indian context are not very structured and the destination addresses are written using the location description. The postal addresses generally make use of well known land marks, houses of famous personalities and popular names of roads for describing the addressee and mail delivery point. All these give an unstructured nature to the postal addresses. People also use synonyms like street/ road for cross, avenue for road etc when writing destination addresses and may some times use wrong spellings. After studying a large number of postal addresses, the various components that may be present in a typical Indian postal address are found to be about twenty. Every address will not contain all the components, and some addresses may contain more than one value for the same component type. The postal addresses in general are approximate/ incomplete/ imprecise descriptions of the mail delivery points (addressee). It is required to identify these components of an address for its proper interpretation. This address component labelling task is not trivial, particularly when the addresses are unstructured and the labelling is to be based on the address information itself. This paper presents a fuzzy symbolic inference system for labelling the address components of an unstructured postal address taking Indian addresses as a case study. The methodology can be adopted in other countries having similar unstructured format.

Address 1:UK Address:	Nildram Ltd	[recipient]
	Ardenham Court	[probably the building name: Not all addresses have this part.]
	Oxford Road	[street name]
	AYLESBURY	[postal town (town/city)]
	BUCKINGHAMSHIRE	[county (not needed)]
	HP19 3EQ	[postal code]
	GREAT BRITAIN	[country name, if posted from outside country]
Address 2:USA Address:	JOHN DOE	[recipient]
	BITBOOST SYSTEMS	[Organization, required if office address]
	SUITE 5A-1204	[Suite name, if available and length on street name line is not sufficient]
	421 E DRACHMAN	[Site no. and street name with direction]
	TUCSON AZ 85705	[Place, state and zip code]
	USA	[country name, if posted from outside country]
Address 3: Indian Address:	Mr. Joseph	[recipient]
	Near Kalika Devi Temple,	[Landmark]
	Behind Govt Hospital	[Landmark]
	Kollur-01	[Place and PIN}
	Karnataka	[State]
	India	[Country name]

Fig. 1. Typical Addresses

3 Symbolic Representation

The symbolic representation of objects is an advantageous one especially for objects which have different and varying number of fields and corresponding data/ knowledge bases [9]. Section 3.1 presents the symbolic representation of postal address and section 3.2 describes the symbolic knowledge base employed in this work.

3.1 Postal Address

Some of the fields of postal addresses are qualitative, such as addressee name, care of name etc, other fields such as house number; road number, postal code (postal index number/ PIN) etc may be numeric, though their use is non numeric in nature. The values taken by most of the fields for a given address, can be distinct or one among the given range or enumerated list of values. A postal address may not contain all the possible fields. This description of the postal address makes it a suitable candidate for representation using symbolic data approach [9].

Symbolic objects offer a formal methodology to represent such variable information about an entity. Symbolic objects are extensions of classical data types. Symbolic objects can be of three different types, Assertion Object, Hoard Object and Synthetic Object. An assertion object is a conjunction of events pertaining to a given object. An event is a pair which links feature variables and feature values. A Hoard object is a collection of one or more assertion objects, whereas a synthetic object is a collection of one or more hoard objects [12]. The postal address object is described as a hoard object consisting of three assertion type objects [10] namely Addressee, Location and Place as described in (1).

$$POSTAL\ ADDRESS\ OBJECT = \{[Addressee],[\ Location],[Place]\} \tag{1}$$

The Addressee specifies the name and other personal details of the mail recipient; the Location specifies the geographical position of the mail delivery point and Place specifies the city/ town or village of the mail recipient. Each of these assertion objects is defined by a collection of events described by the feature variables. The feature variables or postal address fields of the different assertion objects are listed in (2),(3) and (4). Each of the feature describes some aspect of the object and all the features together completely specify the assertions objects. However, certain features remain missing in a typical postal address because they are not available and in some cases the written address may contain more than the required address components (typically more values for one feature, viz two or more landmarks).

$$\begin{aligned}[\text{Addressee} = {} & (\text{Addressee Name})(\text{Care of Name})(\text{Qualification})(\text{Profession})\\ & (\text{Salutation})(\text{Designation})]\end{aligned} \tag{2}$$

$$\begin{aligned}[\text{Location} = {} & (\text{House Number})(\text{House Name})(\text{Road})(\text{Area})(\text{LandMark})\\ & (\text{PBNo})(\text{Firm})]\end{aligned} \tag{3}$$

$$[\text{Place} = (\text{Post})(\text{Taluk})(\text{District})(\text{State})(\text{Place})(\text{PIN})(\text{Via})] \tag{4}$$

A typical postal address and its representation as a symbolic object is given in Table 1.

Table 1. A Typical Postal Address Object

Postal Address	Symbolic Representation
Shri Shankar S Menisinkai, Certified Engineer, "GuruKrupa", 12th Main Road Vidyagiri Bagalkot-587102 Karnataka State	PostalAddressObject={ [Addressee=(Salutation=Shri),(AddresseeName=ShankarSManisinkai),(Designation=Certified Engineer)], [Location=(HouseName=GuruKrupa),(Road=12thMainRoad),(Area=Vidyagiri)], [Place=(place=Bagalkot),(PIN=587102), (State=Karnataka)] }

3.2 Knowledge Base for Address Component Labelling

The symbolic knowledge base employed for postal address component labelling is devised based on the frame structured knowledge base presented in [11] and study of large number of postal addresses. The symbolic knowledge base used in this work provides a systematic approach for address component labelling and an improved performance as compared to the work described in [11]. The symbolic knowledge base, **AD_COMP_KB** is organized as a synthetic object of three hoard objects namely Addressee Knowledge base: **Addresskb**, Location Knowledge base: **Locationkb** and Place Knowledge base: **Placekb** as given in (5).

$$AD\_COMP\_KB = \{[Addresseekb], [Locationkb], [Placekb]\} \tag{5}$$

```
AD_COMP_KB={
  {Addressekb=
        [Salutation]
        [Addressee Name]
        [Care of Name]
        [Qualification]
        [Profession]
        [Designation]
  }
{Locationkb=
        [House No.]
        [House Name]
        [Road No.]
        [Road Name]
        [Area Name]
        [Land Mark]
        [POST BOX]
        [Firm Name]
        [PIN Code]
        [POST]
  }
{Placekb=
        [Place]
        [Taluk]
        [District]
        [VIA]
        [State]
        [Country]
  }
}
```

Fig. 2. Structure of Symbolic Address Component Knowledge Base

The hoard objects are made of assertion objects as detailed in Figure 2. All the assertion objects of the symbolic knowledge base have the events described in Figure 3. The knowledge base is populated with the values extracted by observing large number of postal addresses.

Events of Assertion Object=
{(Number of Words), (Occurring Line), (Number Present), (Inv Comma Present), (ALL CAPITALS), (keywords), (tokens)}

Fig. 3. Events Associated with Assertion Object

4 Fuzzy Symbolic Inference System

The postal address component labelling for unstructured addresses is carried out by the symbolic knowledge base supported fuzzy inference system. The postal address component inference system takes the destination postal address in text form as input, separates the probable components and labels them. The proposed system assumes that different components are on separate lines or on the same line separated by a comma. The fuzzy symbolic inference system for address component extraction and labelling is depicted in Figure 4.

The inference for address component labelling is done at the assertion object level. The labelled components (the identified assertion objects) are then grouped into postal

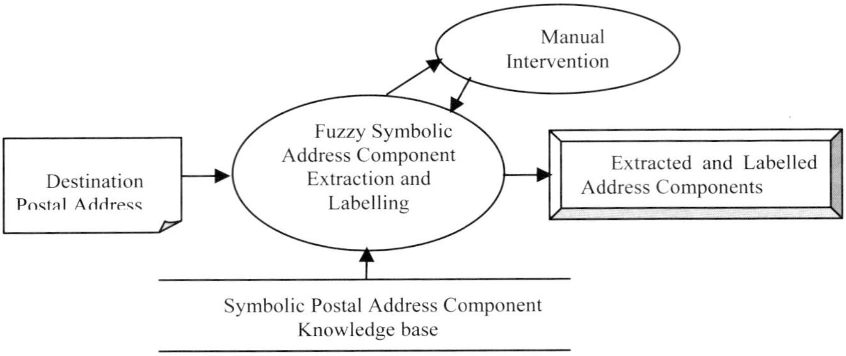

Fig. 4. The Fuzzy Symbolic Inference System for Address Component Extraction and Labelling

hoard object (the symbolic representation of the postal address). The inference mechanism uses symbolic analysis for labelling the address components using the similarity measure, defined in section 5.1 as a fuzzy membership value and fuzzy alpha cut technique for assigning confidence measure for the decision.

4.1 Symbolic Similarity Measure for Address Component Labelling

The problem of address component labelling is not easy and should be ascertained by the information specified by the component only. The presence of some key words and their occurrence relative to the other components helps in identifying the components. The symbolic data analysis for address component labelling needs distance/ similarity measures to map the input to possible candidates. [9,12] describe widely used symbolic data distance measures for similarity. Distance measures for interval type of data, absolute value/ ratio type of data etc are described. The distance/ similarity measure described in [12] is made up of three components, namely similarity due to position, similarity due to content and similarity due to span of the two objects being compared. The position similarity is defined only for interval type of data and describes the distance of one object to the initial position of other object. The span similarity is defined for both interval and absolute type of data and describes the range/fraction of similarity between the objects. The content similarity describes the nearness between the contents of the two objects. The similarity measures defined in [12] have been used for clustering, classification etc, and have been tested on fat oil and iris data. As postal object has only absolute values the span and content similarity measures defined in [12] are modified and used in this fuzzy symbolic inference system for address component labelling.

The similarity measure gives the similarity of the input component with various component labels (assertion objects) of the symbolic synthetic object AD_COMP_KB. The similarity measure between i^{th} input component (IP_i) and j^{th} component label (ct_j) of the knowledge base is found using (6).

$$S(IP_i, ct_j) = \frac{1}{EV} * \sum_{k=1}^{EV} netsim_k \text{ , } for\ 1 \leq i \leq n\ and\ 1 \leq j \leq m \qquad (6)$$

Where,

n is the number of available components in input address and *m* is the number of possible component labels or assertion objects in the knowledge base.

EV takes a value of 7, representing the seven events of the assertion objects The values of $netsim_k$ are calculated for each event of assertion object using the computations implied in (7) for the first five to calculate content similarity and (8) for the last two to calculate span and content similarity.

$$wf_k * \frac{Interse}{Sum\_IP\_KB}, \quad \text{for } 1 \le k \le 5 \tag{7}$$

$$wf_k * \left(\frac{Interse}{Sum\_IP\_KB} + \frac{Comp\_IP + Comp\_KB}{2 * Sum\_IP\_KB} \right), \text{ for } 6 \le k \le 7 \tag{8}$$

Where,

Interse is number of words/elements common to input component and component label under test

Comp_IP is the number of words/ elements in the input component

Comp_KB is the number of words/ elements in the component label (knowledge base) under test and

$$Sum\_IP\_KB = Comp\_IP + Comp\_KB - Interse$$

The weight factors wf_k are pre-defined for every component and the values are assigned based on the importance of the events in different labels. This similarity measure is the fuzzy membership function of the input component in the component label class. The actual decision of the label class is made using the de-fuzzification technique described in section 4.2.

4.2 Fuzzy Symbolic Methodology for Address Component Labelling

The methodology for address component labelling involves separating the components (in separate lines or separated by commas) and extracting the required features. These features are stored in a newly devised data structure called Postal Address Information Structure (PDIS). The structure of PDIS is given in Figure 5. Then the PDIS is used to find the similarity measure with all the component labels.

After the symbolic similarity measure is calculated for the various component labels for an input component using equation (6), the component labels are arranged in the decreasing order of similarity value in a similarity array. Now to make a decision as to which component class, the input component belongs, a de-fuzzification process is taken up. The de-fuzzification is done by defining the fuzzy α-cut set. The α value is calculated using equation (9).

$$\alpha = S_0 - DFC * S_0 \tag{9}$$

Where,

S_0 is the maximum similarity value obtained for the input component *FC* is the de-fuzzification constant and is taken as 0.1, based on the experimentation with postal address components.

The alpha cut set is obtained from the similarity array by taking into the cut set all the members of the similarity array whose value is greater than α. This is depicted pictorially in Figure 6. The α-cut set is used to identify the component label with assigned confidence value for the decision. If the α-cut set has only one member then the component label, ct_0 (corresponding to I_0 and S_0) is assigned to the input component with confidence measure of 100.

```
Postal Address Component
{ Number of words       Integer   // Stores the number of tokens in the component
  Occurring Line        Integer   // The address line where the component occurs
  Number                Boolean   // Flag, set if one of the token is a number
  Inverted Comma        Boolean   // Flag, set if one or more of tokens are in inverted comma
  All Capitals          Boolean   // Flag, set if one of the tokens has all capital characters
  Marked                Boolean   // Flag, set if one of the key words is present
  Category              String    // To store the category of key word/ address component
  Tokens                String    // To store the tokens/ address of the address component
  Confidence            String    // To store the confidence level of the identification
  Component Type        String    // To identify/ label the component
}
```

Fig. 5. Postal Address Information Structure

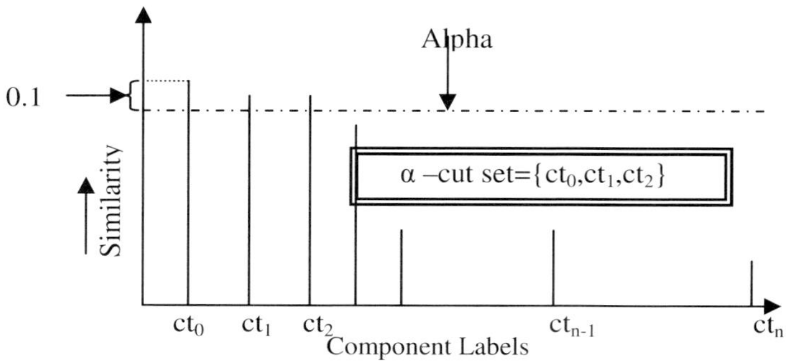

Fig. 6. The De-fuzzification Process and the α cut set

If the α- cut set has more than one component label then the probable component labels are output with the decreasing order of confidence. The confidence of the system in a given component label is evaluated using equation (10). If a particular label has a confidence of above 50% then the component is assigned the label, otherwise manual resolution is resorted to.

$$C_{i,j} = \frac{S_j}{\sum_{k=1}^{p} S_k} * 100 \qquad for\ 1 \leq j \leq p\ and\ 1 \leq i \leq n \tag{10}$$

Where,

$C_{i,j}$= Confidence of assigning j^{th} component label to i^{th} input component

n is the number of input components and p is the number of component labels in α-cut set

S_j is the similarity if i^{th} input component with j^{th} component label in similarity array.

5 Results and Discussions

The fuzzy symbolic inference system for address component labeling is tested on various types of addresses and the results are encouraging. Table 2 summarizes the output of the system for a typical input addresses and lists the highest two similarity values generated with respect to input components and the corresponding identified labels. The overall results are given in Table 3. The total efficiency of the system is about 94% and can be increased by making the symbolic knowledge base much stronger. The developed system is robust enough for use in practical situations. The system has achieved an average component wise address identification efficiency of 94.68%.

Table 2. Result of Address Component Identification

Input Address	Output Address Components					
	Component	Similarity Measure with label	Similarity Measure with label	Alpha cut set	**Assigned Label**	**Confidence of decision**
Mr. Bhosale Chandra, Near Daddennavar Hospital, Extension Area, Bagalkot , 587101	Mr	0.228, Salutation	0.1, Addressee	{Salutation}	Salutation	100
	Bhosale Chandra	0.148, addressee	0.1, Care of Name	{Addressee}	Addressee	100
	Near Daddennavar Hospital	0.228, Landmark	0.1, Care of Name	{Land Mark}	Land Mark	100
	Extension Area	0.278, Area	0.093, Landmark	{Area}	Area	100
	Bagalkot	0.228, Place	0.114, PIN	{Place}	Place	100
	587101	0.114, PIN	0.1, State	{PIN}	PIN	100
Shri, S K Deshpande, "Padmakunja", 15th Cross, Moonlight Bar, Vidyagiri, Bagalkot, 587102	Shri	0.228, Salutation	0.1, Addressee	{Salutation}	Salutation	100
	S K Deshpande	0.123, Addressee	0.1, Care of Name	{Addressee}	Addressee	100
	Padmakunja	0.186, House Name	0.1, Care of Name	{House Name}	House Name	100
	15th Cross	0.119, Road Number	0.1 Care of Name	{Road Number}	Road Number	100
	Moonlight Bar	0.93,Landmark	0.86, Postbox	{Landmark,PostBox}	Landmark	52
	Vidyagiri	0.186, Areaname	0.1,State	{Areaname}	Areaname	100
	Bagalkot	0.2, place	0.1,State	{State}	State	100
	587102	0.126, Pincode	0.107,Post	{Pincode}	Pincode	100

Table 3. Overall Results of Address Component Identification

Sl. No	Particulars	Confidence of Component Labelling			Percentage of Total addresses (=500)
		All 100%	>75% and < 100%	< 75%	
1	Correctly labeled addresses	399	70	01	94
2	Addresses with one incorrectly labeled Component	18	02	03	4.6
3	Addresses with two or more incorrectly labeled components	05	01	01	1.4

6 Conclusions

The fuzzy symbolic methodology for address component labelling presented in this paper has addressed one of the very important sub tasks of integrated postal automation, namely extracting and labelling of postal address components. These labelled address components form a symbolic address object, which can be further used in address interpretation and mapping to the mail delivery point. It employs symbolic similarity measures for address component labelling, which is treated as fuzzy membership function. The fuzzy alpha cut method is employed for de-fuzzification and deciding on the label of components with confidence value. The inference methodology suggested here is an important prior step for postal address interpretation and dynamic optimal route generation for delivery of mail.

References

1. Giovani Garibotto, 2002," *Computer Vision in Postal Automation*" Elsag Bailey-TELEROBOT,2002.
2. P.Nagabhushan, (1998), " Towards Automation in Indian Postal Services : A Loud Thinking", Technovision , Special Volume, pp 128-139
3. M.R.Premalatha and P. Nagabhushan, 2001, "*An algorithmic prototype for automatic verification and validation of PIN code: A step towards Postal Automation*", NCDAR, 13th and 14th July 2001, PESCE Mandya India,pp 225-233
4. Srirangaraj Setlur, A Lawson, Venu Govindaraju and Sargur N Srihari,, 2001," Truthing, Testing and Evaluation Issues in Complex Systems", Sixth IAPR International Conference on Document Analysis and Recognition, Seattle, WA, pp 1205-1214
5. Sargur N. Srihari, Wen-jann Yang and Venugopal Govindaraju, 1999, "Information Theortic Analysis of Postal Address Fields for Automatic Address Interpretation", ICDAR-99, Bangalore India, pp 309-312
6. Fabrizio Sebastiani, 2002, "Machine Learning in Automated Text Categorization", ACM Computing Surveys, Vol 34, No. 1, pp 1-47
7. http://www.bitboost.com/ref/international-address-formats.html
8. Universal Postal Union Address Standard, "FGDC Address Standard Version 2".
9. Bock H.-H. ,Diday E.,2000, "Analysis of symbolic Data", Heidelberg 2000
10. P.Nagabhushan,S.A.Angadi,B.S.Anami,2005, "*A Symbolic Data Structure for Postal Address Representation and Address Validation through Symbolic Knowledge Base*", Premi 05, 18-22 December 2005,Kolkata India, Springer Verlag, LNCS 3776, pp388-393
11. P.Nagabhushan,S.A.Angadi,B.S.Anami,2005, "*A Knowledge -Base Supported Inferencing of AddressComponents in Postal Mail*" NVGIP 05, 2nd and 3rd March 2005, JNNCE,Shimoga, India
12. K.Chidanada Gowda, 2004, "*Symbolic Objects and Symbolic Classification*", Proceedings of International Conference on Symbolic and Spatial Data Analysis :Mining Complex Data Structures *Pisa, September 20th, 2004,pp1-18*

A New Fuzzy MADM Method: Fuzzy RBF Neural Network Model

Hongyan Liu and Feng Kong

North China Electric Power University, Baoding 071003, P.R. China
Liuhongyan000@hotmail.com

Abstract. An RBF neural network model with fuzzy triangular numbers as inputs is set up to solve fuzzy multi-attribute decision making (MADM) problems. The model can determine the weights of attributes automatically so that weights are more objectively and accurately distributed. In this model, decision maker's specific preferences are considered in the determination of weights. It is simple, and can give objective results while taking into decision maker's subjective intensions. A numerical example is given to illustrate the method.

1 Introduction

Weight determination methods in MADM include subjective and objective ones. The former can fully reflect decision makers' subjective intensions, but at the expense of objectivtiy. The latter can yield objective results, but without fully reflecting decision makers' intensions. Since decision makers' preferences for uncertainty: risk-loving, risk-averse or risk-neutral, have an effect on decision results [1], how to combine both subjective and objective information to make results both objective and reflect decision makers' subjective intensions is of theoretical and practical importance [2, 3, 4].

Fuzzy decision making deals with decision making under fuzzy environments [5, 6, 7]. Prevalent fuzzy decision methods are too complicated since they need huge number of calculations. We put forward a fuzzy neural network model which uses triangular fuzzy numbers as inputs, and whose feature is that weights are allocated more objectively and accurately. It has a strong self-learning ability and can also reflect the influence of the decision-maker's preferences for uncertainty on decision results.

2 Neural Network Model with Fuzzy Inputs

Fuzzy RBF neural network method for fuzzy MADM involves a four-leveled network, with the input level being composed of initial uncertain signal sources, the second level being the input revision level which adjusts inputs after considering the decision-maker's specific preferences for uncertainty, the third level being hidden levels, and the fourth level being the output level.

Suppose a decision making problem has M fuzzy attributes and m crisp attributes, then there are $(M+m)$ input neurons. Further suppose there are N hidden units, in the hidden levels, see Fig. 1. We use standardized triangular fuzzy numbers $x_j = (x_{j1}, x_{j2}, x_{j3})$ as inputs of the neural network [1, 2], the output of the network is:

L. Wang et al. (Eds.): FSKD 2006, LNAI 4223, pp. 947 – 950, 2006.

$$y_k = \sum_{i=1}^{N} w_i \varphi(X_k, t_i) = \sum_{i=1}^{N} w_i \exp\left(-\frac{1}{2\sigma_i^2} \sum_{p=1}^{M+m} (t_{kp} - t_{ip}) \right) \tag{1}$$

where w_i represent the weights of the output level, $\varphi(X_k, X_i)$ represent the incentive output functions of the hidden levels, which generally are Gauss functions, and $t_i=(t_{i1}, t_{i2},\ldots, t_{i,M+m})$ is the centre of the Gauss functions, and σ_i^2 the variance.

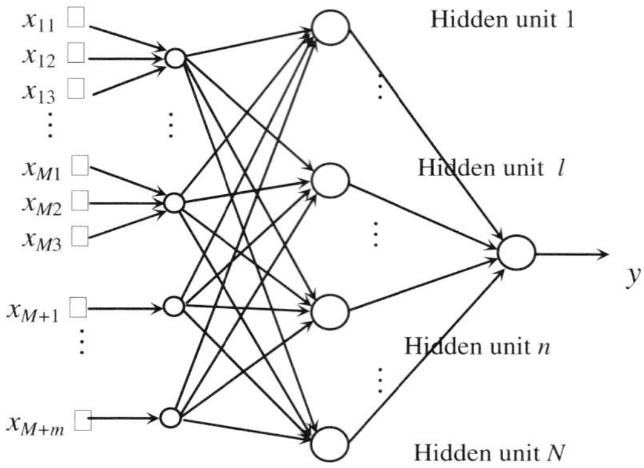

Fig. 1. Fuzzy RBF neural network

We adopt the monitored centre-selection algorithm. The specific learning steps are: Define the objective function to be: $E = \frac{1}{2}\sum_{k=1}^{N} e_k^2 = \frac{1}{2}\sum_{k=1}^{N}\left[d_k - y_k\right]^2$

where d_k represent the expected output of network samples.

The learning of the network is the solving of freedom parameters, t_i, w_i, Σ_i^{-1}, and β_j to minimize the objective function. For weights of the output level, w_i, there is,

$$w_i(n+1) = w_i(n) - \eta_1 \sum_{k=1}^{N} e_k(n)\varphi(X_k, Z_i) \tag{2}$$

$$t_i(n+1) = t_i(n) - \eta_2 2w_i(n)\sum_{k=1}^{N} e_k(n)\varphi'(X_k, t_i(n))\Sigma_i^{-1}(n)(X_k - t_i(n) \tag{3}$$

where $\Sigma_i^{-1} = -\frac{1}{2\sigma_{ii}^2}$. So, for the freedom parameters of the hidden levels, there is,

$$\Sigma_i^{-1}(n+1) = \Sigma_i^{-1}(n) + \eta_3 w_i(n)\sum_{k=1}^{N} e_k(n)\varphi'(X_k, t_i(n))Q_{ki}(n) \tag{4}$$

where $Q_{ki}(n) = (X_k - t_i(n))(X_k - t_i(n))^{\mathrm{T}}$, and for weights of the input levels, there is

$$\beta_j(n+1) = \beta_j(n) - \eta_4 - 2\sum_{k=1}^{N} e_k(n)\varphi'(X_k, t_i(n))\Sigma_i^{-1}(n)(X_k - t_i(n))S \tag{5}$$

$$S_{kj} = \int[\mu(x_{kj}^L) - \mu(0^L)]dx - \int[\mu(x_{kj}^R) - \mu(0^R)]dx$$

$$\mu(x_j^L) = \sup_{x_j^L = y+z, z \le 0} \mu(y) \quad \mu(x_j^R) = \sup_{x_j^R = y+z, z \ge 0} \mu(y)$$

where β_j represents the coefficient of the decision maker's preference for uncertainty, or the decision maker's uncertainty preference weight for the j-th attribute. and with $\eta_1, \eta_2, \eta_3, \eta_4$ being the learning rate.

3 Fuzzy RBF Neural Network MADM Method and Its Application

If there are K alternatives, with M fuzzy attributes and m crisp attributes, then, the decision matrix is: $C = \{c_{ij}\}, i = 1,\dots,K; j = 1,\dots,M+m.$ The corresponding evaluation results, or output samples, are: $D = (d_1, d_2, \dots, d_k)$. To take into account the decision-maker's specific preferences, positive and negative ideal solutions are introduced. The attribute scales of the ideal and negative ideal solutions respectively are (use benefit type scales as examples): $c_j^+ = \sup \max_i \{c_{ij}\}, c_j^- = \inf \min_i \{c_{ij}\}$. Let the expected output of the positive and negative ideal solutions be 0.95 and 0.05 respectively.

Suppose a firm has four new product development alternatives, A_1, A_2, A_3, and A_4, which are to be evaluated from eight aspects: production cost, operational cost, performance,noise,maintenance, reliability, flexibility and safety.The firm makes decisions according to the overall market performances of 15 similar products in the market,whose attribute indices and overall market performances are shown in Table1.

Table 1. Attribute scaless and overall market performances of similar products in the market

	1	2	3	4	5	6	7	8	9	10	11	12	13	14	15	IS	NIS
Production cost ($)	42	20	35	40	30	63	64	84	35	75	49	44	80	41	57	20	48
Operational cost ($)	64	52	47	50	55	65	40	60	40	41	68	35	31	45	68	35	65
Noise (db)	35	70	65	40	55	79	40	54	88	50	79	90	46	42	53	19	70
Function	VG	A	A	G	RG	G	RG	RG	RG	A	G	RG	G	VG	RG	1	0.4
Maintenance	RB	RB	RG	G	G	RG	RG	VG	G	RG	A	G	RG	RG	A	0.9	0.25
Reliability	VG	RG	G	G	G	VB	VG	A	RG	G	G	A	A	VG	RG	1	0.4
Flexibility	RB	G	G	RG	A	G	G	VG	G	RG	G	VG	RG	A	A	1	0.4
Safety	RG	A	RG	G	VG	RG	RG	RG	G	VG	VG	A	A	VG	G	1	0.4
Overall performances	0.78	0.56	0.73	0.92	0.8	0.57	0.87	0.82	0.76	0.69	0.76	0.73	0.74	0.87	0.58	0.95	0.05

(IS , NIS represent the ideal solution and Negative ideal solution respectively).

Table 2. Transformation rules for fuzzy linguistic words[8,9]

Order	Linguistic words	Corresponding triangular fuzzy numbers
1	Very good (VG)	(0.85,0.95,1.00)
2	Good (G)	(0.70,0.80,0.90)
3	Relatively good (RG)	(0.55,0.65,0.75)
4	Average (A)	(0.40,0.50,0.60)
5	Relatively bad (RB)	(0.25,0.35,0.45)
6	Bad (B)	(0.10,0.20,0.30)
7	Very bad (VB)	(0.00,0.05,0.15)

In Table 1, the former 3 are crisp attributes and the latter 5 are fuzzy attributes. For the fuzzy attributes, we can transform them into fuzzy numbers according to Table 2.

Having the network trained, input the data in Table 3 into the network, we will get the outputs (see Table 3).

Table 3. Alternative index values of the product being developed

	Production cost ($)	Operational cost ($)	Noise (db)	Function	Maintenance	Reliability	Flexibility	Safety	Overall performances
A_1	45	35	25	G	A	G	G	RG	0.83
A_2	25	50	60	RG	A	G	A	VG	0.72
A_3	35	45	50	A	G	A	VG	RG	0.75
A_4	48	65	19	RG	RB	RG	A	G	0.71

The ordering of the alternatives are: $A_1 \succ A_3 \succ A_2 \succ A_4$.

4 Conclusion

This Paper set up a RBF neural network model with fuzzy triangular numbers as inputs to solve MADM problems. The model can automatically give weights distributed objectively and accurately . It also has a great self-learning ability so that calculations are greatly reduced and simplified. Further, decision maker's specific preferences for uncertainty are considered in the determination of weights.

References

1. Li, R.-J.: Fuzzy Multi-attribute Decision Making Theory and Its applications. Beijing: Science Press (2002).
2. Song, R.: Multi-attribute Decision Making Method Based on Wavelet Neural Networks. Computer Engineering and Applications, Vol. 35 (2000) 46-48.
3. Qiu, C, Liu, Y.: Multi-attribute Decision Making Based on Artificial Neural Network. Journal of Beijing Science and Engineering University, Vol. 20 (2000) 65-68.
4. Hdgan, M. T., Demuth, H. B., Beale, M.: Neural Network Design. PWS Publishing Company (1996).
5. Song, G., Zou, P.: Weight Determination in Multi-attribute Decision Making. Systems Engineering, Vol. 19 (2001) 84-89.
6. Fuller, R., Carlsson, C.: Fuzzy Multiple Criteria Decision Making: Recent Development. Fuzzy sets and Fuzzy systems, Vol. 78 (1996) 139-153.
7. L, Jie, Hou, Z.: Weight Determination Method Based on a Combination of AHP, Delphi method and Neural Network. Systems Engineering Theories and Practices, Vol. 20 (2001) 59-63.
8. Herrera, F., Herrera E.: Linguistic decision analysis: steps for solving decision problems under linguistic information. Fuzzy Sets and Systems, Vol. 115 (2000) 67-82.
9. Cheng, C.-H., Lin, Y.: Evaluating the best main battle tank using fuzzy decision theory with linguistic criteria evaluation. European Journal of Operational Research, Vol. 142 (2002) 174-186.

A Fuzzy Contrast Model to Measure Semantic Similarity Between OWL DL Concepts

Ming Qiu, Gang Chen, and Jinxiang Dong

Institute of Artificial Intelligence
Zhejiang University, Hangzhou 310027, P.R. China
mingqiu@zju.edu.cn

Abstract. On the basis of psychological studies about similarity, we propose a model, called the fuzzy contrast model, to measure the semantic similarity between concepts expressed by OWL DL. By transforming an OWL DL concept to a set of axioms in description logic $\mathcal{SHOIN}(\mathcal{D})$, the fuzzy contrast model computes the similarity of concepts from their semantic descriptions in $\mathcal{SHOIN}(\mathcal{D})$. In order to imitate human perception of sameness and difference, fuzzy set is introduced to built intersection and set difference of feature set in our model. An iterative method is proposed to compute the similarity of concepts. Two experimental results are provided to show the effectiveness of fuzzy contrast model.

1 Introduction

Semantic similarity has become a very interesting subject in the field of Artificial Intelligence and Psychology. Many computational models of similarity have been proposed, such as distance based approaches [1, 2], information theoretic approaches [3], corpus based approaches [4] and dictionary based approaches [5]. Among them, distance based approaches and information theoretic approaches are widely accepted today. Distance based approaches count edge numbers along the shortest path between two concepts. The shorter the distance the more similar the concepts are semantically. Information theoretic approaches use hybrid approaches that utilize both information content and lexical taxonomy. Information content, obtained statistically from corpora, measures concepts' specification. While lexical taxonomy determines shared information between concepts. By combining information content and taxonomy structure, the information theoretic approaches provide a way of adapting a static knowledge structure to multiple contexts [3].

However, one common deficiency about the approaches mentioned above is that they are all developed based on intuitions about similarity, not based on theoretical foundations, nor experimental data. There are quite a lot of psychological phenomena that can't be represented by these approaches. Meantime, psychologists have been studying human perception of similarity for decades. Many theories and related experimental data can be found in psychological literatures. Artificial intelligence researchers will be benefited from psychological achievements.

In this paper, we propose a fuzzy contrast model to compute semantic similarity between concepts expressed by OWL(Web Ontology Language). The remainder of this paper is organized as follows. Section 2 is an overview of psychology theory about similarity. Section 3 analyzes the semantics of the concepts expressed by OWL DL, which

L. Wang et al. (Eds.): FSKD 2006, LNAI 4223, pp. 951–960, 2006.

is a sublanguage of OWL. Section 4 presents the fuzzy contrast model. Checking of the fuzzy contrast model based on experimental data is provided in Section 5. Conclusions are drawn in Section 6.

2 Similarity Theory

Generally, psychological models of similarity fall into two categories, geometric models and non-geometric models. Geometric models had dominated the theoretical analysis of similarity relations ever since Aristotle (384–332 BC). Such models explain similarity as a distance in some feature space, which is assumed to be a metric space. The validity of geometric models has been experimentally challenged by several researchers [7, 8, 9, 10]. Therefore, many non-geometric models have been proposed. The contrast model, which proposed by Amos Tversky in 1977, is the most famous one [11].

2.1 The Contrast Model

Assuming that functions $\Psi(A)$ and $\Psi(B)$ denote the sets of features relevant to objects A and B, and the similarity between A and B is a function of both common and different features of A and B, Tversky proposes the following equation to describe similarity:

$$sim_{tvr}(A, B) = \theta f(\Psi(A) \cap \Psi(B)) - \alpha f(\Psi(A) - \Psi(B)) - \beta f(\Psi(B) - \Psi(A)) \tag{1}$$

where $\theta, \alpha, \beta \geqslant 0$. f is a function that reflects the salience of features. Asymmetric similarity($sim_{tvr}(A, B) \neq sim_{tvr}(B, A)$) is one of the desirable properties of the contrast model. In equation 1, A is the subject of the comparison and B is the referent. Naturally one focuses on the subject of the comparison. So, the features of the subject are weighted more heavily than those of the referent($\alpha > \beta$). The similarity between toy and real train is a good example of asymmetric similarity [12].

2.2 R and E Model

Equation 1 isn't a normal form. Therefore, Rodriguez and Egenhofer extended the contrast model to a normalized equation as following [13]:

$$sim_{R\&E}(A, B) = \frac{f(\Psi(A) \cap \Psi(B))}{f(\Psi(A) \cap \Psi(B)) + \alpha f(\Psi(A) - \Psi(B)) + (1 - \alpha) f(\Psi(B) - \Psi(A))} \tag{2}$$

where α is a parameter for relative salience($0 \leqslant \alpha \leqslant 1$). The model defined in equation 2 is called R&E(Rodriguez & Egenhofer) model. Both the contrast model and R&E model can imitate human perceptual process of objects' similarity and have many applications [6, 13].

3 OWL DL Concept

OWL is the ontology language developed by W3C Web Ontology working group, and is set to become a W3C Recommendation. The concept expressed by OWL DL ,called OWL DL concept, is the object to be evaluated in the fuzzy contrast model.In this section, we will analyze the semantics of OWL DL concept.

3.1 OWL DL and Description Logic

OWL provides three increasingly expressive sublanguages, OWL Lite, OWL DL and OWL Full. Horrock and Patel-Schneider proved that an OWL DL concept can be translated into a set of axioms in $\mathcal{SHOIN}(\mathcal{D})$ [14]. So we use $\mathcal{SHOIN}(\mathcal{D})$ to analyze the semantics of OWL DL concept.

Description logic is a formalism that supports the logical description of concepts and roles. It is a subset of first-order logic [15]. Complex concepts and roles are constructed from atomic concepts and roles using a variety of concept constructors.Atomic concepts, atomic roles and concept constructors compose the Description Language.The more constructors Description Languages contain, the more expressive they are. For the sake of simplicity, only a portion of constructors in $\mathcal{SHOIN}(\mathcal{D})$ is taken into account to evaluate semantic similarity. They are listed in Table 1.

Table 1. Selected OWL DL constructors and their correspondent semantics in $\mathcal{SHOIN}(\mathcal{D})$

OWL DL	$\mathcal{SHOIN}(\mathcal{D})$	Semantics
C	C	$C^{\mathcal{I}} \sqsubseteq \Delta^{\mathcal{I}}$
ObjectProperty	R	$R^{\mathcal{I}} \sqsubseteq \Delta^{\mathcal{I}} \times \Delta^{\mathcal{I}}$
intersectionOf	$C_1 \cap \cdots \cap C_n$	$(C_1 \cap \cdots \cap C_n)^{\mathcal{I}} = C_1^{\mathcal{I}} \cap \cdots \cap C_n^{\mathcal{I}}$
unionOf	$C_1 \cup \cdots \cup C_n$	$(C_1 \cup \cdots \cup C_n)^{\mathcal{I}} = C_1^{\mathcal{I}} \cup \cdots \cup C_n^{\mathcal{I}}$
complementOf	$\neg C$	$(\neg C)^{\mathcal{I}} = \Delta^{\mathcal{I}} \backslash C^{\mathcal{I}}$
allValuesFrom	$\forall R.C$	$(\forall R.C)^{\mathcal{I}} = \{x \mid \forall y.\langle x, y\rangle \in R^{\mathcal{I}} \rightarrow y \in C^{\mathcal{I}}\}$
someValuesFrom	$\exists R.C$	$(\exists R.C)^{\mathcal{I}} = \{x \mid \exists y.\langle x, y\rangle \in R^{\mathcal{I}} \wedge y \in C^{\mathcal{I}}\}$
subclassOf	$C_1 \sqsubseteq C_2$	$C_1^{\mathcal{I}} \sqsubseteq C_2^{\mathcal{I}}$
equivalentClass	$C_1 \equiv C_2$	$C_1^{\mathcal{I}} \equiv C_2^{\mathcal{I}}$
disjointWith	$C_1 \sqsubseteq \neg C_2$	$C_1^{\mathcal{I}} \sqsubseteq \Delta^{\mathcal{I}} \backslash C_2^{\mathcal{I}}$

In Table 1, $(\cdot)^{\mathcal{I}}$ denotes a set of interpretations to define the formal semantics of concept. A Description Logic(DL) knowledge base is divided into two part, TBox and ABox. TBox is a set of terminology axioms that state facts about concepts and roles. While ABox is a set of assertional axioms that state facts about individual instance of concepts and roles. In this paper, only terminology axioms(TBox) are taken into consideration.

3.2 Semantics of DL Concepts

Suppose that $\mathcal{T}$ is a TBox. Let $\mathscr{R}_{\mathcal{T}}$ and $\mathscr{C}_{\mathcal{T}}$ be the sets of all roles and concepts defined in $\mathcal{T}$ respectively. Then, $\mathscr{C}_{\mathcal{T}}$ can be divided into three subsets, the named concepts $\mathscr{N}_{\mathcal{T}}$ that occur on the left-hand side of axioms in $\mathcal{T}$, the base concepts $\mathscr{B}_{\mathcal{T}}$ that only occur on the right-side of axioms and the composite concepts $\mathscr{X}_{\mathcal{T}}$ that aren't explicitly defined in $\mathcal{T}$ and are built from named or base concepts with constructors $\cap$,$\cup$ and $\neg$. Obviously, $\mathscr{C}_{\mathcal{T}} = \mathscr{N}_{\mathcal{T}} \cup \mathscr{B}_{\mathcal{T}}$.

It is proved that any concept description in TBox $\mathcal{T}$ can be transformed into an equivalent description that is a SSNF(Structural Subsumption Normal Form), which groups the concept description with respect to role names [15]. Therefore, we have following definition.

Definition 1. *Let A be a concept defined in $\mathcal{SHOIN}(\mathcal{D})$ TBox $\mathcal{T}$. Assume its SSNF is of the form* $D_1 \sqcap \cdots \sqcap D_m \sqcap \square R_1.C_1 \sqcap \cdots \sqcap \square R_n.C_n$, *where* $\square$ *stands for* $\leqslant n$, $\geqslant n$, $\forall$ *and* $\exists$, $R_1, \ldots, R_n$ *are distinct roles,* $D_1, \ldots, D_m$ *and* $C_1, \ldots, C_n$ *are concepts in* $\mathcal{T}$. *Then*

- **(i)** $D_i (1 \leqslant i \leqslant m)$ *is called the Explicit-Inclusion Item(ECItem) of A in* $\mathcal{T}$, $\mathscr{D}_{\mathcal{T}}(A) = \{D_1, \ldots, D_m\}$, $\mathscr{D}^*_{\mathcal{T}}(A)$ *is transitive closure of* $\mathscr{D}_{\mathcal{T}}(A)$*;*
- **(ii)** $C_j (1 \leqslant j \leqslant n)$ *is called the Role-Restricted Concept(RCConcept) of A in* $\mathcal{T}$, $\mathscr{C}_{\mathcal{T}}(A) = \{C_1, \ldots, C_n\}$*;*
- **(iii)** $H_j \equiv \square R_j.C_j (1 \leqslant j \leqslant n)$ *is called the Implicit Concept Item(ICoItem) of A in* $\mathcal{T}$, $\mathscr{H}_{\mathcal{T}}(A) = \{H_1, \ldots, H_n\}$.

By Definition 1, given a concept A defined in $\mathcal{SHOIN}(\mathcal{D})$ TBox $\mathcal{T}$, we have that if $C \in \mathscr{D}^*_{\mathcal{T}}(A)$, then $A \sqsubseteq C$. But the reverse is not always true. It is because that there could be a concept $D' \in \mathscr{N}_{\mathcal{T}}$ that satisfies $D' \notin \mathscr{D}^*_{\mathcal{T}}(A)$, but $A \sqsubseteq D'$. It is called D' implicitly includes A.

Definition 2. *Let A, B be concepts defined in $\mathcal{SHOIN}(\mathcal{D})$ TBox $\mathcal{T}$. Assume* $A \sqsubseteq B$, *if not exist concept* $C \in \mathscr{C}_{\mathcal{T}}$ *which satisfies* $A \sqsubseteq C$ *and* $C \sqsubseteq B$, *then we say B directly includes A, denoted by* $A \hat{\sqsubseteq} B$.

Definition 3. *Let D' be a concept defined in $\mathcal{SHOIN}(\mathcal{D})$ TBox $\mathcal{T}$ and for all* $D_i \in \mathscr{D}_{\mathcal{T}}(A)$, *satisfies* $D_i \not\sqsubseteq D'$. *If* $A \hat{\sqsubseteq} D'$, *then D' is called the Implicit-Inclusion Item(ICItem) of A in* $\mathcal{T}$, $D'_{\mathcal{T}}(A)$ *is the set that consist of ICItems of A.*

Definition of ICItem is more rigorous than that of implicit inclusion. Assume a concept $D' \notin \mathscr{D}^*_{\mathcal{T}}(A)$ and a concept $D_i \in \mathscr{D}^*_{\mathcal{T}}(A)$. If $D_i \sqsubseteq D'$ and $A \sqsubseteq D'$, then D' implicitly includes A. It is easy to proved that D' is an ICItem of a concept in $\mathscr{D}^*_{\mathcal{T}}(A)$, but not that of A.

By the Definition 1, we have

$$A^I \sqsubseteq \bigcap_{D_i \in \mathscr{D}_{\mathcal{T}}(A)} D_i^{\mathcal{I}} \cap \bigcap_{H_j \in \mathscr{H}_{\mathcal{T}}(A)} H_j^{\mathcal{I}} \tag{3}$$

The above equation shows that the semantics of concept A, which is defined in $\mathcal{SHOIN}(\mathcal{D})$ TBox $\mathcal{T}$, is explicitly constrained by its ECItems and ICoItems in $\mathcal{T}$. According to Definition 3, it follows that $A^I \sqsubseteq D'^I$. Thus, A is implicitly constrained by its ICItems. Therefore, the semantics of A is described by its ECItems, ICoItems and ICItems. ECItems and ICItems describe the explicit-inclusion and implicit-inclusion relations that constrain the semantics of A. While ICoItems describe the role-restricted relations to A.

3.3 Influences on Similarity Measure

The three kinds of relations mentioned above have different influences upon similarity measure. Explicit-inclusion relations are the taxonomic relations that are manually constructed by ontology engineers. They represent the asserted similarity that ontology engineers are aware of. Role-restricted relations are to constrain the definition of concept. If a pair of concepts are constrained by similar role-restricted relation, it implies that they are similar in semantics. Implicit-inclusion relations are the implicit taxonomic relations inferred from role-restricted relation. They represent the inferred similarity ontology engineers are unaware of. Taxonomic relation magnifies both the similarity of the concepts in same group and the dissimilarity of concepts in different group. Hence, explicit-inclusion and implicit-inclusion relations play an important role in similarity measure.

4 Fuzzy Contrast Model

4.1 Basic Equations of Fuzzy Contrast Model

One problem for the adoption of the contrast model in similarity measure of OWL DL concepts is its characterization of features. In the contrast model, a feature set is the set of logic predicates which are true for the stimulus in question. However, it is difficult to define a set of logic predicates for an OWL DL concept. We propose a method to obtain OWL DL concepts' feature sets from their semantics.

An OWL DL concept is first transformed into a set of axioms in $\mathcal{SHOIN}(\mathcal{D})$ TBox $\mathcal{T}$, then each axiom is normalized to a SSNF. As indicated in Section 3, ECItems, ICoItems and ICItems, which are extracted from the SSNF, represent relations that build the axiom. Therefore, they are considered as features of the concept. For inclusion relations and role-restricted relations have different influences upon similarity measure, concept's feature set is divided into two subsets, inclusion feature set and role-restricted feature set. Hence, we have Definition 4.

Definition 4. *Suppose C is a concept defined in $\mathcal{SHOIN}(\mathcal{D})$ TBox $\mathcal{T}$. Let us denote the inclusion feature set of C by $\mathscr{I}_C$, the role-restricted feature set of C by $\mathscr{R}_C$ respectively. Then $\mathscr{I}_C = \mathscr{D}_\mathcal{T}(C) \cup \mathscr{D}'_\mathcal{T}(C)$ and $\mathscr{R}_C = \mathscr{H}_\mathcal{T}(C)$*

Another problem is how to define intersection and set difference of feature sets. The sameness or difference for features is a pervasive fuzzy relation, which comes from empirical perception. Therefore, we introduce fuzzy set to define intersection and set difference of feature sets. The use of fuzzy set allow us to represent the individual judgement about sameness and difference for features.

We denote the fuzzy intersection of inclusion feature sets of concepts A and B by $\tilde{\mathscr{I}}_{A\cap B}$ and that of role-restricted feature sets of concepts A and B by $\tilde{\mathscr{R}}_{A\cap B}$. $\tilde{\mathscr{I}}_{A\cap B}$ and $\tilde{\mathscr{R}}_{A\cap B}$ are fuzzy sets in space $\mathscr{I}_A$ and $\mathscr{R}_A$ respectively. The membership function of $\tilde{\mathscr{I}}_{A\cap B}$ is defined as

$$\mu_{\tilde{\mathscr{I}}_{A\cap B}}(X) = \max\left(sim\left(X, Y\right)\right) \tag{4}$$

for all $Y \in \mathscr{I}_B$, where $sim(X, Y)$ is a function to measure semantic similarity between concepts X and Y. Assume that concepts $X \in \mathscr{H}_\mathcal{T}(A)$ and $Y \in \mathscr{H}_\mathcal{T}(B)$, with X is of the form $\square R_i.C_i$ and Y is of the form $\square R_j.C_j$. Then the membership function of $\tilde{\mathscr{R}}_{A\cap B}$ is defined as

$$\mu_{\tilde{\mathscr{R}}_{A\cap B}}(X) = \max\left(simr\left(X, Y\right)\right) \tag{5}$$

for all $Y \in \mathscr{R}_B$, where $simr(X, Y)$ is given by

$$simr(X,Y) = \begin{cases} r \times sim(C_i, C_j) & R_i = R_j \wedge (\square \text{ of X is the same as that of } Y), \\ \frac{r}{1+\alpha} \times sim(C_i, C_j) & R_i = R_j \wedge X = \exists R_i.C_i \wedge Y = \forall R_j.C_j, \\ \frac{r}{2-\alpha} \times sim(C_i, C_j) & R_i = R_j \wedge X = \forall R_i.C_i \wedge Y = \exists R_j.C_j, \\ 0 & \text{else.} \end{cases} \quad (6)$$

where r is the coefficient for Role $R_i(R_j)$, α is the parameter for relative salience($0 \leqslant r, \alpha \leqslant 1$).

The fuzzy set differences of inclusion feature sets and role-restricted feature sets of A and B are denoted by $\tilde{\mathscr{I}}_{A-B}$ and $\tilde{\mathscr{R}}_{A-B}$ respectively. They are also fuzzy sets in spaces $\mathscr{I}_A$ and $\mathscr{R}_A$. The membership function of $\tilde{\mathscr{I}}_{A-B}$ is defined as

$$\mu_{\tilde{\mathscr{I}}_{A-B}}(X) = 1 - \mu_{\tilde{\mathscr{I}}_{A\cap B}}(X) \quad (7)$$

Similarly, we define the membership function of $\tilde{\mathscr{R}}_{A-B}$ as

$$\mu_{\tilde{\mathscr{R}}_{A-B}}(X) = 1 - \mu_{\tilde{\mathscr{R}}_{A\cap B}}(X) \quad (8)$$

With these definitions, we propose a computational model of similarity based on R&E model. Let A and B be named concepts or base concepts defined in $\mathcal{SHOIN}(\mathcal{D})$ TBox $\mathcal{T}$, then the similarity function between A and B is defined as

$$\begin{aligned} sim(A,B) = \mu \times & \frac{f\left(\left(\tilde{\mathscr{I}}_{A\cap B}\right)_\lambda\right)}{f\left(\left(\tilde{\mathscr{I}}_{A\cap B}\right)_\lambda\right) + \alpha f\left(\left(\tilde{\mathscr{I}}_{A-B}\right)_\lambda\right) + (1-\alpha) f\left(\left(\tilde{\mathscr{I}}_{B-A}\right)_\lambda\right)} \\ + \nu \times & \frac{f\left(\left(\tilde{\mathscr{R}}_{A\cap B}\right)_\lambda\right)}{f\left(\left(\tilde{\mathscr{R}}_{A\cap B}\right)_\lambda\right) + \alpha f\left(\left(\tilde{\mathscr{R}}_{A-B}\right)_\lambda\right) + (1-\alpha) f\left(\left(\tilde{\mathscr{R}}_{B-A}\right)_\lambda\right)} \end{aligned} \quad (9)$$

where $(\cdot)_\lambda$ denotes the λ-cut of a fuzzy set($0 \leqslant \lambda \leqslant 1$), μ and ν are weights for similarity of inclusion relations and that of role-restricted relations respectively($\mu + \nu = 1$), f is the salience function and will be presented in a separated paper, α is parameters for relative salience($0 \leqslant \alpha \leqslant 1$).

We refer the model defined in Equation 9 as the fuzzy contrast model. As pointed out in Section 3, there are three kinds of concepts defined in TBox $\mathcal{T}$, named concepts, based concepts and composite concepts. The above equations say that to compute the similarity of a pair of concepts, we iterate all the relations that build their axioms in TBox $\mathcal{T}$, and make the similarity of these relations as the membership functions of fuzzy intersection and set difference. Then the similarity of the pair of concepts is got from the fuzzy intersection and set difference of the relations. For base concepts haven't descriptions in $\mathcal{T}$, the similarities between them are specified by ontology engineers. They are the source of similarity, and can be thought of as "propagating" to other concepts through the relations. If $\mathcal{T}$ is acyclic, the propagation of similarity will stop at the leaf concepts in ontology, which aren't used by any named concept in $\mathcal{T}$.

4.2 Similarity of Composite Concepts

For the similarity of composite concepts can't be computed from equations proposed in the previous section, we need to define a set of equations to compute it. For the sake of simplicity, we first define a lower limit to similarity of composite concepts.

Definition 5. *Let $A, B_i(1 \leqslant i \leqslant n)$ be concepts TBox $\mathcal{T}$. Then*

$$sim(A, B_1 \cup \cdots \cup B_n) \geqslant \min_{1\leqslant i\leqslant n} (sim(A, B_i)) \tag{10}$$

$$sim(A, B_1 \cap \cdots \cap B_n) \geqslant \max_{1\leqslant i\leqslant n} (sim(A, B_i)) \tag{11}$$

Next, we state the theorem on similarity of concepts, which is the basis of computing similarity of composite concepts.

Theorem 1. *Let C, D and E be concepts defined in TBox $\mathcal{T}$. If C and D satisfy $C \sqsubseteq D$, then E is no less similar to C than D, denoted by $sim(E, C) \geqslant sim(E, D)$.*

From Definition 5 and Theorem 1, it is easy to prove that Corollaries 1 and 2 hold.

Corollary 1. *Let $A, B_i(1 \leqslant i \leqslant n)$ be concepts defined in TBox $\mathcal{T}$. Then*

$$sim\,(A, B_1 \cup \cdots \cup B_n) = \min_{1\leqslant i\leqslant n} (sim\,(A, B_i)) \tag{12}$$

$$sim\,(B_1 \cup \cdots \cup B_n, A) = \min_{1\leqslant i\leqslant n} (sim\,(B_i, A)) \tag{13}$$

$$sim\,(A, B_1 \cap \cdots \cap B_n) = \max_{1\leqslant i\leqslant n} (sim\,(A, B_i)) \tag{14}$$

$$sim\,(B_1 \cap \cdots \cap B_n, A) = \max_{1\leqslant i\leqslant n} (sim\,(B_i, A)) \tag{15}$$

Corollary 2. *Let $A_i, B_j(1 \leqslant i \leqslant m, 1 \leqslant j \leqslant n)$ be concepts defined in TBox $\mathcal{T}$. Then*

$$sim\,(A_1 \cup \cdots \cup A_m, B_1 \cup \cdots \cup B_n) = \min_{\substack{1\leqslant i\leqslant m\\ 1\leqslant j\leqslant n}} \left(sim\left(A_i, B_j\right)\right) \tag{16}$$

$$sim\,(A_1 \cap \cdots \cap A_m, B_1 \cap \cdots \cap B_n) = \max_{\substack{1\leqslant i\leqslant m\\ 1\leqslant j\leqslant n}} \left(sim\left(A_i, B_j\right)\right) \tag{17}$$

$$\begin{aligned} sim\,((A_1 \cap \cdots \cap A_m, B_1 \cup \cdots \cup B_n) &= \max_{1\leqslant i\leqslant m} \left(\min_{1\leqslant j\leqslant n} \left(sim\left(A_i, B_j\right)\right)\right) \\ &= sim\,((A_1 \cup \cdots \cup A_m, B_1 \cap \cdots \cap B_n) \end{aligned} \tag{18}$$

Finally, we introduce the definition about the similarity from a concept to a complementary concept.

Definition 6. *Let A and B be concepts defined in TBox $\mathcal{T}$. Then*

$$sim(A, \neg B) = sim(\neg A, B) = 1 - sim(A, B) \tag{19}$$

By above equations, the similarity of composite concepts can be obtained from that of the concepts of which they consist.

4.3 Computing Fuzzy Contrast Model

It is natural to compute fuzzy contrast model iteratively. The initial similarity between concepts A and B is defined as

$$sim^0(A,B) = \begin{cases} 1 & A = B \\ 0 & A \neq B \end{cases} \tag{20}$$

To compute the similarity of A and B on iteration $k+1$, we define the equation as

$$\begin{aligned} sim^{k+1}(A,B) = \mu \times & \frac{f\left(\left(\tilde{\mathcal{I}}^k_{A\cap B}\right)_\lambda\right)}{f\left(\left(\tilde{\mathcal{I}}^k_{A\cap B}\right)_\lambda\right) + \alpha f\left(\left(\tilde{\mathcal{I}}^k_{A-B}\right)_\lambda\right) + (1-\alpha) f\left(\left(\tilde{\mathcal{I}}^k_{B-A}\right)_\lambda\right)} \\ + \nu \times & \frac{f\left(\left(\tilde{\mathcal{R}}^k_{A\cap B}\right)_\lambda\right)}{f\left(\left(\tilde{\mathcal{R}}^k_{A\cap B}\right)_\lambda\right) + \alpha f\left(\left(\tilde{\mathcal{R}}^k_{A-B}\right)_\lambda\right) + (1-\alpha) f\left(\left(\tilde{\mathcal{R}}^k_{B-A}\right)_\lambda\right)} \end{aligned} \tag{21}$$

Equation 21 says that on each iteration $k+1$, we update the similarity of A and B using the fuzzy set from the previous iteration k. It can be proved that If $\mathcal{T}$ is acyclic, then the similarity values will stabilize within few iterations.

5 Experimental Results

In this section, we report on two experiments. The first experiment analyzes how well fuzzy contrast model performs for finding similar concepts in an ontology. The second experiment illustrates the effects of varying the parameter λ of the model. We ran the experiments on a bearing ontology, which was built with the Protégé-OWL Plugin.

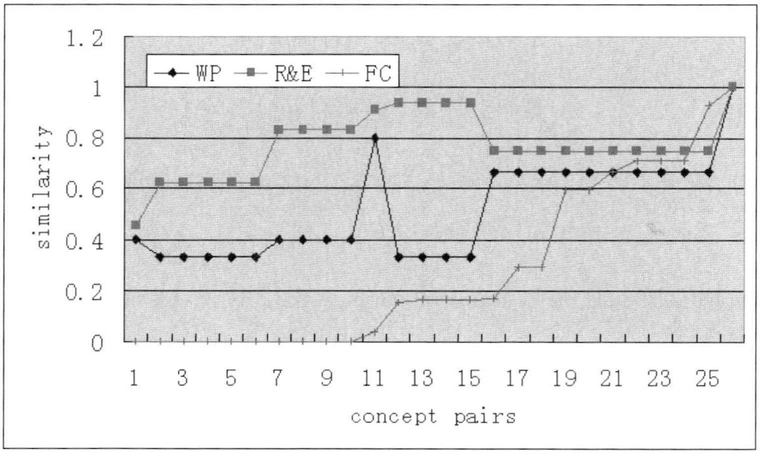

Fig. 1. Ranks of similarity obtained with Fuzzy Contrast model(FC), Wu_Palmer model(WP) and R&E model

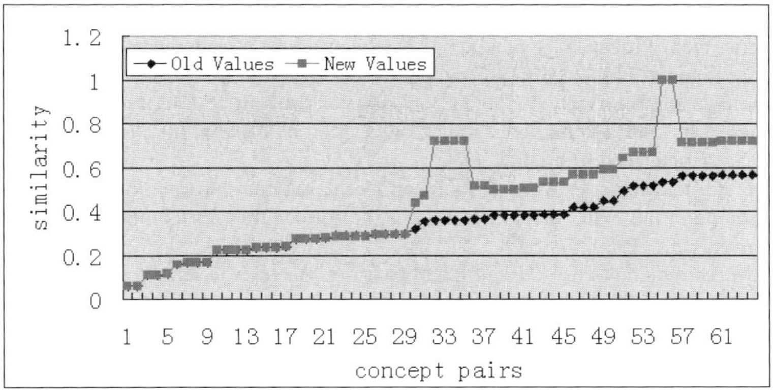

Fig. 2. Ranks of similarity obtained with fuzzy contrast model for different values of λ

RACER was used as the Description Logic Reasoner to compute implicit-inclusion relations.

The first experiment was to compare the fuzzy contrast model with Wu_Palmer model [1] and R&E model [13]. Figure 1 shows that the fuzzy contrast model is highly sensitive to the semantics of concepts. For both implicit-inclusion relations and role-restricted relations are taken into account, the fuzzy contrast model can distinguish the similarity of the concepts in the same category, which neither Wu_Palmer model nor R&E model can do.

In the second experiment, we found parameter λ has great effect on relative ranking. The results shown in Figure 2 are the similarity values computed with parameter λ = 0.6(Old Values) and λ = 0.3(New Values). When λ was decreased from 0.6 to 0.3, some concept pairs obtained higher ranks of similarity than before. It is due to the fact that some concepts were considered to be the same, after λ was changed to 0.3. This experiment shows that fuzzy contrast model can imitate individual judgement of sameness and difference by setting λ to different values.

6 Conclusions

We have presented a model, called the fuzzy contrast model, to compute similarity from the semantics of OWL DL concepts. The model is based on R&E model, which extends the contrast model. The fuzzy contrast model takes into consideration all the relations that build the axioms of the concept. Hence, it has better performance on assessment of the similarity. Another advantage of the fuzzy contrast model is that it can imitate human perception of sameness and difference by the use of fuzzy set.

References

1. Wu, Z., Palmer M.:Verb Semantics and Lexical Selection. Proceedings of 32nd Annual Meeting of Association for Computational Linguistics. Las Cruces, New Mexico (1994) 133–138

2. Song, J., Zhang, W., Xiao, W., et al.:Ontology-based Information Retrieval Model for the Semantic Web. Proceedings of 2005 IEEE International Conference on e-Technology, e-Commerce and e-Service. Hong Kong, China (2005) 152–155
3. Resnik, P.:Using Information Content to Evaluate Semantic Similarity in a Taxonomy. Proceedings of the 14th International Joint Conference on Artificial Intelligence. Morgan Kaufmann, San Francisco (1995) 448–453
4. Church, K. W., Hanks, P.:Word Association Norms, Mutual Information, and Lexicography. Proceedings of the 27th Annual Meeting of the Association for Computational Linguistics. Vancouver, B.C. (1989) 76–83
5. Lesk, M.:Automatic Sense Disambiguation using Machine Readable Dictionaries: How to Tell a Pine Cone from an Ice Cream Cone. Proceedings of the 5th Annual International Conference on Systems Documentation. Toronto,Canada, (1986) 24–26
6. Santini, S., Jain, R.:Similarity Measures. IEEE Transactions on Pattern Analysis and Machine Intelligence. **21(9)** (1999) 871–883
7. Attneave, F.:Dimensions of Similarity. American Journal of Psychology. **63** (1950) 516–556
8. Ashby, F. G., Perrin, N.A.:Toward a Unified Theory of Similarity and Recognition. Psychological Review. **95(1)** (1988) 124–150
9. Rosh, E.:Cognitive Reference Points. Cognitive Psychology. **7** (1975) 532–547
10. Rothkopf, E. Z.:A Measure of Stimulus Similarity and Errors in Some Paired-Associate Learning Tasks. Journal of Experimental Psychology. **53** (1957) 4–101
11. Tversky, A.:Features of Similarity. Psychological Review, **84(4)** (1977) 327–352
12. Tversky, A., Gati, I.:Studies of Similarity. Cognition and Categorization. Rosch,E., Lloyd, B.B. (Ed.), Lawrence Eribaum Associates, New York, (1978) 79–98
13. Rodriguez, A. M., Egenhofer, J. M.:Determining Semantic Similarity among Entity Classes from Different Ontologies. IEEE Transactions on Knowledge and Data Engineering. **15(2)** (2003) 442–456
14. Horrocks, I., Patel-Schneider, P. F.:Reducing OWL Entailment to Description Logic Satisfiability. Proceedings of the 2003 International Semantic Web Conference. Berlin, German, (2003) 17–29
15. Baader, F., McGuinness, D. L., Nardi D. et al,eds.:The Description Logic Handbook: Theory, Implementation and Applications. Cambridge University Press, Cambridge, (2002)

A Fuzzy Trust Model Using Multiple Evaluation Criteria

Keon Myung Lee[1], KyoungSoon Hwang[1], Jee-Hyong Lee[2,⋆], and Hak-Joon Kim[3]

[1] School of Electrical and Computer Engineering, Chungbuk National University, Korea
[2] School of Information and Communication Engineering SungKyunKwan University, Korea
[3] Division of Multimedia Information, Howon University, Korea
kmlee@cbnu.ac.kr

Abstract. This paper is concerned with a fuzzy trust model which takes into account both evaluations from multiple criteria and the recommendations from others in order to set the trust degrees on entities. In the proposed trust model, the entity's preference degrees on the outcomes of the interactions are expressed in fuzzy sets and the trust degrees are determined by aggregating the satisfaction degrees with respect to evaluation criteria with Sugeno's fuzzy integral. In addition, the reputation information is incorporated into the trust degree determination.

1 Introduction

Along with the widespread Internet applications such e-commerce, P2P services and so on, the users have no choice but to take some risks in doing transactions with unknown users or systems over the Internet. In everyday life, we estimate the trust degree on the others by considering the past interaction experience with them and sometimes by referring to the reputation, i.e., word of mouth. In the same token, an on-line entity could reduce the risks to run with the help of the trust information for the interacting entities. Even though there have been proposed various trust models[3-10], there are no models yet generally accepted. Some models are qualitative models[3] and others are quantitative models[4,8-10]. Some models depend only on users' ratings to compute the trust value, and others get the trust values by observing the behaviors of the entity over some period. The trust has been defined in various ways because there are no consensus on what constitutes the trust[3-10]. The following is the Gambetta's[6] which is a well-known definition of trust: *Trust (or, symmetrically, distrust) is a particular level of the subjective probability with which an agent will perform a particular action, both before [we] can monitor such action (or independently of his capacity of ever to be able to monitor it) and in a context in which it affects*

⋆ Corresponding author. This work was supported by the Regional Research Centers Program of the Ministry of Education & Human Resources Development in Korea.

L. Wang et al. (Eds.): FSKD 2006, LNAI 4223, pp. 961–969, 2006.

own action. In the trust models, the following three types of trust are usually considered: situational trust, dispositional trust, and general trust. Situational trust (a.k.a., interpersonal trust) is the trust that an entity has for another entity in a specific situation. Dispositional trust (a.k.a, basic trust) is the dispositional tendency of an entity to trust other entities. General trust is the trust of an entity in another entity regardless of situations. On the meanwhile, the reputation is valuable information for estimating the trust of an entity. The trust of an entity can be differently measured according to which aspects we evaluate. Therefore we propose a trust model to consider multiple evaluation criteria and to enable entities to reflect their preference on the outcomes.

The remainder of this paper is organized as follows: Section 2 briefly presents several related works on the trust models. Section 3 briefly explains the λ-fuzzy measure and the Sugeno's fuzzy integral which are used in the proposed model. Section 4 introduces the proposed fuzzy trust model and Section 5 shows how to apply the model with an example. Finally, Section 6 draws the conclusions.

2 Related Works

Trust and reputation have gained great attention in various fields such as economics, distributed artificial intelligence, agent technology, and so on. Various models for trust and reputation have been suggested as a result[3,4,8-10]. Some models just give theoretical guidelines and others provide computational models.

Abul-Rahman et al.[3] proposed a qualitative trust model where trust degrees are expressed in four levels such as *very trustworthy, trustworthy, untrustworthy,* and *very untrustworthy.* The model has somewhat ad-hoc nature in defining the trust degrees and the weights. Azzedin et al.[4] proposed a trust model for a peer-to-peer network computing system, which maintains a recommender network that can be used to obtain references about a target domain. The model is specialized for the well-structured network computing system and thus there are some restrictions on applying the model to general cases. Derbas et al.[8] proposed a model named TRUMMAR which is a reputation- based trust model that mobile agent systems can use in order to protect agents from malicious systems. The model pays special attention to use reputation for trust modeling, but does not consider the multiple evaluation criteria. Shi et al.[9] proposed a trust model which uses the statistical information about the possible outcome distribution for actions. In the model, trust is described as an outcome probability distribution instead of a scalar value. When choosing a candidate, it is used to compute the expected utility value for the candidate entities' actions. Wang et al.[10] proposed a trust model based on Bayesian networks for peer-to-peer networks. In the model, a Bayesian network is used to represent the trust between an agent and another agent. Such a Bayesian network represents the probabilities to trust an entity in various aspects. The recommendation values from other entities also are incorporated into the model.

3 Fuzzy Integral for Information Aggregation

The ability of the fuzzy integral to combine the evaluation results from various perspectives has been shown in several works[1,2]. In order to use fuzzy integral for information aggregation, we should have importance degrees assigned to each powerset element of evaluation criteria. These importance degrees are required to preserve the properties of fuzzy measure[1]. The widely used λ-fuzzy measure g_λ satisfies the following property along with the fuzzy measure properties[2]: For a finite set $X = \{x_1, x_2, \ldots, x_k\}$, $g_\lambda(\{x_1, \ldots, x_l\}) = \frac{1}{\lambda}\prod_{i=1}^{l}(1 + \lambda g_i) - 1]$, where $g_i = g_\lambda(\{x_i\})$.

When all g_is are given, λ can be derived from the following equation[1]:

$$g_\lambda(X) = \frac{1}{\lambda}[\prod_{i=1}^{t}(1 + \lambda g_i) - 1] = 1 \tag{1}$$

Sugeno's fuzzy integral is a Lebesque integral which has the role of aggregating partial evaluations for an entity in consideration of importance degrees of evaluation criteria[1]. Let X be a set of evaluation items and $g(E)$ the importance degree of evaluation criteria set $E \subset X$ with the properties of fuzzy measure. $g(x)$ denotes the evaluation value on the standpoint of evaluation criterion x, and A denotes the interest focus of evaluation criteria. The Sugeno's fuzzy integral $\oint_A h(x) \circ g(\cdot)$ over the set $A \subset X$ of the function h with respect to a fuzzy measure g is defined as follows:

$$\oint_A h(x) \circ g(\cdot) = \sup_{E \subset X} \{\min\{\min_{x \in E} h(x), g(A \cap E)\}\} \tag{2}$$

$$= \sup_{E \subseteq A} \{\min\{\min_{x \in E} h(x), g(E)\}\} \tag{3}$$

Due to the operation $\min_{x \in E} h(x)$, the fuzzy integral has a tendency to produce pessimistic evaluation. Some decision making problem shows that although an item has poor evaluation, the item can be compensated by other good items. Thus to provide the same effect for the fuzzy integral, we can use a compensatory operator $\psi(\{h(x)|x \in E\})$ instead of the minimum operator in the operation $\min_{x \in E} h(x)$.

4 The Proposed Fuzzy Trust Model

In the literature, there is no consensus on the definition of trust and on what constitutes trust management. In our trust model, however, we take the following definition on the situational trust: *Situational trust is the expectation for an entity to provide satisfactory outcomes with respect to the evaluation criteria in a given situation.* This section presents how to evaluate the situational trust based on the above definition, how to handle the dispositional trust and the general trust, and how to use recommandation from others and to adjust the recommanders' trust.

4.1 Situational Trust

The situational trust is the trust assigned to an entity for a specific situation. Most existing approaches have interest in how much the considered entity's behaviors are satisfactory[3-10]. It is assumed in their methods to rate the satisfaction degree in a single perspective and it is somewhat ad-hoc on how to rate the satisfaction degrees for an entity to other entities in a specific situation. In the proposed method, the situational trust is estimated as follows: First, an entity α accumulatively constructs the empirical probability distribution of possible outcomes for the interacting entities β with respect to each evaluation criterion in the given situation. Each time the entity α needs to measure the trust in another entity β, she computes the satisfaction degrees with β over each evaluation criterion in the situation. After that, the situational trust of α in β is determined by aggregating the satisfaction degrees in the perspective of evaluation criteria.

Let $TS_\alpha(\beta, \delta; EC)$ be the situational trust of entity α in entity β in the situation δ with respect to evaluation criteria $EC = \{ec_1, ec_2, ..., ec_n\}$, where ec_i is an evaluation criterion. It expresses the degree of expectation for trusted entity β to yield satisfactory actions with respect to the evaluation criteria in the given situation. In order to get the situational trust, whenever entity α has an interaction with β, α keeps the records about the evaluation outcomes with respect to the evaluation criteria. The evaluation outcomes are given in either continuous values or categorical attributes. In the case of continuous outcomes, the outcome domain is quantized into several prespecified intervals and outcome values are expressed in the corresponding interval labels.

Empirical Outcome Probability Computation. The entity α's empirical outcome probability for entity β to make outcome o_i in the situation δ up to time t with respect to an evaluation criterion ec_k is computed as follows:

$$P^t(\alpha, \beta, \delta, o_i; ec_k) = \frac{p^t(\alpha, \beta, \delta, o_i; ec_k)}{\sum_{o_j} p^t(\alpha, \beta, \delta, o_j; ec_k)} \tag{4}$$

$$p^t(\alpha, \beta, \delta, o_i; ec_k) = \rho * \frac{N^t_{\alpha\beta}(\delta, o_i; ec_k)}{n_{\alpha\beta}} + (1-\rho) * \frac{N^{[t-dt,t]}_{\alpha\beta}(\delta, o_i; ec_k)}{n^{[t-dt,t]}_{\alpha\beta}} \tag{5}$$

In the above equation, $N^t_{\alpha\beta}(\delta, o_i; ec_k)$ indicates the number of outcome o_i for β to produce with respect to ec_k up to time t, $n_{\alpha\beta}$ is the number of total interactions of α with β, $N^{[t-dt,t]}_{\alpha\beta}(\delta, o_i; ec_k)$ is the number of outcome o_i for β to produce with respect to ec_k within the recent time window $[t - dt, t]$, $n_{\alpha\beta}$ is the number of outcome o_i for β to produce with respect to ec_k within the window $[t - dt, t]$, and ρ indicates the weighting factor to control the ignorance effect on the past experience.

Satisfaction Degree Computation. The trusting entity α makes her mind on which outcomes are satisfactory for her own preference with respect to each evaluation criterion. For an evaluation criterion ec_i, suppose that its possible outcome is $PO(ec_i) = \{o_{1i}, o_{2i}, ..., o_{ni}\}$. Then, the entity α specifies earlier on her satisfactory outcome set $SO(\alpha, ec_i)$ along with the relative preference for each outcome which is expressed in a fuzzy set as follows: $SO(\alpha, ec_i) =$

$\{(o_1, ow_1), ..., (o_j, ow_j)\}$ where $o_i \in PO(ec_i)$ and $wo_k \in [0, 1]$ is the membership degree to indicate the α's relative preference to the outcome o_k. The satisfaction degree $SD_\alpha(\beta, \delta; ec_i)$ of α with β in the perspective of ec_i is determined as follows:

$$SD_\alpha(\beta, \delta; ec_i) = \sum_{(o_k, wo_k) \in SO(\alpha, ec_i)} wo_k \cdot P^t(\alpha, \beta, \delta, o_k; ec_i) \tag{6}$$

Situational Trust Computation. In the proposed method, the situational trust is measured by the satisfaction degrees of an entity α with other entity β with respect to multiple evaluation criteria EC. For example, when a user determines the trust in a restaurant, she considers her satisfaction degrees for it in the point of her own criteria such as taste of food, waiting time to take a table, availability of her favorite food, and so on. The situational trust of α in β in situation δ with respect to evaluation criteria EC is computed as follows: Here, $\psi(SD_\alpha(\beta, \delta; A))$ is the value obtained after the application of a compensatory operator to the situational trust values $SD_\alpha(\beta, \delta; ec_i)$ for $ec_i \in A$, and $WC(A)$ is the relative importance that α weighs for the evaluation criteria set A.

$$TS_\alpha(\beta, \delta; EC) = \oint_{EC} SD_\alpha(\beta, \delta; \cdot) \circ WC(\cdot) = \sup_{A \subset EC} \min\{\psi(SD_\alpha(\beta, \delta; A)), WC(A)\} \tag{7}$$

4.2 Dispositional Trust

The dispositional trust TD_α represents the dispositional tendency for a trusting entity α to trust other entities. Each entity is supposed to assign her own dispositional trust value. It could be used as the initial general trust when an entity starts an interaction with a new entity.

4.3 General Trust

The general trust $TG_\alpha(\beta)$ of entity α in entity β is the trust that α has on β regardless of situations. It plays the role of the initial situational trust for β in a new situation. It can be used as the reputation weight for β at the beginning. It can be also used as the situational trust value while enough interactions have not yet made. The general trust of α in β is obtained by averaging the situational trusts for the experienced situations Φ as follows:

$$TG_\alpha(\beta) = \frac{\sum_{\delta \in \Phi} TS_\alpha(\beta, \delta; EC)}{|\Phi|} \tag{8}$$

4.4 Reputation

When an entity decides whether it starts an interaction with another entity, it is valuable to refer to available reputation information about the entity. A reputation is an expectation about an entity's behavior which is formed by the community having interacted with the entity based on the information about or the observations of its past behaviors.

When the recommenders γ_j are available for an entity β in a situation δ, the entity α might take into account the recommendations for β. Each entity could have different preference on the outcomes with respect to the evaluation criteria and thus it is not so meaningful to directly use the trust values for β from the recommenders γ_j. Therefore we take the approach to take the recommender γ_j's empirical outcome probabilities $P^t(\gamma_j, \beta, \delta, o_k; ec_i)$ for β instead of γ_j's trust value $TS_{\gamma_j}(\beta, \delta; EC)$ on β. With the received outcome probabilities, the satisfaction degrees $SD^r_{\gamma_j}(\beta, \delta; ec_i)$ from γ_j are computed as follows: Here wo_k^α is the preference degree of the entity α for the outcome o_k.

$$SD^r_{\gamma_j}(\beta, \delta; ec_i) = \sum_{o_k \in SO(\alpha, ec_i)} wo_k^\alpha \cdot P^t(\gamma_j, \beta, \delta, o_k; ec_i) \tag{9}$$

Then the situational trust value $TS^r_{\gamma_j}(\beta, \delta; EC)$ from the recommender γ_j is computed as follows: Here $\psi(SD^r_{\gamma_j}(\beta, \delta; A))$ is the value obtained by the application of a compensatory operator to the satisfaction degrees $SD^r_{\gamma_j}(\beta, \delta, ec_k)$ for $ec_k \in A$ from γ_j, and $WC(A)$ is the relative importance that α weighs for the evaluation criteria set A.

$$TS^r_{\gamma_j}(\beta, \delta; EC) = \oint_{EC} SD^r_{\gamma_j}(\beta, \delta; \cdot) \circ WC(\cdot) = \sup_{A \subset EC} \min\{\psi(SD^r_{\gamma_j}(\beta, \delta; A)), WC(A)\} \tag{10}$$

The reputation value $TR_\alpha(\beta, \delta; EC)$ of β for α is computed by taking the weighted sum of the situational trust $TS^r_{\gamma_j}(\beta, \delta; EC)$ from recommenders γ_j as follows: Here the weighting factor wr_j is the recommendation trust value for the recommender γ_j. That is, wr_j is the degree to which α believes the recommendation from γ_j. These weights are updated through the interaction with the entities.

$$TR_\alpha(\beta, \delta; EC) = \frac{\sum_j wr_j \cdot TS^r_{\gamma_j}(\beta, \delta; EC)}{\sum_j wr_j} \tag{11}$$

4.5 Combination of Situational Trust and Reputation

When an entity starts to work in a community, she assigns her own dispositional trust value. The dispositional trust is used as the initial general trust when she interacts with an entity for the first time. Until sufficient number of interactions has made for a given situation, the general trust is used as the situational trust. Once the situational trust $TS_\alpha(\beta, \delta; EC)$ and the reputation $TR_\alpha(\beta, \delta; EC)$ are obtained, the final trust value $TS_\alpha(\beta, \delta; EC)$ is computed by their weighted aggregation as follow: Here, w is the relative weighting factor for the situational trust, $w \in [0, 1]$.

$$TS_\alpha(\beta, \delta; EC) = w \cdot TS_\alpha(\beta, \delta; EC) + (1 - w) \cdot TR_\alpha(\beta, \delta; EC) \tag{12}$$

4.6 Update of the Recommender Trust

The recommender's trust wr_i is updated according to how much their recommendation score is close to the final computed trust value $TS_\alpha(\beta, \delta; EC)$. If the recommendation score is similar to the final trust value, the recommender's recommendation trust is increased by a small amount. Otherwise, the recommender's recommendation trust is decreased by an exponential factor term. The following shows the update rule for the recommender trust wr_i.

Let $\Delta = |TS_\alpha(\beta, \delta; EC) - TS^r_{\gamma_i}(\beta, \delta; EC)|$. If $\Delta < \epsilon$, $wr_i(t+1) = \min\{wr_i(t) \cdot (1+\eta), 1\}$ where ϵ and η are small values such that $0 \le \epsilon, \eta \le 1$. Otherwise, $wr_i(t+1) = wr_i(t)(1 - e^{-\lambda\Delta})$ where λ is a small value such that $0 \le \lambda \le 1$.

5 An Application Example

In order to show how the proposed model works, the section gives an example to apply the model. Suppose that an entity P_1 has interest in the trust of a restaurant R_1 in terms of EC = {taste ts, waiting time wt, favorite food availability fa} with the help of the recommenders P_2 and P_3. Suppose that the satisfying outcome sets of P_1 for EC are given in the following fuzzy sets:

$SO(ts)$ = {(bad bd, 0), (moderate md, 0.5), (good gd, 0.7), (excellent ex, 1)}
$SO(wt)$ = {([0,15] t_1, 1), ((15,30] t_2, 0.7), ((30,50] t_3, 0.5), ((50, ∞) t_4, 0)}
$SO(fa)$ = {(available av, 1), (not available na, 0)}

Let the situation δ be going out to the restaurant R_1 on the weekends. Suppose that the empirical outcome probability distributions $P^t(P_i, R_1, \delta, o_i; ec_k)$ are given as in Table 1:

Table 1. The empirical outcome probability distributions

	o_i									
	taste ts				waiting time wt				availability fa	
	bd	md	gd	ex	t_1	t_2	t_3	t_4	av	na
$P^t(P_1, R_1, \delta, o_i; EC)$	0	0.2	0.3	0.5	0.7	0.1	0.2	0	0.6	0.4
$P^t(P_2, R_1, \delta, o_i; EC)$	0.2	0.2	0.5	0.1	0.4	0.5	0.1	0	0.5	0.5
$P^t(P_3, R_1, \delta, o_i; EC)$	0	0.1	0.2	0.7	0.3	0.4	0.2	0.1	0.7	0.3

Computation of the Trust. Then the satisfaction degree of P_1 and the satisfaction degrees from P_2 and P_3 in perspective of P_1 is obtained using Eq.(6) and Eq.(9) respectively, as in Table 2.

Table 2. The satisfaction degrees

	ec_i		
	taste ts	waiting time wt	availability fa
$SD_{P_1}(R_1, \delta; ec_i)$	0.84	0.87	0.6
$SD^r_{P_2}(R_1, \delta; ec_i)$	0.6	0.8	0.5
$SD^r_{P_3}(R_1, \delta; ec_i)$	0.91	0.68	0.7

Suppose that the relative importance degrees for the evaluation criteria are given as 0.5, 0.3, and 0.2 to ts, wt, and fa, respectively, and the importance degrees satisfy the properties of the λ-fuzzy measure. From Eq.(9), we can get a parameter λ of the λ-fuzzy measure satisfying $0.06\lambda^2 + 0.47\lambda + 0.2 = 0$. The unique root greater than -1 for this equation is $\lambda = -0.45$ which produces the following fuzzy measure on the power set of EC.

subset A of EC	$g_{-0.45}(A)$
ϕ	0
$\{ts\}$	0.5
$\{wt\}$	0.3
$\{fa\}$	0.4
$\{ts, wt\}$	0.87
$\{wt, fa\}$	0.75
$\{fa, ts\}$	0.99
$\{ts, wt, fa\}$	1

For Eq.(7), let us use the compensatory operator $\psi(A) = \tau \cdot \min_{x_i \in A}\{x_i\} + (1-\tau) \cdot \max_{x_i \in A}\{x_i\}$ where $\tau = 0.4$. Then the situational trust $TS_{P_1}(R_1, \delta; EC)$ of P_1 in R_1 is computed by Eq.(7) as follows:

$$\begin{aligned} TS_{P_1}(R_1, \delta; EC) &= \sup_{A \subset EC} \min\{\psi(SD_{P_1}(R_1, \delta; A)), WC(A)\} \\ &= \sup\{\min\{0.84, 0.5\}, \min\{0.6, 0.3\}, \min\{0.91, 0.4\}, \min\{0.744, 0.87\}, \\ &\quad \min\{0.786, 0.75\}, \min\{0.882, 0.99\}, \min\{0.84, 1\}\} = 0.882 \end{aligned}$$

The situational trusts $TS^r_{P_i}(R_1, \delta; EC)$ from the recommenders P_2 and P_3 are computed by Eq.(10) as follows:

$$TS^r_{P_2}(R_1, \delta; EC) = 0.89 \qquad TS^r_{P_3}(R_1, \delta; EC) = 0.80$$

If the recommender trust values wr_{P_i} for P_2 and P_3 are 0.8 and 0.7, respectively, then the reputation of R_1 for P_1 is computed as follows:

$$TR_{P_1}(R_1, \delta; EC) = (0.8 * 0.89 + 0.7 * 0.80)/(0.8 + 0.7) = 0.837$$

If the weighting factor w for the situation trust is 0.7, then the final trust value $TS_{P_1}(R_1, \delta; EC)$ is computed by Eq.(12) as follows:

$$TS_{P_1}(R_1, \delta; EC) = 0.7 * 0.882 + 0.3 * 0.837 = 0.867$$

Based on this trust value, the entity P_1 would decide whether to do business with the entity R_1.

6 Conclusions

The trust information for the online entities are valuable in reducing the risks to take on doing some transactions. We proposed a fuzzy trust model which has the following characteristics: The model allows to look at entity's trust in the point of multiple evaluation criteria. It maintains the empirical outcome distributions for evaluation criteria and enables the trusting entities to express their fuzzy preference on the outcomes when estimating trust in other entities. In addition, the model makes it possible for the entities to put different weights on the evaluation criteria, which are aggregated by using Sugeno's fuzzy integral. When it makes use of the recommendations from others, it takes the outcome distributions instead of their recommending trust values. Thereby, it allows to reflect the trusting entity's preference and her own weighting on the evaluation criteria in the evaluation of the recommendation.

References

[1] H.-J. Zimmermann. *Fuzzy Set Theory and its Applications.* Kluwer-Nijhoff Publishing. 364p. (1985).

[2] K. M. Lee, H. Lee-Kwang. Information Aggregating Networks based on Extended Sugeno's Fuzzy Integral. *LNCS* 1011. 56-66. (1995).

[3] A. Abdul-Rahman, S. Hailes. Supporting Trust in Virtual Communities. *Proc. of the Hawaii Int. Conf. on System Sciences.* (Jan.4-7, 2000, Maui Hawaii).

[4] F. Azzedin, M. Maheswaran. Trust Modeling for Peer-to-Peer based Computing Systems. *Proc. of the Int. Parallel and Distributed Processing Symposium.* (2003).

[5] U. Hengartner, P. Steenkiste. Implementing Access Control to People Location Information. *Proc. of SACMAT'04.* (Jun.2-4, 2004. New York).

[6] D. Gambetta. Can We Trust Trust?. In *Trust: Making and Breaking Cooperative Relations.*(Gambetta. D (ed.)). Basil Blackwell. Oxford. (1990).

[7] D. H. McKnight, N.L. Chervany. The Meanings of Trust. *Technical Report 94-04. Carlson School of Manangement, University of Minnesota.* (1996).

[8] G. Derbas, A. Kayssi, H. artial, A. Cherhab. TRUMMAR - A Trust Model for Mobile Agent Systems Based on Reputation. In *Proc. of ICPS2004.* IEEE. (2004).

[9] J. Shi, G. v. Bochmann, C. Adams. A Trust Model with Statistical Foundation. In *FAST'04.* Academic Press. (2004).

[10] Y. Wang, J. Vassileva. Bayesian Network Trust Model in Peer-to-Peer Network. In *Proc. of WI'03.* IEEE. (2003).

A Context-Aware Music Recommendation System Using Fuzzy Bayesian Networks with Utility Theory

Han-Saem Park, Ji-Oh Yoo, and Sung-Bae Cho

Department of Computer Science, Yonsei University
134 Shinchon-dong, Sudaemoon-ku, Seoul 120-749, Korea
{sammy, taiji391}@sclab.yonsei.ac.kr, sbcho@cs.yonsei.ac.kr

Abstract. As the World Wide Web becomes a large source of digital music, the music recommendation system has got a great demand. There are several music recommendation systems for both commercial and academic areas, which deal with the user preference as fixed. However, since the music preferred by a user may change depending on the contexts, the conventional systems have inherent problems. This paper proposes a context-aware music recommendation system (CA-MRS) that exploits the fuzzy system, Bayesian networks and the utility theory in order to recommend appropriate music with respect to the context. We have analyzed the recommendation process and performed a subjective test to show the usefulness of the proposed system.

Keywords: context-awareness, music recommendation system, fuzzy system, Bayesian networks, utility theory.

1 Introduction

Information recommendation has become an important research area since the first papers on collaborative filtering published in the 1990s [1]. Extensive work has been done in both industry and academia on developing new approaches on recommendation systems over the last decades [2]. Recently, the interests have been increased due to the abundance of practical applications such as recommendation system of books, CDs, and other products at Amazon.com, and movies by MovieLens.

Music recommendation is also an area where this recommendation system is required. As the World Wide Web becomes the source and distribution channels of diverse digital music, a large amount of music is accessible to people. In this situation, music recommendation gets required for each person since it becomes a difficult and time-consuming job to search and change the music whenever he wants to.

There is already a commercial product like iTunes by Apple Computer even though they have used simple rules described by the users [3]. Previously, H. Chen and A. Chen presented the music recommendation system for website, and Kuo and Shan proposed a personalized music filtering system considering user preference [4]. These studies considered the user preference fixed in their recommendation models. However, a user's preference on music changes according to the context. It varies so dynamically that the recommendation system should consider this information.

L. Wang et al. (Eds.): FSKD 2006, LNAI 4223, pp. 970–979, 2006.

This paper proposes a context-aware music recommendation system (CA-MRS) using the fuzzy Bayesian networks and utility theory. CA-MRS exploits the fuzzy system to deal with diverse source information, Bayesian networks to infer the context, and the utility theory to consider the user preference by context. In experiments, CA-MRS with the proposed method provides better recommendations than the original Bayesian networks.

2 Related Works

2.1 Music Recommendation

Generally, there are two approaches for the recommendation system: content-based and collaborative recommendations [2]. The former analyzes the content of objects that user has preferred in the past and recommends the one with relevant content. The latter recommends objects that the user group of similar preference has liked.

Cano *et al.* presented the MusicSurfer in order to provide the content-based music recommendation. They extracted descriptions related to instrumentation, rhythm and harmony from music signal using similarity metrics [5]. H. Chen and A. Chen presented the music recommendation system for website. They clustered the music data and user interests in order to provide collaborative music recommendation. Kuo and Shan proposed a personalized music filtering system which learned the user preference by mining the melody patterns from users' music access behavior [4]. These studies did not consider the user preference which changed by the context. The proposed system, CA-MRS, attempts to work out this problem by reflecting the context sensitively using fuzzy Bayesian networks with utility theory.

2.2 Context Inference Using Bayesian Networks

Dey defined context as any information that can be used to characterize the situation of an entity such as a person, place, or object that is considered relevant to the interaction between a user and an application, including the user and applications themselves [6]. Context is an important factor when one provides services such as music recommendation to the users since user preferences to a service (music in this work) could vary due to context where the user is. There have been many studies on context inference [7, 8].

Bayesian networks (BNs), which constitute a probabilistic framework for reasoning under uncertainty in recent years, have been representative models to deal with context inference [7]. Korpipaa *et al.* in VTT used naïve BNs to learn the contexts of a mobile device user [7], and Horvitz *et al.* in Microsoft research presented the notification system that sense and reason about human attention under uncertainty using BNs [8]. However, context inference using BNs has a limitation that it cannot deal with the diverse information effectively. Since BNs require the discrete input, the loss of information might happen and it cannot reflect the context appropriately. This limitation has been overcome by utilizing the fuzzy system.

3 CA-MRS Using Fuzzy Bayesian Networks and Utility Theory

Overall recommendation process in CA-MRS is as shown in Fig. 1. First, various information is obtained from sensors and internet. This information is pre-processed with the fuzzy system, where fuzzy membership vector is generated. It enters into fuzzy Bayesian network inference module, and FBN module infers context with the probability. Scoring module computes the final score of music in music database considering user preference by context, and then recommendation is conducted based on the final score. User preference can be stored by users.

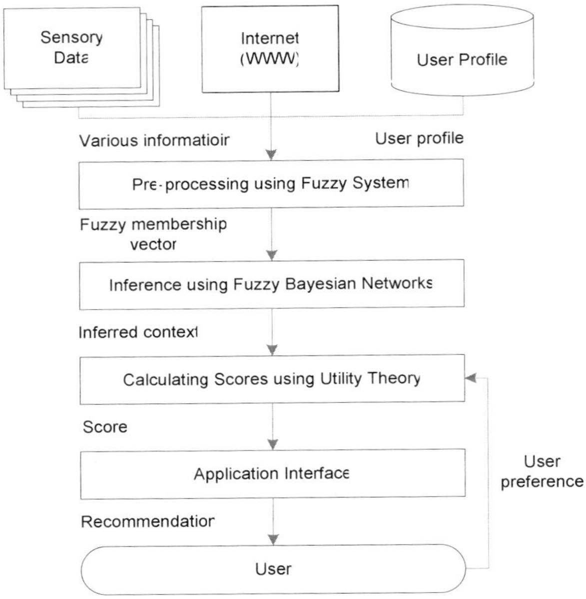

Fig. 1. The recommendation process in CA-MRS

3.1 Data Pre-processing Using Fuzzy System

Since Bayesian networks require discrete data, they generally have disadvantages though they are promising tools for reasoning context. They cannot deal with various types of information effectively because discretization can lose information compared with the original one. Context inference module can use several types of sensor information. Some are continuous, and others are discrete. Besides, it is possible for one data to be categorized into several states at the same time. Usually, a state with the largest value is selected as its state, but this method has a problem when the value is near the criteria or the value belongs to several categories. We have used the fuzzy system for pre-processing step for Bayesian network inference since the fuzzy system is relevant in dealing with diverse information and uncertainty [9].

The input data are pre-processed so that they are represented as a fuzzy membership vector. If the membership degree of $state_k$ of an observed node is μ_{state_k}, fuzzy membership vector is defined as follows.

$$FMV_{node} = (\mu_{state_1}, \mu_{state_2}, \dots, \mu_{state_n}), k = 1,2,\dots,n \tag{1}$$

Here, the membership degree of each state is calculated considering the type of data. If the data are continuous values, pre-defined membership function is used. We have used the trapezoidal function, which is simple and widely used [10].

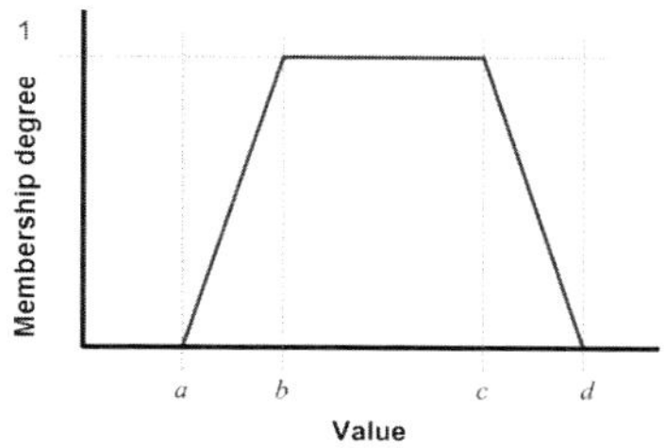

$$\mu_{state_k} = \begin{cases} 0, & x < a \\ \dfrac{x-a}{b-a}, & a \le x < b \\ 1, & b \le x < c \\ \dfrac{x-d}{c-d}, & c \le x < d \\ 0, & d \le x \end{cases}$$

Fig. 2. Trapezoidal fuzzy membership function

When the data are discrete, the membership degree of each state is calculated after normalization. If the membership degree of $state_k$ of an observed node is c_{state_k}, fuzzy membership vector is calculated as follows.

$$\mu_{state_k} = \frac{c_{state_k}}{\arg\max_{c_{state_k}}(k)}, k = 1,2,\dots,n \tag{2}$$

By pre-processing with this fuzzy system, the problems mentioned above are solved. For example, if current temperature is 27.9 degrees, we can represent this information as a fuzzy membership vector of (0, 0.65, 0.81), which means (cold, moderate, hot), using the pre-defined membership function. The data source and type information used in this paper is summarized in Table 1.

Table 1. Data source and type for context inference

Data	Source	Data Type
Temperature	Temperature sensor	Continuous
Humidity	Humidity sensor	Continuous
Noise	Noise sensor (microphone)	Continuous
Illuminance	Illuminance sensor	Continuous
Current weather	Meteorological office website	Discrete
Weather forecast	Meteorological office website	Discrete
Gender	User profile	Discrete
Age	User profile	Continuous
Season	System information	Discrete
Time	System information	Continuous

3.2 Context Inference Using Fuzzy Bayesian Networks

There have been studies to combine the fuzzy system and Bayesian networks. Yang proposed fuzzy Bayesian approach that estimates the density function from the conditional probabilities of the fuzzy-supported values in order to use continuous value in Bayesian framework. Pan and Liu proposed fuzzy Bayesian network and inference algorithm using virtual nodes and Gaussian functions [11]. However, these methods have constraints where observed information should be only one and membership degree sum should be one [11]. Our proposed model includes simple and effective fuzzy Bayesian network without those constraints.

The fuzzy Bayesian networks presented in this paper are extensions of original Bayesian networks. When fuzzy membership vectors are input, fuzzy Bayesian network inference is performed using Eq. (3).

$$\begin{aligned}\text{Fuzzy Evidence} &= FMV_{node_1} \times FMV_{node_2} \times \cdots \times FMV_{node_n} \\ &= (\mu_{s_1}, \mu_{s_2}, \ldots) \times (\mu'_{s_1}, \mu'_{s_2}, \ldots) \times \cdots \\ &= ((\mu_{s_1} \times \mu'_{s_1} \times \cdots), (\mu_{s_2} \times \mu'_{s_2} \times \cdots), \cdots)\end{aligned} \tag{3}$$

Finally, the probability of state s_k^{target} is calculated as follows:

$$P(x_{target} = s_k^{target} \mid \text{Fuzzy Evidence}) = \sum_{\forall E_r} \frac{P(x_{target} = s_k^{target} \mid E_r)\mu_{E_r}(e)}{\sum_{\forall E_r} \mu_{E_r}(e)} \tag{4}$$

where $P(x_{target} = s_k^{target} \mid E_r)$ can be calculated by general Bayesian network inference.

3.3 Application of Utility Theory

After fuzzy Bayesian networks infer the context, the final score of the music is calculated based on user preference by this context. User preference is input by the users and it is represented as a combination of attribute and state of music. Table 2 shows the attributes of music and their possible states, and Table 3 provides an example of user preference.

Table 2. Attribute and states of music

Attribute		States
Genre		Rock, Ballad, Jazz, Dance, Classic
Tempo		Fast, A Little Fast, Moderate, A Little Slow, Slow
Mood	Cheerful – Depressing	Cheerful, A Little Cheerful, Normal, A Little Depressing, Depressing
	Relaxing – Exciting	Relaxing, A Little Relaxing, Normal, A Little Exciting, Exciting
	Disturbing – Comforting	Disturbing, A Little Disturbing, Normal, A Little Comforting, Comforting

Table 3. An example of user preference

	$u^{depressing}$	$u^{content}$	$u^{exuberant}$	$u^{anxious / frantic}$
Genre::Rock	4	3	2	3
Genre::Ballad	2	5	4	1
Genre::Jazz	4	3	4	1
Genre::Dance	3	5	5	2
Genre::Classic	1	2	3	2
Tempo::Fast	4	3	3	2
Tempo::Moderate	2	4	4	2
Tempo::Slow	3	2	5	4
Mood1::Cheerful	1	5	4	1
Mood1::Normal	3	3	2	1
Mood1::Depressing	5	1	2	3
Mood2::Relaxing	2	2	4	3
Mood2::Normal	4	3	4	4
Mood2::Exciting	2	4	2	3
Mood3::Disturbing	2	1	2	3
Mood3::Normal	3	4	4	2
Mood3::Comforting	3	5	4	1

Based on the user preference, the score of each music for a certain attribute is calculated with fuzzy evidence of current context as shown in Eq. (5).

$$Score_{attribute_i} = \sum_{\forall Mood_k} P(Mood_k \mid FuzzyEvidence) \times u_{attribute_i}^{Mood_k} \tag{5}$$

Subsequently, the scores of all music in DB are calculated based on this score. When an attribute saved in DB is $attribute_i$, the recommendation score of $music_k$ is as follows.

$$RecommendationScore_{music_k} = \sum_{\forall attribute_i} Score_{attribute_i} \tag{6}$$

Using these scores, the top n music are selected for recommendation.

4 Experimental Results

We have analyzed the recommendation process and conducted the subjective test so as to show the usefulness of CA-MRS. As analyzing the recommendation process, we have compared the fuzzy Bayesian networks and original Bayesian networks, and also compared the model with the utility theory and the model without one. After that, we have confirmed that user satisfaction increased when CA-MRS was used with the subjective test.

4.1 Analyses of Recommendation Process

1) Experimental Environment

CA-MRS was implemented in Windows XP platform with MFC, and a desktop PC with Pentium IV 2.4GHz CPU was used. In total, 322 music pieces have been collected from music streaming web site [12]. Information on music is also obtained from the same web site, but tempo and mood are set by hand after listening to music.

Data for experiments were created as follows. Weather information was collected from Meteorological Office website during one week [13]. Illumination was generated considering sunrise and sunset time, and noise was generated randomly. The user was a man of 24-years old. Situations in data proceeded from Monday to Sunday with one minute interval, and the total number of situations is 9,803. Music recommendation has been also performed at the same period.

2) Experiments and Analyses

Fig. 3 shows the probability change of 'Mood' node in BN. Two graphs provide the similar tendency, but they are different when the probabilities are changed. Probability by BN shows a sudden change, but that by FBN shows a gradual change. Since the data are continuous values, they usually change gradually: The model with FBN infers more realistic context from data. Fig. 4 shows the change of recommendation score in attribute 'Genre'. The result is similar to Fig. 3. When using context inferred by FBN (See Fig. 4. (b)), the score changes gradually, but that with context inferred by original BN does not.

Fig. 5 compares the number of changed music in top 30 recommended ones at every time point. When using fuzzy Bayesian network, they changed more often and the number is smaller when they are changed. Considering gradual change of actual context, it reflects the context more nicely.

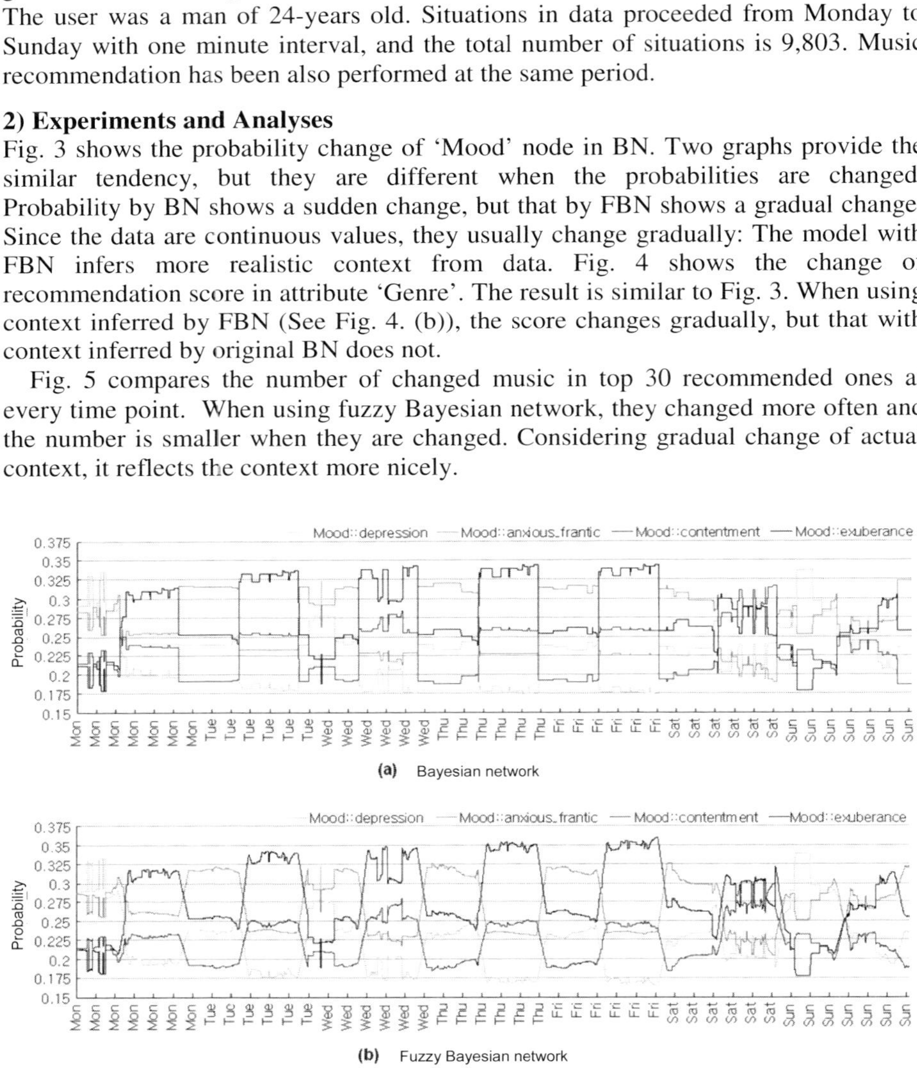

Fig. 3. Probability change comparison of 'Mood' node in Bayesian networks

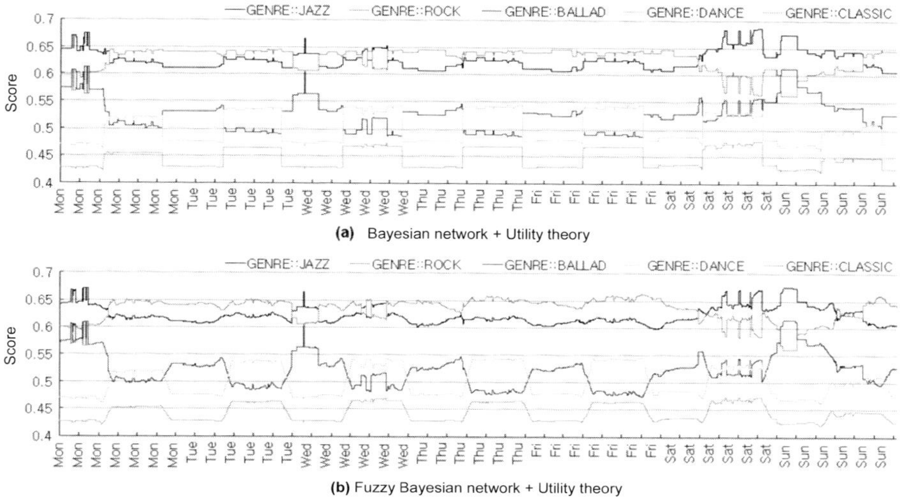

Fig. 4. Score change comparison in music attribute 'Genre'

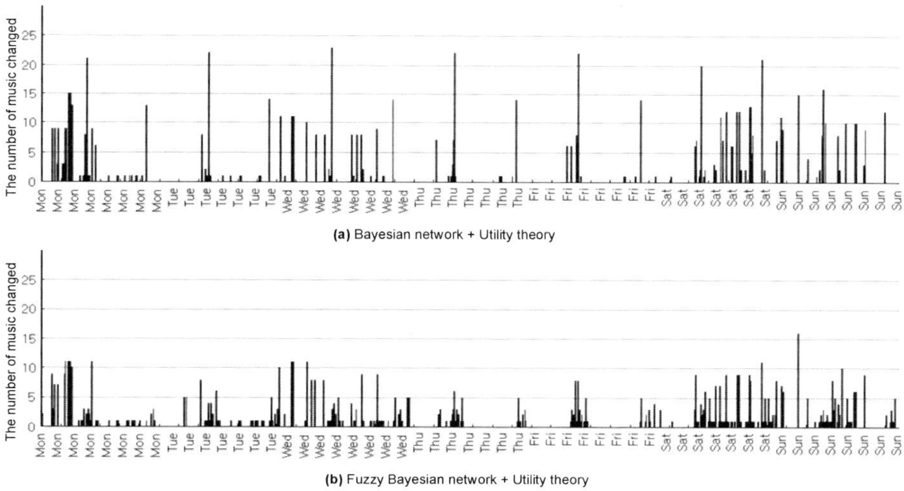

Fig. 5. Comparison of the number of changed music in top 30 recommended

4.2 Subjective Test

In order to test the satisfaction degree of the user, we have used Sheffe's paired comparison [14]. It prevents subjects from evaluating the object too subjectively by requesting them to compare the object relatively.

First, four situations are provided to subjects as shown in Table 4. We have requested answers from 10 college students who often listen to music. For each situation, subjects listen to 10 music pieces selected at random and 10 music pieces recommended by CA-MRS. After listening to all music, subjects evaluate them considering context. Score has 7 degrees from -3 (First recommended music is better) to 3 (Second recommended music is better). The order was not known to subjects.

Table 4. Situations with similar context for experiment

Situation	Context	Collected information
1	End of November, rainy Monday morning	Late fall, Monday, 9 a.m., 4.8 degree C, 82.7% humidity, 500lux, 50Db, Rainy
2	Mid-August, sunny Saturday afternoon	Summer, Saturday, 3 p.m., 30.5 degree C, 65% humidity, 550lux, 65Db, Sunny
3	Early April, cloudy Wednesday evening	Spring, Wednesday, 7 p.m., 16.6 degree C, 40% humidity, 200lux, 65Db, Cloudy
4	End of January, Sunday night with moon	Winter, Sunday, 11 p.m., -7.3 degree C, 57% humidity, 50lux, 30Db, Sunny

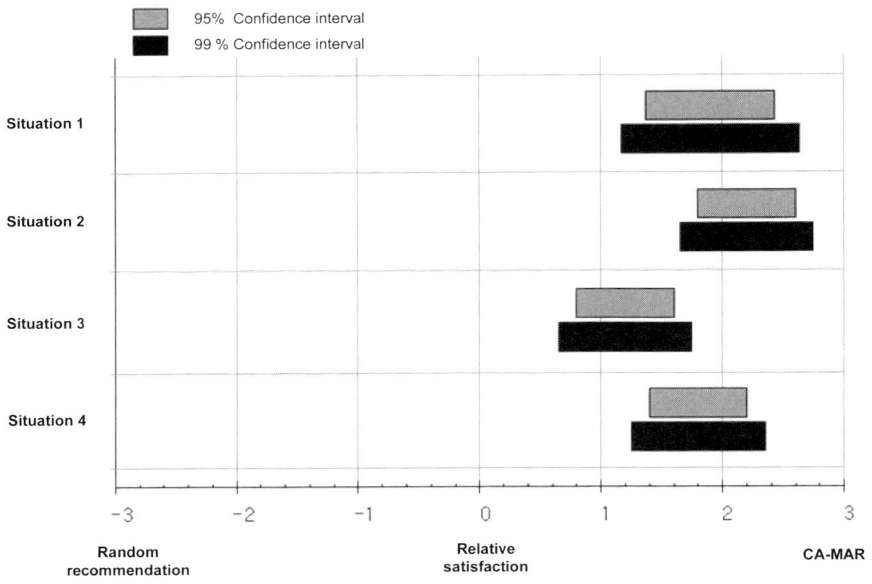

Fig. 6. Probability changes by temperature in case of using original BN

Fig. 6 shows the users satisfaction degree analyzed statistically. It means the recommended music by CA-MAR is better because the confidence intervals do not include 0, and they are closer to 3 for all situations.

5 Conclusion and Future Works

This paper proposes a context-aware music recommendation system using fuzzy Bayesian networks. CA-MRS provides music recommendation considering the sensitive change of context as combining the fuzzy system, Bayesian networks, and the utility theory. In experiments, we have confirmed the usefulness of CA-MRS by analyzing the recommendation process and performing subjective test.

As future works, it is required to determine the user preference with respect to context automatically, and it is also interesting to apply CA-MRS to mobile devices so that the recommendation service can be provided in more dynamic environment with more diverse information.

Acknowledgments. This research was supported by the Ministry of Information and Communication, Korea under the Information Technology Research Center support program supervised by the Institute of Information Technology Assessment, IITA-2005-(C1090-0501-0019).

References

1. U. Shardanand and P. Maes, "Social information filtering: Algorithms for automating 'word of mouth'," *Proc. Conf. on Human Factors in Computing Systems*, vol. 1, pp. 201-217, 1995.
2. G. Adomavicius and A. Tuzhilin, "Toward the next generation of recommender systems: A survey of the state-of-the-art and possible extensions," *IEEE T Knowl Data En*, vol. 17, no. 6, pp. 734-749, 2005.
3. http://www.apple.com/itunes/playlists/
4. F.-F. Kuo and M.-K. Shan, "A personalized music filtering system based on melody style classification," *Proc. IEEE Int. Conf. on Data Mining*, pp. 649-652, 2002.
5. P. Cano, *et al.*, "Content-based music audio recommendation," *Proc. ACM Multimedia*, pp. 212-212, 2005.
6. A. K. Dey, "Understanding and Using Context," *Personal and Ubiquitous Computing*, vol. 5, pp. 20-24, 2001.
7. P. Korpipaa, *et al.*, "Bayesian approach to sensor-based context awareness," *Personal and Ubiquitous Computing*, vol. 7, pp. 113-124, 2003.
8. E. Horvitz, *et al.*, "Models of attention in computing and communications: From principles to applications," *Commun ACM*, vol. 46, no. 3, pp. 52-59, 2003.
9. D. Dubois and H. Prade, "An introduction to fuzzy systems," *Clin Chim Acta*, vol. 270, pp. 3-29, 1998.
10. S. Lertworasirikul, *et al.*, "Fuzzy data envelopment analysis (DEA): A possibility approach," *Fuzzy Set Syst*, vol. 139, no. 2, pp. 379-394, 2003.
11. H. Pan and L. Liu, "Fuzzy Bayesian networks: A general formalism for representation, inference and learning with hybrid Bayesian networks," *Int J Pattern Recogn*, vol. 14, pp. 941-962, 2000.
12. http://www.jukeon.com/
13. http://www.weather.go.kr/
14. H. A. David, *The Method of Paired Comparison*, Charles Griffin and Co. Ltd., 1969.

Application of Fuzzy Logic in Safety Computing for a Power Protection System

Mariana Dumitrescu[1], Toader Munteanu[1], Ion Voncila[1], Gelu Gurguiatu[1], Dan Floricau[2], and Anatoli Paul Ulmeanu[2]

[1] "Dunarea de Jos" Galati University, Electrical Engineering Department, Stiintei Street 2, Galati-800146, Romania
{Mariana.Dumitrescu, Toader.Munteanu, Ion Voncila, Gelu.Gurguiatu}@ugal.ro
[2] Politechnica University Bucharest, Splaiul Independentei 133, Bucharest Romania
{danfl, paul}@pub.ro

Abstract. The paper proposes and exemplifies how to use the fuzzy logic in the critical analysis of the faults including abnormal workings, for the most important elements in Power Systems. An original fuzzy logic-system enables us to analyze the qualitative evaluation of an Electric Transformer Protection System. A fuzzy event-tree allows the use of verbal statement for probabilities and consequences, such as very high, moderate and low probability. The technique is used for quantitative results computing, as "General Safety Degree" associated to all the paths in the tree. The paper focuses on the "General Safety Degree" algorithm. The application of fuzzy logic system is further demonstrated for a case study. A complex software tool named "Fuzzy Event Tree Analysis" had to be elaborated on this purpose.

1 Introduction

Reliability information can be best expressed using fuzzy sets, because seldom it can be crispy, and the use of natural language expressions about reliability offers a powerful approach to handling the uncertainties more effectively [1], [2], [8]. Fuzzy - set logic is used to account for imprecision and uncertainty in data while employing a safety analysis. Fuzzy logic provides an intuitively appealing way of handling this uncertainty by treating the probability of failure as a fuzzy number. This allows the analyst to specify a range of values with an associated possibility distribution for the failure probabilities. If it is associated a triangular membership function with the interval, this implies that the analyst is "more confident" that the actual parameter lies near the center of the interval than at the edges [2], [8].

In a qualitative analysis event trees (ET) give the sequences of events and their probabilities of occurrence. They start with some initiate event (say a failure of some kind) and then develop the possible sequences of events into a tree. At the end of each path of events the result (safe shutdown, the damage of equipment) is obtained. The probability of each result is computed using the probabilities of the events in the sequence leading to it [2], [6], [7].

L. Wang et al. (Eds.): FSKD 2006, LNAI 4223, pp. 980 – 989, 2006.

The fuzzy probability of an event can be put into subcategories based on a range of probability, high-if probability is greater than 0,6 but less than 1,0; very low-if probability is greater than 0 but less than 0,2; etc. The fuzzy event-tree (FET) allows the use of verbal statement for the probabilities and consequences, such as very high, moderate and low probability. The occurrence probability of a path in the event tree is than calculated as the product of the event probabilities in the path [6], [7].

A first direction in event-tree analysis [7] uses fuzzy-set logic to account for imprecision and uncertainty in data while employing this analysis. The fuzzy event-tree allows: uncertainty in the probability of failure and verbal statements for the probabilities/consequences (such as low, moderate and high) concerning the impact of certain sequences of events (such as normal, alert and abnormal). In this case, for a simple 2-component parallel system, let $\tilde{P}_A$ ("low") and $\tilde{P}_B$ ("high") be the component fuzzy failure probability. Then $\tilde{P}_{sys}$ will be the fuzzy probability $\tilde{P}_A \times \tilde{P}_B$ illustrated in Fig.1a. [2].

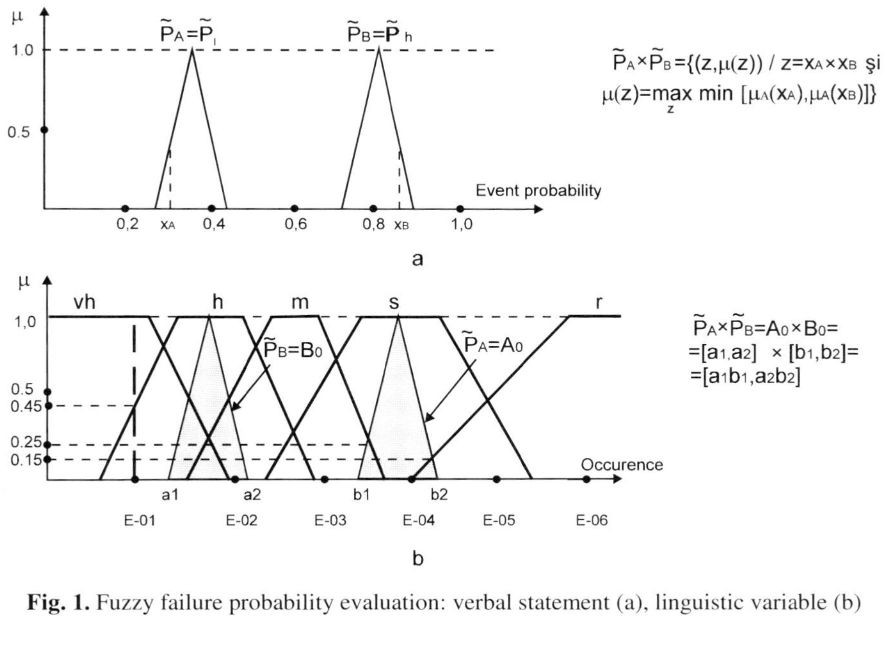

Fig. 1. Fuzzy failure probability evaluation: verbal statement (a), linguistic variable (b)

A second direction in the event- tree analysis [2] uses a linguistic variable to evaluate the fuzzy failure probability. The linguistic values, being assigned to the variable, called "term sets", are generally defined as fuzzy sets that act as restrictions on the base values that them represents. Each of these fuzzy sets means a possibility distribution over the domain of the base variable. For example (Fig. 1.b) "probability of failure" might be a linguistic variable whose values are from the set "remote" ("r"), "low"("l"),"high" ("h") and very high ("vh"). Fuzzy failure probability of the simple 2-component parallel system $\tilde{P}_{sys}$ will be the fuzzy probability $\tilde{P}_A \times \tilde{P}_B$ illustrated in Fig.1.b. The paper uses the linguistic variable in fuzzy failure probability evaluation, because this concept is combined with the possibility theory which becomes an especially powerful tool for working with uncertainly.

Generally fuzzy logic (FL) provides a more flexible and meaningful way of describe the risk. One common procedure for evaluating risk (called the RPN or "Risk priority Number" method) is to provide a numerical ranking for each path of the event-tree. Higher numbers imply a greater risk.

We propose an original fuzzy logic-system (FLS), which enables us to analyze the qualitative evaluation of the event -tree. The technique allows us to develop a fuzzy-event algorithm and to gain quantitative results, as the fuzzy number "General Safety" (GSF) and the crisp value "General Safety Degree" (GSD) associates to all the paths in the tree. It is necessary to say that the calculation of the failure probability is not enough. Another important consideration is the severity of the effect of the failure. The risk, associated with failure, increases as either the severity of the effect of the failure or the failure probability increases. Including the "Severity" in the FLS, ranked according to the seriousness of the failure effect, allow modelling this judgment, by its very nature, highly subjective [3], [4], [5].

The paper focuses on the proposed FLS and its application. The application of fuzzy event-tree algorithm is further demonstrated by using a power protection system, to assess the viability of the method. We needed to elaborate an efficient software tool "Fuzzy Event Tree Analysis" (FETA) to help us in the independent qualitative analysis. Section 2 introduces the fuzzy event tree method. Section 3 gives information about the proposed FLS. Section 4 shows how to apply the algorithm to Electric Transformer Protection System (ETPS). Section 5 presents the conclusions of the paper.

2 Fuzzy Fault-Tree Analysis for a Power Protection System

Event-trees examine sequences of events and their probability of occurrence. They start with some initiating event (for example a failure) and then develop the possible sequences of events into a tree. For example; is the failure detected?, does a safety relay activate?. The result at the end of each chain of events is then evaluated and the probability of each result computed using the probabilities of the events in sequence leading to it.

The procedure for analyzing a fault tree is precise, but the probabilities on which the methodology is based, are not [1], [2], [9]. We illustrate the technique of using fuzzy probabilities in an event- tree analysis, for a simple case of a power protection system.

Usually, the fuzzy event-tree analysis has the following steps:

1. fuzzy "failure" probability and fuzzy "safe" probability evaluation, for all reliability block diagram elements;
2. fuzzy "occurrence" probability evaluation for each path (sequence of events) of the tree;
3. fuzzy "consequence" on power system evaluation, after the events sequence achievement;

4. fuzzy “risk” on power system for each path of the tree evaluation, depending on the path “occurrence” and path “consequence”;
5. the tree-paths ranking, depending on the path “risk”.

2.1 Example

The power system presented in Fig. 2a has two electric transformers protected by a differential schema. Both circuit breakers, protecting each of the transformers, are operated by the same fault detector (FD), a combined relay (R) and trip signal (TS) Fig. 2b. Supposing that a fault occurs on the transformer T1, it is desirable to evaluate the probability of successful operation of the protection system. Generally power protection system involves the sequential operation of a set of components and devices. Event- tree are useful because they recognize the sequential operational logic of a system. Fig. 3a shows the fuzzy event-tree for the network presented in Fig.2.

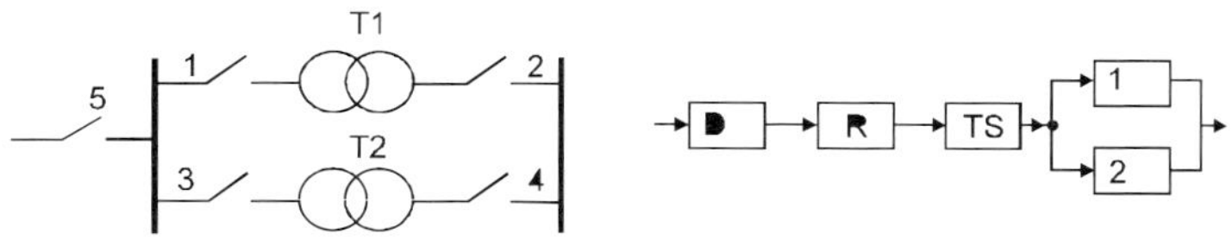

a b

Fig. 2. An electric power- system network (a). Reliability block diagram of differential transformer T1 protection system.

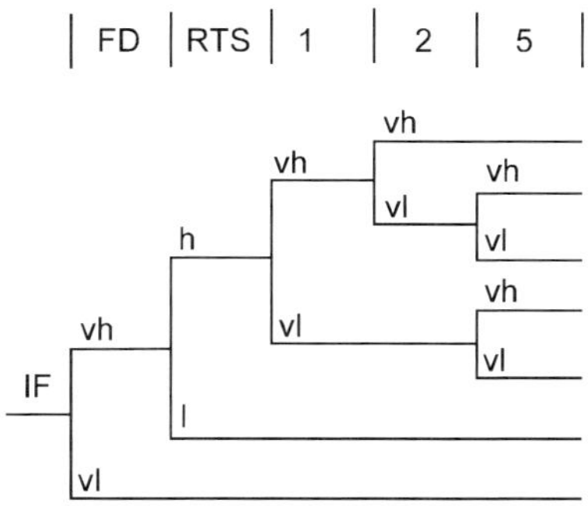

	Path- fuzzy probability	Path- fuzzy consequence
1	$\tilde{P}_1 = \tilde{P}_{vh} \cdot \tilde{P}_h \cdot \tilde{P}_{vh} \cdot \tilde{P}_{vh}$	$\tilde{C}_1 = \tilde{C}_l$
2	$\tilde{P}_2 = \tilde{P}_{vh} \cdot \tilde{P}_h \cdot \tilde{P}_{vh} \cdot \tilde{P}_{vl} \cdot \tilde{P}_{vh}$	$\tilde{C}_2 = \tilde{C}_m$
3	$\tilde{P}_3 = \tilde{P}_{vh} \cdot \tilde{P}_h \cdot \tilde{P}_{vh} \cdot \tilde{P}_{vh} \cdot \tilde{P}_{vl}$	$\tilde{C}_3 = \tilde{C}_h$
4	$\tilde{P}_4 = \tilde{P}_{vh} \cdot \tilde{P}_h \cdot \tilde{P}_{vl} \cdot \tilde{P}_{vh}$	$\tilde{C}_4 = \tilde{C}_m$
5	$\tilde{P}_5 = \tilde{P}_{vh} \cdot \tilde{P}_h \cdot \tilde{P}_{vl} \cdot \tilde{P}_{vl}$	$\tilde{C}_5 = \tilde{C}_h$
6	$\tilde{P}_6 = \tilde{P}_{vh} \cdot \tilde{P}_h$	$\tilde{C}_6 = \tilde{C}_h$
7	$\tilde{P}_1 = \tilde{P}_{vl}$	$\tilde{C}_7 = \tilde{C}_h$

a b

Fig. 3. Fuzzy event- tree for differential transformer T1 protection system

Starting with an initiating fault IF, a very low failure probability is considered for it, because the transformers are rigid systems with no moving parts. The fault detecting system consists of current transformers, and high impedance relays (which from experience are reliable) excepting the case of faults with high currents. The relay/ trip signal device, consists of relay, trip coil, pneumatic systems, coming along with many moving parts, whose high probability is assumed to have a successful operation. Finally, since high technologies have been used in design and manufacturing of the circuit breakers (CB) their successful operation probability is considered to be very high.

As can be seen from Fig., seven consequences are considered for this schema.

- Consequence 1 is rated low, due to the fact that all the components related to the protection schema are operated successfully when a fault appears in the upper trans- former T1. Then, the consequence is the lower transformer T2 supplying the load.
- Consequence 2 results when COB is in failure state and the successful operation of COB saves the generating system, so their consequence is moderate by assuming that the load is not a critical one and can be out for few minutes to allow manual operation.
- Consequence 3 results when both COB and COB are in failure states, thus there is damage to the transformers as well as to the generator along with the instability of the power system, so its consequence is high.
- Consequence 4 results when COB is in failure and COB is safe, therefore the consequence is the same as consequence 3.
- Consequence 5 is the case where both COB and COB are in failure, thus the consequence will be same as consequence 3.
- Consequence 6 results when RATS is in failure due to circuit-breaker tripping system, so it can damage the generator having high consequence.
- Consequence 7 results when the fault-detecting system fails probably due to the CT's saturation, and the result is the loss of supply and, eventually, the generator as well as the transformers is damaged, so its consequence is high.

2.2 The Fuzzy Tree Paths Ranking

In order to see which one of these outcomes has the highest possibility some applications rank the outcomes on the basis of the maximum probabilities associated with outcomes, or on the basis of the probabilities having maximum degree of membership in the fuzzy probabilities. But both of these approaches may lead to improper ranking of the outcomes. The proper approach of the different outcomes is to consider maximum probability, associated with various outcomes and the degree of membership of the rating [7]. Thus the maximizing set can be used, where the maximizing set is defined as set M(Y) of a set to be fuzzy such that the degree of membership for a point $y \in Y$ in M(Y) represent to sup (Y) in some specified sense.

Let the fuzzy probability of each outcome $\tilde{P}_i$ be:

$$\tilde{P}_i = \left\{ \left(p, \mu_{Pi}(p) \right) \right\} \quad , \qquad i = \overline{1,n} \tag{1}$$

and let S be all possible probabilities of the various outcomes,

$$S = \bigcup_{i=1}^{n} S(\tilde{P}i) \tag{2}$$

where S is the union of the supports of the fuzzy probabilities associated with various outcomes. Then the maximizing set of S can be obtained as follows [5], [6]:

$$M(S) = \left\{ \left(p, \mu_M(p) \right) \right\} \tag{3}$$

where:

$$\mu_M(p) = [p / P_{\max}]^n \text{ and } P_{\max} = \sup(S). \tag{4}$$

A fuzzy set P_{io} for each alternative is obtained by taking the intersection of fuzzy sets P_i and M(S):

$$P_{io} = \{(p, \mu_{Pio}(p))\} \tag{5}$$

where:

$$\mu_{Pio}(p) = \min[\mu_{Pi}(p), \mu_M(p)]. \tag{6}$$

The most possible outcome fuzzy set C_i is defined as:

$$C_o = \{(C_i, \mu_{Co}(C_i))\} \tag{7}$$

where:

$$\mu_{Co}(C_i) = \max(\mu_{Pio}(p)) \tag{8}$$

The most possible outcome is therefore:

$$C_o = \{C_1, C_6, C_2, C_4, C_7, C_5, C_3\} \tag{9}$$

where outcome 1 (all components in the system are working) has the highest possibility and outcome 3 and 5 have the lowest possibilities.

3 Fuzzy Logic System for Safety Computing

The tree - paths ranking evaluation is not enough for the power system protection reliability calculation. A methodology to calculate a qualitative index for this kind of systems, is necessary to be developed. On this goal, an adequate fuzzy - logic system (FLS) having the following steps, was created:

1. linguistic variables for FLS parameters construction;
2. FLS inputs "Occurrence" (OC) and "Severity" (SV) evaluation;
3. FLS rule base proposal;
4. FLS rule base evaluation and FLS outputs tree - paths "Safety" (SF) evaluation;
5. fuzzy general conclusion "General Safety" (GSF), for all tree-paths, evaluation;
6. GSF defuzzification and "General Safety Degree" (GSFD) crisp value computing.

3.1 Fuzzy Logic System Input Parameters

The FLS input parameters are the “Occurrence” of the path tree event and the “Severity” of the path tree event. Another input parameters could be introduced also in the future approaches. The FLS output parameter is “Safety” of the analyzed system according to the path tree event.

How the “Occurrence” linguistic variable and the “Severity” linguistic variable were elaborated is presented in table 1, respectively table 2. The “Occurrence” uses five fuzzy sets: very low (Vl), low (L), moderate (M), high (H) and very high (Vh), developed according to the failure probability.

"Severity" FLS input parameters is ranked according to the failure effect associated to the path tree event. It uses five fuzzy sets: very low (Vl), low (L), moderate (M), high (H) and very high (Vh), developed according to the failure effects.

Table 1. Occurrence FLS input parameter elaboration

Rank	Occurrence	Effects	Failure probability
1	Very low "Vl"	Failure not possible	< 1 of 10^5
2 3	Low "L"	Low failure rate	1 of10000 1 of 4000
4 5 6	Moderate "M"	Seldom failures	1 of 1000 1 of 80
7 8	High "H"	Usual failures	1 of 40 1 of 20
9 10	Very high "VH"	Very possible failure	1 of 8 1 of 2

Table 2. Severity FLS input parameter elaboration

Rank	Severity	Effects
1	Very low "Vl"	Minor failure. Not real effects of system performance
2 3	Low "L"	Small deterioration of system performance
4 5 6	Moderate "M"	Perturbation in system functions
7 8	High "H"	High degree of perturbation in system functions, but it still works
9 10	Very high "Vh"	The failure affects the system function

3.2 Fuzzy Logic System Rules Base

The rules base gives the FLS out parameter Safety for each one of the FLS input parameters combination. The rules, presented in table 3, have the following statement: If the Occurrence is "Low" and the Severity is "High" then the Safety is "Moderate".

Table 3. Rules base

		Severity				
		" Vl"	" L "	" M "	" H "	" Vh "
Occurrence	"Vl"	-	(1)"Vh"	(2)"H"	(3)"Im"	(4)"M"
	"L"	-	(5)"H"	(6)"Im"	(7)"M"	(8)"L"
	"M"	(9)"Im"	(10)"Im"	(11)"M"	(12)"L"	-
	"H"	(13)"Hî"	(14)"M"	(15)"Im"	(16)"M"	-
	"V "	(17)"Vh"	(18)"Vh"	(19)"H"	(20)"Im"	-

4 Application of FLS for Electric Transformer Protection System

This paper proposes an original GSF algorithm methodology to obtain a quantitative result for the qualitative analysis. The algorithm uses all tree- paths "conclusions". Because tree- paths "conclusions" have not the same importance for GSF calculation it is necessary to use OC inputs to multiply the SF output.

The GSF general conclusion for all tree-paths is:

Step 1: The tree- paths OC_i $i=\overline{1,N}$ input support is changed into discrete vector.

$$OC_i = \{OC_{i1}...OC_{in}\}, \quad i=\overline{1,N} \tag{10}$$

Step 2: The following vectors are evaluated:

$$OC_j = \{OC_{j1}...OC_{ji}...OC_{jN}\}, \quad j=\overline{1,c} \tag{11}$$

in the following condition

$$\sum_{i=1}^{N} OC_{ji} \cong 1, \tag{12}$$

Step 3: Using SF_i fuzzy numbers, the SF_j fuzzy numbers associated to OC_j are:

$$SF_j = \sum_{i=1}^{N} OC_{ji} \times SF_i', \quad j=\overline{1,c} \tag{13}$$

Step 4: The general fuzzy conclusion GSF is obtained in the following way:

$$SF_{gen} = \bigcup_{j=1}^{c} SF_j \tag{14}$$

For the fuzzy event- tree analysis of a power protection system, an adequate software tool FETA (Fuzzy Event-Tree Analysis) was created. For the proposed power protection system, the following event-tree and minimal cuts set are generated (see table 4, where OP, UN, FA, MC, <MC means operational element state, unoperational element state, failure element state, minimal cut, respectively less than a minimal cut).

The minimal cuts set is {(FD), (RTS), (CB1, CB5), (CB2, CB5)}. As the table 5 shows, the fuzzy event-tree analysis results are the following: the fuzzy number "Safety" associated to the tree-path (SFi); the centered value "Safety Degree" associated to the tree-path (SFDi), the fuzzy set "General Safety" for all the tree-paths (GSF), the centered value "General Safety Degree" for all the tree-paths (GSFD).

Table 4. Event tree and minimal cutset

Path	FD	RTS	CB1	CB2	CB5	Cut
1	OP	OP	OP	OP	OP	<MC
2	OP	OP	OP	FA	OP	<MC
3	OP	OP	OP	FA	FA	MC 1
4	OP	OP	FA	UN	OP	<MC
5	OP	OP	FA	UN	FA	MC 2
6	OP	FA	UN	UN	UN	MC 3
7	FA	UN	UN	UN	FA	MC 4

Table 5.

Path	Possibilities		Centered/ Uncertainty		Active rules	Possibilities	SFDi/ Uncertainty
i	OCi	SVi	OCi	SVi		SFi	SGi
1	l 0.4425	h 1.0000	2.5000E-5	8.1758	4, 4; 4, 5	vl 0.3136	3.6972
	vl 0.7633	vh0.3136	5.2500E-6	0.4000	5, 4; 5, 5	l 0.4425	3.1618
						m 0.7633	
2	l 0.9248	h 0.8589	7.9600E-5	8.6046	4, 4; 4, 5	vl 0.5392	3.0003
	vl 0.3127	vh0.5392	3.5380E-5	0.4000	5, 4; 5, 5	l 0.8589	3.2456
						m 0.3127	
3	l 0.5686	h 0.9328	3.4268E-5	8.4974	4, 4; 4, 5	vl 0.4828	3.3892
	vl 0.6126	vh0.4828	6.3449E-6	0.4000	5, 4; 5, 5	l 0.5686	3.4201
						m 0.6126	
4	0.2409	h 0.6079	1.4583E-5	8.9686	4, 4; 4, 5	vl 0.2409	3.6108
	vl 1.0000	vh0.7308	4.2460E-6	0.4000	5, 4; 5, 5	l 0.7308	3.3306
						m 0.6079	
5	m 1.0000	h 1.0000	4.8465E-3	2.3900	3, 2; 4, 2	m 1.0000	5.6021
	l 0.2911		5.1983E-4	0.4000		im 0.2911	2.6111
6	l 0.9006	h 0.7850	7.4625E-5	8.7117	4, 4; 4, 5	vl 0.5956	3.0106
	vl 0.3475	vh0.5956	3.4880E-5	0.4000	5, 4; 5, 5	l 0.7850	3.3099
						m 0.3475	
7	l 1.0000	h 0.7850	2.3761E-4	8.7117	4, 4; 4, 5	vl 0.5956	2.3297
		vh0.5956	1.7862E-4	0.4000		l 0.7850	2.6236
GSF fuzzy set possibilities: im:0.2537649; h:0.9197336; vh:0.6552803; GSFD crisp value :9.25386							

5 Conclusions

In this paper an application of fuzzy logic for safety computing of an Electric Transformer Protection System was presented. The authors elaborated a fuzzy logic system adequate to fuzzy event-tree analysis. Event-trees are often used in power protection system quality computing, but the paper introduces fuzzy sets and fuzzy logic also to realize a proper model of the analyzed system. An efficient software toll FETA ("Fuzzy Event-Tree Analysis"), for independent analyzing of the power protection system, was elaborated and used to achieve the proposed goal.

The FETA software uses four analyzing methodology steps and gives a global qualitative index for all the paths of the fuzzy event-tree. The adequate rules base, proposed in the paper, allows computing the electric transformer protection system "Safety" using the fuzzy logic. Fuzzy logic system elements, used for the fuzzy event-tree analysis of the electric transformer protection are adequate to the proposed application. The FLS inputs "Occurrence" and "Severity" are associated to the tree- path and the "Safety" of the protection system is obtained as the FLS output.

The proposed FLS uses as an output element the power protection system "Safety" instead of the usually "Risk" parameter, used in engineering applications (with, or without fuzzy logic elements). The introduction of the "Safety" FLS output is necessary and may be used for the power protection system independent analysis. Also it may be used for the combined qualitative analysis of the protected power (electric transformer for example) together with its protection system. This type of analysis implies the hybrid modelling of the combined system. A limit of the proposed method could be the field of use, only for different kinds of protection systems.

References

1. Bastani , F.B., Chen, I.R.: Reliability of Systems with Fuzzy-Failure Criterion. Proc.Ann. Reliability and Maintainability Symp (1994) 265-270
2. Bowles, J.B., Pelaez, C.E.: Application of Fuzzy Logic to Reliability Engineering. Proceedings of the IEEE 3 (1995) 99-107
3. Dumitrescu, M., Munteanu, T.: Fuzzy Probability and Power System Safety. LectureNotes in Computer Science, Vol. 2206. Springer-Verlag, Berlin Heidelberg (2001) 886-889
4. Dumitrescu, M., Munteanu T, Floricau, D.: A Software for Electric-Power Fuzzy Critical Analysis. IEEE Intenational Conference on Fuzzy Systems 25-29 July (2004) 168-172
5. Jain, R.: Decision making in the presence of fuzzy variable. IEEE Trans. System, Man, Cybernetics 6 (1976) 698-703
6. Kenarangui, R.: Verbal rating of alternative sites using multiple criteria weights. Trans. ANS33 (1979) 617-619
7. Kenarangui, R.: Event - tree Analysis by Fuzzy Probability. IEEE Trans. on Reliability 1(1991). 45-52
8. Kumar, A., Ragade, R.: An Extended Reliability Evaluation Framework for Computer Systems Using Fuzzy Logic. Proc.Ann. Reliability and Maintainability Symp (1994) 317-323
9. Mendel, J.M.: Fuzzy Logic Systems for Engineering: A Tutorial. Proceedings of the IEEE 3(1995) 390-295

Fuzzy Multiple Attributive Group Decision-Making for Conflict Resolution in Collaborative Design*

Xinyu Shao, Li Zhang, Liang Gao, and Rui Chen

Department of Industrial & Manufacturing System Engineering,
Huazhong University of Science & Technology, 430074, Wuhan, China
gaoliang@mail.hust.edu.cn

Abstract. Most conflicts in collaborative design are categorized as the problem of fuzzy multiple attributive group decision-making (FMAGDM). Both fuzzy assessments and the aggregation of multiple experts' opinions should be considered in the conflict resolution process. This paper presents a new approach for the problem, where cooperation degree (CD) and reliability degree (RD) are introduced for aggregating the vague experts' opinions. Furthermore, a fuzzy multiple attributive group decision-making expert system (FMAGDMES) is proposed to provide an interactive way to solve conflicts in collaborative environment. It is an intelligent integrated system because it combines fuzzy set theory with the method of group opinion aggregation. The vehicle performance evaluation as a real case is used to validate the efficiency of the proposed expert system, which is implemented by using c++.

1 Introduction

With stepping into the age of global competitive economy, decision-makers must realize that collaboration is playing an important role to improve the efficiency of product development in most of design tasks. In collaborative design process, a large variety of fuzzy numbers and linguistic information are introduced by experts or decision-makers. Therefore, multiple attribute decision-making (MADM) is presented, where these attributes may be assigned as fuzzy, crisp or linguistic valuations. At present, MADM problems are of importance in a variety of fields including engineering, economics, etc [1].

The classical MADM problem is usually resolved by simple additive weighting method, analytic hierarchical process method, outranking relation method, implied conjunction techniques, fuzzy linguistic approaches, miscellaneous techniques, and so on. Above all MADM methods are analyzed by Chen and Hwang in 1992 [2]. Afterward, Chen made use of interval-valued fuzzy sets to represent the characteristics of the alternatives [3], and presented a new method to assess the importance weights and the ratings of multiple attribute alternatives in linguistic terms [4]. Liang and Wang proposed a decision algorithm and applied it to the facility site and robot selection problems [5] [6]. Karsak proposed a two-phase decision method for robot selection procedure [7]. Yeh et al. proposed a new method for the performance evaluation

* Supported by the National Basic Research Program 973 of China (No. 2004CB719405).

L. Wang et al. (Eds.): FSKD 2006, LNAI 4223, pp. 990 – 999, 2006.

problem of urban public transport systems [8]. Yu proposed an AGP-AHP method for solving group decision-making fuzzy AHP problems [9]. Fiordaliso and Kunsch constructed a decision supported system based on the combination of fuzzy expert estimates to assess the financial risks in high-level radioactive waste projects [10]. Zeng proposed an expected value method for fuzzy multiple attribute decision-making (FMADM) problems [11]. Wang and Parkan investigated the MADM problem with fuzzy preference information on alternatives and propose an eigenvector method to rank them [12].

Existing FMADM methods have many remarkable capabilities. However, they also have some shortcomings. Firstly, the group expert opinions are not considered in FMADM problems. But in reality, only one decision-maker cannot present right cooperative resolution, multiple expert opinions frequently come forth in progress of collaborative design. Although, Olcer and Odabasi [1] presented this problem and proposed a new method to solve it with the consensus degree coefficient, they have still not considered the level of reliability. If an expert is not enough reliability, his opinions would lead the final solution inefficacy. Secondly, a suitable expert system is lacked in the existing literature, which provides convenience for participators to enter their opinions and achieve an excellent solution with computer.

On the other hand, in many different areas involving uncertainty and vagueness, fuzzy expert systems have been successfully applied to a wide range of problems [13-21]. Moreover, fuzzy rule based systems are the most suitable system for FMAGDM problems in collaborative design. A fuzzy rule based system consists of three main components: the inference system, the knowledge base and the user interface [22]. So, we must focus our attention on fuzzy multiple attribute group decision-making (FMAGDM), considering how to construct these three main components of the expert system, and finding an effective way to generate high quality estimations for the tasks of conflict resolution in collaborative design.

Accordingly, a new FMAGDM method with cooperation degree (CD) and reliability degree (RD) is introduced in this paper. Furthermore, FMAGDMES is constructed for solving conflict problems in collaborative design. This paper is organized as follows. Section 2, 3 and 4 introduce resolve methods, the architecture and the flow chart of FMAGDMES. In Section 5 a real case vehicle performance evaluation system, validates the efficiency of FMAGDMES. Finally, Section 6 makes some concluding remarks and put forward some improvable directions in future.

2 Basic Methods

This section presents some definitions, formulas and basic solution outline.

2.1 Converting Fuzzy Data to Standardized

As we all know, the decision matrix of the problem is expressed in crisp, fuzzy numbers and linguistic terms. A lot of methods which can convert linguistic terms to fuzzy numbers have already been presented in the past years. In 1992, Chen and Hwang proposed a numerical approximation system to systematically convert linguistic terms to their corresponding fuzzy numbers [2]. It is generally described as Table 1:

Table 1. Linguistic terms and their corresponding fuzzy numbers

	Linguistic terms	Scale 1	Scale 2	Scale 3
1	None	(0,0,0.1)		
2	Very Low	(0,0.1,0.2)	(0,0,0.1,0.2)	(0,0,0.2)
3	Low	(0.1,0.2,0.3)	(0.1,0.2,0.3)	(0,0.2,0.4)
4	Fairly Low	(0.2,0.3,0.4)		(0.2,0.4,0.6)
5	Mol. Low	(0.3,0.4,0.5)	(0.2,0.3,0.4,0.5)	
6	Medium	(0.4,0.5,0.6)	(0.4,0.5,0.6)	
7	Mol. Good	(0.5,0.6,0.7)	(0.5,0.6,0.7,0.8)	
8	Fairly Good	(0.6,0.7,0.8)		(0.4,0.6,0.8)
9	Good	(0.7,0.8,0.9)	(0.7,0.8,0.9)	(0.6,0.8,1)
10	Very Good	(0.8,0.9,1)	(0.8,0.9,1,1)	(0.8,1,1)
11	Excellent	(0.9,1,1)		

For other fuzzy opinions of the k-th expert, such as "approximately equal to z ", we first convert it to a trapezoidal fuzzy number $R_k = (a_k, b_k, c_k, d_k) = (z-m, z, z, z+m)$, where $a_k \le b_k \le c_k \le d_k$, and m is a suitable data to account the lower and upper bounds of the available area for the number z . If the scale of the fuzzy number R_k is not within the interval $[0,1]$, translate it into a standardized trapezoidal fuzzy number. The standardizing formula is given as follows:

$$R_k' = \left(\frac{a-a^*}{d^*-a^*}, \frac{b-a^*}{d^*-a^*}, \frac{c-a^*}{d^*-a^*}, \frac{d-a^*}{d^*-a^*}\right), \text{ when } R_k \text{ is a benefit factor.} \tag{1}$$

$$R_k' = \left(\frac{d^*-d}{d^*-a^*}, \frac{d^*-c}{d^*-a^*}, \frac{d^*-b}{d^*-a^*}, \frac{d^*-a}{d^*-a^*}\right), \text{ when } R_k \text{ is a cost factor.} \tag{2}$$

Here a^* is the smallest scale value of non-standardized trapezoidal fuzzy numbers given by experts for the same attribute, and d^* is the largest scale value.

2.2 Weighting Analysis

In some cases, the relative importance (RI) of experts and attributes are assigned by the most important person. Assume that the RI of expert $E_k\ (k=1,2,\cdots,M)$ is re_k, and the maximal re_k equals to 1. And if there are N unit attributes, we assign the RI of attribute $ra_j\ (0 \le j \le N)$ on a zero to 100 according weighted evaluation technique (WET). So we get the weights of experts and attributes

$$\omega e_k = \frac{re_k}{\sum_{k=1}^{M} re_k} \text{ and } \omega a_j = \frac{ra_j}{\sum_{j=1}^{N} ra_j}. \tag{3}$$

Here $\omega e_k \in [0,1]$, $\sum_{k=1}^{M} \omega e_k = 1$, and $\omega a_j \in [0,1]$, $\sum_{j=1}^{N} \omega a_j = 1$.

2.3 CD and RD

In aggregation phase, we must consider both the degree of cooperation and level of proficiency. CD and RD are the best standards to estimate these two degrees.

CD. The formulas to account CD as follows:
Firstly, we calculate the degree of similarity $S_{uv}(R_u, R_v)$ of the opinions between each pair of experts E_u and E_v, where $S_{uv} \in [0,1]$, $1 \le u \le M$, $1 \le v \le M$, and $u \ne v$. Assume $R_u = (a_u, b_u, c_u, d_u)$ and $R_v = (a_v, b_v, c_v, d_v)$, where $0 \le a_u \le b_u \le c_u \le d_u \le 1$ and $0 \le a_v \le b_v \le c_v \le d_v \le 1$, then

$$S_{uv} = 1 - \frac{|a_u - a_v| + |b_u - b_v| + |c_u - c_v| + |d_u - d_v|}{4}. \tag{4}$$

Secondly, we get the average degree of similarity

$$AA(E_u) = \frac{1}{M-1} \sum_{v=1, u \ne v}^{M} S(R_u, R_v). \tag{5}$$

Calculate the relative degree of similarity

$$RA(E_u) = \frac{AA(E_u)}{\sum_{u=1}^{M} AA(E_u)}. \tag{6}$$

Let $\gamma (0 \le \gamma \le 1)$ is a relaxation factor, which shows the importance of the ωe_u over $RA(E_u)$. If none of decision-makers is the manager to assign the weight of others, a homogenous group of experts problem is considered and γ=0. When $\gamma \ne 0$, a heterogeneous group of expert problem is considered. The CD is calculated that

$$CD(E_u) = \gamma \cdot \omega e_u + (1-\gamma) \cdot RA(E_u). \tag{7}$$

RD. It is represented by the distance between an opinion value and the ideal opinion value of all opinions for the corresponding attribute.

Firstly, we defuzzificate the fuzzy numbers of an expert's opinion. If a fuzzy numbers is $R_k = (a, b, c, d)$, where $0 \le a \le b \le c \le d \le 1$, we get the corresponding crisp opinion value of expert E_k is

$$CE_k = RC_k \cdot \omega e_k = \frac{a+b+c+d}{4} \cdot \omega e_k \,. \tag{8}$$

RC_k is a crisp number after normalize R_k.

If the attribute is a benefit factor, the maximal CE_k is the ideal opinion value. Oppositely, if the attribute is a cost factor, the minimal CE_k is the ideal opinion value.

So we get the distance between an opinion and the ideal opinion

$$DD_k = 1 - \left|\max(CE_k) - CE_k\right| \text{ or } DD_k = 1 - \left|\min(CE_k) - CE_k\right| . \tag{9}$$

According the definition of RD, the formula to account RD as follows:

$$RD_k = \frac{DD_k}{\sum_{k=1}^{M} DD_k} \,. \tag{10}$$

where M is the amount of experts.

The Aggregation Result. We added the parameters α and β to express the important degree of CD and RD separately in aggregate state, and $\alpha + \beta = 1$. So the aggregation result of the experts' opinions for an attribute as:

$$R = \frac{\alpha CD_1 + \beta RD_1}{2} \otimes R_1 \oplus \cdots \oplus \frac{\alpha CD_M + \beta RD_M}{2} \otimes R_M \,. \tag{11}$$

These aggregation results compose a new normalized decision matrix *NR*.

2.4 TOPSIS

TOPSIS is presented by Hwang and Yoon [23], which is quite effective in identifying the best alternative quickly. According to it, following steps to select the best alternative are to be performed.

Firstly, according the formula (8), an element of the weighted normalized decision matrix is calculated as

$$v_{ij} = \omega a_i \cdot r_{ij} = \frac{a+b+c+d}{4} \cdot \omega a_i, \text{ where } i = 1,2,\cdots,N,\ j = 1,2,\cdots,K. \tag{12}$$

Here ωa_i is the weight of the i-th attribute, r_{ij} is an element of *NR*, and K is the amount of alternatives.

Then, the positive-ideal solution sets $A^* = \{v_1^*, v_2^*, \cdots, v_i^*, \cdots, v_N^*\}$ and the negative-ideal solution sets $A^- = \{v_1^-, v_2^-, \cdots, v_i^-, \cdots, v_N^-\}$ can be defined. If the attribute is benefit, v_i^* is the maximal v_{ij}, v_i^- is the minimal v_{ij}. Oppositely, v_i^* is the minimal v_{ij}, v_i^- is the maximal v_{ij}.

Finally, according the n-dimensional Euclidean distance, we calculate similarities to positive-ideal solution and define the similarity of A_j with respect to A^* as

$$C_j^* = \frac{S_j^-}{S_j^* + S_j^-} = \frac{\sqrt{\sum_{i=1}^{N}\left(v_{ij} - v_i^*\right)^2}}{\sqrt{\sum_{i=1}^{N}\left(v_{ij} - v_i^-\right)^2} + \sqrt{\sum_{i=1}^{N}\left(v_{ij} - v_i^*\right)^2}}, 0 < C_j^* < 1; j = 1, 2, \cdots, K . \quad (13)$$

S_j^* is the distance between each alternative to the positive-ideal solution, and S_j^- is the distance between each alternative to the negative-ideal solution. Choose the alternative with the maximum C_j^* as the best alternative.

3 The Architecture of FMAGDMES

In this paper, fuzzy numbers and linguistic valuations are adopted to represent the fuzzy knowledge. Methods in the section 2 are methods of fuzzy inference. The architecture of FMAGDMES is illustrated in Fig 1.

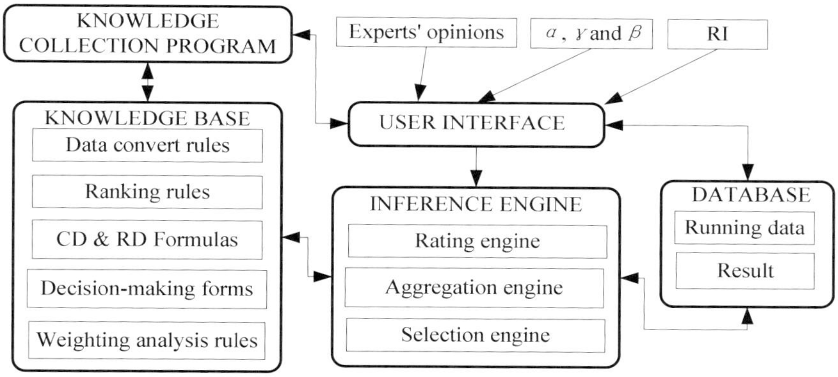

Fig. 1. The architecture of FMAGDMES

The users interface is a dialog module, which helps users communicate with computer. The database deposits running data and final result. The inference engine uses the known information in the database to match the rules in the knowledge base through a lot of mechanisms, strategies and control. If successful matching, the operation will go on with the rules or formulas. In the proposed system, the inference engine is used to get the best alternative by manipulating and applying the knowledge in the knowledge base. The knowledge base is used to storage all kinds of rules, formulas and forms. The decision-making form is used to register all decision-makings. It is convenient to following calculation. The knowledge collection program is a module, which collects experts' opinions and inputs these opinions to the knowledge base.

4 The Flow Chat of FMAGDMES

According the whole process of inference, the proposed system can be summarized to three phases: rating phase, aggregating phase and selection phase. The flow chart of the expert system is described as shown in figure 2.

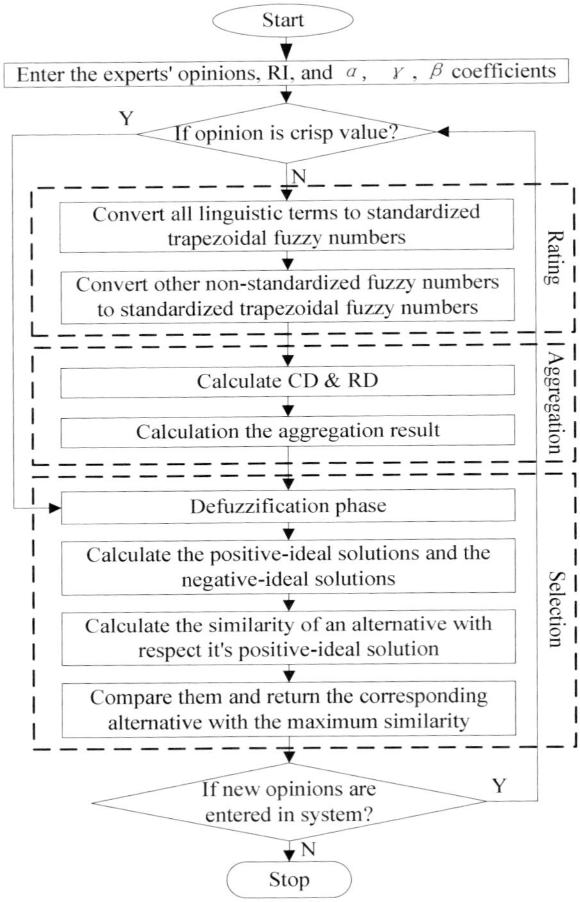

Fig. 2. The flow chart of FMAGDMES

5 A Vehicle Design Case Study

In a vehicle design system, performance evaluation is an important part of whole design process. They involve vagueness and fuzziness. Moreover, most of participators, such as manufacturers, designers and consumers, become design-makers to evaluate these vehicle performances in the collaborative design process. It is difficult for decision-makers to choose and specify the best one among the many alternatives. Hence, we greatly require a useful method and FMAGDMES to provide the ability to handle both multiple attribute data and group decision-makers.

Vehicle performances include develop cost, accelerating time, maximum speed, fuel consumption, ride comfort performance, braking performance, comfort performance, handling performance and yawp [24]. So we assume these 9 performances are input attributes (A_1, A_2, …, A_9). A_1 is the only objective and its type of assessment is crisp, others are subjective. The type of assessment of A_2, A_3, A_4 and A_6 are fuzzy values, A_5, A_7, A_8, A_9 are linguistic values. Except these, A_1, A_2, A_4, $A_6$and A_9 are cost attribute, and others are benefit attribute. We will choose an appropriate style of vehicle among three alternatives X_1, X_2 and X_3. The decision-makers are three experts named manufacturer (E_1), marketer (E_2) and user (E_3). These experts are homogeneous group of experts, so none of them could assign the weights of others and the parameter γ equals 0. Assume the importance degree of RD is equal to the importance degree of CD, so $\alpha = \beta$ =0.5, and $\alpha + \beta = 1$. The opinions of experts are given in Figure 3.

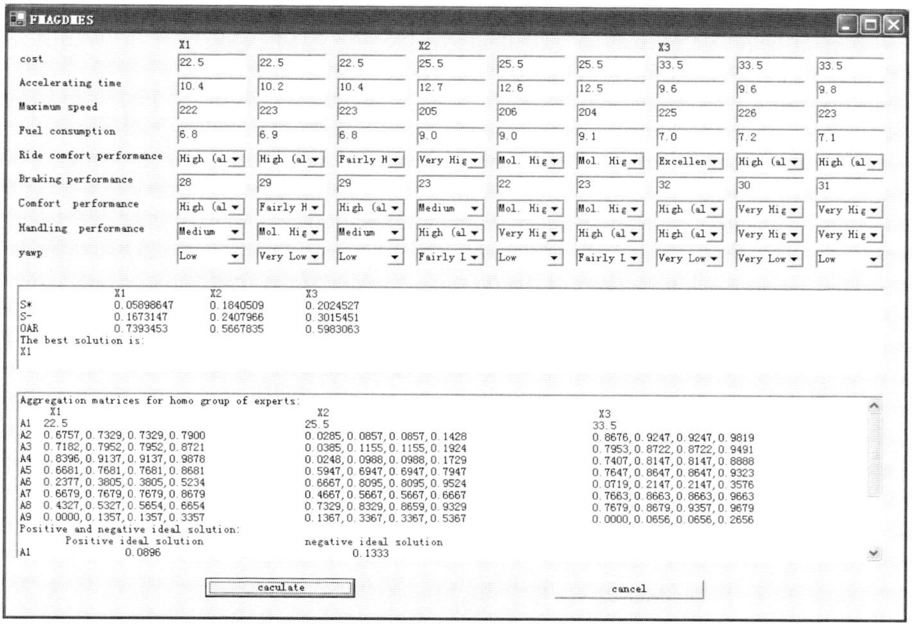

Fig. 3. The interface of the expert system

The RI and weights of attributes and experts are numerated according the formula (3). A_1 is the only objective attribute, so it need not be aggregated.

In the rating phase, linguistic valuations are matched with Scale 1 (for A_5, A_7), Scale 2 (for A_8) and Scale 3 (for A_9) in Table 1. After transformation of original opinions, we get the standardized fuzzy numbers.

In the aggregating phase, we get the aggregation matrix *NR* for homogenous group of experts by CD and RD. It is shown in the analysis interface of FMAGDMES as above Figure 3.

In the last selection phase, we use TOPSIS method to ranking the similarity C_j^* and select the best alternative. The important data of this phase are also shown in the output interface of FMAGDMES as Figure 3. According it, we get the preference order (OAR) of these three alternatives is $X_1 > X_2 > X_3$. So we should choose the best alternative X_1 for our vehicle design task.

6 Conclusion

In our research, we have introduced the definitions of CD and RD for aggregation of group experts' opinions, and constructed a new FMAGDMES, which is very suitable for solving FMAGDM problems in collaborative design. In addition, the measure of converting multiple attribute valuations to fuzzy numbers is also discussed. For the vehicle collaborative design, we used a vehicle performance evaluation case to structure and illustrate the proposed expert system. The case study even illustrated its practicability and validity. The excellences of FMAGDMES include the ability of aggregating group decision-makings and solving FMAGDM problems with computer, clear user interface, elaborate analysis engine, better expanded performance and generalization.

For the future, the method of RD's evaluation may develop further. It is important to adopt a suitable method of RD's evaluation for aggregation of group experts' opinions. It is also important to combine the existed methods to create a new expert system. Future expert system for FMAGDM will be used in more tasks from different areas and the develop directions focus on the accuracy of the results, intelligent inference engines and integration of methods.

References

1. Olcer, A. I., Odabasi, A. Y.: A New Fuzzy Multiple Attributive Group Decision Making Methodology and Its Application to Propulsion/Manoeuvring System Selection Problem. European Journal of Operational Research. Elsevier (2005) 93-114
2. Chen, S. J., Hwang, C. L.: Fuzzy Multiple Attribute Decision-Making: Methods and Applications. Springer-Verlag, New York (1992)
3. Chen, S. M.: A New Method for Handling Multicriteria Fuzzy Decision Making Problems, Vol.25. Cybernetics and Systems (1994) 409-420
4. Chen, S. M.: A New Method for Tool Steel Materials Selection under Fuzzy Environment, Vol. 92. Fuzzy Sets and Systems (1997) 265-274
5. Liang, G. S., Wang, M. J.: A Fuzzy Multi-criteria Decision-Making Method for Facility Site Selection, Vol. 29. International Journal of Production Research (1991) 2313-2330
6. Liang, G. S., Wang, M. J.: A Fuzzy Multi-criteria Decision-Making Approach for Robot Selection, Vol. 10. Robotics and Computer-Integrated Manufacturing (1993) 267-274
7. Karsak, E. E.: A Two-phase Robot Selection Procedure, Vol.9. Production Planning and Control (1998) 675-684

8. Yeh, C. H., Deng, H., Chang, Y. H.: Fuzzy Multi-criteria Analysis for Performance Evaluation of Bus Companies, Vol. 126. European Journal of Operational Research (2000) 459-473
9. Chian-son, Y.: AGP-AHP Method for Solving Group Decision-Making Fuzzy AHP Problems, Vol. 29. Computers and Operations Research (2002) 1969-2001
10. Fiordaliso, A., Kunsci, P.: A Decision Supported System based on The Combination of Fuzzy Expert Estimates to Assess The Financial Risks in High-level Radioactive Waste Projects, Vol. 46. Progress in Nuclear Energy (2005) 374-387
11. Ling, Z.: Expected Value Method for Fuzzy Multiple Attribute Decision Making, Vol. 11. Tsinghua Science and Technology (2006) 102-106
12. Ying-Ming, W., Celik P.: Multiple Attribute Decision Making based on Fuzzy Preference Information on Alternatives: Ranking and weighting, Vol. 153. Fuzzy Sets and Systems (2005) 331-346
13. Buckley, J. J., Siler, W., Tucker, D.: A Fuzzy Expert System, Vol. 20. Fuzzy Sets and Systems (1986) 1-16
14. Jones, P. L. Graham, I.: Expert Systems: Knowledge, Uncertainty, and Decision Making, Chapman & Hall, London (1988)
15. Busse, G.: Managing Uncertainty in Expert Systems. Kluwer Academic Publichers, Dordrecht (1991)
16. Torella, G.: Expert Systems and Neural Networks for Fault Isolation in Gas-turbines. International Society of Air-Breathing Engines (1997)
17. Chang, Y. H., Yeh, C. H., Cheng, J. H.: Decision Support for Bus Operations under Uncertainty: A Fuzzy Expert System Approach, Vol. 26. Omega. Int. J. Mgmt Sci. (1998) 367-380
18. DePold, H. R., Gass, F.D.: The Application of Expert Systems and Neural Networks to Gas Turbine Prognostics and Diagnostics, Vol. 121. J. Eng. Gas Turb. Power (1999) 607–612
19. Wong, S.V., Hamouda, A.M.S.: A Fuzzy Logic based Expert System for Machinability Data-on-demand on the Internet, Vol. 124. Journal of Materials Processing Technology (2002) 57–66
20. William, W. L. C., Tony, J. P., Daniel, P.: A Fuzzy Logic Expert System to Estimate Intrinsic Extinction Vulnerabilities of Marine Fishes to Fishing, Vol. 124. Biological Conservation (2005) 97-111
21. Zbigniew, K., Maria, M. K., Stefan, Z., Marcin, D.: CBR Methodology Application in An Expert System for Aided Design Ship's Engine Room Automation, Vol. 29. Expert Systems with Applications (2005) 256–263
22. Rasmy M. H. etc.: An Expert System for Multi-objective Design Making: Application of Fuzzy Linguistic Preferences and Goal Programming, Vol. 127. Fuzzy Sets and Systems (2002) 209-220
23. Hwang, C. L., Yoon, K. P.: Multiple Attribute Decision-Making: Methods and Applications. Springer-Verlag, Berlin (1981)
24. Ning, L.: Vehicle Design. Engineering and Industry publish company (1999) 10-14

Fuzzy Performance Modeling Aligned with Process and Organization Model of Integrated System in Manufacturing*

Jian Zhou[1,2], Qing Li[1], DaFeng Xu[1], and TianYuan Xiao[1]

[1] Department of Automation, Tsinghua University, Beijing, 100084, P.R. China
j-zhou02@mails.tsinghua.edu.cn
[2] National University of Defense Technology, Changsha, 410073, P.R. China
zhoujiandragon@gmail.com

Abstract. Based on long-term industrial practice, our research team addressed economic view to complement the existing CIM system architecture, which is used to measure the relevant performance and has become the annex C of ISO 15704. However the work on economic view issued before mainly focused on the performance framework, in this paper a fuzzy performance modeling process and method is proposed as the further research. Considering the numerous dynamic and subjective indicators of performance, the related fuzzy method that the research largely relies on here is the best choice for its natural advantages in this area. Before the method for fuzzy performance modeling is put forward, the fuzzy performance modeling framework aligned with process and organization model is discussed firstly. Afterwards, a theorem is presented for conflict and redundancy validation of fuzzy rules that is necessary for performance modeling. Finally the fuzzy performance modeling and measurement method are suggested.

1 Introduction

In recent years, performance measurement is getting increasingly important while many unfinished and rigorous work still remain for us to improve [2]. In long-term industrial practice and the implementation process of a large number of computer integrated manufacturing (CIM) projects in China, it is found that, in a system integration project, demands of the project target are reflected by demands of the economic characteristics, and influences on the system are realized by the integration strategy and the technology project. Therefore our research team addressed Economic View [1] to complement the CIM system architecture. It establishes the relations between economic target and engineering project. It describes the scale indices and influence factors of the economic characteristic in an integrated system, the effect that the economic characteristic indices impose on economic targets in the system integration project, and the function genes of engineering project for the influence factors. It supports for Enterprise Managers, Enterprise Model Developers and

* Sponsored by China 863 Program, No. 2001AA415340; Sponsored by China National Natural Science Fund, No. 60474060.

L. Wang et al. (Eds.): FSKD 2006, LNAI 4223, pp. 1000 – 1009, 2006.

Analyzers, and support for System Developers as well. Because of the significance and applicability of it, Economic View has become the annex C of ISO 15704 since the last year. However the work on Economic View issued before mainly focused on the performance framework and it is lack of modeling methods that support it from concept to reality. In this paper, as a complement of it, a fuzzy performance modeling method are proposed as our further research, which has also gained considerable improvements in comparison with those present methods for performance measurement.

The rest of this paper is organized as follows: The second section provides an overview of the related literature on enterprise performance measurement and a problem statement. The third section is about the fuzzy performance modeling framework aligned with process and organization model, which outlines the major idea of proposed method. The fourth section introduces the fuzzy performance modeling process and the method for fuzzy performance measurement. The final section provides a discussion of limitations and conclusions together with an outlook on future research.

2 Related Work of Performance Measurement

Related research of performance measurement, from the modeling perspective, can be depicted in four categories as follow [2]:

- Performance model consists of some performance indicators, and analysis methods as well as recommendations related to the measurement are presented, such as the work of Kaplan and Norton [5].
- Frameworks as Performance model have undoubtedly made the largest impact on the performance measurement circle. For instance, AHP and ANP developed by Prof. Thomas Saaty [10].
- Performance measurement system, as the most mature model, is basically composed of frameworks and necessary performance management tools. For example, the balanced scorecard performance measurement system [4] is typical academic examples.
- Besides, there are a large number of other researches related to performance of enterprises. Such as, certain researchers apply the dynamic system theory to analyze the mutual causal relations of indicators.

With the efforts of many researchers, research in this field is significantly developed however still has not fully stepped up to meet the actual requirements. There still remains following insufficiencies and obstacles of present study in this research circle:

- Much of the performance information is inherently uncertain--lack of precision, vague concepts, more or less reliable information, etc. The corresponding indicators are numerous and a large number of them are intangible and qualitative. No existing method has provided widely accepted solution about how to analyze these complex and dynamic indicators to measure the performance.

- The simple performance model can be constructed with rigorous logical description, while the complex model is always lack of this crucial support.
- There is no good way to solve the problem that how to balance the contradiction between universality of measurement methods and the individual customized application requirements by different users.
- Performance measurement is supposed to be computer processible and automatic to make the advantage of IT applied in enterprise however methods can not come into use without the participation of people today.
- The multidisciplinary character of the research is hindering developments in the field of Performance measurement [8]. Researchers are always not willing to outstrip traditional functional boundaries of their research, which will result in the fact that their research may be isolated, duplicated and/or contradictory.

Intending to solve the above problems partly, the fuzzy performance modeling method here is the significant complement of Economic View.

3 Fuzzy Performance Modeling Framework Aligned with Process and Organization Model

As above mentioned, the former research of Economic View is largely about the conceptual framework and is lack of concrete modeling method. In the dissertation [11] of one of our team fellows in this year, as a further work, why the performance model is constructed as a structure of multi-indicators hierarchy is discussed from the view point of the depth/span of performance indicator and the information delivery distortion. In his opinion, among the multi-indicators, those with high depth form the bottom layer and are employed to gather professional and special information from real system, while those with wide span compose the top layer and provide the universal performance information of model. To establish the performance model, it is believed that the key problem is to decide what kinds of indicators should be selected and how to bridge the relationships between these indicators. From industrial practice and literature, the selection always depends on experiences, while the research on relationships between them are always configured directly by some experts or related people. Actually, without the special context, the dispersed information in the complex integrated system, denoted by the indicators, is hard even to describe and understand. In this case, the relationships between them are definitely more difficult to analyze and form. So performance modeling is undoubtedly supposed to make full use of all sorts of related knowledge and information. In this paper, the performance model will be constructed mainly with the support of process model and the organization model will also be taken into account. On the one hand, reasons to align with process model are convincing. Firstly the hierarchy of performance model can be constructed associated with the process hierarchy. Secondly process model is map or image of the logical and temporal order of business activities performed on a process object and also contains both behavior and structure of all objects that may be used during process execution, so a large sum of diverse information is embedded in it, which is a great help for model establishment and performance analysis. Thirdly the research on process model is relatively mature and abundant. Furthermore some experts also argued that the measurement system should be focused on processes, not on whole organizations or organizational units [9]. On the other hand the cause to relate the performance modeling to organization model is mainly because process model can not provide relevant organizational information that is also important for modeling. In addition, the related management actions after performance analysis will benefit a lot from the correlation between performance model and process or organization model.

What should also be mentioned about the method proposed in this paper is the application of fuzzy logic. It is widely accepted that fuzzy set theory provides a prominent paradigm in modeling and reasoning with uncertainty [12]. Fuzzy logic is based on fuzzy set theory and it has two major industrial applications in the history: in the field of fuzzy controllers and information systems. In the integrated system, there are an enormous amount of electronic accessible information and knowledge which is inherently uncertain--lack of precision, vague concepts, etc [6]. And it is supposed to make full use of all of them for the performance measurement despite its uncertainties and imperfection. In accordance, fuzzy logic can not only be used to represent this knowledge exactly, but also to utilize it to its full extent. And the complex knowledge presented by fuzzy logic still can be computer processible and allow fast mining and processing even there are a large bodies of them. In view of these facts, the performance modeling in this paper is chiefly based on fuzzy rules and related fuzzy analysis methods.

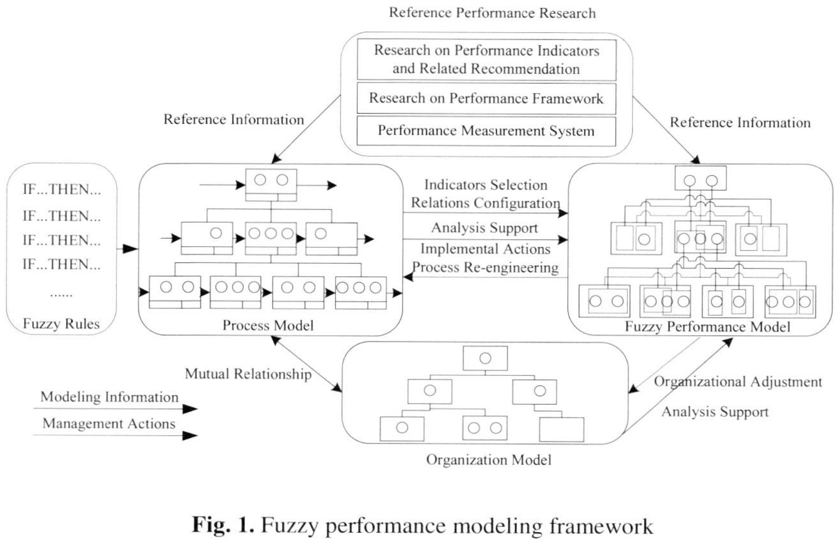

Fig. 1. Fuzzy performance modeling framework

To illustrate the whole idea, a framework for fuzzy performance modeling is put forward as Fig 1. Firstly a large number of fuzzy rules are extracted and selected from integrated system and related business process as well as fuzzy rules base, which will then be correlated to some process activities. Secondly, based on the process model context, fuzzy rules and related performance research, performance indicators are selected and relationships between them are configured. Accordingly the fuzzy performance model is formed. Thirdly, making reference to the context knowledge and information of process and organization model, the performance of integrated system is analyzed based on the performance model. Finally, with the performance analysis conclusions, management actions are implemented, such as process re-engineering and organizational adjustment. What is more, to make the management actions more feasible and effective, it is necessary to set mapping between the performance indicators involving organizational adjustment and corresponding organizational units as well as to set mapping between the key indicators concerning process re-engineering and the related process activities. The major modeling process and mentioned method in this framework will be detailed in the next section.

4 Fuzzy Performance Modeling Process and Measurement Method

4.1 The Preparative Research for the Application of Fuzzy Rules

In this paper, the performance modeling is based on and begins with fuzzy rules. In general such rules are obtained from human experts and are formulated as a linguistic IF THEN paradigm. However because of the insufficient and shortcoming of this kind of approach for knowledge acquisition, many methods have been presented later for automatic extraction from numerical data [3][7]. To make the fuzzy rules more sufficient and reliable, no matter those extracted by certain automatic algorithm or those obtained from human experts and so on, all of them will be adopted to support the performance modeling. In this case, there must exist many unexpected and undesired conflicts and redundancies between those rules. Thus before applying these fuzzy rules, a theorem with logical criteria is suggested for partial conflict and redundancy validation of them that are necessary for our method.

The form of the fuzzy rule f_j is always shown as:

$$\begin{aligned} &if(x_{f_j 1} \in or \notin A_{f_j 1})...and(x_{f_j i} \in or \notin A_{f_j i})...and(x_{f_j n} \in or \notin A_{f_j n}); \\ &then(y_{f_j} \in or \notin B_{f_j}) \qquad\qquad 1 \le i \le n, \quad i, j \in N \end{aligned} \tag{1}$$

Where $A_{f_j i}$ is one of the possible intervals of the i-th premise that is corresponding with certain performance indicator $pi_{f_j i}$ of fuzzy rule f_j. Let us suppose PI_{f_j} is the set of indicators associated with premises of f_j, then $PI_{f_j} = \{pi_{f_j 1}, pi_{f_j 2} ... pi_{f_j i} ... pi_{f_j n}\}$. B_{f_j} denotes one of possible intervals of the relevant conclusion that is associated with certain performance indicator $PI_{f_j c}$,. The alphabet n represents the total number of premises and N denotes the natural number. $x_{f_j i}$ is the value of the i-th premise of f_j and y_{f_j} is the value of associated conclusion.

Before proposing the theorem, it is imperative to construct the corresponding logical value, actually two-value logic here, for every premise and the conclusion of fuzzy rule f_j , which is denoted as $Logic(f_j) = (L_{f_j 1}, L_{f_j 2},..., L_{f_j i},..., L_{f_j n}, L_{f_j c})$. If $x_{f_1 i} \in or \notin A_{f_1 i}$ and $y_{f_1} \in or \notin B_{f_1}$, then $Logic(f_1) = (L_{f_1 1}, L_{f_1 2},..., L_{f_1 i},..., L_{f_1 n}, L_{f_1 c})$. Where if $x_{f_1 i} \in A_{f_1 i} (or \notin A_{f_1 i})$, then $L_{f_1 i} = 1\ (or\ 0)$, and if $y_{f_1} \in B_{f_1} (or \notin B_{f_1})$, then $L_{f_1 c} = 1\ (or\ 0)$. Because of the limitation of paper length, the proof of the theorem has to be omitted.

Theorem 1: If PI_{f_1} is the same with PI_{f_2}, $PI_{f_1 c} = PI_{f_2 c}$, $A_{f_1 i} \cap A_{f_2 i} \neq \phi$, and $Logic(f_1) = (L_{f_1 1}, L_{f_1 2}, ..., L_{f_1 i}, ..., L_{f_1 n}, L_{f_1 c})$, $Logic(f_2) = (L_{f_2 1}, L_{f_2 2}, ..., L_{f_2 i}, ..., L_{f_2 n}, L_{f_2 c})$, and if $B_{f_1} \cap B_{f_2} = \phi$, then $L_{f_1 c} \oplus L_{f_2 c} \neq 0$, then:

Criterion 1:

$$\begin{aligned} &if\ L_{f_1 1} \oplus L_{f_2 1} + L_{f_1 2} \oplus L_{f_2 2} + ... + L_{f_1 n} \oplus L_{f_2 n} = 0 \\ &and\ L_{f_1 c} \oplus L_{f_2 c} = 0 \end{aligned}$$

, then f_1 and f_2 are redundant.

Criterion 2:

$$\begin{aligned} &if\ L_{f_1 1} \oplus L_{f_2 1} + L_{f_1 2} \oplus L_{f_2 2} + ... + L_{f_1 n} \oplus L_{f_2 n} = 0 \\ &and\ L_{f_1 c} \oplus L_{f_2 c} \neq 0 \end{aligned}$$

, then f_1 and f_2 are conflicting.

To expand the range of application of presented theorem, two lemmas are proposed as below, which are easy to be proved.

Lemma 1: If the intervals of the premises corresponding with the same indicator 'pi' involved in different fuzzy rules are the same and actually every indicator 'pi' is correlated with only one possible interval. It is also the same about the associated intervals of the conclusions. Namely if $pi_{f'' i} = pi_{f' i}$ and $PI_{f'' c} = PI_{f' c}$, then $A_{f'' i} = A_{f' i}$ and $B_{f''} = B_{f'}$. And if the value of the premise and the conclusion has been logical value already such as two-value or multi-value logic, then the conclusion of the theorem, namely the logic criteria, can be used directly to validate the conflict and redundancy.

Lemma 2: If $PI_{f''} \subset PI_{f'}$, with the mentioned reasoning method, the conflict and redundancy can be validated only based on the analysis of those premises associated with($pi_{f'' i} = pi_{f' i}$) $\in (PI_{f''} \cap PI_{f'} = PI_{f''})$ and the conclusions corresponding with $PI_{f'' c}$ and $PI_{f' c}$, because, in this case, fuzzy rule f' may be contained inside fuzzy rule f''.

4.2 Selection of Performance Indicators Aligned with Process Model

Based on the theorem for conflict and redundancy validation of fuzzy rules presented above, the selection of performance indicators aligned with process model is an important part of the fuzzy performance modeling process and method.

As discussed in the section 3, the structure of performance model is always a multi-indicators hierarchy, however, the method is lacking and deficient to establish the performance hierarchy independently without the support of other related models, and, to be specific, both the selection of indicators and the configuration of relationship between indicators involving the hierarchy are hard to analyze and determine. To solve these problems and take the related advantages of process model described before, multi-indicators hierarchy of fuzzy performance model will be constructed mainly aligned with process model and actually the hierarchy of performance model will be matched to the hierarchy of relevant process model.

Before assigning the performance indicators to process model, it should be mentioned that the indicators base and the fuzzy rules base, which serve as the most important basis of selection of indicators, are supposed to be built in order to obtain favorable and sufficient performance indicators. The indicators base consists of two major parts, basic indicators base that comprises a large number of generic performance indicators and application indicators base that contains those popular indicators as well as the information of their usual locations in the process model. The fuzzy rules base has the same structure with indicators base, one part is about the generic fuzzy rules for choice and the other is about those frequently used fuzzy rules and relevant relationship between them and the process activities.

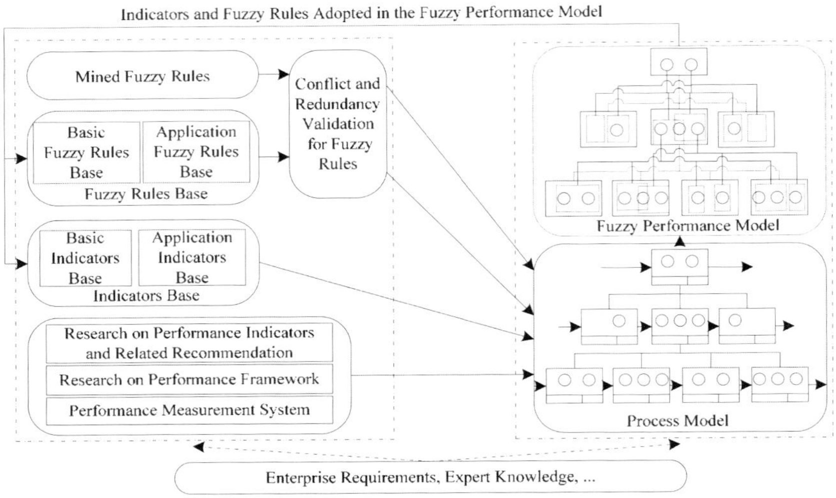

Fig. 2. Selection of indicators aligned with process model

As shown in Fig 2, the core of the selection of indicators is the related process model and actually what we do is to put those adopted indicators into the appropriate process activities. To make this selection process more reliable and dependable, it is encouraged to scratch indicators from various possible aspects, for there is not generally accepted list of performance indicators and fuzzy rules, and also the ones selected in those two knowledge bases are not definitely sufficient. Firstly it will be based on both of the knowledge bases, especially their application sub-bases that can provide not only the popularly applied indicators and fuzzy rules but also the relationship between them and the process model, to make the full use of history experience and knowledge. Secondly, according to the concrete enterprise characteristics and requirements, fuzzy rules mining is getting more convincing and dependable with the rapid development of related research and technology in this area. It can provide sufficient and necessary fuzzy rules for special performance modeling and actually this step is the most important one to obtain discriminative fuzzy performance model for different measurement and application objects. At last other corresponding aspects such as research and outcomes of the performance indicators and framework should also taken into account. By the way the enterprise requirements and expert knowledge are used respectively as the guideline and reference in the whole process of selection of indicators. With these steps, which indicator to be selected and which process activity they will be located in is confirmed. About the indicators adopted, it is important that only those favorable and reasonable indicators are supposed to be selected and each indicator can be put into different process activities in the hierarchy if it is necessary. In accordance with the discussion in the section 3, indicators in the top layer of the hierarchy mainly represent information with wide span, which are always qualitative, while those in the bottom layer depict information with high depth, which are largely quantitative. In the fuzzy performance modeling process, both qualitative and quantitative indicators are concerned and qualitative ones are decomposed into measurable and quantifiable quantitative ones from top to bottom layer until the total indicators in the bottom layer guaranteed to be quantitative with the associated decomposition of the process model.

In addition, three more things should be mentioned. Firstly, with the fuzzy rules and the relationship between them and process activities, it is easy to correlate the process activity with the performance indicators associated with fuzzy rules and then it is also easy to locate these indicators in appropriate process activity. Because the performance grounded on fuzzy performance model in this paper is actually measured through a set of fuzzy rules at last, this kinds of indicators related to certain fuzzy rules are strongly supposed to be adopted. Secondly, the conflict and redundancy validation of fuzzy rules is indispensable, no matter where the fuzzy rules come from. Thirdly, those indicators and fuzzy rules adopted in the fuzzy performance model at the last stage of modeling process will be fed back respectively to both knowledge bases to enrich the resources for the reuse in the future.

4.3 Fuzzy Performance Model and Related Performance Measurement

Based on the process hierarch and performance indicators selected, the method in this sub-section is to measure the fuzzy performance of integrated system based on the fuzzy performance model that will be constructed as below.

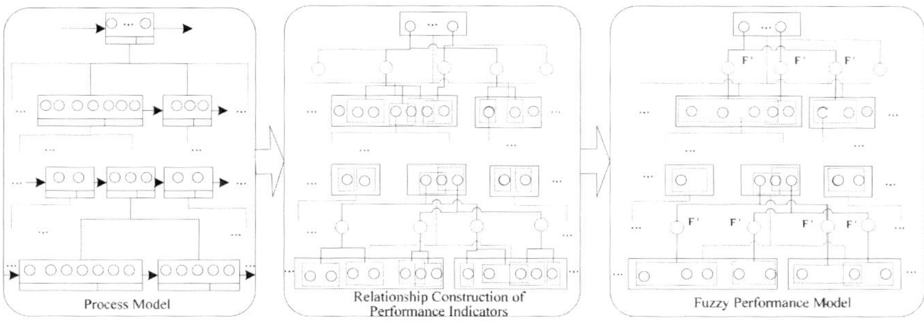

Fig. 3. Relationship construction of performance indicators and fuzzy performance model

The fuzzy performance model will be built with the following five steps. At first, with reference to process context, related research, enterprise experience and expert knowledge, performance indicators in every process activity are respectively categorized into different indicator subsets according to the premises of fuzzy rules and also taking the aggregate characteristics and functions into account as shown in the middle part of Fig 3. To be clarified, fuzzy rules used here are extracted from real enterprise integration system and business process as well as fuzzy rules base, and the subsets can be formed in terms of any significative and meaningful combination. There are no completely identical subsets in the same process activity however different subsets can comprise several same indicators. At second, also taking into account those relevant knowledge mentioned before, the relationship between every subset in different child process activities and their parent indicators in the parent process activities are established respectively. At third, indicators selected in the above procedures are usually numerous, so method should be proposed to prune those relatively unnecessary indicators if it is necessary. Actually, based on the former research, a non-linear programming model is formulated by us to simplify the performance hierarchy, which is not detailed here because of the limitation of paper length. At fourth, the relationship between the indicators and fuzzy rules is checked up and if some indicator subsets are not related to any fuzzy rule, it need mine new corresponding fuzzy rules. At last, those indicators in a child process activity and related to a certain indicator in corresponding parent process activity are put together as a single set. In this case the fuzzy performance model can be finally formed as shown in the right part of Fig 3.

Afterwards, with the associated context of process model and organization model, performance indicators are adjusted appropriately if it is necessary. Then the fuzzy performance of every indicator in bottom layer of hierarchy is designated and fuzzy performance of other indicators can be measured with the following formulas. At last, management actions can be implemented correspondingly.

ψ'_{ij_ik} : If $PI_{(i+1)j_{(i+1)}} \in C'_{ij_ik}$ and ($PI_{(i+1)j_{(i+1)}}$ and $F_{(i+1)j_{(i+1)}}$ in correspondence with some fuzzy term) Then (F'_{ij_ik} in correspondence with relevant fuzzy term), $\forall j_{(i+1)}$.

ψ_{ij_i} : If (C'_{ij_ik} and F'_{ij_ik} in correspondence with some fuzzy term) Then (F_{ij_i} in correspondence with relevant fuzzy term), $\forall k$.

$$C'_{ij_ik} = \left\{ PI_{(i+1)j_{(i+1)}} \middle| PI_{(i+1)j_{(i+1)}} \in S_{ij_ikl}, \forall l, \forall j_{(i+1)} \right\} \tag{2}$$

$$F'_{ij_ik} = \left\{ F_{(i+1)j_{(i+1)}} \middle| PI_{(i+1)j_{(i+1)}} \in C'_{ij_ik} \right\} \otimes \psi'_{ij_ik} \tag{3}$$

$$F_{ij_i} = \left\{ F'_{ij_ik} \middle| C'_{ij_ik} \subseteq C_{ij_i} \right\} \otimes \psi_{ij_i}, \forall k \tag{4}$$

Where PI_{ij_i} is the j-th indicator in the i-th layer of performance hierarchy, and S_{ij_ikl} is the l -th related indicator subset corresponding with the k-th child process activity of PI_{ij_i} . C'_{ij_ik} denotes the whole set of those related indicators associated with the k-th child process activity of PI_{ij_i} , and C_{ij_i} is the set of those related indicators in all of the associated child process activities of PI_{ij_i} . $F_{(i+1)j_{(i+1)}}$ is the fuzzy performance value of $PI_{(i+1)j_{(i+1)}}$ and F_{ij_i} is the corresponding value of PI_{ij_i} . F'_{ij_ik} is about the fuzzy performance value of C'_{ij_ik} , which reflects the synthetic performance of an indicator set. ψ'_{ij_ik} and ψ_{ij_i} are the fuzzy performance relationship functions that are mainly based on fuzzy rule set.

5 Conclusions

To replenish the research of economic view, the fuzzy performance modeling method is put forward. In this proposed method, amounts of related knowledge, information and data are employed and researches and outcomes in various corresponding realms are also made full use of. In addition, based on fuzzy rules, the numerous dynamic and subjective indicators can be analyzed effectively, and we needn't consider the assumption of indicator independence, and the relationship between indicators will be more flexible. In this case, based on knowledge mining and management, our research actually intends to make performance measurement computer processible.

Some problems associated with our work deserve further research, such as Representation of causal relationship to show the causal links between performance conclusion and the management actions, and the further computer aided performance analysis, measurement and management to take the advantage of IT and system.

Anyway, it is expected that a consequence of this research, and the development and application of the model, will be a significant contribution to the performance measurement body of knowledge, particularly in the area of increasing understanding of performance measurement maturity and computer processible method.

References

1. Chen, Y.L., and Tseng, M.M.: A Stair-Like CIM System Architecture, IEEE Trans. on CPMT Part C, (1997) 101-110.
2. Folan, P., Browne, J.: A review of performance measurement: Towards performance management, Computers in Industry 56 (2005) 663-680
3. Gang Leng., Thomas Martin McGinnity, Girijesh Prasad: An approach for on-line extraction of fuzzy rules using a self-organising fuzzy neural network. Fuzzy Sets and Systems 150 (2005) 211-243
4. Kaplan, R., Norton, D.: The balanced scorecard-measures that drive performance, Harvard Business Review (January-February) 70 (1) (1992) 71-79.
5. Kaplan, R., Norton, D.: Putting the balanced scorecard to work, Harvard Business Review (September-October) 71 (5) (1993) 134-147.
6. Larsen, H.L., and Yager, R.R.: Query Fuzzification for Internet Information retrieval. In Dubois, D., Prade H., and Yager, R.R., Eds., Fuzzy Set methods in Information Engineering: A Guided Tour of Applications, John Wiley & Sons, (1997) 291-310
7. Lekova, A., Mikhailov, L., Boyadjiev, D., Nabout, A.: Redundant fuzzy rules exclusion by genetic algorithms. Fuzzy Sets and Systems 100 (1998) 235-243
8. Neely, A.: Business Performance Measurement: Theory and Practice, Cambridge University Press, Cambridge, 2002.
9. Peter Kueng.: Process performance measurement system: a tool to support process-based organizations. Total Quality Management. 11(1) (2000) 67-85.
10. Thomas L. Saaty.: Creative Thinking, Problem Solving & Decision Making. RWS Publications, 4922 Ellsworth Avenue, Pittsburgh, PA 15213. (2001)
11. Wang, Q.: Study on the enterprise performance modeling method supporting the research of economic view. PH.D. Dissertation, Tsinghua University, P.R.China, (2006)
12. Zadeh, L.A.: Fuzzy sets as a basis for a theory of possibility. Fuzzy Sets and Systems (2004) 159-179

Chance Constrained Programming with Fuzzy Parameters for Refinery Crude Oil Scheduling Problem

Cuiwen Cao and Xingsheng Gu

Research Institute of Automation, East China University of Science and Technology, 200237,130 Meilong Road, Shanghai, China
ccw03@mails.tsinghua.edu.cn

Abstract. The main objective of this work is to put forward a chance constrained mixed-integer nonlinear fuzzy programming model for refinery short-term crude oil scheduling problem under demands uncertainty of distillation units. The model studied has characteristics of discrete events and continuous events coexistence, multistage, multiproduct, uncertainty and large scale. Firstly, the model is transformed into its equivalent fuzzy mixed-integer linear programming model by using the method of Quesada & Grossmann [5]. Then the fuzzy equivalent model is changed into its crisp MILP model relies on the theory presented by Liu & Iwamura [12] for the first time in this area. Finally, the crisp MILP model is solved in LINGO 8.0 based on time discretization. A case study which has 265 continuous variables, 68 binary variables and 318 constraints is effectively solved with the proposed solution approach.

1 Introduction

If a refinery is located along seashore, its crude oil scheduling problem includes crude oil unloading process from vessels to storage tanks, transferring process from the storage tanks to charging tanks (where several crude oils are mixed), and charging process from the charging tanks to crude-oil distillation units (CDUs). In literature, mathematical programming technologies have been developed in this area under certainty by Shah [1], Lee et al.[2], Jia, Ierapetritou, & Kelly [3] , Pinto & Joly [4] , and Quesada & Grossmann [5] etc. When the whole process systems are confronted with uncertainty, the fuzzy programming method has been used in different ways in Bellman & Zadeh [13], Zimmermann [7], Liu & Sahinidis [9][10], and Sahinidis [11]in the past. Here we use the theory of Liu [6][12] for the first time to solve our problem in this area which define fuzzy programming analogous to chance constrained programming with stochastic parameters. The paper is organized as follow: Section II states the definition of short-term crude oil scheduling problem. In section III, the chance constrained fuzzy programming methods are introduced and the detail mathematical formulations for crude oil short-term scheduling problem under demands uncertainty are presented. Then the above algorithm is applied to a case study in section IV. In the last section, we put forward the conclusions.

L. Wang et al. (Eds.): FSKD 2006, LNAI 4223, pp. 1010 – 1019, 2006.

2 Problem Definition and Operation Rules

Under the given arrival times of vessels, equipment capacity limitations, key component concentration ranges and the fuzzy demands for CDUs, the problem studied is to determine the following variables to minimize the total operating costs: Waiting time of each vessel at sea; Unloading duration and flow rates from vessels to storage tanks; Crude-oil transfer duration and flow rates from storage tanks to charging tanks; Inventory variation in storage tanks and charging tanks; Concentration levels of key components in storage tanks and charging tanks.

The operation rules that have to be obeyed are as follows: A vessel has to wait at sea if a preceding vessel does not leave; While a charging tank is charging CDU, crude oil cannot be fed into the charging tank and vice versa; Each charging tank can only charge one CDU at one time interval; Each CDU can only be charged by one charging tank at one time interval; CDUs must be operated continuously throughout the scheduling time horizon.

3 Mathematical Model

3.1 Chance Constrained Programming in a Fuzzy Environment

A chance constrained programming we studied with fuzzy parameters can be written as (a1) in Liu & Iwamura's [12] method, where α_i are predetermined confidence levels, Pos{•} denotes the possibility of the events in {•}. So a point x is feasible if and only if the possibility measures of the set $\{\, \xi \mid g_i(x,\xi) \le 0,\ i=1,2,\ldots,p \}$ are at least α_i.

$$\max \quad f(x), \quad s.t. \quad Pos\{\, \xi \mid g_i(x,\xi) \le 0 \,\} \ge \alpha_i,\ i=1,2,\ldots,p \tag{a1}$$

One way of solving chance constrained programming with fuzzy parameters is to convert the chance constraints to their respective crisp equivalents. This process is only successful for some special cases. Some results are as follows.

Case I: Assume that the chance constraints in (a1 can be written as following form:

$$Pos\{\, \xi_i \mid h_i(x) \le \xi_i \,\} \ge \alpha_i,\ i=1,2,\ldots,p \tag{a2}$$

where $h_i(x)$ are functions of decision vector x (linear or nonlinear) and ξ_i are fuzzy numbers with membership functions $\mu_i(\xi_i), i=1,2,\ldots,p$, respectively.

It is clear that, for any given confidence levels $\alpha_i (0 \le \alpha_i \le 1)$, there exist some values K_{α_i} for $Pos\{\, \xi_i \mid K_{\alpha_i} \le \xi_i \,\} = \alpha_i,\ i=1,2,\ldots,p$. If K_{α_i} are replaced by smaller numbers K'_{α_i}, then $Pos\{\, \xi_i \mid K_{\alpha_i} \le \xi_i \,\} \le Pos\{\, \xi_i \mid K'_{\alpha_i} \le \xi_i \,\}$.

Notice that $Pos\{\, \xi_i \mid K_{\alpha_i} \le \xi_i \,\} = \mu_i(\xi_i)$ if the membership function μ_i are unimodal and K_{α_i} are greater than respective modes. Because we have announced that ξ_i are fuzzy members, their membership functions are indeed unimodal, we have $K_{\alpha_i} = \mu_i^{-1}(\alpha_i)$, where μ_i^{-1} are the inverse of μ_i, respectively. For the values K_{α_i} satisfying (a2) are not unique, the functions μ_i^{-1} are multi-valued. If we define K_{α_i} as the maximum values of all potential values, i.e., $K_{\alpha_i} = \sup\{K \mid K = \mu_i^{-1}(\alpha_i)\},\ i = 1,2,\ldots,p$. Thus, the crisp equivalents of chance constraints (a2) are obtained as (a3).

$$h_i(x) \le K_{\alpha_i},\ i = 1,2,\ldots,p \tag{a3}$$

Case II: Assume that the chance constraints in (a1) can be written in the form (a4).

$$Pos\{\ \xi_i \mid h_i(x) \ge \xi_i\ \} \ge \alpha_i,\ i = 1,2,\ldots,p \tag{a4}$$

If we define K_α as the minimum values ($K_{\alpha_i} = \inf\{K \mid K = \mu_i^{-1}(\alpha_i)\},\ i = 1,2,\ldots,p$) of all potential values, thus the crisp equivalents of chance constraints (a5) are obtained.

$$h_i(x) \ge K_{\alpha_i},\ i = 1,2,\ldots,p \tag{a5}$$

With the above method, our crude-oil scheduling problem under CDUs' demands uncertainty can be described as a chance constrained mixed integer nonlinear programming with fuzzy parameters. Now, let's describe the model in detail.

Indices

$i(=1,\ldots,N_{ST})$: Number of crude oil storage tanks; $j, j'(=1,\ldots,N_{BT})$: Number of crude oil blending or charging tanks; $k(=1,\ldots,N_{CE})$: Number of key components; $l(=1,\ldots,N_{CDU})$: Number of crude distillation units; $t = 1,\ldots,SCH$: Time intervals; $v = 1,\ldots,N_V$: Number of crude vessels.

Variables

$D_{j,l,t} = 0-1$: Variable to denote if the crude oil mix in charging tank j charges CDU l at time t; $X_{F,v,t}, X_{L,v,t} = 0-1$: Variable to denote if vessel v starts, complete unloading at time t; $X_{W,v,t} = 0-1$: Variable to denote if vessel v is unloading its crude oil at time t; $q_{SB,i,j,k,t}, q_{BC,j,l,k,t}$: Flow rate of component k from storage tank i to charging tank j, from charging tank j to CDU l at time t; $Q_{VS,v,i,t}, Q_{SB,i,j,t}, Q_{BC,j,l,t}$: Flow rate of crude oil from vessel v to storage tank i, from storage tank i to charging tank j, from charging tank j to CDU l at time t; $T_{F,v}, T_{L,v}$: Vessel v unloading initiation time, completion time; $v_{B,j,k,t}$: Volume of component k in charging tank j at time t; $V_{V,v,t}, V_{S,i,t}$: Volume of crude oil in vessel v, storage tank i at time t; $V_{B,j,t}$: Volume of mixed oil in charging tank j at time t; $Z_{j,j',l,t} = 0-1$: Variable to denote transition from crude mix j to j' at time t in CDU l.

Parameters

C_{Total}: Total operation cost; $C_{UNLOAD,v}$: Unloading cost of vessel v per unit time interval; $C_{SEA,v}$: Sea waiting cost of vessel v per unit time interval; $C_{INVst,i}$: Inventory cost

of storage tank i per unit time per unit volume; $C_{INVbt,j}$: Inventory cost of charging tank j per unit time per unit volume; $C_{SETup,j,j',l}$: Changeover cost for transition from crude mix j to j' in CDU l. $\tilde{d}m_j$: Fuzzy demands of crude mix j by CDUs during the scheduling horizon. $Q_{VS,v,i,\min}$, $Q_{VS,v,i,\max}$: Minimum or maximum crude oil transfer volumetric flow rate from vessel v to storage tank i; $Q_{SB,i,j,\min}, Q_{SB,i,j,\max}$: Minimum or maximum crude oil transfer volumetric flow rate from storage tank i to charging tank j; $Q_{BC,j,l,\min}, Q_{BC,j,l,\max}$: Minimum or maximum crude oil transfer volumetric flow rate from charging tank j to CDU l. $T_{ARR,v}$: Crude vessel v arrival time around the docking station. $V_{V,v,0}$: Initial volume of crude oil in crude vessel v; $V_{S,i,0}, V_{S,i,\min}, V_{S,i,\max}$: Initial, minimum, maximum crude oil volume of storage tank i; $V_{B,j,0}, V_{B,j,\min}, V_{B,j,\max}$: Initial, minimum, maximum mixed crude oil volume of charging tank j. $\xi_{S,i,k}$: Concentration of component k in the crude oil storage tank i;
$\xi_{B,j,k,0}, \xi_{B,j,k,\min}, \xi_{B,j,k,\max}$: Initial, minimum, maximum concentration of component k in the crude mix of charging tank j.

3.2 Mathematical Formulation

Minimize: The total operating cost. It includes unloading cost for the crude vessels, cost for vessels waiting in the sea, inventory cost for storage tanks and charging tanks and changeover cost between different mixed crude oils for CDUs.

$$\begin{aligned} C_{Total} = {} & C_{UNLoad,v}\sum_{v=1}^{Nv}(T_{L,v}-T_{F,v}) + C_{SEA,v}\sum_{v=1}^{Nv}(T_{F,v}-T_{ARR,v}) + \\ & C_{INVst,i}\sum_{i=1}^{N_{ST}}\sum_{t=1}^{SCH}\left(\frac{V_{S,i,t}+V_{S,i,t-1}}{2}\right) + C_{INVbt,j}\sum_{j=1}^{N_{BT}}\sum_{t=1}^{SCH}\left(\frac{V_{B,i,t}+V_{B,i,t-1}}{2}\right) + \\ & \sum_{t=1}^{SCH}\sum_{j=1}^{N_{BT}}\sum_{j'=1}^{N_{BT}}\sum_{l=1}^{N_{CDU}}(C_{SETup,j,j',l}Z_{j,j',l,t}) \end{aligned} \quad (1)$$

Subject to

1) Vessel arrival and departure operation rules: Each vessel arrives at the docking station for unloading only once throughout the scheduling horizon (2a). Each vessel leaves the docking station only once throughout the scheduling horizon (2b).

$$\sum_{t=1}^{SCH} X_{F,v,t} = 1 \quad v = 1,\ldots,N_V \quad (2a)$$

$$\sum_{t=1}^{SCH} X_{L,v,t} = 1 \quad v = 1,\ldots,N_V \quad (2b)$$

Equations for unloading initiation time (2c) and completion time (2d).

$$T_{F,v} = \sum_{t=1}^{SCH} tX_{F,v,t} \quad v = 1,\ldots,N_V \quad (2c)$$

$$T_{L,v} = \sum_{t=1}^{SCH} tX_{L,v,t} \quad v = 1,\ldots,N_V \quad (2d)$$

Each crude vessel's unloading time should start after arrival time set in the planning level (2e). Duration of the vessel unloading is bounded by the initial volume of oil in the vessel divided by maximum unloading rate (2f).

$$T_{F,v} \geq T_{ARR,v} \quad v = 1,\dots,N_V \tag{2e}$$

$$T_{L,v} - T_{F,v} \geq \left\lceil \frac{V_{V,v,0}}{(Q_{VS,v,i,\max})} \right\rceil \quad v = 1,\dots,N_V \tag{2f}$$

Vessel in the sea cannot arrive at the docking station for unloading unless the preceding vessel leaves (2g). Unloading is possible between time $T_{F,v}$ and $T_{L,v}$ (2h).

$$T_{F,v+1} \geq T_{L,v} \quad v = 1,\dots,N_V \tag{2g}$$

$$X_{W,v,t} \leq \sum_{m=1}^{t} X_{F,v,m} \quad X_{W,v,t} \leq \sum_{m=t}^{SCH} X_{L,v,m} \quad v = 1,\dots,N_V; t = 1,\dots,SCH \tag{2h}$$

2) Material balance equations for the vessels: Crude oil in vessel v at time t =initial crude oil in vessel v –crude oil transferred from vessel v to storage tanks up to time t .

$$V_{V,v,t} = V_{V,v,0} - \sum_{i=1}^{N_{ST}} \sum_{m=1}^{t} Q_{VS,v,i,m} \quad v = 1,\dots,N_V, \quad t = 1,\dots,SCH \tag{3a}$$

Operating constraints on crude oil transfer volumetric rate from vessel v to storage tank i at time t (3b).

$$Q_{VS,v,i,\min} X_{W,v,t} \leq Q_{VS,v,i,t} \leq Q_{VS,v,i,\max} X_{W,v,t} \quad v = 1,\dots,N_V; i = 1,\dots,N_{ST}; t = 1,\dots,SCH \tag{3b}$$

The volume of crude oil transferred from v vessel to storage tanks during the scheduling horizon equals to the initial crude oil volume of vessel v (3c).

$$\sum_{i=1}^{N_{ST}} \sum_{t=1}^{SCH} Q_{VS,v,i,t} = V_{V,v,0} \quad v = 1,\dots,N_V \tag{3c}$$

3) Material balance equations for storage tank: Crude oil in storage tank i at time t =initial crude oil in storage tank i + crude oil transferred from vessels to storage tank i up to time t – crude oil transferred from storage tank i to charging tanks up to time t (4a).

$$V_{S,i,t} = V_{S,i,0} + \sum_{v=1}^{N_V} \sum_{m=1}^{t} Q_{VS,v,i,m} - \sum_{j=1}^{N_{BT}} \sum_{m=1}^{t} Q_{SB,i,j,m} \quad i = 1,\dots,N_{ST}; t = 1,\dots,SCH \tag{4a}$$

Operating constraints on crude oil transfer rate from storage tank i to charging tank j at time t (4b). The term $(1 - \sum_{l=1}^{N_{CDU}} D_{j,l,t})$ denotes that if charging tank j is charging any CDU, there is no oil transfer from storage tank i to charging tank j .

$$Q_{SBi,j,\min}(1 - \sum_{l=1}^{N_{CDU}} D_{j,l,t}) \leq Q_{SBi,j,t} \leq Q_{SBi,j,\max}(1 - \sum_{l=1}^{N_{CDU}} D_{j,l,t}) \quad i = 1,\dots,N_{ST}; j = 1,\dots,N_{BT}; t = 1,\dots,SCH \tag{4b}$$

Volume capacity limitations for storage tank i at time t (4c).

$$V_{S,i,\min} \leq V_{s,i,t} \leq V_{S,i,\max} \quad i = 1,\dots N_{ST}, t = 1,\dots,SCH \tag{4c}$$

4) Material balance equations for charging tank: Crude oil mix in changing tank j at time t =initial mixed oil in storage tank j +crude oil transferred from storage tanks to charging tank j up to time t – crude oil mix j charged into CDUs up to time t (5a).

$$V_{B,j,t} = V_{B,j,0} + \sum_{i=1}^{N_{ST}} \sum_{m=1}^{t} Q_{SB,i,j,m} - \sum_{l=1}^{N_{CDU}} \sum_{m=1}^{t} Q_{BC,j,l,m} \quad j = 1,\dots,N_{BT}; t = 1,\dots,SCH \tag{5a}$$

Operating constraints on mixed oil transfer rate from charging tank j to CDU l at time t (5b). Volume capacity limitations for storage tank j at time t (5c).

$$Q_{BC,j,l,\min} D_{j,l,t} \le Q_{BC,j,l,t} \le Q_{BC,j,l,\max} D_{j,l,t} \quad j = 1,\dots,N_{BT}; l = 1,\dots,N_{CDU}; t = 1,\dots,SCH \tag{5b}$$

$$V_{B,j,\min} \le V_{B,j,t} \le V_{B,j,\max} \quad j = 1,\dots,N_{BT}; t = 1,\dots,SCH \tag{5c}$$

Total production amount of crude oil mix j should meet the fuzzy demands of crude mix j for CDUs during the scheduling horizon (5d'), where $\tilde{d}m_j$ are fuzzy members with membership functions μ_j, α_i are predetermined confidence levels to the respective constraints. If we define K_{α_j} as the minimum values of all potential values,

$$Pos\{\sum_{l=1}^{N_{CDU}} \sum_{t=1}^{S_{CH}} Q_{BC,j,l,t} \ge \tilde{d}m_j\} \ge \alpha_j \quad j = 1,\dots,N_{BT} \tag{5d'}$$

$$\sum_{l=1}^{N_{CDU}} \sum_{t=1}^{S_{CH}} Q_{BC,j,l,t} \ge K_{\alpha_j} \quad j = 1,\dots,N_{BT} \tag{5d}$$

i.e., $K_{\alpha_j} = \inf\{ K \mid K = \mu_j^{-1}(\alpha_j) \},\ j = 1,2,\dots,N_{BT}$, then we change the formulation (5d') to (5d).

5) Material balance equations for component k in the charging tank: Volume of component k in charging tank j at time t = initial component k in charging tank j + component k in crude oil transferred from storage tanks to charging tank j up to time t – component k in crude oil mix j transferred to CDUs up to time t (6a).

$$v_{B,j,t} = v_{B,j,0} + \sum_{m=1}^{t} (\sum_{i=1}^{N_{ST}} q_{SB,i,j,m} - \sum_{l=1}^{N_{CDU}} q_{BC,j,l,m}) \quad j = 1,\dots,N_{BT}; k = 1,\dots,N_{CE}; t = 1,\dots,SCH \tag{6a}$$

Operating constraints on volumetric flow rate of component k from storage tank i to charging tank j (6b), from charging tank j to CDU l (6c).

$$q_{SB,i,j,k,t} = Q_{SB,i,j,t} \xi_{S,i,k} \quad i = 1,\dots,N_{ST}; j = 1,\dots,N_{BT}; k = 1,\dots,N_{CE}; t = 1,\dots,SCH \tag{6b}$$

$$Q_{BC,j,l,t} \xi_{B,j,k,\min} \le q_{BC,j,l,k,t} \le Q_{BC,j,l,t} \xi_{B,j,k,\max} \quad l = 1,\dots,N_{CDU}; j = 1,\dots,N_{BT}; k = 1,\dots,N_{CE}; t = 1,\dots,SCH \tag{6c}$$

Volume capacity limitations for component k in charging tank j at time t (6d).

$$V_{BC,j,l,t} \xi_{B,j,k,\min} \le v_{BC,j,l,k,t} \le V_{BC,j,l,t} \xi_{B,j,k,\max} \quad l = 1,\dots,N_{CDU}; j = 1,\dots,N_{BT}; k = 1,\dots,N_{CE}; t = 1,\dots,SCH \tag{6d}$$

Equations (6c) and (6d) are linear formulations translated from non-convex bilinear equations (6'), (6''), (6''') through the method of Quesada and Grossmann[5].

$$\xi_{B,j,k,\min} \le \xi_{B,j,k,t} \le \xi_{B,j,k,\max} \quad j = 1,\dots N_{BT}; k = 1,\dots,N_{CE}; t = 1,\dots,SCH \tag{6'}$$

$$q_{BC,j,l,k,t} = \xi_{B,j,k,t} Q_{BC,j,l,t} \quad j = 1,\dots N_{BT}; k = 1,\dots,N_{CE}; t = 1,\dots,SCH; l = 1,\dots,N_{CDU} \tag{6''}$$

$$v_{B,j,k,t} = \xi_{B,j,k,t} V_{B,j,t} \quad j = 1,\dots N_{BT}; k = 1,\dots,N_{CE}; t = 1,\dots,SCH \tag{6'''}$$

6) Operating rules for crude oil charging: Charging tank j can charge at most one CDU at any time t (7a). CDU l can be charged only by one charging tank at any time t (7b).

$$\sum_{l=1}^{N_{CDU}} D_{j,l,t} \le 1 \quad j = 1,\dots,N_{BT}, t = 1,\dots,SCH \tag{7a}$$

$$\sum_{l=1}^{N_{CDU}} D_{j,l,t} \le 1 \quad j = 1,\dots,N_{BT}, t = 1,\dots,SCH \tag{7b}$$

If CDU l is charged by crude oil mix j at time $(t-1)$ and charged by j' at time t, changeover cost is involved (7c).

$$Z_{j,j',l,t} \ge D_{j',l,t} + D_{j,l,t-1} - 1 \quad j, j'(j \ne j') = 1,\dots,N_{BT}; l = 1,\dots,N_{CDU}; t = 2,\dots,SCH \tag{7c}$$

4 A Case Study

We use the basic data of the deterministic model in Lee et al. [2]. The scheduling horizon of equal duration time intervals is 8 days. The number of vessels is 2. Vessel 1's arrival time is day 1, vessel 2's arrival time is day 5. Vessel 1 contains crude oil 100 (10^4 bbl barrel), the key concentration in it is 0.01; vessel 2 contains crude oil 100 (10^4 bbl), the key concentration in it is 0.06. The number of storage tanks is 2. Storage tank 1's capacity is 0~100 (10^4 bbl), initial crude amount in it is 25 (10^4 bbl), the key concentration in it is 0.01; storage tank 2's capacity is 0~100 (10^4 bbl), initial crude amount in it is 75 (10^4 bbl barrel), the key concentration in it is 0.06. The number of charging tanks is 2. Charging tank 1's capacity is 0~100 (10^4 bbl), initial crude amount in it is 50 (10^4 bbl), the key concentration in it is around 0.015~0.025; charging tank 2's capacity is 0~100 (10^4 bbl), initial crude amount in it is 50 (10^4 bbl), the key concentration in it is around 0.045~0.055. The number of CDU is 1. Unit costs for vessel unloading cost and sea waiting cost are all 8000[$/day]. Unit costs for storage tank inventory is 50[$/(day · bbl)], for charging tank inventory is 80[$/(day · bbl)]. Once unit changeover cost is 50000[$]. The demand of mixed oil 1 and mixed oil 2 by the CDU is $_1$ and $_2$, respectively. The whole model of the case study contains 265 continuous variables, 68 binary variables and 318 constraints.

The fuzzy parameter demand for oil mix $1(\eta_1)$ is assumed to have triangular membership function as follows:

$$\mu_1(\eta_1) = \begin{cases} (\eta_1 - 90)/5 & 90 \le \eta_1 \le 95 \\ (100 - \eta_1)/5 & 95 \le \eta_1 \le 100 \quad (\times 10^4 bbl) \\ 0 & others \end{cases}$$

The fuzzy parameter demand for oil mix $2(\eta_2)$ is assumed to have membership function as follows: $\mu_2(\eta_2) = \exp[-\frac{1}{5}|\eta_2 - 95|]$, $90 \le \eta_2 \le 100$, $(\times 10^4 bbl)$

From the results of Table 1 we can see that if we increase (α_1, α_2) which represent the probability degree of satisfaction of fuzzy demands constraints, the optimal minimum objective cost decrease and vice versa.

Table 1. Main results for case study

$\alpha_1(K_{\alpha 1})$	**0.9(94.5)**	**0.8(94.0)**	**0.7(93.5)**	**0.6(93.0)**	**0.5(92.5)**	**0.4(92.0)**
$\alpha_2(K_{\alpha 2})$	**0.9(94.47)**	**0.8(93.88)**	**0.7(93.22)**	**0.6(92.45)**	**0.5(91.53)**	**0.4(90.42)**
C_{TOTAL}	217.092	217.350	217.629	217.934	218.276	218.671
C_{UNLOAD}	32	32	32	32	32	32
C_{SEA}	10	10	10	10	10	10
C_{INVst}	48.866	49.168	49.493	49.850	50.248	50.708
C_{INVbt}	26.226	26.182	26.135	26.085	26.028	25.963
C_{SETup}	100	100	100	100	100	100
$(T_{F,1},T_{L,1})$	(2,3)	(2,3)	(2,3)	(2,3)	(2,3)	(2,3)
$(T_{F,2},T_{L,2})$	(6,7)	(6,7)	(6,7)	(6,7)	(6,7)	(6,7)
$\alpha_1(K_{\alpha 1})$	**0.9(94.5)**	**0.8(94.0)**	**0.7(93.5)**	**0.6(93.0)**	**0.5(92.5)**	**0.4(92.0)**
$\alpha_2(K_{\alpha 2})$	**0.4(90.42)**	**0.5(91.53)**	**0.6(92.45)**	**0.7(93.22)**	**0.8(93.88)**	**0.9(94.47)**
C_{TOTAL}	218.146	217.961	217.829	217.734	217.665	217.617
C_{UNLOAD}	32	32	32	32	32	32
C_{SEA}	10	10	10	10	10	10
C_{INVst}	50.083	49.873	49.725	49.618	49.543	49.491
C_{INVbt}	26.063	26.088	26.105	26.115	26.122	26.126
C_{SETup}	100	100	100	100	100	100
$(T_{F,1},T_{L,1})$	(2,3)	(2,3)	(2,3)	(2,3)	(2,3)	(2,3)
$(T_{F,2},T_{L,2})$	(6,7)	(6,7)	(6,7)	(6,7)	(6,7)	(6,7)

* C_{XXX} - cost(x10^3\$), T_{XX} – day xx, α_X -predetermined confidence levels, $K_{\alpha x}$ - the minimum demands of all potential values.

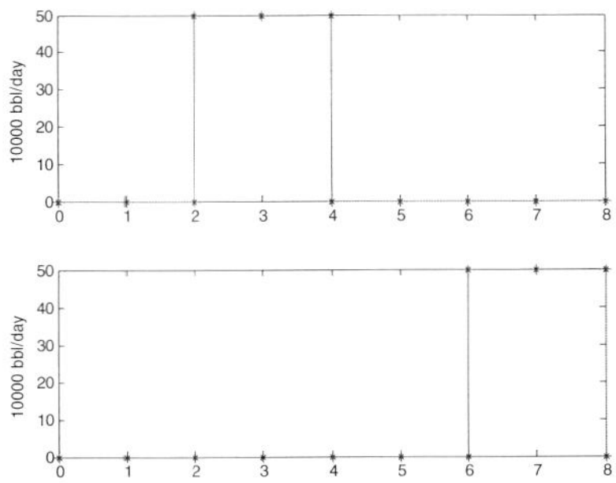

Fig. 1. Flow rate of crude oil from vessels to storage tanks

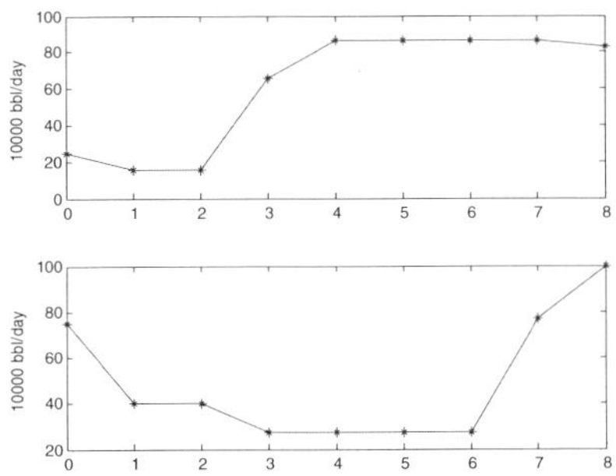

Fig. 2. Inventory status of storage tanks

Main results of once simulation $(\alpha_1 = 0.4, \alpha_2 = 0.9)$ can be found from Fig. 1 to Fig. 5 during the time scheduling horizon. Fig. 1 shows the unloading rate of crude oil in vessel 1 and vessel 2 (Upper for vessel 1, lower for vessel 2); Fig. 2 shows the inventory variation status of storage tank 1 and storage tank 2 (Upper for storage tank 1, lower for storage tank 2); Fig. 3 shows the inventory status of charging tank 1 and 2 (Upper for charging tank 1, lower for charging tank 2); Fig. 4 shows the component concentration in charging tank 1 and charging tank 2 (Upper for charging tank 1, lower for charging tank 2); Fig. 5 shows the charging rate of mixed oil from charging tank 1 and charging tank 2 to one CDU (Upper for charging tank 1, lower for 2).

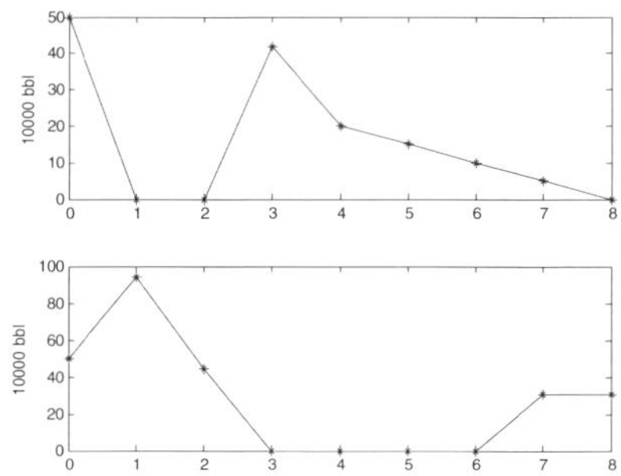

Fig. 3. Inventory status of charging tanks

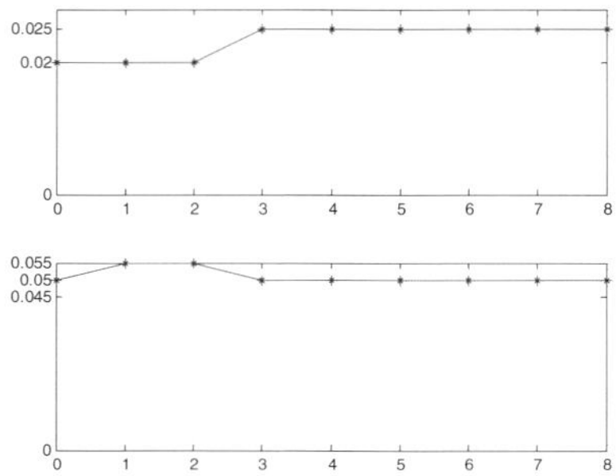

Fig. 4. Component concentration in charging tanks

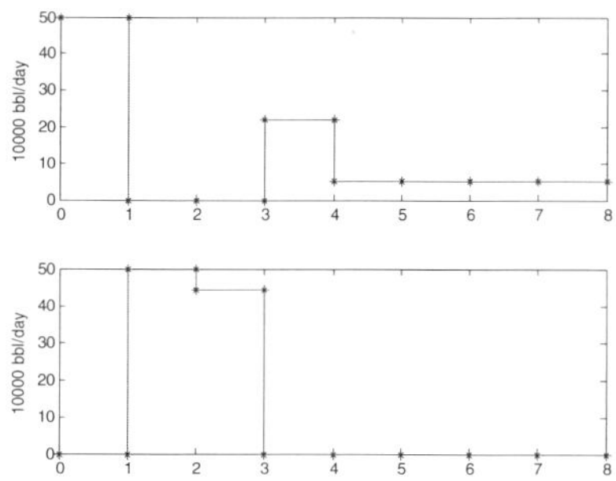

Fig. 5. Charging rate of mixed oil from charging tank 1 and charging tank 2 to the CDU

5 Conclusions

In this paper, a chance constrained mixed-integer nonlinear fuzzy programming model was presented to solve the problem of crude oil short-term scheduling under demands uncertainty of CDUs. After the model was converted into crisp equivalent MILP model, the calculation process was simplified greatly. The branch and bound method in LINGO 8.0 was used to solve the proposed equivalent formulation in a case study. The global optimal results show that this model can work with feasibility and have the possibility to be further used to solve industrial size problem.

Acknowledgment

Financial support from the National Natural Science Foundation of China (Grant No.60474043) and the Key Technology Program of Shanghai Municipal Science and Technology Commission (Grant No.04dz11008) is gratefully appreciated.

References

1. Shah, N.: Mathematical programming technique for crude oil scheduling. Computers and Chemical Engineering. 20 (1996) S1227–S1232
2. Lee, H., Pinto, J.M., Grossmann, I.E., Park, S.: Mixed-integer linear programming model for refinery short-term scheduling of crude oil unloading with inventory management. Industrial & Engineering Chemistry Research. 35 (1996) 1630–1641
3. Jia, Z., Ierapetritou, M., Kelly, J.D.: Refinery short-term scheduling using continuous time formulation: Crude-oil operations. Industrial & Engineering Chemistry Research. 42 (2003) 3085–3097
4. Pinto, J.M., Joly, M., Moro, L.F.L.: Planning and scheduling models for refinery operations. Computers and Chemical Engineering. 24 (2000) 2259–2276
5. Quesada, I., Grossmann, I.E.: Global optimization of bilinear process networks with multicomponent flows. Computers and Chemical Engineering. 19 (1995) 1219–1242
6. Liu, B D., Zhao, R Q., Wang, G.: Uncertain Programming with Applications. Tsinghua University Press (2003) (in Chinese)
7. Zimmermann, H.J.: Fuzzy set theory and its application, Boston: Kluwer Academic Publishers (1991)
8. Xie, J X., Xue, Y.: Optimization Modeling and LINDO/LINGO Software. Tsinghua University Press (2005) (in Chinese)
9. Liu, M L., Sahinidis, N.V.: Optimization in process planning under uncertaity. Industrial & Engineering Chemistry Research. 35 (1996) 4154–4165
10. Liu, M L., Sahinidis, N.V.: Process planning in a fuzzy environment. European Journal of Operational Research. 100 (1997b) 142–169
11. Sahinidis, N.V.: Optimization under uncertainty: state-of-the-art and opportunities. Computers and Chemical Engineering. 28 (2004) 971–983
12. Liu, B D., Iwamura, K.: Chance constrained programming with fuzzy parameters. Fuzzy Sets and Systems. 94 (1998) 227–237
13. Bellman, R.E., Zadeh, L.A.: Decision-making in a fuzzy environment. Management Science. 17 (1970) 141–161

Fuzzy Random Chance-Constrained Programming for Quantifying Transmission Reliability Margin*

Wu Jiekang[1,2,3], Zhou Ju[1], Wu Qiang[4], and Liang Ying[5]

[1] Department of Electrical Engineering, Gunagxi University, 100 University Rd., Nanning, 530004 Guangxi, China
wujiekang@163.com

[2] Department of Electrical Engineering, Zhejiang University, 38 ZheDa Rd., Hangzhou, 310027 Zhejiang, China

[3] Sanxing Science and Technology Co., Ltd., AUX Group, 566 Yinxian Rd., Ningbo, 315100 Zhejiang, China

[4] Department of Computer and Information Engineering, Gunagxi University, 100 University Rd., Nanning, 530004 Guangxi, China
liangying@163.com

[5] Guangxi Water Resources & Electric Power Ltd., 39 North Yuanhu Street, Nanning, 530023 Guangxi, China
wuqiang@163.com

Abstract. A fuzzy random chance-constrained programming method for quantifying transmission reliability margin is presented in this paper. Uncertainty problem of power systems is modeled as a fuzzy stochastic optimization problem, solved the hybrid intelligence algorithm based on chance constrained programming model. The simulation results of IEEE30 system demonstrate the advantages of the proposed approach.

1 Introduction

Transmission reliability Margin (TRM) is defined as the amount of transmission capability necessary to ensure that the interconnected network is secure under a reasonable range of uncertainties in system conditions [1].To take the uncertainties of power system operation conditions into account in computing ATC, several approaches have been proposed to assess the TRM, such as repeated computation method, Monte Carlo statistical approach, probabilistic approach[2], [3] and two-point estimate method [4]. In these methods, some uncertainty factors are not taken into consideration. In this paper, taking some uncertainty factors into account, a mix intelligence algorithm using fuzzy random chance-constrained programming is presented to evaluating transmission reliability margin.

---

* This work is financially supported by Guangxi Nature Science Fundation(#0640028), Guangxi Education Administration and Ningbo Nature Science Fundation(#2005A610013).

L. Wang et al. (Eds.): FSKD 2006, LNAI 4223, pp. 1020 – 1024, 2006.

2 Determination of TRM

Transfer capability can be expressed as

$$P = A(x_1, x_2, x_3, ..., x_m) \tag{1}$$

where x_i is the ith uncertain factor influencing the transfer capability, such as generation dispatch, customer demand, network parameters and topology [4]. In this study uncertainties of line parameters and bus injections are considered.

The problem of TRM allocation can be cast as a problem of supply versus demand. For this purpose, two random variables are defined: R, the reliability margin allocated, and U, the uncertainty of transfer capability due to the uncertainty in the parameters. The objective of reliability analysis is to ensure that R is greater than U for the reserved transmission service. To assess the unreliability probability $P(R \le U)$, the safety margin defined as the difference between reliability margin and transfer capability uncertainty is used[5], [6]:

$$M = R - U \tag{2}$$

If R and U are random variables, M is also a random variable. The probability of unreliability is given by

$$P_{unreliability} = P[(R-U) \le 0] = P[M \le 0] \tag{3}$$

If M is normally distributed, the reliability index β is defined as the number of standard deviation of $M\,(\sigma_M)$ from the mean of the safety margin μ_M :

$$\beta = \frac{\mu_R - \mu_U}{\sqrt{\sigma_R^2 + \sigma_U^2 - 2\operatorname{cov}(R,U)}} \tag{4}$$

To allocate a fixed reliability margin μ_R and assume that R and U are uncorrelated, then $\sigma_R = 0$, $\mu_U = 0$ and $\operatorname{cov}(R,U) = 0$ $\operatorname{cov}(R,U)$ =0. The reliability index β can be expressed as:

$$\beta = \frac{\mu_R}{\sigma_U} = \frac{TRM}{\sigma_U} \tag{5}$$

Transmission reliability Margin is computed by:

$$TRM = \beta\sigma_U \tag{6}$$

3 Fuzzy Random Chance-Constrained Programming (FRCCP)

Fuzzy random chance-constrained programming proposed by Liu[7]-[8] is adopted. Supposes x is a decision-making vector, the parameter ξ is the fuzzy random vector, $f(x,\xi)$ is the objective function, $g_j(x,\xi)$ is the restraint function, $j = 1,2,3,...,m$. In this case, chance constrained problem can be expressed as:

$$ch\{g_j(x,\xi) \le 0, j = 1,2,...,m\}(\alpha) \ge \beta \tag{7}$$

where α and β are confidence level, $ch\{g_j(x,\xi) \le 0\}(\alpha)$ is the primitive chance of fuzzy random event $g_j(x,\xi) \le 0$, It is a function from $(0,1]$ to $[0,1]$.

The standard fuzzy random chance constrained programming model is as follows:

$$\max \overline{f} \tag{8}$$

s.t.

$$ch\{f(x,\xi) \ge \overline{f}\}(\gamma) \ge \delta \tag{9}$$

$$ch\{g_j(x,\xi) \le 0\}(\alpha) \ge \beta,\ j = 1,2,...,m \tag{10}$$

where γ , δ , α_j and β_j are respectively confidence level.

4 Test Results

In order to confirm this algorithm validity, this section tests the proposed method in the IEEE30 system, comparing it with Monte Carlo simulations. The diagram and the data parameter come from [11]. It is assumed that unit 2 and unit 5 is used for base load; unit 1 is used for frequency modulation. The expectation accident and the load of the power systems is respectively given as table 2 and table 3.

Table 1. Contingency set

No.	Fault type
1	The line 19-20 three-phase short-circuits
2	The line 9-10 three-phase short-circuits
3	Units 11 cutting machine

Table 2. Distribution parameters of real power at load buses (p.u.)

Bus	Mean	Standard Deviation	Bus	Mean	Standard Deviation	Bus	Mean	Standard Deviation
1	0.560	0.000	10	0.059	0.010	19	0.094	0.017
2	0.116	0.011	12	0.162	0.012	20	0.025	0.018
3	0.024	0.023	14	0.076	0.021	21	0.178	0.053
4	0.65	0.032	15	0.054	0.015	23	0.032	0.011
5	0.912	0.234	16	0.856	0.047	24	0.089	0.023
7	0.245	0.038	17	0.784	0.033	26	0.035	0.010
8	0.309	0.027	18	0.032	0.010	29	0.028	0.010

For the uncertainty function $V : x \rightarrow \left(V_1(x), V_2(x), V_3(x)\right)$, we have:

$$V_1 = \min\left\{\overline{R} \middle| ch\left\{A(x,\xi) \le \overline{R}\right\}(0.95) \ge 0.85\right\}$$

$$V_2 = \min\left\{\overline{R} \middle| ch\left\{g_1(x,\xi) \le P_{l,\max}\right\}(0.90) \ge 0.95\right\}$$

$$V_3 = \min\left\{\overline{R} \middle| ch\left\{g_2(x,\xi) \le P_{g,\max}\right\}(0.85) \ge 0.80\right\}$$

$$V_4 = \min\left\{\overline{R} \middle| ch\left\{g_3(x,\xi) \le P_{L,\max}\right\}(0.95) \ge 0.90\right\}$$

where 0.95, 90, 0.85 and 0.80 are respectively confidence level. Based on these training samples, we train the neuron network (4 inputs neurons, 6 hide level neurons, 2 outputs neurons) to obtain the uncertainty function V . When the mix intelligence algorithm is carried out (simulates 6,000 generations, 2,000 trainings samples, 400 heredities iterations), the optimal solution is obtained, as shown in table3.

Table 3. TRM computation in IEEE30 system

Probability of Reliability	90%	95%	99%	99.5%
FRCCP	0.6223	0.7981	1.1542	1.2586
Monte Carlo	0.6017	0.8123	1.2281	1.2543

5 Conclusion

This paper proposed an electric transmission reliability margin mathematical model based on the fuzzy random chance constrained programming. Compared with the Monte-Carlo simulation method, this method has more precise in computation and is faster in speed.

References

1. North American Electric Reliability Council (NERC): Available Transfer Capability Definitions and Determinnation.(1996)
2. Sauel P.W.: Alternatives for calculating transmission reliability margin (TRM) in available transfer capability (ATC). In Proc. 31st Hawaii INt. Conf. Syst. SCi., Kohala Coast, HI. January 6-9, (1998) 89.
3. Zhang J., Dobson I., Alvarado F. L.: Quantifying transmission reliability margin. Int. J. Elect.Power Energy Syst., 26 (2004)697-702
4. Chun L. S., Chan N. L.: Two-point estimate method for quantifying transfer capability uncertainty. IEEE Transactions on Power Systems, 20(2) (2005)573-579.
5. Harr M.E.: Reliability-based Design in Civil Engineering. New York: McGraw-Hill(1987)
6. Ang A.. H.–S., W. H.: Probability Concepts in Engineering Planning And Design. New York: Wiley(1975)
7. Liu B.: Fuzzy random chance-constrained programming. IEEE Transactions on Fuzzy Systems, 9(5)(2001)713-720

8. Liu B.: Uncertain Programming with Applications. Tsinghua university Press(2003)
9. Guo Y.J.: Power System Reliability Analysis. Tsinghua university Press(2003)
10. Liu Y. K., Liu B.: A class of fuzzy random optimization: Expected value models. Information Science, to be published.
11. Pai M A.: Computer techniques in power system Analysis. McGraw Hill, New Delhi(1980)

The Fuzzy Weighted k-Cardinality Tree and Its Hybrid Genetic Algorithm

Linzhong Liu[1,2], Ruichun He[1], and Yinzhen Li[1]

[1] School of Traffic and Transportation, Lanzhou Jiaotong University, Lanzhou 730070, China
[2] Dept. of Mathematical Sciences, Tsinghua University, Beijing 100084, China
lzliu@orsc.edu.cn

Abstract. The k-tree problem is to find a tree with k vertices in a given graph such that the total cost is minimum and is known to be NP-hard. In this paper, the k-tree problem with fuzzy weights is firstly formulated as the chance-constrained programming by using the possibility measure and the credibility measure. Then an oriented tree and knowledge-based hybrid genetic algorithm is designed for solving the proposed fuzzy programming models.

Keywords: Fuzzy set, credibility measure, spanning tree, genetic algorithm, quadratic tree, k-tree.

1 Introduction

The problem with respect to trees has a long history in the combinatorial optimization. Since the MST problem was initialized by Boruvka [13] in 1926, it was studied by many researchers and has been applied in wide varieties such as the telecommunication network design, the distribution systems and so on. For example, Ahuja, Orlin and Faster [10] proposed an algorithm for the inverse spanning tree problem; Fujie [11] designed an exact algorithm for the maximum leaf spanning tree problem; Kaneko [22] studied the spanning trees with constraints on the leaf degree; Katagiri and Ishii [14] presented the chance constrained bottleneck spanning tree problem with fuzzy random edge costs; Hassin and Levin [16] studied the minimum restricted diameter spanning trees; Galbiati, Morzenti and Maffioli [19] designed the approximately algorithm of the maximum spanning tree problems. Some other works of the MST problems are introduced in [10]–[23].

The k-cardinality problems such as the k-shortest path problem, the k-matching were studied by many researchers in recent years and many algorithms were designed. Additionally, many optimizing problems such as the fuzzy shortest path problem(Boulmakoul[1],Okada[6] and Okada and Soper[7]) and the fuzzy max-flow problem([3]) have been studied in fuzzy environments. Therefore, it is natural to extend the k-tree problem to fuzzy environments. Since the fuzzy set theory was initialized by Zadeh[8] in 1965, it has been well developed by many researchers, for example, the possibility measure and the credibility measure, in

L. Wang et al. (Eds.): FSKD 2006, LNAI 4223, pp. 1025–1034, 2006.

recent years. In this paper, the models of fuzzy k-tree problem are formulated by using the possibility measure and the credibility measure.

This paper is organized as follows. Firstly, we recall some preliminaries with respect to fuzzy set theory in Section 2. Then we formulate the fuzzy k-tree problem as chance-constrained programming models and induce the crisp equivalents of the models in Section 3. Finally, an oriented tree and knowledge-based hybrid genetic algorithm is given in Section 4 by using heuristic approach and a numerical example is given in Section 5.

2 The Definition and Notations

In this section, we shall introduce some preliminaries and notations. Let ξ be a fuzzy variable with membership function $\mu(x)$ and r a real number. Then the possibility measure and the necessity measure of fuzzy event $\{\xi \le r\}$ were defined as

$$\mathrm{Pos}\{\xi \le \mathrm{r}\} = \sup_{\mathrm{x}\le\mathrm{r}} \{\mu(\mathrm{x})\} \text{ and } \mathrm{Nec}\{\xi \le \mathrm{r}\} = 1 - \mathrm{Pos}\{\xi > \mathrm{r}\},$$

respectively. The credibility measure of the fuzzy event $\{\xi \le r\}$ was defined by Liu and Liu[5] as

$$\mathrm{Cr}\{\xi \le \mathrm{r}\} = \frac{1}{2}(\mathrm{Pos}\{\xi \le \mathrm{r}\} + \mathrm{Nec}\{\xi \le \mathrm{r}\}).$$

One important mathematical property of the credibility measure is the duality: $\mathrm{Cr}\{\xi \le \mathrm{r}\}+\mathrm{Cr}\{\xi > \mathrm{r}\} = 1$ for any fuzzy event $\{\xi \le r\}$. Liu [5] also presented two critical values which serve as the roles to rank fuzzy variables and are defined as $\xi_{\sup}(\alpha) = \sup\{\mathrm{r}|\mathrm{Cr}\{\xi \ge \mathrm{r}\} \ge \alpha\}$ and $\xi_{\inf}(\alpha) = \inf\{\mathrm{r}|\mathrm{Cr}\{\xi \le \mathrm{r}\} \ge \alpha\}$, respectively, where $\alpha \in (0, 1]$.

Throughout this paper, all the graphs are simple graph. For a set S, $|S|$ denotes the cardinality of the set S. Let $G(V, E)$ be a graph with vertices set $V = \{v_1, v_2, \cdots, v_n\}$ and edges set $E = \{e_1, e_2, \cdots, e_m\}$. Sometimes, we quote vertices pair (u, v) to denote the edge with u and v as its end vertices and $E(G)$ and $V(G)$ to denote the edges set and the vertices set of graph G for convenience, respectively. Let S be a subset of E or V. We denote $G[S]$ the induced subgraph of the subset S. Assume that ξ_i is the cost of edge $e_i \in E$ and $\boldsymbol{\xi}$ the vector consists of ξ_i, $i = 1, 2, \cdots, |E|$. For a vertex $v \in V$, let $N_e[v] = \{(u, v)|(u, v) \in E\}$ and $N(v) = \{u|(u, v) \in E\}$. For $u, v \in V(G)$, $G - (u, v)$ and $G + (u, v)$ denote the graphs $G(V, E \setminus \{(u, v)\})$ and $G(V, E \bigcup \{(u, v)\})$, respectively.

A k-tree is a tree with k-vertices. Let T be a k-tree of the graph $G(V, E)$ and $x_i = 1$ if $e_i \in E(T)$ and $x_i = 0$ otherwise. Then the k-tree T can also be denoted by such a vector $\boldsymbol{x} = (x_1, x_2, \cdots, x_{|E|})$. The cost function of the k-tree $\boldsymbol{x}$ can be defined as $f(\boldsymbol{x}, \boldsymbol{\xi}) = \sum_{i=1}^{|E|} \xi_i x_i$.

When the vector $\boldsymbol{\xi}$ is crisp values, then the k-tree problem can be formulated as

$$\begin{aligned} &\min f(\boldsymbol{x}, \boldsymbol{\xi}) = \sum_{i=1}^{|E|} \xi_i x_i \\ &\text{s.t.} \sum_{i=1}^{|E|} x_i = k-1, \\ &\qquad \sum_{e_i \in S} x_i \le |S| - 1, \text{for } S \subset T, \\ &\qquad |\{(u,v)|u \in V(S), v \in V(T \setminus S)\}| \ge 1, \text{for } S \subset T, \\ &\qquad x_i = 0 \text{ or } 1, i = 1, 2, \cdots, |E|, \end{aligned} \tag{1}$$

where $T = \{e_i | x_i = 1\}$.

The second constraint of the model (1) ensures that $\boldsymbol{x}$ is cycle free and the third ensures that $\boldsymbol{x}$ is connected. Therefore, $\boldsymbol{x}$ is a connected and cycle free subgraph, so the first to third constraints of the model (1) ensure that $\boldsymbol{x}$ denotes a k-tree.

It is clear that the cost function $f(\boldsymbol{x}, \boldsymbol{\xi})$ is also a fuzzy variable when the vector $\boldsymbol{\xi}$ is fuzzy vectors. Then the following concept is proposed when ξ_i is fuzzy variables, $i = 1, 2, \cdots, |E|$.

Definition 1. *A k-tree x^* is called α-(k-tree) if*

$$\min \left\{ \bar{f} \middle| \mathrm{M}\left\{ \mathrm{f}(\boldsymbol{x}^*, \boldsymbol{\xi}) \le \bar{\mathrm{f}} \right\} \ge \alpha \right\} \le \min \left\{ \bar{f} \middle| \mathrm{M}\left\{ \mathrm{f}(\boldsymbol{x}, \boldsymbol{\xi}) \le \bar{\mathrm{f}} \right\} \ge \alpha \right\}$$

for any k-tree $\boldsymbol{x}$, where α is a predetermined confidence level between 0 and 1 and $\mathrm{M}\{\cdot\}$ is the possibility measure or credibility measure. The value

$$\min \left\{ \bar{f} \middle| \mathrm{M}\left\{ \mathrm{f}(\boldsymbol{x}, \boldsymbol{\xi}) \le \bar{\mathrm{f}} \right\} \ge \alpha \right\}$$

is called the α-cost of the cost function $f(\boldsymbol{x}, \boldsymbol{\xi})$ at $\boldsymbol{x}$.

3 The Model of Fuzzy k-Tree

In this section, we formulate the fuzzy k-tree problem as chance-constrained programming by using the possibility measure and credibility measure, respectively. In the following process, we assume that all ξ_i are fuzzy variables, $i = 1, 2, \cdots, |E|$.

The chance-constrained programming was initialized by Charnes and Cooper[2]. As an extension of the chance-constrained programming, Liu and Iwamura[4] proposed the idea of fuzzy chance-constrained programming. In some situations, the decision maker hopes to minimize the value $\bar{f}$ with Pos $\left\{\mathrm{f}(\boldsymbol{x}, \boldsymbol{\xi})) \le \bar{\mathrm{f}}\right\} \ge \alpha$ or Cr $\left\{\mathrm{f}(\boldsymbol{x}, \boldsymbol{\xi})) \le \bar{\mathrm{f}}\right\} \ge \alpha$, namely, to find the α-cost, then the fuzzy k-tree problem can be formulated as the chance-constrained programming. The following two models are the chance-constrained programming models formulated by using the possibility measure and the credibility measure, respectively.

$$\begin{aligned}
&\min U_1(\boldsymbol{x}) = \bar{f} \\
&\text{s.t. Pos}\left\{\sum_{i=1}^{|E|} \xi_i x_i \le \bar{f}\right\} \ge \alpha \\
&\sum_{i=1}^{|E|} x_i = k-1, \\
&\sum_{e_i \in S} x_i \le |S|-1, \quad \text{for any } S \subset T, \\
&|\{(u,v)|u \in V(S), v \in V(T \setminus S)\}| \ge 1, \text{for } S \subset T, \\
&x_i = 0 \text{ or } 1, i = 1, 2, \cdots, |E|
\end{aligned} \tag{2}$$

and

$$\begin{aligned}
&\min U_2(\boldsymbol{x}) = \bar{f} \\
&\text{s.t. Cr}\left\{\sum_{i=1}^{|E|} \xi_i x_i \le \bar{f}\right\} \ge \alpha \\
&\sum_{i=1}^{|E|} x_i = k-1, \\
&\sum_{e_i \in S} x_i \le |S|-1, \quad \text{for any } S \subset E(T), \\
&|\{(u,v)|u \in V(S), v \in V(T \setminus S)\}| \ge 1, \text{for } S \subset T, \\
&x_i = 0 \text{ or } 1, i = 1, 2, \cdots, |E|,
\end{aligned} \tag{3}$$

where $T = \{e_i | x_i = 1\}$.

If the costs of the graph are triangular fuzzy variables or trapezoidal fuzzy variables, the models (2) and (3) can be converted to their crisp equivalents. We just take the trapezoidal fuzzy variables as the example to illustrate this idea.

Let $\xi = (r_1, r_2, r_3, r_4)$ be a trapezoidal fuzzy variable with membership function $\mu(x)$, where $r_1 < r_2 < r_3 < r_4$. Then the possibility measure and the credibility measure of fuzzy event $\{\xi \le x\}$ are defined as follows:

$$\text{Pos}\{\xi \le x\} = \begin{cases} 1, & \text{if } r_2 \ge x, \\ \dfrac{x - r_1}{r_2 - r_1}, & \text{if } r_2 \le x \le r_3, \\ 0, & \text{otherwise.} \end{cases} \tag{4}$$

$$\mathrm{Cr}\{\xi \leq \mathrm{x}\} = \begin{cases} 0, & \text{if } x \leq r_1, \\ \dfrac{x - r_1}{2(r_2 - r_1)}, & \text{if } r_1 \leq x \leq r_2, \\ \dfrac{1}{2}, & \text{if } r_2 \leq x \leq r_3, \\ \dfrac{x + r_4 - 2r_3}{2(r_4 - r_3)}, & \text{if } r_3 \leq x \leq r_4, \\ 1, & \text{if } x \geq r_4. \end{cases} \tag{5}$$

Theorem 1. *Assume that all ξ_i are independent trapezoidal fuzzy variables defined as $\xi_i = (a_i^1,\ a_i^2,\ a_i^3,\ a_i^4)$, $i = 1, 2, \cdots, |E|$. If $\alpha > 0.5$, then the models (2) and (3) are equivalent to the following models, respectively:*

$$\begin{aligned} &\min\, (1-\alpha)c_1(\boldsymbol{x}) + \alpha c_2(\boldsymbol{x}) \\ &s.t. \sum_{i=1}^{|E|} x_i = k - 1, \\ &\qquad \sum_{e_i \in S} x_i \leq |S| - 1, \quad \textit{for any } S \subset E(T), \\ &\qquad |\{(u,v)|u \in V(S), v \in V(T \setminus S)\}| \geq 1, \textit{for } S \subset T, \\ &\qquad x_i = 0 \textit{ or } 1, i = 1, 2, \cdots, |E|\,. \end{aligned} \tag{6}$$

and

$$\begin{aligned} &\min\, 2(1-\alpha)c_3(\boldsymbol{x}) + (2\alpha - 1)c_4(\boldsymbol{x}) \\ &s.t. \sum_{i=1}^{|E|} x_i = k - 1, \\ &\qquad \sum_{e_i \in S} x_i \leq |S| - 1, \quad \textit{for any } S \subset E(T), \\ &\qquad |\{(u,v)|u \in V(S), v \in V(T \setminus S)\}| \geq 1, \textit{for } S \subset T, \\ &\qquad x_i = 0 \textit{ or } 1, i = 1, 2, \cdots, |E|\,, \end{aligned} \tag{7}$$

where $T = \{e_i|x_i = 1\}$ and

$$c_1(\boldsymbol{x}){=}\sum_{i=1}^{|E|} a_i^1 x_i, c_2(\boldsymbol{x}){=}\sum_{i=1}^{|E|} a_i^2 x_i, c_3(\boldsymbol{x}){=}\sum_{i=1}^{|E|} a_i^3 x_i, c_4(\boldsymbol{x}){=}\sum_{i=1}^{|E|} a_i^4 x_i.$$

Proof. Firstly, we proof the model (6). Due to the independence property of weights ξ_i and $x_i \geq 0$, it is clear that

$$\sum_{i=1}^{|E|} \xi_i x_i = (c_1(\boldsymbol{x}), c_2(\boldsymbol{x}), c_3(\boldsymbol{x}), c_4(\boldsymbol{x}))$$

is also a trapezoidal fuzzy variable. It follows from equation (4) and $\alpha > 0.5$ that the chance-constraint

$$\mathrm{Pos}\left\{\sum_{\mathrm{i}=1}^{|\mathrm{E}|} \xi_{\mathrm{i}}\mathrm{x}_{\mathrm{i}} \geq \bar{\mathrm{f}}\right\} \geq \alpha$$

of model (2) is equivalent to $\dfrac{\bar{f} - c_1(\boldsymbol{x})}{c_2(\boldsymbol{x}) - c_1(\boldsymbol{x})} \geq \alpha$. Hence, we have $(1-\alpha)c_1(\boldsymbol{x}) + \alpha c_2(\boldsymbol{x}) \leq \bar{f}$. This implies that the α-cost of cost function is just $\alpha c_2(\boldsymbol{x}) + (1-\alpha)c_1(\boldsymbol{x})$.

Similarly, we can prove that the model (3) is equivalent to the model (7). The theorem is thus proved.

4 Hybrid Genetic Algorithm

As stated in Section 1, the algorithms associated with various tree problems have been well studied. In this Section, a heuristic genetic algorithm is designed for solving the proposed models and their crisp equivalent. In order to describe the problem conveniently, we briefly take k to denote the kth edge e_k in the edges set E. In addition, we assume that all the oriented trees in the following process are rooted at vertex 1.

Representation. There are many ways to represent a solution of the optimization problem in genetic algorithm. For a spanning tree, Zhou and Gen[9] employed the Prufer number to encode the chromosome and designed the Prufer number based crossover operator and mutation operator in the genetic algorithm for solving the quadratic minimum spanning tree problem. It is clear that a k-tree can't be encoded by this ways. In this paper, we encode a k-tree by employing the edges indices in a k-tree. By such an encode method, we know that the length of each chromosome is exactly equal to $k-1$ and each encode of chromosome is uniquely correspond to a k-tree. In order to find a cycle when an edge is added to a tree, these operators need us to decode the operated chromosomes to oriented trees firstly, then perform the crossover and mutation operations by using the oriented trees.

Initialization Process. The initialization is an important process in genetic algorithm serving as the role of initializing the chromosomes. We take *pop_size* to denote the number of chromosomes. We initialize chromosomes X_1, X_2, $\cdots$, X_{pop_size} by repeating the following algorithm *pop_size* times, where T denotes a k-tree of the graph $G(V, E)$.

The initialization algorithm

Step 1. Set $T = \emptyset$. Randomly select a vertex $u_0 \in V(G)$ and a vertex $v_0 \in N(u_0)$. Set $T \leftarrow T + \{(u_0, v_0)\}$.

Step 2. Randomly select a vertex $u \in V(T)$ with $N(u) \setminus V(T) \neq \emptyset$. Randomly select a vertex $v \in N(u) \setminus V(T)$. Set $T \leftarrow T + \{(u, v)\}$.

Step 3. Repeat Step 2 until $|E(T)| = k - 1$.

Theorem 2. *The obtained tree in the initialization algorithm is a k-tree.*

Proof. Because the edge (u, v) in the Step 2 has a common vertex u with $V(T)$, the obtained graph T is obviously a connected graph. Therefore, a k-tree is obtained when $|E(T)| = k - 1$.

Crossover Operation. Let $P_c \in (0,1)$ be the crossover probability. In order to determine the parents for the crossover operation, we repeat the following process *pop_size* times: randomly generating a real number r from interval $(0,1)$, the chromosome X_i is selected to crossover if $r < P_c$. We denote the selected parents by $X_1', X_2', \cdots$ and divide them into the following pairs: $(X_1', X_2'), (X_3', X_4'), (X_5', X_6'), \cdots$

Lemma 1. *[24] Let G be a connected simple graph and T be a k-tree of the graph G, $k \le |V(G)| - 1$. If $(u,v) \in E(G)$ satisfies $(u,v) \notin E(T)$ and $(u,v) \notin V(T)$, then there is a unique cycle in $T + \{(u,v)\}$.*

Based on Lemma 1, a heuristic crossover algorithm is designed. Assume that (X_1', X_2') is the chromosomes pair of crossover. The algorithm is summarized as follows.

Crossover algorithm

Step 1. Decode X_1' and X_2' to the oriented k-trees T_1 and T_2.

Step 2. Set $t = |E(T_1) \setminus (E(T_1) \bigcap E(T_2))| = |E(T_1) \setminus (E(T_2) \bigcap E(T_2))|$.

Step 3. Randomly generate an integer s from interval $[1, t]$.

Step 4. Randomly select two sets S_1 and S_2, $|S_1| = |S_2| = s$:

$$S_1 \subset E(T_1) \setminus (E(T_1) \bigcap E(T_2)), \quad S_2 \subset E(T_2) \setminus (E(T_1) \bigcap E(T_2)).$$

Step 5. For every edge $e = (u,v) \in S_1$, if $(u,v) \in V(T_2)$ (or $V(T_1)$), find the unique cycle C_e in $T_2 + \{e\}$. Randomly select an edge $e' \in E(C_e)$, $e' \neq e$. Set $T_2 \leftarrow T_2 - \{e'\} + \{e\}$; if $|\{u,v\} \bigcap V(T_2)| = 1$, then set $T_2 \leftarrow T_2 + e$ and randomly select an leaf edge $e' \in E(T)$, $e' \neq e$ and remove it from T_2.

Step 6. Perform the similar operations as Step 5 between S_2 and T_1.

Step 7. Encode the oriented k-trees T_1 and T_2 to X_1' and X_2'.

From the above crossover operation, we know that the crossover operation is actually an one-point crossover operation. Such an operation can ensure that all new chromosomes are also feasible.

The algorithm to find the cycle C_e

Step 1. Start from the vertex u, along the predecessor vertex of u, find the path P^u from the root to the vertex u. If $v \in V(P^u)$, the cycle C_e is found and set $C_e = P^u_{[u,v]} + \{e\}$, stop. Otherwise, go to next step.

Step 2. Start from the vertex v, along the predecessor vertex of v, find the path P^v from the root to the vertex v. If $u \in V(P^v)$, the cycle C_e is found and set $C_e = P^v_{[u,v]} + \{e\}$, stop. Otherwise, go to next step.

Step 3. Start from the root vertex, along the path P^v or P^u to find the the last common vertex s of the path P^v and P^u, set $C_e = P^u_{[u,s]} + P^v_{[s,v]} + \{e\}$.

Mutation Process. In this section, we will design an oriented tree and knowledge-based heuristic mutation operator. Let $P_m \in (0,1)$ be the mutation probability. We employ the following operator to select the chromosome to be mutated: for $i = 1$ to *pop_size*, randomly generate a random number r from interval $(0, 1)$; if $r \leq P_m$, then the chromosome X_i is selected to be mutated. The idea of mutation operation is also originated from Lemma 1. The mutation algorithm is summarized as follows.

Mutation algorithm

Step 1. For $i = 1$ to *pop_size*, repeat Step 2 to Step 3.

Step 2. Randomly generate a random number r from interval $(0, 1)$. If $r \leq P_m$, then go to Step 3. Otherwise, go to Step 1.

Step 3. Decode the mutated chromosome X_i to an oriented tree T and randomly select an integer s from interval $[1, |E(G) \setminus E(T)|]$ and randomly select k different edges $e_1, e_2, \cdots, e_s$ from set $E(G) \setminus E(T)$. For $j = 1$ to s, repeat Step 4.

Step 4. For every edge $e_i = (u_i, v_i)$, if $(u_i, v_i) \in V(T)$, find the unique cycle C_{e_i} in $T + \{e_i\}$. Randomly select an edge $e' \in E(C_{e_i})$, $e' \neq e_i$. Set $T \leftarrow T - \{e'\} + \{e_i\}$; if $|\{u_i, v_i\} \bigcap V(T)| = 1$, then set $T \leftarrow T + e_i$ and randomly select an leaf edge $e' \in E(T)$, $e' \neq e_i$ and remove it from T.

Step 5. Encode the oriented tree T to the chromosome X_i.

Clearly, such a mutation can also ensure that all new chromosome is also feasible.

Evaluation. The evaluation process is to calculate the fitness of the chromosomes so as to evaluate the chromosomes. We take the functions $U_1(\boldsymbol{x})$ and $U_2(x)$ of models (2) and (3) to calculate the fitness of chromosomes in solving models (2) and (3), respectively. To calculate the fitness, the fuzzy simulations are employed. For more details of fuzzy simulations, the reader can consult to [5].

The fitness f_i of the ith chromosome X_i is defined as the reciprocal of the value of the objective function at X_i, $i = 1, 2, \cdots, pop\_size$. The evaluating function is defined as

$$Eval(X_i) = f_i \Bigg/ \sum_{k=1}^{pop\_size} f_k \ , i = 1, 2, \cdots, pop\_size.$$

Selection Process. The selection process is to select the offsprings of the chromosomes in the genetic algorithm. Let $P_i = \sum_{k=1}^{i} Eval(X_k)$, $i = 1, 2, \cdots, pop\_size$ and $P_0 = 0$, then we employ the spanning roulette wheel to select the chromosomes: randomly generate a number $p \in (0, 1)$; if $p \in [P_{i-1}, P_i)$, then the chromosome X_i is selected. The following is the algorithm.

Genetic Algorithm

Step 1. Randomly initialize *pop_size* chromosomes.

Step 2. Update the chromosomes by crossover process and mutation process.

Step 3. Calculate the objective values $U_1(\boldsymbol{x})$ and $U_2(\boldsymbol{x})$ of the chromosomes by fuzzy simulation.

Step 4. Calculate the fitness of each chromosome according to the objective values.

Step 5. Select the chromosomes by spanning the roulette wheel.

Step 6. Repeat Step 2 to Step 5 for a given number times.

Step 7. Report the best chromosome as the optimal solution.

The two functions $U_1(\boldsymbol{x})$ and $U_2(\boldsymbol{x})$ in the Step 3 of the above algorithm correspond to solve the model (2) and model (3), respectively.

5 The Numerical Experiment and Conclusions

Some numerical experiments have been done with different evolution parameters. We found that the best evolution parameters are that the crossover probability is $P_c = 0.4$ and the mutation probability $P_m = 0.3$. The optimal solution of a numerical example with 21 nodes and 55 edges are found at $400th$ generation, where the number of chromosomes $pop\_size = 30$ and the evolution generation $Gen = 2000$.

In this paper, the fuzzy weighted k-tree was formulated as chance-constrained programming by using possibility measure and credibility measure, respectively. Furthermore, a knowledge-based hybrid genetic algorithm was designed for solving the problem. Actually, the designed algorithm is not only available for the k-tree with fuzzy parameters, but also available for the k-tree with crisp parameters.

Acknowledgements

This research is supported by National Natural Science Foundation of China (No.60425309) and the Qinglan Talent Funds of Lanzhou Jiaotong University.

References

1. A.Boulmakoul, Generalized path-finding algorithms on semirings and the fuzzy shortest path problem. *Journal of Computational and Applied Mathematics*, 162 (2004) 263–272.
2. A.Charnes, W.W.Copper. Chance-constrained programming. *Management Science*, 6(1959),73-79.
3. P.Diamond. A fuzzy max-flow min-cut theorem. *Fuzzy Sets and Systems*, 119(2001), 139–148.
4. B.Liu, and K.Iwamura. A note on chance constrained programming with fuzzy coefficients, *Fuzzy Sets and Systems*, 100(1998), 229-233.
5. B.Liu. Theory and Practice of Uncertain Programing, Physica-Verlag, New York: 2002.
6. S.Okada. Fuzzy shortest path problems incorporating interactivity among paths. *Fuzzy Sets and Systems*, 142(2004), 335–357.

7. S.Okada, T.Soper. A shortest path problem on a network with fuzzy arc lengths. *Fuzzy Sets and Systems*, 109(2000), 129–140.
8. L.A.Zadeh. Fuzzy sets, *Information and Control*, 8(1965), 338–353.
9. G.Zhou,M.Gen. An efficient genetic algorithm approach to the quadratic minimum spanning problem, *Computers ops. Res.*, 25(1998), 229–237.
10. R. K. Ahuja, J. B. Orlin. A Faster Algorithm for the inverse spanning tree problem. *Journal of Algorithms*, 2000(34), 177–193.
11. T. Fujie, An exact algorithm for the maximum leaf spanning tree problem. *Computers & Operations Research*, 30 (2003) 1931–1944.
12. B. Wu, K. Chao, C. Tang. Approximation algorithms for some optimum communication spanning tree problems. *Discrete Applied Mathematics*, 102 (2000) 245–266.
13. R. Graham, P. Hell. On the history of minimum spanning tree problem. *Annals of the History of Computing*, 7(1985), 43–57.
14. H. Katagiri, H. Ishii. Chance constrained bottleneck spanning tree problem with fuzzy random edge costs. *J. of the Operations Research Society of Japan*, 43(2000), 128–137.
15. H. Katagiri, M. Sakawa, H. Ishii. Fuzzy random bottleneck spanning tree problems using possibility and necessity measures. *European Journal of Operational Research*, 152 (2004), 88–95.
16. R.Hassin, A. Levin. Minimum restricted diameter spanning trees. *Discrete Applied Mathematics*, 137(2004), 343–357.
17. B. Boldon, N. Deo, N. Kumar. Minimum-weighted degree-constrained spanning tree problem: heuristics and complementation on an SIMD parallel machine. *Parallel Computing*, 22(1996), 369–382.
18. K. A. Andersen, K. Jornsten, M. Lind. On bi-criterion minimal spanning trees: an approximation. *Computers Operation Research*, 23(1996), 1171–1182.
19. G. Galbiati, A. Morzenti, F. Maffioli. On the approximability of some maximum spanning tree problems. *Theoretical Computer Science*, 181(1997), 107–118.
20. R. Hassia, A. Tamir. On the minimum diameter spanning tree problem. *Information Processing Letters*, 53(1995), 109–111.
21. S. O. Krumke, H. Wirth. On the minimum label spanning tree problem. I*nformation Processing Letters*, 53(1998), 81–85.
22. A. Kaneko. Spanning trees with constraints on the leaf degree. Discrete Applied Mathematics, 115(2001), 73–76.
23. R. Chang, S. Leu. The minimum labeling spanning trees. *Information Processing Letters*, 53(1997), 277–282.
24. J.A.Bondy, U.S.R.Murty. *Graph Theory with Application*, The Macmillan Press Ltd.,1976.

A Fuzzy Method for Evaluating Suppliers

Hsuan-Shih Lee

Department of Shipping and Transportation Management
National Taiwan Ocean University
Keelung 202, Taiwan
Republic of China

Abstract. A key objective of procurement is to purchase the right product from right supplier at right price in due time. In this paper, we are going to propose a fuzzy method for the evaluation of suppliers under fuzzy environment. First we identify the strength and weakness of suppliers from their ratings. Based on the strength and weakness of suppliers, a fuzzy preference relation is constructed. Then we propose a linear programming model to derive the priorities of suppliers.

1 Introduction

Many processes are involved in managing the supply chain to expedite the flow of information and materials. The Supply Chain Council developed a Supply Chain Operations Reference Model (SCOR) as a cross-industry reference model for supply chain management, which identifies five major supply chain process: plan, source, make, deliver, and return. One of the subprocesses in the source process is identifying suppliers. In modern management, a company tries to establish a long-term relationship with its supplier to ensure its stable source and therefore evaluating supplier has become even more critical than ever in gaining strategic advantage in supply chain management [1,2,4,13,14,18].

In this paper, we are going to propose a fuzzy method to evaluate suppliers under the fuzzy environment in which both the weights of criteria and ratings of suppliers may be given in fuzzy numbers. Our method consists of three major steps. The first step is to identify the strength and weakness of suppliers. Since the use of preference relations is usual in decision making [3,5,6,8,9,11,17], the second step is to construct a fuzzy preference relation for suppliers from the strength and weakness of suppliers. The third step "prioritization" is to derive the weight of each supplier from the preference relation. These three steps are described in section 2, 3 and 4 respectively, and an illustrative example is given in section 5.

2 Identifying Strength and Weakness of Suppliers

Definition 1. *The α-cut of fuzzy set A, A^α, is the crisp set $A^\alpha = \{x \mid \mu_A(x) \geq \alpha\}$. The support of A is the crisp set $Supp(A) = \{x \mid \mu_A(x) > 0\}$. A is normal iff $sup_{x \in U} \mu_A(x) = 1$, where U is the universe set.*

L. Wang et al. (Eds.): FSKD 2006, LNAI 4223, pp. 1035–1043, 2006.

Definition 2. *A is a fuzzy number iff A is a normal and convex fuzzy subset of real number.*

Definition 3. *A triangular fuzzy number A is a fuzzy number with piecewise linear membership function μ_A defined by*

$$\mu_A(x) = \begin{cases} \frac{x-a_1}{a_2-a_1}, & a_1 \le x \le a_2, \\ \frac{a_3-x}{a_3-a_2}, & a_2 \le x \le a_3, \\ 0, & otherwise, \end{cases}$$

which can be denoted as a triplet (a_1, a_2, a_3).

Definition 4. *Let A be a fuzzy number. Then A^L_α and A^U_α are defined as $A^L_\alpha = \inf_{\mu_A(z)\ge\alpha}(z)$ and $A^U_\alpha = \sup_{\mu_A(z)\ge\alpha}(z)$ respectively.*

Definition 5. *[12] An extended fuzzy preference relation R on fuzzy numbers is an extended fuzzy subset of the product of fuzzy numbers with membership function $-\infty \le \mu_R(A, B) \le \infty$ being the preference degree of fuzzy number A to fuzzy number B.*

1. *R is reciprocal iff $\mu_R(A, B) = -\mu_R(B, A)$ for all fuzzy numbers A and B.*
2. *R is transitive iff $\mu_R(A, B) \ge 0$ and $\mu_R(B, C) \ge 0 \Rightarrow \mu_R(A, C) \ge 0$ for all fuzzy numbers A, B and C.*
3. *R is additive iff $\mu_R(A, C) = \mu_R(A, B) + \mu_R(B, C)$*
4. *R is a total ordering iff R is reciprocal, transitive and additive.*

In [12], we have defined an extended fuzzy preference relation on fuzzy numbers, which is reciprocal, transitive and additive. For convenience, some results in [12] are reiterated here.

Definition 6. *[12] For any fuzzy numbers A and B, we define the extended fuzzy preference relation F(A, B) by the membership function*

$$\mu_F(A, B) = \int_0^1 ((A - B)^L_\alpha + (A - B)^U_\alpha)d\alpha \tag{1}$$

Lemma 1. *[12] F is reciprocal, i.e., $\mu_F(B, A) = -\mu_F(A, B)$.*

Lemma 2. *[12] F is additive, i.e., $\mu_F(A, B) + \mu_F(B, C) = \mu_F(A, C)$.*

Lemma 3. *[12] F is transitive, i.e., $\mu_F(A, B) \ge 0$ and $\mu_F(B, C) \ge 0 \Rightarrow \mu_F(A, C) \ge 0$.*

Lemma 4. *[12] Let $A = (a_1, a_2, a_3)$ and $B = (b_1, b_2, b_3)$ be two triangular fuzzy numbers. Then $\mu_F(A, B) = (a_1 + 2a_2 + a_3 - b_1 - 2b_2 - b_3)/2$.*

Assume there are m suppliers under evaluation against n criteria. Let fuzzy number A_{ij} be the rating of the ith supplier on the jth criterion and fuzzy number W_j be the weight of jth criterion. To facilitate the computation of strength and weakness of the suppliers, we define the preference intensity function of one fuzzy number A over another number B as follows:

$$Q(A, B) = \max\{\mu_F(A, B), 0\}.$$

Let J be the set of benefit criteria and J' be the set of cost criteria where

$$J = \{1 \leq j \leq n \text{ and } j \text{ belongs to benefit criteria}\}$$

$$J' = \{1 \leq j \leq n \text{ and } j \text{ belongs to cost criteria}\},$$

and

$$J \cup J' = \{1, \ldots, n\}.$$

By benefit criteria, we mean that the larger their value is and the better the supplier is, whereas the cost criteria are on the contrary. The advantage of the ith supplier on the jth criterion is given by

$$a_{ij} = \begin{cases} \sum_{k \neq i} Q(A_{ij}, A_{kj}) & \text{if } j \in J \\ \sum_{k \neq i} Q(A_{kj}, A_{ij}) & \text{if } j \in J'. \end{cases} \tag{2}$$

Similarly, we can define the disadvantage of ith supplier on the jth criterion to be

$$d_{ij} = \begin{cases} \sum_{k \neq i} Q(A_{kj}, A_{ij}) & \text{if } j \in J \\ \sum_{k \neq i} Q(A_{ij}, A_{kj}) & \text{if } j \in J'. \end{cases} \tag{3}$$

Both a_{ij} and d_{ij} are crisp numbers. The fuzzy strength of the ith supplier is defined as

$$FS_i = \sum_j a_{ij} W_j \tag{4}$$

and the fuzzy weakness of the ith supplier is defined as

$$FW_i = \sum_j d_{ij} W_j, \tag{5}$$

where $1 \leq i \leq m$.

The strength of the ith supplier in crisp value can be obtained by

$$S_i = \sum_{k \neq i} Q(FS_i, FS_k) + \sum_{k \neq i} Q(FW_k, FW_i) \tag{6}$$

and the weakness of the ith supplier can be defined similarly as

$$I_i = \sum_{k \neq i} Q(FS_k, FS_i) + \sum_{k \neq i} Q(FW_i, FW_k). \tag{7}$$

3 Building Fuzzy Preference Relation of Suppliers

Definition 7. *[7,15] A fuzzy preference relation P on a set of alternatives X is a fuzzy set on the product set $X \times X$, i.e., it is characterized by a membership function*

$$\mu_P : X \times X \to [0,1].$$

The fuzzy preference relation may be conveniently represented by the matrix $P = (p_{ik})$, being $p_{ik} = \mu_P(x_i, x_k)$ interpreted as the preference degree or intensity of alternative x_i over x_k: $p_{ik} = 0.5$ indicates indifference between x_i and x_k, $p_{ik} > 0.5$ indicates that x_i is preferred to x_k, and vice versa.

Definition 8. *A fuzzy preference relation $P = (p_{ik})$ is called an additive consistent fuzzy preference relation if the following additive transitivity (given by Tanion [16]) is satisfied:*

$$\forall i, j, k \; p_{ik} = p_{ij} + p_{jk} - 0.5$$

Given the strength $S = (S_i)$ and weakness $I = (I_i)$ of the suppliers, the fuzzy preference relation for the suppliers can be constructed as follows. Let the function f measure the intensity of "a being larger than b":

$$f(a, b) = \max\{a - b, 0\}.$$

The preference intensity of the ith supplier over the kth supplier can be measured as the sum of two intensities:

$$f(S_i, S_k) + f(I_k, I_i). \tag{8}$$

Note that the preference intensities thus derived are all nonnegative. The fuzzy preference relation $P = (p_{ik})$ for the suppliers can be defined by normalizing the preference intensity in (8) to [0,1] with

$$p_{ik} = \left(\frac{f(S_i, S_k) + f(I_k, I_i)}{\max_{i,k}(f(S_i, S_k) + f(I_k, I_i))}\right)/2 + 0.5. \tag{9}$$

4 Prioritization of Suppliers

Let w_i be the priority of the ith supplier. Assume all priorities sum to 1. That is,

$$\sum_{i=1}^{m} w_i = 1.$$

An additive consistent fuzzy preference relation based on the suppliers' priorities can be constructed. Let $B = (b_{ik})$, where

$$b_{ik} = (w_i - w_k)/2 + 0.5.$$

Apparently, B is an additive consistent fuzzy preference relation. Since the priorities $w_1, \ldots, w_m$ are unknown, they should be derived in a way that $B = (b_{ik})$ is as close to $P = (p_{ik})$ as possible. That is, $w_1, \ldots, w_m$ are found so that

$$\sum_{i=1}^{m} \sum_{k=1}^{m} |b_{ik} - p_{ik}|$$

is minimized. The problem of finding $w_1, \ldots, w_m$ can be formulated as the following optimization problem:

$$\begin{aligned} &\min \textstyle\sum_{i=1}^{m} \sum_{k=1}^{m} |(w_i - w_k)/2 + 0.5 - p_{ik}| \\ &s.t. \textstyle\sum_{i=1}^{m} w_i = 1 \\ &\quad w_1, \ldots, w_m \geq 0 \end{aligned} \tag{10}$$

Model (10) can be transformed into the following linear programming:

$$\begin{aligned} &\min \textstyle\sum_{i=1}^{m} \sum_{k=1}^{m} (q_{ik} + r_{ik}) \\ &s.t.\ w_i - w_k - 2(p_{ik} - 0.5) = q_{ik} - r_{ik} \\ &\quad q_{ik}, r_{ik} \geq 0 \\ &\quad \textstyle\sum_{i=1}^{m} w_i = 1 \\ &\quad w_i \geq 0 \end{aligned} \tag{11}$$

The priorities of the suppliers can be found by solving (11).

5 Numerical Example

Suppose a company attempts to identify suitable suppliers for establishing a long-term relationship. The evaluation is done by a committee of three decision-makers D_1, D_2, and D_3. After preliminary screening, there are three suppliers A_1, A_2, and A_3 under further evaluation. Assume the linguistic variables employed to represent weights and ratings are respectively shown in Table 1. The evaluation committee then undergoes the proposed evaluation procedure:

Table 1. Linguistic variables for the importance weights of criteria and the ratings

Importance of the criteria		Performance ratings	
linguistic variable	Fuzzy number	Linguistic variable	Fuzzy number
Very low (VL)	(0,0,0.1)	Very poor(VP)	(0,0,1)
Low (L)	(0,0.1,0.3)	Poor (P)	(0,1,3)
Medium low (ML)	(0.1,0.3,0.5)	Medium poor(MP)	(1,3,5)
Medium (M)	(0.3,0.5,0.7)	Fair (F)	(3,5,7)
Medium high (MH)	(0.5,0.7,0.9)	Medium good (MG)	(5,7,9)
High (H)	(0.7,0.9,1.0)	Good (G)	(7,9,10)
Very high (VH)	(0.9,1.0,1.0)	Very good (VG)	(9,10,10)

Table 2. The importance weights of the criteria

	D_1	D_2	D_3
C_1	H	VH	VH
C_2	H	H	H
C_3	MH	H	MH
C_4	MH	MH	MH
C_5	H	H	H

Table 3. The collective weights of the criteria

	C_1	C_2	C_3	C_4	C_5
Weight	(0.83,0.97,1)	(0.7,0.9,1)	(0.57,0.77,0.93)	(0.5,0.7,0.9)	(0.7,0.9,1)

Table 4. The ratings of the suppliers given by the decision makers

Criteria	Suppliers	D_1	D_2	D_3
C_1	S_1	6×10^6	8×10^6	7×10^6
	S_2	3×10^6	4×10^6	5×10^6
	S_3	4×10^6	5×10^6	6×10^6
C_2	S_1	G	VG	F
	S_2	VG	VG	VG
	S_3	MG	G	VG
C_3	S_1	F	G	G
	S_2	G	G	G
	S_3	G	MG	VG
C_4	S_1	VG	G	G
	S_2	G	G	G
	S_3	G	VG	VG
C_5	S_1	F	F	F
	S_2	G	F	G
	S_3	G	G	G

Table 5. The collective ratings of the suppliers

	C_1	C_2	C_3	C_4	C_5
S_1	7×10^6	(6.3,8,9)	(5.7,7.7,9)	(7.7,9.3,10)	(3,5,7)
S_2	4×10^6	(9,10,10)	(7,9,10)	(7,9,10)	(5.7,7.7,9)
S_3	5×10^6	(7,9,10)	(7,9,10)	(8.3,9.7,10)	(7,9,10)

Table 6. The normalized collective ratings of the suppliers

	C_1	C_2	C_3	C_4	C_5
S_1	(1,1,1)	(0.62,0.8,0.9)	(0.57,0.77,0.9)	(0.77,0.93,1)	(0.3,0.5,0.7)
S_2	(0.57,0.57,0.57)	(0.9,1,1)	(0.7.0.9,1)	(0.7,0.9,1)	(0.57,0.77,0.9)
S_3	(0.71,0.71,0.71)	(0.7,0.87,0.97)	(0.7,0.87,0.97)	(0.83,0.97,1)	(0.7,0.9,1)

Step 1: Five evaluation criteria of suppliers are identified. They are
(1) Price (C_1),
(2) Quality (C_2),
(3) Delivery (C_3),
(4) Flexibility (C_4),
(5) Service (C_5),
among which C_2, C_3, C_4, and C_5 are benefit criteria, whereas C_1 is a cost criterion. The weights of criteria given by each decision maker are shown in Table 2 and the collective weights of the criteria for the whole committee are shown in Table 3.

Step 2: The ratings of the suppliers given by the three decision makers are shown in Table 4. There are many methods to aggregate fuzzy ratings into a collective rating [10]. For simplicity, here the collective ratings for suppliers are obtained by averaging the ratings given by the three decision makers, which are shown in Table 5. The normalized collective ratings of the suppliers are obtained by dividing the collective ratings with the largest value in the supports of the fuzzy numbers in the same criterion. The normalized collective ratings of the suppliers are shown in Table 6.

Step 3: Calculate the advantage of the suppliers with respect to each criterion by (2). The advantage of the suppliers with respect to each criterion is shown in Table 7.

Step 4: Calculate the disadvantage of the suppliers with respect to each criterion by (3) . The results are shown in Table 7.

Step 5: Calculate the fuzzy strength of the suppliers by (4). The results are shown in Table 8.

Step 6: Calculate the fuzzy weakness of the suppliers by (5). The results are shown in Table 8.

Step 7: Calculate the strength of the suppliers by (6). The results are shown in Table 9.

Step 8: Calculate the weakness of the suppliers by (7). The results are shown in Table 9.

Step 9: Construct the fuzzy preference relation for the suppliers by (9). The result is shown in Table 10.

Step 10: Calculate the priorities of the suppliers by (11). The priorities of the suppliers are shown in Table 11.

According to the priorities of the suppliers, we find that supplier 2 is the best and supplier 3 is in the second place.

Table 7. The advantage and disadvantage of suppliers

	Advantage						Disadvantage				
	C_1	C_2	C_3	C_4	C_5		C_1	C_2	C_3	C_4	C_5
S_1	0	0	0	0.067	0	S_1	1.43	0.52	0.45	0.067	1.25
S_2	1.14	0.63	0.3	0	0.5	S_2	0	0	0	0.2	0.25
S_3	0.57	0.13	0.2	0.2	1	S_3	0.29	0.25	0.05	0	0

Table 8. The fuzzy strength and fuzzy weakness of the suppliers

Fuzzy strength	Fuzzy weakness
S_1 (0.033,0.278,0.339)	S_1 (2.712,3.369,3.674)
S_2 (1.913,2.283,2.462)	S_2 (0.275,0.365,0.43)
S_3 (1.482,1.766,1.947)	S_3 (0.441,0.541,0.582)

Table 9. The strength and weakness of the suppliers

	Strength		Weakness
S_1	0	S_1	18.376
S_2	11.176	S_2	0
S_3	8.526	S_3	1.325

Table 10. The fuzzy preference relation of the suppliers

$$P = \begin{pmatrix} 0.5 & 0.5 & 0.5 \\ 1.0 & 0.5 & 0.567 \\ 0.933 & 0.5 & 0.5 \end{pmatrix}$$

Table 11. The priorities of the suppliers

Supplier	Priority
1	0.057305
2	0.538602
3	0.404093

6 Conclusions

In this paper, we have proposed a new fuzzy method to evaluate the performance of the suppliers, which is an important issue in supply chain management. Our method enables decision makers to assess suppliers with linguistic variables so that vagueness can be encompassed in the assessment of the suppliers. The proposed fuzzy method consists of three main stages. In the first stage, we identify the strength and weakness of the suppliers. Based on the strength and weakness, we then build a fuzzy preference relation for the suppliers. In the last stage, we formulate the prioritization of fuzzy preference relation as a linear programming problem. By solving the linear programming problem, the priorities of the suppliers can be obtained.

Acknowledgement

This research work was partially supported by the National Science Council of the Republic of China under grant No. NSC94-2416-H-019-006-.

References

1. D.J. Bowersox, D.J. Closs, *Logistical Management - The Integrated Supply Chain Process,* (McGraw-Hill, Singapore, 1996).
2. S.N. Chapman, Just-in-time supplier inventory: an empirical implementation model, *International Journal of Production Research* 29 (1993) 1993-2007.
3. F. Chiclana, F. Herrera, E. Herrera-Viedma, Integrating three representation models in fuzzy multipurpose decision making based on fuzzy preference relations, *Fuzzy Sets and Systems* 97 (1998) 33-48.
4. T.Y. Choi and J.L. Hartley, An exploration of supplier selection practices across the supply chain, *Journal of Operations Management* 14 (1996) 333-343.
5. J. Fodor, M. Roubens, *Fuzzy Preference Modelling and Multicriteria Decision Support,* Kluwer, Dordrecht, 1994.
6. E. Herrera-Viedma, F. Herrera, F. Chiclana, M. Luque, Some issues on consistency of fuzzy preference relations, *European Journal of Operational Research* 154 (2004) 98-109.
7. J. Kacprzyk, Group decision making with a fuzzy linguistic majority, *Fuzzy Sets and Systems* 18 (1986) 105-118.
8. S.H. Kim, B.S. Ahn, Interactive group decision making procedure under incomplete information, *European Journal of Operational Research* 116 (1999) 139-152.
9. S.H. Kim, S.H. choi, J.K. Kim, An interactive procedure for multiple attribute group decision making with incomplete information: Range-based approach, *European Journal of Operational Research* 118 (1999) 139-152.
10. H.-S. Lee, Optimal consensus of fuzzy opinions under group decision making environment, *Fuzzy Sets and Systems* 132 (2002) 303-315.
11. H.-S. Lee, On fuzzy preference relation in group decision making, *International Journal of Computer Mathematics* 82(2) (2005) 133-140.
12. H.-S. Lee, A fuzzy multi-criteria decision making model for the selection of distribution center, *Lecture Notes in Computer Science* 3612 (2005) 1290-1299.
13. H. Min, International supplier selection: a multi-attribute utility approach, *Internation Journal of Physical Distribution and Logistics Management* 24/25 (193) 24-33.
14. C. Muralidharan, N. Anantharaman and S.G. Deshmukh, *Journal of Supply Chain Management* 38 (2002) 22-33.
15. S.A. Orlovski, Decision-making with fuzzy preference relations, *Fuzzy Sets and Systems* 90 (1978) 155-167.
16. T. Tanino, Fuzzy preference orderings in group decision-making, *Fuzzy Sets and Systems* 12 (1984) 117-131.
17. T. Tanino, Fuzzy preference relations in group decision making, in: J. Kacprzyk, M. Roubens (Eds.), Non-Conventional Preference Relations in Decision Making, Springer-Verlag, Berlin, 1988, pp. 54-71.
18. E. Timmerman, An approach to vendor evaluation, *Journal of Purchasing and Materials Management* 1 (1986) 2-8.

Hierarchical σ–Octree for Visualization of Ultrasound Datasets

Sukhyun Lim and Byeong-Seok Shin

Inha University, Dept. Computer Science and Information Engineering
253 Yonghyun-dong, Nam-gu, Inchon, 402-751, Rep. of Korea
slim@inhaian.net, bsshin@inha.ac.kr

Abstract. There are two important factors to visualize ultrasound datasets using volume ray casting method. Firstly, efficient methods to skip over empty space are required. Secondly, adequate noise-detection methods are necessary because ultrasound datasets contain lots of speckle noises. In general, space-leaping and noise-filtering methods are exploited to solve the problems. However, it increases the preprocessing time to generate the filtered datasets, and interesting (meaningful) objects could be affected by a filtering operation. We propose a hierarchical octree containing min-max values and standard deviation for each block, named a hierarchical σ–octree. In rendering step, our method refers to min-max values of a block. If the block is regarded as nontransparent, it also checks its standard deviation value to detect speckle noises. Our method reduces rendering time compared with the method using only the min-max values because most blocks containing speckle noises are considered as transparent.

1 Introduction

Volume visualization is a research area that deals with various techniques to extract meaningful and visual information from volume data [1], [2]. The visualization of ultrasound datasets is a technology for imaging ultrasonic echo information. It is mostly used in obstetrics and gynecology [3], and for visualization of vessels and tumors in soft tissue. The main advantages are that they are non-radiative and relatively inexpensive. In addition, the acquisition procedure is faster than other medical imaging methods such as CT, MR and PET. However, the visualization is difficult since they typically contain a lot of noises [4], [5], [6], [7]. To solve the problem, several filtering techniques are used to separate useful information from the noises [4], [5], [6], [7]. However, if high-density speckle noises still exist in transparent region, we increase the amount of filter size. It causes increasing the preprocessing time to generate filtered datasets, and interesting object can also be influenced (over-blurring).

Volume ray casting is a well-known direct volume rendering method [1]. It is composed of two steps [1]: after a ray advances through a transparent region, it integrates colors and opacities as it penetrates an object boundary. Although it produces high-quality images, the rendering speed is too slow. An octree is one of the data structures

L. Wang et al. (Eds.): FSKD 2006, LNAI 4223, pp. 1044 – 1053, 2006.

to skip over transparent region [8], [9], [10], [11], [12], [13], [14], [15], [16], [17], [18]. Since the region does not contribute to the final image [1], [8], [19], we reduce the rendering time without deteriorating image quality. In this paper, an octant of an octree at an arbitrary level is called a *block*. If an octree stores the minimum and maximum values to determine the transparency of blocks, we call it as a *min-max octree*. After determining the transparency, when all voxels in a block are estimated as transparent, the block is denoted as a *transparent block*.

When we exploit a min-max octree, we can reduce the rendering time because rays skip over transparent blocks without compositions. However, if speckle noises exist in one block, the block is regarded as nontransparent one. It increases the rendering time, and we acquire rendering results containing lots of noises. To solve the problems, we propose a *hierarchical σ–octree*. In preprocessing step, we apply a filtering operation to the original volume dataset, and we make an octree. Although the previous methods store only the minimum and maximum values to determine the transparency of blocks, we compute an addition standard deviation value for each block to detect speckle noises. In rendering step, when a ray reaches a block, we exploit stored min-max values to skip over the block. If the block is regarded as nontransparent, we also check its standard deviation value. If the value is greater than a user-defined threshold (that is, speckle noises exist in transparent region intermittently), the block is skipped. By using three values (min-max and standard variation values) simultaneously, we generate noise-insensitive results while reducing the rendering time.

In Sect. 2, we look into the problems of a hierarchical octree structure when we use it to visualize ultrasound datasets. Then, we explain our data structure in detail in Sect. 3. Experimental results are shown in Sect. 4. Finally, we conclude our work.

2 Problems of a Hierarchical Octree for Visualizing Ultrasound Datasets

One of the main obstacles to visualize ultrasound datasets is speckle noises [4], [5], [6], [7]. When the noises lie on the path of a ray, two problems may arise. The first problem is that unnecessary comparison and level shift between blocks occurs frequently due to the noises according to increase the hierarchy of an octree. Of course, this is the fundamental problem of an octree [13], [15], [16]. However since it is difficult to recognize the speckle noises, we have only to interpret them as meaningful (interesting) objects. Therefore, the rendering time is increased (see Fig. 1).

The second problem occurs if the density (scalar) value of noises on the path of a ray is confined within opaque range. In this case, the color computation step of a pixel can be terminated in the noise region because the opacity value can reach 1.0. Although the density value of the noises is confined to translucent range, color and opacity values are affected by the noises because rays already composite them (see Fig. 2 (b) and (c)). This is a basic problem of volume rendering. However, the number of it is increased since ultrasound datasets contain lots of noises compared with the high-resolution datasets such as CT or MR. As a result, due to two problems, deteriorated results are acquired while requiring long rendering time.

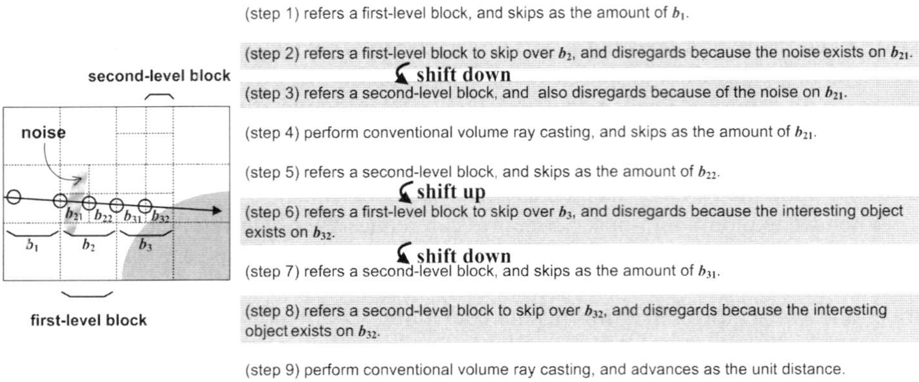

Fig. 1. When we apply a hierarchical octree for a space leaping of ultrasound datasets, unnecessary comparison (shaded box) and level shift (thick arrow) between blocks are increased due to speckle noises. Assume that the octree hierarchy is the second level.

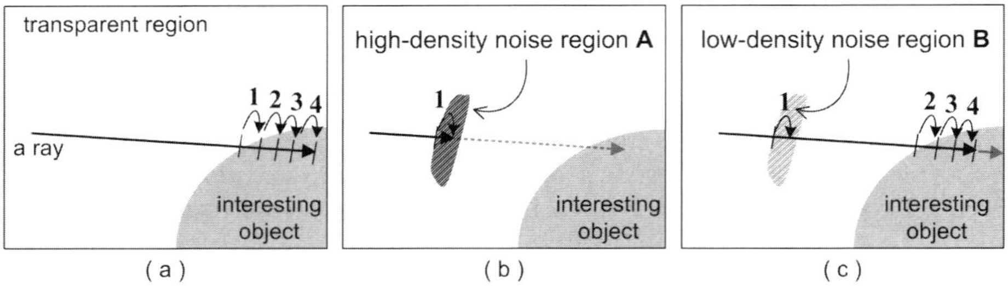

Fig. 2. (a) Ideal case. Assume that four times compositions are performed from an interesting object boundary if there are no noises on the path of a ray. (b) When the density value of the noise is confined to opaque range, the color composition can be ended in the high-density noise region **A**. In this case, only one composition occurs at irrelevant part. (c) Even if the density value is confined within translucent range, we acquire unexpected color and opacity values since the accumulated values are already affected by the low-density noise region **B**.

3 A Hierarchical σ–Octree

To solve two problems as mentioned in Sect. 2, we exploit a filtered dataset. In this case, the amount of noises is reduced because they are spread over neighboring voxels. However, since the filtering can affect to the entire volume dataset, it produces over-blurred results since even interesting objects could be affected to the filtering. If we apply the filtering operation to only noise regions, the final image cannot be deteriorated. However, since the density values between the noises and interesting object are nearly identical [4], [5], [6], it is difficult to distinguish them. As a result, we require a smart method to reduce speckle noises efficiently while reducing the rendering time. To solve it, we propose a hierarchical σ–octree storing standard deviation values as well as min-max values.

The structure containing standard deviation values is proposed in [9]. However, we store three values (min-max and standard deviation values) simultaneously and the method to compute the standard deviation values is different. Although the previous method calculates the deviation from a mean value, we modify it against the position τ where the transparency is changed from transparent to nontransparent region because our concern is the deviation against τ. Eq. (1) represents a method to store our standard deviation value. We assume that a volume dataset is composed of N^3 voxels, and the size of an octree block is B^3. Each voxel $\boldsymbol{v}$ is indexed by $\boldsymbol{v}(x, y, z)$ where $x,y,z = 0,\ldots,N\text{-}1$. Fig. 3 shows an example of a standard deviation when speckle noises lie on transparent region.

$$\sigma = \sqrt{\left(\sum_{i,j,k=0}^{B-1} (v(i,j,k) - \tau)^2 \right) / B^3} \ . \tag{1}$$

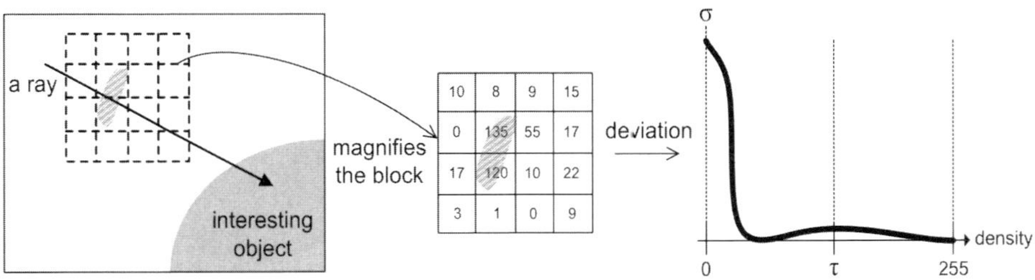

Fig. 3. If speckle noises exist in a block, we detect them by exploiting a standard deviation value (τ=125)

However, it is difficult to detect speckle noised by using only Eq. (1) because it implies two cases. The first is that most voxels in a block are transparent and some speckle noises exist. The second is that the opposite case of the first. That is, meaningful (interesting) voxels and some noises coexist. Fig. 4 (a) shows the cases. Although our concern is only the first case, we cannot distinguish them because the results of standard deviations are almost identical. Therefore, by regarding the voxels whose values are bigger than τ as voxels that have τ values, we complete an equation adequate to visualize ultrasound datasets (see Eq. (2)). In this case, we determine the noises correctly (see Fig. 4 (b)).

$$\sigma = \sqrt{\left(\sum_{i,j,k=0}^{B-1} (v(i,j,k) - \tau)^2 \right) / B^3} \ . \qquad if\,(v(i,j,k) > \tau)\ \ v(i,j,k) = \tau \tag{2}$$

In rendering step, after firing a ray from each pixel, the minimum and maximum values of a block are referred to determine whether to skip over the block or not. When the min-max values are confined within transparent range (that is, current block is a transparent block), the ray jumps over the block. If the block is estimated as

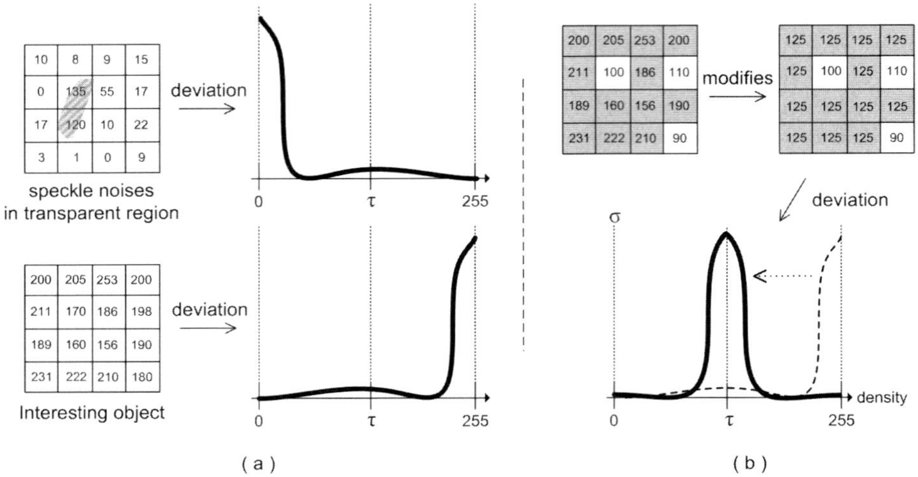

Fig. 4. (a) Two cases occur when we use the previous standard deviation approach. In this case, we cannot determine speckle noises correctly because the results of standard deviations in two cases are almost identical. (b) We regard the voxels whose values are greater than τ (=125) as voxels that have τ values (the voxel marked as gray).

nontransparent, we check stored standard deviation value. When the value is greater than a user-defined threshold (that is, most transparent voxels and some noises coexist), the ray also skips over the block because it involves that speckle noises exist in transparent region. If the min-max values are confined within nontransparent range and its standard deviation value is nearly zero, the ray refers to the child blocks.

To skip over transparent blocks efficiently, we use the *distance template* scheme [15], [17] proposed by Lim and Shin. The distance template is a precomputed distancemap to reduce the cost for determination of the distance between entry point and exit point of a ray with respect to a block. It quickly and reliably leaps to the boundary of a transparent block since it can directly access the distance value to reach the boundary. In addition, we can reuse it without concerning viewing conditions because

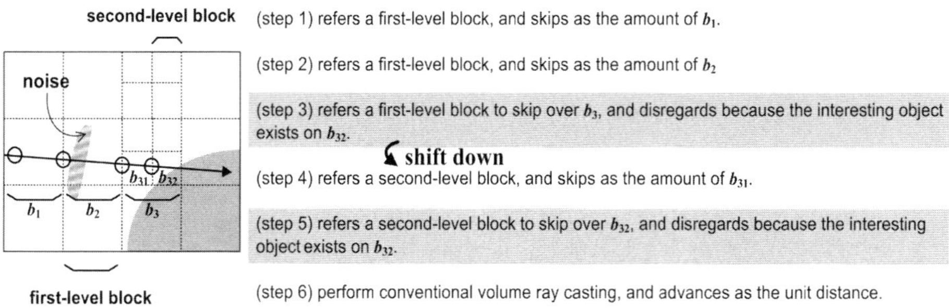

Fig. 5. Since the speckle noise is estimated as transparent with our method, the number of unnecessary comparison (shaded box) is reduced from four to two, and the number of shift (thick arrow) between blocks is reduced from three to one

it is view-independent structure. As a result, we exploit three values (min-max and standard deviation values) simultaneously to determine the transparency of blocks.

Fig. 5 depicts a ray-traversal procedure when applying our approach. Compared with the case of Fig. 1, total number of traversal steps is reduced from nine to six. This results in reducing the rendering time. Fig. 6 shows a ray-traversal example using our method. The traversal condition is identical to Fig. 2. In transparent region, even the speckle noise exists on the path of a ray, our method leaps over the block through a standard deviation value. Therefore, we acquire noises-insensitive rendering results.

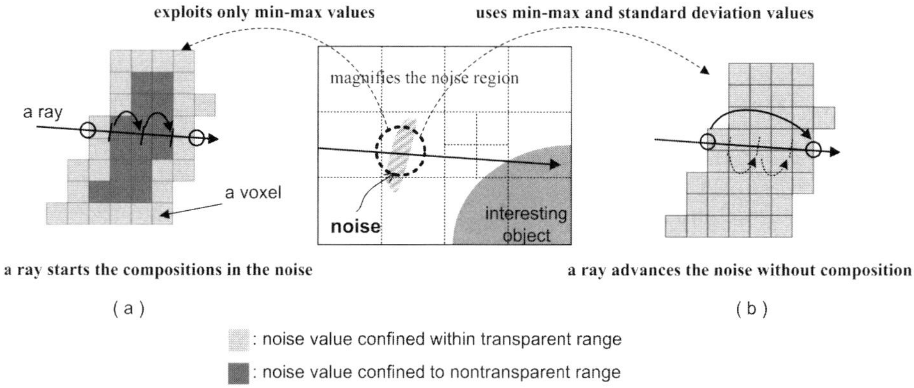

Fig. 6. A ray-traversal example before and after applying standard deviation value. (a) When we exploit only min-max values, a ray starts compositions in the noise region because the noise value is confined to nontransparent range. In this case, we acquire unexpected result. (b) After referring a standard deviation value, the block is estimated as a transparent one even the min-max values are confined to nontransparent region. Therefore, the ray traverses the noise region without compositions. Dotted arrow represents the disregarded sample point.

4 Experimental Results

Our method was implemented on a PC equipped with a Pentium™ IV 3.4 GHz CPU and 2 GB of main memory. Volume datasets were obtained by scanning a fetus with an ultrasound of which the resolutions are 199x123x199, 199x109x199, and 199x89x199. Those datasets are supported by Medison, co., ltd. in Rep. of Korea [20]. We call the datasets as *data A*, *data B*, and *data C*, respectively. We set the maximum hierarchy of an octree as three since the preprocessing time is increased according to increase the hierarchy of it. In addition, we experimented on the Euclidean distance-map method [21], [22] to compare with other space-leaping methods.

Table 1 shows the rendering time under fixed viewing conditions. We exploit an average (box) filter because preprocessing time of it is shorter than that of other filters. By experiment, we set the τ as 70, and a user-defined threshold for standard deviation as 20. When we exploit an octree, root-level blocks are referred to skip transparent region. Therefore, we store standard deviation values for only in root-level blocks because our concern is to detect speckle noises in transparent region. Rendering speed of our method is about 16% faster than that of the previous min-max octree.

Compared with the distancemap-based method [21], [22], the rendering speed of our method is about 10% slow. However, since the distancemap-based approach requires long preprocessing time (about 70 times slower) and large amount of memory (the size of it is identical to that of the volume dataset) as shown in the Table 2 and 3. Therefore, it is not suitable for applications using ultrasound datasets. Compared with the conventional min-max octree, additional processing time to compute standard deviation values is almost disregarded because voxel values are already loaded to cache memory.

Table 1. Comparison of rendering time. Image size is 256 x 256 pixels (seconds)

method	data A	data B	data C
Euclidian distancemap	0.70	0.58	0.67
only min-max values (A)	0.94	0.78	0.86
our method (B)	0.79	0.64	0.74
improvement	**16.0 %**	**17.9 %**	**14.0 %**

Table 2. Comparison of preproceesing time (seconds)

method	data A	data B	data C
Euclidian distancemap	17.23	15.41	12.50
Only min-max values	0.24	0.22	0.18
our method	**0.27**	**0.24**	**0.20**

Table 3. Comparison of required memory. In our method, the time to generate the previous min-max values is included (bytes).

method	data A	data B	data C
Euclidian distancemap	4,871	4,317	3,524
only min-max values	686	608	496
our method	**687**	**609**	**497**

Fig. 7 shows image quality with fixed viewpoint and viewing direction. The images of the first row are the results when we apply the 5^3 average filter and only min-max values to determine the transparency of blocks. The results in second row are the images using min-max and standard deviation values, simultaneously. The images in the upper row contain lots of noises. However, the images generated with our method show natural and clear results.

Along with the increase of the filter size, a few noises are reduced. However, speckle noises still remain in rendered image. To show that we compare the images using the 5^3 filter with our approach and using the 7^3 filter with only min-max values (see Fig. 8). We can verify that our method reduces speckle noises efficiently. The right image appears blurry and non-sharply because even the interesting voxels ar modified by the filtering operation. Moreover, as increasing the filter size from 5^3 to7^3,

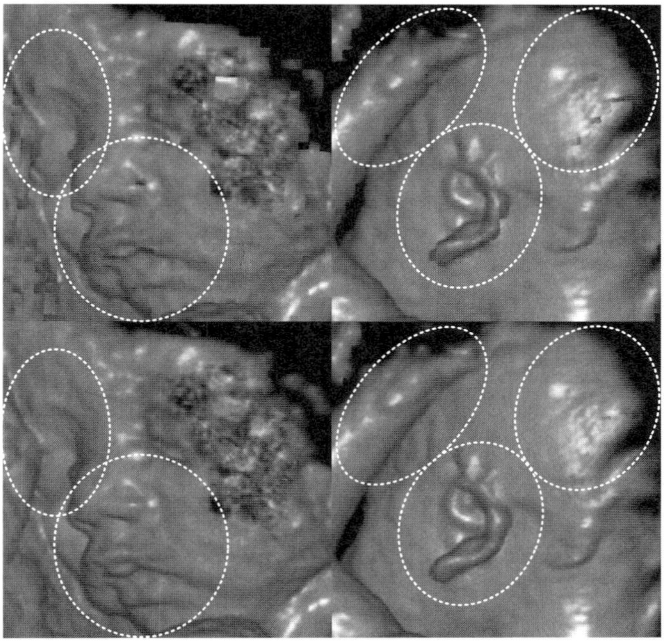

Fig. 7. Comparison of image quality for data A and data B. The images in the first row show the rendering results using the previous method (that is, only exploits min-max values), and the images in second row are the results using min-max and standard deviation values.

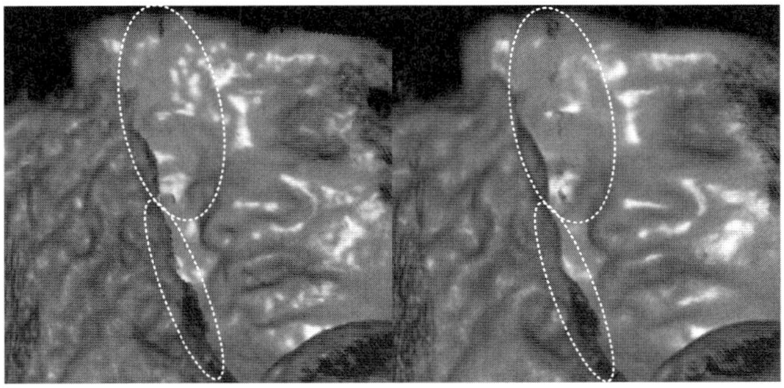

Fig. 8. Comparison of image quality for data A. The left image is generated from our method with 5^3 filter. The right one is the result using only the min-max values with 7^3 filter.

the preprocessing time is increased by 215%. This is critical in visualization of ultrasound datasets.

To confirm that our method is still efficient when we apply other filtering approaches, we implement the [7] method. This method exploits two filtered datasets for the resampling and for the gradient estimation. That is, low-sized filter is used for space leaping, and large-sized filter is exploited to estimate gradient vectors to take less noise image and better depth perception. Even when we exploit the [7] method, the rendering performance is increased approximately 15%. Fig. 9 shows two comparisons when we use our approach with [7] method (left) and only min-max values with [7] (right). Speckle noises are reduced in the left image generated from our method.

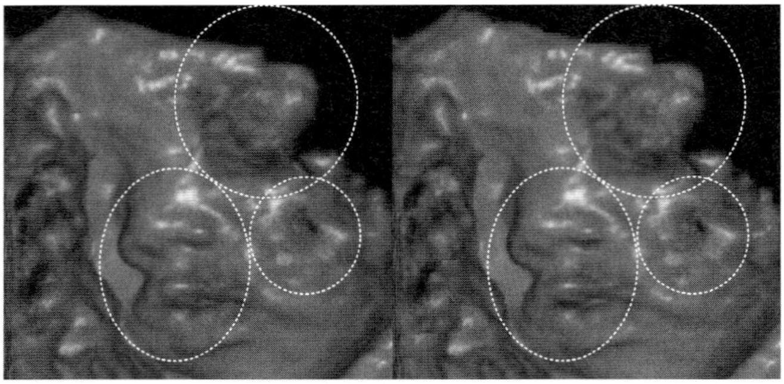

Fig. 9. Comparison of image quality for data C. The left image is generated from our method with [7] method, and the right is the result using only min-max values in [7]. We apply the 5^3 and 7^3 filters for space leaping and gradient estimation, respectively.

5 Conclusion

In order to achieve interactive speed for ultrasound datasets contained lots of speckle noises, we proposed a hierarchical σ–octree. When a ray reaches a block, our method exploits min-max values to check the transparency. If the block is estimated as transparent, it is skipped. When the block is nontransparent, stored standard deviation value is also referred to determine whether it is caused by speckle noises. When it is regarded as nontransparent due to not interesting object but the noises, it is skipped. As a result, by using three values (min-max and standard deviation values) simultaneously, our method produces nature and clear rendering results while reducing the rendering time compared with the method using only min-max values. One of the advantages of our method is that it can be incorporated with any kind of methods using a hierarchical octree since it is independent on them.

Acknowledgment

This work was supported by IITA through IT Leading R&D Support Project.

References

1. Krishnamurthy, B., Bajaj, C.: Data Visualization Techniques. 1st ed., Ed. John Wiley & Sons (1999)
2. Kaufman, A.: Volume Visualization. 1st ed., IEEE Computer Society Press (1991)

3. Baba, K., Jurkovic, D.: Three-Dimensional Ultrasound in Obstetrics and Gynecology. Progress in Obstetric and Gynecological Sonography Series. The Parthenon Publishing Group 1997
4. Fattal, R., Lischinski, D.: Variational Classification for Visualization of 3D Ultrasound Data. Proc. Visualization 2001 (2001) 403-410
5. Sakas, G., Walter, S.: Extracting Surfaces from Fuzzy 3D-Ultrasound Data. In R. Cook, editor. Proc. SIGGRAPH 1995 (1995) 465-474
6. Sakas, G., Schreyer, L.A., Grimm, M.: Preprocessing and Volume Rendering of 3D Ultrasonic Data. IEEE Computer Graphics and Applications, Vol. 15, No. 4 (1995) 47-54
7. Shamdasani, V., Bae, U., Managuli, R., Kim, Y.: Improving the Visualization of 3D Ultrasound Data with 3D Filtering. Proc. SPIE, Vol. 5744 (2005) 455-461
8. Levoy, M.: Efficient Ray Tracing of Volume Data. ACM Transactions on Graphics, Vol. 9 (1990) 245-261
9. Danskin, J., Hanrahan, P.: Fast Algorithms for Volume Ray Tracing. Proc. Volume visualization 1992 (1992) 91-98
10. Wilhelms, J., Gelder, A.V.: Octree for Faster Isosurface Generation. ACM Transactions on Graphics, Vol. 11, No. 3 (1992) 201-227
11. Parker, S., Shirley, P., Livnat, Y., Hansen, C., Sloan, P.: Interactive Ray Tracing for Isosurface Rendering. Proc. IEEE Visualization 1998 (1998) 233-238
12. Knittel, G.: The UltraVis System. Proc. IEEE Volume Visualization 2000 (2000) 71-79
13. Mora, B., Jessel, J., Caubet, R.: A New Object Order Ray-casting Algorithm. Proc. IEEE Volume Visualization 2002 (2002) 203-210
14. Grimm, S., Bruckner, S., Kanitsar, A., GrÖller, E.: Memory Efficient Acceleration Structures and Techniques for CPU-based Volume Raycasting of Large Data. Proc. IEEE Volume Visualization 2004 (2004) 1-8
15. Lim, S., Shin, B.: Reliable Space Leaping Using Distance Template. Lecture Notes in Computer Science, Vol. 3337 (2004) 60-66
16. Hadwiger, M., Sigg, C., Scharsach, H., Buhler, K., Gross, M.: Real-time Ray-casting and Advanced Shading of Discrete Isosurfaces. Graphics Forum, Vol. 24, No. 3 (2005) 303-312
17. Lim, S., Shin, B.: RPO: A Reverse-Phased Hierarchical Min-Max Octree for Efficient Space-Leaping. Proc. Pacific Graphics 2005 (2005) 145-147
18. Lim, S., Shin, B.: Efficient Space-Leaping Using Optimal Block Sets. IEICE Transactions on Information and Systems, Vol. E88-D, No. 12 (2005) 2864-2870
19. Yagel, R., Cohen, D., Kaufman, A.: Discrete Ray Tracing. IEEE Computer Graphics and Applications, Vol. 12, No. 5 (1992) 19-28
20. Medison, co., ltd, Rep. of Korea, http://www.medison.com/eng/index.asp
21. Cohen, D., Sheffer, Z.: Proximity Clouds: An Acceleration Technique for 3D Grid Traversal. The Visual Computer, Vol. 11, No. 1 (1994) 27-28
22. Sramek, M., Kaufman, A.: Fast Ray-tracing of Rectilinear Volume Data Using Distance Transforms. IEEE Trans. on Visualization and Computer graphics, Vol. 6, No. 3 (2000) 236-252

Neural Based CAD and CAP Agent System Framework for High Risk Patients in Ubiquitous Environment

Insung Jung[1], Myeong-Ho Lee[2], and Sanghoon Bae[3]

[1] Department of Industrial & Information Systems Engineering
Ajou University, Korea, 442-749
gabriel7@ajou.ac.kr
[2] Division of Internet Information, Semyung University,
579 Sinweol-Dong, Jechon, Chungbuk, Korea, 390-711
mhlee@semyung.ac.kr
[3] Department of Satellite Information Sciences, Pukyong National University,
599-1 Daeyeon 3 dong, Nam gu , Pusan, Korea, 608-737
sbae@pknu.ac.kr

Abstract. The objective of this paper is to describe the diagnosis, detection, and prediction of high risk patient status or level of risk, based on the hospital domain server with agent based intelligent home medical service system. We have proposed a Computer Aided Diagnosis and monitoring system framework for the high risk patient exploiting the ubiquitous artifacts available in the contemporary medical environment. The proposed framework monitors patients using bio signal devices while they are staying at their homes. Furthermore, the agent based learning process is used to make intelligent inferences and adaptive learning of these private data using Computer Aided Diagnosis system. It can monitor acute COPD using integrated home medical server. The services include prevention, emergency call center and ambulance service in a common platform of ubiquitous environment.

1 Introduction

Recently, the medical treatment technique has improved gradually allowing people to enjoy longer lives. As the studies show, more than 70% the senior citizens have got high risk disease such as hypertension, diabetes and acute respiratory diseases. H.S Lee *et al* [1] "if we can do emergency treatment to them, it is possible to prevent a sudden death." However, recent medical systems are quite lacking in terms of offering real time patient monitoring, early detection and diagnosis. Consequently many people are quite interested in ubiquitous system provisioning for hospital domain. There have been recent studies regarding the use of such u-Health systems. For example, Newandee D. has studied COPD severity classification using principal component and cluster analysis on HRV parameters using bio-signals [2]. Dr. K. Karoui discusses about the multilevel diagnosis framework systems which verify the remote diagnosis information based on four level testing [3]. D. Krenchel and J. P. Calabretto *et al* described about the well-organized storage of remote patient's profiling and quick searching facility based on online services [4, 5]. It can be applied in our project for efficient database handling and on-time automatic updating of the patient regular

L. Wang et al. (Eds.): FSKD 2006, LNAI 4223, pp. 1054 – 1057, 2006.

diagnosis reports. Most of the references have focused on single dimension of the medical facility while our approach integrates the complete solutions utilizing ubiquitous artifacts and online services. It provides just-in-time medical diagnosis and treatment to the home based high risk patients.

The objective of this paper is to combine the agent based decision support system with ubiquitous artifacts. It helps the doctors to acquire on-time diagnosis correctly and select appropriate treatment choices. An attempt is made to supervise the dynamic situation by using agent based ubiquitous artifacts and to find out the appropriate solution for emergency circumstances that provides correct diagnosis and appropriate treatment in time.

2 High Risk Patient System Service Scenario

Health monitoring and Computer Aided diagnosis will be useful for high risk patients for prevention of sudden death. As a specific example, the doctors at Ajou University define four clusters of patient level (regular, careful, serious and dangerous) using vital signal data. They just need to take some medicines and follow prescriptions at home. However, serious situation patients need to be given first aid and readily be in contact with their private doctor. When patient's level is dangerous, a phone call to emergency call center is automatically made and the vital data is sent. The agent computes the location for the nearest hospital to call an ambulance. As a result, we are setting up 3 kinds of health care modules which are called home care module, emergency call center module and an ambulance module.

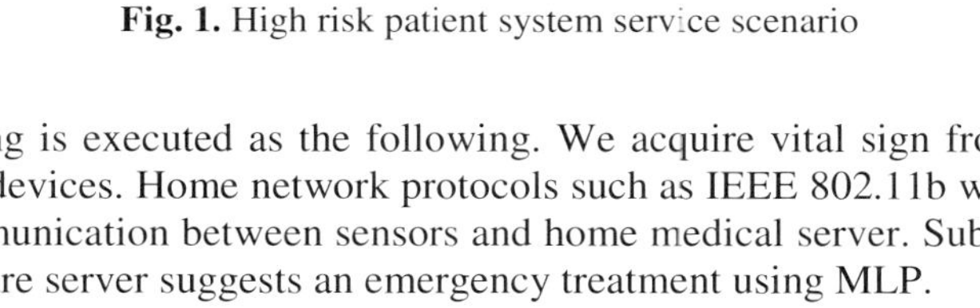

Fig. 1. High risk patient system service scenario

The working is executed as the following. We acquire vital sign from the patient by electronic devices. Home network protocols such as IEEE 802.11b will be used for wireless communication between sensors and home medical server. Subsequently, the home healthcare server suggests an emergency treatment using MLP.

3 System Framework

Our framework of home healthcare system (Fig 2) consists of 4 systems and one knowledge database. The database supports sharing of hospital diagnosis knowledge to CAD system. The vital signal data processing system is used for data normalization, after detecting the vital data. These filtered signals will be used as input data to the neural network based CAD system. With the help of this information, regular monitoring of patient is ensured. It can measure the level of risk by applying regular monitoring and prediction techniques like time series analysis. CAP system predicts vital data and the level of the disease. The last module classifies emergency state of action system for classification of the patient's precarious condition level. If the patient's condition is not normal, the system will react to the emergency situation by suggesting an immediate measure.

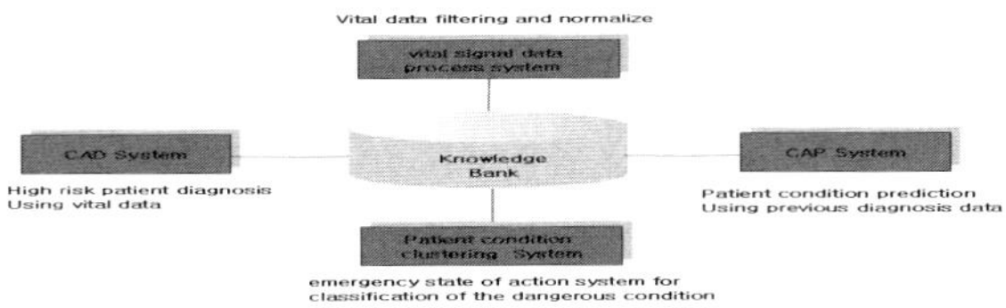

Fig. 2. System framework

3.1 CAD and CAP Framework

Computer Aided Diagnosis (CAD) system (Fig 3-a) uses either of the popular neural network algorithms. We have proposed Multi Layer Process (MLP) for the scope of our paper. It acquires its input data from patients' database. Generally, pulmonary disease patients need to check, SPO_2, Co_2, HP, BP etc, which should be passed as an input data to MLP. The target used by CAD is based on the previous case studies. In addition, the outputs and learning weight results are automatically saved in the database and a message is sent to the server.

Computer Aided Prediction (CAP) system framework (Fig 3-b) works as almost the same as the CAD system. Furthermore, it acquires input vector about patient's previous and current data ($Y_{t-1} \sim Y_{t-n}$) for the prediction of level of risk in the future.

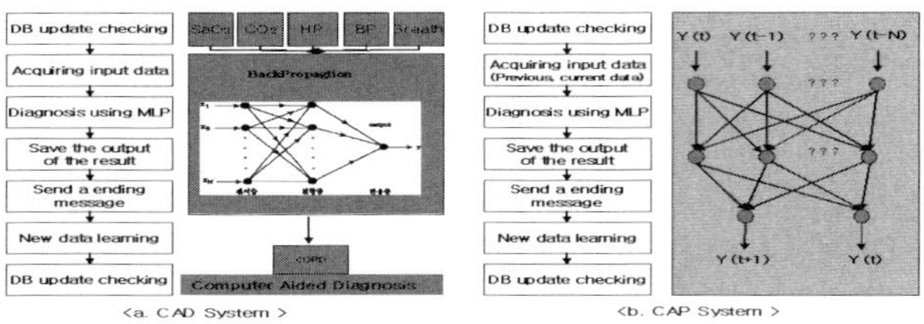

Fig. 3. Computer Aided Diagnosis & Prediction framework

3.2 Classification of Patient's Precarious Level Framework

We are using emergency states to be a measure to trigger an action system for classification of the patient's uncertain level of risk. We have consulted a doctor about important factors for high risk diseases and patient's clustering before using classification of the patients' status. In this paper, four levels of clusters have been defined.

The action will be taken as per the level of risk. For example, Level 1 denotes normal condition. If the risk level rises above level 2 then the result of classification system will send a message to patient's display device like PDA or mobile phone and link to an emergency measurement URL where the data has been stored using XML schema. When the patient receives the message, he can connect on-line to check his feedback action. Level 3 signals dangerous situation, thereby, it involves more interaction between medical server and hospital call center (HCS). When HCS receives a message from medical server, it makes a phone call to the patient's telephone. In case of no response, the HCS sends a message to the doctor and an ambulance service. In addition, the medical server sends a message about these situations to their families ubiquitous devices as well. The fourth and the most perilous level of risk is Level 4. In this stage the medical server automatically sends a message and the data of patient to an ambulance squad and to the private doctor's computer. It assists the doctor and nurses to cut down on their preparatory period. In addition, the medical server does all the previous actions applicable to Level 1~3.

4 Conclusion

Ubiquitous has been one of the hot research topics around the world. One of the best utilization of this technology can be in the field of computer aided diagnosis and prediction. The objective of this paper is to define a framework for the computer aided diagnosis and computer aided prediction of the high risk patients. It will help the medical system to do early detection of high risk diseases, reduce diagnosis error and prevent sudden death situation. One of the limitations in implementation of this system using hospital domain knowledge is different diagnosis pattern of the doctors. It means that the most important factor is their individual experience of diagnosis.

References

1. H.S Lee M, S.J Beak, M.S Chol, C.S Park. : High risk patient death before arriving to hospital J Korean Acad Fam Med Vol. 14, 1993: 601-08.
2. D. A. Newandee, S. S. Reisman, M. N. Bartels, and R. E. De Meersman.: COPD Severity Classification Using Principal Component and Cluster Analysis on HRV Parameters. Bioengineering Conference, 2003 IEEE 29th Annual, Proceedings March 2003:134-35.
3. Dr. K. Karoui, Dr. R. Sammouda, Dr. M. Sammouda.: Framework for a Telemedicine Multilevel Diagnose System, Proceedings of the 23rd annual EMBS International Conference, October 25-28, Istanbul, Turkey.
4. D. Krenchel, R. Bergmann, K. Maximini, A. V. Wangenheim.: A Framework for Radiological Assistant Systems, Proceedings of the 15th IEEE Symposium on Computer-Based Medical Systems (CBMS 2002).
5. J. P. Calabretto et al.: Agent Support for Patients and Community Pharmacists, Proceedings of the 35th Hawaii International Conference on Sysmte Sciences 2002.

Transcription Factor Binding Sites Prediction Based on Sequence Similarity*

Jeong Seop Sim[1] and Soo-Jun Park[2]

[1] School of Computer Science and Engineering, Inha University, Incheon, Korea
jssim@inha.ac.kr

[2] Electronics and Telecommunications Research Institute, Daejeon, Korea
psj@etri.re.kr

Abstract. Sequence algorithms are widely used to study genomic sequences in such fields as DNA fragment assembly, genomic sequence similarities, motif search, etc. In this paper, we propose an algorithm that predicts transcription factor binding sites from a given set of sequences of upstream regions of genes using sequence algorithms, suffix arrays and the Smith-Waterman algorithm.

1 Introduction

There have been vigorous works to study genomic sequences by sequence algorithms in such fields of DNA fragment assembly, genomic sequence similarities, motif search, etc [1,3,4,13]. Motifs are conserved regions in genomic sequences. Some motifs in regulatory regions (or promoter regions) are called transcription factor binding sites because some proteins called transcription factors bind to them. Transcription factor binding sites are extensively studied recently since they are highly related to gene functions.

Transcription factors are proteins involved in regulatations of gene expressions. They bind to some portions of upstream regions of genes and either facilitate or inhibit transcriptions. They recognize very specific parts of genes and bind to specific regions of DNA molecules. Suppose we are given a sequence of a gene whose functions are unknown. If there exists some portions in a regulatory region of the gene whose sequence is very similar to or exactly the same as some binding site of a transcription factor whose function is already known. Then we can assume that the functions of the gene may be related to the functions of the transcription factor. For this, we should have lots of information about the relations between transcription factors and their binding sites. Biologists have vigorously studied to find out the relations via diverse high-cost experiments, such as microarrays [11,12,14,15]. In this paper, we present an algorithm that predicts transcription factor binding sites based on sequence similarity. Our

* This work was supported by INHA UNIVERSITY Research Grant (INHA-32744).

L. Wang et al. (Eds.): FSKD 2006, LNAI 4223, pp. 1058–1061, 2006.

algorithm suggests candidates of transcription factor binding sites from sequences of upstream regions of functionally related genes.

2 Preliminaries

The suffix array SA_T is an array of integers i that represent suffixes S_i of a sequence T [10]. The integers are sorted in lexicograhic order of the corresponding suffixes. To search a pattern efficiently in a suffix array, LCP (longest common prefix) information is used. An LCP array L is an array of the lengths of common prefixes of two adjacent suffixes in a suffix array SA_T for T. In many applications, such as molecular biology, two sequences that have some relations may not be highly similar in their entirety, but they may contain very similar regions. The *local alignment problem* is to find a pair of regions, one from each of the two given sequences, that are very similar. One of the most well-known algorithm to find an optimal local alignment was given by Smith and Waterman [13], which is known as *the Smith-Waterman algorithm*. Many biological programs use some heuristics of this algorithm [3,2,9].

3 Algorithm

Given a set S of DNA sequences $S_1, S_2, \ldots, S_n$, we want to find out all candidates of transcription factor binding sites that satisfy given conditions. The input set of sequences are from regulatory regions of genes whose functions may have some relations, i.e., they can have similar functions, opposite functions, and can be on the same regulatory networks. Some functionally related genes are regulated by the same transcription factors and transcription factors bind to some specific sequences of regulatory regions. Thus, if corresponding input genes have some functional relations, there can be some common specific sequences in the input sequences of regulatory regions. Our algorithm aims to extract such common sequences from the regulatory regions of (potentially) funtionally related genes using a suffix array and the Smith-Waterman algorithm.

Our algorithm mainly consists of three steps. In the first step, we preprocess the input sequences. We make a long sequence T by concatenating all the sequences in S. That is, $T = S_1\#_1S_2\#_2 \ldots S_n\#_n$. Each $\#_i$, $1 \leq i \leq n$, is a special symbol to delimit each sequence.

In the second step, we make a suffix array SA_T of T and make LCP array L for SA_T. There are several linear-time suffix array construction algorithm [5,7,8]. We use Kärkkäinen and Sanders's algorithm [7] for its simplicity. After this process, we make a set of initial candidates. At this time, we use two input parameters of thresholds. The first one denoted by LEN represents the minimum length of the initial candidates of transcription factor binding sites. The initial candidates are longest common prefixes that are longer than LEN. The second input threshold is denoted by RTO, which means that each candidate must appear at least in $n \times RTO$ sequences. Let $C = \{C_1, C_2, \ldots, C_k\}$ be the set of candidates computed in the second step.

Table 1. Results when $LEN = 6, RTO = 0.7, FRQ = 0.8$. We marked a detected candidate motif as known only if a known motif is completely included in the detected candidate motif, i.e., when a known motif is a subsequence of the detected candidate motif.

Group (# of seq.)	Known Motif	Detected Candidate Motif	Actual Frequency (%)	Known or Unknown
1 (24)	ACGCGA	AACGCG	100	unknown
	CGCGAA	ACGCGAAA	100	known
2 (8)	CGCGTC	TGAAAC	100	known
	TGAAAC	TTGAAA	100	unknown
		ACGCGTC	100	known
3 (8)	GTAAACA	AAAAAT	100	unknown
		AGGAAA	100	unknown
	TTAGGAA	GGAAAT	100	unknown
		TTAGGAA	100	known
		GTAAACAA	100	known
4 (8)	TGAAACA	AAAATT	100	unknown
		AAACAA	100	unknown
	CCAGCA	AAACAG	100	unknown
		AAACCA	100	unknown
		CCAGCAA	100	known
		TGAAACA	100	known

Note that the sequences of binding sites may be slightly different. Thus, we use the Smith-Waterman algorithm in the third step to find similar sequences. We perform the Smith-Waterman algorithm with each C_i $(1 \leq i \leq k)$ of C and all the input sequences in $S = \{S_1, S_2, \ldots, S_n\}$. In this step, we use two more threshold values, SV and FRQ, that are used as following. For each C_i to be a final candidate, the number of sequences whose similarity between C_i and some subsequence of each S_j is higher than SV must be larger than $n \times FRQ$.

4 Experimental Results

Kato et al. [6] analyzed transcription factors and binding motifs using various analysis methods together and obtained effective results. We analyze our algorithm based on the result of [6]. To test if our algorithm identifies existing known motifs, we choose 4 groups of sequences each of which contains two known motifs. See Table 1. For each motif in this set, we make input sequences by extracting UTR 300 base pairs of genes where the motif exists. We set threshold values as follows: $LEN = 6, RTO = 0.7, SV = 17, FRQ = 0.8$. The experimental results are shown at Table 1. Our algorithm successfully detects all the motifs in addition to some more candidates of transcription factor binding sites. We are planning to check if these candidates are true positive or not by performing some experiments in the wet lab.

References

1. S. Batzoglou, D. Jaffe, K. Stanley, J. Butler, S. Gnerre, E. Mauceli, B. Berger, J. Mesirov, and E. Lander, *Arachne: Awhole-genome shotgun assembler*, Genome Research, 2002, 12: 177-189.
2. T. Chen and S.S. Skiena, Trie-based data structures for sequence assembly, In *Proc. 8th Annual Symposium on Combinatorial Pattern Matching*, LNCS 1264, 206-223, 1997.
3. P. Green, *Documentation for phrap*, Genome Center, University of Washington, http://www.phrap.org/phrap.docs/phrap.html.
4. D. Gusfield, *Algorithms on Strings, Trees, and Sequences*, Cambridge University Press, 1997
5. P. Ko and S. Aluru, Space efficient linear time construction of suffix arrays, In *Proc. 14th Annual Symposium on Combinatorial Pattern Matching*, LNCS 2676, 200-210, 2003.
6. M. Kato, N. Hata, N. Banerjee, B. Futcher, and M.Q. Zhang, Identifying combinatorial regulation of transcription factors and binding motifs, *Genome Biology*, 5(8), R56, 2004.
7. J. Kärkkäinen and P. Sanders, Simple linear work suffix array construction, *International Colloquium on Automata, Languages and Programming*, LNCS 2676, 943-955, 2003.
8. D.K. Kim, J.S. Sim, H. Park and K. Park, Constructing suffix arrays in linear time, *Journal of Discrete Algorithms*, 3, 126-142, 2005.
9. D. Lipman and W. Pearson, Improved tools for biological sequence comparison, In *Proc. National Academy of Science*, 85, 2444-2448, 1988.
10. U. Manber and G. Myers, Suffix arrays : A new method for on-line string searches, *SIAM Journal on Computing*, 22, 935-938, 1993.
11. V. Matys, E. Fricke, R. Geffers, E. Goling, M. Haubrock, R. Hehl, K. Hornischer, D. Karas, A. E. Kel, O.V. Kel-Margoulis, D.U. Kloos, S. Land, B. Lewicki-Potapov, H. Michael, R. Munch, I. Reuter, S. Rotert, H. Saxel, M. Scheer, S. Thiele, and E. Wingender, TRANSFAC: transcriptional regulation, from patterns to profiles, *Nucleic Acids Research*, 31(1), 374-378, 2003.
12. U. Ohler, H. Niemann, G. Liao, G.M. Rubin, Joint modeling of DNA sequence and physical properties to improve eukaryotic promoter recognition, *Bioinformatics*, 17 Suppl 1, S199-206, 2001.
13. T.F. Smith and M.S. Waterman, Identification of common molecular subsequences, *Journal of Molecular Biology*, 147, 195-197, 1981.
14. G. Stoesser, W. Baker, A. Broek, M. Garcia-Pastor, C. Kanz, T. Kulikova, R. Leinonen, Q. Lin, V. Lombard, R. Lopez, R. Mancuso, F. Nardone, P. Stoehr, M.A. Tuli, K. Tzouvara, and R. Vaughan, The EMBL ncleotide sequence database: major new developments, *Nucleic Acids Research*, 31(1), 17-22, 2003.
15. M.Q. Zhang, Identification of human gene core promoters in silico, *Genome Research*, 8(3), 319-326, 1998.

On the Society of Genome: Social Affiliation Network Analysis of Microarray Data

Jung Hun Ohn[1], Jihoon Kim[1], and Ju Han Kim[1,2,⋆]

[1] Seoul National University Biomedical Informatics (SNUBI)
[2] Human Genome Research Institute, Seoul National University College of Medicine, Seoul 110-799, Korea
jhohn2@snu.ac.kr, hoonie.kim@gmail.com, juhan@snu.ac.kr

Abstract. To investigate the structure of the genomic interaction network built from yeast gene-expression compendium dataset of hundreds of systematic perturbations, social affiliation network analysis methodologies were applied through quantifying various density, closeness and centrality measures and exploring core-periphery structures. Genes affected by a larger number of perturbations were found to be involved in responses to various environmental challenges. Deletion of essential genes was suggested to cause larger number of genes to be significantly up or down regulated. We explored the network structure made up of several sub-networks using core-periphery models to find ancient pathways. Glycolysis and TCA cycle have relatively core positions in the energy-related processes of yeast.

1 Introduction

The current way of describing cellular processes are based on mechanical concepts and each cellular process is regarded as a conveyer belt on which many workers, i.e. proteins, work to give products for the survival of a large factory or a cell. Biology books are full of many such schematic figures, which is, of course, useful for illustrating life phenomena. However, this may mislead. Each gene product or protein has no concept of such processes as DNA replication, apoptosis or signal transduction. They are just interacting with each other without the intention of replicating DNA or transducing signals. These purposeless interactions form the basis of life and may in fact be a better description of life. Complex information exchanges between cellular components keep life go on.

How can we describe this aspect of life? Let us pick the wisdom of social analogy. We endow each gene with its functions from the point of cellular processes like DNA replication and cell cycle control, just as we have our own social roles defined with respect to the social groups like families and jobs. We are in contact with people who share with us the same group memberships, which is the basis of our personal contact and information exchange. One interacts with its group members directly or indirectly and the members are quite important in understanding him: we can know a man by the company he keeps!

⋆ Corresponding author.

L. Wang et al. (Eds.): FSKD 2006, LNAI 4223, pp. 1062–1071, 2006.

Describing the properties of individuals through its social relationship with others has been the subject of study for social network analysts. [1] They try to find social 'stars' in different aspects and to describe the network structure through various centrality measures and navigate its unique structures by graph theoretic approaches. In its graph representation, each node represents an individual and each edge social interaction between two individuals. The presence or absence of interaction between N individuals can be expressed as an N-by-N binary matrix, i.e. 1 for the presence and 0 for the absence of interaction. This matrix is called one-mode matrix. On the other hand, two-mode network represents the affiliation of a set of actors with a set of social occasions. Many social network relations consist of the linkages among actors through their joint participation in social activities or membership in collectivities (i.e. events). Such networks of actors tied to each other through their participation in events and events linked through multiple memberships of actors, are referred to as affiliation networks. [1][2]

Affiliation network is represented as a matrix with binary relationship between actors and events. If an actor is affiliated with an event, the binary relation is given by 1 and otherwise 0 (see methods). Figure 1 shows an example of such affiliation matrices with 18 actors and 12 events.

		1 E	2 E	3 E	4 E	5 E	6 E	7 E	8 E	9 E	10 E	11 E	12 E
1	A1	1	1	1	0	1	1	0	1	1	0	1	0
2	A2	0	0	1	0	1	0	0	1	0	0	0	1
3	A3	0	1	1	1	1	0	1	0	1	0	0	1
4	A4	0	0	0	0	0	1	0	1	1	0	0	0
5	A5	0	0	0	0	1	1	0	0	0	0	0	1
6	A6	0	1	0	0	1	1	1	0	1	0	0	0
7	A7	0	0	0	0	1	1	0	0	0	0	0	1
8	A8	0	0	1	1	1	1	1	0	1	0	1	0
9	A9	0	0	1	0	0	1	0	0	0	0	0	0
10	A10	0	0	1	0	0	1	0	0	1	1	1	1
11	A11	0	0	1	1	1	0	0	0	0	0	0	0
12	A12	0	0	1	0	0	0	1	0	0	0	0	0
13	A13	0	1	0	0	0	0	0	0	1	1	1	1
14	A14	1	0	1	0	0	0	0	0	0	0	1	1
15	A15	0	0	1	1	0	1	1	1	0	0	0	0
16	A16	0	1	1	0	1	1	1	0	0	0	1	0
17	A17	0	0	0	1	1	0	0	0	1	0	0	1
18	A18	1	0	1	1	0	0	0	0	0	0	0	0

Fig. 1. An example of affiliation matrix with 18 actors and 12 events

In the present study, we binarized a yeast microarray dataset to build an affiliation matrix and tried to describe and analyze social behavior of yeast genes. With the classical notation of social network analysis, a gene corresponds to an actor and a group of genes to an event.

Rosetta yeast compendium dataset [3] is hitherto the most systematic approach to profile yeast genes. Gene expression levels were measured in 300 different conditions to investigate the impact of uncharacterized perturbations on the cell like deletion mutations and drug treatments. In the original article authors newly annotated eight deleted genes by hierarchical clustering analysis and confirmed it experimentally.

Cohen *et al.* referred to the 'molecular phenotype' of a gene as the constellation of changes in gene expression profile after deletion of the gene.[4] The molecular phenotype is a group of genes that are significantly up or down regulated by a gene deletion or chemical treatment. Rosetta compendium dataset has gene expression profiles in 300 different gene deletion mutations and drug treatments. A drug treatment works like a gene deletion as it usually blocks the action of several gene products it binds to. Each gene deletion or chemical treatment condition assigns more than 6,000 genes into two groups, molecular phenotype and non-molecular phenotype. This is why the Rosetta compendium dataset of yeast genes is well suited for our purpose. Genes or actors that are differentially transcribed following deletion of one common gene or common chemical treatment belongs to one group and the group is analogous to the events in social network analysis terminology. Genes affected by common perturbations can be assumed to communicate with each other directly or indirectly. This assumption well justifies our approach to analyze social behavior of genes from the perspective of social affiliation network.

This structural uniqueness of the Rosetta dataset led Rung *et al.* to construct, what they called, disruption networks and they analyzed yeast genome graph theoretically and showed that disruption network is scale-free.[5] The social network analysis framework gives additional insights into gene-to-gene communications.

2 Data and Methods

2.1 Data Preprocessing and Determination of Molecular Phenotypes

Rosetta Compendium dataset was downloaded from ExpressDB. [6] It is a compendium of expression profiles corresponding to 300 diverse mutations and chemical treatments (276 deletion mutants, 11 tetracycline regulatable essential genes, 13 chemical treatments) in S. cerevisiae. Excluding genes that have more than 20 missing values left 6,152 genes for analysis. A data matrix containing log expression ratio in each condition was used for analysis. The matrix was normalized with respect to conditions such that mean and standard deviation of each column log ratio value was set to 0 and 1, respectively.

Generally whether a gene is differentially expressed in a condition is determined in a biological sense by its fold ratio. Statistical significance has also been used as a means of selecting differentially expressed gene in a large dataset. [7] We pooled the log ratio values to get a cutoff for binarization process. We obtained 5% quantile (Q0.05) and 95% quantile (Q0.95) (i.e. -1.24 and 1.33, respectively) for the above normalized log ratio values and used them for the cutoff value determining significant log ratio.

2.2 Binarization

Let Eij be the normalized (with respect to condition) log expression ratio of gene i in gene disruption or chemical treatment condition j above. New data matrix A with Aij as its element is given by: A = $<Aij>$,

$$A_{ij} = \begin{cases} 1 \text{ (if } Q0.05 < Eij < Q0.95) \\ 0 \text{ (if } Q0.05 > Eij \text{ or } Q0.95 < Eij) \end{cases}$$

Gene or actor i is affiliated with the molecular phenotype of gene mutation or drug treatment condition j if $Aij = 1$ and is not affiliated if $Aij = 0$. A is the **affiliation matrix** shown in figure 1.

2.3 Analysis of Affiliation Network

Bipartite matrix. Affiliation matrix A is tranformed into a bipartite *square* matrix B given by, (given N actors, M events and O representing zero matrix)

$$\mathbf{B} = \begin{pmatrix} O(NXN) & A(NXM) \\ A(MXN) & O(MXM) \end{pmatrix}$$

Bipartite graph. A graph is bipartite if the vertices are partitioned in two mutually exclusive sets such that there are no ties within either set and every edge in the graph is an unordered pair of nodes in which one node is in one vertex set and the other in the other vertex set. Bipartie graph is very useful in representing two-mode network.

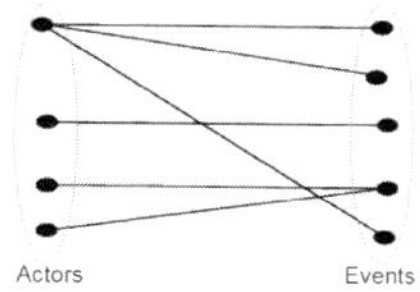

Fig. 2. Bipartite graph representation of an affiliation matrix. Left vertices are actors and right ones events.

Geodesic distance. A shortest path between two nodes is referred to as a geodesic. A geodesic distance matrix G =<Gij> represents geodesic distances between all pairs of nodes in the bipartite graph.

Rates of participation. Rate of participation of actor i is given by

$$\sum_j A_{ij} \quad .$$

which implies how many events an actor participates in. The more sociable an actor is, the more events will he or she participate in.

Size of events. Size of event j is given by

$$\sum_i A_{ij} \quad .$$

which implies how many actors participate in the event j.

Node centrality measures and group centralization measures. For detailed description of the concept of centrality, refer to [1][2][8] and [9]. The origin of this idea in social network analysis can be found in the concept of the 'star'- the person who is the most 'popular' in his or her group or who stands at the center of attention. Group centralization index measures the extent to which the graph is a *star graph - there is one central node with the remaining nodes considerably less central.* Centrality measures were calculated using the UCINET 6.0 software.[10]

Node Degree Centrality. This is the simplest definition of node centrality. The central node must be the one who have the most ties to other nodes in the network. In the two-mode data, actor degree centrality is the number of events an actor attended and event degree centrality is the number of actors participating in the event. Degree centrality of an actor i is given by

$$\sum_j B_{ij} \quad .$$

Node Closeness Centrality. This measures how close a node is to all the other nodes. In two-mode network represented by a bipartite graph, all paths consist of an alternating series of nodes and edges of the form u-v-u'-v' and so on where u and u' are from one vertex set and v and v' from the other. The closeness centrality of a node was defined by Freeman and is inversely proportional to the total geodesic distance from the node to all other nodes in the network.[11] Closeness centrality of an actor i is given by

$$\left[\sum_j G_{ij}\right]^{-1} \quad .$$

Node Betweenness Centrality. This measures the probability that a communication or simply a path from node j to node k takes a particular route through a node i. All lines are assumed to have equal weights. Let g_{jk} be the number of geodesics linking the two nodes j and k. Let g_{jk}(i) be the number of geodesics linking the two nodes that contain node i. In two-mode network, betweenness centrality is a function of paths from actors to actors, events to events, actors to events and vice versa. Betweenness centrality of an actor i is given by

$$\sum_{j<k} g_{jk}(i)/g_{jk} \quad .$$

Group Centralization Measures (Degree, Closeness or Betweenness). Group centralization measure is a group level measure of centrality. Let C(i) be a node centrality index (degree, closeness or betweenness) and C(i)* be the largest value of the indices across all nodes. The general form of group centralization index is given by:

$$C = \frac{\sum_i [C(i)^* - C(i)]}{max \sum_i [C(i)^* - C(i)]} \quad .$$

2.4 Core/Periphery Structures

A common notion in social network analysis is the concept of a core/periphery structure and a dense, cohesive core and a sparse, unconnected periphery are sought. Borgatti *et al.* formalized the notion of core/periphery structure and suggested both discrete and continuous models in detecting core/periphery structure in network data and the computer package UCINET 6 incorporates the model.[12] We adopted the continuous model, which assumes the network has one core and assigns each node a measure of 'coreness'. In UCINET 6, the value of coreness of node i, ci, is obtained so as to maximize the matrix correlation between the data matrix (in affiliation network, the bipartite matrix) and the pattern matrix, P, the element of which is $p_{ij} = c_i{*}c_j$.

3 Results

3.1 Whole Genome View

Rate of participation. Rate of participation of an actor counts the number of events an actor participates in. Actors that participate in a large number of events are regarded as sociable actors. Genes that are differentially expressed in more than 150 out of 300 perturbing conditions are as follows.

YBR072W, YBR145W, YBR296C, YCL018W, YER069W, YFL014W, YFR030W, YFR053C, YGL255W, YHR018C, YHR137W, YHR215W, YIR034C, YJL088W, YJL153C, YJR025C, YML123C, YMR062C, YMR094W, YMR095C, YMR096W, YMR105C, YNL036W, YNL160W, YOL058W, YPL019C, YPR160W, YPR167C, YJR109C, YKL001C, YKL096W, YLR303W

These genes are "social stars" in yeast genome in that they are parts of a large number of molecular phenotypes and in biological sense, are very sensitive to external perturbations. The MIPS functional classifications[13] of these 'star' genes are 1) Stress response, 2) Amino acid biosynthesis, 3) C-compound and carbohydrate biosynthesis, 4) Small molecule transport, 5) Osmoregulation. The functions are important for the survival of the yeast against various environmental challenges. It is natural to suppose that genes that are involved in the processes related to *interaction with cellular environments* will be frequently up or down regulated by external perturbations.

Size of events. Size of an event implies how many actors participate in the event. Examples of gene deletions or drug treatments with large sizes are:

yor078w, erp4, ymr141c, kar2, yef3, cdc42, rpl12a, cla4-haploid, ymr014w, arg5,6, gyp1, dfr1, rps24a, hes1-haploid, idi1, ymr030w, kre1, bub3, yhr011w, ste20, erg11, 2-deoxy-D-glucose, TUNICAMYCIN, she4, yor006c, pac2, mak10, cue1, cat8, hat2, sir1, ymr285c, ade16, phd1-haploid, bub1-haploid, erg4-haploid, yer041w, prb1, aqy2, yml003w, rml2, hir2, msu1, yml011c, top1-haploid, pma1, rnr1-haploid, yor072w, yel033w, sap30

The functional categories are 1) ribosome biogenesis, 2) lipid, fatty-acid and isoprenoid biogenesis, 3) transport, 4) transcriptional control, 5) cell cycle 6) DNA synthesis and replication, 7) budding and pheromone response. The specific kinds of genes giving rise to a large size of perturbation are somewhat different from those found by Featherstone *et al.* and Rung *et al.* because of different normalization process. But the above functional categories lead to the similar conclusion; genes whose deletion strongly 'wiggles' the whole cellular transcriptional system are 'essential' cellular processes that are always switched on irrespective of environmental stimuli.[5][14] The perturbation may be the direct result of the deletion itself or the indirect one of the triggered mechanisms in compensation for the gene disruption to keep one yeast from being lethal. [14][15]

3.2 Analysis of Genes Participating in Energy Related Processes

We explore the structure of a specific network made up of several sub-networks. The MIPS database provides a catalogue of functional categories which groups together genes with similar functions and we explored the network of genes known to participate in 'Energy' related processes.[13] The energy related gene network is composed of 10 subgroups of genes assigned to the following functional categories. A total of 208 genes were included. The number in the parenthesis is the number of genes participating in the process. These genes have no missing values in Rosetta compendium dataset and errors from missing data were excluded.

1. Oxidation of fatty acid (6)
2. Fermentation (28)
3. Glycolysis and gluconeogenesis (28)
4. Glyoxylate cycle (5)
5. Pentose-phosphate pathway (9)
6. Metabolism of energy reserves (glycogen, trehalose) (33)
7. Respiration (70)
8. TCA cycle (20)
9. Other energy generation activities (13)
10. Electron transport (2)

Core/Periphery structure of Energy related genes. In the Energy related affiliation network, the core/periphery structure is investigated. (See methods for details) Figure 3 shows the distribution of coreness scores of genes in each functional categories (Genes with higher coreness scores form the core). Genes participating in fatty acid oxidation and energy transport are mostly placed in the periphery, whereas, glucose metabolism related process (categories 3 and 6) contain core genes in energy process and ATP generating processes (categories 2 and 8) occupy intermediate position. ATP consuming process (Respiration) related genes have relatively peripheral placement. Ancient pathways like Glycolysis and TCA cycle have relatively core positions in the network.[16]

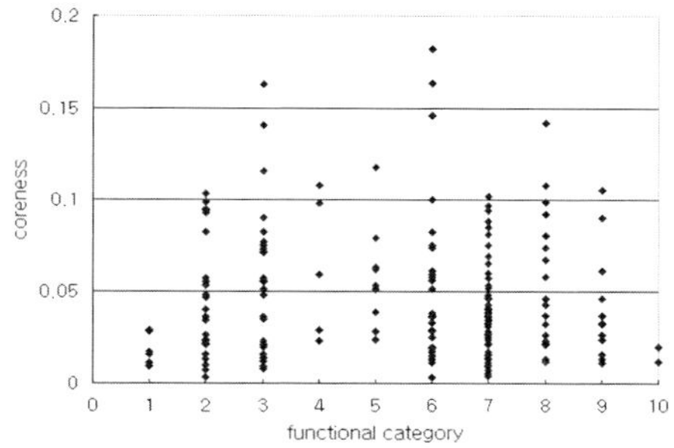

Fig. 3. Core/Periphery structure of energy-related pathways in yeast

Graph centralization index. A graph with higher centralization index is more like a 'star' graph. Table 1 shows the glyoxylate cycle gene group has the highest degree and closeness centralization indices and forms the most 'star' like graph. In contrast, respiration process has the smallest degree, closeness and betweenness centralization indices and has the least star-like structure.

Table 1. Graph centralization index

Process	Degree	Closeness	Betweenness
fatty acid oxidation	26.22	26.99	**66.13**
fermentation	28.32	18.99	16.29
glycolysis	45.81	24.84	24.74
glyoxylate cycle	**52.39**	33.23	49.35
pentose phosphate	45.56	30.84	46.38
energy reserves	49.94	31.23	24.50
respiration	22.78	21.16	8.57
TCA cycle	43.79	26.99	33.68
All actors	26.49	21.59	4.11

Fatty acid oxidation has relatively small degree and closeness centralization indices but it has unusually high betweenness centralization index. YLR284C (ECI1) has the largest betweenness centrality score of all the actors which means other actors depend on this gene to communicate with each other and this gene product might have some control over the interactions.

3.3 TCA Cycle

Now let us focus on one of the sub-networks of energy related processes, or TCA cycle. Borgatti *et al.* pointed out geodesic distance matrix (see methods) as an input for multidimensional scaling gives good visualization results and makes it easy to draw rough conclusions at a glance.[9] Figure 4. shows multidimesional scaling representation of TCA cycle related genes and conditions after geodesic distance matrix is formed from the affiliation matrix. Genes are coded with its enzyme names and 52 conditions (labeled with numbers) were those that contain more than 5 participating genes out of 20 TCA cycle related genes.

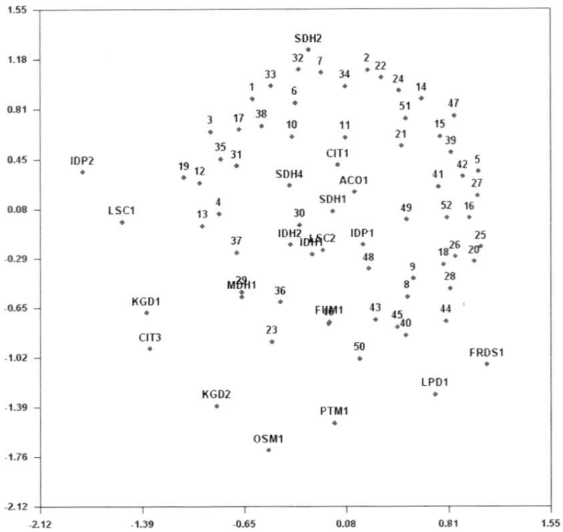

Fig. 4. Multidimesional-scaling representation of TCA cycle-related genes and conditions

We can find a core/periphery structure especially among genes or actors. The core group contains succinate dehydrogenase complex (SDH1, SDH2 and SDH4) and isocitrate dehydrogense complex (IDH1 and IDH2), fumarase, aconitase and citrate synthase. The core group genes are well known TCA cycle related genes and the peripheral genes have hitherto unspecified role in TCA cycle.[17] This might mean the core genes are more exclusively dedicated to a specific process than the peripheral genes.

4 Discussion

The function of a protein is a contextual attribute of strict and quantifiable patterns of interactions between the myriad of cellular constituents.[18] Large-scale gene expression profile was investigated in the context of the social network analysis where genes are regarded as actors, conditions as events, and the network topology as variety of centrality and relatedness indices.

The analysis demonstrated some important features such as core-peripheral players and significant intermediary actors that may be critical for the control of the system and useful for the development of valuable therapeutic substances.

Acknowledgement

This study was supported by a grant from Korea Health 21 R and D Project, Ministry of Health and Welfare, Republic of Korea (A060711).

References

1. Wasserman, S., Faust, K.: Social Network Analysis. Cambridge University Press (1994)
2. Scott J: Social Network Analysis. SAGE publications (2000)
3. Hughes, T.R., Marton, M.J., Jones, A.R., Roberts, C.J., Stoughton, R., Armour, C.D., Bennett, H.A., Coffey, E., Dai, H., He, Y.D., Kidd, M.J., King, A.M., Meyer, M.R., Slade, D., Lum, P.Y., Stepaniants, S.B., Shoemaker, D.D., Gachotte, D., Chakraburtty, K., Simon, J., Bard, M., Friend, S.H.: Functional discovery via a compendium of expression profiles. Cell. 102(2000) 109–126
4. Cohen, B.A., Pilpel, Y., Mitra, R.D., Church, G.M.: Discrimination between Paralogs using Microarray Analysis: Application to the Yap1p and Yap2p Transcriptional Networks. Mol Biol Cell. 13(5) (2002) 1608–1614
5. Rung, J., Schlitt, T., Brazma, A., Freivalds, K., Vilo, J.: Building and analysing genome-wide gene disruption networks. Bioinformatics. suppl. 2 (2002) 202–210
6. Asch, J., Rindone, W., Church, G.M.: Systematic management and analysis of yeast gene expression data. Genome Research 10 (2000) 431–445
7. Ihmels, J., Friedlander, G., Bergmann, S., Sarig, O., Ziv, Y., Barkai, N.: Revealing modular organization in the yeast transcriptional network. Nature Genetics. 31 (2002) 370–377
8. Faust, K.: Centrality in affiliation networks. Social Networks. 19 (1997) 157–191
9. Borgatti, S.P., Everett, M.G.: Network analysis of 2-mode data. Social Networks. 19 (1997) 243–269
10. Borgatti, S.P., Everett, M.G., Freeman, L.C.: Ucinet for Windows: Software for Social Network Analysis. Harvard Analytic Technologies USA (2002)
11. Freeman, L.C.: Centrality in social networks. Social Networks. 1 (1979) 215–239
12. Borgatti, S.P., Everett, M.G.: Models of Core/Periphery Structures. Social Networks. 21 (1999) 375–395
13. Mewes, H.W., Frishman, D., Guldener, U., Mannhaupt, G., Mayer, K., Mokrejs, M., Morgenstern, B., Munsterkotter, M., Rudd, S., Weil, B.: MIPS: a database for genomes and protein sequences. Nucleic Acids Res. 30 (2002) 31–34
14. Featherstone, D.E., Broadie, K.: Wrestling with pleiotropy: genomic and topological analysis of the yeast gene expression network. BioEssays. 24 (2002) 267–274
15. Jeong, H., Mason, S.P., Barabasi, A.L., Oltvai, Z.N.: Lethality and centrality in protein networks. Nature. 411 (2001) 41–42
16. Wagner, A., Fell, D.A.: The small world inside large metabolic networks. Proc. R. Soc. Lond. 268 (2001) 1803–1810
17. Przybyla-Zawislak B, Gadde DM, Ducharme K, McCammon MT: Genetic and biochemical interactions involving tricarboxylic acid cycle (TCA) function using a collection of mutants defective in all TCA cycle genes. Genetics 1999, 152(1): 153–166
18. Barabasi, A.L., Oltvai, Z.N.: Network Biology: Understanding the cell's functional organization. Nature genetics. 5 (2004) 101–113

Prediction of MHC Class I Binding Peptides Using Fourier Analysis and Support Vector Machine

Feng Shi[1,2] and Qiujian Chen[1]

[1] College of Science, Huazhong Agricultural University, Wuhan, P.R. China
[2] School of Mathematics and Statistics, Wuhan University, Wuhan, P.R. China
{shifeng, chenqiujian}@mail.hzau.edu.cn

Abstract. Processing and presentation of major histocompatibility complex class I antigens to cytotoxic T-lymphocytes is crucial for immune surveillance against intracellular bacteria, parasites, viruses and tumors. Identification of antigenic regions on pathogen proteins will play a pivotal role in designer vaccine immunotherapy. We have developed a novel method that identifies MHC class I binding peptides from peptides sequences. For the first time we present a method for MHC class I binding peptides prediction using Fourier analysis and support vector machines (SVM). Using cross-validation, we demonstrate that this novel prediction technique has a reasonable performance.

1 Introduction

Peptides degraded from foreign or self-proteins bind to Major histocompatibility complex (MHC) molecules. The MHC-peptide complex can be recognized by T-cell receptors and trigger an immune response. Identifying characteristic patterns of immunogenic peptide epitopes can provide fundamental information for understanding disease pathogenesis and etiology, and for therapeutics such as vaccine development. According to their different structures and functions, MHC molecules can be classified into two types: MHC class I and MHC class II. Both MHC class I and class II molecules have binding grooves consists of two helices supported by a sheet. The groove of Class I is small, so that those foreign antigenic peptide which has only 9 11 amino acids can bind [2,4,14,15,17], whereas the ends of the MHC Class II binding groove are open, so antigenic peptides as long as 9 30 amino acid residues or even whole protein can be bind [5,20], then be presented to the receptor of CD4+ helper T lymphocytes (HTLs) and consequently stimulate immune response.

It is estimated that only one in 100 to 200 peptides actually binds to a particular MHC [25]. Therefore, a good computational prediction method could significantly reduce the number of peptides that have to be synthesized and tested. Prediction of MHC-peptides can be divided into two groups: sequence based and structure based methods.

Allele specific sequence motifs can be identified by studying the frequencies of amino acids in different positions of identified MHC-peptides. The peptides

L. Wang et al. (Eds.): FSKD 2006, LNAI 4223, pp. 1072–1081, 2006.

that bind to HLA-A* 0201 are often 9 and 10 amino acids long (nonamers and decamers). Nonamers frequently have two anchor residues, a lysine in position 2 and a valine in position 9 [16,18]. Besides the anchor residues, there are also weaker preferences for specific amino acids in other positions. One method to include this information is to use a profile, where a score is given for each type of amino acid in each position[8,13] . The scores can be calculated from observed amino acid frequencies in each position or be set manually. The sum of the scores for a given peptide is then used to make predictions. One frequently used profile based prediction method is SYFPEITHI [17], the matrices in SYFPEITHI were adjusted manually, by assigning a high score (10) for frequently occurring anchor residues, a score of 8 to amino acids that occur in a significant amount and a score of 6 to rarely occurring residues. Preferred amino acids in other positions have scores that range from 1 to 6 and amino acids regarded as unfavorable have scores ranging from -3 to -1. SYFPEITHI prediction can be done for 13 different MHC class I types. A profile based method does not take into account correlations between frequencies in different positions, neither they consider information from peptides that do not bind. This information can be used by machine learning methods. Prediction of MHC-peptides has been made by using machine learning approaches such as artificial neural networks and hidden Markov models [11,22]. It was also showed that one advantage of machine learning algorithms compared to profile methods seems to be that they have a higher specificity[9]. This is possible due to the inclusion of non-binding data in the training. A machine learning approach extracts useful information from a large amount of data and creates a good probabilistic model [1]. Brusic reported a total accuracy of 88% on predictions for the mouse MHC H-2K^d, using artificial neural networks and hidden Markov models which perform 2-15% better than artificial neural networks [2].

Structural approaches for prediction evaluate how well a peptide fit in the binding groove of a MHC molecule. A peptide is threaded through a structural template to obtain a rough estimate of the binding energy. The energy estimation is based on the interactions defined in the binding pocket of a particular MHC molecule [19]. To our knowledge no comparisons of the performance between structural and sequence based methods has been published. Obviously, a structural approach is limited to MHC types with a known structure. However, the advantage of a structural approach is that one known structure alone might be sufficient for creating a prediction model.

In this paper, we incorporated Fourier transform and SVM to predict MHC class I binding peptides. As well known, Fourier analysis is efficient on dealing with numerical signal. It is capable of capturing information about long-range correlations and global symmetries that are completely missed by other approaches, and it has extended into the field of bioinformatics [21]. To our knowledge, this is the first time using Fourier transform in predicting MHC binding peptides.

2 Materials and Methods

2.1 MHC Class I Binding Data

There are two important MHC binding peptides databases. One is the SYFPEITHI [17]; the other is MHCPEP [3]. MHCPEP is a curated database comprising over 13000 peptide sequences known to bind MHC molecules. Entries are compiled from published reports as well as from direct submissions of experimental data. SYFPEITHI (Ver. 1.0) is a database comprising more than 4000 peptide sequences known to bind class I and class II MHC molecules and is supposed to be of a higher quality and is restricted to published data and only contains sequences that are natural ligands to T-cell epitopes. The two databases have different advantages. MHCPEP contains significantly more data (13000 vs 4000), while the quality of the data in SYFPEITHI is assumed to be higher. Therefore, using MHCPEP data for SVM training, it is possible to make predictions for 26 different MHC types. This can be compared with only 6 MHC types when SYFPEITHI data is used for SVM training. However, the predictions from SYFPEITHI might be more reliable and should therefore be used when enough data exists. In order to compare the predict performance, peptide sequences known to bind a MHC class I alleles were extracted from the database SYFPEITHI. There are just three types MHC class I have more than 20 binding peptides: HLA-A2*0201 nonamers(127), HLA-A2*0201 decamers(43), and HLA-A2402 nonamers(26), and some of them are also contained in MHCPEP database. All of these were fetched to compose our binding peptides.

Unfortunately, there are very few experimentally verified examples of peptides that do not bind to a particular MHC. Therefore, the non-binding training examples were extracted randomly from the MitoProteome database of human proteins [6]. Protein sequences from the MitoProteome database were chopped up into the length of interest and known MHC-peptides were removed. Obviously, there is a risk that some of the non-binders actually binds, but since less than 1% of the peptides are expected to bind to a MHC molecule, we do not expect this to cause any major problems [7]. The ratio of binder/non-binders was kept to 1:2 for all MHC types. Conveniently, we denote the dataset of HLA-A2*0201 nonamer 127 binding peptides from SYFPEITHI database and 254 non-binding peptides of nonamer as S0201-9, denote the dataset of HLA-A2*0201 decamer binding peptides(43) from SYFPEITHI database and 86 non-binding peptides of decamer as S0201-10, and denote the dataset of HLA-A2402 nonamer 26 binding peptides from SYFPEITHI database and 52 non-binding peptides of nonamer as S2402-9.

2.2 Discrete Fourier Transform

A common use of the Fourier transform is to identify the frequency components of a weak time-dependent signal which buried in noise. Assume $x = \{x_1, x_2, \cdots, x_n\}$ to be a numerical signal (n is the length of sequence), it is converted to a sequence in the frequency domain with the Fast Fourier Transform (FFT) formula described below:

$$X(f) = \sum_{i=1}^{n} x(t) \exp[i(2\pi t f / n)]$$

and on the contrary,

$$X(t) = \frac{1}{n} \sum_{f=1}^{n} x(f) \exp[-i(2\pi t f / n)]$$

The main advantage of using FFT is that it enhances the characteristics for each localization, and generates a compact set of features [21].

2.3 Support Vector Machine

Support Vector Machine (SVM) is one type of learning machine based on statistical learning theory. It performs a nonlinear mapping of the input vector X from the input space R^d into a higher dimensional Hilbert space by a kernel function. It finds the OSH (Optimal Separating Hyperplane) [24] in the space H corresponding to a non-linear boundary in the input space. Here we briefly state the basic ideas of applying SVM to pattern recognition, especially for the two-class classification problem as follows.

Suppose a series of vectors with corresponding labels are a set of samples, where +1 and -1 represent the positive or negative class respectively. The goal is to construct one binary classifier or derive one decision function from the available samples, which has small probability of misclassifying a future sample. The typical kernel function of SVM is:

$$K(x_i, x_j) = \exp(-\lambda \|x_i - x_j\|^2) \tag{1}$$

This kernel function is called the RBF (radial basis function) kernel with one parameter λ . Finally, for the selected kernel function, the learning task amounts to solving the following convex Quadratic Programming (QP) problem:

$$\max[\sum_{i=1}^{N} \alpha_i - \frac{1}{2} \sum_{i=1}^{N} \sum_{j=1}^{N} \alpha_i \alpha_j \cdot y_i y_j \cdot K(x_i, x_j)]$$

subject to:

$$0 \le \alpha_i \le C \qquad \sum_{i=1}^{N} \alpha_i y_i = 0 \qquad i = 1, 2, \cdots, N$$

Then the form of the decision function is

$$f(x) = sgn(\sum_{i=1}^{N} y_i \alpha_i K(x_i, x_j) + b)$$

For a given sample set, only the kernel function and the regularity parameters C and λ must be selected. A complete description to the theory of SVMs for pattern recognition is in Vapnik's book [24].

In this paper, we apply Vapnik's Support Vector Machine to the prediction of MHC class I binding peptides. We have used the OSU-SVM, a Matlab SVM toolbox (can be download freely from: http://www.ece.osu.edu/~maj/osu_svm), which is an implementation of SVM for the problem of pattern recognition.

2.4 Vector Representation of Peptide

In our experiment, Fourier analysis is used as feature extracting tool and SVMs as the learning framework. Each peptides was encoded into numerical format, the procedure can be described as the following.

Table 1. The correlation coefficients among 7 indices

	HC	NRFB	PK-C	MP	HPLC	NVWV
HV	-0.37	0.47	-0.56	-0.83	0.82	-0.25
HC		-0.58	0.10	0.50	-0.48	0.59
NRFB			-0.37	-0.70	0.68	-0.36
PK-C				0.36	-0.56	-0.01
MP					-0.86	0.23
HPLC						-0.51

Each amino acid in a peptide was encoded by 6 or 7 indexes. These indexes were obtained from amino acid index database AAindex [10]. The database is a collection of published indices together with the result of cluster analysis using the correlation coefficient as the distance between two indexes. Release 6.0 currently contains 494 indices. After done many experiments, we selected 6 indexes for decamers. These indexes included Hydrophilicity value (HV, No: HOPT810101), Heat capacity (HC, No: HUTJ700101), Normalized relative frequency of bend (NRFB, No: ISOY800103), pK-C(No: FASG760105), Mean polarity (MP, No: RADA880108), HPLC parameter (HPLC, No: PARJ860101). We standardized index HV and then transform it by using Fourier transform. The indexes of HC, HPLC, NRFB were standardized. For nonamers, we selected 7 indexes, which including another index: Normalized van der Waals volume (NVWV, No: FAUJ880103). The indexes of HC, HPLC, NRFB were also standardized. The correlation coefficients among these 7 indices are in table 1. For decamer FIASNGVKLV, after standardized index HV and then transform it by using Fourier transform, we get numerical vector (0, 1.23, 0.99, 0.71, 1.09, 1.28, 1.09, 0.71, 0.99, 1.22); standardized index HC, we get (0.27, 0.16, -0.37, -0.26, -0.07, -0.55, 0.00, 0.56, 0.26, 0.00), and do so for other indexes.

Finally, we get a vector of decamer FIASNGVKLV by concatenate all of these vectors, it is a 60 dimension vector. In this paper, we denote our methods as FS.

3 Results

In our experiment, we use the 10 folds cross-validation method to train and test our model FS. We select the RBF kernel function. For dataset S0201-9,

$\lambda = 0.001$, $C = 80$, for dataset S0201-10,$\lambda = 0.008$, $C = 70$, for dataset S2401-9,$\lambda = 0.004$, $C = 15$. Assess the model from such four aspects as accuracy, sensitivity, Matthews Correlation coefficients [12], the area under ROC curve [23]. These concepts are described as follows:

Table 2. Comparison of FS performance to other methods using SYFPEITHI data as test samples

Dataset	Mc			Aroc			Accuracy$^-$(%)			Sensitivity$^-$(%)		
	FS	SVMHC	SYF	FS	SVMHC	SYF	FS	SVMHC	SYF	FS	SVMHC	SYF
*S0201-9	0.88	0.80	0.81	0.97	0.92	0.95	94.0	91.3	91.9	92.1	86.6	83.5
*S0201-10	0.72	0.75	0.74	0.93	0.86	0.92	88.4	89.2	88.4	83.7	72.1	81.4
*S2402-9	0.88	0.72	0.88	0.99	0.81	0.96	94.9	87.2	94.9	92.3	61.5	84.6

*The SVMHC model was train based on A*0201 nonamersdecamers and nonamers restricted MHC binding data from MHCPEP.

Assume TP is the number of correctly predicted binding peptides of test samples, TN is the number of correctly predicted non-binding peptides, FN is the number of under-predicted binding peptides and FP is the number of over-predicted non-binding peptides. Then some definitions may be given by the following equations:

$$accuracy = \frac{TP + TN}{TP + TN + FP + FN}$$

$$Sensitivity = \frac{TP}{TP + FN} \qquad Specificity = \frac{TN}{TN + FP}$$

$$MC = \frac{TP * TN - FP * FN}{\sqrt{(TP + FN)(TP + FP)(TN + FP)(TN + FN)}}$$

In order to compare FS with other methods, we submitted all these data sets into the online prediction systems SYFPEITHI and SVMHC, and obtained their scores. Because there is no control over which sequences are in the train/test sets, the results of SYFPEITHI and SVMHC are just the predicted performance.

Table 2 show the results of this prediction. The generation of relative operating characteristics (ROC) curve allows us for a consideration of the technique's capacity to maximize both specificity and sensitivity. If a high cutoff point is set, then the majority of peptides will be predicted to be non-binders, thus reducing the number of true and false positives so that the sensitivity increases while specificity decrease. The number of true and false negatives is reduced so that sensitivity decreased and specificity increases. A ROC curve may be generated by plotting sensitivity against 1-specifivity for a range of values calculated by varying the cutoff point. The accuracy of the prediction technique can be quantified by measuring Aroc, The Aroc value varies from 0.5 (indicating an entirely random prediction) to 1.0 (indicating a perfect prediction system). Because we use 10-folds cross-validation, each time, we can get a ROC curve, after 10 folds cross-validation, we get 10 curves (denote $f_i(x)$,$i = 1, 2, \cdots, 10$), then we get a new curve $f(x) = \frac{1}{10}\sum_{i=1}^{10} f_i(x)$, we call the curve as average ROC curve. For dataset S0201-9, the Aroc of FS is 0.97, it is higher than the SVMHC and SYF method. The average Aroc values for our method, SVMHC, SYFPEITHI, respectively, are 0.97, 0.92, 0.95. The Mc of FS is 0.88, it is slightly higher than the SVMHC and SYF method. The Mc values for SVMHC, SYFPEITHI, respectively, are 0.80, 0.81. The accuracy of FS is 94%, it is higher than the SVMHC and SYF method. The accuracy values for SVMHC, SYFPEITHI are 91.3% and 91.9% respectively. The sensitivity of FS is 92.1%, it is slightly higher than the SVMHC and SYF method. The sensitivity values for SVMHC, SYFPEITHI are 86.6%and 83.5%respectively.

We have also constructed another data set, test these methods by using MHCPEP data as binding test samples, and the binding train binding were taken from SYFPEITHI database. For HLA-A2*0201 nonamers, it contains 359 binding test peptides and we use 431 non-binding peptides which generated by the way described as above, these compose our dataset P0201-9. For HLA-A2*0201 decamers, it contains 131 test binding peptides. Some of them are contained in SYFPEITHI database(train samples), we used 262 non-binding peptides, these composed our dataset P0201-10. Table 3 show the results according to the best Mc.

We can see that the performance of these methods are not as well as we see in table 2, this is because in this test, the train samples is much less than the test samples. Even then, FS can also achieve a good result.

Table 3. Comparison of FS performance to other methods using MHCPEP data as test samples

Dataset	Mc			Aroc			Accuracy$^-$(%)			Sensitivity$^-$(%)		
	FS	SVMHC	SYF	FS	SVMHC	SYF	FS	SVMHC	SYF	FS	SVMHC	SYF
[a]P0201-9	0.84	0.76	0.77	0.96	0.89	0.94	92.0	88.1	88.2	90.0	81.1	80.8
[a]P0201-10	0.68	0.60	0.70	0.90	0.73	0.94	85.8	81.9	85.8	76.3	46.4	89.3

[a]The SVMHC was train based on A*0201 nonamers and dexamers restricted MHC binding data from SYFPEITHI.

4 Discussion

We have developed a novel method for detecting the binding peptides for MHC class I by using Fourier transform and SVM, and the prediction accuracy of the method described is reasonable performance. Because most binding peptides from SYFPHITEI database are also contained in MHCPEP database, when predict with SVMHC, most test samples are also train samples. At the same time, when use SYFPHITEI (SYF), the test set are also the train set. All these will increase the predict accuracy of SVMHC and SYF, that is to say, the predict accuracy of FS is more reliable than SVMHC and SYF. When use 10-fold cross-validation to train and test our method, the small variance among the test results were occurred. For Mc, the biggest variances are 0.08, 0.05 for Mc and Aroc respectively.

We also extracted randomly another 4 groups of non-binding peptides with the same number to train and test our method, the variance is also small. Only the BBF kernel function parameters should changed little to get almost the same results. The Fourier transform is important in FS method. Due to standardization, the first component of Fourier coefficient is 0, and the others are symmetry. So, in our discussion, we ignore the first and the later half Fourier coefficients. For HLA-A2*0201 nonamers SYFPEITHI binding peptides, the key one is the first component, for most case, it is much less then the others. The average profile of Fourier transform is (0.8907, 0.9777, 1.0630, 0.9700,), while for nonbonding peptides, the average profile of Fourier transform is (0.9913, 0.9284, 1.0006, 0.9551). For decamer binding peptides, the key one is also the first component, for most case, it is larger then the others, the average profile of Fourier transform is (0.9866, 0.9568, 0.9533, 0.9563, 0.9064), and for the non-binding peptides, the average profile of Fourier transform is (0.9457, 1.0019, 0.9978, 0.9965, 0.7890). Maybe this is the reason why Fourier transform can extract some common feather of binding peptides, and we can use these in improving our predict performance.

5 Conclusion

We can see that Fourier transform is useful in binding peptides predict, it can extract some period features of the peptide structure, and these feature is helpful in discriminating binding peptides from non-binding peptides, especially when cooperate with other amino acid index. The Fourier transform may obtain the frequency domain feature of the peptide, and the original thought we used Fourier transform is that the same functional peptides would be share the same frequency domain, and the SVM can grasp these feature efficient.

Acknowledgement

The work was partly supported by the National Natural Science Funds of China (NO: 30170214), and Huazhong Agricultural University.

References

1. Baldi P, Brunak S.: Bioinformatics, the machine learning approach. MIT Press Cambridge Massachusetts, London England (1998).
2. Brusic.V, Rudy G, Harrison LC.: Prediction of MHC binding peptides using artificial neural networks. In: Stonier RJ, Yu XS (eds). Complex Systems: Mechanism of Adaptation. IOS Press, Amsterdam. Holland. (1994) 253-260.
3. Brusic V, Rudy G, Harrison LC.: MHCPEP, a database of MHC-binding peptides: update 1997. Nucleic Acids Res. (1998) 26: 368-371.

4. Buus S, Lauemoller SL, Worning P, Kesmir C, Frimurer T, Corbet S, Fomsgaard A, Hilden J, Holm A, Brunak S.: Sensitive quantitative predictions of peptide-MHC binding by a 'Query by Committee' artificial neural network approach. Tissue Antigens (2003) 62: 378-384.
5. Castellino F, Zhong G, Germain RN.: Antigen presentation by MHC class II molecules: invariant chain function, protein trafficking, and the molecular basis of diverse determinant capture. Hum Immunol. (1997) 54: 159-169.
6. Dawn C, Purnima G, Eoin F, Shankar S.: MitoProteome: Mitochondrial Protein Sequence Database and Annotation System. Nucleic Acids Research.,(2004)32: 463-467.
7. Dönnes P, Elofsson A.: Prediction of MHC I binding peptides, using SVMHC. BMC Bioinformatics. (2002).3: 1-8.
8. Gribskov M, McLachlan AD, Eisenberg D.: Profile analysis: detection of distantly related proteins. Proc. Natl Acad Sci USA (1987) 84: 4355-4358.
9. Gulukota K, Sidney J, Sette A, DeLisi C.: Two complementary methods for predicting peptides binding major histocompatibility complex molecules. J. Mol. Biol. (1997) 267: 1258-1267.
10. Kawashima S, Kanehisa M.: AAindex: amino acid index database. Nucleic Acids Res. (1999) 28: 374
11. Mamitsuka H.: MHC molecules using supervised learning of hidden Markov models. Proteins: Structure, Function and Genetics. (1998) 33: 460-474
12. Matthews B.: Comparison of the predicted and observer secondary structure of T4 phage lysozyme. Biochim Biophys Acta.. (1975) 405: 442-451
13. Nielsen M, Lundegaard C, Worning P, Hvid CS, Lamberth K, Buus S, Brunak S, Lund O.: Improved prediction of MHC class and class epitopes using a novel Gibbs sampling approach. Bioinformatics. (2004) 20: 1388-1397
14. Nielsen M, Lundegaard C, Worning P, Lauemoller SL, Lamberth K, Buus S, Lund O. :Reliable prediction of T-cell epitopes using neural networks with novel sequence representation. Protein Sci. (2003) 12: 1007-1017
15. Parker KC, Bednarek MA, Coligan JE.: Scheme for ranking potential HLA-A2 binding peptides based on independent binding of individual peptide side-chains. J. Immunol. (1994) 152: 163-175
16. Rammensee HG, Friede T, Stevanovic S.: MHC ligands and peptide motifs: first listing. Immunogenetics. (1995) 41: 962-965
17. Rammensee HG, Bachmann J, Emmerich NP, Bachor OA, Stevanovic S.: SYFPEITHI: database for MHC ligands and peptide motifs. Immunogenetics. (1999) 50: 213-219
18. Rotzschke O, Falk K, Stevanovic S, Jung G, Rammensee HG.: Peptide motifs of closely related HLA class I molecules encompass substantial differences. European Journal of Immunology. (1992) 22: 2453-2456
19. Schueler-Furman O, Altuvia Y, Sette A.: Structure-based prediction of binding peptides to MHC class I molecules: application to a broad range of MHC alleles. Protein Science, (2000). 9: 1838-1846
20. Sette A, Adorini L, Colon SM, Buus S, Grey HM.: Capacity of intact proteins to bind to MHC class II molecules. J. Immuno (1989) 143: 1265-1267
21. Shepherd AJ, Gorse D, Thornton JM.: novel approach to the recognition of protein architecture from sequence using Fourier analysis and neural networks. Proteins: Structure, Function, and Genetics.200350: 290-302
22. Srinivasan KN, Zhang GL, Khan AM, August JT, Brusic V.: rediction of calss I T-cell epitopes:evidence os presence of immunogical hot spots inside antigens. Bioinformatics, (2004) 20 (suppl. 1): 297-302

23. Swet J.: Measuring the accuracy of diagnostic systems. Science, (1988) 240: 1285-1293
24. Vapnik V.: Statistical Learning Theory. Wiley-Interscience. New York (1998)
25. Yewdell J, Bennink J.: Immunodominance in major histocompatibility complex class I-restricted T lymphocyte responses. Annu. Rev. Immunol. (1999) 17: 51-81
26. ZhaoYD, Pinilla C, Valmori D, Martin R, Simon R.: Application of Support vector machines for T-cell epitopes prediction. Bioinformatics. (2003) 19: 1978-1984

Clustering and Classification Based Anomaly Detection

Hongyu Yang[1,2], Feng Xie[3], and Yi Lu[4]

[1] Software Research Center, Civil Aviation University of China
Tianjin 300300, China
yhyxlx@hotmail.com
[2] Tianjin Key Lab for Advanced Signal Processing, Civil Aviation University of China
Tianjin 300300, China
[3] Software of Computing Tech., Chinese Academy of Science
Beijing 100080, China
xiefeng@software.ict.ac.cn
[4] Security and Cryptography Laboratory, Swiss Federal Institute of Technologies
(EPFL), CH-1015 Lausanne, Switzerland
yilu@epfl.ch

Abstract. This paper presents an anomaly detection approach based on clustering and classification for intrusion detection (ID). We use connections obtained from raw packet data of the audit trail as basic elements, then map the network connection records into 8 feature spaces typically of high dimension according to their protocols and services. The approach includes two steps, training stage and testing stage. We perform clustering to group training data points into clusters, from which we select some clusters as normal and known-attack profile according to certain criterion. For those training data excluded from the profile, we use them to build a specific classifier. During the testing stage, we utilize influence-based classification algorithm to classify network behaviors. In the algorithm, an influence function quantifies the influence of an object. The experiments on the KDD'99 Intrusion Detection Data Set demonstrate the detection performance and the effectiveness of our ID approach.

1 Introduction

With the exponential growth of the Internet and networked computers, cyber crime has become one of the most important problems in the computer world. Therefore the development of a robust and reliable network intrusion detection system (IDS) is increasingly important.

There are many IDS systems available that are primarily signature-based detectors. Although these are effective at detecting known intrusion attempts and exploits, they fail to recognize new attacks and carefully crafted variants of old exploits[1].

Anomaly Detection systems model normal or expected behavior in a system, and detect deviations of interest that may indicate a security breach or an attempted attack. Anomaly Detection identifies activities that vary from established patterns of users, or groups of users and Misuse Detection involves a

L. Wang et al. (Eds.): FSKD 2006, LNAI 4223, pp. 1082–1091, 2006.

comparison of a user's activities with the known behavior of system penetration[2]. The major benefit of anomaly detection systems is their ability to potentially detect novel attacks. In addition, they may be able to detect new or unusual, but non-intrusive, network behavior that is of interest to a network manager, and needs to be added to the normal profile.

From previous works, we found that maintaining a profile of each authorized user or group is useful in detecting anomaly attacks in systems, but it is difficult to maintain a behavior profile for each legitimate user when the number of user increases. Behavior patterns also change with the mental state of the person and thus detecting anomaly attacks by comparing patterns with user profiles gives rise to a large number of false positive alerts.

It is not possible to observe every possible legitimate pattern in training, so an anomaly detector requires some type of machine learning algorithm in order to generalize from the training set.

2 Related Work

Recently, many researchers have focused on the anomaly detection and some prototypes are also constructed with different methods. Some anomaly detection systems, e.g. SPADE[3], PHAD[4] and ALAD[5], compute statistical models for normal network traffic and generate alarms when there is a large deviation from the normal model.

Lee et al[6] apply RIPPER rule learning algorithm to the labeled data sets and learn the intrusions. Meanwhile, association rules and frequent episodes are used to network connection records to obtain additional features for data mining algorithms. ADAM[7] uses association rules to build a customizable profile of rules of normal behavior, and a classifier (decision tree) that sifts the suspicious activities, classifying them into real attacks (by name), unknown types or false alarms.

MINDS[8] first constructs features from the raw network data, and then the anomaly detection module assigns a degree of being an outlier to each data point, which is called the local outlier factor (LOF). MINDS association pattern analysis module summarizes network connections that are ranked highly anomalous in the anomaly detection module, which could be further used in the known attack detection module as the feedback.

Bernhard[9] used an ensemble of C5 decision trees with cost-sensitive bagged boosting to classify the network connections. Levin[10] used kernel Miner, a data-mining tool based on building the optimal decision forest, to detect the intrusion detections. Using this tool and the KDD data set, Levin created a set of locally optimal decision trees (called the decision forest) from which optimal subset of trees (called the sub-forest) was selected for predicting new cases.

3 Framework Overview

We use a combination of clustering and classification to discover attacks in an audit trail. In our framework, the training set needs to be labeled or

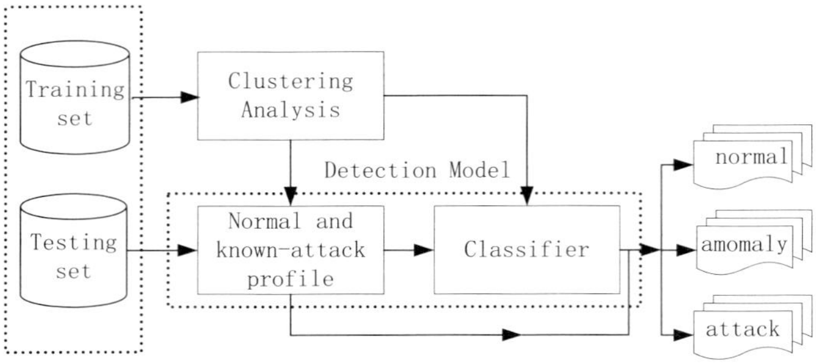

Fig. 1. Architecture of the IDS prototype

attack-free. If the data set includes the labeled attacks, we could get the known attack profile. Otherwise, we only have the normal profile. When training is finished and detection model is built, we can use it to discriminate the new incoming connection on line. The whole framework is shown in Fig.1.

The purpose of clustering is to model the normal and known-attack network behaviors. We think the connections of the same type are more resemble statistically, which means these data are more easily clustered together. Therefore, we can use the centroid of a cluster to represent all members within that cluster, which will reduce the mass of raw data markedly.

For those ambiguous data in sparse space, we need a classifier to deal with them. Other than traditional classifiers, our classifier has the ability to classify a connection record as "anomaly". We let the classifier include a "default" label by which the classifier expresses its inability to recognize the class of the connection as one of the known classes. Of course, the "default" label is "anomaly" in our method.

First, the raw network packets are reconstructed to a connection and correspondingly preprocessed according to its protocol and service. Then it is compared with the profile modeled in the training stage. If it exists in the profile, we will label it as the matched type. Otherwise, it will be fed to the classifier, which will label it as normal, known attack, or anomaly. Finally, when the number of data labeled as known-attack or anomaly surpasses a threshold, an analysis module using the association algorithms will deal with the vast data in order to extract the frequent episodes and rules.

4 Clustering

We map the connection records from the audit stream to a feature space. The feature space is a vector space of high dimension. Thus, a connection is transformed into a feature vector. We adopt 8 feature spaces according to the protocol and service of the connection. That is, we choose the different attributes for the connections with different services. An important reason is that different services usually have the specific security-related features. For example, the attributes of an HTTP connection are different from those of an FTP connection. The eight services include HTTP, FTP, SMTP, TELNET, FINGER, UDP, ICMP and OTHER, in which OTHER service is default. So, even if a new service occurs in the data stream for the first time, it can be simply regarded as OTHER service without reconfiguring the system.

4.1 Distance Function

In order to describe the similarity of two feature vectors, we use the Euclidean Distance as a measure function:

$$d(v_i, v_j) = (\sum_{k=1}^{\|v_i\|} (v_i^{(k)} - v_j^{(k)})^2)^{\frac{1}{2}} \tag{1}$$

where both v_i and v_j are feature vectors in the vector space $\Re^n$ of the same dimensions. $v_i^{(k)}$ represents the $k$the component of vector v_i, and $\|v_i\|$ means the dimensions of v_i, i.e. n.

4.2 Discrete and Continuous Attributes

There are two attribute types in our connection records. One is discrete, i.e. nominal, and the other is continuous. For a discrete attribute, we take the method introduced by Chan[12] that represent a discrete value by its frequency. As a result, discrete attributes are transformed to continuous ones. For a continuous attribute, we adopt the "cosine" normalization to quantize the values. Furthermore, the values of each attribute are normalized to the range [0,1] to avoid potential scale problem. The whole normalization processes include two steps: the first step is normalization of each continuous attribute,

$$v_i^{(k)'} = v_i^{(k)} / \sum_{j=1}^{\|D\|} (v_j^{2(k)})^{\frac{1}{2}} \tag{2}$$

where $\|D\|$ represents the total number of vectors in the training set D. And the second step is the normalization of the feature vector. Note that we don't regard those transformed from discrete attributes in this step.

$$v_i^{(k)'} = v_i^{(k)} / \sum_{k=1}^{\|v_i\|} (v_i^{2(k)})^{\frac{1}{2}} \tag{3}$$

4.3 Clustering and Profile Selection

At present, we use standard k-means algorithm[11] as our clustering approach. K-means is a centroid-based clustering with low time complexity and fast convergence, which is very important in intrusion detection due to the large size of the network traffic audit dataset.

Each cluster in profile can be simply expressed as a centroid and an effect influence radius. So a profile record can be represented as the following format:

$$\langle \ \textbf{\textit{centroid}}, \textbf{\textit{radius}}, \textbf{\textit{type}} \ \rangle$$

Centroid is a centric vector of the cluster, *radius* refers to influence range of a data point (represented as the Euclidean distance from the centroid), and *type* refers to the cluster's category, e.g. normal or attack. We can determine whether a vector is in the cluster or not only by computing the distance between the vector and the centroid and comparing the distance with the radius. If the distance is less than radius, we consider that the vector belongs to the cluster. And then we can label the vector as the cluster's type. Therefore, the whole search in the profile only includes several simple distance calculations, which means we can deal with the data rapidly.

Of course, not all clusters can serve as the profile. Some maybe include both normal and attack examples and not fit for the profile apparently. It is necessary to select some clusters according to a strategy.

A majority example is an example that belongs to the most frequent class in the cluster. The higher the purity is, the better the cluster is served as a profile. A cluster with small purity means that there are many attacks with different types in the cluster, so we don't select such cluster as our profile. Instead, we use them as the training set for classifier.

After the clusters are selected for the profile, we put them into the profile repository. The basic contents include *centroid*, *radius* and *type*. Here, we use the type of majority examples in one cluster as the whole cluster's type regardless of the minority examples.

5 Classification

Apart from the normal and known-attack profile, classifiers for distinguish suspect event are also needed. At the training stage, we use those examples not included in the profile to build a classifier. Our classifier has a "default" label to deal with those unseen or uncertain events. We present a new algorithm, influence-based classification to address this problem.

5.1 Influence Function

We use a function to quantify the influence of an object. This function is called influence function. Informally, the influence function is a mathematical description of the influence a data object has within its neighborhood. In principle, it can be an arbitrary function. In our scheme, we adopt the Gaussian function to measure the influence. We denote the d-dimension feature space by $\Re^D$.

As mentioned above, influence function is a function of position quantifying the influence of a data object in the field. Formally, we can define it as follows.

The influence function of a data object $y \in F^d$ is a function $f_y : F^d \rightarrow R_0^+$ which is defined in terms of a basic influence function ϕ

$$f_y(x) = \phi(x, y) \tag{4}$$

Generally speaking, it is a function of distance function. The distance function is defined as $d : F^D \times F^d \to R_0^+$ which determines the distance of two d-dimensional feature vectors, and it has to be reflexive and symmetric. In our experiment, we assume d as a Euclidean distance function. Examples of basic influence functions include:

1. Square Wave Influence Function

$$\phi(x, y) = \begin{cases} 1 & \text{if } d(x, y) \leq \varepsilon \\ 0 & \text{otherwise} \end{cases} \tag{5}$$

where *d(x,y)* is the distance from x to y, and ε is a parameter indicating the influence radius of point y.
2. Electric Field based Influence Function

$$\phi(x, y) = \frac{1}{d^2(x, y)} \tag{6}$$

3. Gaussian Influence Function

$$\phi(x, y) = e^{-d^2(x,y)/2\sigma^2} \tag{7}$$

where σ is called influence factor determining the influence scope of y. Here, we adopt the Gaussian function to measure the force magnitude of data field. We also call the influence magnitude at a given position p in the field created by point y as the potential energy at that position, or simply potential, represented as $f_y(x)|_{x=p}$.

For convenience of description, we define the field formed only by a data point as the $single - sourcedatafield$. Many little single-source data fields can be emerged into a big *multi-source field* where a test point is affected by many source points other than a single point. The influence function of a multi-source filed D is defined as the sum of the influence function of all data sources in D, i.e., the data field satisfies the super-position principle, which is one of elemental properties of the field. Consequently, the influence function can be represented as follows:

$$f_d(x) = \sum_{y \in D} (f_y(x) = \sum_{y \in D} e^{-d^2(x,y)/2\sigma^2} \tag{8}$$

5.2 Influence-Based Classification

As we know, for a Gaussian distribution, approximately 99.7% of the values fall within a margin. This is the famous " 3 σ criterion". That is, the influence scope of a data object is rough equal to 3 σ. So, in our algorithm, we only focus on those objects inside this range and ignore others. The algorithm is illustrated in Fig. 2.

In the influence-based Classification, we have an important parameter, namely σ. The parameter σ determines the influence of a point on its neighborhood. A bigger σ means the wider influence range of a data point. Generally speaking,

Input: a sample P to be labeled, the influence factor σ, and the training set D
Output: Label P as normal, known-attack or anomaly
Begin
1. normalize P;
2. $f_+ \leftarrow 0$, $f_- \leftarrow 0$;
3. **for each** sample Q in D
4. **if** $d(P,Q) > 3\sigma$ **continue**;
5. compute the influence at P generated by Q and add it to f_+ **if** Q is normal, otherwise add it to f_-;
endfor
6. **if** $f_+/(f_- + f_+) > T_N$ label P as normal;
7. **else if** $f_-/(f_- + f_+) > T_A$ label P as known-attack;
8. **else** label P as anomaly.
End.

Fig. 2. Influence-based Classification Algorithm

since most samples in the training set are normal, a testing point will obtain more positive influences and be liable to being labeled as "normal" for a bigger σ , which will reduce the false alarm rate and meanwhile, unfortunately, reduce the true detection rate. Otherwise, a smaller σ can make the IDS with the high detection rate and high false alarm rate. So we need set a proper σ to get the tradeoff between false alarm rate and true detection rate.

6 Experiments and Results

To validate the claim that our detection approach is accurate, two experiments were conducted. For the first experiment, we ran our intrusion detection prototype on the well-known KDD Cup 1999 Intrusion Detection Data Set[13].

In the experiment, the test data set is the KDD'99 Cup Data Set which is generally used for benchmarking intrusion detection. This database contains a standard set of data to be audited, which includes a wide variety of intrusions simulated in a military network environment. Each sample is labeled as either normal, or as an attack. In our experiment, attack samples fall into four main categories, Probing (PROBE), Denial of Service Attacks (DOS), User to Root Attacks (U2R) and Remote to User Attacks (R2L).

We handled part of the whole KDD'99 data set corresponding to 494019 training connections and 311029 testing connections. Fig. 3 shows the results of this experiment, in which there are 5 ROC curves, 4 curves corresponding to 4 categories of attacks respectively, i.e. PROBE, DOS, U2R and R2L, and the left one corresponding to the overall attacks. "PROBE (4166)" denotes there are 4166 probing examples in the test set. Also, "OVERALL (250436/60593)" means there are total 250436 attacks and 60593 normal examples in the test set, and the corresponding curve describes the overall detection performance of our ID approach.

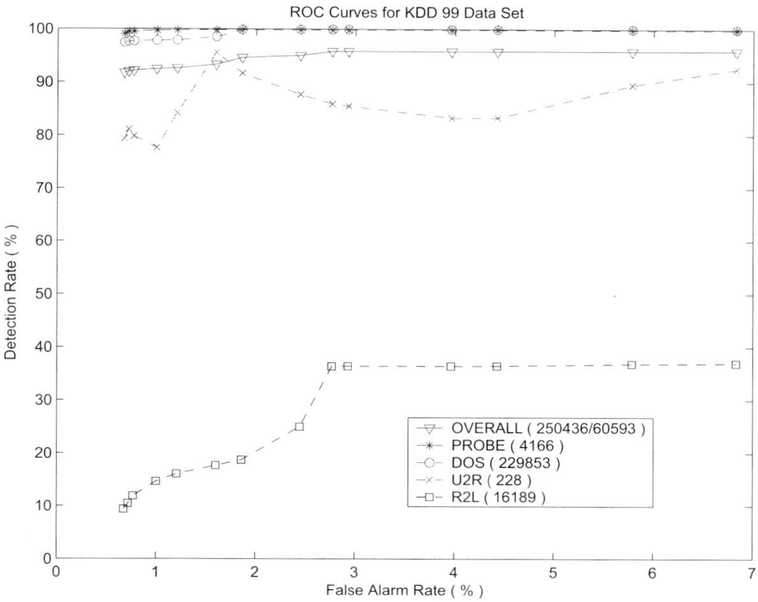

Fig. 3. The performance of proposed system. The curves are obtained by varying the influence factor σ.

The experimental results show that the detection performance of PROBE and DOS attacks of our approach is superior to that of other attacks, especially detection of R2L attacks. The reason for the low detection rate for R2L attacks is that both PROBE and DOS attacks often have the distinct traffic characteristic while U2R and R2L are more similar to normal examples. Especially, there are a lot of *snmpgetattck* and *snmpguess* attacks in R2L dataset, which account up rough 63 percent of all R2L attacks. In fact, they are almost identical with normal examples and hardly detected only by the connection information. This means the detection rate for R2L attacks would reach 37% at most no matter what the false alarm rate is. Therefore, in Fig. 2, the detection rate for R2L attacks keeps stable (about 36.6%) when false positive rate surpasses 2.8%. Except both kinds attacks above, our approach can detect other attacks with the interesting detection and false alarm rates.

Furthermore, we compared our approach with other proposed methods, in which some participated in the task of KDD Cup. Since KDD Cup is concerned with multi-class classification but we are interested only in normal/intrusion detection, we converted the results of those methods into our format. Specifically, the detection rate measures the percentage of intrusive connections in the test set that were labeled as known-attack or anomaly, without considering whether they are classified into the correct intrusion categories. The results are showed in Table 1, in which FAR means false alarm rate. It can be seen that our approach clearly outperforms the others, especially in detection of PROBE and U2R attacks, while its false alarm rate (FAR) is comparable to the others.

Table 1. Comparison of our system with other approaches

METHOD	FAR	PROBE	DOS	U2R	R2L
Our approach	0.7%	**99.5%**	**97.92%**	**81.14%**	10.44%
C5 Bagged Boosting	0.55%	87.73%	97.7%	26.32%	10.27%
Kernel Miner	0.55%	89%	97.57%	22.37%	7.38%
NN	**0.45%**	83.3%	97.3%	8.33%	2.5%
Decision Tree	0.5%	77.92%	97.24%	13.6%	0.52%
Naive Bayes	2.32%	88.33%	96.65%	11.84%	8.66%
PNrule	0.5%	78.67%	97%	14.47%	**10.8%**

7 Conclusion

This paper presents a new ID approach to detecting anomaly attacks. The ID approach maps connection records to different feature spaces in the light of protocol and service of connection and then detects anomalies by clustering and classification. We use the labeled training data to model the network behavior. Differing from ADAM which uses frequent episodes to build the normal profile, we cluster data instances that contain both normal behaviors and attacks and then select some clusters as the normal and known-attack profile according to a criterion. In order to detect novel attacks, we designed a classifier with the "default" label. We present influence-based classification algorithm to classify network behaviors.

Acknowledgement

This work is supported in part by grants from the Research Foundation of Civil Aviation University of China (05YK12M) and the Open Foundation of Tianjin Key Lab for Advanced Signal Processing. We would like to thank those organizations and people for their supports.

References

1. Ke Wang and Salvatore J. Stolfo: Anomalous Payload-Based Network Intrusion Detection. International Symposium on Recent Advances in Intrusion Detection (RAID 2004), LNCS 3224, Springer-Verlag, Berlin Heidelberg (2004) 203-222
2. M. Roesch: Snort: The open source network intrusion detection system. http://snort.sourcefire.com/ (2005)
3. Stanifor, Hoagland and McAlerney: Practical Automated Detection of Stealthy PortScans. Journal of Computer Security Vol. 10(1/2) (2002) 105-136
4. M. V. Mahoney, P.K. Chan: PHAD: Packet Header Anomaly Detection for Identifying Hostile Network Traffic. Technical report, Florida Tech., technical report CS-2001-4, April (2001)

5. M. Mahoney: Network Traffic Anomaly Detection Based on Packet Bytes. Proc. ACM. Symposium on Applied Computing (2003) 346-350
6. Wenke Lee, Stolfo SJ: A Framework for Constructing Features and Models for Intrusion Detection Systems, ACM Transactions on Information and System Security, Vol. 3(4) November (2000) 227-261
7. D. Barbara, Julia Couto, Sushil Jajodia, Ningning Wu: ADAM: A Testbed for Exploring the Use of Data Mining in Intrusion Detection. SIGMOD Record (2001) 15-24
8. L. Ertoz, E. Eilertson, et al: The MINDS - Minnesota Intrusion Detection System. Workshop on Next Generation Data Mining. MIT /AAAI Press (2004)
9. B.Pfahringer : Winning the KDD99 Classification Cup: Bagged Boosting, SIGKDD explorations, Vol. 1(2) (2000) 65-66
10. R. Agarwal and M. V. Joshi: PNrule: A New Framework for Learning Classifier Models in Data Mining (A Case-Study in Network Intrusion Detection). Technical Report TR 00-015, Department of Computer Science, University of Minnesota. (2000)
11. MacQueen: Some methods for classification and analysis of multivariate observations. In Proc. 5th Berkeley Symp. Math. Statistics and Probability (1967) 281-297
12. PK. Chan, MV. Mahoney and MH. Arshad: A machine learning approach to anomaly detection. Technical Report, Department of Computer Science, Florida Institute of Technology (2003)
13. University of California, Irvine: KDD Cup 1999 Data. http://www.ics.uci.edu/kdd/databases/kddcup99/kddcup99.html (2005)

Blind Channel Estimation for Space-Time Block Coded MC-CDMA System*

Aifeng Ren[1,2] and Qinye Yin[1]

[1] School of Electronics and Information Engineering,
Xi'an Jiaotong University, Xi'an 710049, P.R. China
[2] School of Electronic and Engineering, XiDian University, Xi'an 710071, P.R. China
renaifeng@xinhuanet.com, qyyin@xjtu.edu.cn

Abstract. To achieve high system capacity over frequency-selective fading channel, Multicarrier code-division multiple access (MC-CDMA) system with spatial diversity using space-time block coding (STBC) is considered. In this paper, we focus on the downlink frequency-selective fading channel identification problem of STBC multiple-input multiple-output (MIMO) MC-CDMA systems. Based on subspace decomposition technique and the special structure of STBC, we present a blind channel estimation scheme under multiuser environment. Computer simulations illustrate the performance of this algorithm.

1 Introduction

Combined with MC-CDMA, space-time coding (STC) techniques are proposed to exploit both spatial diversity and channel coding in multi-input multi-output (MIMO) systems[1][2]. We consider the orthogonal space-time block code with rate 3/4 constructed in[3]. Three successive symbols for the k-th user, such as $b_k(1), b_k(2), b_k(3)$, are mapped to the following 3×4 matrix

$$\boldsymbol{S}_k^{(i)} = \begin{bmatrix} b_k^{(3i-2)} & b_k^{(3i-1)} & b_k^{(3i)}/\sqrt{2} \\ -b_k^{(3i-1)} & b_k^{(3i-2)} & b_k^{(3i)}/\sqrt{2} \\ b_k^{(3i)}/\sqrt{2} & b_k^{(3i)}/\sqrt{2} & -b_k^{(3i-2)} \\ b_k^{(3i)}/\sqrt{2} & -b_k^{(3i)}/\sqrt{2} & b_k^{(3i-1)} \end{bmatrix} \tag{1}$$
$$\begin{matrix} \downarrow & \downarrow & \downarrow \\ \boldsymbol{s}_k^{(1)}(n) & \boldsymbol{s}_k^{(2)}(n) & \boldsymbol{s}_k^{(3)}(n) \end{matrix}$$

where the mth column of $\boldsymbol{S}_k^{(i)}$ $(i=1,\cdots,N)$ is the transmitted symbol sequence of the kth user at transmit antenna m in the four consecutive symbol periods.

We assume that all K active users transmit the signals through the same downlink dispersive frequency selective fading channel and finally arrive the receiver antenna simultaneously, and all K active users share the same set of subcarriers. The number of subcarriers equals to the spreading factor G. For convenience, we define the assigned frequency-domain spreading code vectors for the k-th user as $\boldsymbol{c}_k^{(m)} = [c_k^{(m)}(1) \quad c_k^{(m)}(2) \quad \dots \quad c_k^{(m)}(G)]^T$ $(m=1,\cdots,M_T)$, where $[\bullet]^T$ denotes transpose.

* Partly supported by the National Natural Sciences Foundation (No. 60572046) of China.

L. Wang et al. (Eds.): FSKD 2006, LNAI 4223, pp. 1092 – 1095, 2006.

2 Signal Model and Blind Channel Estimation

We depict the frequency-domain attenuations on all subcarrier channels between the p-th receive antenna and the m-th transmit antenna as the $G\times 1$ vector $\boldsymbol{\eta}^{(pm)}$, and $\boldsymbol{\eta}^{(pm)}=[\eta^{(pm)}(1)\ \cdots\ \eta^{(pm)}(G)]^T=\boldsymbol{F}_{PRO}\boldsymbol{h}^{(pm)}$ by performing the discrete Fourier transform (DFT) on the time-domain FIR vector $\boldsymbol{h}^{(pm)}=[h^{(pm)}(0)\ \ h^{(pm)}(1)\ \cdots\ h^{(pm)}(L-1)]^T$ $(p=1,\cdots,M_R;m=1,\cdots,M_T)$, where the $G\times L$ matrix $\boldsymbol{F}_{PRO}=\boldsymbol{F}_{DFT}(:,1:L)$ consists of the first L columns of the DFT matrix $\boldsymbol{F}_{DFT}$, whose every entry is defined as $[F_{DFT}]_{i,l}=1/\sqrt{G}\exp(-j2\pi(i-1)(l-1)/G)$, $(i,l=1,\cdots G)$.

Then, we can consider $(G\cdot M_T\cdot M_R)$ parallel subcarrier channels between all inputs of OFDM modulations and all outputs of OFDM demodulations[4]. At the receiver, we can obtain the $4\times M_RG$ data matrix $\boldsymbol{Y}_k^{(i)}(i=1,2,...,N;k=1,2,...,K)$ after OFDM demodulation of the received signals for the k-th user in the i-th group as

$$\boldsymbol{Y}_k^{(i)}=\boldsymbol{S}_k^{(i)}\left[\boldsymbol{h}_k^{(1)T}\quad \boldsymbol{h}_k^{(2)T}\quad \boldsymbol{h}_k^{(3)T}\right]^T=\boldsymbol{S}_k^{(i)}\begin{bmatrix}\boldsymbol{c}_k^{(1)T}\boldsymbol{H}^{(11)} & \boldsymbol{c}_k^{(1)T}\boldsymbol{H}^{(21)} & \cdots & \boldsymbol{c}_k^{(1)T}\boldsymbol{H}^{(M_R1)}\\ \boldsymbol{c}_k^{(2)T}\boldsymbol{H}^{(12)} & \boldsymbol{c}_k^{(2)T}\boldsymbol{H}^{(22)} & \cdots & \boldsymbol{c}_k^{(2)T}\boldsymbol{H}^{(M_R2)}\\ \boldsymbol{c}_k^{(3)T}\boldsymbol{H}^{(13)} & \boldsymbol{c}_k^{(3)T}\boldsymbol{H}^{(23)} & \cdots & \boldsymbol{c}_k^{(3)T}\boldsymbol{H}^{(M_R3)}\end{bmatrix}_{(3\times M_RG)}=\boldsymbol{S}_k^{(i)}\boldsymbol{H}_k\ . \tag{2}$$

where the $1\times M_RG$ vector $\boldsymbol{h}_k^{(i)}(i=1,2,3)$ is composed by random frequency-domain attenuations on subcarriers from the i-th transmit antenna to all receive antennas. $\boldsymbol{H}^{(pm)}[n]=diag(\boldsymbol{\eta}^{(pm)})$. Based on the STBC, the product of the code matrix $\boldsymbol{S}_k^{(i)}$ and the matrix $\boldsymbol{H}_k$ in (2) can be transformed to the product of a $4M_RG\times 3$ matrix $\overline{\boldsymbol{H}}_k$, which is coming from $\boldsymbol{H}_k$, and the symbol vector $\boldsymbol{b}_k^{(i)}$ for the k-th user [5], that is

$$\overline{\boldsymbol{H}}_k\boldsymbol{b}_k^{(i)}=vec(\boldsymbol{H}_k{}^T\boldsymbol{S}_k^{(i)T})\ . \tag{3}$$

where $vec(\boldsymbol{A})$ represents the vectorization of matrix $\boldsymbol{A}$. The matrix $\overline{\boldsymbol{H}}_k$ is given by

$$\overline{\boldsymbol{H}}_k=\begin{bmatrix}\boldsymbol{h}_k^{(1)T} & \boldsymbol{h}_k^{(2)T} & \frac{\boldsymbol{h}_k^{(3)T}}{\sqrt{2}}\\ \boldsymbol{h}_k^{(2)T} & -\boldsymbol{h}_k^{(1)T} & \frac{\boldsymbol{h}_k^{(3)T}}{\sqrt{2}}\\ -\boldsymbol{h}_k^{(3)T} & 0 & \frac{\boldsymbol{h}_k^{(1)T}+\boldsymbol{h}_k^{(2)T}}{\sqrt{2}}\\ 0 & \boldsymbol{h}_k^{(3)T} & \frac{\boldsymbol{h}_k^{(1)T}-\boldsymbol{h}_k^{(2)T}}{\sqrt{2}}\end{bmatrix}_{(4M_RG\times 3)} \tag{4}$$

Making the data matrix $\boldsymbol{Y}_k^{(i)}(i=1,2,...,N)$ for the k-th user become a long column vector and stacking N long vectors, we can construct a $4M_RG\times N$ matrix as

$$\boldsymbol{Y}_k=\left[\overline{\boldsymbol{H}}_k\boldsymbol{b}_k^{(1)}\quad \overline{\boldsymbol{H}}_k\boldsymbol{b}_k^{(2)}\quad \cdots\quad \overline{\boldsymbol{H}}_k\boldsymbol{b}_k^{(N)}\right]=\overline{\boldsymbol{H}}_k\begin{bmatrix}b_k^{(1)} & b_k^{(4)} & \cdots & b_k^{(3N-2)}\\ b_k^{(2)} & b_k^{(5)} & \cdots & b_k^{(3N-1)}\\ b_k^{(3)} & b_k^{(6)} & \cdots & b_k^{(3N)}\end{bmatrix}=\overline{\boldsymbol{H}}_k\overline{\boldsymbol{B}}_k\ . \tag{5}$$

The numerical model of this paper for the multiuser STBC MC-CDMA system is

$$\boldsymbol{Y}=\sum_{k=1}^{K}\overline{\boldsymbol{H}}_k\overline{\boldsymbol{B}}_k+\boldsymbol{V}=\overline{\boldsymbol{H}}\cdot\overline{\boldsymbol{B}}+\boldsymbol{V}\ . \tag{6}$$

where $\overline{\boldsymbol{H}}=[\overline{\boldsymbol{H}}_1\quad \overline{\boldsymbol{H}}_2\quad \cdots\quad \overline{\boldsymbol{H}}_K]$ is $4M_RG\times 3K$. $\boldsymbol{V}$ is a $4M_RG\times N$ thermal noise matrix, whose every entry is the independent identically distributed (i.i.d) complex zero-mean Gaussian noise with variance σ_n^2.

By doing the singular value decomposition (SVD) on $\boldsymbol{Y}$ in (6), we obtain the $4M_RG\times(4M_RG-3K)$ vectors $\boldsymbol{U}_0=[u_{3K+1} \quad u_{3K+2} \quad \cdots \quad u_{4M_RG}]$ that are associated with the smallest singular values span the noise subspace that is the orthogonal complement subspace of the signal subspace defined by the column of $\overline{\boldsymbol{H}}$ [6]. That is, the following matrix equations holds for $k=1,\cdots,K$.

$$\boldsymbol{U}_0^H\overline{\boldsymbol{H}}_k=\boldsymbol{O}_{(4M_RG-3K)\times 3} \tag{7}$$

Separate the matrix $\boldsymbol{U}_0^H$ into four segments and each segment has M_RG columns. That is $\boldsymbol{U}_0^H=[\boldsymbol{W}_1 \quad \boldsymbol{W}_2 \quad \boldsymbol{W}_3 \quad \boldsymbol{W}_4]$, where $\boldsymbol{W}_i\ (i=1,2,3,4)$ is a $(4M_RG-3K)\times M_RG$ matrix. Then, construct a $3(4M_RG-3K)\times 3M_RG$ matrix $\overline{\boldsymbol{W}}$ as

$$\overline{\boldsymbol{W}}=\begin{bmatrix} \boldsymbol{W}_1 & \boldsymbol{W}_2 & -\boldsymbol{W}_3 \\ -\boldsymbol{W}_2 & \boldsymbol{W}_1 & \boldsymbol{W}_4 \\ \frac{\boldsymbol{W}_3+\boldsymbol{W}_4}{\sqrt{2}} & \frac{\boldsymbol{W}_3-\boldsymbol{W}_4}{\sqrt{2}} & \frac{\boldsymbol{W}_1+\boldsymbol{W}_2}{\sqrt{2}} \end{bmatrix} \tag{8}$$

Thus, the product of $\boldsymbol{U}_0^H$ and the matrix $\overline{\boldsymbol{H}}_k$ can be changed to the product of the matrix $\overline{\boldsymbol{W}}$ and the $3M_RG\times 1$ vector $\boldsymbol{h}_k=[\boldsymbol{h}_k^{(1)} \quad \boldsymbol{h}_k^{(2)} \quad \boldsymbol{h}_k^{(3)}]^T$. That is

$$\overline{\boldsymbol{W}}\boldsymbol{h}_k=vec([\boldsymbol{W}_1 \quad \boldsymbol{W}_2 \quad \boldsymbol{W}_3 \quad \boldsymbol{W}_4]\overline{\boldsymbol{H}}_k)=\boldsymbol{0} \tag{9}$$

From (2) and $\boldsymbol{H}^{(pm)}$, the vector $\boldsymbol{h}_k=[\boldsymbol{h}_k^{(1)} \quad \boldsymbol{h}_k^{(2)} \quad \boldsymbol{h}_k^{(3)}]^T$ can be expressed as

$$\boldsymbol{h}_k=\begin{bmatrix} \boldsymbol{I}_{M_R}\otimes(\boldsymbol{C}_k^{(1)}\boldsymbol{F}_{PRO}) & & \boldsymbol{0} \\ & \boldsymbol{I}_{M_R}\otimes(\boldsymbol{C}_k^{(2)}\boldsymbol{F}_{PRO}) & \\ \boldsymbol{0} & & \boldsymbol{I}_{M_R}\otimes(\boldsymbol{C}_k^{(3)}\boldsymbol{F}_{PRO}) \end{bmatrix}\cdot\boldsymbol{h}=\boldsymbol{\Psi}_k\boldsymbol{h}\ . \tag{10}$$

where $\boldsymbol{h}=[\boldsymbol{h}^{(11)T} \ \cdots \ \boldsymbol{h}^{(M_R1)T} \ \boldsymbol{h}^{(12)T} \ \cdots \ \boldsymbol{h}^{(M_R2)T} \ \boldsymbol{h}^{(13)T} \ \cdots \ \boldsymbol{h}^{(M_R3)T}]^T$ is $3M_RL\times 1$, which is stacked by the time-domain FIR channels between all transmit antennas and receive antennas. $\boldsymbol{C}_k^{(m)}=diag(\boldsymbol{c}_k^{(m)})$ is a $G\times G$ matrix, and $\boldsymbol{\Psi}_k$ is a $3M_RG\times 3M_RL$ matrix.

Substituting (10) into (9), we have $\overline{\boldsymbol{W}}\boldsymbol{\Psi}_k\boldsymbol{h}=\boldsymbol{0}$. When $4M_RG-3K\geq M_RL$, that is, $K\leq {(4M_RG-M_RL)}/{3}$, this linear equation set is overdeterminated. With the SVD, we can get the estimation $\hat{\boldsymbol{h}}$ of $\boldsymbol{h}$, which is just the vector associated with the smallest singular value. Note that there is an ambiguous complex coefficient γ between $\hat{\boldsymbol{h}}$ and $\boldsymbol{h}$, that is $\boldsymbol{h}=\gamma\hat{\boldsymbol{h}}$. With the finite alphabet property of the transmitted symbols, γ can be estimated and the accurate estimation of $\boldsymbol{h}$ is obtained [1][7].

3 Simulation Results

In simulation, the DBPSK modulation is used. Spreading factor G is 32. Hadamard multiple access codes are distributed to different users. 50 Monte-Carlo trials are performed for each simulation. The performance is evaluated by the root mean square error (RMSE) which is defined as

$$RMSE=\frac{1}{\|\boldsymbol{h}\|}\sqrt{\frac{1}{N_t}\sum_{i=1}^{N_t}\left\|\hat{\boldsymbol{h}}(i)-\boldsymbol{h}\right\|^2}, \qquad N_t=50 \tag{11}$$

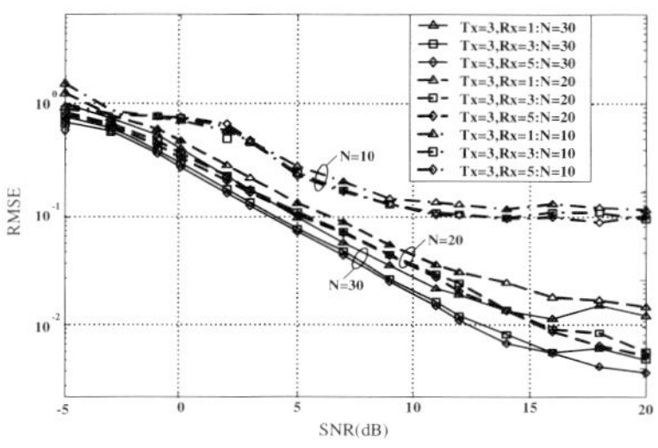

Fig. 1. Performance for different N

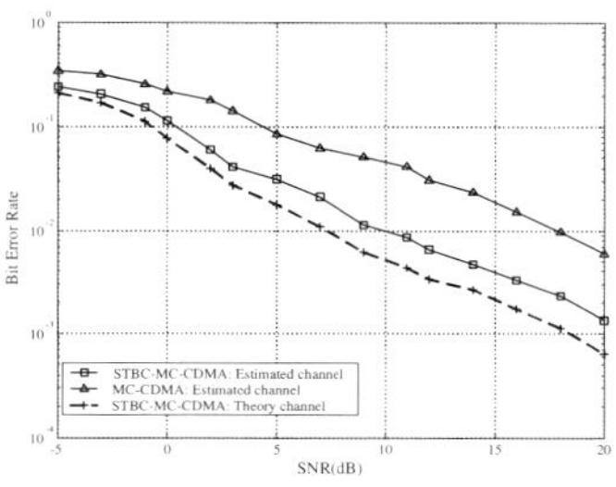

Fig. 2. BER performance comparison

Fig.1 shows the RMSE versus the number of groups N, where M_T=3, K=5, L=6 and M_R=1,3,5. If let N be constant and M_R=1,3,5, the performance of multiple receive antennas is better than single receive antennas obviously, and when M_R=3 and 5, the channel estimation error curve is almost consistent.

Fig.2 shows the BER performance of the decoder with estimated channel and theory channel, respectively, where K=5, L=6, M_T=M_R=3 and N=20. And the BER curve of MC-CDMA system without space-time code is given, where signal is spread and then converted into parallel data stream to be transmitted through different subcarriers.

4 Conclusion

With the special structure of STBC, the characteristic of the STBC is transferred to the vectors combined by frequency-domain attenuations of subcarriers in MC-CDMA system. Based on subspace techniques, we derive a novel blind channel identification scheme for STBC MIMO MC-CDMA system, in which the number of receive antennas is no limited. Compared with known channel information and MC-CDMA system without STC, simulation results prove the BER performance of our algorithm.

References

1. Hu X., Chew Y. H. : Performance of Space-time block cloded MC-CDMA system over frequency selective fading channel using semi-blind channel estimation technique. IEEE Wireless Communications and Networking, Vol. 1. WCNC (2003) 414-419
2. Iraji S., Lilleberg J. : Interference cancellation for space-time block-coded MC-CDMA systems over multipath fading channels. Vehicular Technology Conf., Oct. (2003)
3. Tarokh V., Jafarkhani H., and Calderbank A.R.: Space-time block codes from orthogonal designs. IEEE Trans., Vol. 45. July (1999) 1456-1467
4. Wang, L.P.(Eds.): Soft Computing in Communications. Springer, Berlin Heidelberg New York (2003)
5. Zhao Z.: Blind estimation and reduced-dimension processing based on multiple-input multiple-output antenna systems. Ph.D. Dissertation, xi'an jiaotong university, China, Oct. (2002)
6. Moulines E., Duhamel P., Cardoso J. F., and Mayrargue S. : Subspace methods for the blind identification of multichannel FIR filters. IEEE Trans. Signal Processing, Vol. 43. Feb (1995) 516-525
7. Jiang B., Wu X., and Yin Q. : Blind uplink channel estimation and multiuser detection for wideband CDMA. Proc. WCC-ICSP 2000, Beijing, China, August (2000) 1798-1801

A Polyclonal Selection Clustering for Packet Classification

Fang Liu and Liqi Wei

School of Computer Science and Engineering, Xidian University, Xi'an, 710071, China
f63liu@163.com

Abstract. The process of categorizing packets into "flows" in an Internet router is called packet classification. Packet classification is one of the most difficult problems in the Internet routers. Traditional packet classification algorithms focus on the time complexity and storage complexity of the classification and the rules used for classification are fixed and can not meet the increasing network requirement. In this paper, a polyclonal selection clustering algorithm for packet classification (PSC-PC) is proposed, which can produce the rules for classification automatically. Experimental results show that the rules obtained by PSC-PC are feasible for the packet classification, and the proposed algorithm is self-adaptive and self-learning, which makes it more applicable to the network whose types of application are changeable.

1 Introduction

At present, Internet routers provide only "best-effort" service for servicing packets in a first-come-first-served manner. Routers are now called upon to provide different qualities of service to different applications, therefore, some new mechanisms such as admission control, resource reservation, per-flow queuing and fair scheduling are needed. All of these mechanisms require the router to distinguish packets belonging to different flows.

Flows are specified by rules applied to incoming packets. We call a collection of rules a classifier. Each rule specifies a flow to which a packet may belong based on some criteria applied to the packet header. Some of the header fields, such as the destination/source IP address in network layer and the destination/source port in transport layer, might be used to classify the packets.

The packet classification is considered as an instance of the best matching rule problem in traditional packet classification algorithms, in which, an effective lookup method is the focus. There are many packet classification algorithms with good lookup method available, such as RFC (Recursive Flow Classification) algorithm [1] Modular algorithm [2], Grid of Tries [3], Cross-Product [4] and Tuple Space Search [5]. Generally, a good lookup algorithm should satisfy the follow three conditions: 1) high search speed, 2) low storage requirements, 3) update easily [6]. At present, many available algorithms emphasize the first two conditions, namely, the time efficiency and the space efficiency. But these two factors often restrict each other. Therefore, most traditional algorithms are a compromise of the two factors and have their own advantages. However, traditional algorithms seldom consider the importance of filter

L. Wang et al. (Eds.): FSKD 2006, LNAI 4223, pp. 1096 – 1104, 2006.

set for classification. Classification requires collecting certain behavior rules as the foundation, and its key part is the filter set. The filter set in practice are often disposed artificially, and the number of rules is very limited [7]. However, with the high-speed developing of Internet, the flows of the network are becoming greater and various types of network application are arising. How to extract the rules automatically is important increasingly.

Taylor D.E. and Turner J.S. first proposed the idea of automatic rules generation [8] in 2004, in which, the statistic characteristics of the filter sets are got by studying a large amount of real filter sets, and then generate the corresponding filter set for classification. We were inspired by the idea that the filter set for classification is automatically generated [8], a new polyclonal selection clustering algorithm for packet classification (PSC-PC) is proposed in this paper, in which, the optimal filter set of rules is searched by the Immune Polyclonal Selection Algorithm [9] [10].

2 Mathematical Model of Packet Classification

An address D is a bit string with W bits in length.

A prefix Pr is a bit string of length between 0 and W. We use *length* (Pr) to denote the number of bits in a prefix.

A header H has K fields, which are denoted by $H[1]$, $H[2]$, ..., $H[K]$ respectively. Each field is a binary string.

A rule R also has K fields. Each field $R[i]$ can specify any of the three kinds of matches: exact match, prefix match, or range match.

It is called an exact match *iff* a single value is specified for the ith rule field (i.e. $R[i]$) and the header field $H[i]$ is equal to $R[i]$.

It is called a prefix match *iff* a prefix is specified for the ith rule field of the header field $H[i]$ are the same as those of $R[i]$.

It is called a range match *iff* a range of values $R[i] = val1...val2$ is specified for the ith rule field and the header field $H[i]$ falls into that range, i.e. $val1 <= H[i] <= val2$.

A rule R is said to be a matching rule for a header H *iff* each field $H[i]$ of H matches the corresponding field $R[i]$ of R. The type of match is specified by $R[i]$ and could be an exact match, a prefix match or a range match.

A set of N rules is called a rule database, which is denoted by RS.

Each rule R has a *cost* property denoted by $cost(R)$. For $\forall R_1, R_2 \in RS$, if $cost(R_1)=cost(R_2)$, then $R_1=R_2$. We use the *cost* property to assure that there is at most one matching rule.

In theory, seven fields can be used for the packet classification rules: destination/source IP address (32 bits each), destination/source transport port (16 bits each), type of service (8 bits), protocol type (8 bits) and flag of transport layer (8 bits). The number of bits of these fields is 120 totally (we assume that all the seven fields reside in IP packet header for the sake of convenience, although some fields are in TCP header actually). Statistical results of some actual rule databases used by ISPs show that 17% of the rules specify only one field, 23% specify three fields, and 60% specify four fields [6]. Five fields will be taken into account in this paper: destination/source IP address, destination/ source transport port and protocol type.

A rule is generated by a field or some fields mentioned above, and a filter set is generated by many rules. Each rule *R* defines a flow type (Class) or connects an action. The action shows the corresponding treatment to the packet that meets to the corresponding rule in the filter set. Each type of flow has only a sign, and it corresponds to a kind of specific treatment method. If a packet *p* is matched by a rule *R*, *p* should be ranked the class defined by *R* [11]. Different fields have different value types and match methods, and rules might overlap each other in the same field, that is, a packet may match more than one rule in the filter set. Our purpose is to search for the best one.

3 Polyclonal Selection Clustering for Packet Classification (PSC-PC)

PSC-PC combines the Immune Polyclonal Selection Algorithm [9] [10] with clustering [12] to generate filter set of rules for packet classification automatically. Section 3.1 and 3.2 will introduce the proposed algorithm PSC-PC, which include the polyclonal selection clustering for classification and the packet classification algorithm.

3.1 Polyclonal Selection Clustering for Filter Set Learning

The polyclonal selection algorithm is different from the monoclonal selection in the clonal operator. In the former, the mutation and crossover operations are both carried out, while in the latter, the mutation operation is implemented only.

3.1.1 Encoding

The binary encoding strategy is used. A filter set of rules *RS* composed of $R_1, R_2, R_3, \cdots R_K$ is regarded as an antibody. The *i*th antibody A_i in an antibody population *G* can be described as the formula (1).

$$A_i = \{R_{i1}, R_{i2}, R_{i3}, \cdots R_{iK}\}, i = 1, 2, \cdots, M \tag{1}$$

3.1.2 Affinity Function of Antibody-Antigen

The widely used cost function is the trace of scatter matrix within cluster [11]. We also apply the cost function model here, in which, the similarity measure used is different. We redefine the similarity measure function below.

Assume that each rule sample is described as

$$R_i = \left[r_{i1}^a, \cdots r_{it}^a, r_{i(t+1)}^p, \cdots r_{is}^p, r_{i(s+1)}^c, \cdots r_{ih}^c\right]^T, t = 2, s = 4, h = 5 \tag{2}$$

and a datum sample used is described as

$$X_i = \left[x_{i1}^a, \cdots x_{it}^a, x_{i(t+1)}^p, \cdots x_{is}^p, x_{i(s+1)}^c, \cdots x_{ih}^c\right]^T, t = 2, s = 4, h = 5 \tag{3}$$

where *a* is an address, *p* is a port and *c* is a protocol.

The fields of destination/source transport port and protocol type in the packet sample belong to the categorical values. The fields of destination/source IP address are similar to the numeric values. But they require the longest match for the address, so they cannot be considered as the simple numeric values. Then, the similarity measure function for a rule sample and a datum sample is defined as

$$d(X_i, R_j) = \sum_{l=1}^{t} \sigma(x_{il}^a, r_{jl}^a) + \lambda \sum_{l=t+1}^{s} \delta(x_{il}^p, r_{jl}^p) + \gamma \sum_{l=s+1}^{h} \psi(x_{il}^c, r_{jl}^c) \tag{4}$$

where λ and γ are coefficients, and $\sigma(\cdot)$ is defined as

$$\sigma(x,r) = \begin{cases} 1, & r \neq * \quad and \quad r = x \\ 2 \sim 32, & r \neq * \quad and \quad r \in x \quad and \quad r \neq x \\ 32, & r = * \\ 100, & r \neq * \quad and \quad r \notin x \end{cases}. \tag{5}$$

$\delta(\cdot)$ is defined as

$$\delta(x,r) = \begin{cases} 1, & r \neq * \quad and \quad r = x \\ 16, & r \supset x \\ 31, & r = * \\ 100, & r \neq * \quad and \quad x \not\subset r \end{cases}. \tag{6}$$

$\psi(\cdot)$ is defined as

$$\psi(x,r) = \begin{cases} 1, & r \neq * \quad and \quad r = x \\ 32, & r = * \\ 100, & r \neq * \quad and \quad r \neq x \end{cases}. \tag{7}$$

According to the above formulas, we define the clustering cost function as

$$C(U,R) = \sum_{i=1}^{k} \left(\sum_{j=1}^{n} \mu_{ij}^{\ 2} \sum_{l=1}^{t} \sigma(x_{jl}^a, r_{il}^a) + \lambda \sum_{j=1}^{n} \mu_{ij}^{\ 2} \sum_{l=t+1}^{s} \delta(x_{jl}^p, r_{il}^p) + \gamma \sum_{j=1}^{n} \mu_{ij}^{\ 2} \sum_{l=s+1}^{h} \psi(x_{jl}^c, r_{il}^c) \right), \mu_{ij} \in [0,1] \tag{8}$$

where k is the number of clusters, and n is the number of data samples. R is a prototype of cluster where $R_i = \left[r_{i1}^a, \cdots r_{it}^a, r_{i(t+1)}^p, \cdots r_{is}^p, r_{i(s+1)}^c, \cdots r_{ih}^c \right]^T$, $t = 2, s = 4, h = 5$ is the representative prototype vector of cluster ω_i. U is a partition matrix, and μ_{ij} is the membership degree of x_j belonging to cluster ω_i, which satisfies the probability constraint $\sum_{i=1}^{k} \mu_{ij} = 1, \forall j$. $\sigma(\cdot)$, $\delta(\cdot)$ and $\psi(\cdot)$ are dissimilarity measures which are defined as formula (5), formula (6) and formula (7) respectively.

We define the antibody-antigen affinity function using the clustering cost function.

$$f(Antibody) = \frac{1}{1 + C(U,R)} \tag{9}$$

3.1.3 Filter Set Learning Algorithm by Polyclonal Selection Clustering

Step1. $l = 0$, initialize the antibody population $G(0) = \{A_1(0), A_2(0), \cdots, A_M(0)\}$, set the operation parameters.

Step2. Calculate the affinities of antibodies.

Step3. Judge the iterative termination criterion. If the optimal affinities of successive *m* generations are unchanged, or when the maximal number of evolutionary generation reaches, the algorithm will be halted, otherwise, it continues.

Step4. According to the affinities and the clonal scale, the clonal operation is performed on the original antibodies, the clonal scale N_c is set to 10.

Step5. According to the mutation probability *Pm*, the clonal mutation operation is performed on the antibody population. To reserve the information of the original antibody population, the mutation operation will not be performed on the original antibodies but on the copied ones [13]. The mutation probability in PSC-PC is set to 0.2 before 20 generations, then 0.01 after ones.

Step6. According to the crossover probability *Pc*, the clonal crossover operation is performed on the antibody population. To reserve the information of the original population, the crossover operation will not be performed on the original antibodies but on the mutated ones. The clonal crossover probability *Pc* in PSC-PC is set to 0.75.

Step7. According to the selection probability *Ps*, the clonal selection operation is performed on all the new generated antibodies. The clonal selection probability *Ps* depends on the parameter *Pb,* and *Pb*>0 is the parameter related with the diversity of antibody population [12]. *Pb* is set to 0.6 here.

Step8. According to the death probability *Pd*, the clonal death operation is performed on all the antibodies after the above operations. If there are some antibodies with equal affinity, and the affinity is the highest of all the antibody affinities, then reserve one of the antibodies and the other antibodies with the biggest and equal affinity will be deserted according to the death probability *Pd*. *Pd* in PSC-PC is set to 0.2.

Step9. The antibodies with higher affinity in the population are stored as vaccines [14]. The vaccines will be used in the clonal mutation operation next time. And the antibodies are recoded. Then go to step2.

3.2 Packet Classification Algorithm

The packet set P={p1, p2,…} is classified by using the rules clustered by PSC-PC,

1. If $d(p_i, \omega_j) = \min d(p_i, \omega_h),\ h = 1, 2, \cdots, k$ and $d(p_i, \omega_j) \le \beta$, then $p_i \in \omega_j$.

2. If $d(p_i, \omega_j) > \beta$, p_i is considered an unknown type, in which, β is a predefined threshold, and $d(\cdot,\cdot)$ is defined as formula (4).

4 Experiment Results and Analyses

4.1 Data Set

Since it is difficult to get a large number of data from the real network, the experiment data used in this paper is produced by a random function. The data records used in this paper have five attributes, destination/source IP address, destination/source transport port and protocol type. The details of that the experiment data is generated are as follows:

1. The range of the destination/source IP address is from 1.0.0.1 to 223.255.255.254.
2. The used protocol specifications include TCP, UDP, ICMP, GRE, IGP, EIGRP, ESP, AH and IPE. As shown by many studies, the most commonly used one is TCP. As a whole, TCP and UDP account for more than 70% of data packets in the real network, ICMP is little quantitative, the remaining six protocols are only a few. So protocols of our data records are generated according to the ratio.
3. The range of port is from 0 to 65535. There are three types of ports in the real network: 1) Well Known Ports, from 0 to 1023; 2) Registered Ports, from 1024 to 49151; 3) Dynamic and/or Private Ports, from 49152 to 65535. Their frequencies of occurrence decrease in turn. So the ports of our data records are generated according to this ratio.

4.2 Influence of Parameters on PSC-PC

There are several parameters in the Polyclonal selection algorithm, which will affect the convergence of PSC-PC. Fig. 1 gives the curve of convergence. It can be found that after fluctuating for some generations, the curve becomes smooth ultimately. And that is the process of convergence.

In addition, in PSC-PC, mutation probability and crossover probability are two main factors that affect the precision of the classification. Fig.2 and Fig.3 respectively show the influences of the two factors on the *Accurate rate* and *Default rate*.

It can be found in Fig. 2 (a) that when *Pm* is in the range [0.2, 0.5], the *Accurate rate* is higher, and in Fig. 2 (b), when *Pm* is in the range [0.1, 0.3], the *Default rate* is lower. Because the condition of high *Accurate rate* with low *Default rate* illustrates

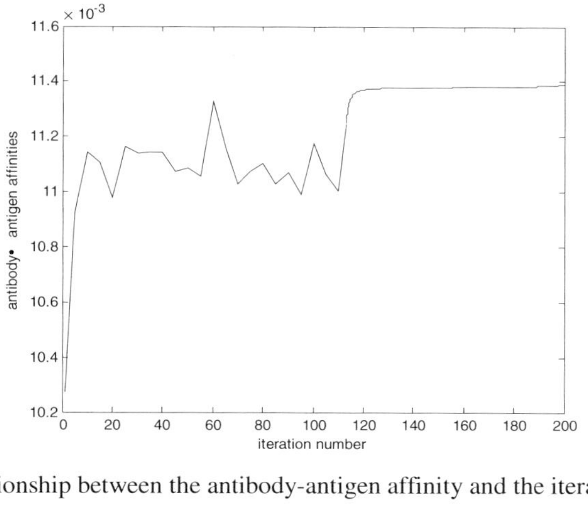

Fig. 1. Relationship between the antibody-antigen affinity and the iteration number

the good classification. So, we set Pm=0.2 in the former 20 generations. However, if the Pm keeps larger in the whole evolutionary process, the algorithm will be unstable and will not converge. Therefore, after 20 generations, we set Pm=0.01.

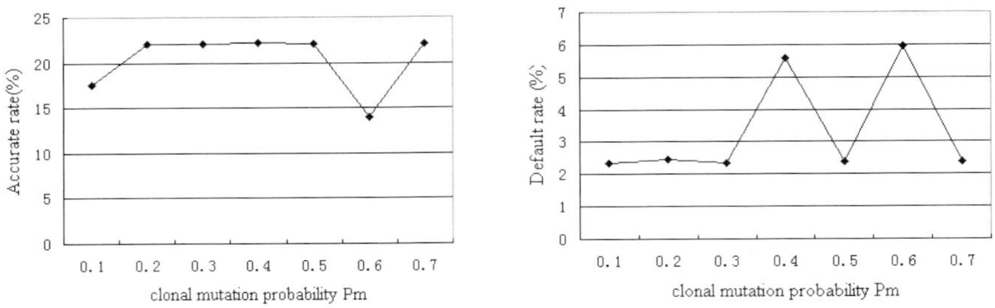

(a) (b)

Fig. 2. Influence of clonal mutation probability *Pm* on *Accurate rate* (a) and *Default rate* (b)

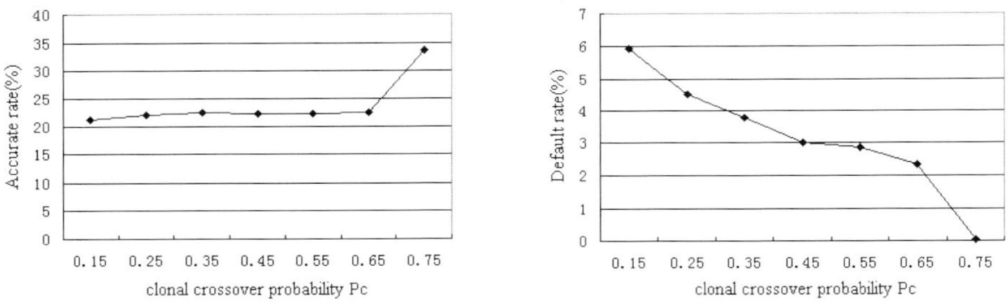

(a) (b)

Fig. 3. Influence of clonal crossover probability *Pc* on *Accurate rate* and *Default rate*

Fig. 3 gives the influences of clonal crossover probability Pc on the *Accurate rate* and *Default rate*. We can see from Fig. 3 (a) that the *Accurate rate* increases with the increase of Pc. And from Fig. 3 (b), it can be found that with the increase of Pc, the *Default rate* decreases. So, we set Pc=0.75 in PSC-PC.

4.3 Performance Test of PSC-PC

First, we will give four evaluation criterions for the measure of the proposed algorithm.

Default rate (D rate) is the ratio of packets mismatched by the rules.

Wildcard rate (W rate) is the ratio of packets matched by the rules whose fields contain many wildcard, and their values of dissimilarity measures functions are between a_1 and b_1.

Fuzzy rate (F rate) is the ratio of packets matched by the rules whose fields are often wide range values, and their values of dissimilarity measures functions are between a_2 and b_2.

Accurate rate (A rate) is the ratio of packets matched by the rules whose fields are often accurate values, and their values of dissimilarity measures functions are between a_3 and b_3.

Where a_1, a_2, a_3, b_1, b_2 and b_3 are thresholds predefined.

The values of *Default rate* and *Accurate rate* are related with the rationality of rules, namely, low *Default rate* with high *Accurate rate* indicate that the rules got are good. The values of *Wildcard rate* and *Fuzzy rate* are related with the agility of rules, namely, high *Wildcard rate* and high *Fuzzy rate* indicate that the rules are agile. And they are restricted by each other at the same time. That is, low *Default rate* whit high *Accurate rate* often result in low *Wildcard rate* and low *Fuzzy rate*, and high *Wildcard rate* and high *Fuzzy rate* often result in low *Accurate rate*.

Table 1 lists the classification results of PSC-PC. They are coincident with the above analysis.

In Table 1, *Tr set* stands for training set and *T set* the test set. The size of the test set is the same as that of the training set, and the test set is generated in the same way as the training set. The training set is used to generate the filter set of rules and test set is used to test the performance of the rules. In PSC-PC, the set with 300 rules is generated by clustering the training set with 6000 packets, the set with 500 rules is got by clustering the training set with 10000 packets, and the set with 700 rules is generated by the training set with 15000 packets. The three sets generated are all tested by the testing set. The numbers of rules are all designated in advance, which can be changed in different experiments.

It can be found from Table 1 that the filter set of rules achieved by PSC-PC is both agile and rational. At the same time, there are very small differences between the classification results on the test set and the training set. It is also shown that the accuracy rate is slightly higher on the test set with the achieved rules from the training set by the PSC-PC, which indicates that PSC-PC has certain learning ability and adaptive capability.

Table 1. The classification results of PSC-PC

Scale n	**Rule size**	**Data set**	**D rate** (%)	**W rate** (%)	**F rate** (%)	**A rate** (%)
6,000	300	*T set*	0.000	0.007	63.350	36.643
		Tr set	0.000	0.017	64.167	35.817
10,000	500	*T set*	0.010	0.346	72.126	27.518
		Tr set	0.000	0.380	72.009	27.611
15,000	700	*T set*	0.007	0.050	63.350	36.593
		Tr set	0.017	0.100	64.217	35.667

5 Conclusions

A new algorithm PSC-PC is proposed in this paper, by which the filter set of rules for the packet classification is generated automatically. Since PSC-PC combines the characteristic of clustering and that of Immune Polyclonal Selection algorithm, namely, the combination of global search and local optimization, it can quickly obtain good rules for the packet classification. Experimental results illustrate that the rules got by PSC-PC are effective for the packet classification. And when the types of network applications are changed, they can be updated easily.

In PSC-PC, however, the parameters are determined experientially. It is still a difficult problem to determine parameters automatically.

Acknowledgement. This work is supported by the National Natural Science Foundation of China under Grant No. 60372045 and No. 60133010, the Defence Pre-Research Project of China under Grant No.51406020104DZ0124, the Key Science-Technology Project of Higher Education of China under Grant No. 0202A022 and the National Research Foundation for the Doctoral Program of Higher Education of China No. 20030701013.

References

1. Gupta, P., McKeown, N.: Packet Classification on Multiple Fields. ACM Computer Review. 29 (1999) 146–160
2. Woo, T.Y.C.: A Modular Approach to Packet Classification: Algorithms and Results. In: Gruein, R. (eds.): Proceedings of IEEE Infocom. 2000. San Francisco, CA: IEEE Computer Society Press (2000) 1210–1217
3. Srinivasan, V.: Fast and Efficient Internet Lookups. Doctor of science degree dissertation, Washington University, August (1999)
4. Srinivasn, V., Varghese, G., Suri, S. et al.: Fast Scalable Level Four Switching. ACM Computer Communication Review. 28 (1998) 191–205
5. Srinivasan, V., Suri, S., Varghese, G.: Packet Classification Using Tuple Space Search. In: Proceedings of ACM Sigcomm99. (1999) 135–1466
6. Yu Zhongchao, Wu Jianping, Xu Ke: Study of IP Classification Technology, Acta Electronica Sinica. 29 (2001) 260–262
7. Yao Xingmiao, Hu Guangming, Li Leming: A Multi-dimensional Packet Classification Algorithm with Trees Divided by Value, Journal of Electronics and Information Technology. 26 (2004) 1413–1419
8. Taylor, D.E., Turner, J.S.: ClassBench: A Packet Classification Benchmark. Tech. Rep. WUCSE2004, Department of Computer Science & Engineering, Washington University in Saint Louis, May (2004)
9. Du Haifeng, Jiao Licheng, Liu Ruochen: Adaptive Polyclonal Programming Algorithm with Applications. In: 5th International Conference on Computational Intelligence and Multimedia Application. (ICCIMA). (2003) 350–355
10. Liu Ruochen, Jiao Licheng, Du Hiafeng: Clonal Strategy Algorithm Based on the Immune Memory. Journal of Computer Science and Technology. 20 (2005) 728–734
11. Gong Xiangyang: High-Speed Data Flow Classification Technology in TCP/IP Network, ZTE Telecommunications. 4 (2001) 15–18
12. Li Jie, Gao Xinbo, Jiao Licheng: A CSA-Based Clustering Algorithm for Large Data Sets With Mixed Numeric and Categorical Values, Acta Electronica Sinica. 32 (2004) 357–362
13. Du Haifeng, Jiao Licheng, Wang Sun'an: Clonal Operator and Antibody Clonal Algorithm. Proceedings of the First International Conference on Machine Learning and Cybernetics. USA: IEEE Press. (2002) 506–510
14. Jiao Licheng, Wang Lei: A Novel Genetic Algorithm Based on Immunity. IEEE Trans. Systems, Man and Cybernetics, Part A. 30 (2000) 552–561

Analyzing Fault Monitoring Policy for Hierarchical Network with MMDP Environment

Xin Zhang[1,2], Yilin Chang[1], Li Jiang[2], and Zhong Shen[1]

[1] State Key Laboratory of Integrated Service Networks,
Xidian University, Xi'an, 710071, China
[2] Dept. of Electronic and Information, Xi'an Institute of Post
and Telecommunications,Xi'an 710061, China
xingerzh@hotmail.com

Abstract. This paper proposed a fault monitoring policy for hierarchical network with multi-manager. It can be used to monitor the network in real time and lightened the burden of the network monitoring management. With the application of the multi-agent Markov Decision Processes in the network management, an appropriate policy model of SNMP polling with the reinforcement learning is given. The simulations results show that the reinforcement-learning model can provide effective fault localization meanwhile decrease the overhead of network management remarkably.

1 Introduction

Real-time network monitoring is required for maintaining the normal network operation. However, the heterogeneity of networks and the variety of the service make the monitoring more difficult especially when the network is in a stressed state. The fault monitoring is a stochastic dynamics process. Markov Decision Processes (MDP) is the only dynamic control method of discrete event dynamic system. Applying the MDP to the policy decision in network, fault monitoring can be treated as a stochastic optimization problem.

In the hierarchical network structure, there are multi-manager under the centre manager, this managerial structure is called multi-manager. This paper proposed a fault-diagnosing model with the multi-manager, combined with the reinforcement learning, analyzed and simulated the polling policy under this model.

2 Multi-manager MDP

The general Markov decision process is made up of following quadruple parameter, $\langle S, A(i), p_{ij}(a), r(i,a), i \in S, j \in S, a \in A(i) \rangle$, S is a finite set of states, $A(i)$ is a finite set of actions, $P_{ij}(a)$ is the State transition probability , Given state i and action $a \in A(i)$, the probability of each possible next state j is $P_{ij}(a)$.

MDP realizes on-line study and decision-making while interacting with the environment. State transition probabilities $P_{ij}(a)$ depend on control actions $a \in A(i)$. For

L. Wang et al. (Eds.): FSKD 2006, LNAI 4223, pp. 1105 – 1108, 2006.

each individual transition, there is an associated immediate reward $r(i,a)$. Costs may be treated as negative rewards; the optimal policy is a control law that yields minimum costs in a finite or infinite horizon. If a certain policy obtain positive reward of system, then Agent will strengthen the trends of this policy, on the contrary, if it get to decrease the reward, it will not select this kind of policy next time, this kind of learning is called Reinforcement Learning, it is suitable for the dynamic system with unknown models.

In the hierarchical network management, the MDP system consists of the manager, managed object (MO) and the monitor. When there are multi middle managers (M/A), there will be a multi-manager MDP (MMDP). Generally speaking, M/As could not obtain other MO's states directly but could exchange information, such as standard Inform message of SNMP, between M/As to get the statistics of other domain. So, in MMDP model, it makes a new policy by treating other whole states' effect as a node of its domain. Therefore, the fault management system can be regarded as a MMDP.

Finding a policy that achieves a lot of reward over the long run is the main task for the MMDP. In MMDP, consider an observing set T, $T=\{t_k \mid k=0,1,......,N\}$, when the agent take the action, there is a reward function R_{ij}^a , $R_{ij}^a = E\{r_j \mid s_t = s_i, a_t = a, s_{t+1} = s_j\}$, and also the corresponding value function $V^a = E\{R_{ij}^a\}$ for each policy to evaluating the decision. The network fault management is a finite MMDP, thus we take the finite expect policy. There is always at least one policy that is better than or equal to all other policies. We call this policy an optimal policy $\pi*$. Although there may be more than one optimal policy, they share the same value function, called the optimal value function, denoted $V*$, defined as $V*(s)=\max_{\pi} V^{\pi}(s)$, for all $s \in S$.

Under the policy π, we define the set $\{X,Y\}$ of two separated domain, and $M^x = (S^x, A^x, p_{ij}^x(s_j^x \mid s_i^x, a^x))$ defines the Markov process in X: its local state space is S^x,local action space is A^x,and the local state transition probability $p^x(s_j^x \mid s_i^x, a^x)$defines the probability of resulting in state s_j^x when taking action a^x in state s_i^x,similarly we can define the Y's process M^y, The joint state space is $S^x \times S^y$, the joint action space is $A^x \times A^y$.In moment t , the cost of SNMP Inform message among domains is $c_t^x(s^x, m^x)$, The reward is $r_t(s_t^x, s_t^y, a_t^x, a_t^y)$.

Clearly, with the Markovian and mutual independent of event, the joint probability density is: $p_{ij}^{a^x,a^y} = \prod_{t=0}^{N-1} p^x(s_{t+1}^x \mid s_t^x, a_t^x) \cdot p^y(s_{t+1}^y \mid s_t^y, a_t^y)$

The reward of the system is :

$R_{ij}^{a^x,a^y} = r_{t_j}(s_{t_j}^x, s_{t_j}^y) + \sum_{t=0}^{t_j-1} r_t(s_t^x, s_t^y, a_t^x, a_t^y) - \sum_{t=0}^{t_j-1}(c_t^x(s_t^x, m_t^x) + c_t^y(s_t^y, m_t^y))$

The value function in policy (π^x, π^y) is:

$$V^{(\pi^x,\pi^y)} = \sum_{ij} p_{ij}^{a^x,a^y} \cdot R_{ij}^{a^x,a^y} \tag{1}$$

If a model is not available, then it is particularly useful to estimate *action* values rather than *state* values. The fault model is often uncertain, generally we adopt the action value function.

Define: Action-value function $Q^{\pi}(s,a)$ in the expected value function under policy π with the action a, $Q^{\pi}(s,a)=E_{\pi}(R_t \mid s_t=s,a_t=a)=E_{\pi}\left\{\sum_{k=0}^{N-1}\gamma^k r_{t+k+1} \mid s_t=s,a_t=a\right\}$.

According to Bellman optimality equation and [2], in practice, the iterative Q-value function is:

$$Q^{t+1}(s_t,a_t)=Q^t(s_t,a_t)+\alpha\left[r_{t+1}+\max_{a_{t+1}}Q^t(s_{t+1},a_{t+1})-Q^t(s_t,a_t)\right] \quad (2)$$

Here α is a study factor.

Define the optimal action-value function as $Q*(s,a)$, and then the optimal policy is:

$$\pi*(s)=\arg\max_a Q*(s,a) \quad (3)$$

3 Applying MMDP Model in the Hierarchical Network Fault Management

Assuming only one MO can be in fault at any time. In each step the fault occurrence is accorded with a certain probability distribution. The communication between the manager and agent is SNMP polling and trap. The system state set S consists of mutual exclusive sub-state, i.e. $S=S^0 \cup S^1$, $S^0=\{s_0\}$ which means all MOs are normal and $S^1=\{s_1,s_2,......s\}$ which means merely i^{th} MO is in fault in M MOs. Action taken by manager consists of polling, repair and inform. $A=A^P \cup A^R \cup A^I$,here $A^P=\{a_1^p,...a_i^p...a_N^p\}$, $A^R=\{a_1^r,...a_i^r...a_N^r\}$, $A^I=\{a_x^i,a_y^i\}$ in which a_j^p is the action to polling i^{th} MO, a_j^r is the action to repair j^{th} MO and a_i^x,a_i^y is the action to inform the other domain's manager.

The optimal solution of value function convergences in the faulty node. Suppose there are M managed objects in the domain, the initial action is polling, after $j^{th}(j\in T)$ step, find the faulty node, and then the $V_t(M)$ represents the value function under the given state. The value function relates to the left stage. According to the reference [3], we get the follow expression:

$$V_t(M)=\max_a\left[R(l,a^p)+\gamma\sum_{l=1}^M P_{il}(a^p)\cdot V_{t\_1}(l)\right]=R(l,a^p)+\max_a\left[\gamma\sum_{l=1}^M P_{il}(a^p)\cdot V_{t\_1}(l)\right] \quad (4)$$

4 Simulation Results and Conclusions

The simulation model adopts the hierarchical management, which includes two sub domains. Each domain includes 6 to 10 MOs. The node of this model include manage node, service node, switch node, which produce different data source according to different distribution. The link is duplex link of above 150kbps, data package include manage package and service package.

Fig.1 is the action-value function for M/A's polling the node. The initial state of MO is in faulty with some probability, then, with reinforcement learning of MMDP, finding the fault source by the decision of Agent, until it has been finding and repaired. This Fig. shows the faulty node is MO 9 through MMDP exactly.

Fig.2 is for comparison of polling step. The solid line is for two divided sub domain and adopt MMDP model, the dashed line is for mixed two domains together and become the centralized management, compare the polling time. It is obvious that fault finding time is decreased remarkably with the MMDP algorithm.

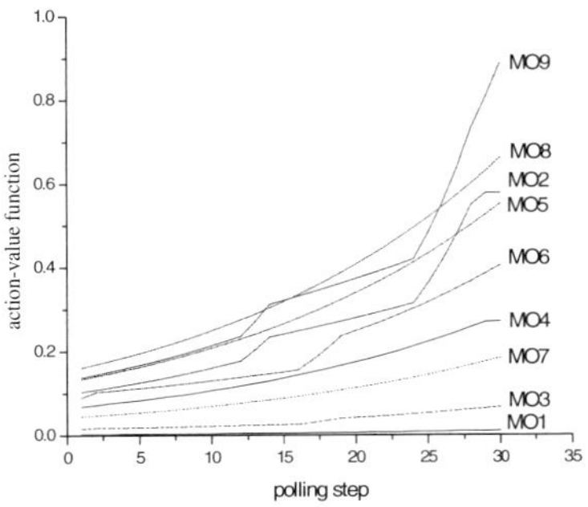

Fig. 1. Value-function of MO with MMDP

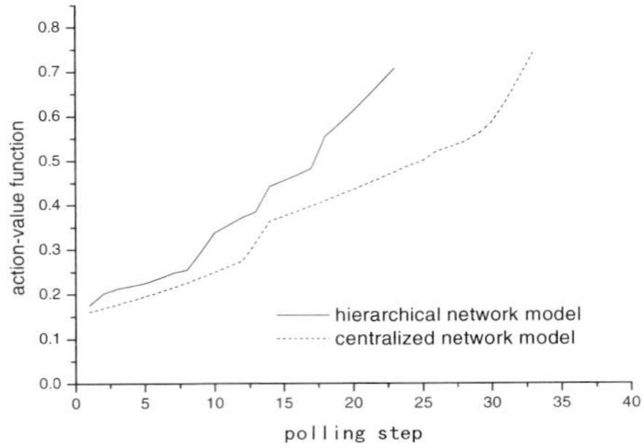

Fig. 2. Polling step of hierarchical network

The study on the MMDP shows that it can decrease the finding time of fault source and improved the link utilization, especially for the manager, shorten the processing time of management package.

References

1. William Stallings.: SNMP,SNMPV2 :the infrastructure for network management.IEEE Communications Magazine, Vol.36 No.3, (1998)37-43.
2. Richard S. Sutton and Andrew G. Barto. :Reinforcement Learning: An Introduction.Cambridge, MA, MIT Press, (1998)
3. Q.Y.Hu,:An Introduction to Markov Decision Processes.Xi'an, Xidian University Press, Ver No.1. (2000)
4. Hood C.S., Ji Chuanyi. :Automated Proactive Anomaly Detection, Integrated Network Management. Proceedings of the Fifth IFIP/IEEE International Symposium on Integrated Network Management. San Diego, USA, IFIP Conference Proceedings, (1997) 688-699.
5. Q. He, MA Shayman, :Using Reinforcement Learning for Proactive Network Fault Management. IEEE Computer, Vol.33 No.9, (2000)515-521

Using Meta-Level Control with Reinforcement Learning to Improve the Performance of the Agents

Daniela Pereira Alves[1], Li Weigang[2], and Bueno Borges Souza[2]

[1] Brazilian Institute of Information in Science and Technology - IBICT, Ministry of Science and Technology - MCT, Brasilia - DF CEP: 70070-914, Brazil
[2] Department of Computer Science, University of Brasilia, Brasilia-DF, CEP: 70910-900, Brazil
{Weigang, Bueno}@cic.unb.br

Abstract. In a complex environment where the messages exchange tensely among the agents, a difficulty task is to decide the best action for new arriving messages during on-line control. The Meta-Level Control model is modified and used to improve the performance of the communication among the agents in this research. During the control process, the decision is made from the experience acquired by the agents with reinforcement learning. The research proposed a Messages Meta Manager (MMM) model for Air Flow Management System (AFMS) with the combination of the Meta-Level Control approach and reinforcement learning algorithms. With the developed system, the cases of initial heuristic (IH), epsilon adaptative (EA) and performance heuristic (PH) were tested. The results from simulation and analyses show the satisfactory to the research purpose.

Keywords: Air Traffic Flow Management, Meta-Level Control, Reinforcement learning.

1 Introduction

A typical form of communication in a distributed system is the exchange of messages [1], especially for processing the immense amount of messages in a complex system. A multi-agent system (MAS) for Air Traffic Flow Management in grid computing - ATFMGC, was developed recently [2]. The research proposed an approach of cooperation and negotiation among agents using grid computing in a real time traffic synchronization problem. In some aspects such as the agent functions, their knowledge representation and inference processes [3] were developed. Standard of Balancing among Agents (SBA) as a criterion was also used to balance and measure the amount of communication among agents and the tolerated delay of the flights [2].

On the other hand, some problems had appeared for the fact of the intense exchange of messages in ATFMGC related with more than 10 airports. For a fixed SBA, it is impossible to efficiently equilibrate the communication. It is necessary to adapt a suitable mechanism to assist the decision process. The idea is to introduce Meta-Control approach [4] in Air Traffic Flow Management (ATFM). During information process, the reinforcement learning [5] is inserted to acquire experience to make Markov decision process more efficiency.

L. Wang et al. (Eds.): FSKD 2006, LNAI 4223, pp. 1109 – 1112, 2006.

Meta-Level Control approach was developed by Raja and Lesser since 2002 [4, 6, 7] for Multi-Agent System. The research proposed a Messages Meta Manager (MMM) model for ATFM with the combination of the Meta-Level Control approach and reinforcement learning algorithms (Q-leanring and SARSA [5]). With the developed system, the cases of initial heuristic (IH), epsilon adaptative (EA) and performance heuristic (PH) were tested. The results from simulation and analyses show the satisfactory to the research purpose.

2 Proposed Model: Messages Meta Manager (MMM)

The model MMM (Messages Meta Manager) is developed as a meta control layer that a system host receives the messages in a more appropriate sequence, set appointments determined messages and prioritizing others. In such a way, Meta-Control uses a set of parameters to associate each message (good utility of the message and maximum state period for the execution) and generate a series of other parameters: the probability of arrive message in the entrance list with high utility, the utility of the messages to set appointments in the agenda list, the period of execution for the message that is in the beginning of the agenda and reason of flow that measures the flow of the messages that they enter and leave in the Module of Decision and Control - MDC.

Exchanging messages in the distributed system with a manner of communication needs to establish a hierarchy according to aspects of this system. The attributes defined for the system are attached to the message of the Meta-Level Controller (in this case: Meta-Control) that the message encapsulates and sends it. The destination of the message is also processed by a Meta-Controller within the Multi-Agent System, which receives the message and analyzes the enclosed attributes to make the decision in the most appropriate. The manager in Meta-Level Control can decide among three actions that are: to set appointments the message for posterior act of receiving; to transfer the message in the system or still to discard it.

The approach of intelligent agent makes use of some aspects of Meta-Control level as to use the parameters in meta level of the messages in the direction of obtaining a process of taking of efficient decision and that it does not overload the performance usual of the system.

3 Case Study

From the case study, we intend to show the results from a simulation of exchange of messages from the developed system. Each knot of the distributed system presents a layer managed in Meta-Level where the meta-parameters were analyzed for each message. With this consideration, we can see the classification of the process messages in the priority by the importance of messages, classification of messages and disrespect of some messages that do not present great damages.

We consider four variants of agents of during learning process. Generally, SARSA algorithm is with a better performance than the Q-learning. As mentioned in the literatures, the exploitation of the agent in SARSA algorithm uses randomly the distribution of probabilities when taking actions. The decided actions are with higher probability values. At the same time, the Q-learning algorithm are focused on to search the maximize rewards, the performance is worse than it by using SARSA.

The simulation considers four cases with 18 periods (every period means a circle of a message in the system) and 125 messages during these periods. In the first case, learning algorithm uses initial heuristic (IH) as the parameters (see Fig. 1); In the second case, IH and epsilon adaptative (EA); In the third, IH and performance heuristic (PH); In the last case, IH, EA and PH. The MMM flow is an index to mean the efficiency of the operation of MMM. Higher MMM flow, better operation performance.

Figure 1 shows the results from Q-learning algorithm. Comparing the four variations of the performance of the communication among agents, the results from case 1 (IH), case 2 (IH and EA) and case 3 (IH and PH) show better performance. In all of these three cases, the index of MMM flow approximated 1 after the simulation in 18 periods. For the last case, the combination of the IH, EA and PH, the simulation shows worse result.

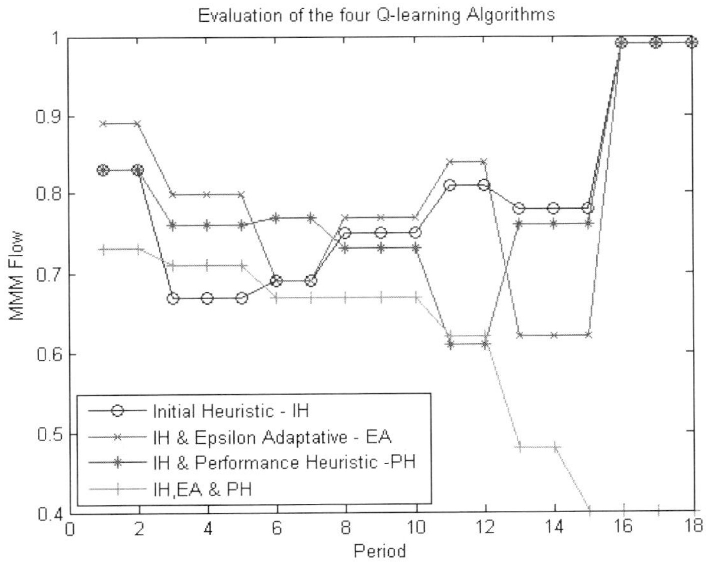

Fig. 1. System performance by Q-learning algorithm

4 Conclusions

This research presented a solution of the application of reinforcement learning in the decision process to better the efficiency for the exchange of messages within distributed systems in Meta-Level Control. The reinforcement learning approach was developed as a module (MRL) with two algorithms: Q-learning and SARSA. The Meta-Level Control was developed as another module (MDC) for making decision to process information.

One of the advantages presented here is the use of reinforcement learning for allowing the agents acquire experience during the process of iterative form the environment. Being a similar form to the learning it uses the performance of the system as criterion to verify the performance of the agents in the environment.

The simulation results from four cases by two reinforcement learning algorithms have shown the correctness and efficiency of the combination of Meta-Level Control and reinforcement learning in a special problem of ATFM. As a stage research report, the simulation is just the part of internal messages process, the part of ATFM is still under the development. In further research, we are also going to apply other reinforcement learning algorithms such as R-learning or Dyna in the system.

References

1. Tanenbaum, A. S.; Steen, M. V. Distributed Systems: Principles and Paradigms. Prentice-Hall, (2002).
2. Weigang, L.; Cardoso, D. A.; Dib, M. V. P.; Melo, A. C. M. A., Method to Balance the Communication among Multi-Agents in Real Time Traffic Synchronization, International Conference on Fuzzy Systems and Knowledge Discovery - FSKD 2005, LNAI, Vol. 3613, pp. 1053-1062, Changsha, (2005).
3. Weigang, L.; Alves, C. J. P.; Omar, N. An Expert System for Air Traffic Flow Management. Journal of Advanced Transportation. Vol. 31, N. 3, (1997), pp. 343-361.
4. Raja, A.; Lesser, V., Meta-Level Reasoning in Deliberative Agents. In Proceedings of the International Conference on Intelligent Agent Technology (IAT 2004), (2004).
5. Sutton, R. S.; Barto, A. G., Reinforcement Learning: An Introduction. The MIT Press, England, (1998).
6. Raja, A.; Lesser, V., Efficient Meta-Level Control in Bounded-Rational. In Proceedings of Autonomous Agents and Multi-Agent System, pp.1104-1105, Melbourne, Australia, (2003).
7. Raja, A.; Lesser, V., Automated Meta-Level Reasoning in Complex Agents. In Proceedings of Eighteenth International Conference on Artificial Intelligence (IJCAI 2003), Workshop on 'Agents and Automated Reasoning", Acapulco Mexico. (2003).
8. Russel, S.; Norvig, P., Artificial Intelligence – A modern Approach. Pearson Education, Inc., Second Edition, New Jersey, (2003).
9. Ribeiro, C. H. C. A Tutorial on Reinforcement Learning Techniques. Division of Computer Science. Departament of Theory of Computation. Technological Institute of Aeronautics. São José dos Campos, Brazil, (2004).
10. Watkins, P. D., Technical note Q-learning. Machine Learning, Vol. 8, 279-292, (1992).
11. Bianchi, R. A. C.; Costa, A. H. R. Uso de Heurísticas para a Aceleração do Aprendizado por Reforço. Artigo apresentado no XXV Congresso da Sociedade Brasileira de Computação. São Leopoldo, RS, Brasil, (2005).

Intrusion Detection Based on Clustering Organizational Co-Evolutionary Classification

Fang Liu and Yun Tian

School of Computer Science and Engineering, Xidian University, Xi'an, 710071, China
f63liu@163.com

Abstract. Organizational Co-Evolutionary Classification (OCEC) is a novel classification algorithm, based on co-evolutionary computation. Differing from Genetic Algorithm, OCEC can work without encoding datasets because introducing "organization" concept. To deal with mass data in intrusion detection effectively, we develop a new algorithm, Clustering Organizational Co-Evolutionary Classification (COCEC) by introducing the clustering method to OCEC. COCEC divides initial data into many sections, and each section is considered as an organization, thus COCEC allows more data to obtain evolutionary learning, so the rule set worked out by COCEC contains fewer rules. In addition to improvement of the initial state in OCEC, some improvements have also been done in the choice strategy of the operators and the rule matching method The experiment results show that COCEC is more accurate and more effective than OCEC and OCEFC (Organizational Co-Evolutionary Fuzzy Classification) with the KDD CUP 99 database, and it greatly reduces the number of rules and testing time.

1 Introduction

Marked by the paper *An Intrusion-Detection Model* [1] delivered by Dorothy E. Denning (1986) as the beginning, the history of intrusion detection technique has developed for more than 20 years. Researchers have applied various pattern classification methods in the field of intrusion detection: methods based on statistics, support vector machine [2], clustering method [3], genetic algorithm based on immune [4] and artificial immune system [5] [6], etc. Owing to the employment of those methods mentioned, promising results have been achieved, but there is still a long way to go to get the satisfactory effects in practice.

In 1995, Wilcox [7] ushered the concept "organization" in economy into GA-based classification and put forward "organizational learning within a learning classifier system". In 2006, Jiao Licheng, Liu Jing, and Zhong Weica [8] put forward the Organization Co-Evolutionary algorithm for Classification (OCEC). OCEC is a recently developed classification method with high accuracy, which is based on Co-Evolutionary computation. In OCEC, they adopted multi-population co-evolutionary scheme, let new evolutionary operators act on organizations comprised of data, and finally drew the rules for classification through evolution step by step. Combining OCEC with fuzzy set theory, Liu Fang and Chen Zhenguo [9] proposed the Organizational Co-Evolutionary Fuzzy Classification (OCEFC), and made the first attempt in the application of OCEC in intrusion detection.

L. Wang et al. (Eds.): FSKD 2006, LNAI 4223, pp. 1113 – 1120, 2006.

However, we find that the classification of mass data by OCEC is not quiet effective through the analysis of OCEC and intrusion detection data set KDDCUP99 [10] in this paper. Therefore, we improve OCEC with the consideration of clustering method, and propose a Clustering Organizational Co-Evolutionary Classification (COCEC). Using a clustering method, COCEC divides initial data into many sections, and each section is considered as an organization which is different from one of the initial stage in OCEC, thus COCEC allows more data to obtain evolutionary learning, so the number of rules extracted by COCEC is fewer than OCEC. In addition to improvement of the initial state in OCEC, some improvements have also been made in the choice strategy of the operators and the rule matching method to satisfy the need of intrusion detection application in this paper. According to the experiments with KDDCUP99 data set and comparison with OCEC and OCEFC, COCEC greatly reduces the number of rules and testing time, and is more accurate and more efficient for the classification.

2 Organizational Co-Evolutionary Algorithms for Classification

The basic idea of OCEC [8] is to utilize different significance of attribute as the leading principle and let the self-defined evolutionary operators act on training data to evolve gradually, then enable us to collect the most similar sample to form organization from which we can draw rules for classification.

In OCEC, the given data are represented in a data table, each row of which is labeled by an object. Each column of the data table is labeled by an attribute. Organizations refer to the set of certain data objects, which are classified into three categories according to the concept of useful attributes.

In the initial stage of OCEC, the objects of every target class are initialized as free organizations to form initial population. Afterwards, two organizations are selected randomly as parent organizations ORG_{p1} and ORG_{p2} for evolutionary operations. But OCEC does not encode like traditional genetic algorithm, new evolutionary operators have to be defined for OCEC.

Add and subtract, exchange and merge operators in OCEC are selected randomly to act on parent organizations. In COCEC, we make a little improvement: when two parent organizations are free, add and subtract operator and exchange operator are useless, therefore those organizations are combined directly.

If there is only one organization left in a population, it would be involved in next generation without any other operations, otherwise, parent organizations would compete with offspring organizations for the chance to go to next generation: If the maximal fitness of the former is bigger than that of the later, the former would go to next generation directly, otherwise the later would. In the organizations, which will go to next generation, if there is any one abnormal, it must be dispersed so as to go to next generation in the form of free organization.

After the evolutionary operation new offspring organizations are produced, the fitness of which have to be calculated for the selection operation. It is pointed out that two factors have to be considered in the fitness of the organizations: the number of objects covered by an organization and the size of useful attribute set [8]. Attribute significance CI_c is introduced in OCEC.

The selection of rules in OCEC takes the way based on the scale of the organization and matching degree. First of all, the rules are ranked the declining sequence according to the scale of the organization. For every sample to test, the first rule with the largest matching degree is chosen to classify in OCEC.

3 Improved OCEC-COCEC

In OCEC [8], each datum is introduced into the population evolution in the form of free organization in the initial state. After the application of OCEC in intrusion detection data set, we find there are some problems in the initial state of OCEC, and the speed of the evolution is quite slow. The reason is that for the three operators, add and subtract operator, exchange operator and merge operator, only the merge operator can reduce the number of organizations. For a data set that contains 10000 objects, although the merge operator is used in every generation, at least 9000 objects are left without being evolved after 1000 generations. Thus, this evolution learning approach to training data is inadequate and incomplete. Although the organizations not involved in the evolution are free ones, finally they need to be extracted into rules, and taken part in the classification of testing data, so the efficiency and accuracy would be decreased.

Given to the reason mentioned above, OCEC has to be improved in order to make it enable to learn large-scale intrusion detection data set thoroughly, so that it would be suitable for intrusion detection. We improved the initial state of OCEC. Data are divided into many sections by using a Fuzzy C-Mean clustering algorithm (FCM) [11]. Then the same attribute sets of every section are found respectively, the corresponding attribute significance and the fitness of every section are defined; finally, the initialization of the organization is completed.

Algorithm 1: The evolution of attribute significance

Presuppose t is the generation of evolution, the organization that is updating the attribute significance is labeled by ORG , c_j is the jth condition attribute in $SAME_{ORG}$, and N is the predefined parameter.

Step1. If $t=0$, then $CI_i \leftarrow 1$, $i=1,2,\cdots,|C|$; $j \leftarrow 1$, $USE_{ORG} \leftarrow \varnothing$;

Step2. If $j > |SAME_{ORG}|$, go to step3, otherwise a population will be randomly selected from other populations, and an organization ORG_1 will also be randomly selected from it. if $c_j \in SAME_{ORG_1}$ and c_j of ORG_1 is different from that of ORG , $USE_{ORG} \leftarrow USE_{ORG} \cup c_j$; otherwise, attribute significance of c_j will be reduced according to formula (1) (*Case* 1), $j \leftarrow j+1$, go to Step2;

Step3. If $USE_{ORG} = \varnothing$, return; otherwise, N objects are randomly selected from other target classes, if the combination of values of USE_{ORG} does not appear in the N objects, each attribute significance in USE_{ORG} will be increased according to formula (1) (*Case* 2); otherwise, $USE_{ORG} \leftarrow \varnothing$, return.

$$CI_c(t+1) = \begin{cases} 0.9CI_c(t) + 0.05 & Case1 \\ 0.9CI_c(t) + 0.2 & Case2 \end{cases} \tag{1}$$

There is some difference between the concept of organization in OCEC and that of organization in COCEC: the organizations of OCEC evolve from free organizations gradually, while those of COCEC come from the evolution of initial data clusters after FCM clustering.

The initial attribute significance, $CI_c(0)$ is updated itself by the algorithm 2, which is different from the definition of attribute significance.

Algorithm 2: Initial clustering division algorithm

Presuppose $|c|$ is the number of data's attribute; c_j is the *j*th attribute of object *O*; initial population *p* is null.

Step1. For initial data set, FCM is employed to divide all sorts of training data into clusters;

Step2. $j \leftarrow 1$, $j < |c|$, attribute c_j of all data in each data cluster is considered as a set, if the set contains only one value of c_j, $SAME_{ORG} \leftarrow SAME_{ORG} \cup c_j$, $j \leftarrow j+1$, go to Step 2;

Step3. For every data cluster, if $SAME_{ORG} = \varnothing$, every object O in it is taken to match $SAME_{ORG}$ of other data clusters one by one. If the matching is successful, O would be added to that data cluster; otherwise, it would be added to SET_{FREE};

Step4. For every data cluster, if $SAME_{ORG} \neq \varnothing$, it would be considered as an organization, and its attribute significance would be defined according to the algorithm1, but the formula of attribute significance would be modified as following:

$$CI_c(0) = \begin{cases} 0.9CI_c(0) + 0.05 & Case1 \\ 0.9CI_c(0) + 0.2 & Case2 \end{cases} \tag{2}$$

The initial value of $CI_c(0)$ is 1, every object in SET_{FREE} would be regarded as a free organization;

Step5. The fitness of every organization is defined as formula (3), and the organization is added to population in order to update the population.

$$fitness_{ORG} = \begin{cases} 0 & ORG \in FREE \\ -1 & ORG \in ABNORMAL \\ |ORG| \prod_{i=1}^{|USE_{ORG}|} CI_i & ORG \in NORMAL \end{cases} \tag{3}$$

Initial data are divided into clusters by the clustering algorithm to learn every object with a larger possibility. In this way, training data could be learned quite thoroughly. Thus, the initial state of OCEC has been improved.

For the low efficiency of the rule matching method in OCEC, a little improvement should be made to satisfy the need of intrusion detection application. In reality, network data often come in some order: majorities of network data are normal data, and the intrusion data always come together in a period of time. The class of current data is mostly the same (or related) with the class of last data (or last several data). The rule matching method should be improved as following: remember the rule R1 which determine the class of the current data D_1, and if the matching degree of the next data D_2 and R1 is not less than that of D_1 and R1, the class of D_2 is same as that of D_1. But in our experiment, the effect of the improvement to the rule matching method is related with the distribution of the test data. The best situation is the data of same class are placed together. In the contrary situation there are not any data the same with the last data and the improvement is not effective.

Algorithm 3: COCEC

Step1. According to algorithm 2 initial training data are divided into the clusters to form the organizations that would be added to population P^0_{Class}, $t \leftarrow 0$, $i \leftarrow 0$;

Step2. If $i > |Class|$, go to Step7;

Step3. If the organizations in P^t_i are more than 1, go to Step4; otherwise, go to Step6;

Step4. Two parent organizations ORG_{p1}, ORG_{p2} are randomly selected from P^t_i, and a random evolutionary operator selected from add and subtract, exchange and merge operators [8] according to the improved choice strategy of operators will act on ORG_{p1}, ORG_{p2}, compute attribute significance and the fitness of child organizations ORG_{c1}, ORG_{c2};

Step5. Selection mechanism [8] will work on ORG_{p1}, ORG_{p2}, and ORG_{c1}, ORG_{c2}, the organizations with higher fitness survive to the next generation;

Step6. The rest organizations in P^t_i to P^{t+1}_i, $i \leftarrow i+1$, go to Step2;

Step7. If stopping criteria are met, go to Step8; otherwise, $t \leftarrow t+1$, $i \leftarrow 1$, go to Step2;

Step8. Organizations in each population are further combined while their evolution has been finished: if the useful attribute set of certain organization is the subset of another one's useful attribute set, these two organizations should be combined together. The useful attribute set of the new organization is the convergent set of those two's useful attribute set.

Step9. Turn the organizations into rules, and let them take the declining sequence according to scale of the organization [8];

Step10. Classify every test object: if the matching degree of the next data D_2 is not less than that of the current data D_1, the class of D_2 is same as that of D_1; otherwise, work out the matching degree of it to the rules, select the first rule with maximal matching degree to perform classification.

The algorithm above is the improved OCEC-Clustering Organization Co-Evolutionary Classification.

4 Experiments and Result Analysis

In order to examine the effectiveness of COCEC, data set KDDCUP99 [10] is employed to do our experiments, and to compare with other algorithms. Three sets of data are selected randomly from KDDCUP99, and all test data are placed randomly in our experiments.

Firstly, two data sets are selected randomly from 10% KDDCUP99, 22 attack types contained in them. The first data set Data 1 contains 14991 data, 8153 data of them are selected randomly as training data, 6838 as test data. Training data contains 5000 normal data, and 3153 attacks (belong to 12 attack types), test data contains 2000 normal data, and 4838 attacks, in which contains 2838 known ones (12 types already exist in training data), and 2000 unknown (10 types don't exist in training data). The second data set Data 2 contains 147605 data, 50000 of them are selected randomly as training data, 97605 as test data. Training data contains 30000 normal data, and 20000 attacks (belong to 12 attack types), test data contains 39588 normal data, and 58017 attacks, in which contains 23113 known ones (12 types already exist in training data), and 34904 unknown (10 types don't exist in training data).

In order to examine the ability of COCEC in learning massive data, a third data set is selected randomly from original KDDCUP99 including 37 attack types. The third data set Data 3 contains 998673 data; 500000 of them are selected randomly as training data, 498673 as test data. Training data contains 300000 normal data, and 200000 attacks (belong to 20 attack types), test data contains 299756 normal data, and 198917 attacks, in which contains 106546 known ones (20 types already exist in training data), and 92371 unknown (17 types don't exist in training data).

The experiment environment is as follows: CPU P4 3.0G, memory 1G, language C. The number of evolutionary generation is 1000 and the number of initial clusters is 1/10 the number of training data.

To evaluate the IDS, there are two major indications of performance: the Detection Rate (DR) is defined as the number of intrusion samples detected by the system divided by the total number of intrusion samples presented in the test set, the False Positive Rate (FPR) is defined as the total number of normal samples that were incorrectly classified as intrusions divided by the total number of normal samples.

The results in Table1, Table2 and Table3 are the means after 10 individual experiments. What has to be mentioned is that the number of rules in the experiments is obtained after the mergence of the organizations. OCEC' is the algorithm OCEC with improved the rule matching method mentioned above.

Table 1. The results of Data1

Algorithm	DR of known(%)	DR of unknown(%)	FPR (%)	Training time (s)	Test time (s)	Number of rules
OCEC	**98.03**	75.50	10.75	**148.623**	2386.334	3090.4
OCEFC	97.85	76.00	9.30	151.454	2509.487	3111.0
OCEC'	97.00	74.95	10.95	149.965	1992.257	3090.6
COCEC	97.32	**79.25**	**3.05**	206.197	**864.954**	**1068.3**

Table 2. The results of Data2

Algorithm	DR of known(%)	DR of unknown(%)	FPR (%)	Training time(s)	Test time (s)	Number of rules
OCEC	97.26	63.78	2.01	301.445	10068.942	7575.2
OCEFC	97.93	64.94	**1.67**	**289.711**	9994.358	7490.6
OCEC'	96.00	65.05	2.07	302.655	8596.951	7571.0
COCEC	**98.98**	**72.19**	1.69	1815.148	**3232.548**	**2250.1**

Table 3. The results of Data3

Algorithm	DR of known(%)	DR of unknown(%)	FPR (%)	Training time(s)	Test time (s)	Number of rules
OCEC	95.38	60.94	**3.42**	**618.421**	153648.684	16281.5
OCEFC	94.62	59.46	3.91	632.997	160025.852	16205.7
OCEC'	95.09	60.49	3.46	618.421	124586.389	16285.3
COCEC	**97.55**	**65.32**	4.32	3568.489	**48817.225**	**4952.0**

At first, the test time of OCEC' is less than that of OCEC, which shows that the improvement to the rule matching method is effective. And the change to the rule matching method will be more effective in reality. From the three tables, we can also find that COCEC performed well in accuracy of classification either for known normal data and attacks in training data, or for unknown attacks. Especially COCEC has more superior detection rate of unknown data, which indicates that it has better generalization than other three algorithms. The reason for the good results is that COCEC can produce fewer organizations after clustering. Therefore, training data can be learned thoroughly to some extent.

In the aspect of training time, the adoption of clustering algorithm by COCEC leads to fewer data contained in every organization, and it may take longer time for every step of evolution. Although the training time of COCEC is a little bit longer than the other two, it's acceptable, because when a classifier is well trained, it wouldn't cost any more training time in practice use. While the shorter the test time the better the classification. When attacks occur, the consequence would be terrifying if the classifier cannot work in time. The test time of the other two algorithms is much longer than that of COCEC. The reason for the difference is that organizations in COCEC contain more data, and the rules for classification, which are extracted after evolution, are less than 1/3 of other two algorithms' rules. Those advantages accelerate classification and improve efficiency to a large extent. The experiment results (especially for Data3 with large quantities of data) show that the sacrifice of training time is worthwhile.

5 Conclusions

To fit for thorough learning of large-scale intrusion detection data set, COCEC divides initial data into many sections, and each section is considered as an organization which is different from one of the initial stage in OCEC, it improves the initial state of OCEC, and allows more data to obtain evolutionary learning, so the rule set worked out by COCEC contains fewer rules. And the rule matching method is changed to make COCEC more effective for Large-scale data in intrusion detection. The experiment results show that DR and FPR of COCEC are satisfying, and especially the DR of unknown data is quite high with KDDCUP99, so COCEC is better generalization ability. At the expense of longer training time, COCEC greatly reduces the number of rules and testing time, and is more accurate and more efficient for the classification.

Acknowledgement. This work is supported by the National Natural Science Foundation of China under Grant No. 60372045 and No. 60133010, the Defence Pre-Research Project of China under Grant No.51406020104DZ0124, the Key Science-Technology Project of Higher Education of China under Grant No. 0202A022 and the National Research Foundation for the Doctoral Program of Higher Education of China No. 20030701013.

References

1. Dorothy E. Denning: An Intrusion-Detection Model. Proceedings of the 1986 IEEE Symposium on Security and Privacy. (1986)
2. Mill, J., Inoue, A.: Support vector classifiers and network intrusion detection. Fuzzy Systems, 2004. Proceedings. 2004 IEEE International Conference. 1 (2004) 407–410
3. Shah, H., Undercoffer, J., Joshi A.: Fuzzy clustering for intrusion detection. Fuzzy Systems, 2003. FUZZ '03. The 12th IEEE International Conference. 2 (2003) 1274–1278
4. Jiao Licheng, Wang Lei: A novel genetic algorithm based on immune, IEEE Trans. on System, Man, and Cybernetics—Part A, 30 (2000) 552–561
5. Liu Fang, Qu Bo, Chen Rongsheng: Intrusion Detection Based on Immune Clonal Selection Algorithms. The Proceedings of 17th Australian Joint Conference on Artificial Intelligence, Lecture Notes in Computer Science, Published by Springer-Verlag Heidelberg, V3339, Cairns, Australia. (2004) 1226–1232
6. Liu Fang, Lin Leping: Unsupervised Anomaly Detection Based On An Evolutionary Artificial Immune Network, Lecture Notes in Computer Science, Published by Springer-Verlag Berlin Heidelberg, 3449 (2005) 166–174
7. Wilcox J R. Organizational learning within a learning classifier system. IlliGAL Report No.95003. (1995)
8. Jiao Licheng, Liu Jing, Zhong Weica: An organizational coevolutionary algorithm for classification. IEEE Trans. Evol. Comput. 10 (2006) 67–80
9. Liu Fang, Chen Zhen-Guo: Intrusion Detection Based on Organizational CoEvolutionary Fuzzy Classifiers. The Proceedings of International Conference on Intelligent Information Processing, IIP2004. (2004)
10. KDD CUP99 dataset: http://kdd.ics.uci.edu/databases/kdd cup99/kdd cup99.html. (1999)
11. Bian Zhao Qi: Patter recognization (in Chinese). Tsinghua University press.(2000)

Clustering Based Stocks Recognition

Yaoyuan Shi[1,2] and Zhongke Shi[1]

[1] The Northwestern Polytechnical University, Xi'an 710068, China
[2] Xidian University, Xi'an, 710071, China
yyshi@xidian.edu.cn

Abstract. A new stocks analysis method based on clustering is presented in this paper, in which, six-dimension feature space is constructed according to the data structure of stock chief-index, and the constructed feature space is analyzed with a new fuzzy kern clustering algorithm. We use the Shanghai and Shenzhen's stock index since 1997 to test our presented method. The results show that the method could intelligently recognizes some rules of essence trends of the stock markets and forecasts essence direction of the stock markets not only in short-term but also in long-term.

1 Introduction

Stock technical analysis is a process that estimates the changing trend of the whole stock market or single stock price and discusses the possible paths of investing behavior in stock markets by analyzing the basic market data (e.g. open price, close price, high price, low price, volume of trade and so on). According to the theory of "All included in market behavior", the changing trend of the market can be grasped after mathematic modeling and further analysis of the essential data. Although there are so many conventional technical analyses, none of them is universal for markets. The main reason is that the mathematical tools cannot excavate the essential rules of the changes in markets, and this is also the reason why the technical analysis was suspected. The outlet of technical analysis is to excavate the essence of stock market by combining the distillate of conventional technical analysis with the modern signal processing technology.

After 1960s, data based enginery learning theory was developed [1], and was swiftly used in excavating stock data, mainly time series forecasting from the aspect of function fitted. As a kind of nonlinear system with an ability of self-study, Neural Network can approach to any nonlinear function at any precision defined in compact set theoretically, therefore, this theory represents remarkable superiority in time series forecasting. Stock market is a system with limited sample data (historical data) and so much noise, and people always get into an unsolvable hobble of over-learning or ill-learning when analyzing it because of the shortage of statistic sense of Neural Network. In paper [2] and [3], data are filtered with wavelet decomposition before fitted. Essentially, these can just smooth data because the probability density function of the noise of stock market is unknown. In 1989, Schekman and Lebaron discovered the phenomenon of chaos in day-income and week-income series of stocks [4], which testified the scientific property of time series forecasting of the stock data from

L. Wang et al. (Eds.): FSKD 2006, LNAI 4223, pp. 1121 – 1129, 2006.

dynamic aspect and also presented the short-term effect of time series instead of long-term effect.

In paper [5], [6], [7], [8], [9] and [10], different methods are proposed for the analysis of the stock market respectively and got satisfactory results. Among them, clustering technique is one of the efficient methods for stock investment analysis and there is great potential in the study of stock investment with it. However, there exist limits to some degree by using these clustering algorithms to analyze the stock market, and the cluster space constructed can not reflect the characteristic of stocks completely. In this paper, we describe the idea of rules mining of stock market using the statistical pattern recognition. With the study of the importance of techno-index in technical analysis, we present a method of constructing a feature space using the techno-index. And based on the study of space structure of stock data, we present an algorithm of fuzzy kern clustering. On the basis of the Dow theory [10], we analyzed the essential trend of stock markets of Shanghai and Shenzhen since 1997 by clustering algorithm and made the experiments in short-term, medium-term and long-term. The results show that the proposed method has an ability of recognizing the stocks in real time or for long-term forecast.

2 Stocks Recognition Based on Clustering

The details of stock analysis under the framework of pattern recognition are as follows:

(1) Information acquisition. It includes measuring, sampling, and quantifying investigated subjects and expressing them by signs that can be operated by computer. All basic data of stock market (open price, close price, high price, low price, volume of trade) are time series data, which are easy to be processed by computer in real time.

(2) Filtering the noise. As a huge complicated system, the development of the stock market is affected by many factors, such as economic function, profitability of enterprises, political environment, and the behavior of "technical bargainer". The distribution of signals and noises in the stock market can not be grasped clearly by far, so it is necessary to preprocess the data of the stock market with smooth technology first. There are three kinds of smooth techniques for stock technical analysis: average smooth technique, exponential smooth technique and wavelet decomposition smooth technique. The Smooth data obtained by average smooth technique have time lag to some extent, while the exponential smooth technique can remove time lag, both of which are classical smooth techniques. Wavelet decomposition smooth technique is a new one developed in the latest 20 years, which obtains smooth data without time lag by using low-frequency signals restructuring of wavelet decomposition. Among these, the exponential smooth technique without time lag is adopted popularly.

The formula of exponential smooth is as follows:

$$\begin{aligned} EXPMA_1 &= C_1 \\ EXPMA_i &= \frac{n-1}{n+1} EXPMA_{i-1} + \frac{2}{n+1} C_i \end{aligned} \tag{1}$$

C_i denotes the close price in the ith day, n is the exponential smooth factor, and n=30 in this paper.

(3) Feature extraction and selection. It is the basis for classification and identification of the data. In this stage, we obtain the features that can reflect the essential of the pattern family by the transformation of the original data, and then construct the feature space. Technical analyses are of two major groups: exponential analysis and graphic analysis. Exponential analysis is based on the techno-index that was built on mathematic transform of the basic data of the stock market. Graphic analysis is a technique by charting some evolution of the basic data of the stock market, and analyzing the charts by experiences. From the viewpoint of signal processing, graphic analysis can be converted to exponential analysis by signifying graphic analysis. Hence, techno-index is the research focus of experts in the technical analysis of the stock market for more than one century. Many techno-indexes have been proposed at present. Although none of these indexes is universal, each of them can reflect some characteristics of the market which are not easy to be perceived to a certain extent in a given period. It follows that techno-index is the essential for the technical analysis of the stock market, and also the quality technique for the extraction of characteristic of the stock market. Generally, it is necessary to compress and extend techno-indexes in order to construct a feature space which is efficient for clustering.

This paper deals with chief-index, and constructs a kind of feature space with six dimensions by combining six techno-indexes [11] of chief-index after proper compression and extension (DMKI, A/D, DEA, MAOSC, MFI, and VRSI) according to the principle of reflecting most possibilities of market features and clustering easily.

(4) Classification decision, which is the classification of identified subjects with statistic techniques in feature space according to the designed classifier. Here, it can be divided into supervised classification and unsupervised one according to if there are training samples for classifier. Because of the continuous evolution of the stock market and there are no fixed patterns and training data, only unsupervised classification technique can be used in stock analysis. Unsupervised technique can be popularly includes two major kinds: direct clustering technique and indirect clustering technique. The former is based on the estimation of probability density function and the latter is based on the measurement of samples similarity. For the stock market, a huge complicated system, estimating the probability density function is very difficult, so indirect clustering technique is proper for stocks recognition. There are many uncertainties in stock analysis with clustering technique because there is no training set of known classes, or even the number of families. The data can be classified according to sample similarity measuring only, for example, the distance information. It is necessary to take three problems into account for effective recognition of the stock market by clustering: the first one is the construction of feature space including the selection of techno-index and the compression and extension of space scale; the second one is the design of clustering algorithm including the distance measurement, the selection of clustering rules and the construction of kern function; the third one is to make full use of professional knowledge in stock analysis to compensate the shortage of information.

The stock market is a complicated system. There exists undefinite in the data of stock market. Fussy system is an efficient technique for its analysis. Fuzzy c-means

(FCM) is a typical algorithm of the fuzzy system in the application of pattern recognition. However, it can not used in the analysis of the stock market directly. First, FCM is suitable for the analysis of data with globularity or like-globularity structure. Second, the number of clusters should be preestablished for FCM, while the number of classes of stock market data is unknown generally.

By studying the constructed feature space of stock data, we presented a fuzzy kern clustering algorithm based on similarity measurement here. In the method, the principal axis kernel function is constructed by K-L transformation, which is served as the kern of FCM. Therefore, the method is suitable for the stock market data wich non-globularity structure. Furthermore, the similarity measurements of samples and kern, the nearest distance, farthest distance, and average distance are used for the preprocessing of the clustering, which supplies the number of clusters objectively and supplies the initial value of the principal axis kernel function. We call the improved method the fuzzy kern clustering. The process of this algorithm is as follows:

Step 1. Clustering the samples y_l, l=1,…, n by three kinds of similarity measurements respectively: nearest distance, farthest distance, and average distance, and get the number of clusters C_i where $C_i \in [2,4]$. Obtain the kern function K_{ij} i=1, 2, 3, j=1,…, C_i by K-L transformation.

Step 2. Fuzzy Clustering.

For i=1, 2, 3

a. C_i seaves as the initial partition, and K_{ij} serves as the initial kern.

b. According to the following rule to implement the clustering.

$$u_j(y_l) = \frac{\left(\Delta\left(y_l, K_{ij}\right)\right)^{-1/(b-1)}}{\sum\left(\Delta\left(y_l, K_{ij}\right)\right)^{-1/(b-1)}}, \qquad b > 1 \tag{2}$$

If $u_j(y_l) = \max(u_j(y_l))$, then $y_l \in \Gamma_{ij}$, namely, samples y_l is classified to the corresponding cluster Γ_{ij}.

c. Modify kern function K_{ij} by K-L transformation and calculate the clustering loss function J_i

$$J_i = \sum_j \sum_i \left(u_j(y_i)\right)^b \Delta\left(y_l, K_{ij}\right) \tag{3}$$

If the change of J_i is less than a threshold value θ, go on the next calculation, else, turn to b.

End

Step 3. Determine the best clustering.

If J_i=min (J_i), Γ_{ij} is the best clustering.

In this algorithm, Δ is the Euclidean distance. Here, the purpose of clustering is to identify the evolvement form of the essence trend that the Dow Theory indicated, so the number of clusters is limited between 2 and 4 generally.

3 Experimental Results and Analysis

Dow Theory indicated that markets form three kinds of trends in the whole evolution process: essence trend, secondary trend, and temporary trend, which is just same as the movement of sea: tide, wave and ripple. Essence trend, reflecting the primary condition of the market, is the most important one and can be divided into essence ascent trend and essence descent trend. In general, essence ascent trend shows three steps by turns: the building of storehouse (accumulation), stable up-trend, or market crest. On the premise that the essence data of the stock market contains all the market information, clustering can be used to analyze stocks market if the essence trend of the stock market is discriminable in the statistic sense.

3.1 Preprocessing

The basic data of 1894 trade days (from Jan. 2, 1997 to Nov. 25, 2004) of Shanghai stock index (000001) and Shenzhen stock index (399001) [12] were preprocessed by exponential smoothing. The curves of daily close points are shown in Fig.1 and Fig.2. The real lines are curves of day-trade close points, and the dashed lines are the exponential smoothing curves of daily close points for 30 days. The horizontal axis denotes days of trade (unit: day), and the vertical ordinate denotes price (unit: yuan).

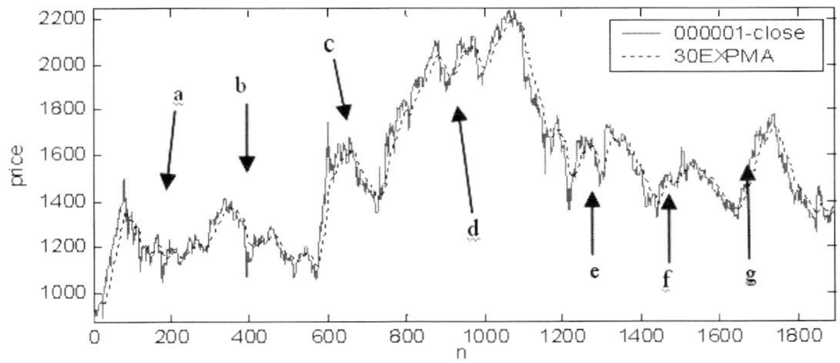

Fig. 1. Graph of daily close points of Shanghai stock index

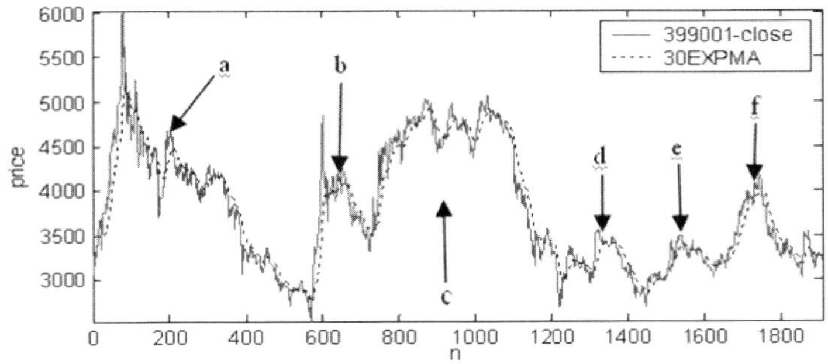

Fig. 2. Graph of daily close points of Shenzhen stock index

It can be shown in Fig.1 and Fig.2 that Shanghai stock index and Shenzhen stock index reflected the basic fluctuation of the stock market without time lag after thirty days' exponential smoothing. For Shanghai stock index, the fluctuation of 300 points between wave crest and wave hollow is regarded as an essence trend. We can see from Fig.1 that Shanghai stock index has formed seven times essence trends since 1997 (represented with a-g in Fig.1). Among these, the fourth one is the biggest one with triple crest structure, and the fifth one has dual crest structure. For Shenzhen stock index, the fluctuation of 600 points between wave crest and wave hollow is regarded as an essence trend. We can see from Fig.2 that Shenzhen stock index has formed six times essence trends since 1997 (represented with a-f in Fig.2). The third one is the biggest one with triple crest structure, and the fourth one has dual crest structure. Comparing the essential trends of Shanghai and Shenzhen stock indexes, we can find that the formation of each essence trend is almost synchronous, which shows that the two markets are of strong relevance and suitable for combined analysis.

3.2 Stock Markets Recognition According to Dow Theory

A test of stock markets recognition was made with the data of former 1450 trade days in Fig.1 and Fig.2. The prior five essential trends of Shanghai stock index are clustered separately in feature space with fuzzy kern clustering algorithm. The results are shown in Fig.3. The clustering results of the former four essential trends of Shenzhen stock index in feature space are shown in Fig.4. (All the classification results of nine essential trends are of three clusters, which were separately represented by '.' 'o' '+' in Figures).

The evolutionary waveforms of nine essence trends of stock markets in Shanghai and Shenzhen are different with each other by referring to Fig. 3 and Fig.4. However, they display a common characteristic by using the clustering technology: there are continuous 'o' signals during the period of steady rise of markets, continuous '.' signals at the crest of markets, and '+' signals will appear swiftly when markets fall

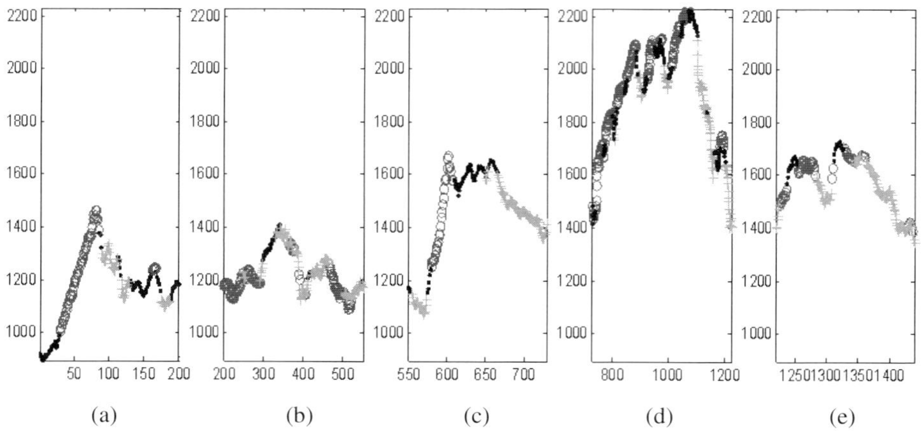

Fig. 3. Results of clustering analysis of the former five essential trends of Shanghai stock market

sharply. This phenomenon sufficiently indicates that there are rules in stock market of China, which can be mined and applied with. It also showed that '.' signals may appear at the beginning of the rise period ((a) and (c) in Fig.3, (a) and (b) in Fig.4), something that is considered as the right expression of the changeability of stock rules and the influence of noise. According to clustering results of Fig.3 and 4, 'o' at the rise phase of markets can be defined as the signal of stable rise of markets, '.' the signal of the crest of markets, and '+' the signal of fall of markets (signals '+' appeared before crest are non-effective). We analyze the later market with data of those 9 essence trends.

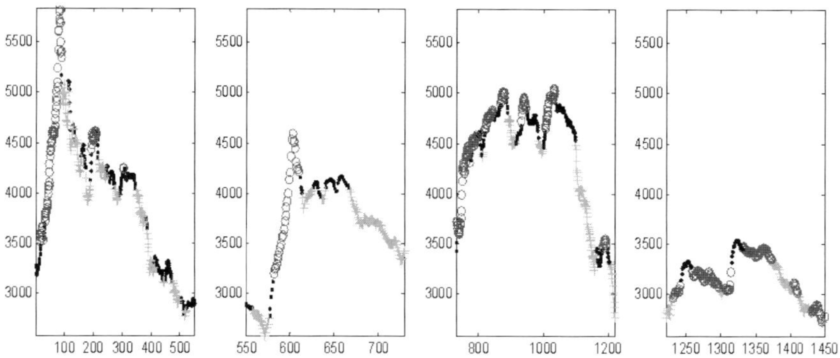

(a) (b) (c) (d)

Fig. 4. Clustering analyses results of the former four essential trends of Shenzhen stock market

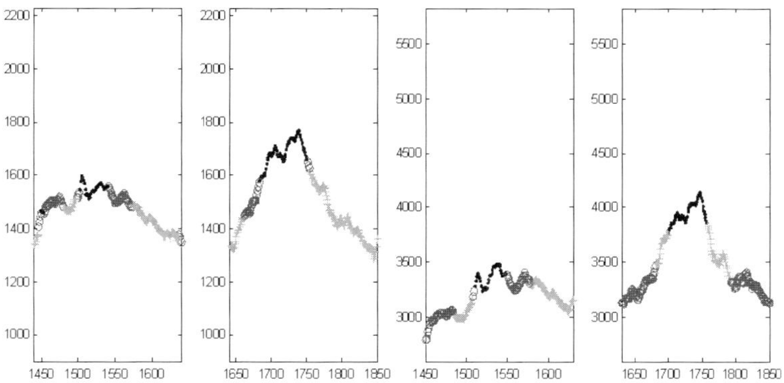

(a) (b) (c) (d)

Fig. 5. The recognition results of later market from Shanghai and Shenzhen stock markets

3.3 The Recognition of Later Market

We select samples from the 9 essence trends of former 1450 trade days in Shanghai and Shenzhen stock markets since January 1st, 1997, and classify the 4 basic trends of later Shanghai and Shenzhen market (there is monitored pattern recognition of sample data), Fig.5 shows the results, in which, (a) and (b) are the recognition results of the sixth and seventh essence trends of Shanghai market, (c) and (d) are the recognition results of the fifth and sixth essence trends of Shenzhen market.

It is shown from Fig.5 that the signal of steady rise of market, 'o', can indicate its stable rise status before the pinnacle of markets, the signal of the market crest, '.', can correctly indicate the market crests of four essential trends, and the signal of fall of market, '+', can send out signals duly after the crest of market. That means that we can recognize the future evolutionary form and forecast the developing direction of markets based on the statistic rules of the prevenient essential trends of markets. It must be pointed that the recognition method based on clustering is suitable for the forecast not only in short-term and median-term, but also in long-term.

4 Conclusions and Discussion

A method for analyzing the rules of stock markets with pattern recognition technique is presented in this paper, in which, a new feature space is designed and a new clustering algorithm is used. From the experimental analysis of Shanghai and Shenzhen's stock markets, we can get the following results.

(1) There are statistic rules in essential trends of the stock market, and implied rules can be recognized to a certain degree by constructing proper feature space and clustering algorithm.

(2) Recognition of short-term, medium-term, or even long-term essential trends evolution of stock markets can be made with pattern recognition techniques in statistic sense.

The method proposed in this paper only offered a few commonplace remarks by way of introduction, and the purpose is to show the feasibility of the idea in excavating the rules of stock markets. The construction of feature space and the design of clustering algorithm are not limited to these, for example, support vector machine is also an effective method for solving the problem of data pattern recognition of stock markets. The key point is how to apply with these advanced learning machines to find the inner rules of stock markets, which is just the developing orientation of the technical analysis of stock markets.

References

1. Cherkassky V., Mulier F.: Learning From Data: Concepts, Theory and Methods. New York: John Wiley & Sons, (1997)
2. Yang Yi-wen: Multivariable Time Series Prediction Based on Neural Networks and Its Application in Stock Market. Information and Control, 30 (2001) 413–417
3. Yang Yi-wen, Liu Gui-zhong, Zhang Zong-ping: Stock Market trend Prediction Based on Neural Networks, Multiresolution Analysis and Dynamical Reconstruction. In: Proceedings of IEEE/IAFE Conference on Computational Intelligence for Financial Engineering, (2000) 155–157.

4. Schekmen J.A., Lebaron B.: Nonlinear Dynamics and Stock Returns. Journal of Business, 62 (1989) 311–317
5. Deng Xiuqin: Application of Cluster Analysis in Stock Market Board Analysis. Application of Statistics and Management, 18 (1999) 1–4 (In Chinese)
6. Chen Zuo, Xie Chi, Chen Hui: Mining Temporal Patterns of Stock Yield Sequences Based on Wave Cluster Method. Systemes Engineering, 23 (2005) 102–107 (In Chinese)
7. Wang, Y.F.: On-demand Forecasting of Stock Prices using a Real-Time Predictor. IEEE Transactions on Knowledge and Data Engineering, 15 (2003): 1033–1037
8. Kim, K.: Financial Time Series Forecasting using Support Vector Machines. Neurocomputing. 55 (2003) 307–319
9. Chung F.L., Fu T.C., Luk, R., Ng, V.: Evolutionary Time Series Segmentation for Stock Data Mining. In: Proceedings of 2002 IEEE International Conference. (2002) 83–90
10. Povinelli, R.J.: Identifying Temporal Patterns for Characterization and Prediction of Financial Time Series Events.Temporal, Spatialand Spatio-Temporal Data Mining. (2000) 46–61
11. Jonnson M.A.: The Random Walk and Beyond: An Inside Guide to the Stock Market. New York: John Willy & Sons, (1988)
12. Yang J.: Handbook of Stock Market Technological Analysis. China Astronautics Publishing House, (2002)
13. http://www.stockstar.com

Stock Time Series Categorization and Clustering Via SB-Tree Optimization

Tak-chung Fu[1,2,*], Chi-wai Law[1], Kin-kee Chan[1], Fu-lai Chung[1], and Chak-man Ng[2]

[1] Department of Computing, The Hong Kong Polytechnic University, Hong Kong
{cstcfu, c1516689, c1922434, cskchung}@comp.polyu.edu.hk
[2] Department of Computing and Information Management, Hong Kong Institute of Vocational Education (Chai Wan), Hong Kong
cmng@vtc.edu.hk

Abstract. SB-Tree is a data structure proposed to represent time series according to the importance of the data points. Its advantages over traditional time series representation approaches include: representing time series directly in time domain (shape preservation), retrieving time series data according to the importance of the data points and facilitating multi-resolution time series retrieval. Based on these benefits, one may find this representation particularly attractive in financial time series domain and the corresponding data mining tasks, i.e. categorization and clustering. In this paper, an investigation on the size of the SB-Tree is reported. Two SB-Tree optimization approaches are proposed to reduce the size of the SB-Tree while the overall shape of the time series can be preserved. As demonstrated by various experiments, the proposed approach is suitable for different categorization and clustering applications.

1 Introduction

A time series is a collection of observations made chronologically. Time series data can be easily obtained from scientific and financial applications, e.g., daily temperatures, daily sale totals, and prices of mutual funds and stocks. The nature of time series data include: large in data size, high dimensionality and update continuously. Indeed, a large set of time series data is from the stock market. Stock time series has its own characteristics over other time series data like ECG. For example, a stock time series is typically characterized by a few critical points and multi-resolution consideration is always necessary for long-term and short-term analyses. In addition, technical analysis is usually used to identify patterns of market behavior, which have high probability to repeat themselves. These patterns are similar in the overall shape but with different amplitudes and/or durations. Moreover, these patterns are characterized by a few data points. Based on such characteristics, a representation of the time series data is needed for manipulating the stock time series effectively and efficiently. Our previous work [1] is proposed to deal with this problem.

---

* Corresponding author.

L. Wang et al. (Eds.): FSKD 2006, LNAI 4223, pp. 1130 – 1139, 2006.

The state-of-the-art time series representation schemes are mainly based on different dimensionality reduction techniques, including Principal Component Analysis (PCA), Discrete Wavelet Transform (DWT), Piecewise Linear Representation (PLR), Piecewise Aggregate Approximation (PAA) and Piecewise Constant Approximation (PCA). Based on these time series representation approaches, various classification and clustering applications have been proposed, e.g., applying PCA [2] and PLR [3,4] to classification and applying PCA [5], DWT [6] and a bit-level representation [7] to clustering. Furthermore, moving average has been proposed for clustering task in [8,9]. Reference [9] compares the clustering result based on different representation schemes, including DFT, DWT, PCA and the proposed ARIMA approach.

In this paper, the time series categorization and clustering tasks which take advantages of our previously proposed time series representation scheme, i.e. specialized binary tree (SB-Tree) [1], are described. SB-Tree is particularly effective in stock time series data. Based on optimizing the size of the SB-Tree, both stock time series categorization and clustering can be facilitated. The paper is organized into five sections. A brief review on SB-Tree, which is based on reordering the time series data points according to their importance, is given in section 2. Two approaches for optimizing the size of SB-Tree are proposed in this section. Section 3 introduces the categorization and clustering processes based on the optimized SB-Trees. The simulation results are reported in section 4 and the final section concludes the paper.

2 A Specialized Binary Tree Representation and Its Optimization

In this section, the SB-Tree structure for financial time series representation is briefly revisited. It is based on determining the data point importance in the time series. Then, the optimization approaches for this time series representation are proposed.

2.1 Specialized Binary Tree Data Structure

In view of the importance of extreme points in stock time series, the identification of perceptually important points (PIP) is firstly introduced in [10]. The frequently used stock patterns are typically characterized by a few critical points. These points are perceptually important in the human identification process and should be considered as more important points. The proposed scheme follows this idea by reordering the sequence *P* based on PIP identification, where the data point identified in an earlier stage is considered as being more important than those points identified afterwards. The distance measurement for evaluating the importance is the vertical distance (VD) [10].

After introducing the concept of data point importance, a binary tree (B-tree) structure has been proposed to store the time series data and is called specialized binary tree (SB-Tree) [1]. To create a SB-Tree, the PIP identification process [10] is adopted. A sample time series and the corresponding SB-Tree built are shown in Fig.1. The arc of the tree represents the VD of the corresponding node (PIP). Detail creating and accessing process of the SB-Tree can be found in [1].

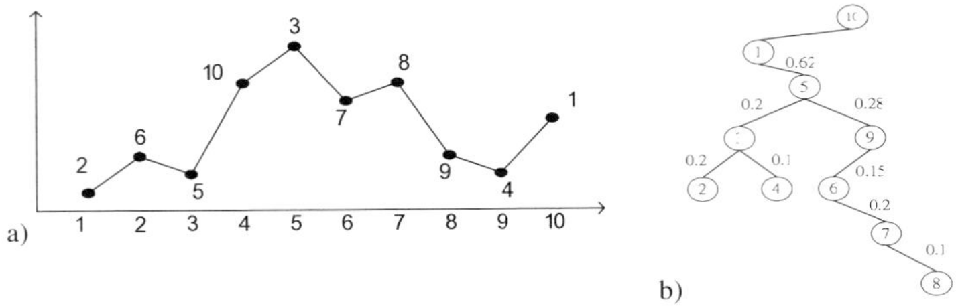

Fig. 1. (a) Sample time series and (b) the SB-Tree built

2.2 SB-Tree Optimization

After transforming the time series data into a SB-Tree, it is possible to further reduce the size of the tree so as to minimize the space consumption of a SB-Tree. It can be done by determining the number of PIP necessarily to represent the time series while the shape of the time series can still be preserved. If only a few more important PIPs are used to represent the whole time series, the error will be very large and the overall shape may also be deformed. Conversely, if all the PIPs are manipulated, the system performance will be very low.

The simplest way to reduce the size of the SB-Tree is applying a lossless pruning approach which only prunes the nodes with distance measured (VD) equal to 0. This kind of nodes has no effect on the shape of the time series because they are the data points located on the straight line formed by other PIPs only.

On the other hand, an acceptable level of error can be specified for which a large number of PIP can be filtered. Thus, a lossy approach is preferred for optimizing the SB-Tree to prune the "unnecessary" nodes of the SB-Tree. Error is defined as the mean square distance between the original time series and the series formed by n PIPs to represent the time series. In other words, the error is calculated by the linear interpolation between retrained points (i.e. PIPs from the optimized SB-Tree) and the original time series. Fig.2 shows the error when only 3 PIPs are used.

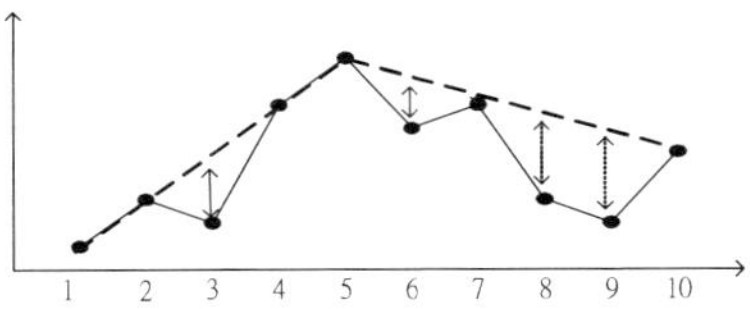

Fig. 2. Error of representing a time series with 3 PIPs compared to the original time series

Two optimization approaches are proposed below:

Tree Pruning Approach: Unimportant signals (i.e. data points) of a time series can be filtered according to a threshold λ. As the tree is accessed from the top and the VD of each node is considered. When the VD of a node is smaller than λ, the fluctuation does not vary a lot and the descendants are considered as less important to the users. Thus, this node and all its descendants should be pruned. Fig.3 shows the pruning of a sample SB-Tree using a threshold equal to 0.15 (i.e. λ =0.15).

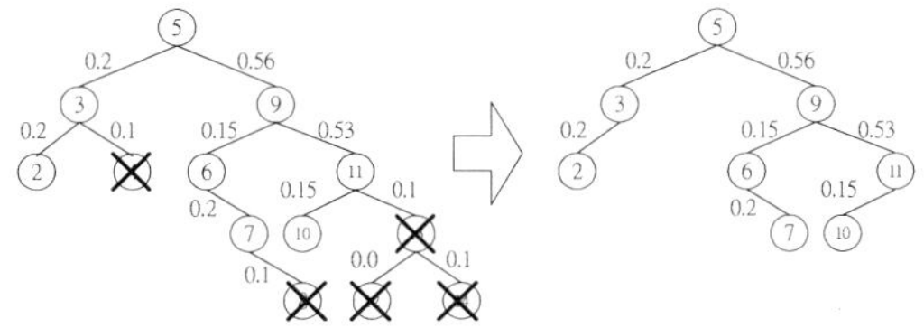

Fig. 3. Example of tree pruning approach using λ =0.15

However, it is difficult for the users to define the threshold λ. In addition, for different time series, different λ may be needed to preserve the general shape of the series. Therefore, an automatic approach for determining this threshold is necessary. It can be achieved by finding the natural gap of the change of VD along with the retrieval of PIPs from the SB-Tree. An example will be shown in section 4.

Error Threshold Approach: The second optimization approach is based on determining the error of the representation compared to the original time series. The PIPs are retrieved from the SB-Tree for representing the time series until the error is lower than a given threshold, α (cf. Fig.2). Again, it is necessary to determine the value of α. A reasonable α value is the point that has no significant decrease in the error. It can be determined by finding the largest decrease in error when adding one more PIP for representing the time series. By including such a PIP, one may come up with an optimized number of PIP for representing the time series as the decrease of error will be at a much lower level compared with the previous stages. Again, an example will be given in section 4.

3 SB-Tree Categorization and Clustering

After representing a set of time series by SB-Trees and determining the optimized tree sizes, it is possible to manipulate the set of optimized SB-Trees for different tasks. In this section, categorization and clustering of the stock time series based on SB-Tree representation are described.

3.1 Categorization

Class generation and time series pattern classification are the two main steps of the categorization process. First, the class generation process is introduced. The generated classes will be used to categorize/index similar time series (or subsequences). Each class is constructed by a class pattern. Two parameters are required to determine the class patterns. The first parameter is the range of number of PIP for building the class patterns, pip_{min} to pip_{max}. In our targeting domain, i.e. stock time series, the common technical patterns are always constructed by 4 to 7 data points. The second parameter is the number of point level, θ. After normalizing the time series, the distribution space (y-axis of the data point) will be equally divided by θ between 0 and 1.

The class patterns are generated by all the combinations of different data point values, such as {0.0, 0.0, 0.0, 0.0}, {0.0, 0.0, 0.0, 0.3}, {0.0, 0.0, 0.0, 0.6} and so on. The generation process carries out for each possible number of PIP, i.e. from pip_{min} to pip_{max}. The total number of classes that can be generated will be equal to $\sum_{n=pip_{min}}^{pip_{max}} \theta^n$.

Fig.4 shows some examples of class pattern using θ=3 for pip_{min}=5 and pip_{max}=7. As the patterns have been normalized before categorization, some class patterns are not necessary such as those class patterns with constant data point values, e.g. {0.00, 0.00, 0.00} and {0.33, 0.33, 0.33}. Therefore, these class patterns can be filtered.

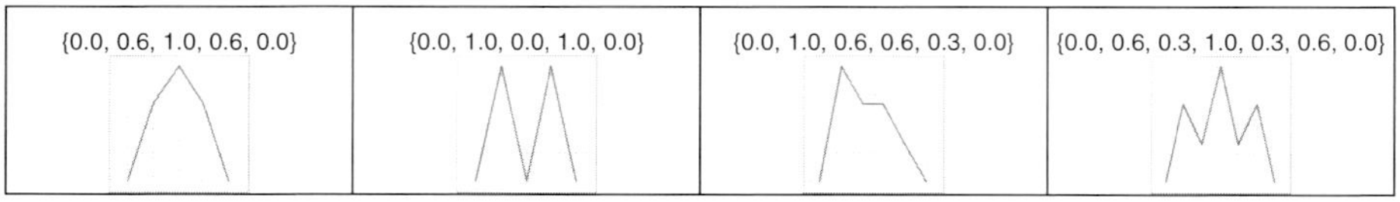

Fig. 4. Examples of class pattern

After a set of class patterns is generated, the time series represented by the optimized SB-Tree can be classified into one of the classes. The time series represented by the optimized SB-Tree will be compared with each class pattern with the same number of data points. The pattern will then be categorized to a class with the minimum distance. The simplest way to measure the distance (*dist*) is by the point-to-point Euclidean distance. Enhanced distance measure approach with horizontal distance consideration can also be applied [10].

Based on this categorization process, different applications can be considered. For example, users can select any interested classes based on the class patterns and investigate on the time series segments belonging to the corresponding classes. Another application is to speed up the time series pattern query process. Given a query time series pattern and a threshold δ, the query pattern will be first compared to all the class patterns, which have the same number of data points. If the distance is greater than δ, the whole class will be ignored; otherwise, the optimized SB-Trees in this class will be compared with the query pattern. If the distance between them is also less than δ, it will be selected as one of the query results.

3.2 Clustering

Clustering is a common approach for finding structure in the given data, in particular for finding structure related to time. There are many popular clustering techniques developed, such as hierarchical, nearest neighbor, and *k*-means algorithms. In the data mining context, the most common one perhaps is the *k*-means algorithm. In the *k*-means clustering, each cluster is represented by the center of the cluster called centroid. The *k*-means algorithm can be implemented with four steps: (1) cluster objects into *k* nonempty subsets, (2) compute seed points as the centroids of the clusters of the current partition, (3) assign each object to the cluster with the nearest seed point and (4) go back to step 2 and stop when no more new assignment is observed or the maximum number of iterations is reached. Suppose that there exist *N* objects, x_1, x_2, ... x_N, in the dataset and they fall into *k* compact clusters, $k << N$. Let M_i be the mean of the vectors, i.e., the cluster centroid, in cluster *i*. If the clusters are well separated, a minimum distance classifier can be used to separate them.

To cluster a set of time series, the time series are the input objects. However, they may have different lengths and one might have to transform the time series data into other representation which provides the same number of features for clustering. By reducing the size of the SB-Trees using the proposed optimization process and based on the assumption that the common technical patterns are always constructed by 4 to 7 data points, the variation in the length of the time series can be greatly reduced. In other words, time series represented by the optimized SB-Tree will provide similar number of feature vectors (from 4 to 7 in this case). Moreover, the number of PIP required to represent the time series after optimization already provides the preliminary information for clustering, i.e., the patterns represented by 5 PIPs are fundamentally different from the patterns represented by 7 PIPs (see Fig.5a). Therefore, a two-step clustering approach is adopted here as shown in Fig.5b. First, the set of time series is clustered by the number of PIP in the corresponding optimized SB-Trees, ranging from pip_{min} to pip_{max}. Then, the subset of time series with the same number of PIP will be further clustered by the *k*-means clustering process with the same number of features and therefore, each of the time series will be grouped into one cluster in one of the *k*-means clustering processes. The point-to-point Euclidean distance can be applied to compute the distance between the time series and the cluster centroid.

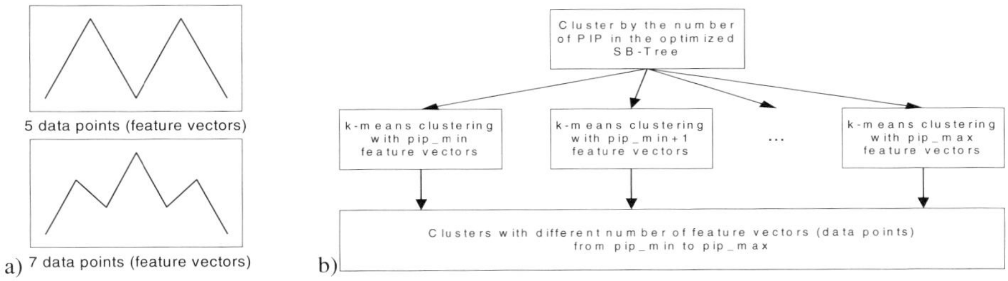

Fig. 5. (a) Time series patterns with different numbers of PIP and (b) A two-step clustering process

In this section, the usages of the optimized SB-Trees in time series categorization and clustering have been described. As the unimportant points of the time series are filtered after the optimization process, time series with different lengths, but with similar shape, will be represented by the same number of PIP for different data analysis tasks.

4 Experimental Results

In this section, three sets of experimental results are reported. First, the performance of the two SB-Tree optimization approaches is shown. Then, the applications of the optimized SB-Tree to categorization and clustering are evaluated. A synthetic time series dataset, which consists of 110 time series with different lengths (25, 43 and 61 data points) was used. Each of them belongs to one of the five technical patterns: head-and-shoulder (H&S), double tops, triple tops, rounded top and spike top. Each technical pattern was used to generate 22 variants by applying different levels of scaling, time wrapping and noise. First, the patterns are uniform time scaling from 7 data points to 25, 43 and 61 data points. Then, each critical point of the patterns can be warp between its previous and next critical points. Finally, noise is added to the set of patterns. Adding noise is controlled by two parameters, namely, the probability of adding noise for each data point and the level of noise being added to such point.

4.1 SB-Tree Optimization

The optimized numbers of data point for representing the five technical patterns are: H&S=7, double tops=5, triple tops=7, rounded top=5 and spike top=5 as the corresponding variants were generated based on the primitive patterns constructed by these numbers of PIP. However, after investigating the generated patterns, it is reasonable to represent the rounded top pattern by either 4 or 5 data points. Therefore, both 4 and 5 PIPs were considered as correct for representing the rounded top pattern in the following experiments.

Taking the H&S pattern as an example, Fig.6a shows the VD values for different numbers of PIP. By using the tree pruning approach, Fig.6b indicates that the optimal number of PIP can be obtained by locating the PIP number with peak change of VD. The dotted lines in the figures show the optimized size of the SB-Tree (i.e. the number of PIP) identified for representing the time series. Fig.7 shows the result using the error threshold approach. Both approaches could identify the correct number of PIP, i.e., 7 PIPs.

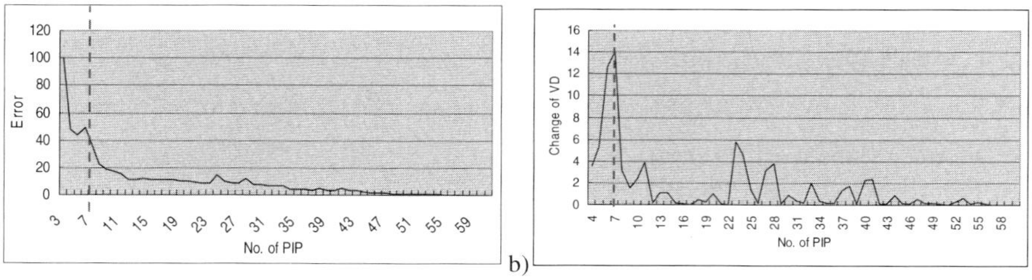

a) b)

Fig. 6. (a) The VD curve with different numbers of PIP retrieved by applying the tree pruning approach and (b) The change of VD with different numbers of PIP. The peak value corresponds to the optimal number of PIP.

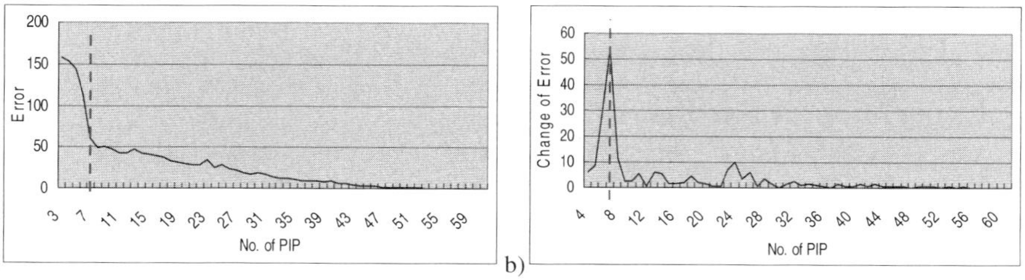

a) b)

Fig. 7. (a) The error curve with different numbers of PIP retrieved by applying the error threshold approach and (b) The change of error with different numbers of PIP. The peak value corresponds to the optimal number of PIP.

Table 1 compares the accuracy and processing time of the two optimization approaches. Accuracy here means the correct number of PIP identified by the optimization process. According to Table 1, the accuracy of the error threshold approach outperformed the tree pruning approach. However, the time consumed by the error threshold approach was 1/3 higher than that of the tree pruning approach. It is because from 3 PIPs till obtaining the correct number of PIP, the error between the original time series and the time series constructed by the selected PIPs has to be calculated correspondingly. On the other hand, the low accuracy of the tree pruning approach was due to the over pruning of the SB-Tree based on determining the largest gap among the VD. Furthermore, the accuracy of the proposed optimization approaches is independent from the length of the time series. Fig.8 shows three sample time series with incorrect optimized size.

Table 1. Comparisons on the two optimization approaches

	Tree Pruning	**Error Threshold**
Length of time series = 25		
Accuracy	72.22%	97.30%
Processing Time	0.09s	0.12s
Length of time series = 43		
Accuracy	62.63%	94.74%
Processing Time	0.28s	0.49s
Length of time series = 61		
Accuracy	72.86%	100.00%
Processing Time	0.65s	1.19s
Overall		
Accuracy	69.09%	97.27%
Processing Time	1.02s	1.80s

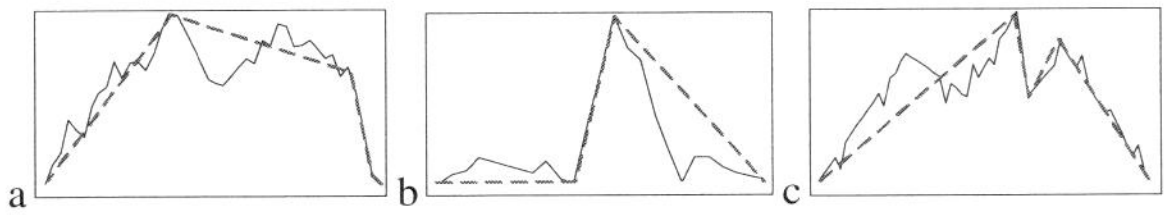

Fig. 8. Wrong optimization cases: (a) resulting from both approaches, (b) resulting from the error threshold approach and (c) resulting from the tree pruning approach

4.2 SB-Tree Categorization

The 110 optimized SB-Trees based on the error threshold approach were used in both the categorization and clustering tasks. During the categorization process, the first step is to generate the class patterns. The range of the number of PIP used for representing the class patterns was 4 to 7 and the point level, θ, was 3. The numbers of class pattern generated after filtering are 110, 570, 2702 and 12138 according to the numbers of PIP are 4, 5, 6 and 7 respectively.

After the categorization process, the 110 optimized SB-Trees or time series took up 20 classes (out of the total 15520 classes), i.e. with an average of 5.5 time series in a class. The implication here is that the subsequent query process can be limited to only one or a few classes in order to save the processing time. At the very beginning of this section, we mentioned that the 110 syntactic time series were generated from 5 technical patterns. They are expected to be assigned to only 5 classes. It is not the case because similar class patterns exist. As exemplified in Fig.9, the 4 class patterns found should belong to the triple tops pattern. However, this is not an undesired behavior as the users can choose to search in both these classes or not during the query process. Furthermore, by investigating these class patterns visually, users can determine which classes they are interested in.

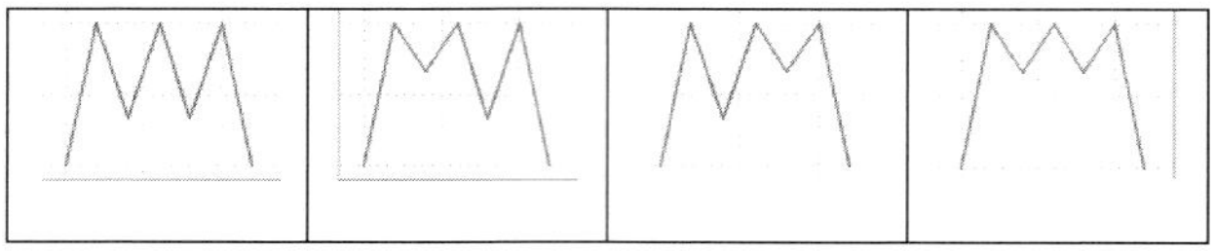

Fig. 9. Class patterns that should belong to the sample technical pattern

4.3 SB-Tree Clustering

In the last experiment, the proposed time series clustering process was simulated. After the first-step clustering, i.e. clustering by the number of PIP in the optimized SB-Tree, the 110 synthetic time series were clustered into 3 groups as shown in Table 2. Both the H&S and triple tops patterns were clustered to the group with 7 PIPs. The correctly optimized SB-Trees of the spike top and double tops were clustered to the group with 5 PIPs. The problem occurred in the rounded top pattern as both 4 and 5 PIPs for representing it was considered as correct. In the first-step clustering process, the accuracy is the same as the optimization process (i.e. 97.27% in this case) and it can be easily seen that the incorrect patterns here come from double tops (1) and spike tops (2) patterns and both were over optimized to 4 PIPs.

Table 2. Number of time series in the clusters with different numbers of PIP

No. of PIP	4	5	6	7
H&S	0	0	0	22
Double Tops	1	21	0	0
Triple Tops	0	0	0	22
Rounded Top	15	7	0	0
Spike Top	2	20	0	0
No. of SB-Tree	18	48	0	44

Then, the *k*-means clustering process was further carried out for each number of PIP. It was benchmarked with the clustering process using other representation scheme, i.e. the Piecewise Aggregate Approach (PAA) [11]. By using PAA, the dimension of the time series will be reduced to the same as the minimum length of the time series in the dataset (i.e. 25 in this experiment). The result shows that there are fewer incorrect groupings of patterns by using the proposed method (i.e. accuracy=91.8%) than the benchmarked approach (i.e. accuracy=75.5%). Based on the finding from the experiment, one may suggest to introduce a redundancy removal process to group similar clusters. First, it is necessary to combine the clusters in each number of PIP when k is larger than the number of actual groups. Also, it is necessary to combine the clusters across different numbers of PIP that are representing the same pattern, such as the rounded top pattern in this experiment. It exists in both 4 and 5 PIPs clusters. This is one of our current works.

5 Conclusion

In this paper, two approaches for optimizing the size of the SB-Tree are proposed. By carrying out the optimization process, the size of the SB-Tree can be reduced under the control of the user preference while the overall shape of the time series can be preserved. In other words, dimensionality reduction of the time series can be achieved. The proposed approach is customized for representing stock time series based on its unique properties. Furthermore, the usages of the optimized SB-Tree are demonstrated by two applications, namely, categorization and clustering. A comprehensive evaluation on the proposed approaches is now conducting on real dataset and will be reported in the near future.

References

1. Fu, T.C., Chung, F.L., Luk, R. and Ng, C.M.: A specialized binary tree for financial time series representation. The 10th ACM SIGKDD Workshop on Temporal Data Mining (2004) 96-104
2. Geurts, P." Pattern extraction for time series classification. In Proc. of the 5th PKDD (2001) 115-127
3. Smyth, P. and Keogh, E.: Clustering and mode classification of engineering time series data. In Proc. of the 3rd Int.l Conf. on KDD (1997) 24-30
4. Keogh, E. and Pazzani, M.: An enhanced representation of time series which allows fast and accurate classification, clustering and relevance feedback. In Proc. of the 4th Int. Conf. on KDD (1998) 239-341
5. Abonyi, J., Feil, B., Nemeth, S. and Arva, P. Principal component analysis based time series segmentation - Application to hierarchical clustering for multivariate process data. In Proc, of the IEEE Int. Conf. on Computational Cybernetics (2003) 29-31
6. Lin, J., Vlachos, M., Keogh, E. and Gunopulos, D.: Iterative incremental clustering of time series. In Proc. of the 9th EDBT (2004) 106-122
7. Ratanamahatana, C.A., Keogh, E., Bagnall, A.J. and Lonardi, S.: A novel bit level time series representation with implications for similarity search and clustering. Technical Report, UCR, TR-2004-93 (2004)
8. Xiong, Y. and Yeung, D.Y.: Mixtures of ARMA models for model-based time series clustering. In Proc. of ICDM (2002) 717-720
9. Kalpakis, K., Gada, D. and Puttagunta, V.: Distance measures for effective clustering of ARIMA time-series. In Proc. of ICDM (2001) 273-280
10. Chung, F.L., Fu, T.C., Luk, R. and Ng, V.: Flexible Time Series Pattern Matching Based on Perceptually Important Points. International Joint Conference on Artificial Intelligence Workshop on Learning from Temporal and Spatial Data (2001) 1-7
11. Keogh, E., Chakrabarti, K., Pazzani, M. and Mehrotra, S.: Dimensionality reduction for fast similarity search in large time series databases. JKIS (2000) 263-286

Similarity Classifier with Generalized Mean; Ideal Vector Approach

Jouni Sampo and Pasi Luukka

Lappeenranta University of Technology,
P.O. Box 20, Lappeenranta,
Finland
Jouni.Sampo@lut.fi, Pasi.Luukka@lut.fi

Abstract. In this paper a study of similarity based classifier with generalized mean and ideal class vector approach is carried out. Before this ideal class vectors in the classifier has been very little investigated area and here focus is changed to study truly 'ideal' vectors to represent class and similarity measure with its power parameters has been taken from best results in our previous studies. To find correct ideal vectors a search using differential evolution algorithm is carried out.

1 Introduction

In this article we study suitability of fuzzy similarity based classification in the Łukasiewicz-structure with generalized mean. Usually mean used in the Łukasiewicz-structure has been arithmetic, but obviously also other means can be used. Here we have tested similarity measure with generalized mean which has proven to be effective in our previous investigations [3] [2], [7]. Łukasiewicz-structure mentioned here is the only MV-structure which holds the fact that the mean of many fuzzy similarities is still a fuzzy similarity.

Same way as notion of fuzzy subset generalizes that of the classical subset, the concept of similarity can be considered as a many-valued generalization of the classical notion of equivalence [1]. As equivalence relation is a familiar way to classify similar objects, fuzzy similarity is an equivalence relation that can be used to classify multi-valued objects. Due to this property, it is suitable for classifying problems that are possible to classify based on clustering by finding similarities in objects.

We also study suitability of proper ideal vector where samples similarity between classes is measured. Main focus on this study is to find best possible ideal vectors for each class. In our previous studies we have simply used mean vector from class samples as ideal vector but here effort is devoted to find best possible ideal vector. We have used differential evolution algorithm to find best possible ideal vectors for each class. Result are very promising and even better classification accuracy has been found using this approach.

2 Mathematical Background

There are several reasons for why Łukasiewicz structure is chocen in defining memberships of objects. One reason is that in the Łukasiewicz structure, it holds that the mean of

L. Wang et al. (Eds.): FSKD 2006, LNAI 4223, pp. 1140–1147, 2006.

many fuzzy similarities is still a fuzzy similarity [9]. Secondly, the Łukasiewicz structure also has a strong connection to first-order logic [10] which is a well studied area in modern mathematics. Thirdly, it also holds the fact that any pseudo-metric induces fuzzy similarity on a given non-empty set X with respect to the Łukasiewicz conjunction [11]. Next, there is shortly introduced the mathematical background concerning fuzzy similarity in the Łukasiewicz structure.

If we examine Łukasiewicz valued fuzzy similarities, $S_i, i = 1, ..., n$ in a set X, we can define a binary relation in L by stipulating $S\langle x, y\rangle = \frac{1}{n}\sum_{i=1}^{n} S_i\langle x, y\rangle$ for all $x, y \in X$. It is easy to prove that this is still a Łukasiewicz-valued fuzzy similarity [9].

In classifier we examine a choice situation where the features of different objects can be expressed in values between [0,1]. Let X be a set of m objects. If we know the similarity value of the n different features between the objects, we can choose the object that has the highest similarity value. The problem is finding, for object x_i a similar object x_j, where $1 \leq i, j \leq m$ and $i \neq j$. By choosing the Łukasiewicz structure for the features of the objects we get n fuzzy similarities for comparing two objects (x, y):

$$S_i\langle x, y\rangle = x(i) \leftrightarrow y(i), \tag{1}$$

where $x, y \in X$ and $i \in \{1, ..., n\}$. As the Łukasiewicz structure is chosen for the membership of objects, we can define the fuzzy similarity as follows:

$$S\langle x, y\rangle = \frac{1}{n}\sum_{i=1}^{n}(x(i) \leftrightarrow y(i)). \tag{2}$$

It is very important to realize that this holds only in the Łukasiewicz structure. Moreover, in the Łukasiewicz structure we can give different non-zero weights $(w_1, \ldots, w_n)$ to the different features and obtain the following equation which again meets the definition of the fuzzy similarity:

$$S\langle x, y\rangle = \sum_{i=1}^{n} w_i(x(i) \leftrightarrow y(i)). \tag{3}$$

In the Łukasiewicz structure, the equivalence relation $a \leftrightarrow b$ is defined as

$$a \leftrightarrow b = 1 - |a - b|. \tag{4}$$

In the generalized Łukasiewicz structure, the equivalence relation can be defined as [3]:

$$a \leftrightarrow b = (1 - |a^p - b^p|)^{\frac{1}{p}}. \tag{5}$$

Weighted similarity measure in the generalized Łukasiewicz structure can be defined as follows [3]:

$$S\langle x, y\rangle = \sum_{i=1}^{n} w_i \sqrt[p]{1 - |(x(i))^p - (y(i))^p|} \tag{6}$$

This formula uses arithmetic mean but it is clear that also other means can be used. If we choose generalized mean it can be stated as following:

$$M_m(a_1, a_2, ..., a_n) = \left(\frac{1}{n}\sum_{i=1}^{n} a_k^m\right)^{1/m} \tag{7}$$

This can be implemented into the similarity measure based on Łukasiewicz structure. The formula of the similarity based on generalized mean and generalized Łukasiewicz structure gets the following form:

$$S\langle x, y\rangle = \left[\frac{1}{n}\sum_{i=1}^{n} w_i \left(\sqrt[p]{1 - |x^p(i) - y^p(i)|}\right)^m\right]^{\frac{1}{m}} . \tag{8}$$

Measure has strong mathematical backround [11], [17] and has proven to be very efficient measure in classification [3].

3 A Classifier

Next we describe the general type of classification problem we are dealing with and structure of our classifier.

We would like to classify set X of objects to the N different classes $C_1, \ldots, C_N$ by their features. We suppose that n is number of different kind of features that we can measure from objects. We suppose that values for magnitude of each feature is normalized so that it can be presented as value between $[0, 1]$. So, the objects we want to classify are basically vectors that belongs to $[0, 1]^n$. This is mandatory if we want to use fuzzy similarity.

Before doing actual classification, all parameters for classifier should be decided. These parameters are

1. The weights w_i, $i = 1, \ldots, n$, for dimensions
2. The ideal vectors $v_i = (v_i(1), \ldots, v_i(n))$ for each classes $i = 1, \ldots, N$
3. The power values p and m.

In this study we set all weights to be equal to one and try to optimize ideal vectors once p and m are fixed. We tested our classifier with several fixed p and m which were found suitable in [24]. We use differential evolution [25] for optimization, the short description of this algorithm is also presented in the next section.

Once ideal vectors have been determined then decision to which class arbitrarily chosen $x \in X$ belongs is made by comparing it to each ideal vector. We have made that comparision by using generalized Łukasiewicz fuzzy similarity which is defined

$$S\langle x, y\rangle = \left(\frac{1}{n}\sum_{i=1}^{n} (1 - |x(i)^p - y(i)^p|)^{m/p}\right)^{1/m} , \tag{9}$$

for $x, y \in [0, 1]^n$. We decide that $x \in C_m$ if

$$S\langle x, v_m\rangle = \max_{i=1,\ldots,N} S\langle x, v_i\rangle . \tag{10}$$

In the algorithmic form, a classifier would be:

Require: $test, learn[1...n], weights, dim$
scale $test$ between $[0, 1]$
scale $learn$ between $[0, 1]$
for $i = 1$ to n **do**
$idealvec[i] = IDEAL[learn[i]]$

$$maxsim[i] = \frac{\left(\sum_{j=1}^{dim} weights[j](1 - |idealvec[i][j]^p - test[j]^p|)^{\frac{m}{p}}\right)^{\frac{1}{m}}}{dim^{\frac{1}{m}}}$$

end for
$class = \arg\max_i maxsim[i]$

---

4 Differential Evolution

The basic idea of evolutionary algorithms is that we create a population V_0 of trial solutions (vectors) for the optimization problem. Next we combine the members of V_0 in a certain way and check if the combined solutions are better than the original trial solutions in V_0. The best solutions continue to population V_1, the next step of the evolution, and then the whole procedure starts again.

We denote the population at evolution step k by set V_k. Our fitness function $f : \mathbb{R}^D \to [0, 1]$ is set so that if $f(\mathbf{v}) < f(\mathbf{u})$ then $\mathbf{v}$ is a better solution for our problem than $\mathbf{u}$. Because we optimize ideal vectors, we have $D = Nn$.

A basic DE contains crossover operation but it has not a real mutation operation. Instead of mutation, DE has *differential variation* operation. In one evolution step we make the following two basic operations for all $\mathbf{v}_k = \{v_1, \ldots, v_D\} \in V_k$:

1. *Differential variation.* The basic idea of differential variation is to add "noise vector" $\mathbf{n}$ to vector $\mathbf{w} \in V_k$. Both $\mathbf{n}$ and $\mathbf{w}$ can be chosen in many ways. Usually $\mathbf{n}$ is either the difference vector between two vectors in V_k or $\mathbf{n}$ is a linear combination of these kind of difference vectors. From $\mathbf{n}$ and $\mathbf{w}$ we form a new vector

$$\mathbf{u} = \mathbf{w} + F\mathbf{n}, \tag{11}$$

where usually parameter F is constant and between 0 and 1.

2. *Crossover.* From $\mathbf{u} = \{u_1, \ldots, u_D\}$ and $\mathbf{v}_k$ we form a trial vector $\mathbf{t} = \{t_1, \ldots, t_D\}$ by setting

$$t_i = \begin{cases} u_i & \text{if } x_i < CR \\ v_i & \text{otherwise} \end{cases} \tag{12}$$

where $x_i \in [0, 1]$ is a uniformly distributed random variable and CR is the crossover probability. Finally the vector $\mathbf{v}_{k+1}$ which is chosen to the next generation V_{k+1} is

$$\mathbf{v}_{k+1} = \begin{cases} \mathbf{t} & \text{if } f(\mathbf{t}) < f(\mathbf{v}) \\ \mathbf{v}_k & \text{otherwise} \end{cases} \tag{13}$$

We chose the same parameters for differential evolution which we found to work best in [26]. Vector $\mathbf{w}$ was chosen as the best member of V_k,

$$f(\mathbf{w}) = \min_{\mathbf{v} \in V_k} \{f(\mathbf{v})\}, \tag{14}$$

and $\mathbf{n}$ was the difference vector

$$\mathbf{n} = \mathbf{y}_1 - \mathbf{y}_2, \tag{15}$$

where $\mathbf{y}_1 \in V_k$ and $\mathbf{y}_2 \in V_k$ were randomly chosen vectors. Other parameters were $F = 0.9$ and $CR = 0.8$.

5 Datasets

The data sets were chosen such that properties of the classifiers would be apparent. The data sets were taken from a UCI Repository of Machine Learning Database (available in [8]) archive. The classifier were implemented with the MATLAB™-software. We used four different data sets. The fundamental properties of the data sets are shown in table 1.

Table 1. Test data sets and their properties

Name	classes	Dimension
Waveform data	3	21
Wisconsin breast cancer data	2	9
Bupa data	2	6
Thyroid data	3	5

Wisconsin Breast Cancer Data: This work was based on Dr. Wolberg's desire to accurately diagnose breast masses based solely on Fine Needle Aspiration (FNA). He identified nine visually assessed characteristics of an FNA sample which he considered relevant to the diagnosis. In collaboration with Prof. Mangasarian and two of his graduate students, Rudy Setiono and Kristin Bennett, a classifier was constructed using the multisurface method (MSM) of pattern separation on these nine features that successfully diagnosed 97% of new cases [19]. The resulting data set is well-known as the Wisconsin Breast Cancer Data (BCW1).

Waveform data set: Waveform data consist of three different classes and has 21 dimensions. There are 21 "attributes" consisting of continuous values between 0 and 6. Each class represents about 1/3 of the data set. Each class is generated from a combination of 2 of 3 "base" waves and each instance is generated by added noise (mean 0, variance 1) in each attribute. There are 5000 rows, which are results of the simulation. [8]

Bupa data set: This data was donated by R. S. Forsyth to [8]. The problem is to predict whether or not a male patient has a liver disorder based on blood tests and alcohol consumption. The attribute information for BUPA data is the following 1) mean corpuscular volume 2) alkaline phosphotase 3) alamine aminotransferase 4) aspartate aminotransferase,5) gamma-glutamyl transpeptidase 6) number of half-pint equivalents of alcoholic beverages drunk per day. The first five variables are all blood tests which are thought to be sensitive to liver disorders that might arise from excessive alcohol consumption.

Thyroid data set: With thyroid diagnosis data is collected to predict whether a patient's thyroid symptoms correspond to the three possible diagnoses: euthyroidism, hypothyroidism or hyperthyroidsim [18]. The diagnosis is based on a complete medical record, including anamnesis, scan etc. The five inputs correspond to: 1) the T3-resin uptake test 2) Total Serum Thyroxin 3) Total Serum triiodothyronine 4) Basal thyroid-stimulating hormone (TSH) 5) Maximal absolute difference of the TSH value after injection of 200 micro grams of thyrotropin-releasing hormone as compared to the basal value.

All data sets were splitted in half; one half was used for training and the other for testing the classifier.

6 Empirical Classification Results

Wave form data: Classification results with waveform data can be seen in Figure 1. Area where best results can be found is with slightly negative mean values and small p values. Best classification result here was 87.69% when previously best results achieved with this classifier was 86.31%. This result was quite good also when comparing to other classifiers (see the table 2).

Table 2. Classification results with different classifiers using waveform data

Method	Classification results
Optimal Bayes	86 %
CART decision tree	72 %
1-Nearest Neighbor	78 %
Similarity classifier 1	86.31 %
Similarity classifier 2	87.69 %

Wisconsin breast cancer data: Classification results with breast cancer data can be seen in Figure 1 (b). Best mean value was around 2.5 and p value 3.5. Best classification accuracy was 98.83%, previously 97.72% was achieved [7] using similarity classifier.

Bupa data: With bupa data best classification results were found when mean value was around $m = 2.5$ and p value $p = 9.3$. Classification results plotted wit several m and p values can be seen in Figure 1 c. Best classification results was 70.52%. which again was higher than usual mean result (66.50%) in [7] with this classifier.

Thyroid data: With thyroid data set we managed to find the area where we got 100.00% as classification accuracy. Previously our best result with this data set was 99.07% (see [24] using similarity classifier. Area where we found the best result was when mean value were around one and p value around two.

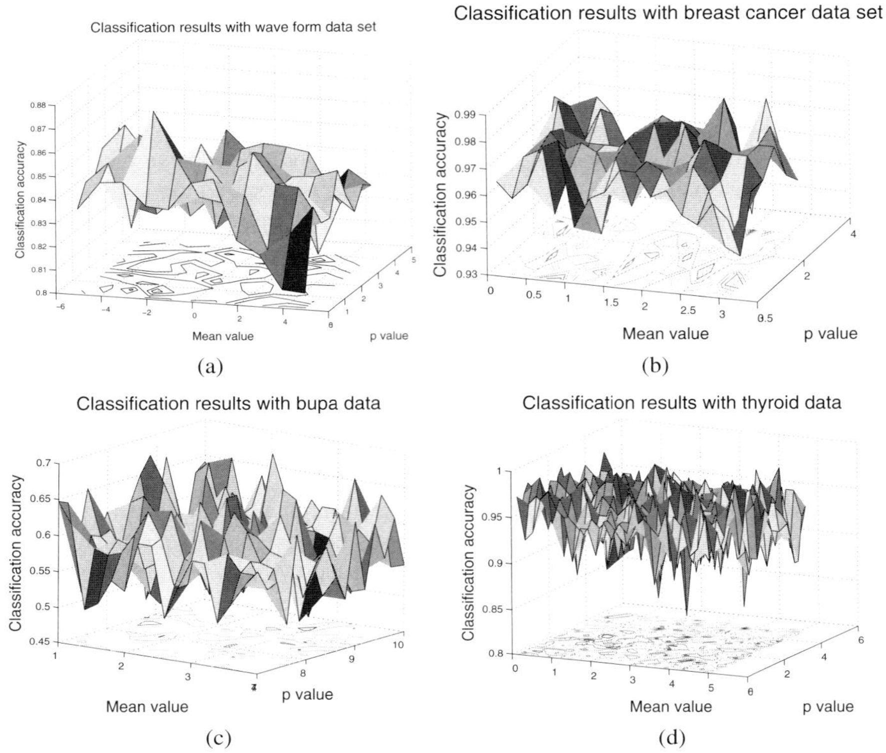

Fig. 1. Classification results for a) wave form b) breast cancer c) bupa d) thyroid data; Result plotted with respect to several p and mean values

7 Conclusion

In this paper we have presented the new way of searching appropriate ideal vectors for our similarity classifier. This is done by optimizing ideal vectors with differential evolution algorithm. Results achieved from the experiments are very promising. With these data set we managed to get better results than before. For future work more experiments will be done with the data sets. Also more work will be done in further development of optimization procedure of parameters used in the classifier.

References

1. Zadeh, L.: *Similarity Relations and Fuzzy Orderings*. Inform Sci, 3, 1971
2. Luukka, P., Saastamoinen, K.: *Similarity Classifier with p-mean and Generalized Łukasiewicz Algebra* Submitted to International Journal of Hybrid Intelligent Systems.
3. Luukka, P., Saastamoinen, K., Könönen, V.: *A Classifier Based on the Maximal Fuzzy Similarity in the Generalized Łukasiewicz-structure*. Paper published in the proceedings of the FUZZ-IEEE 2001 conference, Melbourne, Australia, 2001

4. Saastamoinen, K., Könönen, V., Luukka, P.: *A Classifier Based on the Fuzzy Similarity in the Łukasiewicz-Structure with Different Metrics.* Paper published in the proceedings of the FUZZ-IEEE 2002 Conference, Hawaii, USA.
5. Saastamoinen, K. Luukka, P.: *Classification in the Łukasiewicz Algebra with Different Means* paper published in the proceedings of FUZZ-IEEE 2003 conference, St Louis, USA 2003.
6. Luukka, P., Meyer A.: *Comparison of two different dimension reduction methods in classification by arithmetic, geometric and harmonic similarity measure* paper published in the proceedings of FUZZ-IEEE 2004 conference, Budapest, Hungary, 2004
7. Luukka P., Leppälampi, T.: *Similarity classifier with generalized mean applied to medical data using different preprocessing methods* To appear in the proceedings of the FUZZ-IEEE 2005 conference, Reno, USA.
8. UCI Repository of Machine Learning Databases network document. Referenced 4.11.2004. Available: *ftp://ftp.ics.uci.edu/pub/machine-learning-databases*
9. Turunen, E.: *Mathematics behind Fuzzy Logic.* Advances in Soft Computing, Physica-Verlag, Heidelberg, 1999
10. Novak, V.: *On the Syntactico-semantical Completeness of First-Order Fuzzy Logic.* Kybernetika 26, 1990
11. Klawonn, F., Castro J.L.: *Similarity in Fuzzy Reasoning.* Math Soft Comp, 2, 1995
12. Donoho, D.L.: *High-dimensional data analysis: The curses and blessings of dimensionality.* Lecture at the "Mathematical Challenges of the 21st Century" conference of the American Math. Society. Los Angeles, August 6-11 2000.
13. Jolliffe, I.T.: *Principal Component Analysis*, Springer-Verlag, 1986.
14. Martens, H. Naes, T.: *Multivariate Calibration* John Wiley, UK, 1989
15. Yao,Y.Y., Wong, S.K.M., Butz, C.J.: *On Information-Theoretic Measures of Attribute Importance*, Pasific-Asia Conference on Knowledge Discovery and Data Mining, 1999.
16. Yuan,L., Kesavan, H.K.: *Minimum Entropy and Information Measurement*, IEEE Transaction on System, Man, and Cybernetics, Vol 28, 3, 1998.
17. Formato, F., Gerla, G., Scarpati, L: *Fuzzy Subgroups and Similarities.* Soft Computing, 3, 1999
18. Coomans, D., Broeckaert, M. Jonckheer M. and Massart D.L., *Comparison of Multivariate Discriminant Techniques for Clinical Data - Application to the Thyroid Functional State*, Meth. Inform. Med. 22 (1983) pp. 93-101.
19. O. L. Mangasarian and W. H. Wolberg: *Cancer diagnosis via linear programming*, SIAM News, Volume 23, Number 5, September 1990, pp. 1-18.
20. W.N. Street, W.H. Wolberg and O.L. Mangasarian,: *Nuclear feature extraction for breast tumor diagnosis* IS & T/SPIE 1993 International Symposium on Electronic Imaging: Science and Technology, volume 1905, pages 861-870, San Jose, CA, 1993.
21. J.W. Smith, J.E. Everhart, W.C. Dickson, W.C. Knowler & R.S. Johannes, *Using the ADAP learning algorithm to forecast the onset of diabetes mellitus.* Proceedings of the Symposium on Computer Applications and Medical Care, pp. 261-265, 1988.
22. Bologna, G.: *Symbolic Rule Extraction from the DIMLP Neural Network* To appear in Neural Hybrid Systems, Springer Verlag.
23. Quinlan, J.R.: *C4.5: Programs for Machine Learning.* Morgan Kaufman (1993)
24. Luukka, P.: *Similarity measure based classification*, PhD thesis, Lappeenranta University of Technology, ISBN 952-214-162-3, 2005.
25. Price, K.V.: *Differential Evolution: A Fast and Simple Numerical Optimizer.* Biennial Conference of the North American, 1996.
26. Luukka, P. and Sampo, J.: *Weighted Similarity Classifier Using Differential Evolution and Genetic Algorithm in Weight Optimization.* Journal of Advanced Computational Intelligence and Intelligent Informatics Vol. 8 No.6. Pages 591-598 2004.

A Novel Algorithm for Identification of Body Parts in Medical Images

Jongan Park, Gwangwon Kang, Sungbum Pan, and Pankoo Kim

College of Electronics & Information Engineering
Chosun University, Gwangju, Korea
japark@chosun.ac.kr

Abstract. In this paper, we introduce an algorithm based on energy information obtained from Wavelet Transform for classification of medical images according to imaging modalities and body parts. Various medical image retrieval systems are available today that classify images according to imaging modalities, orientations, body parts or diseases. Generally these are limited to either some specific body part or some specific disease. Further, almost all of them deal with the DICOM imaging format. Our technique, on the other hand, can be applied to any of the imaging formats. The results are shown for JPEG images in addition to DICOM imaging format. We have used two types of wavelets and we have shown that energy obtained in either case is quite distinct for each of the body part.

1 Introduction

Medical image databases are increasingly becoming an essential part of the medical diagnosis. Apart from medical diagnosis, their other applications include therapy control and provision of real world cases as educational examples. Therefore, an efficient as well as precise content based medical image retrieval system has become a necessity.

In clinical practice, the physicians have access to large amount of data. This large amount of data has been made available because of the medical imaging instrumentation. Different anatomical features are captured in various orientations using different imaging procedures. Often physicians must combine the information from different images to fully visualize the imaged anatomical structure.

Hence, a need arises to fully archive the medical images originating from all kind of imaging modalities with their imaged orientations combining with the information of the anatomical structure. Further every individual anatomical structure is labeled and divided in various regions so that region of interests can be specified. However, the complexity increases keeping in mind the following factors:

a) Various imaging modalities and each of them is further sub-divided, for example, X-rays is an imaging modality which further can be divided to include Computer Tomography (CT), Computed Radiography (CR), Plain Radiography, Fluoroscopy etc. Similarly other modalities include Ultrasound, MRI, PET and so on [1].

L. Wang et al. (Eds.): FSKD 2006, LNAI 4223, pp. 1148 – 1158, 2006.

b) Various Imaging Orientations and each of them can further be sub-divided, for example, postero-anterior with sub-classes like inspiration, expiration, supine, prone, micturition etc [1].
c) Large number of anatomical structures and each of the structures can further be subdivided, for example, chest with bones, lungs etc and then further lungs with upper lobe, middle lobe etc and so on [1].
d) And finally an infinite number of diseases and disease pattern make it a very cumbersome task.

On other side, with the rapid development of the Internet and the World Wide Web, the data exchange has become a reality between a large numbers of users. Hence a medical image retrieval system on the internet can have far reaching benefits.

With all these aspects in mind, this paper presents our research related with the classification of medical images according to imaging modalities and according to the anatomical structures, which is the first component of our CBIR system. Our method first recognizes imaging modality based on a lookup table and then different body parts are identified on the bases of their distinct energies.

2 Related Work

The Content Based 3D Neuroradiologic Image Indexing and Retrieval System [2] was concentrated on dealing with multimodal 3D images (MR/CT). Its special characteristics include image similarity based on anatomical structures of the human brain and combining both visual and collateral information for indexing and retrieval.

ASSERT system [3] is based on a human-in-the-loop CBIR system for Medical Images. To pose a query to the database, the physician circles one or more pathology bearing regions (PBR) in the query image. High Resolution Computed Tomography (HRCT) of the lung is targeted in this system.

IBM introduced Ilive (Interactive Lifesciences Imaging Visualization & Exploration) system [4]. ILive was developed for the automatic categorization of medical images according to their modalities. The system is based on semantical set of visual features, their relevance and organization for capturing the semantics of imaging modalities. The main emphasis of IBM ILive is with various imaging modalities.

In IRMA (Image Retrieval in Medical Applications) system [5], an approach for content-based image retrieval in medical applications is presented. The system contains different semantic layers of information modeling, a hierarchical concept for feature representation and uses distributed system architecture for efficient implementation. The IRMA system uses multi-scale segmentation as well as model-based and knowledge-based segmentation. The classification of images is performed by supporting texture analysis.

The method for image classification presented in this paper differs from the methods used in the above mentioned systems for content based Image retrieval.

3 Method

In this paper, an algorithm developed for classification of medical images according to various imaging modalities as well as different body parts is presented. Further, the algorithm is equally applicable to different image formats. The algorithm is tested for both the DICOM images as well as JPEG medical images and has shown satisfactory results.

Generally, most of the medical image retrieval systems use only DICOM images. In DICOM images, all the information regarding imaging modalities, body parts, orientations etc is available in the DICOM header. However, if the information in the DICOM header had been entered incorrectly by mistake or is changed because of some communication error, then it requires that some automated classification be applied to identify the information required for medical image retrieval system. Further, other imaging formats like JPEG etc do not have such elaborate information in their header. And hence requires some automated classification method. With these issues in mind, we have developed automated classification technique for medical images.

The first step is to identify imaging modality. We use two different techniques to identify the imaging modality and the body parts. Each of the technique complements each other hence making the result very accurate and precise.

The first technique checks for the image size, number of bits per pixel, number of rows, number of columns, dimensionality and the color information of the input image. Then this information is compared with the information given in the lookup table, Table 1, yielding the exact imaging modality or narrowing it to a smaller group of imaging modalities. Table 1 does not cover all the modalities and it will be expanded in future. Further, four parameters are shown in Table 1 for each modality.

Table 1. Classification table for Imaging Modalities

Modality	Matrix Size	Bits per pixel	Av. Size (Dicom)	Color Format
Nuclear Medicine	128 x 128	8/16	2MB	Gray/ Pseudo
Ultrasound	512 x 512	8	5~8 MB	Gray/ Pseudo
Doppler Ultrasound	512 x 512	24	15~24 MB	Color
Digital Electron Microscopy	512 x 512	8	Varies	Gray
Digital Color Microscopy	512 x 512	24	Varies	Color
MRI	256 x 256	12	8~12 MB	Gray/ Pseudo
CT	512 x 512	12	20MB	Gray
Computed Radiography	2048 x 2048	12	8~32 MB	Gray
Digitized Mammogram	4096 x 4096	12	64 MB	Gray
Digitized X-Ray Films	2048 x 2048	12	8 MB	Gray
Digital Subtraction Angiography	512 x 512	8	4~10 MB	Gray

In addition to these four parameters, one additional parameter is also used, i.e., dimensionality information telling whether the image is 2D or 3D. And if it's 3D, then the total number of slices in 3D image are also taken into account. So, these facts are exploited to form well defined sets for identification of imaging modalities. The imaging modalities are identified in a tree fashion as depicted in Fig. 1.

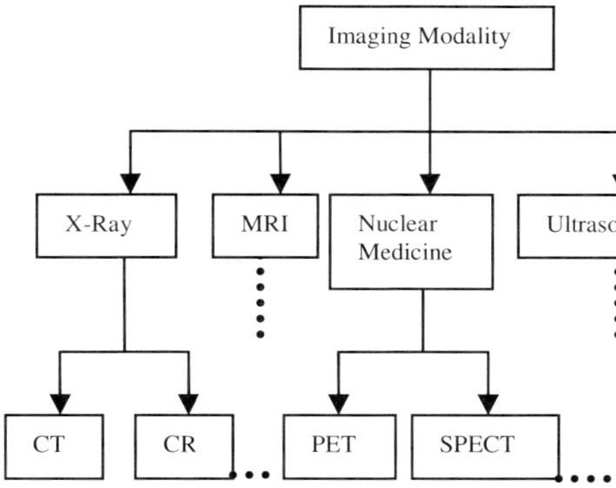

Fig. 1. Identification tree for Imaging Modalities

Once the imaging modality is identified, the next step is to determine the imaged body part. We exploit the fact that all body parts differ in shapes. Hence, each body part could be represented by some unique characteristic. And that unique attribute will differentiate it from rest of the body parts. And hence that attribute or characteristic, unique to that body part, will be an important value for indexing of the medical images according to body parts. It is similar to the concept that every body part has unique absorption characteristic when exposed to X-Rays. X-ray attenuation can be used to distinguish between various body parts. Alternatively other information can be also used; for example, the energy information of a body part.

However, the point that needs to be emphasized is that the attribute vector size should be small in order to enable faster image retrieval in case of a query. At the same time, the attribute vector should also provide accurate result.

Here the energy information is used for identification of each body part. That energy information is obtained using wavelet transforms. We have used two types of wavelets: Biorthogonal and Reversible Biorthogonal wavelet [7, 8] and we have shown that energy obtained in either case is quite distinct for each of the body part.

Fig. 2 shows the decomposition scaling function and the decomposition wavelet function for the Biorthogonal wavelet [7, 8] used. If the same FIR (Finite Impulse Response) filters are used for reconstruction and decomposition then except for the Haar wavelet transform, symmetry and exact reconstruction are incompatible. Therefore two wavelets, instead of just one, are used. Fig. 3 shows the reconstruction scaling function and reconstruction wavelet function.

- One, ψ', is used in the analysis and the coefficients of a signal *s* are:

$$\hat{c}_{j,k} = \int s(x)\, \psi'_{j,k}(x)\, dx \tag{1}$$

- The other, ψ, is used in the synthesis:

$$s = \Sigma_{j,k}\, \hat{c}_{j,k}\, \psi_{j,k} \tag{2}$$

Generally, an infinite number of wavelets are required for complete computational analysis. To solve this problem, the scaling function is used which plays a vital role as it decreases infinite number of wavelets to a finite number. Fig 2. is the visual representation of functions given in equations 1 and 2. Similarly, Fig. 3 depicts the corresponding reconstruction functions.

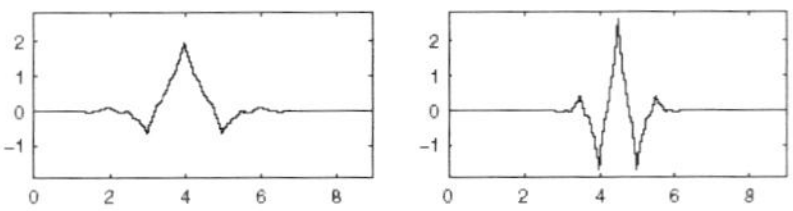

Fig. 2. Decomposition scaling function & Decomposition Wavelet Function

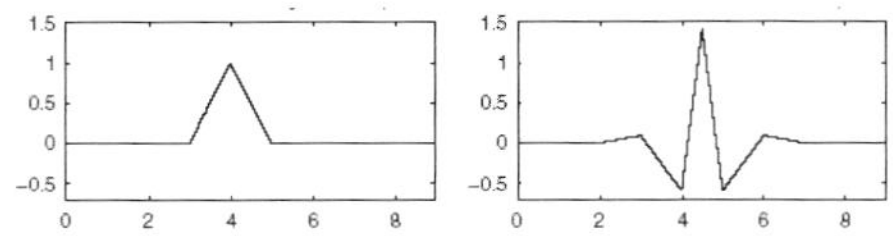

Fig. 3. Reconstruction scaling function & Reconstruction Wavelet Function

We performed 3-level wavelet decomposition. Hence the coefficient matrices for all the three levels were generated. The coefficient matrices were generated for the approximation, horizontal, vertical and diagonal details. Once the coefficient matrices were obtained, we calculated energy corresponding to each of the coefficient matrix, i.e., for a two-dimensional wavelet decomposition, the percentage of energy corresponding to the approximation, horizontal, vertical, and diagonal details.

$$\text{Energy (E)} = \Sigma s^2 \tag{3}$$

Then the energy values for all decomposition levels for each of the coefficient detailed matrix were added to yield a single energy value. i.e., one value depicting the percentage of energy in the horizontal component is generated, one for vertical, one for diagonal and one decimal value for the approximation component. Hence a total of four decimal values per wavelet are calculated for each image.

The same procedure was repeated for Reversible Biorthogonal wavelet [7, 8]. Fig. 4 shows the decomposition scaling and wavelet function in a similar way as described for Biorthogonal case in Fig. 2. While Fig. 5 shows the reconstruction scaling and wavelet function for the corresponding decomposition functions.

Again 3-level wavelet decomposition was applied to obtain the coefficient matrices (i.e., approximation, horizontal, vertical and diagonal details) for all the three levels. Similarly, the energy was calculated corresponding to each of the coefficient matrix. The percentages of energy were calculated for the approximation, horizontal, vertical, and diagonal details according to the formula given earlier in equation 3.

Finally, we obtained a single value by adding the energy values for all decomposition levels for each of the coefficient detailed matrix. Therefore, we obtained four values, i.e., one value each for energy in the horizontal, vertical, diagonal and the approximation component.

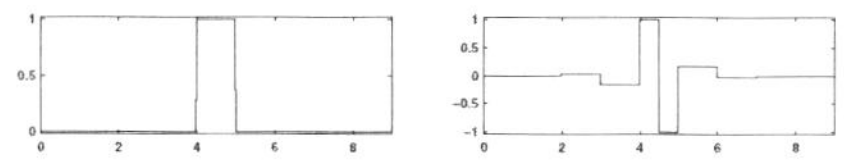

Fig. 4. Decomposition scaling & Wavelet Function respectively

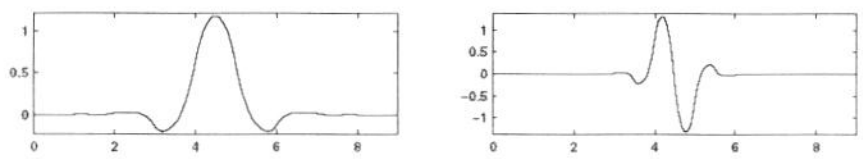

Fig. 5. Reconstruction scaling & Wavelet Function respectively

4 Results and Discussion

Database of 1000 images [9, 10, 11, 12] was used to test the algorithm. Five hundred images were of DICOM format and five hundred were the JPEG images. Further, the images comprised of:

- HRCT lung images
- MRI images of abdomen, brain, spine, heart, knee, shoulder
- CT images of abdomen, brain, chest, ankle
- Ultrasound images of intestine, stomach, kidney

First the images were processed by the method defined in section 3 to identify the imaging modality. Then wavelet transform was applied till third level decomposition. Fig. 6 shows the 3rd level decomposition for Biorthogonal wavelet while Fig. 7 shows the result for Reversible Biorthogonal wavelet.

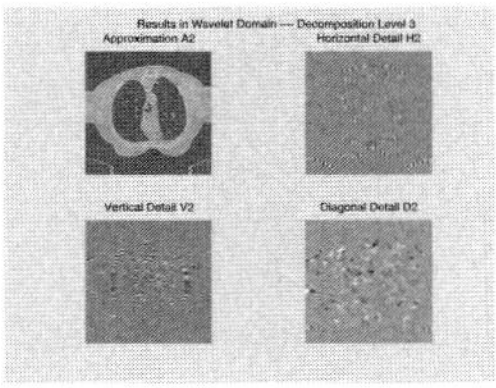

Fig. 6. 3rd level decomposition for HRCT Lung Image using Biorthogonal wavelet

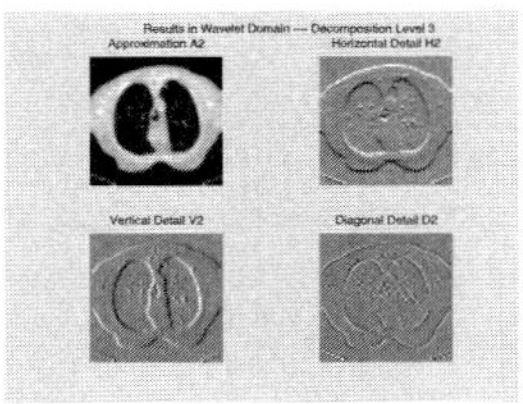

Fig. 7. 3rd level decomposition using Reversible Biorthogonal wavelet

As can be seen from the images above, the detail of HRCT lung image is much more pronounced in Fig. 7 compared to Fig. 6. Hence, the energy values depicted by the Detail Coefficient Matrices (i.e., Horizontal, Vertical and Diagonal) are much higher in Fig. 7 compared to Fig. 6. However, the percentage of energy depicted by the approximation coefficient matrix is higher for Fig. 6 compared to Fig. 7. Hence, we have used both wavelets to balance out the information available in approximation coefficient matrix as well as detail coefficient matrices. Same is true for all other types of images.

The percentage of energy was calculated for each of the detail coefficient matrices including horizontal, vertical and diagonal coefficients as well as for the approximation coefficients. And these values were compared using Euclidean distance measure. This comparison was made for images only within each imaging modality and not across the imaging modalities. For example, the comparison was made between all MRI images of, e.g., brain, heart, knee etc. And it was found that the percentage of energies for each body part within a certain imaging modality is distinct and hence images of a certain modality can be classified according to the body parts based on these energy values.

Below are shown tables which depict energy values for various body parts within certain imaging modality. Table 2 shows percentage of energy values for MRI while table 3 shows percentage of energy values for CT images.

Table 2. Percentage of energy values for MRI DICOM Images

Body Parts	Biorthogonal Wavelet				Reversible Biorthogonal			
	A	H	V	D	A	H	V	D
abdomen	90	.19	2.01	7.7	70	6.1	9	14
spine	81	.67	4.4	13	60	12	13	15
shoulder	93	.53	2.04	3.5	84	4.3	4.07	7.4

Table 3. Percentage of energy values for CT DICOM Images

Body Parts	Biorthogonal Wavelet				Reversible Biorthogonal			
	A	H	V	D	A	H	V	D
ankle	63	1.5	3.6	31	60	15	11	12
brain	99	.001	.04	.81	88	1	3	7
abdomen	74	2	10	13	64	14	10	11

In tables 2 and 3, the first column is for body parts. The next four columns show the result for Biorthogonal wavelet while the last four columns show results for Reversible Biorthogonal wavelet. Further, 'A' stands for Approximation, 'H' stands for

Horizontal, 'V' for Vertical and 'D' for Diagonal. The values are calculated up to 4 decimal places but they are truncated for displaying in the tables above.

Fig. 8 and Fig. 9 shows additional examples of images processed by the algorithm described in section 3. Fig. 8 shows the 3rd level decomposition for a MR Shoulder image while Fig. 9 displays the 3rd level decomposition for a CT chest image. The images are labeled with percentage energy values. For example, consider Fig.8. The approximation image of the MR Shoulder is labeled 84. It means that it contains 84% of the total energy. Similarly, the horizontal image contains 4.3% of the total energy and so on. Please refer to table 2 and table 3 for the corresponding energy values.

It can be seen from the tables as well as the figures that as the approximation value is higher than it results in lower energy values for detail coefficients. And as the energy value for the approximation goes low, the energy values for the detail coefficients become large.

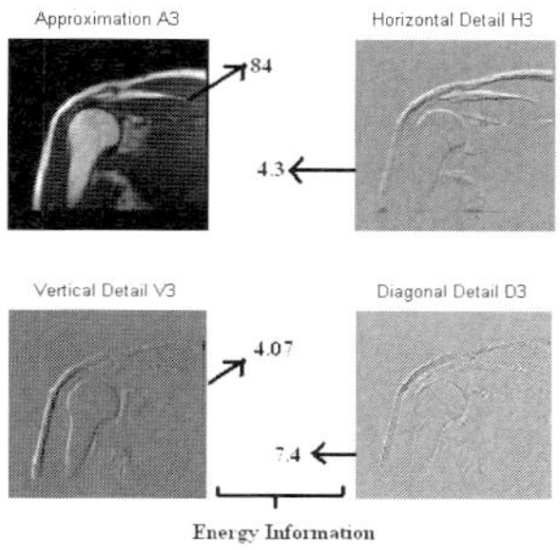

Fig. 8. 3rd level decomposition for a MR Shoulder image

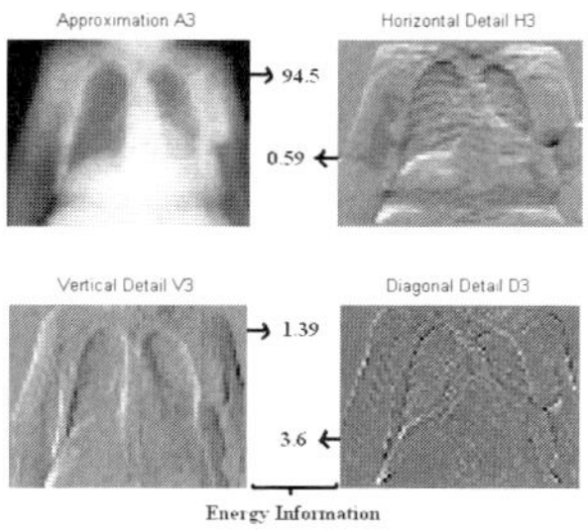

Fig. 9. 3rd level decomposition for a CT chest image

Also, it can be seen from Fig. 10 that there is quite a difference among energy values of different body parts. Hence, it proves the point that energy values can be used to classify body parts within a certain imaging modality. However, it should be noted that these energy values cannot be used for classification of body parts independent of imaging modality. The reason is that the body parts are similar and hence depict very close energy values for same body part in different imaging modality, i.e., brain image by MRI as well as CT provide energy values that are very close to each other.

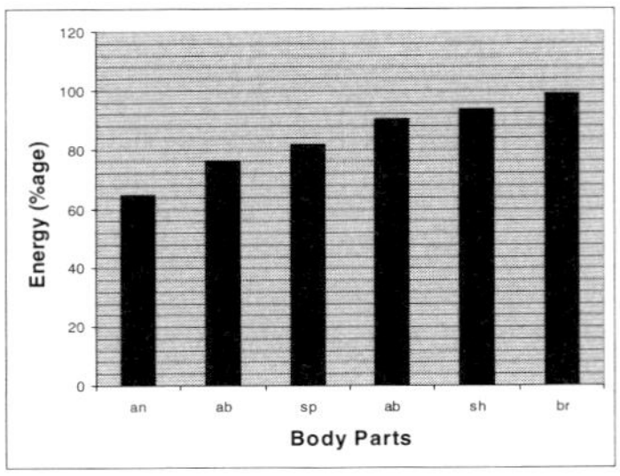

Fig. 10. Energy values from table 2 & 3

Hence, if the imaging modality is not known then it is not possible to find out whether the image of brain is a CT image or MRI.

The tables above are shown only as examples for few body parts in two of the imaging modalities. We have calculated a set of tables for the images available in our database (please see start of this section). Further, the results in Table 2 and 3 are for DICOM images. Table 4 and 5 are provided below for JPEG images. Results for both the wavelets are shown in the table. Again it can be seen from the tables that body parts within a certain imaging modality can be distinctly identified and hence classified.

Table 4. Percentage of energy values for MRI JPEG Images

Body Parts	Biorthogonal Wavelet				Reversible Biorthogonal			
	A	H	V	D	A	H	V	D
abdomen	90	.17	2.04	7.7	70	6.1	9.2	14
spine	80	.7	4.8	14	57	12	14	16
shoulder	93	.52	2.06	3.5	84	4.5	4.05	7.4

Table 5. Percentage of energy values for CT JPEG Images

Body Parts	Biorthogonal Wavelet				Reversible Biorthogonal			
	A	H	V	D	A	H	V	D
ankle	78	0.3	3.5	18	61	8	11	19
brain	85	.28	4.2	9.9	60	9	12	18
abdomen	78	.01	.36	2.01	78	.5	6	14

It is worth noting that the vector size, that will be required for indexing of body parts within certain imaging modality, is quite small, i.e., minimum of 4 decimal values and maximum of 8 decimal values. The time required to compute this vector is fraction of a second on a Pentium-4, 2.66 GHz with 256 MB of RAM. Database used is Microsoft SQL Server 2000.

Initially, the system can take a DICOM Image as well as JPEG image as a query image and the system returns five images in order of similarity. The system was checked with various body part images and it successfully returned images of same body part.

5 Conclusions

In this paper, we have introduced a concept that energy of each body part is distinct and different from the other body part. We have shown the results of calculated energies using wavelet transform. Further, we have used the basic information within the image to identify the imaging modality. We have shown that this technique can be applied to various image formats including DICOM and JPEG. The proposed method offers a novel technique for the classification of medical images in Content based image retrieval system. The vector size is also small for indexing purposes.

In the future, we plan to expand the parameters given in Table 1 so as to eliminate the initial image analysis for identification of imaging modality. Further, we plan to increase the size of our database by incorporating images from different modalities as well as increasing images of various body parts. Also, precision recall curves will be provided once the whole medical image retrieval system is integrated. The system will be able to use queries based on input images, region of interest, Pathology Bearing Regions etc.

Acknowledgement

This study was supported by Ministry of Culture & tourism and Culture & Content Agency in Republic of Korea.

References

1. Lehmann TM, Henning Schubert, Daniel Keysers, Michael Kohnen and Berthold B. Wein, "The IRMA code for unique classification of medical images", Medical Imaging 2003: PACS and Integrated Medical Information Systems: Design and Evaluation, Proceedings of SPIE Vol. 5033, 440-451.
2. Y. Liu, W.E. Rothfus, and T. Kanade, Tech. Report CMU-RI-TR-98-04, Robotics Institute, Carnegie Mellon University, 1998.
3. Shyu, C., Brodley, C., Kak, A., Kosaka, A., Aisen, A. and Broderick, L., ``ASSERT, A physician-in-the-loop content-based image retrieval system for HRCT image databases'', Computer Vision and Image Understanding , Vol. 75, Nos. 1/2, July/August 1999, pp. 111-132.

4. A. Mojsilovic, and J. Gomes, "Semantic based categorization, browsing and retrieval in medical image databases", Proc. Int. Conf. Image Processing, ICIP 2002, Rochester, New York, Sept. 2002.
5. Lehmann TM, Fischer B, Güld MO, Thies C, Keysers D, Deselaers T, Schubert H, Wein BB, Spitzer K, "The IRMA Reference Database and Its Use for Content-Based Image Retrieval in Medical Applications", In: Ammenwerth E, Gaus W, Haux R, Lovis C, Pfeiffer KP, Tilg B, Wichmann HE (Hrsg): GMDS 2004 - Kooperative Versorgung - Vernetzte Forschung - Ubiquitäre Information. Verlag videel OHG, Niebüll, 251-253, 2004.
6. Lehmann TM, Güld MO, Thies C, Fischer B, Spitzer K, Keysers D, Ney H, Kohnen M, Schubert H, Wein BB, "Content-based Image Retrieval in Medical Applications", Methods of Information in Medicine 2004; 43(4): 354-361.
7. Cohen, A., "Ondelettes, analyses multirésolution et traitement numérique du signal," Ph.D. thesis, University of Paris IX, Dauphine, 1992.
8. Daubechies, I., Ten lectures on wavelets, SIAM, 1992.
9. Mallinckrodt Institute of Radiology (MIR), ftp://wuerlim.wustl.edu/pub/dicom/images/version3/
10. ftp://medical.nema.org/medical/dicom/MRMultiframe/
11. Philips: ftp://ftp-wjq.philips.com/medical/interoperability/ out/Medical_Images/
12. http://www.barre.nom.fr/medical/samples/

Improvement of Grey Relation Analysis and Its Application on Power Quality Disturbances Identification

Ganyun Lv[1], Xiushan Cai[1], and Yuanyuan Jin[2]

[1] Department of Information Science and Engineering, Zhejiang Normal University, 299 Beishan Road, 321004 Jinhua, China
{Ganyun_lv, XSCai}@yahoo.com
http://www.zjnu.cn/index.html

[2] Nanjing Micro One Electronics Inc., 11F Huaxin Building, 9 Guanjiaqiao 210005 Nanjin, China
jinyy@microne.com.cn

Abstract. An improved grey relation analysis method was brought forward, based on concept of group relation index and relation index cube. The definition and calculating process was given out in this paper. In contrast to traditional grey relation analysis, the improved grey relation analysis had two advantages over traditional grey relation analysis: A) Greatly strengthened the veracity and reliability of relation analysis; B) Having a much broader range of its application. The improved method was applied to an application of power quality (PQ) disturbance identification in power system. The test result of the application has shows that the improved method has a much better effect than traditional grey relation analysis. The improved method can also be applied to many other applications in a wide range.

1 Introduction

Grey system theory [1, 2] is proposed in 1980's, as a tool for considering with uncertainty in extensive application, such as linear planning, forecasting, system control and identification. Grey relation analysis is an important part of grey system theory. It has been widely used in many applications [3, 4, 5, 6, 7] in recent years. It has even been applied in facing recognition combining with other statistical method [3]. Study of grey relation analysis is going on. Recently an optimal grey relation analysis was brought out in paper [6].

Power quality (PQ) has been an important concern for utility, facility and consulting engineers in recent years. It is necessary to analyze PQ problem and provide reasonable compensating measurements to improve the PQ. Detection and identification of PQ disturbances is an important task in the work. Researches have been adopted many technologies, such as neural networks [8], genetic net [9], expert system [10] and fast match [11], to identify these disturbances. These technologies have a common shortcoming of complex computation, and thus are difficult to be used in real time application. Many PQ disturbances, such as harmonics, transient pulse,

L. Wang et al. (Eds.): FSKD 2006, LNAI 4223, pp. 1159 – 1168, 2006.

low-frequency oscillation, white noise and so on, change in extent and frequency. These PQ disturbances can be taken as variable of grey system. In this paper, we presented an improved grey relation analysis for identification, and realized identification of the PQ disturbances simply and effectively.

2 Improvement of Grey Relation Analysis

Grey relation analysis is an important part of grey system theory. Here, theory of traditional grey relation analysis was introduced first, and then improved grey relation analysis was given out.

2.1 Review of Traditional Grey Relation Analysis

Assuming that there have two groups of sequences, one group is the reference sequences, and the other group is the comparative sequences. The reference sequences are $y_1, y_2, y_3, \cdots, y_m$, and the comparative sequences are $x_1, x_2, x_3, \cdots, x_n$ .They are

$$x_i = \{x_i(1), x_i(2), x_i(3), \cdots, x_i(N)\} \quad (i = 1,2,3,\cdots,n) \ . \tag{1}$$

$$y_j = \{y_j(1), y_j(2), y_j(3), \cdots, y_j(N)\} \quad (j = 1,2,3,\cdots,m) \ . \tag{2}$$

Where $x_i(k)$ and $y_i(k)$ are the k-th characteristic component of x_i and y_i respectively.

Then the grey relevant coefficient is defined in traditional grey relation analysis as follow

$$\xi_{ij}(k) = \frac{\min\limits_i \min\limits_k |y_j(k) - x_i(k)| + 0.5 \max\limits_i \max\limits_k |y_j(k) - x_i(k)|}{|y_j(k) - x_i(k)| + 0.5 \max\limits_i \max\limits_k |y_j(k) - x_i(k)|} \ . \tag{3}$$

In the above equation, the environmental coefficient is set to 0.5.

Then the grey relation value is to be

$$\xi_{ij} = (\xi_{ij}(1), \xi_{ij}(2), \xi_{ij}(3), \cdots \xi_{nj}(N)) \ . \tag{4}$$

Finally the grey relevant index between the reference sequence y_j and the comparative sequence x_i is defined as follow

$$r_{ij} = \frac{1}{N} \sum_{k=1}^{N} \xi_{ij}(k) \ . \tag{5}$$

Where r_{ij} represents the relevant degree between the reference sequence y_j and the comparative sequence x_i. These r_{ij} £¨i=1,2,...,n; j=1,2,...,m£© composed a matrix named Relation Index Matrix, shown as follow

$$R = \begin{bmatrix} r_{11} & r_{12} & r_{13} & \cdots & r_{1N} \\ r_{21} & r_{22} & r_{23} & \cdots & r_{2N} \\ r_{31} & r_{32} & r_{33} & \cdots & r_{3N} \\ \cdots & \cdots & \cdots & \cdots & \cdots \\ r_{m1} & r_{m2} & r_{m3} & \cdots & r_{mN} \end{bmatrix} . \tag{6}$$

The traditional relation analysis is widely used in many applications. However, it relies much on the correctness of reference array. When there are relatively strong characteristic of disperse and random in the reference sequences, the precision and reliability of relation analysis will greatly be worsened. That will be shown in the latter part of the paper.

2.2 Improved Grey Relation Analysis

Assuming that there have m*b reference sequences, as following,

$$\begin{array}{lllll} y_1^1, & y_1^2, & y_1^3, & \cdots, & y_1^b; \\ y_2^1, & y_2^2, & y_2^3, & \cdots, & y_2^b; \\ y_3^1, & y_3^2, & y_3^3, & \cdots, & y_3^b; \\ \cdots, & \cdots, & \cdots, & \cdots, & \cdots; \\ y_m^1, & y_m^2, & y_m^3, & \cdots, & y_m^b; \end{array} . \tag{7}$$

Where $\{y_i^1, \; y_i^2, \; y_i^3, \; \cdots, \; y_i^b;\}$ $(i = 1,2,3,\ldots,\mathrm{m})$ belong to the same group of reference sequence y_i. They are relevant each other, for example, they are the same type of PQ disturbances. On the other hand, they are independent in some degree. The comparative sequences are the same as equation (1) .

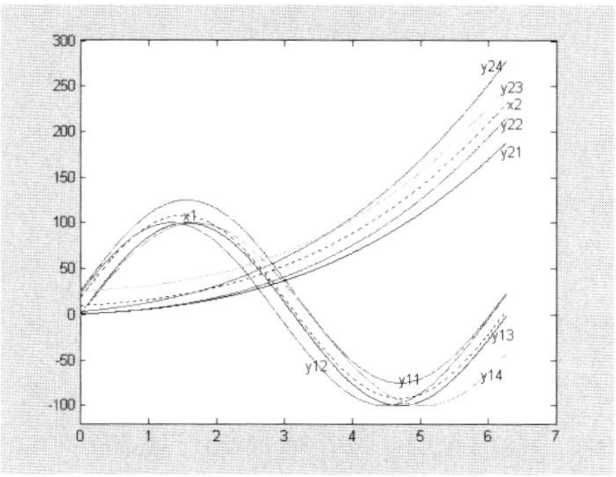

Fig. 1. The sketch map of reference sequences and comparative sequences

The relation sketch of the reference sequences and comparative sequences are shown in Fig.1, in which there have two groups of reference sequences (y_1, y_2) and two comparative sequences (x_1, x_2).

With the assumption above, the Relation Index Cube is defined as follow:

$$V_r = \begin{bmatrix} & & r_{11}^b & & r_{12}^b & \cdots & r_{1n}^b \\ & \iddots & \iddots & & \iddots & \iddots & r_{2n}^b \\ r_{11}^2 & r_{12}^2 & \cdots & & r_{1n}^2 & \iddots & \vdots \\ \begin{bmatrix} r_{11}^1 & r_{12}^1 & \cdots & r_{1n}^1 \\ r_{21}^1 & r_{22}^1 & \cdots & r_{2n}^1 \\ \cdots & \cdots & r_{ij}^1 & \cdots \\ r_{m1}^1 & r_{m2}^1 & \cdots & r_{mn}^1 \end{bmatrix} & & & & \begin{matrix} r_{2n}^2 \\ \vdots \\ r_{mn}^2 \end{matrix} & \iddots & r_{mn}^b \end{bmatrix} . \tag{8}$$

Where r_{ij}^w is the individual relevant coefficient between individual reference sequence y_i^w in group w and comparative sequence x_j . It can be obtained from the following equation.

$$r_{ij}^w = \frac{1}{N}\sum_{k=1}^{N}\xi_{ij}^w(k) . \tag{9}$$

Where the individual grey relevant coefficient $\xi_{ij}^w(k)$ is defined as :

$$\xi_{ij}^w(k) = \frac{\min_j \min_k \left|y_i^w(k) - x_j(k)\right| + 0.5 \max_j \max_k \left|y_i^w(k) - x_j(k)\right|}{\left|y_i^w(k) - x_j(k)\right| + 0.5 \max_j \max_k \left|y_i^w(k) - x_j(k)\right|} . \tag{10}$$

With the Relation Index Cube, the Group Relation Index matrix R_{group} is defined at last as follow,

$$R_{group} = \begin{bmatrix} r_{11} & r_{12} & \cdots & r_{1n} \\ r_{21} & r_{22} & \cdots & r_{2n} \\ \cdots & \cdots & r_{ij} & \cdots \\ r_{m1} & r_{m2} & \cdots & r_{mn} \end{bmatrix} . \tag{11}$$

Where the Group Relation Index $\tilde{r}_{ij}$ is defined as follow

$$\tilde{r}_{ij} = \left(\sum_{w=1}^{b} (r_{ij}^w)^p\right)^{1/p} , \qquad p = 1,2,..,\infty . \tag{12}$$

It represents the relevant degree of the *i*-th comparative sequence and the j-th group reference sequences. The Group Relation Index matrix R_{group} can represent the relationship between the comparative sequences and these groups of reference sequences as a whole.

In contrast to traditional grey relation analysis, the improved grey relation analysis has two following advantages at least. First, it has a much broader range of its application than traditional grey relation analysis because of its lower request of reference sequences data. The improved grey relation analysis can be applied in case of which the reference sequences are strong disperse. Second, it greatly strengthened the veracity and reliability of relation analysis. The efficiency of the improved grey relation analysis relies not on one individual reference sequence but on a group of reference sequences. The total effect of the improved method is much better than that of traditional grey relation analysis.

3 Application of Improved Grey Relation Analysis on Identifying PQ Disturbance

There have many kinds of PQ disturbances in power system, such as harmonics, voltage fluctuations, frequency deviation, sags, swells, over-voltage, under-voltage, transient pulse, low-frequency oscillation, high-frequency oscillation, and white noise. Among them, harmonics, transient pulse, low-frequency oscillation and white noise are similar and most difficult to identify. Many technologies are often relatively complex. This paper gives a new method based on the improved grey relation analysis.

The basic principle of identifying PQ disturbances with improved grey relation analysis is shown as Fig.2.

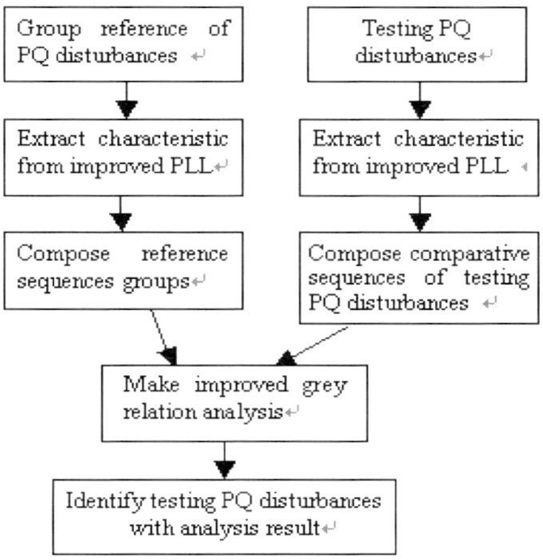

Fig. 2. Identification of PQ disturbances with improved grey relation analysis

Step of identifying PQ disturbances with the improved grey relation analysis is shown as below.

3.1 Construct Reference Signal of PQ Disturbance

In this article, the reference and testing PQ disturbances are generated from power system simulation.

The PQ disturbance is detected through an improved PLL system first, described as Fig. 3.

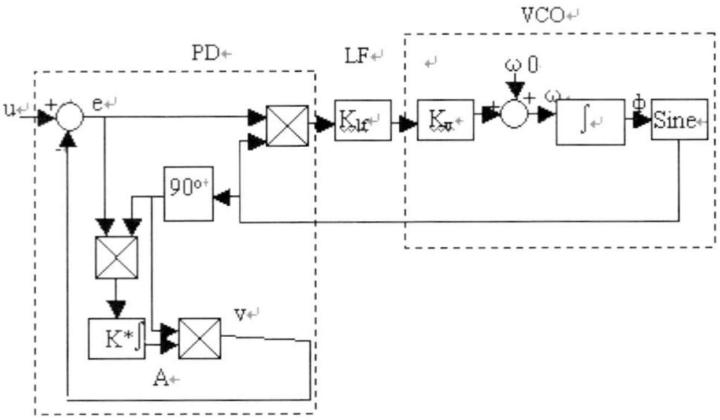

Fig. 3. Block diagram of the improved PLL system

The improved PLL system is a stable phase feedback control system, which could exact out the basic frequency component of the input signal without phase error. It also produces many useful outputs. The output y(t) is not only coherent with but also synchronized with the basic frequency component. The filter also produces instantaneous extent A(t), phase $\phi(t)$ and frequency $\omega(t)$ of the basic frequency component, and the error signal e(t) which represents deviation of input signal from the output signal. With output y(t) and error signal e(t), the filter can exact out the ideal component of input and the total disturbance. With outputs these outputs, the filter can detect out most of PQ disturbance. As an adaptive system, the PLL system is robust with respect to small change of basic frequency of input signal. Because of these characteristics, the PLL system is suited for detecting the PQ disturbance.

3.2 Extract the Features

With the output A(t), $\omega(t)$ and e(t) of PLL system, the characteristic data of the PQ disturbances, including reference signal and detected signal, is obtained. Following features are extracted from PLL output.

1) Time percents of $\Delta f \geq 1\%$, where $\Delta f = 100\% * \dfrac{f - f_0}{f_0}$.

2) Time percents of $\Delta A \geq 10\%$, where $\Delta A = 100\% * \dfrac{A - A_0}{A_0}$.

3) Alternating times of $\left|A-\tilde{A}\right| \geq 1\%$ a, where $\tilde{A}=\frac{1}{N}\sum A_n$.

4) Time percents of $\left|\frac{e(t)}{A_0}\right| \geq 2\%$.

5) Time percents of $\left|\frac{e(t)}{A_0}\right| \geq 2\%$, $\left|\frac{e(t)}{A_0}\right| \geq 4\%$ in 1/4 circle after the disturbance happened.

6) Time percents of $4\% \geq \frac{e(t)}{A_0} \geq 2\%$, $8\% \geq \frac{e(t)}{A_0} \geq 4\%$, $\frac{e(t)}{A_0} \geq 8\%$, $-2\% \geq \frac{e(t)}{A_0} \geq -4\%$, $-4\% \geq \frac{e(t)}{A_0} \geq -8\%$ and $\frac{e(t)}{A_0} \leq -8\%$ in one circle after the disturbance happened.

7) Time percents of $\left|\frac{e(t)}{A_0}\right| \geq 4\%$, $\left|\frac{e(t)}{A_0}\right| \geq 2\%$, $\left|\frac{e(t)}{A_0}\right| \geq 1\%$, $\left|\frac{e(t)}{A_0}\right| \geq 0.5\%$, $\left|\frac{e(t)}{A_0}\right| \geq 0.25\%$, in one circle after the disturbance happened.

8) The four maximum and minimum value of $e(t)$ and its relevant time.

With these features, the PQ disturbances, including harmonics, transient pulse, low-frequency oscillation and noise could be identified. Five samples of reference signal, harmonics, transient pulse, low-frequency oscillation and noise are produced in-order. Five testing signal of harmonics, transient pulse, low-frequency oscillation and noise are produced by the same order.

Five groups of reference sequences obtained from reference signal are shown as following

$$\begin{matrix} y_1^1 & y_1^2 & y_1^3 & y_1^4 & y_1^5; \\ y_2^1 & y_2^2 & y_2^3 & y_2^4 & y_2^5; \\ y_3^1 & y_3^2 & y_3^3 & y_3^4 & y_3^5; \\ y_4^1 & y_4^2 & y_4^3 & y_4^4 & y_4^5; \end{matrix} . \tag{13}$$

Where $\left(y_1^j \quad y_2^j \quad y_3^j \quad y_4^j\right)$, $j=1,2,3,4,5$ is one group of reference sequence, and $\left(y_i^1 \quad y_i^2 \quad y_i^3 \quad y_i^4 \quad y_i^5\right)$ belongs to the same type of reference signal. When *i*=1, it belongs to harmonics; when *i*=2, it belongs to transient pulse; when *i*=3, it belongs to low-frequency oscillation; when *i*=4, it belongs to noise.

Five comparative sequences, including 20 test samples, are gotten from test signals. Each comparative sequences likes as

$$x_1 \quad x_2 \quad x_3 \quad x_4 \ . \tag{14}$$

Where $x_i=\{x_i(1), x_i(2), x_i(3), \cdots, x_i(N)\}$ $(i=1,2,3,\cdots,n)$. $x_i(k)$ is one component of features. In order to better display of analysis, each comparative sequences is arrange as: x_1 is a comparative sequences of harmonics, x_2 is of transient pulse, x_3 is low-frequency oscillation and x_4 is of noise.

3.3 Improved Grey Relation Analysis

Set $p = \infty$ and $p = 1$ in equation (13). Two kinds of Group Relation Index matrixes are obtained in relation analysis. The results of first test are shown as equation (15) and (16):

$$R_{group}\_\max = \begin{bmatrix} 27.3914 & 25.1022 & 25.1639 & 25.3455 \\ 25.3305 & 30.4376 & 28.6053 & 25.2721 \\ 22.9171 & 25.3877 & 30.3629 & 23.8423 \\ 24.0167 & 27.2698 & 27.6393 & 30.7330 \end{bmatrix}, \ p = \infty \ . \tag{15}$$

$$R_{group}\_\text{mean} = \begin{bmatrix} 24.9689 & 24.6388 & 24.1959 & 24.9496 \\ 24.2590 & 29.4873 & 26.3457 & 24.7464 \\ 21.5191 & 24.5295 & 27.8664 & 23.7829 \\ 22.1396 & 24.8720 & 26.2100 & 28.8592 \end{bmatrix}, \ p = 1 \ . \tag{16}$$

The big value of grey relation means the strong relationship. Here we consider the largest relation index as belonging relation. Shown as equation (15) and (16), both two kinds of improved grey relation analysis methods can identify all these testing PQ disturbances correctly in this test. From first line of $R_{group}\_\max$, the test sample x_1 could be identified as the harmonics correctly. From second line of $R_{group}\_\max$, the test sample x_2 could be identified as the transient pulse correctly. From third line of $R_{group}\_\max$, the test sample x_3 could be identified as the low-frequency oscillation correctly. From last line of $R_{group}\_\max$, the test sample x_4 could be identified as the noise correctly. The same results could be obtained in four other tests.

Take each $y_1^j \ \ y_2^j \ \ y_3^j \ \ y_4^j$, $j = 1,2,3,4,5$ as the reference sequences of one samples of reference signal for traditional grey relation analysis. Five times of relation analysis are made. Corresponding to five reference sequences, five Relation Index Matrixes are obtained as follow:

$$\text{R1} = \begin{bmatrix} 24.5356 & 25.1022 & 25.3487 & 25.0411 \\ 25.1514 & 28.4135 & 25.6192 & 24.3341 \\ 22.0076 & 24.0814 & 28.7282 & 23.8423 \\ 21.8743 & 27.2698 & 25.6919 & 28.4574 \end{bmatrix} . \tag{17}$$

$$\text{R2} = \begin{bmatrix} 25.4070 & 24.1566 & 24.0336 & 25.1549 \\ 23.7961 & 30.4376 & 28.6053 & 24.7687 \\ 21.7011 & 25.3877 & 30.3629 & 23.6083 \\ 22.5359 & 24.2295 & 27.3562 & 28.1090 \end{bmatrix} . \tag{18}$$

$$R3 = \begin{bmatrix} 24.9753 & 24.5170 & 25.1103 & 24.1655 \\ 25.3305 & 28.9150 & 24.8934 & 25.0231 \\ 22.9171 & 24.8408 & 24.4128 & 23.7956 \\ 23.2509 & 23.2702 & 26.0330 & 30.7330 \end{bmatrix} . \tag{19}$$

$$R4 = \begin{bmatrix} 27.3914 & 24.9755 & 24.3228 & 25.3455 \\ 24.9985 & 30.0277 & 24.1683 & 25.2721 \\ 22.6025 & 24.2628 & 25.7878 & 23.8260 \\ 24.0167 & 24.5391 & 24.3297 & 28.5390 \end{bmatrix} . \tag{20}$$

$$R5 = \begin{bmatrix} 22.5352 & 24.4425 & 28.1639 & 25.0411 \\ 22.0182 & 29.6428 & 28.4423 & 24.3341 \\ 18.3674 & 24.4746 & 29.6404 & 23.8423 \\ 19.0202 & 25.0516 & 27.6393 & 28.4574 \end{bmatrix} . \tag{21}$$

As shown above, some error results are produced in R1 and R3 because of unmatching between reference signal and test signal. It has been proved that identification with the improved grey relation analysis has better satisfaction with towards dispersed reference signals than that with tradition grey relation analysis. The correct ratio of identification in improved grey relation analysis is much higher than tradition relation analysis method.

The improved grey relation analysis is an exploration method for PQ disturbances identification, it is worth of explore for the future.

4 Conclusion

The main purpose of article is to investigate an improvement of grey relation analysis based on concept of Group Relation Index. Each kind signal of the reference sequences is only one sample in traditional grey relation analysis, while each kind signal of the reference sequences is a group sample in improved grey relation analysis. The improved method has the following properties:

A) Has a much broader range of its application.
B) Greatly strengthened the veracity and reliability of relation analysis.

The verification of the improved relation analysis is asserted through one application on identifying PQ disturbances. It has been proved that identification with the improved grey relation analysis has higher satisfaction with towards dispersed reference signals than that with traditional grey relation analysis. The improved grey relation analysis is a progress of traditional grey relation analysis, and it could be widely applied in many other applications for the future.

References

1. Deng Ju-long: Introduction to grey system theory. Journal of Grey Systems, 1(1), (1989) 24-30
2. Deng Ju-long: Basic methods of grey system. Press of Huazhong University of Science and Technology, Wuhan, China, (1987) 33-48
3. Farn-Shing Chen, Ta-Chun Chang, Hsiu-Hsiang Liao: The application of the grey relation analysis on teacher appraisa. 2000 IEEE International Conference on Systems, Man, and Cybernetics, 5(8), (2000) 6-10
4. Song Bin, Yu Ping, Luo Yunbai, Wen Xishan: Study on the fault diagnosis of transformer based on the grey relational analysis. Power System Technology, 2002 Proceedings, International Conference on Power Con. Vol.4, No.6 (2002)
5. Edwards G.J, Lanitis A, Taylor C.J, Cootes T.F: Face recognition using statistical models. IEE Colloquium on Image Processing for Security Applications, pp. 36-43, 10 March 1997.
6. Bao Rong-chang: An optimal grey relational measurement. 2001. Proceedings of International Joint Conference on IJCNN '01, Vol3, No.6 (2001) 221-227
7. Yang Baoqing, LIU Haiping, Li Yunchen: Model of target identification base on grey relation. Modern Defence Technology, 31(1) (2004) 13-16
8. Shah Baki S.R, Abdullah M.Z, Abidin A.F.: Combination wavelets and artificial intelligent for classification and detection transient overvoltage. Research and Development, (2002) 331-336
9. ZHANG Zhi-yuan, LI Geng-yin, FENG Ren-qin: Auto recognition of power quality disturbance based wavelet and genetic net. Journal of North China Electric University, 29(3), (2003) 24-28
10. Emmanouil Styvaktakis, Math H. J. Bollen, Irene Y. H. Gu: Expert system for classification and analysis of power system events. IEEE Trans. Power Delivery, vol.19 (2004) 423- 428
11. T. K. Abdel-Galil, E. F. EL-Saadany, A.M. Youssef: On-line disturbance recognition utilizing vector quantization based fast match. IEEE Power Engineering Society Summer Meeting, Vol 2 (2002) 1027-1032

Eigen Palmprint Authentication System Using Dimension Reduction of Singular Vector

Jin Soo Noh* and Kang Hyeon Rhee**

Dept. of Electronic Eng., Multimedia & Biometrics Lab.,
Chosun University, Gwangju Metropolitan city, Korea(Daehanminkook)501-759
njinsoo@hanmail.net, multimedia@chosun.ac.kr

Abstract. This paper introduces the palmprint classification and recognition method based on PCA (Principal Components Analysis) using the Palmprint Acquisition Device. And the 75dpi palmprint image which is obtained by the palmprint acquisition device is used for the effectual palmprint recognition system. PCA have been previously applied to other biometric authentication and classification tasks, but not to palmprint images. We describe how the extraction of feature in the palmprint surface using Optimized PCA. The proposed system consists of the palmprint acquisition device and the palmprint authentication algorithm.

1 Introduction

The biometric system is concerned with identifying a person by the physiological characteristics or using some aspect of the behavior such as fingerprint, iris pattern, retina, face, palmprint, voice, gait, signature, and gesture. Fingerprint-based personal authentication has drawn considerable attention for a long time. However, workers and elderly may not provide clear fingerprint because of their problematic skins and physical work. Recently, voice, face and iris-based verifications have been studied and palmprint is also studying extensively [1].

The main characteristics of palmprint images are basically of three types: the principal lines, smaller lines or wrinkles, and fingerprint-like ridges, which, when combined with the skin's background, form specific textures. The skin pattern of a palmprint is completely formulated at birth as like the fingerprints and the pattern that is formed would not change over lifetime thus it could be used as the tool for personal authentication. However, there may be problems affecting the formation or the location of specific patterns generated from the partial damages in the regions of those specific patterns due to the conditions of the hand usage, but principal lines, wrinkles, ridges and minutiae points which are found in palmprint may provide effective means of personal authentication [2]. Although the research history of palmprint is shorter than other method, but various palmprint representations have been proposed for recognition, such as Line features [4], Feature points [5], Fourier spectrum [6], Eigenpalms features [7], Sobel's and morphological features [8], Texture energy [9], Wavelet signatures [10], Gabor phase [11], Fusion code [12], Competitive code [13], etc.

* First author.
** Corresponding author.

L. Wang et al. (Eds.): FSKD 2006, LNAI 4223, pp. 1169 – 1177, 2006.

In this work, in order to resolve the problem of the palmprint authentication, we refer to the palmprint acquisition device [1] to obtain the specific region of the palmprint accurately and suggested the palmprint authentication algorithm using the optimized principal components analysis therefore proposed algorithm reduce the computing time and the database size by extracting the eigen vectors which can generate the palmprint.

The rest of this paper is organized as follows. In section 2, we will be explaining the palmprint acquisition system. In section 3, the database preparation and the authentication process by using the optimized principal components analysis method. And section 4 will demonstrate the performance measurement by calculating the genuine acceptance rate and the false acceptance ratio. In section 5 concludes this paper.

2 The Palmprint Acquisition Device

We designed a palmprint authentication system to minimize changes in the inputted palmprint data at the time of palmprint authentication. To minimize the system operation time of palmprint authentication system, the image resolution of the inputted palmprint data was decided at 75dpi referring to [1]. The size of palmprint image was selected at 135 × 135 pixel after repeated tests.

In order to minimize changes in palmprint acquired, the device was composed of a site to fix the hand, camera and lamp. Fig. 1 shows the palmprint acquisition device. To acquire the wrinkles and palmprint in a certain area to reconstruct the original palmprint, the device was designed with a pin where the hand could be fixed between the middle and ring fingers and palmprint image could be acquired.

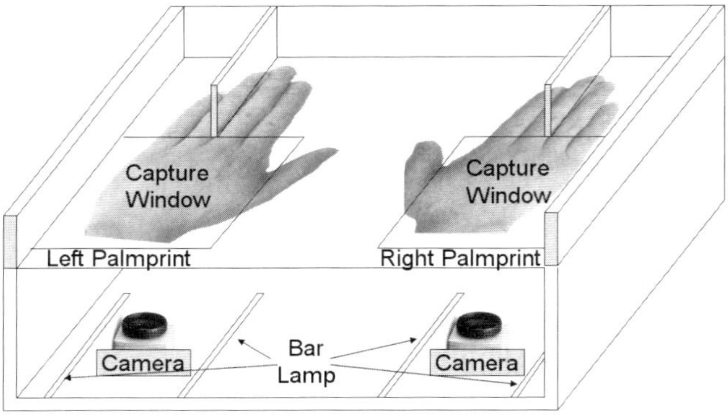

Fig. 1. Palmprint acquisition system

3 Palmprint Authentication Algorithm

The proposed palmprint authentication algorithm in this study was done in 5 steps, ie., palmprint normalization, average palmprint calculation, covariance matrix calculation

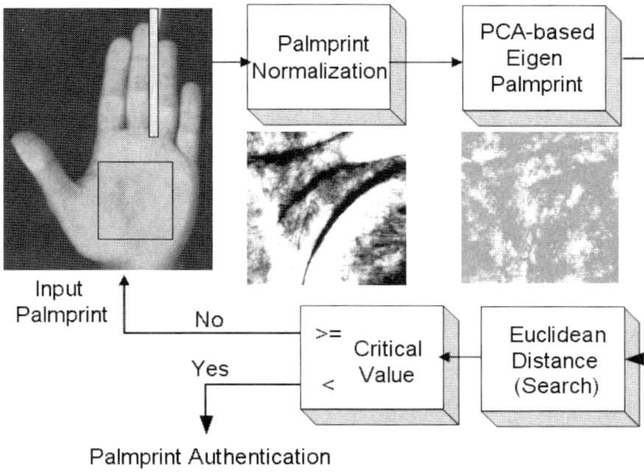

Fig. 2. The proposed algorithm of palmprint authentication

of palmprint, extraction of Eigen palmprint, and searching algorithm. Fig. 2 is the proposed palmprint authentication algorithm.

3.1 PCA-Based Eigen Palmprint

PCA is popular for feature extraction and used to find a low dimensional representation of data. Let us consider a set of *N* sample images $\{x_1, x_2, \cdots, x_n\}$ taking values in an *n*-dimensional image space, and assume that each image belongs to one of *C* classes $\{X_1, X_2, \cdots, X_C\}$. Let us also consider a linear transformation mapping the original *n*-dimensional image space into an *m*-dimensional feature space, where *m*<*n*. The new feature vectors $y_i \in R^m$ are defined by the following linear transformation:

$$y_i = W_{PCA}^t (x_i - \mu), \quad i=1,2, \ldots, M \tag{1}$$

where W_{PCA} is a $n \times m$ matrix. If the total scatter matrix S_T is defined as

$$S_T = \sum_{i=1}^{N} (x_i - \mu)(x_i - \mu)^t \tag{2}$$

where *N* is the number of sample images, and $\mu \in R^n$ is the mean image of all samples, then after applying the linear transformation, W_{PCA}^t, the scatter of the transformed feature vectors $\{y_1, y_2, \cdots, y_N\}$ is $W^t S_T W$. In PCA, the projection W_{PCA} is chosen to maximize the determinant of the total scatter matrix of the projected samples, i.e.,

$$W_{PCA} = \arg\max \left| W^t S_T W \right| = [w_1, w_2, \cdots, w_m] \tag{3}$$

where $[w_1, w_2, \cdots, w_m]$ is the set of n-dimensional eigenvectors of S_T corresponding to the m largest eigenvalues. Distance between $[w_1, w_2, \cdots, w_m]$ is usually measured using the Euclidean distance, but other distance measures also could be used. Fig. 3 is the calculated eigen palmprint using Equation (3).

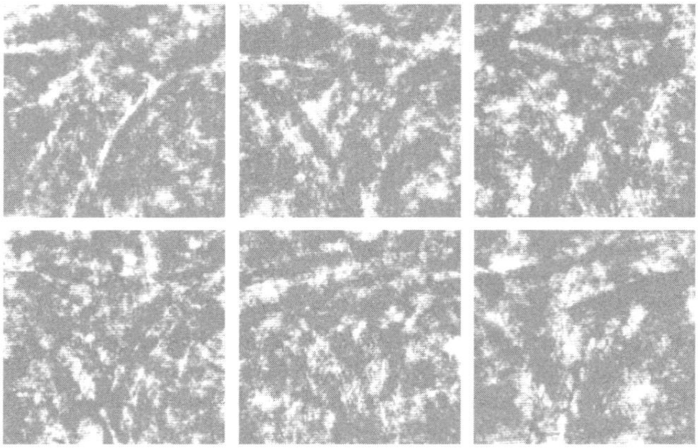

Fig. 3. Eigen Palmprint

3.2 Palmprint Authentication

Once the palmprint is inputted for authentication, the projection for Eigen palmprint is taken to calculate its value. Projection has internal meaning in that it sums the basis vector component of palmprint to distinguish palmprint similarity. Let *X, Y* be eigen palmprint vectors of length *n*. Then we can calculate the following distances between these eigen palmprint vectors. The general distance is introduced from Equation (4) to Equation (8).

-Mnikowski

$$d(X,Y) = \left(\sum_{i=1}^{n} |x_i - y_i|^p \right)^{1/p} \text{, here } p>0 \tag{4}$$

-Manhattan distance

$$d(X,Y) = \sum_{i=1}^{n} |x_i - y_i| \tag{5}$$

-Euclidean distance

$$d(X,Y) = \sqrt{\sum_{i=1}^{n} (x_i - y_i)^2} \tag{6}$$

-Angle-based distance

$$d(X,Y) = -\frac{\sum_{i=1}^{n} x_i y_i}{\sqrt{\sum_{i=1}^{n} x_i^2 \sum_{i=1}^{n} y_i^2}} \tag{7}$$

-Modified Manhattan distance

$$d(X,Y) = \frac{\sum_{i=1}^{n} |x_i - y_i|}{\sum_{i=1}^{n} |x_i| \sum_{i=1}^{n} |y_i|} \tag{8}$$

As in Equation (6), since the shortest Euclidean distance becomes the similar palmprint when comparing the weights of palmprint with the candidate palmprint images and that candidate image is decided the result of authentication.

4 Experiment and Result

The palmprint acquisition device was used in a total of 162 people to acquire the palmprints of both hands. Using 324 palmprints, the capacity of the proposed algorithm was measured. The subjects were 116 men and 46 women between the ages of 21 and 34. The acquisition palmprints had 135 × 135 pixel size. Using a device securing the hand, the rotation and movement of palmprint were minimized.

In order to prepare the palmprint database, one palmprint was measured 3 times. From each palmprint, the basis vector composing the eigen palmprint was calculated to come up with the database composed of average values.

4.1 Performance Measurement

As shown in Fig. 4, palmprint authentication was done as follows. The basis vector of eigen palmprint in the palmprint database and the basis vector coming from the inputted palmprint from the palmprint acquisition device were matched. The Euclidean distance of palmprint was within a certain range, the inputted palmprint same as the palmprint for authentication.

The typical standards used to evaluate the performance of palmprint authentication system are the false acceptance ratio (FAR) and false rejection ratio (FRR). Here, FAR is the ratio of identifying someone's palmprint as the subject's palmprint when the inputted palmprint was compared with the database palmprint. FRR is the ratio of rejecting the subject's palmprint. Equation (9) was used to calculate FAR and FRR in this study. In order to obtain the comparative database for [2, 3], the genuine acceptance rate (GAR) was calculated.

$$FAR = \frac{\gamma_a}{\gamma_b} \qquad FRR = \frac{\delta_a}{\delta_b} \qquad GAR = \frac{\delta_b - \delta_a}{\delta_b} = 1 - FRR \tag{9}$$

γ_a : The number of recognizing someone else as the subject.

γ_b : The number of matching the subject and someone else.

δ_a : The number of recognizing the subject as someone else.

δ_b : The number of matching the subject and the subject.

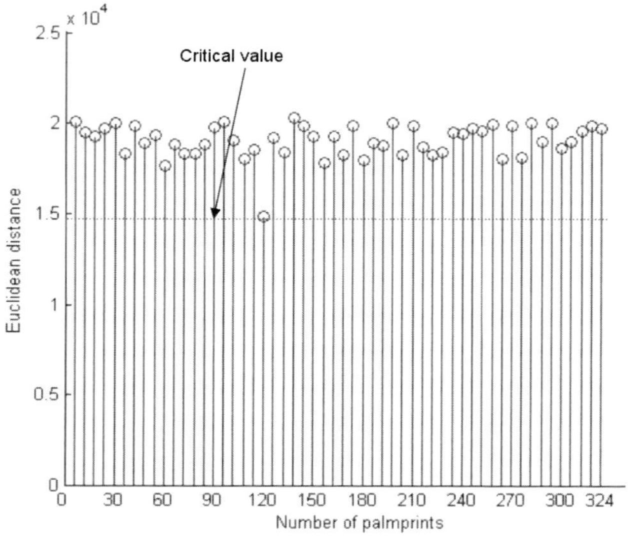

Fig. 4. The procedure of the palmprint authentication

Palmprint authentication was done using the palmprint in the database and that obtained from the palmprint acquisition device in real time. Thus, the palmprint acquisition device was designed to minimize data variance. The total number of palmprint matching in this study was 104,976 (324×324). The number of palmprint matching of the subject and someone else was 104,652. The number of matching the subject and subject was 324.

Among the palmprint authentication system with excellent performance testing ability, the performance of the proposed algorithm was compared using the Hu moment [2] and Gabor filter [3]. Table 1 is FAR and GAR of [2, 3].

Table 1. FAR and GAR of [2, 3]

Hu Moment [2]		Gabor Filter [3]	
FAR[%]	GAR[%]	FAR[%]	GAR[%]
0.038	98.1	0.02	97.7
0.049	98.1	0.03	97.8
0.055	98.4	0.04	98.0
0.087	98.9	0.05	98.2

Fig. 5 is the produced Euclidean distance using the results of matching simulation of 324 subjects vs. other palmprint. Using this result, simulation was done by setting the critical value of palmprint authentication from 1.62×10^4 to 1.65×10^4.

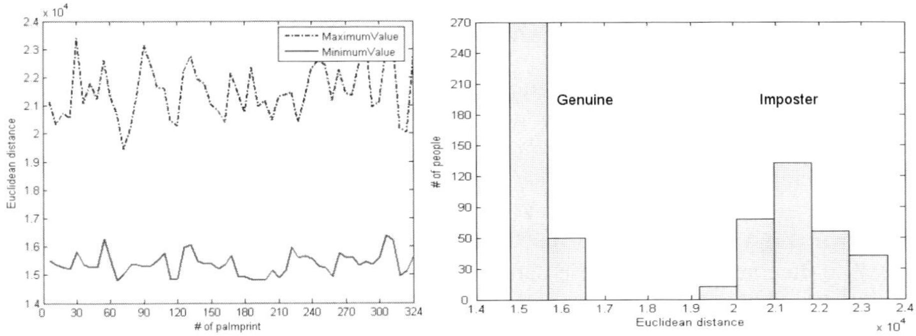

Fig. 5. The euclidean distance of palmprint

Table 2 shows FAR and GAR of the proposed algorithm with changing critical values.

Table 2. The FAR and GAR performances of the first authentication

Critical value	FAR(γ_a / γ_b)[%]	GAR(1- δ_a / δ_b)[%]
1.62×10^4	0.043	96.0
1.63×10^4	0.052	96.6
1.64×10^4	0.059	97.2
1.65×10^4	0.103	98.1

Although FAR and GAR performance was overall high in Table 2, FAR was decreased by 10% compared with [2, 3]. This problem was because palmprint authentication was treated erroneous when the number of data less than the critical values was more than 2. Thus, in order to improve this problem, the system was designed so that the palmprint was acquired again to reconstruct the palmprint when more than 2 values fall into the critical value range. Table 3 is the processed data after the second authentication.

Table 3. The FAR and GAR performances of the second authentication

Critical Value	FAR(γ_a / γ_b)[%]	GAR(1- δ_a / δ_b)[%]
1.62×10^4	0.031	98.1
1.63×10^4	0.033	98.1
1.64×10^4	0.036	98.5
1.65×10^4	0.052	98.8

As shown in Table 3, when the critical value was 1.64 in the processed data after the second authentication, FAR and GAR were 0.036[%] and 98.5[%], respectively.

As a result, the optimized values in this study and those proposed in [2, 3] in which FAR and GAR were improved to 5.3% and 0.4%, respectively.

4.2 Speed of Palmprint Authentication

The proposed palmprint authentication system in this study was realized in Matlab. The specification of the computer was Intel Pentium 4(3.0GHz) process and 2GB RAM. The used camera for palmprint acquisition was PANWEST's web camera.

The overall needed time to process for palmprint authentication was about 1.0 Sec. As shown in Fig. 6, the needed time for acquired palmprint was 0.513 Sec, that for Eigen palmprint extraction was 0.093 Sec, and that seeking the database was 0.313 Sec. The time needed to seek the database changes according to the amount of database. In this study, 324 palmprint images were stored in the database.

Palmprint acquisition...	(0.513Sec.)
Average Palmprint...	(0.063Sec.)
Eigen Palmprint...	(0.093Sec.)
Search...	(0.313Sec.)
Total Time...	(0.982Sec.)

Fig. 6. Total run-time of the proposed algorithm

5 Conclusion

This paper introduces the palmprint classification and recognition method based on Principal Components Analysis. This technique largely constituted of the Palmprint acquisition system using the acquisition device, palmprint authentication system using PCA and data matching algorithm using euclidean distance. The palmprint acquisition system utilized the palmprint fixing device to provide the accuracy in palmprint image acquisition and in order to shorten the time needed for the eigen palmprint calculations, we used the optimized PCA. The palmprint authentication system extracts the basis vector of Eigen palmprint from the inputted palmprint and seeks the Euclidean distance by integrating the database and data weight for palmprint authentication.

Our future work is optimization of the data searching algorithm and implementation.

Acknowledgement. This study was supported by research funds from Chosun University, 2000.

References

1. J. S. Noh and K. H. Rhee, "Palmprint Identification Algorithm Using Hu Invariant Moments," Proc. of the 2nd FSKD, LNAI 3614, pp.91-94, 2005.
2. D. Zhang, W. K. Kong, J. You and M. Wong, "Online Palmprint Identification," IEEE Transactions on Pattern Analysis and Machine Intelligence, vol. 25. NO.9, pp.1041-1050, 2003.
3. D. Zhang and W. Shu, "Two Novel Characteristics in Palmprint Verification: Datum Point Invariance and Line Feature Matching," Pattern Recognition, vol. 32, no. 4, pp. 691-702, 1999.
4. N. Duta, A.K. Jain, and K.V. Mardia, "Matching of Palmprint," Pattern Recognition Letters, vol. 23, no. 4, pp.477-485, 2001.
5. W. Li, D. Zhang, and Z. Xu, "Palmprint Identification by Fourier Transform," International Journal of Pattern Recognition and Artificial Intelligence, vol. 16, no. 4, pp. 417-432, 2002.
6. G. Lu, D. Zhang and K. Wang, "Palmprint Recognition Using Eigenpalms Features," Pattern Recognition Letters, vol. 24, issues 9-10, pp. 1463-1467, 2003.
7. C.C. Han, H.L. Cheng, K.C. Fan and C.L. Lin, "Personal Authentication Using Palmprint Features," Pattern Recognition, vol. 36, no 2, pp. 371-381, 2003.
8. Lei Zhang and David Zhang, "Characterization of Palmprints by Wavelet Signatures via Directional Context Modeling," IEEE Trans. on SMC. B, Vol. 34, No. 3, pp. 1335-1347, June 2004
9. D. Zhang, W. Kong, J. You, and M. Wong, "On-line Palmprint Identification," IEEE Trans. on PAMI, vol. 25, no. 9, pp. 1041-1050, 2003 .
10. J. You, W.K. Kong, D. Zhang and K. Cheung, "On Hierarchical Palmprint Coding with Multi - features for Personal Identification in Large Databases," IEEE Transactions on Circuit Systems for Video Technology, vol.14, no. 2, pp. 234-243, 2004.
11. W.K. Kong and D. Zhang, "Feature-Level Fusion for Effective Palmprint Authentication," Proc. of the 1st ICBA, LNCS 3072, pp.761-767, 2004.
12. W.K. Kong and D. Zhang, "Competitive Coding Scheme for Palmprint Verification," Proc. of the 17th ICPR, vol.1, pp. 520-523, 2004.

Towards Security Evaluation Based on Evidence Collection

Reijo Savola

VTT Technical Research Centre of Finland, P.O. Box 1100,
FIN-90571 Oulu, Finland
Reijo.Savola@vtt.fi

Abstract. Information security evaluation of software-intensive systems typically relies heavily on the experience of the security professionals. Obviously, automated approaches are needed in this field. Unfortunately, there is no practical approach to carrying out security evaluation in a systematic way. We introduce a general-level holistic framework for security evaluation based on security behavior modeling and security evidence collection, and discuss its applicability to the design of security evaluation experimentation set-ups in real-world systems.

1 Introduction

Security evaluation, testing and assessment techniques are needed to be able find adequate solutions. Seeking *evidence* of the actual information security level or performance of systems still remains an ambiguous and undeveloped field. To make progress in the field there is a need to focus on the development of better experimental techniques, better security metrics and models with practical predictive power [3].

The main contribution of this study is to introduce a conceptual approach to security evaluation based on evidence collection and to discuss the evidence collection process in practice. The rest of this paper is organized as follows. Section 2 discusses security metrics and their relationships in general, Section 3 presents our evaluation framework, and finally, last section gives conclusions.

2 Background

The wide majority of the available security metrics approaches offering evidence information have been developed for evaluating security policies and the maturity of security engineering processes. The most widely used of these maturity models is the Systems Security Engineering Capability Maturity Model SSE-CMM (ISO/IEC 21827) [5]. Other well-known models are Trusted Computer Security Evaluation Criteria (TCSEC, The Orange Book) (TCSEC 1985) [8] and Common Criteria (CC) [4]. In connection with policy and process metrics, it is extremely important to evaluate the security functionality of products at the technical level, without forgetting their

L. Wang et al. (Eds.): FSKD 2006, LNAI 4223, pp. 1178 – 1181, 2006.

life cycle management. The goal of the whole process of seeking of security evidence should be targeted at understanding information security threats and vulnerabilities of the product and its usage environment holistically.

Most technical security analysis is currently performed using penetrate-and-patch or "tiger team" tactics. The security level is evaluated by attempting to break into a system under evaluation, exploiting previously known vulnerabilities. If a break-in attempt is successful, the vulnerability is patched. Penetrate-and-patch tactics have been used by special security testing professionals whose methods and tools have not been made public knowledge. There are several problems with penetrate-and-patch: it requires experienced professionals, the actual testing is carried out too late, and the patches are often ignored and even sometimes introduce new vulnerabilities. Most of the technical testing metrics are meant for the unit or source code level. Various methods for system security evaluation and assessment have been proposed in the literature, see e.g. [2] [6] [7] [9]. These frameworks are conceptual and help in understanding the problem area. However, these frameworks do not offer aggregated means for practical security evaluation or the testing process.

3 Framework for Seeking Evidence of Security

The most important task in the whole process of security evaluation is to identify security risks and threats, taking enough assumptions of the attackers' capabilities into account. A holistic and cross-disciplinary threat picture of the system controls the development of security solutions. Threats that are possible during the whole life cycle of the system under evaluation must be considered.

The goal of defining security requirements for a system is to map the results of risk and threat analysis to practical security requirement statements that manage (cancel, mitigate or maintain) the security risks of the system under investigation. The security requirements play a crucial role in the security evaluation. The requirements guide the whole process of security evidence collection. For example, security metrics can be developed based on requirements: If we want to measure security behavior of an entity in the system, we can compare it with the explicit security requirements, which act as a "measuring rod".

All applicable *dimensions* (or *quality attributes*) of security should be addressed in the security requirements definition. See e.g. [1] for a presentation of quality attribute taxonomy. Well-known general dimensions include confidentiality, integrity, availability, non-repudiation and authenticity. Quality attributes like usability, robustness, interoperability, etc., are important requirements too. In fact, an unusable security construct can even turn out to be a security threat.

It must be noted that one cannot easily define a general-level security requirement list that could be used for different kinds of systems. The actual requirements and role of the security dimensions heavily depends on the system itself, and its context and use scenarios.

It is obvious that in order to be able to evaluate security systematically, a model of the security behavior of a system is needed.

3.1 Security Evidence

Security evidence is gathered from various sources as input to the decision process of security evaluation. The evidence collection should be arranged in a way that supports evaluation of security behavior and security actions. We classify the types of security evidence information into three categories:

- **Measured evidence.** The process of gathering measured or assessed information uses security metrics as its basis. Measured evidence can be collected during security testing or in a security audit based on pre-defined metrics.
- **Reputation evidence.** Reputation of software or hardware constructs, or their origin, is an important class of evidence. A software company in charge of implementing a product might have some confidential knowledge of the security of different software components. Reputation evidence can be collected from experience of R&D departments and be based on general-level knowledge.
- **Tacit evidence.** In addition to the measured and reputation evidence, there might be some "silent" or "weak" signals of security behavior. The subjectivity level of tacit evidence might be higher than in the case of measured and reputation evidence. Collection of tacit evidence is typically an ad hoc process. Senior security experts and "tiger teams" play an important role in this kind of evidence.

The objectivity level of the evidence varies a lot. In many cases, even the measurements are arranged in a highly subjective manner. Typically, no single measured value is able to capture the security value of a system. Thus, several pieces of security evidence have to be combined.

3.2 Process for Practical Security Evaluation

We propose the following process to carry out practical-level security evaluation:

1. **Risk and threat analysis.** Carry out risk and threat analysis of the system and its use environment if not carried out before. These are lacking in many practical systems.
2. **Define security requirements** in a way that they can be compared with the security actions of the system. Based on the threat analysis, define the security requirements for the system, if not yet defined. These are lacking in many practical systems.
3. **Prioritize security requirements.** The most critical and most often needed security requirements should be paid the most attention.
4. **Model the security behavior.** Based on the prioritized security requirements, identify the functionality of the system that forms the security actions and their dependencies in a priority order.
5. **Gather evidence** from measured, reputation and tacit security information.
6. **Estimate the probabilities and impacts of security actions** based on the evidence.
7. **Aggregate the results from the probability and impact estimation** to form a clear picture of whether or not the system fulfils the security requirements.

4 Conclusions

We have discussed the problem of information security evaluation in the context of software-intensive systems. There are no systematic means of carrying out security evaluation. In this paper we have presented a conceptual framework for security evaluation with some practical considerations. The framework is based on evidence collection and security requirement centered impact analysis.

This is not a rigorous solution and future work needs to be done on developing a suitable language for expressing security requirements and security behavior in an unambiguous way. A collection of security patterns would be very helpful in modeling the security behavior when carrying out security testing or experimentation.

References

1. Avizienis A., Laprie J.-C., Randell B. and Landwehr C.: Basic Concepts and Taxonomy of Dependable and Secure Computing. In: IEEE Transactions on Dependable and Secure Computing, Vol. 1, No. 1, January/March 2004, pp. 11-33.
2. Brocklehurst S., Littlewood B., Olovsson T. and Jonsson E.: On Measurement on Operational Security. In: IEEE AES Systems Magazine, Oct. 1994. pp. 7-15.
3. Greenwald M., Gunter C., Knutsson B., Seedrov A., Smith J. and Zdancewic S.: Computer Security is not a Science (but it should be). In: Large-Scale Network Security Workshop, Landsdowne, VA, March 13-14, 2003.
4. ISO/IEC 15408: Common Criteria for Information Technology Security Evaluation, Version 2.2, 2004.
5. ISO/IEC 21827: Information Technology – Systems Security Engineering – Capability Maturity Model (SSE-CMM), 2002.
6. McDermid J.A. and Shi, Q.: A Formal Approach for Security Evaluation. In: pp. 47-55, 1992.
7. Nicol D., Sanders W. H., Trivedi K. S.: Model-Based Evaluation: From Dependability to Security. In: IEEE Transactions on Dependable and Secure Computing, Vol. 1, No. 1, January/March 2004, pp. 48-65.
8. Trusted Computer System Evaluation Criteria, "Orange Book", U.S. Department of Defense Standard, DoD 5200.28-std, 1985.
9. Voas J., Ghosh A., McGraw G., Charron F. and Miller K. (1996) Defining an Adaptive Software Security Metric from a Dynamic Software Failure Tolerance Measure. In: Proceedings of the 11th Annual Conference on Computer Assurance, Systems Integrity, Software Safety, Process Security (COMPASS), 1996.

Optical Camera Based Pedestrian Detection in Rainy Or Snowy Weather

Y.W. Xu[1,2], X.B. Cao[1,2], and H. Qiao[3]

[1] Department of Computer Science and Technology, University of Science and Technology of China, Hefei, 230026, P.R. China

[2] Anhui Province Key Laboratory of Software in Computing and Communication, Hefei, 230026, P.R. China

ywxu@mail.ustc.edu.cn, xbcao@ustc.edu.cn

[3] Institute of Automation, Chinese Academy of Sciences, Beijing, 10080, P.R. China

hong.qiao@mail.ia.ac.cn

Abstract. Optical camera based detection method is a popular system to fulfill pedestrian detection; however, it is difficult to be used to detect pedestrians in complicated environment (e.g. rainy or snowy weather conditions). The difficulties mainly include: (1) The light is much weaker than in sunny days, therefore it is more difficult to design an efficient classification mechanism; (2) Since a pedestrian always be partly covered, only using its global features (e.g. appearance or motion) may be mis-detected; (3) The mirror images on wet road will cause a lot of false alarms. In this paper, based on our pervious work, we introduce a new system for pedestrian detection in rainy or snowy weather. Firstly, we propose a cascaded classification mechanism; and then, in order to improve detection rate, we adopt local appearance features of head, body and leg as well as global features. Besides that, a specific classifier is designed to detect mirror images in order to reduce false positive rate. The experiments in a single optical camera based pedestrian detection system show the effeteness of the proposed system.

1 Introduction

Most existing pedestrian detection systems (PDS) [1] [2] [3] [4] [5] are designed for sunny weather; however, a practical PDS should also work well in rainy or snowy weather. One way to solve this problem is to adopt infrared camera based systems [4] [5] or radar/laser based systems [6] which do not affected by rain or snow (it means an infrared camera based PDS works the same in both sunny and rainy/snowy days.); these systems are high cost and maybe affected by other factors. Another way is to design a suitable optical camera based PDS, the system is low cost, simple and having irreplaceable advantages. Compared with an infrared camera based PDS, an optical camera based PDS has much wider detection range and it is not affected by environment temperature.

To an optical camera based PDS, rainy and snowy weather causes following additional difficulties: (1) Brightness of the original images is weaker than in sunny days,

L. Wang et al. (Eds.): FSKD 2006, LNAI 4223, pp. 1182 – 1191, 2006.

and we found that color information can hardly be used according to many experiments, therefore pedestrians are harder to distinguished from the environment; (2) In rainy or snowy weather, a pedestrian might be partly covered by umbrella or raincoat, if take the same way of pedestrian detection in sunny weather which only using global features(e.g. appearance or motion) of pedestrian, a lot of pedestrians will be mis-detected; (3) In rainy or snowy days, the wet road becomes a huge mirror, this makes the images much more complicated, the mirror images of many objects (such as pedestrians and trees) will cause a lot of false alarms.

We have designed a system to detect pedestrians in sunny weather using a two-step method. It firstly used both global appearance features and motion features to select candidates with an AdaBoost classifier; and then a decomposed SVM classifier was used to make accurate classification. [2]

However, to fulfill pedestrian detection in rainy or snowy weather, it is necessary to design a better classification mechanism and efficient classifiers. Based on our previous work for sunny weather [2], new features need to be introduced, and additional measures need to be taken in order to reduce false alarms; however, the detection speed also needs to be guaranteed.

More and more features have been adopted for pedestrian detection; however, most of the existing PDS systems only use pedestrian's global features (features of entire human body) to detect pedestrian. The most widely used global features are appearance or motion features. For instants, Paul Viola et al. took both advantages of global appearance and motion features to detect pedestrian in static scene [1]. Until now, only a few PDS considers using local features to detect pedestrian or to assist detection. For example, Shashua Amnon et al. proposed another PDS which took nine main parts of a human body and their position relations as key features to detect pedestrian [3].

Moreover, since the introduction of new features probably slows down the detection speed, we present a three-stage method.

The remainder of the paper is arranged as follows: Section 2 describes detection and training procedures of the system in detail. Section 3 introduces the validation experiment and shows the results. Section 4 concludes this paper.

2 Procedure of the System

2.1 Detection Procedure

As shown in Figure 1, the whole detection procedure consists of three main stages: candidate selection (stage 1), further confirmation (stage 2) and false alarm reduction (stage 3). Each stage contains one or a group of classifier(s).

To solve the problems caused by rainy or snowy weather, it is necessary to design an efficient classification mechanism and pertinent classifiers. In the system, we improved our two-cascade architecture, adding a new module to solve the false alarm problem caused by specifically weather condition; and more complicated further confirmation layer is designed to solve the problem of being partly covered.

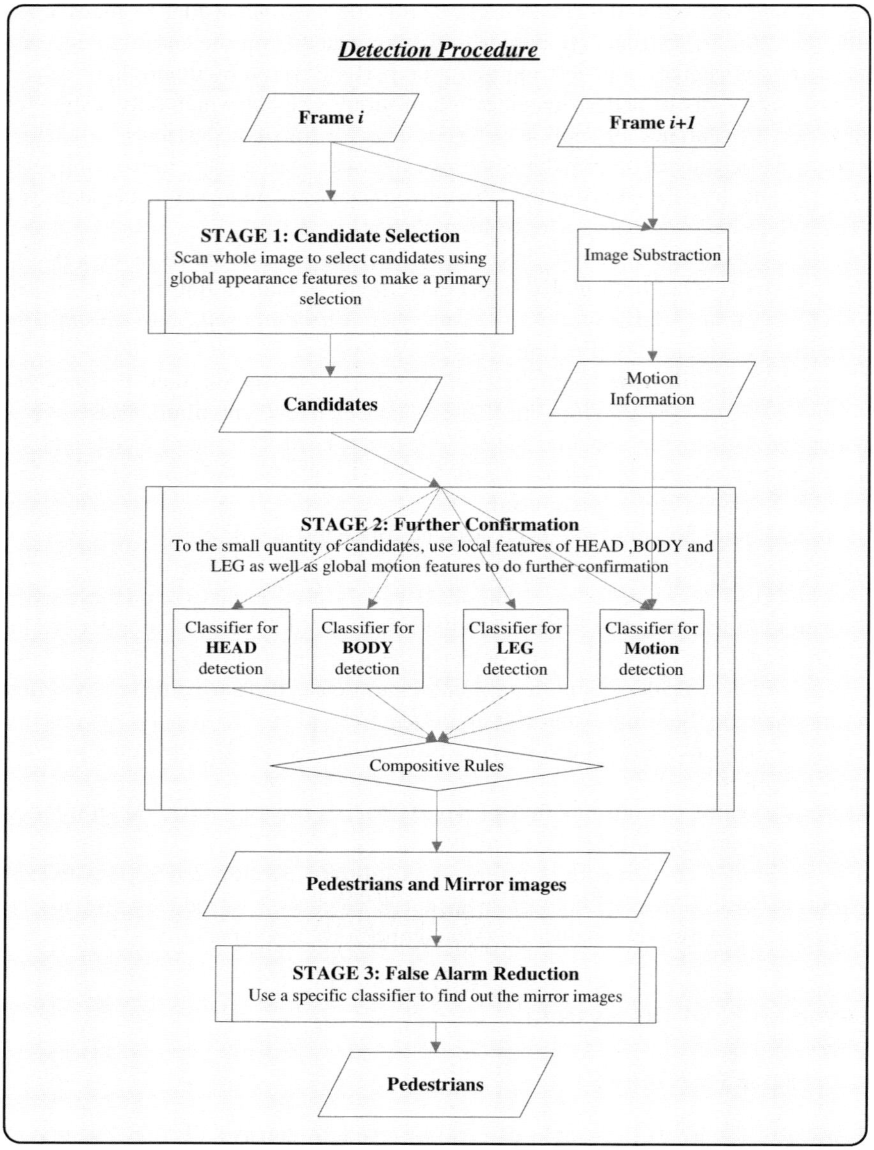

Fig. 1. Detection procedure of the system

2.2 Classification Mechanism and Classifiers

To such a classification based PDS, the classification mechanism and performance of each classifiers are most important. There are six classifiers in the system, the detail of them are listed in Table 1.

Table 1. Detail of all classifiers in the system

Stage	Classifiers / Training algorithm	Features		Training Samples	
		Type	Number	Positive	Negative
1	Candidate selection / AdaBoost	Global Appearance	280	Pedestrians	Non-pedestrians
2	Head detection / AdaBoost	Local appearance	80	Head/ umbrella	Others
	Body detection / AdaBoost	Local appearance	120	Coat/ raincoat	Others
	Leg detection / AdaBoost	Local appearance	160	Legs	Trees and others
	Motion detection / AdaBoost	Global motion	160	Motion	Others
3	Mirror image detection / SVM	Global appearance and motion	160 + 120 (ap. + mo.)	Mirror images	Pedestrians

Candidate selection classifier is designed to quickly wash out most of non-pedestrian objects at first. Hence there is ample time for further confirmation and false alarm reduction. Similar to our previous work [2], zoom-image and slide-window techniques are applied to perform exhaustive search over the whole images at different scale; to each window, the candidate selection classifier is used to decide whether there might be a pedestrian in it.

The classifier is only based on global appearance features in order to get higher detection speed; however, according to our experience and experimental results, only using appearance features is enough if about two times candidates are selected than in our previous system for sunny weather [2].

As shown in Figure 2, four classifiers work in parallel to make further confirmation to the candidates. There are about 100-400 candidates selected, to each candidate, we use local features of a pedestrian's three main parts and global motion features to make further confirmation as following steps:

(1) To get motion information of the candidate region, and then use an AdaBoost classifier to get the probability *P(motion)* that whether the candidates is a pedestrian.

(2) To get the top 1/4 region of the candidate, and then use an 8×8 pixel slide window to scan this area for head with a classifier. The outcome of this classifier is *P(head)* which describes the probability of the region contains a head; it is also a real number between 0 and 1.

(3) To get the middle 1/2 region of the candidate, and then use a well trained classifier to determine whether it contains a human body. The probability is expressed as *P(leg)*.

(4) Similar to previous step, to estimate the probability *P(leg)* that whether the bottom 1/2 region contains a pair of human legs.

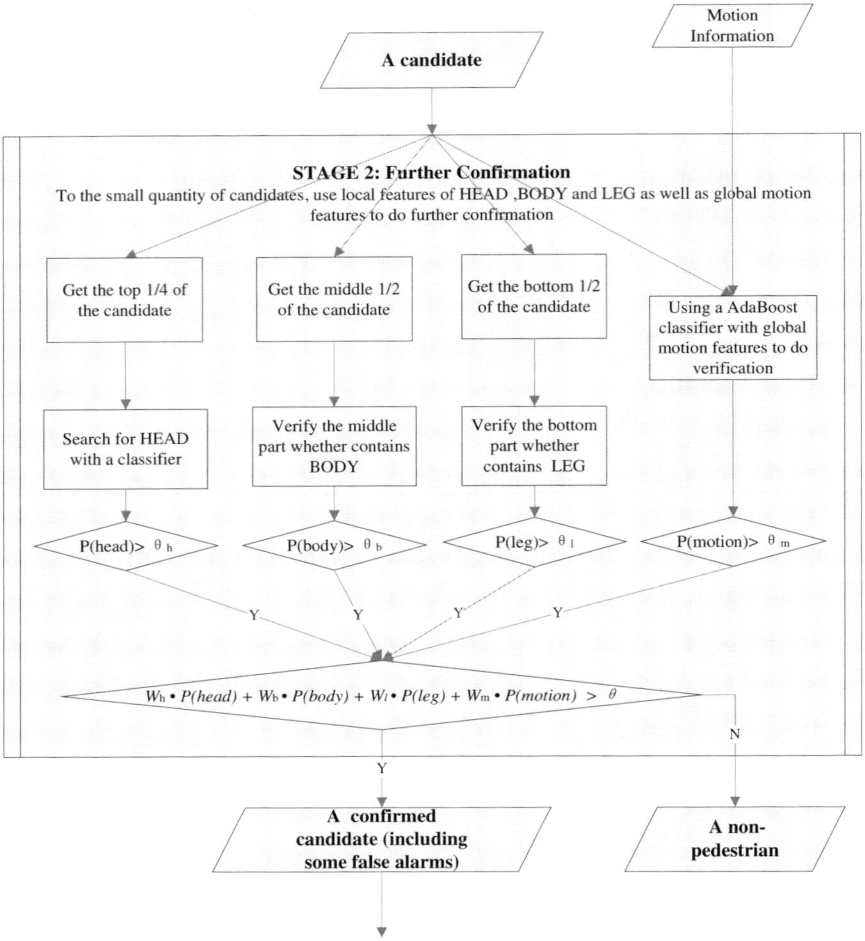

Fig. 2. Detail of further confirmation

(5) To make following computation to make a decision:

① if $P(motion) \le \theta_m$, then $P(motion) = 0$; Others are similar to this.

② if $W_h \bullet P(head) + W_b \bullet P(body) + W_l \bullet P(leg) + W_m \bullet P(motion) > \theta$, then the candidate is judged as a pedestrian in this stage and it will be verified by next stage; Else it is a non-pedestrian.

The four classifiers are all trained by AdaBoost algorithm [1] [7], and all the parameters can be obtained at the same time of training. However, we adjust the eight parameters θ_m, θ_h, θ_b, θ_l, W_m, W_h, W_b, and W_l manually according to the experimental results, because this can increase the training speed great a lot and the performance is also acceptable.

After further confirmation, there are only 10-50 confirmed candidates. Some of them are real pedestrians, and usually more than a half of them are mirror images and other false alarms according to the experimental results, and almost the remainder false alarms are mirror images. To each confirmed candidate, a specific classifier is adopted to judge whether it is a mirror image, if it is not, then it is a pedestrian. (Figure 3)

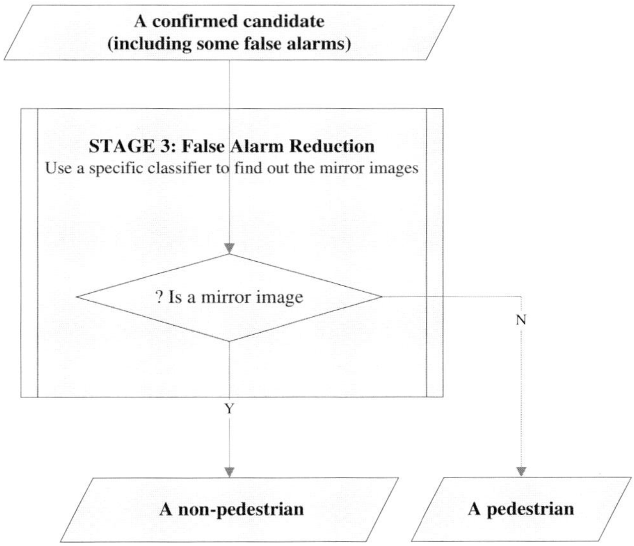

Fig. 3. A classifier to reduce false alarms caused by mirror images

Different from others, mirror image detection classifier takes non-pedestrian (mirror images) as positive samples, and takes pedestrian as negative samples; whilst other classifier take pedestrian as positive one.

2.3 Classifier Training: Samples and Features

As listed in Table 1, we train most of the classifiers with quick and effective AdaBoost algorithm [1] [7]. This algorithm can finish key feature selection and classifier training at the same time. Mirror image detection classifier in stage 3 is trained with SVM ^{light} algorithm proposed by Joachims [8] [9]. Furthermore, different kinds of classifiers aim at different purpose; therefore each classifier is trained separately with its own target. Classifier in stage 1 should be as fast as possible and the false negative rate must be extremely low. Classifiers in stage 2 and 3 should have high positive rate, the detection speed is ignored to them.

To get a well trained classifier, various high quality samples and proper features are most important. Part of the samples of entire pedestrian we used to train the

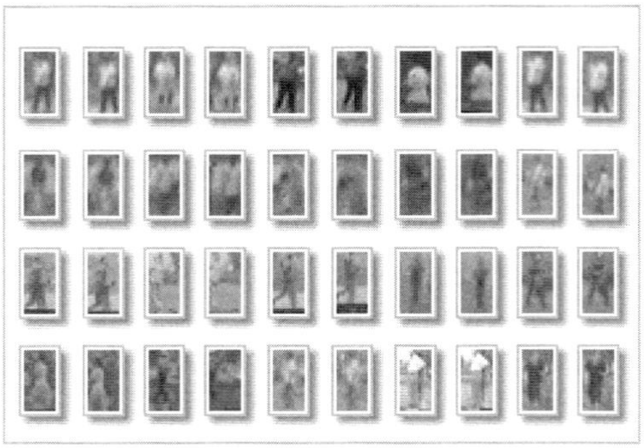

Fig. 4. Some positive samples of entire pedestrian (32×16 pixel)

candidate selection classifier is shown in Figure 4. We totally select 2400 positive samples and 3000 high-quality negative samples.

Definitions of the each main part of a pedestrian is shown in Figure 5, therefore samples of body and leg can be obtained from samples of entire pedestrian automatically. However, we only select clear ones as positive samples. Some of the leg samples are shown in Figure 6. Head samples are manually made from videos captured in rainy or snowy weather; and the head includes human head and umbrella.

A classifier's performance is also strongly depends on its feature set; in the system, we mainly use haar features. Similar to Paul Viola's work [1] and our previous work [2], appearance information can be got from the anterior original gray-scale image. Motion information can be obtained by subtracting two consecutive frames.

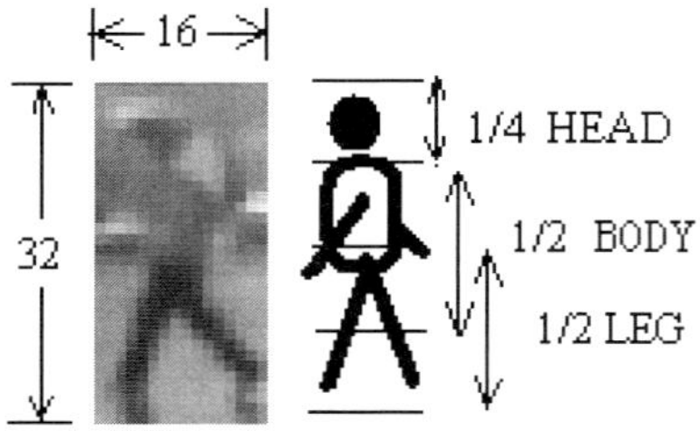

Fig. 5. Definitions of three main parts of a pedestrian

Candidate selection only takes advantages of global appearance features [2] to get higher processing speed. Haar features of head, body and leg are of the similar kinds; however, we choose them according to the character of each part, and only typical features are chosen. Five selected features of leg are shown in Figure 7. We also apply AdaBoost algorithm to choose proper features.

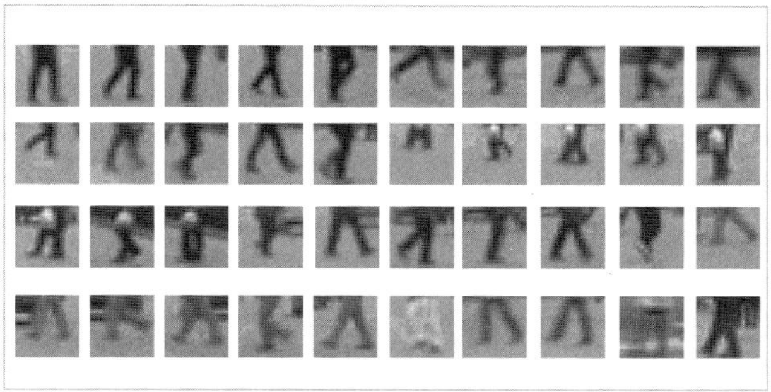

Fig. 6. Some positive samples of leg (16×16 pixel)

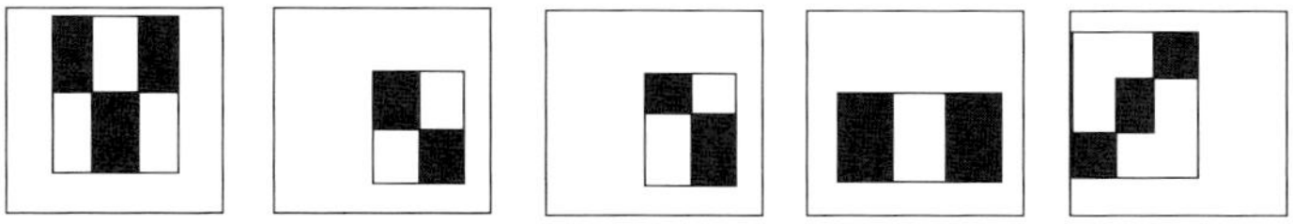

Fig. 7. Five example appearance local features of leg (16×16)

3 Experimental Results

In order to validate the detection ability of our system, we carried out several tests on a high performance PC which equipped a Pentium IV 3.0G CPU and 1G DDR2 RAM. Sixteen test videos (eight of rainy days and eight of snowy days) were captured at a 320×240 resolution with 15fps on a moving vehicle in real city traffic environment and the vehicle speed is 50 km/h in average. Each video has 225 frames (15 seconds). The average result of pedestrian detection for verification videos is listed in Table 2.

Table 2. Average system performance in rainy and snowy weather

Average System performance	Rainy	Snowy
Detection rate	75.2%	83.3%
False positive rate	1. 2%	0.65%
Detection speed	13.3 fps	14.8 fps

Table 2 indicates that:

(1) The system both gets good performance in rainy and snowy weather.

(2) The system performs better in snowy weather than in rainy weather. As shown in Figure 8, the light is weaker in rainy weather because of thick rain cloud, this leads to lower detection rate in rainy weather. The system has a lower false positive rate and a higher speed because of that the road is drier in snowy weather; therefore less mirror image need to be processed.

If other local features are adopted, the system can be easily modified to detect other kind of objects such as bicycle rider (only change local features of leg to features of bicycle). Additionally, we train another PDS for sunny days use similar samples in sunny days and drop body local features. Five random selected test video captured in sunny days show that the average performance is also better than our previous work [2]. Especially the detection speed reaches 15.5fps which is more suitable for real-time vehicular detection in urban area.

Fig. 8. Pedestrian detection in snowy (left) and rainy (right) weather

4 Conclusions

This paper proposed a fast pedestrian detection system for rainy or snowy weather, and the idea and system architecture are also suit for PDS in other weather condition. The experimental results show that the system is suit for real-time detection in city traffic, and if we design a weather judgment module, and for each weather condition train a cascaded classifier which having the same input and output, then a self-reacting PDS can be obtained. Of course this is one of our future works.

Only few optical camera based pedestrian detection system for rainy or snowy weather is proposed; however, infrared (IR) camera based PDS has good performance in rainy or snowy weather, comparing with these systems [4] [5], our system mainly has the following features:

(1) With optical zoom camera (three times max), the system can detect pedestrian in the range of 4.5-90 meters; whilst the infrared camera based system can detect pedestrians within 15 meters.

(2) Bertozzi M. et al. proposed an IR stereo vision based PDS [4]. This PDS worked on single frame and took seven steps to detect pedestrian; it only took shape features to select candidates, and morphological characteristics based head detection was adopted to verify the candidates. The detection speed of our system is much faster than Bertozzi M. et al.'s. (It is about 500 to 600 times faster than theirs if not mention the differences between computers); hence our system is more suitable for a real-time vehicular pedestrian detection.

Acknowledgement

This work was supported by National Natural Science Foundation of China (60204009), and Open Foundation of The Key Laboratory of Complex Systems and Intelligence Science, Chinese Academy of Sciences (20040104).

References

1. Paul Viola, Michael Jones, and Daniel Snow, "Detecting Pedestrians Using Patterns of Motion and Appearance," International Journal of Computer Vision, vol. 63, no. 2, pp. 153-161, 2005
2. Y.W. Xu, X.B. Cao, and H. Qiao, "A low cost pedestrian detection system," IEEE WCICA06, accepted, June 2006
3. Shashua Amnon, Gdalyahu Yoram, and Hayun Gaby, "Pedestrian Detection for Driving Assistance Systems - Single-frame Classification and System Level Performance," IEEE Intelligent Vehicles Symposium, pp. 1-6, 2004
4. Bertozzi M., Broggi A., Lasagni A., and Del Rose M., "Infrared stereo vision-based pedestrian detection," IEEE Intelligent Vehicles Symposium, pp. 24-29, 2005
5. Fengliang Xu, Xia Liu, and Kikuo Fujimura, "Pedestrian Detection and Tracking with Night Vision," IEEE Transaction on Intelligent Transportation Systems, vol. 6, no. 1, pp. 63-71, 2005
6. D.M. Gavrila, J. Giebel, and S. Munder, "Vision-based pedestrian detection: the PROTECTOR system", IEEE Intelligent Vehicles Symposium, pp. 13-18, 2004
7. Yoav Freund, and Robert E. Schapire, "A decision-theoretic generalization of on-line learning and an application to boosting," Computational Learning Theory: Eurocolt '95, pp. 23-37. Springer, 1995.
8. T. Joachims, "Making large-scale SVM learning practical," in Advances in Kernel Methods—Support Vector Learning, B. Schölkopf, C. J. C. Burges, and A. J. Smola, Eds. Cambridge, MA: MIT Press, 1998
9. Zhang XG; "Introduction to statistical learning theory and support vector machines," Acta Automatica Sinica, vol. 26, no. 1, pp. 32-42, 2000

Real Time Face Detection System Based Edge Restoration and Nested K-Means at Frontal View

Hyun Jea Joo, Bong Won Jang, Md. Rezaul Bashar, and Phill Kyu Rhee

Dept. of Computer Science & Engineering
Inha University, Yong-Hyun Dong
Incheon, South Korea
{hj0235, bictory7, bashar}@im.inha.ac.kr, pkrhee@inha.ac.kr

Abstract. Bayesian technique is a popular tool for object detection due to its high efficiency. As it compares pixel by pixel, it takes a lot of execution time. This paper addresses a novel framework for head detection with minimum time and high accuracy. To detect head from motion pictures, motion segmentation algorithm is employed. The novelty of this paper carried out with the following steps: frame differencing, preprocessing, detecting edge lines and restoration, finding the head area and cutting the head candidate. Moreover, nested K-means algorithm is adopted to find head location and statistical modeling is employed to determine face or non-face class, while Bayesian Discriminating Features (BDF) method is employed to verify the faces. Finally, the proposed system is carried out with a lot of experiments and a recognizable success is notified.

1 Introduction

Many visual surveillance systems have adopted camera capture system to detect a lot of typical objects or analyze their motion and patterns. As huge size images, we need only objects active candidate regions. Because, the big sized images take long time to calculate and detect. Therefore, it is the key to find segmentation domains for using real time applications. Because of them, many systems adopt head detection and tracking methods [1][2][3][4][9][10][11]. And they adopt color – based methods to find head regions. [1][3][4][9] However many people have different face color and illumination in various environment. Therefore many systems do not define all face color values. They have solved the restricted condition. And they select the shape modeling like as ellipse. [12][14] But, these applications have disadvantage to solve noise effect. When the original image contains noise, the ellipse model turns to huge space region and image distortion. For overcoming this defect, some methods have used motion segmentation algorithm [7]. Motion segmentation in visual surveillance system falls into three divisions: environmental map, frame differencing and optical flow.

The algorithms using environmental map are very universal. First, they make statistical background. Then, they perform subtraction of current image frame from environment map [9][10]. This method is simple and gives good result in restricted

L. Wang et al. (Eds.): FSKD 2006, LNAI 4223, pp. 1192 – 1200, 2006.

environment. But it is very weak when the light circumstance is changed like as the direct rays of the sun. So, the first consideration in this case is to make accurate environment, which seems difficult to achieve.

Frame differencing approach adopts difference between two or three frames in image sequence. However, an improved performance is achieved by three frame differencing. Frame differencing is very robust method to continuing environmental changes [11]. Because of the differencing result, it is fragile to restore the original edges and to extract the active region. Therefore, the advantages of them are based on making activate candidate regions.

Optical flow selects flow vectors of moving objects in an image sequence. The results apply for gait and activity analysis [12]. But they have defects in complex and sensitive noise effects and do not perform in real time video.

Because of unrestricted surrounding changes in real time, frame-differencing method is used to improve accuracy and to minimize multi-resolution search. So the aim of this paper is to achieve high head detection rate employing minimum multi resolution searches and to reduce overall computational time. Edge detection and restoration challenge is prevailed over by using image evaluation, preprocessing and edge restoration. Then, active head regions are trimmed resulting in reduced search spaces in shorter time for using real time system. Head detection is employed on a predefined threshold value between the people and camera using interpolation method. Moreover, only head detection has been taken into account for this research. So we perform head detection using two clusters K-means algorithm followed by head and body histogram analysis. To detect heads of people, we adopt 13 frames per second frame rate for single person in indoor scene. The camera is fixed 1.6 meter above the ground. The people can move over 25 square meters area in front of camera with their face pointing towards it. Our approaches are divided into several steps. Section 2 presents system architecture. Section 3 deals with image evaluation, image preprocessing, edge detection and restoration. Section 4 presents head detection and face detection using Bayesian method. Section 5 brings up experiments. Finally, sections 6 describe the conclusion and future extensions.

2 Real Time Head Detection System Architecture

Our purposed system architecture is shown in Fig. 1. Input images are of 320 × 240 sizes. Then the difference between the current image (N) frame and previous frame (N-1) is calculated and evaluated. The difference is then compared with certain threshold. The system proceeds towards edge detection only when the difference is significant enough to find the body segment. But when the result is not changed i.e. difference is less than the threshold value (person is not moving), current frame (N) is compared with the second previous (N-2) and difference is evaluated. Therefore the maximum difference is measured between two frames by comparing three image frames. If again the difference is not significant then previously segmented head part is used for face detection. Otherwise, other steps proceeded out which results in new head segmentation. The steps include edge detection or restoration. Here accurate

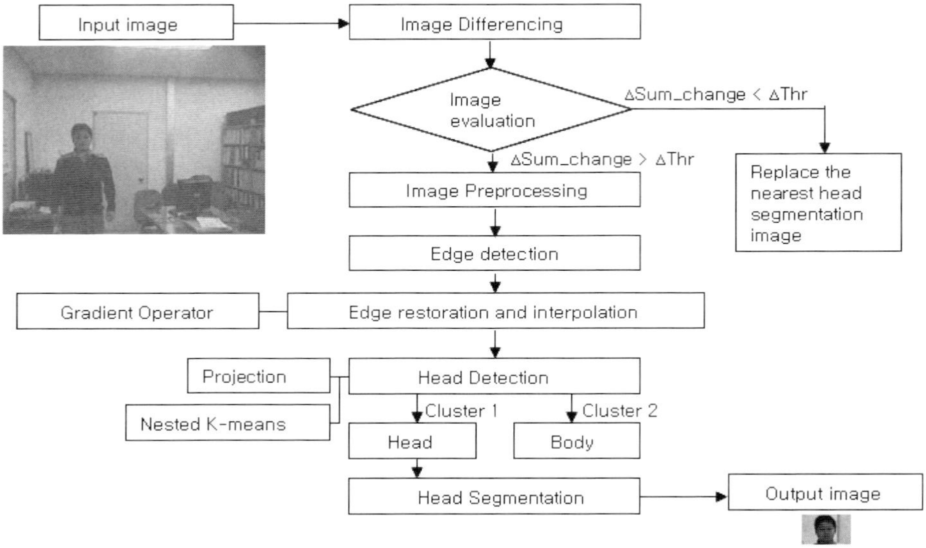

Fig. 1. Architecture of the proposed methodology

edge line across the body is found by performing gradient operators and interpolation, detail of which is discussed in section 3. Next, the projection of horizontal distance between the edges is calculated for each row. Local minima and maxima point is calculated on the projection vector to find the head and body location respectively. To overcome the error due to noise, the projection vector is clustered around the two regions (one for head and one for body) using K-means algorithm. After the head detection region is extracted from the image, Bayesian Discriminating feature method is used to search the target's face in the region to verify.

3 Movement Detection, Edge Detection and Restoration

3.1 Movement Detection

Our proposed method adopts image evaluation for movement detection, i.e. finding people that are moving or static. When the people get far from camera, human's head is very small, for example 25 pixels × 25 pixels, this is why, people can move 25 square meters within target tracking region. However, head size in end point varies on hairstyle, bald headed, man and woman, and other factors. Since we use Bayesian discrimination method for face detection and it uses minimum 16×16 pixel window for statistical analysis. Threshold is calculated assuming that the person should at least move his head for the noticeable change in the successive image frames. The borderline of 16×16 window contains 16×4 pixels and the approach finds the accurate human edges. Therefore the threshold is found to be ΔThr_{total} = 16×3 (i.e. discarding the lower edge of window).

$$\text{Comp(x,y)}=\begin{cases}255 & Set(x,y)<\Delta Thr_{in}\\ 0 & otherwise\end{cases} \tag{1}$$

where

$$\text{Set(x,y)}=\begin{cases}F_N(x,y)-F_{N-1}(x,y) & \Delta Sum_{change}\geq \Delta Thr_{total}\\ F_N(x,y)-F_{N-2}(x,y) & otherwise\end{cases}$$

$$\Delta\text{Sum}_{\text{change}}=\frac{1}{255*Total}\sum_{n=1}^{total}Comp(x,y)$$

Comp(x, y) is the threshold operator and it sets only two extreme gray color values 255 or 0. FN(x, y) is the image sequence indicating the Nth intensity value of (x, y). These values are resolved into white and black for finding edge line. However when ΔSum_{change} value is less than ΔThr_{total} current frame is regarded as static. In this case the previous nearest head segmentation image is used as active region.

3.2 Edge Restoration

Edge is an important issue to detect a particular shape of an object. Human is symmetric shaped especially for frontal view. As frame differencing creates a lot of noise and there is no predefined way to detect noise. To overcome this, edge restoration is employed.

After preprocessing and evaluation, the image edge is searched for the left and right row point to find the first edge pixel .The column points for left and right edge is stored as $Left_{first}(x,y)$ and $Right_{first}(x,y)$ respectively. Some successive points for both columns are considered determining the presence or absence of noise. Noise is detected in two ways: distance and median. For distance method, if there is abrupt change between successive points, noise is detected; otherwise, noise free as shown in equation (2). For median method, the center point between left and right position is determined. If this center point much differs from threshold, we consider this as noise as shown in equation (3). If noise is detected noise-removing technique is applied. Fig2. Shows edge restoration.

$$\text{E}_{\text{left}}(\text{x}_i,\text{y})=\begin{cases}(Left_{first}(x_i,y)=Left_{first}(x_{i-1},y)) & Dev(x,y)\geq Thr_{out}\\ (Left_{first}(x_i,y) & otherwise\end{cases} \tag{2}$$

where

$\text{Dev}_{\text{left}}\text{(x,y)}=(Left_{first}(x_i,y)-Left_{first}(x_{i-1},y)$

$E_{right}(x_i,y)$, $Dev_{right}(x,y)$ are the same ways.

$$\text{Center}_{\text{n-1}}\text{(x,y)}=\frac{1}{n-1}\sum_{i=1}^{n-1}\frac{1}{2}\left(Left_{first}(x_i,y_i)+Right_{first}(x_i,y_i)\right) \tag{3}$$

$J_n(x,y)\geq Thr_{out}$

$$\text{J}_n\text{(x,y)}=\frac{1}{2}\left(Left_{first}(x_n,y_n)+Right_{first}(x_n,y_n)\right)-\text{Center}_{\text{n-1}}\text{(x,y)}$$

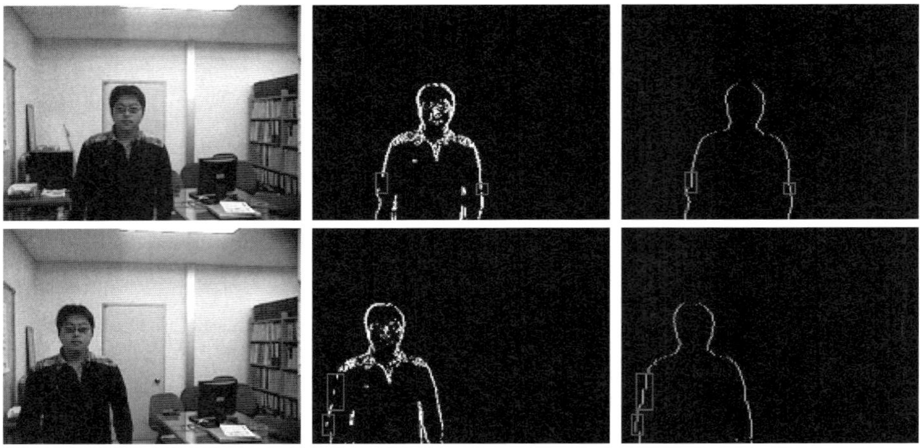

Fig. 2. Original images (left), Differencing images and preprocessed images (center), Edge restoration images (right)

4 Head Region Segmentation Using Nested K-Means Algorithm

To reduce the search space for face detection, first active region of head is detected by the method of head segmentation using vertical projection of edge contours and K-means algorithm.

Here edge contours of the person are extracted by the method described in section 3.2. Then Body width set projection of the edge contours is calculated. Then local minima and maxima are calculated on the projection vector in order to find the expected location of head and body respectively.

$$\mathrm{RP}(\mathrm{x}_{\mathrm{i}}) = \mathrm{Dev}_{\mathrm{right}}(\mathrm{y}) - \mathrm{Dev}_{\mathrm{left}}(\mathrm{y}) \tag{4}$$

The first and second derivative of projection vector is calculated to find local minima and maxima. The definitions about second derivative [5] states that; "If the first derivative *RP*'(xi) is positive (+), then the function RP(xi) is increasing (↑) and RP'(xi) is negative (-), then the function RP(xi) is decreasing (↓). Also, If the second derivative RP'' (xi) is positive (+), then the function RP (xi) is concave up (∪) and RP'' (xi) is negative (-), then the function RP (xi) is concave down (∩)." It is assumed that y=RP(x) is a twice-differentiable function with RP'' (c) =0.

$$\begin{aligned} &f''(x) : y'', \frac{d^2y}{dx^2}, \frac{d^2}{dx^2} f(x) \\ &\mathrm{RP}'(\mathrm{x}_{\mathrm{i}}) = \frac{f(x_i + h) - f(x_i)}{h} \\ &\mathrm{RP}''(\mathrm{x}_{\mathrm{i}}) = \frac{d^2y}{dx^2} = \frac{d}{dx}\left(\frac{dy}{dx}\right) = \lim_{h \to 0} \frac{f'(x_i + h) - f(x_i)}{h} \end{aligned} \tag{5}$$

Therefore if $RP''(d) < 0$, then $RP(x_i)$ has a relative maximum value at x=d. It means that body center ($Max_d(y)$) is in body cluster. In same ways, if $RP''(c) > 0$, then $RP(x_i)$ has a relative minimum value at $x=c$. So, it is near to head center (Minc(y)) and in head cluster, too. Maxd(y), Minc(y) points, that is the center of clusters is used to calculate the similarity:

$$D_b = \|Max_d(y) - RP(x_i)\|^2 \quad\quad (6)$$
$$D_h = \|Min_c(y) - RP(x_i)\|^2$$

$$G(x,y) = \begin{cases} 0 & D_b \geq D_h \quad Head \quad part \\ 1 & otherwise\ Body \quad part \end{cases} \quad\quad (7)$$

$$\mu_{head} = \frac{\sum x}{count(x)} \quad forall x : G(x,y) = 0 \quad\quad (8)$$

$$\mu_{body} = \frac{\sum x}{count(x)} \quad forall x : G(x,y) = 1$$

$$d_b = \|x - \mu_{body}\|^2$$

$$d_h = \|x - \mu_{head}\|^2$$

$$g(x,y) = \begin{cases} 0 & d_b \leq d_h \quad Head \quad part \\ 1 & otherwise\ Body \quad part \end{cases}$$

Nested k-means algorithm is implemented for the purpose of head and body segmentation. Clustering is done in two direction of projection vector. At first local minima and maxima is found in the projection vector RP (x). The operator is defined in equation (4). Then the vector is clustered to search head and body location on the width value using initial centroid as calculated minima and maxima in (5). The distance between the head cluster's center and body cluster's center point is calculated for every entry in projection vector. If distance to the head cluster is greater than that of body cluster, then the point is shifted to body cluster. The first stage of clustering separates head and body parts but error due to noise still exists. That is there are some of the body parts, which are still mis-clustered as head part. To reduce this noise, second level of clustering is needed. Second level of clustering is done on x direction of RP(x), i.e. on the index value of the projection vector taking centroid result of first level clustering as initial centroid. In the same way, the distance between the head cluster's center and body cluster center's point is calculated for every entry in projection vector. If distance to the head cluster is greater than that of body cluster, then the point is shifted to body cluster. So, in second stage of clustering any outline point (x, y) (related to body), which mis-clustered to head part, is corrected. It is found that this method increases the detection rate, because second level of clustering corrects most of the noise occurred in first level of clustering. Fig 4. shows the result of finding head and body regions. The left image is the neck line and body line with the first k-means method using object widths. However, this image contains the noise effect. Because of the nearest Euclidean distance, the first k-means method contains this area to head parts. But it must be a body part. According to my algorithm, the second k-means method eliminates the rectangle using Euclidean distance. The right image is the grouping result about head and body parts using second k-means.

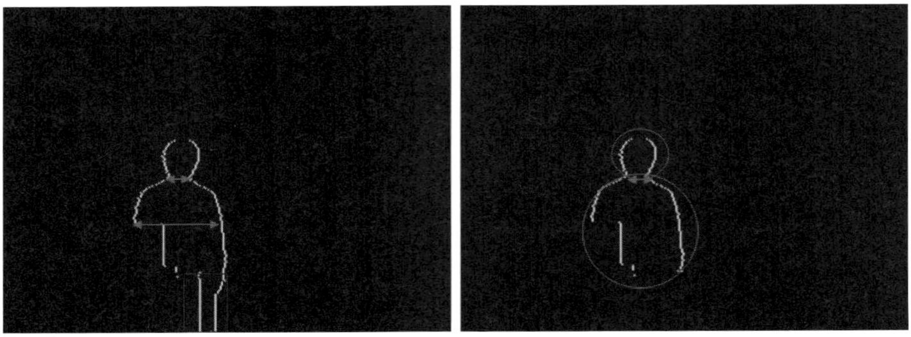

Fig. 3. First k-means method result (left) second k-means method result (right)

5 Experiments

Our proposed method is tested using two series of experiments; first face detection without Nested K-means, second is face detection with head segmentation with or without K means. A cheap LG USB camera is used as a video capturing device. The sequence frames of 320 x 240 pixels were acquired at 13 frames per second (fps) in 2.0GHx Pentium 4 PC computer and stored at the temporary folder in hard disk. Database is separated into 3 groups A, B and C. Group A contains normal people without any spectacles glasses and disguise form. Group B group includes people wearing glass. The last group C contains the people having beard and wearing glass in changing illumination condition. One group consists of 5 people dataset, and each

Table 1. Experiments about face detection rate about head region segmentation methods

	Experiment division	Database A group	Database B group	Database C group	Average total Database
BDF	**Face detection rate**	**92.46%**	**93.48%**	**83.31%**	**89.7%**
Non-including Nested k-means head detection +BDF		**73.29%**	**68.79%**	**65.87%**	**69.32%**
Nested k-means head detection +BDF		**97.29%**	**97.34%**	**97.21%**	**97.28%**
Non-including Nested k-means head detection +BDF	**Accurate Head Location rate**	**93.33%**	**92.14%**	**91.19%**	**92.22%**
Head seg +BDF (proposed method)		**98.07%**	**97.73%**	**99.23%**	**98.34%**

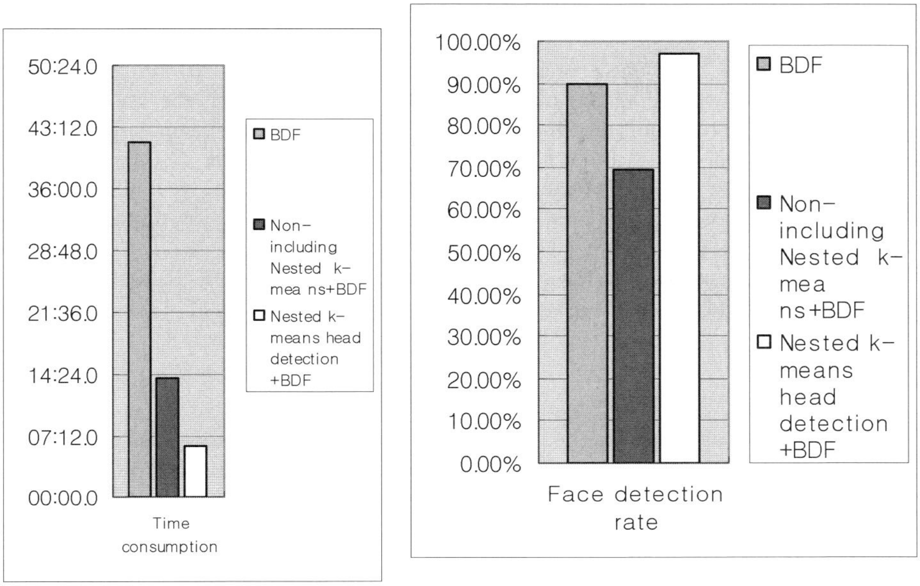

Fig. 4. Time consumption (left), Average face detection rate (right)

person has twenty-second movie clips at 13 fps. So each group has 1300 image sequences. Table 1 shows the experimental results. After the head segmentation, the system finds Bayesian classifier to discriminate face and non-face part of image. FERET database is used to learn the face Bayesian face model where as no-face Bayesian model is generated from the CMU database set.

Table 1 shows the face detection result for different scenario. Here it can see that face detection without segmentation gives high accuracy rate but it is time consuming. The third row of table 1 shows the result of our proposed method, which uses head segmentation and k-means for noise reduction. The result of our proposed method is found to be superior in terms of accuracy 97% and very fast in terms of speed. The experiment was conducted, but that time not using k-means for noise reduction. The recognition rate in this case is found to be very poor. Hence the credit of higher recognition rate goes to the error reduction technique employed by k-means algorithm. And Table 1 shows the result of experiment that shows the accuracy of the head location. Same database and experiments were conducted. The table suggest that our method is almost perfect to find the head location with the accuracy higher than 97%. But, when we do not contain the nested K-means algorithm, the result is very low because our experiment in real-time visual surveillance system contains many noise effects like as changing illumination.

Fig. 4 shows the variation of time consumption and accuracy rates across various methods. It can be seen that the proposed method (Head segmentation +BDF [8]) method based face detection) have high detection rate with less response time.

6 Conclusion

We have proposed real time head region detection based on difference of image frames from active camera. The main component of the system is the use of vertical projection of edge contours to find head location and Nested K-means algorithm for error minimization during differencing and threshold. Overall the system shows low computational cost and high detection rate, which suits for the real time system. Accurate head detection rate of about 97% is achieved while minimizing the computational cost and time.

The future enhancement to the system can be to overcome the occlusion and to find automatic threshold for frames evaluation using multiple cameras.

References

[1] Maolin Chen and Seokcheol Kee, "Head tracking with shape modeling and detection"

[2] G. L. Foresti, "A Real-Time System for Video Surveillance of Unattended Outdoor Environments", IEEE Transactions On Circuits And Systems For Video Technology, Vol. 8, No. 6, October 1998

[3] Tao Yang, Quan Pan, Jing Li, Yongmei Cheng, Chunhui Zhao, "Real-Time Head Tracking System With an Active Camera*, Proceedings of the 5" World Congress on Intelligent Control and Automation, June 15-19, 2004, Hangzhou, PR China

[4] Weimin Huang, Ruijing Luo, Haihong, Beng Hai Lee and Meneka Rajapakse " Real Time Head Tracking and Face and Eyes Detection, Proceedings of IEEE TENCON'02

[5] Jing-Un Won; Yun-Su Chung; In-Soo Kim; Jae-Gark Choi; Kil-Houm Park, "Correlation based video-dissolve detection," Research and Education, 2003. Proceedings. ITRE2003

[6] How-Lung Eng, Junxian Wang, Alvin H. Kam, Wei-Yun Yau, "A Bayesian Framework for Robust Human Detection and Occlusion Handling using Human Shape Model," *icpr*, pp. 257-260, 17th International Conference on Pattern Recognition (ICPR'04) - Volume 2, 2004

[7] Weiming Hu, Tieniu Tan, Fellow, IEEE, Liang Wang, and Steve Maybank "A Survey on Visual Surveillance of Object Motion and Behaviors", Ieee Transactions On Systems, Man, And Cybernetics—Part C: Applications And Reviews, Vol. 34, No. 3, August 2004

[8] Chengjun Liu, Member, IEEE," A Bayesian Discriminating Features Method for Face Detection", IEEE Transactions On Pattern Analysis And Machine Intelligence, Vol. 25, No. 6, June 2003

[9] I. Haritaoglu, D. Harwood, and L. S. Davis, "W : Real-time surveillance of people and their activities," IEEE Trans. Pattern Anal. Machine Intell., vol. 22, pp. 809–830, Aug. 2000.

[10] S. McKenna, S. Jabri, Z. Duric, A. Rosenfeld, and H. Wechsler, "Tracking groups of people," Comput. Vis. Image Understanding, vol. 80, no. 1, pp. 42–56, 2000.

[11] A. J. Lipton, H. Fujiyoshi, and R. S. Patil, "Moving target classification and tracking from real-time video," in Proc. IEEE Workshop Applications of Computer Vision, 1998, pp. 8–14.

[12] D. Meyer, J. Denzler, and H. Niemann, "Model based extraction of articulated objects in image sequences for gait analysis," in Proc. IEEE Int. Conf. Image Processing, 1998, pp. 78–81.

[13] S. Mulassiotis and M. G. Strintzis, "Real-Time Head Tracking And 3d Pose Estimation From Range Data", 2003 IEEE

[14] Harsh Nanda and Kikuo Fujimura, "A Robust Elliptical Head Tracker", Proceedings of the Sixth IEEE International Conference on Automatic Face and Gesture Recognition (FGR'04)

A Context-Aware Music Recommendation Agent in Smart Office

Donghai Guan, Qing Li, Sungyoung Lee*, and Youngkoo Lee

Department of Computer Engineering
Kyung Hee University, Korea
{donghai, sylee}@oslab.khu.ac.kr, yklee@khu.ac.kr,
liqing@icu.ac.kr

Abstract. In this paper, we originally propose a music recommendation agent in a smart office to recommend music for users. Personal tastes are diverse, even the same person may have different preferences in different situations. Hence, our music recommendation is not only based on the user favorite genres but also the current mood of users. By collecting and analyzing the contextual information of users, the agent can automatically senses the mood of users and recommends music.

1 Introduction

In ubiquitous computing, context is any information that can be used to characterize the situation of an entity [1]. A lot of work has been done in trying to make applications in ubiquitous computing environments context aware and many different kinds of agents have been developed [2] [3] [4] [5]. However, to the best of our knowledge, none of them considered entertainment items in ubiquitous environments. Music as a major entertainment item should be considered.

The objective of this paper is to construct an autonomous music recommendation agent which can recommend appropriate music to the users in smart office. This paper sets the stage by building a context-aware agent that can recommend music based on user preference and mood.

Several subtasks should be achieved by our proposed agent. They are music genre classification, user preference and mood deduction.

In the rest of paper, we present the components in our agent in detail. In section 2, we discuss how to classify music into genres and moods. In section 3, we propose the way to deduce user's music preference and mood. Section 4 is conclusions and future works.

2 Music Genre and Mood Classification

To recommend appropriate music to users, first of all, music genre and mood classification is needed.

---

* Dr. Sungyoung Lee is the corresponding author.

L. Wang et al. (Eds.): FSKD 2006, LNAI 4223, pp. 1201 – 1204, 2006.

2.1 Music Genre Classification

This component includes two parts: one is extracting features from music; the other part is using machine learning techniques to classify music based on the extracted features.

1) Feature extraction: three feature sets [6] are used in our paper. They are timbral texture features, rhythmic content features and pitch content features.

2) Classification: Fig. 1 shows the classification accuracy of two classification algorithms (svm and knn) on our dataset. In the experiments, 1,2,3,4,5,6,7 represent mfcc, centroid, rolloff, ssf (sum of scalefactors), flux, pitch and rhythmic content features accordingly. The experiments verify that each of the tradition features contains useful yet incomplete information for characterizing music signals. Here we only show two experiment results. From our experiments, we may draw the conclusion that among all of different combinations of proposed features, the performance is best when all the proposed features used together.

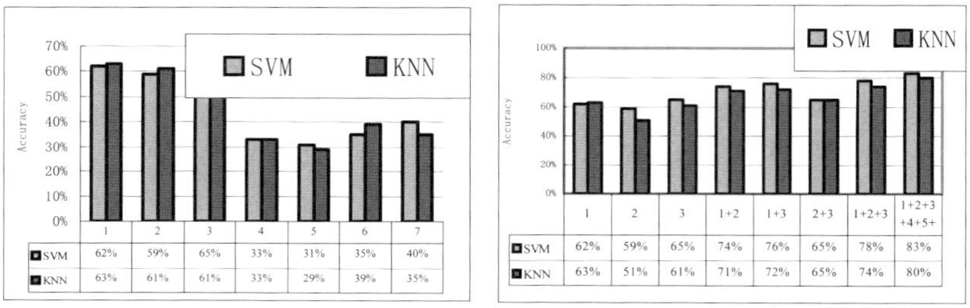

(a) (b)

Fig. 1. Classification performance comparison

2.2 Music Mood Classification

Three features [7] [8], relative tempo, the mean and standard deviation of average silence ratio, are used to classify mood.

1) Tempo: the rate at which notes are played, expressed in score time units per real time unit, i.e. quarter notes per minute or beats per minute.

2) Average silence ration: used to model articulation.

After getting these three features, machine learning techniques can be used to classify the music. Neural network classifier used in [8] showed that music mood could be classified.

3 User Preference and Mood Deduction

User preference and mood are decisive elements in our music recommendation agent. User preference is a long-term activity, while user mood is a state at a given time point which always changes from time to time. User's history profiles are used to deduce user preference and mood.

3.1 User Preference Deduction

To get user preference, training phase is indispensable. In this phase, system provides the genre classified music to each user. Then each user selects his/her favorite music to play. For each genre, the system records the number of music that has been played. Then, we can deduce user's preference by selecting the biggest number. For each user, three most favorite genres are extracted as user preference. For example:

User1 (Classical, Rock, Pop)
User2 (Metal, Pop, Disco)
User3 (Country, Blues, Jazz)
User4 (Metal, Classical, Jazz)

3.2 User Mood Deduction

In our paper, we use the method proposed by Anand [9]. The context used to derive mood includes user's location, the time of day, which other people are in a room with him, the weather outside and his stock portfolio. Native Bayesian algorithm is used to predict user mood.

After the knowledge of music genre, user preference and user mood is available, the recommendation process is quite simple that the music with the matched genre and mood will be selected.

4 Conclusions and Future Work

In this paper, we proposed a novel music recommendation agent, which can provide music to the users according to their preference and mood. Also, we have some issues to solve in the future.

1) Our music recommendation agent is user-centric. In the future, we propose to devise a task-centric music recommendation agent.

2) More multimedia entertainment items in smart office will be considered in the future, such as movie.

Acknowledgement

This work is financially supported by the Ministry of Education and Human Resources Development (MOE), the Ministry of Commerce, Industry and Energy (MOCIE) and the Ministry of Labor (MOLAB) through the fostering project of the Lab of Excellency.

References

1. Dey, A.K., Abowd, G.D., and Salber, D.: A Conceptual Framework and a Toolkit for Supporting the Rapid Prototyping of Context-Aware Applications. In J. of Human-Computer Interaction (HCI), Vol. 16 (2001) 97-166
2. Hong, J. and Landay, J.: An Infrastructure Approach to Context-Aware Computing. In J. of Human-Computer Interaction (HCI), Vol. 16 (2001) 287-303

3. Shafer, S.A.N., Brumitt, B., and Cadiz, J.J.: Interaction Issues in Context-Aware Interactive Environments. In J. of Human-Computer Interaction (HCI), Vol. 16 (2001) 363-378
4. Pascoe, J., Ryan, N.S., and Morse D.R.: Issues in Developing Context-Aware Computing. In Proc. of the International Symposium on Handheld and Ubiquitous Computing, Springer-Verlag, (1999) 208-221
5. Schilit, B.N.: A Context-Aware System Architecture for Mobile Distributed Computing. PhD Thesis, Columbia University, May (1995)
6. Li, Q., Kim, B.M., and Guan, D.H.: Music Recommendation Using Audio Features. In Proc. of the 27th annual international ACM SIGIR conference, (2004) 532-533
7. Juslin, P.N.: Cue Utilization in Communication of Emotion in Music Performance: Relating performance to perception. In J. of Experimental Psychology, Vol. 16, (2000) 1797-1813
8. Feng, Y.Z., Zhuang, Y.T., and Pan, Y.H.: Popular Music Retrieval by Detecting Mood. In Proc. of the 26th annual international ACM SIGIR conference, (2003) 375-376
9. Ranganathan, A. and Campbell, R.H.: A Middleware for Context-Aware Agents in Ubiquitous Computing Environments. In ACM/IFIP/USENIX International Middleware Conference, (2003) 143-161

A Decision Tree-Based Method for Speech Processing: Question Sentence Detection

Vũ Minh Quang, Eric Castelli, and Phạm Ngọc Yến

International research center MICA
IP Hanoi – CNRS/UMI-2954 – INP Grenoble
1,Dai Co Viet - Hanoi – Viet Nam
{minh-quang.vu, eric.castelli, ngoc-yen.pham}@mica.edu.vn

Abstract. Retrieving pertinent parts of a meeting or a conversation recording can help for automatic summarization or indexing of the document. In this paper, we deal with an original task, almost never presented in the literature, which consists in automatically extracting questions utterances from a recording. In a first step, we have tried to develop and evaluate a question extraction system which uses only acoustic parameters and does not need any textual information from a speech-to-text automatic recognition system (called ASR system for Automatic Speech Recognition in the speech processing domain) output. The parameters used are extracted from the intonation curve of the speech utterance and the classifier is a decision tree. Our first experiments on French meeting recordings lead to approximately 75% classification rate. An experiment in order to find the best set of acoustic parameters for this task is also presented in this paper. Finally, data analysis and experiments on another French dialog database show the need of using other cues like the lexical information from an ASR output, in order to improve question detection performance on spontaneous speech.

1 Introduction

There're now increasingly important amount of audio data, including voice, music and various kinds of environmental sounds. The audio segmentation and classification are crucial requirements for a robust information extraction on speech corpus. In these recent years, many researches have been conducted in this domain. Lie Lu [1] in his work shows a success automatic audio segmentation and classification into speech, music, environment sound and silence in a two-stage scheme. The first stage of the classification is to separate speech from non-speech, while the second stage further segments nonspeech class into music, environment sounds and silence with a rule based classification scheme. In [9], an online audio classification and segmentation system is presented where audio recordings are also classified and segmented into speech, music, several types of environmental sounds and silence based on audio content analysis. This last system uses many complicated audio features.

In this paper, a new method of speech classification is presented. We plan to classify speech into two classes: question or nonquestion. Different with previous researches which work on audio in general, we treat in our case only one concrete

L. Wang et al. (Eds.): FSKD 2006, LNAI 4223, pp. 1205 – 1212, 2006.

type of audio: speech. We hope that results of this study will be applied in other researches in DELOC project, or in audio browsing, summarization applications developed in our laboratory. Also, our classification is based more importantly on prosodic characteristic than acoustic characteristic of the speech since we use in this experimentation only one prosodic parameter: it's the F0 (which is the fundamental frequency of a speech signal; it's the inverse of the pitch period length. The pitch period is the smallest repeating unit in this speech signal).

The rest of the paper is organized as follows: speech corpus is presented in section II. The classification scheme is discussed in section III. In Section IV, experiment-tation results of proposed method are given while section V concludes our work.

2 Speech Corpus: Telephone Meeting Recording

Our telephone meeting corpus (called Deloc for delocalized meetings) is made up of 13 meetings of 15 to 60 minutes, involving 3 to 5 speakers (spontaneous speech). The total duration is around 7 hours and the language is French. Different types of meetings were collected which correspond to three categories: recruitment interviews; project discussions in a research team; and brainstorm-style talking. This corpus was manually transcribed in dialog acts. For this experimentation of automatic question detection, we have manually selected and extracted a subset of this corpus. It is composed of 852 sentences: 295 questions (Q) sentences and 557 non-questions (NQ) sentences.

3 Automatic Question Extraction System

3.1 Global System Design

The design of our classification system is illustrated in the following figure 1. Beginning by the F0 calculation for all wave files in our corpus, then for characterizing a wave file, 12 features derived from F0 are calculated for each file. Next, all these features among with the name of corresponding wave file are gathered into a database for facility other management. Then, we apply the 10-folds cross-reference testing method by dividing the corpus randomly into 2 subsets: one subset for training decision tree, another subset for testing the decision tree constructed. This step is repeated 10 times in order to find out the best tree with the best classification performance. These processes are presented in more detail in these following sections.

For this meeting corpus (Deloc), we use 200 Q-utterances and 200 NQ-utterances for training and the remaining utterances for testing (95Q + 357NQ).

For the training data, a decision tree is constructed (the decision-tree algorithm used in our experiments is called "C4.5"[1]) and the classifier obtained is evaluated on the remaining test data.

[1] Ross Quinlan, 1993 C4.5: Programs for Machine Learning, Morgan Kaufmann Publishers, San Mateo, CA.

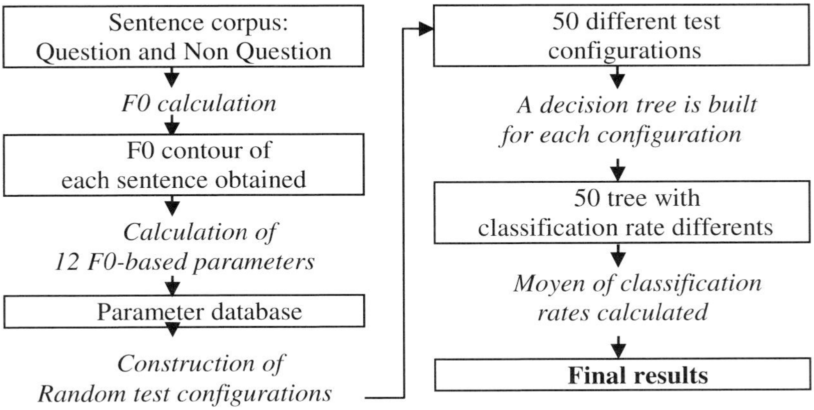

Fig. 1. Global classification system design

The evaluation is either based on a confusion matrix between questions (Q) and non-question (NQ) classes, or on measures coming from information retrieval domain like recall (R), precision (P) and Fratio where :

$$R=\frac{N_{\text{correctly detected questions}}}{N_{\text{total questions in the test set}}} \quad P=\frac{N_{\text{correctly detected questions}}}{N_{\text{total questions detected}}} \quad FRatio=\frac{2P\cdot R}{P+R}$$

3.2 Feature Vectors

In French language, the interrogative form of a sentence is strongly related to its intonation curve. Therefore, we decided to use the evolution of the fundamental frequency to automatically detect questions on an audio input.

From this F0 curve, we derive a set of features which aim at describing the shape of the intonation curve. Some of these features may be found redundant or basic by the reader: however, our methodology is to first evaluate a set of parameters chosen without too much a priori on the intonation curve for both Q and NQ classes. Then, a detailed discussion concerning the usefulness of each parameter will be provided in the experiments of section 4.2. The parameters defined for our work are listed in Table 1. It is important to note here that, contrarily to classical short term feature extraction procedures generally used in speech recognition, a unique long term feature vector is automatically extracted for each utterance of the database.

These features can be divided into 2 main categories: the first 5 features are the statistics on F0 values, but the 7 next features describe the contour of F0 (raising or falling). The F0 contour was extracted using Praat[2] software.

---

[2] http://www.fon.hum.uva.nl/praat/

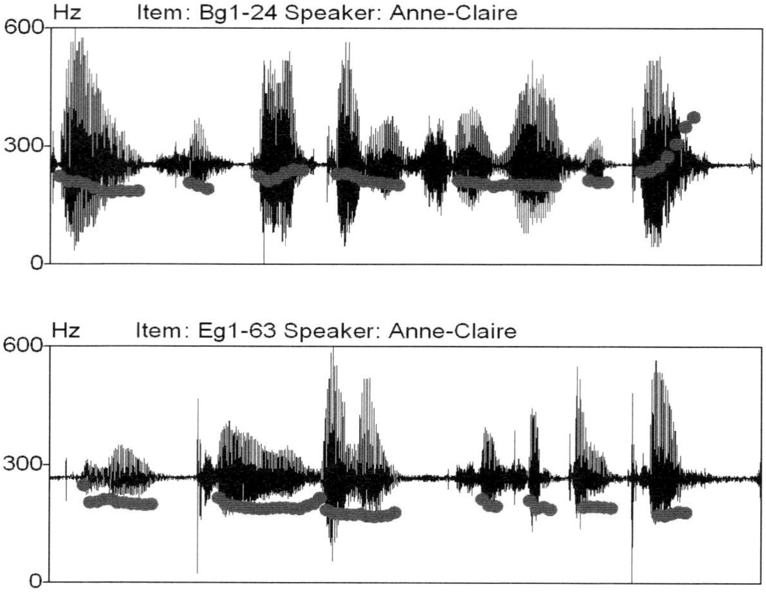

Fig. 2. Signal and F0 contour of French sentences: on top: interrogative sentence; on bottom: affirmative sentence

Table 1. 12-dimensional feature vector derived from F0-curve for each utterance

No	Parameter	Description
1	Min	Minimal value of F0
2	Max	Maximal value of F0
3	Range	Range of F0-values of the whole sentence (Max-Min)
4	Mean	Mean value of F0
5	Median	Median value of F0
6	HighGreaterThan Low	Is sum of F0 values in first half-length smaller than sum of F0 values in last half-length of utterance?
7	RaisingSum	Sum of F0i+1– F0i if F0i+1 > F0i
8	RaisingCount	How many F0i+1 > F0i
9	FallingSum	Sum of F0i+1– F0i if F0i+1 < F0i
10	FallingCount	How many F0i+1 < F0i
11	IsRaising	Is F0 contour rising? (yes/no). Test whether RaisingSum > FallingSum
12	NonZero-FrameCount	How many non-zero F0 values?

3.3 Decision Tree-Based Classifier

Traditionally, statistical-based methods such as Hidden Markov Model (HMM) or Gaussian Mixture Model (GMM) can be used to solve classification problems in speech processing. These statistical methods generally apply on short term features, extracted for instance at a 10ms frame rate. However, in our case, statistical methods are hard to use since we do not use short term feature vectors, as explained in the previous section: one feature vector only is extracted for the whole utterance to be classified, which excludes the use of conventional statistical classifiers. Thus, decision trees, which correspond to another classical machine learning (ML) method [6,7], are a good alternative. In that case, the process is classically divided into two separated stages: training and testing. The training stage is to build a tree-model to represent a set of known-instances while testing stage involves the use of this model to evaluate other unknown-instances. Decision tree is a divide-and-conquer approach to the problem of learning from a set of independent examples (a concrete example is called instance). Nodes in a decision tree involve testing a particular condition, which usually compares an attribute value with a constant. Some other trees compare two attributes with each other, or utilize some functions of one or more attributes. Leaf nodes give a classification for all instances that satisfy all conditions leading to this leaf, or a set of classifications, or a probability distribution over all possible classifications. To classify an unknown instance, it is routed down the tree according to the values of attributes tested in successive nodes, until it reaches a leaf. The instance is classified according to the class assigned to this leaf.

For this work, we have used an implementation of decision-tree algorithms that is included in the open-source toolkit Weka[3] which is a collection of algorithm implementations written in Java for data mining tasks such as classification, regression, clustering, and association rules.

4 Experiments and Results

In this experimentation, we have applied a classic 50-fold cross validation protocol in order to increase the total number of tests made. We repeated 50 times the process of dividing randomly the whole corpus into 2 parts (200Q+200NQ for training; the rest 95Q+357NQ for test). We then obtained 50 decision trees, each with a different classification performance. From these results, we have calculated the mean performance values. Note that the process of training and testing is very fast with decision trees.

Table 2 shows the confusion matrix for Q/NQ classification. The figures correspond to the average performance over all 50 cross-validation configurations. The mean good classification rate is around 75%.

Table 2. Confusion matrix on test data: mean values

Question	Non Question	←classified as
73(77%)	22(23%)	Question
93(26%)	264(74%)	Non Question

[3] http://www.cs.waikato.ac.nz/~ml/weka/

However, it is also interesting to evaluate our question extraction system in terms of precision / recall figures. This is shown in Table 3 where the figures correspond to the average performance over all 50 cross-validation configurations (we give also standard deviations). These results show that our system lead to an acceptable recall (77% of the Q-utterances were retrieved by the system) while the precision is lower, due to a relatively large part of NQ-utterances classified as questions. Looking at the standard deviation calculated over 50 cross-validation configurations, we can say that the system presents a correct stability.

Table 3. Mean performance measures (precision, recall, F-ratio) on test data for the Q-detection task.

	Precision	Recall	Fratio
Mean	44%	77%	56%
Standard deviation	4%	7%	3,5%

This experiment gave us an idea of our first system performance but does not help us to know which acoustic feature is useful for question extraction and which is not. Moreover, since some of these features are correlated with each other, one could try to reduce the original feature set. This is the purpose of the next section experiment.

4.1 Features Ordered by Importance

In order to know how strongly a feature contributes to the classification process, we have performed a "leave-one-out" procedure (as done in [8]). Starting with all N=12 features listed in Table 1, the procedure begins by evaluating the performance of each of the N-1 features. The best subset (in terms of Fratio performance on train data calculated by averaging 50 cross-validation configurations) is determined, and the feature not included in this subset is then considered as the worst feature. This feature is eliminated and the procedure restarts with N-1 remaining features. The whole process is repeated until all features are eliminated. The inverse sequence of suppressed features gives us the list of features ranked from the most effective to the less effective (or most important to less important) for the specific Q/NQ classification task.

This "leave-one-out" procedure lead to the following order of importance (from the most important to the least important feature : 1) isRaising 2) Range 3) Min 4) fallingCount 5) highGreaterThanLow 6) raisingCount 7) raisingSum 8) Median 9) Max 10) nonZeroFramesCount 11) Mean 12) fallingSum.

We observe that isRaising and range parameters are the two most important ones to classify an utterance between Q and NQ classes. The isRaising feature is logically important for detecting questions in French since it corresponds to the case where the F0 contour of a sentence is raising. The second important parameter is the range of the F0 values within the whole sentence. If the decision tree makes use of these only two features isRaising and range, the Fratio obtained is already 54% (to be compared with the 56% obtained with the 12 parameters). Our decision tree can be thus simplified and use two rules only. It is however important to note that some features of the initial parameter set (min, max, RaisingSum, FallingSum) still need to be extracted to calculate these two remaining features.

4.2 Comparison with a "Random" System

In order to really understand what is under the performance obtained in the first experiment, we have compared our system performance to a system that gives a classification output in a basic or random way. For this, we have distinguished three types of "basic" systems: (a) one system which always classifies sentences as Q; (b) one system which always classifies sentences as NQ; and (c) one system that classifies sentences as Q or NQ randomly with a 50% probability. With the data set given in this experimentation, we obtain the following table of Fratio performance for these random systems in comparison with our reference system using 12 parameters (these rates are calculated on test data). In any case, we see that our system is significantly better.

Table 4. Fratio for "basic" or "random" systems and for our system on meeting test data

Always Q (a)	Always NQ (b)	Random 50% Q/NQ (c)	**Our system**
35%	0%	29%	**56%**

5 Conclusion

Retrieving pertinent parts of a meeting or a conversation recording can help for automatic summarization or indexing of the document. In this paper, we have dealt with an original task, almost never presented in the literature, which consists in automatically extracting questions utterances from a recording. In a first step, we have tried to develop and evaluate a question extraction system which uses only acoustic parameters and does not need any ASR output. The parameters used are extracted from the intonation curve and the classifier is a decision tree. Our first experiments on French meeting recordings lead to approximately 75% classification rate. An experiment in order to find the best acoustic parameters for this task was also presented in this paper. Finally, data analysis and experiments on another database have shown the need of using other cues like the lexical information from an ASR output or energy, in order to improve question detection performance on spontaneous speech. Moreover, we plan to apply this Q-detection task to a tonal language like Vietnamese to see if the same parameters can be used or not.

References

1. FERRER L., SHRIBERG E., STOLCKE A., "A Prosody-Based Approach to End-of-Utterance Detection That Does Not Require Speech Recognition", *IEEE Int. Conf. on Acoustics, Speech and Signal Processing (ICASSP)*, vol. I, Hong Kong, 2003, pp. 608-611.
2. SHRIBERG, E., BATES, R. & STOLCKE, A. "A prosody-only decision-tree model for disfluency detection". *Eurospeech 1997*. Rhodes, Greece.

3. STANDFORD V., GAROFOLO J., GALIBERT O., MICHEL M., LAPRUN C., "The NIST Smart Space and Meeting Room Projects: Signal, Acquisition, Annotation and Metrics", *Proc of ICASSP 2003*, Hong-Kong, China, Mai 2003.
4. WANG D., LU L., ZHANG H.J., "Speech Segmentation Without Speech Recognition", *IEEE Int. Conf. on Acoustics, Speech and Signal Processing (ICASSP)*, vol I, april 2003, pp. 468-471.
5. MANA N., BURGER S., CATTONI R., BESACIER L., MACLAREN V., McDONOUGH J., METZE F., "The NESPOLE! VoIP Multilingual Corpora in Tourism and Medical Domains" *Eurospeech 2003*, Geneva, 1-4 Sept. 2003.
6. MARQUEZ L., “Machine learning and Natural Language processing”, Technical Report LSI-00-45-R, Universitat Politechnica de Catalunya, 2000.
7. WITTEN I.H., FRANK E., *Data mining: Pratical machine learning tools and techniques with Java implementations*, Morgan Kaufmann, 1999.
8. L. BESACIER, J.F. BONASTRE, C. FREDOUILLE, “Localization and selection of speaker-specific information with statistical modeling". *Speech Communication, n°31 (2000),* pp 89-106

Application of Chaotic Recurrence Plot Analysis to Identification of Oil/Water Two-Phase Flow Patterns

Jin Ningde, Zheng Guibo, Dong Fang , and Chen Wanpeng

School of Electrical Engineering and Automation,
Tianjin University, Tianjin 300072, China
ndjin@tju.edu.cn

Abstract. Recurrence quantification analysis method was employed to study the identification of oil/water two-phase flow patterns. The results showed recurrence plot appeared different textures for the transitional oil/water two phase flow patterns, and was special for different flow pattern. That indicated the texture of recurrence plot was sensitive to the changes of oil/water two flow patterns, and the recurrence quantification analysis is a useful diagnosis tool in identifying of two phase flow patterns.

Keywords: Oil/water two-phase flow, flow pattern, recurrence plot, recurrence quantification analysis.

1 Introduction

Two phase flow widely exist in many production processes, such as chemical engineering, petroleum industry and other fields. Especially, oil/water two phase flow phenomena are very common in production oil wells. Four flow patterns of oil/water two phase flows can be seen in vertical upward pipe (bubble, slug, froth, mist). The transitional mechanisms between these flow patterns are very complex, and the flow pattern identification is still a very difficult problem. Recently, the recurrence plot (RP) and the recurrence quantification analysis (RQA) methods [1-2] become a new powerful tool to chaotic time series analysis, and many applications in different fields have been developed [3-5]. In this study, we investigated the RP and RQA methods based on the Lorenz chaotic system [6], and then applied the method to conductance fluctuating signals analysis of oil /water two phase flow in order to find a quick look flow pattern identification method.

2 Recurrence Plot and Recurrence Quantification Analysis

2.1 Recurrence Plot (RP)

Phase space reconstruction is a necessary step in chaotic time series analyzing. According to the Takens's embedding theorem, we need to set a time delay to a

L. Wang et al. (Eds.): FSKD 2006, LNAI 4223, pp. 1213 – 1216, 2006.

one-dimension time series to reconstruct its phase space. If the original time series is expressed with $X(x_1, x_2, x_3, \ldots\ldots, x_n)$, after phase space reconstruction with m-dimension embedding and τ -time delay, the vector number is

$$N = n-(m-1)\tau \tag{1}$$

The distance between any two vectors is defined as

$$d_{i,j} = \left\| \boldsymbol{X}_i - \boldsymbol{X}_j \right\| \tag{2}$$

And the threshold is defined as

$$\varepsilon = \alpha \cdot std(\boldsymbol{X}) \tag{3}$$

Where $std(\boldsymbol{X})$ is the standard deviation of time series $\boldsymbol{X}$, α is a constant coefficient which is generally about 0.15. Then the recurrence matrix is expressed by

$$\boldsymbol{R}_{ij} = \Theta(d_{i,j} - \varepsilon) \tag{4}$$

Where:

$$\Theta(x) = \begin{cases} 1, x < 0 \\ 0, x > 0 \end{cases} \tag{5}$$

We define recurrence point at the place where the value is 1 in the recurrence matrix $\boldsymbol{R}_{ij}$, and then dot them on the coordinate plane to get an $N \times N$ plane plot-recurrence plot (RP). The typical recurrence plots of sine signal, Lorenz chaotic signal and random noise signal according to the definition above are showed in Fig. 1, the RP pattern differences are very obvious, and the recurrence plots can sensitively reflect the signals from different physical background.

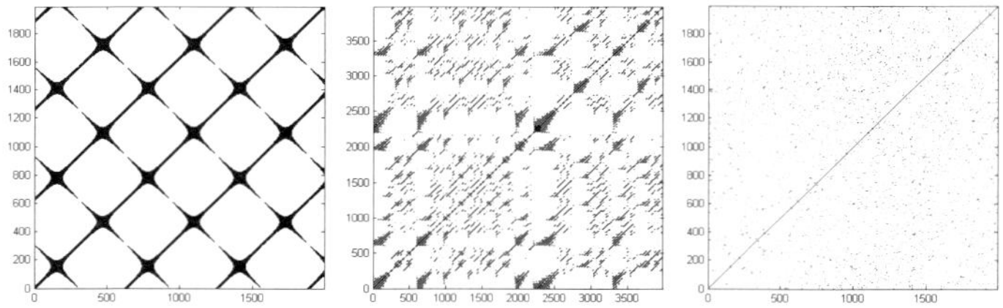

(a) sine signal (b) Lorenz strange attractor (c) random noise signal

Fig. 1. Recurrence plot of three different signals

2.2 Recurrence Quantification Analysis (RQA)

Recurrence quantification analysis was developed based on the recurrence plot, and the detailed texture of recurrence plot can be extracted. From the RP shown in Fig. 1, we can see that the main texture of RP is the recurrence points distributed in the coordinate plane and the parallel lines in the diagonal direction constructed by the points in neighborhood. The feature values of RP can be characterized [2] with Recurrence Rate (RR), Determinism (DET), Ratio, The average diagonal line length (L), the maximum length (Lmax), Divergence (DIV), Entropy (ENTR).

3 Chaotic Recurrence Plot Analysis of Oil/Water Two-Phase Flow Patterns

3.1 Recurrence Plot Analysis of the Transitional Flow Pattern

Because the oil/water two phase flow with water cut of 51% (51.5%) belongs to low dimension chaotic system [7], its RP is not sensitive to the embedding dimension. We choose the embedding dimension $m = 3$; and choose the time delay based on the first minimum of mutual information method [8]. Considering the noise in the experiment data when computing the recurrence matrix, we choose a relatively large threshold coefficient(α=0.2), and the RP are shown in Fig.2. We can see that the total flow rate of oil/water two phase flow has significantly influence on the texture of RP. At the low flow rate (20m^3/day and 40m^3/day), the texture of diagonal line type on the RP almost does not exist and the texture of dispersed points appears on the RP; however, at the high flow rate (50m^3/day), the texture of diagonal line type appears obviously. The phenomenon that two different kind textures appear on the RP corresponds to the character of transitional flow pattern with water cut of 51% (51.5%) that oil-in-water or water-in-oil appears randomly [7].

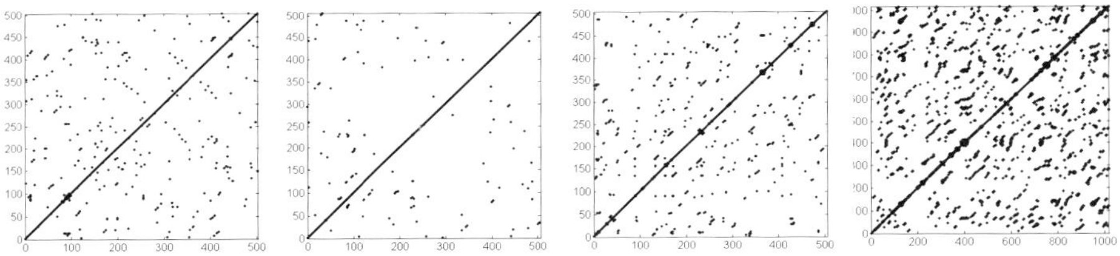

(a) Qt=20m^3/d Kw=51.5% (b) Qt=40m^3/d Kw=51.5% (c) Qt=40m^3/d Kw=51% (d) Qt=50m^3/d Kw=51%

Fig. 2. Recurrence plots for transitional flow patterns

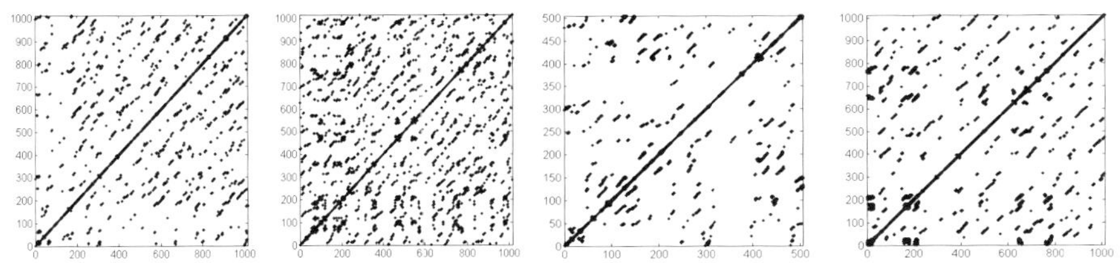

(a) Qt=10m^3/d Kw=91% (b) Qt=20m^3/d Kw=91% (c) Qt=40m^3/d Kw=81% (d) Qt=30m^3/d Kw=71%

Fig. 3. Recurrence plots for oil-in-water flow patterns

3.2 Recurrence Plot Analysis of Oil-in-Water Flow Pattern

Fig.3 shows the RP of oil-in-water flow pattern. They are greatly different from the RP of transitional flow pattern with water cut of 51% (51.5%). On the RP, the diagonal line type texture is only appeared, but not dispersed points. It indicates that the oil-in-water flow pattern is relatively steady, and its flow character is not easily changed with different flows.

4 Conclusions

We used the time series data generated by the Lorenz equation to verify the validity and sensitivity of recurrence quantification analysis, and then applied this method to the identification of oil/water two-phase flow patterns. The research indicated that the texture of dispersed point or diagonal line type was appeared on recurrence plot for the transitional oil/water two phase flow pattern; however, the diagonal line type texture was only appeared for the oil-in-water flow pattern. This phenomenon showed that the texture on recurrence plot was sensitive to the changes of oil/water two flow patterns, and the recurrence quantification analysis is a valid supplementary diagnosis tool for flow patterns identification.

Acknowledgment

The authors wish to thank the support from Natural Science Foundation of China (No. 60374041).

References

1. Eckmann, J. P., Kamphorst, S. O., Ruelle, D., Recurrence plots of dynamical systems, Europhys. Lett., 1987, 4(9): 973–977
2. Zbilut, J. P., Webber, C. L. Jr., Embeddings and delays as derived from quantification of recurrence plots, Phys. Lett. A, 1992, 171(3-4): 199–203
3. Zhong Jikang, Song Zhihuai, Hao Weiqiang, Application of recurrence quantification analysis to EMG, ACTA BIOPHYSICA SINICA, 2002, 18(2), Jun: 241-245
4. Marwan, N., Wessel, N., Kurths, J., Recurrence plot based measures of complexity and its application to heart rate variability data, Phys. Rev. E, 2002, 66(2), 026702
5. Franch, J. B., Contreras, D., Lledo, L. T., Assessing nonlinear structures in real exchange rates using recurrence plot strategies, Physica D, 2002, 171: 249-264
6. Lorenz, E. N., Deterministic nonperiodic flow, J. Atmos. Sci., 1963, 20: 130-141
7. Jin Ningde, Ning Yingnan, Wang Weiwei, Liu Xingbin, Tian Shuxiang, Characterization of oil/water two phase flow patterns in vertical upward flow pipes, Journal of Chemical Industry and Engineering(China), 2001, 52(10): 907-915
8. Guo Baihe, Jin Ningde, Hu Yafan, Hu Yangli, Chaotic time series analysis of oil/water two phase flow patterns, Journal of Dynamics and Control, 2004, 2(3): 48-53

A Clustering Model for Mining Consumption Patterns from Imprecise Electric Load Time Series Data

Qiudan Li, Stephen Shaoyi Liao, and Dandan Li

Laboratory of Complex Systems and Intelligence Science,
Institute of Automation, Chinese Academy of Sciences, Beijing
qiudan.li@ia.ac.cn, qiudanli@gmail.com
Department of Information System, City University of Hong Kong,
School of Economics and Management,
South West Jiao Tong University, China
issliao@cityu.edu.hk
Department of Automation and Computer-Aided Engineering,
The Chinese University of Hong Kong
ddli@acae.cuhk.edu.hk

Abstract. This paper presents a novel clustering model for mining patterns from imprecise electric load time series. The model consists of three components. First, it contains a process that deals with representation and preprocessing of imprecise load time series. Second, it adopts a similarity metric that uses interval semantic separation (Interval SS)-based measurement. Third, it applies the similarity metric together with the k-means clustering method to construct clusters. The model gives a unified way to solve imprecise time series clustering problem and it is applied in a real world application, to find similar consumption patterns in the electricity industry. Experimental results have demonstrated the applicability and correctness of the proposed model.

1 Introduction

With the rapid development of electricity information system, a big amount of electric load time series data are collected, which contain consumption pattern information of electricity consumers. Finding similar consumption pattern groups from these data can help the operators know more about regularities of consumption and make correct decisions. Time series clustering mining can serve for this purpose [1], [2]. However, due to collection or prediction reasons, the load time series data often involves imprecision. Therefore, the clustering mining model has to be designed to deal with this imperfection. Imprecision can be interval valued or fuzzy valued [3].

The purpose of this paper is thus to propose a load time series clustering mining model that can handle the imprecision of load time series efficiently. In [4], the authors discuss the sequence matching problem from a new angle focusing on the imprecise time series data represented by interval values. The method provides an efficient way for mining patterns from imprecise electric load time series.

L. Wang et al. (Eds.): FSKD 2006, LNAI 4223, pp. 1217 – 1220, 2006.

2 A Clustering Model

2.1 Representation and Preprocessing Process

The imprecise electric load time series is a finite sequence X of interval number: X = (x_1, x_2...x_n), where x_i is an interval number $[\underline{x}_i, \overline{x}_i]$ and n is the length of X. From the representation, we notice that X has two boundary sequences, one is lower sequence $\underline{X} = (\underline{x}_1, \underline{x}_2, ..., \underline{x}_n)$ and the other is upper sequence $\overline{X} = (\overline{x}_1, \overline{x}_2, ..., \overline{x}_n)$.

To guarantee the correctness when comparing two load time series, preprocessing procedure is used to handle the differences of baselines and scales [4]. Let $X^* = (x_1^*, x_2^*, \cdots, x_n^*)$ be any boundary sequence, where x_i^* $(1 \le i \le n)$ is any crisp number. In our study, normalization mappings including the Z-score normalization and the max-min linear normalization are used [5].

Definition1. Z-score normalization of X^* is defined as follows:

$$x_i^{*'} = \frac{x_i^* - \mu_{X^*}}{\delta_{X^*}} \tag{1}$$

where μ_{X^*} and δ_{X^*} are mean and standard deviation of the sequence X^*, respectively.

Definition2. Max-min linear normalization of X^* that transforms the values into scale [0,1] is defined as follows:

$$x_i^{*'} = \frac{x_i^* - \min X^*}{\max X^* - \min X^*} \tag{2}$$

where $\min X^*$ and $\max X^*$ are the minimum and maximum values of the sequence X^*, respectively.

2.2 Similarity Metric for Imprecise Electric Load Time Series

The similarity between two imprecise load time series is defined as follows [4]:
Definition3. Let δ > 0 be an integer constant and φ is the normalization mapping. Two load time series *X and Y* have (γ,ε,φ,δ)-similarity if and only if, given 0≤ε≤1, there exist $X_s = (x_{i_1}, ..., x_{i_l})$ and $Y_s = (y_{j_1}, ..., y_{j_l})$ that are subsequences in *X* and *Y*, respectively, and they satisfy the following conditions: 1)for any $1 \le k \le l\text{-}1$, $i_k < i_k+1$ and $j_k < j_k+1$; 2) for any $1 \le k \le l$, $|i_k - j_k| \le \delta$; 3) for any $1 \le k \le$ 1, $SS([\varphi(\underline{x}_{i_k}), \varphi(\overline{x}_{i_k})], [\varphi(\underline{y}_{i_k}), \varphi(\overline{y}_{i_k})]) \le \varepsilon$. Interval SS is the similarity between two interval numbers.

Given ε and φ, the similarity degree between X and Y, Sim(X,Y), is represented as l/n, where l is the length of longest similar subsequences between X and Y.

2.3 Construction of the Cluster

K-means clustering algorithm is widely used in many real world applications [5]. We introduce the proposed similarity metric to k-means to handle imprecise time series clustering. The construction process consists of the following steps:

Step1: Randomly select k objects as initial cluster centers;
Step2: Assign each object to its most similar cluster center, the process can be regarded as a similarity matching process. The similarity degree between each object and cluster center is the length of longest similar subsequences between them, dynamic programming approach can server this purpose;
Step3: Recalculate the cluster center for each cluster;
Step4: The process will continue until the changing value of E in the last two iterations is less than a predefined threshold. E is total sum of dissimilarity degree for object to its cluster center, which can be defined as follows:

$$E=\sum_{i=1}^{k}\sum_{b\in C_i}(1-Sim(b,m_i)) \tag{3}$$

where k is the number of clusters, b is an object, m_i is the center of cluster C_i, and $1-Sim(b,m_i)$ describes the dissimilarity between b and m_i.

3 Use of the Proposed Model in an Electricity Industry

In this section, we apply the proposed model in an electricity industry to demonstrate its correctness and practicability.

We use about one year's electric load time series data of a real electricity system in this study. Each time series has 96 values, which are collected per 15 minutes interval in a day. Cluster patterns are constructed by the proposed model, during the cluster construction process, the total sum of dissimilarity degree decreases at each iteration as objects searching new similar cluster centers. Fig.1 shows the found three representative consumption patterns. Similarity parameters ε, δ in this study are 0.1 and 2, respectively.

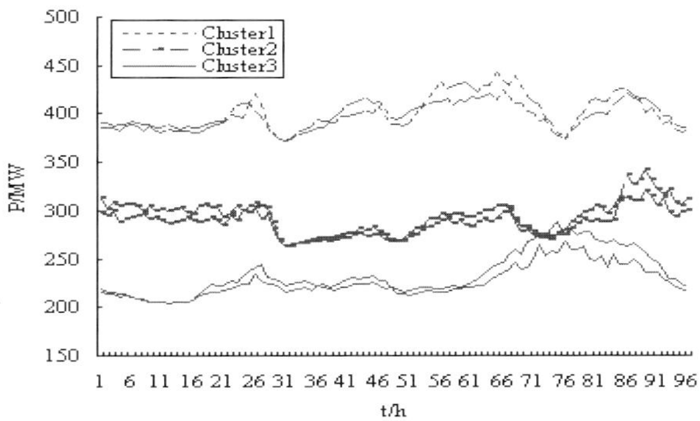

Fig. 1. Three representative cluster patterns

It can be seen from Fig.1 that: 1) Cluster3's load values are relatively low from time points 1 to 20, 85 to 96, while values are relatively high from time points 25 to 44, 65 to 84. By taking time series' date type into account, we can find that cluster3 is a typical holiday's consumption pattern, since some big factories rest and do not consume electricity from time points 1 to 20, 85 to 96. 2) Cluster1 and cluster2's load values' trends at the above time points are opposite to that of cluster3, the reason is that the big factories work on those periods to lessen costs. This trend of cluster2 is more obvious than that of cluster1. Generally speaking, cluster1 and cluster2 can be regarded as normal days' consumption patterns. Since more use of air-condition will leads to higher electricity consumption amount in summer, and some factories will also rest on very hot days, therefore, load values from time points 1 to 20, 85 to 96 are slightly lower, however, the total amount of electricity consumption is still higher than that of other seasons. So cluster1 shows more characteristics of summer while cluster2 shows characteristics of other seasons.

Based on the above analysis, we can find that the proposed model still mines the reasonable consumption patterns from imprecise electric load time series. The found patterns can not only distinguish consumption patterns of holidays from those of normal days, but also show the consumption characteristics of different seasons. Based on the consumption clusters, operators can generate corresponding strategy for different consumer to make the enterprise be competitive.

4 Conclusions and Future Work

We address time series clustering problem from a new angle with a focus on clustering of imprecise time series that are represented by interval numbers. The model is also applicable in other applications to find clustering patterns from imprecise time series data.

Future work will involve the following aspects: First, to further validate the performance of the proposed model. Second, to mine other kinds useful consumption patterns from imprecise electric load time series. Finally, to integrate the model in an electric load time series decision support system.

References

1. Fátima Rodrigues, Jorge Duarte, Vera Figueiredo, Zita A. Vale, Manuel Cordeiro:A comparative analysis of clustering algorithms applied to load profiling. MLDM (2003)73-85
2. Keogh E. J., Kasetty S.: On the need for time series data mining benchmarks: A survey and empirical demonstration. Proc.8th ACM SIGKDD Int. Conf. Knowledge Discovery Data Mining. (2002) 102-111
3. Parsons, S.: Current approaches to handling imperfect information in data and knowledge bases. IEEE Trans. Knowledge Data Eng. (1996)353-372
4. Liao, S. S., Tang, T. H., Liu W.Y.: Finding relevant sequences in time series containing crisp, interval, and fuzzy interval data. IEEE Tran. Syst. Man, Cybern. B (2004)2071-2079
5. Han J., Kamber, M.: Data Mining: Concepts and Techniques. San Mateo, CA: Morgan Kaufmann(2001)

Sequence Outlier Detection Based on Chaos Theory and Its Application on Stock Market

Chi Xie, Zuo Chen, Xiang Yu

College of Business Administration, Hunan University,
Changsha, Hunan, P.R. China

Abstract. There are many observable factors that could influence and determine the time series. The dynamic equations of their interaction are always nonlinear, sometimes chaotic. This paper applied phase space reconstruction method to map time series into multi-dimension space based on chaos theory. Extracted from multi-dimension phase space by the method of sequential deviation detection, outlier set was used to construct a decision tree in order to identify the kinds of outliers. According to the results of decision tree, a trading strategy was set up and applied to Chinese stock market. The results show that, even in bear market, the strategy dictated by decision tree brought in considerable yield.

1 Introduction

The development of information technology improves people's ability of gathering data so that the amounts of data keep increasing rapidly in all trades. A time series database consists of sequences of values or events changing with time. Containing a familiar form of data, time series database has become one of the most important databases for data mining. Naturally, mining time series and sequence data turns into a hotspot of data mining research.

The common methods used to analyze time series are based on statistical models, such as ARMA [1], ARCH [2], GARCH [3] and ARIMA [4]. These basic time series models, which require several special hypotheses, try to describe the system behavior by using a fixed structure. It is inappropriate to apply these models to commercial and financial time series whose structures change with time.

Takens' work on chaos laid a foundation for analyzing the dynamics mechanism of times series [5] and phase space reconstruction is currently the preferred method. It estimates the embedding dimension and delay time for the given time series, and maps the original series into multi-dimensional space in order to fully expose the hidden information.

This paper, which is divided into four sections, presents sequence outlier detection based on chaos theory and applies it to stock market. The first section discusses the basic concept and theory of phase space reconstruction, and introduces the estimating method for embedding dimension and delay time. The second section presents the deviation-based outlier detection according to the outlier set. The third section discusses

L. Wang et al. (Eds.): FSKD 2006, LNAI 4223, pp. 1221 – 1228, 2006.

the problem of identifying outlier by decision tree. Last section presents the results of the application on stock market.

2 Phase Space Reconstruction

2.1 The Basic Concept and Method

A time series is a sequence of observed data, usually ordered in time and could be expressed as $X = \{x_t, t = 1, ..., n\}$, where t is a time index. Formally, time series is a one-dimensional discrete system changing with time. An essential problem in analyzing chaotic time series is how to extract the evolution schema out of a dynamic system.

Let $x(t)$ be an observed time series. Then the state vectors in the reconstructed m-dimensional phase space are defined by

$$Y(t) = (x(t), x(t+\tau), ..., x(t+(m-1)\tau)) \quad Y(t) \in R^m \tag{1}$$

where τ is the delay time.

Takens' research laid a theoretical foundation of using phase space reconstruction for chaotic time series analysis. Takens' theorem states that a dynamical system can be reconstructed from a sequence of observations of the state of the dynamical system. Assume that the original time series is $x(t)$, embedding dimension is m and delay time is τ. According to Takens' theorem, there exists a smooth map $f: R^m \rightarrow R^m$ in the reconstructed phase space R^m which was reconstructed with feasible embedding dimension m and delay time τ. Thus, the state transfer model of m-dimensional phase space could be denoted as:

$$Y(t+1) = f(Y(t)) \tag{2}$$

where $Y(t)$ was the point of reconstructed phase space. Turn (2) into vector form:

$$(x(t+\tau), x(t+2\tau), ..., x(t+m)\tau)) = f(x(t), x(t+\tau), ..., x(t+(m-1)\tau)) \tag{3}$$

For the convenience of computing and simplifying, (3) could be modified as:

$$x(t+m\tau) = F(x(t), x(t+\tau), ..., x(t+(m-1)\tau)) \tag{4}$$

Here, $F(x)$ was a mapping from m-dimensional real space to singular dimensional real number, $F(x)$: $R^m \rightarrow R$. Theoretically, once a function approximation of $F(x)$ is constructed, we could describe the behavior of the chaotic system and make chaotic time series forecasts.

2.2 CC Method

Embedding dimension m and delay time are most important determinants of reconstructing phase space. Generally, the estimate methods of and m could be broadly divided into two major classes according to whether and m are dependent or not. GP

[6], Mutual information [7] and SVF [8] belong to the class of methods holding that and m are independent, while CC method [9] belongs to the other class.

Compared to other methods, CC method is a relatively simple one and easy to be implemented. CC method could estimate both τ and m. It requires relatively small data sets and is not computationally demanding.

The delay time $\tau_d = \tau$ is chosen to ensure the components of x_i are independent and the delay time window $\tau_w = (m\text{-}1)\ \tau_d$ is the entire time spanned by the components of x_i which should be independent of m. CC method is to estimate τ_d and τ_w so as to get delay time and embedding dimension.

The correlation integral introduced by Grassberger and Procaccia is widely used to characterize strange attractors [10]. The correlation integral of embedded time series is defined by:

$$C(m,N,r,t) = \frac{2}{M(M-1)} \sum_{1 \le i < j \le m} \Theta(r - \|X_i - X_j\|), \qquad r > 0 \tag{5}$$

where m is the embedded dimension, N is the size of the data sets, t is the subscript of the time series, M=N-(m-1)t is the number of embedded points in m-dimensional space. Θ is the Heaviside function.

As $N \rightarrow \infty$, to the time series, a statistic S similar to BDS statistics [11] could be constructed as follows:

$$S(m,r,t) = \frac{1}{t} \sum_{s=1}^{t} \left[C_s(m,r,t) - C_s^m(1,r,t) \right] \tag{6}$$

For fixed and m, $S(m, r, t)$ will be identically equal to 0 for all r if $N \rightarrow \infty$. Generally, because the size of real data sets are finite and the data may be serially correlated, $S(m, r, t)$ does not always equal to 0。 The locally optimal times may be either the zero crossings of $S(m, r, t)$ or the times at which $S(m, r, t)$ shows the least variation with r.

In order to measure the variation of $S(m, r, t)$ with r, the quantity is defined by

$$\Delta S(m,t) = \max\{S(m,r_j,t)\} - \min\{S(m,r_j,t)\} \tag{7}$$

The locally optimal times t is then the zero crossings of $S(m, r, t)$ and the minima of $\triangle S(m, t)$.

Brock studied the statistical results of several important asymptotic distributions. The results show that, when $2 \le m \le 5, \sigma/2 \le r \le 2\sigma$, $N \ge 500$, the asymptotic distributions were well approximated by finite time series, where σ is the standard variation of the time series. According to Brock's conclusion, let m=2,3,4,5, $r_i = i\sigma/2$, i=1,2,3,4, computing the following three variables:

$$\bar{S}(t) = \frac{1}{16} \sum_{m=2}^{5} \sum_{j=1}^{4} S(m,r_j,t) \tag{8}$$

$$\Delta \bar{S}(t) = \frac{1}{4} \sum_{m=2}^{5} \Delta S(m,t) \tag{9}$$

$$S_{cor}(t) = \Delta\bar{S}(t) + \left|\bar{S}(t)\right| \tag{10}$$

After looking for the first zero crossing of (8) or the first local minimum of (9) to find the delay time τ_d and looking for the minimum of the quantity of (10) to find delay time window τ_w, we could get τ and m according to the equation $\tau_d = \tau$ and $\tau_w = (m-1)\tau$.

3 Outlier Detection

3.1 Classification of Outlier Detection Methods [12]

Outliers, sometimes called exception, are the data objects that are grossly different from or inconsistent with other objects of the same data sets. There are two possible reasons why outliers do not comply with the general behavior of other data objects. Outliers maybe caused by the noises or the inherent data variability. Thus, there may be important information hidden in outliers and outlier detection and analysis is referred to as outlier mining.

Outlier mining can be described as follows: Given a set of *n* data points or objects and *k*, the expected number of outliers, find the top *k* objects that are considerably dissimilar, exceptional, or inconsistent with respect to the remaining data. Outlier detection methods could be categorized into three approaches: the statistical approach, the distance-based approach and the deviation-based approach.

3.2 Deviation-Based Outlier Detection

Deviation-based approach identifies outliers by examining the main characteristics of objects in a group. If a data object deviate from these characteristics, it would be regarded as an outlier. The sequential exception technique simulates the way in which humans can distinguish unusual objects from among a series of supposedly similar objects. The main idea and details could be found in [13].

To the time series, we reconstructed the phase space after finding the embedding dimension and delay time. Let *Y(t)* be the vector of *m*-dimensional phase space,,we could construct data set *I* according to (4):

$$i_t = F(Y(t)) = x(t + m\tau), \quad i_t \in I \tag{11}$$

The dissimilarity function [13] is used to measure the dissimilarity degree of the subset. Generally, it could be any function that returns a low value if the objects are similar to one another in a given set of objects. The greater the dissimilarity among the objects, the higher the value returned by the function.

Define the dissimilarity function as the variation of data set I：

$$D(I) = \frac{1}{n}\sum_{t=1}^{n}(i_t - \bar{i})^2 \tag{12}$$

Define the cardinality function as the cardinality of data set, that is, the count of the number of objects in a given set. Smoothing factor assesses how much the dissimilarity can be reduced by removing the subset from the original set of objects. The subset, which gets the largest amount of smoothing factors, is the outlier set.

4 Outlier Forecast

Assuming the outlier set is I_x, $i_t \in I_x$, we could create a 2-tuple consisting of the outlier and the corresponding m-dimensional vector

$$< Y(t), F(Y(t)) >=< Y(t), i_t >=< x(t), x(t+\tau), ..., x(t+(m-1)\tau), i_t) > \quad (13)$$

We could set different values or characters to an attribute *flag* according to the value of i_t, so as to categorize outlier into different classes. After mapping m components of vector $Y(t)$ to m attributes: $Y_i(t) \to A_i$, the problem of outlier forecast could be transferred into data classification of (m+1)-tuple $< A_1, A_2, ...A_m, flag >$.

Decision tree is one of the most effective approaches for classification in data mining. It could not only identify the class of all sample set accurately, but also classify the new sample effectively. It is produced by analyzing and inducing the attributes of sample set based on information theory. ID3 is the most famous case-learning algorithm [14], which takes information gain as the attribute selection measure. C4.5 is the improved algorithm derived from ID3. C4.5 takes the ratio of information gain as the attribute measure and is appended with other functions such as pruning, dealing attributes with continuous ranges and extracting rules [14]. We used C4.5 to construct decision tree to identify outliers. Details about the computing method of gain ratio and other functions could be found in [15].

Assuming the forecast point is $x(t)$, we could construct the new m-dimensional vector $X(t)$ from time series according to embedding dimension and delay time:

$$X(t) = (x(t-d\tau), x(t-(d-1)\tau), ..., x(t-\tau)) \quad X(t) \in R^m \quad (14)$$

Use decision tree to identify the outlier class of the new sample $X(t)$. If classification is successful, an outlier will be forecasted and its class could be determined according to the attribute *flag*.

5 Applications

We chose the yield sequence of Shanghai stock composite index as the researching object. Let y_n be the closing price while y_{n-1} be the closing price of last trading day, and the yield sequence could be denoted as $X = \{x_n, x_n = (y_n - y_{n-1}) / y_n\}$. Chinese stock markets have imposed a range limit on the rise and drop in stock market since Dec 16, 1996. We selected the trading data from Jan 2 1997 to Jan 2 2004 to get the yield series with a length of 1685.

We applied CC method to estimate embedding dimension and delay time of the time series. After computing $\Delta\overline{S}(t)$ and $S_{cor}(t)$ according to the equation (9) and (10) respectively, the result was shown as Figure 1.

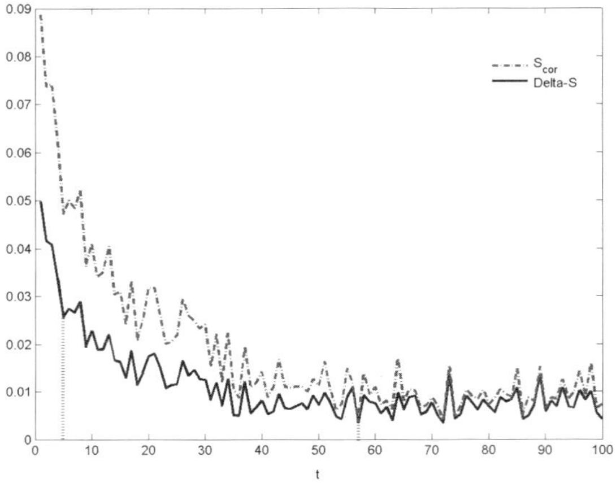

Fig. 1. The result of delay time window estimated. From Figure 1, we could find the first local minimum of $\Delta\bar{S}(t)$ is 5 while the minimum of $S_{cor}(t)$ is 57. We expanded the delay time window to avoid losing information and selected time window τ_w= 60, delay time τ_d= 5, thus got embedding dimension is 13, delay time is 5.

It is necessary to test whether the time series was chaotic or not. A simple method to determine whether a time series is chaotic is to see whether its largest Lyapunov exponent is greater than 0. If it is, the time series is chaotic. We used the practical method [16] to calculate the largest Lyapunov exponent and got the average period p=3.8539, largest Lyapunov exponent λ =0.02381>0.

After phase space reconstruction, we got 1620 points of m-dimensional phase space, i.e. 1620 m-dimensional vectors. Appling sequence exception technique to this data set, we found 342 outliers, almost 21.1% of the total sample set. The count of outliers whose yield was greater than zero is 170, while the ones whose yield was less than zero summed up to 172. The mean of the outlier set is 0.0357% and the variation of it is 3.0466%.

We used P and N to represent the positive outlier and negative outlier according to whether the yield was positive or negative. The interval of Chinese stock market's rise and drop range limit is [-10,10]. In order to dealing attributes with continuous ranges, we divided this interval into 21 subintervals: such as [-10, -9.5], (-9.5:1:9.5), [9.5,10]. We constructed decision tree and took outlier set as training sample set. After pruning and extracting rules, we got 28 rules.

Using constructed decision tree to classify the training set, the result was shown in Table 1.

The classifications accuracy is 73.6%, while 251 cases are correct and 91 cases are incorrect.

In order to verify the effectiveness of the outlier forecast decision tree, we followed the idea of Povinelli [17] and set up a trading strategy, which is directed by the classification result of decision tree. Assuming the forecast point is $x(t)$, we could construct

Table 1. classification result of training set.

	P	N
Class P	136	**34(error)**
Class N	**57(error)**	115

the new m-dimensional vector $X(t)$ from time series according to embedding dimension and delay time and use decision tree to identify the outlier class of the new sample $X(t)$. Taking the T+1 trading mode of Chinese stock market into account, our trading strategy is described as follows, which ignores trading costs.

If classified to P and not holding stock

Buy the stock at opening price of the trading day t and hold;

If classified to N and holding stock

Sell the stock at opening price of the trading day t;

We selected 400 trading days as test set, spanned from Feb 5, 2004 to Sept 19, 2005. The data set was divided into four groups, sizes of which are all 100. SH1: From Feb 5, 2004 to June 30, 2004; SH2: From July 1, 2004 to Nov 24, 2004; SH3: From Nov 25, 2004 to Apr 25, 2005; SH4: From Apr 26, 2005 to Sept 16, 2005.

Table 2 shows the result of comparing the yield of the trading strategy directed by decision tree and the yield of buy-and-hold strategy on different period.

Table 2. Trading result of different strategy

	SH1	SH2	SH3	SH4	AVR	Total (SH1~SH4)
Classification%	4.77	4.18	-1.79	3.47	2.66	10.92
Buy-and-Hold%	-16.72	-1.92	-14.14	5.01	-6.94	-28.24

From Table 2, we could find that the return of our trading strategy was greater than the buy-and-hold's when ignored the trading costs. Although in bear market during SH1, SH2 and SH3, the strategy directed by decision tree brought in considerable yield and the closer the period was to the sample set, the more efficient it was. It didn't work very well at period SH4. The main cause may be that the structures of commercial and economic time series are always changing with time. As a result, the farther the period is, the greater impact it will have on the results. We suggested that it'd be better to recalculate embedding dimension and delay time after a period of time, so as to find the new temporal patterns of the time series.

6 Conclusion

This paper applied CC algorithm to estimate embedding dimension and delay time based on chaos theory, and to reconstruct the phase space of time series. After mapping time series into multi-dimensional phase space, we used sequence derivation technique to extract outlier set. Then, taking the outlier set as training set to construct decision tree, we applied a trading strategy dictated by the decision tree to Chinese stock market. The results show that even in bear market, the strategy dictated by decision tree brought in considerable yield. The structures of commercial and economic time series are always changing with time. As a result, the farther the period is, the greater impact it will have on the estimating result. So it could be more adaptive to short term trading strategy. Because outlier patterns are also changing with time, the focus of our next step would be to change the outlier set automatically for the new sample and modify the structure of decision tree dynamically according to the changed outlier set.

References

1. Box G.E.P, Jenkins G.M.:Time series Analysis: Forecasting and Control[M]. Halden-Day, San Francisco, (1976)
2. Engle R.F.: Autogressive conditional heteroskedasticity with estimates of the variance of UK inflation. Econometrica, Vol.50. (1982) 987-1008
3. Bollerslev T.: Generalized autogressive conditional heteroskedasticity. Journal of Econometrics, Vol.31. (1986) 307-327
4. S. M. Pandit, S. M. Wu.: Time series and system analysis with applications. John Wiley & Sons, New York (1983)
5. Takens F.: Detecting Strange Attractors in Turbulence. Lecture Note in Mathematics, (1980) 366-381
6. Albano A M, et al.: SVD and Grassberger-Procaccia algorithm. Phy Rev A, Vol.38 (1988) 3017-3026
7. Fraser A M.: Information and entropy in strange attractors. IEEE tron IT, Vol.35. (1989) 245-262
8. Kember G, Folwer A C.: A correlation function for choosing time delays in phase portrait reconstructions. Phy Lett A, Vol.179. (1993) 72-80
9. H. S. Kim, R. Eykholt, J. D. Salas.: Nonlinear dynamics, delay times, and embedding windows. Physica D, Vol.127. (1999) 49-59
10. P. Grassberger, I. Procaccia.: Phys. Rev. Lett. Vol.50. (1983) 345
11. W. A. Brock, D. A. Hsieh, B. LeBaron.: Nonlinear Dynamics, Chaos, and Instability: Statistical Theory and Economic Evidence. MIT Press, Cambridge, MA (1991)
12. Jiawei Han, Micheline Kamber.: Data Mining: concepts and techniques. Morgan Kaufmann Publishers, (2001)
13. A. Arning, R. Agrawal, P. Raghavan.: A linear method for deviation detection in large databases. In proc. 1996 int. conf. Data Mining and Knowledge Discovery, Philadelphia, PA, (1999) 164-169
14. J. R. Quinlan.: Induction of Decision Trees. Machine Learning Vol.1. No.1. (1986)
15. J. R. Quinlan.: C4.5: Program of Machine Learning. Margan Kovnfmenn Publishers, (1993)
16. M. T. Rosenstein, J. J. Collins, C. J. De luca.: A practical method for calculating largest Lyapunov exponents in dynamical systems, Physica D, Vol.65. (1992) 117-134
17. Povinelli, R.J.: Identifying Temporal Patterns for Characterization and Prediction of Financial Time Series Events. Temporal, Spatial and Spatio-Temporal Data Mining, (2000), 46-61

Fuzzy-neuro Web-Based Multilingual Knowledge Management

Rowena Chau[1], Chung-Hsing Yeh[1], and Kate Smith-Miles[2]

[1] Clayton School of Information Technology,
Faculty of Information Technology,
Monash University, Clayton, Victoria 3800, Australia
{rowena.chau, chunghsing.yeh}@infotech.monash.edu.au
[2] School of Engineering and Information Technology
221 Burwood Highway, Burwood, Victoria 3125, Australia
katesm@deakin.edu.au

Abstract. This paper presents new methodology towards the automatic development of multilingual Web portal for multilingual knowledge discovery and management. It aims to provide an efficient and effective framework for *selecting* and *organizing* knowledge from voluminous linguistically diverse Web contents. To achieve this, a concept-based approach that incorporates text mining and Web content mining using neural network and fuzzy techniques is proposed. First, a concept-based taxonomy of themes, which will act as the hierarchical backbone of the Web portal, is automatically generated. Second, a concept-based multilingual Web crawler is developed to intelligently harvest relevant multilingual documents from the Web. Finally, a concept-based multilingual text categorization technique is proposed to organize multilingual documents by concepts. As such, correlated multilingual Web documents can be gathered/filtered/organised/ based on their semantic content to facilitate high-performance multilingual information access.

1 Introduction

A portal is a Web site that serves as an entry point into the Web information space. A typical portal has a hierarchical subject catalogue associated with categories of relevant documents, and/or a search engine combined with other services and interactive content. The rapid expansion of the WWW has made electronically accessible information available in almost all natural languages. With majority of this Web data being unstructured text [1], effective multilingual Web portal development technology capable of automatically discovering and organizing relevant multilingual Web documents thus holds the key to exploit the vast human knowledge hidden beneath this largely untapped multilingual text. In this paper, a concept-based approach towards the automatic development of multilingual Web portal, is presented. This approach represents an integration of the text mining and Web content mining paradigms using artificial intelligence techniques.

In the rest of this paper, Section 2 will introduce the concept-based approach and the rationale behind it. This is followed by the generation of a hierarchical

L. Wang et al. (Eds.): FSKD 2006, LNAI 4223, pp. 1229 – 1238, 2006.

concept-based theme taxonomy, which will act as the backbone of the multilingual Web portal in Section 3. Section 4 will give an account the process of concept-based multilingual Web crawling, and Section 5 will explain the development a concept-based multilingual text categorization system for organizing multilingual Web documents by their concept-based thematic content. Finally, Section 6 will conclude the work.

2 The Concept-Based Approach

The focus of this research is to automate the development of a concept-based multilingual Web portal capable of discovering, filtering and organizing multilingual documents relevant to a set of themes underlying a target area of information interest. By associating correlated multilingual documents to their corresponding themes, content-focused information exploration is facilitated. As such, the objective of enabling global knowledge acquisition is achieved.

On discovering and organizing multilingual Web documents relevant to the themes of a target area of interest, multilingual Web portal development is a typical application of text mining and Web content mining that requires an understanding of the natural language text. Text mining concerns the discovery of knowledge from textual data by applying various knowledge discovering techniques. Text clustering and categorization are the two most extensively studied areas in text mining [2]. Using appropriate clustering techniques, terms (i.e. keywords) or documents related to the same concepts can be effectively grouped by their semantic similarity and thus revealing their conceptual content. Emerged as an area of text mining specific to Web documents focusing on analyzing and deriving meaning from textual collection on the Internet [3], Web content mining has attracted much research attention in recent years [4]. Popular Web content mining activities include the exploitation of retrieval results returned by search engines and automatically harvesting relevant Web documents using intelligent Web crawlers [5].

Currently, both text mining and Web content mining technologies are still primarily focused on processing monolingual text. The challenge of discovering knowledge from textual data, which are significantly linguistically diverse, has early been recognised by text mining research [6]. In a monolingual environment, the conceptual content of documents can be discovered by directly detecting patterns of frequent features (i.e. terms) without precedential knowledge of the concept-term relationship. Documents containing an identical known term pattern thus share the same concept. However, in a multilingual environment, *vocabulary mismatch* among diverse languages implies that documents exhibiting similar concept will not contain identical term patterns. This *feature incompatibility* problem thus makes the inference of conceptual contents using term pattern matching inapplicable.

Due to the feature incompatibility problem contributed by the vocabulary mismatch phenomenon across multiple languages, documents describing a common theme in different languages are represented by different sets of features (i.e. terms) in separate feature spaces. This language-specific representation has made multilingual text incomparable. Therefore, monolingual text representation techniques that rely on shared syntactic terms will not work for multilingual text/Web content mining application, such as multilingual Web portal development. To overcome this problem, a concept-based approach is proposed. The idea is: by generating a taxonomy of language-independent concept-based thematic profiles encapsulating all feature variations among multiple languages, semantically-relevant multilingual Web documents can then be scanned, filtered, retrieved and categorized regardless of the syntactic terms they contain. The automatic development of multilingual Web portal typically involves three major tasks, including (1) concept-based theme taxonomy generation, (2) concept-based multilingual Web crawling and (3) concept-based multilingual text categorization.

3 Concept-Based Theme Taxonomy Generation

Web portal development concerns the organization of documents into a pre-defined taxonomy of related themes. In a multilingual environment, documents about a common theme used synonymous terms in different languages. To generate taxonomy of themes capable of accommodating these linguistic variations, concept-based theme profiles, which are linguistically comprehensive as well as semantically rich, are necessary. Each theme profile should be capable of capturing its corresponding semantic content while encapsulating the syntactic variations among term usage. Towards this end, a novel concept-based taxonomy generation approach using a training parallel corpus (i.e. a collection of translated documents) is proposed.

This concept-based taxonomy generation approach involves three constituent tasks: (a) Multilingual term clustering (b) Concept-based document representation (c) Concept-based document clustering. *Multilingual term clustering* aims at forming a set of concepts by grouping related multilingual terms extracted from the parallel documents based on their semantic similarity using self-organizing maps [10]. *Concept-based document representation*, in turn, takes these concepts as a kind of language-independent indexing features to produce a concept-based representation (i.e. a concept-based document vector) for characterizing each training documents. Finally, *concept-based multilingual document clustering*, given the concept-based document vectors as inputs, generates a concept-based theme taxonomy by clustering thematically-related training documents using a hierarchical clustering algorithm. As such, this taxonomy, providing a hierarchical schema, will act as the backbone of the multilingual web portal to guide the collection/filtering/organization of relevant multilingual Web documents for the multilingual Web portal development.

3.1 Multilingual Term Clustering

To cluster multilingual terms with an aim of acquiring a set of concepts by grouping related multilingual terms, parallel corpora containing sets of documents and their translations in multiple languages are ideal sources of multilingual lexical information. Parallel documents basically contain identical concepts expressed by different sets of terms. Therefore, multilingual terms used to describe the same concept tend to occur with very similar inter- and intra-document frequencies across a parallel corpus. An analysis of paired documents has been used to infer the most likely translation of terms between languages in the corpus [7,8,9]. As such, co-occurrence statistics of multilingual terms across a parallel corpus can be used to determine clusters of conceptually related multilingual terms.

To acquire domain-specific concepts, which can effectively characterize the knowledge context of a multilingual Web portal to be developed, a training parallel corpus from the corresponding subject domain is used. Given a parallel corpus D consisting N pairs of parallel documents, meaningful terms from every languages covered by the corpus are extracted. They form the set of multilingual terms for the multilingual term clustering process. Each term is represented by an n-dimensional term vector. Each feature value of the term vector corresponds to the weight of the *nth* document indicating the significance of that document in characterizing the meaning of the term. Parallel documents which are translated versions of one another within the corpus, are considered as the same feature. To determine the significance of each document in characterising the contextual content of a term based on the term's occurrences, the following weighting scheme is used. It calculates the feature value w_{kp} of a document d_p for $p = 1,...,N$ in the vector of term t_k.

$$w_{kp} = \begin{cases} \dfrac{tf_{kp} \cdot log\left(\dfrac{|T|}{|d_p|}\right)}{\sqrt{\sum_{q=1}^{N}\left(tf_{kq} \cdot log\left(\dfrac{|T|}{|d_q|}\right)\right)}} & for\ tf_{kp} > 0 \\ 0 & for\ tf_{kp} = 0 \end{cases} \quad (1)$$

where

tf_{kp} is the occurrence of term t_k in document d_p;

$log\left(\dfrac{|T|}{|d_p|}\right)$ is the inverse term frequency of document d_p; $|T|$ is the number of terms in the whole collection, and $|d_p|$ is the number of terms in document d_p. The longer the document d_p, the smaller the inverse term frequency;

$1 \Big/ \sqrt{\sum_{q=1}^{N}\left(tf_{iq} \cdot log\left(\dfrac{|T|}{|d_q|}\right)\right)}$ is the normalisation factor. With this normalisation factor, the feature value relating a document to a term t_k is reduced according to the total number of documents in which the term occurs.

When contextual contents of every multilingual term are well represented, they are used as the input into the self-organizing map algorithm [10] for multilingual term clustering. Let $\mathbf{x}_i \in R^N$ ($1 \le i \le M$) be the term vector of the ith multilingual term, where N is the number of documents in the parallel corpus for a single language (i.e. the total number of documents in the parallel corpus divided by the number of languages supported by the corpus) and M is the total number of multilingual terms. The self-organizing map algorithm is applied to discover the multilingual term clusters, using these term vectors as the training input to the map. The map consists of a regular grid of nodes. Each node is associated with an N-dimensional model vector. Let $\mathbf{m}_j = \left[m_{jn} \middle| 1 \leq n \leq N\right]$ ($1 \leq j \leq G$) be the model vector of the jth node on the map. The algorithm for multilingual term clustering is given below.

Step 1: Select a training multilingual term vector $\mathbf{x}_i$ at random.
Step 2: Find the winning node s on the map with the vector $\mathbf{m}_s$ which is closest to $\mathbf{x}_i$ such that

$$\left\|\mathbf{x}_i - \mathbf{m}_s\right\| = \min_j \left\|\mathbf{x}_i - \mathbf{m}_j\right\| \tag{2}$$

Step 3: Update the weight of every node in the neighbourhood of node s by

$$\mathbf{m}_t^{new} = \mathbf{m}_t^{old} + \alpha(t)(\mathbf{x}_i - \mathbf{m}_t^{old}) \tag{3}$$

where $\alpha(t)$ is the gain term at time t ($0 \leq \alpha(t) \leq 1$) that decreases in time and converges to 0.
Step 4: Increase the time stamp t and repeat the training process until it converges.

After the training process is completed, each multilingual term is mapped to a grid node closest to it on the self-organizing map. A partition of multilingual term space, represented by a multilingual term cluster map, is thus formed. This process corresponds to a projection of the multi-dimensional term vectors onto an orderly two-dimensional concept space where the proximity of the multilingual terms is preserved as faithfully as possible. Consequently, conceptual similarities among multilingual terms are explicitly revealed by their locations and neighbourhood relationships on the map. Multilingual terms that are synonymous are associated to the same node. In this way, conceptual related multilingual terms are organised into term clusters, representing all existing concepts, within a common semantic space. The problem of feature incompatibility among multiple languages is thus overcome.

3.2 Concept-Based Document Representation

A taxonomy of themes relevant to a particular domain of interest can be generated by finding hierarchy of thematically-related document clusters using a domain-specific training corpus. However, document clustering depends heavily on the document representation (i.e. indexing) scheme. To form the taxonomy of themes that effectively reflects the conceptual content among documents, a suitable method for document indexing must be devised. Contextual contents of documents need to be expressed explicitly in a computationally meaningful way.

In information retrieval, several approaches for document indexing and representation have been suggested. Among them, the vector space model [11] represents documents conveniently as vectors in a multi-dimensional space defined by a set of language-specific index terms. Each element of a document vector corresponds to the weight (or occurrence) of one index term. However, in a multilingual environment, the direct application of the vector space model is infeasible due to the feature incompatibility problem. Multilingual index terms characterising documents of different languages exist in separate vector spaces.

To overcome the problem, a better representation of document contents incurporating information about semantic/conceptual relationships among multilingual index terms is desirable. Towards this end, the multilingual term cluster map obtained in Section 3.1 is used. On the multilingual term cluster map, semantically related multilingual terms have been organised into clusters (i.e. concepts). These concepts, associating semantically related multilingual terms, are thus used to index multilingual documents in place of the documents' original language-specific index terms. As such, a concept-based document vector that explicitly expresses the conceptual context of a document regardless of its language can be obtained. The term-based document vector of the vector space model, which suffers from the feature incompatibility problem, can now be replaced with the language-independent concept-based document vector.

To realize this, every multilingual document is indexed by mapping its text, term by term, onto the multilingual term cluster map. This is done by counting the occurrence of each term on the multilingual term cluster map at the node to which that term has been associated. This statistics of term cluster (i.e. concept) occurrences is then a kind of transformed 'index' of the multilingual document to produce a language-independent concept-based document vector. Concept-based document vectors thus obtained are essential for enabling concept-based multilingual document clustering. Using these concept-based document vectors as input to some appropriate clustering algorithm, multilingual documents, which are originally syntactically incomparable can then be grouped based on the conceptual similarity they convey.

3.3 Concept-Based Multilingual Document Clustering

The hierarchical concept-based theme taxonomy is generated via concept-based multilingual document clustering. With the application of the complete linkage hierarchical clustering algorithm [12] using the concept-based document vectors of the training parallel corpus as inputs, the concept-based multilingual document clustering algorithm is given below.

Step 1: Construct a $K \times Z$ concept-document matrix, *CD*, to represent *K* concepts, c_k ,obtained from the multilingual term clustering process, and the parallel corpus, *D*, of *Z* documents, d_z, as:

$$CD = \begin{array}{c} \\ c_1 \\ c_2 \\ c_3 \\ . \\ . \\ . \\ c_K \end{array} \begin{array}{c} \begin{array}{ccccccc} d_1 & d_2 & d_3 & . & . & . & d_Z \end{array} \\ \begin{bmatrix} w_{11} & w_{12} & w_{13} & . & . & . & w_{1z} \\ w_{21} & w_{22} & w_{23} & . & . & . & w_{2z} \\ w_{31} & w_{32} & w_{33} & . & . & . & w_{3z} \\ . & . & . & . & . & . & . \\ . & . & . & . & . & . & . \\ . & . & . & . & . & . & . \\ w_{k1} & w_{k2} & w_{k3} & . & . & . & w_{kz} \end{bmatrix} \end{array} \tag{4}$$

where w_{kz} is the occurrence of the concept c_k in document d_z

Step 2: Obtain a document association matrix $D \times D$ such that

$$D \times D = \begin{array}{c} \\ d_1 \\ d_2 \\ d_3 \\ . \\ . \\ . \\ d_z \end{array} \begin{array}{c} \begin{array}{cccccc} d_1 & d_2 & d_3 & . \; . \; . & d_Z \end{array} \\ \begin{bmatrix} A_{11} & A_{12} & A_{13} & . \; . \; . & A_{1z} \\ A_{21} & A_{22} & A_{23} & . \; . \; . & A_{2z} \\ A_{31} & A_{32} & A_{33} & . \; . \; . & A_{3z} \\ . & . & . & . \; . \; . & . \\ . & . & . & . \; . \; . & . \\ . & . & . & . \; . \; . & . \\ A_{k1} & A_{k2} & A_{k3} & . \; . \; . & A_{kz} \end{bmatrix} \end{array} \tag{5}$$

where A_{xy} is the coefficient of association between each pair of documents $d_x, d_y \in D$ calculated using cosine similarity measure as defined by the following equation:

$$A_{xy} = \frac{\sum_{k=1}^{K} w_{kx} \cdot w_{ky}}{\sqrt{\sum_{k=1}^{K} w_{kx}^2} \cdot \sqrt{\sum_{k=1}^{K} w_{ky}^2}} \tag{6}$$

Step 3: Apply the complete linkage hierarchical clustering algorithm to the document association matrix $D \times D$ to determine the document clusters. The hierarchy of document clusters resulted from this process is thus a taxonomy of themes representing the contextual content relevant to the domain of the training parallel corpus. This taxonomy, providing a hierarchical schema, will thus act as the backbone of the multilingual Web portal to facilitate organization of theme-related multilingual Web documents collected by the Web crawler.

4 Concept-Based Multilingual Web Crawling

Web crawler plays the significant role in Web portal development by automatically collecting Web documents. Web crawling focusing on a collection of themes requires a set of good seed URLs that point to other potentially relevant documents to guide the process. In this research, a concept-based multilingual Web crawler, incorporating a novel approach to deduce a set of seed URLs using existing search engine is developed.

Related Web documents are often connected by their in- and out- links [13]. Web crawlers traverse the Web to collect related documents by following links. To start crawling the Web for documents relevant to a theme, a set of seed URLs are required by the Web crawler. In a monolingual dimension, it is sufficient to use any documents, written in a particular language, previously considered as relevant to the theme as the seeds for gathering other related documents exist in the same language. However, related documents written in different languages are rarely linked. To make a Web crawler capable of collecting and retrieving relevant documents in various languages, a set of seed URLs in every language must be made available.

To address the above issue, a novel concept-based approach is proposed. The idea is: for every language, we use the term that is most prototypical to a concept to run a Web search. Top-ranked Web documents returned by the search engine are then used as the seed URLs to initialise the concept-based Web crawling. To realise this, identification of the multilingual terms, which are prototypical to each concept becomes the major challenge. Recalling that the multilingual term cluster map obtained in Section 3.1 has already encoded the multilingual concept-term relationships by associating all multilingual terms with the concepts to which they belong, it thus provides the essential clues for effectively inferring such concept-specific prototypical multilingual terms. To identify a term that is prototypical with respect to a concept for every language, we find the term, one for each language, which is closest to the output node of a concept on the self-organizing map. This term is then submitted as a query to the search engine to retrieve a set of the most relevant (e.g. top 10) Web document in each language. These documents then become the seed URLs for the Web crawler to collect other related documents describing similar concepts in various languages on the Web. This approach is to ensure the Web crawler will explore the Web context diversely and yet still remained conceptually focused. Multilingual Web documents collected by the Web crawler are then passed on to a concept-based multilingual text categorization algorithm for organizing into the taxonomy to facilitate global knowledge discovery.

5 Concept-Based Multilingual Text Categorization

Multilingual text categorization algorithm performs the crucial task of Web portal development by organizing documents based on the contextual relevance. As document may belong to multiple themes with different membership degrees, multilingual text categorization is essentially a fuzzy classification process. Towards this end, a fuzzy concept-based multilingual text categorization algorithm is proposed. This algorithm regards the determination of a document's conceptual relevance to a particular theme as a process of fuzzy instance-based classification using the fuzzy nearest-prototype classification algorithm [14]. Multilingual Web document passed on from the Web crawler will then categorized to the corresponding themes where its membership degrees exceeds a certain threshold.

Based on the fuzzy nearest prototype classification algorithm, the fuzzy concept-based multilingual text classifier considered multilingual text categorisation task as a task of determining for every multilingual document a membership value in the range of [0,1] to each of the decision matrix with reference to a set of themes as illustrated in Figure 1.

	d_1	…	…	d_j	…	…	d_q
h_1	$\mu_1(d_1)$	…	…	$\mu_1(d_j)$	…	…	$\mu_1(d_q)$
…	…	…	…	…	…	…	…
h_i	$\mu_i(d_1)$	…	…	$\mu_i(d_j)$	…	…	$\mu_i(d_q)$
…	…	…	…	…	…	…	…
h_p	$\mu_p(d_1)$	…	…	$\mu_p(d_j)$	…	…	$\mu_p(d_q)$

Fig. 1. Decision matrix of fuzzy concept-based text categorization

where $H = \{h_1,\ldots, h_p\}$ is a set of themes, $D = \{d_1,\ldots,d_q\}$ is a set of multilingual documents to be classified and $\mu_i(d_j) \in [0,1]$ is the degree of membership of document d_j in theme h_i. The fuzzy concept-based multilingual text categorization algorithm is as follows:

Let $V = \{ v_1, v_2, \ldots, v_p \}$ be the set of i theme prototypes representing i thematic categories. Such theme prototypes are obtained from the concept-based multilingual document clustering result in Section 3.3 by finding the centroid document in each document cluster. The concept-based document vector of this centroid document thus forms the theme prototype. The membership value $\mu_i(d_x)$ of an unclassified text, d_x, is determined by

$$\mu_i(d_x) = \frac{\left(\dfrac{1}{\left\| d_x - v_i \right\|^{2/(m-1)}} \right)}{\sum_{j=1}^{p} \left(\dfrac{1}{\left\| d_x - v_j \right\|^{2/(m-1)}} \right)} \tag{7}$$

where $m = 2$ is chosen.

As a result of this process, every multilingual document is now assigned a set of theme membership values. Concept-based multilingual text categorisation is then achieved by associating each theme with a set of multilingual documents ranked in decreasing order of relevance as indicated by their theme membership values.

6 Conclusion

This paper has presented novel research towards automatic development of multilingual Web portal. The salient feature of this methodology is the concept-based approach that overcomes the feature incompatibility problem by unifying multilingual terms in a common semantic space. As such, Web documents relevant to the same domain, regardless of language, can then be gathered, filtered and organised based on their comparable semantic content rather than the incomparable syntactic terms they contain. Based on the concept-based approach, a concept-based theme taxonomy representing the hierarchical thematic backbone of a multilingual Web portal is first generated using a training parallel corpus relevant to the domain of a portal to be developed. To collect multilingual documents relevant to the portal's knowledge domain, a concept-based multilingual Web crawler is developed. Multilingual Web documents fetched by the Web crawler, in turn, are organized onto the concept-based theme taxonomy with the fuzzy concept-based multilingual text categorization algorithm. A multilingual Web portal thus developed is particularly significant to multilingual knowledge management where global knowledge relevant to a certain domain of interest need to be scanned, filtered, retrieved and organized to reveal actionable insights. To further evaluate the effectiveness of the proposed approach, a prototype multilingual Web portal is currently being developed, which uses a company's multilingual product catalog as training corpus.

References

1. Chakrabarti, S. (2000) Data mining for hypertext: a tutorial survey. *ACM SIGKDD Exploration*, 1(2), pp. 1 –11.
2. Berry, M. (2003) *Survey of Text Mining: Clustering, Classification, and Retrieval.* Springer-Verlag.
3. Chang, C., Healey, M. J., McHugh, J. A. M. and Wang, J. T. L. (2001) *Mining the World Wide Web: an information search approach.* Kluwer Academic Publishers.
4. Kosala, R. and Blockeel, H. (2000) Web mining research: a survey. *ACM SIGKDD Exploration*, 2(1), pp. 1 –15.
5. Chakrabarti, S. (2002) Mining the Web: Discovering Knowledge from Hypertext Data.Morgan Kaufmann.
6. Tan, A-H. (1999) Text Mining: The state of the art and the challenges. In *Proceedings of PAKDD'99 workshop on Knowledge Disocovery from Advanced Databases*, Beijing, pp. 65-70.
7. Carbonell, J. G., Yang, Y., Frederking, R. E., Brown, R. D., Geng, Y. and Lee, D (1997) Translingual information retrieval: a comparative evaluation. In Pollack, M. E. (ed.) *IJCAI-97 Proceedings of the 15th International Joint Conference on Artificial Intelligence,* pp. 708-714.
8. Davis, M., (1996) New experiments in cross-language text retrieval at nmsu's computing research lab. In Proceedings of the Fifth Retrieval Conference (TREC-5) Gaithersburg, MD: National Institute of Standards and Technology.
9. Landauer, T. K. and Littman, M. L. (1990) Fully automatic cross-language document retrieval. In *Proceedings of the Sixth Conference on Electronic Text Research,* pp. 31-38.
10. Kohonen, T. (1995) *Self-Organizing Maps.* Springer-Verlag, Berlin.
11. Salton, G. (1989) Automatic Text Processing: The Transformation, analysis, and Retrieva of Information by Computer. Addison-Wesley, Reading. MA.
12. Anderberg, M. R. (1973) *Cluster analysis for applications.* Academic Press, Inc., New York.
13. Kumar, R., Raghavan, P., Sridhar Rajagopalan, S., Sivakumar, D., Tompkins, A., and Upfal, E. (2000) The Web as a graph. In Proceedings of the nineteenth ACM SIGMOD SIGACT SIGART symposium on Principles of database systems, pp. 1—10.
14. Keller, J. M., Gray, M R. and Givens, J. A. (1985) A fuzzy *k*-nearest neighbor algorithm. IEEE Transactions of Systems, Man and Cybernetics. Vol. SMC-15, no.4, pp.580-585.

A Maximum Entropy Model Based Answer Extraction for Chinese Question Answering

Ang Sun[1], Minghu Jiang[2], Yanjun Ma[1]

[1] Computational Linguistics Lab, Dept. of Chinese Language, Tsinghua University, Beijing, 100084, China
{sa04, yj-ma03}@mails.tsinghua.edu.cn
[2] Computational Linguistics Lab, Dept. of Chinese Language, Tsinghua University, Beijing, 100084, China
jiang.mh@mail.tsinghua.edu.cn

Abstract. We regard answer extraction of Question Answering (QA) system as a classification problem, classifying answer candidate sentences into positive or negative. To confirm the feasibility of this new approach, we first extract features concerning question sentences and answer words from question answer pairs (QA pair), then we conduct experiments based on these features, using Maximum Entropy Model (MEM) as a Machine Learning (ML) technique. The first experiment conducted on the class-TIME_YEAR achieves 81.24% in precision and 78.48% in recall. The second experiment expanded to two other classes-OBJ_SUBSTANCE and LOC_CONTINENT also shows good performance.

1 Introduction

Normally, the QA system is a combination of three sequential models-Question Analysis Model, Information Retrieval (IR) Model and Answer Extraction Model.

Question Analysis Model analyzes a question sentence, identifies the question type(or answer type) and selects the query words for retrieving information that may contain correct answer to the question.

Information Retrieval Model utilizes the acquired query words to retrieve information related to the question from the Internet or document.

Answer Extraction Model first extracts answer candidates, then ranks, often scores the candidates and finally provides the top 5 answers.

For example, given Question1: 发现大庆油田在哪一年？(In which year did we discover the Daqing oil field?) Based on the analysis of linguistic features and semantic information of the question, Question Analysis Model first recognizes its question type as TIME_YEAR, and then extracts "发现 (discover), 大庆 (Daqing), 油田 (oil field)" as the query words. Secondly, using the query words, IR Model retrieves documents and returns a set of sentences related to the discovery. Thirdly, Answer Extraction Model ranks the sentences according to word-overlap information, answer type, similarity between the question and answer sentence and so forth, and finally gives the correct answer "1959".

L. Wang et al. (Eds.): FSKD 2006, LNAI 4223, pp. 1239 – 1248, 2006.

Focusing on extracting correct answers, two approaches are often applied by QA researchers. One is rule-based answer extraction. By recognizing the question type, the corresponding answer patterns are triggered to extract answers. Take "FALCON", a QA system devised by Harabagiu et al. [4] for example, for the purpose of processing DEFINITION questions, FALCON identifies the question type by its matching result of a set of linguistic question patterns related to definition and then one answer pattern would be applied to extract answer. Although such rules, always linguistic patterns are effective in pinpointing answers, to manually devise useful rules is very costly and time consuming. Moreover, the interference of rules also adds to its inefficiency and inaccuracy in finding answers.

Another approach is based on similarity computation. By mapping questions and answer candidate sentences in different spaces, researchers then compute the "similarity" between them. For example, Moldovan, et al. [1] map questions and answer candidate sentences into logical forms and compute the "similarity" between them using inference rules. The fundamental problem of this approach is that of finding spaces where the distance between questions and correct answer sentences is small and where the distance between questions and incorrect answer sentences is large. Echihabi et al. [3] developed a noisy channel model for computing this distance. They trained a probabilistic model for estimating the conditional probability $P(Q, S_A)$. Using the parameters learned from the model, they can find the sentence $S_i \in \Sigma$ and an answer in it $A_{i,j}$ by searching for the $S_{iA_{i,j}}$ that maximizes the conditional probability $P(Q \mid S_{iA_{i,j}})$.

Motivated by the success of Echihabi's noisy channel approach and other ML techniques that are successfully exploited to QA, we apply MEM to Chinese QA answer extraction. This is a challenge for us because to the best of our knowledge there is no Chinese QA system that exploits ML in extracting answers, although some researchers [6, 7] integrate it into there systems for classifying question types or for calculating the confidence score.

In order to confirm the feasibility of employing MEM to extract answers, we select features including question features and correct answer words features from QA pair and train a MEM model.

Section 2 briefly introduces MEM and its combination with Chinese QA. Section 3 describes the features we extracted from QA pair. Section 4 describes our experiments and analyzes the experimental results. Section 5 summarizes the conclusions.

2 Applying MEM to Chinese QA

The probability model is defined over $X \times Y$, where X is a set of input symbols, or "histories", and Y is a set of class labels. As for QA, X is a set of features extracted from QA pair, and Y reflects the accuracy of the answer, e.g. $Y = 1$ means the answer is correct while $Y = 0$ means incorrect. Thus, we can regard QA as a classification problem. A sample (x, y) is a pair of input $x = \{x_1, \cdots, x_m\} (x_i \in X)$ and output $y \in Y$.

The model's probability of a history x together with a label y is defined as:

$$p(x, y) = \pi\mu \prod_{j=1}^{k} \alpha_j^{f_j(x,y)} \tag{1}$$

where π is a normalization constant, $\{\mu, \alpha_1, \cdots \alpha_k\}$ are the positive model parameters and $\{f_1 \cdots f_k\}$ are known as "features", where $f_j(x, y) \in \{0,1\}$. Note that each parameter α_j corresponds to a feature f_j.

Given data $(x^{(1)}, y^{(1)}), \cdots, (x^n, y^{(n)})(x_i \in X, y_i \in Y)$, let $\bigcup_k (x^{(k)} \times \{y^{(k)}\}) = \{<\tilde{x}_1, \tilde{y}_1>, \cdots, <\tilde{x}_m, \tilde{y}_m>\}$. This means that we enumerate all pairs of an input sample and label and represent them as $<\tilde{x}_i, \tilde{y}_i>$ using index $i(1 \le i \le m)$.

The feature function f_j is often defined as follows.

$$f_i = \begin{cases} 1 & \text{if } x, y \text{ satisfy certain condition} \\ 0 & \text{else} \end{cases}. \tag{2}$$

The Maximum Entropy Principle is to find a model $H(p) = -\sum_{x \in X, y \in Y} p(x \mid y)\log(p(x \mid y))$ which means a probability model $p(x \mid y)$ that maximizes entropy $H(p)$.

The constraints are given by:

$$Ef_i = \tilde{E}f_i, \quad 1 \le j \le k. \tag{3}$$

Where the model's feature expectation is:

$$Ef_j = \sum_{x \in X, y \in Y} p(x, y) f_j(x, y). \tag{4}$$

And the observed feature expectation is:

$$\tilde{E}f_j = \sum_{i=1}^{n} \tilde{p}(x_i, y_i) f_j(x_i, y_i). \tag{5}$$

Where $\tilde{p}(x_i, y_i)$ denotes the observed probability of (x_i, y_i) in the training data. Thus the constraints force the model to match its feature expectations with those observed in the training data. In practice, X is very large and the model's expectation Ef_j can not be computed directly, so the following approximation [8] is used:

$$\tilde{E}f_j = \sum_{i=1}^{n} \tilde{p}(x_i) p(y_{i,} x_i) f_j(x_i, y_i). \tag{6}$$

Where $\tilde{p}(x_i)$ is the observed probability of the history x_i in the training set.

3 Feature Extraction

We use QA pair as our resources for extracting features. Each QA pair consists of one question and its correct answer sentence.

The selected features must reflect the properties of correct answers of a question and can be practically applied to MEM to enable us to extract the sentence containing correct answers. The features used by the three sequential Models of QA: Question Analysis Model, IR Model and Answer Extraction Model, are employed by our approach for its availability and feasibility.

This paper employs three groups of features as features of input data:

- Question Feature (QF) Set
- Sentence Feature (SF) Set
- Combined Feature (CF) Set

3.1 Question Feature (QF) Set

Question Feature Set is extracted from a question sentence.

POS_Q: POS of Question Words, e.g., given Question1, *{发现/v 大庆/ns 油田/n 在/p 哪/r 一/m 年/q}* are features.

Query Words (QW): Based on empirical and linguistic knowledge and QW's contribution to IR Model, we establish three standards for extracting QW. One is selecting all content words such as name entities (locations, organizations, person names…), nouns, verbs, adjectives and so on. Another is selectively extract cardinal number words, ordinal number words and measure words, for those do not immediately follow an interrogative word in the sequence of a question sentence are selected, otherwise are not selected. For example, given POS_Q of Question 2: 第一/m 次/q 世界大战/l 爆发/v 于/p 哪/r 一/m 年/q？/w (When did World War I. break out?), "第一 次" will be selected while "一 年" will not be selected. The last is to remove the interrogative words and all the stop list words(e.g., 是,的,有…).

So, the QW of Question 2 is *{第一 次 世界大战 爆发 于}*

Interrogative Words (IW): e.g., *{谁(who), 哪(where), 多少(how many) …}*
Noun Words (NW): e.g., NW of Question 1 is *{油田}*
Verb Words (VW): e.g., VW of Question 2 is *{爆发}*

3.2 Sentence Feature (SF) Set

Sentence Feature Set is extracted only from answer candidate sentence. Consider, if we define a sentence as S and its words as W_i, then S can be represented as $S = W_1 \cdots W_i \cdots W_n$.

W_i: The enumeration of S 's words, $\{W_1, W_2 \cdots W_{i-1}, W_i, W_{i+1} \cdots W_{n-1}, W_n\}$

For example, given an answer candidate sentence(S_1) of Question 2: 第一次世界大战爆发于1914年，战争的导火索是萨拉热窝事件。(World War I. broke out in 1914, and its fuse was the Sarajevo incident.), W_i of S_1 is *{第一 次 世界大战 爆发 于 1914年 战争 的 导火索 是 萨拉热窝 事件}*

POS Of W_i (POS_ W_i)

$\{W_1 / POS\_TAG_1, \cdots W_i / POS\_TAG_i, \cdots W_n / POS\_TAG_n\}$

e.g., POS_ W_i of S_1 is *{第一/m 次/q 世界大战/l 爆发/v 于/p 1914年/t 战争/n 的/u 导火索/n 是/a 萨拉热窝/ns 事件/n }*

POS Of Correct Answer Word (POS_A): e.g., POS of Question 2's answer word "*1914年*" is *{/t}*

3.3 Combined Feature (CF) Set

Combined Feature contains features created by combining question features and answer sentence features. For each sentence returned by IR Model, the following features are created.

QW_Match_Result: If W_i matches none of QW, then QW_Match_Result= *False*, otherwise QW_Match_Result = *True*.

e.g., the QW_Match_Result of S_1 is *True*, while the QW_Match_Result of another answewer candidate sentence of Question 2: 1914年的这一事件被称为萨拉热窝事件。(The incident occurred in 1914 is called the Sarajevo incident.), is *False*.

IW_Match_Result: If W_i matches one IW, then IW_Match_Result = *False,* otherwise IW_Match_Result = *True.*

e.g., the IW_Match_Result of S_1 is *True*, while the IW_Match_Result of another answewer candidate sentence of Question 2: 第一次世界大战爆发于那一年，1914年还是1915年？(When did World War I. break out, 1914 or 1915?), is *False* because it contains the Interrogative word "哪/r".

POS_A_Match_Result: *True* if POS_TAG_i = POS_A, otherwise *False*.

e.g., the POS_A_Match_Result of S_1 is *True*, while the POS_A_Match_Result of another answewer candidate sentence of Question 2: 这一事件成为第一次世界大战的起源。(This incident serves as the origin of World War I..), is *False*.

NW_Match_Result: If W_i matches one of NW, then NW_Match_Result = *True* , otherwise *False*.

e.g., the NW_Match_Result of S_1 is *True*, while the NW_Match_Result of another answewer candidate sentence of Question 2: 大战的导火线表面上是萨拉热窝事件。(The Sarajevo incident seems as the fuse of World War I..), is *False*.

VW_Match_Result: If W_i matches one of VW, then VW_Match_Result = *True* , otherwise *False*.

e.g., the VW_Match_Result of S_1 is *True*, while the VW_Match_Result of another ansewer candidate sentence of Question 2: 这一事件成为第一次世界大战的导火线。(This incident serves as the fuse of World War I..), is *False*.

4 Experiments

In this section, we first show our Question and Answer set, IR model, training data and testing data, then we describe how we conduct Experiment 1 on the class-TIME_YEAR, and finally present the result of Experiment 2 on the class-OBJ_SUBSTANCE and LOC_CONTINENT.

4.1 Question and Answer Set

We select 400 questions from *HIT-IRLab QA question instances* (a free resource from IRLab of Harbin Institute of Technology), and manually find and collate their answers by searching on the Internet. Table 1 and Table 2 show the details of its class labels and the number of QA pair of each class.

4.2 IR Model

We use a search engine of our Lab called *Web Search* that can automatically acquire information of WebPages's snippets returned by Google. For each question, its query words will be used to search on the Internet, and the result returned is 100 WebPages's snippets. Then we preprocess the acquired snippets by filtering useless information and by cutting them into sentences according to the punctuations such as "。", ".", "！", "!", "？", "?", etc. The result of preprocessing is a group of sentences with number ranging from 200 to 400.

4.3 Training and Testing Data

We select 400 QA pairs belonging to 3 classes. After filtering useless information, we cut the number of answer candidate sentences into 200 for each question.
Table 1 shows the details of our training data.

Table 1. Training Data

Class	Number of QA pairs	Number of WebPages' snippets	Number of answer candidate sentences
TIME_YEAR	150	15,000	30,000
OBJ_SUBSTANCE	50	5,000	10,000
LOC_CONTINENT	50	5,000	10,000

Table 2 shows the details of our testing data.

Table 2. Testing Data

Class	Number of QA pairs	Number of WebPages' snippets	Number of answer candidate sentences
TIME_YEAR	50	5,000	10,000
OBJ_SUBSTANCE	50	5,000	10,000
LOC_CONTINENT	50	5,000	10,000

4.4 Experiment 1

To confirm the feasibility of our approach, we first focus on dealing with only one question type-TIME_YEAR.

4.4.1 Training and Testing

Training Phase:

The training phase trains *MEM* based answer candidate sentence classifier from training data ($(x^{(1)}, y^{(i)})$,..., $(x^{(n)}, y^{(i)})$) . In order to examine the feature's contribution to the efficiency of classifier, we first conduct an experiment only using the first 3 combined features for training Classifier1. Classifier2 is the result of using all the combined features.

1 Given a question q and its correct answer a

2 Use ICTCLAS (a free resource from Institute of Computing Technology, Chinese Academy of Sciences) to get POS_Q and POS_A, generate QW, IW, NW and VW.

3 Input QW to *Web Search* engine, preprocess the candidate sentences returned by the engine.

4 Use ICTCLAS to get W_i and POS_ W_i

5 Compute QW_Match_Result, IW_Match_Result, POS_A_Match_Result, NW_Match_Result and VW_Match_Result.

6 Generate sample $(x^{(i)}, y^{(i)})$: $x^{(i)}$ contains the result obtained by 5, and $y^{(i)}$ is the label 0 or 1.

7 For each question q, execute 1-6.

8 Apply the samples to Maximum Entropy Model to train Classifier1 and Classifier2.

Testing Phase:

The testing phase evaluates the probability of output $y^{(i)}$ given input $x^{(i)}$ using the two classifiers.

1 For each question q, generate its sample $(x^{(i)}, y^{(i)})$

2 Use Classifier 1 to classify the answer candidate sentences.

3 Use Classifier 2 to classify the answer candidate sentences.

4.4.2 Experimental Results

We use precision, recall and F-measure to evaluate the output of the experiment.

The Baseline is computed by the first 3 CF: it regards an example as positive if all the three features are *true,* negative if one feature is *false*.

$$\text{Precision} = \frac{\text{Number of correctly classified candidates}}{\text{Total number of candidates}}. \tag{7}$$

$$\text{Recall} = \frac{\text{Number of correctly classified Positive Examples}}{\text{Number of Positive Examples}}. \tag{8}$$

$$F_\beta\text{-score} = \frac{(\beta^2+1)\cdot \text{Recall} \cdot \text{Precision}}{\beta^2 \cdot \text{Recall} + \text{Precision}}. \tag{9}$$

($\beta = 1$ in all experiments)

Table 3 shows the results on the class-TIME_YEAR.

Table 3. Test result on the Class-TIME_YEAR

Experiment	Precision	Recall	F-measure
Baseline	69.16%	13.04%	21.94%
MEM + 3 CF(Classifier1)	61.54%	97.47%	75.45%
MEM + 5 CF(Classifier2)	81.24%	78.48%	79.83%

By applying the first 3 CF to MEM, the precision of QA system decreases by 7.62% compared to the Baseline, but the recall increases by 84.43%, which means that QA's ability to identify positive candidates is greatly enhanced while its ability to recognize negative candidates is slightly declined.

By applying the 5 CF to MEM, both of the two abilities are improved compared to the Baseline. But the Recall decreases by 18.99% compared to Classifier1, indicating that the NW and VW features of a question is more helpful in estimating negative candidates than positive candidates.

The F-measure shows that both of the two classifiers achieve good performance.

4.5 Experiment 2

For the purpose of testing whether our approach can be successfully applied to other classes, we emulate Experiment 1 and conduct Experiment 2 on the classes OBJ_SUBSTANCE and LOC_CONTINENT.

Table 4. Test result on the Class-OBJ_SUBSTANCE

Experiment	Precision	Recall	F-measure
Baseline	26.52%	8.30%	12.64%
MEM + 3 CF	29.90%	88.42%	44.69%
MEM + 5 CF	84.68%	57.89%	68.77%

Table 5. Test result on the Class-LOC_CONTINENT

Experiment	Precision	Recall	F-measure
Baseline	69.00%	54.29%	60.00%
MEM + 3 CF	69.20%	40.50%	51.10%
MEM + 5 CF	74.00%	50.28%	59.88%

The result on the class-OBJ_SUBSTANCE is in accordance with the result of Experiment 1.

However, the result on the class-LOC_CONTINENT shows that it does not achieve very good performance except for the slightly increased precision. This may be explained by POS_A of this type. Its POS_A comprises several tags, such as "/ns", "/n", "/j", etc.. We select "/ns" as POS_A because it uniquely denotes locations and other tags may mislead our classifier. For instance, we do not select "/n", "/j" because almost all question types can hold these features. The result shows that our classifier can not learn much from the feature-POS_A. Given Question 3: 地球上哪个大陆最大？(Which continent is the largest one on the earth?), because the POS of its answer word "欧亚大陆（Eurasia）" is "/n /j", our classifiers lack the ability of finding correct answers of such types of questions.

Also, several questions of LOC_CONTINENT are *list* questions, e.g., "大象主要生活在哪些大陆上?(Which continents are the main habitats of elephant?)". For those list questions, its answer contains more than one word, and they usually distribute among several sentences. The difficulty of uniting the separated answer words as a single answer and the relatively small number of answer candidate sentences of such questions returned by our IR model result in our classifiers' incapability.

We are informed by this experiment that whether a feature can be useful in constructing a QA answer classifier should be examined, and that we should not only consider the answer types but also consider the forms of the answer.

5 Conclusions

This paper takes the view that answer extraction of QA is actually a problem of classifying answer candidate sentences into positive or negative. Our experimental results confirm the feasibility of this new approach. Meanwhile, it also indicates our future work which includes applying this approach to more classes, selecting features that are more representative of the properties of questions and its correct answers, and taking more answer forms into account.

Acknowledgement

This work was supported by National Natural Science Key Foundation of China (No. 60331010). Tsinghua University 985 research fund; the excellent young teacher program, Ministry of Education, China; and State Key Lab of Pattern Recognition open fund, Chinese Academy of Sciences.

References

1. Moldovan, D., Harabagiu, S., Girju, R., PaulMoraresc, L. F., Novischi, A., Badulescu, A., Bolohan, O.: LCC Tools for Question Answering. In: Proceedings of the Eleventh Text Retrieval Conference, Gaithersburg, Maryland (2002) 144-155

2. Wu, M., Zheng, X., Duan, M., Liu, T., Strzalkowski, T., Albany S.: Question Answering by Pattern Matching, Web-Proofing, Semantic Form Proofing. In Proceedings of the Twelfth Text Retrieval Conference, NIST. Gaithersburg, MD (2003) 578-585
3. Echihabi, A., Marcu, D.: A Noisy-Channel Approach to Question Answering. In: Hinrichs, E., Roth, D., eds.: In:Proceedings of 41st Annual Meeting of the Association for Computational Linguistics (2003) 16–23
4. Harabagiu, S., Moldovan, D., Pasca, M., Mihalcea, R., Surdeanu, M., Bunescu, R., Gîrju, R., Rus, V., Morarescu, P.: FALCON: Boosting Knowledge for Answer Engines. In Proceedings of Ninth the Text Retrieval Conference, NIST. Gaithersburg, MD (2000) 479-488
5. Yutaka, S.: Question Answering as Question-Biased Term Extraction: A New Approach toward Multilingual QA. In: Proceedings of 43rd Annual Meeting of the Association for Computational Linguistics (2005) 215-222
6. You, L., Zhou, Y. Q., Huang, X. J., Wu, L. D.: A maximum entropy model based confidence scoring algorithm for QA. Journal of Software, 16 (8), (2005) 1407-1414
7. Zhang, Q., Chen, Q. X.: Study of Web-Oriented Chinese Q&A System for IT Domain. Master Dissertation of Tsinghua. (2003)
8. Lau, R., Rosenfeld, R., Roukos, S.: Adaptive Language Modeling Using the Maximum Entropy Principle. In: Proceedings of the Human Language Technology Workshop, (1993) 108-113

A Learning Based Model for Chinese Co-reference Resolution by Mining Contextual Evidence

Feifan Liu and Jun Zhao

National Laboratory of Pattern Recognition
Institute of Automation, Chinese Academy of Sciences
{ffliu, jzhao}@nlpr.ia.ac.cn

Abstract. This paper presents a learning based model for Chinese co-reference resolution, in which diverse contextual features are explored inspired by related linguistic theory. Our main motivation is to try to boost the co-reference resolution performance only by leveraging multiple shallow syntactic and semantic features, which can escape from tough problems such as deep syntactic and semantic structural analysis. Also, reconstruction of surface features based on contextual semantic similarity is conducted to approximate the syntactic and semantic parallel preferences in resolution linguistic theories. Furthermore, we consider two classifiers in the machine learning framework for the co-reference resolution, and performance comparison and combination between them are conducted and investigated. We experimentally evaluate our approaches on standard ACE (Automatic Content Extraction) corpus with promising results.

1 Introduction

Co-reference resolution refers to the problem of determining whether discourse references in text correspond to the same real world entities [1]. In the context of ACE (Automatic Context Extraction) we address only specified set of entities [2] for co-reference resolution in Chinese texts here. A mention is a referring expression of an object, and a set of mentions referring to the same object within a document constitute an entity, i.e. an equivalence class of mentions. For example, in the following sentence, mentions are nested bracketed:

"[[微软公司/Microsoft Company]总裁/president][比尔盖茨/Bill Gates]表示/stated[微软/Microsoft]与此事件无关/has nothing to do with this issue, [公司/Company]不需要做任何解释/don't need to give any explanations。"

"微软公司/Microsoft Company", "微软/Microsoft" and "公司/Company" constitute a entity since they refer to the same object. Likewise, "微软公司/Microsoft Company总裁/president" and "比尔盖茨/Bill Gates" also constitute a entity.

Recent research in co-reference resolution has exhibited a shift from knowledge-based approaches to data-driven approaches, yielding learning-based co-reference

* This work was supported by the National Natural Sciences Foundation of China (60372016) and the Natural Science Foundation of Beijing (4052027).

L. Wang et al. (Eds.): FSKD 2006, LNAI 4223, pp. 1249 – 1258, 2006.

systems [3][4][6][9]. These approaches recast the problem as a classification task. Specifically, a pair of mentions is classified as co-referring or not based on a statistical model learned from the training data. Then a separate linking algorithm coordinate the co-referring mentions pairs and partition all the mentions in the document into entities. Soon [4] has been commonly used as a baseline system for comparison under this learning based framework, and many extensions have been conducted at different points. Yang [9] and Strube [10] made improvements in string matching strategy and got good results. Ng [3] proposed a different link-best strategy, and Ng [6] presented a novel ranking approach for partitioning mentions in linking stage.

This paper proposes a Chinese co-reference resolution system employing the statistical framework. Unlike existing work, we focus on exploring the contribution of diverse contextual features inspired by linguistic findings. First, incorporating diverse contextual features try to capture the syntactic structural information, which is inspired by the syntactic constrain rules for anaphora resolution. Second, an information reconstruction method based on contextual similarity is proposed to approximate syntactic and semantic parallel preferences, which plays an important role in co-reference resolution according to linguistic findings. We experimentally evaluate our approaches on standard ACE corpus with promising results.

2 Learning-Based Chinese Co-reference Resolution (Baseline)

Our framework for co-reference resolution is a standard combination of classification and clustering as mentioned above. First, we establish a Chinese co-reference resolution system based on [4] as in figure 1. Note that the dashed part is for offline training.

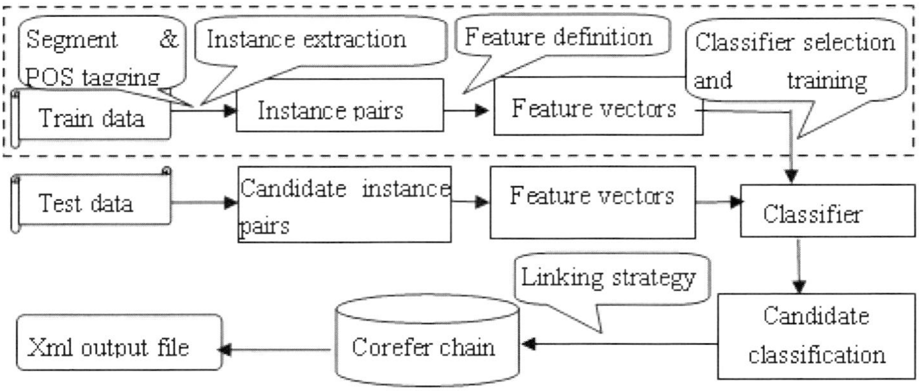

Fig. 1. Statistical Framework for Chinese Co-reference Resolution

2.1 Instance Extraction

Here an instance is a pair of entity mentions (EM) which are either CO-REFERENT or NOT CO-REFERENT. The former is called positive instance, and the latter is called negative instance. We obtained these instances from the ACE training corpus. And we use the extracting strategy in [4] in our baseline system.

Also, since the named and nominal mentions are typed, and we will consider only instances where two mentions belong to the same type, e.g. we will not extract negative instances with two non-pronominal mentions whose types are inconsistent.

2.2 Feature Definition

Every instance is represented by a feature vector. In our baseline system, we try to simulate the feature set in [4] illustrated in Table 1. We use "I" and "J" to denote EM_i and EM_j in an instance respectively. Note that we make some adaptations or modifications according to Chinese characteristics, marked by star symbol.

(1) StringMatch

Since there is no sufficient information in Chinese, such as capitalized information, to determine alias abbreviation, or shorted form of named and nominal mentions, we modify the string match strategy, replacing the binary match feature and alias feature with matching degree feature. A simple matching function is designed as follows.

$$MatchDegree = \frac{\sum_{w_i \in \{C_m\}} len(w_i)}{\max\{len(EM_i), len(EM_j)\}} \tag{1}$$

where C_m is the matched word set of two mentions, $len(.)$ is measured by characters.

Table 1. Feature Set for the Simulated Soon et al. Baseline System

Feature type	Feature	Feature description
Lexical	StringMatch *	real number value between 0 and 1 by equation(1)
Grammatical	MenType_I	"NOM" if EM_i is nominal mentions; "NAM" if named mentions; "PRO" if pronominal mentions.
	MenType_J	same definition to MenType_I
	Definite_I	"1" if EM_i contains words with definitive and demonstrative sense, such as"□/the,this", "那些/those", else "-1"
	Definite_J	Same definition to Definite_I
	Number	"1" if EM_i and EM_j agree in number; "-1" if they disagree; "0" if they can't be determined
	Gender	"1" if EM_i and EM_j agree in gender; "-1" if they disagree; "0" if they can't be determined
	Appositive*	"1" if mentions are in an appositive relationship;else "-1"
Semantic	EntityType*	"1" if EM_i and EM_j are consistent in entity type; "-1" if they are not; "0" if they can't be determined
Positional	Para*	"1" if EM_i and EM_j are in different paragraphs; else"-1"
	SenNum*	see equation (2)
	SubSenNum*	see equstion (3)

(2) Positional Features

Only the sentence number between two mentions is considered in Soon et al. system. Here we extend this type of feature by adding cross paragraph and cross sub-sentence feature. Sentence is delimited by full stop, question mark, or exclamatory mark, while sub-sentence is delimited by colon, semicolon, or comma. We define SenNum and SubSenNum as follows.

$$SenNum = \begin{cases} SenNum & \text{if } SenNum <= 2 \\ \text{"}SenNum > 2\text{"} & \text{if } 2 < SenNum <= 5 \\ \text{"}SenNum > 5\text{"} & \text{if } 5 < SenNum <= 10 \\ \text{"}SenNum > 10\text{"} & \text{if } SenNum > 10 \end{cases} \quad (2)$$

$$SubSenNum = \begin{cases} SubSenNum & \text{if } SubSenNum <= 2 \\ \text{"}SenNum > 2\text{"} & \text{if } 2 < SubSenNum <= 5 \\ \text{"}SenNum > 5\text{"} & \text{if } SubSenNum > 5 \end{cases} \quad (3)$$

2.3 Co-reference Classifier Selection

Diverse machine learning methods have been used for co-reference resolution, such as decision tree(DT) model C4.5[3][4][6], maximum entropy(ME) model[6][11], support vector machine(SVM) model[7][12], and etc. Bryant's work [12] proved experimentally that SVM model (F-value: 72.4) outperform the traditional DT model (F-value: 70.7) in the machine learning framework for co-reference resolution. We consider two learning models in our baseline system: SVM and ME. Our motivation is to compare the two models' performance in the context of co-reference resolution and try some combining strategy on them.

2.4 Linking Strategy

Linking strategy is used to apply the classifier predictions to create co-reference chains. The most popular linking strategy is the link-first strategy [4], which links a mention, EM_j, to the first preceding mention, EM_i, predicated as co-referent. An alternative linking strategy, which can be called link-best strategy [3], links a mention, EM_j, to the most probable preceding mention, EM_i, where the probability is measured by the confidence of the co-reference classifier prediction.

3 Incorporating Multi-level Contextual Evidence

Co-reference is a discourse-level problem, which depends on not only the two candidate mentions themselves but also diverse contextual evidence. Here are two examples.

(1) 小明/Xiaoming 说/said 在学校/in school 数学/math 老师/ teacher 常/often 责怪 /blame他/he。

(2) 李刚/Ligang 常/often 找/go with 刘辉/Liuhui 打篮球/play basketball，徐庆/Xuqing 常/often 找/go with 他/he 去游泳/go swimming。

In the two above sentence, it is very hard for a classifier to predict correct co-referent relations between underlined mentions only using features in the baseline system. But this can be well explained and resolved by linguistic findings on constrains and preferences involved in traditional anaphora resolution theory [13]. Syntactic constrains (他/he≠数学老师/math teacher) should be considered in sentence (1), and syntactic constrains and semantic parallelism preferences (他/he＝刘辉/Liuhui) should be used in sentence (2).

Deep syntactic and semantic knowledge, however, is quite difficult for current Chinese processing technology. So we try to mine multi-level contextual features and investigate their contribution to system performance accordingly. We hope to capture the properties implied in deep syntactic and semantic analysis by incorporating multi-level surface features and reconstruct them using some strategy.

3.1 Word Form and POS (Part of Speech) Evidence

Word forms and POS of them are the most fundamental contextual features at lexical and shallow syntactic level. For each of the two mentions in question, we consider a 5-width window to extract those contextual cues to enrich the feature vector of a training instance, which is expected to be helpful in co-reference resolution.

3.2 Bag of Sememes (BS) Feature

Although deep semantic analysis is not available, we can resort to a Chinese-English knowledge base called HowNet (http://www.keenage.com) to acquire shallow semantic features. HowNet is a bilingual common-sense knowledge base, which uses a set of non-decomposable sememes to define a sense of a word. A total of over 1,600 sememes are involved and they are organized hierarchically. For example, the sense of "研究所/research institute" is defined as "InstitutePlace|场所,*research|研究,#knowledge|知识". Here there are three sememes split by commas, and symbols such as "*" represent specific relations.

So we can acquire a set of sememes for each word in the contextual window. Bag of sememes (BS) is used for modeling the semantic context of each mention, including preceding context BS and post context BS.

3.3 Feature Reconstruction Based on Contextual Semantic Similarity

There are two problems in using BS features in Section 3.2. First, we don't use any disambiguation strategy in extracting word sense from HowNet, which can introduce harmful noises. Second, we just use bag of sememes to model the context, losing the useful associated information between sememes from the same word. We will make it manifest by looking at two different contexts: "校园/schoolyard 歌手/singer" and "学生/student 歌厅/singing hall". Sense for every word and BS features of context are shown in figure 2.

Sense representation of each word in HowNet:

1.校园/schoolyard: {part|部件,%InstitutePlace|场所,education|教育}

2.歌手/singer: {human|人,entertainment|艺,*sing|唱,#music|音乐}

3.学生/student: { human|人,*study|学,education|教育}

4.歌厅/singing hall: {InstitutePlace|场所,@recreation|娱乐,@sing|唱}

Bag of sememes feature of two contexts:

1. 校 园 /schoolyard 歌 手 /singe: {part| 部 件 ,**InstitutePlace**| 场 所 ,**education**|教育,**human**|人,entertainment|艺,**sing**|唱,music|音乐}

2. 学生/student 歌厅/singing hall: {**human**|人,study|学,**education**|教育 ,**InstitutePlace**|场所,recreation|娱乐,**sing**|唱}

Fig. 2. Bag of Sememes of the Two Example Contexts

From Figure 2 we can see that, although the two contexts are semantically discrepant, their BS features are similar to a large extent.

We now have no effective way to resolve the first problem. Motivated by linguistic findings, we can make improvements on modeling the context related to the second question. Undoubtedly, BS features can express the semantic property of the context, which can help us to approximate the semantic parallelism in anaphora theory. But the information in BS is unordered without any relational or structural information. Intuitively, we can use some strategy to reconstruct those unordered BS features to give a better representation of context.

A word similarity computing approach [8] based HowNet is used for reference in our information reconstruction method. We regard context as a set of word and contextual similarity can be acquired by computing the similarity between two word sets.

In our case of computing word similarity, we didn't consider the relation sememe description and the relation symbol description as [8] does. We only consider two aspects: one is the similarity of first basic sememe between two words, denoted as $Sim_1(S_1, S_2)$; one is the similarity of all the other basic sememes in word sense representation from HowNet, denoted as $Sim_2(S_1, S_2)$. Here S_1 and S_2 are sense representations for W_1 and W_2 in consideration. So we compute the similarity of two words using the following equation.

$$Sim(W_1, W_2) = \beta_1 Sim_1(S_1, S_2) + \beta_2 Sim_2(S_1, S_2) \tag{4}$$

where $\beta_1 + \beta_2 = 1$. Since the first basic sememe indicates the most important semantic feature, we set $\beta_1 = 0.7$. Similarity between two sememes is computed according to the distance in the sememe hierarchy of HowNet [8].

Now we turn to the computation of contextual similarity based on word similarity. As mentioned above, contextual similarity can be formulated as a problem of computing similarity between two word sets. We should first find the possible corresponding word pairs between two sets and compute the arithmetical average of the similarity

```
ContextSim ← 0.0
ArrayWordSim ←∅
for each wi∈C1
    for each wj∈C2
        WordSim->value ← Sim(wi,wj), WordSim->index_1←i, WordSim->index_2 ←j
        ArrayWordSim ← ArrayWordSim ∪ {WordSim}
    end for
end for
DeSort(ArrayWordSim)    // sort in decreasing order of similarity value
ProcessedIndex_1←∅, ProcessedIndex_2←∅
for each k = 0,1,2,...,min{size(C1),size(C2)}
    if(WordSimk->index_1∉ProcessedIndex_1 && WordSimk->index_2∉ProcessedIndex_2)
        ContextSim←ContextSim+WordSimk->value
        ProcessedIndex_1 ← ProcessedIndex_1 ∪ {WordSimk->index_1}
        ProcessedIndex_2 ← ProcessedIndex_2 ∪ {WordSimk->index_2}
    end if
end for
ContextSim ← ContextSim / min{size(C1),size(C2)}
```

Fig. 3. Description of Computing Algorithm for Contextual Similarity

values of all the corresponding word pairs. Let C_1 and C_2 denote the word set containing the words in the context of EM_i and EM_j respectively. The algorithm description for contextual similarity computation is illustrated in figure 3.

By calculating contextual similarity, the unordered BS features are reconstructed. The computation is word-based, and the first sememe similarity is given a larger weight, so the BS shortcomings discussed in figure 2 can be overcome to some extent.

4 Experiments and Analysis

In our experiments, two standard classification toolkits are used, namely Maximum Entropy Toolkit (MaxEnt)[1] and Support Vector Machine Toolkit (libSvm)[2]. Parameters in the models are selected by 5-fold cross validation.

4.1 Experiment Data and Evaluation Metric

We now focus on the empirical performance analysis of an implementation of the statistical co-reference model described above. We evaluate the co-reference system on the standard ACE-05 co-reference data. The co-reference classifier is trained on 80% of the data set, and other 20% is used for testing the co-reference resolution systems. Statistics of train data and test data is shown in table 2.

[1] http://homepages.inf.ed.ac.uk/s0450736/maxent_toolkit.html

[2] http://www.csie.ntu.edu.tw/~cjlin/libsvm/

Table 2. Statistics on Experiment Data

	#Doc	#Entity Mention				#Co-reference Chain
		Named	Nominal	Pronominal	Total	
Train data	511	11649	12952	2763	27364	12258
Test data	122	3048	3326	583	6957	3156

Performance of our co-reference system is reported in terms of recall, precision, and F-measure using the model-theoretic MUC scoring program [14].

4.2 Impact of Multi-level Contextual Features on System Performance

This experiment reports the performance of our baseline system and explores the contribution of multi-level contextual features, which is shown in table 3. In table 3, "CR" denotes Co-reference Resolution, "WP" denotes the word and POS features, "BS" denotes the semantic feature represented by Bag of Sememes, and also classification accuracy of classifiers is given for reference.

Table 3. Improved Performances by Incorporating Multi-level Contextual Features (SVM)

	Classification Accuracy	CR recall	CR precision	CR F-value
Baseline System	85.49%	61.06%	93.4%	73.84%
Baseline+WP	85.85%	74.82%	85.37%	79.74%
Baseline+WP+BS	86.27%	74.56%	86.06%	**79.89%**

Table 3 shows that incorporating the contextual lexical and shallow syntactic feature (WP) acquires significant increase in recall, but some drops in precision. The resulting F-value, however, increase non-trivially from 73.84% to 79.74%. The introduction of BS features based on HowNet can further boost the system's performance.

The experimental results are largely consistent with our hypothesis. System performance improves dramatically by applying diverse contextual features. This can be explained that combining multiple contextual features can capture the syntactic constrain information which is definitely helpful for co-reference resolution according to traditional linguistic findings.

4.3 Performance Comparison Between Different Classifiers

We consider two classifiers, ME(Maximum Entropy) and SVM(Support Vector Machine), in our machine learning co-reference resolution framework. Comparison results under three different configurations are demonstrated in table 4.

The results reveal that SVM performs better than ME under all the three configurations. From the principle point of view, ME is a probability model based on log-linear

Table 4. Performance Comparison between ME and SVM

	Baseline			Baseline+WP			Baseline+WP+BS		
	R(%)	P(%)	F(%)	R(%)	P(%)	F(%)	R(%)	P(%)	F(%)
ME	59.46	89.72	71.52	65.43	88.25	75.14	66.27	86.71	75.13
SVM	61.06	93.40	73.84	74.82	85.37	**79.74**	74.56	86.06	**79.89**
SVM +ME	62.12	91.87	**74.12**	70.46	89.24	78.74	72.12	86.97	78.85

regression while SVM is classification model based on large-margin principle. For this reason, SVM can outperform ME when the training data is not much sufficient.

For three systems, only baseline system can benefit from the combination. When adding contextual features, combination doesn't help at all according to the results. We don't know exactly what the reason is now, but we guess it may have something to do with the learning mechanism and confidence measurement of the classifiers.

4.4 Performance Improvement by Similarity Based Information Reconstruction

In this section we try to investigate whether our motivation of modeling context by similarity based information reconstruction is valid or not.

From figure 4, we can see that RC (feature reconstruction) outperforms BS feature and get increase both in recall and precision. The resulting F-value increases from 79.89% to 80.28%. This commendably verifies our analysis in section 3. By contextual similarity, the semantic information can be leveraged in a more reasonable way.

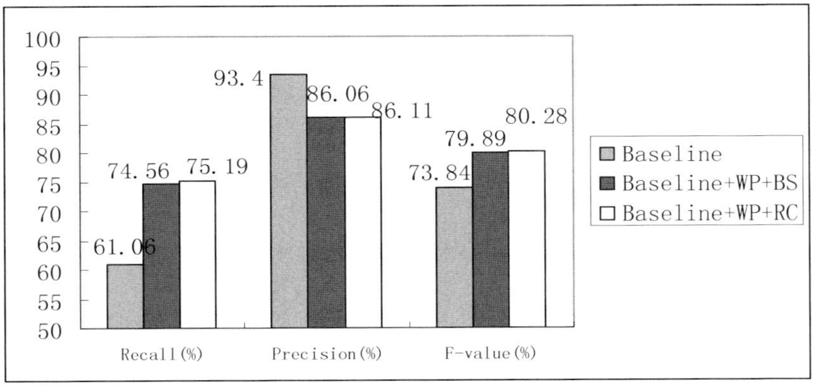

Fig. 4. Performance Improvement by Similarity Based Information Reconstruction

5 Conclusions and Future Work

We propose a learning based model for Chinese co-reference resolution and investigate multiple contextual features to improve the system performance based on related linguistic theory. Experimental results prove that our approach performs very well on the standard ACE data sets without deep syntactic and semantic analysis.

Our future work will focus on the following.

- How to reduce the noise of introducing semantic features more efficiently;
- How to find global features useful for Chinese co-reference resolution.

References

1. R. Mitkov. 2002. Anaphora Resolution. Long-man.
2. NIST. 2005. The Official Evaluation Plan for the ACE 2005 Evaluation.
3. V. Ng and C. Cardie. 2002. Improving Machine Learning Approaches to coreference resolution. In Proc. of the ACL 2002, pages: 104-111.
4. W.M. Soon, H. T. Ng, and D. Lim. 2001. A Machine Learning Approach to Co-reference Resolution of Noun Phrases. Computational Linguistics, 27(4):521-544.
5. M. Vilain, J. Burger, J. Aberdeen, D. Connolly, and L. Hirschman. 1995. A Model-Theoretic Coreference Soring Scheme. In Proc. of MUC-6, pages: 45-52.
6. Vincent Ng. 2005. Machine Learning for Coreference Resolution: From Local Classification to Global Ranking. In Proceedings of ACL 2005, Ann Arbor, MI, June 2005, pp. 157-164.
7. R. Iida, K. Inui, H. Takamura, and Y. Matsumoto. 2003. Incorporating contextual cues in trainable models for coreference resolution. In Proc. of the EACL'03 Workshop on the Computational Treatment of Anaphora, pp: 23–30.
8. Qun Liu, Sujian Li. 2002. Word Similarity Computing Based on How-net. Computational Linguistics and Chinese Language Processing. Vol. 7, No. 2, pp: 59-76.
9. X. Yang, G. Zhou, J. Su & C. L. Tan. 2004. Improving noun phrase co-reference resolution by matching strings, In Proceedings of IJCNLP04, Lecture Notes in Computer Science, Volume 3248, pages 22 - 31. Hainan, China.
10. M. Strube, S. Rapp, and C. Muller. 2002. The Influence of Minimum Edit Distance on Reference Resolution. In Proc. of EMNLP-02, pages:312-319.
11. R Florian, H Hassan, A Ittycheriah, H Jing, N Kambhatla, X Luo, N Nicolov, and S Roukos. 2004. A statistical model for multilingual entity detection and tracking. In Proc. of HLT/NAACL-04, pp: 1-8, Boston Massachusetts, USA.
12. John Bryant. 2004. Combining Feature Based and Semantic Information for Co-reference Resolution. Research report at U.C. Berkeley and ICSI.
13. Daniel Jurafsky, James H. Martin. 2000. Speech and Language Processing: An Introduction to Natural Language Processing, Computational Linguistics, and Speech Recognition. Prentice-Hall.

MFC: A Method of Co-referent Relation Acquisition from Large-Scale Chinese Corpora*

Guogang Tian[1,2], Cungen Cao[1], Lei Liu[1,2], and Haitao Wang[1,2]

[1] Key Laboratory of Intelligent Information Processing
Institute of Computing Technology, Chinese Academy of Sciences
[2] Graduate University of the Chinese Academy of Sciences
Beijing, China, 100080
naitgg@hotmail.com, cgcao@ict.ac.cn, liuliu_leilei@ict.ac.cn, htwang@ict.ac.cn

Abstract. This paper proposes a multi-feature constrained method (MFC) to acquire co-referent relations from large-scale Chinese corpora. The MFC has two phases: candidate relations extraction and verification. The extraction phase uses distribution distance, pattern homogeneity and coordination distribution features of co-referent target words to extract candidate relations from Chinese corpora. In the verification phase, we define an ontology for co-referent token words, and build a relation graph for all candidate relations. Both the ontology and the graph are integrated to generate individual, joint and reinforced strategies to verify candidate relations. Comprehensive experiments have shown that the MFC is practical, and can also be extended to acquire other types of relations.

1 Introduction

In the field of knowledge acquisition from text (KAT), relation acquisition mainly focuses on hyponymy and meronymy. Hearst [1] used lexicon-syntactic patterns to acquire hyponymy from corpora. Cederberg et al. [2] applied latent semantic analysis and coordination patterns to simultaneously increase precision and recall rates of hyponymy acquisition. Girju et al. [3] defined three patterns *Y verb(has) X*, *Y's X* and *X of Y* to acquire meronymy. Berland et al. [4] used statistical method to acquire meronymic relations from large-scale corpora.

Different from the above works, Maedche et al. [5] acquired non-category relations from domain text using association rule algorithm. Lin et al. [6] computed contextual similarity to find similar words from corpora on the basis of dependency tuples.

Co-referent relations are particular relations between words, and they relate two or more words that have identical meanings and designate the same entity in the real world. Based on co-referent relations' semantics and distributions in corpora, this paper proposes a multi-feature constrained method (MFC) to acquire co-referent relations from large-scale Chinese corpora. To our knowledge, there has been little work on the

* This work is supported by the Natural Science Foundation (grant nos. 60273019, 60496326, 60573063, and 60573064), and the National 973 Programme (grants no. 2003CB317008 and G1999032701).

L. Wang et al. (Eds.): FSKD 2006, LNAI 4223, pp. 1259 – 1268, 2006.

co-referent relation acquisition so far. The acquisition is divided into two phases: candidate relations extraction and verification. The applied features include distributive features, semantic features and structural features. The distributive features refer that which position's words in a sentence can constitute a co-referent relation. The semantic features refer to the relations themselves' meaning. The structural features refer to the composition structure while all co-referent relations are linked together.

The rest of this paper is organized as follows. Section 2 and 3 introduce the extraction and verification of the co-referent relations, respectively. The experiments are presented in section 4. Section 5 concludes the paper.

2 A Hybrid Method for Extracting Co-referent Relations

In Chinese, co-referent relations may have a number of token words, e.g. 简称 (Abbreviation), 全称 (Full-Name), 又称 (Other-Name), 俗称 (Colloquialism), 学名 (Academic-Name), and 英文名称 (English-Name). They are called co-referent token words. A token word represents a type of co-referent relation.

Definition 1: Given a word set $W=\{w_1, \ldots, w_n\}$, for any $w_i, w_j \in W$ ($1 \le i, j \le n$), if w_i and w_j nominate an identical entity, then w_i and w_j satisfy a co-referent relation.

Definition 2: Given a word set $W=\{w_1, \ldots, w_n\}$ and a co-referent token word set $\mathcal{R}=\{\mathcal{R}_1, \ldots, \mathcal{R}_m\}$, for any $w_i, w_j \in W$ ($1 \le i, j \le n$), there exists $\mathcal{R}_k \in \mathcal{R}$ ($1 \le k \le m$), if they satisfy a co-referent relation whose token is $\mathcal{R}_k$, then they are represented as $\mathcal{R}_k(w_i, w_j)$, among which $\mathcal{R}_k$ is a token word, and w_i and w_j are called target words (see Fig.2).

To extract candidate co-referent relations from the corpora, we introduce word distribution distance as a fundamental measure, and word pattern homogeneity and coordinate distribution as its supplements.

2.1 Word Distribution Distance

Word distribution distance is based on a hypothesis that co-referent relations often occur in the local sentence, centered by token words, not spread the whole sentence.

Given a sentence $S_k=w_n^- \ldots w_i^- \ldots w_1^-(RT)w_1^+ \ldots w_j^+ \ldots w_m^+$ in the co-referent corpora, where RT is a token word, w_i^- is the ith target word in the left context of RT (from right to left), and w_j^+ is the jth target word in the right context of RT (from left to right). For w_i^- and w_j^+, their relative positions to RT in S_k are denoted as $pos(w_i^-, RT)=i$ and $pos(w_j^+, RT)=j$, respectively. And their relative distances to RT in S_k are formulated in equation (1) and (2). $Len(x)$ represents the length of string x (Note: A Chinese character is 2 long.). For a word w, its distribution distance $Ddis(w, RT)$ in the corpora is defined in equation (3).

$$dis(w_i^-, RT) = \sum_{k=1}^{i} Len(w_k^-) \ . \tag{1}$$

$$dis(w_j^+, RT) = \sum_{k=1}^{j} Len(w_k^+) \ . \tag{2}$$

$$Ddis(w, RT) = \sum_p \sum_d P(pos(w, RT) = p) \times P(dis(w, RT) = d) \times d \ . \tag{3}$$

In equation (3),

$$P(pos(w, RT) = p) = \frac{N(pos(w, RT) = p)}{\sum_k N(pos(w, RT) = k)} \ . \tag{4}$$

$$P(dis(w, RT) = d) = \frac{N(dis(w, RT) = d)}{\sum_l N(dis(w, RT) = l)} \ . \tag{5}$$

$P(pos(w, RT)=p)$ is the probability of that the relative position of w is p. $P(dis(w, RT)=d)$ is the probability of that the relative distance of w is d. $N(pos(w, RT)=p)$ is the occurrence number of w at the relative position p. $N(dis(w, RT)=d)$ is the occurrence number of w whose relative distance is d.

Some explanation is needed for equation (3). Even though w occurs at the same position in different sentences, the relative distance is possibly different in them. Therefore, its contribution to the distribution distance is different. Even though the relative distance of w is the same, it possibly occurs at different positions. Thus, the relative position's contribution to the distribution distance is different. Based on these considerations, we use the probabilities of the relative position and distance as the weight to reveal their contributions to the distribution distance.

In a sentence, a target word w is possibly before or after a token word RT. So w has two distribution distances: *forward distribution distance* and *backward distribution distance*. If the following three conditions are simultaneously satisfied, w_i^- and w_j^+ can form a candidate co-referent relation $\mathcal{R}_k(w_i^-, w_j^+)$:

- $Ddis(w_i^-, RT) \leq -d^l$;
- $Ddis(w_j^+, RT) \leq +d^r$;
- $Ddis(w_i^-, RT) + Ddis(w_j^+, RT) \leq d_t$;

where $-d^l$ is the threshold of the backward distribution distance, $+d^r$ is the threshold of the forward distribution distance, d_t is the threshold of the sum of two distances.

2.2 Pattern Homogeneity of Target Words

Word patterns are classified into composition patterns and context patterns. In the MFC, a composition pattern refers to the POS (part of speech) pattern of a target word, and the context pattern refers to paired symbols surrounding a target word, such as quotation marks and book marks (e.g. 《 and 》 used in Chinese).The word pattern homogeneity means that two target words have the same pattern, i.e. composition or context pattern. For any two target words separated by token words in a sentence,

(1) If they have the same affix (prefix or suffix), or the same POS in the head or tail, then their composition patterns are homogeneous, and they can form a candidate co-referent relation.

For example, in the sentence 囊状淋巴管瘤好发于颈部, 又称囊状水瘤 (*Cystic lymphangioma, also called hygroma, often occurs in the neck*), 囊状淋巴管瘤(*cystic lymphangioma*) and 囊状水瘤(*hygroma*) have the same prefix (i.e. 囊状) and suffix(i.e. 瘤), so they can form a candidate relation 又称(囊状淋巴管瘤, 囊状水瘤), i.e. *Other-Name* (*cystic lymphangioma*, *hygroma*).

(2) If they have identical paired symbols, then their context patterns are homogeneous, and they can form a candidate co-referent relation.

For example, in the sentence "疯羊病"俗称"羊瘙痒病" (*Mad sheep disease has a colloquialism sheep scrapies.*), 疯羊病(*mad sheep disease*) and 羊瘙痒病(*sheep scrapies*) have the same context pattern that are both enclosed by double quotation marks. Thus, they can form a candidate relation 俗称(疯羊病, 羊瘙痒病), i.e. *colloquialism* (*mad sheep disease*, *sheep scrapies*).

2.3 Target Words Coordination

Given a word set W, for any w_i, $w_j \in W$, if they satisfy the following constraints:

- Occurring in the same sentence S_k;
- Being separated by what words, such as "、" (a coordinate sign in Chinese), "和 (*and*)" and "与 (*and*)", can express a coordinate relation;

Then words in W are considered to have coordinate relation. Based on it, we propose a coordinate distribution strategy to generate more candidate relations.

Given a word w and a coordinate set W, if there exist $w_i \in W$ such that $\mathcal{R}'_k(w, w_i)$ holds, then for any other word $w_j \in W$, $\mathcal{R}'_k(w, w_j)$ also holds.

For example, in the sentence 月饼，又称宫饼、小饼、月团、团圆饼等 (*Moon cake is also called palace cake*, *small cake*, *moon paste*, *and Mid-Autumn cake*), 宫饼 (*palace cake*), 小饼(*small cake*), 乐团(*moon paste*) and 团圆饼(*Mid-Autumn cake*) are coordinate in the sentence. If a candidate relation 又称(月饼, 宫饼) (i.e. *Other-Name* (*moon cake*, *palace cake*))is known in advance, the MFC can generate other three candidates, that is

又称(月饼, 小饼), i.e. *Other-Name* (*moon cake*, *small cake*)
又称(月饼, 月团), i.e. *Other-Name* (*moon cake*, *moon paste*)
又称(月饼, 团圆饼), i.e. *Other-Name* (*moon cake*, *Mid-Autumn cake*)

3 Co-referent Relations Verification

Errors may exist in the candidate relations set. The MFC must verify the candidate set to remove incorrect ones. To do this, we define a co-referent relation ontology and graph that the former reflects the semantic features, and the latter reflects the structural features.

3.1 The Ontology of Co-referent Relations

The co-referent relation ontology defines co-referent token words and their interrelations, and constraint axioms derived from them. This section mainly discusses the constraint axioms that are very important to the verification.

I. Word's Length Inequality Axioms

(A) $\forall w_i, w_j \in W, \exists \mathcal{R}_p \in \mathcal{R}, \mathcal{R}_p(w_i, w_j) \rightarrow Len(w_i) > Len(w_j) \vee Len(w_i) < Len(w_j)$;

For example, $Abbreviation(w_i, w_j) \rightarrow Len(w_i) > Len(w_j)$ and $Full\text{-}Name(w_i, w_j) \rightarrow Len(w_i) < Len(w_j)$.

II. Language Distribution Axioms

(B) $\forall w_i, w_j \in W, \exists \mathcal{R}_p \in \mathcal{R}, \mathcal{R}_p(w_i, w_j) \rightarrow ContainLang(w_i, \text{LANGTOKEN}) \vee ContainLang(w_j, \text{LANGTOKEN})$;

$ContainLang(w_i, \text{LANGTOKEN})$ means that the word w_i must contain a certain language. LANGTOKEN is a system-defined token. For example ENG refers to English, and CHN refers to Chinese. For example, $Is\text{-}Chinese\text{-}Name\ (w_i, w_j) \rightarrow ContainLang\ (w_j, \text{ENG})$.

III. Word's Extended Inclusion Axioms

(C) $\exists \mathcal{R}_p \in \mathcal{R}, \forall w_i, w_j \in W, \mathcal{R}_p(w_i, w_j) \wedge ExtInclude(w_i, w_j) \rightarrow \mathcal{R}_p$ is an *Abbreviation* relation;

(D) $\exists \mathcal{R}_p \in \mathcal{R}, \forall w_i, w_j \in W, \mathcal{R}_p(w_i, w_j) \wedge ExtInclude(w_j, w_i) \rightarrow \mathcal{R}_p$ is a *Full-Name* relation;

(C) is a sufficiency axiom of the *Abbreviation* relation, and (D) is sufficiency axiom of the *Full-Name* relation. $ExtInclude(x, y)$ is an extended inclusion relation, which represents the constitutions of y are from x.

IV. Co-referent Relation's Divergence and Convergence Axioms

The co-referent relation has two prominent features: *pointing-from* and *pointing-to* whose meaning are, for a co-referent relation $\mathcal{R}_p(w_i, w_j)$, that $\mathcal{R}_p$ points from w_i and points to w_j.

Definition 3: Given a target word w and a word set $W^o=\{w^o{}_1, \ldots, w^o{}_k\} \subset W$, for any $w^o{}_j \in W^o$ $(1 \le j \le k)$ such that $\mathcal{R}_p(w, w^o{}_j)$ holds, however, for any $w^{o'} \in W \backslash W^o$ that $\mathcal{R}_p(w, w^{o'})$ does not hold anymore, then the *pointing-from degree* of $\mathcal{R}_p$ on w is k, denoted as $P^O(\mathcal{R}_p, w)=k$.

Definition 4: Given a target word w and a word set $W^i=\{w^i{}_1, \ldots, w^i{}_h\} \subset W$, for any $w^i{}_j \in W^i$ $(1 \le j \le h)$ such that $\mathcal{R}_q(w^i{}_j, w)$ holds, however, for any $w^{i'} \in W \backslash W^i$ that $\mathcal{R}_q(w^{i'}, w)$ does not hold anymore, then the *pointing-to degree* of $\mathcal{R}_q$ on w is h, denoted as $P^I(\mathcal{R}_q, w)=h$.

Definition 5: Given a word set $W^O=\{w^o{}_1, \ldots, w^o{}_k\} \subset W$, for any $w^o{}_j \in W^O$ $(1 \le j \le k)$, there exists a word $w_f \in W$ such that $\mathcal{R}_p(w^o{}_j, w_f)$ holds, but there is no any word $w_g \in W \backslash W^O$

such that $\mathcal{R}_p(w_g, w_f)$ holds, then W^O is called the *pointing-from* set of $\mathcal{R}_p$ which means $\mathcal{R}_p$ points from any word of W^O.

Definition 6: Given a word set $W^I=\{w^i_1, \ldots, w^i_h\}\subset W$, for all $w^i_j \in W^I$ $(1\leq j\leq h)$, there exists $w_f\in W$ such that $\mathcal{R}_q(w_f, w^i_j)$ holds, but there is no any word $w_g\in W\backslash W^I$ such that $\mathcal{R}_q(w_f, w_g)$ holds, then W^I is called the *pointing-to* set of $\mathcal{R}_q$ which means $\mathcal{R}_q$ points to any word of W^I.

Given a co-referent relation $\mathcal{R}_p$ and its *pointing-from* set $W^O=\{w^o_1, \ldots, w^o_k\}$ and *pointing-to* set $W^I=\{w^i_1, \ldots, w^i_h\}$,

(E) The *divergence degree* of $\mathcal{R}_p$ is $Div(\mathcal{R}^c{}_p) = \underset{w^o_j\in W^O}{Max} P^O(\mathcal{R}^c{}_p, w^o_j)$, which means the *maximum pointing-from* degree of $\mathcal{R}_p$ on the set W^O.

(F) The *convergence degree* of $\mathcal{R}_p$ is $Cov(\mathcal{R}^c{}_p) = \underset{w^i_j\in W^I}{Max} P^I(\mathcal{R}^c{}_p, w^i_j)$, which means the *maximum pointing-to* degree of $\mathcal{R}_p$ on the set W^I.

V. Co-referent Token Words' Relation Axioms

Given two token words $\mathcal{R}_p$, $\mathcal{R}_q\in \mathcal{R}$, for any target word $w_i, w_j\in W$,

If $\mathcal{R}_p(w_i, w_j) \rightarrow \mathcal{R}_q(w_i, w_j)$ and $\mathcal{R}_q(w_i, w_j) \rightarrow \mathcal{R}_p(w_i, w_j)$ hold, then $\mathcal{R}_p$ and $\mathcal{R}_q$ are semantically equivalent, denoted as $\mathcal{R}_p\equiv\mathcal{R}_q$.

If $\mathcal{R}_p(w_i, w_j) \rightarrow \mathcal{R}_q(w_i, w_j)$ holds, but $\mathcal{R}_q(w_i, w_j)\rightarrow\mathcal{R}_p(w_i, w_j)$ does not hold, then $\mathcal{R}_p$ is contained by $\mathcal{R}_q$, denoted as $\mathcal{R}_p\Rightarrow\mathcal{R}_q$.

If $\mathcal{R}_p(w_i, w_j) \rightarrow \mathcal{R}_q(w_j, w_i)$ and $\mathcal{R}_q(w_j, w_i) \rightarrow \mathcal{R}_p(w_i, w_j)$ hold, then $\mathcal{R}_p$ and $\mathcal{R}_q$ are semantically reversible, denoted as $\mathcal{R}_p\equiv^{-1}\mathcal{R}_q$.

(G) $\exists\mathcal{R}_p, \mathcal{R}_q\in \mathcal{R}, \mathcal{R}_p\equiv\mathcal{R}_q; \forall w_i, w_j\in W, \mathcal{R}_p(w_i, w_j) \leftrightarrow \mathcal{R}_q(w_i, w_j)$;

(H) $\exists\mathcal{R}_p, \mathcal{R}_q\in \mathcal{R}, \mathcal{R}_p\Rightarrow\mathcal{R}_q; \forall w_i, w_j\in W, \mathcal{R}_p(w_i, w_j) \rightarrow \mathcal{R}_q(w_i, w_j)$;

(I) $\exists\mathcal{R}_p, \mathcal{R}_q\in \mathcal{R}, \mathcal{R}_p\equiv^{-1}\mathcal{R}_q; \forall w_i, w_j\in W, \mathcal{R}_p(w_i, w_j) \leftrightarrow \mathcal{R}_q(w_j, w_i)$.

(G) is a semantic equivalence axiom, (H) is a semantic implication axiom, and (I) is a semantic reverse axiom.

3.2 The Co-referent Relation Graph

If a target word is deemed as a vertex, and a co-referent relation as a directed edge, where a token word is a label of the directed edge, all candidate co-referent relation can be organized into a directed graph by their inter-links that is called co-referent relation graph.

Definition 7: Given a node set $V=\{v_1, \ldots, v_n\}$, $v_i, v_j\in V(1\leq i, j\leq n)$, if $<v_i, v_j>\neq<v_j, v_i>$, then $<v_i, v_j>$ is called a ordered node tuple, abbreviated as ordered tuple.

Definition 8: A *semantic association graph* is a 4-tuple $SAG=(V, \mathcal{R}^t, E, f)$, where

- $V=\{v_1, \ldots, v_n\}$ is a set of target words;
- $\mathcal{R}^t=\{\mathcal{R}^t_1, \ldots, \mathcal{R}^t_t\}$ is a set of token words;
- $E=\{e_1, \ldots, e_m\}$ is a set of ordered relations;

– f is a mapping from E to the set of triples composed of the ordered tuples on V and $\mathcal{R}^a$.

Formally, for any directed edge $e_k \in E$ $(1 \leq k \leq m)$, there always exist two nodes v_i, $v_j \in V$ $(1 \leq i,\ j \leq n)$ and a relation token word $\mathcal{R}^a_h \in \mathcal{R}^a$ $(1 \leq h \leq t)$ such that $f(e_k) = <v_i, v_j, \mathcal{R}^a_h>$ holds. That is, e_k is a directed edge, labeled as $\mathcal{R}^a_h$, which points from v_i to v_j. v_i is the *beginning-node* of e_k, and v_j is the *ending-node* of e_k. According to e_k's direction, v_j is the *forward adjacency-node* of v_i. Reversely, v_i is the *backward adjacency-node* of v_j.

Definition 9: Given a semantic association graph $SAG=(V, \mathcal{R}^a, E, f)$, there exists $v \in V$, the number of edges whose ending-node is v is v's *in-degree*, and the number of edges whose beginning-node is v is v's *out-degree*.

Definition 10: Given a semantic association graph $SAG=(V, \mathcal{R}^a, E, f)$, for any $\mathcal{R}^a_h \in \mathcal{R}^a$ $(1 \leq h \leq t)$, if $\mathcal{R}^a_h$ is a co-referent token word, then the SAG is called a *co-referent relation graph*, denoted by $CRG=(V, \mathcal{R}^c, E, f)$, as illustrated in Fig.1.

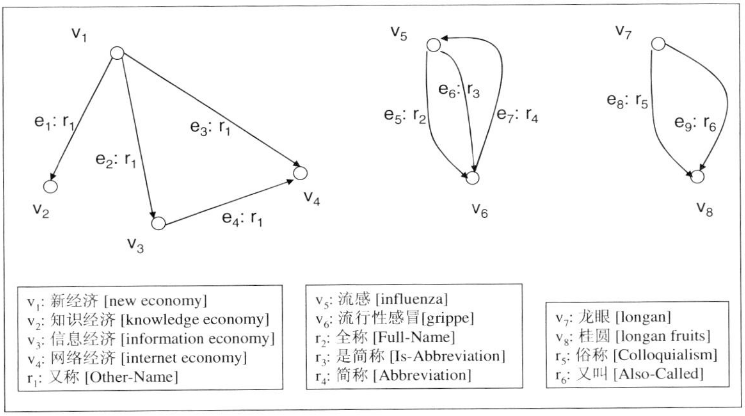

Fig. 1. A sample of co-referent relation graph

Definition 11: Given a co-referent relation graph $CRG=(V, \mathcal{R}^c, E, f)$, there exist a node $v \in V$ and a edge set $E^h \subseteq E$, for any edge $e_k \in E^h$, e_k has the same token word $\mathcal{R}^c_j$ ($\mathcal{R}^c_j \in \mathcal{R}^c$),

If v is e_k's beginning node, then v's out-degree on $\mathcal{R}^c_j$ is called *homogeneous out-degree*, denoted as $OutDegree^H(v, \mathcal{R}^c_j)=|E^h|$, and e_k's ending node constitutes v's *homogeneously forward adjacency-node set*, denoted as $AdV^{P\text{-}H}(v, \mathcal{R}^c_j)$.

If v is e_k's ending node, then v's in-degree on $\mathcal{R}^c_j$ is called *homogeneous in-degree*, denoted as $InDegree^H(v, \mathcal{R}^c_j)=|E^h|$, and e_k's beginning node constitutes v's *homogeneously backward adjacency-node set*, denoted as $AdV^{N\text{-}H}(v, \mathcal{R}^c_j)$.

In Fig.1, for the node v_1 and the label r_1, $OutDegree^H(v_1, r_1)=3$, $AdV^{P\text{-}H}(v_1, r_1)=\{v_2, v_3, v_4\}$. For the node v_4 and the label r_1, $InDegree^H(v_4, r_1)=2$, $AdV^{N\text{-}H}(v_4, r_1)=\{v_1, v_3\}$.

3.3 Verification Strategies of Co-referent Relations

The basic verification idea is that the MFC constructs a co-referent relation graph after inputting all candidate relations, and then applies a series of axioms defined by the co-referent ontology to verify them. The graph is regulated continually till its structure does not change anymore. Finally correct candidates are outputted.

The first three types of axioms in the co-referent ontology are independent verification strategies which can verify a single candidate relation. The other two types are joint verification strategies that can only be used with the co-referent relation graph, and they can verify multiple candidates simultaneously.

Given a co-referent relation graph $CRG=(V, \mathcal{R}, E, f)$:

There exist $w \in V$ and $\mathcal{R}_j \in \mathcal{R}$ such that w's homogeneous out-degree on $\mathcal{R}_j$ is $OutDegree^H(w, \mathcal{R}_j)$, and $\mathcal{R}_j$'s divergence degree is $Div(\mathcal{R}_j)$, or w's homogeneous in-degree on $\mathcal{R}_j$ is $InDegree^H(w, \mathcal{R}_j)$, and $\mathcal{R}_j$'s convergence degree is $Cov(\mathcal{R}_j)$,

1. If there is $OutDegree^H(w, \mathcal{R}_j) > Div(\mathcal{R}_j)$, then $\mathcal{R}_j(w, w_i)$ does not hold for all $w_i \in AdV^{P\text{-}H}(w, \mathcal{R}_j)$;
2. If there is $InDegree^H(w, \mathcal{R}_j) > Cov(\mathcal{R}_j)$, then $\mathcal{R}_j(w_i, w)$ does not hold for all $w_i \in AdV^{N\text{-}H}(w, \mathcal{R}_j)$.

There exist $e_g, e_h \in E$ and $\mathcal{R}_p, \mathcal{R}_q \in \mathcal{R}$ such that $f(e_g)=<w_i, w_j, \mathcal{R}_p>$ and $f(e_h)=<w_i, w_j, \mathcal{R}_q>$ hold,

3. If there is $\mathcal{R}_p \equiv \mathcal{R}_q$ or $\mathcal{R}_p \Rightarrow \mathcal{R}_q$, then $\mathcal{R}_p(w_i, w_j)$ and $\mathcal{R}_q(w_i, w_j)$ hold, otherwise, $\mathcal{R}_p(w_i, w_j)$ and $\mathcal{R}_q(w_i, w_j)$ do not hold.

There exist $e_g, e_h \in E$ and $\mathcal{R}_p, \mathcal{R}_q \in \mathcal{R}$ such that $f(e_g)=<w_i, w_j, \mathcal{R}_p>$ and $f(e_h)=<w_j, w_i, \mathcal{R}_q>$ hold,

4. If there is $\mathcal{R}_p \equiv^{-1} \mathcal{R}_q$, then $\mathcal{R}_p(w_i, w_j)$ and $\mathcal{R}_q(w_j, w_i)$ hold, otherwise, $\mathcal{R}_p(w_i, w_j)$ and $\mathcal{R}_q(w_j, w_i)$ do not hold.

In Fig.1, token words r_2 and r_3 have $r_2 \equiv r_3$, so $r_2(v_5, v_6)$ and $r_3(v_5, v_6)$ are both correct. Because token words r_5 and r_6 have $r_5 \Rightarrow r_6$, $r_5(v_7, v_8)$ and $r_6(v_7, v_8)$ are also correct. Both $r_3(v_5, v_6)$ and $r_4(v_6, v_5)$ are correct because of $r_3 \equiv^{-1} r_4$.

Except for the above verification strategies, the MFC can verify undecided relations using some already verified ones. This is called reinforced verification.

4 Experimental Analysis

From the Chinese open corpora of *2.6G* bytes, we use the predefined co-referent relation patterns (represented by regular expressions composed of token words and target words) to get the co-referent corpora (not limit domain, theme and style) of *18.4M* bytes. Using the MFC method, we acquire more than *60* types of co-referent relations from it. In Fig.2, the left is a group of sentences matched with patterns, and the right is a group of acquired relations from these sentences.

In the extraction phase, we get *66293* pieces of candidate relations, among which the distribution distance gets less than *3/4* of candidates, and the other two strategies get more than *1/4* as its supplements that increase the recall rate of extraction (see Table 1).

The verification orderly executes preprocessing, individual, joint and reinforced verification. Different strategies have different verification capabilities (see Table 2).

Sentence	Relations
南方的老百姓很早就有吃鱼头的习惯，而且大多是吃鳙鱼（又叫花鲢鱼，俗称胖头鱼）。 (The people in south China have habit of eating fish-head for long history. They often eat big-head carp (also called spotted silver carp, colloquialism is fat-head fish))	又叫(鳙鱼, 花鲢鱼) i.e. Also-Called (big-head carp, spotted silver carp) 俗称(鳙鱼, 胖头鱼) i.e. Colloquialism (big-head carp, fat-head fish)
光动力疗法（简称PDT，photodynamic therapy）。 (photodynamic therapy (abbreviated as PDT))	简称(光动力疗法, PDT) i.e. Abbreviation (photodynamic therapy, PDT) 英文名称(光动力疗法, photodynamic therapy) i.e. English-Name(光动力疗法, photodynamic therapy)
科学家们说，严重急性呼吸道综合征（SARS，即非典型肺炎） (Scientists said Severe Acute Respiratory Syndrome (SARS, that is Atypical Pneumonia))	英文名称(严重急性呼吸道综合征, SARS) i.e. English-Name (严重急性呼吸道综合征, SARS) 又称(严重急性呼吸道综合征, 非典型肺炎) i.e. Other-Name (Severe Acute Respiratory Syndrome, Atypical Pneumonia)
目前中国市场对聚氯乙烯（俗称PVC） (At present Polyvinyl Chloride in Chinese market (colloquialism is PVC))	俗称(聚氯乙烯, PVC) i.e. Colloquialism (Polyvinyl Chloride, PVC)
而根据美国风险投资协会（National Venture Capital Association 简称NVCA） (However, according to National Venture Capital Association (abbreviated as NVCA))	英文名称(美国风险投资协会, National Venture Capital Association) i.e. English-Name (美国风险投资协会, National Venture Capital Association) 简称(美国风险投资协会, NVCA) i.e. Abbreviation (National Venture Capital Association, NVCA)
通常所说的失眠症乃是非器质性失眠症的简称。 (The so-called insomnia is in fact the abbreviation of Non-organic insomnia.)	简称(失眠症, 非器质性失眠症) i.e. Abbreviation (insomnia, Non-organic insomnia)
"小灵通"的学名是无线市话，是固定电话业务的延伸。 (*Xiaolingtong*'s academic name is wireless city-phone, and is an extension of fixed phone business.)	学名(小灵通, 无线市话) i.e. Academic-Name (Xiaolingtong, wireless city-phone)
西安古称长安，先后有12个王朝在此建都。 (Xi'an was called Chang'an in ancient times, which had been the capital of twelve dynasties.)	古称(西安, 长安) i.e. Ancient-Name (Xi'an, Chang'an)
囊状淋巴管瘤好发于颈部，又称囊状水瘤。 (Cystic lymphangioma often occurs in the neck, which has a other-name of hygroma.)	又称(囊状淋巴管瘤, 囊状水瘤) i.e. Other-Name (Cystic lymphangioma, hygroma)
对联又称楹联、门联、对子。 (Antithetical couplet's other-name is pillar couplet, gatepost couplet, or couplet.)	又称(对联, 楹联) i.e. Other-Name (antithetical couplet, pillar couplet) 又称(对联, 门联) i.e. Other-Name (antithetical couplet, gatepost couplet) 又称(对联, 对子) i.e. Other-Name (antithetical couplet, couplet)

Fig. 2. Some examples of co-referent sentences and relations

Table 1. Performance of extraction strategies

Extraction Strategy	Number of Relations	Ratio (%)
distribution distance	48723	73.5%
pattern homogeneity	4312	6.5%
coordinate distribution	13258	20.0%
total	66293	100%

Table 2. Performance of verification strategies

Verification Strategy	Number of Relations	R-R (%)	F-R (%)
pre-processing	66293	87.93%	12.07%
individual verification	58293	88.20%	11.80%
joint verification	51414	65.16%	34.84%
reinforced verification	33501	92.24%	7.76%

The retained-ratio (R-R) is the ratio of retained relations after verifying. The filtered-ratio (F-R) is the ratio of filtered relations after verifying.

Table 3 lists different verification strategies' precision and recall rate on the acquisition. The benchmark is *66293* candidate relations. Finally, the precision rate (P) reaches *82.43%*, and the recall rate (R) reaches *92.03%*, and the error verification (correct relations but filtered) have *2206* pieces.

Table 3. Precision and recall rates of verification strategies

Verification Strategy	P (%)	R (%)	Verification Errors
pre-processing	46.70%	98.35%	457
individual verification	51.62%	95.88%	682
joint verification	76.27%	92.32%	987
reinforced verification	82.43%	92.03%	80

5 Conclusions

This paper proposes a multi-feature constrained method (MFC) for acquiring co-referent relations from large-scale Chinese corpora. It also provides some valuable guidance for other types of relation acquisitions. Such guidance is that it should use relations' features as many as possible, which include target words' distributive features and token words' semantic features, and structural features of all candidate relations. However, the acquisition by the MFC is restricted to the patterns. In the future, we will apply a pattern leaning method so that the relation and pattern acquisition would be integrated into one process to acquire more co-referent relations.

References

1. Marti A. Hearst, Automatic Acquisition of Hyponyms from Large Text Corpora, 14th International Conference on Computational Linguistics (COLING 1992), August 23-28, (1992), 539-545
2. Scott Cederberg and DominicWiddows, Using LSA and Noun Coordination Information to Improve the Precision and Recall of Automatic Hyponymy Extraction. Conference on Natural Language Learning (CoNLL-2003), Edmonton, Canada, (2003), 111-118
3. Rosana Girju, Adriana Badulescu and Dan Moldovan, Learning Semantic Constraints for the Automatic Discovery of Part-Whole Relations. In Proceedings of HLT-NAACL 2003, Edmonton, May-June, (2003), 1-8
4. Matthew Berland and Eugene Charniak, Finding Parts in Very Large Corpora. In Proceedings of the 37th Annual Meeting of the Association for the Computational Linguistics (ACL-99), College Park, MD, (1999), 57-64
5. Alexander Maedche and Steffen Staab, Discovering Conceptual Relations from Text. Proceedings of the 14th European Conference on Artificial Intelligence (ECAI 2000), Berlin, Germany, August 20-25, (2000), 321-325
6. Lin D and Pantel P, Induction of Semantic Classes from Natural Language Text. Proceedings of SIGKDD-01, San Francisco, CA, USA, (2001), 317-322

Location-Aware Data Mining for Mobile Users Based on Neuro-fuzzy System*

Romeo Mark A. Mateo[1], Marley Lee[2], Su-Chong Joo[3], and Jaewan Lee[1]

[1] School of Electronic and Information Engineering, Kunsan National University
68 Miryong-dong, Kunsan, Chonbuk 573-701, South Korea
{rmmateo, jwlee}@kunsan.ac.kr
[2] School of Electronic and Information Engineering, Chonbuk National University
664-14, DeokJin-dong, Jeonju, Chonbuk 561-756, South Korea
mrlee@chonbuk.ac.kr
[3] School of Electrical, Electronic and Information Engineering
Wonkwang University, South Korea
scjoo@wonkwang.ac.kr

Abstract. Data mining tools generally deal with highly structured and precise data. However, classical methods fail to handle imprecise or uncertain information. This paper proposes a neuro-fuzzy data mining approach which provides a means to deal with the uncertainty of data. This presents a location-based service collaboration framework and uses the neuro-fuzzy algorithm for data mining. It also introduces the user-profile frequency count (UFC) function to determine the relevance of the information to mobile users. The result of using neuro-fuzzy system provides comprehensive and highly accurate rules.

1 Introduction

Ubiquitous and mobile technologies providing location-awareness and information through location-based services (LBS) have experienced dramatic increase in the world market [1]. Such technologies include radio frequency identifiers (RFID), smart personal devices, and global positioning systems. Researchers investigate the methods of acquiring information using these distributed technologies [2]. Moreover, identifying patterns and rules of locations by using data mining are challenging areas for these researchers. Classical methods provide meaningful information and are use for predictions of data [3]. The rules extracted from classification mining predict the next event from the history of transactions [4]. The knowledge represented by rules is useful for analyzing patterns of data especially in allocating resources from the LBS [3].

On the contrary, classical methods avoid imprecise or uncertain information because it is not considered useful in processing the data. Moreover, the goal of obtaining understandable results is often neglected. In location-based services, the imprecision is due to the errors and inaccuracies of measuring devices.

* This research was supported by grant R01-2006-000-10147-0 from the Basic Research Program of the Korea Science and Engineering Foundation.

L. Wang et al. (Eds.): FSKD 2006, LNAI 4223, pp. 1269 – 1278, 2006.

Fuzzy systems are used to handle uncertainty from the data that cannot be handled by classical methods. It uses the fuzzy set to represent a suitable mathematical tool for modeling of imprecision and vagueness [5]. The pattern classification of fuzzy classifiers provides a means to extract fuzzy rules for information mining that leads to comprehensible method for knowledge extraction from various information sources [6]. The fuzzy algorithm is also a popular tool for information retrieval [7]. Fuzzy *c*-means classifier (FCM) uses an iterative procedure that starts with an initial random allocation of the objects to be classified to *c* clusters. Among other data mining techniques, FCM is a very popular tool for knowledge extraction in the distributed environment like in Ko, et. al [8]. The output of FCM can be substantially improved by means of preprocess filtering [9]. The filtering removes the unnecessary data and consequently increases the processing speed of FCM as well as improves the quality of rules extracted. In the case of neuro-fuzzy, a fuzzy system is used to represent knowledge in an interpretable manner. The algorithm borrows the learning ability of neural networks to determine the membership values. It is among the most popular data mining techniques used in recent research [10, 11].

In this paper, we propose a location-aware data mining approach for mobile users based on neuro-fuzzy system. We present a framework to enhance the location-based service (LBS) by filtering using the user-profile frequent count (UFC) function to select the most relevant object service. To demonstrate our framework, we perform data mining using the neuro-fuzzy algorithm on the location information obtained from the LBS. Also, the proposed system is compared to other classical methods.

2 Related Works

2.1 Data Mining Using Mobile Devices

In location-based services, data mining is used to reveal patterns of services and provide prediction of location. A sequential mining approach for the location prediction is used to allocate resources in a PCS network [3]. This technique can effectively allocate resources to the most probable-to-move cells instead of blindly allocating excessive resources in the cell-neighborhood of a mobile-user. Location-awareness agent using data mining is found in the work of Lee, et. al. [4]. This is done by sending a mobile agent to the LBS and then it performs the classification mining in the database. A guide system combines the positioning technique and location-awareness service to provide the surrounding information for users [12]. The guide system not only accepts the user's search query to find the target but also receives the information from other users who took notes during the tour guide.

2.2 Neuro-fuzzy Systems

Fuzzy classification is based on the concept of fuzzy sets, which was conceived by Lotfi Zadeh [14]. It is presented as a way of processing data by allowing partial set membership rather than crisp set membership or non-membership. Typical fuzzy data analysis discovers rules in large set of data and these rules can be used to describe the dependencies within the data and to classify a new data [6]. Neuro-fuzzy systems are fuzzy classifiers and uses neural networks for learning by performing induction of the structure and adaptation of the connection weights [10, 11]. There are many types of neuro-fuzzy rule generation algorithm [15]. FuNE-I is a neuro-fuzzy model that is based on the architecture of feed-forward neural network with five layers which uses only rules with one or two variables in antecedents [16]. A Sugeno-Type neuro-fuzzy system is used for a scheme to construct an n-link robot manipulator to achieve high-precision position tracking [17]. A scheme of knowledge encoding in a fuzzy multi-layer perceptron using rough set-theoretic concepts [18] that is utilized for encoding the crude domain knowledge. A neuro-fuzzy classification (NEFCLASS) is a fuzzy classifier that creates fuzzy rule from data by a single run through the data set [11]. Figure 1 is a type of neuro-fuzzy system which consists of two input nodes, two output nodes and five hidden node use for the linguistic rules.

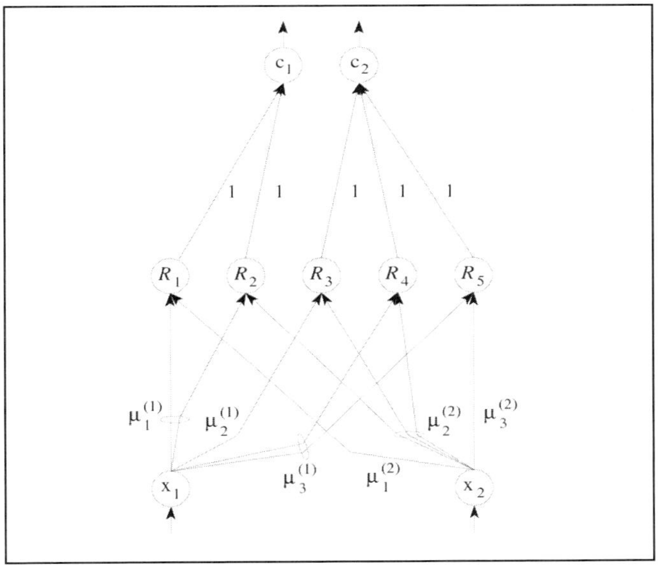

Fig. 1. A NEFCLASS system with two inputs, five rules and two output classes

3 Collaborative Framework for Location-Based Services

The proposed framework of location-based service (LBS) consists of four interactive components namely, location-aware agent, location information service, mapping service and object group services. These components collaborate to produce the necessary location information and use neuro-fuzzy system for classifying the data and extract rules. Figure 2 illustrates the interaction of object service registration and processing the mobile request from the LBS framework. The LBS is implemented in Common Object Request Broker Architecture (CORBA) to support the communication of distributed objects and to provide an efficient service.

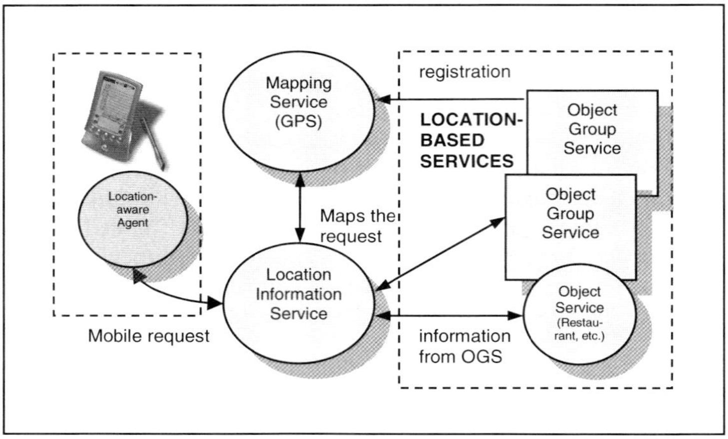

Fig. 2. Location-based services collaboration framework

The interaction of these components starts with a client requesting for information like finding of 30 nearest restaurant. The request is sent to the location information service and then it invokes the services that match the request of client. Mapping service estimates the location of the mobile user among the services. The extracted information from different services is collected by the location information service. After collecting the information, location-aware agent process the data using the neuro-fuzzy algorithm. The processed information is then presented into table of location information and fuzzy rules as outputs. The components from the proposed architecture are defined in more details in the next sub-sections.

3.1 Components of the Location-Based Services

Location-aware Agent. First, the query of the mobile user is processed by the location information service. Before sending the request, user profiles like the interest of mobile user are sent. The request is forwarded to mapping services to locate the service and object services to provide information. Lastly, after gathering the location information, it is returned to the location-aware agent and process data mining.

Location Information Service (LIS). Location information service collects information provided by location-based services. Thus, when a mobile user wants to search the nearest restaurants, then the location information service invokes the mapping service to map the nearest restaurants. The location information are collected and sent to the location-aware agent for processing the neuro-fuzzy algorithm.

Mapping Service. The mapping service accomplishes the inquiry of mobile users to select the nearest services. To be able to map the object service, it has to register first to the mapping service. Mapping service accepts the membership of the object service and the object service agrees to collaborate to the system. Once the location of the object service is change, it notifies the mapping service to update its location.

Object Group Services (OGS). Object group service provides customable implementation of objects and gives challenge to developers. These is done by implementing membership schemes like binding other objects to a group and defined through IDL for accessibility of the object services [19]. In our proposed system, the object group services are represented by location-based services or as simple as web servers. Object services allow information retrieval through location information service.

4 Neuro-fuzzy Data Mining Model

Conventional methods in data mining are geared towards structured and precise data. Lee, et. al [4] propose a location-aware agent to mine the database of library using a J48 classifier to predict the availability of books which can be borrowed. However, it does not deal with imprecise data which can be factor for time estimation and the result is not easy to comprehend. In our study, we use the neuro-fuzzy system to manage imprecise data. Figure 3 presents the overall procedure of the location-aware data mining approach. First, the mobile user sends request to the LIS. The LIS then performs the UFC function to each object services and the data are collected and produce *LI*. Finally, the *LI* is sent to the location-aware agent to perform data mining.

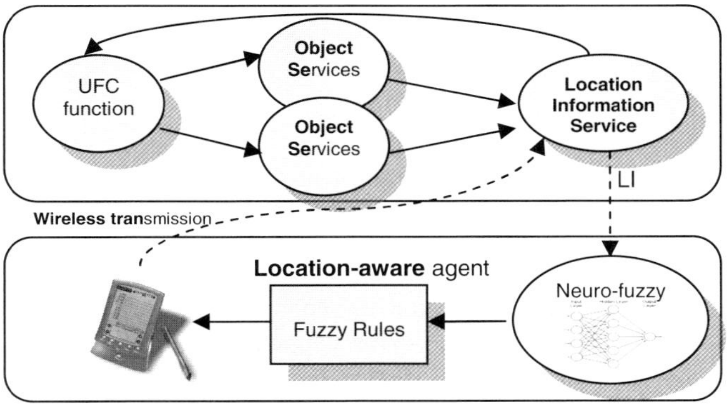

Fig. 3. Neuro-fuzzy data mining using location-aware agent

There are two common problems in information retrieval and these are overflow of information and relevance of information. This can be solved by preprocessing method like filtering based on user-profile. We used the method of user-profile frequency count (UFC) function to select the most relevant object service. Let P as the collected user-profile and p_i as a single profile. A single profile of a mobile user contains subsets of words. These are words describe p_i. In our example, we use American food as a single user-profile and this is described by burgers, steaks, and ice cream that are subset of this profile. The process of calculating P is presented in Equation 1.

$$P = \{ p_1, p_2, ..., p_i \} \tag{1}$$

P is used for user-profile frequency count (UFC) function through the object services. Before executing UFC, the LIS requests the mapping service to query the location of the object services that contains the possible information. The object services then allows information retrieval from their web pages by using UFC. In information retrieval, term frequency [8, 20] is used for classifying documents. This study uses the UFC to determine information relevance where the frequency of p_i is counted over all words from every object service (o_j). Equation 2 shows the calculation of UFC.

$$UFC\,(p_i, o_j) = \sum_{i=1}^{n} \sum_{j=1}^{m} p_{ij} \tag{2}$$

This computation counts the frequency of p_i from each o_j. Thus, this function is done after the location information from the o_j was already invoked by the LIS. The UFC is a separate procedure from the invocation of location information from o_j. The repetition of the word from profile will not be counted at this stage. After getting the UFC value of each object service, it is compared to a threshold value (θ) for filtering. The location information which has a UFC value that is less than the threshold will be filtered or removed. If $UFC\ (p_i, o_j) < \theta$ then it is not relevant to the mobile user and it can be ignored. Moreover, we collect all $UFC\ (p_i, o_j)$ to generate a new attribute from the location information. Equation 3 presents the collected information or I_j from the object services where C is the iteration through k attributes.

$$I_j = \sum_{1}^{m} C_k \tag{3}$$

$$LI = I_j + P_j(UFC) \tag{4}$$

Equation 4 presents the merged value of previous location information attributes and the attribute from UFC value. Here, LI is the merged attribute of I_j, which are previous attributes from location information, and $P_j(UFC)$, which is the new attribute that was obtained by UFC function. It is assumed that this relevant information from the web pages is likely important factor for generating fuzzy rules. The LI is transmitted from LIS to the user's mobile device containing a location-aware agent which processes the data mining. The neuro-fuzzy algorithm is shown below.

1. **Fuzzy Classification.** Fuzzy rules shown in Equation 5 are used in classification of pattern where small numbers of linguistic terms that are shared by all rules. These are linguistic terms which are readable and easy to interpret.

$$If\ \ a_1\ is\ A_{i,1}\ and\ \ a_2\ is\ A_{i,2}\ and\ \ldots and\ \ a_k\ is\ A_{i,K}\ then\ \ B_i \tag{5}$$

2. **Learning Fuzzy Classification Rules.** The domain attributes are mapped to the units of the input layer and output layer of the neural network which contains one unit for each possible value of the class attribute. This procedure creates fuzzy rules and adapts the fuzzy set appearing in the rules of learning.

5 Experimental Evaluation

The components of the proposed location-based service collaboration framework were designed using Java. To evaluate different algorithms for data mining, we used

Weka for classical methods like J48 classifier, FCM and MLP, and NEFCLASS for neuro-fuzzy. The following subsection describes the simulation environment, performance analysis and result.

5.1 Data Mining Environment

The environment OS platform used here were Windows OS, Red Hat Linux and Sun Solaris 8 to simulate the heterogeneity of the location-based services. The data were gathered from the web pages of restaurants in California and used the mapping service of Google to estimate the location points of the restaurants. These data were processed to the location information service using UFC function. We used the favorite foods of the user-profile to perform UFC and set θ to 1 for filtering the object services. The data mining procedure was executed after sending the information to the mobile users.

5.2 Performance Analysis

Precision and recall are two typical measures for evaluating the performance of information retrieval systems [20]. Given a discovered cluster γ and the associated reference cluster Γ, precision ($P\gamma\Gamma$) and recall ($R\gamma\Gamma$) applied to evaluate the performance of clustering algorithms. In classifier algorithm, recall and precision is performed by cross-validation of the classified instances. To evaluate the performance of our experiment, we used these measurements of precision and recall. This is done by calculating the average precisions in Equation 6 where *AvgP* is the summation of precision (P_n) of classes divided by the number of classes.

$$AvgP = \frac{\sum_{i=1}^{n} P_n}{n} \tag{6}$$

Average of recall is computed in Equation 7 where *AvgP* is the summation of recall (R_n) of classes divided by the number of classes.

$$AvgR = \frac{\sum_{i=1}^{n} R_n}{n} \tag{7}$$

A high percentage of precision and recall means that the classification method is more accurate. The processing time is observed to determine the time constraint of processing the algorithm. The number of correctly classified instances was used to determine accuracy. The performance of neuro-fuzzy and classical methods is discussed in Section 5.3. The classical methods that we compared are fuzzy *c* means (FCM), J48 classifier and multilayer perceptron (MLP).

5.3 Results of Data Mining

We used the data environment from Section 5.1 and simulated on computer networks. The result of data mining using neuro-fuzzy algorithm processed by the location-aware agent and generated fuzzy rules from the location information. Fuzzy rules are shown in Table 1 and consist of 9 patterns classified out of 30 restaurant information.

Table 1. Fuzzy rules processed from neuro-fuzzy data mining

Fuzzy rules
if nSea is **N** and nPark is **F** and sCrowd is **S** and fPref is **M** then American
if nSea is **N** and nPark is **M** and sCrowd is **S** and fPref is **H** then Seafood
if nSea is **F** and nPark is **M** and sCrowd is **S** and fPref is **H** then American
if nSea is **N** and nPark is **M** and sCrowd is **S** and fPref is **M** then Seafood
if nSea is **N** and nPark is **M** and sCrowd is **L** and fPref is **H** then American
if nSea is **N** and nPark is **N** and sCrowd is **S** and fPref is **M** then Italian
if nSea is **N** and nPark is **M** and sCrowd is **S** and fPref is **L** then Seafood
if nSea is **M** and nPark is **N** and sCrowd is **S** and fPref is **M** then Italian
if nSea is **F** and nPark is **N** and sCrowd is **S** and fPref is **M** then Italian

The attributes **nSea** and **nPark** determines the distance of the restaurant from the sea and the park, respectively, which have variables for membership given by **N** for near, **M** for medium near and **F** for far. The attribute **sCrowd** determines if there are lots of people within the place, which have variables for membership given by **S** for small, **M** for medium and **L** for large and **fFood** determines the relevant information from favorite food of the mobile user which have variables for membership given by **L** for less-favorite, **M** for medium-favorite and **H** for highly-favorite. These rules classify types of restaurant which are Italian, American and Seafood. In the first rule, it can be explained that American restaurants are near from the sea and far from parks and not crowded and the food are medium favorite of the mobile user. The second, it can be explained that seafood restaurants are near to the sea and medium far from parks and not crowded and the food are highly favorite of the mobile user.

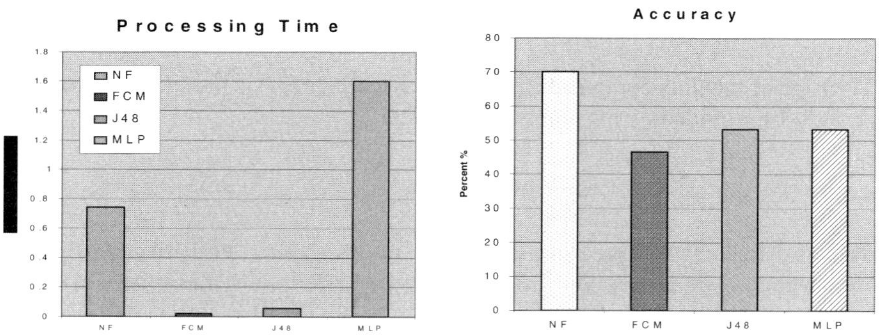

(a) (b)

Fig. 4. Bar graphs showing the comparison of processing time and accuracy of neuro-fuzzy and classical methods

Comparison of classical methods for performance is shown in Figure 4. The bar graphs present the comparison of processing time and accuracy of neuro-fuzzy and other classical methods in. In Figure 4a, the processing time of neuro-fuzzy is much faster than the MLP while FCM and J48 classifier has almost same processing time.

In accuracy, we can justify the performance of neuro-fuzzy is better than the other classical methods in the sense that even though the processing time is slower than the FCM and J48, it is more accurate of classifying patterns shown in Figure 4b. Table 2 shows the result of precision and Table 3 for recall.

Table 2. Precision of each class

Classes	**Neuro-fuzzy**	**FCM**	**J48**	**MLP**
American	0.85	0.29	0.71	0.44
Seafood	0.33	0.36	0.46	0.43
Italian	0.72	0.50	0.50	0.57
Average	0.63	0.38	0.56	0.48

Table 3. Recall of each class

Classes	**Neuro-fuzzy**	**FCM**	**J48**	**MLP**
American	0.60	0.25	0.50	0.40
Seafood	0.70	0.56	0.50	0.60
Italian	0.80	0.25	0.60	0.40
Average	0.70	0.35	0.53	0.47

The result of precision (Equation 6) and recall (Equation 7) are presented in Table 2 and 3, respectively. Neuro-fuzzy has the highest precision which has an average of 0.63 and recall which has an average of 0.70 compared to J48 (0.56, 0.53), MLP (0.48, 0.47), and FCM (0.38, 0.35). Most of these classical methods were able to predict testing data with the number of misclassified patterns between 14 to 16 while neuro-fuzzy has only 9 misclassified patterns. In addition, the results outperformed the mentioned classical methods in terms of simplicity by providing comprehensive rules from the data.

6 Conclusion

In this paper we proposed a location-aware data mining approach for mobile user based on neuro-fuzzy system. We present a collaborative framework for location-based services and enhanced it by filtering using the user-profile frequent count (UFC) function to select the most relevant object service. Neuro-fuzzy algorithm is used for data mining of location information. The proposed system is compared to other classical methods and shows that it provides comprehensive results of rules from the location information. More correctly classified instances are obtained by the neuro-fuzzy which indicates that its classification accuracy is much better than classical methods.

We presented our experiment on restaurant services and our future works will be implementing this work in an on-going project of intelligent home healthcare services.

References

1. Schiller, J., and Voisard, A.: Location-Based Services. Morgan Kaufmann Publishers, Elveiser Inc. (2004) pp. 10-15
2. Patterson, C., Muntz, R., and Pancake, C.: Challenges in Location-Aware Computing. IEEE Pervasive Computing. Vol.2, No. 2 (April-June 2003) pp. 80-89
3. Yavas, G., Katsaros, D., Ulusoy, O., and Manolopoulos, Y.: A Data Mining Approach for Location Prediction in Mobile Environments. Data & Knowledge Engineering Vol. 54 (2005) pp. 121-146
4. Lee, J. W., Mateo, R. M., Gerardo, B. D., and Go, S.: Location-aware Agent using Data mining for the Distributed Location-based Services. LNCS Vol. 3894, Glasgow, Scotland, U.K. (May 2006) pp. 867-876
5. Dumitrescu, D., Lazzerini, B., and Jain, L. C.: Fuzzy Sets and Their Application to Clustering and Training. The CRC Press International Series on Computer Intelligence, CRC Press LLC (2000)
6. Kruse, R., Bolgelt, C., and Nauck, D.: Fuzzy Data Analysis: Challenges and Perspectives. In Proceedings of the 8th IEEE International Conference on Fuzzy Systems, IEEE Press, Piscataway, NJ, USA (1999)
7. Mendes Rodrigues, M.E.S. and Sacks, L.: A Scalable Hierarchical Fuzzy Clustering Algorithm for Text Mining. In Proceedings of the 5th International Conference on Recent Advances in Soft Computing, Nottingham, U. K. (December 2004)
8. Ko, J., Gerardo, B. D., Lee, J. W., and Hwang, J.: The Information Search System Using Neural Network and Fuzzy Clustering Based on Mobile Agents. LNCS Vol. 3481, Singapore (May 2005) pp. 205-214
9. Yi, S., Gerardo, B. D., Lee, Y. S., and Lee, J. W.: Intelligent Information Search Mechanism using Filtering and NFC based on Multi-agents in the Distributed Environment. LNCS Vol. 3982, Glasgow, Scotland, U.K. (May 2006) pp. 867-876
10. Klose, A., Nürnberger, A., Nauck , D., and Kruse R.: Data Mining with Neuro-Fuzzy Models. Data Mining and Computational Intelligence, Springer-Verlag (2001) pp. 1-36
11. Nauck, D., and Kruse, R.: NEFCLASS - A Neuro-Fuzzy Approach for the Classification of Data. In Proceedings of ACM Symposium on Applied Computing, Nashville (1995)
12. Huang, Y. P., Chuang. W. P.: Improving the Museum's Service by Data Mining and Location-aware Approach. In Proceedings of Systems, Man and Cybernetics (2004)
13. Bellavista, P., Corradi, A., and Stenfalli, S.: Mobile Agent Middleware for Mobile Computing. Computer Journal. (March 2001) pp. 73-81
14. Zadeh, L. A.: Fuzzy Sets. Information and Control (1965) pp. 338-353
15. Mitra, S., and Hayashi, Y.: Neuro-Fuzzy Rule Generation: Survey in Soft Computing Framework. IEEE Trans. Neural Networks, Vol. 11 (2000) pp. 748-768
16. Halgamuge, S. K., and Glesner, M.: Neural Networks in Designing Fuzzy Systems for Real World Applications. Fuzzy Sets and Systems, 65 (1994) pp. 1-12
17. Wai, R. J., and Chen, P. C.: Intelligent Tracking Control for Robot Manipulator Including Actuator Dynamics via TSK-type Fuzzy Neural Network. IEEE Trans. Fuzzy Systems Vol. 12 (2004) pp. 552-560
18. Banerjee, M., Mitra, S., and Pal, S. K.: Rough Fuzzy MLP: Knowledge Encoding and Classification. IEEE Trans. Neural Networks, Vol. 9 (1998) pp. 1203-1216
19. Felber, P., Guerraoui, R., Schiper, A.: Evaluating CORBA Portability: The Case of an Object Group Service, Proceedings of the Second International Enterprise Distributed Object Computing Workshop (1998) pp. 164-173
20. Baeza-Yates, R., and Ribeiro-Neto, B.: Modern Information Retrieval. New York: Addison Wesley, ACM Press (1999)

Biomedical Named Entities Recognition Using Conditional Random Fields Model*

Chengjie Sun, Yi Guan, Xiaolong Wang, and Lei Lin

School of Computer Science, Harbin Institute of Technology,
Heilongjiang Province, China
{cjsun, guanyi, wangxl, linl}@insun.hit.edu.cn
http://www.insun.hit.edu.cn

Abstract. Biomedical named entity recognition is a critical task for automatically mining knowledge from biomedical literature. In this paper, we introduce Conditional Random Fields model to recognize biomedical named entities from biomedical literature. Rich features including literal, context and semantics are involved in Conditional Random Fields model. Shallow syntactic features are first introduced to Conditional Random Fields model and do boundary detection and semantic labeling at the same time, which effectively improve the model's performance. Experiments show that our method can achieve an F-measure of 71.2% in JNLPBA test data and which is better than most of state-of-the-art system.

1 Introduction

With the development of computational and biological technology, the amount of biomedical literature is increasing unprecedentedly. MEDLINE database has colleted 11 million biomedical related records since 1965 and is increasing at the rate of 1500 abstracts a day [1]. The research literature is a major repository of knowledge. From them, researchers can find the knowledge, such as connections between diseases and genes, the relationship between genes and specific biological functions and the interactions of different proteins and so on.

The explosion of literature in the biomedical field has provided a unique opportunity for natural language processing techniques to aid researchers and curators of databases in the biomedical field by providing text mining services. Yet typical natural language processing tasks such as named entity recognition, information extraction, and word sense disambiguation are particularly challenging in the biomedical domain with its highly complex and idiosyncratic language.

Biomedical Named Entities Recognition (NER) is a critical task for automatically mining knowledge from biomedical literature. Two special workshops for biomedical named entities recognition BioCreAtIvE [2] (Critical Assessment for Information Extraction in Biology) and JNLPBA [3] (Joint Workshop on Natural Language

* This work is supported by National Natural Science Foundation of China (60504021) and The 863 high Technology Research and Development Programme of China (2002AA117010-09).

L. Wang et al. (Eds.): FSKD 2006, LNAI 4223, pp. 1279 – 1288, 2006.

Processing in Biomedicine and its Applications) were held in 2004 respectively and each of them contained an open evaluation of biomedical named entities recognition technology. The data and guidelines afforded by the two workshops greatly promote the biomedical NER technology. According to the evolution results of JNLPBA2004, the best system can achieve an F-measure of 72.6%. This is somewhat lower than figures for similar tasks from the news wire domain. For example, extraction of organization names has been done at over 0.90 F-measure [2]. Therefore, biomedical NER technology need further study in order to make it applied.

Current research methods for NER can be classified into 3 categories: dictionary-based methods [4], rule-based methods [5] and machine learning based methods. In biomedical domain, dictionary-based methods suffer from low recall due to new entities appear continually with the biology research advancing. Biomedical NEs do not follow any nomenclature, which makes rule-based methods to be helpless. Besides, rule-based method itself is hard to port to new applications. More and more machine learning methods are introduced to solve the biomedical NER problem, such as Hidden Markov Model [6] (HMM), Support Vector Machine [7] (SVM), Maximum Entropy Markov Model [8] (MEMM) and Conditional Random Fields [1, 9] (CRFs). Biomedical NER problem can be cast as a sequential labeling problem. Conditional random fields for sequences labeling offer advantages over both generative models like HMM and classifiers applied at each sequence position[10].

In this research, we utilize Conditional Random Fields model involving rich features to extract biomedical named entities from biomedical literature. The feature set includes orthographical features, context features, word shape features, prefix and suffix features, Part of Speech (POS) features and shallow syntactic features. Shallow syntactic features are first introduced to Conditional Random Fields model and do boundary detection and semantic labeling at the same time, which effectively improve the model's performance. Although some features have been used by some researchers, we show the effect of each kind of features in detail, which can afford valuable reference to other researchers. Our method does not need any dictionary resources and post-processing, so it has strong adaptability. Experiments show that our method can achieve an F-measure of 71.2% in JNLPBA test data and which is better than most of state-of-the-art system.

The remainder of this paper is structured as follows. In section 2, we define the problem of biomedical named entities recognition and its unique characteristics. In section 3, a brief introduction of linear-chain conditional random fields model are given. In section 4 we explain the features involved in our system. Experiment results are shown in section 5. Section 6 is a brief conclusion.

2 Biomedical Named Entity Recognition

Biomedical NER can be addressed as a sequential labeling problem. It is defined as recognizing objects of a particular class in plain text. Depending on required application, NER can extract objects ranging from protein/gene names to disease/virus names. In practice, we regard each word in a sentence as a token and each token is associated with a label. Each label with a form of B-C, I-C or O indicates not only the category of the Named Entity (NE) but also the location of the token within the an NE. In this label denotation, C is the category label; B and I are location labels, standing

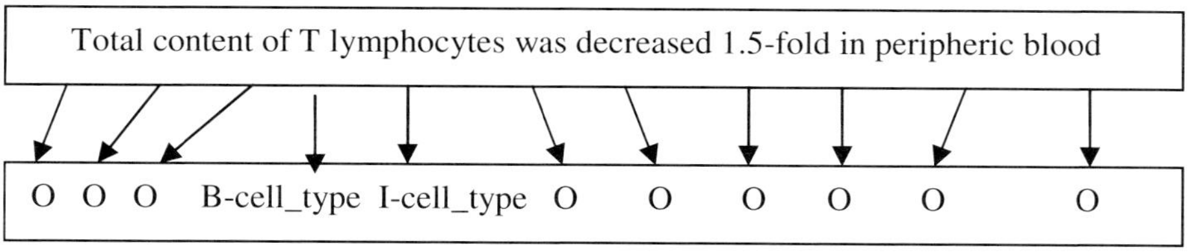

Fig. 1. An example of biomedical NER

for the beginning of an entity and inside of an entity respectively. O indicates that a token is not part of an NE. Fig. 1 is an example of biomedical NER.

Biomedical NER is a challenging problem. There are many different aspects to deal with. In general, biomedical NEs do not follow any nomenclature [11] and can comprise long compound words and short abbreviations. Biomedical NEs are often English common nouns (as opposed to proper nouns, which, are the nouns normally associated with names) and are often descriptions [12]. For example, some Drosophila (fruit fly) gene names are *blistery, inflated, period, punt and midget*. Some NEs contain various symbols and other spelling variations. On average, any NE of interest has five synonyms. An NE may also belong to multiple categories intrinsically; An NE of one category may contain an NE of another category inside it [13].

In natural language processing domain, Generative Models and Discriminative Models are often used to solve the sequential labeling problem, such as NER. Recently, Discriminative Models are preferred due to their unique characteristic and good performance [14]. Generative Models define a joint probability distribution $p(X,Y)$ where X and Y are random variables respectively ranging over observation sequences and their corresponding label sequences. In order to define a joint distribution of this nature, generative models must enumerate all possible observation sequences – a task which, for most domains, is intractable unless observation elements are represented as isolated units, independent from the other elements in an observation sequence. Discriminative Models directly solve the conditional probability $p(Y \mid X)$.The conditional nature of such models means that no effort is wasted on modeling the observations and one is free from having to make unwarranted independence assumptions about these sequences; arbitrary attributes of the observation data may be captured by the model, without the modeler having to worry about how these attributes are related.

Table 1. Biomedical Named Entities label list

Meaning	Label	Meaning	Label
Beginning of protein	B-protein	Inside protein	I-protein
Beginning of DNA	B-DNA	Inside DNA	I-DNA
Beginning of RNA	B-RNA	Inside RNA	I-RNA
Beginning of cell_type	B-cell_type	Inside cell_type	I-cell_type
Beginning of cell_line	B-cell_line	Inside cell_line	I-cell_line
others	O		

This paper utilizes a Discriminative Model – Conditional Random Fields to solve biomedical NER problem. Using the definition in [3], we recognize 5 categories entities. There are 11 labels in all using BIO notation mentioned before. All labels are show in Table 1. Each token in the biomedical text will be assigned one of the 11 labels in the recognition results.

3 Conditional Random Fields Model

Conditional Random Fields (CRFs) model is a kind of undirected graph model [14]. A graphical model is a family of probability distributions that factorize according to an underlying graph. The main idea is to represent a distribution over a large number of random variables by a product of local functions that each depend on only a small number of variables [15]. The power of graph model lies in it can model multi variables, while an ordinary classifier can only predicate one variable.

The result of NER is a label sequence, so linear-chain CRFs model is adopted in this research.

Let $\mathbf{y}$, $\mathbf{x}$ be random vectors, $\Lambda = \{\lambda_k\} \in \Re^K$ be a parameter vector, and $\{f_k(y, y', \mathbf{x}_t)\}_{k=1}^{K}$ be a set of real-valued feature functions. Then a linear-chain CRFs is a distribution $p(\mathbf{y}|\mathbf{x})$ that takes the form

$$p(\mathbf{y}|\mathbf{x}) = \frac{1}{Z(\mathbf{x})}\exp\{\sum_{k=1}^{K}\lambda_k f_k(y_t, y_{t-1}, \mathbf{x}_t)\} , \tag{1}$$

where $Z(\mathrm{x})$ is an instance-specific normalization function.

$$Z(\mathrm{x}) = \sum_{\mathbf{y}}\exp\{\sum_{k=1}^{K}\lambda_k f_k(y_t, y_{t-1}, \mathbf{x}_t)\} . \tag{2}$$

For the application of linear-chain CRFs model, the key problem is how to solve the parameter vector $\theta = \{\lambda_k\}$. This is done during the training process.

Suppose there are iid training data $D = \{\mathbf{x}^{(i)}, \mathbf{y}^{(i)}\}_{i=1}^{N}$, where each $\mathbf{x}^{(i)} = \{x_1^{(i)}, x_2^{(i)}, \cdots, x_T^{(i)}\}$ is a sequence of inputs and each $\mathbf{y}^{(i)} = \{y_1^{(i)}, y_2^{(i)}, \cdots, y_T^{(i)}\}$ is a sequence of corresponding predictions. Then parameter estimation is performed by penalized maximum conditional log likelihood $l(\theta)$,

$$l(\theta) = \sum_{i=1}^{N}\log p(\mathbf{y}^{(i)} \mid \mathbf{x}^{(i)}) . \tag{3}$$

Take formula (1) into formula (3), we get

$$l(\theta) = \sum_{i=1}^{N}\sum_{t=1}^{T}\sum_{k=1}^{K}\lambda_k f_k(y_t^i, y_{t-1}^i, \mathbf{x}_t^{(i)}) - \sum_{i=1}^{N}\log Z(\mathbf{x}^{(i)}) . \tag{4}$$

In order to avoiding overfitting, a penalty term is involved, the formula (4) becomes into

$$l(\theta)=\sum_{i=1}^{N}\sum_{t=1}^{T}\sum_{k=1}^{K}\lambda_k f_k(y_t^i, y_{t-1}^i, \mathbf{x}_t^{(i)})-\sum_{i=1}^{N}\log Z(\mathbf{x}^{(i)})-\sum_{k=1}^{K}\frac{\lambda_k^2}{2\sigma^2}. \tag{5}$$

In formula (5), σ^2 determines the strength of the penalty. Finding the best σ^2 can require a computationally-intensive parameter sweep. Fortunately, according to [15], the accuracy of the final model does not appear to be sensitive to changes in σ^2. In our experiment, the σ^2 is set to 10. Given formula (5), we can use Improved Iterative Scaling (IIS) method or Numerical Optimization Techniques to find its maximum value and solve $\theta = \{\lambda_k\}$. We adopt L-BFGS [16] afforded by MALLET toolbox [17] to do that, which is an Numerical Optimization Techniques with high efficiency compared to IIS method. If $\theta = \{\lambda_k\}$ is available, we can use formula (1) to do NER.

For biomedical NER problem, the input sequence $\mathbf{x}$ is a sentence, the output sequences $\mathbf{y}$ is corresponding labels. The function set $\{f_k(y, y', \mathbf{x}_t)\}_{k=1}^{K}$ contains binary-value functions, which embody the features of the training data. For example $f_k(y, y', \mathbf{x}_t)$ may be defined as $f_k(y, y', \mathbf{x}_t)=\begin{cases}1 \ \textit{if WORD}_t\textit{=T}, \textit{WORD}_{t+1}\textit{=cells, y'=O, y=B-cell\_type}\\ 0 \ \textit{others}\end{cases}$.

4 Features

In order to describe the complexity language phenomena in biomedical literatures, we involve orthographical features, context features, word shape features, prefix and suffix features, Part of Speech (POS) features and shallow syntactic features. Compare to others exist biomedical NER system using CRFs, we first introduce shallow syntactic features in CRFs model. Shallow syntactic features are embodied using chunk labels (Therefore, chunk features and shallow syntactic features are same meaning in this paper). One of the most remarkable advantages of CRFs model is that it is convenient to involve rich features without considering the dependency of features. Also, when new features are added, the model doesn't need modification.

4.1 Shallow Syntactic Features

In order to get shallow syntactic features, we use GENIA Tagger [18] to do text chunking. Text chunking is the techniques of recognizing relatively simple syntactic structures. It consists of dividing a text into phrases in such a way that syntactically related words become member of the same phrase. These phrases are non-overlapping which means that one word can only be a member of one chunk [19]. After chunking, each token will be assigned a chunk label.

The syntactic information contains in chunk labels can afford much more reliable clues for NER than literal information. For example, a noun chunk is more likely to form an entity. In our research, shallow syntactic features include chunk labels with a window of size 5. If we use "c" denote a chunk label, -n denote n position prior to target token, +n denote n position after target token. The chunk features can be denoted as c-2, c-1, c0, c1, c2. Besides, some combined features are used in order to make full use of syntactic features. We employ 3 kinds of combined features: p-1c0, c0t0 and p0c0, where p denotes a POS tag and t denotes a token.

4.2 Other Features

Orthographical features: Orthographical features describe how a token is structured. For example, whether it contains both upper and lower letters, whether it contains digits and whether it contains special character. Orthographical features are important to biomedical NER for its special structures. We use regular expressions to characterize orthographical features which are listed in Table 2. Some of them are also used in [1, 9].

Table 2. Orthographical features

<table>
<tr><th>Feature name</th><th>Regular Expression</th></tr>
<tr><td>ALLCAPS</td><td>[A-Z]+</td></tr>
<tr><td>INITCAP</td><td>^[A-Z].*</td></tr>
<tr><td>CAPSMIX</td><td>.*[A-Z][a-z].*|.*[a-z][A-Z].*</td></tr>
<tr><td>SINGLE CHAR</td><td>[A-Za-z]</td></tr>
<tr><td>HAS DIGIT</td><td>.*[0-9].*</td></tr>
<tr><td>SINGLE DIGIT</td><td>[0-9]</td></tr>
<tr><td>DOUBLE DIGIT</td><td>[0-9][0-9]</td></tr>
<tr><td>NATURAL NUMBER</td><td>[0-9]+</td></tr>
<tr><td>REAL NUMBER</td><td>[-0-9]+[.,]+[0-9.,]+</td></tr>
<tr><td>HAS DASH</td><td>.*-.*</td></tr>
<tr><td>INIT DASH</td><td>-.*</td></tr>
<tr><td>END DASH</td><td>.*-</td></tr>
<tr><td>ALPHA NUMERIC</td><td>(.*[A-Za-z].*[0-9].*)|(.*[0-9].*[A-Za-z].*)</td></tr>
<tr><td>ROMAN</td><td>[IVXDLCM]+</td></tr>
<tr><td>PUNCTUATION</td><td>[,.;:?!-+]</td></tr>
</table>

Word shape features: Tokens with similar word shape may belong to the same category [13]. We come up with a simple way to normalize all similar tokens. According to our method, upper-case characters are all substituted by "X", lower-case characters are all substituted by "x", digits are all substituted by "0" and other characters are substituted by "_". For example, "IL-3", "IL-4" and "IL-5" will be normalized as "XX_d". Thus, there tokens can share the weight of feature "XX_d". To further normalize these tokens, we substitute all consecutive strings of identical characters with one character. For example, "XX_d" is normalized to "X_d".

Prefix and suffix Features: Some prefixes and suffix can provide good clues for NER. For example, tokens ending in "ase" are usually proteins, tokens ending in"RNA" are usually RNAs. In our work, the length range of affix is 3-5. If the length is too short, the distinguishing ability of affix will decrease. The frequency of the affix will be low if the length of affix is too long.

Context Feature: Tokens near the target token may be indicators of its category. For example, "IL-3" may belong to "DNA" or "protein". If we know the next token is "gene", we can decide that it belong to "DNA" category. According to [1, 9], we choose 5 as the context window size, i.e. the target token, the two tokens right prior to target token and the two tokens right after target token.

POS Features: The granule of POS features is larger than context features, which will help to increasing the generalization of the model. GENIA Tagger is used to do POS tagging. GENIA Tagger is a trained on biology literatures, whose accuracy is 98.20% as described in [18]. For POS features, we use the same window size as context features.

5 Experiment

5.1 Experiment Dataset

In the experiment, JNLPBA 2004 dataset is adopted. Its basic statistics is summarized in Table 3 and Table 4. Only 106 abstracts' publish year among 404 in test dataset are same as training dataset [3]. The difference in publish year between training data and test data demands the model should have a good generalization.

Table 3. Dataset of JNLPBA

dataset	#abs	#sen	#tokens
Training set	2,000	18,546	472,006
Test set	404	3,856	96,780

Table 4. Entity distribution in JNLPBA dataset

dataset	protein	DNA	RNA	cell_type	cell_line	All
Training set	30,269	9,533	951	6,718	3,830	51,031
Test set	5,067	1,056	118	1,921	500	8,662

5.2 Experiment Results

We use JNLPBA training set to train our model. Evaluation is done at JNLPBA test set. Training our model with all feature sets in section 4 took approximately 45 hours (3.0G CPU, 1.0G Memory, 400 iterations). Once trained, the model can annotate the test data in less than a minute. The experiment results are shown in Table 5. In Table 5, P, denoting the precision, is the number of NEs a system correctly detected divided by the total number of NEs identified by the system. R, denoting the recall, is the number of NEs a system correctly detected divided by the total number of NEs contained in the input text. $F = 2PR/(P+R)$ stands for the synthetic performance of a system.

Table 5. Experiment results

Entity Category	P (%)	R (%)	F (%)
protein	69.03	78.05	73.27
DNA	70.98	66.48	68.66
RNA	68.91	69.49	69.20
Cell_line	52.21	56.60	54.32
Cell_type	80.23	64.45	71.48
overall	70.16	72.27	71.20

Our system achieves F-measure of 71.20%, which is better than most of the state-of-the-art systems. Especially for protein, the most important entity category, our system's F-measure is 73.27%, which is much closer to the best system with F-measure 73.77% of protein in JNLPBA2004.

Table 6 shows our system's performance with different feature sets. The baseline feature set includes orthographical features, context features, word shape features and prefix and suffix features. These features are literal features and easy to collection. So they are often adopted by most biomedical NER system, such as [1, 9, 13]. POS features contain larger granule knowledge than literal feature. They can increase the model's generalization, so the F-measure increases to 70.33% from 69.52% when adding them into the model. Chunk features contain syntactic information which is more general linguistic knowledge than POS features. Involving shallow syntactic features can increase the performance from 70.33% to 71.20%. From Table 6, we can conclude that features contain large granule linguistic knowledge can prompt the CRFs model's generalization and get better results.

Table 6. The effect of different features set

Feature set	P (%)	R (%)	F (%)
Baseline	69.01	70.03	69.52
+POS features	69.17	71.53	70.33
+chunk features	70.16	72.27	71.20

In order to compare our work with others, Table 7 lists the performance of other systems adopting CRFs model and the state-of-the-art system. All results are tested in the same dataset, so they are comparable.

Table 7. Performance comparison

Number	System name	P (%)	R (%)	F (%)
1	Our system	**70.2**	**72.3**	**71.2**
2	Tzong-han Tsai(CRF)[1]	69.1	71.3	70.2
3	Settles et al., 2004 (CRF)[9]	69.3	70.3	69.8
4	Zhao, 2004[6]	69.4	76.0	72.6

System 3 only involves orthographical features, context features, word shape features and prefix and suffix features. Its performance is near to our baseline system. System 2 adds POS features and lexical features into system 1. Besides, system 2 adopts two post processing methods including Nested NE Resolution and Reclassification based on the rightmost word. But the F-measure of system 2 is still lower than our system with 1 percent. This also shows that syntactic features are effective in prompting the model's performance. System 4 is the state-of-the-art system in JNLPBA2004. But according to [6]□system 4 also need lexical resource and post processing. The F-measure of system 4 will below 70% if post processing is removed. Our system need not any lexical resource and post processing. It achieves good performance with good adaptability.

6 Conclusion

Conditional random fields for sequences labeling offer advantages over both generative models like HMM and classifiers applied at each sequence position. In this paper, we cast biomedical NER as a sequential labeling problem and utilize Conditional Random Fields model involving rich features to solve it.

The main contributions of this research are:

- ➢ First introduce shallow syntactic features to CRFs model and do boundary detection and semantic labeling at the same time. Experiment shows that shallow syntactic features greatly improve the model's performance.
- ➢ Show the effect of POS features and shallow syntactic features in detail; conclude that large granule linguistic knowledge can prompt the CRFs model's generalization, which can afford valuable reference to other researchers.
- ➢ Achieve a biomedical NRE system with an F-measure of 71.2% in JNLPBA test data and which is better than most of state-of-the-art system. The system has strong adaptability because it does not need any dictionary resources and post-processing.

References

1. Tsai, T.H., Chou, W.C., Wu, S.H., Sung, T.Y., Hsiang, J., Hsu, W.L.: Integrating Linguistic Knowledge into a Conditional Random Field Framework to Identify Biomedical Named Entities. Expert Systems with Applications. 30(1) (2006) 117-128.
2. Hirschman, L., Yeh, A., Blaschke, C., Valencia, A.: Overview of BioCreAtIvE: critical assessment of information extraction for biology. BMC Bioinformatics. 6(Suppl 1) (2005)
3. KIM, J.D., OHTA, T., TSURUOKA, Y., TATEISI, Y.: Introduction to the Bio-Entity Recognition Task at JNLPBA. In Joint Workshop on Natural Language Processing in Biomedicine and its Applications. (2004) 70-75
4. Kou, Z., Cohen, W.W., Murphy, R.F.: High-recall protein entity recognition using a dictionary. bioinformatics. 21(Suppl. 1) (2005) i266-i273
5. Cohen, A.M., Hersh, W.R.: A survey of current work in biomedical text mining. BRIEFINGS IN BIOINFORMATICS. 6(1) (2005) 57-71

6. Zhou, G.D., Su, J.: Exploring Deep Knowledge Resources in Biomedical Name Recognition. In Joint Workshop on Natural Language Processing in Biomedicine and its Applications. (2004) 96-99
7. Kazama, J., Makino, T., Ohta, Y., Tsujii, J.: Tuning Support Vector Machines for Biomedical Named Entity Recognition. In Proceedings of the ACL Workshop on Natural Language Processing in the Biomedical Domain. (2002) 1-8
8. Finkel, J., Dingare, S., Nguyen, H., Nissim, M., Manning, C., Sinclair, G.: Exploiting Context for Biomedical Entity Recognition: From Syntax to the Web. In Joint Workshop on Natural Language Processing in Biomedicine and its Applications. (2004) 88-91
9. Burr, S.: Biomedical Named Entity Recognition Using Conditional Random Fields and Novel Feature Sets. In Joint Workshop on Natural Language Processing in Biomedicine and its Application. (2004) 104-107
10. Sha, F., Pereira, F.: Shallow parsing with conditional random fields. In Proceedings of HLT-NAACL. (2003) 213-220
11. Shatkay, H., Feldman, R.: Mining the Biomedical Literature in the Genomic Era: An Overview. JOURNAL OF COMPUTATIONAL BIOLOGY. 10(6) (2003) 821-855
12. Yeh, A.S., Morgan, A., Colosimo, M., Hirschman L.: BioCreAtIvE task 1A: gene mention finding evaluation. BMC Bioinformatics 6(Suppl 1) (2005)
13. Tsai, T.H., Wu, C.W., Hsu, W.L.: Using Maximum Entropy to Extract Biomedical Named Entities without Dictionaries. In Proceedings of IJCNLP2005. (2005) 270-275.
14. Lafferty, J., McCallum, A., Pereira, F.: Conditional Random Fields: Probabilistic Models for Segmenting and Labeling Sequence Data. In Proceedings of the International Conference on Machine Learning. (2001) 282–289
15. Sutton, C., McCallum, A.: An Introduction to Conditional Random Fields for Relational Learning. http://www.cs.umass.edu/~mccallum/papers/crf-tutorial.pdf. (2005)
16. Wallach, H.M.: Efficient training of conditional random fields. Master's thesis. University of Edinburgh (2002)
17. McCallum, A.: MALLET: A Machine Learning for Language Toolkit. http://mallet.cs.umass.edu. (2002)
18. Tsuruoka, Y., Tateishi, Y., Kim, J.D.: Developing a Robust Part-of-Speech Tagger for Biomedical Text. In Advances in Informatics - 10th Panhellenic Conference on Informatics. (2005) 382-392
19. Erik, F., Sang, T.K., Buchholz, S.: Introduction to the CoNLL-2000 Shared Task: Chunking. In Proceedings of CoNLL-2000 and LLL-2000. (2000) 127-132

Spam Behavior Recognition Based on Session Layer Data Mining*

Xuan Zhang, Jianyi Liu, Yaolong Zhang, and Cong Wang

School of Information Engineering,
Beijing University of Posts and Telecommunications, Beijing, China, 100876
{zhangx, liujy, zyl, wangc}@nlu.caai.cn

Abstract. Various approaches are presented to solve the growing spam problem. However, most of these approaches are inflexible to adapt to spam dynamically. This paper proposes a novel approach to counter spam based on spam behavior recognition using Decision Tree learned from data maintained during transfer sessions. A classification is set up according to email transfer patterns enabling normal servers to detect malicious connections before mail body delivered, which contributes much to save network bandwidth wasted by spams. An integrated Anti-Spam framework is founded combining the Behavior Classification with a Bayesian classification. Experiments show that the Behavior Classification has high precision rate with acceptable recall rate considering its bandwidth saving feature. The integrated filter has a higher recall rate than either of the sub-modules, and the precision rate remains quite close to the Bayesian Classification.

1 Introduction

Knowledge Discovery has been broadly applied in Email Management and Spam Detecting [1], but until now applications here mainly concerned with mail body processing and user modelling. In this paper, we propose a framework using Decision Tree learning to find the patterns of spam transfer in session layer. Therefore, a Behavior Classification can be set up against spam flooding while saving the bandwidth.

Spam, officially called unsolicited bulk email (UBE) or unsolicited commercial email (UCE) [2], has long become a social problem and even brought serious threat to the Cyber Security. These annoying mails not only waste the end users' time to deal with them, but also consume large amount of bandwidth as well as enormous volume of server storage.

Several approaches have been proposed and typical server-side techniques exist in following three main categories:

Black/White-list filtering blocks bulk mail delivery during the transport process by checking blackhole list or white list when suspicious mail servers

* This work was supported by the National Natural Science Foundation of China under the Grant No. 60575034.

L. Wang et al. (Eds.): FSKD 2006, LNAI 4223, pp. 1289–1298, 2006.

connect [3]. This kind of techniques are quite simple and effective but somehow too "radical" to accommodate the exceptions.

Content-based Filtering targets the body of a mail message or the message header [3] using keywords matching or statistic methods. However, automatic software, often called "bulk mailers", usually add "innocent" content to outsmart content-based filters [1]. Furthermore, this kind of techniques can't prevent these unsolicited mails from taking up bandwidth.

Traffic monitoring technique, usually known as rate limiting, monitors and limits the mail traffic by observing the traffic flows of various hosts in a network [4], aims to defend mail server from large-scale spam attack. Impacts brought by junk mails mitigate but quantity of received spam not reduced.

Our approach is put in force during the email transfer session to recognize anomalous connections based on Behavior Classification which is set up through session log mining using Decision Tree algorithm. In this way, individual mail server can be more active in defending the spam flow according to its own "communication" history without getting the whole mail message and therefore not wasting bandwidth.

The rest of paper is organized as follows: In Section 2 we introduce related works and investigate whether the behavioral difference between spam and ham (normal mail) can be detected. Section 3 presents our approach in detail. Then behavior patterns discovered in a real mail server is shown in Section 4, effectiveness of the Behavior Classification (and also the integrated filter combined with Bayesian Classification) is evaluated here. Finally, we come to the conclusion in Section 5.

2 Related Works and Research Basis

To counter spam during the email delivery transaction is a new trend for the related research [5,6,7,8,9]. To detect abnormal email behavior in training phase, there are three typical approaches:

Man-Made Rule method: define classification model and rules manually through analyzing the possible characteristics in spam transfer.

Blacklisting method: query both a local black list file as well as Internet databases in real-time to defeat unsolicited traffic [9].

Filtering method: using statistical method to scan the data of mail to calculating spam probability of an email message [9].

These approaches are to some extent either unhandy or empirical. Our approach also focuses in the delivery phase, but applied at the mail server side in session layer and use Decision Tree learning in training phase. And it can be integrated with various server side techniques. Our opinion is based on the following reasons.

Email does not work so differently than it used to when it first appeared [1]. Sending email relies on SMTP(Simple Mail Transfer Protocol), including commands in Table 1 [10] during a typical interaction.

Table 1. Command Message in a typical SMTP Session

Command	Specification
HELO/EHLO	Sender's host name
MAIL	Sender of the messager
RCPT	Intended recipient of the message
DATA	Body of the mail
QUIT	Goodbye message

Unfortunately, SMTP is so simple that it can be easily faked and cheated. Spammers employ anonymous or compromise (spam trojan) server to mass mail junk mails without going through a specific mail server or a particular ISP, instead, they connect to the desination mail server directly or use a so-called open relay [3].

On the other hand, spam abuse the transferring protocol thus has distinct characters in behavior level, difference between bulk mailer and normal server is discriminated, for example, spammers send large amount of similar mails continuously and have to alter the message (e.g. domain info in command "HELO") at intervals to hide the track or spoof anti-spam filter while an ordinary mail server rarely. More details are given in Section 3.1.

3 Spam Behavior Recognition

3.1 Data Collecting and Pre-processing

The data collector works as an MTA (Mail Transfer Agent) so that command messages of each mail connection can be recorded. In fact, we add a Bayesian filter [11] to the collector for data pre-processing. Samples of the session data are listed in Table 2.

These items are almost all we can get during the SMTP session besides the user account and password, in which the item "IP" shows the mailer's IP address, "Domain" records the hostname of the mailer claimed well after the "HELO" command, "Sender" and "Receiver" represent the sender and receiver available in the "MAIL" and "RCPT" command respectively, and "Time" is the time stamp of each email.

The filter mentioned above checks the legitimacy of each mail first. And then we verify the result in manual additionally by looking through the subject of the mails (and only the subject items due to the privacy consideration). Finally, we make all these records clustered according to the "IP" field, and the abnormal behavior set of mail connections is separated from the alternative one.

What should be pointed out is that samples in the spam collections does not mean these spams were sent from the exact domains we list, they were just sent as such by spam trojan.

Table 2. Sample of Session Data

Legitimate Sessions

IP	Domain	Sender	Receiver	Time
202.108.3.174	smtp.vip.sina.com	-@vip.sina.com	-	2005-9-26 18:38:57
202.108.3.174	smtp.vip.sina.com	-@vip.sina.com	-	2005-9-27 16:54:03
202.108.3.174	smtp.vip.sina.com	-@vip.sina.com	-	2005-11-09 9:07:29

Spam Sessions

IP	Domain	Sender	Receiver	Time
59.64.128.128	sina.com	david@sina.com	joe@-	2005-3-25 13:10:28
59.64.128.128	pro.ro	joe@pro.ro	michael@-	2005-3-25 13:23:28
59.64.128.128	tom.com	brenda@tom.com	fred@-	2005-3-25 13:31:26
...	...	...	...	...
202.105.72.231	zx433dan	i8f9d9g1@microsoft.com	-	2005-3-25 16:22:15
202.105.72.231	zx738tuj	n1s2e9w8@msa.hinet.net	-	2005-3-25 17:16:26
202.105.72.231	MISS	Mk2o0l2h7@msn.com	-	2005-3-25 17:55:55

Note: due to the privacy issue, legitimate sender, receiver and domain of the target server is presented as "-" here.

3.2 Feature Selection

The feature selection is very important for building efficient decision tree [12]. Bulk mailers tamper with command messages in SMTP session to pretend innocently, therefore we'd better not learn the patterns of spamming behavior from each entry of the record directly, but clues lie here.

As shown in Table 2, mails from IP "59.64.128.128" were obviously spams for the reason that the mailer altered it's "domain" from "sina.com" to "pro.ro" in such a while, the possibility of a normal server to behavior like this is very little (normal server may modify this but the change is usually tiny), therefore we check the consistency of claimed domain from the same IP and named the feature as "ip2domain".

The feature "ip2domain", "ip2sender", "domain2sender" check the consistency of "domain", postfix of "sender"(e.g. sina.com) and the correlation between each other compared with the previous record of the same IP, as we have pointed out, normal servers behave consistently while bulk mailers send various inconsistent message pretending that they are innocent in every individual connections. It is easy to be cheated once but only once if consistency is checked.

"Open Relay" is a term used to describe mail servers that allow unauthorized users to send email through them [13], now it become a backdoor for spammers relaying bulk mails. Though most mail servers have forbidden open relay for the secure consideration, spammers usually intrude to the normal servers to modify the option to allow open relay. Feature "ip2receiver" is to detect the trick.

"Senderlth" and "Receiverlth" may not be noticeable but the pattern mining proved they are useful. Explanation is given in Section 4.1 .

Table 3. Behavior Features and Specifications

Feature	Specification
ip2domain	y, if "domain" consists with previous records' of the same IP ("consists" for short in following items); n, otherwise
ip2sender	y, if postfix in "sender" consists; n, otherwise
ip2receiver	y, if postfix in "receiver" is correct; n, otherwise (check if "open relay")
domain2sender	y, if "domain" associate with postfix in "receiver" consists; n, otherwise
senderlth	length of characters before "@" in "sender"
receiverlth	length of characters before "@" in "receiver"

In fact, the malicious mailer also behave abnormally that it often send mails to email addresses not exist in the target server, like data in Table 2, all receivers of mails sent from "59.64.128.128" are nonexistent in our mail server. This is one of the ways spammers collect their "Mailing List". We do not take the authenticity of target address as feature in our training phase because the validation of receiver can be accomplished by the mail server and this kind of connections will not take up much bandwidth (but may imply the mail attack). Besides, it is not worth to maintain a list of valid addresses in the Behavior Recognition phase and that will also reduce the flexibility of the Behavior Classification. Servers that attempt to make nonexistent-receiver mailings frequently can be simply added into the black list.

In addition, frequency of connections from the same IP would be taken as a feature in situation that the throughput of mail server maintains in high level.

4 Experimentation

The total of the terms is 27500, 15000 for training, 2500 for pruning and 10000 for testing. For comparison, we also evaluate the integrated effectiveness of our Behavior Classification and Bayesian Filtering [11].

4.1 Pattern Mining and Analysis

Decision tree is chosen because they provide a comprehensible representation of their classification decisions [14]. C4.5 Decision Tree Algorithm, which has been widely used and tested [15], is applied to the training set to generate the decision tree. The Decision Tree retrieved is often a very large, complex that overfits the training data and incomprehensive to experts. It's impossible to avoid noise in the training set, then overfitting the tree to the data in this manner can lead

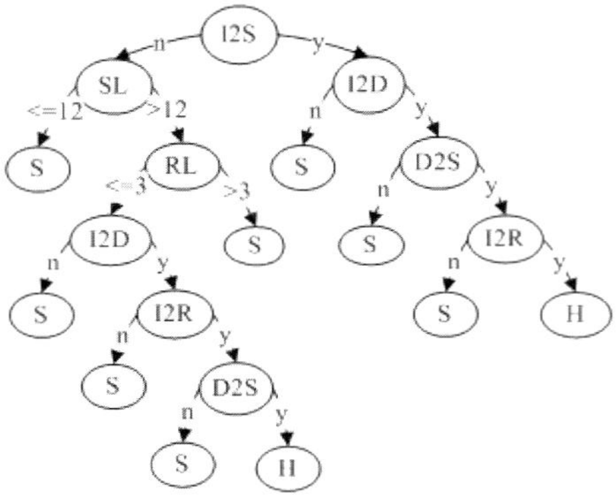

Fig. 1. Decision Tree after pruning. Nodes Specification(detail can be find in Table 3): I2S:ip2sender, SL:senderlth, RL:receivelth, I2D:ip2domain, I2R:ip2receiver, D2S:domain2sender, S:Spam, H:Ham.

to poor performance on unseen data [16]. Therefore, we pruned the tree in C4.5 algorithm firstly and then in manual way. Figure 1 shows the decision tree which was generated and pruned for characterizing the attribute post test.

From the Decision Tree, we can find:

1. Left branch of the Tree is the main branch, which contains approximately 58.13% of the whole entries.

2. The significance of each feature can be sorted as below: ip2sender, senderlth, receiverlth, ip2domain, ip2receiver, domain2sender.

3. The tree generated above may include some phenomena that are not noticeable. Length of email address are rarely taken into consideration in the anti-spam topic, however, the tree shows that length of mail sender and receiver can be valuable for classifying mail connections. For example, "senderlth $\leqslant$ 12", which can be explained that generally legitimate sender don't apply for long-name accounts from their Email Service Provider while spammers like to adopt such long names ("wdjhzfdomkvyotv" e.g.) to "produce" more different email addressed randomly. And "receiverlth $\leqslant$ 3", that's because the mail server under protected provides email accounts according to the initial letter of employees' name (Chinese name usually contains two to three characters). It is implied that the Data Mining approach do help discover unrevealed knowledge hide in behavior record.

4. The tree may differ from one mail server and another, from time to time after long intervals, so that new patterns of spam behavior can be detected dynamically. This approach of spam pattern learning can be "personalized" to every single mail server individually.

4.2 Evaluation

The integrated framework is illustrated in Figure 2, dashed frame presents the Behavior Recognition Process given in Section 3. Mail stream is filtered through Anti-Dos module, IP block module, Behavior Classification and Bayesian

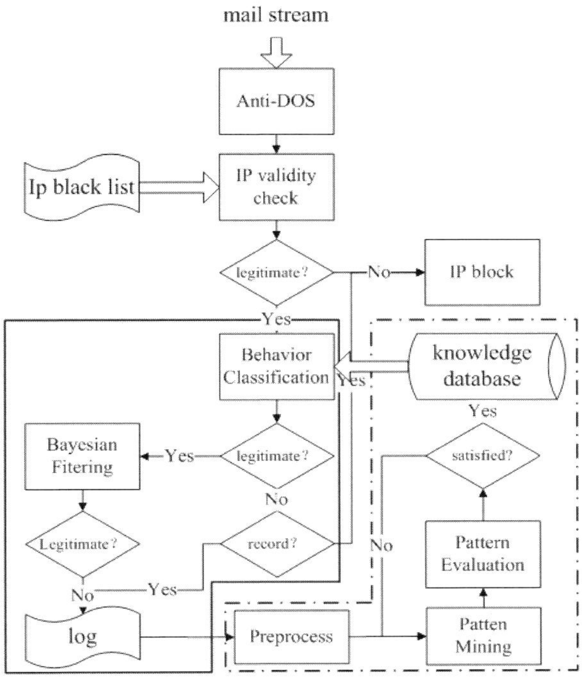

Fig. 2. Integration of Two Classfication

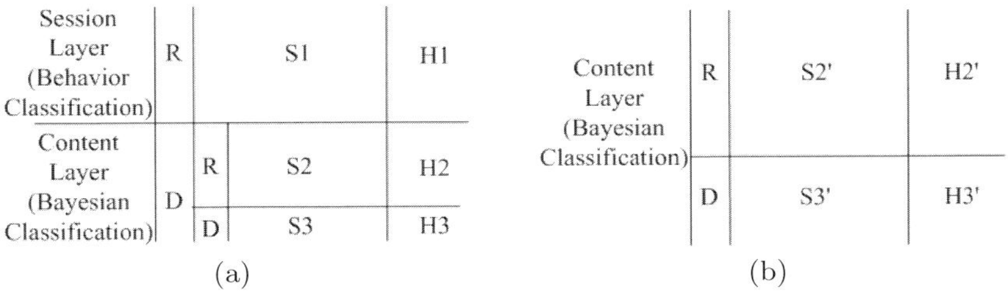

Fig. 3. Recall and Discharge. (a) Integrated Classification (b) Bayesian Classification R denotes Recall, D denotes Discharge while S denotes Spam, H denotes Ham.

Filtering. For evaluation, Anti-Dos and IP block module are turned off, and the Behavior Classification and Bayesian Filtering Module are evaluated individually and integratively.

From Figure 3, recall rate R and precision rate P can be calculated. Results are listed in Table 4. Binary tree traversal algorithm is used to present the Decision Tree generated above in Figure 1, judging whether a mail session is legitimate or not in classification implementation [17].

Table 4. Evaluation comparing with Bayesian Filtering

Items	Behavior Classification(%)	Bayesian Classification(%)	Integration(%)
Precision	98.00	92.03	91.71
Recall	53.27	73.47	78.32

4.3 Behavior Classification Evaluation

According to the experiment, the precision rate (98.00%) of Behavior Classification is quite high while the recall rate (53.27%) is relatively low. As mentioned above, the behavior patterns of spam will alter with time going, and bulk mailer would also adjust its behavior against the anti-spam's strategy. So, to always recall most spam patterns is ideal but hard to realize, and we should always emphasize on the precision rate taking the quality of mail service in consideration.

On the other hand, the advantage of Behavior Recognition in the session layer is that it's time saving and can protect bandwidth resource against junks. In this aspect, more than 50% of the malicious connection be detected meet the requirement in reality. By adopting strategies such as reject connection or rate limiting, the mail server being protected can benefit a lot from Behavior Classification.

4.4 Integrated Evaluation

The evaluation of the integrated filtering focus in the real line frame in Figure 2. We select Bayesian Classification in email content to be integrated with Behavior Recognition, not only because they work in different phases of email transfer, but also for the reason that the Bayesian filtering usually have high recall rate.

As shown in Table 4, the integrated recall rate is higher compared to either of the classification individually, but precision rate is 0.32% lower than the Bayesian Classification (the lower one). In fact, for the two layer filtering, there are (Consist with the Figure 3, subscript 1 for Behavior Classification, 2 for Bayesian Classification individually, 2'for Bayesian Classification works after the Behavior Classification and *intg* for the integrated filter):

$$\max(S_1, S_2') \leqslant S_1 + S_2 \leqslant \min(S_1 + S_2 + S_3, S_1 + S_2') \tag{1}$$

$$\max(H_1, H_2') \leqslant H_1 + H_2 \leqslant H_1 + H_2' \tag{2}$$

So, we can get:

$$\frac{1}{1 + \frac{H1 + H_2'}{\max(S_1, S_2')}} \leqslant P_{intg} \leqslant \frac{1}{1 + \frac{\max(H_1, H_2')}{\min(S_1 + S_2 + S3, S_1 + S_2')}} \tag{3}$$

$$R_{intg} \geqslant \max(R_1, R_2) \tag{4}$$

Taken high precision (98.00% here) of the front Module (Behavior Classification), the integrated precision is bounded from 90.82% to 94.03% according to Equation 3 in our experiment. And the recall rate would be higher than $\max(R_1, R_2)$, helping to save much of time and bandwidth. The performance tally with our expectation that the Behavior Classification can be integrated with current various anti-spam techniques.

5 Conclusion

In this paper, an approach is given to discover different patterns of Spam Behavior in transfer phase using Data Mining method, upon which a Behavior Classification was set up. Evaluated in experiment, the Behavior Classification, with high precision, contributes much to the bandwidth saving during email delivery.

Obviously, no single approach can be expected to expire all those unsolicited mails so far. Behavior Classification can be applied in ordinary mail servers and integrated with currently used spam filtering and blocking techniques.

Nevertheless, for the spammer behavior change dynamically, the Behavior Classification needs to be renewed periodically accordingly.

For future works, incremental learning of Behavior Patterns for the classification need to be realised, pattern incorporation and the frequency of learning are involved. Classifying spam connections in a finer granularity should also be considered, such as to distinguish the virus mail stream from common advertisement mail streams. Further more, it is a exacting work for individual mail server to counter the infestation of spam, cooperating with other servers is an essential trend for all anti-spam systems. By sharing Email Behavior Patterns among spam filters will make the Behavior Classification working more effectively.

References

1. Ioannis Katakis, Grigorios Tsoumakas, I.V.: Email Mining: Emerging Techniques for Email Management. In: Web Data Management Practices: Emerging Techniques and Technologies. Idea Group Publishing (2006) 32
2. Flavio D. Garcia, Jaap-Henk Hoepman, Jeroen van Nieuwenhuizen: Spam Filter Analysis. In: Proc. 19th IFIP International Information Security Conference, WCC2004-SEC, Toulouse, France, Kluwer Academic Publishers (2004)
3. Lueg, C.: Spam and anti-spam measures: A look at potential impacts. In: Proc. Informing Science and IT Education Conference, Pori, Finland (2003) 24–27
4. Anti-Spam Technologies: Anti-Spam Technology Overview http://e-com.ic.gc.ca/epic/Internet/inecic-ceac.nsf/en/gv00297e.html#3.4.3
5. Salvatore J. Stolfo, Shlomo Hershkop, K.W., Nimeskern, O.: Emt/met: Systems for modeling and detecting errant email. In: Proc. DARPA Information Survivability Conference and Exposition. Volume 2. (2003) 290–295
6. Prasanna Desikan, J.S.: Analyzing network traffic to detect e-mail spamming. In: Proc. ICDM Workshop on Privacy and Security Aspects of Data Mining, Brighton UK (2004) 67–76
7. Qiu Xiaofeng, H.J., Ming, C.: Flow-based anti-spam. Proc. IEEE Workshop on IP Operations and Management (2004) 99–103

8. Banit Agrawal, Nitin Kumar, M.M.: Controlling spam emails at the routers. In: Proc. International Conference on Communications. Volume 3., Seoul, South Korea (2005) 1588–1592
9. Tran, M.: Freebsd server anti-spam software using automated tcp connection control. Technical report, CAIA Technical Report 040326A (2004)
10. Forouzan, B.A., Gegan, S.C.: TCP/IP Protocol Suite. McGraw-Hill (2000)
11. Jianyi Liu, Yixin Zhong, Y.G., Wang, C.: Intelligent spam mail filtering system based on comprehensive information. In: Proc. 16th International Conference on Computer Communication. (2004) 1237–1242
12. Tarek Abbes, Adel Bouhoula, M.R.: Protocol analysis in intrusion detection using decision tree. In: Proc. International Conference on Information Technology: Coding and Computing. Volume 1., Las Vegas, Nevada (2004) 404–408
13. Glossary:Open Rlay http://www.viruslist.com/en/glossary?glossid=153949388
14. Mitchell, T.: Machine Learning. McGraw-Hill (1997)
15. Quinlan, J.R.: C4.5: Programs for Machine Learning. Morgan Kaufmann, San Mateo, CA (1993)
16. James P. Early, C.E.B., Rosenberg, C.: Behavioral authentication of server flows. In: Proc. 19th Annual Computer Security Applications Conference, Las Vegas, Nevada (2003) 46–55
17. Zhang, Y.: Research and application of behavior recognition technology in anti-spam system. Master thesis of Beijing University of Posts and Telecommunications (2006)

A Face Detection Using Multiple Detectors for External Environment

Mi Young Nam and Phill Kyu

Dept. of Computer Science & Engineering, Inha University
253, Yong-Hyun Dong, Incheon, Korea
rera@im.inha.ac.kr, pkrhee@inha.ac.kr

Abstract. We propose a method of multiple context fusion based robust face detection scheme, multiple cascade and finally decision using correlation table. It takes advantage of multiple cascade face detector fusion by context. We propose the filtering classifier method for illumination face image. And then we constructed cascade classifier from applied different filtering method. The multiple cascade detectors made from six single context detectors. Six contexts are divided k-means algorithm, and classify illuminant. In this paper, we proposed the classifier fusion method by using correlation between face images. The proposed face detection achieves the capacity of the high level attentive process by taking advantage of the context-awareness using the information from illumination. We achieved very encouraging experimental results having varying illuminant.

1 Introduction

The face detection technology involves automatically detecting the position and the area of face from a random scanned picture or digital video data. The algorithm of this technology searches for a limited set of facial features, but the features appear to be different from a case to case due to various conditions of the real world [1, 2, 3, 4]. Context, in this paper, is modeled as the effect of the change of application working environment. The context information used here is illumination. As environment context changes, it is identified by the multiple context fusion, and the detection scheme is restructured [5, 6, 7, 8]. The goal of this paper is to explore the possibility of environment-insensitive face detection by adopting the concept of multiple context fusion. According to the environmental change, the face image is variously expressed [3, 5]. Therefore, in this paper, we propose the face detecting method using several contexts for robust face detection in the various environmental changes. And, we proposed the multiple face detectors to fuse each the context face recognition result. The multiple face detectors are comprised of several detectors. And the single face detector has the cascade face detection structure. The cascade face detector is comprised of two face detection units. We employ K-means for the context modeling and awareness. It provides the capacity of the high level attentive process by the environment context-awareness in face detection. We achieve very encouraging results from extensive experiments.

L. Wang et al. (Eds.): FSKD 2006, LNAI 4223, pp. 1299 – 1302, 2006.

2 Face Detection Architecture Using Multi-context

2.1 Learning

In this session, we make the correlation table and cascade classifier from training face images. The correlation tables decide final detection result from cascade face detector. Clustering method is used K-means algorithm, we divided six class. And we make the classifier for context. First classifier practice histogram equalization and second classifier practice the contrast stretching. Because we divided the face for illumination, we applied different preprocessing filter.

2.2 Testing

Fig. 1 shows our face detection flowchart. In testing step, we apply cascade classifier and correlation table. Therefore, we improve the reliability by using correlation table between contexts. Environment context is analyzed and identified using the K-means. Input pattern is vectorized for image size of 20x20 sizes; input node had size of 10x10 pixels. That image is converted 1x100 dimension vectors. Context shows images of three clusters various illuminant face dataset, we define 3 step illuminant environment EQ(1). We use probability of face is true and probability of face is false.

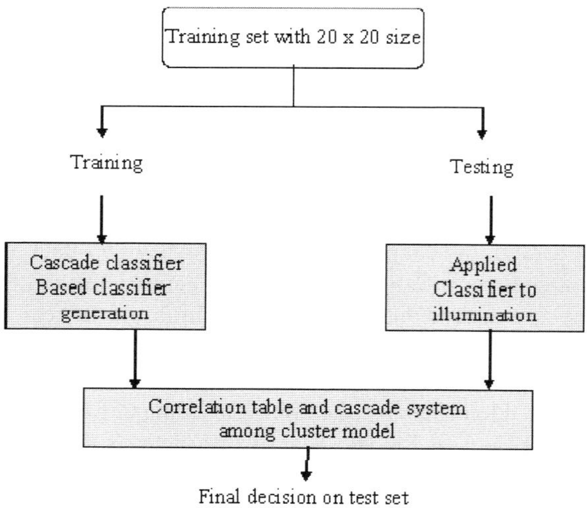

Fig. 1. The proposed face detection architecture

$$P(X \mid True) = \exp(-0.5 * ((x - \mu \mathbf{T})/\delta \mathbf{T})^{\wedge}2) / (\sqrt{2\pi} * \delta \mathbf{T}) \quad (1)$$

3 Experiment

The experiment of the proposed method has been performed with images captured in various environments 100 images are captured and used in the experiment. Inha data set include 100 of person. In CMU database, total image number is 130 images including 507 faces. Tables compare the detection rates (AR) for the context-based color system and the umbers of context-based Bayesian classifiers for the three boosting-based systems, given the number of false alarms (FA). And Table 1 shows that the result of face detection between multiple context based Detector and context based multiple cascade detector fusion in CMU testset1. Table 2 shows that the result of face detection between multiple context based Detector and context based multiple cascade detector fusion in CMU testset2. We know that propose method is good face detection performance other method. Also, number of context very closes, and different face detection ratio. We could improve illuminant face detection performance by using cascade detector combination.

Table 1. Face detection result of proposed face detector in CMU testset1

Face detection according from face context number					
Nine face context		Five face context		Proposed architecture	
AR	FAR	AR	FAR	AR	FAR
88.75	59	90.0	62	92.80	24
91.70	68	92.3	70	94.08	35

Table 2. Face detection result of proposed face detector in CMU testset2

Face detection according from face context number					
Nine face context		Five face context		Proposed architecture	
AR	FAR	AR	FAR	AR	FAR
84.6	3	90.3	7	94.9	2
86	14	91.2	15	96.1	10

The combined cascade face detector is also investigated. In this experiment, the factor illumination was considered and experimental images were classified by the actor of illumination. We classified bad illumination images into the image including a partially lighted face, good images into that including a nearly uniformly lighted face.

4 Conclusion

The context information used here is illumination. We generated cascade face detector for each context. And then we combine face detection result from each face detector. We could improve the face detection by using cascade face detector combination. As environment context changes, it is reliability. The detection scheme aims at robustness as well as fast execution way under dynamically changing context. In this paper, we has been resolved by employing K-means for divided context illuminant group. The proposed face detection achieves the capacity of the high level attentive process by taking advantage of the illumination context-awareness. Experimental result has shown that the proposed system has detected the face successfully 96.1% of CMU testset2.

References

[1] Ludmila I. Kuncheva, Switching Between Selection and Fusion in Combining Classifiers: An Experiment, IEEE Transactions on Systems, Man, and Cybernetics - part B: cybernetics, Vol.32, Vo.2, (2002) 146-156

[2] Ludmila I. Kuncheva, ,A Theoretical Study on Six Classifier Fusion Strategies IEEE S on PAMI, Vol. 24, No. 2, (2002)

[3] L. I. Kuncheva, J. C. Bezdek, and R. P. W. Duin, "Decision templates for multiple classifier fusion: An experimental comparison," Pattern Recognit., Vol. 34, No. 2, (2001) 299–314

[4] Y.S. Huang and C.Y. Suen, "A Method of Combining Multiple Classifiers—A Neural Network Approach," Proc. 12th Int'l Conf. Pattern Recognition

[5] S. Yau, F. Karim, Y. Wang, B. Wang, and S. Gupta, Reconfigurable Context-Sensitive Middleware for Pervasive Computing, IEEE Pervasive Computing, Vol.1, No.3, (2002) 33-40

[6] M.Y. Nam and P.K. Rhee, An Efficient Face Recognition for Variant Illumination Condition, ISPACS2005, Vol.1, (2004) 111-115

[7] M.Y. Nam and P.K. Rhee, A Novel Image Preprocessing by Evolvable Neural Network, LNAI3214, Vol.3, (2004) 843-854

[8] Kuncheva L.I. and L.C. Jain, Designing classifier fusion systems by genetic algorithms, IEEE Transactions on Evolutionary Computation, Vol.4, No.4, (2000) 327-336

An Intelligent Decision Support System for IT Outsourcing

Gülçin Büyüközkan* and Orhan Feyzioğlu

Department of Industrial Engineering, Galatasaray University,
Çırağan Caddesi No: 36, Ortaköy 34357 İstanbul-Turkey
Phone: +90 212 227 4480; Fax: +90 212 259 5557
{gbuyukozkan, ofeyzioglu}@gsu.edu.tr

Abstract. Outsourcing information technology (IT) is a major contemporary strategic decision. This paper proposes an intelligent decision support framework for effective IT outsourcing management. The proposed framework uses case-based reasoning as the main intelligent technique and integrates rule-base reasoning and compromise programming techniques in fuzzy environment for a real-time decision-making. While integrating different methodologies, our motaivation is to take the advantage of their strengths and cancel out each other's weaknesses. The framework potentially leads to more accurate, flexible and efficient retrieval of alternatives that are most similar and most useful to the current decision situation. Finally, a real-life case is given to validate the feasibility of the proposed framework.

1 Introduction

Throughout the past 25 years of rapid advances in information technology (IT), most businesses have realized the strategic role and competitive advantage IT can provide to an organization. Many companies see opportunity to cut IT costs while still maintaining the benefits of technology by downsizing the IT function and outsourcing it to firms that specialize in operating IT efficiently and reliably. Outsourcing IT can include mainframe and data centers, wide area networks, applications development and maintenance functions, end-user computing and business processing [10]. A decision to outsource is driven by numerous factors, from a desire for cost-cutting or managing legacy systems, to the desire to focus on the core business, enable rapid business change or expansion, or obtain strategic advantage by keeping up with ever changing technology [9, 16, 17, 27]. Two themes in the IT outsourcing research have attracted interest among the researchers: (1) the reasons for, the benefits and risks of outsourcing decision (make or buy), (2) the selection of a partner for the outsourcing relationship. This study focuses on the latter. Prior research discusses partner or supplier selection in various ways. The most common approaches and methods for supplier evaluation include different multi criteria decision-making (MCDM) methods such as analytic hierarchy process [14] and analytic network process [15], statistical techniques such as principal components analysis and factor analysis [4],

* Corresponding author.

L. Wang et al. (Eds.): FSKD 2006, LNAI 4223, pp. 1303 – 1312, 2006.

data analysis techniques such as cluster analysis, discriminant analysis, data envelopment analysis [20] and simulation [24]. In this study, we develop an intelligent decision support (IDS) framework based on case-based reasoning (CBR) in fuzzy environment for effective IT supplier evaluation and selection. The proposed framework also integrates rule-base reasoning (RBR) and compromise programming techniques in fuzzy environment to deal with uncertain and imprecise decision situations. CBR is a good concept to transform the evaluation know-how from experts into a computer-assessment method to make the evaluation process more convenient and efficient [6]. It can capture all 'memory' of human being without losing them due to lapse of time and carelessness. Data processing procedure is the heart of CBR system, which guides the way to succession. Lack of a good information flow could let the system misunderstand how to store the cases and make use of them for decision-making [5]. In addition, CBR method alone in context-aware systems may not work as expected because it is highly possible to have a high number of items or variables to consider. That is, as the system becomes more realistic, the number of items, consequently the volume of the context tends to increase. Moreover, as the system increases the sensitivity, more contextual information can be newly involved. In that case, the number of criteria will exponentially increase, which will adversely affect system feasibility and hence performance. Thus, some supplemental methods to determine weights among the items need to be combined with the CBR method [7, 8]. For this reason, fuzzy logic and RBR concepts are merged into proposed CBR system. Fuzzy logic is a formal tool eminently suited for solving problems with imprecision inherent in empirical data [11]. RBR is a natural knowledge representation, in the form of 'If…Then.' structures [19]. It consists of an inference engine and assertion, which is employed for interpreting sets of facts and rules. In order to simultaneously improve searching speed and accuracy, we also integrate a particular multi criteria decision-making technique, more precisely compromise programming, with learning methodologies. Common tasks involved in these methodologies are considered and the manner of how to combine different techniques to build an IDS model is explored. The remainder of the paper is organized as follows. Section 2 describes the main points of the proposed IT outsourcing decision support model while Section 3 presents its implementation in a Turkish company. Concluding remarks are given in Section 4.

2 The Proposed IT Outsourcing Decision Support System

2.1 The Proposed Framework and Utilized Techniques

In the IT outsourcing decision support system, the description of previous IT outsourcing evaluation and selection processes and results, serves as the central link of decision support for the decision maker. The framework is constructed to assist IT customers on formulating IT outsourcing decisions involves three main sub-systems. Thus, customers define their service and strategic preferences through a hybrid CBR, RBR and fuzzy compromise programming approach. The IDS system's initial input is the IT service request of customer. It aims to eliminate the unnecessary data, which is not valuable for the given request. The usefulness of data is defined by its similarities to the customer requirements. Thus, our framework primarily requires that the customer defines the degree of importance of the strategic level evaluation criteria. This is the distinguishing part of our model, since many CBR and RBR applications [5, 7] require case features in this stage of the decision-making problem instead of customer preferences. The importance weights of the strategic level criteria are evaluated by the customer and then carried as input to the first rule-based sub-system. This sub-system enables an initial elimination of the cases in the case base regarding their similarities to the requested case considering their strategic ratings.

In a second elimination, the customer's IT service specifications, namely the requested case features are considered. In a similar fashion, the remaining cases in the case base are reexamined to remove the ones that are irrelevant. For instance, this second rule-based sub-system prevents dealing with mainframe service cases when the customer looks for an IT management and support service.

The final assignment before entering to the CBR cycle is the determination of importance of IT business service levels, which will be the input of the CBR sub-system. The first stage of the CBR cycle is the representation of the cases and, the next stage involves the measurement of similarities between the requested (new) case and each case already in the case base. Our framework makes use of the fuzzy compromise programming approach to calculate these similarities. The case(s) with the minimum distance value is (are) the most similar case(s) to the new case; therefore recently proposed solutions can be applied to the new problem. The CBR cycle ends with the adaptation of this case, if confirmed, into the case base.

Before giving the evaluation criteria of the proposed framework, techniques used in IDS system are briefly presented in the following sub-sections.

2.1.1 Rule-Based Reasoning (RBR) and Case-Based Reasoning (CBR)

RBR and CBR are two of the problem solving methodologies of Artificial Intelligence (AI). CBR, rooted in early 1980s, provides a theoretical basis and application method for human analogy, with many distinguishing features from other major AI methods. The main characteristic is that CBR enables to make use of the specific knowledge by remembering a previous similar situation and by reusing information and knowledge of that situation. It is an alternative approach to RBR where the knowledge is expressed with the rules. On contrary, major AI techniques rely solely on general knowledge of a problem domain or they establish generalized relationships between problem descriptors and conclusions [1]. As a result, CBR is recommended to developers trying to reduce the knowledge acquisition task, avoid repeating mistakes, learn over time, and maybe most significantly reason with incomplete or imprecise data [21]. It is difficult to assume that CBR is a concept completely different from RBR. In fact, some researchers present CBR as a specific type of RBR [5], while others define RBR as an alternative to CBR [18]. RBR represents knowledge using "If-Then" rules. This characteristic of RBR systems renders them poor at dealing with vague nature inherent in the IT service provider selection. The most significant difference between the CBR and RBR processes is their performance in learning duration curve. Since in CBR, the decisions are made according to previous cases, the system requires an amount of time for accumulating a sufficient number of cases. However, the performance of the RBR is always the same, as the decisions are made concerning only the predetermined rules [5]. The initial step of a CBR system is the representation of cases. A case should contain the composition of the problem, its solution and the outcome; namely both the content and the context. The case is a data format containing words, numbers and symbols to represent solutions and a state of affairs [8]. The cases are stored in a case base and each case is represented with a set of attributes for an effective storage during matching process. Another advantage of using discrete and standard attributes is that they preclude the subjectivity when defining a case. These attribute values are useful during the case retrieval stage when analyzing the similarities between the new case and the old ones. Among the steps of the CBR cycle, case retrieval is the key challenge, since a CBR system is not a valuable methodology without effective case retrieval. The similarity measure used to quantify the degree of resemblance between a pair of cases plays a very important role in case retrieval. Hence, CBR systems are sometimes called similarity searching systems [18]. In this study we propose to use the compromise programming technique as a distance-based approach for similarity identification.

2.1.2 Compromise Programming

Initially proposed by Zeleny [28], compromise programming is a distance-based MCDM approach and it is viewed as an effort to approach or emulate the ideal solution as closely as possible. It defines a metric as the distance of the alternatives from an ideal solution, where each alternative under consideration reaches its optimum value. Consequently, a rational decision maker should select the alternatives that are closer to the defined ideal solution than those that are farther. Recently, a compromise ranking method (known as VIKOR) is introduced as one applicable technique to implement within MCDM [23]. VIKOR method provides a maximum group utility for the majority and a minimum of an individual regret for the opponent. It introduces the multi-criteria ranking index based on the particular measure of closeness to the ideal solution. The details of this method can be found in [23, 26]. Within the VIKOR method, the distance is formulated as below

$$S_j = \sum_{i=1}^{n} w_i \left(f_i^+ - f_{ij}\right) / \left(f_i^+ - f_i^-\right), \tag{1}$$

$$R_j = \max_i \left[w_i \left(f_i^+ - f_{ij}\right) / \left(f_i^+ - f_i^-\right) \right], \tag{2}$$

$$Q_j = v\left(S_j - S^+\right) / \left(S^- - S^+\right) + (1-v)\left(R_j - R^+\right) / \left(R^- - R^+\right) \tag{3}$$

with $S^+ = \min_j S_j$, $S^- = \max_j S_j$, $R^+ = \min_j R_j$, $R^- = \max_j R_j$. Here, w_i are the associated weights of each of the objectives i; f_{ij} is the objective value of the jth alternative in ith objective; f_i^+ and f_i^- are the best and worst possible solution of the alternatives in the objective space; S_j and R_j values represent the average and the worst group scores for alternative j respectively. Finally, v is the weight of the decision-making strategy "the majority of criteria" (or "the maximum group utility"). The compromise Q_j can be selected with "voting by majority" ($v > 0.5$), with "consensus" ($v = 0.5$), or with "veto" ($v < 0.5$). Incorporating fuzzy arithmetic within the general framework of composite programming necessitates the use of fuzzy numbers for the ideal and anti-ideal points, as well as, to the outcomes of the objective functions. Consequently, the equations 1-3 are written as follows,

$$\tilde{S}_j = \sum_{i=1}^{n} \tilde{w}_i \left(\tilde{f}_i^+ - \tilde{f}_{ij}\right) / \left(\tilde{f}_i^+ - \tilde{f}_i^-\right), \tag{4}$$

$$\tilde{R}_j = \max_i \left[\tilde{w}_i \left(\tilde{f}_i^+ - \tilde{f}_{ij}\right) / \left(\tilde{f}_i^+ - \tilde{f}_i^-\right)\right], \tag{5}$$

$$Q_j = v\left(\tilde{S}_j - \tilde{S}^+\right) / \left(\tilde{S}^- - \tilde{S}^+\right) + (1-v)\left(\tilde{R}_j - \tilde{R}^+\right) / \left(\tilde{R}^- - \tilde{R}^+\right). \tag{6}$$

In our case, we try to find the alternative with the minimum distance to the ideal point, thus, we will choose the smallest Q_j value among the ones obtained by equation 6. The fuzzy subtractions used in this formula are calculated using the distance formulation proposed in [3].

2.2 IT Supplier Evaluation Criteria

A set of evaluation criteria has to be defined in advance to determine the IT service provider who offers the best all-around package of products and services for the customer. Traditionally suppliers focused on a technical output evaluation, in terms of quality, delivery speed and reliability, price offered, but when the relationship becomes closer and longer, the number of selection criteria increase, and suppliers are selected on their global performances. Global evaluations range from total costs analysis to the consideration of supplier's capacity, their future service capability or the closeness of the relation and continuous improvement capabilities. In strategic evaluations technological, financial and organisational capabilities are considered together with technological and strategic coherence [2, 12, 13, 22, 25].

Based on these emphasized different factors, two groups of evaluation criteria are determined and used in our proposed framework. The first group focuses on the strategic aspects of the IT service provider companies and identifies them as follows [25]: financial stability, successful track record, similar size, comparable culture, similar values and goals, and fit to develop a sustainable relationship. The second group of evaluation criteria is developed to measure important aspects of the supplier's business in five main groups: technical expertise, performance, quality, total cost and intangibles. By using these two groups of criteria, we created an IT service provider evaluation system as explained in detail in the case study.

3 A Case Study

The XYZ Company is a part of an important group in Turkey, which consists of 14 companies, each spread over some sectors. In 2005, the board of directors decided that the companies should outsource their IT service needs to reap various benefits, from cost savings to increased flexibility, and from improved quality of services to better access to state-of-the-art technology. For this reason, a prototype tool is developed to solve the IT outsourcing decision problem of this company. We proposed to establish a common knowledge base, which will benefit from the past experiences of each company. Such a knowledge base would provide companies the information necessary when signing contracts with IT firms.

3.1 Case Representation

Case representation and organization is the key part of a CBR and RBR system in each stage of the methodology. A case includes not only the concrete features such as its category, its region or its duration but also strategic and business related features. Each case should be defined in respect to three main categories of features:

- *General case features:* Main service category, Sub service category, Service scale, Technical capacity, Price, Duration, Personal qualification, Region.
- *Strategic case features:* Similar values and goals, Similar size, Financial stability, Comparable structure, Successful track record, Fit to develop a sustainable relationship.
- *IT business case features:* Technical expertise, Performance, Total cost, Quality, Intangibles.

3.2 Identification of Strategic Evaluation Criteria and Case Features

In the following part of the section, we have chosen an illustrative example to clearly show the application of the model with the developed tool. The tool is made as user-friendly as possible, to facilitate its use to personnel from different backgrounds. Figure 1 depicts the initial front end where the customer is expected to give his/her IT service requests' importance degree. At this point, we assume that the decision makers (customers in our case) use linguistic terms in the following set to express the importance of criteria: W = {Absolutely Important, Very Strongly Important, Strongly Important, Moderately Important, Important}. These weights are then quantified with fuzzy numbers.

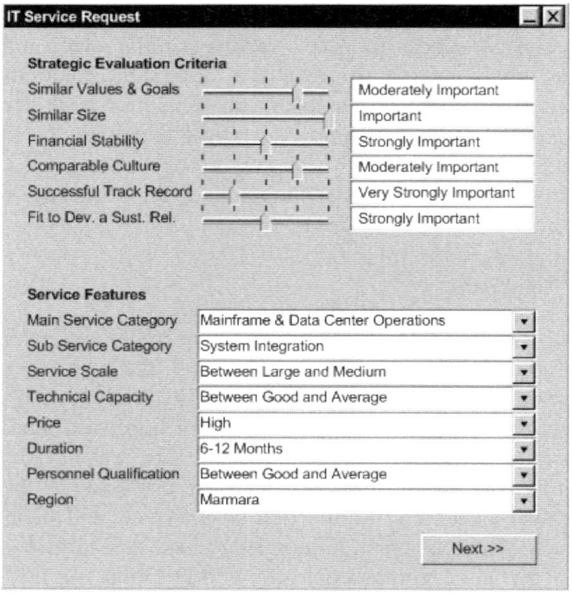

Fig. 1. The first user interface of the developed decision support tool

As an illustrative case, we have chosen a system integration service request for mainframe and data center operations of a high price and between large and medium scale, with between good and average personnel qualification and technical capacity, whose duration is between six-twelve months in the region Marmara. Besides for this service, the successful track record is defined as very strongly important, the financial stability and the development of a sustainable relationship is defined as strongly important, having comparable culture and similar values and goals as moderately important, while being similar size with the outsourcing company as important.

3.3 Case Retrieval in RBR

The interviews performed with several company experts have directed us to integrate a rule-based elimination stage in respect to strategic criteria; since the majority of the experts have stated that the strategic criteria are the most significant indicators in the selection process. The criteria values evaluated in the previous stage are utilized to compose the first rule:

If ((OldCase(i)..SimilarValuesGoalsValue < Moderately important) and
(OldCase(i).SimilarSize < Important) and
(OldCase(i).FinancialStabilityValue < Strong important) and
(OldCase(i).ComparableCultureValue < Moderately important) and
(OldCase(i).SuccessfulTrackRecordValue < Very strongly important) and
(OldCase(i).FitToDevelopRelationshipsValue < Strongly important))
Then Disregard (NewCase)

For the second rule-base sub-system, the cases are again eliminated, but this time in a more strict way according to their features. The second rule written for the illustrative example is

If ((OldCase(i).MainCategory ≠ Mainframe & Data Center Operations) and
(OldCase(i).SubCategory ≠ System Integration) and
(OldCase(i).Scale ≠ Between Large and Medium) and
(OldCase(i).TechnicalCapacity ≠ Between Good and Average) and
(OldCase(i).Price ≠ High) and
(OldCase(i).Duration ≠ (6-12 months)) and
(OldCase(i).PersonnelQualification ≠ Between Good and Average) and
(OldCase(i).Region ≠ Marmara))
Then Disregard (NewCase);

3.4 Case Retrieval in CBR

The intelligent decision support tool, developed using SWI Prolog environment, enabled to diminish the search space to find the matching cases by means of these two consequent RBR sub-systems. In the next stage, namely the CBR sub-system, the customer is expected to determine the performance values that s/he required. Hence, the next sub-system illustrated in Figure 2 demands from the customer the degree of importance and the ratings of each IT business evaluation criterion.

In the example, the customer have stated that when deciding on a IT service, the technical expertise is strongly important, the performance and the intangible factors are moderately important, the total cost is very strongly important and the quality of the service is strongly important for her/him. These evaluations enable the tool to calculate the importance weights of each IT business criteria as given in Table 1.

Table 1. The importance weights of IT business criteria

	Linguistic assessment	Logistics business criteria weights
Technical Expertise	Strongly important	(0.122, 0.219, 0.409)
Performance	Moderately important	(0.098, 0.188, 0.364)
Total cost	Very strongly important	(0.146, 0.250, 0.409)
Quality	Strongly important	(0.122, 0.219, 0.409)
Intangibles	Moderately important	(0.098, 0.188, 0.364)

Following the weighting of the business criteria, the customer is expected to evaluate the performance of these criteria. For our example, these values are determined as, total cost should be in very strong level, technical expertise and intangibles should be in strong level, the quality should be in average level and the performance should be in sufficient level. This is the ideal solution for the MCDM. The aim of the CBR sub-system is to find the most similar cases to this ideal solution using the importance weights when measuring the distance.

In the last part, the intelligent decision support tool measures similarities using the fuzzy compromise programming and it gives a performance value to each one of the cases. Figure 2 shows the most similar cases, in other words the cases with the lowest

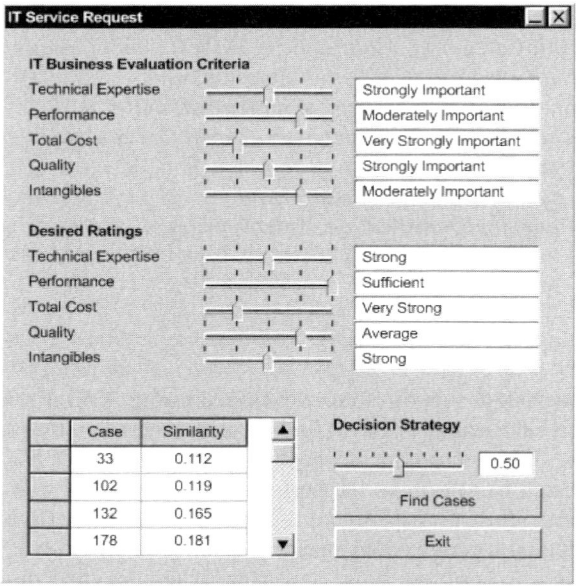

Fig. 2. The second user interface of the decision support tool

distances from the desired (ideal) case. In our application, the case # 33 is lowest value, which implies the highest similarity.

4 Conclusions

This paper proposes an IDS model for an effective IT supplier evaluation and selection. The proposed framework use CBR as the main intelligent technique and integrated RBR and fuzzy compromise programming techniques for real-time decision-making involving uncertain and imprecise decision situations. The details of the methodology are given within an application of IT outsourcing supplier selection for a Turkish company. In conclusion, the results and benefits of the proposed framework can be summarized as follows: faster reaction to a change in the IT activity requests and in the importance degrees of supplier evaluation criteria, a decrease in the decision-making time in IT outsourcing, the right choice of suppliers and the retention of supplier intelligence in IT outsourcing management. Further development of the framework in the connection of other AI techniques (such as neural network) to become a hybrid AI approach is worthwhile direction in this field.

References

1. Aamodt, A., Plaza, E. (1994). Case-Based Reasoning: Foundational Issues, Methodological Variations, and System Approaches, AI Communications. IOS Press, 7 (1) 39-59.
2. Akomode, O.J., Lees, B., Irgens, C. (1998). Constructing customized models and providing information to support IT outsourcing decisions, Logistics Information Management, 11 (2) 114-127
3. Bojadziev, G., Bojadziev, M. (1995). Fuzzy Sets, Fuzzy Logic, Applications. Advances in Fuzzy Systems & Applications and Theory Vol. 5 (Singapore: World Scientific).
4. Carr, A.S., Pearson, J.N. (2002). The impact of purchasing and supplier involvement on strategic purchasing and its impact on firm's performance. International Journal of Operations and Production Management, 22 (9), 1032-1055.
5. Chan, F. (2005). Application of a hybrid case-based reasoning approach in electroplating industry. Expert Systems with Applications, 29, 121-130.
6. Chang, C.L., Cheng, B.W., Su, J.L. (2004). Using case-based reasoning to establish a continuing care information system of discharge planning. Expert Systems with Application, 26, 601–613.
7. Changchien, S.W., Lin, M-C. (2005). Design and implementation of a case-based reasoning system for marketing plans. Expert Systems with Applications, 28, 43–53.
8. Chow H.K.H., Choy K.L., Lee W.B., Chan F.T.S. (2005). Design of a knowledge-based logistics strategy system. Expert Systems with Applications, 1-19.
9. Cullen, S., Willcocks, L.P. (2003). Intelligent IT Outsourcing: Eight Building Blocks to Success. Elsevier, Chichester.
10. Cullen, S., Seddon, P.B., Willcocks, L.P. (2005). IT outsourcing configuration: Research into defining and designing outsourcing arrangements, Journal of Strategic Information Systems, 14 (4), 357-387.
11. Entemann, C.W. (2002). Fuzzy logic: Misconceptions and clarifications. Artificial Intelligence Review, 17, 65–84.

12. Feeny, D., Lacity, M., Willcocks, L. (2005). Taking the measure of outsourcing service providers. Sloan Management Review 46 (3), 41–48.
13. Fink, D., Shoeib, A. (2003). Action: the most critical phase in outsourcing information technology, Logistics Information Management. 16 (5), 302-311.
14. 14.Handfield, R., Walton, S.V., Stroufe, R., Melnyk, S.A. (2002). Applying environmental criteria to supplier assessment: a study in the application of the AHP. European Journal of Operational Research, 141 (1), 70-87.
15. Jharkharia, S., Shankar, R. (2005). Selection of logistics service provider: An analytic network process (ANP) approach. Omega, Article in Press, Available through (www.sciencedirect.com).
16. Klepper, R., Jones, W.O., (1998). Outsourcing Information Technology Systems and Services. Prentice Hall, New Jersey.
17. Lacity, M.C., Willcocks, L.P. (2001). Global IT Outsourcing: In Search of Business Advantage. Wiley, Chichester.
18. Liao, T.W., Zhang, Z., Mount, C.R. (1998). Similarity Measures for retrieval in case-based reasoning systems, Applied Artificial Intelligence 12, 267-288.
19. Lin, Y.T., Tseng, S.S., Tsai, C.F. (2003). Design and implementation of new object-oriented rule-base management system. Expert Systems with Application, 25, 369–385.
20. Liu, J., Ding, F.Y, Lall, V. (2000). Using data envelopment analysis to compare suppliers for supplier selection and performance improvement. Supply Chain Management: an International Journal, 5 (3), 143-150.
21. Main, J., Dillon, T.S., Shiu, S.C.K. (2000). A tutorial on case-based reasoning. In S. K. Pal, T. S. Dillon, D. S. Yeung (Eds.), Soft computing in case based reasoning. London: Springer.
22. Michell, V., Fitzgerald, G. (1997). The IT outsourcing market-place: vendors and their selection, Journal of Information Technology 12, 223-237
23. Opricovic, S., Tzeng, G.H. (2004). Compromise solution by MCDM methods: a comparative analysis of VIKOR and TOPSIS, European Journal of Operational Research, 156, 445–455.
24. Paisittanand, S., Olson D.L., A simulation study of IT outsourcing in the credit card business, European Journal of Operational Research, Article in Press, Available through (www.sciencedirect.com)
25. Poisson, J.F. (2004). Managing Business Process Outsourcing for Business Results, Presentation of Turkey Outsourcing Conference. 14-15 December, İstanbul.
26. Tzeng, G.H., Lin, C.W., Opricovic, S. (2005). Multi-criteria analysis of alternative-fuel buses for public transportation, Energy Policy, 33, 1373–1383.
27. Venkatraman, N., (1997). Beyond outsourcing: managing IT resources as a value center. Sloan Management Review 38 (3), 51–64.
28. Zeleny, M. (1974). Linear Multiobjective Programming, New York: Springer Verlag, 197-220.

Fuzzy Support Vector Machines Regression for Business Forecasting: An Application

Yukun Bao[12], Rui Zhang[1], and Sven F. Crone[2]

[1] Department of Management Science & Information System, School of Management, Huazhong University of Science and Technology, Wuhan 430074, China
yukunbao@mail.hust.edu.cn

[2] Department of Management Science, Management School, Lancaster University, Lancaster LA1 4YX, United Kingdom

Abstract. This study proposes a novel method for business forecasting based on fuzzy support vector machines regression (FSVMR). By an application on sales forecasting, details of proposed method are presented including data preprocessing, kernel selection, parameters tuning and so on. The experimental result shows the method's validity.

1 Introduction

Business forecasting has consistently been a critical organizational capability for both strategic and tactical business planning [1]. Time series forecasting methods such as exponential smoothing have been widely used in practice, but it always doesn't work when the market fluctuates frequently and at random [2]. Research on novel business forecasting techniques have evoked researchers from various disciplines such as computational intelligence.

Recently, support vector machines (SVMs) have been extended to solve nonlinear regression estimation problems and they have been shown to exhibit excellent performance in time series forecasting [3, 4, 5].

One of the key issues encountered in training support vector is the data preprocessing. Some raw data points corrupted by noises are less meaningful and they make different senses to later training process. But standard SVMs algorithm lacks this ability. To solve this problem, Fuzzy support vector machines regression(FSVMR) apply a fuzzy membership to each input points so that different input points can make different contributions to the learning of decision surface and can enhances the SVM in reducing the effect of outliers and noises in data points. Details on the principal and application of FSVMR can be found in ref. [6, 7, 8]

2 Experimental Setting and Algorithms

2.1 Data Sets

We selected 5 goods with 430 daily sales data from a manufacturing firm's management information system. We used the former 400 data points as training

L. Wang et al. (Eds.): FSKD 2006, LNAI 4223, pp. 1313–1317, 2006.

Table 1. Details of Data Sets

Goods	Mean	SD	Min	Max	Train	Test
A	6.95	3.58	0	14.7	400	30
B	171	81.24	23	285	400	30
C	79.17	28.31	12	167	400	30
D	7.52	5.24	1	13.6	400	30
E	18.19	10.11	12.33	41.56	400	30

data sets and the rest 30 data points as testing data.More details of the data sets are listed in Table 1.

2.2 Embedding Dimension

Given a time-series $\{x_1, x_2, \ldots, x_n\}$ generated by a dynamical system. We assume that $\{x_{t+\tau}\}(\tau \geq 1)$ is a projection of dynamics operation in a high-dimensional state space [3]. In order to make prediction, we must reconstructing the input time series data into state space. That is to say if $\{x_t\}$ is the goal value of prediction, the previous values $\{x_{t-(d-1)\tau}, x_{t-(d-2)\tau}, \ldots, x_{t-\tau}\}$ should be the corrected state vector. We call d the embedding dimension or the sliding window, τ the prediction period. In this experiment, we only analyze forecasting $\{x_{t+1}\}$, thus, $\tau = 1$. After transformation, we get the samples in matrix form:

$$\mathbf{X} = \begin{pmatrix} x_1 & x_{1+\tau} & \cdots & x_{1+(d-1)\tau} \\ x_2 & x_{2+\tau} & \cdots & x_{2+(d-1)\tau} \\ \vdots & \vdots & \ddots & \vdots \\ x_{n-(d-1)\tau} & x_{n-(d-2)\tau} & \cdots & x_{n-\tau} \end{pmatrix} \quad \mathbf{Y} = \begin{pmatrix} x_{1+d\tau} \\ x_{2+d\tau} \\ \vdots \\ x_n \end{pmatrix} \tag{1}$$

The value of embedding dimension of a time series data set affects prediction performance. In following experiments, embedding dimension is fixed at 5 balancing error and training cost.

2.3 Defining Fuzzy Membership

It is easy to choose the appropriate fuzzy membership. First, we choose $\sigma > 0$ as the lower bound of fuzzy membership. Second, we make fuzzy membership s_i be a function of time t_i

$$s_i = f(t_i) \tag{2}$$

We suppose the last point x_n be the most important and choose $x_n = f(t_n) = 1$, and the first point x_1 be the most least important and choose $s_1 = f(t_1) = \sigma$. If we want to let fuzzy membership be a linear function of the time, we can select

$$s_i = f(t_i) = \alpha t_i + b = \frac{1-\sigma}{t_n - t_1} t_i + \frac{t_n \sigma - t_1}{t_n - t_1} \tag{3}$$

If we want to make fuzzy membership be a quadric function of the time, we can select

$$s_i = f(t_i) = \alpha(t_i - b)^2 + c = (1 - \sigma)\left(\frac{t_i - t_1}{t_n - t_1}\right)^2 + \sigma \tag{4}$$

2.4 Performance Criteria

The prediction performance is evaluated using the normalized mean squared error (NMSE). NMSE is the measures of the deviation between the actual and predicted values. The smaller the values of NMSE, the closer are the predicted time series values to the actual values. The NMSE of the test set is calculated as follows:

$$\text{NMSE} = \tfrac{1}{\delta^2 n} \sum_{i=1}^{n} (y_i - \hat{y}_i)^2 , \tag{5}$$

$$\delta^2 = \tfrac{1}{n-1} \sum_{i=1}^{n} (y_i - \overline{y})^2 , \tag{6}$$

where n represents the total number of data points in the test set. $\hat{y}_i$ represents the predicted value. $\overline{y}$ denotes the mean of the actual output values.

2.5 Kernel Function Selection and Parameters Tuning

The literatures [9, 10] show that RBF kernel usually get better results than others and use it as the default kernel in predicting time series data. In our experiment,We use general RBF as the kernel function. Comparative results of Goods A between different kernels are shown in Table 2.

There are two parameters while using RBF kernels: C and γ. We use a grid-search on C and γ using cross-validation. We found that trying exponentially growing sequences of C and γ is a practical method to identify good parameters.

Table 2. Results of forecasting with different kernels on Goods A. $\varepsilon = 0.1$.

Kernels	Parameter	NMSE Training	NMSE Testing	Time
Poly	$C = 8, d = 1$	0.012	0.401	0.211
RBF	$C = 8, \gamma = 0.25$	0.008	0.317	0.102
Sigmoid	$C = 8, \gamma = 0.0625$	0.021	0.474	0.176

3 Experimental Results

Table 3 shows the averaged NMSE values of EMA, standard SVM compared with FSVMR. Figure 1 illustrates the predicted and actual values of Goods A in testing. By computing of standard deviation, FSVMR's accuracy is 6.6% and 14.3% higher than standard SVM and EMA respectively.

Table 3. Averaged Forecasting Results for All 5 Goods

Methods	EMA	Standard SVMs	FSVMR
NMSE	0.3610	0.3313	0.3095

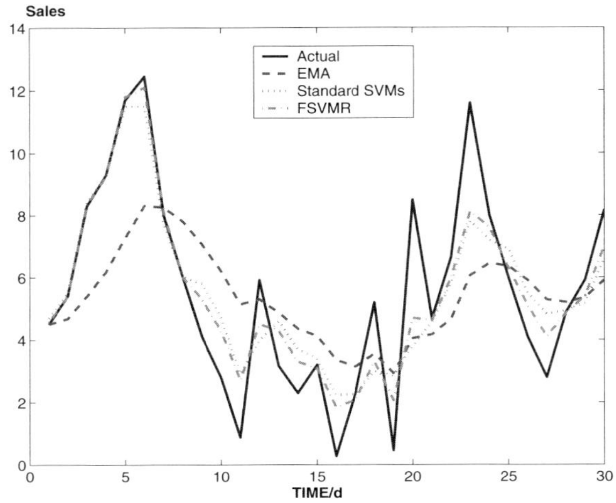

Fig. 1. Forecasting results comparison for Goods A

Acknowledgements

This research is granted by NSFC Project No.70401015 and Hubei Provincial Key Social Science Research Center of Information Management.

References

1. Fildes R, Hastings R. The organization and improvement of market forecasting. Journal of Operation Research Society, 45(1994) 1-16.
2. Lawrence M, O'Connor M. Sales forecasting updates: how good are they in practice? International Journal of Forecasting,16(2000) 369-382.
3. Müller K.R., Smola A.J., Rätsch G., Schölkopf B., Kohlmorgen J., and Vapnik V. Prediction time series with support vector machies. Lecture Notes in Computer Science Vol.1327,1997,999-1004.
4. Francis, E.H.T. and Cao L. Application of support vector machines in financial time series forecasting. Omega: International Journal of Management Science, 29(2001) 309-317.
5. Bao YK, Lu YS and Zhang JL. Forecasting stock price by SVMs regression. Lecture Notes in Artifical Intelligence, Vol.3192,2004:295-303.
6. Chun-Fu Lin, Sheng-De Wang, Fuzzy Support Vector Machines, IEEE Trans on Neural Networks. 13(2002)464-471.

7. Dug H. H, Changha H.. Support vector fuzzy regression machines.Fuzzy Sets and Systems, 138 (2003) 271-281.
8. Bao YK, Zou H and Liu ZT. Forecasting intermittent demand by fuzzy support vector machines. Lecture Notes in Artifical Intelligence, Vol.4031,2006:1080-1089.
9. T. B. Trafalis and H. Ince. Support vector machine for regression and applications to financial forecasting. In Proceedings of the IEEE-INNS-ENNS International Joint Conference, 2000:348-353.
10. Kim K.J. Financial time series forecasting using support vector machines. Neurocomputing, 55(2003) 307-319.

Applying Sensitivity Analysis in Structure Damage Identification

Huazhu Song[1], Luo Zhong[1], and Bo Han[2,3]

[1] School of Computer Science and Technology, Wuhan University of Technology, Wuhan, Hubei 430070, China
[2] School of Computer Science, Wuhan University, Wuhan, Hubei 430072, China
[3] Dept. of Computer and Information Science, Temple University, Philadelphia, PA 19122, U.S.A.
shuazemail@yahoo.com

Abstract. Structure health monitoring aims to detect the nature of structure damage by using a network of sensors, whose sensor signals are highly correlated and mixed with noise, it is difficult to identify direct relationship between sensors and abnormal structure characteristics. In this study, we apply sensor sensitivity analysis on a structure damage identifier, which integrates independent component analysis (ICA) and support vector machine (SVM) together. The approach is evaluated on a benchmark data from University of British Columbia. Experimental results show sensitivity analysis not only helps domain experts understand the mapping from different location and type of sensors to a damage class, but also significantly reduce noise and improve the accuracy of different level damages identification.

1 Introduction

Structural stiffness decreases due to aging, damages, and other harmful effects. These adverse changes lead to abnormal dynamic characteristics in natural frequencies and mode shapes. By instrumenting structures with a network of sensors, structural health monitoring (SHM) aims to provide reliable and economical approaches to detect the nature of structure damage in an early stage so as to prevent catastrophic failures[1,2]. The technology of machine learning has been used, such as independent component analysis (ICA) or principal component analysis (PCA) for feature extraction, artificial neural networks (ANN) or support vector machines (SVM) for classification. However, the complicated data transformation and classification make it difficult to identify direct relationship between sensors and abnormal structure characteristics. Structure engineers are keen to explore the relationship because different type and location sensors have empirically been proved to provide varied quality information.

In this paper, firstly ICA with SVM is combined together to construct a structure damage classifier. Next, the classifier is regarded as a black box and apply ICA-SVM based first-order sensitivity analysis to select most important sensors. Our experiments, based on the benchmark data from University of British Columbia, showed sensitivity analysis can clearly reveal the relationship between selected sensors and

L. Wang et al. (Eds.): FSKD 2006, LNAI 4223, pp. 1318 – 1321, 2006.

specific damages, and the ICA-SVM classifier significantly improves the identification accuracy with the most sensitive signals.

2 Methodology

2.1 Architecture of Structure Damage Classifier on Sensitive Sensors

The architecture of sensitive information prediction based on ICA and classifiers SVM is shown in Fig.1, where Fast-ICA algorithm is used with a non-quadratic function g(y) = tanh (a1×y) to measure nongaussianity, and linear kernel function is used in SVM [3].

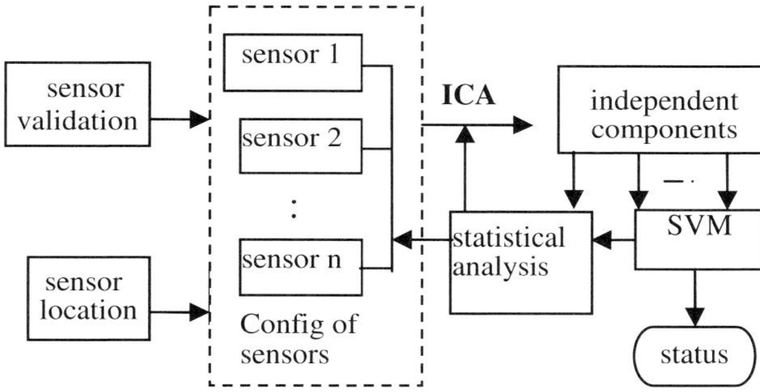

Fig. 1. Architecture of sensitive information prediction

2.2 ICA-SVM Based Sensitivity Analysis

By sensitivity analysis, the classifier is regarded as an ICA-SVM black box, whose inputs are sensor signals x_1, x_2, …, x_h and output is status label Y. We assume each signal x_i (i=1,2,…,h) observers normal distribution with $N(\overline{x_i}, \sigma_i)$. By perturbing a sensor signal with a small value Δx_i, we explore how much difference a new predictor Y_i will make, comparing with the predictor Y_{full} constructed by full set of original sensor signals. Thereby, the normalized stimulation sensitivity $S_i = \frac{\Delta Y_i / \sigma_{Y_i}}{\Delta x_i / \sigma_i} = \frac{\sigma_i (Y_i - Y_{full})}{\sigma_{Y_i} \Delta x_i}$, where σ_{Y_i} is the standard derivation of predictor Y_i. Given all sensor signals have the same standard derivation, $\sigma_i = \sigma_j$, $\sigma_{Y_i} = \sigma_{Y_i}$ (here i,j=1,2,…, h and $i \neq j$), S_i is simplified as the first-order derivative $\frac{(Y_i - Y_{full})}{\Delta x_i}$. Sorting the S_i, we will rank the sensors signals by their sensitivity. The top features play the most important roles in the damage detection. The detailed algorithm is listed in [4].

3 Experiments

A popular benchmark to testify the classification accuracies is used, which was set up by the IASC-ASCE SHM task Group at University of British Columbia. The structure is a 4-story, 2-bay by 2-bay steel-frame scale-model structure, which has a 2.5 m × 2.5m plane and is 3.6m tall[5]. In our experiments, seven data sets in the ambient data was served, where C01 is an undamage dataset, C02-C07 are different type of damaged datasets. There are 15 attributes in each dataset. They correspond to the signals from 15 sensors located in this steel-frame, and the 16 attribute is noise attribute.

3.1 Experimental Results

(1) Sensitive sensor list
For each undamaged or damaged dataset, 6000 samples are randomly chosen. According to sensitive information algorithm, we obtain a sorted attribute list shown in Table 1. The bold attributes denotes they have been selected into sensitive sensor list SL. The table also helps domain experts to explore different location and type of sensors to a specific damage class.

Table 1. Sorted attributes list

Data	1st	2nd	3rd	4th	5th	6th	7th	8th	9th	10th	11th	12th	13th	14th	15th
C01	**4**	**12**	**6**	**11**	**2**	**1**	**15**	13	3	14	8	5	9	7	10
C02	**11**	**4**	**13**	**5**	**15**	**6**	**1**	2	3	8	7	12	14	10	9
C03	**15**	**6**	**3**	**2**	**9**	**4**	**11**	12	7	1	13	5	14	8	10
C04	**15**	**9**	**10**	**13**	**6**	**14**	**12**	2	1	7	11	3	5	4	8
C05	**12**	**10**	**11**	**1**	**13**	**3**	**4**	2	8	5	15	14	9	6	7
C06	**7**	**10**	**3**	**2**	**9**	**12**	**1**	14	15	4	11	6	5	13	8
C07	**8**	**15**	**14**	**7**	**12**	**2**	**4**	6	5	13	3	1	10	11	9

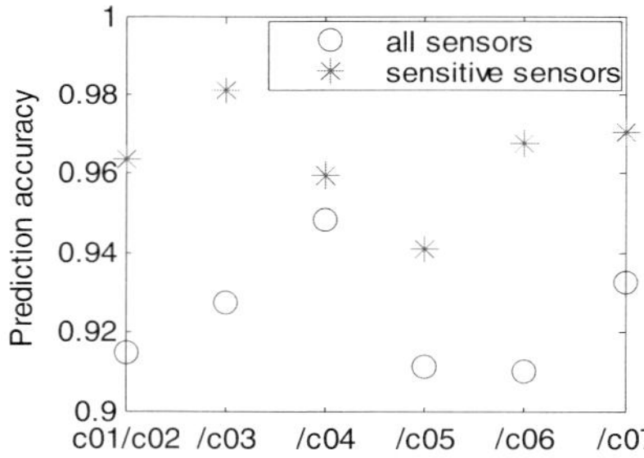

Fig. 2. Two damage identification

For all seven data sets in Table 1, we counted the total occurring frequency for each selected sensitive attribute, and get 7 attributes 4,12,15,1,2,6,11 occurring more than 3 times in all datasets.

(2) Identification of two kinds of damage level
For comparing the classification on accuracy by using all signals or using the most sensitive 7 signals, two damage levels experiment is done. 70% of C01 work as for training, the remaining 30% of C01 as test; and the same number of samples from another damaged dataset in C02-C07 for test, the result is shown in Fig.2, which shows sensitive sensors significantly improve the prediction accuracy.

(3) Identification of multi-damage level
Further comparing the classification on accuracy by using all signals or using the most sensitive 7 signals, multi-damage levels experiments are done.For multi-damage level experiment, C01 is treated as undamage data, its output is '1' ; C02 is regarded

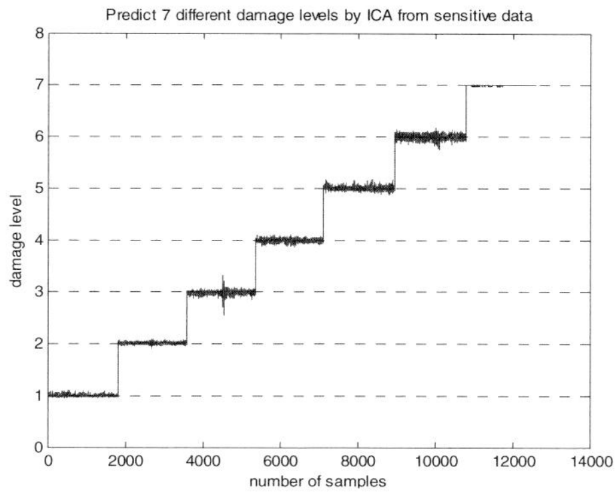

Fig. 3. Prediction with sensitive sensors

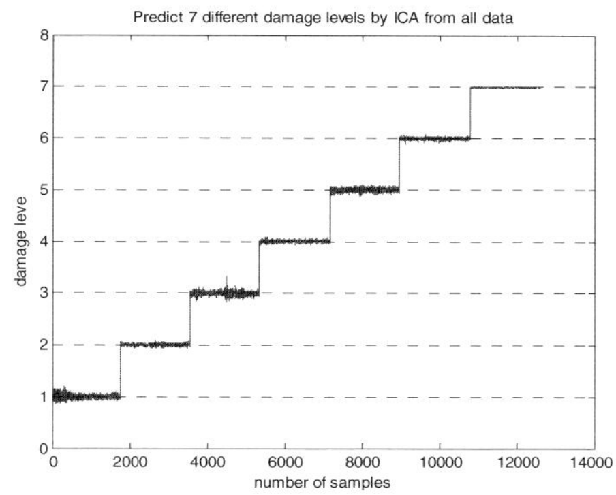

Fig. 4. Prediction with all sensors

as damage data whose output is '2', and so on. 70% of C01-C07 are training data, the rest 30% of C01-C07 are test data, predict the damage value. The results is shown in Fig.3 and Fig.4.

The above experiments show that the sensitive sensors can get an accuracy prediction for different damage levels .Compared with used all sensors, the number for sensitive sensors is reduced nearly half of all sensors, but it can performs damage identification well. They show the validity of the architecture in Fig.1.

4 Conclusions

In this paper, sensitivity analysis is applied in a structure damage classifier, whose architecture combines ICA with SVM, and it is evaluated by the benchmark data from University of British Columbia. The damage detection accuracy using sensitive attributes is significantly better than those obtained by using full sensor signals in two damage level identification; it can perform well for multi-damage level identification.

References

[1] S.W. Doebling, C.R. Farrar, (et): A Summary Review of Vibration-Based Damage Identification Methods, The Shock and Vibration Digest, 30 (2), (1998) 91-105

[2] T. Pothisiri, K. D. Hjelmstad: Structural Damage Detection and Assessment from Modal Response, J. of Engineering Mechanics, 129 (2) , (2003)135-145

[3] H. Song, L. Zhong, (et): Structural Damage Detection by Integrating Independent Component Analysis and Support Vector Machine. ADMA, Springer LNAI3584, (2005) 670-677

[4] B. Han, L. Kang, (et): Improving Structrue Damage Identification by Using ICA-ANN Based Sensitivity Analysis. ICIC 2006, Accepted.

[5] http://www.bc.cityu.edu.hk/asce.shm

Evaluation Function for Siguo Game Based on Two Attitudes

ZhengYou Xia, YongPing Zhu, and Hui Lu

Department of computer science, Nanjing University of Aeronautics and Astronautics
zhengyou_xia@yahoo.com

Abstract. Siguo game is a fascinating imperfect information game that provides a new testbed for AI. We have written a computer program that plays Siguo game. This paper reveals and discusses the method that is based on optimistic and cautious attitudes to construct evaluation function of our system. We also present the worth and rank of material and analyze several features of evaluation function, which are piece capture, position domain of piece and oriflamme guard. Each feature of evaluation function is evaluated by general and optimistic algorithm, respectively.

1 Introduction

Games can model some elements of the real world, and offer more exploring methods for dealing with uncertainty. Indeed, the difficulties associated with handling incomplete or uncertain information are now receiving an increasing amount of attention in many other computer science research domains.The study of games like poker, bridge and Siguo game [1][2] could be highly valuable. Siguo, Poker and Bridge provide an excellent test bed for studying decision-making under conditions of uncertainty. There are many benefits to be gained from designing and experimenting with poker and Bridge programs [3][4]. The Siguo game can be classified two kinds (i.e. 1vs1 and 2 vs 2 model). Siguo game is an imperfect-information game, which is different to Poker or Bridge game. Siguo game can obtain less information than poker acquires during playing game. The player of Siguo game cannot get exact type information of the confederate and opponents' piece from the previous rounds and only get results of “>”,”<”and “maybe equal”. It involves very lacking information: the premises, process and the consequences of a decision are vagueness, imprecision, and uncertainty. Therefore, Siguo game is a new test bed for AI, which is a game of imperfect information, where competing players must deal with possible knowledge, risk assessment, and possible deception and leaguing players have to deal with cooperation and information signal transmission.

As the conventional game, the evaluation function is very important for designing the senior intelligent Siguo game system and is used to estimate the players' winning chance in positions at the leaves of game-trees. The conventional games (e.g. go, chess, amazons, and Othello games, etc.) are based on the perfect information and the score of position evaluation can be exact gotten. Related works about evaluation function of these games [5][6][7][8][9][10] have been well achieved. However, since

L. Wang et al. (Eds.): FSKD 2006, LNAI 4223, pp. 1322 – 1331, 2006.

the Siguo game is different to traditional game, the method to construct evaluation function may be different to previous methods. In this paper, we reveal and discuss the evaluation function of our Nhope system. We define the worth and rank of material and analyze several features of evaluation function, which are piece capture, position domain of piece and oriflamme guard, respectively. We use these features to introduce the method to construct evaluation function of Siguo game. The method is based on two attitudes (i.e. optimism and pessimism) to evaluate the score of strategy.

2 Worth and Worth Rank of Material in the Siguo Game

In Siguo game[1], every player has twenty-five pieces. To express conveniently, we denote the Sapper, Second Lieutenant, Captain, Major, Lieutenant Colonel, Colonel, Senior Colonel , Lieutenant General, Marshal ,Mine, Bomb as x_1, x_2, x_3, x_4, x_5, x_6, x_7, x_8, x_9, x_{10}, x_{11}, respectively. In the Siguo game, every material worth should be abode the experience rules of equation 1.

$$\begin{aligned}&x_9 > (x_7 + x_{11}); x_9 < (x_8 + x_{11}); x_8 \le (x_7 + x_{11}); x_7 \ge x_{11}\\&x_{11} \ge x_6; x_8 \ge (x_6 + x_{11}); x_9 \ge (2x_6 + x_{11}); x_1 \ge x_{10}; x_1 \ge x_5\\&x_1 \le x_6; x_6 \le 2x_5; x_5 \le 2x_4; x_4 \le 2x_3; x_3 \le 2x_2\end{aligned} \tag{1}$$

For example, in the equation 1, $x_9 > (x_7 + x_{11})$ denotes that the value of marshal material is more than the sum value of Senior Colonel and bomb material. Similarly, $x_9 < (x_8 + x_{11})$ denotes that the value of marshal material is less than the sum value of Lieutenant General and bomb material. In the Siguo game, worth of the Second Lieutenant material is suggested to be 1 (i.e. x1=1) and Worth of other materials are inferred and suggested as the follows.

$$\begin{aligned}&worth(x_1) = 9, worth(x_2) = 1, worth(x_3) = 2, worth(x_4) = 4,\\&worth(x_5) = 7, worth(x_6) = 12, worth(x_7) = 20, worth(x_8) = 33,\\&worth(x_9) = 45, worth(x_{10}) = 8, worth(x_{11}) = 19\end{aligned} \tag{2}$$

To analyze characteristic of these materials, we give the worth rank according to characteristic of materials as follows.

$$\begin{aligned}&rank(x_1) = 1, rank(x_2) = 2, rank(x_3) = 3, rank(x_4) = 4, rank(x_5) = 5, rank(x_6) = 6\\&, rank(x_7) = 8, rank(x_8) = 9, rank(x_9) = 10, rank(x_{10}) = 6.5, rank(x_{11}) = 7\end{aligned} \tag{3}$$

3 Features of Evaluation Function

In our Nhope Siguo system, evaluation function is composed of nine features and evaluated at the leaf node of game tree. In this paper, we only discuss four features and use these features to present the method of constructing evaluation function for

Siguo game. The four features include the piece capture, position domain of piece, oriflamme guard of opponent and oriflamme guard of computer.

3.1 Piece Capture

When piece α fights with piece β, let piece α be winner and piece β be loser. If piece α and piece β are not bombs or oriflamme, the piece α can get following score shown as equation 4 and piece β can get score shown as equation 5.

$$score(\alpha, \beta) = worth(\beta) - \frac{|worth(\alpha) - worth(\beta)|}{e^{|rank(\alpha) - rank(\beta)|}} \tag{4}$$

$$score(\beta, a) = -score(\alpha, \beta) \tag{5}$$

In the Siguo game, computing of piece capture is different to chinese chess, chess, and other games that the winner can get the score that is worth of loser' piece during fight. However, in Siguo game, when the piece of winner kills piece of opponent, the type information about piece of the winner is leaked and opponent of the winner can know the exact type scope of winner' piece. To detect type scope of the opponent' piece by using suitable piece to fight with very fuzzy and uncertain opponent' piece is general method for players of Siguo game. Therefore, loser can get some worth from fight because loser can get some type information about piece of winner. Similarly, the winner loses some worth because winner expose type of itself piece to opponent during fight. In equation 4, $\frac{|worth(\alpha) - worth(\beta)|}{e^{|rank(\alpha) - rank(\beta)|}}$ is the worth that the loser can get from fight and the winner loses during fight.

When piece α fights with piece β, let piece α be bomb and piece β be non-bomb or non-oriflamme piece. The piece α can get the following score shown as equation 6 and piece β get score shown as equation7.

$$score(\alpha, \beta) = (worth(\beta) - worth(\alpha)) - \frac{|\, worth(\alpha) - worth(\beta)\,|}{e^{|rank(\alpha) - rank(\beta)|}} \tag{6}$$

$$score(\beta, \alpha) = -score(\alpha, \beta) \tag{7}$$

If piece α and piece β are the same type, $score(\beta, \alpha) = score(\alpha, \beta) = 0$.

If any pieces can capture oriflamme of opponent, the piece will get score $+\infty$ and the game is over.

In our Nhope system, we use two attitudes to compute piece capture, which are optimistic and cautious attitudes. The algorithm about piece capture is illustrated as Fig1. Let ψ_{oca} be the score by using optimistic capture algorithm and ψ_{gca} be the score by using general capture algorithm. We order ψ_{oca} and ψ_{gca} by score (i.e. $\min(\psi_{oca}, \psi_{gca}), \max(\psi_{oca}, \psi_{gca})$). We can assume that the score of piece capture is

General Capture algorithm

Let α be piece of computer system and β be piece of human player

1. Make the type membership set of piece β to do unitary process and $\sum_{i=1}^{12} w_i = 1$.
2. General capture score (ψ_{gca}) $= \sum_{i=1}^{12} w_i * score(\alpha, \beta)$.

Optimistic Capture algorithm

1. Search the max membership value from type fuzzy set of piece β.
2. If only one maximal type membership value can be found and the type is β_T_i, the optimistic capture= $score(\alpha, \beta\_T_i)$.
3. If there are l ($l > 1$) different types Φ that have the same membership value and are the maximal value in the type fuzzy set of piece β, the optimistic capture score (ψ_{oca})= $\sum_{T_i \in \Phi} score(\alpha, \beta\_T_i) \Big/ l$

Fig. 1. Algorithms for piece capture

in $[\min(\psi_{oca}, \psi_{gca}), \max(\psi_{oca}, \psi_{gca})]$. In other words, we can use the two algorithms to compute the score of piece capture and the score is between $\min(\psi_{oca}, \psi_{gca})$ and $\max(\psi_{oca}, \psi_{gca})$. $R_{capture} = [r_1^1 = \min(\psi_{oca}, \psi_{gca}), r_1^2 = \max(\psi_{oca}, \psi_{gca})]$ is called the score of piece capture.

3.2 Position Domain of Piece

The goal of the Siguo game is to capture opponent' oriflamme or kills out opponent' pieces that can move. To achieve the goal, player often moves piece to fight with the opponent' piece or occupy good position during playing Siguo game. To evaluate the piece capture has been discussed in the above section. In this section, we discuss evaluation about position of piece.

Definition **conjoint piece** $cp(a,b), a\ and\ b\ denote\ piece$: if the position of piece a is empty and piece b can be moved to the position of piece a by next move, then piece b is conjoint to piece a.

Definition **position domain of piece** a , $pdp(a,\Theta), \Theta\ denotes\ position\ domain$: $\exists b_i$piece, $i \in (1, \cdots n)$, if b_i satisfies $cp(a,b_i)$, then $b_i \in \Theta$. If

$\{a, b_i\} \subseteq \Theta$ *and* $\{a, b_i\} \supseteq \Theta, i \in (1, \ldots \cdots n)$,then $pdp(a, \Theta)$ is called position domain of piece a .

Fig2a and Fig2b show position domain of senior colonel piece of player 1. In the domain, black piece with unknown type is piece of player 2. In the Siguo game, more fifty moving strategies can be generated for every piece of sapper type. If every piece is considered to be possible sapper, the moving strategies will increase explosively. Therefore, in our Nhope Siguo system, if the membership value of sapper type is maximal in the type fuzzy set of opponent' piece, we consider the piece as piece of sapper type to avoid every piece as the piece of sapper type. We assume the computer system moves SC to the position that shown as in Fig2a and Fig2b. Therefore, in the Fig2a, the position domain of SC is $pdp(SC, \Theta) = \{\text{B,SC,1st piece,3rd piece,4th piece}\}$. In the Fig2b, since membership value of sapper type in the 1st piece is maximal, we consider the piece as sapper, which can fly to position of SC.Therefore, we consider the 1st piece is neighboring to SC and the position domain of SC is $pdp(SC, \Theta) = \{\text{B,SC,1st piece,4th piece}\}$.

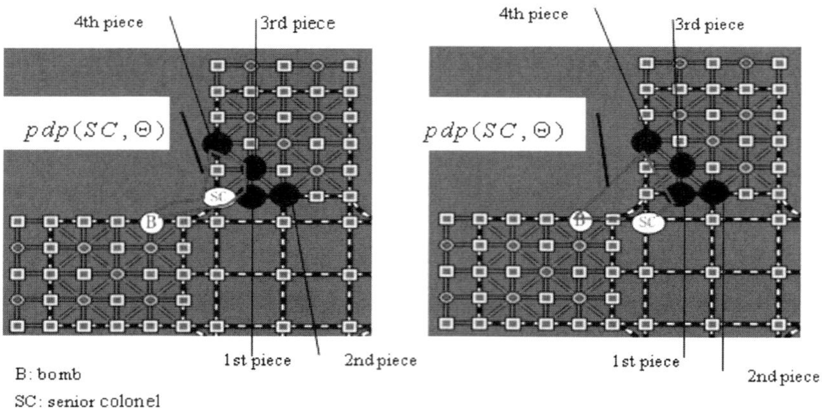

(a) (b)

1st piece type fuzzy set :(0.484,0.167,0.177,0.054,0.056,0.054,0.008,0,0,0,0,0)
2nd piece type fuzzy set :(0.137,0.116,0.145,0.089,0.083,0.191,0.038,0.089,0.113,0,0,0)
3rd piece type fuzzy set: (0,0,0,0,0,0.16,0.2,0.1,0.11,0,0.43,0)
4th piece type fuzzy set :(0,0,0.051,0.17,0.2,0.147,0.21,0.139,0.083,0,0,0)

Fig. 2.(a) Piece position domain (b) Piece position domain

When one player moves a piece to one position, we evaluate the quality of the position by analyzing the position domain of the piece according to several features such as safety, joint defense, etc. We present an algorithm for evaluation of position domain of piece based on two attitudes that are optimistic and cautious attitudes. The general algorithm for evaluating position domain of piece is presented in the Fig3. If we use Optimistic Capture algorithm instead of General Capture algorithm in the Fig3, general algorithm for evaluating position domain of piece in the Fig3 becomes optimistic algorithm for evaluating position domain of piece. Similarly, let $pdp(a,\Theta)_{general}.sq$ be the score of position domain of piece a by using general algorithm for evaluating position domain of piece and $pdp(a,\Theta)_{optimistic}.sq$ be the score of position domain of piece a by using Optimistic algorithm for evaluating position domain of piece. The score of position domain of piece a is $R_{pdp} = [r^{1}_{2} = \min(pdp(a,\Theta)_{optimistic}.sq, pdp(a,\Theta)_{general}.sq)$ $, r^{2}_{2} = \max(pdp(a,\Theta)_{optimistic}.sq, pdp(a,\Theta)_{general}.sq)]$

1. Check conjoint piece of piece of player1 and build position domain of piece $\alpha : pdp(\alpha,\Theta)$.The pieces of player 1 in the $pdp(\alpha,\Theta)$ are marked as $o\alpha_i.sq = 0, i \in (1...n)$.
2. If there is not any piece of opponent in the $pdp(\alpha,\Theta)$,return.
3. If the piece $o\alpha_i$ is in the arm camp, then set the security quality of piece $o\alpha_i.sq$=0.25
4. Else,use the General Capture algorithm to compute, $\psi^{\beta_j}_{gcc}.\alpha_i = \sum_{k=1}^{12} w_i * score(o\alpha_i,\beta_j), j \in (1...l)$,$\beta_j$ satisfy cp($o\alpha_i,\beta_j$), $\beta_j \in pdp(\alpha,\Theta) and \beta_j$ is piece of player 2 .
5. Choose the minimal value from $\psi^{\beta_j}_{gcc}.\alpha_i$ and $o\alpha_i.sq$=min($\psi^{\beta_j}_{gcc}.\alpha_i$), $j \in (1...l)$
6. until the computation of $o\alpha_i$ pieces have been finished
7. If $o\alpha_i, i \in (1,...n)$ are all positive or negative, choose the minimal value from $o\alpha_i.sq$ and $pdp(\alpha,\Theta).sq = min(o\alpha_i.sq)$. Return pdp$(\alpha,\Theta).sq)$
8. Choose pieces of player 1, and they are marked as $l\alpha_k \in, k \in (1,...m)$,$l\alpha_k \in pdp(\alpha,\Theta), l\alpha_k \subseteq (o\alpha_i) and l\alpha_k.sq \geq 0$.
   (a) Search $s\alpha_j$,it satisfies cp($l\alpha_k, s\alpha_j$),$s\alpha_j \neq bomb, s\alpha_j \subseteq o\alpha_i and s\alpha_j.sq < 0, j \in (1,...n_1)$
   8.1.1 Search piece β_k of player2, the β_k satisfies min($\psi^{\beta_k}_{gcc}.\alpha_j$), $k \in (1,...l)$ /* for optimistic algorithm, $\beta_k \neq bomb$ */
   8.1.2 If $l\alpha_k.sq > o and l\alpha_k = bomb$,then goto (b)
   8.1.3 Compute $F = \sum_{i=1}^{12} w_i * score(l\alpha_k,\beta_k)$,compute $G^{l\alpha_k}_{s\alpha_j} = F + s\alpha_j.sq$
   8.1.4 If the check of $s\alpha_j$ is not finished ,then goto 8.1.1
   (b) If the check of $l\alpha_k$ is not finished ,then goto (a)
9. $s\alpha_j.sq = max(G^{l\alpha_k}_{s\alpha_j}), k \in (1,...m), j \in (1,...n_1)$
10. $pdp(\alpha,\Theta).sq = min(s\alpha_j.sq), and return(pdp(\alpha,\Theta).sq)$

Fig. 3. Algorithm for analyzing position domain of piece

3.3 Oriflamme Guard

In the Fig4, the 2k, 3j and 2i positions around the oriflamme are called external guard position, and the 1k, 2j and 1i positions around oriflamme are called interior guard position. When no any self-pieces are on external guard position or interior guard position and these positions are exposed to opponent, we consider guard position danger.

Two methods can be used to analyze the dangerous degree about guard position of oriflamme, which are methods of static and dynamic analysis. For static analysis, only six guard positions of oriflamme are considered in constructing evaluation function. However, dynamic method is too complex and performance of system will be worse because plenty of pieces that can arrive to guard position of oriflamme in next move

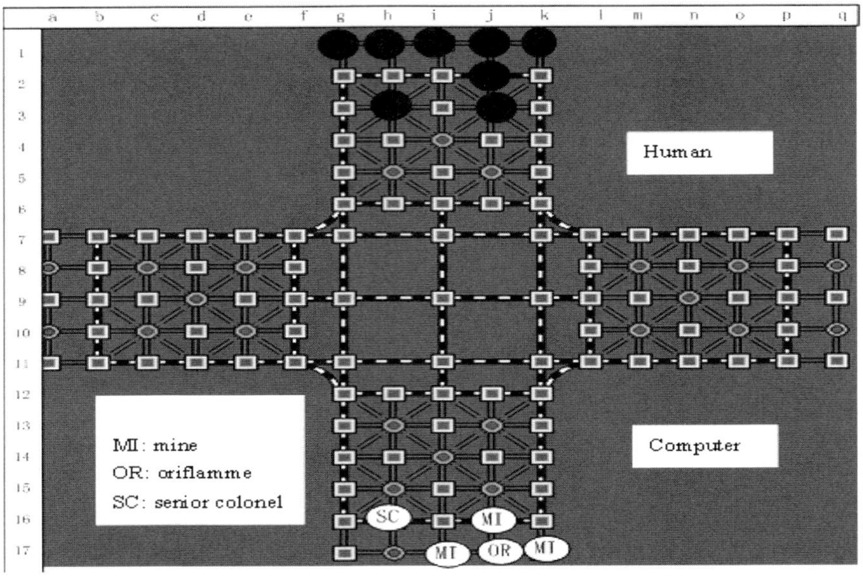

Fig. 4. Oriflamme guard

must be considered. In our previous version of Nhope system, the result of system performance is not good by using dynamic analysis method. Therefore, in this paper, we only discuss the static method to analyze guard position. The equation8 and 9 are experience equations for evaluating dangerous degree about external guard and interior guard position. If there are no any dangerous guard positions, we consider the guard position safety and set score of oriflamme guard zero. To compute oriflamme guard feature for computer is illustrated as in Fig5 and to compute oriflamme guard feature for human player is illustrated as in Fig6.

$$\text{external\_guard}=\frac{\mathrm{m}}{1+x^{-2}},\ x\in(1,2,3),\ \mathrm{m}\ \textit{is suggested as}\ \text{in}\ [8,20] \tag{8}$$

x denotes the number of dangerous guard positions (i.e. 2k, 3j and 2i positions).

$$\text{interior\_guard}=\frac{\mathrm{n}}{1+y^{-2}},\ y\in(1,2,3),\ \mathrm{m}\ \text{is}\ \textit{suggested as}\ \text{in}\ [20,45] \tag{9}$$

y denotes the number of dangerous guard positions (i.e. 2k, 3j and 2i positions).

4 Experiment About Problem of Evaluation Function

For perfect information, the evaluation function can evaluate exactly the strategy of player. However, for imperfect information game, since information about opponent is imperfect or very fuzzy and uncertain, evaluation function cannot evaluate validly

1. Check the six guard positions around computer' oriflamme
2. Check external guard positions, if the three positions are all occupied by self-pieces, then external_guard =0.

   Else check the number of dangerous external guard positions and get x .

   Compute external_guard= $\frac{m}{1+x^{-2}}$, $x \in (1, 2, 3)$.
3. Check interior guard positions, if the three positions are all occupied by self-piece, then interior_guard =0.

   Else checks the number of dangerous external guard positions and gets y .compute interior_guard= $\frac{n}{1+y^{-2}}$, $y \in (1, 2, 3)$ ·
4. Ori_guard = external_guard+interior_guard , and return (Ori_guard).

Fig. 5. Algorithm for oriflamme guard feature of computer

1. If the oriflamme position of human has been revealed, then

/* under this situation, general algorithm and optimistic algorithm is the same as the following steps*/

1.1. Check external guard positions, if the three positions are all occupied by self-pieces, then external_guard =0. Else check the number of dangerous external guard positions and get x .

Compute external_guard= $\frac{m}{1+x^{-2}}$, $x \in (1, 2, 3)$.

1.2 Check interior guard positions, if the three positions are all occupied by self-piece, then interior_guard =0. Else check the number of dangerous external guard positions and gets y .compute interior_guard= $\frac{n}{1+y^{-2}}$, $y \in (1, 2, 3)$.

1.3 Ori_guard = external_guard+interior_guard , and return (Ori_guard).

2. There are two positions that contain oriflamme piece.

2.1 If it is optimistic attitude, then choose the position that is maximal membership value of oriflamme type between two positions. Go to 1.1 for computing oriflamme guard by using the position.

2.2 Check the two oriflamme base camp, respectively and operations are same as the above steps

$$Ori\_guard_{general} = w_{first} * Ori\_guard_{first} + w_{secnd} * Ori\_guard_{second}$$

Return ($Ori\_guard_{general}$)

Fig. 6. Algorithm for oriflamme guard feature of human player strategy of player in few situations. The situation is that factual information about opponent is entirely different to that we get from model or experience data. Under the situation, evaluation function cannot well give proper score about strategy. We called the phenomena invalid problem of evaluation function.We use our Nhope system to fight with human player of 1st level, 2nd level, 3rd level and 4th level in internet Siguo game club. The rate of invalid problem of evaluation function is increased with level of human player shown as in Fig7.

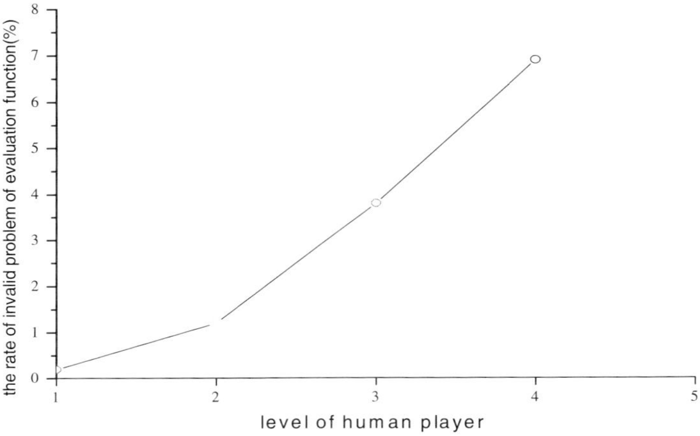

Fig. 7. Invalid problem level of human player

The first reason is that our system analyzes and infers type information of pieces based on statistical data that is mined from more than one thousand different lineups of Siguo game. Statistical data only shows general rules and information about opponent. The second reason is that our system assumes that behavior of human player is rational. However, good human player can disturb the judgment of our computer system by cheat. Generally, human player of high level can easily apperceive the intention of opponent of low level and very exact infer the type of piece of opponent based on psychology of opponent that is concealed in action of moving piece by opponent. For human player of low level, the ability to apperceive the intention of opponent is weak and they often move piece according to principle of piece force balance and joint defense, etc. Therefore, we can validly construct model to fight with human player of low level by using previous statistical data. Presently, the force rank of our Nhope system is equal to the second rank of human player in the two-person competition (i.e.1V1) in the internet club. However, the force rank of our system is equal to novice of human player in four-person competition (i.e.2V2).

5 Conclusion

In this paper, we present a novel method to construct evaluation function of our Siguo game system. An important idea in the method of constructing evaluation function is based on two attitudes, which are optimistic and cautious attitudes. To use the method, we can make score of evaluation function in rational interval and decrease invalid problem of evaluation function. However, it is not very valid to evaluate the human player of high level because sporadic invalid problem of evaluation is occurred. To improve the exact guessing type of opponent' piece can decrease validly problem of evaluation function, which is the research that we are implementing now.

References

1. Z. Y. Xia, Y. A. Hu, J.Wang, Analyze and Guess Type of Piece in the Computer Game Intelligent System. FSKD (2) 2005, Lecture Notes In Artificial Intelligent (LNAI), Vol. 3339, pp.1174-1183, Springer-Verlag Heidelberg.
2. Lianzhong, Introduciton to SiGuo. Available at http://www.ourgame.com/srvcenter/game-intro/junqi.html.
3. Frank,I.;Basin,D. search in games with incomplete information: A case study using bridge card play. Artificial intelligent 100:87-123.
4. Schaeffer, J., Culberson, J., Treloar, N., Knight, B., Lu, P., Szafron, D, A World Championship Caliber Checkers Program. Articial Intelligence 53, 273-289, 1982.
5. M.Buro, Statistical feature combination for the evaluation of game positions,Journal of Arti¯cial Intelligence Research 3, 373-382.1995.
6. Marsland, T.A., Evaluation Function Factors, ICCA Journal 8(2), 1985.
7. M. Buro.,A Small Go Board Study of Metric and Dimensional Evaluation Functions, Computers and Games,2002.
8. Jack van Rijswijck,A Data Mining Approach to Evaluation Function Learning in Awari, Computers and Games, 2000.
9. T.Hashimoto,Y.Kajihara,N.Sasaki,H.Iida,et al,. An evaluatin function for amazons advances in computer games 9,the Netherlands.
10. Aviezri S. Fraenkel, Michal Ozery,Adjoining to Wytho®'s Game its P-Positions as Moves. Theor. Comput. Sci. 205(1-2): 283-296 (1998).

Author Index

Lecture Notes in Artificial Intelligence (LNAI)

Vol. 4223: L. Wang, L. Jiao, G. Shi, X. Li, J. Liu (Eds.), Fuzzy Systems and Knowledge Discovery. XXVIII, 1337 pages. 2006.

Vol. 4213: J. Fürnkranz, T. Scheffer, M. Spiliopoulou (Eds.), Knowledge Discovery in Databases: PKDD 2006. XXII, 660 pages. 2006.

Vol. 4212: J. Fürnkranz, T. Scheffer, M. Spiliopoulou (Eds.), Machine Learning: ECML 2006. XXIII, 851 pages. 2006.

Vol. 4211: P. Vogt, Y. Sugita, E. Tuci, C. Nehaniv (Eds.), Symbol Grounding and Beyond. VIII, 237 pages. 2006.

Vol. 4203: F. Esposito, Z.W. Ras, D. Malerba, G. Semeraro (Eds.), Foundations of Intelligent Systems. XVIII, 767 pages. 2006.

Vol. 4201: Y. Sakakibara, S. Kobayashi, K. Sato, T. Nishino, E. Tomita (Eds.), Grammatical Inference: Algorithms and Applications. XII, 359 pages. 2006.

Vol. 4196: K. Fischer, I.J. Timm, E. André, N. Zhong (Eds.), Multiagent System Technologies. X, 185 pages. 2006.

Vol. 4188: P. Sojka, I. Kopeček, K. Pala (Eds.), Text, Speech and Dialogue. XIV, 721 pages. 2006.

Vol. 4183: J. Euzenat, J. Domingue (Eds.), Artificial Intelligence: Methodology, Systems, and Applications. XIII, 291 pages. 2006.

Vol. 4180: M. Kohlhase, OMDoc – An Open Markup Format for Mathematical Documents [version 1.2]. XIX, 428 pages. 2006.

Vol. 4160: M. Fisher, W.v.d. Hoek, B. Konev, A. Lisitsa (Eds.), Logics in Artificial Intelligence. XII, 516 pages. 2006.

Vol. 4155: O. Stock, M. Schaerf (Eds.), Reasoning, Action and Interaction in AI Theories and Systems. XVIII, 343 pages. 2006.

Vol. 4149: M. Klusch, M. Rovatsos, T.R. Payne (Eds.), Cooperative Information Agents X. XII, 477 pages. 2006.

Vol. 4139: T. Salakoski, F. Ginter, S. Pyysalo, T. Pahikkala, Advances in Natural Language Processing. XVI, 771 pages. 2006.

Vol. 4133: J. Gratch, M. Young, R. Aylett, D. Ballin, P. Olivier (Eds.), Intelligent Virtual Agents. XIV, 472 pages. 2006.

Vol. 4130: U. Furbach, N. Shankar (Eds.), Automated Reasoning. XV, 680 pages. 2006.

Vol. 4120: J. Calmet, T. Ida, D. Wang (Eds.), Artificial Intelligence and Symbolic Computation. XIII, 269 pages. 2006.

Vol. 4114: D.-S. Huang, K. Li, G.W. Irwin (Eds.), Computational Intelligence, Part II. XXVII, 1337 pages. 2006.

Vol. 4108: J.M. Borwein, W.M. Farmer (Eds.), Mathematical Knowledge Management. VIII, 295 pages. 2006.

Vol. 4106: T.R. Roth-Berghofer, M.H. Göker, H. A. Güvenir (Eds.), Advances in Case-Based Reasoning. XIV, 566 pages. 2006.

Vol. 4099: Q. Yang, G. Webb (Eds.), PRICAI 2006: Trends in Artificial Intelligence. XXVIII, 1263 pages. 2006.

Vol. 4095: S. Nolfi, G. Baldassarre, R. Calabretta, J.C. T. Hallam, D. Marocco, J.-A. Meyer, O. Miglino, D. Parisi (Eds.), From Animals to Animats 9. XV, 869 pages. 2006.

Vol. 4093: X. Li, O.R. Zaïane, Z. Li (Eds.), Advanced Data Mining and Applications. XXI, 1110 pages. 2006.

Vol. 4092: J. Lang, F. Lin, J. Wang (Eds.), Knowledge Science, Engineering and Management. XV, 664 pages. 2006.

Vol. 4088: Z.-Z. Shi, R. Sadananda (Eds.), Agent Computing and Multi-Agent Systems. XVII, 827 pages. 2006.

Vol. 4087: F. Schwenker, S. Marinai (Eds.), Artificial Neural Networks in Pattern Recognition. IX, 299 pages. 2006.

Vol. 4068: H. Schärfe, P. Hitzler, P. Øhrstrøm (Eds.), Conceptual Structures: Inspiration and Application. XI, 455 pages. 2006.

Vol. 4065: P. Perner (Ed.), Advances in Data Mining. XI, 592 pages. 2006.

Vol. 4062: G. Wang, J.F. Peters, A. Skowron, Y. Yao (Eds.), Rough Sets and Knowledge Technology. XX, 810 pages. 2006.

Vol. 4049: S. Parsons, N. Maudet, P. Moraitis, I. Rahwan (Eds.), Argumentation in Multi-Agent Systems. XIV, 313 pages. 2006.

Vol. 4048: L. Goble, J.-J.C.. Meyer (Eds.), Deontic Logic and Artificial Normative Systems. X, 273 pages. 2006.

Vol. 4045: D. Barker-Plummer, R. Cox, N. Swoboda (Eds.), Diagrammatic Representation and Inference. XII, 301 pages. 2006.

Vol. 4031: M. Ali, R. Dapoigny (Eds.), Advances in Applied Artificial Intelligence. XXIII, 1353 pages. 2006.

Vol. 4029: L. Rutkowski, R. Tadeusiewicz, L.A. Zadeh, J.M. Zurada (Eds.), Artificial Intelligence and Soft Computing – ICAISC 2006. XXI, 1235 pages. 2006.

Vol. 4027: H.L. Larsen, G. Pasi, D. Ortiz-Arroyo, T. Andreasen, H. Christiansen (Eds.), Flexible Query Answering Systems. XVIII, 714 pages. 2006.

Vol. 4021: E. André, L. Dybkjær, W. Minker, H. Neumann, M. Weber (Eds.), Perception and Interactive Technologies. XI, 217 pages. 2006.

Vol. 4020: A. Bredenfeld, A. Jacoff, I. Noda, Y. Takahashi (Eds.), RoboCup 2005: Robot Soccer World Cup IX. XVII, 727 pages. 2006.

Vol. 4013: L. Lamontagne, M. Marchand (Eds.), Advances in Artificial Intelligence. XIII, 564 pages. 2006.

Vol. 4012: T. Washio, A. Sakurai, K. Nakajima, H. Takeda, S. Tojo, M. Yokoo (Eds.), New Frontiers in Artificial Intelligence. XIII, 484 pages. 2006.

Vol. 4008: J.C. Augusto, C.D. Nugent (Eds.), Designing Smart Homes. XI, 183 pages. 2006.

Vol. 4005: G. Lugosi, H.U. Simon (Eds.), Learning Theory. XI, 656 pages. 2006.

Vol. 3978: B. Hnich, M. Carlsson, F. Fages, F. Rossi (Eds.), Recent Advances in Constraints. VIII, 179 pages. 2006.

Vol. 3963: O. Dikenelli, M.-P. Gleizes, A. Ricci (Eds.), Engineering Societies in the Agents World VI. XII, 303 pages. 2006.

Vol. 3960: R. Vieira, P. Quaresma, M.d.G.V. Nunes, N.J. Mamede, C. Oliveira, M.C. Dias (Eds.), Computational Processing of the Portuguese Language. XII, 274 pages. 2006.

Vol. 3955: G. Antoniou, G. Potamias, C. Spyropoulos, D. Plexousakis (Eds.), Advances in Artificial Intelligence. XVII, 611 pages. 2006.

Vol. 3949: F. A. Savacı (Ed.), Artificial Intelligence and Neural Networks. IX, 227 pages. 2006.

Vol. 3946: T.R. Roth-Berghofer, S. Schulz, D.B. Leake (Eds.), Modeling and Retrieval of Context. XI, 149 pages. 2006.

Vol. 3944: J. Quiñonero-Candela, I. Dagan, B. Magnini, F. d'Alché-Buc (Eds.), Machine Learning Challenges. XIII, 462 pages. 2006.

Vol. 3930: D.S. Yeung, Z.-Q. Liu, X.-Z. Wang, H. Yan (Eds.), Advances in Machine Learning and Cybernetics. XXI, 1110 pages. 2006.

Vol. 3918: W.K. Ng, M. Kitsuregawa, J. Li, K. Chang (Eds.), Advances in Knowledge Discovery and Data Mining. XXIV, 879 pages. 2006.

Vol. 3913: O. Boissier, J. Padget, V. Dignum, G. Lindemann, E. Matson, S. Ossowski, J.S. Sichman, J. Vázquez-Salceda (Eds.), Coordination, Organizations, Institutions, and Norms in Multi-Agent Systems. XII, 259 pages. 2006.

Vol. 3910: S.A. Brueckner, G.D.M. Serugendo, D. Hales, F. Zambonelli (Eds.), Engineering Self-Organising Systems. XII, 245 pages. 2006.

Vol. 3904: M. Baldoni, U. Endriss, A. Omicini, P. Torroni (Eds.), Declarative Agent Languages and Technologies III. XII, 245 pages. 2006.

Vol. 3900: F. Toni, P. Torroni (Eds.), Computational Logic in Multi-Agent Systems. XVII, 427 pages. 2006.

Vol. 3899: S. Frintrop, VOCUS: A Visual Attention System for Object Detection and Goal-Directed Search. XIV, 216 pages. 2006.

Vol. 3898: K. Tuyls, P.J. 't Hoen, K. Verbeeck, S. Sen (Eds.), Learning and Adaption in Multi-Agent Systems. X, 217 pages. 2006.

Vol. 3891: J.S. Sichman, L. Antunes (Eds.), Multi-Agent-Based Simulation VI. X, 191 pages. 2006.

Vol. 3890: S.G. Thompson, R. Ghanea-Hercock (Eds.), Defence Applications of Multi-Agent Systems. XII, 141 pages. 2006.

Vol. 3885: V. Torra, Y. Narukawa, A. Valls, J. Domingo-Ferrer (Eds.), Modeling Decisions for Artificial Intelligence. XII, 374 pages. 2006.

Vol. 3881: S. Gibet, N. Courty, J.-F. Kamp (Eds.), Gesture in Human-Computer Interaction and Simulation. XIII, 344 pages. 2006.

Vol. 3874: R. Missaoui, J. Schmidt (Eds.), Formal Concept Analysis. X, 309 pages. 2006.

Vol. 3873: L. Maicher, J. Park (Eds.), Charting the Topic Maps Research and Applications Landscape. VIII, 281 pages. 2006.

Vol. 3864: Y. Cai, J. Abascal (Eds.), Ambient Intelligence in Everyday Life. XII, 323 pages. 2006.

Vol. 3863: M. Kohlhase (Ed.), Mathematical Knowledge Management. XI, 405 pages. 2006.

Vol. 3862: R.H. Bordini, M. Dastani, J. Dix, A.E.F. Seghrouchni (Eds.), Programming Multi-Agent Systems. XIV, 267 pages. 2006.

Vol. 3849: I. Bloch, A. Petrosino, A.G.B. Tettamanzi (Eds.), Fuzzy Logic and Applications. XIV, 438 pages. 2006.

Vol. 3848: J.-F. Boulicaut, L. De Raedt, H. Mannila (Eds.), Constraint-Based Mining and Inductive Databases. X, 401 pages. 2006.

Vol. 3847: K.P. Jantke, A. Lunzer, N. Spyratos, Y. Tanaka (Eds.), Federation over the Web. X, 215 pages. 2006.

Vol. 3835: G. Sutcliffe, A. Voronkov (Eds.), Logic for Programming, Artificial Intelligence, and Reasoning. XIV, 744 pages. 2005.

Vol. 3830: D. Weyns, H. V.D. Parunak, F. Michel (Eds.), Environments for Multi-Agent Systems II. VIII, 291 pages. 2006.

Vol. 3817: M. Faundez-Zanuy, L. Janer, A. Esposito, A. Satue-Villar, J. Roure, V. Espinosa-Duro (Eds.), Nonlinear Analyses and Algorithms for Speech Processing. XII, 380 pages. 2006.

Vol. 3814: M. Maybury, O. Stock, W. Wahlster (Eds.), Intelligent Technologies for Interactive Entertainment. XV, 342 pages. 2005.

Vol. 3809: S. Zhang, R. Jarvis (Eds.), AI 2005: Advances in Artificial Intelligence. XXVII, 1344 pages. 2005.

Vol. 3808: C. Bento, A. Cardoso, G. Dias (Eds.), Progress in Artificial Intelligence. XVIII, 704 pages. 2005.

Vol. 3802: Y. Hao, J. Liu, Y.-P. Wang, Y.-m. Cheung, H. Yin, L. Jiao, J. Ma, Y.-C. Jiao (Eds.), Computational Intelligence and Security, Part II. XLII, 1166 pages. 2005.

Vol. 3801: Y. Hao, J. Liu, Y.-P. Wang, Y.-m. Cheung, H. Yin, L. Jiao, J. Ma, Y.-C. Jiao (Eds.), Computational Intelligence and Security, Part I. XLI, 1122 pages. 2005.